Strategic Applications of Named Reactions in Organic Synthesis

Strategic Applications of Named Reactions in Organic Synthesis

Background and Detailed Mechanisms

by

László Kürti and Barbara Czakó

UNIVERSITY OF PENNSYLVANIA

250 Named Reactions

ELSEVIER
ACADEMIC
PRESS

AMSTERDAM • BOSTON • HEIDELBERG • LONDON • NEW YORK • OXFORD • PARIS
SAN DIEGO • SAN FRANCISCO • SINGAPORE • SYDNEY • TOKYO

Senior Publishing Editor Jeremy Hayhurst
Project Manager Carl M. Soares
Editorial Assistant Desiree Marr
Marketing Manager Linda Beattie
Cover Printer Phoenix Color
Interior Printer RR Donnelley

Elsevier Academic Press
30 Corporate Drive, Suite 400, Burlington, MA 01803, USA
525 B Street, Suite 1900, San Diego, California 92101-4495, USA
84 Theobald's Road, London WC1X 8RR, UK

This book is printed on acid-free paper. ♾

Library of Congress Cataloging-in-Publication Data
Application Submitted

British Library Cataloguing in Publication Data
A catalogue record for this book is available from the British Library

ISBN-13: 978-0-12-369483-6 Casebound Edition
ISBN-10: 0-12-369483-3 Casebound Edition

Printed in the United States of America
05 06 07 08 09 10 9 8 7 6 5 4 3 2 1

ISBN-13: 978-0-12-429785-2 Paperback Edition
ISBN-10: 0-12-429785-4 Paperback Edition

Printed in the United States of America
05 06 07 08 09 10 9 8 7 6 5 4 3 2

For all information on all Elsevier Academic Press Publications
visit our Web site at www.books.elsevier.com

This book is dedicated to

Professor Madeleine M. Joullié

for her lifelong commitment

to mentoring graduate students

ABOUT THE AUTHORS

Barbara Czakó was born and raised in Hungary. She received her Diploma from Lajos Kossuth University in Debrecen, Hungary (now University of Debrecen) where she conducted research in the laboratory of Dr. Sándor Berényi. She obtained her Master of Science degree at University of Missouri-Columbia working for Professor Shon R. Pulley. Currently she is pursuing her Ph.D. degree in synthetic organic chemistry under the supervision of Professor Gary A. Molander at the University of Pennsylvania. She will join the research group of Professor E.J. Corey at Harvard University as a postdoctoral fellow in 2006.

László Kürti was born and raised in Hungary. He received his Diploma from Lajos Kossuth University in Debrecen, Hungary (now University of Debrecen) where he conducted research in the laboratory of Professor Sándor Antus. He obtained his Master of Science degree at University of Missouri-Columbia working for Professor Michael Harmata. Currently he is pursuing his Ph.D. degree in synthetic organic chemistry under the supervision of Professor Amos B. Smith III at the University of Pennsylvania. He will join the research group of Professor E.J. Corey at Harvard University as a postdoctoral fellow in 2006.

ACKNOWLEDGEMENTS

The road that led to the completion of this book was difficult, however, we enjoyed the support of many wonderful people who guided and helped us along the way. The most influential person was **Professor Madeleine M. Joullié** whose insight, honest criticism and invaluable suggestions helped to mold the manuscript into its current form.

When we completed half of the manuscript in early 2004, **Professor Amos B. Smith III** was teaching his synthesis class "Strategies and Tactics in Organic Synthesis" and adopted the manuscript. We would like to thank him for his support and encouragement. We also thank the students in his class for their useful observations that aided the design of a number of difficult schemes.

Our thanks also go to **Professor Gary A. Molander** for his valuable remarks regarding the organometallic reactions. He had several excellent suggestions on which named reactions to include.

Earlier this year our publisher, Academic Press/Elsevier Science, sent the manuscript to a number of research groups in the US as well as in the UK. The thorough review conducted by the professors and in some cases also by volunteer graduate students is greatly appreciated.
They are (in alphabetical order):

Professor Donald H. Aue (University of California Santa Barbara)
Professor Ian Fleming (University of Cambridge, UK)
Professor Rainer Glaser (University of Missouri-Columbia)
Professor Michael Harmata (University of Missouri-Columbia)
Professor Robert A. W. Johnstone (University of Liverpool, UK)
Professor Erik J. Sorensen (Princeton University)
Professor P. A. Wender (Stanford University) and two of his graduate students **Cindy Kan** and **John Kowalski**
Professor Peter Wipf (University of Pittsburgh)

We would like to express our gratitude to the following friends/colleagues who have carefully read multiple versions of the manuscript and we thank them for the excellent remarks and helpful discussions. They were instrumental in making the manuscript as accurate and error free as possible:

James P. Carey (Merck Research Laboratories)
Akin H. Davulcu (Bristol-Myers Squibb/University of Pennsylvania)
Dr. Mehmet Kahraman (Kalypsys, Inc.)
Justin Ragains (University of Pennsylvania)
Thomas Razler (University of Pennsylvania)

There were several other friends/colleagues who reviewed certain parts of the manuscript or earlier versions and gave us valuable feedback on the content as well as in the design of the schemes.

Clay Bennett (University of Pennsylvania)
Prof. Cheon-Gyu Cho (Hanyang University, Korea/University of Pennsylvania)
Dr. Shane Foister (University of Pennsylvania)
Dr. Eugen Mesaros (University of Pennsylvania)
Dr. Emmanuel Meyer (University of Pennsylvania)
David J. St. Jean, Jr. (University of Pennsylvania)
Dr. Kirsten Zeitler (University of Regensburg, Germany)

Finally, we would like to thank our editor at Elsevier, **Jeremy Hayhurst**, who gave us the chance to make a contribution to the education of graduate students in the field of organic chemistry. He generously approved all of our requests for technical support thus helping us tremendously to finish the writing in a record amount of time. Our special thanks are extended to editorial assistants **Desireé Marr** and previously, **Nora Donaghy**, who helped conduct the reviews and made sure that we did not get lost in a maze of documents.

CONTENTS

FOREWORD

This book on "Strategic Applications of Named Reactions in Organic Synthesis" is destined to become unusually useful, valuable, and influential for advanced students and researchers in the field. It breaks new ground in many ways and sets an admirable standard for the next generation of texts and reference works. Its virtues are so numerous there is a problem in deciding where to begin. My first impression upon opening the book was that the appearance of its pages is uniformly elegant and pleasing – from the formula graphics, to the print, to the layout, and to the logical organization and format. The authors employ four-color graphics in a thoughtful and effective way. All the chemical formulas are exquisitely drawn.

The book covers many varied and useful reactions for the synthesis of complex molecules, and in a remarkably clear, authoritative and balanced way, considering that only two pages are allocated for each. This is done with unusual rigor and attention to detail. Packed within each two-page section are historical background, a concise exposition of reaction mechanism and salient and/or recent applications. The context of each example is made crystal clear by the inclusion of the structure of the final synthetic target. The referencing is eclectic but extensive and up to date; important reviews are included.

The amount of information that is important for chemists working at the frontiers of synthesis to know is truly enormous, and also constantly growing. For a young chemist in this field, there is so much to learn that the subject is at the very least daunting. It would be well neigh impossible were it not for the efforts of countless authors of textbooks and reviews. This book represents a very efficient and attractive way forward and a model for future authors. If I were a student of synthetic chemistry, I would read this volume section by section and keep it close at hand for reference and further study.

I extend congratulations to László Kürti and Barbara Czakó for a truly fine accomplishment and a massive amount of work that made it possible. The scholarship and care that they brought to this task will be widely appreciated because they leap out of each page. I hope that this wonderful team will consider extending their joint venture to other regions of synthetic chemical space. Job well done!

E. J. Corey

January, 2005

INTRODUCTION

The field of chemical synthesis continues to amaze with its growing and impressive power to construct increasingly complex and diverse molecular architectures. Being the precise science that it is, this discipline often extends not only into the realms of technology, but also into the domains of the fine arts, for it engenders unparallel potential for creativity and imagination in its practice. Enterprises in chemical synthesis encompass both the discovery and development of powerful reactions and the invention of synthetic strategies for the construction of defined target molecules, natural or designed, more or less complex. While studies in the former area –synthetic methodology– fuel and enable studies in the latter –target synthesis– the latter field offers a testing ground for the former. Blending the two areas provides for an exciting endeavor to contemplate, experience, and watch. The enduring art of total synthesis, in particular, affords the most stringent test of chemical reactions, old and new, named and unnamed, while its overall reach and efficiency provides a measure of its condition at any given time. The interplay of total synthesis and its tools, the chemical reactions, is a fascinating subject whether it is written, read, or practiced.

This superb volume by László Kürti and Barbara Czakó demonstrates clearly the power and beauty of this blend of science and art. The authors have developed a standard two-page format for discussing each of their 250 selections whereby each named reaction is concisely introduced, mechanistically explained, and appropriately exemplified with highlights of constructions of natural products, key intermediates and other important molecules. These literature highlights are a real treasure trove of information and a joy to read, bringing each named reaction to life and conveying a strong sense of its utility and dynamism. The inclusion of an up-to-date reference listing offers a complete overview of each reaction at one's fingertips.

The vast wealth of information so effectively compiled in this colorful text will not only prove to be extraordinarily useful to students and practitioners of the art of chemical synthesis, but will also help facilitate the shaping of its future as it moves forward into ever higher levels of complexity, diversity and efficiency. The vitality of the enduring field of total synthesis exudes from this book, captivating the attention of the reader throughout. The authors are to be congratulated for the rich and lively style they developed and which they so effectively employed in their didactic and aesthetically pleasing presentations. The essence of the art and science of synthesis comes alive from the pages of this wonderful text, which should earn its rightful place in the synthetic chemist's library and serve as an inspiration to today's students to discover, invent and apply their own future named reactions. Our thanks are certainly due to László Kürti and Barbara Czakó for a splendid contribution to our science.

K.C. Nicolaou

January, 2005

PREFACE

Today's organic chemist is faced with the challenge of navigating his or her way through the vast body of literature generated daily. Papers and review articles are full of scientific jargon involving the description of methods, reactions and processes defined by the names of the inventors or by a well-accepted phrase. The use of so-called "named reactions" plays an important role in organic chemistry. Recognizing these named reactions and understanding their scientific content is essential for graduate students and practicing organic chemists.

This book includes some of the most frequently used named reactions in organic synthesis. The reactions were chosen on the basis of importance and utility in synthetic organic chemistry. Our goal is to provide the reader with an introduction that includes a detailed mechanism to a given reaction, and to present its use in recent synthetic examples. This manuscript is not a textbook in the classical sense: it does not include exercises or chapter summaries. However, by describing 250 named organic reactions and methods with an extensive list of leading references, the book is well-suited for independent or classroom study. On one hand, the compiled information for these indispensable reactions can be used for finding important articles or reviews on a given subject. On the other hand, it can also serve as supplementary material for the study of organic reaction mechanisms and synthesis.

This book places great emphasis on the presentation of the material. Drawings are presented accurately and with uniformity. Reactions are listed alphabetically and each named reaction is presented in a convenient two-page layout. On the first page, a brief introduction summarizes the use and importance of the reaction, including references to original literature and to all major reviews published after the primary reference. When applicable, leading references to modifications and theoretical studies are also given. The introduction is followed by a general scheme of the reaction and by a detailed mechanism drawn using a four-color code (red, blue, green and black) to ensure easy understanding. The mechanisms always reflect the latest evidence available for the given reaction. If the mechanism is unknown or debatable, references to the relevant studies are included. The second page contains 3 or 4 recent synthetic examples utilizing the pertinent named reaction. In most cases the examples are taken from a synthetic sequence leading to the total synthesis of an important molecule or a natural product. Some examples are taken from articles describing novel methodologies. The synthetic sequences are drawn using the four-color code, and the procedures are described briefly in 2-3 sentences. If a particular named reaction involves a complex rearrangement or the formation of a polycyclic ring system, numbering of the carbon-skeleton is included in addition to the four-color code. In the depicted examples, the reaction conditions as well as the ratio of observed isomers (if any) and the reported yields are shown. The target of the particular synthetic effort is also illustrated with colors indicating where the intermediates reside in the final product.

The approach used in this book is also unique in that it emphasizes the clever use of many reactions that might otherwise have been overlooked.

The almost 10,000 references are indexed at the end of the book and include the title of the cited book, book section, chapter, journal or review article. The titles of seminal papers written in a foreign language were translated to English. The name of the author of a specific synthetic example was chosen as the one having an asterisk in the reference.

In order to make the book as user-friendly as possible, we have included a comprehensive list of abbreviations used in the text or drawings along with the structure of the protecting groups and reagents. Also in an appendix, the named organic reactions are grouped on the basis of their use in contemporary synthesis. Thus the reader can readily ascertain which named organic reactions effect the same synthetic transformations or which functional groups are affected by the use of a particular named reaction. Finally, an index is provided to allow rapid access to desired information based on keywords found in the text or the drawings.

László Kürti & Barbara Czakó

University of Pennsylvania

Philadelphia, PA

January 2005

IV. EXPLANATION OF THE USE OF COLORS IN THE SCHEMES AND TEXT

The book uses four colors (black, red, blue, and green) to depict the synthetic and mechanistic schemes and highlight certain parts of the text. In the "**Introduction**" and "**Mechanism**" sections of the text, the title named reaction/process is highlighted in blue and typed in italics:

> "The preparation of ketones *via* the C-alkylation of esters of 3-oxobutanoic acid (acetoacetic esters) is called the *acetoacetic ester synthesis*. Acetoacetic esters can be deprotonated at either the C2 or at both the C2 and C4 carbons, depending on the amount of base used."

All other named reactions/processes that are mentioned are typed in italics:

> "Dilute acid hydrolyzes the ester group, and the resulting β-keto acid undergoes decarboxylation to give a ketone (mono- or disubstituted acetone derivative), while aqueous base induces a *retro-Claisen reaction* to afford acids after protonation."

In the "**Synthetic Applications**" section, the name of the target molecule is highlighted in blue:

> "During the highly stereoselective total synthesis of epothilone B by J.D. White and co-workers, the stereochemistry of the alcohol portion of the macrolactone was established by applying *Davis's oxaziridine oxidation* of a sodium enolate."

In the schemes, colors are applied to highlight the changes in a given molecule or intermediate (formation and breaking of bonds). It is important to note that due to the immense diversity of reactions, it is impossible to implement a strictly unified use of colors. Therefore, **each scheme has a unique use of colors specifically addressing the given transformation**. By utilizing four different colors the authors' goal is to facilitate understanding. The authors hope that the readers will look up the cited articles and examine the details of a given synthesis. The following sample schemes should help the readers to understand how colors are used in this book.

- In most (but not all) schemes the starting molecule is colored blue, while the reagent or the reaction partner may be of any of the remaining two colors (red and green). **The newly formed bonds are always black.**

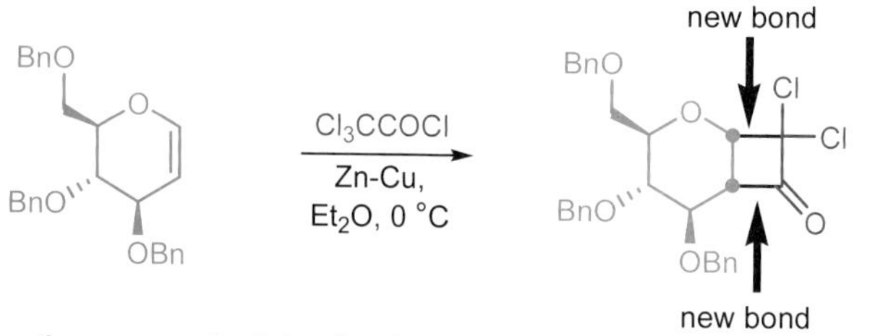

- The general schemes follow the same principle of coloring, and where applicable the same type of key reagents are depicted using the same color. (In this example the two different metal-derived reagents are colored green.)

Simmons & Smith (1958):

Furukawa modification (1966):

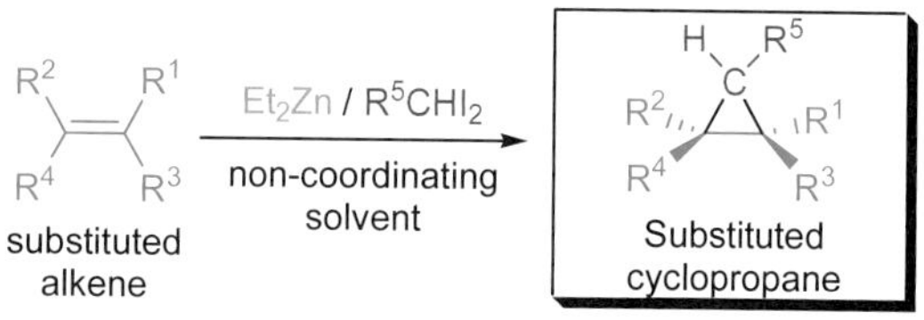

Charette asymmetric modification (1994):

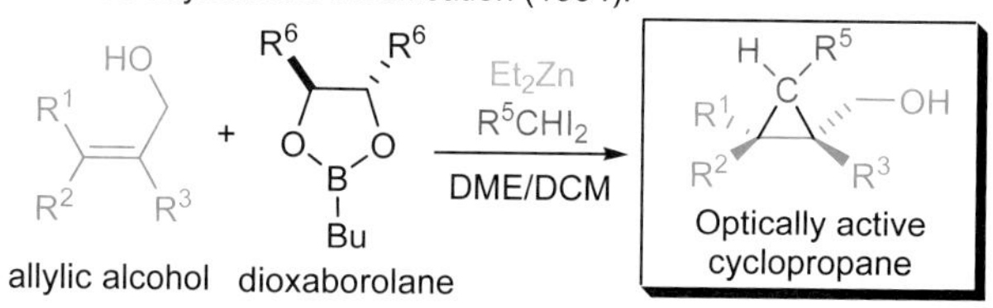

R^{1-4} = H, substituted alkyl and aryl; R^5 = H, Me, phenyl; R^6 = $CONMe_2$; non-coordinating solvent: toluene, benzene, DCM, DCE

- The mechanistic schemes benefit the most from the use of four colors. These schemes also include extensive arrow-pushing. The following two schemes demonstrate this point very well.
  - The catalytic cycle for the *Suzuki cross-coupling*:

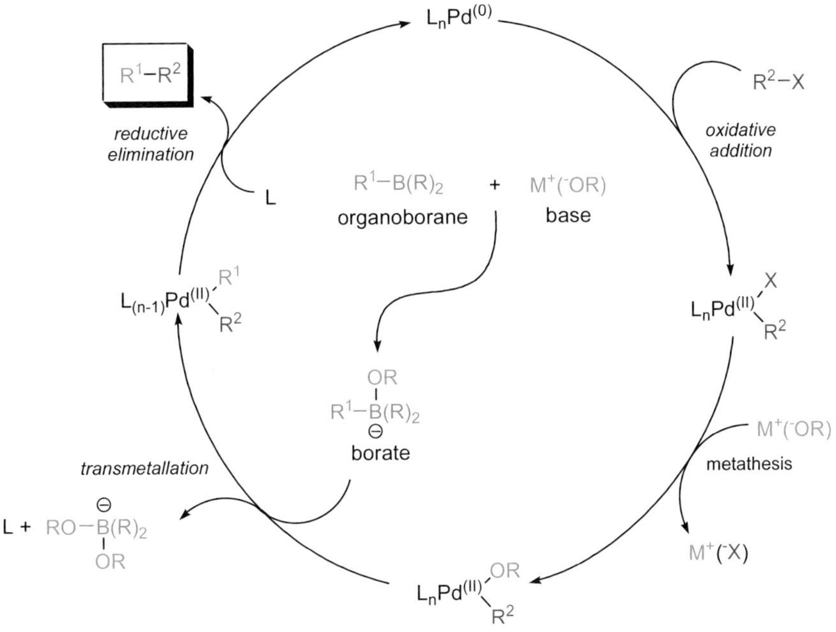

  - The mechanism of the *Swern oxidation*:

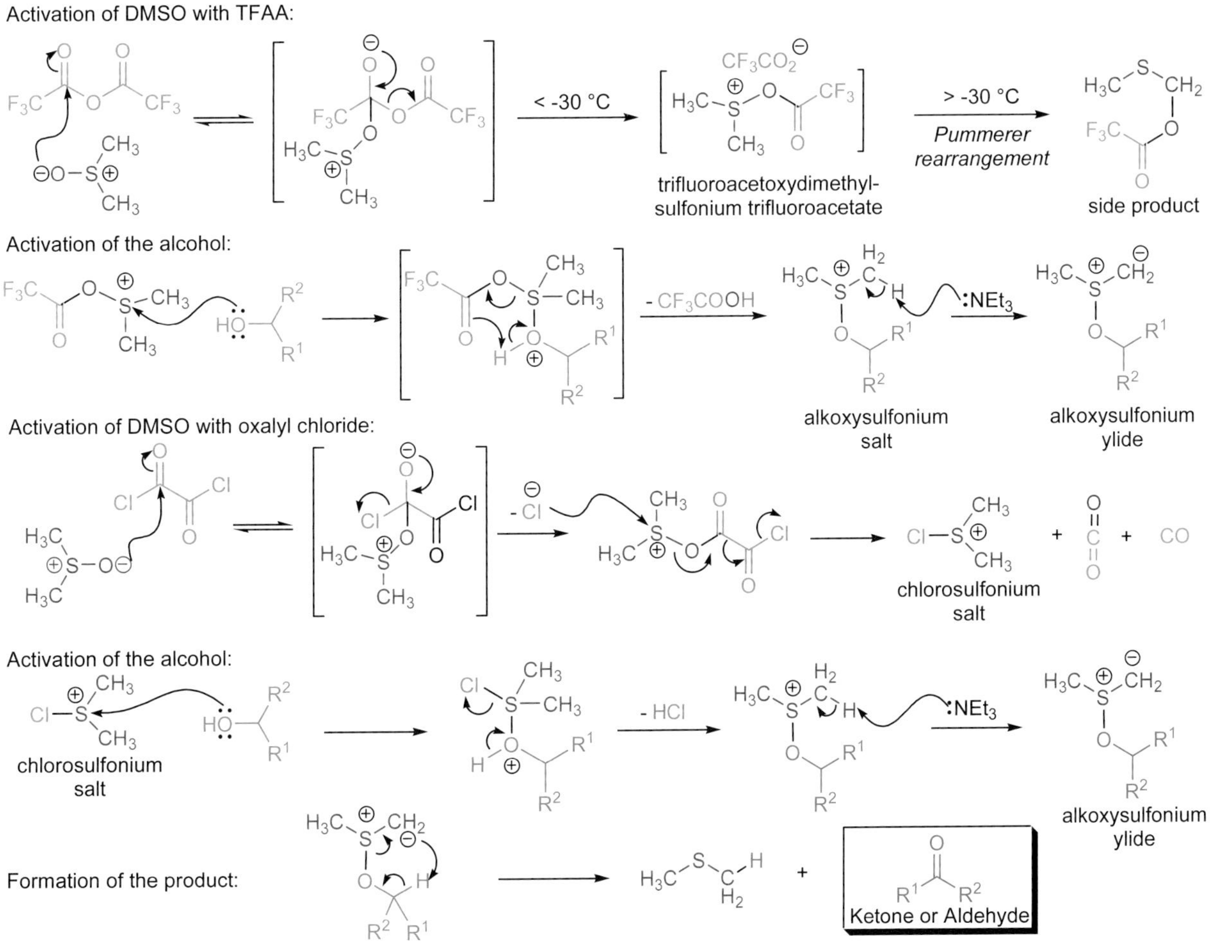

- In the case of complex rearrangements, numbering of the initial carbon skeleton has been applied in addition to the colors to facilitate understanding. Again, the newly formed bonds are black.

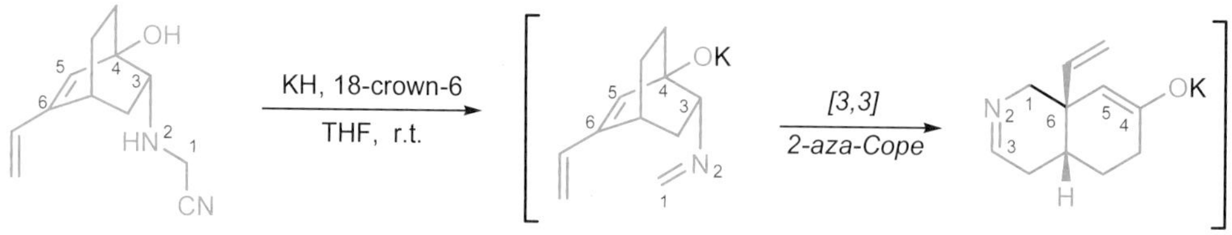

- In most instances, the product of a given named reaction/process will be part of a larger structure (e.g., natural product) at the end of the described synthetic effort. For pedagogical reasons, the authors decided to indicate where the building block appears in the target structure. It is the authors' hope that the reader will be able to put the named reaction/process in context and the provided synthetic example will not be just an abstract one.

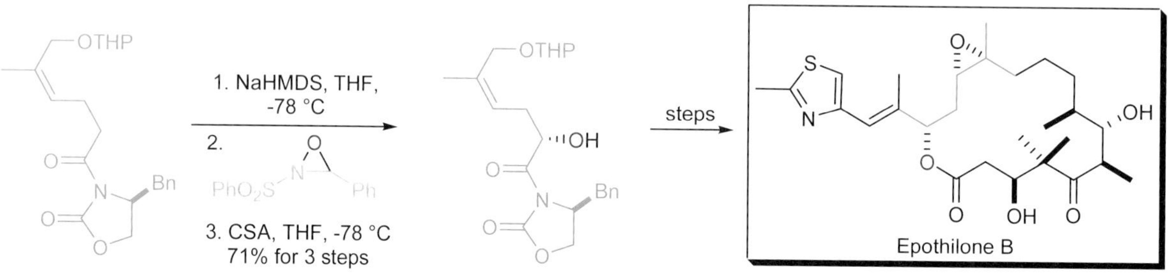

- The references at the end of the book are listed in alphabetical order, and the named reaction for which the references are listed is typed in blue and with boldface (see *Dakin oxidation*). **Important: the references are listed in chronological order when they appear as superscript numbers in the text** (e.g., reference 10 is a more recent paper than reference 12, but it received a smaller reference number because it was cited in the text earlier).

Mechanism: [12,10,15-17]

The mechanism of the *Dakin oxidation* is very similar to the mechanism of the *Baeyer-Villiger oxidation*.

- For the *Dakin oxidation* example, the references at the end of the book will be printed in the order they have been cited, but within a group of references (e.g., 15-17) they appear in chronological order.

Dakin oxidation

10. Hocking, M. B. Dakin oxidation of o-hydroxyacetophenone and some benzophenones. Rate enhancement and mechanistic aspects. *Can. J. Chem.* **1973**, 51, 2384-2392.
11. Matsumoto, M., Kobayashi, K., Hotta, Y. Acid-catalyzed oxidation of benzaldehydes to phenols by hydrogen peroxide. *J. Org. Chem.* **1984**, 49, 4740-4741.
12. Ogata, Y., Sawaki, Y. Kinetics of the Baeyer-Villiger reaction of benzaldehydes with perbenzoic acid in aquo-organic solvents. *J. Org. Chem.* **1969**, 34, 3985-3991.
13. Boeseken, J., Coden, W. D., Kip, C. J. The synthesis of sesamol and of its β-glucoside. The Baudouin reaction. *Rec. trav. chim.* **1936**, 55, 815-820.
14. Kabalka, G. W., Reddy, N. K., Narayana, C. Sodium percarbonate: a convenient reagent for the Dakin reaction. *Tetrahedron Lett.* **1992**, 33, 865-866.
15. Hocking, M. B., Ong, J. H. Kinetic studies of Dakin oxidation of o- and p-hydroxyacetophenones. *Can. J. Chem.* **1977**, 55, 102-110.
16. Hocking, M. B., Ko, M., Smyth, T. A. Detection of intermediates and isolation of hydroquinone monoacetate in the Dakin oxidation of p-hydroxyacetophenone. *Can. J. Chem.* **1978**, 56, 2646-2649.
17. Hocking, M. B., Bhandari, K., Shell, B., Smyth, T. A. Steric and pH effects on the rate of Dakin oxidation of acylphenols. *J. Org. Chem.* **1982**, 47, 4208-4215.

V. LIST OF ABBREVIATIONS

Abbreviation	Chemical Name	Chemical Structure
18-Cr-6	18-crown-6	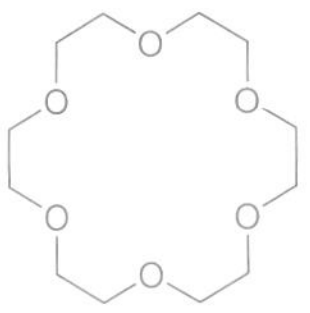
Ac	acetyl	
acac	acetylacetonyl	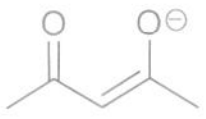
AA	asymmetric aminohydroxylation	NA
AD	asymmetric dihydroxylation	NA
ad	adamantyl	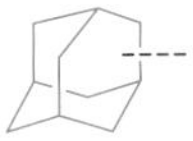
ADDP	1,1'-(azodicarbonyl)dipiperidine	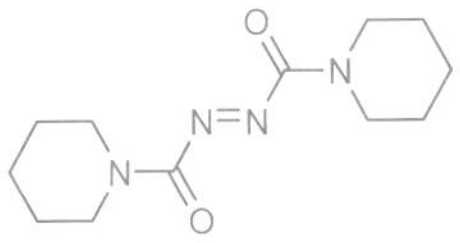
ADMET	acyclic diene metathesis polymerization	NA
acaen	*N,N'-bis*(1-methyl-3-oxobutylidene)ethylenediamine	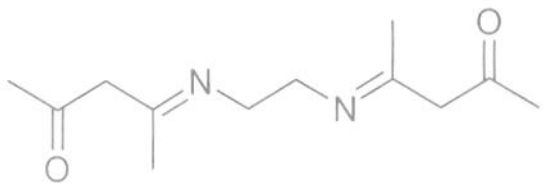
AIBN	2,2'-azo *bis*isobutyronitrile	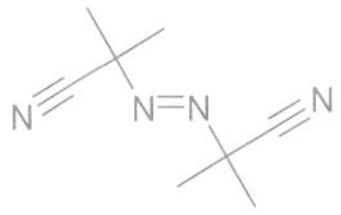
Alloc	allyloxycarbonyl	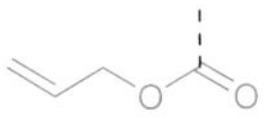
Am	amyl (*n*-pentyl)	
An	*p*-anisyl	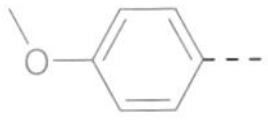
ANRORC	anionic ring-opening ring-closing	NA
aq	aqueous	NA
AQN	anthraquinone	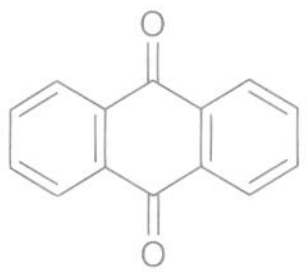
Ar	aryl (substituted aromatic ring)	NA

Abbreviation	Chemical Name	Chemical Structure
ATD	aluminum tris(2,6-*di-tert*-butyl-4-methylphenoxide)	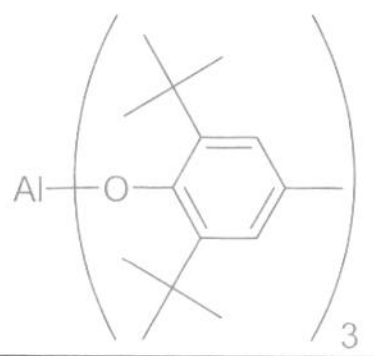
atm	1 atmosphere = 10^5 Pa (pressure)	NA
ATPH	aluminum tris(2,6-diphenylphenoxide)	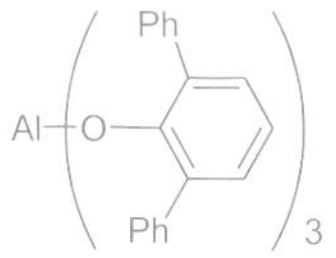
BBN (9-BBN)	9-borabicyclo[3.3.1]nonane (9-BBN)	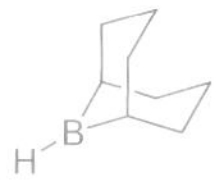
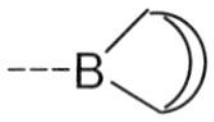	9-borabicyclo[3.3.1]nonyl	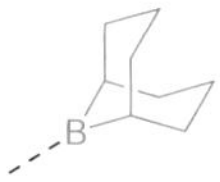
BCME	*bis*(chloromethyl)ether	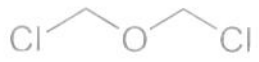
BCN	*N*-benzyloxycarbonyloxy-5-norbornene-2,3-dicarboximide	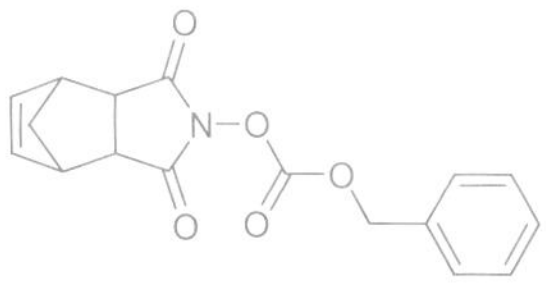
BDPP	(2*R*, 4*R*) or (2*S*, 4*S*) *bis*(diphenylphosphino)pentane	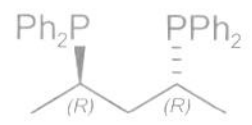
BER	borohydride exchange resin	NA
BHT	2,6-*di*-*t*-butyl-*p*-cresol (butylated hydroxytoluene)	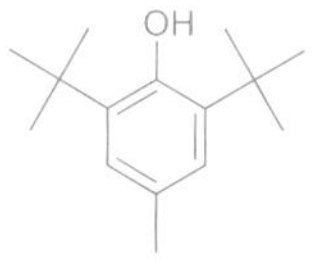
BICP	2(*R*)-2’(*R*)-*bis*(dipenylphosphino)-1(*R*),1’(*R*)-dicyclopentane	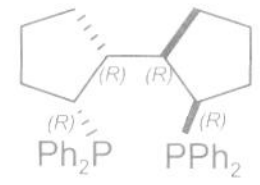
BINAL-H	2,2'-dihydroxy-1,1'-binaphthyl lithium aluminum hydride	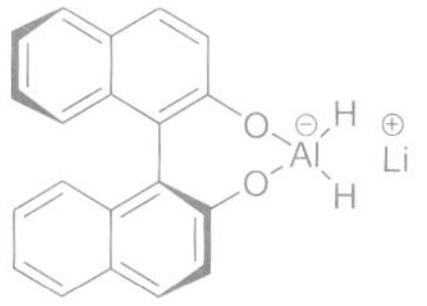
BINAP	2,2'-*bis*(diphenylphosphino)-1,1'-binaphthyl	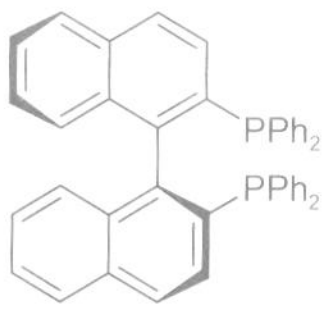

Abbreviation	Chemical Name	Chemical Structure
BINOL	1,1'-bi-2,2'-naphthol	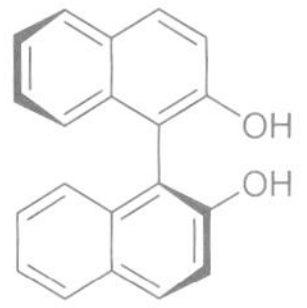
Bip	biphenyl-4-sulfonyl	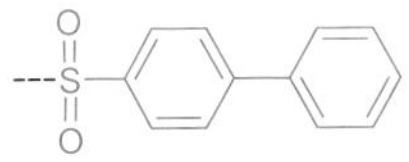
bipy	2,2'-bipyridyl	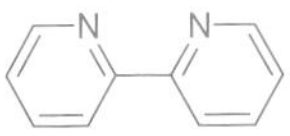
BLA	Brönsted acid assisted chiral Lewis acid	NA
bmin	1-butyl-3-methylimidazolium cation	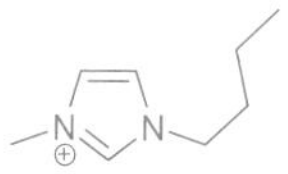
BMS	Borane-dimethyl sulfide complex	$H_3B \bullet SMe_2$
Bn	benzyl	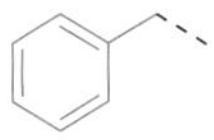
BNAH	1-benzyl-1,4-dihydronicotinamide	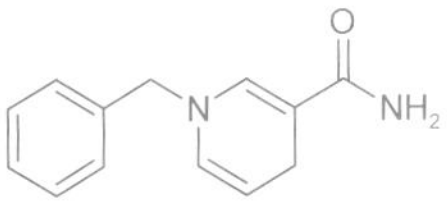
BOB	4-benzyloxybutyryl	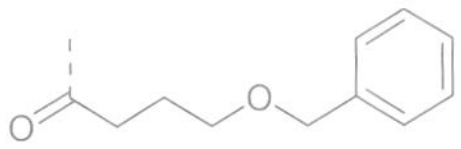
Boc	*t*-butoxycarbonyl	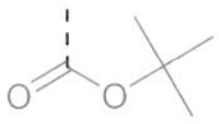
BOM	benzyloxymethyl	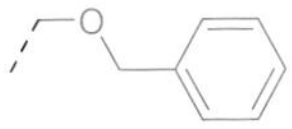
BOP-Cl	*bis*(2-oxo-3-oxazolidinyl)phosphinic chloride	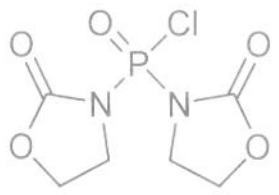
bp	boiling point	NA
BPD	*bis*(pinacolato)diboron	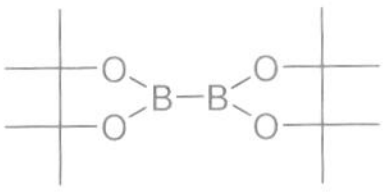
BPO	benzoyl peroxide	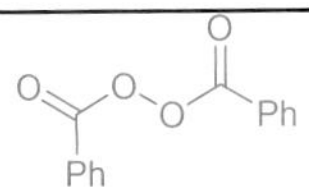
BPS (TBDPS)	*t*-butyldiphenylsilyl	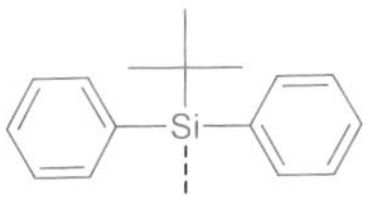

Abbreviation	Chemical Name	Chemical Structure
BQ	benzoquinone	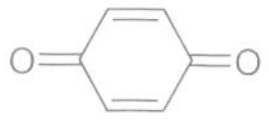
Bs	brosyl = (4-bromobenzenesulfonyl)	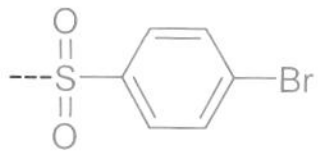
BSA	*N,O-bis*(trimethylsilyl)acetamide	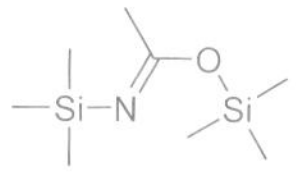
BSA	Bovine serum albumin	NA
Bt	1- or 2-benzotriazolyl	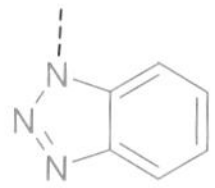
BTAF	benzyltrimethylammonium fluoride	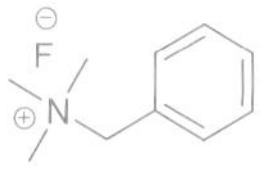
BTEA	benzyltriethylammonium	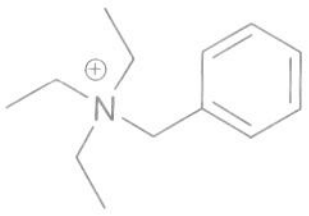
BTEAC	benzyltriethylammonium chloride	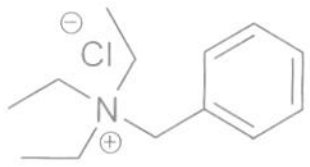
BTFP	3-bromo-1,1,1-trifluoro-propan-2-one	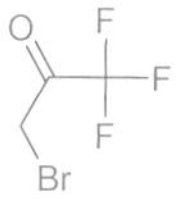
BTMA	benzyltrimethylammonium	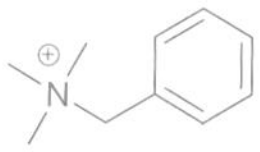
BTMSA	*bis*(trimethylsilyl) acetylene	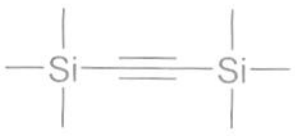
BTS	*bis*(trimethylsilyl) sulfate	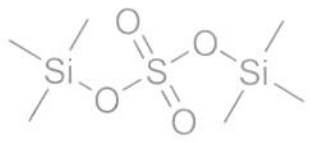
BTSA	benzothiazole 2-sulfonic acid	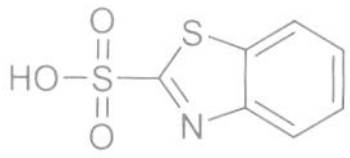
BTSP	*bis*(trimethylsilyl) peroxide	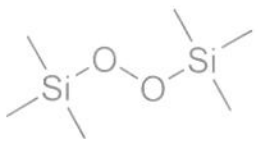
Bz	benzoyl	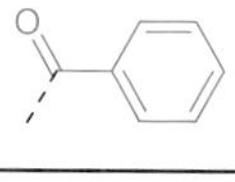
Bu (^{n}Bu)	*n*-butyl	
c	cyclo	NA

Abbreviation	Chemical Name	Chemical Structure
ca	circa (approximately)	NA
CA	chloroacetyl	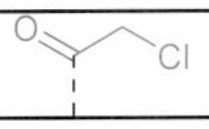
CAN	cerium(IV) ammonium nitrate (cericammonium nitrate)	$Ce(NH_4)_2(NO_3)_6$
cat.	catalytic	NA
CB	catecholborane	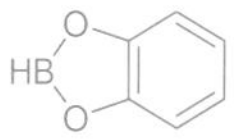
CBS	Corey-Bakshi-Shibata reagent	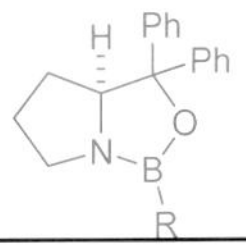 R = H, alkyl
Cbz (Z)	benzyloxycarbonyl	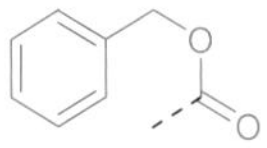
cc. or conc.	concentrated	NA
CCE	constant current electrolysis	NA
CDI	carbonyl diimidazole	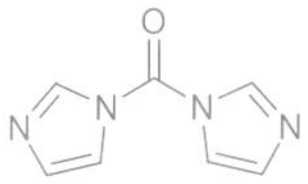
CHD	1,3 or 1,4-cyclohexadiene	 1,3-CHD 1,4-CHD
CHIRAPHOS	2,3-*bis*(diphenylphosphino)butane	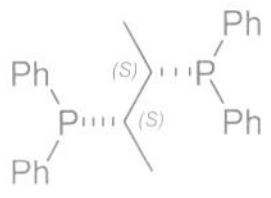
Chx (Cy)	cyclohexyl	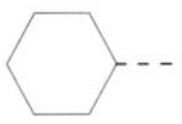
CIP	2-chloro-1,3-dimethylimidazolidinium hexafluorophosphate	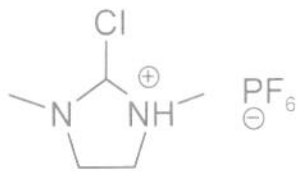
CM (XMET)	cross metathesis	NA
CMMP	cyanomethylenetrimethyl phosphorane	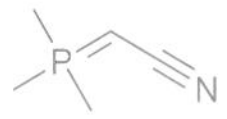
COD	1,5-cyclooctadiene	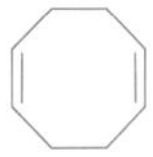
COT	1,3,5-cyclooctatriene	
Cp	cyclopentadienyl	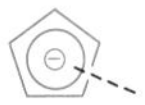
CPTS	collidinium-*p*-toluenesulfonate	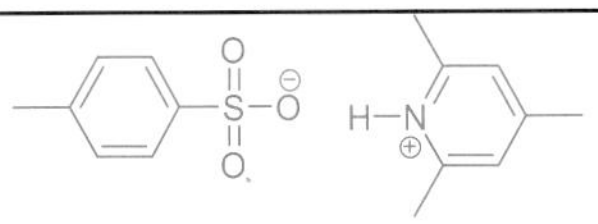

Abbreviation	Chemical Name	Chemical Structure
CRA	complex reducing agent	NA
Cr-PILC	chromium-pillared clay catalyst	NA
CSA	camphorsufonic acid	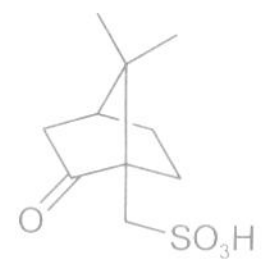
CSI	chlorosulfonyl isocyanate	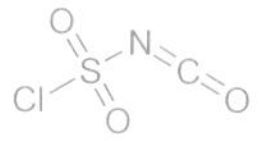
CTAB	cetyl trimethylammonium bromide	
CTACl	cetyl trimethylammonium chloride	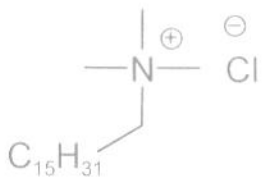
CTAP	cetyl trimethylammonium permanganate	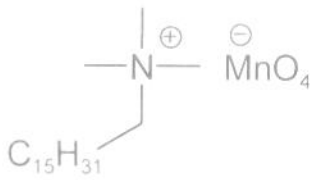
Δ	heat	NA
d	days (length of reaction time)	NA
DABCO	1,4-diazabicyclo[2.2.2]octane	
DAST	diethylaminosulfur trifluoride	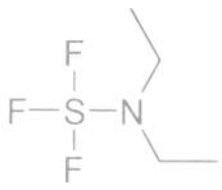
DATMP	diethylaluminum 2,2,6,6-tetramethylpiperidide	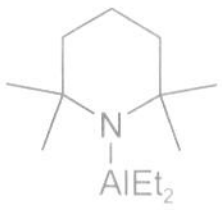
DBA (dba)	dibenzylideneacetone	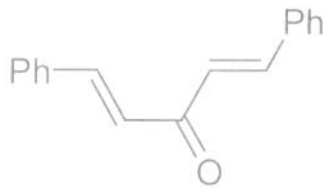
DBAD	*di-tert*-butylazodicarboxylate	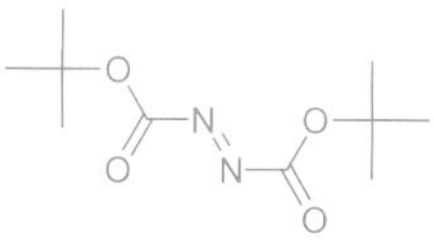
DBI	dibromoisocyanuric acid	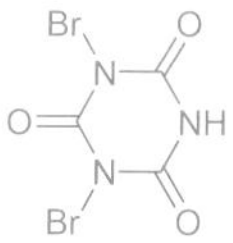

Abbreviation	Chemical Name	Chemical Structure
DBM	dibenzoylmethane	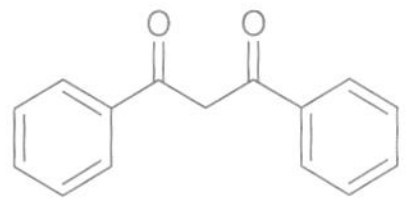
DBN	1,5-diazabicyclo[4.3.0]non-5-ene	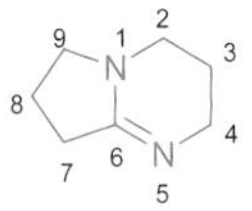
DBS	dibenzosuberyl	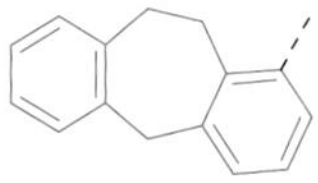
DBU	1,8-diazabicyclo[5.4.0]undec-7-ene	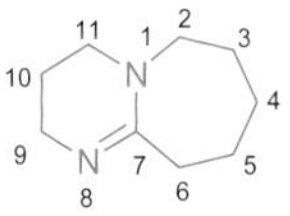
DCA	9,10-dicyanoanthracene	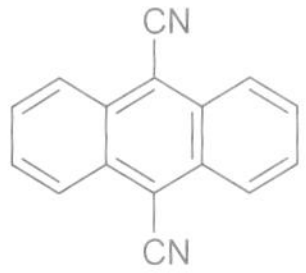
DCB	1,2-dichlorobenzene	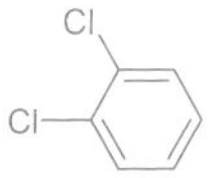
DCC	dicyclohexylcarbodiimide	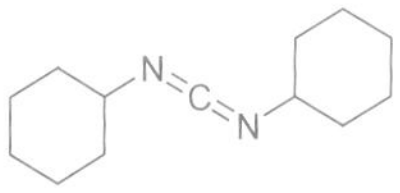
DCE	1,1-dichloroethane	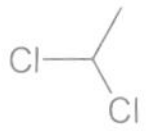
DCM	dichloromethane	CH_2Cl_2
DCN	1,4-dicyanonaphthalene	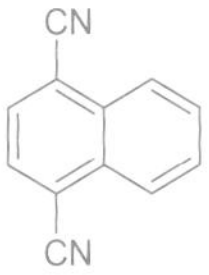
Dcpm	dicyclopropylmethyl	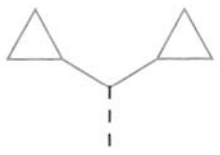
DCU	*N,N'*-dicyclohexylurea	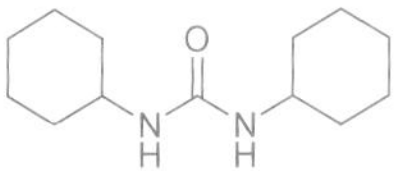
DDQ	2,3-dichloro-5,6-dicyano-1,4-benzoquinone	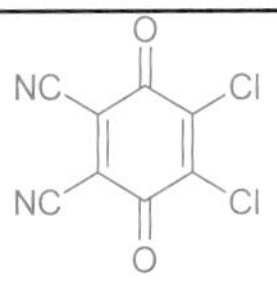
de	diastereomeric excess	NA

Abbreviation	Chemical Name	Chemical Structure
DEAD	diethyl azodicarboxylate	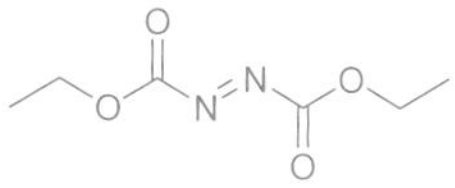
DEIPS	diethylisopropylsilyl	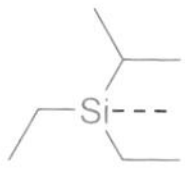
DEPBT	3-(diethoxyphosphoryloxy)-1,2,3-benzotriazin-4(3*H*)-one	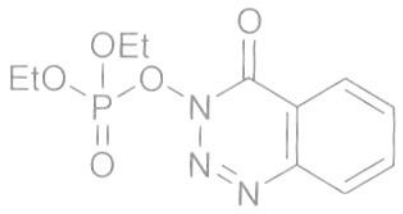
DET	diethyl tartrate	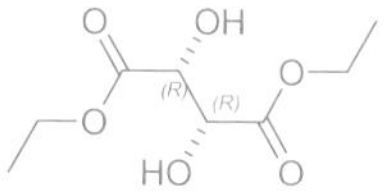
DHP	3,4-dihydro-2*H*-pyran	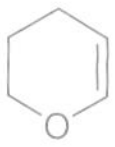
DHQ	dihydroquinine	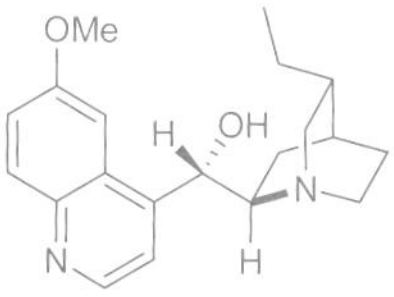
$(DHQ)_2$PHAL	*bis*(dihydroquinino)phthalazine	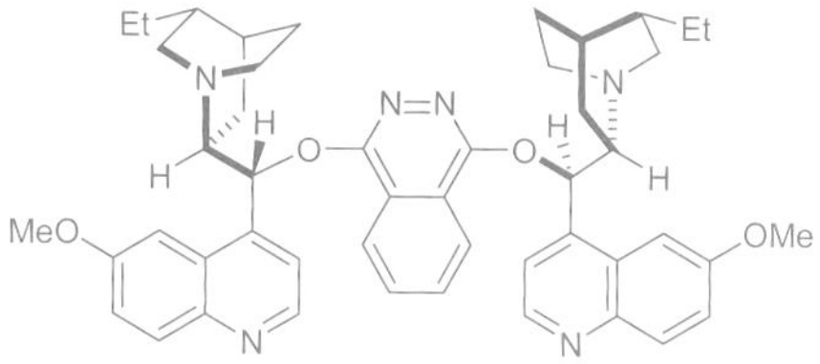
DHQD	dihydroquinidine	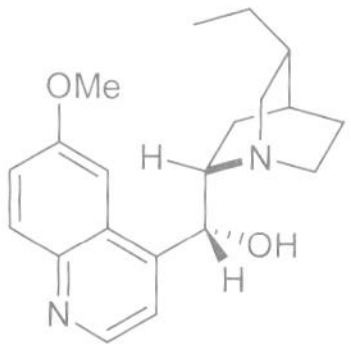
$(DHQD)_2$PHAL	*bis*(dihydroquinidino)phthalazine	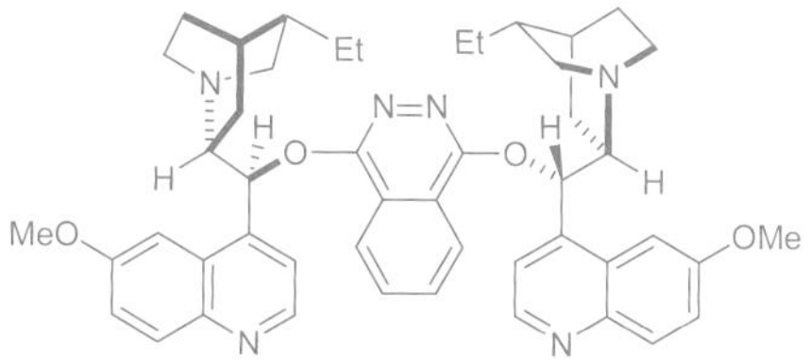
DIAD	diisopropyl azodicarboxylate	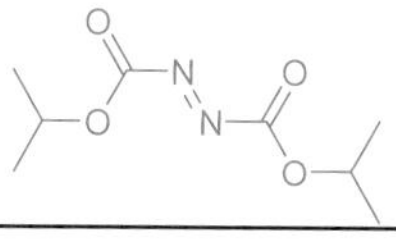
DIB (BAIB or PIDA)	(diacetoxyiodo)benzene	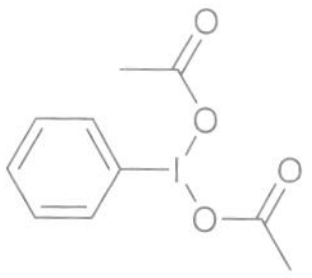

Abbreviation	Chemical Name	Chemical Structure
DIBAL (DIBAH) DIBAL-H	diisobutylaluminum hydride	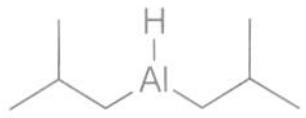
DIC	diisopropyl carbodiimide	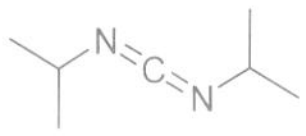
diop	4,5-*bis*-[(diphenylphosphanyl)methyl]-2,2-dimethyl-[1,3]dioxolane	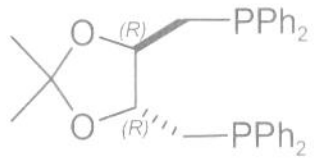
DIPAMP	1,2-*bis*(*o*-anisylphenylphosphino)ethane	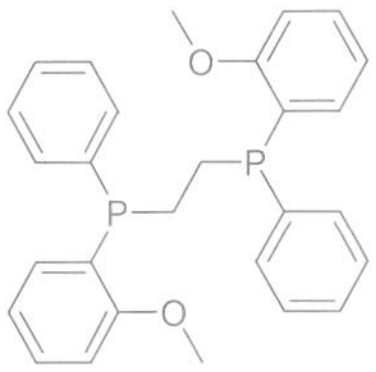
DIPEA (Hünig's base)	diisopropylethylamine	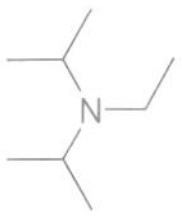
DIPT	diisopropyl tartrate	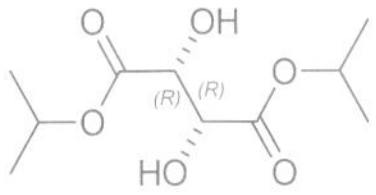
DLP	dilauroyl peroxide	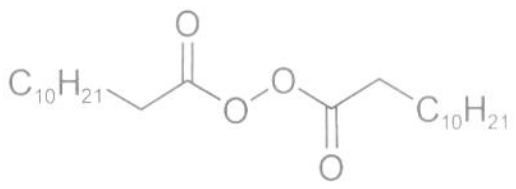
DMA (DMAC)	*N,N*-dimethylacetamide	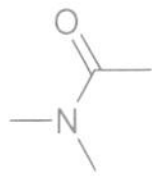
DMAD	dimethyl acetylene dicarboxylate	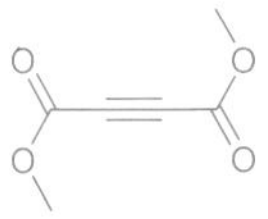
DMAP	*N,N*-4-dimethylaminopyridine	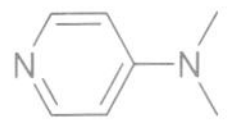
DMB	*m*-dimethoxybenzene	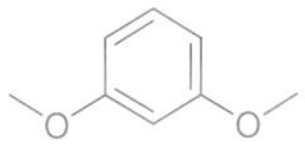
DMDO	dimethyl dioxirane	
DME	1,2-dimethoxyethane	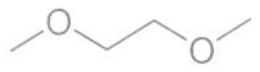

Abbreviation	Chemical Name	Chemical Structure
DMF	*N,N*-dimethylformamide	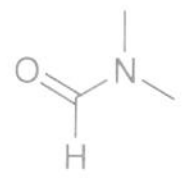
DMI	1,3-dimethylimidazolidin-2-one	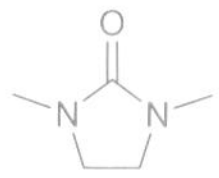
DMP	Dess-Martin periodinane	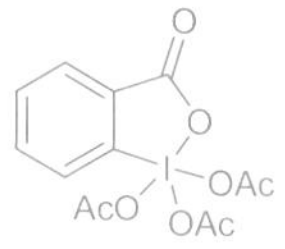
DMPS	dimethylphenylsilyl	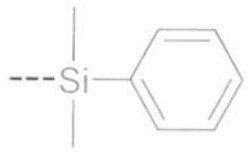
DMPU	1,3-dimethyl-3,4,5,6-tetrahydro-2(1*H*)-pyrimidone (*N,N*-dimethyl propylene urea)	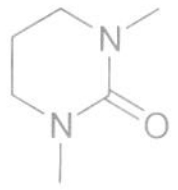
DMTSF	dimethyl(methylthio)sulfonium tetrafluoroborate	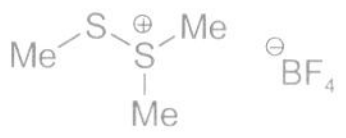
DMS	dimethylsulfide	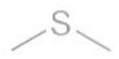
DMSO	dimethylsulfoxide	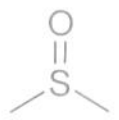
DMT	4,4'-dimethoxytrityl	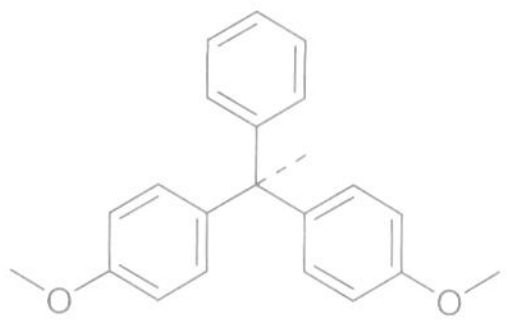
DMTMM	4-(4,6-dimethoxy[1,3,5]triazin-2-yl)-4-methylmorpholinium chloride	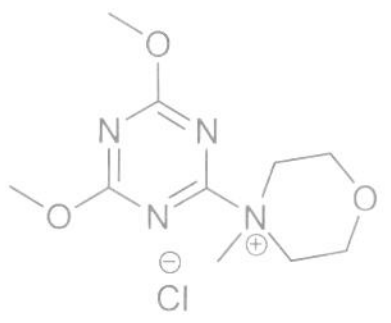
DMTr	4,4'-dimethyltrityl	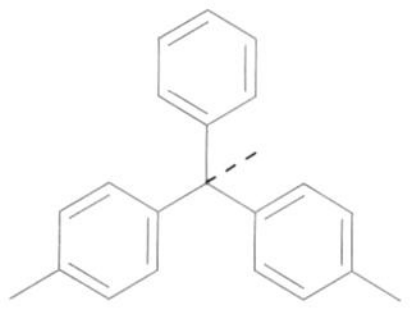
DMTST	(dimethylthio)methylsulfonium trifluoromethanesulfonate	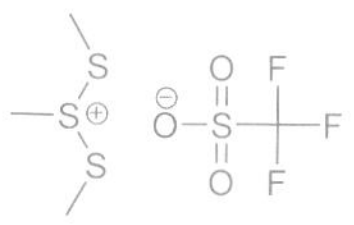
DNA	deoxyribonucleic acid	NA

Abbreviation	Chemical Name	Chemical Structure
DPA (DIPA)	diisopropylamine	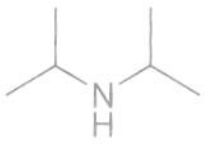
DPBP	2,2'-*bis*(diphenylphosphino)biphenyl	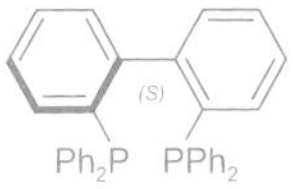
DPDC	diisopropyl peroxydicarbonate	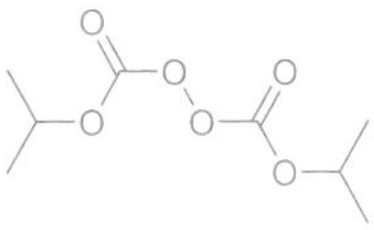
DPDM	diphenyl diazomethane	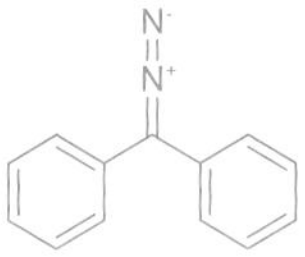
DPEDA	1,2-diamino-1,2-diphenylethane	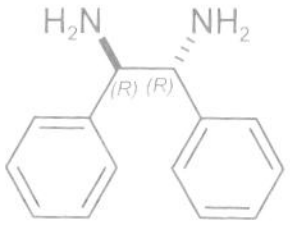
DPIBF	diphenylisobenzofuran	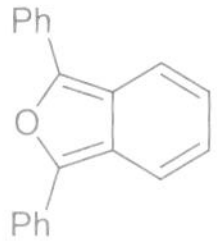
DPPA	diphenylphosphoryl azide (diphenylphosphorazidate)	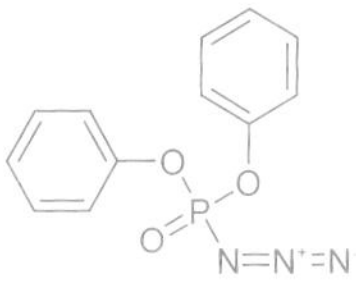
Dppb (ddpb)	1,4-*bis*(diphenylphosphino)butane	
dppe	1,2-*bis*(diphenylphosphino)ethane	
dppf	1,1'-*bis*(diphenylphosphino)ferrocene	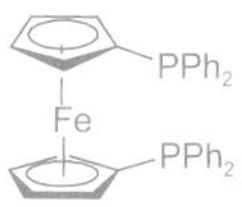
dppm	*bis*(diphenylphosphino)methane	
dppp	1,3-*bis*(diphenylphosphino)propane	
DPS (also TBDPS or BPS)	*t*-butyldiphenylsilyl	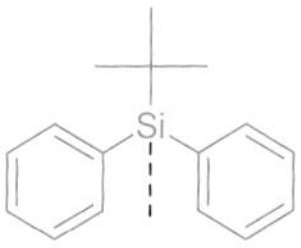

Abbreviation	Chemical Name	Chemical Structure
DPTC	*O,O'*-di(2'-pyridyl)thiocarbonate	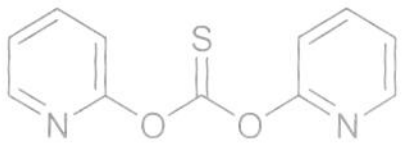
dr	diastereomeric ratio	NA
DTBAD (DBAD)	*di-tert*-butyl azodicarboxylate	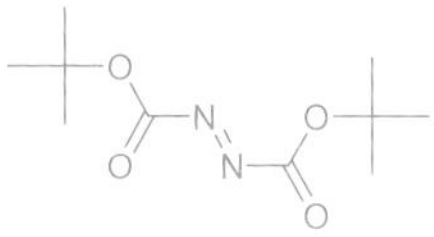
DTBB	4,4'-*di-tert*-butylbiphenyl	
DTBP	2,6-*di-tert*-butylpyridine	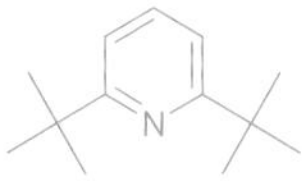
DTBMP	2,6-*di-tert*-butyl-4-methylpyridine	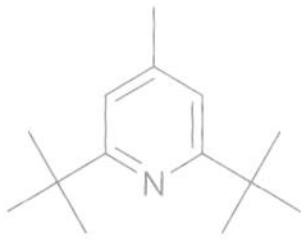
DTE	1,4-dithioerythritol	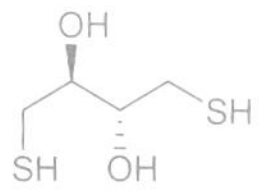
DVS	1,3-divinyl-1,1,3,3-tetramethyldisiloxane	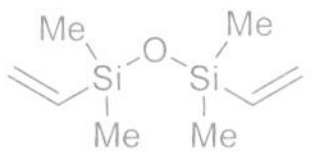
E^+	electrophile (denotes any electrophile in general)	NA
E2	bimolecular elimination	NA
ED	effective dosage	NA
EDA	ethyl diazoacetate	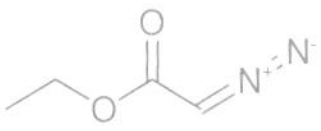
EDDA	ethylenediamine diacetate	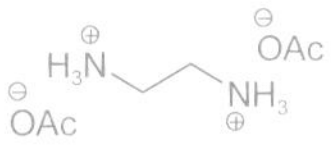
EDC (EDAC)	1-ethyl-3-(3-dimethylaminopropyl)carbodiimide (ethyldimethylaminopropylcarbodiimide)	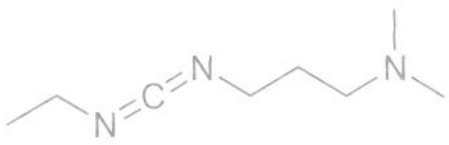
EDCI	1-ethyl-3-(3-dimethylaminopropyl)carbodiimide hydrochloride	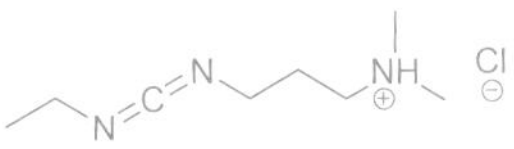
EDCP	2,3-*bis*-phosphonopentanedioic acid (ethylene dicarboxylic 2,3-diphosphonic acid)	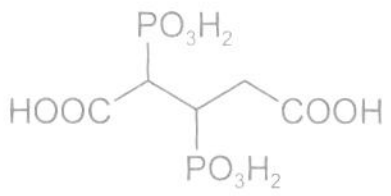
EDG	electron-donating group	NA
EDTA	ethylenediamine tetraacetic acid	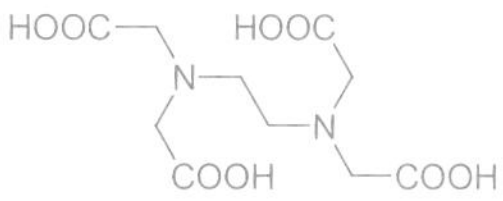

Abbreviation	Chemical Name	Chemical Structure
ee	enantiomeric excess	NA
EE	ethoxyethyl	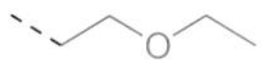
E$_i$	intramolecular *syn* elimination	NA
en	ethylenediamine	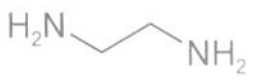
EOM	ethoxymethyl	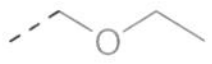
ESR	electron spin resonance (spectroscopy)	NA
Et	ethyl	
ETSA	ethyl trimethylsilylacetate	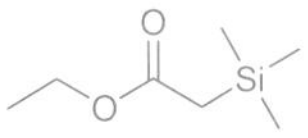
EVE	ethyl vinyl ether	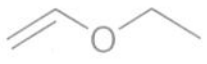
EWG	electron-withdrawing group	NA
Fc	ferrocenyl	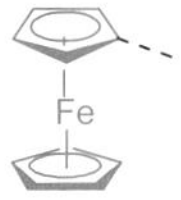
FDP	fructose-1,6-diphosphate	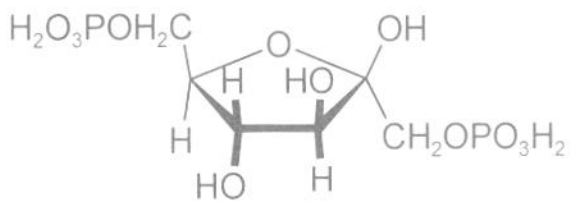
FDPP	pentafluorophenyl diphenylphosphinate	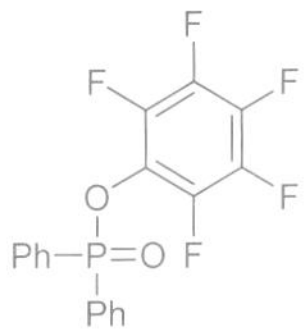
Fl	fluorenyl	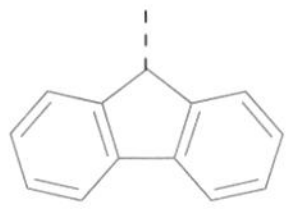
FMO	frontier molecular orbital (theory)	NA
Fmoc	9-fluorenylmethoxycarbonyl	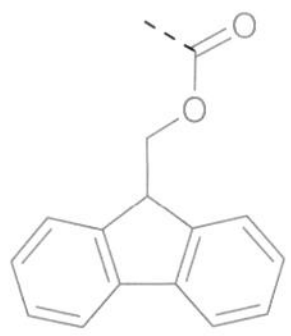
fod	6,6,7,7,8,8,8-heptafluoro-2,2-dimethyl-3,5-octanedione	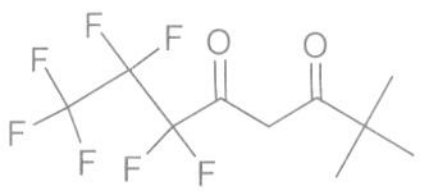
fp	flash point	NA
FSM	Mesoporous silica	NA
FTT	1-fluoro-2,4,6-trimethylpyridinium triflate	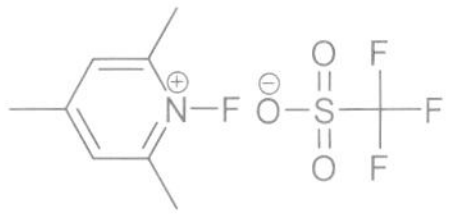

Abbreviation	Chemical Name	Chemical Structure
FVP	flash vacuum pyrolysis	NA
GEBC	gel entrapped base catalyst	NA
h	hours (length of reaction time)	NA
hν	irradiation with light	NA
HATU	*O*-(7-azabenzotriazol-1-yl)-*N*,*N*,*N'*,*N'*-tetramethyluronium hexafluorophosphate	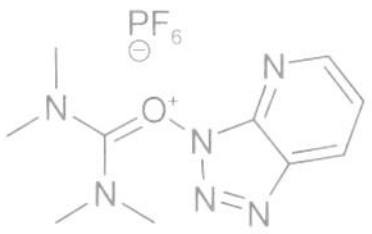
Het	heterocycle	NA
hfacac	hexafluoroacetylacetone	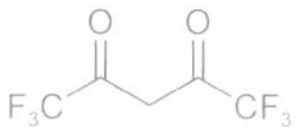
HFIP	1,1,1,3,3,3-hexafluoro-2-propanol (hexafluoroisopropanol)	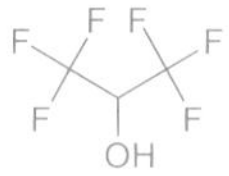
HGK	4-hydroxy-2-ketoglutarate	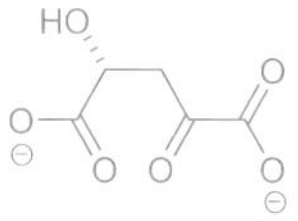
Hgmm	millimeter of mercury (760 Hgmm = 1 atm = 760 Torr)	NA
HLE	horse liver esterase	NA
Hmb	2-hydroxy-4-methoxybenzyl	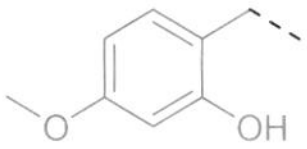
HMDS	1,1,1,3,3,3-hexamethyldisilazane	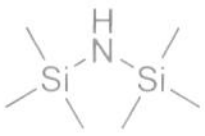
HMPA	hexamethylphosphoric acid triamide (hexamethylphosphoramide)	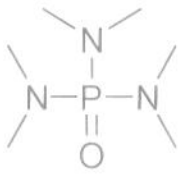
HMPT	hexamethylphosphorous triamide	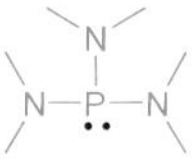
HOAt	1-hydroxy-7-azabenzotriazole	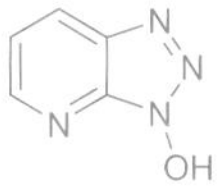
HOBt (HOBT)	1-hydroxybenzotriazole	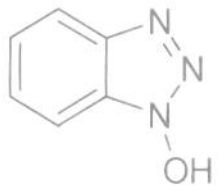
HOMO	highest occupied molecular orbital	NA
HOSu	*N*-hydroxysuccinimide	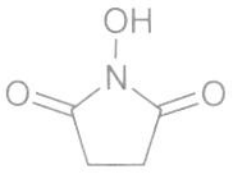
HPLC	high-pressure liquid chromatography	NA
HWE	Horner-Wadsworth-Emmons	NA
i	iso	NA

Abbreviation	Chemical Name	Chemical Structure
IBA	2-iodosobenzoic acid	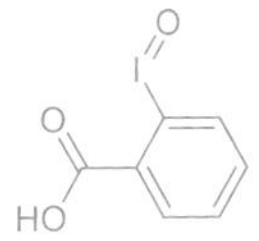
IBX	*o*-iodoxybenzoic acid	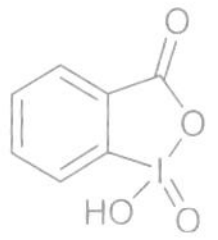
IDCP	*bis*(2,4,6-collidine)iodonium perchlorate	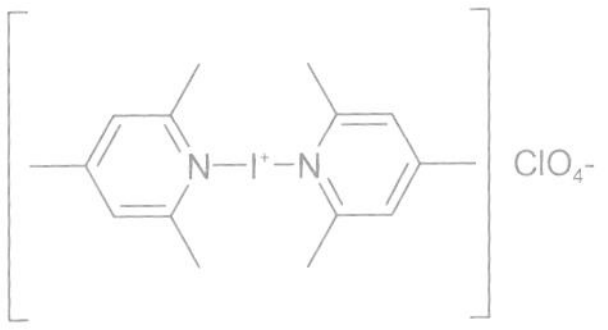
Imid (Im)	imidazole	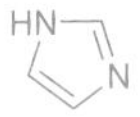
INOC	intramolecular nitrile oxide cycloaddition	NA
IPA	isopropyl alcohol	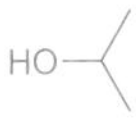
Ipc	isopinocamphenyl	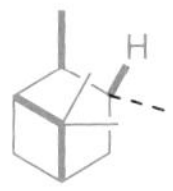
IR	infrared spectroscopy	NA
K-10	a type of Montmorillonite clay	NA
KDA	potassium diisopropylamide	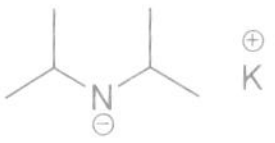
KHMDS	potassium *bis*(trimethylsilyl)amide	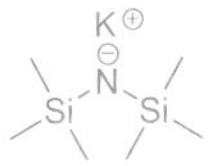
KSF	a type of Montmorillonite clay	NA
L	ligand	NA
L.R.	Lawesson's reagent (2,4-*bis*-(4-methoxyphenyl)-[1,3,2,4]dithiadiphosphetane 2,4-dithion)	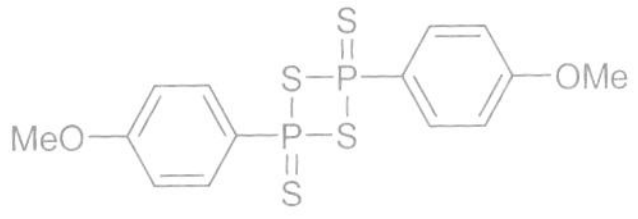
LA	Lewis acid	NA
LAB	lithium amidotrihydroborate	LiH_2NBH_3
LAH	lithium aluminum hydride	$LiAlH_4$
LD_{50}	dose that is lethal to 50% of the test subjects (cells, animals, humans etc.)	NA
LDA	lithium diisopropylamide	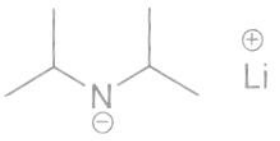
LDBB	lithium 4,4'-*t*-butylbiphenylide	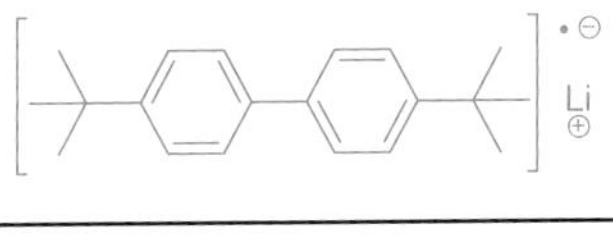

Abbreviation	Chemical Name	Chemical Structure
LDE	lithium diethylamide	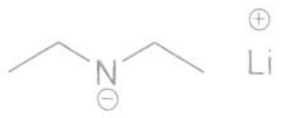
LDPE	lithium perchlorate-diethyl etherate	$LiClO_4$ - Et_2O
LHMDS (LiHMDS)	lithium *bis*(trimethylsilyl)amide	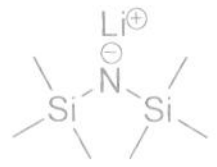
LICA	lithium isopropylcyclohexylamide	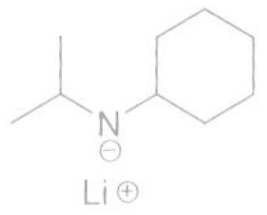
LICKOR (super base)	butyllithium-potassium *tert*-butoxide	BuLi - KO*t*-Bu
liq.	liquid	NA
LiTMP (LTMP)	lithium 2,2,6,6-tetramethylpiperidide	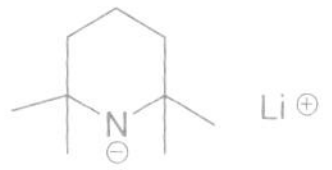
LPT	lithium pyrrolidotrihydroborate (lithium pyrrolidide-borane)	$Li(CH_2)_4NBH_3$
L-selectride	lithium tri-*sec*-butylborohydride	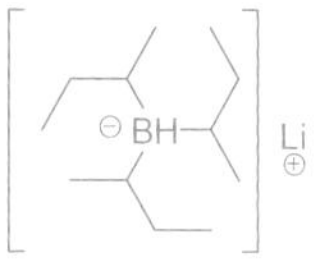
LTA	lead tetraacetate	$Pb(OAc)_4$
LUMO	lowest unoccupied molecular orbital	NA
lut	2,6-lutidine	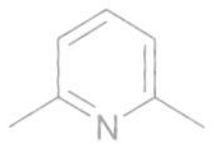
m	meta	NA
MA	maleic anhydride	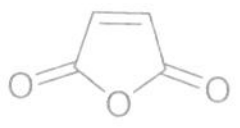
MAD	methyl aluminum *bis*(2,6-di-*t*-butyl-4-methylphenoxide)	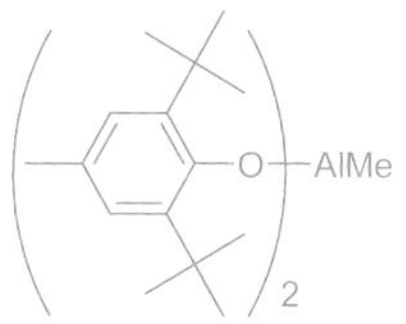
MAT	methyl aluminum *bis*(2,4,6-tri-*t*-butylphenoxide)	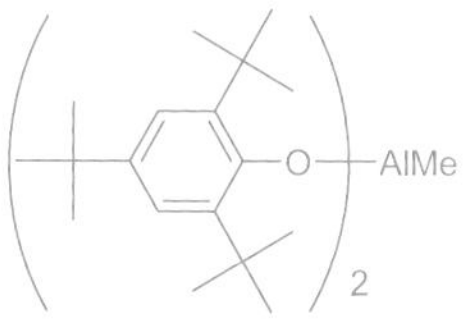

Abbreviation	Chemical Name	Chemical Structure
MBT	2-mercaptobenzothiazole	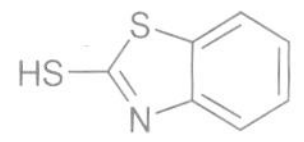
***m*-CPBA**	*meta* chloroperbenzoic acid	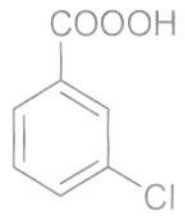
Me	methyl	$- -CH_3$
MEM	(2-methoxyethoxy)methyl	
MEPY	methyl 2-pyrrolidone-5(*S*)-carboxylate	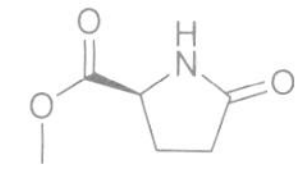
Mes	mesityl	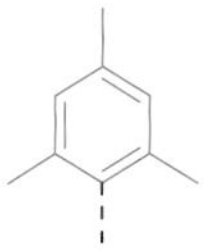
mesal	*N*-methylsalicylaldimine	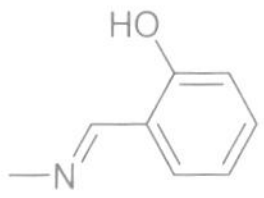
MIC	methyl isocyanate	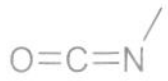
MMPP (MMPT)	magnesium monoperoxyphthalate	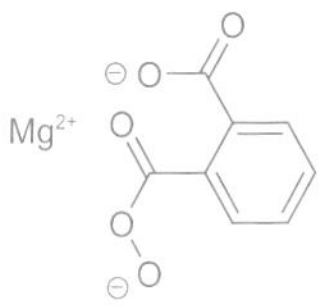
MOM	methoxymethyl	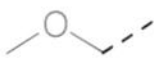
MoOPH	oxodiperoxomolybdenum(pyridine)-(hexamethylphosphoric triamide)	NA
mp	melting point	NA
MPa	megapascal = 10^6 Pa = 10 atm (pressure)	
MPD (NMP)	N-methyl-2-pyrrolidinone	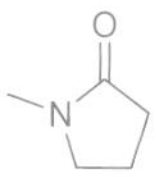
MPM	methoxy(phenylthio)methyl	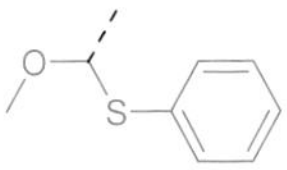
MPM (PMB)	*p*-methoxybenzyl	

Abbreviation	Chemical Name	Chemical Structure
MPPC	*N*-methyl piperidinium chlorochromate	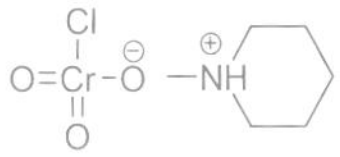
Ms	mesyl (methanesulfonyl)	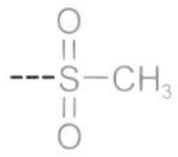
MS	mass spectrometry	NA
MS	molecular sieves	NA
MSA	methanesulfonic acid	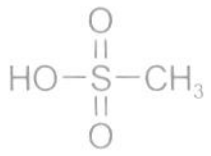
MSH	*o*-mesitylenesulfonyl hydroxylamine	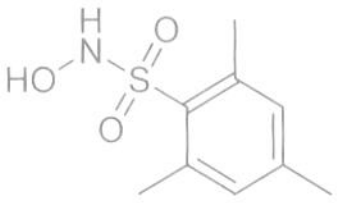
MSTFA	*N*-methyl-*N*-(trimethylsilyl) trifluoroacetamide	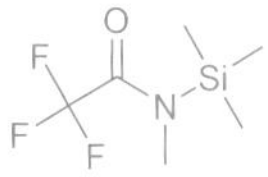
MTAD	*N*-methyltriazolinedione	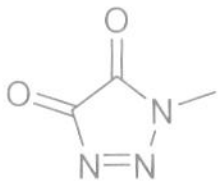
MTEE (MTBE)	methyl *t*-butyl ether	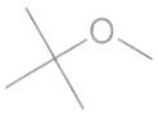
MTM	methylthiomethyl	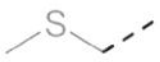
MTO	methyltrioxorhenium	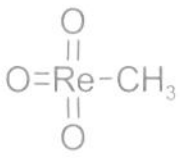
Mtr	(4-methoxy-2,3,6-trimethylphenyl)sulfonyl	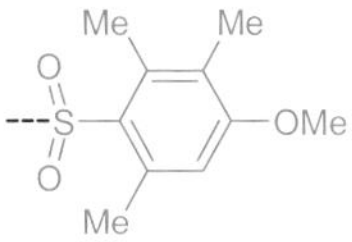
MVK	methyl vinyl ketone	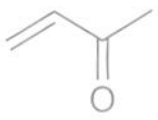
mw	microwave	NA
n	normal (e.g. unbranched alkyl chain)	NA
NADPH	nicotinamide adenine dinucleotide phosphate	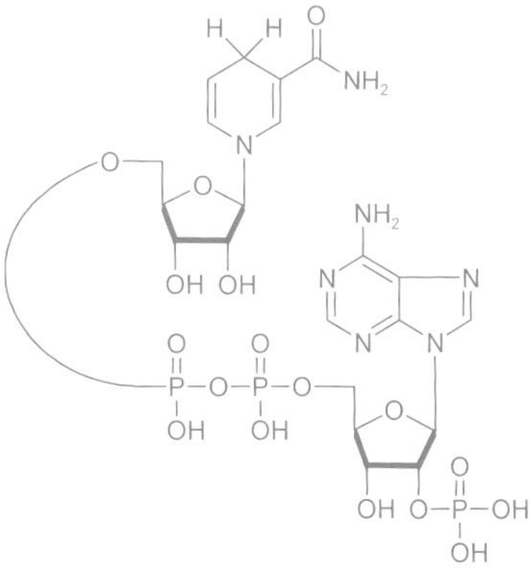

Abbreviation	Chemical Name	Chemical Structure
NaHMDS	sodium *bis*(trimethylsilyl)amide	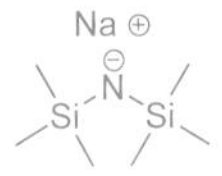
Naph (Np)	naphthyl	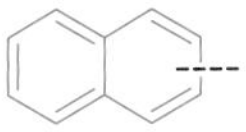
NBA	*N*-bromoacetamide	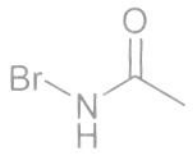
NBD (nbd)	norbornadiene	
NBS	*N*-bromosuccinimide	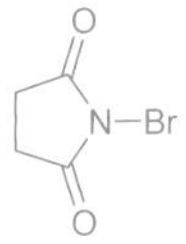
NCS	*N*-chlorosuccinimide	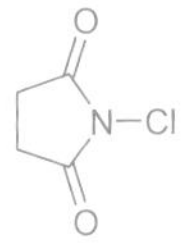
N_f	nonafluorobutanesulfonyl	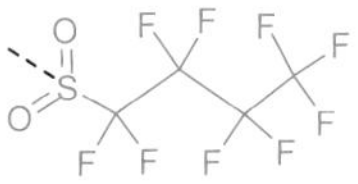
NHPI	*N*-hydroxyphthalimide	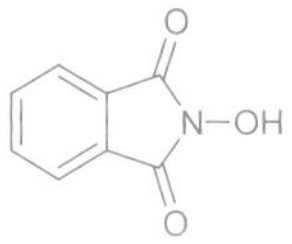
NIS	*N*-iodosuccinimide	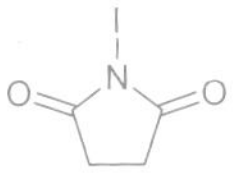
NMM	*N*-methylmorpholine	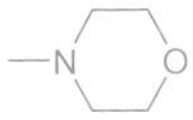
NMO	*N*-methylmorpholine oxide	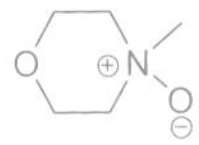
NMP	*N*-methyl-2-pyrrolidinone	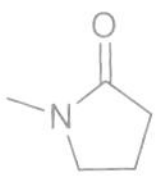
NMR	nuclear magnetic resonance	NA
NORPHOS	*bis*(diphenylphosphino)bicyclo[2.2.1]-hept-5-ene	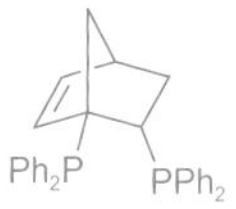
Nos	4-nitrobenzenesulfonyl	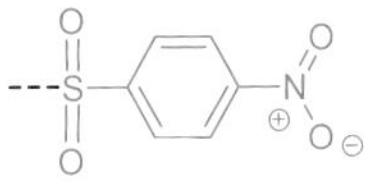

Abbreviation	Chemical Name	Chemical Structure
NPM	*N*-phenylmaleimide	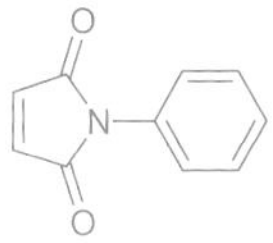
NR	no reaction	NA
Ns	2-nitrobenzenesulfonyl	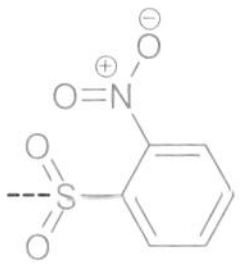
NSAID	non steroidal anti-inflammatory drug	NA
Nuc	nucleophile (general)	NA
o	ortho	NA
Oxone	potassium peroxymonosulfate	$KHSO_5$
p	para	NA
PAP	2,8,9-trialkyl-2,5,8,9-tetraaza-1-phospha-bicyclo[3.3.3]undecane	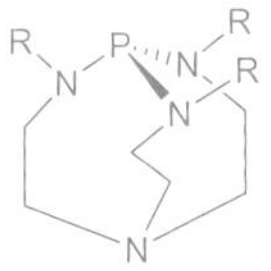
PBP	pyridinium bromide perbromide	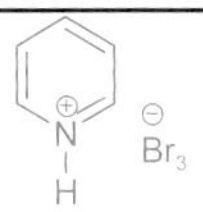
PCC	pyridinium chlorochromate	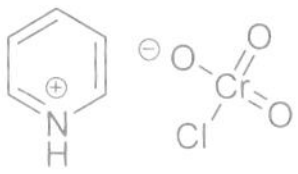
PDC	pyridinium dichromate	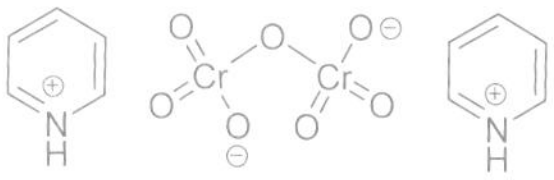
PEG	polyethylene glycol	NA
Pf	9-phenylfluorenyl	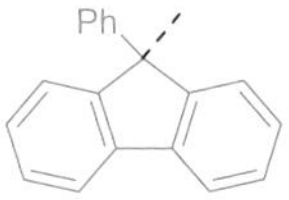
pfb	perfluorobutyrate	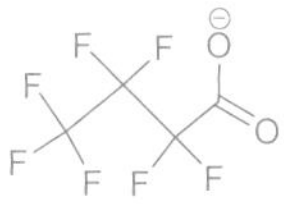
Ph	phenyl	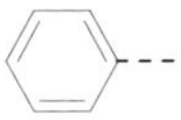
PHAL	phthalazine	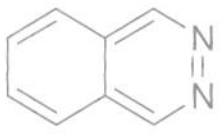
phen	9,10-phenanthroline	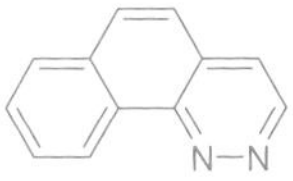

Abbreviation	Chemical Name	Chemical Structure
Phth	phthaloyl	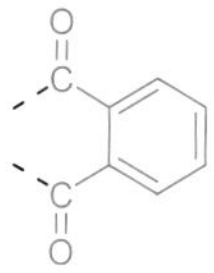
pic	2-pyridinecarboxylate	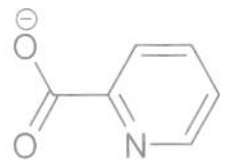
PIDA (BAIB or DIB)	phenyliodonium diacetate	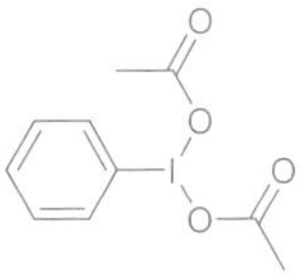
PIFA	phenyliodonium *bis*(trifluoroacetate)	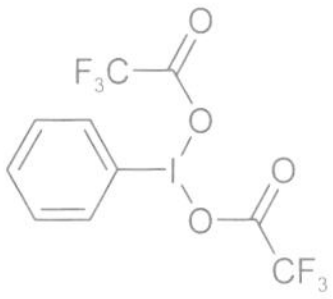
Piv	pivaloyl	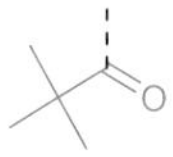
PLE	pig liver esterase	NA
PMB (MPM)	*p*-methoxybenzyl	
PMP	4-methoxyphenyl	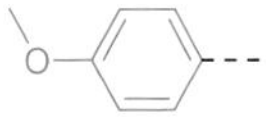
PMP	1,2,2,6,6-pentamethylpiperidine	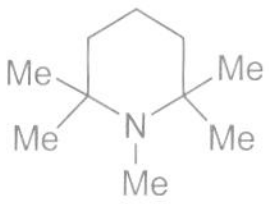
PNB	*p*-nitrobenzyl	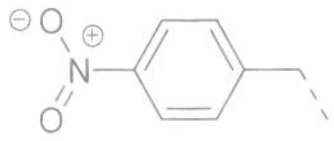
PNZ	*p*-nitrobenzyloxycarbonyl	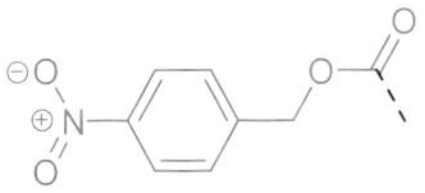
PPA	polyphosphoric acid	NA
PPI	2-phenyl-2-(2-pyridyl)-2*H*-imidazole	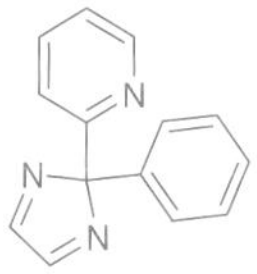
PPL	pig pancreatic lipase	NA
PPO	4-(3-phenylpropyl)pyridine-N-oxide	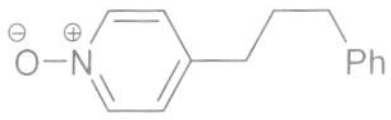
PPSE	polyphosphoric acid trimethylsilyl ester	NA

Abbreviation	Chemical Name	Chemical Structure
PPTS	pyridinium *p*-toluenesulfonate	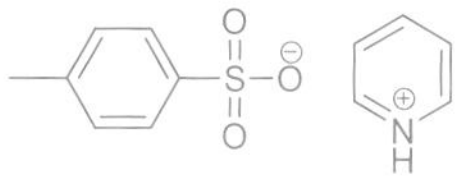
Pr	propyl	
psi	pounds per square inch	NA
PT	1-phenyl-1*H*-tetrazol-yl	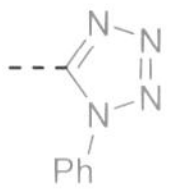
P.T.	proton transfer	NA
PTAB	phenyltrimethylammonium perbromide	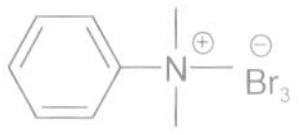
PTC	Phase transfer catalyst	NA
PTMSE	(2-phenyl-2-trimethylsilyl)ethyl	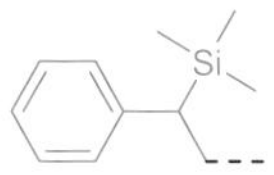
PTSA (or TsOH)	*p*-toluenesulfonic acid	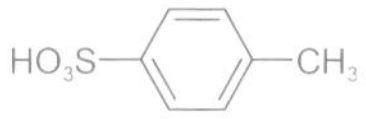
PVP	poly(4-vinylpyridine)	NA
Py (pyr)	pyridine	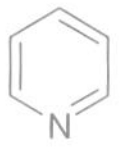
r.t.	room temperature	NA
rac	racemic	NA
RAMP	(*R*)-1-amino-2-(methoxymethyl)pyrrolidine	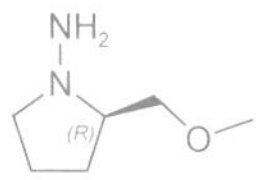
RaNi	Raney nickel	NA
RB	Rose Bengal	See Rose bengal
RCAM	ring-closing alkyne metathesis	NA
RCM	ring-closing metathesis	NA
Rds (or RDS)	rate-determining step	NA
Red-Al	sodium *bis*(2-methoxyethoxy) aluminum hydride	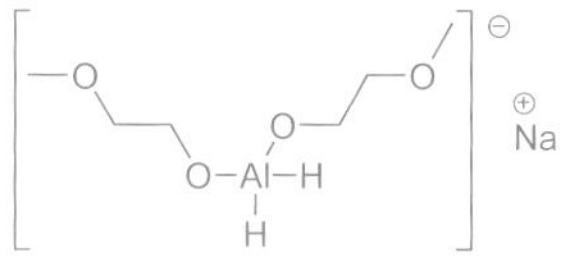
Rham	rhamnosyl	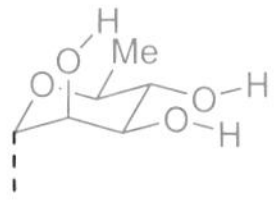
R_f	perfluoroalkyl group	C_nF_{2n+1}
R_f	retention factor in chromatography	NA
ROM	ring-opening metathesis	NA
ROMP	ring-opening metathesis polymerization	NA

Abbreviation	Chemical Name	Chemical Structure
Rose Bengal (RB)	2,4,5,7-tetraiodo-3',4',5',6'-tetrachlorofluorescein disodium salt (a photosensitizer)	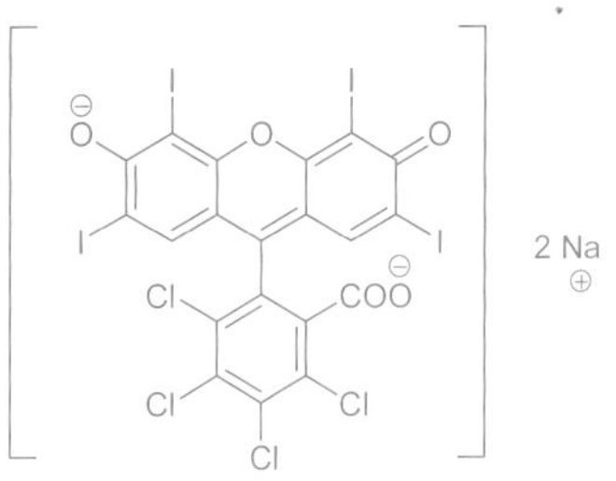
s	seconds (length of reaction time)	NA
S,S,-chiraphos	(*S,S*)-2,3-*bis*(diphenylphosphino)butane	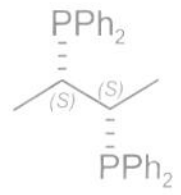
Salen	*N,N'*-ethylene*bis*(salicylideneiminato) *bis*(salicylidene)ethylenediamine	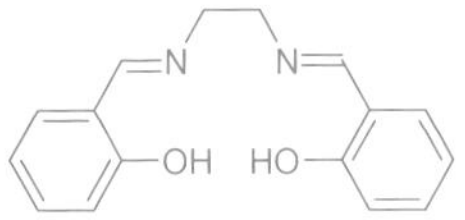
salophen	*o*-phenylene*bis*(salicylideneiminato)	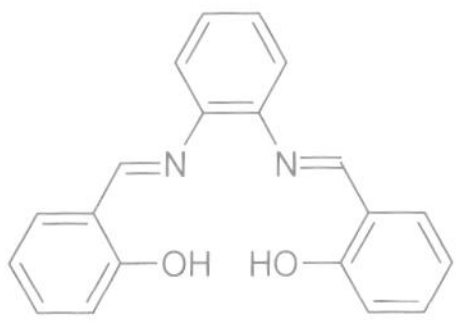
SAMP	(*S*)-1-amino-2-(methoxymethyl)pyrrolidine	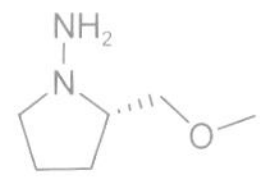
SC CO_2	supercritical carbon-dioxide	NA
SDS	sodium dodecylsulfate	
sec	secondary	NA
SEM	2-(trimethylsilyl)ethoxymethyl	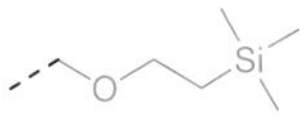
SES	2-[(trimethylsilyl)ethyl]sulfonyl	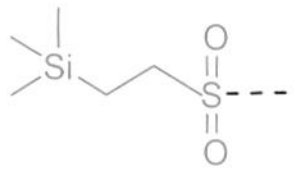
SET	single electron transfer	NA
Sia	1,2-dimethylpropyl (secondary isoamyl)	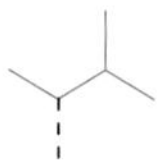
SPB	sodium perborate	$Na^{\oplus}\ BO_3^{\ominus}$

Abbreviation	Chemical Name	Chemical Structure
TADDOL	2,2-dimethyl-α,α,α^1, α^1-tetraaryl-1,3-dioxolane-4,5-dimethanol	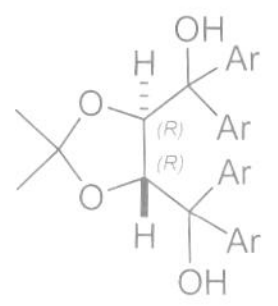
TASF	tris(diethylamino)sulfonium difluorotrimethylsilicate	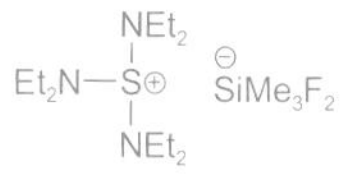
TBAB	tetra-*n*-butylammonium bromide	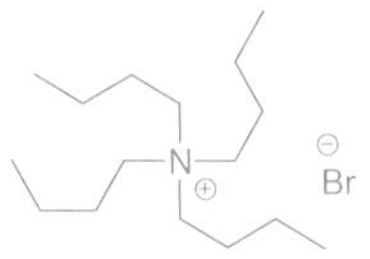
TBAF	tetra-*n*-butylammonium fluoride	
TBAI	tetra-*n*-butylammonium iodide	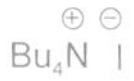
TBCO	tetrabromocyclohexadienone	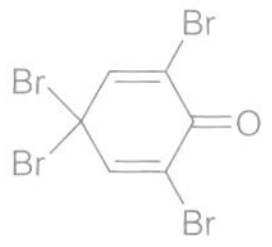
TBDMS (TBS)	*t*-butyldimethylsilyl	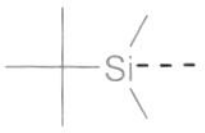
TBDPS (BPS)	*t*-butyldiphenylsilyl	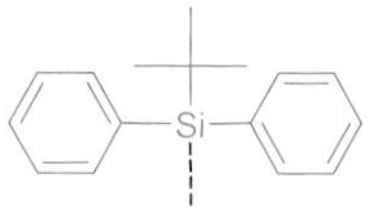
TBH	*tert*-butyl hypochlorite	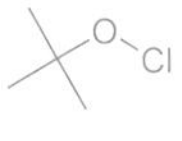
TBHP	*tert*-butyl hydroperoxide	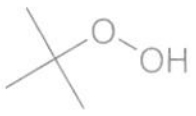
TBP	tributylphosphine	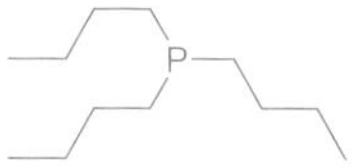
TBT	1-*tert*-butyl-1*H*-tetrazol-5-yl	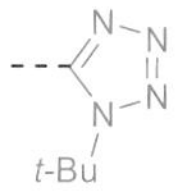
TBTH	tributyltin hydride	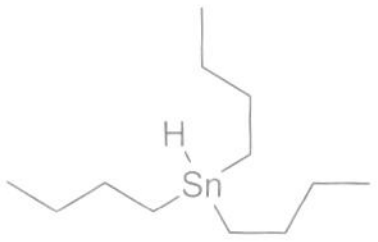
TBTSP	*t*-butyl trimethylsilyl peroxide	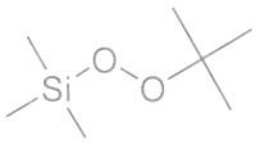

Abbreviation	Chemical Name	Chemical Structure
TCCA	trichloroisocyanuric acid	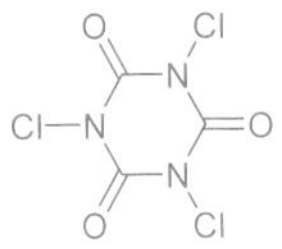
TCDI	thiocarbonyl diimidazole	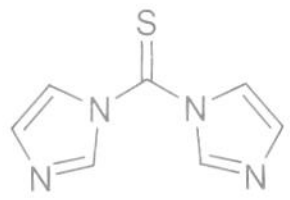
TCNE	tetracyanoethylene	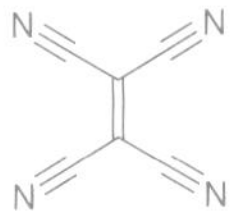
TCNQ	7,7,8,8-tetracyano-para-quinodimethane	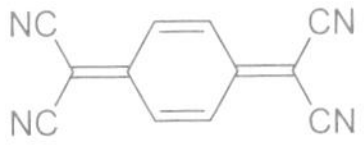
TDS	dimethyl thexylsilyl	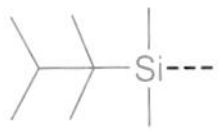
TEA	triethylamine	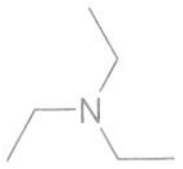
TEBACl	benzyl trimethylammonium chloride	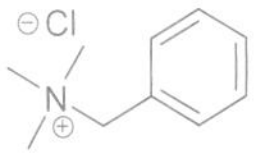
TEMPO	2,2,6,6-tetramethyl-1-piperidinyloxy free radical	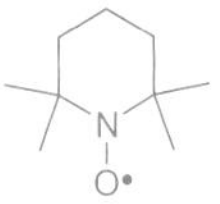
Teoc	2-(trimethylsilyl)ethoxycarbonyl	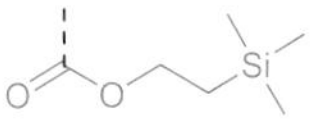
TEP	triethylphosphite	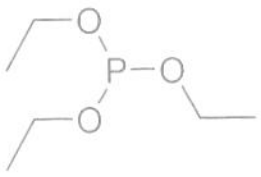
TES	triethylsilyl	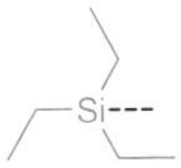
Tf	trifluoromethanesulfonyl	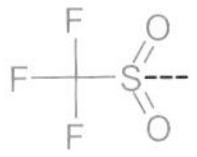
TFA	trifluoroacetic acid	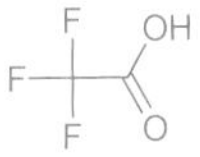
Tfa	trifluoroacetamide	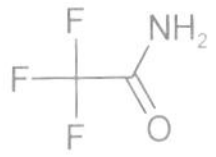

Abbreviation	Chemical Name	Chemical Structure
TFAA	trifluoroacetic anhydride	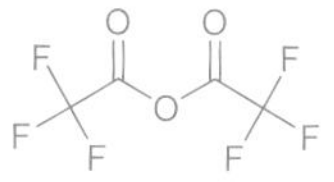
TFE	2,2,2-trifluoroethanol	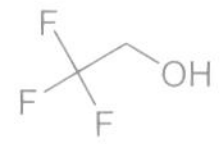
TFMSA	trifluoromethanesulfonic acid (triflic acid)	
TFP	tris(2-furyl)phosphine	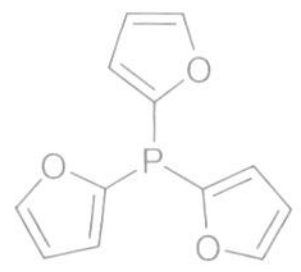
Th	2-thienyl	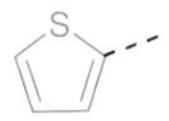
thexyl	1,1,2-trimethylpropyl	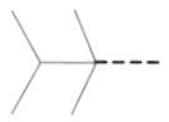
THF	tetrahydrofuran	
THP	2-tetrahydropyranyl	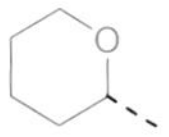
TIPB	1,3,5-tri*iso*propylbenzene	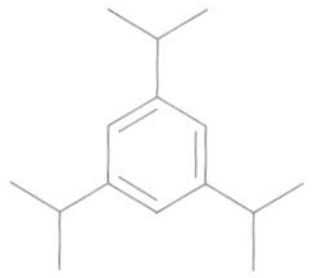
TIPS	triisopropylsilyl	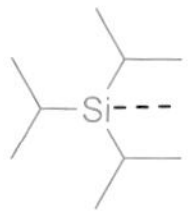
TMAO (TMANO)	trimethylamine *N*-oxide	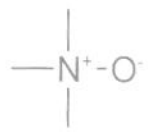
TMEDA	*N,N,N',N'*-tetramethylethylenediamine	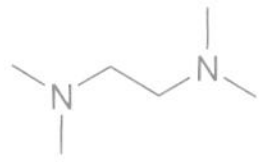
TMG	1,1,3,3-tetramethylguanidine	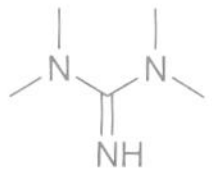
TMGA	tetramethylguanidinium azide	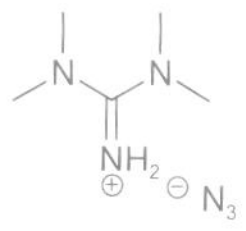

Abbreviation	Chemical Name	Chemical Structure
Tmob	2,4,6-trimethoxybenzyl	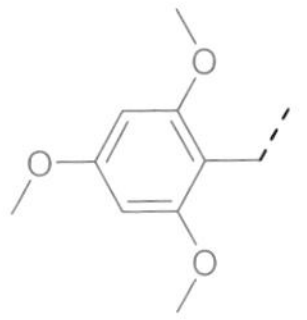
TMP	2,2,6,6-tetramethylpiperidine	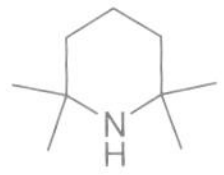
TMS	trimethylsilyl	
TMSA	trimethylsilyl azide	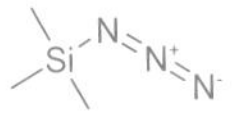
TMSEE	(trimethylsilyl)ethynyl ether	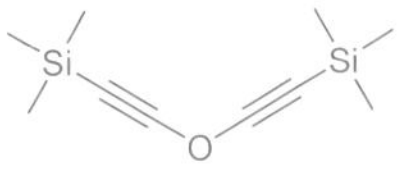
TMU	tetramethylurea	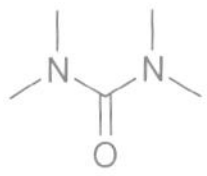
TNM	tetranitromethane	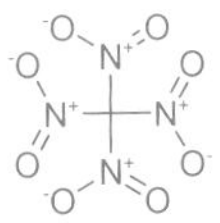
Tol	*p*-tolyl	
tolbinap	2,2'-*bis*(di-*p*-tolylphosphino)-1,1'-binaphthyl	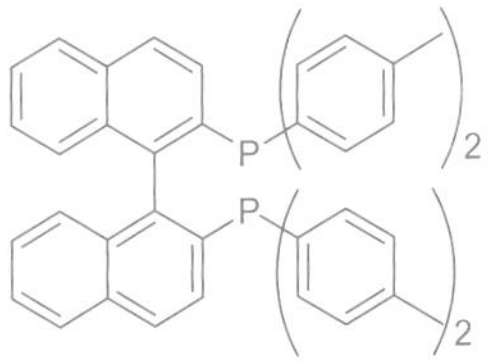
TPAP	tetra-*n*-propylammonium perruthenate	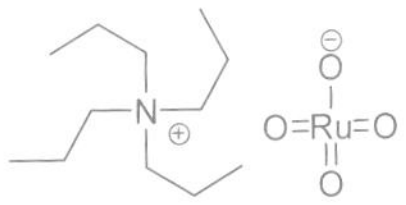
TPP	triphenylphosphine	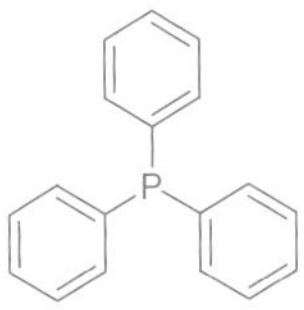
TPP	5,10,15,20-tetraphenylporphyrin	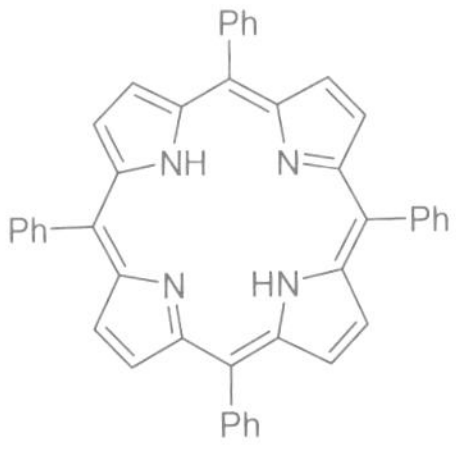

Abbreviation	Chemical Name	Chemical Structure
TPS	triphenylsilyl	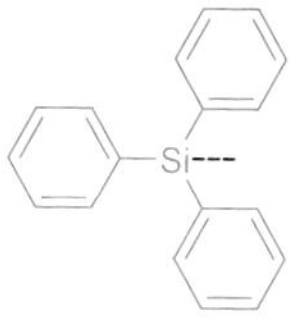
Tr	trityl (triphenylmethyl)	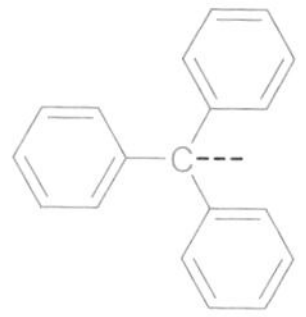
Trisyl	2,4,6-triisopropylbenzenesulfonyl	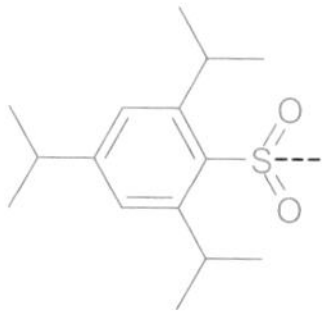
Troc	2,2,2-trichloroethoxycarbonyl	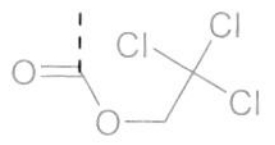
TS	transition state (or transition structure)	NA
Ts (Tos)	*p*-toluenesulfonyl	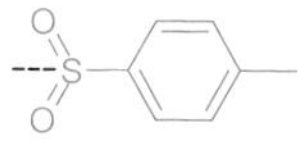
TSE (TMSE)	2-(trimethylsilyl)ethyl	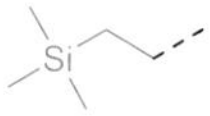
TTBP	2,4,5-tri-*tert*-butylpyrimidine	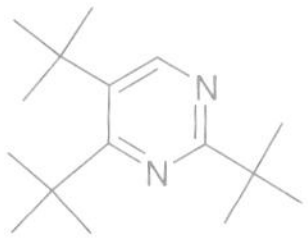
TTMSS	tris(trimethylsilyl)silane	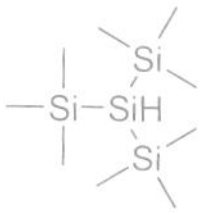
TTN	thallium(III)-trinitrate	$Tl(NO_3)_3$
UHP	urea-hydrogen peroxide complex	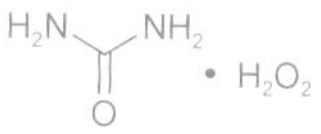
Vitride (Red-Al)	sodium *bis*(2-methoxyethoxy)aluminum hydride	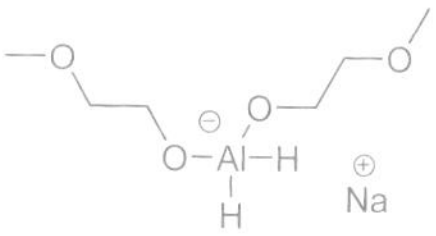
wk	weeks (length of reaction time)	NA
Z (Cbz)	benzyloxycarbonyl	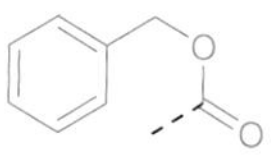

VI. LIST OF NAMED ORGANIC REACTIONS

VII. NAMED ORGANIC REACTIONS IN ALPHABETICAL ORDER

ACETOACETIC ESTER SYNTHESIS

(References are on page 531)

Importance:

[*Seminal Publications*[1-4]; *Reviews*[5-9]; *Modifications & Improvements*[10-19]]

The preparation of ketones *via* the *C*-alkylation of esters of 3-oxobutanoic acid (acetoacetic esters) is called the *acetoacetic ester synthesis*. Acetoacetic esters can be deprotonated at either the C2 or at both the C2 and C4 carbons, depending on the amount of base used. The C-H bonds on the C2 carbon atom are activated by the electron-withdrawing effect of the two neighboring carbonyl groups. These protons are fairly acidic (pK_a ~11 for C2 and pK_a ~24 for C4), so the C2 position is deprotonated first in the presence of one equivalent of base (sodium alkoxide, LDA, NaHMDS or LiHMDS, etc.). The resulting anion can be trapped with various alkylating agents. A second alkylation at C2 is also possible with another equivalent of base and alkylating agent. When an acetoacetic ester is subjected to excess base, the corresponding dianion (extended enolate) is formed.[13-15,18,19] When an electrophile (e.g., alkyl halide) is added to the dianion, alkylation occurs first at the most nucleophilic (reactive) C4 position. The resulting alkylated acetoacetic ester derivatives can be subjected to two types of hydrolytic cleavage, depending on the conditions: 1) dilute acid hydrolyzes the ester group, and the resulting β-keto acid undergoes decarboxylation to give a ketone (mono- or disubstituted acetone derivative); 2) aqueous base induces a *retro-Claisen reaction* to afford acids after protonation. The hydrolysis by dilute acid is most commonly used, since the reaction mixture is not contaminated with by-products derived from ketonic scission. More recently the use of the *Krapcho decarboxylation* allows neutral decarboxylation conditions.[11,12] As with malonic ester, monoalkyl derivatives of acetoacetic ester undergo a base-catalyzed coupling reaction in the presence of iodine. Hydrolysis and decarboxylation of the coupled products produce γ-diketones. The starting acetoacetic esters are most often obtained *via* the *Claisen condensation* of the corresponding esters, but other methods are also available for their preparation.[5,8]

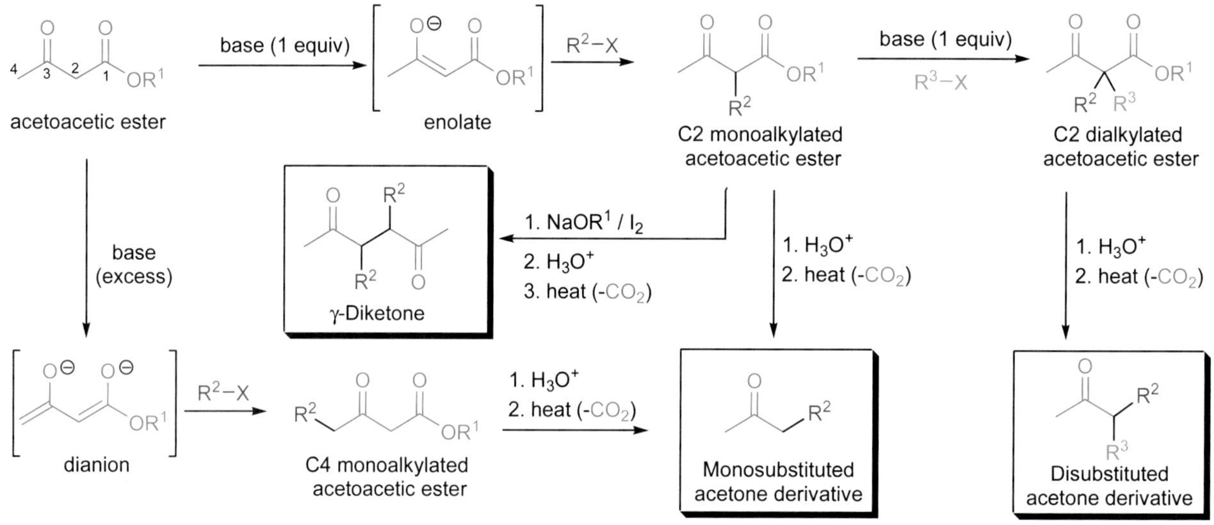

R^1 = 1°, 2° or 3° alkyl, aryl; R^2 = 1° or 2° alkyl, allyl, benzyl; R^3 = 1° or 2° alkyl, allyl, benzyl; base: NaH, NaOR^1, LiHMDS, NaHMDS

Mechanism:[3,20]

The first step is the deprotonation of acetoacetic ester at the C2 position with one equivalent of base. The resulting enolate is nucleophilic and reacts with the electrophilic alkyl halide in an S_N2 reaction to afford the C2 substituted acetoacetic ester, which can be isolated. The ester is hydrolyzed by treatment with aqueous acid to the corresponding β-keto acid, which is thermally unstable and undergoes decarboxylation *via* a six-membered transition state.

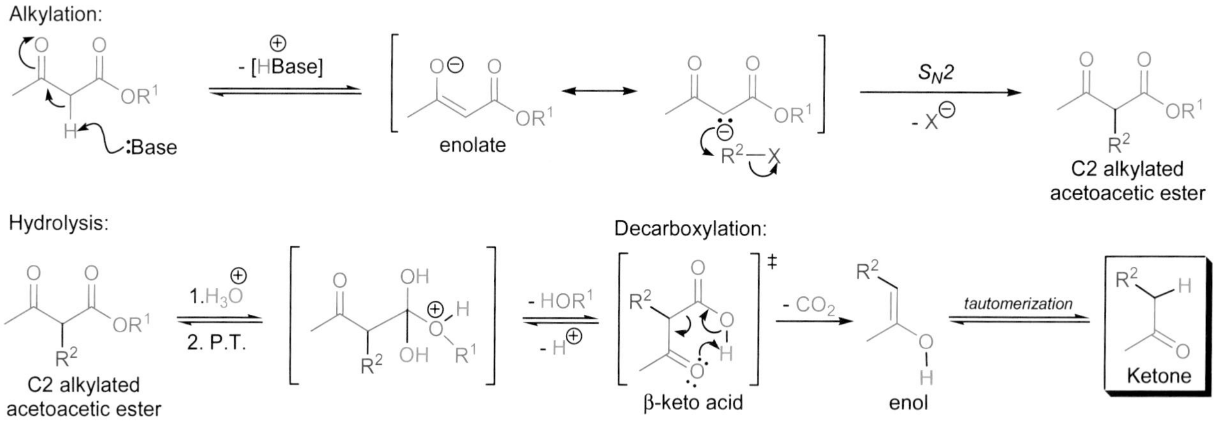

Synthetic Applications:

In the laboratory of H. Hiemstra, the synthesis of the bicyclo[2.1.1]hexane substructure of solanoeclepin A was undertaken utilizing the intramolecular photochemical dioxenone-alkene [2+2] cycloaddition reaction.[21] The dioxenone precursor was prepared from the commercially available *tert*-butyl acetoacetate using the *acetoacetic ester synthesis*. When this dioxenone precursor was subjected to irradiation at 300 nm, complete conversion of the starting material was observed after about 4h, and the expected cycloadduct was formed in acceptable yield.

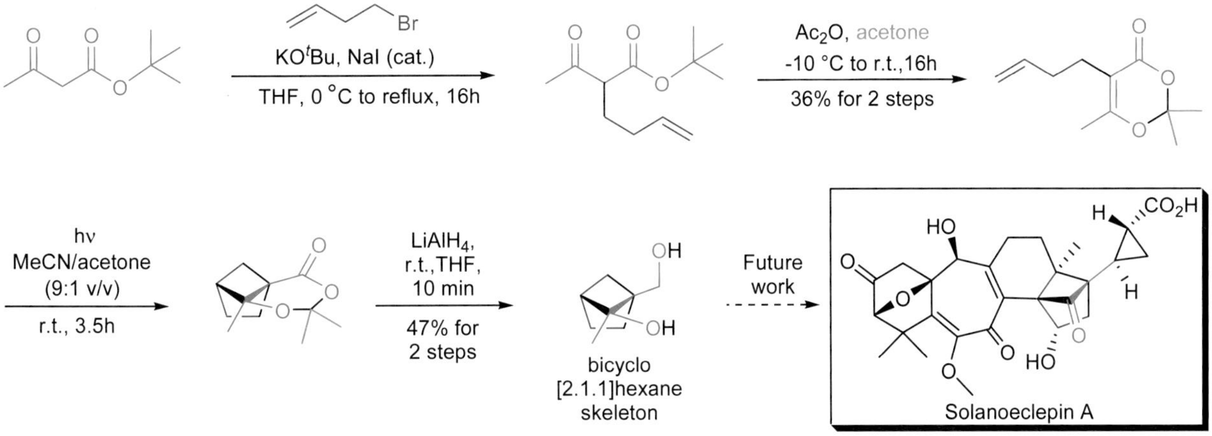

R. Neier et al. synthesized substituted 2-hydroxy-3-acetylfurans by the alkylation of *tert*-butylacetoacetate with an α-haloketone, followed by treatment of the intermediate with trifluoroacetic acid.[22] When furans are prepared from β-ketoesters and α-haloketones, the reaction is known as the *Feist-Bénary reaction*. A second alkylation of the C2 alkylated intermediate with various bromoalkanes yielded 2,2-disubstituted products, which upon treatment with TFA, provided access to trisubstituted furans.

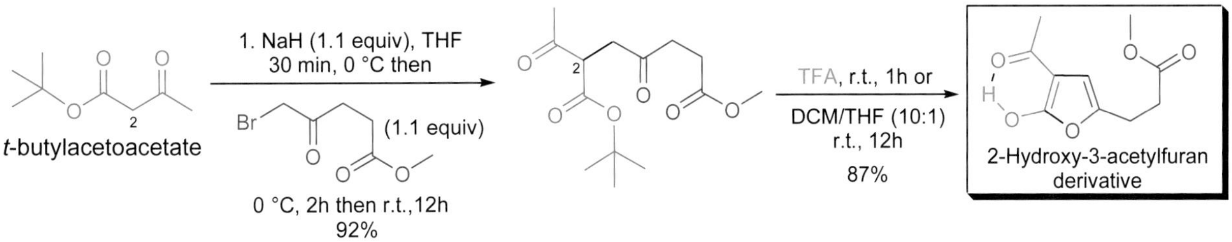

M. Nakada and co-workers developed a novel synthesis of tetrahydrofuran and tetrahydropyran derivatives by reacting dianions of acetoacetic esters with epibromohydrin derivatives.[23] The selective formation of the tetrahydrofuran derivatives was achieved by the use of $LiClO_4$ as an additive.

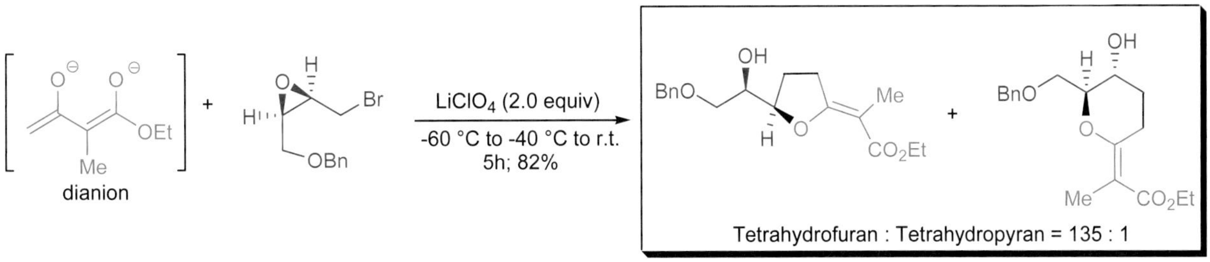

A synthetic strategy was developed for the typical core structure of the *Stemona* alkaloids in the laboratory of C.H. Heathcock.[24] The precursor for the 1-azabicyclo[5.3.0]decane ring system was prepared *via* the successive double alkylation of the dianion of ethyl acetoacetate.

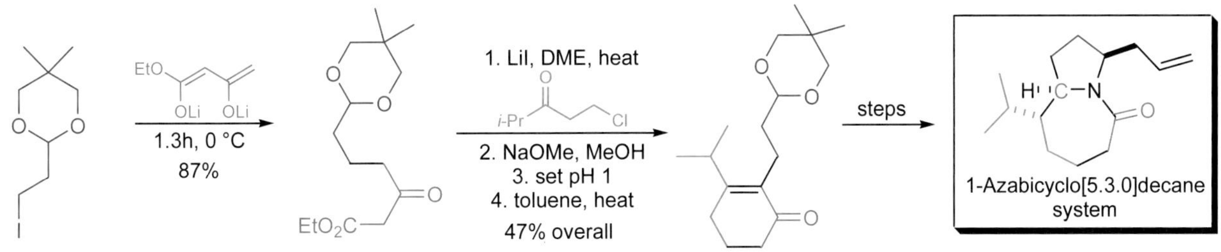

ACYLOIN CONDENSATION

(References are on page 531)

Importance:

[*Seminal Publications* [1-4]; *Reviews* [5-9]; *Modifications & Improvements* [10-22]]

The *acyloin condensation* affords acyloins (α-hydroxy ketones) by treating aliphatic esters with molten, highly dispersed sodium in hot xylene.[8] The resulting disodium acyloin derivatives are acidified to liberate the corresponding acyloins, which are valuable synthetic intermediates. Aliphatic monoesters give symmetrical compounds, while diesters lead to cyclic acyloins. The *intramolecular acyloin condensation* is one of the best ways of closing rings of 10 members or more (up to 34 membered rings were synthesized).[6] For the preparation of aromatic acyloins (R=Ar), the *benzoin condensation* between two aromatic aldehydes is applied. The acyloin condensation is performed in an inert atmosphere, since the acyloins and their anions are readily oxidized. For small rings (ring size: 4-6), yields are greatly improved in the presence of TMSCl and by the use of ultrasound.[11,13] The addition of TMSCl increases the scope of this reaction by preventing base-catalyzed side reactions such as β-elimination, *Claisen* or *Dieckmann condensations*. The resulting bis-silyloxyalkenes are either isolated or converted into acyloins by simple hydrolysis or alcoholysis.

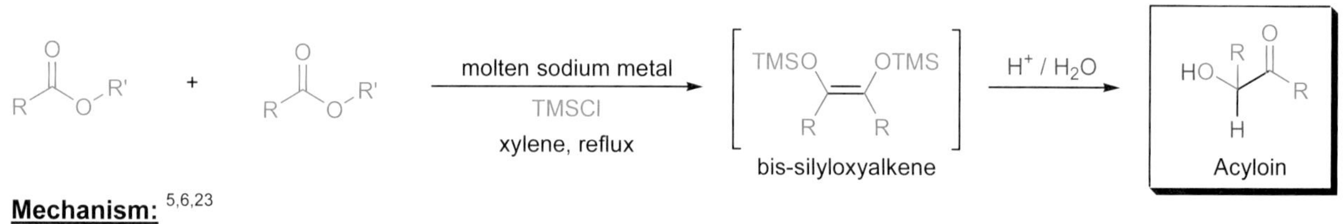

Mechanism: [5,6,23]

There are currently two proposed mechanisms for the *acyloin ester condensation* reaction. In mechanism **A** the sodium reacts with the ester in a single electron transfer (SET) process to give a radical anion species, which can dimerize to a dialkoxy dianion.[5,6] Elimination of two alkoxide anions gives a diketone. Further reduction (electron transfer from the sodium metal to the diketone) leads to a new dianion, which upon acidic work-up yields an enediol that tautomerizes to an acyloin. In mechanism **B** an epoxide intermediate is proposed.[23]

Mechanism **A**:

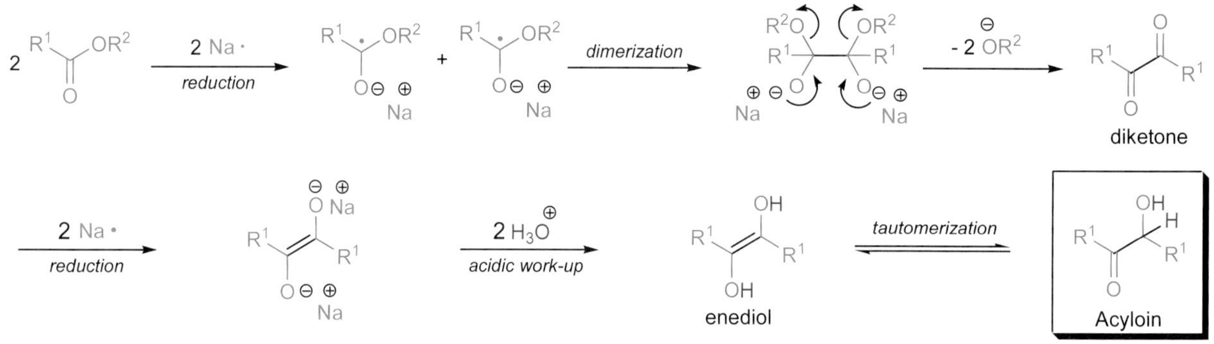

Mechanism **B**:

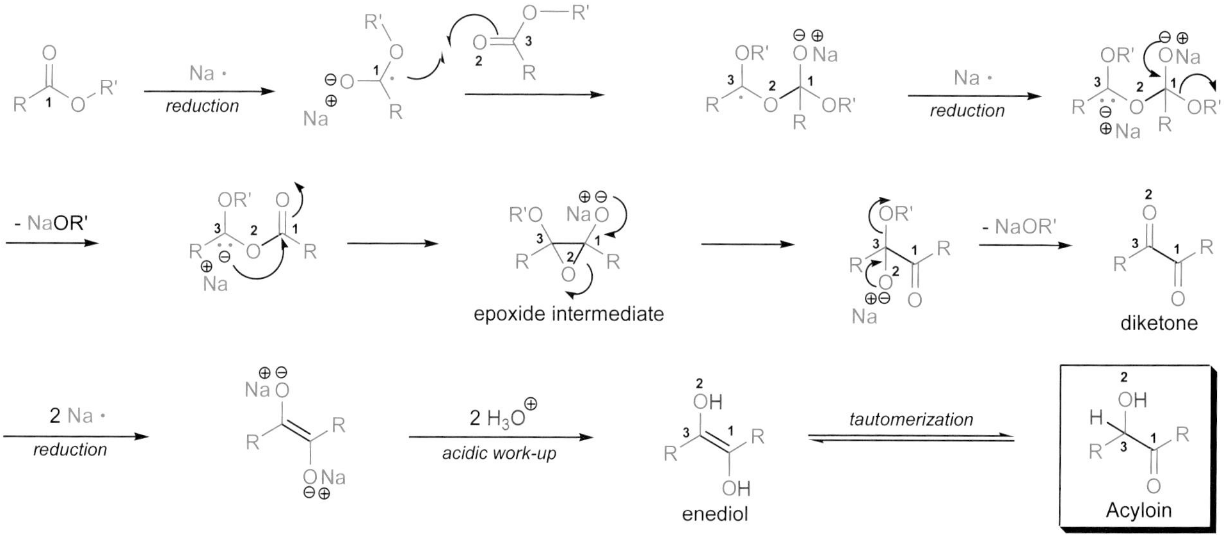

Synthetic Applications:

J. Salaün and co-workers studied the ultrasound-promoted *acyloin condensation* and cyclization of carboxylic esters.[13] They found that the acyloin coupling of 1,4-, 1,5-, and 1,6-diesters afforded 4-, 5- and 6-membered ring products. The cyclization of β-chloroesters to 3-membered ring products in the presence of TMSCl, which previously required highly dispersed sodium, was simplified and improved under sonochemical activation.

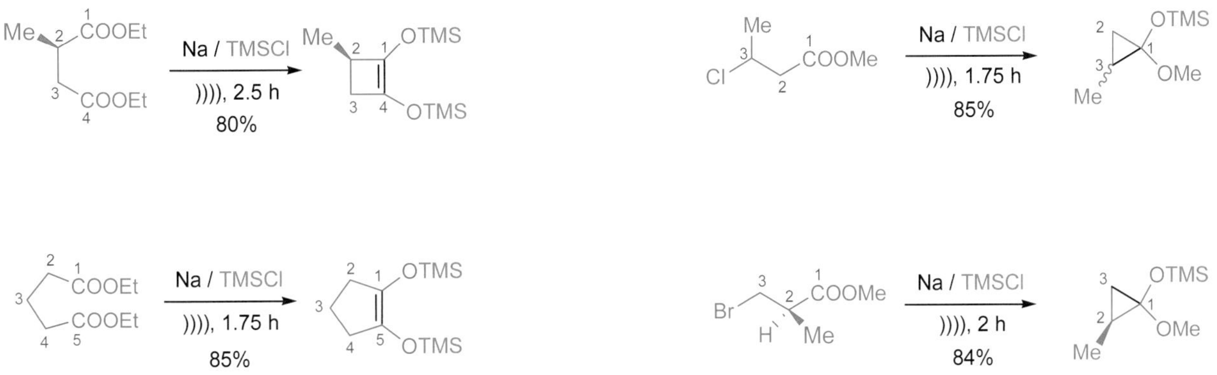

The diterpene alkaloids of the *Anopterus* species, of which anopterine (R=tigloyl) is a major constituent, are associated with a high level of antitumor activity. All of these alkaloids contain the tricyclo[3.3.2[1,4].0]decane substructure. S. Sieburth et al. utilized the *acyloin condensation* as a key step in the short construction of this tricyclic framework.[24]

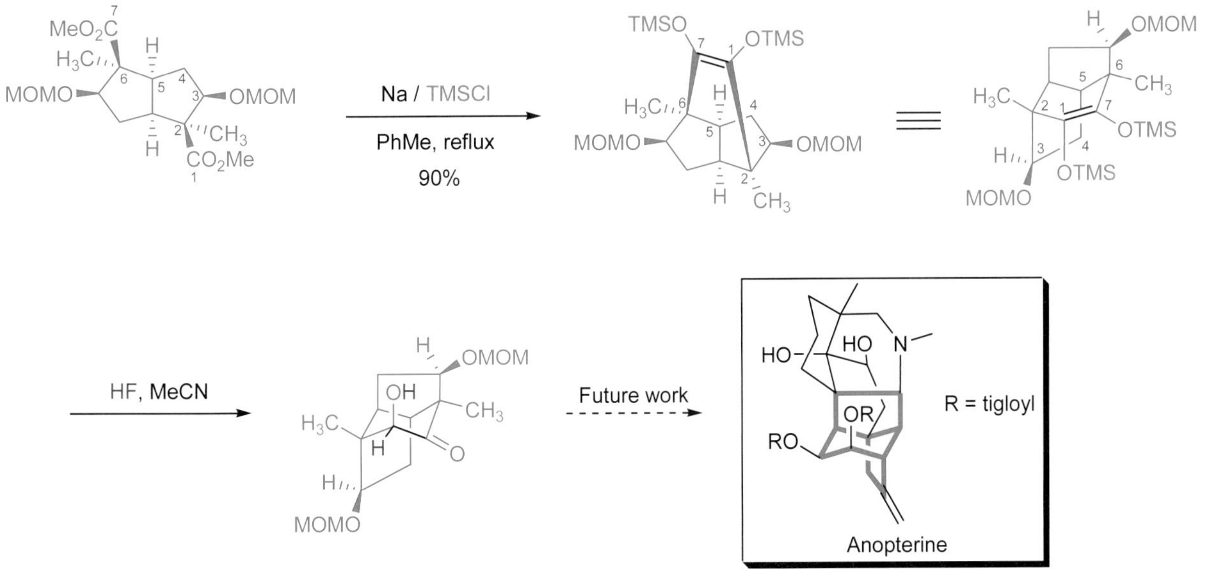

D.J. Burnell et al. synthesized bicyclic diketones by Lewis acid-promoted geminal acylation involving cyclic acyloins tethered to an acetal. The required *bis*-silyloxyalkenes were prepared by using the standard *acyloin condensation* conditions.[25]

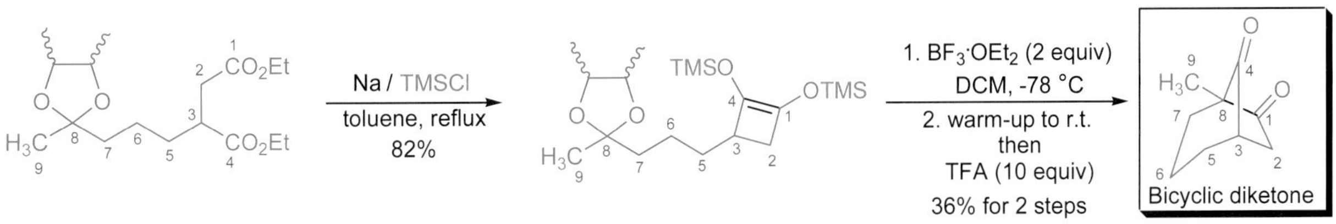

ALDER (ENE) REACTION
(HYDRO-ALLYL ADDITION)

(References are on page 532)

Importance:

[*Seminal Publications*[1-6]; *Reviews*[7-33]; *Theoretical Studies*[34-44]]

In 1943, K. Alder systematically studied reactions that involved the activation of an allylic C-H bond and the allylic transposition of the C=C bond of readily available alkenes.[4-6] This reaction is known as the *ene reaction.* Formally it is the addition of alkenes to double bonds (C=C or C=O), and it is one of the simplest ways to form C-C bonds. The *ene reaction* of an olefin bearing an allylic hydrogen atom is called "*carba-ene reaction*". For the reaction to proceed without a catalyst, the alkene must have an electron-withdrawing (EWG) substituent. This electrophilic compound is called the *enophile*. The *ene reaction* has a vast number of variants in terms of the enophile used.[7-9,11,12,45,14-16,46,18-20,24,47,27-30] Olefins are relatively unreactive as enophiles, whereas acetylenes are more enophilic. For example, under high pressure acetylene reacts with a variety of simple alkenes to form 1,4-dienes. When the enophile is a carbonyl compound, the *ene reaction* leads exclusively to the corresponding alcohol instead of the ether (*carbonyl-ene reaction*). However, thiocarbonyl compounds react mainly to give allylic sulfides rather than homoallylic thiols. Schiff bases derived from aldehydes afford homoallylic amines (*aza-ene, imino-ene* or *hetero-ene reaction*).[19] *Metallo-ene reactions* with Pd, Pt, and Ni-catalyzed versions have been successful in intramolecular systems. The *ene reaction* is compatible with a variety of functional groups that can be appended to the ene and enophile. The *ene reaction* can be highly stereoselective and by adding Lewis acids ($RAlX_2$, $Sc(OTf)_3$, $LiClO_4$, etc.), less reactive enophiles can also be used. The regioselectivity of the reaction is determined by the steric accessibility of the hydrogen. Usually primary hydrogens are abstracted faster than secondary hydrogens and tertiary hydrogens are abstracted last. Functionalization of the reacting components by introduction of a silyl, alkoxy, or amino group, thus changing the steric and electronic properties, affords more control over the regioselectivity of the reaction.

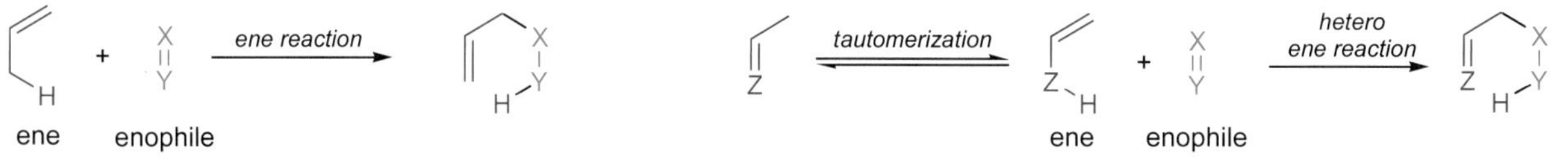

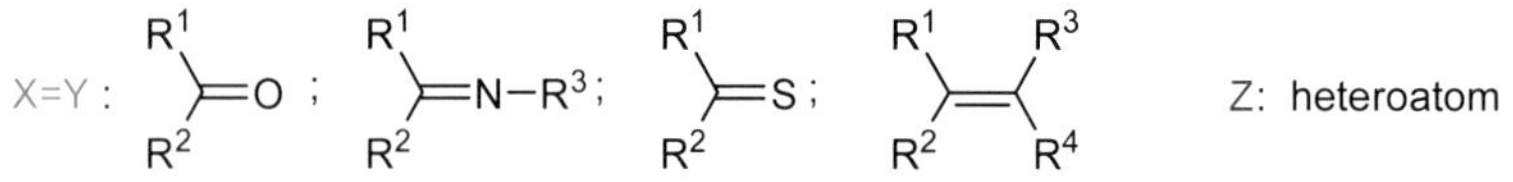

Mechanism:[48-52,31]

The *ene reaction* is mechanistically related to the better-known *Diels-Alder reaction* and is believed to proceed *via* a six-membered aromatic transition state.[50,51] Thermal *intermolecular ene reactions* have high negative entropies of activation, and for this reason the *ene reaction* requires higher temperatures than the *Diels-Alder reaction*. The forcing conditions were responsible for the initial paucity of *ene reactions*. However, *intramolecular ene reactions* are more facile. The enophile reacts with the ene component in a "*syn*-fashion" and this observation suggests a *concerted mechanism*. There is a frontier orbital interaction between the HOMO of the ene component and the LUMO of the enophile. The *ene-reaction* is favored by electron-withdrawing substituents on the enophile, by strain in the ene component and by geometrical alignments that position the components in a favorable arrangement. Some *thermal ene reactions*, such as the ene reaction between cyclopentene and diethyl azodicarboxylate (DEAD), are catalyzed by free radical initiators, so for these processes a *stepwise biradical pathway* had been suggested.[48,49] The mechanism of the *Lewis acid-promoted ene reaction* is believed to involve both a concerted and a cationic pathway.[53] Whether the mechanism is concerted or stepwise, a partial or full positive charge is developed at the ene component in Lewis acid-promoted reactions.

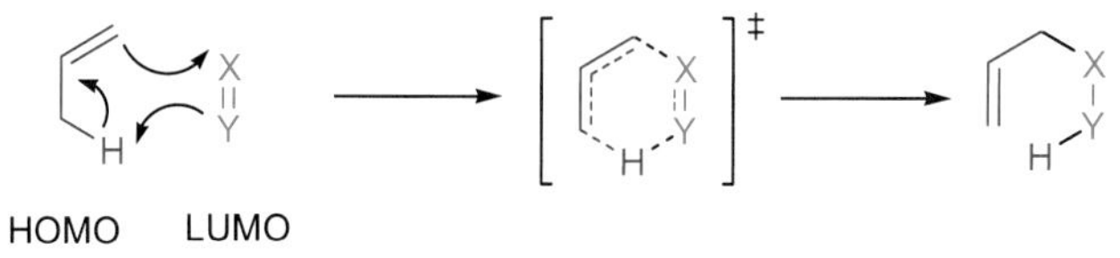

ALDER (ENE) REACTION
(HYDRO-ALLYL ADDITION)

Synthetic Applications:

The *aza-ene* reaction recently found application in the synthesis of imidazo[1,2-*a*]pyridine and imidazo[1,2,3-*ij*][1,8]naphthyridine derivatives in the laboratory of Z.-T. Huang.[54] The reaction of heterocyclic ketene aminals with enones such as MVK proceeded *via* an *aza-ene addition*, followed by intramolecular cyclization to afford the products. The aroyl-substituted heterocyclic ketene aminals (Ar=Ph, 2-furyl, 2-thienyl) underwent two subsequent *aza-ene reactions* when excess MVK was used.

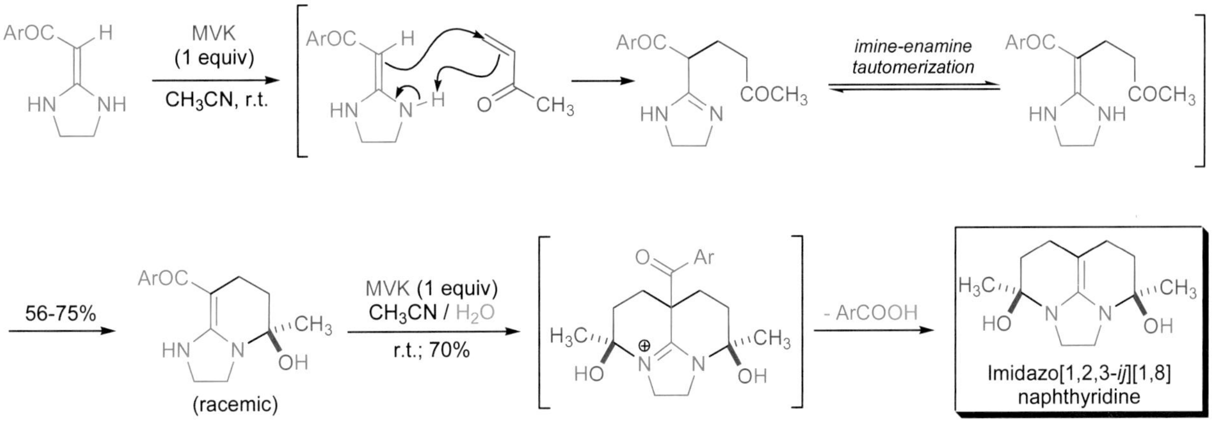

B. Ganem and co-workers accomplished the asymmetric total synthesis of (–)-α-kainic acid using an enantioselective, *metal-promoted ene cyclization*.[52] The chiral bis-oxazoline-magnesium perchlorate system strongly favored the formation of the *cis*-diastereomer in the cyclization. Enantiomerically pure kainic acid was synthesized from readily available starting materials on a 1-2 g scale in six steps in an overall yield of greater than 20%.

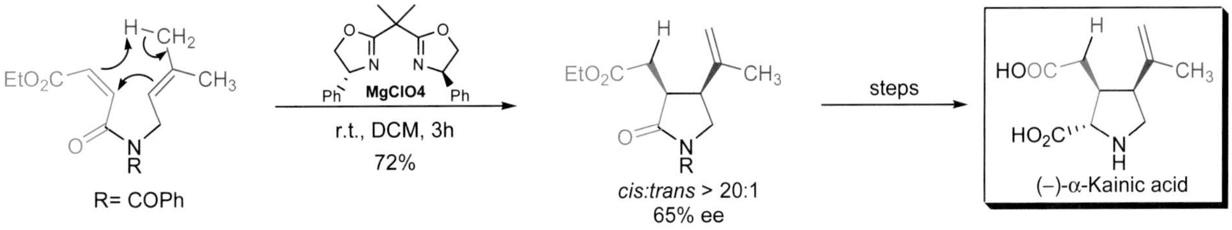

The first total synthesis of (+)-arteannuin M was completed by L. Barriault et al. using a tandem *oxy-Cope/transannular ene reaction* as the key step to construct the bicyclic core of the natural product.[55] The tandem reaction proceeded with high diastereo- and enantioselectivity.

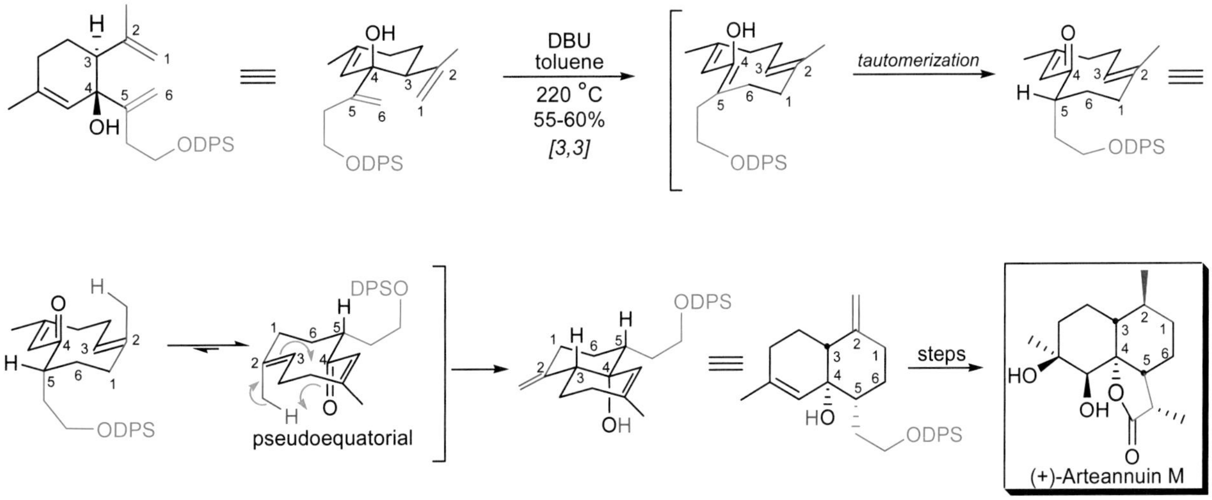

ALDOL REACTION

(References are on page 533)

Importance:

[*Seminal Publications*[1,2]; *Reviews*[3-46]; *Theoretical Studies*[47-74]]

The *aldol reaction* involves the addition of the enol/enolate of a carbonyl compound (nucleophile) to an aldehyde or ketone (electrophile). The initial product of the reaction is a β-hydroxycarbonyl compound that under certain conditions undergoes dehydration to generate the corresponding α,β-unsaturated carbonyl compound. The transformation takes its name from 3-hydroxybutanal, the acid-catalyzed self-condensation product of acetaldehyde, which is commonly called aldol. Originally the *aldol reaction* was carried out with Brönsted acid[1,2] or Brönsted base catalysis,[75,76] but these processes were compromised by side reactions such as self-condensation, polycondensation, and dehydration followed by *Michael addition*. Development of methods for the formation and application of preformed enolates was a breakthrough in the aldol methodology. Most commonly applied enolates in the aldol reaction are the lithium-,[12] boron-,[14] titanium-,[15] and silyl enol ethers, but several other enolate derivatives have been studied such as magnesium-,[12] aluminum-,[14] zirconium-,[15] rhodium-,[15] cerium-,[15] tungsten-,[15] molybdenum-,[15] rhenium-,[15] cobalt-,[15] iron-,[15] and zinc enolates.[16] Enolate formation can be accomplished in a highly regio- and stereoselective manner. The *aldol reaction* of stereodefined enolates is highly diastereoselective.[3,13] (*E*)-Enolates generally yield the *anti* product, while (*Z*)-enolates lead to the *syn* product as the major diastereomer. Lewis acid mediated *aldol reaction* of silyl enol ethers (*Mukaiyama aldol reaction*) usually provides the *anti* product.[77,78] Control of the absolute stereochemical outcome of the reaction can be achieved through the use of enantiopure starting materials (reagent control) or asymmetric catalysis.[6,7,79,8,9,22,41] Reagent control can be realized by: 1) utilizing chiral auxiliaries in the enol component, such as oxazolidinones (also see *Evans aldol*), bornanesultams, pyrrolidinones, arylsulfonamido indanols, norephedrines and *bis*(isopropylphenyl)-3,5-dimethylphenol derivatives;[80] 2) applying chiral ligands on boron enolates such as isopinocampheyl ligands, menthone derived ligands, tartrate derived boronates, and C_2-symmetric borolanes;[24,25,80] 3) using chiral aldehydes.[7,17,29,41] Direct *asymmetric catalytic aldol reactions* can be achieved *via* 1) biochemical catalysis applying enzymes or catalytic antibodies;[11,18,20,81,27] 2) chiral metal complex mediated catalysis; and 3) organocatalysis utilizing small organic molecules.[21,28,29,33,82,35-37,39]

Classical aldol reaction:

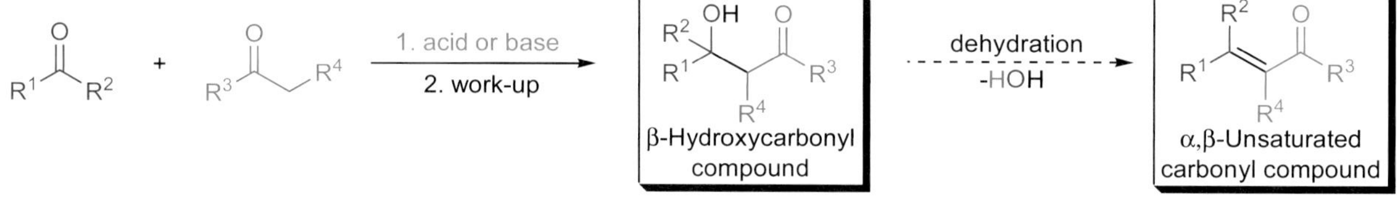

Aldol reaction through the use of preformed enolate:

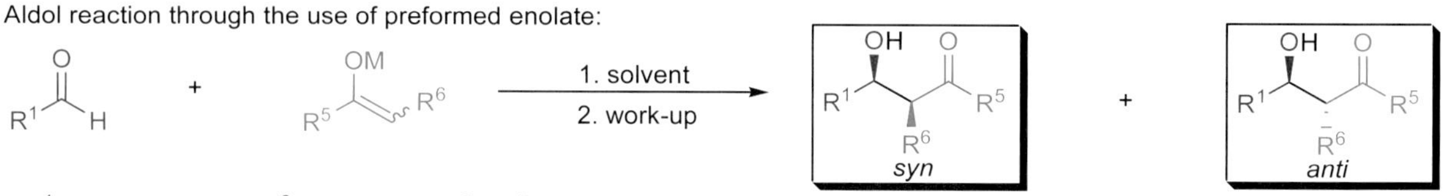

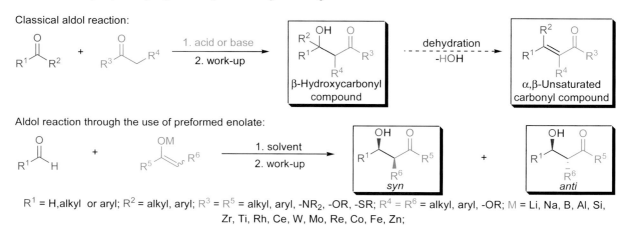

R^1 = H,alkyl or aryl; R^2 = alkyl, aryl; R^3 = R^5 = alkyl, aryl, -NR_2, -OR, -SR; R^4 = R^6 = alkyl, aryl, -OR; M = Li, Na, B, Al, Si, Zr, Ti, Rh, Ce, W, Mo, Re, Co, Fe, Zn;

Mechanism: [7,12,13]

The mechanism of the classical *acid catalyzed aldol reaction* involves the equilibrium formation of an enol, which functions as a nucleophile. The carbonyl group of the electrophile is activated toward nucleophilic attack by protonation. In the base catalyzed reaction, the enolate is formed by deprotonation followed by the addition of the enolate to the carbonyl group. In both cases, the reaction goes through a number of equilibria, and the formation of the product is reversible. *Aldol reaction* of preformed enolates generally provides the products with high diastereoselectivity, (*Z*)-enolates yielding the *syn* product, (*E*)-enolates forming the *anti* product as the major diastereomer. The stereochemical outcome of the reaction can be rationalized based on the Zimmerman-Traxler model, according to which the reaction proceeds through a six-membered chairlike transition state. The controlling factor according to this model is the avoidance of destabilizing 1,3-diaxial interactions in the cyclic transition state.

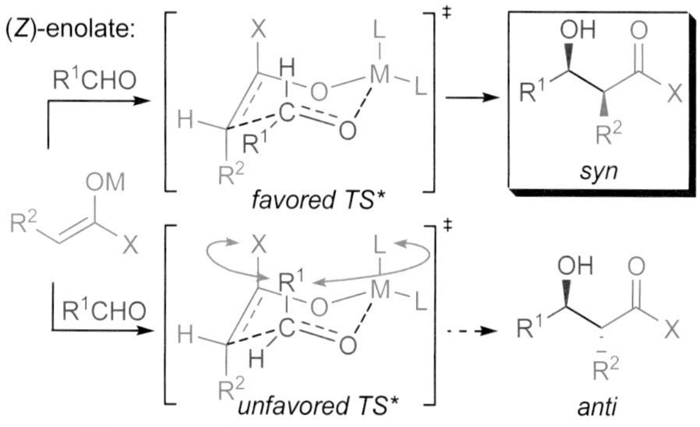

Zimmerman-Traxler model for (Z)- enolate

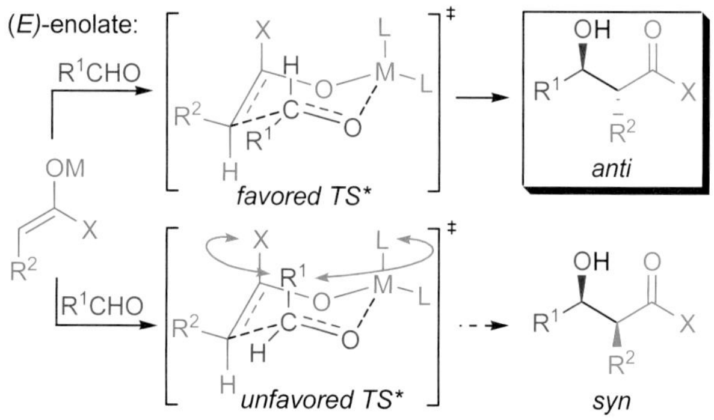

Zimmerman-Traxler model for (E)- enolate

Synthetic Applications:

The first enantioselective total synthesis of (–)-denticulatin A was accomplished by W. Oppolzer.[83] The key step in their approach was based on enantiotopic group differentiation in a *meso* dialdehyde by an *aldol reaction*. In the *aldol reaction* they utilized a bornanesultam chiral auxiliary. The enolization of *N*-propionylbornane-10,2-sultam provided the (*Z*)-borylenolate derivative, which underwent an *aldol reaction* with the *meso* dialdehyde to afford the product with high yield and enantiopurity. In the final stages of the synthesis they utilized a second, *double-diastereodifferentiating aldol reaction*. Aldol reaction of the (*Z*)-titanium enolate gave the *anti*-Felkin *syn* product. The stereochemical outcome of the reaction was determined by the α-chiral center in the aldehyde component.

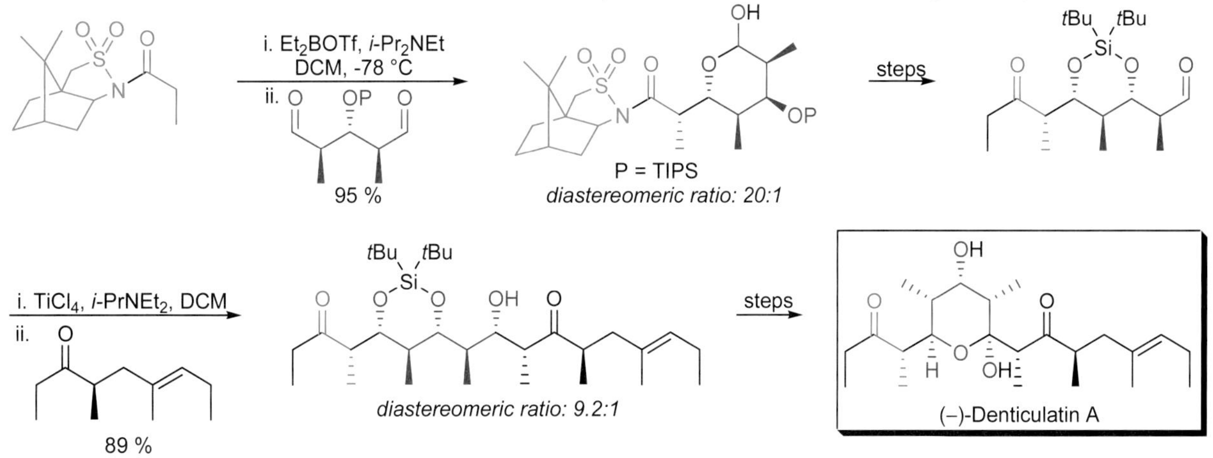

During the total synthesis of rhizoxin D by J.D. White et al., an *asymmetric aldol reaction* was utilized to achieve the coupling of two key fragments.[84] The *aldol reaction* of the aldehyde and the chiral enolate derived from (+)-chlorodiisopinocampheylborane afforded the product with a diastereomeric ratio of 17-20:1 at the C13 stereocenter. During their studies, White and co-workers also showed that the stereochemical induction of the chiral boron substituent and the stereocenters present in the enolate reinforce each other thus representing a "*matched*" *aldol reaction*.

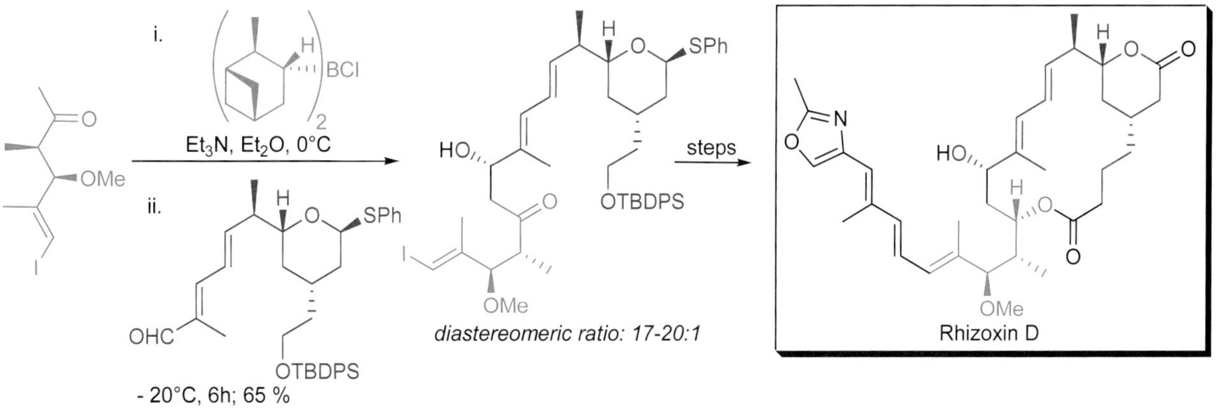

A possible way to induce enantioselectivity in the *aldol reaction* is to employ a chiral catalyst. M. Shibasaki and co-workers developed a bifunctional catalyst, (*S*)-LLB (L=lanthanum; LB=lithium binaphthoxide), which could be successfully applied in direct *catalytic asymmetric aldol reactions*.[85] An improved version of this catalyst derived from (*S*)-LLB by the addition of water and KOH was utilized in the formal total synthesis of fostriecin.[86]

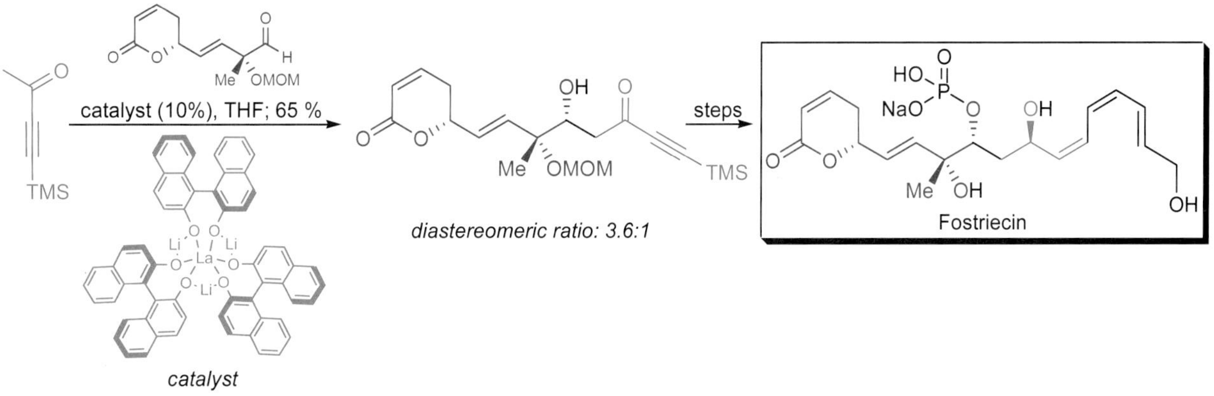

ALKENE (OLEFIN) METATHESIS

(References are on page 534)

Importance:

[*Seminal Publications*[1,2]; *Reviews*[3-61]; *Modifications & Improvements*[62-70]; *Theoretical Studies*[71-76]]

The metal-catalyzed redistribution of carbon-carbon double bonds is called *alkene (olefin) metathesis*. The first report of double-bond scrambling was published in 1955[1] but the term "*olefin metathesis*" was introduced only thirteen years later by N. Calderon[2] and co-workers. There are several different *olefin metathesis* reactions: *ring-opening metathesis polymerization* (ROMP), *ring-closing metathesis* (RCM), *acyclic diene metathesis polymerization* (ADMET), *ring-opening metathesis* (ROM), and *cross-metathesis* (CM or XMET). These various olefin metathesis reactions give access to molecules and polymers that would be difficult to obtain by other means. ROMP makes it possible to prepare functionalized polymers, while the application of RCM provides easy entry into medium and large carbocycles as well as heterocyclic compounds. The application of *olefin metathesis* for the synthesis of complex organic molecules did not appear until the beginning of the 1990s because the available catalysts had low performance and little functional group tolerance. In the past 10 years *olefin metathesis* has become a reliable and widely used synthetic method. The currently used $L(L')X_2Ru{=}CHR$ catalyst system is highly active, and it has sufficient functional group tolerance for most applications. However, new catalysts are still needed, because the current ones do not always perform well in several demanding transformations. Some of the problems still encountered are: 1) incompatibility with basic functional groups (nitriles and amines); 2) cross metathesis to form tetrasubstituted olefins; and 3) low stereoselectivity in CM and macrocyclic RCM reactions.

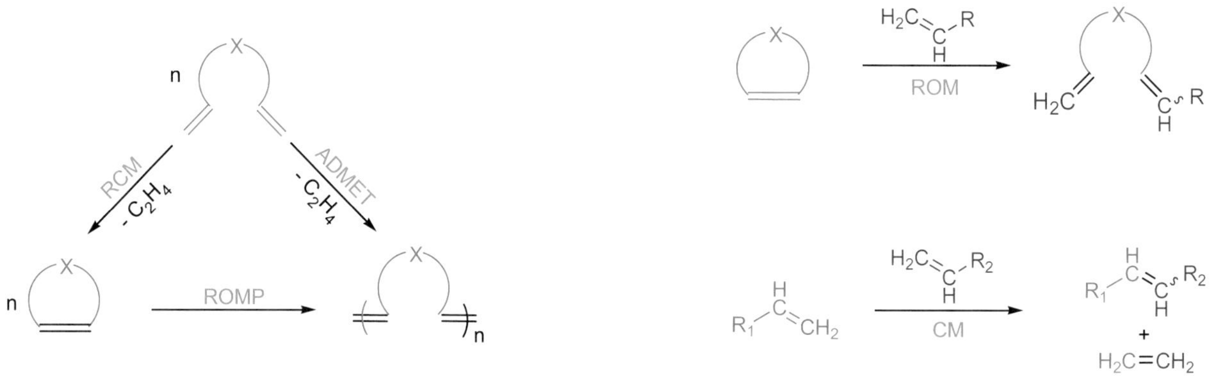

Mechanism:[77-86]

Crystal structures of the $L_2X_2Ru{=}CHR$ carbene complexes reveal that they have a distorted square pyramidal geometry with the alkylidene in the axial position and the *trans* phosphines and halides in the equatorial plane.[87,88] R. H. Grubbs and co-workers have conducted extensive kinetic studies on $L_2X_2Ru{=}CHR$ complexes and proposed a mechanism that is consistent with the observed activity trends.[89] There are two possible mechanistic pathways (**I** & **II**):

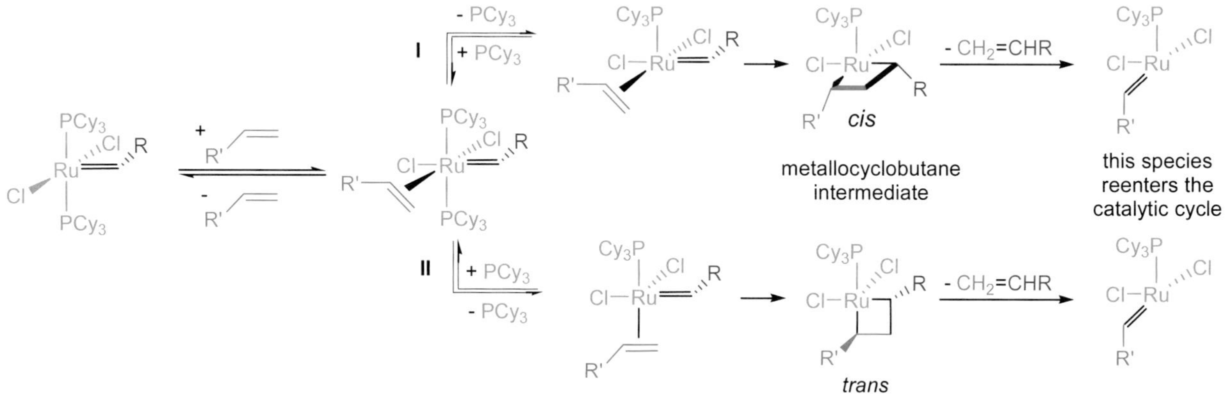

Synthetic Applications:

A.B. Smith and co-workers have devised an efficient strategy for the synthesis of the cylindrocyclophane family of natural products.[90,91] Olefin *ring-closing metathesis* was used for the assembly of the [7,7]-paracyclophane skeleton. During their investigations they discovered a remarkably efficient *CM dimerization* process, that culminated in the total synthesis of both (–)-cylindrocyclophane A and (–)-cylindrocyclophane F. They established that the *cross metathesis dimerization* process selectively led to the thermodynamically most stable member of a set of structurally related isomers. Out of three commonly used RCM catalysts, Schrock's catalyst proved to be the most efficient for this transformation.

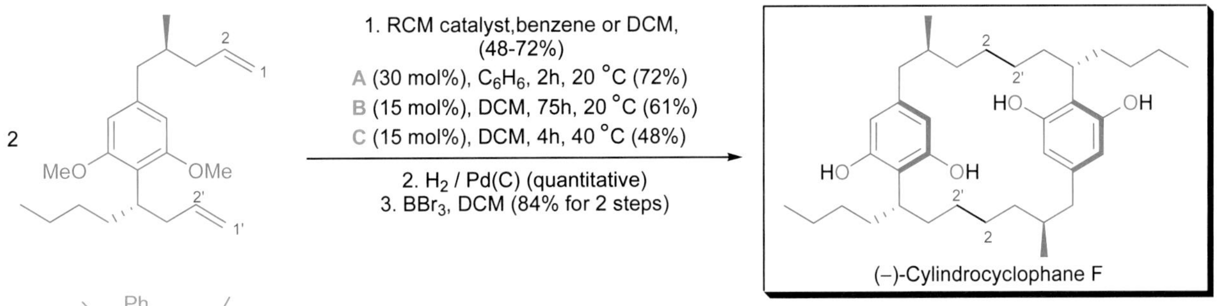

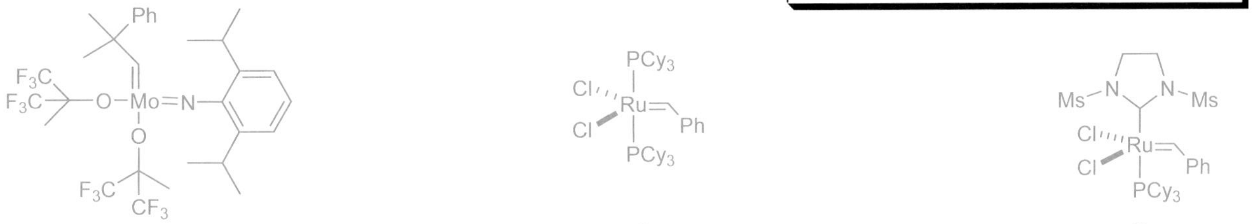

A Schrock's catalyst

B Grubb's catalyst

C Grubb's modified perhydroimidazolidine catalyst

The streptogramin antibiotics are a family of compounds that were isolated from a variety of soil organisms belonging to the genus *Streptomyces*. They are active against bacteria resistant to vancomycin. In the laboratory of A.I. Meyers the first total synthesis of streptogramin antibiotics, (–)-griseoviridin and its C8 epimer, featuring a 23-membered unsaturated ring, was accomplished using a novel *RCM* that involved a highly diastereoselective triene to diene macrocyclic ring formation.[92] The metathesis was performed in 37-42% yield using 30 mol% of Grubbs catalyst. The natural product was obtained as a single diastereomer; no other olefin isomers were formed in the ring-closing step.

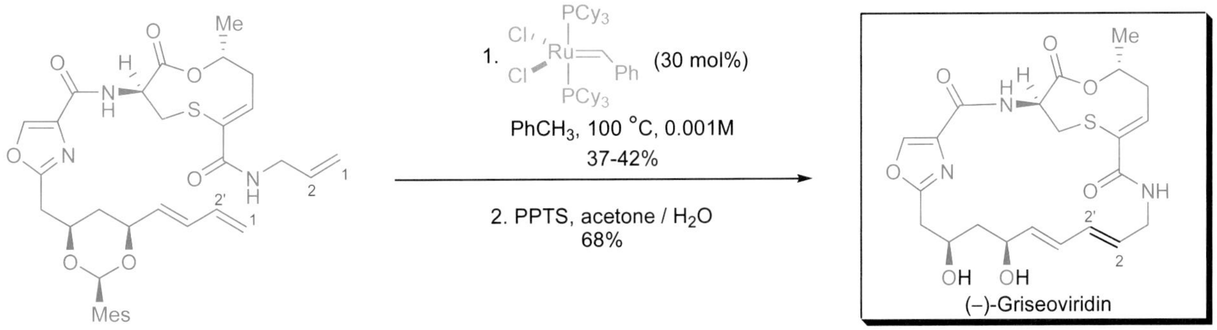

The first enantioselective total synthesis of (+)-prelaureatin was achieved by M.T. Crimmins et al.[93] The oxocene core of the natural product was constructed in high yield by a *RCM* reaction using the first generation Grubbs catalyst.

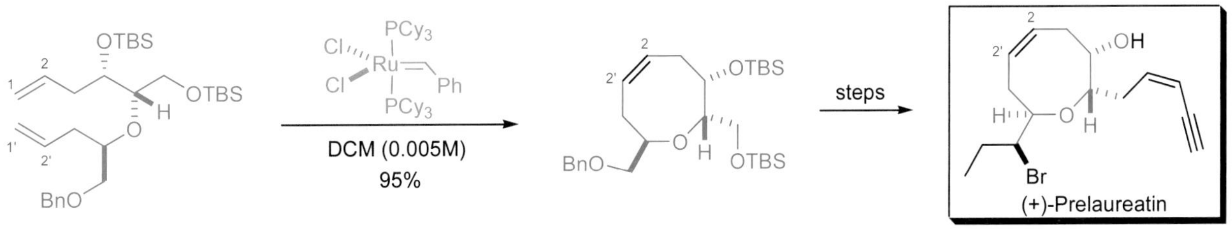

ALKYNE METATHESIS

(References are on page 536)

Importance:

[*Seminal Publications*[1-3]; *Reviews*[4-11]]

The metal-catalyzed redistribution of carbon-carbon triple bonds is called *alkyne metathesis*. In the beginning of the 1970s, A. Mortreux and co-workers were the first to achieve the homogeneously catalyzed metathesis reaction of a C-C triple bond in which they statistically disproportionated *p*-tolylphenylacetylene to tolan (diphenyl acetylene) with an *in situ* formed [$Mo(CO)_6$]/resorcinol catalyst at 110 °C.[1] However, all attempts to convert terminal alkynes by metathesis failed with this catalyst. Cyclotrimers and complex polymers were isolated instead. A decade later, in the 1980s, the well-defined Schrock tungsten carbyne complex [(t-BuO)$_3$W≡C-t-Bu] was shown to catalyze the metathesis of terminal alkynes accompanied by the evolution of gaseous acetylene.[12] This reaction also suffered from substantial polymerization of the substrate to polyacetylenes. In the 1990s research efforts intensified to find suitable catalysts. M. Mori and co-workers successfully cross-metathesized internal alkynes in the presence of a Mortreux-type catalyst,[13,14] while in the laboratory of A. Fürstner the conditions for RCAM (ring-closing alkyne metathesis) were developed.[15] The cycloalkynes obtained by the RCAM can be stereoselectively converted into the corresponding (*Z*)- or (*E*)-alkenes by catalytic hydrogenation,[16-18] hydroboration, and subsequent protonation, as well as by other methods.[19] In the years to come *alkyne metathesis* will probably become a useful tool for organic synthesis as well as for the synthesis of polymers.

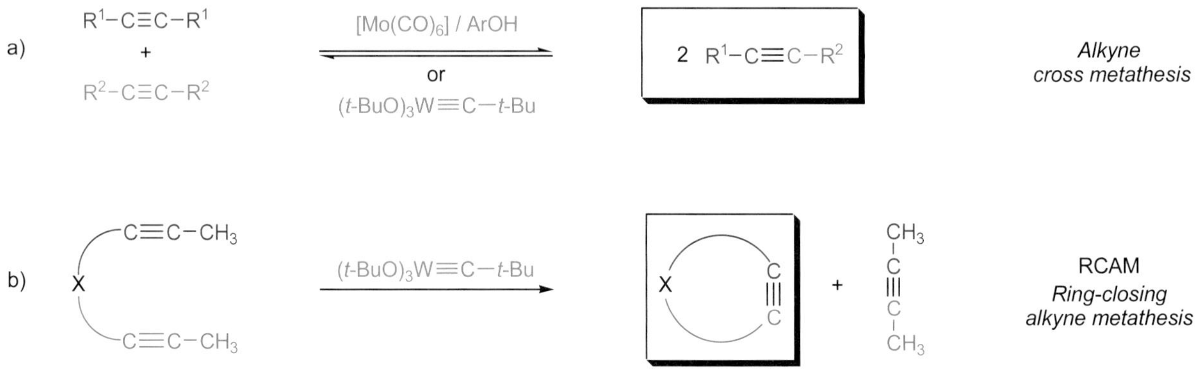

Mechanism: [20-27]

The *alkyne cross metathesis* and *metathesis polymerization* can be carried out both thermally and photochemically. The nature of the catalytically active species in the thermally and photochemically activated systems is unknown. The mechanism shown below accounts for the formation of the *alkyne cross metathesis* products, but none of the currently proposed mechanisms are supported by solid experimental evidence.

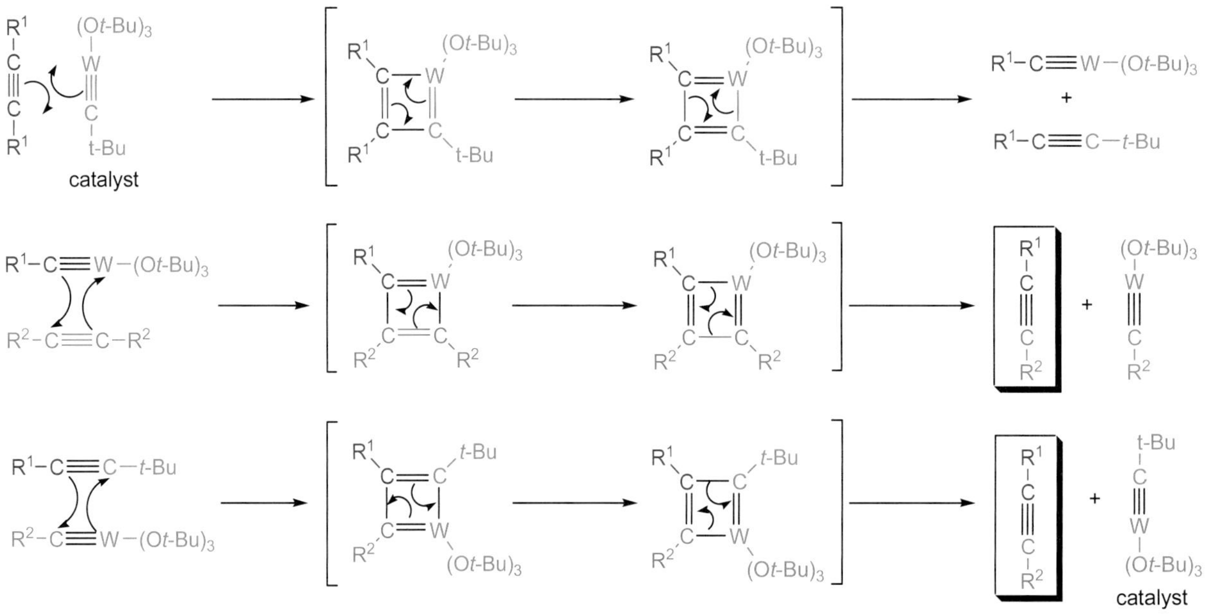

Synthetic Applications:

The total synthesis of the recently discovered azamacrolides was undertaken in the laboratory of A. Fürstner.[16] These compounds are the defense secretions of the pupae of the Mexican beetle *Epilachnar varivestis*, and they are the first examples of naturally occurring macrolactones containing a basic nitrogen atom in the tether that do not ring-contract to the corresponding amides. RCAM followed by Lindlar reduction provided a convenient, high-yielding, and stereoselective way to introduce the (*Z*)-double bond. (The usual RCM approach using Grubbs carbene only yielded a mixture of alkenes (*Z*) : (*E*) = 1:2.)

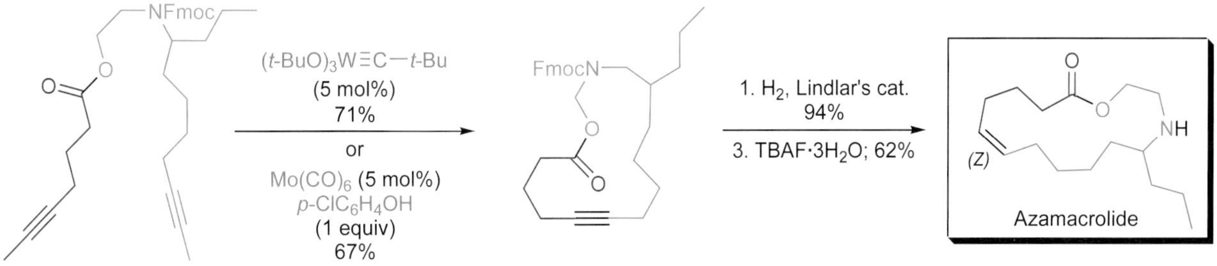

A. Fürstner and co-workers also showed that RCAM is indeed a very mild method, because during their stereoselective total synthesis of prostaglandin E_2-1,15-lactone, the Mo[N-(*t*-Bu)(Ar)$_3$]-derived catalyst tolerated a preexisting double bond and a ketone functionality.[17] Chromatographic inspection of the reaction mixture revealed that no racemization took place before or after the ring closure, and the *ee* of the substrate and the product were virtually identical.

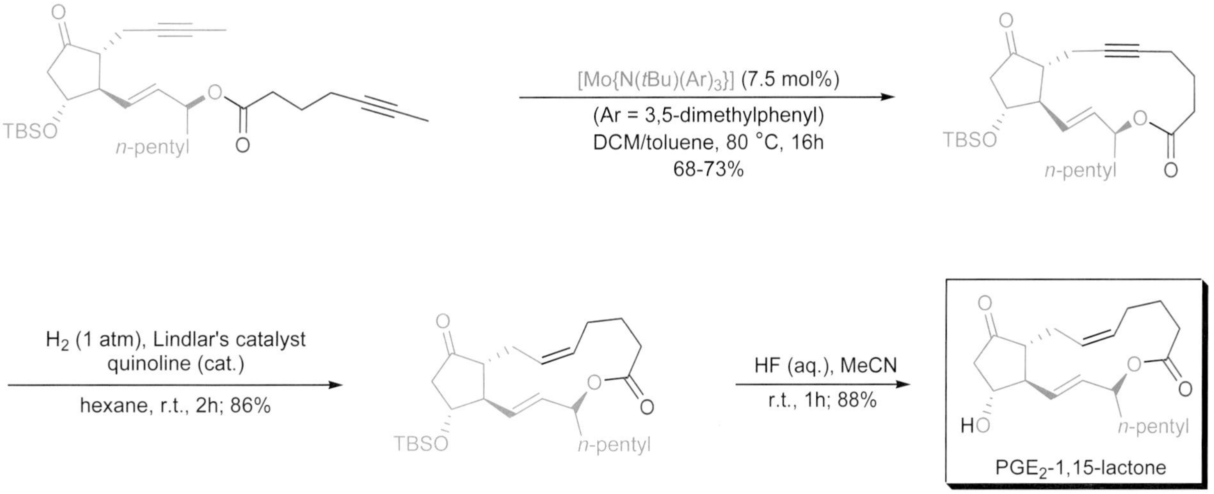

The first total synthesis of three naturally occurring cyclophane derivatives belonging to the turriane family of natural products was also described by A. Fürstner et al.[28] These natural products have a sterically hindered biaryl moiety and saturated as well as unsaturated macrocyclic tethers. Stereoselective entry to this class of compounds is possible using RCAM followed by Lindlar reduction of the resulting cycloalkynes.

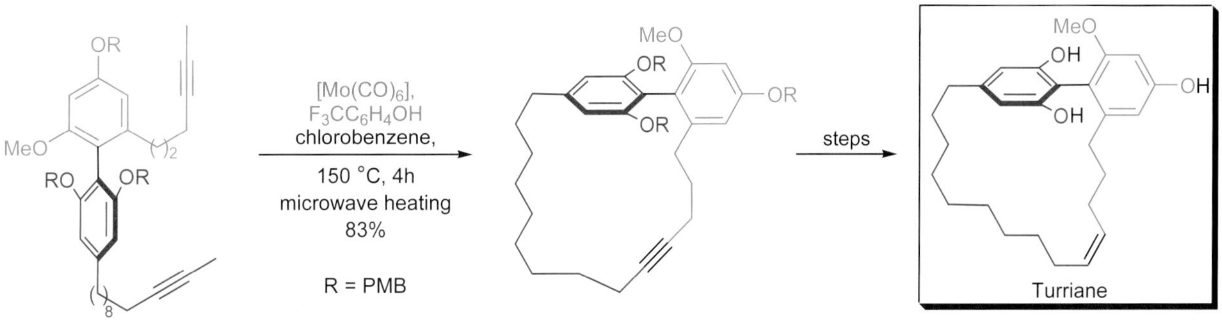

AMADORI REACTION / REARRANGEMENT

(References are on page 537)

Importance:

[*Seminal Publications*[1,2]; *Reviews*[3-9]]

The acid- or base-catalyzed isomerization of *N*-glycosides (glycosylamines) of aldoses to 1-amino-1-deoxyketoses is called the *Amadori reaction/rearrangement*. Both the substrates and the products are referred to as "Amadori compounds". A variety of Lewis acids ($CuCl_2$, $MgCl_2$, $HgBr_2$, $CdCl_2$, $AlCl_3$, $SnCl_4$, etc.) have been employed as catalysts to induce this rearrangement. The rearrangement takes place if an aldose is reacted with an amine in the presence of a catalytic amount of acid. The amine component can be primary, secondary, aliphatic, or aromatic. Glycosylamine derivatives are implicated in the complex *Maillard reaction*, whereby sugars, amines, and amino acids (proteins) condense, rearrange, and degrade often during cooking or preservation of food.[10] The dark-colored products formed in this reaction are responsible for the non-enzymatic browning observed with various foodstuffs.

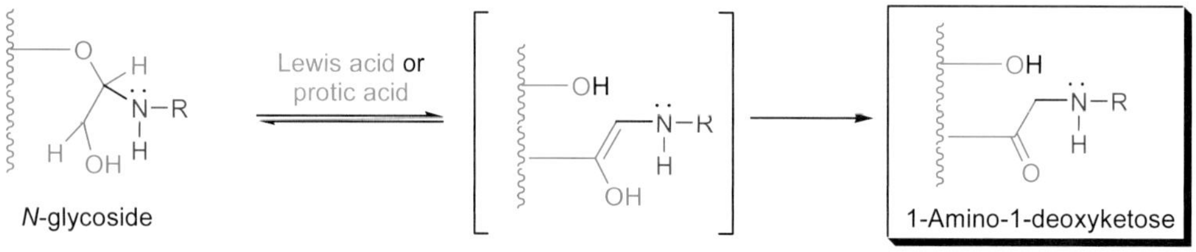

Mechanism: [11,12]

The first step of the mechanism is the coordination of the Lewis acid (proton in the scheme) to the ring oxygen atom of the *N*-glycoside. Subsequently the ring is opened, and the loss of a proton gives rise to an enolic intermediate, which in turn undergoes tautomerization to the corresponding 1-amino-1-deoxyketose.

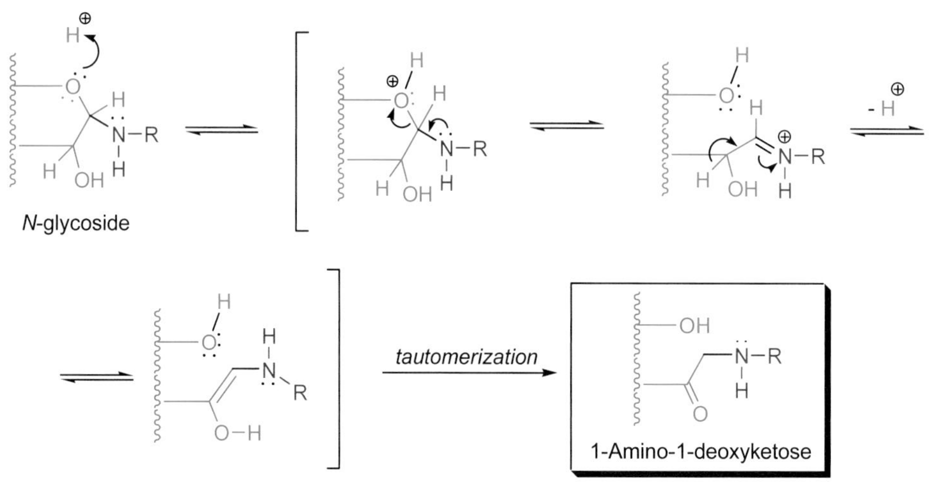

Synthetic Applications:

C. Blonski and co-workers utilized the *Amadori rearrangement* in the synthesis of various D-*fructose analogs* that were modified at C1, C2, or C6 positions.[13] The key intermediate, 1-deoxy-1-toluidinofructose, was obtained from D-glucose quantitatively by reacting D-glucose with *p*-toluidine in acetic acid.

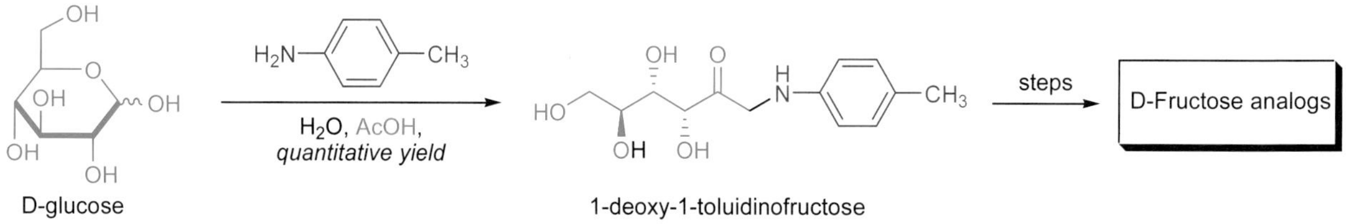

Synthetic Applications:

The synthesis of novel DNA topoisomerase II (topo II) inhibitors was undertaken in the laboratory of T.L. Macdonald.[14] Their research program dealt with the synthesis of piperidin-3-one derivatives, which were needed as synthetic intermediates for a variety of potential topo II-directed agents. The key step in their approach was the *Amadori reaction* for the preparation of highly functionalized piperidin-3-ones under mild conditions. Upon treatment with a catalytic amount of *p*-toluenesulfonic acid in toluene at reflux, the desired rearrangement took place in high yield.

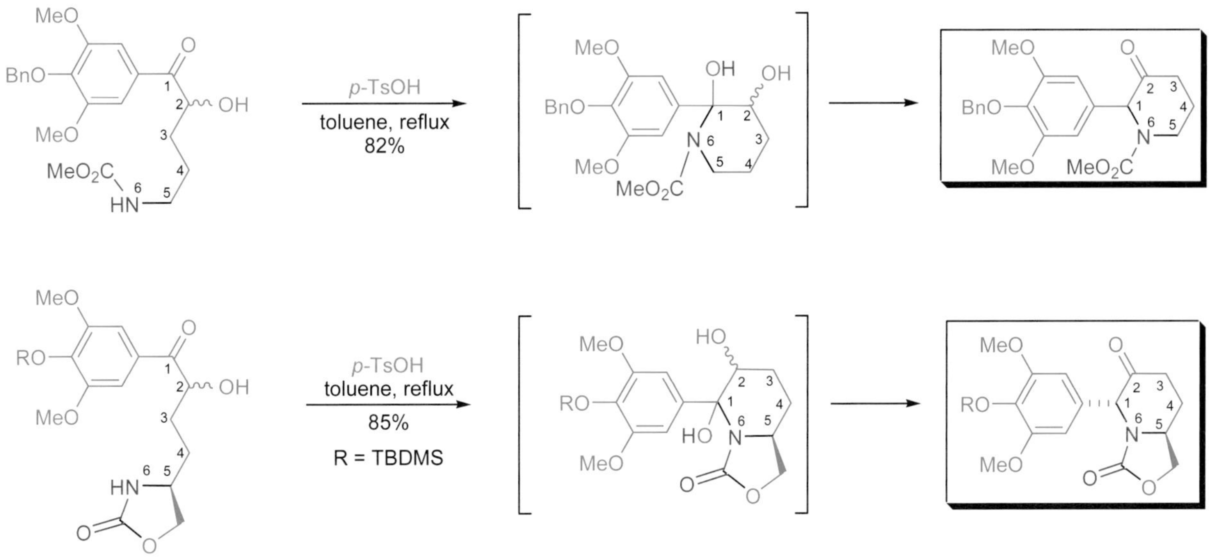

S. Horvat and co-workers conducted studies on the *intramolecular Amadori rearrangement* of the monosaccharide esters of the opioid pentapeptide leucine-enkephaline.[15] The esters were prepared from either D-glucose, D-mannose or D-galactose by linking their C6 hydroxy group to the C-terminal carboxy group of the endogenous opioid pentapeptide leucine-enkephaline (H-Tyr-Gly-Gly-Phe-Leu-OH). Exposure of these monosaccharide esters to dry pyridine-acetic acid (1:1) mixture for 24h at room temperature, resulted in the desired *Amadori rearrangement* to afford novel bicyclic ketoses that are related to the furanose tautomers of 1-deoxy-D-fructose (**I**) and 1-deoxy–D-tagatose (**II**).

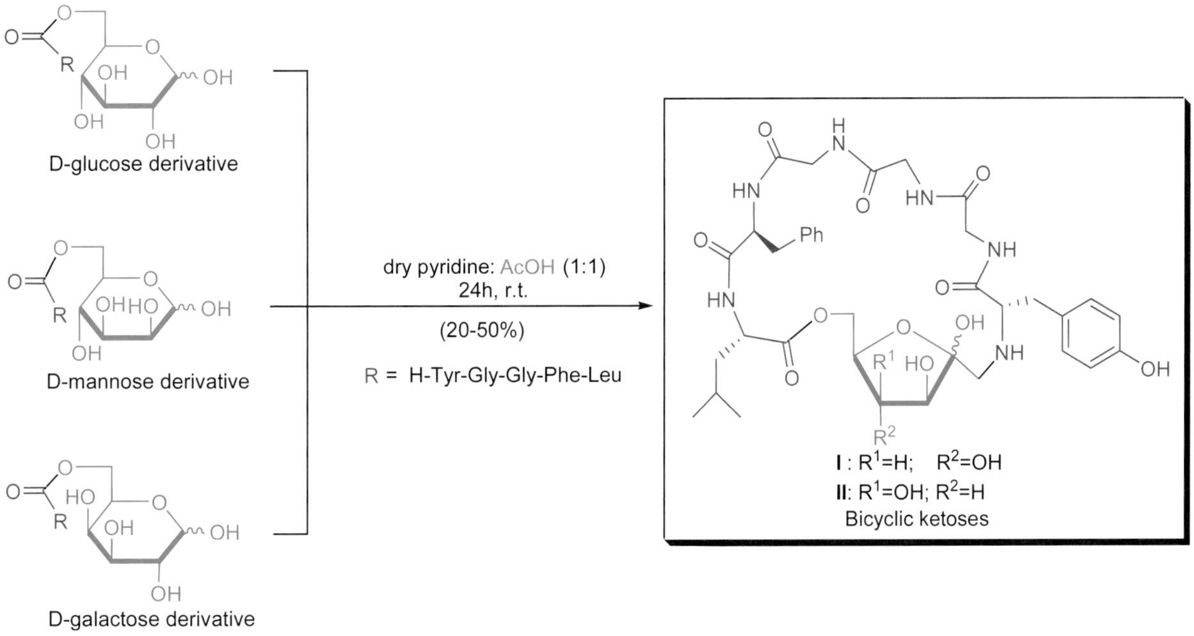

ARBUZOV REACTION
(MICHAELIS-ARBUZOV REACTION)

(References are on page 537)

Importance:

[*Seminal Publications*[1-4]; *Reviews*[5-14]; *Modifications & Improvements*[15-22]]

In 1898, A. Michaelis and R. Kaehne reported that, upon heating, trialkyl phosphites reacted with primary alkyl iodides to afford dialkyl phosphonates.[2] A few years later, A.E. Arbuzov investigated the reaction in great detail and determined its scope and limitations.[3] The synthesis of pentavalent alkyl phosphoric acid esters from trivalent phosphoric acid esters and alkyl halides is known as the *Arbuzov reaction* (also known as *Michaelis-Arbuzov reaction*). The general features of this transformation are:[9] 1) it usually proceeds well with primary alkyl halides (mainly iodides and bromides); 2) certain secondary alkyl halides such as *i*-PrI or ethyl α-bromopropionate do react, but with most secondary and tertiary alkyl halides the reaction does not take place or alkenes are formed; 3) besides simple alkyl halides, other organic halides are also good substrates for the reaction including benzyl halides, halogenated esters, acyl halides, and chloroformic acid esters; 4) aryl and alkenyl halides do not undergo S_N2 substitution, so they are unreactive under the reaction conditions; 5) activated aryl halides (e.g., heteroaryl halides: isoxazole, acridine, coumarin) do react; 6) the alkyl halides may not contain ketone or nitro functional groups, since these usually cause side reactions; 7) α-chloro- and bromo ketones undergo the *Perkow reaction* with trialkyl phosphites to afford dialkyl vinyl phosphates, but α-iodo ketones give rise to the expected Arbuzov products; 8) the trivalent phosphorous reactant can be both cyclic and acyclic; 9) in most cases the reaction takes place in the absence of a catalyst, but for certain substrates the presence of a catalyst is needed; and 10) catalysts can be various metals, metal salts, and complexes (e.g., Cu-powder, Ni-halides, $PdCl_2$, $CoCl_2$), protic acids (e.g., AcOH), or light. Phosphonates are of great importance in organic synthesis, agriculture, and chemical warfare. Organophosphoric acid esters are produced on the multiton scale and used as insecticides (e.g., methidathion, methyl-parathion, etc.). Organophosphonates also found application in chemical warfare (nerve gases such as VX, Sarin, etc.). They are potent inhibitors of the enzyme acetyl cholinesterase *via* phosphorylation and therefore extremely toxic to the parasympathetic nervous system. The *Horner-Emmons-Wadsworth modification of the Wittig reaction* (synthesis of alkenes from carbonyl compounds) utilizes phosphonates instead of phosphoranes. Phosphonates are easily deprotonated to yield ylides that are more reactive than the corresponding phosphoranes (phosphorous ylides). Phosphonates react with ketones that are unreactive toward phosphoranes.

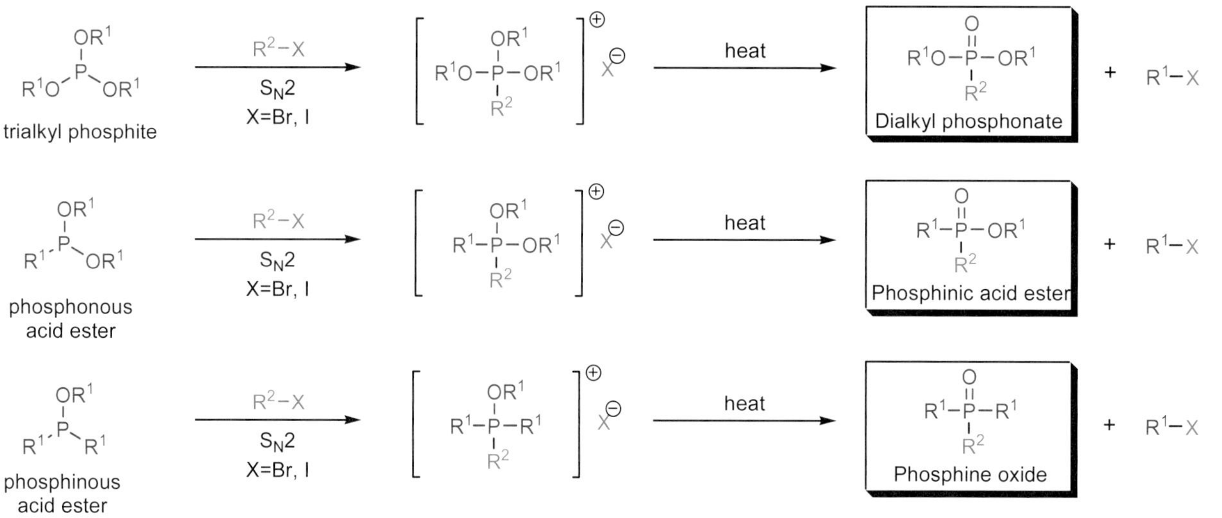

Mechanism: [23-30,15,18]

The first step of the mechanism is the nucleophilic attack (S_N2) of the alkyl halide by the phosphorous to form a phosphonium salt **A**. Under the reaction conditions (heat) the phosphonium salt **A** is unstable and undergoes a C-O bond cleavage (the halide ion (X^-) acts as a nucleophile and attacks one of the alkyl groups in an S_N2 reaction) to afford the phosphonate ester.

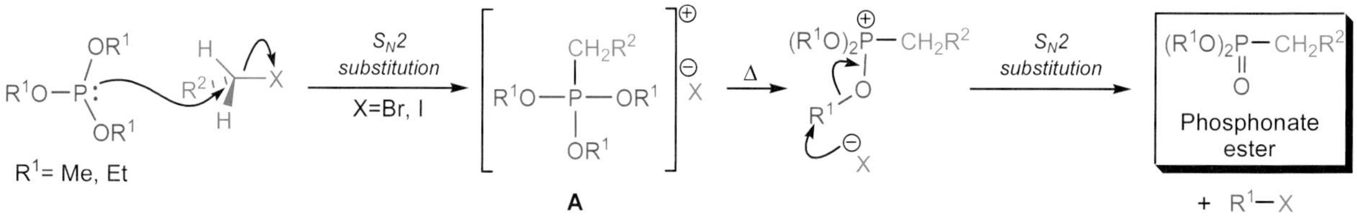

ARBUZOV REACTION
(MICHAELIS-ARBUZOV REACTION)

Synthetic Applications:

The phosphonic acid analog of NSAID (Non-Steroidal Anti-Inflammatory Drug) diclofenac® was successfully synthesized in the laboratory of B. Mugrage using a novel acid catalyzed *Arbuzov reaction* as the key step followed by a TMSBr promoted dealkylation.[31] It needs to be pointed out that the nucleophilic attack takes place on the *ortho*-quinonoid intermediate in a non-S_N2 process.

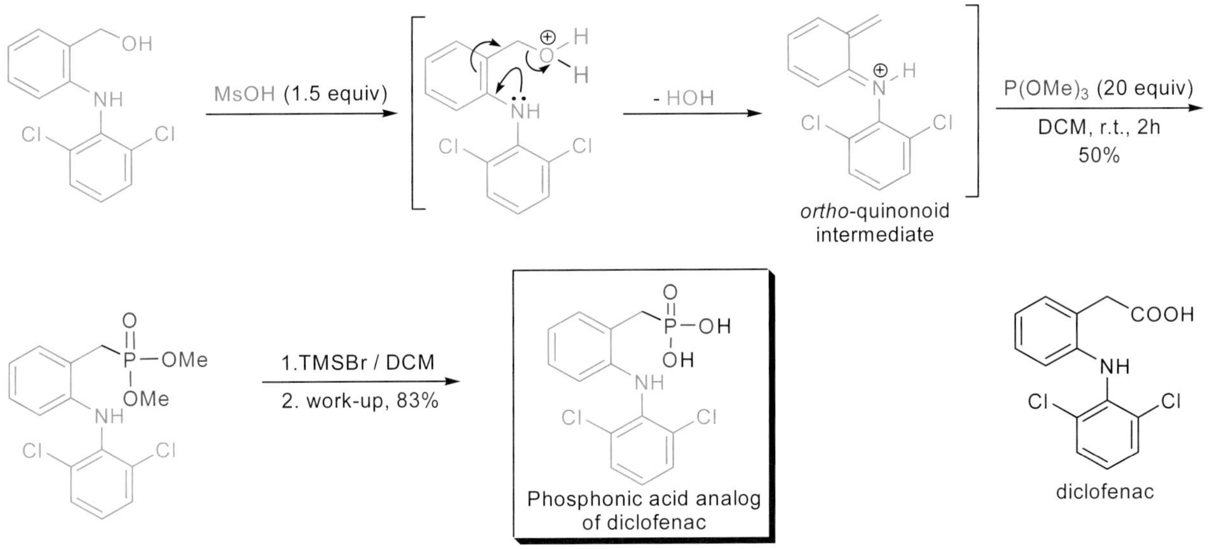

R.R. Schmidt and co-workers designed and synthesized a novel class of glycosyltransferase inhibitors.[32] The key synthetic steps involve an *Arbuzov reaction* followed by a coupling with uridine-5'-morpholidophosphate as the activated derivative.

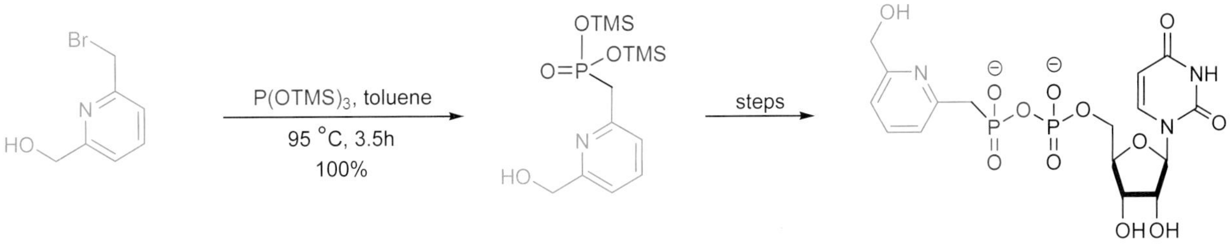

A novel enantioselective synthesis of an antagonist of the NMDA receptor, *cis*-perhydroisoquinoline LY235959, was achieved in 13% overall yield and 17 steps from (*R*)-pantolactone in the laboratory of M.M. Hansen.[33] The phosphoric acid portion of the target was introduced by a high-yielding *Arbuzov reaction*.

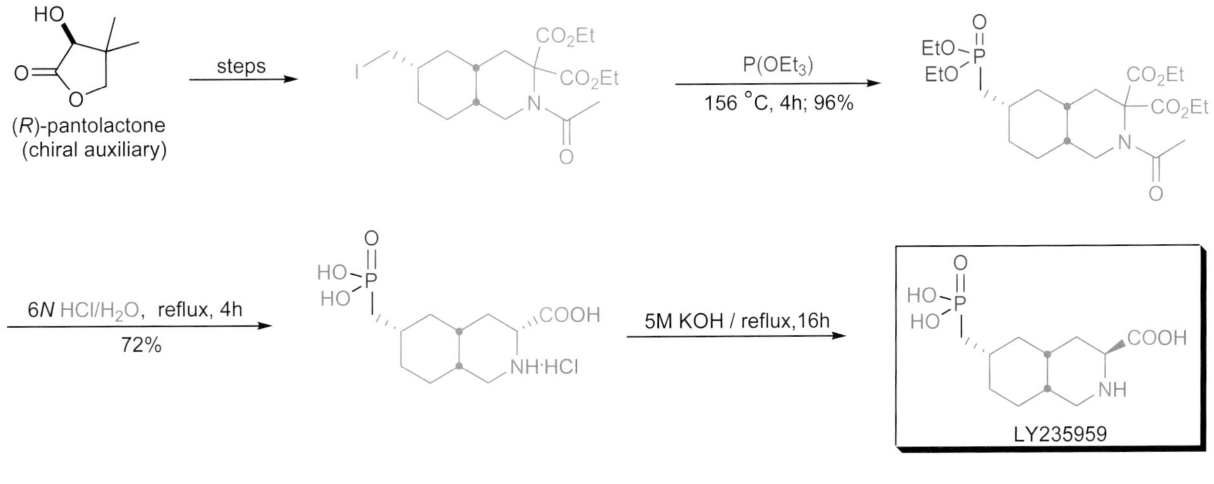

ARNDT-EISTERT HOMOLOGATION / SYNTHESIS

(References are on page 538)

Importance:

[*Seminal Publication*[1]; *Reviews*[2-4]; *Modifications & Improvements*[5-10]]

The conversion of a carboxylic acid to its homolog (one CH_2 group longer) in three stages is called the *Arndt-Eistert synthesis*. This homologation is the best preparative method for the chain elongation of carboxylic acids. In the first stage of the process the acid is converted to the corresponding acid chloride. The second stage involves the formation of a α-diazo methylketone, followed by a *Wolff rearrangement* in the third stage.[4] The third stage is conducted either in the presence of solid silver oxide/water or silver benzoate/triethylamine solution. The yields are usually good (50-80%). If the reaction is conducted in the presence of an alcohol (ROH) or amine (RNHR'), the corresponding homologated ester or amide is formed. Other metals (Pt, Cu) also catalyze the decomposition of the diazo ketones. An alternative method is to heat or photolyze the diazo ketone in the presence of a nucleophilic solvent (H_2O, ROH, or RNH_2), and in these cases no catalyst is required. The reaction tolerates a wide range of non-acidic functional groups (alkyl, aryl, double bonds). Acidic functional groups would react with diazomethane or diazo ketones.

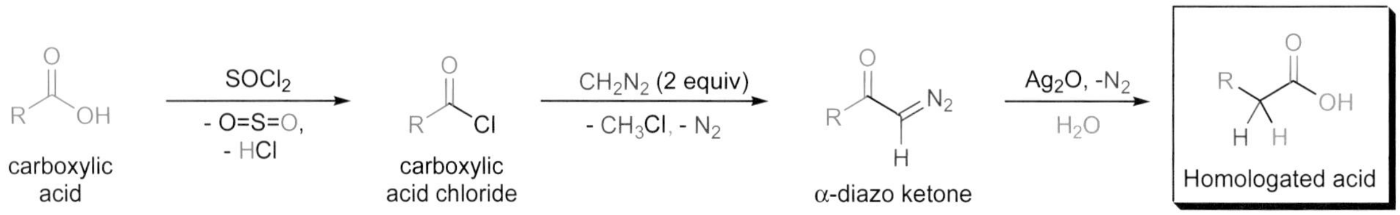

Mechanism: [2,11,4]

Since hydrogen chloride (HCl) is the by-product of the reaction between the acid chloride and diazomethane, two equivalents of diazomethane are needed so that the presence of HCl does not give side products (e.g. chloroketones). The HCl reacts with the second equivalent of diazomethane to form methyl chloride and dinitrogen. The role of the catalyst is not well understood. The diazo ketone can exist in two conformations, namely the *s-(E)* and *s-(Z)* conformations, which arise from the rotation about the C-C single bond. It has been shown that the *Wolff rearrangement* takes place preferentially from the *s-(Z)* conformation. With the loss of a molecule of nitrogen, the decomposition of the diazo ketone involves the formation of a carbene, followed by a carbene rearrangement with the intermediacy of an oxirene. The carbene undergoes a rapid *[1,2]*-shift to afford a ketene that reacts with the nucleophilic solvent to give the homologated acid derivative.

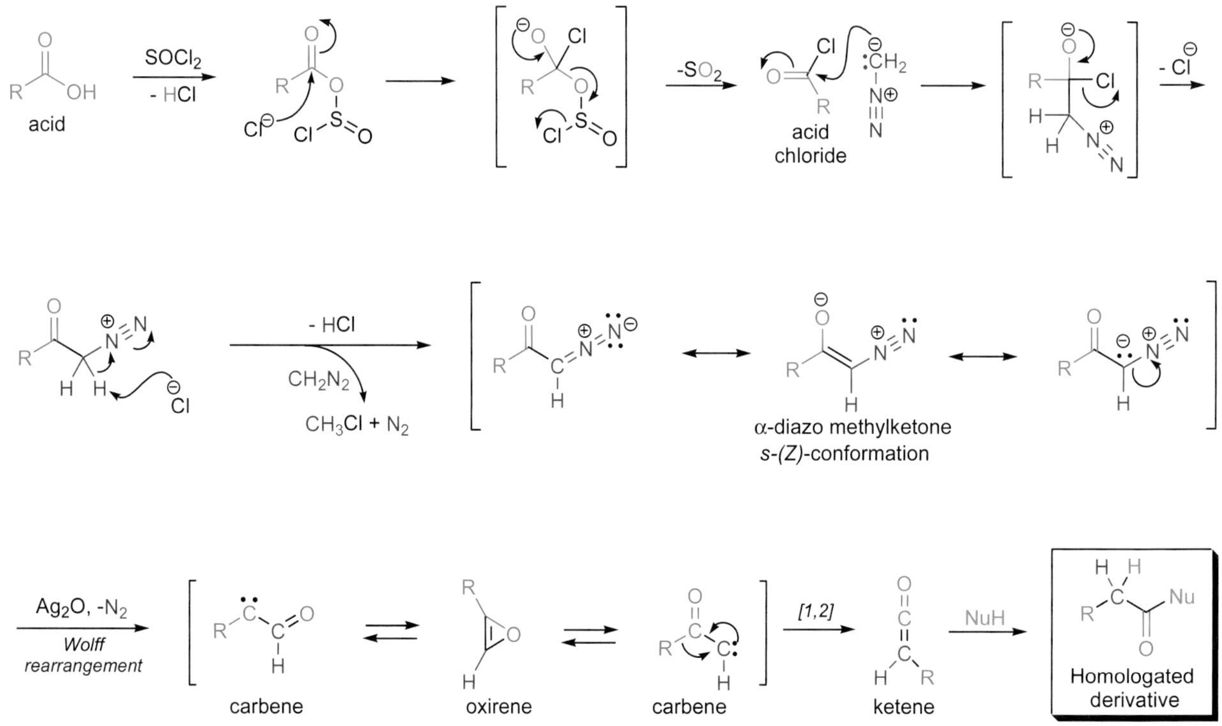

Synthetic Applications:

The oligomers of β-amino acids, as opposed to α-peptides, show a remarkable ability to fold into well-defined secondary structures in solution as well as in the solid state. The β-amino acid building blocks were synthesized from α-amino acids using the *Arndt-Eistert homologation reaction* in the laboratory of D. Seebach.[12]

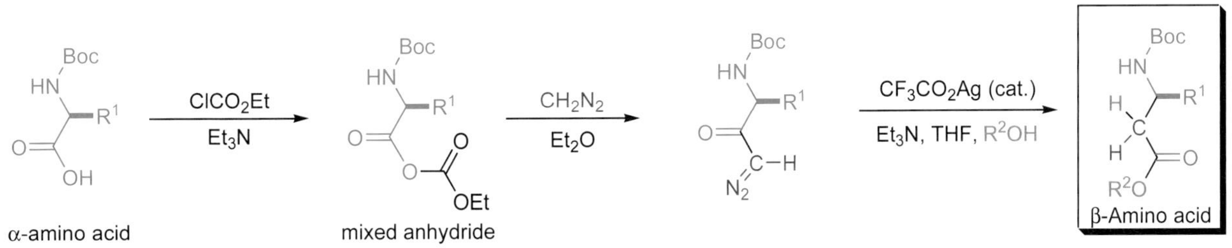

During the total synthesis of the CP molecules, K.C. Nicolaou et al. homologated a sterically hindered carboxylic acid, which was part of an advanced intermediate.[13] Due to the sensitive nature of this intermediate, the diazo ketone was prepared *via* the acyl mesylate rather than the acid chloride. The diazo ketone then was immediately dissolved in DMF:H_2O (2:1) and heated to 120 °C in the presence of excess Ag_2O for one minute to generate the homologated acid in 35% yield.

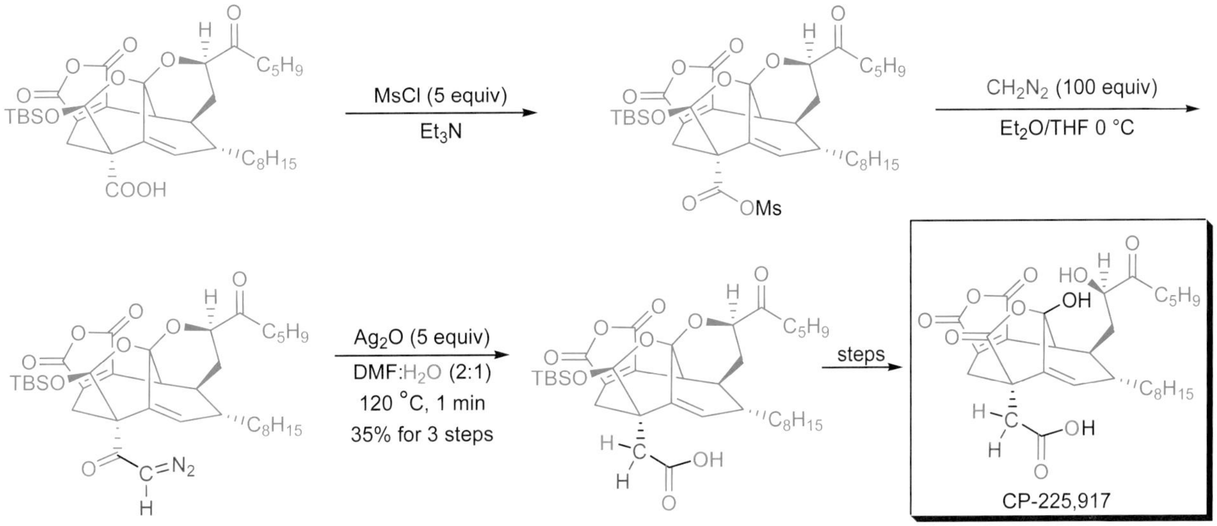

A.T. Russell and co-workers synthesized (*R*)-(–)-homocitric acid-γ-lactone in multigram quantities starting from a citric acid derivative and using the *Arndt-Eistert homologation* as the key step.[14]

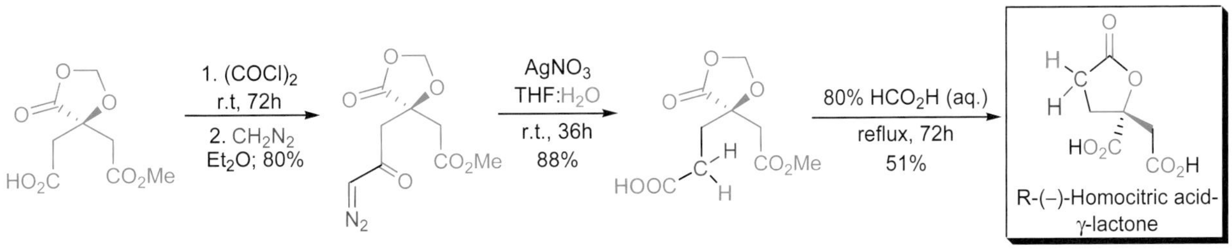

In the laboratory of B.M. Stolz, the first total synthesis of the bis-indole alkaloid (±)-dragmacidin D was accomplished.[15] During the endgame, a carboxylic acid was homologated to the corresponding α-bromo ketone by treating the diazo ketone intermediate with hydrobromic acid.

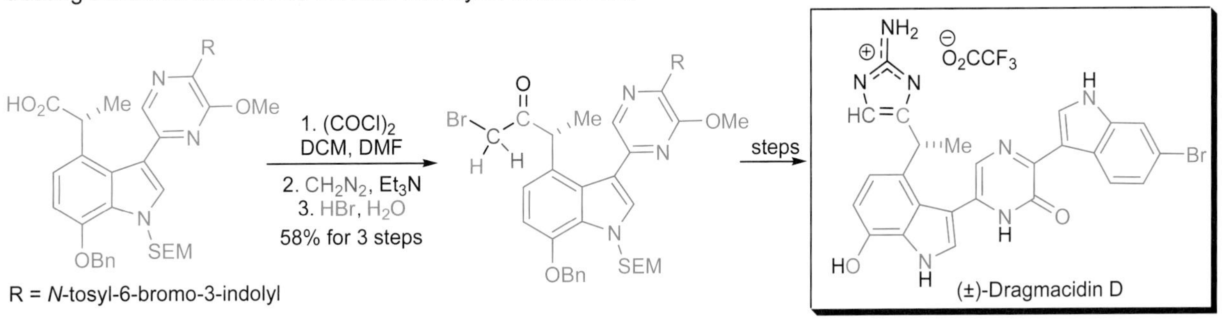

AZA-CLAISEN REARRANGEMENT
(3-AZA-COPE REARRANGEMENT)

(References are on page 538)

Importance:

[*Seminal Publications*[1]; *Reviews*[2,3]; *Modifications & Improvements*[4-11]; *Theoretical Studies*[12]]

The thermal *[3,3]-sigmatropic rearrangement* of allyl vinyl ethers is called the *Claisen rearrangement.*[13,14] Its variant, the thermal *[3,3]-sigmatropic rearrangement* of *N*-allyl enamines, is called the *aza-Claisen rearrangement* (*3-aza-Cope* or *amino-Claisen rearrangement*). There are several known variations of the *aza-Claisen rearrangement*, and each one belongs to a subclass of this type of reaction. The rates of the rearrangement depend mainly on the structural features of the specific system, which can be: 1) *3-aza-1,5-hexadienes;* 2) *3-azonia-1,5-hexadienes*; and 3) *3-aza-1,2,5-hexatrienes*. The observed temperature trend for these reactions is that milder temperatures are required as one progresses from the "*neutral*" to the "*charged*" and finally to the *keteneimine* rearrangement. The rearrangement generally occurs between 170-250 °C for the neutral species, and between room temperature and 110 °C for the Lewis acid coordinated or quaternized molecules.

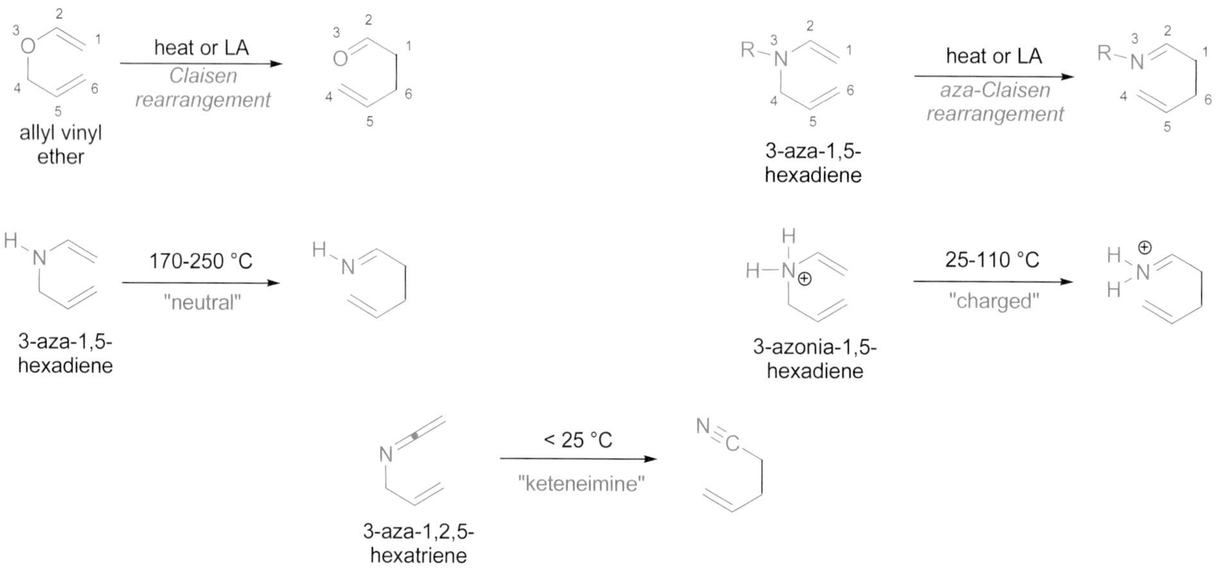

Mechanism:

The *aza-Claisen rearrangement* is a concerted process, and it usually takes place *via* a chairlike transition state where the substituents are arranged in quasi-equatorial positions. (See more details in *Claisen rearrangement*.)

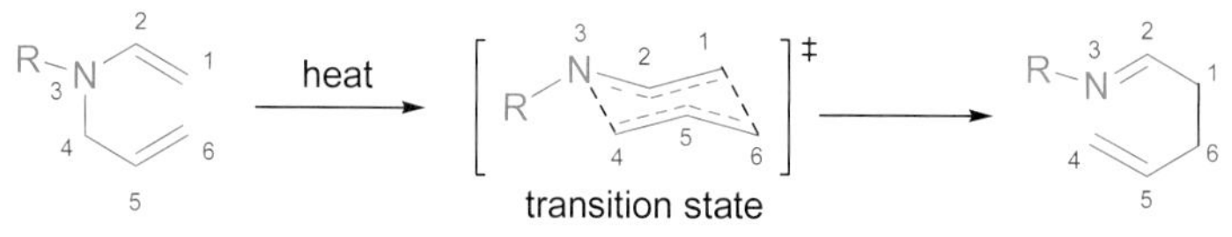

Synthetic Applications:

S. Ito et al. utilized the *aza-Claisen rearrangement* of carboxamide enolates for the enantioselective total synthesis of (–)-isoiridomyrmecin, which is a constituent of *Actinidia polygama* and exhibits unique bioactivity.[2] The rearrangement of the (*S,S*) stereoisomer was conducted under standard conditions, and the product was isolated as a single (*R,R*) stereoisomer in 77% yield.

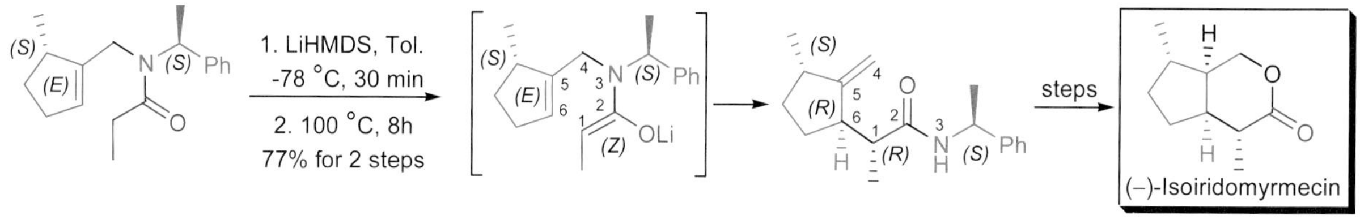

AZA-CLAISEN REARRANGEMENT
(3-AZA-COPE REARRANGEMENT)

Synthetic Applications:

The first asymmetric synthesis of fluvirucinine A1 was accomplished in the laboratory of Y.-G. Suh.[15] Key steps of the synthesis involved a diastereoselective vinyl addition to the amide carbonyl group as well as an *amide enolate induced aza-Claisen rearrangement.*[15]

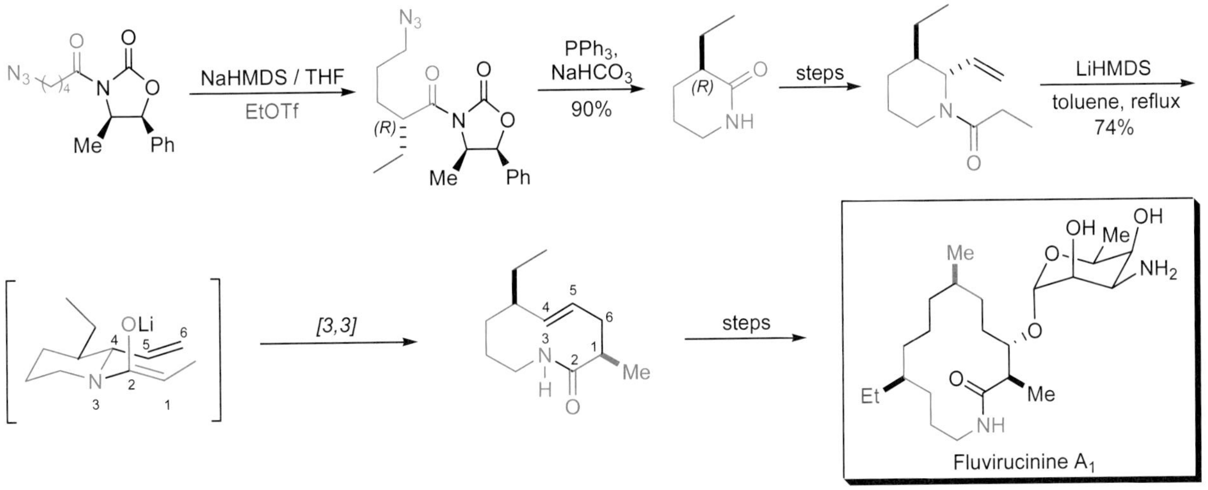

T. Tsunoda and co-workers synthesized the antipode of natural antibiotic antimycin A3b starting from (*R*)-(+)-methylbenzylamine and utilizing the *asymmetric aza-Claisen rearrangement.*[16] The amide precursor was deprotonated with LiHMDS at low temperature then the reaction mixture was refluxed for several hours to bring about the sigmatropic rearrangement.

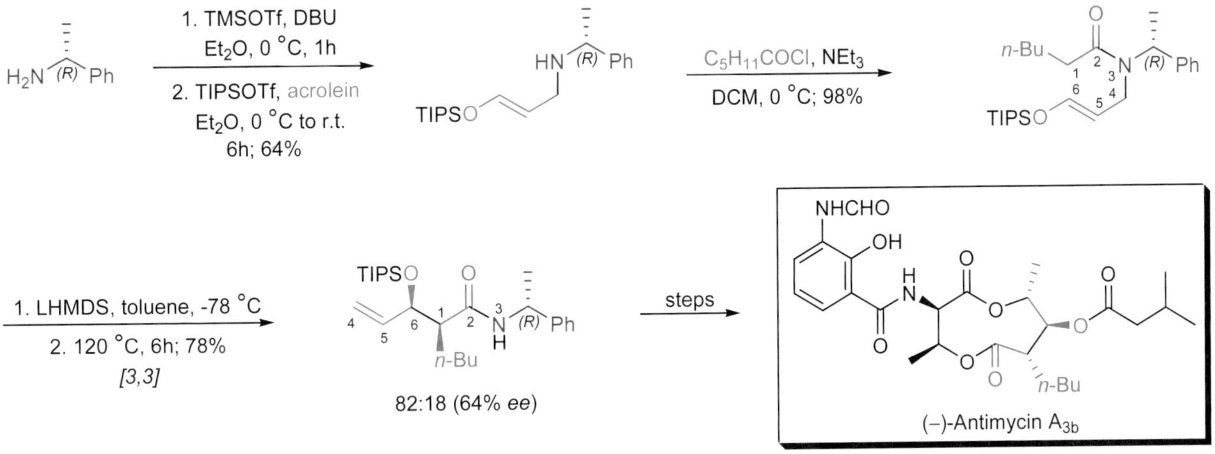

U. Nubbemeyer et al. achieved the enantioselective total synthesis of the bicyclic tetrahydrofuran natural product (+)-dihydrocanadensolide *via* a key step utilizing the *diastereoselective zwitterionic aza-Claisen rearrangement* of an *N*-allylpyrrolidine.[17]

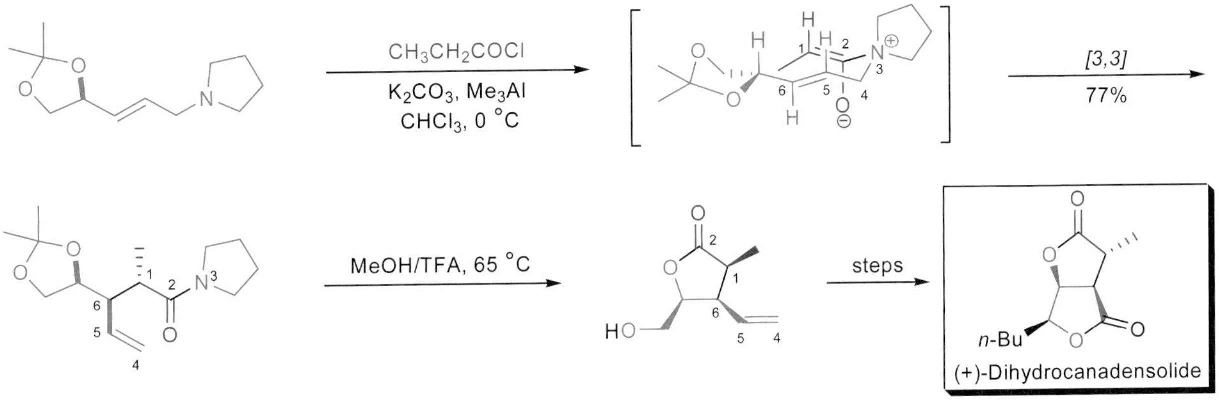

AZA-COPE REARRANGEMENT

(References are on page 538)

Importance:

[*Seminal Publications*[1-3]; *Reviews*[4-6]; *Modifications & Improvements*[7-28]; *Theoretical Studies*[18,21,29]]

When 1,5-dienes are heated, they isomerize *via* a *[3,3]-sigmatropic rearrangement* known as the *Cope rearrangement*. The rearrangement of *N*-substituted 1,5-dienes is called the *aza-Cope rearrangement*. This reaction has many variants, namely *1-aza-*, *2-aza-*, *3-aza-* and *1,3-*, *2,3-*, *2,5-*, *3,4- diaza-Cope rearrangements*.[7,8] The *3-aza-Cope rearrangement* is also known as the *aza-Claisen rearrangement*. The rearrangement of *cis*-2-vinylcyclopropyl isocyanates to 1-azacyclohepta-4,6-dien-2-ones (*2-aza-divinylcyclopropane rearrangement*) is analogous to the well-known and highly stereospecific *cis-divinylcyclopropane rearrangement*. It is well established that the presence of an oxygen atom adjacent to the π-bond accelerates the *Cope rearrangement*. When there is a group attached to C3 or C4 with which the newly formed double bond can conjugate, the reaction takes place at a lower temperature than in the unsubstituted case. As with all *[3,3]-sigmatropic rearrangements*, the activation energies are significantly lowered when the starting diene is charged.

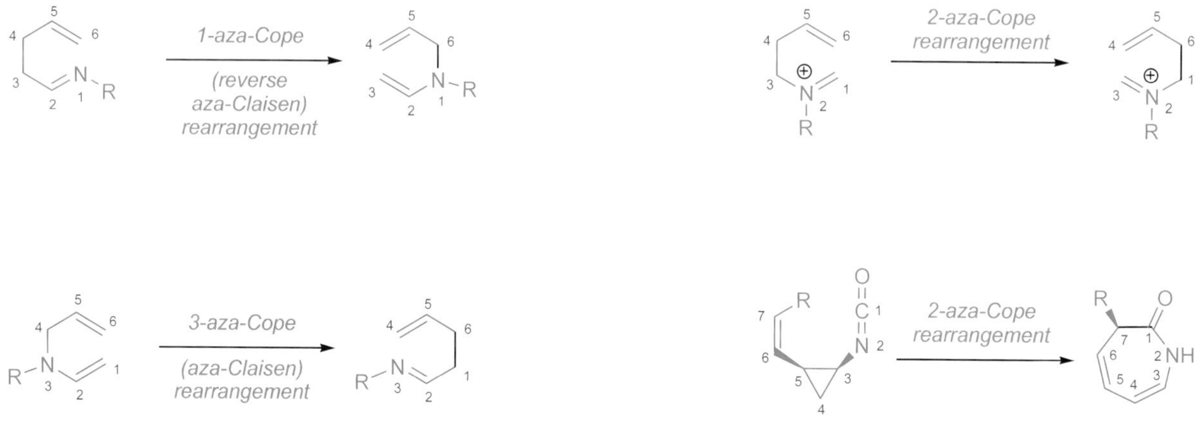

Mechanism: [30-35,18,36,21,37,38,29]

The *aza-Cope rearrangement* is a concerted process, and it usually takes place *via* a chairlike transition state where the substituents are arranged in a quasi-equatorial position. (*See more detail in Cope rearrangement.*)

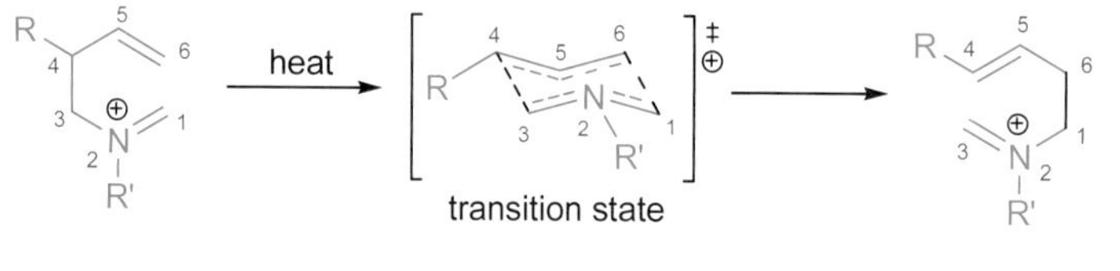

Synthetic Applications:

The *tandem cationic aza-Cope rearrangement* followed by a *Mannich cyclization* was applied in the synthesis of the novel tricyclic core structure of the powerful immunosupressant FR901483 in the laboratory of K. Brummond.[39] Their approach was the first synthetic example in which this tandem reaction passes through a bridgehead iminium ion.

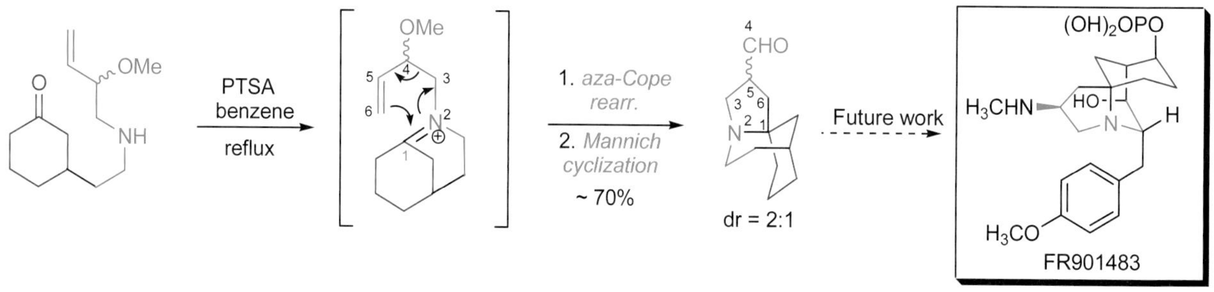

Synthetic Applications:

D.J. Bennett et al. developed a facile synthesis of *N*-benzylallylglycine based on a *tandem 2-aza-Cope/iminium ion solvolysis reaction*.[40] *N*-Benzylallylglycine can be prepared in good yield through a one-pot reaction of *N*-benzylhomoallylamine with glyoxylic acid monohydrate in methanol.

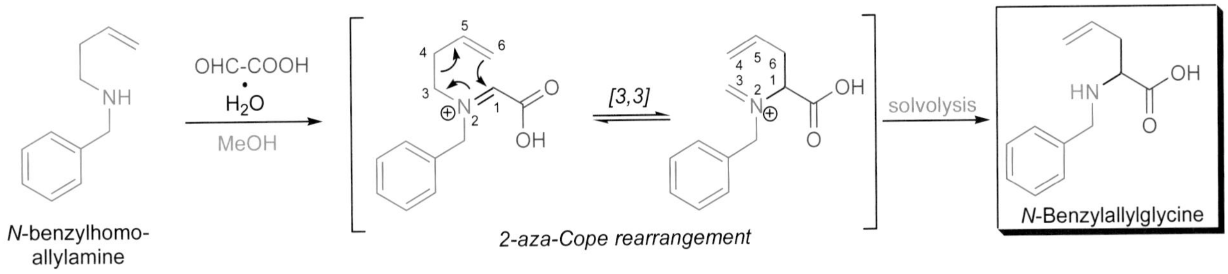

L.E. Overman and co-workers accomplished a total synthesis of (±)-gelsemine by a sequence where the key strategic steps are a sequential anionic 2-*aza-Cope rearrangement* and *Mannich cyclization*, an *intramolecular Heck reaction*, and a *complex base-promoted molecular reorganization* to generate the hexacyclic ring system.[41] The exposure of the bicyclic substrate to potassium hydride in the presence of 18-crown-6 initiated the anionic aza-Cope rearrangement of the bicyclic formaldehyde-imine alkoxide. The rearrangement product was quenched with excess methyl chloroformate then was treated with base to afford the desired *cis*-hexahydroisoquinolinone.

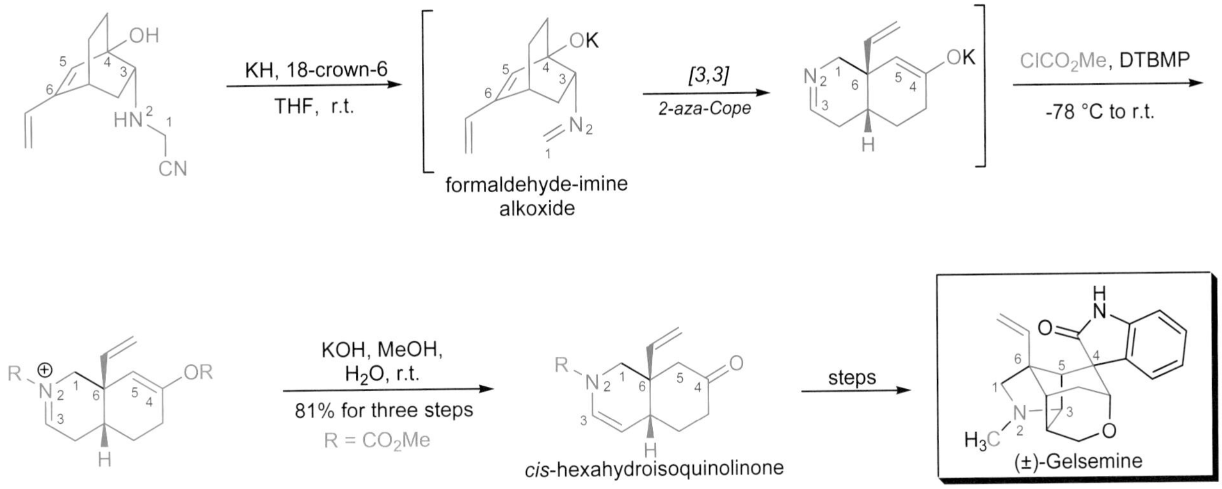

During the enantioselective total syntheses of (–)- and (+)-strychnine and the Wieland-Gumlich aldehyde, L.E. Overman and co-workers used the tandem *aza-Cope rearrangement/Mannich reaction* as a key step.[42] This central *aza-Cope/Mannich reorganization* step proceeded in 98% yield.

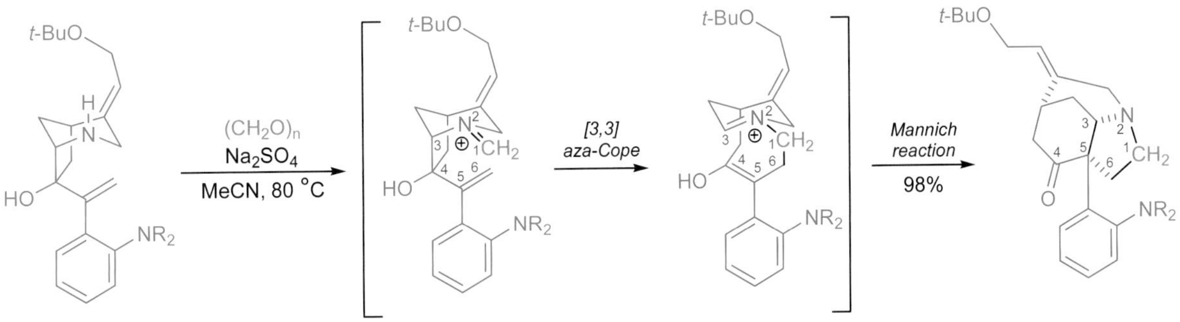

AZA-WITTIG REACTION

(References are on page 539)

Importance:

[*Seminal Publications*[1]; *Reviews*[2-11]; *Theoretical Studies*[12-17]]

In 1919, H. Staudinger and J. Meyer prepared $PhN{=}PPh_3$, an aza-ylide which was the first example of an aza-Wittig reagent.[1] By definition an ylide is "a substance in which a carbanion is attached directly to a heteroatom carrying a substantial degree of positive charge and in which the positive charge is created by the sigma bonding of substituents to the heteroatom".[4] The reaction of aza-ylides (iminophosphoranes) with various carbonyl compounds is called the *aza-Wittig reaction*. The product of the reaction is a Schiff base. Just as in the regular *Wittig reaction*, the by-product is triphenylphosphine oxide. Over the last decade, the *aza-Wittig* methodology has received considerable attention because of its utility in the synthesis of C=N double bond containing compounds, in particular, nitrogen heterocycles. The *intramolecular aza-Wittig reaction* is a powerful tool for the synthesis of 5-, 6-, 7-, and 8 membered heterocycles.

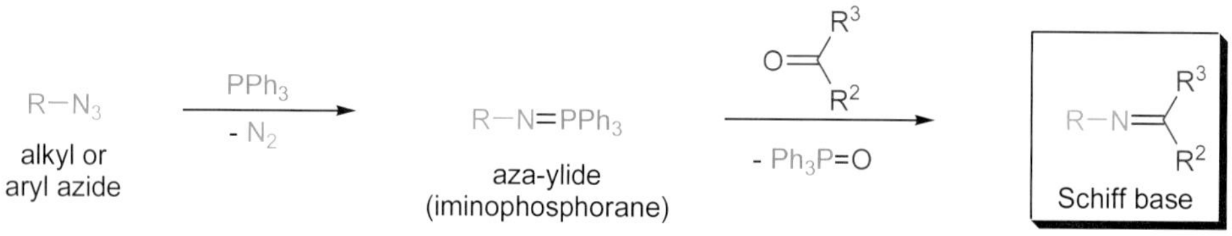

Mechanism:[18,15]

In the first step, the triphenylphosphine reacts with an alkyl azide to form an iminophosphorane with loss of nitrogen (*Staudinger reaction*). In the second step, the nucleophilic nitrogen of the iminophosphorane attacks the carbonyl group to form a four-membered intermediate (oxazaphosphetane) from which the product Schiff base and the by-product triphenylphosphine oxide are released.

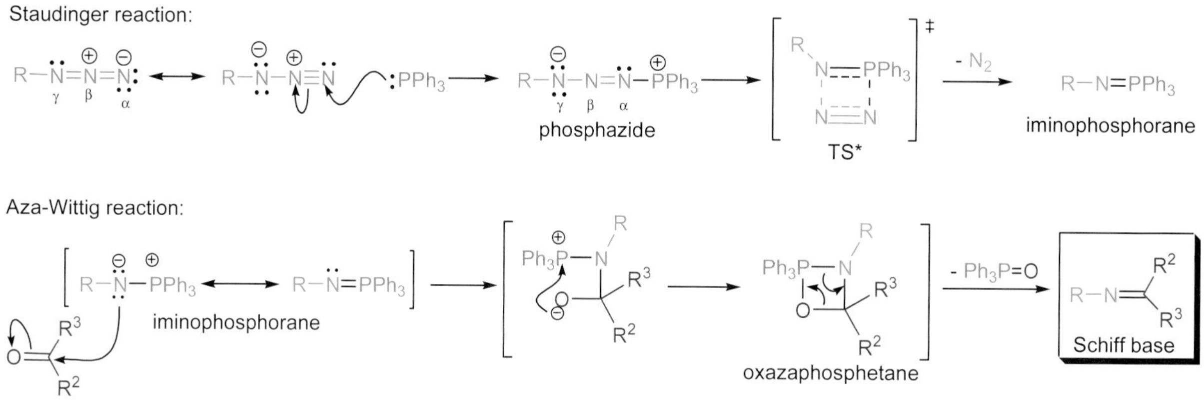

Synthetic Applications:

The solid phase synthesis of trisubstituted guanidines was achieved in the research group of D.H. Drewery by utilizing the *aza-Wittig reaction*. The reaction of solid-supported alkyl iminophosphorane and aryl or alkyl isothiocyanates afforded carbodiimides, which upon treatment with primary or secondary amines provided the trisubstituted guanidines.[19]

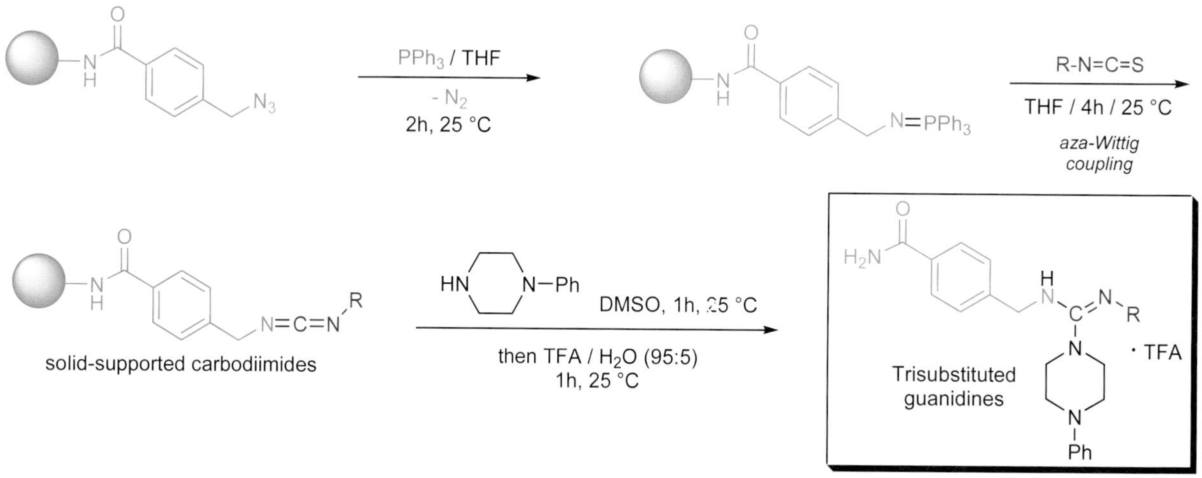

Synthetic Applications:

D.R. Williams and co-workers have accomplished the stereocontrolled total synthesis of the polycyclic *Stemona* alkaloid, (–)-stemospironine.[20] Key transformations included the use of a *Staudinger reaction* leading to the *aza-Wittig ring closure* of the perhydroazepine system. The *Staudinger reaction* was initiated by the addition of triphenylphosphine, leading to an aza-ylide for intramolecular condensation providing a seven-membered imine. An *in situ* reduction yielded the azepine system. Finally, (–)-stemospironine was produced by the iodine-induced double cyclization reaction in which the vicinal pyrrolidine butyrolactone was formed *via* the stereoselective intramolecular capture of an intermediate aziridinium salt.

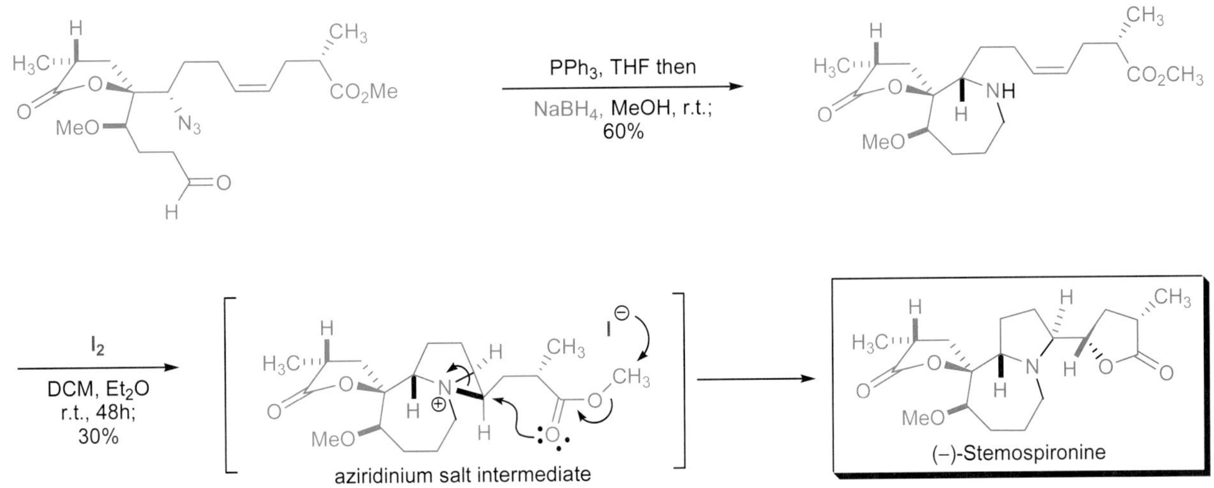

The first total synthesis of (–)-benzomalvin A, which possesses a 4(3H)-quinazolinone and 1,4-benzodiazepin-5-one moiety, was accomplished in the laboratory of S. Eguchi.[21] Both 6- and 7-membered ring skeletons were efficiently constructed by the *intramolecular aza-Wittig reaction*. The precursors were prepared from L-phenylalanine. The reaction of the azide derivative with tributylphosphine formed the corresponding iminophosphorane intermediate, which spontaneously underwent the *aza-Wittig cyclization* to give the 7-membered ring. Finally the 6-membered ring of (–)-benzomalvin A was constructed by another *intramolecular aza-Wittig cyclization reaction*.

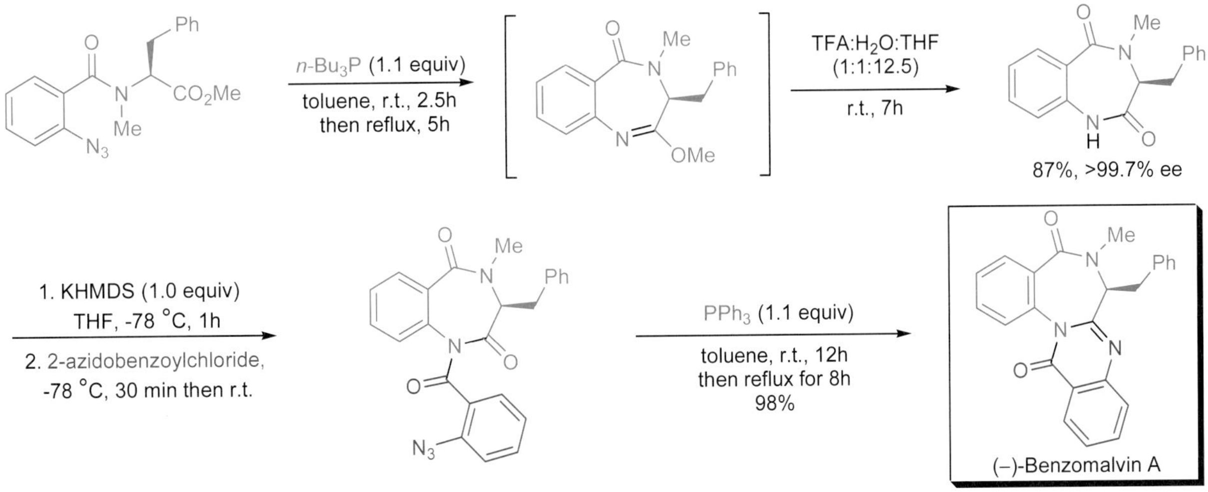

In the total synthesis of antitumor antibiotic (±)-phloeodictine A1 by B.B. Snider and co-workers, the key step was an *aza-Wittig reaction* followed by a *retro-Diels-Alder reaction* to afford the desired bicyclic amidine.[22] The polystyrene-supported PPh_3 made it easy to separate the product from by-products with a simple filtration.

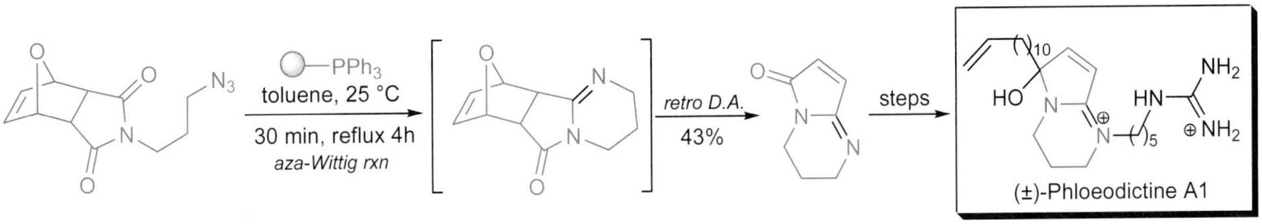

AZA-[2,3]-WITTIG REARRANGEMENT

(References are on page 540)

Importance:

[*Seminal Publications*[1,2]; *Review*[3]; *Modifications & Improvements*[4-14]; *Theoretical Studies*[9,15]]

The highly stereoselective rearrangement of α-metalated ethers to metal alkoxides is called the *Wittig rearrangement* and was first reported by G. Wittig and L. Löhmann in 1942.[16] The product is a secondary or tertiary alcohol after hydrolytic work-up. The nitrogen analog of this reaction is the isoelectronic *aza-Wittig rearrangement* that involves the isomerization of α-metalated tertiary amines to skeletally rearranged metal amides. The corresponding homoallylic secondary amines are obtained upon work-up. It was shown that the *aza-[2,3]-Wittig rearrangement* proceeds with the inversion of configuration of the lithium bearing carbon[17] as it occurs in the oxygen series. The *aza-Wittig rearrangement* should not be confused with the *Stevens* or *Sommelet-Hauser rearrangement* that both require quaternary ammonium salts as starting materials. These two rearrangements may lead to side products (e.g., when a quaternary ammonium salt is treated with a strong base, a rearranged tertiary amine may be formed by the *Stevens rearrangement* through a vicinal alkyl migration). In the case of a benzyltrialkylammonium salt the *Sommelet-Hauser rearrangement* may also compete; it is favored at low temperatures and yields an *o*-substituted benzyldialkylamine through a *[2,3]-sigmatropic rearrangement*.[3] In general, the *aza-[2,3]-Wittig rearrangement* of α-metalated amines is considerably slower (due to the lack of a thermodynamic driving force) and less selective than that of α-metalated ethers. Exceptions are noted when the rate of rearrangement is increased due to the relief of ring strain.

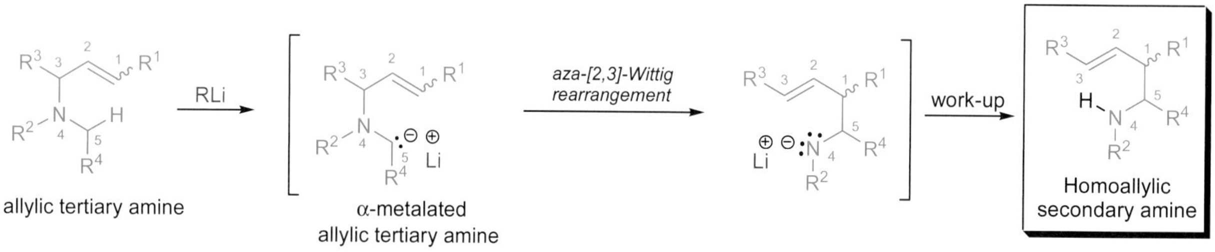

Mechanism: [18-22,9]

The *aza-[2,3]-Wittig rearrangement* proceeds by a concerted process through a six-electron, five-membered cyclic transition state of envelope-like geometry. According to the Woodward-Hoffmann rules, the *[2,3]-sigmatropic rearrangement* is a thermally allowed, concerted sigmatropic rearrangement that proceeds in a suprafacial fashion with respect to both fragments. Therefore, the *aza-[2,3]-Wittig rearrangement* is a one-step S_Ni-reaction, which results in a regiospecific carbon-carbon bond formation by suprafacial allyl inversion in which the heteroatom function gets transposed from allylic to homoallylic. The driving force for these rearrangements is the transfer of a formal negative charge from the less electronegative α-carbon to the more electronegative heteroatom.

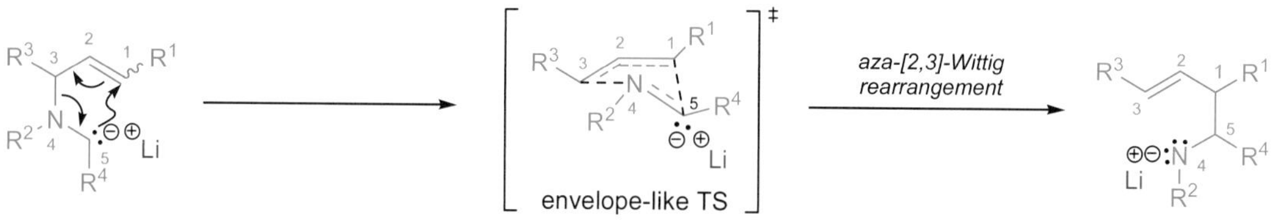

Synthetic Applications:

In the laboratory of J.C. Anderson, the total synthesis of (±)-kainic acid was accomplished relying on a route that utilized an *aza-[2,3]-Wittig rearrangement* as the key step to install the correct relative stereochemistry between C2 and C3.[23] The C4 stereocenter was established *via* an *iodolactonization reaction*.

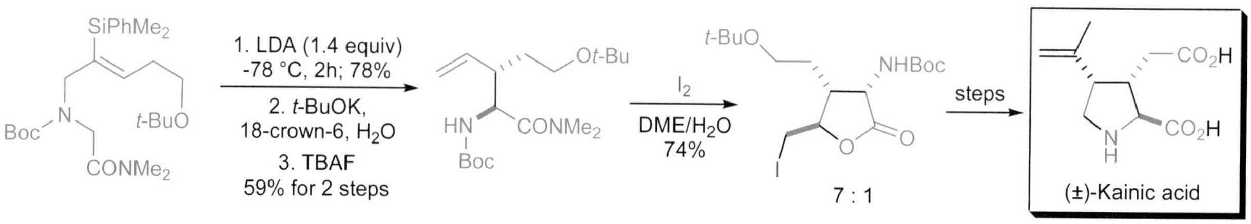

Synthetic Applications:

The *aza-Wittig rearrangement* of appropriately substituted vinylaziridines leads to the stereoselective formation of tetrahydropyridines, which are key intermediates in the synthesis of piperidines. A one-pot, two-step synthesis of unsaturated piperidines from 2-ketoaziridines utilizing the *aza-[2,3]-Wittig rearrangement* was reported by I. Coldham and co-workers.[24] Treatment of 2-ketoaziridines with two equivalents of a phosphonium ylide generates vinylaziridines that rearrange by a *[2,3]*-sigmatropic shift with the concomitant ring opening of the aziridines to give unsaturated piperidines.

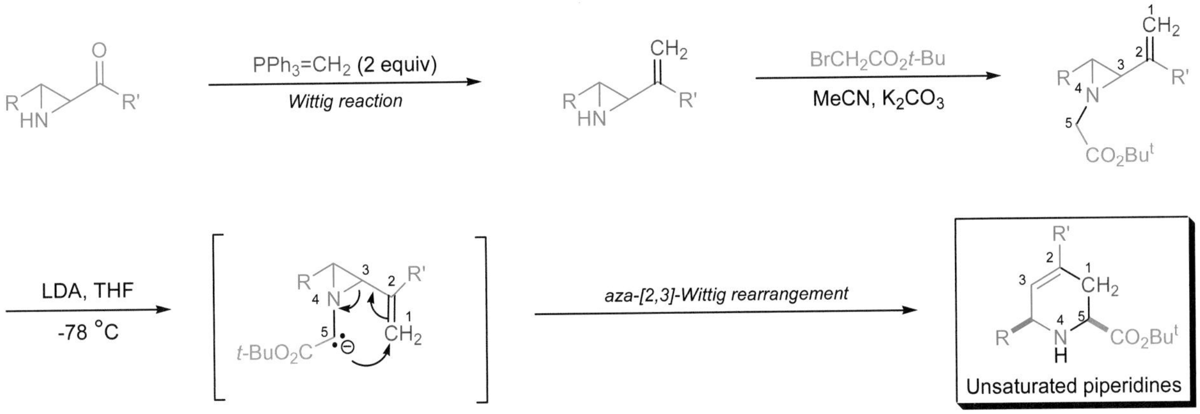

Research by J.C. Anderson et al. has shown that the inclusion of a C2 trialkylsilyl substituent into allylic amine precursors allows the base-induced *aza-[2,3]-sigmatropic rearrangement* to proceed in excellent yield and diastereoselectivity.[14] The rearrangement precursors require a carbonyl-based nitrogen protecting group that must be stable to the excess strong base required for the reaction. The *N*-Boc and *N*-benzoyl groups are very good at stabilizing the product anion and initiating deprotonation. The migrating groups need to stabilize the initial anion by resonance and a pK_a>22 is required for the rearrangement to occur. Products are formed with high *anti* diastereoselectivity (10:1-20:1).

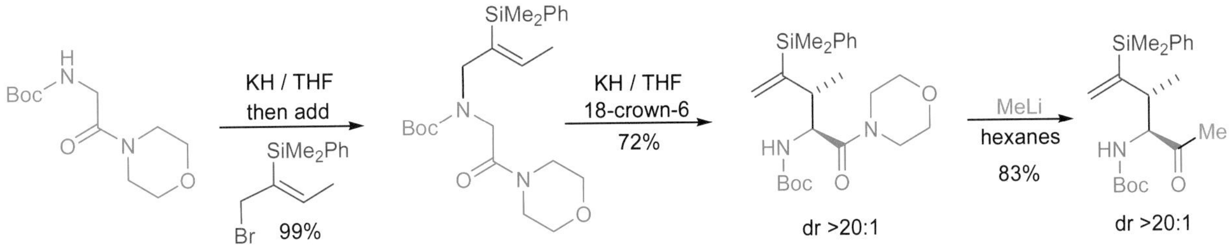

Tertiary amines are generally reluctant to undergo the *[2,3]-aza-Wittig rearrangement* and promotion of the rearrangement leads to unreacted starting material or *[1,2]*-rearranged products. However, in certain cases the addition of Lewis acids can lead to successful *aza-[2,3]-Wittig rearrangements*. In the laboratory of I. Coldham, the *aza-[2,3]-Wittig rearrangement* of *N*-alkyl-*N*-allyl-α-amino esters to *N*-alkyl-*C*-allyl glycine esters was investigated in detail.[25] It was reported that instead of using Lewis acids, the addition of iodomethane or benzyl bromide to tertiary amines promoted quaternary ammonium salt formation. *In situ*, these salts underwent spontaneous *[2,3]*-sigmatropic rearrangement when DMF was used as the solvent along with K_2CO_3 and DBU at 40 °C. In all cases when R=Me, a 60:40 *anti:syn* ratio of diastereomers was obtained.

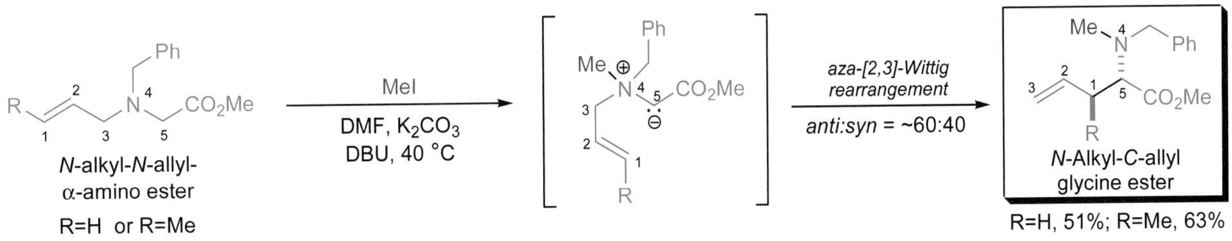

BAEYER-VILLIGER OXIDATION/REARRANGEMENT

(References are on page 540)

Importance:

[*Seminal Publication*[1]; *Reviews*[2-25]; *Modifications & Improvements*[26-36]; *Theoretical Studies*[37-48]]

The transformation of ketones into esters and cyclic ketones into lactones or hydroxy acids by peroxyacids was discovered as early as 1899 by A. Baeyer and V. Villiger when they were investigating the ring cleavage of cyclic ketones. This reaction was later named after them as the *Baeyer-Villiger oxidation*. The oxidation of ketones using this method has the following features: 1) it tolerates the presence of many functional groups in the molecule, for example, even with α,β-unsaturated ketones, the oxidation with peroxyacids generally occurs at the carbonyl group and not at the C=C double bond; 2) the regiochemistry depends on the migratory aptitude of different alkyl groups. For acyclic compounds, R' must usually be secondary, tertiary, or vinylic. For unsymmetrical ketones the approximate order of migration is tertiary alkyl > secondary alkyl > aryl > primary alkyl > methyl, and there are cases (e.g., bicyclic systems) in which various stereoelectronic aspects can influence which group migrates; 3) the rearrangement step occurs with retention of the stereochemistry at the migrating center; 4) a wide variety of peroxyacids can be used as oxidants for the reaction; and 5) the oxidation can also be performed asymmetrically on racemic or prochiral ketones using enzymes or chiral transition metal catalysts. A wide range of oxidizing agents can be used to perform the *Baeyer-Villiger oxidations* and their activity is ranked as follows: CF_3CO_3H > monopermaleic acid > monoperphthalic acid > 3,5-dinitroperbenzoic acid > *p*-nitroperbenzoic acid > *m*CPBA ~ performic acid > perbenzoic acid > peracetic acid » H_2O_2 > *t*-BuOOH.[34] Recently there has been considerable effort to make the *B.-V. oxidation* catalytic and at the same time preserve the high regio- and stereoselectivity of the reaction. Some of the most promising catalysts are substituted seleninic acids that are usually generated *in situ* from diaryl diselenides with H_2O_2 (*Syper method* of activation).[28,34]

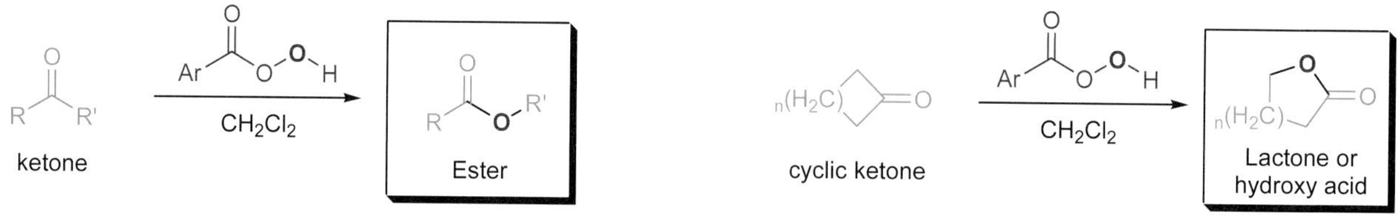

Mechanism:[49-61]

In 1953 Doering and Dorfman clarified the mechanism by performing a labeling experiment. Their experimental results confirmed Criegee's hypothesis, which he presented in 1948. In the first step, the carbonyl group is protonated to increase its electrophilicity, then the peroxyacid adds to this cationic species to form the so-called *Criegee intermediate* (adduct). When the carboxylic acid (R^1COOH) departs from this intermediate, an electron-deficient oxygen substituent is formed, which immediately undergoes an alkyl migration. This alkyl migration and the loss of the carboxylic acid both take place in a concerted process. It is assumed that the migrating group has to be in a position antiperiplanar to the dissociating oxygen-oxygen single bond of the peroxide. The FMO (frontier molecular orbital) theory states that this antiperiplanar arrangement allows the best overlap of the C-R^2 σ bond with the O-O σ* orbital (primary stereoelectronic effect). In 1998, Y. Kishi and co-workers showed that in allylic hydroperoxides the bond antiperiplanar to the dissociating peroxide bond is always and exclusively the bond that migrates, even when this migration is electronically disfavored.[57] Despite the numerous investigations of the mechanism of the *Baeyer-Villiger oxidation*, the factors that control the migratory aptitude are still not completely understood. Electron density and steric bulk strongly influence the migration ability, but the exact nature of these influences remains obscure.

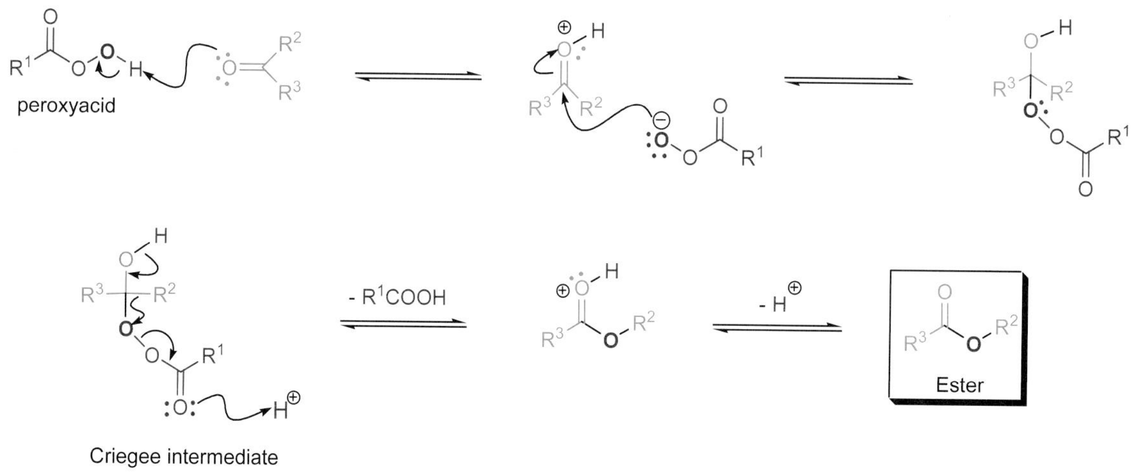

Synthetic Applications:

Investigations by J. Oh showed that the cycloaddition of dichloroketene to glucal followed by *Baeyer-Villiger oxidation* afforded a bicyclic γ-lactone, an α-D-*C*-glucoside, which was further transformed to a C1-methyl glucitol derivative.[62]

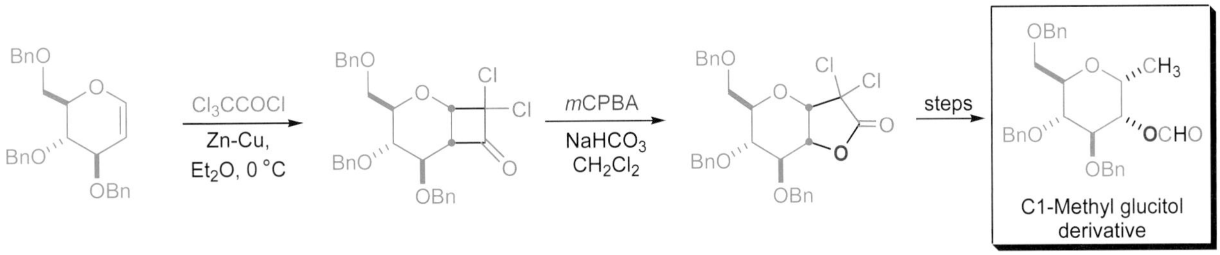

In the laboratory of T.K.M. Shing, the functionalized CD-ring of Taxol® was synthesized in 21 steps starting out from (*S*)-(+)-carvone.[63] The key steps were *Baeyer-Villiger oxidation*, *Oppenhauer oxidation*, *Meerwein-Ponndorf-Verley reduction*, a stereospecific *Grignard addition*, and an intramolecular S_N2 reaction.

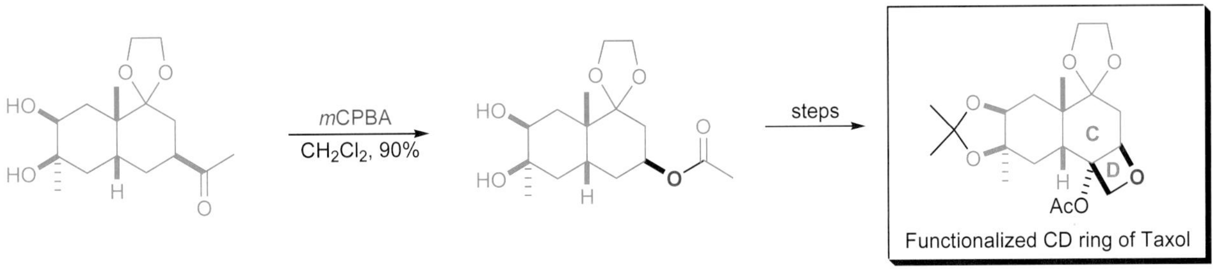

Only a few methods are known for the preparation of cage-annulated ethers. A.P. Marchand and co-workers have used the *Baeyer-Villiger oxidation* for the synthesis of novel cage heterocycles and developed a general procedure that can be used to synthesize cage ethers by replacing the carbonyl group in a cage ketone by a ring oxygen atom or by a CH_2O group.[64]

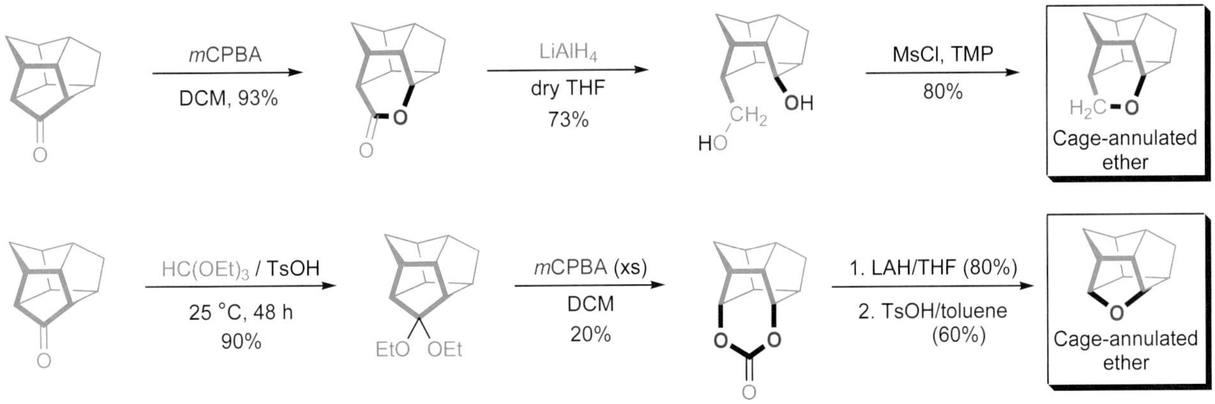

An unexpected rearrangement was observed in the peroxytrifluoroacetic acid-mediated *Baeyer-Villiger oxidation* of *trans*-3β-hydroxy-4,4,10β-trimethyl-9-decalone by F.W.J. Demnitz and co-workers.[65] The initially formed ring-expanded lactone product underwent a trifluoroacetic acid-catalyzed cleavage of the lactone C-O bond, and the resulting tertiary carbocation was trapped by the free hydroxyl group to afford a 7-oxabicyclo[2.2.1]heptane derivative. This compound was then used for the total synthesis and structure proof of the sesquiterpene (±)-farnesiferol C.

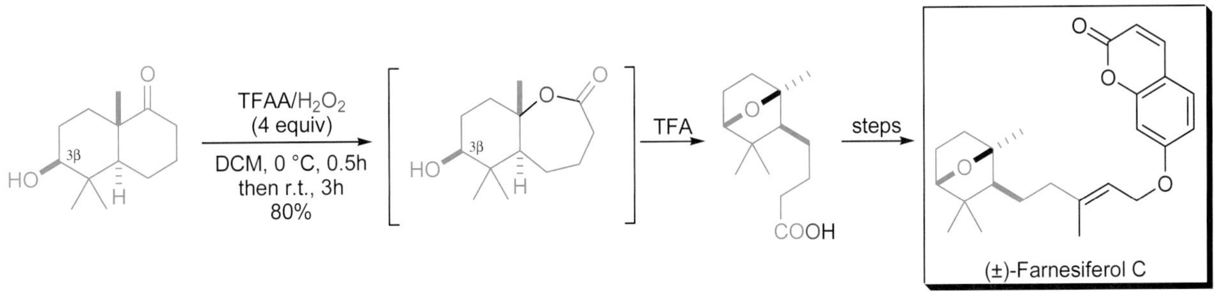

BAKER-VENKATARAMAN REARRANGEMENT

(References are on page 542)

Importance:

[*Seminal Publications*[1-4]; *Reviews*[5-7]; *Modifications & Improvements*[8-17]]

The base-catalyzed rearrangement of aromatic *ortho*-acyloxyketones to the corresponding aromatic β-diketones is known as the *Baker-Venkataraman rearrangement*. β-Diketones are important synthetic intermediates, and they are widely used for the synthesis of chromones, flavones, isoflavones, and coumarins. The most commonly used bases are the following: KOH, potassium *tert*-butoxide in DMSO, Na metal in toluene, sodium or potassium hydride, pyridine, and triphenylmethylsodium.

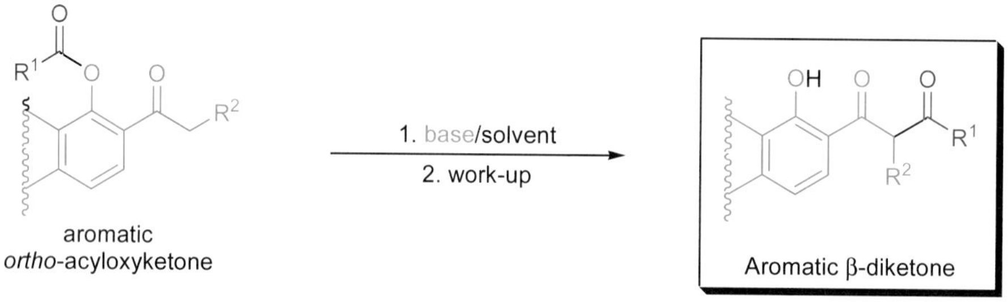

R^1 = alkyl, aryl, NH_2; R^2 = alkyl, aryl; base: KOH, KO*t*-Bu, NaH, Na metal, KH, C_5H_5N

Mechanism: [18-22]

In the first step of the mechanism, the aromatic ketone is deprotonated at the α-carbon and an enolate is formed. This nucleophile attacks the carbonyl group of the acyloxy moiety intramolecularly to form a tetrahedral intermediate that subsequently breaks down to form the aromatic β-diketone.

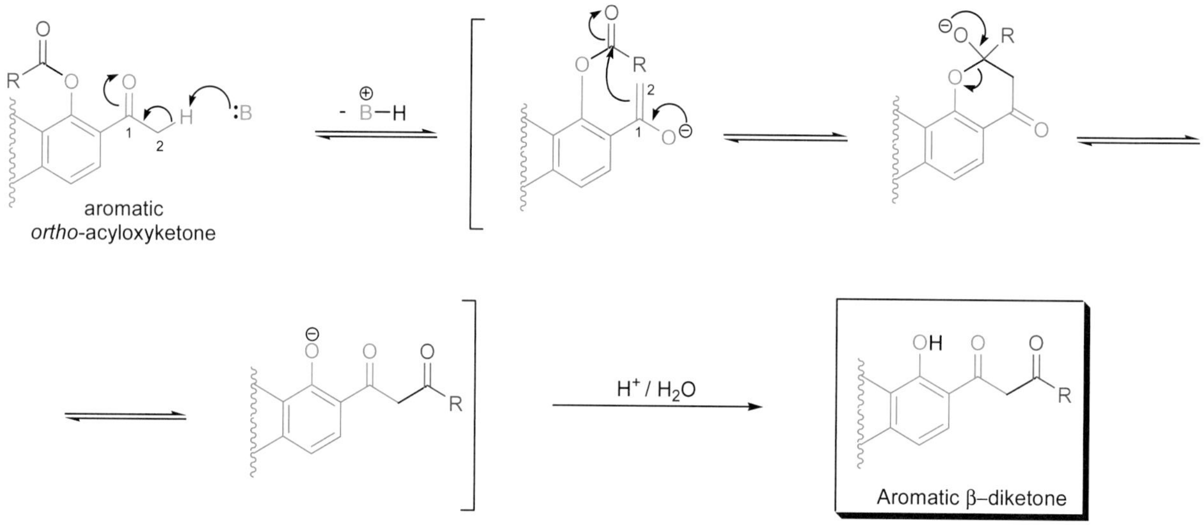

Synthetic Applications:

In the laboratory of K. Krohn, the total synthesis of aklanonic acid and its derivatives was undertaken, utilizing the *Baker-Venkataraman rearrangement* of *ortho*-acetyl anthraquinone esters in the presence of lithium hydride.[23] Using this method, it was possible to introduce ketide side-chains on anthraquinones in a facile manner.

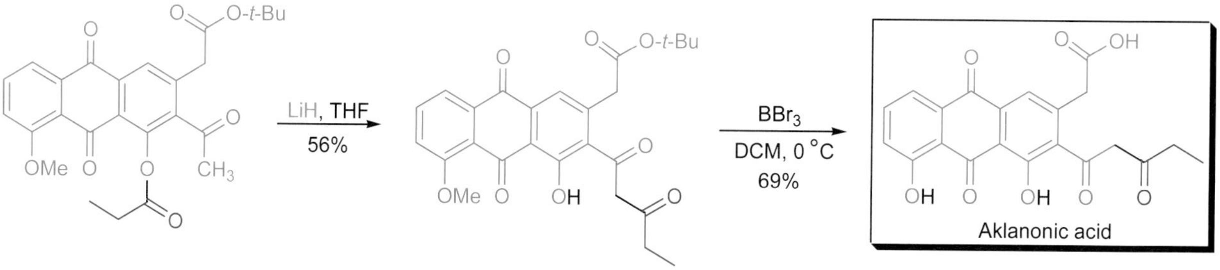

Synthetic Applications:

V. Snieckus and co-workers developed a new *carbamoyl Baker-Venkataraman rearrangement,* which allowed a general synthesis of substituted 4-hydroxycoumarins in moderate to good overall yields.[16] The intermediate arylketones were efficiently prepared from arylcarbamates *via directed ortho metallation* and *Negishi cross coupling*. The overall sequence provided a regiospecific anionic *Friedel-Crafts* complement for the construction of *ortho*-acyl phenols and coumarins.

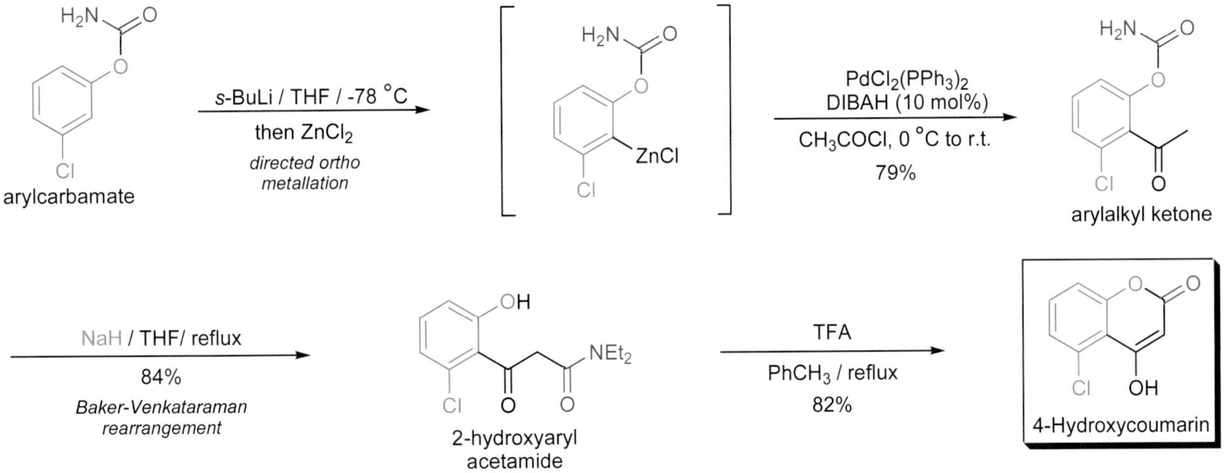

Stigmatellin A is a powerful inhibitor of electron transport in mitochondria and chloroplasts. During the diastereo- and enantioselective total synthesis of this important natural product, D. Enders et al. utilized the *Baker-Venkataraman rearrangement* for the construction of the chromone system in good yield.[24]

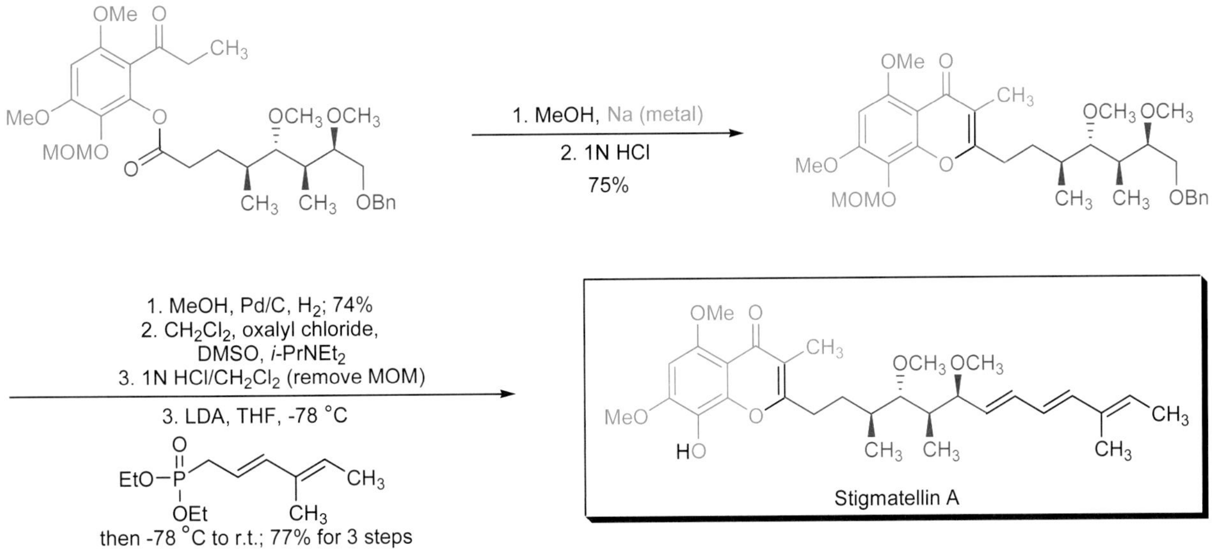

A highly efficient and operationally simple domino reaction was developed in the laboratory of S. Ruchiwarat for the synthesis of benz[*b*]indeno[2,1-*e*]pyran-10,11-diones.[25] The initial aroyl-transfer was achieved by the *Baker-Venkataraman rearrangement* by subjecting the starting material to KOH in pyridine under reflux for 30 minutes.

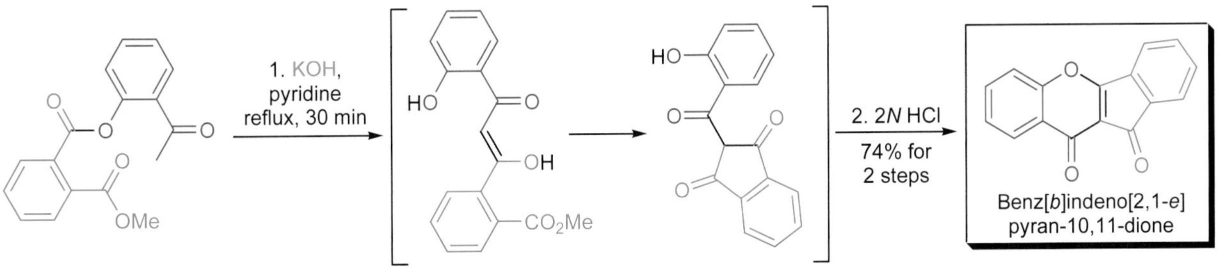

BALDWIN'S RULES / GUIDELINES FOR RING-CLOSING REACTIONS

(References are on page 542)

Importance:

[*Seminal Publication*[1]; *Reviews*[2,3]; *Related Publications*[4-14]]

In 1976, J.E. Baldwin formulated a set of rules/guidelines governing the ease of intramolecular ring-closing reactions, the so-called *Baldwin's rules* or *Baldwin's guidelines*.[1] Baldwin used these rules/guidelines to gain valuable insight into the role of stereoelectronic effects in organic reactions and predict the feasibility of these reactions in synthetic sequences. A few years later in 1983, J.D. Dunitz and co-workers demonstrated that there are favored trajectories for the approach of one reactant molecule toward another.[15] We must note, however, that there is substantial limitation on these rules/guidelines; a large number of examples are known for which they do not apply.

Summary of most important ring closures:

(F=favored, D=disfavored)

Ring size	Exo-dig	Exo-trig	Exo-tet	Endo-dig	Endo-trig	Endo-tet
3	D	F	F	F	D	-
4	D	F	F	F	D	-
5	F	F	F	F	D	D
6	F	F	F	F	F	D
7	F	F	F	F	F	-

Disfavored Processes

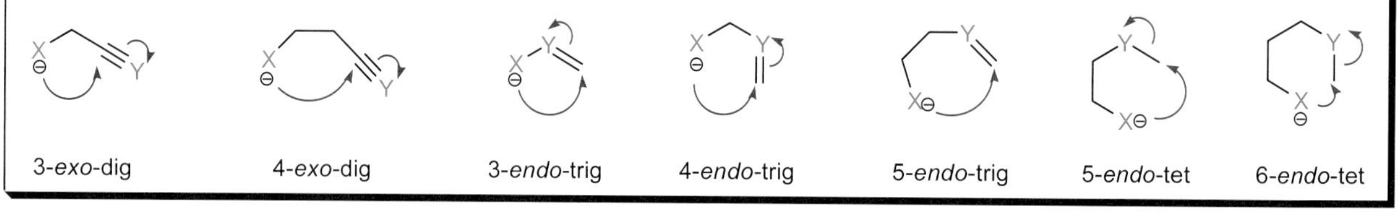

Favored Processes

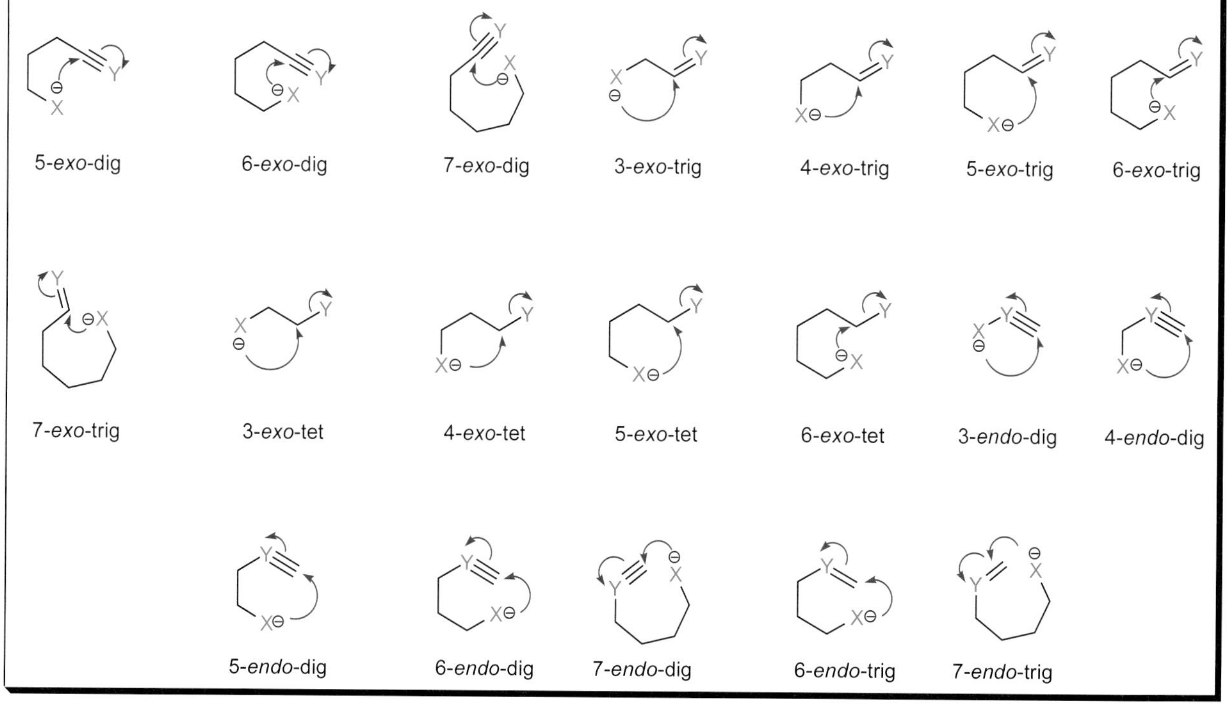

Synthetic Applications:

D.L Boger and co-workers reported an asymmetric total synthesis of *ent*-(–)-roseophilin, the unnatural enantiomer of a naturally occurring antitumor antibiotic.[16] Their approach featured a *5-exo-trig* acyl radical-alkene cyclization to construct the fused cyclopentanone unit. To this end, the hindered methyl ester functionality was hydrolyzed and the resulting acid was transformed to the corresponding phenyl selenoester *via* a two-step sequence. The *5-exo-trig* acyl radical-alkene cyclization was achieved by using AIBN and Bu_3SnH to provide the tricyclic *ansa*-bridged azafulvene core.

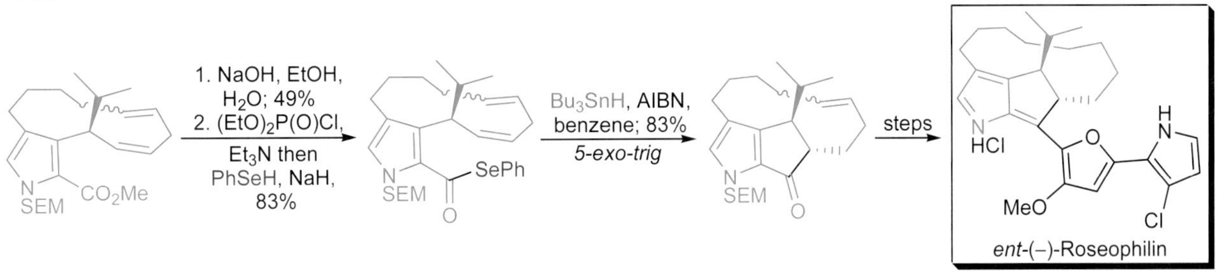

The total synthesis of balanol, a fungal metabolite was accomplished by K.C. Nicolaou *et al.*[17] For the construction of the central hexahydroazepine ring, they have utilized a *7-exo-tet* cyclization. The substitution reaction between the mesylate of the primary alcohol and the Cbz-protected amine was effected by a slight excess of base to produce the desired 7-membered ring in high yield.

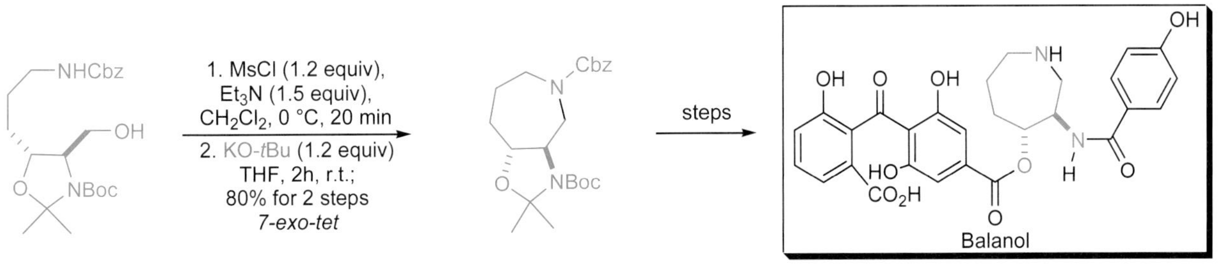

The total synthesis of pyrrolidinol alkaloid, (+)-preussin was achieved in five efficient transformations from commercially available *tert*-Boc-(*S*)-phenylalanine in the laboratory of S.M. Hecht.[18] The key step involved the $Hg^{(II)}$-mediated *5-endo-dig* cyclization of ynone substrate affording the desired pyrrolidinone which, in two more steps, was converted into the natural product.

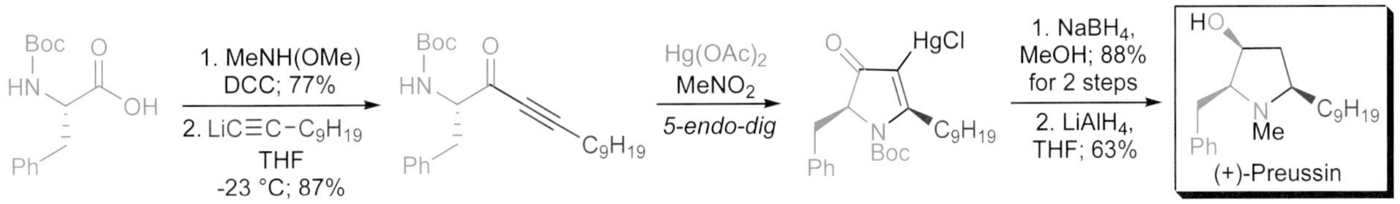

In the laboratory of K. Nacro, a cyclization process leading stereoselectively to *six- and/or five-membered ring lactones* and *lactone ethers* from optically active epoxy- or diepoxy β-hydroxyesters or diastereomeric epoxy lactones was developed.[19] The diastereomeric lactones were prepared from nerol and geraniol. The acid catalyzed cyclization of epoxyalcohols is one of the most effective methods for constructing cyclic ethers. The cyclization proceeds in the *exo* mode giving cyclic ethers with a hydroxyl group in the side chain. The regioselectivity of the cyclization is predicted by the *Baldwin's rules*; in the case shown below the ether formation takes place *via* a *5-exo-tet* cyclization.

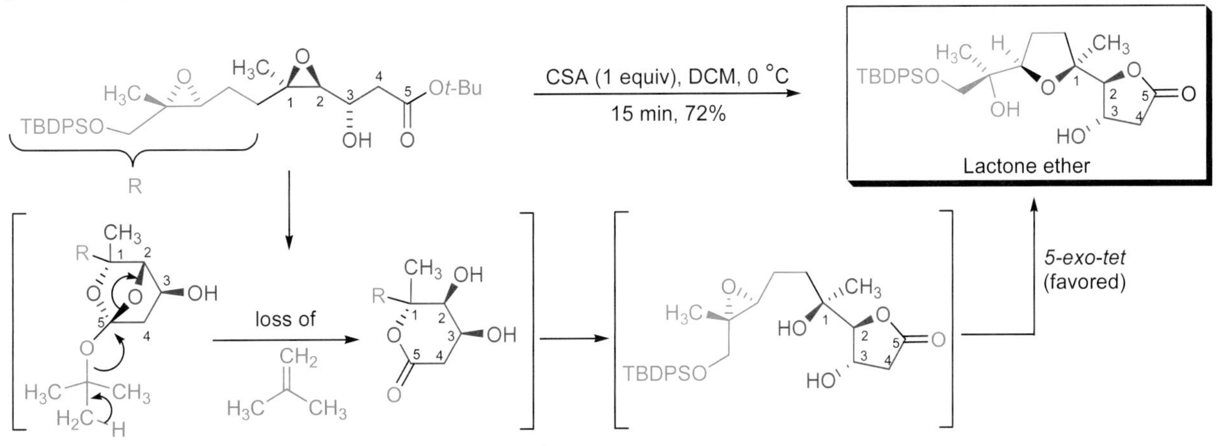

BALZ-SCHIEMANN REACTION
(SCHIEMANN REACTION)

(References are on page 543)

Importance:

[*Seminal Publication*[1]; *Reviews*[2-6]; *Modifications & Improvements*[7-14]]

The thermal decomposition of aromatic diazonium tetrafluoroborates ($ArN_2^+BF_4^-$) to give aromatic fluorides is called the *Balz-Schiemann reaction*. Normally diazonium salts are unstable but diazonium tetrafluoroborates are fairly stable and may be obtained in high yields. Aromatic heterocyclic diazonium tetrafluoroborates may also be used. The diazonium salts are obtained from the diazotization of aromatic amines in the presence of hydrogen tetrafluoroborate (HBF_4). Improved yields of aryl fluorides may be achieved when instead of tetrafluoroborates, hexafluorophosphates (PF_6^-) or hexafluoroantimonates (SbF_6^-) are used as counterions.[7,8] One drawback of the reaction is the potential danger of explosion when large-scale thermal decomposition of the aromatic diazonium tetrafluoroborates is attempted. However, when the decomposition is carried out, either thermally or photolytically, in pyridine·HF solution, the reaction proceeds smoothly even on a larger scale. This approach is especially useful for the preparation of aryl fluorides having polar substituents (OH, OMe, CF_3, etc.).[15]

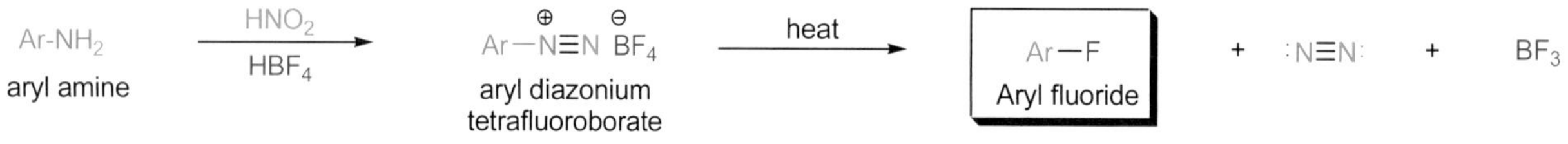

Mechanism: [16-24]

The mechanism involves a positively charged intermediate,[21] which is attacked by BF_4^- rather than the fluoride ion.[20] Both the thermal and photochemical decomposition of diazonium tetrafluoroborates afford the same product ratio, which suggests the intermediacy of the aryl cation. The decomposition follows a first-order rate law, so it is probably of S_N1 type.

Formation of the aryldiazonium salt:

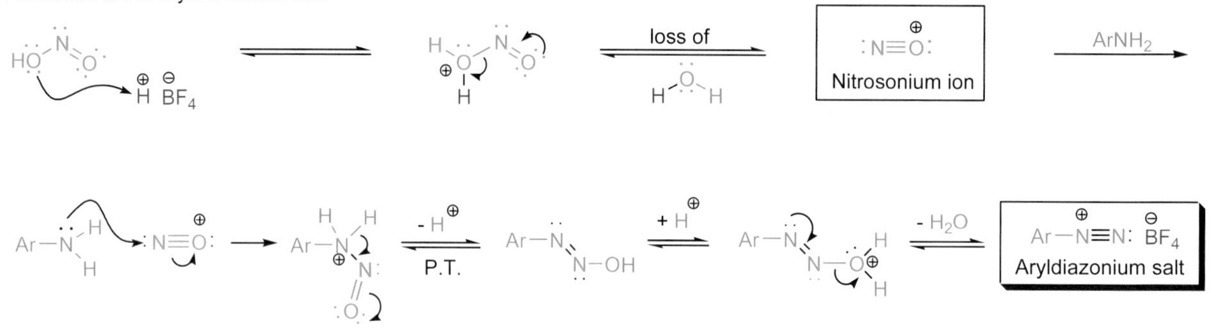

Decomposition of the aryldiazonium salt:

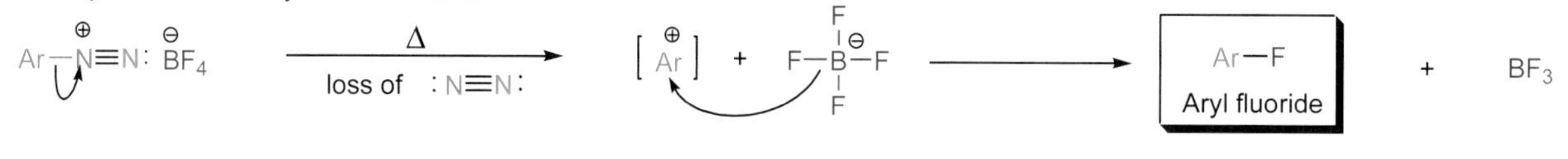

Synthetic Applications:

In the laboratory of D.A. Holt, the synthesis of a new class of steroid 5α-reductase inhibitors was undertaken.[25] They found that unlike the steroidal acrylates, steroidal A ring aryl carboxylic acids exhibit greatly reduced affinity for rat liver steroid 5α-reductase. The tested steroidal A ring carboxylic acids were synthesized from estrone; in one example, fluorine was incorporated into the 4-position of estrone *via* the *Balz-Schiemann reaction*.

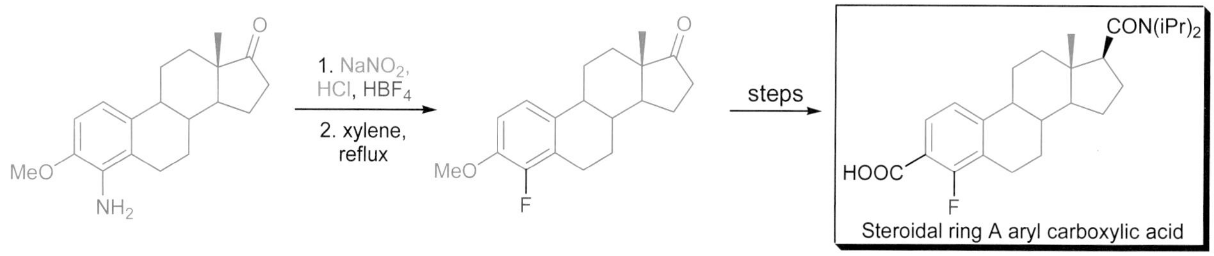

Synthetic Applications:

C. Wiese and co-workers have synthesized 5-fluoro-D/L-dopa and the corresponding [^{18}F]5-Fluoro-L-dopa starting from 5-nitrovanillin *via malonic ester synthesis*, the *Balz-Schiemann reaction*, and the separation of the racemic mixture [^{18}F]5-fluoro-D/L-dopa utilizing a chiral HPLC system.[26] The inactive 5-fluoro-D/L-dopa was obtained in an eight-step synthesis with an overall yield of 10%.

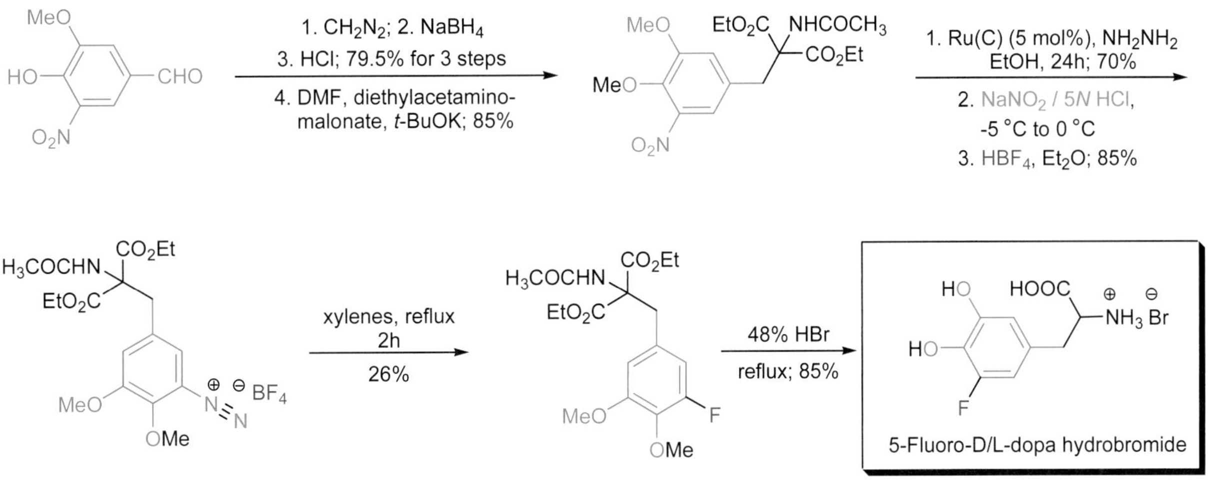

D.R. Thakker synthesized K-region monofluoro- and difluorobenzo[c]phenanthrenes using the *Balz-Schiemann reaction* in order to elucidate the metabolic activation and detoxification of polycyclic aromatic compounds.[27]

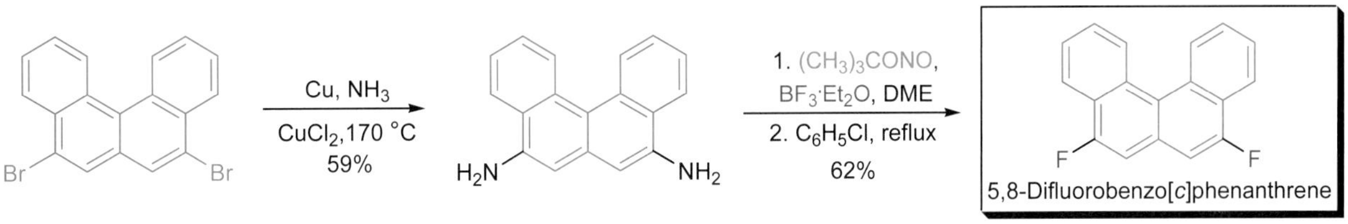

Dibenzo[*a,d*]cycloalkenimines were synthesized and pharmacologically evaluated as *N*-methyl-D-aspartate antagonists by P.S. Anderson et al.[28] A symmetrical 3,7-difluoro derivative was accessed by applying the *Balz-Schiemann reaction* on the corresponding 3,7-diamino analog.

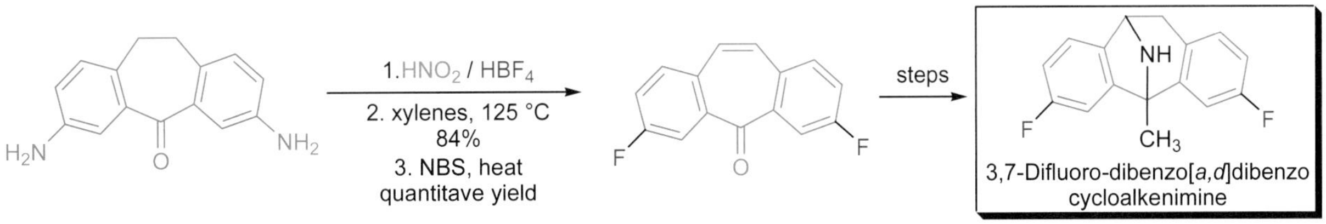

The synthesis of 7-azaindoles is a challenging task and there are few efficient routes to substituted derivatives. In the laboratory of C. Thibault, the concise and efficient synthesis of 4-fluoro-1*H*-pyrrolo[2,3-*b*]pyridine was achieved.[29] The fluorination was carried out using the *Balz-Schiemann reaction*. The aromatic amine precursor was prepared *via* the *Buchwald-Hartwig coupling* of the aryl chloride with *N*-allylamine followed by deallylation. The diazonium tetrafluoroborate intermediate was generated at 0 °C and it decomposed spontaneously in 48% HBF_4 solution to afford the desired aromatic fluoride.

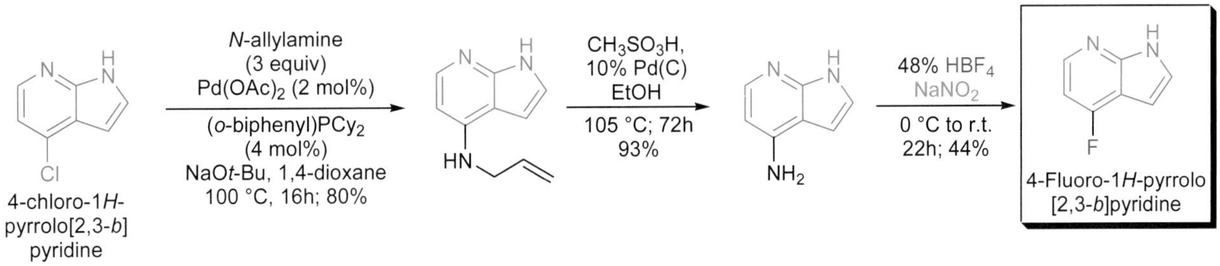

BAMFORD-STEVENS-SHAPIRO OLEFINATION

(References are on page 543)

Importance:

[*Seminal Publication*[1]; *Reviews*[2-4]; *Modifications & Improvemens*[5-18]]

The base catalyzed decomposition of arylsulfonylhydrazones of aldehydes and ketones to provide alkenes is called the *Bamford-Stevens reaction*. When an organolithium compound is used as the base, the reaction is termed the *Shapiro reaction*. The most synthetically useful protocol involves treatment of the substrate with at least two equivalents of an organolithium compound (usually MeLi or BuLi) in ether, hexane, or tetramethylenediamine. The *in situ* formed alkenyllithium is then protonated to give the alkene. The above procedure provides good yields of alkenes without side reactions and where there is a choice, the less highly substituted alkene is predominantly formed. Under these reaction conditions tosylhydrazones of α,β-unsaturated ketones give rise to conjugated dienes. It is also possible to trap the alkenyllithium with electrophiles other than a proton.

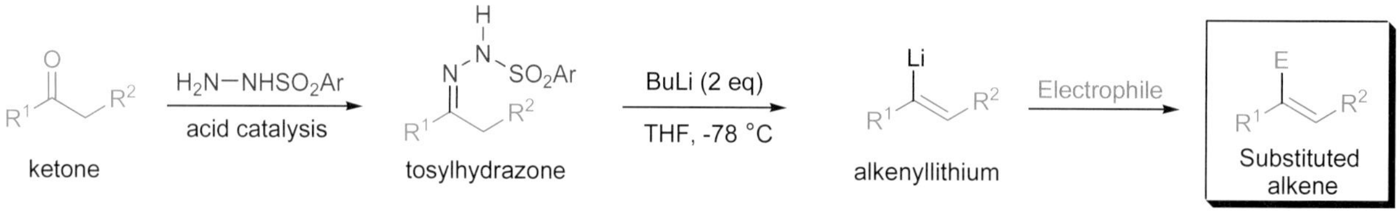

Mechanism: [19,7,20]

The reaction mechanism depends on the reaction conditions used. The reaction of tosylhydrazone with a strong base (usually metal-alkoxides) in protic solvents results in the formation of a diazo compound that in some cases can be isolated.[20] The diazo compound gives rise to a carbocation that may lose a proton or undergo a *Wagner-Meerwein rearrangement*. Therefore, a complex mixture of products may be isolated. When aprotic conditions are used, the initially formed diazo compound loses a molecule of nitrogen and a carbene intermediate is formed, which either undergoes a *[1,2]-H shift* or various carbene insertion reactions. In the case of the *Shapiro reaction*, two equivalents of alkyllithium reagent deprotonate the tosylhydrazone both at the nitrogen and the α-carbon and an alkenyllithium intermediate is formed *via* a carbanion mechanism. Subsequently, the protonation of the alkenyllithium gives rise to the alkene.

Carbene and Carbocation Mechanism:

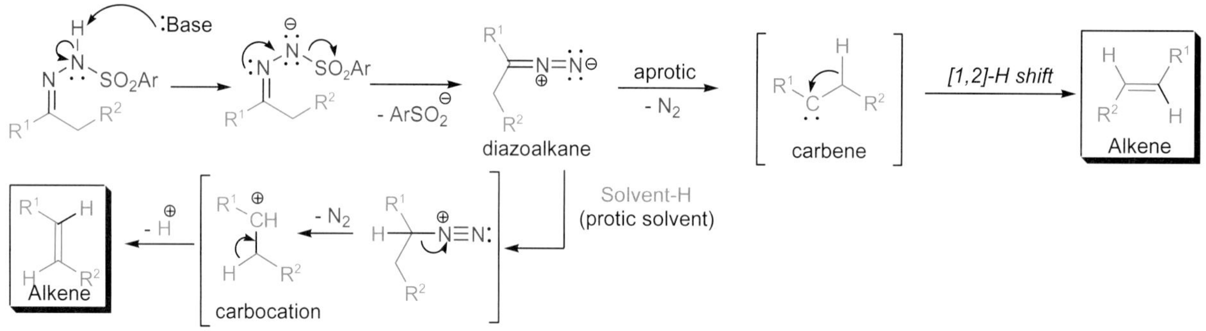

Carbanion Mechanism (*Shapiro reaction*):

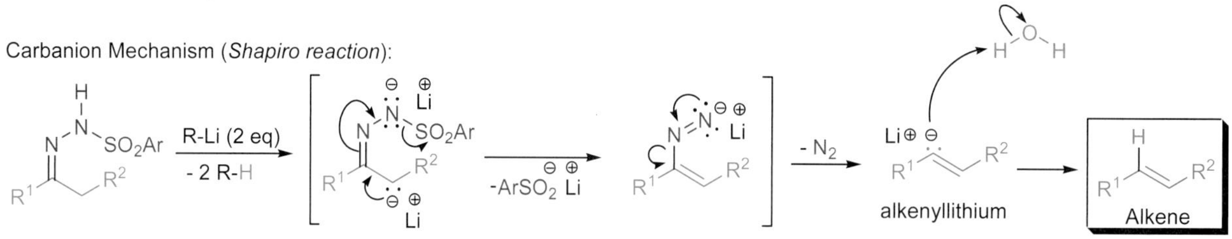

Synthetic Applications:

The first enantioselective total synthesis of (–)-myltaylenol was achieved in the laboratory of E. Winterfeldt.[21] The authors used an *intramolecular Diels-Alder cycloaddition* and the *Shapiro reaction* as key transformations to construct the unusual carbon framework of this sesquiterpenoid alcohol natural product, which contains three consecutive quaternary carbon atoms.

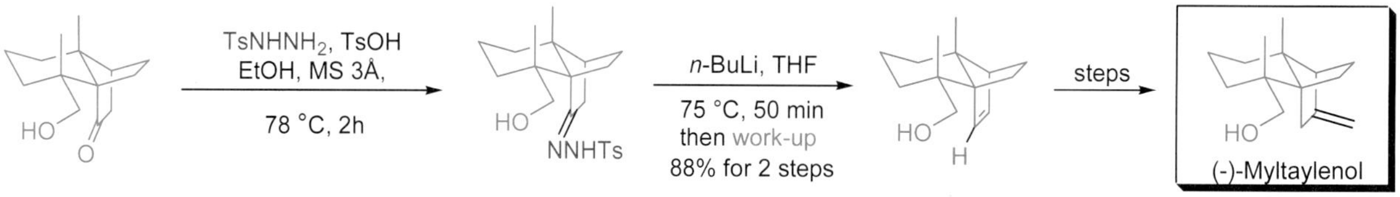

Synthetic Applications:

In the laboratory of K. Mori the task of determining the absolute configuration of the phytocassane group of phytoalexins was undertaken. To this end, the naturally occurring (–)-phytocassane D was synthesized from (*R*)-Wieland-Miescher ketone.[22] During the synthesis, a tricyclic ketone intermediate was subjected to the *Shapiro olefination* reaction to give the desired cyclic alkene in good yield.

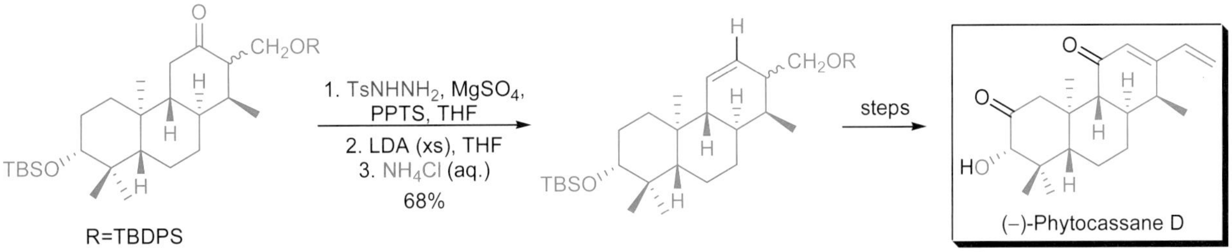

L. Somsák et al. developed a one-pot reaction to prepare exo-glycals from glycosyl cyanides.[23] In this one-pot reaction, acylated glycosyl cyanides were first converted to the corresponding aldehydes with Raney nickel-sodium hypophosphite, and then converted into 2,5- and 2,6-anhydroaldose tosylhydrazones to give *exo*-glycals under aprotic *Bamford-Stevens* conditions. During the reaction *C*-glycosylmethylene carbenes are formed and spontaneously rearrange to give the observed *exo*-glycals.

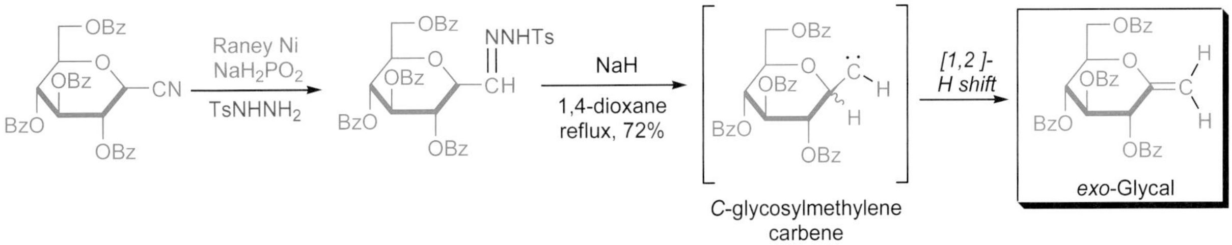

A novel class of chiral indenes (verbindenes) was prepared from enantiopure verbenone by K.C. Rupert and co-workers who utilized the *Shapiro reaction* and the *Nazarov cyclization* as the key transformations.[24] The bicyclic ketone substrate was treated with triisopropylbenzenesulfonyl hydrazide to prepare the trisyl hydrazone that was then exposed to *n*-BuLi. The resulting vinyllithium intermediate was reacted with various aromatic aldehydes to afford the corresponding allylic alcohols.

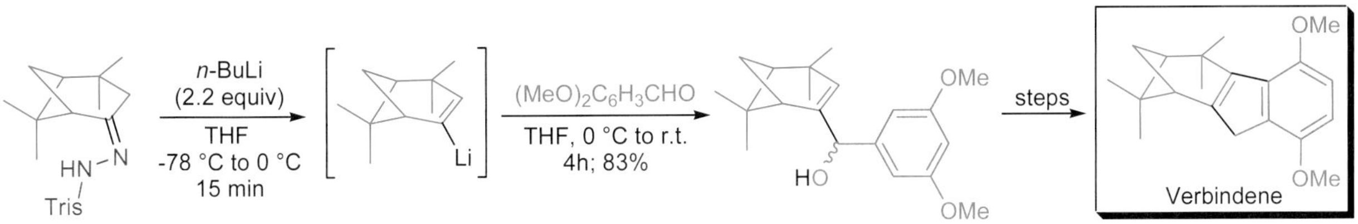

During the total synthesis of (–)-isoclavukerin A by B.M. Trost et al., the introduction of the diene moiety was achieved by the use of the *Bamford-Stevens reaction* on a bicyclic trisylhydrazone compound.[25] Interestingly, the strongly basic Shapiro conditions (e.g., alkyllithiums or LDA) led only to uncharacterizable decomposition products. However, heating of the trisylhydrazone with KH in toluene in the presence of diglyme gave good yield of the desired diene. It was also shown that the olefin formation and the following decarboxylation could be conducted in one pot. According to this procedure, excess NaI was added and the temperature was elevated to bring about the *Krapcho decarboxylation*.

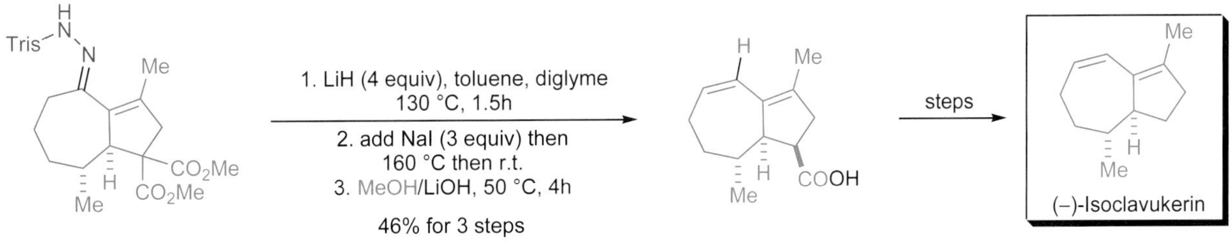

BARBIER COUPLING REACTION

(References are on page 544)

Importance:

[*Seminal Publication*[1]; *Reviews*[2-16]; *Modifications & Improvements*[17-22]; *Theoretical Studies*[23]]

In the case of unstable organometallic reagents, it is convenient to generate the reagent in the presence of the carbonyl compound, to produce an immediate reaction. This procedure is referred to as the *Barbier reaction*. The original protocol with magnesium metal was described by P. Barbier and later resulted in the development of the well-known *Grignard reaction*. Most recently other metals (e.g., Sn, In, Zn, etc.) in aqueous solvents have been used under similar conditions with good results. The obvious advantages of these procedures are their safety and simplicity, as well as the ability to treat unprotected sugars with organometallic reagents.

R^1, R^2 = H, alkyl, aryl; R^3 = alkyl, aryl, allyl, benzyl; X = Cl, Br, I; M = Mg, Sm, Zn, Li, etc.

Mechanism: [24-29]

The mechanism of the formation of the organometallic reagent is identical to the formation of a *Grignard reagent*, presumably involving a single electron transfer (SET) mechanism from the metal surface to the alkyl halide. The mechanism of the addition of Grignard reagents to carbonyl compounds is not understood, but it is thought to take place mainly *via* either a concerted process or a radical pathway (stepwise).[30-32]

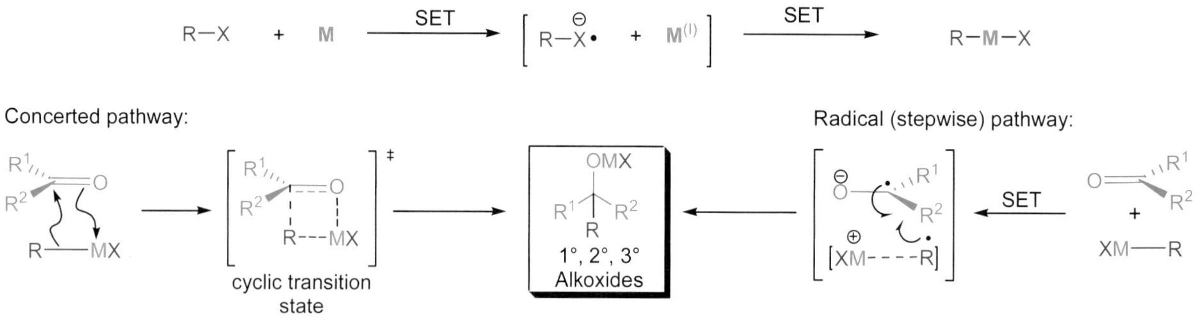

Synthetic Applications:

B.M. Trost and co-workers conducted studies toward the total synthesis of saponaceolide B, an antitumor agent active against 60 human cancer cell lines.[33,34] One of the challenging structural features of this compound was the *cis* 2,4-disubstituted 1-methylene-3,3-dimethylcyclohexane ring. The key steps to construct this highly substituted cyclohexane ring were a diastereoselective *Barbier reaction* to install a vinyl bromide moiety followed by an *intramolecular Heck cyclization reaction*.

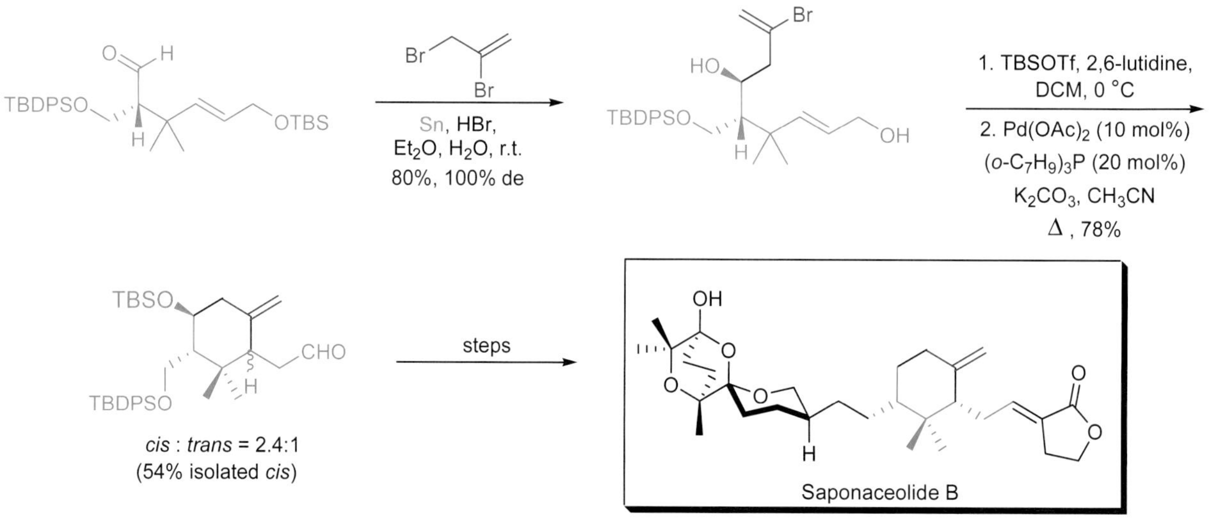

Synthetic Applications:

During the enantioselective total synthesis of the sarpagine-related indole alkaloids talpinine and talcarpine, J.M. Cook and co-workers prepared an important allylic alcohol precursor for an *anionic oxy-Cope rearrangement*.[35] However, the desired allylic carbanion was expected to undergo an undesired allylic rearrangement when stabilized as either a magnesium or lithium species. This problem was overcome by using the *Barbier reaction conditions,* which was a modification of the allylbarium chemistry of Yamamoto.[36,37] The mixture of the allylic bromide and the aldehyde was added to freshly prepared barium metal at -78 °C to generate the desired allylic carbanion. The resulting barium-stabilized species then added to the aldehyde, affording the 1,2-addition product in high yield, without allylic rearrangement.

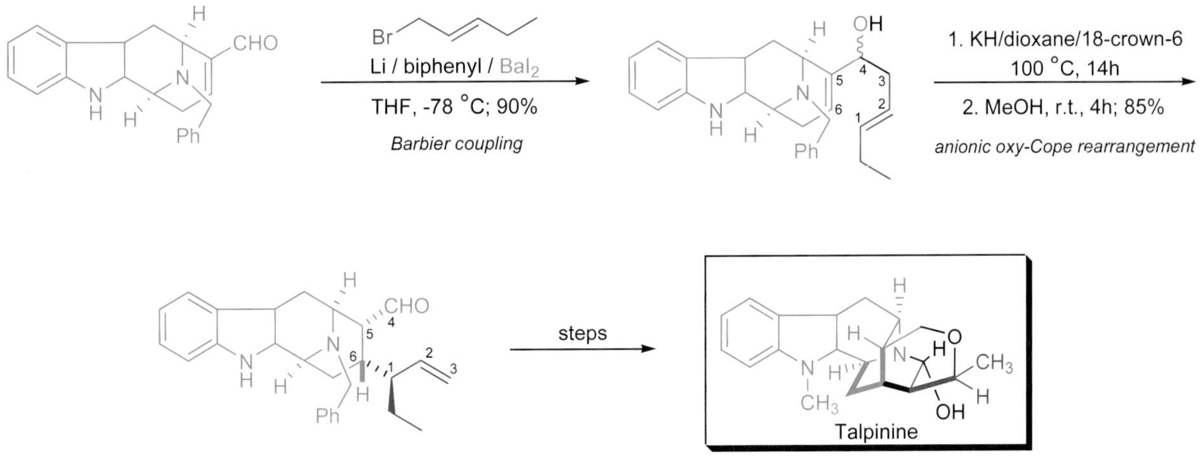

Stypodiol, epistypodiol and stypotriol are secondary diterpene metabolites produced by the tropical brown algae *Stypopodium zonale*. These compounds display diverse biological properties, including strong toxic, narcotic, and hyperactive effects upon the reef-dwelling fish. In the laboratory of A. Abad an efficient stereoselective synthesis of stypodiol and its C14 epimer, epistypodiol, was accomplished from (*S*)-(+)-carvone.[38] The key transformations in the synthesis of these epimeric compounds were an *intramolecular Diels-Alder reaction*, a *sonochemical Barbier reaction* and an *acid-catalyzed quinol-tertiary alcohol cyclization.*

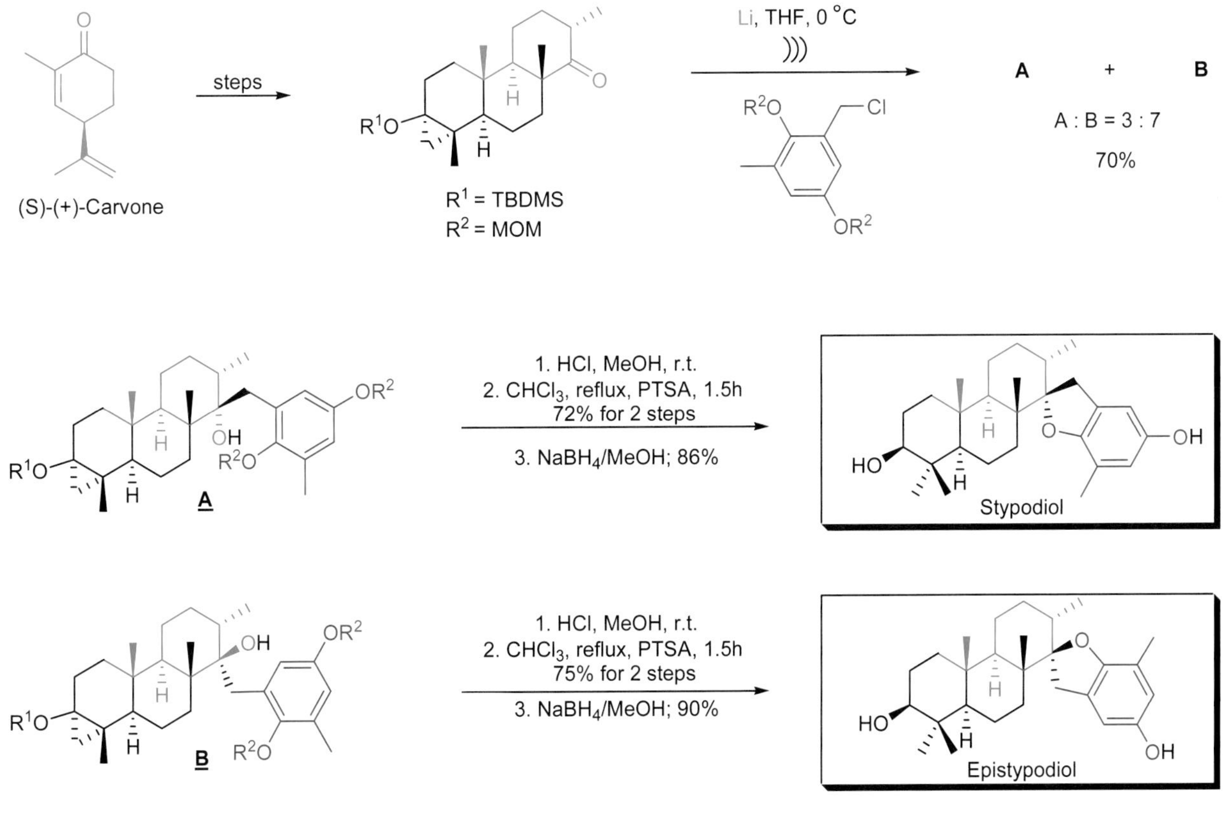

BARTOLI INDOLE SYNTHESIS

(References are on page 545)

Importance:

[*Seminal Publications*[1-3]; *Reviews*[4-7]; *Modifications & Improvements*[8-11]]

In 1989, G. Bartoli et al. described the reaction of substituted nitroarenes with excess vinyl Grignard reagents at low temperature to afford substituted indoles upon aqueous work-up.[2] The authors found that the highest yields were obtained with *ortho*-substituted nitroarenes. According to their procedure three equivalents of vinylmagnesium bromide were added to the cold solution of the nitroarene, which was stirred for 20 minutes, then quenched with a saturated NH_4Cl solution, followed by extraction of the product with diethyl ether. The formation of 7-substituted indoles from *ortho*-substituted nitroarenes (or nitrosoarenes) and alkenyl Grignard reagents is known as the *Bartoli indole synthesis*. The general features of this transformation are: 1) when the nitroarene does not have a substituent *ortho* to the nitro group, the reaction gives low or no yield of the desired indole; 2) the size of the *ortho* substituent also has an effect on the yield of the reaction and the sterically more demanding groups usually give higher yield of the product; 3) most often simple vinylmagnesium bromide is used but substituted alkenyl Grignard reagents can also be applied and they give rise to the corresponding indoles with substituents at the C2 or C3 positions; and 4) when nitrosoarenes are the substrates, only two equivalents of the Grignard reagent are necessary.

Bartoli (1989):

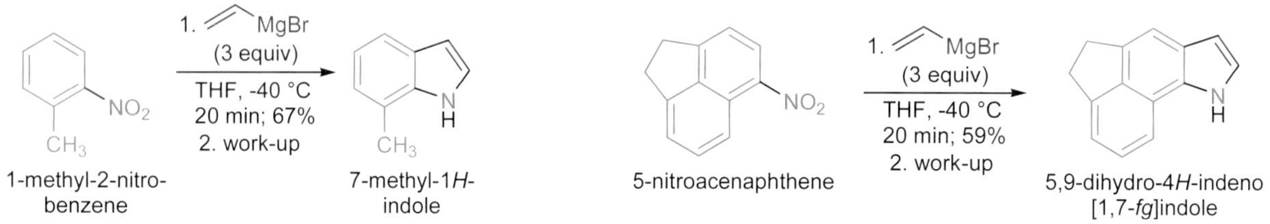

Bartoli indole synthesis:

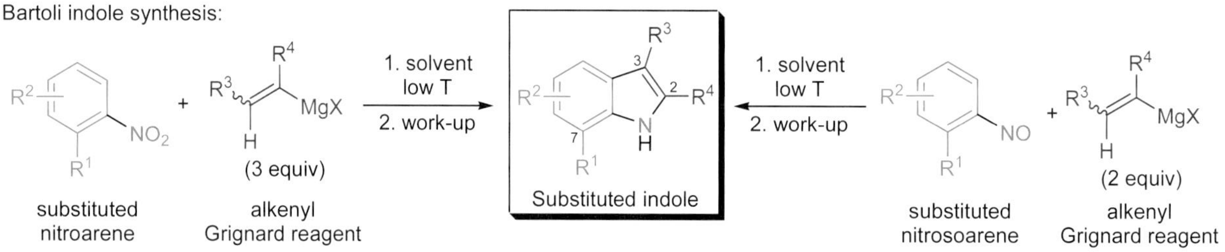

R^1 = Me, alkyl, aryl, F, Cl, Br, I, $OSiR_3$, *O*-benzyl, *O-sec*-alkyl, $CH(OR)_2$; R^2 = H, alkyl, aryl, *O*-alkyl, etc.; R^{3-4} = H, alkyl, aryl, SiR_3

X = Cl, Br, I; solvent: Bu_2O, Et_2O, THF

Mechanism: [12,7]

The mechanism of the *Bartoli indole synthesis* is not clear in every detail, but G. Bartoli and co-workers successfully elucidated the main steps in the process. The first step is the addition of Grignard reagent to the oxygen atom of the nitro group followed by the rapid decomposition of the resulting *O*-alkenylated product to give a nitrosoarene. The nitrosoarene is much more reactive than the starting nitroarene, and it is attacked by the second equivalent of Grignard reagent to give an *O*-alkenyl hydroxylamine derivative, which rearranges in a facile *[3,3]-sigmatropic process*. The rearranged product then undergoes intramolecular nucleophilic attack, and the proton in the ring junction is removed by the third equivalent of the Grignard reagent. Finally, acidic work-up affords the indole.

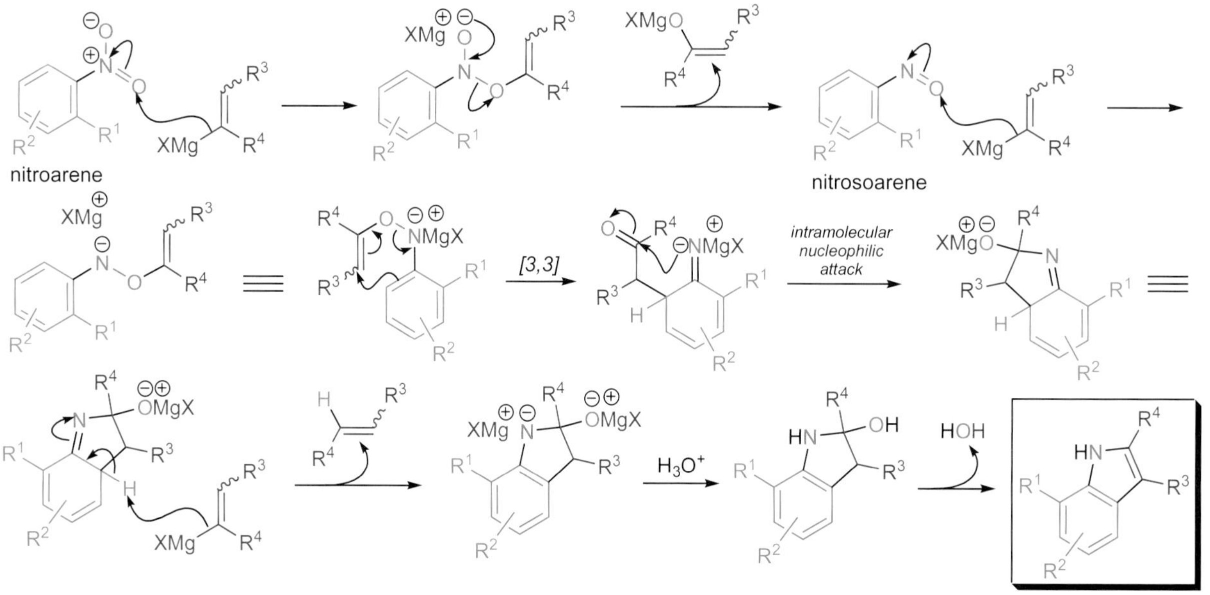

Synthetic Applications:

In the laboratory of T. Wang a general method for the preparation of 4- and 6-azaindoles from substituted nitropyridines based on the *Bartoli indole synthesis* was developed.[13] The substrates were treated with excess vinylmagnesium bromide according to the original procedure described by Bartoli et al. The yields were usually moderate and similarly to the simple nitroarenes, the larger the ortho substituent was, the higher yields were obtained. Interestingly, it was noted that the presence of a halogen atom at the 4-position of the pyridine ring resulted in significantly improved product yields.

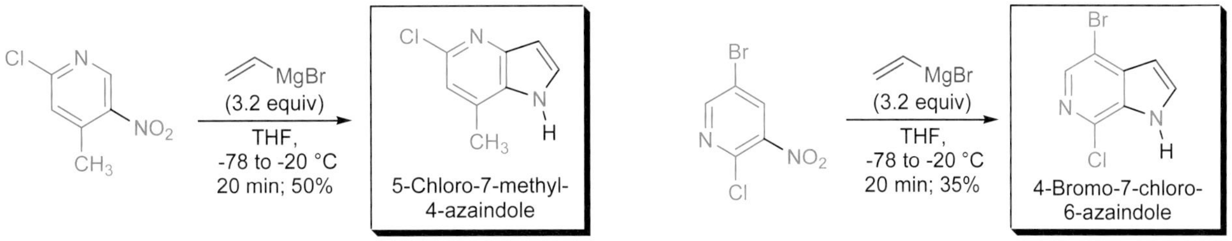

The short synthesis of the pyrrolophenanthridone alkaloid hippadine was accomplished by D.C. Harrowven and co-workers.[14] The key step of the synthetic sequence was the *Ziegler modified intramolecular Ullmann biaryl coupling* between two aryl bromides. One of the aryl halides was 7-bromoindole which was prepared using the *Bartoli indole synthesis*. The second aryl bromide was connected to 7-bromoindole *via* a simple *N*-alkylation.

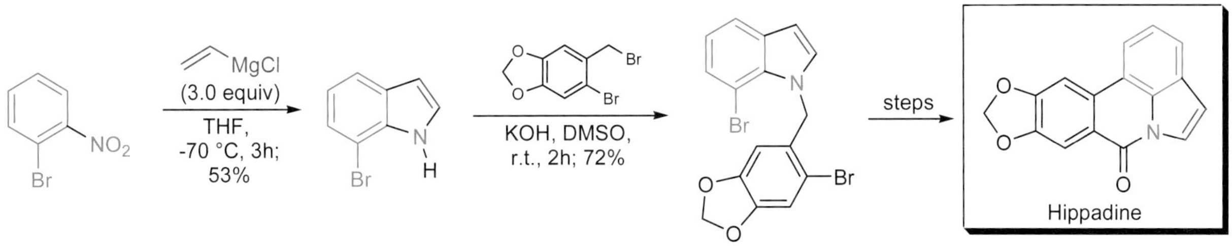

The research team of T.A. Engler and J.R. Henry identified and synthesized a series of potent and selective glycogen synthase kinase-3 (GSK3) inhibitors.[15] One of the targets required the preparation of 5-fluoro-7-formylindole, which was achieved by the *Bartoli indole synthesis*. Since the unprotected formyl group is incompatible with the Grignard reagent, a two-step protocol was implemented. First, the formyl group of 5-fluoro-2-nitrobenzaldehyde was protected as the corresponding di-*n*-butyl acetal, then excess Grignard reagent was added at low temperature, and finally the acetal protecting group was removed by treatment with aqueous HCl.

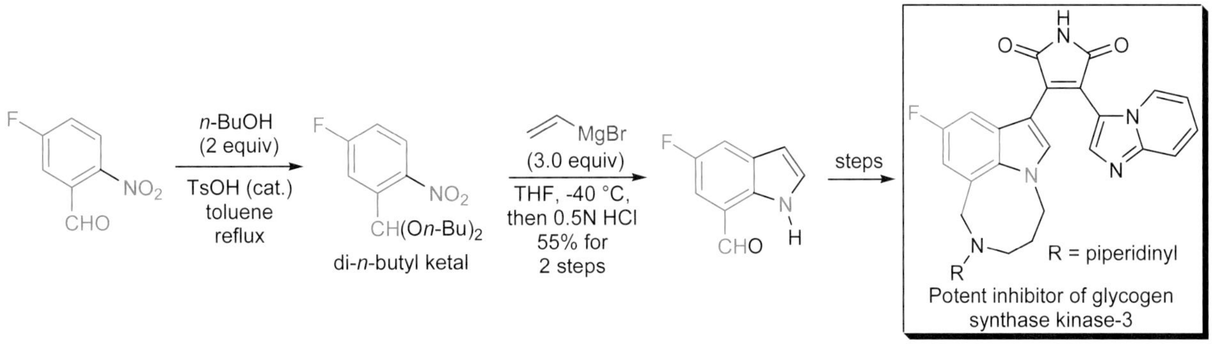

Several heterocycles were prepared from dehydroabietic acid, and their antiviral properties were evaluated in the laboratory of B. Gigante.[16] Dehydroabietic acid was first esterified, then brominated. Nitrodeisopropylation was achieved using a mixture of nitric acid and sulfuric acid. The resulting *o*-bromo nitroarene was treated with excess vinyl Grignard reagent to obtain the corresponding methyl-12-bromo-13,14b-pyrrolyl-deisopropyl dehydroabietate.

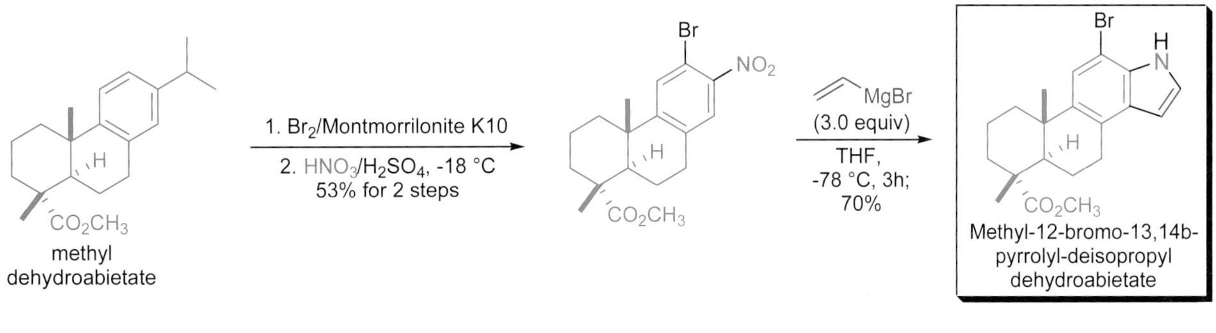

BARTON NITRITE ESTER REACTION

(References are on page 545)

Importance:

[*Seminal Publications*[1-7]; *Reviews*[8-12]; *Theoretical Studies*[13,14]]

The Barton nitrite ester reaction (*Barton reaction*) is a method for achieving remote functionalization on an unreactive aliphatic site of a nitrite ester under thermal or photolytic conditions *via* oxygen-centered radicals. The nitrite esters are converted to the corresponding γ-hydroxy oximes in the reaction. The most common way to generate an oxygen-centered radical is by the thermolysis or photolysis of nitrite, hypochlorite, or hypoiodite esters. Nitrogen-centered radicals are generated by heating the appropriate *N*-haloamines with sulfuric acid to give pyrrolidines or piperidines (*Hofmann-Löffler-Freytag reaction*). The *Barton nitrite ester reaction* was a landmark in the development of free radicals as valuable intermediates for organic synthesis. Most of the synthetic examples are from the steroid field because the *Barton reaction* occurs readily in rigid molecules. Usually skeletons with several fused rings are well-suited for remote functionalizations.

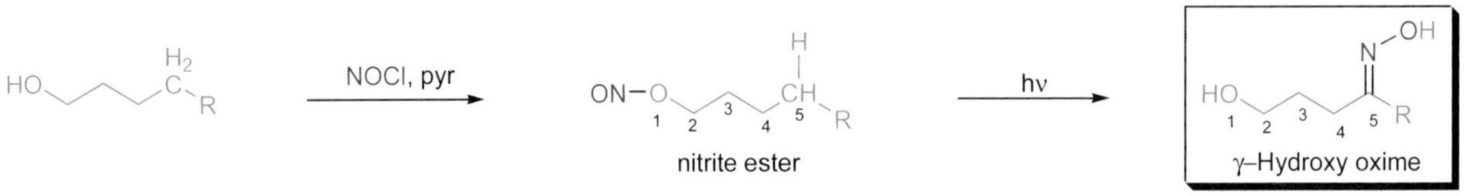

Mechanism:[15,16]

The first step in the mechanism is the homolysis of the O-N bond to form an oxygen-centered radical and a nitrogen-centered free radical. Next, the highly reactive alkoxyl radical abstracts a hydrogen atom from the δ-position (5-position) *via* a quasi chair-like six-atom transition state to generate a new carbon-centered radical that is captured by the initially formed NO free radical. If a competing radical source such as iodine is present, the reaction leads to an iodohydrin, which can cyclize to form a tetrahydrofuran derivative. Occasionally, tetrahydropyran derivatives are obtained in low yields.

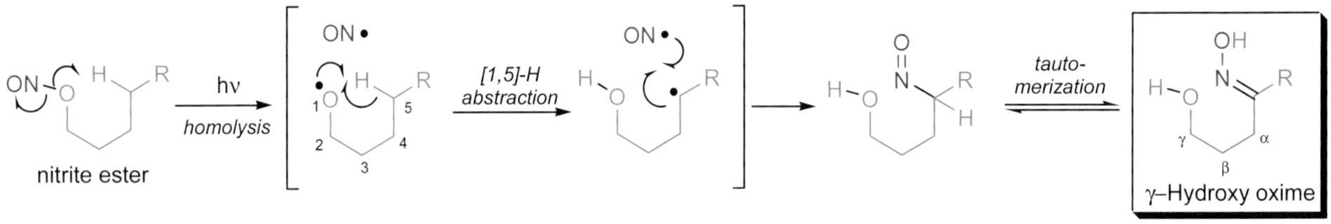

Synthetic Applications:

In the partial synthesis of myriceric acid A by T. Konoike and co-workers, the *Barton nitrite ester reaction* was utilized in a large-scale preparation of one of the intermediates.[17]

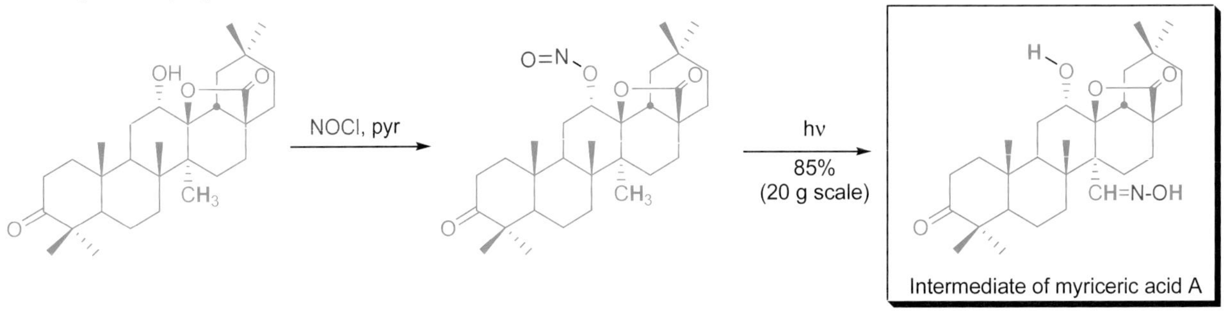

Cephalosporins are important β-lactams, but a number of pathogenic microorganisms have developed resistance to these antibiotic compounds. In order to prepare novel antibiotic cephalosporin analogs, I. Chao and co-workers synthesized 1-dethia-3-aza-1-carba-2-oxacephem, which is not a substrate of the inducible β-lactamase enzyme.[18] The key step of the synthetic sequence was the *Barton nitrite ester reaction* in which regioisomeric oximino β-lactams were generated and transformed into the desired product.

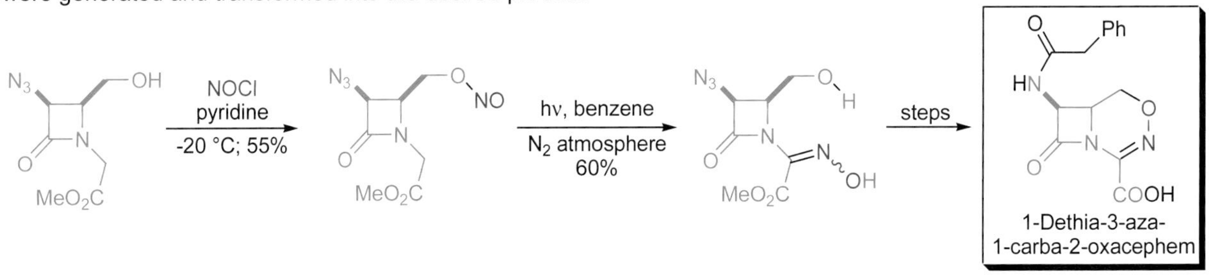

Synthetic Applications:

The *Barton reaction* was utilized during the synthesis of various terpenes and has played a crucial role in the elucidation of terpene structures. The *Barton nitrite ester reaction* was a key step in E.J. Corey's synthesis of azadiradione[19] and perhydrohistrionicotoxin[20]. Even though the yields were low, other ways to access the same intermediates would have been tedious, and afforded lower overall yields than in the applied *Barton reactions*.

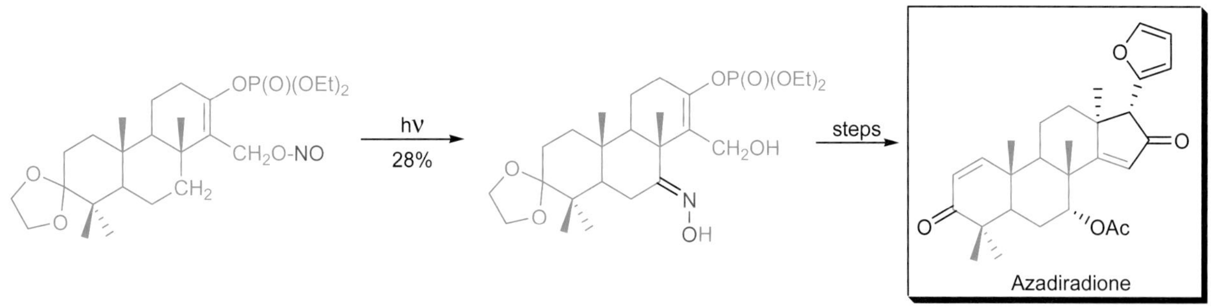

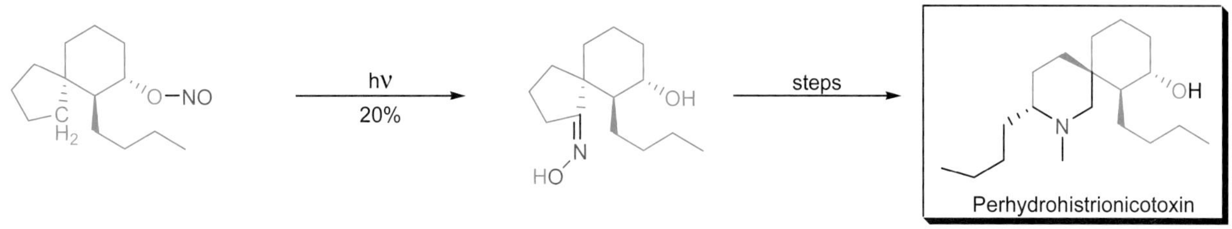

The *Barton reaction* does not always afford only a single major product. J. Sejbal and co-workers isolated two products in a *Barton reaction* on triterpene substrates.[21] In this example, reaction at either (or both) the C4 and C10 methyl groups was expected, but oxidation of the C8 methyl group was not. This remote functionalization occured *via* two consecutive *[1,5]- H-atom transfers*.

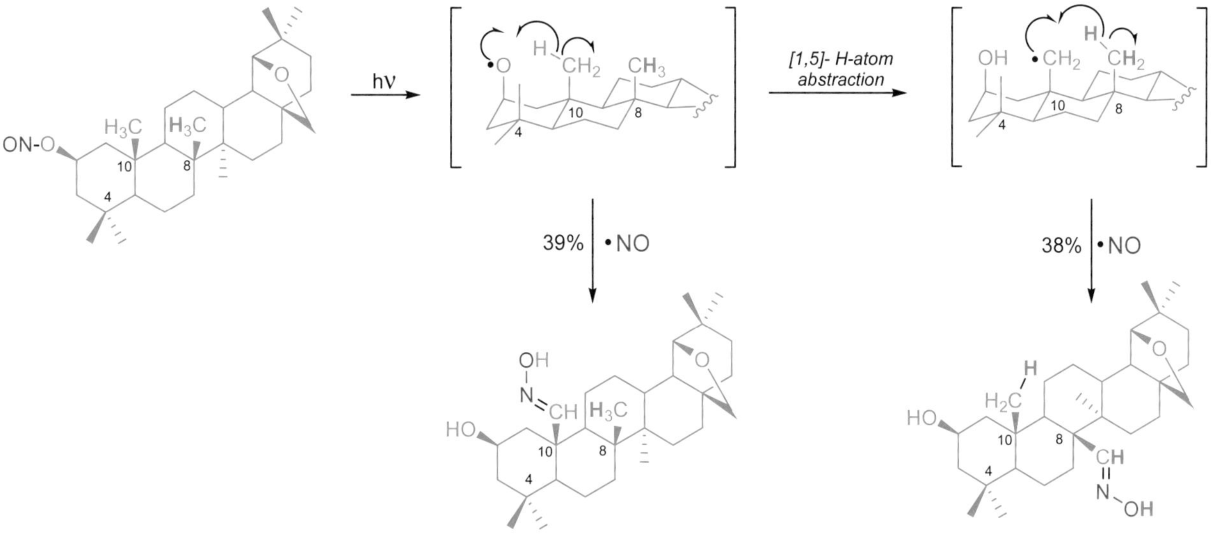

The carbon-centered radical at the δ-position can be reacted by various trapping agents other than the nitrosyl radical. Z. Čekovič and co-workers used electron-deficient olefins (Michael acceptors) such as acrylonitrile to trap the δ-carbon radial and obtain functionalized alkyl chains.[22] In order to maximize the yield of the desired chain-elongated product, a high concentration of the acrylonitrile had to be used. The final radical was trapped by the nitrosyl radical.

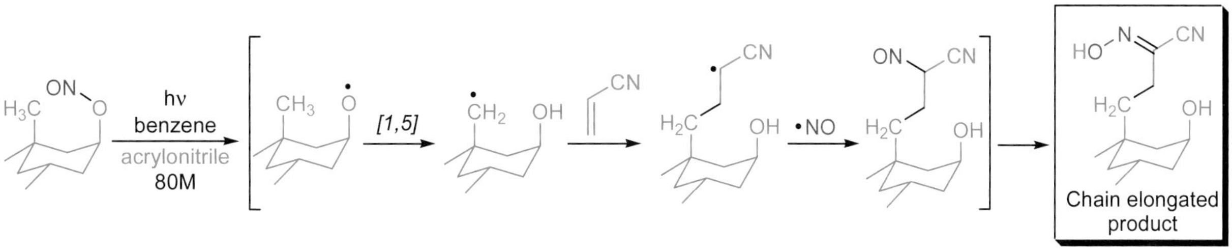

BARTON RADICAL DECARBOXYLATION REACTION

(References are on page 546)

Importance:

[*Seminal Publications*[1-3]; *Reviews*[4]; *Modifications & Improvements*[5-12]]

Conversion of a carboxylic acid to a thiohydroxamate ester, followed by heating the product in the presence of a suitable hydrogen donor such as tri-*n*-butyltin hydride, produces a *reductive decarboxylation*. This sequence of reactions is called the *Barton decarboxylation reaction* and may be used to remove a carboxylic acid and replace it with other functional groups.

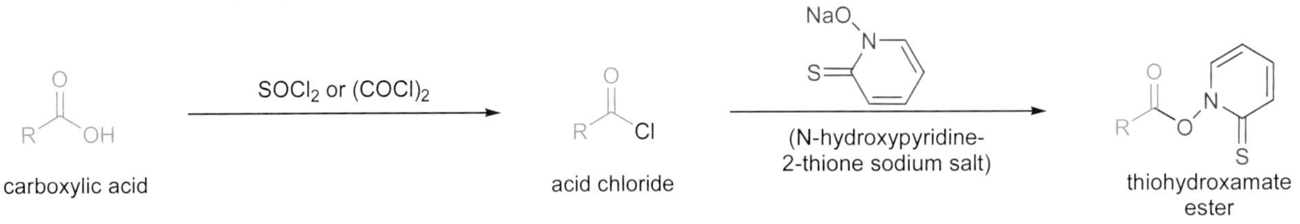

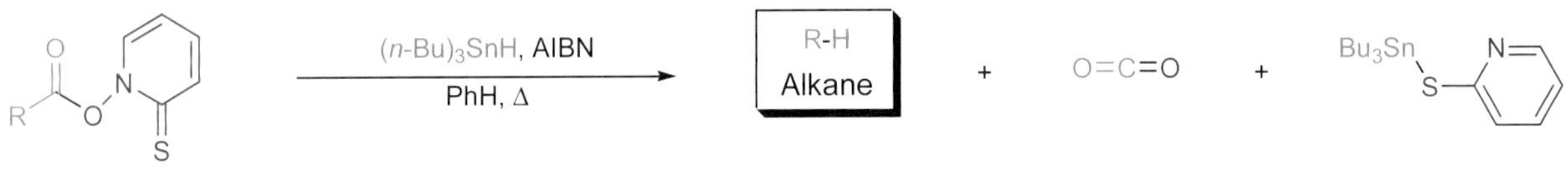

Mechanism: [13]

The first step of the reaction is the homolytic cleavage of the radical initiator AIBN upon heating. This initiation step generates the first radical to start the chain reaction. The initial radical abstracts a hydrogen atom from the tri-*n*-butyltin hydride to afford a tri-*n*-butyltin radical that attacks the sulfur atom of the thiohydroxamate ester, forming a strong Sn-S bond. Next, carbon dioxide is lost, and the released alkyl radical (R·) gets reduced to the product (R-H) by abstracting a hydrogen atom from a tri-*n*-butyltin hydride molecule. The tin radical generated in this last step enters another reaction cycle until all of the starting thiohydroxamate ester is consumed.

Initiation step:

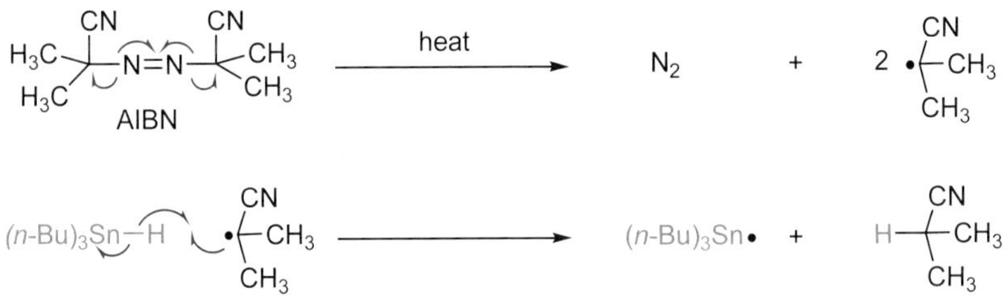

Propagation step:

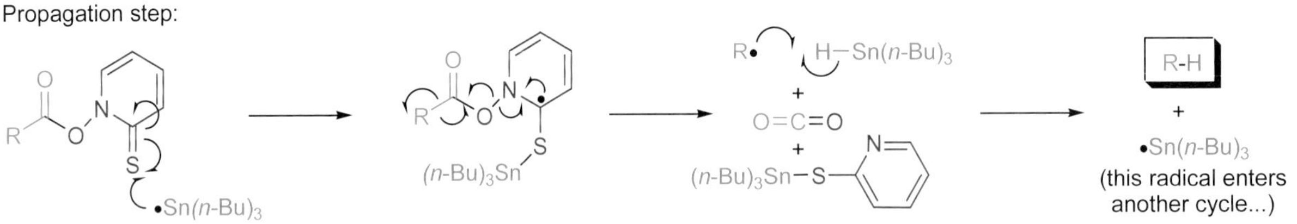

Synthetic Applications:

The *Barton decarboxylation procedure* was used in the total synthesis of (–)-verrucarol by K. Tadano et al. The initially formed thiohydroxamic ester was decarboxylated to leave a methylene radical on the cyclopentyl ring, which was then trapped by molecular oxygen. Reductive work-up in the presence of *t*-BuSH finally provided the hydroxylated product.[14]

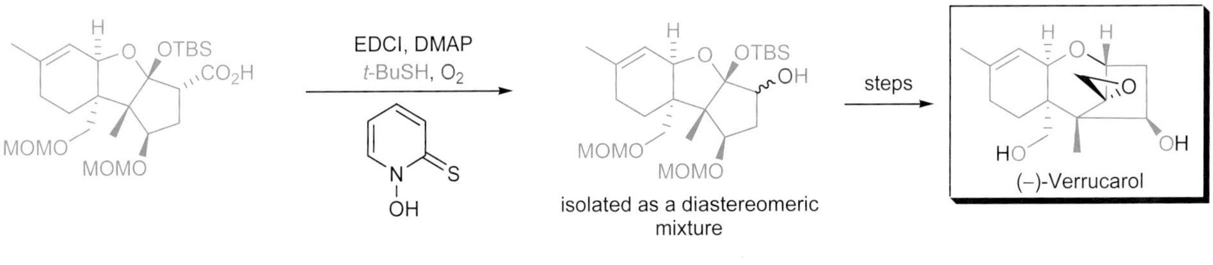

Synthetic Applications:

(–)-Quinocarcin exhibits notable antitumor activity against several strains of solid mammalian carcinomas. In the laboratory of S. Terashima, synthetic studies on quinocarcin and its related compounds were conducted.[15] In an effort to establish structure-activity relationships, the synthesis and *in vitro* cyctotoxicity of C10 substituted quinocarcin congeners was carried out. To prepare 10-decarboxyquinocarcin, the *Barton decarboxylation protocol* was employed. The corresponding acid was esterified with 2-mercaptopyridine-*N*-oxide, and the resulting thiohydroxamate ester was immediately subjected to *Barton radical decarboxylation* using AIBN and tributyltin hydride giving rise to the C10 decarboxylated derivative in 65% overall yield.

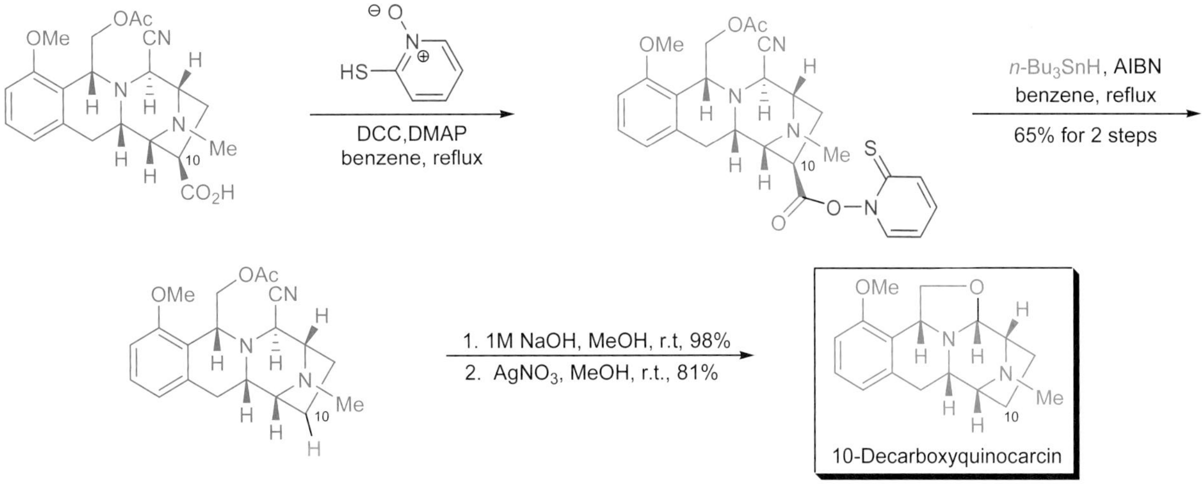

B. Zwanenburg and co-workers synthesized 6-functionalized tricyclodecadienones (*endo*-tricyclo[5.2.1.0^{2,6}]deca-4,8-dien-3-ones) using *Barton's radical decarboxylation reaction* from the corresponding tricyclic carboxylic acid.[16] Their work expanded the chemical scope of the tricyclodecadienone system as a synthetic equivalent of cyclopentadienone. The synthesis of functionalized cage compounds was also undertaken beginning with 1,3-bishomocubanone carboxylic acid, obtained by irradiating the tricyclic ester. After the *bromodecarboxylation* and *phenylselenodecarboxylation* of 1,3-bishomocubanone carboxylic acid under the conditions of the *Barton reaction*, the corresponding bridgehead bromide and phenylselenide were obtained in high yield.

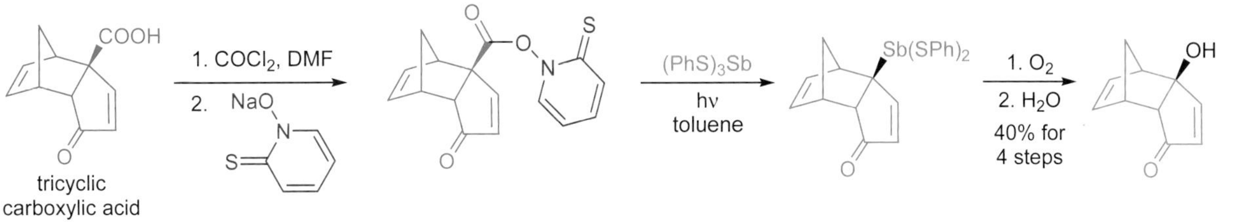

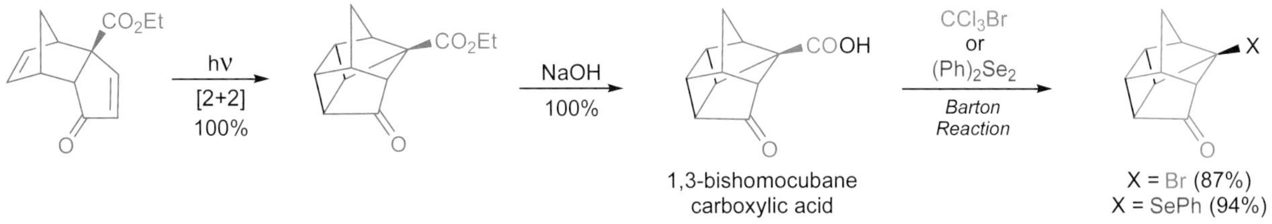

A *double Barton radical decarboxylation* was utilized during the one-step total synthesis of tyromycin A and its analogs by M. Samadi et al.[17] The *bis*-thiohydroxamic ester was irradiated in the presence of citraconic anhydride, and the resulting product was stirred for two days at room temperature to ensure complete elimination.

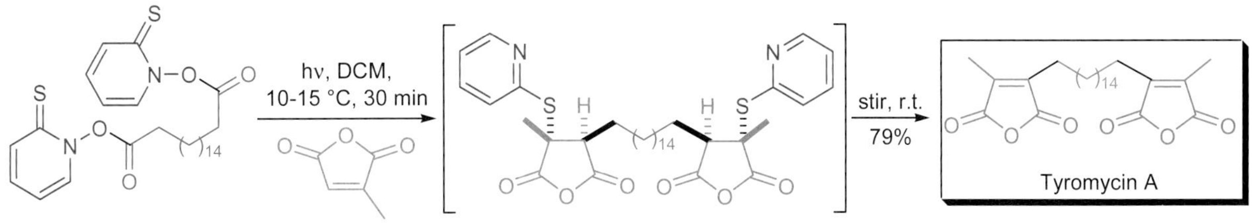

BARTON-McCOMBIE RADICAL DEOXYGENATION REACTION

(References are on page 546)

Importance:

[*Seminal Publications*[1-6]; *Reviews*[7-12]; *Modifications & Improvements*[13-24]]

In the *Barton-McCombie radical deoxygenation* reaction the hydroxyl group of an alcohol is replaced with a hydrogen atom. Even hindered secondary and tertiary alcohols may be deoxygenated by this method. In a typical procedure the alcohol is first converted to a thioxoester derivative, which is then exposed to tri-*n*-butyltin hydride in refluxing toluene.

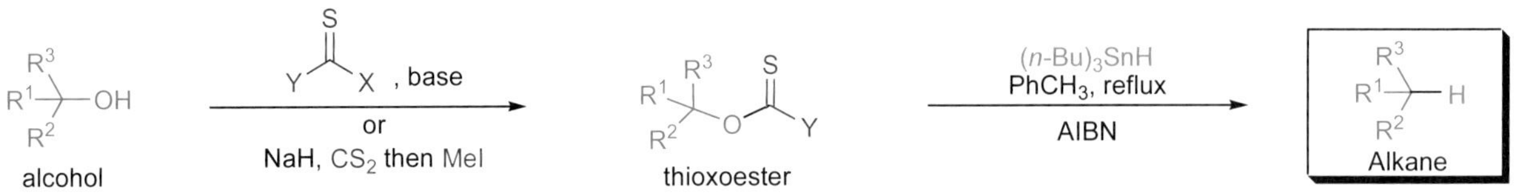

Y = SMe, imidazolyl, OPh, OMe; X = Cl, imidazolyl; base: NaH

Mechanism: [25,13,26]

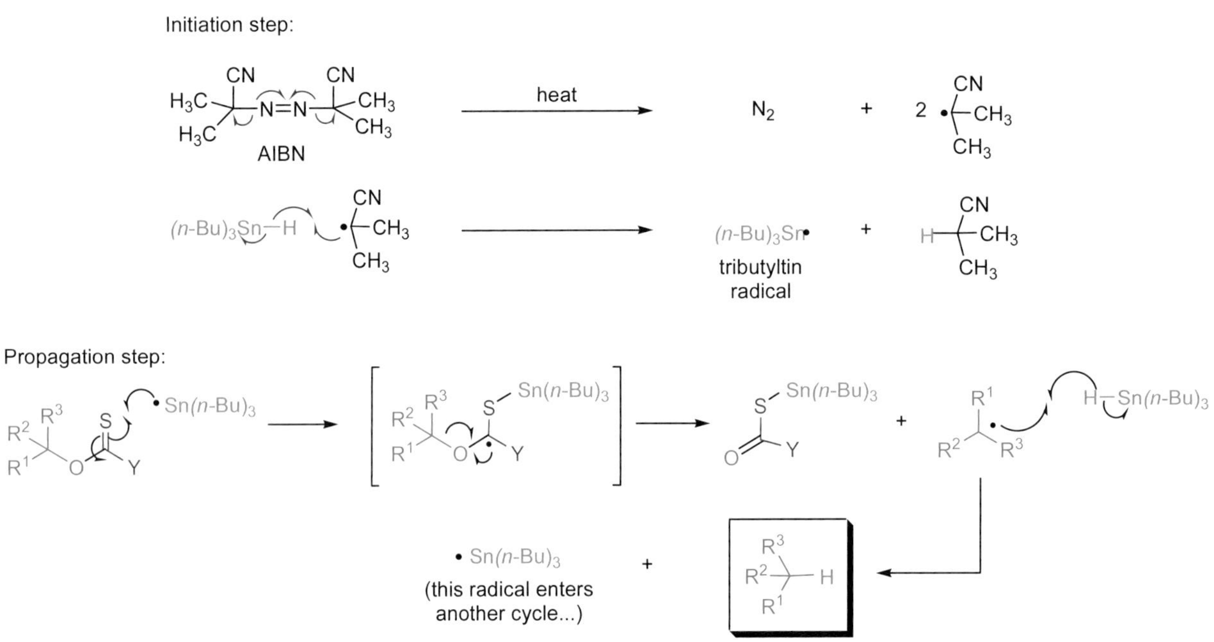

Synthetic Applications:

In the asymmetric synthesis of the C1-C19 fragment of kabiramide C, to complete the stereochemical array, J. Panek and co-workers used, among other methods, *the Barton-McCombie deoxygenation* protocol.[27]

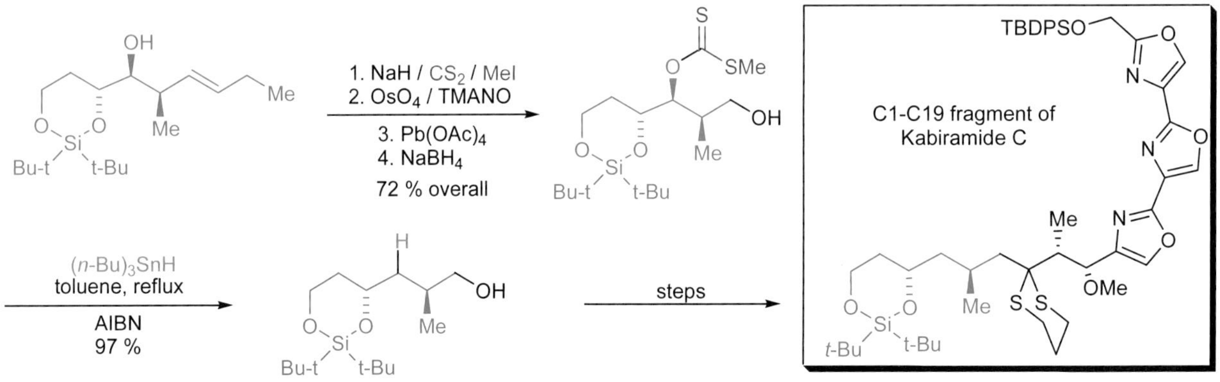

Synthetic Applications:

S.J. Danishefsky and co-workers developed a synthetic route to the neurotrophic illicinones and a total synthesis of the natural product tricycloillicinone.[28] Illicinones were found to enhance the action of choline acetyltransferase, which catalyzes the synthesis of acetylcholine from its precursors. The application of *Corey-Snider oxidative cyclization* and the *Barton-McCombie radical deoxygenation* provided a direct route to tricycloillicinone.

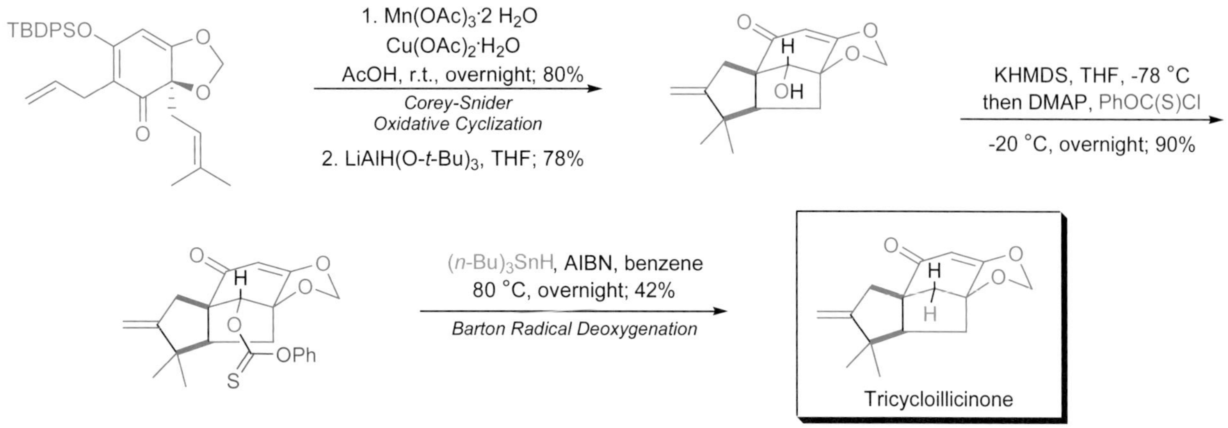

In the laboratory of V. Singh a novel and efficient stereospecific synthesis of the marine natural product (±)-$\Delta^{9(12)}$-capnellene from *p*-cresol was developed.[29] After rapidly assembling the desired carbon framework, it was necessary to remove the carbonyl group from the tricyclic intermediate which was accomplished using *Barton's deoxygenation procedure*.

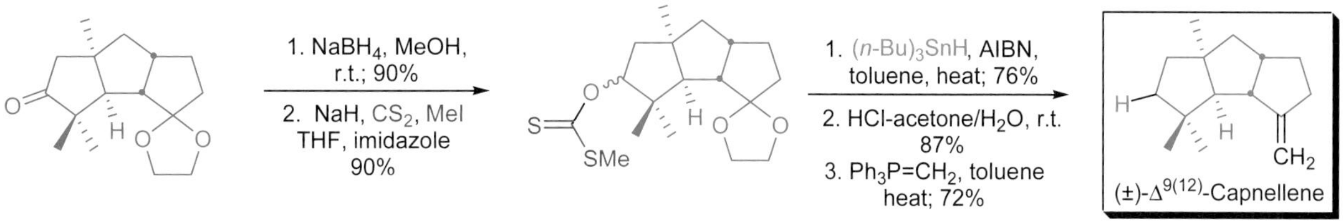

F. Luzzio and co-workers devised a total synthesis for both antipodes of the (–)-Kishi lactam, which is a versatile intermediate for the synthesis of the perhydrohistrionicotoxin (pHTX) alkaloids.[30] In the final stages of the synthesis of the (-)-Kishi lactam, it was necessary to remove one of the secondary alcohol groups. The *Barton radical deoxygenation protocol* was utilized for this operation.

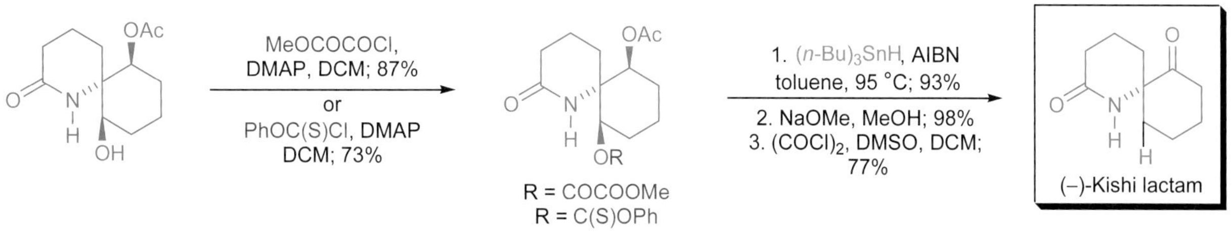

R.H. Schlessinger et al. have successfully synthesized the α,β-unsaturated octenoic acid side chain of zaragozic acid, which contains acyclic "skip" 1,3 dimethyl stereocenters.[31] Their approach utilized the *Barton radical deoxygenation reaction* in the last step of the total synthesis for the removal of the unnecessary hydroxyl group.

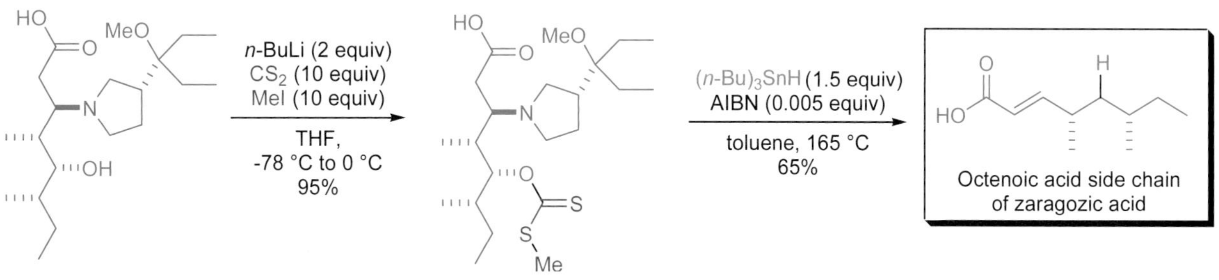

BAYLIS-HILLMAN REACTION

(References are on page 547)

Importance:

[*Seminal Publications*[1,2]; *Reviews*[3-13]; *Modifications & Improvements*[14-31]]

In 1968, K. Morita reported the reaction of acetaldehyde with ethyl acrylate to give α-hydroxyethylated products in the presence of tertiary phosphines.[1] Four years later A.B. Baylis and M.E.D. Hillman carried out the same transformation by using the cheaper and less toxic DABCO as the catalyst.[2] The *Baylis-Hillman reaction* involves the formation of a C-C single bond between the α-position of conjugated carbonyl compounds, such as esters and amides, and carbon electrophiles, such as aldehydes and activated ketones in the presence of a suitable nucleophilic catalyst, particularly a tertiary amine. The most frequently used catalysts are DABCO, quinuclidine, cinchona derived alkaloids and trialkylphosphines. The *asymmetric Baylis-Hillman reaction* can be mediated efficiently by hydroxylated chiral amines derived from cinchona alkaloids. The reaction works with both aliphatic and aromatic aldehydes and results in high enantioselectivities.[7] A catalytic amount of BINAP was also shown to promote the reaction with selected aldehydes.[17] The major drawbacks of the *organocatalytic Baylis-Hillman reaction* are the slow reaction rate (days and weeks) and the limited scope of substrates. However, these shortcomings may be partly overcome by using metal-derived Lewis acids.[15,16]

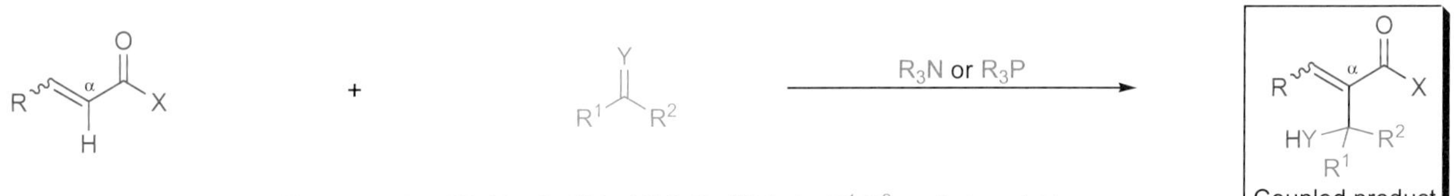

X = NH_2, NR_2, OR; Y = O, NTs, NCO_2R, NSO_2Ar; R^1,R^2 = alkyl, aryl, H

Mechanism:[32-34,17,35-38]

The currently accepted mechanism of the *Baylis-Hillman reaction* involves a *Michael addition* of the catalyst (tertiary amine) at the β-position of the activated alkene to form a zwitterionic enolate. This enolate reacts with the aldehyde to give another zwitterion that is deprotonated, and the catalyst is released. Proton transfer affords the final product.

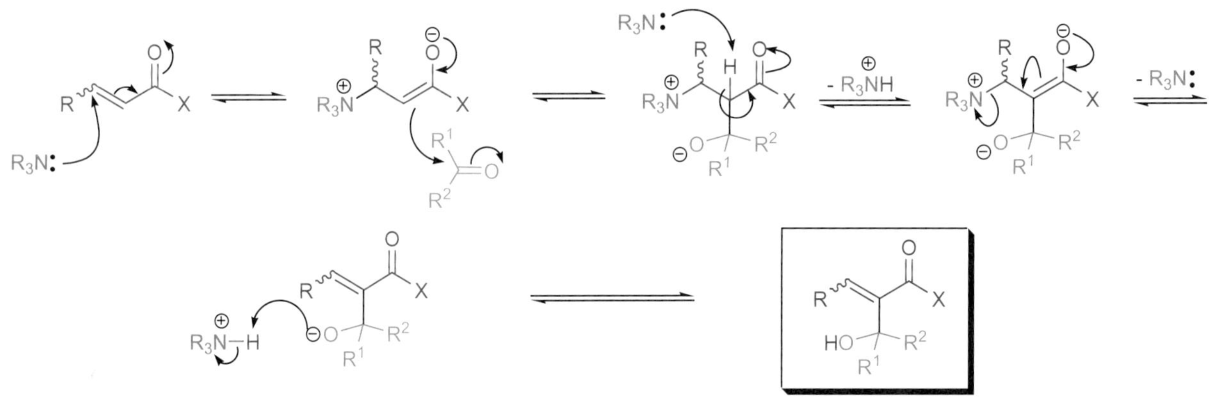

Synthetic Applications:

S. Hatekayama and co-workers developed a highly enantio- and stereocontrolled route to the key precursor of the novel plant cell inhibitor epopromycin B, using a *cinchona-alkaloid catalyzed Baylis-Hillman reaction* of *N*-Fmoc leucinal.[39]

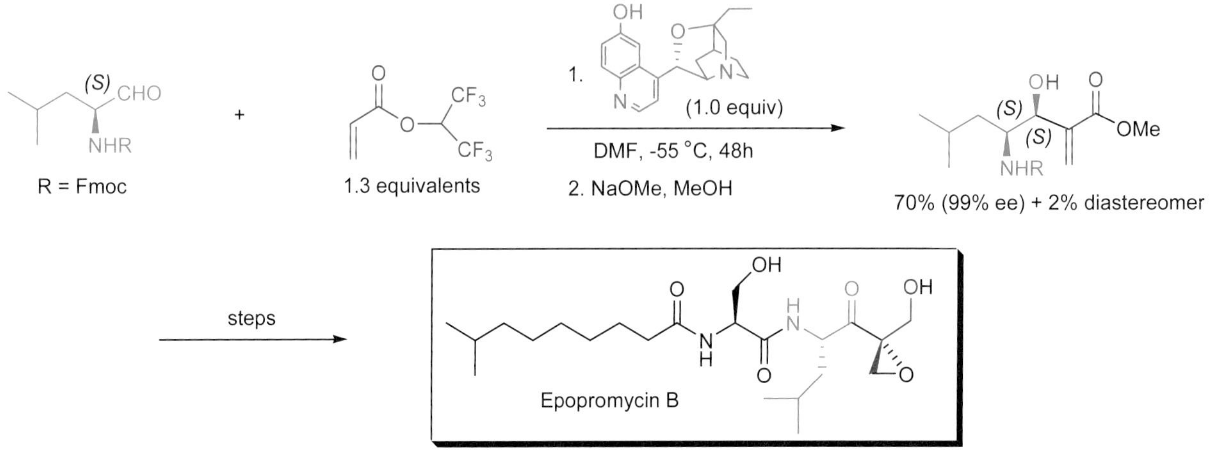

Synthetic Applications:

It was shown in the laboratory of P.T. Kaye that the reactions of 2-hydroxybenzaldehydes and 2-hydroxy-1-naphthaldehydes with various activated alkenes proceeded with regioselective cyclization under *Baylis-Hillman conditions* to afford the corresponding 3-substituted 2H-chromene derivatives in high yields.[40] Previous attempts to prepare 2H-chromenes chemoselectively *via* the cyclization of 2-hydroxybenzaldehyde-derived Baylis-Hillman products had proven unsuccessful. Complex mixtures containing coumarin and chromene derivatives were obtained. Good results were observed after the careful and systematic study of the various reactants and reaction conditions.

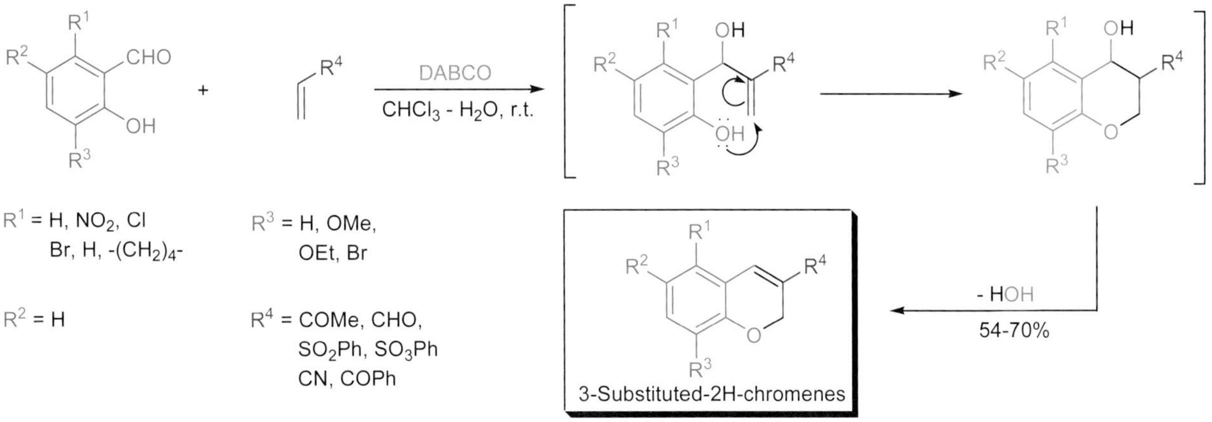

D. Basavaiah and co-workers achieved the simple and convenient stereoselective synthesis of (*E*)-α-methylcinnamic acids *via* the nucleophilic addition of hydride ion from sodium borohydride to acetates of Baylis-Hillman adducts (methyl 3-acetoxy-3-aryl-2-methylenepropanoates), followed by hydrolysis and crystallization.[41] The potential of this methodology was demonstrated in the synthesis of (*E*)-*p*-(myristyloxy)-α-methylcinnamic acid, which is an active hypolipidemic agent.

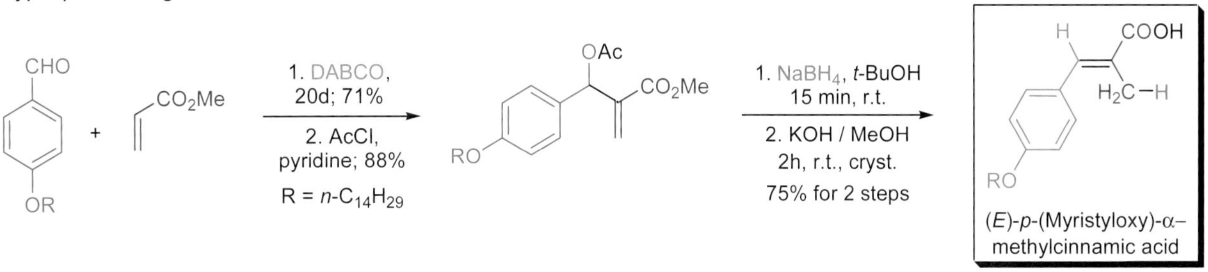

Research by J. Jauch showed that in the case of highly base-sensitive substrates the *Baylis-Hillmann reaction* can be carried out by using lithium phenylselenide, which is a strong nucleophile but only weakly basic. This variant of the reaction is highly diastereoselective and was successfully applied to the total synthesis of kuehneromycin A.[42]

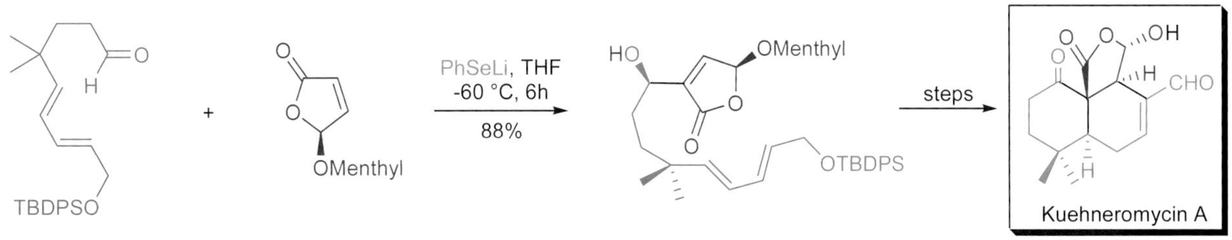

In the simple stereoselective total synthesis of salinosporamide A, E.J. Corey and co-workers applied the *intramolecular Baylis-Hillman reaction* to a ketoamide substrate.[43] The reaction was catalyzed by quinuclidine and the γ-lactam product was formed as a 9:1 mixture of diastereomers favoring the desired stereoisomer.

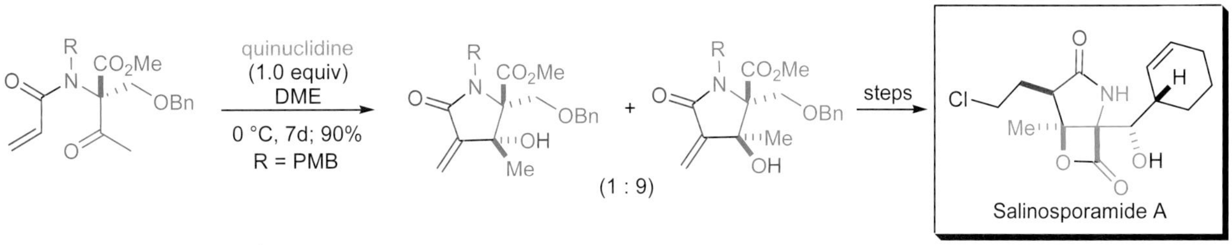

BECKMANN REARRANGEMENT

(References are on page 548)

Importance:

[*Seminal Publication*[1]; *Reviews*[2-5]; *Modifications & Improvements*[6-17]; *Theoretical Studies*[18-27]]

The conversion of aldoximes and ketoximes to the corresponding amides in acidic medium is known as the *Beckmann rearrangement*. It is especially important in the industrial production of ε-caprolactam, which is used as a monomer for polymerization to a polyamide for the production of synthetic fibers. The reaction is usually carried out under forcing conditions (high temperatures >130 °C, large amounts of strong Brönsted acids) and it is non-catalytic. The applied Brönsted acids are: H_2SO_4, $HCl/Ac_2O/AcOH$, etc., which means that sensitive substrates cannot be used in this process. The stereochemical outcome of this rearrangement is predictable: the R group *anti* to the leaving group on the nitrogen will migrate. If the oxime isomerizes under the reaction conditions, a mixture of the two possible amides is obtained. The hydrogen atom never migrates, so this method cannot be used for the synthesis of *N*-unsubstituted amides.

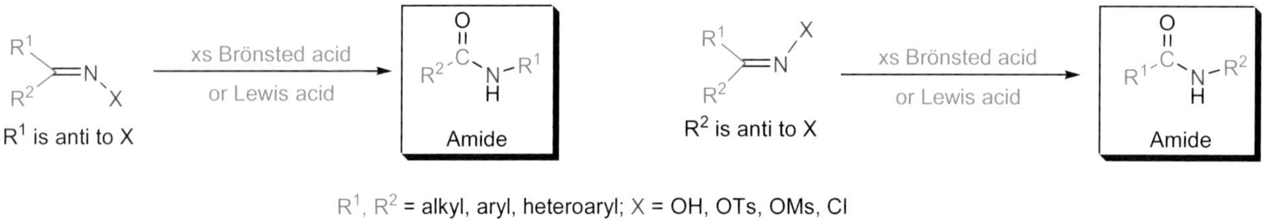

Mechanism: [28,19,22-24,29-31]

In the first step of the mechanism the X group is converted to a leaving group by reaction with an electrophile. The departure of the leaving group is accompanied by the *[1,2]*-shift of the R group, which is *anti* to the leaving group. The resulting carbocation reacts with a nucleophile (a water molecule or the leaving group) to afford the amide after tautomerization.

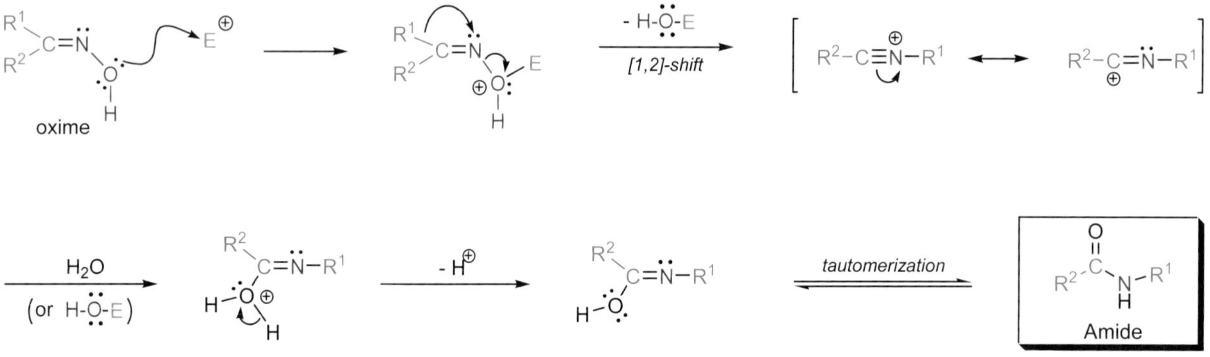

Synthetic Applications:

N.S. Mani and co-workers utilized the *organoaluminum promoted modified Beckmann rearrangement* during their efficient synthetic route to chiral 4-alkyl-1,2,3,4-tetrahydroquinoline. (4*R*)-4-Ethyl-1,2,3,4-tetrahydroquinoline was obtained by rearrangement of the ketoxime sulfonate of (3*R*)-3-ethylindan-1-one.[32] The resulting six-membered lactam product was reduced to the corresponding cyclic secondary amine with diisobutylaluminum hydride.

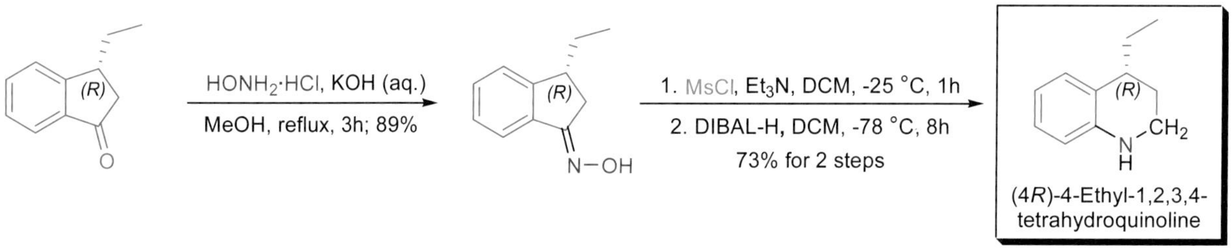

Synthetic Applications:

In the laboratory of J.D. White, the asymmetric total synthesis of the non-natural (+)-codeine was accomplished *via intramolecular carbenoid insertion*.[33] In the late stages of the total synthesis it was necessary to install a 6-membered piperidine moiety. This transformation was accomplished utilizing a *Beckmann rearrangement* of the cyclopentanone oxime portion of one of the intermediates. Later the 6-membered lactam was reduced to the corresponding amine with LAH. To this end, an oxime brosylate (Bs) was prepared, which underwent a smooth *Beckmann rearrangement* in acetic acid to provide a 69% yield of two isomeric lactams in an 11:1 ratio in favor of the desired isomer.

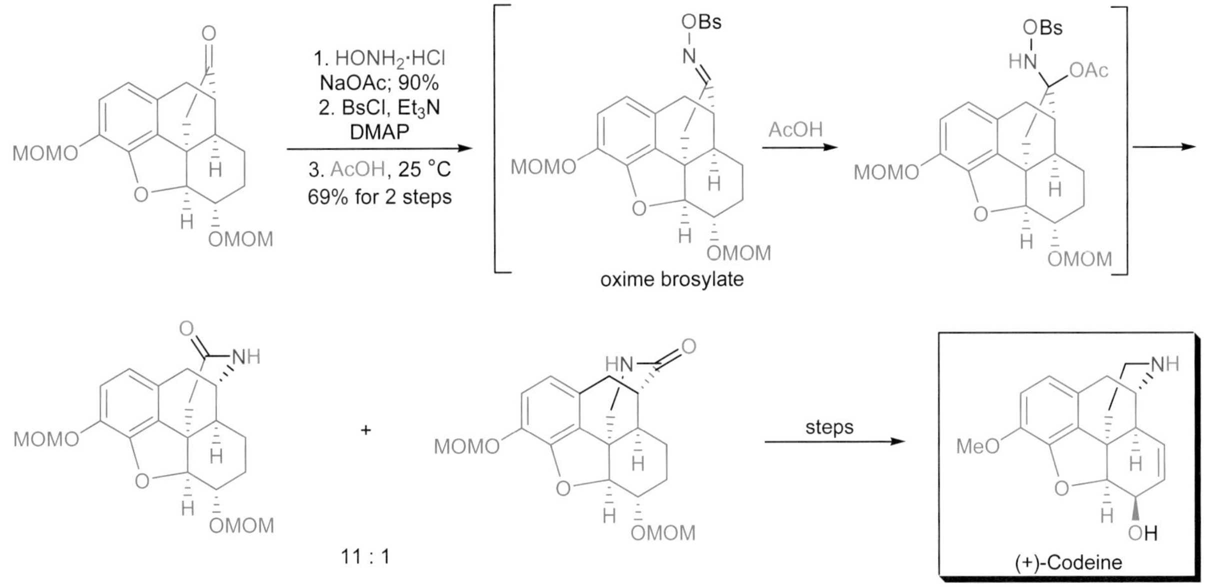

J.D. White et al. reported the total synthesis of (–)-ibogamine *via* the catalytic *asymmetric Diels-Alder reaction* of benzoquinone.[34] The azatricyclic framework of the molecule was established by converting the bicyclic ketone to the *anti* oxime and then subjecting it to a *Beckmann rearrangement* in the presence of *p*-toluenesulfonyl chloride to afford the 7-membered lactam. Elaboration of this lactam into the azatricyclic core of ibogamine and later to the natural product itself was accomplished in a few additional steps.

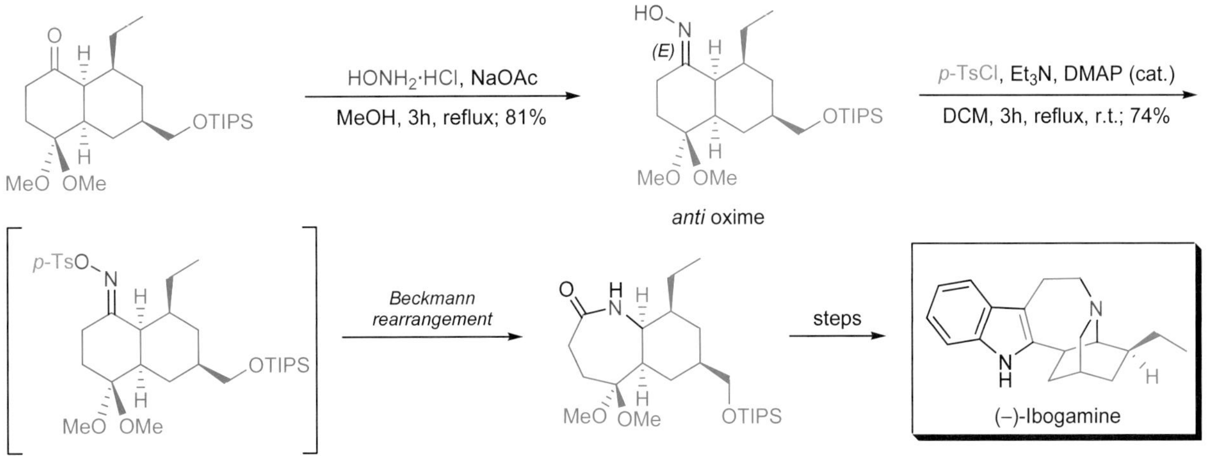

A novel variant of the *photo-Beckmann rearrangement* was utilized by J. Aubé and co-workers in the endgame of the total synthesis of (+)-sparteine.[35] The hydroxylamine was generated *in situ*, and reacted intramolecularly with the ketone to form a nitrone . Photolysis of the nitrone afforded the desired lactam in good yield.

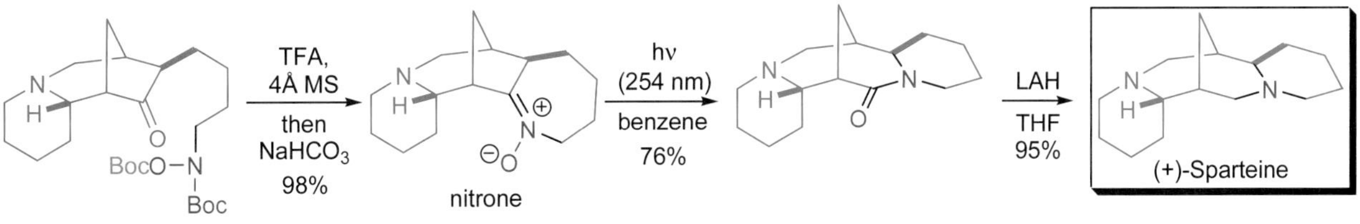

BENZILIC ACID REARRANGEMENT

(References are on page 548)

Importance:

[Seminal Publications[1,2]; *Reviews*[3-6]; *Modifications & Improvements*[7-10]; *Theoretical Studies*[11]]

Upon treatment with base (e.g., NaOH), α-diketones rearrange to give salts of α-hydroxy acids. This process is called the *benzilic acid rearrangement*. The reaction takes place with both aliphatic and aromatic α-diketones and α-keto aldehydes. Usually diaryl diketones undergo *benzilic acid rearrangements* in excellent yields, but aliphatic α-diketones that have enolizable α-protons give low yields due to competing *aldol condensation* reactions. Cyclic α-diketones undergo the synthetically useful *ring-contraction benzilic acid rearrangement* reaction under these conditions. When alkoxides or amide anions are used in place of hydroxides, the corresponding esters and amides are formed. This process is called the *benzilic ester rearrangement*. Alkoxides that are readily oxidized such as ethoxide (EtO^-) or isopropoxide (Me_2CHO^-) are not synthetically useful, since these species reduce the α-diketones to the corresponding α-hydroxy ketones. Aryl groups tend to migrate more rapidly than alkyl groups. When two different aryl groups are available, the major product usually results from migration of the aromatic ring with the more powerful electron-withdrawing group(s).

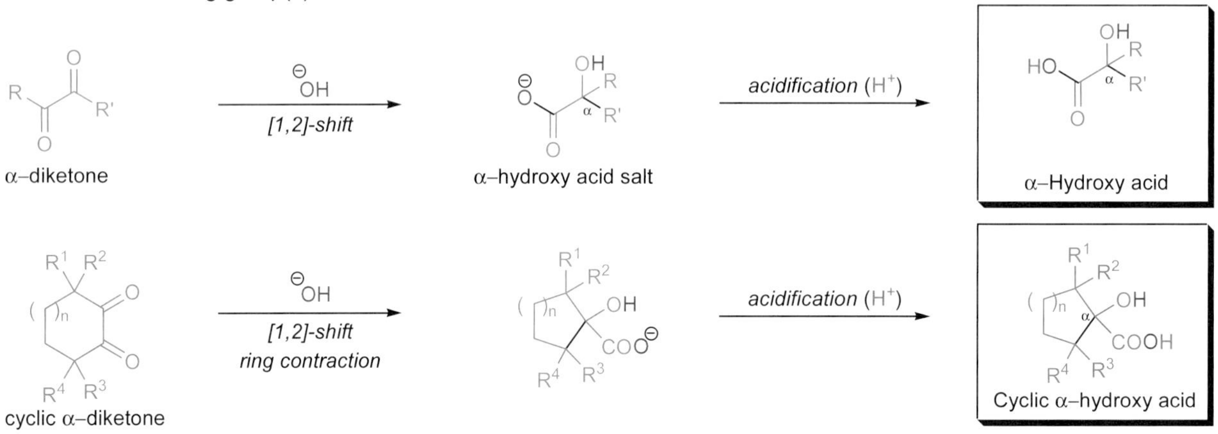

Mechanism: [12-16,11,17,18,8,6]

The *benzilic acid rearrangement* is an irreversible process. The first step of the mechanism is the addition of the nucleophile (HO^-, alkoxide, or amide ion) across the C=O bond to give a tetrahedral intermediate. The next step is aryl or alkyl migration to form the corresponding α-hydroxy acid salt.

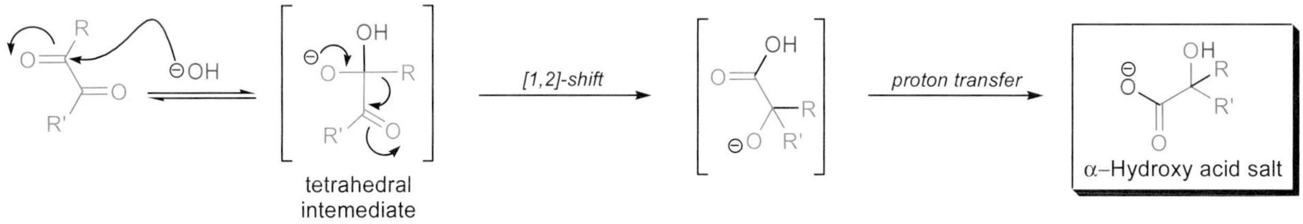

Synthetic Applications:

J.L. Wood et al. were able to convert a pyranosylated indolocarbazole to the carbohydrate moiety of (+)-K252a utilizing the stereoselective *ring-contraction benzilic acid rearrangement*.[19] This reaction suggested a possible biosynthetic link between furanosylated and pyranosylated indolocarbazoles.

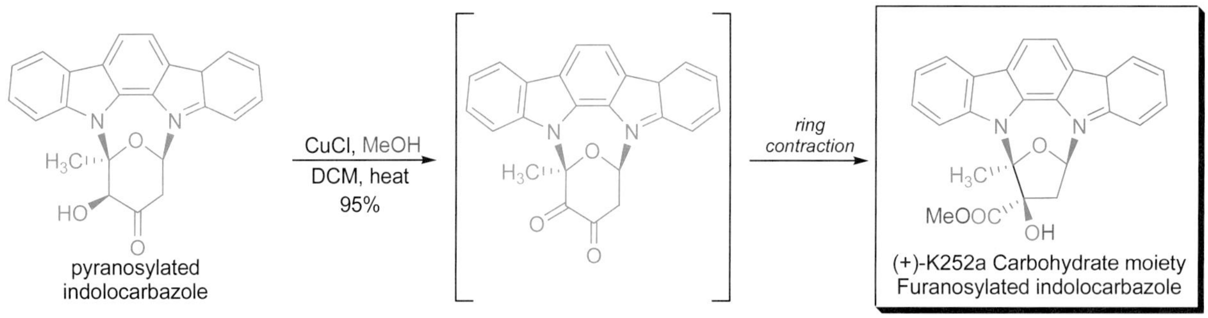

Synthetic Applications:

In an attempt to isolate 16α,17α-dihydroxyprogesterone by the stereoselective *cis*-dihydroxylation of 16-dehydroprogesterone using cetyltrimethylammonium permanganate (CTAP) as an oxidant, J.A. Katzenellenbogen and co-workers isolated a novel 5-ring D-homosteroid instead of the desired diol.[20] The mechanism of the final step was similar to the *benzilic acid rearrangement*. Under reaction conditions in which the permanganate concentration was high, the C21 enolate of the diketone attacked the aldehyde to form the 5-membered ketolactol. The final ring contraction was accomplished by the *benzilic acid rearrangement*.

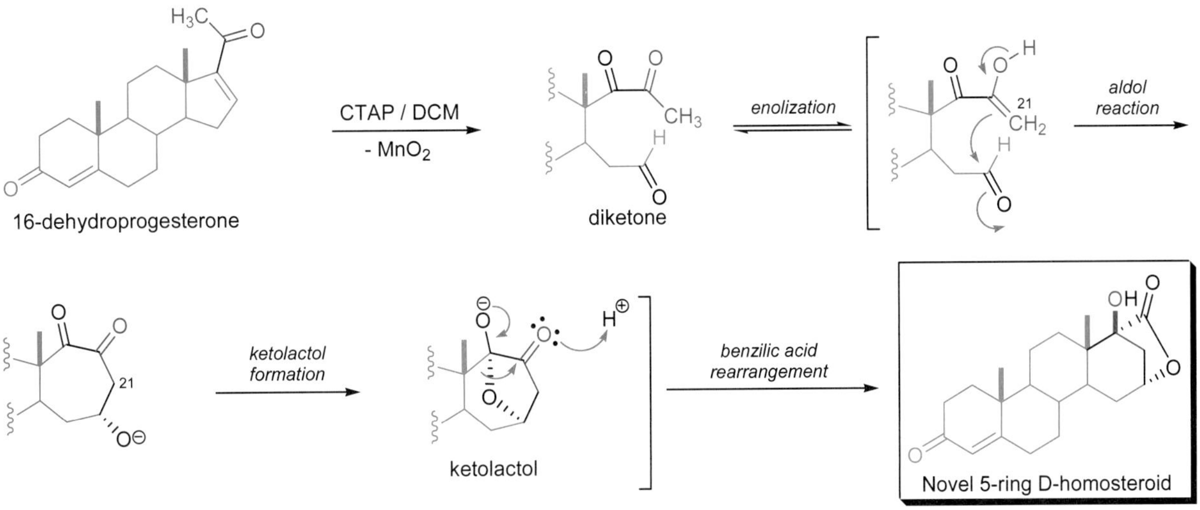

H. Takeshita and co-workers devised a short synthesis of (±)-hinesol and (±)-agarospirol *via* a mild base-catalyzed *retro-benzilic acid rearrangement* of *proto*-[2+2] photocycloadducts to the desired spiro[4,5]decanedione framework.[21]

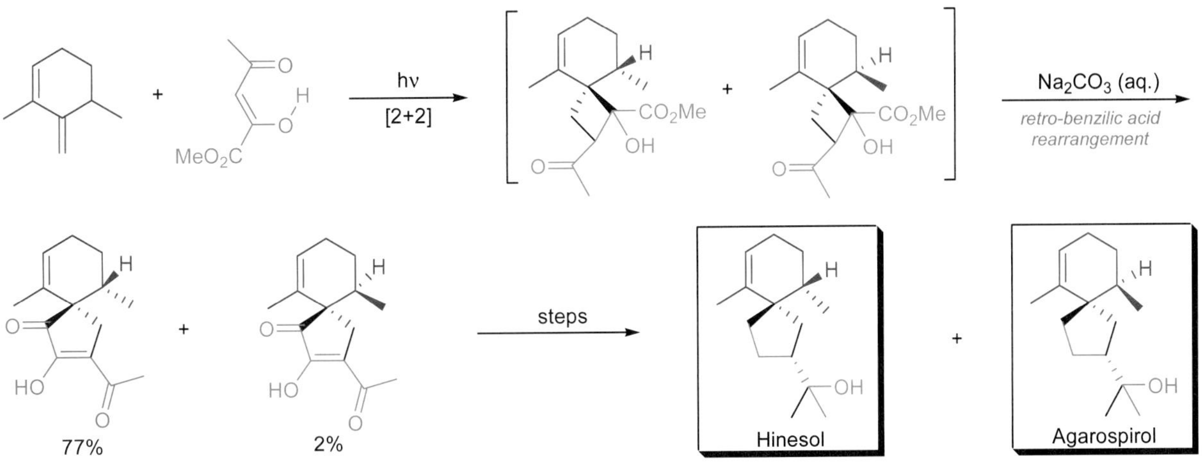

P.A. Grieco et al. accomplished the total synthesis of (±)-shinjudilactone and (±)-13-*epi*-shinjudilactone *via* a *benzilic acid-type rearrangement*.[22] The substrate was exposed to basic conditions and the two desired products were obtained as a 1:1 mixture. Interestingly, when the C1 position was methoxy substituted, the rearrangement failed to take place under a variety of acidic and basic conditions.

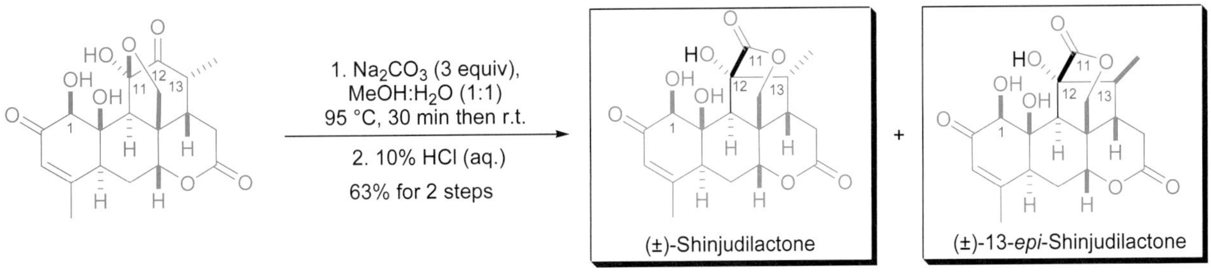

BENZOIN AND RETRO-BENZOIN CONDENSATION

(References are on page 549)

Importance:

[*Seminal Publication*[1-3]; *Reviews*[4-8]; *Modifications & Improvements*[9-16]; *Theoretical Studies*[17,18]]

Upon treating certain (but not all) aromatic aldehydes or glyoxals (α-keto aldehydes) with cyanide ion (CN^-), benzoins (α-hydroxy-ketones or acyloins) are produced in a reaction called the *benzoin condensation*. The reverse process is called the *retro-benzoin condensation*, and it is frequently used for the preparation of ketones. The condensation involves the addition of one molecule of aldehyde to the C=O group of another. One of the aldehydes serves as the donor and the other serves as the acceptor. Some aldehydes can only be donors (e.g. *p*-dimethylaminobenzaldehyde) or acceptors, so they are not able to self-condense, while other aldehydes (benzaldehyde) can perform both functions and are capable of self-condensation. Certain thiazolium salts can also catalyze the reaction in the presence of a mild base.[11,12,19] This version of the benzoin condensation is more synthetically useful than the original procedure because it works with enolizable and non-enolizable aldehydes and asymmetric catalysts may be used. Aliphatic aldehydes can also be used and mixtures of aliphatic and aromatic aldehydes give mixed benzoins. Recently, it was also shown that thiazolium-ion based organic ionic liquids (OILs) promote the benzoin condensation in the presence of small amounts of triethylamine.[12] The stereoselective synthesis of benzoins has been achieved using chiral thiazolium salts as catalysts.[11]

R = aryl, heteroaryl, 3° alkyl, C(=O)-alkyl; catalyst: NaCN, KCN, thiazolium salt, NHC (*N*-heterocyclic carbenes)

Mechanism: [20,21,17,22-24,18,25-30,19,31]

All the steps of *the cyanide ion catalyzed benzoin condensation* are completely reversible, and the widely accepted mechanism involves the loss of the aldehydic proton in the key step. This deprotonation is possible due to the increased acidity of this C-H bond caused by the electron-withdrawing effect of the CN group. The cyanide ion is a very specific catalyst of the reaction. Cyanide is a good nucleophile, a good leaving group, and its electron-withdrawing effect enhances the acidity of the aldehyde hydrogen.

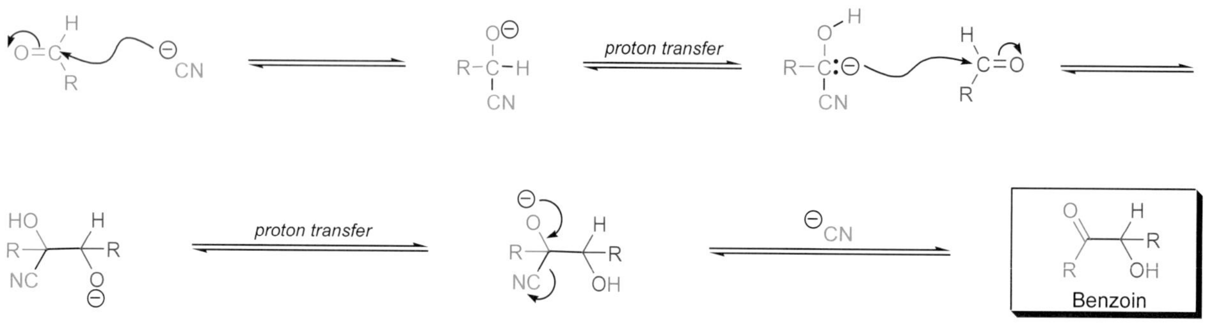

The generally accepted mechanism of the *thiazolium salt-catalyzed benzoin condensation* was first proposed by R. Breslow.[26]

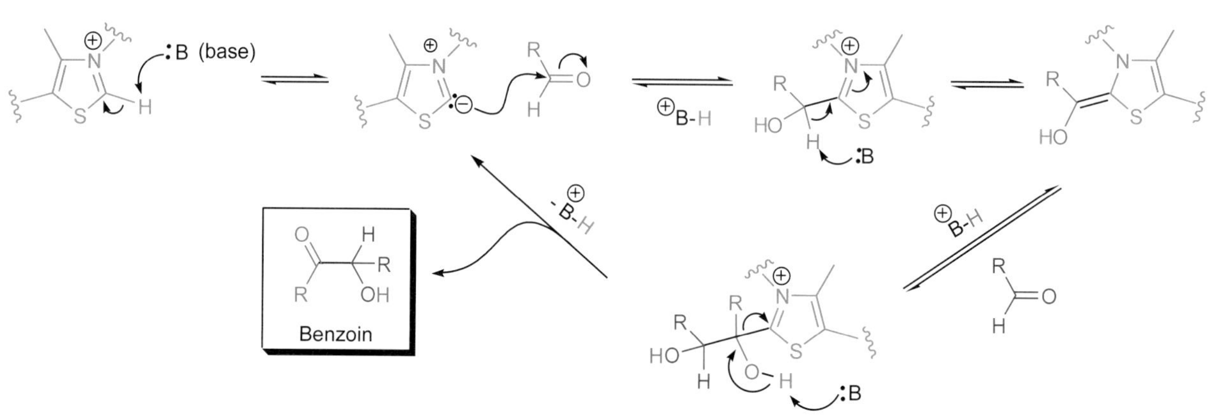

Synthetic Applications:

A. Miyashita and co-workers have developed a new method for the synthesis of ketones based on the concept that the *benzoin condensation* is reversible (*retro-benzoin condensation*) and affords the most stable product.[32] When α-benzylbenzoin was treated with KCN in DMF, the C-C bond was cleaved, resulting in the formation of deoxybenzoin and benzaldehyde. This method of synthesizing ketones has been applied to several α-substituted benzoins, and the corresponding ketones were formed in good yields. The authors also realized, based on the known analogy between the chemical behavior of the C=O double bond of ketones and the C=N double bond of nitrogen-containing heteroarenes, that a cyanide ion catalyzed *retro-benzoin condensation* of α-hydroxybenzylheteroarenes would also be possible.[33]

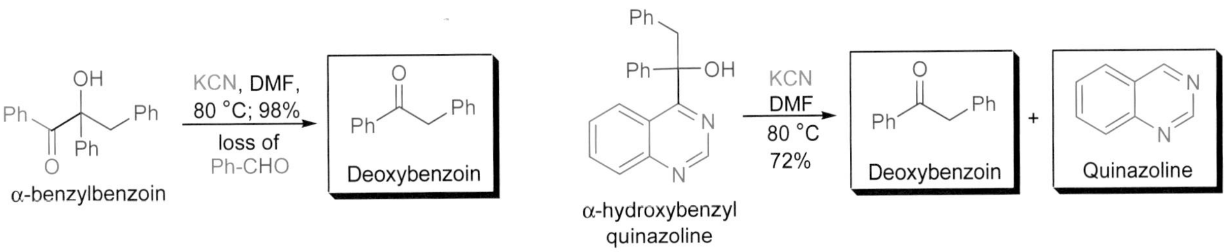

The *retro-benzoin condensation* methodology was used to synthesize 2-substituted quinazolines in good overall yield from 2,4-dichloroquinazoline. 2-Substituted quinazolines are obtained by substitution of 2-chloroquinazoline with nucleophiles, though it is difficult to prepare the starting 2-chloroquinazoline. These results indicate that the aroyl group, which may be introduced onto nitrogen-containing heteroarenes at the α-position, can be used as a protecting group. Later it can be easily removed by conversion to an α-hydroxybenzyl group, followed by a *retro-benzoin condensation*.[33]

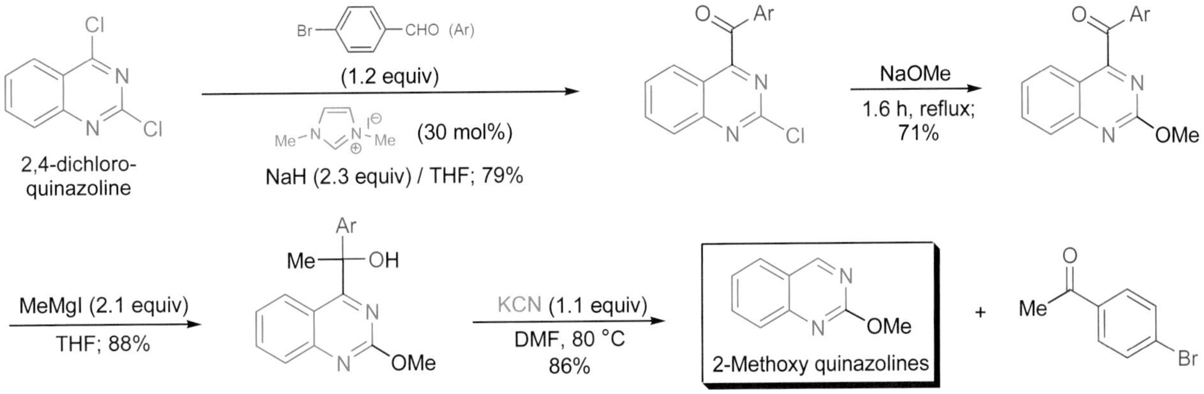

In the laboratory of K. Suzuki, a *catalytic crossed aldehyde-ketone benzoin condensation* was developed and applied to the synthesis of stereochemically defined functionalized preanthraquinones.[15]

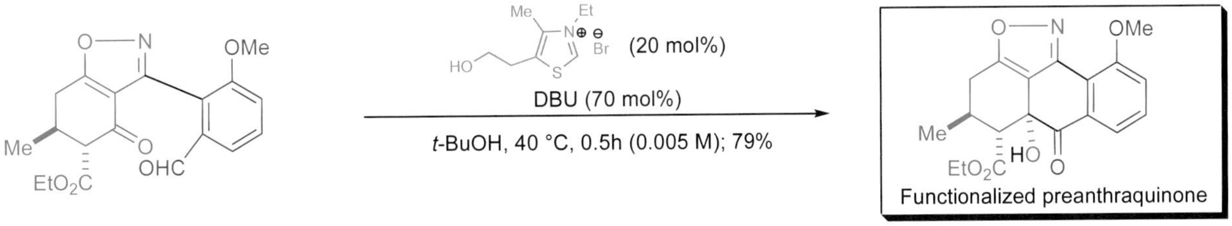

The *benzoin condensation* was the key carbon-carbon bond forming step during the synthesis of anti-inflammatory 4,5-diarylimidazoles by T.E. Barta and co-workers.[34] The benzaldehyde was first converted to the cyanohydrin using TMSCN. Deprotonation was followed by the addition of 4-(MeS)-benzaldehyde to afford the benzoin.

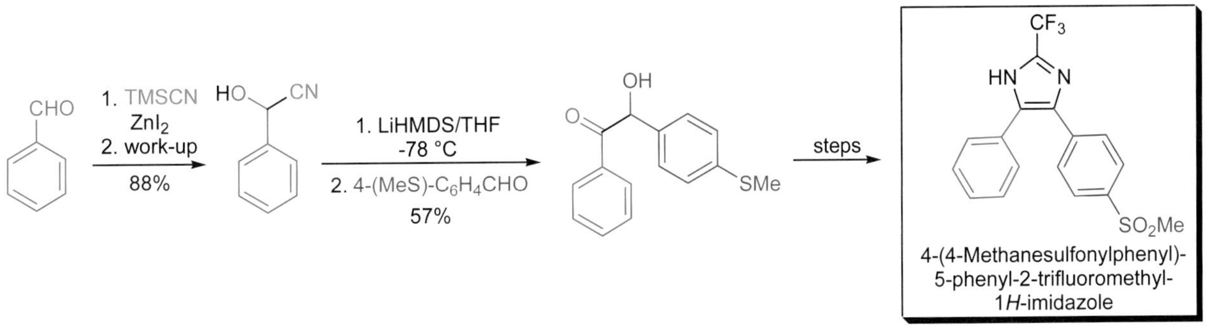

BERGMAN CYCLOAROMATIZATION REACTION

(References are on page 550)

Importance:

[*Seminal Publications*[1,2]; *Reviews*[3-14]; *Modifications & Improvements*[15-18]; *Theoretical Studies*[19-33]]

The thermal cycloaromatization of enediynes, which proceeds *via* the formation of benzenoid diradicals, is known as the *Bergman cycloaromatization reaction*. It received little attention in the 1970s when it was first reported, but it became the subject of intense research in the 1990s when certain marine natural products containing the enediyne moiety showed remarkable antitumor activity *via* the cleavage of double stranded DNA. Synthetically the *Bergman cyclization* was exploited to prepare fused ring systems by tethering alkenes to an enediyne unit and allowing the alkenes to react with the cycloaromatized species to form additional saturated rings. It is also possible to make fused aromatic ring systems, such as acenaphthenes or perylene derivatives. The *Bergman cyclization* tolerates a wide range of functional groups, many of which also increase the yield of the cycloaromatization reaction.[34] The distance between the triple bonds is crucial: the further away the triple bonds are, the higher the temperature required for the cyclization to occur. In order to observe cyclization at physiological temperatures, the enediyne unit should be part of a 10-membered ring.

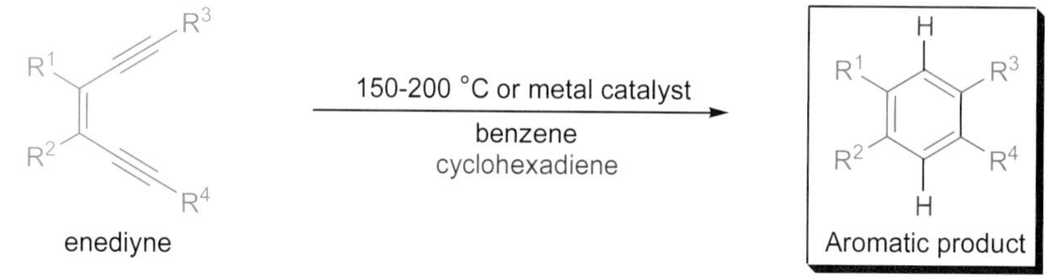

Mechanism: [35-40,26-28]

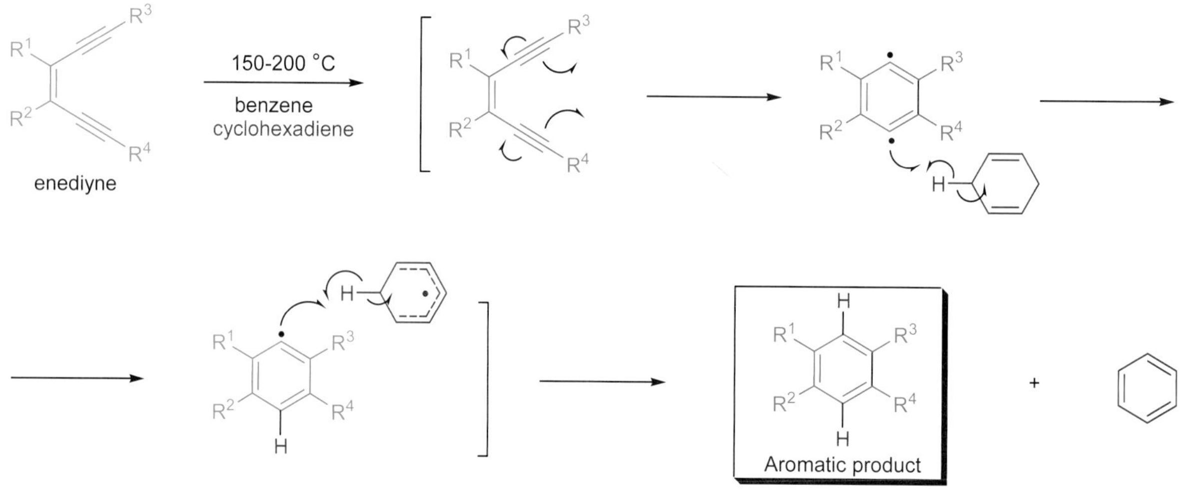

Synthetic Applications:

To make the *Bergman cyclization* synthetically more appealing, the reaction temperature had to be lowered significantly. J.M. Zaleski and co-workers developed a *Mg^{2+}-induced thermal Bergman cyclization* at ambient temperature.[29]

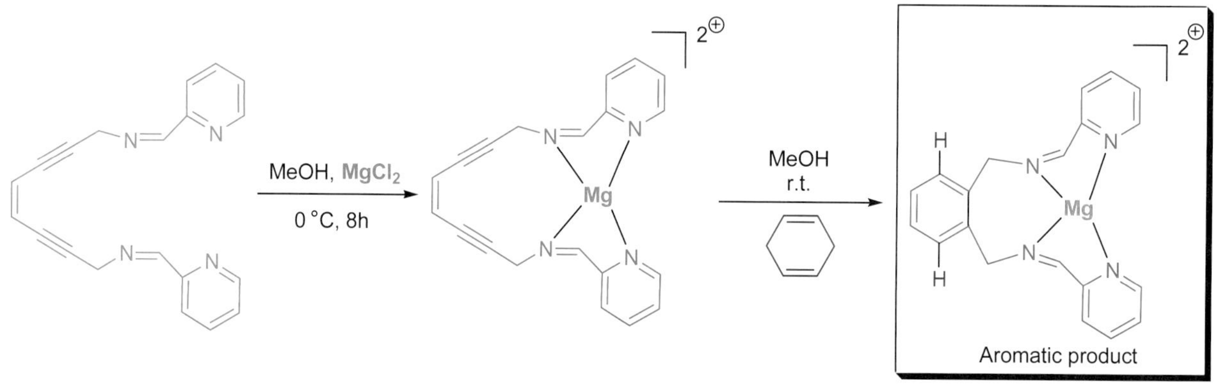

Synthetic Applications:

In the laboratory of K.C. Russell novel 10-membered pyrimidine enediynes were synthesized in seven and eight steps, respectively.[41] These compounds were tested for their ability to undergo the *Bergman cyclization* both thermally and photochemically. Where X=OH, the enediynol readily cyclizes both thermally and photochemically in isopropanol, while when X=O, the enediynone only cyclizes under thermal conditions to give excellent yield of the corresponding aromatic compound. The difference in reactivity between the alcohol and the ketone was assumed to arise from different excited states. Ketones are well-known to possess different excited states and different reactivity from triplet excited states that can undergo hydrogen- and electron-abstraction processes. If the *photochemical Bergman cyclization* is favored by a singlet excited state, then a triplet state ketone could interfere with the normal cyclization process and result in poor yield and conversion.

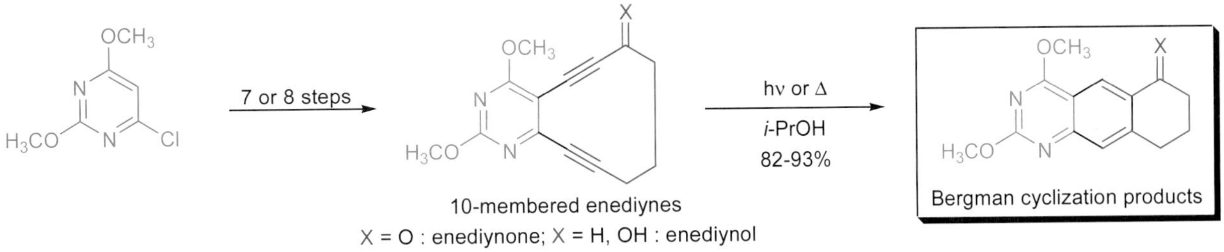

Porphyrin chromophores have received much attention, particularly as photoelectric devices and molecular wires. Efficient π-electronic communication between porphyrin macrocycles is pivotal in various complex functions. K.M. Smith et al. showed that neighboring acetylenic units on porphyrins provide a means for the efficient construction of aromatic superstructures triggered by the *Bergman cyclization* reaction conditions and give rise to novel [n]phenacenoporphyrins, which belong to a new class of highly π-extended porphyrins.[42]

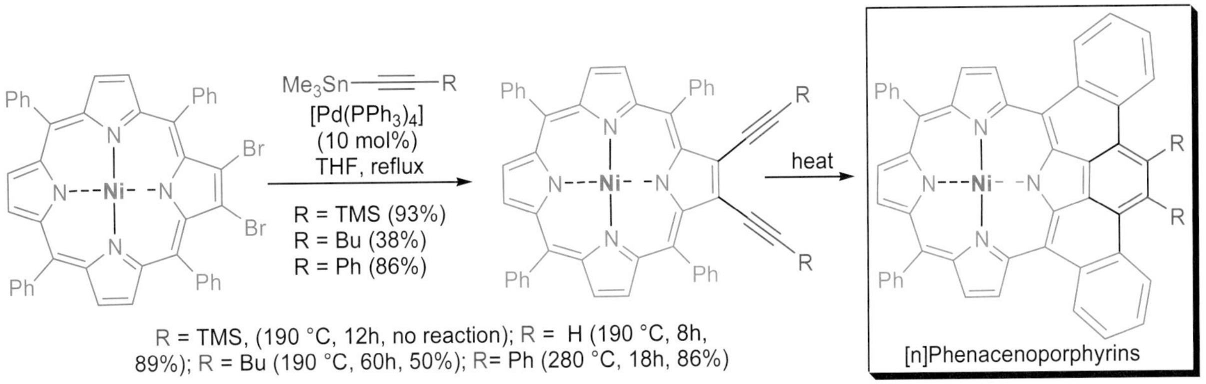

Research by S.J. Danishefsky et al. has shown that calicheamicin/esperamicin antibiotics containing an allylic trisulfide trigger can undergo a mild *Bergman cyclization* when treated with benzyl mercaptan.[43]

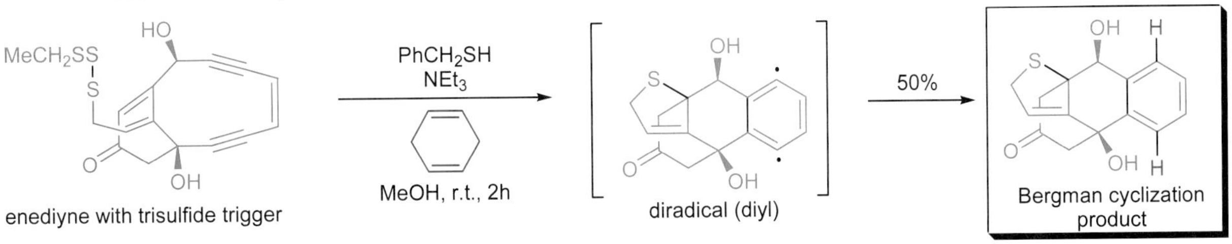

When the enediyne substrate has functional groups that can trap the initially formed Bergman diradical, the rapid construction of complex fused ring systems becomes feasible. J.E. Anthony and co-workers prepared an acenaphthene derivative as well as a substituted perylene using this concept.[34]

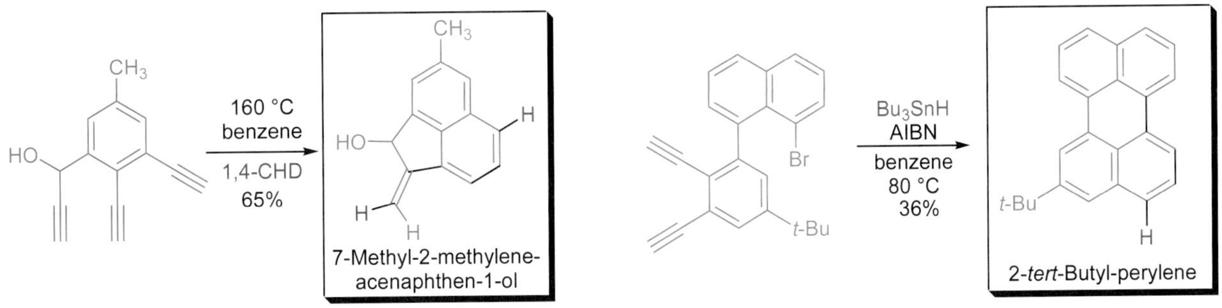

BIGINELLI REACTION

(References are on page 551)

Importance:

[*Seminal Publications*[1,2]; *Reviews*[3-14]; *Modifications & Improvements;*[15-38] *Theoretical Studies*[39,40]]

In 1893, P. Biginelli was the first to synthesize functionalized 3,4-dihydropyrimidin-2(1H)-ones (DHPMs) by the one-pot three-component condensation reaction of an aromatic aldehyde, urea, and ethyl acetoacetate in the presence of catalytic HCl in refluxing ethanol.[2] This process is called the *Biginelli reaction* and the products are referred to as Biginelli compounds.[5] The *Biginelli reaction* was not utilized widely until the early 1990s when the growing demand for biologically active compounds made multicomponent reactions attractive.[8] The general features of the reaction are: 1) it is usually carried out in alcohols as solvents containing a small amount of catalyst; 2) several Lewis and Brönsted acids catalyze the process: HCl, H_2SO_4, TsOH,[31] TMSI,[36] LiBr,[35], $InBr_3$,[30] $BF_3{\cdot}OEt_2$, $FeCl_3$,[21] $Yb(OTf)_3$, $Bi(OTf)_3$,[26,37] VCl_3[41] and PPE;[19] 3) the structure of all three components can be widely varied; 4) aliphatic, aromatic, or heteroaromatic aldehydes are used but with aliphatic or hindered aromatic aldehydes (*ortho*-substituted) the yields are moderate; 5) a variety of β-keto esters, including ones with chiral centers at R^2 as well as tertiary acetoacetamides have been utilized; 6) monosubstituted ureas and thioureas give exclusively *N*-1 substituted dihydropyrimidines while *N*-3 alkylated products are never formed; 7) *N,N'*-disubstituted ureas do not react under the standard *Biginelli reaction* conditions; and 8) the preparation of enantiomerically pure Biginelli compounds is currently easiest *via* resolution, and a true intermolecular asymmetric version does not yet exist.[42] There are several variations of the *Biginelli reaction*: 1) the most significant variant is called the *Atwal modification* in which an enone is reacted with a protected urea or thiourea derivative under neutral conditions to first give a 1,4-dihydropyrimidine, which is converted to the corresponding DHPM upon deprotection with acid;[15-17] 2) in the *Shutalev modification* α-tosyl substituted ureas and thioureas are reacted with enolates of 1,3-dicarbonyl compounds to afford hexahydropyrimidines, that are readily converted to DHPMs;[20] 3) solid phase synthesis with Wang resin-bound urea derivatives or with PEG-bound acetoacetate allows the preparation of DHPMs in high yield and high purity;[43,44] 4) a fluorous phase variant was developed using a fluorous urea derivative;[18] and 5) microwave-assisted and solvent-free conditions were also successfully implemented.[23,25,32]

Biginelli reaction:

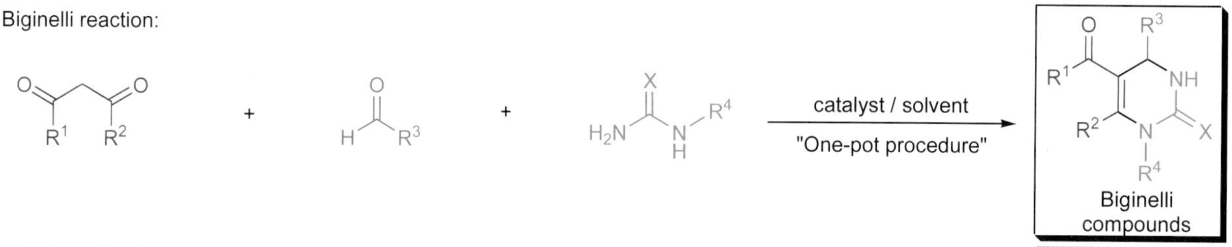

Atwal modification:

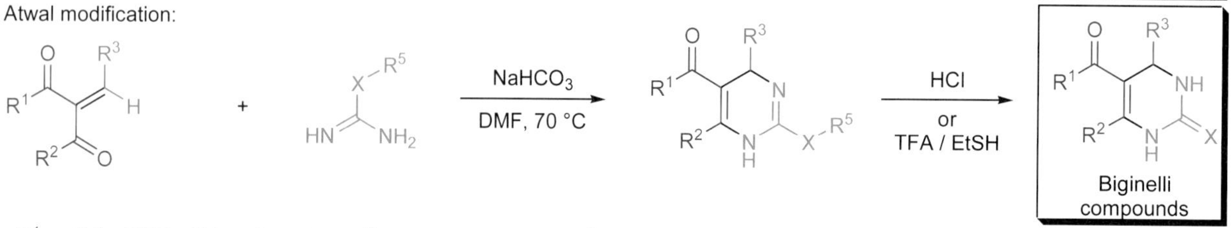

R^1 = -OEt, -NHPh, NEt_2, alkyl, -SEt; R^2 = alkyl, aryl, $-CH_2Br$; R^3 = aryl, heteroaryl, alkyl; R^4 = Me, Ph, H; X = O, S; R^5 = Me when X = O; R^5 = *p*-$OMeC_6H_4$ when X = S; <u>catalyst</u>: HCl, $FeCl_3$, $InCl_3$, PPE, $BF_3{\cdot}OEt_2$

Mechanism: [45-50,40,9]

The first step in the mechanism of the *Biginelli reaction* is the acid-catalyzed condensation of the urea with the aldehyde affording an aminal, which dehydrates to an *N*-acyliminium ion intermediate. Subsequently, the enol form of the β-keto ester attacks the *N*-acyliminium ion to generate an open chain ureide, which readily cyclizes to a hexahydropyrimidine derivative.

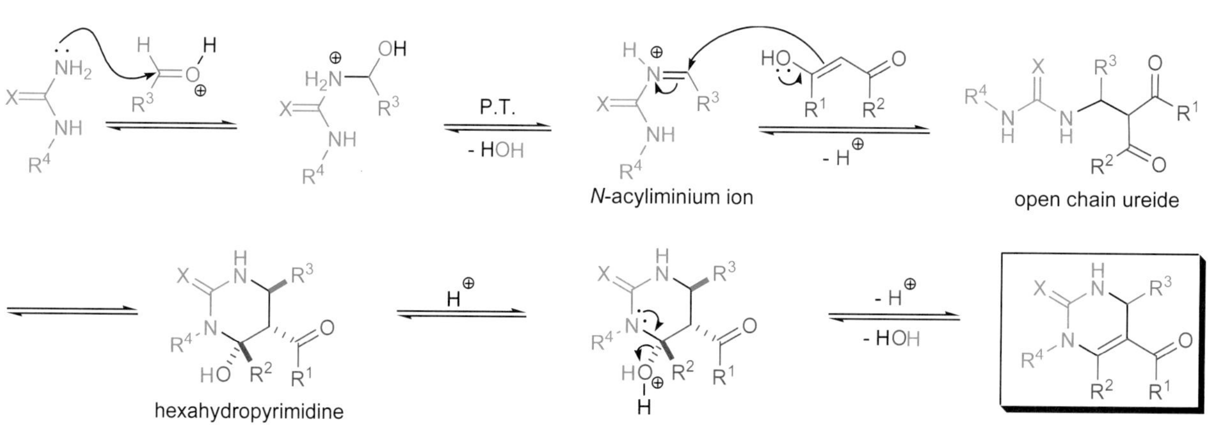

Synthetic Applications:

The only way to realize an enantioselective *Biginelli reaction* is to conduct it intramolecularly where the enantiopure urea and aldehyde portions are tethered.[51] This reaction was the key step in L.E. Overman's total synthesis of guanidine alkaloid 13,14,15-Isocrambescidin 800.[52] An optically active guanidine aminal was reacted with an enantiopure β-keto ester in trifluoroethanol to afford 1-iminohexahydropyrrolo[1,2-*c*]pyrimidine carboxylic ester with a 7:1 *trans* selectivity between C10 and C13 positions.

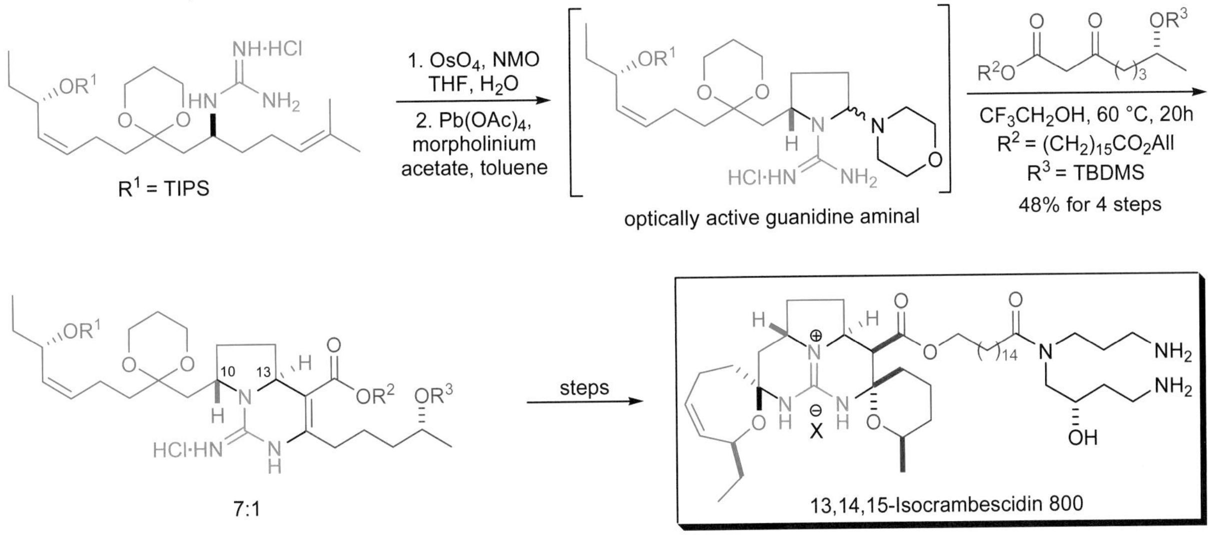

The traditional intermolecular three-component version of the *Biginelli reaction* was utilized for the improved synthesis of racemic monastrol by A. Dondoni and co-workers.[28] The one-pot $Yb(OTf)_3$ catalyzed reaction took place between 3-hydroxybenzaldehyde, ethyl acetoacetate, and thiourea. Racemic monastrol was isolated in 95% yield and was resolved on a preparative scale using diastereomeric *N*-3-ribofuranosyl amides.

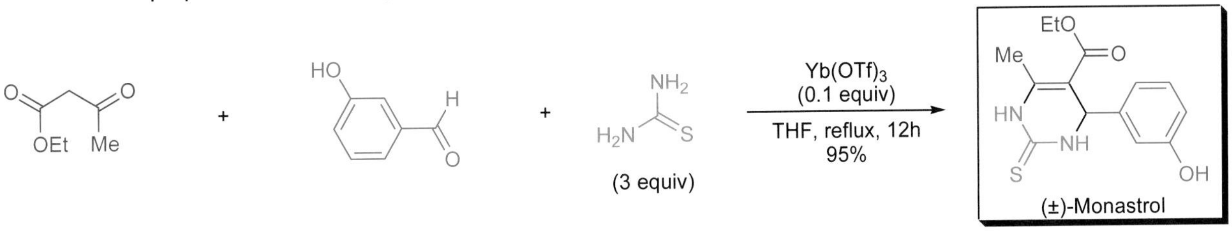

The first total synthesis of batzelladine F was accomplished using the tethered version of the *Biginelli reaction* in the laboratory of L.E. Overman.[53] The assembly of complex bisguanidines was achieved by reacting an enantiopure β-keto ester with 3 equivalents of a guanidine derivative in trifluoroethanol in the presence of morpholinium acetate. The product pentacyclic bisguanidine was isolated in 59% yield after HPLC purification. To complete the total synthesis, the trifluoroacetate counterions were exchanged for BF_4^-, the final ring was closed by converting the secondary alcohol to the corresponding mesylate followed by treatment with base, and the vinylogous amide was reduced by catalytic hydrogenation. Interestingly, the choice of counterion was crucial since model studies indicated the formation of complex product mixtures when the counterion was formate, acetate or chloride.

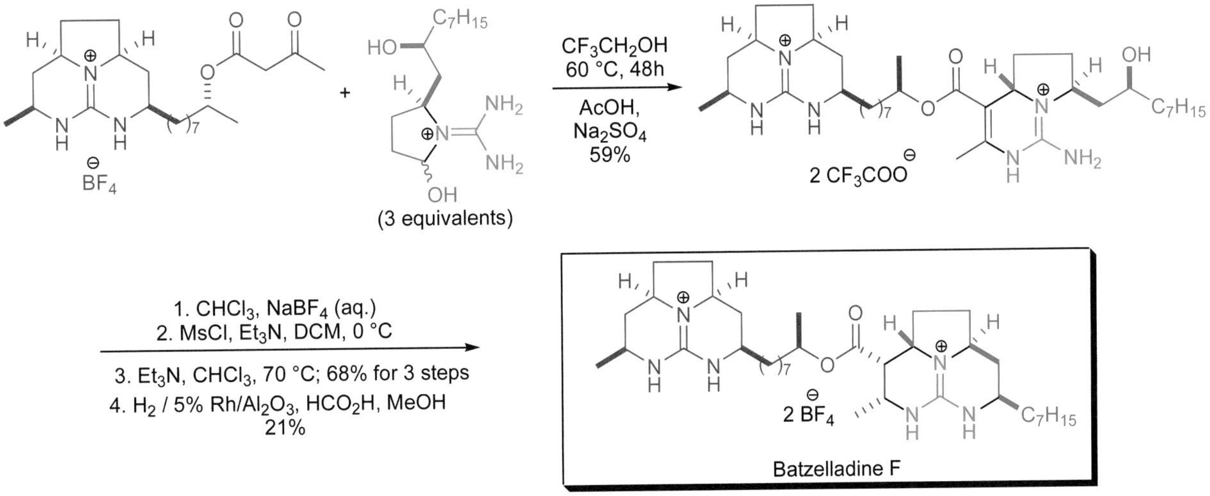

BIRCH REDUCTION

(References are on page 552)

Importance:

[*Seminal Publication*[1]; *Reviews*[2-15]; *Modifications & Improvements*[16-21]; *Theoretical Studies*[22-33]]

The 1,4-reduction of aromatic rings to the corresponding unconjugated cyclohexadienes and heterocycles by alkali metals (Li, Na, K) dissolved in liquid ammonia in the presence of an alcohol is called the *Birch reduction*. Heterocycles, such as pyridines, pyrroles, and furans, are also reduced under these conditions. When the aromatic compound is substituted, the regioselectivity of the reduction depends on the nature of the substituent. If the substituent is electron-donating, the rate of the reduction is lower compared to the unsubstituted compound and the substituent is found on the non-reduced portion of the new product. In the case of electron-withdrawing substituents, the result is the opposite. Ordinary alkenes are not affected by the *Birch reduction* conditions, and double bonds may be present in the molecule if they are not conjugated with an aromatic ring. However, conjugated alkenes, α,β-unsaturated carbonyl compounds, internal alkynes, and styrene derivatives are reduced under these conditions. There are some limitations to the *Birch reduction*: electron-rich heterocycles need to have at least one electron-withdrawing substituent, so furans and thiophenes are not reduced unless electron-withdrawing substituents are present.

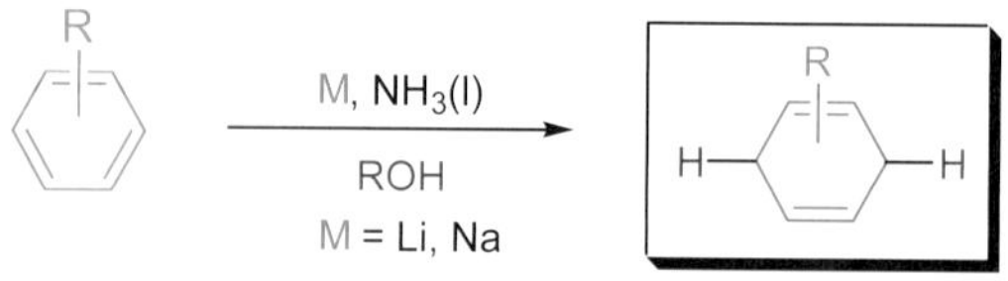

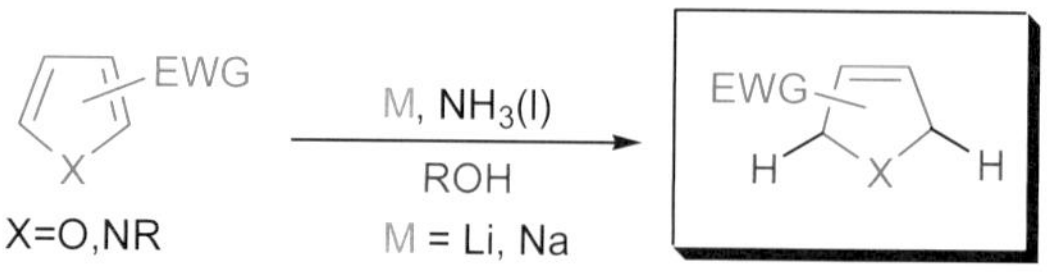

Mechanism:[34-38]

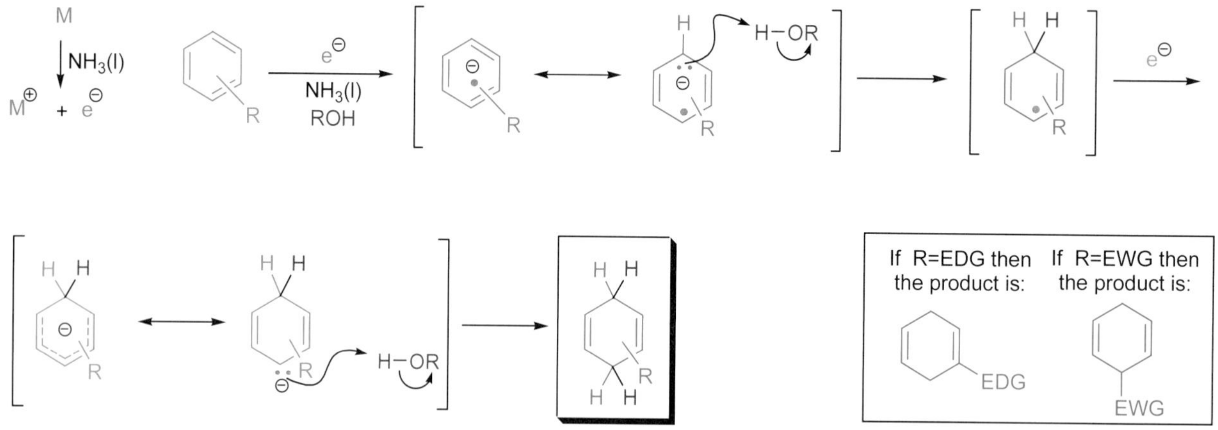

Synthetic Applications:

In the first example (**I**) T.J. Donohoe et al. utilized the *Birch reduction* to reduce then alkylate electron-deficient 2- and 3-substituted pyrroles.[39,40] This reductive alkylation method proved to be very efficient for the synthesis of substituted 3- and 2-pyrrolines, respectively. An alcohol as a proton source was not necessary for the reduction to occur. In the second example (**II**) the same researchers performed a *stereoselective Birch reduction* on a substituted furan during the enantioselective total synthesis of (+)-nemorensic acid.[41]

I.

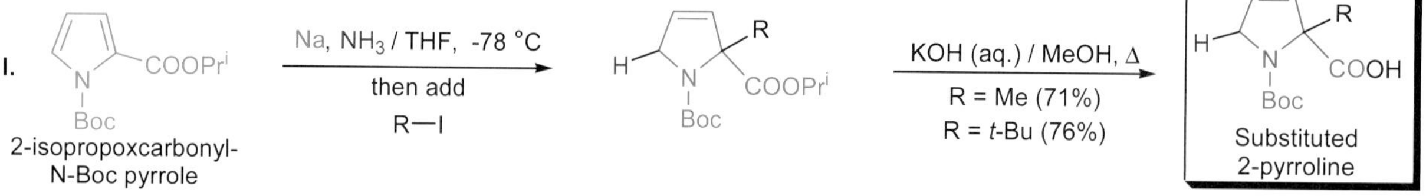

II.

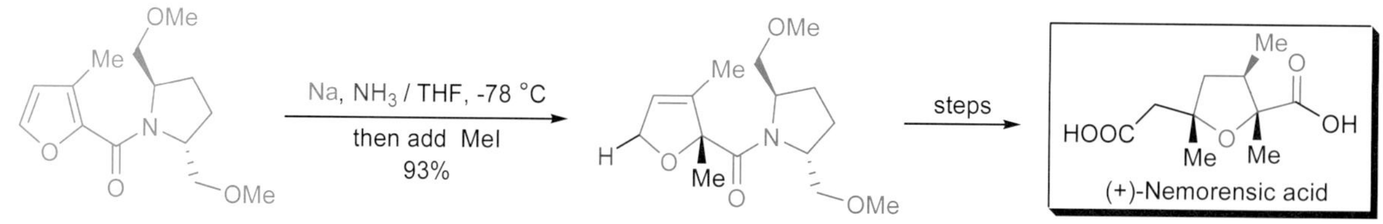

BIRCH REDUCTION

Synthetic Applications:

During the enantioselective total synthesis of (–)-taxol, I. Kuwajima and co-workers used the *Birch reduction* to elaborate an array of functional groups on the C-ring of the natural product.[42] The originally 1,2-disubstituted benzene ring was subjected to typical *Birch reduction* conditions (K/liquid ammonia/*t*-BuOH), and the resulting 1,3-cyclohexadiene (**I**) was oxygenated by singlet oxygen from the convex β-face to give the desired $C_4\beta$-$C_7\beta$ diol. The side product benzyl alcohol (**II**) was recycled as starting material *via Swern oxidation* in excellent yield providing a total conversion that was acceptable for synthetic purposes.

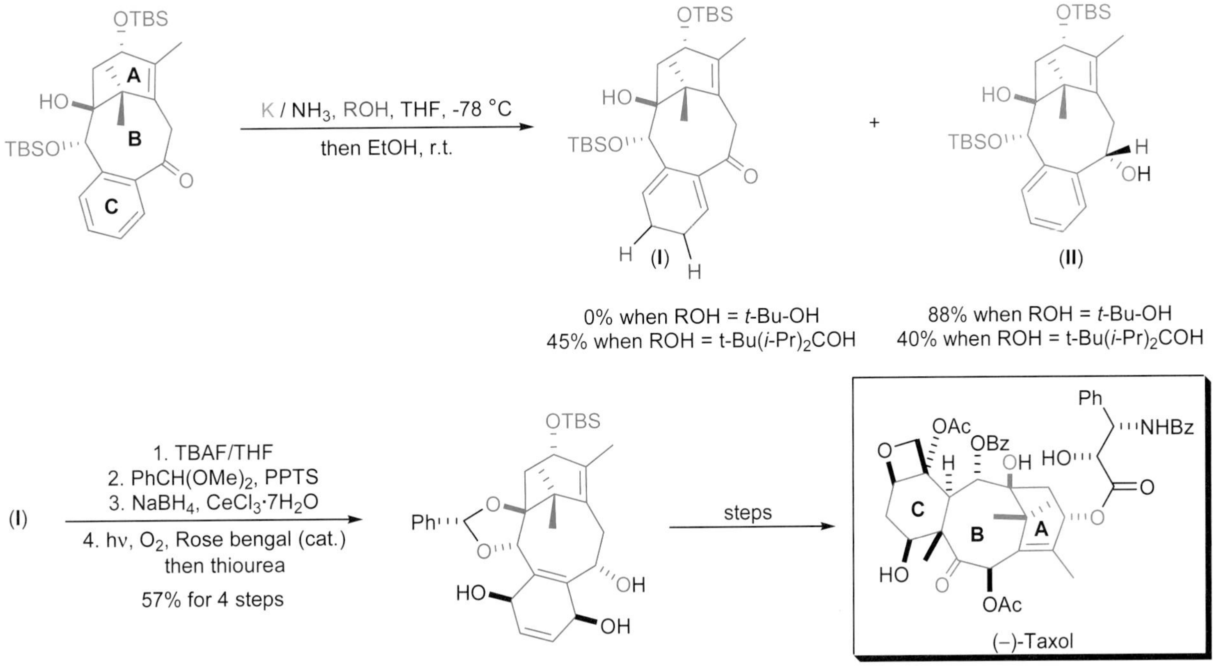

In the laboratory of A.G. Schultz during the asymmetric total synthesis of two vincane type alkaloids, (+)-apovincamine and (+)-vincamine, it was necessary to construct a crucial *cis*-fused pentacyclic diene intermediate.[43] The synthesis began by the *Birch reduction-alkylation* of a chiral benzamide to give 6-ethyl-1-methoxy-4-methyl-1,4-cyclohexadiene in a >100:1 diastereomeric purity. This cyclohexadiene was first converted to an enantiopure butyrolactone which after several steps was converted to (+)-apovincamine.

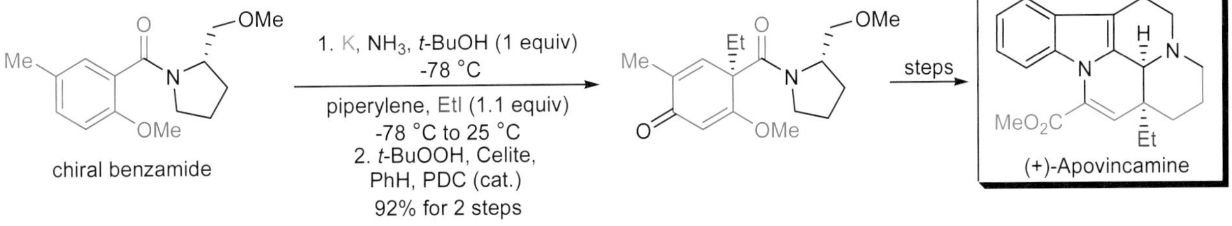

The total synthesis of galbulimima alkaloid GB 13 was accomplished by L.N. Mander and co-workers. The *Birch reduction* of a complex intermediate was necessary in order to prepare a cyclic α,β-unsaturated ketone.[44] The treatment of the substrate with lithium metal in liquid ammonia first resulted in a quantitative *reductive decyanation* of the C6a cyano group. The addition of excess ethanol to the reaction mixture reduced the aromatic ring to the corresponding enol ether that was hydrolyzed in a subsequent step to afford the unsaturated ketone.

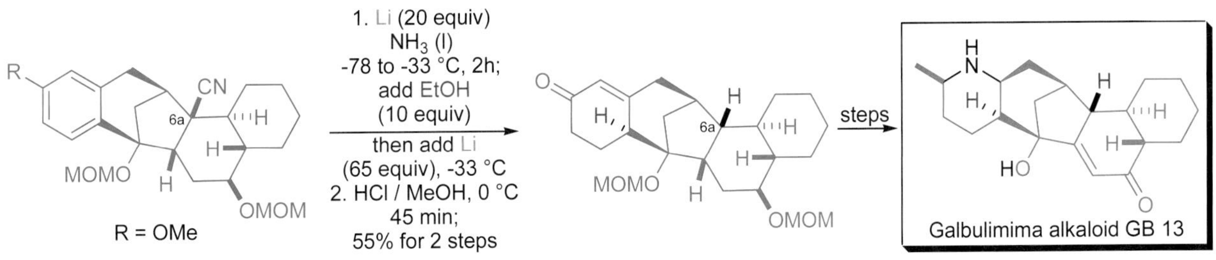

BISCHLER-NAPIERALSKI ISOQUINOLINE SYNTHESIS

(References are on page 553)

Importance:

[*Seminal Publication*[1]; *Reviews*[2-4]; *Modifications & Improvements*[5-15]]

One of the *Friedel-Crafts acylation* routes toward the synthesis of isoquinolines is the Bischler-Napieralski synthesis. When an acyl derivative of a phenylethylamine is treated with a dehydrating agent ($POCl_3$, P_2O_5, PPA, TFAA, or Tf_2O)[6] a *cyclodehydration reaction* takes place to form a 3,4-dihydroisoquinoline derivative. If the starting compound contains a hydroxyl group in the α-position, an additional dehydration takes place yielding an isoquinoline.

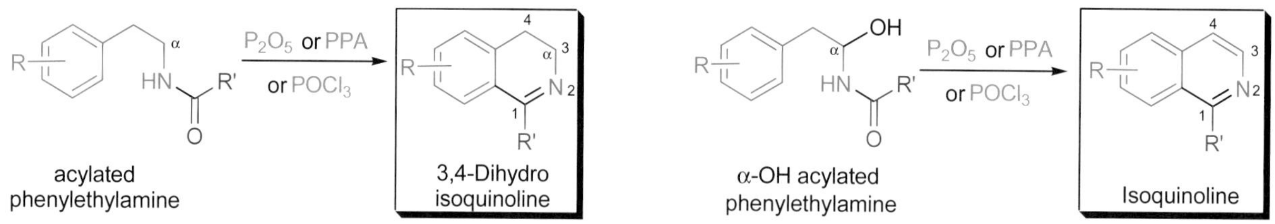

Mechanism: [16,5]

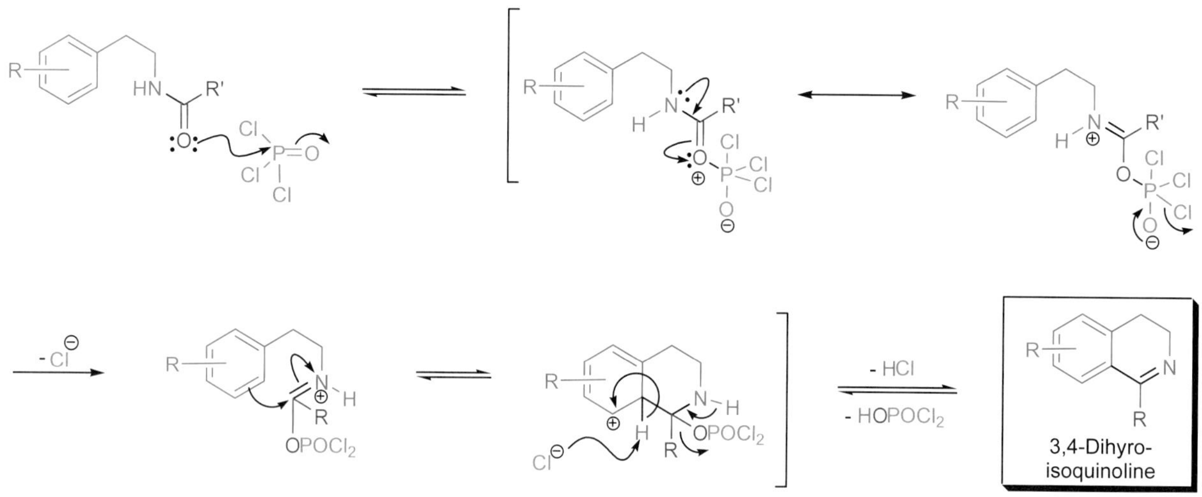

Synthetic Applications:

In the laboratory of J. Bonjoch the first total syntheses of the pentacyclic (±)-strychnoxanthine and (±)-melinonine-E alkaloids were accomplished using a *radical carbocyclization via* α-carbamoyldichloromethyl radical followed by the *Bischler-Napieralski cyclization*, as the two key cyclization steps.[17]

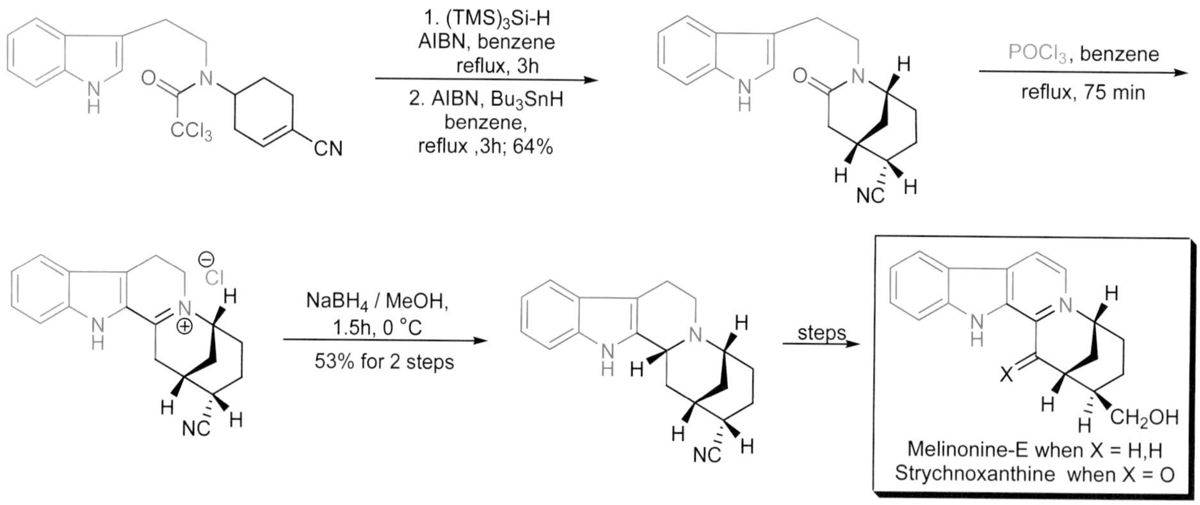

Synthetic Applications:

The first total synthesis of annoretine, an alkaloid containing the 1,2,3,4-tetrahydronaphtho [2,1-*f*]isoquinoline moiety was achieved by J.C. Estevez and co-workers.[18] The total synthesis had two key steps: first a *Bischler-Napieralski reaction* to form the 5-styrylisoquinoline unit followed by a *photocyclization* to provide the desired naphthoisoquinoline skeleton.

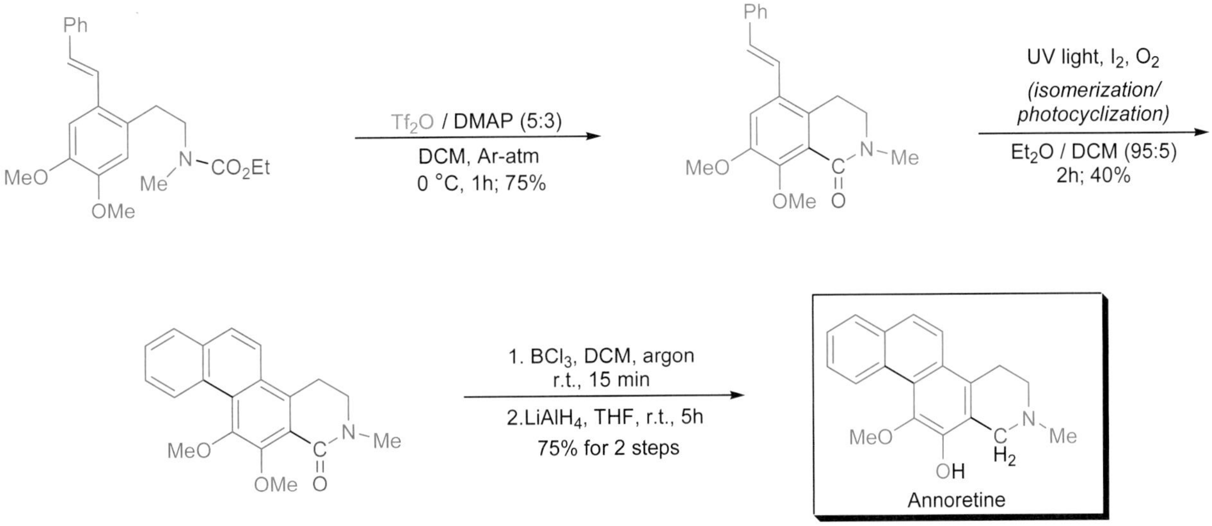

The asymmetric total synthesis of rauwolfia alkaloids (–)-yohimbane and (–)-alloyohimbane was carried out by S.C. Bergmeier et al. by utilizing a *novel aziridine-allylsilane cyclization* and the *Bischler-Napieralski isoquinoline synthesis* as key steps.[19] These alkaloids have a characteristic pentacyclic ring framework and exhibit a wide range of interesting biological activities such as antihypertensive and antipsychotic properties.

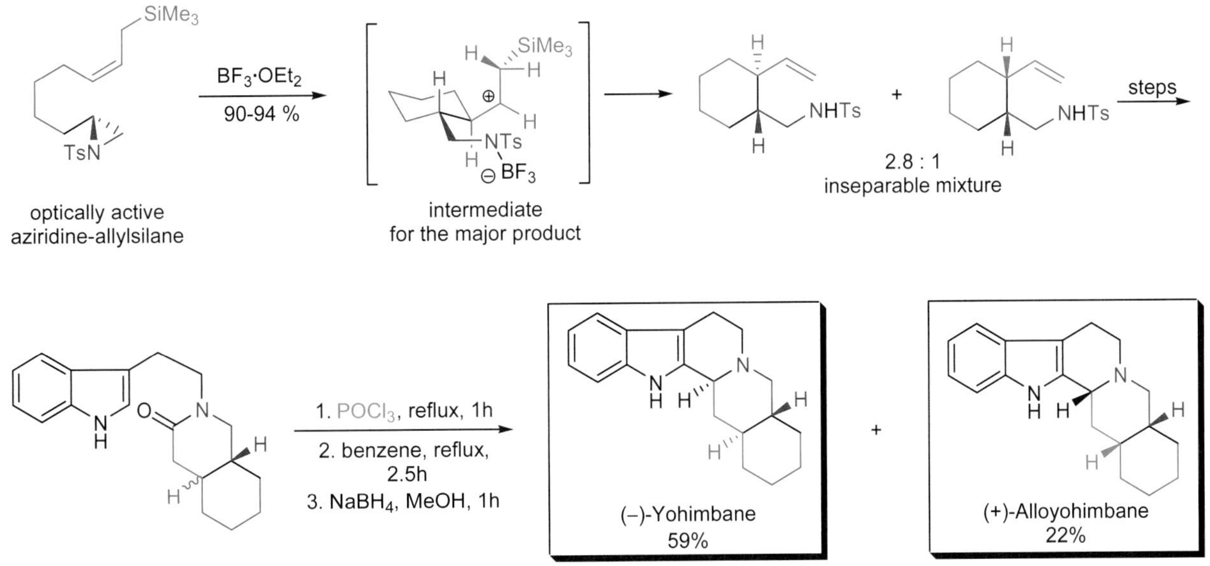

The first enantioselective total synthesis of the 7,3'-linked naphthylisoquinoline alkaloid (–)-ancistrocladidine was accomplished by J.C. Morris and co-workers.[20] The key steps of the synthesis were the *Pinhey-Barton ortho-arylation* and the *Bischler-Napieralski cyclization*. The natural product was isolated from the 1:1 mixture of atropisomers by recrystallization from toluene/petroleum ether.

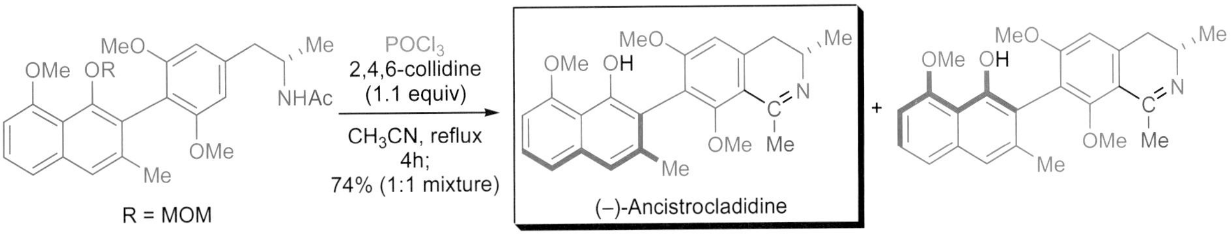

BROOK REARRANGEMENT

(References are on page 553)

Importance:

[*Seminal Publications*[1-4]; *Reviews*[5-12]; *Modifications & Improvements*[13-19]; *Theoretical Studies*[20-25]]

In the late 1950s, A.G. Brook observed the intramolecular anionic migration of silyl groups from a carbon to an oxygen atom.[1,2] This migratory aptitude of the silyl group was later found to be more general. Therefore, all the *[1,n]*-carbon to oxygen silyl migrations are referred to as *Brook rearrangements*. The reaction is based on the great susceptibility of silicon toward a nucleophilic attack and the formation of a strong silicon-oxygen bond (Si-O) from the relatively weak silicon-carbon bond. The reverse process is called the *retro-Brook rearrangement* and was first reported by J.L. Speier.[26,27]

[1,2]-Silyl migrations: *[1,n]*-Silyl migrations:

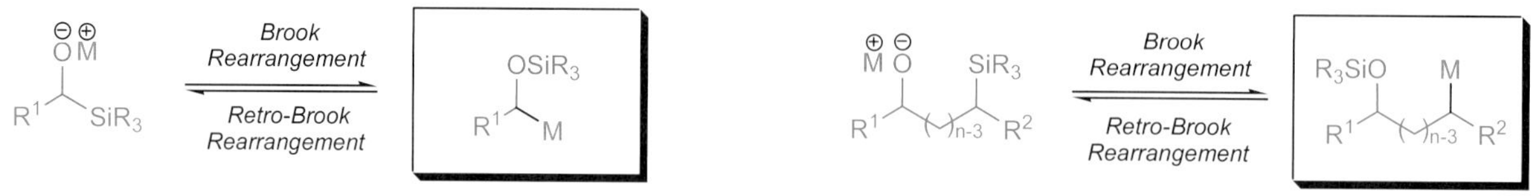

R^{1-2} = alkyl, aryl; SiR_3 = $SiMe_3$, $SiEt_3$, $SiMe_2t$-Bu, etc.; n = 2-5

Mechanism: [28,29,13,30-32,25]

The mechanism is believed to involve a pentacoordinate-silicon atom.[30]

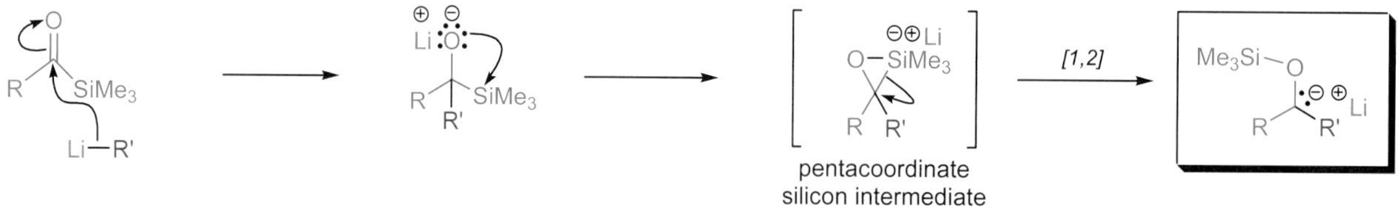

Synthetic Applications:

In the laboratory of K. Takeda, a new synthetic strategy was developed for the stereoselective construction of eight-membered carbocycles utilizing a *Brook rearrangement-mediated [3+4] annulation.*[33] The unique feature of this methodology is the generation, in two steps, of eight-membered ring systems containing useful functionalities from readily available compounds containing three- and four-carbon atoms.

General Scheme:

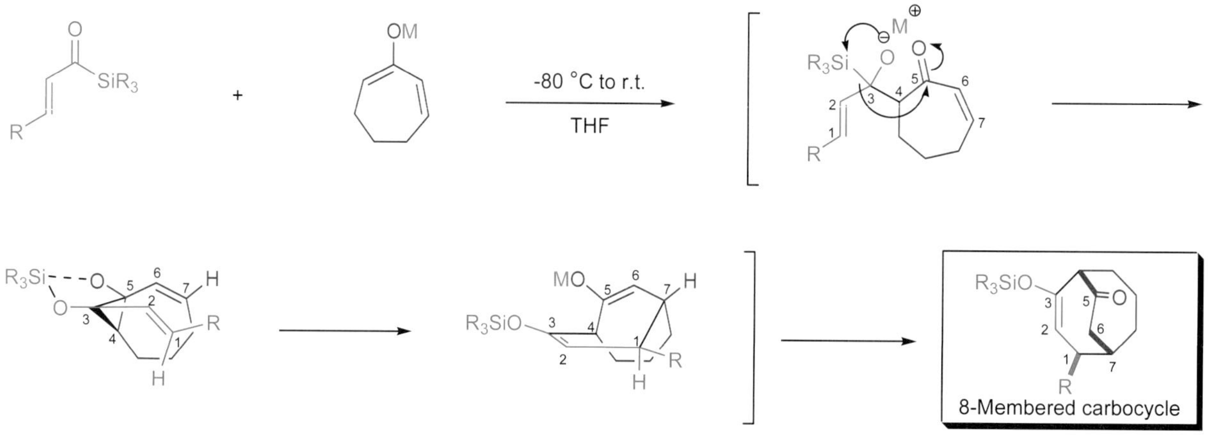

Specific Example:

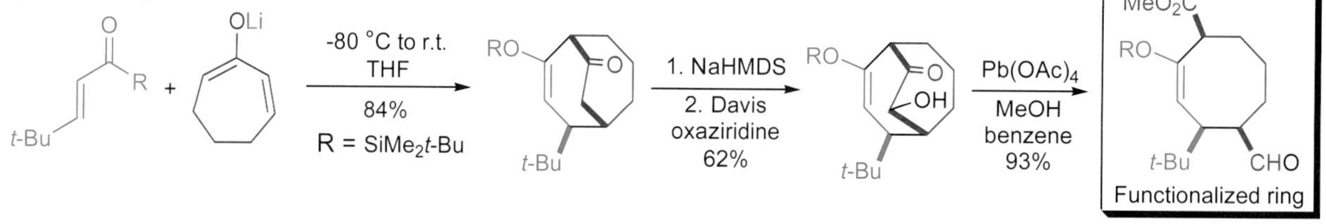

Synthetic Applications:

W.H. Moser and co-workers developed a new and efficient method for the stereocontrolled construction of spirocyclic compounds, including the spirocyclic core of the antitumor agent fredericamycin A.[34] The strategy involved a one-pot *aldol addition/Brook rearrangement/cyclization* sequence beginning from arene chromium tricarbonyl complexes and can formally be described as a [3+2] annulation.

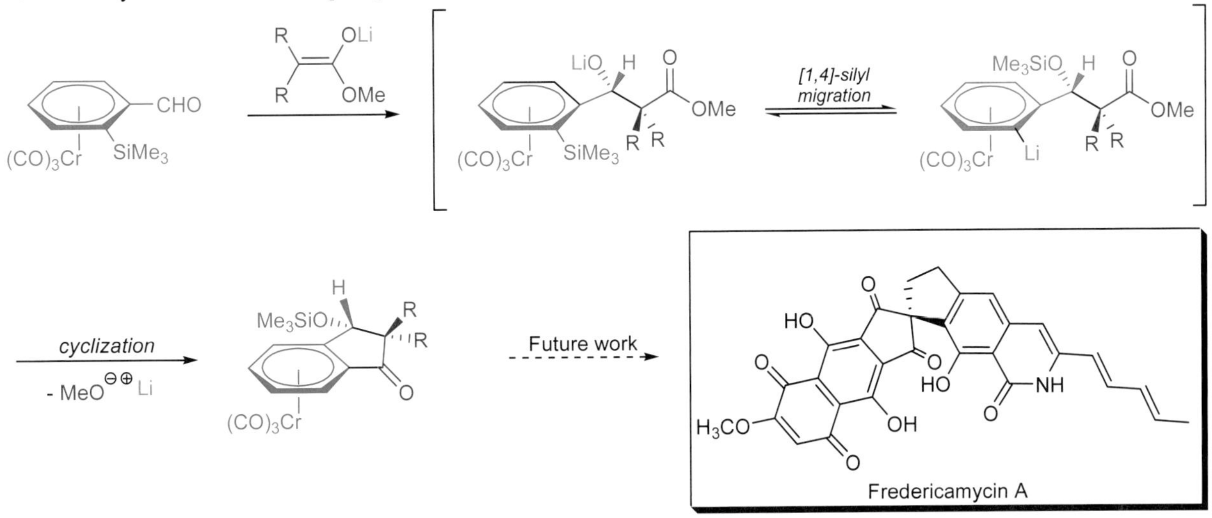

Cyathins, isolated from bird nest fungi, are interesting compounds because of their unusual 5-6-7 tricyclic ring system and their important biological activities. K. Takeda and co-workers synthesized the tricyclic core of the cyathins using a *Brook rearrangement mediated-[4+3] annulation* reaction.[35] The seven-membered ring was formed *via* the *oxy-Cope rearrangement* of a divinylcyclopropane intermediate.

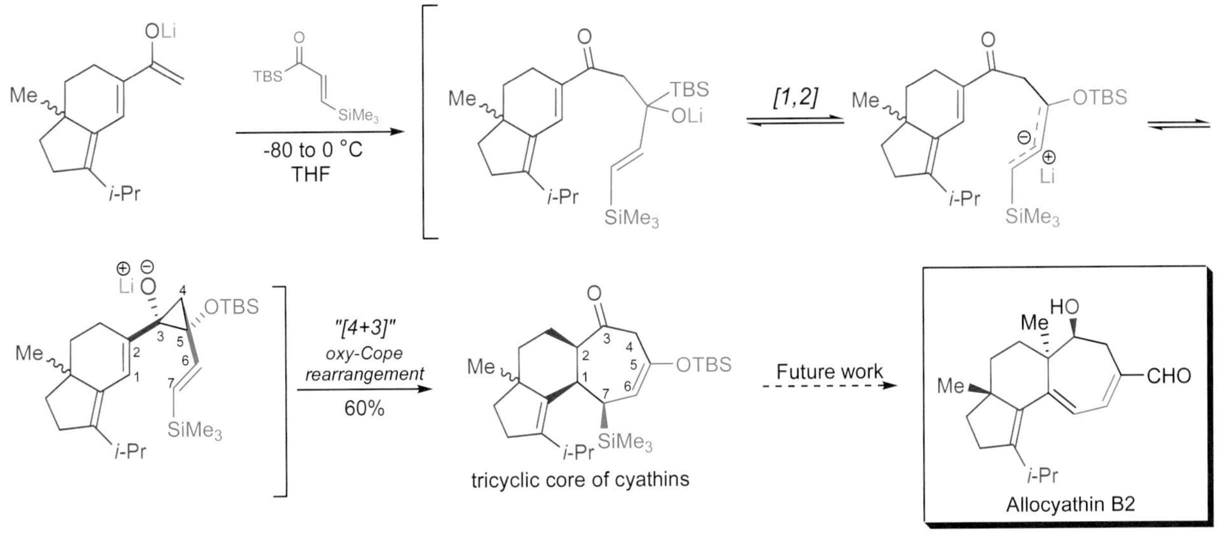

The total synthesis of (+)-α-onocerin *via* four-component coupling and tetracyclization steps was achieved in the laboratory of E.J. Corey.[36] The farnesyl acetate-derived acyl silane was treated with vinyllithium, which brought about the stereospecific formation of a (*Z*)-silyl enol ether as a result of a spontaneous *Brook rearrangement*. In the same pot, the solution of I_2 was added to obtain the desired diepoxide *via oxidative dimerization*.

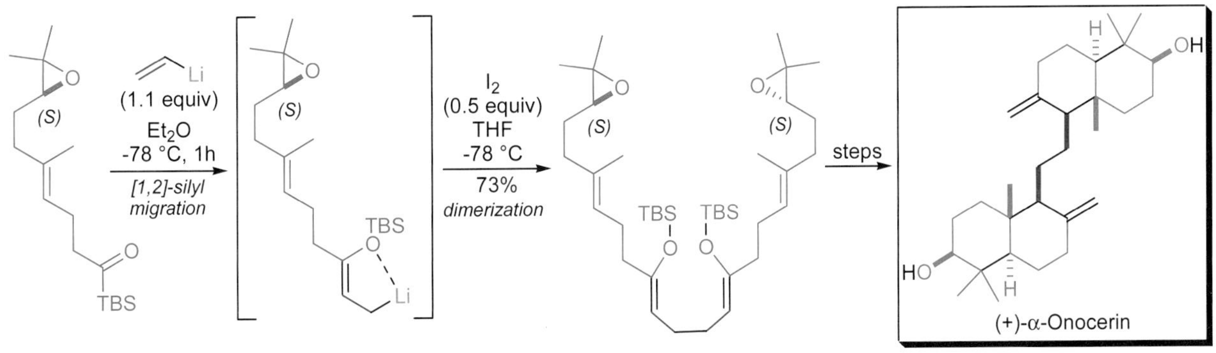

BROWN HYDROBORATION REACTION

(References are on page 554)

Importance:

[*Seminal Publication*[1]; *Reviews*[2-21]; *Modifications & Improvements*[22-28]; *Theoretical Studies*[29-34]]

The addition of a B-H bond across a carbon-carbon double or triple bond is called the *Brown hydroboration reaction*. This process is highly regioselective and stereospecific (*syn*). The boron becomes bonded primarily to the less substituted carbon atom of the alkene (*anti-Markovnikoff product*). The resulting organoboranes are very useful intermediates in organic synthesis. The boron can be replaced for hydroxyl (*hydroboration/oxidation*), halogen, or amino groups (*hydroboration/amination*). If BH_3 is used in the hydroboration reaction, it will react with three molecules of alkenes to yield a trialkylborane (R_3B). Transition metal complexes catalyze the addition of borane to alkenes and alkynes and significantly enhance the rate of the reaction. This variant may alter the chemo-, regio-, and diastereoselectivity compared to the uncatalyzed hydroboration.[27] In the presence of a chiral transition metal complex, enantioselectivity can be achieved.[25]

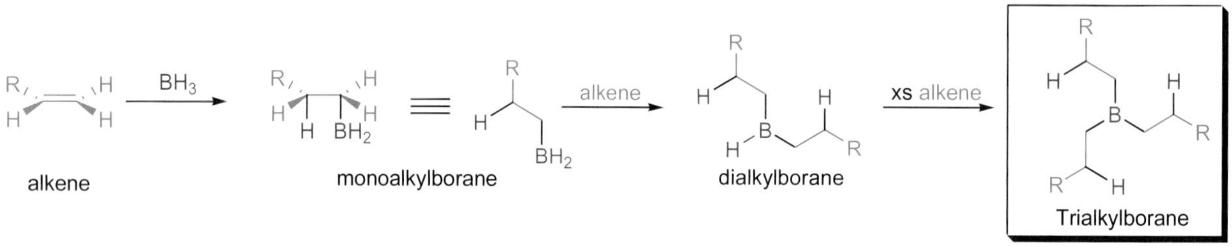

Mechanism: [35-43]

Boron has only three electrons in the valence shell, and therefore its compounds are electron deficient and there is a vacant *p*-orbital on the boron atom. Borane (BH_3) exists as a mixture of B_2H_6/BH_3, as dimerization partially alleviates the electron deficiency of the boron. This equilibrium is fast, and most reactions occur with BH_3. The addition of borane to a double bond is a concerted process going through a four-centered transition state. The formation of the C-B bond precedes the formation of the C-H bond so that the boron and the carbon atoms are partially charged in the four-centered transition state.

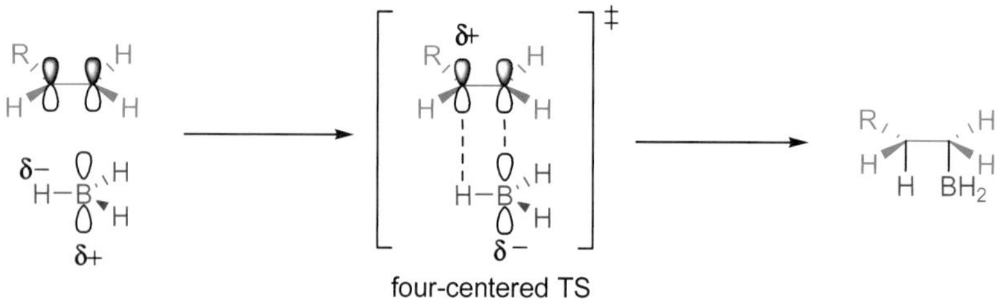

In the Cp_2TiMe_2-catalyzed hydroboration of alkenes, a titanocene bis(borane) complex is responsible for the catalysis.[43] This bis(borane) complex initially dissociates to give a monoborane intermediate. Coordination of the alkene gives rise to the alkene-borane complex, which is likely to be a resonance hybrid between an alkene borane complex and a β-boroalkyl hydride. An intramolecular reaction extrudes the trialkylborane product, and coordination of a new HBR_2 regenerates the monoborane intermediate.

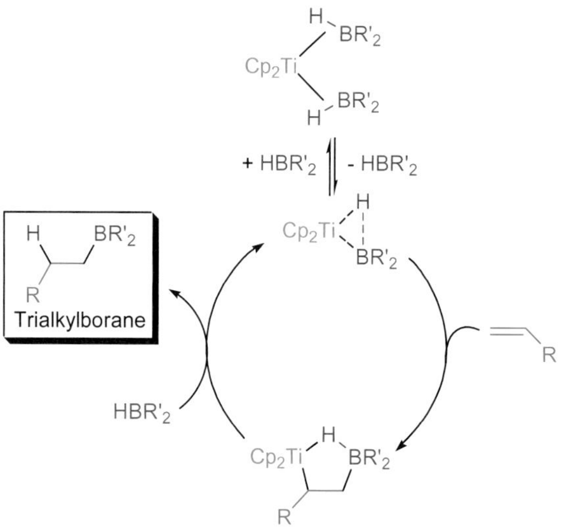

Synthetic Applications:

In the enantiospecific total synthesis of the indole alkaloid trinervine, J.M. Cook and co-workers used the *hydroboration/oxidation* sequence to functionalize the C19-C20 *exo* double bond with excellent regioselectivity.[44]

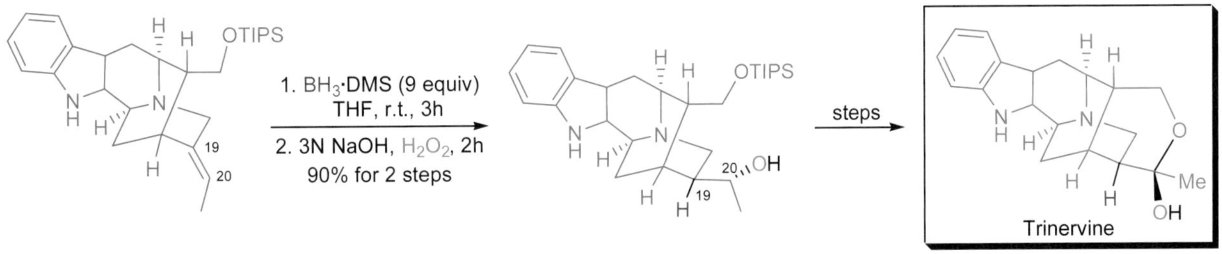

During the enantioselective synthesis of (3a*R*,4*R*,7a*S*)-4-hydroxy-7a-methylperhydro-1-indenone, a suitable CD-ring fragment for vitamin D-analogs, M. Vandewalle et al. realized that the *hydroboration/oxidation* of (1,1)-ethylenedioxy-8a-methyl-1,2,3,4,6,7,8,8a-octahydronaphtalene led to a *cis*-decalin structure instead of the literature reported *trans*-fusion.[45]

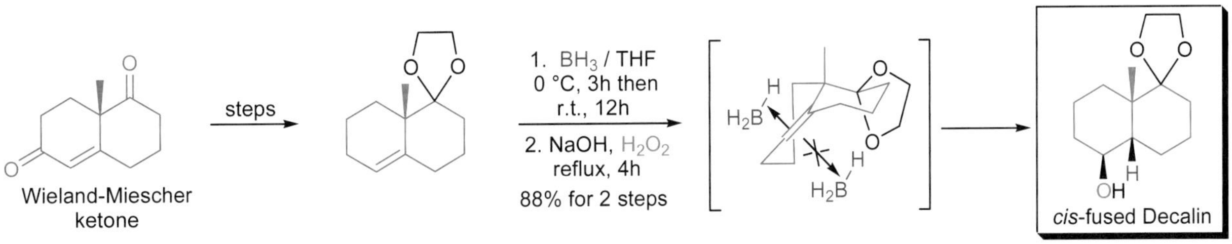

P. Knochel and co-workers used diphosphines as ligands in the rhodium-catalyzed asymmetric hydroboration of styrene derivatives.[46] The best results were obtained with the very electron rich diphosphane, and (*S*)-1-phenylethanol was obtained in 92% ee at –35 °C, with a regioselectivity greater than 99:1 (Markovnikoff product). A lower reaction temperature resulted in no reaction, while a higher temperature resulted in lower enantioselectivity and regioselectivity. The regioselectivity was excellent in all cases. Irrespective of the electronic nature of the substituents, their position and size had a profound effect on the enantioselectivity.

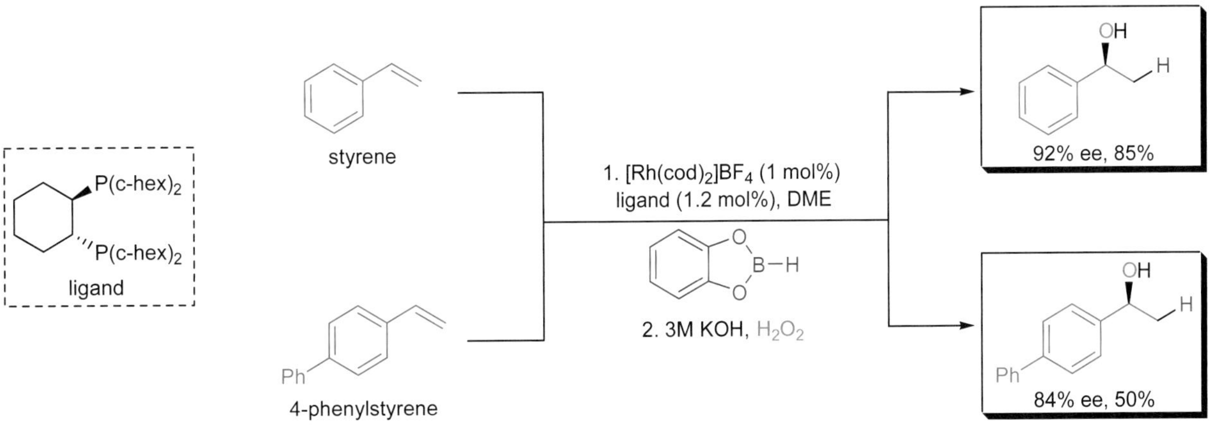

The enantioselective total synthesis of (–)-cassine was accomplished in the laboratory of H. Makabe.[47] The synthetic sequence involved a key, highly diastereoselective *PdCl*$_2$*-catalyzed cyclization* of an amino allylic alcohol. The cyclic product was then subjected to *hydroboration* with 9-BBN followed by oxidation to afford the desired primary alcohol, which was converted to (–)-cassine.

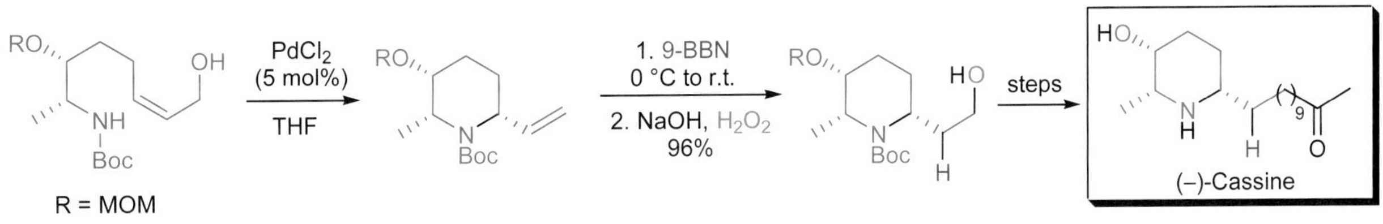

BUCHNER METHOD OF RING EXPANSION

(References are on page 555)

Importance:

[*Seminal Publications*[1-3]; *Reviews*[4-9]; *Modifications & Improvements*[10-20]]

The thermal or photochemical reaction of ethyl diazoacetate with benzene and its homologs to give the corresponding isomeric esters of cycloheptatriene carboxylic acid (*via* the corresponding esters of norcaradienic acid) is called the *Buchner reaction*. This transformation was first reported by E. Buchner and T. Curtius in 1885, when they synthesized cycloheptatrienes from thermal and photochemical reactions of ethyl diazoacetate with benzene *via* arene cyclopropanation, followed by the electrocyclic ring opening of the intermediate norcaradiene.[1] The reaction offers a convenient entry to seven-membered carbocycles both inter- and intramolecularly. The complexity of the product mixture was significantly reduced or completely eliminated with the advent of modern transition metal catalysts: at first it was copper-based, and then in the 1980s it became almost exclusively rhodium-based (e.g., $RhCl_3 \cdot 3H_2O$, $Rh_2(OAc)_4$, Rh(II)-trifluoroacetate). For example, rhodium(II)-trifluoroacetate catalysis provides a single isomer of the cycloheptatriene in 98% yield.[11] Synthetically, it is convenient that chromium tricarbonyl-complexed aromatic rings do not undergo the *Buchner ring expansion* either inter- or intramolecularly.[20]

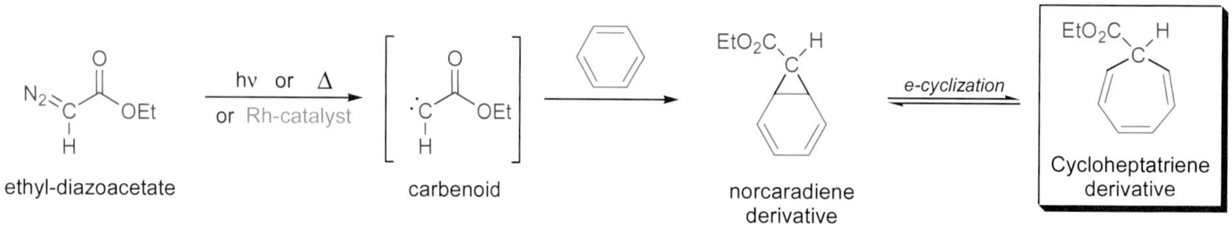

Mechanism:[21-23]

In the first step of the *Buchner reaction*, one of the π-bonds of the aromatic ring undergoes cyclopropanation catalyzed by a metal-carbenoid complex, which is the reactive intermediate. Metal carbenoids are formed when transition-metal catalysts [e.g., $Rh_2(OCOR)_4$] react with diazo compounds to generate transient electrophilic metal carbenes. The catalytic activity of the transition-metal complexes depends on the coordinative unsaturation at their metal center that allows them to react with diazo compounds as electrophiles. Electrophilic addition causes the loss of N_2 and the formation of the metal-stabilized carbene. Transfer of the carbene to electron-rich substrates completes the catalytic cycle. There are two possible scenarios for the first step: a) the intermediate can be represented as a metal-stabilized carbocation where the carbenoid α-carbon atom is the electrophilic center, that undergoes nucleophilic attack by the electron-rich double bond of olefins on route to cyclopropane; and b) the metal-carbenoid intermediate forms a rhodium-based metallocycle resulting from the nucleophilic attack of the negative charge on the rhodium atom onto one of the carbon atoms of the double bond. In the second step, the norcaradiene derivative undergoes an electrocyclization to afford the corresponding cycloheptatriene.

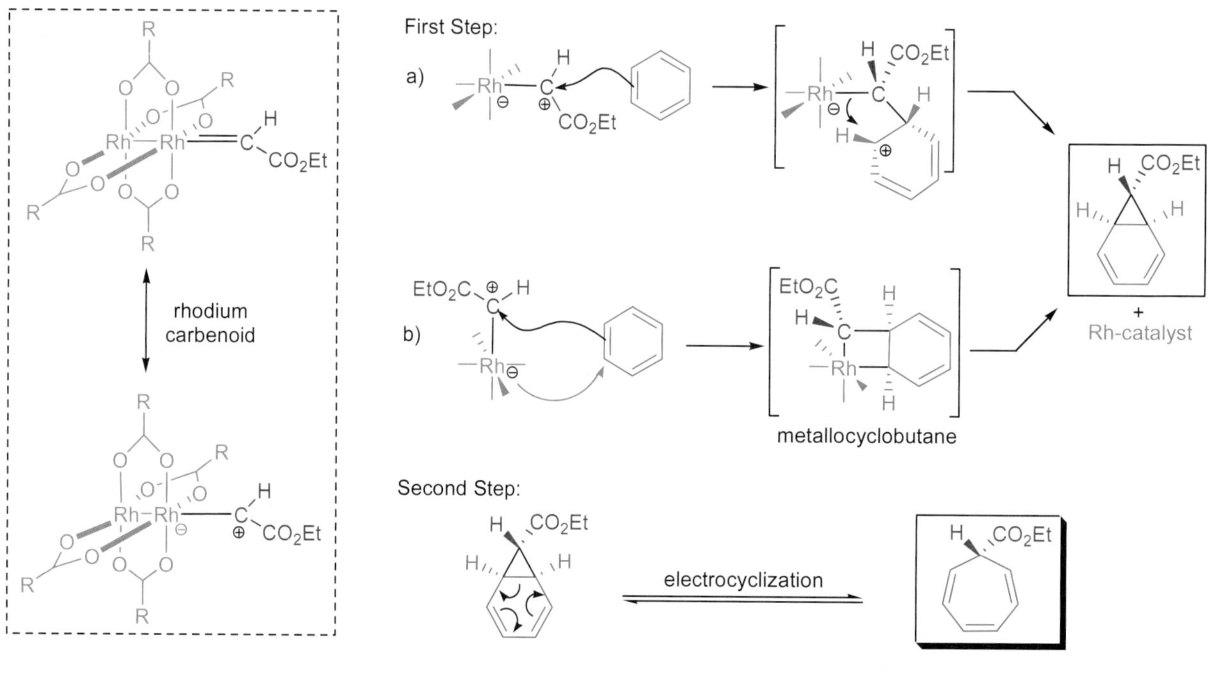

Synthetic Applications:

R.L. Danheiser and co-workers developed a new strategy for the synthesis of substituted azulenes, which is based on the reaction of β-bromo-α-diazo ketones with rhodium carboxylates.[24] The key transformation involves the following steps: *intramolecular addition of rhodium carbenoid* to an arene double bond, *electrocyclic ring opening*, *β-elimination*, *tautomerization*, and trapping to produce 1-hydroxyazulene derivatives. The advantage of this method over previous approaches is the ability to prepare a variety of azulenes substituted on both the five- and seven-membered rings from readily available benzene starting materials. The synthetic utility of the method was demonstrated in the total synthesis of the antiulcer drug egualen sodium (KT1-32).

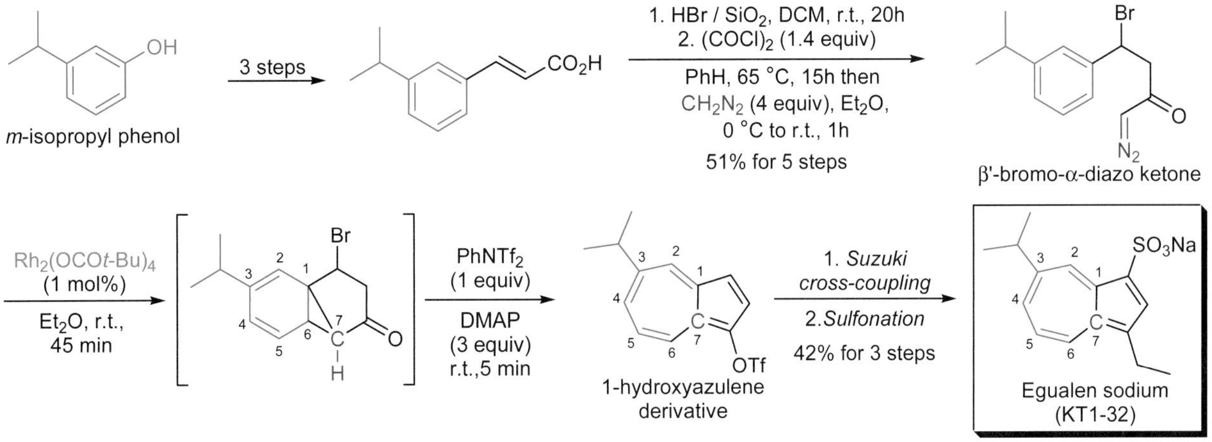

The need to prepare fullerene derivatives for possible applications to medicine and material sciences resulted in the development of novel synthetic methods for the functionalization of C_{60}. R. Pellicciari et al. reacted C_{60} with carboalkoxycarbenoids generated by the $Rh_2(OAc)_4$-catalyzed decomposition of α-diazoester precursors.[25] This reaction was the first example of a transition metal carbenoid reacting with a fullerene and the observed yields and product ratios were better than those obtained by previously reported methods. The reaction conditions were mild and the specificity was high for the synthesis of carboalkoxy-substituted[6,6]-methanofullerenes. When the same reaction was carried out thermally, the rearranged product (the [6,5]-open fullerene) was the major product.

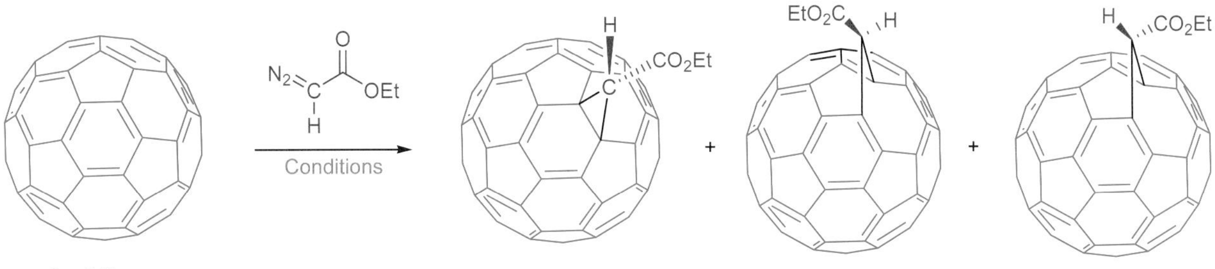

C60-fullerene	[6,6]-closed		[6,5]-open		[6,5]-open
Thermal: (110 °C, 7h; toluene; 35%)	1	:	4	:	2
$Rh_2(OAc)_4$ (stoichiometric): (r.t., 8h; 1-MeNap; 42%)	52	:	1	:	0

The total synthesis of the diterpenoid tropone, harringtonolide was accomplished in the laboratory of L.N. Mander.[26] The key step to form the seven-membered ring was the *Buchner reaction* of a complex polycyclic diazo ketone intermediate. Upon treatment with rhodium mandelate, an unstable adduct was formed and was immediately treated with DBU to afford the less labile cycloheptatriene.

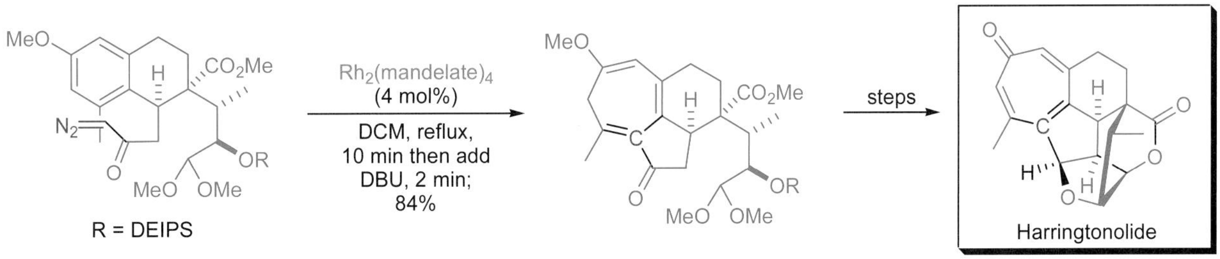

BUCHWALD-HARTWIG CROSS-COUPLING

(References are on page 556)

Importance:

[*Seminal Publications*[1-4]; *Reviews*[5-15]]

The direct Pd-catalyzed C-N and C-O bond formation between aryl halides or trifluoromethanesulfonates and amines (1° and 2° aliphatic or aromatic amines; imides, amides, sulfonamides, sulfoximines) or between aryl halides or triflates and alcohols (aliphatic alcohols and phenols) in the presence of a stoichiometric amount of base is known as the *Buchwald-Hartwig cross-coupling*. The coupling can be both inter- and intramolecular. The first palladium-catalyzed formation of aryl C-N bonds was reported by T. Migita and co-workers in 1983.[1] More than a decade later, in the laboratory of S. Buchwald, a new catalytic procedure was developed based on Migita's amination procedure.[2] The great disadvantage of these early methods was that both procedures called for the use of stoichiometric amounts of heat- and moisture-sensitive tributyltin amides as coupling partners. In 1995, S. Buchwald[16] and J. Hartwig[17] concurrently discovered that the aminotin species can be replaced with the free amine if one uses a strong base (e.g., sodium *tert*-butoxide or LHMDS), which generates the corresponding sodium amide *in situ* by deprotonating the Pd-coordinated amine. The typical procedure calls for either an aryl bromide or iodide, while the $Pd^{(0)}$-catalyst is usually complexed with chelating phosphine type ligands such as BINAP, DPPF, XANTPHOS, and DPBP or bidentate ligands such as DBA (*trans,trans*-dibenzylideneacetone). The base has to be present in stoichiometric amounts and the temperature for the reaction can be sometimes as low as 25 °C. Since the mid-1990s the reaction conditions for this coupling have gradually become milder, and by applying the appropriate ligand, even the otherwise unreactive aryl chlorides can be coupled with amines or alcohols.[9]

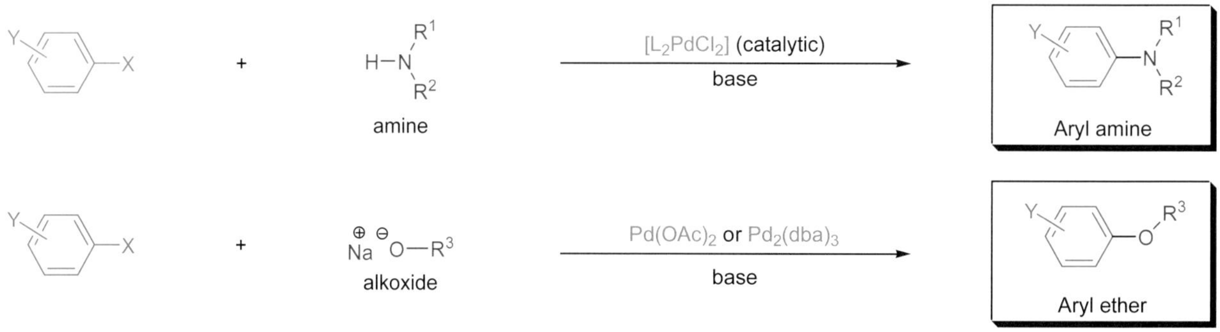

X = Cl, Br, I, OTf; Y = *o, m* or *p*-alkyl, phenacyl, amino, alkoxy; R^{1-2} = 1° or 2° aromatic or aliphatic; R^3 = 1°, 2°, or 3° aliphatic or aromatic; L = $P(o\text{-}Tol)_3$, BINAP, dppf, dba; base: NaO*t*-Bu, LHMDS, K_2CO_3, Cs_2CO_3

Mechanism: [3,17,4,7,9,11,14]

The first step in the catalytic cycle is the oxidative addition of $Pd^{(0)}$ to the aryl halide (or sulfonate). In the second step the $Pd^{(II)}$-aryl amide can be formed either by direct displacement of the halide (or sulfonate) by the amide *via* a $Pd^{(II)}$-alkoxide intermediate. Finally, reductive elimination results in the formation of the desired C-N bond and the $Pd^{(0)}$ catalyst is regenerated. Below is the catalytic cycle for the formation of an arylamine.

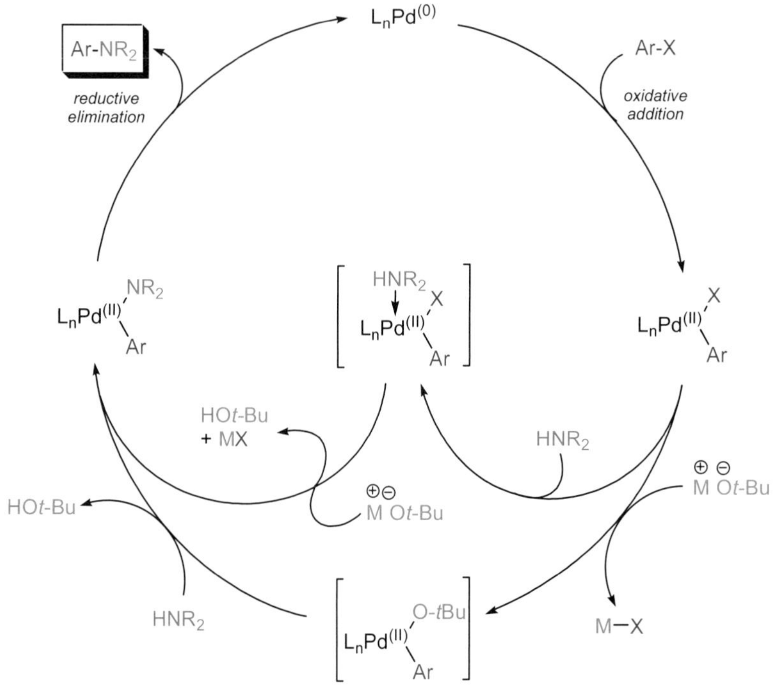

Synthetic Applications:

The opioid (±)-cyclazocine is known to be an analgesic and in the 1970s it was thought to prevent relapse in post-addicts of heroin. Unfortunately cyclazocine is *O*-glucuronidated in humans, and therefore it has a short duration of action. M.P. Wentland and co-workers synthesized analogues by replacing the prototypic 8-OH substituent of cyclazocine by amino and substituted amino groups using the *Buchwald-Hartwig cross-coupling reaction.*[18]

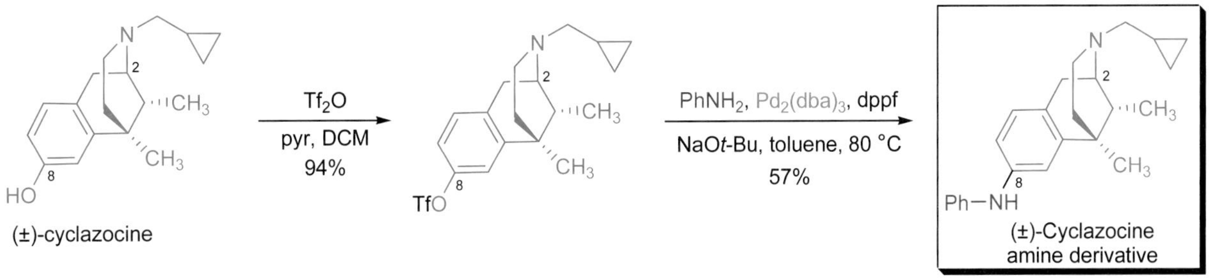

In the laboratory of G.A. Sulikowski, an enantioselective synthesis of a 1,2-aziridinomitosene, a key substructure of the mitomycin antitumor antibiotics, was developed.[19] Key transformations in the synthesis involved the *Buchwald-Hartwig cross-coupling* and chemoselective *intramolecular carbon-hydrogen metal-carbenoid insertion reaction*.

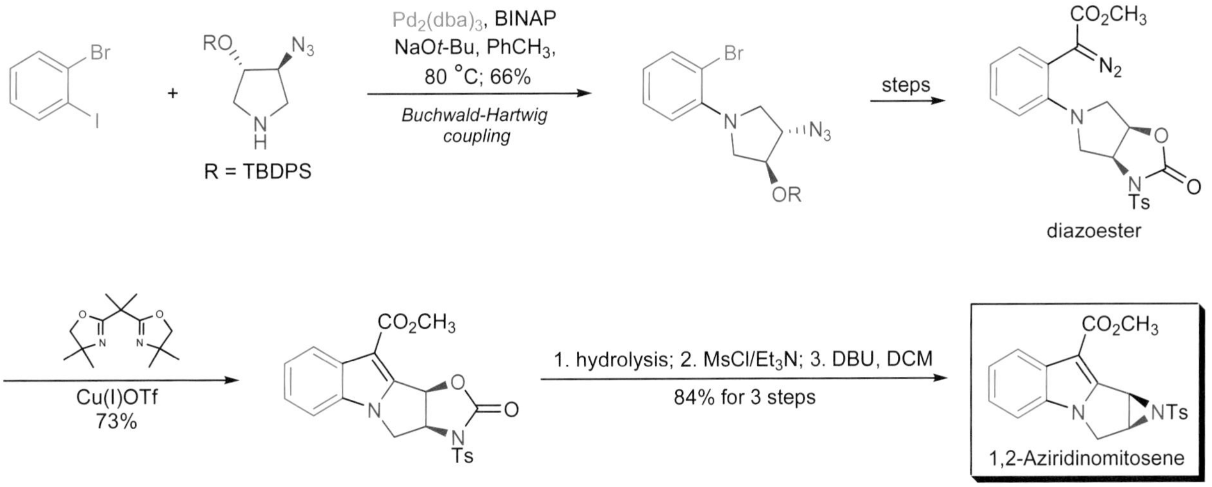

Naturally occurring phenazines have interesting biological activities but the available methods for their preparation offer only poor yields. T. Kamikawa et al. prepared polysubstituted phenazines by a new route involving two subsequent $Pd^{(II)}$*-catalyzed aminations* of aryl bromides using the conditions developed by Buchwald and Hartwig.[20]

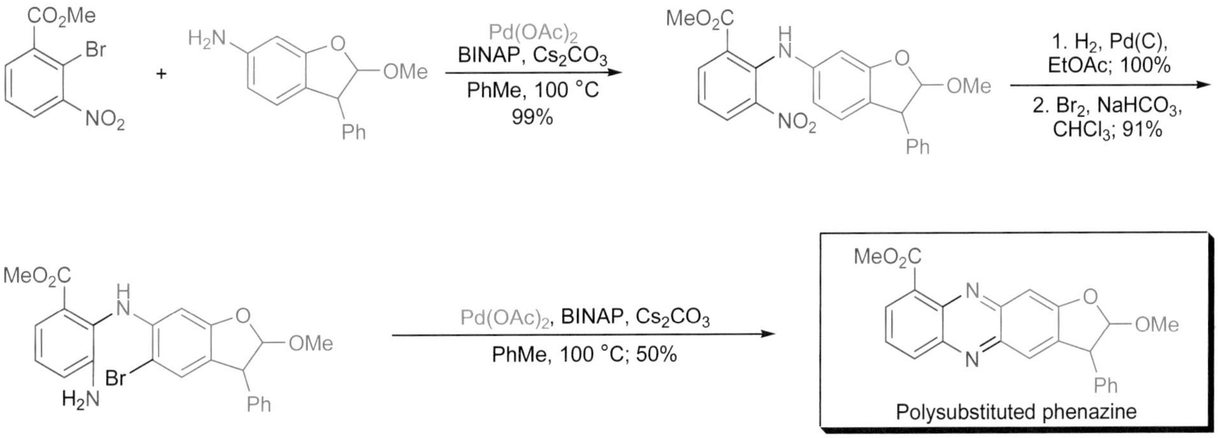

BURGESS DEHYDRATION REACTION

(References are on page 556)

Importance:

[*Seminal Publications*[1,2]; *Reviews*[3,4]; *Modifications & Improvements*[5-17]]

In the early 1970s, E.M. Burgess and co-workers discovered that secondary and tertiary alcohols could be dehydrated with the inner salt of (methoxycarbonylsulfamoyl)triethylammonium hydroxide to afford the corresponding olefins.[1] This process is now known as the *Burgess dehydration reaction* and the reagent is called the Burgess reagent. The *Burgess dehydration reaction* offers an advantage over other dehydration methods, namely it takes place under mild conditions (low temperature and neutral medium). Therefore, excellent yields can be achieved even with acid-sensitive substrates that are prone to rearrange. The elimination is *syn*-selective, but the *syn*-selectivity is higher for secondary alcohols. Tertiary alcohols tend to react faster and under milder conditions; E1 elimination products are observed when stabilized carbocations are formed. In most cases the elimination leads to the formation of the conjugated product, if conjugation with other C=C or C=O double bonds is possible. Primary alcohols are converted to the corresponding carbamates, which in turn give primary amines after hydrolysis.[5] The Burgess reagent is compatible with a wide range of functional groups, such as epoxides, alkenes, alkynes, aldehydes, ketones, alkyl halides, acetals, amides, and esters, and this enables the efficient dehydration of highly functionalized molecules. In the second half of the 1980s, the Burgess reagent was also used for dehydrating primary amides[6] and oximes[13] to the corresponding nitriles at room temperature. Other functional groups can also be dehydrated, so formamides give isonitriles,[10] ureas are converted to carbodiimides,[8] and primary nitro alkanes yield nitrile oxides[9] upon treatment with the Burgess reagent.

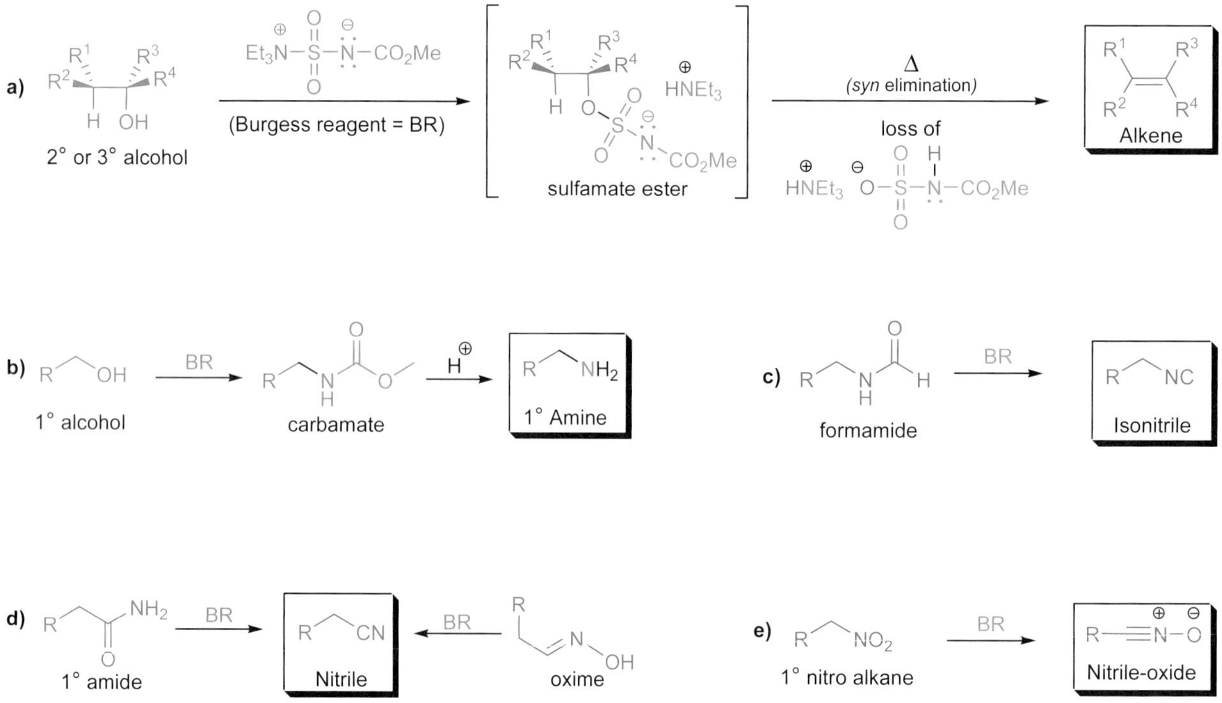

Mechanism:[1,2]

The mechanism involves a stereospecific *syn*-elimination *via* ion-pair formation from the intermediate sulfamate ester (comparable to the *Chugaev elimination* of xanthate esters). Kinetic and spectroscopical data are consistent with an initial rate-limiting formation of an ion-pair followed by a fast *cis*-β-proton transfer to the departing anion.

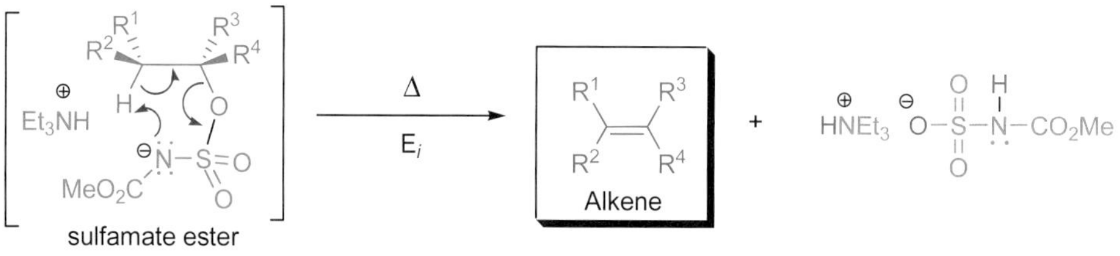

Synthetic Applications:

During the first total synthesis of taxol®, R. Holton and co-workers installed an *exo*-methylene group on the C ring in order to set the stage for the D ring (oxetane) formation.[18] The *Burgess dehydration reaction* was applied to a complex tricyclic tertiary alcohol intermediate (ABC rings) and the desired exocyclic alkene was isolated in 63% yield.

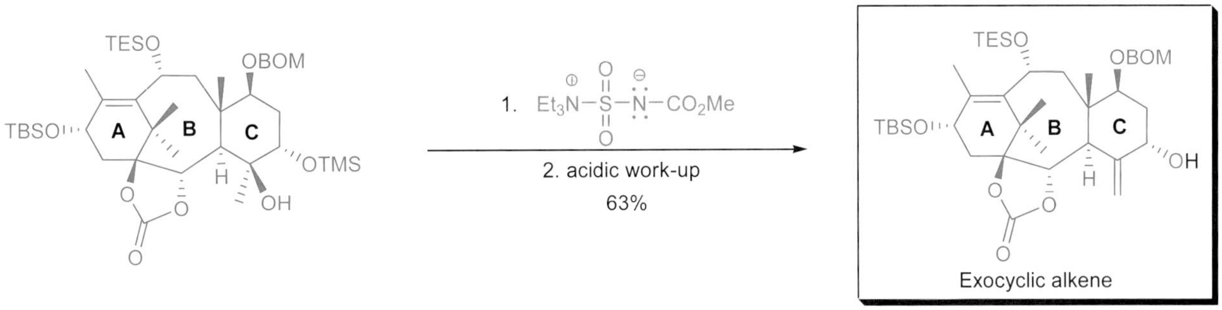

In the laboratory of A.I. Meyers, the first enantioselective total synthesis of the streptogramin antibiotic (–)-madumycin II was achieved.[19] The natural product contains an oxazole moiety, which may be considered a masked dehydropeptide. The oxazole moiety was introduced in two steps: by the *Burgess cyclodehydration reaction* followed by oxidation of the resulting oxazoline to the corresponding oxazole.

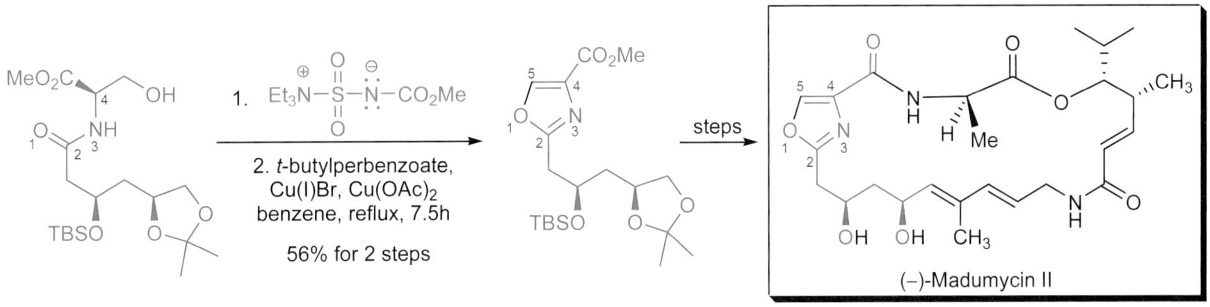

The first total synthesis of the nucleoside antibiotic herbicidin B was achieved by A. Matsuda et al. using a SmI_2 promoted *novel aldol-type C-glycosidation reaction* as the key step.[20] After the key step, the resulting secondary alcohol functionality was removed with the Burgess reagent. The corresponding α,β-unsaturated ketone was isolated in good yield. Hydrogenation of the enone double bond followed by the removal of protecting groups and cyclic ketal formation afforded herbicidin B.

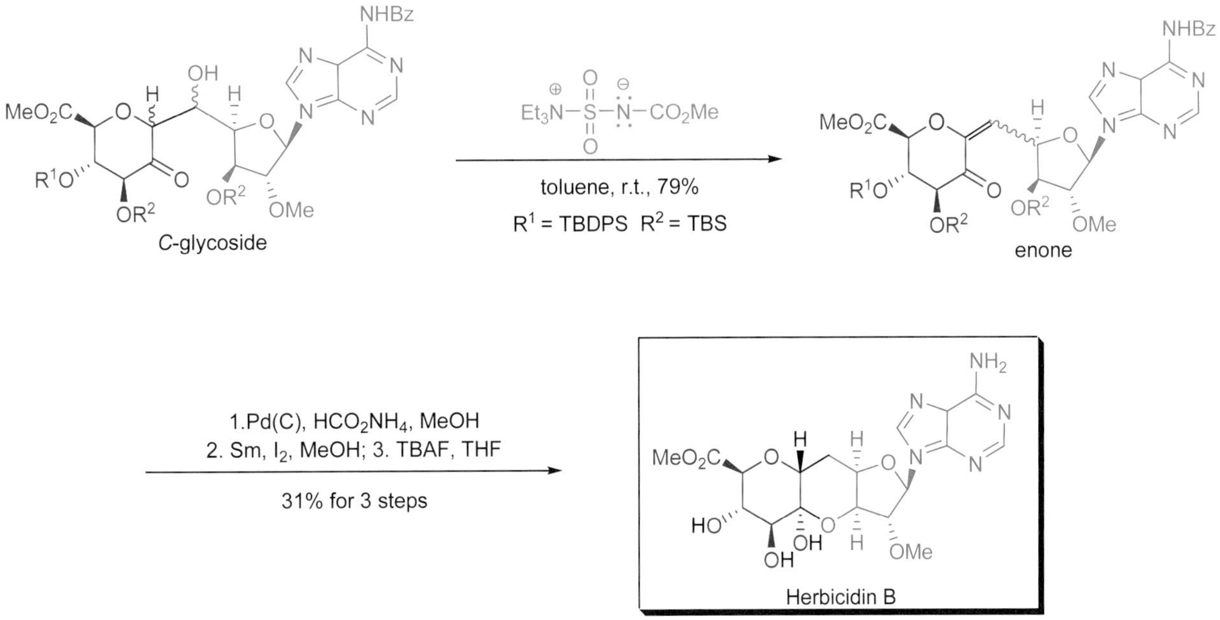

CARROLL REARRANGEMENT
(KIMEL-COPE REARRANGEMENT)

(References are on page 557)

Importance:

[*Seminal Publications*[1,2]; *Reviews*[3,4]; *Modifications & Improvements*[5-12]]

The *[3,3]-sigmatropic rearrangement* of allylic β-keto esters to γ,δ-unsaturated ketones is known as the *Carroll rearrangement* (*Kimel-Cope rearrangement* or *decarboxylative Claisen rearrangement*). Although discovered in 1940, this reaction was not applied in drug synthesis until the early 1990s.[6] The reaction has found limited use in synthetic organic chemistry, since harsh conditions (130-220 °C) were needed to induce the *[3,3]-sigmatropic rearrangement*. However, these thermal barriers were lowered through modifications of the precursor β-keto ester.[6-12] Many different variations of the *Carroll rearrangement* are known, but most of them proceed with decarboxylation of the initially formed β-keto acid. The decarboxylation can be avoided by esterification or intramolecular lactonization of the β-keto acids at low temperatures, leading to the rearranged products with excellent *syn/anti* selectivities.

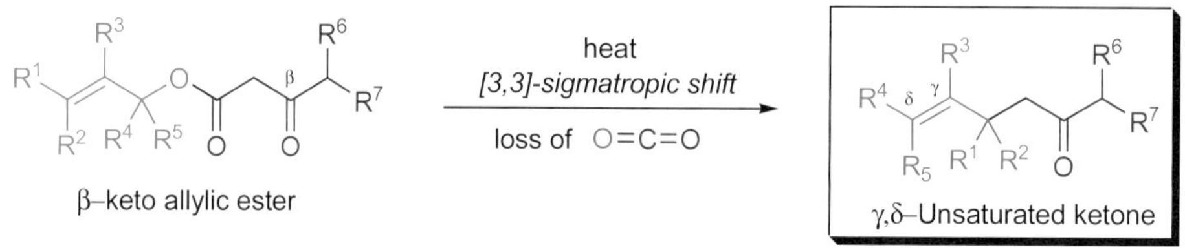

Mechanism: [13,14]

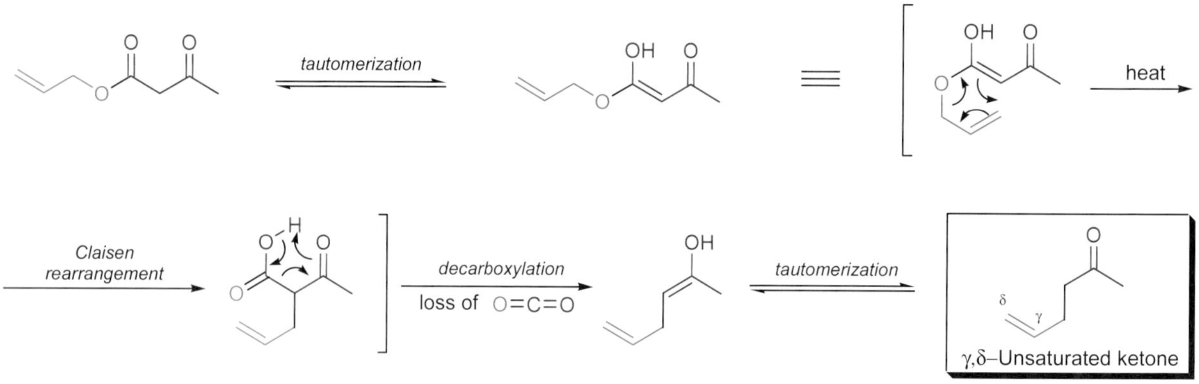

Synthetic Applications:

D. Enders and co-workers have achieved the enantioselective total synthesis of antibiotic (–)-malyngolide by using the *asymmetric Carroll rearrangement* as the key step.[11]

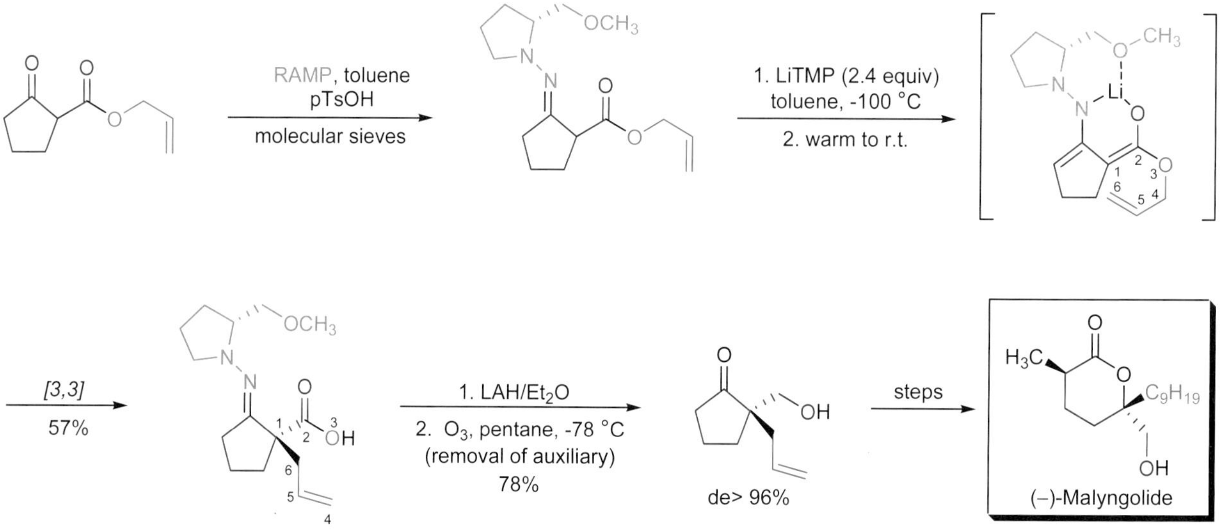

Synthetic Applications:

In the laboratory of A.M. Echavarren, the total synthesis of the antibiotic (±)-4-*epi*-acetomycin was completed by using the stereoselective *ester enolate Carroll rearrangement* of (*E*)-butenyl-2-methylacetoacetate as the key step, followed by ozonolysis and acetylation. The stereochemistry of the major isomer resulted from the rearrangement of the (*E*)-enolate through a chair-like transition state.[6]

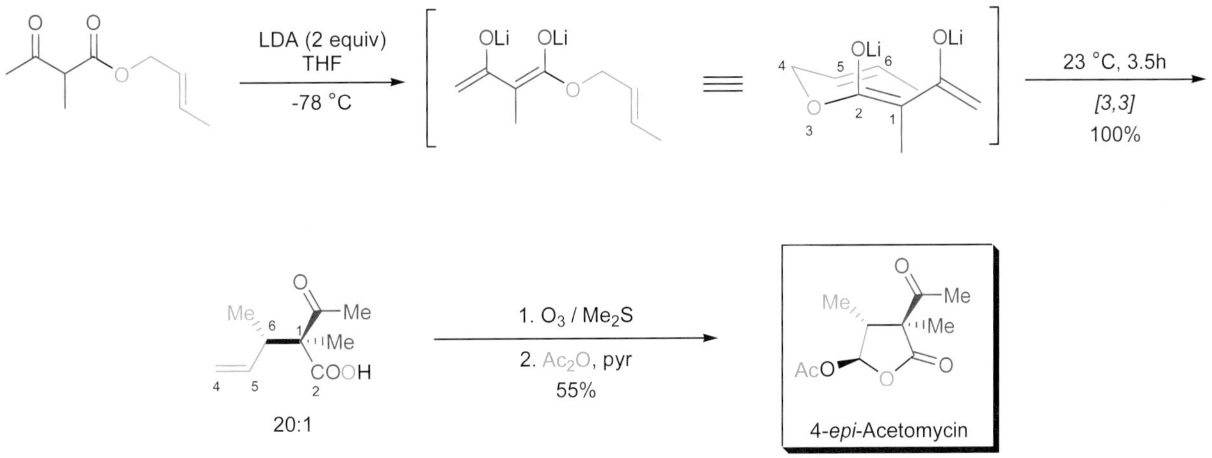

J. Rodriguez et al. have investigated the stereoselective *ester dienolate Carroll rearrangement* of (*E*)- and (*Z*)-allylic β-keto esters and found a new, attractive approach to the synthesis of the Prelog-Djerassi lactone and related compounds.[7]

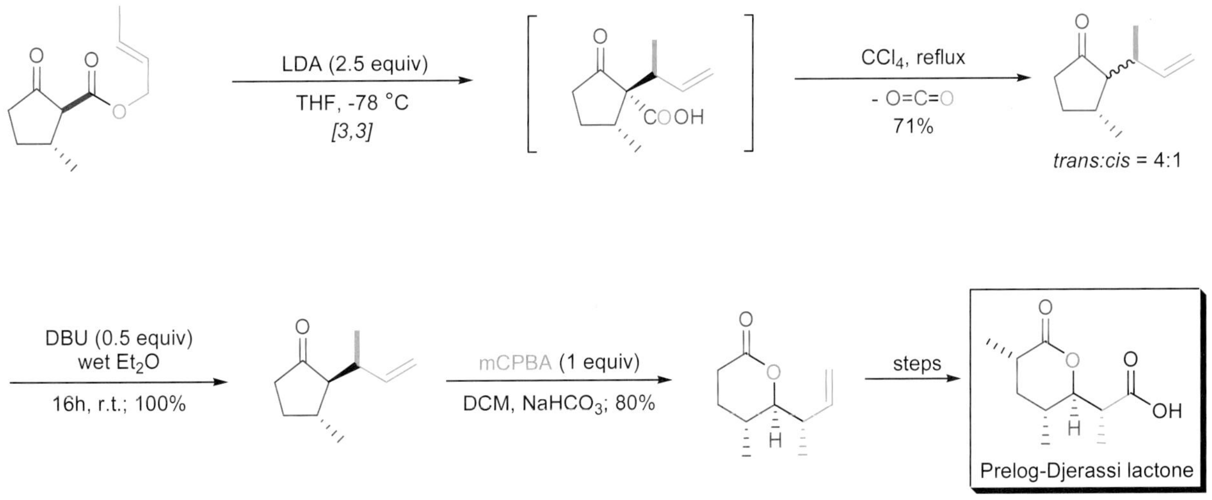

K.L. Sorgi and co-workers prepared acetoacetates from substituted *p*-quinols and found that they underwent the *Carroll rearrangement* at room temperature to afford substituted arylacetones and related derivatives in moderate to good yields.[10]

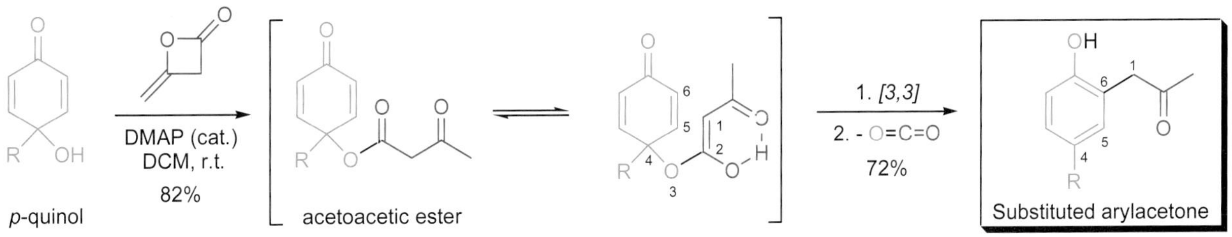

CASTRO-STEPHENS COUPLING

(References are on page 558)

Importance:

[*Seminal Publications*[1,2]; *Reviews*[3-8]; *Modifications & Improvements*[9-13]]

The copper(I) mediated coupling of aryl or vinyl halides with aryl- or alkyl-substituted alkynes is known as the *Castro-Stephens coupling*. In the early 1960s, R.D. Stephens and C.E. Castro discovered that disubstituted (diaryl or arylalkyl) acetylenes were produced in good yield upon treatment of aryl iodides with stoichiometric amounts of copper(I) acetylides under a nitrogen atmosphere in refluxing pyridine (a).[1] The best results are obtained with electron-poor aryl halides. When aryl iodides bear a nucleophilic substituent in the *ortho* position, cyclization to the corresponding heterocycles occurs exclusively (b).[2] Vinyl iodides and bromides are also suitable partners affording enynes. Traditional copper-mediated aryl coupling reactions have several drawbacks compared to the currently used Pd-catalyzed reactions (e.g., *Sonogashira coupling*). The problems encountered are: 1) most copper(I) salts are insoluble in organic solvents, so the reactions are often heterogeneous and require high reaction temperatures; and 2) the reactions are sensitive to functional groups on the aryl halides, and the yields are often irreproducible. Recent modifications allow the use of catalytic amounts of copper(I) complexes and milder conditions for the couplings.[11,13]

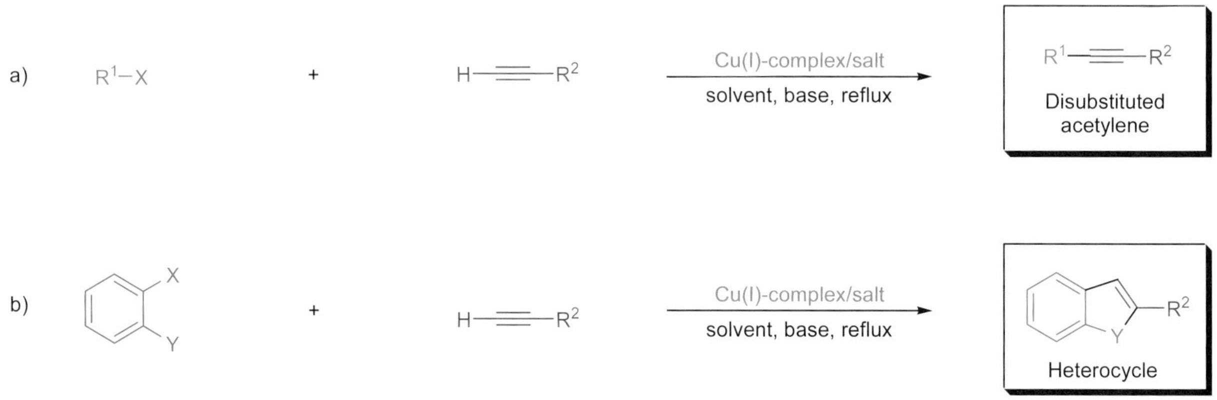

R^1 = aryl, vinyl; X = I, Br; Y = OH, NH_2; R^2 = alkyl, aryl; base: pyridine, KO*t*-Bu, NEt_3

Mechanism:[4]

The reaction is believed to proceed *via* a four-centered transition state.

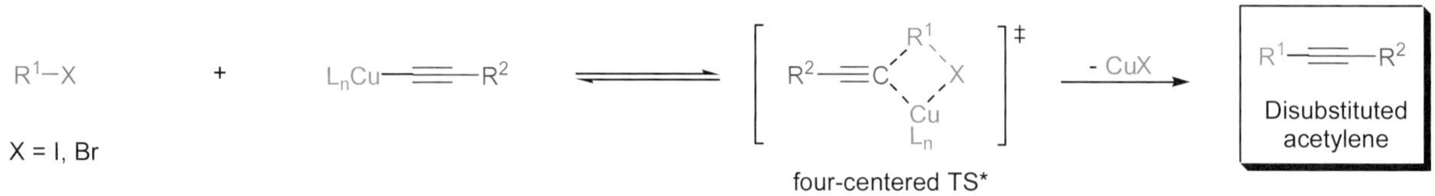

four-centered TS*

Synthetic Applications:

In the laboratory of M. Nilsson, a facile one-pot synthesis of isocumestans (6*H*-benzofuro[2,3-*c*][1]benzopyran-6-ones) was developed *via* a novel extension of the *Castro-Stephens coupling* utilizing *ortho*-iodophenols and ethyl propiolate.[14] The reaction can be regarded as an extended *Castro-Stephens coupling* where an intermediate cuprated benzofuran couples with a second equivalent of *ortho*-iodophenol, and the product lactonizes to isocumestan.

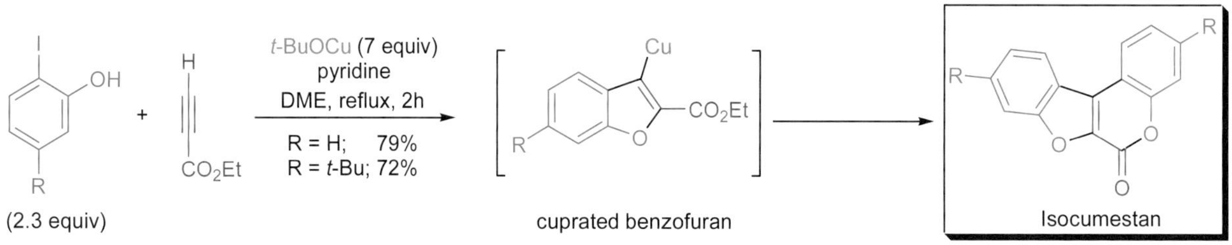

Synthetic Applications:

Tribenzocyclotriyne (TBC) is a planar, anti-aromatic, annelated dehydroannulene. The cavity of TBC is of sufficient size to form complexes with low oxidation state first-row transition metals. When the complex of Ni(TBC) is partially reduced with alkali metals, the complex increases its conductivity by four orders of magnitude. This remarkable property was the reason for the synthesis of cyclotriynes by W.J. Youngs et al. as precursors to conducting systems.[15] The synthesis of a methoxy-substituted tribenzocyclotriyne was accomplished starting from (2-iodo-3,6-dimethoxyphenyl)ethyne using the *Castro-Stephens coupling*. The copper acetylide was prepared by dissolving the alkyne in ethanol and adding it to an equal volume of CuCl in ammonium hydroxide. Refluxing the copper acetylide in pyridine under anaerobic conditions produced the cyclotriyne in 80% yield.

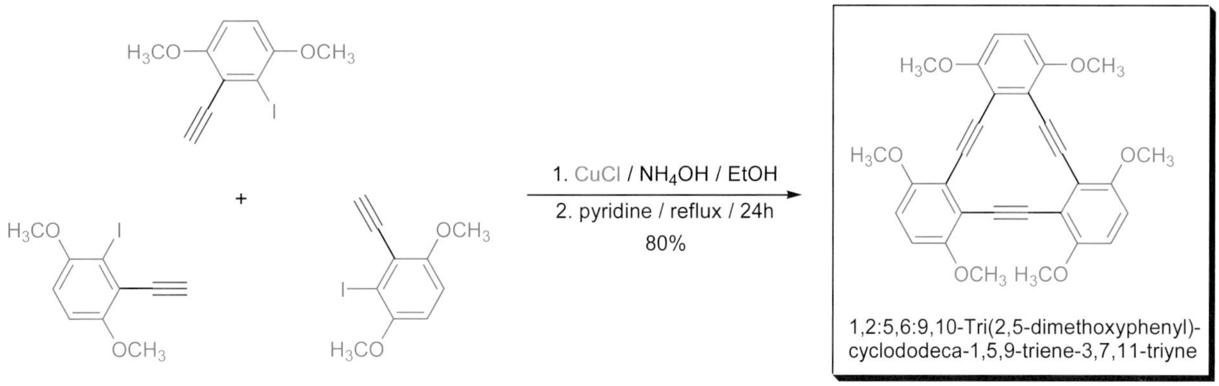

R.S. Coleman and co-workers have developed a stereoselective synthesis of the 12-membered diene and triene lactones characteristic of the antitumor agent oximidines I and II, based on an *intramolecular Castro-Stephens coupling*.[16] The effectiveness of this protocol rivals the efficiency of standard macrolactonization. The stereoselective reduction of the internal alkyne afforded the 12-membered (*E*,*Z*)-diene lactone in good yield.

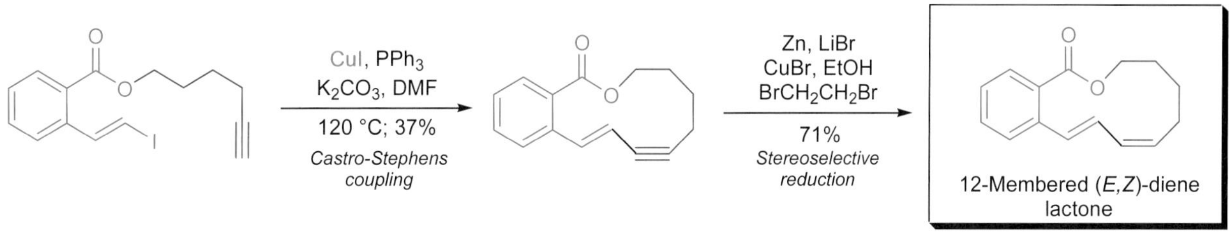

During the total synthesis of epothilone B, J.D. White et al. used the *modified Castro-Stephens reaction* instead of a *Wittig reaction* for the coupling of two important subunits (**A** & **B**) to avoid strongly basic conditions.[17]

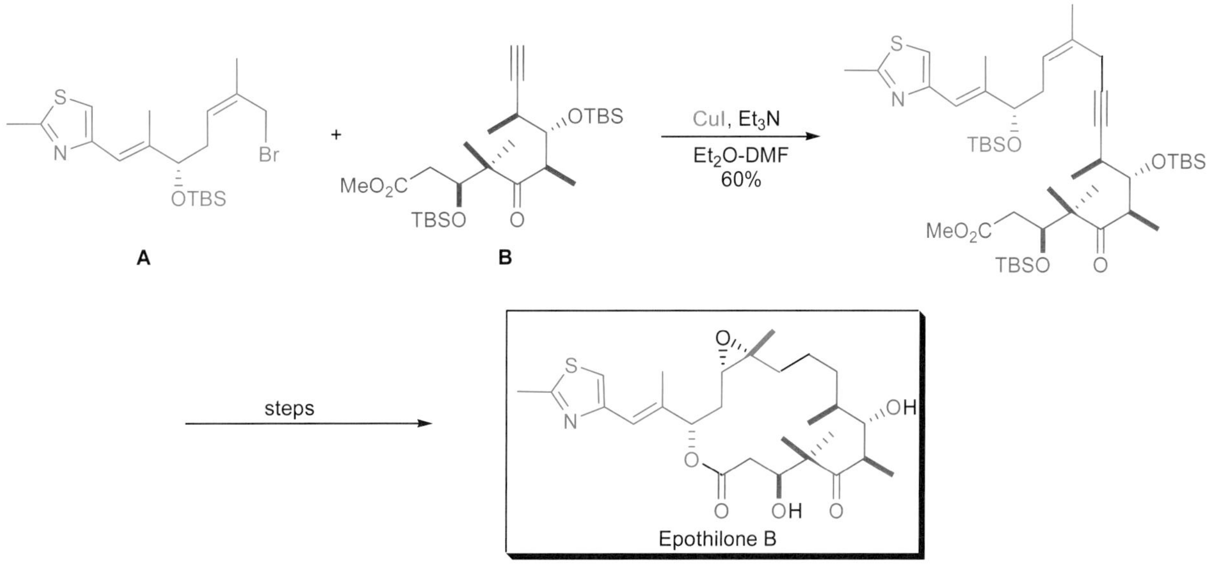

CHICHIBABIN AMINATION REACTION

(References are on page 558)

Importance:

[*Seminal Publications*[1,2]; *Reviews*[3-8] ; *Modifications & Improvements*[9,10]]

In the early 1900s, A.E. Chichibabin reacted pyridine with sodium amide ($NaNH_2$) in dimethylamine at high temperature (110 °C). After aqueous work-up, he isolated 2-aminopyridine in 80% yield.[1] A decade later, he added pyridine to powdered KOH at 320 °C, and after aqueous work-up 2-hydroxypyridine was isolated.[2] Similar reactions take place when pyridine or its derivatives are treated with strong nucleophiles such as alkyl- and aryllithiums to give 2-alkyl and 2-arylpyridines.[11] The direct amination of pyridine and its derivatives at their electron-deficient positions *via* nucleophilic aromatic substitution (S_NAr) is known as the *Chichibabin reaction*. This reaction is also widely used for the direct introduction of an amino group into the electron-deficient positions of many azines and azoles (e.g., quinoline is aminated at C2 & C4, isoquinoline at C1, acridine at C9, phenanthridine at C6, quinazoline at C4). Both inter- and intramolecular[12-14] versions are available, but investigations have mainly focused on intermolecular reactions. There are two procedures for conducting the *Chichibabin reaction*: **A**) the reaction is carried out at high temperature in a solvent that is inert toward $NaNH_2$ (e.g., *N,N*-dialkylamines, arenes, mineral oil, etc.) or without any solvent; or **B**) the reaction is run at low temperature in liquid ammonia with KNH_2 (more soluble than $NaNH_2$). Procedure **A** proceeds in a heterogeneous medium and the reactions effected under these conditions show strong dependence on substrate basicity, while procedure **B** proceeds in a homogeneous medium and there is no substrate dependence. Frequently, an oxidant such as KNO_3 or $KMnO_4$ is added during procedure **B** to facilitate the amination by oxidizing the hydride ion (poor leaving group) in the intermediate σ-complex.[9,6] The low temperature conditions make it possible to aminate substrates such as diazines, triazines, and tetrazines, which are destroyed at high temperatures, but pyridine itself does not undergo amination in liquid ammonia because it is not sufficiently electron-deficient.

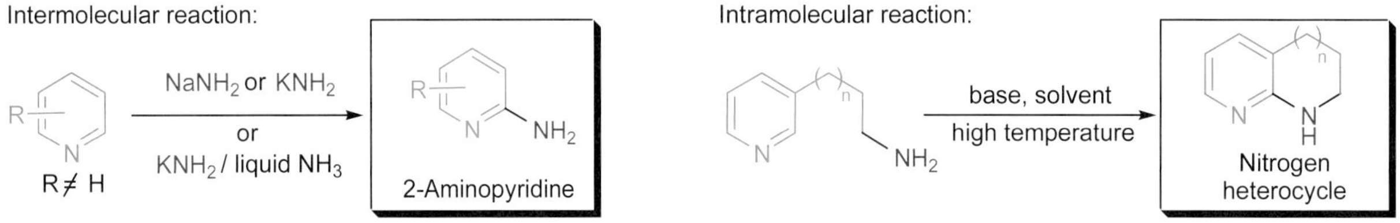

Mechanism: [15-26,7]

The *Chichibabin reaction* is formally the nucleophilic aromatic substitution of hydride ion (H^-) by the amide ion (NH_2^-). In the first step, an *adsorption complex* is formed with a weak coordination bond between the nitrogen atom in the heterocycle and the sodium ion (Na^+); this coordination increases the positive charge on the ring α-carbon atom, and thus facilitates the formation of an *anionic σ-complex* that can be observed by NMR in liquid ammonia solution. This σ-complex is then aromatized to the corresponding sodium salt while hydrogen gas (H_2) is evolved (a proton from an amino group reacts with the leaving group hydride ion). It is possible to monitor the progress of the reaction by the volume of the hydrogen gas evolved. However, this mechanism may not be the only one operating, since indirect evidence (formation of heterocyclic dimers) suggests that under heterogeneous conditions there is a single-electron-transfer (SET) from the amide nucleophile to the substrate.

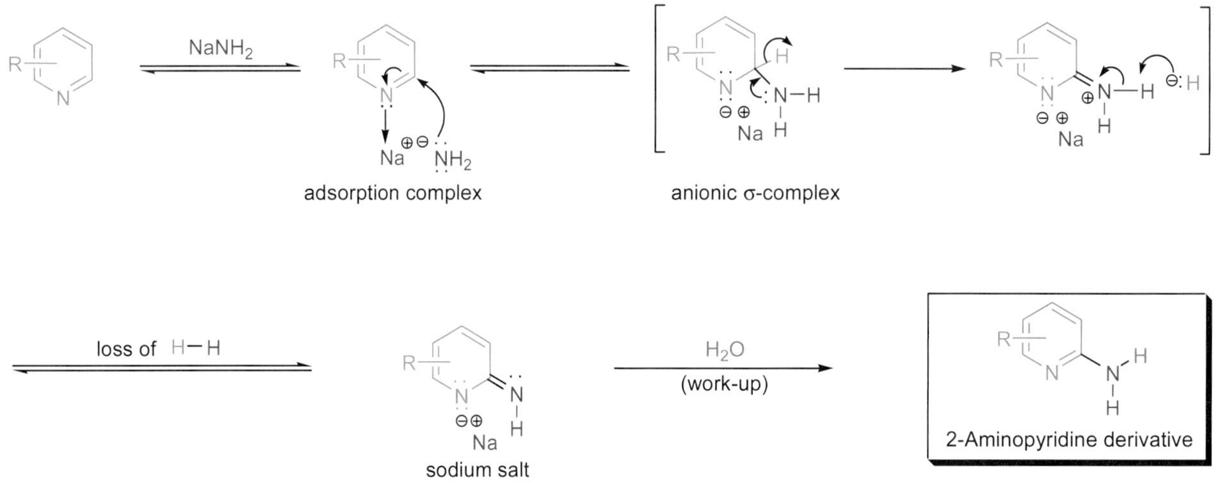

Synthetic Applications:

In the laboratory of J.S. Felton, the synthesis of 2-amino-1-methyl-6-phenyl-1*H*-imidazo[4,5-*b*]pyridine (PHIP), a mutagenic compound isolated from cooked beef, and its 3-methyl isomer have been accomplished.[27] The synthesis of PHIP began with the commercially available 3-phenylpyridine, which was aminated at the 6-position with sodium amide in toluene by the *Chichibabin reaction* in 58% yield.

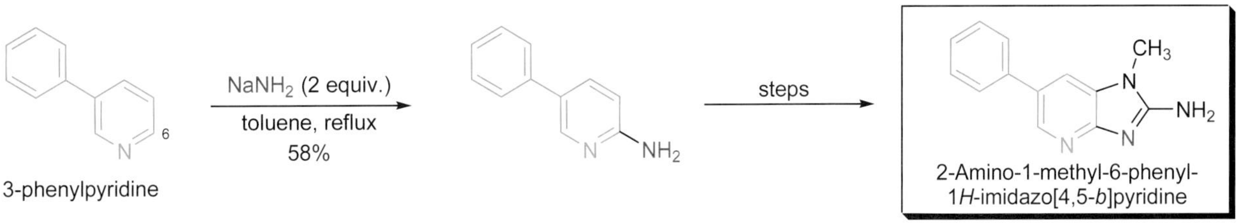

M. Palucki and co-workers synthesized 2-[3-aminopropyl]-5,6,7,8-tetrahydronaphthyridine in large quantities for clinical studies *via* a one-pot *double Suzuki reaction* followed by deprotection and a highly regioselective *intramolecular Chichibabin cyclization*.[14] This approach was amenable to scale-up unlike the traditional methods such as the *Skraup* and *Friedländer reactions* that involve carbon-carbon bond forming steps. The *Chichibabin reaction* was optimized and afforded the desired product in high yield, excellent regioselectivity, and a significant reduction in reaction time compared to literature precedent.

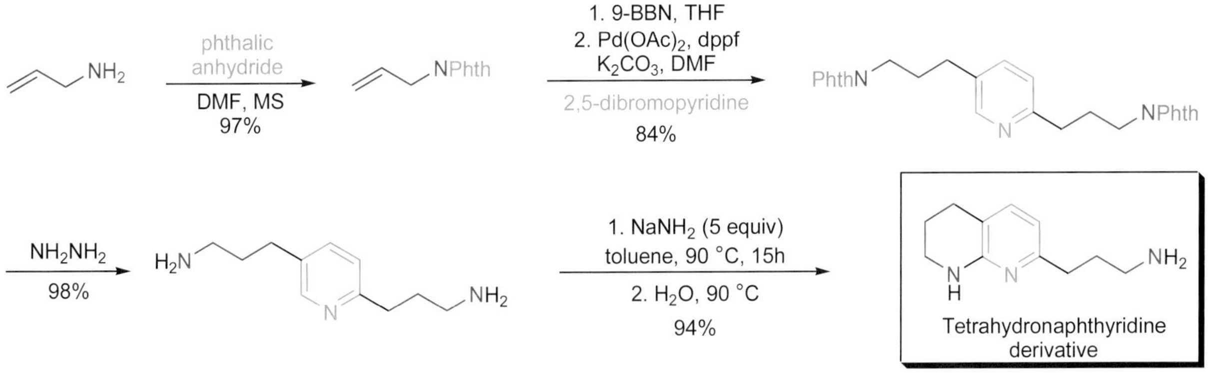

T.R. Kelly et al. have synthesized bisubstrate reaction templates utilizing the *Chichibabin amination reaction* during the preparation of one precursor.[28] This reaction template was designed to use hydrogen bonding to bind two substrates simultaneously but transiently, giving rise to a ternary complex, which positions the substrates in an orientation that facilitates their reaction.

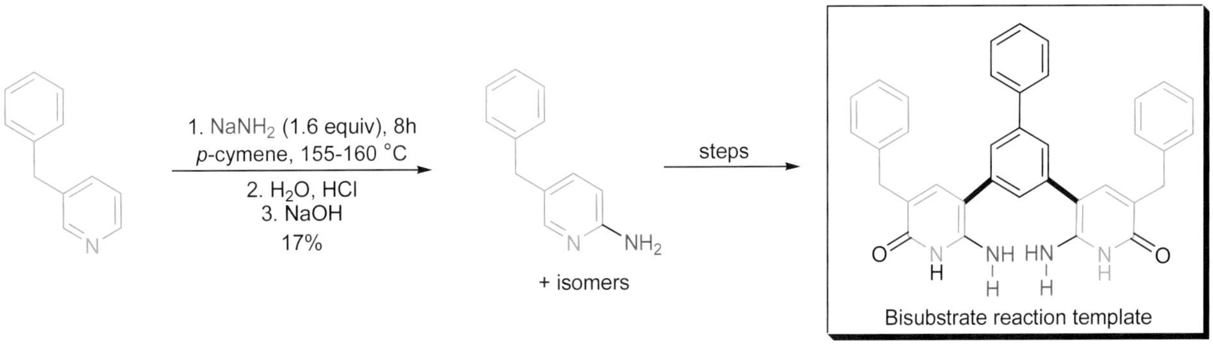

A.N. Vedernikov and co-workers designed and synthesized tridentate facially chelating ligands of the [2.n.1]-(2,6)-pyridinophane family.[29] The key step in their synthesis of these tripyridine macrocycles was a *double Chichibabin-type condensation* of 1,2-*bis*(2-pyridyl)ethanes with lithiated 2,6-dimethylpyridines.

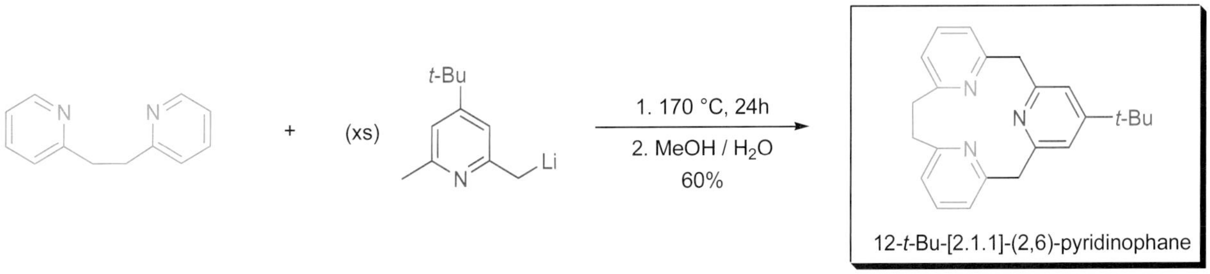

CHUGAEV ELIMINATION REACTION
(XANTHATE ESTER PYROLYSIS)

(References are on page 559)

Importance:

[*Seminal Publications*[1,2]; *Reviews*[3,4]; *Modifications & Improvements*[5-7]]

The formation of olefins by pyrolysis (100-250 °C) of the corresponding xanthates (containing at least one β-hydrogen atom) *via cis*-elimination is known as the *Chugaev elimination reaction*. This transformation was discovered by L. Chugaev in connection with his studies on the optical properties of xanthates[1] in 1899. Xanthates are prepared from the corresponding alcohols (1°, 2°, and 3°) by first deprotonating the alcohol with a base (e.g., NaH, NaOH, or KOH) and reacting the resulting alkoxide with carbon disulfide. The metal xanthate is then trapped with an alkyl iodide (often methyl iodide). Primary xanthates are usually more thermally stable than secondary and tertiary xanthates and therefore undergo elimination at much higher temperatures (>200 °C). The *Chugaev elimination* reaction of xanthates is very similar to the ester (acetate) pyrolysis, but xanthates eliminate at lower temperatures than esters and the possible isomerization of alkenes is minimized. The by-products (COS, R^4-SH) of the *Chugaev reaction* are very stable, thus making the elimination irreversible. The reaction is especially valuable for the conversion of sensitive alcohols to the corresponding olefins without rearrangement of the carbon skeleton. If the elimination of the xanthate can occur in two directions, when more than one β-hydrogen is available on each carbon atom, the utility of the *Chugaev reaction* is greatly diminished by the formation of complex mixtures of olefins.

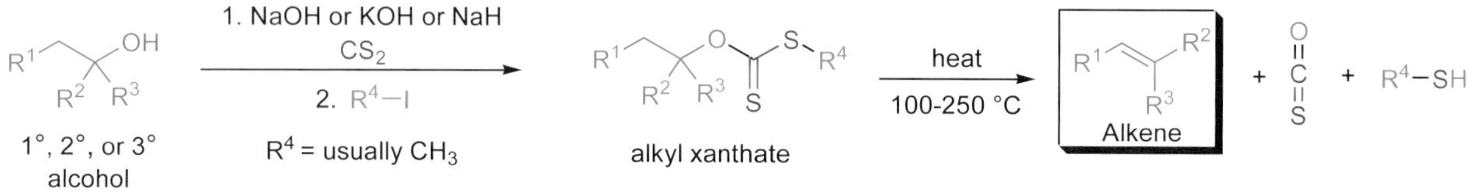

Mechanism:[8-12]

The *Chugaev reaction* is an intramolecular *syn* elimination (E_i), and it proceeds through a six-membered transition state involving a *cis*-β-hydrogen atom of the alcohol moiety and the thione sulfur atom of the xanthate. Isotopic studies involving ^{34}S and ^{13}C showed that the C=S, and not the thiol sulfur atom, closes the ring in the transition state.[12] The β-hydrogen and the xanthate group must be coplanar in the cyclic transition state.

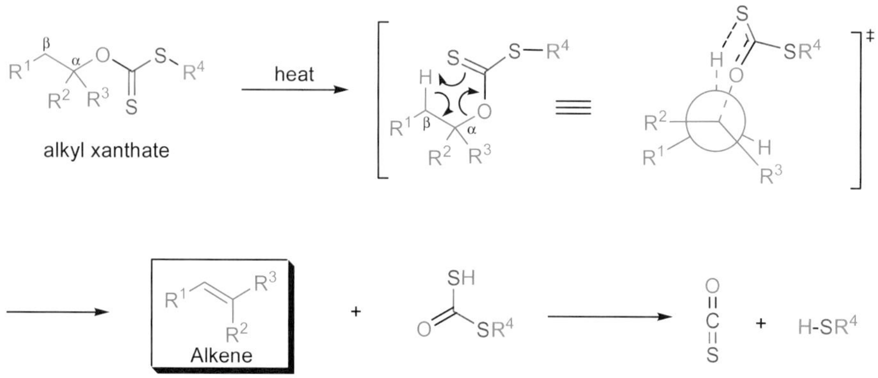

Synthetic Applications:

A concise route to (–)-kainic acid was developed by K. Ogasawara and co-workers by employing sequentially a *Chugaev syn-elimination* and an *intramolecular ene reaction* as the key steps.[13] After preparing the xanthate under standard conditions, the compound was heated to reflux in diphenyl ether in the presence of sodium bicarbonate. The desired tricyclic product bearing the trisubstituted pyrrolidine framework was formed as a single diastereomer in 72% yield.

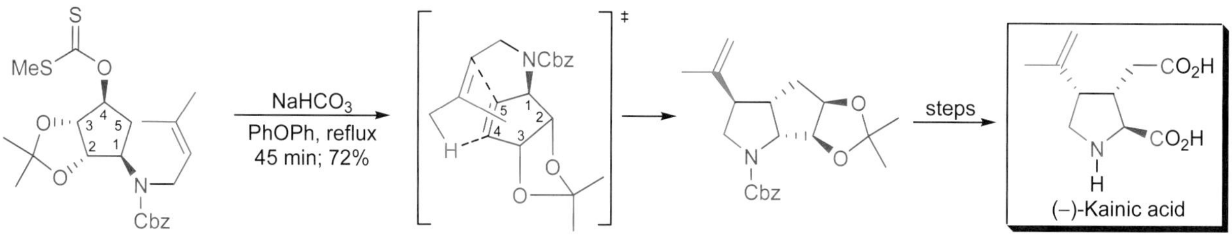

CHUGAEV ELIMINATION REACTION
(XANTHATE ESTER PYROLYSIS)

Synthetic Applications:

In the late stages of the total synthesis of dihydroclerodin, A. Groot and co-workers used the *Chugaev elimination reaction* to install an exocyclic double bond on ring A.[14] Before employing the xanthate ester pyrolysis, the authors tried several methods that failed to convert the primary alcohol to the exocyclic methylene functionality. The corresponding xanthate ester was prepared followed by heating to 216 °C in *n*-dodecane for 2 days to afford the desired alkene in 74% yield.

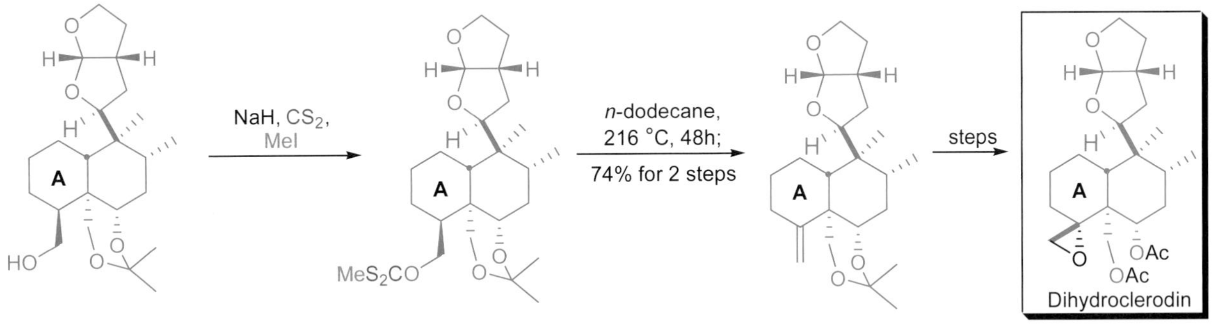

During the first total synthesis of (–)-solanapyrone E by H. Hagiwara et al., it was necessary to install the C3-C4 double bond in the decalin ring of the natural product by eliminating the C4 secondary alcohol.[15] Since the stereochemistry of the *xanthate pyrolysis* is *syn*, it was possible to install this double bond regioselectively, without observing any of the undesired C4-C5 double bond. The C4 alcohol was first converted to the xanthate in 91% yield using *t*-BuOK as a base. The double bond at C3 was then selectively introduced by heating the xanthate at 190 °C in 1-methylnaphthalene.

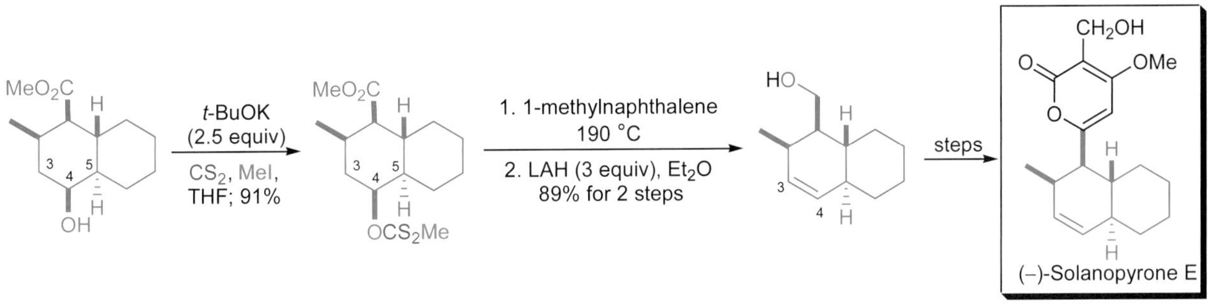

J.M. Cook and co-workers accomplished the total synthesis of ellacene (1,10-cyclododecanotriquinancene) by utilizing the *Weiss reaction* and the *Chugaev elimination* as key steps.[16] The elimination of the *tris*-xanthate was performed in HMPA at 220-230 °C in very high yield. This pyrolysis was superior to the elimination conducted under neat conditions.

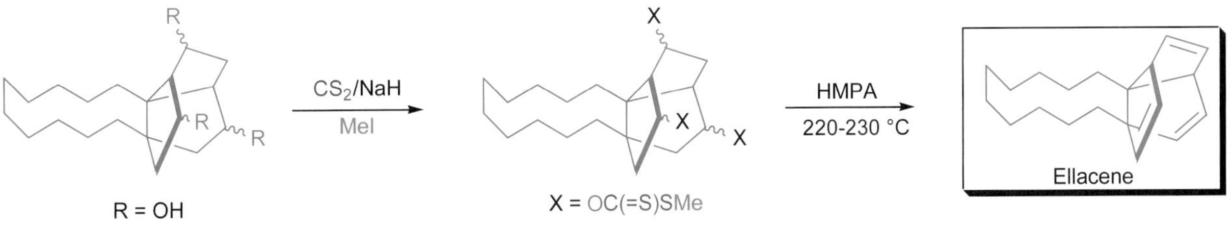

Synthetic studies on kinamycin antibiotics in the laboratory of T. Ishikawa resulted in the elaboration of the highly oxygenated D ring with all the required stereocenters for the kinamycin skeletons.[17] The tricyclic tertiary alcohol was converted to the corresponding xanthate and then smoothly pyrolyzed under reduced pressure to yield the desired tetrahydrofluorenone system.

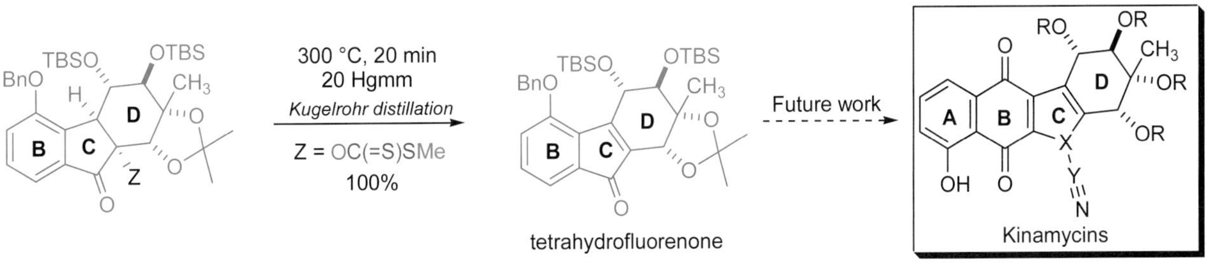

CIAMICIAN-DENNSTEDT REARRANGEMENT

(References are on page 559)

Importance:

[*Seminal Publications*[1-4]; *Theoretical Studies*[5]]

The rearrangement of pyrroles to 3-halo-pyridines upon treatment with haloforms (CHX_3 where X = Cl, Br, I) in the presence of a strong base was first described by G.L. Ciamician.[1] Its synthetic utility was later extended by M. Dennstedt to the sodium methoxide catalyzed reaction of pyrrole with methylene iodide to give pyridine.[6] Soon after the initial discovery, the methodology was also extended for the indole series to prepare substituted quinolines.[7-9] The reaction is known as the *Ciamician-Dennstedt rearrangement*, but it is also referred to as the *"abnormal" Reimer-Tiemann reaction*.

Mechanism: [10-19]

The mechanism starts with the generation of a dihalocarbene *via* an α-elimination, followed by insertion into the most electron rich π-bond of the pyrrole. The 6,6-dihalo-2-azabicyclo[3.1.0]hexane intermediate then undergoes a ring expansion to give the 3-halogen-substituted pyridine derivative triggered by the deprotonation of the pyrrole nitrogen. In the case of indoles the dihalocyclopropane intermediate interconverts with an open-ring indolyldihalomethyl anion, and therefore two different products, 3*H*-indole and quinoline, are formed.[19]

Carbene formation:

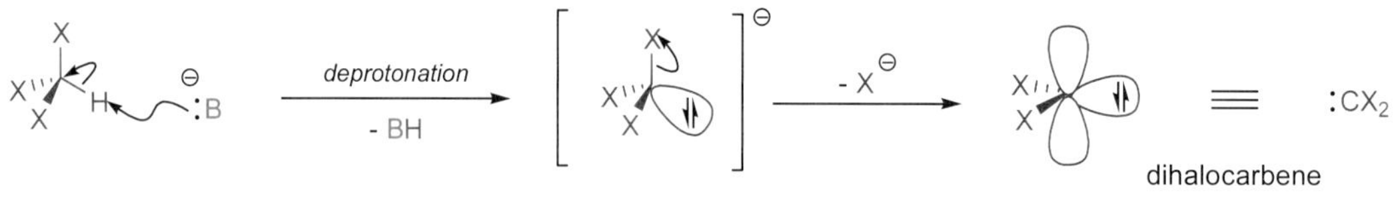

Insertion of carbene:

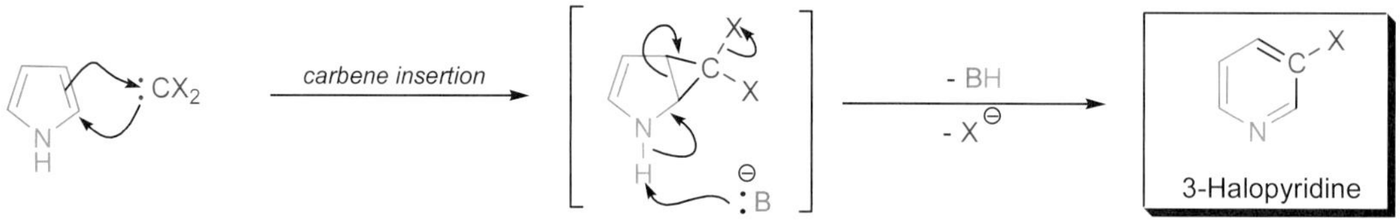

Synthetic Applications:

In an effort to expand the available synthetic tools for the preparation of various metacyclophanes and pyridinophanes, C.B. Reese and co-workers prepared [6](2,4)pyridinophane derivatives by treating 4,5,6,7,8,9-hexahydro-1*H*-cyclo-octa[*b*]pyrrole with dichloro- and dibromocarbene respectively.[20] The dihalocarbenes predominantly inserted into the most substituted (more electron rich) double bond of the pyrrole ring in modest to poor yields.

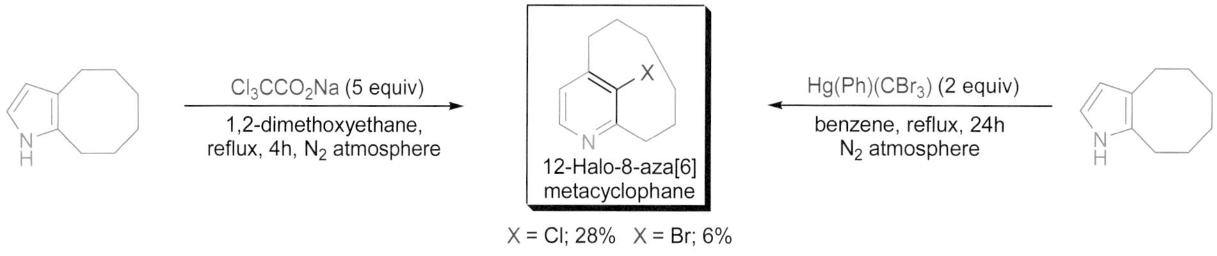

Synthetic Applications:

For a long time heterocyclic analogues of calix[4]arene such as calix[4]pyridines were unknown. In the laboratory of J.L. Sessler, a universal and easy synthetic protocol was devised for the preparation of calix[m]pyridine-[n]pyrrole (m+n=4) and calix[4]pyridine systems based on the nonmetal mediated ring expansion of pyrrole.[21] The reaction of dichlorocarbene with *meso*-octamethyl-calix[4]pyrrole brought about a pyrrole ring expansion to give chlorocalixpyridinopyrroles and chlorocalixpyridines. Using 15 equivalents of sodium trichloroacetate as the carbene source and 1,2-dimethoxyethane as the solvent afforded a 1:1:1 ratio of calix[3]pyridine[1]pyrrole : calix[4]pyridine : calix[2]pyridine[2]pyrrole. Only monochlorinated pyridines were formed but each pyridine ring gave rise to two regioisomers. Yields were between 26-65%.

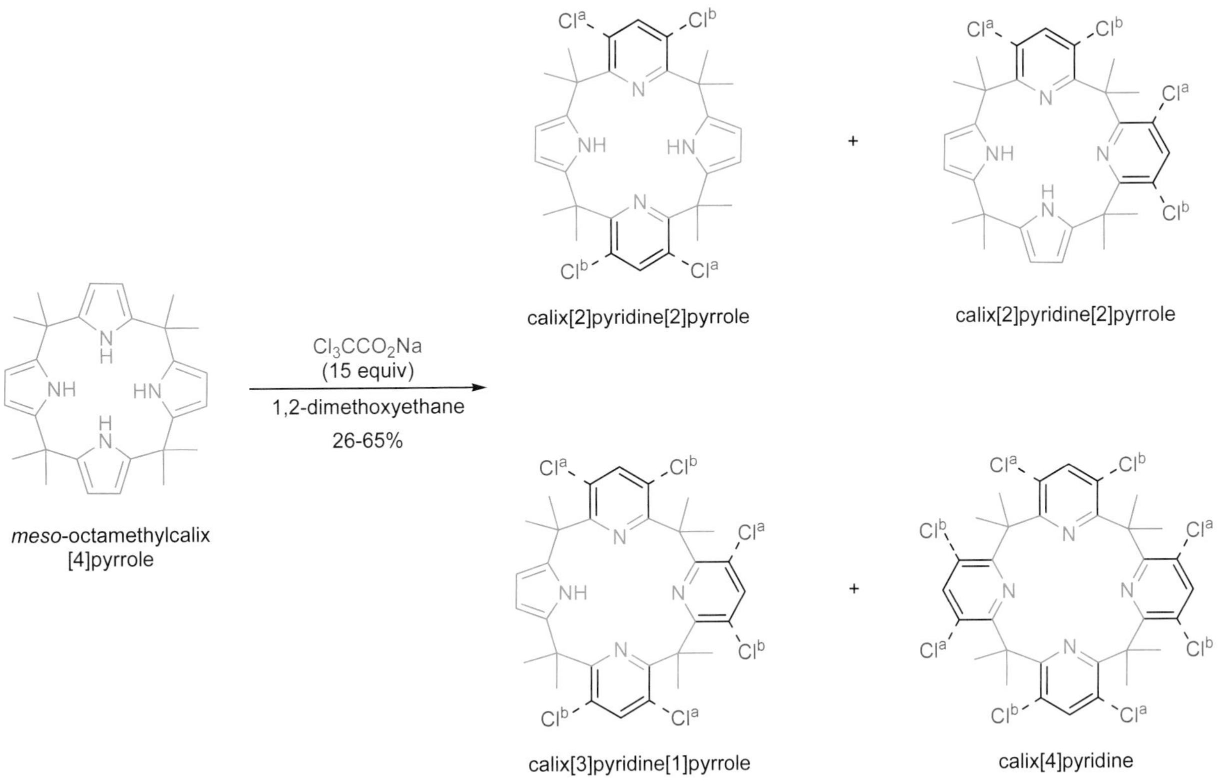

The first example for the insertion of an electrogenerated dichlorocarbene into substituted indoles was described by F. De Angelis and co-workers.[19] The dichlorocarbene was generated by reduction of CCl_4, followed by fragmentation of the resulting trichloromethyl anion. Under these conditions, 2,3-dimethylindole was converted to 3-chloro-2,4-dimethylquinoline and 3-(dichloromethyl)-2,3-dimethyl-3*H*-indole in moderate yield. The study revealed that the reaction mechanism and product formation are determined by the acidity of the solvent.

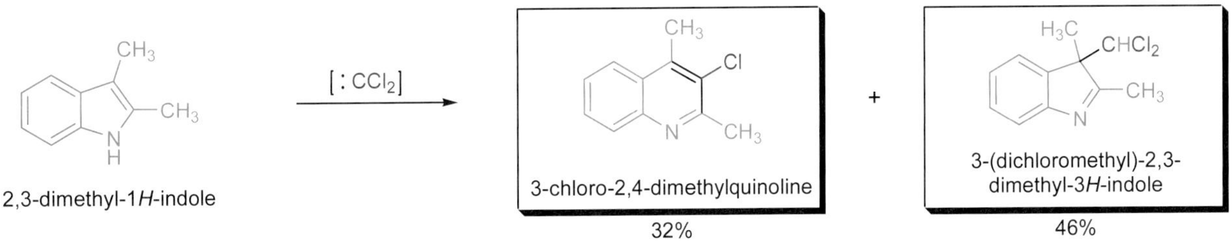

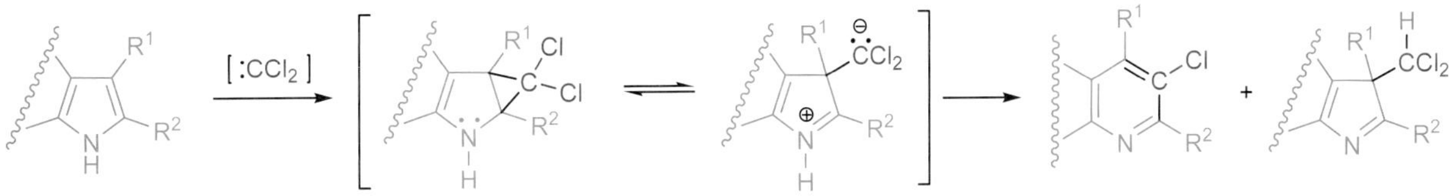

CLAISEN CONDENSATION / CLAISEN REACTION

(References are on page 559)

Importance:

[*Seminal Publication*[1]; *Reviews*[2-8]; *Modifications & Improvements*[9-11]; *Theoretical Studies*[12-14]]

The base mediated condensation of an ester containing an α-hydrogen atom with a molecule of the same ester to give a β-keto ester is known as the *Claisen condensation*. If the two reacting ester functional groups are tethered, then a *Dieckmann condensation* takes place. The reaction between two different esters under the same conditions is called *crossed (mixed) Claisen condensation*. Since the *crossed Claisen condensation* can potentially give rise to at least four different condensation products, it is a general practice to choose one ester with no α-protons (e.g., esters of aromatic acids, formic acid and oxalic acid). The ester with no α-proton reacts exclusively as the acceptor and this way only a single product is formed. A full equivalent of the base (usually an alkoxide, LDA or NaH) is needed and when an alkoxide is used as the base, it must be the same as the alcohol portion of the ester to prevent product mixtures resulting from ester interchange. There are two other variants of this process: a) an ester enolate reacts with a ketone or aldehyde to give an β-hydroxyester, and b) a ketone or aldehyde enolate reacts with an ester to give a 1,3-diketone, both of these are referred to as the *Claisen reaction*. A useful alternative to the *Claisen condensation* is the reaction of an ester enolate with an acid chloride to generate a β-ketoester.

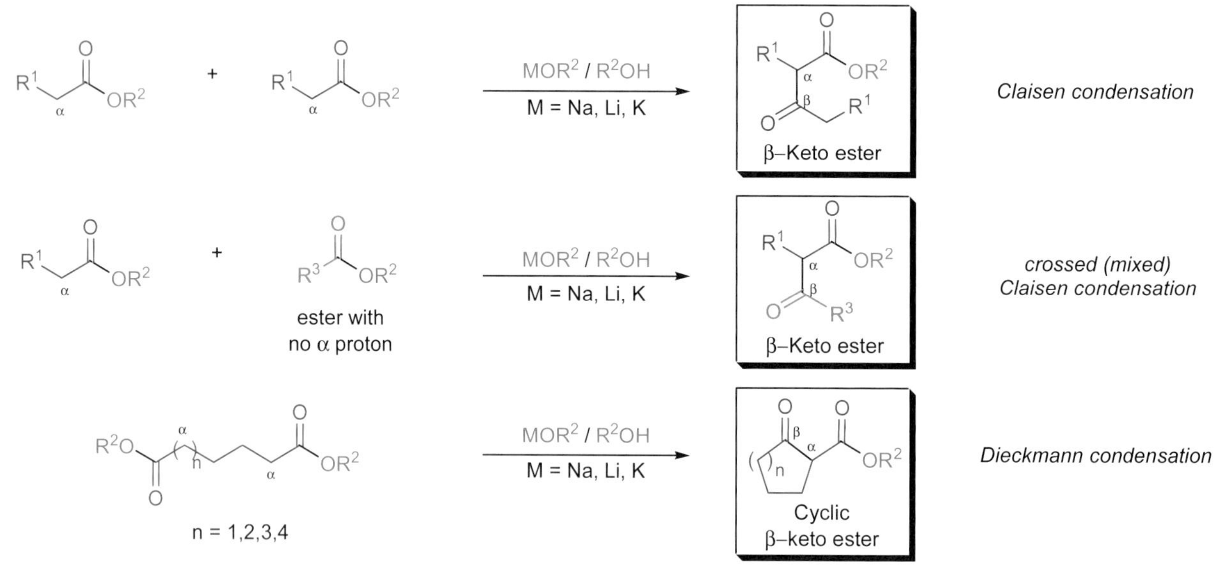

Mechanism: [15-23]

In the first step the base (usually an alkoxide, LDA, or NaH) deprotonates the α-proton of the ester to generate an ester enolate that will serve as the nucleophile in the reaction. Next, the enolate attacks the carbonyl group of the other ester (or acyl halide or anhydride) to form a tetrahedral intermediate, which breaks down in the third step by ejecting a leaving group (alkoxide or halide). Since it is adjacent to two carbonyls, the α-proton in the product β-keto ester is more acidic than in the precursor ester. Under the basic reaction conditions this proton is removed to give rise to a resonance stabilized anion, which is much less reactive than the ester enolate generated in the first step. Therefore, the β-keto ester product does not react further.

First step:

Second step:

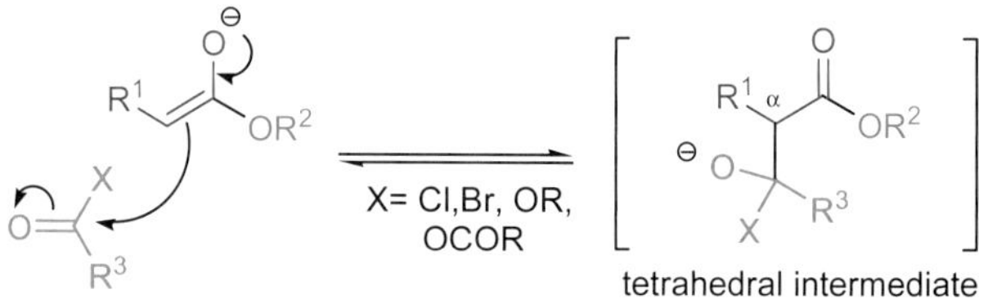

Third step:

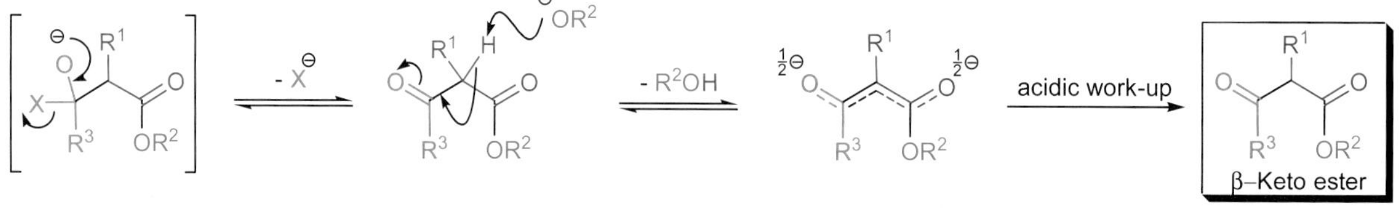

Synthetic Applications:

C.H. Heathcock and co-workers devised a highly convergent asymmetric total synthesis of (–)-secodaphniphylline, where the key step was a *mixed Claisen condensation*.[24] In the final stage of the total synthesis, the two major fragments were coupled using the *mixed Claisen condensation*; the lithium enolate of (–)-methyl homosecodaphniphyllate was reacted with the 2,8-dioxabicyclo[3.2.1]octane acid chloride. The resulting crude mixture of β-keto esters was subjected to the *Krapcho decarboxylation* procedure to afford the natural product in 43% yield for two steps.

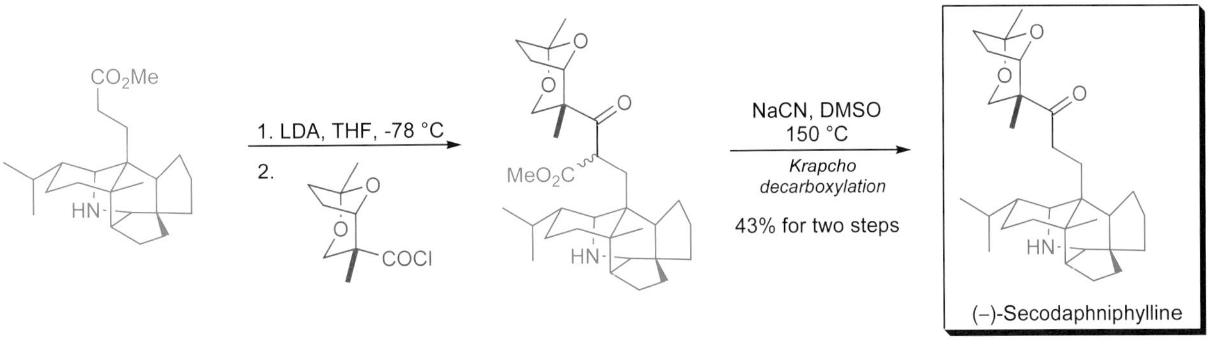

The short total syntheses of justicidin B and retrojusticidin B were achieved in the laboratory of D.C. Harrowven.[25] A novel tandem *Horner-Emmons olefination/Claisen condensation* sequence was used between an aldehyde and a phosphonate tetraester to prepare the highly substituted naphthalene core of the natural products. Simultaneous addition of the aromatic ketoaldehyde and phosphonate to a cooled solution of sodium ethoxide in THF-ethanol effected the desired annulation in 73% yield. The resulting diester was then converted to justicidin B and retrojusticidin B.

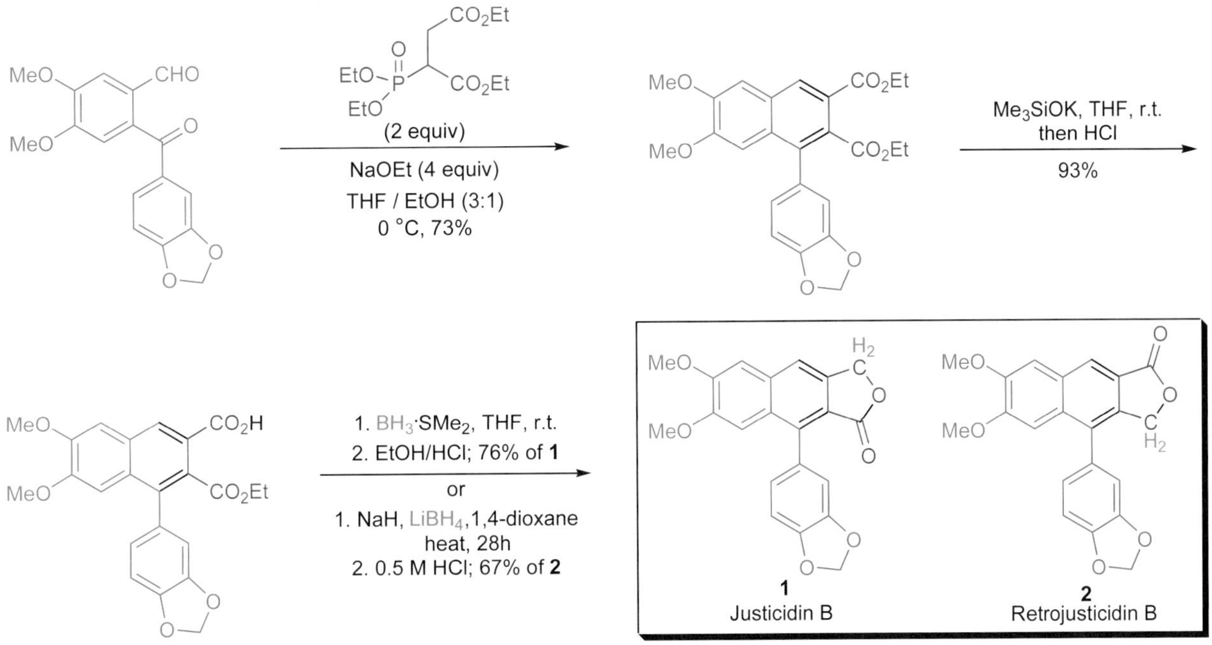

T. Nakata et al. developed a simple and efficient synthetic approach to prepare (+)-methyl-7-benzoylpederate, a key intermediate toward the synthesis of mycalamides.[26] The key steps were the *Evans asymmetric aldol reaction*, *stereoselective Claisen condensation* and the *Takai-Nozaki olefination*. The *diastereoselective Claisen condensation* took place between a δ-lactone and the lithium enolate of a glycolate ester.

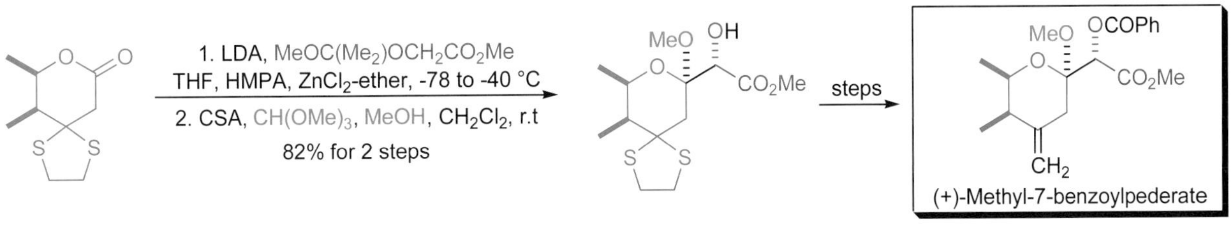

CLAISEN REARRANGEMENT

(References are on page 560)

Importance:

[*Seminal Publications*[1,2]; *Reviews*[3-32]; *Modifications & Variants*[33-48]; *Theoretical Studies*[49-55]]

In 1912, L. Claisen described the rearrangement of allyl phenyl ethers to the corresponding *C*-allyl phenols and also described the transformation of *O*-allylated acetoacetic ester to its *C*-allylated isomer in the presence of ammonium chloride upon distillation.[1] Named after its discoverer, the thermal *[3,3]-sigmatropic rearrangement* of allyl vinyl ethers to the corresponding γ,δ-unsaturated carbonyl compounds is called the *Claisen rearrangement*. The allyl vinyl ethers can be prepared in several different ways: 1) from allylic alcohols by mercuric ion–catalyzed exchange with ethyl vinyl ether;[56,57] 2) from allylic alcohols and vinyl ethers by acid catalyzed exchange;[58,59] 3) thermal elimination;[60,61] 4) *Wittig olefination* of allyl formates and carbonyl compounds;[62,63] and 5) *Tebbe olefination* of unsaturated esters;[64,65]. It is usually not necessary to isolate the allyl vinyl ethers, since they are prepared under conditions that will induce their rearrangement.

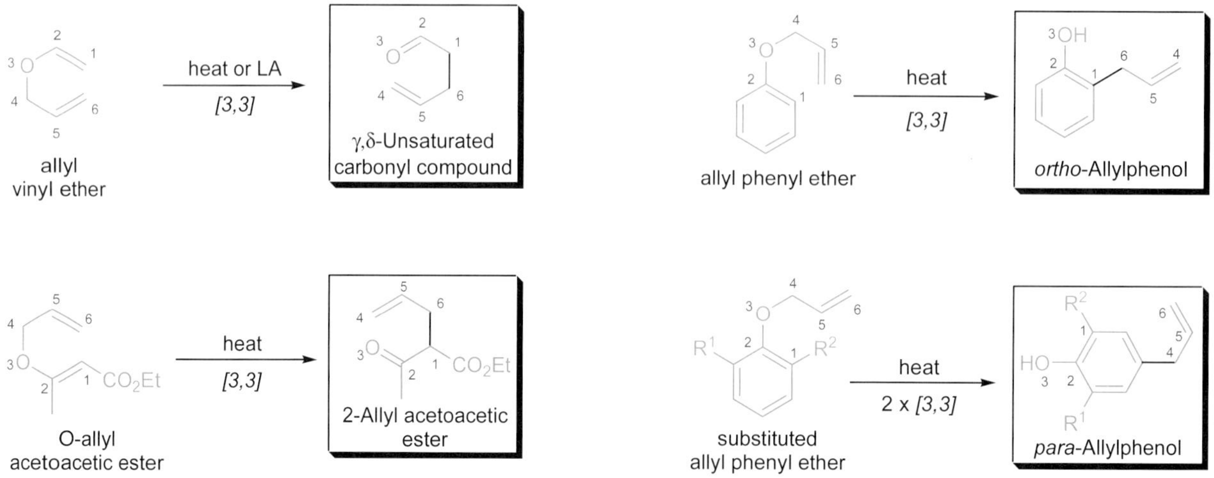

Mechanism:[66,67,6,68-70,18,20,31]

Mechanistically the reaction can be described as a suprafacial, concerted, nonsynchronous *[3,3]-sigmatropic rearrangement*. The *Claisen rearrangement* is a unimolecular process with activation parameters (negative entropy and volume of activation) that suggest a constrained transition state.[20] Studies revealed that the stereochemical information is transferred from the double bonds to the newly formed σ-bond. Based on this observation, an early six-membered chairlike transition state is believed to be involved. There are several transition state extremes possible. The actual transition state depends on the nature of substituents at the various positions of the starting allyl vinyl ether. If a chiral allylic alcohol is used to prepare the starting allyl vinyl ether, then the chirality is transferred to the products; the stereoselectivity will depend on the energy difference between diastereomeric chairlike transition states. In acyclic systems, the observed stereoselectivity can usually be rationalized by assuming that the unfavorable 1,3-diaxial interactions are minimized in the chairlike transition state with the large groups adopting an equatorial position. When the geometry of the ring or other steric effects preclude or disfavor a chairlike structure, the reaction can proceed through a boatlike transition state.[71,72]

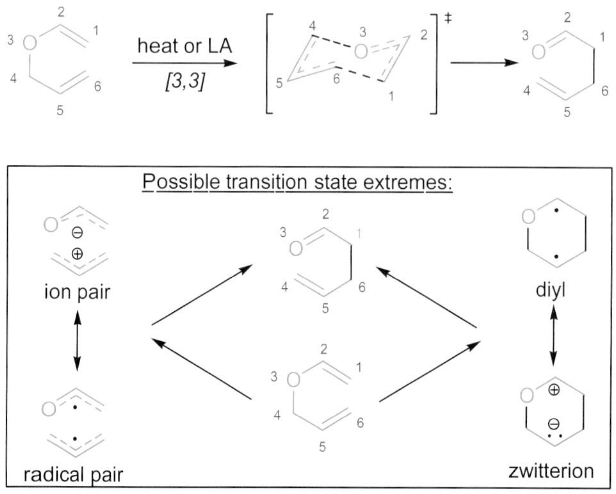

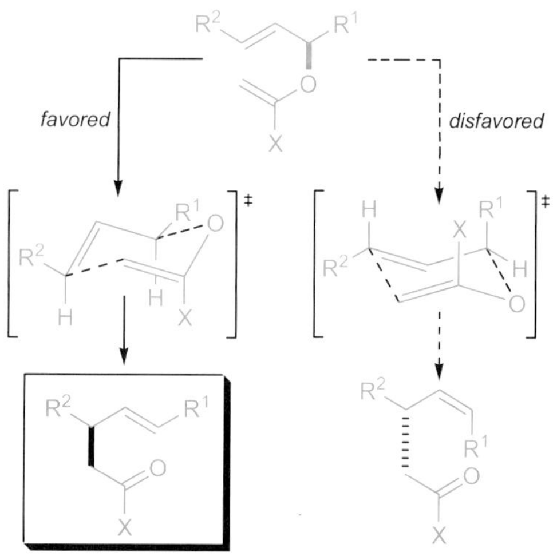

Synthetic Applications:

The asymmetric total synthesis of the putative structure of the cytotoxic diterpenoid (–)-sclerophytin A was realized *via* a *Tebbe-Claisen rearrangement* of a tricyclic lactone precursor in the laboratory of L.A. Paquette.[73] The tricyclic lactone was subjected to the *Tebbe methylenation* protocol to provide the allyl vinyl ether that was then heated to 130-140 °C in *p*-cymene to undergo the *Claisen rearrangement* in good yield.

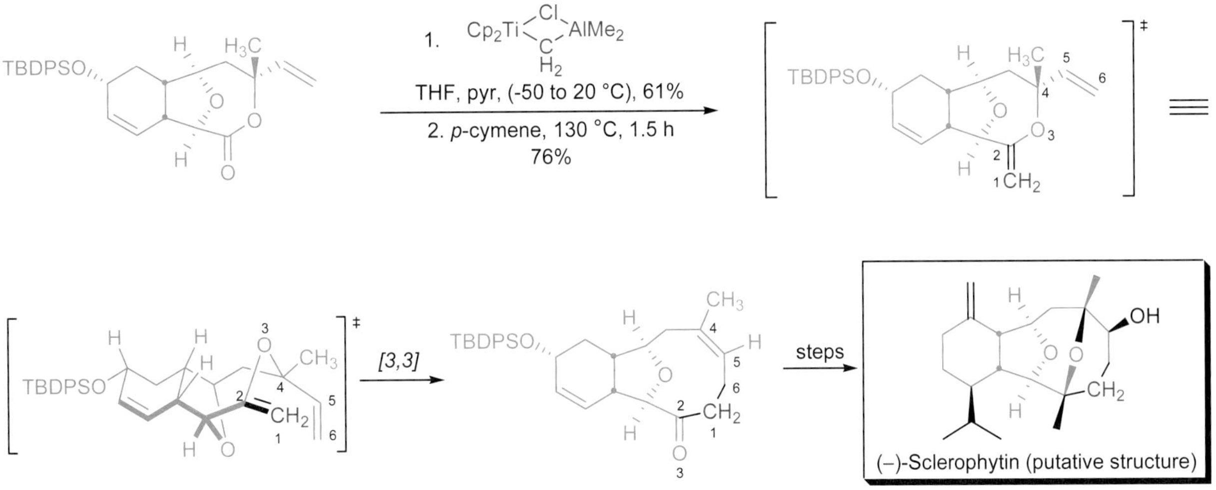

In the enantioselective total synthesis of (+)- and (–)-saudin, the core of the synthetic strategy was a *Lewis acid mediated stereoselective Claisen rearrangement* to establish the correct relative stereochemistry between the C1 and C6 stereocenters.[74] R.K. Boeckman Jr. and co-workers had to overcome the stereochemical preference of the thermal rearrangement by using a bidentate Lewis acid promoter ($TiCl_4$) that coordinated to both the oxygen of the vinyl ether and the ester. This coordination enforced a boatlike conformation for the existing six-membered ring in the transition state. The rearrangement itself took place *via* a chairlike transition state.

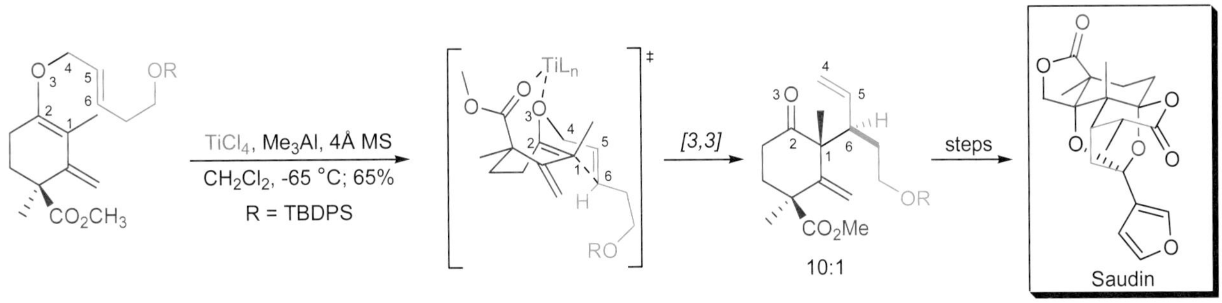

In K.C. Nicolaou's biomimetic synthesis of 1-*O*-methylforbesione, the construction of the 4-oxatricyclo[4.3.1.0]decan-2-one framework was achieved by using a *double Claisen rearrangement* that was followed by an *intramolecular Diels-Alder reaction*.[75] This one-pot biomimetic *double Claisen rearrangement/intramolecular Diels-Alder reaction cascade* afforded the natural product in 63% yield.

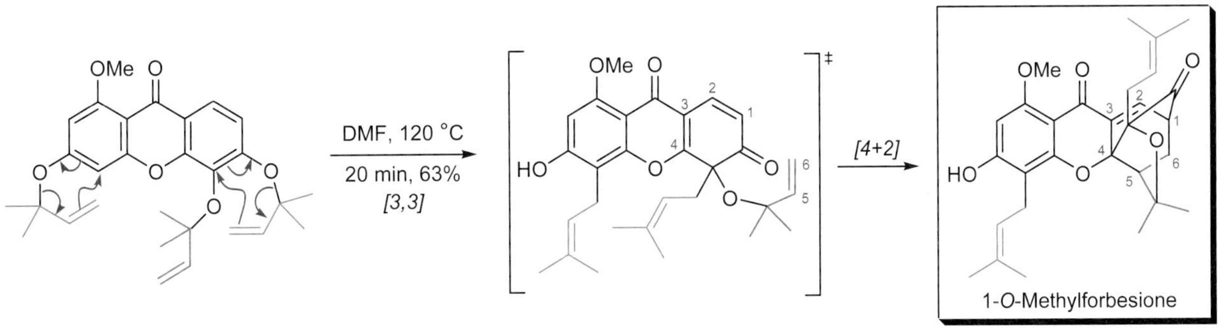

CLAISEN-IRELAND REARRANGEMENT

(References are on page 561)

Importance:

[*Seminal Publications*[1-8]; *Reviews*[9-20]; *Modifications & Improvements*[21-25]; *Theoretical Studies*[26]]

The *[3,3]-sigmatropic rearrangement* of *O*-trialkylsilylketene acetals to γ,δ-unsaturated carboxylic acids was first reported by R.E. Ireland in 1972, and it is referred to as the *Claisen-Ireland rearrangement* or *ester enolate Claisen rearrangement*.[6] Silylketene acetals are readily available by preparing the lithium enolate of allylic esters and trapping the enolate with a trialkylsilyl halide. The *Claisen-Ireland rearrangement* takes place under much milder conditions (room temperature and above) than the regular *Claisen rearrangement*. The ease of rearrangement is attributed to the highly nucleophilic enolate that generally accelerates sigmatropic processes (see *oxy-Cope rearrangement*). The reaction is very versatile, since it allows the assembly of highly functionalized structures. The conversion of a carbon-oxygen bond into a carbon-carbon bond affords a convenient way to assemble contiguous quaternary centers. Due to the highly ordered cyclic transition state, high levels of stereocontrol can be achieved. The high product stereoselectivities can be realized by efficient control of the ketene acetal geometry; deprotonation with LDA/THF leads to the kinetically favored (*Z*)-ester enolates, whereas the (*E*)-ester enolates are formed in the presence of THF/HMPA.[27,28] The rearrangement of the (*Z*)-ester enolates of (*E*)-allyl esters affords *anti*-products, whereas *syn*-products are obtained by the rearrangement of the (*E*)-ester enolates of (*E*)-allyl esters. The first asymmetric enantioselective version of the *Claisen-Ireland rearrangement* using a chiral boron reagent was reported by E.J. Corey et al.[21,23] It is also possible to achieve high levels of enantioselectivity by using chiral auxiliaries or chiral catalysts.[15,25]

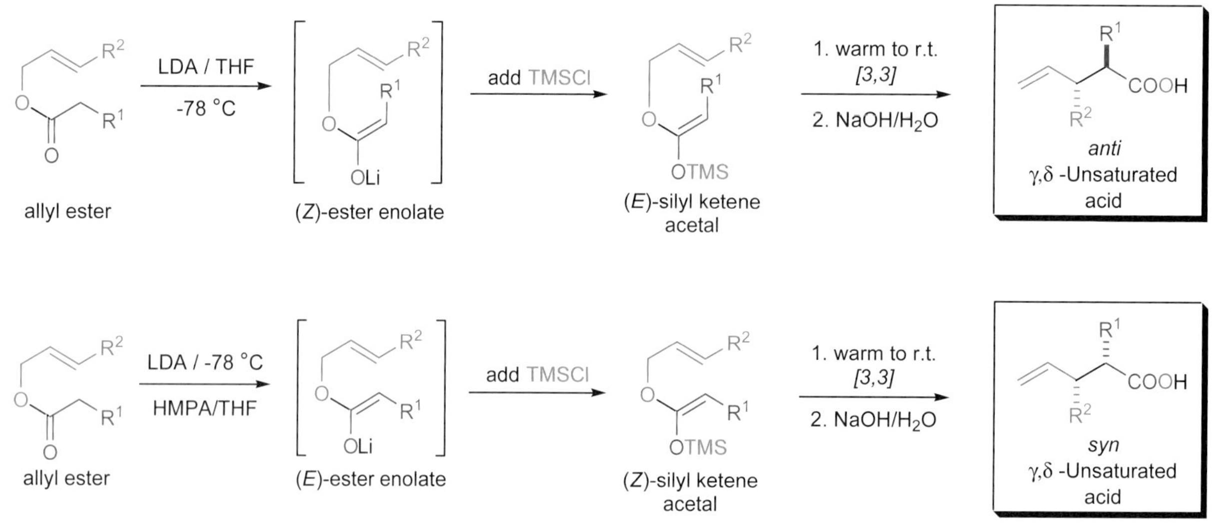

Mechanism:[29,27,28,25]

In acyclic systems the *Claisen-Ireland rearrangement* proceeds *via* a chairlike transition state (TS*). However, in cyclic systems conformational constraints can override the inherent preference for chairlike TS* and the boatlike TS* becomes dominant. One explanation for the preference of boatlike transition states in cyclic systems is the destabilizing steric interactions of the silyloxy substituent and the ring atoms in a chairlike TS*.

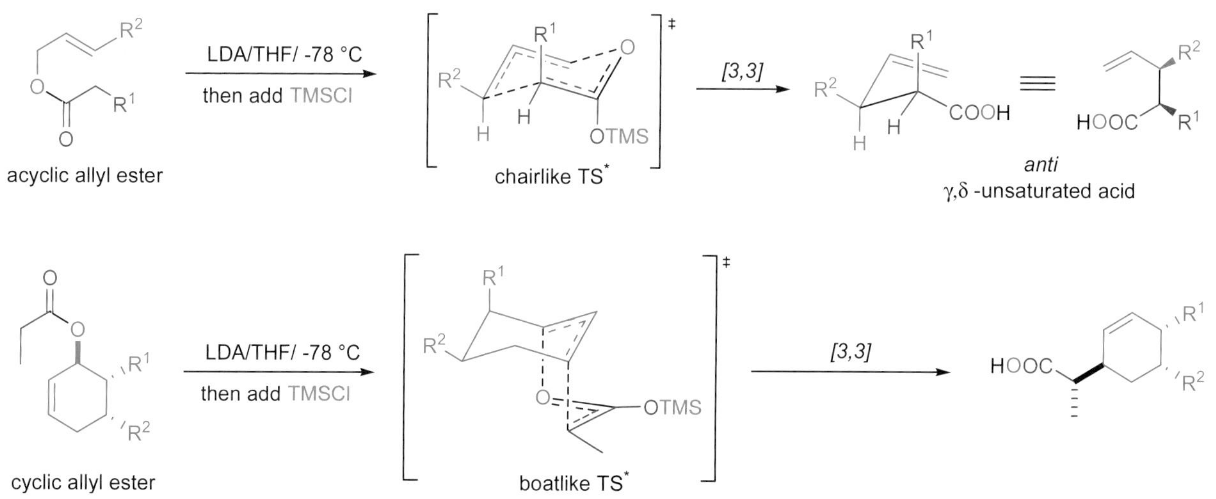

Synthetic Applications:

In the enantioselective total synthesis of β-lactone enzyme inhibitor (–)-ebelactone A and B, I. Paterson and co-workers constructed seven stereocenters and a trisubstituted alkene plus a very sensitive β-lactone ring.[30] The backbone of their strategy applied an *aldol reaction / Ireland-Claisen rearrangement* sequence and used minimal functional group manipulation. The *Ireland-Claisen rearrangement* was performed in the presence of an unprotected ketone moiety and set a precedent for this protocol. The diastereoselectivity was 96:4, indicating highly (*E*)-selective silylketene acetal formation.

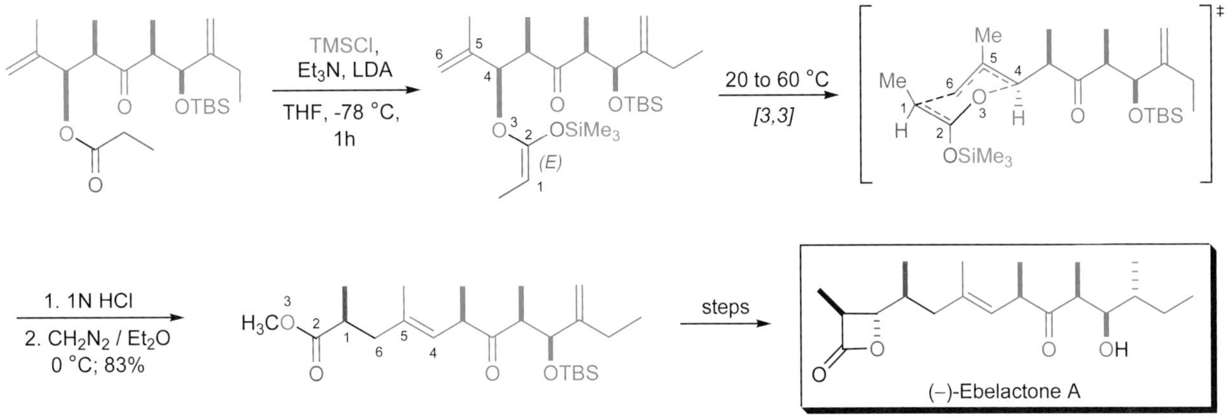

It was nearly a quarter century after the structure determination of aspidophytine that its first convergent enantioselective total synthesis was accomplished in the laboratory of E.J. Corey.[31] The *Claisen-Ireland rearrangement* was used to construct one of the key intermediates.

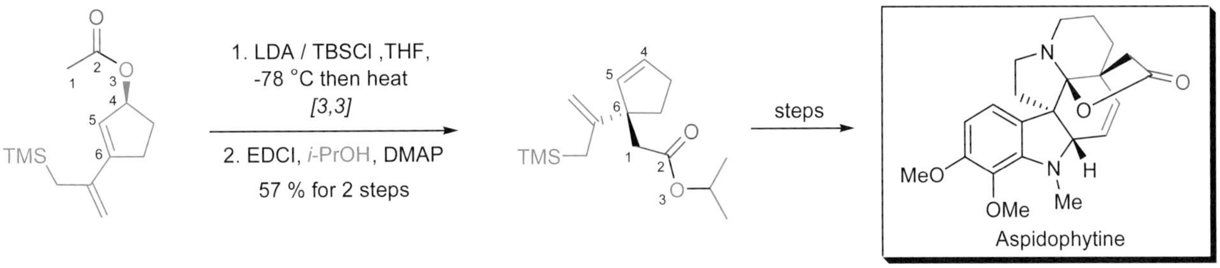

The first chemical synthesis of an optically active trichodiene, (–)-trichodiene involved a *Claisen-Ireland rearrangement* as the key step to connect the vicinal quaternary centers.[32] J.C. Gilbert and co-workers found that the rearrangement occurred with complete facial selectivity and excellent diastereoselectivity to afford an advanced intermediate that was directly converted to (–)-trichodiene.

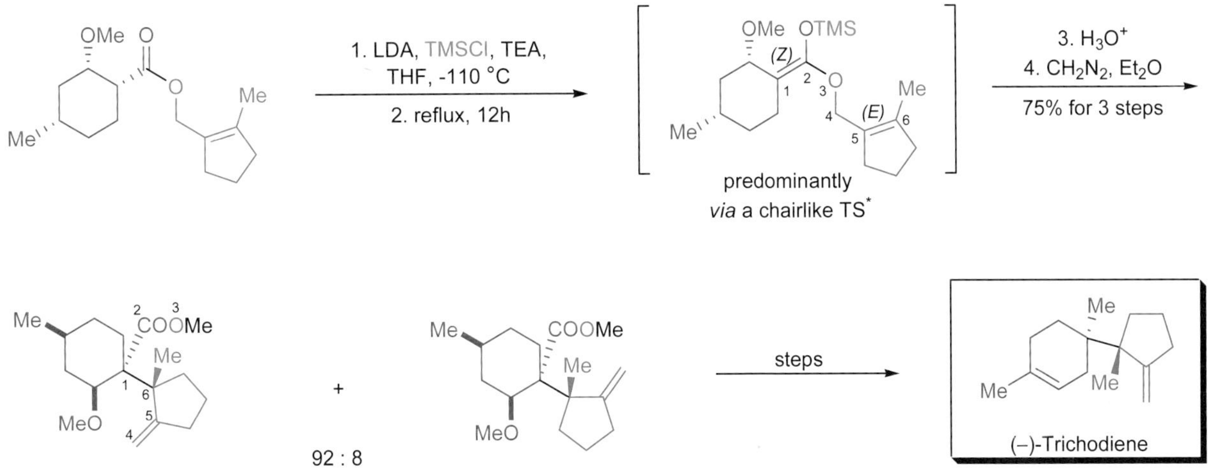

CLEMMENSEN REDUCTION

(References are on page 562)

Importance:

[*Seminal Publications*[1-3]; *Reviews*[4-9]; *Modifications & Improvements*[10-13]]

In 1913, E. Clemmensen reported that simple ketones and aldehydes were converted to the corresponding alkanes upon refluxing for several hours with 40% aqueous hydrochloric acid, amalgamated zinc (Zn/Hg), and a hydrophobic organic co-solvent such as toluene.[1] This method of converting a carbonyl group to the corresponding methylene group is known as the *Clemmensen reduction*. The original procedure is rather harsh so not surprisingly the *Clemmensen reduction* of acid-sensitive substrates and polyfunctional ketones is rarely successful in yielding the expected alkanes. The *Clemmensen reduction* has been widely used in synthesis and several modifications were developed to improve its synthetic utility by increasing the functional group tolerance. Yamamura and co-workers have developed a milder procedure which uses organic solvents (THF, Et_2O, Ac_2O, benzene) saturated with dry hydrogen-halides (HCl or HBr) and activated zinc dust at ice-bath temperature.[10-13] Compared to the original Clemmensen procedure[1] these modified reductions are complete within an hour at 0 °C and are appropriate for acid- and heat-sensitive compounds. Certain carbonyl compounds, however, have very low solubility in the usual solvents used for the *Clemmensen reduction*, so in these cases a second solvent (acetic acid, ethanol, or dioxane) is added to the reaction mixture to increase the solubility of the substrate and allow the reduction to take place. The *Clemmensen reduction* of polyfunctional ketones such as 1,2-, 1,3-, 1,4-, 1,5-diketones, α,β-unsaturated ketones and ketones with α-heteroatom substituents is less straightforward than the reduction of monofunctional substrates.[6] Usually complex mixtures are formed in these reactions, which contain a substantial amount of rearranged products.

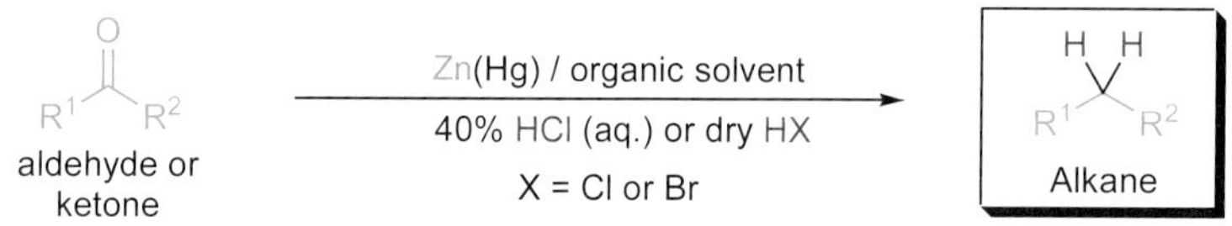

Mechanism: [14-20,6,21-25,8,26,27]

The mechanism of the *Clemmensen reduction* is not well understood. The lack of a unifying mechanism can be explained by the fact that the products formed in the various reductions are different when the reaction conditions (e.g., concentration of the acid, concentration of zinc in the amalgam) are changed. It was shown that the reduction occurs with zinc but not with other metals of comparable reduction potential. The early mechanistic papers came to the conclusion that the *Clemmensen reduction* occurs stepwise involving organozinc intermediates.[15-17] It was also established that simple aliphatic alcohols are not intermediates of these reductions, since they do not give alkanes under the usual Clemmensen conditions. However, allylic and benzylic alcohols undergo the *Clemmensen reduction*.[14,21] Currently, there are two proposed mechanisms for the *Clemmensen reduction*, and they are somewhat contradictory. In one of the mechanisms the rate determining step involves the attack of zinc and chloride ion on the carbonyl group[17] and the key intermediates are carbanions, whereas in the other heterogeneous process, the formation of a radical intermediate and then a zinc carbenoid species is proposed.[20,22]

Carbanionic mechanism:

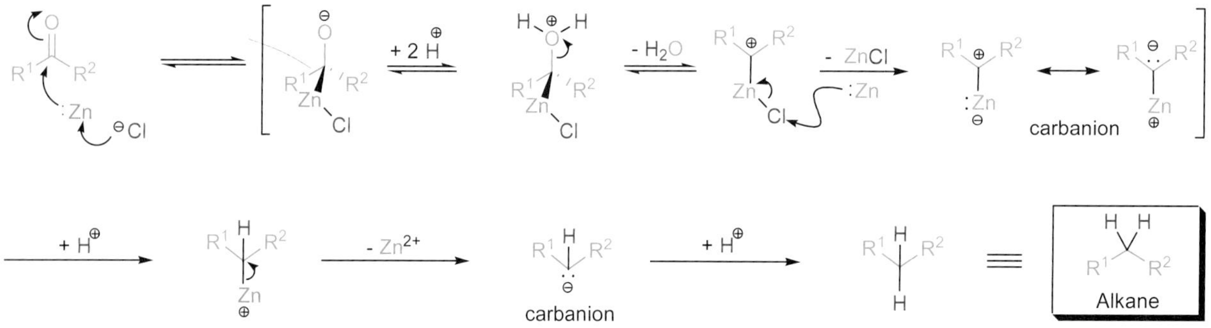

Carbenoid mechanism:

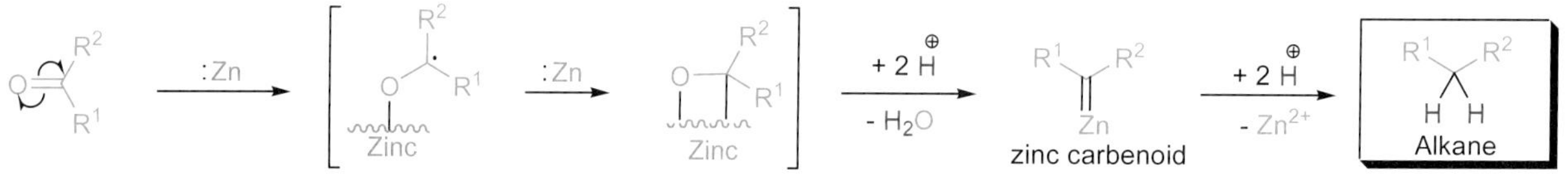

Synthetic Applications:

Numerous heterocyclic 1,3-dicarbonyl compounds possessing alkyl substituents at the electronegative 2-position exhibit interesting biological properties. The synthesis of these compounds is either cumbersome or calls for expensive starting materials. T. Kappe and co-workers have found a simple and effective method for the reduction of acyl substituted 1,3-dicarbonyl compounds to the corresponding alkyl derivatives.[28] For example, 3-acyl-4-hydroxy-2(1H)-quinolones and 3-acyl-4-hydroxy-6-methypyran-2-ones were reduced in good yields to 3-alkyl-4-hydroxy-2(1H)-quinolinones and 3-alkyl-4-hydroxy-6-methylpyran-2-ones, respectively, using zinc powder in acetic acid/hydrochloric acid.

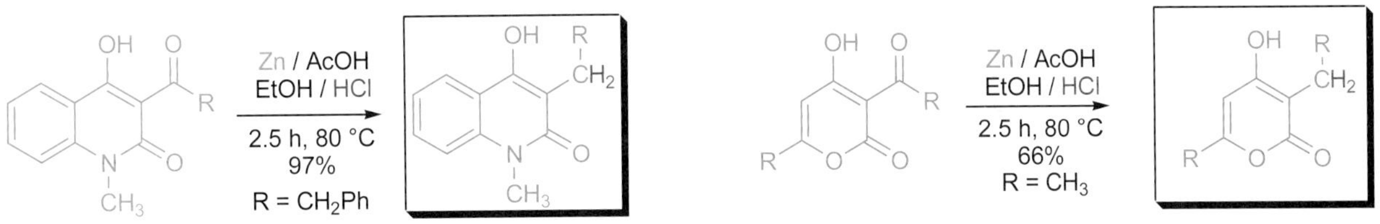

During the enantioselective total synthesis of denrobatid alkaloid (–)-pumiliotoxin C by C. Kibayashi et al., an aqueous *acylnitroso Diels-Alder cycloaddition* was used as the key step.[29] In the endgame of the total synthesis, the *cis*-fused decahydroquinolone was subjected to the *Clemmensen reduction* conditions to give a 2:1 epimeric mixture of deoxygenated products in 57% yield. Subsequent debenzylation converted the major isomer into 5-*epi*-pumiliotoxin C.

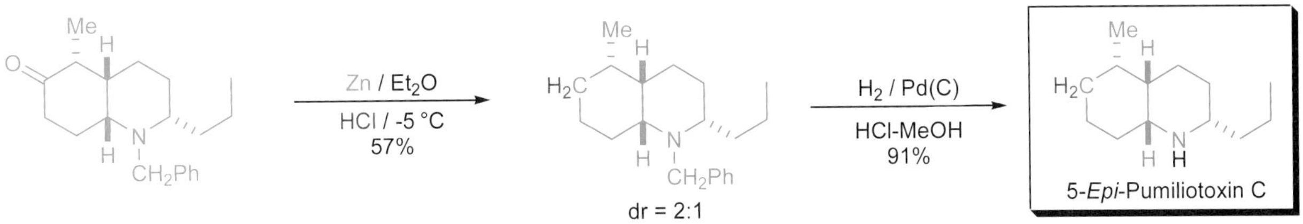

S.M. Weinreb and co-workers were surprised to find that the convergent stereoselective synthesis of marine alkaloid lepadiformine resulted in a product that gave a totally different NMR spectra than the natural product.[30] This finding led to the revision of the proposed structure of lepadiformine. In the final stages of the synthesis, they exposed a tricyclic piperidone intermediate to *Clemmensen conditions* to remove the ketone functionality. Under these conditions the otherwise minor elimination product (alkene) was formed predominantly; however, it was possible to hydrogenate the double bond to give the desired alkane.

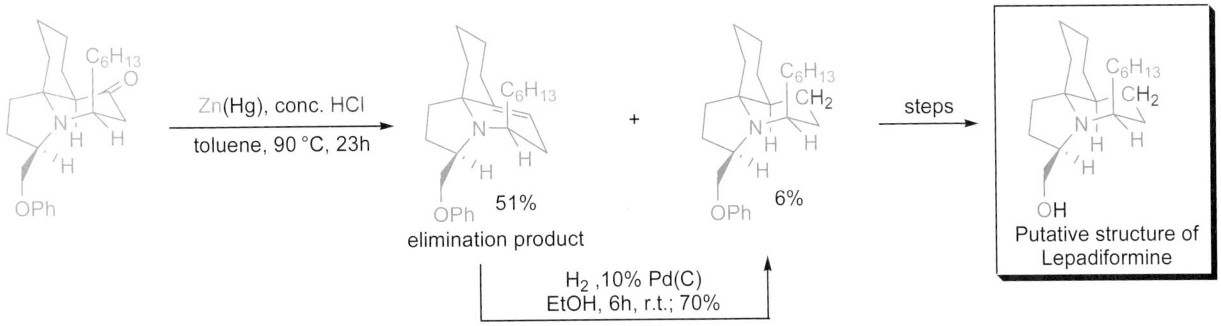

In the laboratory of F.J.C. Martins the synthesis of novel tetracyclic undecane derivatives was undertaken. In one of the synthetic sequences the *Clemmensen reduction* was used to remove a ketone functionality in good yield.[31]

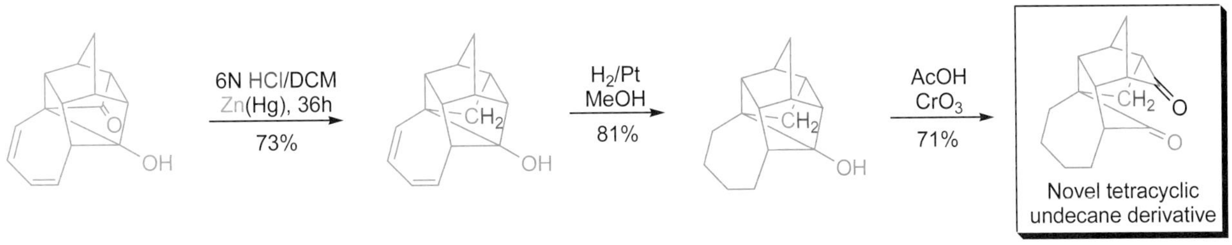

COMBES QUINOLINE SYNTHESIS

(References are on page 563)

Importance:

[*Seminal Publication*[1]; *Reviews*[2-4]; *Modifications & Improvements*[5]]

The formation of quinolines and benzoquinolines by the condensation of primary aryl amines with β-diketones followed by an acid catalyzed ring closure of the Schiff base intermediate is known as the *Combes quinoline synthesis*. The closely related reaction of primary aryl amines with β-ketoesters followed by the cyclization of the Schiff base intermediate is called the *Conrad-Limpach reaction* and it gives 4-hydroxyquinolines as products.[6-8]

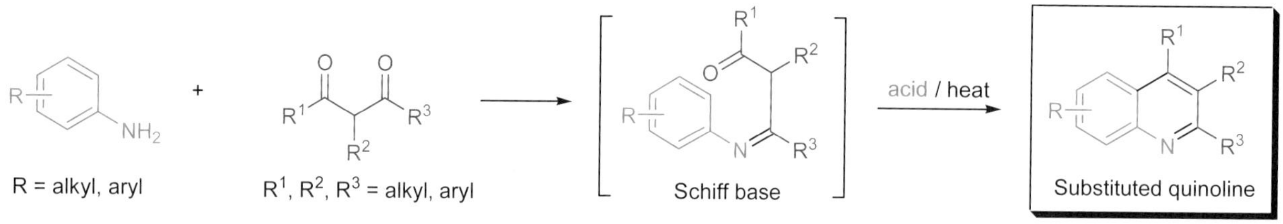

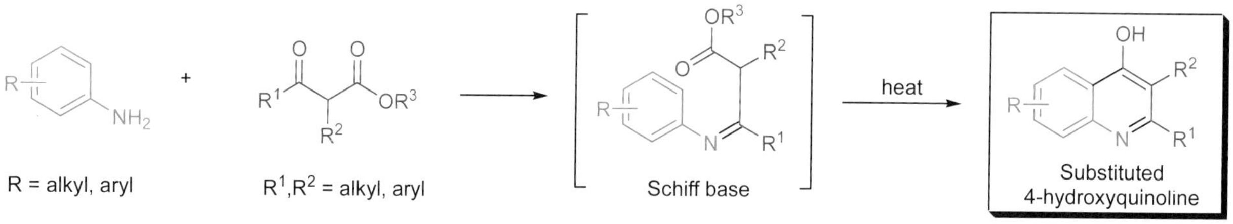

Mechanism: [9]

The first step in the *Combes reaction* is the acid-catalyzed condensation of the diketone with the aromatic amine to form a Schiff base (imine), which then isomerizes to the corresponding enamine. In the second step, the carbonyl oxygen atom of the enamine is protonated to give a carbocation that undergoes an electrophilic aromatic substitution. Subsequent proton transfer, elimination of water and deprotonation of the ring nitrogen atom gives rise to the neutral substituted quinoline system.

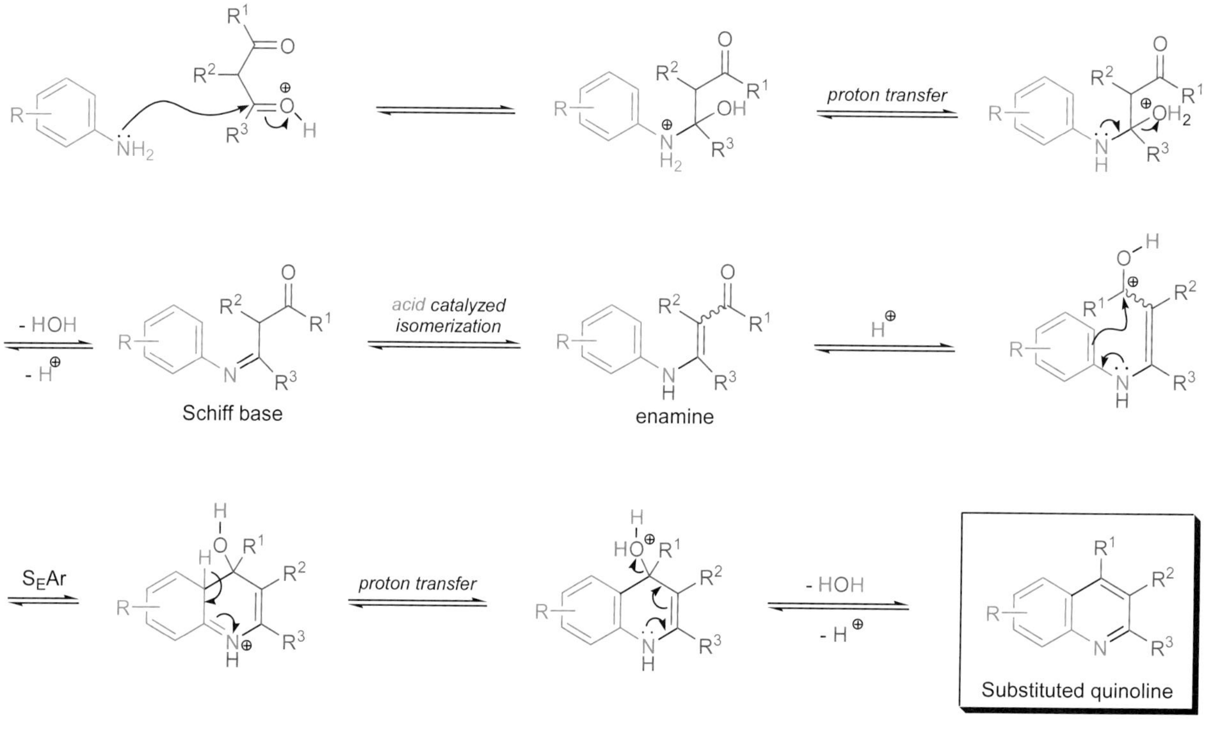

Synthetic Applications:

In the laboratory of S. Gupta, the synthesis of novel heterocyclic ring systems was accomplished utilizing the *Combes reaction.*[10] The condensation of 1-naphthylamine with 2-acylindan-1,3-diones produced the corresponding anils in good yield. The anils were cyclodehydrated to benz[*h*]indeno[2,1-*c*]quinoline-7-ones in the presence of polyphosphoric acid. Subsequent *Wolff-Kishner reduction* gave rise to the novel 7*H*-benzo[*h*]indeno[2,1-*c*]quinolines.

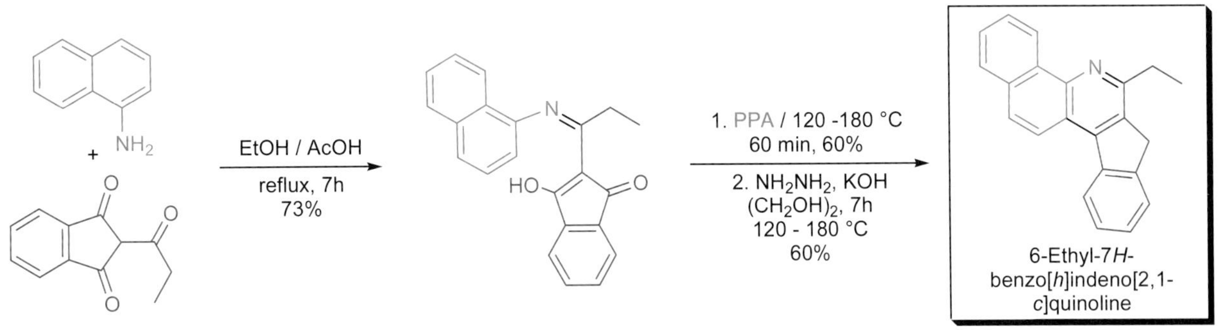

During a study of the reactivity of 4(7)-aminobenzimidazole as a bidentate nucleophile, C. Avendano and co-workers obtained 7*H*-imidazo[1,5,4-*e*,*f*][1,5]benzodiazepine-4-ones by using β-ketoesters as electrophiles.[11] The reactions were regioselective and took place with equimolar amounts of the β-ketoesters without the use of a catalyst. Isolated yields were around 50%. However, when the benzimidazole was treated with 2,4-pentanedione in a 1:5 ratio in the presence of an acid catalyst, an 1*H*-imidazo[4,5-*h*]quinoline was formed and no traces of imidazobenzodiazepines were observed.

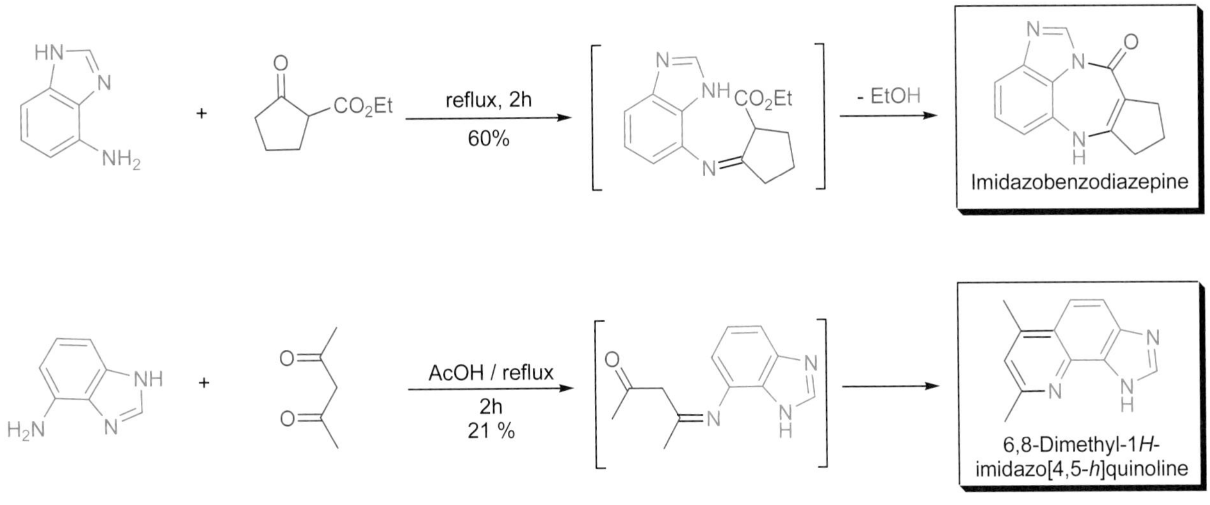

In the attempted synthesis of twisted polycycle 1,2,3,4-tetraphenylfluoreno[1,9-*gh*]quinoline, R.A. Pascal Jr. et al. used the *Combes quinoline synthesis* to assemble the azaaceanthrene core.[12] Oxidation with DDQ was followed by a Diels-Alder reaction with tetracyclone (tetraphenylcyclopentadienone) to afford the corresponding cycloadduct. However, the last decarbonylation step of the sequence failed to work even under forcing conditions, presumably due to steric hindrance.

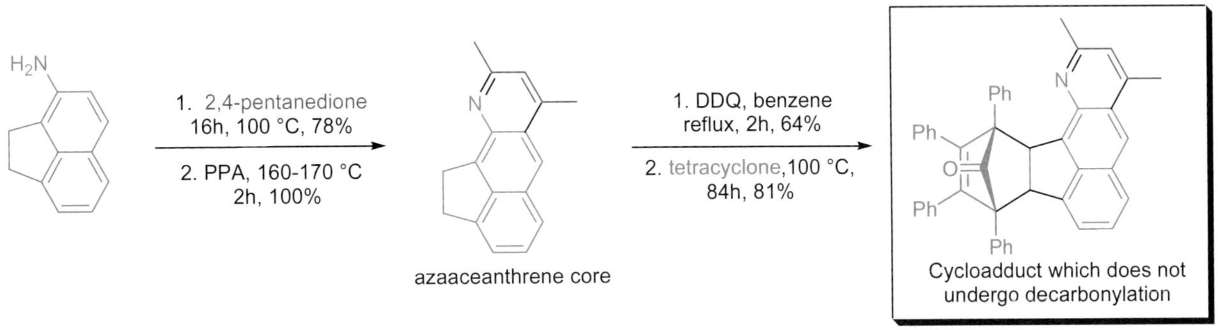

COPE ELIMINATION / COPE REACTION

(References are on page 563)

Importance:

[*Seminal Publications*[1-3]; *Reviews*[4-6]; *Modifcations & Improvements*[7-11]; *Theoretical Studies*[12,13]]

In 1949, A.C. Cope and co-workers discovered that by heating trialkylamine-*N*-oxides having hydrogens in the β-position, an olefin and *N,N*-dialkylhydroxylamine are formed.[1] The transformation involving the stereoselective *syn* elimination of tertiary amine oxides is now referred to as the *Cope elimination* or *Cope reaction*. The substrates, tertiary amine oxides, are easily prepared by the oxidation of the corresponding tertiary amine with hydrogen peroxide or peroxycarboxylic acids such as *m*CPBA. Isolation of the *N*-oxides is usually not necessary; the amine is mixed with the oxidizing agent and heated. Amine oxides are very polar compounds and the oxygen serves as a base to remove the β-hydrogen atom via a *syn* conformation. The synthetic utility of the *Cope elimination* is comparable to the *Hofmann elimination* of quaternary ammonium hydroxides, but it takes place at lower temperatures (100-150 °C). The *Cope elimination* is almost free of side reactions due to the intramolecular nature of the elimination (the base is part of the molecule). However, in certain cases, the product alkene may isomerize[14] to the more stable conjugated system, and allyl- or benzyl migration[2] is sometimes observed to give O-allyl or benzyl-substituted hydroxylamines. Cyclic amine oxides (5, 7-10-membered rings, where the nitrogen is part of the ring) can also be pyrolysed but with 6-membered rings the reaction is usually low-yielding or does not occur.[15,16] The direction of the *Cope elimination* is governed almost entirely by the number of hydrogen atoms at the various β-positions, and therefore there is no preference for the formation of the least substituted alkene unlike in the *Hofmann elimination* reaction. Upon pyrolysis, *N*-cycloalkyl-substituted amine oxides give mainly the thermodynamically more stable endocyclic olefins. Cyclohexyl derivatives, however, form predominantly exocyclic olefins, since the formation of the endocyclic double bond would require the cyclohexane ring to be almost planar in the transition state.

Acyclic systems:

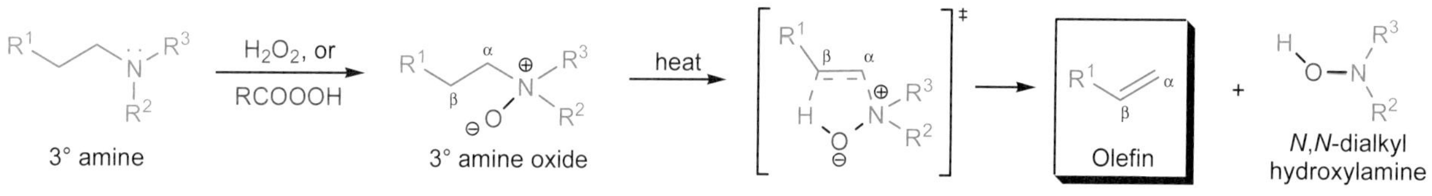

Cyclic systems:

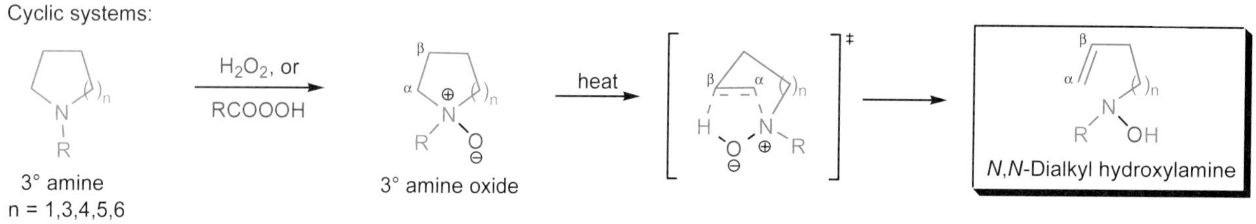

Mechanism: [17,16,5,18-21]

The *Cope elimination* is a stereoselective *syn* elimination and the mechanism involves a planar 5-membered cyclic transition state. There is strong resemblance to the mechanism of *ester pyrolysis* and the *Chugaev elimination*. The first evidence of the stereochemistry of the elimination was the thermal decomposition of the *threo* and *erythro* derivatives of *N,N*-dimethyl-2-amino-3-phenylbutane.[17] The *erythro* isomer gives predominantly the *(Z)*-alkene (20:1), while the *threo* isomer forms the *(E)*-olefin almost exclusively (400:1). Two decades later deuterium-labeling evidence confirmed the mechanism of the *Cope elimination* to be 100% *syn*.[18]

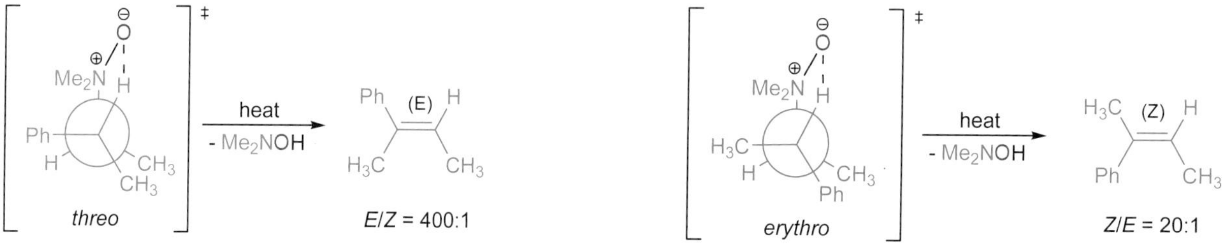

Synthetic Applications:

In their search for conformationally biased mimics of mannopyranosylamines, A. Vasella and co-workers planned to synthesize compounds that would inhibit β-mannosidases.[22] In order to construct the bicyclo[3.1.0]hexane framework, a five-membered *O*-silylated *N,N*-dimethyl-amino alcohol was synthesized. Oxidation of the tertiary amine with *m*CPBA yielded 83% of the *N*-oxide, which was subsequently subjected to the *Cope elimination* to give 69% of the desired benzyl enol ether. Cyclopropanation of this enol ether gave rise to the highly functionalized bicyclic skeleton.

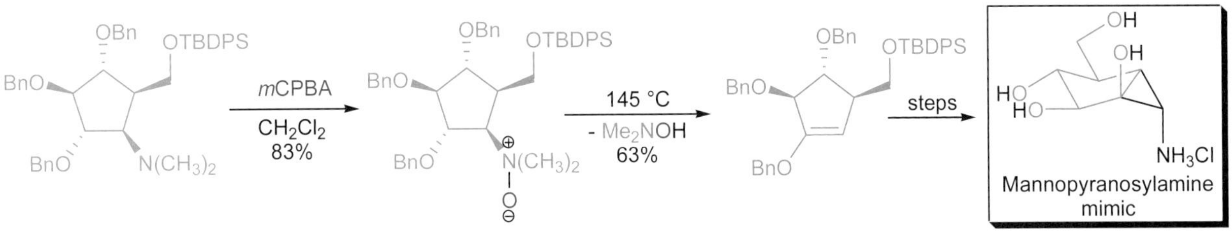

A convenient synthesis of secondary hydroxylamines using secondary amines as starting material was developed in the laboratory of I.A. O'Neil.[10] Secondary amines were treated with a Michael acceptor such as acrylonitrile in methanol to give tertiary β-cyanoethyl amines in excellent yield. These tertiary amines were then oxidized with *m*CPBA to give the corresponding *N*-oxides, which underwent the *Cope elimination in situ* to generate the hydroxylamine in excellent yield. The great advantage of this method is that it works for both cyclic and acyclic systems.

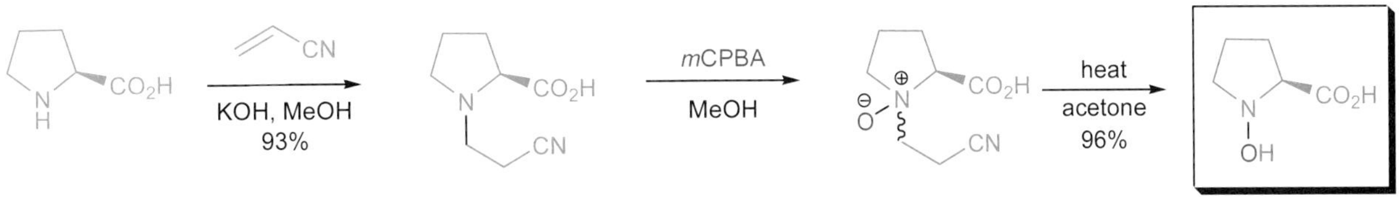

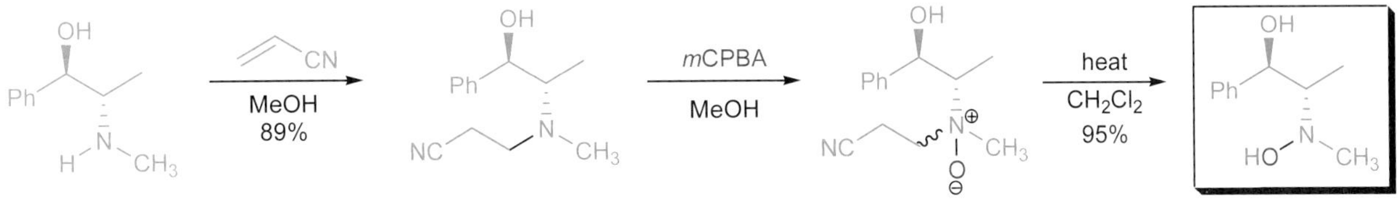

A new enantiospecific synthesis of taxoid intermediate (1*S*)-10-methylenecamphor was described by A.G. Martinez utilizing the *Cope elimination* to generate the vinyl group at the bridgehead norbornane position.[23] (1*R*)-3,3-Dimethyl-2-methylenenorbornan-1-ol was treated with Eschenmoser's salt, to initiate a tandem *electrophilic carbon-carbon double bond addition/Wagner-Meerwein rearrangement* to give (1*S*)-10-dimethylaminomethylcamphor. This tertiary amine was oxidized to the corresponding *N*-oxide in 95% yield, and subsequent *Cope elimination* gave the desired taxoid intermediate in 80% yield.

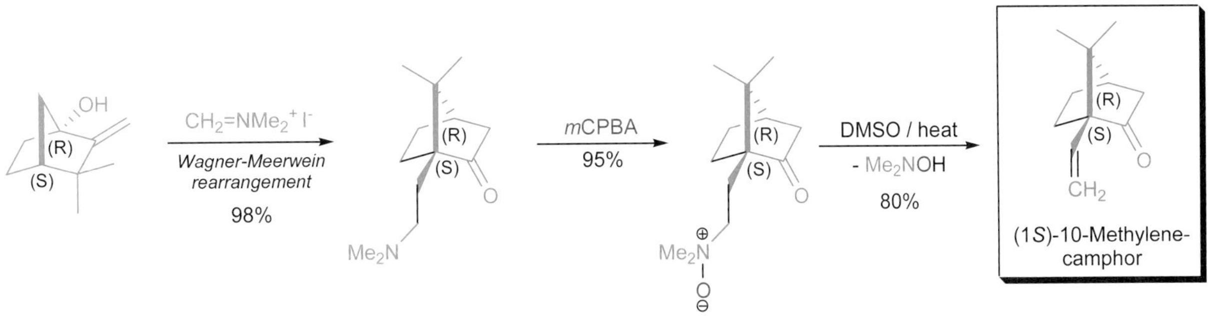

COPE REARRANGEMENT

(References are on page 564)

Importance:

[*Seminal Publications*[1]; *Reviews*[2-14]; *Theoretical Studies*[15-28]]

In 1940, A.C. Cope observed the rearrangement of (1-methylpropenyl)allylcyanoacetate into the isomeric (1,2-dimethyl)-4-pentylidinecyanoacetate upon distillation, and he recognized that this rearrangement was similar in type to the known *Claisen rearrangement.*[1] The thermal *[3,3]-sigmatropic rearrangement* of 1,5-dienes to the isomeric 1,5-dienes is called the *Cope rearrangement,* and it can only be detected when the 1,5-diene substrate is substituted. The rearrangement is reversible because there are no changes in the number or types of bonds, and the position of the equilibrium is determined by the relative stability of the starting material and the product. When the product is stabilized by conjugation or the resulting double bond is more highly substituted, the equilibrium will be shifted toward the formation of the product. The reaction is both stereospecific and stereoselective as a result of a cyclic chairlike transition state. The typical temperature required to induce *Cope rearrangement* in acyclic dienes is 150-260 °C. The required temperature is significantly lower (room temp. or below) when: 1) the dienes are substituted in positions C3 or C4; 2) the dienes are cyclic and ring strain is relieved; or 3) the *Cope rearrangement* is catalyzed by transition metal complexes.[4] The *Cope rearrangement* of strained 1,2-divinyl cycloalkanes (cyclopropane and cyclobutane) gives convenient access to synthetically useful seven- and eight-membered carbocycles.

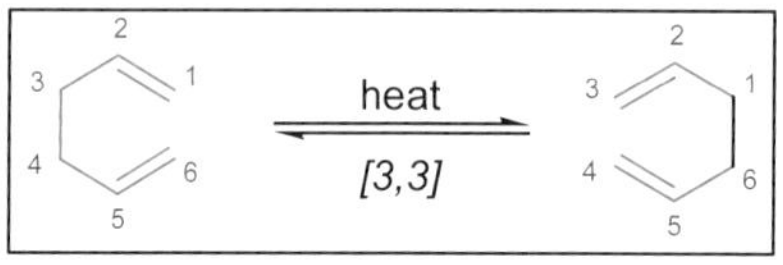

Mechanism:[29-43]

Soon after its discovery, the *Cope rearrangement* was investigated in great detail in order to establish its mechanism. In the classical sense, *[3,3] sigmatropic rearrangements* do not have observable intermediates. Therefore, in the 1960s these rearrangements were dubbed "no mechanism reactions".[29] The *Cope rearrangement* predominantly proceeds *via* a chairlike transition state where there is minimal steric interaction between the substituents.[29,32] The exact nature of the transition state depends on the substituents and varies between two extreme forms: from two independent allyl radicals to a cyclohexane-1,4-diradical depending on whether the bond making or bond breaking is more advanced. In most cases the reactions are concerted with a relatively late transition state where the bond between C1 and C6 is well-developed.

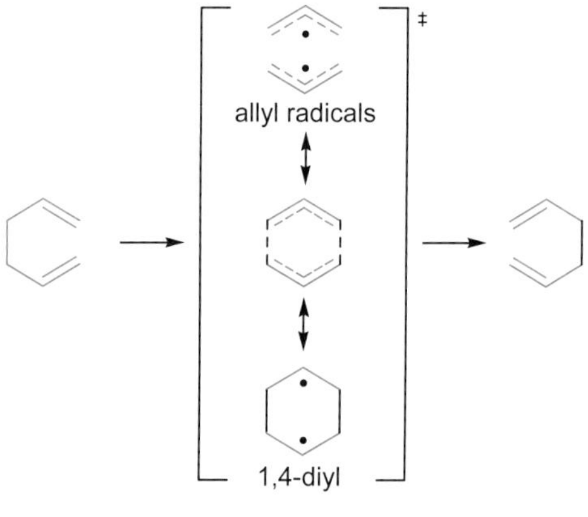

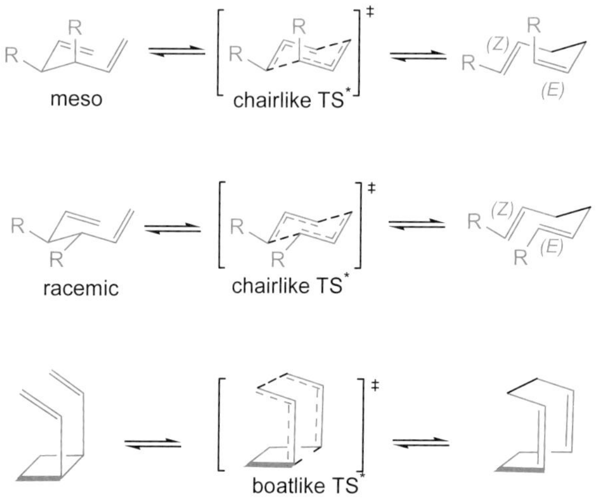

Synthetic Applications:

The enantioselective total synthesis of (+)- and (–)-asteriscanolide was accomplished in the laboratory of M.L. Snapper utilizing a sequential *intramolecular cyclobutadiene cycloaddition*, *ring-opening metathesis* and *Cope rearrangement*.[44] The key cycloadduct was treated with Grubbs's catalyst under an ethylene atmosphere to generate a divinylcyclobutane intermediate in a selective *ring-opening metathesis* of a strained trisubstituted cyclobutene. The divinylcyclobutane intermediate subsequently underwent a facile *Cope rearrangement* under mild conditions to afford the 8-membered carbocycle of (+)-asteriscanolide.

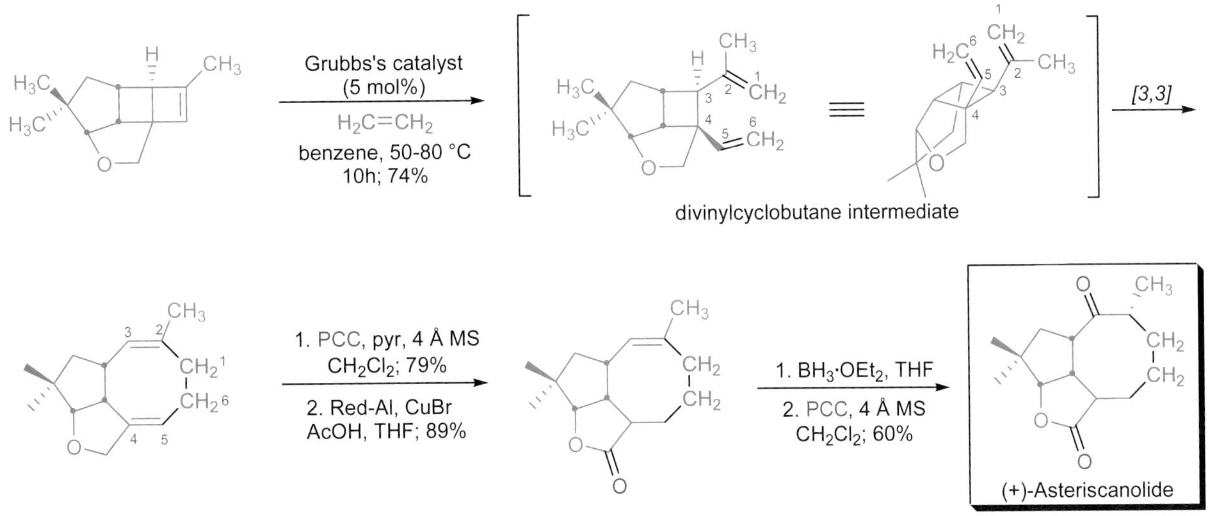

The Cope rearrangement of a divinylcyclopropane intermediate was the key step in the total synthesis of (±)-tremulenolide A by H.M.L. Davies et al.[45] The divinylcyclopropane intermediate was obtained by a Rh-*catalyzed stereoselective cyclopropanation* of a hexadiene. Usually the *Cope rearrangement* of divinylcyclopropanes occurs at or below room temperature, In this case, a congested boat transition state was required for the rearrangement so forcing conditions were necessary. The product cycloheptadiene was obtained by Kugelrohr distillation at 140 °C as a single regioisomer in 49% yield.

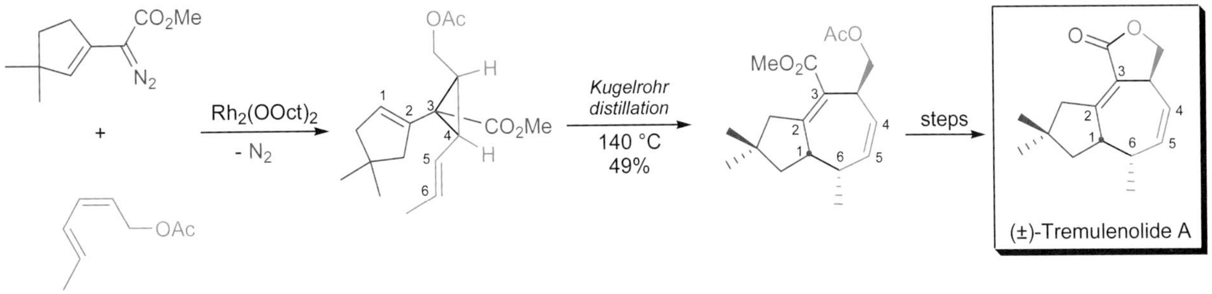

A tricyclic ring system containing all the stereogenic centers of the nonaromatic portion of (–)-morphine was prepared by T. Hudlicky and co-workers using an *intramolecular Diels-Alder cycloaddition* followed by a *Cope rearrangement*.[46] Interestingly, the initial Diels-Alder cycloadduct did not undergo the *Cope rearrangement* even under forcing conditions. However, when the hydroxyl group was oxidized to the corresponding ketone, the *[3,3]*-sigmatropic shift took place at 250 °C in a sealed tube. The driving force of the reaction was the formation of an α,β-unsaturated ketone.

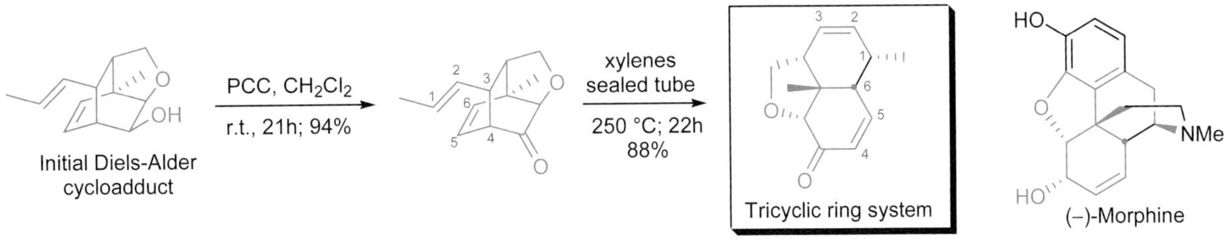

COREY-BAKSHI-SHIBATA REDUCTION (CBS REDUCTION)

(References are on page 565)

Importance:

[*Seminal Publications*[1-4]; *Reviews*[5-12]; *Modifications & Improvements*[13,14]; *Theoretical Studies*[15-23]]

In 1981, S. Itsuno and co-workers were the first to report that stoichiometric mixtures of chiral amino alcohols and borane-tetrahydrofuran complex (BH_3·THF) reduced achiral ketones to the corresponding chiral secondary alcohols enantioselectively and in high yield.[1] Several years later, E.J. Corey and co-workers showed that the reaction of borane (BH_3) and chiral amino alcohols leads to the formation of oxazaborolidines, which were found to catalyze the rapid and highly enantioselective reduction of achiral ketones in the presence of BH_3·THF.[2,3] The enantioselective reduction of ketones using catalytic oxazaborolidine is called the *Corey-Bakshi-Shibata reduction* or *CBS reduction*. Research in the Corey group showed that the methyl-substituted oxazaborolidines (*B*-Me) were more stable and easier to prepare than the extremely air and moisture-sensitive original *B*-H analogs. The systematic study of oxazaborolidine-catalyzed reductions revealed that high enantiomeric excess (*ee*) is achieved when the oxazaborolidine has a rigid bicyclic (proline based) or tricyclic structure. More flexible ring systems resulted in lower enantioselectivities. The advantages of the CBS catalysts are: 1) ease of preparation; 2) air and moisture stability; 3) short reaction times (high catalyst turnover); 4) high enantioselectivity; 5) typically high yields; 6) recovery of catalyst precursor by precipitation as the HCl salt; and 7) prediction of the absolute configuration from the relative steric bulk of the two substituents attached to the carbonyl group.

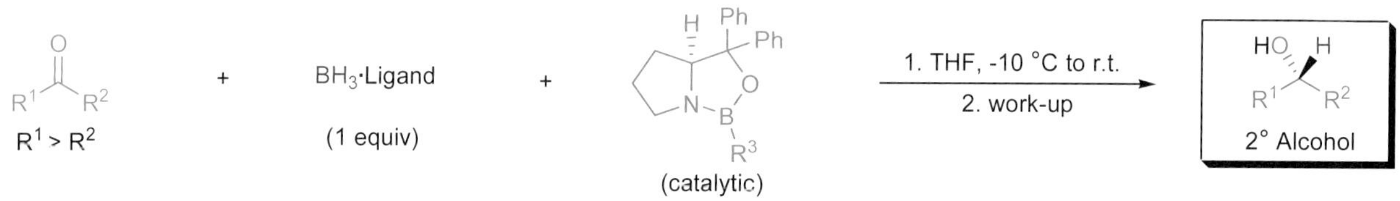

R^{1-2} = alkyl, aryl; <u>Ligand</u>: THF, Me_2S, 1,4-thioxane, diethylaniline; R^3 = H, alkyl

Mechanism: [2-4,24-27]

The first step of the mechanism is the coordination of BH_3 (Lewis acid) to the tertiary nitrogen atom (Lewis base) of the CBS catalyst from the α-face.[27] This coordination enhances the Lewis acidity of the endocyclic boron atom and activates the BH_3 to become a strong hydride donor. The CBS catalyst-borane complex then binds to the ketone at the sterically more accessible lone pair (the lone pair closer to the smaller substituent) *via* the endocyclic boron atom. At this point the ketone and the coordinated borane in the vicinal position are *cis* to each other and the unfavorable steric interactions between the ketone and the CBS catalyst are minimal. The face-selective hydride transfer takes place *via* a six-membered transition state.[24,26] The last step (regeneration of the catalyst) may take place by two different pathways (**Path I** or **II**).[25,19,21]

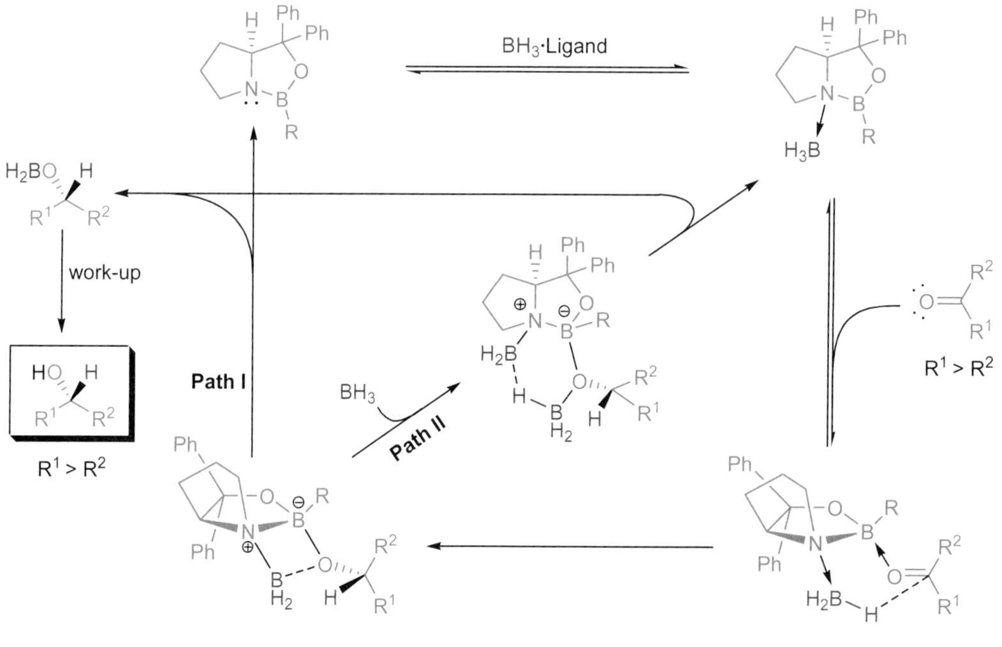

Synthetic Applications:

The asymmetric total synthesis of prostaglandin E_1 utilizing a *two-component coupling process* was achieved in the laboratory of B.W. Spur.[28] The hydroxylated side-chain of the target was prepared *via* the *catalytic asymmetric reduction* of a γ-iodo vinyl ketone with catecholborane in the presence of Corey's CBS catalyst. The reduction proceeded in 95% yield and >96% *ee*. The best results were obtained at low temperature and with the use of the *B-n*-butyl catalyst. The *B*-methyl catalyst afforded lower enantiomeric excess and at higher temperatures the *ee* dropped due to competing non-catalyzed reduction.

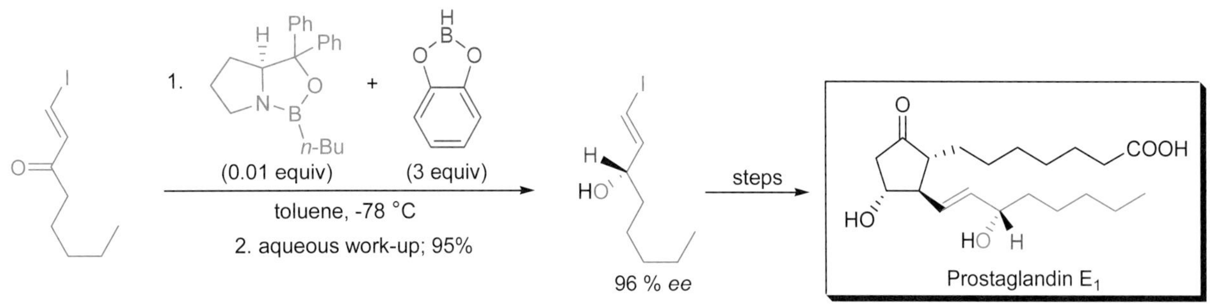

E.J. Corey and co-workers synthesized the cdc25A protein phosphatase inhibitor dysidiolide enantioselectively.[29] In the last phase of the total synthesis, the secondary alcohol functionality of the side-chain was established with a highly diastereoselective *oxazaborolidine-catalyzed reduction* using borane-dimethylsulfide complex in the presence of the (*S*)-*B*-methyl CBS catalyst. Finally, a *photochemical oxidation* generated the γ-hydroxybutenolide functionality. This total synthesis confirmed the absolute stereochemistry of dysidiolide.

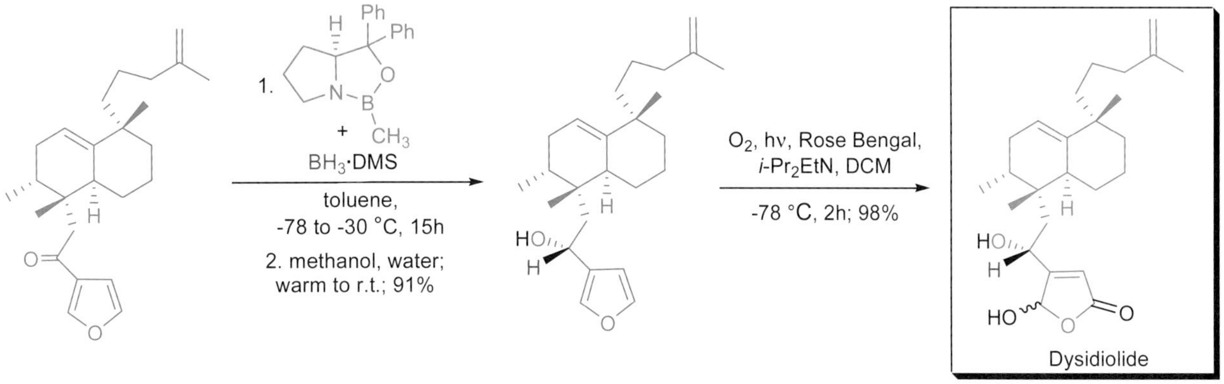

In the final stages of the total synthesis of okadaic acid by C.J. Forsyth et al., the central 1,6-dioxaspiro[4,5]decane ring system was introduced by the enantioselective reduction of the C16 carbonyl group using (*S*)-CBS/BH_3, followed by *acid-catalyzed spiroketalization*.[30]

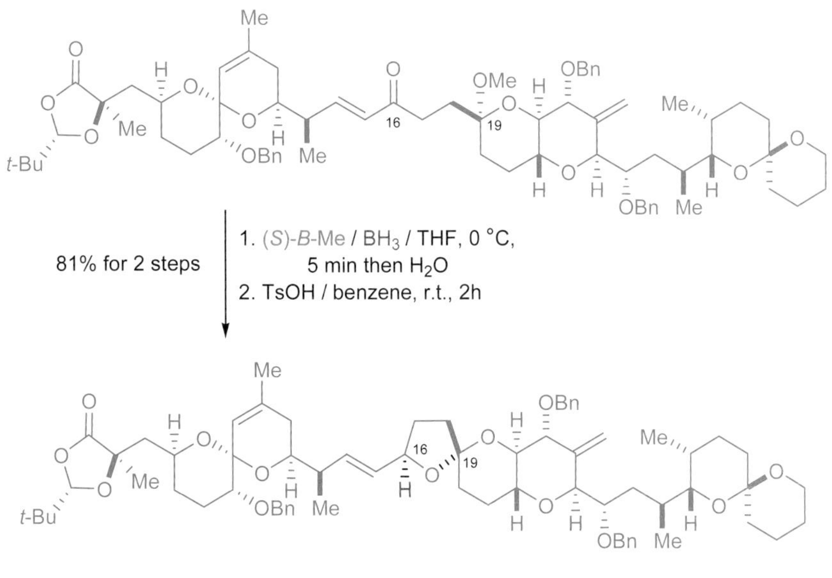

CORETY-CHAYKOVSKY EPOXIDATION AND CYCLOPROPANATION

(References are on page 565)

Importance:

[*Seminal Publications*[1,2]; *Reviews*[3-11]; *Modifications & Improvements*[12-14]; *Theoretical Studies*[15-17]]

In 1962, E.J. Corey and M. Chaykovsky deprotonated trimethylsulfoxonium halides using powdered sodium hydride under nitrogen at room temperature to form a reactive compound, dimethylsulfoxonium methylide (**I**).[1] When simple aldehydes and ketones were mixed with **I**, the formation of epoxides was observed. Likewise, the reaction of dimethylsulfonium methylide (**II**) with aldehydes and ketones also resulted in epoxide formation.[2] Compounds **I** and **II** are both sulfur ylides and are prepared by the deprotonation of the corresponding sulfonium salts. The preparation of epoxides (oxiranes) from aldehydes and ketones using sulfur ylides is known as the *Corey-Chaykovsky epoxidation*. When **I** is reacted with α,β-unsaturated carbonyl compounds, a conjugate addition takes place to produce a cyclopropane as the major product. This reaction is known as the *Corey-Chaykovsky cyclopropanation*.[1] Sulfur ylide **II** is more reactive and less stable than **I**, so it is generated and used at low temperature. The reaction of substituted sulfur ylides with aldehydes is stereoselective, leading predominantly to *trans* epoxides. Asymmetric epoxidations are also possible using chiral sulfides.[12,6] The use of various substituted sulfur ylides allows the transfer of substituted methylene units to carbonyl compounds (isopropylidene or cyclopropylidene fragments) to prepare highly substituted epoxides. Since the S-alkylation of sulfoxides is not a general reaction, it is not practical to obtain the precursor salts in the trialkylsulfoxonium series. This shortcoming limits the corresponding sulfur ylides to the unsubstituted methylide. However, sulfur ylide reagents derived from sulfoximines offer a versatile way to transfer substituted methylene units to carbonyl compounds to prepare oxiranes and cyclopropanes.[12]

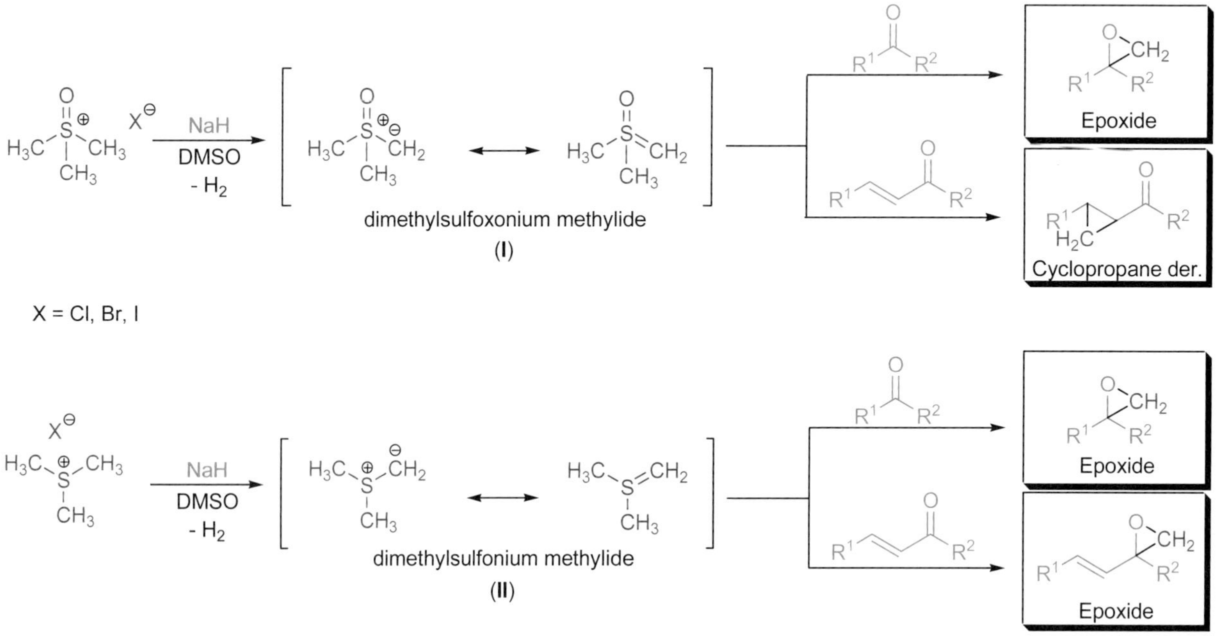

Mechanism: [18-25]

Epoxide Formation:

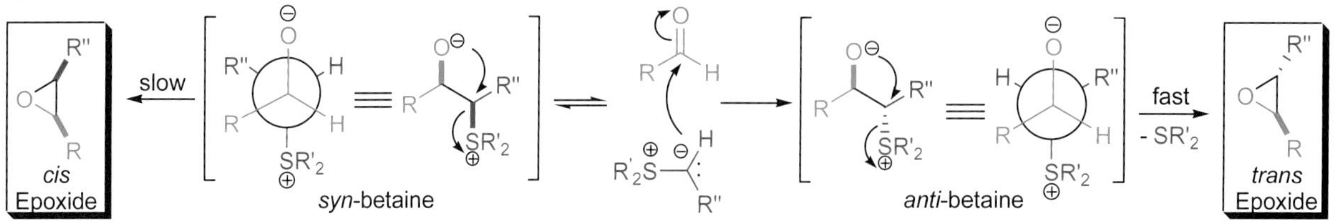

Cyclopropane Formation:

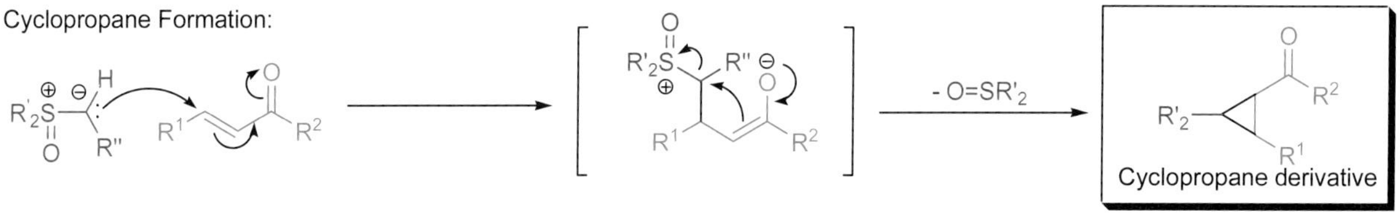

Synthetic Applications:

During the total synthesis of (+)-phyllanthocin, A.B. Smith and co-workers installed the epoxide functionality *chemo-* and *stereoselectively* at the C7 carbonyl group of the intermediate diketone by using dimethylsulfoxonium-methylide in a 1:1 solvent mixture of DMSO-THF at 0 °C.[26] The success of this chemoselective methylenation was attributed to the two α-alkoxy substituents, which render the C7 carbonyl group much more electrophilic than C10.

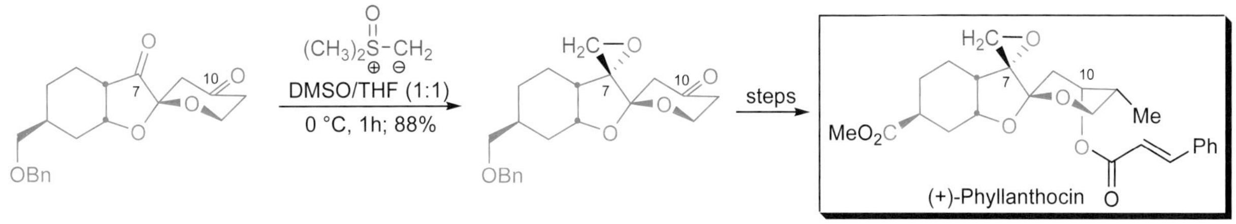

A short enantiospecific total synthesis of (+)-aphanamol I and II from limonene was achieved and the absolute stereochemistry of I and II established in the laboratory of B. Wickberg.[27] The key steps were a *de Mayo photocycloaddition*, a *Corey-Chaykovsky epoxidation* and finally a *base-catalyzed fragmentation* of the γ,δ-epoxyalcohol intermediate. Upon treating the photocycloadduct with dimethylsulfoxonium methylide, only the *endo* epoxide diastereomer was formed due to the steric hindrance provided by the methyl and isopropyl groups.

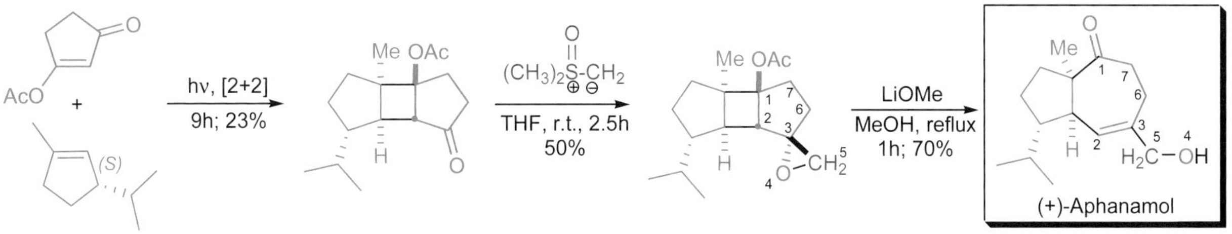

The conversion of a bicyclo[2.2.1]octenone derivative to the corresponding bicyclo[3.3.0]octenone, a common intermediate in the total synthesis of several iridoid monoterpenes, was achieved by N.C. Chang et al. The target was obtained by sequential application of the *Corey-Chaykovsky epoxidation*, *Demjanov rearrangement* and a *photochemical [1,3]-acyl shift.*[28]

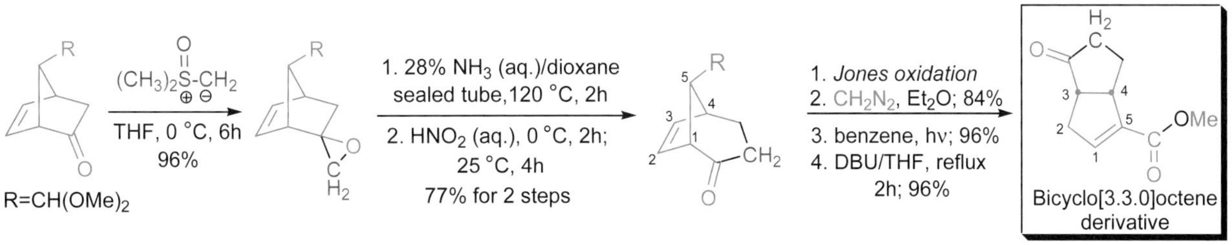

One of the steps in the highly stereoselective total synthesis of (±)-isovelleral involved the cyclopropanation of an α,β-unsaturated ketone using dimethylsulfoxonium methylide.[29] C.H. Heathcock and co-workers studied this transformation under various conditions and they found that THF at ambient temperature gave superior results to DMSO, which is the most common solvent for the *Corey-Chaykovsky cyclopropanation*.

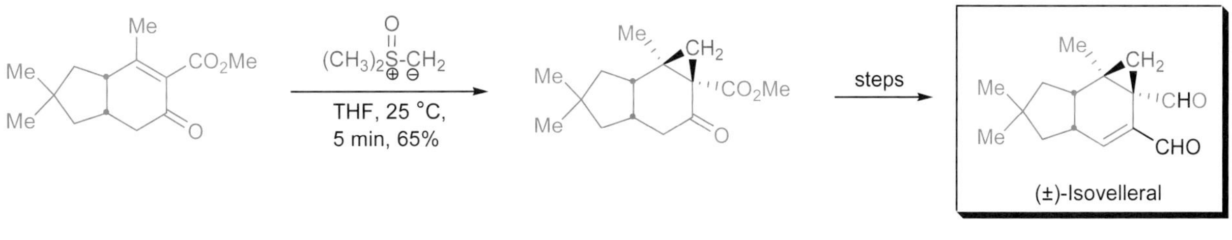

COREY-FUCHS ALKYNE SYNTHESIS

(References are on page 566)

Importance:

[*Seminal Publication*[1]; *Reviews*[2]; *Modifications and Improvements*[3-5]]

The one-carbon homologation of aldehydes to the corresponding terminal alkynes using carbon tetrabromide and triphenylphosphine is known as the *Corey-Fuchs alkyne synthesis*. In 1972, E.J. Corey and P.L. Fuchs examined the synthetic possibility of transforming aldehydes to the corresponding one-carbon chain-extended alkynes.[1] The first step of their procedure involved the conversion of the aldehyde to the corresponding homologated dibromoolefin in two possible ways: **I**) addition of the aldehyde (1 equivalent) to a mixture of triphenylphosphine (4 equivalents) and carbon tetrabromide (2 equivalents) in CH_2Cl_2, at 0 °C in 5 minutes;[6] or **II**) addition of the aldehyde to a reagent, which is prepared by mixing zinc dust (2 equivalents) with Ph_3P (2 equivalents) and CBr_4 (2 equivalents) in CH_2Cl_2 at 23 °C for 24-30h (the reaction time to form the alkyne is 1-2h). Yields are typically 80-90% for this first Wittig-type step. Procedure **II**, using zinc dust and less Ph_3P, tends to give higher yields of dibromoolefins and simplifies the isolation procedure. In the second step, the conversion of the prepared dibromoolefins to the corresponding terminal alkynes is accomplished by treatment with 2 equivalents of *n*-butyllithium at -78 °C (*lithium-halogen exchange* and elimination), followed by simple hydrolysis. The intermediate is a lithium acetylide, which can be treated with a number of electrophiles to produce a wide variety of useful derivatives. Recently, a one-pot modified procedure using *t*-BuOK/(Ph_3PCHBr_2)Br followed by the addition of excess *n*-BuLi was published.[5]

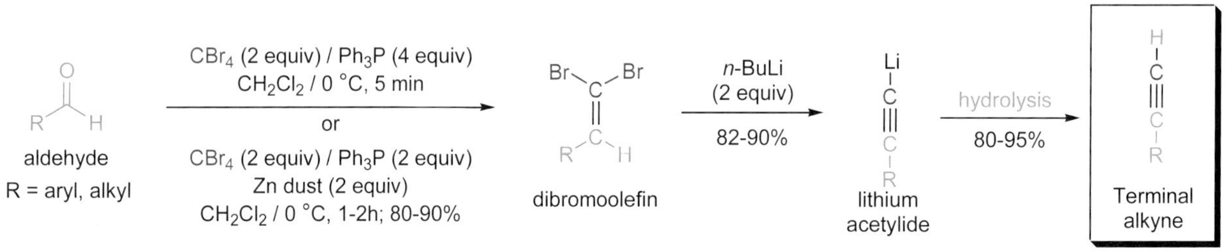

Mechanism:[6,1]

The mechanism of dibromoolefin formation from the aldehyde is similar to the mechanism of the *Wittig reaction*. However, there is very little known about the formation of the alkyne from the dibromoolefin. The mechanism below is one possible pathway to the observed product.

Generation of the phosphorous ylide:

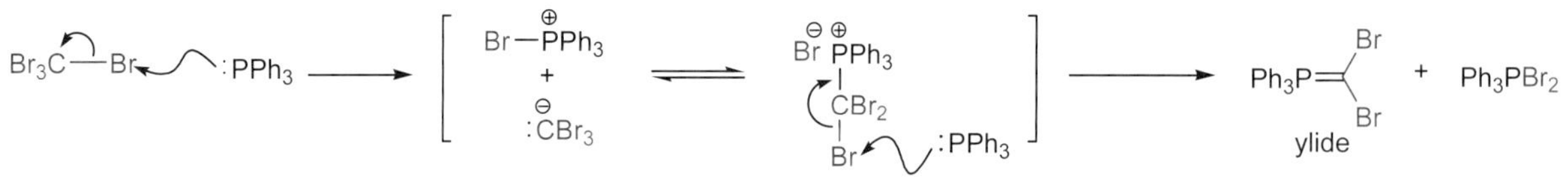

Reaction of the phosphorous ylide with the aldehyde:

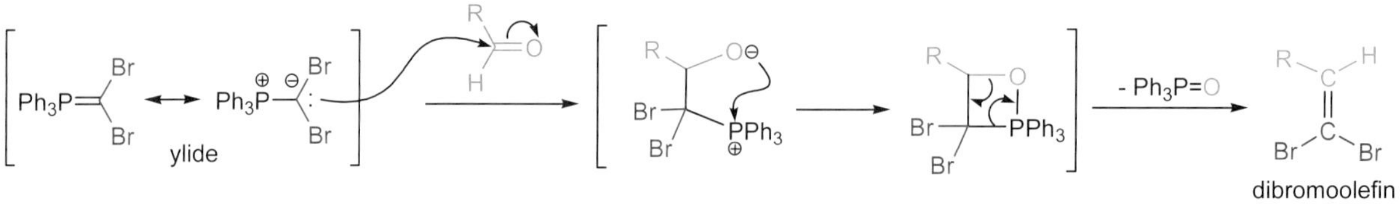

Conversion of dibromoolefin to terminal alkyne:

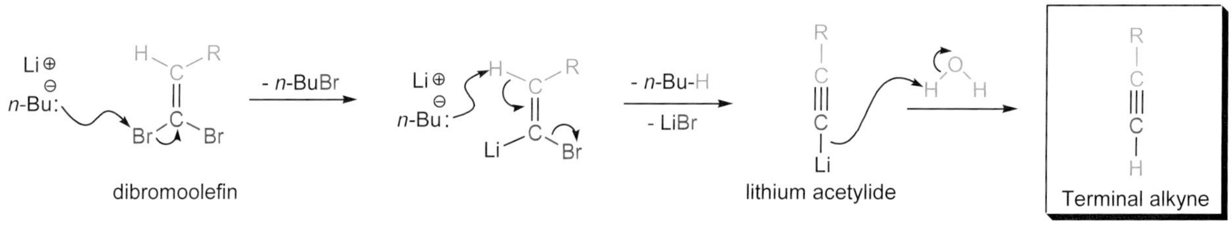

Synthetic Applications:

In the laboratory of J.H. van Boom, the synthesis of highly functionalized *cis-* and *trans*-fused polycyclic ethers of various ring sizes *via* radical cyclization of carbohydrate-derived β-(alkynyloxy)acrylates was developed.[7] The radical cyclization precursors were prepared iteratively and the terminal alkyne moieties were installed using the *Corey-Fuchs procedure*.

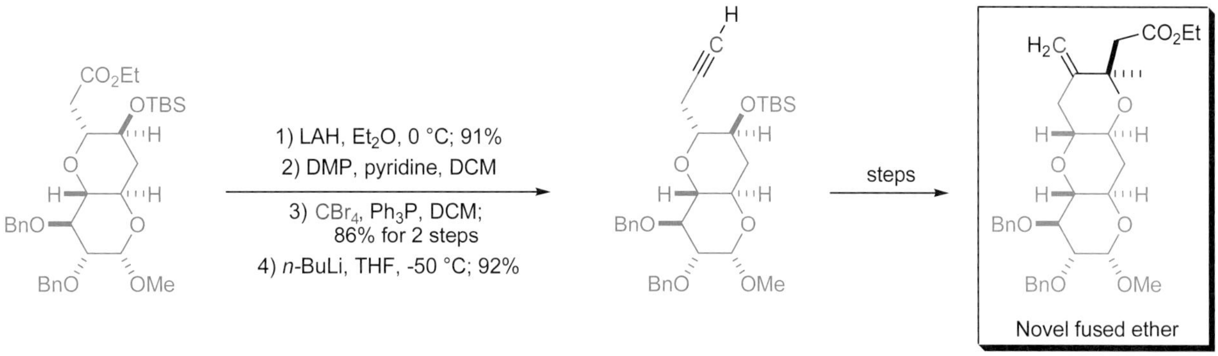

The total synthesis of Galubulimima alkaloid 4,4*a*-didehydrohimandravine, using an *intramolecular Diels-Alder reaction* and a *Stille coupling* as the key steps, was accomplished in the laboratory of M.S. Sherburn.[8] The required vinylstannane intermediate for the *Stille coupling* was prepared *via* the *one-pot Corey-Fuchs reaction*,[5] followed by *radical hydrostannylation*.

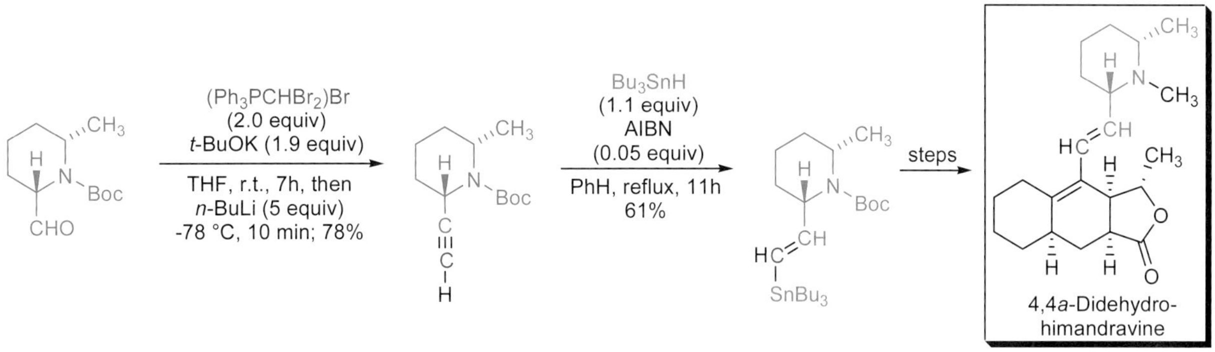

W.J. Kerr and co-workers carried out the total synthesis of (+)-taylorione starting from readily available (+)-2-carene and using a *modified Pauson-Khand annulation* with ethylene gas as the key step.[9] The key terminal alkyne intermediate was prepared by the *Corey-Fuchs reaction*. Interestingly, the ketal protecting group was sensitive to the excess of CBr_4, so the addition of this reagent had to be monitored carefully to cleanly transform the aldehyde to the desired dibromoolefin.

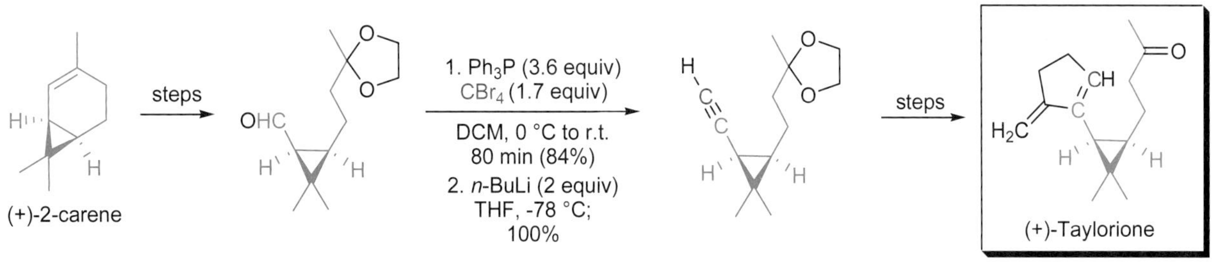

W. Oppolzer et al. utilized the *Corey-Fuchs alkyne synthesis* for the preparation of a key acyclic enynyl carbonate during the total synthesis of (±)-hirsutene.[10]

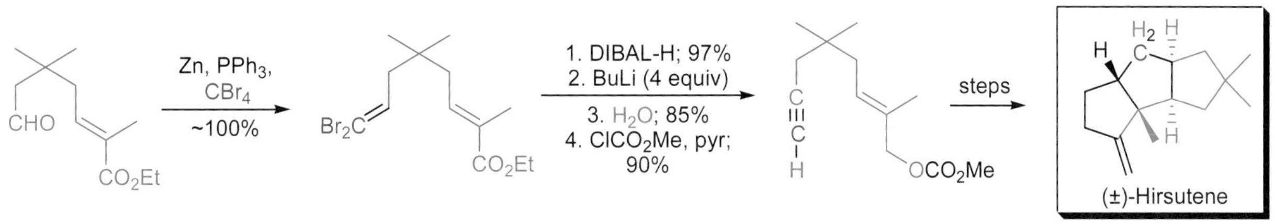

COREY-KIM OXIDATION

(References are on page 566)

Importance:

[*Seminal Publications*[1-3]; *Modifications & Improvements*[4,5]]

In 1972, E.J. Corey and C.U. Kim developed a new process for the efficient conversion of alcohols to aldehydes and ketones using *N*-chlorosuccinimide (NCS), dimethylsulfide (DMS) and triethylamine (TEA).[2] The oxidation of primary and secondary alcohols with NCS/DMS is known as the *Corey-Kim oxidation*. The active reagent, S,S-dimethylsuccinimidosulfonium chloride, is formed *in situ* when NCS and DMS are reacted and is called the *Corey-Kim reagent*.[1] This protocol can be used for the oxidation of a wide variety of primary and secondary alcohols except for allylic and benzylic alcohols, where the substrates are predominantly converted to the allylic and benzylic halides. In polar solvents, a side-reaction may occur in which the alcohol forms the corresponding methylthiomethyl ether ($ROCH_2SCH_3$). The reaction conditions for the *Corey-Kim oxidation* are mild and tolerate most functional and protecting groups. Therefore, the protocol can be applied to the oxidation of polyfunctionalized molecules. Recent modifications of the original procedure led to the development of the fluorous[4] and odorless[5] *Corey-Kim oxidations*. In addition to being an effective oxidant for alcohols, the *Corey-Kim reagent* has also been used to dehydrate aldoximes to nitriles,[6] convert 3-hydroxycarbonyl compounds to 1,3-dicarbonyls,[7] synthesize stable sulfur ylides from active methylene compounds[8] and to prepare 3(*H*)-indoles from 1(*H*)-indoles.[9]

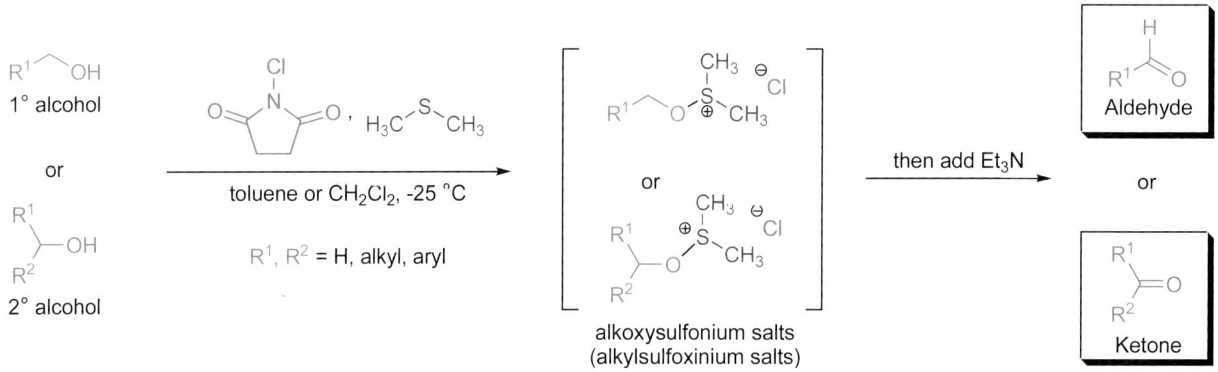

Mechanism:[2,4]

The first step of the mechanism of the *Corey-Kim oxidation* is the reaction of dimethylsulfide with *N*-chlorosuccinimide to generate the electrophilic active species, *S,S*-dimethylsuccinimidosulfonium chloride (*Corey-Kim reagent*) *via* dimethylsulfonium chloride. The sulfonium salt is then attacked by the nucleophilic alcohol to afford an alkoxysulfonium salt. This alkoxysulfonium salt is deprotonated by triethylamine and the desired carbonyl compound is formed. The dimethylsulfide is regenerated, and it is easily removed from the reaction mixture *in vacuo*. In the *odorless Corey-Kim oxidation*[5] instead of dimethylsulfide, dodecylmethylsulfide is used. This sulfide lacks the unpleasant odor of DMS due to its low volatility.

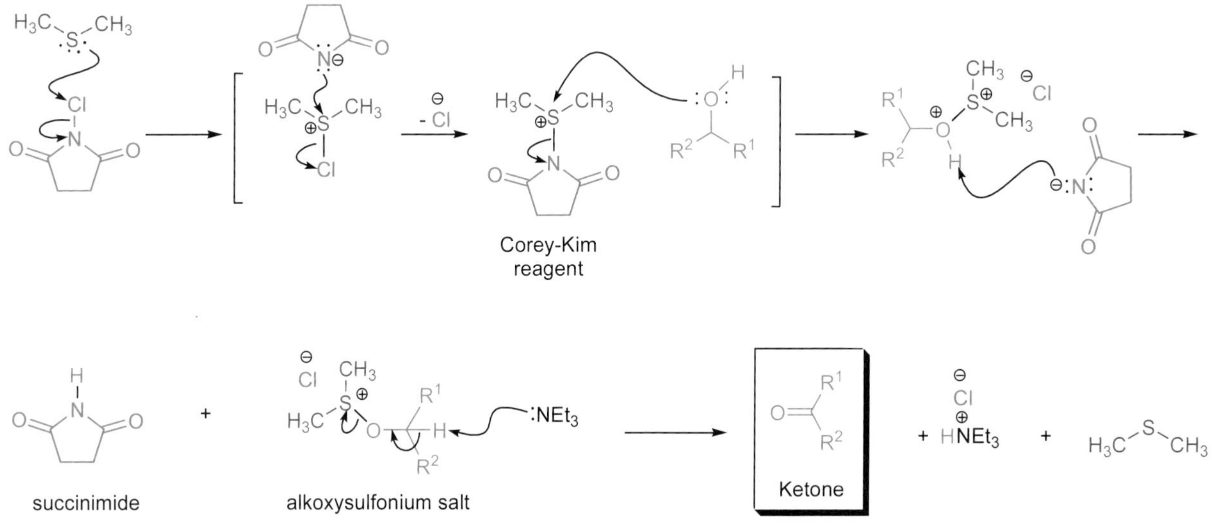

Synthetic Applications:

During the total synthesis of (±)-ingenol by I. Kuwajima and co-workers, an advanced tricyclic diol intermediate was selectively converted to the corresponding α-ketol utilizing the *Corey-Kim oxidation*.[10] The diol was oxidized only at the less hindered C6 hydroxyl group.

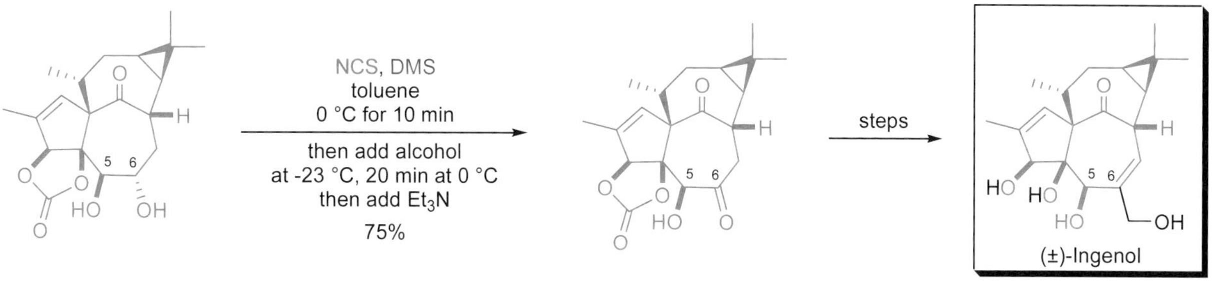

In the laboratory of L.S. Hegedus, the total synthesis of (±)-*epi*-jatrophone was accomplished using a *palladium-catalyzed carbonylative coupling* as the key step.[11] In the endgame of the synthesis, a β-hydroxy ketone moiety was oxidized in excellent yield to the corresponding 1,3-dione using the mild *Corey-Kim protocol*.

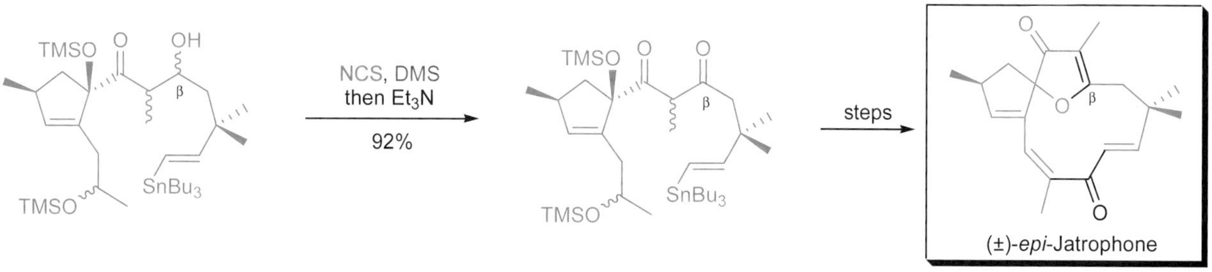

In the final stages of the total synthesis of (±)-cephalotaxine by M.E. Kuehne et al., a tetracyclic *cis*-vicinal diol was oxidized to the α-diketone.[12] Using PCC, pyridine/SO_3 or the Swern protocol did not yield the desired product. However, by applying the *Corey-Kim protocol*, NCS-DMS in dichloromethane at -42 °C, afforded the diketone in 89% yield.

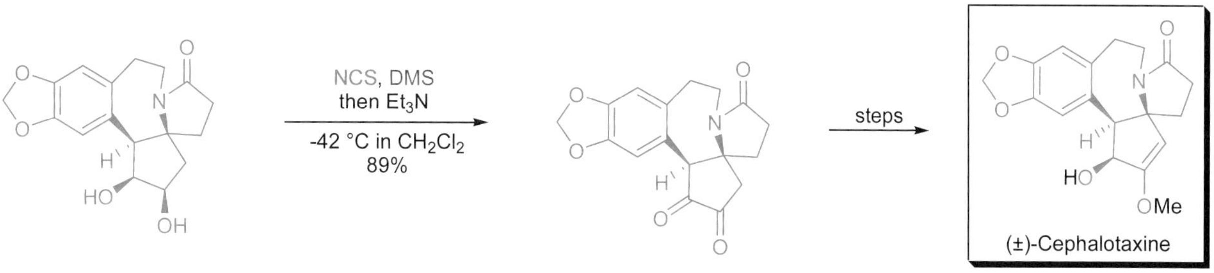

The serotonin antagonist LY426965 was synthesized using *catalytic enantioselective allylation* with a chiral biphosphoramide in the laboratory of S.E. Denmark.[13] In order to prepare the necessary 3,3-disubstituted allyltrichlorosilane reagent, the (*E*)-allylic alcohol was first converted by the *Corey-Kim procedure* to the corresponding chloride.

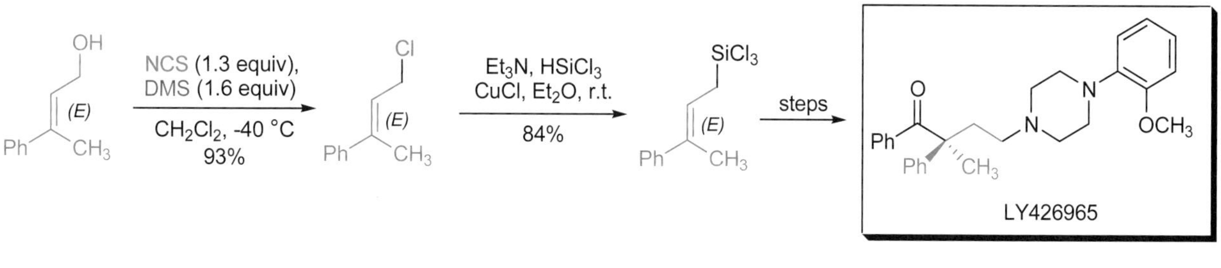

COREY-NICOLAOU MACROLACTONIZATION

(References are on page 567)

Importance:

[*Seminal Publications*[1,2]; *Reviews*[3-13]; *Modifications & Improvements*[14-16]; *Theoretical Studies*[17]]

Before the 1970s there was no general way to efficiently prepare medium- and large-ring lactones from highly functionalized hydroxy acids under mild conditions. When the ring size of the target lactone is large, the probability of the hydroxyl group reacting with the carboxylic acid moiety within the same molecule is very low, and mainly intermolecular coupling occurs unless the concentration of the substrate is very low (high-dilution conditions). In 1974, E.J. Corey and K.C. Nicolaou reported a novel and mild method for the formation of macrolactones from complex hydroxy acid precursors.[1] A series of ω-hydroxy acids were lactonized by first converting them to the corresponding 2-pyridinethiol esters, which were then slowly added to xylene at reflux. The formation of lactones from hydroxy acids *via* their 2-pyridinethiol esters is known as the *Corey-Nicolaou macrolactonization*. The power of the method was first demonstrated by the total synthesis of (±)-zearalenone in which the functionalized hydroxy acid was first treated with 2,2'-dipyridyl disulfide and the resulting 2-pyridinethiol ester was heated to reflux in benzene.[1] Removal of the protecting groups furnished the natural product. The general features of this macrolactonization strategy are: 1) the reaction is conducted under neutral and aprotic conditions, so substrates containing acid- and base-labile functional groups are tolerated; 2) the formation of the 2-pyridinethiol ester is conducted in the presence of a slight excess of PPh_3 and 2,2-dipyridyl disulfide;[18] 3) the actual cyclization is usually conducted in refluxing benzene or toluene under high-dilution conditions to keep the undesired intermolecular ester formation at a minimum; 4) the lactonization is not catalyzed by acids, bases, or by-products; and 5) lactones with ring sizes 7-48 have been successfully prepared, but reaction rates strongly depend on ring-size and the functionality of the substrate. Over the past three decades several modifications of the method were introduced: 1) the use of silver perchlorate (or $AgBF_4$) to activate the 2-pyridinethiol esters by complexation; significant reduction of reaction time is observed (*Gerlach-Thalmann modification*);[14] and 2) the development of other bis-heterocyclic disulfide reagents by Corey et al.[15]

Corey & Nicolaou (1974):

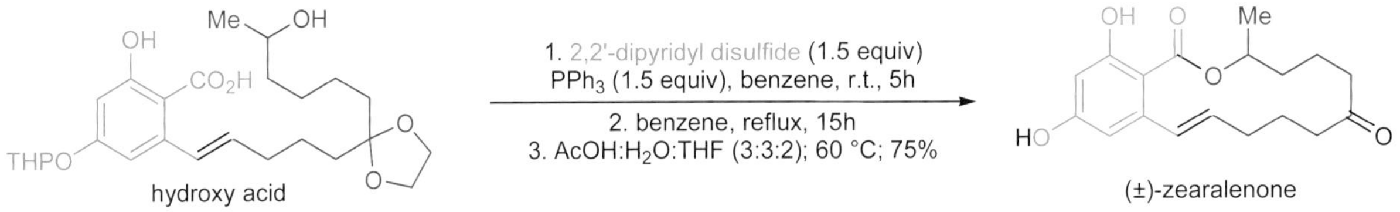

Corey-Nicolaou macrolactonization (double activation):

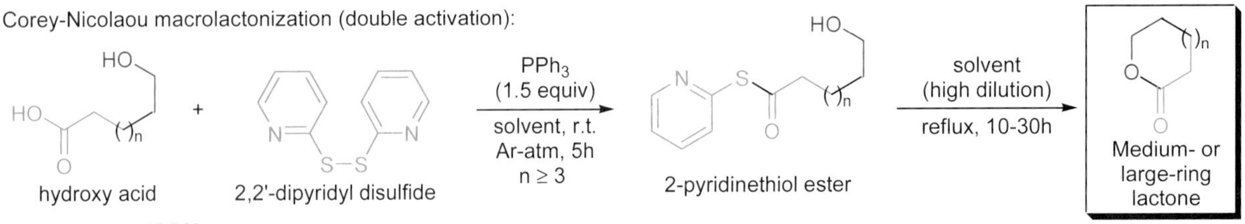

Mechanism: [19,5,20]

The 2-pyridinethiol ester undergoes an intramolecular proton transfer to give rise to a dipolar intermediate in which the carbonyl group is part of a six-membered ring held by hydrogen bonding. In this dipolar intermediate both the carbonyl group and the oxygen atom of the alcohol are activated because the carbonyl group is more electrophilic but the oxygen is more nucleophilic than before. The intramolecular attack of the alkoxide ion onto the carbonyl group is electrostatically driven and the tetrahedral intermediate collapses to yield the desired lactone as well as 2-pyridinethione. This mechanistic picture is supported by the observation that thiolesters in which the intramolecular hydrogen bonding was not possible did not undergo lactonization upon heating.

Formation of the 2-pyridinethiol ester:

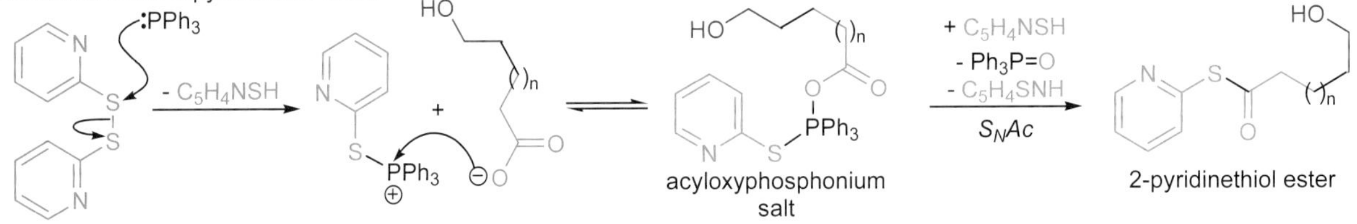

Formation of the lactone by double activation:

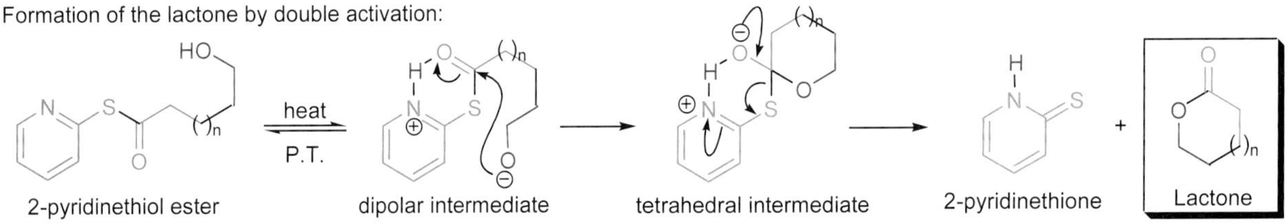

Synthetic Applications:

The *modified Corey-Nicolaou macrolactonization* was applied for the construction of the BCD ring system of brevetoxin A by K.C. Nicolaou and co-workers.[21] The dihydroxy dicarboxylic acid substrate was subjected to a one-pot bis-lactonization. After the formation of the bis-2-pyridinethiol ester, the lactonization was conducted at low substrate concentration (0.013 M) in toluene at reflux temperature.

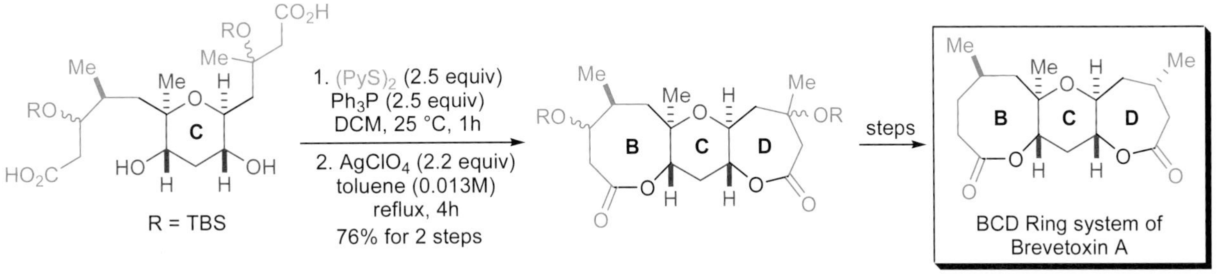

The research team of M. Hirama conducted synthetic studies toward the C-1027 chromophore, which contains a highly unsaturated 17-membered macrolactone.[22] The authors investigated several macrolactonization protocols including the *Mukaiyama-*, *Corey-Nicolaou-*, and *Yamaguchi protocols*. The Mukaiyama and *Yamaguchi macrolactonization* conditions gave dimers as the major product, but the *Corey-Nicolaou procedure* yielded the desired macrolactone as the only product, albeit in modest yield. The modification of the protecting groups in the hydroxy acid precursor helped to optimize the yield of the macrolactone which was obtained as a 1:1.1 mixture of inseparable atropisomers.

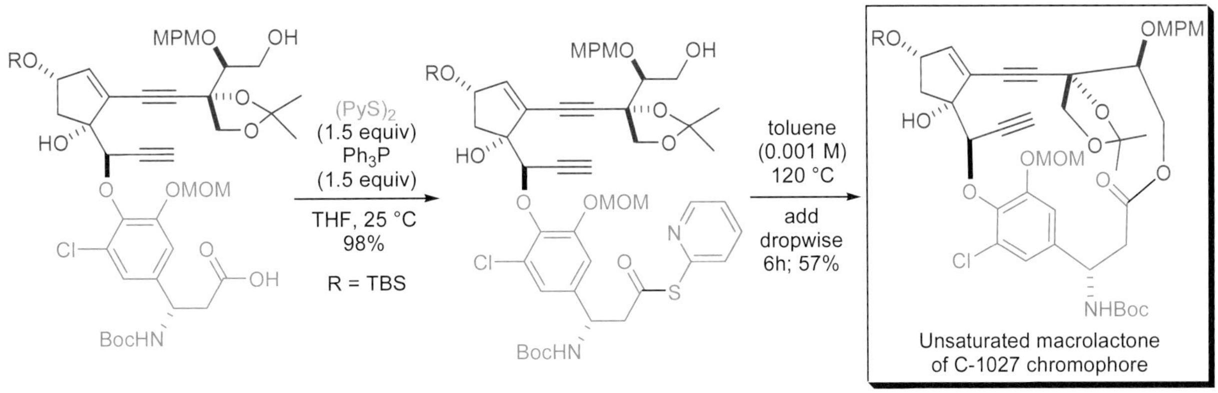

The first total synthesis of the ichthyotoxic marine natural product (–)-aplyolide A was accomplished by Y. Stenstrøm and co-workers.[23] The compound has a 16-membered lactone ring, four (*Z*)-double bonds, as well as a stereogenic center. Numerous macrolactonization protocols were tested, but most of them gave the diolide (dimer) except for the *Corey-Nicolaou procedure*.

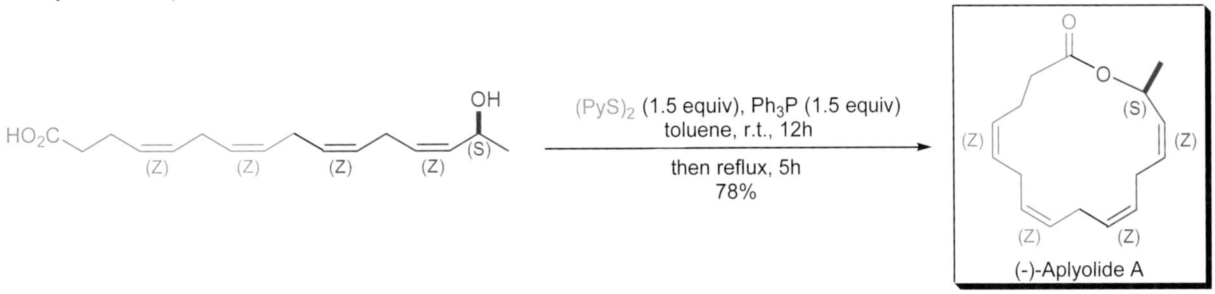

M.B. Andrus and T.-L. Shi achieved the total synthesis of the 10-membered lactone (–)-tuckolide (decarestrictine D), which potentially inhibits cholesterol biosynthesis.[24] The lactonization was only successful under the *Corey-Nicolaou conditions*. Interestingly, the unsubstituted 9-hydroxynonanoic acid did not lactonize under these conditions.

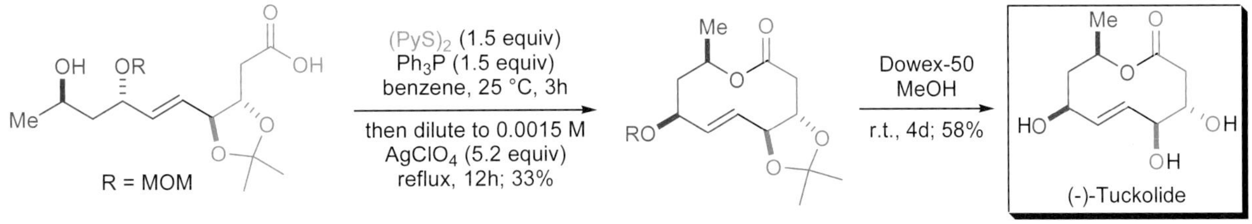

CORREY-WINTER OLEFINATION

(References are on page 567)

Importance:

[*Seminal Publications*[1,2]; *Review*[3]]

In 1963, E.J. Corey and R.A.E. Winter described a new two-step method for the stereospecific synthesis of alkenes from 1,2-diols *via* cyclic 1,2-thionocarbonates and 1,2-trithiocarbonates.[1,2,4] This method of alkene synthesis is called the *Corey-Winter olefination*. In the first step, the 1,2-diol is converted quantitatively to the corresponding cyclic thionocarbonate derivative using thiocarbonyldiimidazole. In the second step, the thionocarbonate is treated with excess trialkylphosphite [$P(OR')_3$, where R'=Me, Et or alkyl] at reflux, and a *cis*-elimination reaction takes place to yield the alkene and by-products [CO_2 and $(OR)_3P=S$]. The reaction is completely *stereospecific* and high-yielding. Even highly substituted and strained olefins (e.g., *trans*-cycloheptene)[2] can be prepared. However, no elimination is observed in those cases in which the *cis*-elimination would lead to an extremely strained structure (e.g., *trans*-cyclohexene). The stereochemistry of the product olefin is only determined by the stereochemistry of the starting 1,2-diol (*cis* or *trans*) and usually under the reaction conditions, no isomerization of the product is observed. A *cis* olefin, may be converted to *trans*-1,2-diol and subjected to the *Corey-Winter procedure* to afford the corresponding *trans* olefin. Similarly, *trans* olefins can be converted to the corresponding *cis* olefins.

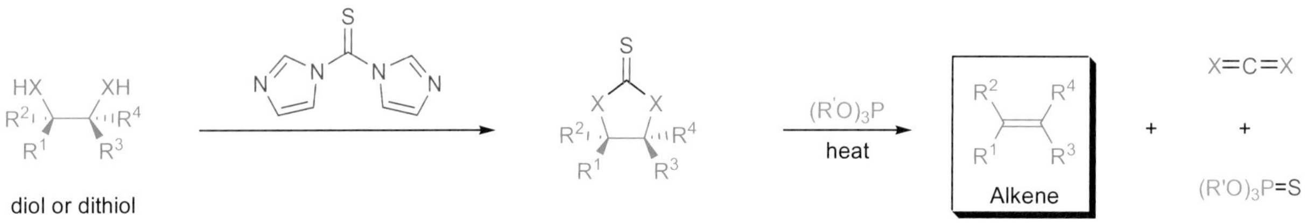

R^1, R^2, R^3, R^4 = H, alkyl, aryl; R' = Me, Et; substrate: X = O (1,2-diol), X = S, 1,2-dithiol; cyclic intermediate: X = O (cyclic 1,2-thionocarbonate), X = S (cyclic 1,2-trithiocarbonate)

Mechanism: [2,5-7]

The exact mechanism of the reaction between the thionocarbonate and the trialkylphosphite is not known. There are two possible pathways (**I** and **II**) and both of them are presented. In pathway **I**, the formation of an ylide intermediate is postulated based on inhibition studies,[4] while in pathway **II** the generation of a carbenoid intermediate is assumed. There is direct experimental evidence that the elimination of the cyclic 1,2-thionocarbonate involves the formation of a carbenoid intermediate.[6]

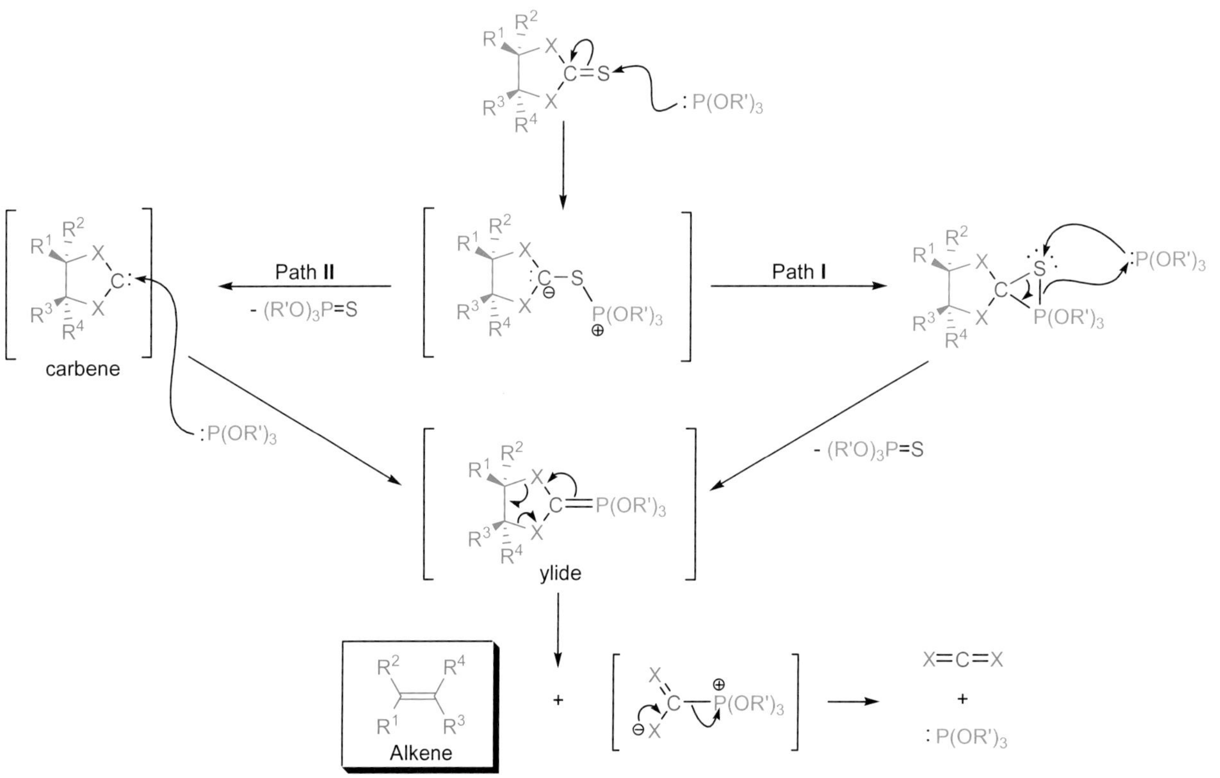

Synthetic Applications:

The enantiospecific synthesis of naturally occurring cyclohexane epoxides such as (+)-crotepoxide and (+)-boesenoxide was accomplished by T.K.M. Shing et al.[8] The key intermediate 1,3-cyclohexadiene was prepared using the *Corey-Winter protocol* on a *cis*-vicinal diol. The resulting diene was then converted to the natural product after several steps.

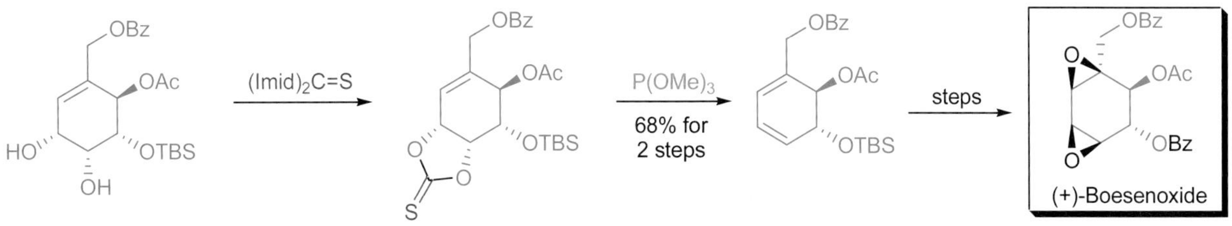

The absolute configuration of radiosumin, a novel potent trypsin inhibitory dipeptide, was determined by T. Shioiri and co-workers by carrying out the first enantioselective total synthesis of the natural product.[9] The *s-trans* 1,3-diene in one of the key synthetic intermediates was installed by the *Corey-Winter olefination* using the Corey-Hopkins reagent (1,3-dimethyl-2-phenyl-1,3,2-diazaphospholidine).

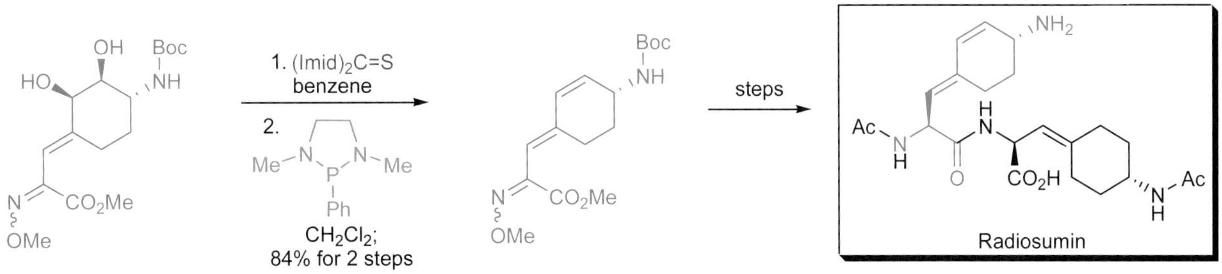

In the laboratory of J.H. Rigby, synthetic studies were undertaken on the ingenane diterpenes.[10] During these studies, it was necessary to investigate the ring opening reactions of a structurally complex allylic epoxide intermediate. This allylic epoxide was prepared from a 1,3-diene in three steps: *dihydroxylation*, *epoxidation* and *Corey-Winter olefination*.

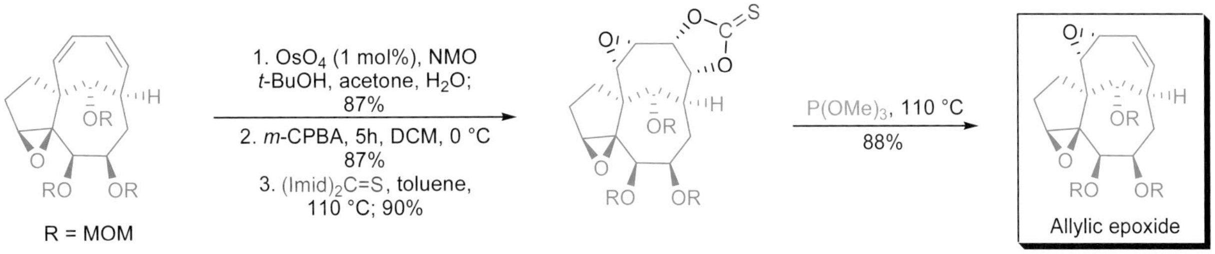

G.W.J. Fleet and co-workers synthesized L-(+)-swainsonine and other more highly oxygenated monocyclic structures that exhibited inhibitory activity toward naringinase (L-rhamnosidase).[11] In order to remove a *cis*-vicinal diol moiety in the endgame of the synthesis, the *Corey-Winter olefination* was utilized.

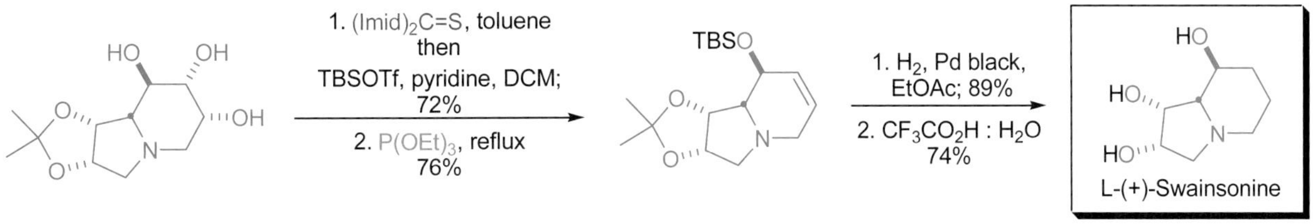

CORNFORTH REARRANGEMENT

(References are on page 567)

Importance:

[*Seminal Publication*[1]; *Reviews*[2-7]; Modifications[8]; *Theoretical Studies*[9-11]]

In 1949 J.W. Cornforth observed that upon heating, 2-phenyl-5-ethoxyoxazole-4-carboxamide (R^1=Ph, R^2=OEt, and R^3=NH_2) rearranged to ethyl 2-phenyl-5-aminooxazole-4-carboxylate.[1] The thermal rearrangement of 4-carbonyl substituted oxazoles to their isomeric oxazoles is known as the *Cornforth rearrangement*. The extent of the rearrangement depends on the thermodynamic stability of the starting material versus the product. When R^2=R^3, the *Cornforth rearrangement* is degenerate and leads to a 1:1 equilibrium mixture.[12] In the early 1970s, the scope and limitations of the reaction were investigated in depth by M.J.S. Dewar and co-workers.[13,12] They found that the rearrangement was general and that secondary and tertiary alkyl and aryl oxazole-4-carboxamides were converted to the corresponding secondary and tertiary 5-aminooxazoles.[13] When the amide nitrogen is part of a heterocycle (R^3=N-heterocycle), the rearrangement occurs in typically excellent (>90%) yield. The *Cornforth rearrangement* was also found to be a general method for the synthesis of 5-thiooxazole-4-carboxylic esters from 5-alkoxyoxazole-4-thiocarboxylates (R^3=SAr). A special case of the rearrangement is the base-induced or pyrolytic isomerization of 4-hydroxymethylene-5-oxazolones or their potassium salts to the corresponding oxazole-4-carboxylic acids.[14]

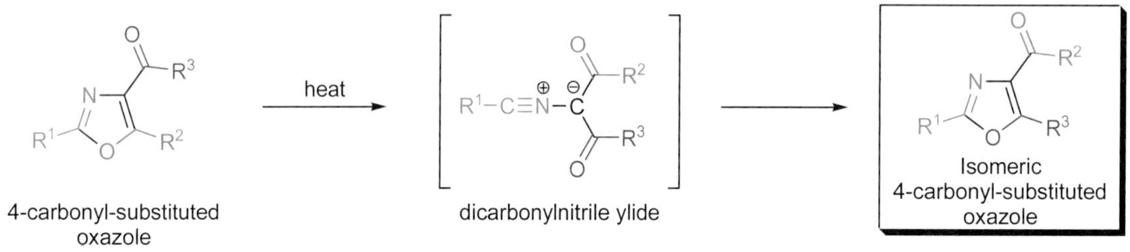

Mechanism: [15,13,3]

The mechanism involves the electrocyclic opening of the oxazole ring to a dicarbonylnitrile ylide intermediate, which undergoes a *[1,5]-dipolar electrocyclization*[3,11] to give the rearranged oxazole. The intermediate nitrile ylide cannot be isolated. To prove that the mechanism involves this intermediate, G. Höfle and W. Steglich generated carbonylnitrile ylides by a thermally induced *[1,3]-dipolar cycloreversion reaction* of 4-acyl-2-oxazolin-5-ones and found that the resulting ylides readily cyclized to oxazoles in preparatively useful yields.[16] Whether or not the rearrangement occurs depends solely on the free energy difference between the starting material and product, or more precisely on the nature of R^2 and R^3 substituents.[13,12] In aprotic solvents the rate of isomerization increases with increasing solvent polarity suggesting that only a small positive charge builds up in the transition state.[15] However, there is a substantial rate increase when the solvent is changed from an aprotic ($PhNO_2$) to a protic solvent ($PhCH_2OH$), suggesting that the negative charge in the transition state is stabilized *via* hydrogen bonding.[13]

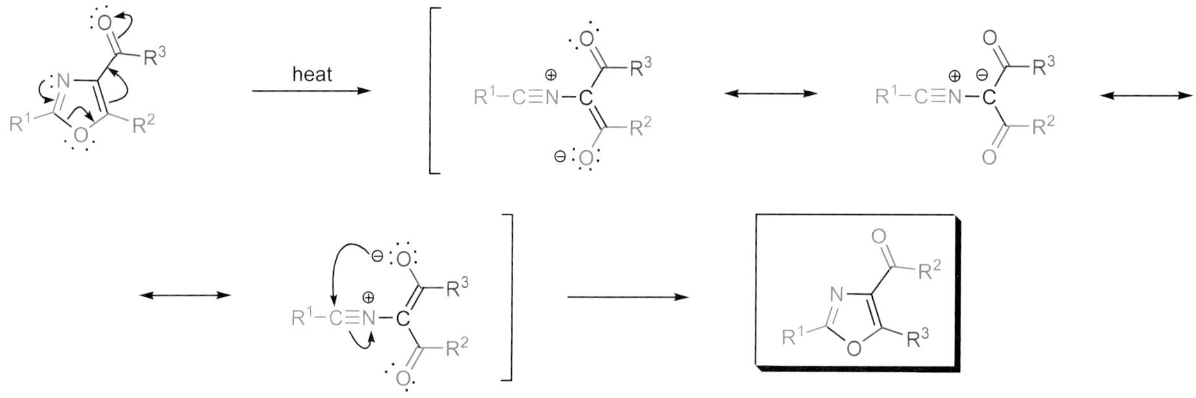

Preparation of carbonylnitrile ylide:

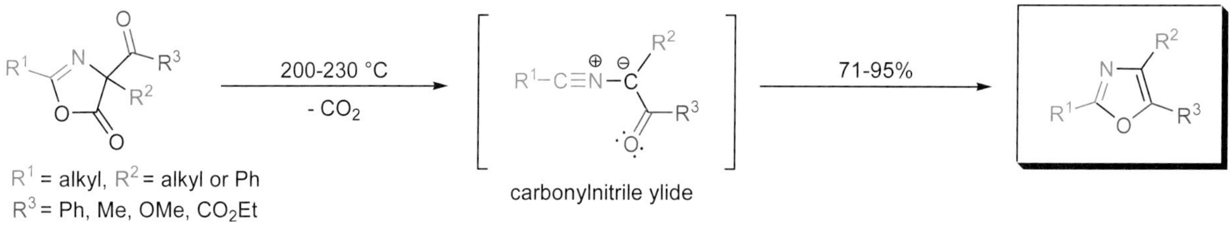

Synthetic Applications:

Substituted oxazoles are attractive starting materials for a variety of heterocyclic ring transformations due to their reactivity toward acids, bases, heat, dienophiles, and dipolarophiles. Despite the numerous ring transformations of oxazoles, the oxazole to thiazole interconversion was mainly unexplored until I.J. Turchi and co-workers examined the thermal *Cornforth rearrangement* of 4-(aminothiocarbonyl)-5-ethoxyoxazoles to 5-aminothiazoles.[17] The reaction turned out to be a simple and relatively general route to thiazoles from readily available starting materials, and the procedure is applicable to the synthesis of any 2-alkyl- or 2-aryl-4-(alkoxycarbonyl)-5-aminothiazoles.

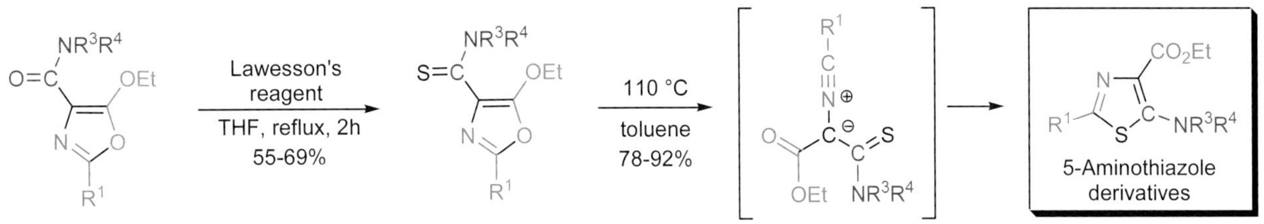

In the laboratory of D.R. Williams, a carbanion methodology for the alkylations and acylations of substituted oxazoles was investigated.[8] The study showed that the monoalkylation of the dianion generated from 2-(5-oxazolyl)-1,3-dithiane exclusively led to the substitution of the carbon adjacent to sulfur. However, acylation reactions of the dianion afforded 4,5-disubstituted oxazoles. These new products presumably arose from carbonylnitrile ylide intermediates, which were generated by the selective C-acylation of a ring-opened dianion tautomer. This is the first example of a base-induced, low-temperature *Cornforth rearrangement*.

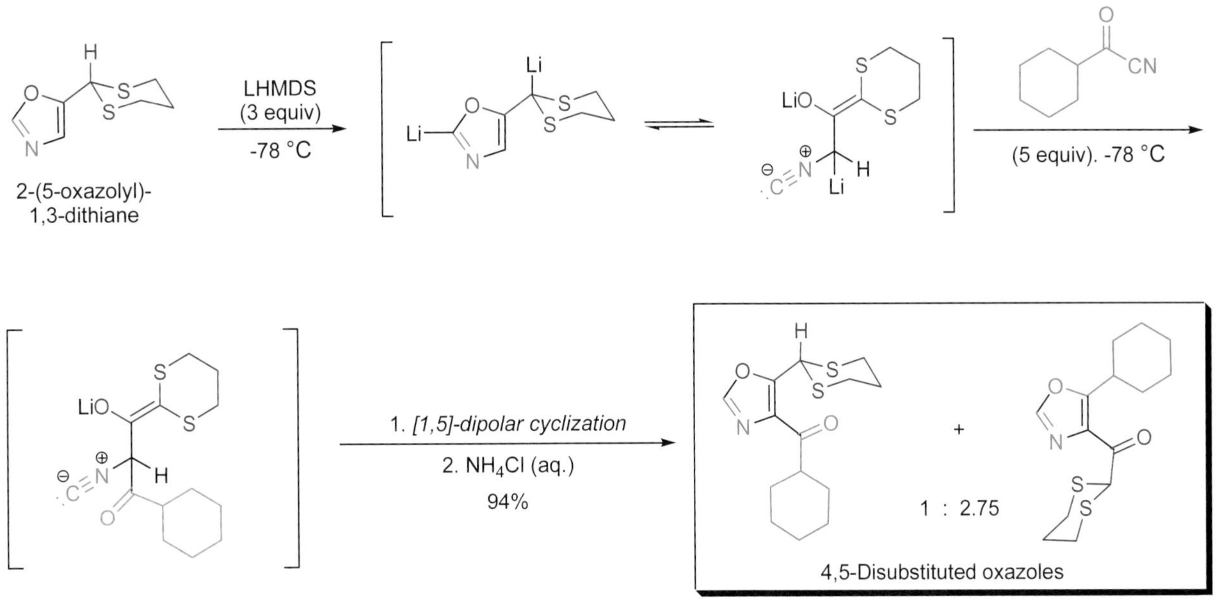

During the investigation of the scope and limitations of the *Cornforth rearrangement*, M.J.S. Dewar and co-workers treated 2-phenyl-5-ethoxyoxazole-4-aziridinylcarboxamide with sodium iodide in acetone (*Heine reaction*) to prepare 2-(2-phenyl-5-ethoxyoxazolyl)-Δ^2-oxazoline in 60% yield.[12] This oxazoline was a *Cornforth rearrangement* precursor, which upon thermolysis in boiling toluene gave 5-phenyl-7-carboethoxyimidazo[5,1-*b*]-2,3-dihydrooxazole in 97% yield.

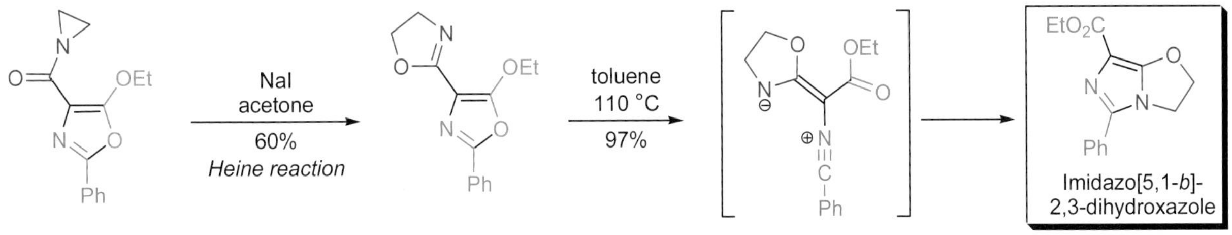

CRIEGEE OXIDATION

(References are on page 568)

Importance:

[*Seminal Publication*[1]; *Reviews*[2-5]; *Modifications & Improvements*[6-8]]

The cleavage of 1,2-diols (glycols) to the corresponding carbonyl compounds by lead tetraacetate [$Pb(OAc)_4$, LTA] in an organic solvent is known as the *Criegee oxidation*. Glycols are cleaved with ease under mild conditions and in good yield with periodic acid (HIO_4) or LTA. Other functional groups, such as β-amino alcohols, 1,2-diamines, α-hydroxy aldehydes and ketones, α-diketones and α-keto aldehydes undergo similar cleavage upon treatment with LTA. Several oxidizing agents (e.g., sodium bismuthate, manganese(III) pyrophosphate, PIDA, cerium(IV) salts, vanadium(V) salts, chromic acid, nickel peroxide, silver(I) salts, etc.) also cleave glycols, but these oxidizing agents are synthetically much less efficient. *Cis*-vicinal diols and *threo* diols are cleaved much faster than the corresponding *trans*-vicinal diols and *erythro* diols. *Cis* diols can be titrated using LTA without the interference of aliphatic glycols and *trans*-glycols on five-membered rings.[9] The *Criegee oxidation* is complementary to the *ozonolysis* of double bonds, since alkenes are easily dihydroxylated and then cleaved to afford the desired carbonyl compounds. During the past decade, the oxidative cleavage of bicyclic unsaturated diols led to the development of a new *ring-expansion/rearrangement* methodology for the preparation of densely functionalized six- and seven-membered rings from simple and well-known building blocks.[6-8]

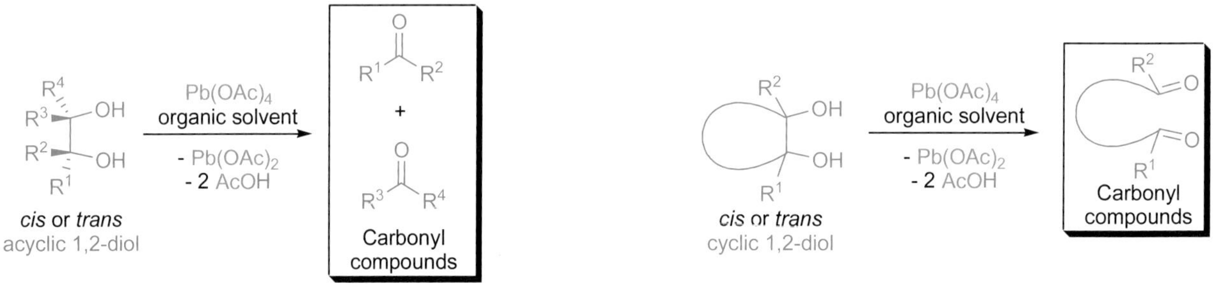

R^{1-4} = H, alkyl

R^{1-2} = H, alkyl

Mechanism:[10-20,8]

The mechanism of the *Criegee oxidation* most likely involves the formation of a bidentate metal - 1,2-glycol five-membered complex (**Path I**), which then breaks down to products *via* a two-electron process. The breakdown of the cyclic intermediate is the rate-determining step and the driving force is the electronegativity of $Pb^{(IV)}$, which abstracts the bonded electron pair of one of the O-atoms adjacent to the C-C bond and is reduced to $Pb^{(II)}$. The kinetics of the reaction is overall second order, first order in each reactant. It was found that the addition of acetic acid retards the reaction by shifting the equilibrium to the left. For substrates where the formation of the cyclic five-membered intermediate is not possible (e.g., bicyclic *trans* diols), an alternative concerted electron displacement is proposed (**Path II**) involving one of the acetate groups attached to the metal.[13]

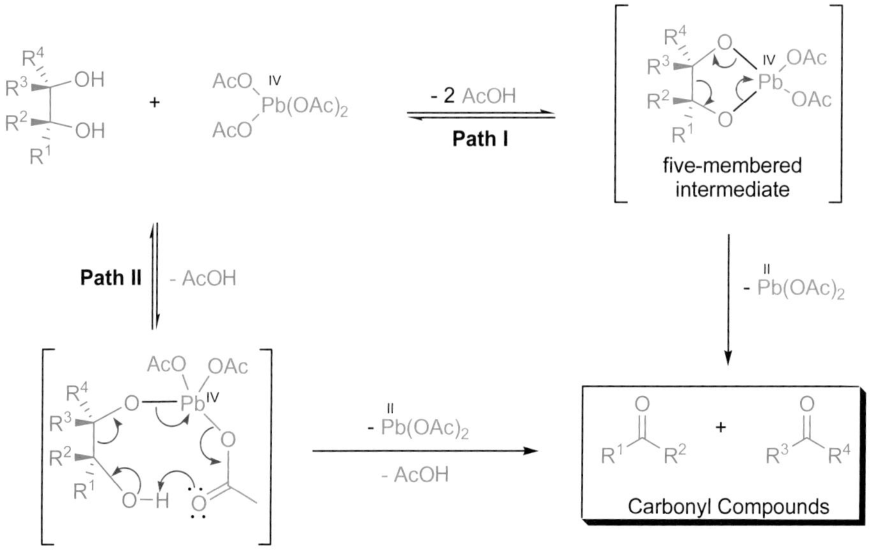

Synthetic Applications:

G.S.R. Rao and co-workers described the conversion of aromatic compounds to linear and angular triquinanes, which involved a *5-exo-trig allyl radical cyclization* as the key step.[21] To install the third five-membered ring of the linear triquinane, the tricyclic 1,2-diol intermediate was cleaved using the *Criegee oxidation* to afford a diketone. The remaining double bond was cleaved by ozonolysis and the resulting triketone was treated with PTSA in refluxing benzene to give the desired linear triquinane.

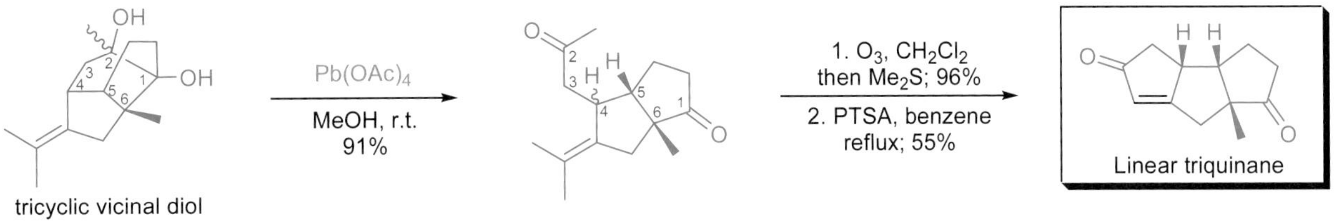

In the laboratory of Y. Takemoto, the asymmetric total synthesis of the marine metabolite, halicholactone was accomplished.[22] One advanced intermediate contained a 1,2-vicinal diol moiety which was cleaved under mild conditions to afford the corresponding aldehyde. The *Criegee oxidation* was chosen to effect this transformation at low temperature, followed by the *stereoselective allylation* of the resulting aldehyde with tetraallyltin.

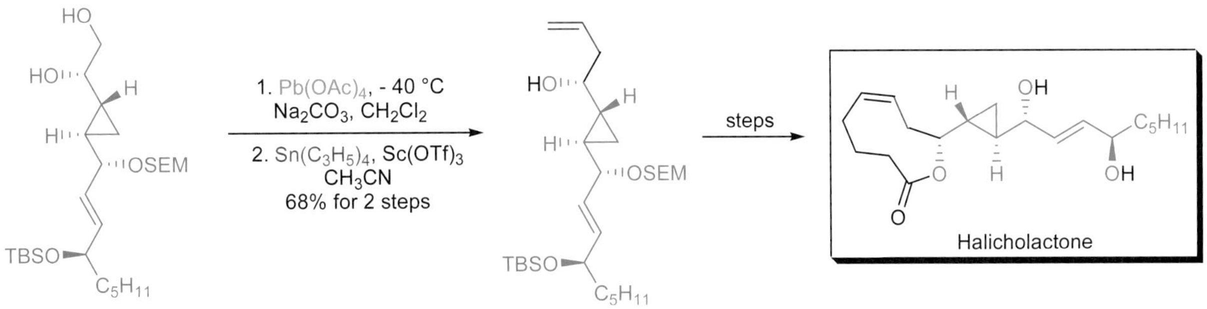

M. Hesse and co-workers synthesized (±)-pyrenolide B, a macrocyclic natural product isolated from a phytopathogenic fungus.[23] The key transformation of the synthesis was the *ring enlargement reaction* of a bicyclic enol ether intermediate to the corresponding oxolactone. The ring enlargement was performed using a two-step procedure: dihydroxylation of the enol ether double bond, followed by oxidation of the resulting diol with $Pb(OAc)_4$ to quantitatively afford the ring-expanded product.

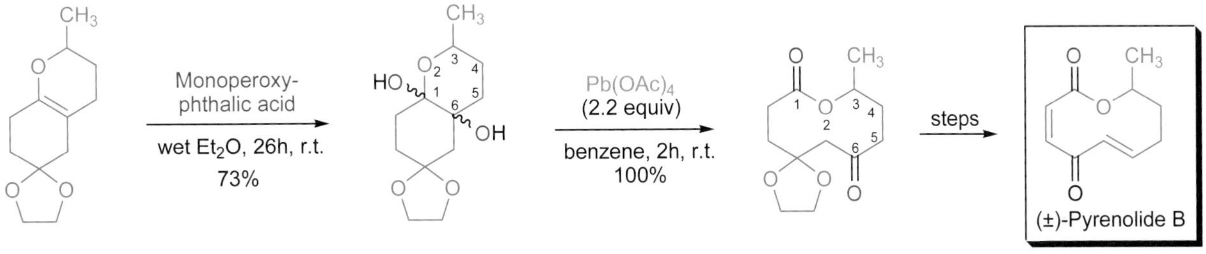

In the synthesis of angular triquinane (±)-silphinene by S. Yamamura et al., the *Criegee oxidation* was used to obtain a key bicyclic intermediate.[24]

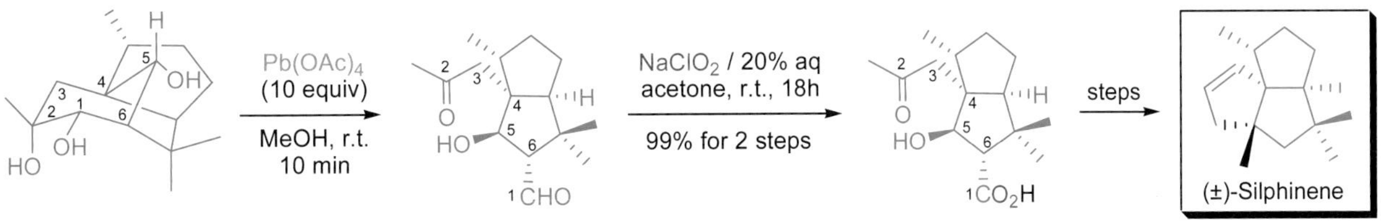

CURTIUS REARRANGEMENT

(References are on page 568)

Importance:

[*Seminal Publications*[1-5]; *Reviews*[6-10]; *Modifications & Improvements*[11-14]; *Theoretical Studies*[15-17]]

The thermal decomposition (pyrolysis) of acyl azides to the corresponding isocyanates is known as the *Curtius rearrangement*. The rearrangement is catalyzed by both protic[18] and Lewis acids and the decomposition temperature is significantly lowered compared to the uncatalyzed reaction.[19] Acyl azides can be prepared in several different ways: 1) reacting acid chlorides or mixed anhydrides[11] with alkali azide[13] or trimethylsilyl azide;[20] 2) treating acylhydrazines with nitrous acid or nitrosonium tetrafluoroborate;[21] and 3) treating carboxylic acids with diphenyl phosphoryl azide (DPPA).[12] The product isocyanate can be isolated if the pyrolysis is conducted in the absence of nucleophilic solvents. If the reaction is carried out in the presence of water, amines ($R-NH_2$), or alcohols (R-OH), the corresponding amines, ureas, and carbamates are formed. The *Curtius rearrangement* is a very general reaction and can be applied to carboxylic acids containing a wide range of functional groups. It is also possible to induce a *Curtius rearrangement* under photochemical conditions, but this pathway gives rise to several side-products in addition to the desired isocyanate.[22] The *photochemical Curtius rearrangement* of phosphinic azides is also known as the *Harger reaction*.[23-25]

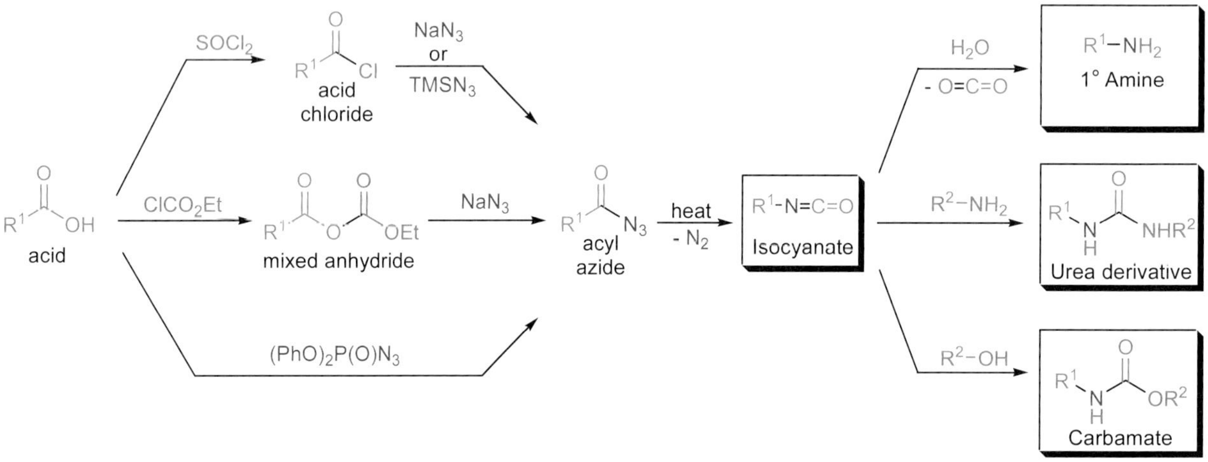

Mechanism: [26-30]

Nitrene intermediates are formed in the pyrolysis of most alkyl azides, aryl azides, sulfonyl azides, and azidoformates. However, the mechanism of the *Curtius rearrangement* under thermal conditions is most likely a concerted process.[27] This hypothesis is based on the lack of any evidence indicating the formation of a free acyl nitrene species.[15] For example, neither insertion, addition, nor amide products are isolated in the *thermal Curtius rearrangement*, which would be expected if a nitrene intermediate is involved.[6] The values of the entropy of activation are also in good agreement with a synchronous mechanism.[28] The *photochemical Curtius rearrangement* on the other hand proceeds by the formation of nitrenes, which undergo typical nitrene reactions. This is not surprising, since the energy of the photon is high enough to break the $N-N_2$ bond without alkyl or aryl participation.

Thermal rearrangement:

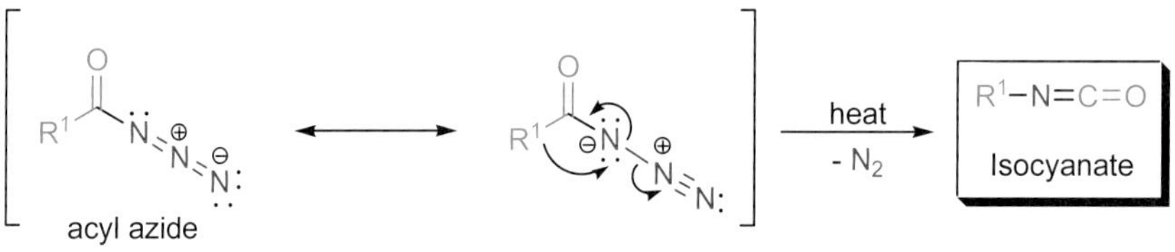

Photochemical rearrangement:

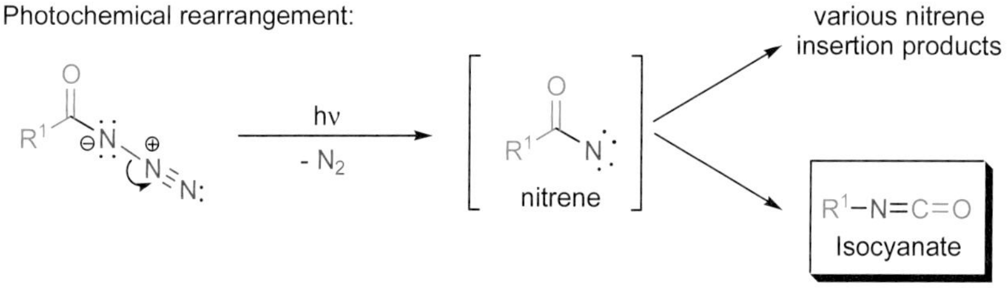

Synthetic Applications:

The enantioselective total synthesis of the cytokine modulator (–)-cytoxazone using a *syn-stereoselective aldol addition* and a *Curtius rearrangement* as key steps was described by J.A. Marco et al.[31] The key intermediate acid was treated with DPPA and triethylamine in toluene at reflux. This step furnished the oxazolidinone directly and in good yield through an *in situ* capture of the isocyanate group by the free secondary alcohol functionality. Removal of the protecting group led to the formation of the natural product.

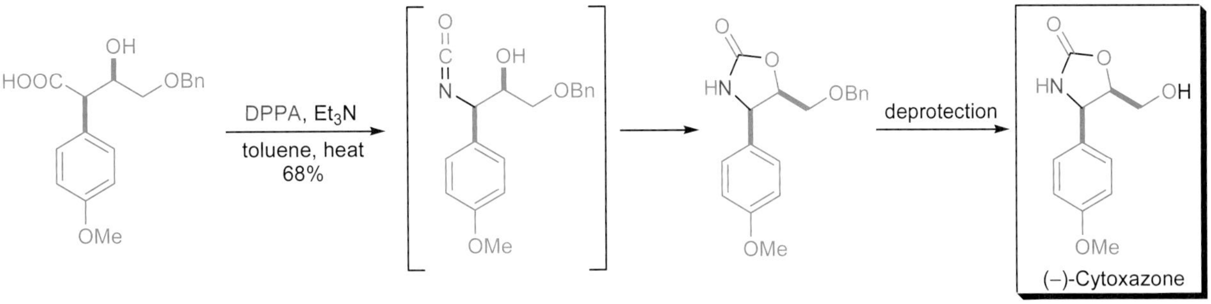

The first total synthesis of streptonigrone utilizing an *inverse electron demand Diels-Alder reaction* was accomplished in the laboratory of D.L. Boger.[32] In order to introduce the C5 pyridone amine functionality, the carboxylic acid was exposed to the Shioiri-Yamada reagent (DPPA) in benzene-water. Subsequent hydrolysis with lithium hydroxide in THF/water was necessary to complete the conversion to the primary amine.

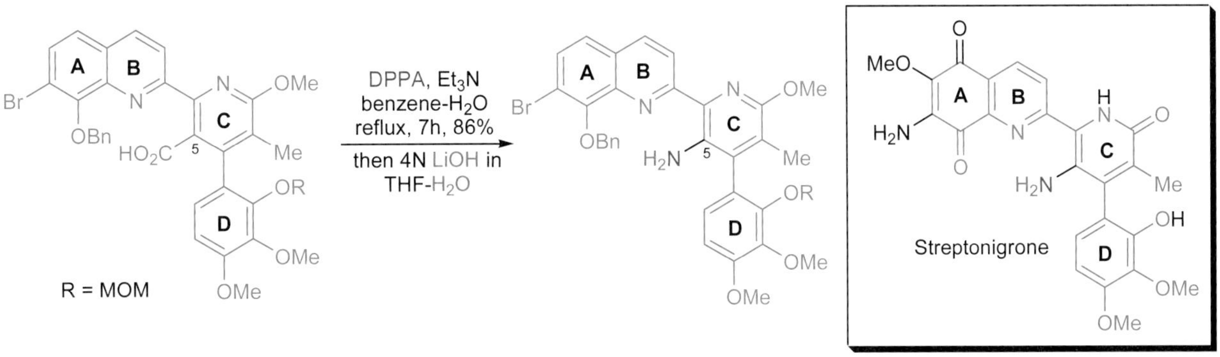

The antimuscarinic alkaloid (±)-TAN1251A possesses a unique tricyclic skeleton that consists of a 1,4-diazabicyclo[3.2.1]octane ring and a cyclohexanone ring bonded through a spiro carbon atom. K. Murashige and co-workers introduced the nitrogen connected to the spiro carbon atom by applying the *Curtius rearrangement*.[33]

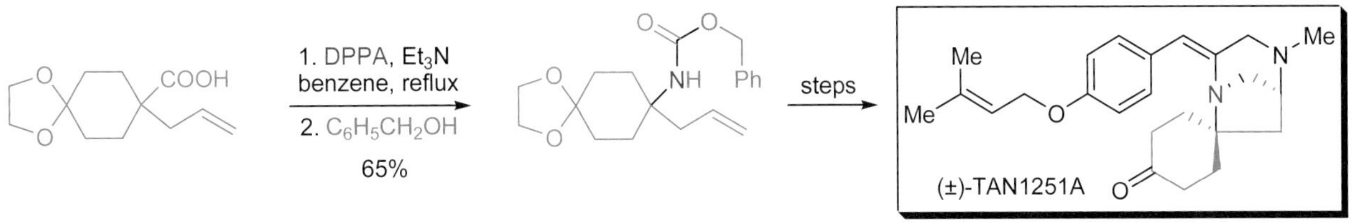

A key carbamate intermediate during the total synthesis of pancratistatin was prepared *via* the *Curtius rearrangement* of the corresponding carboxylic acid by S. Kim et al.[34] The isocyanate intermediate was rather stable and was converted to the desired carbamate in 82% overall yield by treatment with NaOMe/MeOH.

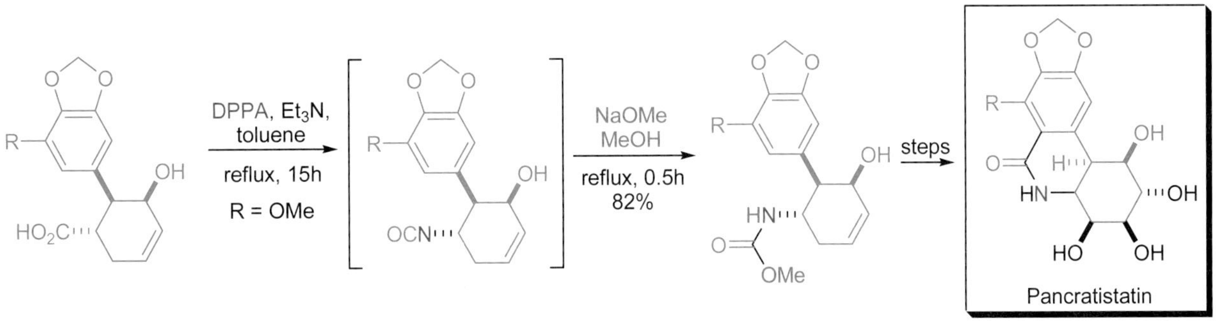

DAKIN OXIDATION

(References are on page 569)

Importance:

[*Seminal Publications*[1,2]; *Reviews*[3-7]; *Modifcations & Improvements*[8,9]]

When treated with organic peracids (RCO_3H) or hydrogen peroxide (H_2O_2), aliphatic aldehydes are smoothly oxidized to carboxylic acids. Aromatic aldehydes, however, undergo a more complex reaction in which the aldehyde group is converted to the acylated phenolic hydroxyl group. In 1909, H.D. Dakin obtained high yields of pyrocatechol (1,2-dihydrohybenzene) when he oxidized *ortho*-hydroxybenzaldehyde with perbenzoic acid.[1] The oxidation of aromatic aldehydes and ketones to the corresponding phenols is known as the *Dakin oxidation*, and this transformation is very similar to the well-known *Baeyer-Villiger oxidation*. The reaction works best if the aromatic aldehyde or ketone is electron rich (-R, -OH, -OR, $-NH_2$, or -NHR substituents in the *ortho* or *para* positions). When the aromatic ring is substituted with electron-withdrawing groups, the product of the oxidation is usually the carboxylic acid. The *Dakin oxidation* is usually performed using the following reagents: alkaline H_2O_2,[5,10] acidic H_2O_2,[11] peroxybenzoic acid,[12] peroxyacetic acid,[13] sodium percarbonate,[14] 30% H_2O_2 with arylselenium compounds as activators (*Syper process*),[8] and urea-H_2O_2 (UHP) adduct.[9]

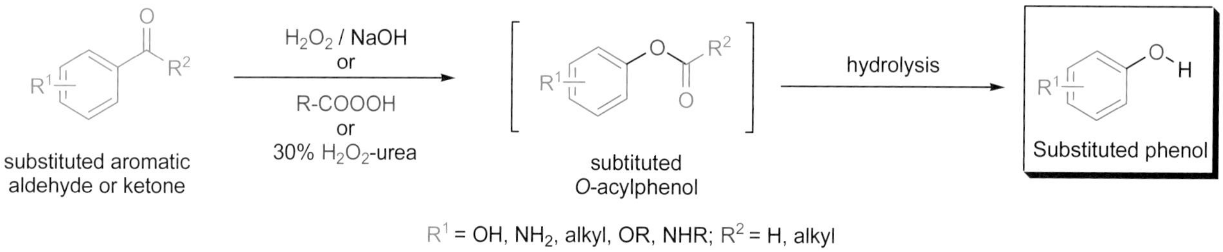

Mechanism: [12,10,15-17]

The mechanism of the *Dakin oxidation* is very similar to the mechanism of the *Baeyer-Villiger oxidation*. Under basic conditions (H_2O_2/NaOH) the hydrogen-peroxide is deprotonated to give the hydroperoxide anion (HO_2^-), which adds across the carbonyl group of the substituted aromatic aldehyde or ketone. The resulting tetrahedral intermediate undergoes a *[1,2]-aryl shift* to afford an *O*-acylphenol, which is hydrolyzed to the corresponding phenolate anion under the reaction conditions. Finally, the work-up liberates the substituted phenol from the phenolate salt.

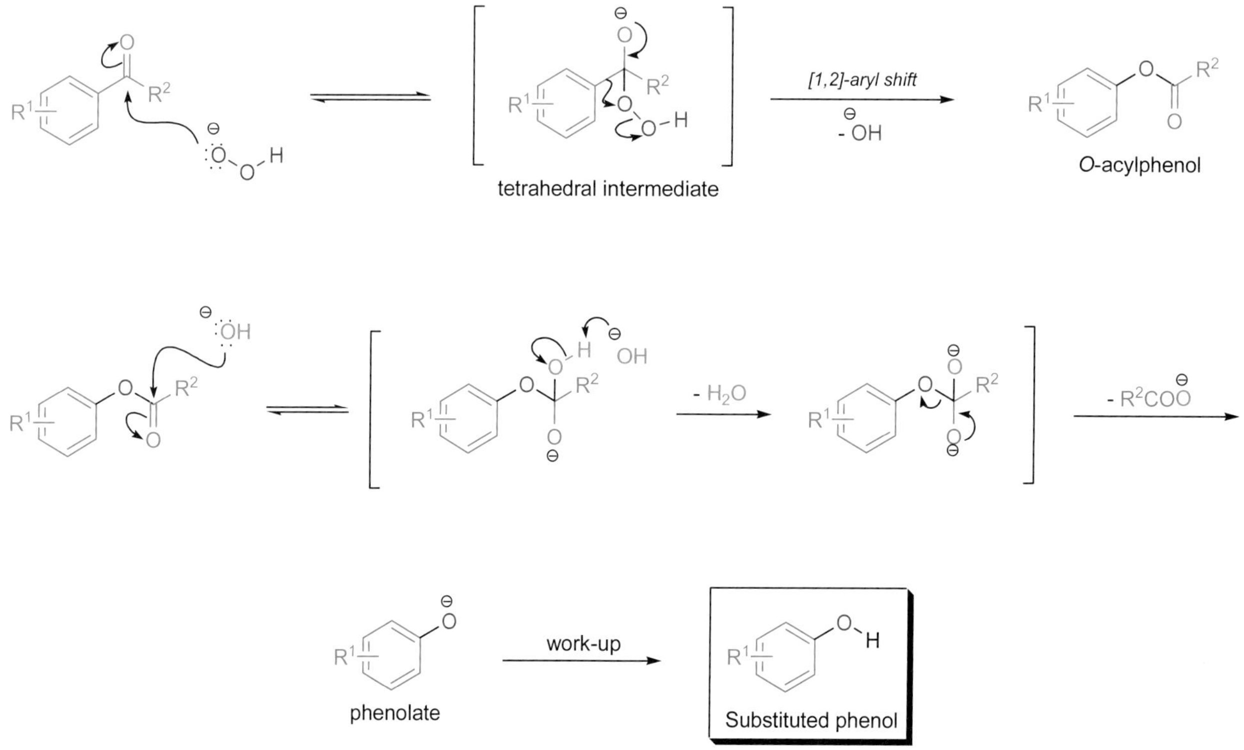

Synthetic Applications:

The total synthesis of vineomycinone B_2 methyl ester was accomplished in the laboratory of C. Mioskowski using a *double Bradsher cyclization*, a *modified Dakin oxidation*, and a *singlet oxygen oxidation* as key steps.[18] The substituted anthracene-dialdehyde derivative was treated under *modified Dakin oxidation* conditions, that is, with phenylselenic acid and hydrogen peroxide at 20 °C for 20h, to introduce the phenolic oxygens. This was followed by a singlet oxygen addition across the central aromatic ring with reductive work-up and air oxidation to generate the desired anthraquinone functionality.

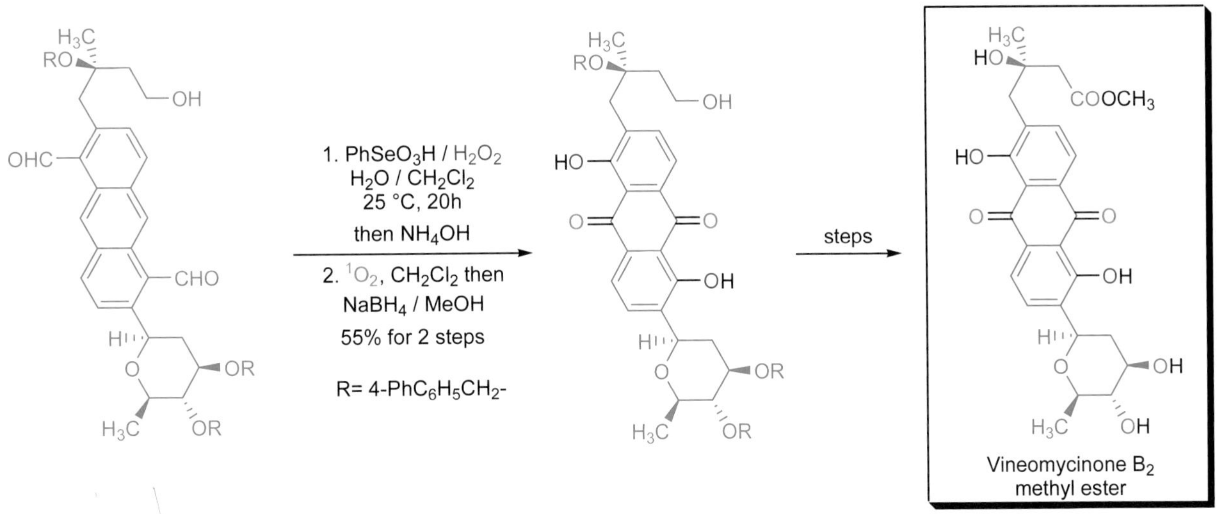

M.E. Jung and co-workers have developed a synthesis of selectively protected L-Dopa derivatives from L-tyrosine *via* a *Reimer-Tiemann reaction* followed by the *modified Dakin oxidation*.[19] The formyl group introduced by the *Reimer-Tiemann reaction* had to be converted to the corresponding phenol. After trying many sets of conditions, the *Syper process* was chosen, which uses arylselenium compounds as activators for the oxidation. Treatment of the aromatic aldehyde with 2.5 equivalents of 30% hydrogen peroxide in the presence of 4% diphenyl diselenide in dichloromethane for 18h gave the aryl formate in excellent yield. This ester was cleaved by treatment with methanolic ammonia for 1h to afford the desired phenol in good yield.

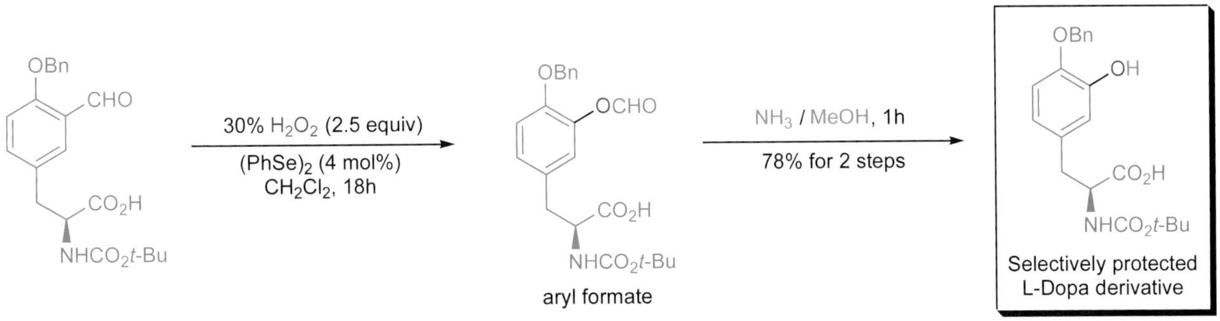

Carboxy-functionalized fluorescein dyes are important as conjugated fluorescent markers of biologically active compounds. M.H. Lyttle et al. have used the *Dakin oxidation* on 4-methoxy-3-hydroxy-2-chloro-benzaldehyde to obtain the desired resorcinol derivative that served as an intermediate in their improved synthesis.[20]

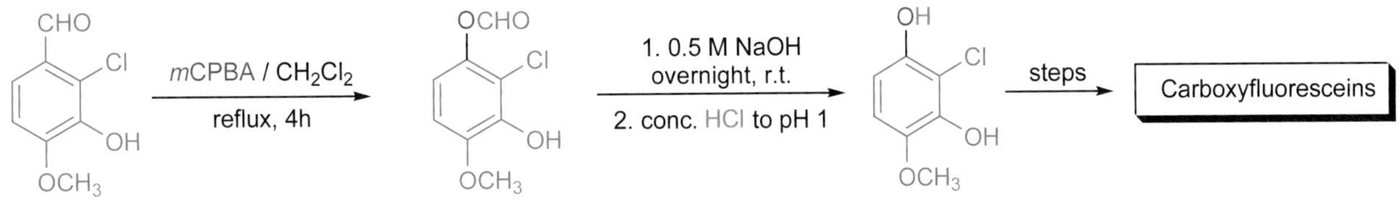

DAKIN-WEST REACTION

(References are on page 569)

Importance:

[*Seminal Publications*[1-3]; *Reviews*[4-6]; *Modifications & Improvements*[7-10]]

The conversion of carboxylic acids to ketones has been known for centuries.[6] It is therefore interesting that since the mid-1800s several chemists have claimed to have discovered this transformation (e.g., W.H. Perkin, Sr., W. Heintz, etc.).[1] In 1928, H.D. Dakin and R. West reported that when certain amino acids, such as aspartic acid and histidine, were heated in acetic anhydride in the presence of pyridine, the corresponding α-acetamido methyl ketones were formed in high yield.[2,3] The formation of α-acylamino alkyl ketones from α-amino acids and symmetrical carboxylic acid anhydrides in the presence of a base is known as the *Dakin-West reaction*. The general features of this transformation are: 1) both primary and secondary α-amino acids undergo this transformation, but β-amino acids only afford the corresponding *N*-acylated derivatives; 2) the α-amino acids need to have a proton at their α-position, otherwise they simply undergo *N*-acylation; 3) the anhydride component is most often acetic anhydride, but other anhydrides such as propionic anhydride can also be used; 4) when acetic anhydride is used, the product is an α-acetylamino methyl ketone, whereas with propionic anhydride the corresponding α-propionylamino ethyl ketone is obtained; 5) the base is usually pyridine, but various alkylpyridines and sodium acetate have been successfully employed; 6) primary α-amino acids react with anhydrides at around 100 °C, but secondary α-amino acids require significantly higher reaction temperatures; and 7) the addition of a nucleophilic catalyst such as DMAP allows the reaction to take place at room temperature.[8]

Dakin & West (1928):

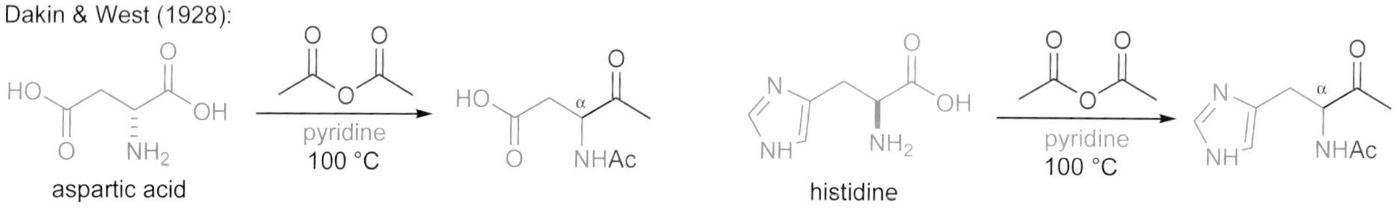

Dakin-West reaction:

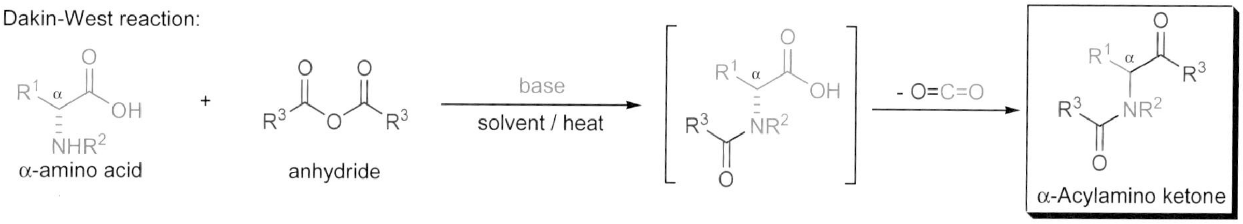

R^1 = H, alkyl, substituted alkyl; R^2 = H, alkyl, substituted alkyl, aryl, heteroaryl; R^3 = Me, Et, *n*-Pr; base: pyridine, alkylpyridine, NaOAc; solvent: pyridine, Et_3N

Mechanism: [11-26,6]

Formation of *N*-acetyl-α-amino acid:

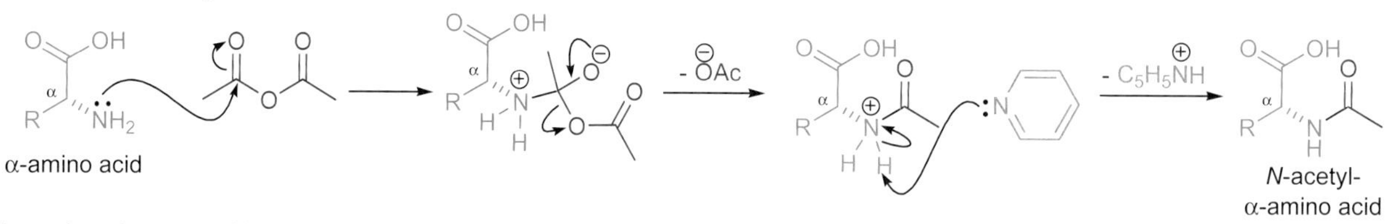

Formation of α-acetamido ketone from *N*-acetyl-α-amino acid:

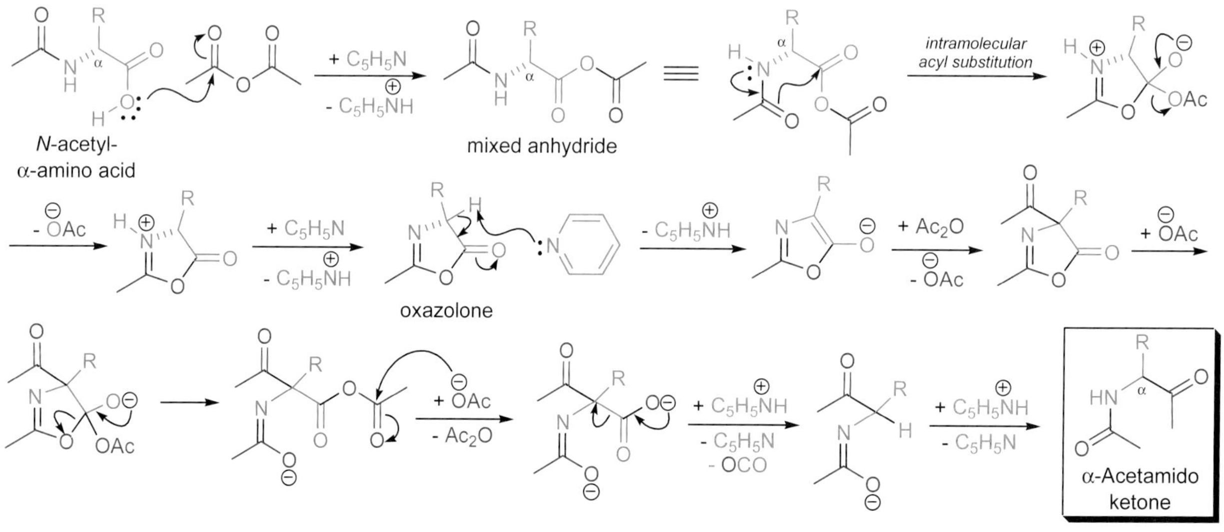

Synthetic Applications:

In the laboratory of E.B. Pedersen, several 2-methylsulfanyl-1*H*-imidazoles were prepared and tested for their activity against HIV-1.[27] These compounds can be regarded as novel non-nucleoside reverse transcriptase inhibitors. The required α-aminoketone hydrochloride building blocks were prepared using the *Dakin-West reaction*. L-Cyclohexylalanine was dissolved in excess pyridine and propionic anhydride and was kept at reflux overnight. The resulting α-propionylamino ethyl ketone was hydrolyzed with concentrated hydrochloric acid and the α-aminoketone hydrochloride was heated with one equivalent of potassium thiocyanate in water to afford 4-cyclohexylmethyl-5-ethyl-1,3-dihydroimidazole-2-thione. This material was then advanced to 4-cyclohexylmethyl-1-ethoxymethyl-5-ethyl-2-methylsulfanyl-1*H*-imidazole.

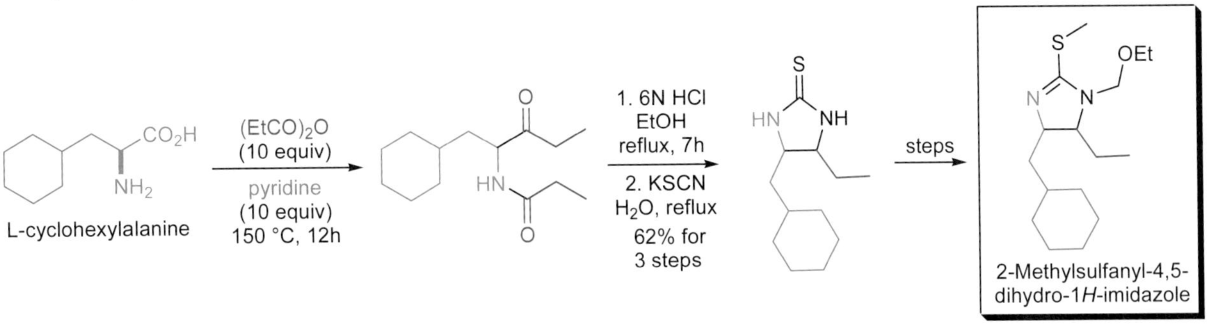

The synthesis of ketomethylene pseudopeptide analogues was accomplished by L. Cheng et al., and their biological activity as thrombin inhibitors was tested.[28] These analogues were prepared through a *modified Dakin-West reaction* under mild conditions and in almost quantitative yield. The required anhydride was prepared from monomethyl succinate, and a large excess of it was mixed with the tripeptide substrate in pyridine in addition to triethylamine and catalytic amounts of DMAP. The reaction mixture was heated for one hour at 40-50 °C.

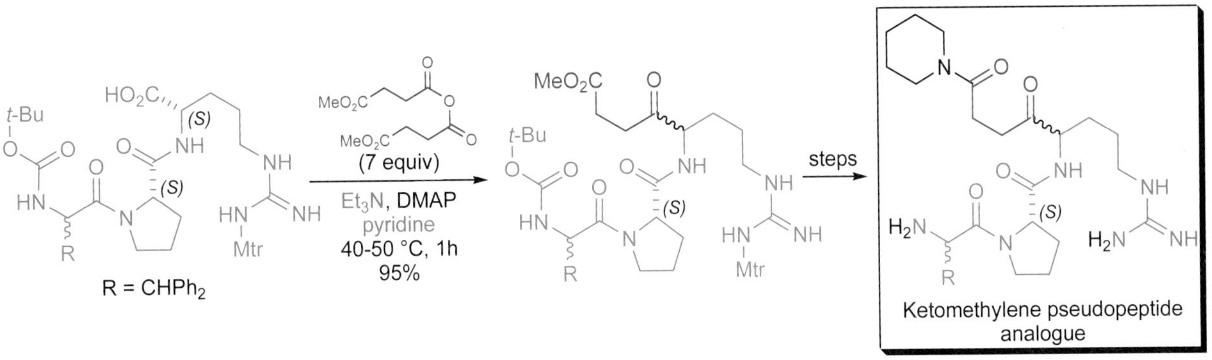

The efficient solution and solid phase synthesis of a 3,9-diazabicyclo[3.3.1]non-6-en-2-one scaffold was developed by R. Giger and co-workers from L-tryptophan using a novel sequential *Dakin-West/intramolecular Pictet-Spengler reaction*.[10]

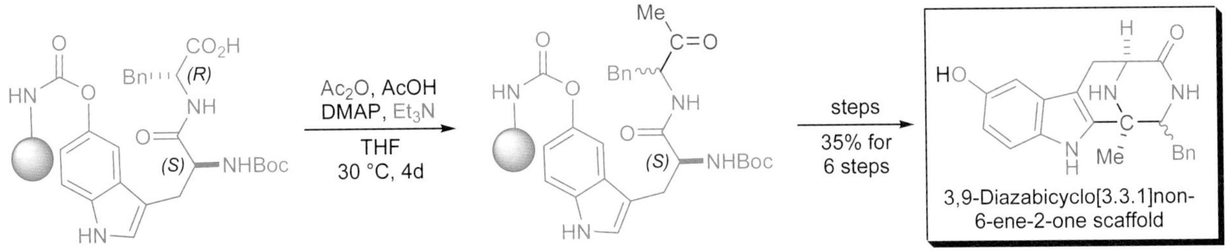

An improved method for the preparation of a series of oxazole-containing dual PPARα/γ agonists was reported by A.G. Godfrey et al.[29] The synthesis utilized the *Dakin-West reaction* which allowed the introduction of a phenyl ketone moiety. This ketone was subsequently converted to the corresponding oxazole using $POCl_3$/DMF.

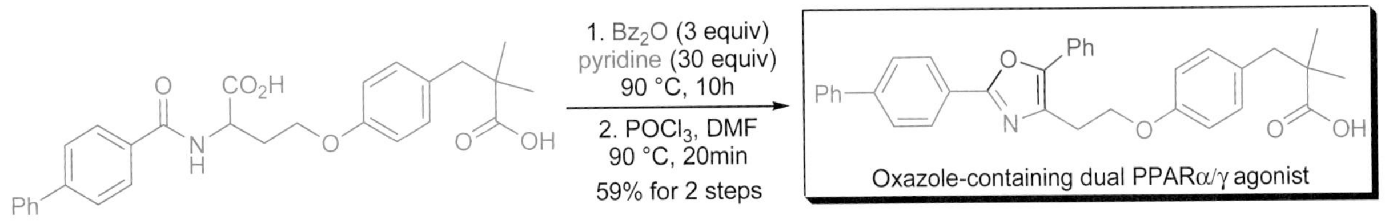

DANHEISER BENZANNULATION

(References are on page 570)

Importance:

[*Seminal Publication*[1]; *Modifications & Improvements*[2]]

In 1984, R.L. Danheiser and co-workers developed a new, one-step method for the regiocontrolled synthesis of highly substituted aromatic compounds by heating cyclobutenone derivatives with activated (heterosubstituted)[1,3-5] or unactivated acetylenes. This convergent annulation process is referred to as the *Danheiser benzannulation*, and it proceeds *via* a vinylketene intermediate. Alkoxyacetylenes were found to be the best partners for this annulation, but the relatively harsh conditions required to cleave the aryl ether moiety in the products led to the use of trialkylsilyloxyalkynes instead.[4] In the typical annulation procedure, the solution of the cyclobutenone component (in $CHCl_3$, benzene, or toluene) in the presence of a slight excess of the heterosubstituted acetylene is heated to 80-160 °C in a sealed Pyrex tube.[1] Modification of the original strategy involves the generation of the vinyl- or arylketene intermediate *via* the *photochemical Wolff-rearrangement* of an unsaturated (vinyl or aryl) α-diazo ketone.[2] This new two-step *modified Danheiser benzannulation* allows the synthesis of polycyclic aromatic and heteroaromatic systems (e.g., substituted naphthalenes, benzofurans, benzothiophenes, indoles, carbazoles, etc.), which cannot be accessed using the original methodology. The advantage of this new procedure is that the various functionalized aryl and vinyl α-diazo ketones are easily accessible from a wide range of available simple ketones and carboxylic acid derivatives. The best yields are obtained when 3-alkoxy phenol derivatives are formed, and in this respect the *modified Danheiser benzannulation* complements the *Dötz benzannulation* reaction, which results in the formation of 4-alkoxy phenol derivatives.

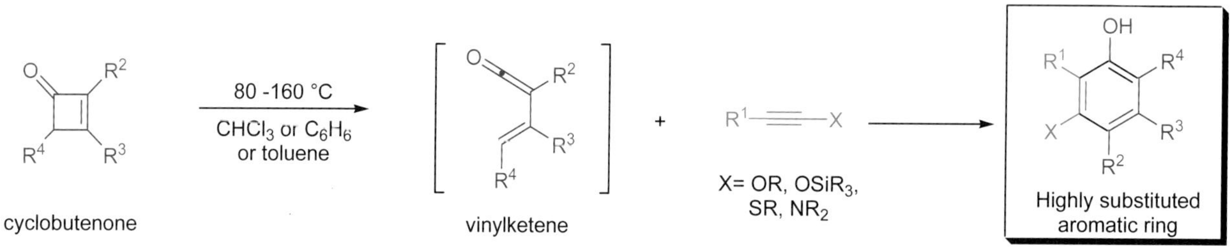

Mechanism: [1,2]

In the original version of the annulation, the vinylketene[6] intermediate is generated in a reversible *4π electrocyclic ring opening* of the cyclobutenone followed by a cascade of three more pericyclic reactions. The ketenophilic alkyne reacts with the vinylketene in a regiospecific *[2+2] cycloaddition.* The resulting 2-vinylcyclobutenone then undergoes a reversible *4π electrocyclic cleavage* to give a dienylketene, which immediately rearranges in a *six-electron electrocyclization* to afford a cyclohexadienone. The highly substituted phenol is formed after tautomerization. The *photochemical Wolff rearrangement* of the unsaturated α-diazoketone also yields the vinylketene, and most likely proceeds *via* carbene and oxirene intermediates.[7]

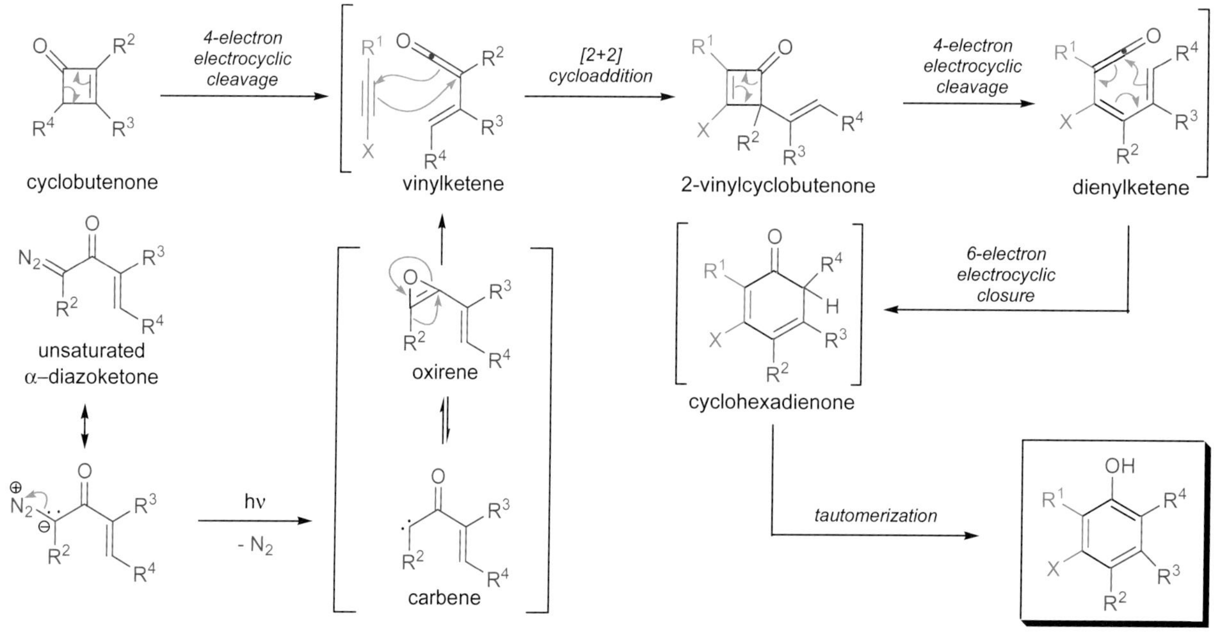

Synthetic Applications:

R.L. Danheiser and co-workers have used the *modified Danheiser benzannulation* for the synthesis of the marine carbazole alkaloid hyellazole.[2] The required diazoketone was prepared from the *N*-Boc derivative of 3-acetylindole using a *diazo transfer reaction*. The diazoketone was irradiated in the presence of the alkyne to afford the desired carbazole annulation product in 56% yield. Finally, in order to install the phenyl group of hyellazole at C1, the phenolic hydroxyl group was converted to the corresponding triflate and a *Stille cross-coupling* was performed.

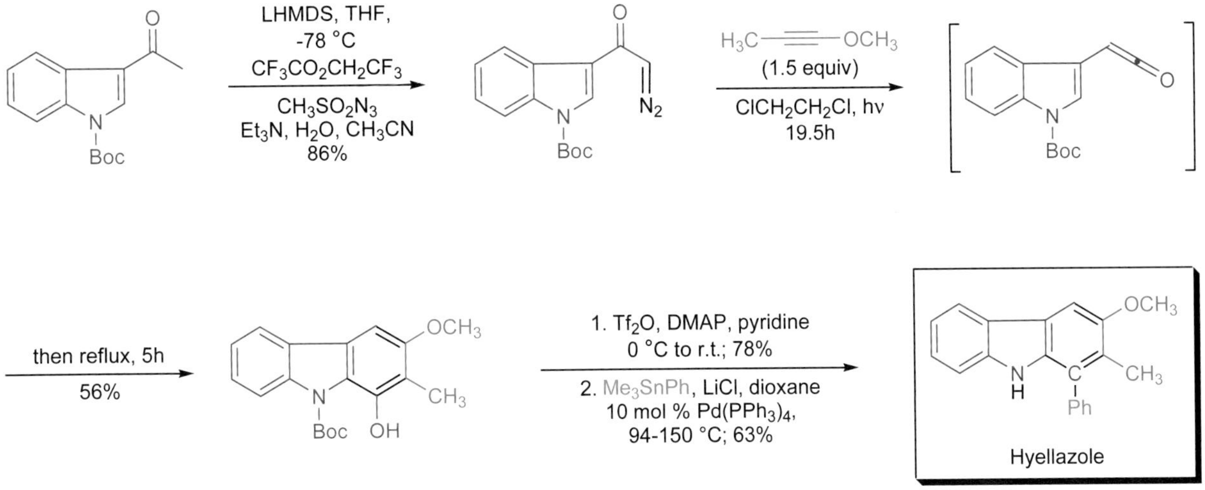

The use of substituted alkoxyacetylenes in synthesis is fairly limited due to the lack of simple, general methods for their preparation. However, silyloxyacetylenes are easier to make and can be prepared from esters in a one-pot operation.[8] In the laboratory of C.J. Kowalski, research has shown that silyloxyacetylenes could be successfully used in the *Danheiser benzannulation*.[5] This modification was used in the total synthesis of Δ-6-tetrahydrocannabinol.

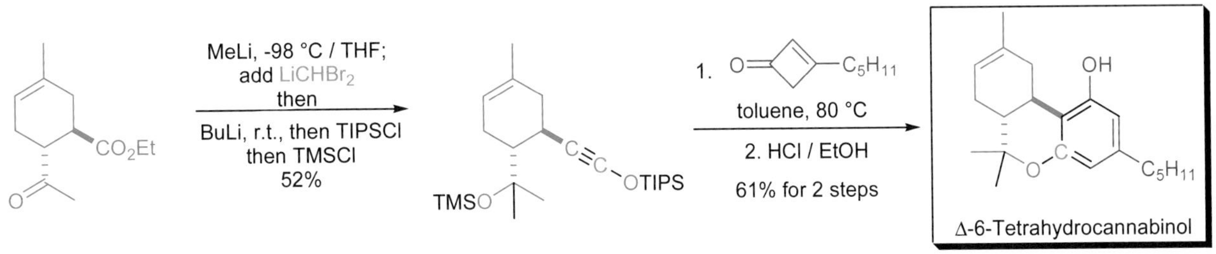

During the total synthesis of (–)-cylindrocyclophane F, A.B. Smith et al. used the *Danheiser benzannulation* to construct the advanced aromatic intermediate for an *olefin metathesis dimerization* reaction.[9] The starting material triisopropylsilyloxyalkyne was synthesized from the corresponding ethyl ester using the *Kowalski two-step chain homologation*.[8]

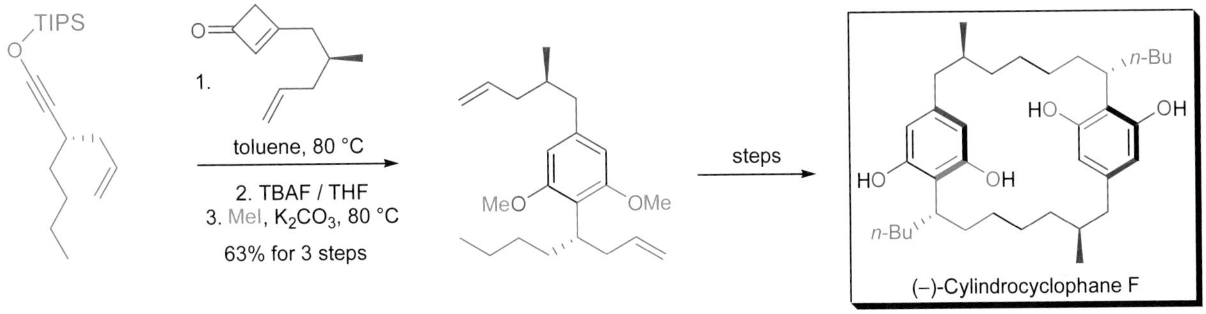

DANHEISER CYCLOPENTENE ANNULATION

(References are on page 570)

Importance:

[*Seminal Publications*[1,2]; *Modifications and Improvements*[3-8]]

The one-step regio- and stereoselective [3+2] annulation of (trimethylsilyl)allenes and electron-deficient alkenes (allenophiles) in the presence of titanium tetrachloride ($TiCl_4$) to produce highly substituted cyclopentene derivatives is referred to as the *Danheiser cyclopentene annulation*. The typical annulation involves rapid addition of 1.5 equivalents of distilled $TiCl_4$ to a methylene chloride solution containing the allenophile and 1.0-1.5 equivalents of (trimethylsilyl)allene at -78°C.[1,2] The required (trimethylsilyl)allenes are relatively easy to prepare, and the allenophiles are usually readily available α,β-unsaturated ketones. Both cyclic and acyclic enones are good reaction partners. However, other allenophiles such as α-nitro olefins only react with allenes in a *Michael type process*. α,β-Unsaturated aldehydes give complex reaction mixtures, whereas α,β-unsaturated esters react sluggishly to afford the desired cyclopentene derivative in moderate yields. The annulation works most efficiently using 1-substituted (trimethylsilyl)allenes. The addition of the allene to the allenophile is predominantly suprafacial, and as a result, the annulation is highly stereoselective. The reaction of allenylsilanes with other electrophiles results in the formation of heterocycles.[4,5,8]

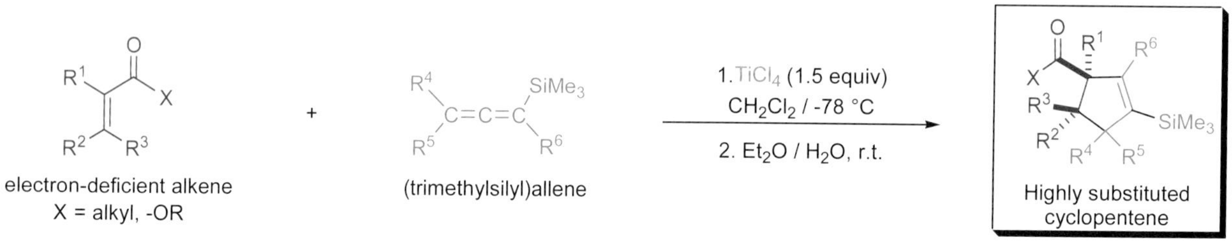

Mechanism: [1,2]

The first step of the mechanism involves the initial complexation of titanium tetrachloride to the carbonyl group of the electron-deficient alkene (enone) to give an alkoxy-substituted allylic carbocation. The allylic carbocation attacks the (trimethylsilyl)allene regiospecifically at C3 to generate vinyl cation **I**,[9] which is stabilized by the interaction of the adjacent C-Si bond. The allylic π-bond is only coplanar with the C-Si bond in (trimethylsilyl)allenes, so only a C3 substitution can lead to the formation of a stabilized cation.[1] A *[1,2]-shift* of the silyl group follows to afford an isomeric vinyl cation (**II**), which is intercepted by the titanium enolate to produce the highly substituted five-membered ring.[10,11] Side products (**III** – **V**) may be formed from vinyl cation **I**.

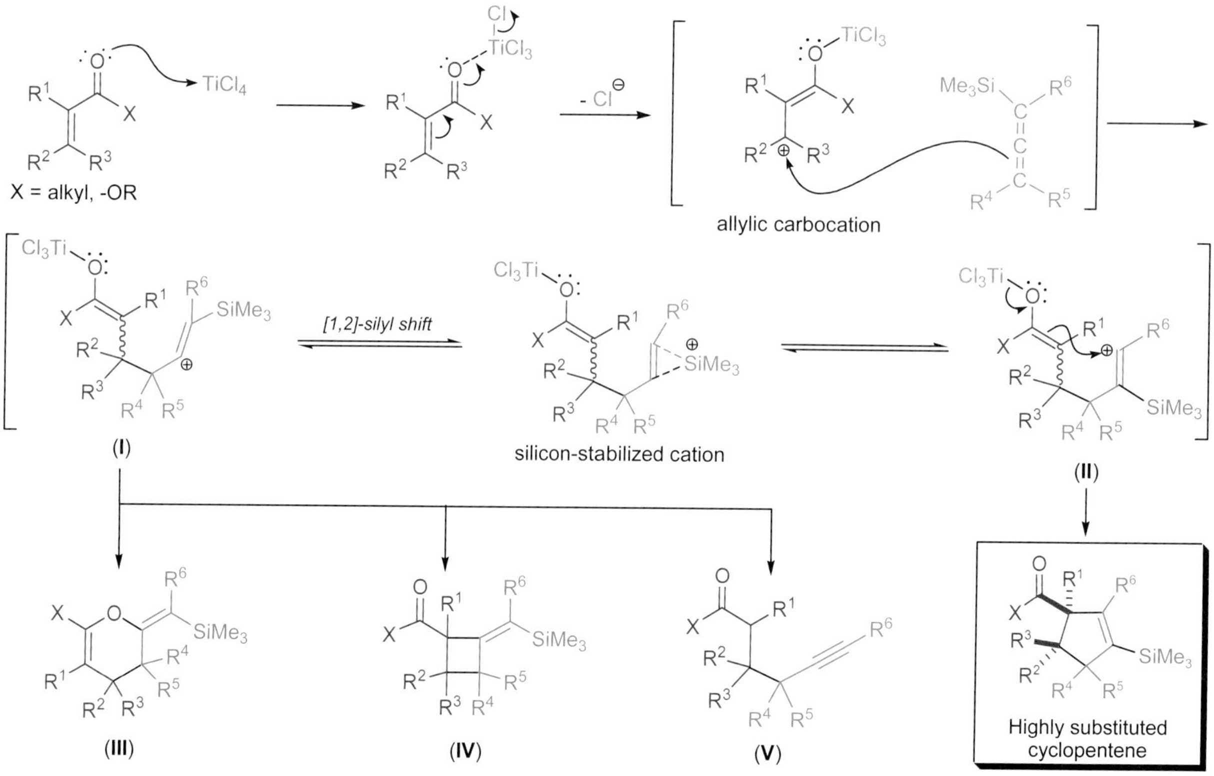

Synthetic Applications:

Disilanyl groups are considered the synthetic equivalent of the hydroxyl group. These groups can be easily converted in a one-pot reaction to the corresponding hydroxyl group by treatment with TBAF in THF followed by $H_2O_2/KHCO_3$ oxidation.[12] Y. Ito and co-workers have demonstrated the synthetic usefulness of the disilanyl groups in the *disilane version of the Danheiser cyclopentene annulation*. In the presence of 1.5 equivalents of $TiCl_4$, allenyldisilanes reacted with 1-acetylcycloalkenes to give bicyclic alkenyldisilanes in moderate to good yields. Then the bicyclic alkenyldisilanes were converted to the corresponding bicyclic ketones *via* oxidation.

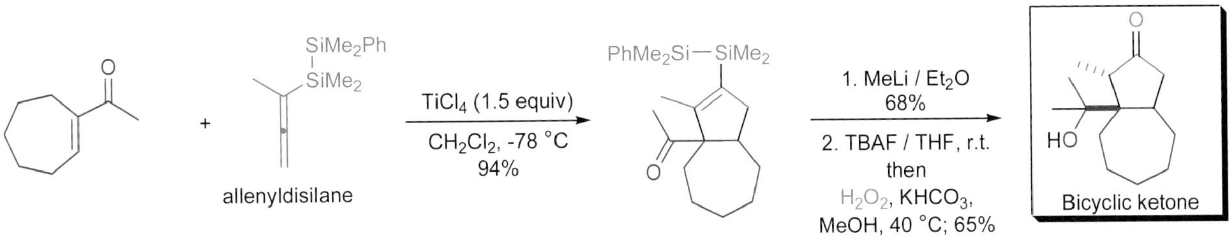

H.J. Schäfer et al. achieved the formal total synthesis of the trinorguaiane sesquiterpenes (±)-clavukerin and (±)-isoclavukerin by using the *Danheiser cyclopentene annulation* as the key step.[13] Racemic 4-methylcyclohept-2-en-1-one was reacted with (trimethylsilyl)allene in the presence of 1.7 equivalents of $TiCl_4$ in dichloromethane at -78 °C to afford a 1:1 mixture of the *cis*-fused diastereomers, which were easily separated by HPLC. The diastereomers were then converted to key fragments of earlier total syntheses of the above mentioned natural products.

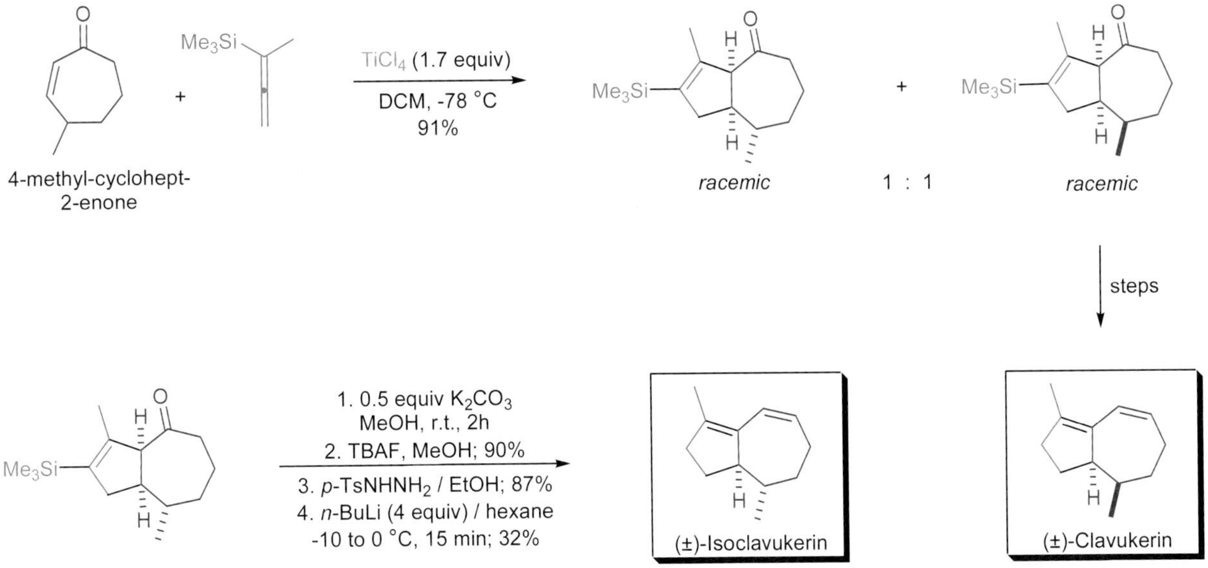

Research in the laboratory of R.L. Danheiser has shown that allenylsilanes can be reacted with electrophiles other than enones, such as aldehydes and *N*-acyl iminium ions to generate oxygen and nitrogen heterocycles.[4] Aldehydes can function as heteroallenophiles and the reaction of C3 substituted allenylsilane with the achiral cyclohexane carbaldehyde afforded predominantly *cis*-substituted dihydrofurans.

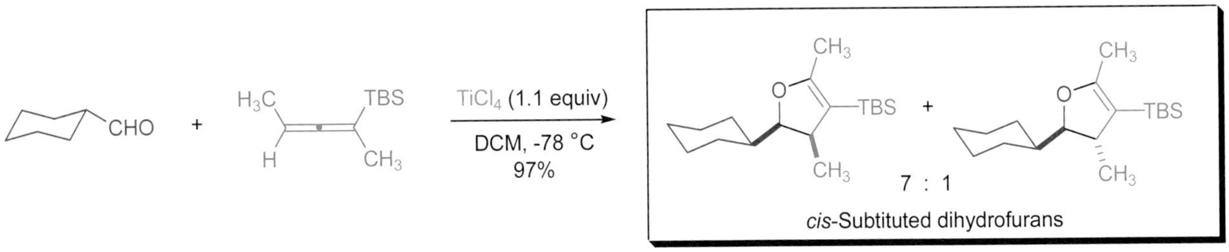

DANISHEFSKY'S DIENE CYCLOADDITION

(References are on page 570)

Importance:

[*Seminal Publication*[1]; *Reviews*[2-5]; *Modifications and Improvements*[6-16]]

Following the discovery of the *Diels-Alder cycloaddition reaction* in 1928, a wide variety of functional groups were incorporated into the dienophile component, while the variation of substituents on the diene component was fairly limited. In 1974, S.J. Danishefsky et al. prepared an electron-rich heteroatom substituted diene, (*E*)-1-methoxy-3-(trimethylsilyloxy)-1,3-diene, which was later successfully used in *normal and hetero Diels-Alder cycloaddition* reactions.[1] Cycloaddition reactions involving this particular diene are referred to as *Danishefsky's diene cycloadditions*. Danishefsky's diene readily reacts with imines,[6,16] aldehydes,[4,11,14] alkenes, alkynes, and even with certain electron-deficient aromatic rings[17] to afford the corresponding heterocyclic and carbocyclic rings. In general, heteroatom substituents with lone pairs of electrons have the following effects on the diene component: 1) the diene becomes more electron-rich, making it more reactive toward dienophiles; 2) regioselectivity of the cycloaddition is improved when unsymmetrical dienophiles are used; and 3) the heteroatom serves as a handle for post-cycloaddition modifications (e.g., the β-alkoxy enol silyl ether is converted to the corresponding enone under acidic conditions).[18] The increase in reactivity can be explained by the FMO theory, namely that the electron-rich heteroatom increases the HOMO energy level of the diene thereby decreasing the energy difference between the diene's HOMO and the dienophile's LUMO. As a result, the transition state is stabilized, and the reaction rate is increased. Over the years, structural modifications to *Danishefsky's diene* improved the reactivity and selectivity as well as the acid and heat sensitivity of these electron-rich dienes.[7,9,10,12] The *Danishefsky's diene cycloaddition* reactions are catalyzed by various Lewis acids, and asymmetric versions have also been developed.[11,13-15]

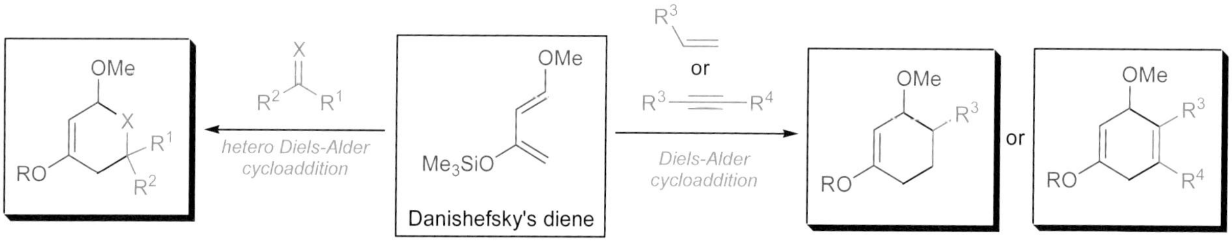

R = TMS; X = O, NH, NR; R^1 = alkyl, aryl; R^2 = H, alkyl; R^3, R^4 = alkyl, aryl, EWG

Mechanism: [19,11,20,13]

There are two different modes of cyclizations in hetero [4+2] cycloadditions involving *Danishefsky's diene*: 1) concerted (pericyclic) and 2) stepwise. When carbonyl compounds are reacted with Danishefsky's diene, the stepwise pathway is often referred to as the *Mukaiyama aldol reaction pathway*. The concerted process is called the *Diels-Alder pathway*. The mode of cyclization in the case of Lewis acid catalyzed reactions depends on the Lewis acid itself and whether it is present in stoichiometric or catalytic amounts.[19] The *Mukaiyama aldol pathway* has been observed only with titanium[21] and boron[22,23] complexes, while the *Diels-Alder pathway* occurred when aluminum,[11] chromium,[24] europium,[25] rhodium,[14] zinc,[19] and ytterbium[26] complexes were used. The scheme below shows that the intermediates of both mechanistic pathways give the same product upon treatment with acid.

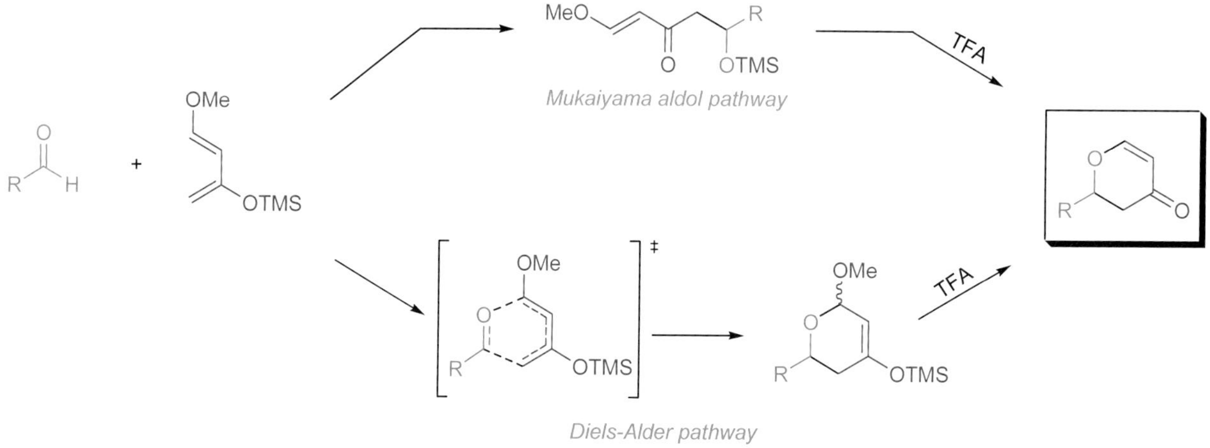

Synthetic Applications:

The first total synthesis of the marine furanosesquiterpenoid tubipofuran was accomplished in the laboratory of K. Kanematsu.[27] The *cis*-fused furanodecalin system was constructed by the *regioselective Diels-Alder cycloaddition* reaction of benzofuran quinone and *Danishefsky's diene* in refluxing toluene. The reaction gave an 11:1 mixture of the desired *ortho-endo* adduct versus the undesired *para-endo* product in 98% isolated yield. The major isomer then was subjected to sequential *radical deoxygenation reactions* before it was finally converted to the natural product.

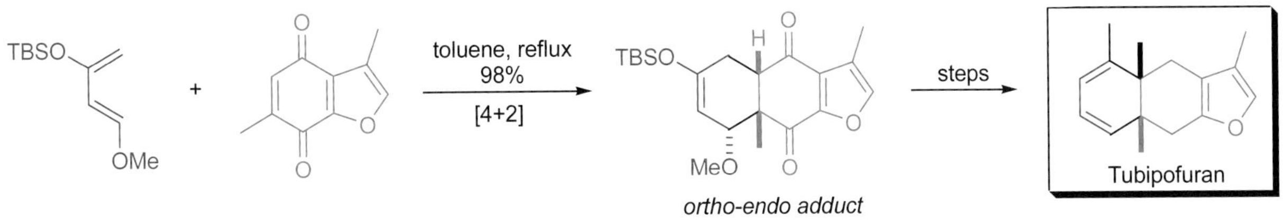

The enantioselective total synthesis of the Securienega alkaloid (–)-phyllanthine by S.M. Weinreb et al. involved a stereoselective $Yb(OTf)_3$-promoted *hetero Diels-Alder reaction* between a cyclic imine dienophile and *Danishefsky's diene*.[26] This was the first example of using an unactivated cyclic imine in this type of cycloaddition. Commonly used Lewis acid catalysts (e.g., $SnCl_2$, $TiCl_4$, etc.) produced only low yields of the desired cycloadduct. However, it was discovered that ytterbium triflate catalyzed the cycloaddition and afforded the product in 84% yield. Later they also found that the cyclization could occur at high pressure and in the absence of the catalyst, although a slightly lower yield (71%) of the product was obtained.

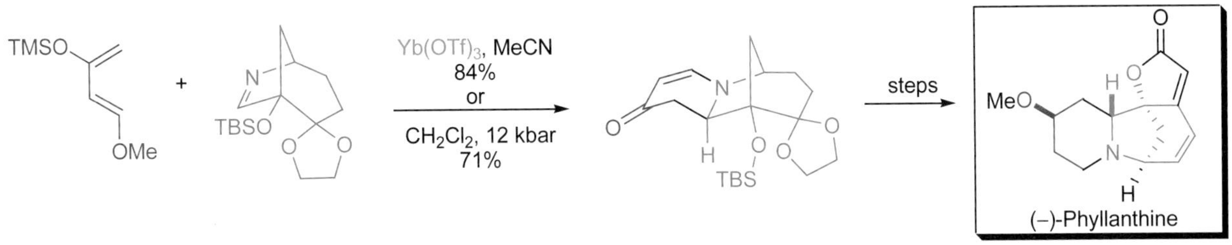

(±)-A80915G is a member of the napyradiomycin family of antibiotics. Its concise total synthesis was published by M. Nakata and co-workers using sequential *Stille cross-coupling* of aryl halides with allyltins and the *Diels-Alder reaction* of a chloroquinone with the *Danishefsky-Brassard diene*.[28]

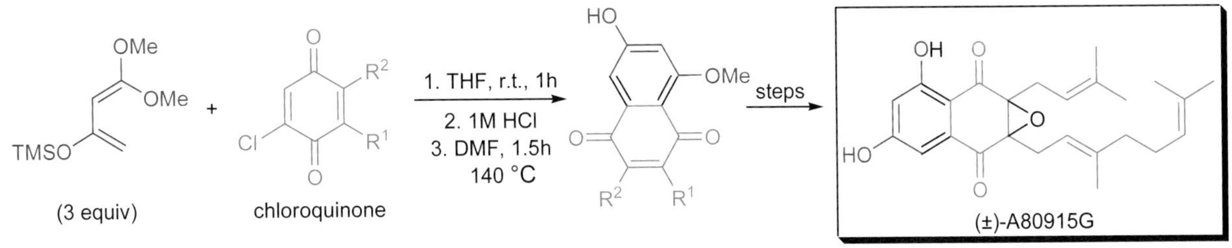

A versatile C_4-building block, *difluorinated Danishefsky's diene*, was developed for the construction of fluorinated six-membered rings in the laboratory of K. Uneyama. The diene was prepared by the selective C-F bond cleavage of trifluoromethyl ketones. The reaction of this novel diene with benzaldehyde afforded the corresponding difluoro dihydropyrone in 92% *ee* in the presence of equimolar Ti(IV)-(*R*)-BINOL.[12]

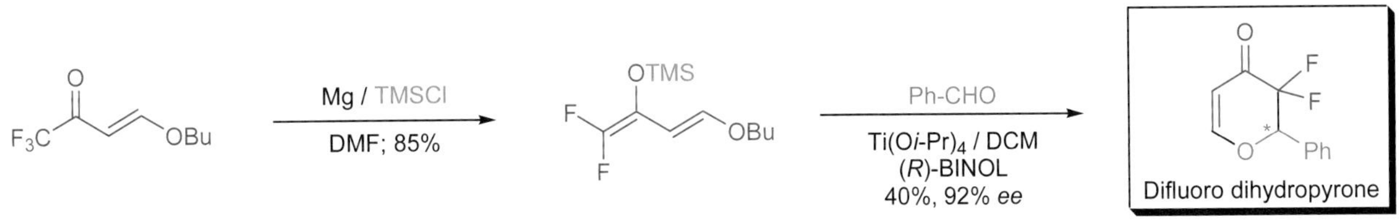

DARZENS GLYCIDIC ESTER CONDENSATION

(References are on page 571)

Importance:

[*Seminal Publications*[1-3]; *Reviews*[4-7]; *Modifications and Improvements*[8-16]]

The formation of α,β-epoxy esters (glycidic esters) from aldehydes and ketones and α-halo esters under basic conditions is known as the *Darzens glycidic ester condensation*. The first report of this transformation was published by E. Erlenmeyer, and he described the condensation of benzaldehyde with ethyl chloroacetate in the presence of sodium metal.[1] During the early 1900s G. Darzens developed and generalized the reaction and found that sodium ethoxide (NaOEt) was a very efficient condensing agent.[3] Sodium amide[2] and other bases such as *N*-ethyl-*N*-(tributylstannyl)carbamate[17] can also be used to bring about the *Darzens condensation*. The reaction is general, since aromatic aldehydes and ketones, aliphatic ketones as well as α,β-unsaturated and cyclic ketones react smoothly and give good yields of the expected glycidic esters. Aliphatic aldehydes usually give lower yields, but the deprotonation of the α-halo ester with a strong kinetic base prior to the addition of the aldehyde results in acceptable yields.[18] α-Chloro esters are preferable to bromo or iodo esters, since they give higher yields. In addition to α-halo esters, α-halo sulfones,[19,15] nitriles,[20,16] ketones,[17] ketimines,[21] thiol esters,[22] or amides[14,16] can also be used to obtain the corresponding glycidic derivatives. A useful extension of the reaction is the *Darzens aziridine synthesis* (*aza-Darzens reaction*) when the α-halo esters are condensed with imines.[8] Newer versions of the *aza-Darzens reaction* allow the preparation of aziridines in optically pure form.[11,12] Glycidic esters are versatile synthetic intermediates: the epoxide functionality can be opened with various nucleophiles and upon thermolysis the intermediates undergo decarboxylation to afford the corresponding one carbon homologue of the starting aldehyde or ketone.[23]

R^1 = alkyl, aryl; X = Cl, Br, I; EWG = CO_2R, CN, SO_2R, $CONR_2$, C(=O), C(=NR); R^2 = alkyl, aryl, H; R^3 = alkyl, aryl; Y = O, NR;
base = Na, NaOEt, $NaNH_2$, NaOH, K_2CO_3, NaO*t*-Bu;
when Y = O and EWG = CO_2R then the product is called glycidic ester

Mechanism:[24-26,6,27-29]

The first step of the mechanism is an *aldol reaction*: the base deprotonates the α-halo ester in a rate-determining step and the resulting carbanion (enolate) attacks the carbonyl group of the reactant aldehyde or ketone. The resulting intermediate is a halohydrin that undergoes an S_Ni reaction in the second step to form the epoxide ring. The strereochemical outcome of the *Darzens condensation* is usually in favor of the *trans* glycidic derivative. However, changing the solvents, bases, and the substituents can give either the *cis* or *trans* diastereomers. The stereochemistry of the product is determined by the initial enolate geometry and the steric requirements of the transition state.[29]

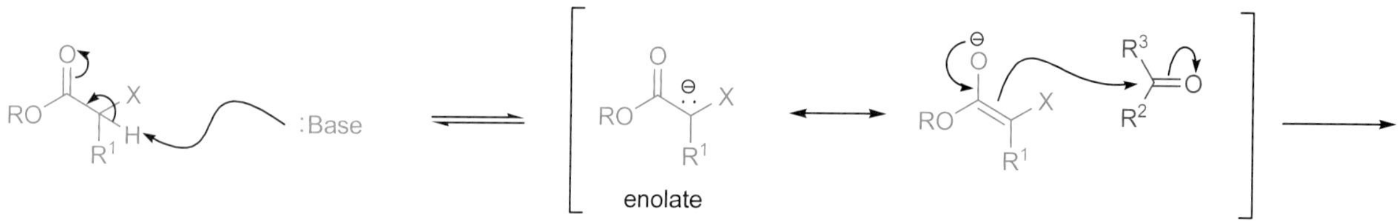

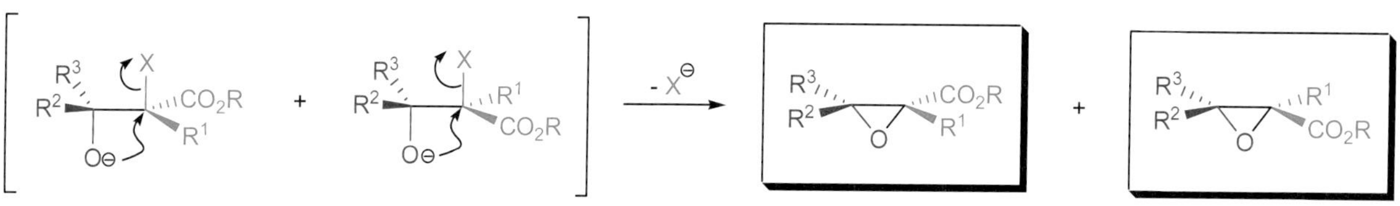

Synthetic Applications:

During the enantioselective total synthesis of (–)-coriolin, I. Kuwajima and co-workers used a *Darzens-type reaction* to construct the spiro epoxide moiety on the triquinane skeleton.[30] Interestingly, the usual *Darzens condensation* where the α-bromoketone was condensed with paraformaldehyde yielded a bromohydrin in which the hydroxymethyl group was introduced from the concave face of the molecule. This bromohydrin upon treatment with DBU gave the undesired stereochemistry at C3 (found in 3-*epi*-coriolin). To obtain the correct stereochemistry at C3, the substituents were introduced in a reverse manner. It was also necessary to enhance the reactivity of the enolate with potassium pinacolate by generating a labile potassium enolate in the presence of NIS. The *in situ* formed iodohydrin, then cyclized to the spiro epoxide having the desired stereochemistry at C3.

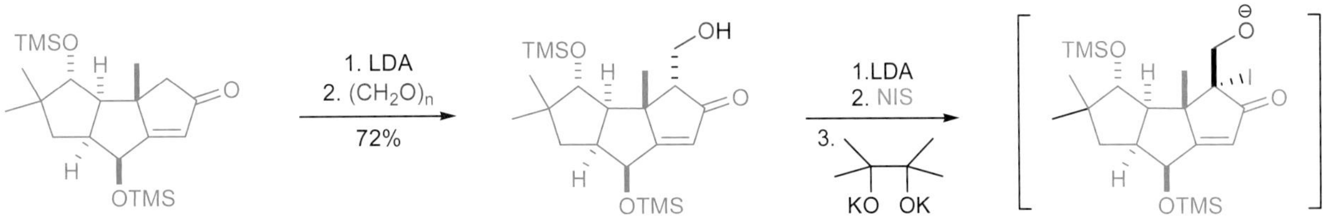

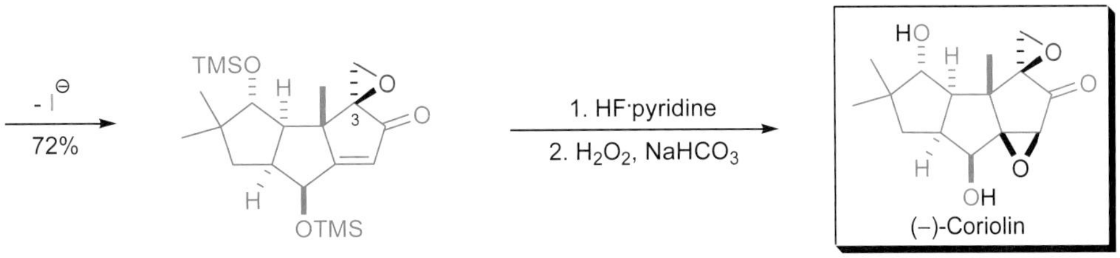

In the laboratory of P.G. Steel, a five-step synthesis of (±)-epiasarinin from piperonal was developed.[31] The key steps in the sequence involved the *Darzens condensation*, *alkenyl epoxide-dihydrofuran rearrangement* and a Lewis acid mediated cyclization. The desired vinyl epoxide intermediate was prepared by treating the solution of (*E*)-methyl-4-bromocrotonate and piperonal with LDA, then quenching the reaction mixture with mild acid (NH_4Cl).

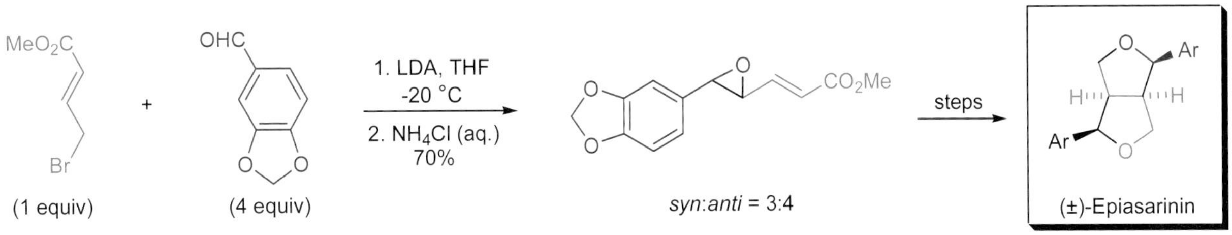

A. Schwartz et al. synthesized several calcium channel blockers of the diltiazem group enantioselectively by using an auxiliary-induced asymmetric *Darzens glycidic ester condensation*.[32] The condensation of *p*-anisaldehyde with an enantiopure α-chloro ester afforded a pair of diastereomeric glycidic esters that possessed significantly different solubility. The major product was crystallized directly from the reaction mixture in 54% yield and in essentially enantiopure form. This major glycidic ester was then converted to diltiazem in a few more steps.

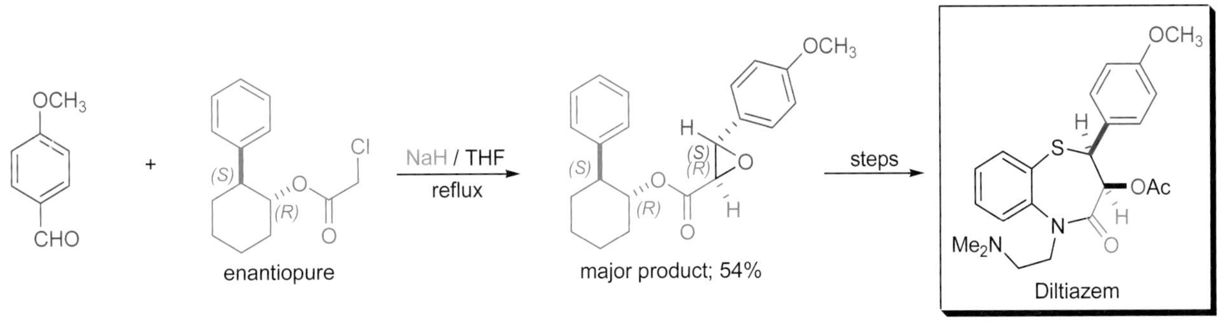

DAVIS' OXAZIRIDINE OXIDATIONS

(References are on page 572)

Importance:

[*Seminal Publications*[1-9]; *Reviews*[10-17]; *Modifications & Improvements*[18-22]; *Theoretical Studies*[23-25]]

Three-membered heterocyclic compounds containing oxygen, nitrogen, and carbon atoms are called oxaziridines. The first oxaziridines were prepared by treating imines with peroxyacids in the second half of the 1950s.[26,27] Oxaziridines are highly reactive compounds due to the ring strain and the relatively weak N-O bond, and they can serve as both aminating and oxygenating agents. Nucleophiles attack at the nitrogen atom if the substituent attached to the aziridine nitrogen is small (R^1 = H, Me). However, in the case of larger substituents, the nucleophilic attack takes place at the oxygen atom instead. In the late 1970s, F.A. Davis prepared *N*-sulfonyloxaziridines, which act exclusively as oxidizing agents with nucleophiles and their rate of oxidation is comparable to peracids.[2] The oxidation reactions involving 2-arylsulfonyl-3-aryloxaziridines (Davis' reagents) are called *Davis' oxaziridine oxidations*. *N*-sulfonyloxaziridines offer two major advantages: they are highly chemoselective and also neutral, aprotic oxidizing agents. The following oxidative transformations are easily carried out: 1) sulfides and selenides to sulfoxides[28,29] and selenoxides[7] without overoxidation; 2) alkenes to epoxides;[4,6,22] 3) amines to hydroxylamines and amine oxides;[30] and 4) organometallic compounds to alcohols or phenols.[31] The most widespread application of *N*-sulfonyloxaziridines is the oxidation of enolates to α-hydroxy carbonyl compounds (acyloins).[13] Recently, the synthetic utility of a new class of oxaziridines, perfluorinated oxaziridines, is being investigated due to the unique reactivity profile of these oxidizing agents.[19]

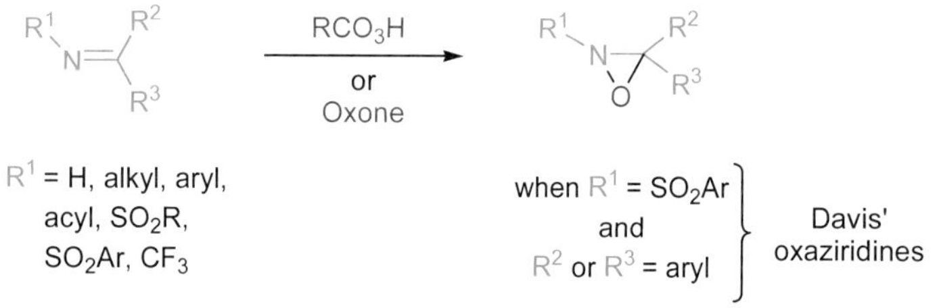

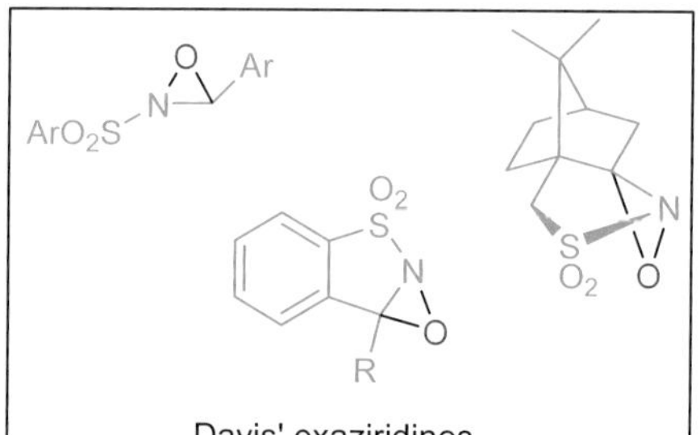

Davis' oxaziridines

Mechanism: [6,9,32-34,13,20,35]

The mechanism of oxygen transfer from oxaziridines to nucleophiles is believed to involve an S_N2 type reaction and this assumption is supported by theoretical[23-25] and experimental[9] studies. When sulfides are oxidized to the corresponding sulfoxides and sulfones, the molecular recognition is steric in origin, and it is determined by the substituents on both the substrate and the oxaziridine.[9] For the oxidation of enolates, the molecular recognition is explained with an S_N2 mechanism as well as by an open (non-chelated) transition state where the nonbonded interactions are minimized.[33,36,20] The mechanism of oxygen transfer to an enolate to form the corresponding acyloin is shown below.[13]

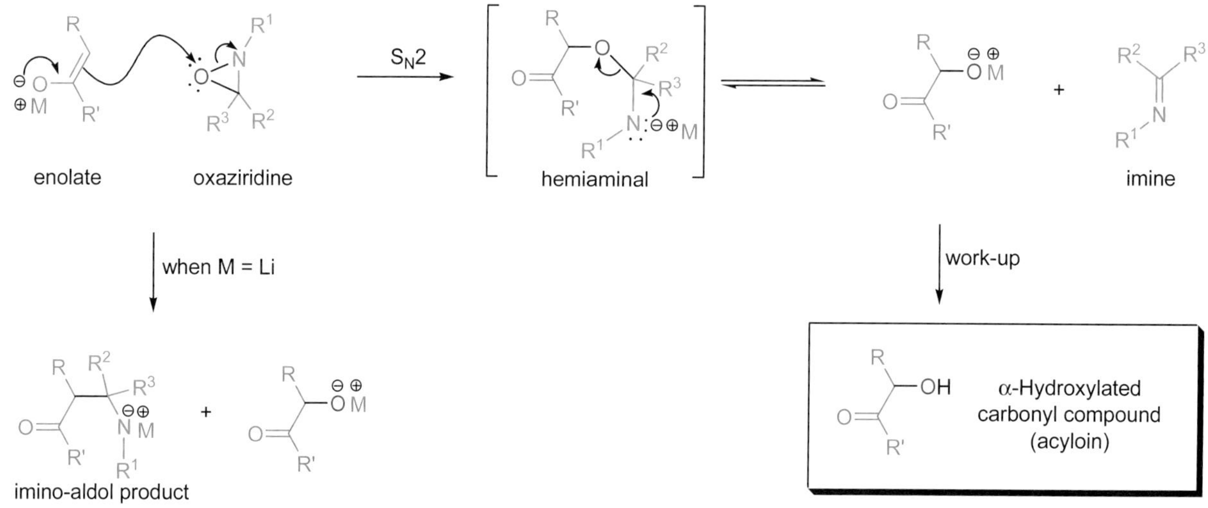

Synthetic Applications:

During the highly stereoselective total synthesis of epothilone B by J.D. White and co-workers, the stereochemistry of the alcohol portion of the macrolactone was established by applying *Davis' oxaziridine oxidation* of a sodium enolate.[37] The sodium enolate was generated from the corresponding chiral oxazolidinone derivative, which upon oxidation gave 71% yield of α-hydroxylated compound.

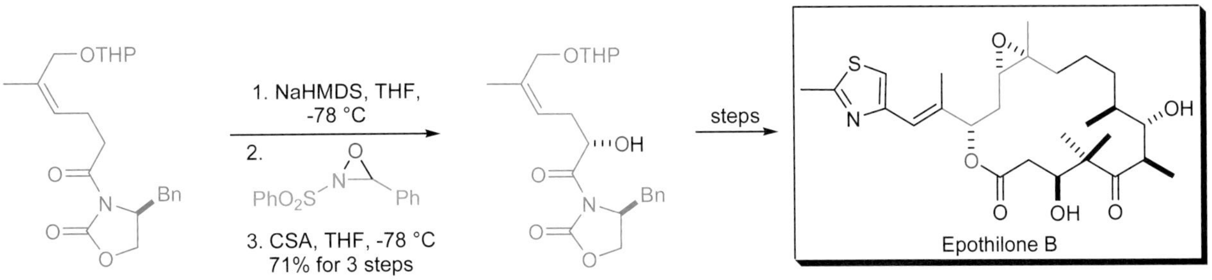

An abbreviated synthesis of a substituted 1,7-dioxaspiro[5.5]undec-3-ene system constituting the C3-C14 portion of okadaic acid was developed in the laboratory of C.J. Forsyth.[38] The C3-C8 fragment, a substituted valerolactone, was prepared in three steps. The diastereoselective α-hydroxylation of this lactone was accomplished by using *Davis' chiral camphorsulphonyl oxaziridine* on the corresponding lithium enolate at -78 °C. The isolated yield was 61% and the ratio of diastereomers was 10:1.

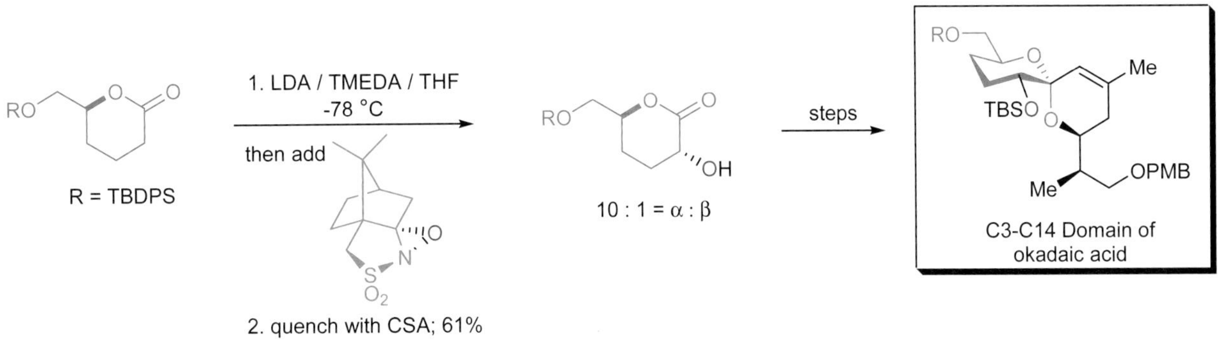

The first total synthesis of (–)-fumiquinazoline A and B was accomplished by B.B. Snider and co-workers using a *Buchwald-Hartwig Pd-catalyzed cyclization* of an iodoindole carbamate to construct the imidazoindolone moiety.[39] In order to set up the stereochemistry at the benzylic position of the indole fragment, the double bond was oxidized with the *saccharine-derived Davis' oxaziridine* in the presence of methanol to give the major diastereomer in 65% yield.

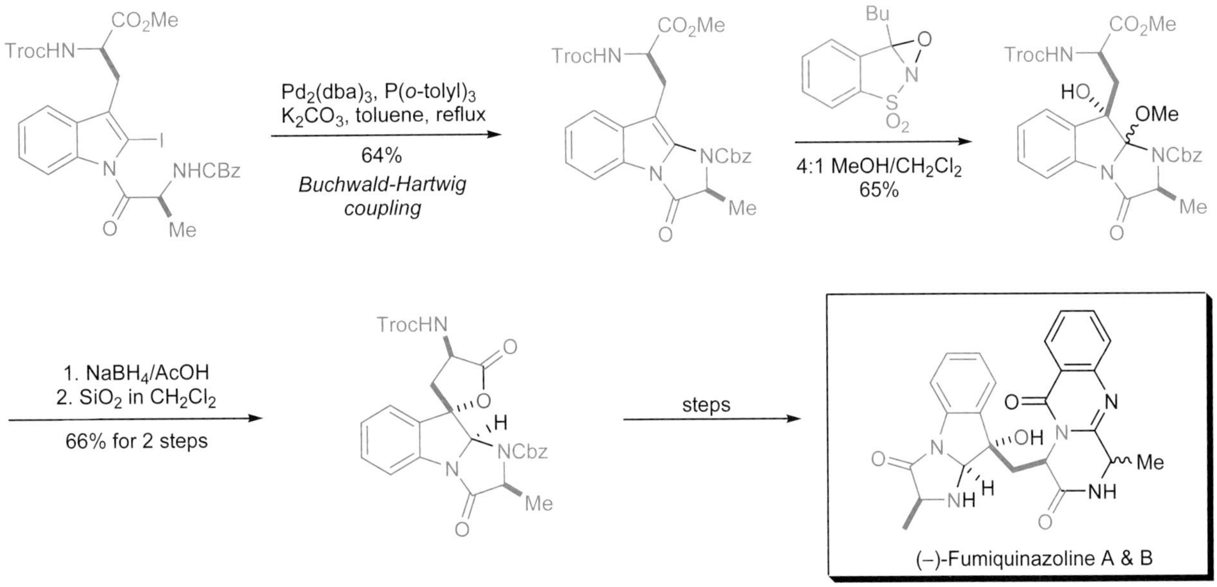

DE MAYO CYCLOADDITION (ENONE-ALKENE [2+2] PHOTOCYCLOADDITION)

(References are on page 573)

Importance:

[*Seminal Publications*[1-4]; *Reviews*[5-14]; *Modifications & Improvements*[15-18]]

The photochemical [2+2] cycloaddition of enones (α,β-unsaturated carbonyl compounds) with alkenes is known as the *de Mayo cycloaddition*. A substituted cyclobutane is formed in the process. The first example of this transformation was the "Italian sunlight-induced" *intramolecular photoisomerization* of carvone to carvoncamphor published by G.L. Ciamician in 1908.[19] Ciamician's finding was verified by G. Büchi 50 years later.[20] It was not until the early 1960s when P. de Mayo, P.E. Eaton, and E.J. Corey demonstrated that the intermolecular *enone-alkene photocycloaddition* was possible as well.[1-4] De Mayo's first paper described the intermolecular [2+2] cycloaddition of enolized 1,3-diketones (enone) and olefins. The cycloadducts (β-hydroxy ketones) underwent a spontaneous *retro-aldol reaction* to afford 1,5-diketones.[1] The alkene (olefin) and enone reaction partners can vary widely; cycloadditions with enol esters of β-diketones, dioxolenones, vinylogous esters and amides, and with cycloalkenones have been successfully carried out. The *de Mayo cycloaddition* is highly stereo- and regioselective, but there are no simple rules available to predict the stereo- and regiochemistry of the products. In intermolecular processes, the stereochemical information carried by the alkene component is often scrambled in the product indicating that the mechanism of the cycloaddition is not concerted.[12] The cycloadducts of cyclic enones are most often *cis*-fused.[4] The regiochemical outcome of intermolecular reactions is determined by orbital coefficients. In intramolecular processes, the number of atoms connecting the two double bonds (the enone and alkene double bonds) also has an effect: two-atom tethers give rise to a mixture of regiosiomers, while tethers of three or more atoms generally yield single products.[10]

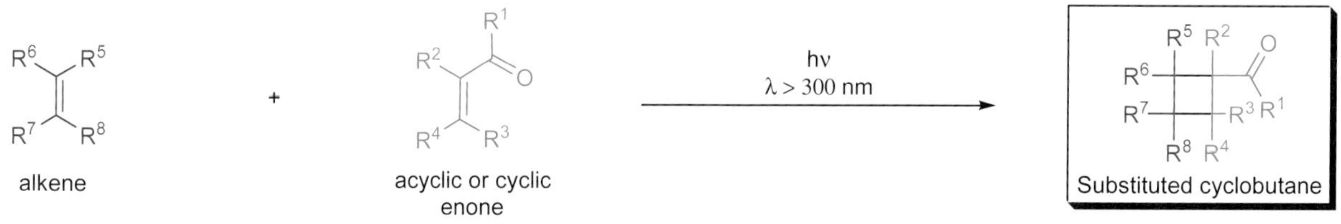

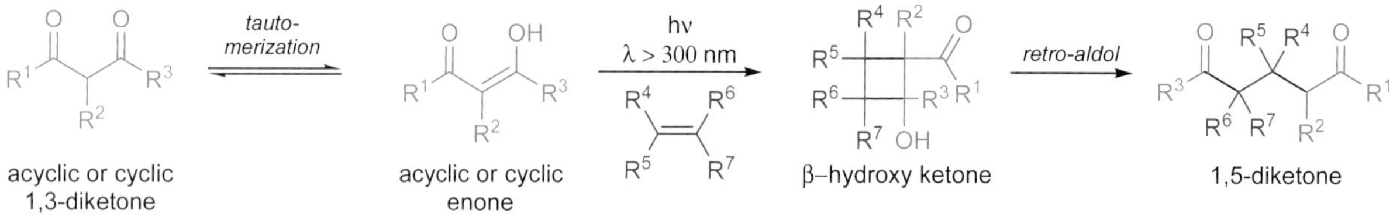

Mechanism: [2,4,21,7,22-26,12,27,28,14]

The mechanism of the *enone-alkene [2+2] photocycloaddition* presumably follows the scheme below. Upon irradiation: 1) a triplet exciplex is irreversibly formed from the triplet enone and ground state alkene; 2) the triplet exciplex collapses to one or more 1,4-biradicals.; 3) the biradicals either cyclize to the cyclobutane or revert to starting materials; and 4) the biradical reversion decreases the overall efficiency of the process.

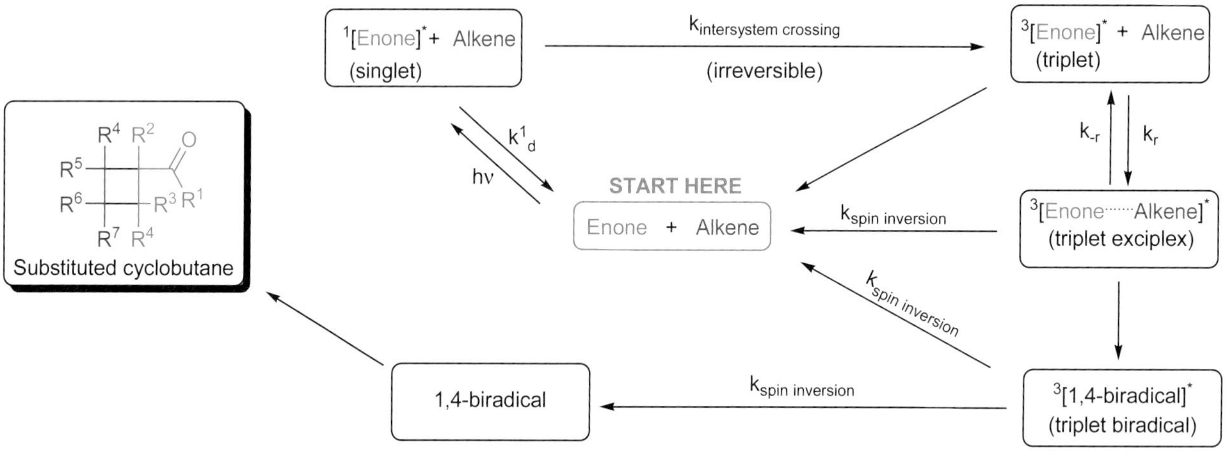

Synthetic Applications:

During the early 1990s, the research group of M. Fetizon was developing novel methods for the synthesis of taxane diterpenes.[29] Their goal was to construct the AB ring skeleton of taxol. The construction of the bicyclo[5.2.1]decane system was realized by the *intermolecular de Mayo cycloaddition* of an enolized bicyclic 1,3-dione and vinyl acetate followed by a *Lewis acid catalyzed ring opening reaction*. The methanolic solution of the β-diketone and vinyl acetate was irradiated (λ>245 nm) at 0 °C and a mixture of diastereomers was formed in excellent yield. The *retro-aldol reaction* was effected by treatment with BF_3 etherate in dichloromethane to afford good yields of the desired bicyclic ring system.

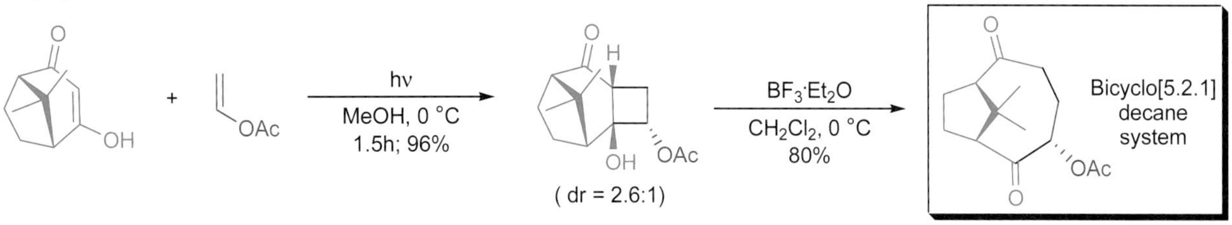

E.J. Sorensen and co-workers have synthesized the tricyclic carbon framework of guanacastepenes by applying an *intramolecular [2+2] photocycloaddition* followed by a SmI_2-induced fragmentation as key steps.[30] The enone was irradiated to effect an *intramolecular enone-olefin [2+2] cycloaddition* to afford the desired cyclobutyl ketone in 76% yield. The cyclobutane fragmentation was achieved by treatment with SmI_2 and the resulting Sm(III) enolate was trapped with a selenium electrophile. The double bond in the seven-membered ring was introduced by the oxidation of the selenium with *m*CPBA.

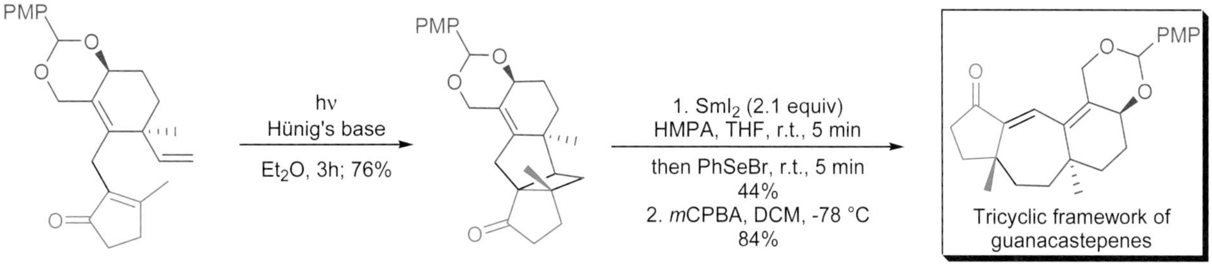

The first total synthesis of (±)-ingenol was accomplished in the laboratory of J.D. Winkler.[31] In order to establish the highly unusual C8 / C10 *trans* ("inside-outside") intrabridgehead stereochemistry of the BC ring system of the natural product, a *dioxenone-alkene intramolecular [2+2] photocycloaddition-fragmentation* sequence was employed. The photocycloaddition of the allylic chloride with the tethered dioxenone proceeded in 60% yield. The fragmentation was induced by methanolic potassium carbonate, followed by LAH reduction of the ester, elimination of the chloride with DBU, and silylation of the primary alcohol with TBSCl. The yield was 35% over four steps and the product was a 7:1 mixture of epimers at C6.

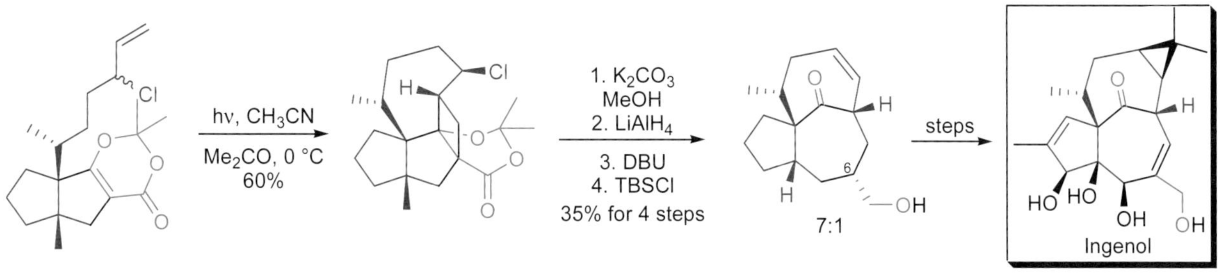

The total synthesis of the naturally occurring guaiane (±)-alismol was accomplished by G.L. Lange and co-workers using a *free radical fragmentation/elimination* sequence of an initial [2+2] de Mayo photocycloadduct.[32]

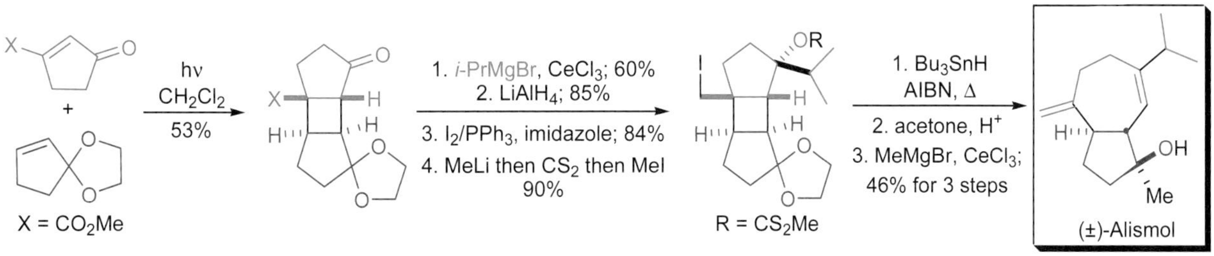

DEMJANOV AND TIFFENEAU-DEMJANOV REARRANGEMENT

(References are on page 573)

Importance:

[*Seminal Publications*[1-3]; *Reviews*[4-6]; *Modifications & Improvements*[7]]

The ring enlargement of aminomethylcycloalkanes upon treatment with nitrous acid (HNO_2) to the corresponding homologous cycloalkanols is called the *Demjanov rearrangement*. This name is also given to the rearrangement of acyclic primary amines with nitrous acid. The first rearrangement of this type was observed and reported in the early 1900s.[1,2] Synthetically, the *Demjanov rearrangement* is best applied for the preparation of five-, six-, and seven-membered rings, but it is not well-suited for the preparation of smaller or larger rings due to low yields. In 1937, M. Tiffeneau observed that the treatment of 1-aminomethyl cycloalkanols (β-aminoalcohols) with nitrous acid led to the formation of the ring-enlarged homolog ketones.[3] This transformation can be regarded as a variant of the *pinacol rearrangement* (*semipinacol rearrangement*) and is known as the *Tiffeneau-Demjanov rearrangement*. This transformation can be carried out on four- to eight-membered rings, and the yields of the ring-enlarged products are always better than for the *Demjanov rearrangement*. However, the yields tend to decrease with increasing ring size.[8,9,6] If the aminomethyl carbon atom is substituted, the *Demjanov rearrangement* is significantly retarded and mostly unrearranged alcohols are formed, but the *Tiffeneau-Demjanov rearrangement* readily occurs. Substrates with substitution on the ring carbon atom to which the aminomethyl group is attached undergo facile *Demjanov rearrangement*.

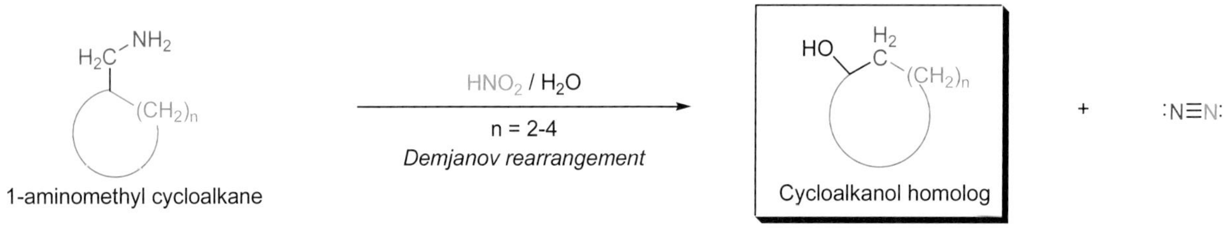

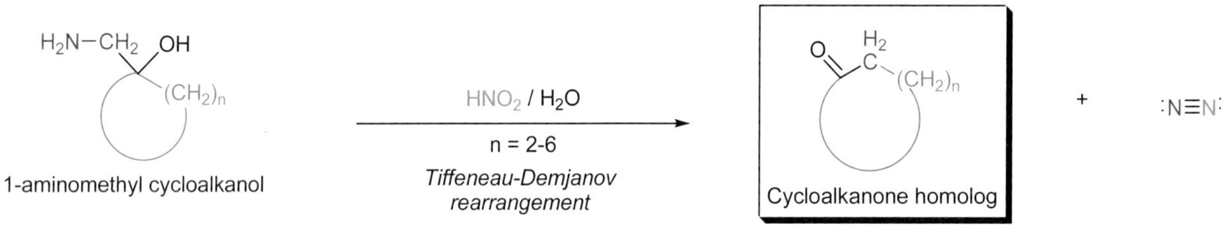

Mechanism:[10-12]

The mechanism of both the *Demjanov* and *Tiffeneau-Demjanov rearrangements* is essentially the same. The first step is the formation of the nitrosonium ion or its precursor (N_2O_3) from nitrous acid. This electrophile is attacked by the primary amino group and in a series of proton transfers the diazonium ion is formed. This diazonium ion is very labile due to the lack of stabilization and it readily undergoes a *[1,2]-alkyl shift* accompanied by the loss of nitrogen. The rearrangement is competitive with the substitution of the diazonium leaving group by the solvent (e.g., water) or with the formation of carbocations that may undergo other rearrangements (e.g., hydride shift). The ring expansion is favored in the *Demjanov rearrangement*, since the entropy of activation for *hydride shift* is higher.

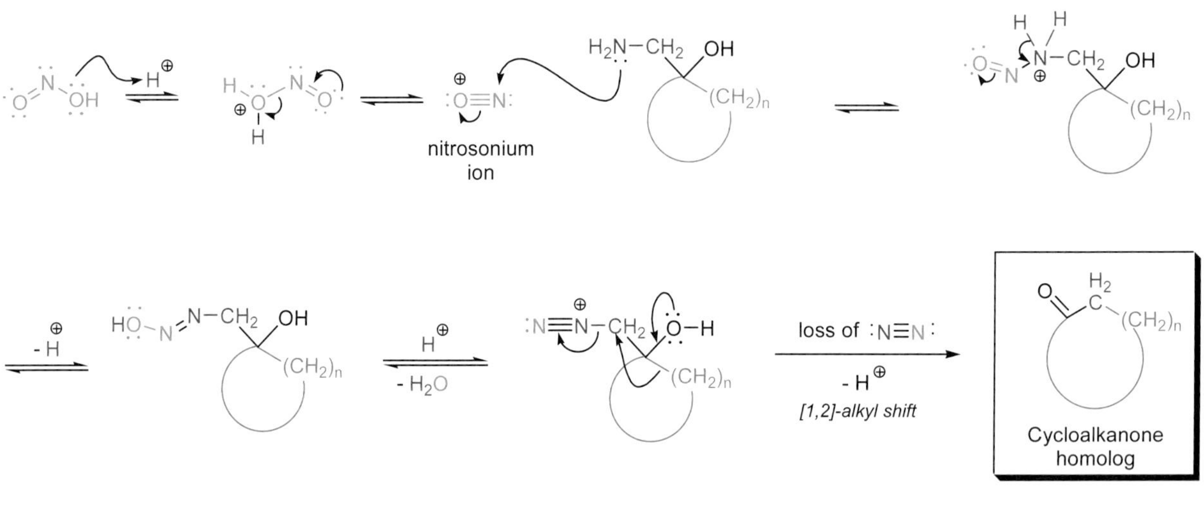

Synthetic Applications:

In the laboratory of A. Nickon, the syntheses of brexan-2-one (tricyclo[4.3.0.0^{3,7}]nonane-2-one) and the ring-expanded homolog (homobrexan-2-one) were undertaken.[13] Brexanes are frequently used in mechanistic studies, so an efficient and versatile method for the preparation of these molecules was necessary. The key step leading to the brexane-2-one parent molecule was an *endo-selective intramolecular Diels-Alder cycloaddition*, while the ring-expansion to the homolog was achieved using the *Tiffeneau-Demjanov rearrangement*. Toward this end the tricyclic ketone was efficiently converted to the corresponding aminoalcohol by treatment with TMSCN followed by LAH reduction. Upon treatment with HNO_2, the rearrangement proceeded in excellent yield to afford homobrexan-2-one.

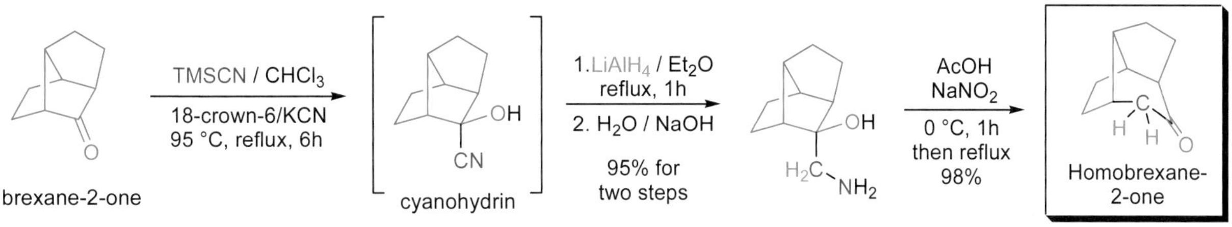

To explore the biological activity of spectinomycin analogs, E. Fritzen and co-workers prepared the ring-expanded homospectinomycins containing a seven-membered carbohydrate ring.[14] The *Tiffeneau-Demjanov* ring expansion was attempted on two epimeric aminoalcohols. Surprisingly, only the (*R*)-epimer gave the desired ring-expanded ketone, while the (*S*)-epimer afforded the corresponding epoxide as the only product. Upon treatment with nitrous acid, the (*R*)-epimer gave rise to three products in equal amounts. Only one of the products was the desired ring-expanded ketone, whereas the other two products were the (*R*)-epoxide and the corresponding vicinal diol.

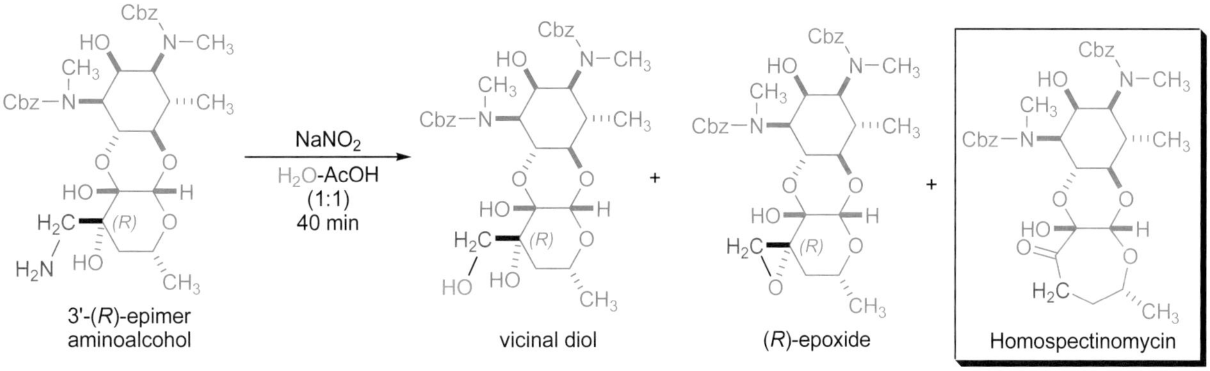

The stereochemistry of cyclic primary amines or aminoalcohols dramatically influences the product distribution of their respective *Demjanov* and *Tiffeneau-Demjanov rearrangements*. P. Vogel and co-workers have studied the ring-expansion of 2-aminomethyl-7-oxabicyclo[2.2.1]heptane derivatives upon treatment with nitrous acid. Some of their findings are shown below.[6]

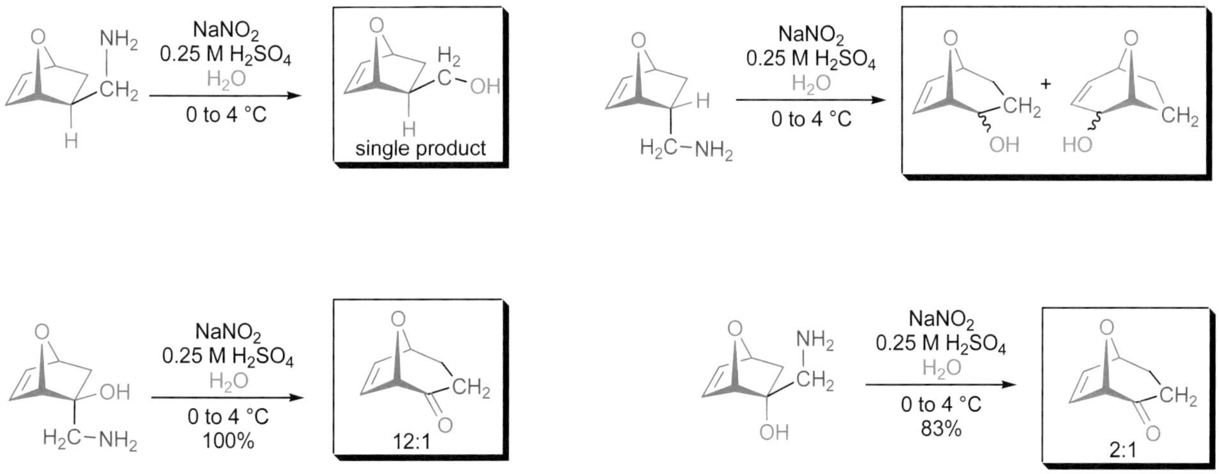

DESS-MARTIN OXIDATION

(References are on page 574)

Importance:

[*Seminal Publication*[1]; *Reviews*[2-8]; *Modifications & Improvements*[9-15]]

Since the early 1980s, hypervalent iodine reagents have emerged as selective, mild, and environmentally friendly oxidizing agents in organic synthesis.[7] One class of these reagents encompasses the organic derivatives of pentacoordinate iodine(V), which are called periodinanes.[16,17] The best-known members of this class are 2-iodoxybenzoic acid (IBX)[18] and *Dess-Martin periodinane* (DMP).[1] IBX has been known since 1893, but its almost complete insolubility in most organic solvents prevented its widespread use in organic synthesis.[19] In 1983, D.B. Dess and J.C. Martin reported the preparation of 1,1,1-tris(acetoxy)-1,1-dihydro-1,2-benziodoxol-3-(1*H*)-one (DMP) *via* the acylation of IBX.[19] This new periodinane is far more soluble in organic solvents than IBX; since its discovery it has emerged as the reagent of choice for the oxidation of alcohols to the corresponding carbonyl compounds.[2] Oxidations using DMP are called *Dess-Martin oxidations*. Currently, DMP is commercially available, but it is rather expensive. Therefore, it is usually prepared by the oxidation of 2-iodobenzoic acid to IBX, followed by the acylation of IBX to DMP. The oxidation of 2-iodobenzoic acid can be done with potassium bromate ($KBrO_3$)[1,10,12] in aqueous sulfuric acid or with Oxone ($2KHSO_5$-$KHSO_4$-K_2SO_4)[13] in water. During the 1990s, modifications[10,13] to the original procedure were necessary because the morphology[12] and purity of the IBX strongly influenced the quality of DMP and therefore the reproducibility of DMP oxidations. The advantages of the *Dess-Martin oxidation* over the conventional oxidation of alcohols are: 1) mild reaction conditions (room temperature, neutral pH); 2) high chemoselectivity; 3) tolerance of sensitive functional groups on complex substrates; and 4) long shelf-life and thermal stability (unlike IBX, which has been found to be explosive[20]). Besides the conversion of alcohols to carbonyl compounds, the *DMP oxidation* was also successfully utilized for the oxidation of functional groups for which traditional mild oxidants failed to work: 1) allylic alcohols to α,β-unsaturated carbonyls;[21] 2) cleavage of aldoximes and ketoximes to aldehydes and ketones;[22] 3) *N*-acyl hydroxylamines to acyl nitroso compounds;[23] 4) 4-substituted anilides to *p*-quinones;[24] 5) β-amino alcohols to α-amino aldehydes without epimerization;[25] and 6) γ,δ-unsaturated aromatic amides to complex heterocycles.[26]

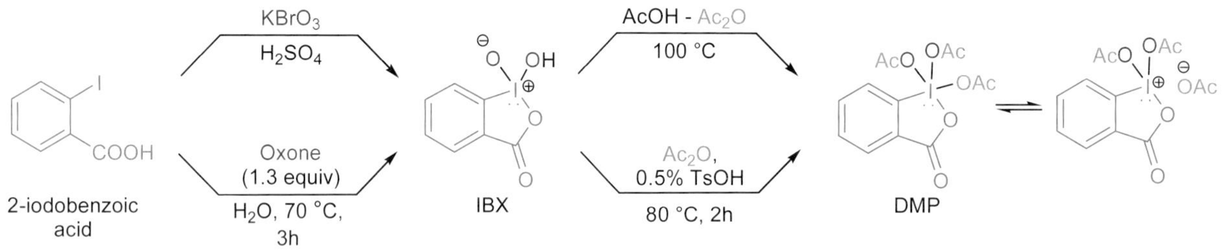

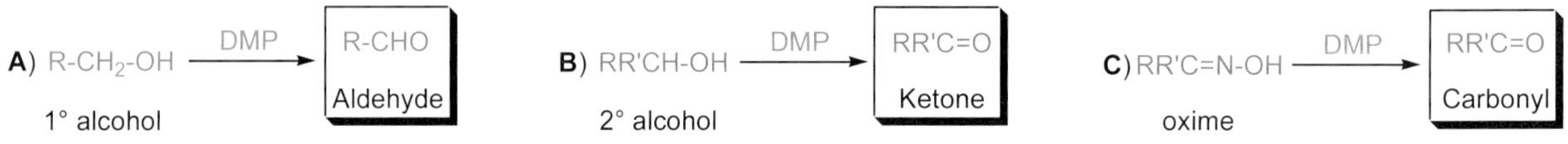

Mechanism: [9,11,27,28]

It has been shown by ^{1}H-NMR that DMP reacts rapidly with 1 equivalent of alcohol (1° or 2°) to give diacetoxyalkoxyperiodinanes, while in the presence of 2 equivalents of alcohol (or diol) a double displacement takes place to produce acetoxydialkoxyperiodinanes. Next, the α-proton of the alcohol is removed by a base (acetate), and the carbonyl compound is released along with a molecule of iodinane. When excess alcohol is present, the oxidation is much faster due to the especially labile nature of acetoxydialkoxyperiodinanes.[9] It has also been shown that added water accelerates DMP oxidations.[11]

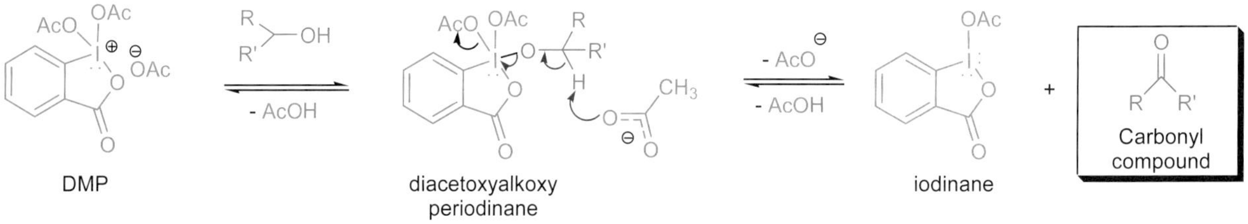

Synthetic Applications:

In the final stages of the total synthesis of ustiloxin D, M.M. Joullié and co-workers had to install the amide side-chain onto the already assembled macrocycle.[29] To achieve this goal, the macrocyclic primary alcohol was treated with the Dess-Martin periodinane to generate the corresponding aldehyde, which was subsequently treated with sodium chlorite to afford the carboxylic acid. The carboxylic acid was then coupled with the benzyl ester of glycine to complete the installation of the side-chain in 66% yield for three steps.

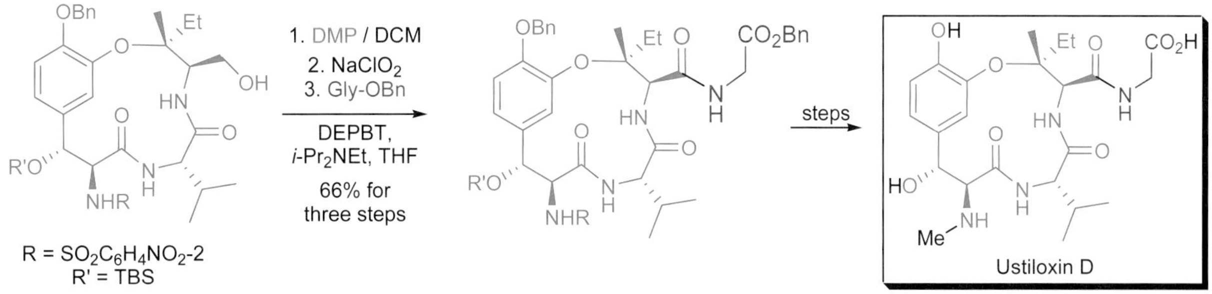

A novel one-pot *Dess-Martin oxidation* was developed for the construction of the γ-hydroxy lactone moiety of the CP-molecules in the laboratory of K.C. Nicolaou.[30] Bicyclic 1,4-diol was treated with 10 equivalents of DMP in dichloromethane for 16h to promote a tandem reaction: first, the bridgehead secondary alcohol was selectively oxidized to the ketone, followed by a ring closure to afford the isolable hemiketal, which was further oxidized by DMP to give a keto aldehyde. Trace amounts of water terminated the cascade to give a stable diol, which was not further oxidized with DMP. Subsequent TEMPO oxidation furnished the desired γ-hydroxy lactone.

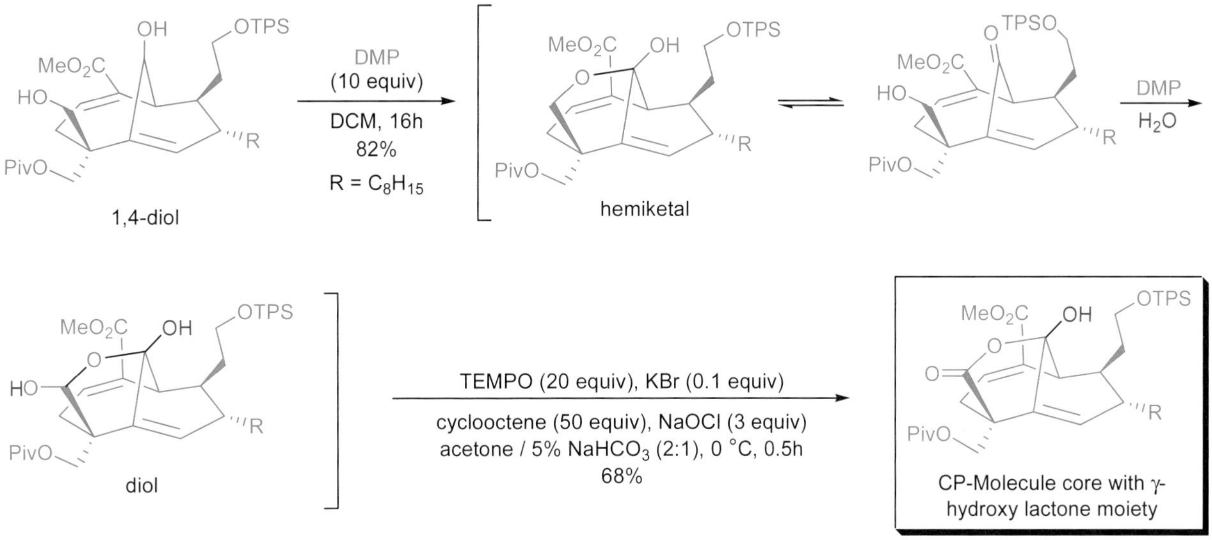

For the elaboration of the dienyl side-chain of the E-F fragment of (+)-spongistatin 2, A.B. Smith et al. oxidized the sensitive primary allylic alcohol moiety using the *Dess-Martin oxidation*.[31] The resulting α,β-unsaturated aldehyde was treated with a Wittig reagent to obtain the desired 1,3-dienyl side chain.

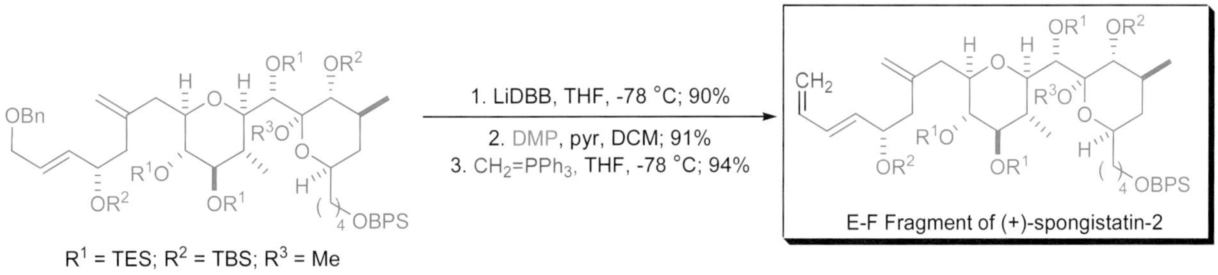

DIECKMANN CONDENSATION

(References are on page 574)

Importance:

[*Seminal Publications*[1,2]; *Reviews*[3-8]; *Modifications & Improvements*[9-17]]

The base mediated condensation of an ester, containing an α-hydrogen atom, with a molecule of the same ester to give a β-keto ester is known as the *Claisen condensation*. When the two reacting ester functional groups are tethered the reaction is called the *Dieckmann condensation*, and a cyclic β-keto ester is formed. In the related *Thorpe-Ziegler condensation* the intramolecular base-catalyzed cyclization of dinitriles affords enaminonitriles.[18,19] The commonly used procedure involves prolonged treatment of the diesters with at least one equivalent of a strong base (alkoxide, sodium amide, or alkali metal hydrides)[20] in dry solvent under reflux in an inert atmosphere. The *Dieckmann condensation* forms 5-, 6-, 7-, and 8-membered rings in high yield but gives very low yields for larger rings.[21,4] It is possible, however, to effect the cyclization at high-dilution so the intramolecular reaction dominates and in certain cases the preparation of large rings (>12) is possible.[22,23] If the product β-keto ester does not have an acidic α-hydrogen, the reaction is sluggish and the *retro-Dieckmann cyclization* predominates; the equilibrium is shifted to the right if one equivalent of an alcohol-free base is used. With the *Thorpe-Ziegler cyclization* it is possible to assemble 5- to 33-membered rings and this method is superior to the *Dieckmann condensation* for the formation of 7- and 8-membered rings. Modifications of the original *Dieckmann procedure* made it possible to use mild reaction conditions: 1) dithiols (dithioesters) are treated with sodium hydride so the cyclizations take place in only 2h at room temperature;[9] 2) environmentally friendly solvent-free conditions allow the presence of air and the reaction proceeds in high yield at room temperature in 1h;[16] and 3) the use of $TiCl_4/Bu_3N$ with catalytic amounts of TMSOTf in toluene gives high yields in 2-3h at room temperature.[14,24,17]

Dieckmann condensation:

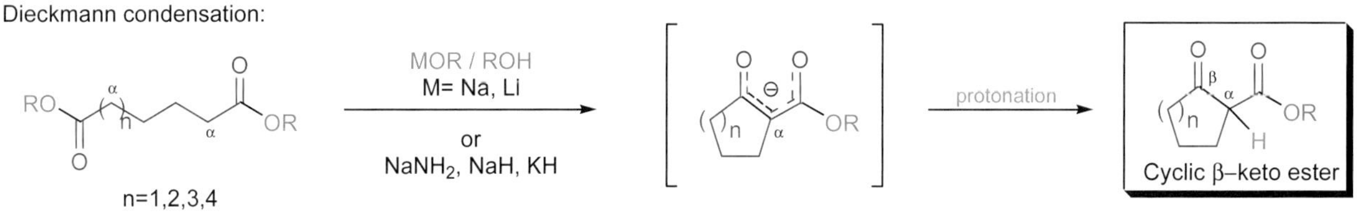

Thorpe-Ziegler cyclization:

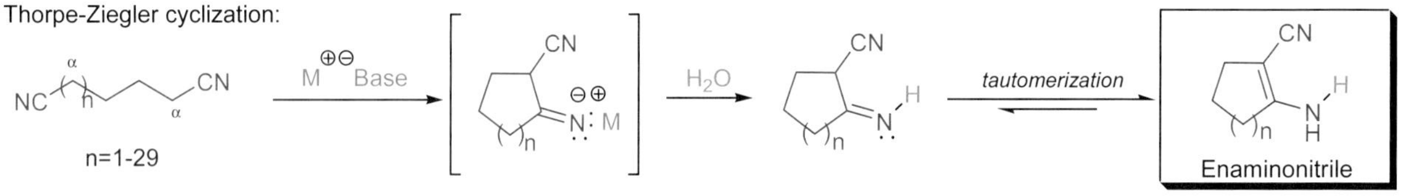

Mechanism:[25-31]

Each step of the *Dieckmann condensation* is completely reversible. The driving force of the reaction is the generation of the resonance-stabilized enolate of the product β-keto ester. As stated above, the condensation usually fails if it is not possible to generate this stable intermediate. The mechanism of the *Dieckmann condensation* is almost identical to the mechanism of the *Claisen condensation*. The rate-determining step, however, is the ring formation in which the ester enolate attacks the carbonyl group of the second ester functional group.[25,26] The resulting tetrahedral intermediate then rapidly breaks down to the enolate of the β-keto ester. Protonation of the enolate affords the final product.

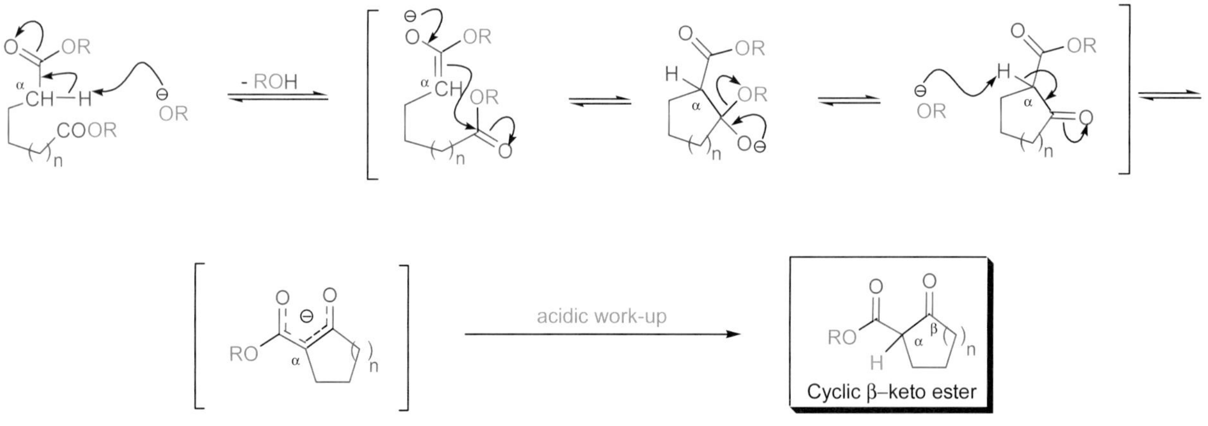

Synthetic Applications:

Mycophenolic acid is one of the highly substituted phthalide natural products and possesses many *in vitro* and *in vivo* biological activities. The synthetic strategy toward its convergent total synthesis by A. Covarrubias-Zúñiga was based on a ring annulation sequence involving a *Michael addition* and a *Dieckmann condensation* as key steps.[32] The deprotonation of 2-geranyl 1,3-acetonedicarboxylate with sodium hydride was followed by the addition of a protected alkynal to give rise to the enolate *in situ*, which cyclized to the hexasubstituted aromatic ring of the natural product in 33% yield.

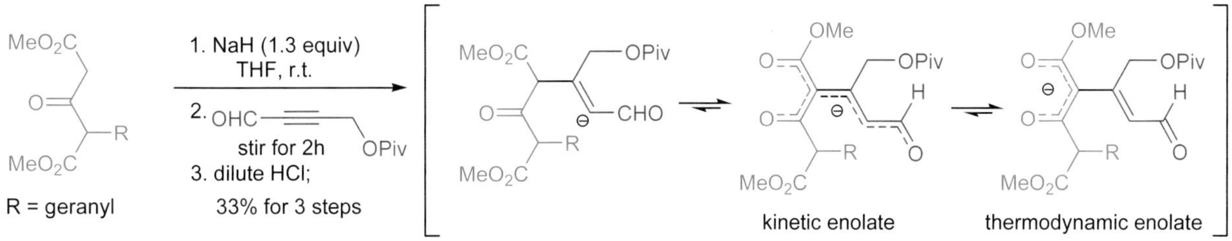

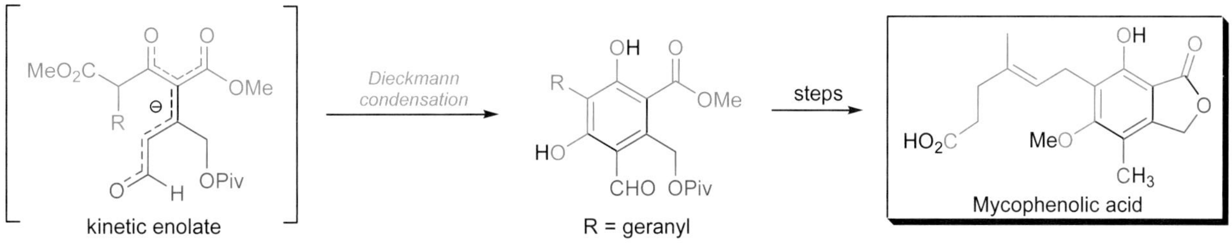

The 14-membered macrocyclic ring of (–)-galbonolide B was formed utilizing a novel *macro-Dieckmann cyclization* which was developed in the laboratory of B. Tse.[33] In order to bring about the desired macrolactonization, the secondary acetate was treated with LiHMDS in refluxing THF under high-dilution conditions to afford the desired lactone in 75% yield. It is important to note, however, that the analogous secondary propionate failed to cyclize under identical conditions.

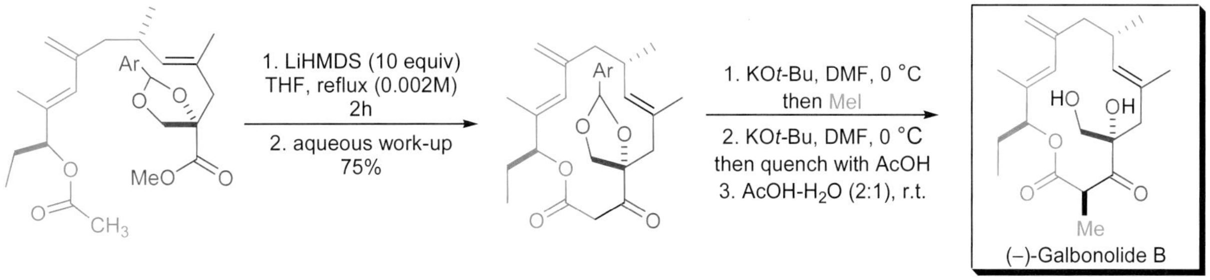

The naturally occurring clerodane diterpenoid (±)-sacacarin has been synthesized by R.B. Grossman and co-workers in only 10 steps using a double annulation of a tethered diacid and 3-butyn-2-one.[34] The second ring of sacacarin was prepared by an *intramolecular Dieckmann condensation* of an ester and a methyl ketone in excellent yield. The resulting enol was then immediately converted to the corresponding ethyl enol ether using ethanol and an acid catalyst.

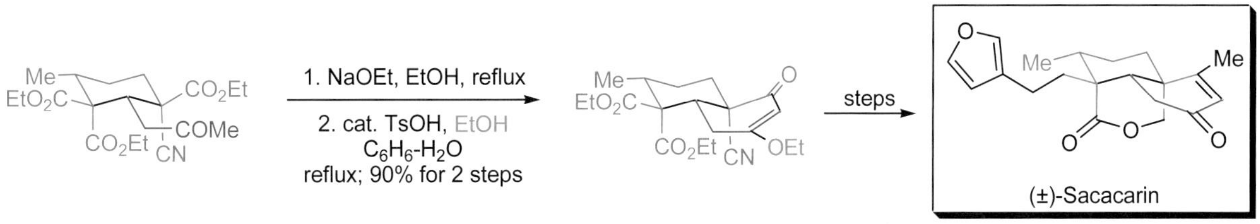

DIELS-ALDER CYCLOADDITION

(References are on page 575)

Importance:

[*Seminal Publications*[1-4]; *Reviews*[5-47]; *Theoretical Studies*[48-66]]

The [4π + 2π] cyclization of a diene and alkene to form a cyclohexene derivative is known as the *Diels-Alder cycloaddition* (*D-A cycloaddition*). Reports of such cyclizations were made by H. Wieland,[67] W. Albrecht,[68] Thiele, H. Staudinger, and H.V. Euler[69] in the early 1900s, but the structures of the products were misassigned. It was not until 1928 when O. Diels and K. Alder established the correct structure of the cycloadduct of *p*-quinone and cyclopentadiene.[2] Since its discovery, the *D-A cycloaddition* has become one of the most widely used synthetic tools. The diene component is usually electron rich, while the alkene (dieneophile) is usually electron poor and the reaction between them is called the *normal electron-demand D-A reaction*. When the diene is electron poor and the dienophile is electron rich then an *inverse electron demand D-A cyclization* takes place. Besides alkenes, substituted alkynes, benzynes, and allenes are also good dienophiles. If one or more of the atoms in either component is other than carbon, then the reaction is known as the *hetero-D-A reaction*.[70] In the *retro-D-A reaction* unsaturated six-membered rings break down to yield dienes and dienophiles.[8] The synthetic value of the *D-A cycloaddition* is due to the following features: 1) it can potentially set four stereocenters in one step; 2) if unsymmetrical dienes and dienophiles react it is highly *regioselective* and *stereospecific;* 3) the regioisomers are predominantly the "*ortho*" and "*para*" products over the "*meta*" product; 4) if a disubstituted *cis* (*Z*) alkene is used, the stereochemistry of the two substituents in the product will be *cis* and when an (*E*) alkene is used, the stereochemistry in the product will be *trans*; 5) the stereochemical information (*E* or *Z*) in the diene is also transferred to the product; 6) the predominant product is the *endo* cycloadduct; 7) by using appropriate chiral catalysts the cycloaddition can be made enantioselective;[71-73,41] and 8) multiple rings can be created in one step with defined stereochemistry.

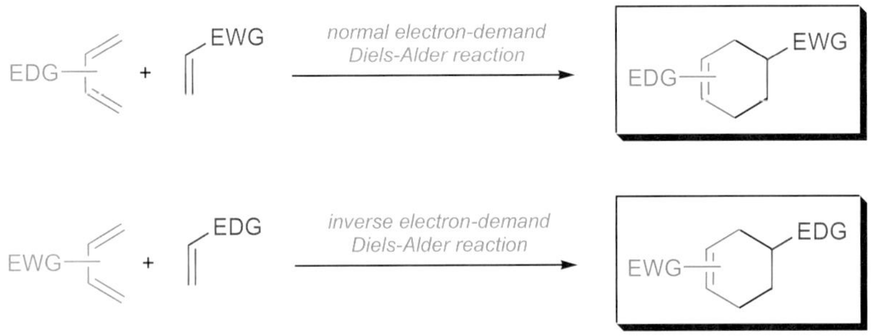

EDG (electron-donating group) = alkyl, *O*-alkyl, *N*-alkyl, etc.

EWG (electron-withdrawing group) = CN, NO_2, CHO, COR, COAr, CO_2H, CO_2R, COCl etc.

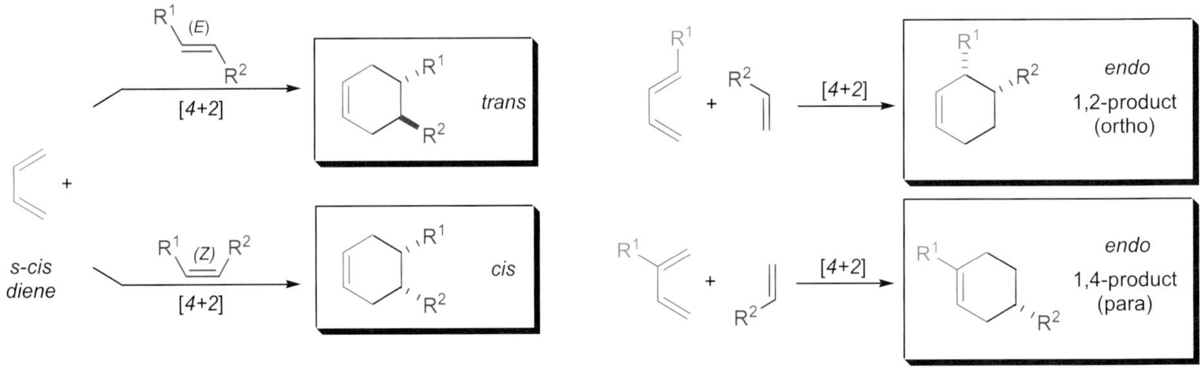

Mechanism: [23,74-78,29,79-82,38,83-85,44,86]

Mechanistically the D-A reaction is considered a concerted, pericyclic reaction with an aromatic transition state. The driving force is the formation of two new σ-bonds. The *endo* product is the kinetic product and its formation is explained by *secondary orbital interactions*.[80] Some of the mechanistic studies suggested that a diradical[79] or a di-ion mechanism may be operational in certain cases.[82] It was also shown that solvents and salts can influence reaction kinetics.[38]

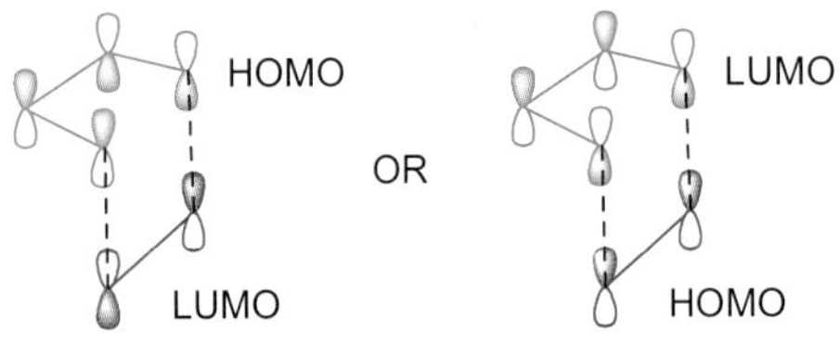

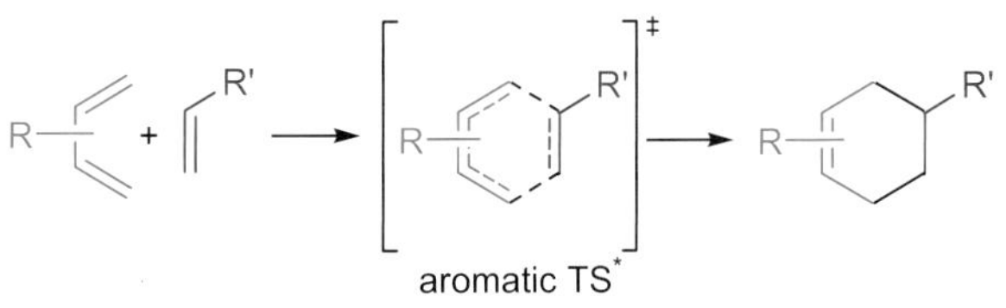

Synthetic Applications:

The *intramolecular Diels-Alder cycloaddition* is a very powerful synthetic tool, since it can generate molecular complexity in a single step. S. Antus and co-workers obtained reactive cyclohexa-2,4-dienones by dearomatizing *o*-methoxyphenols with hypervalent iodine reagents (e.g., PIDA). These dienones rapidly dimerized to give heavily substituted complex tricyclic compounds. The dearomatization of 2,6-dimethoxy-4-allyllphenol with PIDA/methanol resulted in the formation of the natural product asatone in a single step.[87]

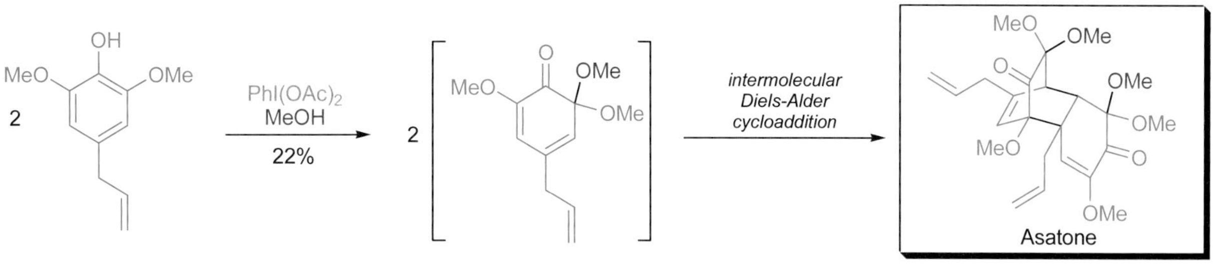

The total synthesis of the rubrolone aglycon was accomplished in the laboratory of D.L. Boger as part of the ongoing research to explore the cycloaddition reaction of cyclopropenone ketals.[88] The key step in the production of the seven-membered C-ring was the *intermolecular Diels-Alder reaction* of an electron-rich diene with the very strained dienophile. The cycloaddition took place in excellent yield (97%) and with complete disastereoselectivity.

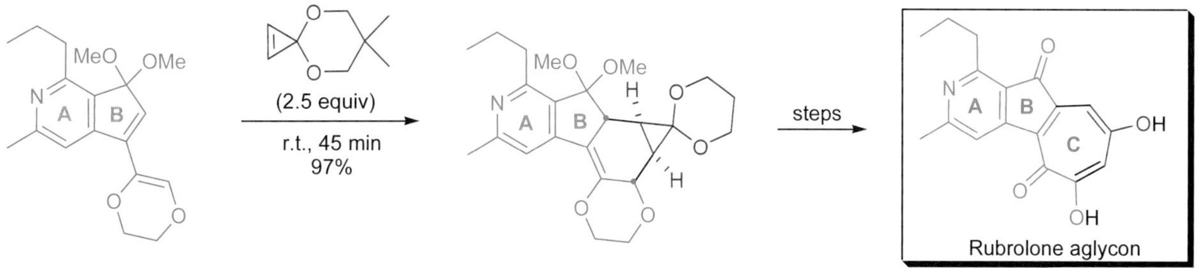

The critical step in the enantioselective and stereocontrolled total synthesis of eunicenone A by E.J. Corey et al. was the highly efficient chiral Lewis acid catalyzed *intermolecular Diels-Alder cycloaddition reaction*.[89] The diene component was mixed with 5 equivalents of 2-bromoacrolein and 0.5 equivalents of the chiral oxazaborolidine catalyst in CH_2Cl_2 at -78 °C for 48h. The reaction gave 80% of the desired cycloadduct in 97% *ee* and the *endo*/*exo* selectivity was 98:2.

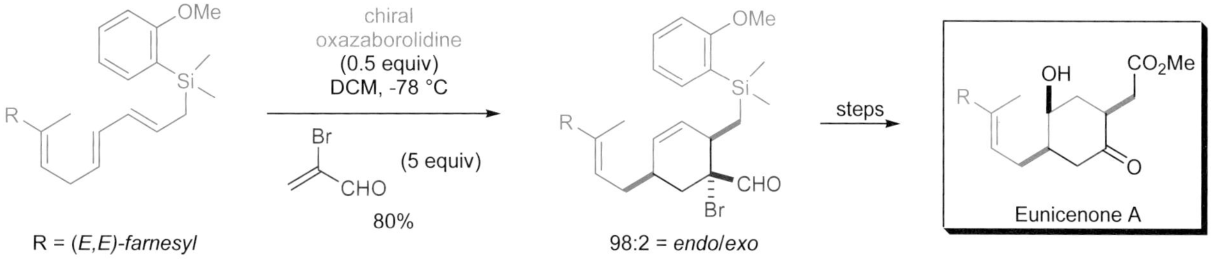

Certain functional groups can direct through hydrogen-bonding the outcome of the *intermolecular Diels-Alder cycloaddition*. This was the case in the key *Diels-Alder cycloaddition* step during the total synthesis of (±)-rishirilide B in the laboratory of S.J. Danishefsky.[90] The diene was thermally generated *in situ*.

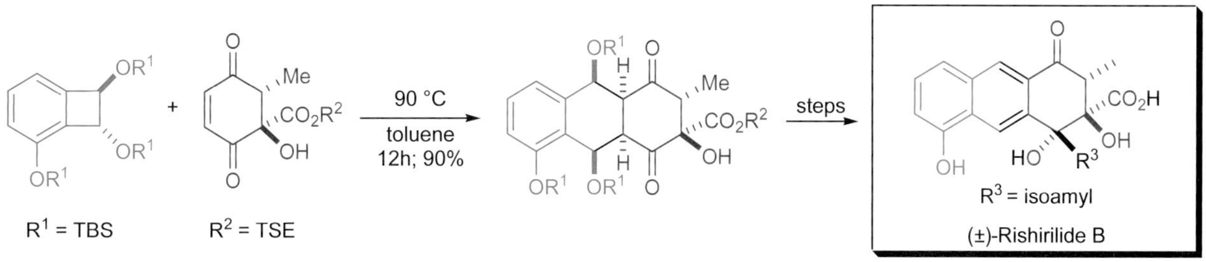

DIENONE-PHENOL REARRANGEMENT

(References are on page 577)

Importance:

[*Seminal Publications*[1-3]; *Reviews*[4-8]; *Modifications & Improvements*[9-11]; *Theoretical Studies*[12]]

The acid- and base-catalyzed or photochemically-induced migration of alkyl groups in cyclohexadienones is known as the *dienone-phenol rearrangement*, and is widely used for the preparation of highly substituted phenols. In 1893, A. Andreocci described the rearrangement of santonin to desmotroposantonin upon acidic treatment, but it was only in 1930 that the starting material and the product of this rearrangement were carefully characterized.[1,3] The term "*dienone-phenol rearrangement*" was introduced by A.L. Wilds and C. Djerassi.[13] Cyclohexadienones (both *ortho* and *para*) can be considered as "blocked aromatic molecules" in which the migration of an alkyl group converts the non-aromatic substrate into an aromatic one.[6] *Dienone-phenol rearrangements* require only moderately strong acidic media (e.g., H_2SO_4 in acetic acid, acetic anhydride, Lewis acidic clay,[9] etc.), and they are considerably exothermic due to the formation of very stable aromatic compounds.

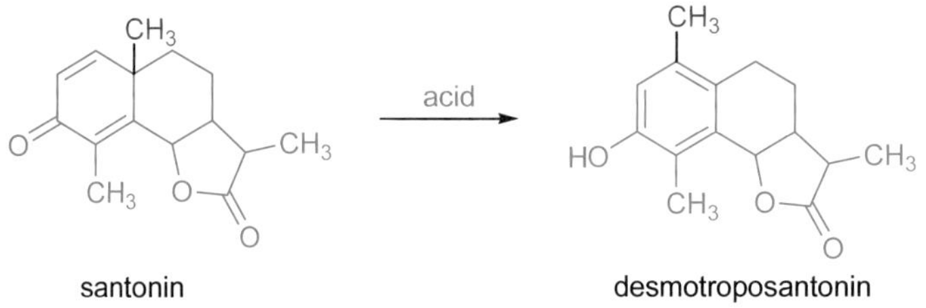

Mechanism: [14-28]

Most *dienone-phenol rearrangements* involve acid catalysis and the products appear to be the result of sigmatropic *[1,3]-migrations* of C-C bonds. The *[1,3]-alkyl migrations* are actually the result of two subsequent *[1,2]-alkyl shifts* as was demonstrated by ^{14}C isotope labeling studies.[14,15] Depending on the nature of the migrating groups, other rearrangements such as *[1,2]*, *[1,3]*, *[1,4]*, *[1,5]*, *[3,3]*, *[3,4]*, and *[3,5]* can also take place.[6,7] When the migrating group is benzyl, the products predominantly arise from [1,5]-migrations, and the rate of these rearrangements is several orders of magnitude greater than for simple alkyl groups. If the migrating group is allyl, crotyl, or propargyl, then the main course of the rearrangement takes place *via [3,3]-shifts* rather than *[1,2]-shifts*. The scheme below depicts the mechanism of the acid-catalyzed rearrangement of *p*-cyclohexadienone to the corresponding 3,4-disubstituted phenol as well as the rearrangement of a bicyclic dienone *via* two subsequent *[1,2]-shifts*.

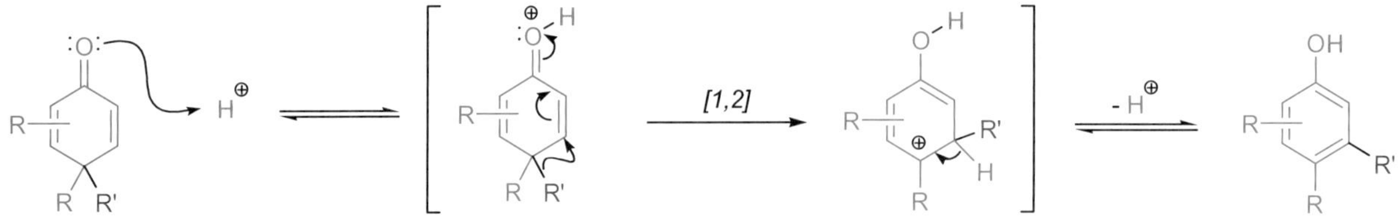

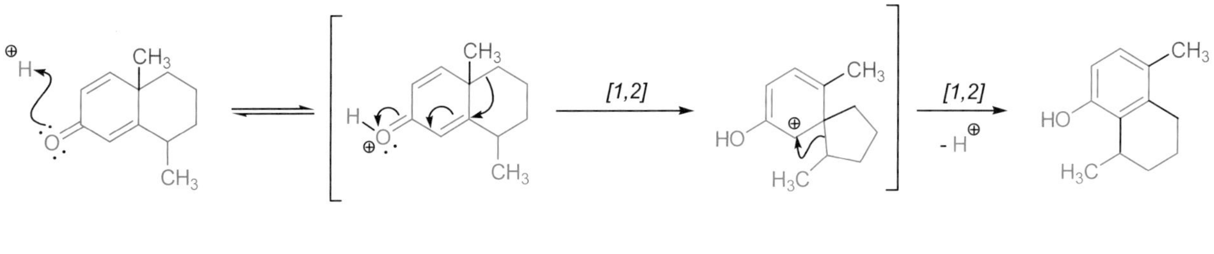

Synthetic Applications:

An efficient synthetic route to tetra- and pentasubstituted phenols was developed in the laboratory of A.G. Schultz by the *photochemical dienone-phenol rearrangement* of 4,4-disubstituted 2-phenyl-2,5-cyclohexadienones.[29] The photorearrangement substrates were conveniently prepared by the *Birch reduction-alkylation* of the corresponding aromatic compounds followed by the *bis allylic oxidation* of the initial diene products using *t*-butyl hydroperoxide and catalytic amounts of PDC. Upon irradiation with 366 nm light, the dienones underwent a regioselective *dienone-phenol rearrangement* to afford the phenols in high yield.

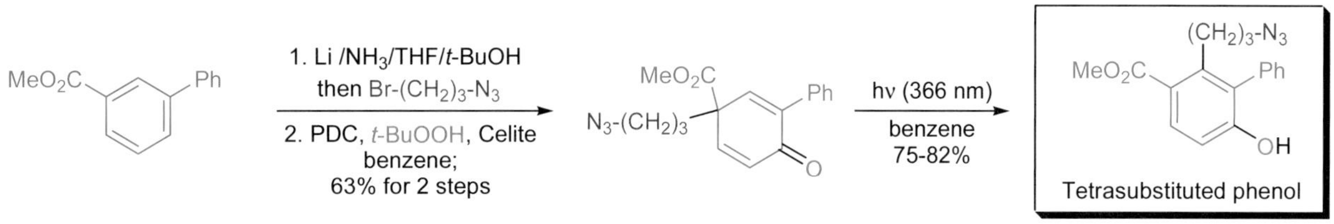

During model studies toward kidamycins, K.A. Parker and co-workers developed a methodology for the synthesis of *bis C*-aryl glycosides.[30] Phenolic *bis* glycosides were synthesized using the regiocontrolled Lewis acid mediated *dienone-phenol type rearrangement* as the key step in which a glycal undergoes a *[1,2] shift*. The resulting *bis C*-aryl glycal was first hydrogenated over PtO_2 to give the *bis* glycoside followed by global desilylation to afford the desired kidamycin model.

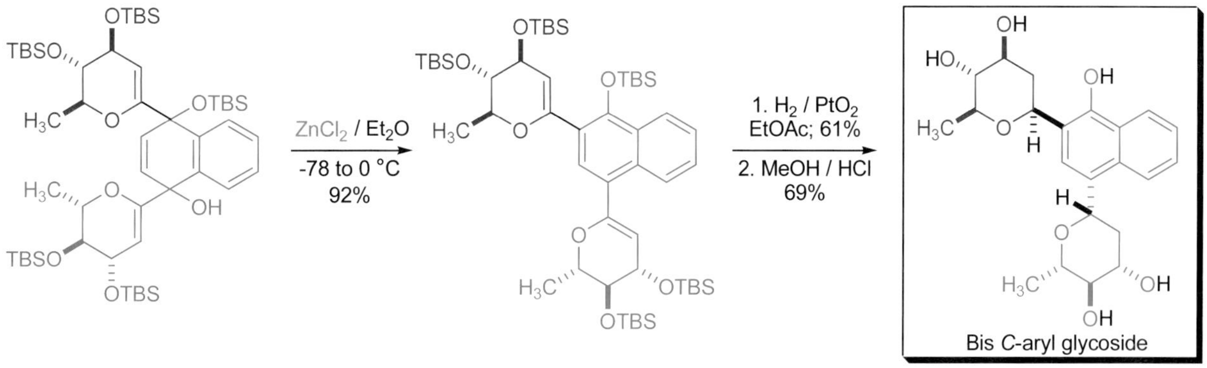

Rearrangement of spirodienones under a variety of conditions (both acidic and basic) afforded substituted 6*H*-dibenzo[*b*,*d*]pyran-6-ones.[31] D.J. Hart et al. showed that rearrangements in aqueous sulfuric acid gave products of formal *O*-migration, whereas rearrangements in trifluoroacetic anhydride (TFAA)/trifluoroacetic acid (TFA)/sulfuric acid mostly resulted in *C*-migration products. The *dienone-phenol rearrangement* also worked well for highly substituted spirodienone systems and afforded either the *C*- or *O*-migration products depending on the applied reaction conditions.

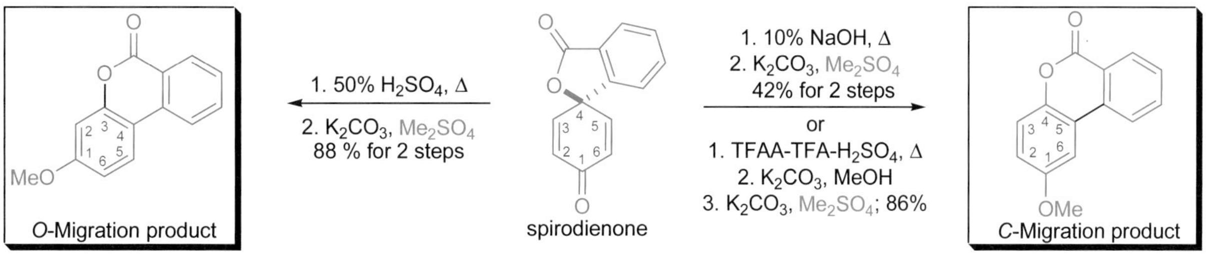

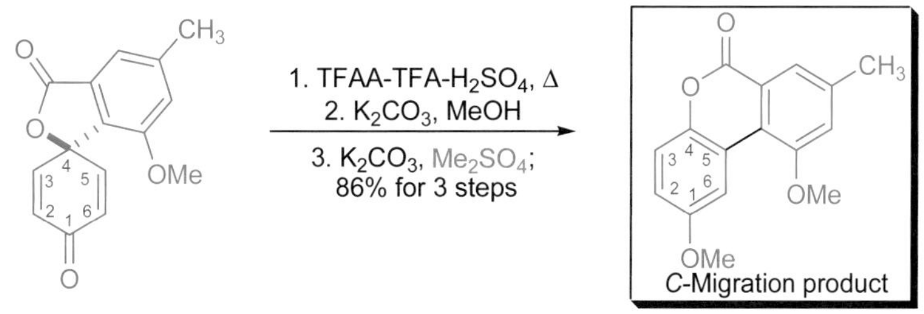

DIMROTH REARRANGEMENT

(References are on page 578)

Importance:

[*Seminal Publications*[1-3]; *Reviews*[4-10]; *Theoretical Studies*[11]]

The isomerization of heterocycles in which endocyclic or exocyclic heteroatoms and their attached substituents are translocated *via* a ring-opening-ring-closure sequence is known as the *Dimroth rearrangement*. The first observation of this type of rearrangement was made by B. Rathke on a triazine derivative but no rationalization was provided to explain the findings.[1] In 1909, O. Dimroth proposed the correct mechanism for the rearrangement of a triazole derivative.[2] The generality of the process was first recognized in the pyrimidine series[12,13] in the mid-1950s and later proved to be even more general; it was shown to occur in many nitrogen-containing heterocyclic systems.[10] It was in 1963 when the term *Dimroth rearrangement* was coined by D.J. Brown and J.S. Harper.[14] The rearrangement may be divided into two types: 1) translocation of heteroatoms within rings of fused systems (***Type I***) and 2) translocation of exo- and endocyclic heteroatoms in a heterocyclic ring (***Type II***). The second type of rearrangement is more common than the first. The *Dimroth rearrangement* can be catalyzed by acids,[15,16] bases (alkali),[17,18] heat, or light.[19,20] Numerous factors influence the course of the *Dimroth rearrangement* in heterocyclic systems: 1) degree of aza substitution in the rings (more nitrogen atoms in the ring lead to more facile nucleophilic attack);[21] 2) pH of the reaction medium (affects the rate of the rearrangement);[22] 3) presence of electron-withdrawing groups (give rise to more facile ring-opening); and 4) the relative thermodynamic stability of the starting material and the product.

Type I rearrangement:

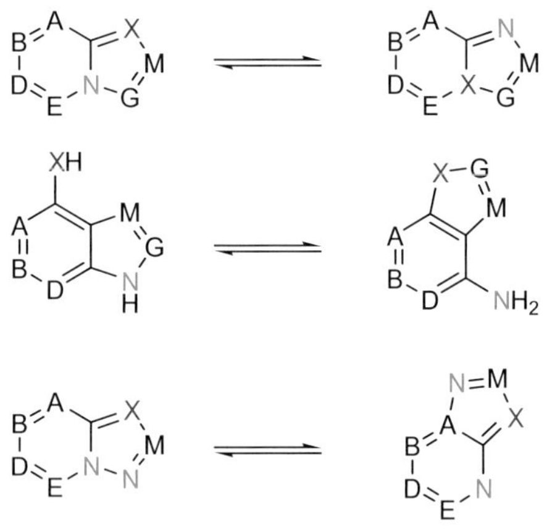

X = heteroatom; N = nitrogen

Type II rearrangement:

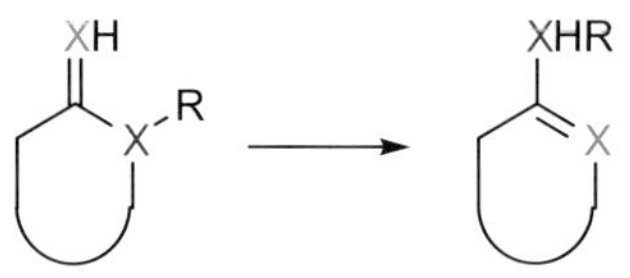

Mechanism: [23,24]

The exact pathway by which the *Dimroth rearrangement* takes place in a given heterocycle depends on many factors (see above). However, in general there are three distinct steps: 1) attack of the heterocyclic ring by a nucleophile; 2) electrocyclic ring opening followed by rotation about a single bond; and 3) ring closure. These steps are known collectively as the ANRORC mechanism. If the rearrangement takes place as a result of heat or irradiation, then the first step is the electrocyclic ring opening followed by the ring closure. The mechanism illustrates the rearrangement of 2-amino-5-nitropyridine to 2-methylamino-5-nitropyridine.

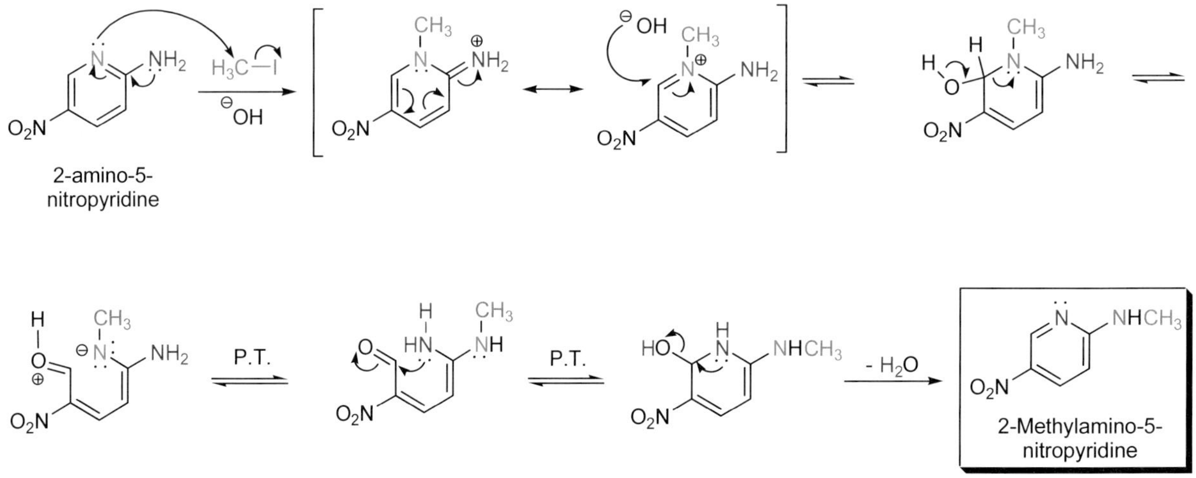

Synthetic Applications:

The marine ascidian metabolite purine aplidiamine-9-β-D-ribofuranoside was prepared by T. Itaya et al. by alkylation of 8-oxoadenosine with 4-benzyloxy-3,5-dibromobenzyl bromide followed by a *Dimroth rearrangement* and acid hydrolysis.[25] The rearrangement was induced by treating the nucleoside in boiling 1N NaOH for 1h. The desired rearranged nucleoside was formed in 58% overall yield.

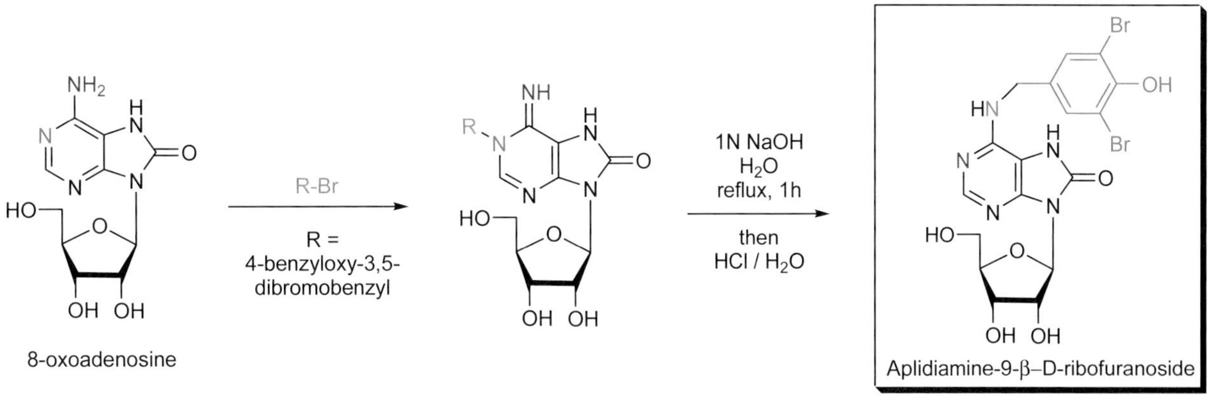

In the laboratory of R.A. Jones, N^1-methoxy derivatives of adenosine and 2'-deoxyadenosine were found to undergo a facile *Dimroth rearrangement*.[26] The high-yielding process allowed the efficient synthesis of [1,7-$^{15}N_2$]- and [1,7, NH_2-$^{15}N_3$] adenosine and 2'-deoxyadenosine that are important tools in the NMR studies of nucleic acid structure and interactions. The rearrangement was carried out in weakly acidic refluxing methanol.

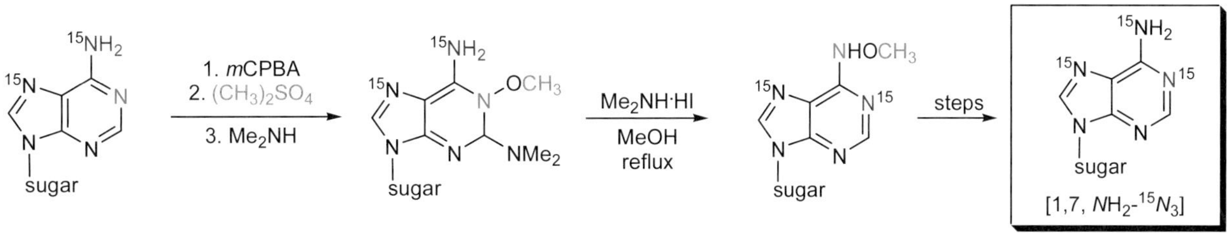

A new synthetic approach to tricyclic 1,3,6-thiadiazepines was developed by V.A. Bakulev and co-workers.[27] The synthetic sequence involved a base-catalyzed *Smiles rearrangement* followed by an *in situ Dimroth rearrangement*. The starting substituted 1,2,3-thiadiazole was treated with triethylamine in refluxing ethanol. In the first step, the thiadiazole ring was transposed from the sulfur to the nitrogen atom (*Smiles rearrangement*). In the second step, the 5-amino-1,2,3-thiadiazole underwent a *Dimroth rearrangement* to form the bis(triazole) intermediate, which immediately formed the tricyclic 1,3,6-thiadiazepine accompanied by the loss of hydrogen sulfide anion.

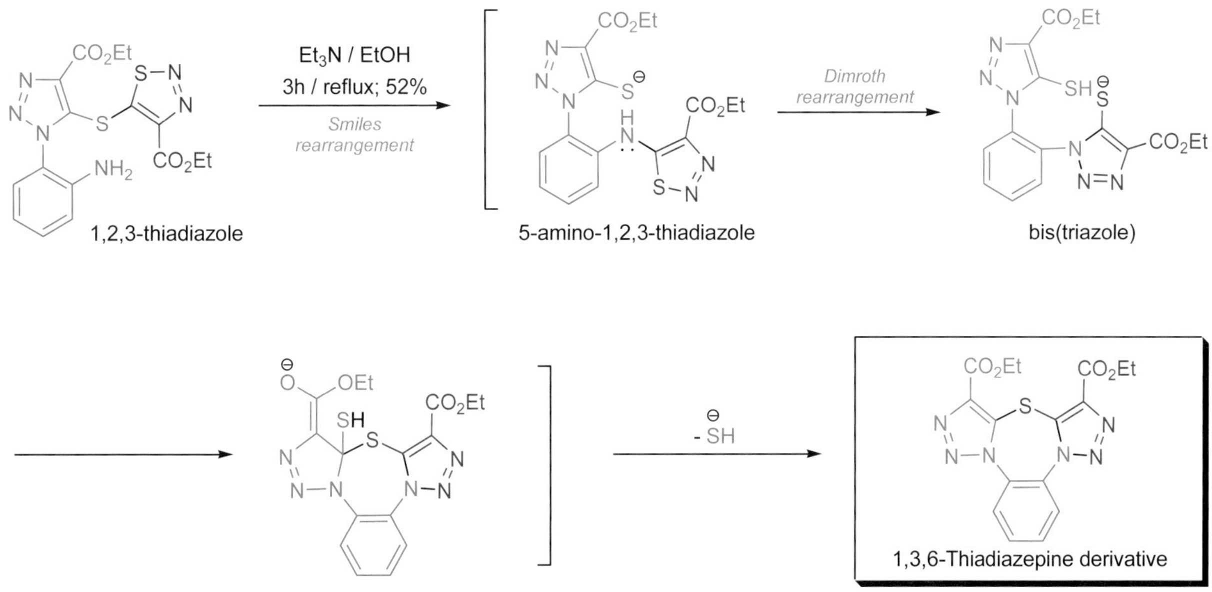

DOERING-LAFLAMME ALLENE SYNTHESIS

(References are on page 578)

Importance:

[*Seminal Publications*[1-3]; *Reviews*[4-7]; *Modifications & Improvements*[8,9]]

In 1958, W. Doering and P.M. LaFlamme developed a two-step one-carbon homologation procedure to prepare allenes from alkenes.[3] The first step of the synthesis involves the addition of dibromocarbene to an olefin. Then, in the second step, the 1,1-dibromocyclopropane derivative is reduced with an active metal (high surface area Na or Mg) to afford the allene in moderate to good yield. The method was shown to be general and today the preparation of allenes from olefins *via* dihalocyclopropanes is known as the *Doering-LaFlamme allene synthesis*. Geminal dihalocyclopropanes are readily available from the reaction of dihalocarbene with an olefin, as described by Doering and Hoffmann in 1954.[1] Drawbacks of the original allene synthesis are: 1) isomerization of unsaturated compounds is common with sodium metal; 2) sluggish reaction and the formation of allene-cyclopropane mixtures with magnesium metal; and 3) dichlorocyclopropanes are less reactive than the dibromo analogs. The modification of the original procedure by reacting the dihalocarbene with alkyllithiums[8,9] or Grignard reagents[10] results in higher yields of allenes. For example ethyl- and isopropylmagnesium bromide can be used at room temperature to convert dibromocyclopropanes into aliphatic and non-strained cyclic allenes.[10]

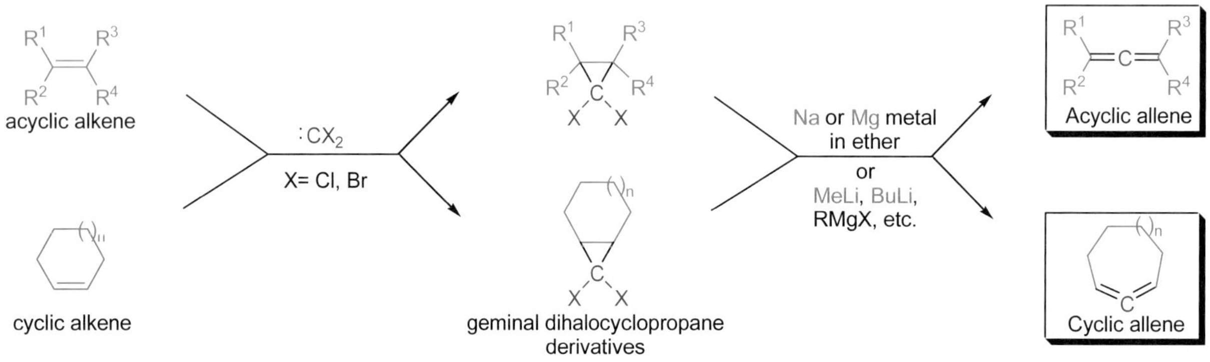

Mechanism: [3,11-14,10]

The first step of the *Doering-LaFlamme allene synthesis* is the generation of a dihalocarbene that reacts with the olefin *in situ*. First, the haloform is deprotonated by a strong base to form an unstable trihalomethyl carbanion, which undergoes a facile α-elimination to the dihalocarbene. The dihalocarbene then quickly inserts into the double bond of the olefin to afford a geminal dihalocyclopropane. In the second step, the alkyllithium performs a lithium-halogen exchange with the dihalocyclopropane to form lithiobromocyclopropane, which in turn loses lithium halide to generate a cyclopropylidene or a related carbenoid. The cyclopropylidene undergoes rearrangement to the corresponding allene.

Dihalocarbene formation and insertion:

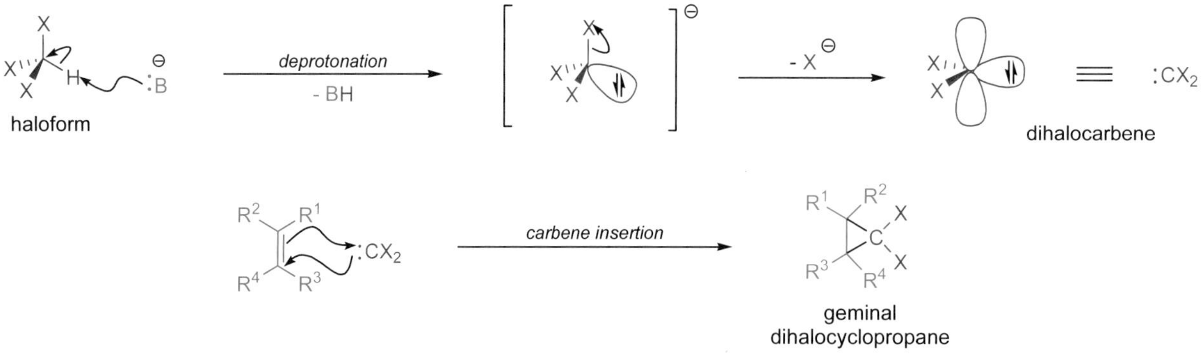

Metal-halogen exchange and rearrangement:

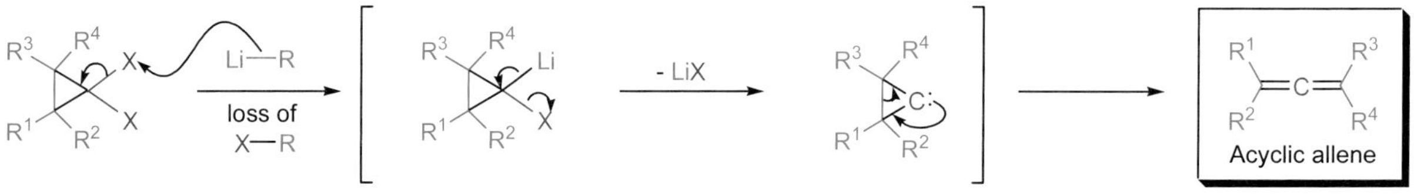

Synthetic Applications:

The synthesis of novel 4α-substituted sterols was undertaken in the laboratory of C.H. Robinson.[15] These compounds are potential inhibitors of sterol 4-demethylation. To prepare the desired 4-allenyl-5α-cholestan-3β-ol, the exocyclic olefin precursor was first reacted with bromoform/potassium *t*-butoxide to afford the geminal dibromo-substituted cyclopropane derivative. Next, methyllithium was used to bring about the rearrangement to afford the allene, and finally acidic conditions were applied for the removal of the THP protecting group.

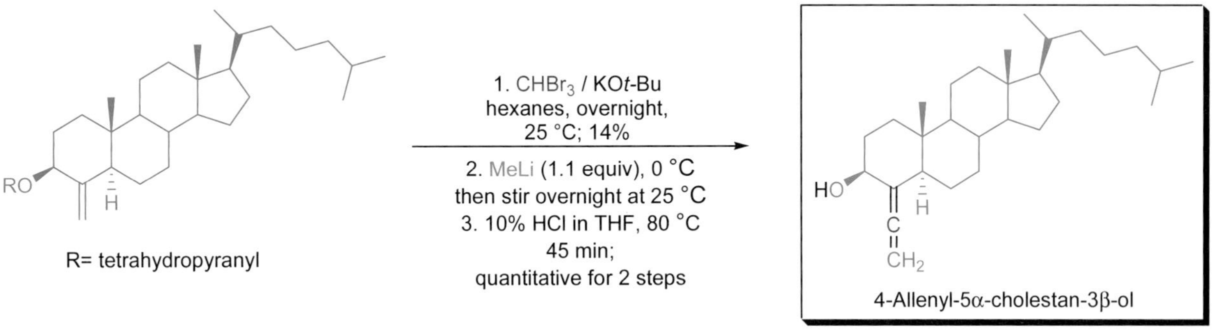

During studies of the preparation and chemical behavior of spirocyclopropanated bicyclopropylidenes, A. de Meijere and co-workers successfully synthesized a branched [8]triangulane from 7-cyclopropylidenespiro[2.0.2.1]heptane.[16] The key transformation in their approach was the *Doering-LaFlamme allene synthesis*. The 7-cyclopropylidene spiro[2.0.2.1]heptane was first dibromocyclopropanated and then treated with methyllithium to afford the key intermediate allene in good yield. Upon reaction with diazocyclopropane (generated *in situ* from *N*-nitroso-*N*-cyclopropylurea), the allene gave the desired branched [8]triangulane in modest yield.

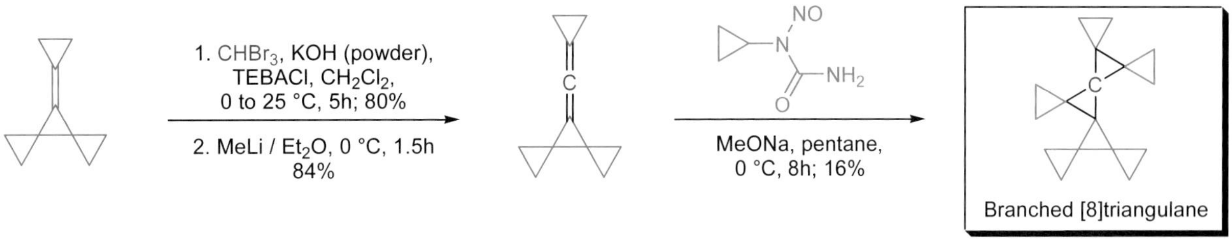

M. Santelli et al. developed a general synthesis of β-silylallenes from allylsilanes utilizing the *Doering-LaFlamme allene synthesis*.[17]

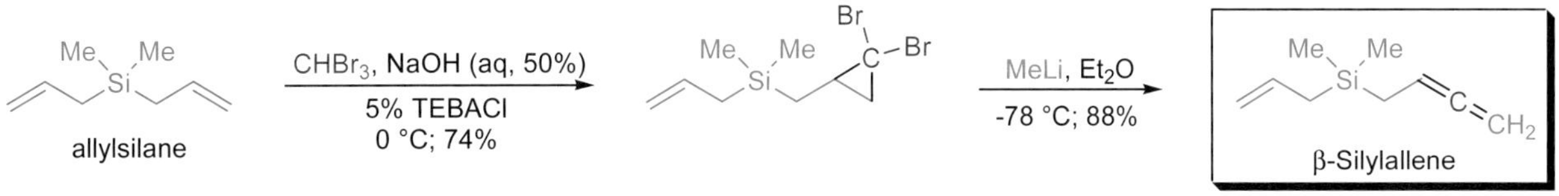

The synthesis and thermal rearrangement of π and heteroatom bridged diallenes was investigated by S. Braverman and co-workers.[18] Bis(γ,γ-dimethylallenyl)ether was generated by the addition of dibromocarbene to diisobutenyl ether and treating the resulting dibromocyclopropane derivative with methyllithium. However, the allene proved to be impossible to isolate, since it underwent spontaneous cyclization to give 3-isopropenyl-4-isopropylfuran in high yield.

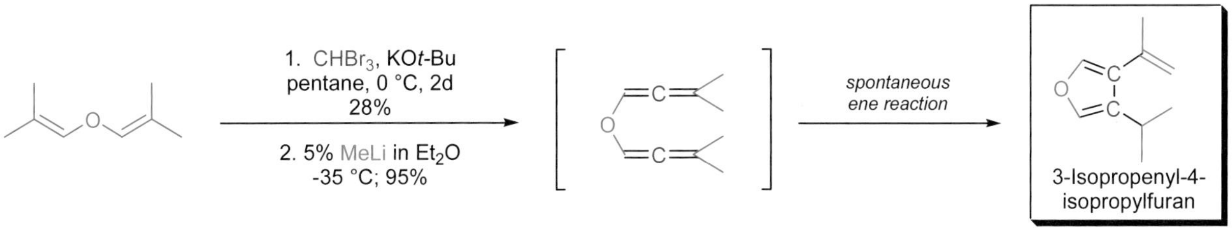

DÖTZ BENZANNULATION REACTION

(References are on page 579)

Importance:

[*Seminal Publications*[1,2]; *Reviews*[3-20]; *Modifications & Improvements*[21-24]; *Theoretical Studies*[25-28]]

In 1975, K.H. Dötz reported the formal [3+2+1] cycloaddition of a chromium phenylmethoxycarbene complex with diphenylethyne that yielded primarily a chromium tricarbonyl-complexed 4-methoxy-1-naphthol upon heating in dibutyl ether at 45 °C.[1] The reaction of an α,β-unsaturated pentacarbonyl chromium carbene complex (Fischer-type carbene) with an alkyne to afford a substituted hydroquinone (1,4-dihydroxybenzene) derivative is called the *Dötz benzannulation reaction*. Since its initial discovery, the transformation has become one of the most studied reactions of chromium complexes. The nature of the products depends largely on the nature of the substituents on the carbene and the reaction conditions (solvents, temperature, concentration, etc.).[29,30] The required Fischer chromium carbenes can be prepared with ease by treating $Cr(CO)_6$ with an organolithium nucleophile followed by the *O*-alkylation of the resulting acyl metalate with a strong alkylating agent (e.g., Meerwein's salt, alkyl triflates, etc.). This process allows the preparation of a wide variety of unsaturated chromium-carbenes and is limited only by the availability of the organolithium reagent. Advantages of the *Dötz benzannulation reaction* are: 1) access to densely functionalized aromatic compounds with excellent chemo- and regioselectivity (the large alkyne substituent, R_L, always ends up *ortho* to the phenolic OH group); 2) compatibility with a variety of substituents on both the alkyne and the unsaturated carbene side chain; 3) aryl carbene complexes with electron-withdrawing or electron-donating substituents work as well as unsubstituted aryl- or heteroaryl carbenes; 4) alkynes bearing electron-donating groups give moderate to excellent yields; 5) the hydroquinone products can be oxidized to give highly substituted quinones; and 6) the annulation is also possible intramolecularly with a reversal of the regioselectivity. The disadvantages are: 1) toxicity of chromium complexes; 2) alkynes with electron-withdrawing groups give poor yields or do not react at all; 3) heterosubstituted alkynes generally give low yields; and 4) the benzannulation is often accompanied by the formation of indenes and cyclobutenones.

Preparation of α,β–unsaturated Fischer-type carbenes:

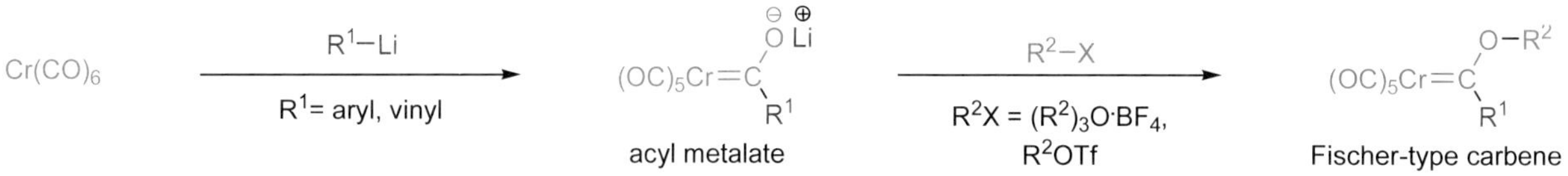

Dötz benzannulation reaction:

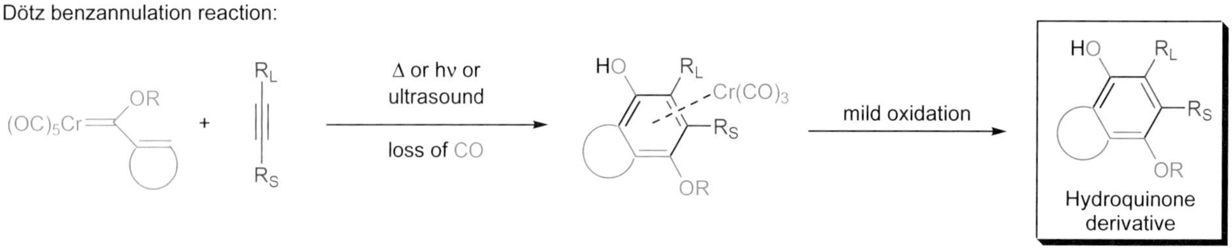

Mechanism: [31-34,17,35]

The mechanism of the *Dötz benzannulation reaction* has not been fully elucidated. The first step is the rate-determining dissociation of one carbonyl ligand from the Fischer carbene complex, which is *cis* to the carbene moiety. Subsequently, the alkyne component coordinates to the coordinatively unsaturated carbene complex, and then it inserts into the metal-carbon bond. After the alkyne insertion, a vinylcarbene is formed that can lead to the product by two different pathways (**Path A** or **Path B**).[36-39]

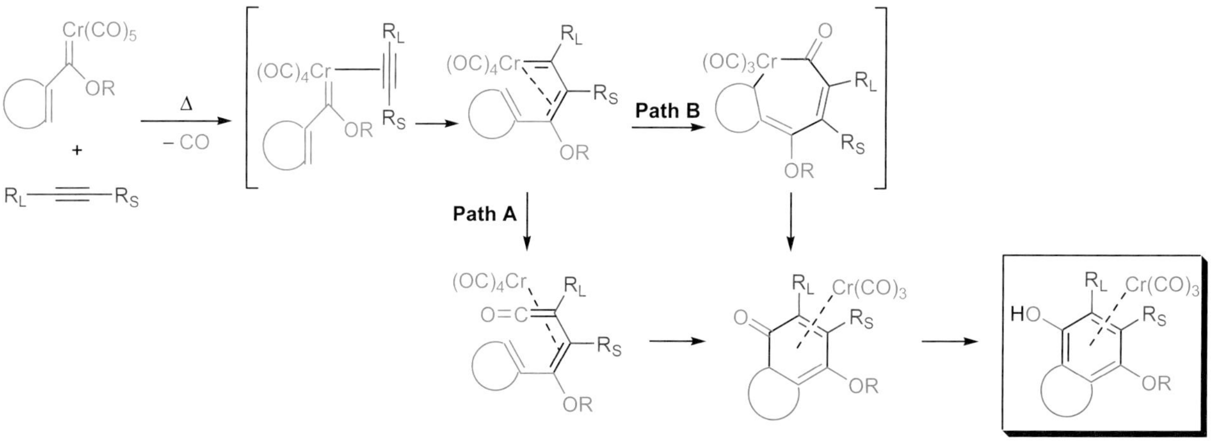

Synthetic Applications:

The architecturally interesting and biologically significant protein kinase C inhibitor calphostins (A-D), and their analogs were synthesized in the laboratory of C.A. Merlic.[40] The key steps in their approach were a *Dötz aminobenzannulation* utilizing an enantiopure Fischer carbene complex to prepare a pentasubstituted naphthylamine, followed by a *biomimetic oxidative dimerization* to produce the perylenequinone skeleton.

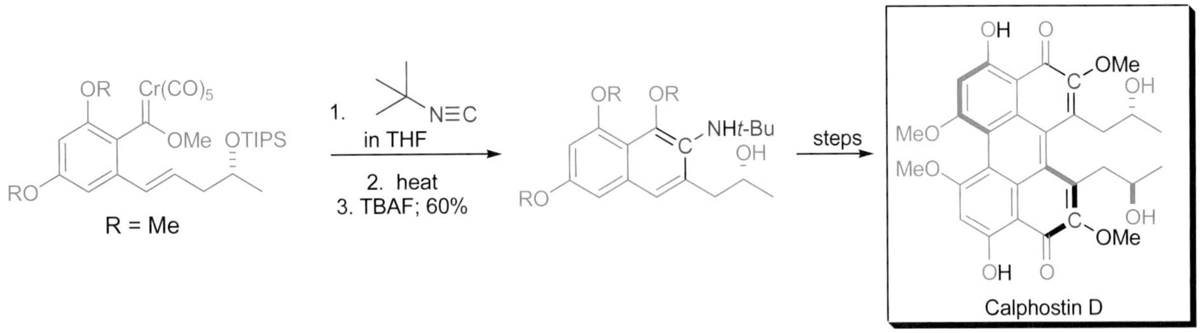

P. Quayle and co-workers utilized the *Dötz benzannulation reaction* for the synthesis of diterpenoid quinones.[41] The authors developed a novel synthetic approach to 12-*O*-methyl royleanone using a simple vinyl chromium carbene complex along with a disubstituted oxygenated acetylene. The bicyclic hydrazone was converted to the corresponding vinyllithium derivative by the *Shapiro reaction* and then functionalized to give the desired crude Fischer chromium carbene complex. The benzannulation took place in refluxing THF with excellent regioselectivity, and the natural product was obtained in 37% overall yield from the hydrazone.

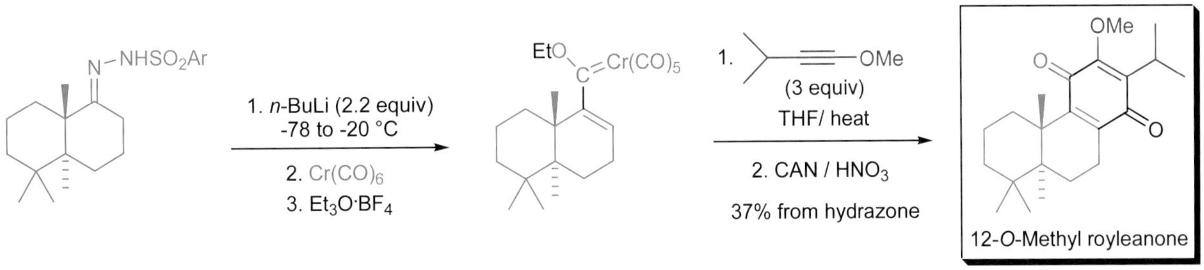

C-Arylglycosides possess a stable C-C glycosidic linkage and exhibit a broad range of useful antitumor, antifungal and antibiotic properties. S.R. Pulley et al. developed a novel method for the synthesis of this important class of compounds by using the *Dötz benzannulation* reaction between alkynyl glycosides and alkoxy phenyl chromium carbenes.[42]

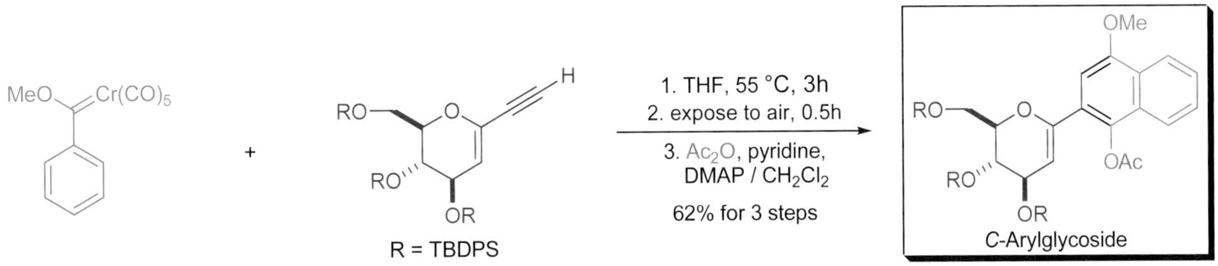

An exceptionally mild *Dötz benzannulation* was used by W.J. Kerr and co-workers for the total synthesis of a natural insecticide, 2-(1,1-dimethyl-2-propenyl)-3-hydroxy-1,4-naphthalenedione, by utilizing dry adsorption (DSA) techniques.[24]

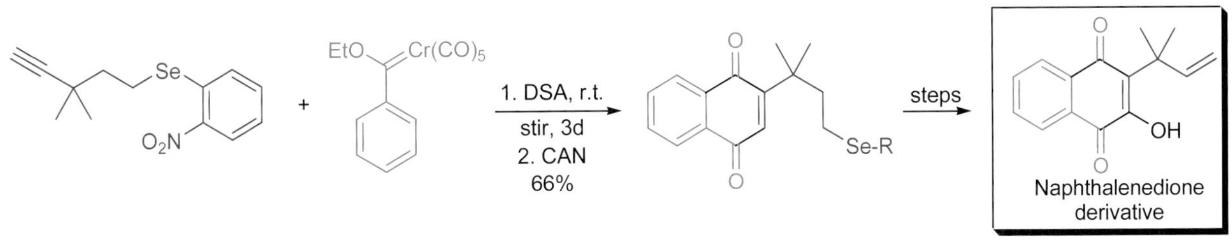

ENDERS SAMP/RAMP HYDRAZONE ALKYLATION

(References are on page 579)

Importance:

[*Seminal Publications*[1-5]; *Reviews*[6-15]; *Modifications & Improvements*[16-18]]

In 1976, D. Enders reported the asymmetric α-alkylation of ketones *via* the corresponding (*S*)-1-amino-2-methoxymethylpyrrolidine (SAMP) hydrazone derivatives.[1] According to the general procedure, the SAMP hydrazone was deprotonated with lithium diisopropylamide in tetrahydrofuran, and the corresponding lithium derivative was reacted with an alkyl halide. The product was ozonized to provide the α-alkylated ketone with high enantioselectivity. The opposite enantiomer can be obtained by using (*R*)-1-amino-2-methoxymethylpyrrolidine (RAMP) as the chiral auxiliary. This transformation can also be carried out on aldehydes.[2,3] The asymmetric alkylation of ketones and aldehydes *via* their SAMP/RAMP hydrazone derivatives is referred to as the *Enders SAMP/RAMP hydrazone alkylation*. General features of the reaction are: 1) the SAMP/RAMP hydrazones of aldehydes can be formed by mixing the aldehyde with the hydrazone derivative at 0 °C, while ketones need to be heated to reflux in the presence of a catalytic amount of acid in benzene or cyclohexane under Dean-Stark conditions;[1,2] 2) the hydrazones can be purified by distillation or chromatography, although purification is not always necessary, and they can be stored at -20 °C under inert atmosphere without decomposition;[1,2] 3) cyclic and acyclic ketones and aldehydes undergo the transformation;[1,2,14] 4) deprotonation can be effected with lithium bases, most commonly with lithium diisopropylamide;[1,2,14] 5) the alkylating reagents are alkyl-, benzyl-, and allyl bromides and iodides; 6) upon completion of the alkylation, the ketone can be regenerated by ozonolysis or methylation with methyl iodide and subsequent acidic hydrolysis;[1,2] and 7) the hydrazones can be transformed into various functionalities such as nitrile,[19,20] dithiane,[21] or amine.[19,13] The SAMP chiral auxiliary can be obtained from (*S*)-proline in four steps in a 58% overall yield, while RAMP is available from (*R*)-glutamic acid in six steps in 35%.[22-24] Several related chiral auxiliaries were also developed such as SADP, SAEP, SAPP, and RAMBO.[16-18] In addition to alkyl halides, the deprotonated SAMP/RAMP hydrazones react with Michael acceptors, ketones, α-halogen substituted esters, oxiranes, and aziridines.[14]

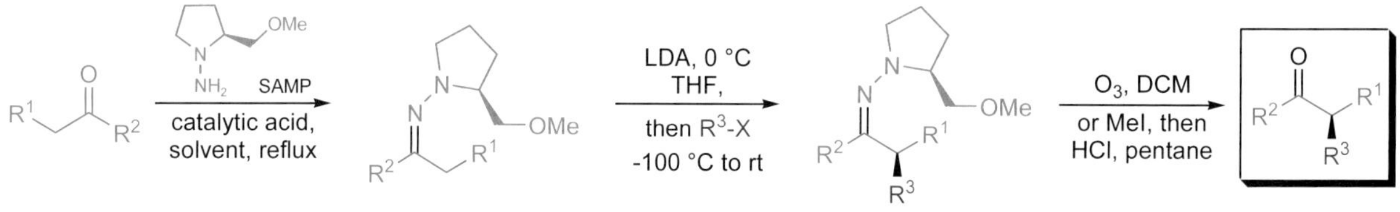

R^1 = alkyl, aryl; R^2 = H, alkyl, R^1 = R^2 = -$(CH_2)_3$-, -$(CH_2)_4$-, -$(CH_2)_5$-, -$(CH_2)_6$-, -CH=CH$(CH_2)_2$-; R^3 = alkyl, benzyl, allyl; X = I, Br;
solvent: benzene, cyclohexane

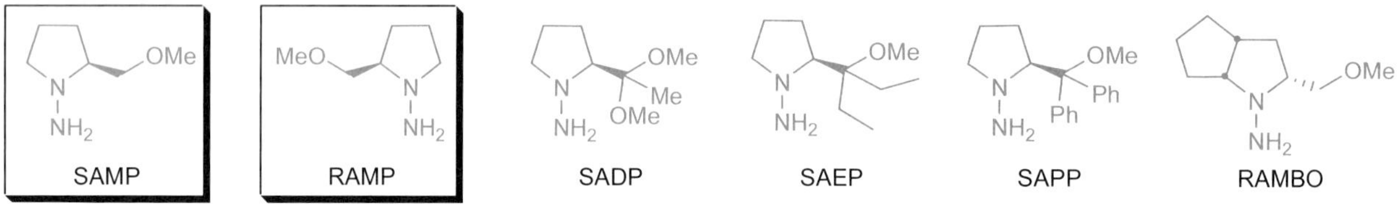

Mechanism:[25,1,2,26,3,22,27,5,28]

The deprotonation of the SAMP/RAMP hydrazone derivatives leads to the formation of azaenolates that can be trapped by the alkyl halide. In theory, four isomeric azaenolates can form in the deprotonation step, but it was shown that around the C-C double bond *E* stereochemistry is dominant, while around the C-N bond *Z* stereochemistry ($E_{CC}Z_{CN}$) is dominant for cyclic- and acyclic ketones. This observation was confirmed by trapping experiments,[1,2,22,27,5] MNDO calculations,[25] spectroscopic investigations,[26,3] and X-ray analysis.[28] It was also shown by freezing point depression experiments that the lithiated SAMP hydrazones exist in a monomeric form.[29] Electrophilic attack by the electrophile on this system proceeds from the sterically more accessible face with high diastereoselectivity.

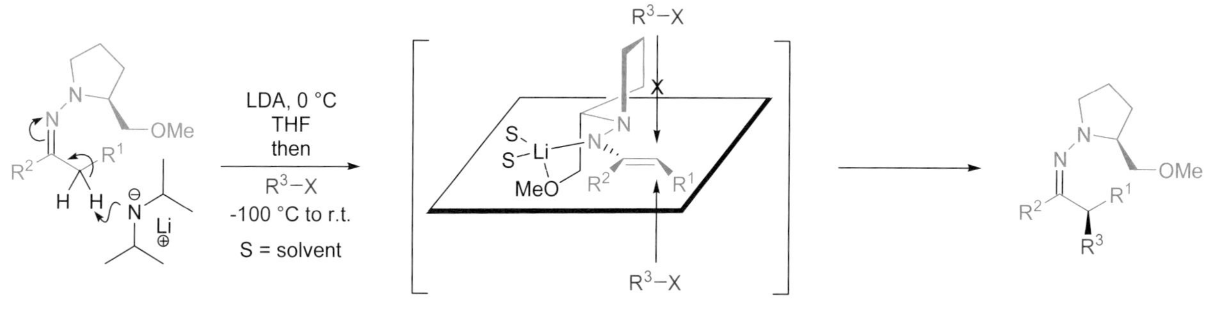

Synthetic Applications:

The synthesis of (–)-C_{10}-desmethyl arteannuin B, a structural analog of the antimalarial artemisinin, was developed by D. Little et al.[30] In their approach, the absolute stereochemistry was introduced early in the synthesis utilizing the *Enders SAMP/RAMP hydrazone alkylation* method. The sequence begins with the conversion of 3-methylcyclohexenone to the corresponding (*S*)-(–)-1-amino-2-(methoxymethyl)pyrrolidine (SAMP) hydrazone. Deprotonation with lithium diisopropylamide, followed by alkylation in the presence of lithium chloride at -95 °C afforded the product as a single diastereomer. The SAMP chiral auxiliary was removed by ozonolysis.

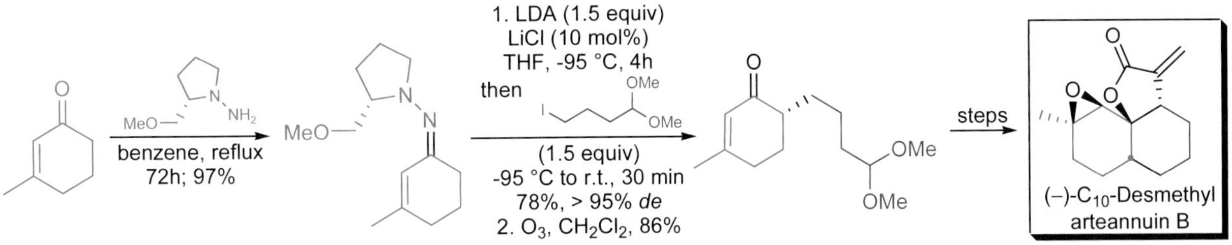

The total synthesis of (–)-denticulatin A, a polypropionate metabolite, was accomplished in the laboratory of F.E. Ziegler.[31] To establish the absolute stereochemistry at C12, they utilized the *Enders SAMP/RAMP hydrazone alkylation.* To this end, the RAMP hydrazone of 3-pentanone was successfully alkylated with 1-bromo-2-methyl-2(*E*)-pentene. Hydrolysis of the hydrazone under standard acidic conditions led to loss of the enantiomeric purity. This problem was avoided by using cupric acetate for the cleavage.

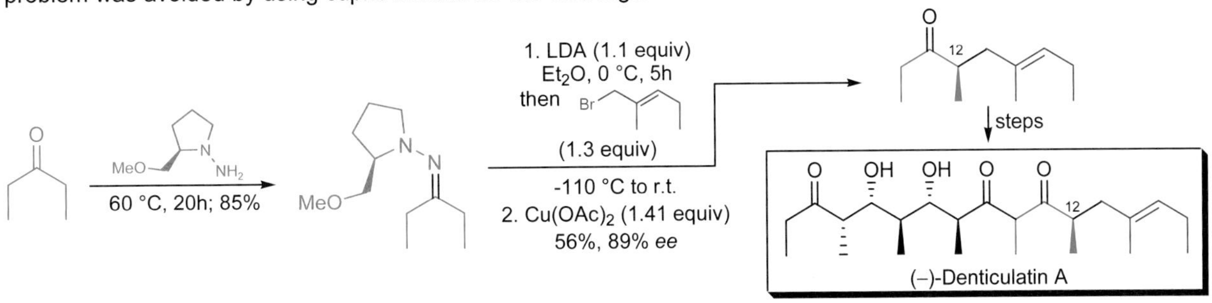

The first asymmetric total synthesis of (+)-maritimol, a diterpenoid natural product that possesses a unique tetracyclic stemodane framework was accomplished by P. Deslongchamps.[32] To introduce the C12 stereocenter, the *Enders SAMP/RAMP hydrazone alkylation* was used. This stereocenter played a crucial role in controlling the diastereoselectivity of the key transannular Diels-Alder reaction later in the synthesis. The required SAMP hydrazone was formed under standard conditions using catalytic *p*-toluenesulfonic acid. Subsequent protection of the free alcohol as a *t*-butyldiphenylsilyl ether, deprotonation of the hydrazone with LDA and alkylation provided the product in high yield and excellent diastereoselectivity. The hydrazone was converted to the corresponding nitrile by oxidation with magnesium monoperoxyphthalate.

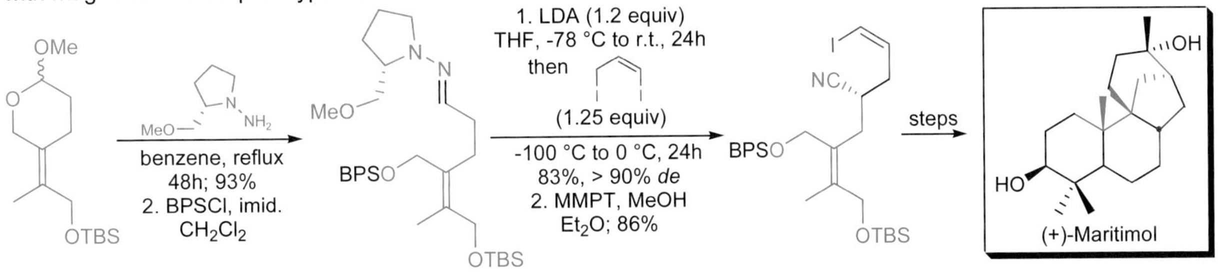

Application of the *Enders SAMP/RAMP hydrazone alkylation method* on 1,3-dioxan-5-one derivatives leads to versatile C_3 building blocks.[33] To demonstrate the usefulness of the above method, the research group of D. Enders applied it during the first asymmetric total synthesis of both enantiomers of streptenol A.[34] To obtain the natural isomer, the RAMP hydrazone of 2,2-dimethyl-1,3-dioxan-5-one was used as starting material. This compound was deprotonated with *t*-butyllithium and alkylated with 2-bromo-1-*tert*-butyldimethylsilyloxyethane. The chiral auxiliary could be hydrolyzed under mildly acidic conditions to provide the ketone in excellent yield and enantioselectivity.

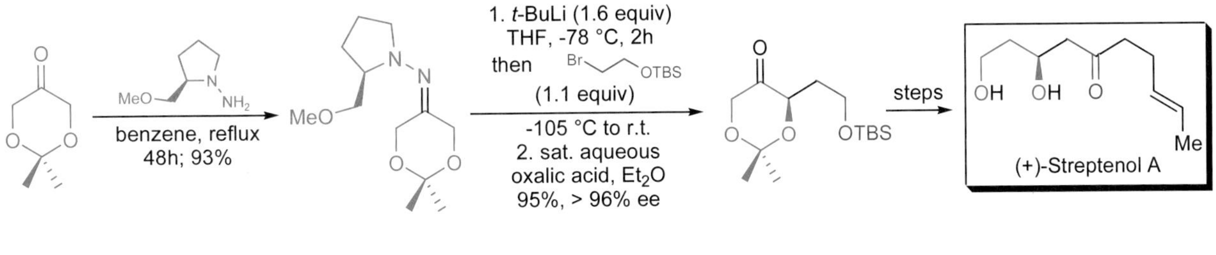

ENYNE METATHESIS

(References are on page 580)

Importance:

[*Seminal Publications*[1-5]; *Reviews*[6-16]; *Modifications & Improvements*[17-23]; *Theoretical Studies*[24,25]]

In 1985, T.J. Katz reported an intriguing methylene migration reaction when a biaryl 1,7-enyne was exposed to 1 mol% of a tungsten Fischer carbene complex to give a 1,3-diene as the product in 31% yield.[1] This was the first example of the metal carbene catalyzed intramolecular redistribution of carbon-carbon multiple bonds between an alkene and an alkyne. The transition metal catalyzed cycloisomerization of 1,*n*-enynes to the corresponding 1,3-dienes is known as the *intramolecular ring-closing enyne metathesis*. Another variant is the *cross enyne metathesis* between independent molecules of an alkene and an alkyne.[26] Soon after Katz's report, molybdenum, and chromium Fischer carbene complexes were also successfully utilized, but the catalysts were often required in stoichiometric amounts, and the yields were generally low due to side reactions. Besides metal carbene complexes, the *enyne metathesis* can be catalyzed by the following low-valent transition metals: Pd(II),[7] Pt(II),[27] Ru(II),[28] and Ir(I)[29] complexes.[15] The most widely used and most efficient *enyne metathesis* catalysts are ruthenium benzylidene complexes such as Grubbs first and second generation catalysts, which were originally developed for *olefin metathesis* reactions. The general features of the *ring-closing enyne metathesis* are: 1) the substituents of the olefin have a profound influence on the reaction rate, the number of different products, and their distributions;[30,31] 2) monosubstituted alkenes react faster than di- or trisubstituted ones; 3) enynes with monosubstituted olefins form exclusively the smallest possible ring size; 4) the substitution of the alkyne partner also has an influence on the reaction rate: terminal alkynes react slower than internal ones; 5) alkyl substituents on the alkyne tend to give high yields, whereas electron-withdrawing substituents usually result in lower yields; 6) the presence of ethylene gas (instead of the usual argon) may substantially increase the rate of the reaction in certain cases;[32,33] 7) reactions are usually conducted in dichloromethane, toluene, or benzene either at ambient temperature or at reflux; 8) a wide range of functional groups (esters, amides, ethers, ketones, acetals, etc.) are tolerated under the reaction conditions, but amines and alcohols need to be protected to obtain high yields; 9) the formation of a five- and six-membered ring is easily achieved, whereas 7-, 8-, and 9-membered carbocycles are not formed as readily unless the enyne tether contains a heteroatom;[12] and 10) the *enyne metathesis* in combination with other metathesis reactions and cycloadditions leads to powerful tandem reactions.

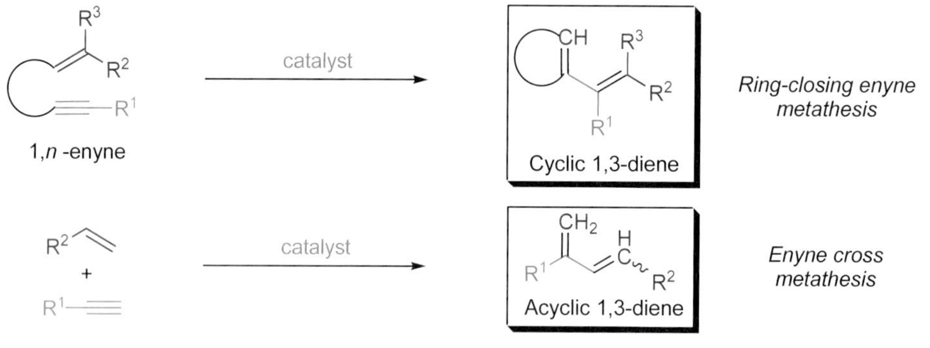

Mechanism:[34-36,17,37-39,30,27,40,31,41]

The mechanism of the *enyne metathesis* depends on the type of catalyst used while the fine details of the process are much less understood than in the case of olefin metathesis. Since the most widely used catalyst is the Grubbs's second-generation ruthenium carbene complex, the reaction mechanism of a *ring-closing enyne metathesis* employing this carbene is discussed. Two different mechanistic pathways may operate depending on whether the metal carbene first reacts with the alkene or alkyne. In **Path I** the alkene forms a metallacyclobutane intermediate that subsequently undergoes several ring openings and closures to give the final diene product. However, in **Path II** the metal carbene first reacts with the alkyne and two different dienes can be formed *via* two regioisomeric metallacyclobutenes (only one is shown).

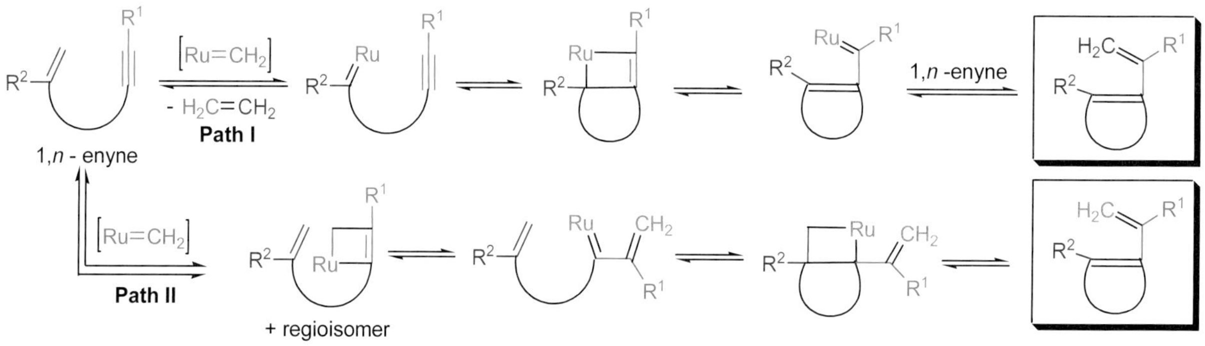

Synthetic Applications:

The short total synthesis of (±)-differolide based on a tandem *enyne metathesis* / [4+2] cycloaddition was accomplished by T.R. Hoye et al.[42] The *enyne metathesis* was carried out on allyl propynoate using Grubbs's first-generation metathesis catalyst. The catalyst was added to the substrate slowly to maintain high substrate and low ruthenium carbene concentrations. The initially formed 2-vinylbutenolide readily dimerized *via* a *Diels-Alder cycloaddition* in which the vinyl group participated as the dienophile to afford the natural product.

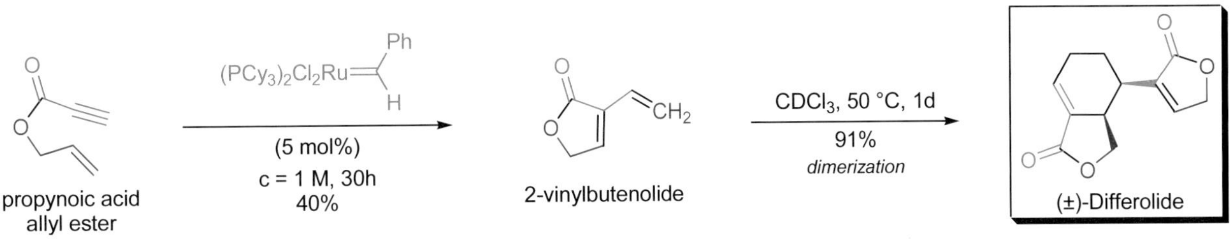

The total synthesis of polycyclic alkaloid (–)-stemoamide was achieved in the laboratory of M. Mori *via* a ruthenium carbene catalyzed *enyne metathesis*.[43] The cyclization was effected by 5 mol% of catalyst in benzene at 50 °C. After 11h of stirring under these conditions, 87% of the 5,7-fused bicyclic system was formed.

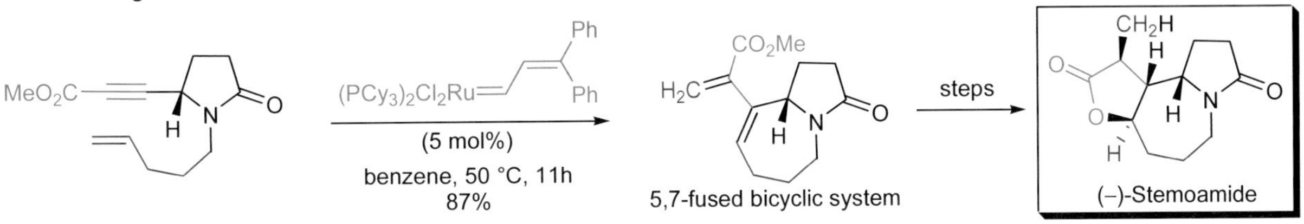

A platinum- and Lewis acid catalyzed *enyne metathesis* was used as the key step in the formal total synthesis of antibiotics streptorubin B and metacycloprodigiosin by A. Fürstner.[37] The electron-deficient enyne was cyclized with either a platinum halide or a hard Lewis acid (e.g., $BF_3{\cdot}OEt_2$) to the desired *meta*-pyrrolophane core of the target molecules. A few more steps completed the formal synthesis.

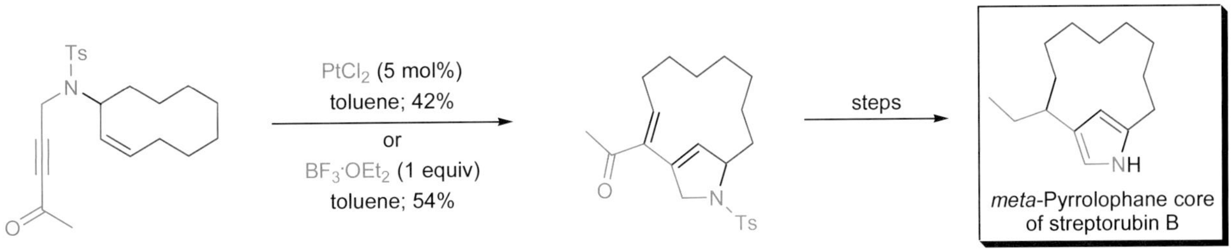

M. Shair and co-workers were the first to apply the *enyne metathesis* for macrocyclization during the biomimetic synthesis of (–)-longithorone A.[44] The two 16-membered paracyclophane building blocks, one diene and one dienophile component, were prepared using 50 mol% Grubbs's first-generation catalyst under 1 atm ethylene gas pressure. These components, after several additional steps, underwent two facile *Diels-Alder cycloaddition reactions* to afford the natural product.

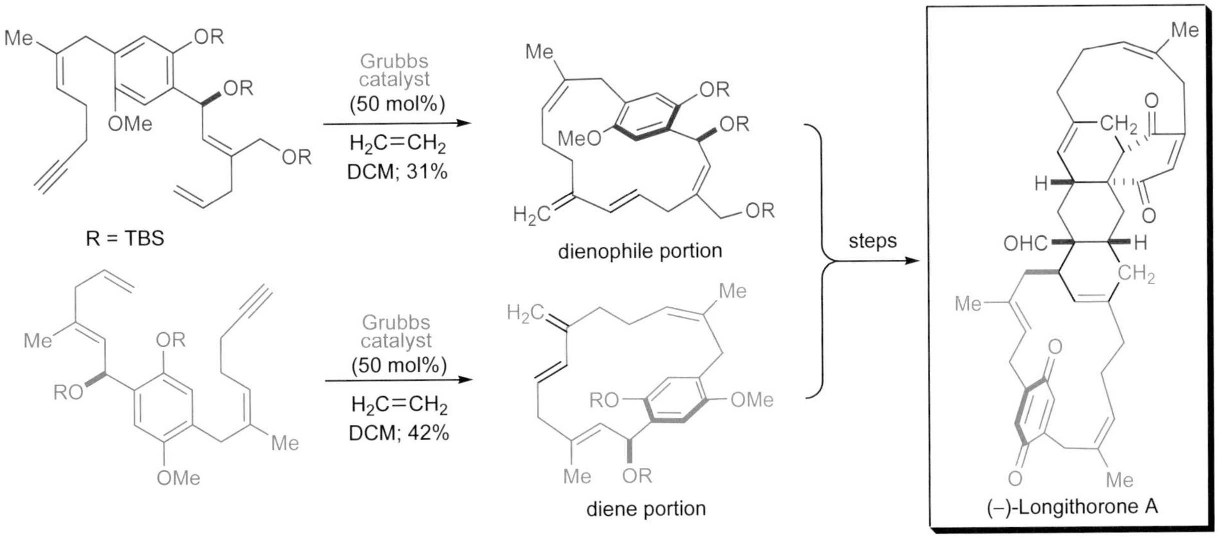

ESCHENMOSER METHENYLATION

(References are on page 581)

Importance:

[*Seminal Publications*[1]; *Reviews*[2]; *Modifications & Improvements*[3,4]]

The introduction of a (dimethylamino)methyl group ($-CH_2NMe_2$) into the α-position of a carbonyl group (ketone, ester, lactone, etc.) using dimethyl(methylene)ammonium iodide, $[CH_2{=}NMe_2]^+I^-$ (Eschenmoser's salt), followed by an elimination to the corresponding α-methylene carbonyl compound is known as the *Eschenmoser methenylation*. The first step of the methenylation procedure can be regarded as a modified *Mannich reaction*, in which the enolizable carbonyl compound is reacted with a preformed iminium ion (Eschenmoser's salt). Next, the resulting α-(dimethylamino)methyl carbonyl compound can be eliminated by using one of the following methods: 1) heat; 2) conversion to the corresponding quaternary ammonium salt, which is then heated (*Hoffmann elimination*);[5] 3) conversion to the corresponding *N*-oxide to induce a *Cope elimination* upon heating;[6] or 4) treatment with base.[7] The methenylation process is most efficient when the substrates are symmetrical ketones or ketones that have only one available enolizable α-position. In the case of unsymmetrical ketones (in which both the α and α' positions are available), regioselective methenylation is possible by the use of modified versions of Eschenmoser's salt.[3,4]

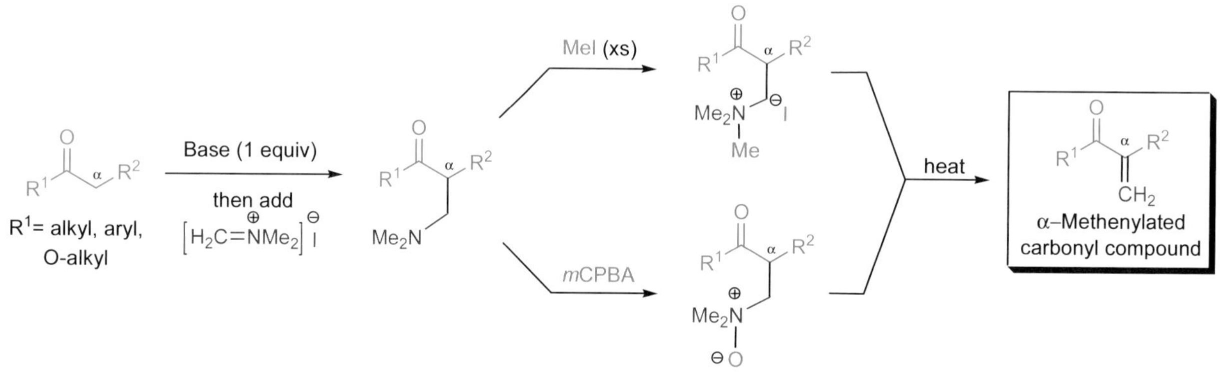

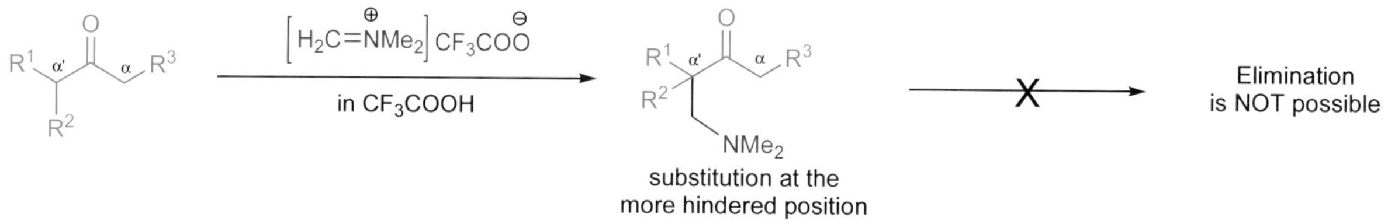

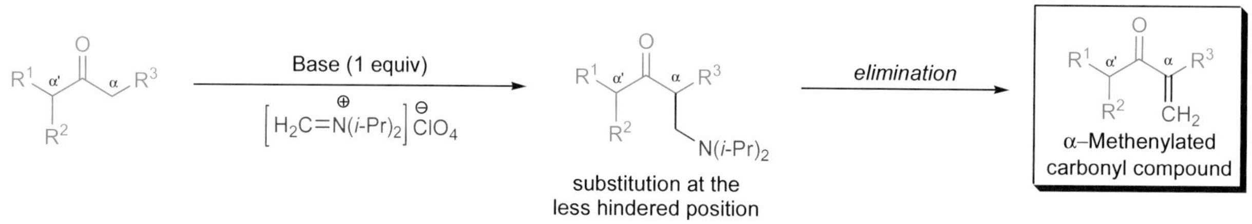

Mechanism:

The first step of the mechanism of the *Eschenmoser methenylation* is the deprotonation of the substrate at the α-position. The resulting enolate ion then reacts with the electrophilic iminium salt to afford the α-(dimethylamino)methyl carbonyl compound. In a second operation, the elimination is carried out in one of four ways as mentioned above. The scheme shown below depicts the *Cope elimination* of the tertiary amine *N*-oxide.

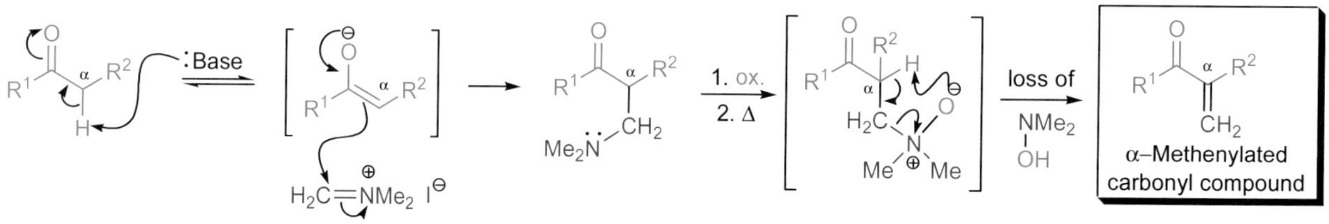

ESCHENMOSER METHENYLATION

Synthetic Applications:

S.J. Danishefsky and co-workers identified an *exo*-methylene hydroazulenone as a versatile intermediate in efforts directed toward the total synthesis of guanacastepene.[6] The *exo*-methylene group was introduced on the hydroazulene by the two-step *Eschenmoser methenylation* procedure. The substrate was deprotonated with LiHMDS followed by the addition of 3 equivalents of Eschenmoser's salt. The resulting α-(dimethylamino)methyl ketone was treated with *m*CPBA to form the *N*-oxide, which spontaneously underwent a *Cope elimination* to afford the desired *exo*-methylene hydroazulenone.

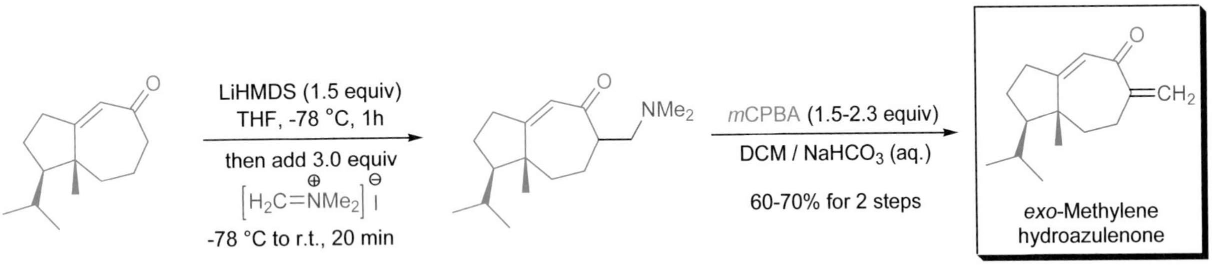

In the laboratory of J.L. Wood, an expeditious approach to the densely functionalized isotwistane core of CP-263,114 was developed.[7] For the proposed *radical cyclization*, an *exo*-methylene group was installed on a five-membered lactone ring. It was discovered that both the formation of the lactone ring and the *Eschenmoser methenylation* could be conducted in a one-pot operation by simply treating the α-acetoxy ketone with excess amounts of LiTMP and then with Eschenmoser's salt.

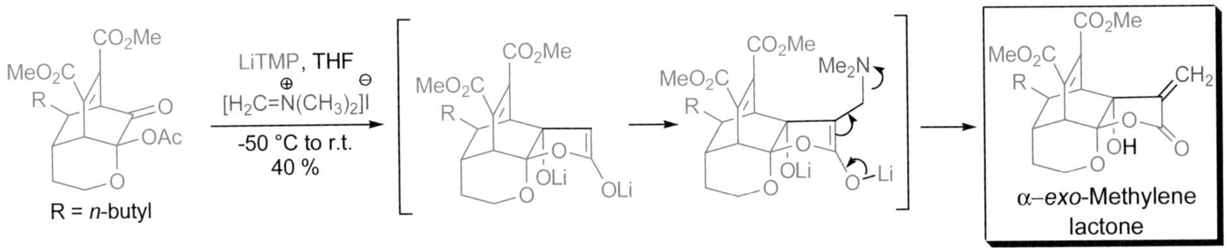

The total synthesis of the cembranoid diterpene (±)-crassin acetate methyl ether was accomplished by W.G. Dauben et al.[8] In the final stages of the total synthesis, the sensitive α-methylene group was introduced onto the six-membered lactone by using the *Eschenmoser methenylation* procedure. The lactone was deprotonated with LDA and then treated with Eschenmoser's salt. In the second step, the dimethylamino group was exhaustively methylated and the quaternary ammonium salt underwent a smooth *Hofmann elimination* upon deprotonation with DBU.

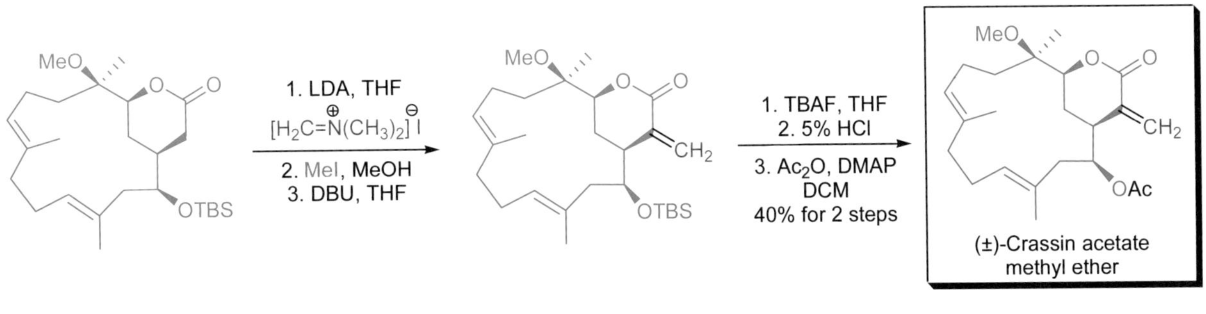

During the early stages of the total synthesis of (±)-gelsemine, S.J. Danishefsky et al. wanted to install a key oxetane ring on a bicyclic ketone intermediate.[9] The *Eschenmoser methenylation* was chosen to prepare the required bicyclic α-methylene ketone which was later converted to the oxetane in a few steps.

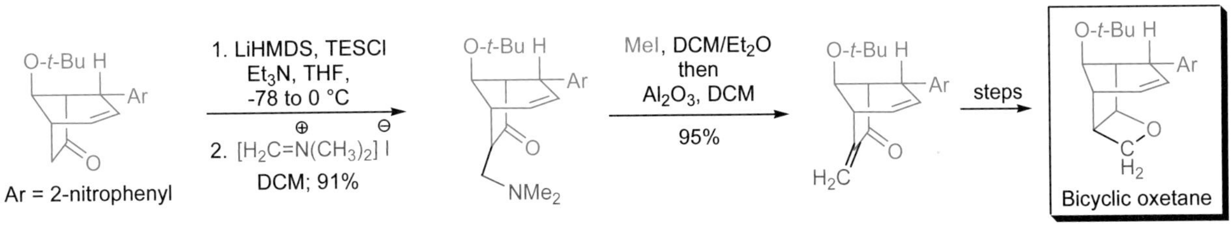

ESCHENMOSER-CLAISEN REARRANGEMENT

(References are on page 581)

Importance:

[*Seminal Publications*[1,2]; *Reviews*[3,4]; *Modifications & Improvements*[5]]

In 1964, A. Eschenmoser reported a reaction in which allylic or benzylic alcohols underwent a *Claisen-type rearrangement* when heated with *N,N*-dimethylacetamide dimethyl acetal in xylenes.[1] The rearrangement took place with a high degree of sterospecificity and generated a γ,δ-unsaturated amide as the product. Today this transformation is referred to as the *Eschenmoser-Claisen rearrangement*. The rearrangement is more (*E*)-selective and usually takes place at lower temperature (100-150 °C) than the other variants such as the *Claisen* and *Johnson-Claisen rearrangements*. Allylic alcohols substituted at the 2-position afford trisubstituted alkene products with significant levels of diastereoselection, just as in the case of the *Johnson-Claisen rearrangement*.[6] This selectivity is explained by 1,3-diaxial nonbonding interactions in the chairlike transition state.

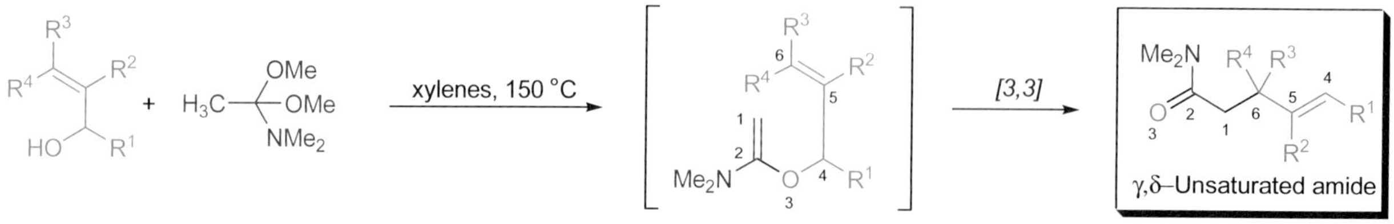

Mechanism: [6]

The reaction does not require the presence of an acid catalyst, the allylic alcohol readily exchanges one of the alkoxy groups of *N,N*-dimethylacetamide dimethyl acetal. The resulting mixed acetal loses methanol and the ketene aminal intermediate undergoes a *[3,3]-sigmatropic shift via* a chairlike transition state in acyclic systems. In certain cases, cyclic systems may prefer a boatlike transition state due to conformational constraints. The ratio of the products will depend on the energy difference between the transition states. Generally the *Eschenmoser-Claisen rearrangement* of secondary allylic alcohols proceeds with very high (*E*)-selectivity due to destabilizing 1,3-diaxial interactions in the transition state that would lead to the (*Z*)-isomer.[6]

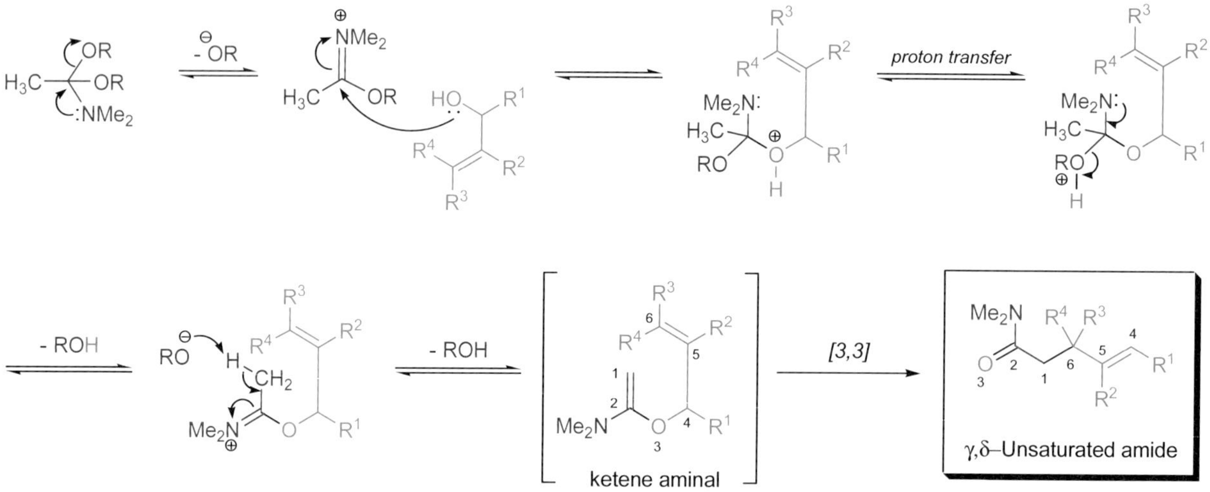

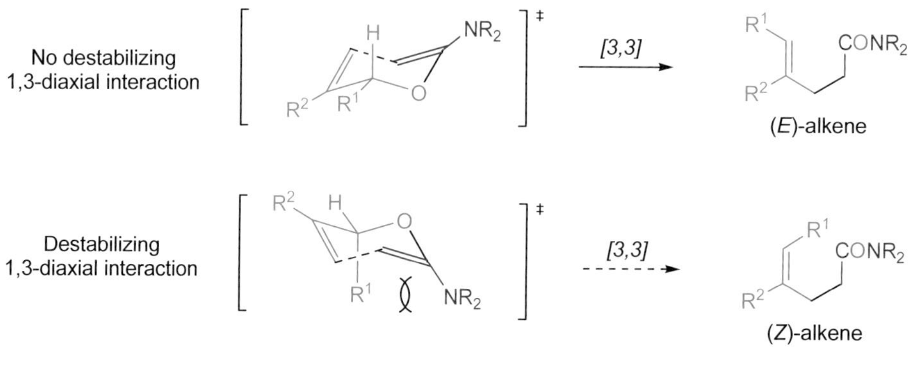

Synthetic Applications:

The first total synthesis of (±)-stenine has been accomplished in the laboratory of D.J. Hart.[7] The key steps were an *intramolecular Diels-Alder reaction*, an amidine variant of the *Curtius rearrangement*, an *Eschenmoser-Claisen rearrangement*, a *halolactonization*, and a *Keck allylation*. The allylic alcohol precursor and *N,N*-dimethylacetamide dimethyl acetal was heated to reflux in xylenes for 4h to afford the desired amide in 93% isolated yield. The transition state most likely adopted a boatlike conformation.

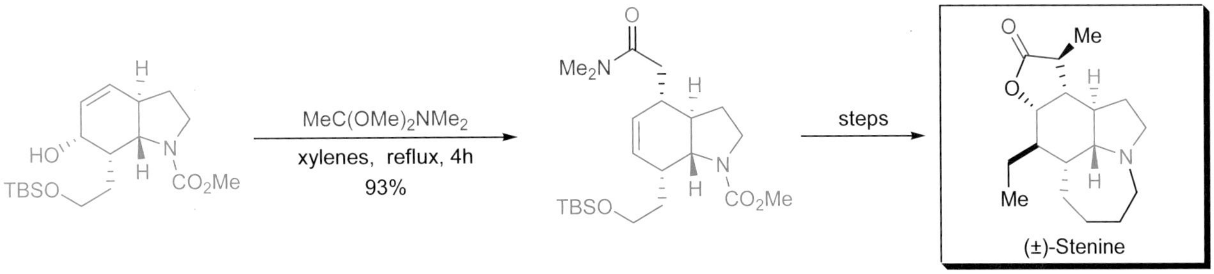

During the asymmetric total synthesis of (+)-pravastatin by A.R. Daniewski et al., one of the stereocenters was introduced with the *Eschenmoser-Claisen rearrangement*.[8] The tertiary alcohol intermediate was heated in neat *N,N*-dimethylacetamide dimethyl acetal at 130 °C for 48h, during which time the by-product methanol was distilled out of the reaction mixture to afford the desired amide in 92% yield.

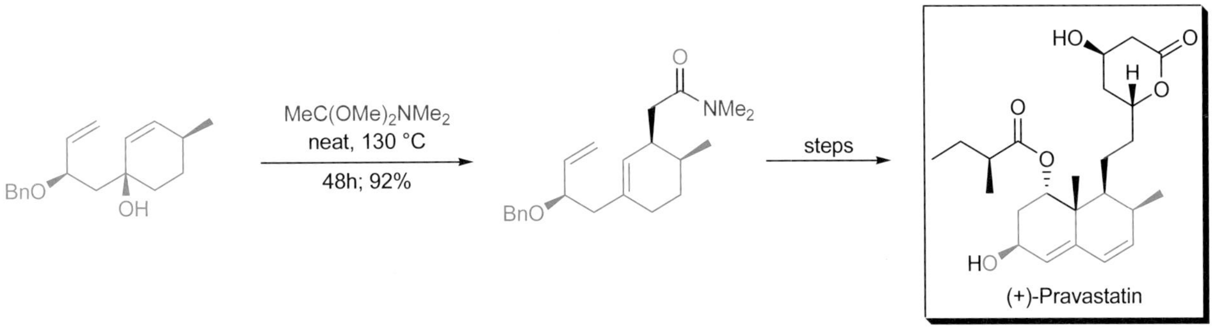

In order to construct the sterically congested C7a quaternary chiral center in the natural product anisatin, T.P. Loh and co-workers developed an efficient strategy by way of an *Eschenmoser-Claisen rearrangement*.[9] The resulting amide was converted to an ε-lactone (reported by A.S. Kende) in four steps, thereby completing a concise formal synthesis of (±)-8-deoxyanisatin. Other attempted *[3,3]-sigmatropic rearrangements* to construct C7a stereocenter resulted in re-aromatized products.

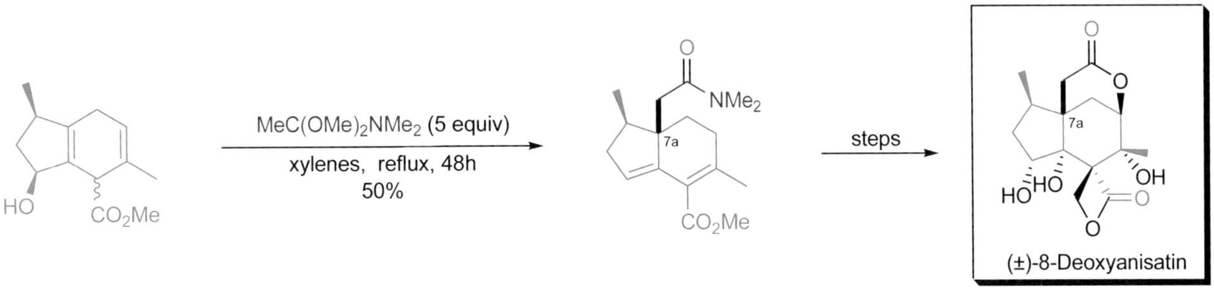

D.R. Williams et al. successfully synthesized the AB ring system of norzoanthamine by the *intramolecular Diels-Alder cyclization* of an (*E*)-1-nitro-1,7,9-decatriene.[10] The key transformation for establishing the quaternary stereocenter at C12 in the cycloaddition precursor was the *Eschenmoser-Claisen rearrangement*.

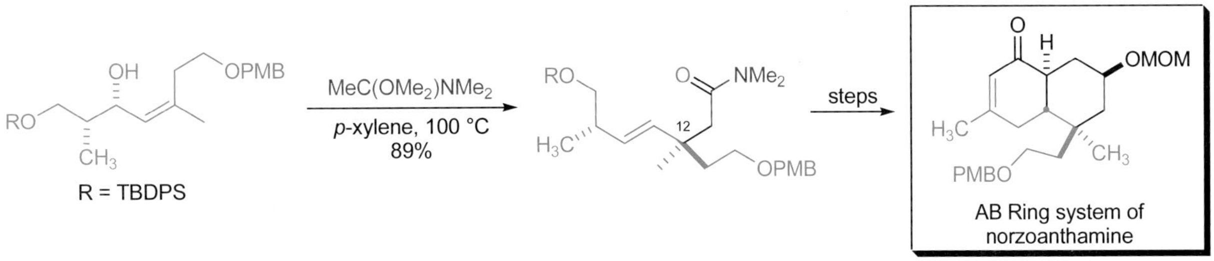

ESCHENMOSER-TANABE FRAGMENTATION

(References are on page 582)

Importance:

[*Seminal Publications*[1-4]; *Reviews*[5,6]; *Modifications & Improvements*[7-12]]

In 1967, A. Eschenmoser was the first to describe the fragmentation of the tosylhydrazone of an α,β-epoxy ketone to the corresponding acetylenic ketone (alkynone).[1] Soon after this initial report, M. Tanabe and J. Schreiber published their independent findings of the same fragmentation generating medium-sized cyclic alkynones.[2-4] Today, the preparation of cyclic alkynones, acyclic alkynals, and alkynones *via* cyclic epoxy ketone hydrazones is known as the *Eschenmoser-Tanabe fragmentation*. Although the fragmentation readily occurs on acyclic epoxy ketone hydrazones, from a synthetic point of view one needs to start from a cyclic epoxy ketone in order to isolate the desired cyclic or acyclic alkynones. The starting cyclic epoxy ketones are not always easy to synthesize, especially when they are sterically hindered. They are usually prepared by the epoxidation of the corresponding α,β-unsaturated ketones.[13] Next, the epoxy ketone hydrazones are exposed to base (or acid in certain cases) or heated to bring about the fragmentation, which is accompanied by the evolution of nitrogen gas. Acyclic acetylenic aldehydes can be efficiently prepared by using the 2,4-dinitrophenylhydrazone derivatives of the epoxy ketones.[10] When the preparation of the epoxy ketone is not possible or has to be avoided, treatment of the unsaturated hydrazones with excess NBS in methanol leads directly to the desired alkynones.[12] Over the last few decades, the scope of the reaction was extended, and improvements have been implemented by the use of the following epoxy ketone derivatives: 1) oximes;[14] 2) aminoaziridines;[8,9] 3) 2,4-dinitrobenzenesulfonyl hydrazones;[10] 4) 1,3,4-oxadiazolines;[11] and 5) diazirines.[7] Advantages of the *Eschenmoser-Tanabe fragmentation* are the following: 1) easy access to medium-sized cyclic ketones; 2) both terminal and disubstituted alkynes can be prepared; and 3) the fragmentation is not limited to the use of aromatic sulfonylhydrazones. Besides the fragmentation of epoxy ketone derivatives, there are only very few examples in the literature for the "*nitrogen*- and *carbon*-analogue" of the *Eschenmoser-Tanabe fragmentation*.[15-17]

Synthesis of cyclic alkynones:

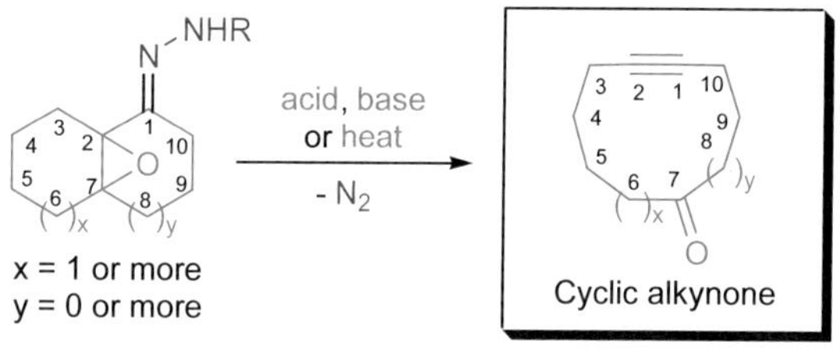

Synthesis of acyclic alkynones and alkynals:

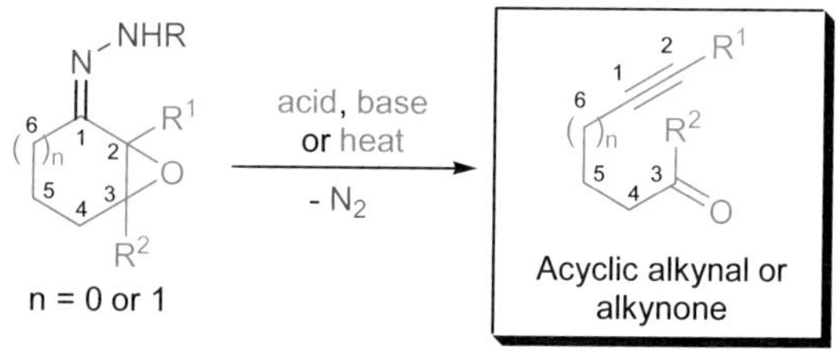

R = tosyl, 2,4-dinitrophenyl; R^{1-2} = H, alkyl; when R^2 = H, then the product is an alkynal, and when R^2 = alkyl, then it is an alkynone

Mechanism: [1,3,6,18,19]

The *Eschenmoser-Tanabe fragmentation* is basically a seven-center *Grob-type fragmentation* in which the starting molecule breaks into three fragments. The mechanism is concerted for epoxy ketone hydrazones and oxadiazolinones, while the thermal decomposition of epoxy-diazirines involves a free oxiranylcarbene intermediate.[18,19] The deprotonation of the starting epoxy ketone arylhydrazone leads to the formation of an alkoxide, which rapidly undergoes fragmentation to give an alkyne, ketone, nitrogen gas, and a leaving group (usually arylsulfinate).

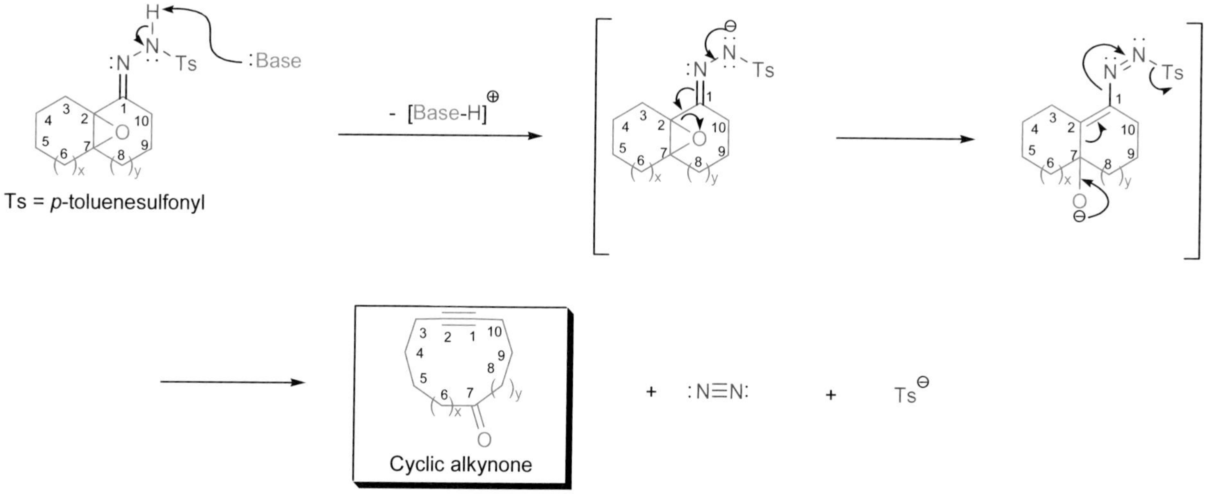

Synthetic Applications:

The first total synthesis of the Galbulimima alkaloid GB 13 was accomplished in the laboratory of L.N. Mander.[20] In the late stages of the synthesis, the plan was to convert the pentacyclic α,β-unsaturated ketone to the corresponding tetracyclic alkynone using the *Eschenmoser-Tanabe fragmentation*. Interestingly, the direct epoxidation of the enone was unsuccessful. Therefore, a sequence of reduction-epoxidation-oxidation gave the desired epoxy ketone in 77% yield. The treatment of this epoxy ketone with *p*-nitrobenzenesulfonylhydrazide afforded the alkynone in good yield.

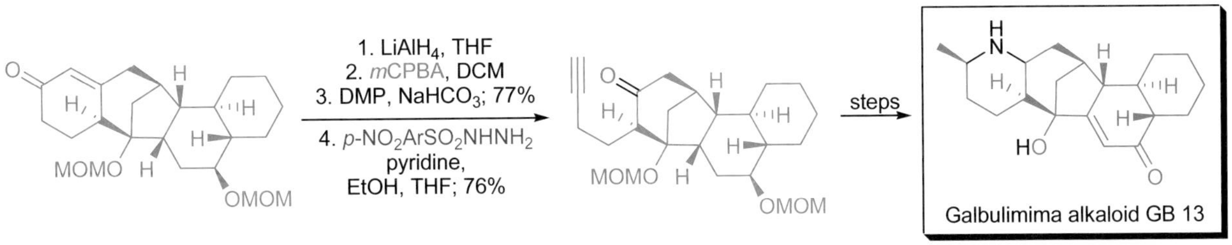

J.A. Katzenellenbogen et al. developed an efficient method for the synthesis of alkyl-substituted enol lactones that are potent inhibitors of the serine protease elastase.[21] The precursors for the enol lactones were α- and β-alkyl-substituted 5-hexynoic acids, which were prepared by the *bromoform reaction* of the corresponding alkynoic methyl ketones. These alkynones were synthesized by an *Eschenmoser-Tanabe fragmentation* of suitably substituted cyclohexenones.

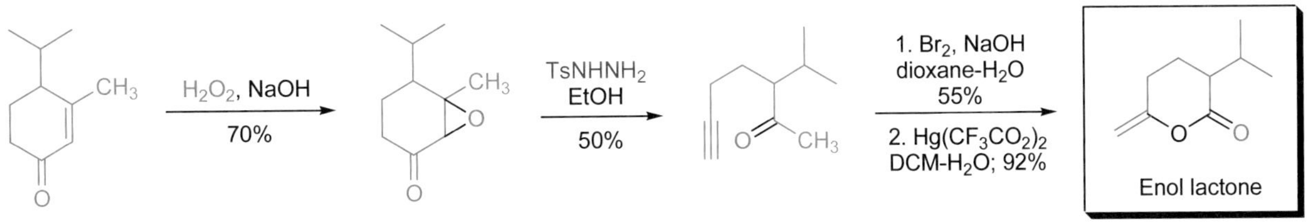

During model studies for the synthesis of botrydiane sesquiterpene antibiotics, B.M. Trost and co-workers prepared a complex 1,6-enyne precursor for transition metal catalyzed *enyne metathesis* reactions.[22] The 1,6-enyne was prepared from a heavily substituted alkynal, which was synthesized *via* the *Eschenmoser-Tanabe fragmentation* of an epoxy ketone. The resulting alkynal was unstable, so it was immediately subjected to a *Wittig olefination* to afford the desired 1,6-enyne.

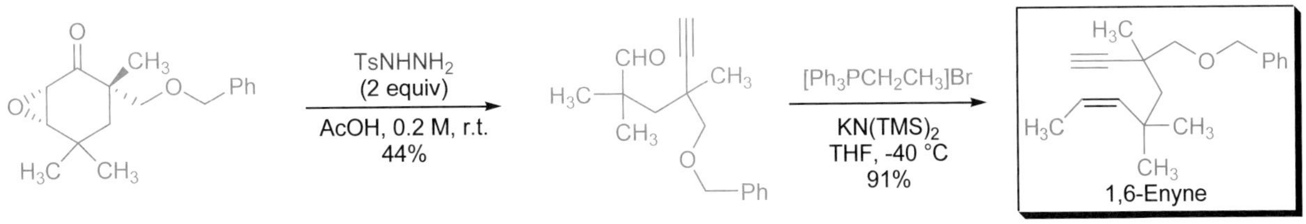

In the laboratory of S.J. Danishefsky, the synthesis of antibiotics containing the benz[*a*]anthracene core structure was investigated using the *Dötz benzannulation* of a cycloalkynone.[23] The required cycloalkynone was prepared from azulenone using the *Eschenmoser-Tanabe fragmentation*.

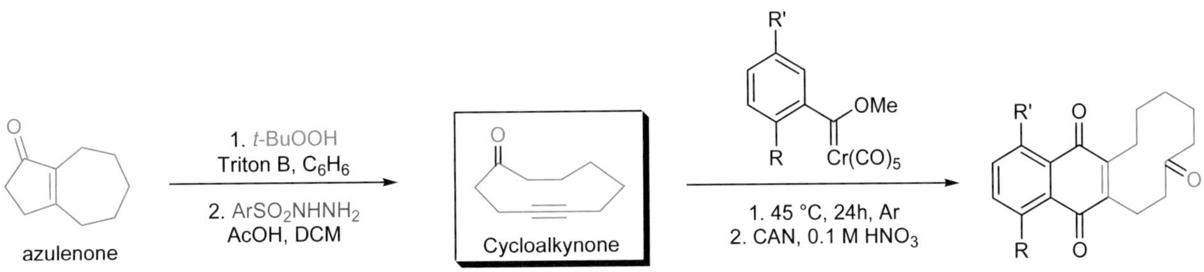

ESCHWEILER-CLARKE METHYLATION
(REDUCTIVE ALKYLATION)
(References are on page 582)

Importance:

[*Seminal Publications*[1-4]; *Reviews*[5-7]; *Modifications & Improvements*[8-13]]

The one-pot reductive methylation of primary and secondary amines to the corresponding tertiary amines is known as the *Eschweiler-Clarke methylation*. This reaction falls into the category of *reductive alkylation* of amines by carbonyl compounds (aldehydes and ketones), and it is considered as a modification of the *Leuckart-Wallach reaction*.[14] The first reductive alkylation of an amine was reported by R. Leuckart in 1885, and a few years later the scope of the reaction was explored by Wallach and co-workers.[15,16] In 1905, W. Eschweiler and then in 1933, H.T. Clarke demonstrated that formaldehyde could be used along with formic acid to introduce methyl groups to primary and secondary amines to obtain tertiary amines.[1,2] Formic acid serves as a reducing agent (hydride donor), which reduces the Schiff base intermediate to the corresponding amine. Today, other reducing agents, such as sodium borohydride, sodium cyanoborohydride,[8] sodium cyanoborohydride-titanium(IV)isopropoxide [$NaBH_3CN$-Ti(Oi-Pr)$_4$)],[17] sodium triacetoxyborohydride [NaBH(OAc)$_3$],[18] borohydride exchange resin (BER),[19] formic acid derivatives (formamide, ammonium formate, etc.), or hydrogen gas/catalyst[20] are used in place of formic acid. When the amine substrate is unsaturated, it is possible to obtain a cyclic amine product under the *Eschweiler-Clarke methylation* conditions, and the process is referred to as the *Eschweiler-Clarke cyclization*.[3,4]

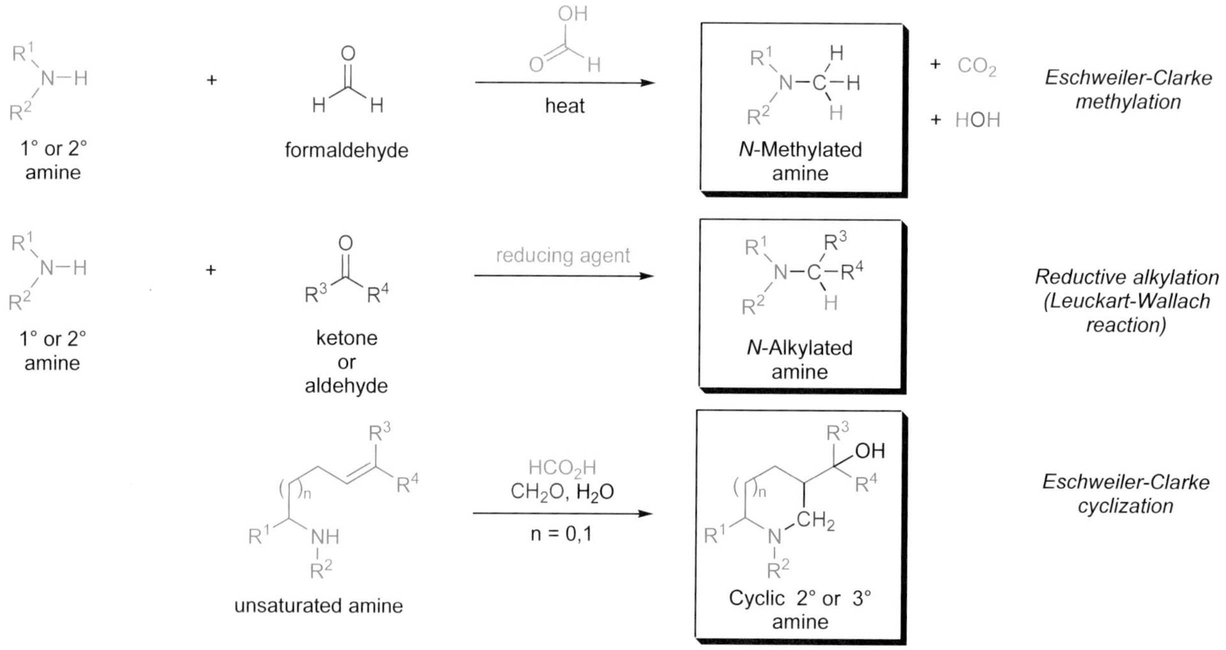

Mechanism:[21-26,13]

The mechanism of all of the above mentioned reactions is essentially the same. However, some steps in the mechanism are still not fully understood. The following steps are believed to be involved in the *Eschweiler-Clarke methylation*: 1) formation of a Schiff-base (imine) from the starting primary or secondary amine and formaldehyde *via* an aminoalcohol (aminal) intermediate; 2) hydride transfer from the reducing agent (e.g., formic acid, cyanoborohydride, etc.) to the imine to get the corresponding *N*-methylated amine along with the loss of CO_2; and 3) if the starting amine was primary, then steps 1 and 2 are repeated.

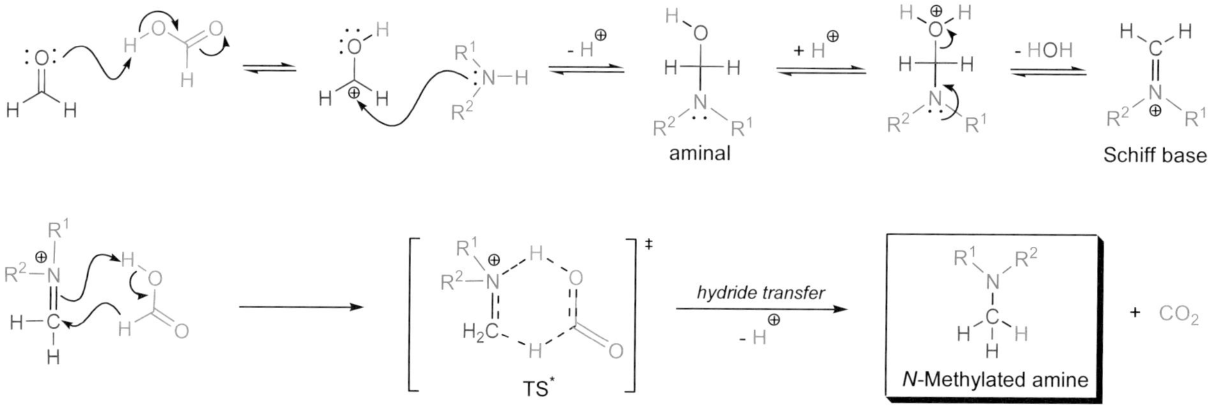

Synthetic Applications:

During the total synthesis of (–)-calyculin A and B, A.B. Smith and co-workers utilized a modified *Eschweiler-Clarke methylation* to convert a complex primary amine to the corresponding *N,N*-dimethylamino derivative.[27] The *N*-Boc protected primary amine was first deprotected using TMSOTf, followed by introduction of the two methyl groups using HCHO/$NaBH_3CN$ in AcOH/CH_3CN solvent mixture. The acetonide protecting group was subsequently removed, and the resulting diol was silylated.

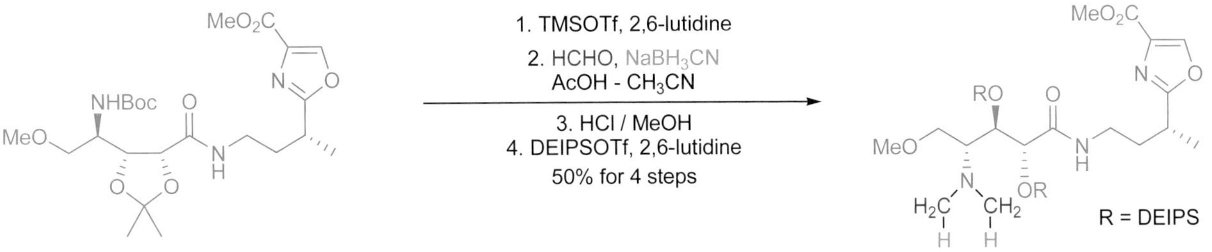

The enantioselective total syntheses of several piperidine and pyrrolidine alkaloids of tobacco were accomplished in the laboratory of J. Lebreton.[20] In the final stage of the total synthesis of (*S*)-*N*-methylanabasine, a one-pot *Cbz-deprotection-hydrogenation-Eschweiler-Clarke methylation* was carried out using a HCHO/MeOH/Pd(C)/H_2 system at room temperature with an overall 88% yield.

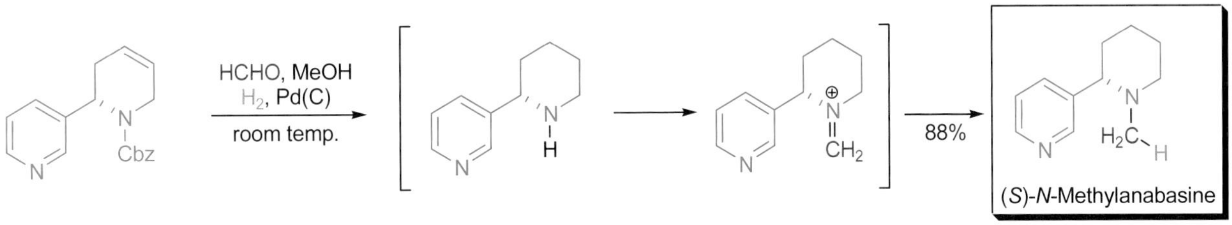

The oxindole alkaloid (–)-horsfiline was synthesized by K. Fuji et al. using an *asymmetric nitroolefination* as the key step.[28] During the endgame of the total synthesis, an *N*-methylation was performed on the five-membered secondary amine using the original *Eschweiler-Clarke methylation* conditions (HCO_2H/HCHO/reflux). Unfortunately, these harsh methylation conditions led to the racemization of the quaternary stereocenter. Therefore, milder modified conditions were applied ($NaBH_3CN$ as the reducing agent) to retain the optical activity of the substrate.

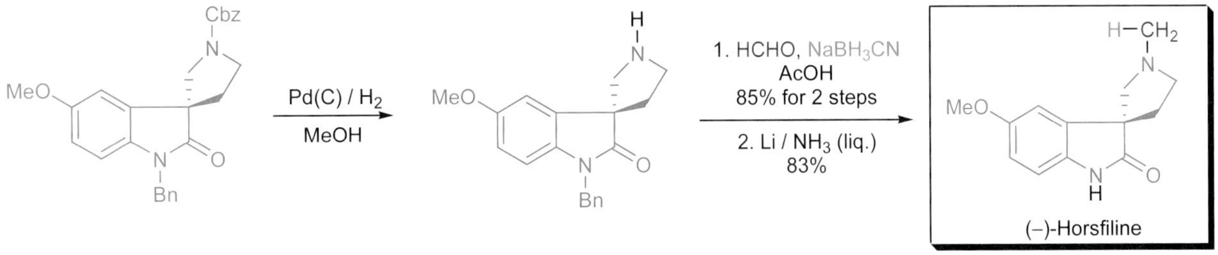

C.L. Gibson and co-workers developed an efficient synthesis for chiral ring annulet 2,6-disubstituted 1,4,7-trimethyl-1,4,7-triazamacrocycles. This class of molecules is capable of stabilizing transition metals in their high oxidation states and therefore can be used as oxidation catalysts.[29] The *N*-methylation of the three nitrogens in the last step was conducted using the original *Eschweiler-Clarke methylation conditions*.

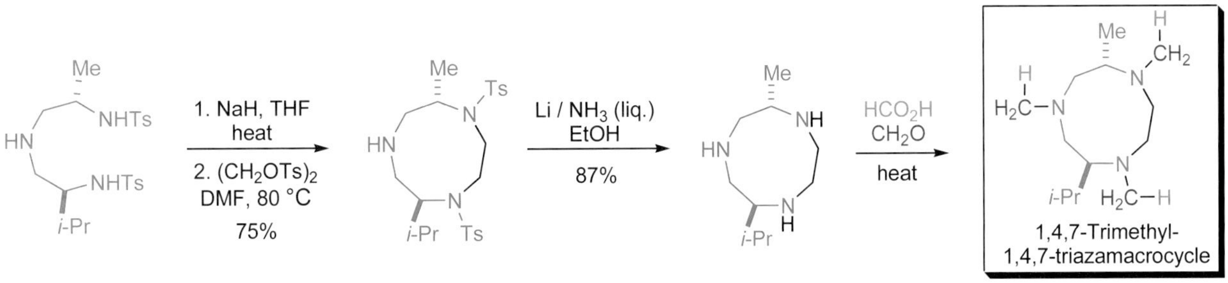

EVANS ALDOL REACTION

(References are on page 583)

Importance:

[*Seminal Publication*[1]; *Reviews*[2-10]; *Modifications & Improvements*[11-22]; *Theoretical Studies*[23-28]]

The boron mediated *aldol reaction* is a powerful method for highly stereoselective carbon-carbon bond formation. The high diastereoselectivity of this process can be attributed to the relatively short boron-oxygen bond length (1.36-1.47 Å) in the boron enolate,[29] which upon reacting with an aldehyde leads to a tight, six-membered chairlike transition state. Reaction of (*Z*)-boron enolates with aldehydes gives the *syn* aldol product while, (*E*)-boron enolates lead to formation of the *anti* aldol product with high diastereoselectivity.[30,31] Control of the absolute stereochemistry can be achieved through the application of covalently attached chiral auxiliaries in the enol component. D.A Evans and his co-workers developed a pair of oxazolidinone based chiral auxiliaries, which could be obtained from (*S*)-valinol and (1*S*,2*R*)-norephedrine with excellent enantiopurity.[1] *Asymmetric aldol reactions* relying on the application of these chiral auxiliaries are called the *Evans aldol reaction*. General features of the *Evans aldol reaction* are: 1) enolization of the *N*-acyl oxazolidinones under standard conditions (1.1 equiv Bu_2BOTf, 1.2 equiv diisopropylamine, 0 °C, 30 min) affords the (*Z*)-enolates with excellent selectivity;[1] 2) *aldol reaction* of the resulting (*Z*)-boron enolates with a wide variety of aldehydes yields the *syn* aldol product with very high diastereo- and enantioselectivity;[1] 3) when a chiral aldehyde is used, the facial bias of the enolate overrides the π-facial selectivity of the chiral aldehyde;[32] 4) aldol reaction of boron enolates derived from *N*-acetyloxazolidinone (R^1=H) provide the products with low stereoselectivity, but this can be overcome by the incorporation of a heteroatom substituent in the α-position, such as a thioalkyl group (R^1=SR), which can be reductively removed;[1] and 5) there are several methods for the nondestructive removal and recovery of the chiral auxiliary: hydrolysis and transesterification (LiOH, LiOOH, LiOR, LiSEt),[33-35] reductive removal ($LiAlH_4$),[33,36] and transamination to Weinreb amide (Me(OMe)NH, Me_3Al).[37] Since the introduction (S)-4-isopropyl-oxazolidin-2-one and (1S,2*R*)-4-methyl-5-phenyl-oxazolidin-2-one chiral auxiliaries by D.A. Evans, several modifications have been reported.[11-22] Besides the *aldol reaction*, the Evans chiral auxiliaries were successfully applied in enolate alkylation,[33] enolate acylation,[33] enolate amination,[38-41] and hydroxylation[42] processes.

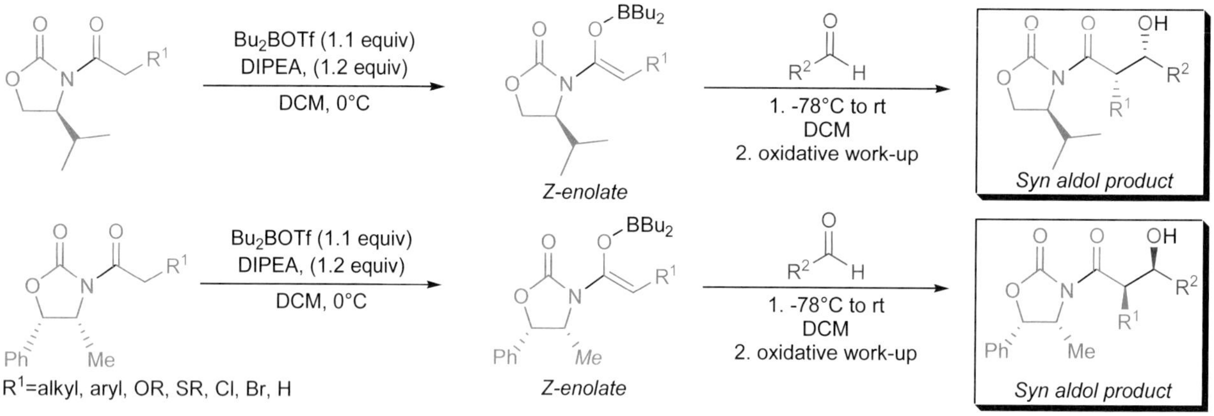

Mechanism:[2]

The observed stereoselectivity in the *Evans aldol reaction* can be explained by the *Zimmerman-Traxler* transition state model.[2] There are eight possible transition states, four of which would lead to the *anti* aldol product. These, however, are disfavored due to the presence of unfavorable 1,3-diaxial interactions (not depicted below). The possible transition states leading to the *syn* aldol product are shown below. The preferred transition state leading to the product is transition state **A**, where the dipoles of the enolate oxygen and the carbonyl group are opposed, and there is the least number of unfavored steric interactions.

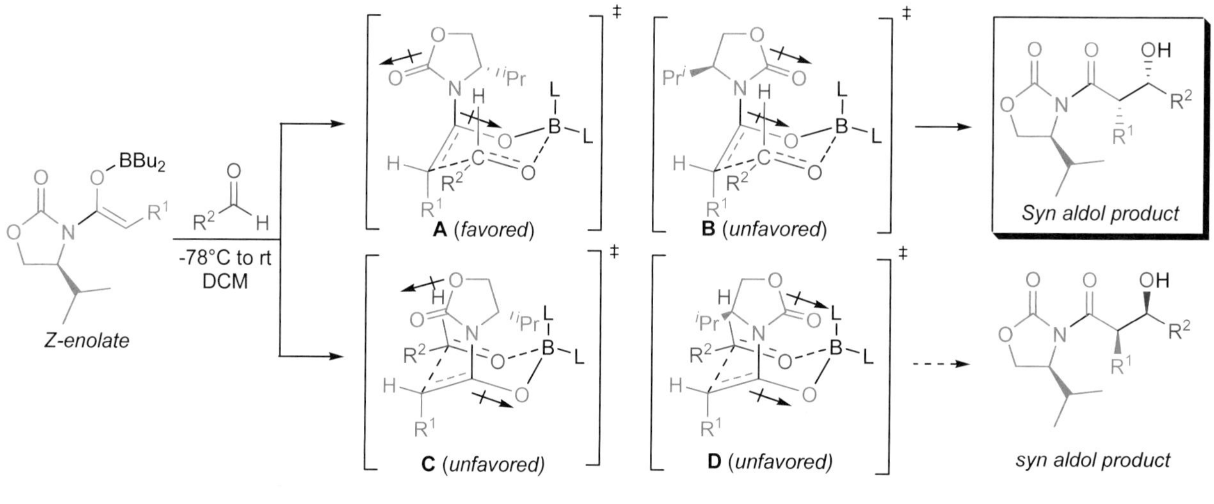

Synthetic Applications:

Glucolipsin A, a glycolipid possessing glycokinase-activating properties, was discovered at Bristol-Myers Squibb, but the absolute stereochemistry of the natural product remained elusive. A. Fürstner and co-workers elucidated the absolute stereochemistry *via* synthesis and spectroscopic analysis of the natural macrolide and its C_2-symmetric stereoisomers.[43] In their approach, they utilized the *Evans aldol reaction* that provided the *syn* aldol product with good yield and excellent diastereoselectivity.

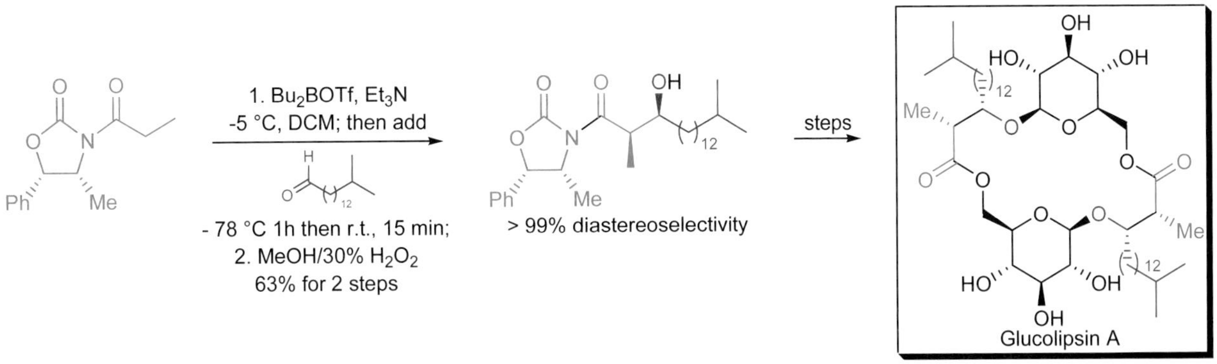

D.L Boger et al. reported the total synthesis of bleomycin A_2. They devised an efficient synthesis for the construction of the tripeptide S, tetrapeptide S, and pentapeptide S subunits of the natural product.[44,45] In their strategy, they utilized an *Evans aldol* reaction between the (*Z*)-enolate derived from (*S*)-4-isopropyl-3-propionyl-oxazolidin-2-one and *N*-Boc-D-alaninal. In order to synthesize one of the diastereomers of the pentapeptide S subunit, they carried out an *Evans aldol reaction* between the same aldehyde and the (*Z*)-enolate of (*R*)-4-isopropyl-3-propionyl-oxazolidin-2-one. The formation of the diastereomeric *syn* aldol product in this reaction clearly shows that the stereochemical outcome of the transformation is determined by the chiral auxiliary.

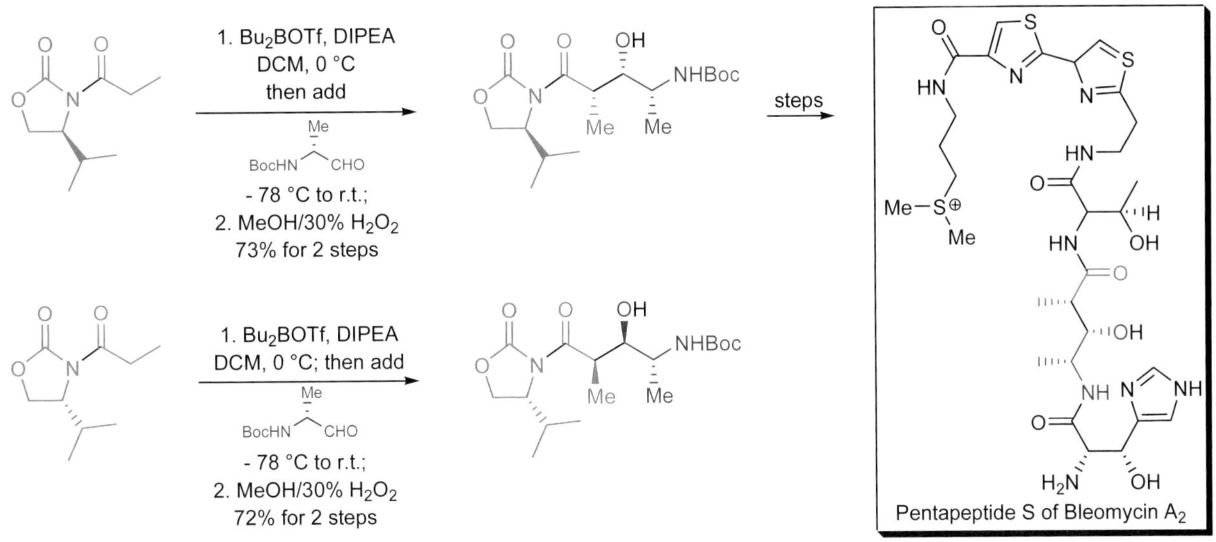

The asymmetric total synthesis of cytotoxic natural product (–)-FR182877 was accomplished by D.A. Evans and co-workers.[46,47] To establish the absolute stereochemistry, a boron mediated aldol reaction was utilized applying (*R*)-4-benzyl-*N*-propionyl-2-oxazolidinone[48] as a chiral auxiliary to yield the *syn* aldol product.

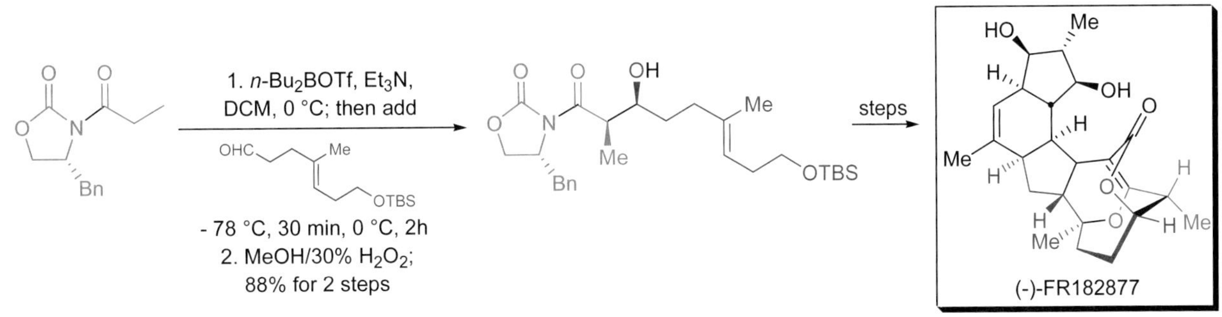

FAVORSKII AND HOMO-FAVORSKII REARRANGEMENT

(References are on page 584)

Importance:

[*Seminal Publications*[1,2]; *Reviews*[3-8]; *Modifications & Improvements*[9-12]; *Theoretical Studies*[13,14]]

Treatment of α-halo ketones possessing at least one α-hydrogen with base in the presence of a nucleophile (alcohol, amine, or water) results in a skeletal rearrangement *via* a cyclopropanone intermediate to give carboxylic acids or carboxylic acid derivatives (esters or amides). This reaction is known as the *Favorskii rearrangement,* and it is widely used for the synthesis of highly branched carboxylic acids. The halogen substituent can be a chlorine, bromine or iodine, while the base is usually an alkoxide or hydroxide. Upon rearrangement, acyclic α-halo ketones give acyclic carboxylic acid derivatives, while cyclic α-halo ketone substrates undergo a ring-contraction reaction to afford one-carbon smaller cyclic carboxylic acid derivatives. The reaction is both regio- and stereoselective.[15,16,12] The rearrangement of unsymmetrical α-halo ketones leads to the product, which is formed through the cleavage of the cyclopropanone intermediate to usually give the thermodynamically more stable of the two possible carbanions. Besides α-halo ketones, other α-substituted ketones such as α-hydroxy,[9] α-tosyloxy,[17] and α,β-epoxy ketones[18,19] can undergo the rearrangement upon treatment with base. When the starting ketone is α,α'-dihalogenated, the product is an α,β-unsaturated carboxylic acid derivative and in analogous fashion trihaloketones give rise to α,β-unsaturated-α-halo acids. α-Halo ketimines are also suitable substrates for the *Favorskii rearrangement*, although they are less reactive than the corresponding α-halo ketones.[10,11] General features of the *Favorskii rearrangement* are: 1) sensitivity to structural factors (bulkiness of substituents, degree of alkyl substitution) and reaction conditions (base, solvent, temperature); 2) alkyl or aryl substitution on the halogen-bearing carbon increases the rate of rearrangement; 3) in cyclic α-halo ketones, the rearrangement is general in rings from 6-10; and 4) yields are widely varied from moderate to good. There are two important variations of the *Favorskii rearrangement*: 1) when β-halo ketones are treated with base in the presence of a nucleophile, the *homo-Favorskii rearrangement* takes place *via* a cyclobutanone intermediate;[20,21] and 2) if the α-halo ketone does not have any enolizable hydrogens ($R^{3-5}\neq H$), then the *quasi-Favorskii rearrangement* is operational.

Favorskii-rearrangement:

Homo-Favorskii-rearrangement:

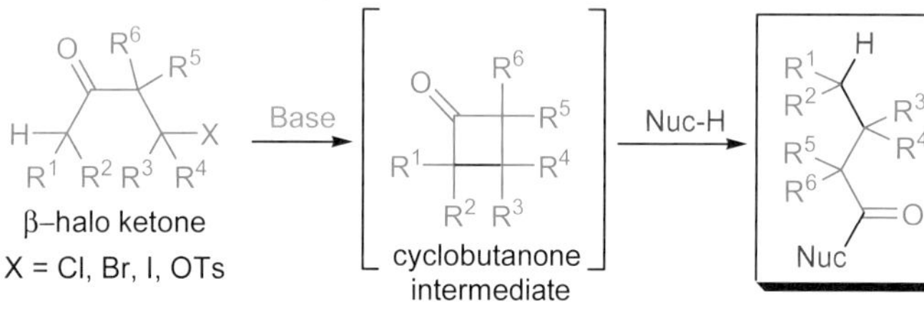

Quasi-Favorskii-rearrangement:

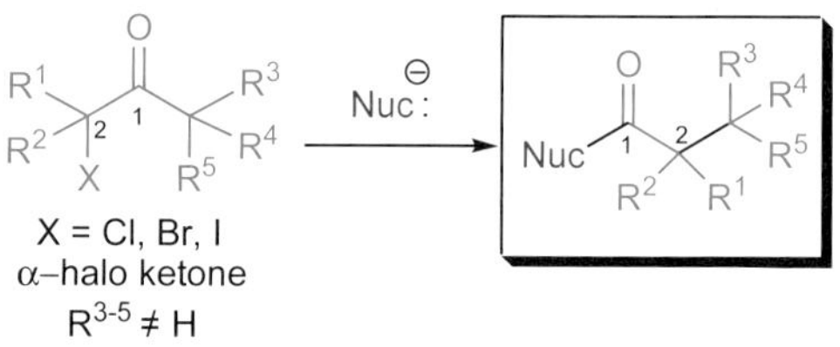

Mechanism: [22-26,9,27-31,10,11]

During the last century there have been numerous proposals for the mechanism of the *Favorskii rearrangement*. Currently the widely accepted mechanism involves the following steps: 1) deprotonation at the α-carbon and formation of an enolate; 2) intramolecular attack by the enolate on the α'-carbon bearing the leaving group to form a cyclopropanone intermediate; 3) regioselective opening of the intermediate to give the most stable carbanion; and 4) proton transfer to the carbanion to afford the product.

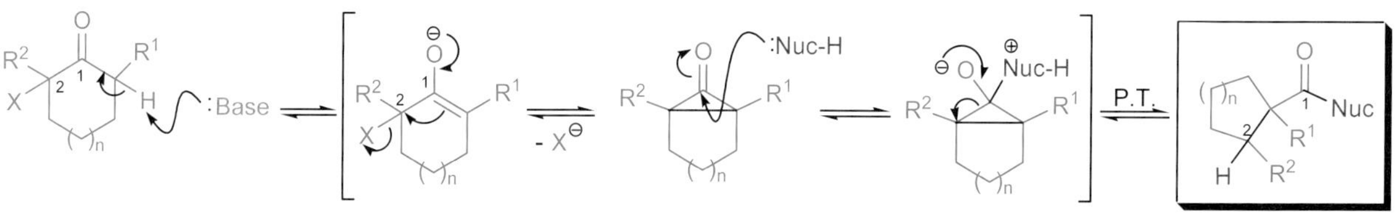

Synthetic Aplications:

The total synthesis of the symmetrical cage compound hexacyclo[6.4.2.0^{2,7}.0^{3,11}.0^{6,10}.0^{9,12}]tetradecene was accomplished in the laboratory of H. Takeshita by using sequential *Diels-Alder cycloaddition*, *Favorskii rearrangement* and *[2π+2π] photocycloaddition* as key steps.[32] The *Favorskii rearrangement* of a bridgehead α-halo ketone afforded the anticipated bridgehead carboxylic acid in 88% yield. Next, the acid was converted to the corresponding *tert*-butyl peroxy ester, which was subsequently photocyclized. The final step was the removal of the bridgehead carboxylic acid functionality by heating the perester in *p*-diisopropylbenzene for 2h at 150 °C.

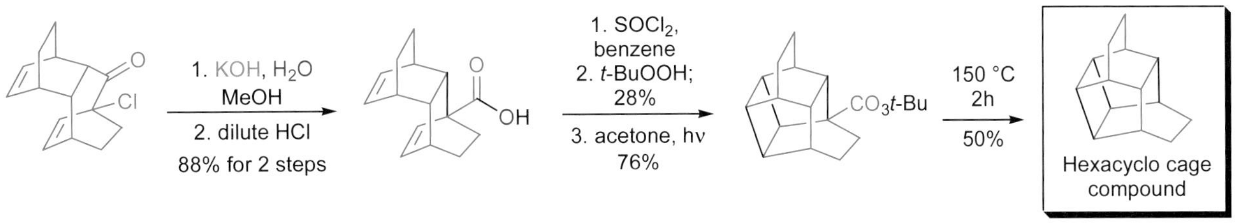

E. Lee and co-workers demonstrated that the chlorohydrin derived from (+)-carvone undergoes a *stereoselective Favorskii rearrangement* to afford a highly substituted cyclopentane carboxylic acid derivative.[33] This intermediate was then converted to (+)-dihydronepetalactone. When the THP-protected chlorohydrin was treated with sodium methoxide in methanol at room temperature, the rearrangement took place with excellent stereoselectivity (10:1) and high yield. Interestingly, the major product was the thermodynamically less stable cyclopentanecarboxylate.

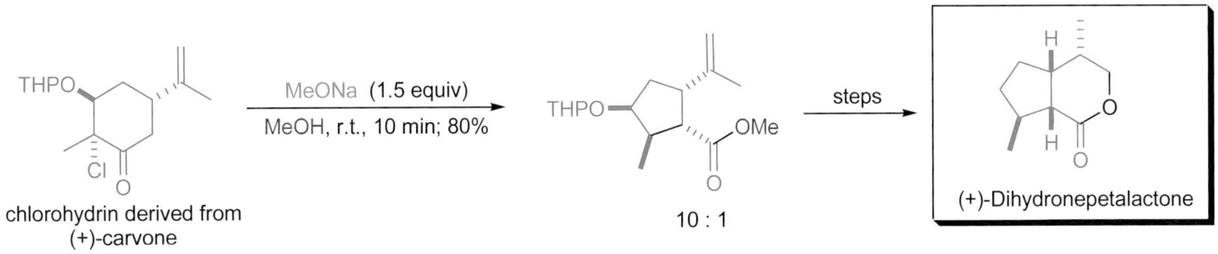

The key step in the stereocontrolled total synthesis of the tricyclic (±)-kelsoene by M. Koreeda et al. was a base-catalyzed *homo-Favorskii rearrangement* of a γ-keto tosylate to elaborate the 4-5 fused ring portion of the target molecule.[34] The bicyclic 5-6 fused γ-keto tosylate was treated with excess potassium *tert*-butoxide, which effected the desired rearrangement in less than 2 minutes at room temperature. The nucleophilic solvent was too bulky to effect the opening of the cyclobutanone intermediates, making their isolation possible. The mixture of isomeric cyclobutanones was converted to a separable 1:1 mixture of cyclobutanones with *p*-TsOH, and the ketone functionality was then removed *via* the corresponding tosylhydrazone.

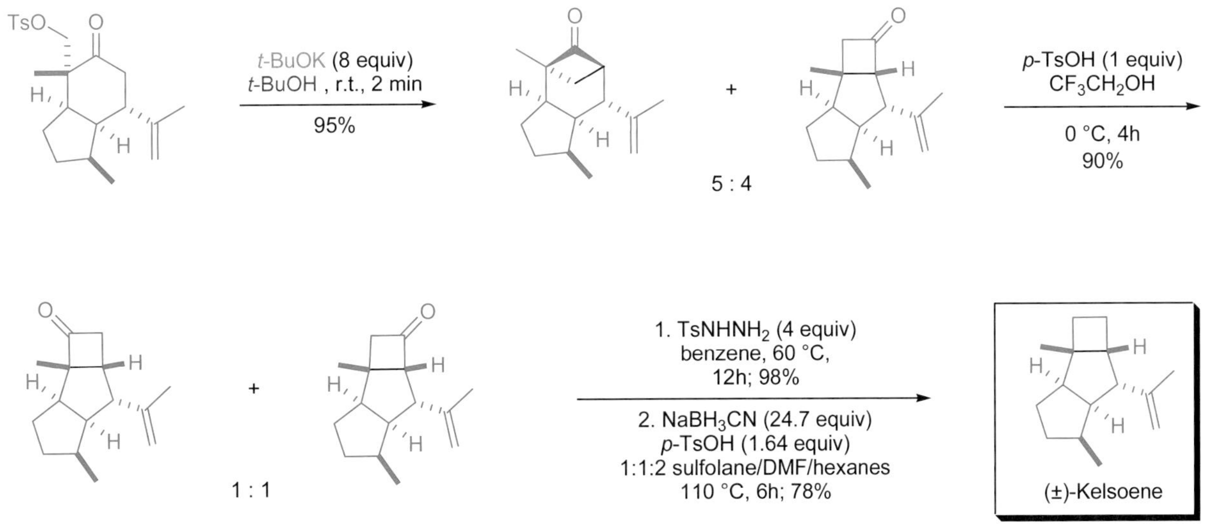

FEIST-BÉNARY FURAN SYNTHESIS

(References are on page 585)

Importance:

[*Seminal Publications*[1,2]; *Reviews*[3]; *Modifications & Improvements*[4-6]]

The synthesis of furans from β-keto esters and α-halogenated carbonyl compounds (aldehydes and ketones) under basic conditions is known as the *Feist-Bénary furan synthesis*. The general features of this reaction are: 1) the yields are strongly dependent on the substrates and are often moderate; 2) the initially isolated product of the reaction is usually the substituted dihydrofuranol ("*interrupted Feist-Bénary reaction*"), which is dehydrated under acid-catalyzed conditions to isolate the substituted furan;[7] 3) the regiochemical outcome depends on the reactivity of the α-halogenated carbonyl compound: α-halogenated aldehydes (R^1=H) tend to first undergo an *aldol reaction* followed by an *O*-alkylation, while α-halogenated ketones (R^1=alkyl) first *C*-alkylate the β-keto ester and then acid treatment is necessary to obtain the substituted furan;[8] 4) the following bases are often used to deprotonate the β-keto esters: NaH, NaOMe, NaOEt, aqueous NaOH, or Et_3N; 5) the reaction is general with respect to the nature of the β-dicarbonyl compound: in addition to β-keto esters, β-oxopropionates, β-diketones and β-dialdehydes can also be used;[7] and 6) the diastereoselectivity of the *interrupted Feist-Bénary reaction* depends on the basicity of the nucleophile: mainly the *cis* isomer is formed when nucleophiles derived from moderately acidic β-dicarbonyl compounds are used, while nucleophiles derived from highly acidic β-dicarbonyl compounds mainly yield the *trans* isomer.[7] There are several modifications of the original *Feist-Bénary synthesis* and they use more complex α-halogenated carbonyl compounds as reaction partners: 1) β-keto esters were condensed with 1,2-dibromoacetate to afford high yields of 2,3-disubstituted furans;[5] 2) alkylation of the sodium salts of β-keto esters with 3-halogenated alkynes (propargyl halides) in the presence of Cu(II)-salts yielded alkylidenefurans, which were isomerized to tetrasubstituted furans upon treatment with acid;[9] and 3) heating of β-keto esters with 5-hydroxy-5*H*-furan-2-one in the presence of Et_3N gave 3-alkoxy carbonylfurans.[10]

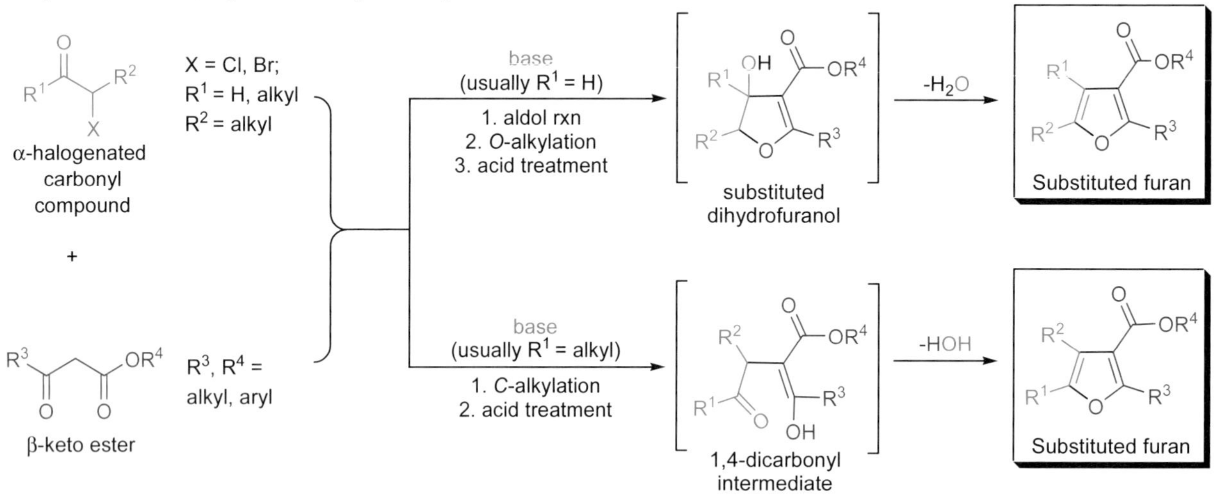

Mechanism: [11,12]

The first step of the *Feist-Bénary furan synthesis* is the deprotonation of the β-keto ester at the α-carbon atom. The resulting stabilized enolate undergoes an *aldol reaction* with the α-halogenated carbonyl compound by attacking the carbonyl group. Subsequent proton transfer generates a stable enolate anion that displaces the α-halogen atom in an intramolecular S_N2 reaction. The resulting dihydrofuranol, which often can be isolated, is treated with aqueous acid to generate the substituted furan.

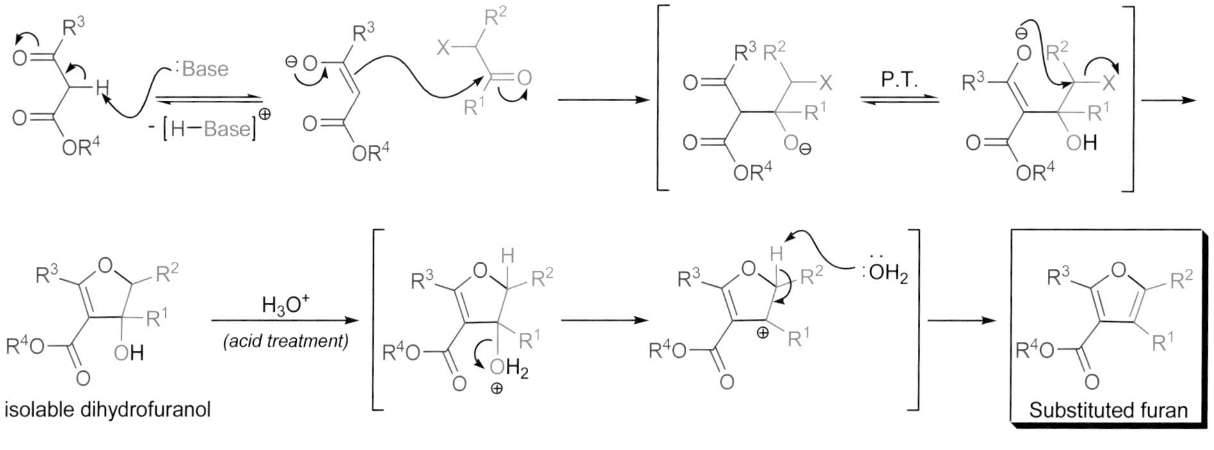

Synthetic Applications:

An efficient synthesis of the 7-deoxy zaragozic acid core was developed by M.A. Calter and co-workers.[13] The assembly of this complex structure was based on the *"interrupted" Feist-Bénary reaction*, which produces highly oxygenated dihydrofuranols that can be isolated. To this end, the sodium enolate of malondialdehyde was reacted with 2-bromo-3-oxo-diethyl succinate in benzene at room temperature to afford 29% of the *cis*-dihydrofuranol. This product was converted to the zaragozic acid core in four steps.

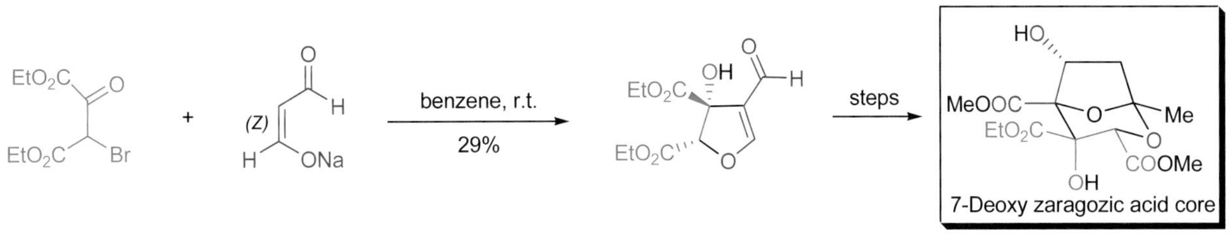

An efficient synthetic sequence for the preparation of 2,4-*bis*(trifluoromethyl)furan was developed by R. Filler and co-workers.[14] The potassium enolate of ethyl 4,4,4-trifluoroacetate was reacted with 3-bromo-1,1,1-trifluoroacetate in DMSO to afford 2,4-bis (trifluormethyl)-4-hydroxydihydro-3-furoate as a result of *O*-alkylation. Interestingly, under these conditions usually *C*-alkylation is preferred. Next, dehydration was performed to give the corresponding 2,4-*bis* (trifluoromethyl)-3-furoate in good yield. Finally, decarboxylation by heating with quinoline and $CuSO_4$ yielded the target furan in excellent yield.

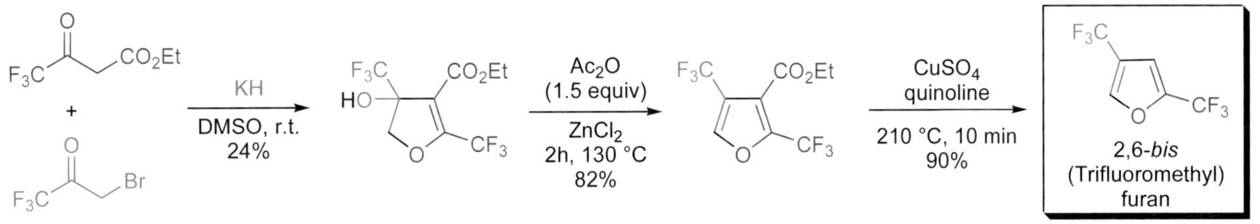

Research by P. Xinfu et al. has shown that the *Feist-Bénary furan synthesis* is well-suited for the construction of furolignans having two different aryl groups.[15] 3,4-Dimethyl-2-piperonyl-5-veratrylfuran was prepared by first reacting the sodium enolate of a β-keto ester derived from piperonal with an α-bromo β-keto ester derived from vanillin. The resulting 1,4-diketone was then subjected to acid-catalyzed cyclization with TsOH to the corresponding tetrasubstituted furan. The desired furolignan was obtained in two more steps.

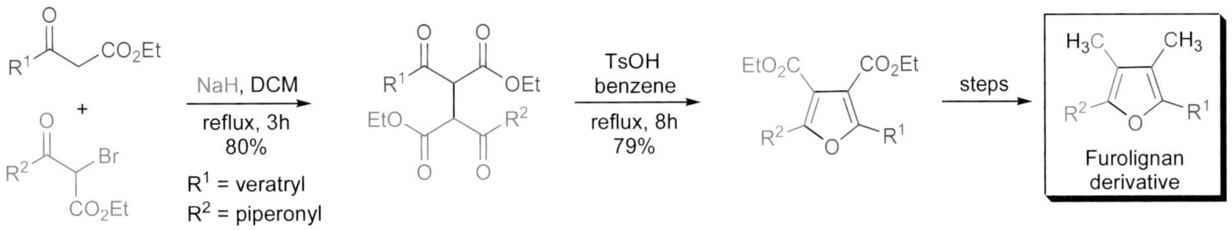

The mycotoxin patulin was synthesized via the oxidation of a disubstituted furan in the laboratory of M. Tada.[16] The required 2,3-disubstituted furan was conveniently prepared *via* the *Feist-Bénary reaction* of acetonedicarboxylic acid dimethyl ester and chloroacetaldehyde in the presence of pyridine. Subsequent functional group modification and oxidation of this furan finally gave the natural product.

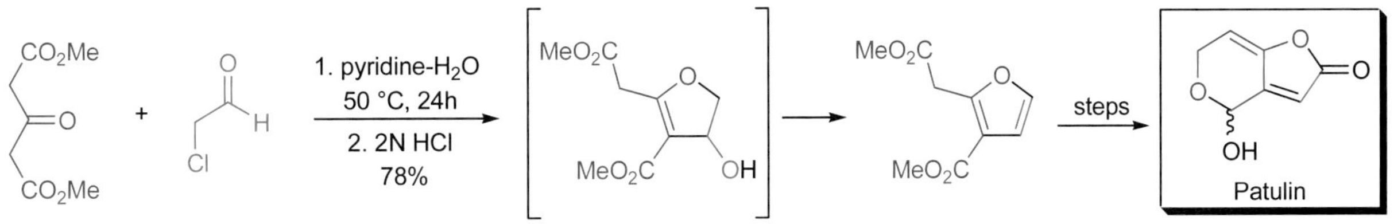

FERRIER REACTION / REARRANGEMENT

(References are on page 585)

Importance:

[*Seminal Publications*[1-5]; *Reviews*[6-16]; *Modifications & Improvements*[17-24]]

The Lewis acid promoted rearrangement of unsaturated carbohydrates is known as the *Ferrier reaction/ rearrangement*. The first report was made in 1914 by E. Fischer when he observed the allylic rearrangement of tri-*O*-acetyl-D-glucal to the corresponding 2,3-unsaturated hemiacetal upon heating with water.[1,25] The synthetic utility of this transformation was recognized by R.J. Ferrier during the early 1960s when he successfully prepared *O*-, *S*-, and *N*-linked unsaturated glycosyl compounds from 1,2-glycals and nucleophiles in the presence of Lewis acids.[2-4] This reaction is the *Type I Ferrier reaction* and its general features are: 1) substrates with good leaving groups, for example, acyloxy groups, in the 3-position (sugar nomenclature) successfully undergo the rearrangement upon heating in the presence of strong nucleophiles, such as alcohols and phenols, even in the absence of a catalyst; 2) commonly used Lewis acids are: $BF_3{\cdot}OEt_2$, $SnCl_4$, I_2,[21] $FeCl_3$,[24] TMSOTf-$AgClO_4$ [23]; 3) the hydroxyl group at C3 in the glycal can be activated under *Mitsunobu reaction* conditions without the use of a Lewis or protic acid;[20] and 4) the stereochemistry of the 2,3-unsaturated glycosyl product at the anomeric center depends on the relative stereochemistry of the groups at C3 and C4 in the starting material, but the α-anomer is usually predominant. The *Type II Ferrier rearrangement* was first reported in 1979 when exocyclic enol ethers were converted to substituted cyclohexanones upon treatment with mercury(II) salts.[5] The *Type II rearrangements* also became synthetically significant for the following reasons: 1) the precursors are readily available from carbohydrates, so the synthesis of chiral, highly-substituted cyclohexanone derivatives is possible; 2) in most reactions, single diastereomers are isolated in high yield;[7] and 3) the Lewis acid can be used in catalytic amounts and complex targets having acid sensitive functionalities can be prepared.[18] It was established that there is a strong correlation between the stereochemistry of the group at C3 and the stereochemistry of the group β to it: the newly generated OH groups and the C3 substituents are generally *trans* disposed in the product.[26]

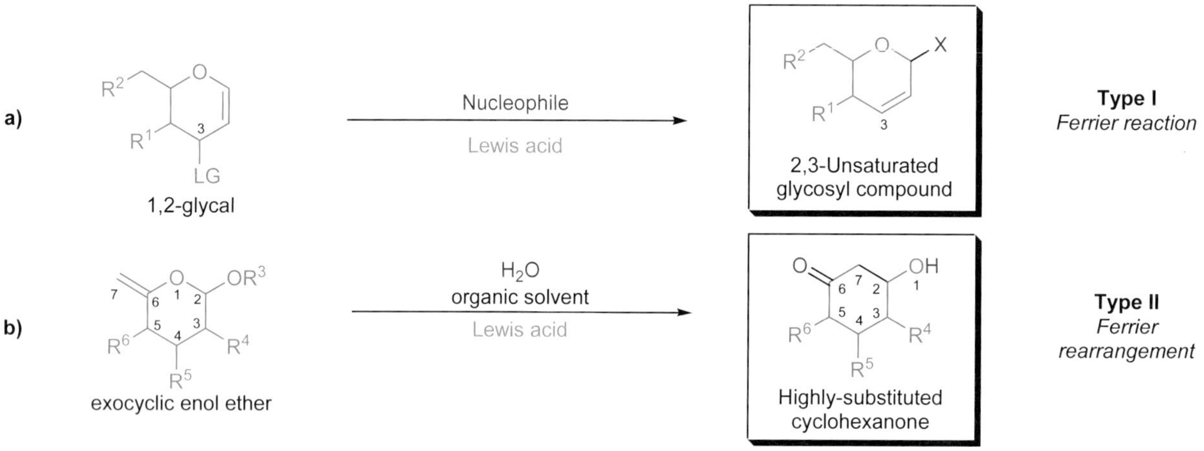

R^1, R^2 = *O*-acyl; LG = *O*-acyl, OTs, etc.; Lewis acid: $BF_3{\cdot}OEt_2$, $SnCl_4$, I_2, H_3O^+, TMSOTf, $FeCl_3$, etc.; X = OR, SR, NR_2, CR_3
R^3 = alkyl; R^4, R^5, R^6 = *O*-alkyl, *O*-acyl; Lewis acid: $HgCl_2$, $HgSO_4$, $Hg(OCOCF_3)_2$, $PdCl_2$, $Pd(OAc)_2$, etc.

Mechanism:[27-33,15]

The first step of the mechanism in the *Type I Ferrier reaction* is the departure of the leaving group from the C3 position of the glycal to give an allyloxocarbenium ion upon treatment with Lewis acid. The allyloxycarbenium ion is then captured by the nucleophile to give the corresponding glycoside. In the *Type II Ferrier rearrangement*, the enol ether first undergoes regiospecific hydroxymercuration to give a ketoaldehyde. This ketoaldehyde intermediate then undergoes an *aldol-like intramolecular cyclization* to afford the product cyclohexanone.

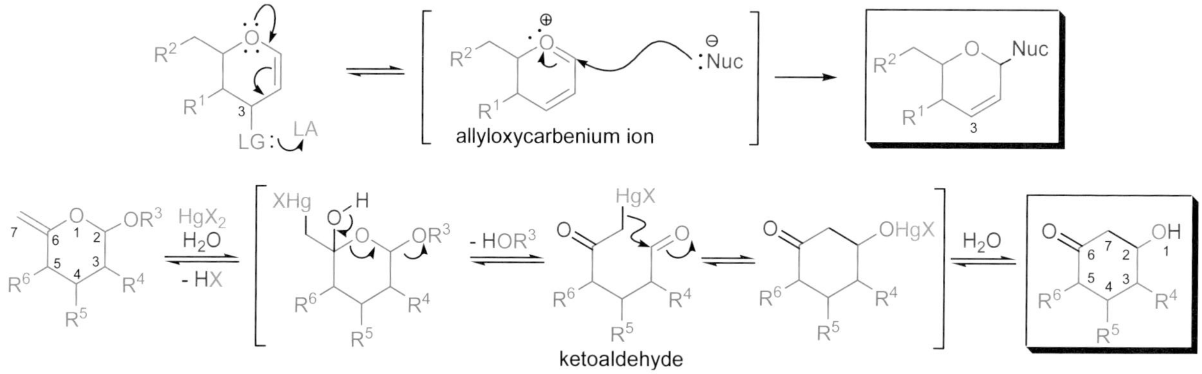

Synthetic Applications:

Research in the laboratory of H.M.I. Osborn showed that the use of cyclohexene derivatives as nucleophiles in the Lewis acid-mediated *Type I carbon-Ferrier reaction* of 3-*O*-acetylated glycals can be used to prepare unsaturated β-linked *C*-disaccharides.[34] The incorporation of the alkene took place with one equivalent of glucal in the presence of boron-trifluoride etherate in 33% yield. The desired *C*-disaccharide was obtained by selective hydrogenation of the exocyclic double bond in the presence of an endocyclic one.

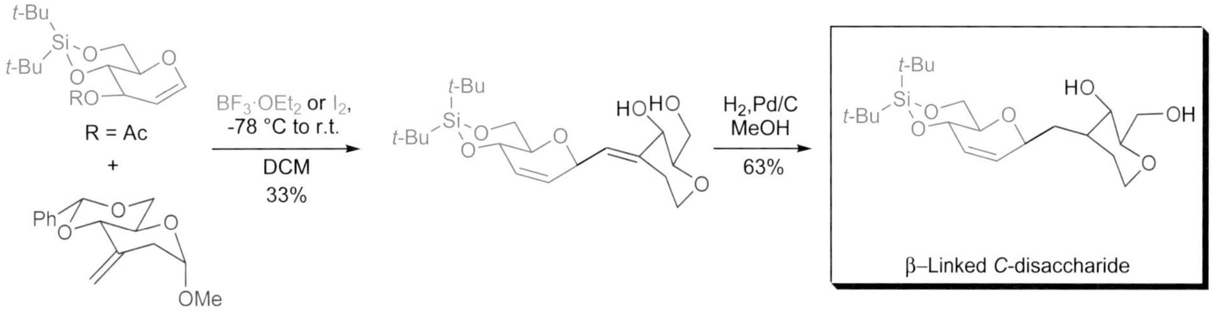

D.R. Williams and co-workers accomplished the first total synthesis of marine dolabellane diterpene (+)-4,5-deoxyneodolabelline.[35] The *Type I carbon-Ferrier reaction* was utilized to assemble the key *trans*-2,6-disubstituted dihydropyran with complete stereoselectivity (α-anomer). The macrocyclization was carried out with a vanadium-based *pinacol coupling*.

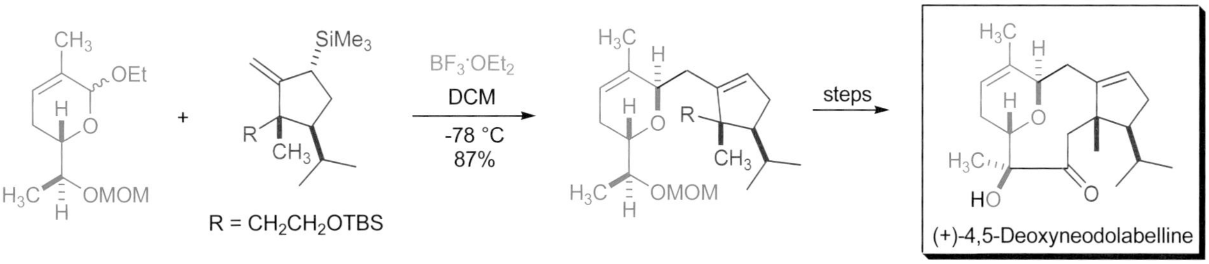

The highly oxygenated sesquiterpene paniculide A was synthesized by N. Chida et al. starting from D-glucose.[36] The key step to construct the substituted cyclohexane subunit of the natural product involved the *Type II Ferrier rearrangement*.

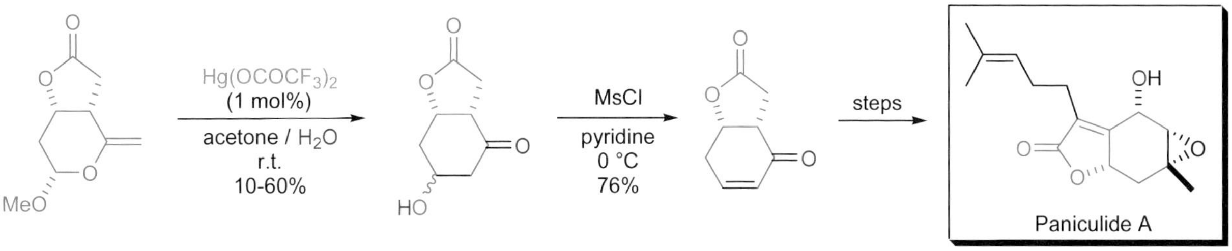

The stereoselective total synthesis of antimitotic alkaloid (+)-lycoricidine was accomplished by S. Ogawa and co-workers by utilizing the catalytic version of the *Type II Ferrier rearrangement* for the synthesis of the optically active substituted cyclohexenone fragment.[37] The rearrangement was effected with 1 mol% of mercuric(II)trifluoroacetate in acetone-water solvent system.

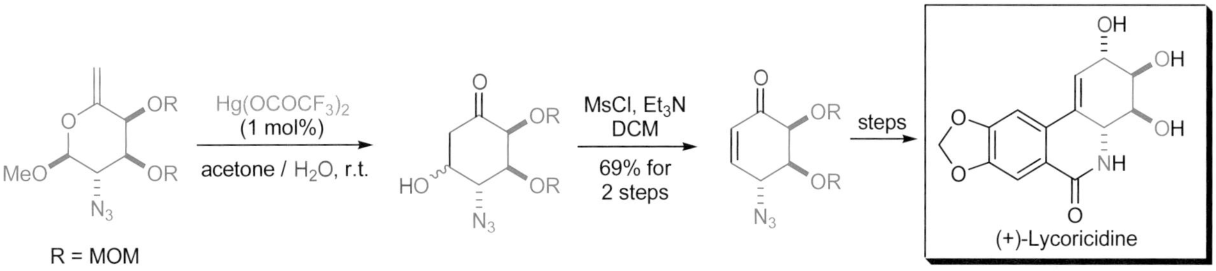

FINKELSTEIN REACTION

(References are on page 586)

Importance:

[*Seminal Publication*[1]; *Review*[2]; *Modifications & Improvements*[3-9]; *Theoretical Studies*[10-14]]

The equilibrium exchange of the halogen atom in alkyl halides for another halogen atom is known as the *Finkelstein reaction*. The first example of a halogen-exchange reaction was reported in the mid 1800s by W.H. Perkin,[15] but the systematic study of the reaction was conducted several decades later by H. Finkelstein in 1910.[1] Finkelstein observed that when various alkyl chlorides and bromides (1°, 2°, 3°, benzylic, etc.) were boiled with a 15 wt% solution of NaI in acetone, the corresponding alkyl iodides were formed in good yield. He also noted that the reaction time varied greatly, being the shortest for primary, allylic, and benzylic halides and the longest for tertiary alkyl halides. The *Finkelstein reaction* is an equilibrium process and capitalizes on the substantial solubility difference of sodium-halides in organic solvents (acetone, 2-butanone, etc.). While NaI dissolves readily in acetone, the solubility of NaBr and NaCl in organic solvents is very low. Therefore, the equilibrium can be shifted toward the direction of halogen-exchange according to the *Le Chatelier principle*: the formed NaBr and NaCl precipitates from the solution. Even today, the preparatively most important *Finkelstein reactions* are the conversion of alkyl bromides, chlorides, tosylates and mesylates to the corresponding alkyl iodides which are often difficult to prepare by other methods. Other halogenated compounds such as α-halogenated ketones and acids also undergo the *Finkelstein reaction* with ease. There are numerous modifications of the reaction: 1) solid-phase supported KI avoids the use of large excess of the reagent;[5] 2) microwave irradiation at high pressure considerably increases the rate of the reaction;[6,8] 3) alkyl fluorides can be prepared from other alkyl halides with lipophilic quaternary ammonium fluorides (TBAF) even in aprotic solvents of low polarity;[7] 4) the alkyl halide to alkyl fluoride conversion can also be done by using KF/18-crown-6 in dipolar aprotic solvents;[16] 5) the displacement of fluorine in alkyl fluorides with iodide is possible with the use of TMSI;[4] and 6) sterically hindered secondary and tertiary alkyl halides can be converted to the alkyl iodides by treatment with NaI/CS_2 in the presence of various Lewis acids ($AlMe_3$, $ZnCl_2$ $FeCl_3$, etc.).[3]

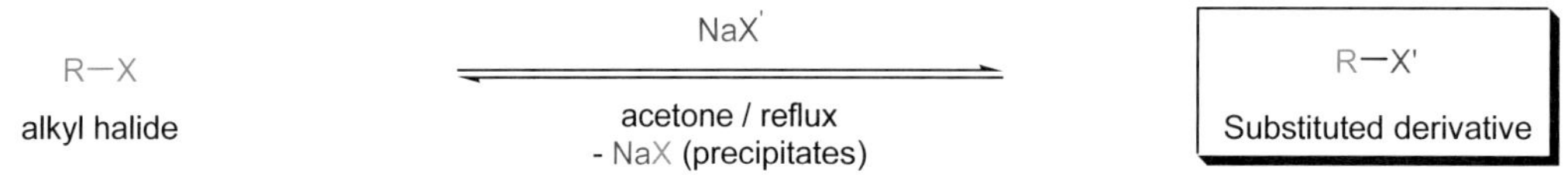

X = Cl, Br, OMs, OTs; R = 1° and 2°alkyl, allyl, benzyl; when X = Cl then X' = Br or I; when X = Br then X' = I

Mechanism: [17-27]

The mechanism of the *Finkelstein reaction* is often described as a typical S_N2 reaction where the filled orbital of the nucleophile (halide ion) interacts with the σ* orbital of the carbon-halogen bond, and the reaction proceeds with an overall inversion of configuration. This mechanistic picture depicts most transformations involving primary and secondary alkyl, allylic or benzylic halides. The driving force for the reaction is the removal of one of the nucleophiles from the equilibrium as an insoluble salt. Usually alkyl fluorides are very stable, and therefore they are sluggish to participate in nucleophilic displacement reactions unless the fluoride ion can be tied up in a stronger bond (such as Si-F) to compensate for the cleavage of the strong C-F bond. In certain cases, however, the *Finkelstein reaction* gave rise to dimeric and rearranged products, which were isolated and characterized; detailed mechanistic studies concluded that a sequential cation-free radical mechanism was operational.[22]

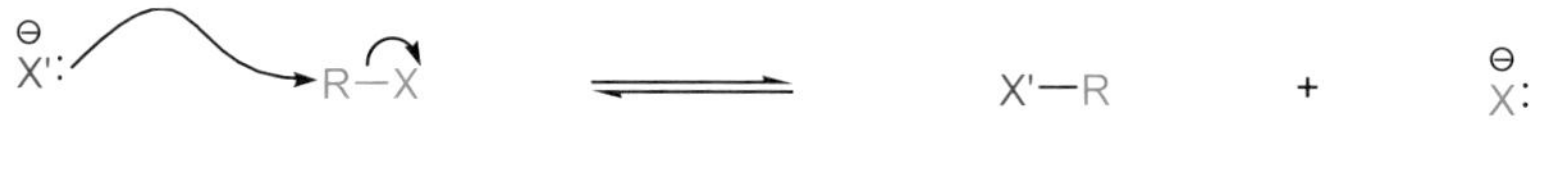

Description of the process with molecular orbitals:

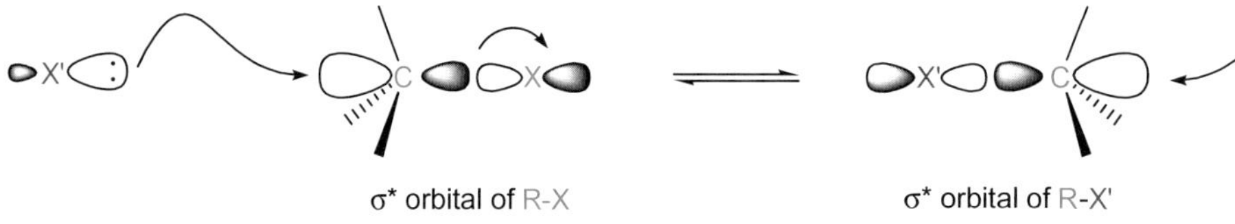

Synthetic Applications:

During the endgame of the total synthesis of the stemona alkaloid (–)-stenine, Y. Morimoto and co-workers utilized the *Finkelstein reaction* to prepare a primary alkyl iodide from a primary alkyl mesylate.[28] The mesylate was prepared from the corresponding primary alcohol with MsCl/Et_3N. The resulting primary alkyl iodide was used in the subsequent *intramolecular N-alkylation* to construct the final perhydroazepine C-ring of the natural product.

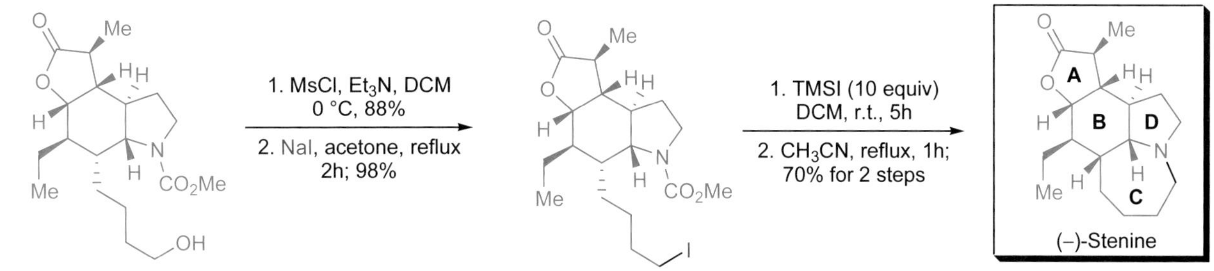

In the laboratory of J. Zhu, the synthesis of the fully functionalized 15-membered biaryl-containing macrocycle of RP 66453 was accomplished.[29] One of the key steps in their approach was *Corey's enantioselective alkylation of a glycine template* with a structurally complex biaryl benzyl bromide. This benzyl bromide was prepared from the corresponding benzyl mesylate *via* the *Finkelstein reaction* using lithium bromide in acetone.

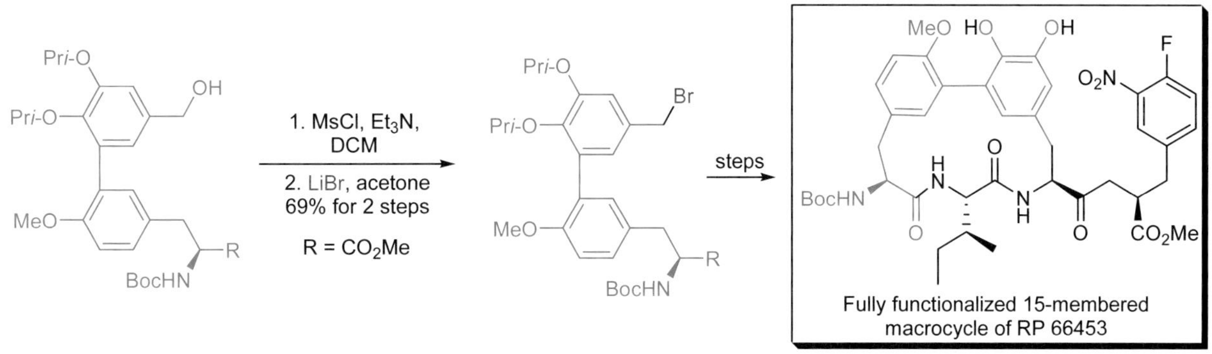

The marine sesquiterpene nakijiquinones were synthesized and biologically evaluated by H. Waldmann et al.[30] The core structure of the natural product was assembled *via* a *reductive alkylation* of a bicyclic enone with tetramethoxybenzyl iodide. This aryl iodide was obtained in a two-step procedure: treatment of the corresponding 1,2,4,5-tetramethoxybenzene with HBr/paraformaldehyde/AcOH followed by the *Finkelstein reaction* to replace the bromide with iodide.

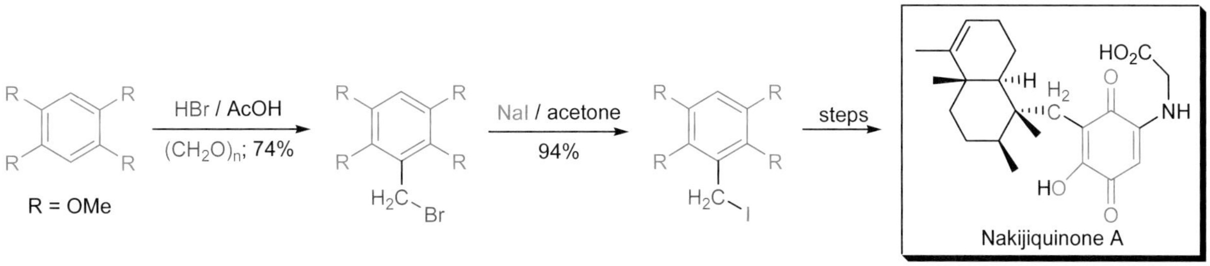

The key step in D. Kim's total synthesis of (–)-brefeldin A was an *intramolecular nitrile-oxide cycloaddition*.[31] In order to prepare the substrate for this cycloaddition, a *double Finkelstein reaction* was performed; first an alkyl tosylate was replaced with iodide; then the iodide was exchanged with a nitrite ion to afford the desired alkyl nitro compound.

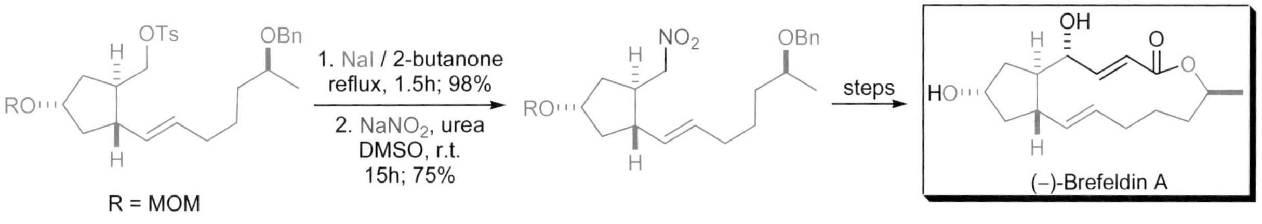

FISCHER INDOLE SYNTHESIS

(References are on page 587)

Importance:

[*Seminal Publications*[1,2]; *Reviews*[3-9]; *Modifications & Improvements*[10-13]; *Theoretical Studies*[14-21]]

In 1883, E. Fischer and F. Jourdan[1] treated pyruvic acid 1-methylphenylhydrazone with alcoholic hydrogen chloride, and the product of this reaction was later identified as 1-methylindole-2-carboxylic acid.[2] The preparation of indoles by heating arylhydrazones of ketones or aldehydes in the presence of a protic acid or a Lewis acid catalyst is known as the *Fischer indole synthesis*. Since its discovery, it has become the most important method to prepare substituted indoles. The catalysts that successfully lead to indolization are: 1) strong acids (e.g., PTSA, PPA, HCl, H_2SO_4); 2) weak acids (e.g., pyridinium chloride, AcOH); 3) solid acids (e.g., montmorillonite KSF clay, Mordenite, Zelotite Y, ion-exchange resins); and 4) Lewis acids (PCl_3, polyphosphoric acid trimethylsilyl ester, $ZnCl_2$). The Lewis acid catalyzed reactions often proceed under milder conditions (room temperature rather than high temperature) than the reactions catalyzed by protic acids. In the case of heteroaromatic arylhydrazones, however, the use of any acid is problematic (due to the protonation of the heteroatom), and for these compounds simple heating at high temperatures (thermal non-catalytic method) can also lead to indolization. The acid catalyzed cyclizations are usually 7 to 30 times faster than the thermal reactions. The main features of the *Fischer indole synthesis* are the following: 1) it is not necessary to isolate the arylhydrazones, the indole formation can be conducted by mixing the aldehyde and hydrazine and carrying out the indolization in one-pot; 2) unsymmetrical ketones give two regioisomeric 2,3-disubstituted indoles, and the regioselectivity depends on a combination of factors: acidity of the medium, substitution of the hydrazine, steric effects in the ketone and in the ene-hydrazines; 3) with unsymmetrical ketones indolization usually occurs at the least substituted α-carbon atom in strongly acidic medium, whereas weak acids give rise to the other regioisomer; 4) indolization of α,β-unsaturated ketones is generally unsuccessful due to the formation of unreactive pyrazolines; 5) 1,2-diketones can give both mono- and *bis*-indoles and the mono-indoles are usually formed with strong acid catalysts in refluxing alcohols; 6) 1,3-diketones and β-keto esters are not ideal substrates, since their arylhydrazones form pyrazoles and pyrazol-3-ones, respectively; 7) due to their sensitivity, aldehydes are used in their protected forms (acetal, aminal, or bisulfite addition product), and they give rise to 3-substituted indoles; 8) hydrazines are often used as their HCl salt or in their Boc protected form (they are not very stable in their free base form); 9) electron-withdrawing substituents on the aromatic ring of the hydrazine causes the indolization to become low-yielding and slow; 10) *ortho*-substituted arylhydrazines generally react much slower than the *meta*-substituted ones; and 11) the *Japp-Klingemann reaction* provides an easy way to obtain the starting arylhydrazones from β-dicarbonyls and arenediazonium salts.

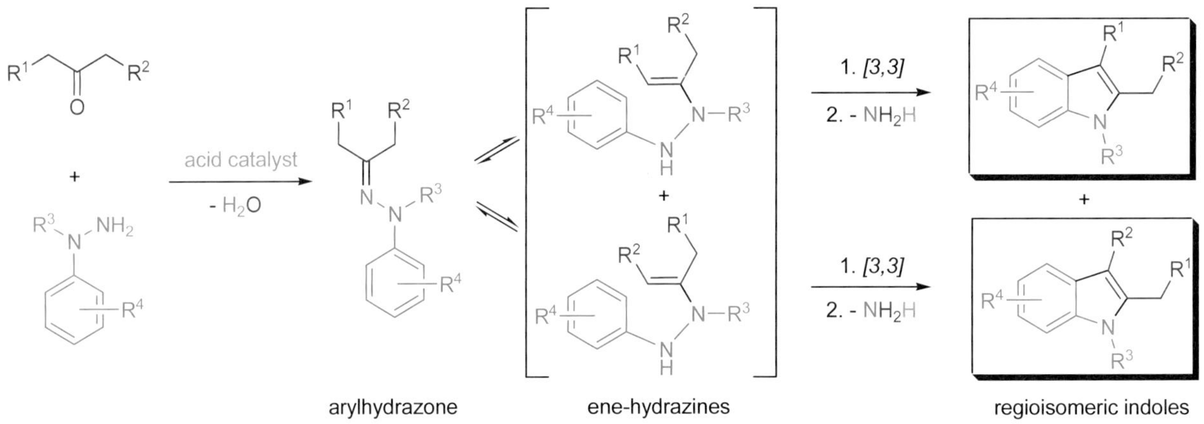

Mechanism: [22-39]

The currently accepted mechanism of the *Fischer indole synthesis* was originally proposed by R. Robinson in 1924.[22] There are five distinct steps: 1) coordination of the Lewis acid (e.g., proton) to the imine nitrogen; 2) tautomerization of the hydrazone to the corresponding ene-hydrazine; 3) disruption of the aromatic ring by a *[3,3]-sigmatropic rearrangement*; 4) rearomatization *via* a proton shift and formation of the 5-membered ring by a favored *5-exo-trig* cyclization; and 5) the loss of a molecule of ammonia to finally give rise to the indole system.

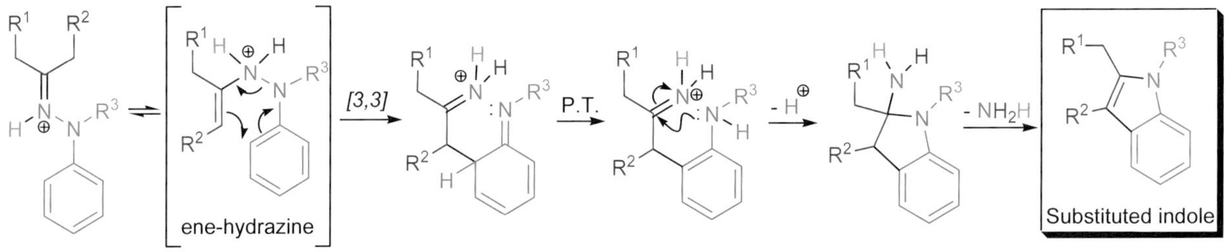

Synthetic Applications:

The total synthesis of (±)-deethylibophyllidine was accomplished by J. Bonjoch and co-workers, who applied a regioselective *Fischer indole synthesis* as one of the key steps to obtain octahydropyrrolo[3,2-*c*]carbazoles.[40] The indole formation was followed by a tandem *Pummerer rearrangement-thionium ion cyclization* to generate the quaternary spiro stereocenter.

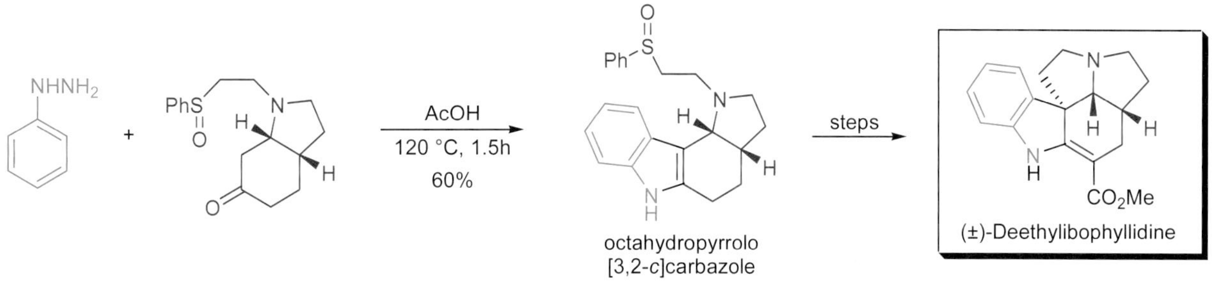

During the total synthesis of (+)-aspidospermidine by J. Aubé et al., the final steps involved an efficient *Fischer indolization* of a complex tricyclic ketone.[41] This ketone was unsymmetrical and the indole formation occurred regioselectively at the most substituted α-carbon in a weakly acidic medium (glacial AcOH).

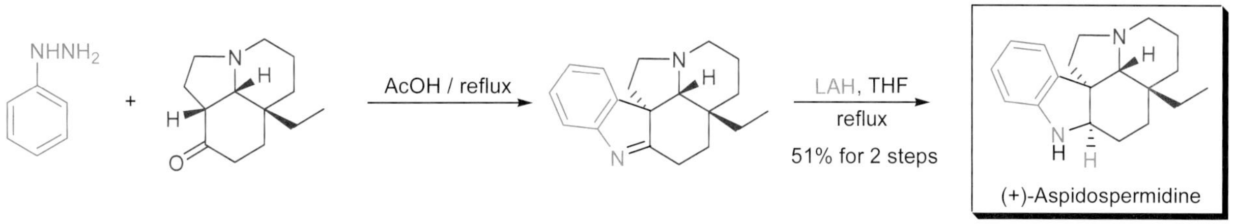

The unusual 6-azabicyclo[3.2.1]oct-3-ene core of the alkaloid (±)-peduncularine was assembled using the [3+2] annulation of an allylic silane with chlorosulfonyl isocyanate by K.A. Woerpel and co-workers.[42] In the endgame of the total synthesis, the bicyclic aldehyde was masked as the acetal, and an efficient *Fischer indole synthesis* was performed using phenylhydrazine hydrochloride along with 4% H_2SO_4. Several subsequent steps led to the natural product.

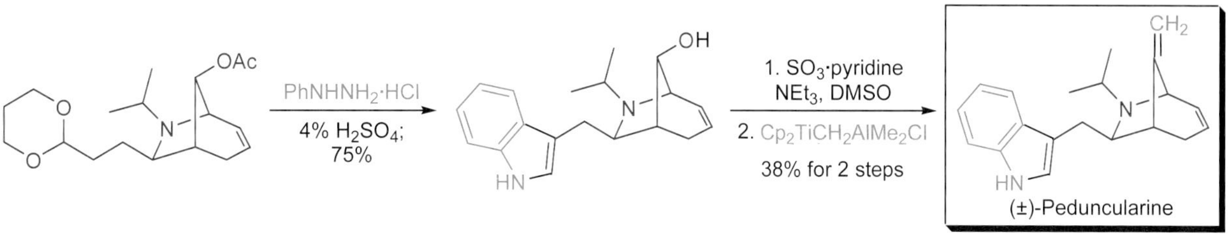

J.M. Cook et al. accomplished the enantiospecific total synthesis of the indole alkaloid tryprostatin A.[43] The substituted indole nucleus was assembled at the beginning of the synthesis, and the necessary arylhydrazone was prepared *via* the *Japp-Klingemann reaction* using the diazonium salt derived from *m*-anisidine and the anion of ethyl-α-ethylacetoacetate. The regioselectivity of the *Fischer indole synthesis* favored the 6-methoxy-3-methylindole-2-carboxylate regioisomer in a 10:1 ratio.

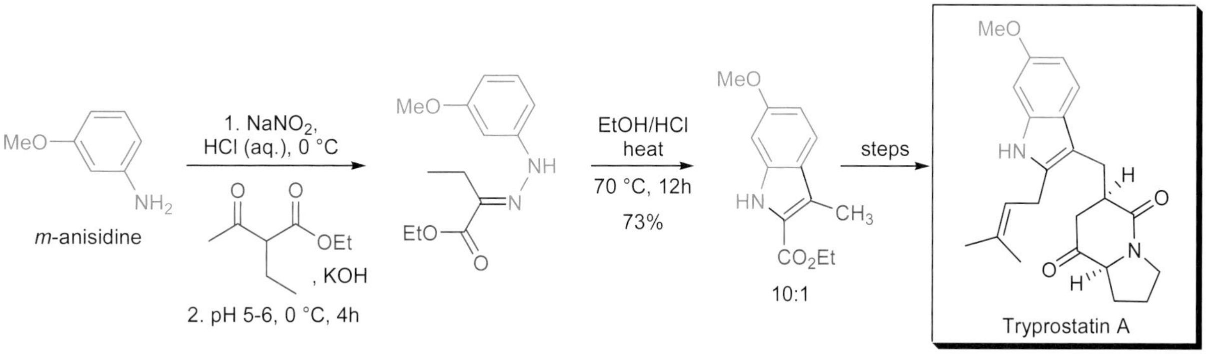

FLEMING-TAMAO OXIDATION

(References are on page 588)

Importance:

[*Seminal Publications*[1-7]; *Reviews*[8-12]; *Modifications & Improvements*[13-17]; *Theoretical Studies*[18,19]]

In 1983, K. Tamao and M. Kumada reported that silicon-carbon bonds can be cleaved by hydrogen peroxide, under basic conditions in the presence of bicarbonate salts, to afford the corresponding alcohols, provided that the silicon atom had at least one electron-withdrawing substituent.[3] A year later, I. Fleming and co-workers discovered that the dimethylphenylsilyl-carbon bond ($PhMe_2Si$-C) can be oxidatively cleaved in two steps to the corresponding alcohol with retention of configuration at the carbon atom to which the silicon is attached.[5] The two steps were: 1) protodesilylation of the phenyl ring using HBF_4 or BF_3·AcOH complex; and 2) treatment of the resulting silyl fluoride with a peracid (e.g., *m*CPBA, AcOOH). These early discoveries paved the way to the development of a large number of silicon-based reagents and the use of various silyl groups as the masked form of the hydroxyl group.[16] The mild, stereospecific oxidation of silicon-carbon bonds to yield the corresponding carbon-oxygen bonds (alcohols) is called the *Fleming-Tamao oxidation*. In terms of laboratory execution of the oxidation, the following facts are noteworthy: 1) phenylsilanes are more robust than alkoxysilanes, so they can be removed at the end of a long synthetic sequence; 2) aryl, heteroaryl and allyl substituents on the silicon atom behave the same way as the phenyl group, and they are all replaced by the fluoride in the first step of the oxidation; 3) in the second step fluoride additives are often needed in addition to the oxidizing agent; and 4) usually more than one equivalent of oxidizing agent is necessary for each silicon-carbon bond. Advantages of the *Fleming-Tamao oxidation* are: 1) carbon-silicon bonds can be introduced stereospecifically, and therefore the preparation of substrates is straightforward (e.g., *via* the regioselective transition metal catalyzed *hydrosilylation of olefins*); 2) by carefully choosing the substituents on the silicon atom, the oxidation of a specific silyl group is possible in the presence of other silyl groups; 3) unlike the oxygen atom, the silicon does not have lone pairs of electrons, so it does not coordinate to electrophiles or Lewis acids; 4) in the case of optically active substrates, the reaction is stereospecific, that is, there is a retention of configuration; 5) the oxidation conditions are mild enough to tolerate a wide range of functional groups even in complex substrates; 6) the two-step reaction can also be conducted in one-pot by using Hg^{2+} or Br^+ as electrophiles;[7] and 7) the isolation of the product alcohol is straightforward, since the by-products of the oxidation are usually water-soluble. There are some disadvantages as well: 1) the oxidation of silyl groups attached to tertiary carbons of cyclic systems do not always proceed with ease;[14] and 2) in the presence of tertiary amines, special conditions are required to avoid *N*-oxide formation.[19]

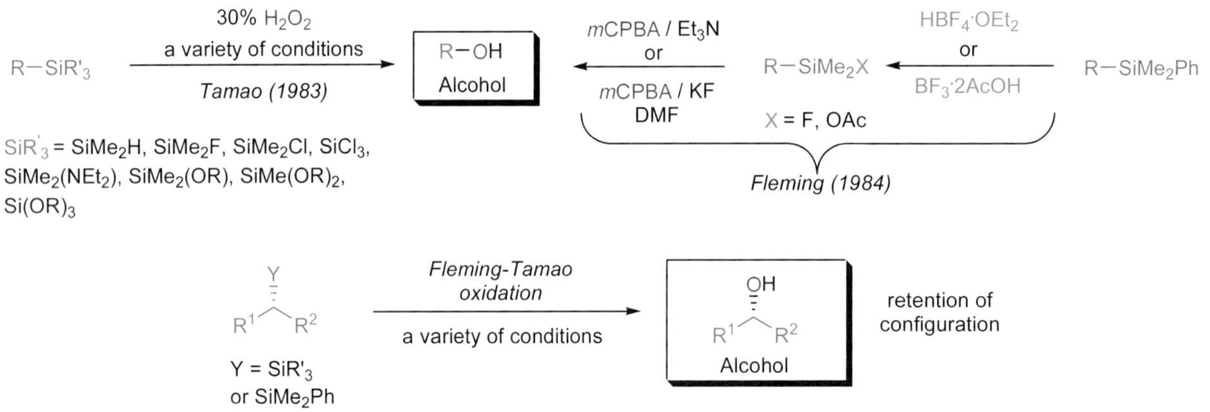

Mechanism: [1,11,18]

The mechanism of the *Fleming-Tamao oxidation* has four distinct steps when the silyl group is $-SiMe_2Ph$: 1) S_EAr by the electrophile on the phenyl ring in the *ipso* position affords the heteroatom-substituted silane ($-SiMe_2X$) derivative; 2) attack of the heterosilane by the peroxide to give tetracoordinated silyl peroxide; 3) *[1,2]*-alkyl shift to give a dialkoxy silane (analogous to the step in *Baeyer-Villiger oxidation*), followed by conversion to a siloxane; and 4) hydrolysis of the siloxane to the desired alcohol.

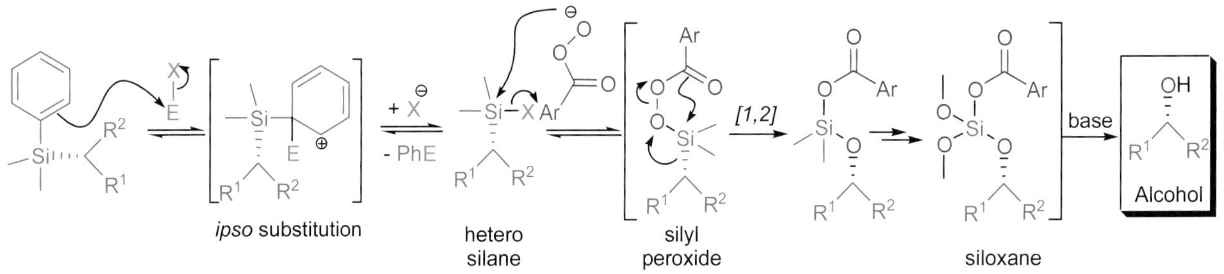

Synthetic Applications:

In the laboratory of F.G. West, the stereoselective *silyl-directed [1,2]-Stevens rearrangement* of ammonium ylides was investigated as a potential key step toward the enantioselective synthesis of various hydroxylated quinolizidines.[19] The dimethylphenylsilyl group served as a surrogate for one of the hydroxyl groups in the product. The *Fleming-Tamao oxidation* was performed under Denmark's conditions to avoid oxidation of the tertiary amine to the corresponding *N*-oxide, and the desired quinolizidine diol was obtained in 81% yield.[17]

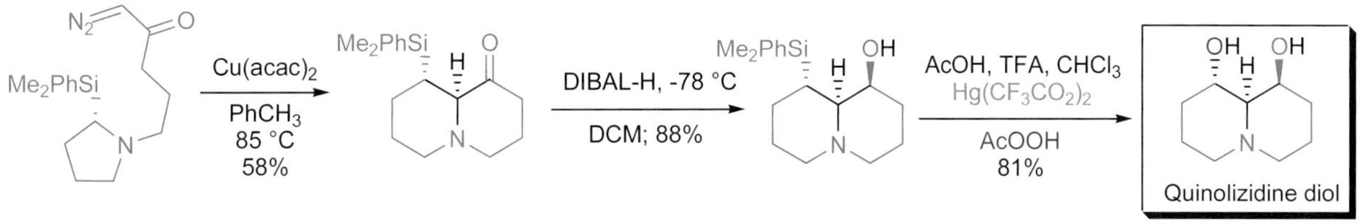

During the total synthesis of the marine alkaloid (±)-lepadiformine by S.M. Weinreb et al., one of the key bicyclic *N*-acyliminium salt intermediates was subjected to a nucleophilic attack by an organocuprate.[20] The resulting allyldimethylsilyl derivative was then treated under the *Fleming-Tamao oxidation* conditions to afford the corresponding hydroxymethyl compound in excellent yield.

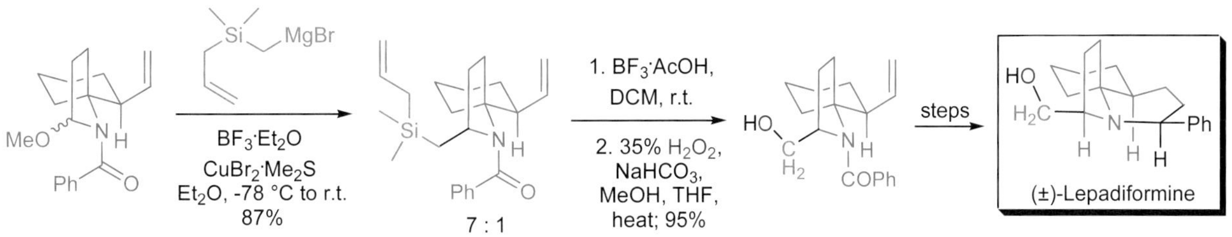

M. Shibasaki and co-workers reported a concise stereocontrolled synthesis of the 18-*epi*-tricyclic core of garsubellin A.[21] In the endgame, the unmasking of an α,β-unsaturated ketone became necessary just prior to the cyclization of the third ring. The latent β-hydroxyl group was best carried through several steps as a pentamethyldisilyl substituent, which was removed by a *modified Fleming-Tamao oxidation*.[15]

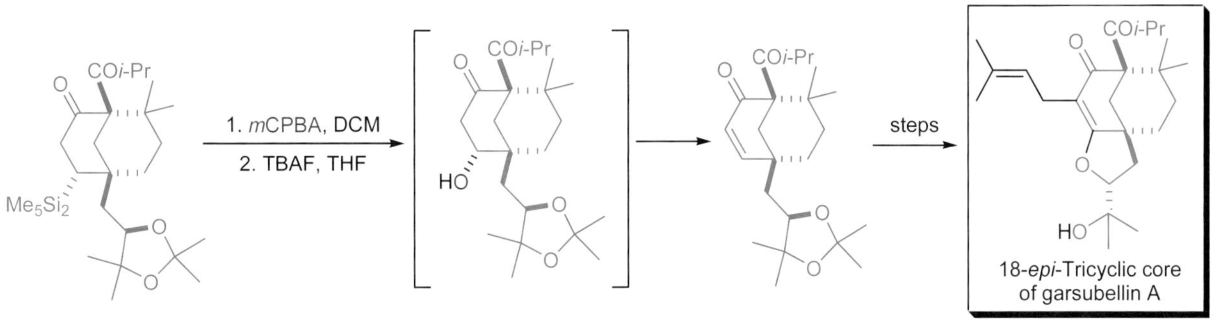

The synthesis of the C1-C21 subunit of the protein phosphatase inhibitor tautomycin was accomplished by J.A. Marshall et al.[22] During the last steps of the synthetic sequence, the *hydrosilylation* of a terminal alkyne afforded a five-membered siloxane that was oxidized by the *Fleming-Tamao oxidation*. The initially formed enol tautomerized to the corresponding methyl ketone.

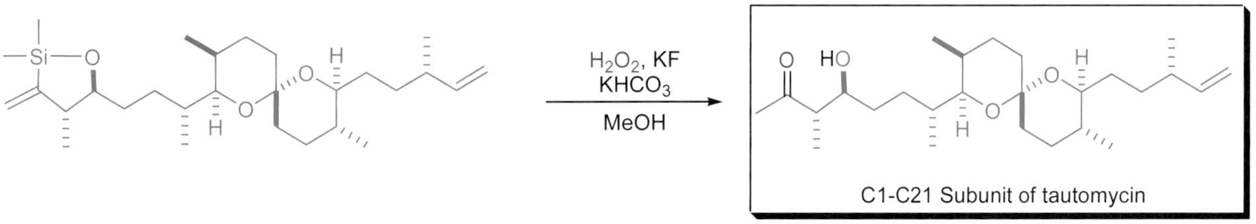

FRIEDEL-CRAFTS ACYLATION

(References are on page 588)

Importance:

[*Seminal Publications*[1,2]; *Reviews*[3-18]; *Modifications & Improvements*[19-29]; *Theoretical Studies*[30-40]]

The introduction of a keto group into an aromatic or aliphatic substrate by using an acyl halide or anhydride in the presence of a Lewis acid catalyst is called the *Friedel-Crafts acylation*. The reaction is closely related to the *Friedel-Crafts alkylation*, which introduces alkyl groups into aromatic and aliphatic substrates. General features of the *Friedel-Crafts acylations* are the following: 1) substrates that undergo the *Friedel-Crafts alkylation* are also easily acylated and in most cases electron-rich substrates (R^1 = -OH, -NR_2, alkyl, etc.) are needed to obtain the desired ketone in good yield; 2) aromatic substrates with strongly electron-withdrawing groups (R^1 = -NO_2, -CX_3, etc.) and certain heteroaromatic compounds (e.g., quinolines, pyridines) do not undergo the acylation at all, and they may be used as solvents (these unreactive substrates, however, are efficiently acylated by the *Minisci reaction*); 3) acylating agents besides acyl halides are: aromatic and aliphatic carboxylic acids, anhydrides, ketenes and esters, as well as polyfunctional acylating agents (oxalyl halides); 4) acyl iodides are usually the most reactive, while acyl fluorides are the least reactive (I > Br > Cl > F); 5) unlike in the alkylations, *Friedel-Crafts acylations* require substantial amounts of catalyst (slightly more than one equivalent), since the acylating agent itself coordinates one equivalent of Lewis acid, and therefore excess is needed to observe catalysis; 6) most often used catalysts are: AlX_3, lanthanide triflates, zeolites, protic acids (e.g., H_2SO_4, H_3PO_4), $FeCl_3$, $ZnCl_2$, PPA; 7) in the case of very reactive acylating agents (e.g., acyloxy triflates) or very electron-rich substrates there is little or no catalyst required;[8] 8) no polyacylated products are observed, since, after the introduction of the first acyl group, the substrate becomes deactivated; 9) rearrangement of the acylating agent under the reaction conditions is rarely observed and this feature allows the preparation of straight chain alkylated aromatic compounds in a two-step process (acylation followed by reduction); 10) unprotected Lewis basic functional groups (e.g., amines) are poor substrates, since the acylation will preferentially take place on these functional groups instead of the aromatic ring; 11) the *intramolecular Friedel-Crafts acylation* is well-suited for the closure of 5-, 6- and 7-membered rings with a tendency for the formation of the 6-membered ring. One drawback of the *Friedel-Crafts acylation* is that the Lewis acid catalyst usually cannot be recovered at the end of the reaction, since it is destroyed in the work-up step. However, recent studies showed that the use of heterogeneous catalysts (mainly zeolites) makes this important reaction more feasible on an industrial scale.[41]

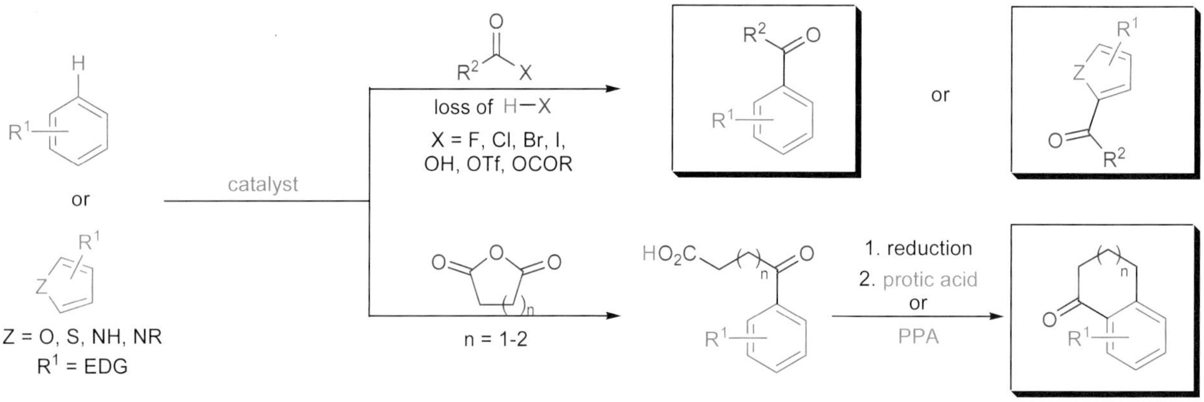

Mechanism: [4,42-47]

The initial step of the mechanism is the coordination of the first equivalent of the Lewis acid to the carbonyl group of the acylating agent. Next, the second equivalent of Lewis acid ionizes the initial complex to form a second donor-acceptor complex which can dissociate to an acylium ion in ionizing solvents. The typical S_EAr reaction gives rise to an aromatic ketone-Lewis acid complex that has to be hydrolyzed to the desired aromatic ketone.

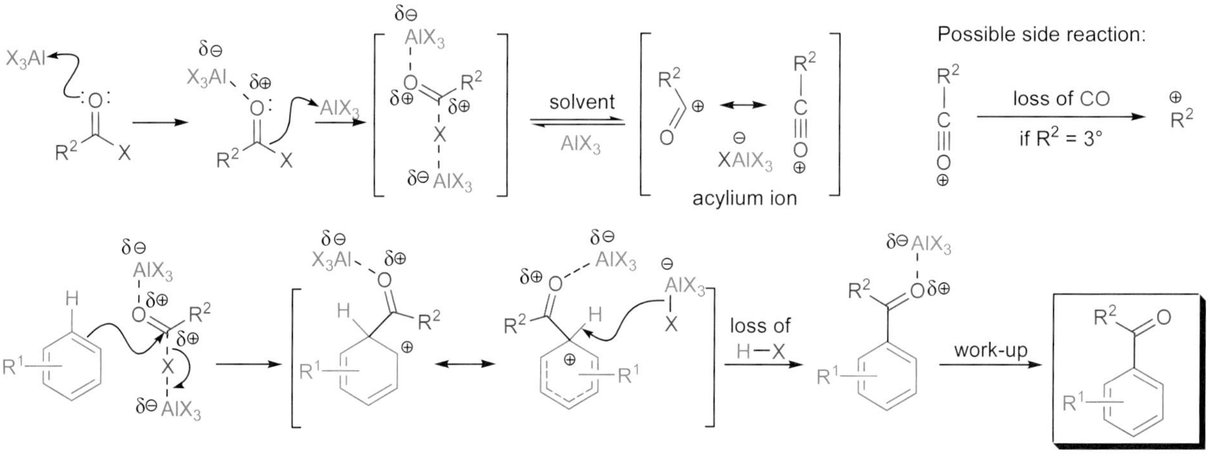

Synthetic Applications:

L.E. Overman et al. accomplished the enantioselective total synthesis of (–)-hispidospermidin by utilizing an *aliphatic intramolecular Friedel-Crafts acylation* as the key step to assemble the rigid tricyclic core.[48] The bicyclic acid precursor was first converted to the corresponding acid halide followed by treatment with one equivalent of titanium tetrahalide (TiX_4). Interestingly, upon cyclization with $TiCl_4$, the acid chloride gave substantial quantities of a side-product arising from a facile *[1,2]-hydride shift*. The extent of this unwanted hydride shift was greatly suppressed by first preparing the acid bromide followed by a $TiBr_4$ mediated cyclization. The authors attributed this improvement to the increased nucleophilicity of the bromide ion *vs.* chloride ion.

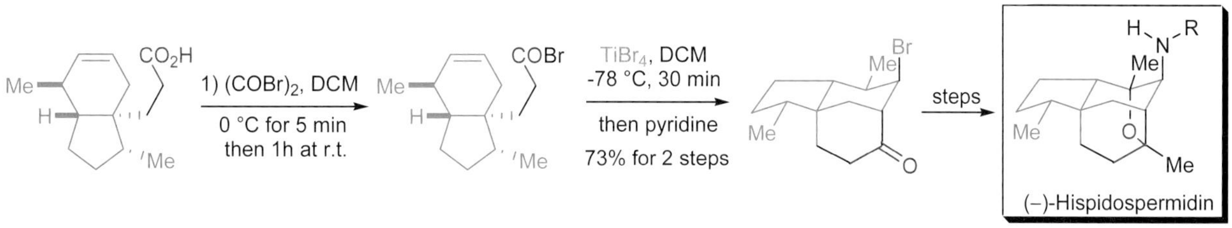

During the total synthesis of phomazarin, D.L. Boger and co-workers closed the B ring of the natural product with a *Friedel-Crafts acylation* reaction.[49] This key step provided the fully functionalized phomazarin skeleton. The carboxylic acid precursor was exposed to trifluoroacetic anhydride at 50 °C for 72h. The initial product was a C5 trifluoroacetate, which was subsequently hydrolyzed in the presence of air, which oxidized the phenol to the corresponding B-ring quinone.

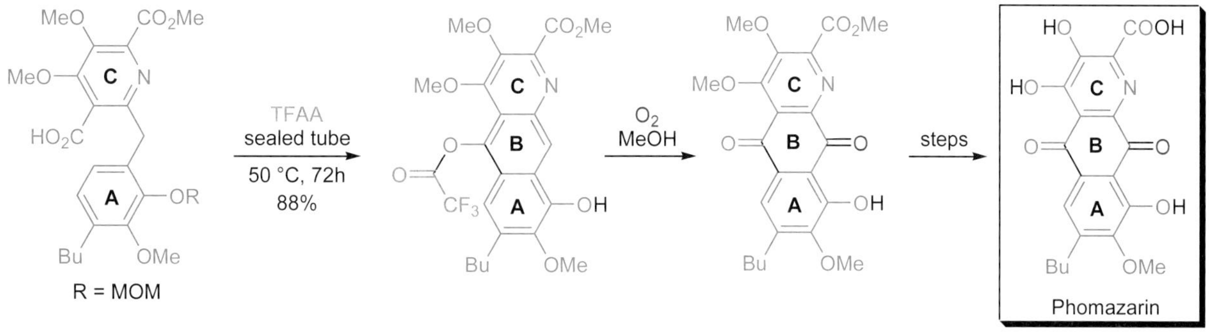

In the laboratory of K. Krohn, the total synthesis of phytoalexine (±)-lacinilene C methyl ether was completed.[50] In order to prepare the core of the natural product, an *intermolecular Friedel-Crafts acylation* was carried out between succinic anhydride and an aromatic substrate, followed by an *intramolecular acylation*. After the first acylation, the 4-keto arylbutyric acid was reduced under *Clemmensen reduction* conditions (to activate the aromatic ring for the intramolecular acylation).

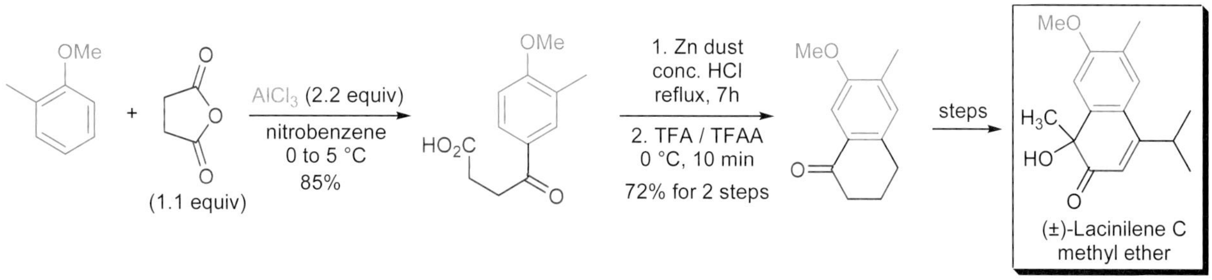

The first synthesis of the macrotricyclic core of roseophilin was carried out by A. Fürstner and co-workers.[51] An *intramolecular Friedel-Crafts acylation* was used to close the third ring of the macrotricycle.

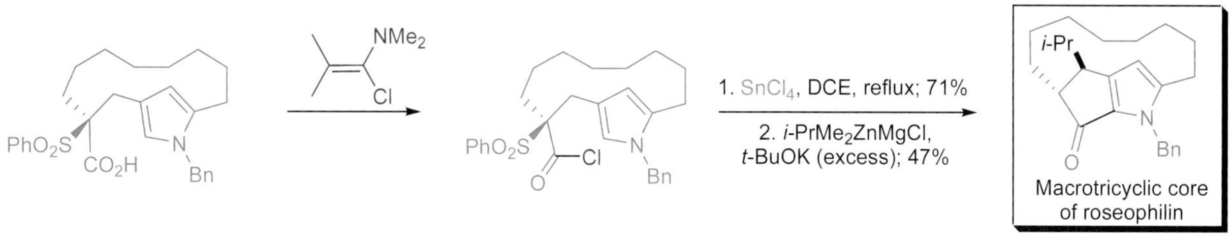

FRIEDEL-CRAFTS ALKYLATION

(References are on page 589)

Importance:

[*Seminal Publications*[1,2]; *Reviews*[3-13]; *Modifications & Improvements*[14-36]; *Theoretical Studies*[37-45]]

In 1877, C. Friedel and J.M. Crafts treated amyl chloride with thin aluminum strips in benzene and observed the formation of amylbenzene.[1,2] The reaction of alkyl halides with benzene was found to be general, and aluminum chloride ($AlCl_3$) was identified as the catalyst. Since their discovery, the substitution of aromatic and aliphatic substrates with various alkylating agents (alkyl halides, alkenes, alkynes, alcohols, etc.) in the presence of catalytic amounts of Lewis acid is called the *Friedel-Crafts alkylation*. Until the 1940s the alkylation of aromatic compounds was the predominant reaction, but later the alkylation of aliphatic systems also gained considerable importance (e.g., isomerization of alkanes, polymerization of alkenes and the reformation of gasoline). In addition to aluminum chloride other Lewis acids are also used for *Friedel-Crafts alkylations*: $BeCl_2$, $CdCl_2$, BF_3, BBr_3, $GaCl_3$, $AlBr_3$, $FeCl_3$, $TiCl_4$, $SnCl_4$, $SbCl_5$, lanthanide trihalides, and alkylaluminum halides ($AlRX_2$). The most widely employed catalysts are $AlCl_3$ and BF_3 for alkylations with alkyl halides. When the alkylating agent is an alkene or an alkyne, in addition to the catalyst, a cocatalyst (usually a proton-releasing substance such as water, an alcohol, or a protic acid) is also necessary for the reaction to occur. Other efficient catalysts are: 1) aluminum trialkyls (e.g., AlR_3) and alkoxides [$Al(OPh)_3$]; 2) acidic oxides and sulfides; 3) modified zeolites; 4) acidic cation-exchange resins (e.g., Dowex 50); 5) Brönsted acids (e.g., HF, H_2SO_4, H_3PO_4); 6) Brönsted and Lewis superacids (e.g., $HF{\cdot}SbF_5$, $HSO_3F{\cdot}SbF_5$); 7) clay-supported metal halides;[18] and 8) enzymes.[22] The general features of the *Friedel-Crafts alkylations* are: 1) the reactivity of alkyl halides is the highest for alkyl fluorides and the lowest for alkyl iodides (F > Cl > Br > I); 2) the branching of the alkyl group has a dramatic influence, since tertiary alkyl halides are the most reactive: tertiary, benzyl > secondary > primary; 3) if the alkyl halide is polyfunctional (it has more than one halogen atom (e.g., $RCHX_2$) or has a double bond besides the halogen), a wide range of products can be formed, and the product ratio mainly depends on the type of catalyst used; 4) 1° and 2° alkyl groups tend to rearrange and therefore product mixtures are formed; 5) if the aromatic substrate is substituted, electron-donating substituents are required, and electron-poor substrates do not undergo the alkylation (e.g., $C_6H_5NO_2$); and 6) the orientation of substitution is catalyst dependent; in addition to the expected *o-* and *p-*disubstituted products, substantial amounts of *meta-*derivatives can be obtained under harsh conditions (e.g., with $AlCl_3$ at high temperature). The reaction also has disadvantages: 1) only electron-rich (usually alkyl substituted) aromatic rings can be used as substrates; 2) after the first alkyl group is introduced, the aromatic ring becomes more reactive and polyalkylation often occurs; 3) catalysts and alkylating agents that are too reactive may degrade the substrate; 4) nucleophilic functional groups (-OH, -OR, -NH2) coordinate to the Lewis acid catalyst, thereby deactivating it; and 5) the *Friedel-Crafts alkylation* reaction is reversible, and therefore alkyl groups that are already in the substrate may migrate, rearrange, or be removed under the reaction conditions.

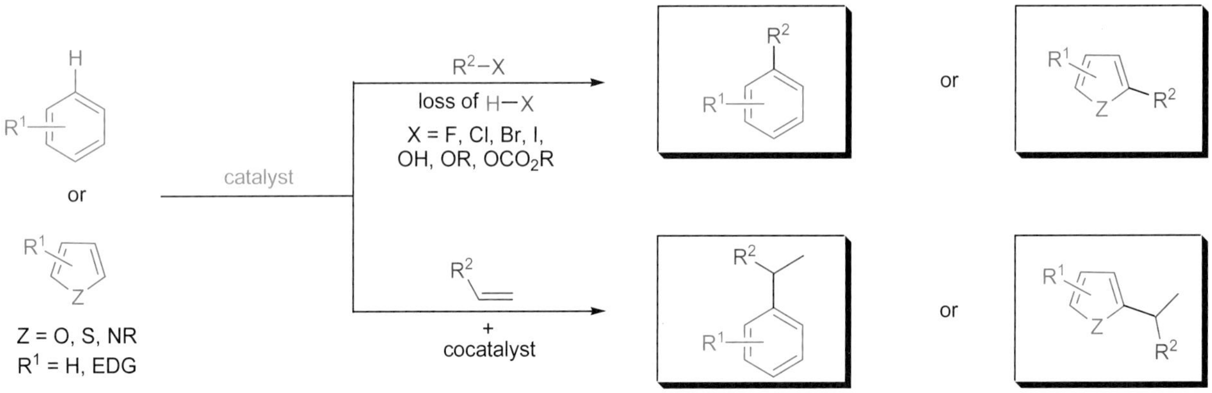

Mechanism:[46-54]

The first step of the *Friedel-Crafts alkylation* is the coordination of the Lewis acid to the alkylating agent (e.g., alkyl halide) to give a polar addition complex. The extent of polarization in this complex depends on the branching of the alkyl group and almost total dissociation is observed in the case of tertiary and benzylic compounds. The rate determining step is the formation of the σ-complex by the reaction of the initial complex (electrophile) and the aromatic ring; this step disrupts the aromaticity of the substrate. In the last step of the mechanism a proton is lost and the aromaticity is reestablished.

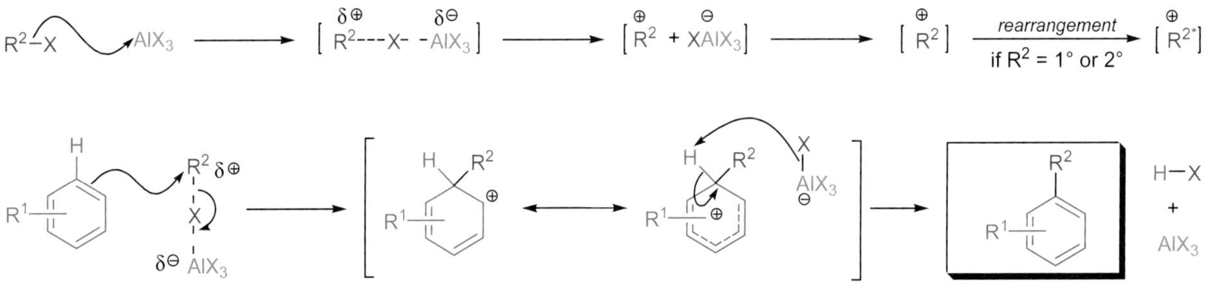

Synthetic Applications:

S. L. Schreiber et al. carried out the total synthesis of the potent cytotoxin (±)-tri-*O*-methyl dynemicin A methyl ester.[55] The key step was a *regioselective Friedel-Crafts alkylation* of an extremely sensitive aromatic enediyne with 3-bromo-4,7-dimethoxyphthalide. The coupling of these two fragments took place in the presence of silver triflate at 0 °C in 1 minute, and after methylation, gave a 1:1 mixture of diastereomers in 57% yield.

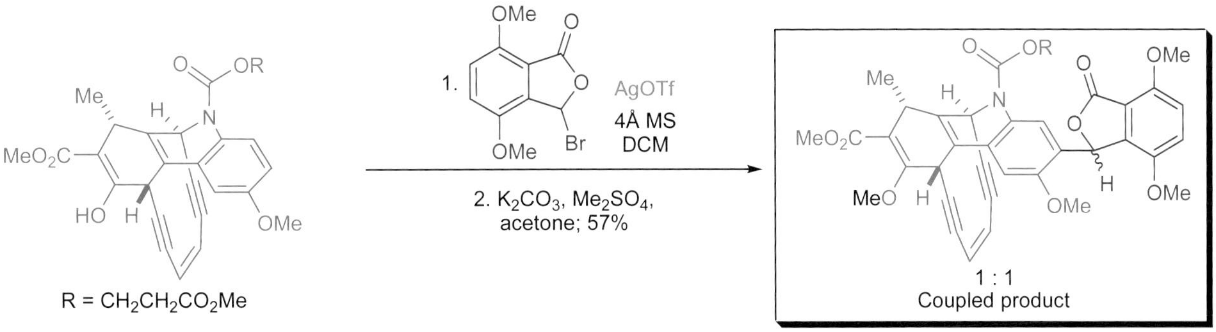

In the laboratory of G.A. Posner, semisynthetic antimalarial trioxanes in the artemisinin family were prepared *via* an efficient *Friedel-Crafts alkylation* using a pyranosyl fluoride derived from the natural trioxane lactone artemisinin.[56] The alkylating agent, pyranosyl fluoride, was prepared from the lactone in two steps: reduction to the lactol followed by treatment with diethylaminosulfur trifluoride. The highly chemoselective alkylation was promoted by $BF_3 \cdot OEt_2$ and several electron-rich aromatic and heteroaromatic compounds were alkylated in moderate to high yield using this method.

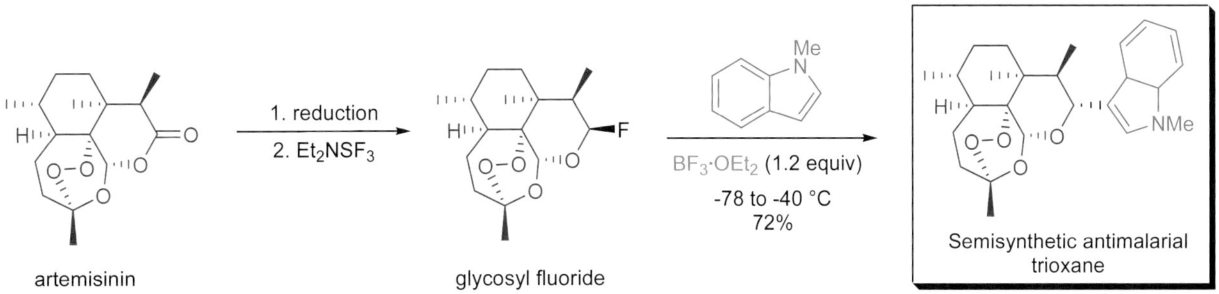

The first total synthesis of (±)-brasiliquinone B was accomplished by V.H. Deshpande and co-workers starting from 7-methoxy-1-tetralone.[57] The key step of their synthesis was the *Friedel-Crafts alkylation* of 2-ethyl-7-methoxytetralin with 3-bromo-4-methoxyphthalide in the presence of tin tetrachloride.

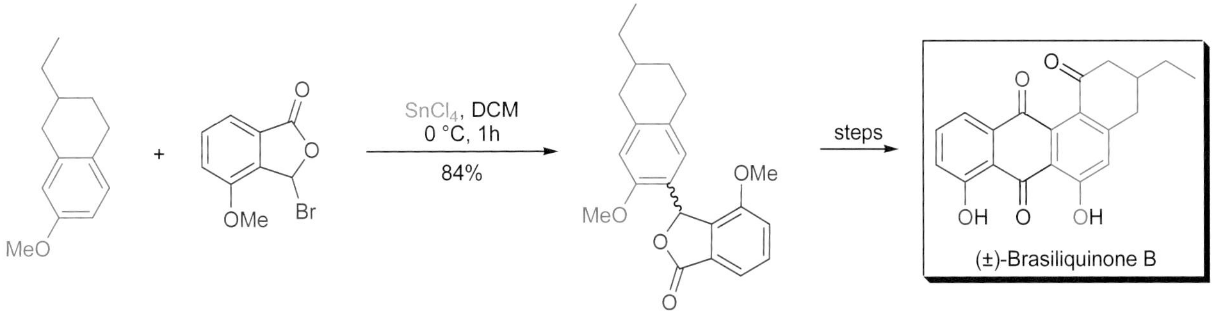

During the synthesis of anti-HIV cosalane analogues, M. Cushman et al. attached substituted benzoic acid rings to the pharmacophore through methylene and amide linkers.[58] In order to assemble a complex highly substituted benzophenone derivative, 3-chlorosalicylic acid had to be benzylated. A substituted benzyl alcohol was chosen as the alkylating agent and the benzylation proceeded smoothly in methanol using sulfuric acid as the catalyst.

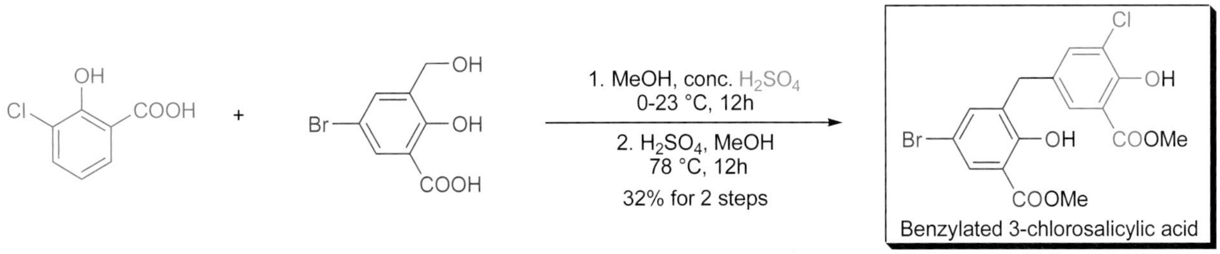

FRIES-, PHOTO-FRIES, AND ANIONIC ORTHO-FRIES REARRANGEMENT

(References are on page 590)

Importance:

[*Seminal Publications*[1-4]; *Reviews*[5-14]; *Modifications & Improvements*[15-31]; *Theoretical Studies*[32-38]]

In the early 1900s, K. Fries and co-workers reacted phenolic esters of acetic and chloroacetic acid with aluminum chloride and isolated a mixture of *ortho*- and *para*-acetyl- and chloroacetyl phenols.[3,4] Reports in the literature described similar rearrangements in the presence of Lewis acids during the late 1800s,[1,2] but Fries was the one who recognized that the rearrangement of phenolic esters was general. In his honor the conversion of phenolic esters to the corresponding *ortho* and/or *para* substituted phenolic ketones and aldehydes, in the presence of Lewis or Brönsted acids is called the *Fries rearrangement*. The *Fries rearrangement* has the following general features: 1) usually it is carried out by heating the phenolic ester to high temperatures (80-180 °C) in the presence of at least one equivalent of Lewis acid or Brönsted acid (e.g., HF, $HClO_4$, PPA); 2) the reaction time can vary between a few minutes and several hours; 3) Lewis acids that catalyze the *Friedel-Crafts acylation* are all active but recently solid acid catalysts (e.g., zeolites, mesoporous molecular sieves) and metal triflates have also been used;[12,30] 4) the rearrangement is general for a wide range of structural variation in both the acid and phenol component of phenolic esters; 5) yields are the highest when there are electron-donating substituents on the phenol, while electron-withdrawing substituents result in very low yields or no reaction; 6) with polyalkylated phenols alkyl migration is often observed under the reaction conditions; 7) the *Friedel-Crafts acylation* of phenols is usually a two-step process: formation of a phenolic ester followed by a *Fries rearrangement*; 8) the selectivity of the rearrangement to give *ortho*- or *para*- substituted products largely depends on the reaction conditions (temperature, type, and amount of catalyst, solvent polarity, etc.); 9) at high temperatures without any solvent the *ortho*-acylated product dominates while low temperatures favor the formation of the *para*-acylated product; 10) with increasing solvent polarity the ratio of the *para*-acylated product increases; and 11) optically active phenolic esters rearrange to optically active phenolic ketones. There are two main variants of the *Fries rearrangement*: 1) upon irradiation with light phenolic esters undergo the same transformation, which is known as the *photo-Fries rearrangement*;[8,11] and 2) an *anionic ortho-Fries rearrangement* takes place when *ortho*-lithiated *O*-aryl carbamates undergo a facile *intramolecular [1,3]-acyl migration* to give substituted salicylamides at room temperature.[17,27]

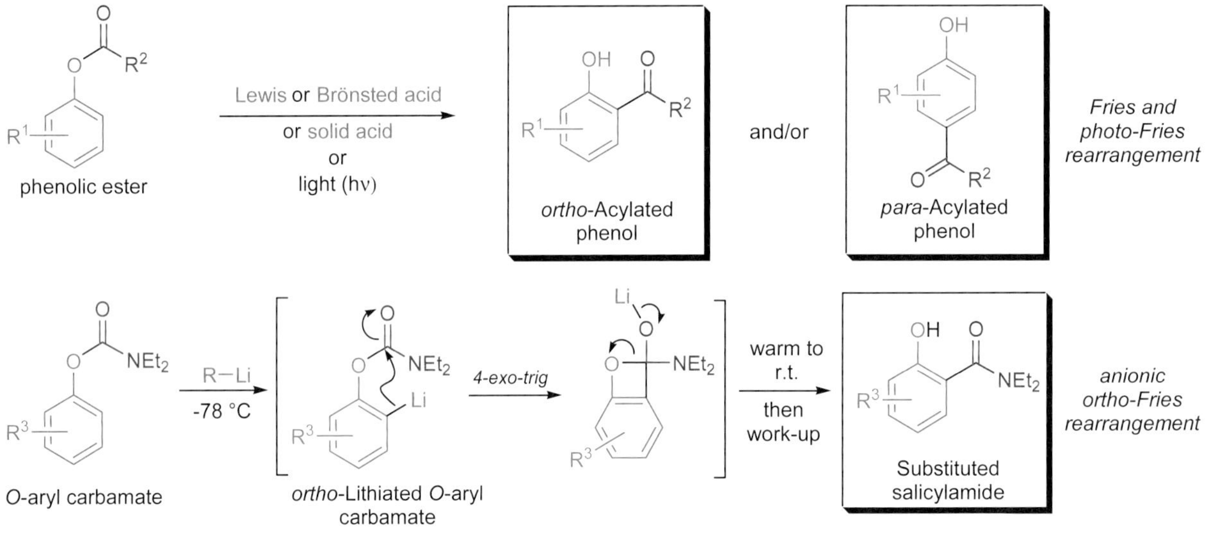

R^1 = alkyl, -OR,-NR_2, -aryl; R^2 = alkyl, aryl; R^3 = alkyl, -OR, Cl

Mechanism: [39-49,11,50]

The *Fries rearrangement* proceeds *via* ionic intermediates but the exact mechanistic pathway (whether it is inter- or intramolecular) is still under debate. There are many reports in the literature that present evidence to support either of the pathways, but it appears that the exact route depends on the structure of the substrates and the reaction conditions. The scheme depicts the formation of an *ortho*-acylated phenol from a substituted phenolic ester in the presence of aluminum trihalide catalyst. The *photo-Fries rearrangement* proceeds *via* radical intermediates.[11,50,13]

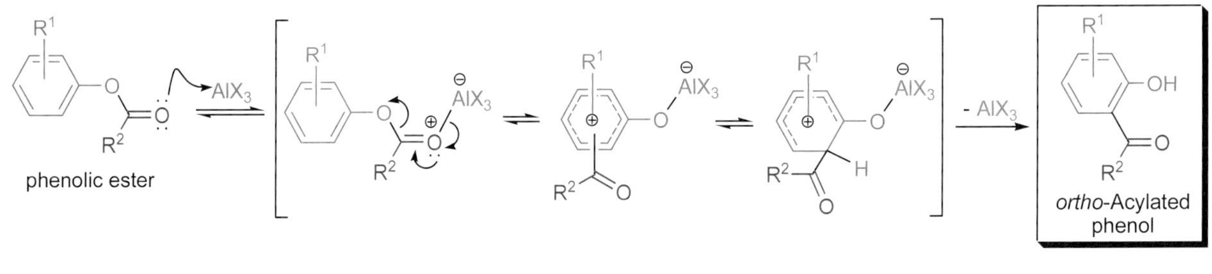

Synthetic Applications:

The first atropo-enantioselective total synthesis of a phenylanthraquinone natural product (*M*)-knipholone was reported by G. Bringmann et al.[51] In the late stages of the synthesis, an acetyl group had to be introduced under mild conditions. The advanced substituted anthraquinone intermediate was first deprotected with $TiCl_4$ and then acylated with Ac_2O in the presence of $TiCl_4$. A spontaneous *Fries-rearrangement* took place to afford the *ortho*-acylated product in high yield. The natural product was obtained by a mono *O*-demethylation at C6 with $AlBr_3$.

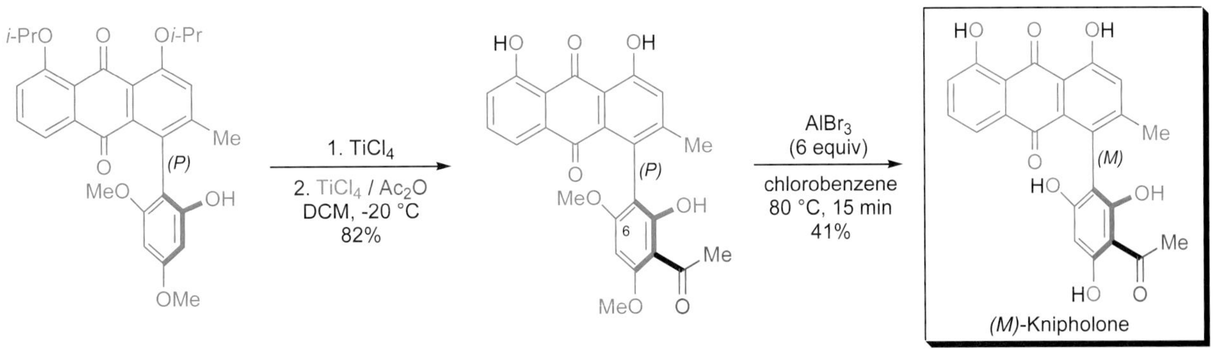

The total synthesis of the potent protein kinase C inhibitor (–)-balanol was accomplished by J.W. Lampe and co-workers.[52] They took advantage of the *anionic homo-Fries rearrangement* to prepare the sterically congested benzophenone subunit. To this end, 2-bromo-3-benzyloxy benzyl alcohol was first acylated with a 1,3,5-trisubstituted benzoyl chloride to obtain the ester precursor in 84% yield. Next, the ester was treated with *n*-BuLi at -78 °C to perform a metal-halogen exchange. The resulting aryllithium rapidly underwent the *anionic homo-Fries rearrangement* to afford the desired tetra *ortho*-substituted benzophenone in 51% yield.

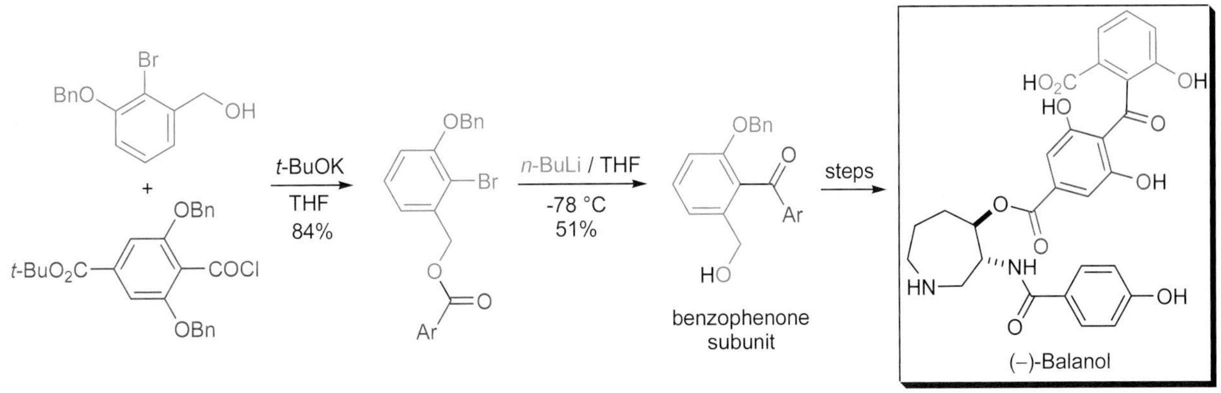

Research in the laboratory of P. Magnus showed that the macrocyclic skeleton of diazonamide could be synthesized with the use of *macrolactonization* followed by a *photo-Fries rearrangement*.[53] First, the aromatic carboxylic acid and the phenol were coupled with EDCI to form the macrolactone (phenolic ester), which was then exposed to light at high-dilution to cleanly afford the macrocyclic *ortho*-acylated phenol skeleton of diazonamide.

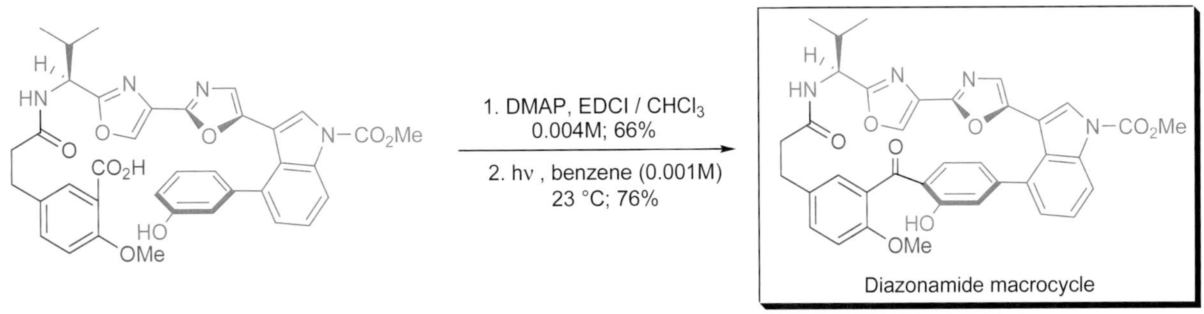

GABRIEL SYNTHESIS

(References are on page 592)

Importance:

[*Seminal Publication*[1]; *Reviews*[2-4]; *Modifications & Improvements*[5-16]]

The mild, two-step preparation of primary amines from the corresponding alkyl halides, in which potassium phthalimide is first alkylated and the resulting *N*-alkylphthalimide is subsequently hydrolyzed, is known as the *Gabriel synthesis*. Alkylation of phthalimide with simple alkyl halides was first reported in 1884,[17] but it was not until 1887 when S. Gabriel recognized the generality of the process and came up with the two-step procedure for the synthesis of primary amines.[1] The alkylation reaction can be conducted in the absence or in the presence of a solvent.[2] The best solvent is DMF (good for S_N2 reactions), but DMSO, HMPA, chlorobenzene, acetonitrile, and ethylene glycol can also be used. The following alkylating agents give good to excellent yields during the preparation of the required *N*-alkylphthalimides: 1) sterically unhindered 1°and 2° alkyl halides give the best results with alkyl iodides being the most reactive (I > Br > Cl) followed by allylic, benzylic, and propargylic halides; 2) alkyl sulfonates (mesylates, tosylates) often give higher yields than the alkyl halides and are easier to obtain; 3) α-halo ketones, esters, nitriles, and β-keto esters (e.g., diethyl bromomalonate);[18,19] 4) *O*-alkylisoureas;[20] 5) alkoxy- and alkylthiophosphonium salts;[21] 6) 1°and 2° alcohols under the *Mitsunobu reaction* conditions (DEAD/Ph_3P/phthalimide);[12] 6) aryl halides with several electron-withdrawing groups (S_NAr reaction to prepare 1° arylamines); 7) aryl halides in the presence of Cu(I) catalysts;[6,9] 8) epoxides and aziridines (preparation of amino alcohols and diamines);[22,23] and 9) α,β-unsaturated compounds undergo facile *Michael-addition* by the phthalimide anion.[24] The original *Gabriel synthesis* had the following problems that limited its widespread application: 1) when the potassium phtalimide and the alkyl halide required high temperatures (120-240 °C) without a solvent, heat sensitive substrates could not be used; 2) the hydrolysis was usually carried out with a strong acid (e.g., H_2SO_4, HBr, HI) at high temperatures therefore substrates containing acid-sensitive functionalities were excluded; and 3) strong alkaline hydrolysis was also used and was incompatible with base-sensitive functional groups. In 1926, H.R. Ing and R.H.F. Manske came up with a modification by introducing hydrazine hydrate in refluxing ethanol for the cleavage of the *N*-alkylphthalimide under mild and neutral conditions (*Ing-Manske procedure*).[5] During the past century, several other modifications of the original procedure were introduced: 1) novel Gabriel reagents (replacement of phthalimide with other nitrogen sources) to achieve milder deprotection conditions;[4] 2) addition of catalytic amounts of a crown ether or a cryptand to the reaction mixture of alkyl halides with potassium phthalimide gives almost quantitative yields;[8,10] and 3) the use of $NaBH_4$ in isopropanol for the exceptionally mild cleavage of the phthalimide.[11] A related process is the *Gabriel-malonic ester synthesis* in which the anion of diethyl phthalimidomalonate is alkylated and after hydrolysis/decarboxylation an amino acid is obtained.[19]

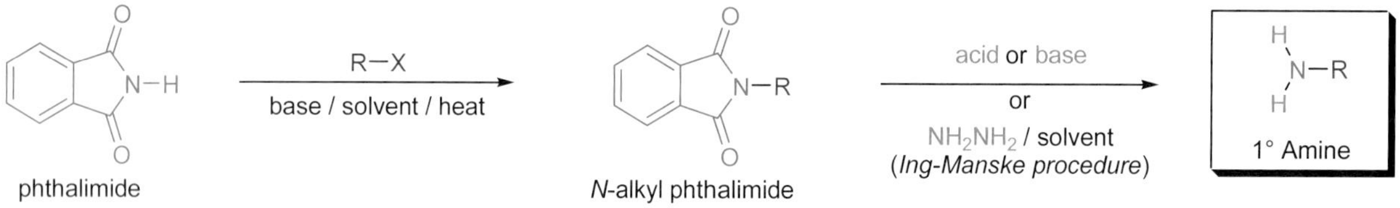

X = halogen, OTf, OMs, etc.; R = 1°, 2° alkyl, allylic, benzylic, etc.

Mechanism: [2,15]

The first step of the *Gabriel synthesis*, the alkylation of potassium phthalimide with alkyl halides, proceeds *via* an S_N2 reaction. The second step, the hydrazinolysis of the *N*-alkylphthalimide, proceeds by a nucleophilic addition of hydrazine across one of the carbonyl groups of the phthalimide. Subsequently, the following steps occur: ring-opening then proton-transfer followed by an intramolecular S_NAc reaction, another proton-transfer and finally, the breakdown of the tetrahedral intermediate to give the desired primary amine and the side product phthalyl hydrazide.

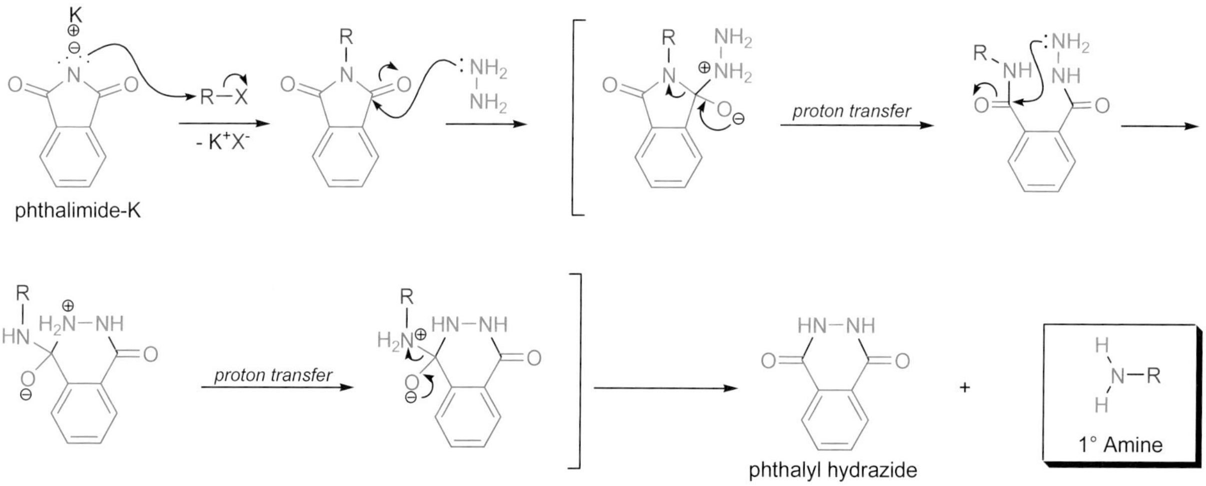

Synthetic Examples:

The total synthesis of the insect feeding deterrent peramine was accomplished by D.J. Dumas at du Pont laboratories.[25] The *Gabriel synthesis* was successfully employed in the last steps of the synthesis. The primary alkyl chloride was treated with potassium phthalimide in DMF at 77-82 °C for 1.5h. The resulting *N*-alkylphthalimide was cleaved in high yield using the *Ing-Manske procedure*.

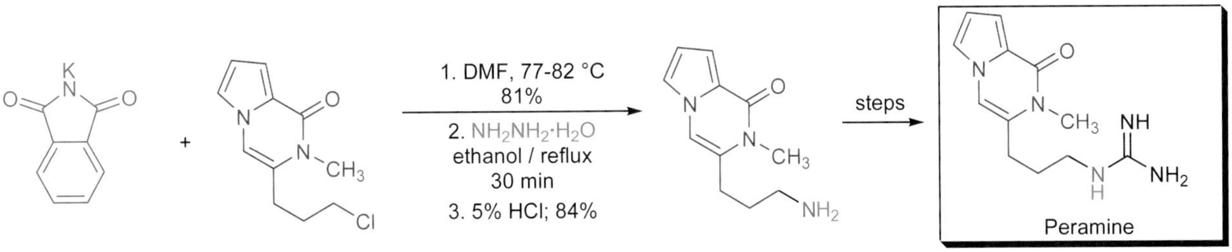

During the synthesis of swainsonine- and castanospermine analogues (amino sugars), K. Burgess et al. introduced the nitrogen atom by replacing a primary hydroxyl group using phthalimide under the *Mitsunobu reaction conditions*.[26] The phthalyl group was not immediately removed but carried over several steps. Interestingly, deprotection with hydrazine was not compatible with the terminal alkene functionality due to significant hydrogenation of the double bond by the *in situ* formed diimide. Using methylamine instead of hydrazine cleanly afforded the deprotected primary amine that readily displaced a secondary mesylate to form a substituted pyrrolidine ring.

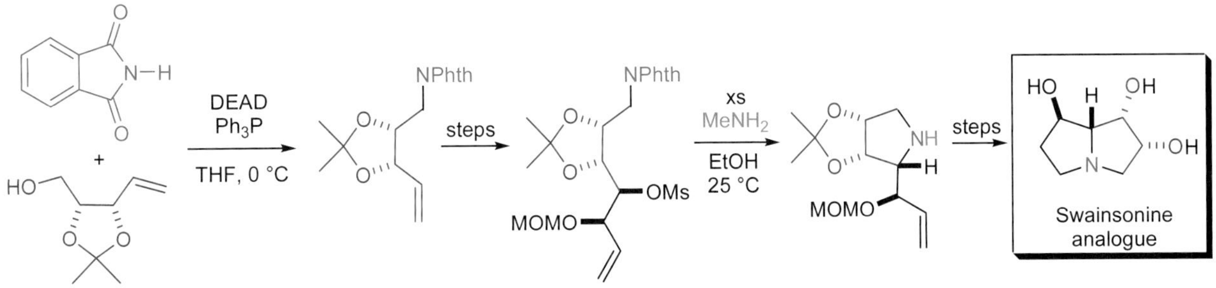

A dynamic kinetic resolution was utilized for the highly stereoselective *Gabriel synthesis* of α-amino acids by K. Nunami and co-workers.[27] The substrate, *t*-butyl-(4*S*)-1-methyl-3-2-(bromoalkanoyl)-2-oxoimidazolidine-4-carboxylate, smoothly reacted with potassium phthalimide at room temperature to give only one diastereomer in good yield. The removal of the chiral auxiliary afforded an *N*-phthaloyl-L-α-amino acid.

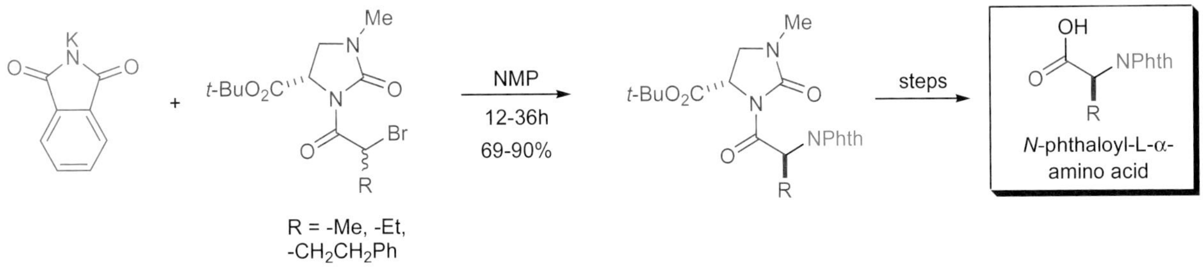

The preparation of vicinal diamines in an enantioselective fashion is a challenging task. F.M. Rossi et al. undertook the synthesis of a β-benzoylamino-phenylalanine (2,3-diamino acid), which is an analogue of the taxol side chain.[28] During their synthetic studies, the secondary alcohol of an enantiopure oxazolidinone was mesylated and displaced by potassium phthalimide in DMF. Interestingly, there was a net retention of configuration due to neighboring group participation by the oxazolidinone nitrogen atom. For this reason, the authors later decided to displace the mesylate with NaN_3 and to protect the oxazolidinone nitrogen with a TMS group to avoid participation.

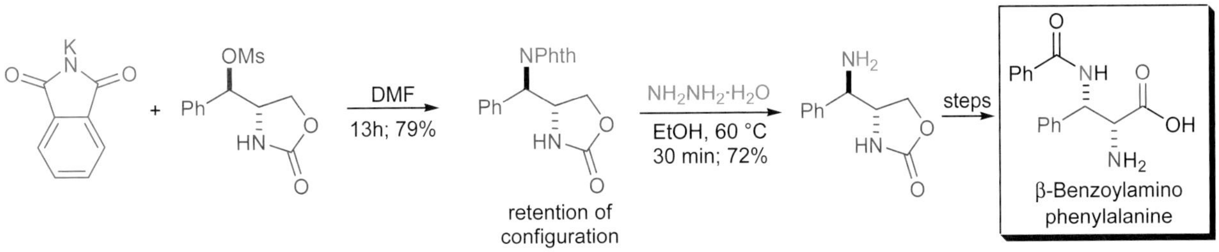

GATTERMANN AND GATTERMANN-KOCH FORMYLATION

(References are on page 592)

Importance:

[*Seminal Publications*[1-3]; *Reviews*[4-8]; *Modifications & Improvements*[9-13]; *Theoretical Studies*[14,15]]

In 1897, L. Gattermann and J.A. Koch successfully introduced a formyl group (CHO) on toluene by using formyl chloride (HCOCl) as the acylating agent under *Friedel-Crafts acylation* conditions.[1] Although the researchers were not able to prepare the acid chloride, they assumed that by reacting carbon monoxide (CO) with hydrogen chloride (HCl), formyl chloride would be formed *in situ*, and in the presence of catalytic amounts of $AlCl_3$-Cu_2Cl_2 formylation of the aromatic ring would occur. The introduction of a formyl group into electron rich aromatic rings by applying CO/HCl/Lewis acid catalyst (AlX_3, FeX_3, where X = Cl, Br, I) to prepare aromatic aldehydes is known as the *Gattermann-Koch formylation*. The general features of this formylation reaction are: 1) at atmospheric pressure activated aromatic compounds can be used as substrates (e.g., alkylbenzenes); 2) at high CO pressure (100-250 atm) the reaction rate increases significantly and even non-activated aromatics (chlorobenzene, benzene) can be formylated; 3) deactivated aromatic compounds (having *meta*-directing substituents) cannot be formylated with this method; 4) a carrier/activator (Cu_2Cl_2, $TiCl_4$ or $NiCl_2$) for the catalyst is necessary at atmospheric pressure; however, no activator is needed at high pressure; 5) the amount and purity of the catalyst is very important and often a full equivalent of catalyst is needed; 6) monosubstituted substrates are formylated almost exclusively at the *para* position, but when there is already a *para* substituent present in the substrate, the formyl group is introduced at the *ortho* position; 7) just as in the *Friedel-Crafts reactions*, alkyl migration occurs with highly alkylated aromatic substrates; and 8) the need for high pressures renders this method mainly useful to industrial applications. The scope of the *Gattermann-Koch reaction* in terms of suitable substrates is also limited, since it is mostly restricted to alkylbenzenes. Gattermann introduced a modification where HCN is mixed with HCl in the presence of $ZnCl_2$ to formylate phenols, phenolic ethers and heteroaromatic compounds (e.g., pyrroles and indoles). This modification is called the *Gattermann formylation* (or *Gattermann synthesis*).[2,3] The main drawback of the *Gattermann formylation* was that it called for the use of anhydrous HCN, which is a very toxic compound. To avoid the handling of HCN, R. Adams generated it *in situ* along with $ZnCl_2$ by reacting $Zn(CN)_2$ with HCl in the presence of the aromatic substrate (*Adams modification*).[10] This method has since become the most widely used variant in organic synthesis. Other modifications used NaCN and CNBr successfully instead of HCN.[9] A serious limitation of both title reactions is that they cannot be used for the formylation of aromatic amines due to numerous side reactions.

Gattermann-Koch Formylation:

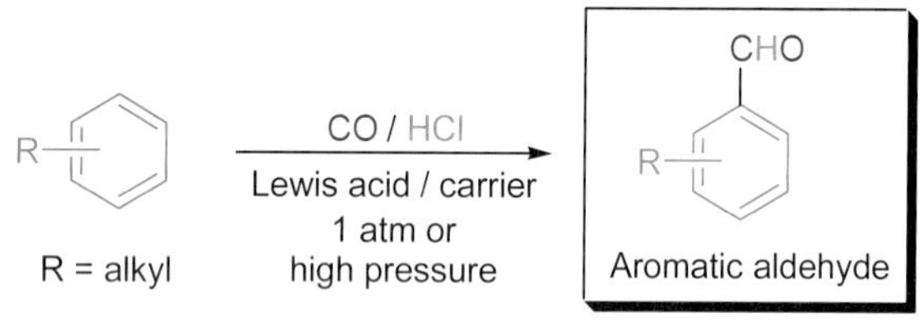

Gattermann Formylation:

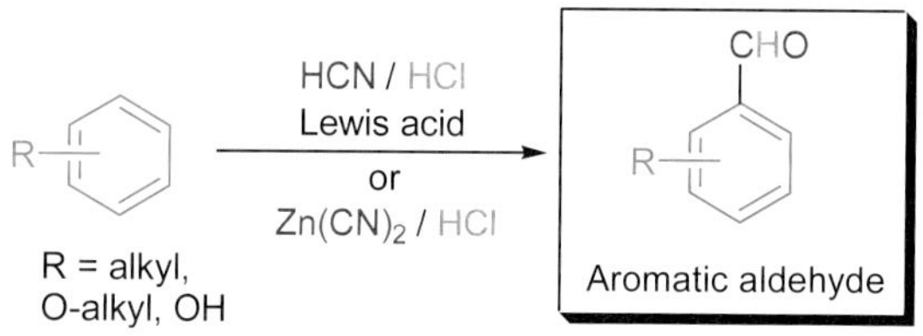

Mechanism: [16-23]

The mechanisms of the *Gattermann* and *Gattermann-Koch formylation* belong to the category of electrophilic aromatic substitution (S_EAr) but are not known in detail, since they have a tendency to vary from one substrate to another, and the reaction conditions may also play a role. When carbon monoxide is used, the electrophilic species is believed to be the formyl cation, which is attacked by the aromatic ring to form a σ-complex. This σ-complex is then converted to the aromatic aldehyde upon losing a proton. When HCN is used, the initial product after the S_EAr reaction is an imine hydrochloride, which is subsequently hydrolyzed to the product aldehyde.

Gattermann-Koch Formylation:

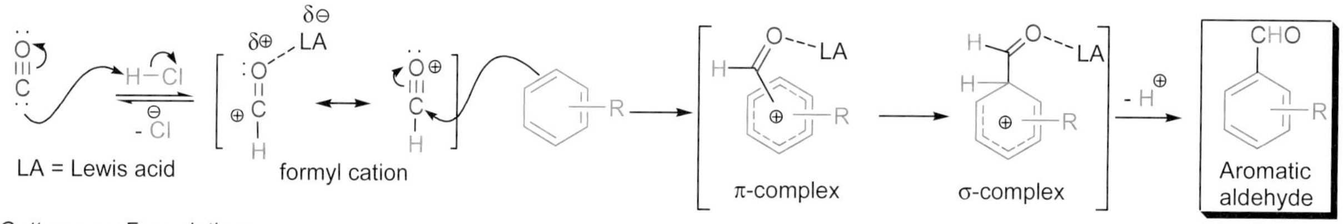

Gattermann Formylation:

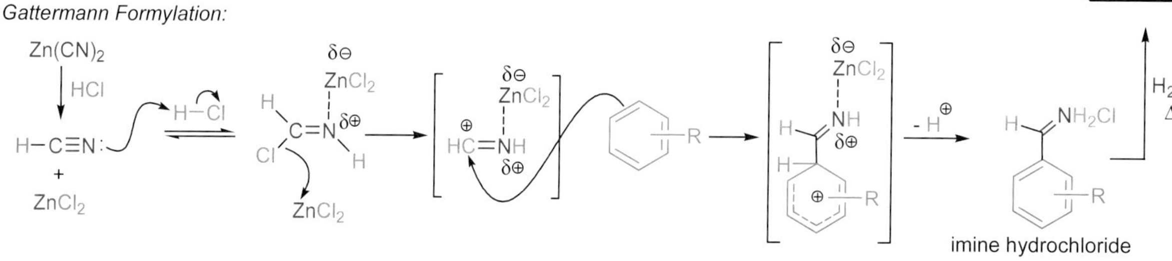

Synthetic Applications:

The benzofuran-derived natural product caleprunin A was synthesized by R. Stevenson et al. using the *Gattermann formylation* as the key step.[24] The starting 3,4,5-trimethoxyphenol was suspended with $Zn(CN)_2$ in ether and dry HCl gas was bubbled through the reaction mixture at room temperature for 2h. The solvent was decanted, water was added and the mixture was heated for 15 minutes. The natural product was obtained by reacting the benzaldehyde derivative with chloroacetone in DMF in the presence of anhydrous K_2CO_3.

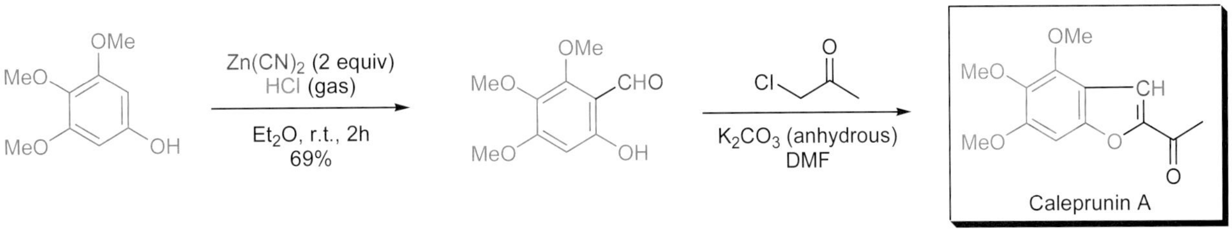

The regiospecific introduction of the formyl group into the C3 postion of 2,5-dialkyl-7-methoxy-benzo[*b*]furans was achieved by H.N.C. Wong and co-workers by using the *Adam's modification of the Gattermann formylation.*[25] A potential ligand for adenosine A_1 receptors was prepared from 2-cyclopentyl-5-(3-hydroxypropyl)-7-methoxy-benzo[*b*]furan in 50% yield by bubbling HCl gas through its etheral solution containing $Zn(CN)_2$ at -10 °C for 1h. The resulting imine hydrochloride was hydrolyzed with a water-ethanol mixture at 50 °C.

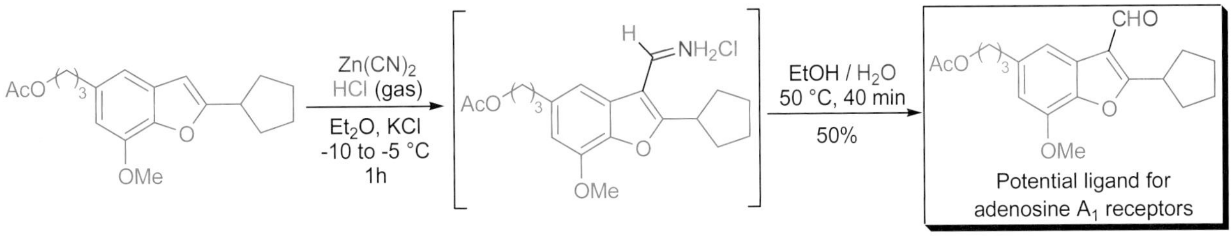

Compounds containing the pyridocarbazole ring are known to have DNA intercalating properties and therefore they are potent antitumor agents. For example, several syntheses of pyrido[2,3-*a*]carbazole derivatives have been published, but these methods are often lengthy and low-yielding. R. Prasad and co-workers synthesized 2-hydroxypyrido[2,3-a]carbazoles starting from 1-hydroxycarbazoles.[26] The key transformation was the *Gattermann formylation* of 1-hydroxycarbazoles to obtain 1-hydroxycarbazole-2-carbaldehydes, from which the target compounds could be obtained *via* a *Perkin reaction*.

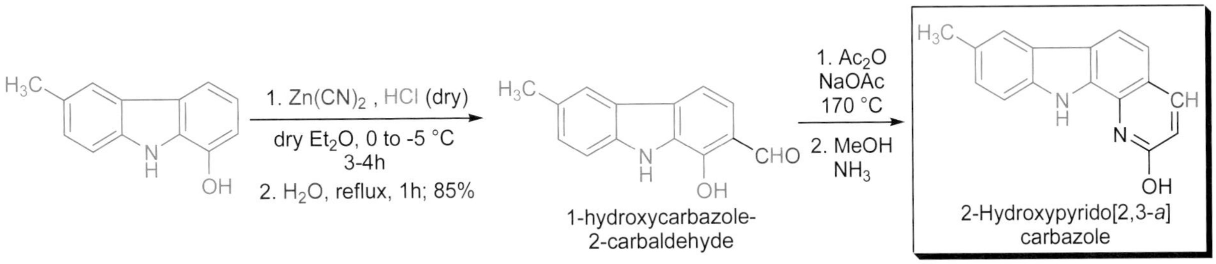

Certain aromatic analogues of natural amino acids can be used as potential fluorescent probes of peptide structure and dynamics in complex environments. The research team of M.L. McLaughlin undertook the gram scale synthesis of racemic 1- and 2-naphthol analogues of tyrosine.[27] The synthesis of the 1-naphthol tyrosine analogue started with the *Gattermann formylation* of 1-naphthol using the *Adams modification* to afford the formylated product 4-hydroxy-1-naphthaldehyde in 67% yield.

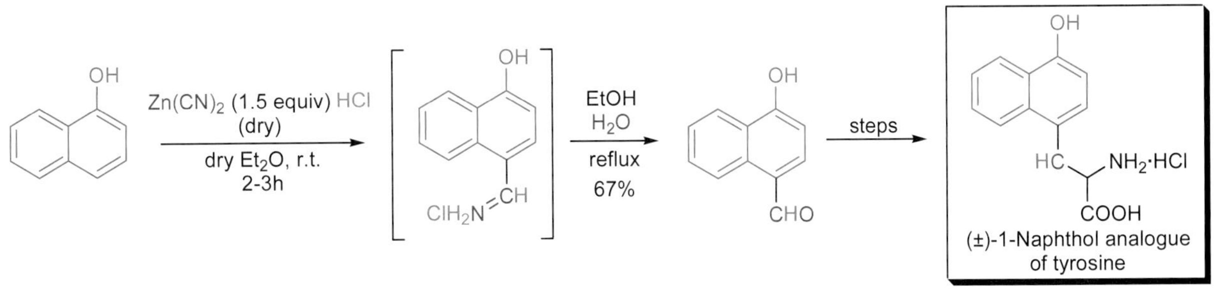

GLASER COUPLING

(References are on page 593)

Importance:

[*Seminal Publication*[1]; *Reviews*[2-9]; *Modifications & Improvements*[10-16]]

In 1869, C. Glaser discovered that when phenylacetylene was treated with a copper(I)-salt in the presence of aqueous ammonia, a precipitate formed, which after air oxidation yielded a symmetrical compound, 1,4-diphenyl-1,3-butadiyne (diphenyldiacetylene).[1] The preparation of symmetrical conjugated diynes and polyynes (linear or cyclic) by the oxidative homocoupling of terminal alkynes in the presence of copper salts is known as the *Glaser coupling*. There are numerous versions of the original procedure developed by Glaser, and these differ mainly in the type and amount of oxidants used: 1) besides oxygen and air, $CuCl_2$ and $K_3Fe(CN)_6$ are used most often as oxidizing agents; 2) Glaser's procedure was heterogeneous and slow, but G. Eglinton and A.R. Galbraith showed that using $Cu(OAc)_2$ in methanolic pyridine made the process homogeneous and faster (*Eglinton procedure*). This method was successfully applied to the synthesis of macrocyclic diynes;[10] and 3) A.S. Hay used tertiary amines such as pyridine or the bidentate ligand TMEDA as complexing agents to solubilize the $Cu^{(I)}$-salt. Next, oxygen gas was passed through this solution to give the homocoupled product in a few minutes at room temperature in almost quantitative yield (*Hay coupling conditions*).[11,12] General features of the *Glaser coupling* and related methods are: 1) it works well for acidic terminal alkynes, but the yield tends to drop when the alkyne is less acidic (e.g., alkyl- or silicon-substituted terminal alkynes); 2) the reaction rate is often increased when a small amount of DBU, which most likely serves as a strong base to deprotonate the alkyne, is added to the reaction mixture;[7] 3) the reaction conditions tolerate a wide range of functional groups as the oxidation is mostly restricted to the triple bond; 4) if the reactants or the product is oxygen sensitive, side reactions can be minimized by either running the reaction for shorter periods of time or applying an inert atmosphere and using large amounts of the $Cu^{(II)}$-salt; 5) the yield of the coupling of heterocyclic alkynes strongly depends on the solvent used, and DME was found to be the best; 6) for oligomerization reactions, *o*-dichlorobenzene is the best solvent; and 7) besides using common solvents, recent modifications employed supercritical CO_2 and ionic liquids for the couplings.[13,16] The *Glaser coupling* is not well-suited for the preparation of unsymmetrical diynes. Therefore, other methods were developed using both oxidative and non-oxidative conditions: 1) the *Chodkiewitz-Cadiot reaction* couples a terminal alkyne with a 1-bromoalkyne in the presence of a copper(I)-salt and an aliphatic amine (e.g., $EtNH_2$);[17-19] 2) copper(I)- and cobalt(I)-salts are efficient catalysts for the coupling of alkynyl Grignard derivatives with 1-haloalkynes;[4] and 3) $Pd^{(0)}$-catalyzed coupling of terminal alkynes with 1-iodoalkynes in the presence of a $Cu^{(I)}$-salt is also successful.[20]

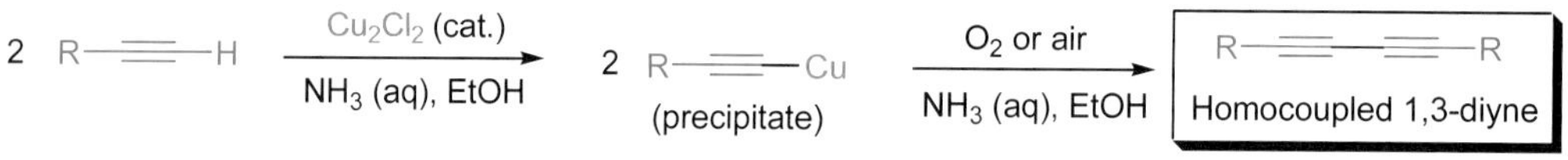

Eglinton procedure:

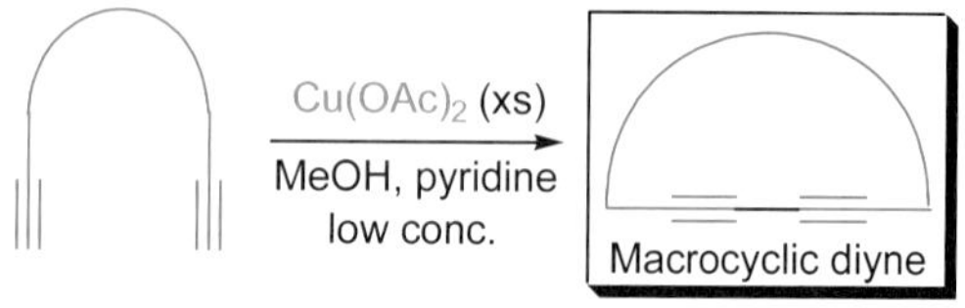

Hay's conditions:

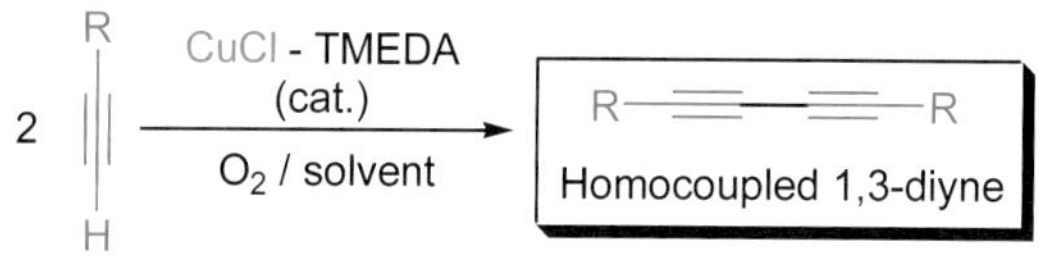

Chodkiewitz-Cadiot heterocoupling:

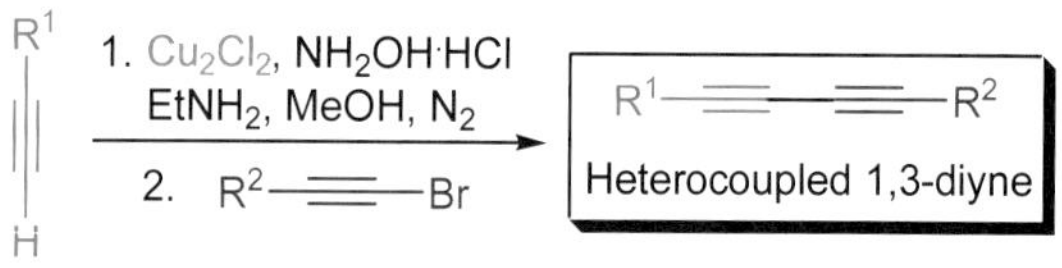

Pd-catalyzed heterocoupling:

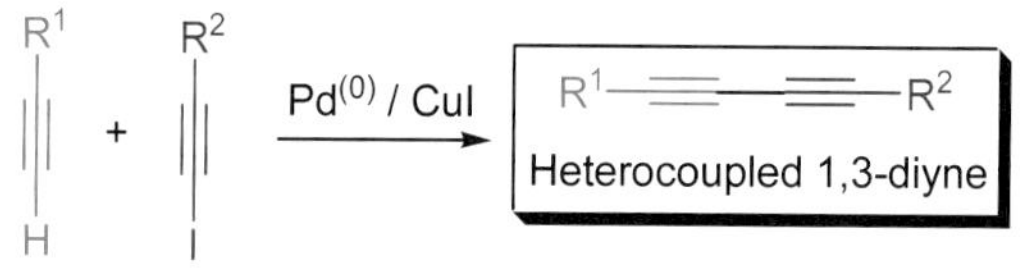

Mechanism: [21-29]

The mechanism of the *Glaser coupling* and related methods is very complex and is not fully understood. Studies revealed that the mechanism is highly dependent on the experimental conditions. The early proposal involving a radical mechanism has been rejected. The currently accepted mechanism involves dimeric copper(II)acetylide complexes.

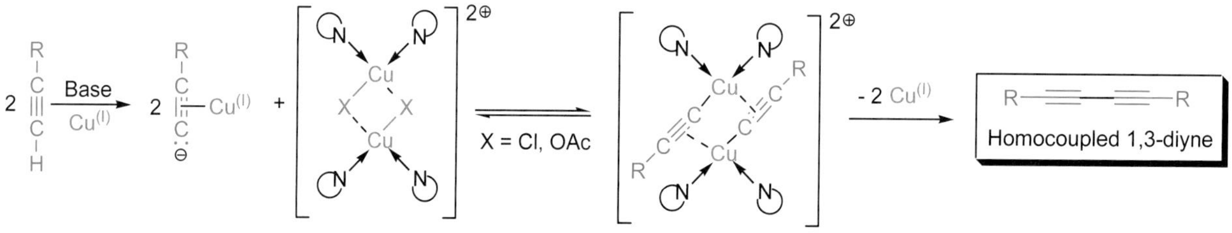

Synthetic Applications:

Novel polymerizable phosphatidylcholines were successfully synthesized by the research team of G. Just.[30] To prepare a 32-membered macrocyclic diyne, the *Eglinton modification* of the *Glaser coupling* was utilized. The diester-diyne starting material was slowly added to a refluxing solution containing 10 equivalents of cupric acetate in dry pyridine. The macrocycle was isolated in 54% yield after column chromatography.

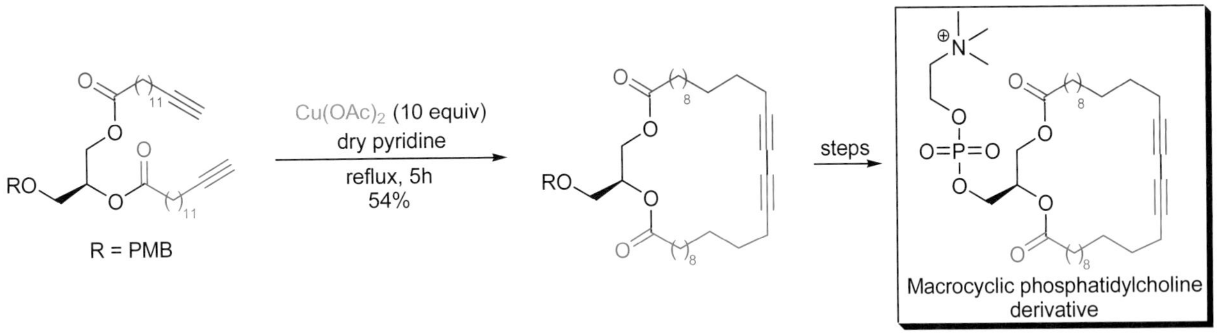

During the biomimetic total synthesis of endiandric acids A-G by K.C. Nicolaou and co-workers, the key polyunsaturated precursor was assembled *via* the *Glaser coupling* of two different terminal alkynes.[31-34] One of the alkynes was used in excess so the yield of the heterocoupled diyne could be maximized. In a solvent mixture of pyridine:methanol (1:1), the two reactant alkynes were treated with $Cu(OAc)_2$ at 25 °C to provide the desired diyne in 70% yield.

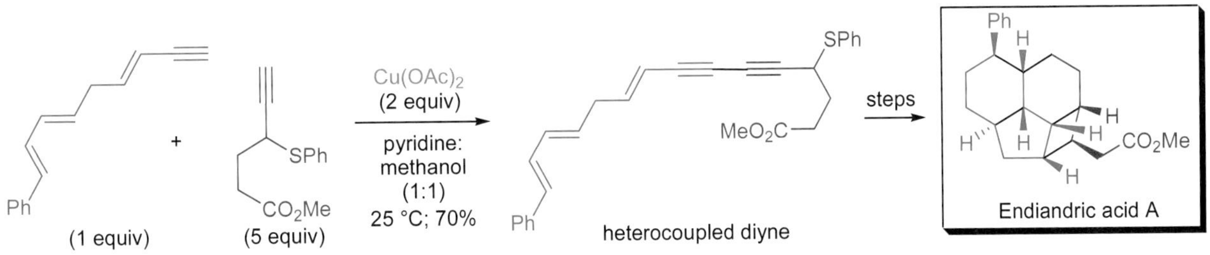

C.S. Wilcox and his research team designed and synthesized chiral water-soluble cyclophanes based on carbohydrate precursors.[35] These compounds are also dubbed as "glycophanes" and they are potentially valuable enzyme models. The key macrocyclization step utilized the *Glaser coupling* and the reaction was carried out in a thermal flow reactor at 80 °C in 67% yield.

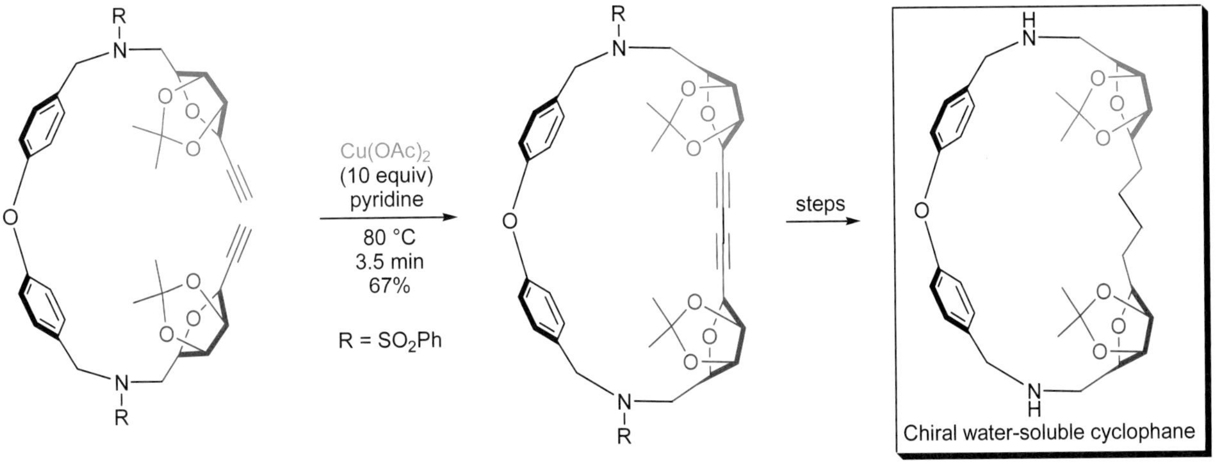

Nucleoside dimers linked by the butadiynediyl group were prepared by A. Burger et al. using the *Eglinton modification* of the *Glaser coupling via* dimerization of 3'β-C-ethynyl nucleosides.[36]

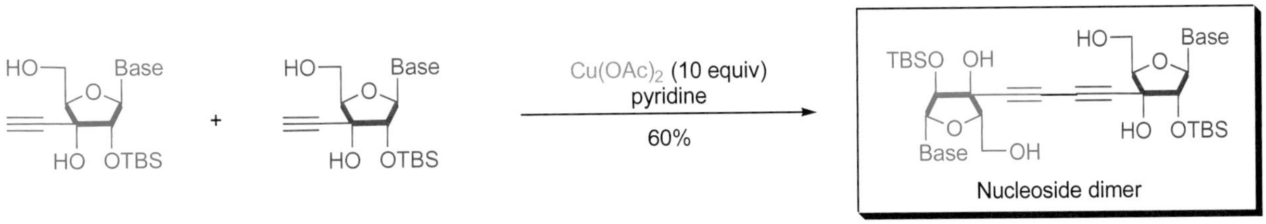

GRIGNARD REACTION

(References are on page 593)

Importance:

[*Seminal Publications*[1,2]; *Reviews*[3-17]; *Modifications and Improvements*;[18-20] *Theoretical Studies*[21-26]]

In 1900, V. Grignard reported that an alkyl halide (RX) reacts with magnesium metal (Mg) in diethyl ether to give a cloudy solution of an organomagnesium compound (RMgX), which upon reaction with aldehydes and ketones afforded secondary and tertiary alcohols, respectively.[1] These organomagnesium compounds are called *Grignard reagents*, and their addition across carbon-heteroatom multiple bonds is referred to as the *Grignard reaction*. Soon after its discovery, the *Grignard reaction* became one of the most versatile C-C bond forming tools. The general features of Grignard reagents and their reactions are: 1) the reagents are predominantly prepared by reacting alkyl, aryl, or vinyl halides with magnesium metal in aprotic nucleophilic solvents (e.g., ethers, tertiary amines); 2) the reagents are usually thermodynamically stable but air and moisture sensitive and incompatible with acidic functional groups (e.g., alcohols, thiols, phenols, carboxylic acids, 1°, 2° amines, terminal alkynes); 3) the C-Mg bond is very polar and the partial negative charge resides on the carbon atom, so Grignard reagents are excellent carbon nucleophiles (in the precursor halides the carbon has a partial positive charge so overall a reversal of polarity known as *umpolung* takes place upon formation of the reagent); 4) in most carbon-heteroatom multiple bonds the carbon atom is partially positively charged so the formation of C-C bonds with the nucleophilic Grignard reagents is straightforward; 5) addition of one equivalent of Grignard reagent followed by a work-up converts aldehydes to secondary alcohols (formaldehyde to primary alcohols), ketones to tertiary alcohols, nitriles to ketones and carbon-dioxide to acids; 6) acid derivatives react with two equivalents of Grignard reagent: esters and acyl halides (RCOX) are converted to tertiary alcohols; 7) prochiral aldehydes and ketones give rise to racemic mixtures of the corresponding alcohols upon reacting with achiral Grignard reagents, since the addition takes place on both faces of the carbonyl group; 8) chiral substrates, however, lead to diastereomeric mixtures with the predominant formation of one diastereomer as predicted by the *Felkin-Anh* or *chelation-control* models; and 9) alkyl halides can couple with Grignard reagents in a *Wurtz reaction* to give alkanes, while epoxides are opened in an S_N2 reaction at the less substituted carbon to give two-carbon homologated alcohols. Grignard reactions are often accompanied by certain side-reactions: 1) the generation of the Grignard reagent from alkyl halides can lead to undesired *Wurtz coupling* products; 2) the presence of oxygen (air) and moisture can consume some of the reagent to give alkoxides and alkanes, respectively; 3) if the carbonyl compound has a proton at the α-position, the Grignard reagent can act as a base and enolize the substrate (alkyllithium or organocerium reagents offer a solution to this problem, because they are more covalent and therefore less basic); and 4) if the reagent has a β-hydrogen and the substrate is hindered, reduction of the carbonyl group may occur by an intermolecular hydride transfer.

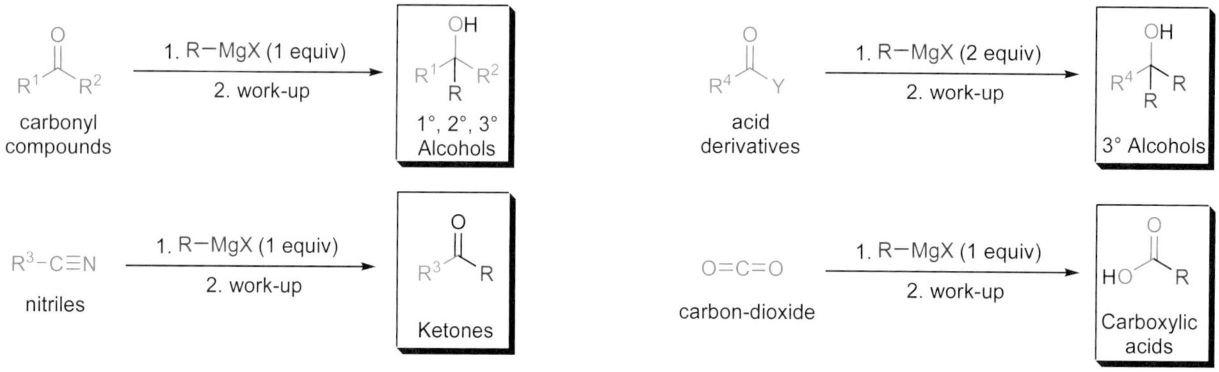

R¹, R² = alkyl, aryl, H; R³ = alkyl, aryl; R⁴ = alkyl, aryl; Y = OR, Cl, Br, I; R = alkyl, aryl; X = Cl, Br, I

Mechanism:[5,27-33,18,34]

The mechanism of the formation of the Grignard reagent is most likely a single-electron-transfer (SET) process, and it takes place on the metal surface.[33] The mechanism of the addition of Grignard reagents to carbonyl compounds is not fully understood, but it is thought to take place mainly *via* either a concerted process or a radical pathway (stepwise).[5,27,29] It was found that substrates with low electron affinity react in a concerted fashion passing through a cyclic transition state. On the other hand, sterically demanding substrates and bulky Grignard reagents with weak C-Mg bonds tend to react through a radical pathway, which commences with an electron-transfer (ET) from RMgBr to the substrate.[34]

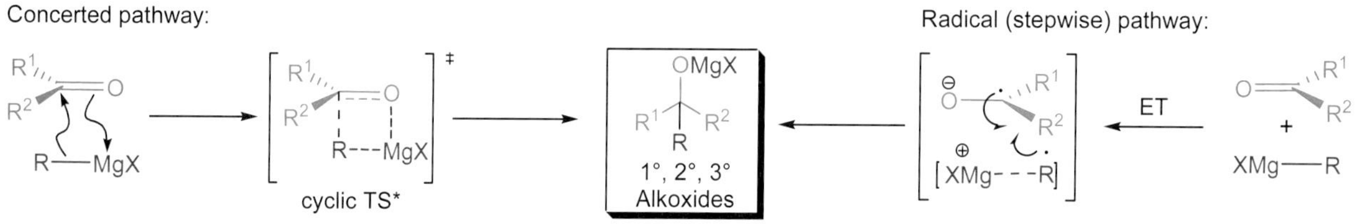

Synthetic Applications:

The stereoselective total synthesis of (±)-lepadiformine was accomplished in the laboratory of S.M. Weinreb.[35] The introduction of the hexyl chain in a stereoselective fashion was achieved by a *Grignard reaction* to an iminium salt during the last steps of the synthetic sequence. The iminium salt was generated *in situ* from an α-amino nitrile with boron trifluoride etherate, and the addition of hexylmagnesium bromide gave a 3:1 mixture of alkylated products favoring the desired stereoisomer. Removal of the benzyl group completed the total synthesis.

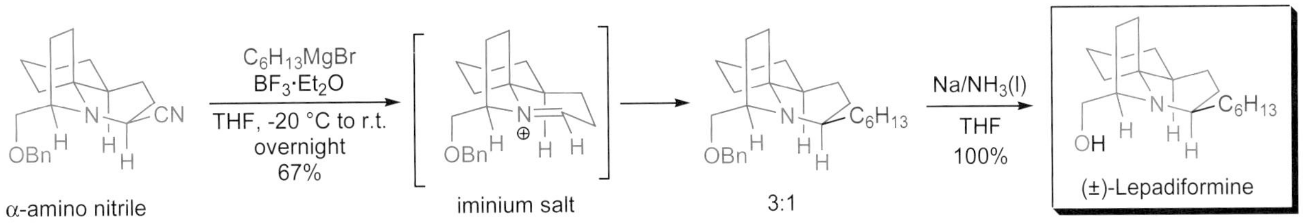

The conjugate addition of Grignard reagents to cyclic α,β-unsaturated ketones can be efficiently directed by an alkoxy substituent in the γ-position. This was the case in J.D. White's total synthesis of sesquiterpenoid polyol (±)-euonyminol in which an isopropenyl group was introduced to a bicyclic substrate *via* a chelation-controlled conjugate *Grignard addition*.[36] The γ-hydroxy unsaturated cyclic ketone was first treated with LDA and 15-crown-5 and then with isopropenylmagnesium bromide, which led to the formation of a reactive ate complex through a *Schlenk equilibrium*. From the ate complex, the isopropenyl group was intramolecularly transferred to the β-carbon of the enone.

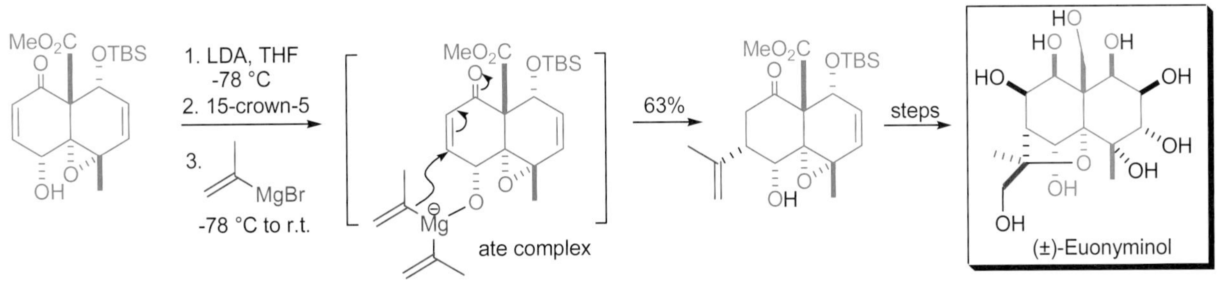

The addition of Grignard reagents to complex molecules sometimes results in side reactions that may destroy the substrate. These side reactions are often attributed to the basicity of the reagent. Therefore, more nucleophilic derivatives must be prepared. This was the case during the total synthesis of (–)-lochneridine by M.E. Kuehne et al., when the attempted conversion of a pentacyclic ketone to the corresponding tertiary alcohol with ethylmagnesium bromide failed.[37] However, the formation of an organocerium reagent by adding the Grignard reagent to anhydrous $CeCl_3$ increased its nucleophilicity, therefore the reaction afforded the desired tertiary alcohol in 73% yield with complete diastereoselection.

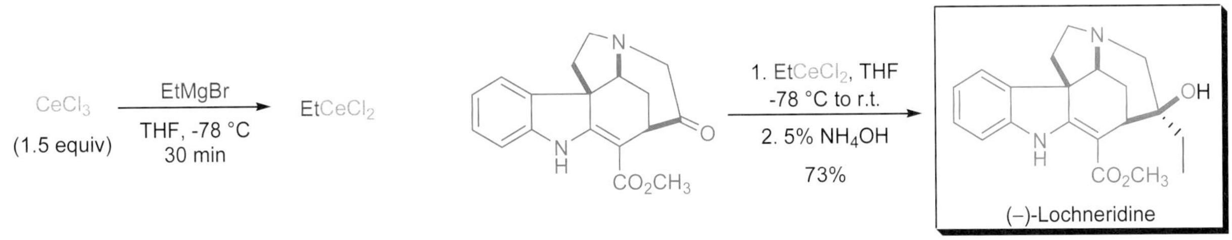

During the synthesis of natural and modified cyclotetrapeptide trapoxins, S.L. Schreiber and co-workers prepared a fully functionalized nonproteinogenic amino acid surrogate *via* the ring-opening of Cbz serine β-lactone with an organocuprate derived from a Grignard reagent.[38]

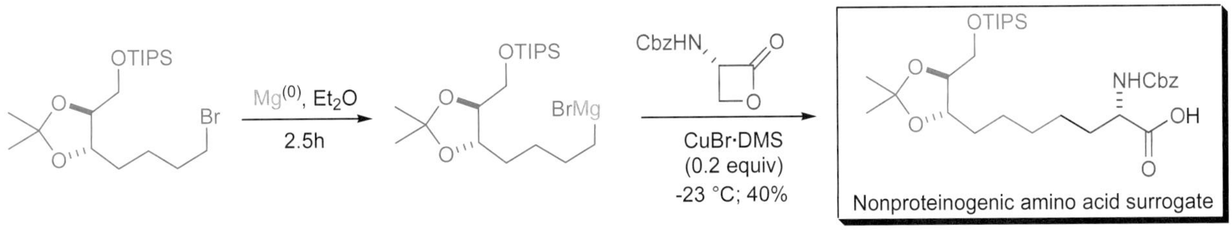

GROB FRAGMENTATION

(References are on page 594)

Importance:

[*Seminal Publications*[1-3]; *Reviews*[4-6]; *Theoretical Studies*[7-9]]

In the 1950s, C.A. Grob was the first to systematically investigate the regulated heterolytic cleavage reactions of molecules containing certain combinations of carbon and heteroatoms (e.g., B, O, N, S, P, halogens). Cleavage reactions of this type are referred to as *Grob fragmentations*, and as a result, three fragments (products) are formed. The general formula of "*a-b-c-d-X*" represents three embedded components: 1) "*a-b*" is the electrofuge, which leaves without the bonding electron pair and becomes the electrofugal fragment; 2) "*c-d*" will become the unsaturated fragment at the end of the reaction; and 3) "*X*" is the nucleofuge, which leaves with a bonding electron pair. Typical electrofugal fragments are carbonyl compounds, carbon dioxide, imonium-, carbonium- and acylium ions, olefins, and dinitrogen. Stabilization of the incipient positive charge on atom "*b*" and the inductive effect of atom "*a*" together determine how facile the formation of the electrofugal fragment is. The unsaturated fragment is usually an olefin, alkyne, imine, or nitrile while the nucleofugal fragment is often a halide, carboxylate, or sulfonate ion. The nucleofuge can have a charge (e.g., diazonium ion) before the fragmentation occurs, and that can accelerate the cleavage of the *b-c* and *d-X* bonds. The *Grob fragmentation* is often accompanied by side reactions such as substitution, elimination, or ring closure. It is most synthetically useful when it takes place in rigid bi- or polycyclic systems in a concerted and highly stereoselective fashion, so the stereochemical outcome of the product is predictable.

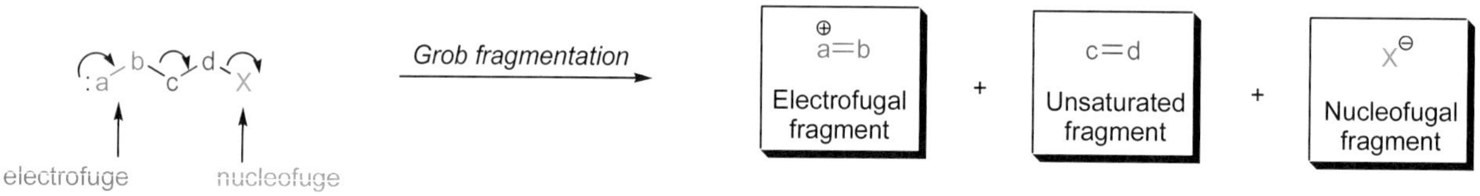

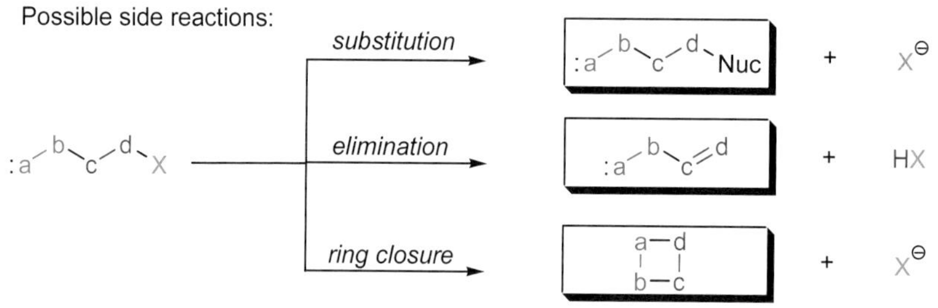

Mechanism:[10,5,11-13]

Heterolytic cleavage reactions such as the *Grob fragmentation* can take place by several different mechanisms, and the exact pathway depends on the structural, steric and electronic factors present in the substrate. There are three main mechanistic pathways: 1) one-step synchronous (concerted) cleavage in which the *a=b* and *X* fragments depart from the middle *c=d* group simultaneously; 2) two-step cleavage starting with the loss of *X* and the departure of the a=b fragment from the carbocationic intermediate; and 3) two-step cleavage starting with the departure of *a=b* and the loss of *X* from the carbanionic intermediate (this is rare). The synchronous mechanism has very strict structural and stereochemical requirements, since five atoms are involved in the transition state: all five atomic orbitals need to overlap. These requirements are best met in rigid polycyclic systems and the *Grob fragmentation* of these rigid molecules exhibits a significant increase in reaction rates compared to the non-concerted fragmentations (frangomeric effect). When the stereochemical arrangement for the concerted process cannot be achieved due to strain, then the so-called *syn fragmentation* or side reactions (e.g., elimination) take place.[6]

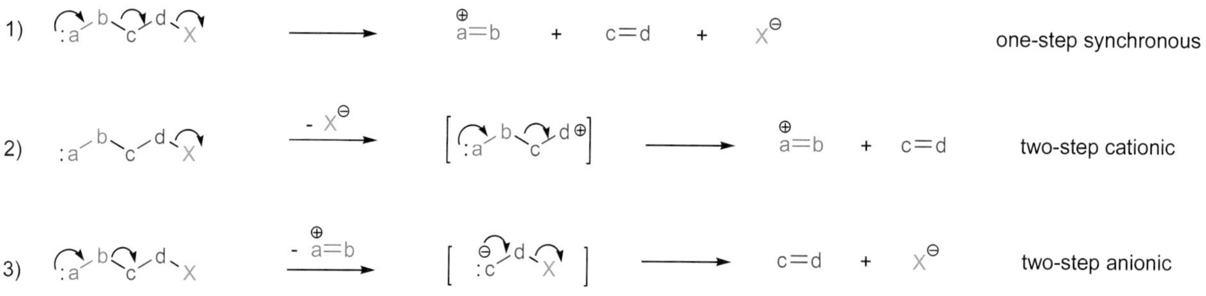

Synthetic Applications:

L.A. Paquette and co-workers accomplished the first total synthesis of the antileukemic agent jatrophatrione.[14] This natural product has a [5.9.5] fused tricyclic skeleton with a *trans*-B/C ring fusion. The key step in their approach was the *Grob fragmentation* to obtain the tricyclo[5.9.5] skeleton. The tetracyclic 1,3-diol was monomesylated on the less hindered hydroxyl group and then treated with potassium *tert*-butoxide, triggering the concerted fragmentation to afford the desired tricyclic product in almost quantitative yield.

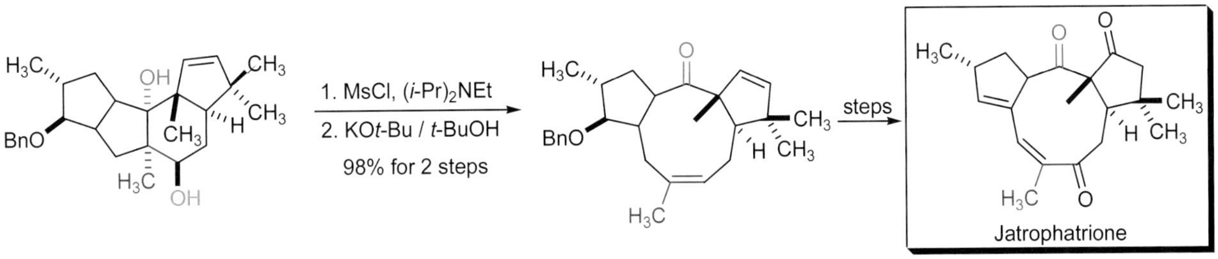

In the laboratory of J.D. Winkler, the synthesis of the carbon framework of the eleutherobin aglycon was developed using a *tandem Diels-Alder reaction* and a *Grob fragmentation* as key steps.[15] The tricyclic fragmentation precursor was subjected to potassium carbonate in DMF at 75 °C to afford the fragmentation product in 68% yield *via* a dianion intermediate that underwent a spontaneous hemiketalization.

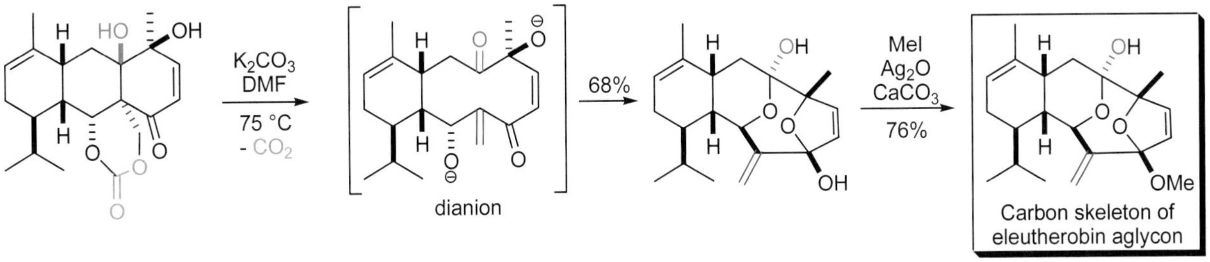

G.A. Molander et al. used samarium(II) iodide to prepare *highly functionalized stereodefined medium sized (8-, 9-, and 10-membered) carbocycles via* a domino reaction composed of a cyclization/fragmentation process.[16] The method involved the reduction of substituted keto mesylates bearing iodoalkyl, allyl, or benzyl side chains under Barbier-type conditions. The *intramolecular Barbier reaction* occurred between the iodoalkyl chain and the ketone of the cycloalkanone and generated a bicyclic alkoxide that underwent *Grob fragmentation*. The reaction proceeded in a stereoselective manner with high yields under mild conditions. The cyclization of cycloalkanediones under similar conditions was also observed, yielding functionalized polycyclic hydroxyl ketones in high yields with complete diastereoselectivity.

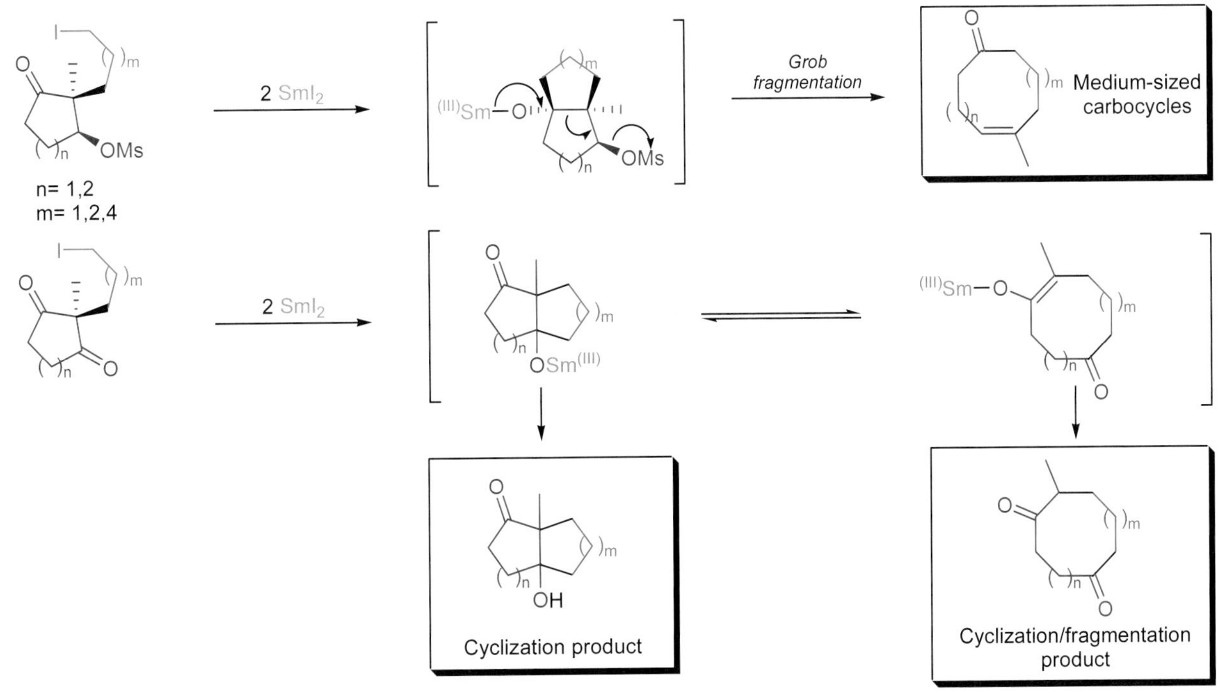

HAJOS-PARRISH REACTION

(References are on page 595)

Importance:

[*Seminal Publications*[1-4]; *Reviews*[5-12]; *Modifications & Improvements*[13-18]; *Theoretical Studies*[19-22]]

In the early 1970s, two industrial groups independently examined the *asymmetric intramolecular aldol reaction* of 2-alkyl-2-(3-oxoalkyl)-cyclopentane-1,3-diones using amino acids. Z.G. Hajos and D.R. Parrish at Hoffmann-LaRoche found that a catalytic quantity of (*S*)-(–)-proline was sufficient to furnish the cyclization of 2-methyl-2-(3-oxobutyl)-cyclopentane-1,3-dione and induce enantioselectivity.[3,4] Best results were obtained when the reaction was carried out in polar aprotic solvents such as DMF at room temperature in the presence of 3 mol% (*S*)-(–)-proline yielding the product quantitatively with 93.4% *ee*. *p*-Toluenesulfonic acid catalyzed dehydration to the corresponding bicyclic enone (*Hajos-Parrish ketone*) could be realized without the loss of optical purity. R. Wiechert and co-workers showed that the enone product could be formed directly when the cyclization was performed in the presence of (*S*)-(–)-proline (10-200 mol%) and an acid co-catalyst such as $HClO_4$.[1,2] The amino acid catalyzed *intramolecular aldol reaction* of prochiral 2-alkyl-2-(3-oxoalkyl)-cyclopentane-1,3-diones is known as the *Hajos-Parrish reaction*, but it is also referred to as the *Hajos-Parrish-Eder-Sauer-Wiechert reaction*. (*S*)-(–)-Proline catalyzed *intramolecular aldol reaction* of 2-methyl-2-(3-oxobutyl)-cyclohexane-1,3-dione leading to 8*a*-methyl-3,4,8,8*a*-tetrahydro-2*H*,7*H*-naphthalene-1,6-dione (*Wieland-Miescher ketone*) could also be realized in high yields, although the optical purity of the product was moderate (70%) and further recrystallization was required to obtain the product in optically pure form.[23] Since its invention, the *Hajos-Parrish reaction* was applied to the synthesis of several differently substituted hexahydroindene-1,5-dione-, 2,3,7,7*a*-tetrahydro-6*H*-indene-1,5-dione- and 3,4,8,8*a*-tetrahydro-2*H*,7*H*-naphthalene-1,6-dione derivatives.[1-5,16,18] The most general catalyst is (*S*)-(–)-proline, but in certain cases (*S*)-(–)-phenylalanine proved to be more efficient.[24] The reaction was also studied applying polymer bound (*S*)-(–)-proline as catalyst.[15] Precursors for the *Hajos-Parrish reaction* can be easily obtained by the *Michael addition* of cyclopentane-1,3-dione and cyclohexane-1,3-dione derivatives to α,β-unsaturated ketones.

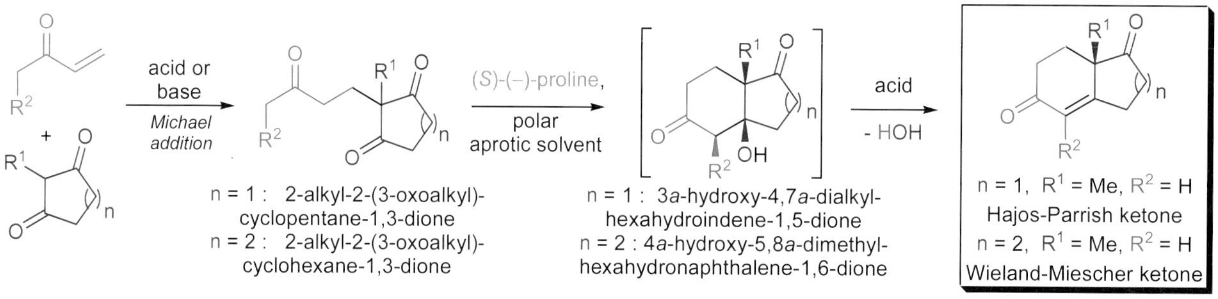

Mechanism:[4,25-28,20-22]

The originally proposed stereochemical model by Hajos and Parrish[4] was rejected by M.E. Jung[25] and A. Eschenmoser.[26] They proposed a one-proline aldolase-type mechanism involving a side chain enamine. The most widely accepted transition state model to account for the observed stereochemistry was proposed by C. Agami et al. suggesting the involvement of two (*S*)-(–)-proline molecules.[14,27-29] Recently, K.N. Houk and co-workers reexamined the mechanism of the intra- and intermolecular (*S*)-(–)-proline catalyzed *aldol reactions*. Their theoretical studies, kinetic, stereochemical and dilution experiments support a one-proline mechanism where the reaction goes through a six-membered chairlike transition state.[19-22]

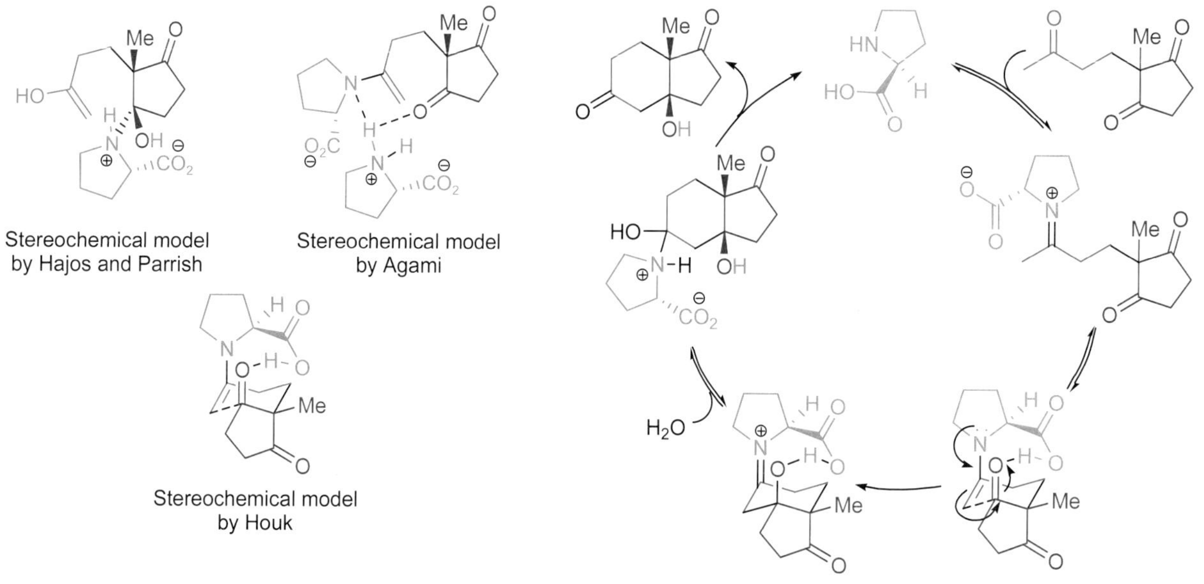

Synthetic Applications:

A short, enantioselective total synthesis of (+)-desogestrel, the most prescribed third-generation oral contraceptive, was accomplished by E.J. Corey et al.[30] They started out from a *Hajos-Parrish ketone* analogue (*S*)-(+)-7*a*-ethyl-2,3,7,7*a*-tetrahydro-6*H*-indene-1,5-dione, which was readily available by the original procedure by Hajos and Parrish.[4] The desired enone could be synthesized starting out from 2-ethylcyclopentane-1,3-dione that underwent *Michael addition* with methyl vinyl ketone. *Intramolecular aldol reaction* in the presence of 30 mol% (*S*)-(–)-proline followed by dehydration gave the product in high yield and excellent enantioselectivity. The product enone could be converted to desogestrel in 16 consecutive steps.

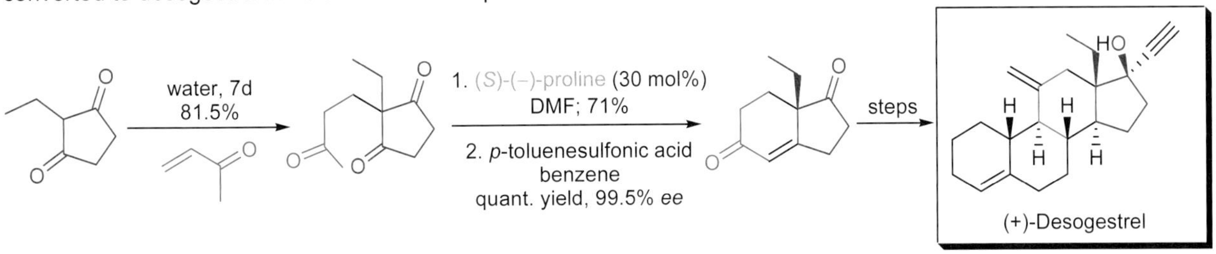

The first enantioselective total synthesis of tetracyclic sesquiterpenoid (+)-cyclomyltaylan-5α-ol, isolated from a Taiwanese liverwort, was accomplished by H. Hagiwara and co-workers.[31] They started out from *Hajos-Parrish ketone* analogue, (*S*)-(+)-4,7*a*-dimethyl-2,3,7,7*a*-tetrahydro-6*H*-indene-1,5-dione, that could be synthesized from 2-methylcyclopentane-1,3-dione and ethyl vinyl ketone in an acetic acid-catalyzed *Michael addition* followed by an *intramolecular aldol reaction*. The *intramolecular aldol reaction* was carried out in the presence of one equivalent (*S*)-(–)-phenylalanine and 0.5 equivalent D-camphorsulfonic acid. The resulting enone was recrystallized from hexane-diethyl ether to yield the product in 43% yield and 98% *ee*. Since the absolute stereochemistry of the natural product was unknown, the total synthesis also served to establish the absolute stereochemistry.

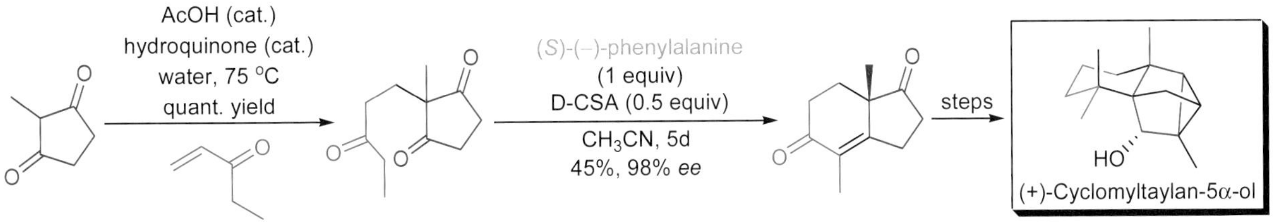

J. Wicha and co-workers reported the enantioselective synthesis of the CD side-chain portion of *ent*-vitamine D_3.[18] The key step in their approach was the amino acid mediated *asymmetric Robinson annulation* between 2-methyl-cyclopentane-1,3-dione and 1-phenylsulfanyl-but-3-en-2-one. During their optimization studies they found that the annulation is most efficient if the reaction is carried out in the presence of (S)-(–)-phenylalanine and D-camphorsulfonic acid, giving the product in 69% yield and 86.2% *ee*. The optical purity of the enone could be improved to 95.6% by recrystallization from methanol.

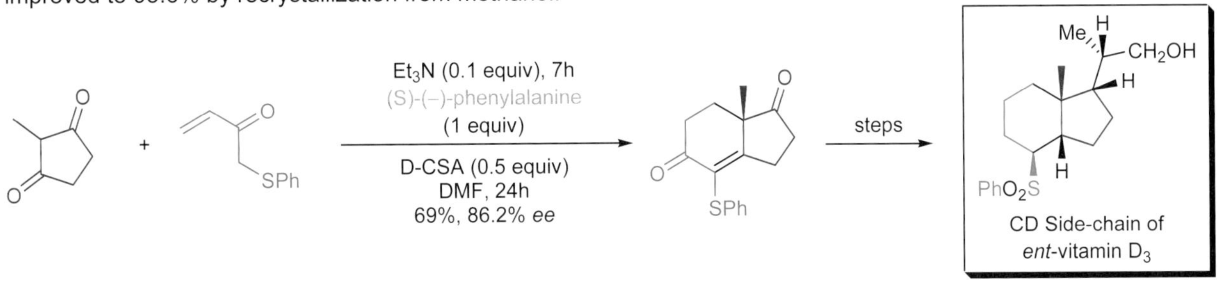

The first total synthesis of barbacenic acid, a bisnorditerpene containing five contiguous stereocenters, was achieved by A. Kanazawa et. al.[32,33] They started out from a *Wieland-Miescher ketone* analogue that could be synthesized with high yield and excellent enantioselectivity by the procedure of S. Takahashi. According to this procedure, the *Michael addition* product 2-methyl-2-(3-oxo-pentyl)-cyclohexane-1,3-dione was cyclized in the presence of (*S*)-(–)-phenylalanine and D-camphorsulfonic acid.

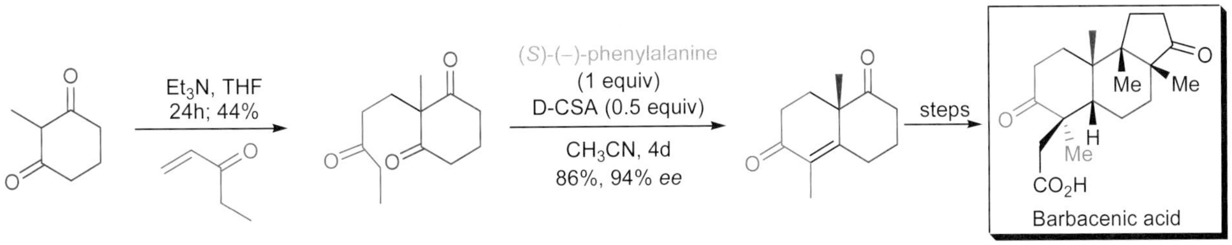

HANTZSCH DIHYDROPYRIDINE SYNTHESIS

(References are on page 595)

Importance:

[*Seminal Publication*[1]; *Reviews*[2-13]; *Modifications & Improvements*[14-22]]

In 1882, A. Hantzsch condensed two moles of ethyl acetoacetate with one mole of acetaldehyde and ammonia to obtain a fully substituted symmetrical dihydropyridine.[1] He initially assigned the structure as a 2,3-dihydropyridine, but it was later shown to be a 1,4-dihydropyridine. The one-pot condensation of a β-keto ester or a 1,3-dicarbonyl compound with an aldehyde and ammonia to prepare 1,4-dihydropyridines is known as the *Hantzsch dihydropyridine synthesis*. Frequently, the 1,4-dihydropyridine products are spontaneously oxidized to the corresponding substituted pyridines, but in the case of stable dihydropyridines, the use of an oxidizing agent [e.g., HNO_2, HNO_3, $(NH_4)_2Ce(NO_3)_6$, MnO_2, $Cu(NO_3)_2$] is necessary.[23-30] General features of the reaction are: 1) aliphatic, aromatic, heterocyclic, and α,β-unsaturated aldehydes can be used as the aldehyde component; 2) ammonia or primary amines are suitable as the amine component; 3) the dicarbonyl component is usually an acyclic or cyclic β-keto ester, β-keto aldehyde, or a 1,3-diketone; 4) the product of the reaction is a symmetrical dihydropyridine, which is formed in good or excellent yield; 5) if the C3 and C5 substituents are electron-withdrawing (e.g., acyl, nitro, sulfonyl) the dihydropyridine is stable enough to be isolated; 6) the reaction conditions can range from basic media all the way to strongly acidic solutions, and the choice of conditions needs to be optimized for the given system; 7) good yields are obtained with substrates having electron-withdrawing groups; and 8) sterically congested aldehydes generally give low yields (e.g., *o*-substituted benzaldehyde). The original procedure only affords symmetrical products, but there are several modifications that allow the preparation of unsymmerical dihydropyridines: 1) one equivalent of a β-keto ester is condensed with an aldehyde of choice to give an α,β-unsaturated carbonyl compound (alkylidene), which in turn is treated with another β-keto ester and a nitrogen source; 2) an α,β-unsaturated carbonyl compound (derived from the condensation of active methylene compounds and aldehydes) is condensed with an enamine;[31-33] and 3) in the *Knoevenagel modification* various substituted 1,5-dicarbonyl compounds can be prepared (e.g., *Michael addition* of a 1,3-dicarbonyl compound to an α,β-unsaturated carbonyl compound under basic conditions) and reacted with a nitrogen source (usually ammonium acetate-acetic acid).[34,35]

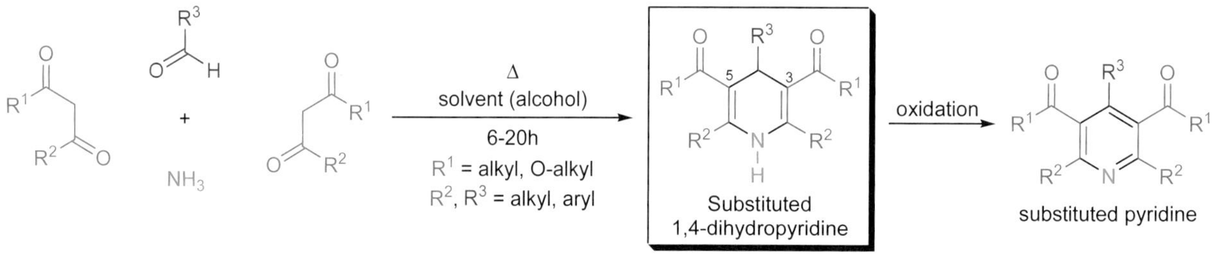

Mechanism:[36-38]

There have been many studies aiming to determine the exact mechanistic pathway of the *Hantzsch dihydropyridine synthesis*, but the ^{13}C and ^{15}N-NMR experiments conducted by A.R. Katritzky et al. were the only ones that confirmed the existence of certain intermediates.[37] All of the investigated reactions had two common intermediates: an enamine and an α,β-unsaturated carbonyl compound. The initial steps of the reaction involve a *Knoevenagel condensation* of the 1,3-dicarbonyl compound with the aldehyde to give an α,β-unsaturated carbonyl compound and a condensation of ammonia with another equivalent of the 1,3-dicarbonyl compound to give an enamine. The rate determining step is the *Michael addition* of the enamine to the α,β-unsaturated carbonyl compound. Subsequently, the addition product undergoes an intramolecular condensation of the amino and carbonyl groups to afford the desired substituted 1,4-dihydropyridine.

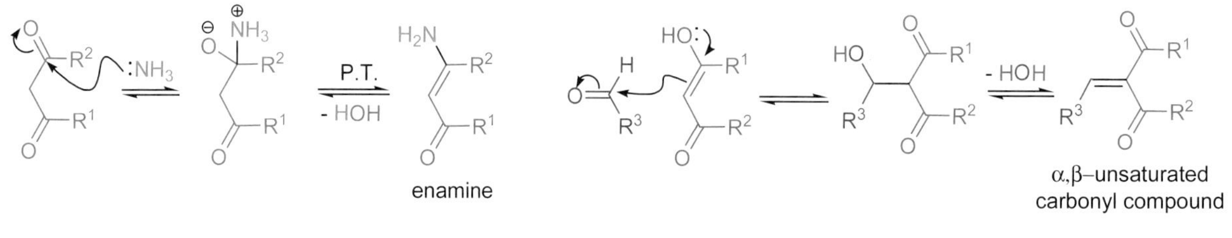

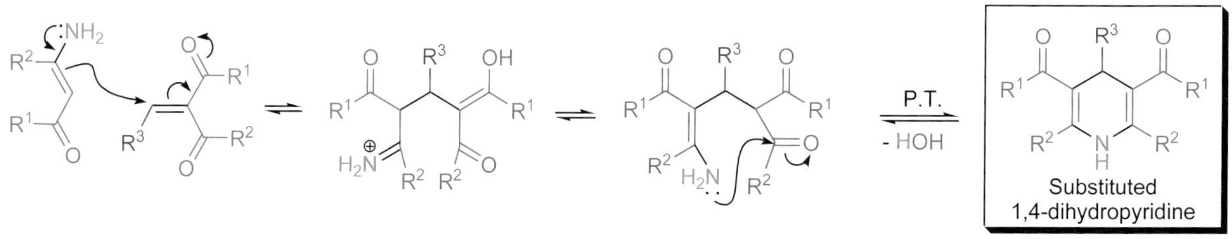

Synthetic Applications:

F. Dollé and co-workers synthesized (–)-S12968, an optically active 1,4-dihydropyridine that is a calcium channel antagonist.[39] The key step in their synthetic approach was a *modified Hantzsch dihydropyridine synthesis* and the resulting racemic mixture was separated by chiral HPLC. The starting β-keto ester was condensed with 2,3-dichlorobenzaldehyde under slightly acidic conditions to obtain the corresponding benzylidene derivative in 50% yield. Next, the second β-keto ester was heated in ethanol along with ammonium formate, which was the source of ammonia, to give the racemic 1,4-dihydropyridine. Finally, HPLC separation of the enantiomers followed by deprotection and esterification gave (–)-S12968.

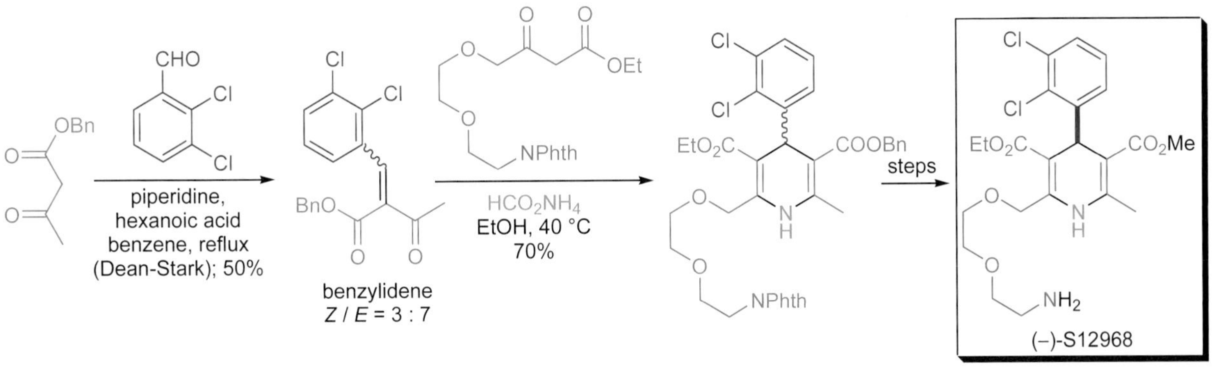

A new strategy for the synthesis of heterocyclic α-amino acids utilizing the *Hantzsch dihydropyridine synthesis* was developed in the laboratory of A. Dondoni.[40] The enantiopure oxazolidinyl keto ester was condensed with benzaldehyde and *tert*-butyl amino crotonate in the presence of molecular sieves in 2-methyl-2-propanol to give a 85% yield of diastereomeric 1,4-dihydropyridines. The acetonide protecting group was removed and the resulting amino alcohol was oxidized to the target 2-pyridyl α-alanine derivative.

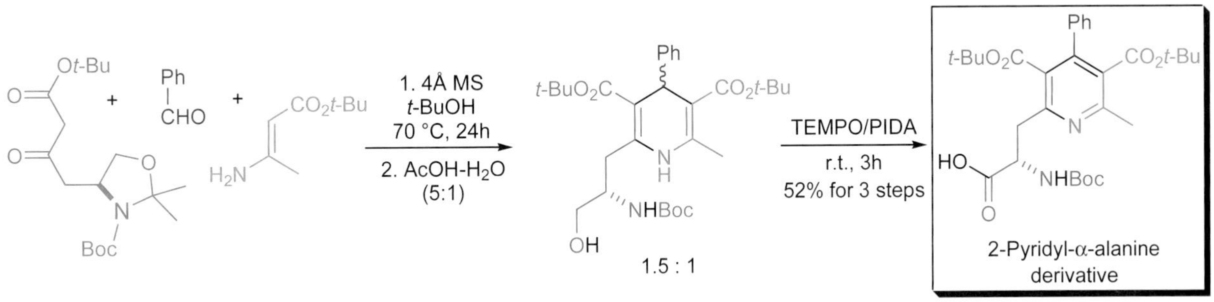

Lipophilic 1,4-dihydropyridines, such as 4-aryl-1,4-dihydropyridines, exhibit significant calcium channel antagonist activity. N.R. Natale et al. have synthesized a series of 4-isoxazolyl-1,4-dihydropyridines bearing lipophilic side chains at the C5 position of the isoxazole ring.[41] The *Hantzsch synthesis* was carried out in an aerosol dispersion tube at 110 °C in ethanol in the presence of 2 equivalents of ethyl acetoacetate and aqueous ammonia solution.

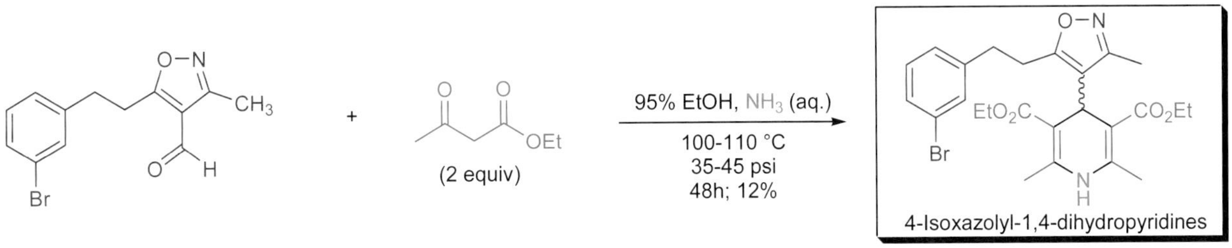

M. Baley reported the first synthesis of an unsymmetrical 2,2'-6'2''-terpyridine containing two carboxylic acids using the *Hantzsch dihydropyridine synthesis* followed by an oxidation.[42] The furan ring served as a latent carboxylic acid functional group.

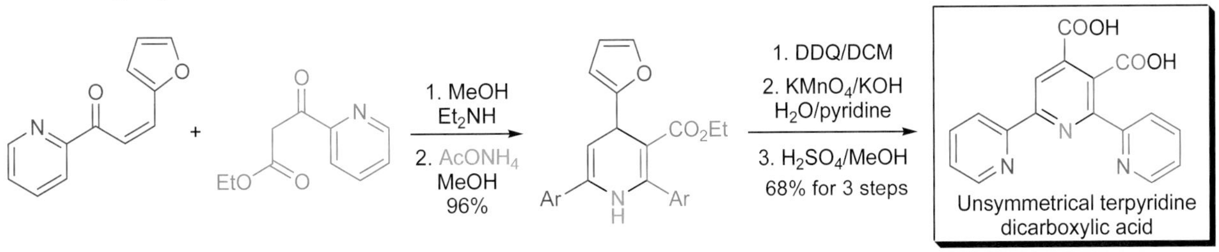

HECK REACTION

(References are on page 596)

Importance:

[*Seminal Publications*[1-4]; *Reviews*[5-39]; *Modifications & Improvements*[40-47]; *Theoretical Studies*[48-54]]

In the early 1970s, T. Mizoroki and R.F. Heck independently discovered that aryl, benzyl and styryl halides react with olefinic compounds at elevated temperatures in the presence of a hindered amine base and catalytic amount of $Pd^{(0)}$ to form aryl-, benzyl-, and styryl-substituted olefins.[1-3] Today, the palladium-catalyzed arylation or alkenylation of olefins is referred to as the *Heck reaction*. Since its discovery, the *Heck reaction* has become one of the most widely used catalytic carbon-carbon bond forming tools in organic synthesis. The general features of the reaction are: 1) it is best applied for the preparation of disubstituted olefins from monosubstituted ones; 2) the electronic nature of the substituents on the olefin only has limited influence on the outcome of the reaction; it can be either electron-donating or electron-withdrawing but usually the electron poor olefins give higher yields; 3) the reaction conditions tolerate a wide range of functional groups on the olefin component: esters, ethers, carboxylic acids, nitriles, phenols, dienes, etc., are all well-suited for the coupling, but allylic alcohols tend to rearrange; 4) the reaction rate is strongly influenced by the degree of substitution of the olefin and usually the more substituted olefin undergoes a slower *Heck reaction*; 5) unsymmetrical olefins (e.g., terminal alkenes) predominantly undergo substitution at the least substituted olefinic carbon; 6) the nature of the X group on the aryl or vinyl component is very important and the reaction rates change in the following order: I > Br ~ OTf >> Cl; 7) the R^1 group in most cases is aryl, heteroaryl, alkenyl, benzyl, and rarely alkyl (provided that the alkyl group possesses no hydrogen atoms in the β-position), and these groups can be either electron-donating or electron-withdrawing; 8) the active palladium catalyst is generated *in situ* from suitable precatalysts (e.g., $Pd(OAc)_2$, $Pd(PPh_3)_4$) and the reaction is usually conducted in the presence of monodentate or bidentate phosphine ligands and a base; 9) the reaction is not sensitive to water, and the solvents need not be thoroughly deoxygenated; and 10) the *Heck reaction* is stereospecific as the migratory insertion of the palladium complex into the olefin and the β-hydride elimination both proceed with *syn* stereochemistry. There are a couple of drawbacks of the *Heck reaction*: 1) the substrates cannot have hydrogen atoms on their β-carbons, because their corresponding organopalladium derivatives tend to undergo rapid β-hydride elimination to give olefins; and 2) aryl chlorides are not always good substrates because they react very slowly. Several modifications were introduced during the past decade: 1) asymmetric versions;[23,36] 2) generation of quaternary stereocenters in the *intramolecular Heck reaction*;[17,55,34] 3) using water as the solvent with water-soluble catalysts;[56,57,47] and 4) heterogeneous palladium on carbon catalysis.[40]

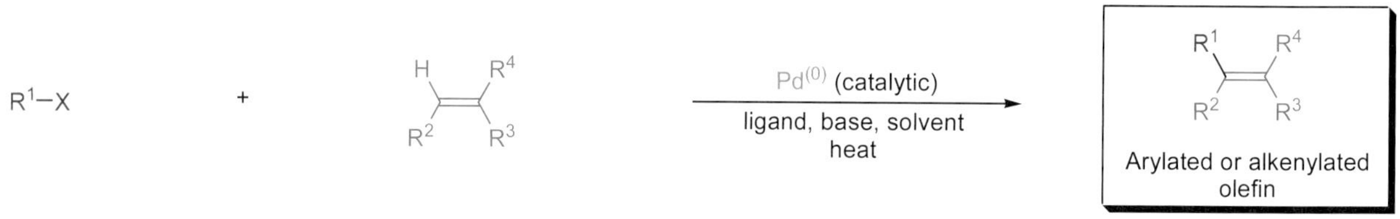

R^1 = aryl, benzyl, vinyl (alkenyl), alkyl (no β hydrogen); R^2, R^3, R^4 = alkyl, aryl, alkenyl; X = Cl, Br, I, OTf, OTs, N_2^+;
<u>ligand</u> = trialkylphosphines, triarylphosphines, chiral phosphines; <u>base</u> = 2° or 3° amine, KOAc, NaOAc, $NaHCO_3$

Mechanism: [58,59,21,22,51,53]

The mechanism of the *Heck reaction* is not fully understood and the exact mechanistic pathway appears to vary subtly with changing reaction conditions. The scheme shows a simplified sequence of events beginning with the generation of the active $Pd^{(0)}$ catalyst. The rate-determining step is the *oxidative addition* of $Pd^{(0)}$ into the C-X bond. To account for various experimental observations, refined and more detailed catalytic cycles passing through anionic, cationic or neutral active species have been proposed.[21,36]

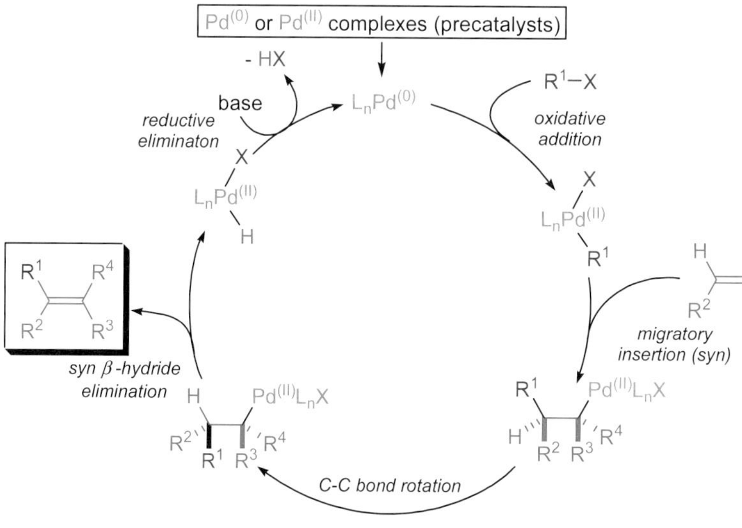

Synthetic Applications:

Ecteinascidin 743 is a potent antitumor agent that was isolated from a marine tunicate. T. Fukuyama et al. applied the *intramolecular Heck reaction* as the key step in the assembly of the central bicyclo[3.3.1] ring system.[60] Toward this end, the cyclic enamide precursor was exposed to 5 mol% of palladium catalyst and 20 mol% of a phosphine ligand in refluxing acetonitrile to afford the desired tricyclic intermediate in 83% isolated yield.

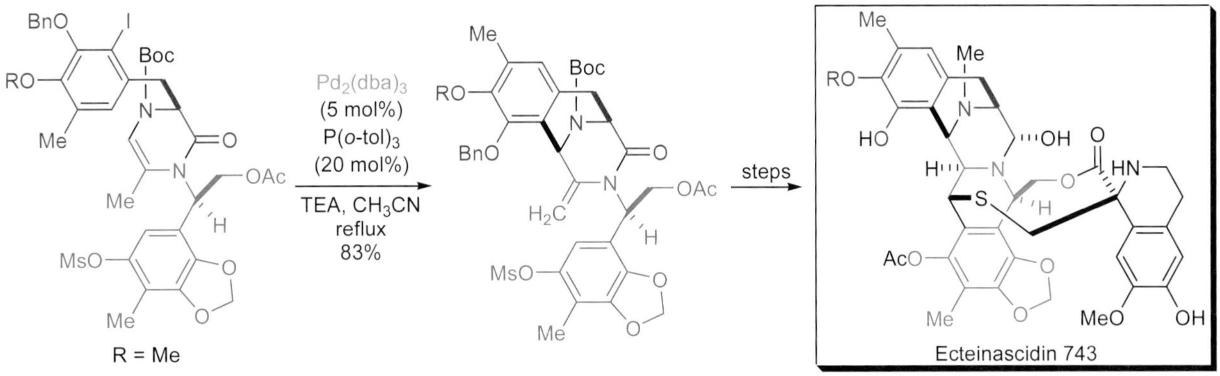

The introduction of the C3 quaternary center was the major challenge during the total synthesis of asperazine by L.E. Overman and co-workers.[61] To address this synthetic problem, a diastereoselective *intramolecular Heck reaction* was used. The α,β-unsaturated amide precursor was efficiently coupled with the tethered aryl iodide moiety in the presence of 20 mol% $Pd_2(dba)_3{\cdot}CHCl_3$ and one equivalent of $(2\text{-furyl})_3P$ ligand. The desired hexacyclic product was obtained as a single diastereomer in 66% yield.

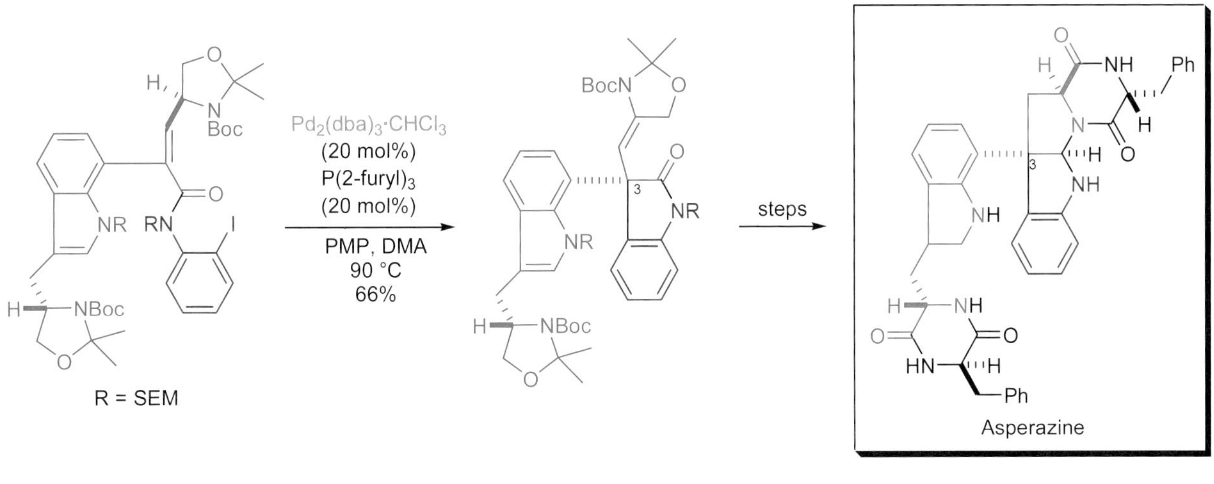

The total synthesis of the potent anticancer macrocyclic natural product lasiodiplodin was achieved in the laboratory of A. Fürstner.[62] The key macrocyclization step was carried out by the *alkene metathesis* of a styrene derivative, which was prepared in excellent yield *via* an *intermolecular Heck reaction* between an aryl triflate and high-pressure ethylene gas.

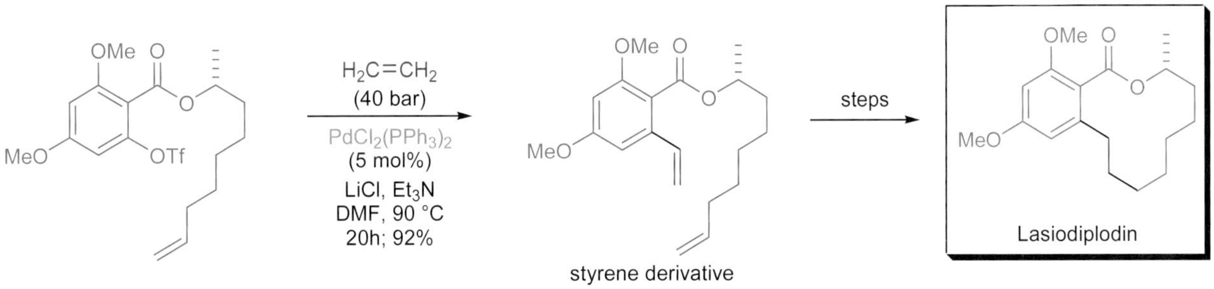

HEINE REACTION

(References are on page 597)

Importance:

[*Seminal Publications*[1-4]; *Reviews*[5-9]]

In 1959, H.W. Heine described the isomerization of 1-aroylaziridines to the corresponding 2-aryl-2-oxazolines in the presence of excess sodium iodide in acetone at room temperature or at reflux.[1] The isomerizations took place in almost quantitative yields. The *intramolecular ring expansion* of substituted *N*-acylaziridines by nucleophilic reagents (e.g., NaI or KSCN) to the corresponding substituted oxazolines is known as the *Heine reaction*. The isomerization of various substituted aziridines to oxazolines under acidic and thermal conditions are very well known, but the *Heine reaction* is the only reaction that induces these isomerizations under mild and neutral conditions.[5,6,9] The main features of the *Heine reaction* are: 1) iodide ion and thiocyanate ion were found to be the only nucleophiles to induce isomerizations;[2] 2) the course of the reaction is greatly influenced by the choice of solvent and acetone, acetonitrile, and 2-propanol give the best results;[2] 3) the *Heine reaction* is stereospecific; when non-racemic aziridines are used as substrates, the stereochemical outcome is a net retention of configuration; 4) 3-aryl substituted *N*-acyl aziridinecarboxylic esters (R^2 = aryl) or aryl disubstituted C_2-symmetric *N*-acyl aziridines are the best substrates, since it is essential to open the aziridine ring regiospecifically; 5) substrates for which the aziridine ring-opening is not regiospecific give rise to a mixture of products; and 6) aziridines that are substituted at C1 with electron-withdrawing groups often undergo dimerization when treated with sodium iodide.[5,6] The ring expansion of *N*-substituted aziridines (X = O, S, N) with iodide or thiocyanate ions is quite general and can lead to other five-membered heterocycles such as thiazolines, imidazolines and triazolines.

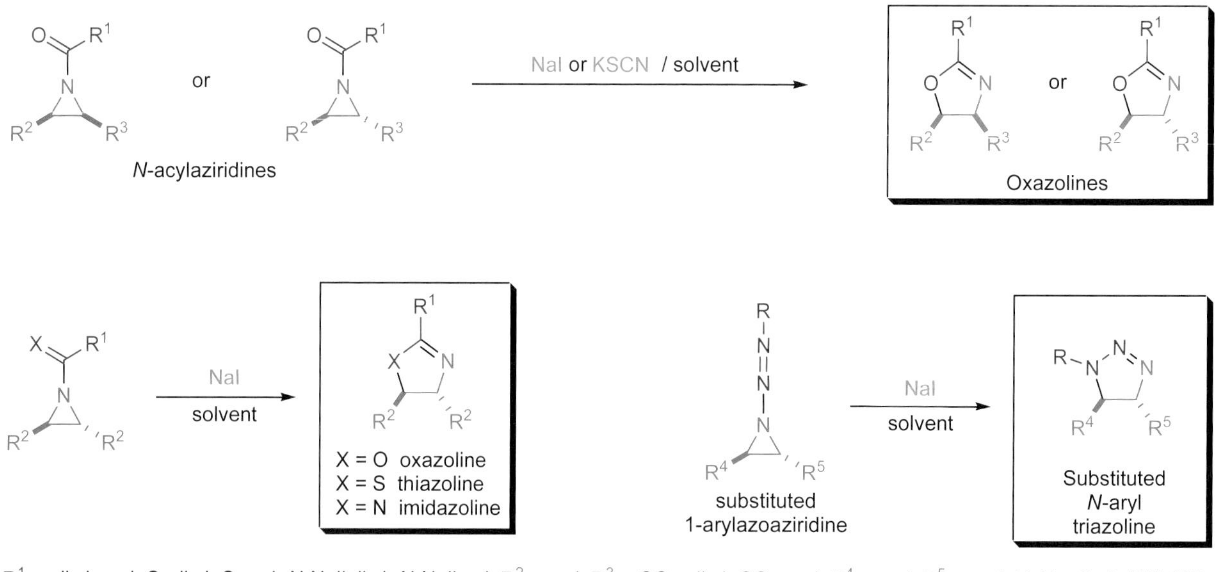

R^1 = alkyl, aryl, O-alkyl, O-aryl, *N,N*-dialkyl, *N,N*-diaryl; R^2 = aryl; R^3 = CO_2-alkyl, CO_2-aryl; R^4 = aryl; R^5 = aryl, H; X = O, S, NH, NR; solvent = 2-propanol, acetone, acetonitrile

Mechanism: [5,6,9]

The first step of the *Heine reaction* is the regiospecific S_N2 attack of the iodide ion at the C3 carbon resulting in the ring-opening of the aziridine and the inversion of stereochemistry at C3. Next, the secondary alkyl iodide is attacked by the negatively charged oxygen atom in an S_N2 reaction causing the stereochemistry to invert once again at C3. Since two consecutive inversions (double inversion) take place at C3, the stereochemical outcome of the *Heine reaction* is a net retention.

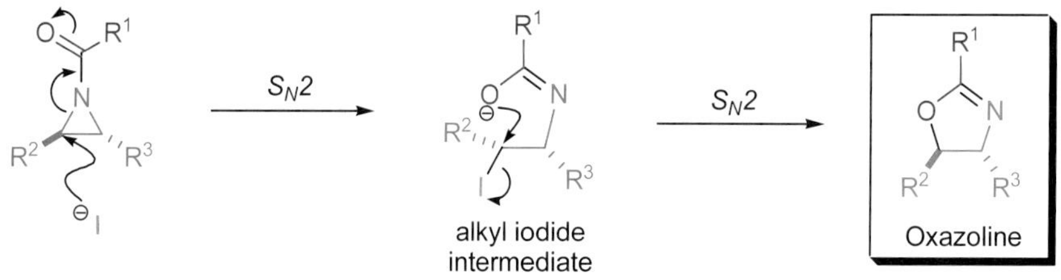

Synthetic Applications:

The synthesis of ferrocenyl oxazolines was accomplished in the laboratory of B. Zwanenburg using the *Heine reaction* as the key step to form the oxazoline rings.[10] *N*-Ferrocenoyl-aziridine-2-carboxylic esters were prepared by the acylation of optically active aziridines with either ferrocenecarbonyl chloride or ferrocene-1,1'-dicarbonyl dichloride and treated with catalytic amounts of NaI in boiling acetonitrile. The ring expansions proceeded in good yields affording the expected ferrocenyl oxazolines and ferrocenyl *bis*-oxazolines. The ester functionality provided a convenient handle for further modifications of the ligands by the addition of a Grignard reagent to form the corresponding ferrocenyl oxazoline carbinols.

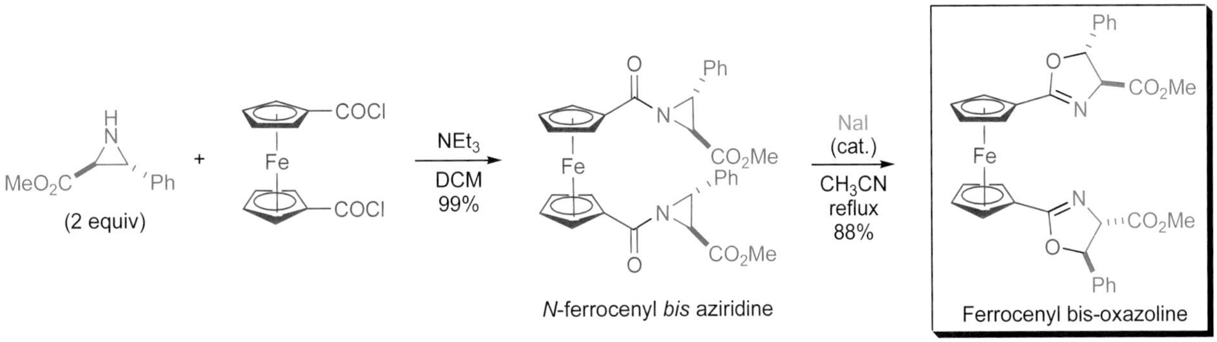

J.M.J. Tronchet and co-workers prepared functionalized octenopyranoses to investigate the synthetic utility of glycosylaziridine derivatives.[11] The authors found that by treating bromoenoses with methanolic ammonia at room temperature, the corresponding disubstituted glycosylaziridines were formed with an *E*/*Z* ratio of 16:5. The aziridines were acylated, and the resulting *N*-acyl glycosylaziridines were subjected to a nucleophilic ring-expansion to afford oxazolines in excellent yield. As expected, the overall stereochemical outcome was a net retention of configuration.

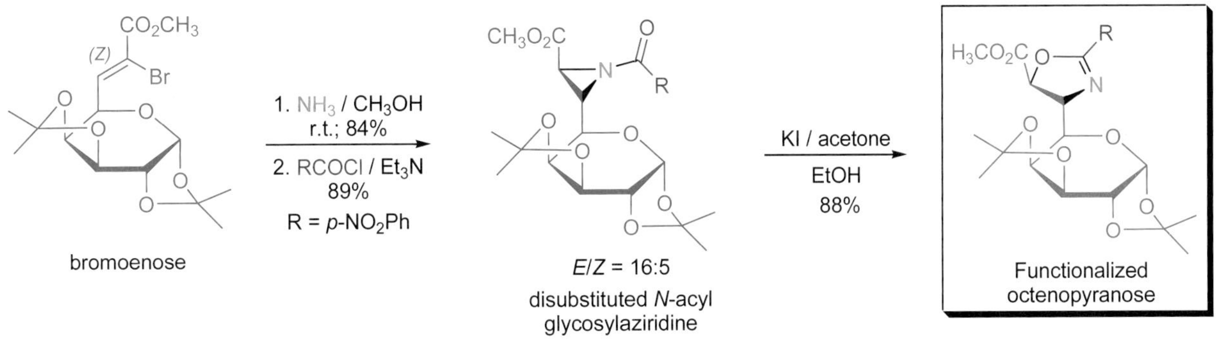

The synthesis of proline containing tripeptides constrained with phenylalanine-like aziridine and dehydrophenylalanine residues was accomplished in the laboratory of J. Iqbal.[12] These tripeptides show β-turn structure in solution and are good models for studying the mechanism of HIV protease. The aziridine rings in these tripeptides were stereoselectively transformed *via* the *Heine reaction* in two steps to the corresponding dehydrophenylalanine containing tripeptides, which also prefer to form β-turn structures in solution.

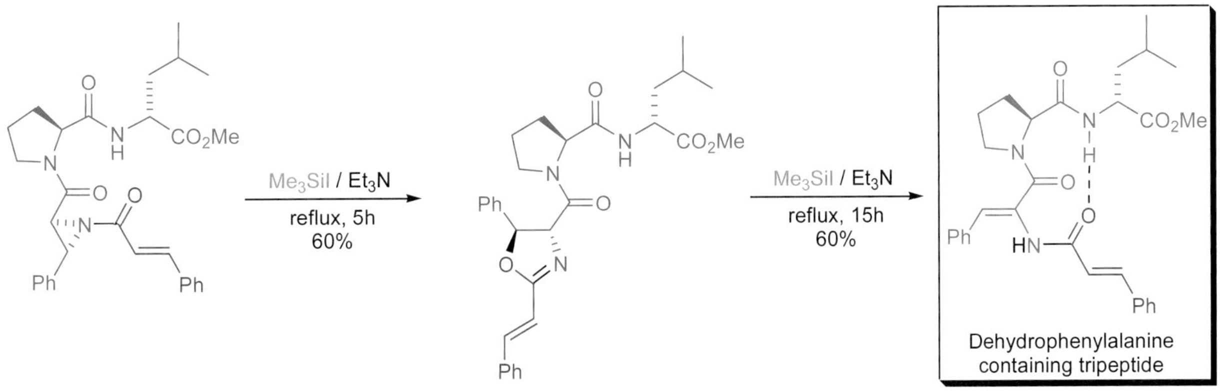

HELL-VOLHARD-ZELINSKY REACTION

(References are on page 598)

Importance:

[*Seminal Publications*[1-3]; *Reviews*[4-6]; *Modifications & Improvements*[7-11]]

The preparation of α-halo carboxylic acids by treating the corresponding carboxylic acid with elemental halogen (Cl_2 or Br_2) at elevated temperatures in the presence of catalytic amounts of red phosphorous (P) or phosphorous trihalide (PCl_3 or PBr_3) is known as the *Hell-Volhard-Zelinsky reaction* (*HVZ reaction*). The reaction was first described by C. Hell[1] and was slightly modified by J. Volhard[2] and N. Zelinsky[3] a few years later. The initial product of the *HVZ reaction* is an α-halo acyl halide, which usually is hydrolyzed to the corresponding α-halo acid during the aqueous work-up. However, when the work-up is conducted in the presence of nucleophiles such as alcohols, thiols, and amines, the corresponding α-halo esters, thioesters, and amides are formed, respectively. General features of the *HVZ reaction* are: 1) reaction conditions are relatively harsh, involving high temperatures (usually above 100 °C) and extended reaction times; 2) usually less than one equivalent of P or PX_3 catalyst is needed; 3) certain activated carboxylic acids and acid derivatives (e.g. anhydrides, acyl halides, 1,3-diesters) that are readily enolized can be halogenated in the absence of a catalyst; 4) α-bromination of substrates with long alkyl chains is completely selective; however, α-chlorination competes with random free radical chlorination processes so a mixture of mono- and polychlorinated products are obtained;[12,13] 5) attempts to bring about the fluorination or iodination of carboxylic acids under HVZ conditions have not been successful (however, there are other means of introducing these elements directly into carboxylic acids);[14] and 6) conducting the reaction at too high a temperature may result in the elimination of hydrogen halide from the product resulting in the formation of α,β-unsaturated carboxylic acids.[12] To improve the low selectivity of chlorination, certain modifications were introduced: 1) passing chlorine gas through the neat aliphatic acid (chains are no longer than C_8) at 140 °C in the presence of a strong acid catalyst and a free radical inhibitor;[7,8] 2) using TCNQ as the radical initiator gives monochlorinated products of acids of any chain length;[9] and 3) treatment of acylphosphonates with SO_2Cl_2 and subsequent hydrolysis of the α-chloro acylphosphonates to the corresponding α-halo acids.[10,11]

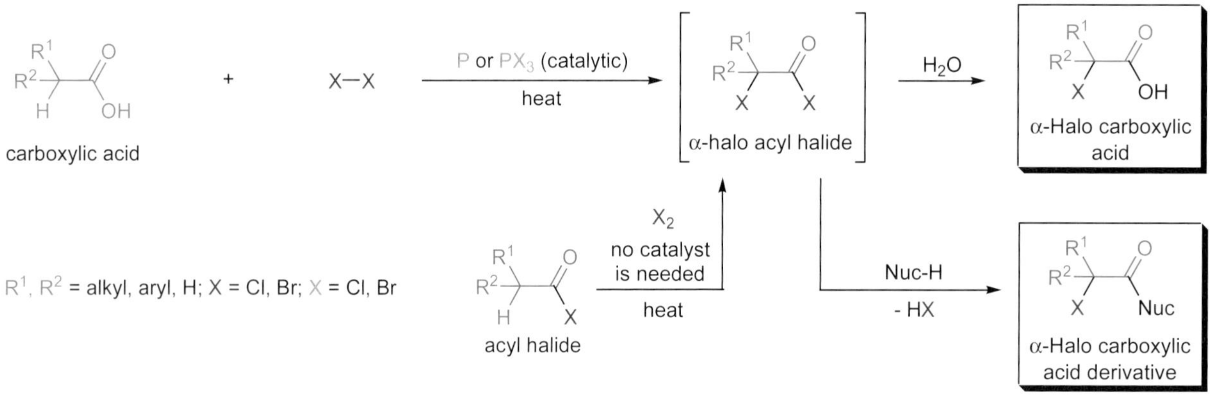

Mechanism:[15-17,4,18,19]

The first part of the mechanism includes the conversion of the carboxylic acid functionality to the acyl halide by the phosphorous trihalide. The acyl halide easily tautomerizes to the corresponding enol in the presence of a catalytic amount of acid.[15,4] The halogen subsequently reacts with the enol to afford the α-halo acyl halide, accompanied by the loss of a hydrogen halide. The halogen atoms in the PX_3 catalyst/reagent are not incorporated in the α-position of the acid.

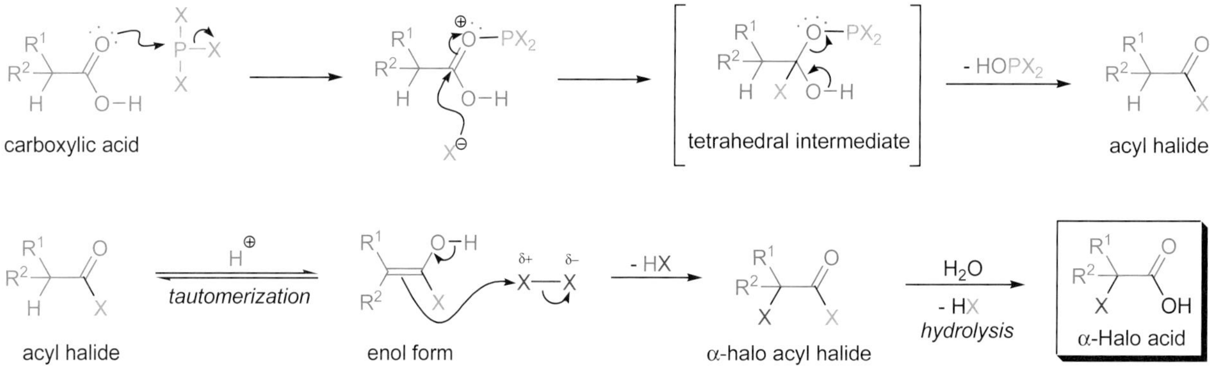

Synthetic Applications:

A convenient one-pot procedure for the preparation of α-bromo thioesters from carboxylic acids based on the *HVZ reaction* was developed by H.-J. Liu and co-workers.[20] The neat carboxylic acid was mixed with 0.4 equivalents of PBr_3, the resulting mixture was heated to 100-120 °C in an oil bath and 1.2 equivalents of liquid bromine was added in 1.5h. In the same flask, now containing the α-bromo acyl bromide, the solution of the thiol in dichloromethane was added to give the desired α-bromo thioesters in high yield.

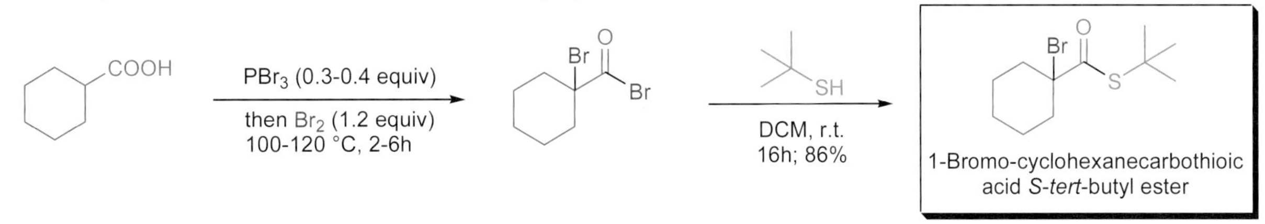

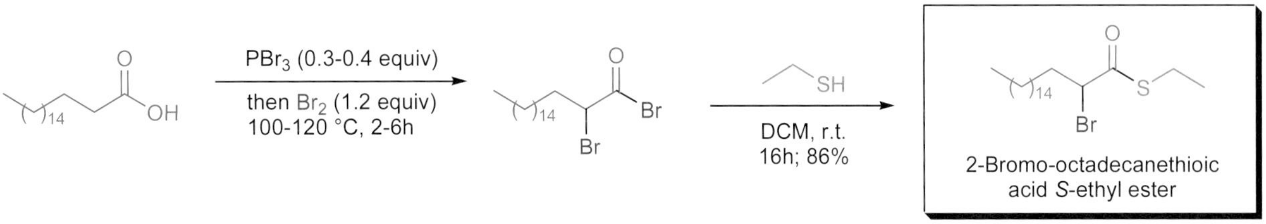

The preparation of C_2-symmetric 2,5-disubstituted pyrrolidines (utilized as chiral auxiliaries) often calls for *meso*-2,5-dibromoadipic esters as starting materials. An improvement in the synthesis of the *meso* stereoisomer was published by T. O'Neill and co-workers.[21] The authors began with the α-bromination of adipoyl chloride followed by esterification with ethanol to obtain a complex mixture of dibromo adipates (racemic + *meso*) in quantitative yield. The racemic and *meso*-dibromoadipates have very different crystalline properties, and these stereoisomers were found to be in equilibrium in an alcohol solution. Crystallizing the higher melting *meso* isomer and removing it from the equilibrium caused the remaining racemic mixture to convert to the *meso* isomer by shifting the equilibrium to the right, according to *Le Chatelier*'s *principle*.

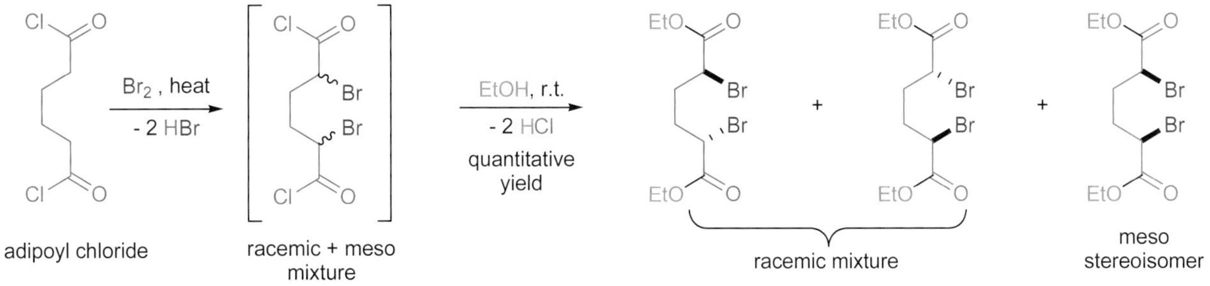

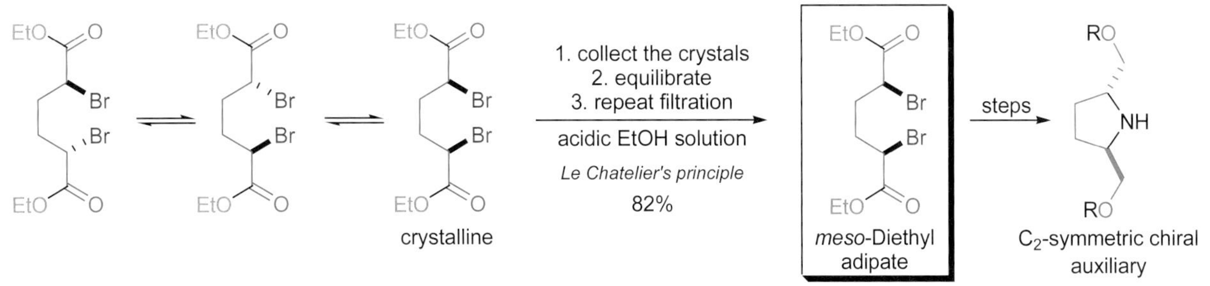

In order to determine the structure of the photochemical rearrangement product of carvone camphor in methanol, and to prove its structure, the research team of T. Gibson subjected the bicyclic carboxylic acid product to a degradation sequence, which commenced with the *HVZ reaction*, followed by dehydrohalogenation, dihydroxylation and glycol cleavage.[22]

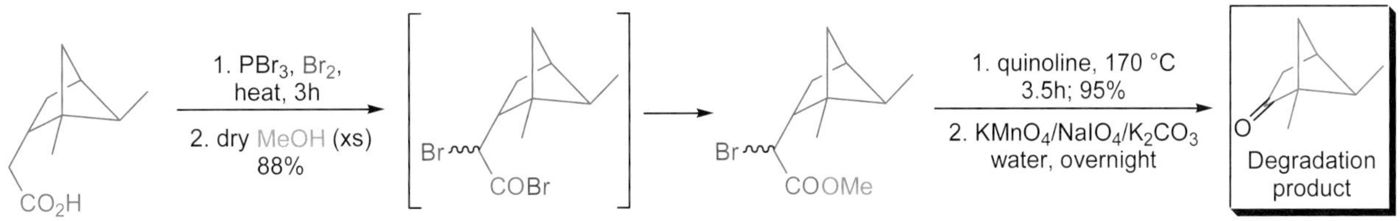

HENRY REACTION

(References are on page 598)

Importance:

[*Seminal Publications*[1,2]; *Reviews*[3-18]; *Modifications & Improvements*[19-38]; *Theoretical Studies*[39]]

In 1895, L. Henry discovered that nitroalkanes were easily combined with aldehydes and ketones to give β-nitro alcohols in the presence of a base.[1,2] Since its discovery, the *aldol condensation* between nitroalkanes and carbonyl compounds (*nitro-aldol reaction*) has become a significant tool in the formation of C-C bonds and is referred to as the *Henry reaction*. The β-nitro alcohols are easily converted to other useful synthetic intermediates: 1) upon dehydration, nitroalkenes are formed that may be used as: a) dienes and dienophiles;[40-42] b) Michael acceptors;[43] or c) masked ketones (since the *Nef reaction* converts them to the corresponding ketones); 2) oxidation of the secondary alcohol functionality affords α-nitro ketones; 3) reduction of the nitro group gives β-amino alcohols; and 4) radical denitration affords secondary alcohols. General features of the *Henry reaction* are: 1) only a catalytic amount of base is necessary; 2) both ionic and nonionic bases may be used such as alkali metal hydroxides, alkoxides, carbonates, sources of fluoride ion (e.g., TBAF,[44] KF,[45] Al_2O_3-supported KF[46]), solid supported bases,[47] rare earth metal salts,[48] transition metal complexes[31,33,34] and nonionic organic nitrogen bases (e.g., amines,[49] TMG,[50] DBU,[51] DBN,[52] PAP[27]); 3) the solvents and bases do not have significant influence on the outcome of the reaction; 4) the steric properties of the reactants play an important role: hindered substrates (usually ketones) react slowly and side reactions often occur; 5) usually the β-nitro alcohols are formed as a mixture of diastereomers (*syn* and *anti*) but by modification of the reaction conditions high levels of diastereoselectivity can be achieved;[6,17] and 6) the stereocenter to which the nitro group is attached to is easy to epimerize. The *Henry reaction* is often accompanied by side reactions: 1) the β-nitro alcohols undergo dehydration, especially when aromatic aldehydes are used as substrates; however, by carefully chosen conditions this can be supressed; 2) with sterically hindered carbonyl compounds, a base-catalyzed self-condensation or *Cannizzaro reaction* may take place; and 3) the *retro-Henry reaction* may prevent the reaction from going to completion. Several modifications have been developed: 1) unreactive alkyl nitro compounds are converted to their corresponding dianions which react faster with carbonyl compounds;[19,20] 2) reactions of ketones are accelerated by using PAP as the base;[27] 3) high-pressure and solvent-free conditions improve chemo- and regioselectivity; 4) aldehydes react with α,α-doubly deprotonated nitroalkanes to give nitronate alkoxides that afford mainly *syn*-nitro alcohols upon kinetic protonation;[6] 5) nitronate anions on which the alcohol oxygen atom is silyl-protected give predominantly *anti*-β-nitro alcohols upon kinetic protonation;[6] 6) nitronate anions in which one oxygen atom of the nitro group is silyl-protected give mainly anti-β-nitro alcohols when reacted with aldehydes in the presence of catalytic amounts of fluoride ion;[6] 7) in the presence of chiral catalysts the *asymmetric Henry reaction* can be realized;[13,15,17,18,34] and 8) when imines are used instead of carbonyl compounds as substrates, the *aza-Henry reaction* takes place to afford nitroamines; upon the reduction of nitroamines, vicinal diamines are obtained.[28,37]

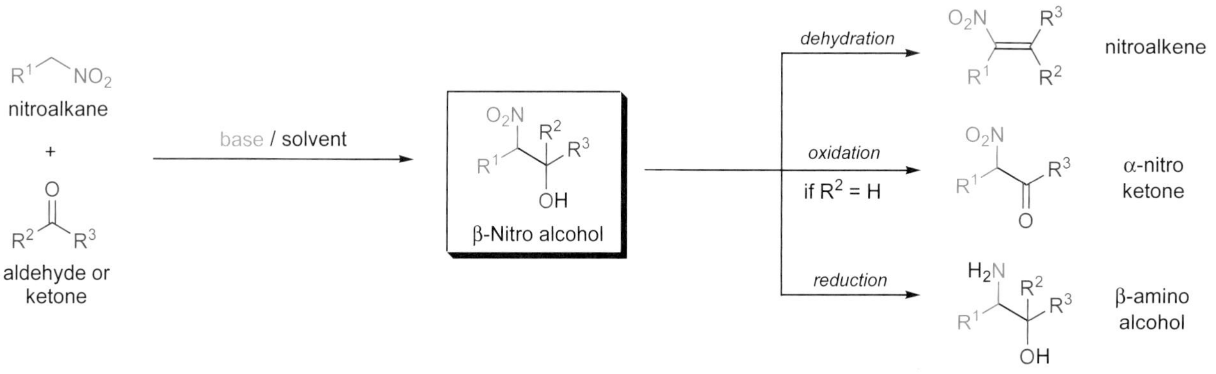

R^1 = alkyl, aryl, CO_2R, alkenyl; R^2, R^3 = alkyl, aryl, H; base = NR_3, DBU, DBN, PAP, TMG, KF, TBAF, Al_2O_3, $La_3(OR)_9$, NaOH, NaOR, amberlyst A-21, etc.

Mechanism: [53,51]

All the steps in the *Henry reaction* are completely reversible. The first step of the mechanism is the deprotonation of the nitroalkane by the base at the α-position to form the corresponding resonance stabilized anion. Next, an *aldol reaction* (*C*-alkylation of the nitroalkane) takes place with the carbonyl compound to form diastereomeric β-nitro alkoxides. Finally the β-nitro alkoxides are protonated to give the expected β-nitro alcohols.

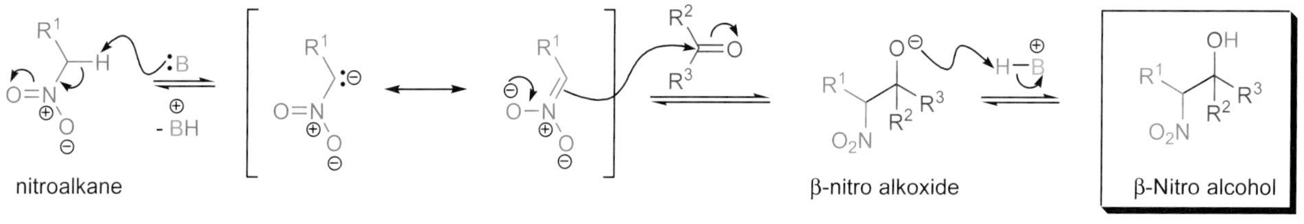

Synthetic Applications:

R.J. Estévez and co-workers utilized the *intramolecular Henry reaction* in their synthetic strategy to convert nitroheptofuranoses into deoxyhydroxymethylinositols.[54] The starting nitroheptofuranoses were prepared as a mixture of diastereomers from a D-glucose derivative and 2-nitroethanol using the *intermolecular Henry reaction*. The key *intramolecular Henry reaction* was brought about by treating this diastereomeric mixture with 2% aqueous sodium bicarbonate solution to afford an enantiomerically pure six-membered carbocycle. Removal of the nitro group and cleavage of the protecting groups gave the desired 1D-3-deoxy-3-hydroxymethyl-*myo*-inositol.

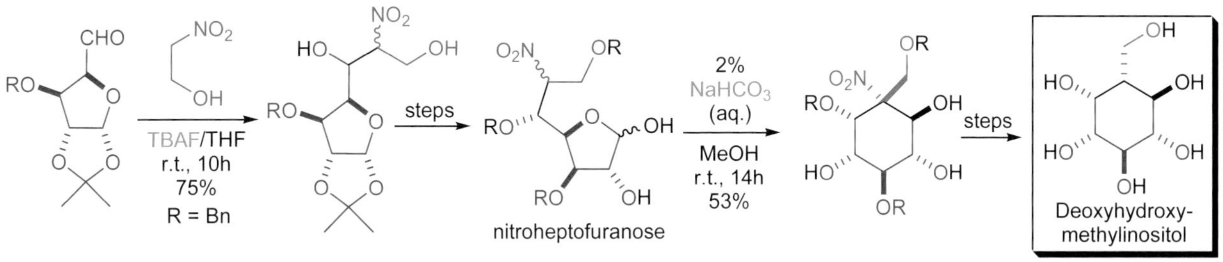

The first total synthesis of the 14-membered para ansa cyclopeptide alkaloid (–)-nummularine F was accomplished in the laboratory of M.M. Joullié.[55] The *N*3 nitrogen atom was introduced by using the *Henry reaction* between the 4-formylphenoxy group and the anion of nitromethane, followed by reduction of the nitro group to the corresponding amine. The epimeric benzyl alcohols did not pose a problem since they were dehydrated at the end of the synthetic sequence to give the C1-C2 double bond.

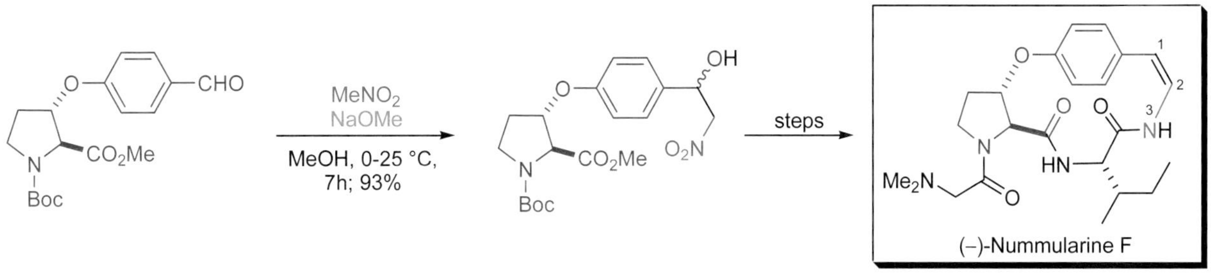

The bone collagen cross-link (+)-deoxypyrrololine has potential clinical utility in the diagnosis of osteoporosis and other metabolic bone diseases. Intrigued by its novel structure and its promise to allow the early discovery of various bone diseases, the research team of M. Adamczyk developed a convergent total synthesis for this 1,3,4-trisubstituted pyrrole amino acid.[56] The key step of the synthesis was the union of the nitroalkane and aldehyde fragments to obtain a diastereomeric mixture of the expected β-nitro alcohol in good yield. This new functionality served as a handle to install the pyrrole ring.

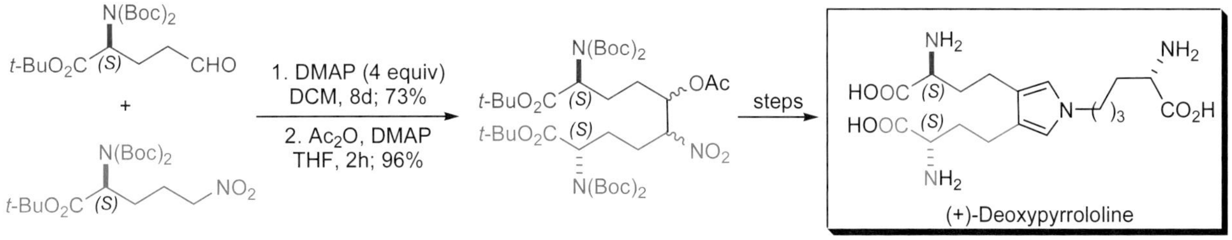

The total synthesis of (+)-cyclophellitol containing a fully oxygenated cyclohexane ring was accomplished by T. Ishikawa and co-workers.[57] The synthetic strategy was based on the *intramolecular silyl nitronate [3+2] cycloaddition* reaction. The cycloaddition precursor was prepared by the *Henry reaction* starting from a D-glucose-derived aldehyde.

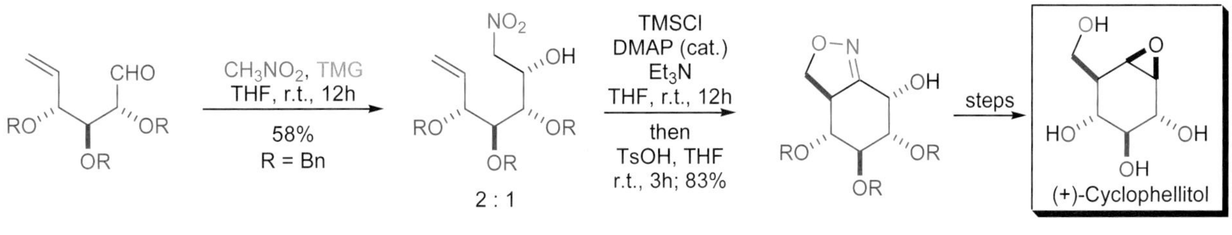

HETERO DIELS-ALDER CYCLOADDITION

(References are on page 599)

Importance:

[*Seminal Publications*[1-3]; *Reviews*[4-43]; *Theoretical Studies*[44-59]]

The [$4\pi + 2\pi$] cyclization of a diene and a dienophile to form a cyclohexene derivative is known as the *Diels-Alder cycloaddition* (*D-A cycloaddition*), but if one or more of the atoms in either component is other than carbon, then the reaction is referred to as the *hetero D-A cycloaddition (HDA)*. The first example of an imine participating as a heterodienophile was reported by K. Alder in 1943.[1] Since this initial report, the utilization of the *HDA reaction* in the synthesis of heterocyclic compounds has become pervasive. The general features of these reactions are: 1) high levels of regio- and diastereocontrol are observed and the outcome of the reaction can be predicted to the same extent as in the case of the *all-carbon D-A reaction*; 2) when the diene component does not contain a heteroatom and the heterodienophile is electron-deficient because of the heteroatom(s), the cycloaddition proceeds as a *normal electron-demand D-A reaction* (diene HOMO interacts with the LUMO of the heterodienophile); 3) when the diene contains one or more heteroatoms and/or electron-withdrawing substituents, it becomes electron-deficient, and therefore an electron-rich dienophile is needed and the reaction proceeds as an *inverse electron-demand D-A reaction* (heterodiene LUMO interacts with the HOMO of the dienophile); 4) when the heterodiene is substituted with one or more strongly electron-donating groups, the electron-deficient nature of the diene can be reversed and a *normal electron-demand hetero D-A reaction* can take place with a suitably electron-deficient dienophile; 5) *HDA reactions* can be catalyzed by Lewis acids, usually exhibiting higher regio- and stereoselectivities than uncatalyzed processes; and 6) by using a chiral auxiliary or catalyst the *asymmetric HDA reaction* can be realized.[22,31,38]

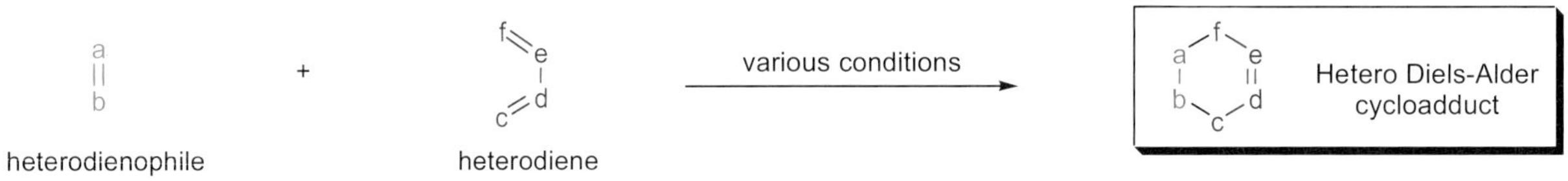

Most common heterodienophiles

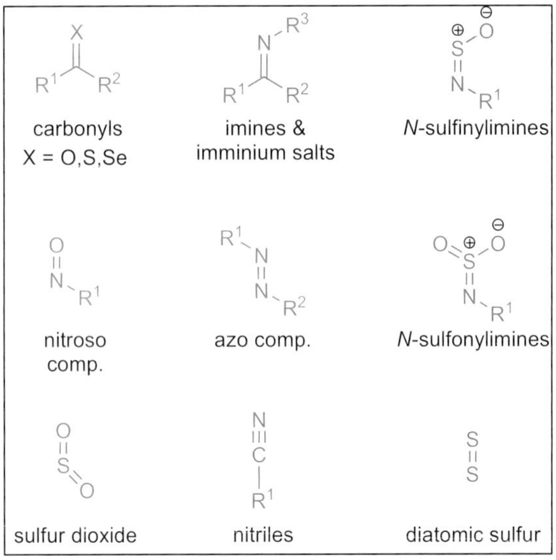

Most common heterodienes

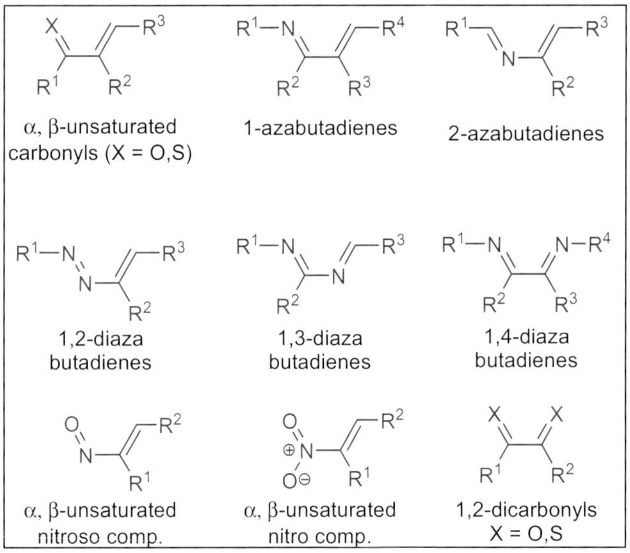

Mechanism: [60-69,51,53]

Mechanistically the *all-carbon Diels-Alder reaction* is generally considered a concerted, pericyclic reaction with an aromatic transition state, but there is also evidence for a stepwise (diradical or diion) process. For *HDA reactions*, theoretical studies revealed that the transition states are usually concerted, but less symmetrical. Depending on the reaction conditions and the number and type of substituents on the reactants, the *HDA reaction* can become stepwise, exhibiting a polar transition state.

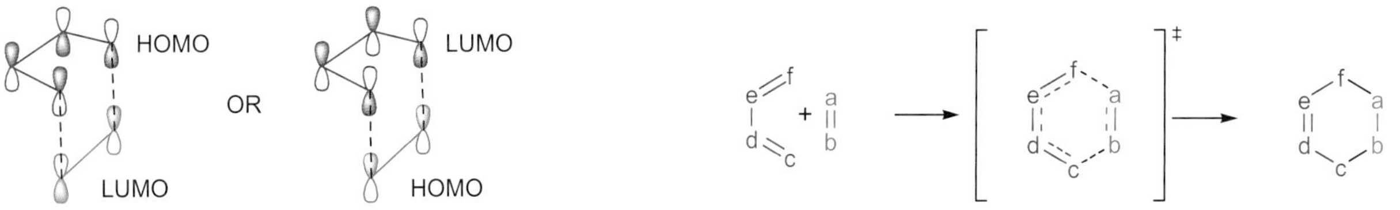

Synthetic Applications:

The enantioselective total synthesis of the epidermal growth factor inhibitor (–)-reveromycin B was completed by M.A. Rizzacasa and co-workers.[70] The key step to assemble the 6,6-spiroketal moiety was the *HDA reaction* between an α,β-unsaturated aldehyde (butylacrolein) and an enantiopure methylene pyran. The desired 6,6-spiroketal was obtained as a single enantiomer after heating the neat reactants in the absence of solvents at 110 °C for 2 days.

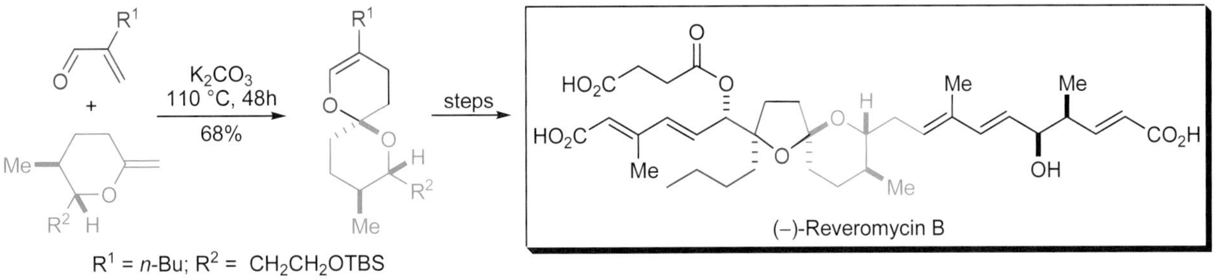

In the laboratory of S.F. Martin, a biomimetic approach toward the total synthesis of (±)-strychnine was developed by using tandem *vinylogous Mannich addition* and *HDA reaction* to construct the pentacyclic heteroyohimboid core of the natural product.[71] The commercially available 4,9-dihydro-3*H*-β-carboline was first converted to the corresponding *N*-acylium ion and then reacted with 1-trimethylsilyloxybutadiene in a *vinylogous Mannich reaction*. The resulting cycloaddition precursor readily underwent the expected *HDA reaction* in 85% yield.

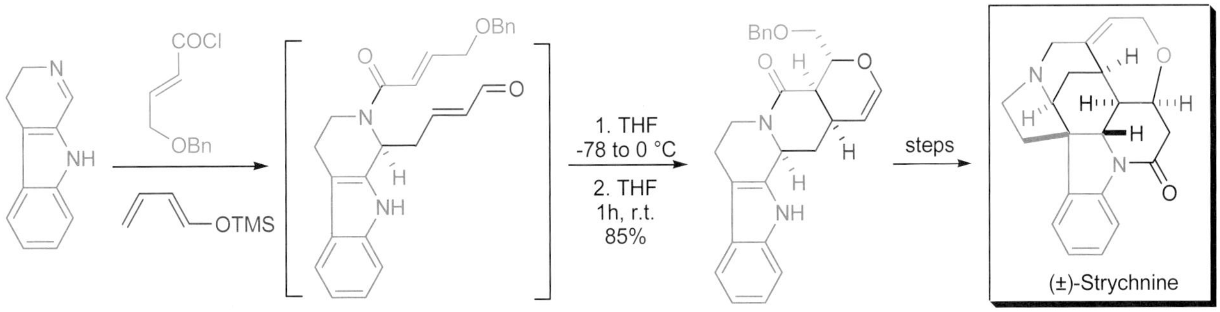

The first total synthesis of the decahydroquinoline alkaloid (–)-lepadin A was reported by C. Kibayashi et al.[72] The authors' approach was based on the *intramolecular HDA reaction* of an *in situ* generated acylnitroso compound. The precursor hydroxamic acid was oxidized with $Pr_4N(IO_4)$ in water-DMF (50:1) to form an acylnitroso compound that smoothly underwent the *[4+2] cycloaddition*. The *trans* bicyclic oxazino lactam product was formed as a 6.6:1 mixture of diastereomers; a result of the hydrophobic effect.

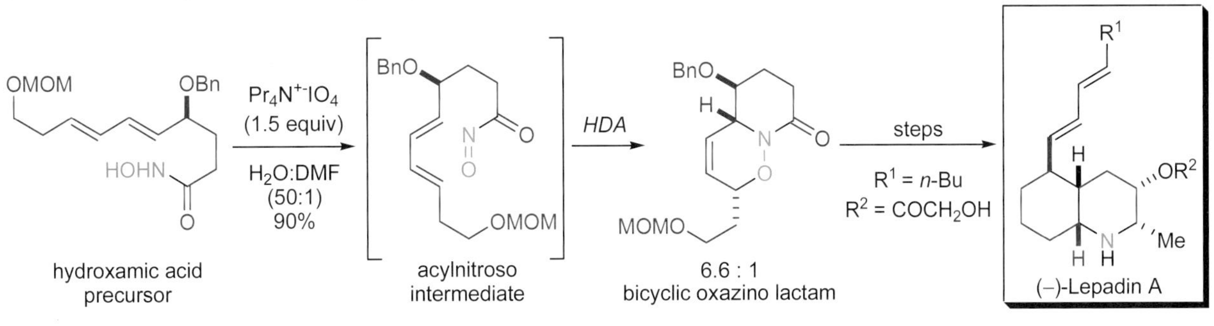

C.H. Swindell and co-workers enantioselectively prepared the Taxol A-ring side chain by using a *thermal inverse electron-demand HDA reaction* as the key step.[73] The (*Z*)-ketene acetal was attached to a chiral auxiliary and reacted with the *N*-benzoylaldimine to give the desired dihydrooxazine in 75% yield with good diastereoselectivity.

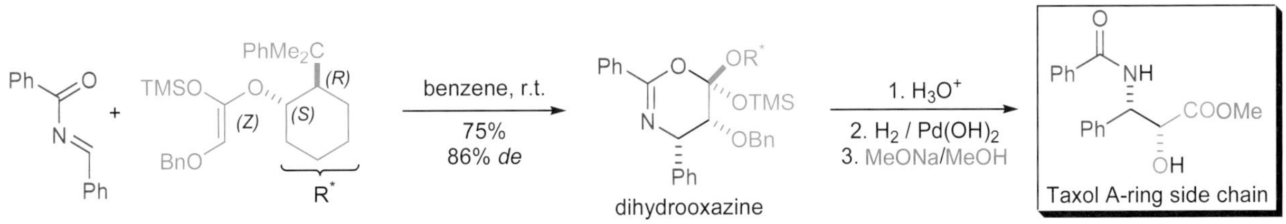

HOFMANN ELIMINATION

(References are on page 601)

Importance:

[*Seminal Publications*[1-4]; *Reviews*[5-9]; *Modifications & Improvements*[10,11]; *Theoretical Studies*[12]]

In 1851, A.W. Hofmann discovered that when trimethylpropylammonium hydroxide is heated, it decomposes to form a tertiary amine (trimethylamine), an olefin (propene), and water.[1,2] Widespread use of this transformation did not occur until 1881, when Hofmann applied this method to the study of the structure of piperidines and nitrogen-containing natural products (e.g., alkaloids).[3,4] The pyrolytic degradation of quaternary ammonium hydroxides to give a tertiary amine, an olefin and water is known as the *Hofmann elimination*. The process involves three steps: 1) exhaustive methylation of the primary, secondary or tertiary amine with excess methyl iodide to yield the corresponding quaternary ammonium iodide; 2) treatment with silver oxide and water (the iodide counterion is exchanged with hydroxide ion); and 3) the aqueous or alcoholic solution of the quaternary ammonium hydroxide is concentrated under reduced pressure and heated between 100-200 °C to bring about the elimination. Under reduced pressure, the elimination tends to take place at lower temperatures with higher yields. When the substrate is heterocyclic or the nitrogen is at a ring junction or at the bridgehead, the above steps need to be repeated multiple times to completely eliminate the nitrogen from the molecule. In the old days the number of repetitions indicated the position of the nitrogen atom in the original molecule and gave valuable structural clues about the unknown substance. The *Hofmann elimination* is a β-elimination, that is, the hydrogen is abstracted by the base (hydroxide ion) from the β-carbon atom. In the case of unsymmetrical compounds (in which more than one alkyl group attached to the nitrogen has β-hydrogen atoms), the β-hydrogen located at the least substituted carbon is abstracted by the base to form the less substituted alkene (Hofmann's rule).[1] The *Hofmann elimination* has few side reactions: occasionally the base can act as a nucleophile and substitution products are isolated. When the substrate does not have any alkyl groups with β-hydrogen, the main product of the pyrolysis is the substitution product (alcohol when water is the solvent or ether when no solvent is used).[13] An important variant of the *Hofmann elimination* is the *Wittig modification* in which the quaternary ammonium halide is treated with strong bases (alkylithiums, KNH_2/liquid NH_3, etc.) to afford an olefin and tertiary amine *via* an E_i mechanism.[11]

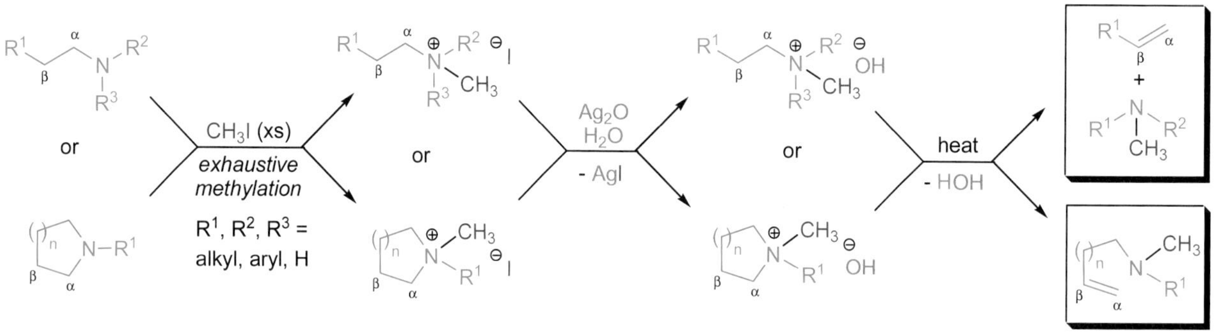

Mechanism: [14-27,11,28-30,12,31-34]

Generally the mechanism of the *Hofmann elimination* is E2, and it is an *anti* elimination (the leaving groups have to be *trans*-diaxial/antiperiplanar). However, in the case of certain substrates, the mechanism can be shifted in the carbanionic $E1_{cb}$ direction when the *trans* elimination process is unfavorable and the compounds contain sufficiently acidic allylic or benzylic β-hydrogen atoms. In acyclic substrates, the elimination gives rise to the least substituted alkene (Hofmann product). There are three factors which play a role in determining the outcome of the elimination: 1) the extent to which the double bond is developed in the transition state; 2) the acidity of the β-hydrogen atom; and 3) the influence of steric interactions in the transition state (this is the most widely accepted argument). In cycloalkyl ammonium salts, the most important factor in the elimination process is the availability of the *trans* β-hydrogen atoms. When both the β and β' trans hydrogens atoms are available in cyclic substrates, the elimination gives the most substituted alkene (*Saytzeff's rule*).

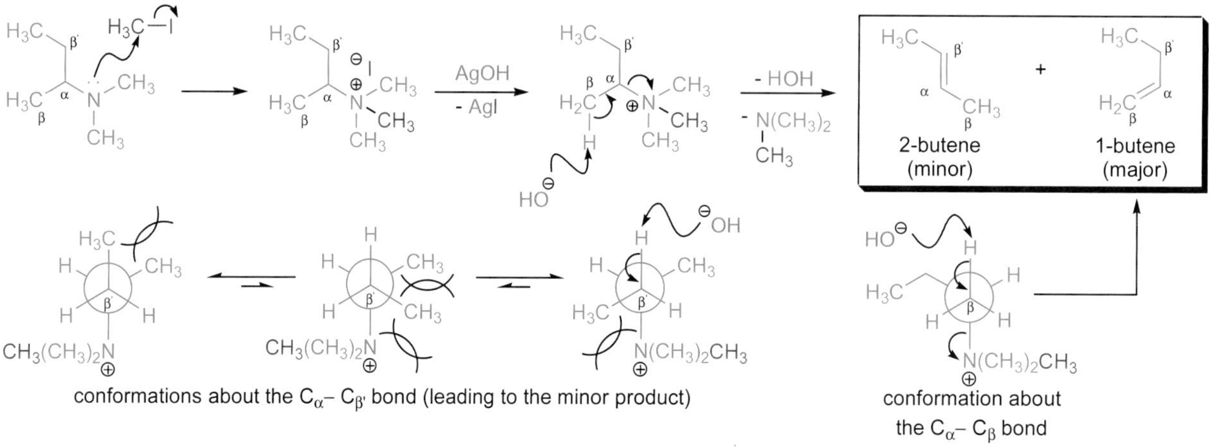

Synthetic Applications:

The enantioselective formal total synthesis of 4-demethoxydaunomycin was accomplished in the laboratory of M. Shibasaki.[35] The key intermediate was prepared from an enantiomerically enriched *trans*-β-amino alcohol, which was first exhaustively methylated to the corresponding quaternary ammonium salt. This salt was then treated with excess *n*-BuLi to afford the desired allylic alcohol in moderate yield.

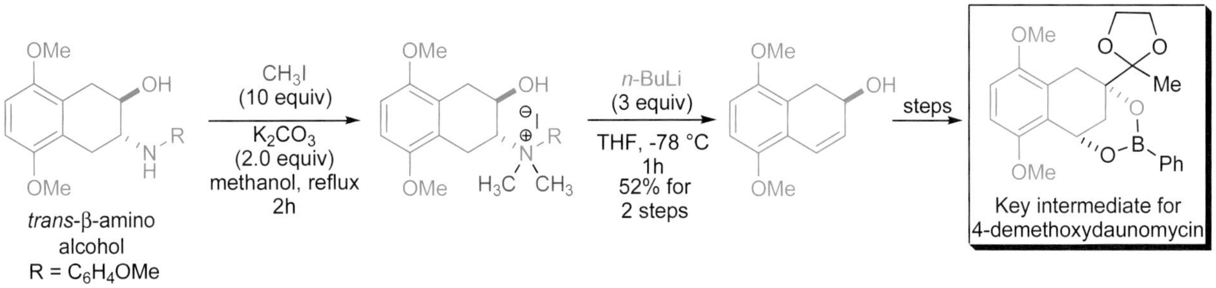

During the total synthesis of fungal metabolite (–)-cryptosporin, R.W. Franck and co-workers developed an efficient method for the regiospecific synthesis of naturally occurring naphtho[2,3-*b*]pyrano- and [2,3-*b*]furanoquinones using the *Bradscher cycloaddition* as the key step.[36] The *Hofmann elimination* of a primary amine located at the benzylic position, was carried out in the last steps of the synthesis. Interestingly, exhaustive methylation of the primary amine with excess MeI in MeOH/K_2CO_3 resulted in spontaneous elimination of the quaternary ammonium salt at room temperature.

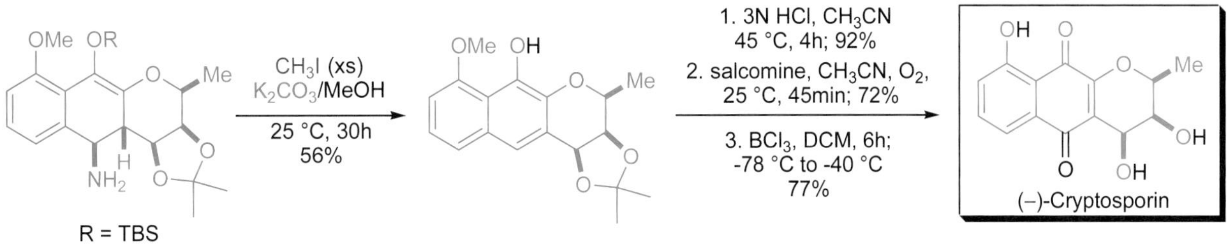

The ABCD ring system of the diterpene alkaloid atisine was constructed by T. Kametani et al using an *intramolecular Diels-Alder cycloaddition* reaction as the key step.[37] The dienophile was obtained by the traditional *Hofmann degradation* of the corresponding dimethylamino precursor. The diene was prepared by the kinetic enolization of the cyclohexenone system with LDA.

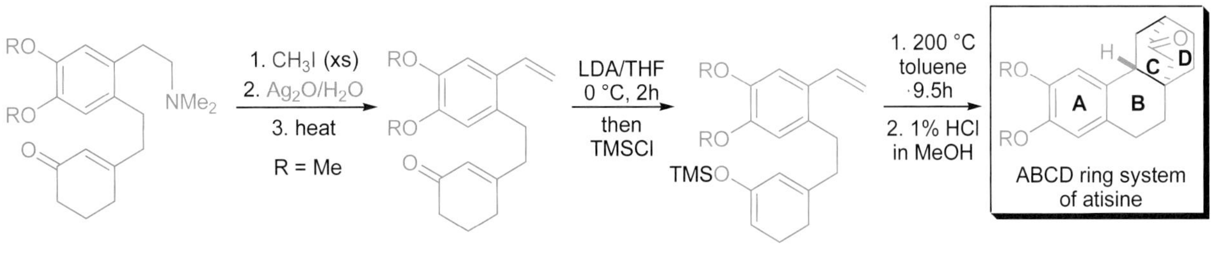

In the laboratory of D.S. Watt, the enantioselective total synthesis of (+)-picrasin B was achieved from (–)-Wieland-Miescher ketone.[38] At the early stages of the synthetic effort, an exocyclic double bond was introduced in a two-step procedure by first alkylating the bicyclic conjugated TMS enol ether with Eschenmoser's salt at the γ-position, followed by *Hofmann elimination* of the dimethylamino group.

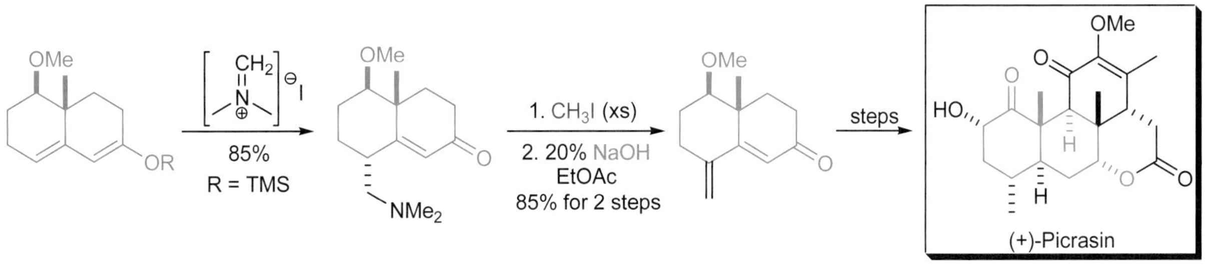

HOFMANN-LÖFFLER-FREYTAG REACTION
(REMOTE FUNCTIONALIZATION)

(References are on page 602)

Importance:

[*Seminal Publications*[1-4]; *Reviews*[5-14]; *Modifications & Improvements*[15-22]; *Theoretical Studies*[23]]

In the early 1880s, A.W. Hofmann was trying to determine if piperidine, whose structure was unknown at the time, was unsaturated by exposing it to hydrohalic acids or bromine. During these investigations he prepared various *N*-haloamines and *N*-haloamides and studied their reactions under acidic and basic conditions. The treatment of 1-bromo-2-propylpiperidine with hot sulfuric acid, followed by basic work-up, yielded octahydroindolizine, a bicyclic tertiary amine.[1-3] In 1909, K. Löffler and C. Freytag applied this transformation to simple secondary amines and realized that it was a general method for the preparation of pyrrolidines.[4] The formation of cyclic amines from *N*-halogenated amines *via* an *intramolecular 1,5-hydrogen atom transfer* to a nitrogen radical is known as the *Hofmann-Löffler-Freytag reaction* (*HLF reaction*). General features of the reactions are: 1) it may be carried out in acidic solutions, but neutral and even weakly basic reaction conditions have been applied successfully;[24,25] 2) it can be conducted under milder conditions if the intermediate alkyl radical is stabilized by a heteroatom (e.g., nitrogen);[24] 3) initiation of the radical process can be done by heating, irradiation with light or with radical initiators (e.g., dialkyl peroxides, metal salts); 4) the initially formed nitrogen-centered radical abstracts a H-atom mostly from the δ-position (or 5-position) and predominantly 5-membered rings are formed; and 5) rarely, in rigid cyclic systems, the formation of 6-membered rings is possible.[24,15] The original strongly acidic reaction conditions are often not compatible with the sensitive functional and protecting groups of complex substrates, therefore several modifications were introduced: 1) photolysis of *N*-bromoamides proceeds under neutral conditions;[26] 2) in the presence of persulfates and metal salts, sulfonamides undergo remote γ- and δ-halogenation under neutral conditions;[27] 3) the most important variant of this reaction is the *Suárez modification* in which *N*-nitroamides,[20] *N*-cyanamides,[18] and *N*-phosphoramidates[22] react with hypervalent iodine reagents in the presence of iodine (I_2) under neutral conditions to generate nitrogen-centered radicals *via* the hypothetical iodoamide intermediate. The *HLF reaction* is closely related to the well-known *Barton nitrite ester reaction*, which proceeds *via* alkoxyl radicals and has been extensively used for remote functionalization in steroid synthesis.

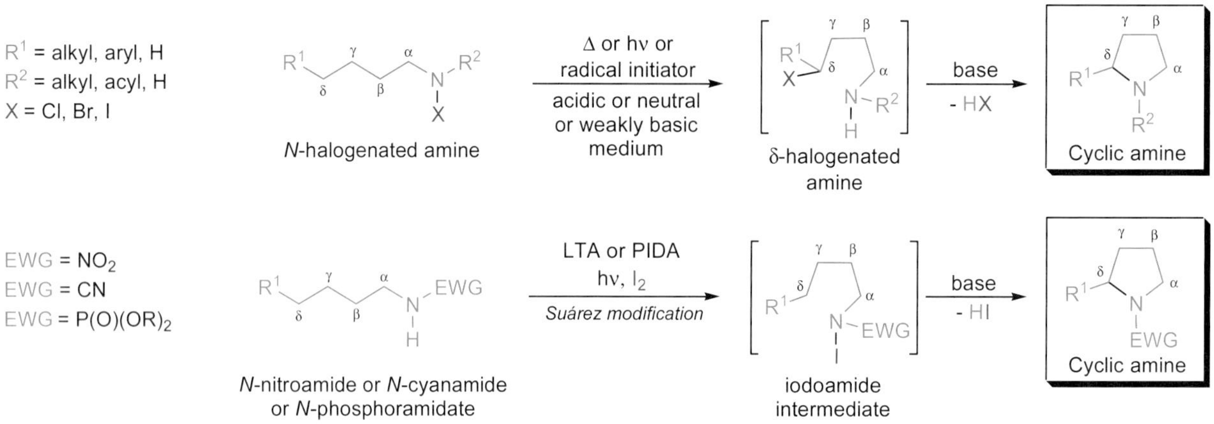

Mechanism: [28-31]

The mechanism of the *HLF reaction* is a radical chain reaction. When the reaction is conducted in acidic medium, the first step is the protonation of the *N*-halogenated amine to afford the corresponding *N*-halogenated ammonium salt. Heat, irradiation with light or treatment with radical initiators generates the nitrogen-centered radical, *via* the homolytic cleavage of the *N*-halogen bond, which readily undergoes an *intramolecular 1,5-hydrogen abstraction*. Next, the newly formed alkyl radical abstracts a halogen atom intermolecularly. Treatment of the δ-halogenated amine with base gives rise to the desired cyclic amine product.

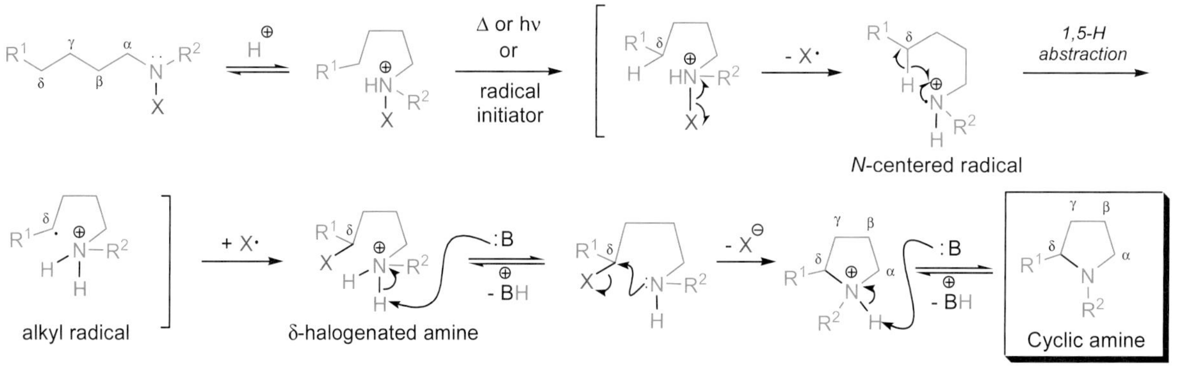

Synthetic Applications:

In the laboratory of Y. Shibanuma, a novel synthetic approach was developed to construct the bridged azabicyclic ring system of the diterpene alkaloid kobusine.[32] The bridged nitrogen structure of the target (±)-6,15,16-imino-podocarpane-8,11,13-triene was synthesized by means of a *Hofmann-Löffler-Freytag reaction* from a bicyclic chloroamine. First the bicyclic amine was converted to the corresponding *N*-chloro derivative in good yield by treatment with NCS in dichloromethane. The solution of the bicyclic *N*-chloroamine in trifluoroacetic acid was then irradiated with a 400 W high pressure Hg-lamp under nitrogen atmosphere at r.t. for several hours to afford a moderate yield of the product.

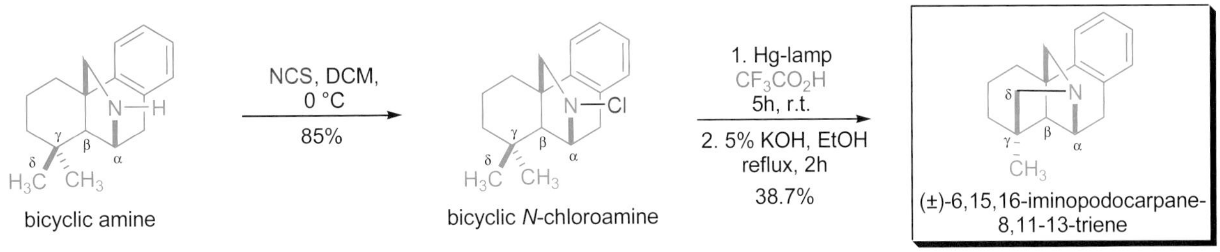

E. Suárez and co-workers prepared chiral 7-oxa-2-azabicyclo[3.2.1]octane and 8-oxa-6-azabicyclo[3.2.1]octane ring systems derived from carbohydrates *via* an intramolecular hydrogen abstraction reaction promoted by *N*-centered radicals.[22] The *N*-centered radicals were obtained under mild conditions (*Suárez modification*) from phenyl and benzyl amidophosphates and alkyl and benzyl carbamate derivatives of aminoalditols by treatment with PIDA/I_2 or PhIO/I_2. The initial *N*-radical undergoes a 1,5-hydrogen abstraction to form an alkyl radical, which is oxidized to the corresponding stabilized carbocation (oxocarbenium ion) under the reaction conditions. The overall transformation may be considered as an *intramolecular N-glycosidation* reaction.

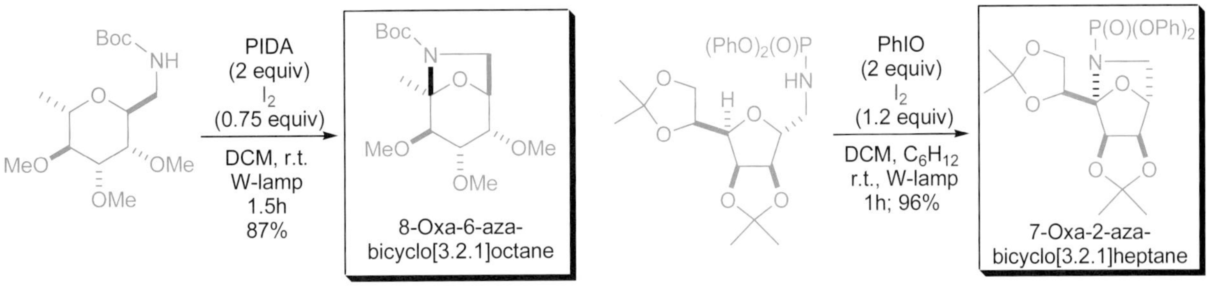

The *Suárez modification of the HLF reaction* was the basis of the new synthetic method developed by H. Togo et al.[33] The authors prepared *N*-alkyl-1,2-benzisothiazoline-3-one-1,1-dioxides (*N*-alkylsaccharins) from *N*-alkyl(*o*-methyl)-arenesulfonamides using (diacetoxyiodo)arenes in the presence of iodine *via* sulfonamidyl radicals. The transformations did not work in the dark, indicating the radical nature of the reaction. The yields varied from moderate to excellent and the nature of the aromatic substituents on both the substrate and the (diacetoxyiodo)arenes were important. It should be noted that the oxygen atom at the C3 position most likely arises from the hydrolysis of a C3 diiodo intermediate (not isolated).

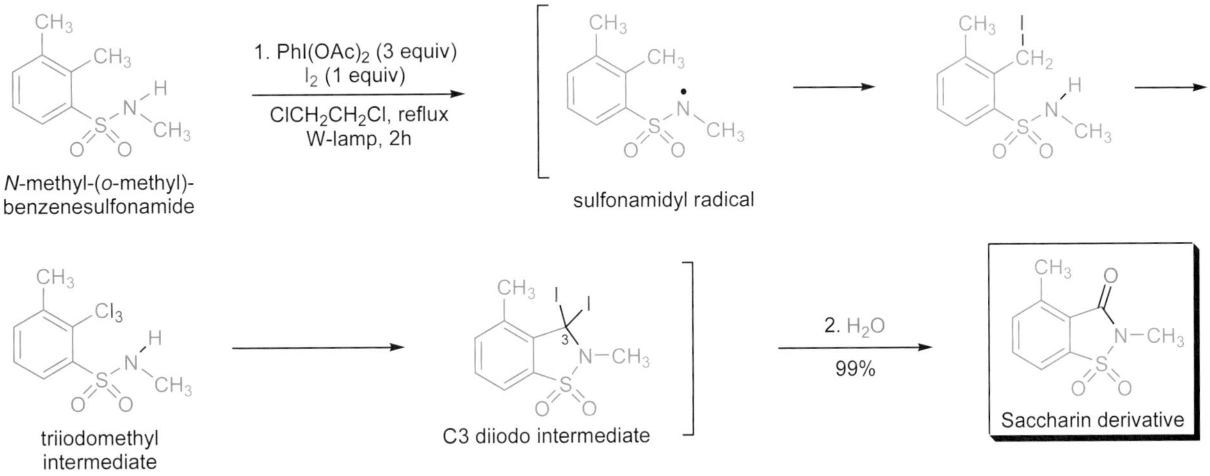

HOFMANN REARRANGEMENT

(References are on page 602)

Importance:

[*Seminal Publications*[1-5]; *Reviews*[6-15]; *Modifications & Improvements*[16-30]]

In 1881, A.W Hofmann found that by treating acetamide with one equivalent of bromine (Br_2) and sodium or potassium hydroxide it afforded *N*-bromoacetamide. Upon further deprotonation and heating, *N*-bromoacetamide gave an unstable salt that in the absence of water readily rearranged to methyl isocyanate.[1] However, in the presence of water and excess base the product was methylamine. The conversion of primary carboxamides to the corresponding one-carbon shorter amines is known as the *Hofmann rearrangement* (also known as the *Hofmann reaction*). According to the standard procedure, the amide is dissolved in a cold solution of an alkali hypobromite or hypochlorite and the resulting solution is heated to ~70-80 °C to bring about the rearrangement. The general features of this transformation are: 1) the hypohalite reagents are freshly prepared by the addition of chlorine gas or bromine to an aqueous solution of KOH or NaOH; 2) the amides cannot contain base-sensitive functional groups under the traditional basic reaction conditions, but acid-sensitive groups (e.g., acetals) remain unchanged; 3) the isocyanate intermediate is not isolated, since under the reaction conditions it is readily hydrolyzed (or solvolyzed) to the corresponding one-carbon shorter amine *via* the unstable carbamic acid; 4) when the reaction is conducted under phase-transfer catalysis conditions, the isocyanates may be isolated;[31,25] 5) if the starting amide is enantiopure (the carbonyl group is directly attached to the stereocenter), there is a *complete retention of configuration* in the product amine; 6) the *Hofmann rearrangement* gives high yields for a wide variety of aliphatic and aromatic amides but the best yields for aliphatic amides are obtained if the substrate has no more than 8 carbons (hydrophilic amides); and 7) α,β-unsaturated amides and amides of α-hydroxyacids rearrange to give aldehydes or ketones.[32,33] Since the discovery of the *Hofmann rearrangement*, several modifications were introduced: 1) for hydrophobic amides, the use of methanolic sodium hypobromite (bromine added to sodium methoxide in methanol) results in high yields of the corresponding methylurethanes;[6] 2) for acid- and base-sensitive substrates the use of neutral *electrochemically induced Hofmann rearrangement* was developed;[18,26,28] 3) in order to extend the scope of the reaction for base-sensitive substrates, the *oxidative Hofmann rearrangement* may be carried out with LTA or hypervalent iodine reagents (PIDA, PIFA, PhI(OH)OTs, etc.) under mildly acidic conditions;[16,23,14,29] and 4) when hypervalent iodine reagents or LTA are used in the presence of an amine or an alcohol, the generated isocyanate is *in situ* converted to the corresponding carbamate or urea derivative.[17]

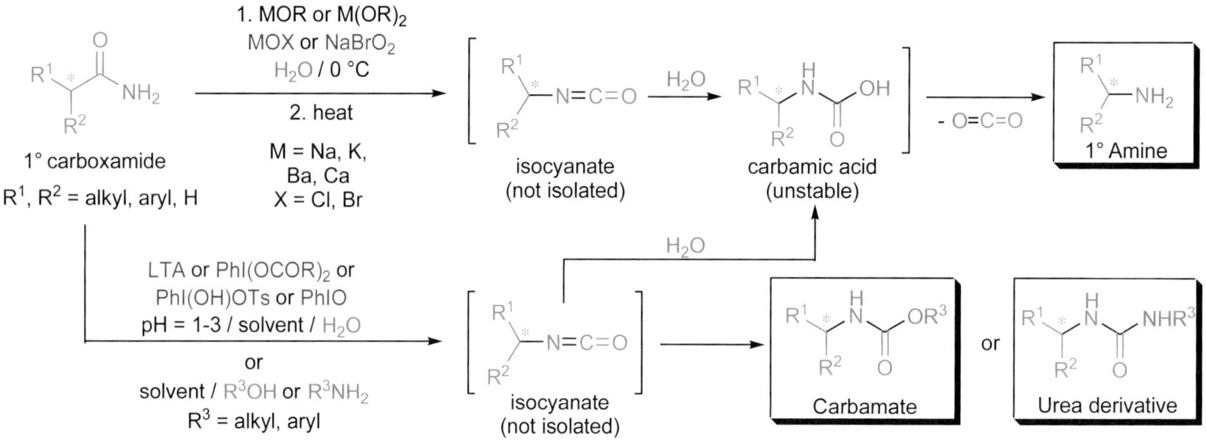

Mechanism:[34-40,19,41]

The mechanism of the *Hofmann rearrangement* is closely related to the *Curtius*, *Lossen* and *Schmidt rearrangements*. The first step is the formation of an *N*-halogen substituted amide. Next, the *N*-haloamide is deprotonated by the base to the corresponding alkali salt that is quite unstable and quickly undergoes a concerted rearrangement to the isocyanate *via* a bridged anion. This mechanistic picture is strongly supported by kinetic evidence.[36-39] As a result, the *Hofmann rearrangement* proceeds with complete retention of configuration.

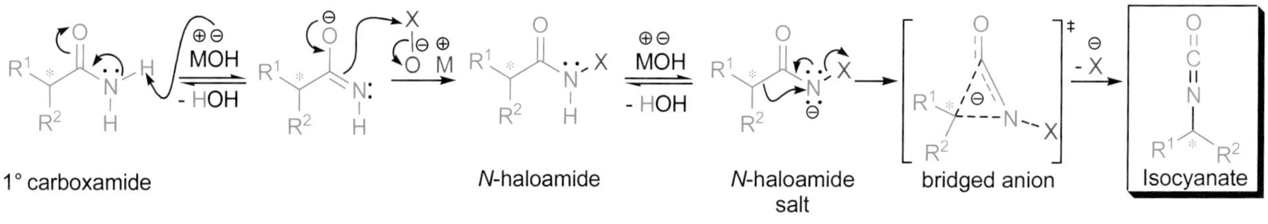

Synthetic Applications:

The enantioselective total synthesis of (–)-epibatidine was accomplished in the laboratory of D.A. Evans.[42] The key steps in the synthetic sequence included a *hetero Diels-Alder reaction* and a modified *Hofmann rearrangement*. The primary carboxamide was subjected to lead tetraacetate in *tert*-butyl alcohol that brought about the rearrangement and gave the corresponding *N*-Boc protected primary amine in good yield. A few more steps from this intermediate led to the completion of the total synthesis.

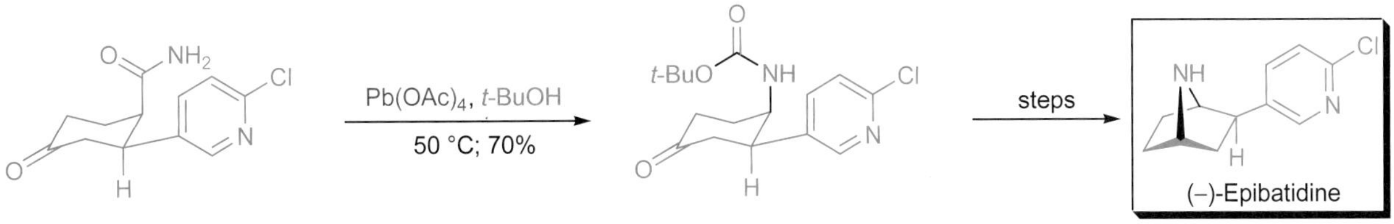

The first asymmetric total synthesis of the hasubanan alkaloid (+)-cepharamine was completed by A.G. Schultz et al.[43] In order to construct the *cis*-fused *N*-methylpyrrolidine ring, the advanced tetracyclic lactone was first converted to the primary carboxamide by treatment with sodium amide in liquid ammonia. Next the *Hofmann rearrangement* was induced with sodium hypobromite in methanol initially affording the isocyanate, which upon reacting with the free secondary alcohol intramolecularly gave the corresponding cyclic carbamate in excellent yield.

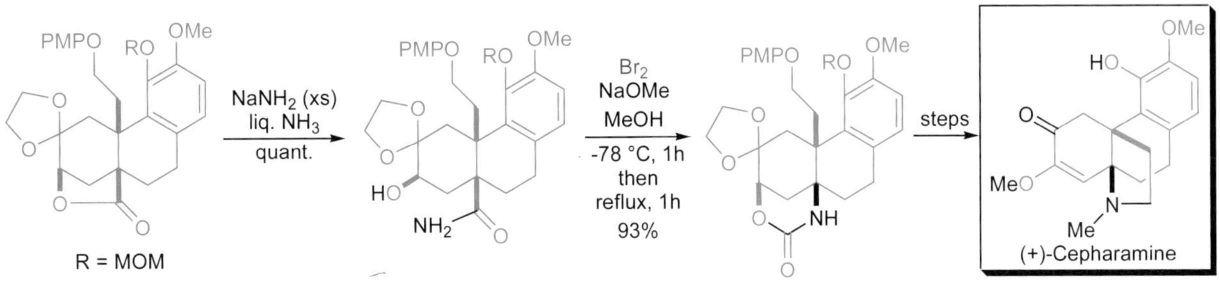

R. Verma and co-workers developed a silicon-controlled total synthesis of the antifungal agent (+)-preussin using a *modified Hofmann rearrangement* as one of the key steps in the final stages of the synthetic sequence.[44] The primary carboxamide was exposed to LTA in DMF in the presence of benzyl alcohol, which resulted in an efficient *Hofmann rearrangement* to afford the Cbz-protected primary amine. As expected, there was no loss of optical activity in the product. The silicon group was finally converted to the corresponding secondary alcohol by the *Fleming-Tamao oxidation*.

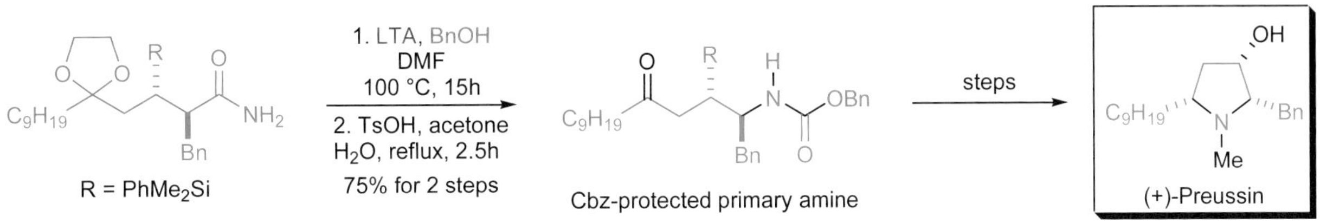

During the late stages of the asymmetric total synthesis of capreomycidine IB it was necessary to transform an asparagine residue into a diaminopropanoic acid residue.[45] R.M. Williams et al. employed a chemoselective *Hofmann rearrangement*, thereby avoiding protection and deprotection steps that would have been necessary had the diaminopropanoic acid been introduced directly. The complex pentapeptide was treated with PIFA and pyridine in the presence of water to afford the primary amine in high yield.

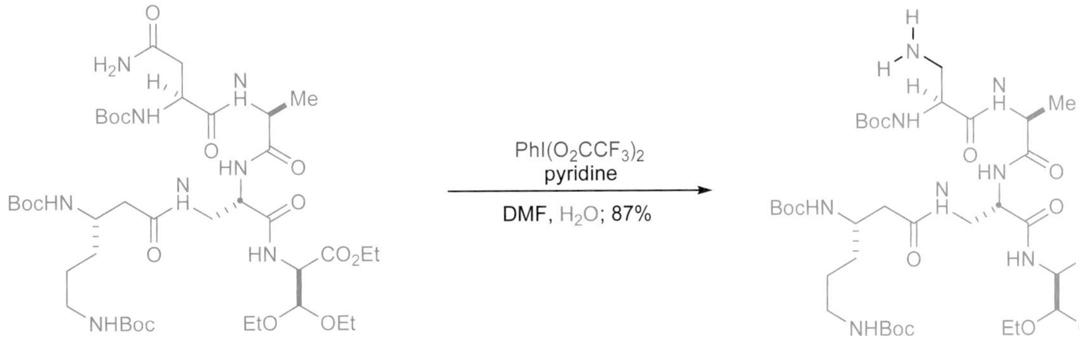

HORNER-WADSWORTH-EMMONS OLEFINATION

(References are on page 603)

Importance:

[*Seminal Publications*[1-4]; *Reviews*[5-24]; *Modifications & Improvements*[25-39]; *Theoretical Studies*[40-46]]

In 1958, L. Horner utilized the carbanions of alkyl diphenyl phosphine oxides (R^1=Ph) to prepare alkenes from aldehydes and ketones.[1,2] This modification of the *Wittig reaction* is known as the *Horner-Wittig reaction* (or *Horner reaction*) but its widespread use in organic synthesis became a reality only in the early 1960s when W.S. Wadsworth and W.D. Emmons studied the synthetic utility of phosphonate carbanions (R^1=O-alkyl) for the preparation of olefins.[3] In this detailed study, Wadsworth and Emmons revealed the significant advantages these phosphonate carbanions had over the traditional triphenyl phosphorous ylides used in *Wittig reactions*. The stereoselective olefination of aldehydes and ketones using phosphoryl-stabilized carbanions (most often R^1=O-alkyl and R^2=CO_2-alkyl) is referred to as the *Horner-Wadsworth-Emmons olefination* (or *HWE olefination*). The *HWE olefination* has the following advantages over the traditional *Wittig olefination*: 1) the preparation of the starting alkyl phosphonates is easier (usually the *Arbuzov reaction* is used) and cheaper than the preparation of phosphonium salts; 2) the phosphonate carbanions are more nucleophilic than the corresponding phosphorous ylides, so they readily react with practically all aldehydes and ketones under milder reaction conditions; 3) hindered ketones that are unreactive in *Wittig reactions* react readily in *HWE olefinations*; 4) the α-carbon of the phosphonate anions can be further functionalized with various electrophiles (e.g., alkyl halides) prior to the olefination, but phosphorous ylides usually do not undergo smooth alkylation; 5) the by-product dialkyl phosphates are water-soluble, so it is much easier to separate them from the alkene products than from the water-insoluble triphenylphosphine oxide. General features of the *HWE olefination* are: 1) high (*E*)-selectivity for disubstituted alkenes under much milder conditions than normally used in *Wittig reactions* (R^2 needs to be able to conjugate with the incipient double bond); 2) the (*E*)-selectivity is maximized by increasing the size of the alkyl group of the R^1 or R^2 substituents (e.g., R=isopropyl is best); and 3) the stereoselectivity is strongly substrate dependent but can be reversed to form predominantly (*Z*)-olefins by using smaller alkyl groups (e.g., methyl) in the R^1 and R^2 substituents and a strongly dissociating base (e.g., KO*t*-Bu). There are a couple of important modifications of the *HWE olefination*: 1) in the *Still-Gennari modification* R^1=OCH_2CF_3 and the reaction affords (*Z*)-olefins exclusively;[27] 2) for base-sensitive substrates, the use of a metal salt (LiCl or NaI) and a weak amine base (e.g., DBU) has proven effective to avoid epimerization;[28,30,35] 3) asymmetric *HWE olefinations*;[29,36,23] 4) the *Corey-Kwiatkowski modification* uses phosphoric acid bisamides to prepare (*Z*)-alkenes stereoselectively, $(Me_2N)_2P(O)CH_2R$, where R=aryl.[25,26]

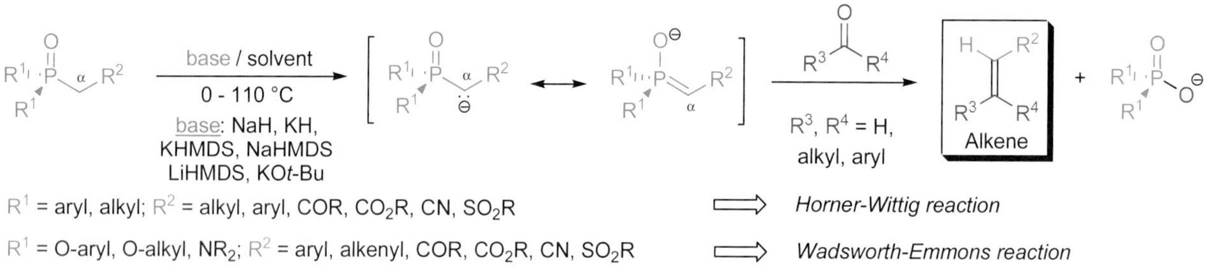

Mechanism:[47,9,48,11]

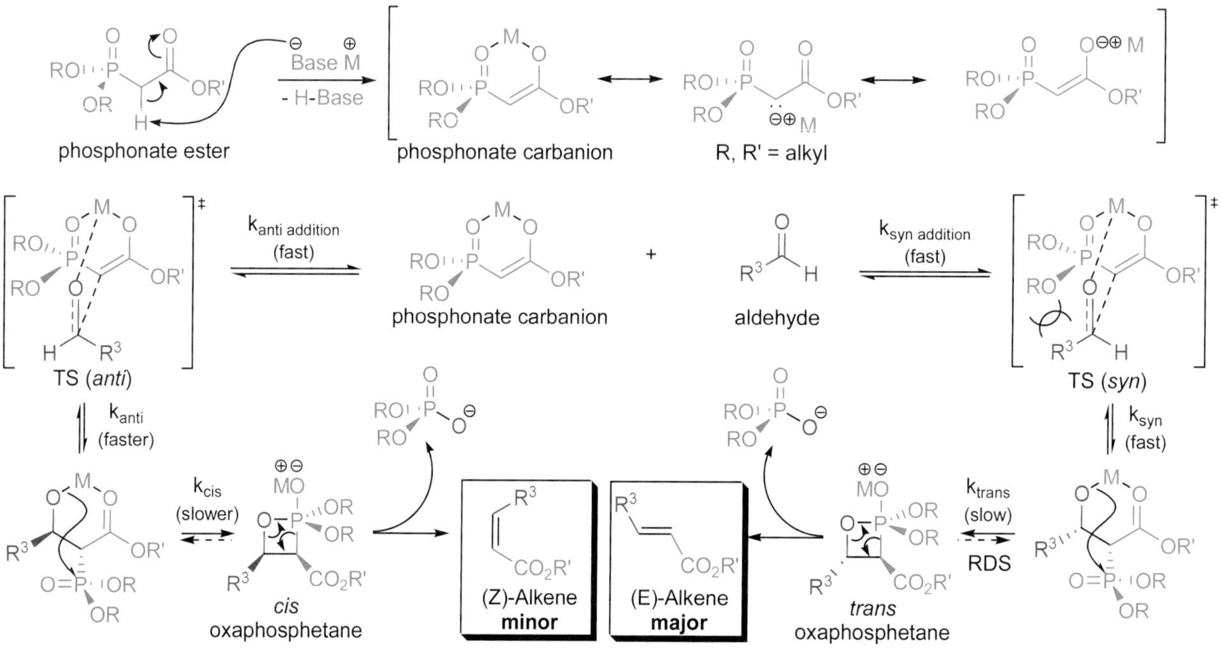

Synthetic Applications:

In the laboratory of T.R. Hoye, a *HWE macrocyclic head-to-tail dimerization* was used to construct the C_2-symmetric macrocyclic core of (–)-cylindrocyclophane A.[49] The monomer phosphono ester aldehyde was subjected to sodium hydride in benzene containing a catalytic amount of 15-crown-5 ether and 55% of the (*E,E*)-macrocyclized product was obtained. None of the (*Z,Z*) stereoisomer was observed. Macrocyclization reactions usually require high-dilution conditions but even relatively concentrated solutions (0.02M) did not decrease the yield of the product in this case.

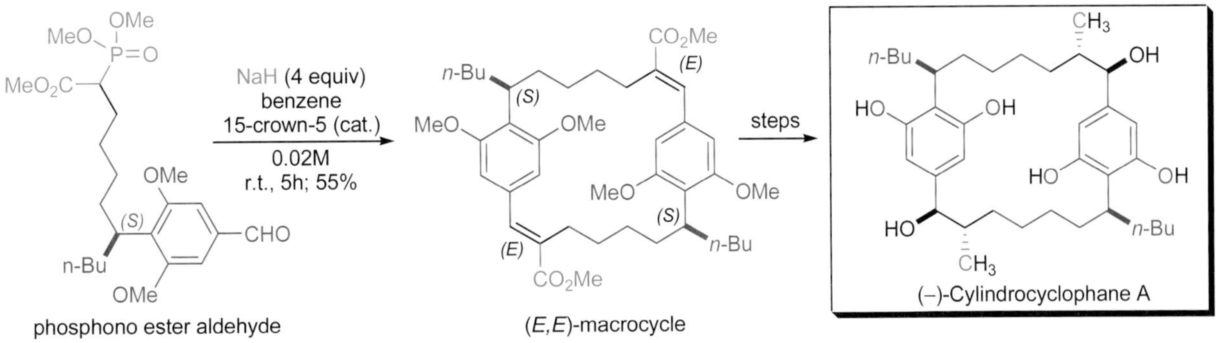

A short, asymmetric total synthesis of an important 3-(hydroxymethyl)carbacephalosporin antibiotic was achieved by M.J. Miller and co-workers.[50] The β-lactam ring was formed *via a Mitsunobu cyclization*, while the six-membered unsaturated ring was constructed by a *HWE cyclization*. This intramolecular olefination afforded a single diastereomer in 85% yield.

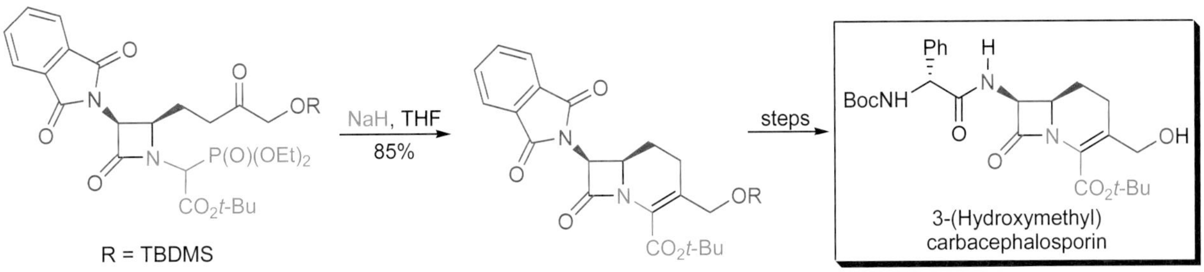

In order to assign the absolute stereochemistry and relative configuration of *callipeltoside A*, B.M. Trost et al. devised a highly convergent total synthesis by which several stereoisomers were prepared.[51] The key steps in the synthetic sequence were a *ruthenium-catalyzed Alder-ene alkene-alkyne coupling*, a *Pd-catalyzed asymmetric allylic alkylation* and a late-stage coupling of the side chain by the *HWE olefination*. The olefination step gave the coupled product in a moderate yield and with moderate stereoselectivity (*E:Z* = 4:1).

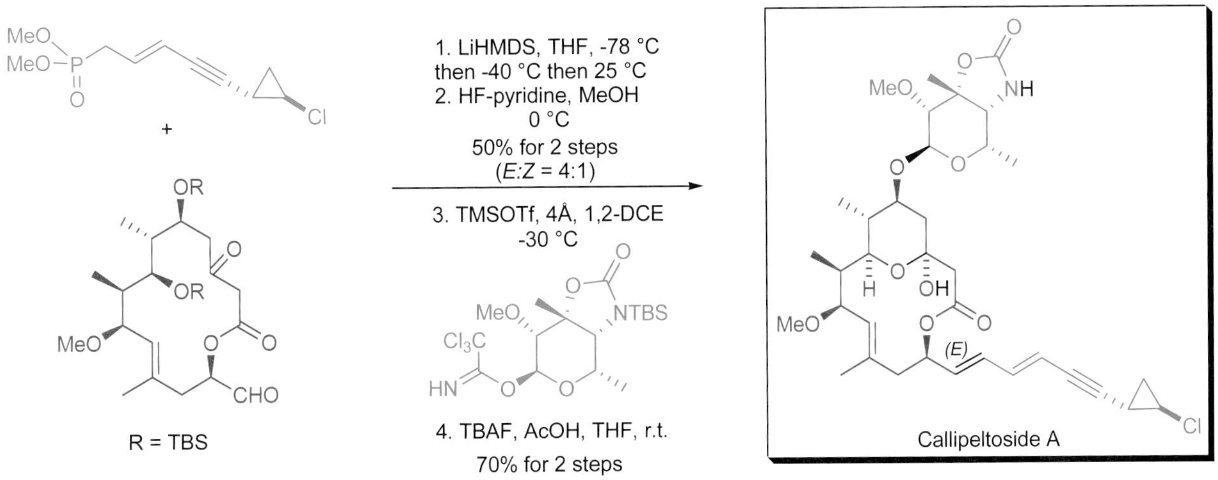

HORNER-WADSWORTH-EMMONS OLEFINATION – STILL-GENNARI MODIFICATION

(References are on page 604)

Importance:

[*Seminal Publication*[1]; *Reviews*[2,3]; *Modifications & Improvements*[4-10]; *Theoretical Studies*[11]]

The *Horner-Wadsworth-Emmons olefination* and the *Wittig reaction* of stabilized ylides with aldehydes are the two most widely used methods for the preparation of (*E*)-alkenes. The *HWE olefination* gives rise to (*E*)-α,β-unsaturated ketones and esters, while the *trans*-selective *Wittig reaction* affords simple, unconjugated (*E*)-alkenes. In 1983, W.C. Still and C. Gennari introduced the first general way to prepare (*Z*)-olefins from aldehydes by the modification of the phosphonate reagent used in the *HWE olefination*.[1] The preparation of (*Z*)-α,β-unsaturated ketones and esters by coupling electrophilic *bis*(trifluoroalkyl) phosphonoesters in the presence of strong bases with aldehydes is known as the *Still-Gennari modification of the HWE olefination*. General features of this process are: 1) the necessary *bis*(trifluoroethyl)phosphonoesters are easily prepared from the commercially available trialkylphosphonoesters and trifluoroethanol; 2) (*Z*)-stereoselectivity is observed not only for 1,2-disubstituted but for trisubstituted alkenes as well; 3) the phosphonate reagent must have an electron-withdrawing (carbanion-stabilizing) group at its α-position, otherwise the phosphonate carbanion decomposes; 4) a well-dissociating base must be used in which the metal cation is not coordinating (this is usually achieved by adding 18-crown-6 into the reaction mixture); and 5) when R^2=CN, the (*Z*)-selectivity is high as opposed to the poor (*E*)-selectivity of α-cyano-stabilized regular phosphonates.

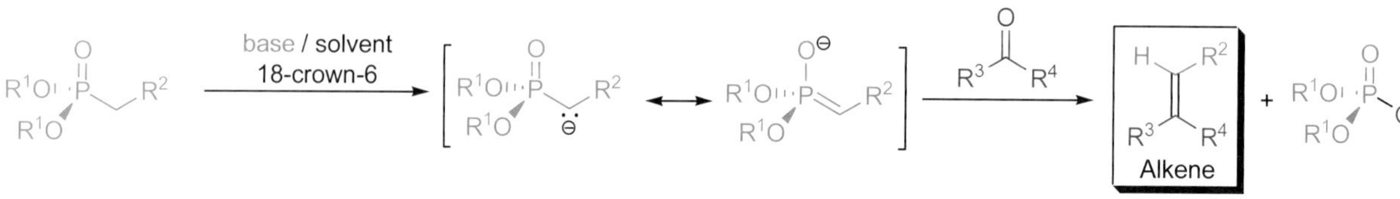

R^1 = CH_2CF_3, trifluoroalkyl; R^2 = COR, CO_2R, CN, SO_2R; R^{3-4} = H, alkyl, aryl; base = KH, KHMDS

Mechanism: [12]

The mechanism of the *HWE olefination* is not fully understood. In the *Still-Gennari modified HWE olefination* the phosphorous has two electron-withdrawing trifluoroalkoxy groups. In this case the rearrangement from the chelated adduct to form the oxaphosphetane is favored and the elimination step is faster than the initial addition, which essentially becomes irreversible (unlike in the case of the regular HWE olefination). As a result, the formation of the (*Z*)-stereoisomer is predominant.

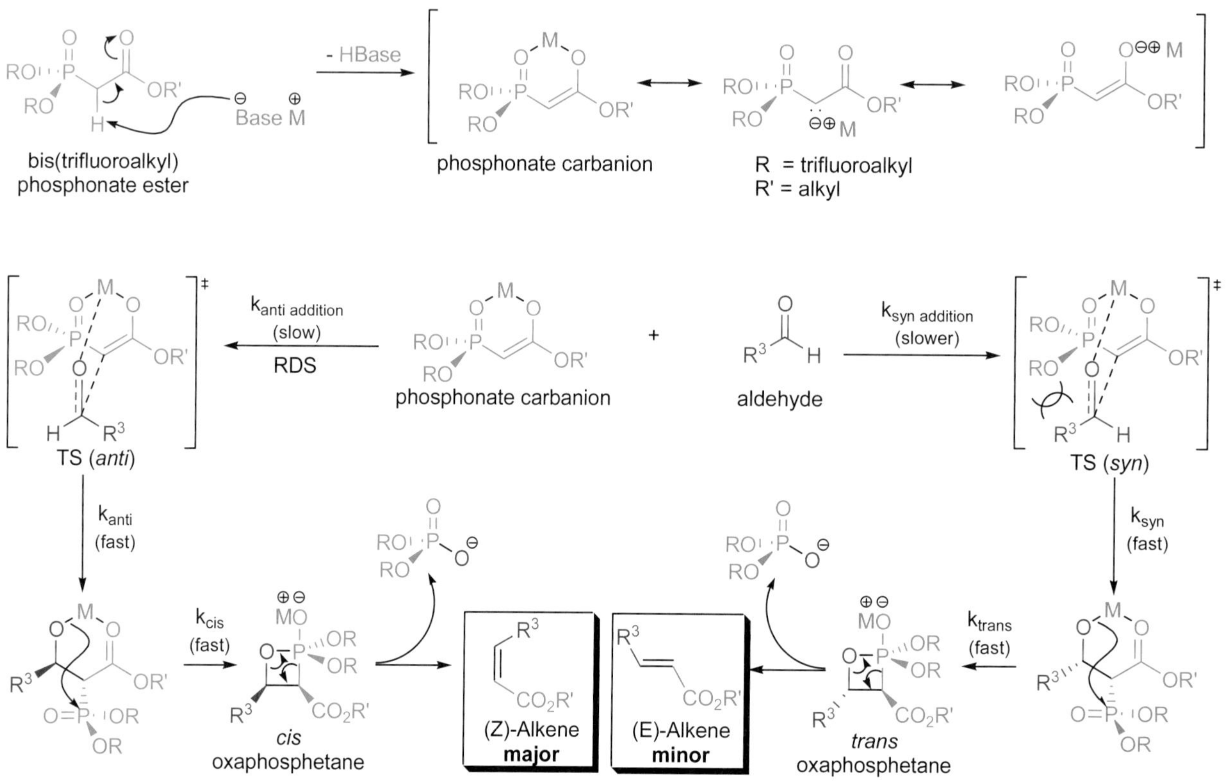

Synthetic Applications:

In C.J. Forsyth's total synthesis of phorboxazole A, the intramolecular version of the *Still-Gennari modified HWE olefination* was used to affect the macrocyclization of a complex *bis*(trifluoroethoxy) phosphonate-aldehyde precursor.[13] The precursor was dissolved in toluene and was exposed to K_2CO_3 in the presence of 18-crown-6. The desired C1-C3 (*Z*)-acrylate moiety was formed in 77% yield with a 4:1 (*Z*:*E*) ratio. Interestingly, when the same cyclization was carried out with the regular *bis*(dimethoxy) phosphonate, the macrocyclization was markedly slower, but the stereoselectivity was the same (4:1).

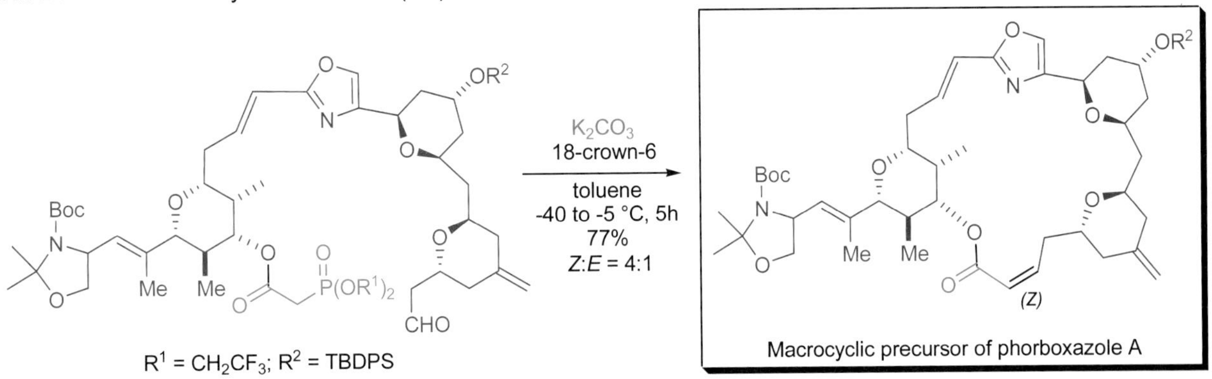

In the laboratory of S.V. Ley, the total synthesis of the β-lactone cholesterol synthase inhibitor 1233A was achieved by using the *oxidative decomplexation* of a (π-allyl)tricarbonyliron lactone as the key step.[14] The (*Z*)-alkene present in the target was introduced using the *S-G modified HWE olefination* of an aldehyde with *bis*(2,2,2-trifluoroethyl) (methoxycarbonylmethyl)phosphonate to give the desired α,β-unsaturated methyl ester in excellent yield.

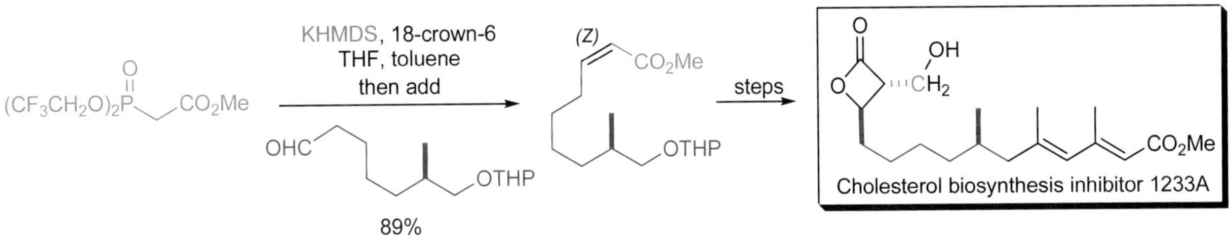

The stereoselective synthesis of the anti-ulcer 3,4-dihydroisocoumarin AI-77B was accomplished by E.J. Thomas and co-workers.[15] The key transformation was the *stereoselective dihydroxylation* of 4-(*Z*)-alkenylazetidinones that were prepared from 4-formylazetidinone *via* the *Still modified HWE olefination*. The benzyl *bis*(trifluoroethyl) phosphonoacetate was prepared from phosphonic dichloride and 2,2,2-trifluoroethanol and was alkylated using benzyl bromoacetate.

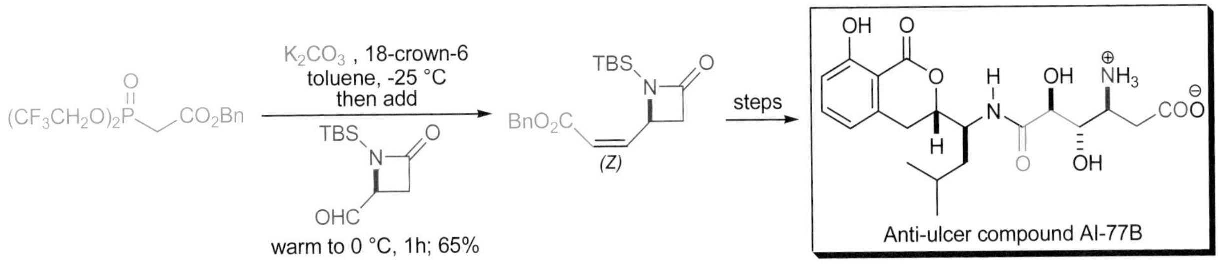

The key tricyclic intermediate toward the total synthesis of spinosyn A was assembled by W.R. Roush et al. featuring a one-pot tandem *intramolecular Diels-Alder reaction* and an *intramolecular vinylogous Baylis-Hillman cyclization*.[16] The cyclization precursor was prepared *via* the *S-G modified HWE reaction*.

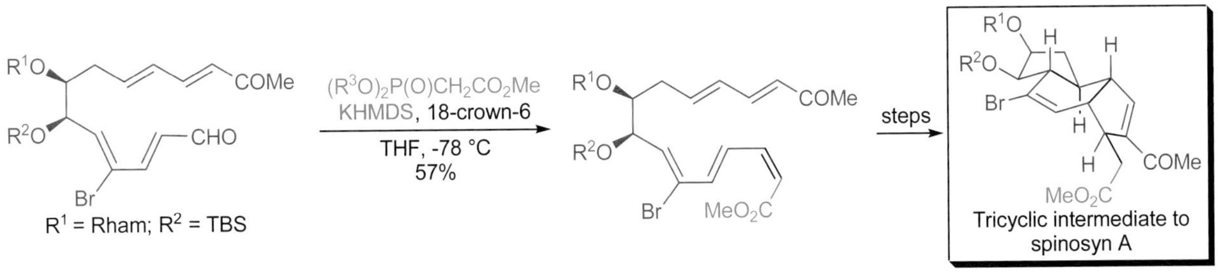

HOUBEN-HOESCH REACTION / SYNTHESIS

(References are on page 605)

Importance:

[*Seminal Publications*[1-5]; *Reviews*[6-10]; *Modifications & Improvements;*[11-17] *Theoretical Studies*[18]]

By the early 1900s the *Friedel-Crafts acylation* and the *Gattermann formylation* were widely used to prepare aromatic ketones and aldehydes, respectively. The preparation of monoacylated derivatives of highly activated (electron rich) substrates (e.g., polyphenols) was not possible, since usually more than one acyl group was introduced using the standard *Friedel-Crafts acylation* conditions. In 1915, K. Hoesch reported the extension of the *Gattermann reaction* for the preparation of aromatic ketones by using nitriles instead of hydrogen cyanide and replaced the aluminum chloride with the milder zinc chloride.[1,2] A decade later the scope and the limitation of this novel ketone synthesis was examined in great detail by J. Houben, who showed that the procedure principally worked for polyphenols or polyphenolic ethers.[3] The condensation of nitriles with polyhydroxy- or polyalkoxyphenols to prepare the corresponding polyhydroxy- or polyalkoxyacyloxyphenones is known as the *Houben-Hoesch reaction*. The general features of this reaction are: 1) only highly activated disubstituted aromatic compounds undergo the transformation (at least one of the substituents should be a hydroxy or an alkoxy group); 2) the aromatic compound can be heterocyclic so pyrroles, indoles, and furans are also substrates of this transformation; 3) the structure of the nitrile is freely variable: alkyl, aryl, and substituted alkyl groups (e.g., α-halogenonitriles, α-hydroxynitriles, and their ethers and esters) are all compatible with the reaction conditions; 4) aliphatic nitriles tend to give higher yields than aromatic nitriles; 5) the aromatic nitrile cannot have a strongly electron-withdrawing group in its *ortho*-position (no reaction is observed), but these groups in the *meta*-position have no effect on the reactivity of the aromatic nitrile; 6) the nitriles are often introduced as their hydrochloride salts;[11] 7) zinc chloride is the most widely used Lewis acid but for very electron rich substrates (e.g., phloroglucinol) no Lewis acid is needed; and 8) the initial product of the reaction is the imine hydrochloride that is hydrolyzed to afford the final product aromatic ketone. The most important modifications of the *Houben-Hoesch reaction* are: 1) by using trichloroacetonitrile, even non-activated aromatics can be acylated; and 2) switching the Lewis acid to BCl_3 the acylation of aromatic amines can be realized with high *ortho* regioselectivity.[13]

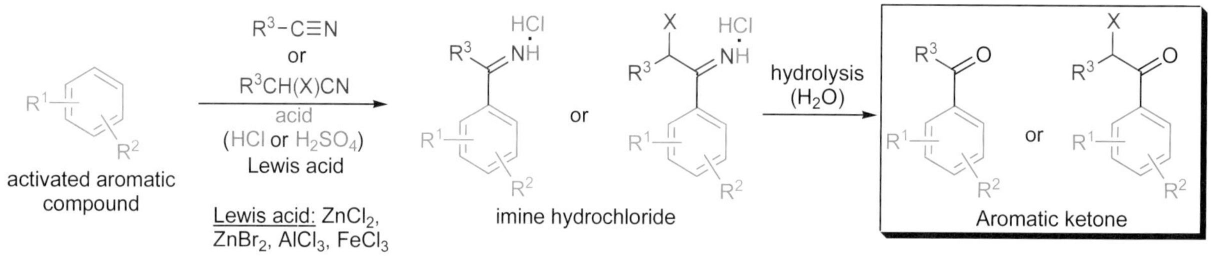

R^1 = OH, *O*-alkyl; R^2 = OH, *O*-alkyl, alkyl, Cl, Br, I; R^3 = alkyl, aryl, substituted alkyl or aryl; X = H, OH, *O*-alkyl, Cl, Br, I

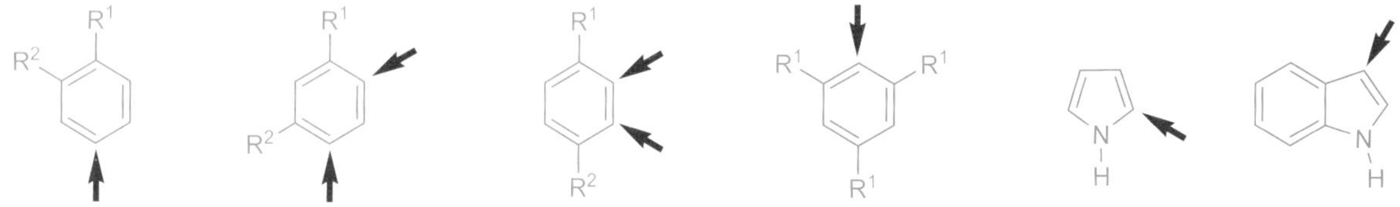

1,2-disubstituted 1,3-disubstituted 1,4-disubstituted 1,3,5-trisubstituted pyrrole indole

Mechanism: [19,15,20,21]

The mechanism is not fully understood, but it is very similar to the mechanism of the *Gattermann-Koch formylation*. The first step is the formation of a nitrilium chloride that is subsequently transformed to an imino chloride from which the reactive species, the iminium ion is generated.

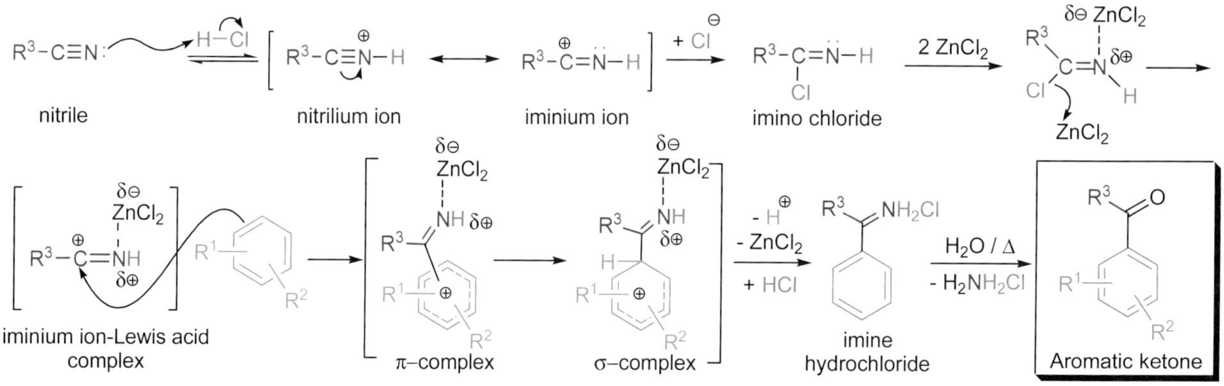

Synthetic Applications:

In the laboratory of D.W. Cameron the total synthesis of the azaanthraquinone natural product bostrycoidin was undertaken using the *Minisci reaction* and the *intramolecular Houben-Hoesch reaction* as the key steps.[22] It is worth noting that the synthesis of specific di- and trihydroxyazaanthraquinones by the *Friedel-Crafts acylation* is very limited due to the lack of orientational specificity and the lack of reactivity of pyridine derivatives in acylation reactions.

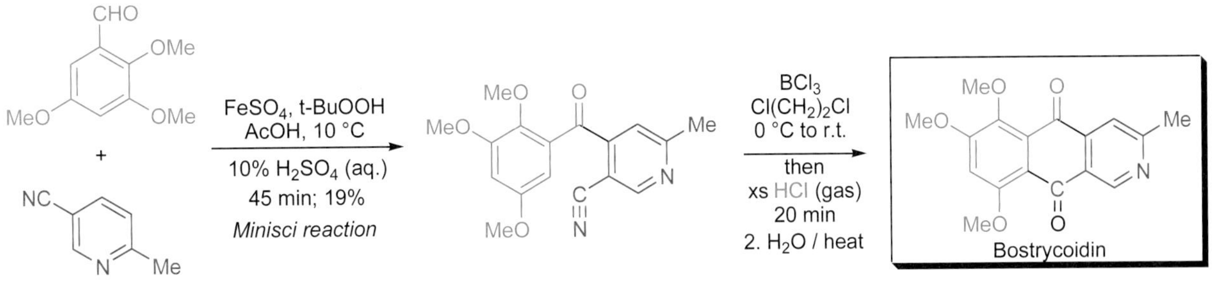

Genistein (4',5,7-trihydroxyflavone) is an important nutraceutical molecule found in soybean seeds, and it has a wide range of pharmacological effects.[23] The two-step total synthesis of genistein was achieved by M.G. Nair et al. using the *Houben-Hoesch reaction* to acylate phloroglucinol with *p*-hydroxyacetonitrile.[24] The resulting deoxybenzoin was treated with DMF/PCl_5 in the presence of $BF_3{\cdot}OEt_2$ to give genistein in 90% yield. The DMF/PCl_5 mixture was the source of the $[(Me_2N{=}CHCl)^+]Cl^-$ reagent. This synthetic sequence was suitable for the large scale (~1 metric ton) one-pot preparation of the natural product.

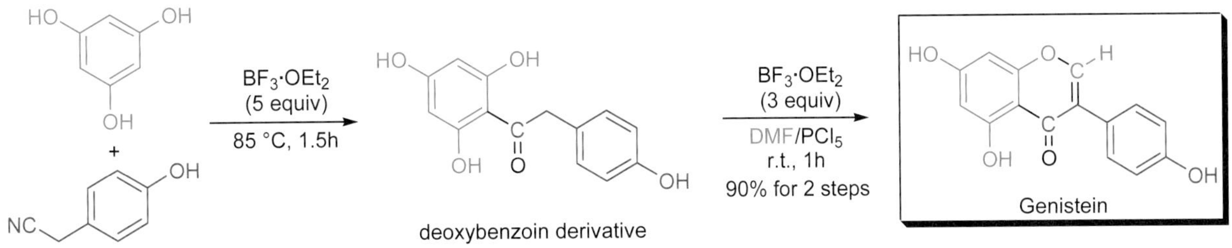

Nitriles having electrophilic or leaving groups in their α- or β-postions often lead to so-called "abnormal" Houben-Hoesch products besides the expected "normal" acylation products. Especially notorious is the reaction of β-oxonitriles with phenols that afford exclusively 2*H*-1-benzopyran-2-one derivatives instead of the expected 1,2-diketones. α-Halogenonitriles react with phenols to give the expected 3-benzofuranone and also the abnormal 2-benzofuranone. R. Kawecki and co-workers found that the condensation of phenols with aromatic α-hydroxyiminonitriles or α-oxonitriles under the Hoesch conditions leads to benzofuro[2,3-*b*]benzofuran derivatives.[25]

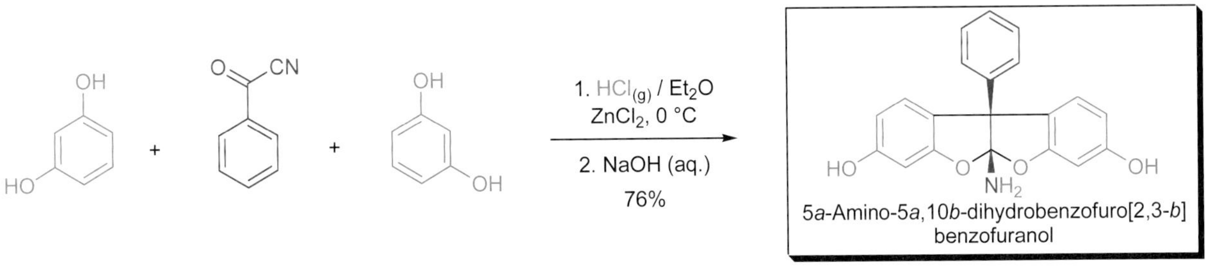

The synthesis of 11-hydroxy *O*-methylsterigmatocystin (HOMST) was carried out in the laboratory of C.A. Townsend by utilizing the *alkylnitrilium ion variant of the Houben-Hoesch reaction.*[17] The alkylnitrilium salt was prepared by reacting the aryl nitrile with 2-chloropropene in the presence of $SbCl_5$. Next, the phenol was added in a 2.5:1 excess. Alkaline hydrolysis then afforded the xanthone, which was subsequently converted to HOMST in few more steps.

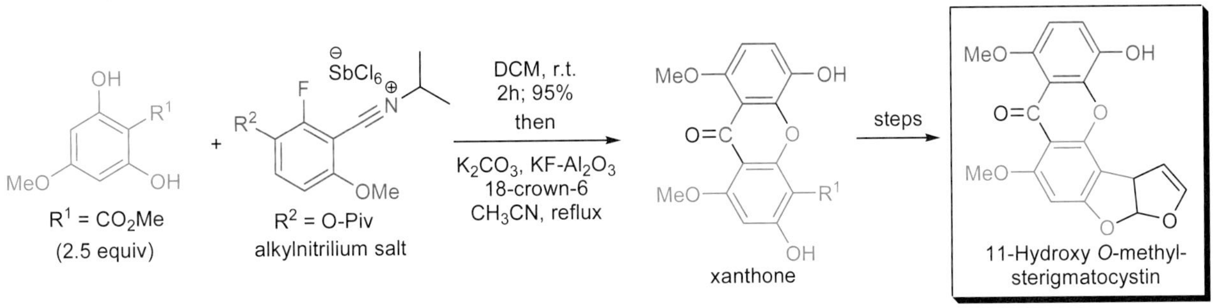

HUNSDIECKER REACTION

(References are on page 605)

Importance:

[*Seminal Publications*[1-3]; *Reviews*[4-7]; *Modifications & Improvements*[8-30]]

In 1939, H. Hunsdiecker reported that when the dry silver salts of aliphatic carboxylic acids were treated with bromine, the corresponding one-carbon shorter alkyl bromides were obtained.[2,3] The halogenative decarboxylation of aliphatic-, α,β-unsaturated-, and certain aromatic carboxylic acids to prepare the one-carbon shorter alkyl halides is referred to as the *Hunsdiecker reaction*. The general features of this transformation are: 1) the silver salts are prepared from the corresponding carboxylic acids with silver oxide; 2) the slurry of the silver salt in carbon tetrachloride is treated with one equivalent of the halogen, and carbon-dioxide is evolved as rapidly as the halogen is added; 3) in order to obtain high yields, the silver salts must be pure and scrupulously dry, which is not easy to achieve, since the silver salt is often heat sensitive; 3) aliphatic carboxylic acids are the best substrates, but aromatic carboxylic acids with electron-withdrawing substituents are also suitable; 4) electron-rich (activated) aromatic carboxylic acids undergo electrophilic aromatic substitution under the reaction conditions; 5) instead of silver salts, the much more stable thallium(I)- and mercury(I)-salts can be used;[13] 6) functional groups that react with halogens are incompatible (e.g., alkenes, alkynes) under the reaction conditions; and 7) if optically active silver carboxylates are used, there is a significant loss of optical activity in the product alkyl halides. Due to the technical difficulties with the preparation of the silver carboxylates, numerous modifications were introduced to simplify the procedure: 1) the preparation of the silver carboxylate is avoided and higher yields are observed if one adds the solution of the acid chloride to a slurry of dry silver oxide/CCl_4/bromine at reflux temperature;[8,9] 2) the use of crystallizable thallium(I)-carboxylates instead of silver salts improve the yield;[13] 3) the *Cristol-Firth modification* uses excess red HgO and one equivalent of halogen in one-pot;[10] 4) in the *Suárez modification*, the acid is treated with a hypervalent iodine reagent in CCl_4 with remarkable functional group tolerance;[20] 5) LTA can be used directly with iodine or with lithium halides (chlorides and bromides) to produce the corresponding alkyl halides (*Kochi modification*);[11,6] 6) the *Barton modification* exploits the thermal or photolytic decomposition of thiohydroxamate esters in halogen donor solvents (e.g., $BrCCl_3$, CHI_3) and this modification is compatible with almost all functional groups;[17,19] 7) if AIBN is used in the *Barton modification*, any kind of aromatic acid (both activated and deactivated) can be decarboxylated in high yield;[31] and 8) the reaction can be made metal-free and catalytic, but this reaction probably follows a non-radical mechanistic pathway.[24,27,29]

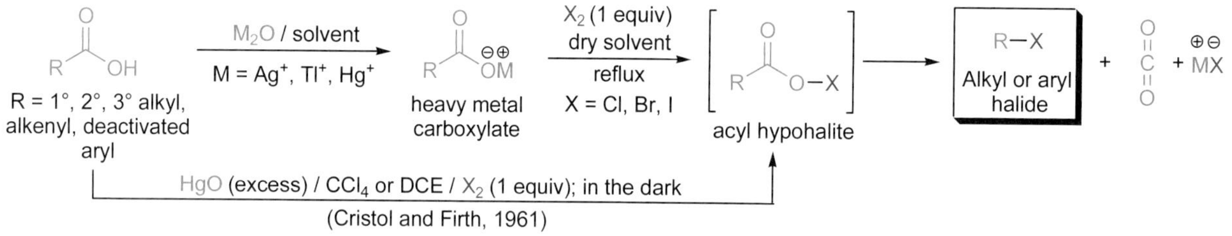

Hunsdiecker-type reactions:

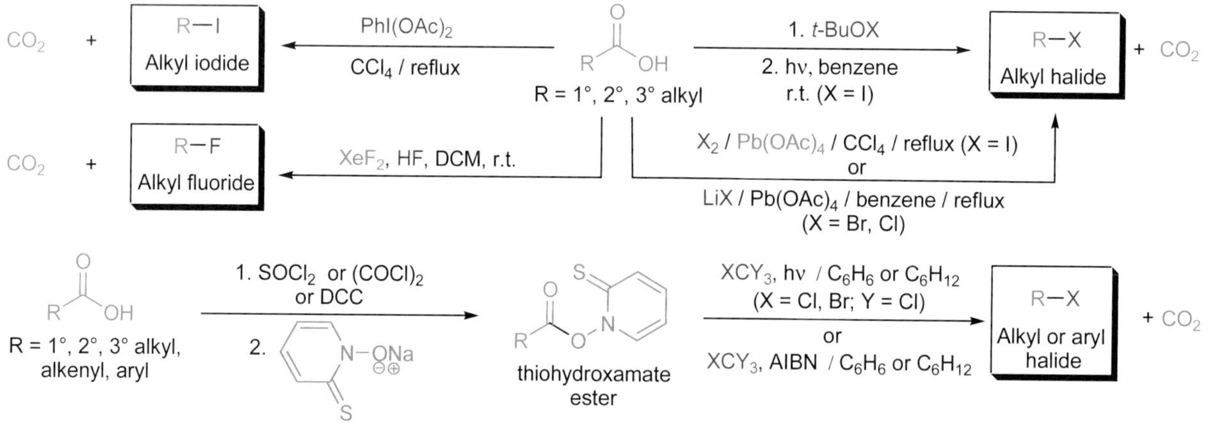

Mechanism: [4,32-37,29]

Classical Hunsdiecker reaction:

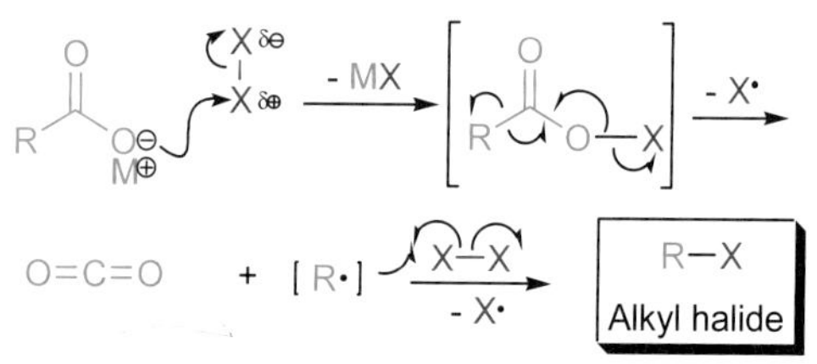

Cristol-Firth modified Hunsdiecker reaction:

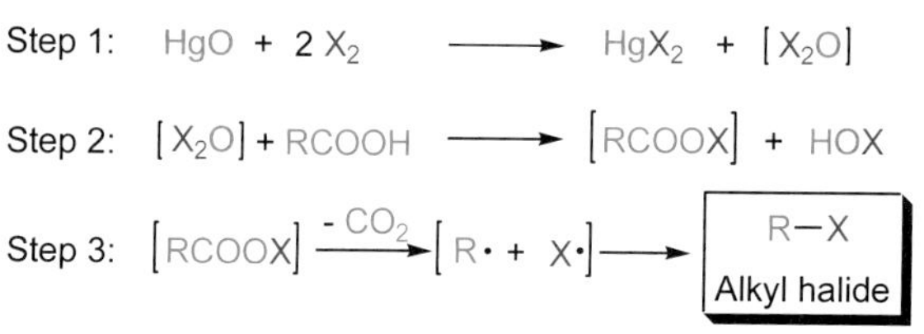

Synthetic Applications:

There are a few efficient methods for the stereoselective synthesis of vinyl halides, and this transformation remains a synthetic challenge. Research by S. Roy showed that the *Hunsdiecker reaction* can be made metal free and catalytic (*catalytic Hunsdiecker reaction*) and can be used to prepare (*E*)-vinyl halides from aromatic α,β-unsaturated carboxylic acids.[27] The unsaturated aromatic acids were mixed with catalytic amounts of TBATFA and the *N*-halo-succinimide was added in portions over time at ambient temperature. The yields are good to excellent even for activated aromatic rings which do not undergo the *classical Hunsdiecker reaction*. The fastest *halodecarboxylation* occurs with NBS, but NCS and NIS are considerably slower. The nature of the applied solvents is absolutely critical, and DCE proved to be the best. This strategy was extended and applied in the form of a *one-pot tandem Hunsdiecker reaction-Heck coupling* to prepare aryl substituted (2*E*,4*E*)-dienoic acids, esters, and amides.

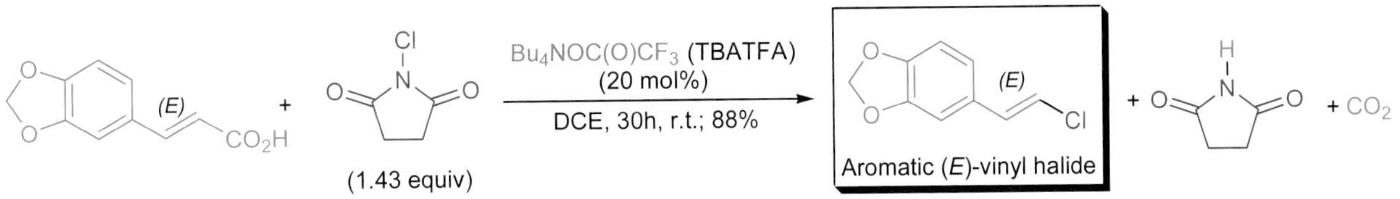

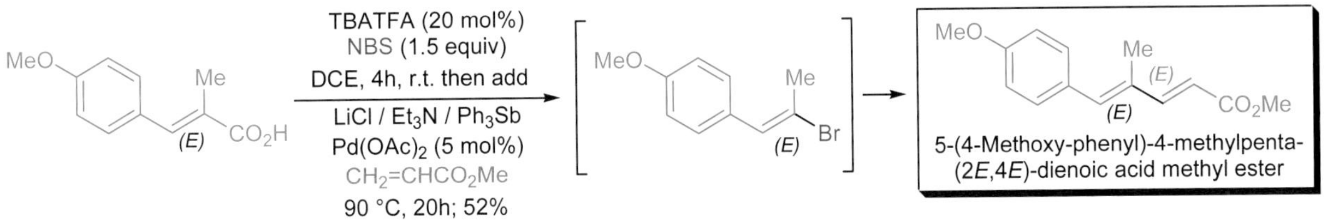

The *classical Hunsdiecker reaction* was utilized in the laboratory of P.J. Chenier for the preparation of a highly strained cyclopropene, tricyclo[3.2.2.0^{2,4}]non-2(4)-ene.[38] The *Diels-Alder cycloaddition* was used to prepare the bicyclic 1,2-diacid, which surprisingly failed to undergo the *Cristol-Firth modified Hunsdiecker reaction*, most likely due to the unreactive nature of the diacid mercuric salt. However, the classical conditions proved to work better to afford the bicyclic 1,2-dibromide in modest yield. Treatment of this dibromide with *t*-BuLi generated the desired strained cyclopropene, which was trapped with diphenylisobenzofuran (DPIBF).

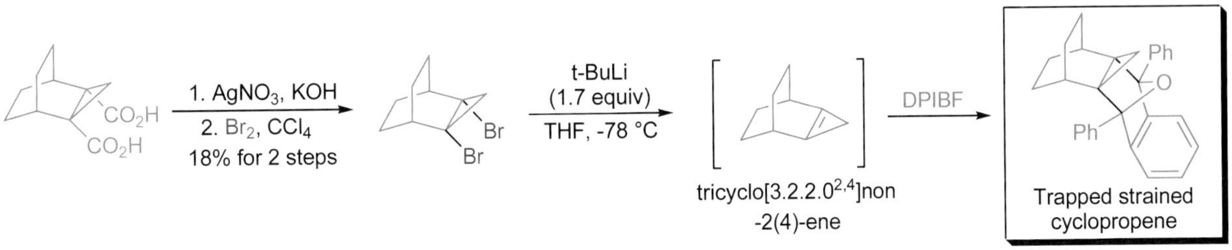

During the final stages of the asymmetric total synthesis of antimitotic agents (+)- and (-)-spirotryprostatin B, the C8-C9 double bond had to be installed, and at the same time the carboxylic acid moiety removed from C8. R.M. Williams et al. found that the *Kochi- and Suárez modified Hunsdiecker reaction* using LTA or PIDA failed and eventually the *Barton modification* proved to be the only way to achieve this goal.[39] After the introduction of the bromine substituent at C8, the C8-C9 double bond was formed by exposing the compound to sodium methoxide in methanol. This step not only accomplished the expected elimination but also epimerized the C12 position to afford the desired natural product as a 2:1 mixture of diastereomers at C12. The two diastereomers were easily separated by column chromatography.

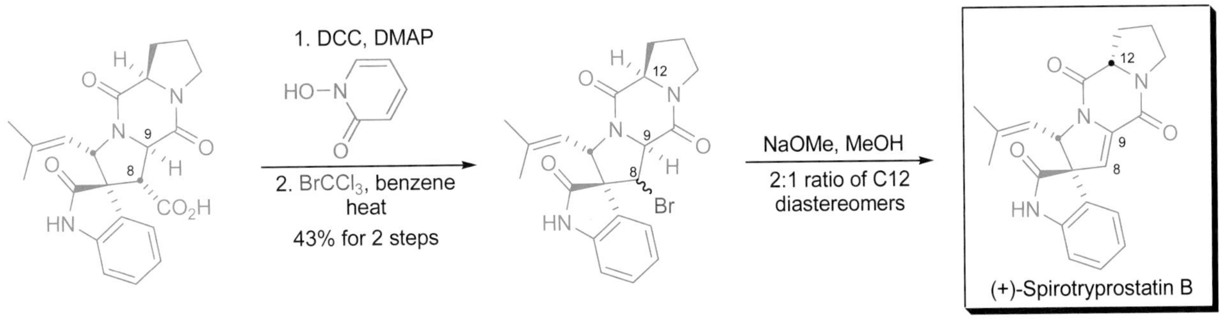

JACOBSEN HYDROLYTIC KINETIC RESOLUTION

(References are on page 606)

Importance:

[*Seminal Publications*[1-5]; *Reviews*[6-15]; *Modifications & Improvements*[16-24]]

In 1995, a few years after the discovery of the enantioselective epoxidation of unfunctionalized olefins (*Jacobsen-Katsuki epoxidation*), E.N. Jacobsen and co-workers discovered that *meso* epoxides undergo *asymmetric ring-opening* (*ARO*) by various nucleophiles (e.g., $TMSN_3$) in the presence of catalytic amounts of chiral Cr(III)(salen) complexes.[3] Although several enantioselective ring-opening reactions of epoxides were known at the time,[1,2] it was shown that the chromium(III)-salen complex catalyzed these ring-opening reactions with an unprecedented high level of enantioselectivity. In 1997, it was discovered that Co(III)salen complexes catalyzed the reaction of racemic terminal epoxides with water to afford highly enantiomerically enriched terminal epoxides and diols. This method is known as the *Jacobsen hydrolytic kinetic resolution* (*HKR*).[5] The general features of this reaction are: 1) racemic terminal epoxides are readily available and inexpensive substrates; 2) water is the most environmentally benign reactant possible; 3) catalyst loadings are low (0.5-5 mol%); 4) both enantiomers of the catalyst are readily available; 5) the scale of the reaction has no effect on the yield and enantiomeric excess (mg to ton scale); 6) the enantioselectivity of the ring-opening is extremely high (k_{rel} = >100); 7) the scope of substrates is completely general and practically every terminal epoxide undergoes *HKR*; 8) both products of the *HKR* are isolated in a highly enantio-enriched form (>99% *ee*); 9) separation of the products is straightforward based on the large difference of boiling points and solubility of epoxides and diols; 10) the yields are generally high considering that the theoretical maximum yield for each of the products is 50%; 11) solvent-free conditions can be achieved in many cases (unless the epoxide is too hydrophobic) and generally the volumetric productivity is very high; and 12) the catalyst can be recovered and reused many times without noticeable decrease of its activity.

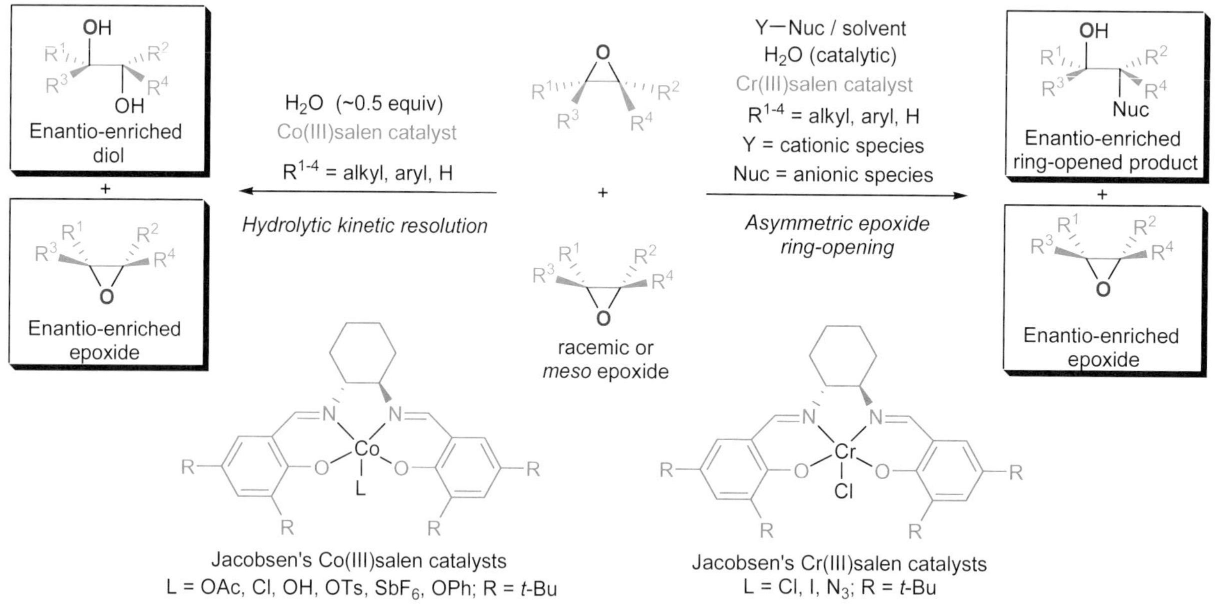

Mechanism:[25-28]

The mechanism of the *Jacobsen HKR* and *ARO* are analogous. There is a second order dependence on the catalyst and a cooperative bimetallic mechanism is most likely. Both epoxide enantiomers bind to the catalyst equally well so the enantioselectivity depends on the selective reaction of one of the epoxide complexes. The active species is the Co(III)salen-OH complex, which is generated from a complex where L≠OH. The enantioselectivity is counterion dependent: when L is only weakly nucleophilic, the resolution proceeds with very high levels of enantioselectivity.

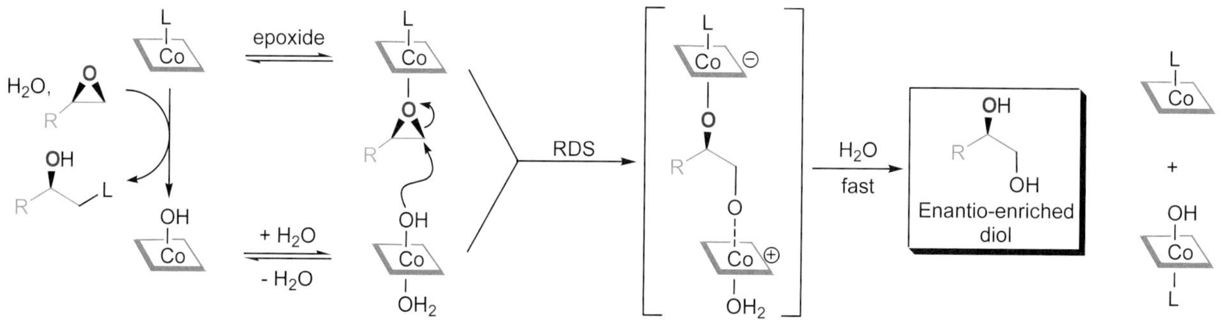

Synthetic Applications:

In the laboratory of J. Mulzer, the total synthesis of laulimalide, a microtubule stabilizing antitumor agent, was accomplished.[29] The C9 stereochemistry of the natural product was introduced using the *Jacobsen HKR* on a diastereomeric mixture of a terminal epoxide. The epoxide mixture was prepared *via* the *Corey-Chaykovsky epoxidation* of citronellal. The *HKR* proceeded in high yield and high selectivity at room temperature, and the products were easily separated by flash chromatography. The diol was converted into the diastereomerically pure epoxide in three steps.

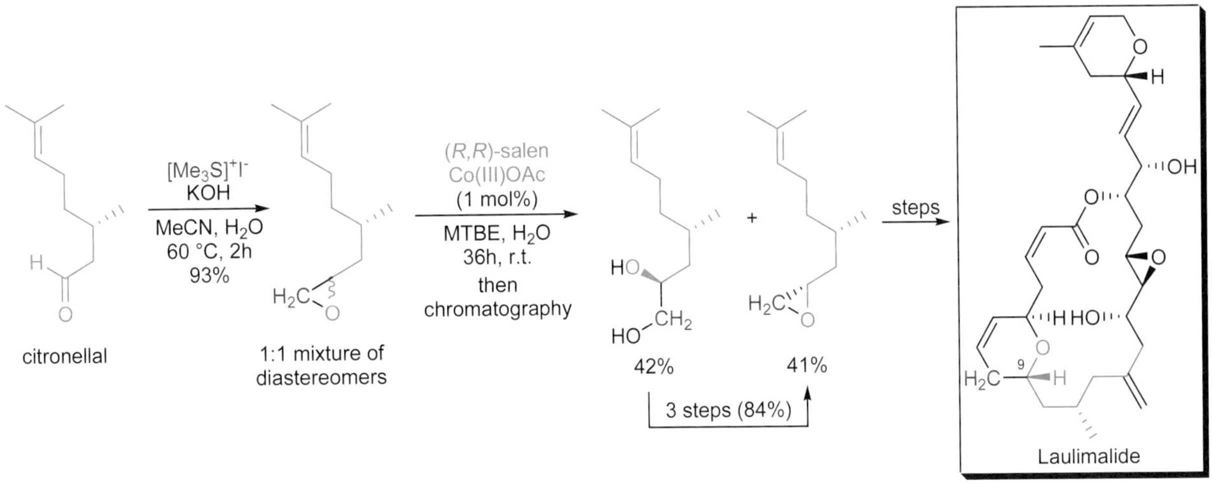

The highly convergent total synthesis of the antitumor agent fostriecin (CI-920) was achieved by E.N. Jacobsen and co-workers.[30] The goal was to make the synthetic route flexible enough to prepare structural analogs of the natural product. One of the key building block terminal epoxides was prepared in enantio-enriched form by the *Jacobsen HKR*. The racemic epoxide was readily available by the epoxidation of the inexpensive methyl vinyl ketone. However, the *HKR* catalyst was easily reduced to its Co(II) form and precipitated with low substrate conversion. This problem was resolved by carrying out the reaction in the presence of oxygen, which reoxidized the inactive Co(II)salen complex to the catalytically active Co(III)salen complex. The enantiopure epoxide was the source for the C9 stereocenter of the product.

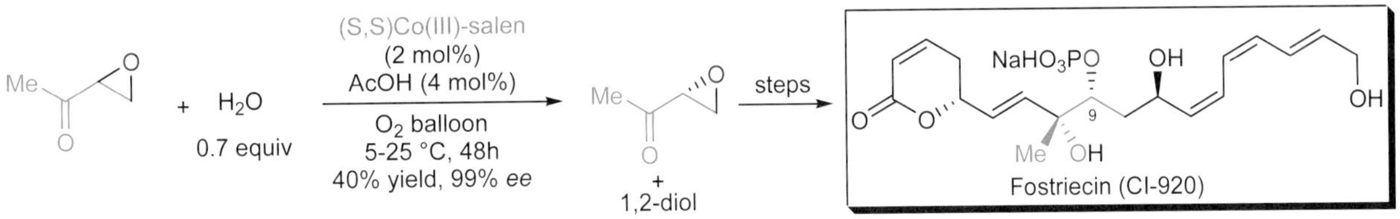

Annonaceous acetogenins have shown potent activity as inhibitors of certain tumor cells. The (4*R*)-hydroxylated analogue of the naturally occurring annonaceous acetogenin bullatacin was synthesized by Z.-J. Yao et al., and it showed enhanced cytotoxicity compared to other analogues.[31] This compound combines the advantages of bullatacin, one of the most potent naturally occurring acetogenins, and the previous analogues. The (4*R*)-hydroxylated butenolide subunit was introduced by the ring opening of a diastereomerically pure epoxide, which was prepared by the *Jacobsen HKR* in high yield and with almost perfect diastereoselectivity. This approach will allow the synthesis of other (4*R*)-hydroxylated analogs of annonaceous acetogenins.

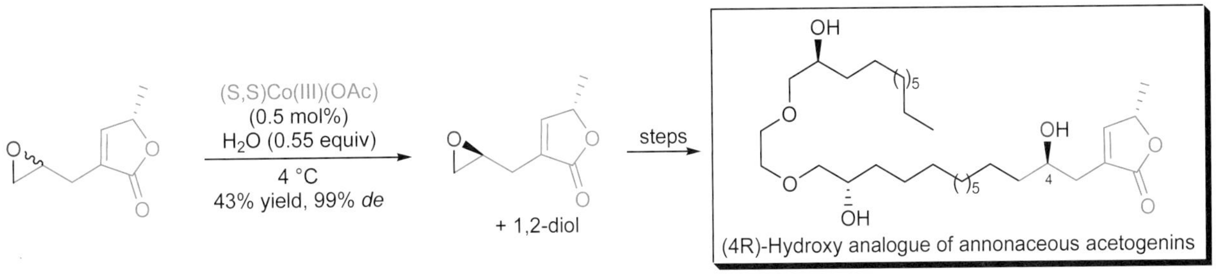

JAPP-KLINGEMANN REACTION

(References are on page 608)

Importance:

[*Seminal Publications*[1-4]; *Reviews*[5-7]; *Modifications & Improvements*[8-11]]

In 1887, F.R. Japp and F. Klingemann attempted to prepare an azo ester by coupling benzenediazonium chloride with the sodium salt of ethyl-2-methylacetoacetate.[1] However, the isolated product turned out to be the phenylhydrazone of ethyl pyruvate, which contained two carbon atoms less than the expected azo ester.[2-4] Subsequent experiments showed that the reaction was general and the initial coupling product was the azo ester, which was unstable under the reaction conditions and it rapidly rearranged to the phenylhydrazone with loss of the aliphatic acyl group. The coupling reaction between aryldiazonium salts and 1,3-dicarbonyl compounds to yield arylhydrazones is known as the *Japp-Klingemann reaction*. The general features of the reaction are: 1) the substituted arenediazonium salts are prepared from the corresponding *o*-, *m*-, and *p*-substituted anilines *via diazotization* (treatment with HNO_2); 2) the reaction works for compounds having an acidic C-H bond between two or three electron-withdrawing groups (e.g., substituted β-diketones, β-keto esters, malonic esters, cyanoacetic esters, or alkali salts of their corresponding acids); 3) if the coupling is carried out with the alkali metal salt of a β-keto acid, the carboxylate anion will undergo decarboxylation (CO_2 is lost) to give the arylhydrazone of the corresponding 1,2-diketone; 4) when a mixed β-diketone (having both an aliphatic and an aromatic acyl group) is used, the aliphatic acyl group will be cleaved preferentially; 5) when acyl derivatives of acetoacetic esters are used (R^2=acyl), the products are the monoarylhydrazones of α,β-diketo esters; 6) cyclic β-keto esters undergo ring-opening in the second stage of the reaction; 7) alkali metal salts of cyclic β-keto acids are not opened, but rather they undergo decarboxylation to give 1,2-diketone monoarylhydrazones; 8) the coupling is usually carried out in acidic or basic aqueous medium at 0 °C and if solubility of the substrate is poor, ethanol or methanol is added; 9) under basic conditions both stages of the reaction take place, whereas under acidic conditions the azo compound can be isolated, and it has to be treated with a mild base to bring about the rearrangement; 10) the rate of the reaction depends on the C-H acidity of the 1,3-dicarbonyl compound and the more activated compounds tend to react faster; 11) excess diazonium salt leads to numerous decomposition products, so the use of one equivalent is advised; 12) the reaction is easy to monitor visually, since the intermediate azo compounds are more highly colored than the product arylhydrazones; and 13) the main use of arylhydrazones is as substrates for the *Fischer indole synthesis* as well as for the synthesis of enantiopure amino acids.

Japp and Klingemann, 1877:

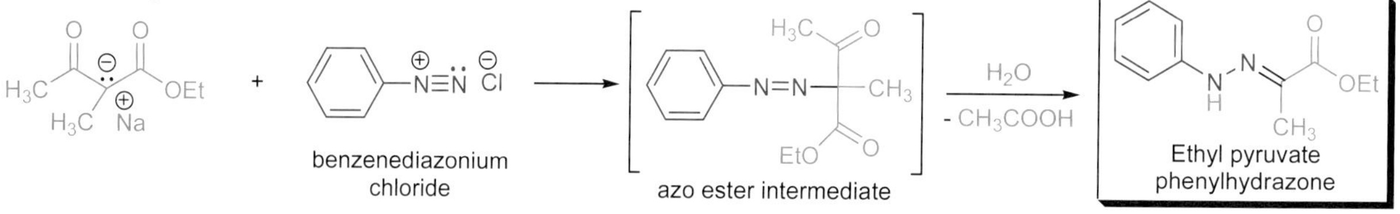

General equation:

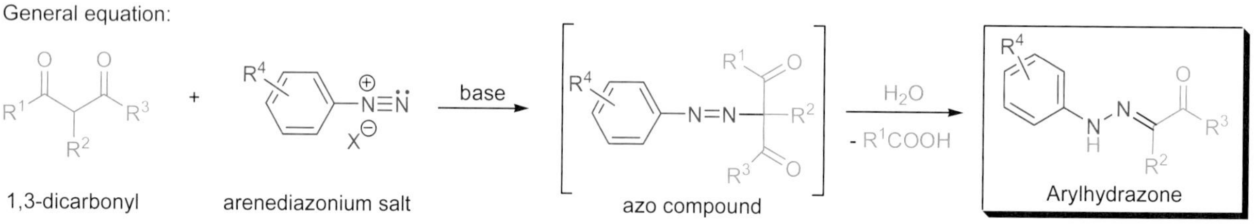

R^1 = alkyl, aryl; R^2 = H, alkyl, aryl, acyl, CN, Cl, Br; R^3 = *O*-alkyl, *O*-aryl, OH; R^4 = electron-withdrawing or electron-donating groups

Mechanism: [12-21]

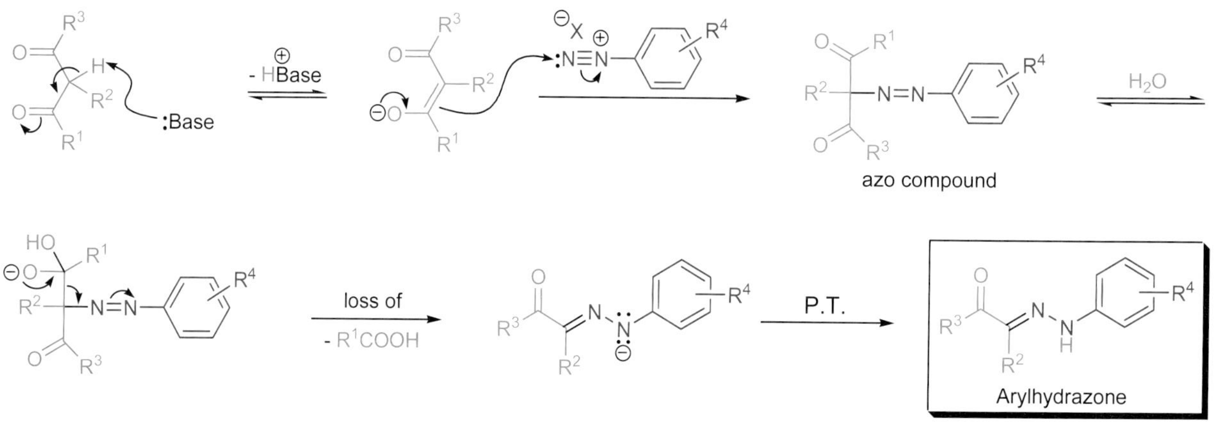

Synthetic Applications:

The first enantioselective total synthesis of (–)-gilbertine was accomplished by S. Blechert and co-workers using a *cationic cascade cyclization* as the key step.[22] The indole moiety was introduced by first applying the *modified Japp-Klingemann reaction* on a substituted formylcyclohexanone precursor followed by the *Fischer indole synthesis* of the resulting phenylhydrazone. The benzenediazonium chloride was prepared prior to the reaction by treating aniline with concentrated HCl/ aqueous $NaNO_2$. Then the strongly acidic solution was buffered by the addition of NaOAc before the formylcyclohexanone derivative was added. The buffering increased the yield of the phenylhydrazone from 10% to 90%!

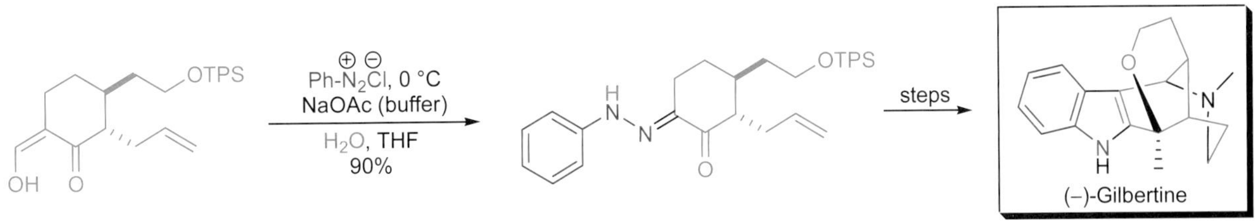

The *Japp-Klingemann reaction* was the key step during the first synthesis of the pentacyclic pyridoacridine marine cytotoxic alkaloid arnoamine A by E. Delfourne et al.[23] The diazonium salt was added to a vigorously stirred solution of ethyl-2-methyl-3-oxobutyrate in ethanol containing KOH, NaOAc and water. The resulting hydrazone was exposed to polyphosphoric acid to form the indole ring.

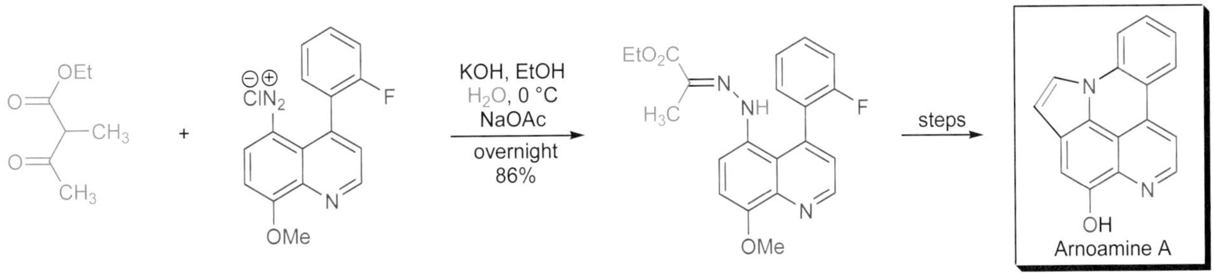

The macrolide soraphen A was shown to exhibit potent fungicidal activity against a variety of plant pathogenic fungi. In the laboratory of J.-L. Sinnes, a new approach was undertaken in which the natural product was degraded to a key lactone, which was used to build several simplified analogs of soraphen A.[24] The key degradation step was the *Japp-Klingemann reaction* of the macrocyclic β-keto ester in its enol form. Treatment of this enol with 4-(methoxyphenyl)diazonium tetrafluoroborate under mildly basic conditions resulted in the quantitative cleavage of the C-C bond of the macrocycle. Since the natural product was very sensitive to strong acids and bases, this approach was a mild alternative to a *retro-Claisen reaction*, which would have required the use of strongly acidic or basic conditions.

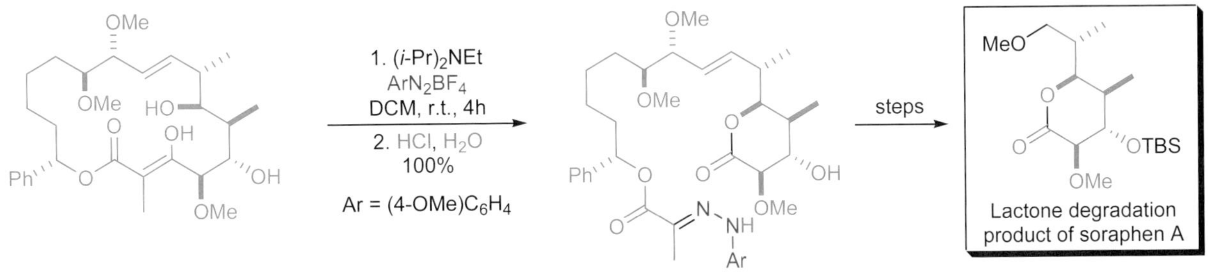

A new heterocyclic ring system, *5H,12H*-[1]Benzoxepino[4,3-*b*]indol-6-one, was prepared by the *Fischer indole cyclization* of a substituted benzoxepin-5b-one phenylhydrazone by G. Primofiore and co-workers.[25] The phenylhydrazone precursor was prepared *via* the *Japp-Klingemann reaction* of the corresponding 3,4-dihydro-4-hydroxymethylene[1]benzoxepin-5(2*H*)-one.

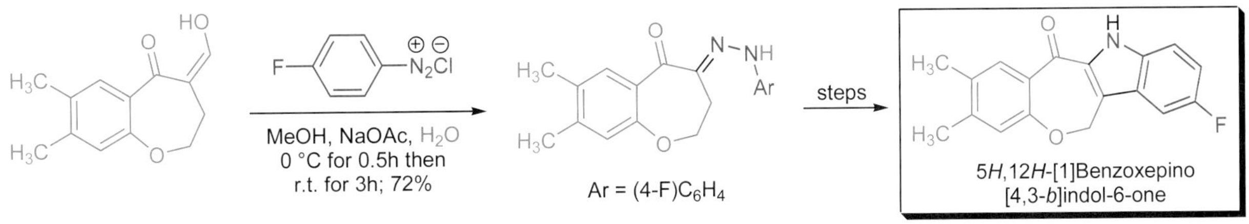

JOHNSON-CLAISEN REARRANGEMENT

(References are on page 609)

Importance:

[*Seminal Publication*[1]; *Reviews*[2-6]]

In 1970, W.S. Johnson reported a reaction in which allylic alcohols were heated in the presence of excess triethyl orthoacetate under weakly acidic conditions (e.g., catalytic amounts of propionic acid).[1] The initial product was a ketene acetal that underwent a facile *[3,3]-sigmatropic rearrangement* to afford γ,δ-unsaturated esters. This method is a modification of the original *Claisen rearrangement*, and is referred to as the *Johnson-Claisen-* or *ortho ester Claisen rearrangement.* The reaction is highly stereoselective and is well-suited for the synthesis of *trans*-disubstituted olefinic bonds. The temperature required for the transformation is usually 100-180 °C. The rearrangement can be significantly accelerated by clay-catalyzed microwave thermolysis.[7] While the traditional *Claisen rearrangement* has excellent acyclic stereocontrol, the *Johnson-Claisen rearrangement* exhibits only modest levels of acyclic stereoselection when the double bond is disubstituted. However, using allylic alcohols substituted at the 2-position affords trisubstituted alkene products with significant levels of diastereoselection.[8] This is explained by 1,3-diaxial nonbonding interactions in the chairlike transition state. Therefore, the *Johnson-Claisen rearrangement* of *(E)*-allylic alcohols mainly give *syn* products while *(Z)*-allylic alcohols predominantly give *anti* products.

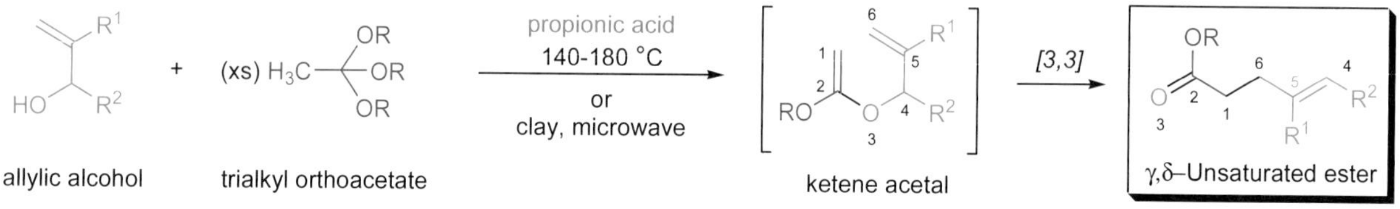

Mechanism: [1,8]

The reaction starts with the exchange one of the alkoxy groups of the ortho ester for the allylic alcohol under acid catalysis. The resulting mixed ortho ester then eliminates a molecule of alcohol to afford an unstable ketene acetal, which undergoes a *[3,3]-sigmatropic shift*. In all of the known Claisen rearrangements, acyclic systems prefer chairlike transition states, whereas cyclic systems may prefer boatlike transition states due to conformational constraints. The ratio of the products will depend on the energy difference between the transition states. The *Johnson-Claisen rearrangements* of secondary allylic alcohols proceed with high *(E)*-selectivity due to the destabilizing 1,3-diaxial interactions in the transition state, which would lead to the (*Z*)-isomer.

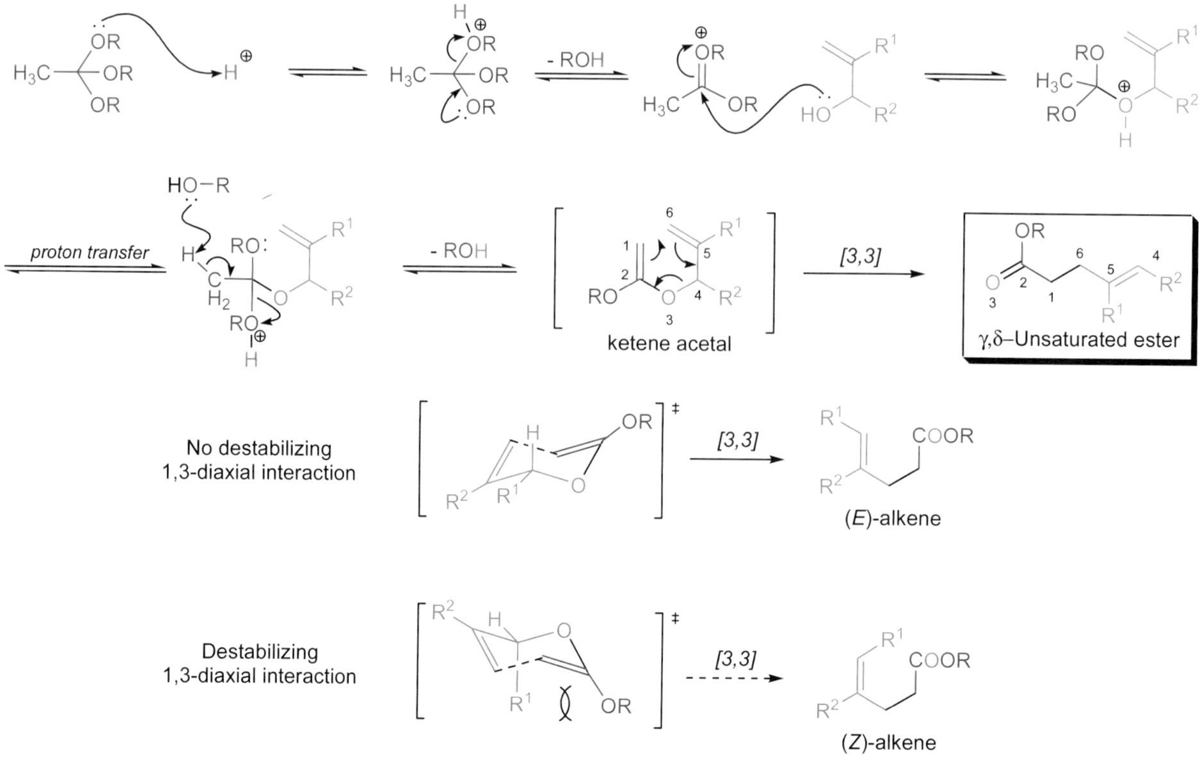

Synthetic Applications:

The potent antitumor agent halomon has a tertiary chlorinated carbon stereocenter at C3, which also contains an α-chlorovinyl group. C. Mioskowski and co-workers developed a strategy that enabled them to prepare a wide range of analogs and establish the correct stereochemistry at C3.[9] These operations were achieved by using a *Johnson-Claisen rearrangement* of a *trans*-dichlorinated allylic alcohol. The reaction was carried out in trimethyl orthoacetate as the solvent and using *p*-toluenesulfonic acid instead of the usual propionic acid as the catalyst. Interestingly, no other *[3,3]-sigmatropic rearrangements* (*Cope*, *Stevens*, *Claisen* or *Ireland-Claisen*) were successful to bring about the same transformation. Halomon was synthesized in 13 steps starting from 2-butyne-1,4-diol with an overall yield of 13%.

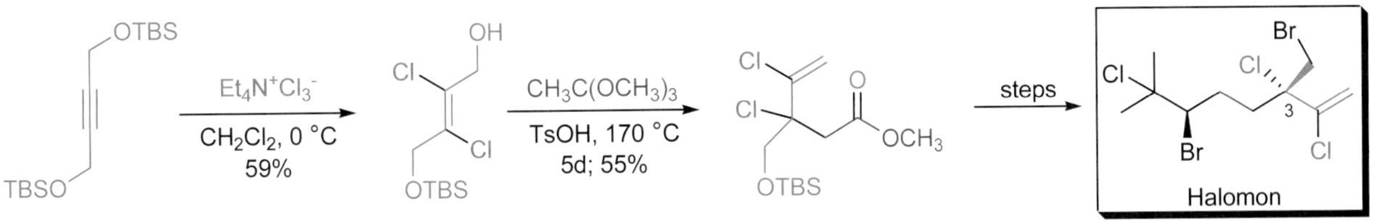

During the total synthesis of the pentacyclic sesquiterpene dilactone (±)-merrilactone A by S.J. Danishefsky et al., a two-carbon unit was introduced at C9 by a *Johnson-Claisen rearrangement.*[10] This high yielding transformation was carried out in the presence of catalytic 2,2-dimethyl propanoic acid at 135 °C using mesitylene as the solvent. A mixture of diastereomeric esters were formed, which were later hydrolyzed and subjected to *iodolactonization* to form the second lactone ring present in merrilactone A. The natural product was synthesized in 20 steps with an overall yield of 10.7%.

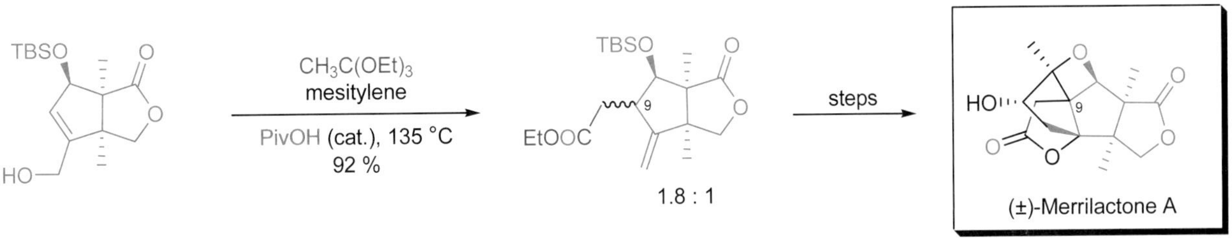

The enantioselective total synthesis of the 13-membered macrolide fungal metabolite (+)-brefeldin A was accomplished using a *triple chirality transfer process* and *intramolecular nitrile oxide cycloaddition* in the laboratory of D. Kim.[11] To set the correct stereochemistry at C9, the stereoselective *ortho ester Claisen rearrangement* was applied on a chiral allylic alcohol precursor. The rearrangement was catalyzed by phenol and it took place at 125 °C in triethyl orthoacetate to give 84% isolated yield of the desired diester.

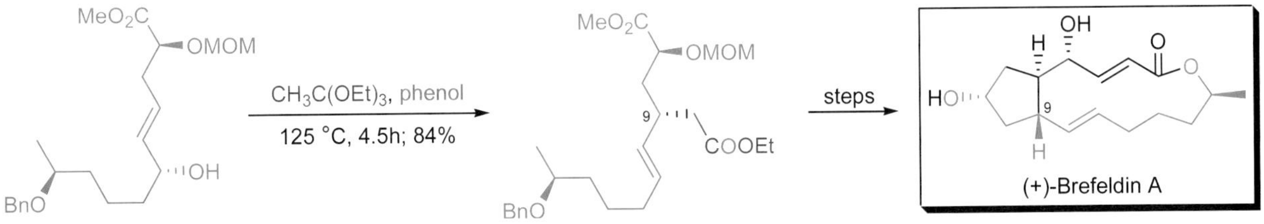

The C7 quaternary stereocenter of (±)-gelsemine was established utilizing a *Johnson-Claisen rearrangement* by S.J. Danishefsky and co-workers.[12] The starting stereoisomeric allylic alcohols were individually subjected to the rearrangement conditions, and each gave rise to the same γ,δ-unsaturated ester.

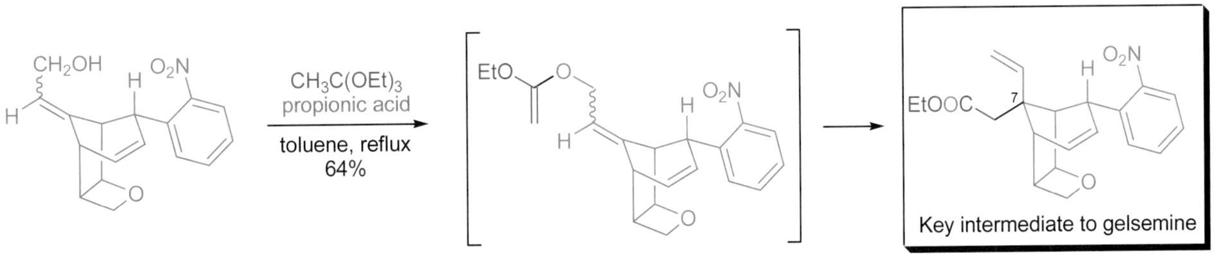

JONES OXIDATION / OXIDATION OF ALCOHOLS BY CHROMIUM REAGENTS

(References are on page 609)

Importance:

[*Seminal Publications*[1,2]; *Reviews*[3-7]; *Modifications & Improvements*[8-20]]

In 1946, E.R.H. Jones and co-workers successfully converted alkynyl carbinols with chromic acid (CrO_3 mixed with dilute sulfuric acid) to the corresponding alkynyl ketones without oxidizing the sensitive triple bond.[1] The reaction was carried out in acetone by slowly adding the aqueous chromic acid to the substrate at ambient temperature, and the product was isolated in high yield. The oxidation of primary and secondary alcohols with chromic acid is referred to as the *Jones oxidation*. The general features of the reaction are: 1) the chromic acid (H_2CrO_4) can be prepared by dissolving chromic trioxide (CrO_3) or a dichromate salt ($Cr_2O_7^{2-}$) in acetic acid or in dilute sulfuric acid; 2) the oxidation is usually carried out in acetone, which serves a dual purpose: it dissolves most organic substrates, and it reacts with any excess oxidant so it protects the product from overoxidation; 3) in practice the alcohol substrate is titrated with the aqueous solution of the oxidant; 4) excess of the reagent should be avoided because other functional groups of the substrate may be oxidized; 5) the process is amenable to large-scale oxidations; 6) primary alcohols are converted to carboxylic acids with the intermediacy of aldehydes that sometimes can be isolated by distillation if the aldehyde is volatile; 7) secondary alcohols are converted to the corresponding ketones; 8) allylic and benzylic alcohols are efficiently oxidized to the corresponding aldehydes with little or no over-oxidation; 9) glycols and acyloins often suffer C-C bond cleavage under the reaction conditions, but in certain cases the addition of Mn^{2+} or Ce^{3+} salts prevents this side reaction;[10] 10) isolated double and triple bonds remain unchanged, but α,β-unsaturated aldehyde products may undergo double bond isomerization; 11) in rigid cyclic systems axial alcohols tend to react faster than the equatorial alcohols; 12) acid sensitive protecting groups are easily removed under the reaction conditions; and 13) free amines are often incompatible with the *Jones oxidation*, and they need to be protected as the corresponding perchlorate salts prior to the oxidation. For particularly acid sensitive or otherwise delicate substrates the use of the strongly acidic *Jones reagent* is clearly not the best method of oxidation, so several mildly acidic CrO_3-derived oxidizing agents were developed: 1) Sarett prepared CrO_3-(pyridine)$_2$ and carried out the oxidations in pyridine as the solvent;[8] 2) due to difficulties during work-up and with the isolation of products, the *Sarett oxidation* was modified by Collins by using the macrocrystalline form of the reagent that was soluble in dichloromethane and made the oxidations very fast at room temperature (*Collins oxidation*) and highly tolerant toward a wide range of functional groups;[11] 3) Corey et al. developed the mildly acidic pyridinium chlorochromate (PCC) and the neutral pyridinium-dichromate (PDC) reagents that rapidly oxidize 1° and 2° alcohols, as well as allylic and benzylic alcohols in dichloromethane to the corresponding aldehydes and ketones;[12,16] and 4) a large number of other very mild CrO_3-amine reagents have been developed.[5,7]

Jones oxidation (1946):

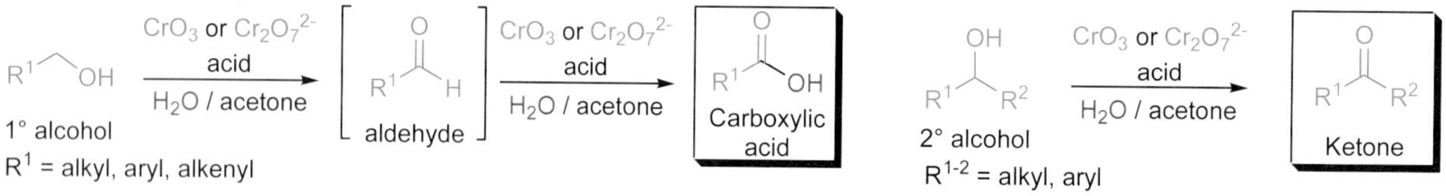

Sarett and Collins oxidations (1953 & 1968):

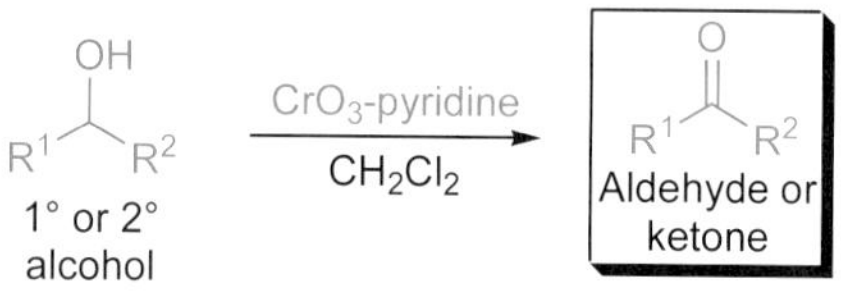

PCC and PDC oxidations (Corey, 1975 & 1979):

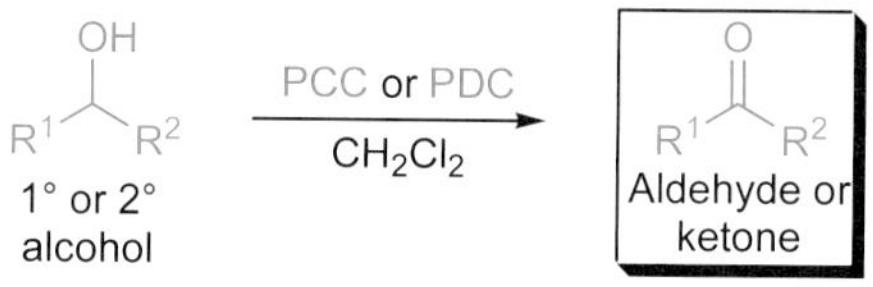

Mechanism: [21,9,22-24]

The concentration and the pH determines the form of $Cr^{(VI)}$ in aqueous solutions: in dilute solution the monomoeric form ($HCrO_4^-$) dominates while in concentrated solution the dimeric form ($HCr_2O_7^-$) is prevalent. The alcohol substrate is first converted to the corresponding chromate ester, which suffers a rate-determining deprotonation by a base to release the $Cr^{(IV)}$ species. This mechanism is supported by a large kinetic isotope effect observed during the oxidation of an α-deuterated alcohol substrate.[21]

Complete mechanism which accounts for the observed stoichiometry:

$$R^1R^2CHOH + Cr^{(VI)} \longrightarrow R^1R^2C{=}O + Cr^{(IV)} + 2\,H^+$$
$$R^1R^2CHOH + Cr^{(IV)} \longrightarrow R^1R^2C{=}O + Cr^{(II)} + 2\,H^+$$
$$Cr^{(II)} + Cr^{(VI)} \longrightarrow Cr^{(III)} + Cr^{(V)}$$
$$R^1R^2CHOH + Cr^{(V)} \longrightarrow R^1R^2C{=}O + Cr^{(III)} + 2\,H^+$$

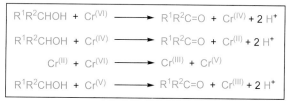

Step #1:

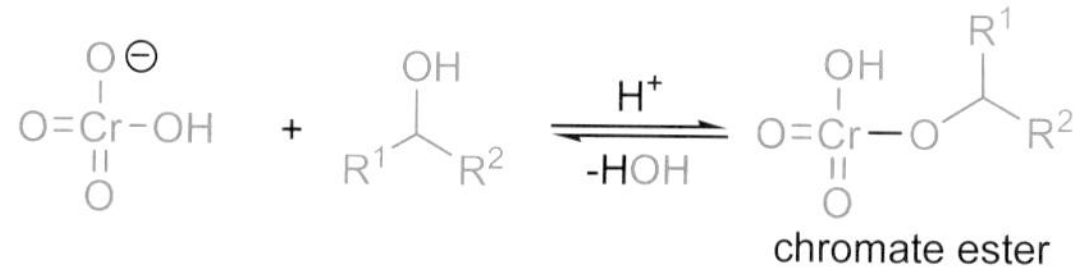

Step #2:

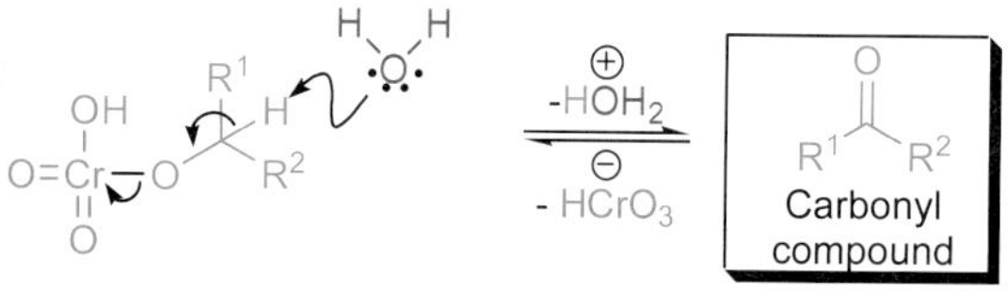

Synthetic Applications:

The *Jones oxidation* was used during the endgame of the total synthesis of (–)-CP-263,114 (Phomoidride B) by T. Fukuyama and co-workers.[25] The secondary alcohol functionality of the side chain on the fully elaborated carbon skeleton was exposed to excess CrO_3 in H_2SO_4 for 20 minutes to afford the corresponding ketone in quantitative yield. The last step was the removal of the *tert*-butyl ester with formic acid to give the natural product in 96% yield.

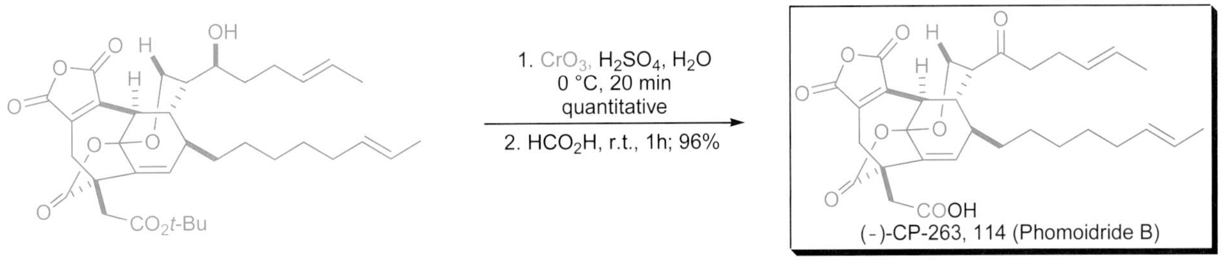

The total synthesis of (±)-bilobalide, a C15 ginkgolide, was accomplished in the laboratory of M.T. Crimmins using a *[2+2] photocycloaddition* as the key step to secure most of the stereocenters.[26] In the final stages of the total synthesis the *Jones oxidation* was used twice. First, the five-membered acetal moiety was oxidized with Jones reagent to the corresponding lactone in refluxing acetone. Next, the five-membered enol ether was epoxidized with excess DMDO and the resulting epoxide was treated with Jones reagent to afford the natural product.

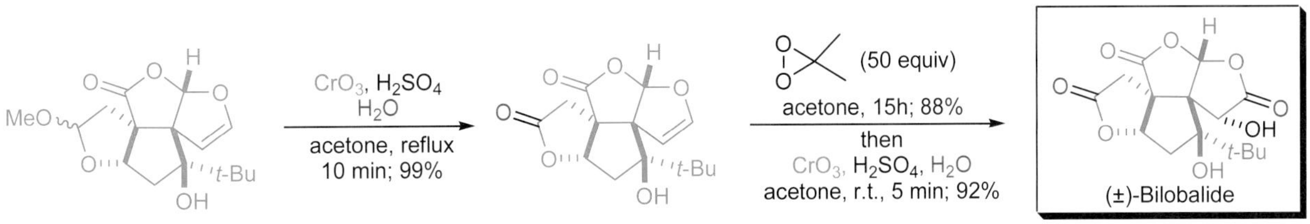

An *α-carbonyl radical cyclization* was the key step in C.-K. Sha's enantioselective total synthesis of the alkaloid (–)-dendrobine.[27] The five-membered nitrogen heterocycle was installed during the final stages of the synthetic effort. The bicyclic azido alcohol intermediate was oxidized using the Jones reagent to give the corresponding azido ketone, which was converted in three steps to the natural product.

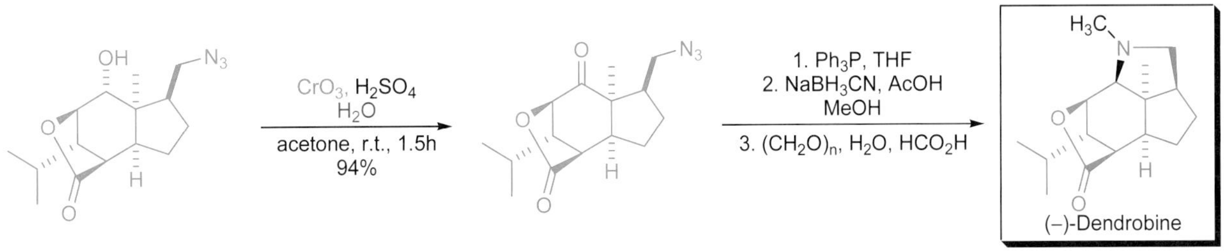

In the laboratory of H. Hagiwara, the first total synthesis of the polyketide natural product (–)-solanapyrone E was achieved.[28] The installation of the pyrone moiety required the addition of the *bis*(trimethylsilyl) enol ether of methyl acetoacetate to a bicyclic aldehyde precursor in the presence of titanium tetrachloride. The resulting δ-hydroxy-β-ketoester was oxidized with the Jones reagent to afford the corresponding β,δ-diketoester in good yield.

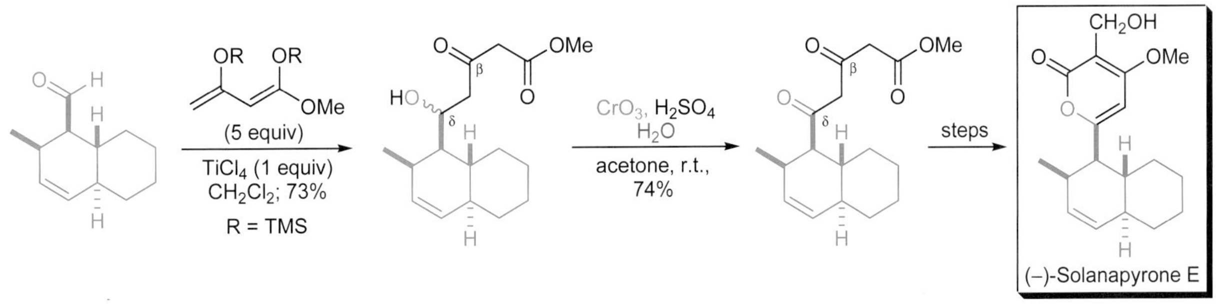

JULIA-LYTHGOE OLEFINATION

(References are on page 610)

Importance:

[*Seminal Publication*[1]; *Reviews*[2-9]; *Modifications & Improvements*[10-22]]

In 1973, M. Julia and J.-M. Paris reported a novel olefin synthesis in which β-acyloxysulfones were reductively eliminated to the corresponding di-, tri-, or tetrasubstituted alkenes.[1] This olefin synthesis requires the following steps: 1) addition of an α-metalated phenylsulfone to an aldehyde or ketone; 2) acylation of the resulting β-alkoxysulfone; and 3) reductive elimination of the β-acyloxysulfone with a single-electron donor to yield the desired alkene. Not long after the seminal publication, B. Lythgoe and P.J. Kocienski explored the scope and limitation, and today this olefination method is known as the *Julia-Lythgoe olefination*.[10-13] The classical *Julia-Lythgoe olefination* has the following general features: 1) high (*E*)-stereoselectivity; 2) the (*E*)-selectivity is increased with increasing chain branching around the newly formed double bond; and 3) the relative stereochemistry in the intermediate β-acyloxysulfones does not influence the geometry of the alkene product. Since the classical procedure was quite tedious (3 steps) to carry out in the laboratory, a more convenient *one-pot modification* was developed by S.A. Julia and co-workers who added α-metalated heteroarylsulfones to carbonyl compounds instead of the traditional phenylsulfones.[15] The initial intermediate β-alkoxy heteroarylsulfone is very labile, and it quickly undergoes the *Smiles rearrangement* in which the heterocycle is transferred from the sulfur to the oxygen atom to afford yet another unstable intermediate, a sulfinate salt. This sulfinate salt readily decomposes to the desired (*E*)-alkene, sulfur dioxide and the metal salt of benzothiazol-2-ol. Several heteroaromatic activators were examined, and it was revealed that not all heteroarylsulfones worked equally well in terms of product yield and stereoselectivity.[8] The BT-sulfones react with α,β-unsaturated or aromatic aldehydes to give conjugated 1,2-disubstituted (*E*)-alkenes. Kocienski found that the PT-sulfone (1-phenyl-1*H*-tetrazol-5-yl sulfone) provides nonconjugated 1,2-disubstituted alkenes with high (*E*)-selectivity if no significant electronic or steric bias is present (*Kocienski-modified Julia olefination*).[17] For the preparation of conjugated 1,2-disubstituted (*Z*)-alkenes, the use of allylic or benzylic TBT-sulfones (1-*t*-butyl-1*H*-tetrazol-5-yl sulfones) is recommended.[18]

Classical Julia-Lythgoe olefination:

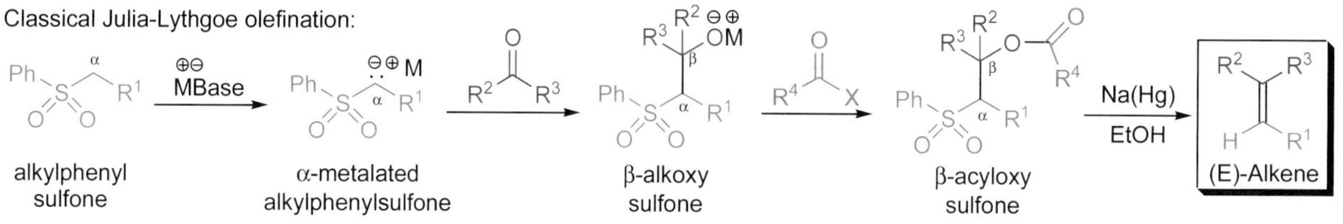

R^1 = H, alkyl, aryl; R^2, R^3 = H, alkyl, aryl, alkenyl; R^4 = alkyl, aryl; X = Cl, Br, OCOR

Modified (One-pot) Julia olefination:

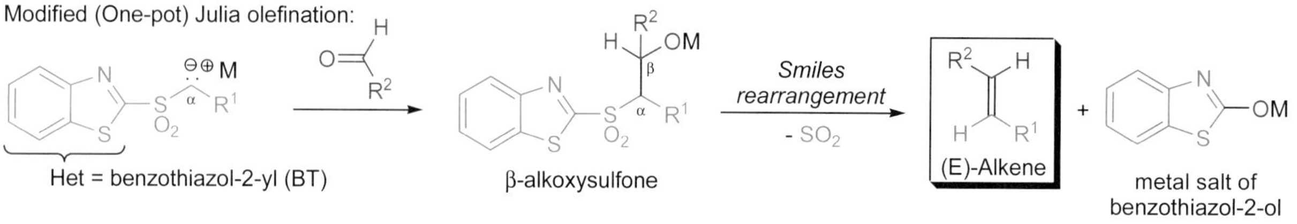

R^1 = H, alkyl, aryl; R^2 = alkyl, aryl,alkenyl; Het = benzothiazol-2-yl (BT), pyridin-2-yl (PYR), 1-phenyl-1*H*-tetrazol-5-yl (PT)

Mechanism: [11,13,3,16]

The exact mechanistic pathway of the *classical J-L olefination* is unknown. Deuterium-labeling studies showed that the nature of the reducing agent (sodium amalgam or SmI_2) determines what type of intermediate (vinyl radical or secondary alkyl radical) is involved.[16] Both intermediates are able to equilibrate to the more stable isomer before conversion to the product. The high (*E*)-selectivity of the *Kocienski-modified reaction* is the result of kinetically controlled irreversible diastereoselective addition of metalated PT-sulfones to nonconjugated aldehydes to yield *anti*-β-alkoxysulfones which stereospecifically decompose to the (*E*)-alkenes.

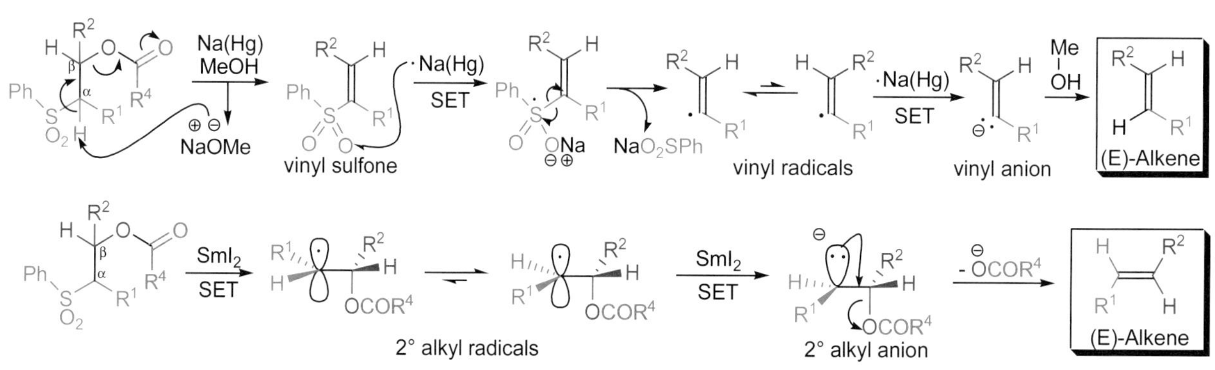

Synthetic Applications:

The first total synthesis of racemic indolizomycin was accomplished by S.J. Danishefsky et al.[23] The natural product's trienyl side chain was elaborated using the classical *J-L olefination*. The macrocyclic α,β-unsaturated aldehyde was treated with an (*E*)-allylic lithiated sulfone to give epimeric acetoxy sulfones upon acetylation. The mixture of epimers was exposed to excess sodium amalgam in methanol to afford the desired (*E,E,E*) triene stereospecifically.

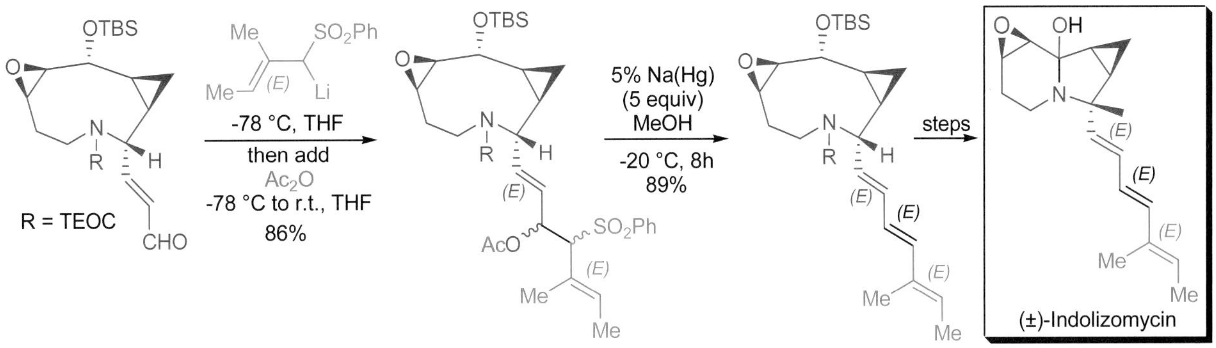

In the asymmetric total synthesis of (–)-callystatin A by A.B. Smith and co-workers, two separate *Julia olefinations* were used to install two (*E*)-alkene moieties.[24] The C6-C7 (*E*)-alkene was installed using the *Kocienski-modified process* in which the PT-sulfone was dissolved along with the α,β-unsaturated aldehyde in DME and treated with NaHMDS in the presence of HMPA. The (*E*)-olefin was the only product but due to the relative instability of the starting PT-sulfone, the isolated yield of the product was only modest.

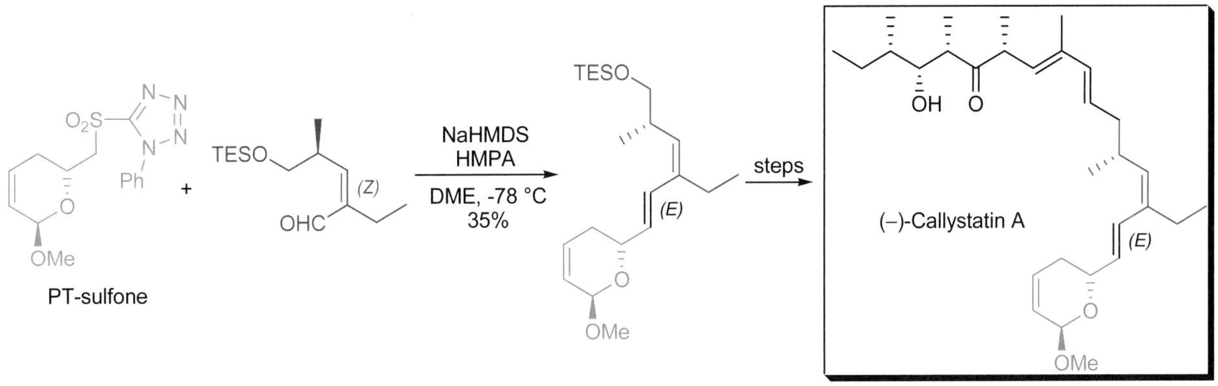

The novel antifungal agent (+)-ambruticin was synthesized in the laboratory of E.N. Jacobsen.[25] The key coupling step in this convergent synthesis was the formation of the C8-C9 (*E*)-alkene *via* the *Kocienski modified Julia olefination*. Interestingly, the coupling showed great selectivity for either the (*E*)- or (*Z*)-stereoisomers depending on the base or solvent used. When NaHMDS was used in THF, the (*Z*)-olefin was formed predominantly (8:1), whereas when LiHMDS was used in DMF/DMPU, the (*E*)-olefin was formed with very high stereoselectivity.

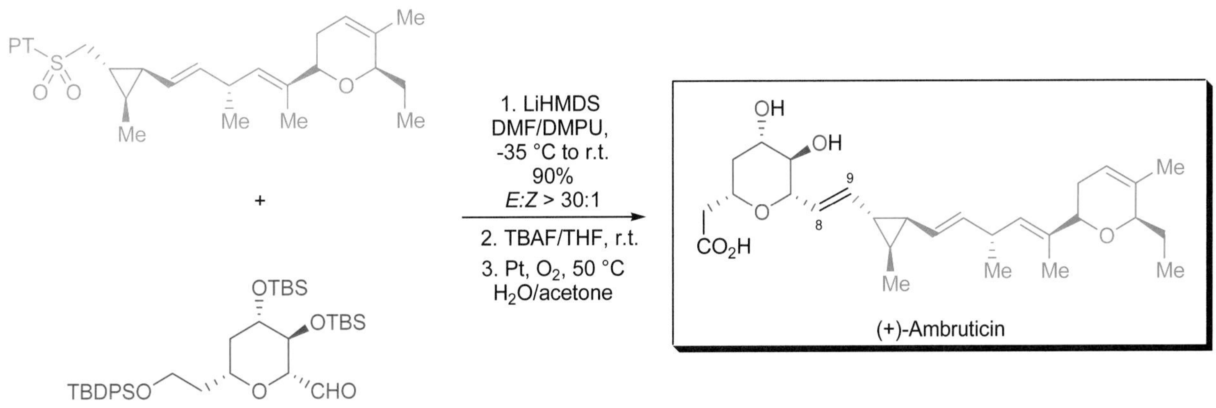

KAGAN-MOLANDER SAMARIUM DIIODIDE-MEDIATED COUPLING

(References are on page 610)

Importance:

[*Seminal Publications*[1-3]; *Reviews*[4-26]]

During the late 1970s, H. Kagan systematically examined the reducing properties of lanthanide(II) iodides. During his studies he found that in the presence of two equivalents of samarium diiodide, alkyl bromides, iodides and tosylates react with aldehydes and ketones to provide the corresponding alcohols.[2] The original transformation was carried out in tetrahydrofuran at room temperature for 24 hours or at reflux for a few hours. Kagan also noted that the addition of catalytic amounts of ferric choride significantly decreased the reaction time. This method was later extensively studied by G.A. Molander. In 1984, he reported the first intramolecular version of this transformation.[3,27] He also discovered that ω-iodoesters undergo intramolecular acyl substitution in the presence of samarium diiodide and catalytic amounts of iron(III) salts.[28] Tandem reactions leading to complex carbocycles were also developed.[29] Today, these transformations are referred to as the *Kagan-Molander samarium diiodide-mediated coupling*. The reaction can be performed in two different ways: 1) adding the ketone to a preformed solution of the organosamarium that is prepared by treating the alkyl halide with two equivalents of samarium diiodide (*samarium Grignard reaction*); and 2) reacting the alkyl halide with samarium diiodide in the presence of the ketone (*samarium Barbier reaction*). The most common method for the preparation samarium diiodide is to react the finely ground samarium metal with diiodomethane, diiodoethane or iodine in tetrahydrofuran.[30,5,10] The general features of the reaction are:[19] 1) it is usually carried out in tetrahydrofuran by employing two equivalents of samarium diiodide in the presence of additives or catalysts; 2) in some cases, tetrahydropyran, alkylnitriles, and benzene were used as solvents; 3) under standard conditions, alkyl bromides and iodides undergo the transformation, but alkyl chlorides are unreactive; 4) reaction of alkyl choride under visible light irradiation was reported; 5) the substrate scope of organic bromides and iodides is wide: primary alkyl-, secondary alkyl, allylic and benzylic halides, iodoalkynes, α-heterosubstituted alkyl halides, and α-halogeno carbonyl compounds (*samarium Reformatsky reaction*) undergo the reaction; 6) aryl, vinyl, and tertiary halides are not viable substrates; they are reduced to the radical stage but are usually not reduced further by samarium diiodide; they instead abstract a hydrogen atom from tetrahydrofuran; and 7) the reaction of aryl chlorides with ketones was reported in benzene as a solvent, where hydrogen abstraction is not feasible. The reactions in most cases are relatively slow in tetrahydrofuran, and the addition of co-solvents or catalysts is necessary. The most commonly used co-solvent is HMPA, which dramatically improves the reducing ability of samarium diiodide ($E°_{(Sm(II)/Sm(III)\ in\ THF)}$ = -1.33V; $E°_{(Sm(II)/Sm(III)/4\ equiv\ HMPA\ in\ THF)}$ = -2.05V).[31] DMPU is also often used as an additive.[19] Several transition metal salts proved to be efficient catalysts for this transformation: iron(III) salts, copper(I)- and copper(II) salts, nickel(II) salts, vanadium trichloride, silver(I) halides, cobalt dibromide, zirconium tetrachloride, and Cp_2ZrCl_2.[19]

Intermolecular reaction (Kagan 1980)

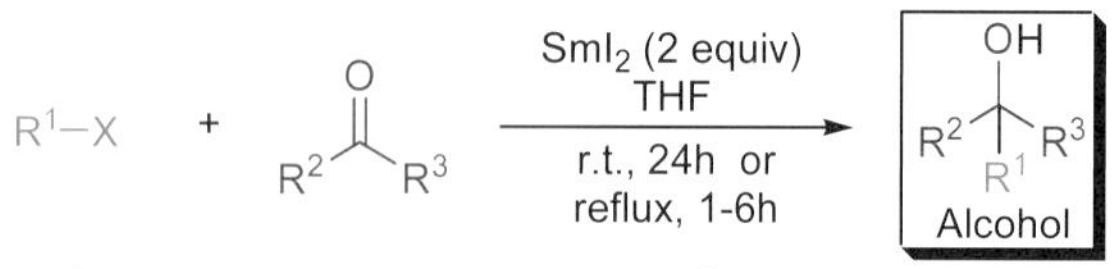

R^1 = alkyl, allyl, benzyl, propargyl; R^2 = H, alkyl, aryl;
R^3 = alkyl aryl; X=Br, I, OTs

Intramolecular reaction (Molander 1984)

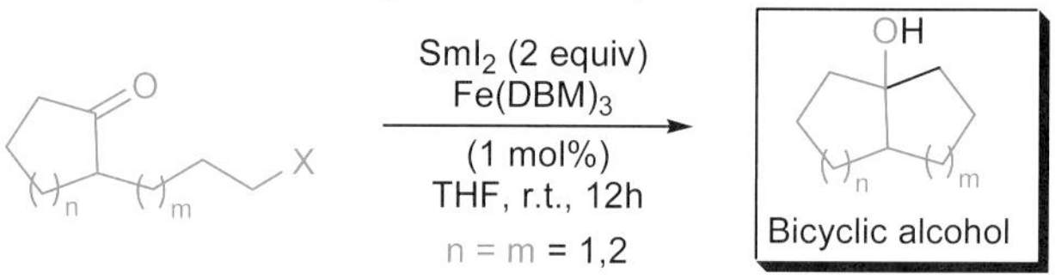

Nucleophilic acyl substitution (Molander 1993):

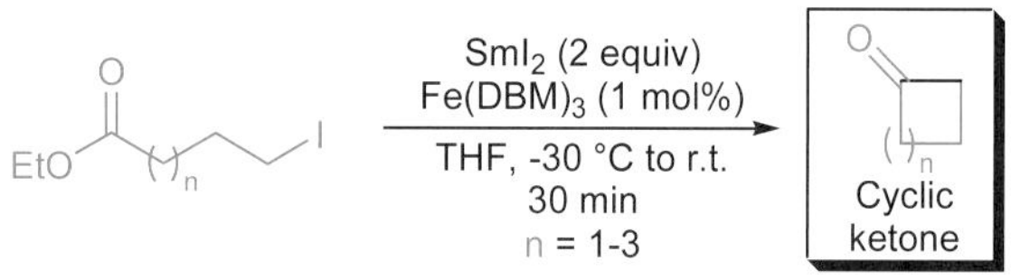

Tandem reactions (Molander 1995):

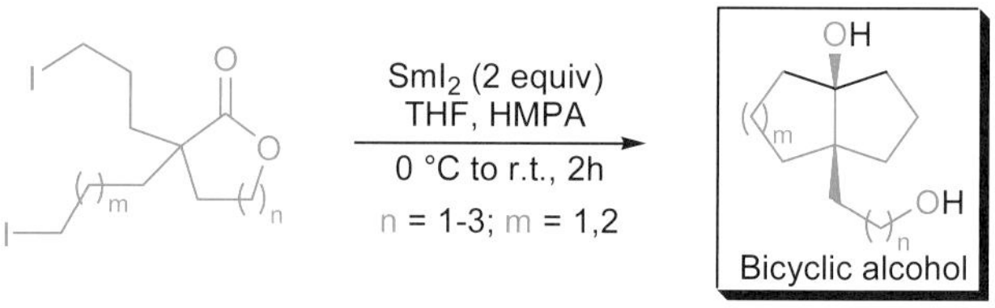

Mechanism:[32-34,7,35-37,31,38-45]

Samarium diiodide is a one electron reductant that is capable of reducing both alkyl halides and carbonyl compounds. The rate of the reduction depends on the nature of the substrate and the reaction conditions. The mechanism of the addition of alkyl halides to carbonyls was extensively studied.[33,7,35] In case of the *samarium Grignard processes*, it was concluded that the reaction proceeds through an organosamarium intermediate. However, the mechanism of the samarium Barbier processes is not fully understood and there is no unambiguous evidence in favor of any of the possible pathways.

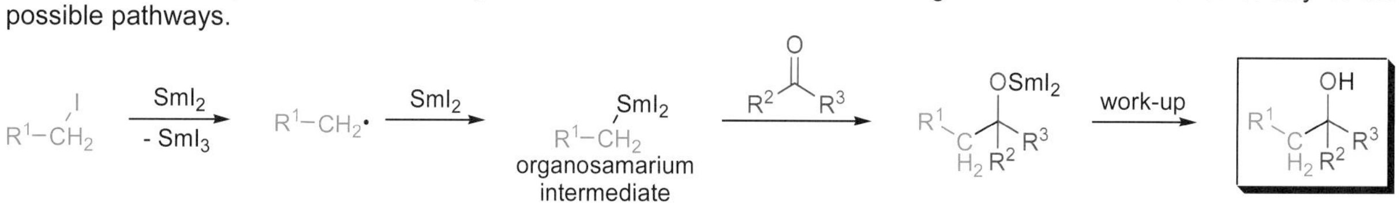

Synthetic Applications:

The ABC ring system of the carbocyclic skeleton of variecolin, a sesterterpenoid natural product was accomplished by G.A. Molander and co-workers.[46] In their approach, they utilized two samarium diiodide mediated processes. First, a primary alkyl iodide was reacted with a ketone substrate in the presence of two equivalents of samarium diiodide and catalytic nickel(II) iodide under *samarium Grignard* conditions. Subsequent oxidation and lactone formation provided the chlorolactone substrate. As alkyl chlorides are less reactive than alkyl bromides and iodides, the second samarium diiodide mediated process, an intramolecular nucleophilic acyl substitution, required visible light irradiation.

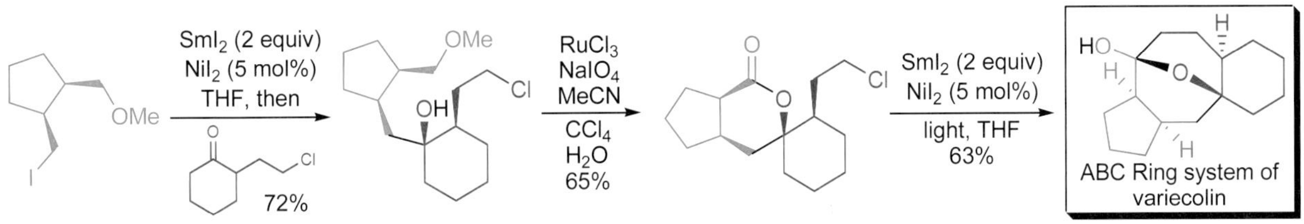

Vinigrol is a tricyclic diterpene with interesting biological activity such as antihypertensive activity and platelet aggregation inhibition property. The eight-membered framework of this natural product was synthesized by F. Matsuda et al. utilizing an *intramolecular Kagan-Molander coupling reaction*.[47] The substrate for the cyclization was prepared starting out from chlorodihydrocarvone in six steps. The samarium diiodide mediated cyclization took place within minutes in tetrahydrofuran using HMPA as the co-solvent.

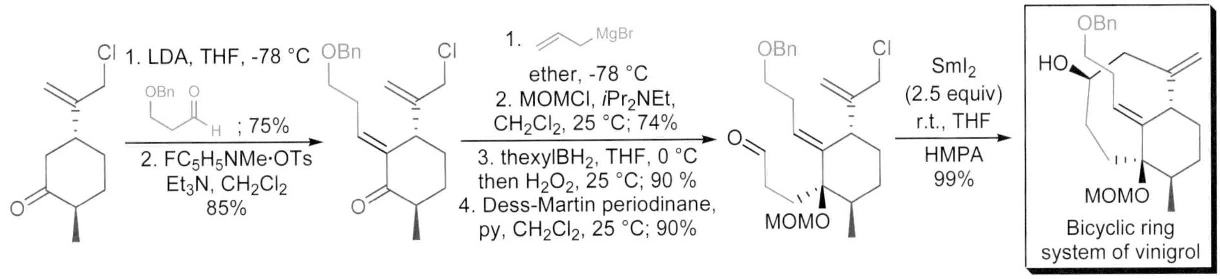

The research group of T. Nakata developed a convergent synthesis for the construction of a *trans*-fused 6-6-6-6-membered tetracyclic ether ring system, a subunit, which is present in several polycyclic marine ether natural products.[48] Late in their synthesis, they utilized a *samarium diiodide mediated nucleophilic acyl substitution* as the key step to form one of the tetrahydropyran rings.

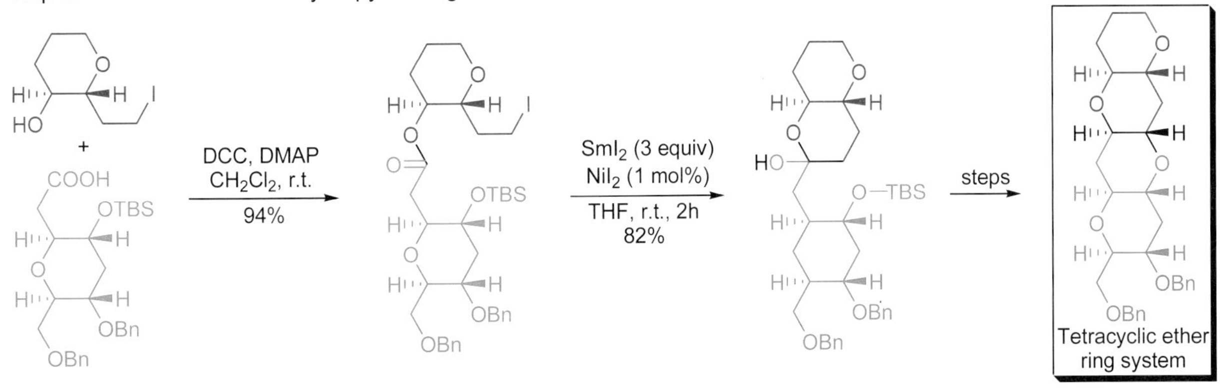

The total synthesis of pederin, a potent insect toxin was achieved by T. Takemura and co-workers.[49] One of the key steps of the synthesis was an *intramolecular samarium diiodide induced Reformatsky reaction* to construct the lactone subunit of the molecule. The transformation was carried out in tetrahydrofuran at 0 °C without the use of additives or catalysts.

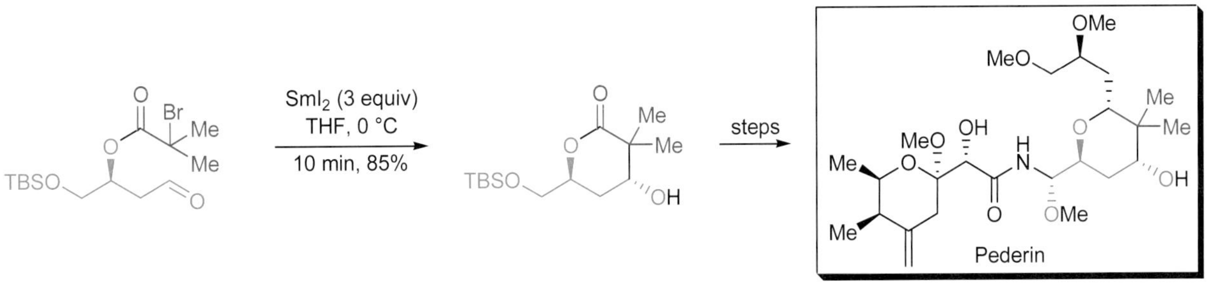

KAHNE GLYCOSIDATION

(References are on page 611)

Importance:

[*Seminal Publications*[1]; *Reviews*[2-9]; *Modifications & Improvements*[10-19]; *Theoretical Studies*[20,21]]

The efficient preparation of glycosides from sterically hindered or otherwise unreactive substrates using standard glycosidation methods (e.g., *Koenigs-Knorr glycosidation*, thioglycoside method, etc.) was a significant challenge until the late 1980s. In 1989, D. Kahne and co-workers developed a novel glycosidation method in which they treated glycosyl sulfoxides with trifluoromethanesulfonic anhydride in toluene at low temperature and to the resulting reaction mixture they added the solution of the nucleophile (alcohols, phenols, or amides) and a base also in toluene.[1] The products were the corresponding *O*- or *N*-glycosides with predominantly α-stereochemistry in the absence of neighboring group participation and with predominantly β-stereochemistry when anchimeric assistance was involved. The highly stereoselective preparation of *O*-, *S*-, or *N*-glycosides *via* the activation of glycosyl sulfoxides is known as the *Kahne glycosidation* (*sulfoxide method*). The general features of this transformation are:[8] 1) the sulfoxides are usually prepared *via* the oxidation of the corresponding thioglycosides (axial thioglycosides are oxidized to give a single sulfoxide diastereomer while equatorial thioglycosides give rise to a mixture of diastereomeric sulfoxides);[22-24] 2) the most common oxidizing agents are *m*CPBA and MMPP; 3) both alkyl and aryl sulfoxides can be used as substrates; 4) the reactivity of aryl glycosyl sulfoxides can be modulated by placing electron-donating or electron-withdrawing substituents on the aromatic ring (multicomponent couplings are possible this way[25]); 5) primary-, secondary and tertiary alcohols, phenols, trialkylstannylated phenols, silylated amides can be used as nucleophiles; 6) the method is especially well-suited for the glycosidation of sterically hindered alcohols, which are unreactive under other glycosidation methods; 7) the most common activating agent is triflic anhydride (Tf_2O) and trimethylsilyl triflate (TMSOTf), but occasionally Lewis acids (e.g., $Cp_2ZrCl_2/AgClO_4$)[16] and mineral acids[14,15] can be used as activating agents; 8) since triflic acid or phenylsulfenyl triflate is generated in the reaction, the use of a hindered, non-nucleophilic base to buffer the reaction mixture is recommended (sometimes the use of a base results in the formation of an orthoester instead of a glycoside, a problem that is resolved by simply omitting the base); 9) the reaction is conducted at low temperatures and is usually complete in a matter of minutes or a few hours; and 10) the stereochemical outcome of the coupling is a function of the solvent and the protecting groups in both the glycosyl donor and acceptor.

Kahne (1994):

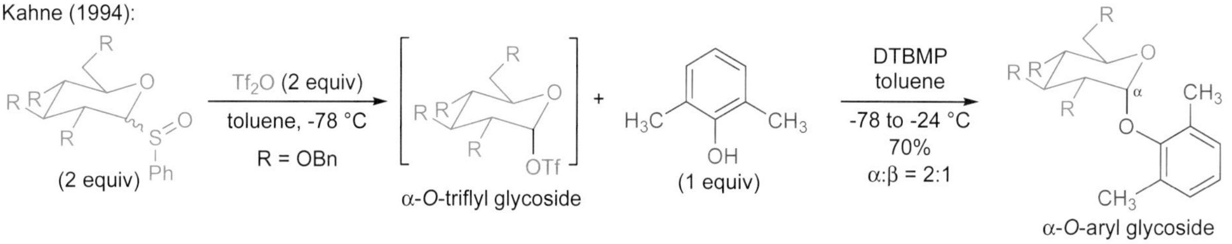

Kahne glycosidation (sulfoxide method):

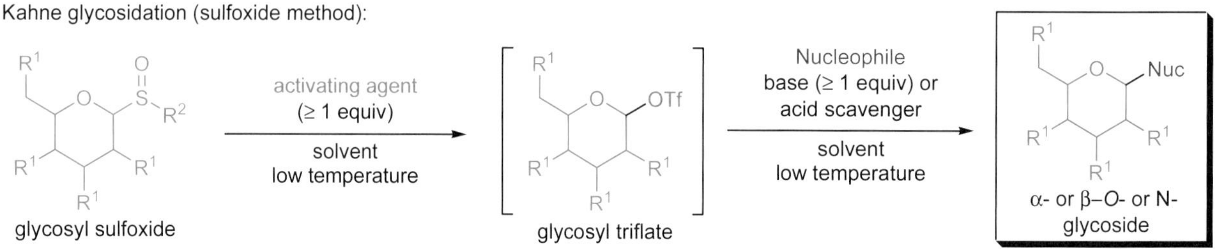

R^1 = *O*-alkyl, *O*-aryl, *O*-acyl; R^2 = alkyl, aryl; triflate activator: Tf_2O, TMSOTf, TfOH; solvent: toluene, CH_2Cl_2, Et_2O, EtOAc, EtCN; base: DTBMP, DTBP, TTBP; acid scavenger: methyl propiolate, allyl-1,2-dimethoxybenzene, $P(OMe)_3$, $P(OEt)_3$; Nucleophile: 1°, 2° and 3° alcohols, phenols, thiols, silylated amides, *O*-trialkylstannyl phenols

Mechanism: [26,20,27,21,8]

The precise mechanism of the glycosidic bond formation in the *Kahne glycosidation* is not known. NMR studies have revealed that when the activating agent is a triflate, glycosyl triflates are formed and act as glycosyl donors.[26] It is not clear whether the nucleophile displaces the leaving group in an S_N2 reaction or oxocarbenium/triflate contact ion pairs trap it stereoselectively. There is no structural information on the active species, which are generated upon activation by Lewis acids.

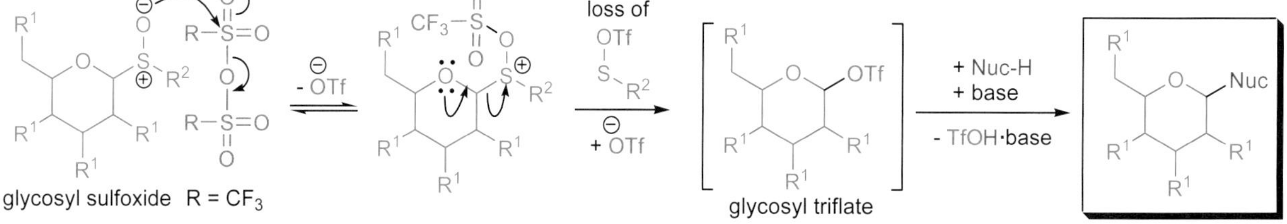

Synthetic Applications:

The first enantioselective total synthesis of the potent antiulcerogenic glycoside (–)-cassioside was accomplished in the laboratory of R.K. Boeckman Jr.[28] The natural product features a β-glycosidic bond to the extremely hindered neopentyl alcohol functionality of the aglycon. The *Kahne glycosidation* proved to be well-suited for the challenging glycosidation step at the final stages of the synthetic effort. The choice of the protecting groups proved to be important, since the authors found that after the coupling the removal of the benzyl groups failed in the presence of the unsaturations present in the coupled product. The tetrakis(MPM)glucosylphenyl sulfoxide was activated with Tf_2O at -90 °C. The resulting reactive intermediate was unstable at -78 °C, so the addition of the nucleophile was performed at -90 °C.

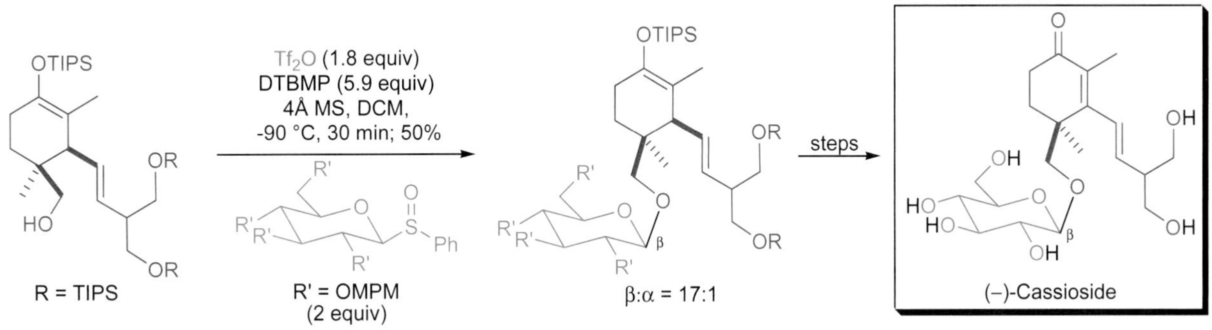

When the alkyl or aryl sulfoxide functionality is placed on the aglycon, a useful variant of the *Kahne glycosidation* arises which is known as the *reverse Kahne glycosidation*. D.B. Berkowitz and co-workers utilized this method for the total synthesis of etoposide, a semisynthetic glucoconjugate of epipodophyllotoxin, which has been used as an antineoplastic agent.[12] The activation of the phenyl sulfoxide occurred at low temperature, and after the addition of excess glycosyl acceptor, the reaction mixture was warmed to -40 °C in 5 hours and quenched. The coupled product was exclusively the β anomer, which was isolated in good yield. The final step was the removal of the benzyl and Cbz groups.

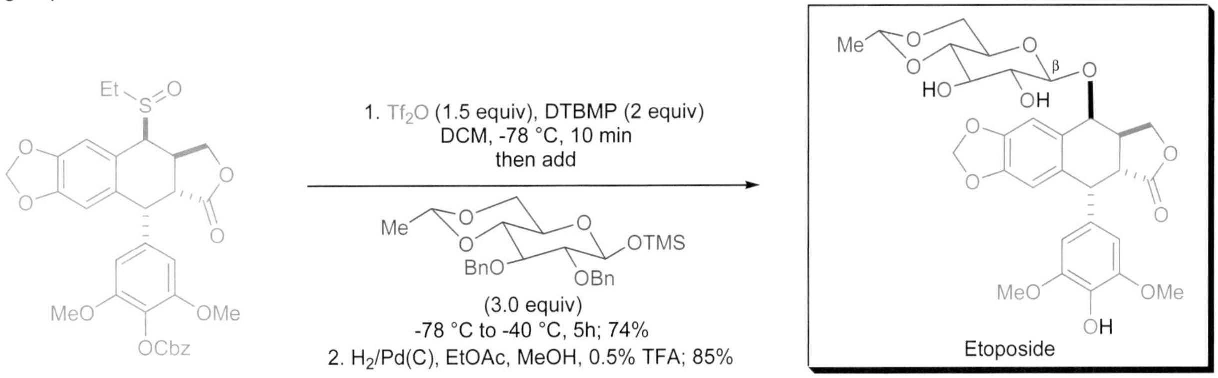

D. Kahne et al. developed a one-pot multicomponent stereoselective synthesis for the trisaccharide portion of cyclamycin 0 using the *Kahne glycosidation*.[25,29] The reactivity of the glycosyl donor was tuned (the rate limiting step is the triflation of the sulfoxide) and the *p*-methoxyphenyl sulfoxide was activated first. The trisaccharide was obtained in an overall 25% yield with complete α-selectivity.

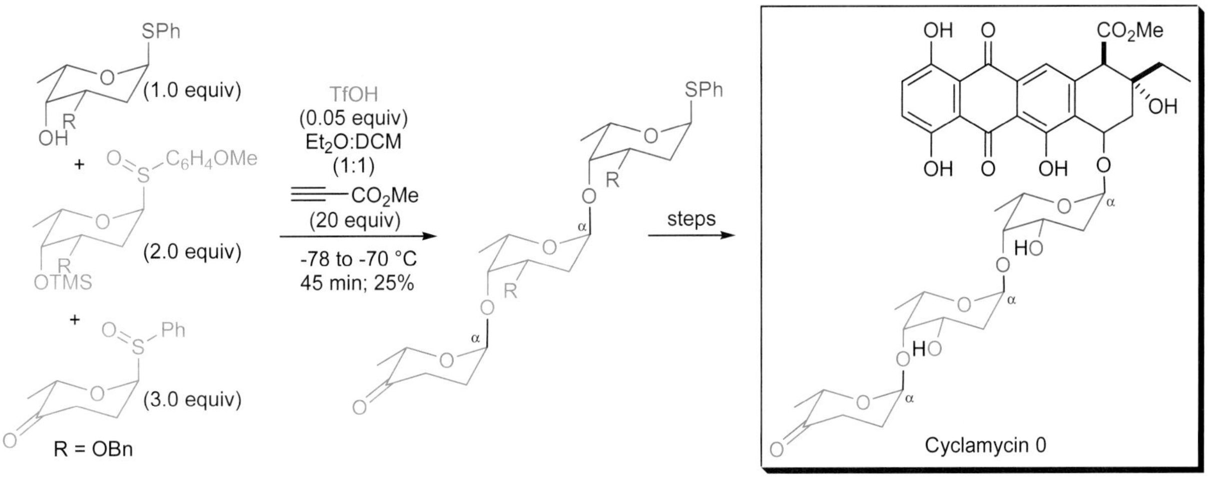

KECK ASYMMETRIC ALLYLATION

(References are on page 612)

Importance:

[*Seminal Publication*[1-3]; *Reviews*[4-11]; *Modifications & Improvements*[12-19]]

The formation of chiral secondary homoallylic alcohols *via* the enantioselective addition of allylic nucleophiles to aldehydes is an important tool in organic synthesis. An efficient way to achieve this transformation is to use allylic organometallic reagents in the presence of chiral Lewis acid catalysts. The most widely studied catalysts in the area are the 1,1'-binaphthalene-2,2'-diol (BINOL) complexes of titanium$^{(IV)}$. The first application of a Ti$^{(IV)}$-BINOL complex for enantioselective allylation was reported by K. Mikami in 1993.[20] According to this procedure, the catalyst was prepared from $TiCl_2(Oi\text{-}Pr)_2$ and (*S*)-BINOL in the presence of 4Å molecular sieves *in situ*. The addition of allylsilanes and allylstannanes to glyoxylate in the presence of 10% of the catalyst provided the products with low enantio- and diastereoselectivity. The same year, G.E. Keck independently reported the application of the BINOL/Ti$^{(IV)}$ catalyst system for asymmetric allylation.[1-3] He utilized allyltributylstannane as the nucleophile, and reacted it with aliphatic, aromatic, and unsaturated aldehydes in the presence of 10 mol% catalyst. The catalyst was prepared by combining two equivalents of the (*R*)- or (*S*)-BINOL ligand with one equivalent of $Ti(Oi\text{-}Pr)_4$ in dichloromethane, and the mixture was kept at room temperature for five minutes to an hour. The reaction of unbranched aliphatic, aromatic and unsaturated aldehydes with allyltributylstannane in the presence of 10% catalyst provided the homoallylic alcohols with high yields and enantioselectivity; α-branched aldehydes gave the products with lower yields and enantioselectivity. Today, this reaction is referred to as the *Keck asymmetric allylation*. About the same time, the research group of E. Tagliavini reported similar results using BINOL/Ti$^{(IV)}$ complexes for asymmetric allylation.[21] His procedure for the preparation of the catalyst system was similar to Mikami's original method, except that they used a slight excess of the BINOL ligand. The high selectivity and wide applicability of the above method stimulated further studies and several modifications of the original catalyst system were reported: 1) instead of the original BINOL ligand, derivatives of BINOL were utilized;[16,17] 2) dendritic BINOL ligands were applied for easy separation of the reaction mixture from the catalyst;[15] 3) racemic BINOL and enantiopure diisopropyl tartrate was combined to prepare the catalyst;[12] 4) bidentate catalysts prepared by mixing $Ti(Oi\text{-}Pr)_4$, BINOL, and aromatic diamines showed improved reactivity and selectivity;[18,19] and 6) rate enhancement could be achieved by the addition of stoichiometric amounts of additives such as *i*-PrSSiMe, *i*-$PrSBEt_2$, *i*-$PrSAlEt_2$, and $B(OMe)_3$.[13,14] The scope of the reaction was extended to β-substituted allylic stannanes.[22-25]

R^1 = alkyl, aryl, alkenyl; R^2 = alkyl, O-alkyl; Mikami's catalyst: $TiCl_2(Oi\text{-}Pr)_2$ + (*S*)-BINOL (0.3 equiv) + 4Å MS in CH_2Cl_2, toluene, 1h, r.t.; Keck's catalyst: $Ti(Oi\text{-}Pr)_4$ + (*R*)-BINOL (2 equiv) + 4Å mol sieves in CH_2Cl_2, 1h, r.t.; Tagliavini's catalyst: $TiCl_2(Oi\text{-}Pr)_2$ + (*S*)-BINOL (slight excess) + 4Å mol. sieves in CH_2Cl_2, 2h, r.t.;

Mechanism:[2,26,12,27]

The exact course of the mechanism of the allylation is not fully understood. The chiral Lewis acid presumably activates the aldehyde toward nucleophilic attack by the allyltributyltin. After loss of the tributyltin group, the homoallylic titanium$^{(IV)}$ alkoxide forms. Subsequently, the Ti$^{(IV)}$ Lewis acid is regenerated through transmetallation. This process can be facilitated by additives such as *i*-$PrSSiMe_3$.[13] Investigation of the mechanism of the enantioselective process revealed a positive nonlinear effect that suggests the involvement of a dimeric titanium complex $(BINOL)_2Ti_2X_4$.[2,12] To account for the absolute stereochemistry, a stereochemical model was proposed by E.J. Corey and co-workers. They postulated that a C-H···O hydrogen bond in the transition state assembly is a key factor in determining the absolute stereochemistry.[27]

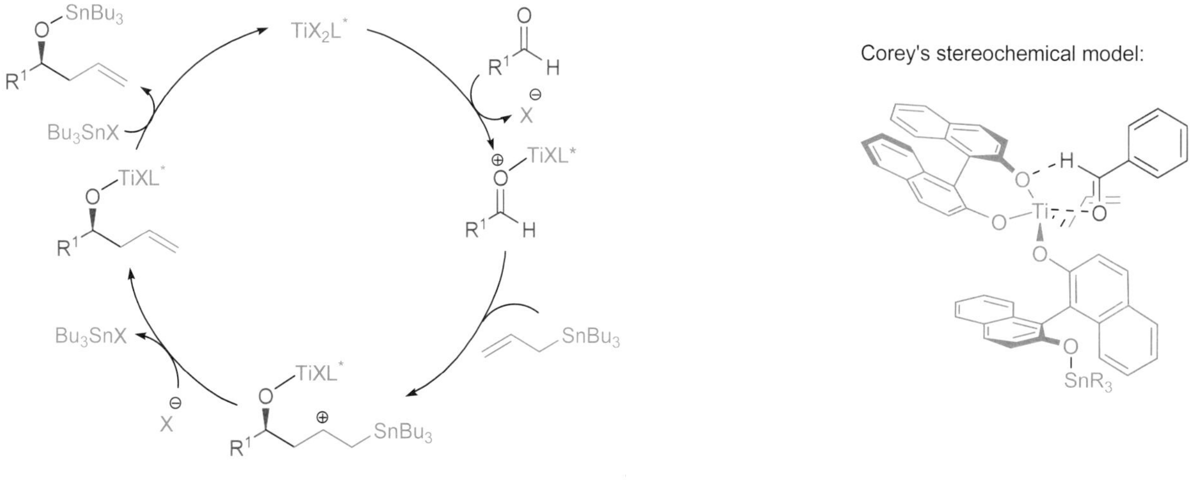

Synthetic Applications:

A. Fürstner and co-workers devised an efficient synthesis of (–)-gloeosporone, a fungal germination inhibitor.[28] They utilized the *Keck asymmetric allylation* method to create the 7(*R*)-homoallylic alcohol subunit. The reaction of the substrate aldehyde in the presence of the *in situ* generated catalyst provided the product with high yield and as the only diastereomer. It is important to note that it was essential to use freshly distilled Ti(*i*-OPr)$_4$ for the preparation of the catalyst in order to get high enantioselectivity and reproducible results.

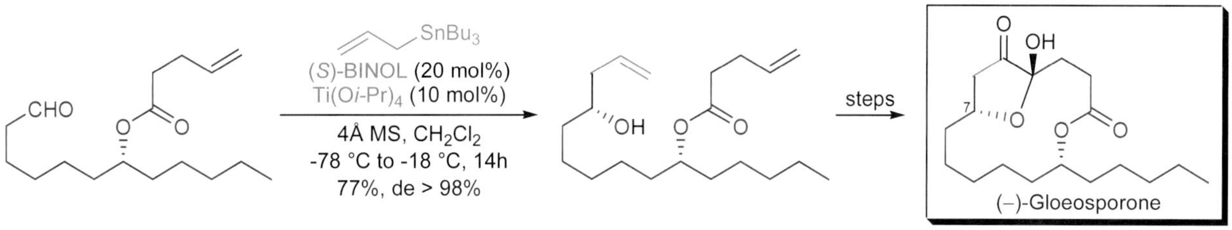

A convergent, stereocontrolled total synthesis of the microtubule-stabilizing macrolides, epothilones A *and* B was achieved in the laboratory of S.J. Danishefsky.[29] During their investigations, they examined several approaches to construct these natural products. One possible strategy to introduce the C15-hydroxyl group in an enantioselective fashion was to use Keck's asymmetric allylation method. Under standard conditions, the reaction provided the desired homoallylic alcohol in good yield and excellent enantioselectivity.

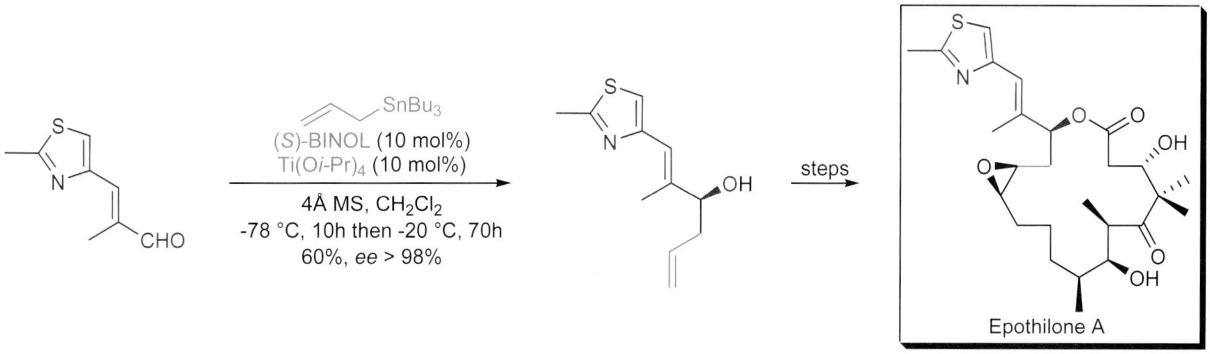

The spongistatins are a family of architecturally complex bisspiroketal macrolides, which display extraordinary cytotoxicity. During the second generation synthesis of the ABCD subunit of *spongistatin 1*, A.B. Smith and co-workers utilized the Keck allylation to construct the Kishi epoxide.[30] The allylation was carried out under standard conditions, using tributyl-(2-ethylallyl)-stannane as the allylstannane reactant. The desired product was formed in high yield and a diastereomeric ratio greater than 10:1.

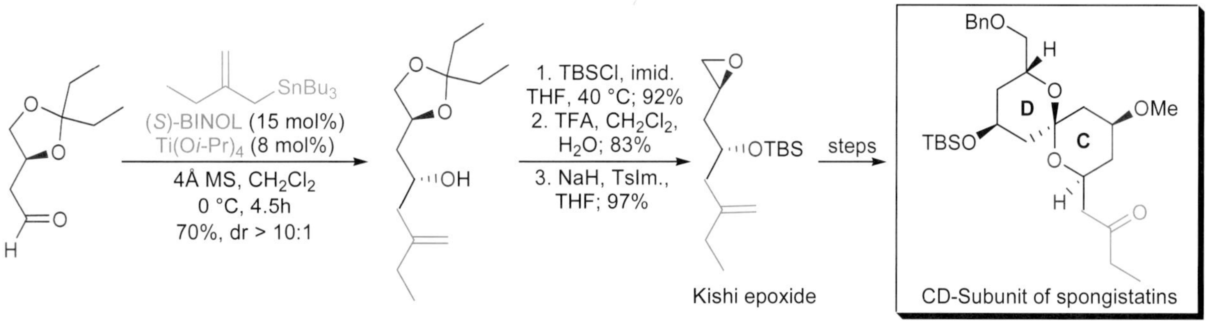

Rhizoxin is a macrocyclic natural product possessing antibiotic and antifungal properties, and it also exhibits antitumor activity. G.E. Keck and co-workers described a synthetic approach for the construction of this natural product, where they utilized the catalytic asymmetric allylation method as a key strategic element to establish the C13 stereochemistry.[31]

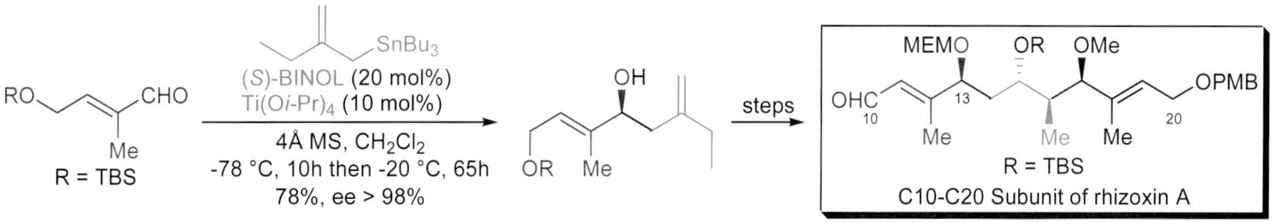

KECK MACROLACTONIZATION

(References are on page 613)

Importance:

[*Seminal Publications*[1]; *Reviews*[2,3]; *Modifications & Improvements*[4]]

The introduction of the *Corey-Nicolaou macrolactonization* in the mid-1970s had a tremendous impact in the field of natural product total synthesis, and it was followed by numerous other macrolactonization procedures.[2] By the early 1980s the total synthesis of several very complex macrolide antibiotics was achieved. In 1985, G.E. Keck and E.P. Boden were trying to develop a new macrolactonization protocol in which the activated ester derived from the hydroxy acid substrate is generated *in situ* and does not need to be isolated.[1] At the outset of their studies they attempted to use the conditions of the *Steglich esterification* (DMAP/DCC)[5] for the formation of macrolactones, but even in the presence of excess reagents the experiments failed. However, when a proton source such as the hydrochloride salt of dimethylamino pyridine (DMAP•HCl) was added to mediate the crucial proton-transfer step, the macrocyclizations occurred in good to excellent yields. The formation of medium- and large-ring lactones from hydroxy acids using a combination of a dialkyl carbodiimide, an amine hydrochloride, and an amine base is known as the *Keck macrolactonization*. The general features of this transformation are: 1) as with other macrolactonization procedures the reaction requires high-dilution conditions (≤ 0.03 M); 2) the substrate is usually dissolved in an aprotic solvent and added to the refluxing solution of the reagents *via* a syringe pump over several hours; 3) the activating agent is a *N,N'*-dialkyl carbodiimide (DCC or EDCI) that prevents small amounts of water from destroying the activated acyl derivative; the process is essentially self-drying; 4) the carbodiimide reagent is typically used in several fold excess to ensure high conversion of the starting material; and 5) the use of DMAP•HCl prolongs the lifetime of the activated acyl intermediate and suppresses the formation of the undesired *N*-acyl urea by-product. The main disadvantage of the method is the need to use excess amounts of the carbodiimide reagent. At the end of the reaction, the excess carbodiimide must be destroyed with AcOH/MeOH and the product has to be separated from large amounts of dialkylurea. The most important modification of the *Keck macrolactonization* utilizes polymer-bound DCC to simplify the work-up.[4]

Keck & Boden (1985):

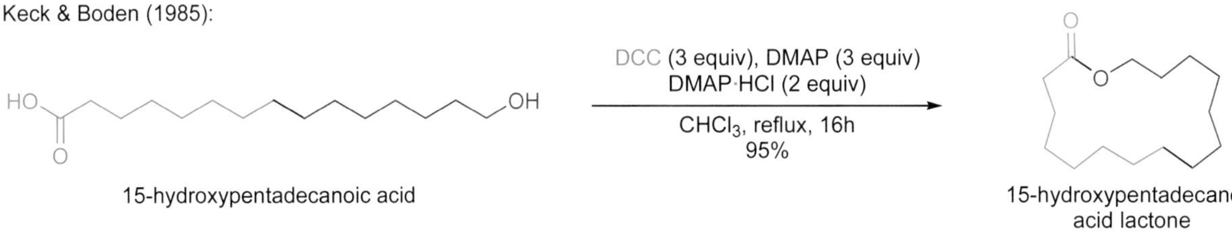

Keck macrolactonization:

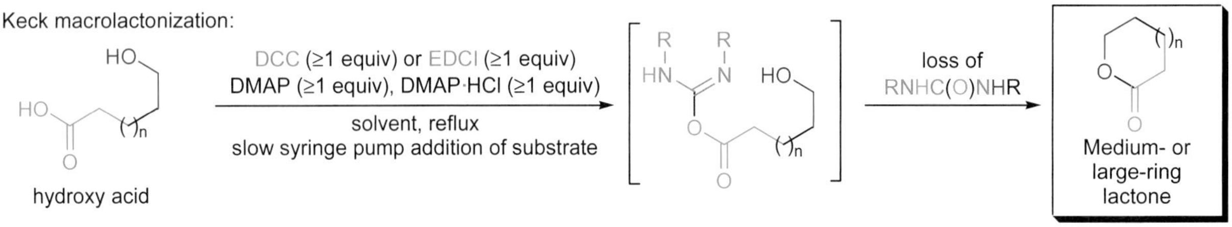

Mechanism:

Formation of the activated ester intermediate:

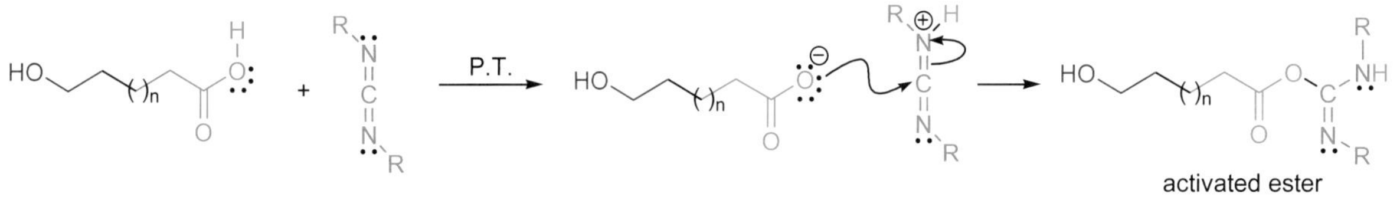

Formation of the macrolactone and the N,N'-dialkylurea by-product:

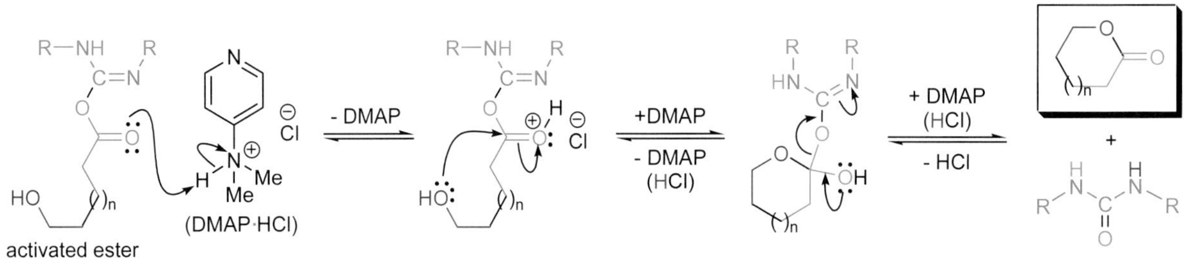

Synthetic Applications:

The total synthesis of a novel fungicidal natural product, (–)-hectochlorin, was accomplished by J.R.P. Cetusic and co-workers.[6] The final step in their synthetic route was the *Keck macrolactonization* under the original conditions developed by Keck et al. The substrate hydroxy acid was dissolved in ethanol-free $CHCl_3$ and was slowly added to a chloroform solution of DCC, DMAP and DMAP·HCl at reflux temperature.

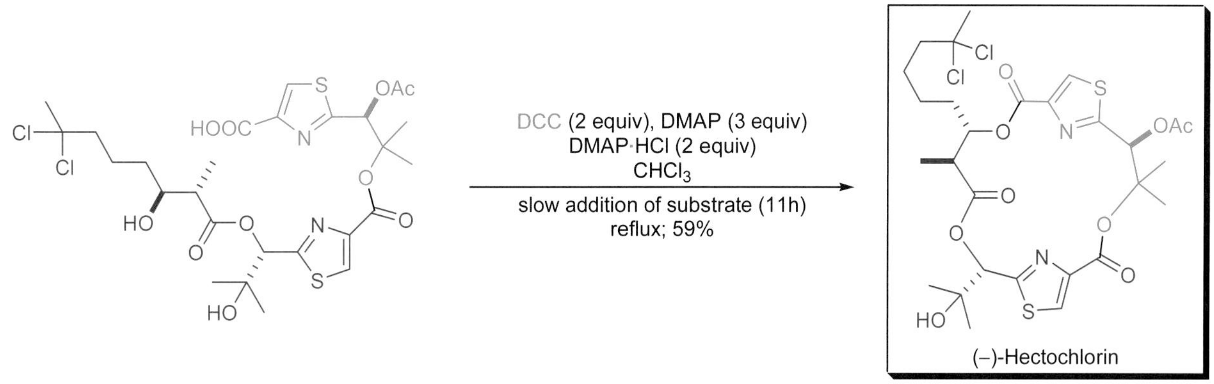

The 16-membered tetraenic macrolactone (–)-bafilomycin A_1 was synthesized in the laboratory of S. Hanessian.[7] The key macrolactonization step was conducted under the *modified Keck conditions* using EDCI instead of DCC. Interestingly, model studies on the macrocyclization of the hydroxy acid containing the entire bafilomycin A_1 carbon framework yielded a mixture of products. However, if the hydroxy acid did not contain the pseudosugar moiety, the macrolactonization took place uneventfully, and the thermodynamically more stable 16-membered lactone ring (with the C15 hydroxyl group) was formed exclusively.

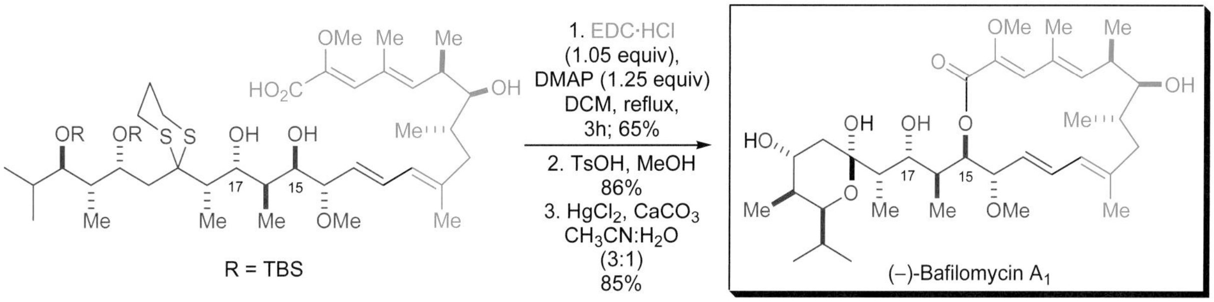

The *Keck macrolactonization* was used by R.J.K. Taylor et al. to close the 10-membered ring of (+)-apicularen A.[8] The lactonization was attempted using both the Yamaguchi and Mitsunobu procedures and neither gave even a trace of the cyclic product. However, when the *Keck conditions* were applied, the desired lactone was isolated in moderate yield.

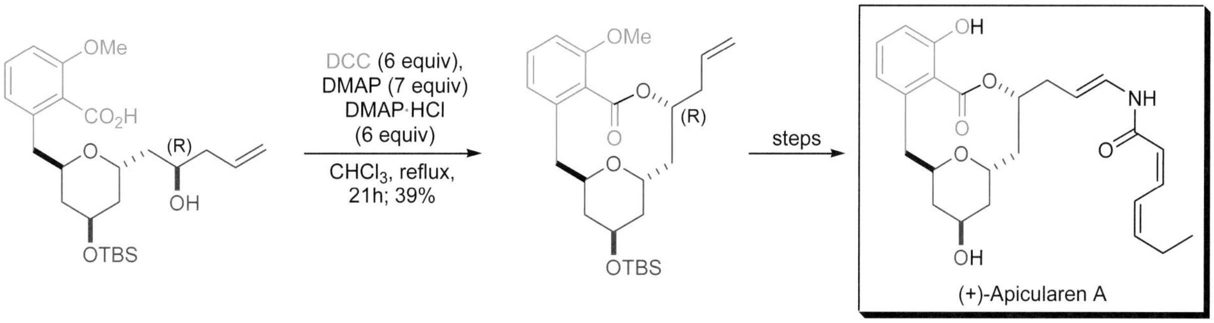

The total synthesis of the microtubule stabilizing antitumor drug epothilone B was achieved by J. Mulzer et al. who cyclized the 16-membered macrocycle using the *Keck macrolactonization*.[9]

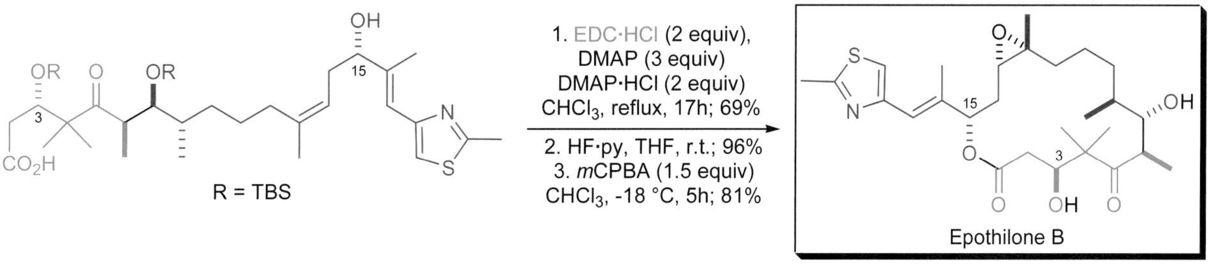

KECK RADICAL ALLYLATION

(References are on page 613)

Importance:

[*Seminal Publications*[1-4]; *Reviews*[5-8]]

During the total synthesis of perhydrohistrionicotoxin, G.E. Keck and co-workers faced a challenge to replace a halogen with an allyl moiety.[9] They solved this problem by applying a free radical chain process, namely reacting an alkyl halide with allyltributyltin. The reaction was carried out in benzene, at 80 °C in the presence of catalytic amounts of AIBN as a radical initiator. Since this report, the coupling of an alkyl halide with allyltributyltin under radical conditions to introduce the allyl functionality is referred to as the *Keck radical allylation*. Keck examined the scope of the reaction and he found the following:[4] 1) the reaction is general for primary-, secondary-, and tertiary alkyl bromides: 2) it tolerates a wide range of functional groups such as free hydroxyl groups, esters, ethers, epoxides, acetals, ketals, and sulfonate esters; 3) the reaction is highly chemoselective: aldehydes that readily undergo allylation with allyltributyltin under acidic conditions, were not affected under the reaction conditions; 4) the process is tolerant of steric hindrance; 5) in addition to alkyl bromides, alkyl chlorides, phenylselenides, and thioacylimidazole derivatives also react; and 6) to initiate the process, a catalytic amount of AIBN proved to be the most efficient, but photoinitiation can also be used. Although this transformation was studied and extended by Keck, it should be noted that the first example of such a reaction was reported independently by M. Kosugi[2] and J. Grignon[1,3] in 1973. For the initiation, they utilized benzoyl peroxide, pyrolysis, or photoinitiation, and the isolated yields of the products were low to moderate.

Kosugi's procedure (1973):

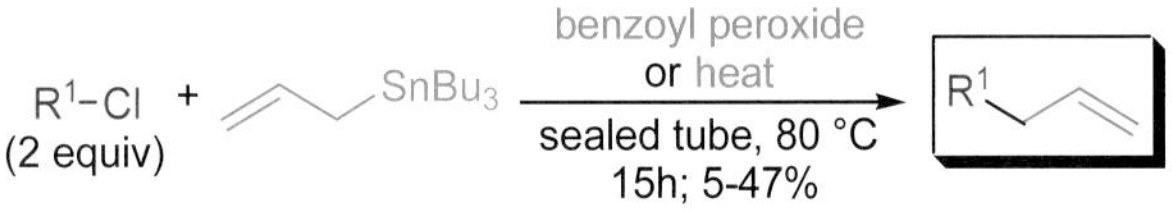

R^1 = $-CCl_3$, $-CHCl_2$, $-CH_2CO_2Me$, $-CH_2CCl_3$

Grignon's procedure (1973, 1975):

R^2 = $-CCl_3$, $-CHCl_2$, - CH_2Cl, $-CH_2CO_2Et$, $-CCl_2CO_2Et$, - CCl_2CHO, $-CBr_2CHO$, $-CH(CH_3)CO_2Me$, *n*-propyl, *i*-Pr, *t*-Bu, allyl, *t*-butylcyclohexyl, aryl; X = Cl, Br

Keck's general process (1982):

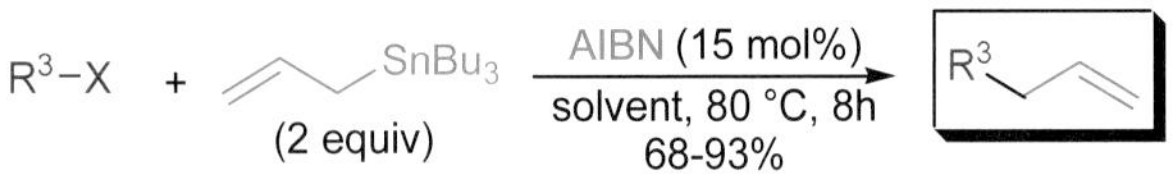

R^3 = 1°, 2°, and 3° alkyl; X = Cl, Br, SePh, thioacylimidazole;
solvent = benzene, toluene;

Keck's specific example (1982):

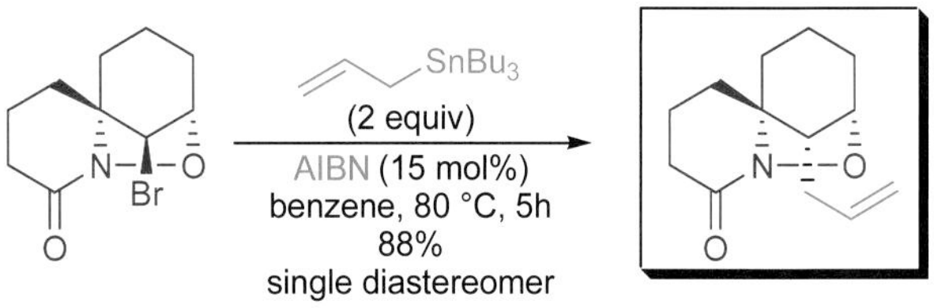

Mechanism:[1-3]

The mechanism of this transformation was examined by M. Kosugi.[2] He found that the reaction was promoted by benzoyl peroxide, a radical initiator and was retarded by *p*-quinone, a radical scavenger. These results are in accordance with a free radical chain mechanism. The initiation of the reaction may take place *via* a variety of possible pathways, one possibility is depicted below.

Initiation step:

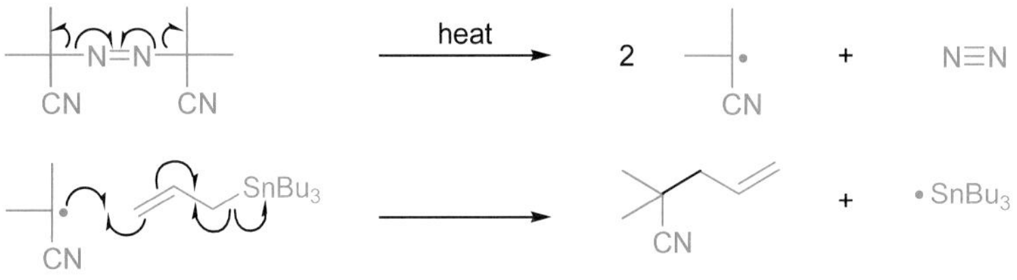

Propagation step:

Termination steps

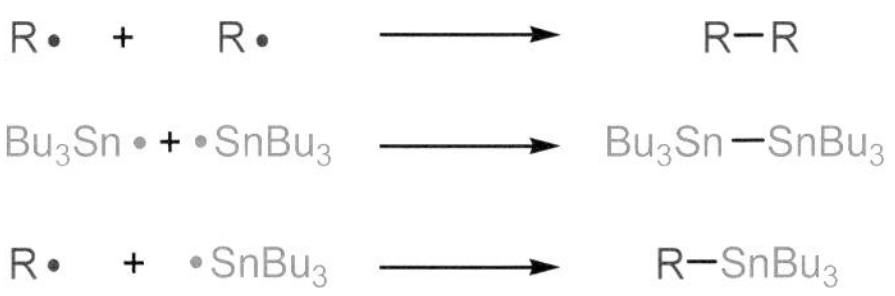

Synthetic Applications:

S.J. Danishefsky and co-workers reported the total synthesis of pentacyclic sesquiterpene dilactone, merrilactone A.[10] In their approach, they utilized *Keck's radical allylation* method to achieve the required chain extension. This sidechain was later used to construct one of the cyclopentane rings of the natural product.

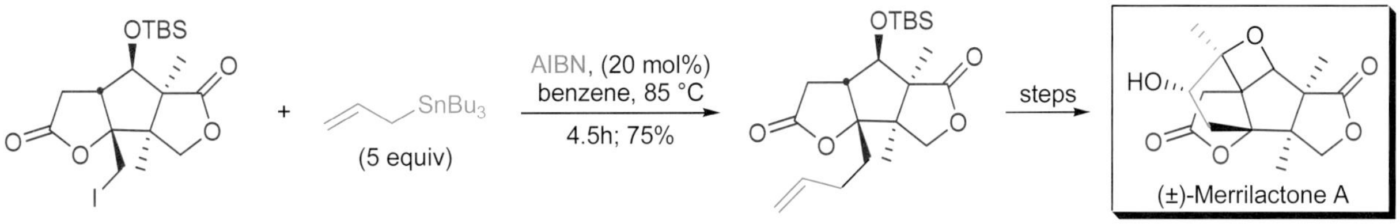

The total synthesis of *Stemona* alkaloid (–)-tuberostemonine was accomplished by P. Wipf.[11] Late in the synthesis, the introduction of an ethyl sidechain was required. This could be achieved in a novel four-step sequence. First, the allyl sidechain was introduced by the *Keck radical allylation*. To this end, the secondary alkyl phenylselenide substrate was treated with allyltriphenyltin in the presence of catalytic amounts of AIBN. This was followed by the introduction of a methyl group onto the lactone moiety. The allyl group then was transformed into the desired ethyl group as follows: the terminal double bond was isomerized to the internal double bond by the method of R. Roy. This was followed by ethylene cross metathesis and catalytic hydrogenation to provide the desired ethyl sidechain.

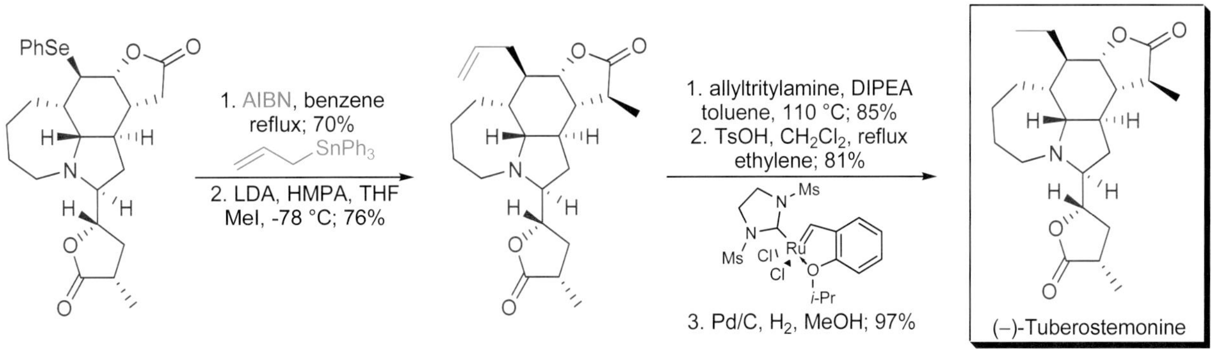

Oligosaccharides are structurally diverse biopolymers that play an important role in many biological processes. To examine the biological function of these compounds and develop therapeutic agents, the construction of synthetic polysaccharides is essential. Carbon-linked glycosides, called *C-glycosides*, are hydrolytically stable carbohydrate mimetics that were widely studied for their biological activity. C.R. Bertozzi and co-workers reported the synthesis of *β-C*-glycosides of N-acetylglucosamine *via* the *Keck radical allylation*.[12] This transformation was carried out on the corresponding bromide- and chloride derivatives, using a large excess of allyltributyltin. In case of the chloride substrate, higher temperature (110 °C) was required to effect the transformation.

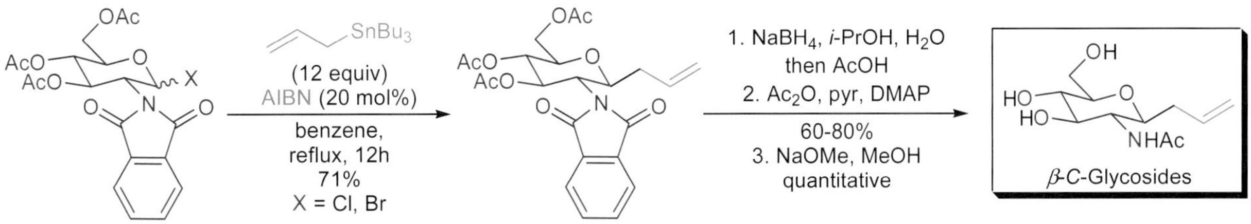

Manzamine A is an alkaloid that was shown to inhibit the growth of P-388 mouse leukemia cells. The synthesis of the tetracyclic substructure of this natural product was reported by D.J. Hart.[13] For the construction of the perhydroisoquinoline moiety, he utilized the *Keck radical allylation*. This transformation was carried out under standard conditions, reacting a secondary alkyl iodide with allyltributyltin.

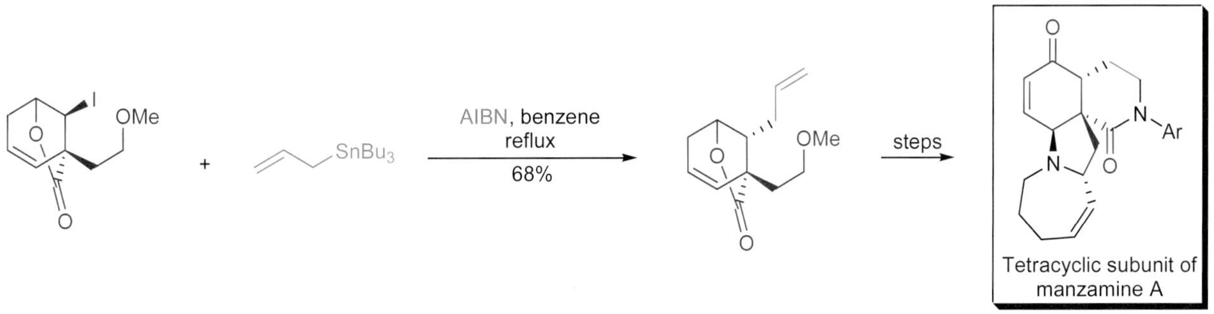

KNOEVENAGEL CONDENSATION

(References are on page 613)

Importance:

[*Seminal Publications*[1,2]; *Reviews*[3-10]; *Modifications & Improvements*[11-41]]

In 1894, E. Knoevenagel reported the diethylamine-catalyzed condensation of diethyl malonate with formaldehyde in which he isolated the *bis* adduct.[1] He found the same type of *bis* adduct when formaldehyde and other aldehydes were condensed with ethyl benzoylacetate or acetylacetone in the presence of primary and secondary amines. Two years later in 1896, Knoevenagel carried out the reaction of benzaldehyde with ethyl acetoacetate at 0 °C using piperidine as the catalyst and obtained ethyl benzylidene acetoacetate as the sole product.[2] The reaction of aldehydes and ketones with active methylene compounds in the presence of a weak base to afford α,β-unsaturated dicarbonyl or related compounds is known as the *Knoevenagel condensation*. The general features of the reaction are: 1) aldehydes react much faster than ketones; 2) active methylene compounds need to have two electron-withdrawing groups and typical examples are malonic esters, acetoacetic esters, malonodinitrile, acetylacetone, etc.; 3) the nature of the catalyst is important, usually primary, secondary, and tertiary amines and their corresponding ammonium salts, certain Lewis acids combined with a tertiary amine (e.g., $TiCl_4/Et_3N$), potassium fluoride, or other inorganic compounds such as aluminum phosphate are used; 4) the by-product of the reaction is water and its removal from the reaction mixture by means of azeotropic distillation, the addition of molecular sieves, or other dehydrating agents shifts the equilibrium toward the formation of the product; 5) the choice of solvent is crucial and the use of dipolar aprotic solvents (e.g., DMF) is advantageous, since protic solvents inhibit the last 1,2-elimination step; 6) the dicarbonyl product can be hydrolyzed and decarboxylated to afford the corresponding α,β-unsaturated carbonyl compounds; 7) when R^3 and R^4 or R^5 and R^6 are different, the product is obtained as a mixture of geometrical isomers, and the selectivity is dictated by steric effects; and 8) usually the thermodynamically more stable compound is formed as the major product.

Knoevenagel (1894):

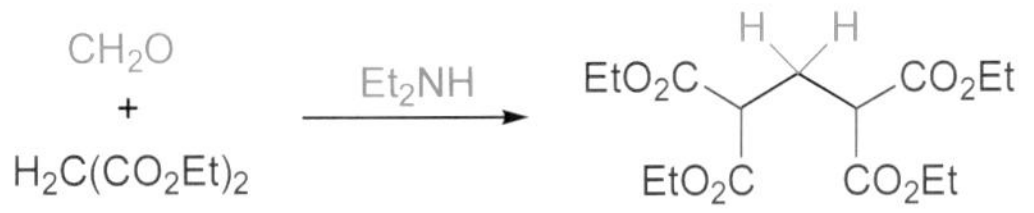

Knoevenagel (1896):

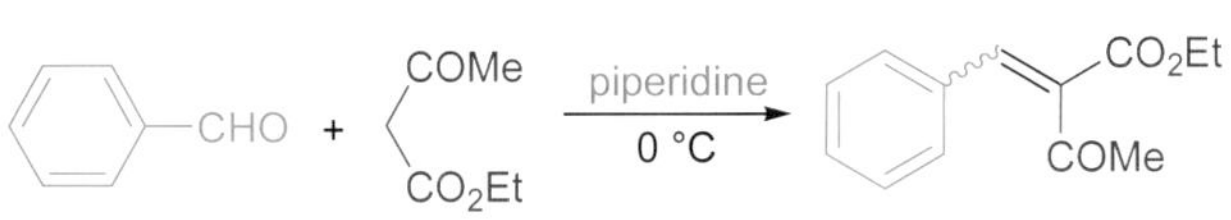

Knoevenagel condensation:

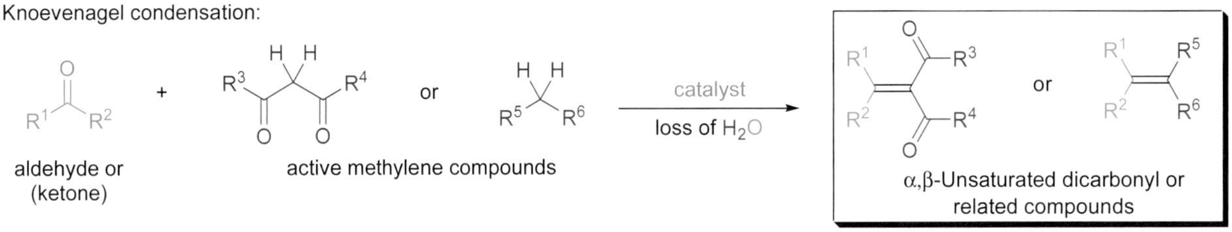

R^1 = H, alkyl, aryl; R^2 = H, alkyl, aryl; R^{3-4} = alkyl, aryl, OH, *O*-alkyl, *O*-aryl, NH-alkyl, NH-aryl *N*-dialkyl, *N*-diaryl; R^{5-6} = CO_2H, CO_2-alkyl, CO_2-aryl, C(O)NH-alkyl, C(O)NH-aryl, C(O)N-dialkyl, C(O)N-diaryl, C(O)-alkyl, C(O)-aryl, CN, $CNNR_2$, $PO(OR)_2$, SO_2OR, SO_2NR_2, SO_2R, SOR, SiR_3; catalyst: 1°, 2° or 3° amines, R_3NHX such as $[H_3NCH_2CH_2NH_3](OAc)_2$, piperidinium acetate/AcOH, NH_4OAc, KF, CsF, RbF, $TiCl_4/R_3N$ (*Lehnert modification*), pyridine/piperidine (*Doebner modification*), dry alumina (*Foucaud modification*), $AlPO_4/Al_2O_3$, xonotlite with KO*t*-Bu, $Zn(OAc)_2$

Mechanism: [42,4,43-49,7,50-55]

The *Knoevenagel condensation* is a base-catalyzed *aldol-type reaction*, and the exact mechanism depends on the substrates and the type of catalyst used. The first proposal for the mechanism was set forth by A.C.O. Hann and A. Lapworth (*Hann-Lapworth mechanism*) in 1904.[42] When tertiary amines are used as catalysts, the formation of a β-hydroxydicarbonyl intermediate is expected, which undergoes dehydration to afford the product. On the other hand, when secondary or primary amines are used as catalyst, the aldehyde and the amine condense to form an iminium salt that then reacts with the enolate. Finally, a 1,2-elimination gives rise to the desired α,β-unsaturated dicarbonyl or related compounds. The final product may undergo a *Michael addition* with the excess enolate to give a *bis* adduct.

Hann-Lapworth mechanism with tertiary amines as catalysts:

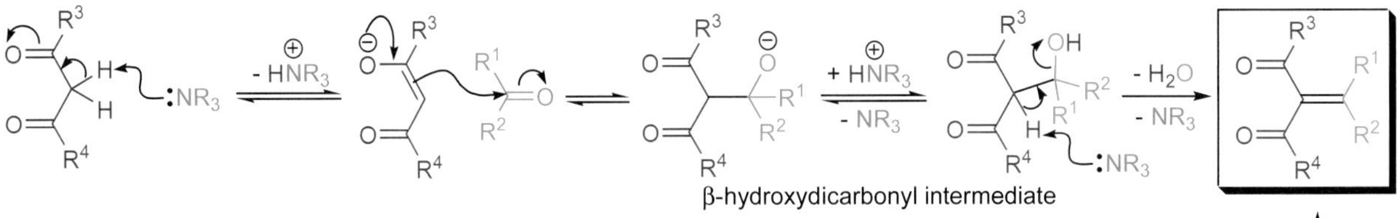

Mechanism with primary or secondary amines as catalysts:

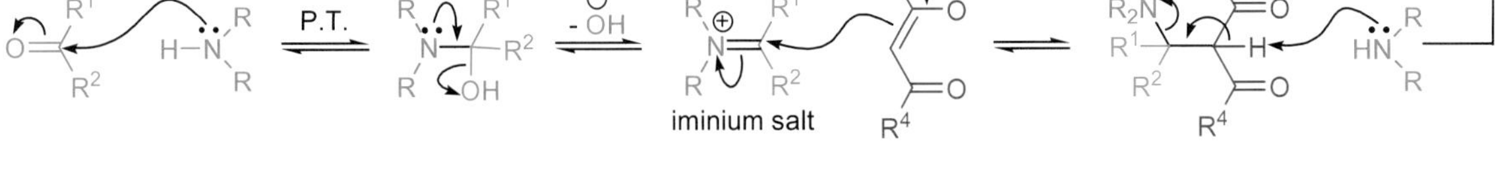

Synthetic Applications:

The total synthesis of the marine-derived diterpenoid sarcodictyin A was accomplished in the laboratory of K.C. Nicolaou.[56] The most challenging part of the synthesis was the construction of the tricyclic core, which contains a 10-membered ring. This macrocycle was obtained by the intramolecular 1,2-addition of an acetylide anion to an α,β-unsaturated aldehyde. This unsaturated aldehyde moiety was installed by utilizing the *Knoevenagel condensation* catalyzed by β-alanine. The Knoevenagel product was exclusively the (*E*)-cyanoester.

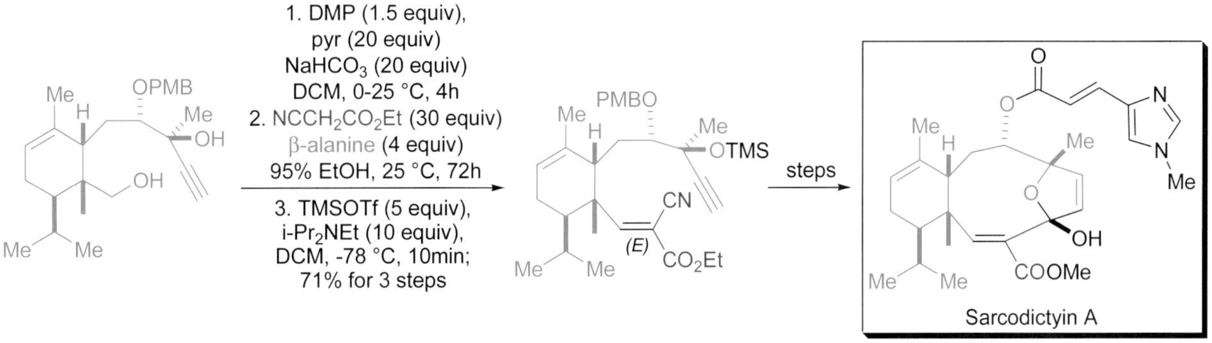

The domino *Knoevenagel condensation/hetero-Diels-Alder reaction* was used for the enantioselective total synthesis of the active anti-influenza A virus indole alkaloid hirsutine and related compounds by L.F. Tietze and co-workers.[57] The *Knoevenagel condensation* was carried out between an enantiopure aldehyde and Meldrum's acid in the presence of ethylenediamine diacetate. The resulting highly reactive 1-oxa-1,3-butadiene underwent a *hetero-Diels-Alder reaction* with 4-methoxybenzyl butenyl ether (*E*/*Z* = 1:1) *in situ*. The product exhibited a 1,3-asymmetric induction greater than 20:1.

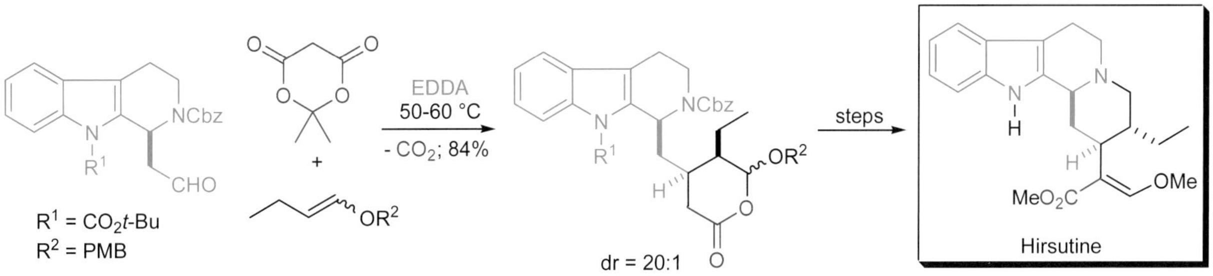

During the total synthesis of (±)-leporin A, a tandem *Knoevenagel condensation/inverse electron demand intramolecular hetero-Diels-Alder reaction* was employed by B.B. Snider et al. to construct the key tricyclic intermediate.[58] The condensation of pyridone with the enantiopure acyclic aldehyde in the presence of triethylamine as catalyst afforded an intermediate that underwent a [4+2] cycloaddition to afford the tricyclic core of the target.

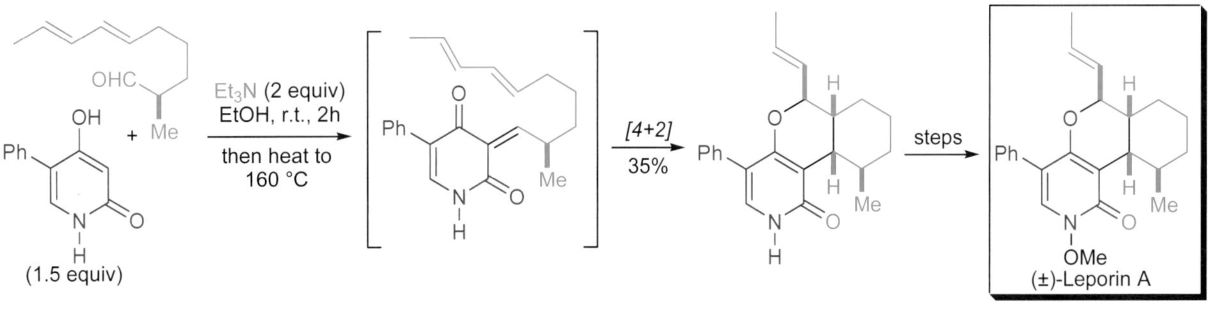

The stereocontrolled total synthesis of (±)-gelsemine was accomplished by T. Fukuyama and co-workers using the *Knoevenagel condensation* to prepare a precursor for the key *divinylcyclopropane-cycloheptadiene rearrangement*.[59] The use of 4-iodooxindole as the active methylene component allowed the preparation of the (*Z*)-alkylidene indolinone product as a single stereoisomer.

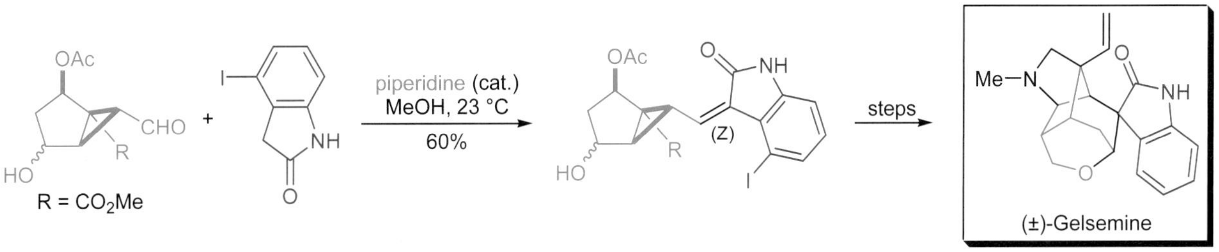

KNORR PYRROLE SYNTHESIS

(References are on page 614)

Importance:

[*Seminal Publications*[1,2]; *Reviews*[3-8]; *Modifications & Improvements*[9-17]]

In 1886, L. Knorr reported that by heating the mixture of α-nitroso ethyl acetoacetate and ethyl acetoacetate in glacial acetic acid with zinc dust, a tetrasubstituted pyrrole is formed. The nitroso compound underwent reduction under the reaction conditions, and the resulting α-amino-β-ketoester reacted with the acetoacetic ester to afford the highly substituted pyrrole product. The condensation of an α-amino ketone or an α-amino-β-ketoester with an active methylene compound is known as the *Knorr pyrrole synthesis*. The general features of the transformation are: 1) the reaction can be conducted under both acidic and basic conditions; 2) α-amino ketones are often quite labile and tend to undergo self-condensation (to form the corresponding pyrazines), so it is common to prepare them by first nitrosating the ketone and then reducing the resulting α-nitroso ketone *in situ*; 3) the reduction of the α-nitroso ketone (or α-oximino ketone in its tautomerized form) is conducted using zinc powder in acetic acid, aqueous solution of sodium dithionate ($Na_2S_2O_4$), or catalytic hydrogenation under which conditions ketones and esters are not reduced; 4) the hydrochloride salts of α-amino ketones are stable, and they can be used directly and the HCl can be neutralized *in situ*; 5) carbonyl-protected (e.g., acetal) derivatives of α-amino ketones are often utilized to avoid self-condensation; 6) alternatively the required α-amino ketones can be prepared by the *Neber rearrangement* of *O*-acylated ketoximes; 7) *N*-substituted pyrroles are formed when a secondary amino ketone is used; 8) the active methylene component is usually a 1,3-diketone, β-ketoester or a β-cyanoester; 9) if the active methylene compound is not reactive enough, the formation of the pyrrole will be slow and the self-condensation of the α-amino ketone becomes predominant; and 10) when non-symmetrical ketones are used, there is a modest regioselectivity favoring the regioisomer in which the bulkier group is part of the acyl substituent at C4.

Knorr (1886):

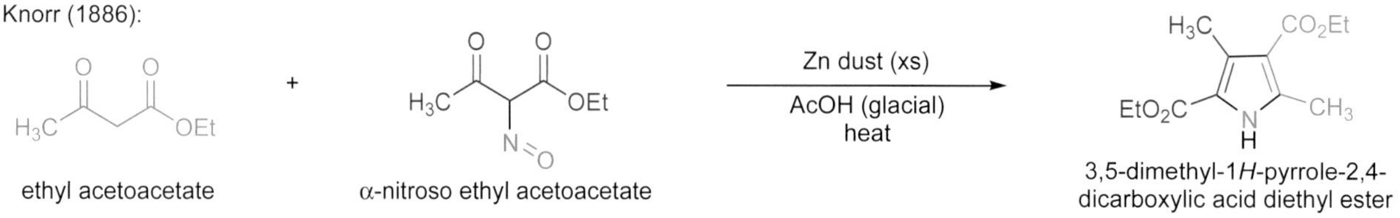

Knorr pyrrole synthesis:

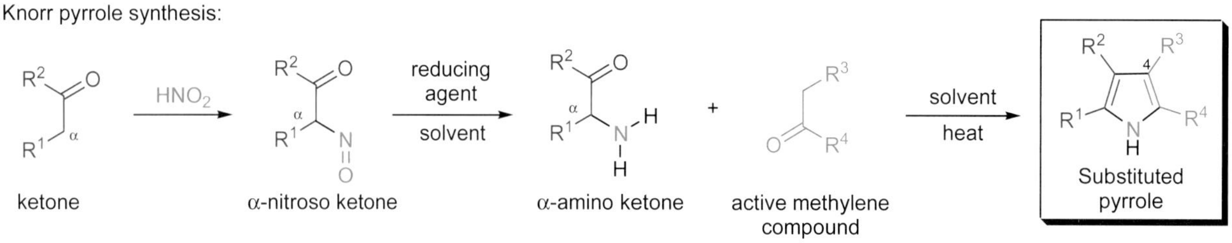

R^1 = H, aryl, CO_2R; R^2 = alkyl, aryl; R^3 = electron-withdrawing group (EWG) = COR, CO_2R, CN, SO_2R; R^4 = H, alkyl, aryl, CO_2R;
reducing agent: Zn/AcOH, $Na_2S_2O_4$, Pd(C)/H_2; solvent: AcOH, H_2O

Mechanism:[18-20]

Condensation of the amino ketone and ketone to give an imine:

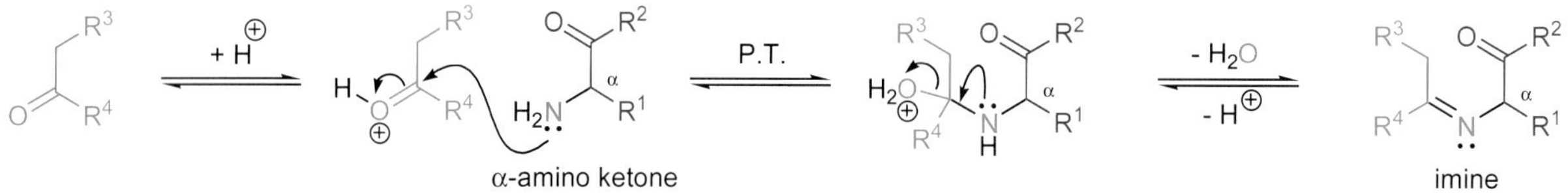

Tautomerization of the imine to the enamine and cyclization:

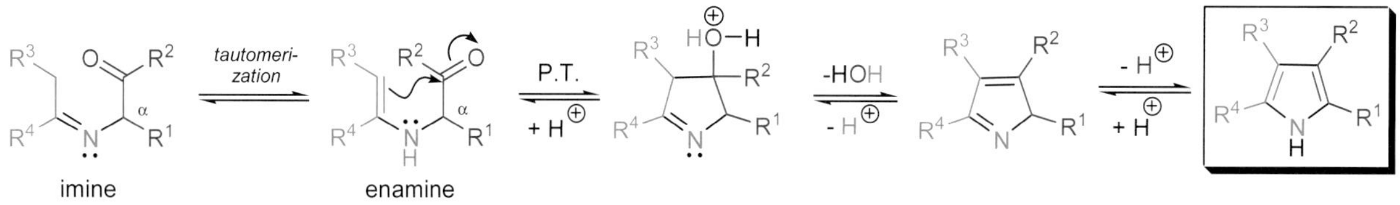

Synthetic Applications:

A new anti-inflammatory/analgesic agent, 4,5,8,9-tetrahydro-8-methyl-9-oxothieno[3',3':5,6]cyclohepta[1,2-*b*]-pyrrole-7-acetic acid, was synthesized by H.E. Rosenberg and R.W. Ward et al. using the *Knorr pyrrole synthesis* for the construction of the highly substituted pyrrole ring.[21] The starting β-ketoamide was first nitrosated under standard conditions in acetic acid/water to afford the corresponding α-oximino ketone. This was followed by the addition of diethyl acetone-1,3-dicarboxylate, zinc powder, and sodium acetate, and the resulting mixture was heated at reflux. The cyclization to obtain the desired tricyclic ketone was achieved under Vilsmeier-Haack conditions using $POCl_3$.

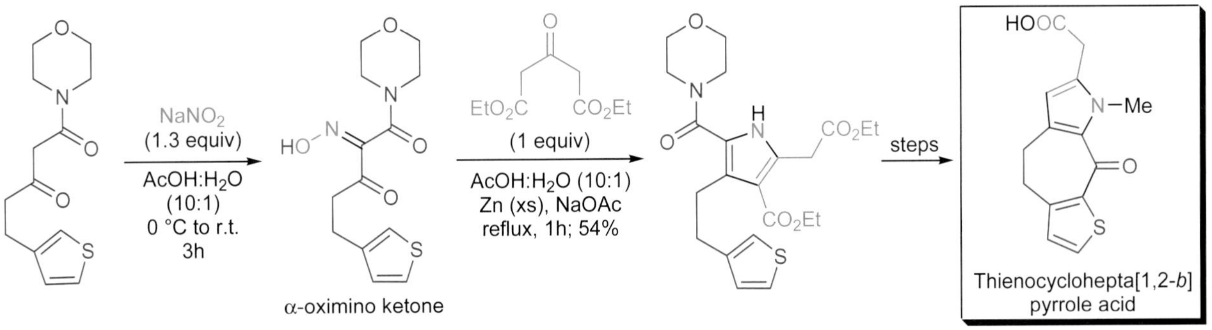

A useful modification of the *Knorr pyrrole synthesis* was developed in the laboratory of J.M. Hamby for the construction of tetrasubstituted pyrroles. The necessary α-amino ketones were prepared from *N*-methoxy-*N*-methylamides of amino acids (Weinreb amides).[13] These Weinreb amides were prepared by the mixed anhydride method and treated with excess methylmagnesium bromide in ether to afford the corresponding Cbz-protected α-amino ketones in excellent yield. The Cbz group is removed by catalytic hydrogenation in the presence of the active methylene compound (e.g., acetoacetic ester), the catalyst is then filtered and the resulting solution is heated to reflux to bring about the condensation.

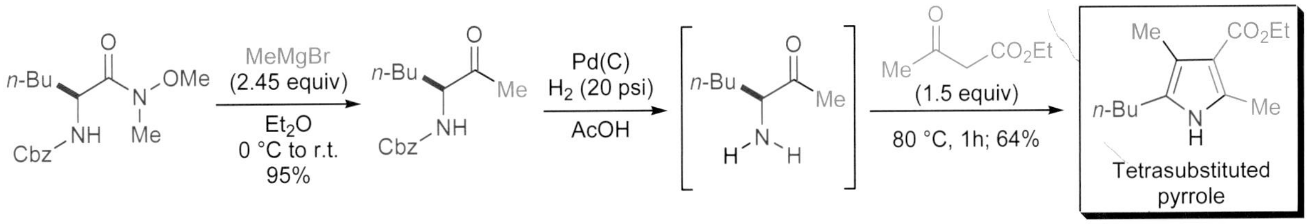

The large-scale synthesis of a potent δ-opioid antagonist, SB-342219, was accomplished by the research team of J.S. Carey.[22] The route developed by medicinal chemists could not be fully adapted for the large-scale preparation, since it required the addition of finely divided zinc powder in portions to a hot and flammable solvent containing a phenylhydrazone and a low concentration of the resulting α-amino ketone had to be maintained. Therefore, a procedure was sought that avoided the use of zinc metal altogether. The tricyclic ketone was mixed with an excess of the amino ketone hydrochloride in acetic acid and heated. Only one regioisomer of the pyrrole was formed in good yield, which was then converted to the final compound in a few steps.

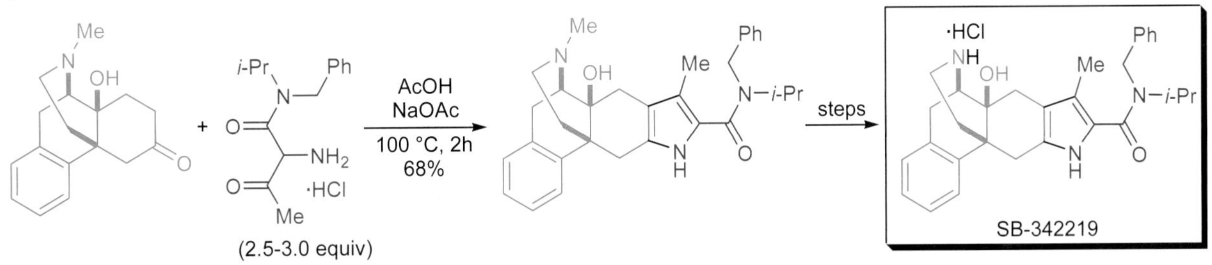

The two-step one-pot total synthesis of Ro 22-1319, an antipsychotic agent featuring a rigid pyrrolo[2,3-*g*]isoquinoline skeleton, was accomplished by D.L. Coffen and co-workers.[23] The cyclic 1,3-diketone precursor was prepared from arecoline and dimethyl malonate, and in the same reaction vessel an amino ketone hydrochloride was added. The pH of the reaction mixture was adjusted to 4 in order to initiate the formation of the pyrrole.

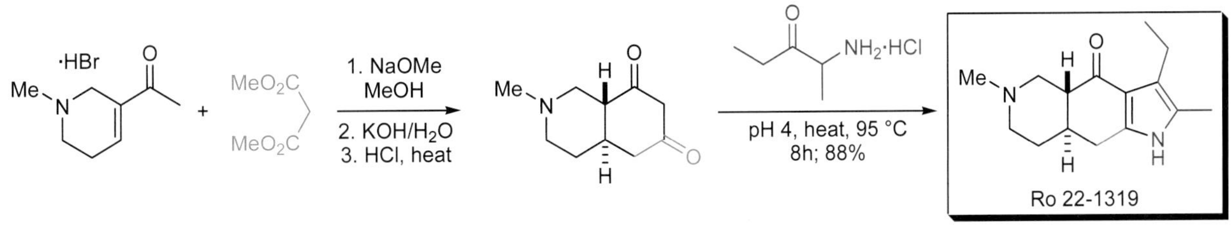

KOENIGS-KNORR GLYCOSIDATION

(References are on page 615)

Importance:

[*Seminal Publications*[1,2]; *Reviews*[3-21]; *Modifications & Improvements*[22-34]; *Theoretical Studies*[35-37]]

The first synthesis of a glycoside was reported by A. Michael in 1879, when he treated 2,3,4,6-tetra-*O*-acetyl-α-D-glucopyranosyl chloride with the potassium salt of 4-methoxy phenol in absolute ethanol.[1] The product was the corresponding β-D-*O*-phenyl glycoside, but the acetyl groups were hydrolyzed under the strongly basic reaction conditions. This procedure could only be used for the synthesis of aryl glycosides, and the integrity of the acetyl functionality could not be preserved. Two decades later, in 1901, W. Koenigs and E. Knorr modified the procedure and by taking tetra-*O*-acetyl-α-D-glucopyranosyl bromide and treating it with excess silver carbonate in methanol they isolated the corresponding β-D-*O*-methyl glycoside with all the acetyl groups intact.[2] The synthesis of alkyl- and aryl *O*-glycosides from glycosyl halides and alcohols or phenols in the presence of heavy metals salts or Lewis acids is known as the *Koenigs-Knorr glycosidation*. The general features of this transformation are: 1) the preparation of glycosyl halides can be achieved typically by the exchange of the anomeric hydroxyl group with halogenating agents; 2) the various glycosyl halide substrates may have very different reactivities and stabilities, and these mainly depend on the nature of the halogen atom and the substituents on the carbohydrate scaffold: chlorides are more stable than bromides, while iodides are usually very unstable and electron-withdrawing protecting groups tend to increase the stability; 3) the reactivity of the glycosyl halide is also influenced by the choice of solvent, the temperature and the nature of the coactivator (Lewis acid or heavy metal salt); 4) the reaction is regiospecific, since the substitution always takes place at the anomeric carbon (C1) and can be highly diastereoselective; 5) formation of α-*O*-glycosides can be aided by the anomeric effect when neighboring group participation is not operational (if R^4=*O*-alkyl); 6) formation of β-*O*-glycosides is usually achieved from α-glycosyl halides when neighboring group participation is operational (e.g., R^4=*O*-acyl); 7) the coactivator or catalyst is typically a silver- or mercury salt dissolved in an aprotic solvent and the by-product acid is usually trapped by a base (e.g., Ag_2CO_3, collidine); and 8) due to the relatively low thermodynamic stability of glycosyl halides, reactions are conducted at or below room temperature. Disadvantages of the procedure are: 1) harsh conditions are needed for the preparation of the glycosyl halides, which are thermally not very stable; 2) glycosyl halides can undergo hydrolysis or 1,2-elimination; and 3) the coactivators are usually required in equimolar quantities, and they are often toxic and sometimes explosive. Numerous significant modifications and variants of the reaction exist.[22-34]

Michael (1879):

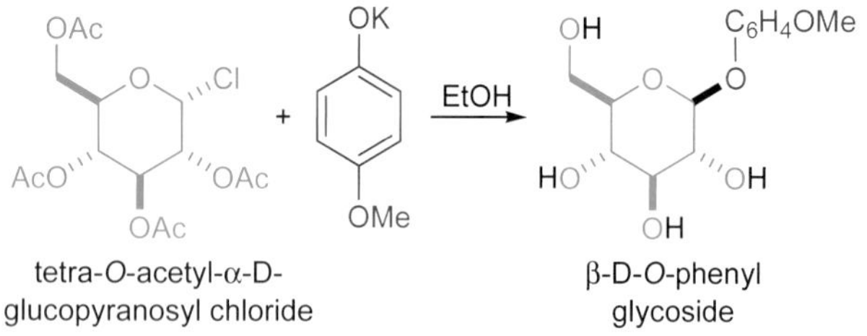

Koenigs & Knorr (1901):

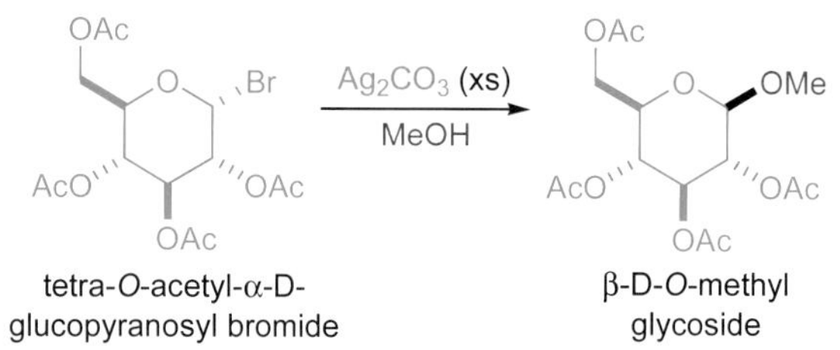

Koenigs-Knorr glycosidation:

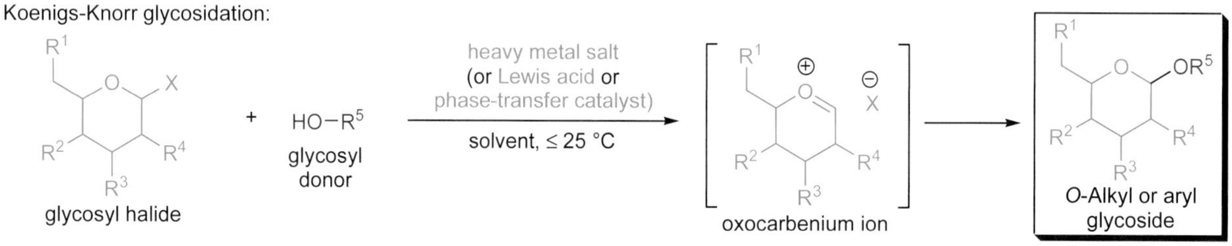

R^{1-4} = *O*-alkyl, *O*-acyl, alkyl ,aryl; X = Cl, Br, I; R^5 = alkyl, aryl, heteroaryl; heavy metal salts: AgOTf, Ag_2O, Ag_2CO_3, $AgNO_3$, $AgClO_4$, HgI_2, $HgCl_2$, $HgBr_2$, $Hg(CN)_2$; Lewis acids: $Sn(OTf)_2$, $Sn(OTf)_2$-collidine, $Sn(OTf)_2$-TMU, $SnCl_4$, TrCl-$ZnCl_2$; Phase-transfer catalysts: $(Bu_4N)Br$, $(Et_3NCH_2Ph)Br$, $(Et_3NCH_2Ph)Cl$; solvent: DCM, cyclohexane, petroleum ether, etc.

Mechanism:[38-42]

In order to achieve high levels of diastereoselectivity, the attack of the nucleophile should proceed *via* an S_N2 type mechanism. This is the case when the acyloxy group at C2 forms a dioxolanium ion with the oxocarbenium ion.

Glycosidation with neighboring group participation:

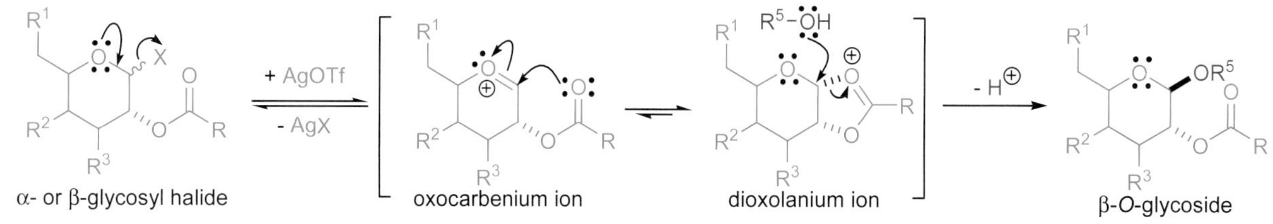

Synthetic Applications:

The first total synthesis of the major component of the microbial biosurfactant sophorolipid, sophorolipid lactone, was accomplished in the laboratory of A. Fürstner.[43] The natural product features a 26-membered ring, a (Z)-double bond, and two β-glycosidic linkages. The macrocyclization was achieved *via ring-closing alkyne metathesis* followed by hydrogenation of the alkyne in the presence of Lindlar's catalyst and finally the glycosidic linkages were installed using a *modified Koenigs-Knorr glycosidation*. In order to preserve the labile *p*-methoxybenzaldehyde dimethylacetal functionality, the anomeric hydroxyl group was converted to the corresponding glycosyl bromide under neutral conditions. The glycosidation was performed in the presence of excess silver triflate and base to afford an excellent yield of the desired β-*O*-glycoside. Interestingly, coactivators other than AgOTf gave inferior results.

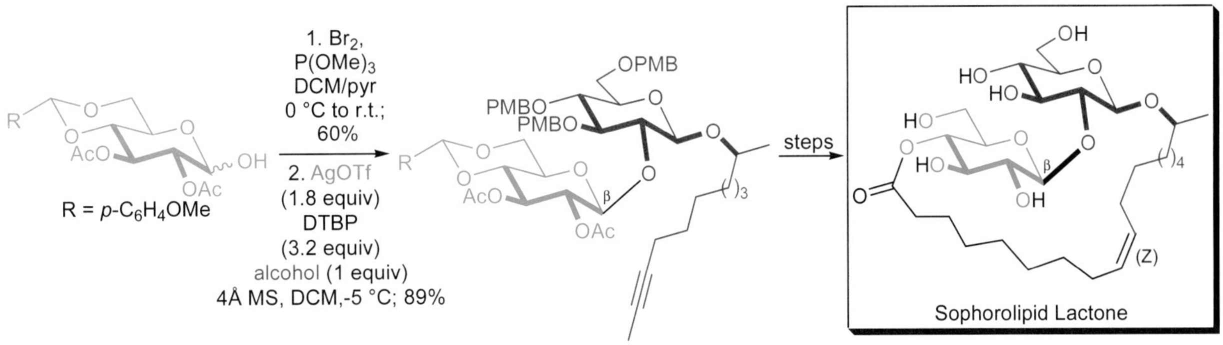

The macrolide insecticide (+)-lepicidin A (or (+)-A83543A) was first synthesized by D.A. Evans and co-workers. In the final stages of the total synthesis, the β-selective glycosidation of the C17 alcohol was required. The task was made even more difficult by the fact that a 2''-deoxy-β-glycosidic linkage had to be formed. The strategy was to take an α-glycosyl halide and its S_N2 inversion would afford the desired β-glycoside. The glycosyl bromide was generated prior to the reaction from the corresponding glycosyl acetate, but it was not purified due to its instability. NMR spectra confirmed that it was exclusively the α-anomer. The α:β selectivity was poor, and the yield could only be improved by using as much as 4 equivalents of the glycosyl bromide. The reaction was conducted several times and the anomers were separated providing enough β-glycoside to complete the total synthesis. The last two steps were the removal of the Fmoc protecting group under mildly basic conditions (Et_2NH) and reductive alkylation of the free amino group under the *Eschweiler-Clarke methylation* conditions.

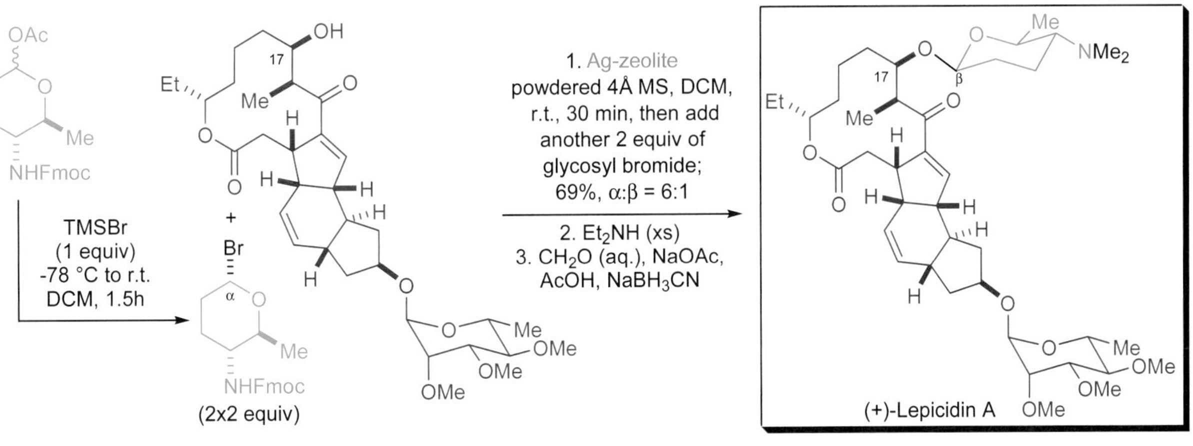

The naturally occurring noncyanogenic cyanoglucoside (–)-lithospermoside was prepared by C. Le Drian et al.[44] The key *Koenigs-Knorr glycosidation* step was very sensitive to steric hindrance, so the protecting groups on the aglycon had to be carefully chosen to obtain a reasonable yield.

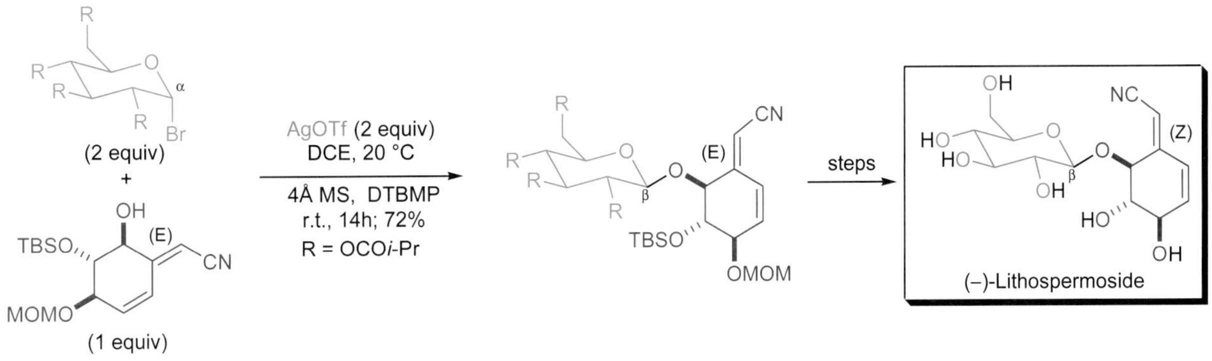

KOLBE-SCHMITT REACTION

(References are on page 616)

Importance:

[*Seminal Publications*[1-4]; *Reviews*[5,6]; *Modifications & Improvements*[7-15]; *Theoretical Studies*[16,17]]

In 1860, J. Kolbe and E. Lautemann reported the successful synthesis of salicylic acid (2-hydroxybenzoic acid) by heating phenol and sodium metal in an atmosphere of carbon dioxide.[1-3] The same year, they published similar transformation of *p*-cresol and thymol to obtain the corresponding *p*-cresotinic acid and *o*-thymotic acid, respectively.[4] This initial procedure was capricious and the yields varied greatly. In 1884, R. Schmitt found that exposing dry sodium phenoxide to a high-pressure of CO_2 in a sealed tube and heating it above 100 °C gave quantitative yields of the corresponding salicylic acid derivatives.[7,8] These conditions worked equally well for substituted phenols and naphthols. The preparation of *ortho*- or *para*-substituted aromatic hydroxy acids from the corresponding phenols under basic conditions using gaseous CO_2 is referred to as the *Kolbe-Schmitt reaction*. The general features of this transformation are: 1) phenols, substituted phenols, naphthols, and electron-rich heteroaromatic compounds (e.g., hydroxypyridine, carbazole, etc.) are good substrates; 2) monohydric phenols are first converted to the corresponding alkali or alkali earth phenoxides (e.g., Na, K, Mg, Ca, Ba), dried and then heated in the presence of pressurized CO_2 (5-100 atm); 3) di- or polyhydric phenols (with more than two hydroxyl groups) can be carboxylated with carbon dioxide at atmospheric pressure; 4) simple acidification of the reaction mixture affords the desired aromatic hydroxy acid; 5) the size of the alkali metals greatly influences the position of attack, the use of large alkali metal ions such as Rb^+ or Cs^+ gives rise to *p*-hydroxybenzoic acid derivatives, whereas smaller alkali metal ions (Na^+ or K^+) afford salicylic acid derivatives;[14] and 6) the presence of even trace amounts of water significantly decreases the yield of the product, so the reactants, reagents, and the solvents should be thoroughly dried before use.

Kolbe & Lautemann (1860):

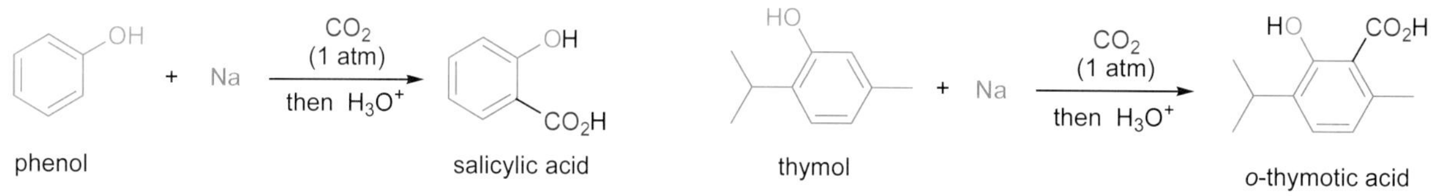

Kolbe-Schmitt reaction:

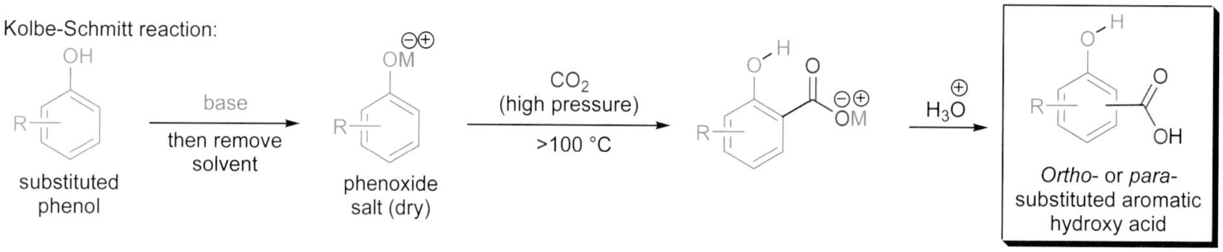

R = H, alkyl, aryl, OH, *O*-alkyl, NR_2; base: alkali metal hydroxides (e.g., NaOH, KOH, CsOH), K_2CO_3, $KHCO_3$

Mechanism: [8,18,5,19-24,15]

The mechanism of the *Kolbe-Schmitt reaction* was investigated since the late 1800s, but the mechanism of the carboxylation could not be elucidated for more than 100 years. For a long time, the accepted mechanism was that the carbon dioxide initially forms an alkali metal phenoxide-CO_2 complex, which is then converted to the aromatic carboxylate at elevated temperature.[8,18] The detailed mechanistic study conducted by Y. Kosugi et al. revealed that this complex is actually not an intermediate in the reaction, since the carefully prepared phenoxide-CO_2 complex started to decompose to afford phenoxide above 90 °C.[17] They also demonstrated that the carboxylated products were thermally stable even at around 200 °C.[17] The CO_2 electrophile attacks the ring directly to afford the corresponding *ortho*- or *para*-substituted products. (When the counterion is large (e.g., cesium) the attack of CO_2 at the *ortho*-position is hindered; therefore, the *para*-substituted product is the major product.)

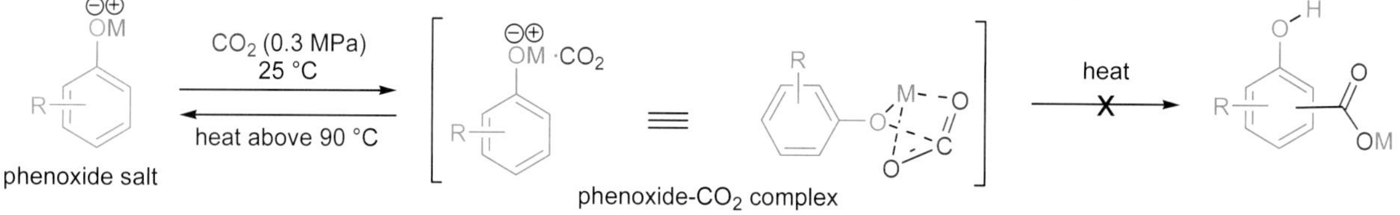

Direct attack of CO_2 on the aromatic ring:

Synthetic Applications:

In the laboratory of S. Blechert, the large-scale synthesis of a new and highly efficient alkene metathesis catalyst was achieved.[25] The catalyst was a biphenyl-based ruthenium alkylidene complex, and it was ideal for the ring-opening cross-metathesis of substrates that contain unprotected chelating atoms. The starting material 2-hydroxybiphenyl was first deprotonated, and the resulting dry sodium salt subjected to the *Kolbe-Schmitt reaction* conditions. The crude carboxylation product was alkylated with excess isopropyl bromide to afford the corresponding isopropyl ester that in three steps was converted to a vinyl derivative and finally to the desired ruthenium alkylidene complex.

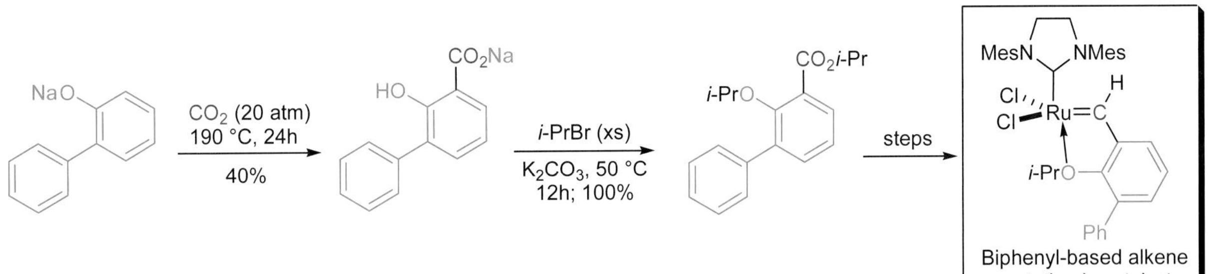

Biphenyl-based alkene metathesis catalyst

Phenols that have more than one hydroxyl group may be carboxylated with CO_2 at atmospheric pressure under basic conditions. The research team of Y.-C. Gao synthesized 3,5-di-*tert*-butyl-γ-resorcylic acid from 4,6-di-*tert*-butyl resorcinol using the *Kolbe-Schmitt reaction* under these conditions.[26] The resorcylic acid derivative was needed in order to prepare ternary complexes of lanthanide(III)-3,5-di-*tert*-butyl-γ-resorcylate with substituted pyridine-*N*-oxide.

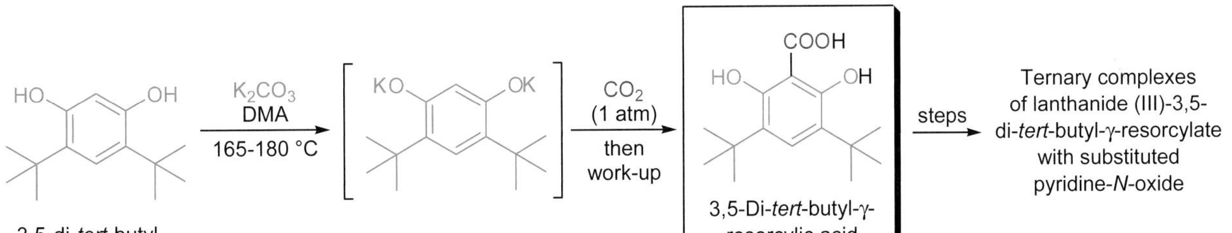

B.S. Green and co-workers developed an improved preparation of the clathrate host compound tri-*o*-thymotide (TOT) and other trisalicylide derivatives.[27] The synthesis began with the preparation of *ortho*-thymotic acid from thymol using the *Kolbe-Schmitt reaction*. The authors found that the yield of the product was dramatically increased when the reactants, solvents, and reagents were dried before use. Thus, thymol was dissolved in dry xylene, sodium metal was added and the temperature was kept at 130 °C for 20h in a dry carbon dioxide atmosphere. The desired carboxylated product was isolated in good yield. Finally, cyclodehydration with $POCl_3$ afforded TOT in almost quantitative yield.

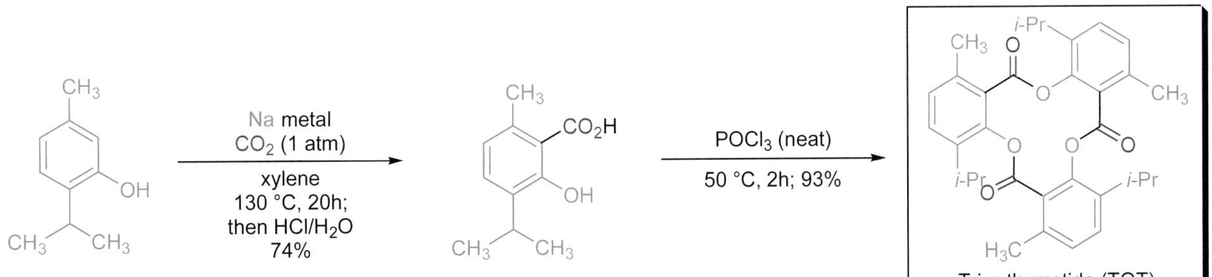

Tri-*o*-thymotide (TOT)

The first enantioselective total synthesis of the fungal metabolite (+)-pulvilloric acid was accomplished by H. Gerlach et al.[28] At the final stages of the synthetic effort the carboxylic acid moiety was installed *via* the *Kolbe-Schmitt reaction* using CO_2 at atmospheric pressure. The final formylation and ring-closure were achieved with triethyl orthoformate.

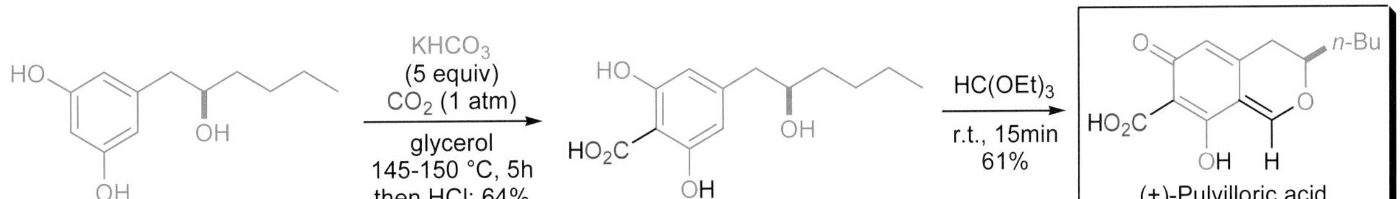

(+)-Pulvilloric acid

KORNBLUM OXIDATION

(References are on page 616)

Importance:

[*Seminal Publications*[1,2]; *Reviews*[3-8]; *Modifications & Improvements*[9-16]]

In 1957, N. Kornblum and co-workers discovered that activated primary benzyl bromides and α-bromo aromatic ketones are efficiently oxidized to the corresponding aldehydes and phenylglyoxals by simply dissolving the substrates in dimethyl sulfoxide (DMSO).[1] The drawback of this procedure was that it gave low yields for benzyl bromides having no electron-withdrawing groups, and less reactive halides, such as aliphatic alkyl halides, did not get oxidized at all. It was quickly recognized that the unreactive alkyl halides first had to be converted to the more reactive tosylates, which were oxidized readily in hot DMSO in the presence of a base (e.g., Na_2CO_3).[2] The oxidation of alkyl halides to the corresponding carbonyl compounds using DMSO as the oxidant is known as the *Kornblum oxidation*. The general features of the reaction are: 1) the typical procedure calls for the heating of the activated primary or secondary alkyl halide in DMSO in the presence of a base; 2) for unactivated alkyl halides the process requires two steps: first the addition of silver tosylate forms the tosylate, which is heated in DMSO in the presence of a base; 3) for primary alkyl halides the oxidation usually gives high yield of the carbonyl product, but with secondary alkyl halides, elimination of HX to form olefins is often a side reaction; 4) for sterically hindered substrates the yields are only moderate; 5) tertiary alkyl halides do not react; 6) the relative reactivity of the substrates is the following: tosylate>iodide>bromide>chloride; 7) the base plays a dual role: it neutralizes the hydrogen halide to avoid the oxidation of HX by DMSO (X_2 can lead to side reactions), as well as facilitates the deprotonation of the alkoxysulfonium intermediate; and 7) for substrates that dissolve poorly in DMSO a co-solvent is needed (e.g., DME). There are a number of variants and alternatives of the *Kornblum oxidation*: 1) silver-assisted DMSO oxidations;[11] 2) the use of amine oxides as oxidants (occasionally called the *Ganem oxidation*);[13] 3) the use of pyridine *N*-oxide or 2-picoline *N*-oxide and a base;[17,18] 4) the use of metal nitrates;[19,9,20,21] 5) *Sommelet oxidation*;[22] and 6) *Kröhnke oxidation*.[23]

Original procedure (Kornblum, 1957):

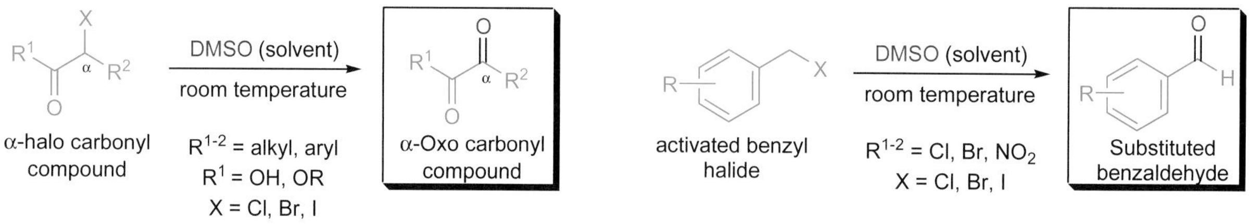

Improved procedure (Kornblum, 1959):

Improved procedure (Ganem, 1974):

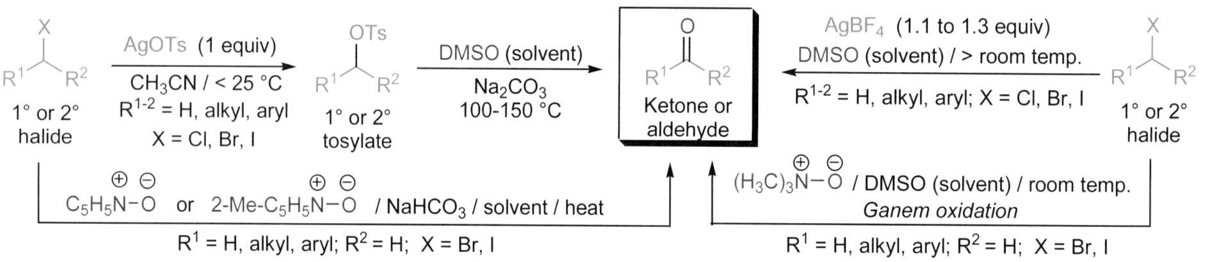

Mechanism:[4,6-8]

With alkyl halide substrates, the first step of the oxidation is the S_N2 displacement of the halide with tosylate anion. Next the alkyl tosylate undergoes a second S_N2 reaction with the nucleophilic oxygen atom of the DMSO to form the alkoxysulfonium salt that undergoes deprotonation to give the alkoxysulfonium ylide, which upon a *[2,3]-sigmatropic shift* affords the carbonyl compound. In the case of α-halo carbonyl substrates, the deprotonation takes place at the more acidic α-carbon instead of the methyl group attached to the sulfur atom of the alkoxysulfonium salt.

Oxidation of an alkyl halide:

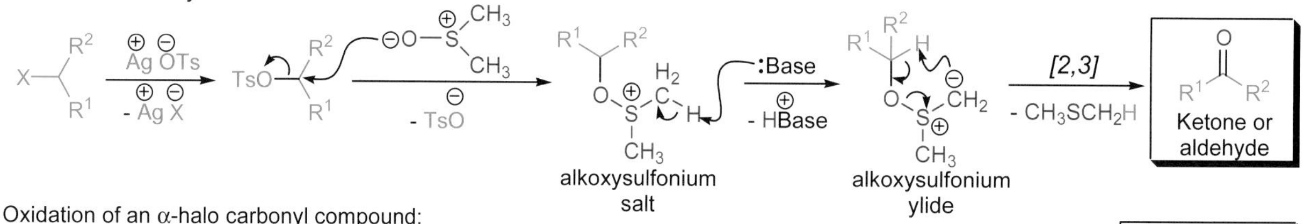

Oxidation of an α-halo carbonyl compound:

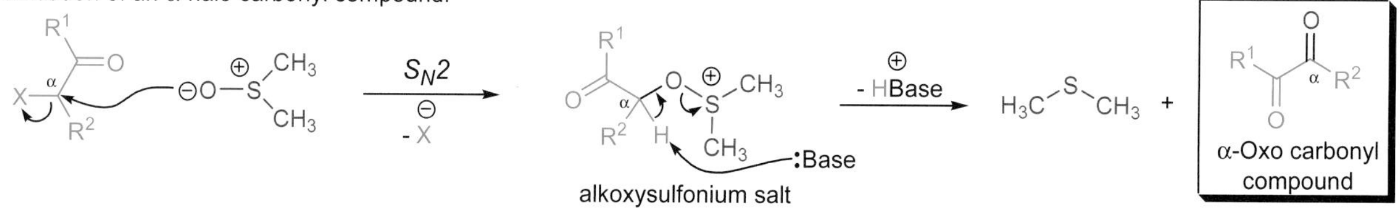

Synthetic Applications:

A tandem *Kornblum oxidation*/imidazole formation reaction was used during the preparation of new fluorescent nucleotides by B. Fischer and co-workers.[24] The adenosine monophosphate free acid was mixed with 10 equivalents of 2-bromo-(*p*-nitro)-acetophenone and dissolved in DMSO. The required pH value was maintained with the addition of DBU which also served as a base. The *Kornblum oxidation* of the alkyl halide yielded the glyoxal, which reacted *in situ* with the aromatic amine to form the desired imidazole derivative.

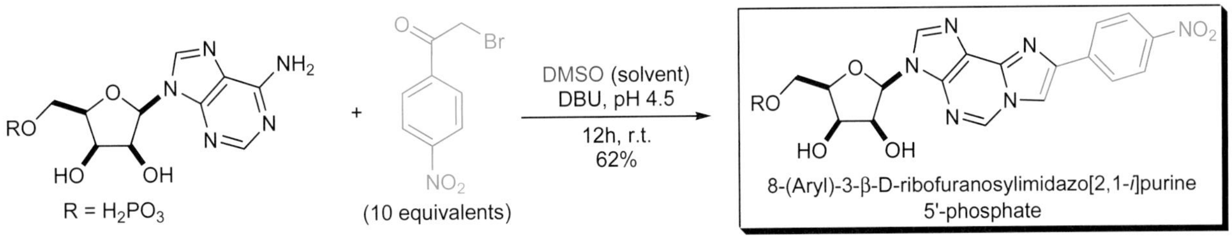

The first total synthesis of the clerodane alkaloid solidago alcohol was achieved in the laboratory of H.-S. Liu, using a highly diastereoselective *Diels-Alder cycloaddition* as the key step.[25] The installation of the 3-furyl side chain required the conversion of the bicyclic primary alkyl bromide to the corresponding aldehyde. This was accomplished by the *modified Kornblum oxidation*, which employed silver tetrafluoroborate to activate the substrate.

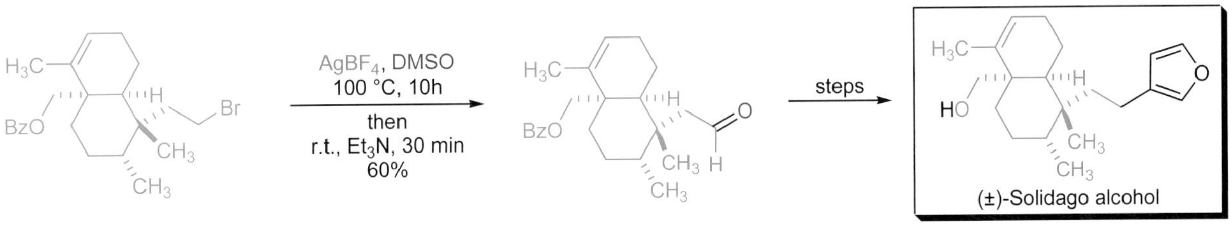

A number of simple analogs of the antipsoriatic agent anthralin (dithranol) were prepared by K. Müller and co-workers by changing the positions of the hydroxyl groups as well as adding new functional groups into various positions of the anthracenone nucleus.[26] The benzyl bromide functionality was converted to the corresponding aldehyde by the *Kornblum oxidation* in fair yield.

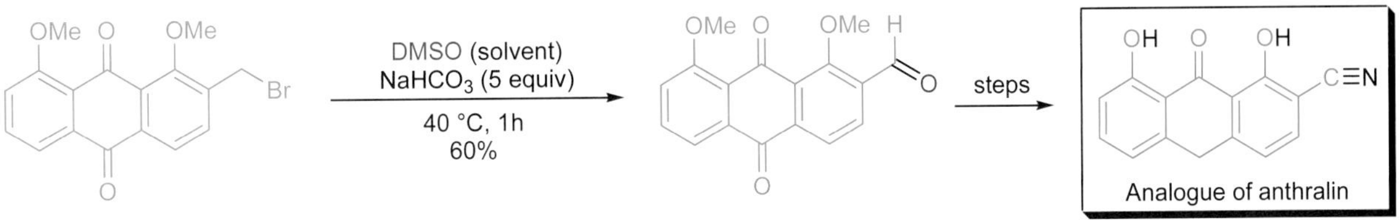

A novel synthetic approach was developed by R.E. Taylor et al. for the preparation of the triene portion of the biologically active polyketide apoptolidin.[27] The allylic chloride substrate was prepared from an allylic alcohol *via* a *thionyl chloride mediated rearrangement*. Next, the allylic chloride was subjected to the *Ganem oxidation* by treating it with five equivalents of trimethylamine *N*-oxide (*TMANO*) in DMSO at room temperature to obtain the desired α,β-unsaturated aldehyde. Interestingly, the original *Kornblum oxidation* conditions were not well suited for this system because of the required high reaction temperature.

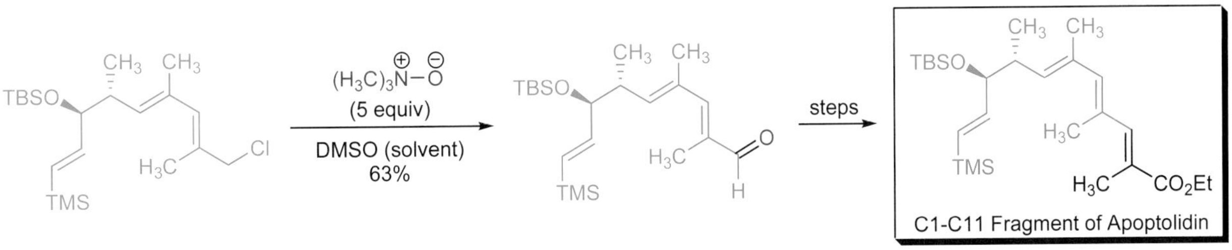

KRAPCHO DEALKOXYCARBONYLATION (KRAPCHO REACTION)

(References are on page 617)

Importance:

[*Seminal Publications*[1-6]; *Reviews*[7-9]; *Modifications & Improvements*[10-15]]

In 1967, A.P. Krapcho reported that upon heating geminal dicarbethoxy compounds with sodium cyanide in dimethyl sulfoxide, the corresponding ethyl esters were obtained in high yield.[5] The products could be purified by distillation following an aqueous work-up. The discovery of this transformation happened serendipitously during an attempted conversion of a ditosylate to the corresponding dinitrile with potassium cyanide in hot DMSO, and the product of the reaction was the demethoxycarbonylated dinitrile (this result was reported only in 1970).[6] The dealkoxycarbonylation of β-keto esters, α-cyano esters, malonate esters, and α-alkyl- or arylsulfonyl esters to the corresponding ketones, nitriles, esters, and alkyl- or arylsulfones is known as the *Krapcho dealkoxycarbonylation* (also *Krapcho reaction* or *Krapcho decarboxylation*). The general features of this reaction are:[8,9] 1) this nucleophilic dealkoxycarbonylation process is general for methyl- or ethyl esters of carboxylic acids, which have an electron-withdrawing group (CO_2-alkyl, CN, CO-alkyl, SO_2-alkyl, etc.) at their α-position; 2) this one-pot procedure obviates the need to perform the multistep decarboxylation of geminal diesters to the corresponding monoesters, which involves the following steps: basic or acidic hydrolysis of the ester followed by the decarboxylation of the resulting diacid and the esterification of the final carboxylic acid to obtain the desired monoester; 3) the reaction conditions are essentially neutral, so both acid- and base-sensitive substrates can be used and the otherwise frequent acid-catalyzed rearrangements are avoided; 4) the chemoselectivity and the functional group tolerance of the method is very high; 5) double bonds are not isomerized and in the overwhelming majority of cases labile stereocenters are not racemized; 6) the choice of specific reaction conditions is always dependent on the substitution pattern of the substrate; 7) monosubstituted malonic esters are dealkoxycarbonylated in hot dipolar aprotic solvent containing at least one equivalent of water; 8) as a rule of thumb when the substrate has at least one proton at the α-position, the dealkoxycarbonylation can be achieved with wet DMSO at reflux in the absence of a salt; 9) disubstituted malonic esters, however, are dealkoxycarbonylated only in the presence of at least one equivalent of a salt (e.g., KCN, LiCl, etc.) in wet DMSO at reflux; 10) the presence of a salt tends to accelerate the rate of the dealkoxycarbonylation of many (but not all) substrates; 11) besides DMSO, other dipolar aprotic solvents can be used such as dimethylacetamide, HMPT and DMF; 12) methyl esters are dealkoxycarbonylated faster than ethyl esters; and 13) vinylogous β-keto esters are also dealkoxycarbonylated in high yield.

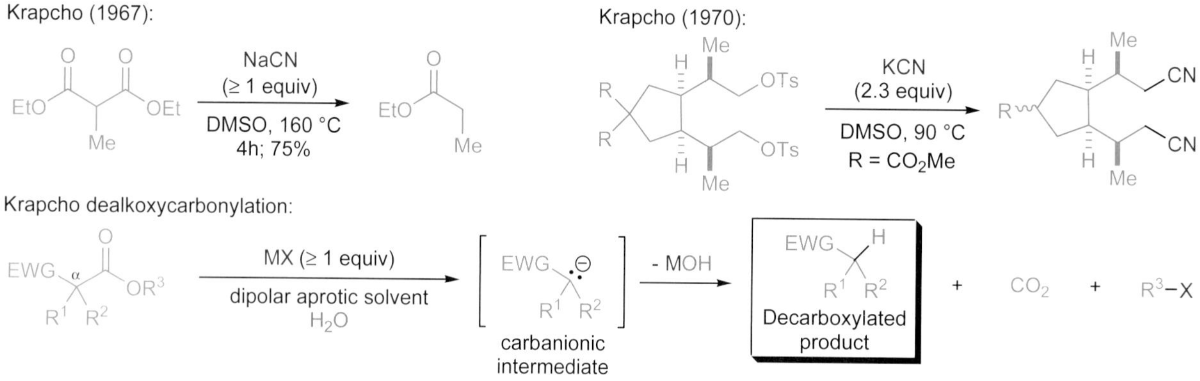

EWG = CO_2-alkyl, CO_2-aryl, CN, CO-alkyl, SO_2-alkyl, SO_2-aryl; R^{1-2} = H, alkyl, aryl; R^3 = Me, Et; MX = NaCN, KCN, LiCl, NaCl, NaBr, NaI, LiI·H_2O, Na_2CO_3·H_2O, Na_3PO_4·$12H_2O$, Me_4NOAc ; solvent: DMSO, DMF, DMA, HMPT

Mechanism:[16,17,9,18,19]

The mechanism of the *Krapcho dealkoxycarbonylation* is dependent on the structure of the substrate ester and the type of anion used. In the case of α,α-disubstituted diesters (especially the methyl esters), the anion from the salt (cyanide ion in the scheme) attacks the alkyl group of the ester in an S_N2 fashion and the decarboxylation results in the formation of a carbanionic intermediate that is quenched by the water. In the case of α-monosubstituted diesters the cyanide attacks the carbonyl group to form a tetrahedral intermediate, which breaks down to give the same carbanionic intermediate and a cyanoformate, which is hydrolyzed to give carbon dioxide and an alcohol.

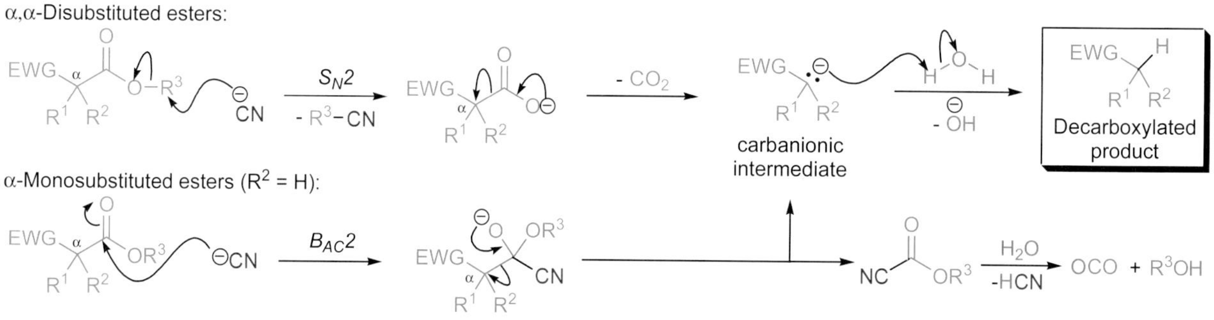

Synthetic Applications:

In the laboratory of A. Fürstner, a practical synthesis of the immunosuppressive alkaloid metacycloprodigiosin and its functional derivatives was developed.[20] Toward the end of the synthetic sequence a *meta*-pyrrolophane β-keto ester was decarboxylated under standard Krapcho conditions. The substrate was dissolved in wet DMSO, and two equivalents of sodium chloride were added and the reaction mixture was heated to 180 °C to afford the desired *meta*-pyrrolophane ketone in excellent yield. This ketone functionality was first converted to an ethyl group and then the product was advanced to metacycloprodigiosin.

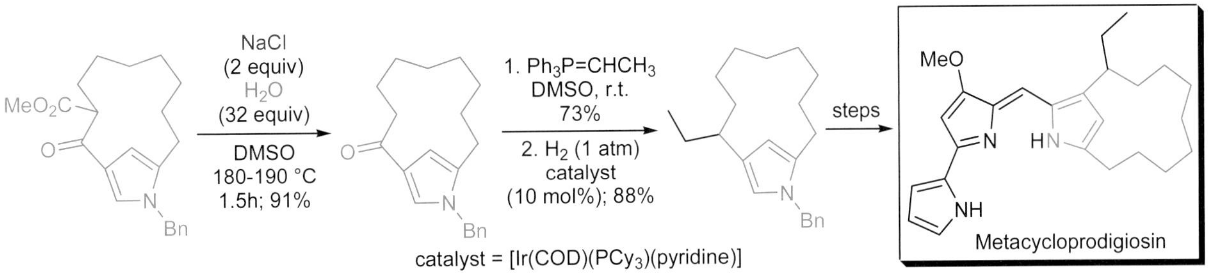

A highly *exo*-selective asymmetric *hetero Diels-Alder reaction* was the key step in D.A. Evans' total synthesis of (–)-epibatidine.[21] The bicyclic cycloadduct was then subjected to a fluoride-promoted fragmentation that afforded a β-keto ester, which was isolated exclusively as its enol tautomer. The removal of the ethoxycarbonyl functionality was achieved using the *Krapcho decarboxylation*. Interestingly, the presence of a metal salt was not necessary in this transformation. Simply heating the substrate in wet DMSO gave rise to the decarboxylated product in quantitative yield.

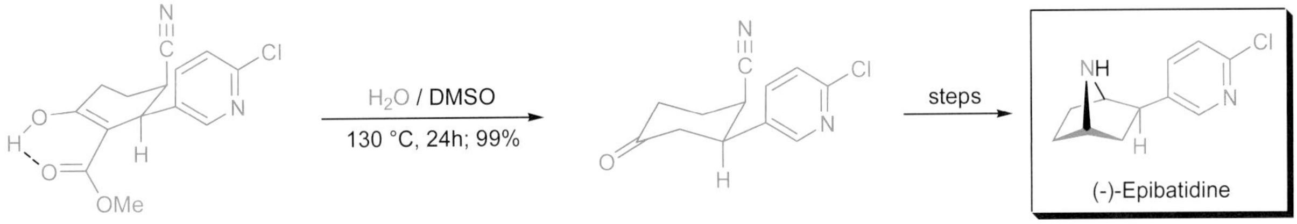

A general synthetic route toward the marine metabolite eunicellin diterpenes was developed by G.A. Molander and co-workers.[22] The power of this method was demonstrated by the completion of the asymmetric total synthesis of deacetoxyalcyonin acetate. A tricyclic β-keto ester intermediate was methylated in the γ-position with complete diastereoselectivity using dianion chemistry and the crude product was subjected to the *Krapcho decarboxylation*. This was one of the rare cases when the transformation did not only remove the methoxycarbonyl group, but at the same time epimerized the newly formed stereocenter to yield a separable mixture of methyl ketones.

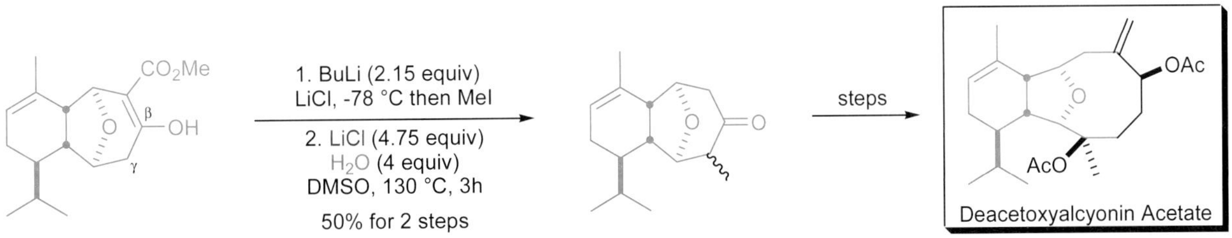

The first enantioselective formal total synthesis of paeonilactone A was reported by J.E. Bäckvall who used a palladium(II)-catalyzed 1,4-oxylactonization of a conjugated diene as the key step.[23] The lactonization precursor diene acid was obtained from an enantiopure dimethyl malonate derivative *via* sequential *Krapcho decarboxylation* and ester hydrolysis.

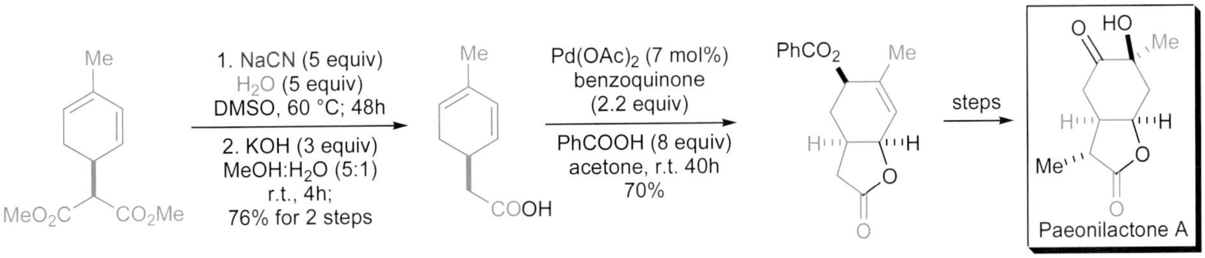

KRÖHNKE PYRIDINE SYNTHESIS

(References are on page 617)

Importance:

[*Seminal Publications*[1,2]; *Reviews*[3]; *Modifications and Improvements*[4,5]]

In 1961, F. Kröhnke and W. Zeher reported that phenacyl isoquinolinium bromide reacted with benzalacetophenone under basic conditions to afford an isoquinolinium betaine, which upon treatment with ammonium acetate in acetic acid at reflux temperature yielded 2,4,6-triphenylpyridine in moderate yield.[1,2] This synthetic sequence was a new and efficient way to access highly substituted pyridines. The condensation of acylmethylpyridinium salts with α,β-unsaturated ketones and ammonia to give substituted pyridines is known as the *Kröhnke pyridine synthesis*. The general features of the transformation are:[3] 1) α-haloketones are prepared from the corresponding methyl ketones using standard halogenation conditions (e.g., Br_2, Bu_4NBr_3, etc.); 2) α-haloketones are mixed with pyridine to afford the required acylmethylpyridinium salts that are considered 1,3-diketone equivalents; 3) treatment of the acylmethylpyridinium salt with ammonium acetate (or other ammonia equivalents) in acetic acid in the presence of an α,β-unsaturated ketone gives rise to a Michael adduct (a 1,5-diketone), which undergoes cyclization with ammonia to produce the substituted pyridine; 4) the great advantage of the method is that unlike in the *Hantzsch dihydropyridine synthesis*, oxidation (dehydrogenation) is not necessary, since the pyridine is formed directly; 5) the substitution pattern of the two components can be varied widely ranging from simple alkyl all to way to substituted aryl and heteroaryl groups; 6) the α,β-unsaturated ketones can be used directly or in the form of the corresponding Mannich bases, which undergo cleavage under the reaction conditions to afford the α,β-unsaturated ketones; 7) in most cases the reaction is used to prepare 2,4,6-trisubstituted pyridines, but occasionally higher substitution (at C3 and C5) can be achieved; 8) if R^4=CO_2H, 2-carboxypyridines are formed that can be decarboxylated thermally to afford 2,4-disubstituted pyridines; and 9) the preparation of symmetrically or unsymmetrically substituted bi- and oligopyridines (up to seven pyridine units) is accomplished with ease unlike with other methods that are less straightforward and require many steps.

Kröhnke (1961):

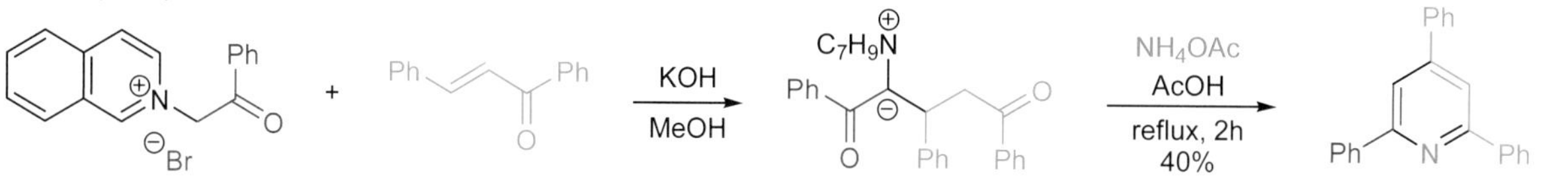

N-phenacyl isoquinolinium bromide — benzalacetophenone — isoquinolinium betaine — 2,4,6-triphenylpyridine

Kröhnke pyridine synthesis:

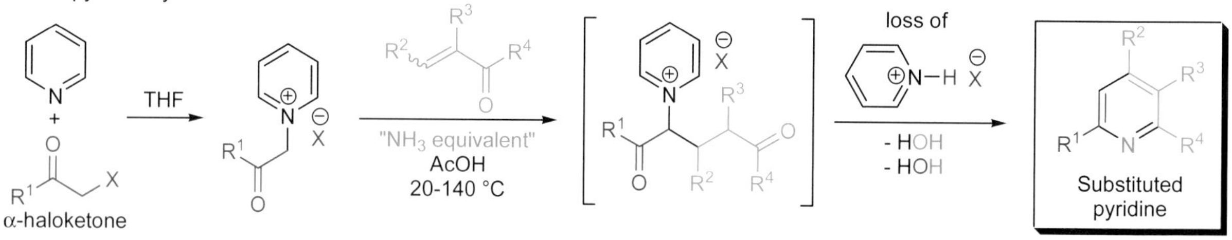

R^1 = alkyl, substituted aryl, heteroaryl; X = Cl, Br, I; R^2 = H, alky, aryl, heteroaryl; R^3 = H, (alkyl, aryl, heteroaryl); R^4 = alkyl, aryl, heteroaryl, CO_2^-, CO_2-alkyl; NH_3 equivalent: NH_4OAc, $HCONH_2$, CH_3CONH_2

Mechanism:[6-9]

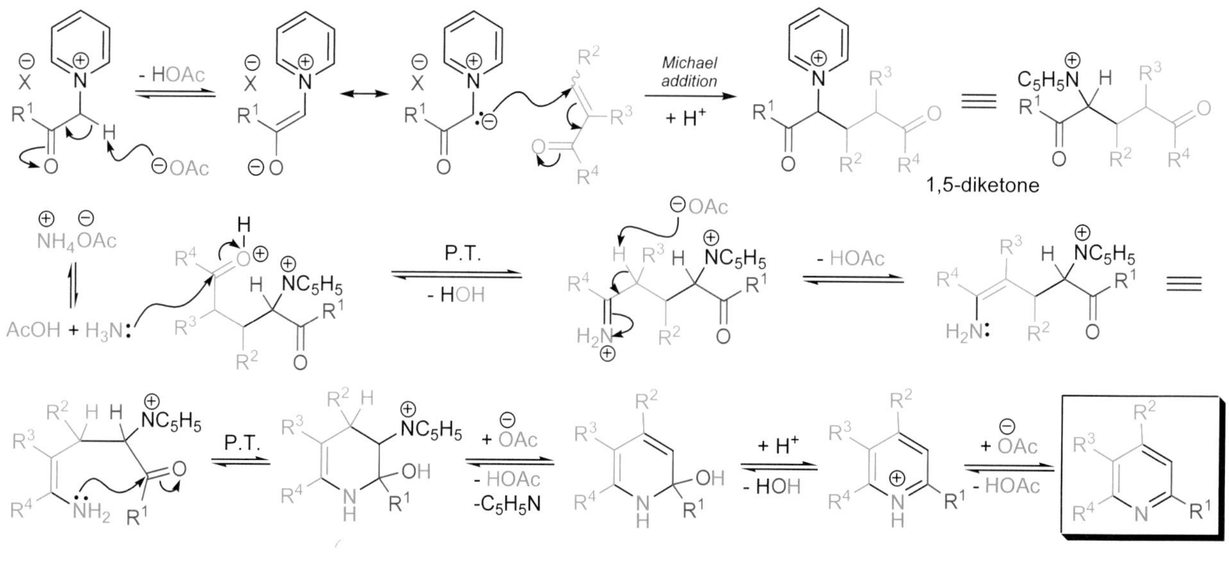

Synthetic Applications:

In the laboratory of P. Kočovský, novel pyridine-type *P,N*-ligands were prepared from various monoterpenes.[10] The key step was the *Kröhnke pyridine synthesis*, and the chirality was introduced by the α,β-unsaturated ketone component, which was derived from enantiopure monoterpenes. One of these ligands was synthesized from (+)-pinocarvone which was condensed with the acylmethylpyridinium salt under standard conditions to give good yield of the trisubstituted pyridine product. The benzylic position of this compound was deprotonated with butyllithium, and upon addition of methyl iodide the stereoselective methylation was achieved. The subsequent nucleophilic aromatic substitution (S_NAr) gave rise to the desired ligand.

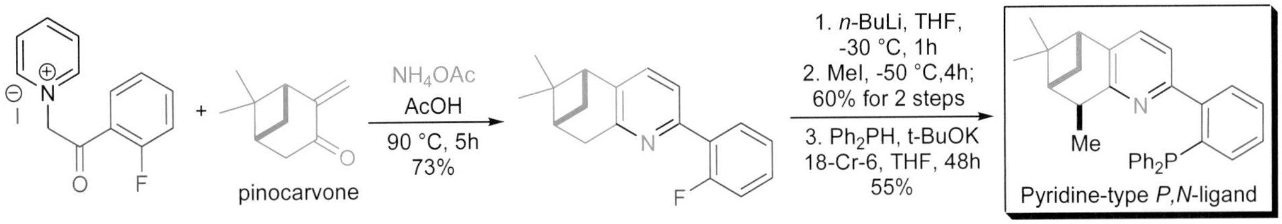

The synthesis of cyclo-2,2':4',4'':2'',2''':4''',4'''':2'''',2''''':4'''''-sexipyridine was accomplished by T.R. Kelly and co-workers by using multiple *Stille cross-couplings* and the *Kröhnke pyridine synthesis* for the final macrocyclization.[11] The bromination of the quinquepyridine was conducted with wet *N*-bromosuccinimide in THF, and the resulting α-bromoketone was immediately converted to the corresponding acylmethylpyridinium salt by strirring it with excess pyridine in acetone overnight. The crucial macrocyclization took place in the presence of excess ammonium acetate in acetic acid at reflux. Interestingly, other macrocyclization attempts using the *Ullmann biaryl coupling* or the *Glaser coupling* all failed.

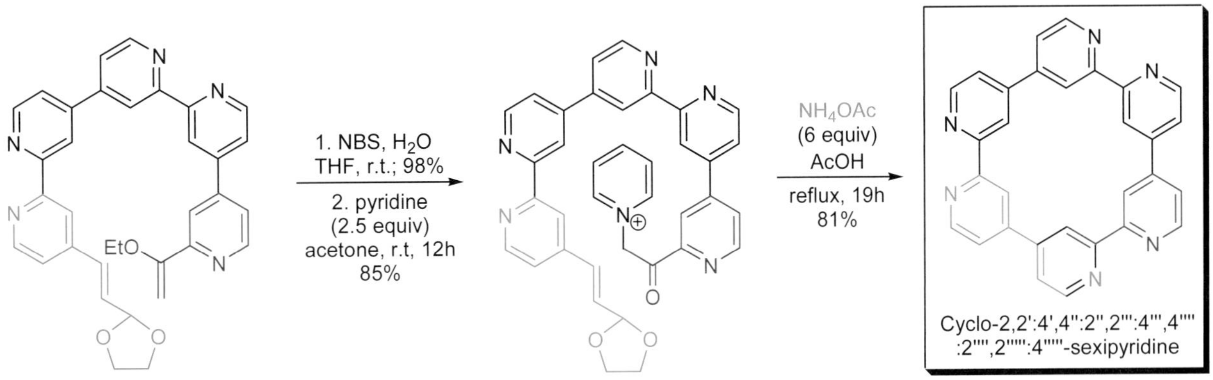

The research team of E.-S. Lee synthesized and evaluated several 2,4,6-trisubstituted pyridine derivatives as potential topoisomerase I inhibitors.[12] One of these compounds, 4-furan-2-yl-2-(2-furan-2-yl-vinyl)-6-thiophen-2-yl-pyridine, was prepared by the *Kröhnke pyridine synthesis* and showed strong topoisomerase I inhibitory activity.

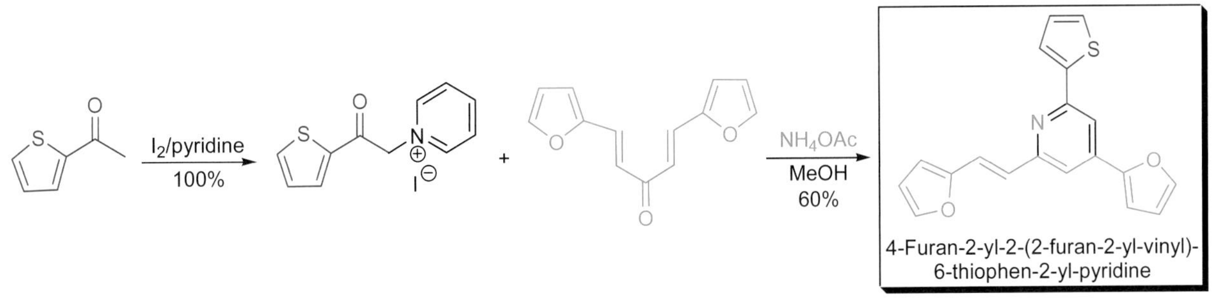

Novel, tetrahydroquinoline-based *N,S*-type ligands were prepared by the *Kröhnke pyridine synthesis* and their catalytic activity was assessed by G. Chelucci et al.[13] The acylmethylpyridinium iodide was reacted with a cyclic α,β-unsaturated ketone derived from 2-(+)-carene.

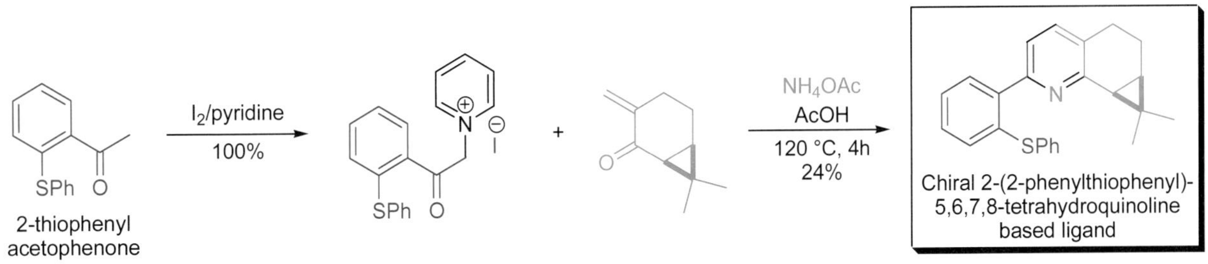

KULINKOVICH REACTION

(References are on page 618)

Importance:

[*Seminal Publications*[1,2]; *Reviews*[3-15]; *Modifications & Improvements*[16-23]; *Theoretical Studies*[24]]

In 1989, O.G. Kulinkovich reported that 1-alkylcyclopropanols were formed when an excess of ethylmagnesium bromide was added to simple carboxylic esters in the presence of one equivalent of titanium tetraisopropoxide.[1] The reaction could also be carried out with catalytic amounts of $Ti(O\textit{i}\text{-}Pr)_4$ and only two equivalents of Grignard reagent was necessary. The titanium(II)-mediated one-pot conversion of carboxylic esters and amides to the corresponding 1-alkylcyclopropanols and 1-alkylcyclopropylamines is known as the *Kulinkovich reaction*. The general features of the reaction are: 1) the active species is a titanacyclopropane intermediate that acts as a 1,2-dicarbanion equivalent and doubly alkylates the carbonyl group; 2) more complex Grignard reagents yield 1,2-*cis* disubstituted cyclopropanols with good diastereoselectivity; 3) the observed *cis*-selectivity is lower for the formation of 1,2-disubstituted cyclopropylamines from amides; 4) the reaction is sensitive to the nature of the R^1 group (aromatic esters do not react) and steric crowding (α-branched R^1 groups and too bulky R^2 groups) give lower yields); 5) when terminal alkenes are added into the reaction mixture, these are incorporated into the cyclopropane products. There are several important modifications of the procedure, which helped to expand the scope of the reaction.[16-23]

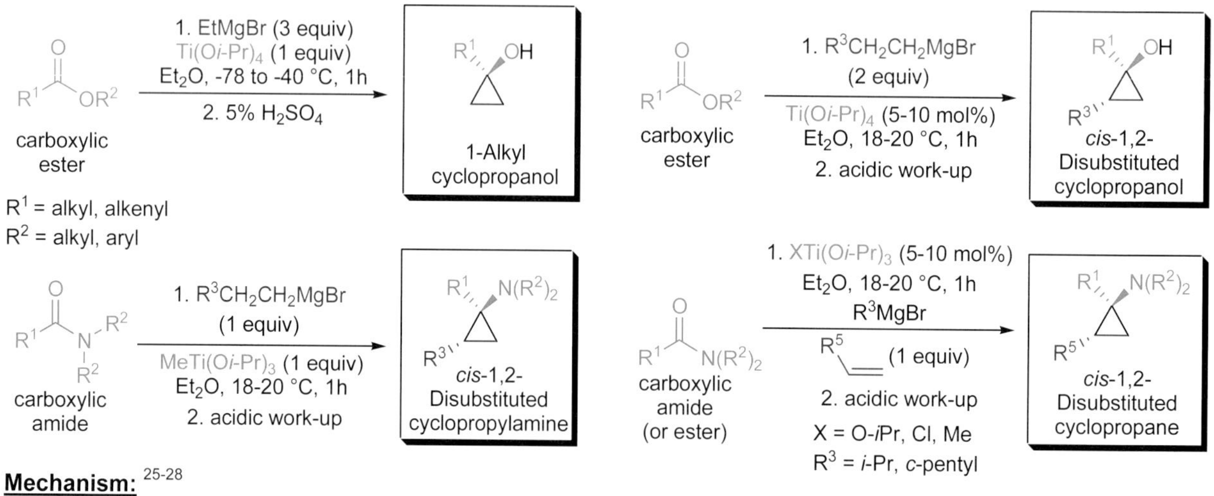

Mechanism:[25-28]

The catalytic cycle of the *Kulinkovich reaction* begins with the dialkylation of the $Ti(O\textit{i}\text{-}Pr)_4$ with two equivalents of ethylmagnesium bromide to form the thermally unstable diethyltitanium intermediate, which quickly undergoes a β-hydride elimination to give ethane and titanacyclopropane. This titanacyclopropane acts as a 1,2-dicarbanion equivalent when it reacts with the carboxylic ester, and it performs a double alkylation. The addition of ethylmagnesium bromide to the titanium in the titanacyclopropane-ester complex triggers the formation of the first C-C bond formation and leads to the oxatitanacyclopentane ate-complex. At this point, the alkoxy group of the original ester is eliminated as its magnesium salt and the second C-C bond is formed to generate the cyclopropane ring. The resulting titanium cyclopropoxide undergoes alkylation at the titanium by ethylmagnesium bromide, and thus the diethyltitanium intermediate is regenerated and the product magnesium cyclopropoxide is formed. Upon aqueous/acidic work-up, the 1-cyclopropanol is isolated. For carboxylic amides the mechanism is slightly different.

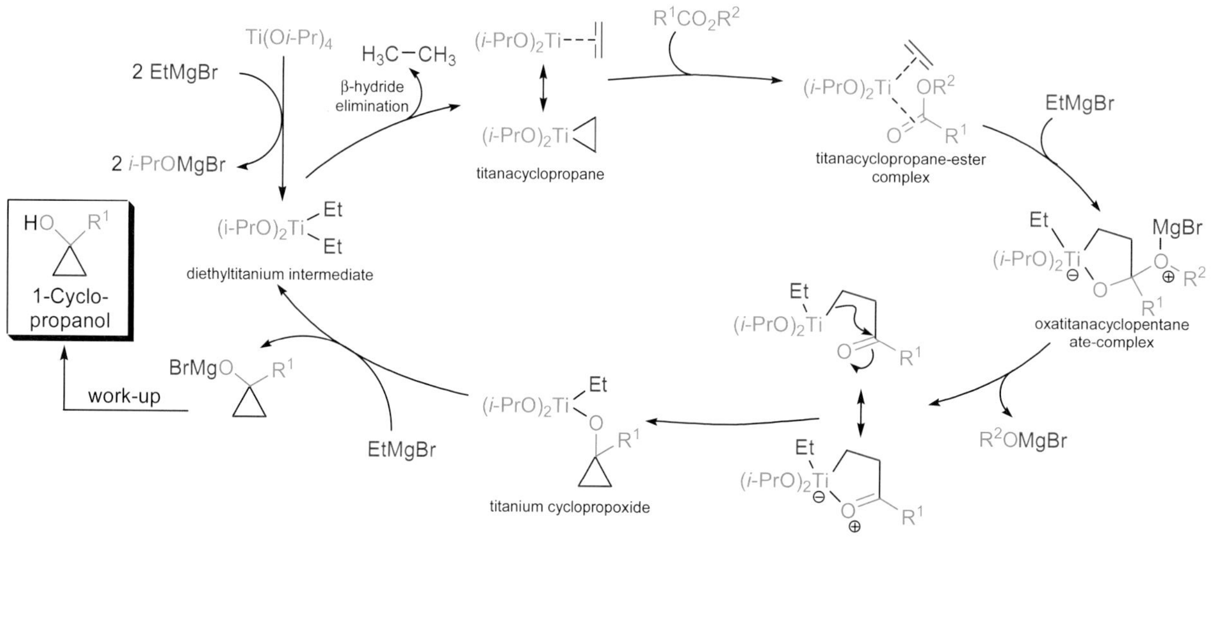

Synthetic Applications:

The key component of the antitumor antibiotic cleomycin, (*S*)-cleonin, was prepared from (*R*)-serine using the *Kulinkovich reaction* as the key step in the laboratory of M. Taddei.[29] The methyl ester of *N*-Cbz serine acetonide was exposed to freshly prepared ethylmagnesium bromide in the presence of substoichiometric amounts of titanium tetraisopropoxide to afford the desired cyclopropylamine in good yield. Subsequent functional group manipulations gave (*S*)-cleonin.

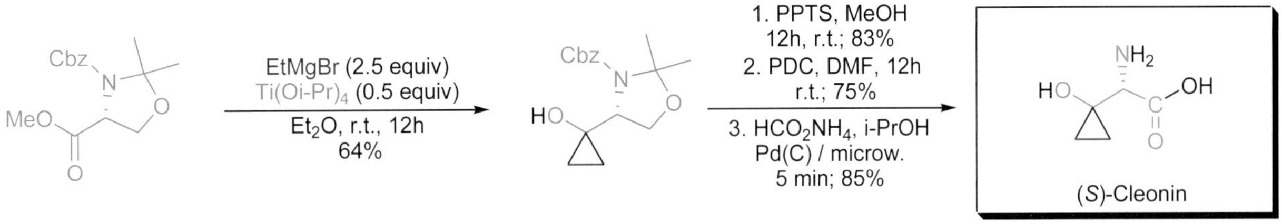

Cyclopropylamines and their substituted derivatives are important building blocks in a large number of biologically active compounds. The synthesis of potentially biologically active *N,N*-dimethyl bicyclic cyclopropylamines from *N*-allylamino acid dimethylamides by the *intramolecular variant of the Kulinkovich reaction* was accomplished by M.M. Joullié and co-workers.[30]

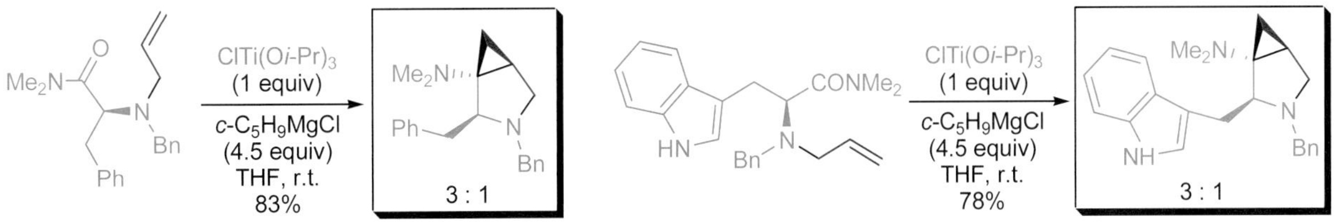

J.K. Cha et al. developed a stereocontrolled synthesis of bicyclo[5.3.0]decan-3-ones from readily available acyclic substrates.[31] Acyclic olefin-tethered amides were first subjected to the *intramolecular Kulinkovich reaction* to prepare bicyclic aminocyclopropanes. This was followed by a *tandem ring-expansion-cyclization sequence* triggered by *aerobic oxidation*. The reactive intermediates in this tandem process were aminium radicals (radical cations). The *p*-anisidine group was chosen to lower the amine oxidation potential. This substituent was crucial for the generation of the aminium radical (if Ar = phenyl, the ring aerobic oxidation is not feasible).

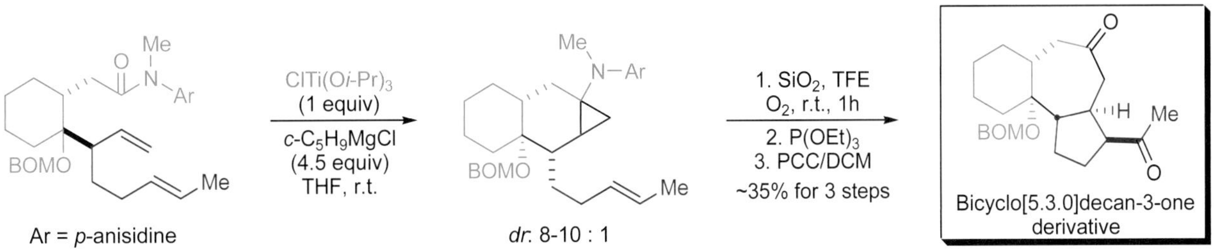

A general diastereoselective synthesis of fused bicyclic compounds using a sequential *Kulinkovich cyclopropanation* and an *oxy-Cope rearrangement* was achieved by J.K. Cha and co-workers.[32] *cis*-1,2-Divinylcyclopropanes have found significant synthetic utility as substrates for *[3,3]-sigmatropic rearrangements*. The *Kulinkovich reaction* offered a straightforward and facile synthesis of *cis*-1,2-dialkenylcyclopropanols that gave fused bicyclic carbocycles upon *oxy-Cope rearrangement*.

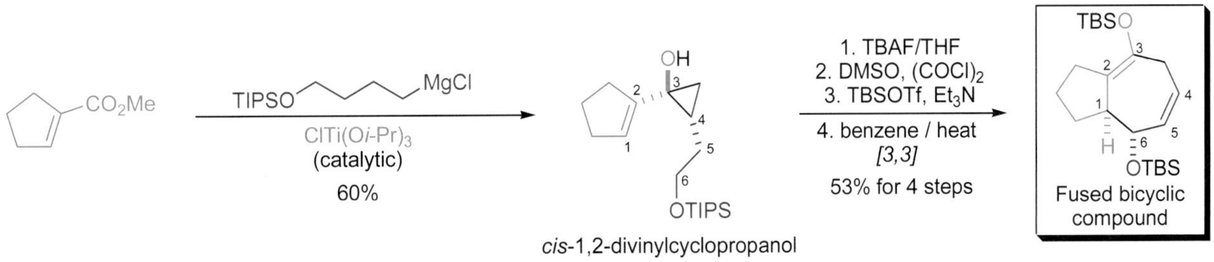

KUMADA CROSS-COUPLING

(References are on page 619)

Importance:

[*Seminal Publications*[1-4]; *Reviews*[5-18]; *Modifications & Improvements*[19-30];]

During the 1970s a great deal of research effort was focused on the transition metal-catalyzed carbon-carbon bond forming reactions of unreactive alkenyl and aryl halides.[31,32] In 1972, M. Kumada and R.J.P. Corriu independently discovered the stereoselective cross-coupling reaction between aryl- or alkenyl halides and Grignard reagents in the presence of a catalytic amount of a nickel-phosphine complex. In the following years, Kumada explored the scope and limitation of the reaction. Consequently, this transformation is now referred to as the *Kumada cross-coupling*. Nickel catalysis only worked for Grignard reagents and excluded the highly versatile organolithium reagents. Therefore, the use of alternative catalysts such as various palladium complexes was explored.[19-24,26] The characteristic features of the *Kumada cross-coupling* are: 1) in the Ni-catalyzed process the catalytic activity depends largely on the nature of the phosphine ligand, and the following reactivity trend is observed: $Ni(dppp)Cl_2$ > $Ni(dppe)Cl_2$ > $Ni(PR_3)_2Cl_2$ ~ $Ni(dppb)Cl_2$; 2) even alkyl (sp^3) Grignard reagents having β-hydrogens can selectively undergo cross-coupling reactions without any undesired β-hydride elimination; 3) with *sec*-alkyl Grignard reagents the alkyl group tends to isomerize to the corresponding primary alkyl group, and this isomerization is dependent on the basicity of the phosphine ligand and the nature of the aromatic halide; 4) the use of the *dppf* ligand slows the β-hydride elimination considerably and accelerates the reductive elimination, thereby allowing the coupling of *sec*-Grignard reagents without isomerization;[24] 5) chlorinated aromatic compounds react with ease and even fluorobenzene can undergo Ni-catalyzed cross-coupling;[16] 6) the coupling is stereoselective and the stereochemistry of the starting vinyl halides is preserved; 7) the Pd-catalyzed process is more chemo- and stereoselective and has a much broader scope with carbanions than the Ni-catalyzed reaction. However, the coupling does not take place with aryl chlorides, only with aryl bromides and iodides; 8) organomagnesium and organolithium reagents are used most often. However, the coupling will take place with organosodium (RNa), organocopper (R_2CuLi), organoaluminum, organozinc, organotin, organozirconium, and organoboron compounds;[14] 9) organolithiums are by far the most versatile, since these reagents can be prepared in many different ways including the direct lithiation of hydrocarbons;[9] and 10) functional groups that are base-sensitive are not tolerated because of the polar nature of the organomagnesium and organolithium compounds (this tolerance is greatly improved in the *Negishi cross-coupling* by using much less basic organozinc compounds). There are not many side-reactions except for the occasional isolation of homocoupled and reduction products that can be avoided by observing the following precautions: 1) the organolithium should be added slowly because fast addition produces α-bromo alkenyllithiums that undergo rearrangement to give lithium acetylides, thus lowering the overall yield; 2) the $Pd^{(0)}$ catalyst should be clean to ensure high activity; and 3) no reagents should be added in excess.[33,14]

R^{1-3} = H, alkyl, aryl, alkenyl; X = F, Cl, Br, I. OTf; R^4 = alkyl, aryl, alkenyl; X = Br, I; L = PPh_3 or L_2 = dppp, dppe, dppb

Mechanism: [34-38]

Ni-catalyzed process:

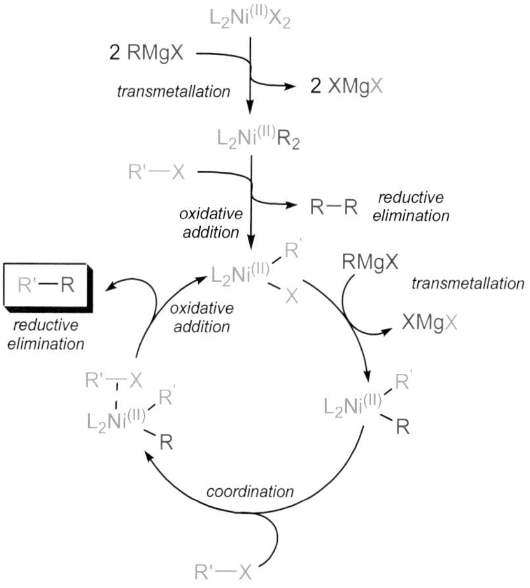

Pd-catalyzed process:

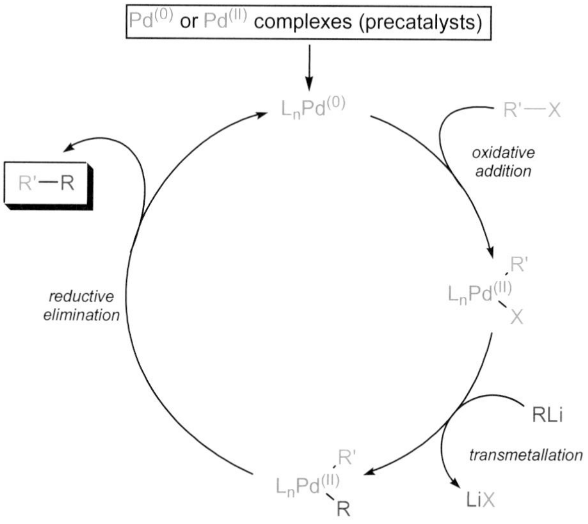

Synthetic Applications:

The enantioselective total synthesis of (+)-ambruticin was accomplished in the laboratory of E.N. Jacobsen. The *Kumada cross-coupling* was utilized to convert an (*E*)-vinyl iodide intermediate to the corresponding conjugate diene in good yield.[39] The stereochemistry of the vinyl iodide was completely preserved.

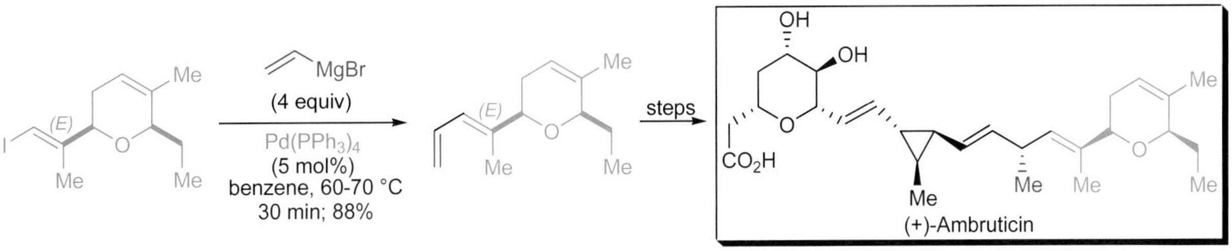

The highly concise synthesis of [18]dehydrodesoxyepothilone B, the 18-membered ring homologue of 10,11-dehydro-12,13-desoxyepothilone B, was based on a convergent *RCM* strategy.[40] S.J. Danishefsky et al. assembled the metathesis precursor by first converting a (*Z*)-vinyl iodide precursor to the corresponding 1,5-diene *via* the *Kumada cross-coupling*.

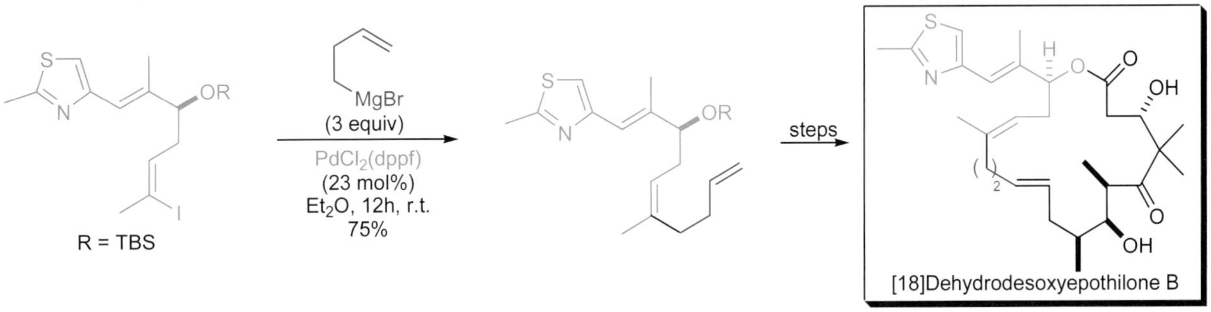

Enol phosphates were used as substrates for the *Kumada cross-coupling* during the final stages of the total synthesis of tetrahydrocannabinol and several of its analogs.[41] Y. Kobayashi and co-workers developed an indirect three-step 1,4-addition strategy to functionalize α-iodinated cyclohexanones with the addition of cuprates. The resulting enolates were trapped as corresponding phosphates, which underwent facile *Kumada cross-coupling* with methylmagnesium chloride in the presence of Ni(acac)$_2$.

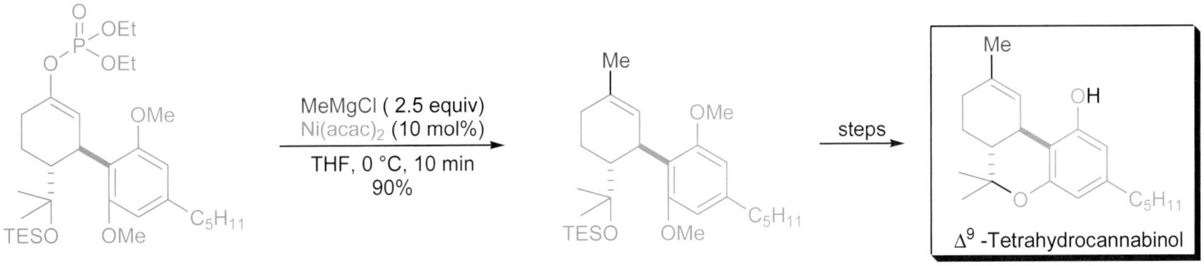

Research by M. Ikunaka showed that C$_2$-symmetrical chiral quaternary ammonium salts can serve as asymmetric phase-transfer catalysts.[42] To prepare significant quantities of (*R*)-3,5-dihydro-4*H*-dinaphth[2,1-*c*:1',2'-*e*]azepine, a novel short and scalable synthetic approach was undertaken. The synthesis commenced with the triflation of (*R*)-binol to give the *bis-O*-triflate. The *Kumada cross-coupling* was used to install two methyl groups in good yield.

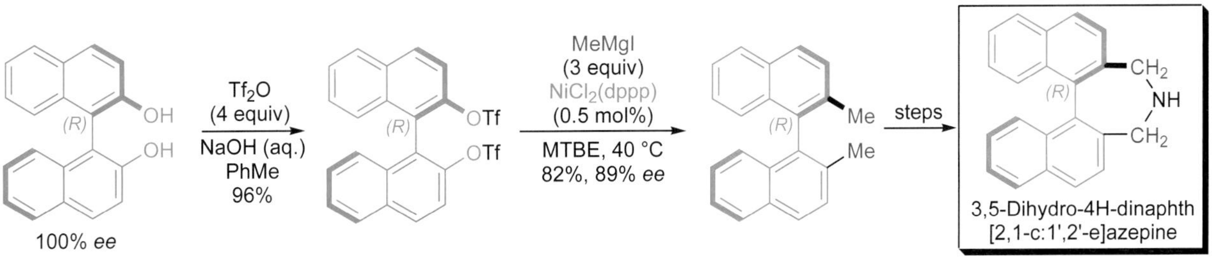

LAROCK INDOLE SYNTHESIS

(References are on page 620)

Importance:

[*Seminal Publications*[1]; *Reviews*[2-11]; *Modifications & Improvements*[12-20]]

In 1991, R.C. Larock reported the synthesis of indoles *via* the Pd-catalyzed coupling of 2-iodo anilines and disubstituted alkynes.[1] In the following years, the scope and limitation of the method were further explored by Larock and co-workers.[21] The one-pot Pd-catalyzed heteroannulation of *o*-iodoanilines and internal alkynes to give 2,3-disubstituted indoles is known as the *Larock indole synthesis* (*Larock heteroannulation*). The general features of the reaction are: 1) a wide variety of disubstituted alkynes can be used as coupling partners, and the substitution pattern of R^2 and R^3 groups does not have a marked effect on the efficiency of the reaction; 2) the nitrogen atom on the aniline can also be diversely substituted; 3) only *o*-iodoanilines are good substrates for the coupling and *o*-bromoanilines were found to be unreactive under the reaction conditions; 4) the coupling is highly regioselective: the larger alkyne substituent (R^2) almost always becomes located at the 2-position of the indole;[17] 5) when R^2=SiR_3, 2-silylindoles are obtained that can be protodesilylated, halogenated, or coupled with alkenes *via* a Pd-catalyzed reaction; 6) usually an excess (1.5-2 equivalents) of the alkyne coupling partner is needed. However, in the case of volatile alkynes, multiple equivalents are needed to achieve high yields; 7) the use of a full equivalent of LiCl and excess base was found to be necessary for the reproducibility of the reaction; and 8) typically DMF is used as the solvent at 100 °C. There are several modifications of the *Larock indole synthesis*: 1) the coupling of imines derived from *o*-iodoanilines with disubstituted alkynes gives rise to isoindolo[2,1-*a*]indoles;[14,15] 2) the *o*-iodoanilines can be replaced with vicinal iodo-substituted heterocyclic amines to prepare 5-,6- or 7-azaindoles,[13] pyrrolo[3,2-c]quinolines, tetrahydroindoles and 5-(triazolylmethyl)tryptamine analogs;[5] and 3) the coupling partner alkynes can be replaced with substituted allenes to synthesize 3-methyleneindolines.[12]

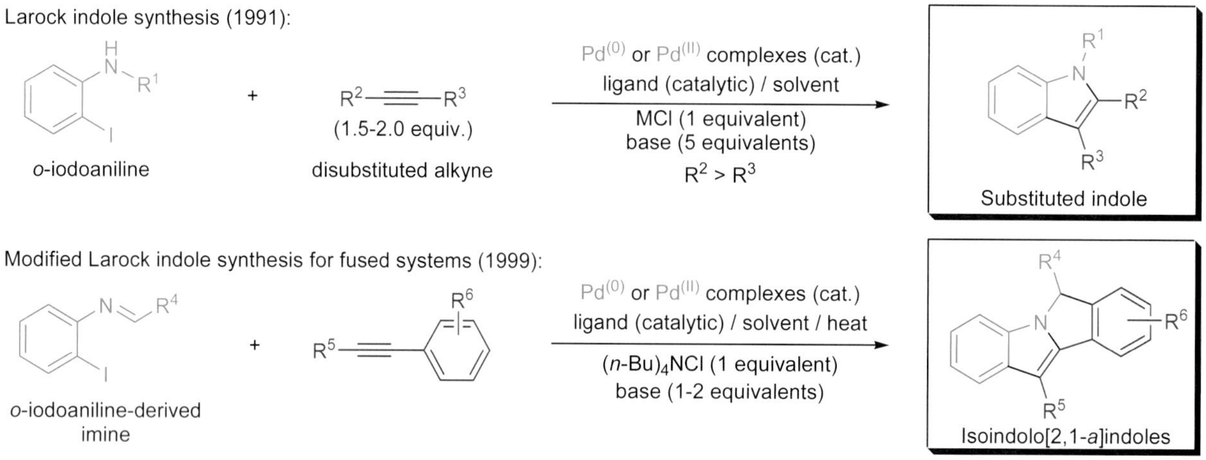

R^1 = alkyl, acyl, SO_2Ar; R^{2-3} = 1°, 2°, 3° alkyl, aryl, alkenyl, CH_2OH, SiR_3; M = $(n\text{-}Bu)_4N^+$, Li; base = Na_2CO_3, K_2CO_3, KOAc

R^4 = alkyl, aryl; R^5 = 1°, 2°, 3° alkyl, aryl, CH_2OH, CO_2R; R^6 = EWG or EDG; base = Na_2CO_3, *i*-Pr_2NEt

Mechanism:[21]

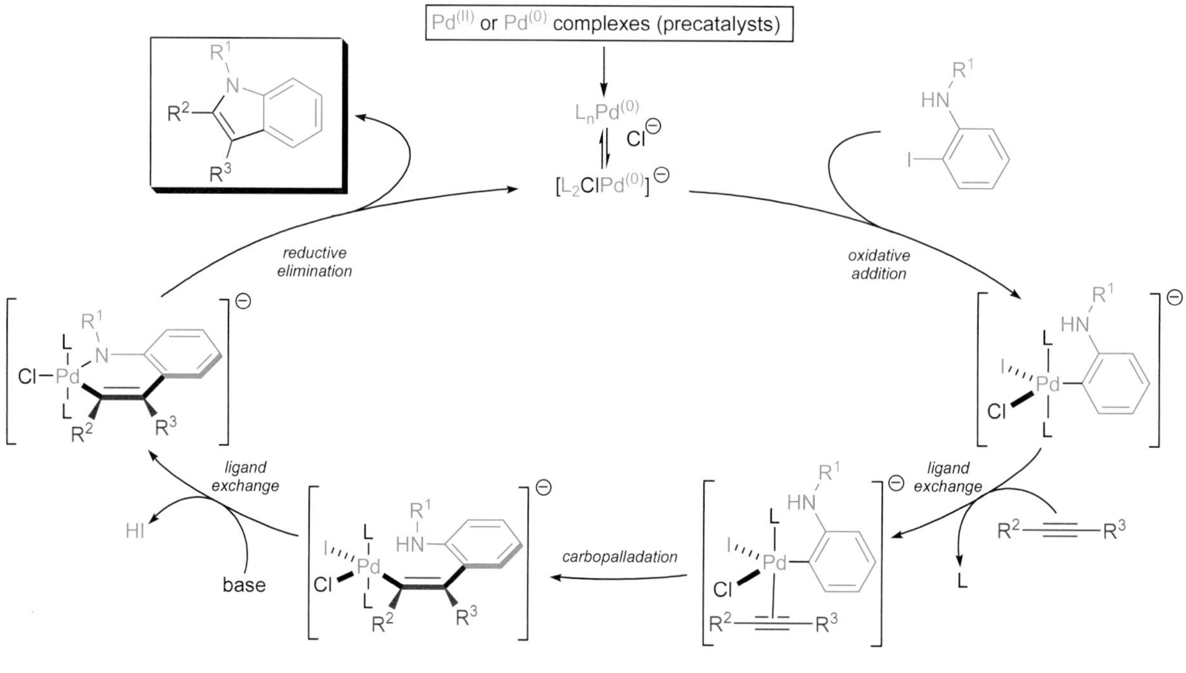

Synthetic Applications:

The total synthesis of (–)-fuchsiaefoline was accomplished in the laboratory of J.M. Cook using the *Larock indole synthesis* to prepare the key precursor 7-methoxy-D-tryptophan in enantiopure form.[22] The propargyl-substituted Schöllkopf chiral auxiliary was reacted with 2-iodo-6-methoxyaniline in the presence of 2 mol% $Pd(OAc)_2$ to give the expected indole in good yield. Interestingly, the *Bartoli indole synthesis* gives 7-substituted indoles only in moderate yield.

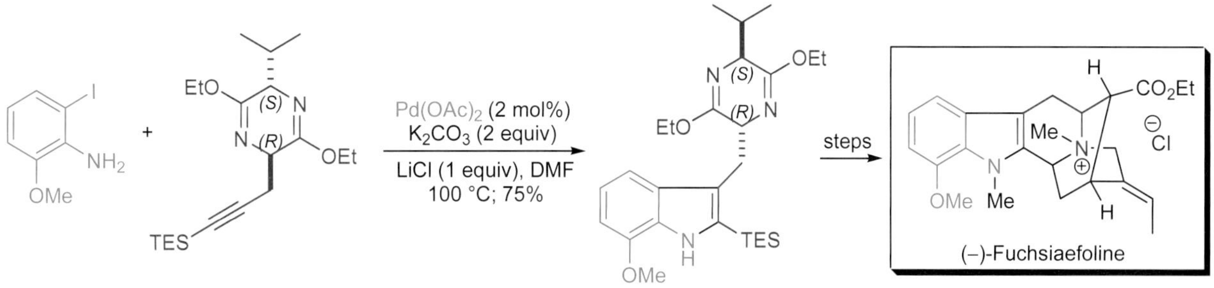

T.F. Walsh and co-workers synthesized two (*S*)-β-methyl-2-aryltryptamine based gonadotropin hormone antagonists *via* a consecutive *Larock indole synthesis* and *Suzuki cross-coupling*. The required (*S*)-β-methyltryptophol derivatives were prepared by coupling 4-substituted *o*-iodoanilines with optically active internal alkynes under standard conditions. The resulting 2-trialkylsilyl substituted indoles were then subjected to a *silver-assisted iododesilylation reaction* to afford the 2-iodo-substituted indoles that served as coupling partners for the *Suzuki cross-coupling* step.

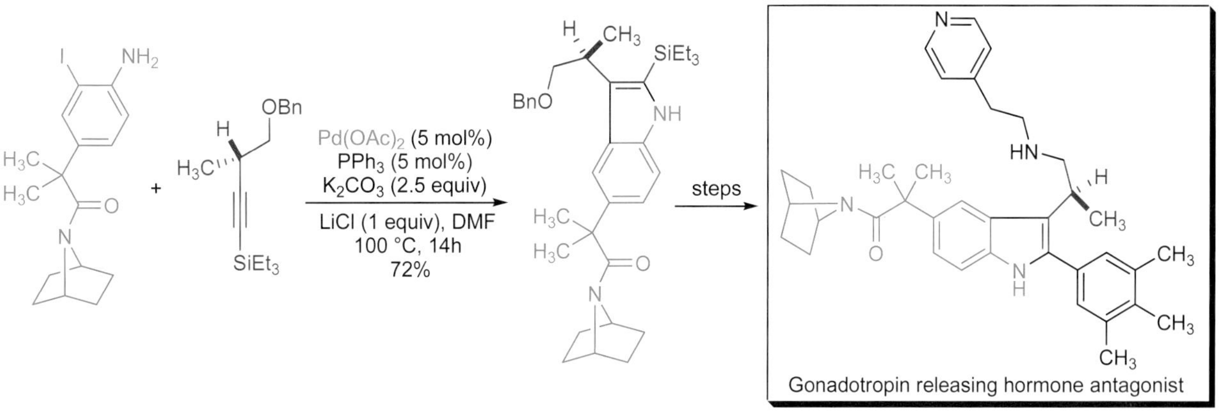

The preparation of diversely substituted azaindoles is fairly difficult, and there are no generally applicable strategies in the literature. Research by L. Xu et al. showed that 2-substituted-5-azaindoles could be synthesized by the Pd-catalyzed coupling of aminopyridyl iodides with terminal alkynes.[13] The coupling reaction proceeded in good yield under the conditions originally developed by Larock. Therefore, this example can be considered an extension of the *Larock indole synthesis*. By stopping the reaction early it was shown that the intermediate was an internal alkyne.

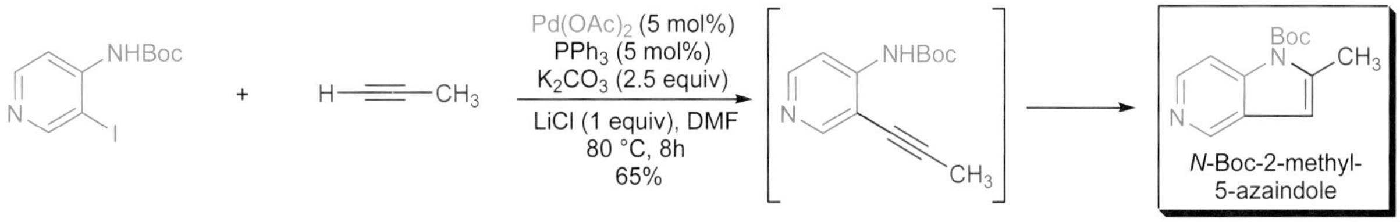

A complete reversal of regioselectivity was observed by M. Isobe and co-workers during the *Larock heteroannulation* of *o*-iodoaniline with α-C-glucosylpropargyl glycine in an attempt to prepare C-glycosyltryptophan.[14]

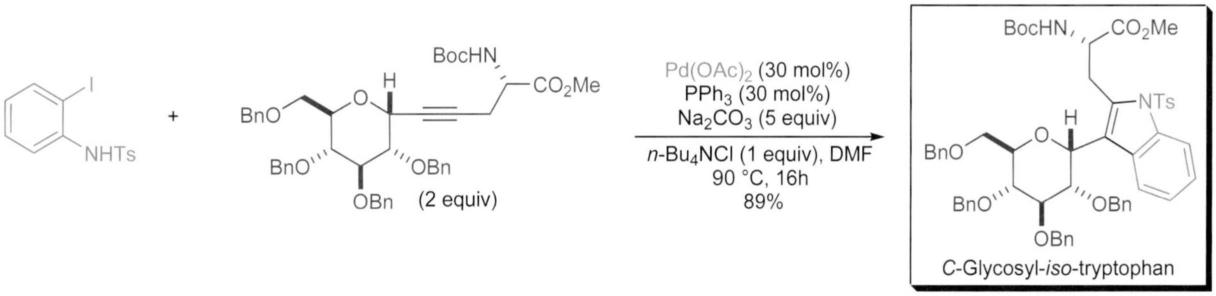

LEY OXIDATION

(References are on page 620)

Importance:

[*Seminal Publications*[1,2]; *Reviews*[3-12]; *Modifications & Improvements*[13-18]]

There are only two elements in the periodic table, ruthenium (Ru), and osmium (Os), which can sustain the uniquely high +8 oxidation state in their complexes containing strongly σ- and π-donating oxo (O^{2-}) ligands. Both metals can have eleven different oxidation states (d^0 to d^{10}), and any of these oxidation states can be stabilized with the appropriate choice of ligands. At any given oxidation state ruthenium complexes are more potent oxidizing agents than the corresponding osmium complexes (e.g., OsO_4 does not cleave double bonds, while RuO_4 does).[6] The greater lability of ruthenium complexes makes it possible to participate in catalytic processes. Despite the unselective nature of RuO_4 as an oxidant, it was possible to design ruthenium complexes with lower oxidation states which were less reactive and therefore more selective toward organic substrates containing several different functional groups.[5] The organic salts of perruthenate ion with large cations, R[RuO_4], (R=Pr_4N^+ or Bu_4N^+) are soluble in organic solvents and are milder oxidizing agents than RuO_4.[1,2,13] In 1987, S.V. Ley and co-workers introduced tetrapropylammonium perruthenate (TPAP) as a selective and mild oxidant of primary and secondary alcohols without the undesired cleavage of double bonds. The oxidation takes place with catalytic amounts (5-10 mol%) of TPAP when a co-oxidant such as *N*-methylmorpholine *N*-oxide (NMO) is used. The catalytic process to convert primary and secondary alcohols to the corresponding aldehydes and ketones with TPAP/NMO is referred to as the *Ley oxidation*. The general features of the reagent and reaction are: 1) TPAP is an air stable and non-volatile dark green solid and can be stored indefinitely when kept in the freezer (it decomposes when heated over 150 °C); 2) TPAP is soluble in a wide range of organic solvents, but in practice dichloromethane or acetonitrile (or their mixture) are used almost exclusively; 3) in a typical procedure, 5 mol% of TPAP is added to the solution of the substrate alcohol (1 equivalent) and NMO (1.5 equivalent) in CH_2Cl_2/MeCN in the presence of finely ground 4Å molecular sieves (0.5 g/mmol of substrate); 4) oxidations take place at room temperature in a few minutes or a couple of hours and the isolated yield of products is usually good or excellent (the catalyst turnover number is ~250); 5) the oxidations are vigorous, especially when the co-oxidant is not NMO (e.g., TMAO) and in these cases the TPAP should be added slowly to the reaction mixture in small portions; 6) the process works well on both small and large scale (e.g., *Swern oxidation* is difficult to run on a scale of a few milligrams); 7) due to the rapid nature of this oxidation, there is a danger of a runaway reaction (explosion) on multigram scale, so adequate cooling is necessary and the TPAP should be added to the reaction mixture slowly and portionwise; 8) the reaction rate and efficiency is improved when finely ground 4Å molecular sieves are added to the reaction mixture; 9) if pure CH_2Cl_2 is used as the solvent, the oxidations may not go to completion on a large scale but the addition of 10% (by volume) acetonitrile to the reaction mixture drives the oxidation to full conversion; 10) the work-up is very simple when the solvent is pure dichloromethane: the reaction mixture is filtered through a pad of silica-gel (or a short column), the silica-gel is washed with EtOAc and the filtrate is evaporated in vacuo; and 11) when the reaction is carried out in a mixture of CH_2Cl_2/acetonitrile, the solvent is first removed on a rotary evaporator, the residue is dissolved in dichloromethane or EtOAc and filtered through a pad of silica-gel (this is necessary, since acetonitrile can co-elute some residual TPAP, which contaminates the product).

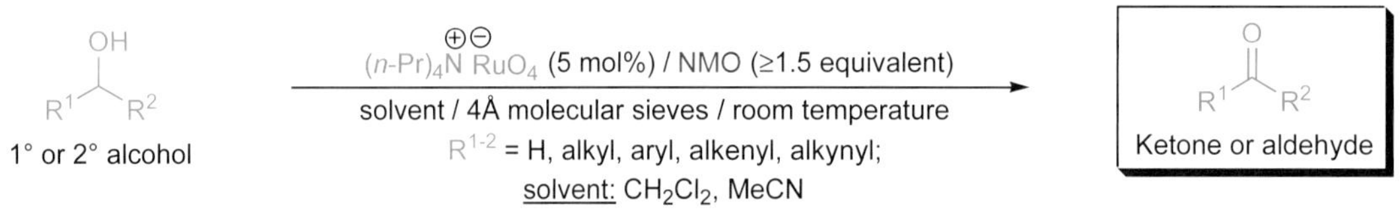

Mechanism: [19,6,20-25]

The mechanism of the *Ley oxidation* is complex and the exact nature of the species involved in the catalytic cycle is unknown. The difficulty in establishing an exact mechanism arises from the fact that the complexes of $Ru^{(VIII)}$, $Ru^{(VII)}$, $Ru^{(VI)}$, $Ru^{(V)}$ and $Ru^{(IV)}$ are all capable of stoichiometrically oxidizing alcohols to carbonyl compounds.[6] The TPAP reagent can oxidize alcohols stoichiometrically as a three-electron oxidant and can also be used as a catalyst when a co-oxidant is present (e.g., NMO, TMAO, or hydroperoxides). Data suggests that the oxidation proceeds *via* the formation of a complex between the alcohol and TPAP (ruthenate ester).[21] It was also found that the stoichiometric oxidation of isopropyl alcohol with TPAP is autocatalytic and the catalyst is suspected to be colloidal RuO_2. Small amounts of water decrease the degree of autocatalysis. This observation is supported by the finding that the addition of molecular sieves improves the efficiency of the reaction.

Step #1: $Ru^{(VII)} + RCH_2OH \longrightarrow Ru^{(V)} + RCHO + 2H^+$

Step #2: $Ru^{(VII)} + Ru^{(V)} \longrightarrow 2\,Ru^{(VI)}$

Step #3: $Ru^{(VI)} + RCH_2OH \longrightarrow Ru^{(IV)} + RCHO + 2H^+$

Step #4: $Ru^{(IV)} + NMO \longrightarrow Ru^{(VI)} + NMM$

Synthetic Applications:

The total synthesis of the immunosuppressant (–)-pironetin (PA48153C) was accomplished by G.E. Keck and co-workers.[26] The six-membered α,β-unsaturated lactone moiety was installed using a *lactone annulation reaction* by reacting the advanced aldehyde intermediate with the lithium enolate of methyl acetate. The aldehyde was prepared by the *Ley oxidation* of the corresponding primary alcohol and was used without purification in the subsequent annulation step.

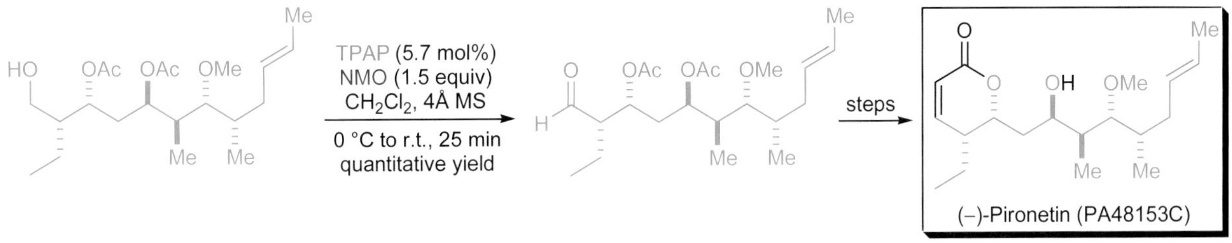

D.E. Ward et al. reported a general approach to cyathin diterpenes and the total synthesis of allocyathin B_3. The tetracyclic secondary alcohol was converted to the corresponding ketone using TPAP/NMO in good yield.[27]

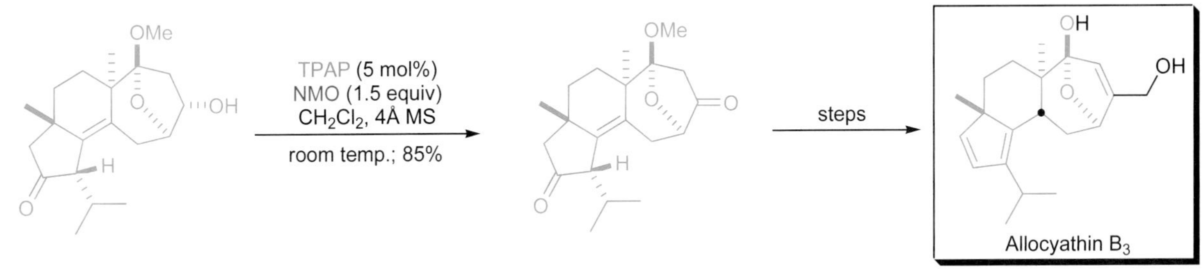

In the laboratory of D. Tanner, a novel method was developed for the stereoselective synthesis of (*E*)-tributylstannyl-α,β-unsaturated ketones in two steps from secondary propargylic alcohols.[28] The first step was the highly regio- and stereoselective *Pd-catalyzed hydrostannylation* of the triple bond followed by a mild *Ley oxidation*. This method was utilized for the construction of a key intermediate for the total synthesis of zoanthamine.

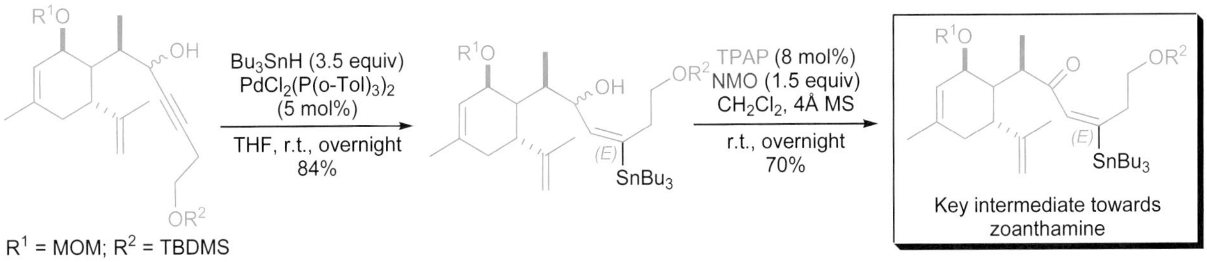

During the total synthesis of (–)-motuporin by J.S. Panek et al., the *modified Ley oxidation* was utilized in the preparation of the key *N*-Boc-valine-Adda fragment.[29] In order to obtain the carboxylic acid, the TPAP and NMO were administered twice, and the second portion of TPAP/NMO was accompanied by the addition of water. The water formed aldehyde hydrate which was oxidized to the carboxylic acid. The oxidation is so mild that the labile α-stereocenter was left intact.

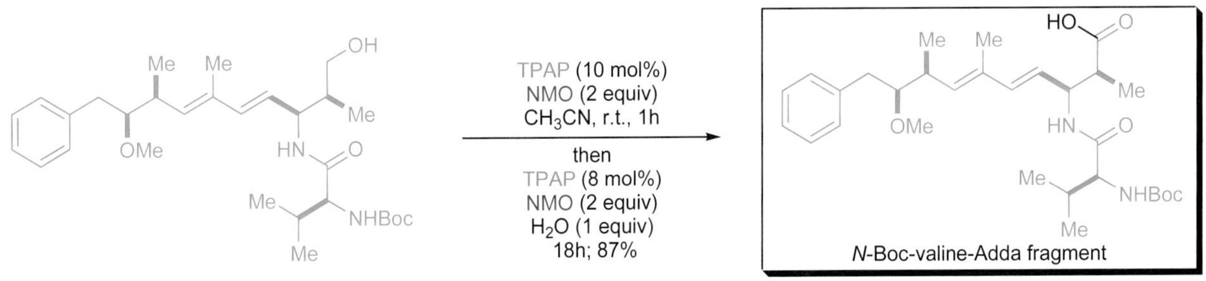

LIEBEN HALOFORM REACTION

(References are on page 621)

Importance:

[*Seminal Publications*[1-3]; *Reviews*[4-8]; *Modifications & Improvements*[9-13]]

In 1822, Serullas discovered that when iodine crystals were added to the mixture of an alkali and ethyl alcohol, a yellow precipitate was formed that he identified as "hydroiodide of carbon", but it was actually iodoform (CHI_3).[1] The discovery of chloroform ($CHCl_3$) came a decade later, when J. Liebig reacted chloral (trichloroacetaldehyde) with aqueous calcium hydroxide solution.[2] The reaction did not attract attention until 1870 when A. Lieben studied the action of iodine and alkali on many carbonyl compounds and formulated the rules that provide the basis of the so-called iodoform test.[3] Before spectroscopic methods became widely available for structural elucidation, the use of the iodoform test provided crucial information regarding the structure of organic compounds.[14] Presently, the reaction is more useful as a method of synthesizing carboxylic acids with one less carbon atom. The formation of haloforms from organic compounds upon treatment with hypohalites is known as the *Lieben haloform reaction* (or *haloform reaction*). The general features of this reaction are: 1) compounds containing the methyl ketone (CH_3-CO) functional group or compounds that get oxidized under the reaction conditions to methyl ketones will undergo the transformation; 2) in addition to methyl ketones and methyl carbinols, mono-, di-, and trihalogenated methyl ketones also give rise to haloforms; 3) the reaction is usually conducted in aqueous alkali, but for compounds that are insoluble in water the addition of a co-solvent such as dioxane or THF is necessary; 4) the halogen can be chlorine, bromine, and iodine, but elemental fluorine gas cannot be used due to its immense reactivity; 5) the reaction is sensitive to steric hindrance, so when the R^1 group is bulky, the hydrolysis of the trihalomethyl ketone usually does not take place, and the reaction stops; 6) certain side reactions such as the α-halogenation and subsequent cleavage of the other alkyl group is possible.

Serullas (1822): Lieben (1831):

Lieben haloform reaction:

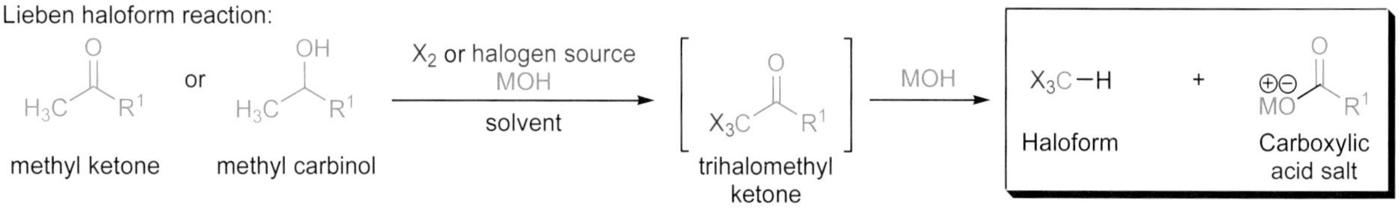

R^1 = H, alkyl, aryl; X_2 = Cl_2, Br_2, I_2; halogen source: NaOCl, NaOBr, NaOI, ICN; X_3C = F_3C, Cl_3C, Br_3C, I_3C; MOH = NaOH, KOH; solvent: H_2O, dioxane/H_2O, THF/MeOH

Mechanism: [15,4,16,5,17]

The mechanism of the *haloform reaction* has been extensively studied, and it can be concluded that it is a very complex process. The exact mechanistic pathway is dependent on the structure of the substrate and the specific reaction conditions.[17] The scheme depicts the oxidation of a methyl carbinol to the corresponding methyl ketone *via* an organic hypohalite. The methyl ketone then undergoes deprotonation, and three sequential α-halogenations take place to afford the trihalomethyl ketone. This compound undergoes rapid hydrolysis to afford the haloform and a carboxylate.

Oxidation of the carbinol to the methyl ketone:

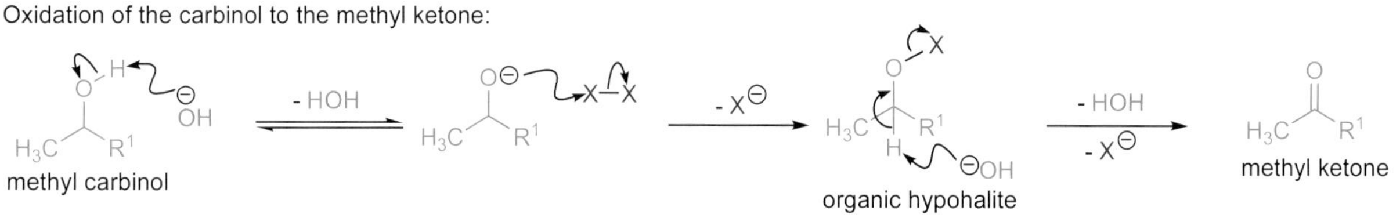

Sequential halogenation of the methyl group:

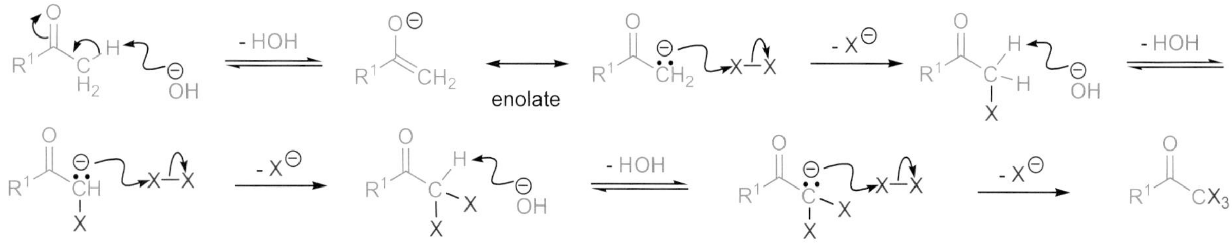

Hydrolysis of the trihalomethyl ketone:

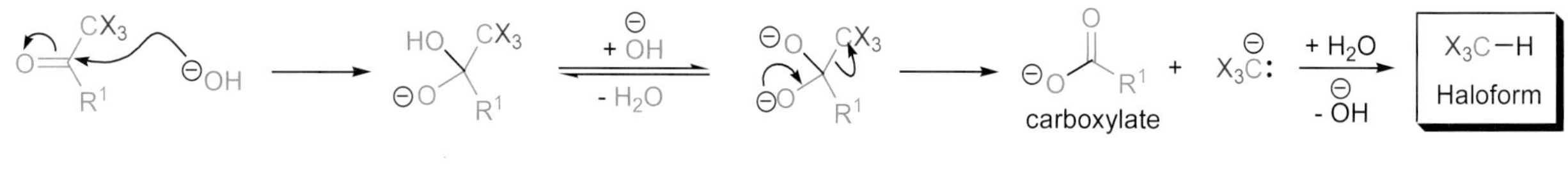

Synthetic Applications:

The blossoms of many flowers contain methyl jasmonates that are frequently used as ingredients in perfumes. It is noteworthy that the methyl epi-isomers have greater biological activity, and they play a role in inducing gene expression, mediate plant defense mechanisms, and signal transmission. The total synthesis of (±)-methyl epijasmonate was undertaken by H.C. Hailes and co-workers, who used a highly regioselective *Diels-Alder reaction* to install the required 2,3-*cis* stereochemistry.[18] After the ozonolysis of the cyclohexene double bond, the resulting methyl ketone moiety had to be transformed to a methyl ester, which was accomplished by using the *Lieben haloform reaction*. The aqueous solution of sodium hypobromite (prepared by adding Br_2 to sodium hydroxide) was slowly added to the solution of substrate in dioxane. The resulting carboxylate salt was converted to the methyl ester using *Fischer esterification* conditions under which the silyl protecting group was also removed. A final *Dess-Martin oxidation* furnished the natural product.

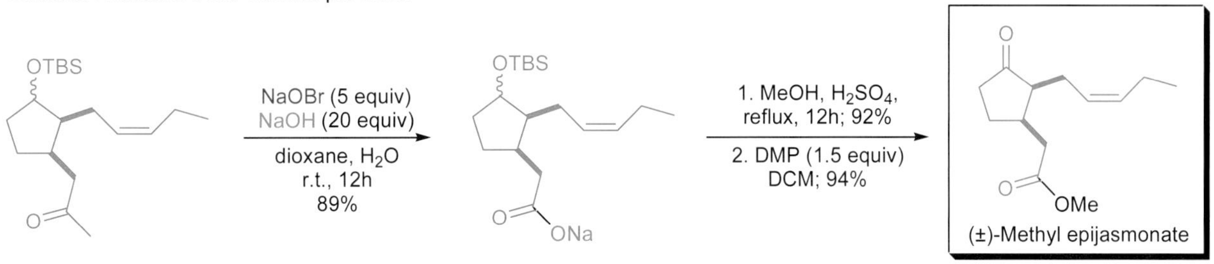

A novel synthetic route for the preparation of unsymmetrically substituted benzophenones was developed in the laboratory of C.-M. Andersson utilizing an iron-mediated aromatic substitution as one of the key steps.[19] The power of this method was demonstrated by the formal synthesis of the benzophenone moiety of the protein kinase C inhibitor balanol. In the late stages of the synthesis, it became necessary to convert the aromatic methyl ketone functionality of the highly substituted benzophenone substrate to the corresponding carboxylic acid. Bromine was added to sodium hydroxide solution, and the resulting sodium hypobromite solution was slowly added to the substrate at low temperature. Upon acidification the desired carboxylic acid was obtained in fair yield.

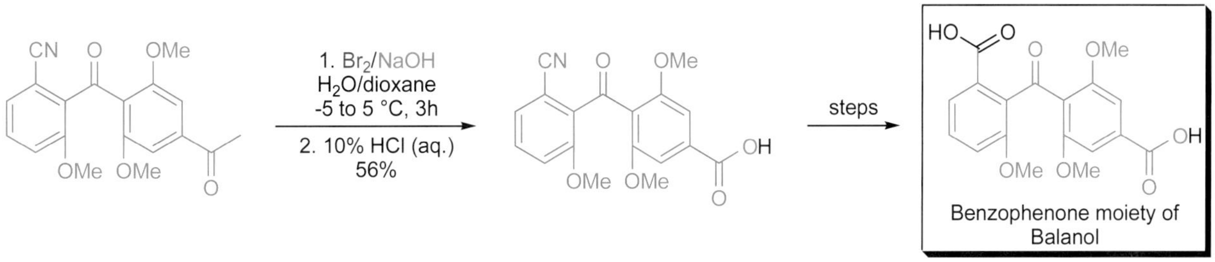

The biomimetic total synthesis of (±)-20-epiervatamine was accomplished by J. Bosch et al.[20] The authors used the addition of 2-acetylindole enolate to a 3-acylpyridinium salt as akey step to connect the two main fragments. The *in situ* formed 1,4-dihydropyridine was trapped with trichloroacetic anhydride to afford the corresponding trichloroacetyl-substituted 1,4-dihydropyridine derivative. The conversion of the trichloroacetyl group to a methyl ester was achieved by treatment with sodium methoxide. This transformation can be regarded as the second step of the *haloform reaction*.

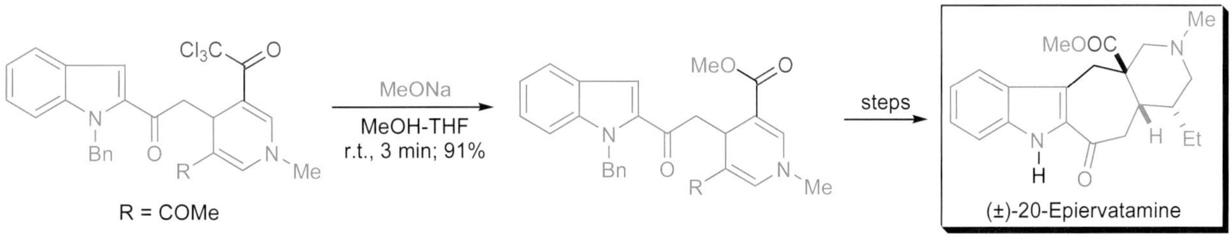

During the total synthesis of (±)-anthoplalone by K. Fukumoto et al. one of the intermediates was a cyclopropyl methyl ketone, and the synthetic sequence required the conversion of this functionality to the corresponding cyclopropane carboxylic acid methyl ester.[21] This transformation was accomplished via the *haloform reaction* using bleach in methanol. The methyl ester and some carboxylic acid was obtained after this step, so the product mixture was treated with diazomethane to convert the acid side product to the methyl ester.

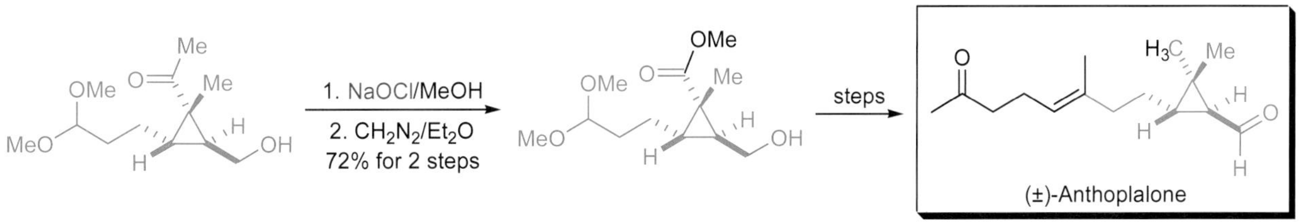

LOSSEN REARRANGEMENT

(References are on page 621)

Importance:

[*Seminal Publications*[1-3]; *Reviews*[4-9]; *Modifications & Improvements*[10-19]]

In 1872, W. Lossen reported that the pyrolysis of benzoyl benzohydroxamate (the mixed anhydride derived from phenylhydroxamic acid and benzoic acid) gave phenyl isocyanate and benzoic acid.[2] A few years later, he observed that the potassium salt of anisoyl benzohydroxamate was readily converted to diphenylurea, potassium anisoate, and carbon dioxide in boiling water. In this latter transformation the initial product was phenyl isocyanate, half of which reacted with water to afford aniline and carbon dioxide, and the other half reacted with aniline to form diphenylurea. The conversion of *O*-acyl hydroxamic acids to the corresponding isocyanates is known as the *Lossen rearrangement*. The general features of the reaction are: 1) hydroxamic acids can be readily prepared in several different ways:[4,5,7] a) from the corresponding carboxylic acids by first conversion to acid chlorides or mixed anhydrides then reaction with hydroxylamine; b) from esters with hydroxylamine; c) from aliphatic and aromatic carboxamides with hydroxylammonium chloride; 2) the free hydroxamic acids do not undergo the *Lossen rearrangement* under any condition, so the activation of the oxygen atom is necessary for the rearrangement to take place; 3) the acylation of the hydroxyl group of hydroxamic acids can be carried out with the following types of reagents: anhydrides,[4,5] acyl halides,[4,5] $SOCl_2$, $SO_3{\cdot}Et_3N$,[11] dialkylcarbodiimides,[10] activated aromatic halides[14] (e.g., 2,4-dinitrochlorobenzene), under *Mitsunobu reaction* conditions[12] (PPh_3, DEAD, ROH) and silylation;[13] 4) the rearrangement is usually initiated by heating the *O*-activated hydroxamic acids with bases (e.g., NaOH, DBU) in the presence of water or other nucleophiles (e.g., amines, alcohols); 5) the more active *O*-sulfonyl and *O*-phosphoryl derivatives, however, tend to rearrange spontaneously; 6) the initial product of the rearrangement is an isocyanate that after reacting with water gives an unstable carbamic acid, which breaks down to give a primary amine and carbon dioxide; 7) when an amine is present as the nucleophile, the product of the reaction is a substituted urea; 8) when there is a neighboring nucleophilic functional group (e.g., NH_2, OH, COOH) within the molecule, it will react with the isocyanate; and 9) the stereocenter adjacent to the hydroxamic acid functional group remains intact during the rearrangement (optical activity is unchanged). The *Lossen rearrangement* is closely related to the *Hofmann and Curtius rearrangements*, but its main advantage over the other methods is the mild reaction conditions, since it does not require the use of concentrated strong bases or intense heat.

W. Lossen (1872):

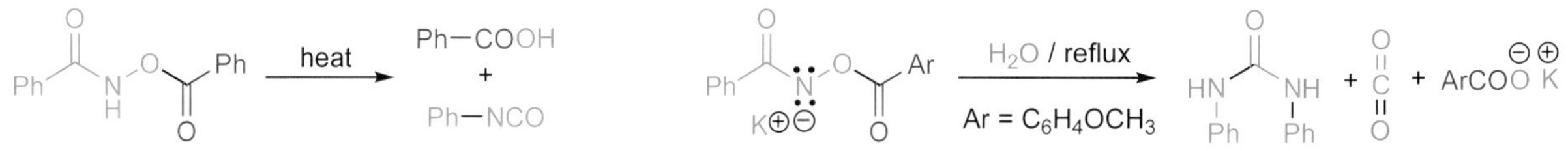

benzoyl benzohydroxamate

potassium salt of anisoyl benzohydroxamate

Lossen rearrangement:

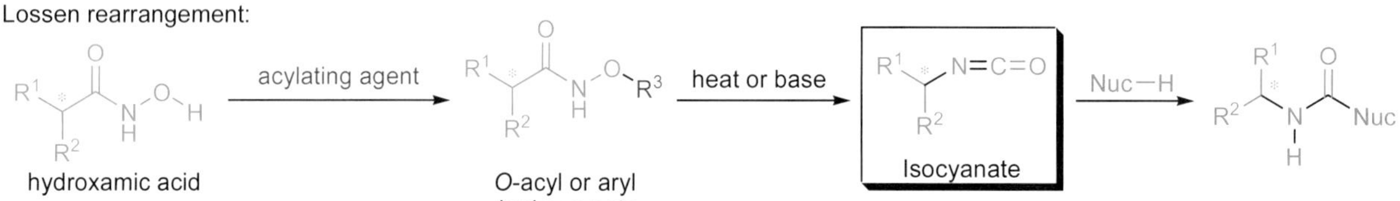

hydroxamic acid

O-acyl or aryl hydroxamate

R^{1-2} = alkyl, aryl ; acylating agent: anhydrides, acyl halides (RCOCl, RSO_2Cl, RPO_2Cl), $SOCl_2$, activated aromatic halides, RNCNR (carbodiimides); R^3 = CO-alkyl, CO-aryl, Cl, SiR_3, $C_6H_3(EWG)_2$(O-aryl), PO_2R, SO_2R, C=NR(NHR); base: NaOH, KOH, DBU, $(i\text{-}Pr)_2NEt$; nucleophile: H_2O, ROH, RNH_2

Mechanism: [20,10,21-23]

The mechanism of the *Lossen rearrangement* is closely related to the *Curtius-*, *Hofmann-,* and *Schmidt rearrangements*. The first step is the deprotonation of the *O*-acyl hydroxamate at the nitrogen atom by the base to the corresponding alkali salt, which is quite unstable and quickly undergoes a concerted rearrangement to the isocyanate *via* a bridged anion. The rate of the rearrangement strongly depends on the electronic nature of the substituents: the more electron-withdrawing R^3 is and the more electron-donating R^1 and R^2 are, the higher the rate is.

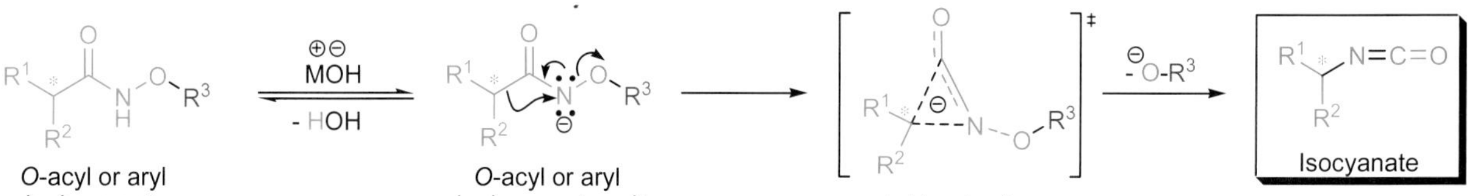

O-acyl or aryl hydroxamate

O-acyl or aryl hydroxamate salt

bridged anion

Synthetic Applications:

An improved synthesis of ONO-6818, a new nonpeptidic inhibitor of human neutrophil elastase, was developed by K. Ohmoto and co-workers.[24] The main difference between this new synthesis and the previous ones is that a dangerous (explosive) *Curtius rearrangement* of an acyl azide was replaced with a safer *Lossen rearrangement*. The required hydroxamic acid was prepared from a carboxylic acid by first converting it to the mixed anhydride with isobutyl chloroformate followed by the addition of hydroxylamine. The hydroxamic acid then was acetylated using acetic anhydride and the resulting *O*-acetyl hydroxamate was exposed to DBU in the presence of water. The intermediate isocyanate reacted with water to give the corresponding amine and CO_2.

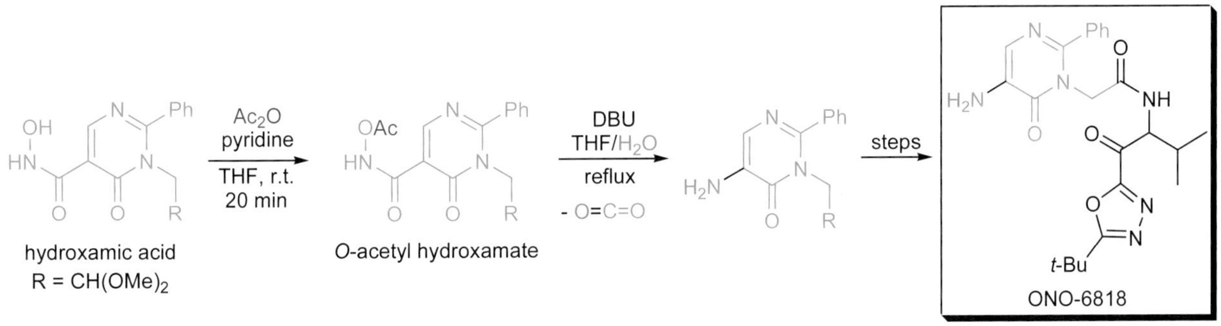

5,6-Disubstituted benz[*cd*]indoles have been shown to be effective inhibitors of the enzyme thymidylate synthase. The improved large scale synthesis of 5-methylbenz[*cd*]indol-2(1*H*)-one was accomplished by G. Marzoni et al.[25] The *Lossen rearrangement* was the key step to set up the ring system of the target compound. The cyclic hydroxamic acid (*N*-hydroxynaphthalimide) was deprotonated and used in a nucleophilic aromatic substitution with 2,4-dinitrochlorobenzene to afford *N*-(2,4-dinitrophenoxy)naphthalimide. The rearrangement took place under basic conditions with complete regioselectivity so that the amine was formed on the more electron rich aromatic ring. The cyclization of the resulting γ-amino acid to the amide was achieved by adjusting the pH to 3 with concentrated sulfuric acid.

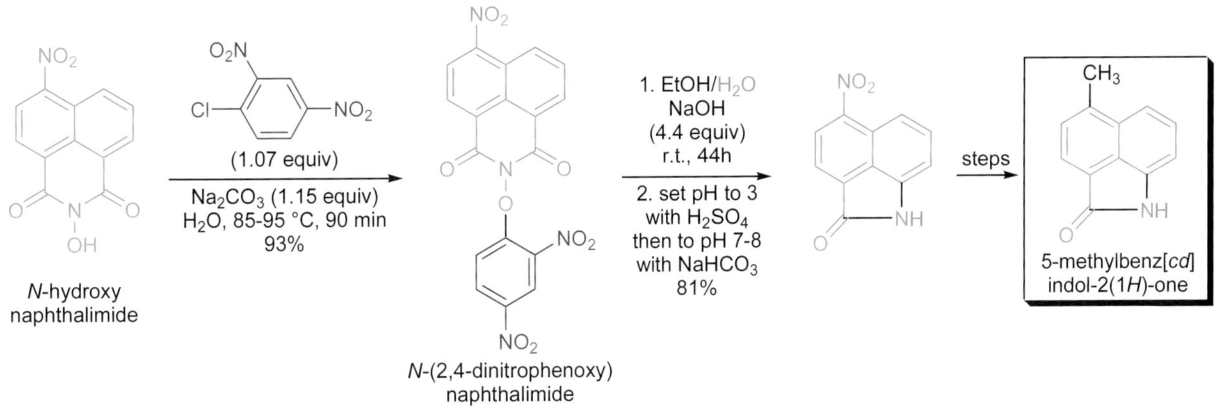

Pectins are important in cell wall assembly and detailed information of their structure will help to elucidate the relationship between the structures and physical properties. One possible approach is the chemical degradation of pectins. The specific degradation of the methyl esterified galacturonic acid residues of pectin to the corresponding oligogalacturonic acids bearing an arabitol residue was carried out in the laboratory of P.W. Needs.[26] The esters were first converted to the hydroxamic acids then reacted with EDC to give isoureas that upon the *Lossen rearrangement* and hydrolysis afforded 5-aminoarabinopyranose derivatives.

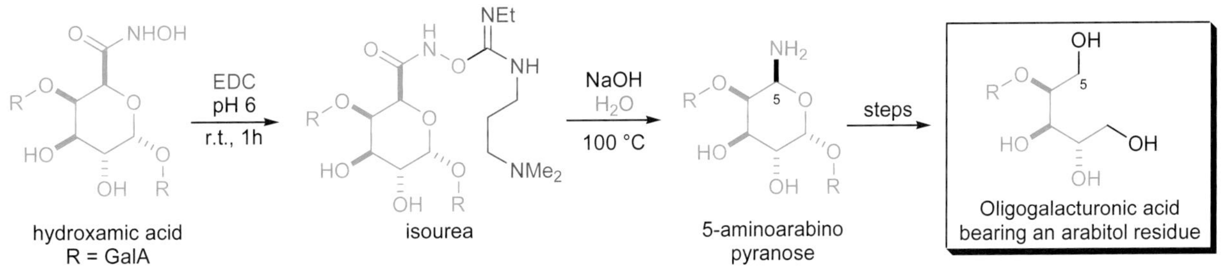

LUCHE REDUCTION

(References are on page 622)

Importance:

[*Seminal Publications*[1-4]; *Reviews*[5-11]; *Modifications & Improvements*[12-15]; *Theoretical Studies*[16]]

In 1978, J.L. Luche reported the selective conversion of α,β-unsaturated ketones to allylic alcohols using a mixture of lanthanide chlorides and sodium borohydride ($NaBH_4$).[1,2] Later the scope and limitation of the reaction was determined, and it was found that the 1,2-reduction of enones was best achieved by the use of $CeCl_3 \cdot 7H_2O/NaBH_4$ in ethanol or methanol.[4] The transformation of enones to the corresponding allylic alcohols using the combination of cerium chloride/sodium borohydride is known as the *Luche reduction*. The discovery by Luche was a breakthrough in the reduction of unsaturated carbonyl compounds, since metal hydrides usually give a mixture of 1,2- and 1,4-reduction products, and it was rare to obtain the 1,2-reduction product exclusively and in good yield. Usually hard metal hydrides (containing more ionic metal-H bonds) deliver the hydride mostly to the carbonyl group (1,2-addition), whereas soft metal hydrides (containing a more covalent metal-H bonds) favor conjugate addition. Alkali metal borohydrides are softer reducing agents than aluminum hydrides, so they are expected to favor the conjugate reduction of enones. Borohydrides can be made harder by the replacement of some of the hydride ligands with alkoxy groups so that the 1,2-selectivity will be larger. The general features of the *Luche reduction* are: 1) both acyclic and cyclic enones are reduced to the corresponding allylic alcohols in high yield with no or little 1,4-reduction by-product; 2) among various lanthanide salts, the heptahydrate of $CeCl_3$ was found to give the highest 1,2-selectivity; 3) under the reaction conditions most functional groups (such as carboxylic acids, esters, amides, alkyl halides, tosylates, acetals, sulfides, azides, epoxides, nitriles, nitro compounds) are unaffected; 4) the reactions are usually conducted at or below room temperature, and the reduction is complete within 5-10 minutes; 5) the reaction vessel and the solvents do not need to be dried, the regioselectivity and the yield is unaffected by water content up to 5% by volume; 6) the cerium chloride can be used directly as its heptahydrate and no drying is needed; 7) no inert atmosphere is required as the reaction is not sensitive to the presence of oxygen; 7) the best solvent is methanol, since the reaction rates are the highest, but occasionally ethanol and isopropanol are used, even though the reduction is slower in these solvents; 8) steric hindrance has little or no effect on the regioselectivity; 8) the combination of $CeCl_3/NaBH_4$ is excellent for the chemoselective reduction of ketones in the presence of aldehydes, since under these conditions aldehydes undergo rapid *acetalization*, which prevents their reduction; 9) substituted cyclohexenones undergo mainly an axial attack of hydride, so equatorial alcohols are obtained; 10) in rigid cyclic or polycyclic systems the hydride delivery occurs from the least hindered face of the carbonyl group; 11) conjugated or aromatic aldehydes are reduced preferentially in the presence of isolated aliphatic aldehydes; and 12) the lowering of the reaction temperature well below zero (e.g., -78 °C) usually increases the diastereoselectivity of the reduction of chiral substrates.

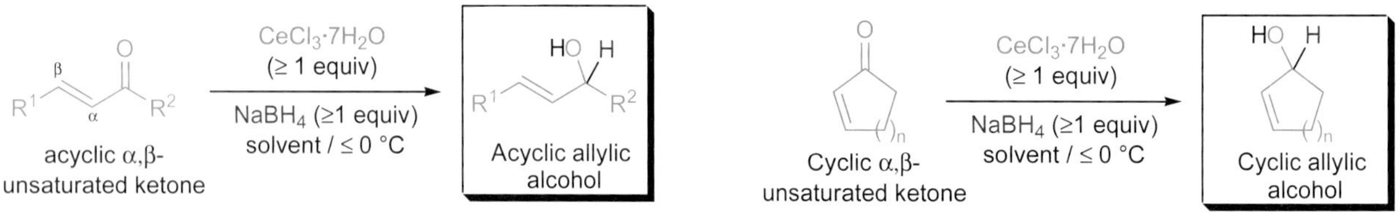

R^{1-2} = H, alkyl, aryl; n = 1-3; solvent = methanol, ethanol, isopropanol

Mechanism: [17,4,18]

As mentioned above, $NaBH_4$ is a soft reducing agent and it has a tendency to reduce enones at the β-position of the double bond. The active species during the *Luche reduction* is believed to be an alkoxy borohydride, which in combination with the hard cerium cation acts as a hard reducing agent. The involvement of cerium borohydrides have been discounted based on experimental evidence.[19] The mechanism is complicated by the fact that more than one type of borohydride is formed. The role of the cerium is twofold: 1) catalysis of the formation of alkoxyborohydrides; and 2) increasing the electrophilicity of the carbonyl carbon atom. By coordinating to the oxygen atom of the solvent, cerium increases the acidity of the medium and helps activating the carbonyl of the enone indirectly (lanthanoid ions were shown to preferentially coordinate to alcohols rather than carbonyl groups by NMR spectroscopy).[20]

Formation of alkoxyborohydrides:

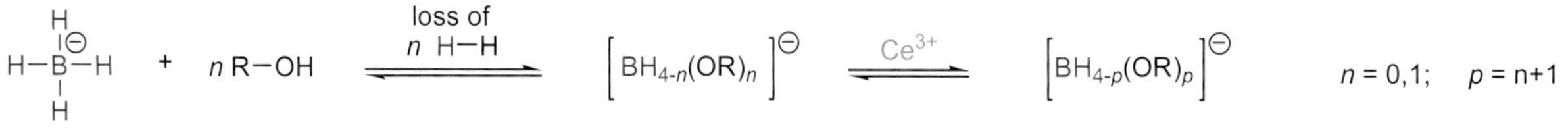

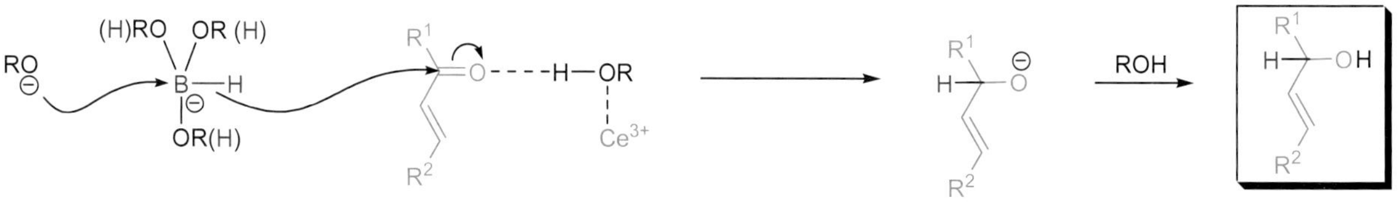

Synthetic Applications:

The total synthesis of several *amaryllidaceae alkaloids* including that of narciclasine was accomplished in the laboratory of T. Hudlicky.[21] The C2 stereochemistry was established by a two-step sequence: *Luche reduction* of the α,β-unsaturated cyclic ketone followed by a *Mitsunobu reaction*. The ketone was first mixed with over one equivalent of $CeCl_3$ in methanol and then the resulting solution was cooled to 0 °C, and the sodium borohydride was added. In 30 minutes the reaction was done, and the excess $NaBH_4$ was quenched with AcOH. The delivery of the hydride occurred from the less hindered face of the ketone and the allylic alcohol was obtained as a single diastereomer.

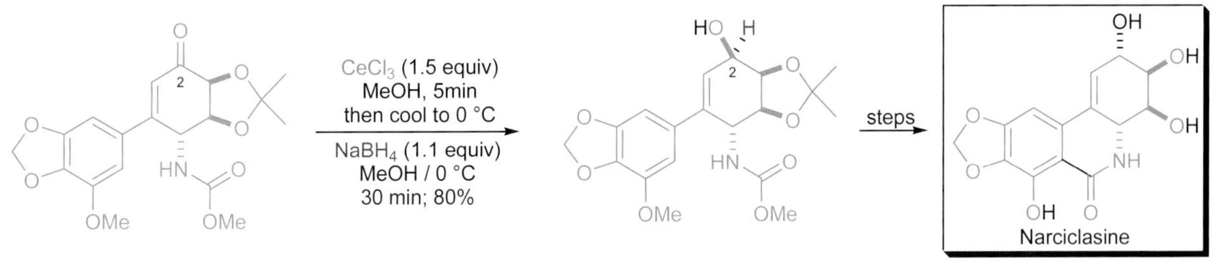

During the final stages of the total synthesis of (–)-subergorgic acid by L.A. Paquette and co-workers, the transposition of a tricyclic enone was needed.[22] The enone was exposed to the *Luche conditions* and an 85:15 mixture of diastereomers was obtained. In order to achieve this level of diastereoselectivity, the reaction temperature had to be lowered to -50 °C instead of the usual 0 °C. The major product was formed *via* the *exo* attack of the carbonyl group by the hydride. The allylic alcohol was later converted to the corresponding sulfoxide followed by a *Mislow-Evans rearrangement* to the isomeric allylic alcohol.

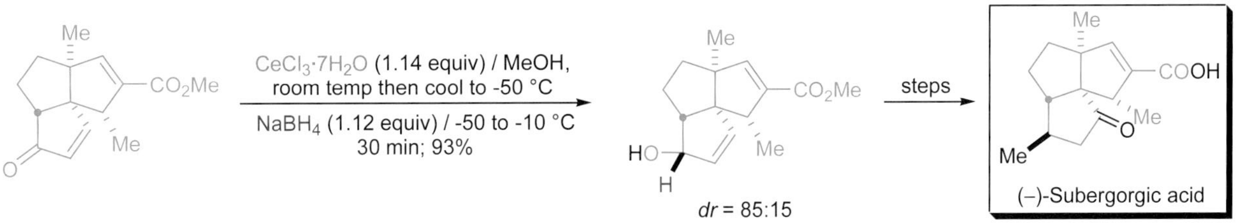

A general synthetic route to several polyhydroxylated agarofurans was developed by J.D. White and co-workers and the total synthesis of (±)-euonyminol was achieved.[23] The key intermediate was prepared *via* a *Diels-Alder reaction* between a diene and a substituted benzoquinone. The resulting bicyclic homoannular diene was reduced under the Luche conditions with excellent regio- and stereoselectivity at C6. The substrate was mainly in the boat conformation and the β-face of the ketone was more exposed to hydride attack. The C6 ketone was also more sterically accessible and more basic than the C9 ketone functionality.

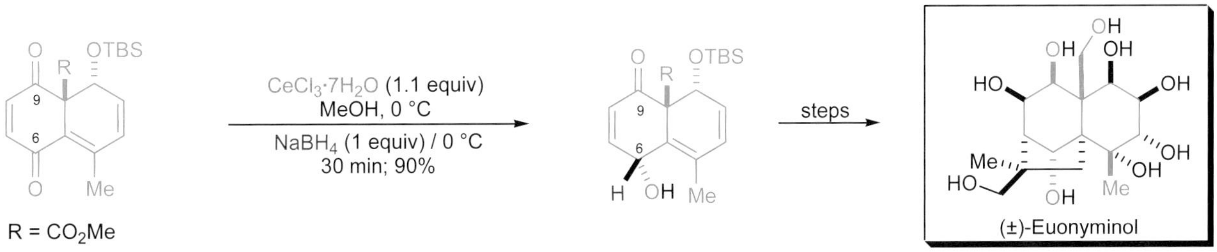

The deoxygenation of the C6 position of an advanced intermediate was accomplished in a two-step procedure by Y. Kishi et al. in their synthesis of (±)-batrachotoxinin A.[24] The *Luche reduction* was followed by the formation of the C6 pyridylthioether, which was desulfurized using Raney nickel.

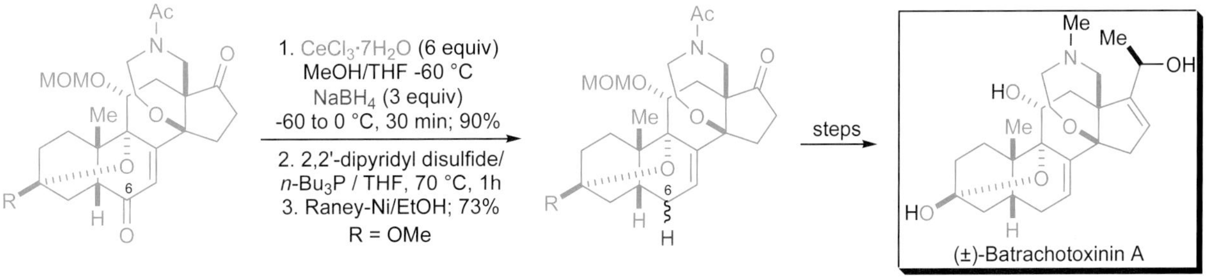

MADELUNG INDOLE SYNTHESIS

(References are on page 622)

Importance:

[*Seminal Publications*[1,2]; *Reviews*[3-7]; *Modifications & Improvements*[8-20]]

In 1912, W. Madelung reported that *N*-benzoyl-*o*-toluidine was converted to the corresponding 2-phenylindole when heated with two equivalents of sodium ethoxide at high temperatures in the absence of air.[2] Madelung also showed that the yields could be improved by using the alkoxides of higher aliphatic alcohols such as *n*-amyl alcohol. The intramolecular cyclization of *N*-acylated-*o*-alkylanilines to the corresponding substituted indoles in the presence of a strong base is known as the *Madelung indole synthesis*. A decade later in 1924, A. Verley demonstrated that sodium amide ($NaNH_2$) was a more general reagent and a wide range of *N*-acylated-*o*-toluidines could be converted to the corresponding 2-substituted indoles.[8,9] The general features of the transformation are: 1) when $NaNH_2$ or sodium alkoxides are used as bases, usually temperatures over 250 °C are required; 2) the use of alkyllithiums allows the reaction to take place at ambient or slightly higher temperatures; 3) high yields are observed when the aromatic ring has electron-donating substituents, while electron-withdrawing substituents tend to give lower yields; 4) the efficiency of the reaction is dependent of the steric bulk of the R^2 substituent; and 5) when the methyl group is substituted with an electron-withdrawing group (e.g., CN), the cyclization takes place at lower temperatures.[13] One of the most important modifications of the *Madelung indole synthesis* was introduced by A.B. Smith et al. who metalated substituted *N*-TMS-o-toluidines with *n*-BuLi. The resulting benzylic anion was reacted with non-enolizable esters or lactones to afford *N*-lithioketamine intermediates that first underwent *intramolecular heteroatom Peterson olefination* to give indolinines, and then tautomerized to the corresponding 2-substituted indoles. This modification is referred to as the *Smith indole synthesis*.[6]

Madelung (1912):

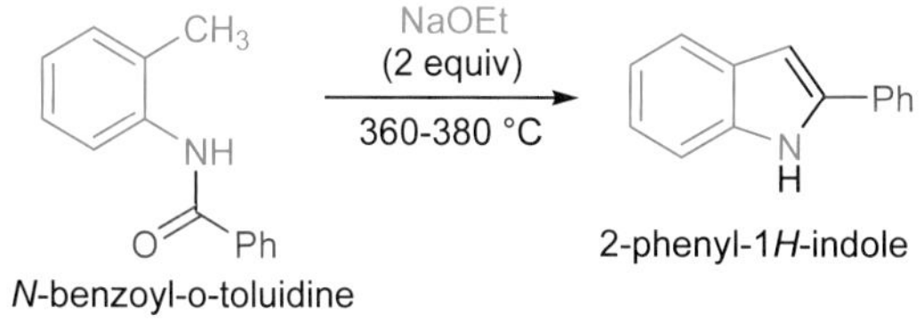

Verley (1924 & 1925):

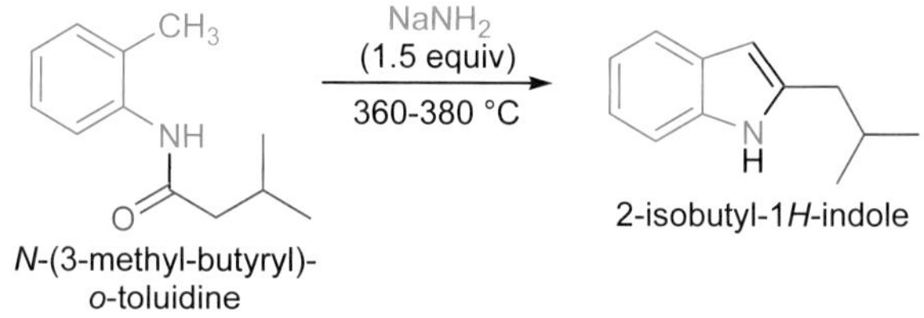

Madelung indole synthesis:

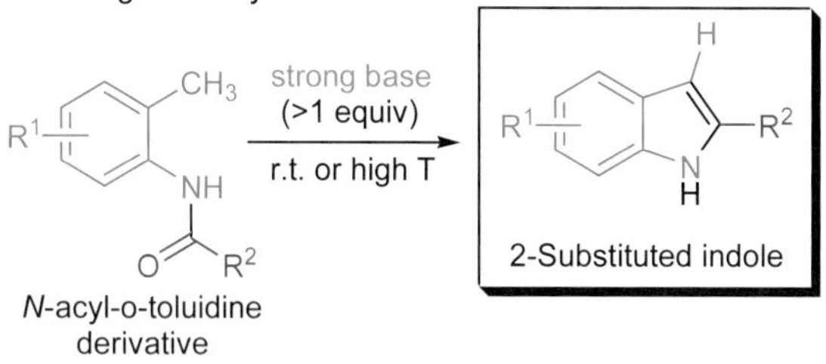

Modified Madelung indole synthesis:

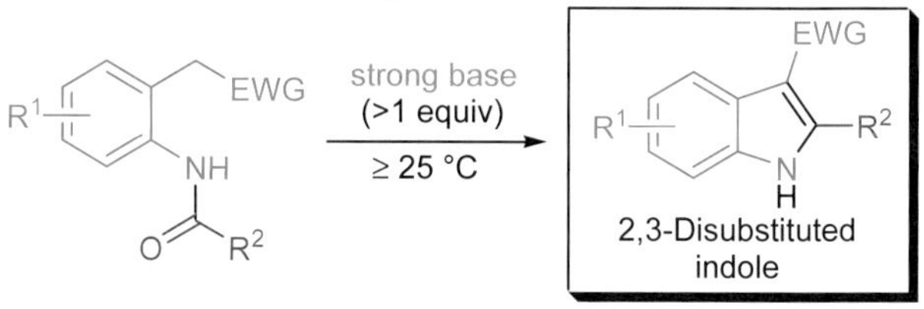

Smith-modified Madelung indole synthesis (Smith indole synthesis):

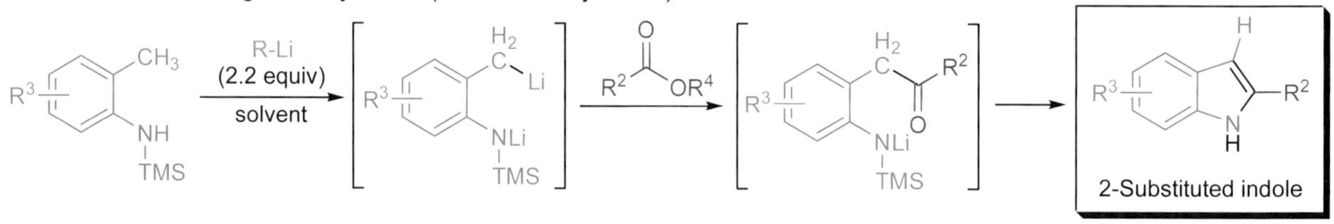

R^1 = H, alkyl, aryl, typically EDG; R^2 = alkyl, aryl; R^3 = alkyl, O-alkyl, O-aryl, Cl, F; R^4 = Me, Et; EWG = CN, CO_2R
strong base: KOEt, NaOEt, $NaNH_2$, Na(O-alkyl); alkyllithium, aryllithium; solvent: hexanes, THF

Mechanism:[4,11]

Mechanism of the Madelung indole synthesis:

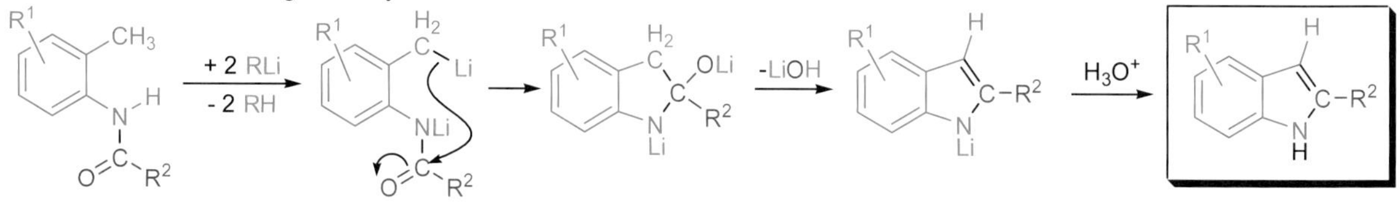

Mechanism of the Smith indole synthesis:

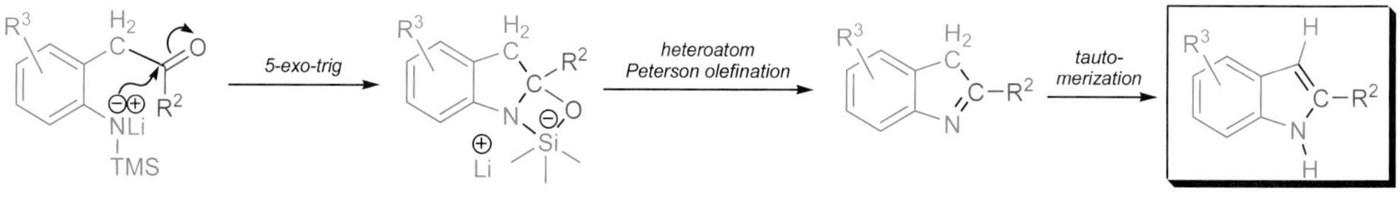

Synthetic Applications:

In the laboratory of A.B. Smith, the total synthesis of (–)-penitrem D, one of the most architecturally complex indole alkaloids, was accomplished.[21] The *Smith-modified Madelung indole synthesis* was utilized for the coupling of the two main fragments to form the desired 2-substituted indole ring. The *o*-toluidine derivative was first *N*-silylated and then treated with 2.1 equivalents of *sec*-BuLi. In the same pot, the addition of the lactone furnished an initial coupled product. In order to facilitate the final *heteroatom Peterson olefination*, exposure to silica gel was necessary and the indole was formed in high yield. It is worth noting that the use of large excess of the lithiated *o*-toluidine fragment was necessary to achieve the full conversion of the lactone.

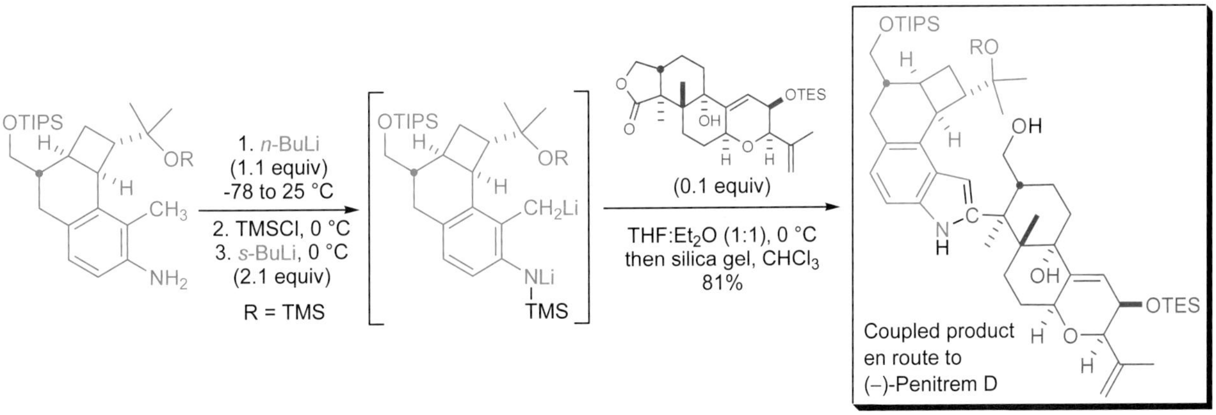

The synthesis of a novel indacene (2,6-diphenyl-1,5-diaza-1,5-dihydro-*s*-indacene) was completed by H.J. Geise and co-workers.[22] This compound had a great potential to be used as an organic light-emitting diode based on its optical and electroluminescent properties. The authors chose the conditions of the original high-temperature *Madelung indole synthesis*. First, 2,5-dimethyl-4-amino aniline was benzoylated then mixed with a large excess of potassium-*tert*-butoxide and heated to high temperatures in a preheated oven.

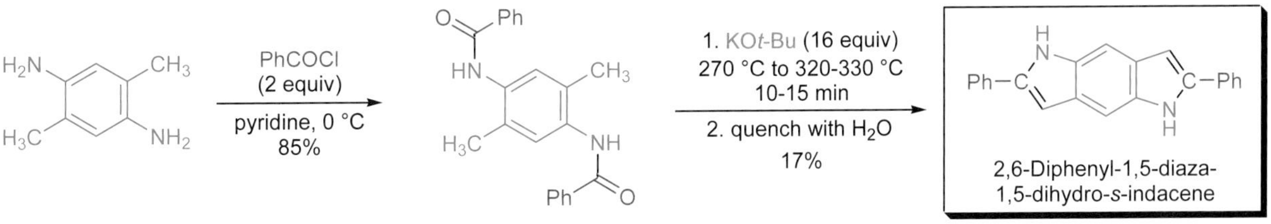

The *solid-phase version of the Madelung indole synthesis* was developed by D.A. Wacker et al. for the preparation of 2,3-disubstituted indoles.[20] The *ortho*-substituted aniline substrate was first attached to the Bal resin using reductive amination. The resin-bound aniline was then acylated and the cyclization was brought about with a variety of bases to afford high yields of the disubstituted indoles. The products were quantitatively removed from the resin with TFA:Et_3SiH (95:5).

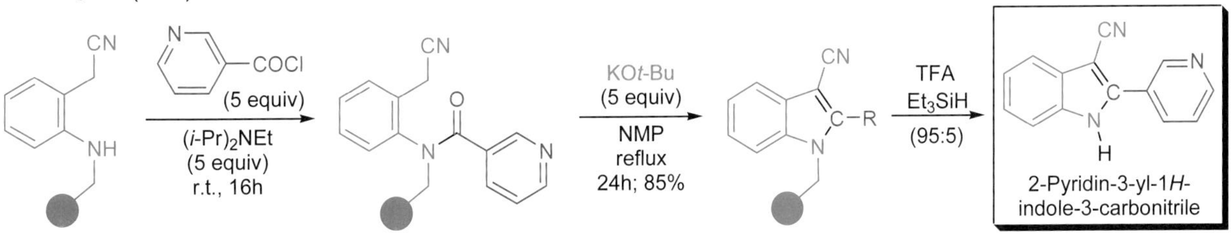

A practical synthetic route to the spiro analogues of triketinins was devised by V. Kouznetsov and co-workers utilizing the *Madelung indole synthesis* in the final step.[23] The starting *N*-acetylated spiroquinolines were rearranged to 4-*N*-acetylaminoindanes, which were finally converted to the desired indoles.

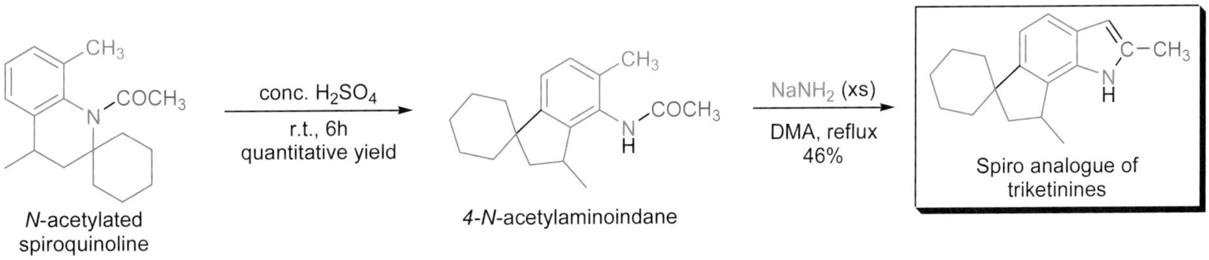

MALONIC ESTER SYNTHESIS

(References are on page 623)

Importance:

[*Seminal Publications*[1,2]; *Reviews*[3-6]; *Modifications & Improvements*[7-16]]

In 1863, Geuther was the first investigator to perform the *C*-alkylation of an enolate derived from an active methylene compound (a methylene or methine group with two electron-withdrawing groups attached to it). Namely, he deprotonated ethyl acetoacetate and reacted the resulting sodium enolate with ethyl iodide and isolated the corresponding ethyl α-ethyl acetoacetate.[17] More than a decade later, J. Wislicenus investigated the reaction between the sodium enolates of malonic esters and primary and secondary alkyl halides and made the observation that primary alkyl halides reacted faster than secondary ones.[1,2] The alkylation of malonic ester enolates with various organic halides and the subsequent decarboxylation of the alkylated products to yield substituted acetic acid derivatives is known as the *malonic ester synthesis*. The general features of the transformation are:[4] 1) the alcohol component of the malonic ester substrates is primarily derived from aliphatic alcohols (e.g., OMe, OEt, O*t*-Bu); 2) the pK_a of the methylene group is usually between 9-11, so relatively weak bases are sufficient for the generation of the reactive ester enolate; 3) the base most often corresponds to the alcohol component of the substrate to avoid the generation of mixtures of esters (e.g., dimethyl malonate is deprotonated with NaOMe in MeOH); 4) the applied solvent can vary from hydroxylic solvents (e.g., alcohols) all the way to dipolar aprotic solvents (e.g., DMF) and nonpolar aprotic solvents (e.g., benzene); 5) the reaction is bimolecular (S_N2) for 1° and 2° alkyl halides, especially in dipolar aprotic solvents, so high concentration of both the enolate and the organic halide results in faster alkylation; 6) allylic and benzylic halides may also react in a monomolecular fashion (S_N1); 7) 1° and 2° alkyl halides and allylic and benzylic halides react the fastest, while tertiary alkyl halides mainly give elimination products; 8) the order of reactivity of the halides is I ~ OTs > Br > Cl; 9) *C*-monoalkyl malonic esters are less acidic than unsubstituted ones, so the use of a stronger base is needed to effect the second deprotonation, and the alkylation of the corresponding enolates is slower; 10) when α,ω-dihalides are used as the alkylating agents, cycloalkanes are obtained and the formation of five-, six-, and seven-membered rings is favored; and 11) saponification of the mono- or disubstituted malonic ester with base affords a 1,3-diacid, which undergoes decarboxylation upon heating with an acid to give substituted acetic acids.

Monoalkylation of malonic ester: Dialkylation of malonic ester:

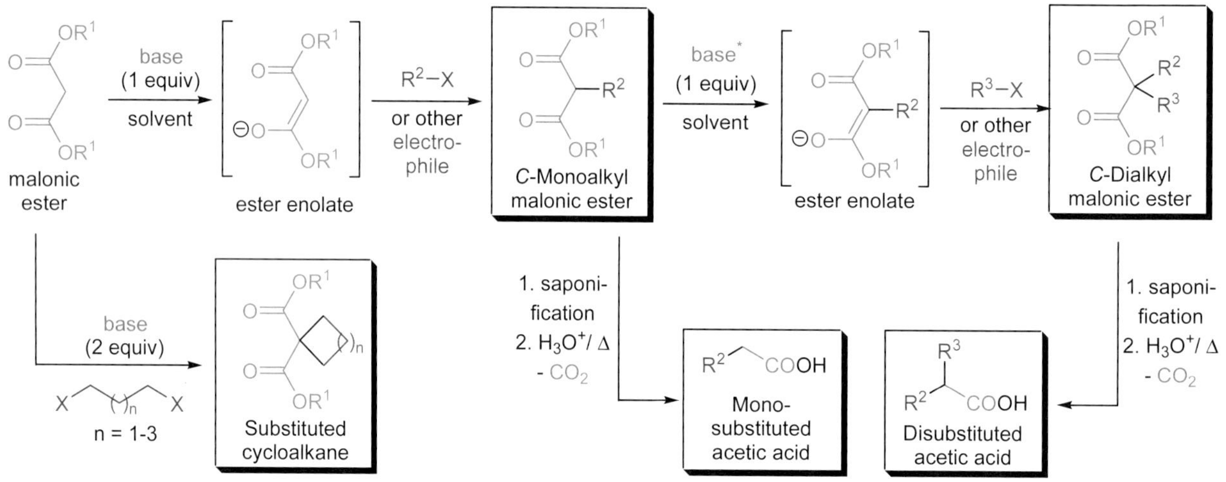

R^1 = alkyl, aryl; R^{2-3} = 1° or 2° alkyl, allyl, benzyl, activated aryl, acyl; X = Cl, Br, I, OTs; electrophile: epoxide, dialkyl sulfate, alkyl sulfonate, alkyl nitrate; base = $NaOR^1$, NaH; base* = KO*t*-Bu, conc. NaOEt, NaH; solvent = R^1OH, *t*-BuOH, benzene, ether, DMF

Mechanism: [18,4]

Mechanism of mono- and dialkylation:

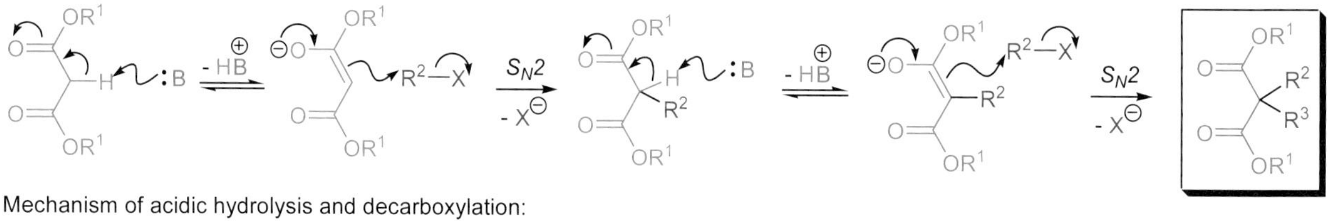

Mechanism of acidic hydrolysis and decarboxylation:

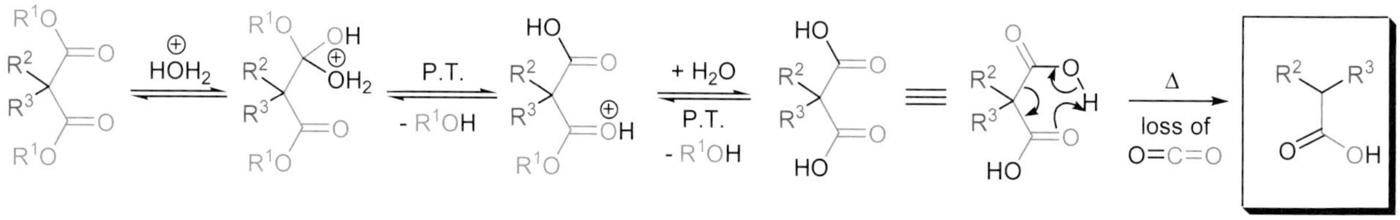

Synthetic Applications:

The first enantioselective total synthesis of (+)-macbecin I was accomplished by R. Baker and co-workers.[19] A key vinyl iodide precursor was prepared stereoselectively using the *malonic ester synthesis*. Diethyl methylmalonate was treated with *in situ* generated diiodocarbene in ether at reflux to afford diiodomethylmethylmalonate in good yield. This dialkylated malonic ester then was converted to (*E*)-3-iodo-2-methyl-2-propenoic acid by reacting it with aqueous KOH. The saponification was accompanied by a concomitant decarboxylation.

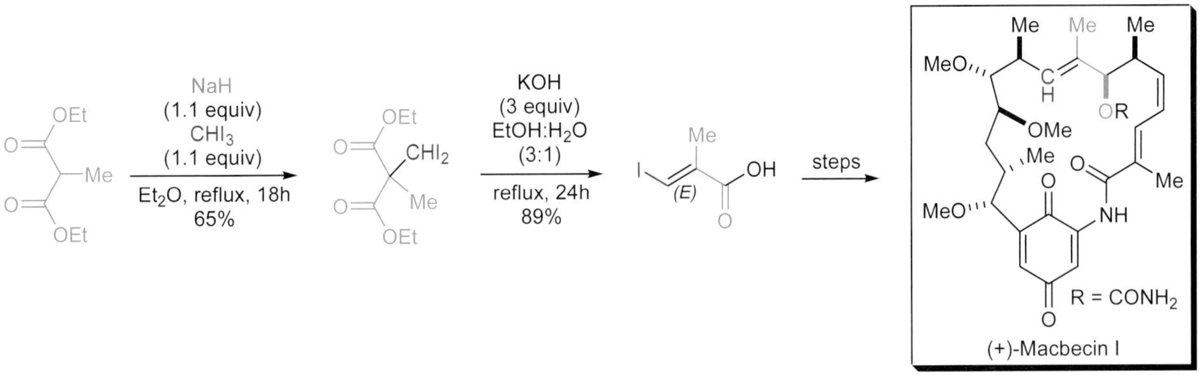

The novel humulane-type sesquiterpene (+)-bicyclohumulenone was synthesized for the first time in the laboratory of M. Kodama.[20] The natural product features a cyclodecenone ring fused to a cyclopropane ring, having two stereocenters at the ring junction. The cyclopropane moiety was installed using a stereoselective *Simmons-Smith cyclopropanation* reaction, while the 10-membered ring was formed *via* an intramolecular alkylation of an α-sulfenyl carbanion with an epoxide. The two main fragments were united by the *malonic ester synthesis* in which the monosubstituted dimethyl malonate was alkylated with an allylic chloride.

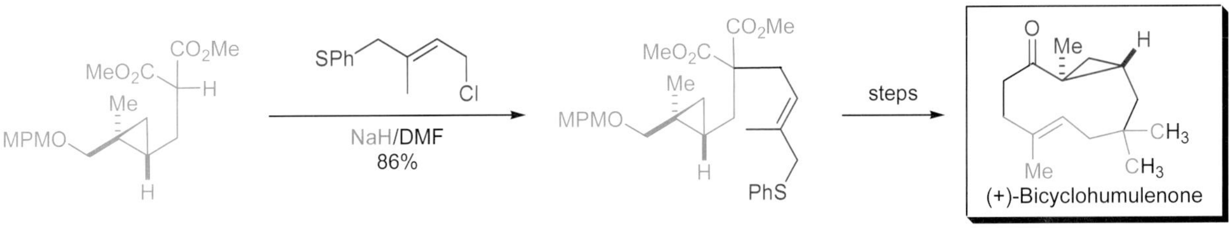

The structural elucidation of the secondary metabolites of *Dictyostellium* cellular slime molds was achieved by Y. Oshima et al.[21] The total synthesis of a novel compound, dictyopyrone A, which possesses a unique α-pyrone moiety with a side-chain at the C3 position, was successfully carried out using the *malonic ester synthesis*. Meldrum's acid was acylated and the product was subjected to transesterification with an optically active diol. Specific rotation of the final product was identical with that of the natural product, so the absolute configuration was established as (S).

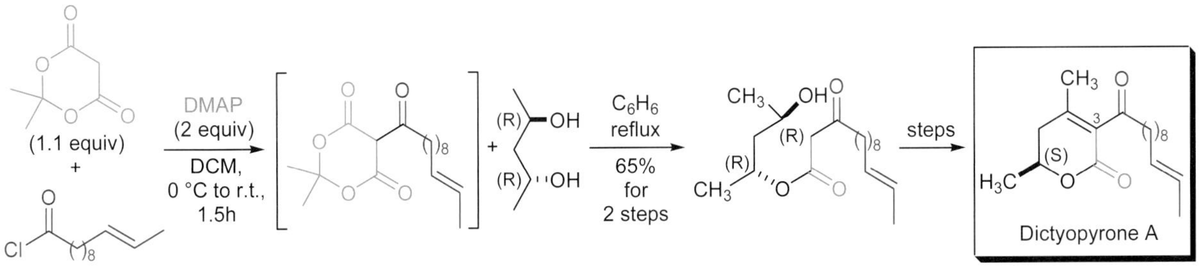

The key step in total synthesis of (+)-juvabione by G. Helmchen and co-workers was the *Pd-catalyzed allylic substitution* with the anion of (pivaloyloxy)malonate.[22] The substitution proceeded with very high regio- and stereoselectivity.

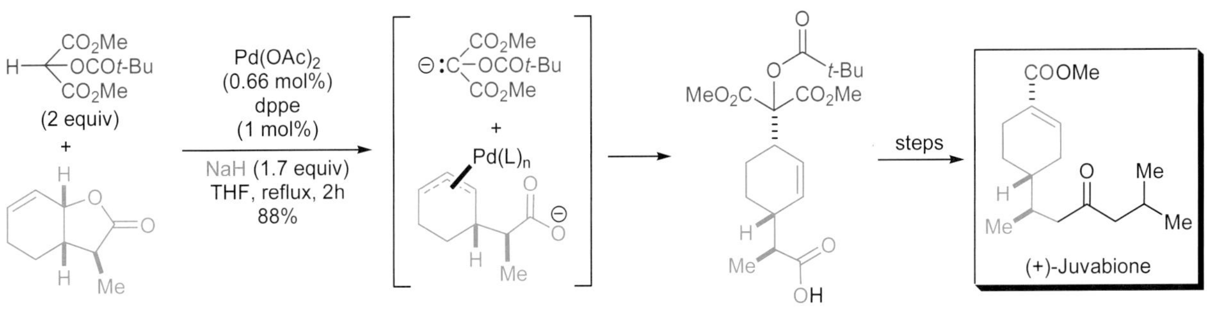

MANNICH REACTION

(References are on page 623)

Importance:

[*Seminal Publications*[1-3]; *Reviews*[4-23]; *Modifications & Improvements*[24-36]; *Theoretical Studies*[37-49]]

In 1903, B. Tollens and von Marle made the observation that the reaction of acetophenone with formaldehyde and ammonium chloride led to the formation of a tertiary amine.[1] In 1917, C. Mannich also isolated a tertiary amine by exposing antipyrine to identical conditions and recognized the generality of this reaction.[2,3] The condensation of a CH-activated compound (usually an aldehyde or ketone) with a primary or secondary amine (or ammonia) and a non-enolizable aldehyde (or ketone) to afford aminoalkylated derivatives is known as the *Mannich reaction*. More generally, it is the addition of resonance-stabilized carbon nucleophiles to iminium salts and imines. The product of the reaction is a substituted β-amino carbonyl compound, which is often referred to as the Mannich base. The general features of the reaction are: 1) the CH-activated component (activated at their α-position) is usually an aliphatic or aromatic aldehyde or ketone, carboxylic acid derivatives, β-dicarbonyl compounds, nitroalkanes, electron-rich aromatic compounds[12] such as phenols (activated at their *ortho* position) and terminal alkynes;[13] 2) only primary and secondary aliphatic amines or their hydrochloride salts can be used since aromatic amines tend not to react; 3) the non-enolizable carbonyl compound is most often formaldehyde; 4) when the amine component is a primary amine, the initially formed β-amino carbonyl compound can undergo further reaction to eventually yield a *N,N*-dialkyl derivative (a tertiary amine); however, with secondary amines overalkylation is not an issue; 5) the reaction medium is usually a protic solvent such as ethanol, methanol, water, or acetic acid to ensure sufficiently high concentration of the electrophilic iminium ion, which is responsible for the aminoalkylation; 6) unsymmetrical ketones usually give rise to regioisomeric Mannich bases, but the product derived from the aminoalkylation of the more substituted α-position tends to be dominant; and 7) Mannich bases are useful synthetic intermediates, since they can undergo a variety of transformations: β-elimination to afford α,β-unsaturated carbonyl compounds (Michael acceptors), reaction with organolithium, or Grignard reagents to yield β-amino alcohols and substitution of the dialkylamino group with nucleophiles to generate functionalized carbonyl compounds. There have been several improvements to the original three-component *Mannich reaction.* The use of preformed iminium salts is the most significant modification because it allows faster, more regioselective, and even stereoselective transformations under very mild conditions.[18]

Tollens and von Marle (1903):

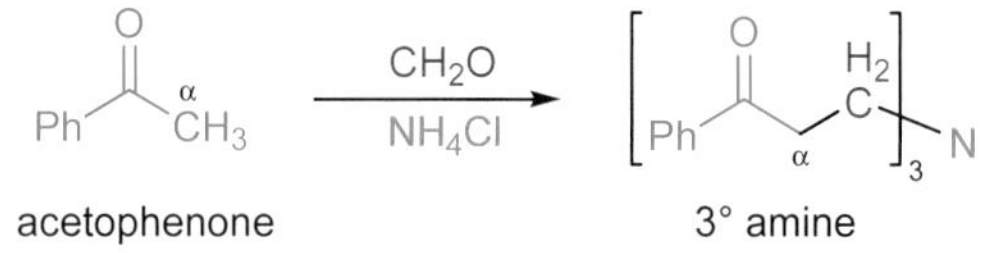

Mannich (1917):

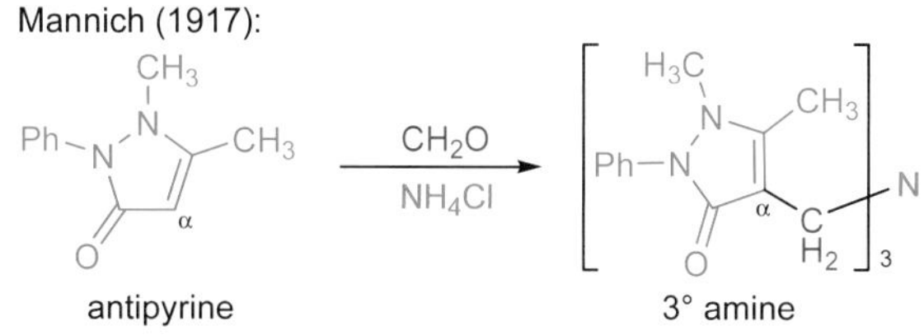

Mannich reaction:

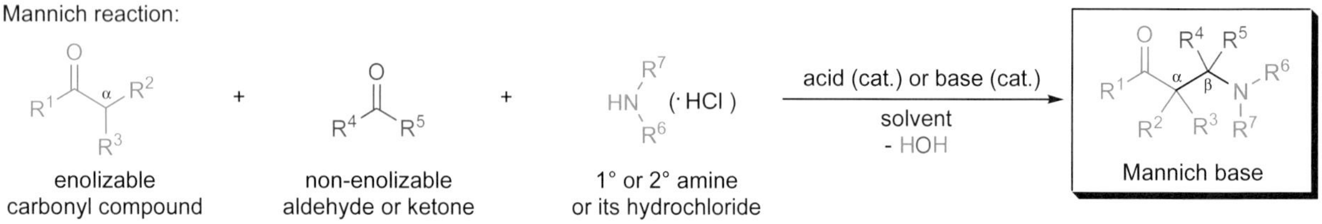

R^1 = H, alkyl, aryl, OR; R^{2-3} = H, alkyl, aryl; R^{4-5} = H, alkyl, aryl; R^6 = H, alkyl, OH, NH_2; R^7 = H, alkyl; solvent = ROH, H_2O, AcOH

Mechanism: [6,50,12-14]

The mechanism of the *Mannich reaction* has been extensively investigated. The reaction can proceed under both acidic and basic conditions, but acidic conditions are more common. Under acidic conditions the first step is the reaction of the amine component with the protonated non-enolizable carbonyl compound to give a hemiaminal, which after proton transfer loses a molecule of water to give the electrophilic iminium ion.[50] This iminium ion then reacts with the enolized carbonyl compound (nucleophile) at its α-carbon in an *aldol-type reaction* to give rise to the Mannich base.

Formation of the reactive iminium ion under acidic conditions:

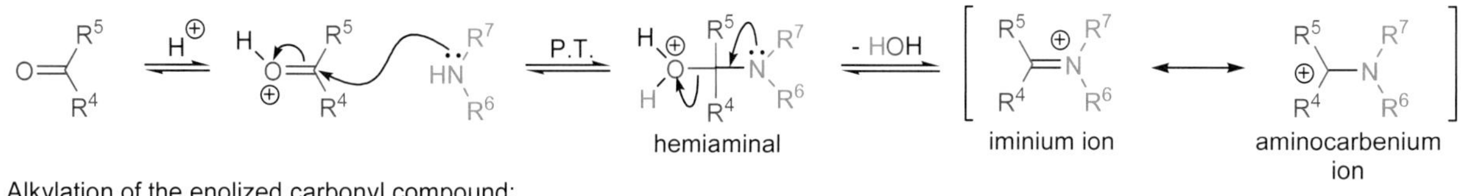

Alkylation of the enolized carbonyl compound:

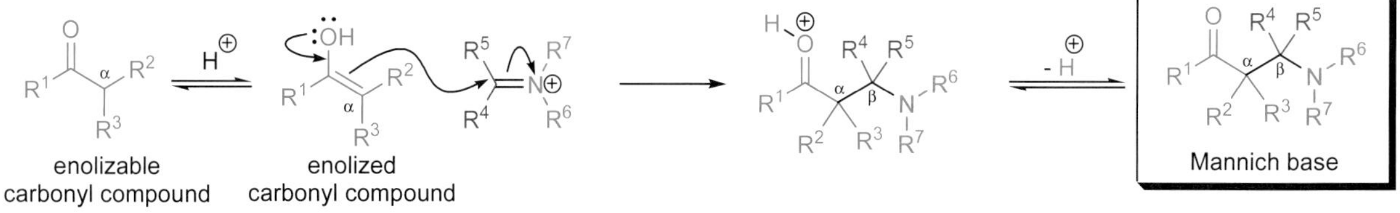

Synthetic Applications:

The total synthesis of (±)-aspidospermidine was accomplished by C.H. Heathcock and co-workers.[51] The synthetic strategy relied on an intramolecular cascade reaction, which simultaneously formed the B, C, and D rings of the natural product. As we mentioned previously, the CH-activated component of the *Mannich reaction* can also be an electron-rich aromatic ring such as an indole. The starting material was subjected to TFA in dichloromethane which first resulted in the formation of an indole (B ring) and an acylammonium ion (D ring) that *in situ* underwent an *intramolecular Mannich-type cyclization* giving rise to the C ring.

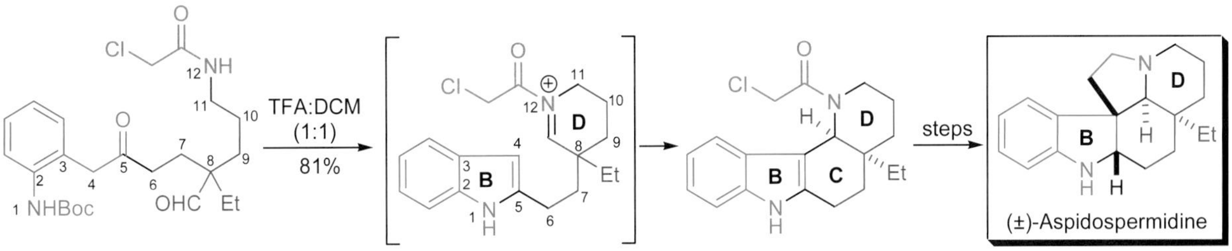

When preformed iminium salts are utilized in *Mannich reactions*, the reaction medium no longer needs to be a protic solvent, so the use of aprotic solvents allows the transformation of sensitive intermediates such as metal enolates. L.A. Paquette et al. carried out the highly regioselective introduction of an *exo*-methylene functionality during the total synthesis of (–)-*O*-methylshikoccin by reacting a potassium enolate with the Eschenmoser salt.[52] The resulting β-*N,N*-dimethylamino ketone was converted to the corresponding quaternary ammonium salt and elimination afforded the desired α,β-unsaturated ketone (*Eschenmoser methenylation*).

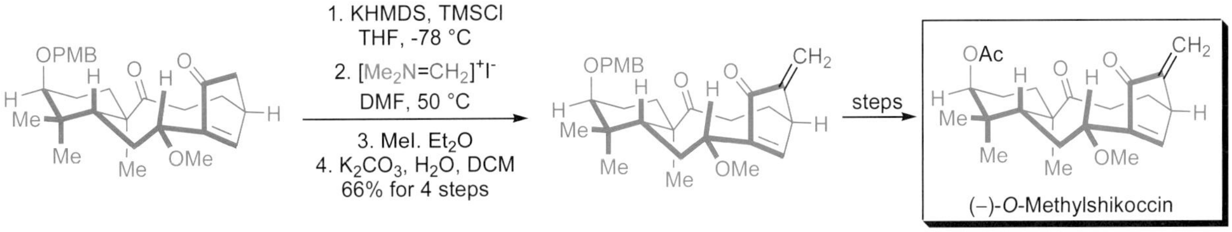

One of the most well-known applications of the *Mannich reaction* is its use in a tandem fashion with the *aza-Cope rearrangement* to form heterocycles. This reaction was the cornerstone of the strategy in the research group of L.E. Overman during the total synthesis of (±)-didehydrostemofoline (asparagamine A).[53] The bicyclic amine hydrogen iodide salt was exposed to excess paraformaldehyde, which led to the formation of the first iminium ion intermediate that underwent a facile *[3,3]-sigmatropic rearrangemen*t. The resulting isomeric iminium ion spontanaeously reacted with the enol in an *intramolecular Mannich cyclization*.

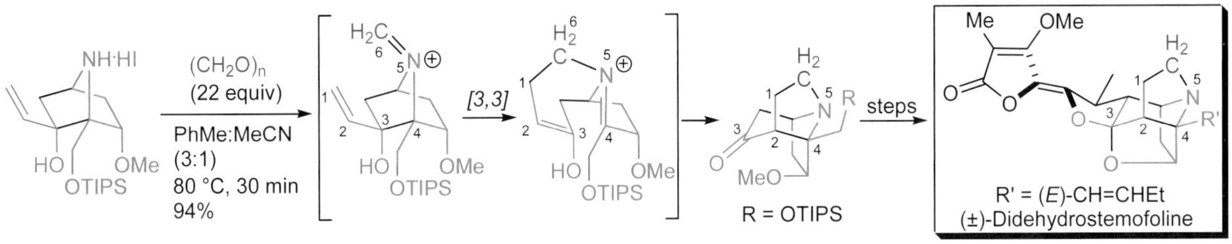

In the laboratory of S.F. Martin, the *vinylogous Mannich reaction* (VMR) of a 2-silyloxyfuran with a regioselectively generated iminium ion was utilized as the key step in the enantioselective construction of (+)-croomine.[54,55] The carboxylic acid moiety of the starting material was converted to the acid chloride which spontaneously underwent decarbonylation to give the corresponding iminium ion. Reaction of this iminium ion with the 2-silyloxyfuran afforded the desired *threo* butenolide isomer as the major product.

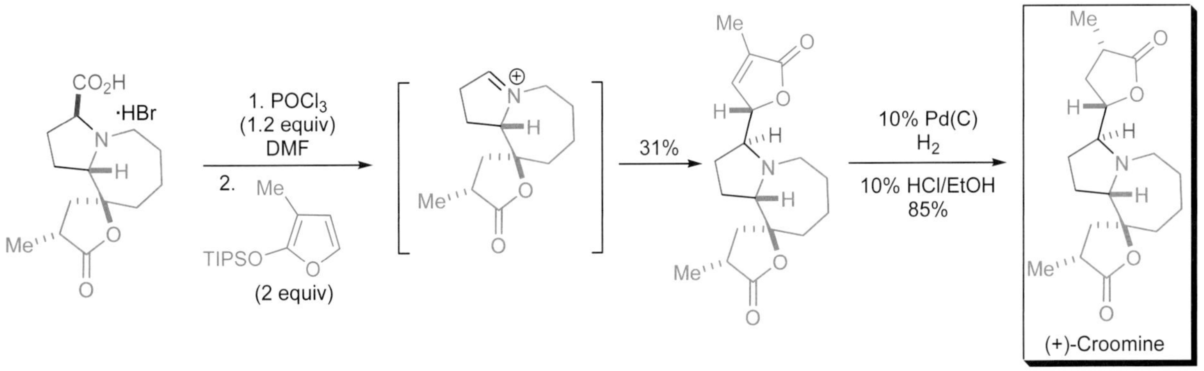

McMURRY COUPLING

(References are on page 624)

Importance:

[*Seminal Publications*[1-5]; *Reviews*[6-19]; *Modifications & Improvements*[20-28]; *Theoretical Studies*[29]]

In the early 1970s, the research groups of T. Mukaiyama,[3] S. Tyrlik,[4] and J.E. McMurry[5] independently discovered that the treatment of carbonyl compounds with low-valent titanium led to olefinic coupled products. In the following years, McMurry investigated the scope and limitation of the process,[20] and today the reductive coupling of carbonyl compounds using low-valent titanium complexes to form the corresponding alkenes is known as the *McMurry coupling*. The general features of this coupling reaction are: 1) it is used most often for the homocoupling of aldehydes and ketones to afford alkenes. However, mixed coupling is feasible if one component is used in excess or one of the coupling partners is a diaryl ketone; 2) the low-valent titanium reducing agent can be prepared in many ways but the most common is the reduction of $TiCl_3$ with a zinc-copper couple (Zn-Cu) in DME;[20] 3) if the reaction is conducted at low temperature, the pinacol intermediate may be isolated; 4) at high temperature the alkenes are formed directly; 5) sterically hindered and/or strained olefins, which cannot be prepared by other means, are formed in high yield; 6) even sterically hindered tetrasubstituted alkenes can be prepared; 7) macrocyclization under high-dilution conditions is successful for the synthesis of medium and large rings and the yields are independent of the ring size unlike in other macrocyclizations (e.g., *acyloin condensation*); 8) intramolecular reactions are the fastest for the formation of five- and six-membered rings and the formation of eight- or higher-membered rings is considerably slower; 9) the reaction conditions do not tolerate the presence of easily reducible functional groups (e.g., epoxides, α-halo ketones, unprotected 1,2-diols; allylic and benzylic alcohols, quinones, halohydrins, aromatic and aliphatic nitro compounds, oximes, and sulfoxides), but most other functional groups are compatible; 10) aldehydes react much faster than ketones so the coupling of two aldehydes in the presence of a ketone can be performed chemoselectively; 11) the alkene product is formed with poor stereoselectivity, although there is a slight preference for the formation of (*E*)-alkenes in intermolecular reactions; and 12) in the presence of a chlorosilane the *McMurry reaction* becomes catalytic.[18]

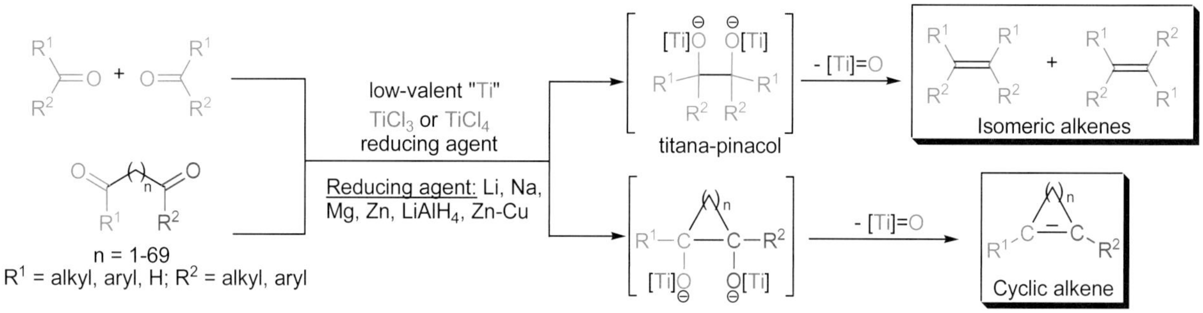

Mechanism:[30,20,31-38,13,39,40]

The mechanism of the *McMurry coupling* is not entirely clear, but it is composed of two distinct steps: 1) pinacol formation and 2) deoxygenation to the alkene. Extensive research showed that the low-valent titanium is most likely a mixture of $Ti^{(II)}$ and $Ti^{(0)}$, and the ratio of these species depends on the method of preparation (solvent, temperature, reducing agent, etc.). Recent findings suggest that the reaction possibly involves the formation of a carbene or a metal carbenoid.[34-36,13] The nature of the intermediates is strongly dependent on the structure of the carbonyl substrate and the reaction conditions, which is why the reaction is “tricky” and yields are difficult to reproduce in the laboratory.

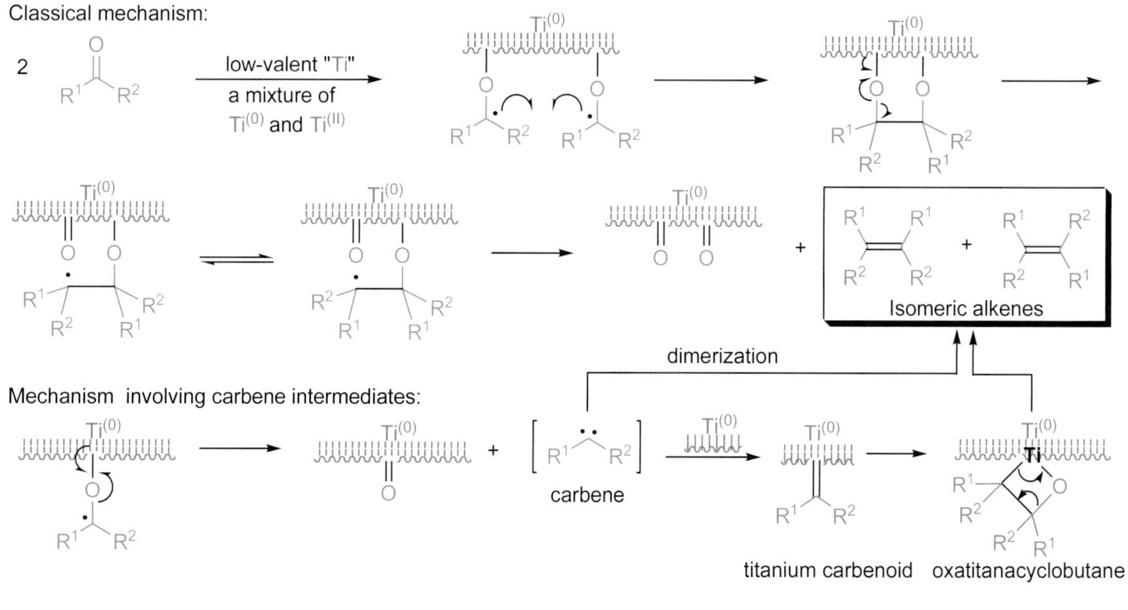

Synthetic Applications:

The first enantioselective total synthesis of (–)-13-hydroxyneocembrene using an *intramolecular McMurry coupling* as the key macrocyclization step was accomplished by Y. Li and co-workers.[41] To avoid any intermolecular coupling, high-dilution conditions were used. The cyclization precursor was added slowly *via* a syringe pump to a suspension of low-valent titanium reagent ($TiCl_4$/Zn) in refluxing DME. The reaction favored the formation of the (*E*)-olefin, the *E/Z* ratio was 2.5:1. The final step was the removal of the silyl protecting group with TBAF.

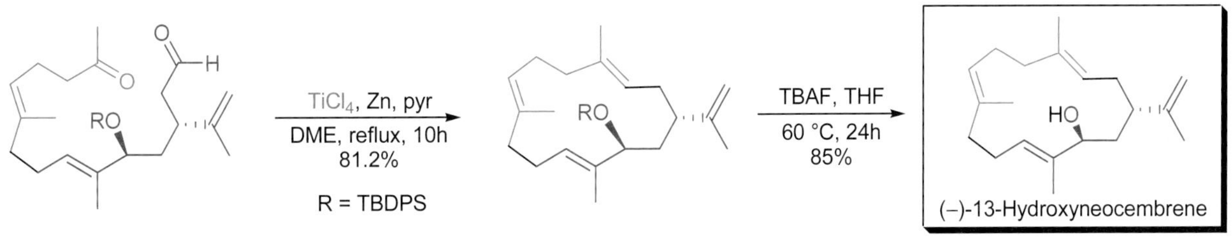

In the laboratory of T. Nakai, the asymmetric tandem *Claisen-rearrangement-ene reaction* sequence followed by a *modified McMurry coupling* was used to access (+)-9(11)-dehydroestrone methyl ether.[42] The Claisen-ene product was subjected to ozonolysis and epimerization to the 8,14-*anti* configuration. The C-ring was constructed by treating the tricyclic diketo aldehyde with $TiCl_3$-Zn(Ag) in DME to afford the desired final product in 56% yield.

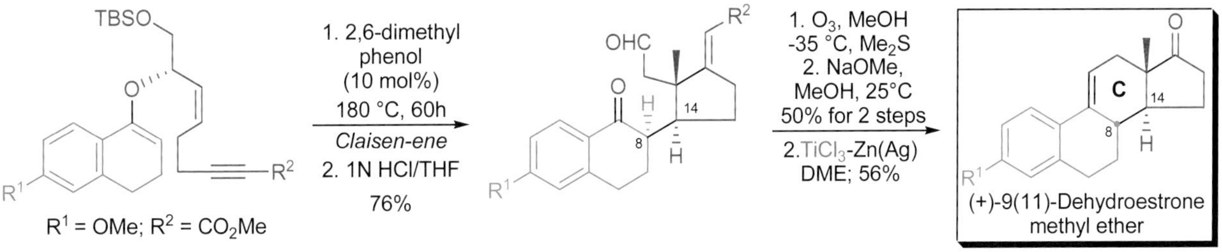

Several ADAM (alkenyldiarylmethane) II non-nucleoside reverse transcriptase inhibitors were prepared by M. Cushman and co-workers.[43] The *McMurry reaction* was the key transformation that enabled the coupling of the diaryl ketone with a variety of aldehydes in good yield. The commercially available $TiCl_4$-THF (2:1) and zinc dust was used to prepare the low-valent titanium reagent in refluxing THF. To this suspension was added the diaryl ketone and the aldehyde successively.

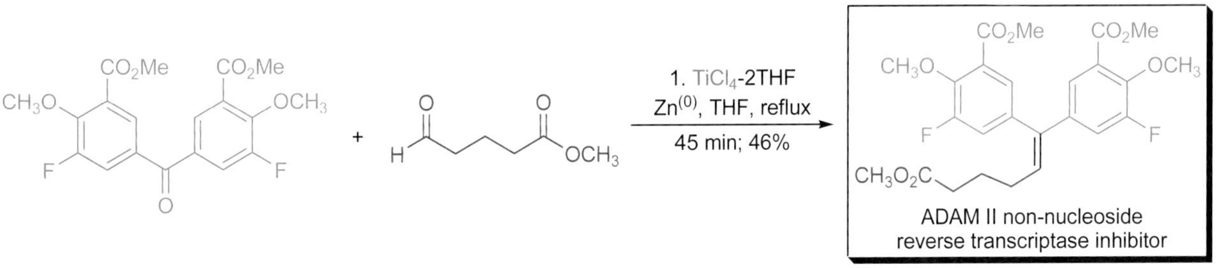

The impressive synthetic power of the *McMurry coupling* was demonstrated by K. Kakinuma et al. when they synthesized archaeal 72-membered macrocyclic lipids.[44] The final macrocyclization between the dialdehyde proceeded in 66% yield, giving rise to a single diastereomer.

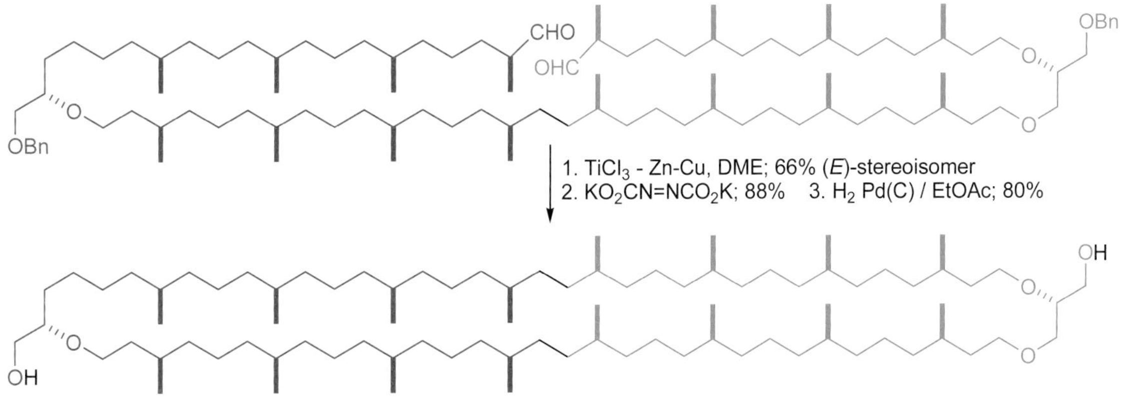

MEERWEIN ARYLATION

(References are on page 625)

Importance:

[*Seminal Publications*[1]; *Reviews*[2-7]; *Modifications & Improvements*[8-18]; *Theoretical Studies*[19-21]]

In 1939, H. Meerwein and co-workers reported in an extensive study that aromatic diazo compounds reacted with α,β-unsaturated carbonyl compounds in which the aryl group added across the double bond and a molecule of nitrogen was lost.[1] In one experiment, coumarin was reacted with *p*-chlorodiazonium chloride in the presence of catalytic amounts of copper(II)chloride, and the corresponding 3-(*p*-chlorophenyl)coumarin was isolated in moderate yield. When the unsaturated reaction partner was cinnamic acid, a molecule of carbon dioxide was lost in addition to nitrogen and the product was the corresponding styrene derivative. The arylation of substituted alkenes with aryldiazonium halides (formally the addition of an aryl halide to a carbon-carbon double bond) in the presence of a metal salt catalyst is known as the *Meerwein arylation*. The general features of this reaction are: 1) the procedure is simple; no special laboratory equipment is needed; 2) the aryldiazonium halides are prepared by the diazotization of aromatic amines using sodium nitrite and aqueous hydrohalic acids and are not isolated, rather immediately reacted with the alkenes in the presence of an organic solvent (e.g., acetone, acetonitrile); 3) the presence of electron-withdrawing substituents on the aromatic ring tends to increase the yield, whereas electron-donating groups often give lower yields; 4) the alkene component usually has an electron-withdrawing substituent and mostly α,β-unsaturated carbonyl compounds are used; 5) if there are two electron-withdrawing substituents on the double bond, and they are attached to the same carbon and then the aryl group will add to the other sp^2 hydbridized carbon atom; 6) when each of the olefin carbon atoms has an electron-withdrawing substituent, regioisomeric products may be formed; however, the major product will arise from the most resonance stabilized radical intermediate; 7) cinnamic acids and maleic acids are arylated at the α-carbon, and the reaction is accompanied by decarboxylation which is a pH-dependent process; 8) alkynes with electron-withdrawing substituents also react, but the yields are often poor; 9) furan derivatives are arylated with ease under the reaction conditions; and 10) the initial product of the reaction is a substitution product (alkyl halide), which can be dehydrohalogenated under basic conditions to afford the corresponding aryl substituted olefin. The *Meerwein arylation* is not free of side reactions (e.g., *Sandmeyer reaction,* formation of azo compounds, etc.), which are the primary cause of the often moderate product yields.

Meerwein (1939):

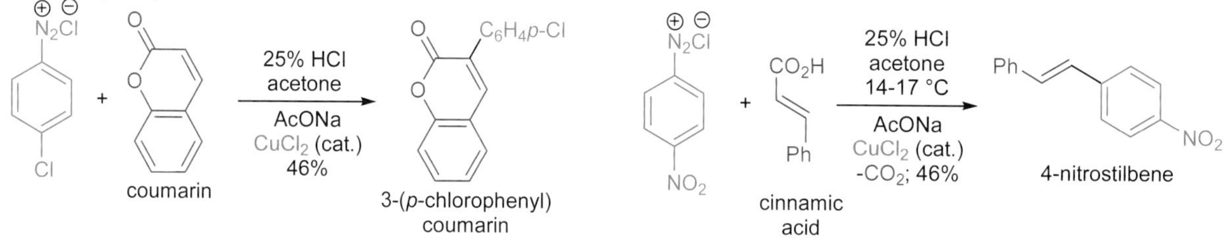

Meerwein arylation:

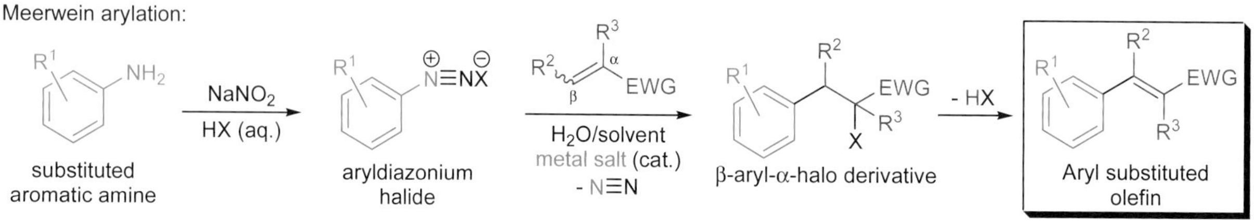

R^1 = H, alkyl, aryl, *O*-alkyl, Cl, Br, I, CO_2-alkyl, CONHR, SO_2R, NO_2, CF_3; R^{2-3} = H, alkyl, aryl; EWG = CHO, CO-alkyl, CO_2-alkyl, CO_2H, CO_2NH_2, CO_2NR_2, CN, alkenyl, Cl, Br; HX: HCl, HBr; solvent: acetone, acetonitrile; metal salt: $CuCl_2$, $CuBr_2$

Mechanism: [22-24,4,21]

The mechanism of the *Meerwein arylation* is not completely understood. In his seminal paper, Meerwein proposed the involvement of aryl cations, however, this hypothesis was soon eliminated when J.K. Kochi suggested that aryl radicals are formed under the reaction conditions.[22] The actual catalyst is a copper(I) species, which is formed *in situ* from copper(II) salts and carbonyl compounds (e.g., acetone which is often used as a solvent).[23]

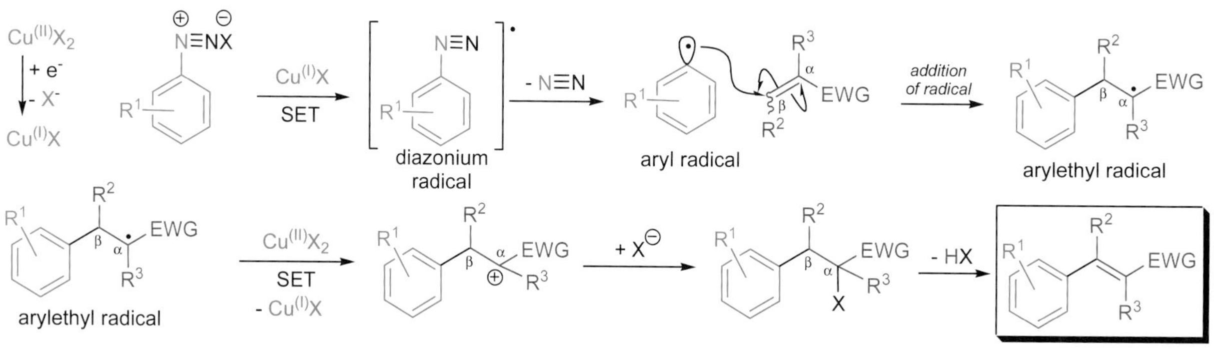

Synthetic Applications:

In the laboratory of R. Bihovsky, a series of peptide mimetic aldehyde inhibitors of calpain I was prepared in which the P_2 and P_3 amino acids were replaced with substituted 3,4-dihydro-1,2-benzothiazine-3-carboxylate-1,1-dioxides.[25] The synthesis began with the diazotization of the substituted aniline substrate using sodium nitrite and hydrochloric acid. The aqueous solution of the corresponding diazonium chloride product was added dropwise to the solution of acrylonitrile in a water-acetone mixture, which contained catalytic amounts of copper(II) chloride. This *Meerwein arylation* step afforded the chloronitrile derivative, which was subjected to sulfonation with chlorosulfonic acid, and the resulting sulfonyl chloride was treated with the solution of ammonia in dioxane to give the desired 3,4-dihydro-1,2-benzothiazine-2-carboxamide.

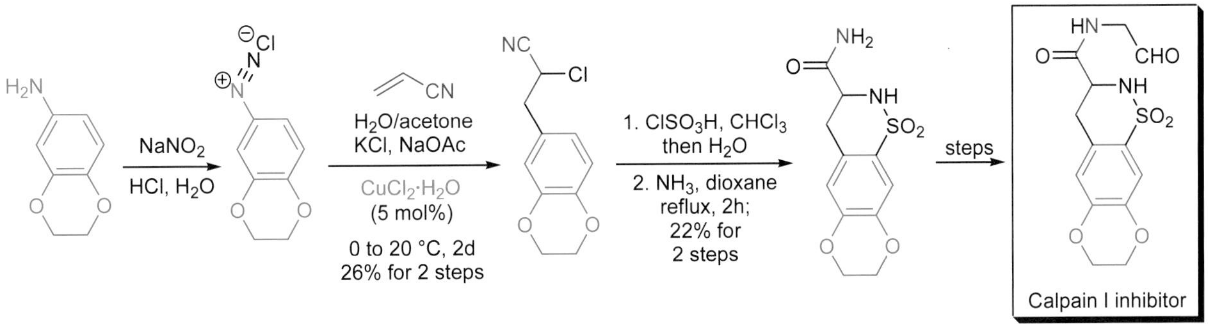

The research team of J.E. Baldwin developed the first synthetic sequence for the preparation of N(5)-ergolines.[26] The key step was a *hetero-Diels-Alder reaction* of a substituted phenyl butadiene to form the piperidine ring. The phenyl butadiene substrate was prepared *via* the *Meerwein arylation* of 1,4-butadiene and a diazonium salt derived from 2,6-dinitrotoluene. The initially formed chlorinated product was subjected to dehydrochlorination using DBU as the base.

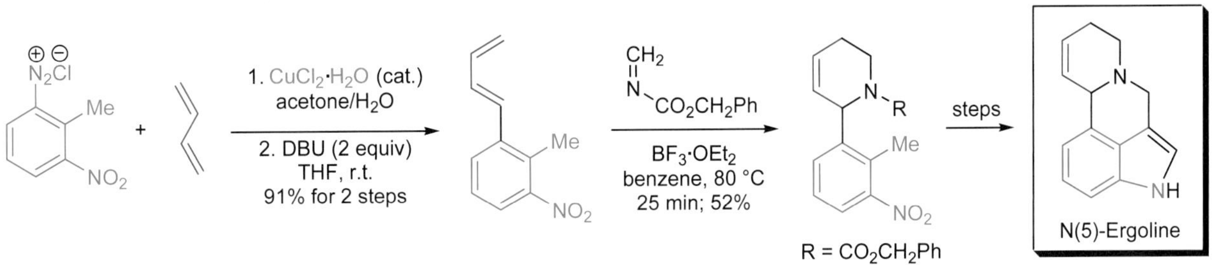

The synthesis of the aglycone of the antibiotic gilvocarcin-M was accomplished by T.C. McKenzie et al. by a sequential *Meerwein arylation-Diels-Alder cycloaddition*.[27] The anthranilic methyl ester substrate was first subjected to diazotization and then the resulting diazonium chloride was coupled to 2,6-dichlorobenzoquinone in water to afford the quinone product in moderate yield. It is important to mention that the *Meerwein arylation* was conducted in water at 80 °C in the absence of a catalyst.

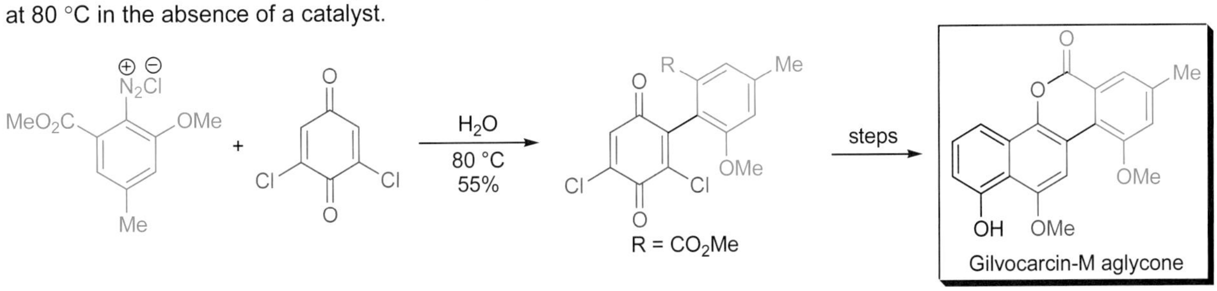

T. Sohda and co-workers prepared a series of novel thiazolidinedione derivatives of the potent antidiabetic pioglitazone (AD-4833, U-72, 107).[28] The *para*-substituted aniline was diazotized with $NaNO_2$/HBr, and the diazonium bromide was used to arylate methyl acrylate in the presence of copper(II) oxide. The bromopropionate product was first treated with thiourea, and the resulting iminothiazolidinone hydrolyzed with aqueous hydrochloric acid to afford the desired thiazolidinedione derivative.

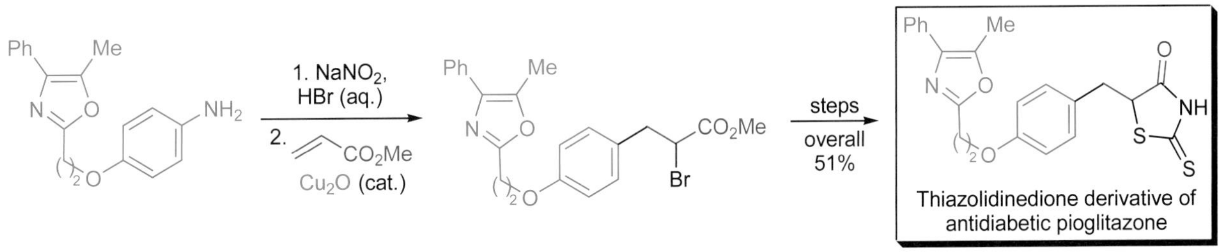

MEERWEIN-PONNDORF-VERLEY REDUCTION

(References are on page 626)

Importance:

[*Seminal Publications*[1-3]; *Reviews*[4-18]; *Modifications & Improvements*[19-31]; *Theoretical Studies*[32,33]]

In the mid-1920s, three researchers independently described reduction of carbonyl compounds with the use of aluminum alkoxides: 1) in 1925, H. Meerwein successfully reduced aldehydes with ethanol in the presence of aluminum ethoxide;[1] 2) during the same year, A. Verley reduced ketones with aluminum ethoxide as well as aluminum isopropoxide but found that sterically hindered ketones (e.g., camphor) reacted very slowly;[2] and 3) in 1926, W. Ponndorf demonstrated that the reduction of aldehydes and ketones was general for a variety of metal alkoxides (e.g., alkali metal and aluminum alkoxides) derived from secondary alcohols, and he found the process completely reversible.[3] The reduction of aldehydes and ketones by metal alkoxides (mainly by aluminum isopropoxide) is known as the *Meerwein-Ponndorf-Verley reduction* (*MPV reduction*).[34] The reverse reaction, the oxidation of alcohols to aldehydes and ketones, is referred to as the *Oppenauer oxidation*. The general features of the *MPV reduction* are: 1) the reaction is completely reversible and the removal of the low boiling ketone or the addition of excess isopropyl alcohol shifts the equilibrium to the right according to *Le Chatelier's principle*; 2) the reduction takes place in boiling isopropanol under mild conditions, and it is very chemoselective for aldehydes and ketones, whereas other functional groups (e.g., double bond, esters, acetals, etc.) remain unchanged, and this is the greatest advantage over the use of metal hydride reducing agents; 3) the most popular metal alkoxides are aluminum alkoxides, and these are often used in stoichiometric amounts (one or more equivalents for ketones), but Ln(III) alkoxides (e.g., $Sm(Ot\text{-}Bu)I_2$) can be applied in catalytic amounts;[21,22] 4) aluminum alkoxides are readily soluble in both alcohols and hydrocarbon solvents, whereas other metal alkoxides have limited solubility; 5) aldehydes react faster than ketones; 6) keto aldehydes are reduced to hydroxy ketones, whereas α,β-unsaturated aldehydes and ketones give the corresponding allylic alcohols; 7) cyclic diketones usually give rise to hydroxyl ketones unless an aromatic ring can be formed *via hydrogen transfer*; 8) β-diketones or β-keto esters cannot be reduced due to the formation of stable β-enolate chelate complexes with metal alkoxides, but when these compounds do not have enolizable hydrogens at the α-position, the reduction proceeds smoothly; 9) the method is sensitive to steric hindrance, so sterically hindered ketones and aldehydes are reduced more slowly than unhindered ones; 10) to increase the rate of reduction for slow reactions, the alcohol solvents may be mixed with higher boiling solvents (e.g., toluene, xylene) or multiple equivalents of aluminum alkoxide should be applied; 11) the reaction rate is significantly increased by the addition of protic acids (e.g., TFA, HCl, propionic acid);[19,24,25] 12) in rigid cyclic substrates, the reduction proceeds with high diastereoselectivity; 13) catalytic asymmetric versions are known, but currently only the *intramolecular asymmetric MVP reduction* gives high *ee*'s;[15] and 14) both small-, and large-scale reduction can be carried out with ease (few milligrams to several hundred grams). The most important side reactions are: 1) *aldol condensation* of aldehyde substrates, which have an α-hydrogen atom to form β-hydroxy aldehydes and/or α,β-unsaturated aldehydes, but with ketones this side reaction is not common; 2) *Tishchenko reaction* of aldehyde substrates with no α-hydrogen atom, but this can be suppressed by the use of anhydrous solvents; 3) dehydration of the product alcohol to an olefin, especially at high temperature; and 4) the migration of the double bond during the reduction of α,β-unsaturated ketones.

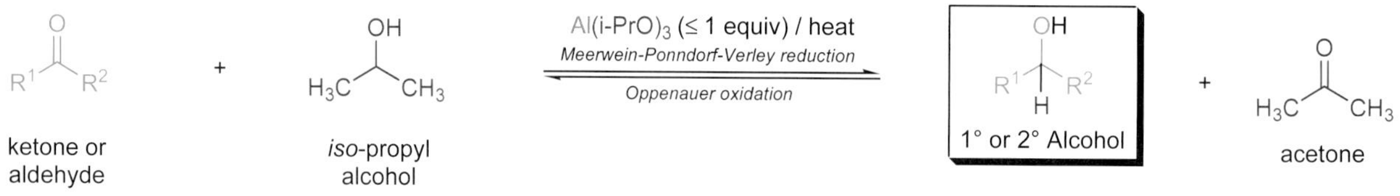

R^1 = alkyl, aryl, alkenyl; R^2 = H, alkyl, aryl, alkenyl

Mechanism: [35-40,19,41-48]

The currently accepted concerted mechanism that goes through a chairlike six-membered transition state was first proposed by Woodward.[35] The special activity of aluminum alkoxides for the *MVP reduction* can be explained as a result of the activation of both the hydride donor and the hydride acceptor. For aromatic ketones the involvement of radicals was suggested, but for aliphatic carbonyl compounds there is no evidence for a SET mechanism.[44]

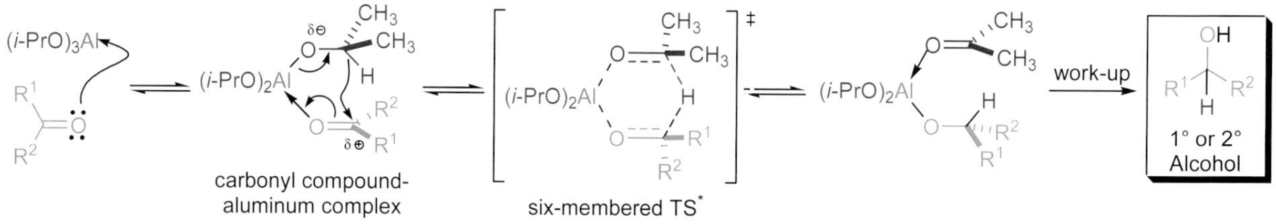

Synthetic Applications:

The highly stereoselective formal total synthesis of GA_{111} and GA_{112} methyl esters was accomplished using the combination of a *Pd-catalyzed cycloalkenylation reaction* and *inverse-electron demand Diels-Alder cycloaddition* in the laboratory of M. Ihara.[49] The final step of the synthesis was the reduction of the tetracyclic ketone to obtain both diastereomers of the corresponding secondary alcohols. It was found, however, that the hydride reduction of this ketone gave GA_{112} methyl ester exclusively as a single diastereomer. When the reduction was carried out in the presence of large excess of aluminum isopropoxide, both diastereomers were formed, but the GA_{111} methyl ester was the major product.

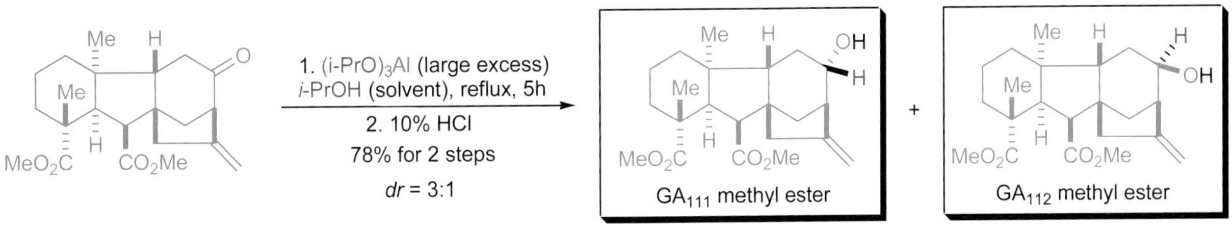

The *MPV reduction* was used in a highly stereoselective fashion during the final stages of the total synthesis of *dl*-coccuvinine and *dl*-coccolinine by T. Sano et al.[50] The α,β-unsaturated ketone moiety was selectively reduced in the presence of an α,β-unsaturated lactam to give the β-allylic alcohol in good yield. The methylation of the allylic alcohol under phase-transfer conditions (*Williamson ether synthesis*) was followed by the reduction of the lactam carbonyl group to the corresponding methylene group with excess allane to afford the natural product.

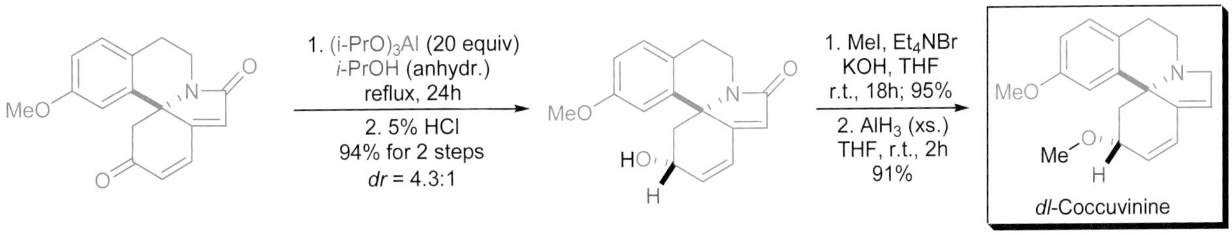

The absolute stereochemistry of the rutamycin antibiotics was established through asymmetric synthesis of the known bicyclic degradation product by D.A. Evans and co-workers.[51] The introduction of the equatorial secondary alcohol functionality turned out to be problematic when traditional metal hydrides were used for the reduction of the ketone. For example, $LiAlH_4$ gave only a 1:1 mixture of axial and equatorial diastereomers. The use of the *samarium(II)-catalyzed MVP reduction* gave a 98:1 mixture of diastereomers favoring the equatorial alcohol. Subsequent examination of this highly stereoselective reduction revealed that the reaction operated under kinetic control, and the observed product was formed due to the coordination of the reducing agent to the axial spiroketal oxygen atom.

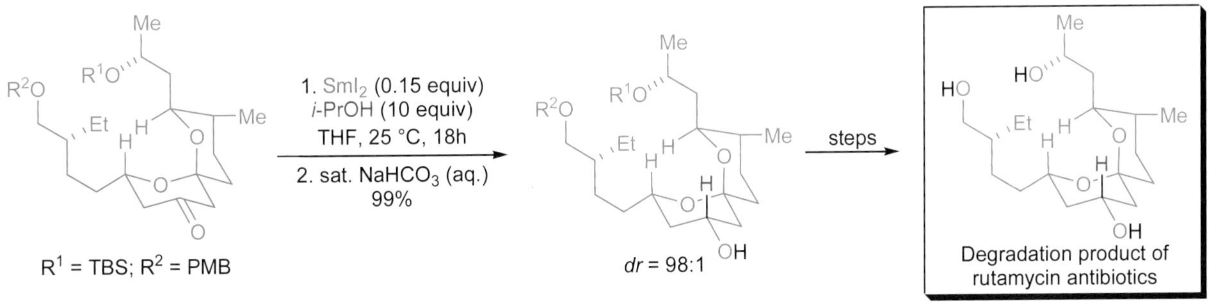

The synthesis of the rare furochromone ammiol was achieved by R.B. Gammill starting from (methylthio)furochromone in four steps.[52] The last step was the selective conversion of the aldehyde moiety of a six-membered 1,4-dicarbonyl compound using the *MVP reduction*.

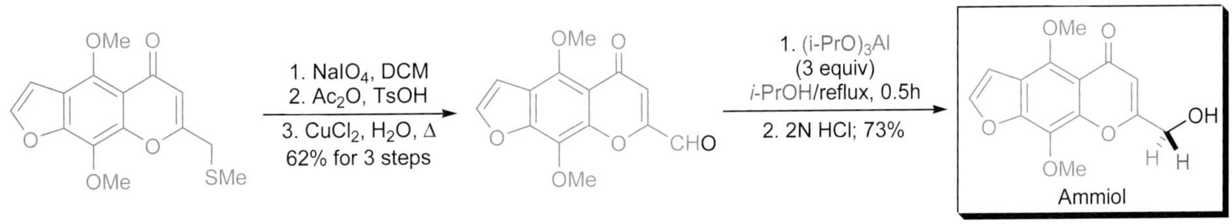

MEISENHEIMER REARRANGEMENT

(References are on page 627)

Importance:

[*Seminal Publications*[1,2]; *Reviews*[3-5]; *Modifications & Improvements*[6-13]; *Theoretical Studies*[14-17]]

In 1919, J. Meisenheimer reported that upon heating in an aqueous sodium hydroxide solution, *N*-benzyl-*N*-methyl aniline-*N*-oxide underwent a facile isomerization to afford *O*-benzyl-*N*-methyl-*N*-phenyl hydroxylamine.[1] Three decades later, A.C. Cope and co-workers reinvestigated the rearrangement to explore its mechanism.[18] They discovered that the isomerization of *N*-crotyl-*N*-methyl aniline *N*-oxide occurred with the inversion of the allylic system to give *N*-methyl-*O*-(1-methyl-allyl)-*N*-phenylhydroxylamine. This result suggested that the isomerization occurred *via* a five-membered cyclic transition state analogous to the mechanism of the *Claisen rearrangement*. The thermal rearrangement of certain tertiary amine *N*-oxides to the corresponding *O*-substituted-*N,N*-disubstituted hydroxylamines is known as the *Meisenheimer rearrangement*. The general features of the reaction are: 1) the rearrangement takes place in both open-chain and cyclic systems; 2) the *[1,2]*- and *[2,3]-shift* of substituents are the two different modes of the transformation; 3) the *[1,2]-shift* occurs when one of the substituents is capable of stabilizing radicals (R^1 = benzyl, diphenylmethyl, etc.); 4) the *[2,3]-shift* is common when one of the substituents is allylic; 5) during the *[1,2]-shift*, the stereocenter on the migrating group suffers extensive racemization;[3] 6) the *[2,3]-shift* usually takes place much faster than the *[1,2]-shift* and the transfer of chirality of the migrating group is possible; 7) when any of the R^2,R^3 or R^6,R^7 are alkyl groups that have a hydrogen atom at their β-position, the *Cope elimination* becomes competitive; 8) the *N*-oxides of *N*-benzyl and *N*-allyl cyclic amines mainly undergo *[1,2]-shifts* to afford the corresponding *O*-benzyl and *O*-allyl hydroxylamines, respectively; 9) the *N*-oxides of 2-aryl-, 2-heteroaryl, and 2-vinyl cyclic amines predominantly undergo ring-enlargement to give 1,2-oxazaheterocycles; and10) the ring-enlargement is general for four- to ten-membered cyclic amine *N*-oxides.[5]

Meisenheimer (1919):

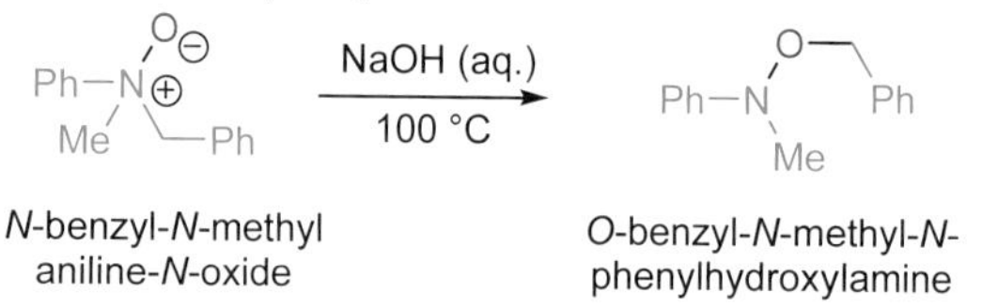

Cope and Kleinschmidt (1944):

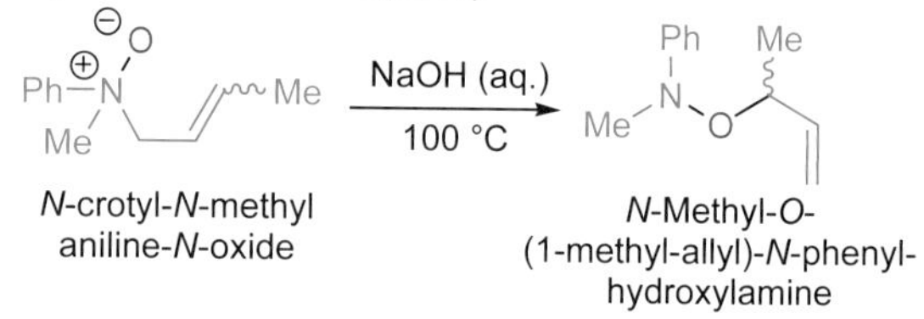

[1,2]-Meisenheimer rearrangement in acyclic systems:

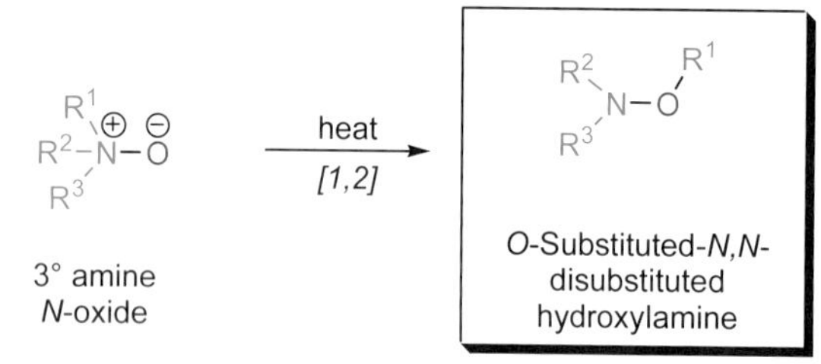

[2,3]-Meisenheimer rearrangement in acyclic systems:

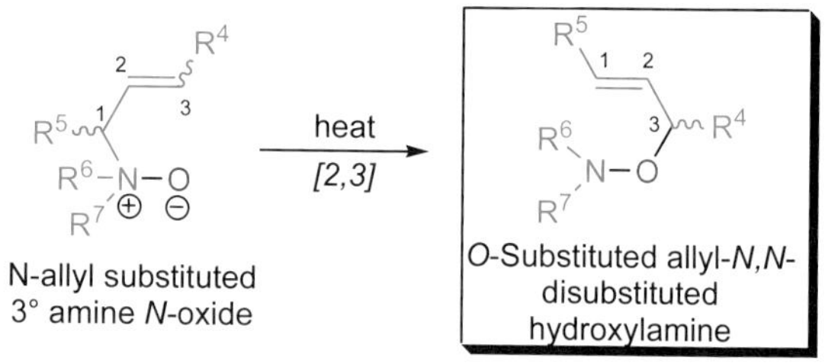

[1,2]-Meisenheimer rearrangement in cyclic systems:

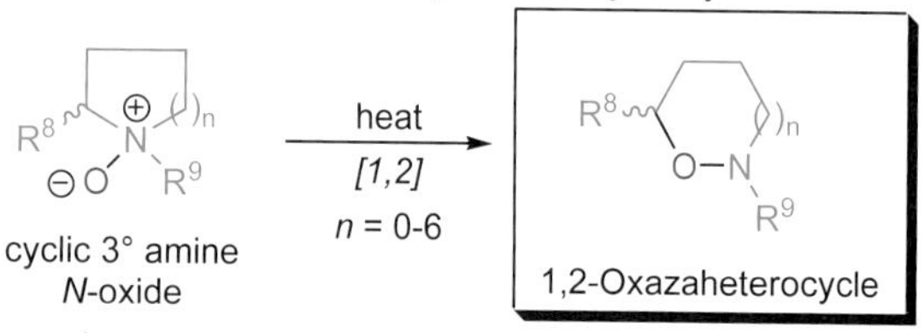

[2,3]-Meisenheimer rearrangement in cyclic systems:

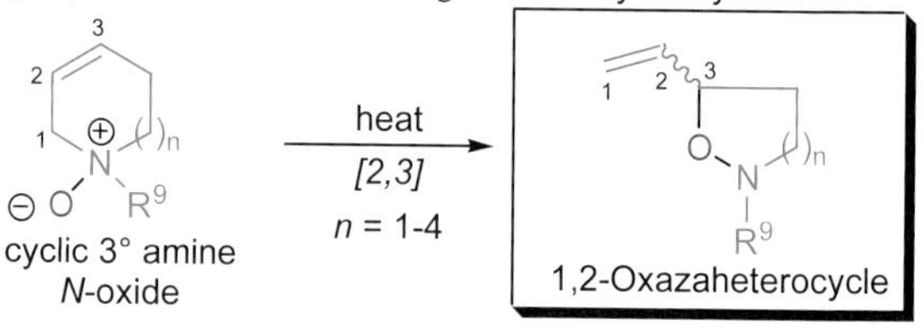

R^1 = CH_2Ph, $CHPh_2$, CH_2Ar, allyl; R^{2-3} = alkyl with no β-hydrogen, aryl; R^{4-5} = H, alkyl, aryl; R^{6-7} = alkyl with no β-hydrogen, aryl; R^8 = alkenyl, aryl; R^9 = alkyl with no β-hydrogen

Mechanism: [18,3,19,20,6,13]

The *[1,2]-Meisenheimer rearrangement* most likely proceeds *via* a *homolytic dissociation-recombination* mechanism,[19] whereas the *[2,3]-Meisenheimer rearrangement* is a concerted sigmatropic process that goes through a five-membered envelopelike transition state.

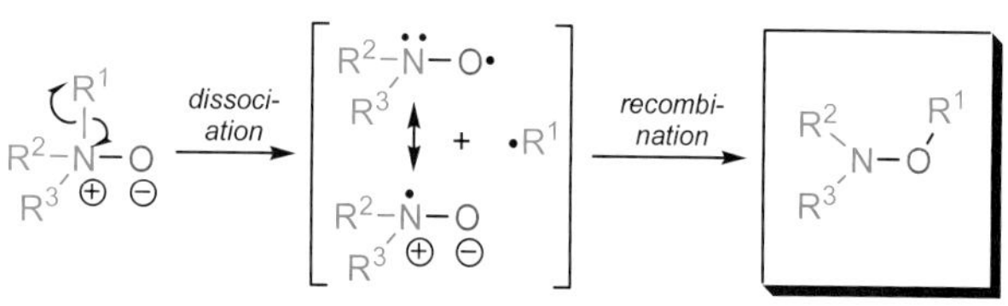

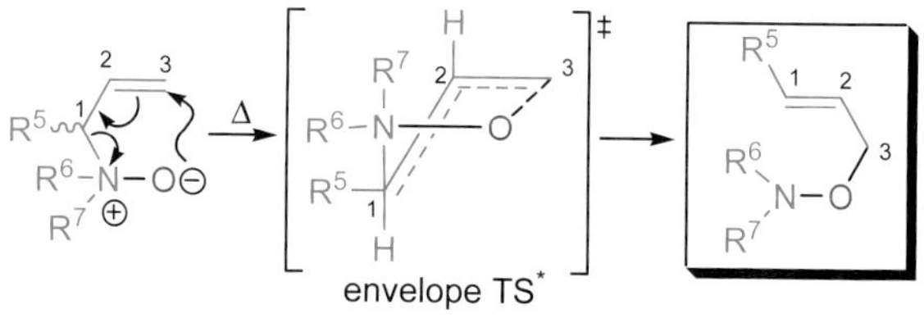

Synthetic Applications:

The natural product (*R*)-sulcatol is a male-produced aggregation pheromone of the ambrosia beetle. This insect can devastate entire forests when its population is out of control.[21] Various studies revealed that different species respond to the compound in different enantiomeric excess. The asymmetric synthesis of (*R*)-sulcatol was accomplished in the laboratory of S.G. Davies using a *stereospecific [2,3]-Meisenheimer rearrangement* as the key step. The treatment of the allylic amine substrate with *m*CPBA followed by the filtration of the reaction mixture through deactivated basic alumina afforded the desired hydroxylamine as a single diastereomer.

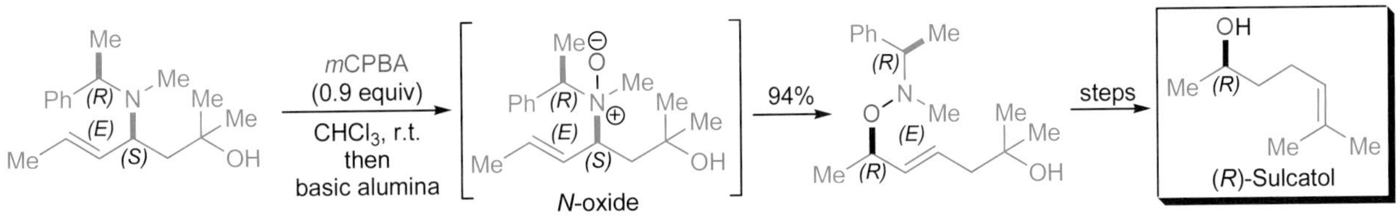

A new route to the 12(S)carba-eudistomin skeleton was developed by T. Kurihara et al.[22] The key substrate for this new route was a 1,2-*cis*-2-ethenylazetopyridoindole, which was readily oxidized at 0 °C to afford the corresponding *N*-oxide. This *N*-oxide spontaneously underwent a *[2,3]-Meisenheimer rearrangement* to afford the desired oxazepine derivative. Interestingly, when the 1,2-*trans*-2-ethenylazetopyridoindole was subjected to identical conditions, the *[1,2]-Meisenheimer rearrangement* occurred exclusively and gave rise to an isoxazolidine derivative.

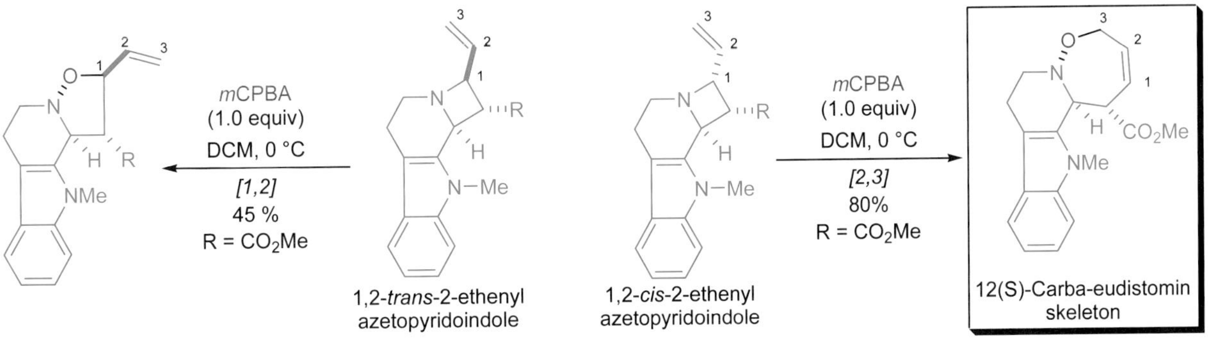

In the laboratory of H. Kondo, various prodrugs of the clinically effective antibacterial agent norfloxacin (NFLX) were synthesized.[23] The *N*-masked derivatives of NFLX were efficiently unmasked *in vivo*, and they exhibited equal or higher activity than NFLX itself. In order to reveal the mode of action of these prodrugs, the *N*-allylic derivative of NFLX was subjected to *m*CPBA at low temperatures. The resulting *N*-oxide was then heated to bring about a *[2,3]-Meisenheimer rearrangement* to afford the corresponding *O*-allyl-hydroxylamine derivative. This hydroxylamine derivative also acted as a prodrug, since it liberated a higher concentration of NFLX in plasma and had a higher activity than NFLX itself.

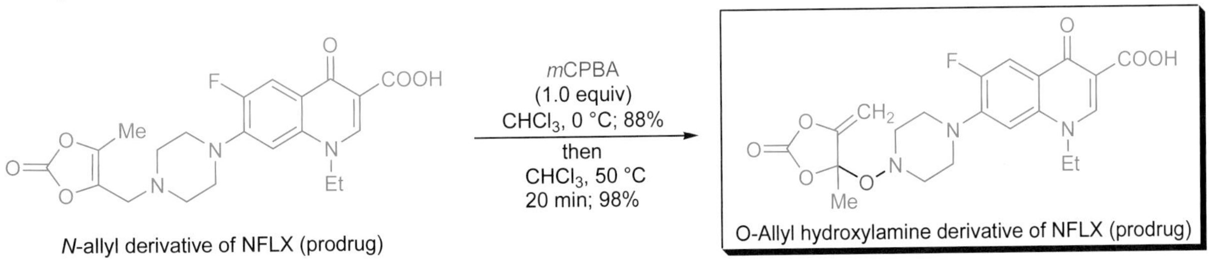

The *[1,2]-Meisenheimer rearrangement* and a *Heck cyclization* were the key steps in T. Kurihara's synthesis of magallanesine.[24] The azetidine was exposed to H_2O_2, and the resulting azetidine *N*-oxide was refluxed in THF to afford the desired azocine derivative. Other usual oxidants such as *m*CPBA or MMPP gave rise to complex mixtures.

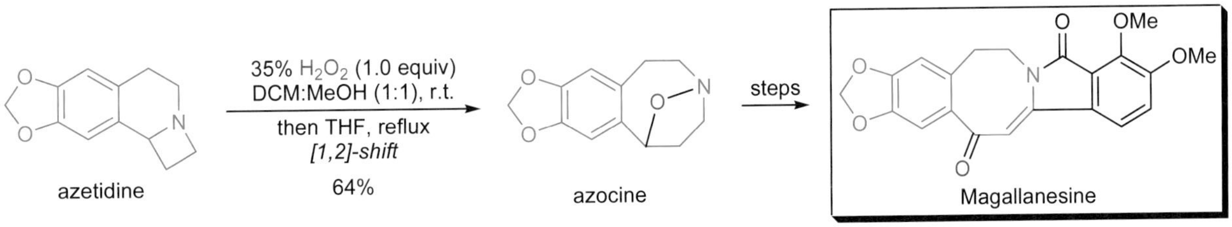

MEYER-SCHUSTER AND RUPE REARRANGEMENT

(References are on page 627)

Importance:

[*Seminal Publications*[1-7]; *Reviews*[8]; *Modifications & Improvements*[9-15]; *Theoretical Studies*[16-21]]

In 1922, K.H. Meyer and K. Schuster reported that the attempted conversion of 1,1,3-triphenyl-2-propynol to the corresponding ethyl ether with concentrated sulfuric acid and ethanol afforded 1,3,3-triphenyl propenone, an α,β-unsaturated ketone.[1] The authors showed that the use of other reagents such as acetic anhydride and acetyl chloride also brought about the same reaction. A few years later, H. Rupe and co-workers investigated the acid-catalyzed rearrangement of a large number of α-acetylenic (propargylic) alcohols.[2-7] The acid-catalyzed isomerization of secondary and tertiary propargylic alcohols, *via* a *[1,3]*-shift of the hydroxyl group, to the corresponding α,β-unsaturated aldehydes or ketones is known as the Meyer-Schuster rearrangement. The general features of this transformation are: 1) when the substrate contains a terminal alkyne, the product is an aldehyde, whereas substrates containing disubstituted alkynes yield ketones; 2) the substrates, 2° or 3° propargylic alcohols, may not have a proton at their α-position so that the initial propargylic cation can isomerize to an allenyl cation, which provides the product carbonyl compound; 3) the rearrangement can be catalyzed by both protic and Lewis acids under anhydrous or aqueous conditions. The related acid-catalyzed rearrangement of tertiary propargylic alcohols, *via* a formal *[1,2]*-shift of the hydroxyl group, yielding the corresponding α,β-unsaturated ketones is called the Rupe rearrangement. The most important features of this reaction are: 1) the product is always the α,β-unsaturated ketone regardless of the substitution of the triple bond; 2) the substrates are tertiary propargylic alcohols that have hydrogen atoms available at their α-position; 3) most often strong protic acids mixed with alcohol solvents are used to bring about the rearrangement, but certain Lewis acid such as mercury(II)-salts and even dehydrating agents ($SOCl_2$, P_2O_5, etc.) were shown to be effective; 4) the nature of the acid catalyst does not affect the course of the rearrangement. The disadvantages of the above two rearrangements are: 1) certain substrates may give rise to a mixture of *Rupe and Meyer-Schuster rearrangement* products; 2) low yields are observed when the product (especially aldehydes) undergoes self-condensation, or is readily oxidized under the reaction conditions; 3) acid-sensitive functionalities in the substrate may give undesired elimination products; and 4) the initial propargylic cation occasionally undergoes *Wagner-Meerwein or Nametkin rearrangement*.

Meyer and Schuster (1922): Rupe (1924-1928):

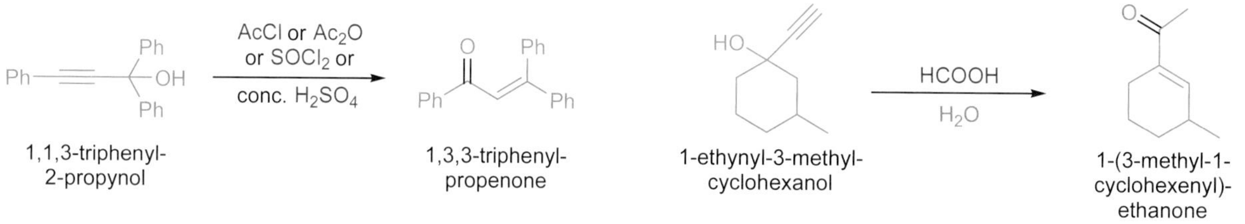

Meyer-Schuster rearrangement: Rupe rearrangement:

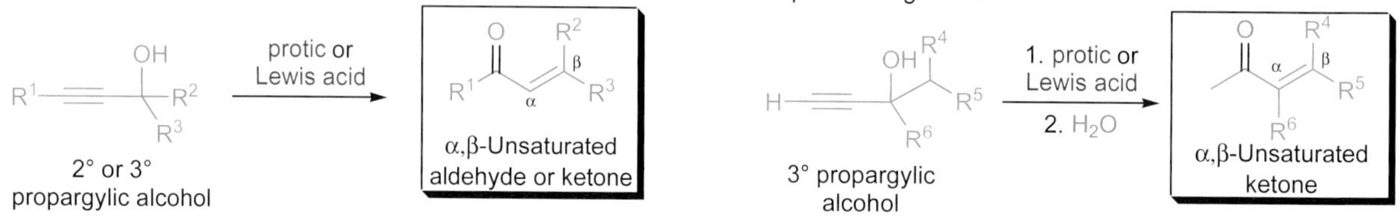

R^1 = H, alkyl, aryl; R^{2-3} = H, aryl or alkyl with no H atoms adjacent to the α-carbon; R^{4-6} = alkyl, aryl; protic acid: H_2SO_4, AcOH, HCO_2H, Dowex-50/HCO_2H, HCl/2-propanol, HCl/Et_2O, *p*-TsOH. etc.; Lewis acid: $HgSO_4$/EtOH, $POCl_3$/pyridine, $HgSO_4$/H_2SO_4

Mechanism: [22,8,23]

Meyer-Schuster rearrangement:

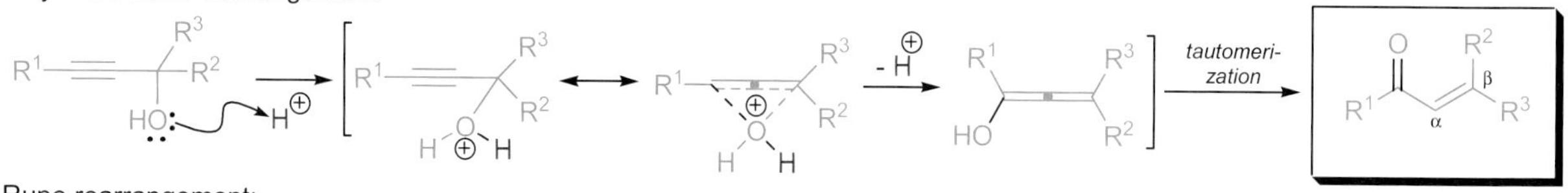

Rupe rearrangement:

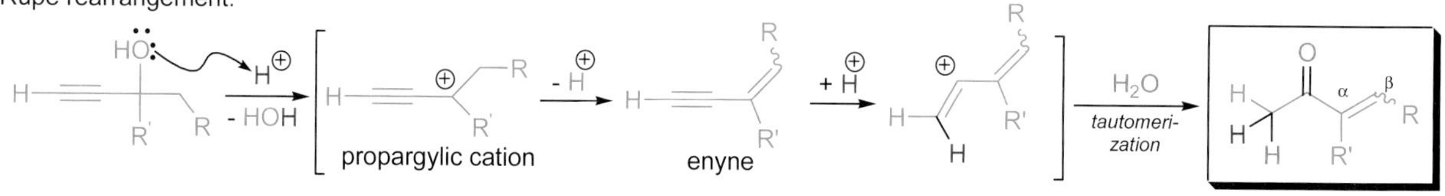

Synthetic Applications:

The first fully stereoselective total synthesis of the linear triquinane sesquiterpene (±)-capnellene was achieved by L.A. Paquette et al.[24] The C-ring is a fused cyclopentenone moiety, and the authors tried to assemble it using the *Nazarov cyclization*. However, the dienone precursor failed to undergo the cyclization under a variety of conditions, so an alternative strategy was sought that was based on the *Rupe rearrangement*. The treatment of the bicyclic tertiary propargylic alcohol substrate with formic acid and trace amounts of sulfuric acid afforded high yield of the α,β-unsaturated methyl ketone product. Interestingly, the double bond of the enone did not end up in the most substituted position as it is expected in most cases.

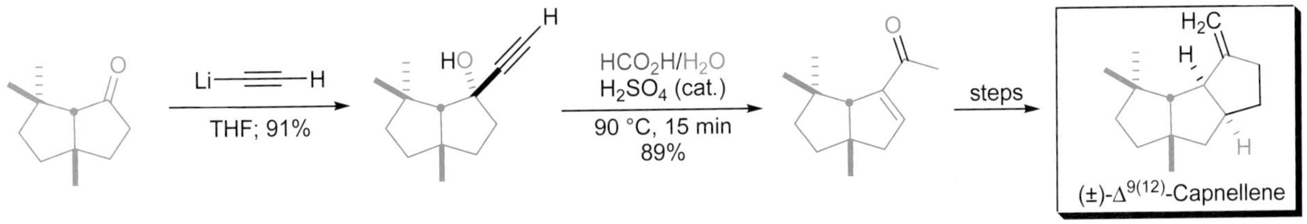

H. Stark and co-workers prepared novel histamine H_3-receptor antagonists with carbonyl-substituted 4-[(3-phenoxy)propyl]-1*H*-imidazole structures.[25] The *Meyer-Schuster rearrangement* was used for the synthesis of one of the compounds. The *p*-hydroxybenzaldehyde derivative was reacted with ethynylmagnesium bromide to afford a secondary propargylic alcohol. Upon hydrolysis with 2N HCl in a refluxing ethanol/acetone mixture, the corresponding *p*-hydroxy cinnamaldehyde was obtained.

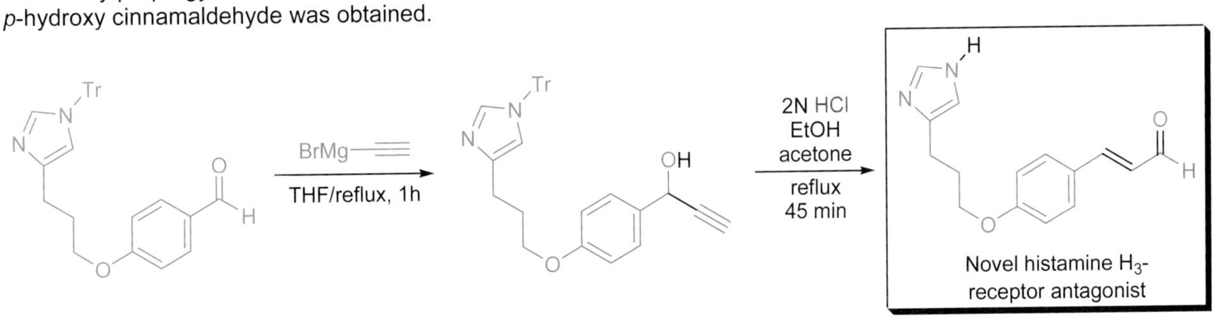

One of the disadvantages of the *Rupe rearrangement* is the harsh reaction conditions needed, making it very difficult to adapt the reaction to large-scale synthesis of unsaturated ketones. The research team of H. Weinmann investigated the rearrangement of a steroidal tertiary propargylic alcohol using a variety of acid catalysts.[15] They found that the macroporous Amberlyst-type resin A-252C in refluxing ethyl acetate containing 2 equivalents of water were ideal for the rearrangement in a pilot plant on a 64 kg scale.

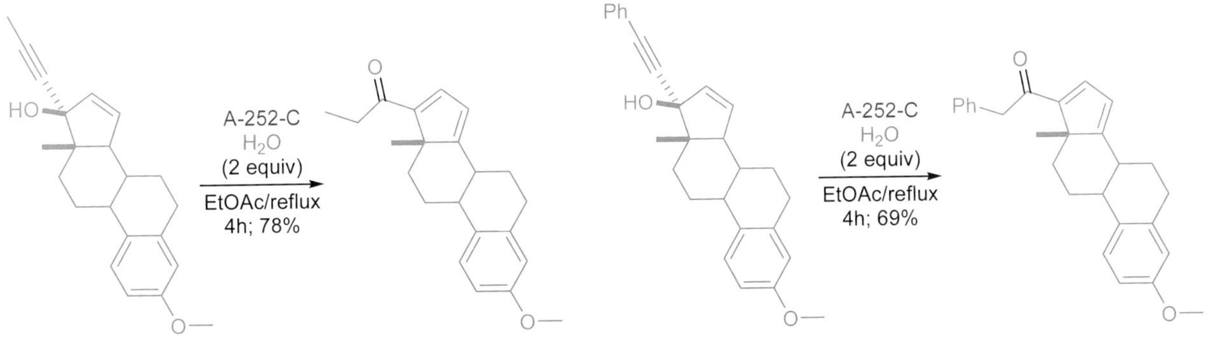

In the laboratory of S.C. Welch, the *Meyer-Schuster rearrangement* was the key step in the stereoselective total synthesis of the antifungal mold metabolite (±)-LL-Z1271α.[26] A tricyclic enone acetal was treated with lithium ethoxyacetylide, and the crude product was exposed to H_2SO_4 in anhydrous methanol, which brought about the rearrangement and afforded the desired product in 30% yield along with 12% of an epimer.

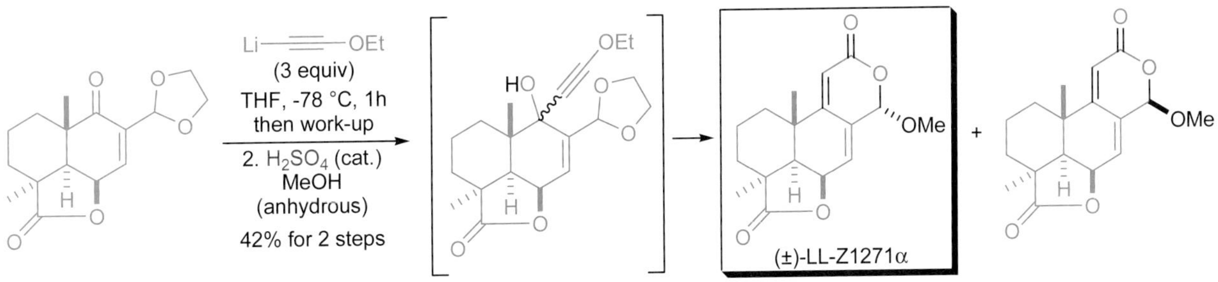

MICHAEL ADDITION/REACTION

(References are on page 628)

Importance:

[*Seminal Publications*[1-4]; *Reviews*[5-26]; *Modifications & Improvements*[27-46]; *Theoretical Studies*[47-66]]

The first example of a carbon nucleophile adding to an electron-deficient double bond was published in 1883 by T. Komnenos, who observed the facile addition of the anion of diethyl malonate to ethylidene malonate.[1] However, it was not until 1887 that A. Michael systematically investigated the reaction of stabilized anions with α,β-unsaturated systems; during this study he found that diethyl malonate added across the double bond of ethyl cinnamate in the presence of sodium ethoxide to afford a substituted pentanedioic acid diester.[2] A few years later, in 1894, he demonstrated that not only electron-deficient double bonds but also triple bonds can serve as reaction partners for carbon nucleophiles.[4] This method of forming new carbon-carbon bonds became exceedingly popular by the early 1900s and today the addition of stabilized carbon nucleophiles to activated π-systems is known as the *Michael addition* (or *Michael reaction*) and the products are called Michael adducts. Currently, however, all reactions that involve the 1,4-addition (conjugate addition) of virtually any nucleophile to activated π-systems are also referred to as the *Michael addition.* The general features of this reaction are: 1) the nucleophile (Michael donor) can be derived by the deprotonation of CH-activated compounds such as aldehydes, ketones, nitriles, β-dicarbonyl compounds, etc. as well as by the deprotonation of heteroatoms; 2) depending on the type and strength of the electron-withdrawing group (negative charge stabilizing group), the use of even relatively weak bases is possible (e.g., NEt_3); 3) it is possible to carry out the reaction using only catalytic amount of base, so when a full equivalent base is used, the product is an anion that can be reacted further with various electrophiles; 4) the structure of the activated alkene or alkyne (Michael acceptor) can be varied greatly; virtually any electron-withdrawing group could be used; 5) the reaction may be conducted in both protic and aprotic solvents; 6) both inter- and intramolecular versions exist; 7) the reaction can be highly diastereoselective when both the Michael donor and acceptor have defined stereochemistry; and 8) asymmetric versions have been developed.[28,30,31,41,25] The main drawback of the *Michael addition* is that other processes may compete with the desired 1,4-addition such as 1,2-addition and self-condensation of the carbon nucleophile, but the careful choice of reaction medium and the use of additives can suppress these undesired reactions.

Michael (1887):

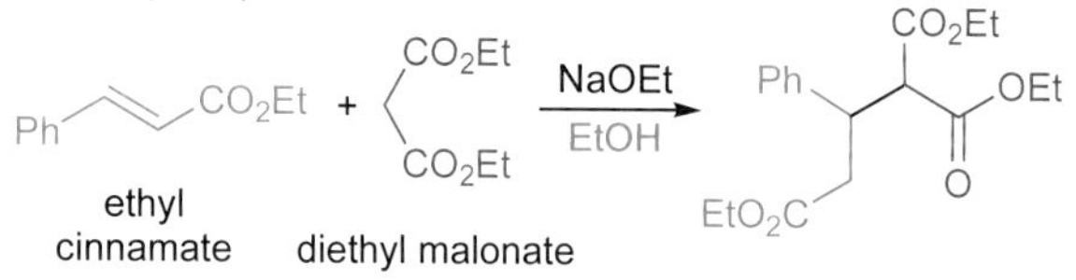

Michael (1894):

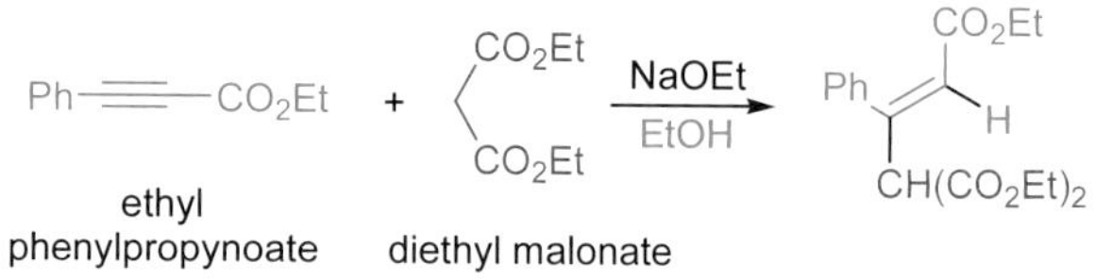

Michael addition of carbon nucleophiles:

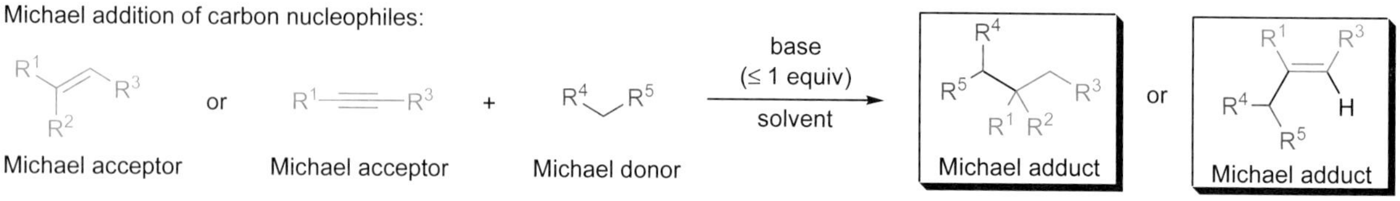

Michael addition of heteroatom nucleophiles:

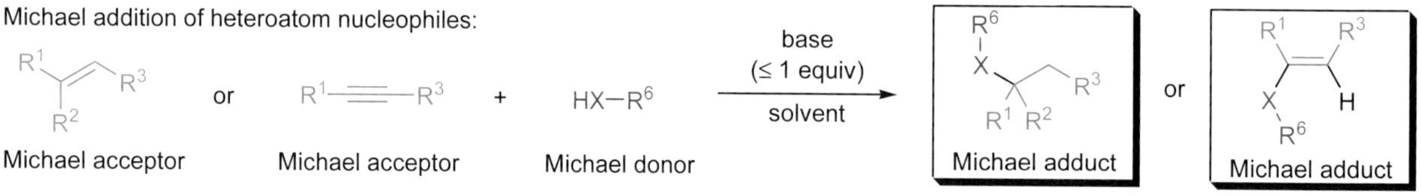

R^{1-2} = H, alkyl, aryl; R^3 = C(=O)-alkyl, C(=O)-aryl, CO_2-alkyl, CO_2-aryl, C(=O)NR_2, CN, CHO, NO_2, S(=O)R, $[PR_3]^+$, $PO(OR)_2$, heteroaryl (e.g. pyridine); R^4 = H, alkyl, aryl, C(=O)-alkyl, C(=O)-aryl, CO_2-alkyl, CO_2aryl, C(=O)NR_2, CN, CHO, NO_2; R^5 = C(=O)-alkyl, C(=O)-aryl, CO_2-alkyl, CO_2-aryl, CN, CHO; R^6 = H, alkyl, aryl; X = O, S, NH, NR, etc.; base: piperidine, NEt_3, NaOH, KOH, NaOEt, KO*t*-Bu; solvent: EtOH, *t*-BuOH, etc. or aprotic solvents such as THF, acetonitrile, benzene, etc.

Mechanism: [9,11,67,17]

The mechanism is illustrated with the addition of a malonate anion across the double bond of ethyl cinnamate. The reaction is reversible in protic solvents and the thermodynamically most stable product usually predominates. When organometallic reagents are used as Michael donors (e.g., copper-catalyzed organomagnesium additions) SET-type mechanisms may be operational.

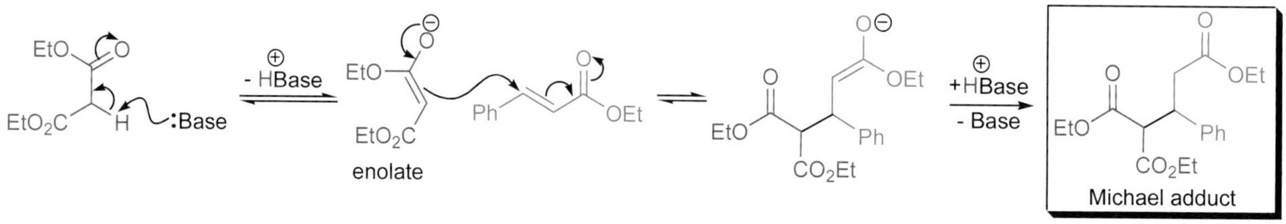

Synthetic Applications:

A unique class of steroidal alkaloids, the batrachotoxinins, is isolated in small quantities from the skins of poison arrow frogs and also from the feather of a New Guinea bird. One of the key steps during the total synthesis of (±)-batrachotoxinin A by Y. Kishi et al. was a *Michael addition* to form a seven-membered oxazapane ring.[68] The removal of the primary TBS protecting group was achieved by treatment with TASF and the resulting alkoxide attacked the enone at the β-position to afford an enolate as the Michael adduct. The enolate was trapped with phenyl triflimide as the enol triflate.

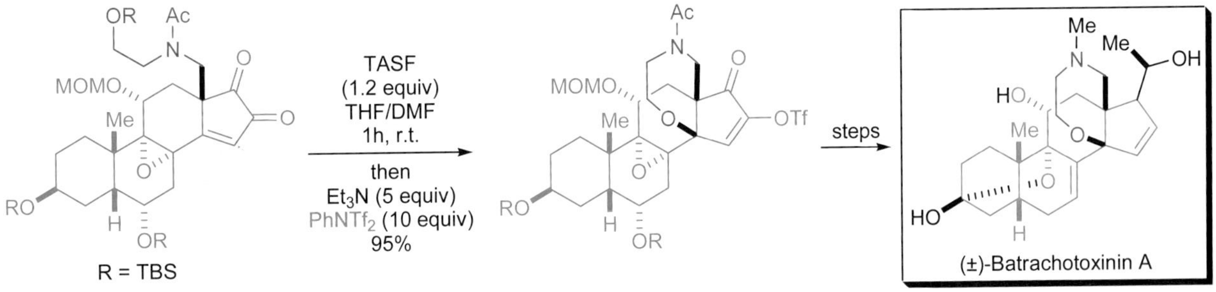

The synthesis of both enantiomers of the antitumor-antibiotic fredericamycin A was achieved in the laboratory of D.L. Boger.[69] The DE ring system of the natural product was assembled *via* a tandem *Michael addition-Dieckmann condensation*. The highly substituted 4-methylpyridine precursor was treated with excess LDA followed by the addition of the Michael acceptor cyclopentenone. The Michael adduct underwent an intramolecular acylation with the ester functionality *in situ* to afford the desired DEF tricycle.

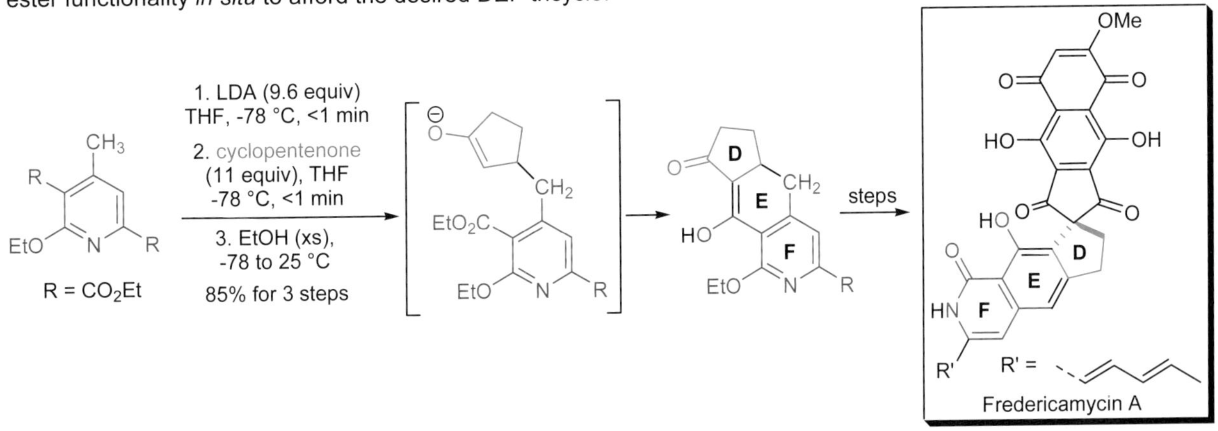

M. Ihara and co-workers utilized an *intramolecular double Michael addition* for the efficient and completely stereoselective construction of the tricyclo[6.3.0.0^{3,9}]undecan-10-one framework during the total synthesis of (±)-longiborneol.[70] The substituted cyclopentenone precursor was exposed to several different reaction conditions, and the highest yield was obtained when LHMDS was used as the base. The first deprotonation took place at C11; the resulting enolate added to C9, and the ester enolate (negative charge located at C10) in turn added to the cyclopentenone at C3.

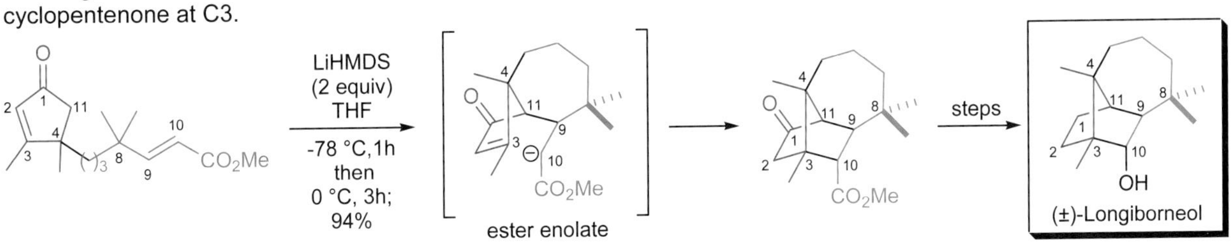

The potent neurotoxin (–)-dysiherbaine was synthesized by S. Hatekayama et al. who assembled the central pyran ring *via* an *intramolecular Michael addition* of a primary alcohol to an α,β-unsaturated ester.[71] The sole product of this key cyclization was a tricyclic lactone, which was isolated in good yield.

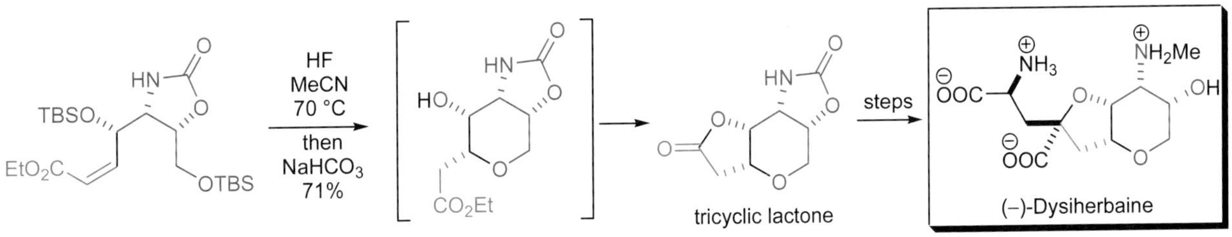

MIDLAND ALPINE-BORANE® REDUCTION (MIDLAND REDUCTION)

(References are on page 630)

Importance:

[*Seminal Publications*[1-7]; *Reviews*[8-14]; *Modifications & Improvements*[15-20]; *Theoretical Studies*[21]]

In the late 1970s, M.M. Midland and co-workers reported a surprising observation that certain B-alkyl-9-borabicyclo[3.3.1]nonanes reduced benzaldehyde to benzyl alcohol in THF solution at reflux.[3] The rate of the reaction was strongly dependent on the structure of the B-alkyl group, and it was found that increasing substitution at the β-position significantly increased the rate of reduction. Soon after this initial communication, the asymmetric version was developed by the same authors using B-3α-pinanyl-9-BBN as the reducing agent, which was easily available by reacting (+)-α-pinene with 9-BBN.[2] The asymmetric induction was comparable to that of an enzyme catalyzed reduction. This new reducing agent was later commercialized by Aldrich Co. under the name Alpine-Borane®. The asymmetric reduction of carbonyl compounds (mostly ketones) using either enantiomer of Alpine-Borane® is known as the *Midland reduction* (or *Midland Alpine-Borane reduction*). The general features of this transformation are: 1) since both enantiomers of α-pinene are available, the corresponding chiral reducing agents are readily available by reaction with 9-BBN; 2) suitable substrates are prochiral ketones and aldehydes (e.g., deutero aldehydes); 3) by using one enantiomer of Alpine-Borane® the carbonyl compounds are reduced consistently to give the same absolute configuration of the corresponding alcohol; 4) alcohols of the opposite absolute configuration may be obtained by using the other enantiomer of Alpine-Borane®; 5) the reduction takes place under mild conditions at room temperature or slightly above using 40-100% excess of the reducing agent; 6) the rate of reduction is the greatest for aldehydes, whereas ketones are reduced at significantly slower rates depending on the steric bulk of the substituents; 7) when the reaction is conducted under high-pressure conditions, the rate is increased as well as the level of asymmetric induction; 8) the level of asymmetric induction is usually very high (>90% *ee*), and existing stereocenters in the substrates usually do not influence the outcome of the reduction; 9) Alpine-Borane® exhibits a remarkable degree of chemoselectivity for aldehydes and ketones. Other functional groups remain unchanged unless forcing condition induce a dehydroboration process to form 9-BBN and α-pinene.

Midland (1977):

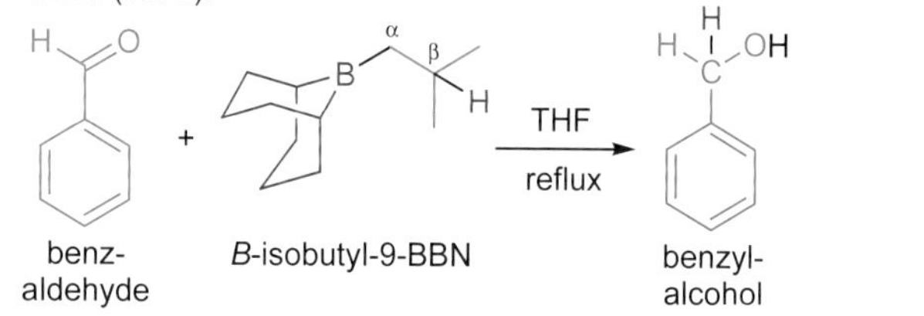

Midland (1977):

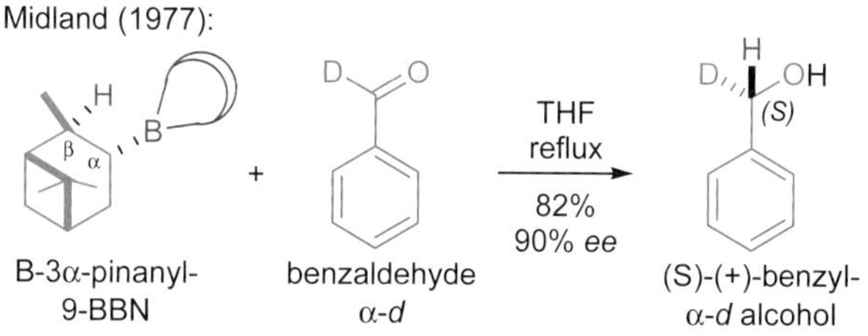

Preparation of B-3α-pinanyl-9-BBN (Alpine-Borane®):

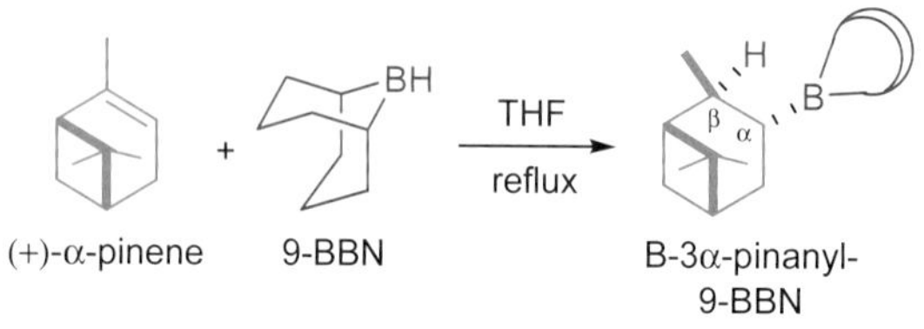

Midland reduction:

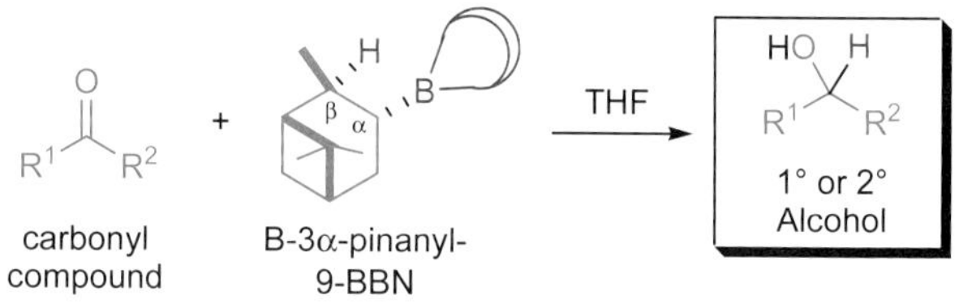

R^1 = alkyl, aryl, alkenyl, alkynyl; R^2 = H, D, alkyl, aryl, CO_2-alkyl

Mechanism: [22,23,9,24,25]

Kinetic studies of the *Midland reduction* confirmed that the reduction of aldehydes is a bimolecular process and the changes in ketone structure have a marked influence on the rate of the reaction (e.g., the presence of an EWG in the *para* position of aryl ketones increases the rate compared to an EDG in the same position).[23] However, when the carbonyl compound is sterically hindered, the rate becomes independent of the ketone concentration and the structure of the substrate. The mechanism with sterically unhindered substrates involves a cyclic boatlike transition structure (similar to what occurs in the *Meerwein-Ponndorf-Verley reduction*). The favored transition structure has the larger substituent (R_L) in the equatorial position, and this model correctly predicts the absolute stereochemistry of the product.

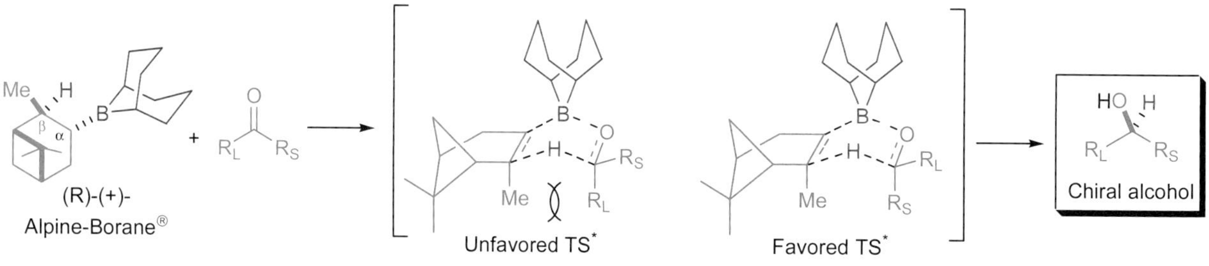

Synthetic Applications:

The first total synthesis of the neuritogenic spongean polyacetylene lembehyne A was accomplished by M. Kobayashi and co-workers.[26] The single stereocenter of the molecule was introduced *via* the *Midland reduction* of a propargylic ketone using an Alpine-Borane®, which was prepared from (+)-α-pinene and 9-BBN.

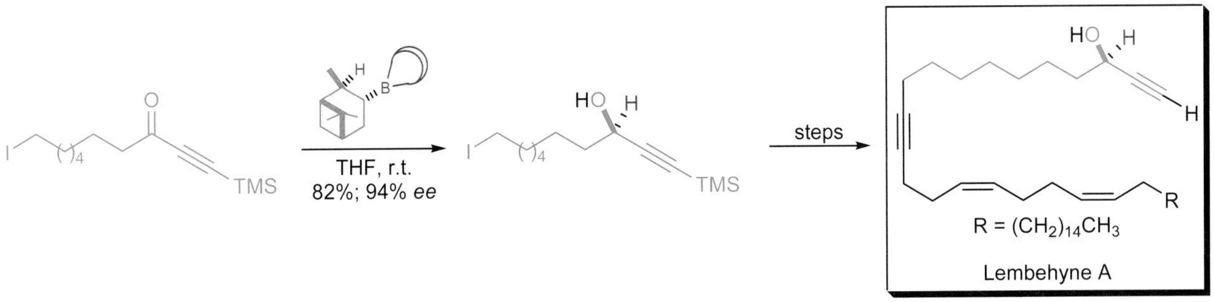

Chirally deuterated sugars are useful in elucidating mechanisms of biosynthesis and chemical reactions. In the laboratory of N.P.J. Price, the stereoselective synthesis of chirally deuterated (*S*)-D-(6-$^2H_1$)glucose was achieved utilizing (*R*)-(+)-Alpine-Borane to reduce a deutero aldehyde precursor stereoselectively.[27] The substrate was dissolved in dichloromethane, and at room temperature the solution of the reducing agent was added in THF in excess. When all the starting material was consumed, the excess reagent was destroyed with acetaldehyde and the reaction mixture was worked-up oxidatively using $NaOH/H_2O_2$.

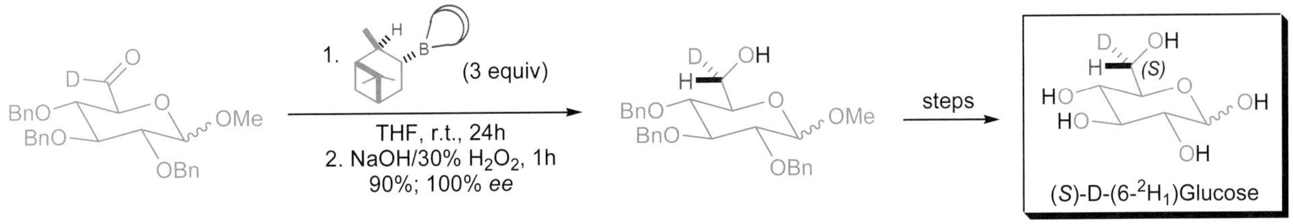

The cyclic peroxide natural product (+)-chondrillin was prepared by P.H. Dussault and co-workers using a *singlet oxygenation/radical rearrangement* sequence as the key step.[28] The first stereocenter was introduced *via* the *Midland reduction* of an ynone substrate.

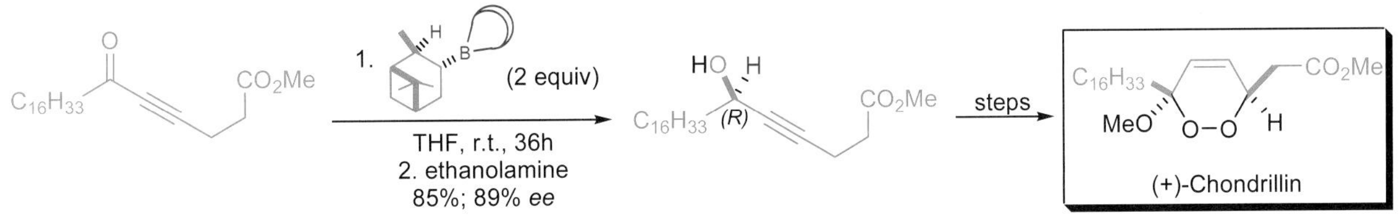

Stable, isotope-labeled amino acids are often utilized in the elucidation of protein structures and in probing the mechanism of enzyme catalyzed processes as well as revealing the metabolic pathways of amino acids. When deuterium is introduced, the protein in which the labeled amino acids are incorporated can be studied by NMR techniques. For instance, the absence of signal in the ^{1}H-NMR spectrum simplifies the assignment of peaks. An improved synthesis of the doubly labeled (*R*)-glycine-d-^{15}N was developed by R.W. Curley Jr. et al.[29] The current synthetic sequence introduced chirality by reducing a deutero aldehyde with (*R*)-(+)-Alpine-Borane. The resulting benzyl alcohol was subjected to a *Mitsunobu reaction* using ^{15}N-phthalimide, which inverted the stereochemistry and introduced the labeled nitrogen atom (overall a *Gabriel synthesis*).

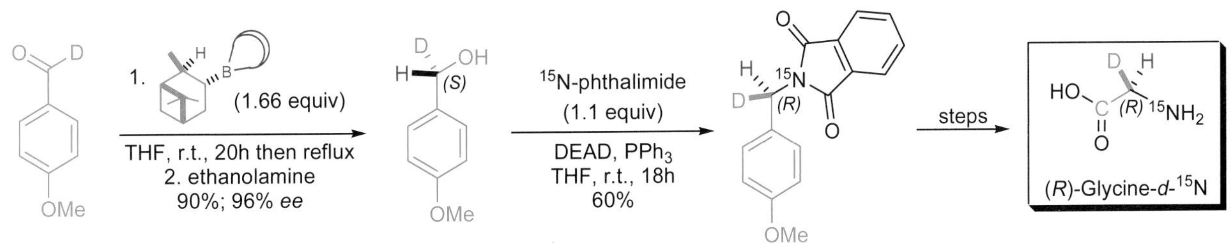

MINISCI REACTION

(References are on page 630)

Importance:

[*Seminal Publications*[1-11]; *Reviews*[12-16]]

The substitution of protonated heteroaromatic bases by nucleophilic carbon-centered radicals is known as the *Minisci reaction*, named after its discoverer F. Minisci. In the late 1960s, radical processes were generally not considered selective, and their synthetic use was limited to simple molecules. In 1968, Minisci demonstrated that selective substitutions could be realized by reacting nucleophilic carbon-centered radicals with electron-deficient substrates (olefin conjugated with EWG, protonated heteroaromatic bases, quinines, etc.).[1] This transformation was especially important because it resembled the *Friedel-Crafts aromatic substitution*, but with opposite reactivity and selectivity. The *Minisci reaction* introduces acyl groups directly into heteroaromatic rings, a reaction that would be impossible under the ususal Friedel-Crafts reaction conditions. Pyridines,[5,17,18] pyrazines,[19] quinolines,[4,5] diazines,[20] imidazoles,[21] benzothiazoles[22] and purines were shown to selectively react with a wide range of nucleophilic radicals, at the positions α- and γ to the nitrogen. All heteroaromatic bases in which at least one α- or γ position is free undergo this reaction. The reactivity and the selectivity generally increase with the number of heteroatoms in the aromatic rings or polycyclic heterocycles. The observed high selectivity is due to polar effects, and is strictly related to the nucleophilic character of carbon-centered radicals. The radicals may be generated from a wide range of compounds (alkanes, alkenes, alkylbenzenes, alcohols, ethers, aldehydes, ketones, carboxylic acids, esters, amides, amines, alkyl halides, peroxides, *N*-chloroamines, oxaziridines, etc.), making the reaction synthetically useful.[1,23-26,15,27,16,28,29] Most of the Minisci substitution reactions occur in aqueous or mixed aqueous media (e.g., methanol-water) under acidic conditions at room temperature. The reactions are immediate, and isolation of the organic products is convenient.

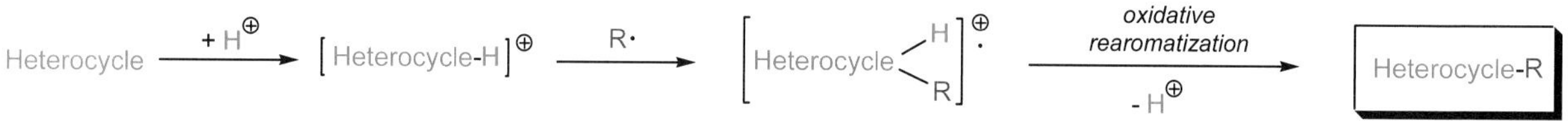

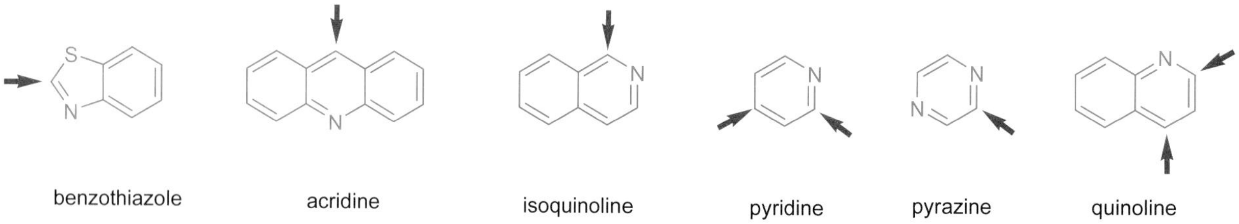

Mechanism: [1,5,6,23,30-33,24,34-36]

In the first step, the carbon centered radical is generated. The second step involves the addition of this radical to the protonated ring. The third step consists of the rearomatization of the radical adduct by oxidation. The rates of addition of alkyl and acyl radicals to protonated heteroaromatic bases are much higher than those of possible competitive reactions, particularly those with solvents. Polar effects influence the rates of the radical additions to the heteroaromatic ring by decreasing the activation energy as the electron deficiency of the heterocyclic ring increases.

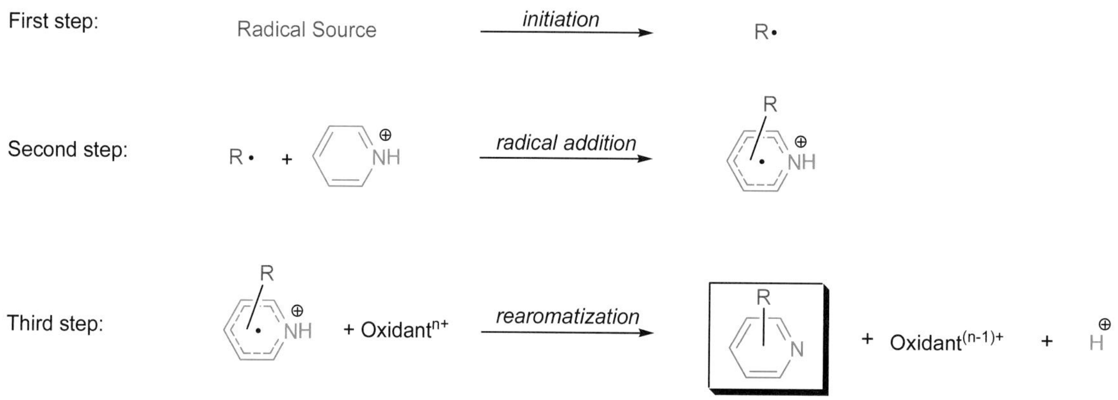

Synthetic Applications:

F. Minisci and co-workers generated alkyl radicals from alkyl iodides under simple conditions (thermal decomposition of dibenzoyl peroxide) and used it for selective C-C bond formation on protonated heterocycles.[37] The method was successfully applied to complex substrates, such as 6-iodo-1,2,3,4-diisopropylidene-α-galactose, which was reacted with protonated 2-methylquinoline to give the corresponding C-nucleoside in excellent yield.

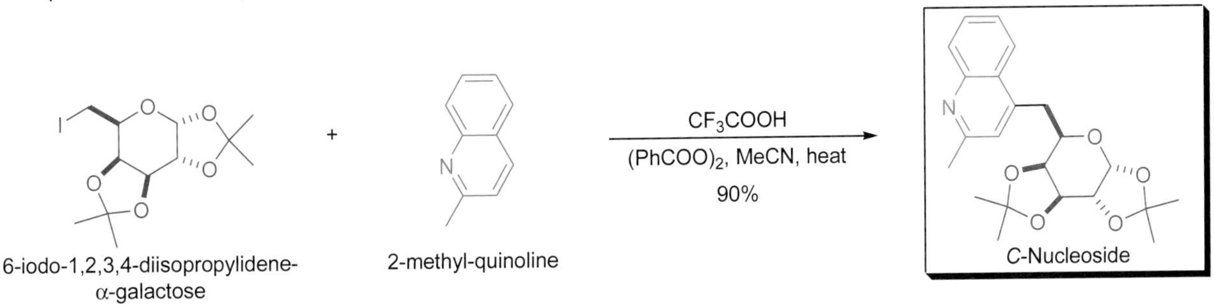

In the course of synthetic and pharmacological investigations, some non-natural azaergoline analogs were efficiently synthesized in the laboratory of M.K.H Doll.[38] Previous syntheses of these analogs were too long to be practical. Therefore, an *intramolecular tandem decarboxylation-cyclization Minisci reaction* was developed to achieve a short synthesis of the 8-azaergoline ring system. Starting from simple, commercially available precursors, the target tetracycle was obtained in four steps with an overall yield of 28%.

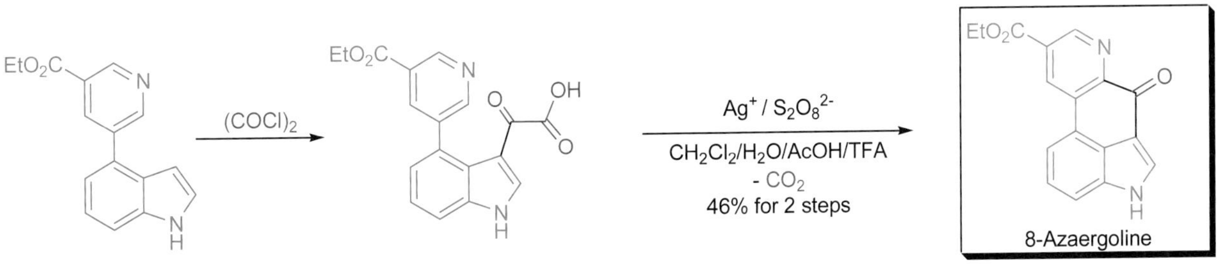

In order to evaluate fluoroheteroaromatic compounds as intracellular pH probes, R.A.J. Smith and co-workers prepared monofunctionalized polymethylated pyridines.[28] To this end, *radical Minisci-type substitution reactions* were used on substituted pyridines. Reaction of hydroxymethyl radicals with *N*-methoxy 2,4- and 2,6-dimethylpyridinium salts gave 2,4,6-substituted hydroxymethylpyridines. Similar reactions with 2,3,5,6-tetramethylpyridine and derivatives failed, but substitution at the 4-position could be achieved using a carbamoyl radical to yield 2,3,5,6-tetramethyl isonicotinamide, which suggested that steric and reactivity restrictions can be overcome by appropriate choice of the reactive radical intermediate.

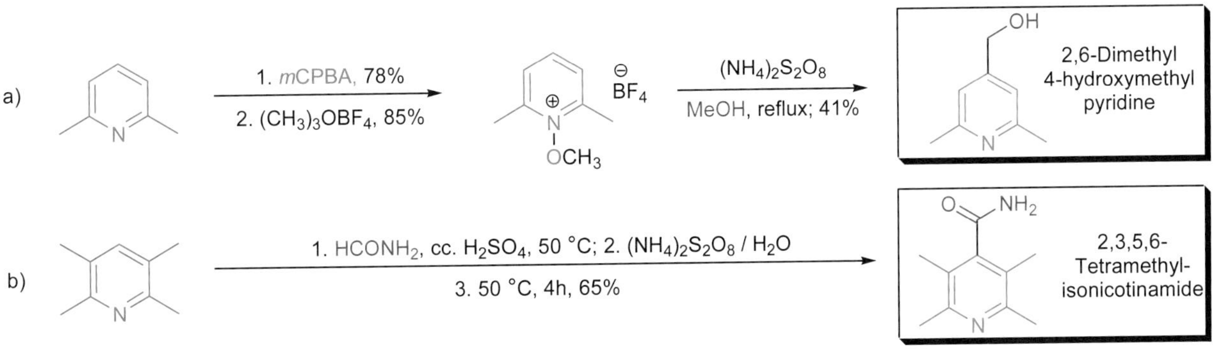

Commercially available glycine derivatives were used by C.J. Cowden to generate 1-amidoalkyl radicals for the alkylation of 3,6-dichloropyridazine in moderate to good yields.[39]

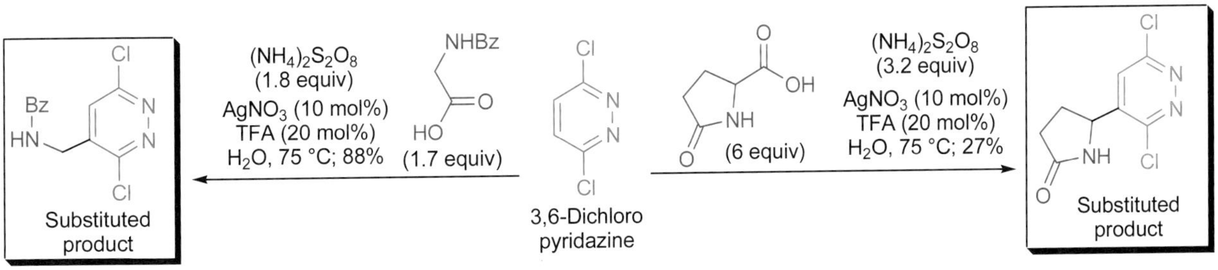

MISLOW-EVANS REARRANGEMENT

(References are on page 631)

Importance:

[*Seminal Publications*[1-3]; *Reviews*[4-8]; *Modifications & Improvements*[9-12]; *Theoretical Studies*[13,14]]

In 1968, K. Mislow and co-workers reported that upon heating, enantiomerically pure allylic sulfoxides underwent facile thermal racemization, while enantiopure allylic sulfenates afforded optically active sulfoxides.[1] Mechanistic studies revealed that these transformations were closely related, reversible, and concerted intramolecular processes that could be classified as *[2,3]*-sigmatropic rearrangements.[2] Soon after this discovery, D.A. Evans et al. recognized the synthetic potential of this rearrangement by converting allylic sulfoxides to allylic alcohols in the presence of a sulfenate ester trapping agent (thiophile) and demonstrated that it was general for a wide range of substrates.[3] The reversible 1,3-transposition of allylic sulfoxide and allylic alcohol functionalities is known as the *Mislow-Evans rearrangement*. The general features of the reaction are: 1) it is used mainly for the stereoselective synthesis of allylic alcohols from sulfoxides; 2) sulfoxides can be synthesized in variety of ways for example from the corresponding sulfides *via* oxidation or by the thermal rearrangement of sulfenate esters and can be obtained in enantiomerically pure form;[3] 3) allylic sulfoxides are regioselectively deprotonated at the α-position, and the resulting sulfoxide-stabilized allylic carbanion can be alkylated regioselectively α to the sulfur; 4) the formation of the allylic carbanion is achieved by the use of a strong base such as *n*-BuLi or LDA at low temperatures; 5) the alkylation of the allylic carbanion is conducted also at low temperatures with a variety of alkyl, allylic, and benzylic halides; 6) in the presence of a thiophile the allylic sulfoxides are cleanly transformed into the rearranged allylic alcohol products; 7) when heated in the absence of a thiophile, α,α'-disubstituted allylic sulfoxides may undergo rearrangement to afford the thermodynamically more stable isomers;[4] 8) the reaction is stereoselective, the chirality of the sulfur atom can be transferred to the carbon and *vice versa* allowing the preparation of allylic alcohols with defined double bond geometries; 9) the choice of thiophile can alter the stereochemical outcome of the rearrangement depending on the relative rates of the *sulfoxide-sulfenate ester rearrangement* and *sulfenate ester cleavage* by the thiophile;[15] 10) phosphite and amine thiophiles favor the almost exclusive formation of the (*E*) stereoisomer; 11) usually no purification of the intermediate products is required; after work-up the allylic alcohol product is isolated in good to excellent yield; and 12) propargyl sulfenates also undergo the rearrangement to give allenic sulfoxides.[6]

The thermal racemization of allylic sulfoxides (Mislow, 1968):

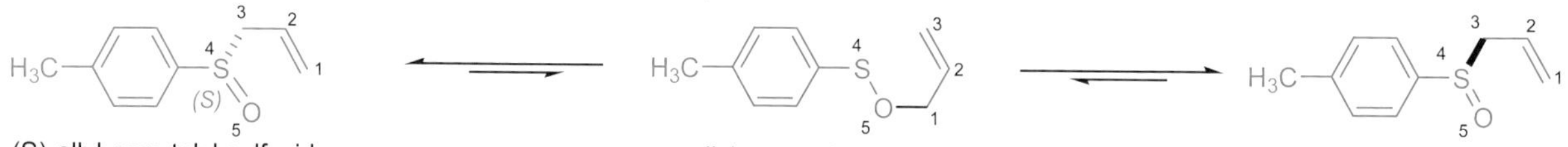

(S)-allyl *para*-tolyl sulfoxide — allyl *para*-toluenesulfenate — (*R*)-allyl *para*-tolyl sulfoxide

Conversion of allylic alcohols to allylic sulfoxides and allylic sulfoxides to allylic alcohols (Evans, 1971):

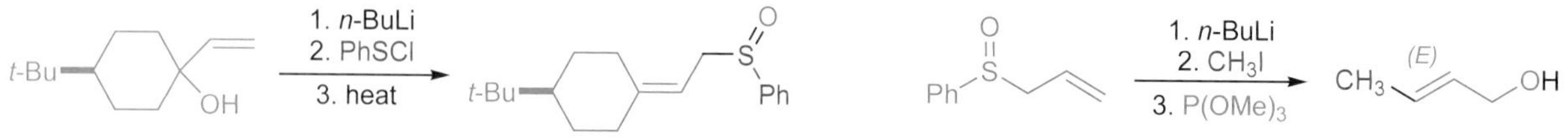

Mislow-Evans rearrangement:

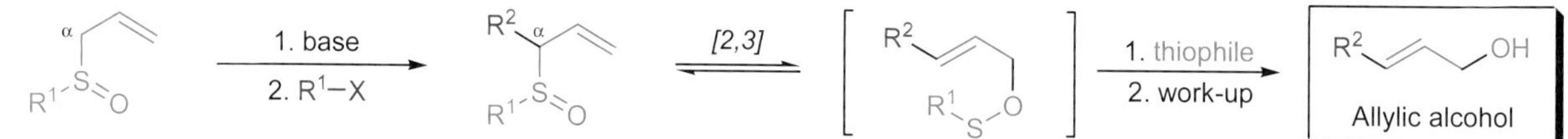

unsubstituted allylic sulfoxide — α-substituted allylic sulfoxide — sulfenate ester

R^1 = alkyl, aryl; R^2 = alkyl, allyl, propargyl, benzyl; base: alkyllithiums, LDA; thiophile: PhSNa, $P(OMe)_3$, $P(OEt)_3$, $P(NEt_2)_3$, Et_2NH

Mechanism: [1,2,15,5,16]

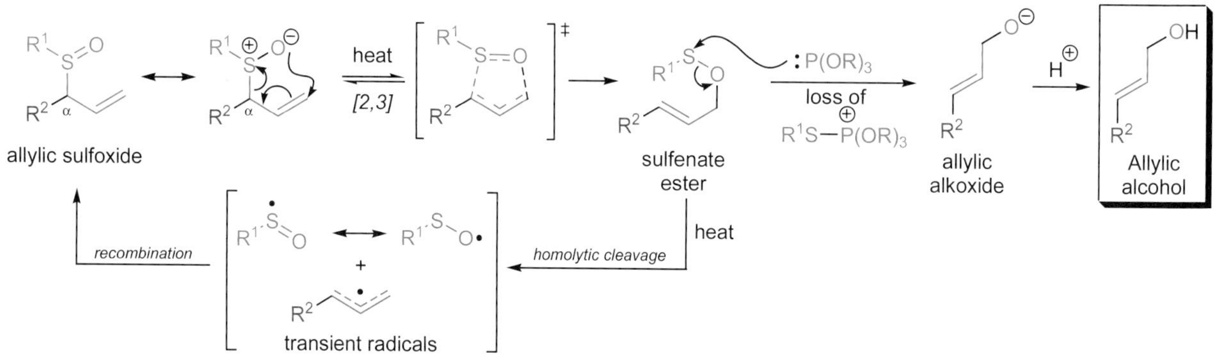

Synthetic Applications:

Prostaglandin E_2 is one of the most important members of the mammalian hormone prostaglandins that exhibit a wide range of biological activity. The quantification of the total amount of prostaglandin E_2 produced in humans is best achieved by assessing the accumulation of the major urinary metabolite PGE_2U_m. Since the supply of this material for assays has been depleted, the total synthesis of the ethyl ester of the major urinary metabolite of prostaglandin E_2 (PGE_2U_m) was undertaken by D.F. Taber et al.[17] In order to ensure the (*E*) stereochemistry of the double bond, the *Mislow-Evans rearrangement* was utilized. The phenyl sulfide substrate was first oxidized to the corresponding sulfoxide with *m*CPBA, and without purification, it was treated with trimethyl phosphite to produce the desired (*E*)-allylic alcohol in excellent yield.

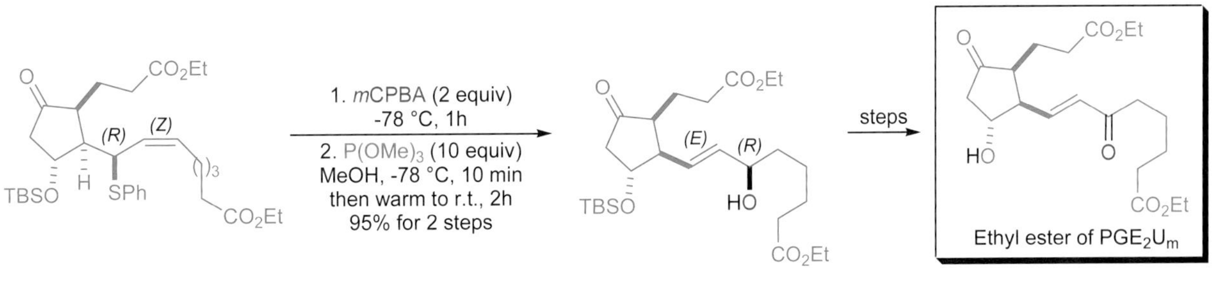

The first asymmetric total synthesis of the macrocyclic lactone metabolite (+)-pyrenolide D was accomplished in the laboratory of D.Y. Gin.[18] The natural product has a densely functionalized polycyclic structure and its absolute configuration had to be established. The key step of the synthesis was a *stereoselective oxidative ring-contraction* of a 6-deoxy-D-gulal, which was prepared from anomeric allylic sulfoxide *via* the *Mislow-Evans rearrangement*.

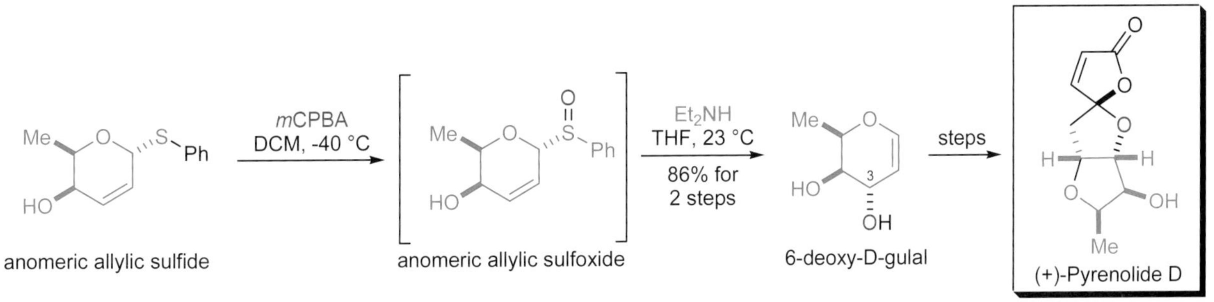

In the stereoselective total synthesis of (±)-14-deoxyisoamijiol by G. Majetich et al., the last step was the epimerization of the C2 secondary allylic alcohol functionality.[19] The *Mitsunobu reaction* resulted only in a poor yield (30%) of the inverted product, so the well-established *sulfoxide-sulfenate rearrangement* was utilized. The allyic alcohol was first treated with benzenesulfenyl chloride, which afforded the thermodynamically more stable epimeric sulfenate ester *via* an allylic sulfoxide intermediate. The addition of trimethyl phosphite shifted the equilibrium to the right by consuming the desired epimeric sulfenate ester and produced the natural product.

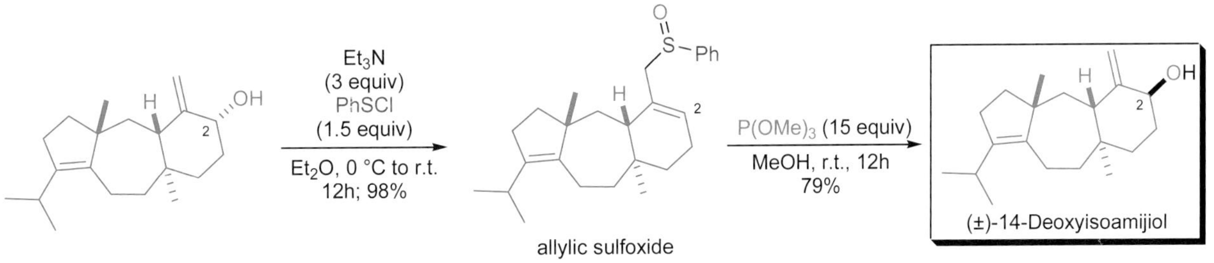

The *Mislow-Evans rearrangement* was chosen by T. Tanaka and co-workers to create the C12 stereocenter of halicholactone and ensure the (*E*) stereochemistry of the C9-C11 double bond.[20]

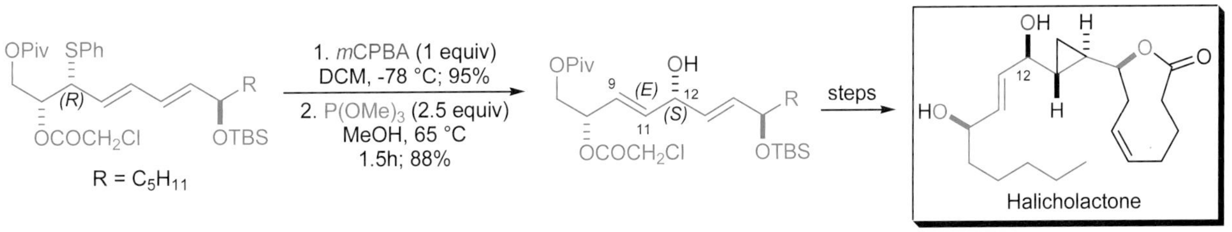

MITSUNOBU REACTION

(References are on page 632)

Importance:

[*Seminal Publications*[1,2]; *Reviews*[3-12]; *Modifications & Improvements*[13-24]]

In 1967, O. Mitsunobu et al. reported that secondary alcohols could be efficiently acylated with carboxylic acids in the presence of diethyl azodicarboxylate (DEAD) and triphenylphosphine.[1,2] A few years later it was shown that optically active secondary alcohols underwent complete inversion of configuration under the reaction conditions. Later the procedure was found to be general for the synthesis of optically active amines, azides, ethers, thioethers, and even alkanes. The substitution of primary and secondary alcohols with nucleophiles in the presence of a dialkyl azodicarboxylate and a trialkyl- or triaryl phosphine is known as the *Mitsunobu reaction*. The general features of this transformation are:[3-5] 1) primary and secondary alcohols are the best substrates and secondary alcohols undergo complete inversion of configuration; 2) tertiary alcohols do not undergo the reaction, but certain tertiary propargylic alcohols have been successfully converted; 3) the nucleophile is a relatively acidic compound ($pK_a \leq 15$); 4) among oxygen nucleophiles carboxylic acids give rise to esters, alcohols, and phenols to ethers, while thiols and thiophenols afford thioethers; 5) common nitrogen nucleophiles include imides, hydroxamates, nitrogen heterocycles, and hydrazoic acid; 6) the formation of carbon-carbon bonds is also possible, but the nucleophiles in this case are mainly active methylene compounds (β-diketones, β-keto esters, etc.); however, β-diesters are not reactive enough; 7) the reaction is also feasible intramolecularly, 3-,4-,5-,6-, and 7-membered cyclic ethers and cyclic amines can be prepared; 8) when halide ion sources (e.g., alkyl and acyl halides, zinc halides) are used along with DEAD/PPh_3, the alcohol substrates are converted to the corresponding primary and secondary alkyl halides;[4] 9) the reaction is usually conducted in THF, but dioxane and DCM are also used; 10) PPh_3 or $P(n\text{-}Bu)_3$ are the most commonly used phosphines; 11) the azodicarboxylate reagents are most often DEAD and DIAD, which can be used interchangeably; 11) the reaction temperature is usually between 0 °C and 25 °C, but certain sterically hindered substrates may require higher temperatures; and 12) in the typical procedure the mixture of the phosphine, alcohol, and the nucleophile are dissolved and the solution of the azodicarboxylate is added dropwise; alternatively, the azodicarboxylate is first reacted with the phopshine, and the solution of the alcohol and the nucleophile is added drowpwise. An important variant of the *Mitsunobu reaction* was developed by T. Mukaiyama, who described the preparation of inverted *tert*-alkyl carboxylates from chiral tertiary alcohols *via* alkoxydiphenylphosphines formed *in situ* using 2,6-dimethyl-1,4-benzoquinone.[20,23]

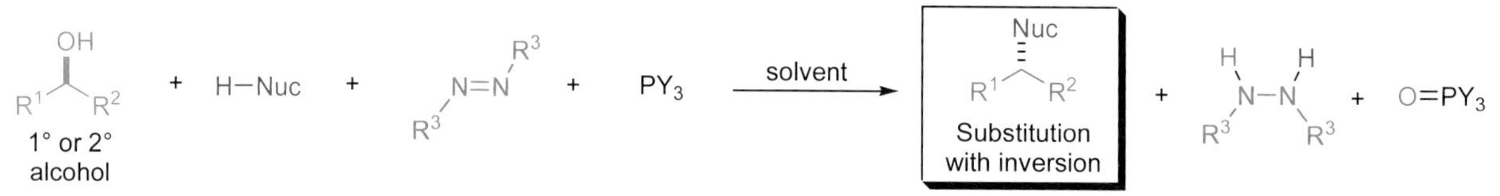

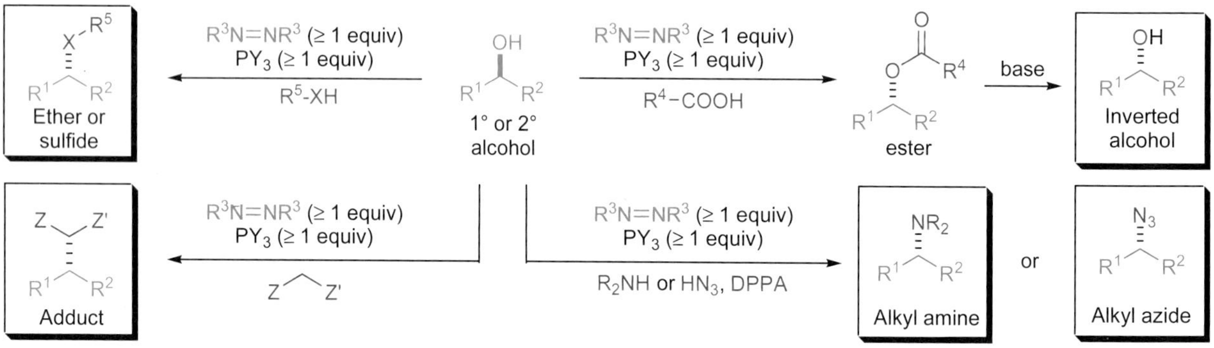

R^{1-2} = alkyl ,aryl, heteroaryl, alkenyl; H-Nuc: *O-*, *S-*, *N-* and *C*-nucleophiles; R^3 = CO_2Et (DEAD), $CO_2i\text{-}Pr$ (DIAD), $CON(CH_2)_5$ (ADDP), $CONMe_2$ (TMAD); Y = alkyl, aryl, heteroaryl, O-alkyl; solvent: THF, dioxane, DCM, $CHCl_3$, DMF, toluene, benzene, HMPA; R^4 = H, CH_3, Ph, 4-$NO_2C_6H_4$, 3,5-$(NO_2)_2C_6H_3$, alkyl, aryl; R^5 = alkyl, aryl, heteroaryl; X = O, S; Z & Z' = CO-alkyl, CO-aryl, CO_2-alkyl, CO_2-aryl, CN

Mechanism: [25-45]

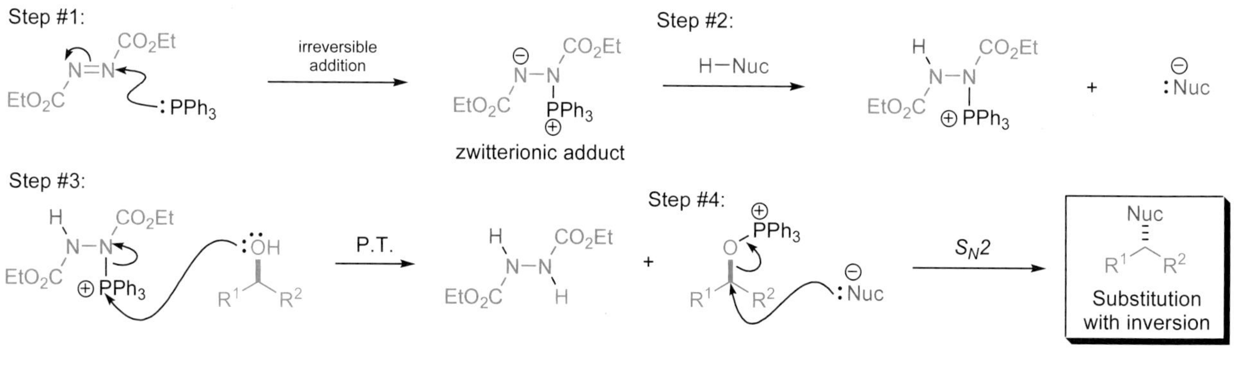

Synthetic Applications:

The architecturally novel macrolide (+)-zampanolide was synthesized in the laboratory of A.B. Smith.[46] The C8-C9 (*E*)-olefin moiety was constructed using the *Kocienski-modified Julia olefination*. The required PT-sulfone was prepared from the corresponding primary alcohol *via* a two-step protocol employing sequential *Mitsunobu reaction* and sulfide-sulfone oxidation. The primary alcohol and two equivalents of 1-phenyl-1*H*-tetrazolo-5-thiol was dissolved in anhydrous THF at 0 °C and treated sequentially with triphenylphosphine and DEAD. The desired tetrazolo sulfide was isolated in nearly quantitative yield.

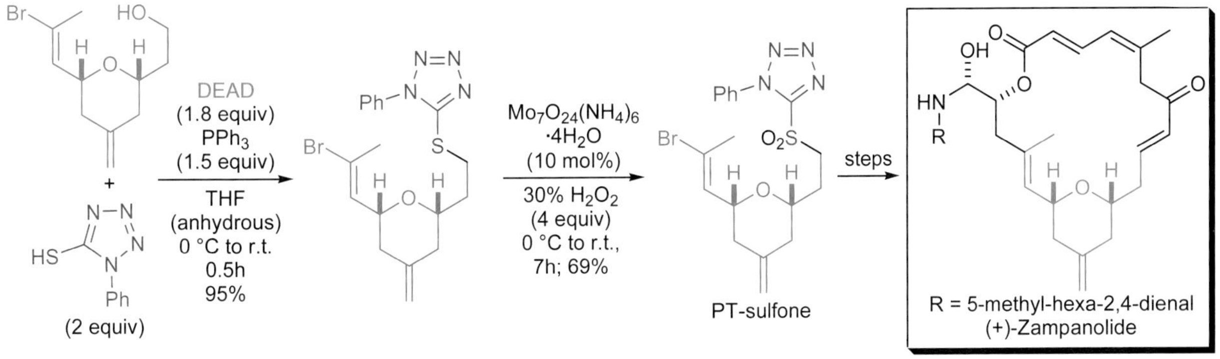

The enantioselective total synthesis of the complex bioactive indole alkaloid *ent*-WIN 64821 was accomplished by L.E. Overman and co-workers.[47] This natural product is a representative member of the family of the C_2-symmetric bispyrrolidinoindoline diketopiperazine alkaloids. The stereospecific incorporation of two C-N bonds was achieved using the *Mitsunobu reaction* to convert two secondary alcohol functionalities to the corresponding alkyl azides with inversion of configuration. The azides subsequently were reduced to the primary amines and cyclized to the desired *bis*-amidine functionality.

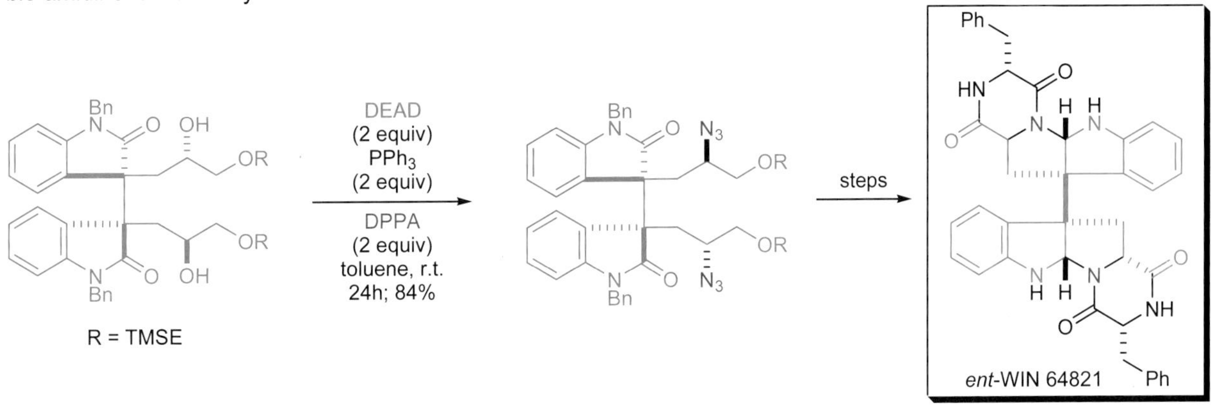

The naturally occurring potent antitumor antibiotic (+)-duocarmycin A, its epimer, and unnatural enantiomers were prepared by D.L. Boger et al.[48] The last step of the synthesis was the elaboration of the reactive cyclopropane moiety, which was carried out *via* a *transannular spirocyclization* using Mitsunobu conditions. This is a special case when the *Mitsunobu reaction* is utilized to create new carbon-carbon bonds.

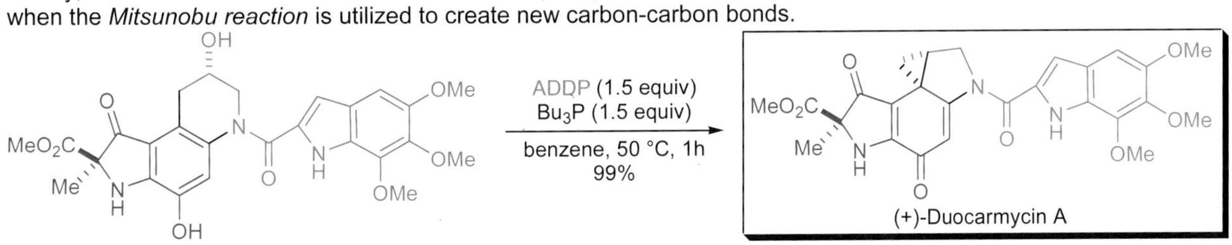

The first total synthesis of the tricyclic marine alkaloid (±)-fasicularin was completed by the research team of C. Kibayashi.[49] The secondary alcohol functionality was inverted using the Mitsunobu protocol. The resulting *p*-nitro benzoate was readily hydrolyzed under basic conditions.

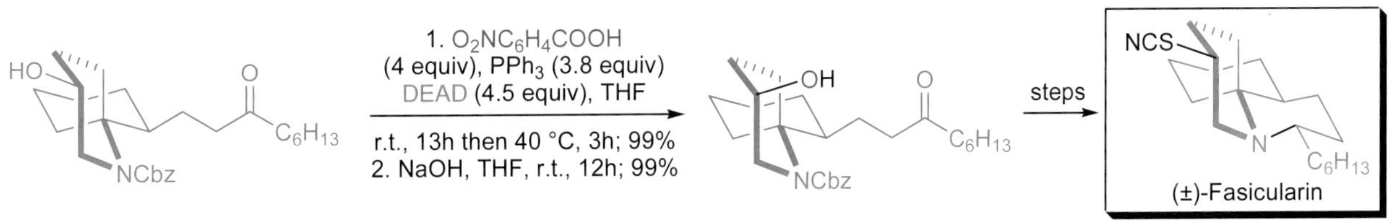

MIYAURA BORATION

(References are on page 633)

Importance:

[*Seminal Publications*[1,2]; *Reviews*[3-12]; *Modifications & Improvements*[13-31]; *Theoretical Studies*[32]]

In 1993, N. Miyaura and co-workers found that alkynes could be efficiently *cis*-diborated with the pinacol ester of diboronic acid (abbreviated as B_2pin_2 or pinB-Bpin) in the presence of catalytic amounts of platinum tetrakistriphenylphosphine.[1] Later, in 1995, the same authors discovered that tetraalkoxydiboron compounds could be coupled with aromatic halides in the presence of catalytic amounts of $PdCl_2$(dppf) to afford arylboronic esters, which are important substrates for the *Suzuki cross-coupling* and *Ullmann biaryl ether synthesis*.[2] Surprisingly, only Pd-based catalysts were effective; other metal complexes did not catalyze the reaction at all. The palladium-catalyzed cross-coupling reaction of aromatic and heteroaromatic halides or triflates with tetraalkoxyboron compounds to give arylboronic and heteroarylboronic esters is referred to as the *Miyaura boration*. The general features of this transformation are: 1) the one-pot coupling proceeds under mild conditions, which is a significant improvement over the traditional synthesis of arylboronic esters and acids (the reaction between trialkyl borates and arylmagnesium halides or aryllithiums); 2) most functional groups are tolerated under the mildly basic reaction conditions; 3) the best substrates are aryl bromides and iodides, but recently aryl triflates[15,3] and aryldiazonium tetrafluoroborates[21] have also been used; 4) the aryl group may have either electron-donating or electron-withdrawing substitutents; 5) electron-rich aryl bromides tend to react slower than electron-rich aryl iodides, and the chemoselective boration of an aryl iodide in the presence of an aryl bromide can be achieved in high yield; 6) the use of palladium(0)-tricyclohexylphosphine as the catalyst allows the coupling of the much less reactive aryl chlorides;[23] and 7) the presence of potassium acetate (KOAc) as the base in the reaction mixture is critical for the successful coupling of aryl halides, and it not only accelerates the reaction but also prevents the formation of biaryl by-products (*Suzuki cross-coupling*). A number of synthetically useful variants of this reaction have been developed.[13-31]

Miyaura (1993):

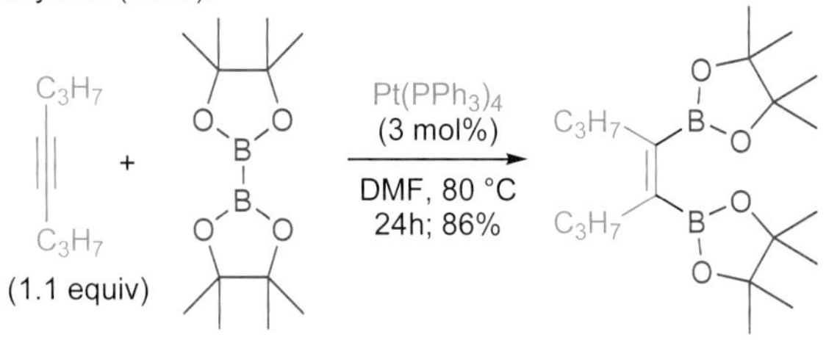

Miyaura (1995):

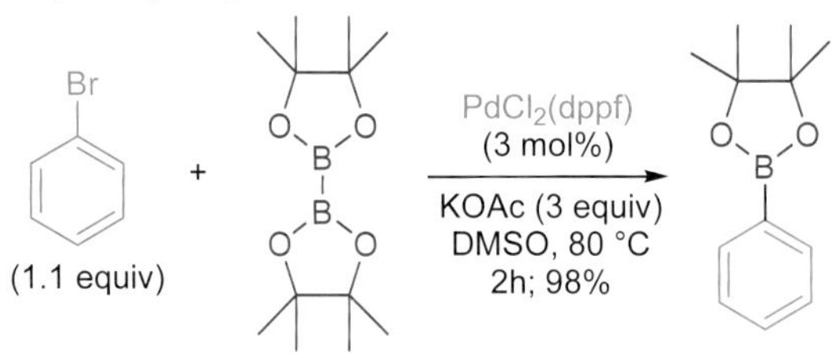

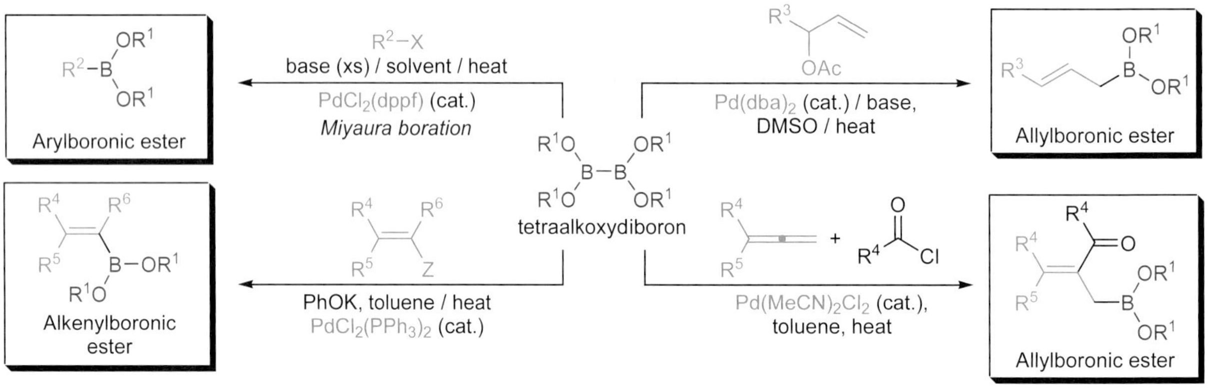

R^1 = alkyl; R^2 = aryl, heteroaryl; R^3 = H, alkyl, aryl; R^{4-6} = alkyl, aryl; X = Br, I, (or Cl) OTf, N_2BF_4; Z = I, Br, OTf; base: KOAc; solvent: DMF, DMSO, dioxane, toluene

Mechanism: [2,32]

The first step of the *Miyaura boration* is the oxidative addition of the $Pd^{(0)}$-complex into the C-X bond of the aryl halide. Next, a transmetallation takes place, the exact mechanism of which depends on the nature of the substrate, and finally the reductive elimination affords the product.

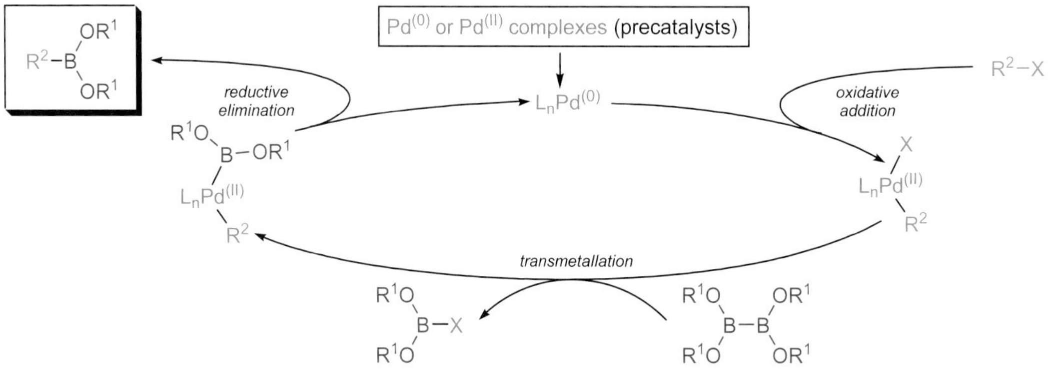

Synthetic Applications:

The total synthesis of the proteasome inhibitor cyclic peptide TMC-95A was accomplished by. S.J. Danishefsky and co-workers.[33] The biaryl linkage in the natural product was constructed by a *Suzuki cross-coupling* between an aryl iodide and an arylboronic ester derived from L-tyrosine. The required arylboronic pinacolate substrate was prepared using the *Miyaura boration*. The aryl iodide was exposed to *bis*(pinacolato)diboron in the presence of a palladium catalyst and potassium acetate in DMSO. The coupling proceeded in high yield and no symmetrical biaryl by-product was observed.

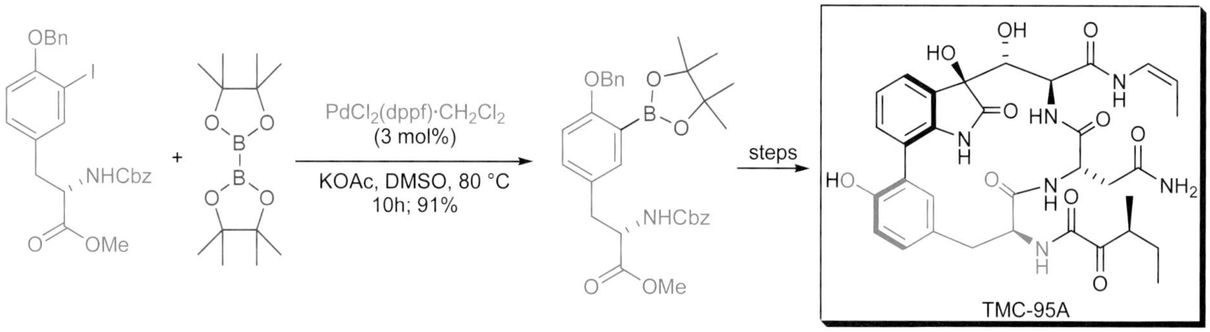

A novel macrocyclization reaction was developed based on a domino *Miyaura boration/intramolecular Suzuki cross-coupling* sequence in the laboratory of J. Zhu.[34] This strategy was applied in the synthesis of biaryl-containing macrocycles. The diiodide substrate was dissolved in degassed DMSO, and then the catalyst and the base were added. Successful macrocyclization required extensive experimentation, and the authors determined that the concentration and the nature of the base were the two most important factors. Interestingly, potassium carbonate is not suitable as a base in the *Miyaura boration*, since it tends to give biaryl by-products, but in this particular macrocyclization reaction it proved to be completely ineffective because the reaction failed to take place.

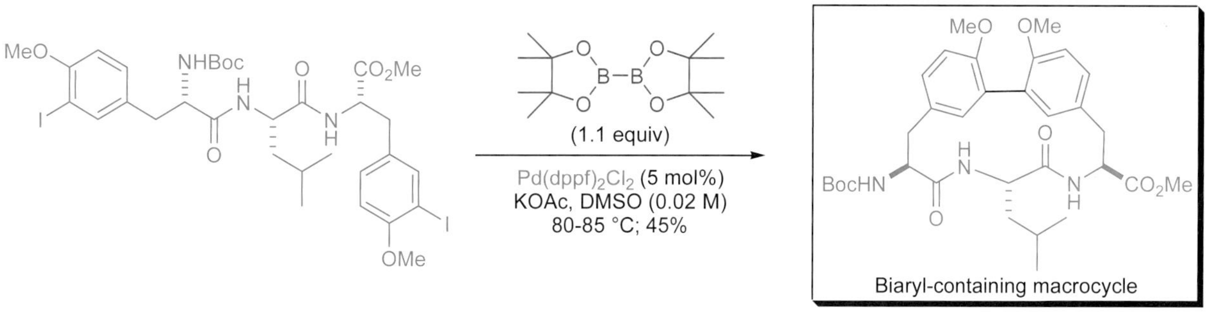

The first total synthesis of the potent antibiotic marine natural product (±)-spiroxin C was completed by T. Imanishi et al., who employed a *TBAF-activated Suzuki cross-coupling* as the key step to form the biaryl linkage.[35] The coupling partner naphthylborate ester was prepared using the *Miyaura boration*.

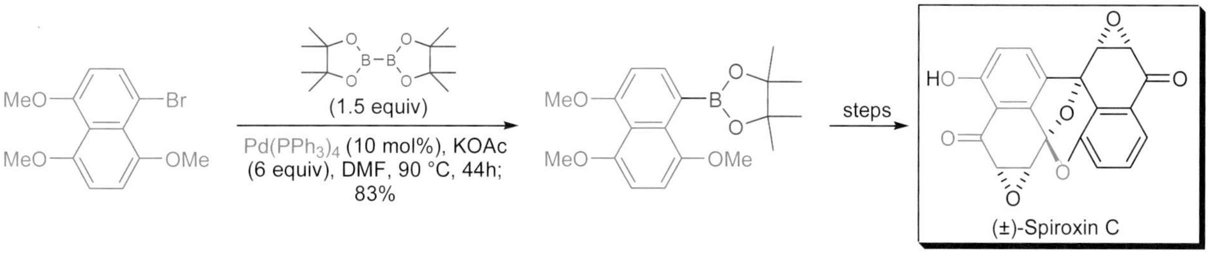

The efficient synthesis of a potent topoisomerase I poison terbenzimidazole was developed in the laboratory of P.J. Smith.[36] The desired aryl-aryl bonds were created *via* iterative *Suzuki-cross couplings*. The arylboronic ester was derived from 1-benzyl-5-iodo-1*H*-benzimidazole using the *Miyaura boration*.

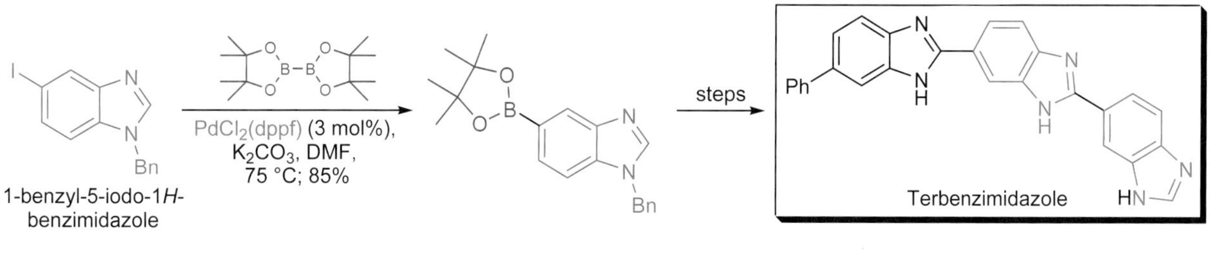

MUKAIYAMA ALDOL REACTION

(References are on page 633)

Importance:

[*Seminal Publications*[1,2]; *Reviews*[3-22]; *Theoretical Studies*[23]]

The *crossed aldol reaction* between preformed enolates and carbonyl compounds is among the most important carbon-carbon bond forming reactions. A powerful version of this transformation is the Lewis acid mediated addition of enol silanes to carbonyl compounds, a process that was discovered by T. Mukaiyama in the early 1970's[1,2] and today is referred to as the *Mukaiyama aldol reaction*. The general features of the reaction are: 1) according to the original procedure, stoichiometric quantities of the Lewis acid such as $TiCl_4$, $SnCl_4$, $AlCl_3$, $BCl_3{\cdot}OEt_2$, and $ZnCl_2$ were required to effect the transformation;[1,2] 2) lately, several catalytic versions were developed utilizing Lewis acids such as $Sn^{(IV)}$, $Sn^{(II)}$, $Mg^{(II)}$, $Zn^{(II)}$, $Li^{(I)}$, $Bi^{(III)}$, $In^{(III)}$, $Ln^{(III)}$, $Pd^{(II)}$, $Ti^{(IV)}$, $Zr^{(IV)}$, $Ru^{(II)}$, $Rh^{(II)}$, $Fe^{(II)}$, $Al^{(III)}$, $Cu^{(II)}$, $Au^{(I)}$, R_3SiX, Ar_3C^+, and clay as catalyst;[6,10,11,13,14,16] 3) several Lewis base catalyzed transformations were also developed;[24-27] 4) the enol silane component can be derived from aldehydes, ketones, esters, and thioesters;[4] 5) the reactions of unsubstituted, mono- and disubstituted enol silanes were examined;[4] and 6) the most commonly used carbonyl reactants are aldehydes, but ketones and acetals also react under appropriate reaction conditions.[4] The diastereoselectivity of the *Mukaiyama aldol reaction* can be controlled if substrates and conditions are carefully chosen. The diastereochemical outcome of monosubstituted enol silanes is usually as follows: 1) when R^2 is small and R^3 is bulky, the reaction leads to the *anti* product independent of the double bond geometry; 2) when R^2 is large, *syn* diastereoselection predominates independent of the enol silane geometry; and 3) when the aldehyde is capable of chelation, the formation of the *syn*-diastereomer is preferred.[14] Control of the absolute stereoselectivity can be achieved by utilizing chiral enol silanes or chiral aldehydes.[14] The fastest-growing area in the *Mukaiyama aldol* methodology is the development of asymmetric catalytic processes, utilizing chiral Lewis acid complexes and Lewis bases.[13]

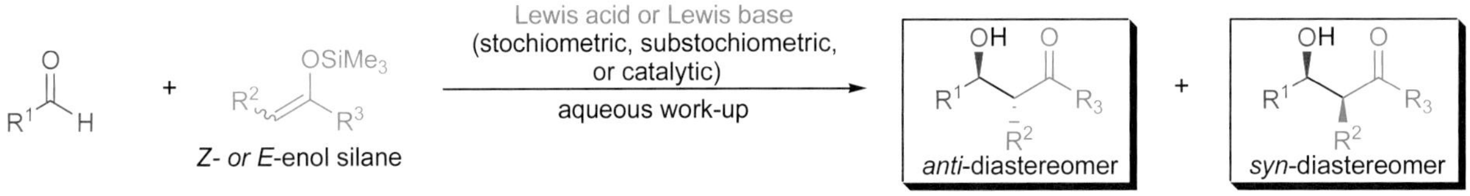

R^1 = alkyl, aryl, -OR; R^2 = H, alkyl, aryl; R^3 = alkyl, aryl, -OR, -SR, H; Lewis acid = $Sn^{(IV)}$, $Sn^{(II)}$, $Mg^{(II)}$, $Zn^{(II)}$, $Li^{(I)}$, $Bi^{(III)}$, $In^{(III)}$, $Ln^{(III)}$, $Pd^{(II)}$, $Ti^{(IV)}$, $Zr^{(IV)}$, $Ru^{(II)}$, $Rh^{(II)}$, $Fe^{(II)}$, $Al^{(III)}$, $Cu^{(II)}$, $Au^{(I)}$, R_3SiX, Ar_3C^+, clay; Lewis base = F^-, $(R_2N)_3PO$;

Mechanism:[3,28-39,14]

The mechanism of the *Mukaiyama aldol reaction* largely depends on the reaction conditions, substrates, and Lewis acids. Under the classical conditions, where $TiCl_4$ is used in equimolar quantities, it was shown that the Lewis acid activates the aldehyde component by coordination[30,31,35] followed by rapid carbon-carbon bond formation. Silyl transfer may occur in an intra- or intermolecular fashion. The stereochemical outcome of the reaction is generally explained by the open transition state model, and it is based on steric- and dipolar effects.[14] For *Z*-enol silanes, transition states **A**, **D**, and **F** are close in energy. When substituent R^2 is small and R^3 is large, transition state **A** is the most favored and it leads to the formation of the *anti*-diastereomer.[34] In contrast, when R^2 is bulky and R^3 is small, transition state **D** is favored giving the *syn*-diastereomer as the major product. When the aldehyde is capable of chelation, the reaction yields the *syn* product, presumably *via* transition state **H**.[29,32,36]

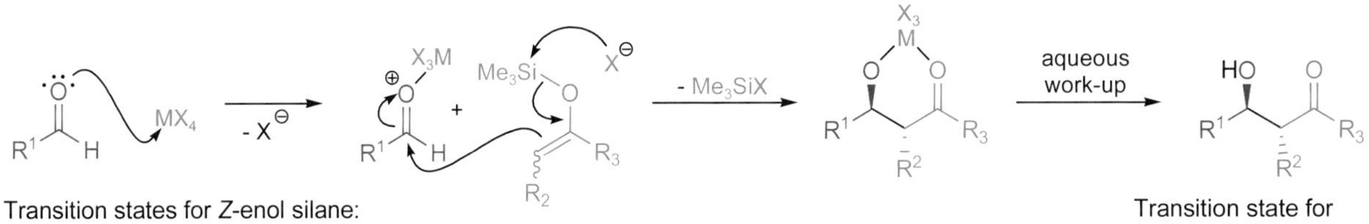

Transition states for *Z*-enol silane:

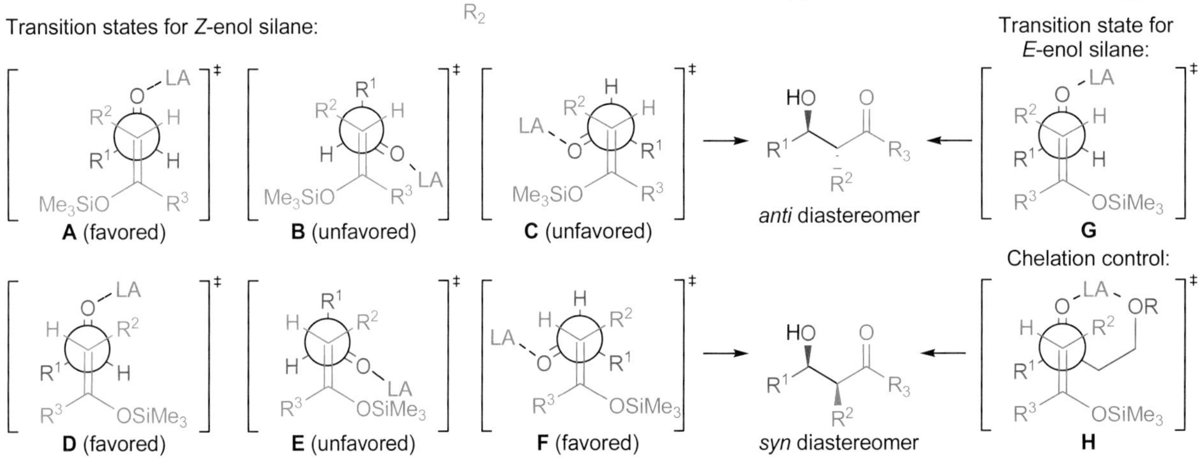

Synthetic Applications:

The asymmetric total syntheses of rutamycin B and oligomycin C was accomplished by J.S. Panek et al.[40] In the synthesis of the C3-C17 subunit, they utilized a *Mukaiyama aldol reaction* to establish the C12-C13 stereocenters. During their studies, they surveyed a variety of Lewis acids and examined different trialkyl silyl groups in the silyl enol ether component. They found that the use of $BF_3 \cdot OEt_2$ and the sterically bulky TBS group was ideal with respect to the level of diastereoselectivity. The stereochemical outcome was rationalized by the open transition state model, where the orientation of the reacting species was *anti* to each other, and the absolute stereochemistry was determined by the chiral aldehyde leading to the *anti* diastereomeric Felkin aldol product.

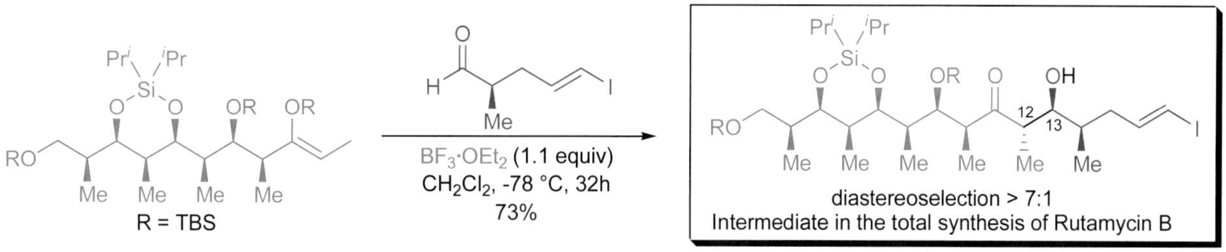

Tin(II) mediated asymmetric *aldol reactions* are among the first chiral Lewis acid controlled *Mukaiyama aldol reactions*.[41,42] A catalytic version of this method was utilized during the total syntheses of sphingofungins B and F by S. Kobayashi.[43] The *asymmetric tin catalyzed Mukaiyama aldol reaction* provided the two main fragments of the molecule with excellent enantio- and diastereoselectivities. Combination of the two fragments and subsequent steps led to the total synthesis of sphingofungins B and F.

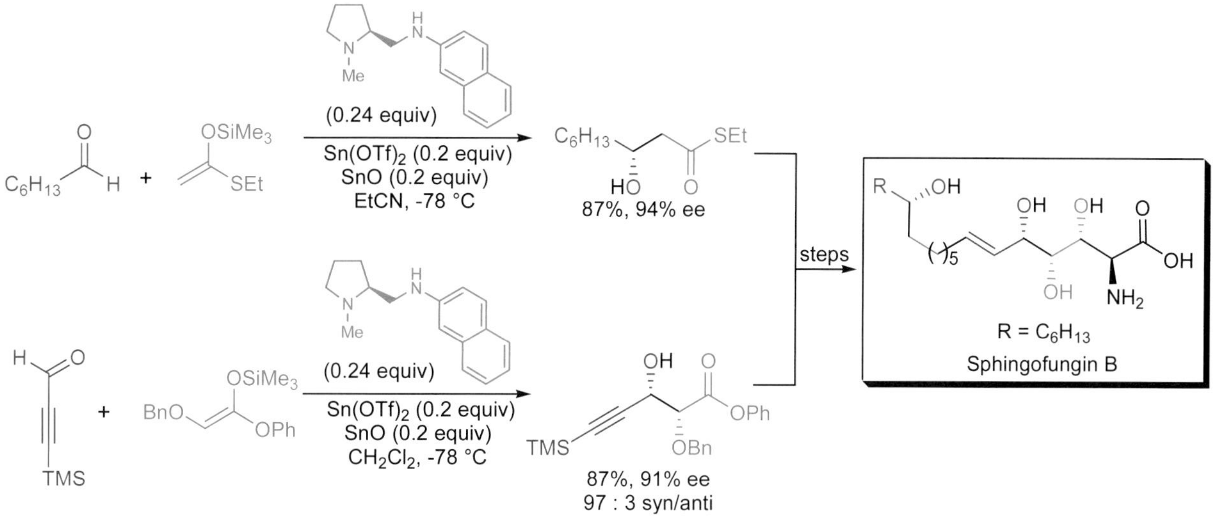

A convergent total synthesis of polyene macrolide roflamycoin was achieved by S.D. Rychnovsky and co-workers.[44] In their approach, they introduced the C25 stereocenter *via* an asymmetric catalytic *Mukaiyama aldol reaction* utilizing Carreira's chiral titanium catalyst.[45]

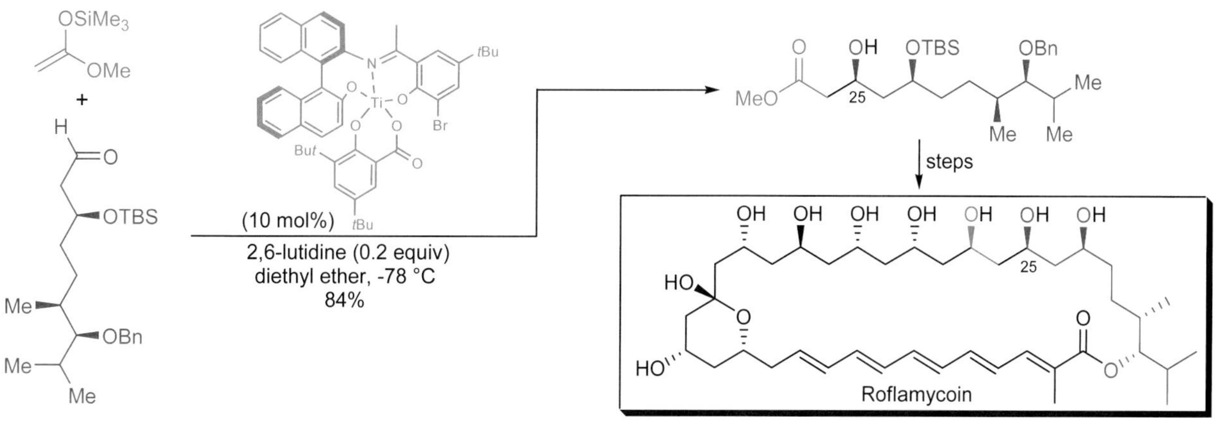

MYERS ASYMMETRIC ALKYLATION

(References are on page 634)

Importance:

[*Seminal Publications*[1-4]; *Reviews*[5-7]; *Modifications & Improvements*[8]]

In 1978, M. Larcheveque and co-workers reported that the *N*-acylation of commercially available D- or L-ephedrine led to highly crystalline *N,N*-disubstituted amides that could be alkylated in high yield and with good diastereoselectivity.[1,2] The alkylated products were easily converted to the corresponding optically active α-substituted ketones and carboxylic acids. Almost two decades later, A.G. Myers et al. developed an efficient alkylation of *N*-acylated pseudoephedrines (the diastereomers of ephedrine) to obtain enantiomerically enriched α-alkylated, aldehydes, ketones, carboxylic acids, and alcohols.[3] This transformation is known as the *Myers asymmetric alkylation*. The general features of this alkylation are:[4] 1) both enantiomers of pseudoephedrine are inexpensive and commercially available commodity chemicals; 2) the *N*-acylation can be achieved in almost quantitative yields using symmetrical and mixed anhydrides or acid chlorides and the *N*-acyl derivatives (tertiary amides) are usually highly crystalline materials; 3) the alkylation of these tertiary amides is achieved by first deprotonation with lithium diisopropylamide and the resulting (*Z*)-enolate undergoes highly diastereoselective alkylation at the α-position; 4) allylic, benzylic, as well as the less reactive alkyl halides (including β-branched alkyl iodides and β-oxygenated *n*-alkyl halides) are all good alkylating agents, since the enolates are highly nucleophilic (unlike imide enolates that react only with highly reactive halides); 5) the α-alkylated products are often crystalline and can be enriched by recrystallization to get >99% *de*; 6) there are two general procedures to conduct the alkylation: in the first the alkylating agent is used in slight excess, while in the second the enolate is used in excess; 7) in order to obtain high yields and high levels of diastereoselectivity, the use of a large excess (6-10 equivalents) of anhydrous lithium chloride is necessary; 8) the role of the LiCl is twofold: it accelerates the rate of the alkylation and suppresses the *O*-alkylation of the pseudoephedrine hydroxyl group; 9) when β-branched alkyl iodides are used as alkylating agents the transformation leads to 1,3-dialkyl substituted alkyl chains (*syn* or *anti*), a common motif in a large number of natural products; 10) the 1,3-*syn* products represent matched cases demonstrating that the diastereofacial bias exerted by this chiral auxiliary overrides the secondary effects originating from the existing stereocenter in the alkyl iodide; 11) substrates that are both β-alkyl branched and β-alkoxy substituted react very slowly, albeit the diastereoselectivity of the alkylation remains high; and12) the removal of the chiral auxiliary from the alkylated tertiary amide products gives rise to the following useful functionalities: simple acidic, basic or Lewis acid catalyzed hydrolysis affords carboxylic acids, reduction with lithium pyrrolidide-borane (LPT) or with lithium amidotrihydroborate (LAB) gives primary alcohols, reduction with lithium triethoxyaluminum hydride results in aldehydes, while the addition of alkyllithium reagents followed by an aqueous work-up leads to ketones.

Myers (1994):

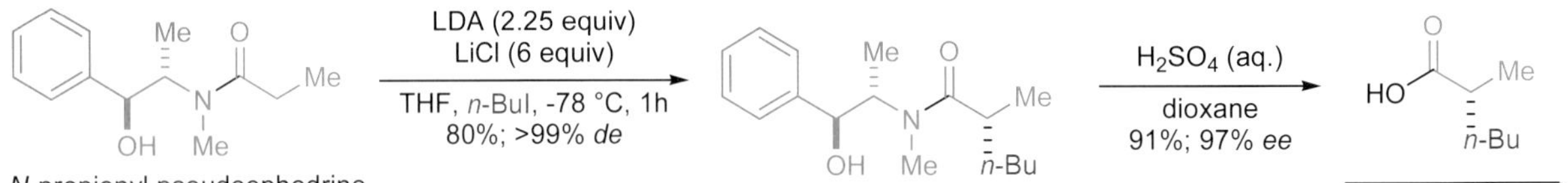

N-propionyl pseudoephedrine

Myers asymmetric alkylation:

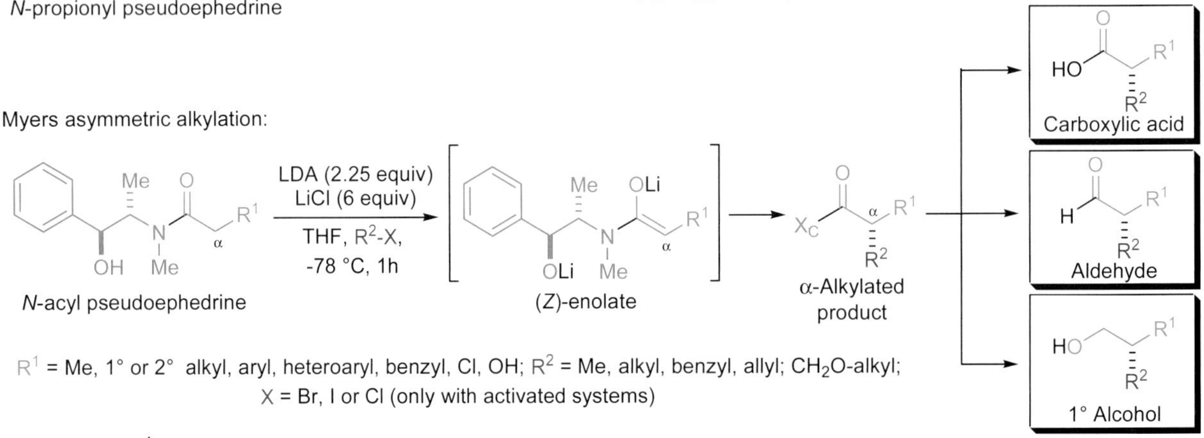

R^1 = Me, 1° or 2° alkyl, aryl, heteroaryl, benzyl, Cl, OH; R^2 = Me, alkyl, benzyl, allyl; CH_2O-alkyl;
X = Br, I or Cl (only with activated systems)

Mechanism:[4]

The origin of the high diastereoselectivity in this alkylation is not fully understood. The stereochemical outcome is consistent with a model in which the (*Z*)-enolate is alkylated from the α-face while the β-face is blocked by the solvated lithium alkoxide.

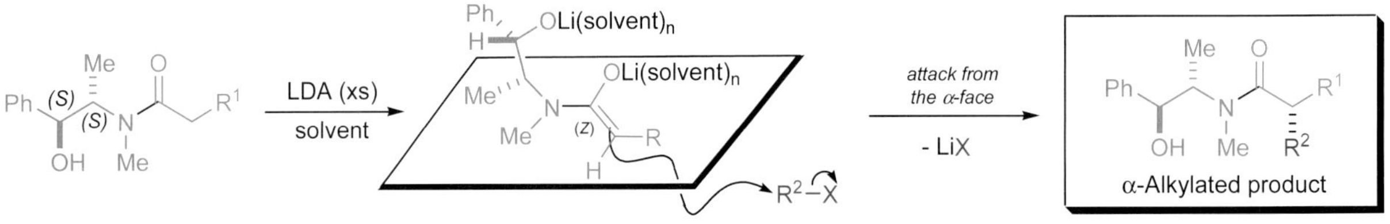

Synthetic Applications:

The enantioselective total synthesis of borrelidin, a structurally unique macrolide with angiogenesis inhibitory activity, was completed by J.P. Morken and co-workers.[9] The *Myers asymmetric alkylation* was used to set the C8 stereocenter. *N*-Propionyl pseudoephedrine was deprotonated with LDA in the presence of excess lithium chloride. It is well known that when excess enolate is used and the alkyl halide is the limiting reagent, yields tend to be higher than in those cases where the enolates are used as the limiting reagent. The authors used twice as much enolate as the alkyl iodide, and the product was isolated in excellent yield and with complete diastereoselectivity. Subsequently, the auxiliary was removed reductively using LAB as the reducing agent.

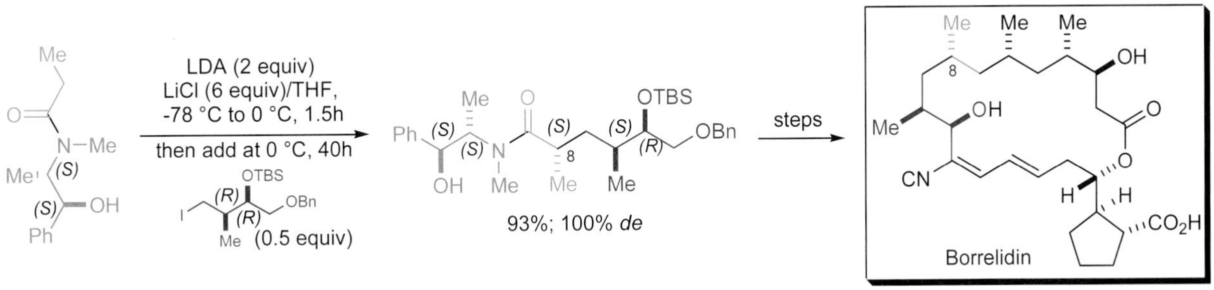

In the laboratory of T.F. Jamison, the synthesis of amphidinolide T1 was accomplished utilizing a catalytic and stereoselective macrocyclization as the key step.[10] The *Myers asymmetric alkylation* was chosen to establish the correct stereochemistry at the C2 position. In the procedure, the alkyl halide was used as the limiting reagent and almost two equivalents of the lithium enolate of the *N*-propionyl pseudoephedrine chiral auxiliary was used. The alkylated product was purified by column chromatography and then subjected to basic hydrolysis to remove the chiral auxiliary.

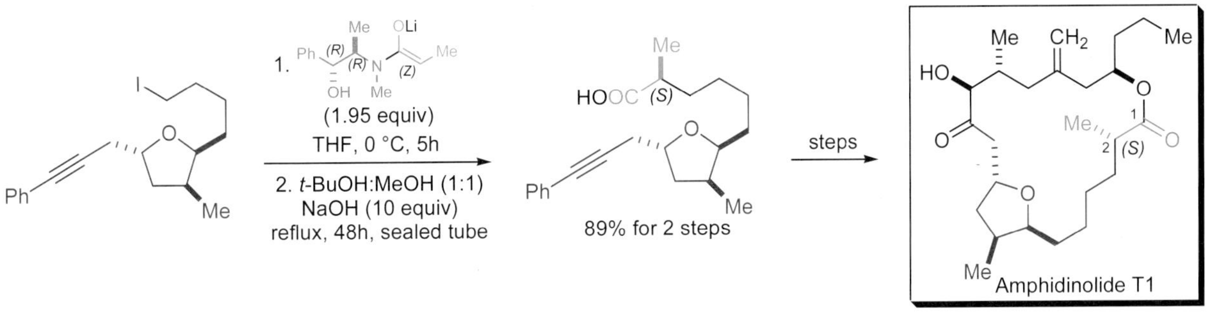

The total synthesis of the potent cytotoxic macrolide (–)-dictyostatin was accomplished by I. Paterson et al.[11] This natural product exhibits a powerful growth-inhibitory activity against a number of human cancer cell lines at nanomolar concentrations and it is active against Taxol-resistant cancer cells that express active P-glycoprotein. In order to create the C16 stereocenter, the *Myers asymmetric alkylation* was chosen as the method that achieved the desired three-carbon homologation of the β-branched alkyl iodide substrate. The alkylated product was removed reductively using LAB and the resulting primary alcohol was oxidized to the corresponding aldehyde by the *Dess-Martin oxidation*.

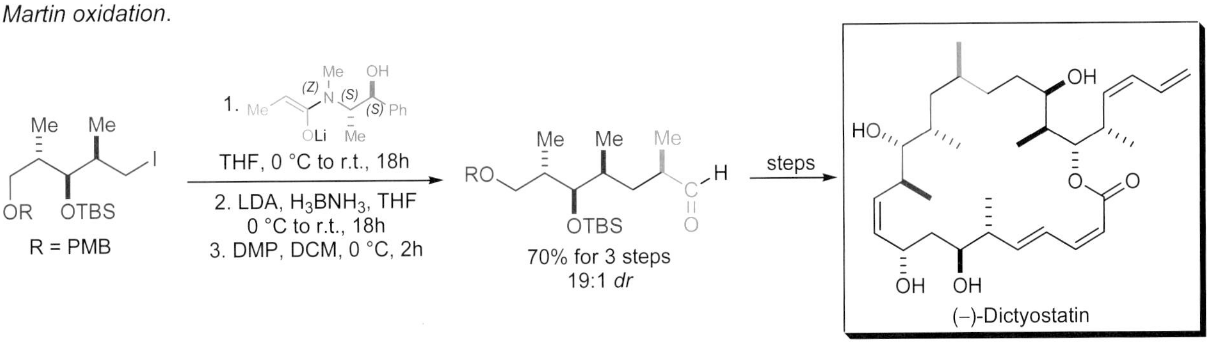

The neurotoxic lipopeptide (+)-kalkitoxin was prepared by J.D. White et al., who installed one of the stereocenters *via* the *Myers asymmetric alkylation* followed by reductive workup to obtain the enantiopure primary alcohol.[12]

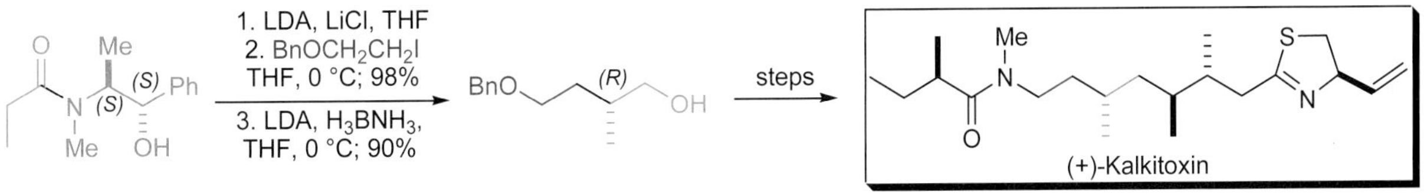

NAGATA HYDROCYANATION

(References are on page 635)

Importance:

[*Seminal Publications*[1-6]; *Reviews*[7]; *Modifications & Improvements*[8,9]]

In 1873, A. Claus reported that tricarballic acid (1,2,3-propanetricarboxylic acid) was isolated when an ethanolic solution of 2,3-dichloropropene was heated with excess potassium cyanide in a sealed tube.[1] Although Claus did not realize at that time, this observation was the first example of a 1,4-addition of hydrogen cyanide to an activated alkene. Two decades later in 1896, J. Bredt and J. Kallen observed that the treatment of benzalmalonate with KCN and sulfuric acid followed by hydrolysis and decarboxylation (see *malonic ester synthesis*) gave rise to 2-phenylsuccinic acid.[2] In the first half of the 1900s, many research groups used the conjugate hydrocyanation in organic synthesis, but the method was far from being efficient, and it was plagued by numerous side reactions. A breakthrough in efficiency occurred in 1962, when W. Nagata and co-workers developed a new hydrocyanation method in which they added α,β-unsaturated ketones to a mixture of triethylaluminum and HCN in THF and observed that the reaction took place considerably faster and with much higher selectivity than under the original reaction conditions.[3] Later it was found that dialkylaluminum cyanides were even better reagents for conjugate hydrocyanations. The formation of β-cyano ketones and esters from α,β-unsaturated ketones and esters using dialkylaluminum cyanides (or HCN with Al-based Lewis acids) is known as the *Nagata hydrocyanation*. The general features of this transformation are: 1) in the overwhelming majority of cases the carbonyl compound is a ketone and rarely the aldehyde (since it undergoes 1,2-addition); 2) α,β-unsaturated esters and nitriles are also good substrates; 3) when HCN is used in conjunction with aluminum trialkyls or with alkylaluminum halides, the order of reactivity is as follows: $EtAlCl_2$>Me_3Al>Et_3Al>Et_2AlCl; 4) the reaction is almost exclusively conducted in a dipolar aprotic solvent such as THF, hydrocarbon solvents are not suitable, since the HCN reacts with AlX_3 immediately in nonpolar media; 5) a small amount of water in the reaction medium accelerates the hydrocyanation when HCN/$AlMe_3$ is used (substrates with free hydroxyl groups have the same effect); 6) in the case of less reactive substrates, the trialkylaluminum and the HCN are both used in excess, but always more of the aluminum reagent is applied to prevent the polymerization of HCN; 7) the reactivity of the dialkylaluminum cyanide reagent is strongly dependent on the basicity of the solvent and increases with decreasing solvent basicity: THF>dioxane>*i*-Pr_2O>benzene>toluene; and 8) the rate of hydrocyanation is much faster with the dialkylaluminum cyanide than with the other reagent combination (HCN/AlX_3).

Claus (1873):

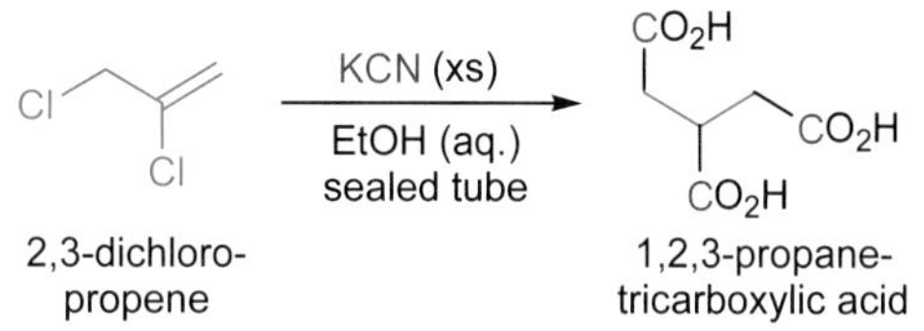

Bredt & Kallen (1896):

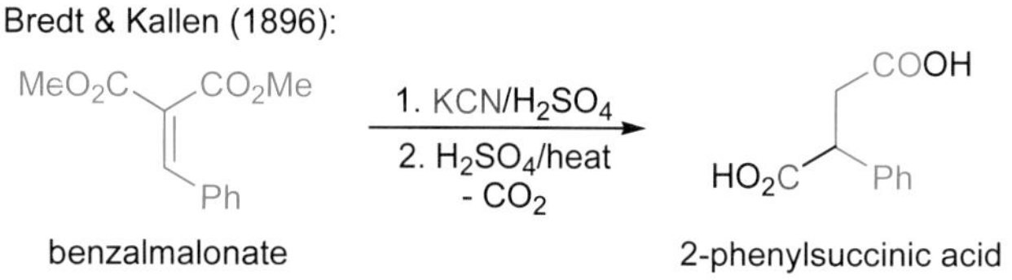

Nagata (1962):

Nagata (1966):

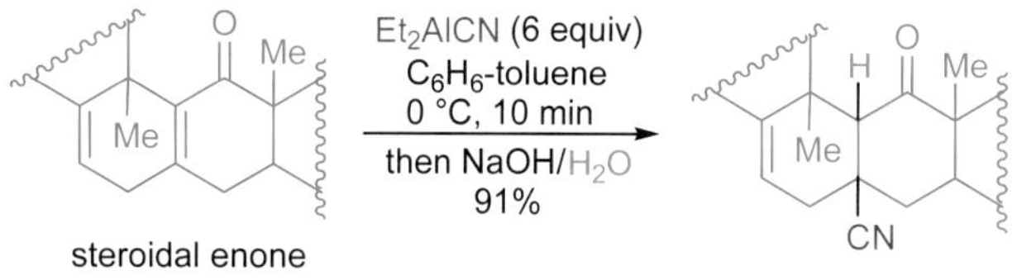

Nagata hydrocyanation:

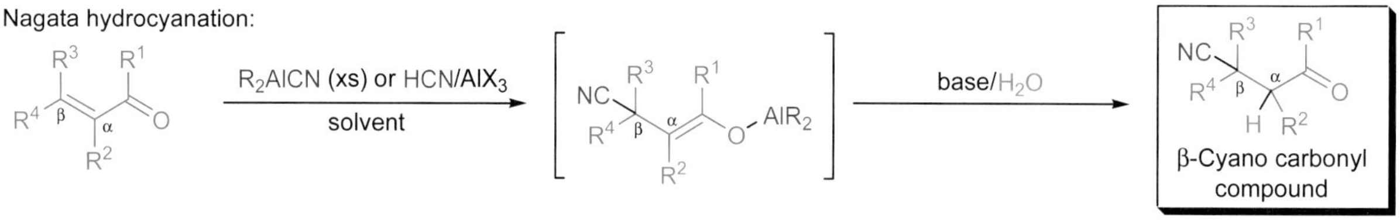

R^1 = alkyl, aryl, *O*-alkyl, *O*-aryl, (NR_2); R^{2-4} = H, alkyl, aryl; R = Me, Et, *i*-Bu; $\underline{AlX_3}$ = $EtAlCl_2$, Me_3Al, Et_3Al, Et_2AlCl

Mechanism:

Reversible 1,4-addition using dialkylaluminum cyanide (S = solvent):

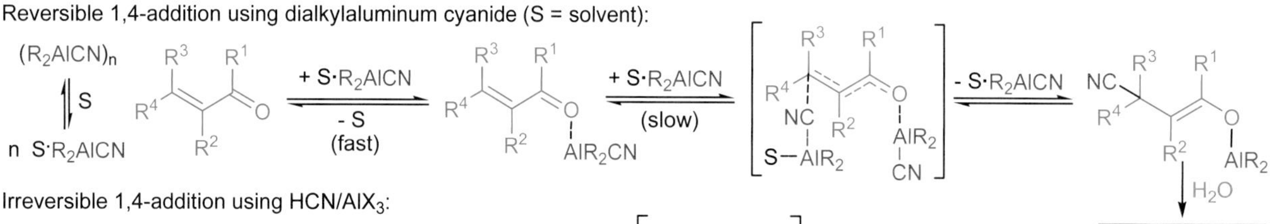

Irreversible 1,4-addition using HCN/AlX_3:

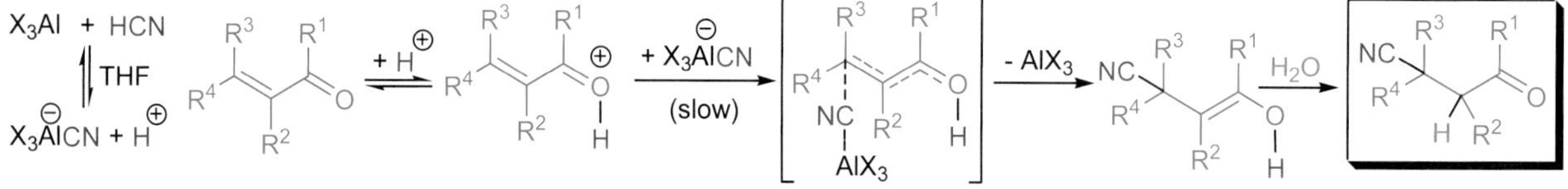

Synthetic Applications:

The total synthesis of (±)-scopadulcic acid B was completed by L.E. Overman et al. who used a *double-Heck cyclization* as the key step.[10] In the endgame of the synthetic effort, the stereoselective introduction of the quaternary methyl group at the C10 position was required. The authors anticipated that the pentacyclic β,β-disubstituted enone would be a poor Michael acceptor. However, they were surprised that virtually none of the standard conjugate addition procedures worked, giving rise only to 1,2-adducts and large amounts of recovered starting material. Fortunately, the *Nagata hydrocyanation* protocol using diethylaluminum cyanide was able to effect the desired conjugate addition. Since in rigid bicyclic systems the cyano group is usually delivered from the axial position, the stereochemical outcome of the *Nagata hydrocyanation* was first assigned tentatively. Later it was confirmed that indeed the addition occurred from the axial position. Subsequently, the cyano group was reduced to the corresponding methyl group in two steps.

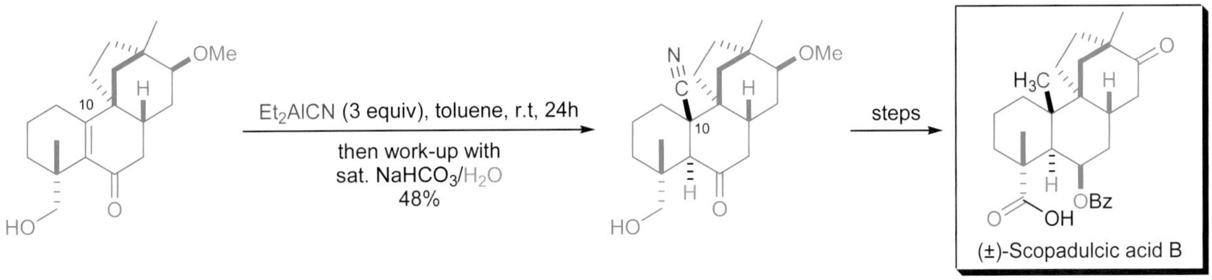

Termite soldiers produce a large number of different chemical defense agents. Several of these molecules are unusual bioactive terpenoids such as the secotrinervitanes that have been isolated and their structure elucidated. In the laboratory of T. Kato, the total synthesis of (±)-3α-acetoxy-7,16-secotrinervita-7,11-dien-15β-ol was accomplished.[11] The *Nagata hydrocyanation* was used to introduce a carbon at the β-position of a macrocyclic enone intermediate. The substrate was treated with excess diethylaluminum cyanide in dry toluene and the addition resulted in the formation of a 1:1 mixture of diastereomers, which could be readily separated by column chromatography. The cyano group was later converted to the corresponding methyl ester.

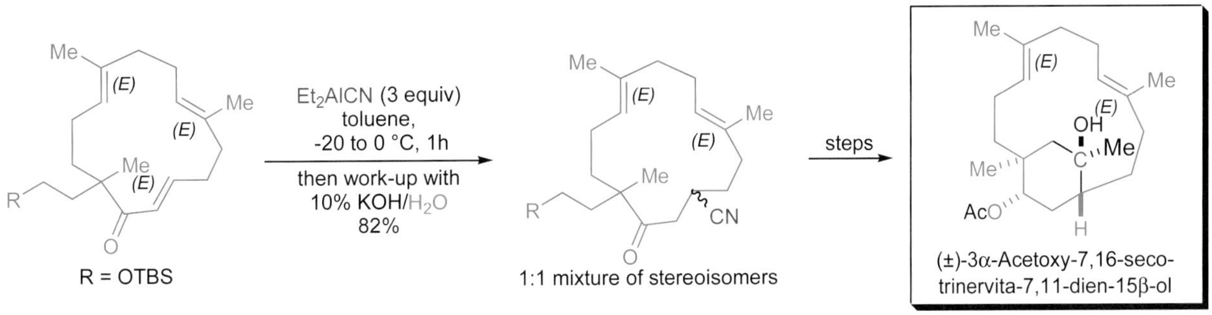

The highly stereoselective synthesis of the tricyclic diterpene moiety of radarins was achieved by K. Fukumoto and co-workers, who utilized an *intramolecular Diels-Alder cycloaddition* to construct the B and C rings simultanaeously.[12] Radarins are indole alkaloids that exhibit potent cytotoxicity against solid tumor cells. The preparation of the Diels-Alder cycloaddition precursor commenced with the *Nagata hydrocyanation* of a known bicyclic enone. The ring fusion of the decalin system had to be *trans*, so the hydrocyanation was conducted under thermodynamic conditions using diethylaluminum cyanide. The choice of the cyano group at the ring junction was strategic for two reasons: 1) a cyano group can be easily converted to the corresponding methyl group, which is actually required in the natural product; and 2) in the cycloaddition, the small steric bulk of the cyano group avoids the substantial 1,3-diaxial interactions that would have occurred if a methyl group was present. The enolate formed in the hydrocyanation step was trapped as the silyl enol ether, and was subsequently halogenated at the α-position with NBS. The resulting α-bromo ketone was converted to the corresponding α-hydroxy ketone, which was subsequently cleaved in a *Criegee-type oxidation*.

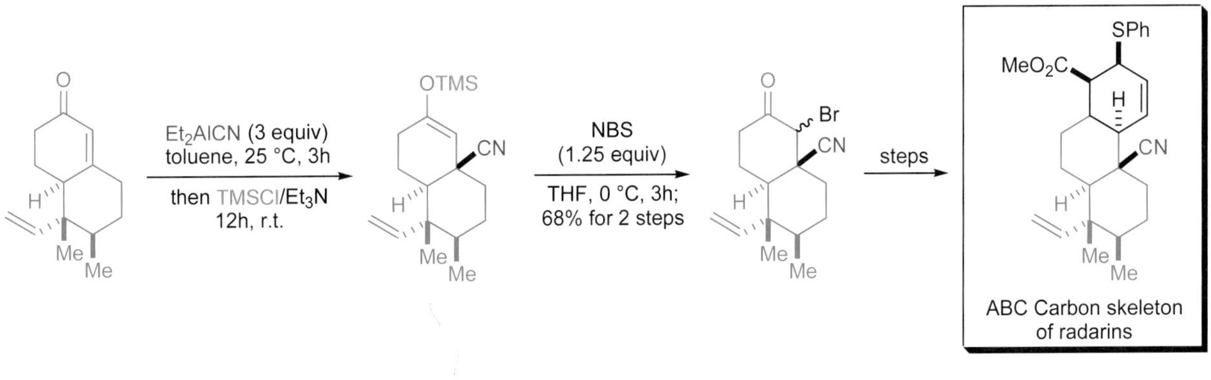

NAZAROV CYCLIZATION

(References are on page 635)

Importance:

[*Seminal Publications*[1-6]; *Reviews*[7-11]; *Modifications & Improvements*[12-28]; *Theoretical Studies*[29-31]]

In 1903, D. Vorländer and co-workers found that treatment of dibenzylideneacetone with concentrated sulfuric acid and acetic anhydride followed by hydrolysis by sodium hydroxide yielded a cyclic ketol, the structure of which was unknown at the time.[1] In the 1930s, the research group of C.S. Marvel examined the acid-catalyzed hydration of dienynes.[2,3] Later, in the 1940s and 1950s, I.N. Nazarov et al. revisited the topic and extensively studied the cyclization of the intermediate allyl vinyl ketones to the corresponding 2-cyclopentenones.[4-6] The protic- or Lewis acid catalyzed ring-closure of divinyl ketones (and their acid-labile precursors) *via* pentadienylic cations is known as the *Nazarov cyclization*. The general features of the reaction are:[8,10] 1) in a broader sense, any compound that affords the key pentadienylic cation or its equivalent is a viable substrate for the transformation; 2) allyl vinyl ketones are isomerized *in situ* to the corresponding divinyl ketones; 3) electron-donating substituents in the α and α' positions accelerate the cyclization, whereas rate retardation is observed when they are in the β and β'-positions; 4) fused cyclic systems are formed when one or both of the groups attached to the ketone are cyclic; and 5) the introduction of trialkylsilyl (or trialkylstannyl) groups in the β or β'-position ensures the controlled collapse of the cyclopentenylic cation thus undesired *Wagner-Meerwein rearrangements* are avoided, the final double bond is formed regioselectively, and the stereocenters at the ring fusion are preserved (*silicon-directed Nazarov cyclization*).[13]

Vorlander & Schroeter (1903): Nazarov (1940-1950s):

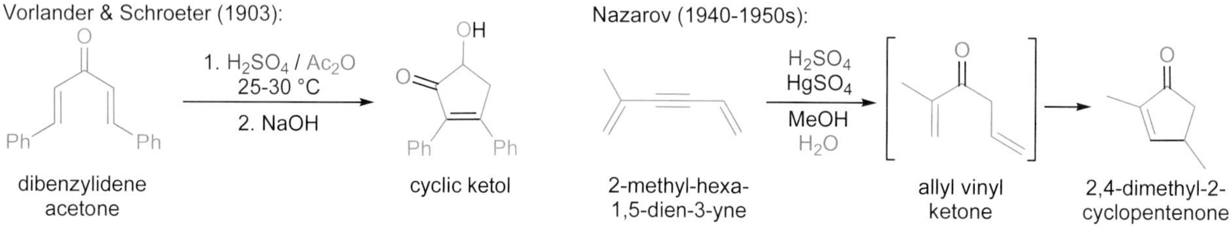

General scheme for Nazarov cyclization:

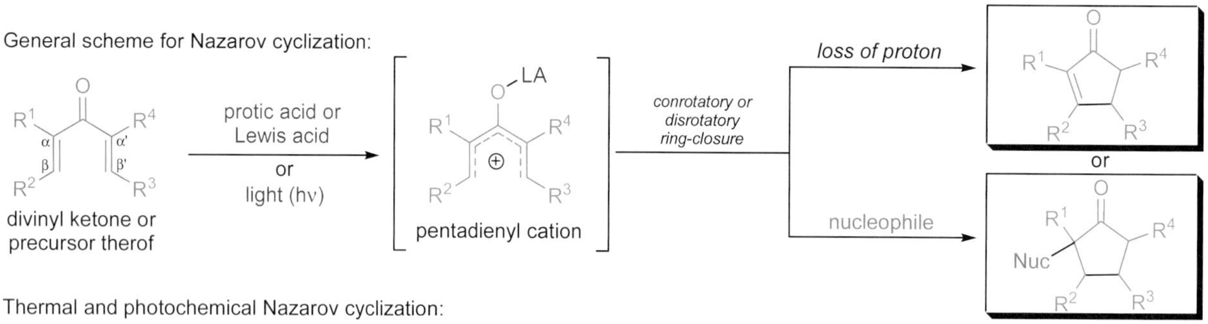

Thermal and photochemical Nazarov cyclization:

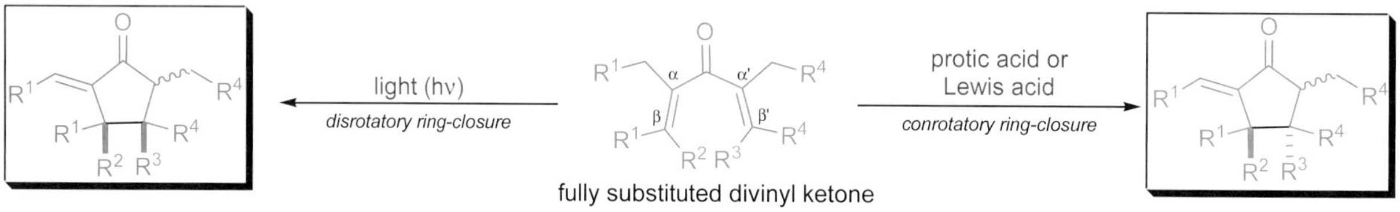

Mechanism: [32-37,15,10]

The mechanism of the *Nazarov cyclization* was not clarified until 1952, when it was realized that the cyclization proceeded *via* carbocation intermediates.[32] The *Nazarov cyclization* is a pericyclic reaction that belongs to the class of 4π electrocyclizations. The first step is the coordination of the Lewis acid to the carbonyl group of the substrate and the formation of the pentadienylic cation, which undergoes a conrotatory ring closure to give a cyclic carbocation that may be captured by a nucleophile, may undergo deprotonation, or further rearrangement may take place. The electrocyclization step may proceed in a clockwise or counterclockwise fashion (torquoselectivity) generating two diastereomers when the divinyl ketone substrate is chiral. The sense of torquoselection is primarily controlled by steric factors such as the torsional and nonbonding interactions between the substituents in the vicinity of the newly forming bond. Under photochemical conditions the cyclization proceeds in a disrotatory fashion.

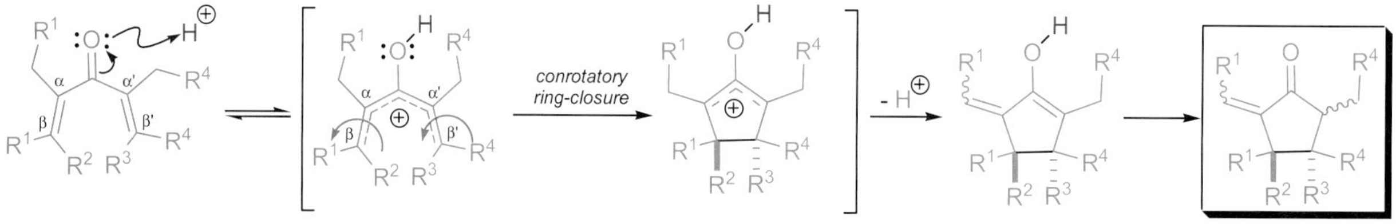

Synthetic Applications:

The stereoselective synthesis of (±)-trichodiene was accomplished by K.E. Harding and co-workers.[38] The synthesis of this natural product posed a challenge, since it contains two adjacent quaternary stereocenters. For this reason, they chose a stereospecific electrocyclic reaction, the *Nazarov cyclization*, as the key ring-forming step to control the stereochemistry. The cyclization precursor was prepared by the *Friedel-Crafts acylation* of 1,4-dimethyl-1-cyclohexene with the appropriate acid chloride using $SnCl_4$ as the catalyst. The *Nazarov cyclization* was not efficient under protic acid catalysis (e.g., TFA), but in the presence of excess boron trifluoride etherate high yield of the cyclized products was obtained. It is important to note that the mildness of the reaction conditions accounts for the fact that both of the products had an intact stereocenter at C2. Under harsher conditions, the formation of the C2-C3 enone was also observed.

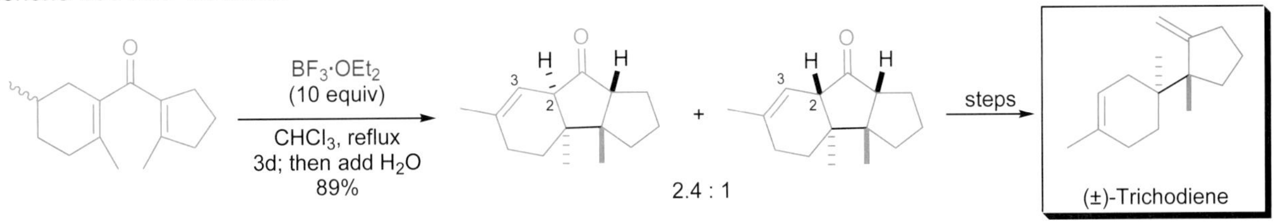

In the laboratory of M. Miesch, the *silicon-directed Nazarov cyclization* was utilized in the synthesis of the angular triquinane (±)-silphinene.[39] The cyclization precursor was prepared by the addition of a Grignard reagent derived from bromovinylsilane to the α,β-unsaturated aldehyde on the A ring, followed by MnO_2 oxidation of the resulting allylic alcohol. The addition of large excess of boron trifluoride etherate in refluxing ethylbenzene brought about the *Nazarov cyclization* to form the C ring of the natural product. The benzyloxy group on the B ring was also eliminated during the cyclization step.

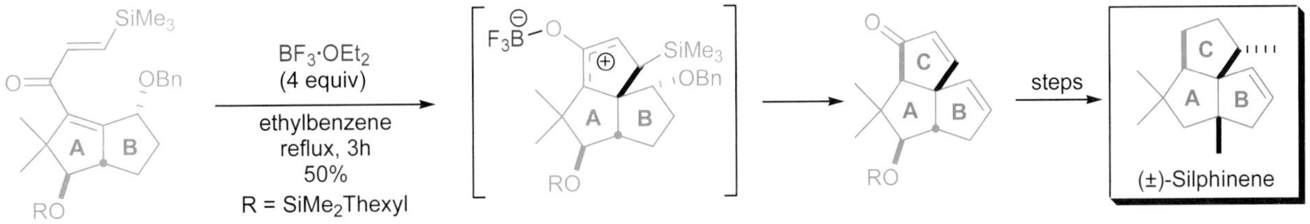

The six-step synthesis of (±)-desepoxy-4,5-didehydromethylenomycin A methyl ester from diethyl methanephosphonate was reported by M. Mikolajczyk et al.[40] The key ring-forming step was the *Nazarov cyclization* of an α-phosphoryl dienone to afford the corresponding cyclopentenone in high yield. In the product, the phosphoryl and carboxymethyl groups were exclusively *trans* disposed to each other. The double bond was found to be also exclusively at the C2 and C3 positions. The final step of the synthetic sequence was the introduction of the exo-methylene functionality by using the *Horner-Wittig reaction*.

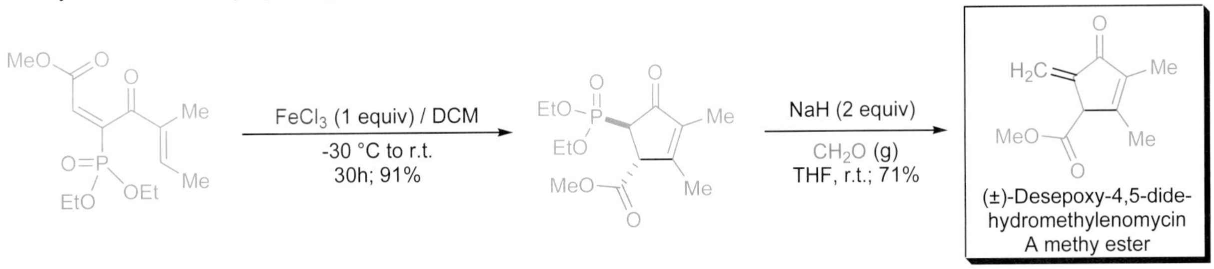

The naturally occurring *bis*-indole yuehchukene is considered the dimer of 3-didehydroprenylindole and exhibits potent anti-implantation activity in rats. Part of an SAR study, K.-F. Cheng and co-workers synthesized inverto-yuehchukene, which can be considered as the dimer of 2-didehydroprenylindole.[41] The five-membered ring of the target was constructed by the *Nazarov cyclization* of the corresponding divinyl ketone in a refluxing dioxane solution containing concentrated hydrocholoric acid.

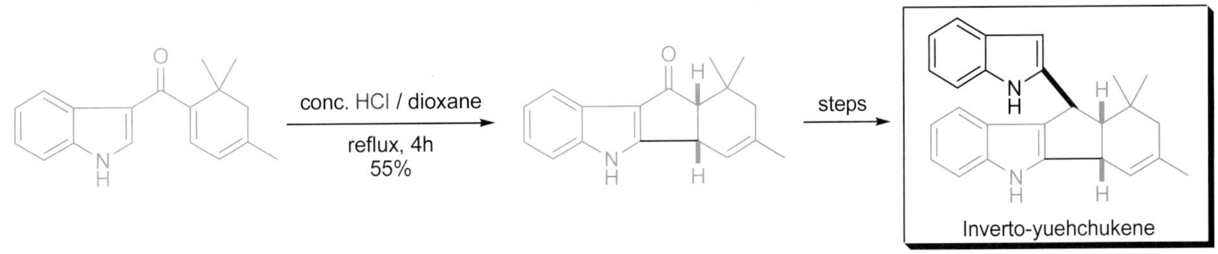

NEBER REARRANGEMENT

(References are on page 636)

Importance:

[*Seminal Publications*[1-5]; *Reviews*[6-11]; *Modifications & Improvements*[12-19]]

In 1926, during the investigation of the *Beckmann rearrangement*, P.W. Neber and A. Friedolsheim reported that the successive treatment of ketoxime tosylates with potassium ethoxide, acetic acid, and hydrochloric acid yielded the hydrochloride salts of α-amino ketones.[1] The base-induced rearrangement of *O*-acylated ketoximes to the corresponding α-amino ketones is known as the *Neber rearrangement*. Since its discovery, the rearrangement has become an important synthetic tool in the synthesis of heterocycles in which amino ketones are used as key intermediates. The general features of the reaction are: 1) acylated ketoximes derived from both acyclic and cyclic ketones can be used; 2) the required oximes are readily prepared from the ketones by reacting them with hydroxylamine under acidic conditions; 3) *O*-acylation of the oximes is conducted using acyl halides or anhydrides in the presence of a mild base (e.g., pyridine); 4) the rearrangement is usually carried out in an alcohol solution containing equimolar quantities of an alkali alkoxide; 5) when two methylene groups are available at the α- and α'-positions, the rearrangement mainly gives rise to a product in which the amino group is located on the more electrophilic carbon; 6) the rearrangement is not stereospecific, since the stereochemistry of the substrate (*syn* or *anti*) usually does not influence the outcome of the reaction, and this is in sharp contrast with the stereospecificity of the *Beckmann rearrangement*; and 7) the product amino ketones have a tendency to dimerize, so they often need to be prepared in a protected form as their amino acetals or hydrochloride salts (e.g., the amino acetals are prepared from the 2*H*-azirine intermediates by treatment with acidic alcohols). There are a few limitations to the *Neber rearrangement*: 1) *O*-acylated aldoximes do not yield α-amino ketones upon treatment with base, but rather undergo *E2 elimination* to afford the corresponding nitriles or isonitriles; and 2) the substrate must have a methylene group in the α-position in the overwhelming majority of the cases. Other types of compounds having at least one α-hydrogen atom also undergo the *Neber rearrangement* upon treatment with base: 1) ketone dimethylhydrazonium halides;[20] 2) *N,N*-dichloro-*sec*-alkyl amines;[21,22] 3) *N*-chloroimines;[12] and 4) *N*-chloroimidates.[13,23]

Rearrangement of ketoxime tosylates (Neber, 1926):

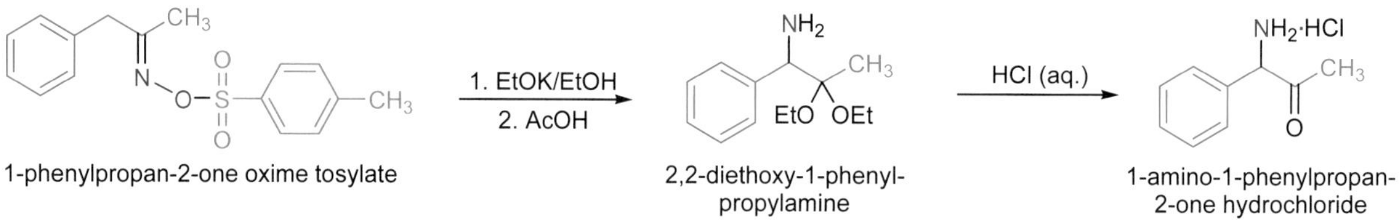

Neber rearrangement of O-acylated ketoximes:

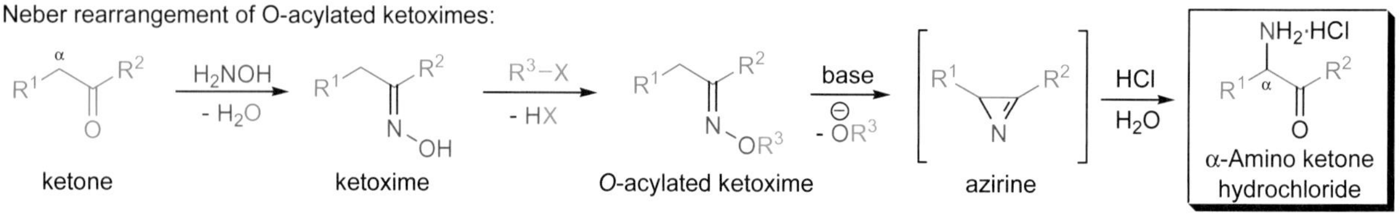

R^1 = H, alkyl, aryl; R^2 = alkyl, aryl, O-alkyl, NH_2, NH-alkyl; R^3 = $SO_2C_6H_4CH_3$, SO_2CH_3; base: NaOEt, KOEt

Mechanism: [24,25,22,26-28,19]

The first step of the mechanism is the deprotonation of the *O*-acylated ketoxime at its α-position, which gives rise to the corresponding enolate. This enolate then can react *via* two possible pathways: 1) a concerted anionic pathway in which the leaving group is directly displaced to give the isolable 2*H*-azirine or 2) a nitrene pathway that leads to the same 2*H*-azirine intermediate *via nitrene insertion*. The nitrene pathway has not been disproved experimentally.

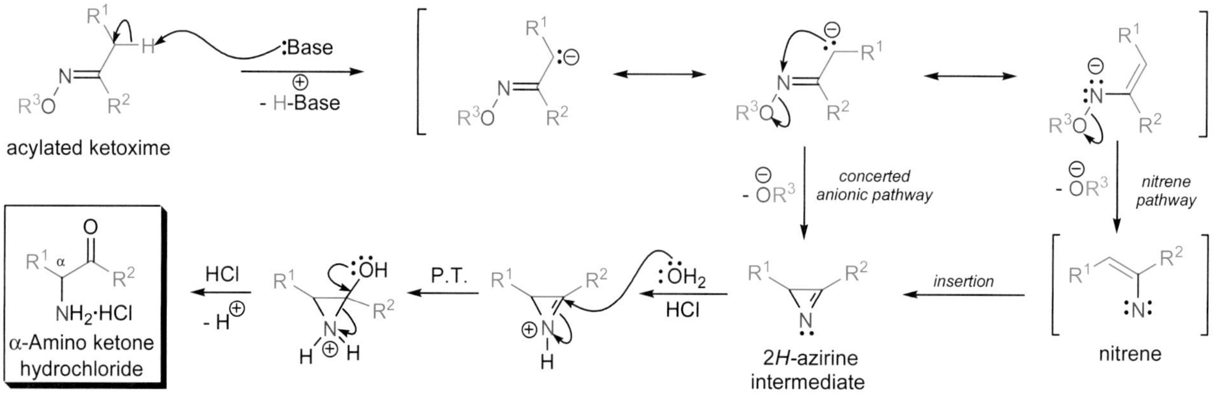

Synthetic Applications:

The chemoenzymatic synthesis of a β3 adrenergic receptor agonist was developed by J.Y.L. Chung and co-workers.[29] The key chiral 3-pyridylethanolamine intermediate was prepared *via* the *Neber rearrangement* of the ketoxime tosylate derived from 3-acetylpyridine. The oxime formation and the tosylation were carried out in a one-pot process using pyridine as the solvent. The solution of the ketoxime tosylate in ethanol was then cooled to 10 °C and potassium ethoxide was added. After the TsOK salt was removed from the reaction mixture, HCl gas was bubbled through the solution until the pH reached ~2 and the 3-pyridylaminomethyl ketal was isolated as its di-HCl salt.

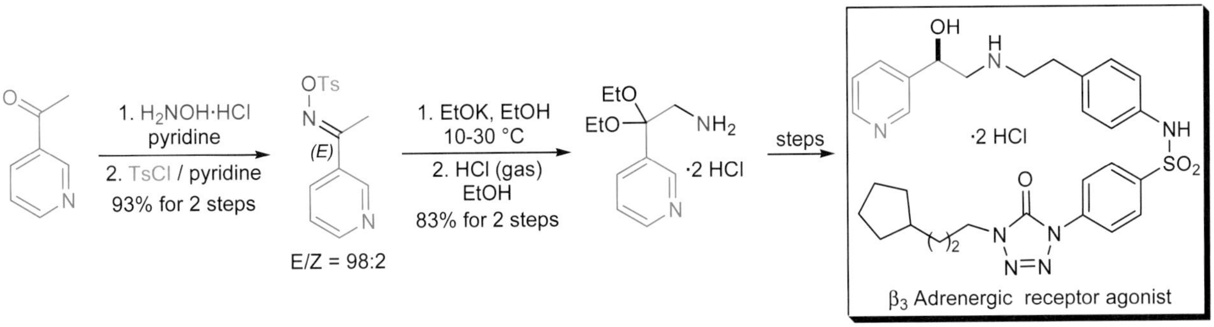

In the laboratory of M. Rubiralta, the general synthesis of the potential substance P antagonist 3-aminopiperidines was accomplished.[30] The (*E*)-oxime of 2-phenyl-2-piperidone was first tosylated and the resulting ketoxime tosylate was immediately subjected to KOEt/EtOH in the presence of anhydrous $MgSO_4$. The resulting regioisomeric aminopiperidines were formed in a 4:1 ratio. The major regioisomer was identified as the 2,3-*cis* diastereomer. Interestingly, when the (*Z*)-oxime was rearranged under identical conditions, the other regioisomer was the major product. This finding suggested that the intermediate 2*H*-azirine was formed *via* the *anti* displacement of the tosyl group.

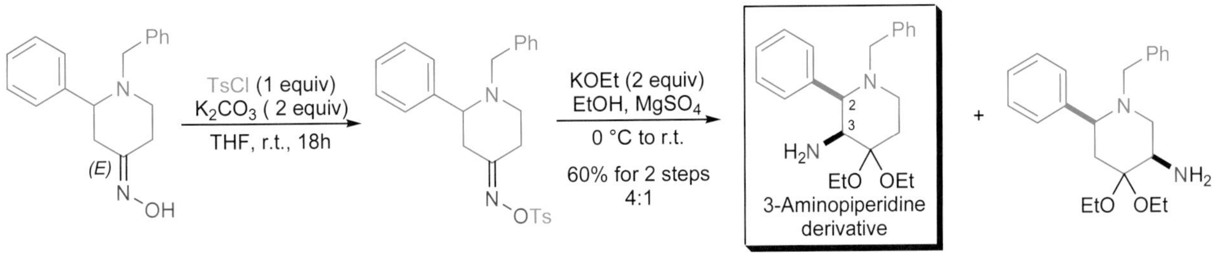

The short synthesis of L- and D-vinylglycine was achieved by D.H.G. Crout and co-workers using the *Neber rearrangement* of an *N*-chloroimidate prepared from but-3-enenitrile.[23] The synthesis started with the *Pinner reaction*, which gave rise to the imino ether in quantitative yield. Oxidation of the imino ether with sodium hypochlorite afforded the *N*-chloroimidate, which was then exposed to aqueous NaOH to induce the *Neber rearrangement*. The racemic vinylglycine was isolated in 53% yield using a cation exchange resin. The resolution of this racemic product was carried out by a *papain-catalyzed enantioselective esterification* in a two-phase system.

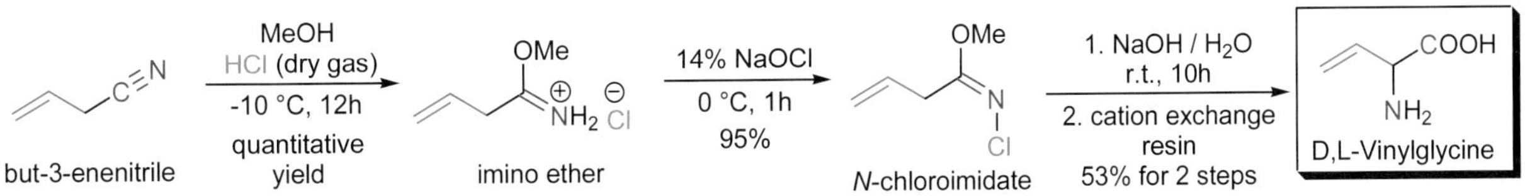

The synthesis of optically active 3-amino-2*H*-azirines was carried out using a *modified Neber rearrangement* in the laboratory of I.P. Piskunova.[16] The optically active amidoximes were acylated using mesyl chloride to give *O*-mesyl derivatives that upon treatment with sodium methoxide afforded the desired product with high diastereoselectivity.

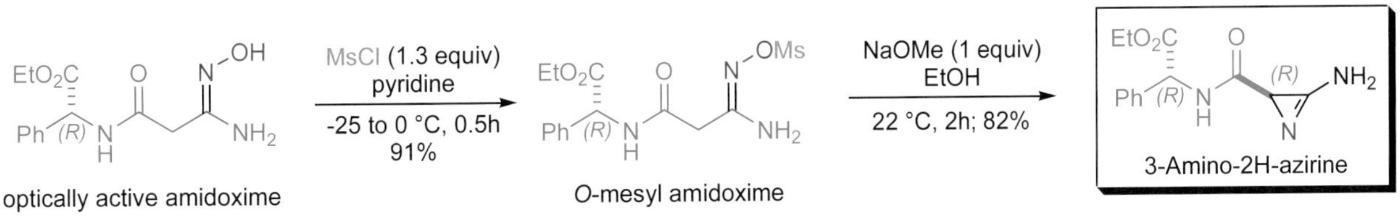

NEF REACTION

(References are on page 636)

Importance:

[*Seminal Publications*[1,2]; *Reviews*[3-9]; *Modifications & Improvements*[10-30]]

In 1893, M. Konovalov observed that the treatment of the potassium salt of 1-phenylnitroethane with dilute acid (AcOH, H_2SO_4) led to the formation of 1-phenylnitroethane and acetophenone.[1] In 1894, J.U. Nef systematically studied the acidic hydrolysis of several nitroparaffin sodium salts, while he was completely unaware of Konovalov's experiments, and showed that the major product of all these reactions were the corresponding carbonyl compounds.[2] Since Nef demonstrated the generality of this transformation, which he discovered independently, the conversion of nitroalkanes into the corresponding carbonyl compounds is known as the *Nef reaction*. The general features of the reaction are: 1) the product distribution is strongly influenced by the acid concentration, and for best results the pH need to be smaller than unity; 2) when the pH>1, a number of by-products such as oximes and hydroxynitroso compounds can be formed; and 3) original reaction conditions required the addition of the nitronate salt to the solution of the acid to avoid the formation of undesired products. To make the reaction more chemoselective and tolerant toward many functional groups, several modifications have been developed during the past three decades: 1) oxidative methods allow the conversion of primary nitroalkanes into aldehydes or carboxylic acids, while secondary nitroalkanes are converted to ketones;[11,13,21,23,24,27] 2) reductive methods are available for the direct preparation of nitroalkanes to aldehydes, ketones, or oximes;[10,12,17,26] 3) carbonyl compounds and oximes can also be prepared from nitroolefins (nitroalkenes) using various reducing agents.[14,15,18,25]

Konovalov (1893): Nef (1894):

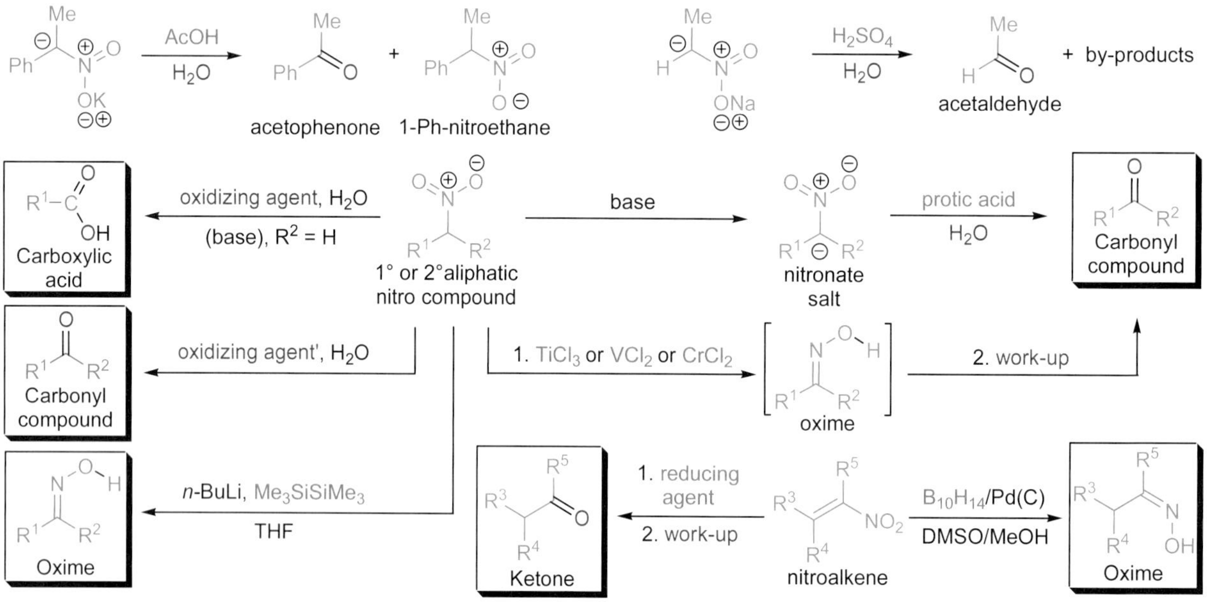

R^{1-2} = H, alkyl, aryl; R^{3-5} = H, aryl, alkyl; oxidizing agent: $KMnO_4$ (at pH~11), Oxone, $(OTMS)_2$, TPAP/NMO, $Cu(OAc)_2/O_2$, $NaNO_2$/AcOH/DMSO; oxidizing agent': to get aldehydes (R^2 = H) use DMDO, $Na_2CO_3 \cdot 1.5\ H_2O_2$, $KMnO_4$ while for ketones use any of the above oxidants; reducing agent: Al powder/$NiCl_2 \cdot 6H_2O$, Zn dust/TFA, Mg powder/$CdCl_2$; protic acid: HCl, H_2SO_4, AcOH

Mechanism: [31-39,4,5,40,41]

The mechanism of the *Nef reaction* has been extensively studied. Under the original reaction conditions, the nitronate salt is first protonated to give the nitronic acid, which after further protonation is attacked by a molecule of water. The process is strongly dependent on the pH of the reaction medium. Weakly acidic conditions favor the regeneration of the nitro compound and by-product formation (oximes and hydroxynitroso compounds), whereas strongly acidic medium (pH<1) promotes the formation of the carbonyl compound. The most popular reductive method ($TiCl_3$) proceeds *via* a nitroso compound that tautomerizes to form an oxime and finally upon work-up the desired product is obtained.

Nef reaction under acidic conditions:

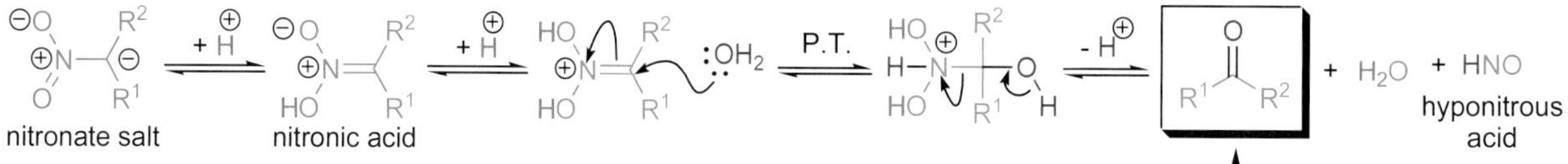

Nef reaction under reductive conditions:

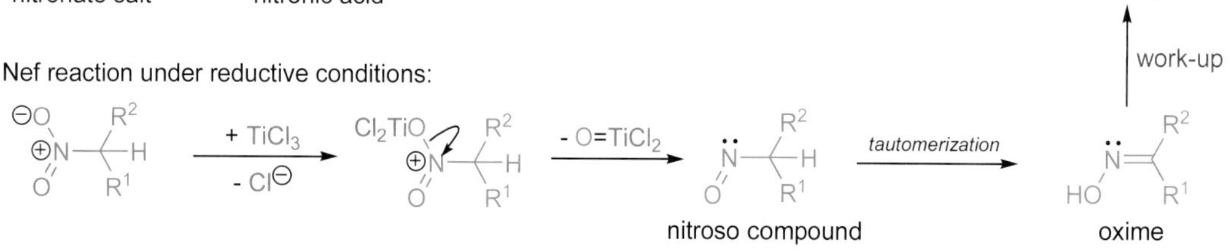

Synthetic Applications:

The synthesis of the bisbenzannelated spiroketal core of the γ-rubromycins was achieved by the research team of C.B. de Koning.[42] The key step was the *Nef reaction* of a nitroolefin, which was prepared by the *Henry reaction* between an aromatic aldehyde and a nitroalkane. The nitroolefin was a mixture of two stereoisomers, and it was subjected to catalytic hydrogenation in the presence of hydrochloric acid. The hydrogenation accomplished two different tasks: it first converted the nitroalkene to the corresponding oxime and removed the benzyl protecting groups. The oxime intermediate was hydrolyzed to a ketone that underwent spontaneous spirocyclization to afford the desired spiroketal product.

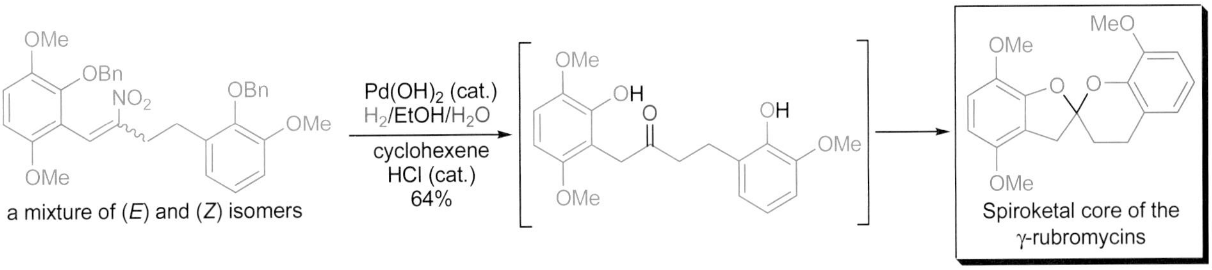

The total synthesis of spirotryprostatin B was accomplished by K. Fuji et al using an *asymmetric nitroolefination* to establish the quaternary stereocenter.[43] The conversion of the nitroolefin to the corresponding aldehyde was carried out under reductive conditions using excess titanium(III) chloride in aqueous solution. The initially formed aldehyde oxime was hydrolyzed *in situ* by the excess ammonium acetate.

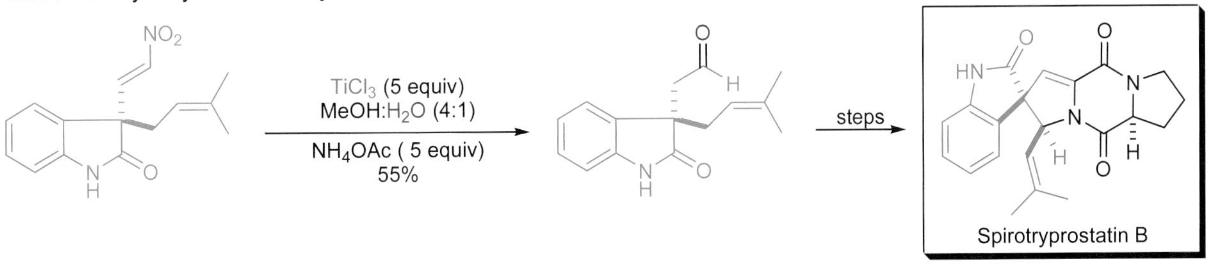

In the laboratory of B.M. Trost, the second generation asymmetric synthesis of the potent glycosidase inhibitor (–)-cyclophellitol was completed using a *Tsuji-Trost allylation* as the key step.[44] The synthetic plan called for the conversion of the α-nitrosulfone allylation product to the corresponding carboxylic acid or ester. Numerous oxidative *Nef reaction* conditions were tested, but most of them caused extensive decomposition of the starting material or no reaction at all. Luckily, the nitrosulfone could be efficiently oxidized with dimethyldioxirane under basic conditions (TMG) to afford the desired carboxylic acid in high yield.

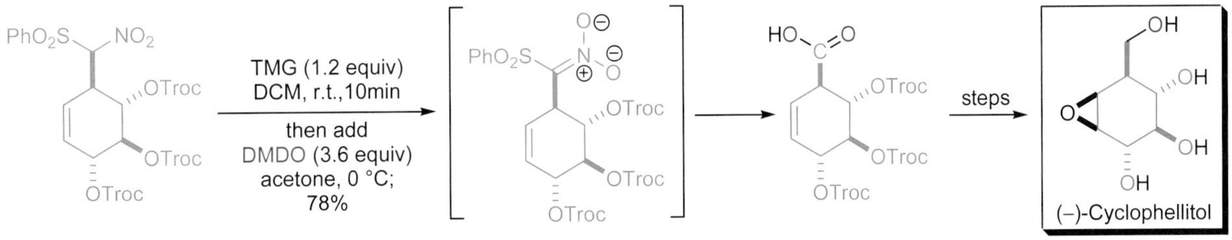

In order to treat influenza infections, the development of neuraminidase inhibitors is required. The currently available compounds are not potent enough, and they have a number of side effects. The stereoselective total synthesis of one potent inhibitor, BXC-1812 (RWJ-270201), was achieved by M.J. Müller and co-workers.[45] The key intermediate substituted nitromethane was prepared *via* a Pd-catalyzed allylation of nitromethane under basic conditions. The transformation of this nitroalkane to the corresponding carboxylic acid methyl ester was carried out in two steps. The *Nef reaction* was conducted in DMF instead of the usual DMSO because DMSO as the solvent caused extensive epimerization of the product. The initially formed carboxylic acid was then esterified.

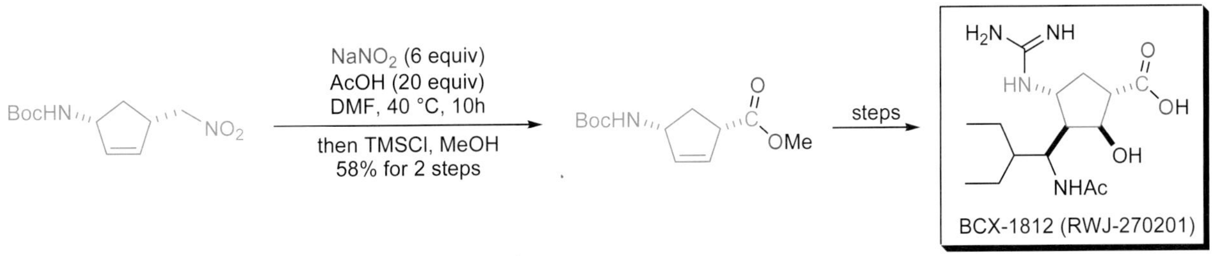

NEGISHI CROSS-COUPLING

(References are on page 637)

Importance:

[*Seminal Publications*[1-6]; *Reviews*[7-24]; *Modifications & Improvements*[25-32]]

In 1972, after the discovery of Ni-catalyzed coupling of alkenyl and aryl halides with Grignard reagents (*Kumada cross-coupling*), it became apparent that in order to improve the functional group tolerance of the process, the organometallic coupling partners should contain less electropositive metals than lithium and magnesium. In 1976, E. Negishi and co-workers reported the first stereospecific Ni-catalyzed alkenyl-alkenyl and alkenyl-aryl cross-coupling of alkenylalanes (organoaluminums) with alkenyl- or aryl halides.[1,2] Extensive research by Negishi showed that the best results (reaction rate, yield, and stereoselectivity) are obtained when organozincs are coupled in the presence of $Pd^{(0)}$-catalysts.[3,4,7] The Pd- or Ni-catalyzed stereoselective cross-coupling of organozincs and aryl-, alkenyl-, or alkynyl halides is known as the *Negishi cross-coupling*. The general features of the reaction are: 1) both Ni- and Pd-phosphine complexes work well as catalysts. However, the Pd-catalysts tend to give somewhat higher yields and better stereoselectivity, and their functional group tolerance is better; 2) the active catalysts are relatively unstable $Ni^{(0)}$- and $Pd^{(0)}$-complexes but these can be generated *in situ* from more stable $Ni^{(II)}$- and $Pd^{(II)}$-complexes with a reducing agent (e.g., 2 equivalents of DIBAL-H or *n*-BuLi); 3) in the absence of the transition metal catalyst, the organozinc reagents do not react with the alkenyl halides to any appreciable extent; 4) the most widely used ligand is PPh_3, but other achiral and chiral phosphine ligands have been successfully used; 5) the various organozinc reagents can be prepared by either direct reaction of the organic halide with zinc metal or activated zinc metal or by transmetallation of the corresponding organolithium or Grignard reagent with a zinc halide (ZnX_2);[33,34] 6) the use of organozinc reagents allows for a much greater functional group tolerance in both coupling partners than in the *Kumada cross-coupling* where organolithiums and Grignard reagents are utilized as coupling partners; 7) other advantages of the use of organozincs include: high reactivity, high regio-, and stereoselectivity, wide scope and applicability, few side reactions and almost no toxicity; 8) the reaction is mostly used for the coupling of two $C(sp^2)$ carbons but $C(sp^2)$-$C(sp)$ as well as $C(sp^2)$-$C(sp^3)$ couplings are well-known; 9) besides organozincs, compounds of Al and Zr can also be utilized; 10) if the organoaluminum and organozirconium derivatives are not sufficiently reactive, they can be transmetallated by the addition of zinc salts, and this protocol is referred to as the *double metal catalysis*;[35] and 11) of all the various organometals (Al, Zr, B, Sn, Cu, Zn), organozincs are usually the most reactive in Pd-catalyzed cross-coupling reactions and do not require the use of additives (e.g., bases as in *Suzuki cross-couplings*) to boost the reactivity;[20] Some of the limitations of the *Negishi cross-coupling* are: 1) propargylzincs do not couple well but homopropargylzincs do; 2) secondary and tertiary alkylzincs may undergo isomerization, but cross-couplings of primary alkyl- and benzylzincs give satisfactory results; and 3) due to the high reactivity or organozincs, CO insertion usually does not happen unlike in the case of less reactive organotins (see *carbonylative Stille cross-coupling*).

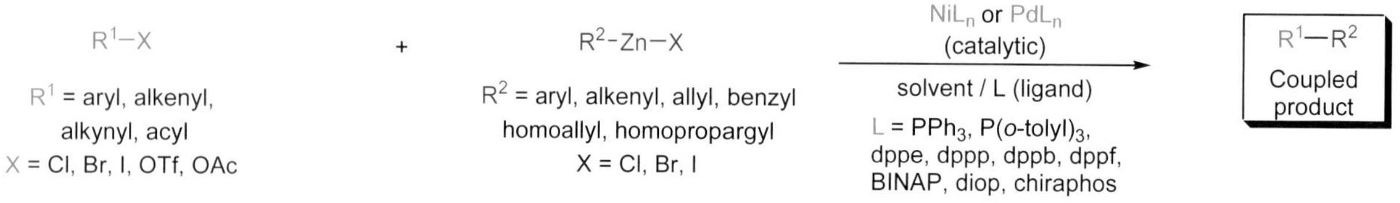

Mechanism:[10]

Ni-catalyzed process:

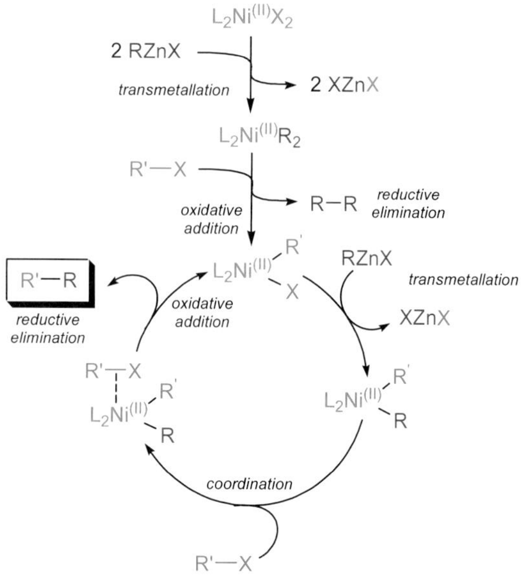

Pd-catalyzed process:

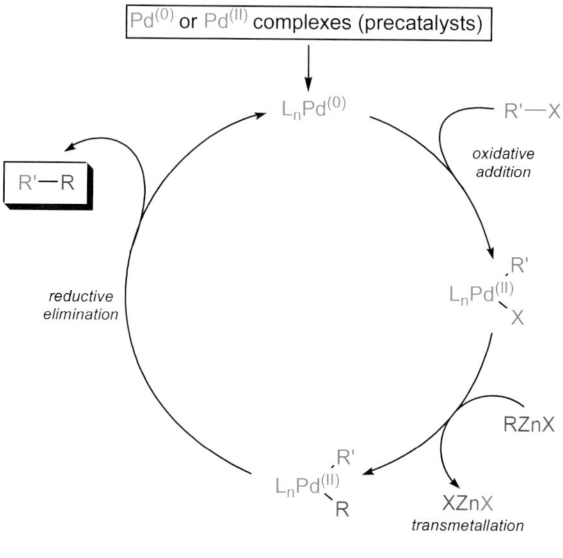

Synthetic Applications:

The *Negishi cross-coupling* was utilized during the final stages of the total synthesis of caerulomycin C for the preparation of the bipyridyl system by T. Sammakia et al.[36] The highly substituted 6-bromopyridine was coupled, in the presence of $Pd_2(dba)_3/PPh_3$ catalyst system, with 2-lithiopyridine, which was transmetallated by $ZnCl_2$ *in situ* to the corresponding organozinc reagent. Interestingly, the analogous *Stille cross-coupling* using 2-tributylstannyl pyridine was far less efficient and gave a low yield of the desired product.

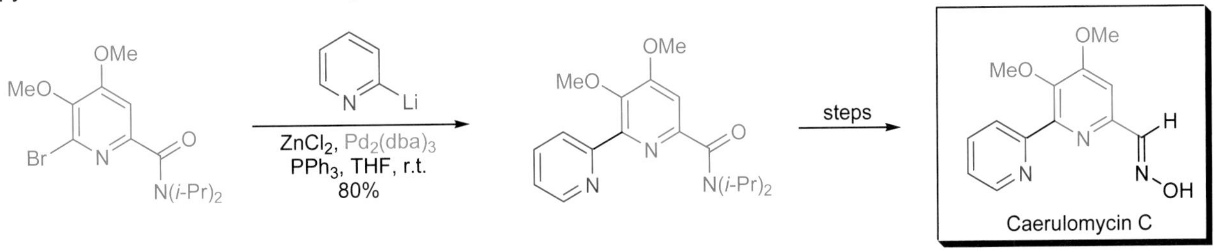

The *modified Negishi protocol* was used in J.S. Panek's total synthesis of (–)-motuporin to couple the left-hand subunit organozinc compound with the right-hand subunit (*E*)-vinyl iodide.[37] The left-hand subunit was prepared by the *Schwartz hydrozirconation* of a disubstituted alkyne to give an (*E*)-trisubstituted zirconate, which was subsequently transmetalated with anhydrous $ZnCl_2$. The resulting vinylzinc species was immediately treated with one equivalent of the (*E*)-vinyl iodide in the presence of 5 mol% $Pd(PPh_3)_4$ to afford the (*E,E*)-diene coupled product with complete stereoselectivity.

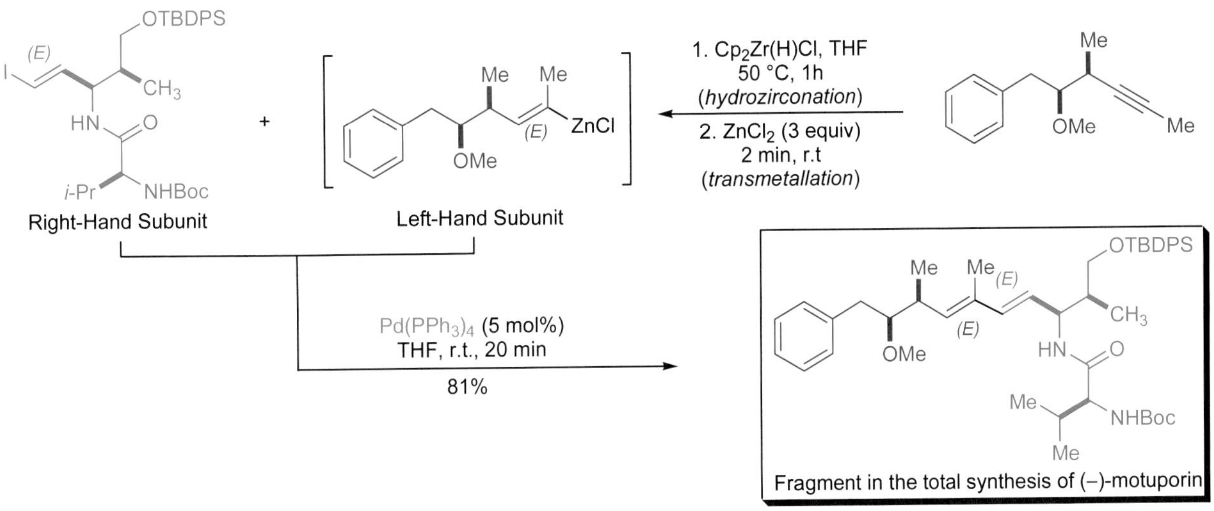

The convergent and stereocontrolled synthesis of (+)-amphidinolide J was achieved in the laboratory of D.R. Williams.[38] To install the (*E*)- C7-C8 double bond stereoselectively, a homoallylic alkylzinc reagent was coupled with an (*E*)-vinyl iodide using the *Negishi reaction*. The very stable homoallylic alkylzinc species was prepared in one pot from the corresponding homoallylic iodide by treatment with two equivalents of *t*-BuLi followed by transmetallation with $ZnCl_2$. The addition of the (*E*)-vinyl iodide in the presence of catalytic amounts $Pd(PPh_3)_4$ gave the coupled 1,5-diene product in high yield.

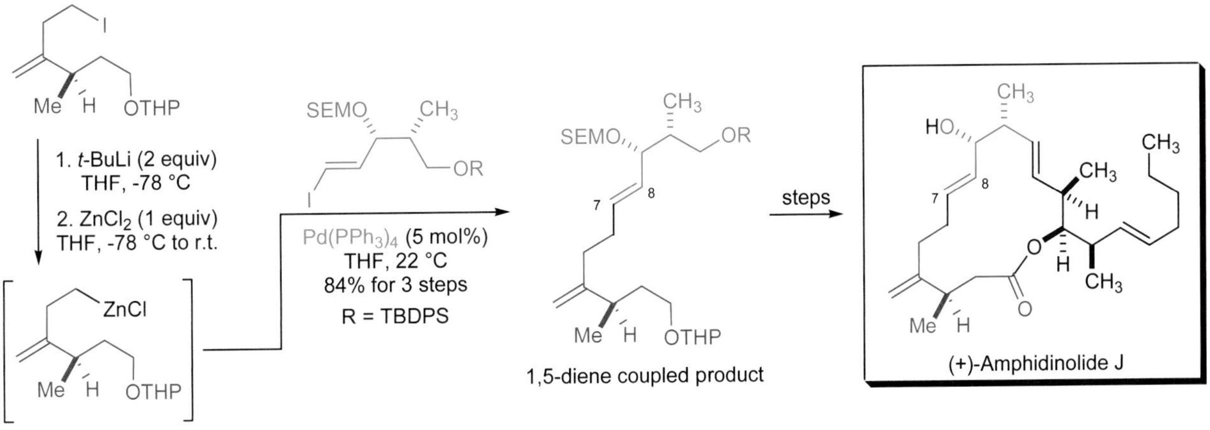

NENITZESCU INDOLE SYNTHESIS

(References are on page 638)

Importance:

[*Seminal Publications*[1]; *Reviews*[2-6]; *Modifications & Improvements*[7-20]]

In 1929, C.D. Nenitzescu described the reaction of *p*-benzoquinone with 3-aminocrotonate in acetone at reflux temperature from which he isolated a 2-methyl-5-hydroxyindole derivative.[1] For the next two decades, the reaction was not explored further, but during the 1950s the scope and limitation of the transformation was thoroughly investigated and applied to the synthesis of melanin-related compounds. The condensation of a 1,4-benzoquinone with enamines to afford substituted 5-hydroxyindole derivatives is known as the *Nenitzescu indole synthesis*. The general features of the reaction are:[3] 1) the benzoquinone component can be unsubstituted, mono-, di-, or trisubstituted; 2) the degree of substitution does not have a significant effect on the rate of the reaction; 3) the structure of the enamine component may be varied widely: β-aminocrotonates (R^3=Me and R^4=O-alkyl), β-aminoacrylates, β-aminoacrylamides (R^4=NH_2 or NR_2), and even β-amino-α,β-unsaturated ketones can be used; 4) when R^4=O-alkyl, the resulting 3-alkoxycarbonyl indoles can be easily decarboxylated; 5) in most instances the R^3 substituent should be other than hydrogen; 6) yields can be very high, but occasionally low yields are observed (varies from substrate to substrate); 7) the reaction is regioselective, and the regioselectivity is strongly influenced by the nature of the substituents on the quinone component; 8) an electron-donating group (e.g., R^1=OH, O-alkyl) at the C2 position deactivates the C3 position and directs the attack of the nucleophile to C5; 9) an electron-withdrawing group (e.g., R^1=CO_2-alkyl, CF_3) at C2 directs the attack of the nucleophile preferentially to the C3 position; 10) a small substituent at C2, which is moderately electron-donating (e.g., R^1=Me, Cl) results in possible nucleophilic attack at either C5 or C6 and the formation of a mixture of regioisomeric indoles is expected; 11) when the C2 substituent is sterically demanding (e.g., R^1=*t*-Bu), the nucleophile is expected to attack at C5 preferentially; 12) besides 5-hydroxyindoles, other heterocycles such as benzofurans can be prepared using the *Nenitzescu reaction* between *N,N*-dialkylaminocrotonates and benzoquinones;[17] and 13) instead of the *p*-benzoquinone, the corresponding quinone imides and quinone diimides can also be used.[7,9]

Nenitzescu (1929):

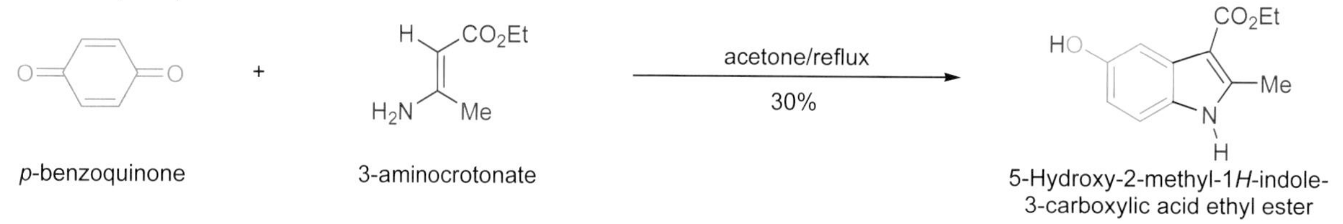

Nenitzescu indole synthesis:

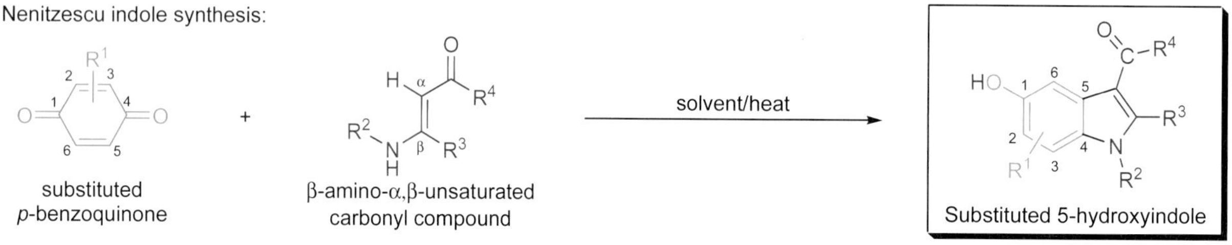

R^1 = H, alky, aryl, OH, *O*-alkyl, *O*-aryl, Cl, Br, CF_3, CO_2-alkyl, etc.; R^2 = H, alkyl, cycloalkyl, aryl, benzyl; R^3 = alkyl, aryl, CO_2-alkyl, *O*-alkyl; R^4 = alkyl, aryl, *O*-alkyl, NH_2, NR_2; solvent: acetone, EtOH, $MeNO_2$, AcOH, $CHCl_3$

Mechanism:[21-27]

The mechanism of the *Nenitzescu indole synthesis* is not fully understood. The most likely first step is a *Michael addition* of the enamine to the *p*-quinone. In the resulting Michael adduct, the imine nitrogen attacks the proximal carbonyl group of the quinone and the bicyclic hemiaminal and then undergoes dehydration to give the 5-hydroxyindole product. In an alternative mechanism, an oxidation-reduction mechanism is proposed: the Michael adduct tautomerizes to the corresponding hydroquinone, which is oxidized by the starting *p*-quinone to another *p*-quinone, which undergoes intramolecular cyclization to qive a quinonimmonium intermediate. This intermediate in turn is a viable oxidant of the hydroquinone and itself gets reduced to give the 5-hydroxyindole product.[27]

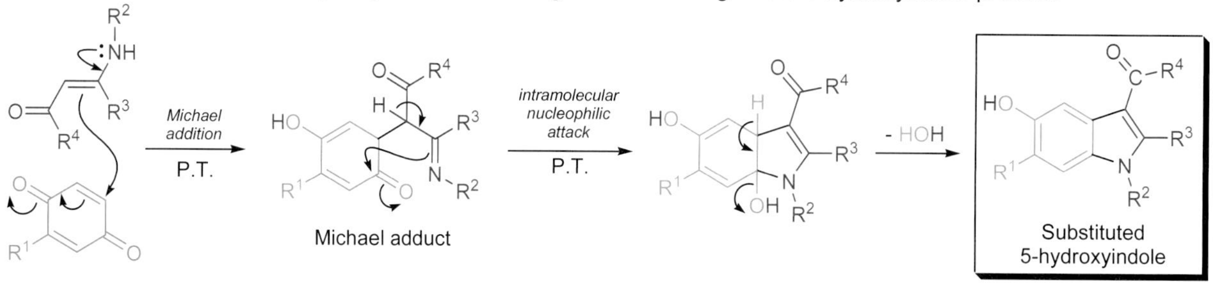

Synthetic Applications:

A facile synthesis of the key intermediate of EO 9, a novel and fully synthetic bioreductive alkylating indolequinone, was accomplished by M. Kasai et al.[28] The authors' goal was to develop a short and efficient synthesis in order to prepare large quantities of the target. The highly functionalized indole nucleus was constructed in one step using the *Nenitzescu indole synthesis*. The benzoquinone and the enamine were dissolved in the solvent mixture and heated to afford the desired methyl-5-hydroxy-2-methoxymethylindole-3-carboxylate in moderate yield. In the work-up step, the excess benzoquinone was destroyed with sodium dithionate ($Na_2S_2O_4$) and the product was crystallized thus obviating the need for chromatographic separation.

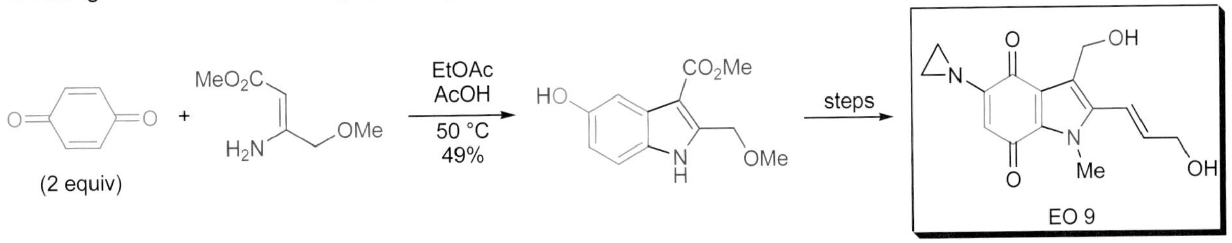

The synthesis of the first potent and selective secretory phospholipase A_2 (s-PLA_2) inhibitor, LY311727, was carried out in the laboratory of M.J. Martinelli.[29] The indole core of the target was prepared by the *Nenitzescu indole synthesis*, which proceeded in high yield. The enamine component was readily prepared from methyl propionylacetate (3-oxo-pentanoic acid methyl ester) and benzylamine in the presence of catalytic amounts of TsOH. A thorough screening of various solvents pinpointed nitromethane as the optimal solvent for the transformation, since the product crystallized from the reaction mixture and was easily removed by filtration.

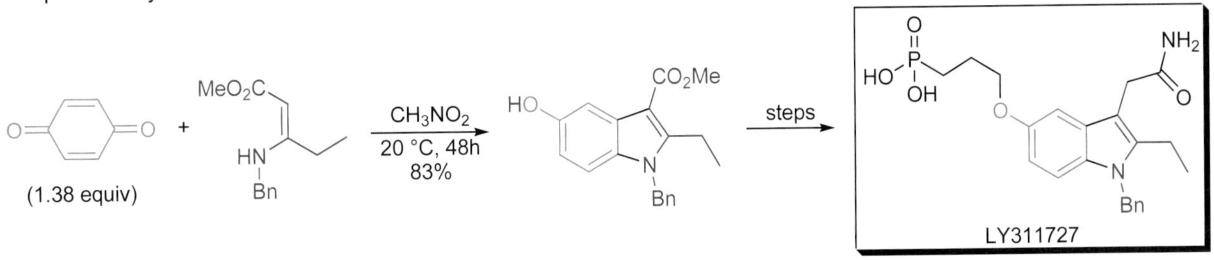

The *Nenitzescu indole synthesis* can be formally regarded as a one-pot three-component condensation where all the components are readily available: β-keto esters, primary amines, and *p*-benzoquinones. This observation prompted the research team of D.M. Ketcha to develop the *solid-phase version of the Nenitzescu indole synthesis* for the preparation of 5-hydroxyindole-3-carboxamides.[30] The process began with the acetoacetylation of ArgoPore®-Rink-NH_2 resin with diketene to obtain a polymer-bound acetoacetamide, which was then converted to the corresponding enamine upon condensation with primary amines and in the presence of trimethyl orthoformate (dehydrating agent). The indole formation generally took place in nitromethane much more efficiently than in acetone, and it was completely regioselective, giving rise exclusively to the C6 regioisomer.

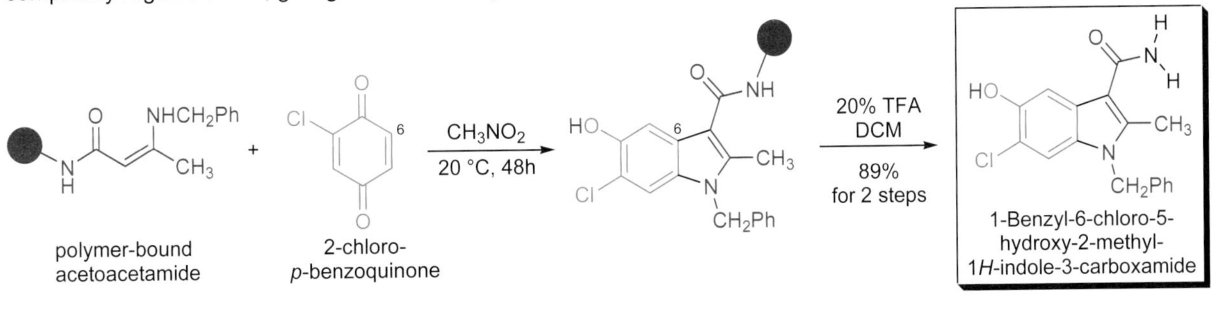

An interesting variant of the *Nenitzescu indole synthesis,* involving the Lewis acid-directed coupling of enol ethers with benzoquinone mono- and *bis*-imides, was developed by T.A. Engler et al. for the synthesis of substituted β- and γ-tetrahydrocarbolines.[31]

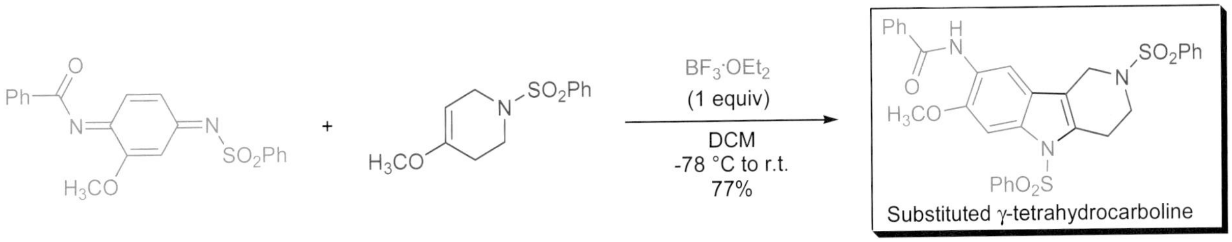

NICHOLAS REACTION

(References are on page 639)

Importance:

[*Seminal Publications*[1-4]; *Reviews*[5-14]; *Modifications & Improvements*[15-19]; *Theoretical Studies*[20,21]]

In 1972, K.M. Nicholas and R. Pettit reported that dicobalt hexacarbonyl-complexed propargylic alcohols were easily dehydrated upon treatment with acid to form the corresponding 1,3-enynes. However, uncomplexed propargylic alcohols did not react under identical conditions.[2] This finding suggested that the intermediates of these reactions were the dicobalt hexacarbonyl-stabilized propargylic cations, which in fact could be isolated and were shown to have significant stability.[3] The trapping of dicobalt hexacarbonyl-stabilized propargylic cations with various nucleophiles is known as the *Nicholas reaction*. The alkyne functionality of the resulting substituted products can be regenerated by a mild oxidation. The general features of the *Nicholas reaction* are: 1) propargylic alcohols are easily prepared by the addition of acetylides to ketones and aldehydes and readily converted to various derivatives; 2) the alkyne complexes are obtained in almost quantitative yields by reacting the propargyl derivatives with $Co_2(CO)_8$ in an appropriate solvent (ether, pentane, hexane, benzene, etc.);[1] 3) the cobalt-alkyne complexes are red, brown, or purple solids or oils that are moderately air stable and can be purified with flash chromatography; 4) the stabilized propargylic cations are either generated by the addition of Brönsted or Lewis acids to propargylic derivatives or by the addition of electrophiles to 1,3-enyne-cobalt hexacarbonyl complexes; 5) a wide range of nucleophiles reacts with the resulting propargylic cations including C-, O-, N-, and S-nucleophiles (see scheme); 6) after the substitution the cobalt complexes can be decomplexed either oxidatively (most common) or reductively; 7) *oxidative decomplexation* regenerates the triple bond, while *reductive decomplexation* (e.g., Li/liquid ammonia, H_2/Rh-catalyst, or Wilkinson catalyst) yields the corresponding alkene; 8) when the cobalt complex is not removed, it can be used in a subsequent Pauson-Khand reaction; 9) the reaction can be both inter- and intramolecular, and even macrocyclization can be achieved; and 10) there are no allene side products that often complicates the reactions of uncomplexed propargylic substrates.

Nicholas & Pettit (1972):

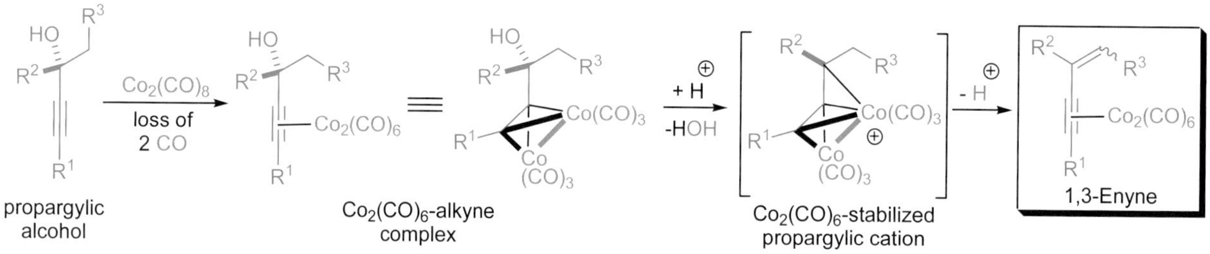

Reaction of $Co_2(CO)_6$-stabilized propargylic cations with nucleophiles (Nicholas, 1977):

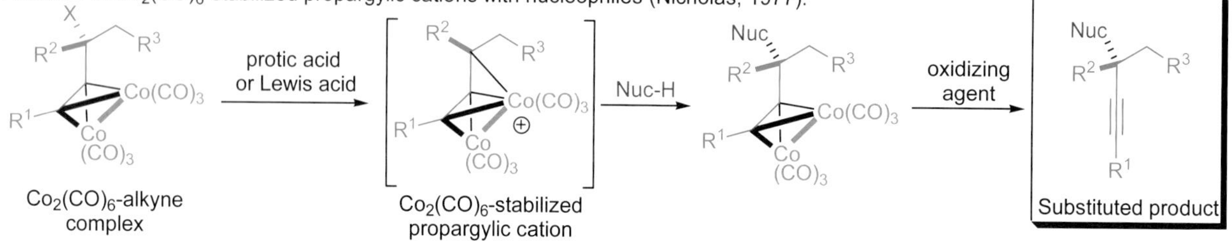

R^{1-3} = H, alkyl, aryl; X = OH, O-alkyl, O-benzyl, O-silyl, acetal, OAc, OCOAr, OCO*t*-Bu, OMs, OTf, Cl; Nuc-H = e-rich aromatics, simple alkenes, allylsilanes, allylstannanes, enol ethers, silylketene acetals, ROH, N_3^-, RNH_2, RR'NH, RSH, HS(R)SH, F^-; oxidizing agent: CAN, $Fe(NO_3)_3$, NMO, TMANO, TBAF, C_5H_5N/air/ether, DMSO/H_2

Mechanism: [2,22,20,23]

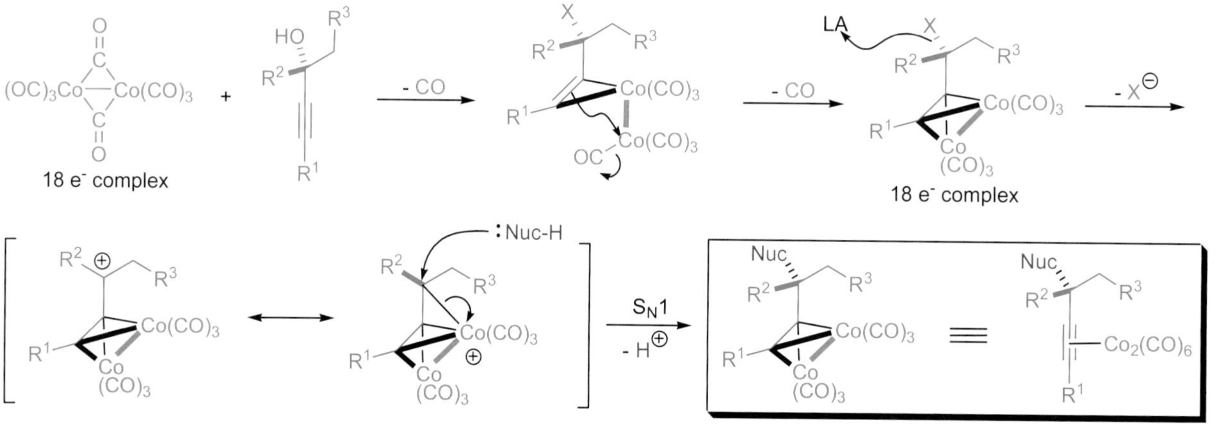

Synthetic Applications:

The *Nicholas reaction* was used to synthesize the β-lactam precursor of thienamycin in the laboratory of P.A. Jacobi and thereby accomplish its formal total synthesis.[24] The necessary β-amino acid was prepared by the condensation of a boron enolate (derived from an acylated oxazolidinone) with the cobalt complex of an enantiopure propargylic ether. The resulting adduct was oxidized with ceric ammonium nitrate (CAN) to remove the cobalt protecting group from the triple bond, and the product was obtained with a 17:1 *anti:syn* selectivity and in good yield.

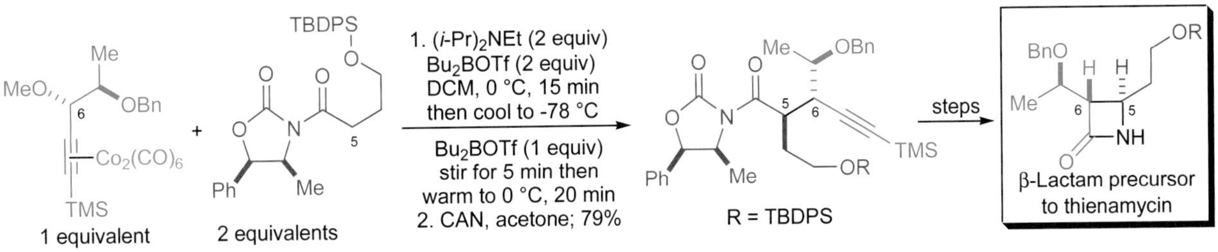

The total syntheses of (+)-secosyrins 1 and 2 was achieved and their relative and absolute stereochemistry was unambiguously established by C. Mukai and co-workers.[25] To construct the spiro skeleton of these natural products, the *intramolecular Nicholas reaction* was utilized. The alkyne substrate was first converted to the dicobalt hexacarbonyl complex by treatment with $Co_2(CO)_8$ in ether. Exposure of the resulting complex to boron trifluoride etherate at room temperature brought about the ring closure with inversion of configuration at C5 to afford the expected tetrahydrofuran derivative. The minor product was the C5 epimer which was formed only in 15% yield.

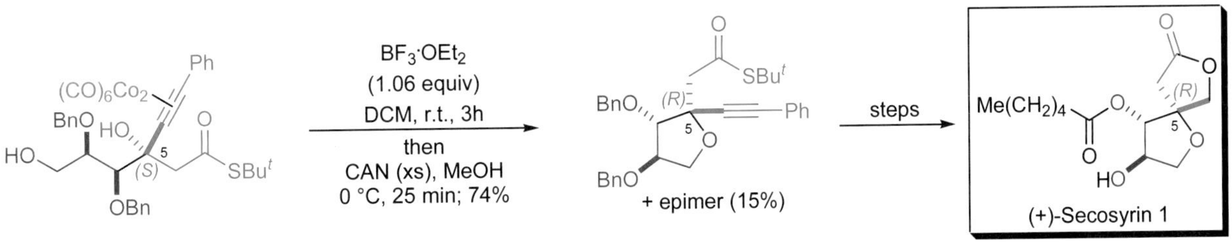

The tandem use of the *intramolecular Nicholas reaction* and the *Pauson-Khand reaction* was featured in S.L. Schreiber's total synthesis of (+)-epoxydictymene.[26] The propargylic acetal, a 1:1 mixture of diastereomers at the acetal carbon, was readily converted to the $Co_2(CO)_6$-complex in excellent yield. The treatment of this complex with a stoichiometric amount of Et_2AlCl afforded the 5-8 fused bicyclic ring system of the natural product as a single diastereomer in 91% yield. The allylsilane served as the nucleophile to capture the stabilized propargylic cation. The alkyne protecting group was not removed as later this cobalt-alkyne complex was utilized in the *Pauson-Khand reaction*.

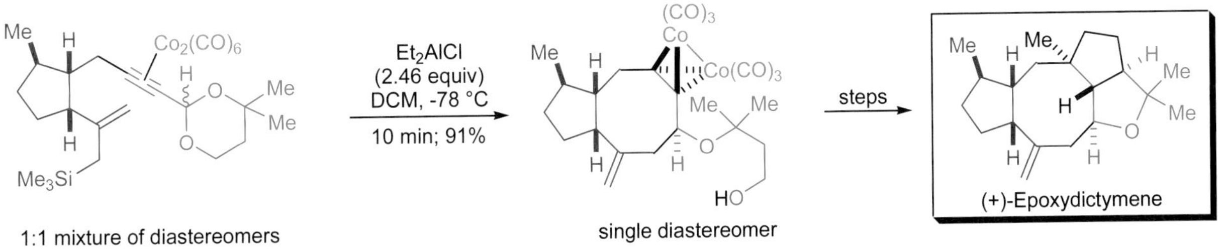

The application of the *intramolecular Nicholas reaction* by C. Mukai et al. made it possible to develop a novel procedure for the construction of oxocane derivatives.[27] Interestingly, several Lewis and Brönsted acids gave rise to complex mixtures. However, the use of mesyl chloride/triethylamine in refluxing DCM afforded the desired oxocane as the sole product.

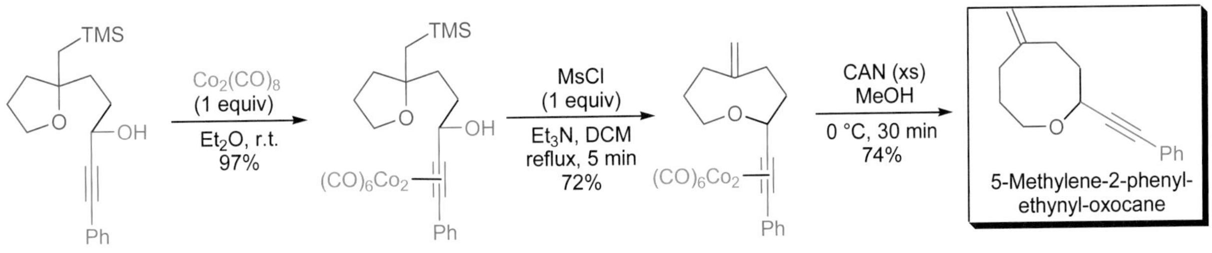

NOYORI ASYMMETRIC HYDROGENATION

(References are on page 640)

Importance:

[*Seminal Publications*[1-4]; *Reviews*[5-32]; *Modifications & Improvements*[33-37]; *Theoretical Studies*[38,39]]

In 1980, T.S.R. Noyori and co-workers reported that cationic BINAP-Rh complexes catalyzed the *asymmetric hydrogenation* of α-(acylamino) acrylic acids or esters to give the corresponding amino acid derivatives in high enantiomeric excess.[1] However, these rhodium catalysts could be used only for the synthesis of amino acids, the rate of hydrogenation was very slow, and the reaction conditions had to be chosen very carefully for each substrate to achieve high enantioselectivity. A few years later, the preparation of BINAP-Ru$^{(II)}$ dicarboxylate complexes proved to be generally applicable for the asymmetric hydrogenation of a wide range of functionalized olefins.[2] Oligomeric halogen-containing BINAP-Ru$^{(II)}$ complexes were found to be efficient catalysts for the asymmetric hydrogenation of functionalized ketones in which coordinative nitrogen, oxygen, and halogen atoms near the C=O functionality direct the reactivity and the absolute stereochemistry of the product.[3,4] The reduction of functionalized olefins and ketones with hydrogen gas (H_2) using BINAP-Ru$^{(II)}$ complexes as catalyst is known as the Noyori asymmetric hydrogenation. The general features of the reaction are: 1) BINAP, a conformationally flexible atropisomeric C_2-symmetric diphosphane ligand is available in both enantiomeric forms;[40,41] 2) the various BINAP-Ru$^{(II)}$ complexes are easily prepared and the catalyst loadings are small; 3) hydrogenation of α,β-unsaturated and β,γ-unsaturated carboxylic acids takes place in alcohol solvents, where the sign and degree of enantioselection are highly dependent on the substitution pattern and hydrogen pressure;[42] 4) allylic and homoallylic alcohols are hydrogenated with high enantioselectivity;[43] 5) substituted enamides give rise to enantio-enriched α- or β-amino acids;[44,45] 6) the sense of chirality is predictable in the hydrogenation of functionalized ketones and preexisting stereogenic centers in the substrate significantly influence the outcome;[3] 7) the double hydrogenation of 1,3-diones *via* chiral β-hydroxy ketones give rise to *anti* 1,3-diols in almost 100% ee;[3] 8) β-keto esters are the best substrates for asymmetric hydrogenation;[46] and 9) racemic β-keto esters with a configurationally labile α-stereocenter can be transformed into a single stereoisomer with high selectivity by undergoing an *in situ* inversion of configuration in the presence of a base (*dynamic kinetic resolution*).[47,48]

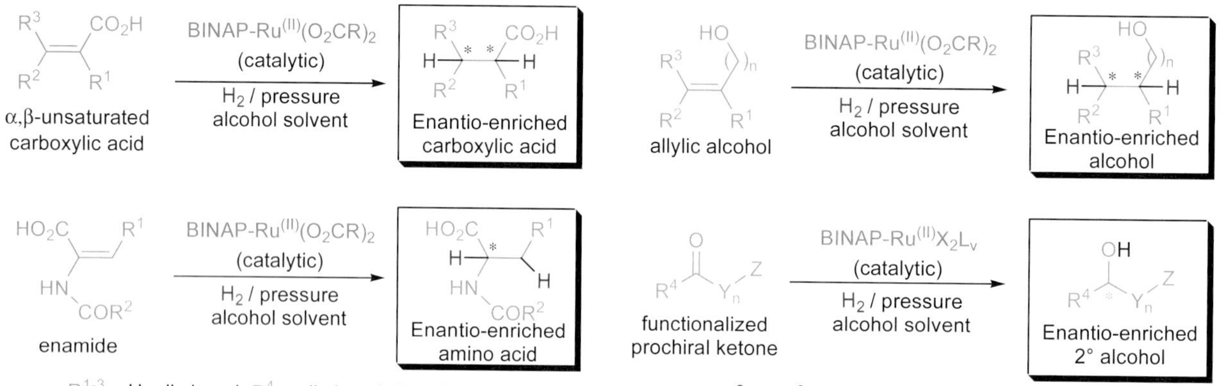

R^{1-3} = H, alkyl, aryl; R^4 = alkyl, aryl; Z = nitrogen, oxygen, halogen; Y = sp^2 or sp^3 hybridized carbon; n =1-3; X = halogen; L = neutral ligand or solvent; R = alkyl, aryl

Mechanism: [1,49-64]

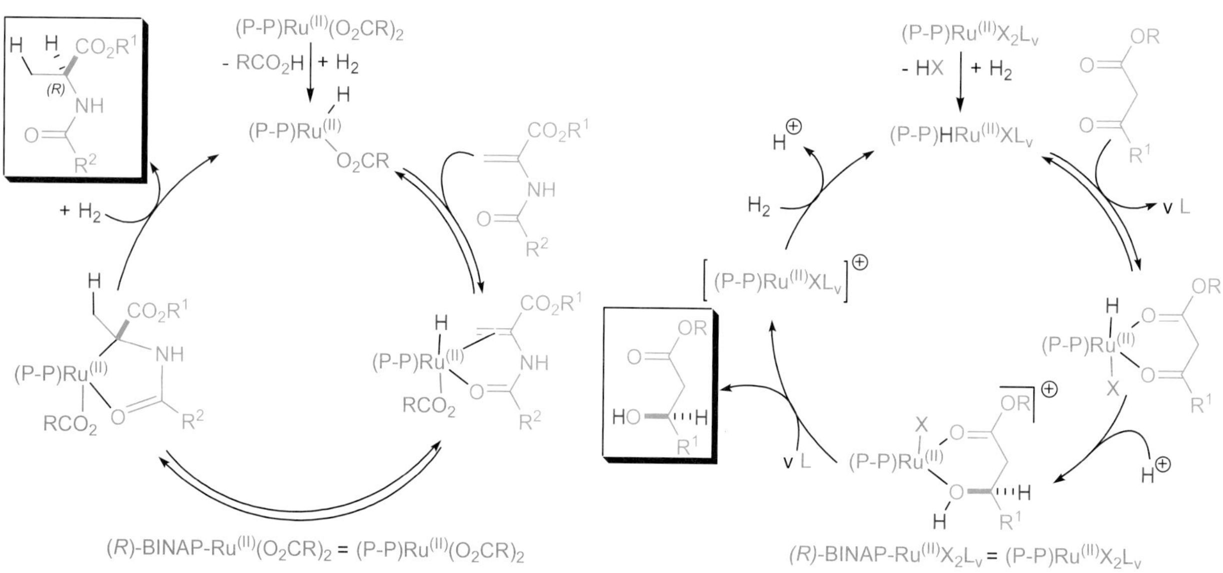

Synthetic Applications:

The total synthesis of pentacyclic alkaloid (–)-haliclonadiamine was accomplished by D.F. Taber and co-workers.[65] The *Noyori asymmetric hydrogenation* was used to prepare a bicyclic β-hydroxy ester intermediate in enantiopure form from a racemic bicyclic β-keto ester *via* kinetic resolution. It was found that the hydrogenation only took place in the presence of added HCl and by optimizing the amount of HCl added, the proportion of the total reduced ketone could be controlled. About 87% of the "matched" ketone was reduced, while the other β-keto ester enantiomer was not significantly converted to the reduced product. Interestingly, the diastereoselectivity of the hydrogenation depended on the nature of the added acid: with HCl, the *trans* diastereomer was the major product, while with AcOH the *cis* diastereomer was dominant.

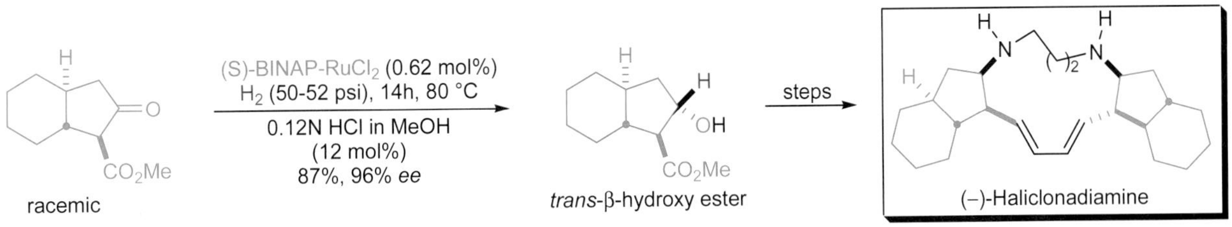

The convergent and stereocontrolled synthesis of the C17-C28 fragment (CD spiroketal unit) of spongistatin 1 was achieved in the laboratory of W.R. Roush.[66] One of the building blocks was prepared by using the *Noyori asymmetric hydrogenation* of a readily available β-keto ester, which gave rise to the corresponding β-hydroxy ester in 81% yield and 95% *ee*.

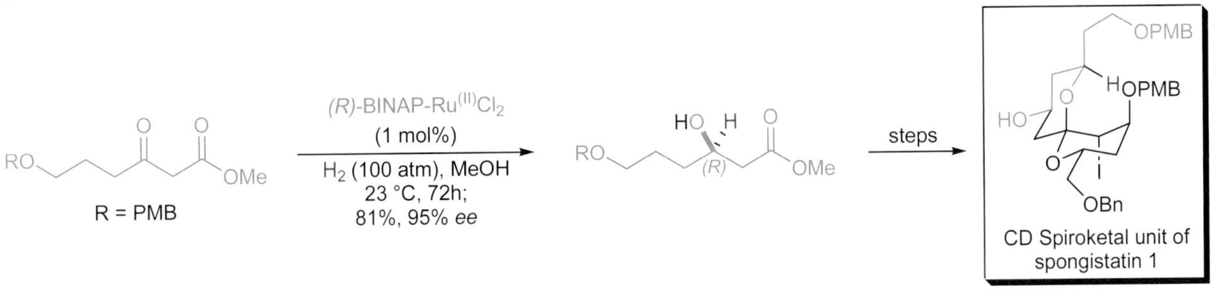

A pronounced enhancement of stereoselectivity was observed in the *asymmetric hydrogenation* of 2-substituted 2-propen-1-ols by *transient acylation* in the laboratory of O. Mitsunobu.[67] The aroylation of the allylic alcohol hydroxyl group prior to the hydrogenation gave the best results.

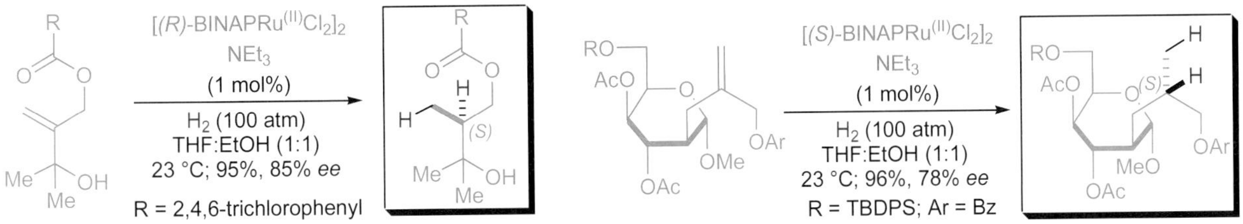

The *Noyori asymmetric transfer hydrogenation* was utilized in the synthesis of the chiral 1,2,3,4-tetrahydroisoquinolines by R.A. Sheldon et al.[68] These compounds are important intermediates in the Rice and Beyerman routes to morphine. The "Rice imine" was exposed to a series of chiral Ru(II) complexes, which was prepared from η^6-arene-Ru(II) chloride dimeric complexes and *N*-sulfonated 1,2-diphenylethylenediamines along with the azeotropic mixture of HCOOH/NEt3. With the best catalyst the desired tetrahydroisoquinoline was isolated in 73% yield and the enantiomeric excess was 99%.

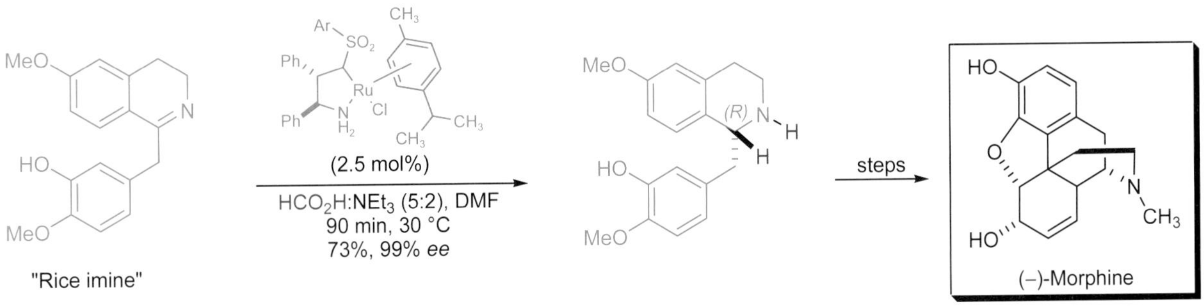

NOZAKI-HIYAMA-KISHI REACTION

(References are on page 641)

Importance:

[*Seminal Publications*[1-7]; *Reviews*[8-16]; *Modifications & Improvements*[17-30]]

In 1977, H. Nozaki and T. Hiyama et al. reacted aldehydes and ketones with organochromium(III) reagents, which were generated *in situ* from allyl and vinyl halides upon treatment with $CrCl_2$ under aprotic and oxygen-free conditions, and obtained the corresponding allylic and homoallylic alcohols with high chemospecificity and stereoselectivity.[1,2] In 1986, Y. Kishi and H. Nozaki independently discovered that traces of nickel salts catalyzed the formation of carbon-chromium(III) bonds, even from otherwise less reactive substrates (e.g., vinyl and aryl halides). This modification helped to make the process more reliable.[6,7] The one-pot *Barbier-type addition* of alkenyl, alkynyl, aryl, allyl, or vinylchromium compounds to aldehydes or ketones is known as the *Nozaki-Hiyama-Kishi (NHK) reaction*. Since its discovery, the *NHK reaction* has become a powerful synthetic tool for the chemoselective formation of carbon-carbon bonds under very mild conditions and has been applied to the total synthesis of a number of complex natural products. The general features of the reaction are: 1) the $CrCl_2$ is either purchased commercially or prepared by the reduction of $CrCl_3$ prior to the reaction; 2) $Cr^{(II)}$ is a one-electron donor, and therefore two moles of the chromium(II) salt are required to reduce one mol of organic halide to the corresponding organochromium(III) reagent; 3) it can take place both inter- and intramolecularly, and the thermodynamic driving force is the formation of a strong O-$Cr^{(III)}$ bond; 4) aldehydes react markedly faster than ketones, so when both functional groups are present, the reaction of the organochromium species with aldehydes proceeds with complete chemoselectivity; 5) because of their low basicity, organochromium reagents are compatible with a wide range of sensitive functional groups; 6) it is possible to maintain the integrity of the various electrophilic functional groups within polyfunctional organochromium reagents; and 7) the addition of crotylchromium(III) reagents to aldehydes is highly diastereoselective and stereoconvergent: in all cases, the *anti* homoallylic alcohol is favored, independent of the configuration of the starting crotyl halide. The drawbacks of the *NHK reaction* are: 1) the nickel and chromium salts are very toxic; 2) the redox potential of $Cr^{(II)}$ shows a significant dependence on the solvents used as the reaction medium and solvent mixtures need to be used for optimum results; 3) usually a large excess of $CrCl_2$ is required, especially in macrocyclization reactions; and 4) the Lewis acidic salts formed during the preparation of $CrCl_2$ may alter the stereochemical outcome of the reaction for polyfunctional substrates where chelation control determines the stereochemical course.

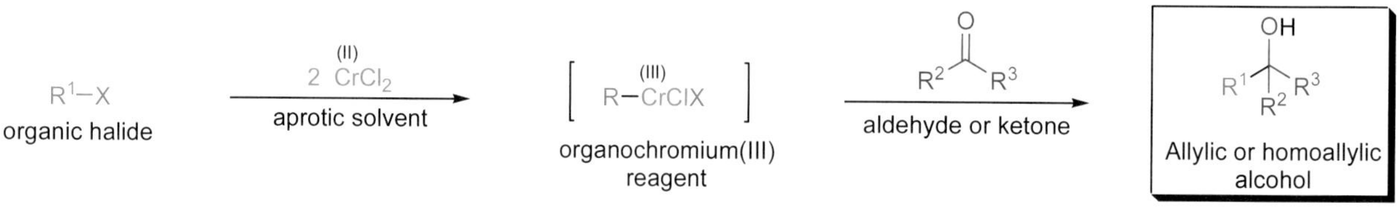

R^1 = alkenyl, aryl, allyl, vinyl, propargyl, alkynyl, allenyl; X = Cl, Br, I, OTf, etc.; R^2, R^3 = alkyl, aryl, alkenyl, H;
solvent: DMF, DMSO, THF

Mechanism:[6,18,19,9,10,13]

In the nickel(II)-catalyzed *NHK reaction*, the first step is the reduction of $Ni^{(II)}$ to $Ni^{(0)}$ that inserts into the halogen-carbon bond *via* an oxidative addition. The *organonickel species* transmetallates with $Cr^{(III)}$ to form the organochromium(III) nucleophile, which then reacts with the carbonyl compound. To make the process environmentally benign, a chromium-catalyzed version was developed where a chlorosilane was used as an additive to silylate the chromium alkoxide species in order to release the metal salt from the product.[18,19] The released $Cr^{(III)}$ is reduced to $Cr^{(II)}$ with manganese powder.

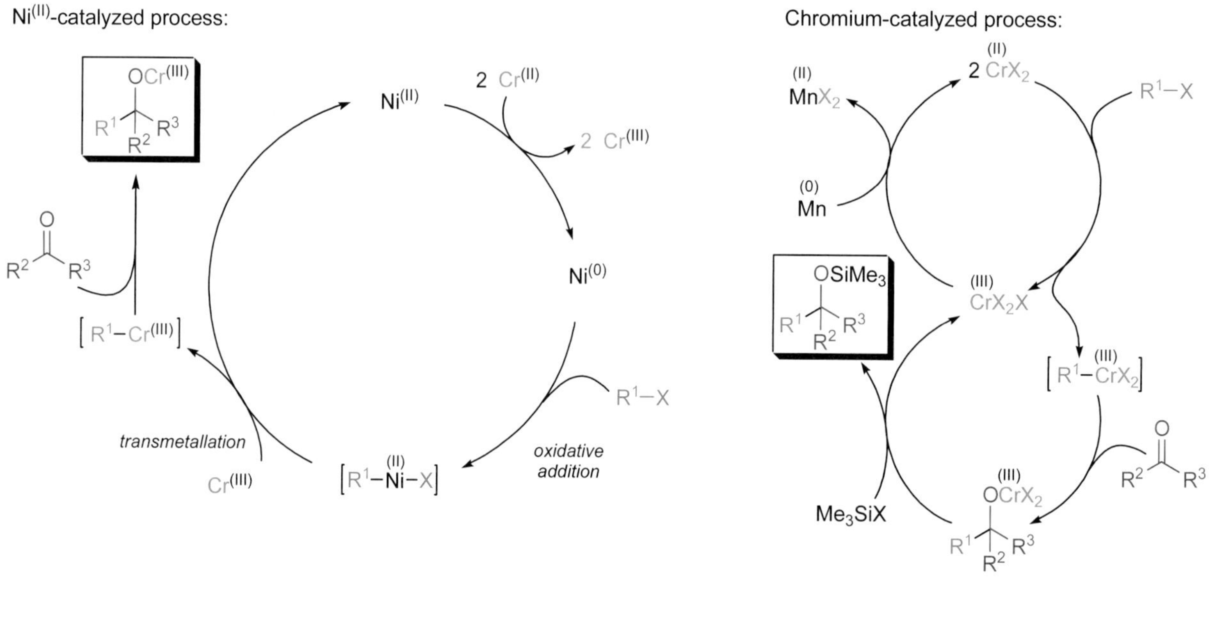

Synthetic Applications:

In the laboratory of G.A. Molander, a general route for the synthesis of eunicellin diterpenes was developed and was applied for the asymmetric total synthesis of deacetoxyalcyonin acetate.[31] One of the key steps was an *inramolecular NHK coupling* reaction between an enol triflate and an aldehyde. The cyclopentenol product was formed in high yield as a 2:1 mixture of diastereomers. The undesired diastereomer could be transformed to the desired one using a *Mitsunobu reaction*.

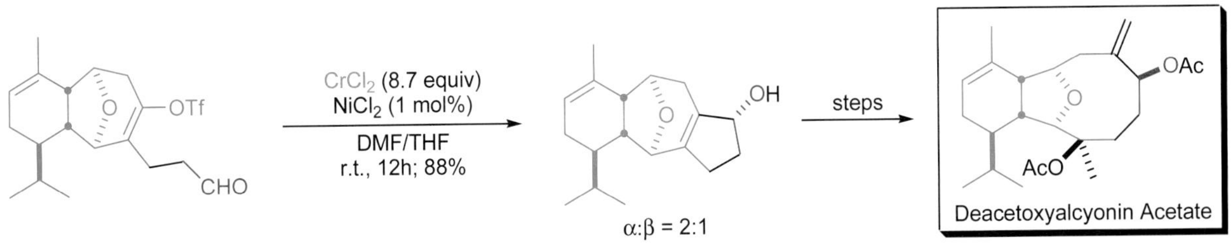

The C1-C19 fragment of (-)-mycalolide was assembled by J.S. Panek et al. *via* the *NHK coupling* between the C1-C6 vinyl iodide and C7-C19 aldehyde subunits.[32] The desired allylic alcohol was obtained as a 1:1 mixture of stereoisomers and was oxidized to the corresponding ketone using Dess-Martin periodinane. The synthesis of the C1-C19 fragment was completed in three more steps.

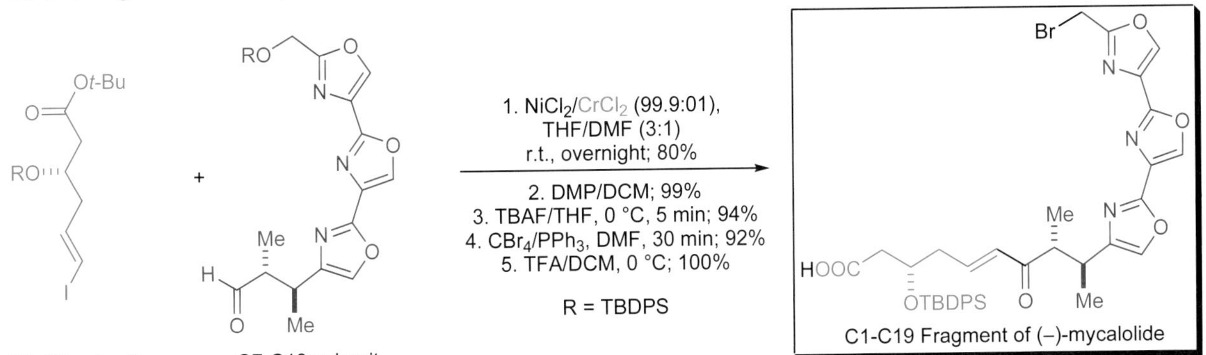

C1-C6 subunit

C7-C19 subunit

One of the key steps during the first total synthesis of (–)-aspinolide B by A. de Meijere and co-workers was the *NHK reaction* to form the ten membered lactone ring.[33] The precursor for this key macrocyclization step was prepared by forming an ester from a three-carbon monoprotected diol fragment and a seven-carbon vinyl iodide fragment. Deprotection of the primary alcohol and its subsequent oxidation afforded the desired vinyl iodide aldehyde precursor. Exposure of this precursor to 15 equivalents of $CrCl_2$ doped with 0.5% of $NiCl_2$ at high dilution in DMF afforded the desired diastereomer in a 1.5:1 ratio.

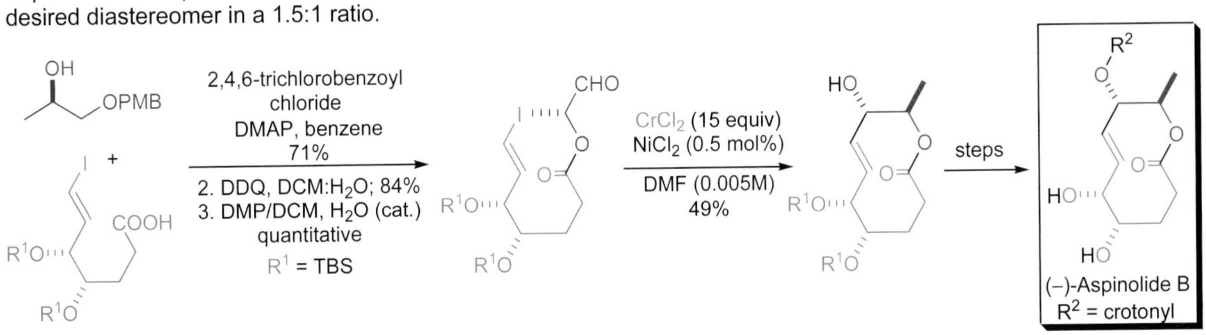

A novel approach to the elaboration of the C12-C13 trisubstituted olefin portion of epothilone D was developed by R.E. Taylor et al.[34] The authors used sequential *NHK coupling* and a thionyl chloride induced *allylic rearrangement* followed by the reductive removal of the chiral auxiliary.

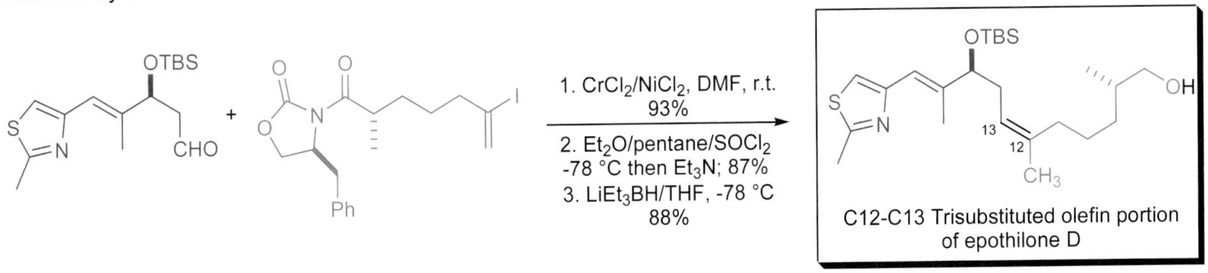

OPPENAUER OXIDATION

(References are on page 642)

Importance:

[*Seminal Publications*[1]; *Reviews*[2-7]; *Modifications & Improvements*[8-17]; *Theoretical Studies*[18,17]]

In 1937, R.V. Oppenauer oxidized steroids with secondary alcohol functionality to the corresponding ketones using acetone in benzene in the presence of catalytic amounts of aluminum *tert*-butoxide.[1] This oxidation proved to be high yielding and superior to other existing oxidation methods due to its mildness. Oppenauer's method came more than a decade after three researchers independently described reduction of carbonyl compounds with the use of aluminum alkoxides: 1) in 1925, H. Meerwein successfully reduced aldehydes with ethanol in the presence of aluminum ethoxide;[19] 2) during the same year A. Verley reduced ketones with aluminum ethoxide as well as aluminum isopropoxide but found that sterically hindered ketones (e.g., camphor) reacted very slowly;[20] and 3) in 1926, W. Ponndorf demonstrated that the reduction of aldehydes and ketones was general for a variety of metal alkoxides (e.g., alkali metal and aluminum alkoxides) derived from secondary alcohols, and he found the process completely reversible.[21] The oxidation of primary and secondary alcohols with ketones in the presence of metal alkoxides (e.g., aluminum isopropoxide) to the corresponding aldehydes and ketones is known as the *Oppenauer oxidation*.[22] The reverse reaction, the reduction of aldehydes and ketones to alcohols, is referred to as the *Meerwein-Ponndorf-Verley reduction*. The general features of the *Oppenauer oxidation* are: 1) the reaction is completely reversible and can be driven to completion according to *Le Chatelier's principle* by adding large excess of the ketone (e.g., acetone) to the reaction mixture; 2) the reaction conditions are mild, since the substrates are usually heated in acetone/benzene mixtures; 3) most functional groups are tolerated (alkenes, alkynes, esters, amides, etc.), but if the substrate contains basic nitrogen atoms, the use of alkali metal alkoxides is necessary in place of aluminum alkoxides;[23] 4) in order to achieve reasonable reaction rates, stoichiometric amounts of the aluminum alkoxide needs to be used; 5) most commonly aluminum isopropoxide, *t*-butoxide, and phenoxide are used; 6) a wide range of primary and secondary alcohols are oxidized under the reaction conditions; 6) secondary alcohols are oxidized much faster than primary alcohols, so complete chemoselectivity can be achieved (this feature makes the *Oppenauer oxidation* unique compared to other oxidations); 7) overoxidation of aldehydes to carboxylic acids never happens; 8) the oxidation of 1,4- and 1,5-diols usually yields lactones; 9) acetone is used most often as the oxidant, but aromatic and aliphatic aldehydes are suitable as oxidants due to their low reduction potentials; 10) addition of protic acids dramatically increases the rate of oxidation;[9] and 11) the oxidation can be conducted using heterogeneous catalysts (e.g., alumina, zeolites), which has one great advantage over the traditional homogeneous variant: the catalyst can be easily separated from the reaction mixture.[12,5] The most important side reactions are: 1) *aldol condensation* of aldehyde products, which have an α-hydrogen atom to form β-hydroxy aldehydes and/or α,β-unsaturated aldehydes, but with ketones this side reaction is not common; 2) *Tishchenko reaction* of aldehyde products with no α-hydrogen atom, but this can be suppressed by the use of anhydrous solvents; and 3) the migration of the double bond during the oxidation of allylic and homoallylic alcohol substrates.[4]

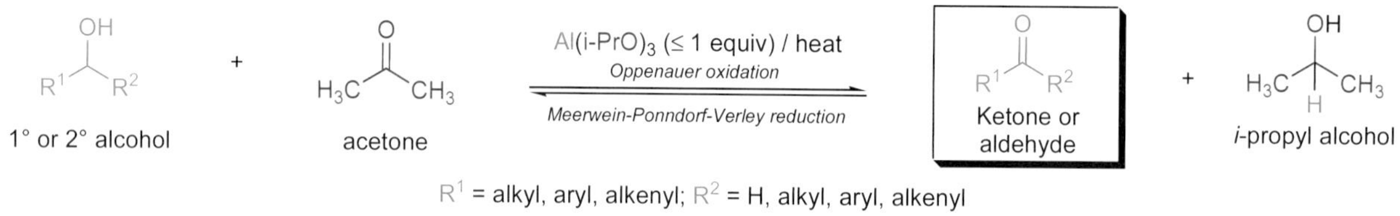

Mechanism:[24-29]

Both the oxidant carbonyl compound (acetone) and the substrate alcohol are bound to the metal ion (aluminum). The alcohol is bound as the alkoxide, whereas the acetone is coordinated to the aluminum which activates it for the hydride transfer from the alkoxide. The hydride transfer occurs *via* a six-membered chairlike transition state. The alkoxide product may leave the coordination sphere of the aluminum *via* alcoholysis, but if the product alkoxide has a strong affinity to the metal, it results in a slow ligand exchange, so a catalytic process is not possible. That is why often stoichiometric amounts of aluminum alkoxide is used in these oxidations.

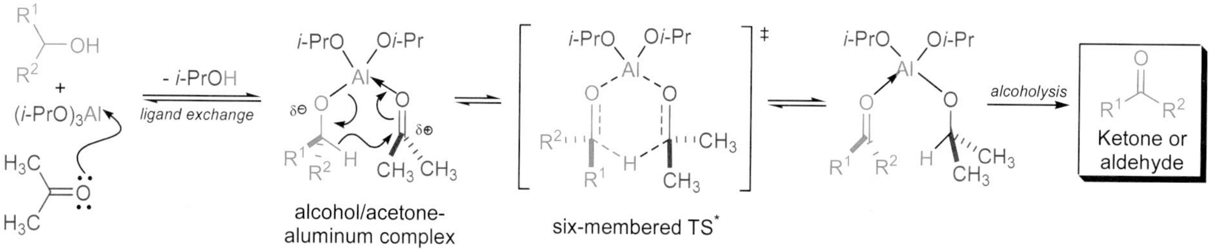

Synthetic Applications:

The *modified Oppenauer oxidation* was used in the synthesis of estrone by P. Kočovský et al.[30] The tetracyclic diol was exposed to aluminum isopropoxide and *N*-methyl-piperidine-4-one (oxidizing agent)[8] to obtain the corresponding enone in good yield. The formation of the enone involved the migration of the initial β,γ-double bond. The treatment of this enone with TsOH overnight in ether led to the formation of estrone by aromatization.

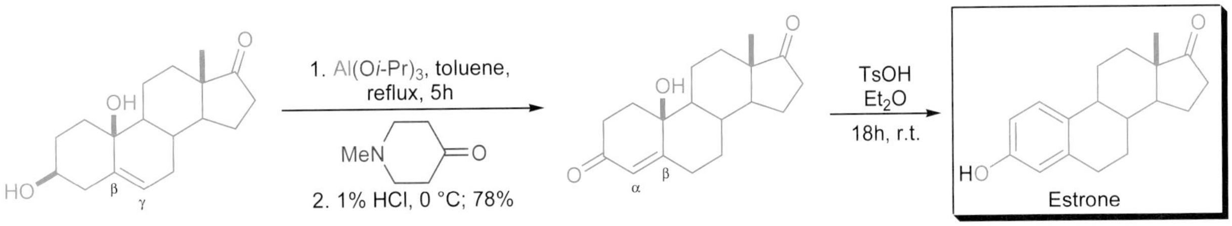

An *intramolecular Diels-Alder reaction* was the key step in D.D. Sternbach's total synthesis of the linearly fused triquinane (±)-hirsutene.[31] The cycloaddition took place between a cyclopentadiene ring and an α,β-unsaturated ketone that was generated *in situ* by using the *Oppenauer oxidation*.

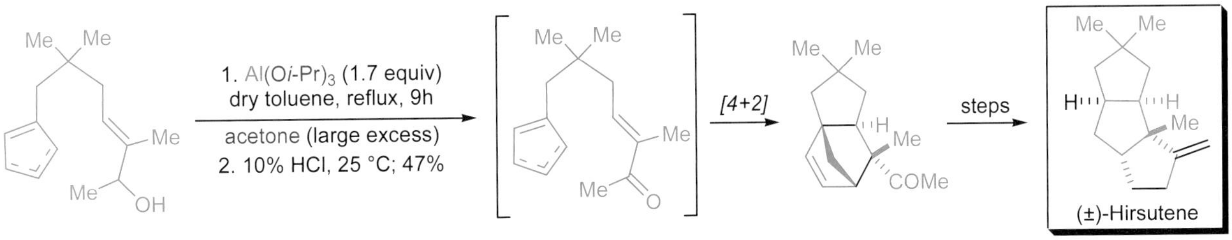

The total synthesis of several lycopodium alkaloids was accomplished by C.H. Heathcock and co-workers.[32] At the final stages of the synthesis of (±)-lycodoline, a *modified Oppenauer oxidation* was planned to carry out the transformation of a primary alcohol to the corresponding aldehyde. However, when the substrate was treated with potassium *t*-butoxide and benzophenone in refluxing benzene, the only product was an *N*-dealkylated tricyclic amino ketone (via *retro Michael reaction*). This problem was resolved by substituting the KO*t*-Bu with potassium hydride which efficiently removed the protons from both the primary and tertiary alcohols, thereby preventing the *retro Michael reaction*. The oxidation product aldehyde quickly underwent a facile *aldol condensation* to form the tricyclic enone.

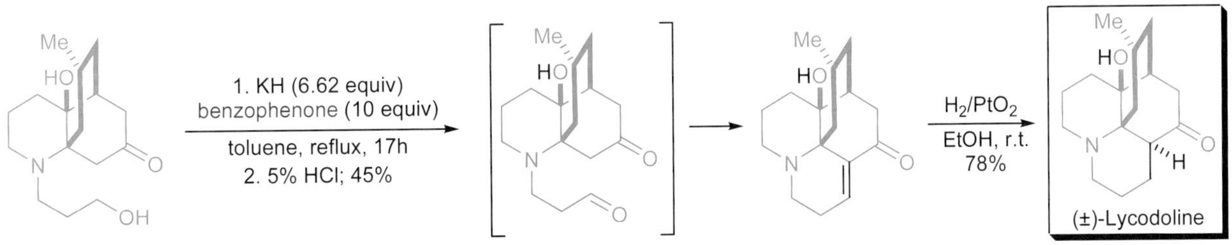

The tricyclic ring system containing the fully functionalized CD ring of taxol was prepared from (*S*)-(+)-carvone by T.K.M. Shing et al.[33] The bicyclic α-hydroxy ketone (4-hydroxy-5-one) was isomerized by an *intramolecular redox reaction* in the presence of catalytic amounts of aluminum isopropoxide. This example was a special case where both reactants were in the same molecule: the ketone was the oxidant for the *Oppenauer oxidation*, whereas the secondary alcohol was the hydride donor for the *MVP reduction*. The conversion to the thermodynamically more stable 5-hydroxy-4-one proceeded in good yield.

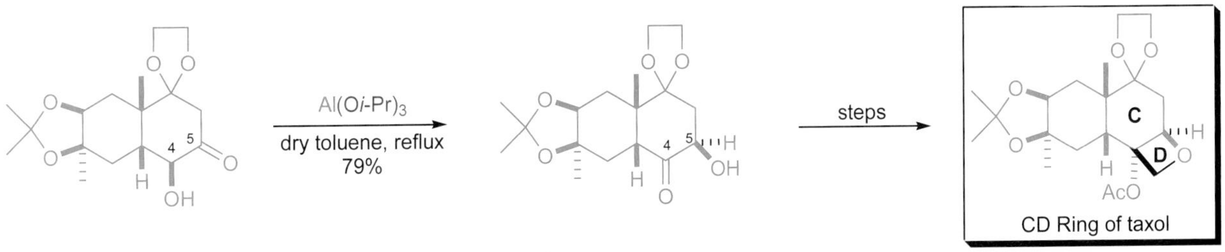

OVERMAN REARRANGEMENT

(References are on page 643)

Importance:

[*Seminal Publications*[1-3]; *Reviews*[4-8]; *Modifications & Improvements*[9-20]; *Theoretical Studies*[21]]

In 1937, O. Mumm and F. Möller, while investigating the mechanism of the *Claisen rearrangement*, observed that the thermal rearrangement of *N*-phenyl-benzimidic acid allyl ester afforded *N*-allyl-*N*-phenyl-benzamide in quantitative yield.[1] They also showed that the termini of the allyl group were switched as a result of the transformation. For the next few decades, several research groups reported similar rearrangements of allylic imidates, but the preparation of the substrates were low yielding, and the relatively harsh conditions did not allow these reactions to become synthetically useful.[4] In 1974, L.E. Overman described the facile thermal and mercuric ion catalyzed rearrangement of allylic trichloroacetimidates to afford the corresponding trichloroacetamides.[2] The 1,3-transposition of alcohol and amine functionalities *via* the *[3,3]*-sigmatropic rearrangement of allylic trichloroacetimidates is known as the *Overman rearrangement*. The general features of the reaction are: 1) the allylic trichloroacetimidates are easily prepared in almost quantitative yield by reacting allylic alcohols with trichloroacetonitrile in the presence of catalytic amounts of base (e.g., NaOR, KOR, DBU);[22,23] 2) heating the crude trichloroacetimidates in a solvent (e.g., xylenes) usually between 25-140 °C for several hours or exposure to certain metal catalysts results in a *[3,3]*-sigmatropic rearrangement;[23,15,18-20] 3) isolated yield of the allylic trichloroacetamides is usually high; 4) the allylic trichloroacetamides can be hydrolyzed under basic conditions (3*M* NaOH solution at room temperature) to afford the corresponding allylic amines; 5) the rearrangement is completely regiospecific, therefore no trichloroacetamide product with an unrearranged carbon skeleton is formed; 6) the rearrangement of trichloroacetimidates derived from secondary allylic alcohols proceeds with a high level of stereoselectivity and preferentially the (*E*)-alkenes are formed; 7) the metal catalysts are usually $Hg^{(II)}$-salts, which are used in 10-20 mol% quantities; 8) the mercury(II)-salts can be removed from the product by flash chromatography or by complexation with pyridine or PPh_3; 9) the metal catalysis, however, usually works well only for imidates derived from 3-substituted primary allylic alcohols and in all other cases the thermal conditions are preferred; 10) the imidates of certain cyclohexenyl allylic alcohols may undergo a competitive elimination;[3] 11) propargylic trichloroacetimidates rearrange to give trichloroacetamido-1,3-dienes;[9] and 12) the trichloroacetamide functionality can be used as a radical precursor or transformed into acylureas or guanidine derivatives.[24,25]

Mumm & Moller (1937):

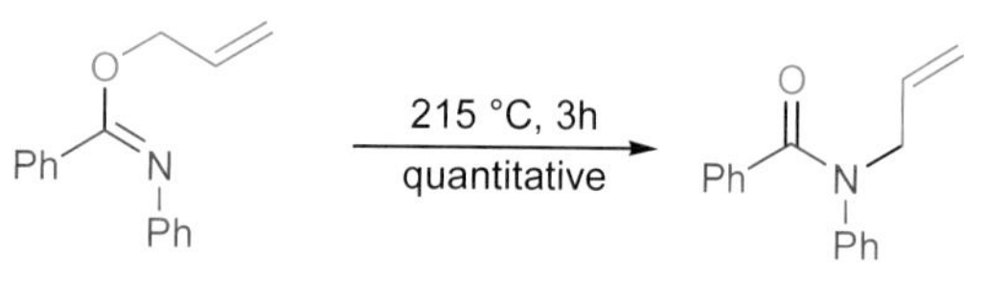

N-Phenyl-benzimidic acid allyl ester — *N*-allyl-*N*-phenyl-benzamide

Overman (1974):

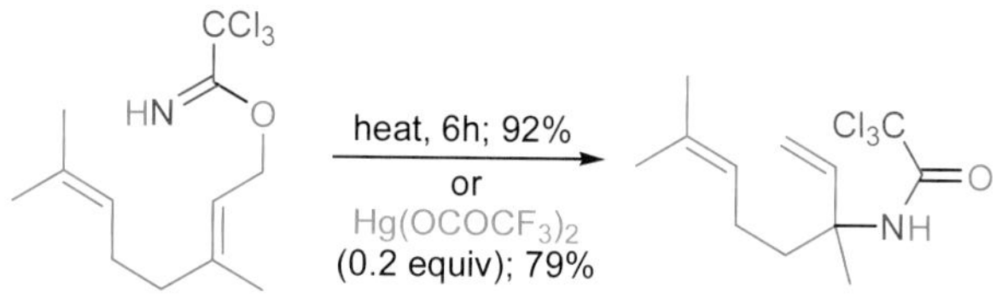

geranyl trichloroacetimidate — geranyl trichloroacetamide

Overman rearrangement:

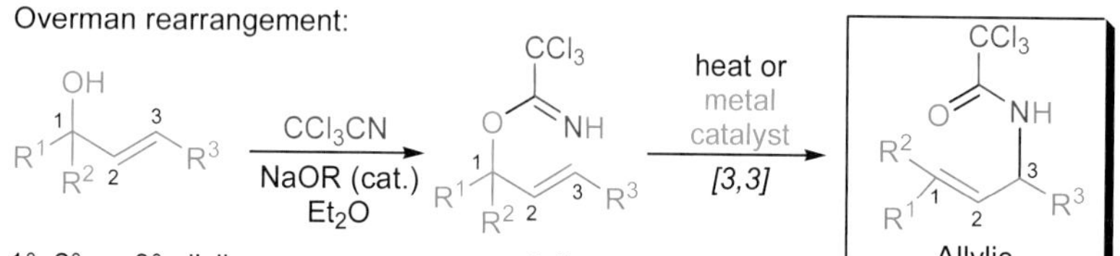

1°, 2°, or 3° allylic alcohol — allylic trichloroacetimidate — Allylic trichloroacetamide

Claisen-type rearrangements:

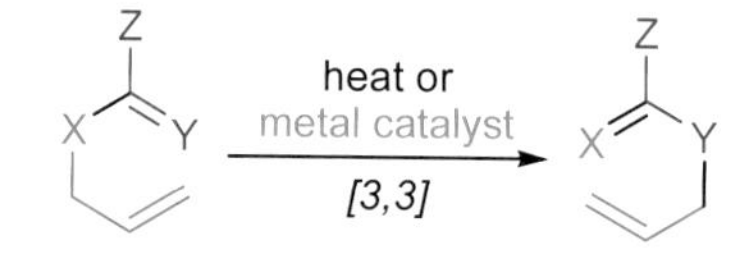

R^{1-3} = H, alkyl, aryl; metal catalyst: $Hg(OCOCF_3)_2$, $Hg(NO_3)_2$, $Pd^{(II)}$-salts; X = O,S, *N*-alkyl, *N*-aryl; Z = CCl_3; Y = NH, *N*-alkyl, *N*-aryl

Mechanism: [2,3,26,27]

Similarly to the mechanism of the *Claisen rearrangement*, the *Overman rearrangement* is a suprafacial, concerted, nonsynchronous *[3,3]*-sigmatropic rearrangement. The reaction is irreversible, which is the result of the significant driving force associated with the formation of the amide functionality. The mechanism of the metal catalyzed reaction is believed to proceed *via* an *iminomercuration-deoxymercuration* sequence and it is only formally a *[3,3]-sigmatropic shift*.

Mechanism of the thermal rearrangement:

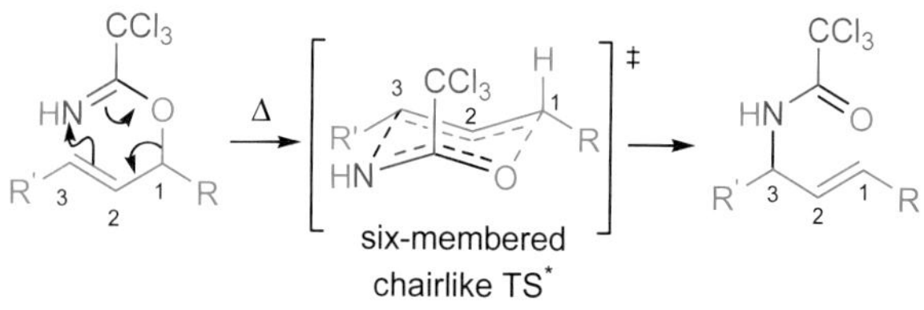

Mechanism of the $Hg^{(II)}$-catalyzed rearrangement:

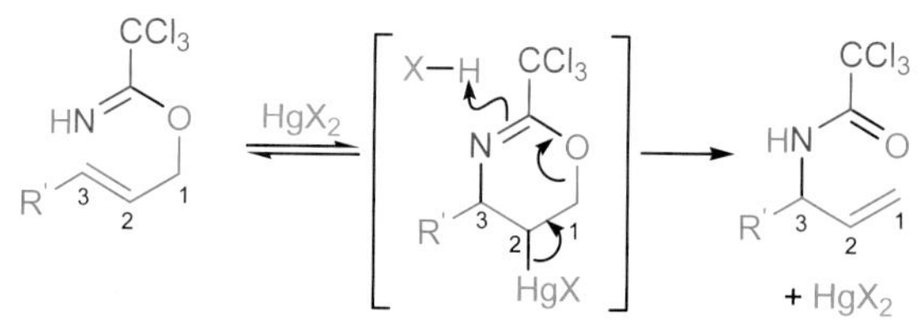

Synthetic Applications:

The total synthesis of sphingofungin E from D-glucose was described by N. Chida and co-workers.[28] The stereocenter at C5 was constructed using the *Overman rearrangement* of an allylic trichloroacetimidate derived from diacetone-D-glucose. The (*Z*)-allylic alcohol was reacted with trichloroacetonitrile in the presence of DBU and the resulting crude trichloroacetimidate was heated in xylenes for six days to afford a 4.3:1 ratio of C5 epimers. Interestingly, the rearrangement of the trichloracetimidate derived from the (*E*)-allylic alcohol gave only moderate yield of the C5 epimers in a 1:4 ratio.

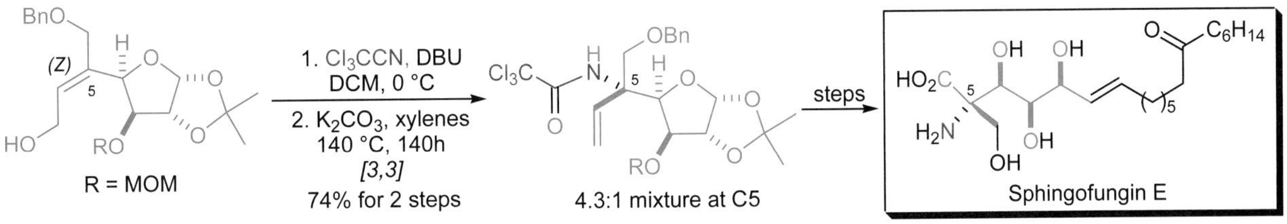

The *Overman rearrangement* was used by S.J. Danishefsky et al. to introduce the nitrogen atom stereoselectively at the C4a position of (±)-pancratistatin.[29] The cyclic allylic alcohol was converted to the trichloroacetimidate in the presence of sodium hydride. The compound was heated as a neat liquid under high vacuum, which afforded the desired rearranged product in reasonable yield.

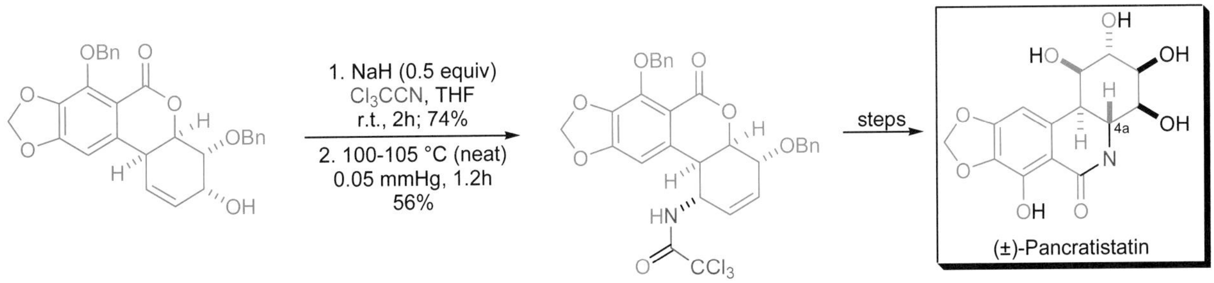

The *transition metal catalyzed Overman rearrangement* allows the reaction to take place at or around room temperature, so thermally sensitive substrates can be used. In the laboratory of M. Mehmandoust, this approach was applied for the synthesis of enantiomerically pure (*E*)-β,γ-unsaturated α-amino acids, which are potent enzyme inhibitors.[30] The trichloroimidate substrates were derived from optically pure monoprotected diallylalcohols and were exposed to 10 mol% of $Pd^{(II)}$-salt. The rearrangements took place rapidly at room temperature with complete transfer of chirality.

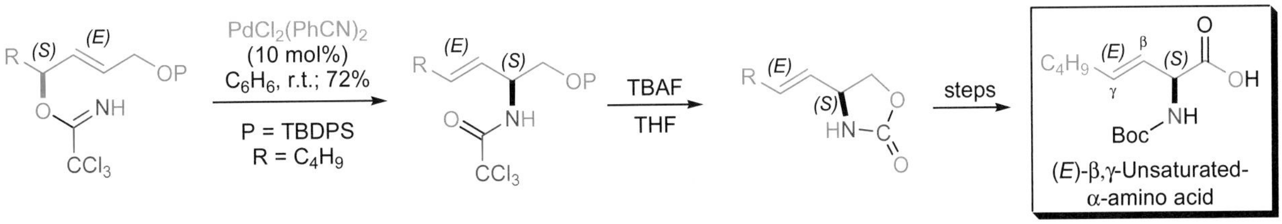

The asymmetric total synthesis of the phenanthroquinolizidine alkaloid (–)-cryptopleurine was reported by S. Kim et al.[31] One of the key steps in the sequence was the *thermal Overman rearrangement* which took place in refluxing toluene in nearly quantitative yield and without any loss of the optical purity of the allyl trichloroimidate substrate.

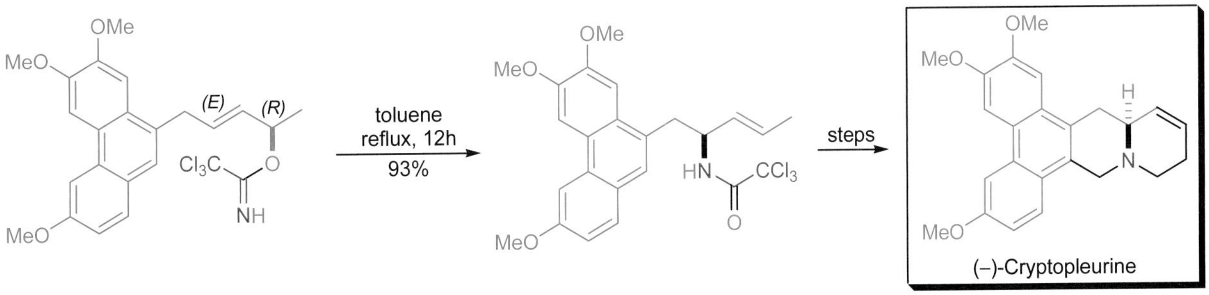

OXY-COPE REARRANGEMENT AND ANIONIC OXY-COPE REARRANGEMENT

(References are on page 643)

Importance:

[*Seminal Publications*[1,2]; *Reviews*[3-14]; *Modifcations & Improvements*[15,16]; *Theoretical Studies*[17,18]]

The thermal *[3,3]-sigmatropic rearrangement* of 1,5-dienes is known as the *Cope rearrangement*. When 1,5-dienes are substituted with a hydroxyl group at the C3 position, they undergo a similar rearrangement to first give enols that are subsequently converted to the corresponding δ,ε-unsaturated carbonyl compounds. The formation of the carbonyl compound is the driving force for the reaction.[19] The *[3,3]-sigmatropic rearrangement* of 1,5-diene-3-ols is called the *oxy-Cope rearrangement*, a term coined by J.A. Berson in 1964.[2] A decade later in 1975, a major improvement in the *oxy-Cope rearrangement* was made when it was found that conversion of the 1,5-diene alcohol to the corresponding potassium alkoxides resulted in 10^{10}-10^{17} rate acceleration of the rearrangement.[15] The *base accelerated oxy-Cope rearrangements* are called *anionic-oxy-Cope rearrangements*. Besides the enormous rate acceleration, there was a considerable drop in the temperature required to bring about the rearrangements. In this anionic rearrangement an enolate anion is first formed, which renders the process irreversible. Potassium bases are used most often along with 18-crown-6 to effect greater charge separation and the maximization of the acceleration. The preparation of the 1,5-diene-3-ol substrates usually involves the 1,2-addition of vinyl organometallics to β,γ-unsaturated aldehydes or ketones or the 1,2-addition of allyl anions to α,β-unsaturated carbonyl compounds. Just as in the parent *Cope rearrangement*, the *oxy-Cope* and *anionic-oxy-Cope rearrangements* are both stereospecific and stereoselective as a result of a cyclic highly ordered transition state. It is worth noting that the use of the *anionic-oxy-Cope rearrangement* in synthesis is advantageous over the *Cope rearrangement* because it does not require high temperature at which side reactions more frequently occur.

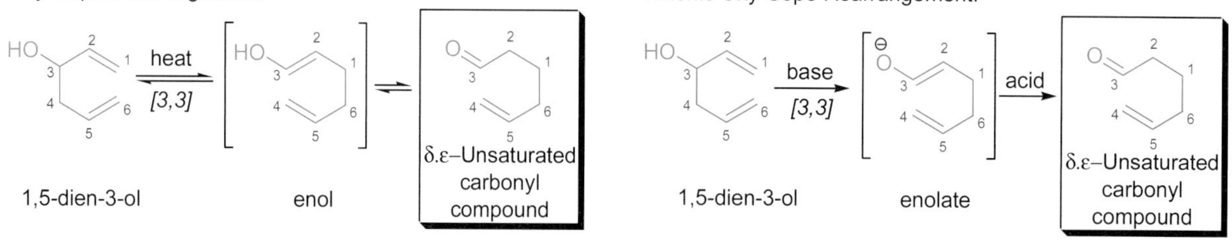

Mechanism: [1,5,9,12,20]

The *oxy-Cope* and *anionic-oxy-Cope rearrangements* involve highly ordered cyclic transition states, so the asymmetry is almost completely transferred from the substrate to the product. Most commonly in acyclic systems as in other *[3,3]-sigmatropic rearrangements*, the transition states are chairlike and the substituents adopt a quasiequatorial position to minimize unfavorable steric interactions. In unsubstituted substrates the diastereoselection is low, but the introduction of an alkyl substituent at C4 improves the diastereoselectivity. In (*Z*)-1-substituted alkenes there is preference for the oxyanionic bond to be pseudo-equatorial, whereas in (*E*)-1-substituted alkenes it tends to be pseudo-axial.[21] Due to conformational constraints in some cyclic substrates, a boatlike transition state may be preferred.

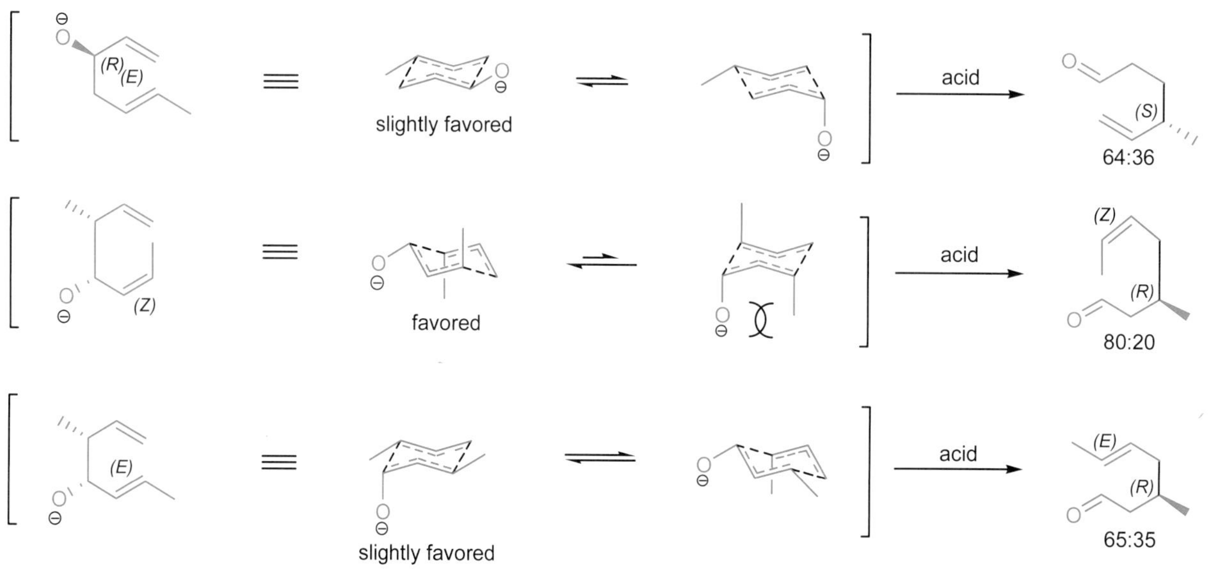

Synthetic Applications:

The enantioselective construction of a key tricyclic intermediate of spinosyn A utilizing a highly stereocontrolled *anionic oxy-Cope rearrangement* was accomplished in the laboratory of L.A. Paquette.[22] The precursor tertiary alcohol was treated with potassium hydride in THF and the *oxy-Cope rearrangement* was complete within 3 hours at room temperature. Interestingly, the yield varied between 77 and 91% depending on the source of KH.

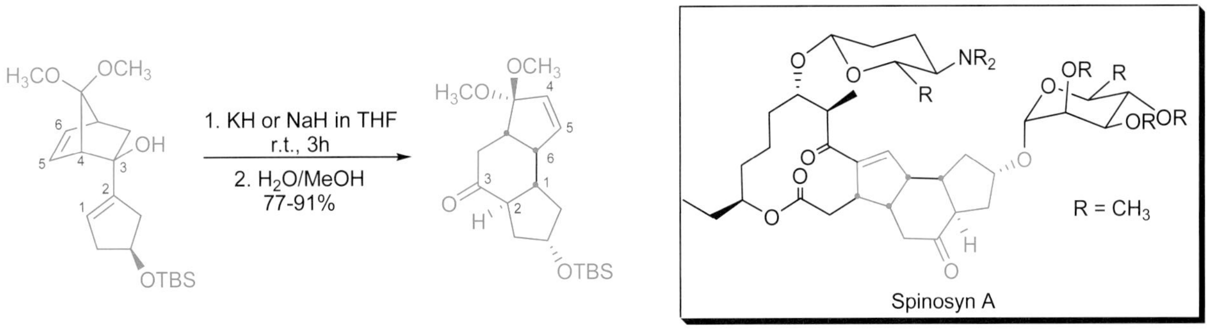

The 1,2-addition of vinyllithium to the carbonyl group of dialkyl squarate-derived bicycloheptenones initiates a low-temperature anion-accelerated *oxy-Cope rearrangement* to afford bicyclo[6.3.0]undecadienone. H.W. Moore and co-workers accomplished the total synthesis of (±)-precapnelladiene using this methodology.[23]

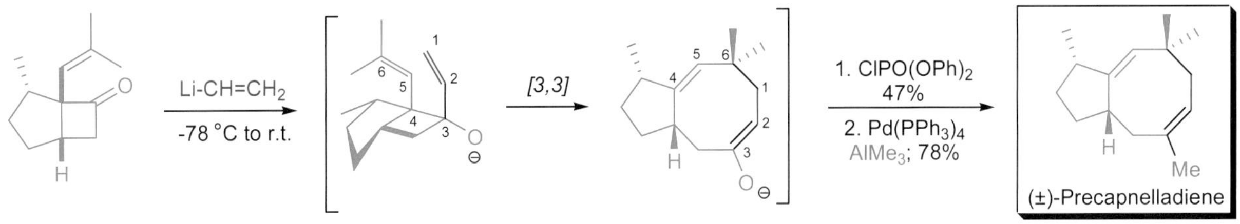

Helicenes are helical compounds consisting of *ortho*-fused aromatic rings. These compounds are potentially useful as catalysts or as platforms for molecular recognition. The currently used syntheses are not practical and do not allow the preparation of helicenes on large scale. M. Karikomi et al. have developed a sequential *double aromatic oxy-Cope rearrangement* strategy for the synthesis of 2-acetoxy[5]helicene.[24] First, 3-phenanthrylmagnesium bromide was synthesized using an *aromatic oxy-Cope rearrangement*. The Grignard reagent was then used to obtain 3-(phenanthrenyl)bicyclo[2.2.2]octanol, which underwent a second *aromatic oxy-Cope rearrangement* upon treatment with KH and one equivalent of 18-crown-6 in THF at 0 °C.

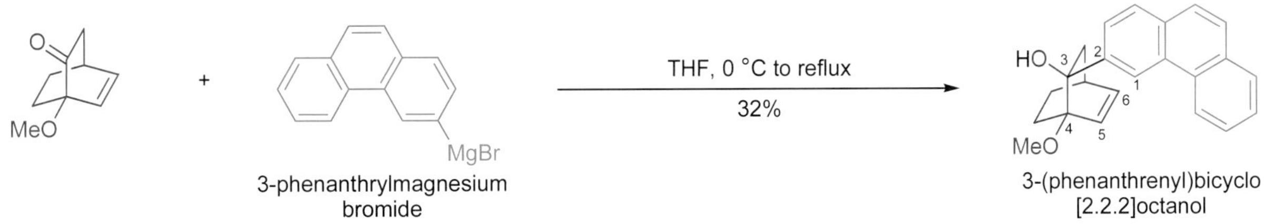

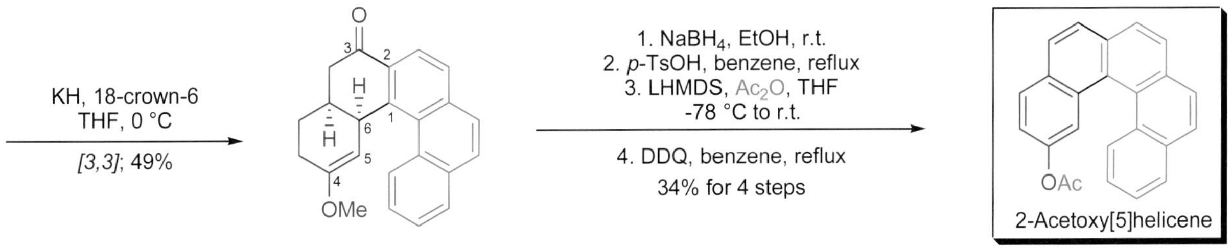

PAAL-KNORR FURAN SYNTHESIS

(References are on page 644)

Importance:

[*Seminal Publications*[1-3]; *Reviews*[4-6]; *Modifications & Improvements*[7,8]]

In 1884, C. Paal and L. Knorr almost simultaneously reported that 1,4-diketones upon treatment with strong mineral acids underwent dehydration to form substituted furans.[1,2] This transformation soon became widely used and now it is referred to as the *Paal-Knorr furan synthesis*. The general features of the method are:[5] 1) virtually any 1,4-dicarbonyl compound (mainly aldehydes and ketones) or their surrogates[9-12] are suitable substrates; 2) the dehydration is affected by strong mineral acids such as hydrochloric acid or sulfuric acid, but often Lewis acids and dehydrating agents (e.g., phosphorous pentoxide, acetic anhydride, etc.) can be used; and 3) the yields are usually moderate to good. The two major drawbacks of the reaction are the relative difficulty to obtain the 1,4-dicarbonyl substrates, and the sensitivity of many functionalities to acidic conditions.

Paal & Knorr (1884):

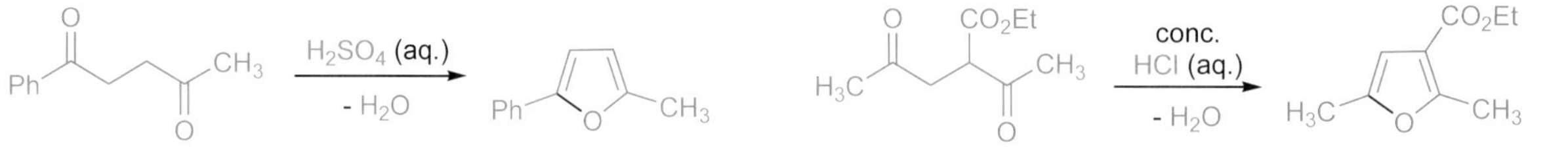

1-phenylpentane-1,4-dione | 2-methyl-5-phenylfuran | 2-acetyl-4-oxopentanoic acid ethyl ester | 2,5-dimethylfuran-3-carboxylic acid ethyl ester

Paal-Knorr furan synthesis:

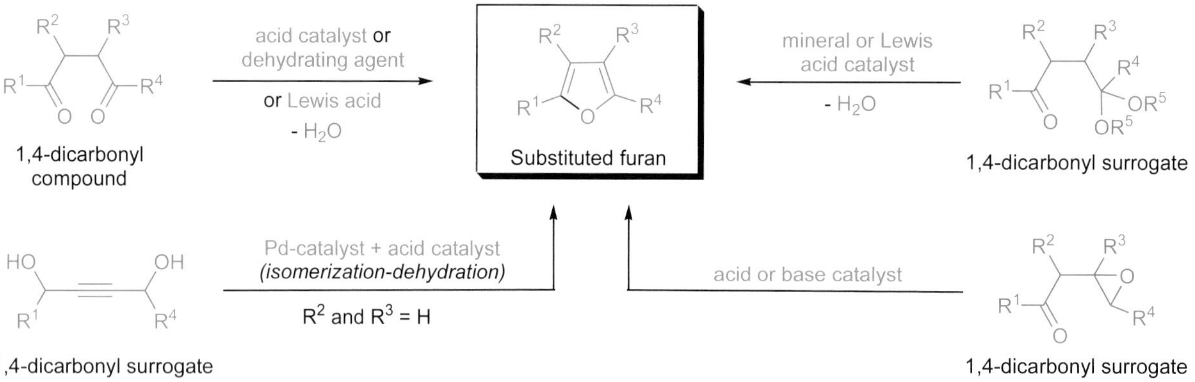

R^1 = H, alkyl, aryl; R^{2-3} = H, alkyl, aryl, CO_2-alkyl, CO_2-aryl; R^4 = H, alkyl, aryl; R^5 = CH_3, C_2H_5; acid catalyst: HCl, H_2SO_4, PPA, *p*-TsOH, $(COOH)_2$, Amberlyst 15; Lewis acid: $ZnBr_2$, ZnC_2, $BF_3 \cdot Et_2O$; dehydrating agent: P_2O_5, Ac_2O

Mechanism: [13,5]

Even though the *Paal-Knorr furan synthesis* has been around for 120 years, its precise mechanism was not known until 1995 when V. Amarnath et al. investigated the intermediates of the reaction and determined the most likely mechanistic pathway.[13] The formation of furans was studied on various racemic and *meso*-3,4-diethyl-2,5-hexanediones. The authors found that the rate of cyclization was different for the racemic and *meso* compounds and that the configuration of the unreacted dione was not affected. This observation strongly suggested that the widely accepted mechanism, involving the ring-closure of a monoenol followed by the loss of water, is incorrect. The most likely pathway involves the rapid protonation of one of the carbonyl groups followed by the attack of the forming enol at the other carbonyl group (rate-determining step). This pathway accounts for the difference in reaction rates for the substrate diastereomers.

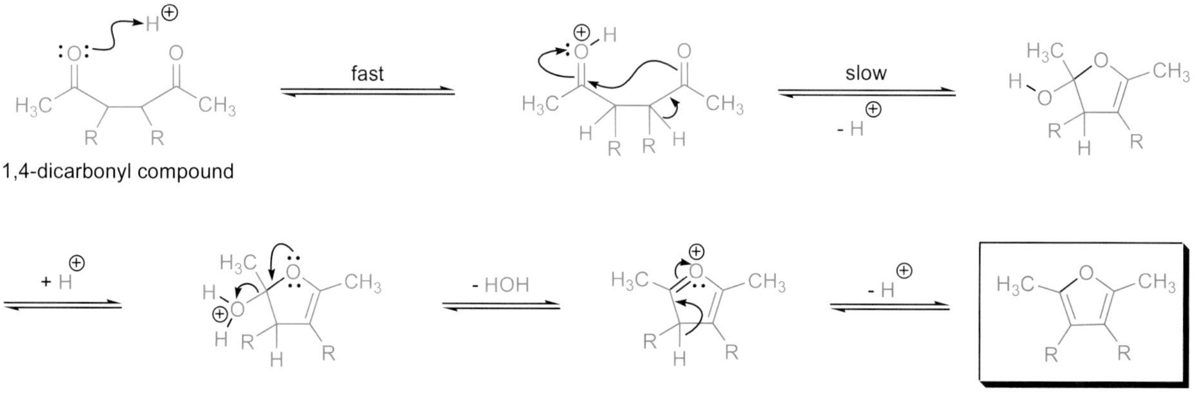

Synthetic Applications:

In the laboratory of H. Hart, the synthesis of various furan macrocycles was accomplished.[14] The preparation of a bisfuran macrocycle, which also contained two naphthalene rings, began with the *Diels-Alder cycloaddition* of the tetraketone substrate with excess benzyne. The benzyne was generated *in situ* from benzenediazonium carboxylate hydrochloride, and it reacted with the two furan rings to afford the corresponding oxabicyclic derivative. The double bond in the newly formed ring was saturated by catalytic hydrogenation. The formation of the desired furan rings was achieved with the *Paal-Knorr furan synthesis* in the presence of *p*-toluenesulfonic acid. Under the reaction conditions the oxabicycles were converted to the naphthalene rings.

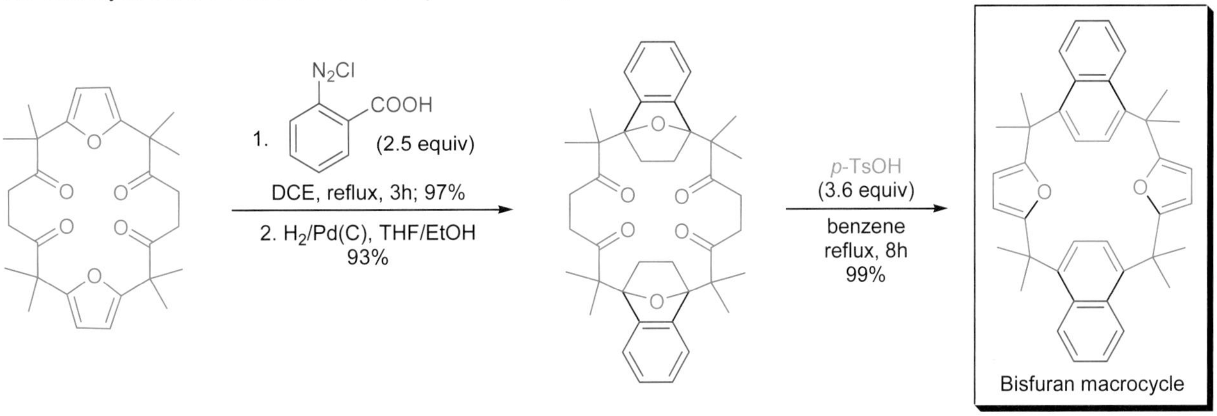

Bisfuran macrocycle

The synthesis of a soluble nonacenetriquinone based on the well-known *Diels-Alder reaction* of 1,3-diarylisobenzofurans was developed by L.L. Miller and co-workers.[15] The preparation of the 1,3-diarylisobenzofuran commenced with the *Paal-Knorr furan synthesis*. The substrate was an aromatic 1,4-diketone, which was treated with excess neat boron trifluoride etherate for almost two days to afford the desired 2,5-diarylfuran in almost quantitative yield. Interestingly, this cyclization could not be achieved efficiently by using the more traditional acid catalysts such as H_2SO_4 or PPA.

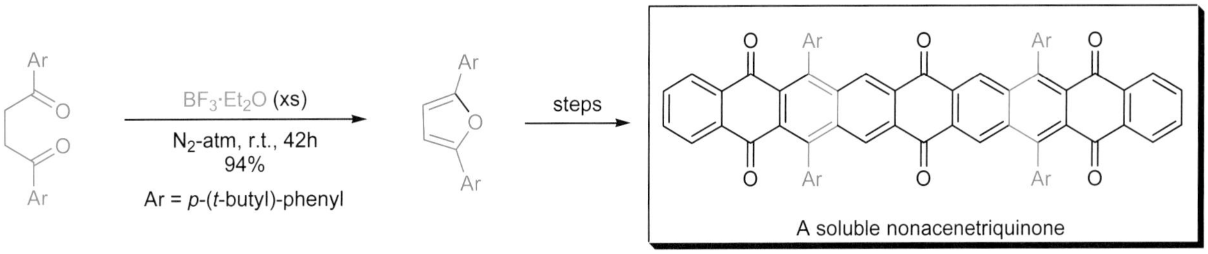

A soluble nonacenetriquinone

The first furan-isoannelated [14]annulene was prepared by Y.-H. Lai et al.[16] The furan moiety was installed by the *Paal-Knorr furan synthesis*. The 1,4-diketone substrate was synthesized *via* an oxidative coupling using MnO_2/AcOH. The dehydration to the furan was effected by phosphorous pentoxide in ethanol.

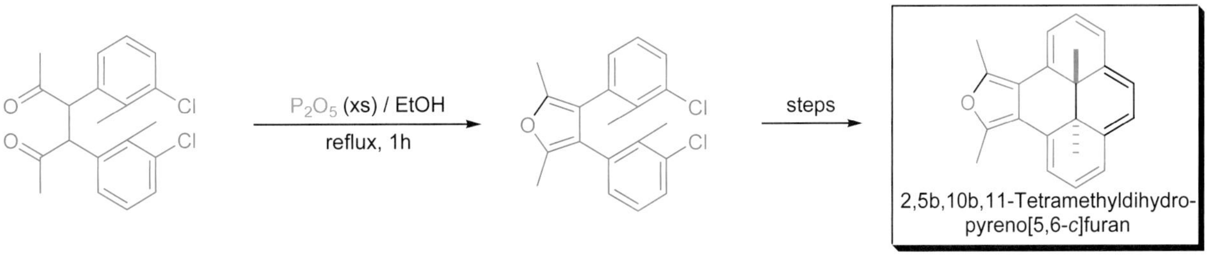

2,5b,10b,11-Tetramethyldihydro-pyreno[5,6-*c*]furan

C.S. Cooper and co-workers synthesized several quinolones containing five- and six-membered heterocyclic substituents at the 7-position and tested their antibacterial activities.[17] The 1,4-diketone substrate was prepared *via* the oxidative coupling of isopropenyl acetate and an acetophenone derivative. The *Paal-Knorr furan synthesis* was conducted in the presence of *p*-TsOH.

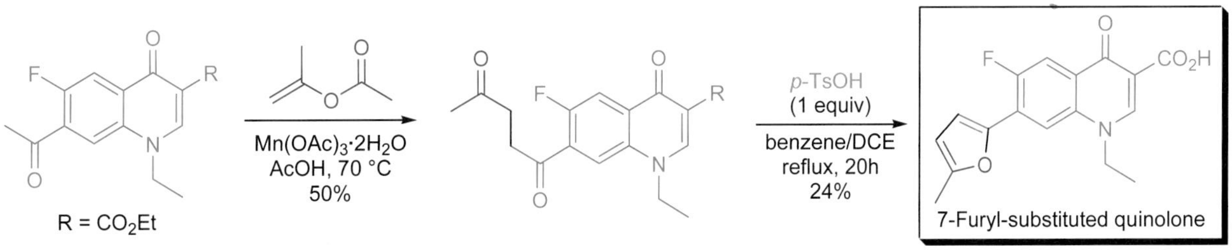

7-Furyl-substituted quinolone

PAAL-KNORR PYRROLE SYNTHESIS

(References are on page 644)

Importance:

[*Seminal Publications*[1,2]; *Reviews*[3-8]; *Modifications & Improvements*[9-20]]

In 1884, C. Paal and L. Knorr almost simultaneously reported that the treatment of 1,4-diketones with concentrated aqueous ammonia or ammonium acetate in glacial acetic acid gave rise to 2,5-disubstituted pyrroles in good yield.[1,2] It was also shown that besides ammonia, primary amines also react with 1,4-diketones to afford *N*-alkyl substituted pyrroles. The preparation of substituted pyrroles by the condensation of 1,4-dicarbonyl compounds with ammonia or primary amines is known as the *Paal-Knorr pyrrole synthesis*. The general features of the transformation are: 1) practically any 1,4-dicarbonyl compound (mainly 1,4-diketones) or their surrogates are good substrates for the reaction; 2) 1,4-dialdehydes or keto aldehydes are used less often mainly because of their relative instability and the lack of general methods for their preparation; 3) the structure of the amine component can be varied widely, since ammonia, aliphatic primary amines, both electron-rich and electron-poor aromatic amines, and heterocyclic amines (e.g., aminopyridines, aminothoazoles, etc.) can be used; 4) α,ω-diamines afford dipyrryl derivatives tethered *via* their nitrogen atoms; 5) ammonia can be introduced either as a concentrated aqueous solution, as ammonium acetate in an alcohol solvent or ammonium carbonate in DMF at high temperature; 6) the relatively basic alkylamines do not react if the acidity of the reaction medium is below pH 5.5, while aromatic amines usually undergo cyclization only when pH<8.2 and the highest yields are observed between pH 4.5 and 5.5; 7) besides protic acids, certain Lewis acids such as $Ti(Oi\text{-}Pr)_4$, as well as layered zirconium phosphate also catalyze the reaction; 8) the solvent of choice depends on the type of amine used, and it can range from polar protic to dipolar aprotic all the way to nonpolar solvents; and 9) yields range from good to excellent and occasionally can be close to quantitative.

Paal (1884):

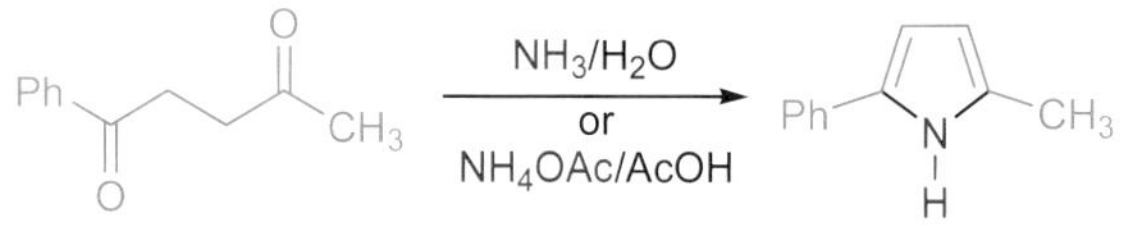

1-phenylpentane -1,4-dione; 2-methyl-5-phenyl-1*H*-pyrrole

Knorr (1884):

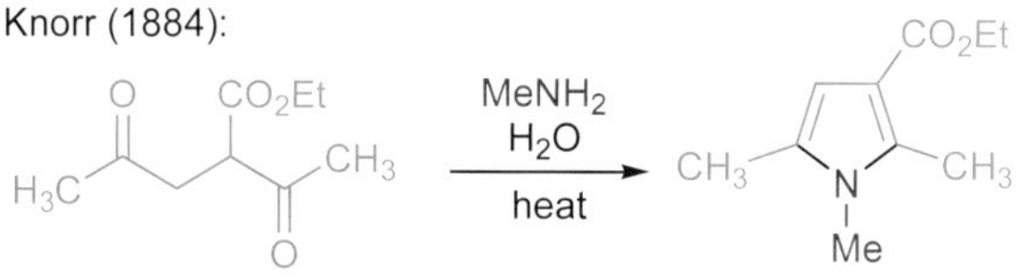

2-acetyl-4-oxopentanoic acid ethyl ester; 1,2,5-trimethyl-1*H*-pyrrole-3-carboxylic acid ethyl ester

Paal-Knorr pyrrole synthesis:

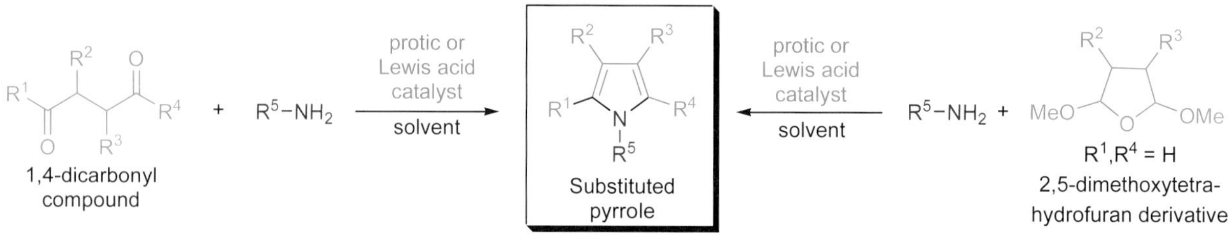

R^1 = H, alkyl, aryl; $R^{2\text{-}3}$ = H, alkyl, aryl, CO_2-alkyl, CO_2-aryl; R^4 = H, alkyl, aryl; R^5 = H, 1°, 2° or 3° alkyl, aryl, heteroaryl, NR_2, NHR, NH_2, OH; ammonia precursors: NH_4OAc, $(NH_4)_2CO_3$; catalyst: zeolite, Al_2O_3, *p*-TSOH, CSA, zirconium phosphate, $Ti(Oi\text{-}Pr)_4$, microwave; solvent: MeOH, EtOH, H_2O, toluene, DMF, ionic liquid

Mechanism: [21,22]

Even though the *Paal-Knorr pyrrole synthesis* has been around for 120 years, its precise mechanism was the subject of debate. In 1991, V. Amarnath et al. investigated the intermediates of the reaction and determined the most likely mechanistic pathway.[23] The formation of pyrroles was studied on various racemic and *meso*-3,4-diethyl-2,5-hexanediones. The authors found that the rate of cyclization was different for the racemic and *meso* compounds and the racemic isomers reacted considerably faster than the *meso* isomers. There were two crucial observations: 1) the stereoisomers did not interconvert under the reaction conditions; and 2) there was no primary kinetic isotope effect for the hydrogen atoms at the C3 and C4 positions. These observations led to the conclusion that the cyclization of the hemiaminal intermediate is the rate-determining (slow) step.

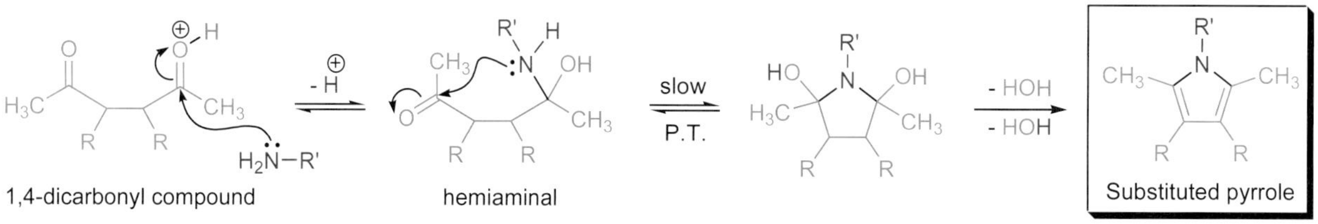

Synthetic Applications:

F.H. Kohnke and co-workers prepared novel heterocyclophanes from cyclic poly-1,4-diketones, which were obtained by the oxidation of calix[6]furan and calix[4]furan.[24] One of the heterocyclophanes, calix[6]pyrrole, was prepared by the *Paal-Knorr pyrrole synthesis* from the corresponding dodecaketone. The substrate was heated with excess ammonium acetate in absolute ethanol. Interestingly, the analogous synthesis of calix[4]pyrrole under identical conditions failed, while calix[5]pyrrole is obtained only in 1% yield.[25,26]

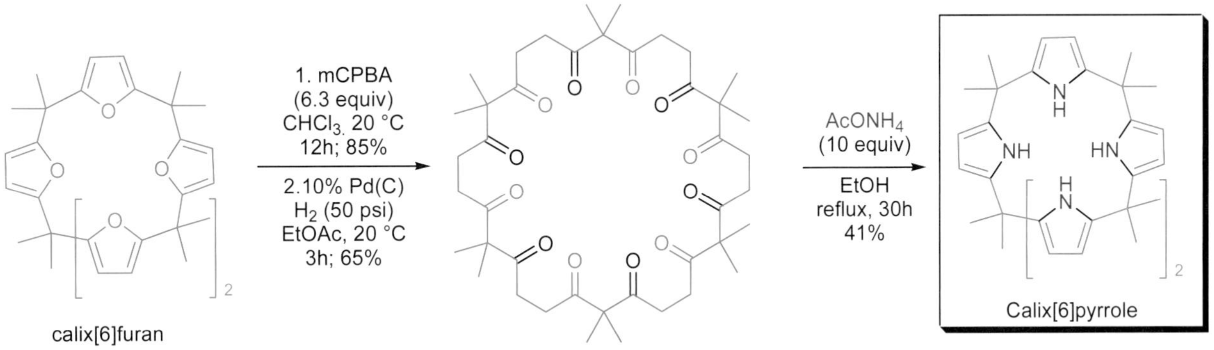

The formal total synthesis of roseophilin was accomplished by B.M. Trost et al. who used the *Paal-Knorr pyrrole synthesis* to install the trisubstituted pyrrole moiety.[27] The 1,4-diketone substrate was reacted with various primary amines to obtain *N*-substituted pyrroles. The best yield was obtained when benzylamine was used as the amine component, but the *N*-deprotection of the product proved to be problematic. This forced the researchers to prepare the otherwise unstable *N*-unprotected pyrrole under carefully controlled conditions and protect it immediately with SEM-chloride.

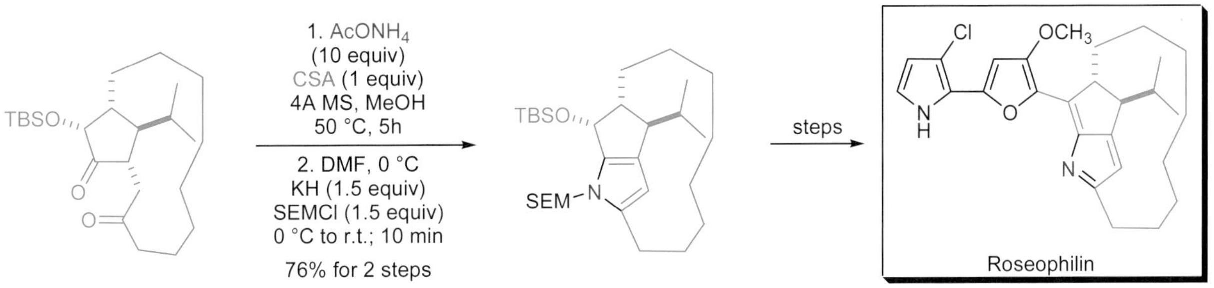

In the laboratory of D.F. Taber, the large-scale preparation of a tetrasubstituted pyrrole, a key precursor for the preparation of hemes and porphyrins, was achieved.[28] The 1,4-dicarbonyl substrate was generated from a ketal *via* hydrolysis and was immediately subjected to the *Paal-Knorr pyrrole synthesis* by heating it with ammonium carbonate in DMF. The resulting 1*H*-pyrrole was formylated with trimethyl orthoformate in trifluoroacetic acid.

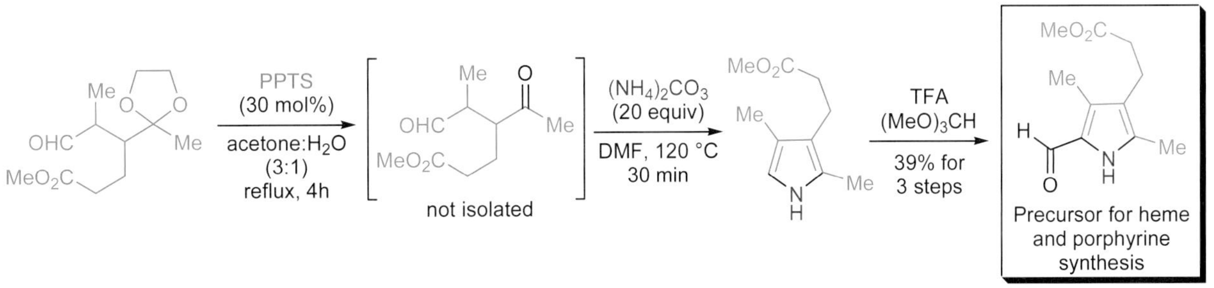

The titanium isopropoxide mediated *Paal-Knorr pyrrole synthesis* was used as the key step in the first total synthesis of magnolamide by W. Le Quesne et al.[29]

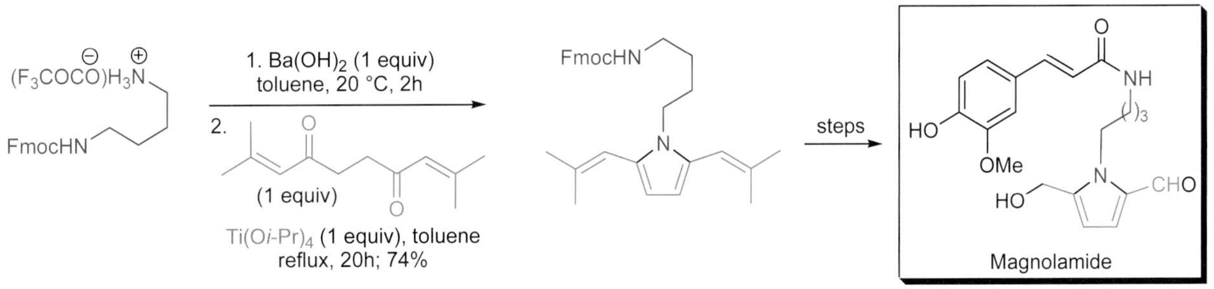

PASSERINI MULTICOMPONENT REACTION

(References are on page 645)

Importance:

[*Seminal Publications*[1]; *Reviews*[2-12]; *Modifications & Improvements*[13-28]]

Isocyanides, also known as isonitriles, are a unique class of organic compounds. The carbon center of the isocyanide group is formally divalent, and it can react with electrophiles and nucleophiles. The first synthetically useful reaction of isocyanides was described by M. Passerini, who reported that isocyanides react with carboxylic acids and carbonyl compounds in one step to provide α-acyloxycarboxamides.[1] This transformation is known today as the *Passerini multicomponent reaction* (MCR). The synthetic power of the *Passerini reaction* is that three reaction partners are combined in one pot under mild conditions (three component reaction or P-3CR) and the product incorporates most atoms of all three starting materials. These types of transformations coupled with combinatorial chemistry and parallel synthesis techniques allow the quick assembly of a wide array of compounds from simple starting materials.[7,9,10] The general features of the classical *Passerini reaction* are: 1) it is carried out at high concentrations of the starting materials in inert solvents at or below room temperature;[1] 2) it is accelerated in apolar solvents;[2] 3) a wide variety of aldehydes and ketones undergo the reaction; 4) there are rare limitations to the carbonyl component, only sterically hindered ketones and α,β-unsaturated ketones are unreactive;[29,16] 5) in addition to C-isocyanides, trimethylsilyl isocyanide also undergoes the reaction;[18] 6) when water is used as the nucleophilic component instead of carboxylic acid, the reaction gives the corresponding α-hydroxycarboxamide under acid catalyzed conditions;[13,15] 7) when hydrazoic acid is combined with the isocyanide and the carbonyl compound under acidic conditions, α-hydroxyalkyltetrazole is the product;[15-17] and 8) catalytic asymmetric variants of the reaction were also developed.[26-28] By choosing the proper starting materials, the *Passerini reaction* often does not stop at the α-acyloxycarboxamide product and it leads to the formation of heterocycles:[20-22,24,25] 1) the reaction of α-oxoaldehydes with carboxylic acids and isocyanides leads to the formation of oxazoles;[20] 2) when cyanoacetic acid is used as the acid component along with α-oxoaldehydes and isocyanides, 2-hydroxyfurans form;[21] and 3) the reaction of β-oxothioamide with isocyanides leads to benzo[c]thiophenes.[22] When bifunctional starting materials incorporating the carbonyl and the carboxylic acid functionality are used, lactones of various ring sizes can be formed.[14] The reaction of α-chloroketones with carboxylic acids and isocyanides under basic conditions leads to the formation of β-lactams.[19,23] In the absence of the carboxylic acid component, this transformation leads to the formation of α-epoxylactones.[19]

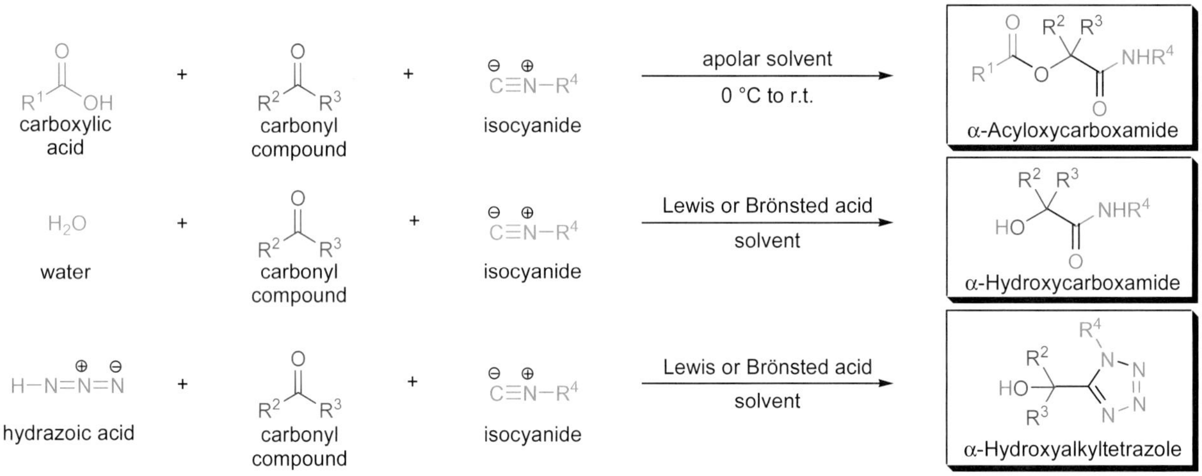

R^1 = alkyl, aryl; R^2 = alkyl, aryl; R^4 = alkyl, aryl; R^3 = H, alkyl, aryl; Brönsted acid: HCl, HNO_3, H_3PO_4, H_2SO_4;
Lewis acids: $BF_3 \cdot OEt_2$, $TiCl_4$

Mechanism:[13,30,31,2,32-37]

The mechanism of the *Passerini reaction* was widely examined. A plausible mechanism that is consistent with experimental data is as follows: First, the carbonyl compound and the carboxylic acid forms a hydrogen bonded adduct. Subsequently, the carbon atom of the isocyanide group attacks the electrophilic carbonyl carbon, and also reacts with the nucleophilic oxygen atom of the carboxylic acid. The resulting intermediate cannot be isolated as it rearranges to the more stable α-acyloxycarboxamide in an intramolecular transacylation.

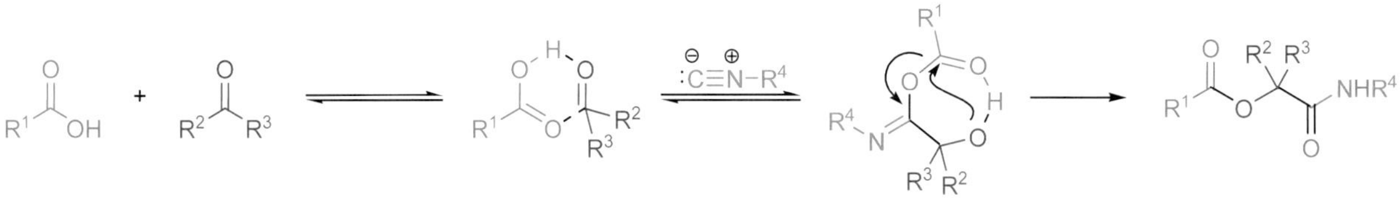

Synthetic Applications:

Eurystatin A is a 13-membered macrocyclic natural product featuring a leucine, ornithine, and an α-ketoalanine amide subunit. This compound exhibits serine protease prolyl endopeptidase inhibition. The total synthesis of this compound was accomplished by E. Semple et al.[38] The key reaction in their approach was the *Passerini reaction* between an *N*-protected ornithine fragment, *N*-α-Fmoc alaninal and a protected leucine isonitrile to give the desired α-acyloxycarboxamide under mild, neutral conditions in high yield and multigram quantities, and as 1:1.2 mixture of diastereomers. Subsequent Fmoc deprotection led to a smooth *O*- to *N*- acyl migration providing the entire acyclic skeleton of the natural product.

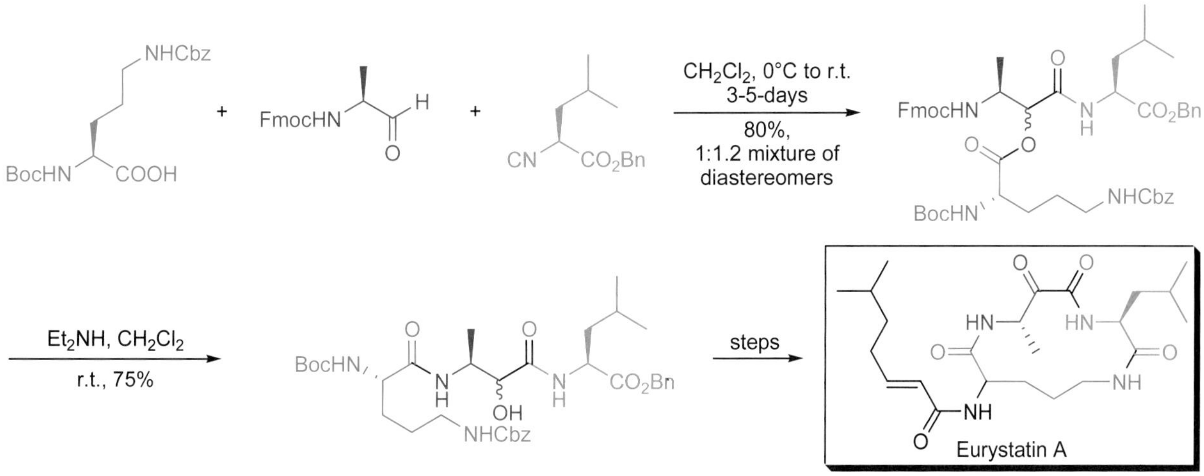

L. Banfi and co-workers utilized the *Passerini three component reaction* to prepare a 9600 member hit generation library of nor-statines.[39,40] These compounds are potential transition state mimetics for the inhibitors of aspartyl proteases. The authors produced the library by starting out from eight *N*-Boc-α-aminoaldehydes, twenty isocyanides and sixty carboxylic acids. The key *Passerini reaction* occurred under mild conditions. This transformation was followed by removal of the Boc protecting group and acyl transfer. Three representative examples of the library are shown.

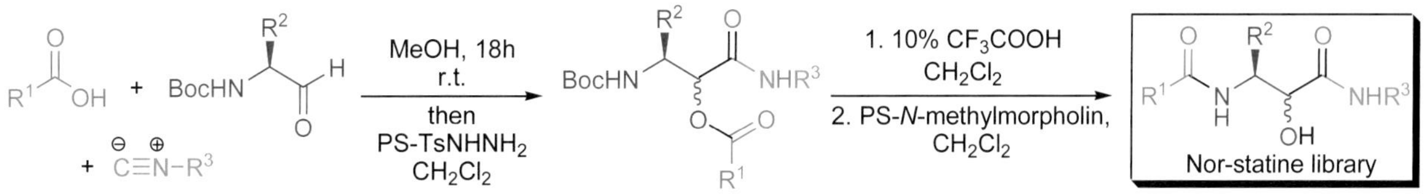

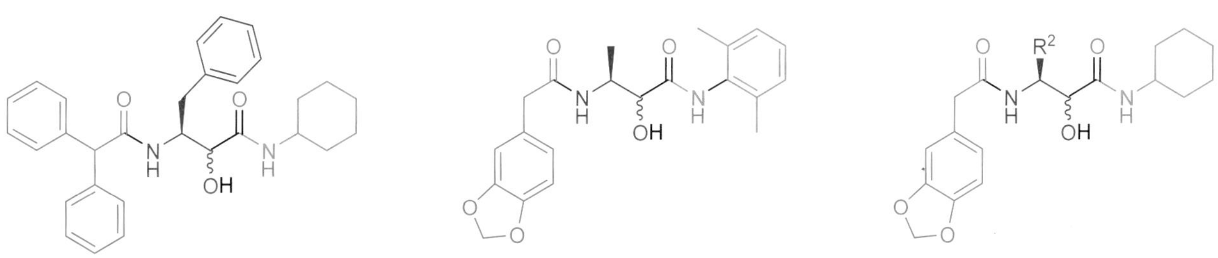

R. Bossio and co-workers developed a novel method for the synthesis of tetrasubstituted furan derivatives.[21] The *Passerini reaction* between arylglyoxals, isocyanides, and cyanoacetic acids led to the formation of *N*-substituted 3-aryl-2-cyanoacetoxy-3-oxopropionamides, which in the presence of amine bases underwent a *Knoevenagel condensation* providing *N*-substituted 3-aryl-cyano-2,5-dihydro-5-oxofuran-2-carboxamides.

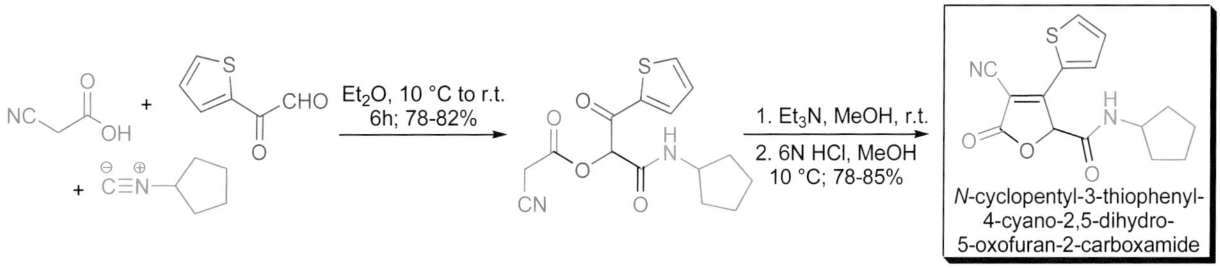

PATERNO-BÜCHI REACTION

(References are on page 646)

Importance:

[*Seminal Publications*[1-4]; *Reviews*[5-18]; *Theoretical Studies*[19-26]]

In 1909, E. Paterno and G. Chieffi reported an interesting reaction that took place between benzaldehyde and 2-methyl-2-butene upon exposure to sunlight.[2] The authors isolated two isomeric compounds that they characterized as trimethylene oxides (oxetanes). The reaction went largely unnoticed until 1954, when G. Büchi decided to reinvestigate Paterno's findings and determine the exact structure of the products.[4] The photochemical cycloaddition between aldehydes and alkenes to form oxetanes is known as the *Paterno-Büchi reaction*. The general features of this transformation are:[7,10] 1) the carbonyl substrate is the energy-absorbing component in the process, and it becomes excited upon irradiation; 2) the carbonyl compound can be either an aldehyde or a ketone; 3) the alkene substrate is most often electron-rich by virtue of one or more electron-donating substituents (e.g., alkoxy, thioalkyl, or alkylamino); 4) the reaction is highly regio- and stereoselective, and the regiochemical outcome can be predicted based on the most stable 1,4-biradical intermediate, which is formed when the excited carbonyl compound adds across the carbon-carbon double bond; 5) the degree of regioselectivity depends on the nature and the position of the substituents on the alkene and, for example, alkenyl sulfides afford the oxetane products with higher regioselectivity than the corresponding enol ethers[27] and also 1,1-disubstituted alkenes give rise to highly regiochemically pure products; 6) when stereochemically pure (*E*) or (*Z*) alkenes are used, the stereochemical information is usually lost and the conformational preference in the 1,4-biradical intermediate results in the predominant formation of the *trans* oxetane product; 7) when cyclic alkenes are used, only the *cis* oxetanes are formed in the case of five- and six-membered alkenes, whereas larger cyclic olefins give rise to a mixture of *cis*- and *trans* oxetanes; 8) conjugated dienes (1,3-dienes and styrenes) and certain five-membered heterocycles (e.g., furans, pyrroles, imidazoles, and indoles) also react to give the corresponding oxetanes; and 9) the facial diastereoselectivity can be induced either with chiral auxiliaries attached to the carbonyl compounds or with the use of chiral alkenes.

Paterno (1909 & 1911):

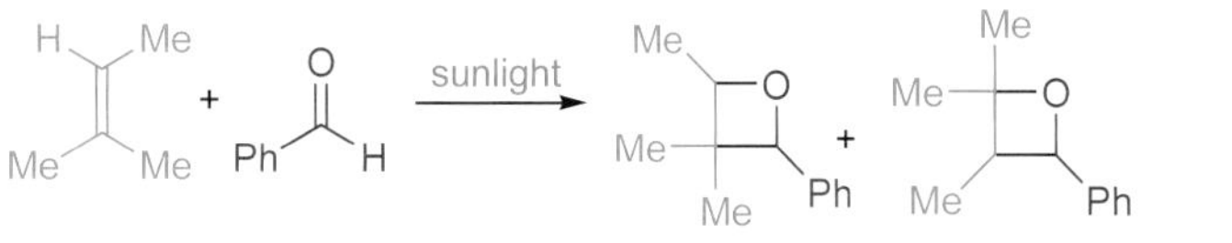

Büchi (1954):

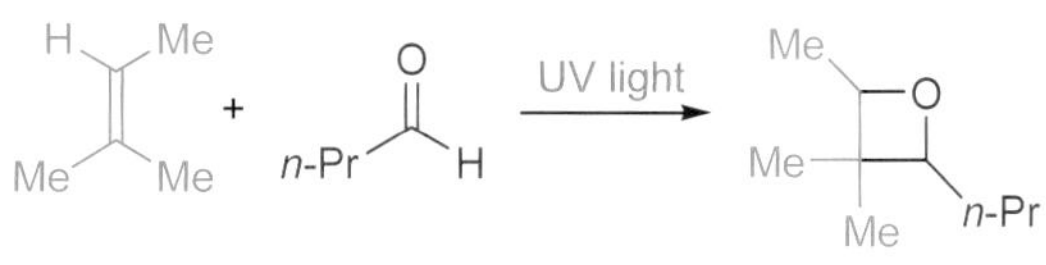

Paterno-Büchi reaction:

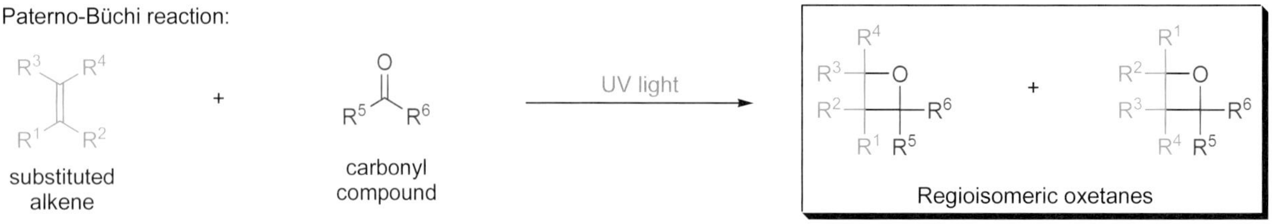

R^{1-4} = H, alkyl ,aryl, O-alkyl, S-alkyl, NR_2; R^{5-6} = H, alkyl ,aryl

Mechanism: [28-51]

The mechanism of the *Paterno-Büchi reaction* has been extensively studied. The current understanding of the process involves the following steps: 1) the carbonyl functionality (S_0) is excited by a UV photon *via* nπ*-absorption to afford the corresponding singlet state (S_1); 2) the carbonyl singlet state can be converted to the carbonyl triplet state (T_1) *via* intersystem crossing (ISC); 3) when the carbonyl singlet reacts with the alkene (mostly in the case of aliphatic aldehydes and ketones and a very high alkene concentration is required in order to quench the singlet state efficiently) the photocycloaddition is stereospecific and the stereochemical information of the alkene substrate is translated into the oxetane product; 4) in the overwhelming majority of the *Paterno-Büchi reactions*, however, the intersystem crossing gives rise to the carbonyl triplet state, which upon addition to the alkene affords a 1,4-biradical (these species have been studied spectroscopically);[33] and 5) finally the most stable 1,4-biradical conformer collapses to the oxetane product.

Start here:

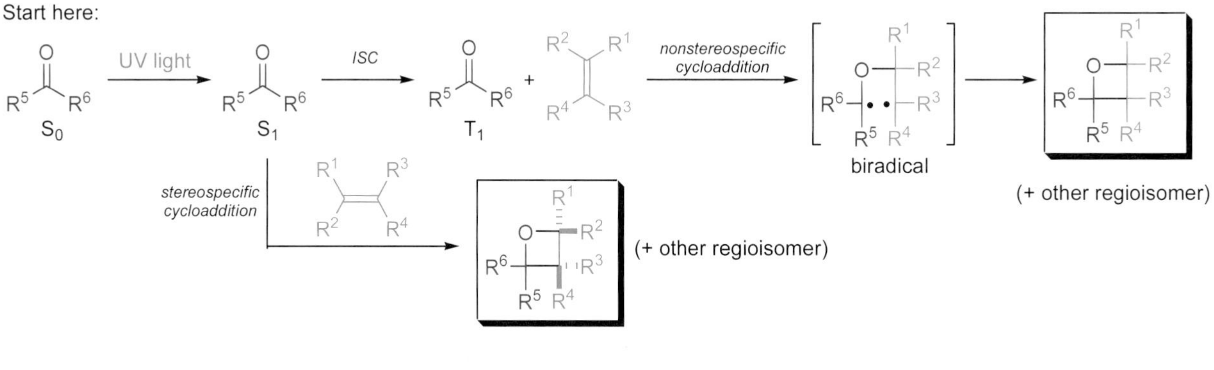

Synthetic Applications:

In the laboratory of T. Bach, the *Paterno-Büchi reaction* of chiral 2-substituted 2,3-dihydropyrroles with benzaldehyde was used in the total synthesis of the antifungal agent (+)-preussin.[52,53] Benzaldehyde was mixed with excess dihydropyrrole substrate in acetonitrile and was irradiated at room temperature at 350 nm UV light. Once all the benzaldehyde was consumed, half an equivalent of benzaldehyde was added and the irradiation continued for another two hours. The addition of the photoexcited benzaldehyde proceeded in a *syn* fashion and the thermodynamically less stable *endo* oxetane was formed as the major product. The oxetane ring was then cleaved under catalytic hydrogenation conditions in the presence of Pearlman's catalyst to form the all-*cis* pyrrolidinol. Finally the reduction of the *N*-carboxymethyl group to the corresponding *N*-methyl group was achieved using lithium aluminum hydride.

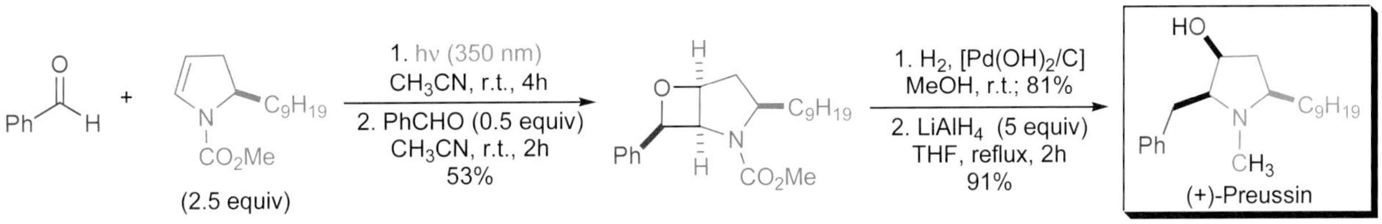

A unique *intramolecular Paterno-Büchi reaction/fragmentation* sequence was utilized during the short total synthesis of the angular triquinane (±)-oxosilphiperfol-6-ene by V.H. Rawal et al.[54] The photocycloaddition substrate was prepared *via* a highly *regio-*, *endo-*, and diastereofacially selective *Diels-Alder cycloaddition* between 1,3-dimethylcyclopentadiene and 1-acetyl-3-methylcyclopentene. The cycloadduct was then irradiated with Corex-filtered light to obtain the strained cage-like product. Reductive cleavage of the oxetane ring with LDBB yielded an allylic alcohol, which was oxidized to the desired α,β-unsaturated ketone with PDC.

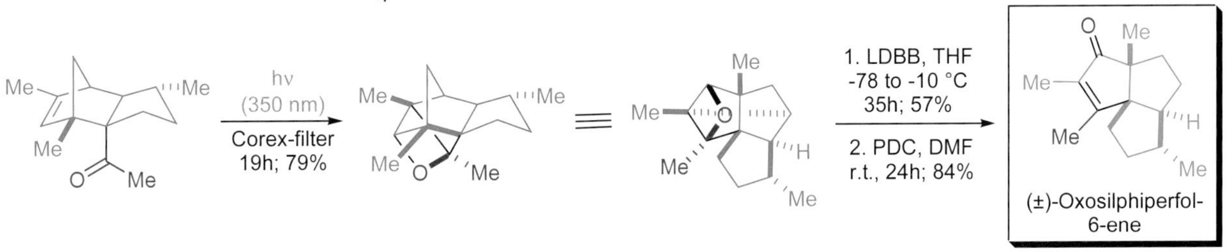

The first total synthesis of the cytotoxic agent (±)-euplotin A was completed by the research team of R.L. Funk.[55] The key step of the synthetic effort was the *intramolecular hetero Diels-Alder cycloaddition* of a 3-acyl oxadiene (generated from 5-acyl-4*H*-1,3-dioxins *via* thermal retrocycloaddition) with a substituted dihydrofuran to afford the tricyclic skeleton of the natural product. The correct relative stereochemistry of the required dihydrofuran substrate was established using the *Paterno-Büchi reaction* between ethyl glyoxylate and furan. Subsequently, the oxetane ring was opened stereoselectively under Lewis acid catalysis.

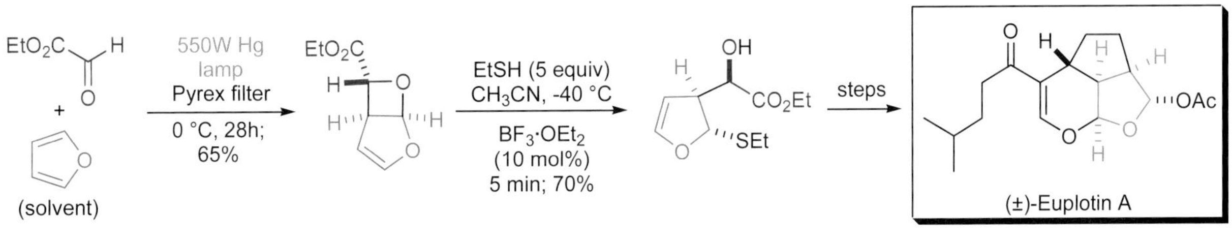

The *Paterno-Büchi reaction* of furan and various aldehydes was shown to be a highly stereoselective photochemical version of the *aldol reaction* by S.L. Schreiber and co-workers in which the furan serves as an enolate equivalent.[56] This strategy was applied to the total synthesis of the antifungal metabolite (±)-avenaciolide.[57] The photocycloaddition of nonanal with excess furan proceeded in nearly quantitative yield, and the two out of the three required stereocenters were created in a single step. The photocycloadduct was first hydrogenated then hydrolyzed under acidic conditions.

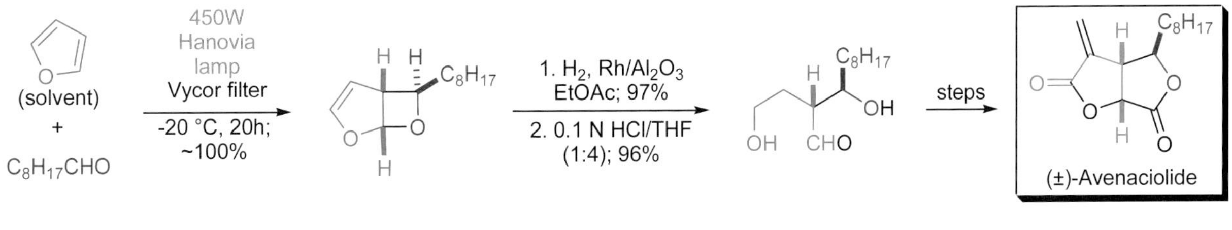

PAUSON-KHAND REACTION
(References are on page 647)

Importance:

[*Seminal Publication*[1]; *Reviews*[2-31]; *Modifications & Improvements*[32-35]; *Theoretical Studies*[36-47]]

In 1973, I.U. Khand and P.L. Pauson reported that various acetylenehexacarbonyl dicobalt complexes reacted with alkenes in hydrocarbon or ether solvents to give cyclopentenones in good yield.[1] The scope and limitation of this reaction was determined by the research group of P.L. Pauson in the 1970s. The transition metal (cobalt) catalyzed formal *[2+2+1]* cycloaddition of alkynes, alkenes, and carbon-monoxide to form substituted cyclopentenones is referred to as the *Pauson-Khand reaction*. The general features of this process are: 1) the reaction is feasible both inter- and intramolecularly; 2) acetylene and terminal as well as internal alkynes are all substrates for the reaction. However, derivatives of propynoic acid do not react; 3) the required alkyne-cobalt complexes are easily prepared by reacting alkynes with dicobalt octacarbonyl; 4) internal alkynes tend to give lower yields of the product than terminal alkynes; 5) a wide range of alkenes are feasible reaction partners and, generally, strained cyclic alkenes react the fastest; 6) the order of reactivity is significantly influenced by the substitution pattern of the alkene substrate: strained cyclic alkene > terminal alkene > disubstituted alkene >> trisubstituted alkene, and tetrasubstituted alkenes do not react; 7) alkenes with strongly electron-withdrawing groups give poor or no reaction; 8) the reaction is highly regioselective: the larger alkyne substituent (R^1) ends up next to the carbonyl group in the product, but the regioselectivity with respect to the alkene is less predictable in intermolecular reactions; 9) with cyclic alkenes the reaction is highly stereoselective and the *exo* product is formed preferentially; 10) intramolecular reactions proceed with excellent regio- and stereoselectivity; 11) with the use of chiral auxiliaries the reaction conditions are compatible with a large number of different functional groups. However there are certain functionalities that are only partially tolerated: alkyl and aryl halides, vinyl ethers, and vinyl esters; 12) the reaction can be accelerated by the addition of various promoters (such as tertiary amine oxides, high-intensity light, etc.), which help to open a coordination site at one of the cobalt atoms for the alkene to coordinate; 13) it is possible to run the cyclization catalytically but only in the presence of a high pressure atmosphere of CO; and 14) besides $Co_2(CO)_8$, other transition metal complexes also efficiently catalyze the cyclization (e.g. $Fe(CO)_5$, $Ru_2(CO)_{12}$, etc.)

Pauson & Khand (1973):

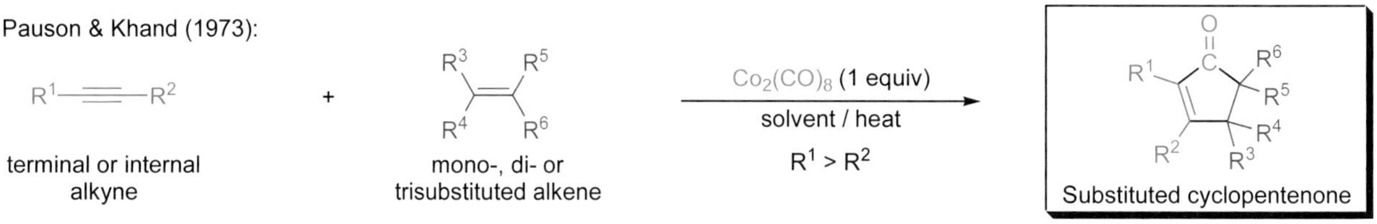

Modified P-K reaction:

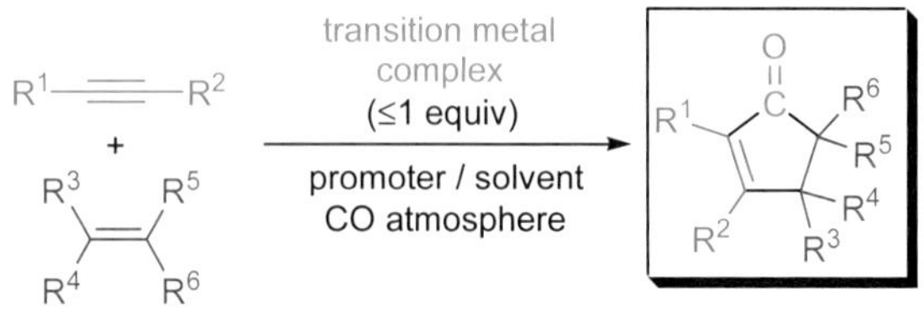

Intramolecular variant:

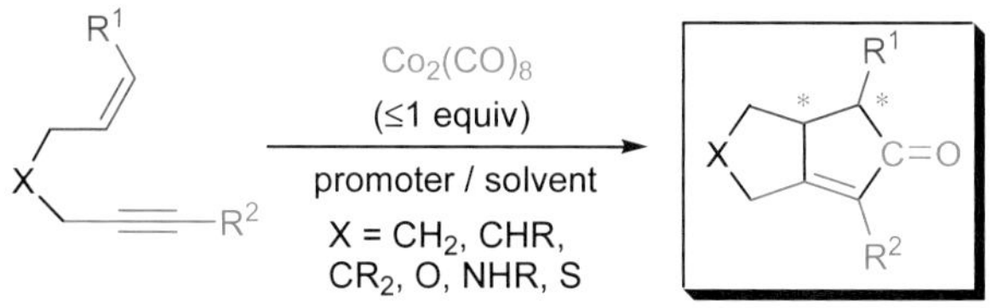

R^{1-6} = H, alkyl, aryl, substituted alkyl and aryl; transition metal complex: $Co_2(CO)_8$, $Fe(CO)_5$, $Ru_2(CO)_{12}$, Cp_2TiR_2, $Ni(COD)_2$, $W(CO)_6$, $Mo(CO)_6$, $[RhCl(CO)_2]_2$; promoter: NMO, TMAO, $RSCH_3$, high-intensity light/photolysis, "hard" Lewis base

Mechanism: [48-62]

The mechanism of the *Pauson-Khand reaction* has not been fully elucidated. However, based on the regio- and stereochemical outcome in a large number of examples, a reasonable hypothesis has been inferred.

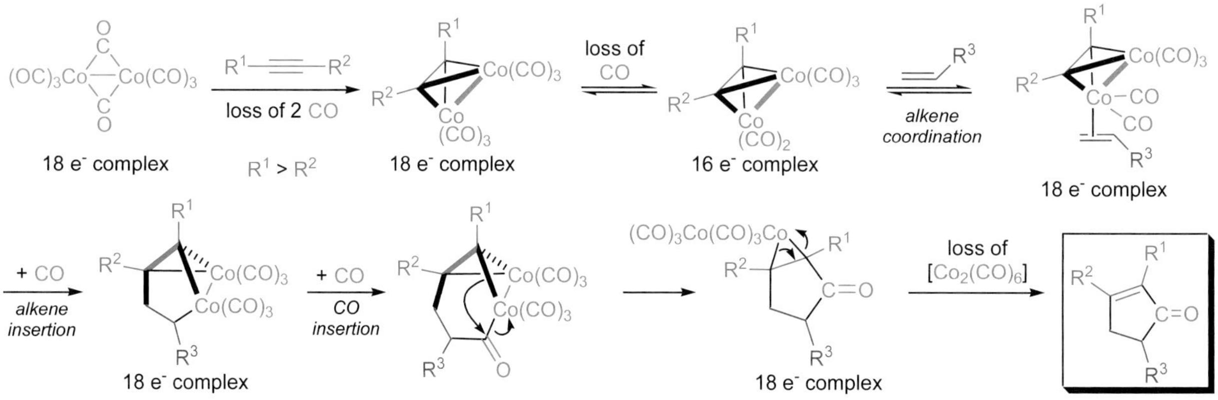

Synthetic Applications:

The total synthesis of the sesquiterpene (+)-taylorione was achieved in the laboratory of J.G. Donkervoort who used the *modified Pauson-Khand reaction* to prepare the five-membered ring of the natural product.[63] The preformed alkyne-cobalt complex was exposed to excess triethylamine-*N*-oxide, which oxidized off two CO ligands to free up a coordination site for the ethylene. The optimum pressure of the ethylene gas had to be at 25 atm, and the reaction was conducted in an autoclave.

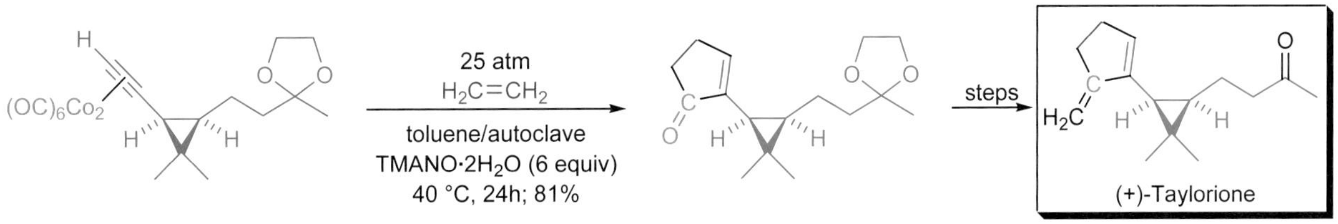

During the synthetic studies toward the natural product kalmanol, L.A. Paquette and co-workers prepared the CD diquinane substructure by using an *intramolecular Pauson-Khand reaction*.[64] The use of an *N*-oxide promoter for the cyclization resulted in very mild conditions and afforded the desired triquinane in good yield and as a single diastereomer.

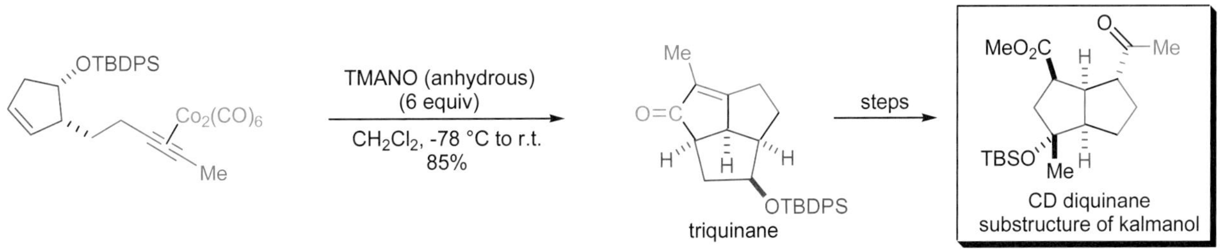

In the laboratory of S.L. Schreiber, the total synthesis of (+)-epoxydictymene was accomplished by the tandem use of cobalt-mediated reactions as key steps.[65] The eight-membered carbocycle was formed *via* a *Nicholas reaction*, while the five-membered ring was annulated by the *Pauson-Khand reaction*. Several *P.-K.* conditions were explored and the best diastereoselectivity was observed when NMO was used as a promoter. The annulated product was isolated as an 11:1 mixture of diastereomers.

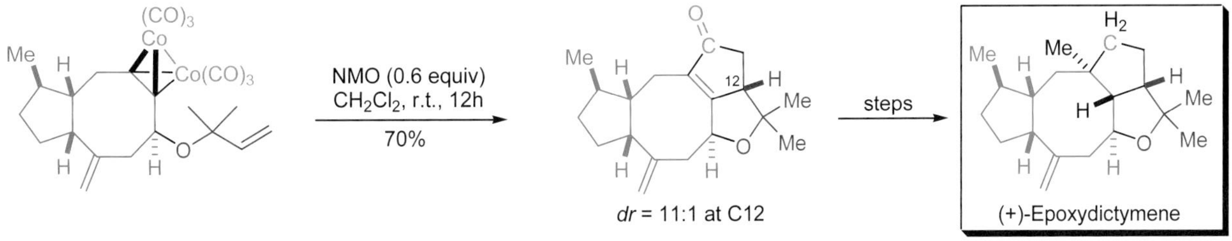

The key bicyclo[4.3.0]nonenone intermediate in the total synthesis of (±)-13-deoxyserratine was prepared by a highly diastereoselective *intramolecular Pauson-Khand reaction* of a functionalized enyne-cobalt complex in the laboratory of S.Z. Zard.[66] The reactive conformation of this complex is one in which the OTBS group occupies the pseudoequatorial position. The observed diastereoselectivity was high as the alternative conformer was significantly higher in energy. The concave shape of the bicyclic product was exploited in controlling the introduction of the remaining three stereocenters.

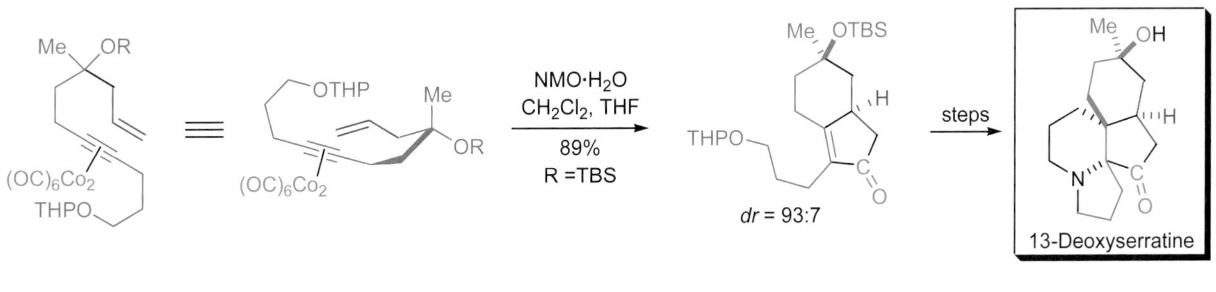

PAYNE REARRANGEMENT

(References are on page 649)

Importance:

[*Seminal Publications*[1-4]; *Reviews*[5-8]; *Modifications & Improvements*[9-12]; *Theoretical Studies*[13-15]]

In 1935, E.P. Kohler and C.L. Bickel described the unusual properties of certain 2,3-epoxy alcohols (β-oxanols), which underwent a rearrangement in the presence of catalytic amounts of a strong base (e.g., alkali hydroxides, barium oxide, magnesium methylate, etc.) to give isomeric 2,3-epoxy alcohols.[1] Three decades later in 1962, G.B. Payne reported that aqueous sodium hydroxide at room temperature was sufficient to bring about the isomerization-equilibration of 2,3-epoxy alcohols, a transformation, which he found to be general and termed as "*epoxide migration*".[4] The base-catalyzed intramolecular nucleophilic displacement of 2,3-epoxy alcohols to give the isomeric 2,3-epoxy alcohols is known as the *Payne rearrangement*. The general features of the reaction are:[7] 1) enantiopure epoxide substrates are accessible most conveniently by the *Sharpless asymmetric epoxidation* of allylic alcohols; 2) the stereochemistry at C2 undergoes inversion; 3) the base needs to be strong and in most cases the use of a protic solvent such as water or an alcohol is necessary; 4) the direction of epoxide equilibration is influenced by both steric and electronic effects; 5) the most substituted epoxide isomer is favored; 6) *trans* epoxides are more stable than *cis* epoxides; 7) vinylyl and phenyl substituents on the oxirane have a stabilizing effect, while EWG are destabilizing; 8) the epoxide isomer with a primary hydroxyl group is favored; and 9) in cyclic systems the favored epoxide is the one that has more pseudoequatorial groups. The two main variants of the reaction are the *aza- and thia-Payne rearrangement* in which aziridines and thiiranes are formed, respectively.[10,5]

Kohler & Bickel (1935):

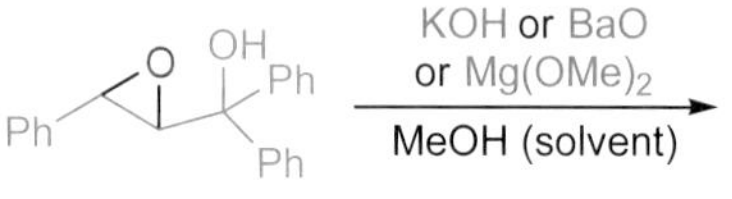

Payne (1962):

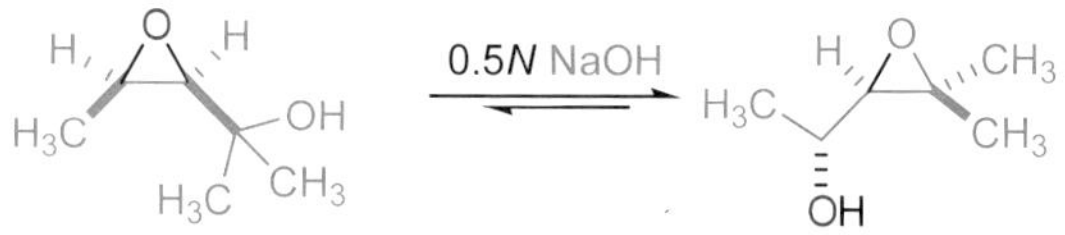

Payne rearrangement followed by nucleophilic epoxide ring-opening:

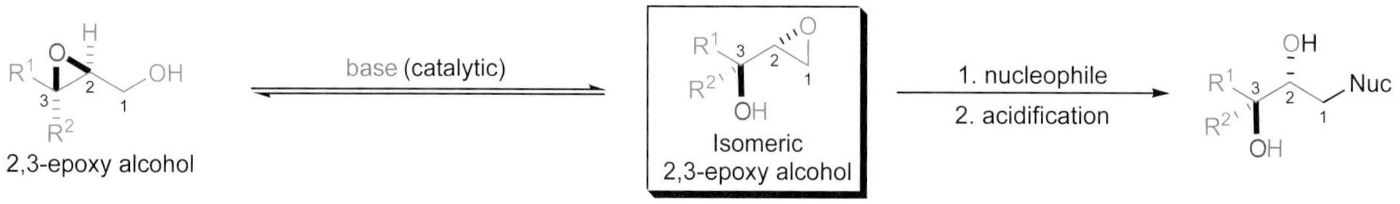

Aza-Payne rearrangement (can be true equilibrium):

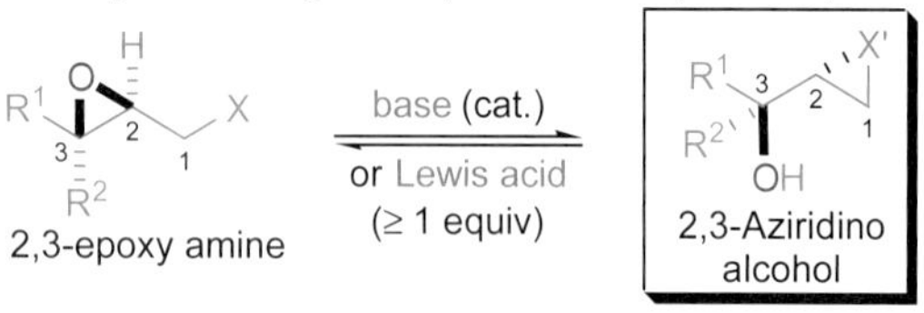

Thia-Payne rearrangement (only forward direction):

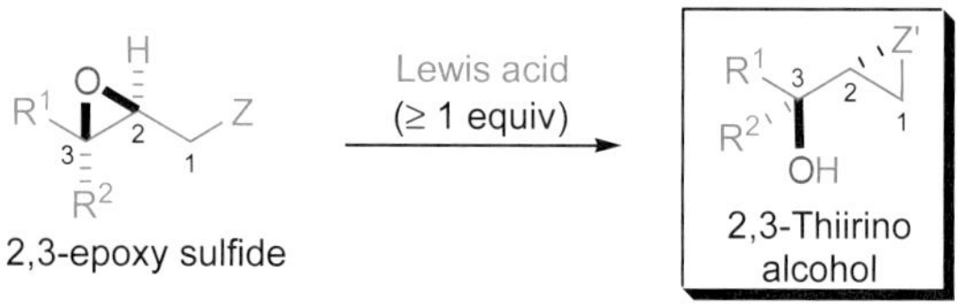

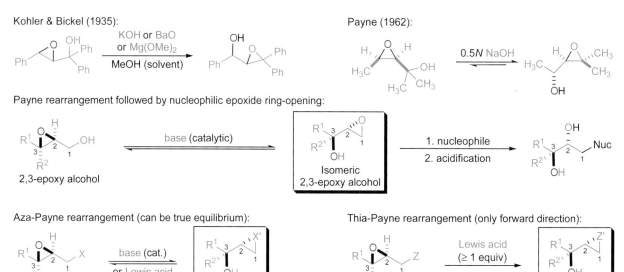

R^{1-2} = H, alkyl, aryl; when X = NR_2, X' = NR_2^+; when X = NHMs, X' = NMs; when Z = SAc, Z' = S; when Z = SR, Z' = SR^+
base: NaOH, KOH, NaOR, NaH, KH; Lewis acid: $AlMe_3$, TMSOTf, $PhB(OH)_2$, $BF_3{\cdot}OEt_2$, $Ti(Oi\text{-}Pr)_4$

Mechanism: [5,7]

The currently accepted mechanism was first proposed by S.J. Angyal and P.T. Gilham in 1957.[3] The first step of the process is the deprotonation of the hydroxyl group at C1 by the strong base and the resulting alkoxide undergoes an S$_N$*i* reaction (*3-exo-tet process*) to open the adjacent epoxide at C2. As a result, a new epoxide is formed at C1 and C2 in which the C2 stereochemistry is inverted. The alkoxide anion at C3 is protonated by the solvent to afford the product. It is worth noting that the success of the rearrangement in most cases depends on the nature of the solvent. Generally, strong bases in aprotic solvents (e.g., NaH/THF) do not affect the reaction, but strong bases in protic solvents (e.g., NaOH/H_2O) do. According to theoretical studies, the product of the *Payne rearrangement* is formed under kinetic control, since the thermodynamically most stable species would be an oxetane, which has never been observed in solution-phase reaction mixtures (only in the mass spectrometer).[13,14] When the reaction is conducted in the presence of a nucleophile so that the equilibrating epoxides are opened *in situ* with a nucleophile (slow step), the product distribution is governed by the Curtin-Hammett principle and exclusive ring-opening at the least substituted carbon of the less substituted epoxide can be achieved. The mechanism of the *aza-Payne rearrangement* is more complex and the outcome is influenced both by the structure of the substrate and the nature of the base or Lewis acid.[5]

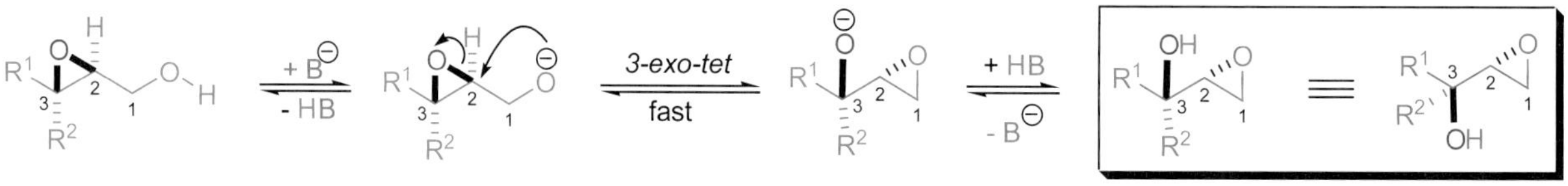

Synthetic Applications:

The Lewis acid-catalyzed *aza-Payne rearrangement* was utilized in the total synthesis of *epi*-7-deoxypancratistatin by T. Hudlicky and co-workers.[16] The 2,3-aziridino alcohol was treated with *t*-BuLi, to generate the epoxy amide that was trapped with piperonyl bromide.

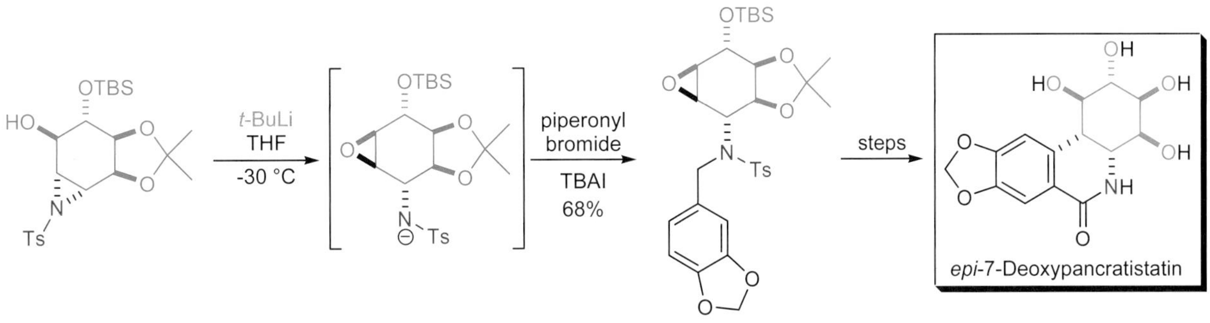

A novel neuroexcitotoxic amino acid, (–)-dysiherbaine, was synthesized starting from a carbohydrate precursor in the laboratory of M. Sasaki.[17] Under benzylation conditions, the cyclic 2,3-epoxy alcohol underwent a facile *Payne rearrangement* and the rearranged alkoxide was trapped with benzyl bromide.

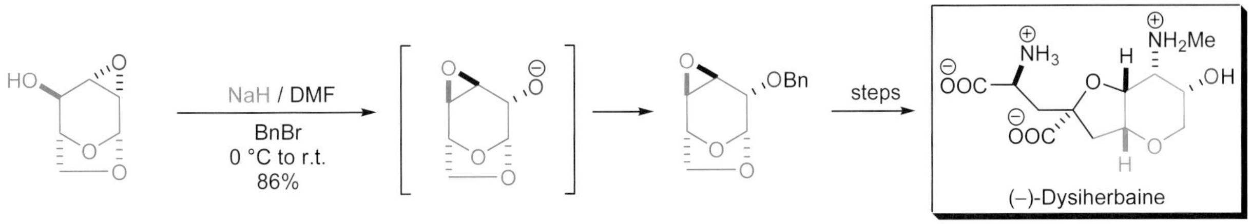

I. Kvarnström et al. prepared novel nucleosides with potential HIV-1 inhibitor acitivity using the *thia-Payne rearrangement* to install the sulfur atom stereoselectively. The 2,3-epoxy alcohol was first converted to the corresponding thioacetate then treated with methanolic ammonia solution to effect the rearrangement to afford the thiirane in excellent yield. As expected, inversion of configuration at C2 occured. The authors also found, that under mild acidic conditions (silica gel), the thioacetate yielded a thiirane with a net retention of configuration at the C2 stereocenter. This result can be explained with the neighboring group participation of the acetate, which opened the protonated epoxide (with inversion at C2) to give a 1,3-oxathiolan-2-ylium ion. This carbocation then rearranged to the more stable 1,3-dioxolan-2-ylium ion. Subsequently, the sulfur nucleophile at C1 attacked C2 for the second time with inversion of configuration to afford the thiirane with a net retention of configuration.

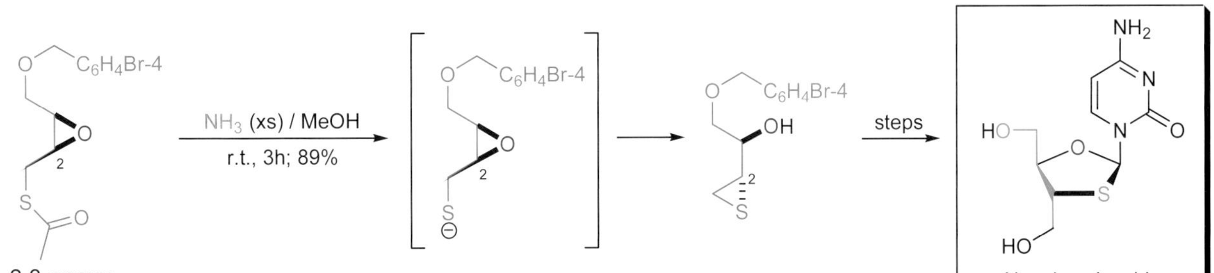

The total synthesis of (±)-merrilactone A was accomplished by S.J. Danishefsky and co-workers.[18] The last step of the sequence was an acid-induced *homo-Payne rearrangement*. The tetracyclic homoallylic alcohol precursor was first epoxidized using *m*CPBA. The epoxidation was expected to occur from the same face as the C7 hydroxyl group, but due to the congested nature of the C1-C2 double bond at its β-face, the epoxide was formed predominantly on the α-face. The epoxide substrate then was exposed to *p*-toluenesulfonic acid at room temperature to afford the desired oxetane ring of the natural product.

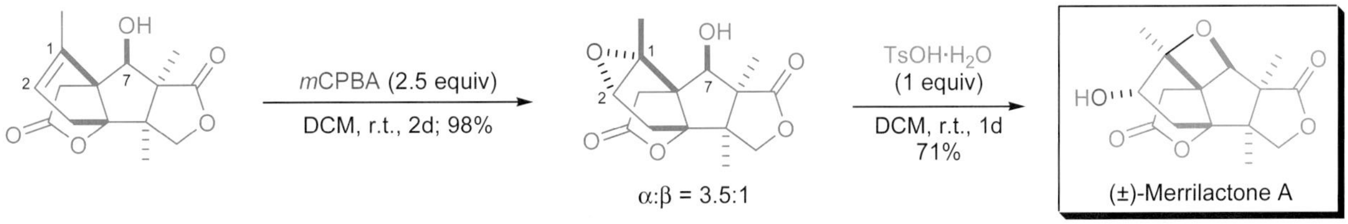

PERKIN REACTION

(References are on page 649)

Importance:

[*Seminal Publications*[1,2]; *Reviews*[3,4]; *Modifications & Improvements*[5-22]]

In 1868, W.H. Perkin described the one-pot synthesis of coumarin by heating the sodium salt of salicylaldehyde in acetic anhydride.[1] After this initial report, Perkin investigated the scope and limitation of the process and found that it was well-suited for the efficient synthesis of cinnamic acids.[2] The condensation of aromatic aldehydes with the anhydrides of aliphatic carboxylic acids in the presence of a weak base to afford α,β-unsaturated carboxylic acids is known as the *Perkin reaction* (or *Perkin condensation*). The general features of the transformation are:[3,4] 1) the aldehyde component is most often aromatic, but aliphatic aldehydes with no α-hydrogens as well as certain α,β-unsaturated aldehydes can also be used;[17] 2) the reaction is more facile and gives higher yield of the product when the aromatic aldehyde has one or more electron-withdrawing substituents; 3) aliphatic aldehydes are not suitable for the reaction, since they often give enol acetates and diacetates when heated with acetic anhydride; 4) the anhydride should be derived from an aliphatic carboxylic acid, which has at least two hydrogen atoms at their α-position (if there is only one α-hydrogen atom, a β-hydroxy carboxylic acid is obtained); 5) the weak base is most often the alkali metal salt of the carboxylic acid corresponding to the applied anhydride or a tertiary amine (e.g., Et_3N); 6) the usual procedure requires heating of the aldehyde in the anhydride (often used as the solvent) at or above 150 °C; and 7) the stereochemistry of the newly formed double bond is typically (*E*). There are two important modifications of the *Perkin reaction*: 1) the condensation of an aromatic aldehyde or ketone with an *N*-acyl glycine in acetic anhydride in the presence of NaOAc to obtain azlactones (oxazolones), which are important intermediates for the synthesis of α-amino acids (*Erlenmeyer-Plöchl azlactone synthesis*);[6-9,15,22] and 2) the condensation of aromatic aldehydes with α-arylacetic acids in acetic anhydride and in the presence of a weak base (proceeding *via* mixed anhydrides generated *in situ*) to obtain α-arylcinnamic acids (*Oglialoro modification*).[5]

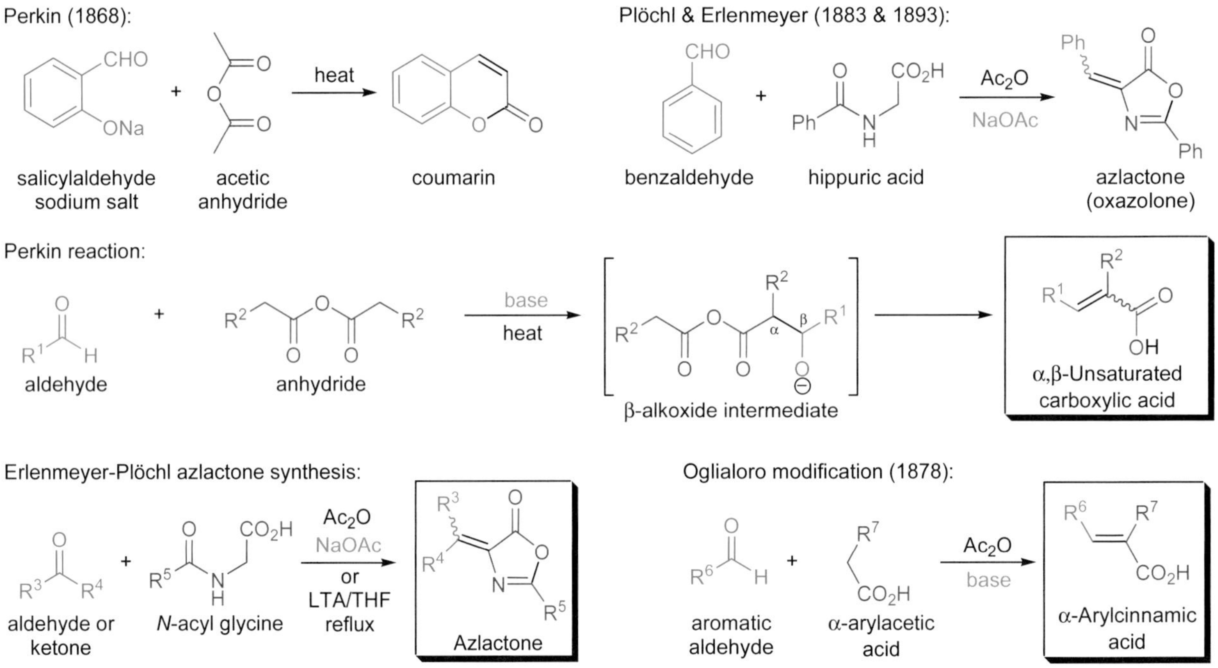

R^1 = aromatic, heteroaromatic, alkenyl, alkyl group with no α-hydrogen atom; R^2 = H, alkyl, aryl; R^{3-4} = H, alkyl, aryl; R^5 = alkyl, aryl; R^6 = aryl, heteroaryl; base: NaOAc, KOAc, CsOAc, Et_3N, pyridine, piperidine, K_2CO_3

Mechanism: [23,3,24-31,4,32-35]

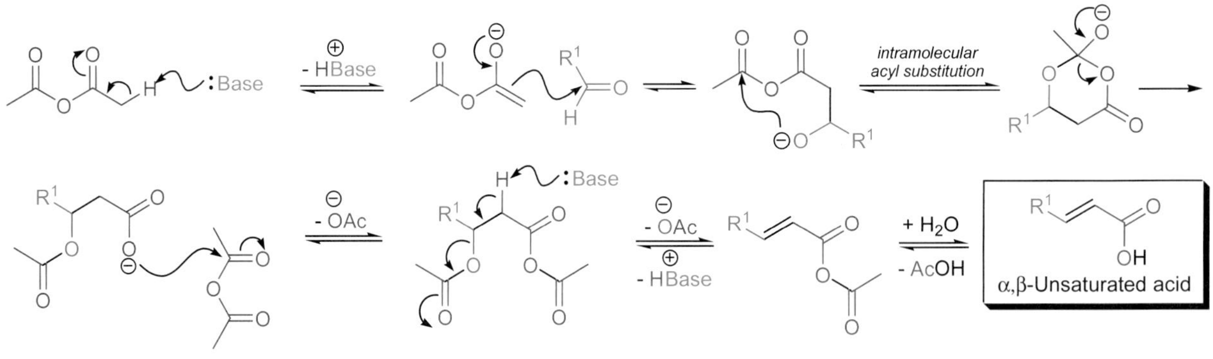

Synthetic Applications:

The combretastatins are a group of antimitotic agents isolated from the bark of the South African tree *Combretum caffrum*. A novel and highly stereoselective total synthesis of both the *cis* and *trans* isomers of combretastatin A-4 was developed by J.A. Hadfield and co-workers.[36] The (*Z*)-stereoisomer was prepared using the *Perkin reaction* as the key step in which 3,4,5-trimethoxyphenylacetic acid and 3-hydroxy-4-methoxbenzaldehyde was heated with triethylamine and acetic anhydride at reflux for several hours. The α,β-unsaturated acid was isolated in good yield after acidification and had the expected (*E*) stereochemistry. Decarboxylation of this acid was effected by heating it with copper powder in quinoline to afford the natural product (*Z*)-combretastatin A-4.

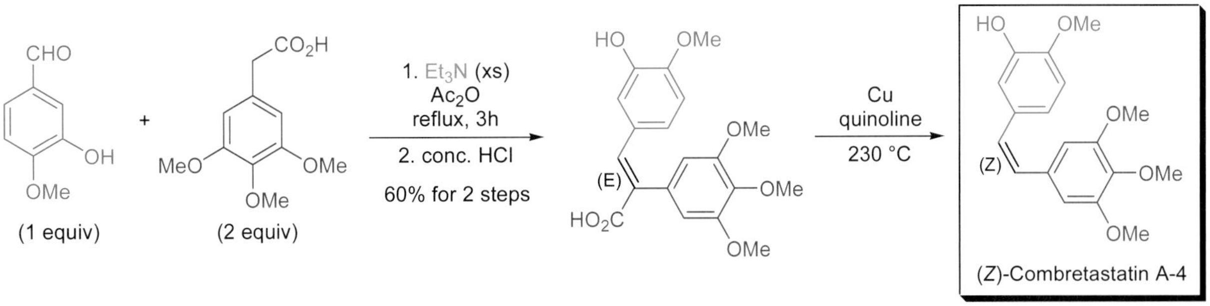

In the laboratory of D. Ma, the asymmetric synthesis of several metabotropic glutamate receptor antagonists derived from α-alkylated phenylglycines was undertaken.[37] The preparation of (*S*)-1-aminoindan-1,5-dicarboxylic acid (AIDA) started with the *Perkin reaction* of 3-bromobenzaldehyde and malonic acid. The resulting (*E*)-cinnamic acid derivative was hydrogenated and the following *intramolecular Friedel-Crafts acylation* afforded the corresponding indanone, which was then converted to (*S*)-AIDA.

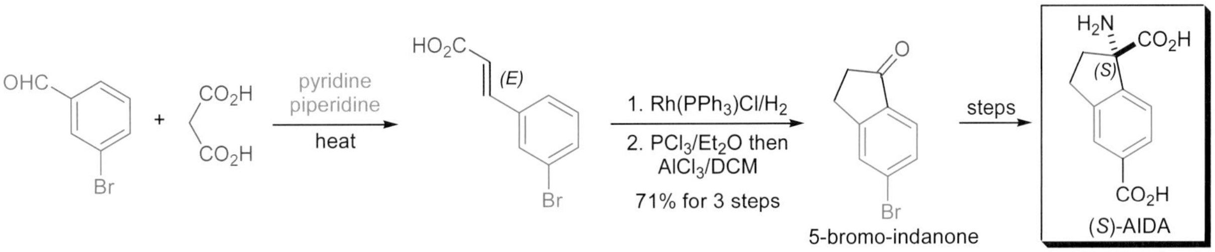

The large-scale pilot plant preparation of the chiral aminochroman antidepressant ebalzotan (also known as NAE-086) was developed by H.J. Federsel and co-workers.[38] The structural features of the target included a disubstituted chroman skeleton, a stereocenter, as well as a non-symmetrical tertiary amine moiety at the C3 position and a secondary carboxamide group at C5. The backbone of the target molecule was constructed using the *Perkin condensation* of 2-hydroxy-6-methoxybenzaldehyde with hippuric acid under mild conditions.

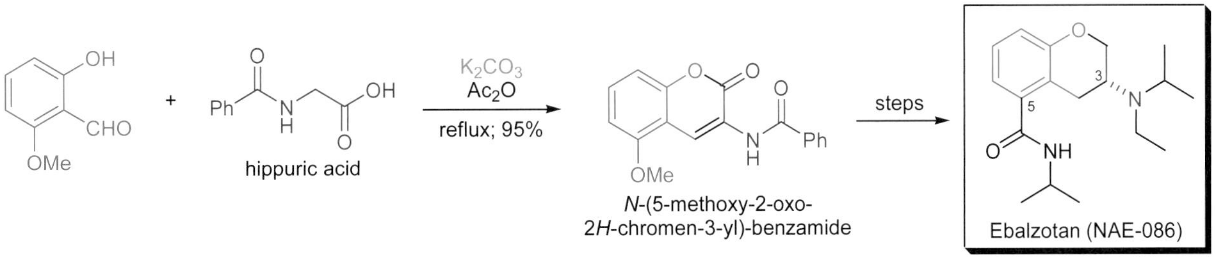

Fluorinated analogs of naturally occurring biologically active compounds, such as amino acids, often exhibit unique physiological properties, and therefore there is substantial interest in their convenient and high-yielding preparation. The research team of K.L. Kirk synthesized 6-fluoro-*meta*-tyrosine and several of its metabolites employing the *Erlenmeyer-Plöchl azlactone synthesis*.[39] Hippuric acid and 2-benzyloxy-5-fluorobenzaldehyde were condensed in the presence of sodium acetate in acetic anhydride to isolate the corresponding azlactone, which was converted to the target fluorinated amino acid in three steps.

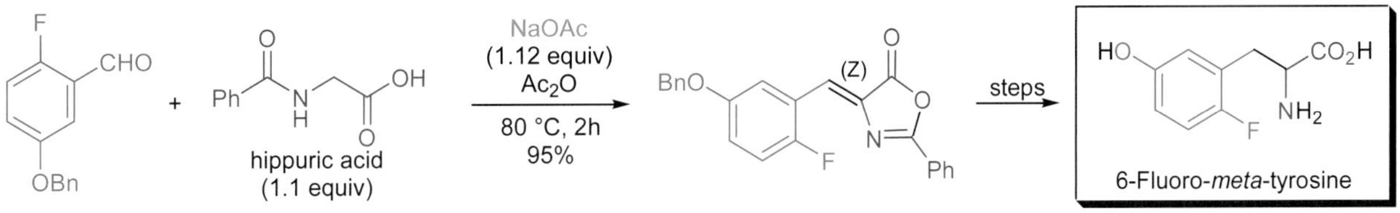

PETASIS BORONIC ACID-MANNICH REACTION

(References are on page 650)

Importance:

[*Seminal Publications*[1]; *Reviews*[2-7]; *Modifications & Improvements*[8-16]]

Allylic amines are synthetically useful building blocks and several derivatives possess diverse biological properties. In 1993, N.A. Petasis and co-workers reported an efficient synthesis of these compounds based on a modified *Mannich reaction* where vinylboronic acids served as the nucleophilic component. This transformation is referred to as the *Petasis boronic acid-Mannich reaction*. The general features of the reaction are: 1) according to the original procedure, the reaction is convenient to carry out: the mixture of paraformaldehyde and a secondary amine are heated to 90 °C in toluene or dioxane for ten minutes followed by the addition of the vinylboronic acid and stirring the reaction mixture at room temperature for several hours or heating to 90 °C for 30 minutes; 2) the work-up procedure includes an acid-base extraction to remove the unreacted vinylboronic acid; 3) the addition of the boronic acid to the amine-paraformaldehyde adduct occurs with complete retention of the geometry of the double bond; 4) the resulting allylamines form with high stereoselectivity; 4) originally, formaldehyde was used as the aldehyde component, but other aldehydes and ketones also undergo the transformation;[8,9] 5) when glyoxylic acid or α-keto acids are used as the carbonyl component, α-amino acids are obtained;[8,9] 5) the boronic acids can be prepared by the condensation of catecholborane with terminal alkynes and subsequent hydrolysis of the vinylboronate esters; 6) vinylboronate esters can also participate in the reaction, but purification of the product is more difficult;[1] 7) arylboronate esters[8] and potassium organotrifluoroborates[15,16] are also viable substrates; 8) in addition to secondary amines, tertiary aromatic amines,[14] substituted hydrazines,[12] substituted hydroxylamines, and sulfinamides[13] undergo the transformation; and 9) upon Lewis acid activation, 2-hydroxy- and 2-alkoxy derivatives of *N*-protected pyrrolidines and piperidines also react.[10] A solid phase version of the reaction was also developed.[11]

Petasis (1993):

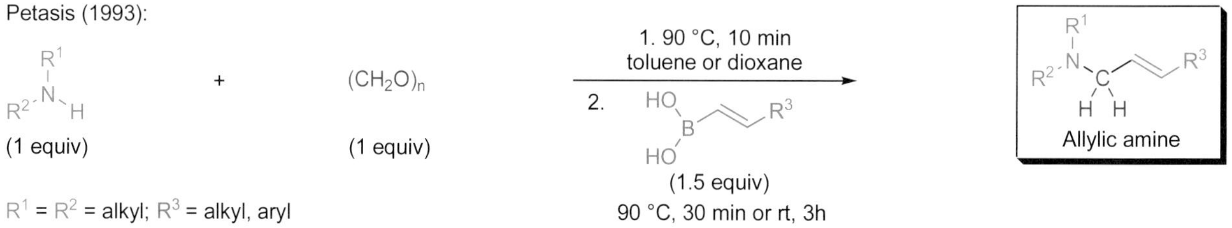

$R^1 = R^2$ = alkyl; R^3 = alkyl, aryl

General scheme:

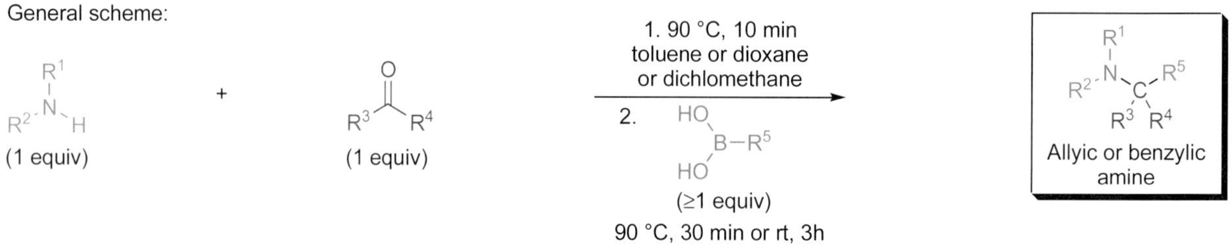

R^1 = alkyl; R^2 = alkyl; OH, *O*-alkyl, -SO*ᵗ*Bu, NHCOO*ᵗ*Bu; R^3 = H, alkyl, R^4 = H, -COOH, aryl; R^5 = alkenyl, aryl, heteroaryl

Mechanism:[1]

The mechanism of the *Petasis boronic acid-Mannich reaction* is not fully understood. In the first step of the reaction, upon mixing the carbonyl and the amine components, three possible products can form: iminium salt **A**, diamine **B**, and α-hydroxy amine **C**. It was shown that preformed iminium salts do not react with boronic acids. This observation suggests that the reaction does not go through intermediate **A**. Both intermediate **B** and **C** can promote the formation of the product. Most likely, the reaction proceeds through intermediate **C**, where the hydroxyl group attacks the electrophilic boron leading to an "ate"-complex. Subsequent vinyl transfer provides the allylic amine along with the boronic acid sideproduct.

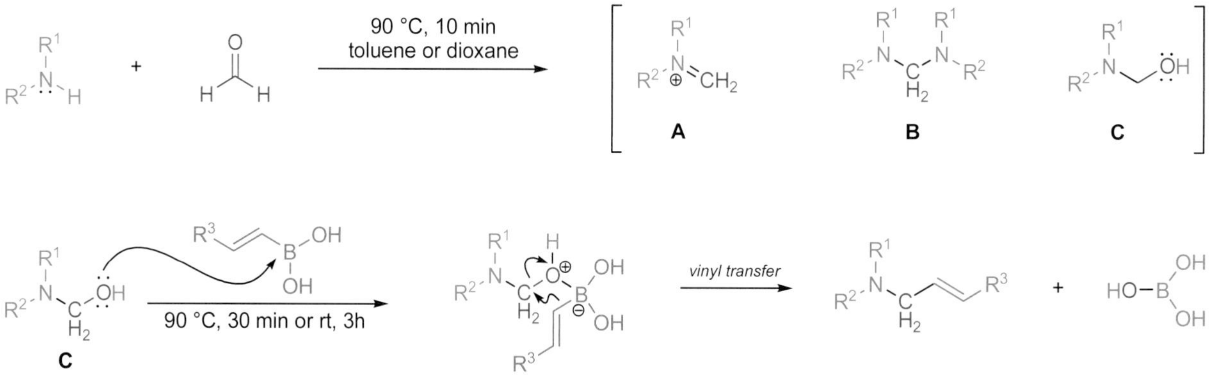

Synthetic Applications:

(–)-Cytoxazone is a novel cytokine modulator. The total synthesis of this natural product and its enantiomer was accomplished by S. Sugiyama.[17] The 3-amino-1,2-propanediol moiety was synthesized by a *Petasis boronic acid-Mannich* reaction between DL-glyceraldehyde, (*R*)-1-(1-naphthyl)ethylamine and 4-methoxyphenylboronic acid to provide a 1:1 mixture of the diastereomeric products. The diastereomers could be separated at a later stage in the synthesis and transformed into (–)- and (+)-cytoxazone.

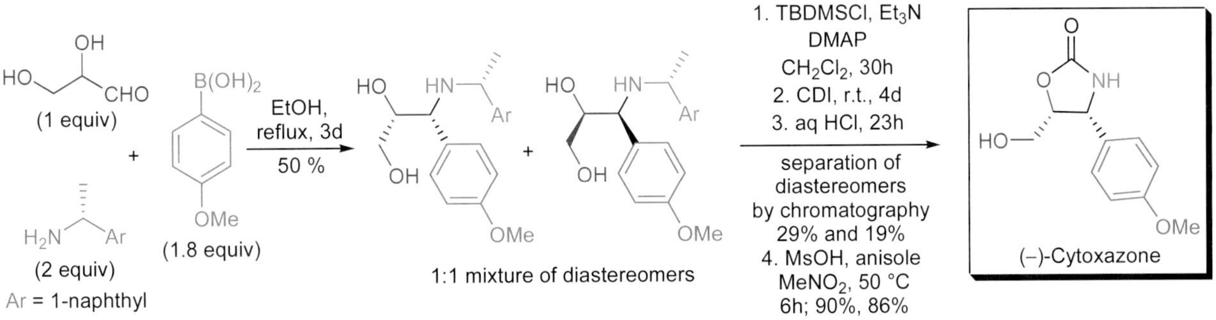

In the laboratory of A. Golebiowski, the high throughput synthesis of diketopiperazines was accomplished.[18] These compounds can serve as β-turn mimetics. The key step in this approach was a *Petasis boronic acid-Mannich reaction* between the Merrifield resin-bound piperazine-2-carboxylic acid, glyoxylic acid, and a wide range of commercially available boronic acids to provide a 1:1 mixture of the products. A specific example is shown below.

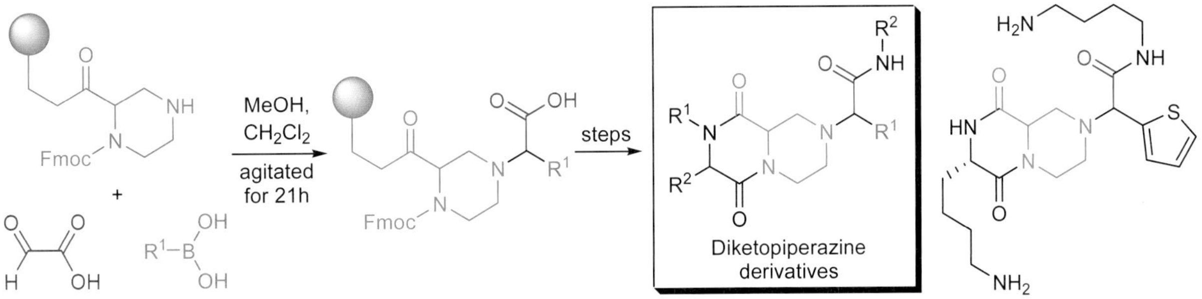

M.G. Finn and co-workers developed a procedure for the preparation of 2*H*-chromene derivatives that includes a Petasis three-component reaction between salicylaldehyde, vinylic- and aromatic boronic acids, and dibenzylamine.[19] The hydroxyl group of the salicylic aldehyde is essential for the activation of the boronic acid. The initially formed allylic amine undergoes a cyclization upon ejecting the dibenzylamine, thus rendering the process catalytic in the amine.

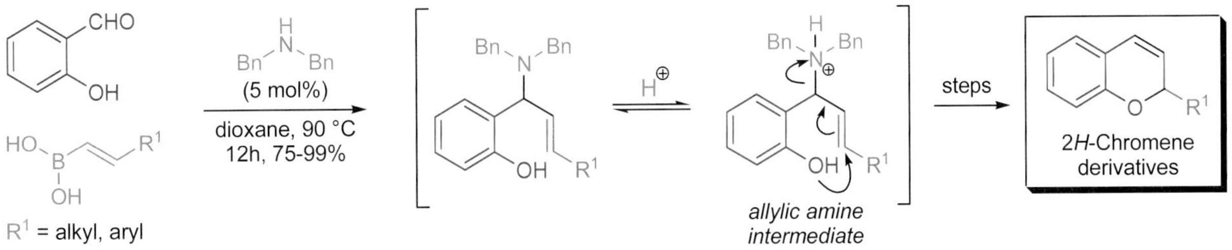

R.A. Batey and co-workers developed a modification of the *Petasis-boronic acid-Mannich reaction* that occurs via *N*-acyliminium ions derived from *N*-protected-2,3-dihydroxypyrrolidine and 2,3-dihydroxypiperidine derivatives.[10] This method was utilized in the total synthesis of (±)-deoxycastanospermine. The formation of the *N*-acyliminium ion was achieved by treating *N*-Cbz-2,3-pyrrolidine with $BF_3{\cdot}OEt_2$.[20] Subsequent vinyl transfer from the alkenylboronic ester provided the product with excellent yield and diastereoselectivity.

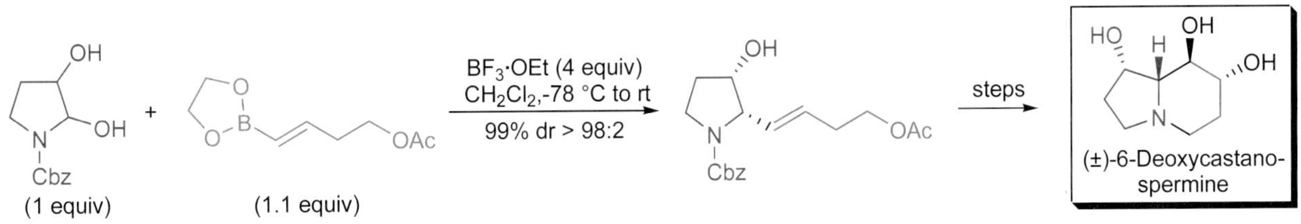

PETASIS-FERRIER REARRANGEMENT

(References are on page 650)

Importance:

[*Seminal Publications*[1,2]; *Modifications & Improvements*[3,4]]

In 1995, N.A. Petasis reported the Lewis acid-promoted rearrangement of five-membered enol acetals to substituted tetrahydrofurans and in 1996, the similar rearrangement of six-membered enol acetals to the corresponding substituted tetrahydropyrans.[1,2] The rearrangement proceeds *via* an oxocarbenium ion intermediate similar to the one which is involved in a *Type II Ferrier rearrangement*. Therefore, the stereocontrolled Lewis acid-promoted rearrangement of cyclic enol acetals to the corresponding substituted tetrahydrofurans and tetrahydropyrans is called the *Petasis-Ferrier rearrangement*. In laboratory practice, the rearrangement is a three-step procedure: 1) highly stereoselective preparation of 1,3-dioxolane-4-ones and 1,3-dioxane-4-ones from α- and β-hydroxy acids and aldehydes, respectively; 2) methenylation of the carbonyl group with dimethyl titanocene (Cp_2TiMe_2) to afford the enol acetals; and 3) treatment of the enol acetals with an aluminum-based Lewis acid to bring about the transposition of an O-atom with a C-atom on the ring. It was not until 1999 that this rearrangement was modified and utilized for the total synthesis of complex natural products by A.B. Smith and co-workers.[3-7] The general features of the *Petasis-Ferrier rearrangement* are: 1) the straightforward construction of the substrate enol acetals allows the stereocontrolled assembly of complex fragments in a linchpin fashion; 2) the configuration of the acetal carbon is retained or enhanced during the rearrangement; 3) the rearrangement of five-membered enol acetals takes place at a much higher temperature than for the six-membered substrates; 4) trialkylaluminums were found to be the most effective reagents to mediate the rearrangement (i-Bu_3Al, Me_3Al, Me_2AlCl being the most common); 5) the stereoselectivity of the aluminum-mediated carbonyl reduction (very last step) depends on the substitution pattern and occurs when i-Bu_3Al is used (the reduction does not take place with Me_2AlCl); and 6) a drawback of the procedure is that the olefination step can lead to a mixture of olefin stereoisomers when the applied titanocene is other than dimethyl titanocene.

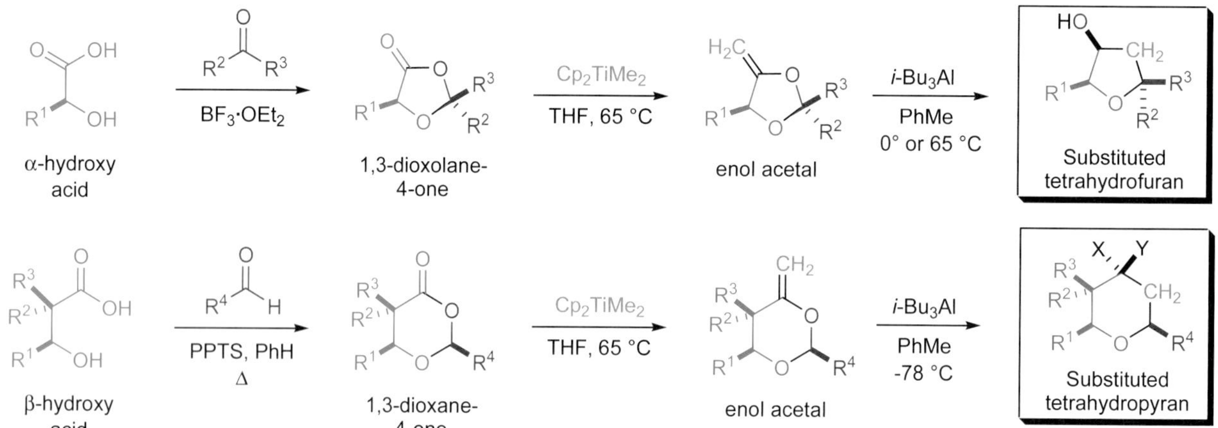

X = OH & Y = H

Mechanism: [1,2]

The aluminum-mediated *Petasis-Ferrier rearrangement* is a stepwise *[1,3]-sigmatropic process*. The first step is the coordination of the Lewis-acid to the O-atom of the enol. Coordination to the ether O-atom is reversible and non-productive. Cleavage of the adjacent C-O-bond, assisted by the antiperiplanar lone pair of the etheral O-atom, stereospecifically gives rise to an oxocarbenium enolate species, which cyclizes to the desired oxacycle. The rate difference in the rearrangement for the five- *versus* six-membered series can be explained by the more facile *6-(enolendo)-endo-trig* cyclization.[8,9] The last step is the intramolecular equatorial hydride delivery.

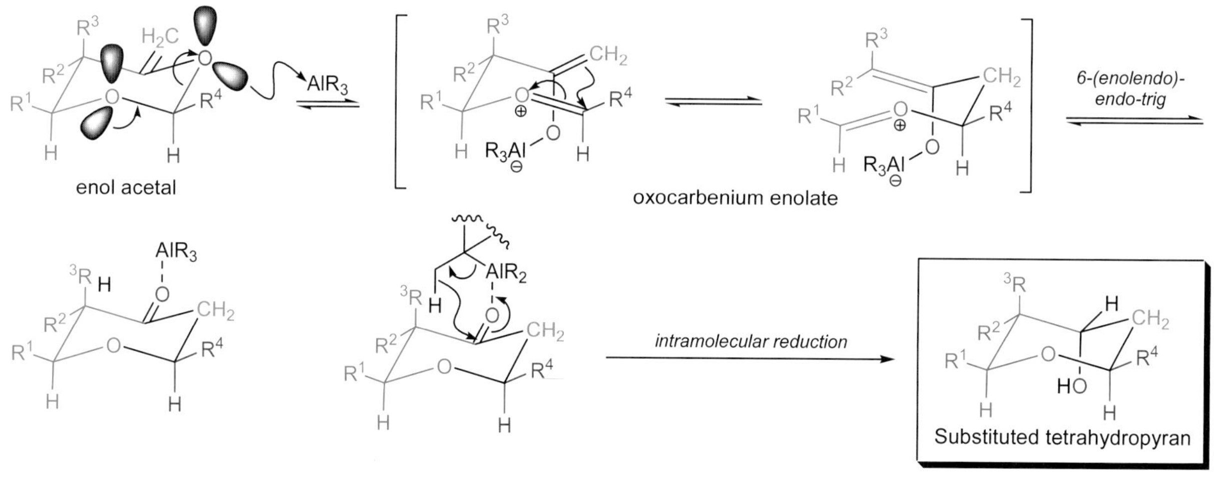

Synthetic Applications:

During the total synthesis of (+)-phorboxazole A by A.B. Smith and co-workers, the *modified Petasis-Ferrier rearrangement* was successfully employed for the preparation of the C11-C15 and C22-C26 *cis*-tetrahydropyran rings.[5] The rearrangement using the conditions prescribed by Petasis (with *i*-Bu_3Al) failed to produce the desired 2,6-*cis*-tetrahydropyran, so Me_2AlCl was investigated. Treatment of the substrate with Me_2AlCl at ambient temperature provided the C3-C19 subtarget of phorboxazole as a single isomer in 89% yield.

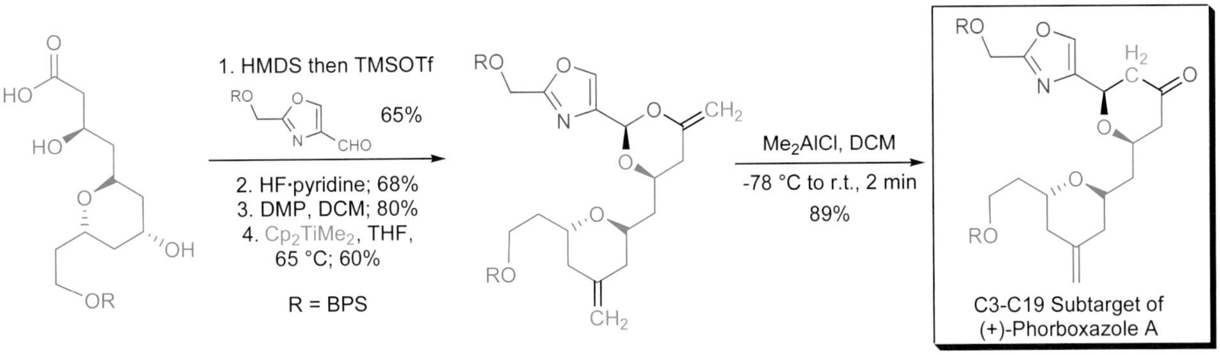

Similarly, the C22-C26 fully substituted central tetrahydropyran ring of phorboxazole was prepared using the *modified Petasis-Ferrier rearrangement*.[5] Based on the known mechanistic model, the enol acetal moiety of the rearrangement substrate required the (*Z*)-configuration. The synthesis of this enol ether was not possible with either the *Takai-* or *Petasis-Tebbe olefinations*. Utilization of the *Type-II Julia olefination* afforded the desired enol acetal, but with no *E*/*Z* selectivity. Upon treatment of these enol ethers with Me_2AlCl, the rearrangement afforded only the desired tetrahydropyran in excellent yield.

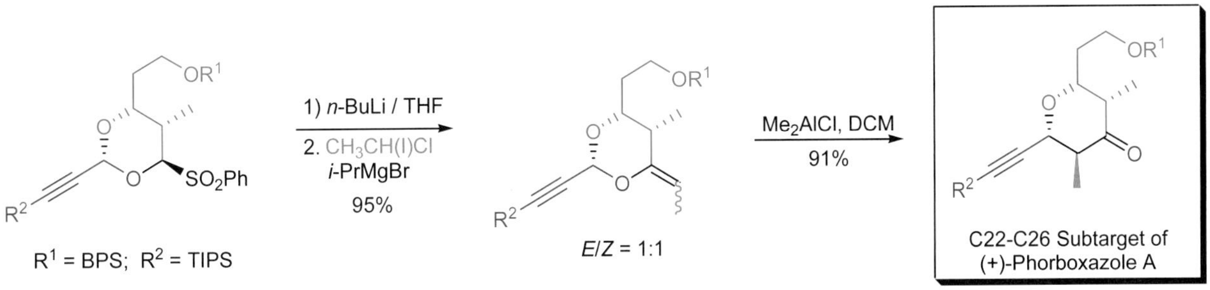

The first total synthesis of (+)-zampanolide and (+)-dactylolide was achieved in the laboratory of A.B. Smith.[6,7] The key step of these syntheses was the application of the *modified Petasis-Ferrier rearrangement* to construct the central *cis*-2,6-disubstituted tetrahydropyran moiety in a stereocontrolled fashion. The treatment of the enol acetal with 1 equivalent of Me_2AlCl at -78 °C effected the rearrangement to furnish the desired *cis*-tetrahydropyranone in 59% yield.

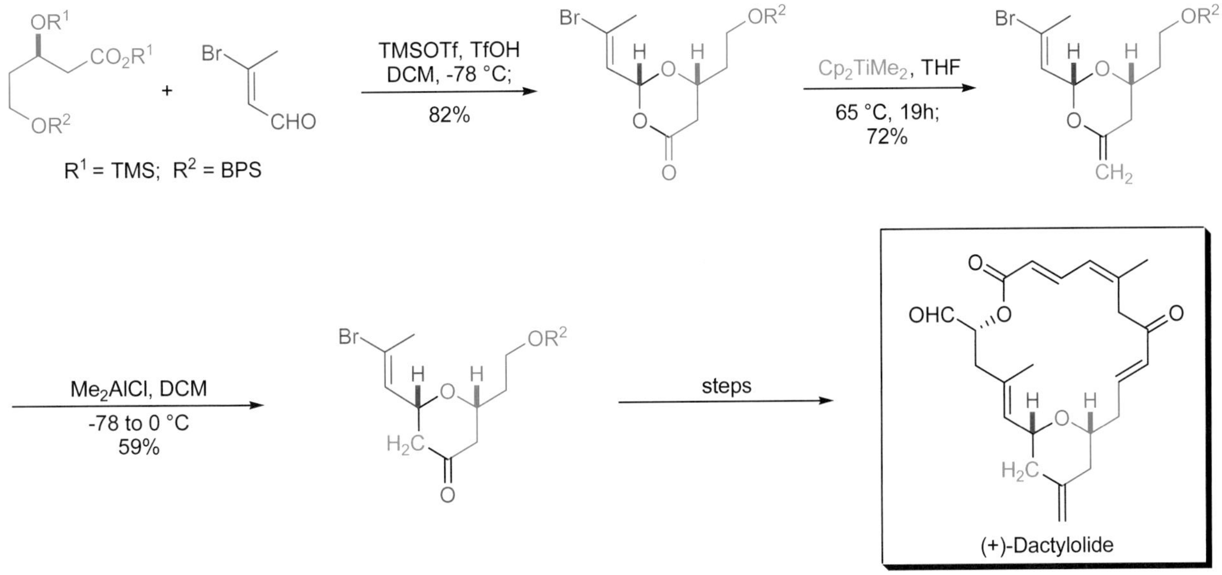

PETERSON OLEFINATION

(References are on page 650)

Importance:

[*Seminal Publications*[1-3]; *Reviews*[4-23]; *Modifications & Improvements*[24-30]; *Theoretical Studies*[31-34]]

In 1968, D.J. Peterson demonstrated in a detailed study that α-trimethylsilyl-substituted organometallic compounds could be used to convert carbonyl compounds *via* β-silylcarbinols to the corresponding olefins.[3] Similar transformations prior to Peterson's publication were reported but the scope of the reaction was not investigated.[1,2] The preparation of alkenes from α-silyl carbanions and carbonyl compounds is known as the *Peterson olefination* and it is considered to be the silicon-variant of the *Wittig-type reactions*. The general features of the reaction are: 1) the α-silyl carbanions are prepared in a variety of ways, including metal-halogen exchange of the α-halogenated alkylsilanes or the direct deprotonation of alkylsilanes at the α-position; 2) the addition of the α-silyl carbanions to carbonyl compounds gives rise to a mixture of diastereomeric β-silylcarbinols, which can be isolated and separated only if the R^2 substituent in the α-silyl carbanion is not electron-withdrawing; 3) when the R^2 substituent is an electron-donating group (e.g., alkyl) the intermediate β-silylcarbinols can be isolated and the diastereomers can be separated by means of chromatography; 4) upon treatment with base (NaH, KH, KO*t*-Bu), the β-silylcarbinols undergo a stereospecific *syn*-elimination, while treatment with dilute acid or a Lewis acid (AcOH, H_2SO_4, $BF_3{\cdot}OEt_2$) results in a stereospecific *anti*-elimination; and 5) either the (E) or (Z)-alkene can be obtained from a diastereomerically pure β-silylcarbinol by choosing acidic or basic conditions, so the stereoselectivity of the reaction depends on the availability of the diastereomerically pure β-silylcarbinol. Since the preparation of a specific α-silyl carbanion is not always possible, a variety of methods were developed to access α-silylcarbinols in a diastereomerically pure form: 1) the stereoselective addition of nucleophiles to α-silyl ketones, aldehydes, and esters;[35,36] 2) ring-opening of α,β-epoxysilanes with nucleophiles;[37,38,22] 3) *aldol reaction* of enolates derived from α-silyl ketones with aldehydes and ketones;[39] and 4) *stereoselective dihydroxylation* of vinylsilanes.[40,17] Related reactions in which the silicon group (SiR_3) has been replaced with groups containing other elements (SbR_2, AsR_2, SnR_3, HgR, etc.) also form alkenes, but usually the corresponding α-carbanions are harder to prepare and the elimination requires special and often harsh conditions.[9]

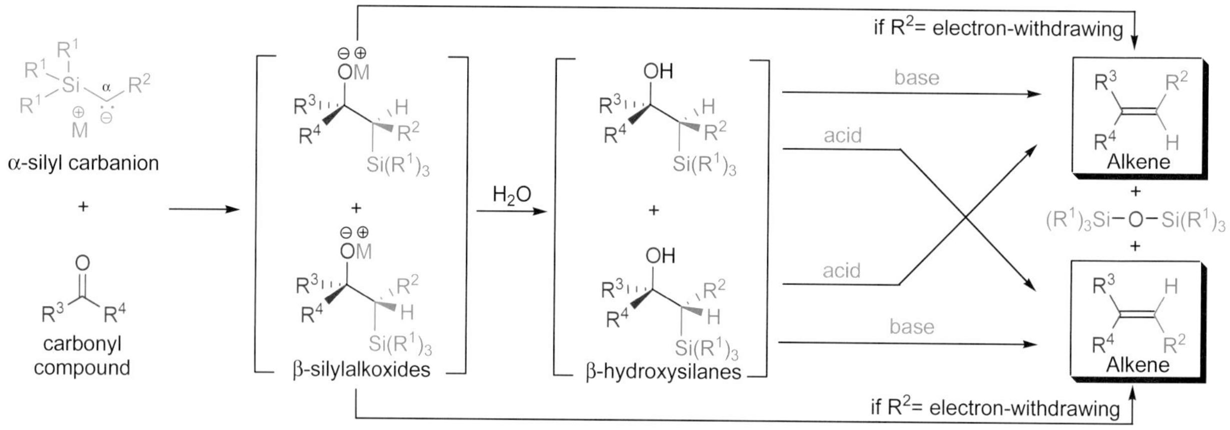

R^1=alkyl, aryl; R^2 = alkyl, aryl, CO_2R, CN, $CONR_2$, CH=NR, SR, SOR, SO_2R, SeR, SiR_3, OR, BO_2R_2; R^3,R^4=alkyl, aryl, H

Mechanism: [41-44,9,45-50,21]

The exact pathway of the *Peterson reaction* is still not clear despite the intensive research effort.[9,21] Most of the mechanistic studies suggest that both the stepwise and concerted pathways are feasible under basic conditions. In the concerted pathway a pentacoordinate 1,2-oxasiletanide is formed. The stepwise pathway is expected when chelation control operates in the reaction. The driving force is the formation of a very strong Si-O bond. Under acidic conditions the β-hydroxysilane undergoes an *E2 elimination* to afford the other alkene isomer.

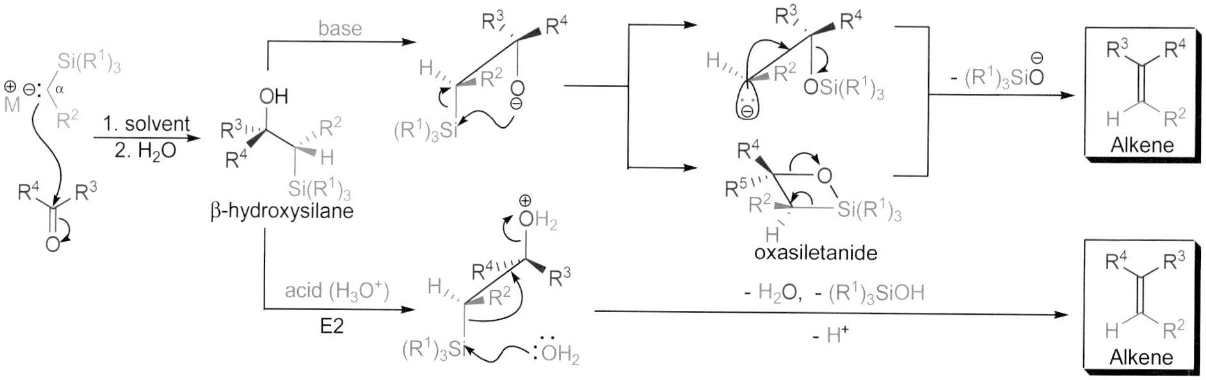

Synthetic Applications:

In the laboratory of P. Deslongchamps, the first asymmetric total synthesis of (+)-maritimol, a member of the stemodane diterpenoids, was accomplished using the *Peterson olefination* as the key step.[51] Close to the end of the synthetic sequence, the D ring of the natural product had to be installed via the *Thorpe-Ziegler annulation* of the corresponding 1,5-dinitrile. This dinitrile was prepared using the *Peterson olefination*. The tricyclic aldehyde was treated with the solution of an α-silyl boronate derived from trimethylsilylacetonitrile. The resulting 6:1 mixture of *cis*- and *trans*-enenitriles was reduced to the desired saturated 1,5-dinitrile.

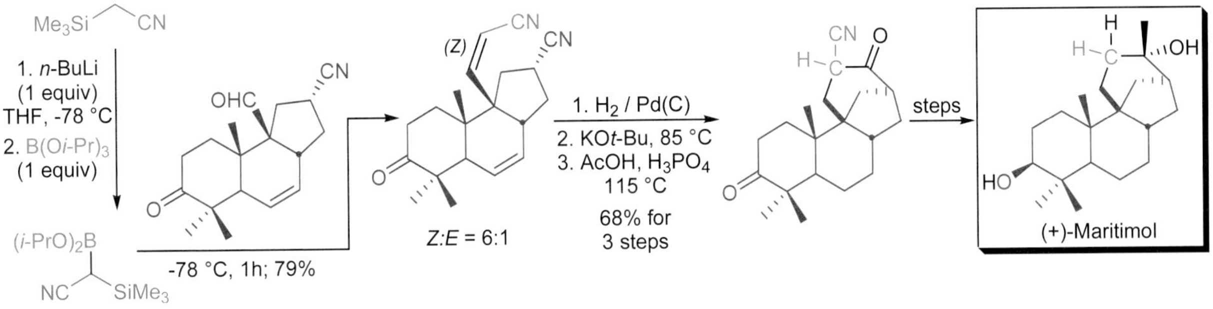

M.A. Tius et al. reported a formal total synthesis of the macrocyclic core of roseophilin.[52] The aliphatic five-membered ring of this core was prepared *via* a variant of the *Nazarov cyclization*. The precursor for this cyclopentannelation reaction is an (*E*)-α,β-unsaturated aldehyde, which was prepared using the *Peterson olefination* on the *t*-butylimine of 5-hexenal. First the α-TMS derivative of the imine was generated; then after a second deprotonation, the additon of isobutyraldehyde gave the (*E*)-α,β-unsaturated imine upon aqueous work-up. Acidic hydrolysis of this imine gave the desired (*E*)-α,β-unsaturated aldehyde in good yield.

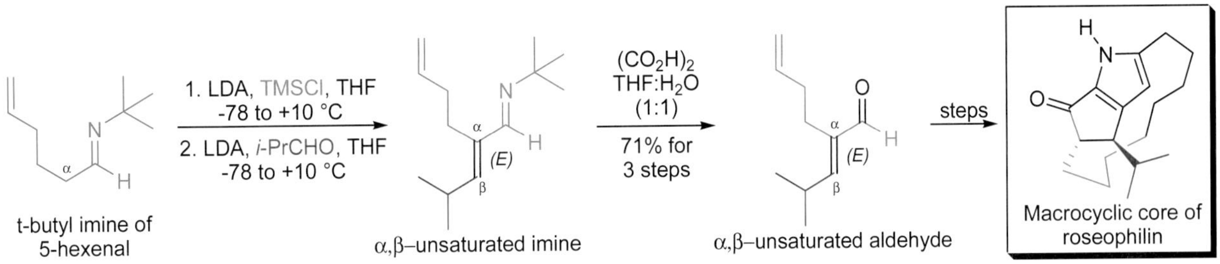

In the final stages of the total synthesis of (+)-brasilenyne by S.E. Denmark and co-workers, the introduction of the (*Z*)-enyne side chain was accomplished with the *Peterson olefination*.[53] The aldehyde was treated with lithiated 1,3-*bis*(triisopropylsilyl)propyne at low temperature followed by slow warming of the reaction mixture to ambient temperature to give a 6:1 (*Z:E*) ratio of the desired enyne.

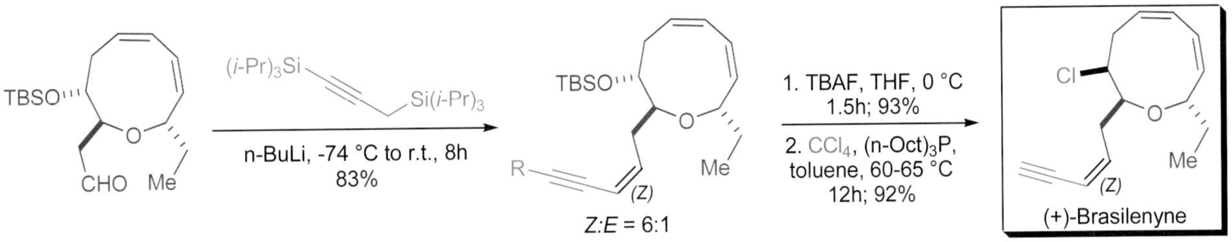

A (Z)-selective *Peterson olefination* was the key step in the first enantioselective total synthesis of both enantiomers of lancifolol in the laboratory of H. Monti.[54] This synthetic approach allowed the correlation of the relationship between absolute configuration and specific rotation. It is important to mention that no other olefination method could be applied successfully in installing this (Z)-alkene moiety.

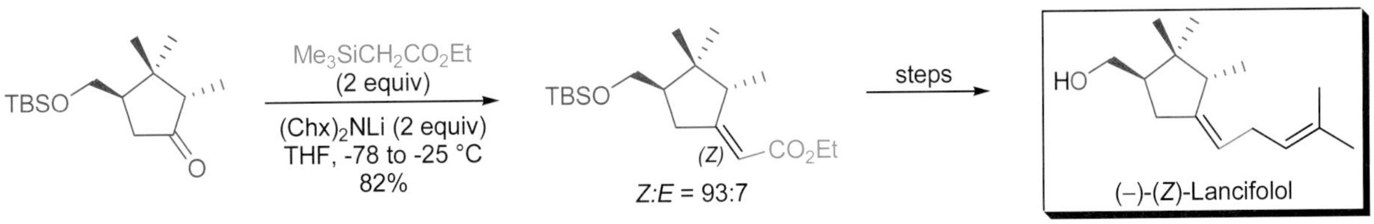

PFITZNER-MOFFATT OXIDATION

(References are on page 652)

Importance:

[*Seminal Publications*[1-3]; *Reviews*[4-9]; *Modifications & Improvements*[10-18]]

In 1963, J.G. Moffatt and K.E. Pfitzner observed that primary and secondary alcohols were efficiently oxidized to the corresponding aldehydes and ketones in a solution of dimethyl sulfoxide (DMSO) upon the addition of dicyclohexyl carbodiimide (DCC) and catalytic amounts of anhydrous phosphoric acid (H_3PO_4).[1] This transformation is known as the *Pfitzner-Moffatt oxidation* (*Moffatt oxidation*) and falls into the general category of activated dimethyl sulfoxide mediated oxidations.[8,9] The scope and limitation of the *P-M oxidation* was quickly established, and it was clear that this procedure was a good alternative to chromium(VI)-based oxidations (using PCC and PDC) to oxidize sensitive alcohol substrates under mild and weakly acidic condition.[2,3] The general features of the reactions are: 1) the necessary reagents are all inexpensive and easy to handle, and the execution of the oxidation does not require special equipment; 2) the yield of the product is generally high on both small and large scale; 3) there are only a few side reactions: the occasional formation of methylthiomethyl ether by-products and the isomerization of β,γ-unsaturated carbonyl compounds under the reaction conditions; 4) most functional groups are tolerated, but unprotected tertiary alcohols are often eliminated; 5) DCC is the most widely used activating agent that needs to be applied in excess (usually 3 equivalents or more); 5) the DMSO can serve as the solvent, but inert co-solvents (e.g., EtOAc, benzene) can also be used to make the isolation of the product easier; 6) the oxidation only works with catalysts that are only moderately acidic compounds such as *ortho*-phosphoric acid (H_3PO_4), dichloroacetic acid and the pyridinium salts of strong acids; and 7) in the presence of strong organic and mineral acids, the oxidation is very slow or it does not take place at all. The main drawbacks of the *P-M oxidation* are: 1) the by-product dialkyl urea is often difficult to remove from the product completely, but the use of water soluble or polymer-bound carbodiimides resolves any purification problem;[15] and 2) the excess DCC has to be removed from the product as well, but this issue can be resolved by the addition of oxalic acid during the work-up. Other well-known ways to activate DMSO involve the use of: 1) acetic anhydride (*Albright and Goldman procedure*);[13] 2) pyridine-SO_3 complex (*Parikh-Doering oxidation*);[14] and 3) oxalyl chloride or trifluoroacetic anhydride (*Swern oxidation*).[16,17]

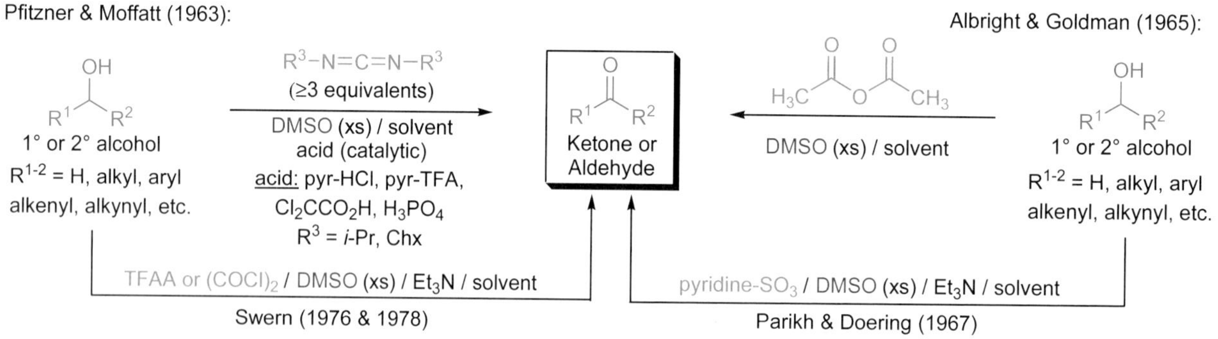

Mechanism: [2,19,20,6,21,8,9]

The mechanism of the *P-M oxidation* consists of three distinct steps: 1) activation of the DMSO by a protonated dialkyl carbodiimide; 2) activation of the alcohol substrate and the formation of the key alkoxysulfonium ylide intermediate; and 3) the intramolecular decomposition of the alkoxysulfonium ylide to afford the product ketone or aldehyde and the dialkyl urea by-product (established by isotopic labeling studies). The alkoxysulfonium ylide is a common intermediate in all other oxidations using activated DMSO.

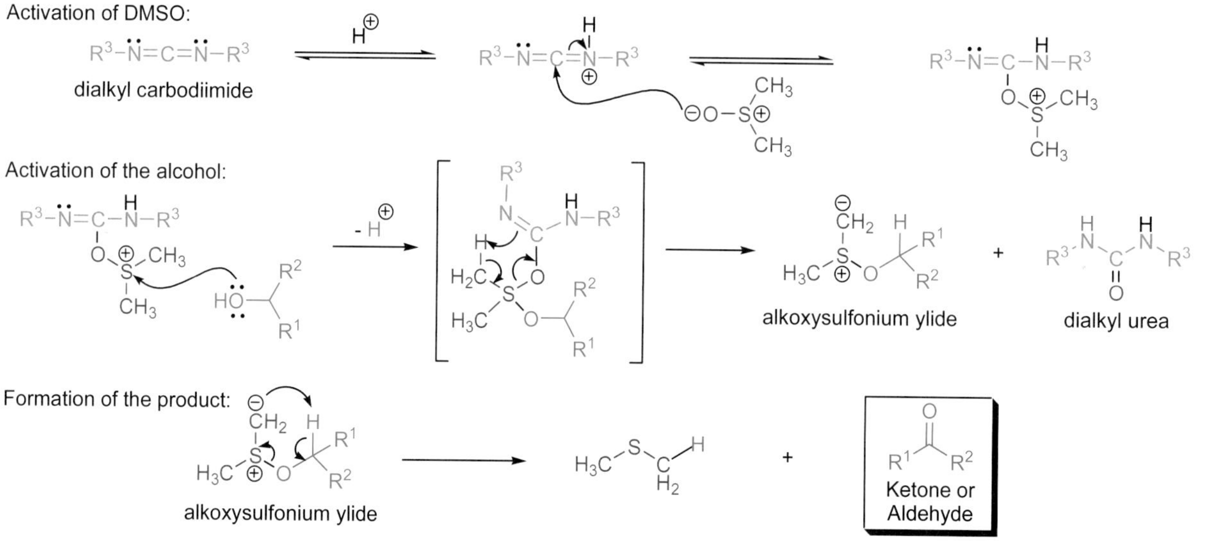

Synthetic Applications:

The first total synthesis of the nucleoside antibiotic herbicidin B was accomplished in the laboratory of A. Matsuda.[22] The key step was a novel *aldol-type C-glycosidation reaction* promoted by SmI_2 between a 1-phenylthio-2-ulose derivative and a 1-β-D-xylosyladenine-5'-aldehyde derivative. During the preparation of the phenylthio sugar subunit, the *Moffatt oxidation* was applied to convert the primary alcohol to the corresponding aldehyde, which was immediately oxidized with PDC in DMF/MeOH to the methyl ester. The reaction conditions were completely compatible with the silyl protecting group as well as the thioacetal functionality.

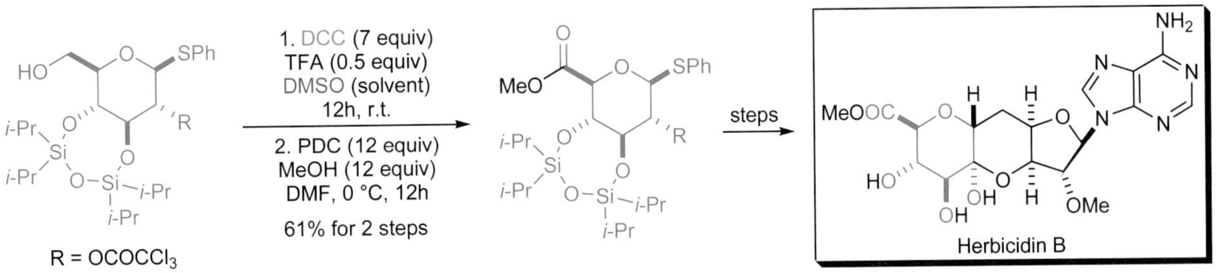

The *Moffatt oxidation* was utilized in the endgame of the total synthesis of (+)-paspalicine by A.B. Smith et al.[23] The advanced intermediate hexacyclic homoallylic alcohol was subjected to the *Moffatt oxidation* conditions using pyridinium trifluoroacetate as the acid catalyst. Under these conditions, the desired β,γ-unsaturated ketone and the rearranged α,β-unsaturated ketone (paspalicine) were formed in a 5:1 ratio. The final step was the *Rh-catalyzed isomerization* of the β,γ-unsaturated ketone to the natural product.

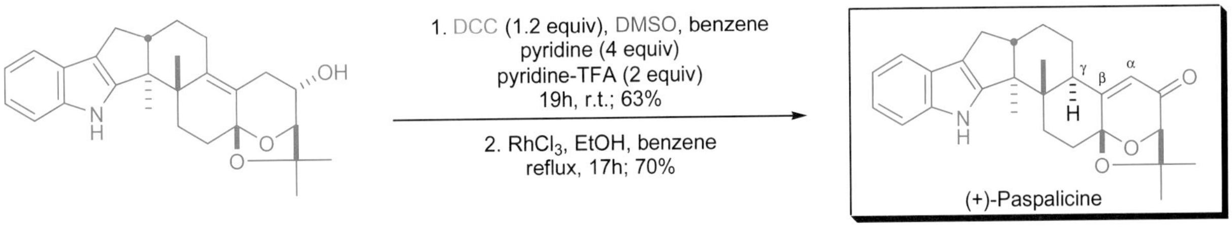

The complex polyene hydroxyl-substituted tetrahydrofuran metabolite (±)-citreoviral was synthesized by G. Pattenden and co-workers.[24] All four carbons on the tetrahydrofuran ring are chiral, and in the final stages of the synthetic effort the stereochemistry of the C3 secondary homoallylic alcohol had to be inverted. This step was best achieved by a *Moffatt oxidation*/$NaBH_4$ reduction sequence.

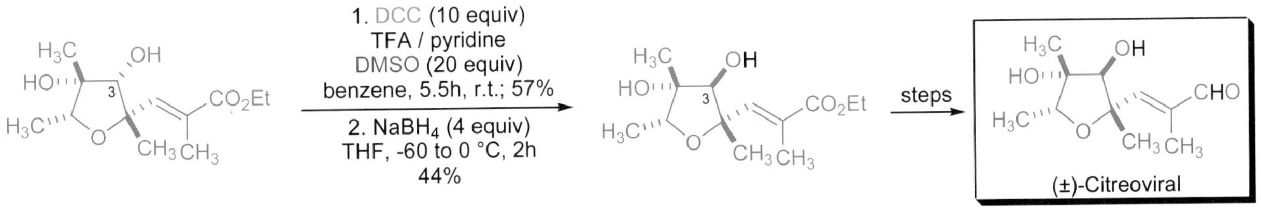

The total synthesis of the antimicrobial drimane-type sesquiterpene (–)-pereniporin A was achieved by the research team of K. Mori.[25] The advanced intermediate bicyclic primary alcohol was first oxidized to the corresponding aldehyde using the *Moffatt oxidation*. Interestingly, the sensitive α-hydroxy aldehyde moiety in the product remained unchanged. The final step was a global deprotection followed by a spontaneous lactol formation.

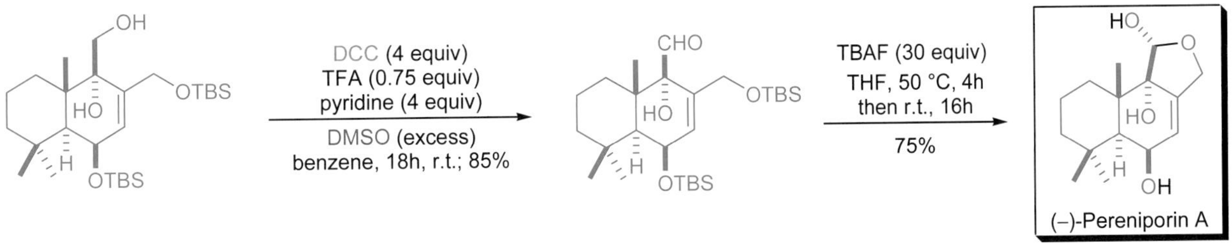

PICTET-SPENGLER TETRAHYDROISOQUINOLINE SYNTHESIS

(References are on page 652)

Importance:

[*Seminal Publications*[1]; *Reviews*[2-13]; *Modifications & Improvements*[14-27]; *Theoretical Studies*[28]]

In 1911, A. Pictet and T. Spengler reported the condensation of phenylethylamine and methylal (dimethoxymethane) in concentrated hydrochloric acid to afford 1,2,3,4-tetrahydroisoquinoline in moderate yield.[1] The authors observed a similar transformation when tyrosine and phenylalanine were subjected to identical conditions. The condensation of a β-arylethylamine with a carbonyl compound in the presence of a protic or Lewis acid to give rise to a substituted tetrahydroisoquinoline is known as the *Pictet-Spengler tetrahydroisoquinoline synthesis* (or *Pictet-Spengler reaction*). The general features of the transformation are: 1) only β-arylethylamines with electron-donating substituents afford high yields; 2) the carbonyl compound can be an aldehyde or a ketone or any acid-labile surrogate; 3) the most frequently used aldehyde is formaldehyde or its dimethyl acetal; 4) the number of electron-donating groups on the aromatic ring influences the ease of the reaction, and, for example, the presence of two alkoxy groups allows the *Pictet-Spengler reaction* to proceed under physiological conditions (this is important in the biosynthesis of alkaloids); 5) the reaction is usually carried out with a slight excess of the carbonyl compound (to ensure the complete consumption of the amine) in either protic or aprotic medium; and 6) since the reaction goes through the intermediacy of a Schiff base, the Schiff base can be prepared separately and subjected to a protic or Lewis acid to afford the cyclized tetrahydroisoquinoline product.

Pictet & Spengler (1911):

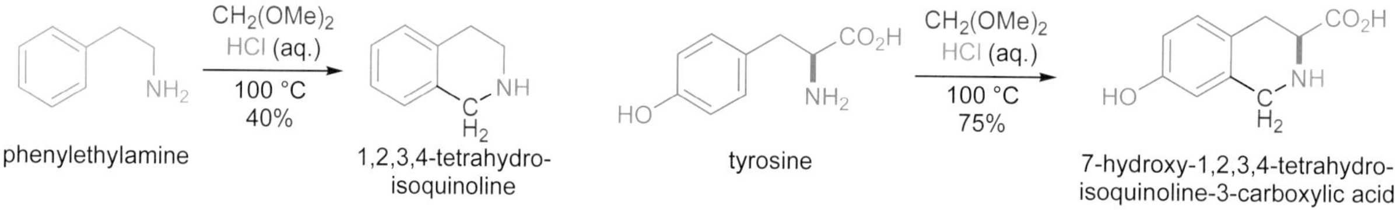

Pictet-Spengler reaction:

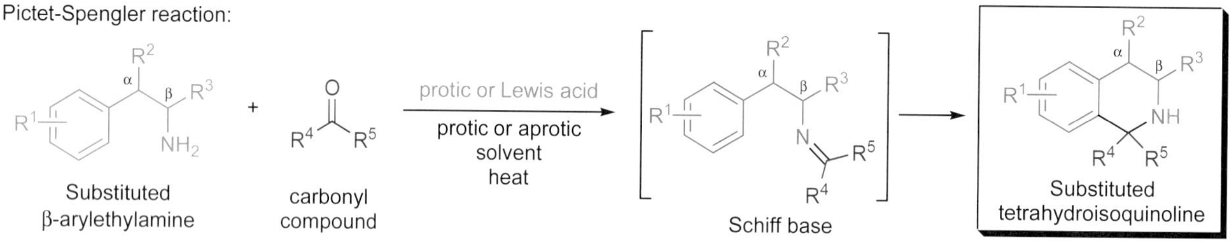

R^1 = H, alkyl , aryl, *O*-alkyl, usually an electron-donating group (EDG); R^{2-3} = H, alkyl ,aryl; R^{4-5} = H, alkyl ,aryl; protic acid: HCl, H_2SO_4, TFA, silica gel; Lewis acid: $BF_3{\cdot}OEt_2$

Mechanism: [2,9]

The first step of the *Pictet-Spengler reaction* is the formation of a Schiff base. The amine and aldehyde give rise to an aminal, which is dehydrated under acidic conditions to afford the corresponding imine. Protonation of the imine results in the formation of an iminium ion, which reacts with the electron-rich aromatic ring in a *6-endo-trig* cyclization to afford the six-membered heterocycle. The same type of reactive intermediate is involved in the *Bischler-Napieralski isoquinoline synthesis*, but that cationic species is more electrophilic and the aromatic ring does not need to be activated to achieve cyclization. The loss of proton restores the aromatic ring, thus giving rise to the product.

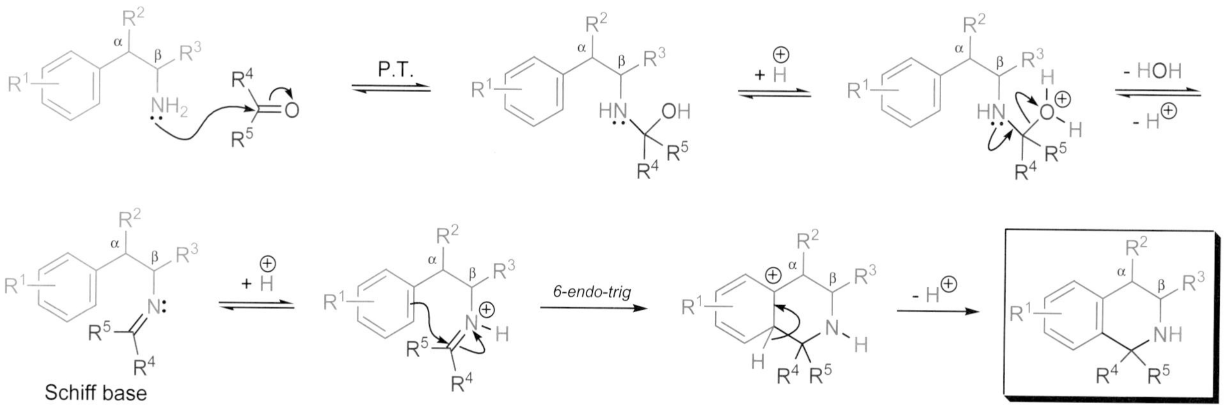

Synthetic Applications:

An important variant of the *Pictet-Spengler reaction* occurs when the aromatic substrate is an indole. In the laboratory of P.D. Bailey the enantioselective total synthesis of the indole alkaloid (–)-suaveoline was accomplished. The authors utilized a *cis*-selective *Pictet-Spengler reaction*.[29] The indole substrate was mixed with an aliphatic aldehyde in dichloromethane in the presence of molecular sieves and stirred for more than two days. Once the formation of the Schiff base was complete, TFA was added at low temperature to bring about the cyclization. Interestingly, no *trans* isomer of the carboline was generated and the *cis* isomer was isolated in high yield. Presumably the aromatic rings of the TBDPS protecting group interacted with the indole ring (π-stacking) during the cyclization causing the high observed *cis*-selectivity.

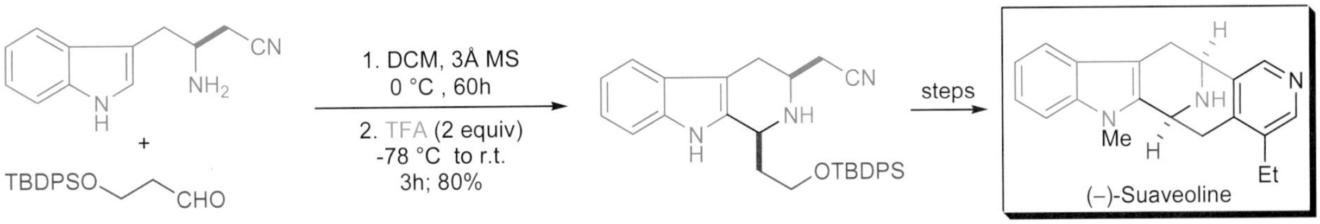

The formal total synthesis of the pyranonaphthoquinone natural product (±)-deoxyfrenolicin was achieved by Y.-C. Xu and co-workers.[30] The naphthopyran intermediate was prepared *via* the *oxa-Pictet-Spengler reaction* between a substituted naphthalene and dimethoxymethane in the presence of $BF_3 \cdot OEt_2$. The natural product has a 1,3-*trans* relationship between the two substituents of the pyran ring, and surprisingly the use of an aliphatic aldehyde only gave rise to the 1,3-*cis* naphthopyran product. For this reason, the stereoselective introduction of the three carbon side chain was accomplished by a *DDQ-induced oxidative carbon-carbon bond formation* using allyltriphenyltin as the source of the allyl group.

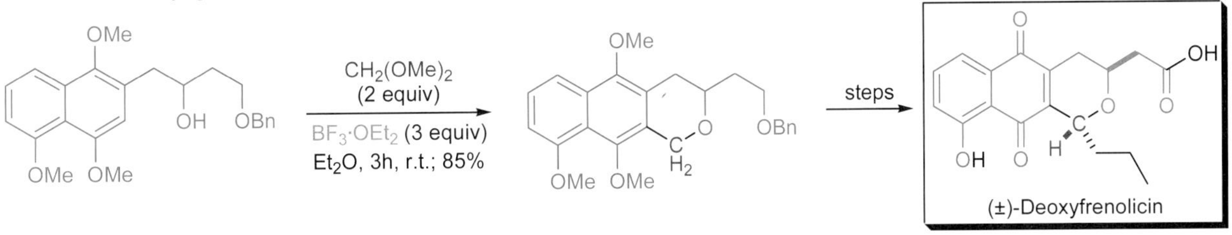

One of the key steps during the enantioselective total synthesis of the montanine-type alkaloid (+)-coccinine by W.H. Pearson et al. was the *Pictet-Spengler reaction* of a highly substituted perhydroindole intermediate.[31] The substrate was exposed to the aqueous solution of formaldehyde in methanol in the presence of 6N hydrochloric acid. The cyclization took place overnight at reflux temperature to afford the pentacyclic product in moderate yield. It is worth noting that under the cyclization conditions the benzyl protecting group was removed.

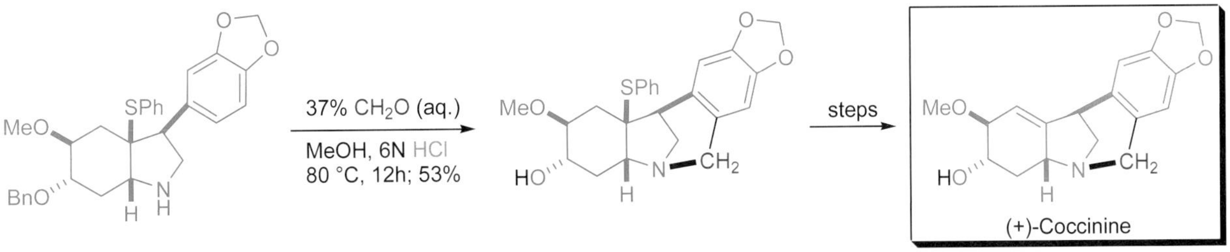

The research group of S.J. Danishefsky investigated model systems in an effort directed toward the total synthesis of ET 743 and its analogues.[32] The stereoselective formation of the spiro stereocenter of the ABFGH subunit of ET 743 was installed *via* a *Pictet-Spengler reaction*. The electron-rich phenylethylamine was mixed with a slight excess of the ketone substrate and the cyclization took place at room temperature in the presence of silica gel.

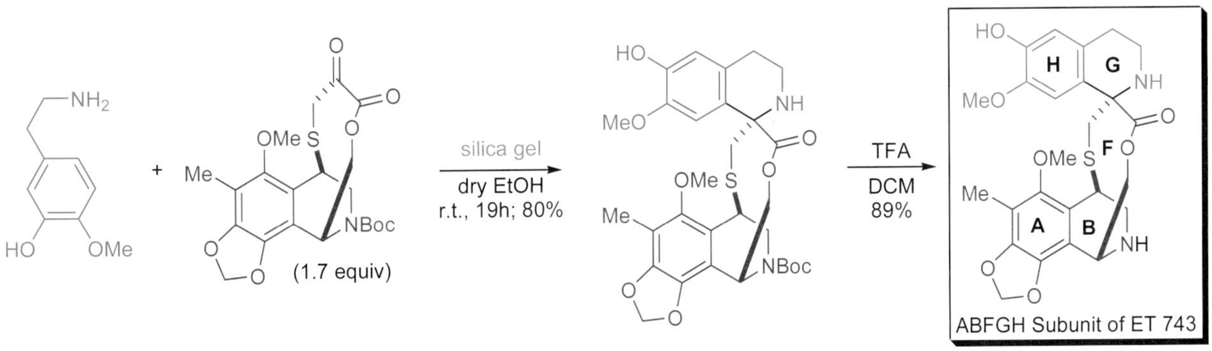

PINACOL AND SEMIPINACOL REARRANGEMENT

(References are on page 653)

Importance:

[*Seminal Publications*[1,2]; *Reviews*[3-15]; *Modifications & Improvements*[16-32]; *Theoretical Studies*[33-38]]

In 1860, R. Fittig reported that treatment of pinacol (2,3-dimethylbutane-2,3-diol) with sulfuric acid gave pinacolone (3,3-dimethylbutane-2-one).[1,39] The reaction was shown to be general for acyclic and cyclic vicinal diols (also known as glycols or 1,2-diols), which, upon treatment with catalytic amounts of acid, undergo dehydration with concomitant *[1,2]*-alkyl,- aryl- or hydride shift to afford ketones or aldehydes. This acid-catalyzed transformation of vicinal diols is known as the *pinacol rearrangement*. The general features of the reaction are: 1) virtually any cyclic or acyclic vicinal glycol can undergo the rearrangement, and, depending on the substitution pattern, aldehydes and/or ketones are formed; 2) when all four substituents are identical, the rearrangement yields a single product; 3) when the four substituents are not identical, product mixtures are formed; 4) the product is usually formed *via* the most stable carbocation intermediate when the glycol substrate is unsymmetrical; 5) the reaction can be highly regioselective and the regioselectivity is determined by the relative migratory aptitudes of the substituents attached to the carbon adjacent the carbocation center; 6) the substituent that is able to stabilize a positive charge better (better electron donor) tends to migrate preferentially; 7) the relative migratory aptitudes are: aryl ~ H ~ vinyl (alkenyl) > *t*-Bu >> cyclopropyl > 2° alkyl > 1° alkyl; 8) the *pinacol rearrangement* can also be stereoselective especially when complex cyclic vicinal diols are involved; 9) cyclic systems may rearrange *via* both ring-expansion and ring-contraction and the course of the rearrangement is strongly influenced by the ring size; 10) most often a cold aqueous solution of sulfuric acid (25% H_2SO_4) is used to effect the rearrangement; however, other acids such as perchloric acid and phosphoric acid have also been utilized;[10] and 11) besides protic acids, Lewis acids (e.g., $BF_3{\cdot}OEt_2$, TMSOTf) are also used. The drawbacks of the *pinacol rearrangement* are: 1) it is generally not easy to prepare complex vicinal diols; 2) in the case of unsymmetrical substrates, the regioselective formation of only one carbocation is usually not trivial, so product mixtures are obtained; 3) side reactions such as β-eliminations yielding dienes and allylic alcohols are often observed; 4) the intermediate carbocations may undergo equilibration; and 5) various conformational effects and neighboring group participation in cyclic systems are complicating factors. When one of the hydroxyl groups is converted to a good leaving group, the regioselective generation of the carbocation intermediate is possible. Similarly selective generation of carbocations can be realized when 2-heterosubstituted alcohols (e.g., halohydrins, 2-amino alcohols, 2-hydroxy sulfides, etc.) are used as substrates. The *pinacol-type rearrangement* of these compounds is referred to as the *semipinacol rearrangement*, a term first coined by M. Tiffeneau.[2] Owing to its predictability and the mild reaction conditions, the *semipinacol rearrangement* is almost exclusively utilized in complex molecule synthesis.

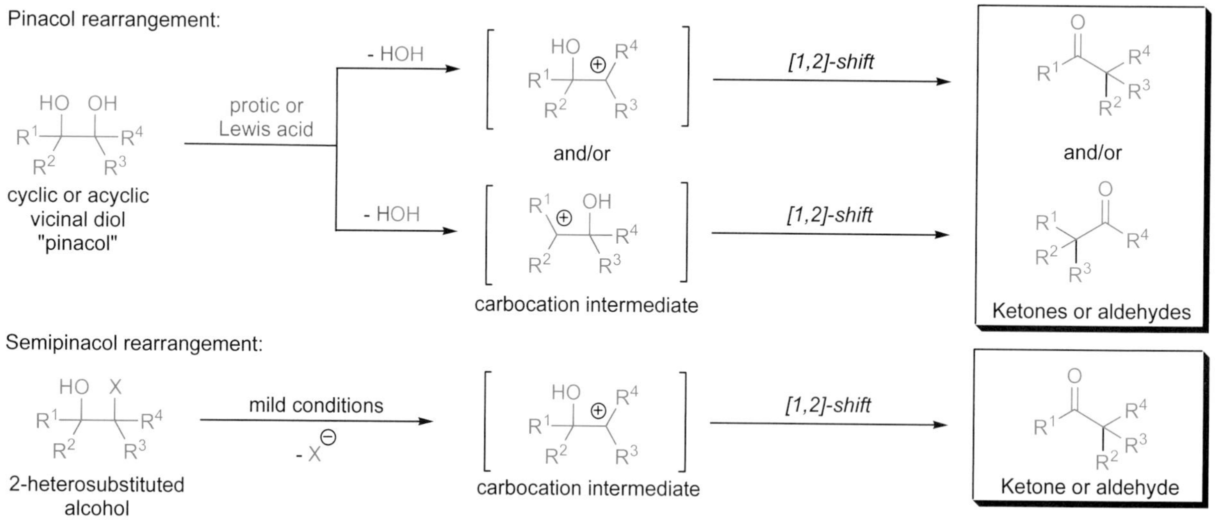

R^{1-4} = H, alkyl, aryl, acyl; X = Cl, Br, I, SR, OTs, OMs, N_2^+ (*Tiffeneau-Demjanov rearrangement*); protic acid: H_2SO_4, $HClO_4$, H_3PO_4, TFA, TsOH; Lewis acid: $BF_3{\cdot}OEt_2$, TMSOTf; mild conditions: $LiClO_4$/THF/$CaCO_3$, Et_3Al/DCM, Et_2AlCl/DCM, etc.

Mechanism:[40-54]

The first step of the process is the protonation of one of the hydroxyl groups, which results in the loss of a water molecule to give a carbocation intermediate. This intermediate undergoes a *[1,2]*-shift to give a more stable carbocation that upon the loss of proton gives the product. The pinacol rearrangement was shown to be exclusively intramolecular, and both inversion and retention were observed at the migrating center.

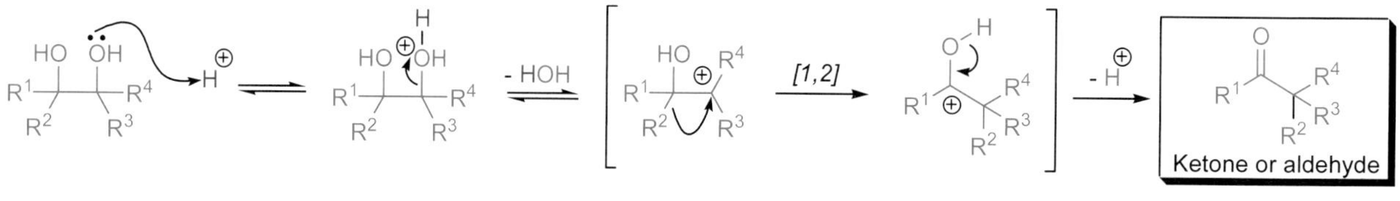

Synthetic Applications:

The total synthesis of (±)-furoscrobiculin B, a lactarane sesquiterpene isolated from basidiomycetes of mushrooms, was accomplished in the laboratory of H. Suemune and K. Kanematsu using a *furan ring transfer reaction* and a *semipinacol rearrangement* as key steps.[55] The secondary hydroxyl group of the tricyclic *cis*-vicinal diol substrate was converted to the corresponding tosylate that *in situ* underwent a ring-expansion reaction to afford an azulenofuran in good yield.

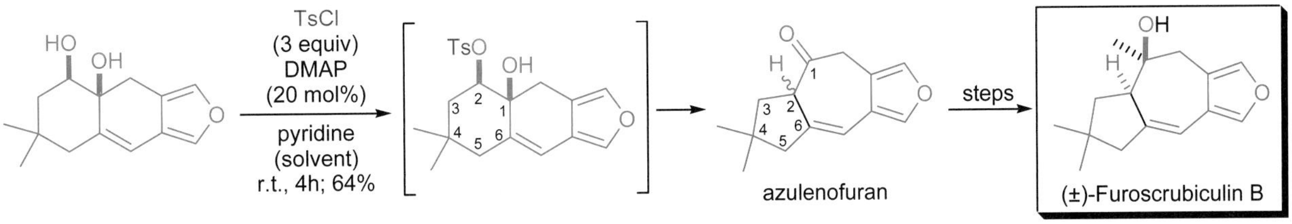

G.R. Pettit and co-workers converted a highly substituted *trans*-stilbene derivative to the strong cancer cell growth inhibitor and antimitotic agent hydroxyphenstatin.[56] The key step of the synthesis was a $BF_3{\cdot}OEt_2$-catalyzed *pinacol rearrangement* of an optically active vicinal diol to afford a substituted diphenylacetaldehyde in racemic form. From this key intermediate, several derivatives were prepared in addition to the target molecule.

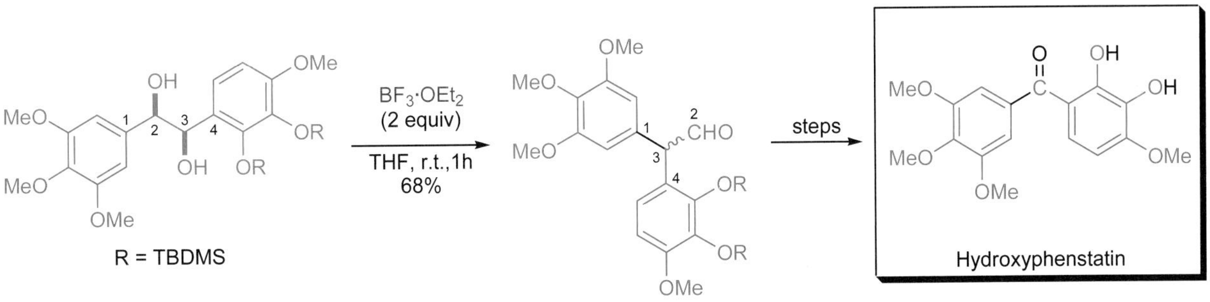

During the total synthesis of (±)-fredericamycin A, the spiro 1,3-dione center was introduced by R.D. Bach et al. utilizing a mild *mercury-mediated semipinacol rearrangement* that involved a *[1,2]*-acyl shift.[57] The indanone dithioacetal was reacted with 1,2-*bis*[(trimethylsilyl)oxy]cyclobut-1-ene in the presence of mercuric trifluoroacetate and the rearrangement took place *in situ*.

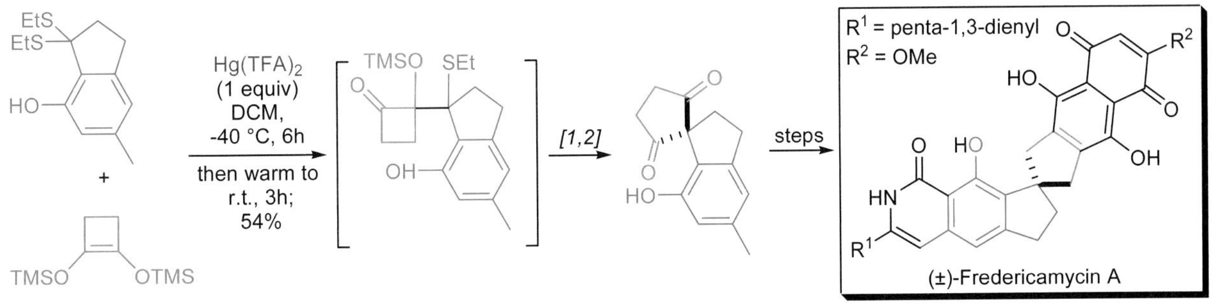

The stereocontrolled asymmetric total synthesis of protomycinolide IV was achieved, based on the *organoaluminum-promoted stereospecific semipinacol rearrangement*, by K. Suzuki and co-workers.[58] The excess DIBALH reduced the C2 carbonyl group to the corresponding aluminum alkoxide, which was immediately treated with one equivalent of Et_3Al to bring about the *[1,2]*-alkenyl shift. The initially formed aldehyde was reduced by the excess reducing agent to afford the primary alcohol upon work-up. There was no E/Z isomerization of the alkenyl group.

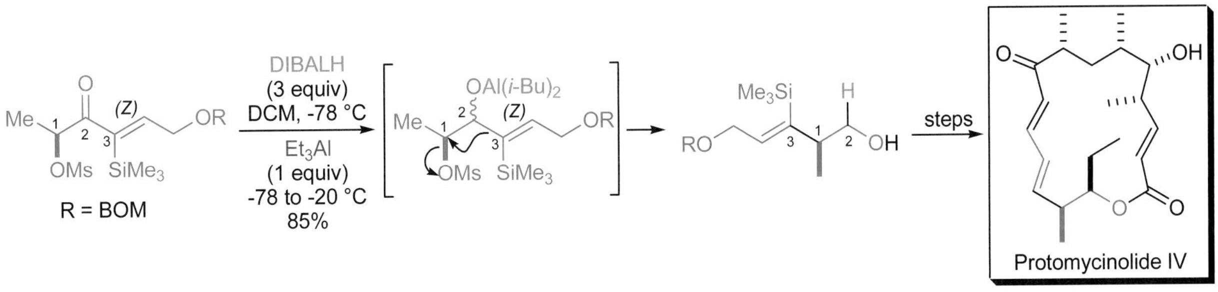

PINNER REACTION

(References are on page 654)

Importance:

[*Seminal Publications*[1-3]; *Reviews*[4-8]; *Modifications & Improvements*[9-15]]

In 1877, A. Pinner and Fr. Klein reported that when dry hydrogen chloride gas was bubbled through the mixture of benzonitrile and isobutanol, a crystalline compound was formed that was characterized as the addition product of all three reactants.[1] A year later in 1878, a similar addition product was isolated by reacting hydrogen cyanide with absolute ethanol and HCl.[2] The condensation of nitriles with alcohols and phenols in the presence of anhydrous hydrogen chloride or hydrogen bromide to afford imino ethers (also referred to as imidates or imino esters) is known as the *Pinner reaction* (or *Pinner synthesis*). The general features of this transformation are:[4-8] 1) the reactants are usually dissolved in an anhydrous solvent (e.g., benzene, chloroform, nitrobenzene, dioxane, etc.), and dry hydrogen chloride gas is bubbled through the solution at 0 C°; 2) if the reaction is conducted at higher than 0 C°, the product imino ether salt may decompose to give an amide and an alkyl halide; 3) in some cases the use of solvent tends to lower the yield of the product, so the neat reactants are simply mixed and treated with dry HCl gas; 4) the structure of the nitrile can vary widely so aliphatic, aromatic, and heteroaromatic nitriles are all good substrates; 5) when the nitrile is sterically hindered (e.g., *ortho*-substituted benzonitrile) the *Pinner reaction* may not take place; 6) the alcohol component is usually methanol and ethanol, but many primary and secondary alcohols have been used successfully; 7) monohydric phenols also react, however, dihydric- or polyhydric phenols may undergo the *Houben-Hoesch reaction* to afford aromatic ketones; 8) thiols and thiophenols also react with nitriles in an analogous fashion to form imino thioethers (thioimidates); 9) the initial product is usually the imino ether hydrohalide salt, which can be easily converted to the corresponding free imino ether by treatment with a weak base; 10) imino ethers are generally not very stable compounds, they undergo rapid hydrolysis to form esters when treated with water and acid (this is especially true for imino ethers generated by the reaction of aliphatic nitriles); 11) if the nitrile and alcohol are treated with aqueous hydrochloric acid, the esters are formed directly; 12) upon treatment with excess alcohol, imino ethers are converted to ortho esters (this can be a side reaction during the preparation when excess alcohol is used); and 13) imino ether hydrohalide salts can be transformed into an amidine hydrohalide salt by treatment with ammonia.

Pinner (1877):

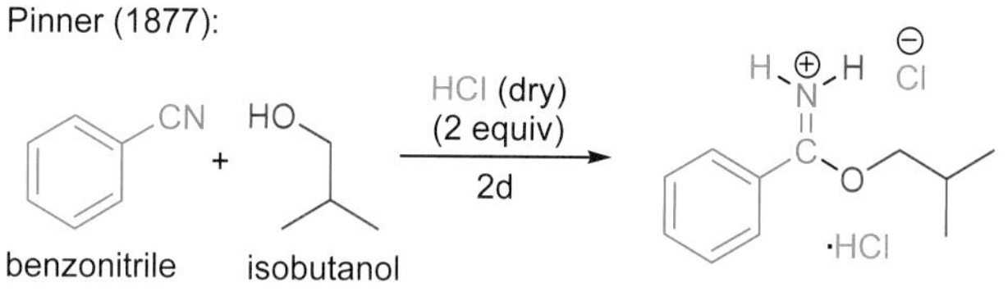

Pinner (1878):

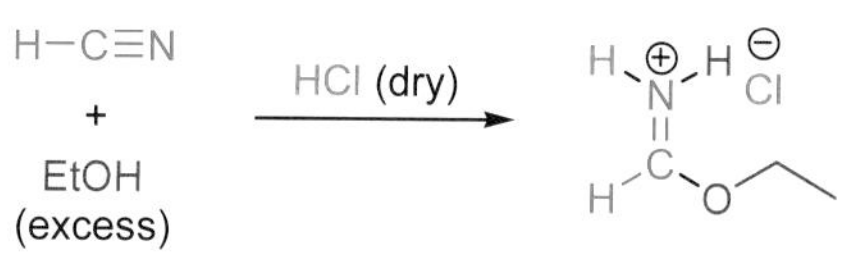

Pinner reaction:

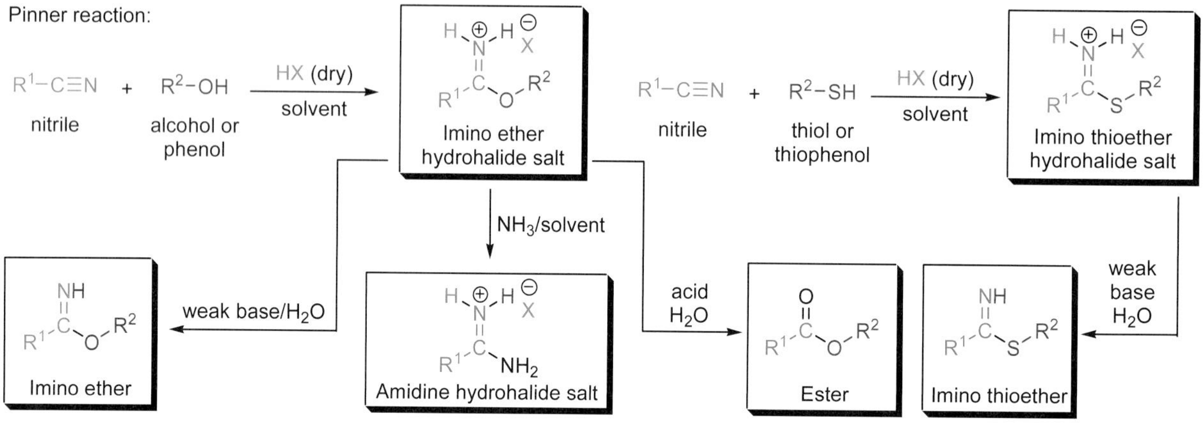

R^1 = H, alkyl, aryl; R^2 = Me, Et, 1° and 2° alkyl, aryl; HX = HCl, HBr; <u>solvent</u>: $CHCl_3$, benzne, nitrobenzene, dioxane, (EtOH, MeOH); <u>base</u>: $NaHCO_3$, Na_2CO_3; <u>acid</u>: HCl, H_2SO_4

Mechanism: [16,17,6,18]

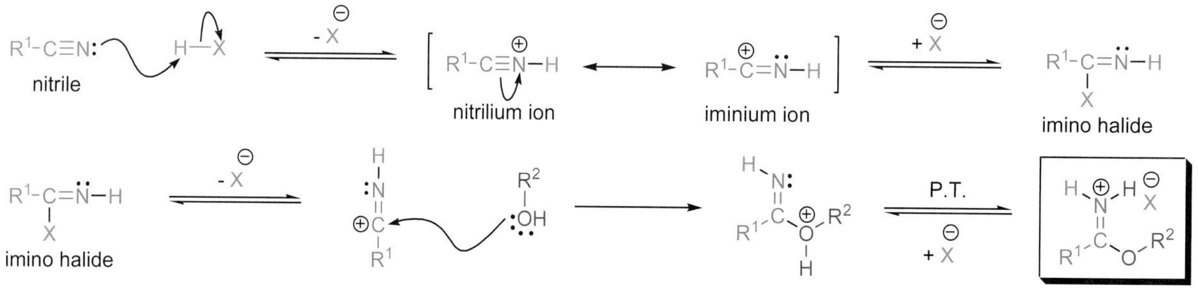

Synthetic Applications:

The first stereoselective total synthesis of AI-77B, a gastroprotective substance, was accomplished by Y. Hamada and co-workers.[19] In the final stages of the synthetic effort, the *intramolecular Pinner reaction* was utilized to convert the cyano group into the corresponding carboxylic acid. The nitrile substrate was dissolved in 5% HCl in methanol, and excess trimethyl orthoformate was added at 5 °C and the reaction mixture was stirred at this temperature for almost two days. Next, the cyclic imino ether hydrochloride salt was treated with water at room temperature followed by basic hydrolysis. Finally, the pH was adjusted with HCl to obtain the natural product.

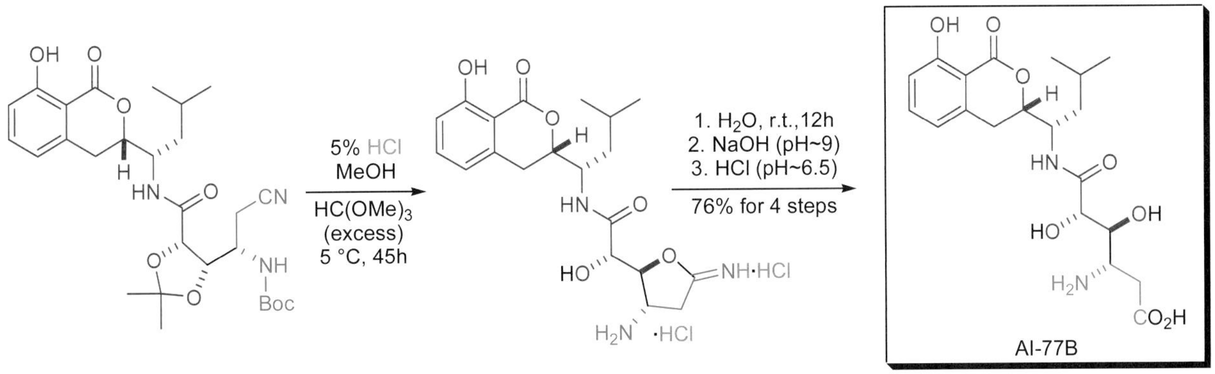

In the laboratory of R.B. Grossman both the putative and the actual structure of the naturally occurring clerodane diterpenoid (±)-sacacarin was prepared.[20] A cyclic geminal dinitrile intermediate was subjected to the conditions of the *Pinner reaction* by passing dry HCl gas through the solution of the substrate in absolute ethanol at room temperature. Under these conditions, only the equatorial cyano group was converted to the imino ethyl ether hydrochloride salt. Most likely the axial cyano group was too sterically hindered, therefore it did not react. The imino ether then was hydrolyzed with concentrated aqueous hydrochloric acid to give the corresponding ethyl ester.

The synthesis of enantiomerically pure nonpeptidic inhibitors of thrombin, a key serine protease in the blood-coagulation cascade, was carried out by F. Diederich et al.[21] These ligands have a conformationally rigid tricyclic core, and the appended substituents fill the major binding pockets at the thrombin active site. The required amidine functionality on the aromatic ring of one of these inhibitors was prepared from the corresponding aromatic nitrile *via* the *Pinner reaction*. The substrate was dissolved in a mixture of dry methanol and chloroform, and dry HCl gas bubbled through the solution for 10 minutes until saturation. The reaction mixture then was stored at 4 °C for one day, and then the imino ether was isolated by filtration. The methanolic solution of ammonia was added to the solution of the imino ether in methanol, and the resulting solution was heated at 65 °C for a few hours to achieve complete conversion to the amidinium salt.

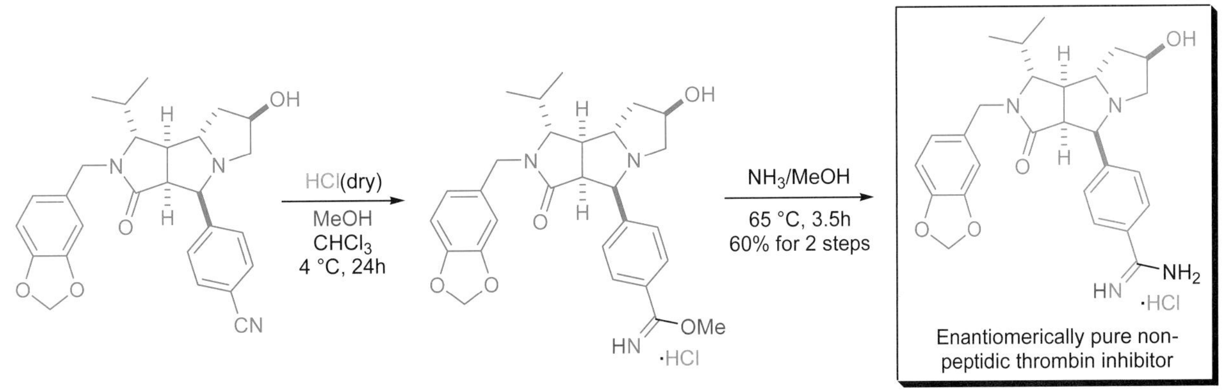

PINNICK OXIDATION

(References are on page 655)

Importance:

[*Seminal Publications*[1-4]; *Reviews*[5]; *Modifications & Improvements*[6,5,7]]

The oxidation of aldehydes to the corresponding carboxylic acids is a very important transformation in organic synthesis. Until the early 1970s most methods required expensive reagents and complex reaction conditions, the functional group tolerance was limited, and the selectivities were low. In 1973, B.O. Lindgren was the first to apply the inexpensive sodium chlorite ($NaClO_2$) in combination with hypochlorous acid (HClO) and scavengers (e.g., sulfamic acid, resorcinol) to convert vanillin to the corresponding vanillic acid under mild conditions.[1] The HClO is formed as a by-product of the oxidation process, and it can cause side reactions such as consumption of the $NaClO_2$ to form chlorine oxide (ClO_2) or reacting with C=C double bonds. A few years later, G.A. Kraus and co-workers were the first to use 2-methyl-2-butene as a scavenger under buffered conditions for the oxidation of an aliphatic- and an α,β-unsaturated aldehyde.[2,3] In 1981, H.W. Pinnick showed that the $NaClO_2$/2-methyl-2-butene system was generally applicable to the oxidation for a wide range of α,β-unsaturated aldehydes without affecting any of the double bonds present. Today, this transformation of aldehydes (aliphatic, aromatic, saturated, or unsaturated) to the corresponding carboxylic acids is referred to as the *Pinnick oxidation*.[4] The general features of the reaction are: 1) in a typical procedure, the aldehyde is dissolved in *tert*-butanol (often in combination with another solvent such as THF) along with the large excess of the scavenger followed by the dropwise addition of the aqueous solution of sodium dihydrogen phosphate buffer (NaH_2PO_4) and $NaClO_2$ at room temperature; 2) the scavenger is most often 2-methyl-2-butene, which has to be added in large excess (caution: the boiling point is low therefore the container should be cold before opening); 3) to ensure a constant pH value, the use of several equivalents of NaH_2PO_4 is recommended; 4) usually slightly more than one equivalent of $NaClO_2$ is necessary, which should be dissolved in water (by itself or together with the phosphate buffer) only prior to the oxidation, since exposure to light or the presence of impurities (e.g., Fe^{2+} and Fe^{3+} complexes) tend to decompose the reagent;[8] 5) with certain substrates the purity of the reagents is crucial, and the oxidation sometimes stops after a few percent of conversion:[9] a) due to the sensitivity/instability of the $NaClO_2$ in acidic medium in the presence of transition metal complexes the use of a steel needle for the addition of the oxidant should be avoided (use a Pasteur pipette instead); b) neat 2-methyl-2-butene or 2M solution in THF should be used instead of the 90% technical grade reagent; 6) when 2-methyl-2-butene is used as the scavenger, none of the double bonds in the substrate will be chlorinated, but with other scavengers, such as H_2O_2, side reactions involving isolated double bonds do occur; 7) stereocenters at the α-position of aldehydes are unaffected; and 8) functional group tolerance is excellent, and hydroxyl groups do not need to be protected.

Lindgren (1973):

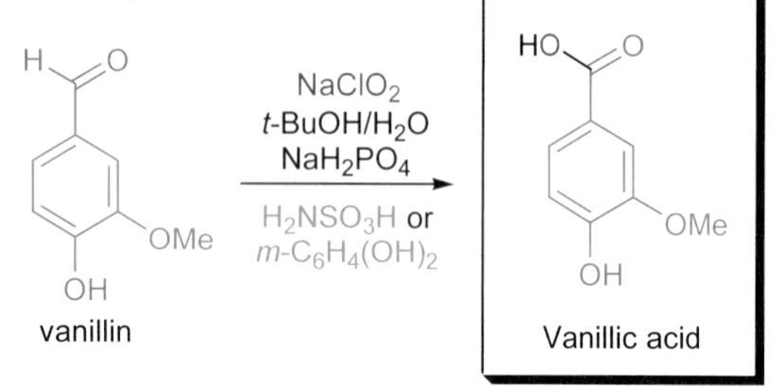

Kraus (1980):

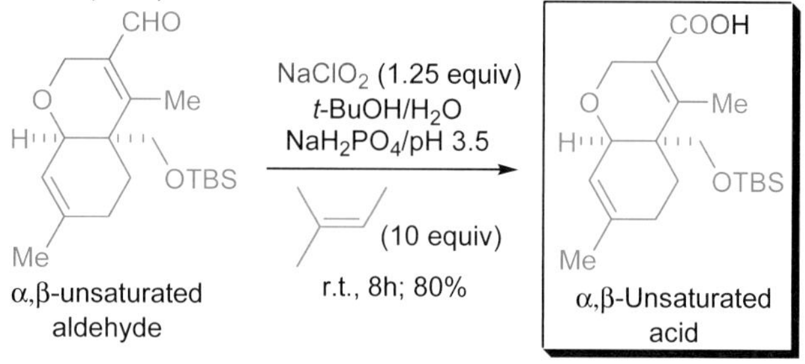

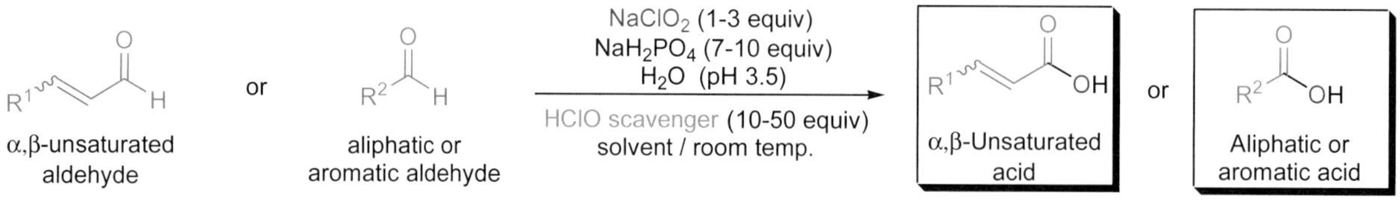

R[1] = H, alkyl, aryl, alkenyl, allyl; R[2] = alkyl, aryl, allyl, homoallyl; <u>scavenger</u> = 2-methyl-2-butene, H_2O_2, H_2NSO_3H, m-$C_6H_4(OH)_2$, DMSO; <u>solvent</u> = *t*-BuOH, *t*-BuOH/THF

Mechanism:[10,6]

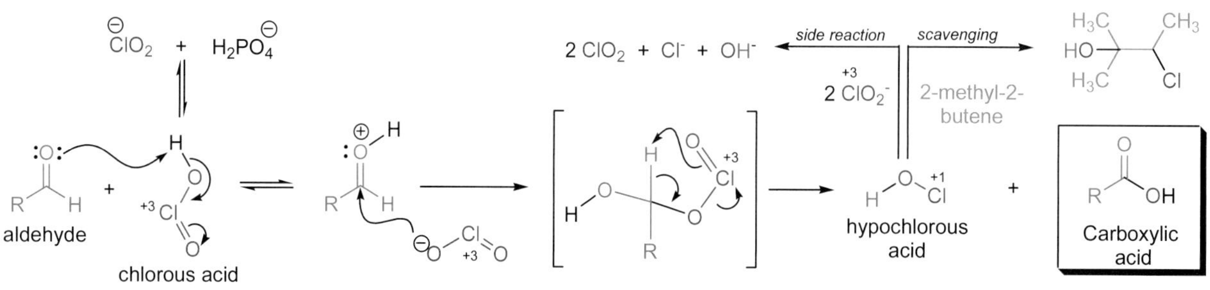

Synthetic Applications:

The total synthesis of the complex bioactive indole alkaloid ditryptophenaline, having two contiguous quaternary stereocenters related by C_2 symmetry, was accomplished in the laboratory of L.E Overman.[11] In the late stages of the synthetic effort the complex diol substrate was oxidized to the dicarboxylic acid using a two-step procedure: first, a *Dess-Martin oxidation* to the dialdehyde followed by the *Pinnick oxidation*. The mild reaction condition ensured that the integrity of the stereocenters at the α-positions was preserved.

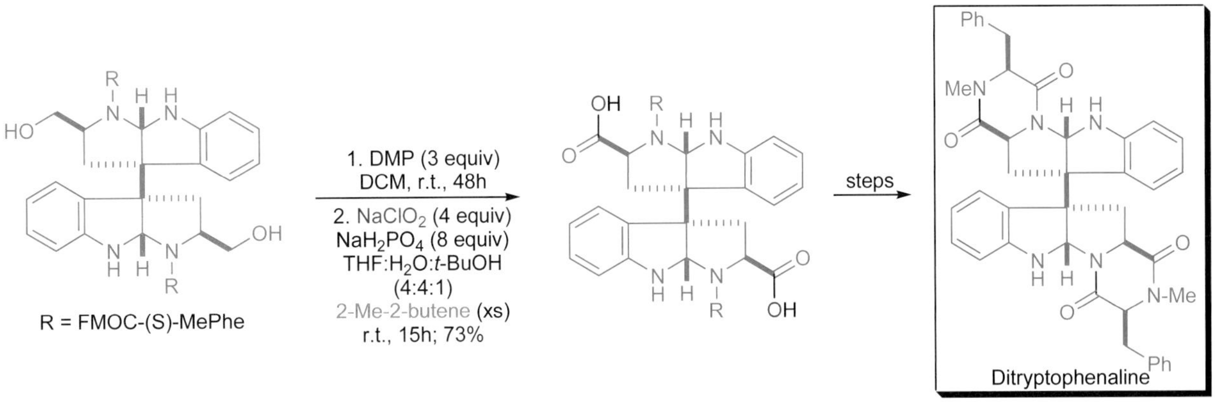

A novel triple oxidation procedure was applied by A. Armstrong et al. to install the tricarboxylic acid moiety during the total synthesis of (+)-zaragozic acid C.[12] The bicyclic triol substrate was first exposed to the *Swern oxidation* conditions to afford the corresponding trialdehyde. Several different oxidations (e.g., *Jones oxidation*, *modified Ley oxidation*) were tried on the crude trialdehyde to convert it to the triacid, but all of these attempts resulted in a complex mixture of products. A clean and high-yielding solution to this problem was to use the *Pinnick oxidation* that gave rise to the desired triacid. Esterification to the tri-*tert*-butyl ester was conducted by using *N,N*-diisopropyl-*O-tert*-butylisourea in dichloromethane.

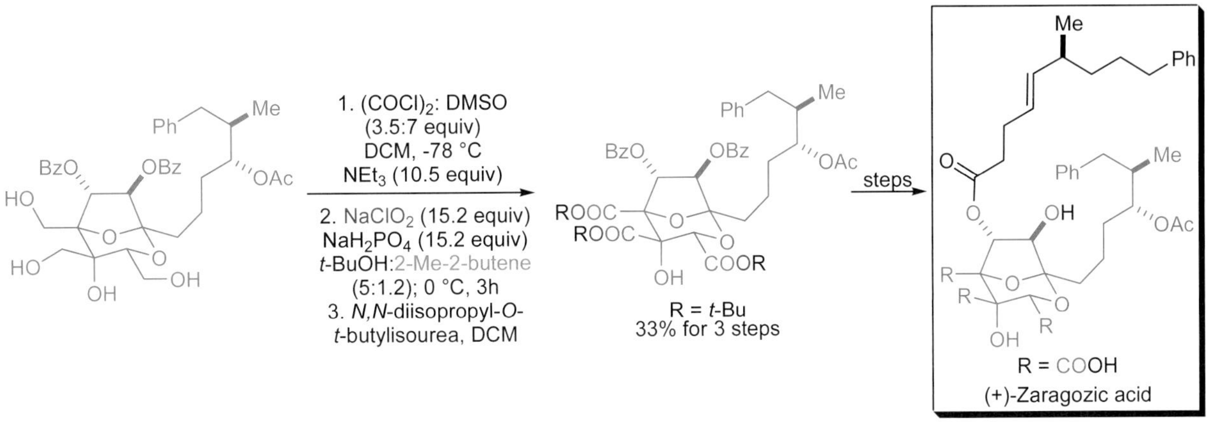

The formal total synthesis of the selective muscarinic receptor antagonist (+)-himbacine was accomplished by M.S. Sherburn and co-workers using an *intramolecular Diels-Alder reaction*, a *Stille cross-coupling*, and a *6-exo-trig acyl radical cyclization* as the key steps.[13] In order to prepare the selenoate ester precursor for the radical cyclization step, the aldehyde-enyne substrate was converted to the carboxylic acid *via* the *Pinnick oxidation* without affecting the delicate enyne moiety.

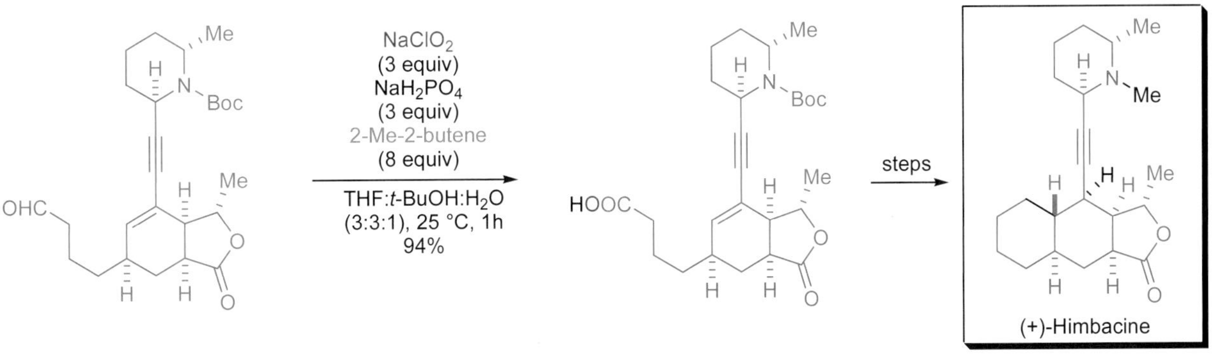

POLONOVSKI REACTION

(References are on page 655)

Importance:

[*Seminal Publications*[1,2]; *Reviews*[3-9]; *Modifications & Improvements*[10-16]]

In 1927, the Polonovski brothers reported that certain alkaloid *N*-oxides, upon treatment with acetic anhydride or acetyl chloride, underwent a rearrangement in which one of the alkyl groups attached to the nitrogen was cleaved and the *N*-acetyl derivative of the alkaloid was obtained.[1] For several decades, the procedure was used, almost exclusively, for the *N*-demethylation of tertiary amines because it took place under much milder conditions than other methods available at the time. The activation of tertiary amine *N*-oxides with acyl halides or anhydrides to form the corresponding iminium ion intermediates is known as the *Polonovski reaction*. The general features of the reaction are: 1) the *N*-oxide substrates are usually prepared from the corresponding tertiary amines by oxidation; 2) the activation of *N*-oxides is effected by acyl halides or anhydrides, but in the majority of the cases acetic anhydride (Ac_2O) is used; 3) when trifluoroacetic anhydride (TFAA) is used, the reaction proceeds under mild reaction conditions (*Polonovski-Potier reaction*) and the reaction can be stopped at the iminium ion stage;[10,6] 4) besides anhydrides, various iron salts and sulfur dioxide can be used as activating agents;[11,12] 5) when formic-acetic or formic-pivalic anhydride is employed as the acylating agent, the *N*-oxide is simply reduced to the amine;[17,18] 6) the initially formed iminium ions are versatile intermediates (e.g., *Mannich* and *Pictet-Spengler reactions*), which can be converted to other important classes of compounds such as enamines, tertiary amides and/or secondary amines, and aldehydes;[8,9] 7) depending on the nature of the activating agent and the reaction conditions, there are two main reaction pathways available for the iminium ions: **A**) reaction with a nucleophile at the α carbon or **B**) *Grob-type* C_α-C_β cleavage to afford alkenes and new iminium ions (only when it is activated by an adjacent electron-donating center and the C_α-C_β bond is antiperiplanar with the N-O bond);[8,9] 8) when more than one group attached to the nitrogen has a hydrogen at the α position, regioisomeric iminium ions are formed; however, the regioselectivity can be controlled, and the thermodynamically more stable iminium ion is formed with TFAA, while with Ac_2O the kinetically more acidic α position is deprotonated; 9) the acidity of the α C-H bond is increased if R^1=EWG; 10) when the 3° amine *N*-oxide is cyclic, the reaction takes place only for five- and six-membered rings, and the endocyclic iminium ions are formed in preference to exocyclic ones; and 11) when the iminium ion is too reactive, the corresponding α-cyanoamines (iminium ion equivalents) can be prepared in high yield.[13,16]

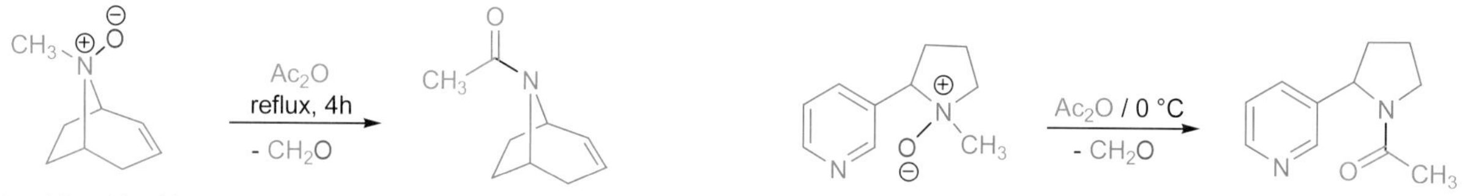

tropidine *N*-oxide — *N*-acetylnortropidine — nicotine *N*-oxide — *N*-acetylnornicotine

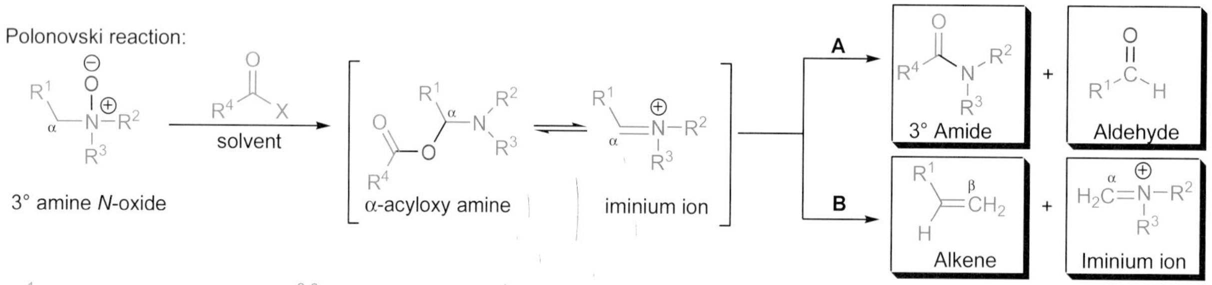

R^1 = H, alkyl, aryl, heteroaryl; R^{2-3} = CH_3, alkyl, aryl; R^4 = CH_3, CF_3 (*Polonovski-Potier rxn*); X = Cl, Br, $OCOR^4$, BF_4^-, ClO_4^-, SbF_6^-

Mechanism: [19-22,8,9,23,24]

The conversion of the *O*-acylimonium salt to the imine proceeds *via* an *E2-type elimination*. The hydrogen that is antiperiplanar to the N-O bond is usually removed preferentially. When the *N*-oxide is activated with iron salts, a SET mechanism is operational, while with SO_2 an intramolecular ionic mechanism is most likely.[11,12]

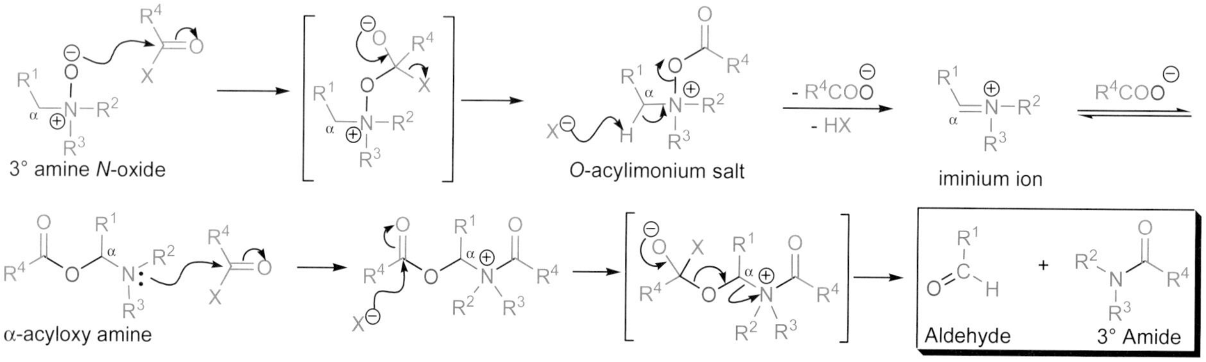

Synthetic Applications:

In the laboratory of J. Kobayashi, the biomimetic one-pot transformation of serratinine into serratezomine A was accomplished using the *Polonovski-Potier reaction*.[25] Serratinine was first treated with *m*-chloroperbenzoic acid to obtain the *N*-oxide, and then excess TFAA was added. The iminium ion was formed in the following fashion: the C13 hydroxyl group formed a hemiacetal with the C5 carbonyl group and simultaneously with the formation of the C5-C13 lactone the C4-C5 bond was broken. The iminium ion was then reduced with sodium cyanoborohydride to afford the tertiary amine functionality. Besides serratezomine A, another lactone (between the C8 hydroxyl and C5 carbonyl) was formed in 27% yield.

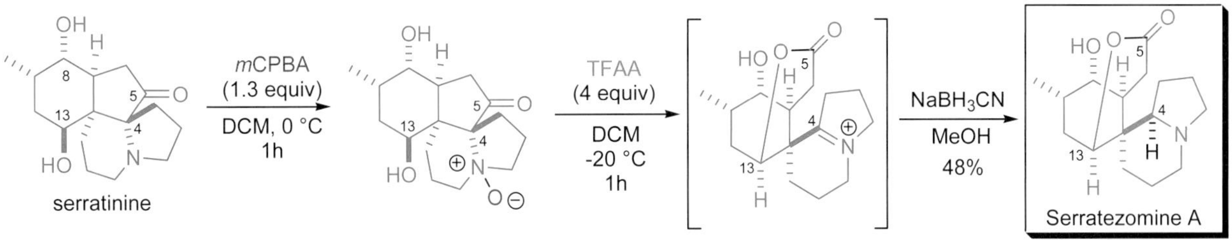

The total synthesis of (±)-dynemicin A was achieved by S.J. Danishefsky et al.[26] As part of the synthetic studies, highly sensitive enediyne containing quinone imine systems were prepared, and their biological properties were evaluated. The first step in the sequence leading to one such quinone imine began with the oxidation of the nitrogen of the phenanthridine substrate, and the resulting *N*-oxide was heated in neat acetic anhydride to induce the *Polonovski reaction*.

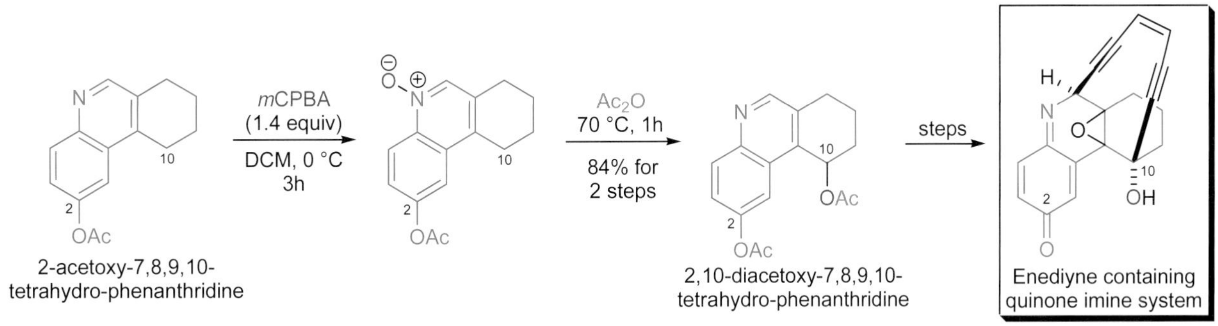

The naturally occurring sulfonamide (–)-altemicidin is the first 6-azaindene monoterpene alkaloid isolated as a metabolite of microorganisms. A.S. Kende utilized the *Polonovski-Potier reaction* in the key step to introduce the carbamoyl enamine functionality.[27] The tertiary amine was oxidized to the *N*-oxide by H_2O_2 followed by treatment with excess TFAA to afford the desired vinylogous trifluoromethyl amide.

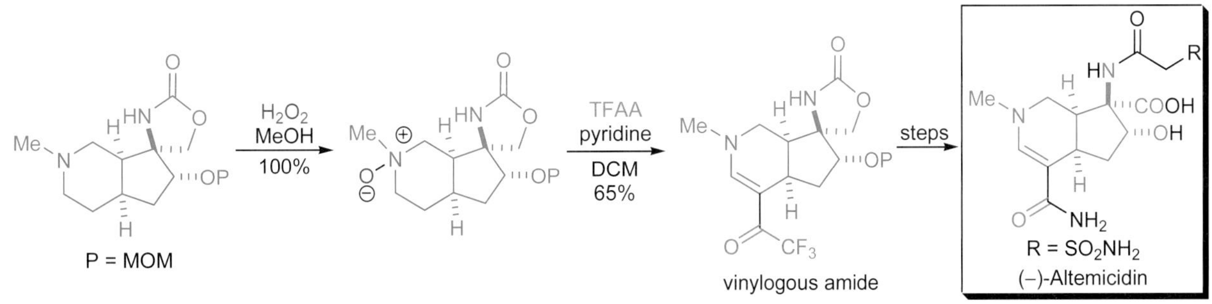

The core nucleus of the mitomycinlike antitumor agent FR-900482 was synthesized by F.E. Ziegler and co-workers.[28] The selective oxygenation of the C9a position was achieved by the *Polonovski reaction*.

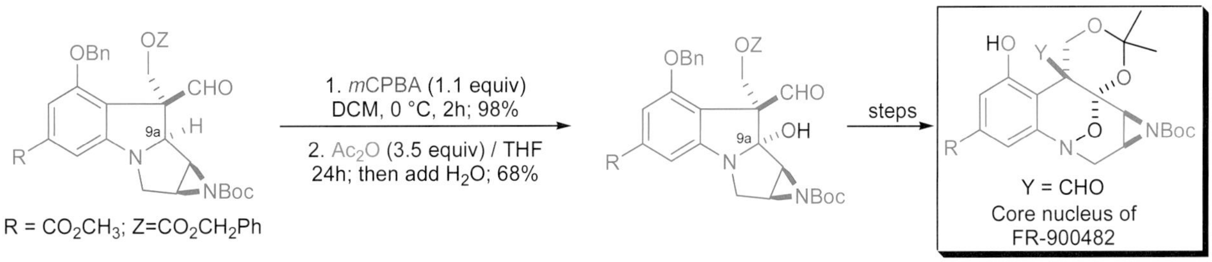

POMERANZ-FRITSCH REACTION

(References are on page 655)

Importance:

[*Seminal Publications*[1-5]; *Reviews*[6-11]; *Modifications & Improvements*[12-19]]

In 1893, C. Pomeranz and P. Fritsch independently reported a new synthesis of isoquinoline by heating a benzalaminoacetal, prepared by the condensation of benzaldehyde and 2,2-diethoxyethylamine, in concentrated sulfuric acid.[1,2] During the 1890s, these authors successfully prepared a wide range of structurally diverse isoquinolines.[3-5] The acid-catalyzed cyclization of benzalaminoacetals (these are Schiff bases) to give substituted isoquinolines is known as the *Pomeranz-Fritsch reaction*. The general features of the transformation are: 1) the benzalaminoacetals are prepared by reacting 2,2-dialkoxyethylamines with substituted aromatic aldehydes or rarely with aromatic ketones; 2) the structural variation of the 2,2-dialkoxyethylamines is very restricted, and, in the overwhelming majority of the cases, the dimethyl or diethyl acetals are used without any substituents on the C1 carbon (C1-substituted analogues tend to fail to undergo the reaction); 3) aromatic aldehydes give rise to C1-unsubstituted isoquinolines, usually in good yield, while aromatic ketones afford C1-substituted isoquinolines albeit in low yield; 4) the highest yields are obtained when the substituents on the aromatic ring are electron-donating; 5) strongly electron-withdrawing substituents (e.g., NO_2) on the aromatic ring prevent the formation of isoquinolines and the corresponding oxazoles are obtained instead;[20] 6) when both of the *ortho*-positions (relative to the carbonyl group) are unoccupied, a regioisomeric mixture of isoquinolines is obtained; 7) the most commonly used protic acids are sulfuric acid and hydrochloric acid, but Lewis acids such as $BF_3{\cdot}OEt_2$, trifluoroacetic anhydride and lanthanide triflates have been used occasionally;[15,17] 8) unless the aromatic ring is highly electron-rich, heating of the reaction mixture is required in order to achieve cyclization. Two of the most important modifications are: 1) when a substituted benzylamine is condensed with glyoxal hemiacetal, the resulting Schiff base is efficiently cyclized to give the corresponding C1-substituted isoquinoline (*Schlittler-Müller modification*);[12] 2) hydrogenation of the benzalaminoacetal and the acid-catalyzed cyclization of the resulting amine gives rise to a tetrahydroisoquinoline (*Bobbitt-modification*).[13,21,19]

Pomeranz and Fritsch (1893-94):

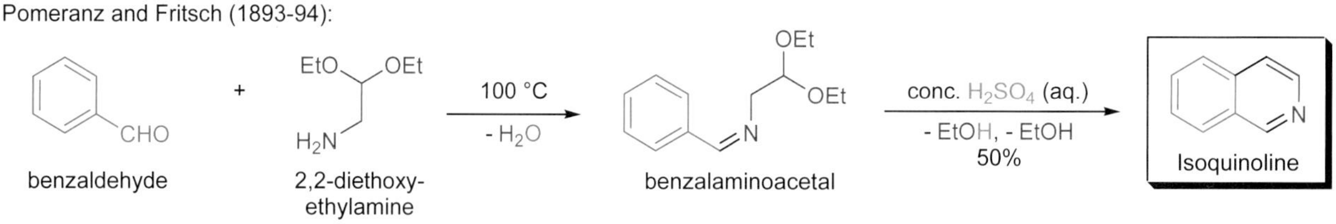

Pomeranz-Fritsch reaction:

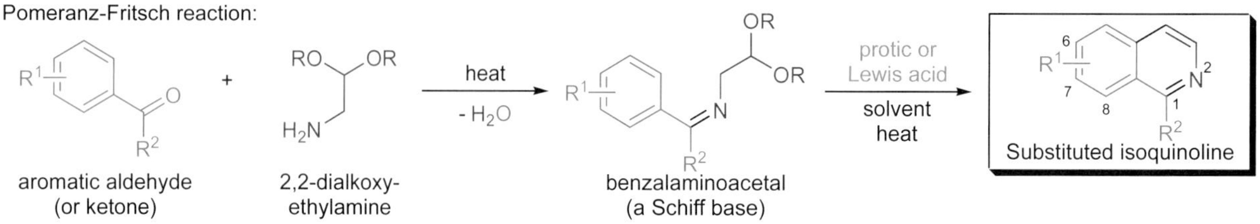

Schlittler-Müller modification (1948):

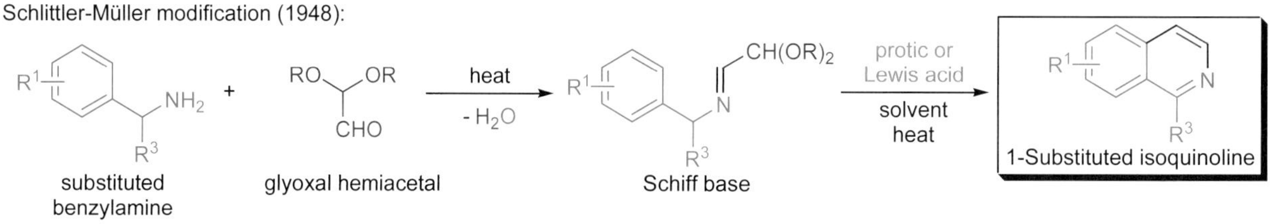

R^1 = usually an electron-donating group (EDG), H, alkyl, aryl, O-alkyl, Cl, Br; R^{2-3} = H, alkyl; R = Me, Et; protic acid: H_2SO_4, HCl, PPA; Lewis acid: $BF_3{\cdot}OEt_2$

Mechanism:[20,7]

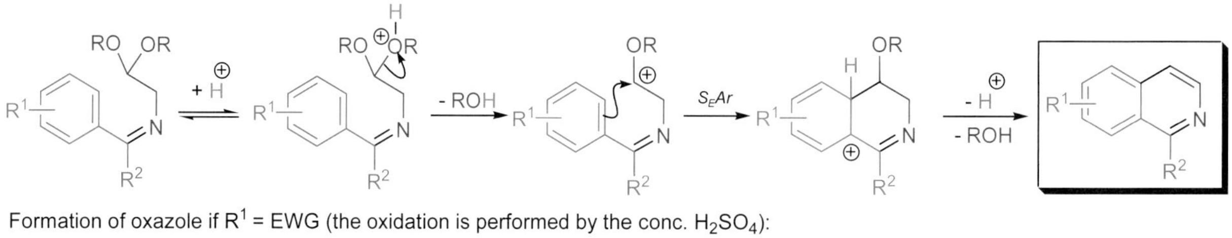

Formation of oxazole if R^1 = EWG (the oxidation is performed by the conc. H_2SO_4):

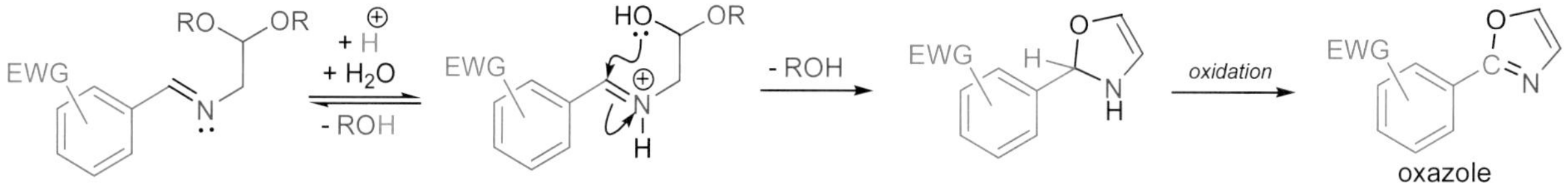

Synthetic Applications:

The *Bobbitt modified Pomeranz-Fritsch reaction* allows the preparation of enantiopure tetrahydroisoquinolines. During the studies directed toward the total synthesis of ET 743 and its analogues, S.J. Danishefsky and co-workers utilized this transformation for the preparation of a key tetrahydroisoquinoline intermediate.[22] The cyclization precursor was efficiently synthesized from the enantiopure benzylamine derivative by *N*-alkylation with excess diethylbromoacetal. The resulting compound was subjected to 6*N* hydrochloric acid at 0 °C and slowly warmed to ambient temperature overnight. The desired tetrahydroisoquinoline was formed as a 4:1 mixture of diastereomers.

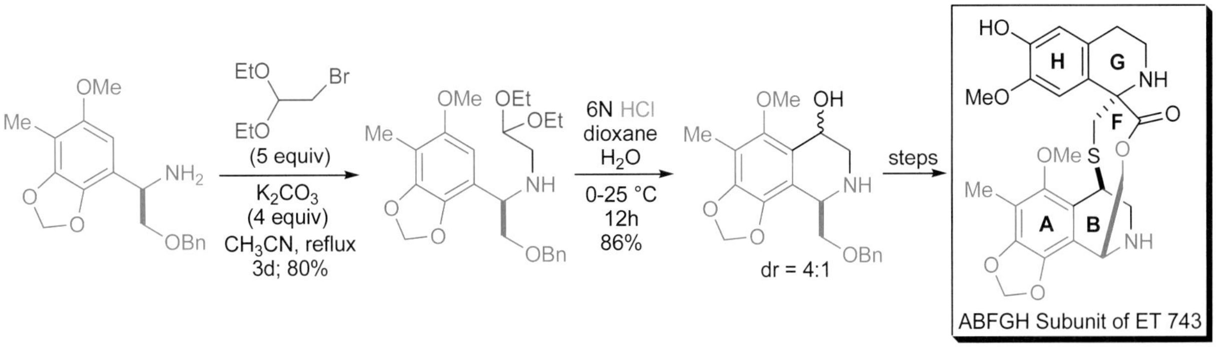

The total synthesis of (±)-4-hydroxycrebanine was accomplished by J.-I. Kunitomo et al., who used the *Bobbitt modification of the Pomeranz-Fritsch reaction* as the key ring-forming step.[23] The aromatic ketone substrate was first condensed with aminoacetaldehyde diethylacetal to afford a Schiff base that was immediately reduced to the corresponding amino compound in high yield. Exposure of this intermediate to concentrated HCl for several days gave rise to the tetrahydroisoquinoline as a mixture of two diastereomers.

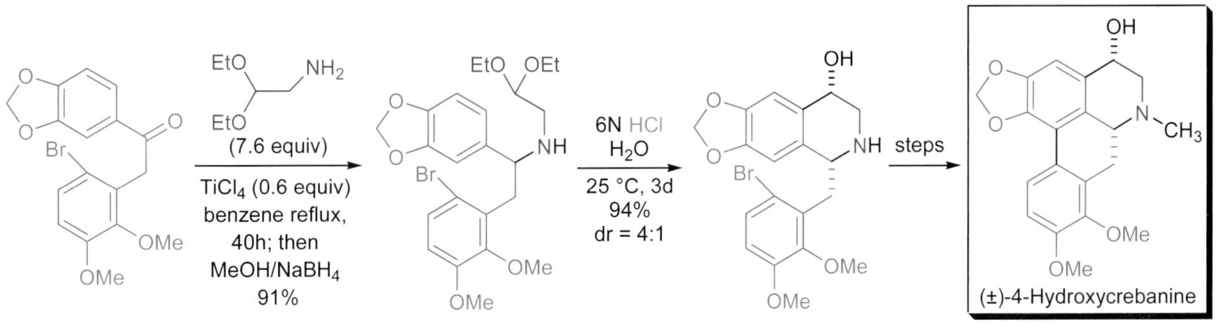

The shortest synthesis of papaverine was achieved in the laboratory of R. Hirsenkorn starting from racemic stilbene oxide and using a *modified Pomeranz-Fritsch reaction*.[24] The aminolysis of the stilbene oxide led to the formation of the cyclization precursor, which upon treatment with excess benzoyl chloride, underwent cyclization to give the *N*-benzoyl 1,2-dihydroisoquinoline derivative. Reduction under Wolff-Kishner conditions afforded papaverine.

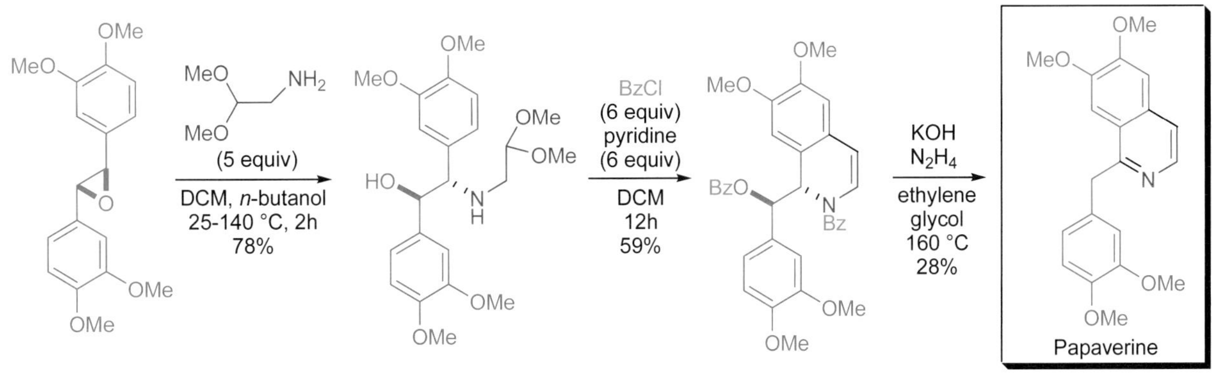

The *asymmetric variant of the Pomeranz-Fritsch reaction* was used by D. Rozwadowska and co-workers in the total synthesis of (–)-salsolidine.[21]

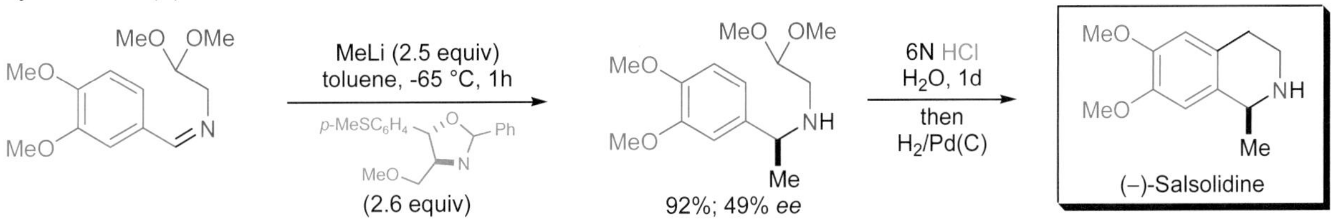

PRÉVOST REACTION

(References are on page 656)

Importance:

[*Seminal Publications*[1-3]; *Reviews*[4-7]; *Modifications & Improvements*[8-16]]

In 1933, C. Prévost reported that the treatment of styrene with silver benzoate and iodine (I_2) in dry benzene gave the dibenzoate ester of the corresponding glycol that upon hydrolysis afforded the 1,2-diol.[1] This two-step transformation of olefins leads to 1,2-*trans* diols, and it is referred to as the *Prévost reaction*. The general features of this reaction are: 1) both acyclic and cyclic alkenes are good substrates; 2) the initial products are diastereomeric *trans*-1,2-dicarboxylates, which are hydrolyzed under basic conditions to the *trans*-1,2-diols *(anti* products); 3) in rigid cyclic systems the reaction is highly diastereoselective; 4) the most commonly used reagent is silver benzoate (R=Ph), but this can be replaced with other silver carboxylates or thallium(I)acetate;[11] 5) when conjugated and isolated double bonds are both present in the molecule, the dihydroxylation usually takes place on the isolated double bond. The most important modification of the *Prévost reaction* was introduced by Woodward and Brutcher, who used wet acetic acid to obtain *cis*-1,2-diols. This modification was based on the observation by Winstein et al., who reported the erosion of *trans* selectivity of the *Prévost reaction* by small amounts of water.[17,18]

Trans-dihydroxylation of olefins (Prévost, 1933):

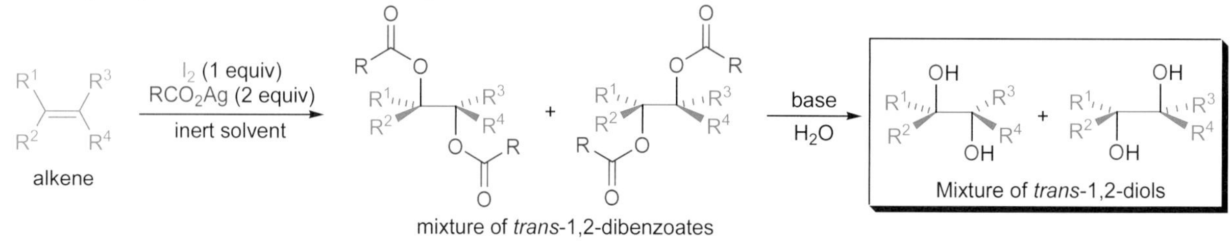

Woodward-Brutcher modification to prepare *cis*-diols (1958):

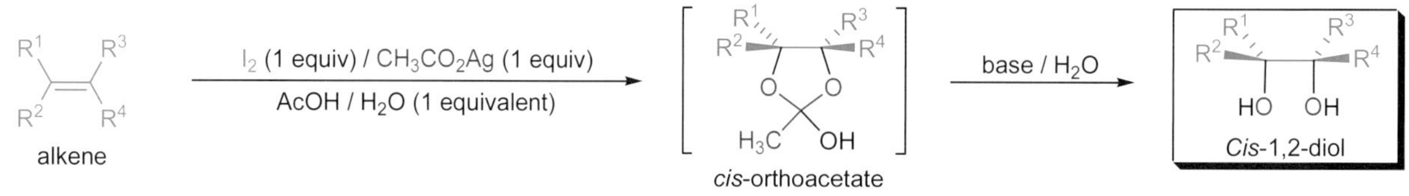

Mechanism:[17-20]

The first step of the *Prévost reaction* is the reaction of the alkene with iodine to form the cyclic iodonium ion. Next, the iodonium ion is stereospecifically opened by the silver carboxylate to form the corresponding *trans*-1,2-iodo carboxylate. The iodine is displaced intramolecularly by the carbonyl group of the carboxylate (anchimeric assistance) to form a cyclic cationic intermediate. In the absence of water, this cation is opened with the inversion of configuration by the second equivalent of silver carboxylate to afford the *trans*-1,2-dicarboxylate. However, in the presence of water (*Woodward-Brutcher modification*) the common intermediate is converted to a *cis*-orthocarboxylate which is hydrolyzed to the corresponding *cis*-1,2-diol.

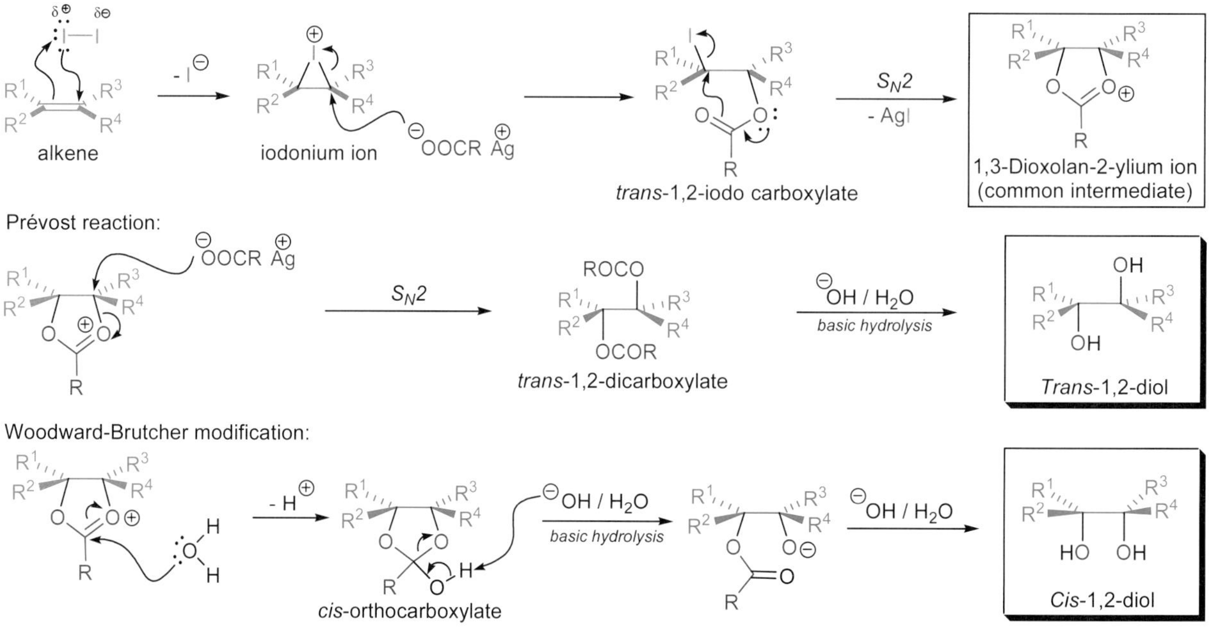

Synthetic Applications:

In the laboratory of S. Kumar, the synthesis of phenolic derivatives of *trans*-7,8-dihydroxy -7,8-dihydrobenzo[*a*]pyrene, a highly tumorigenic compound, was accomplished.[21] The *trans*-vicinal diol functionality was introduced by using the "dry" Prévost conditions. The alkene was subjected to a mixture of iodine and silver benzoate in dry refluxing benzene to give a good yield of the corresponding *trans*-7,8-dibenzoate derivative.

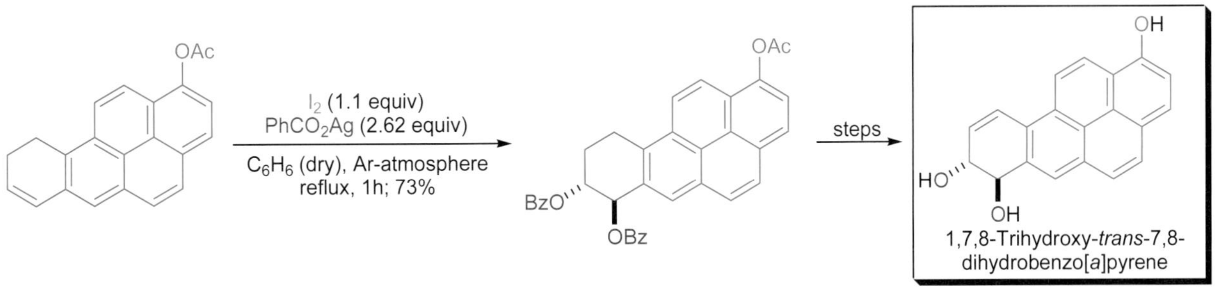

The total synthesis of (–)-SS20846A, a 2-alkyl-4-hydroxypiperidine natural product exhibiting antibacterial and anticonvulsant properties, was achieved by C.R. Johnson and co-workers.[22] The key transformations included an alkene metathesis for the preparation of the piperidine ring and the *Prévost reaction* for the installation of the 4-hydroxy substituent.

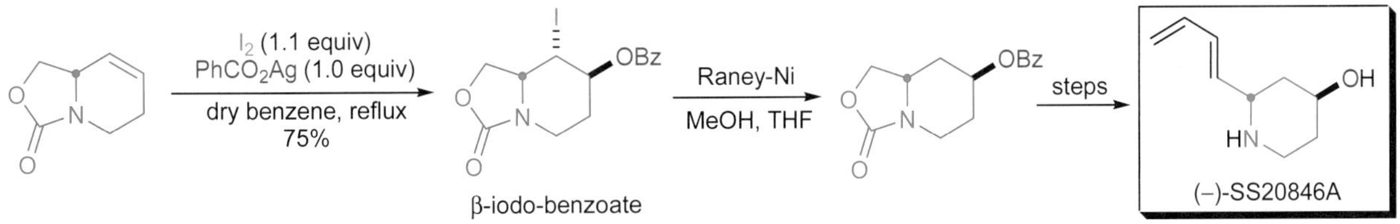

The key steps in the first total synthesis of (±)-momilactone A by P. Deslongchamps et al. were a highly diastereoselective *transannular Diels-Alder cycloaddition* and the *Prévost reaction*.[23] The β-ketolactone moiety was installed by first treating the tricyclic alkene with *N*-bromo acetamide and silver acetate to obtain the *trans* bromoacetate with excellent diastereoselectivity. The *cis* stereochemistry of the lactone was achieved a few steps later by the intramolecular nucleophilic displacement of the bromide with the carboxylate ion on the adjacent six-membered ring.

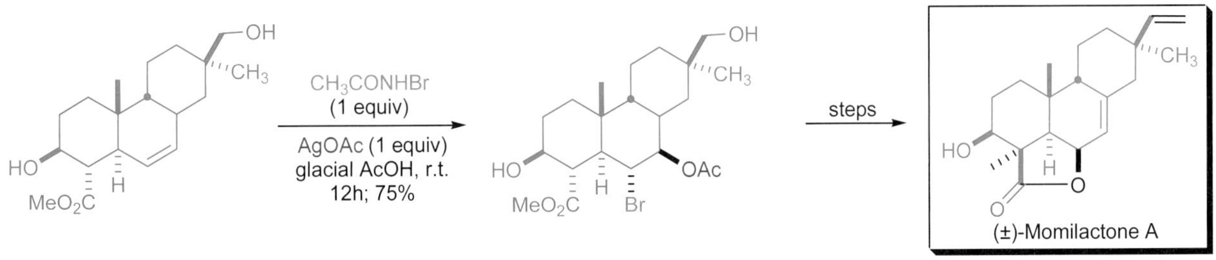

The *Woodward-Brutcher modification of the Prévost reaction* was used by P.T. Lansbury to install the *cis* vicinal diol moiety of (±)-2,3-dihydrofastigilin C.[24] The *cis* vicinal diacetate was formed in high yield and with good diastereoselectivity (5:1) when the reaction was conducted in wet acetic acid.

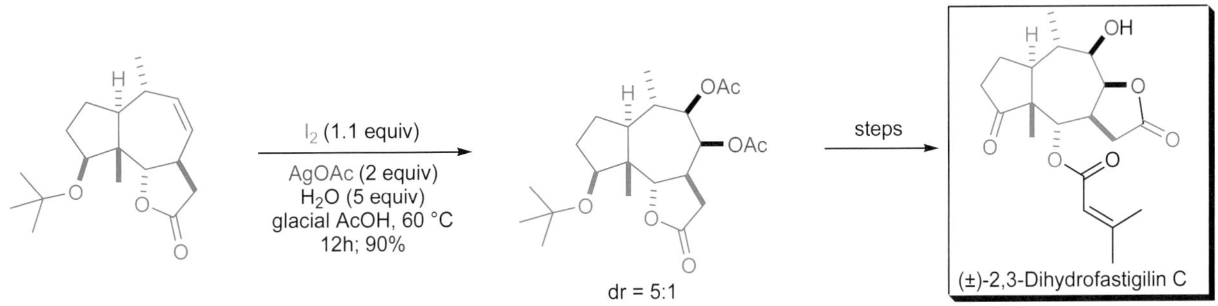

PRILEZHAEV REACTION

(References are on page 656)

Importance:

[*Seminal Publications*[1-3]; *Reviews*[4-12]; *Modifications & Improvements*[13-23]; *Theoretical Studies*[24-33]]

In 1909, N. Prilezhaev was the first to use peroxycarboxylic acids to oxidize isolated double bonds to the corresponding oxiranes (epoxides).[1] This transformation is referred to as the *Prilezhaev reaction*. The use of peroxyacids for the preparation of epoxides is one of the most widely used methods unless the epoxide is needed in an enantiomerically pure form for which other methods are available (e.g., *Sharpless, Jacobsen*, and *Shi asymmetric epoxidation*). The general features of the *Prilezhaev reaction* are: 1) the reaction is stereospecific, since the stereochemistry of the alkene substrate is retained in the epoxide product (*trans* alkene yields the *trans* epoxide, while *cis* alkene affords *cis* epoxide); 2) the reaction rate increases if the substituents on the alkene are electron-donating and decreases if they are electron-withdrawing; 3) an electron-withdrawing substituent (R^5) on the peroxyacid increases the rate of epoxidation; 4) substrates with multiple isolated double bonds can be epoxidized regioselectively, since the more electron-rich double bond reacts faster with the peracid (terminal alkenes are the least reactive, so a disubstituted alkene is selectively epoxidized in the presence of a terminal one); 5) alkenes that have preexisting chiral centers theoretically give rise to two diastereomeric epoxides, but in practice high diastereoselectivities may be achieved by preferentially epoxidizing the less sterically hindered face of the alkene (substrate-directed synthesis);[34] 6) alkenes with no chiral centers give rise to a 1:1 mixture of enantiomeric epoxides (racemic mixture); 7) the steric demand of the peroxyacid is almost negligible, so even very sterically hindered substrates may be epoxidized; 8) cup-shaped molecules are usually epoxidized from the less hindered convex side; 9) if a functional group adjacent to the double bond can coordinate to the peroxyacid, the natural steric bias will be overridden and the epoxidation will occur from that face of the double bond where the coordinating functional group is located (e.g., $OH>CO_2H>CO_2R>OCOR$) and this phenomenon is called the *neighboring group effect*; 10) the reagent peroxyacids can be prepared (by reacting carboxylic acids with hydrogen peroxide) or purchased from commercial sources; 11) most widely used peroxyacid is *m*CPBA, which is a relatively stable solid with good solubility in most organic solvents; 12) less frequently used (and not very stable) peroxyacids are generated *in situ* (e.g., peroxyacetic and performic acid); 13) the peroxyacids are much less acidic than the carboxylic acids, so acid-catalyzed side reactions (e.g., epoxide ring-opening) are rare; 14) when the product is very acid sensitive, the reaction mixture needs to be buffered since the by-product is a strong carboxylic acid; 15) epoxidations with *m*CPBA are usually carried out at or below ambient temperature, and a mildly basic work-up ensures the removal of the benzoic acid by-product from the epoxide product; 16) the reaction tolerates most functional groups, but free amines are readily oxidized, so they must be protected; 17) ketones may undergo a competing *Baeyer-Villiger oxidation*; 18) α,β-unsaturated esters are epoxidized, while α,β-unsaturated ketones remain unchanged under the reaction conditions; and 19) alkynes react 10^3 times slower than alkenes, so alkenes are selectively epoxidized in the presence of alkynes. When the use of peroxyacids is not suitable for the substrates or the products, alternative epoxidizing agents may be applied:[11] 1) peroxycarboximidic acids (by mixing nitriles with H_2O_2);[13,19] 2) magnesium monoperoxyphthalate hexahydrate (MMPP);[16] 3) dimethyldioxirane or dialkyldioxiranes;[17,23] 4) alkyl hydroperoxides in the presence of a transition metal catalyst;[35] 5) molecular oxygen and light (*photoepoxidation*);[15] and 6) inorganic peroxo acids (e.g., peroxoselenic acid).[10,11]

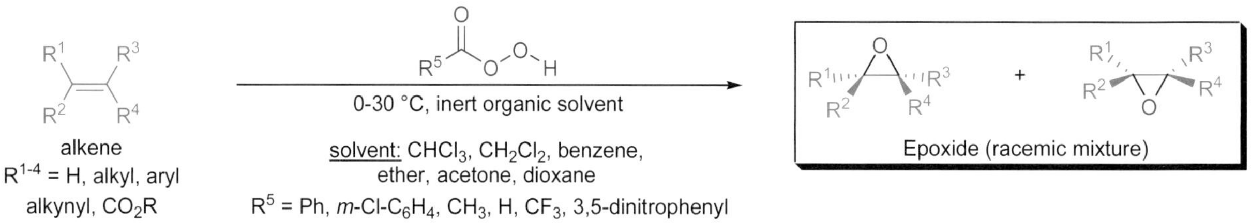

Mechanism: [36,7,37-47]

The *Prilezhaev reaction* is stereospecific, and a *syn* addition of the oxygen to the double bond is observed in all cases. This observation supports the assumption that the epoxidation of alkenes by peroxyacids is a concerted process. The reaction takes place at the terminal oxygen atom of the peroxyacid, and the π HOMO of the olefin approaches the σ^* LUMO of the O-O bond at an angle of 180° (butterfly transition structure).

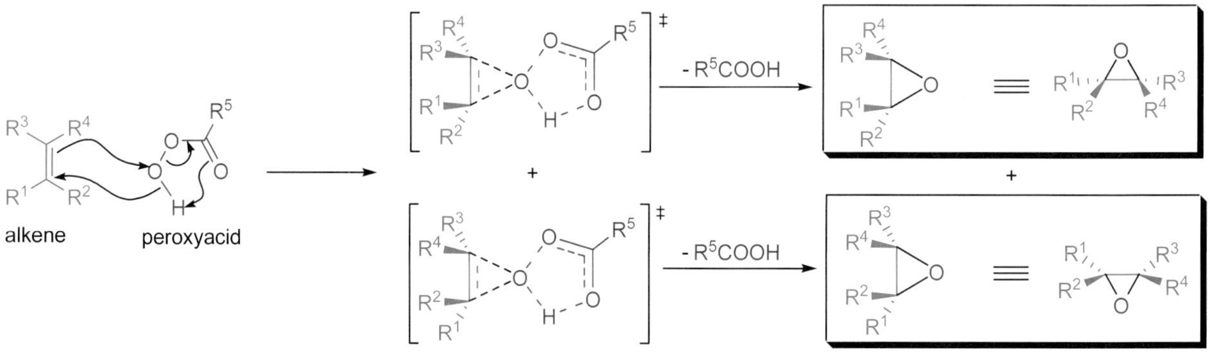

Synthetic Applications:

A *diastereoselective epoxidation* of a tetrasubstituted double bond was accomplished with *m*CPBA in the total synthesis of (–)-21-isopentenylpaxilline by A.B. Smith et al.[48] The tetracyclic lactone substrate containing the tetrasubstituted double bond was exposed to *m*CPBA in toluene at room temperature. The reaction mixture also contained sodium bicarbonate to neutralize the by-product *m*-chloro benzoic acid. The epoxidation exclusively took place from the less hindered α-face of the molecule. At a later stage, this epoxide was converted to the γ-hydroxy enone moiety present in the natural product.

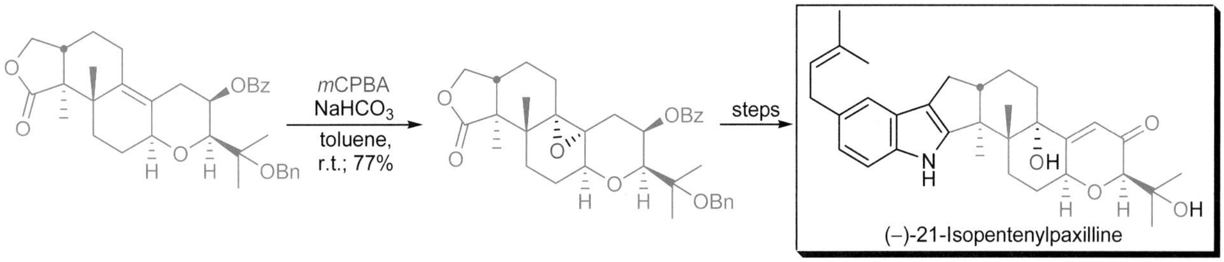

During the first total synthesis of briarellin diterpenes, briarellins E and F, L.E. Overman and co-workers utilized the large reactivity difference between a triple and a double bond in peroxyacid oxidations to selectively epoxidize a trisubstituted double bond in the presence of a terminal alkyne.[49] The epoxidation with *m*CPBA was carried out in DCM in the presence of a base to afford the α-epoxide in a 9:1 diastereomeric ratio.

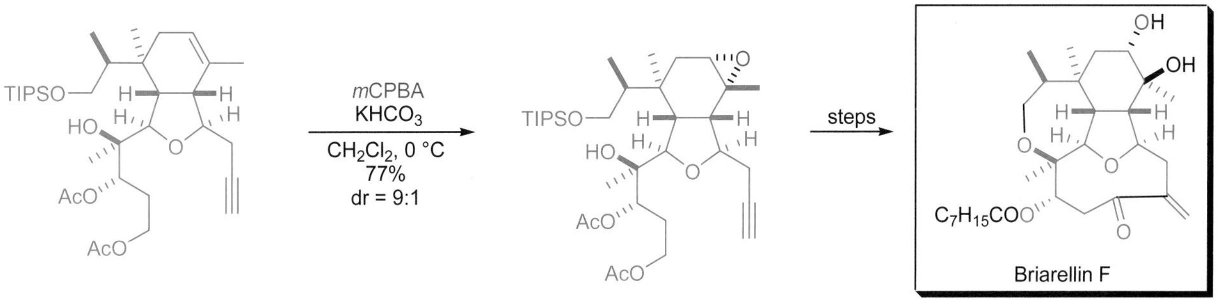

The *hydroxyl group-directed epoxidation* was utilized by M. Isobe et al. in their total synthesis of 11-deoxytetrodotoxin.[50] The six-membered cyclic allylic alcohol was treated with *m*CPBA in the presence of a phosphate buffer to afford an almost quantitative yield of the desired β-epoxide.

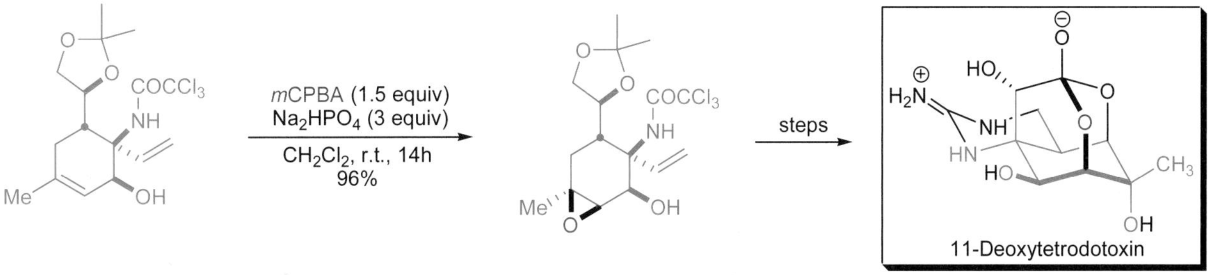

The final step in J. Mulzer's total syntheses of epothilones B and D was the oxidation of the C12-C13 double bond of epothilone D *via* a highly diastereoselective *Prilezhaev reaction* to obtain epothilone B.[51] The same *m*CPBA oxidation endgame was chosen by R. E. Taylor et al. in the total synthesis of these two natural products.[52]

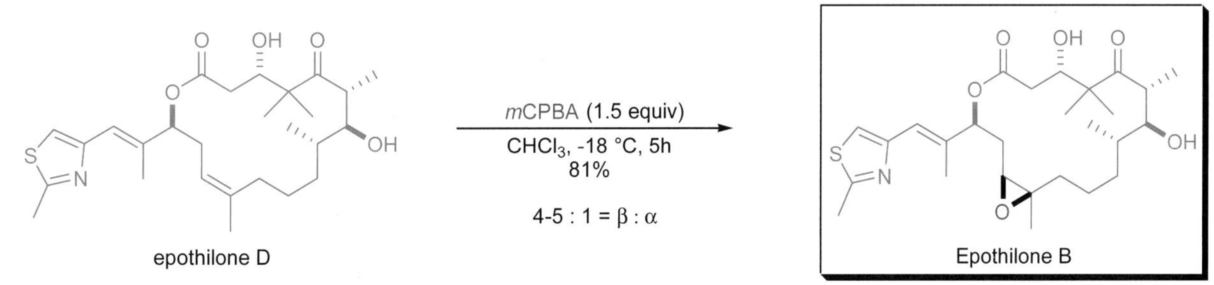

PRINS REACTION

(References are on page 658)

Importance:

[*Seminal Publications*[1-4]; *Reviews*[5-8]; *Modifications & Improvements*[9-14]; *Theoretical Studies*[15]]

In 1899, O. Kriewitz reported that upon heating with paraformaldehyde in a sealed tube, β-pinene gave rise to an unsaturated alcohol (nopol).[1,2] It was not until two decades later that H.J. Prins conducted the first comprehensive study on the sulfuric acid-catalyzed reactions of various alkenes (e.g., styrene, pinene, camphene) with formaldehyde.[3,4] In his honor, the acid-catalyzed condensation of alkenes with aldehydes is referred to as the *Prins reaction*. The general features of the reaction are: 1) potentially a large number of different products can be formed; however, the careful control of the reaction conditions allows the formation of a given product with good selectivity; 2) besides allylic alcohol products, the formation of 3-substituted alcohols, 1,3-diols, and 1,3-dioxanes is possible, depending on what type of nucleophilic species are present in the reaction mixture; 3) a variety of protic and Lewis acids may be employed to catalyze the reaction: H_2SO_4, HCl, HOCl, HNO_3, *p*-TsOH, BF_3, $AlCl_3$, $ZnCl_2$, $TiCl_4$, etc.; 4) when the reaction is conducted under anhydrous conditions, the *carbonyl ene reaction* takes place (*See Ene reaction*), and the corresponding homoallylic alcohols are formed exclusively; 5) the reaction is fastest with formaldehyde and with highly substituted alkenes; 6) both acyclic and cyclic alkenes are substrates for the transformation; 7) the addition of the protonated aldehyde across the double bond of the alkene follows *Marknovnikoff's rule*, and the fate of the resulting carbocation determines what type of products are formed; and 8) with cyclic alkenes, the products often have *anti* stereochemistry due to neighboring group participation.

Kriewitz (1899):

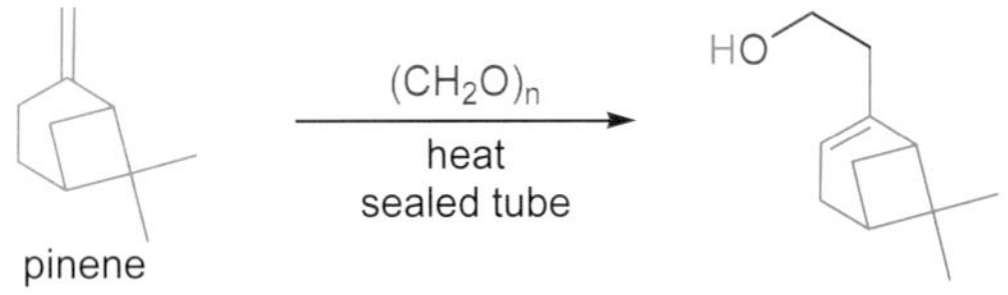

Prins (1919):

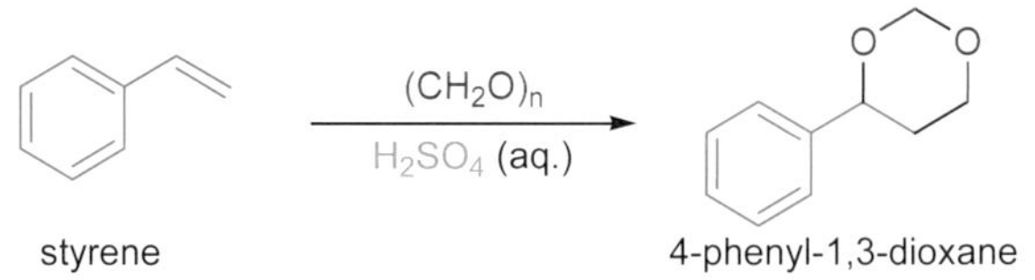

Prins reaction:

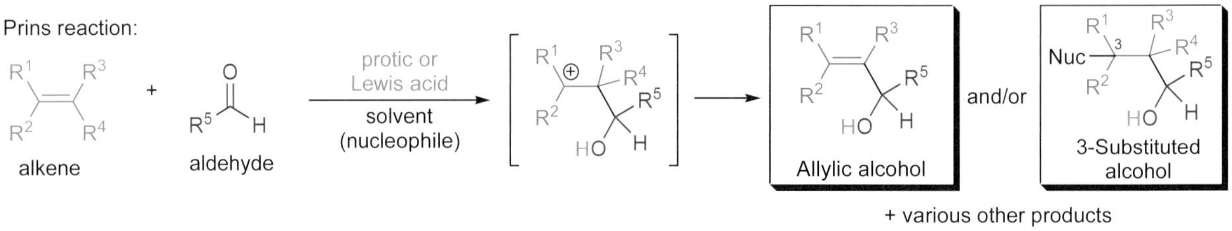

R^{1-4} = H, alkyl, aryl, heteroaryl; R^5 = H, alkyl, aryl; protic acid: dilute aqueous H_2SO_4, HCl, H_3PO_4, HOCl, *p*-TsOH, HNO_3; Lewis acid: BF_3, $AlCl_3$, $ZnCl_2$, $TiCl_4$, cation-exhange resin; solvent: H_2O, ROH, benzene; nucleophile: could be the solvent or the conjugate base derived from the protic acid; other products: dienes, 1,3-dioxanes, 1,3-diols, etc.

Mechanism:[16-29]

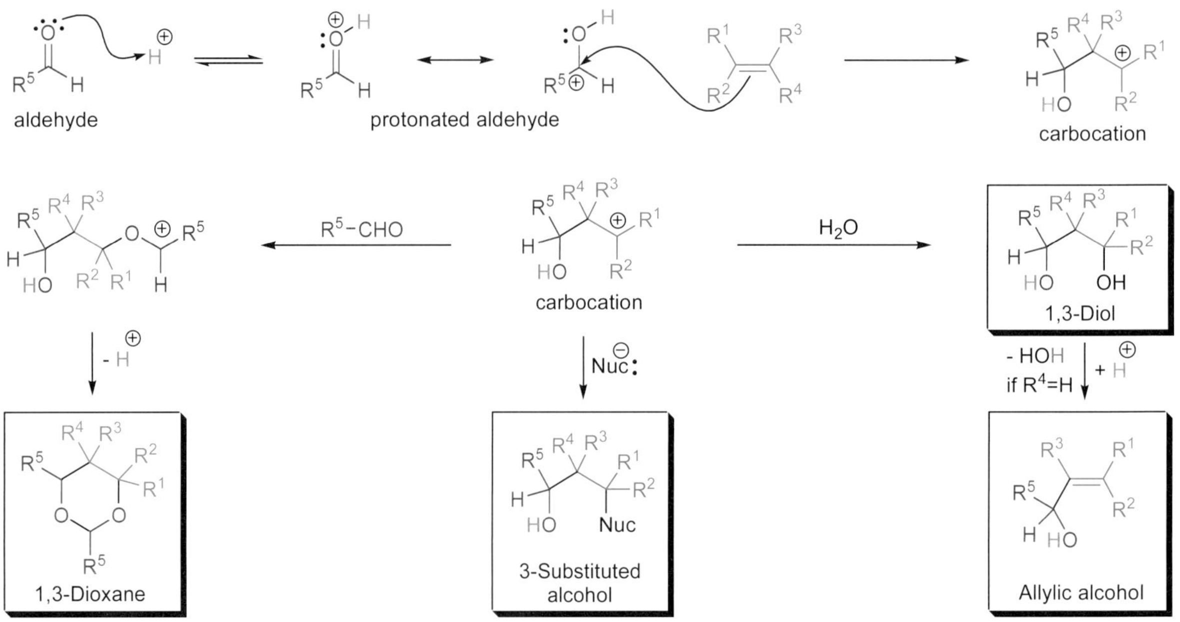

Synthetic Applications:

Studies toward the biomimetic total synthesis of (+)-chatancin were conducted by P. Deslongchamps et al.[30] The authors planned to use a *transannular Diels-Alder reaction* of a pyranophane intermediate as the key ring forming step. The cyclic dienedione precursor for this transformation was prepared using the *Prins reaction* on a substrate derived from *trans-trans* farnesol.

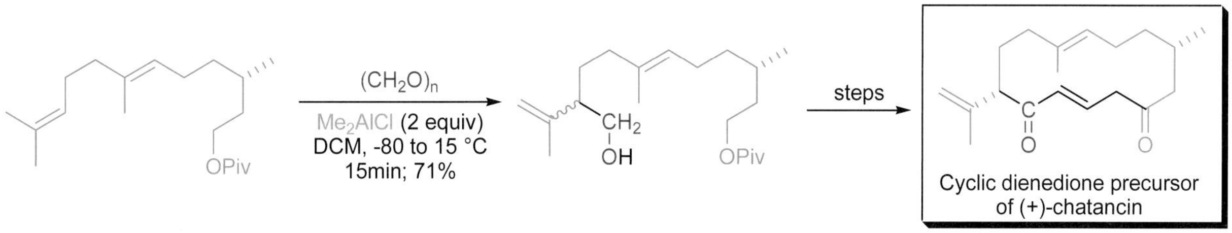

The tandem *Mukaiyama aldol reaction-Prins cyclization* was utilized during the formal total synthesis of leucascandrolide A by S.D Rychnovsky.[31] The addition of the activated aldehyde to the enol ether resulted in the formation of an oxocarbenium ion, which was captured intramolecularly by the allylsilane moiety to form a new tetrahydropyran ring. The reduction of the crude reaction mixture with $NaBH_4$ was performed to remove the unreacted aldehyde starting material, thereby facilitating the chromatographic purification of the product. The product was isolated as a 5.5:1 mixture of epimers at C9.

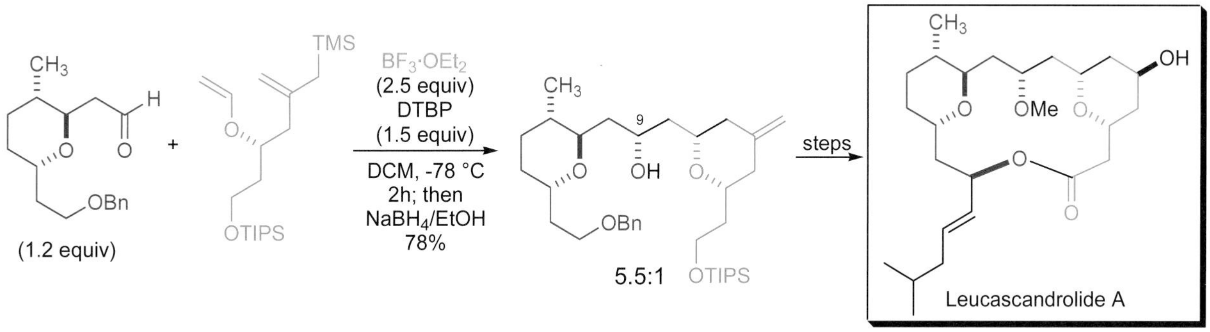

In the laboratory of R.D. Rychnovsky, the *segment-coupling Prins cyclization* was utilized for the total synthesis of (–)-centrolobine.[32] This approach avoided the common side reactions, such as side-chain exchange and partial racemization by reversible *2-oxonia Cope rearrangement*, associated with other *Prins cyclization reactions*. The substrate α-acetoxy ether was subjected to $SnBr_4$ in DCM, which brought about the formation of the all-equatorial tetrahydropyran in good yield.

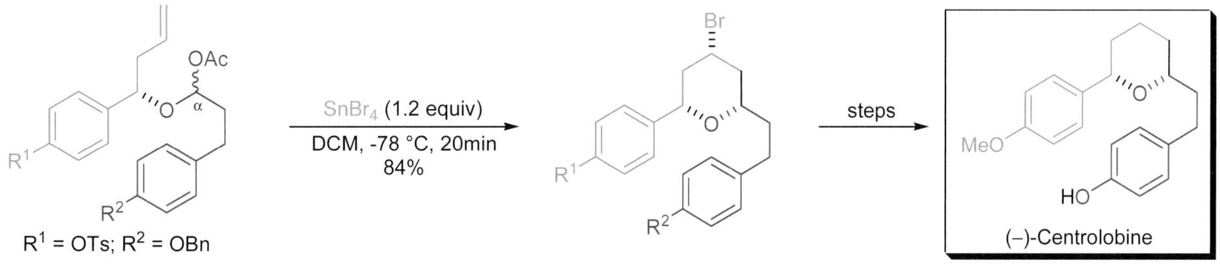

The stereoselective total synthesis of (±)-isocycloseychellene was achieved by S.C. Welch and co-workers.[33] One of the key ring forming reactions was an *oxidative Prins reaction* that took place without the need of a catalyst (*carbonyl ene reaction*) to afford the desired tricyclic ketone.

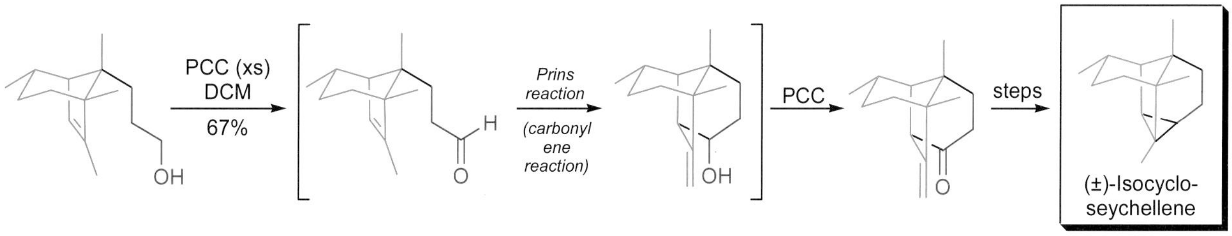

PRINS-PINACOL REARRANGEMENT

(References are on page 658)

Importance:

[*Seminal Publications*[1-6]; *Reviews*[7-9]; *Modifications & Improvements*[10-14]]

In 1969, G. Mousset and co-workers attempted to prepare the acetonide of a *meso* allylic 1,2-diol by refluxing it with acetone in the presence of an acidic clay catalyst.[1] To their surprise, instead of the expected acetal, they isolated a highly substituted tetrahydrofuran derivative. The authors proposed that the acetone condensed with the diol to give an oxocarbenium ion that underwent a *Prins cyclization* to afford a β-hydroxy carbenium ion intermediate, which gave rise to the tetrahydrofuran derivative *via* a *pinacol rearrangement*. Almost two decades later in 1987, L.E. Overman et al. investigated the Lewis acid mediated rearrangement of 4-alkenyl-1,3-dioxolanes (allylic acetals) to afford 3-acyltetrahydrofurans.[3] Subsequent studies conducted by the Overman group demonstrated that the transformation was general and had a broad scope.[9] The formation of oxacyclic and carbocyclic ring systems by terminating *Prins cyclizations* with the *pinacol rearrangement* in a tandem fashion is known as the *Prins-pinacol rearrangement*. The general features of the reaction are:[9] 1) it is completely stereoselective and results in the formation of two C-C bonds, one C-O bond, and two new stereocenters; 2) protic and Lewis acids are the most common in promoting the reaction; 3) most widely used solvents are nitromethane and dichloromethane; 4) alkenyl-substituted cyclic acetals derived from 1,2-diols give rise to highly substituted 3-acyltetrahydrofurans; 5) 1-alkenylcycloalkane-1,2-diols condense with aldehydes and ketones and afford annulated 3-acyltetrahydrofurans accompanied by ring-enlargement; 6) when the double bond of the starting alkenyl diol is part of a ring, a variety of differently annulated polycyclic ethers can be prepared upon condensation with aldehydes and ketones; 7) in the majority of cases, both the *syn* and *anti* acetal stereoisomers afford the same tetrahydrofuran adduct; 8) the acyl substituent at C3 will be preferentially *cis*-disposed to both the C2 and C5 substituents; 9) if the oxocarbenium ion intermediate is external to the ring formed in the *Prins cyclization* step, the formation of a carbocyclic ring takes place;[13] and 10) besides substituted alkenes, terminal alkynes also participate in the rearrangement.[9]

Mousset (1969):

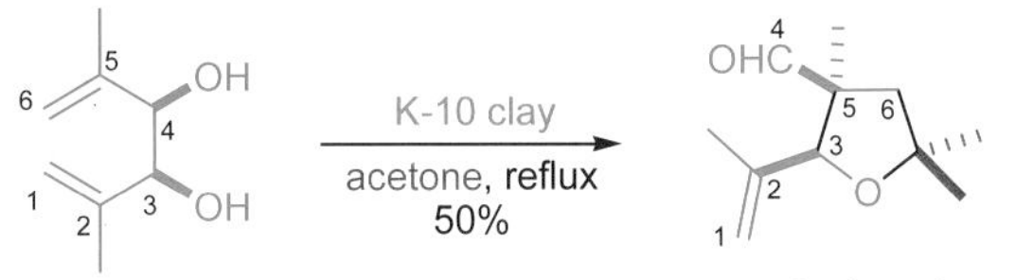

allylic 1,2-diol — substituted tetrahydrofuran

Overman (1987):

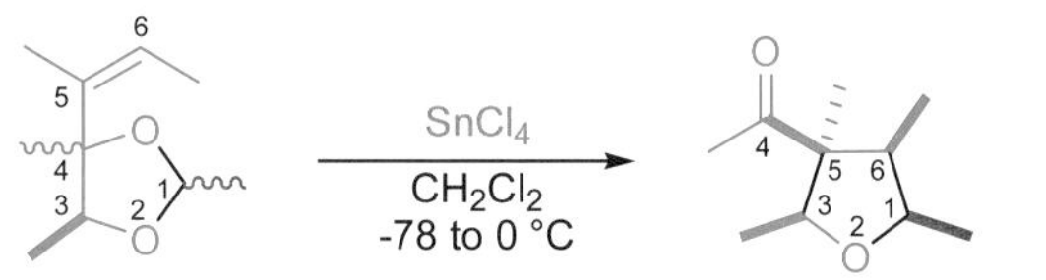

allylic acetal — acyl tetrahydrofuran

Formation of substituted tetrahydrofurans:

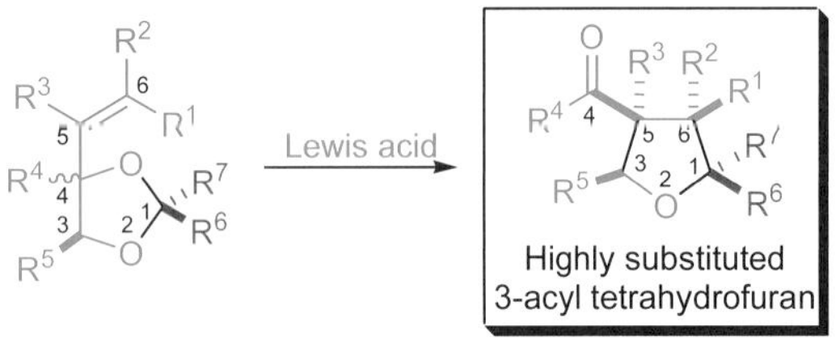

Tetrahydrofuran annulation accompanied by ring-enlargement:

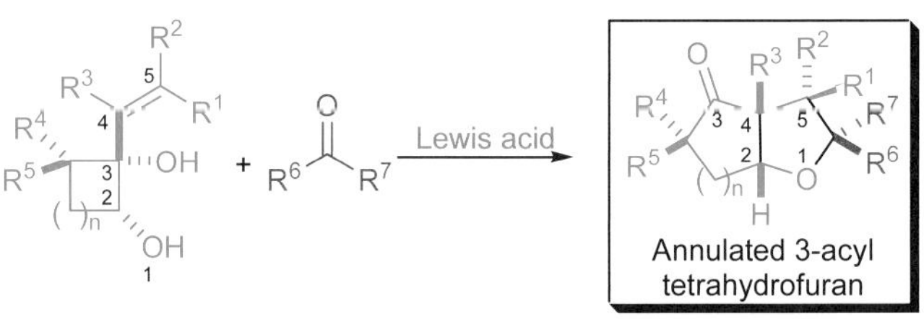

Cyclopentane spiroannulation:

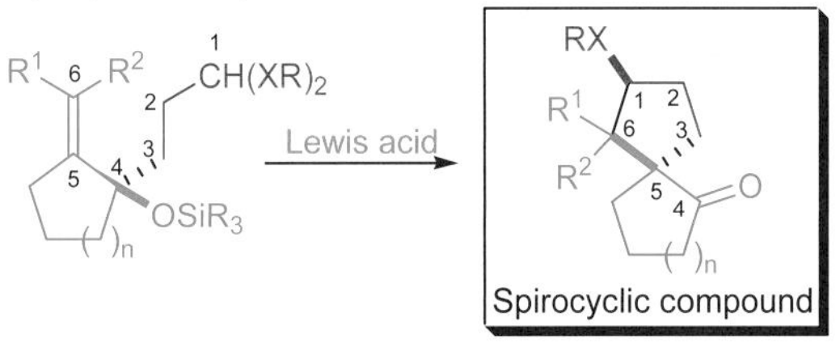

Cyclopentane annulation accompanied by ring-enlargement:

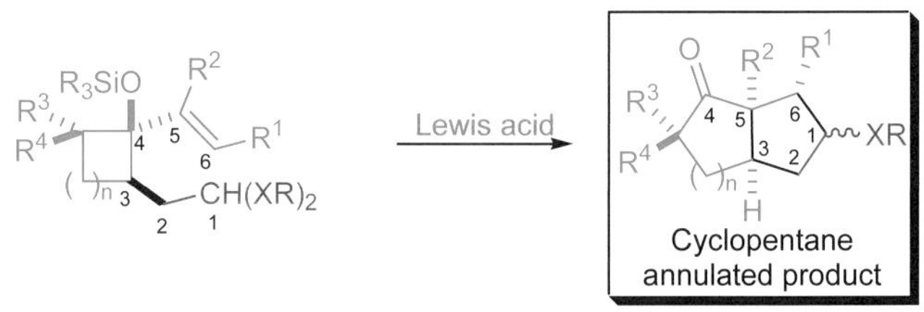

R^{1-5} = H, alkyl, aryl; R^{6-7} = H, alkyl, aryl, alkenyl; n = 1-3; XR = SEt, OMe; SiR_3 = TMS, TES, TBDMS; Lewis acid: BCl_3, $SnCl_4$, BF_3

Mechanism: [10-12,15,16,9]

Originally, the reaction was thought to proceed by an *oxonia-Cope rearrangement* followed by *aldol cyclization*, but this hypothesis was rejected based on the observation that enantiomerically enriched acetals gave rise to tetrahydrofurans of high enantiomeric purity and not a racemic mixture as was expected.[9]

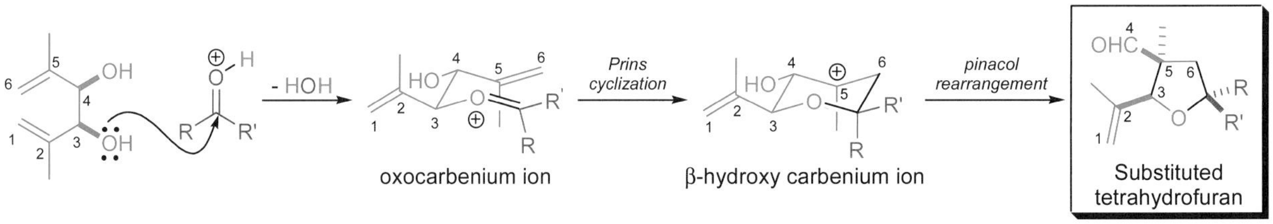

Synthetic Applications:

The *Prins-pinacol rearrangement* was utilized during the first enantioselective total synthesis of briarellin diterpenes by L.E. Overman and co-workers.[17] The cyclohexadienyl diol substrate was condensed with a (*Z*)-α,β-unsaturated aldehyde at low temperature in the presence of catalytic amounts of acid and $MgSO_4$ as dehydrating agent. The initially formed acetal was then exposed to 10 mol% of $SnCl_4$ to afford the desired tetrahydroisobenzofuran as a single stereoisomer that was later converted to briarellin F.

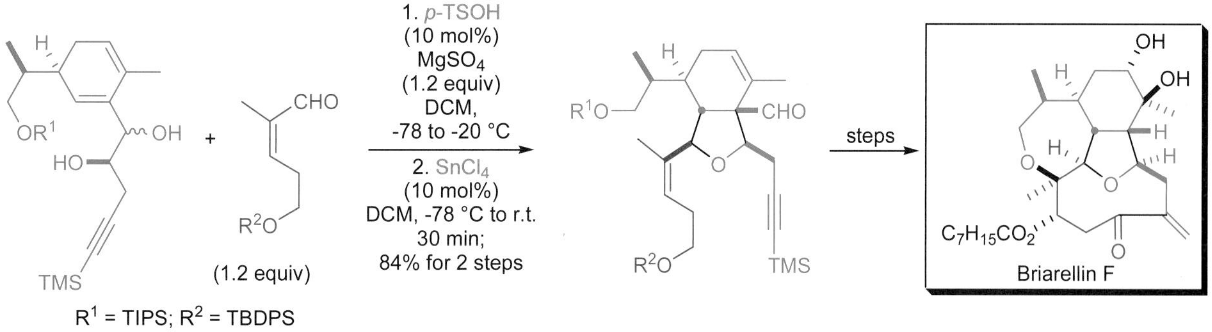

The first total synthesis of *lycopodium* alkaloids of the magellanane group was achieved in the laboratory of L.E. Overman.[18] The angularly fused all-carbon tetracyclic framework of (–)-magellaninone was constructed using the *ring-enlargement Prins-pinacol rearrangement* as the key step. The dienyl acetal substrate was treated with 1.1 equivalents of $SnCl_4$, which gave rise to the desired tetracycle as a mixture of methoxy epimers at C5. The *Prins cyclization* of the oxocarbenium ion took place from the less hindered convex face of the *cis*-bicyclooctadiene moiety and the subsequent *pinacol rearrangement* installed the quaternary stereocenter at C2.

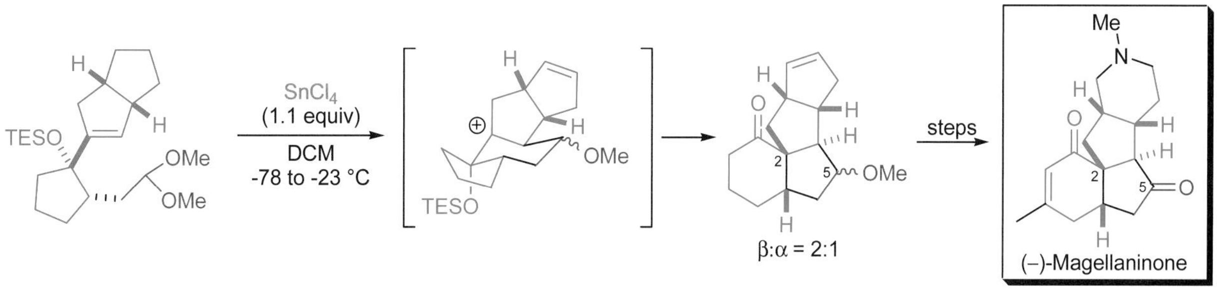

The enantioselective total synthesis of the polysubstituted tetrahydrofuran (–)-citreoviral, the unnatural enantiomer, was synthesized by L.E. Overman et al.[15] The *Prins-pinacol rearrangement* of an allylic 1,2-diol with an unsymmetrical ketone proceeded with high stereoselectivity. The *bis*(trimethylsilyl)-1,2-diol was condensed with the dimethyl acetal of the unsymmetrical ketone in the presence of catalytic amounts of TMSOTf, which yielded a nearly 1:1 mixture of the corresponding acetal and rearrangement product. The acetal was converted to the desired tetrahydrofuran product upon exposure to tin tetrachloride.

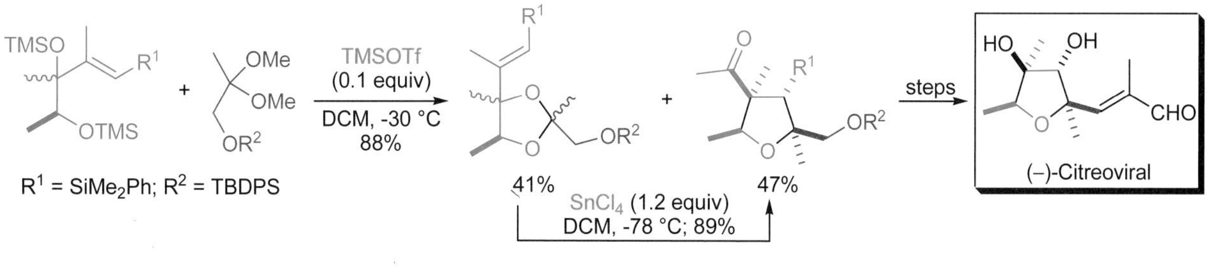

The *thio-Prins-pinacol rearrangement* was the key transformation in L.E. Overman's enantioselective total synthesis of (+)-shahamin K.[19] Treatment of the dithioacetal substrate with DMTSF brought about the rearrangement, which gave rise to the *cis*-hydroazulene core of the natural product.

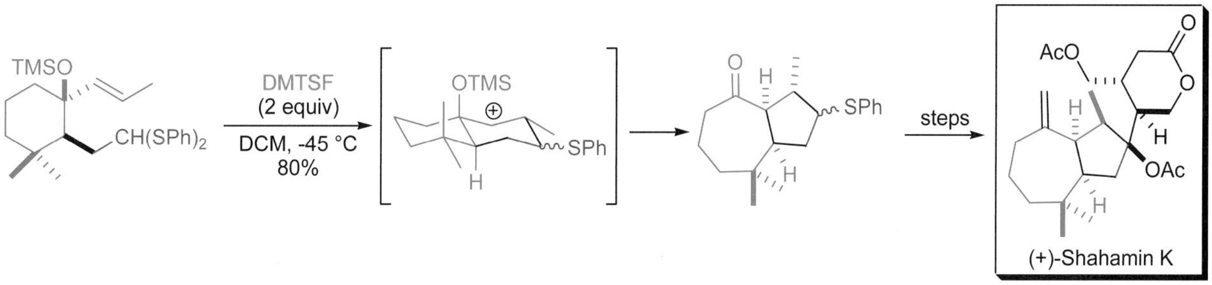

PUMMERER REARRANGEMENT

(References are on page 659)

Importance:

[*Seminal Publications*[1,2]; *Reviews*[3-25]; *Modifications & Improvements*[26-36]; *Theoretical Studies*[37]]

In 1909, R. Pummerer observed that by heating phenylsulfinylacetic acid with mineral acids (e.g., HCl, H_2SO_4), thiophenol and glyoxylic acid were formed.[1] Later this transformation was shown to be general, and today the formation of α-substituted sulfides from the corresponding sulfoxides is referred to as the *Pummerer rearrangement*.[38] The general features of the reaction are: 1) the sulfoxide substrates must have at least one hydrogen atom at their α-position; 2) acetic anhydride (Ac_2O) is the most widely used activating reagent for the rearrangement, and it is often applied as the solvent in combination with other solvents such as benzene or ethyl acetate; 3) the use of acid co-catalysts (e.g., TsOH, AcOH, TFAA) is common to minimize side reactions and increase the product yields; 4) Ac_2O can be replaced with TFAA, which is a stronger reagent and allows for milder reaction conditions;[26] 5) the most common product of the reaction is an α-acetoxy sulfide; 6) upon acidic hydrolysis, the α-acetoxy sulfide affords a thiol and a carbonyl compound that can be easily separated; 7) upon treatment with base, vinyl sulfides are formed *via* a β-elimination; 8) the rearrangement is regioselective when the sulfoxide has hydrogens at both the α- and α'-positions and the more acidic position will get preferentially substituted; 9) the regioselectivity can be altered by steric factors especially in cyclic systems: isomeric sulfoxides often give rise to different products; and 10) the rearrangement can take place both inter- and intramolecularly. Drawbacks of the reaction are: 1) substrates with unprotected hydroxyl or amino groups result in side rections with the activating reagent; 2) unreactive substrates may undergo undesired sulfenic acid elimination if harsh conditions are necessary; 3) fragmentation products are observed when stable carbocations (e.g., allylic, benzylic) can be formed by the heterolytic cleavage of the C-S bond; 4) when the nucleophile is a primary or secondary alcohol, reduction of the sulfoxide to the sulfide may occur along with the oxidation of the alcohol (see *Swern oxidation*). There are several variants of the rearrangement:[12] 1) when selenoxides are the substrates, the *seleno-Pummerer rearrangement* takes place; 2) *sila-Pummerer rearrangement* occurs with sulfoxides bearing a TMS group on the α-carbon, which spontaneously rearrange to α-silyloxy sulfides, and no activating reagents are needed;[39] 3) vinyl sulfoxide substrates may undergo the *additive- and vinylogous Pummerer rearrangement*; 4) chirality transfer from enantiopure sulfoxides to the α-carbon is possible, and it constitutes the *asymmetric Pummerer rearrangement*, but this process is limited in scope.[15]

Pummerer (1909):

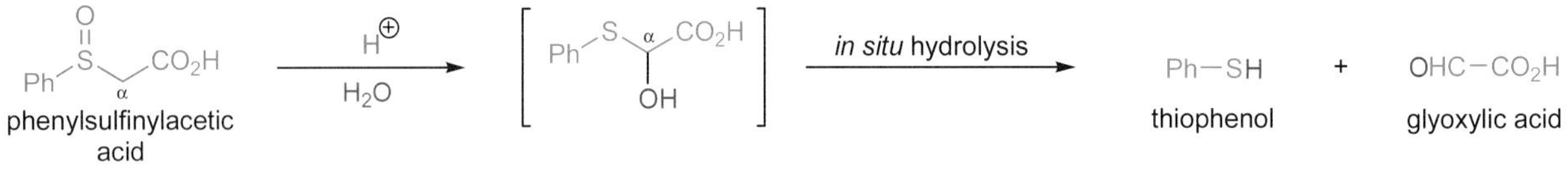

Pummerer rearrangement:

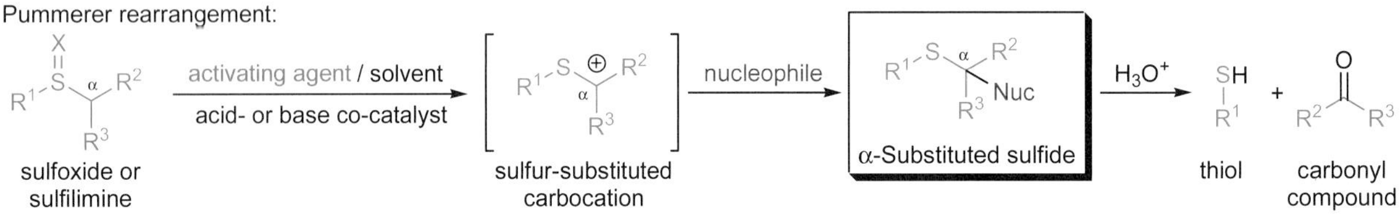

R^1 = alkyl, aryl; R^{2-3} = H, alkyl, aryl; X = O, NR; activating agent: HCl, H_2SO_4, TsOH, I_2/MeOH, Ac_2O, TFAA, *t*-BuBr, Me_3SiX, PCl_3, PCl_5, $Sn(OTf)_2$; nucleophile: H_2O, ROH, RCO_2^-; Nuc: OH, *O*-alkyl, *O*-aryl, O_2CR, F, Cl, Br, SR, NR_2; co-catalysts: AcOH, TsOH, TFAA, NaOAc

Mechanism: [9,15,21]

The mechanism of the *Pummerer rearrangement* consists of four steps: 1) acylation of the sulfoxide oxygen to form an acyloxysulfonium salt; 2) loss of a proton from the α-carbon to afford an acylsulfonium ylide; 3) cleavage of the sulfur-oxygen bond to give sulfur-substituted carbocation (RDS); and 4) capture of the nucleophile by the carbocation.

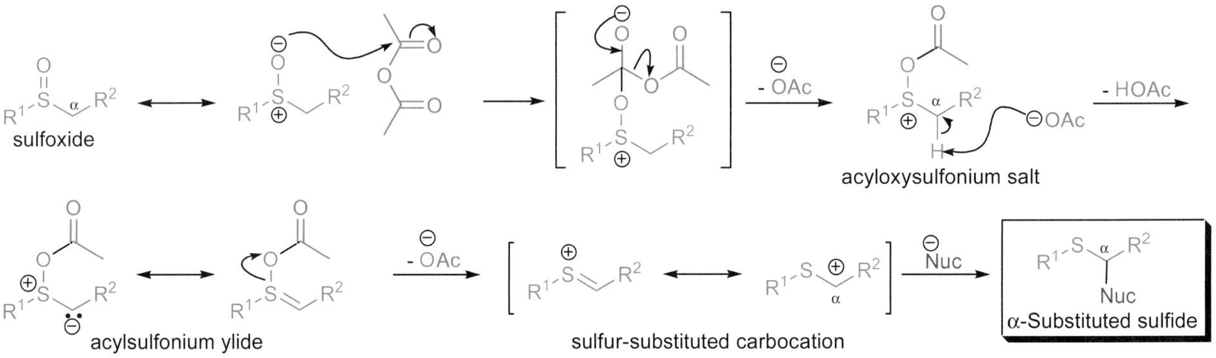

Synthetic Applications:

An enantioselective approach to polyhydroxylated compounds using chiral sulfoxides was developed in the laboratory of G. Solladié and was applied for the synthesis of enantiomerically pure *myo*-inositol and pyrrolidine derivatives.[40] The presence of the chiral sulfoxide directed the reduction of two carbonyl groups in one of the intermediates. In order to form the six-membered ring of *myo*-inositol, the removal of these sulfoxides under mild conditions was necessary. To this end, a one-pot *Pummerer rearrangement*-sodium borohydride reduction was performed using TFAA as the activating reagent. The initially formed thioacetal was reduced with $NaBH_4$ at pH 7 to afford the corresponding diol.

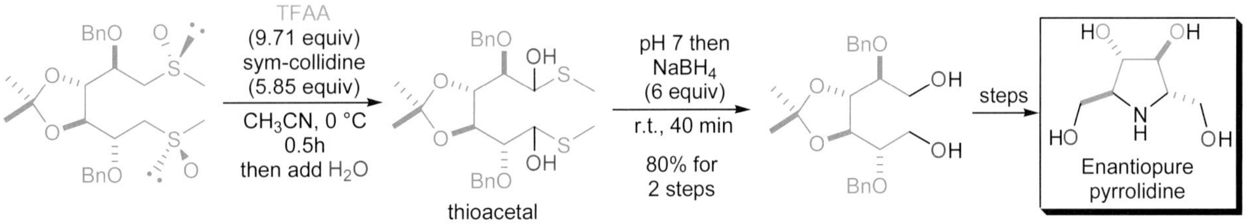

Quartromicins are complex C_2 symmetric macrocyclic natural products that have significant activity against a number of human viral targets.[41] The diastereoselective synthesis of the *endo*- and *exo*-spirotetronate subunits of the quartromicins was accomplished by W.R. Roush and co-workers. The preparation of the *exo*-α-acetoxy aldehyde involved the *Pummerer rearrangement* of a sulfoxide using acetic anhydride as the activating reagent and NaOAc as the co-catalyst. The yield of this transformation was modest and all attempts to improve its efficiency failed.

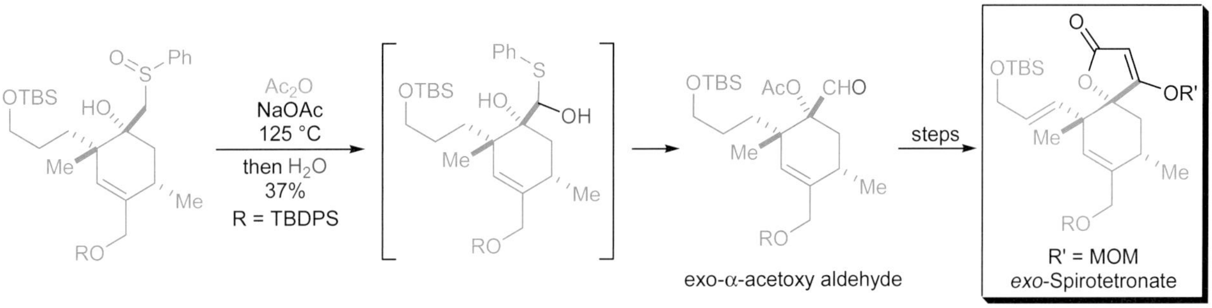

The total synthesis of (±)-deethylibophyllidine was achieved by J. Bonjoch et al. using a tandem *Pummerer rearrangement/thionium ion cyclization* to generate the quaternary spiro center.[42] The sulfoxide was exposed to an equimolar mixture of TFA/TFAA and heated for 2h to form the quaternary stereocenter at C7 with the desired stereochemistry, but at C6 a mixture of epimers were formed. Reductive desulfurization with Raney-Ni followed by photochemical rearrangement afforded the natural product.

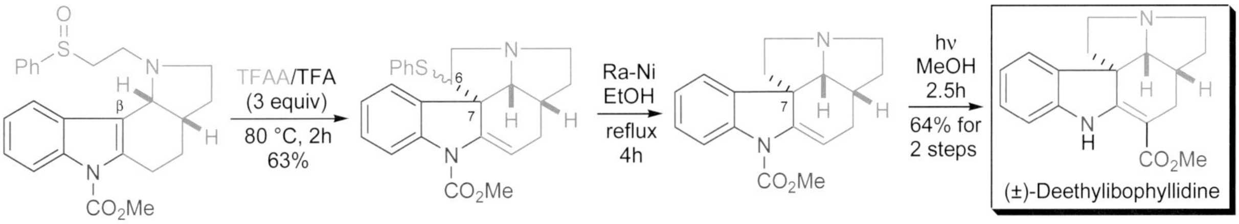

The *Pummerer rearrangement* was utilized to introduce the formyl group into the pyrone ring during H. Hagiwara's total synthesis of solanopyrone D.[43] Extensive screening revealed that the best way to activate the sulfoxide was to use the combination of TMSOTf as the *O*-silylating agent and $TMSNEt_2$ as a mild base. The addition of TBAF in THF afforded the formylated pyrone ring.

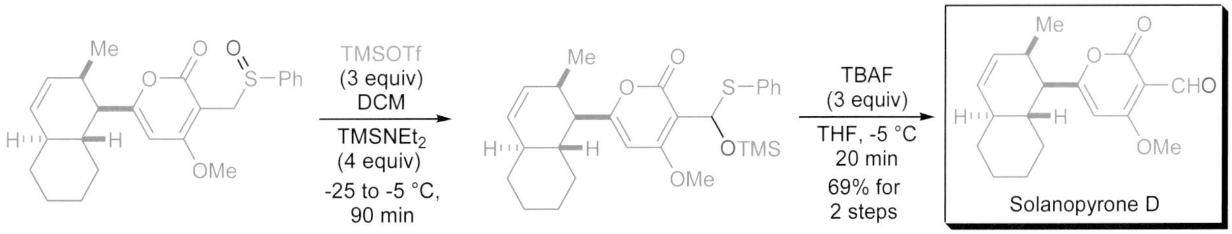

QUASI-FAVORSKII REARRANGEMENT

(References are on page 660)

Importance:

[*Seminal Publications*[1-3]; *Review*[4]]

In 1939, B. Tchoubar et al. reported that upon treatment with powdered sodium hydroxide in ether, α-chlorocyclohexyl phenyl ketone gave a 40% yield of 1-phenylcyclohexanecarboxylic acid *via a semibenzilic type rearrangement*.[1] In 1952, C.L. Stevens and E. Farkas obtained a higher yield when they repeated the same reaction in refluxing xylene. They predicted that the stereochemistry of the rearrangement would involve an inversion at the halogen-bearing carbon.[3] Upon treatment with certain nucleophiles, α-halo ketones with no hydrogen atom at the α'-position or bicyclic α-halo ketones with an α'-hydrogen atom at the bridgehead carbon atom undergo a skeletal rearrangement known as the *quasi-Favorskii rearrangement*. The product of the rearrangement is a carboxylic acid or a carboxylic acid derivative, depending on the nature of the nucleophile. Probably the most well-known example of the *quasi-Favorksii rearrangement* is the key step in the synthesis of cubane by P.E. Eaton et al.[5,6] In addition to nucleophiles, the rearrangement can be initiated by the ionization of the α-halo ketones upon treatment with salts of heavy metals (e.g., $AgNO_3$, $AgSBF_6$, etc.).[2,7] Substrate preparation is primarily carried out in the following three ways: 1) direct α-halogenation of substituted acyclic and cyclic ketones; 2) *Robinson annulation* of cyclic α-halo ketones with methyl vinyl ketone (MVK);[8,9] and 3) *[4+3] cycloaddition* of cyclic α,α'-dihalo ketones with cyclic dienes.[10,11] The analogous reaction of α-halo ketones (having at least one enolizable hydrogen atom in the α'-position) with base in the presence of a nucleophile is called the *Favorskii rearrangement*. The general features of the *quasi-Favorskii rearrangement* are: 1) acyclic and monocyclic α-halo ketones that do not have hydrogens in their α'-positions are good substrates; 2) the reaction is stereospecific (inversion at the carbon to which the halogen is attached); and 3) monocyclic and bicyclic substrates undergo ring-contraction to give the corresponding cyclic or bicyclic homologue.

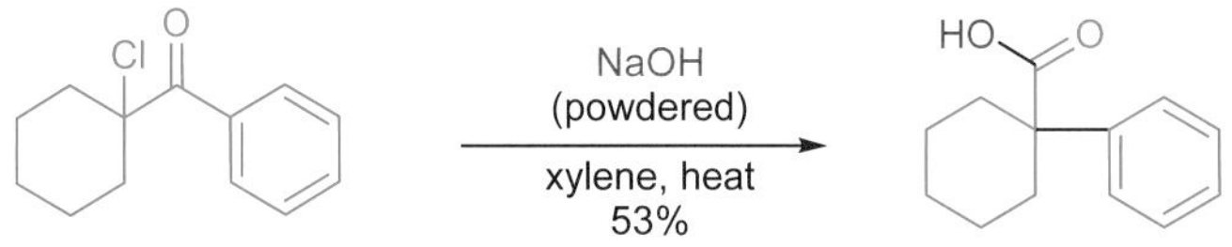

(α-chlorocyclohexyl)-
phenyl ketone

1-phenylcyclohexane-
carboxylic acid

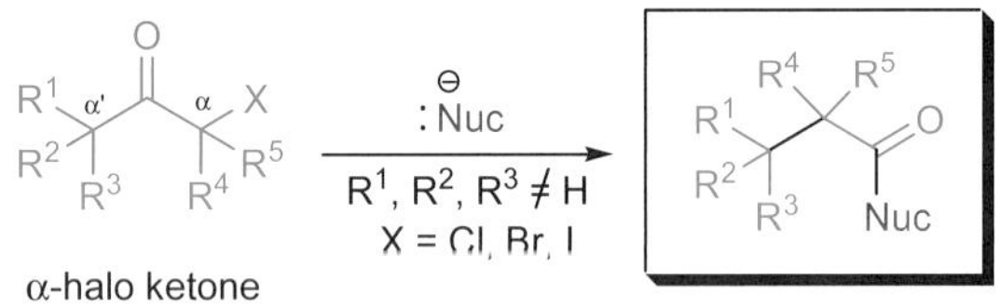

α-halo ketone
(no α' hydrogen)

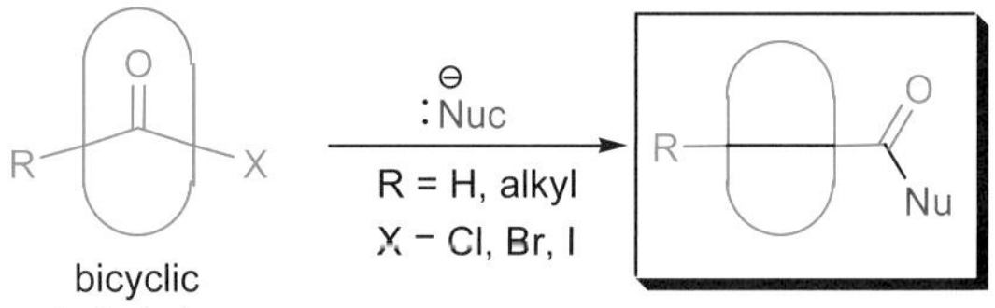

bicyclic
α-halo ketone

Mechanism: [12,13,7,14,15]

The mechanism of the *quasi-Favorskii rearrangement* involves the following steps: 1) attack of the nucleophile on the carbonyl carbon atom to form a tetrahedral intermediate; 2) next, this anionic intermediate undergoes a facile *1,2-alkyl shift*, similar to the mechanism of the *benzilic acid rearrangement*, and as a result, the halogen attached to the α-carbon is displaced with the inversion of configuration. When the substrate is bicyclic and there is a hydrogen in the α'-position, enolization is not possible because the double bond of the enol would be incorporated in the bridgehead and this reaction would violate Bredt's rule. The cyclopropanone intermediate of the *Favorskii rearrangement* would be highly strained (and sterically congested) and therefore its formation is highly disfavored. (This is valid for bicyclic systems in which the *trans* double bond would be part of a ring having less than 8 carbons; however, systems with rings larger than 8 carbons could be enolized.)

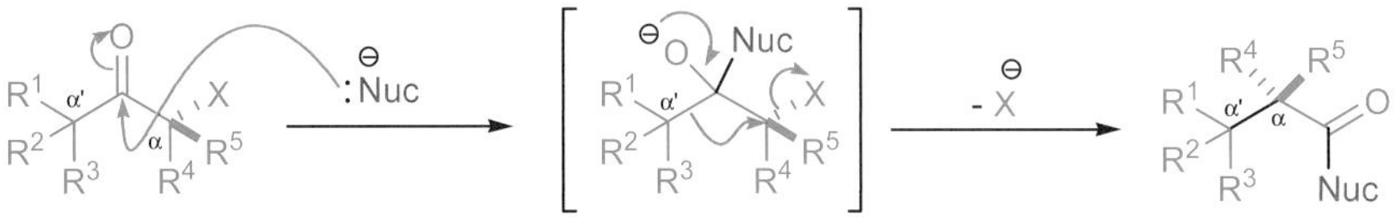

semibenzilic intermediate

inversion took place
at the α–position

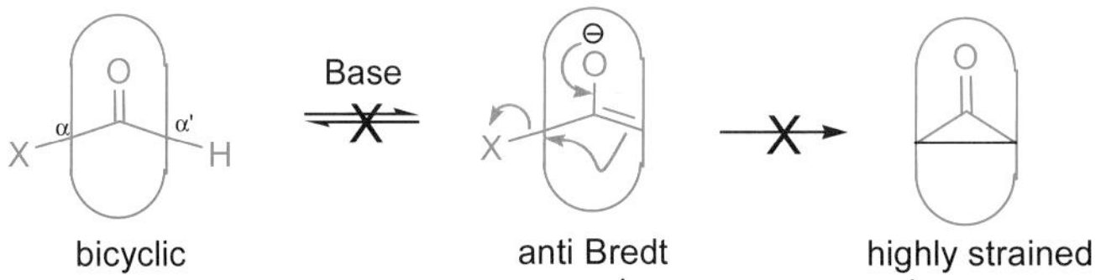

bicyclic
α–halo ketone

anti Bredt
enol

highly strained
cyclopropanone

Synthetic Applications:

G.A. Kraus and co-workers utilized the *quasi-Favorskii rearrangement* of a bicyclic bridgehead bromide as the key step in their formal total synthesis of *epi*-modhephene.[8,9] The required bicyclo[3.3.1]nonenone bridgehead bromide precursor was prepared by a *Robinson annulation* reaction between 3-bromo-2-oxocyclohexanecarboxylate and MVK. Upon treatment with lithiated dimethyl methylphosphonate, the bicyclic bromo ketone underwent a facile *quasi-Favorskii rearrangement* to afford the key intermediate bicyclo[3.3.0]octane derivative.

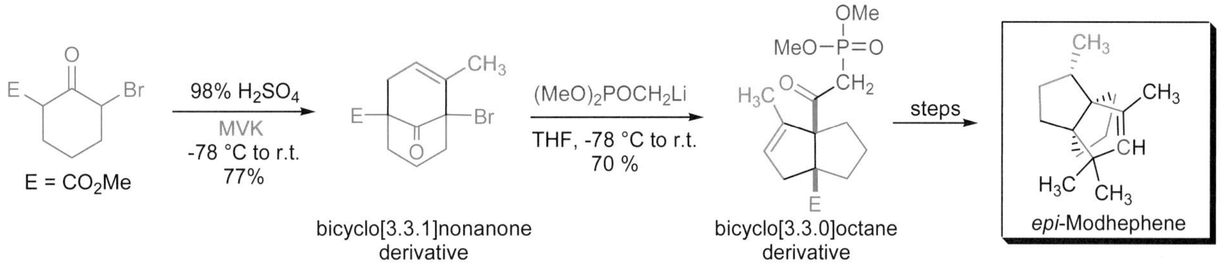

In the laboratory of M. Harmata, a novel methodology utilizing a sequential *[4+3] cycloaddition–quasi-Favorskii rearrangement* was developed for the rapid construction of polycyclic ring systems.[11] The *intramolecular [4+3] cycloaddition* of a halogenated allylic alcohol gave 65% of the expected tricyclic bridgehead α-bromo ketone precursor as a single diastereomer. Upon treating this bromo ketone with LAH in THF, a *quasi-Favorskii rearrangement* took place in nearly quantitative yield to afford a 5-6-5 fused tricyclic product.

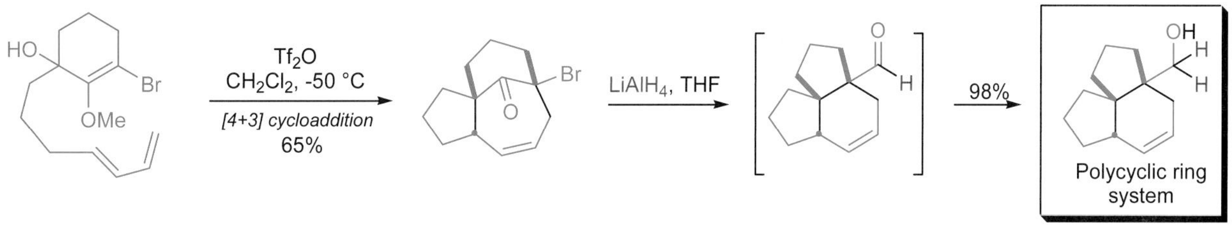

A formal total synthesis of racemic spatol was accomplished by M. Harmata et al. using an *intermolecular [4+3] cycloaddition* of a halogenated cyclopentenyl cation with cyclopentadiene followed by a *quasi-Favorskii rearrangement* as the key steps.[16]

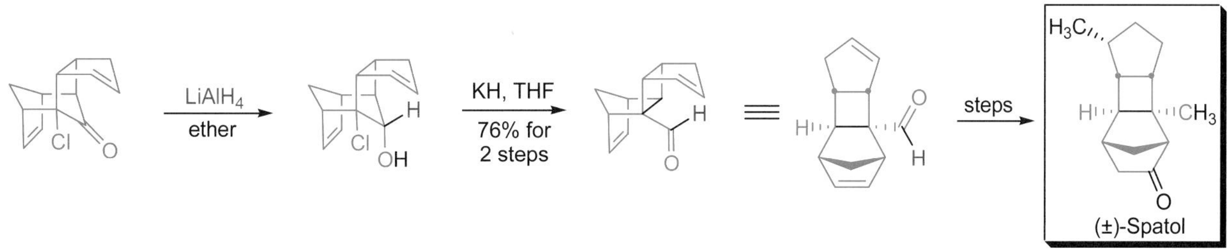

M. Harmata and co-workers successfully synthesized racemic sterpurene using an *intermolecular [4+3] cycloaddition* to prepare the key *quasi-Favorskii rearrangement* precursor.[17] The tricyclic bridgehead α-bromo ketone was first treated with LAH at 0 °C to get the corresponding secondary alcohol. Treatment of this alcohol with KH triggered the expected ring-contraction to afford the 5-6-4 fused tricyclic aldehyde, which was then reduced to the primary alcohol with LAH.

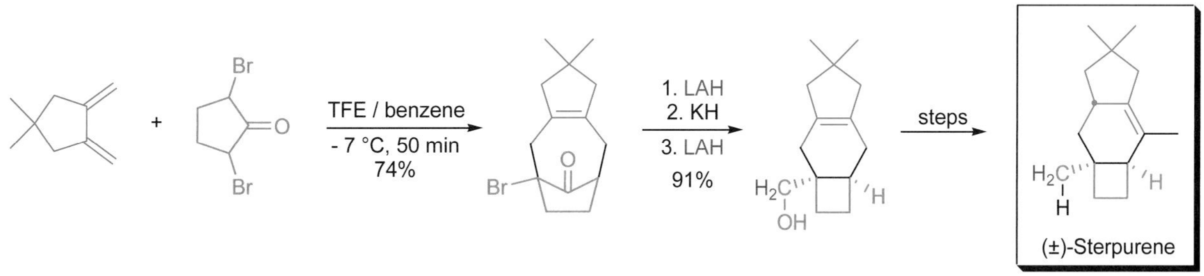

RAMBERG-BÄCKLUND REARRANGEMENT

(References are on page 660)

Importance:

[*Seminal Publications*[1]; *Reviews*[2-9]; *Modifications & Improvements*[10-20]]

In 1940, L. Ramberg and B. Bäcklund described an interesting reaction in which 1-bromo-1-ethanesulfonyl ethane (an α-bromo sulfone) was predominantly converted to (*Z*)-2-butene when treated with a boiling aqueous KOH solution.[1] There was no work published on this transformation until the early 1950s, when F.G. Bordwell and co-workers conducted a thorough kinetic investigation and elucidated the reaction mechanism.[21,22] The base-induced rearrangement of α-halogenated sulfones *via* episulfone intermediates to produce alkenes is referred to as the *Ramberg-Bäcklund rearrangement*. The general features of the reaction are:[5,6,8] 1) the precursor halogenated sulfones can be easily prepared by the halogenation of the corresponding sulfones and the sulfones themselves are usually prepared by the oxidation of sulfides; 2) the reaction is well-suited for the preparation of 1,1- or 1,2-di, tri-, and tetrasubstituted alkenes; 3) the position of the newly formed double bond is unambiguous and under the reaction conditions no double bond migration takes place; 4) both acyclic and cyclic substrates can be used and the reaction is especially useful for the preparation of strained cycloalkenes *via* ring-contraction; 5) the stereochemical outcome of the rearrangement depends on both the base and the solvent, but the temperature is not decisive; 6) aqueous base (e.g., KOH) favors the formation of (*Z*)-alkenes but strong bases in aprotic solvents (e.g., KO*t*-Bu/DMSO) predominantly give rise to (*E*)-alkenes; and 7) base-sensitive functional groups need to be protected.

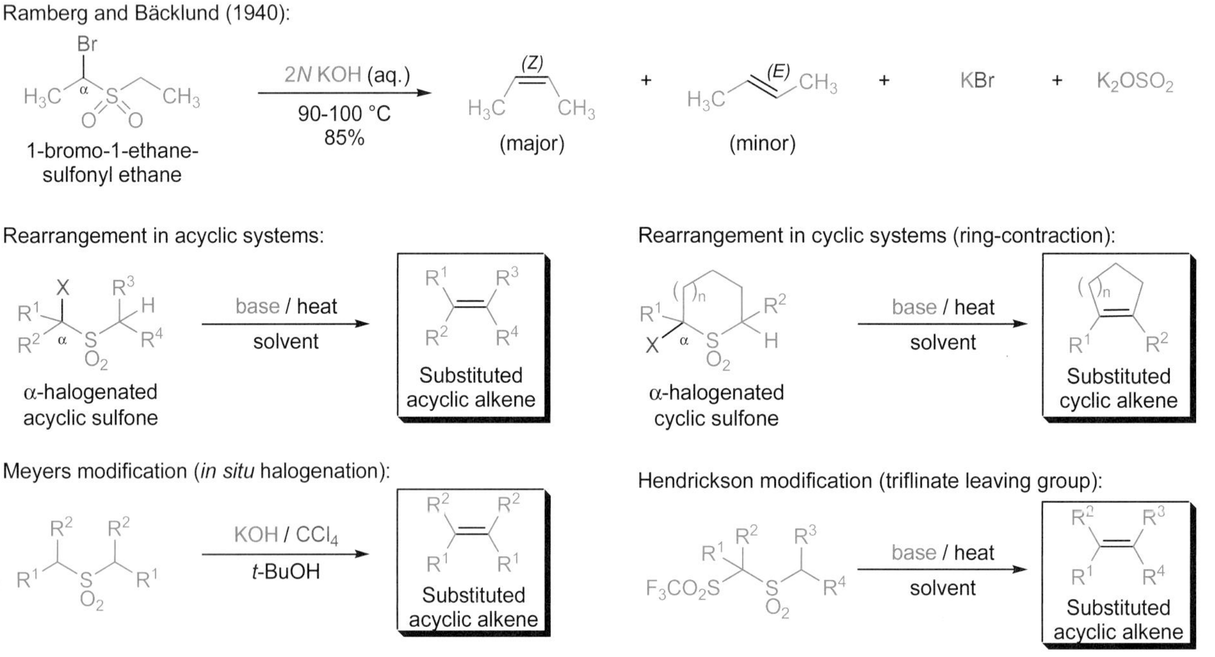

R^{1-4} = H, alkyl, aryl, heteroaryl, CO_2R; n = 0-12; X = Cl, Br, I, OTs; base: KOH, NaOH, KO*t*-Bu; solvent: THF, *t*-BuOH/DCM

Mechanism: [21,22,3,23-31]

The mechanistic details of the rearrangement were investigated in detail predominantly by the research groups of F.G. Bordwell and L.A. Paquette who established that the transformation consists of three distinct steps:[3] 1) the first step of the process is the deprotonation of the sulfone at the α- or α'-position, which undergoes rapid equilibration; 2) only the carbanion at the α'-position results in an intramolecular displacement reaction (S_Ni attack) on the carbon bearing the X group to give the reactive intermediate episulfones (thiirane 1,1-dioxides), which are generally formed as mixtures of *cis*- and *trans* stereoisomers (slow step); and 3) the final step is the loss of SO_2 either thermally or under base catalysis to give a mixture of alkene stereoisomers. The overall stereochemical outcome of the reaction is determined in the second step.

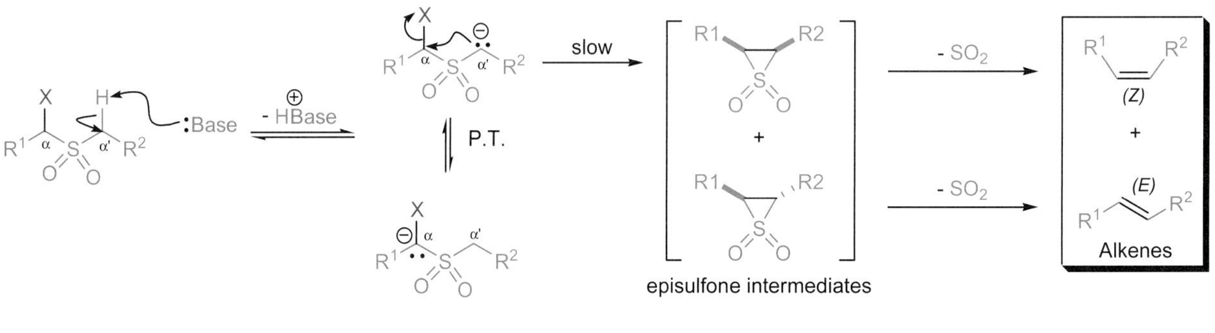

Synthetic Applications:

A concise convergent synthetic strategy was developed by B.M. Trost and co-workers for the synthesis of acetogenins, a class of compounds with a wide breadth of biological activity.[32] The authors chose (+)-solamin as the target to demonstrate the utility of their strategy, which relied on the *Meyers modification of the Ramberg-Bäcklund rearrangement* as the key step. As the chlorination of the sulfone failed, the *in situ* chlorination-rearrangement was attempted and led to the successful conversion of the oxasulfone precursor to the desired 2,5-dihydrofuran core.

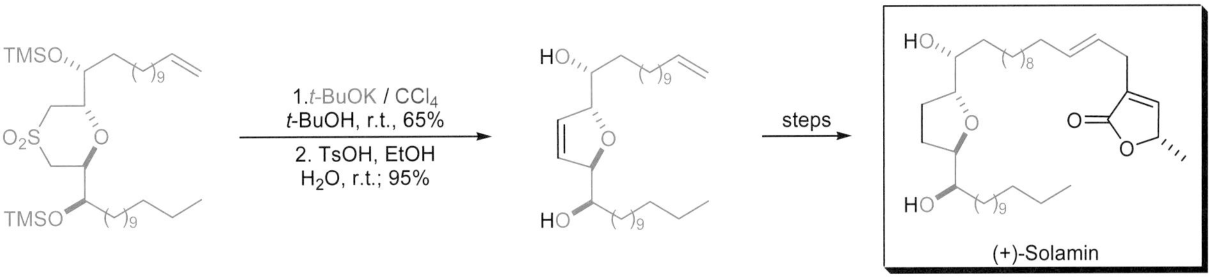

In the laboratory of R.K. Boeckman, the total synthesis of (+)-eremantholide A was accomplished using the *Ramberg-Bäcklund rearrangement* for the crucial ring-contraction step at the end of the synthetic sequence.[33] The nine-membered macrocyclic core of the natural product is highly strained since the C4-C5 double bond is twisted 88° out of the plane of the 3(2*H*)-furanone ring. The ring-contraction precursor 10-membered macrocyclic sulfide was sequentially treated with 6N HCl, Oxone and Amberlyst 15 resin to afford the corresponding sulfone. The chlorination of this sulfone took place exclusively at the more substituted α-position, and upon treatment with a strong base, the rearrangement yielded the desired product in good yield.

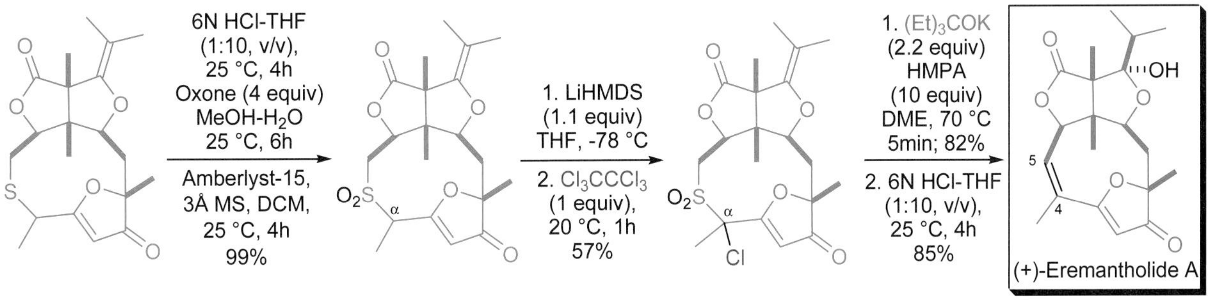

A novel benzannulation strategy featuring a [6+4] cycloaddition followed by *Ramberg-Bäcklund rearrangement* was employed for the total synthesis of (+)-estradiol by J.H. Rigby et al.[34] The higher-order cycloaddition took place between a seven-membered TMS-substituted η^6-thiepin 1,1-dioxide $(CO)_3Cr$-complex and a highly substituted diene to afford directly the bicyclic sulfone rearrangement precursor. The ring-contraction was induced by the sequential treatment with *t*-BuOK and *N*-chlorosuccinimide at very low temperatures followed by the addition of another equivalent of the base.

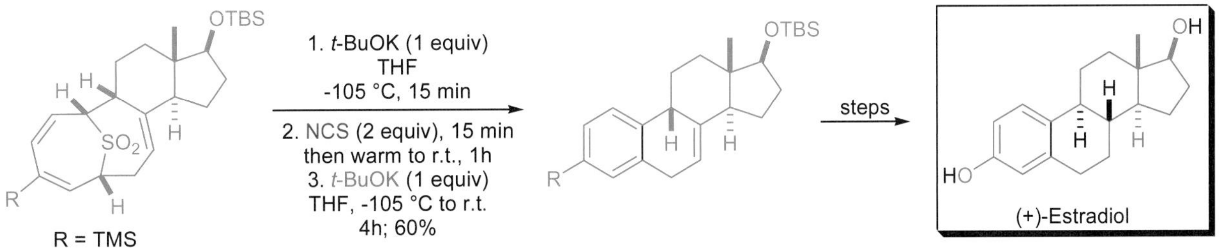

The *Ramberg-Bäcklund rearrangement* was the key step in the total synthesis of the marine alkaloid manzamine C by D.I. MaGee and E.J. Beck.[35] The azacycloundecene ring was stereoselectively formed by exposing the α-chloro sulfone to a strong base. The use of weaker bases either resulted in no reaction or gave rise to mixtures of (*E*)- and (*Z*)-alkenes.

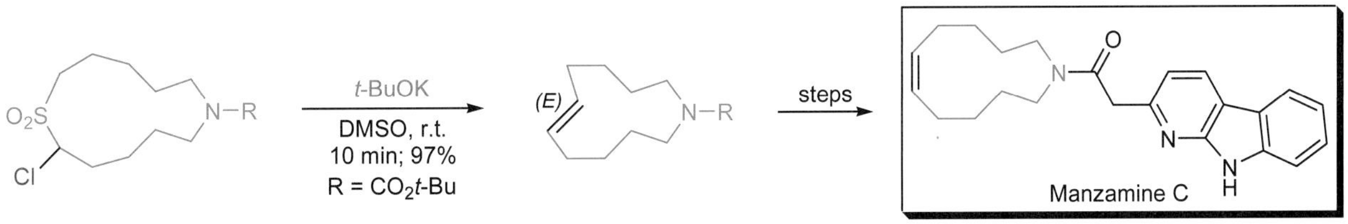

REFORMATSKY REACTION

(References are on page 661)

Importance:

[*Seminal Publication*[1]; *Reviews*[2-19]; *Modifications & Improvements*[20-37]; *Theoretical Studies*[38-40]]

In 1887, S. Reformatsky, reported that in the presence of zinc metal, iodoacetic acid ethyl ester reacted with acetone to yield 3-hydroxy-3-methylbutyric acid ethyl ester.[1] Since this initial report, the classical *Reformatsky reaction* was defined as the zinc-induced reaction between an α-halo ester and an aldehyde or ketone. The scope of the reaction, however, extends far beyond this original definition, and today, the metal-induced reaction of α-carbonyl halides with a wide range of electrophiles are referred to as the *Reformatsky reaction*. The reaction is a two stage process: first the activated zinc metal inserts into the carbon-halogen bond, and this is followed by the reaction of the zinc enolate (Reformatsky reagent) with the carbonyl compound in an *aldol reaction*. The general features of the *Reformatsky reaction* are:[5,7,9] 1) the reaction is most commonly carried out in a single step by addition of the α-halo ester and the carbonyl compound to the suspension of the activated zinc, but preforming the organozinc reagent prior to the addition of the electrophile is also possible; 2) most often ether solvents are used such as diethyl ether, tetrahydrofuran, 1,4-dioxane and dimethoxyethane, but mixtures of these solvents with aromatic hydrocarbons and more polar solvents such as acetonitrile, dimethyl formamide, dimethyl sulphoxide, and hexamethylphosphoric triamide are also used; 3) organozinc reagents can be formed from 2-bromoalkanoates, α-bromo ketones, alkyl 2-bromomethyl-2-alkenoates,[41] and alkyl 4-bromo-2-alkenoates[42]; and 4) in addition to aldehydes and ketones, *Reformatsky reagents* also react with esters,[43] acid chlorides,[44] epoxides,[43] nitrones,[45] aziridines,[46] imines,[47] and nitriles[48] (*Blaise reaction*). The scope of the *Reformatsky reaction* was considerably extended by the development zinc-activation procedures. Activated zinc metal can be formed in two ways:[7] 1) by removal of the deactivating zinc oxide layer from the metal surface employing reagents such as iodine, 1,2-dibromoethane, copper(I) halides, mercuric halides or by using zinc-copper or zinc-silver couple;[2,5,7,9,12] and 2) by reduction of zinc halides in solution by various reducing agents such as potassium[49] (*Rieke zinc*), sodium-[50] or lithium naphthalide[51] and potassium-graphite laminate[52] (C_8K) to form finely dispersed zinc metal. Metals other than zinc were also used including lithium,[22] magnesium,[20] cadmium,[28] barium,[37] indium,[21,34] germanium,[36] nickel,[31] cobalt,[35] and cerium.[24] A major breakthrough in the *Reformatsky reaction* was the application of metal salts with favorable reduction potentials, the most important ones being samarium(II) iodide,[23,32,33] chromium(II) chloride,[29] and titanium(II) chloride.[25] These reactions often can be carried out under mild conditions and afford the products with high stereoselectivity. In addition to these metal salts, cerium(III) halides,[30] disodium telluride,[30] trialkylantimony/iodine,[26,27] and diethylaluminum chloride[26,27] can also be employed. The main advantages of the *Reformatsky reaction* over the classical *aldol reaction* are the following: 1) the reaction succeeds even with highly substituted ketone substrates; 2) the ester enolate can be formed in the presence of highly enolizable aldehyde and ketone functionalities; and 3) the reaction is uniquely suited for intramolecular reactions.

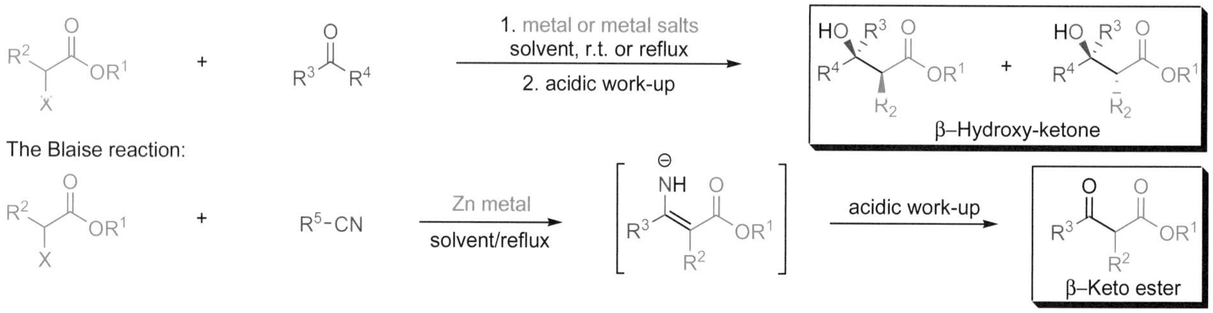

X = Cl, Br, I; R^1 = alkyl; R^2 = H, alkyl, aryl; R^3, R^4 = H, alkyl, aryl; R^5 = alkyl, aryl; solvent: Et_2O, THF, 1,4-dioxane, DME, benzene, toluene, MeCN, DMF, DMSO; metal: Zn, Mg, Cd, Ba, In, Ge, Co, Ni, Ce; metal salt: SmI_2, $CrCl_2$, $TiCl_2$, CeX_3, Na_2Te, R_3SnLi, R_3Sb/I_2, Et_2AlCl;

Mechanism: [53-57]

Spectroscopic[53,56] and crystallographic[54,55] studies of Reformatsky reagents derived from α-halo esters showed that the enolate is present in the *C*-enolate form and in ether solvents they form dimers. Enolates derived from α-halo ketones prefer the *O*-metal enolate form.[57] It is assumed, based on theoretical calculation,[38] that the zinc enolate dimers are dissociated by the action of the carbonyl compound and converted to the corresponding *O*-zinc enolates. Subsequently, the reaction goes through six-membered chairlike transition state.

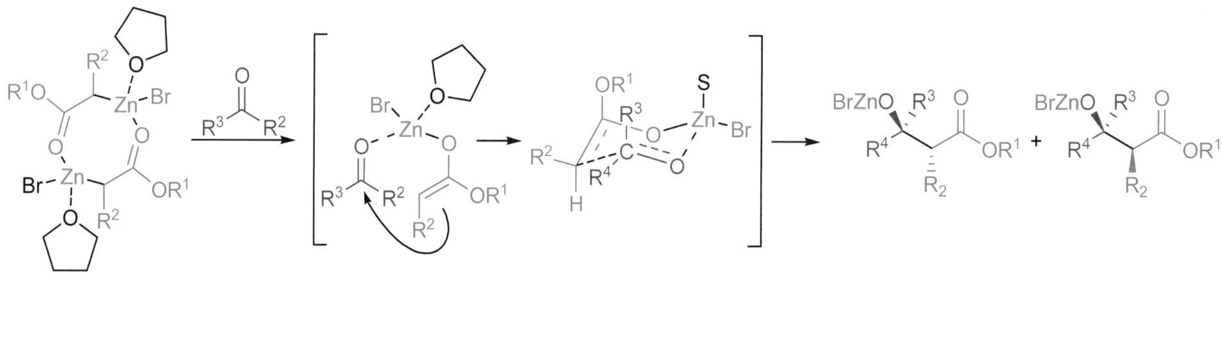

Synthetic Applications:

Cytochalasins are macrocyclic natural products possessing a broad range of biological activity. During the synthesis of C16, C18-bis-*epi*-cytochalasin D, E. Vedejs and co-workers utilized the *Reformatsky reaction* to close the twelve-membered macrocyclic ring.[58] The reaction was induced by finely dispersed zinc metal, which was formed by the reduction of $ZnCl_2$ by sodium naphthalide. The cyclization was carried out at room temperature by the slow addition of the substrate to the above metal suspension. To effect full elimination of the hydroxyl group and hydrolyze the methyl enol ether subunit, the product was treated with 10% H_2SO_4 upon work-up. Subsequent steps led to the formation of C(16),C(18)-bis-*epi*-cytochalasin D, the structure of which was proven by spectroscopic methods and X-ray crystallography.

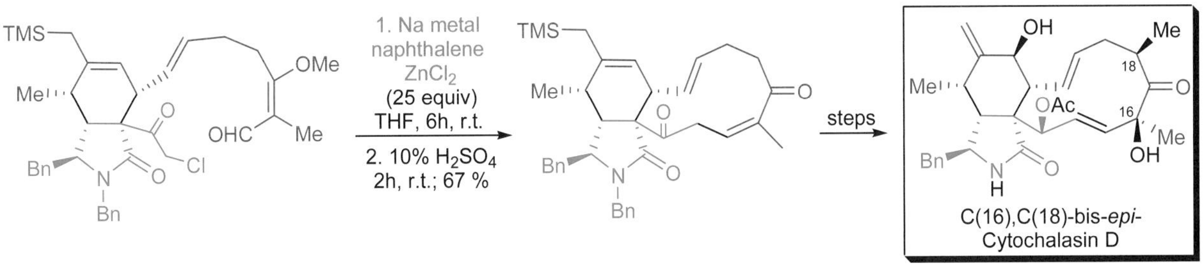

Ciguatoxin and its congeners are naturally occurring polycyclic ethers, which exhibit high affinity binding to voltage-sensitive sodium channels (VSSC). The scarcity of these compounds from natural sources and their structural complexity necessitated the construction of more accessible model systems in order to investigate their interaction with VSSC and conduct structure-activity relationship studies. In the laboratory of M Sakasi, a highly convergent synthesis of the decacyclic ciguatoxin model containing the F-M ring framework was accomplished.[59] To construct the fused oxononane ring system, a SmI_2-mediated intramolecular *Reformatsky reaction* was utilized. The reaction was carried out at -78 °C in THF to give the desired oxacyclic ring with high yield and as a single diastereomer. The resulting hydroxyl group was protected *in situ* as an acetate ester.

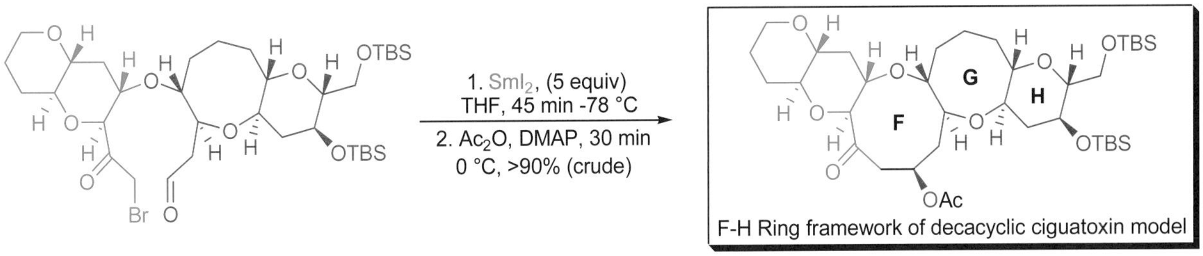

L. Wessjohn and co-workers successfully applied the $CrCl_2$-mediated *Reformatsky reaction* for the synthesis of C1-C6 fragment of epothilones.[60] In their approach, they utilized the Evans (*R*)-4-benzyl-oxazolidinone chiral auxiliary to control the absolute stereochemistry. The chromium-*Reformatsky reaction* between the (R)-4-benzyl-3-(2-bromoacetyl)-oxazolidinone and 2,2-dimethyl-3-oxo-pentanal occurred with complete chemoselection providing the product with 63% yield and as a single diastereomer.

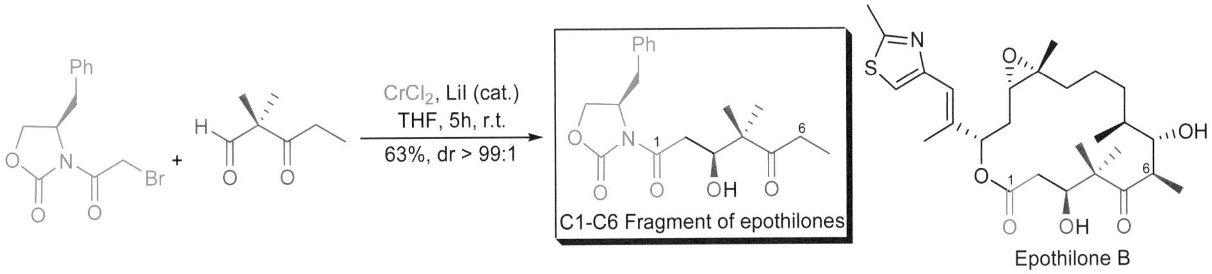

G.R. Pettit and co-workers used a novel *tetrakis*(triphenylphosphine)cobalt(0)-promoted *Reformatsky reaction* for the synthesis of a dolastatin 10 unit, dolaproine in a Boc-protected form.[61]

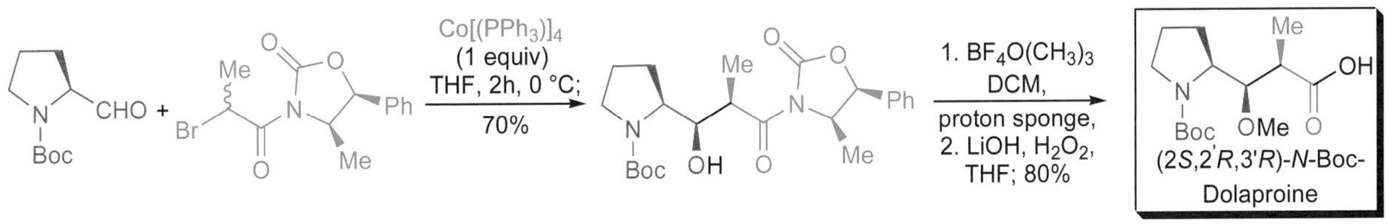

REGITZ DIAZO TRANSFER

(References are on page 662)

Importance:

[*Seminal Publications*[1-6]; *Reviews*[7-13]; *Modifications & Improvements*[14-32]]

In 1910, O. Dimroth reported that the treatment of malonamic acid methyl ester with phenyl azide yielded the corresponding 2-diazomalonamic acid methyl ester.[1] This reaction remained largely unnoticed for more than fifty years until 1964, when M. Regitz et al. investigated the reaction of arylsulfonyl azides with 1,3-diketones to afford α-diazo-β-dicarbonyl compounds.[2] The transfer of a diazo group to active methylene compounds using alkyl- or arylsulfonyl azides is known as the *Regitz diazo transfer*. The general features of the transformation are: 1) both cyclic and acyclic 1,3-diketones and β-keto esters undergo the diazo transfer in the presence of weak bases such as triethylamine, diethylamine, or piperidine, but if the acidity of the methylene group is not sufficient, the use of stronger bases (e.g., NaOEt, KOH) becomes necessary; 2) the azide reagent most often is an arylsulfonyl azide such as *p*-toluenesulfonyl azide, and these reagents can be easily prepared from the corresponding arylsulfonyl halides *via halogen-azide exchange*; 3) simple cyclic and acyclic ketones usually do not react directly with sulfonyl azides, so they need to be activated by formylation (*Claisen reaction*), and the resulting α-formyl ketone is treated with the sulfonyl azide in the presence of a base to give the corresponding α-diazo ketones (*deformylative diazo transfer*);[15] 4) when the substrate is base-sensitive, instead of formylation, trifluoroacetylation can be used, which improves the yield of the diazo ketone considerably;[19] and 5) the side product of the reaction is a sulfonamide which in some cases is fairly difficult to remove from the reaction mixture (especially *p*-$TsNH_2$), so several water-soluble and lipophilic analogues have been developed.[14] The product α-diazo carbonyl compounds are versatile intermediates and can be used in the following applications:[19] 1) *Wolff rearrangement* of α-diazo ketones to give ketenes and products derived from ketenes; and 2) transition metal catalyzed C-H, N-H, O-H insertion reactions and cyclopropanations.

Dimroth (1910):

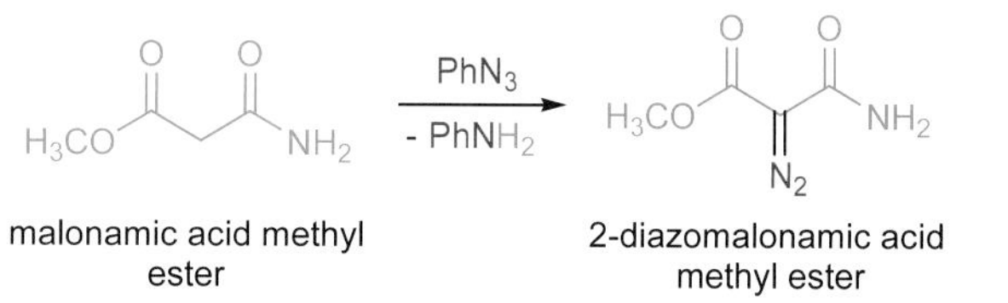

Regitz (1964):

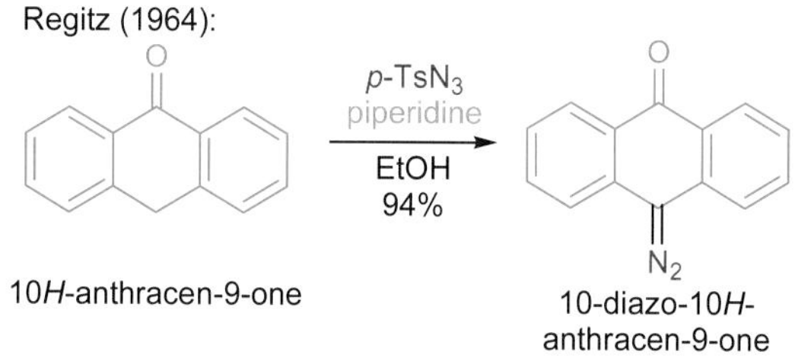

Regitz diazo transfer:

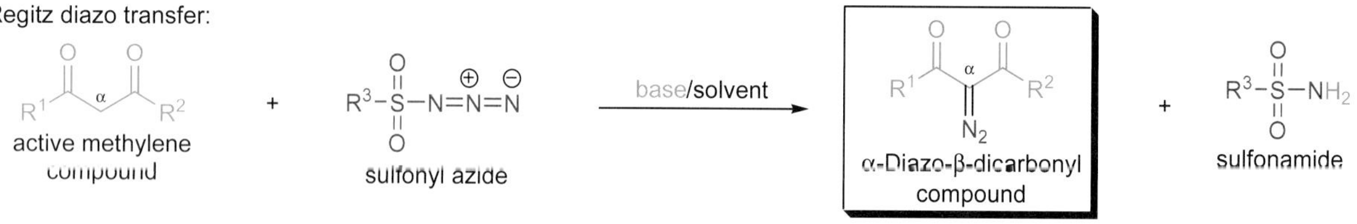

Deformylative diazo transfer:

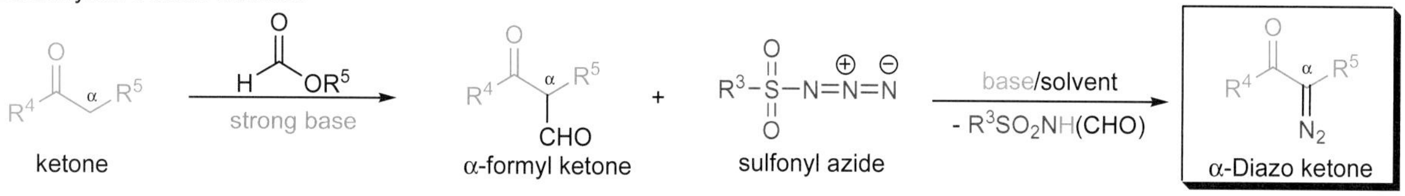

R^{1-2} = aryl, alkyl, *O*-alkyl, *O*-aryl, NH_2, NR_2; R^3 = Me, *p*-tolyl, *p*-CO_2H-phenyl; R^4 = alkyl, aryl; R^5 = H, alkyl, aryl; base: piperidine, NEt_3, $HNEt_2$, KOEt, KOH; solvent: MeOH, EtOH, EtOH-H_2O, Et_2O, CH_2Cl_2, $CHCl_3$, acetonitrile

Mechanism:[7,9,33]

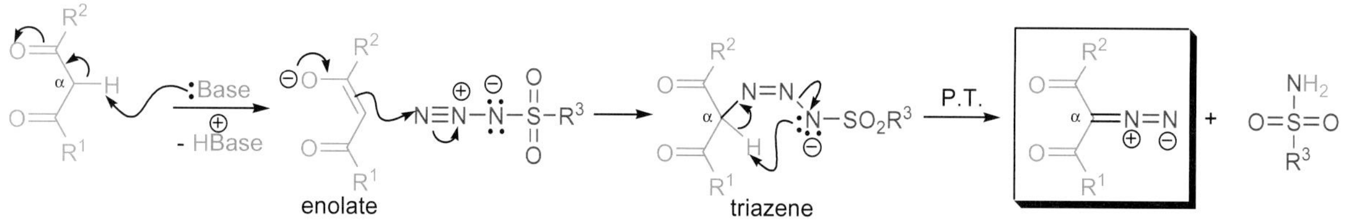

Deformylative diazo transfer:

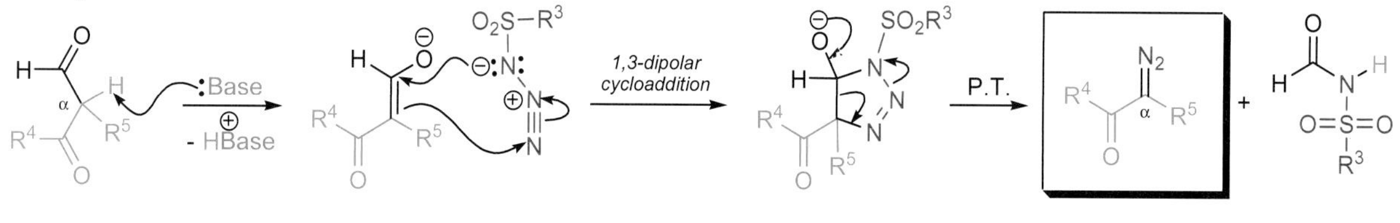

Synthetic Applications:

In the laboratory of A. Padwa, a novel synthetic approach to the fully functionalized core of lysergic acid was developed utilizing an *intramolecular isomünchone cycloaddition* pathway.[34] The key cycloaddition precursor diazo imide was prepared using the standard *Regitz diazo tranfer* conditions. The diazo imide then was heated with catalytic amounts of rhodium(II)-perfluorobutyrate in dichloromethane to afford the desired cycloadduct as a single diastereomer and in excellent yield. The only reason the authors were not able to complete the total synthesis of lysergic acid was that they could not affect the isomerization of the double bond between the two six-membered rings.

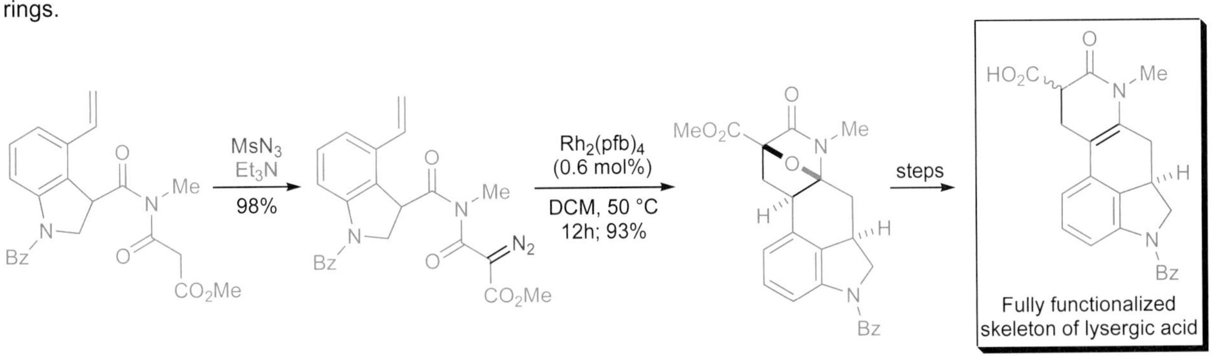

A versatile stereoselective synthesis of *endo,exo*-furofuranones was accomplished by R.C.D. Brown and co-workers.[35] One of the key steps was a Rh(II)-catalyzed C-H insertion reaction and the required diazo lactone was prepared *via* the *Regitz diazo transfer* reaction. The 2-acetyl substituted lactone substrate proved to be recalcitrant toward the *deacylative diazo transfer* under standard conditions. Eventually the authors decided to use the very reactive triflyl azide (TfN_3), which was generated *in situ* under phase-transfer conditions to afford the desired α-diazo lactone. The C-H insertion product was then converted to (+)-methylxanthoxylol.

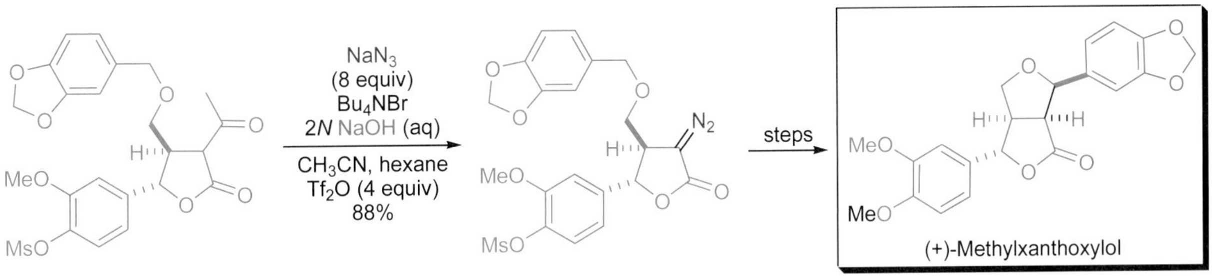

The carbocyclic [6-7] core of guanacastepenes was prepared by. D. Trauner et al. using the intramolecular reaction between carbenoids derived from diazo carbonyl compounds and furans.[36] The required diazo carbonyl substrate was synthesized using *p*-acetamidobenzenesulfonyl azide (*p*-ABSA) as the diazo-donor component in the *Regitz diazo transfer* reaction.

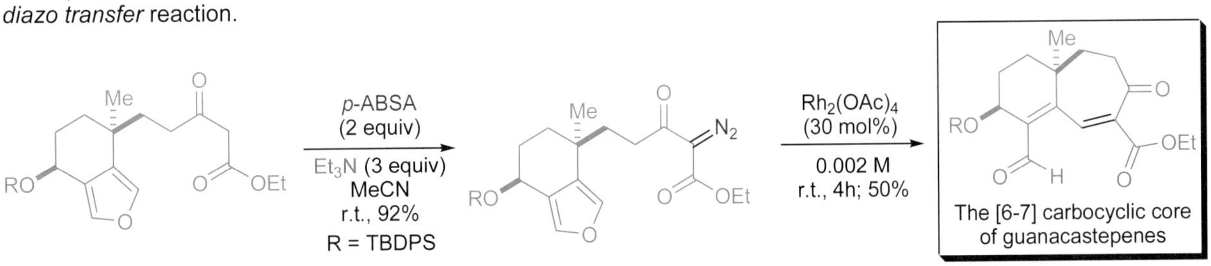

N-Alkyl substituted pyridones are known to exhibit both antibacterial and antifungal activity. The pyridone acid A58365A is a potent angiotensin-converting enzyme inhibitor and it was synthesized in the laboratory of A. Padwa using a *[3+2] cycloaddition* of a phenylsulfonyl substituted isomünchone intermediate with methyl vinyl ketone.[37] The isomünchone intermediate was generated from the corresponding diazo imide which was prepared *via* a *Regitz diazo transfer* reaction.

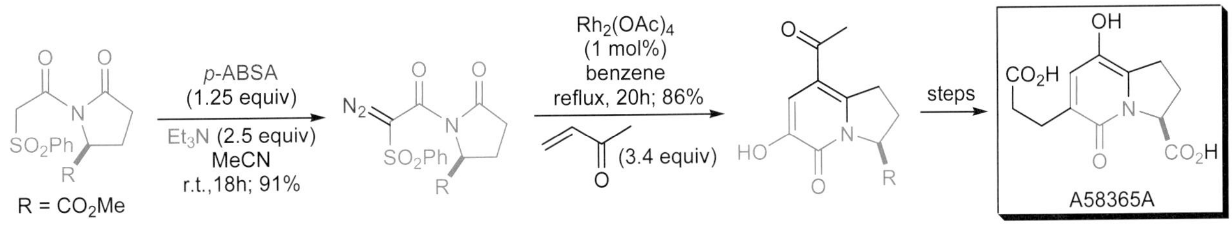

RILEY SELENIUM DIOXIDE OXIDATION

(References are on page 663)

Importance:

[*Seminal Publications*[1,2]; *Reviews*[3-9]; *Modifications & Improvements*[10-22]; *Theoretical Studies*[23]]

In 1932, H.L. Riley and co-workers reported the first general synthetic use of selenium dioxide (SeO_2) as an oxidant of aldehydes and ketones.[1] The various ketones and aldehydes having an α-methylene group were converted to the corresponding 1,2-dicarbonyl compounds in moderate to good yield. Since this initial discovery, the use of SeO_2 rapidly expanded, and it was shown that besides carbonyl compounds, olefinic substrates were oxidized at the allylic position (*allylic oxidation*) to the corresponding allylic alcohols or enones.[2] The oxidation of the methylene group adjacent to a carbonyl group or the double bond of olefins (allylic or benzylic position) with selenium-dioxide is collectively referred to as the Riley oxidation. The general features of these transformations are: 1) ketones and aldehydes with low molecular weights are more reactive than the higher homologs; 2) ketones with available α- and α'-positions will give rise to a mixture of regioisomers; 3) the sterically less hindered α-position is oxidized faster, therefore the methyl group of methyl ketones (R^1=H) is preferentially oxidized over the other available α-position; 4) the allylic positions in acyclic olefins are oxidized at very different rates and the reactivity depends on the substitution pattern of the substrate: a) in 1,2-disubstituted alkenes the trend is: CH > CH_2 > CH_3; b) in geminally disubstituted alkenes the trend is: CH > CH_2 > CH_3; c) in trisubstituted alkenes the oxidation takes place at the more substituted end of the double bond and the trend is CH_2 > CH_3 > CH; d) terminal olefins yield primary allylic alcohols due to the allylic rearrangement of the double bond; 5) the oxidation of acyclic olefins primarily gives rise to (*E*)-allylic alcohols; 6) the oxidation of cyclic olefins occur in the ring and α to the more substituted carbon of the double bond rather than in the side chain; 7) in cyclic olefins where the double bond is unsubstituted the reactivity trend is: CH_2 > CH; 8) for bicyclic olefins in which none of the rings contain more than 7 carbon atoms, the oxidation will not take place at the bridgehead position (Bredt's rule); 9) *gem*-dimethyl olefins exclusively give rise to the (*E*)-allylic alcohols or (*E*)-α,β-unsaturated aldehydes; and 10) rearrangement may occur if the preferred allylic position is adjacent to a quaternary carbon or a cyclopropyl ring.

Selenium dioxide oxidation of ketones and aldehydes (Riley, 1932):

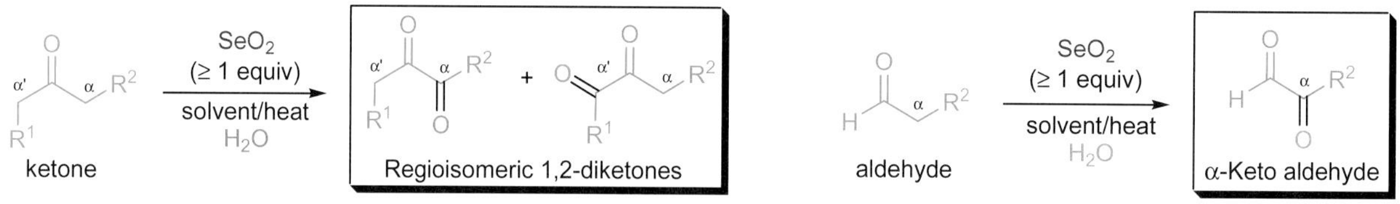

Selenium dioxide oxidation of olefins (Guillemonat, 1939):

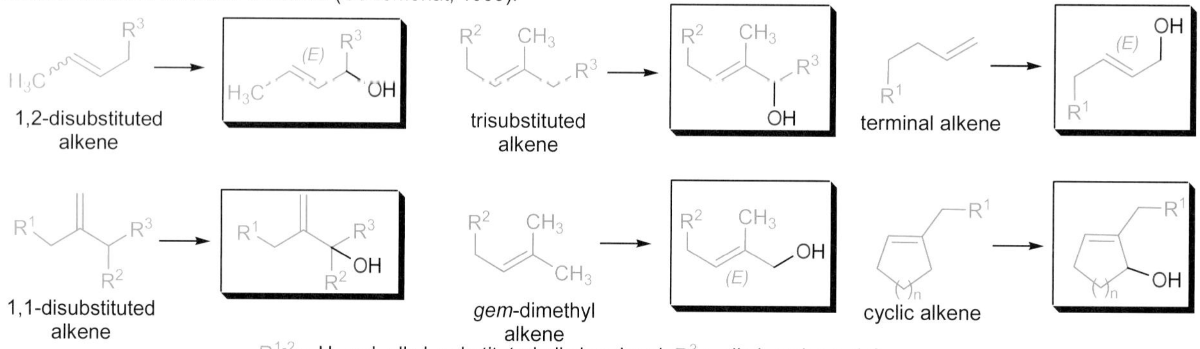

R^{1-2} = H, aryl, alkyl, substituted alkyl and aryl; R^3 = alkyl, aryl; n = 1-3

Mechanism: [24-41]

Oxidation of carbonyl compounds:

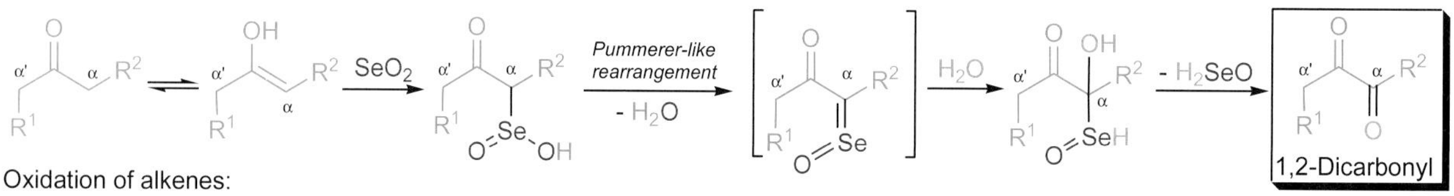

Oxidation of alkenes:

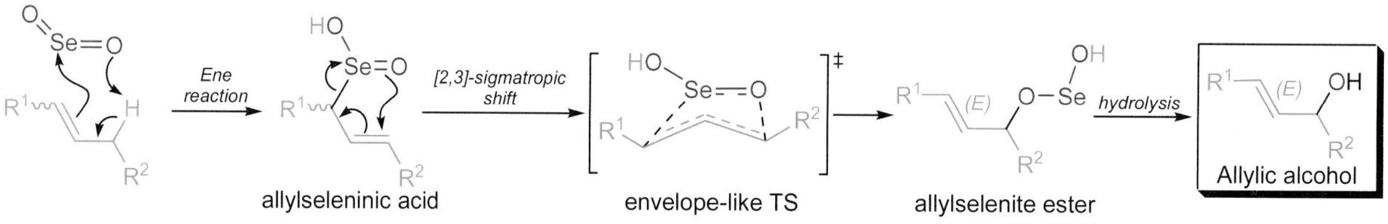

Synthetic Applications:

The antiviral natural product hamigeran B has a unique tricarbocyclic skeleton in which the aromatic nucleus is fused to a hydrindane framework bearing three stereogenic centers. G. Mehta and co-workers accomplished the total synthesis of 6-*epi*-hamigeran B by using an *intramolecular Heck reaction* as the key step to form the six-membered middle ring.[42] At the final stages of the synthesis, the introduction of the 1,2-diketone moiety was performed by using the *Riley oxidation*. The cyclohexanone had only one available α-position, so the oxidation proceeded cleanly and in high yield.

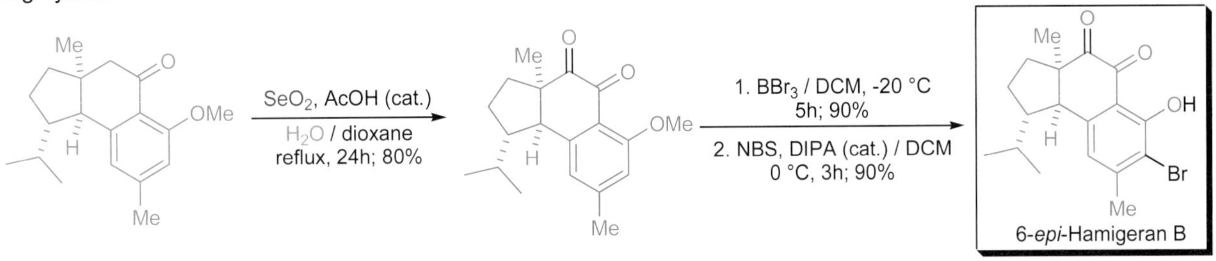

In the laboratory of T.-J. Lu, a highly stereoselective method for the asymmetric synthesis of α-amino acids was developed by the alkylation of a chiral tricyclic iminolactone derived from (+)-camphor.[43] The iminolactone can be considered a glycine equivalent. The synthesis commenced with the *Riley oxidation* of (+)-camphor to obtain the corresponding (+)-camphorquinone. Amino acids are obtained by first alkylating the α-position of the lactone with various alkyl halides and then hydrolyzing the monosubstituted products. The advantage of this technique was that the chiral auxiliary could be fully recovered without the loss of any optical activity.

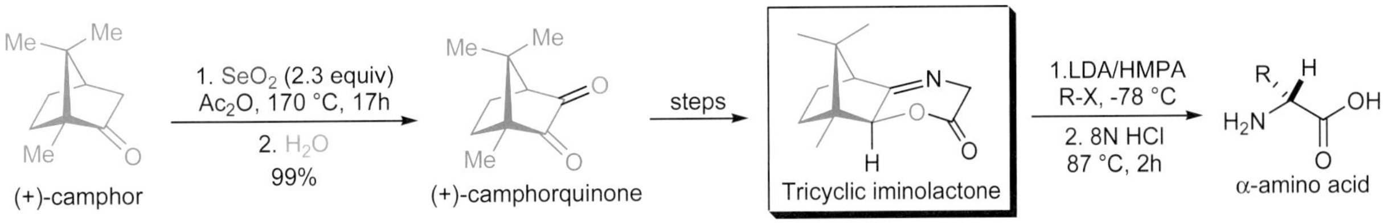

The first total syntesis of cristatic acid, a potent antibiotic against Gram-positive bacteria, was reported by A. Fürstner et al.[44] The prenylated aromatic substrate (trisubstituted *gem*-dimethyl alkene) was subjected to a SeO_2-catalyzed allylic oxidation to obtain stereospecifically the (*E*)-allylic alcohol.

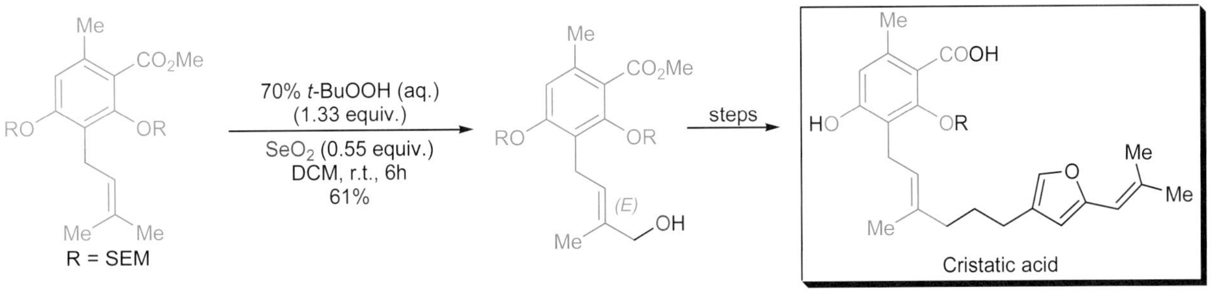

During the enantioselective total synthesis of miroestrol by E.J. Corey and co-workers, the introduction of a hydroxyl group was required at one of the bridgehead positions.[45] This position was α to a ketone and was also the allylic position to a double bond. The oxidation was effected by selenium dioxide/*tert*-butyl hydroperoxide at 25 °C.

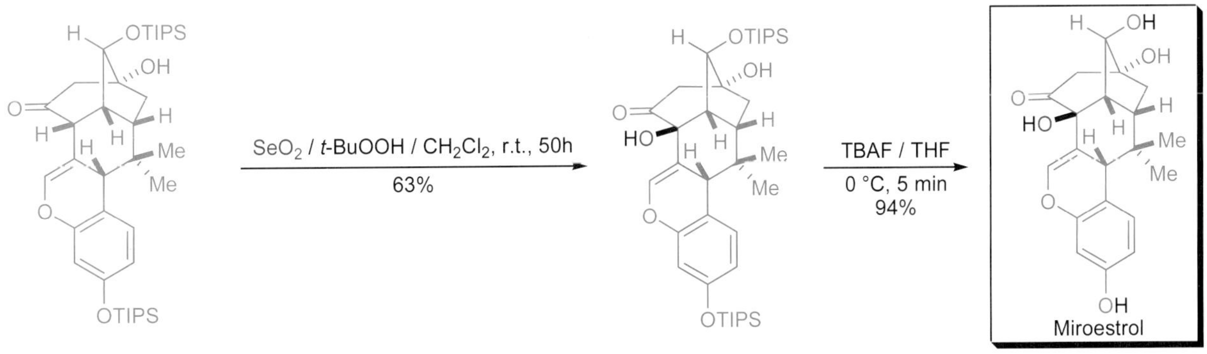

RITTER REACTION

(References are on page 664)

Importance:

[*Seminal Publications*[1,2]; *Reviews*[3-10]; *Modifications & Improvements*[11-26]; *Theoretical Studies*[27,28]]

In 1948, J.J. Ritter and P.P. Minieri reported that treatment of nitriles with alkenes or tertiary alcohols under acidic conditions resulted in the formation of *N-tert*-alkylamides.[1,2] When hydrogen cyanide was used as the nitrile component, *N*-tert-alkyl formamides were obtained, which could be easily hydrolyzed with base to give the corresponding *tert*-alkylamines.[1] The formation of *N*-alkyl carboxamides from aliphatic- or aromatic nitriles and carbocations is known as the *Ritter reaction*. Since its discovery the *Ritter reaction* has enjoyed an enormous success, and it is widely used for the preparation of acyclic amides as well as heterocycles (e.g., lactams, oxazolines, dihydroisoquinolines, etc.). The general features of this transformation are:[5,8] 1) the carbocation can be generated in a variety of ways from tertiary-, secondary, or benzylic alcohols, alkenes or alkyl halides; 2) the classical reaction conditions involve the dissolution of the nitrile substrate in the mixture of acetic acid and concentrated sulfuric acid followed by the addition of the alcohol or alkene at slightly elevated temperatures (50-100 °C); 3) alcohols that are easily ionized (e.g., 2° and 3° alcohols, benzylic alcohols) give the best results; 4) 1,1-disubstituted alkenes give rise to regioisomerically pure products, but with 1,2-disubstituted alkenes a mixture of regioisomers may be formed; 5) the initially formed carbocation (which can be obtained from a large number of different functionalities)[5,8] may undergo a *Wagner-Meerwein rearrangement* to give rise to the most stable carbocation before reacting with the nitrile; 6) besides protic acids, Lewis acids (e.g., $SnCl_4$, $BF_3{\cdot}OEt_2$, $AlCl_3$, etc.) have been successfully employed in the *Ritter reaction* to generate the required carbocations; 7) the structure of the nitrile component can be varied widely and most substrates containing a cyano group will undergo the reaction, so, for example, besides aliphatic and aromatic nitriles, compounds like cyanogen and cyanamide will also react; and 8) the nitrile substrate may not contain acid-sensitive functional groups that would be destroyed under the strongly acidic reaction conditions, but modifications (*Ritter-type reactions*) that proceed under neutral conditions expanded the scope of the substrates.

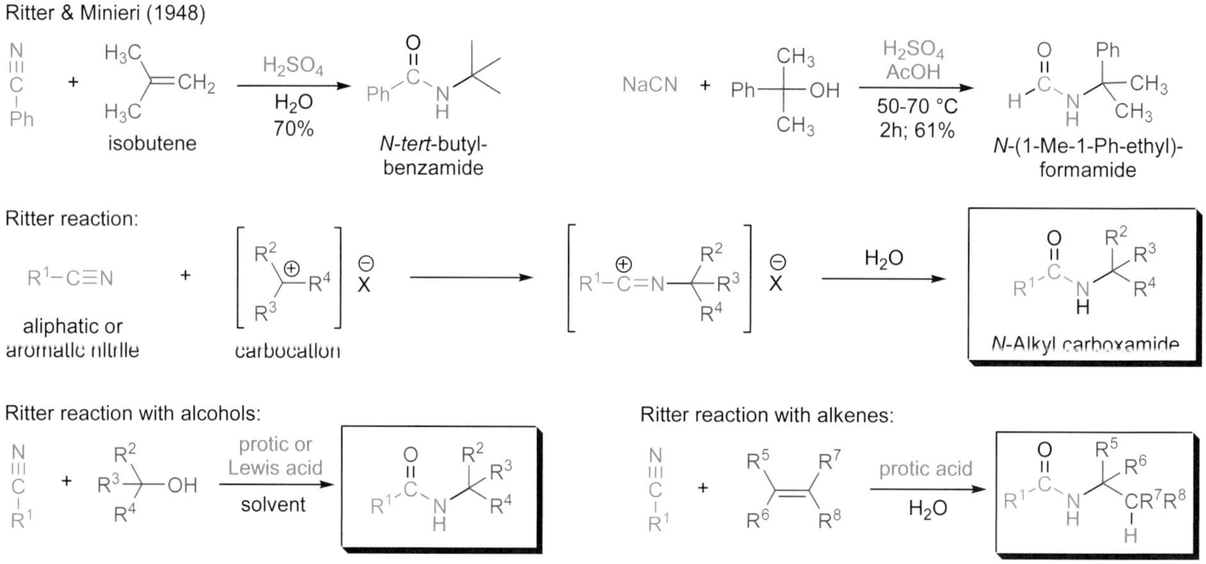

R^1 = H, 1°, 2° or 3° alkyl, alkenyl, alkynyl, aryl, heteroaryl; R^2 = alkyl, aryl, heteroaryl; R^{3-4} = H, alkyl, aryl; R^{5-6} = alkyl, aryl; R^{7-8} = H, alkyl, aryl; protic acid: H_2SO_4, $HClO_4$, PPA, HCO_2H, RSO_3H; Lewis acid: $AlCl_3$, $BF_3{\cdot}OEt_2$, $SbCl_5$; solvent: glacial AcOH, H_2SO_4 (conc.), Ac_2O, $CHCl_3$, CH_2Cl_2, n-Bu_2O, $PhNO_2$

Mechanism: [29-40]

The mechanism of the *Ritter reaction* has been intensely studied. When alcohols are used to generate the carbocation, the hydroxyl group is protonated then under the reaction conditions the C-O bond is heterolytically cleaved to generate a carbocation. This cation is then attacked by the nitrogen atom of the nitrile to form a nitrilium ion, which upon reacting with the conjugate base of the acid (hydrogen sulfate anion in the scheme) gives rise to an imidate. Finally, hydrolysis produces the desired *N*-alkyl carboxamide.

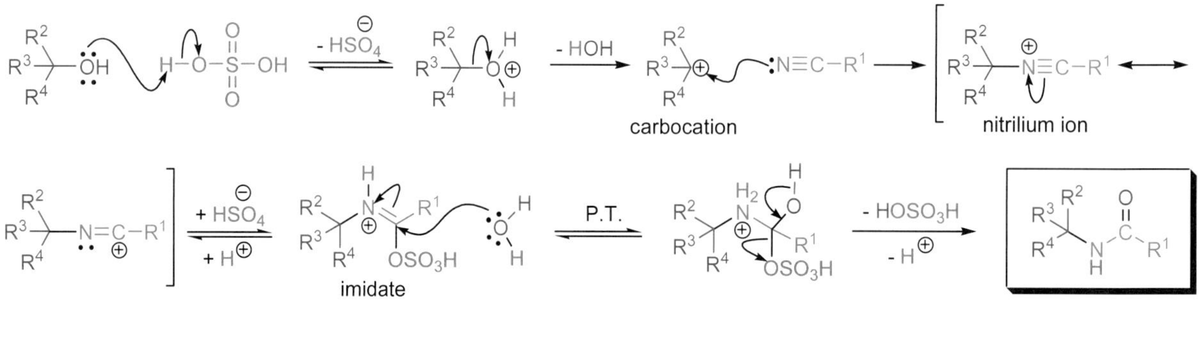

Synthetic Applications:

The enantioselective biomimetic total synthesis of the alkaloid (+)-aristotelone was accomplished by C.H. Heathcock and co-workers.[41] The synthetic sequence commenced with a $Hg(NO_3)_2$-mediated *Ritter reaction* between (1S)-(–)-β-pinene and 3-indolylacetonitrile. Upon protonation, the pinene underwent a *Wagner-Meerwein rearrangement* to generate a tertiary carbocation which reacted with the cyano group. The initially formed imine product was reduced to the corresponding amine by sodium borohydride in methanol.

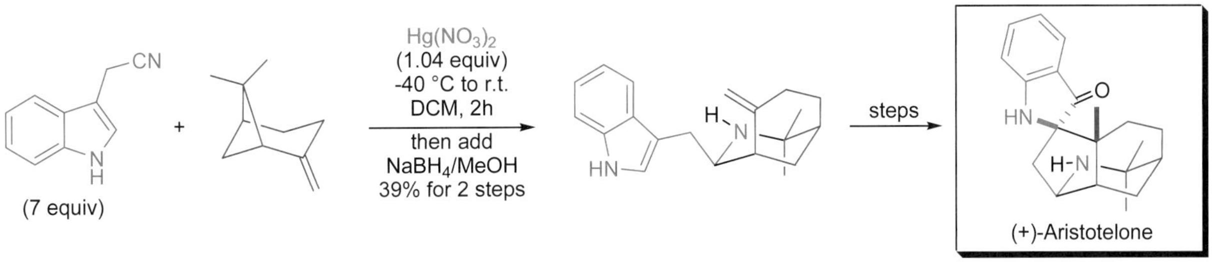

In the laboratory of T.-L. Ho, the total synthesis of the novel marine sesquiterpene (±)-isocyanoallopupukeanane was completed.[42] In the endgame of the synthesis, it was necessary to install the isocyano group onto the tricyclic trisubstituted alkene substrate so that it will occupy the more substituted carbon atom (according to Markovnikov's rule). The *Ritter reaction* was chosen to form the required carbon-nitrogen bond. The alkene substrate was dissolved in glacial acetic acid and first excess sodium cyanide followed by concentrated sulfuric acid was added at 0 °C. The reaction mixture was stirred at ambient temperature for one day and then was subjected to aqueous work-up. The product *N*-alkyl formamide was subsequently dehydrated with tosyl chloride in pyridine to give rise to the desired tertiary isocyanide which indeed was identical with the natural product.

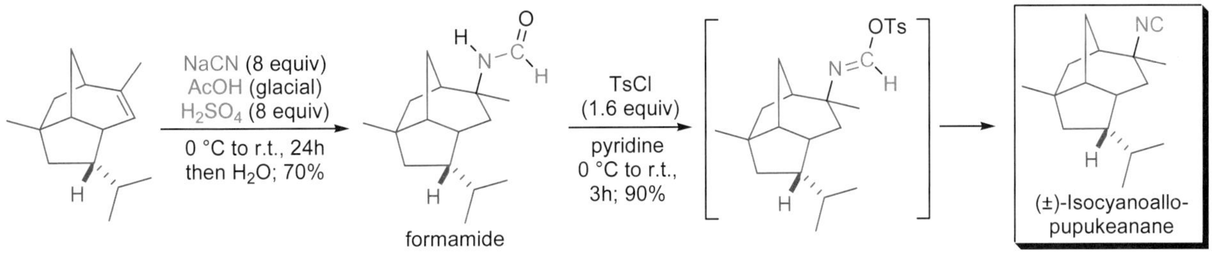

A *modified Ritter reaction* was used by Y.L. Janin et al. for the preparation of electron rich 1-aryl-3-carboxylisoquinolines, which are considered to be the electron-rich analogues of PK 11195, a falcipain-2 inhibitor.[26] Interestingly, the standard *Ritter reaction* conditions (strong acid) led to extensive decomposition of both starting materials, but the use of HBF_4 in ether gave rise to the desired dihydroisoquinoline, albeit in poor yield.

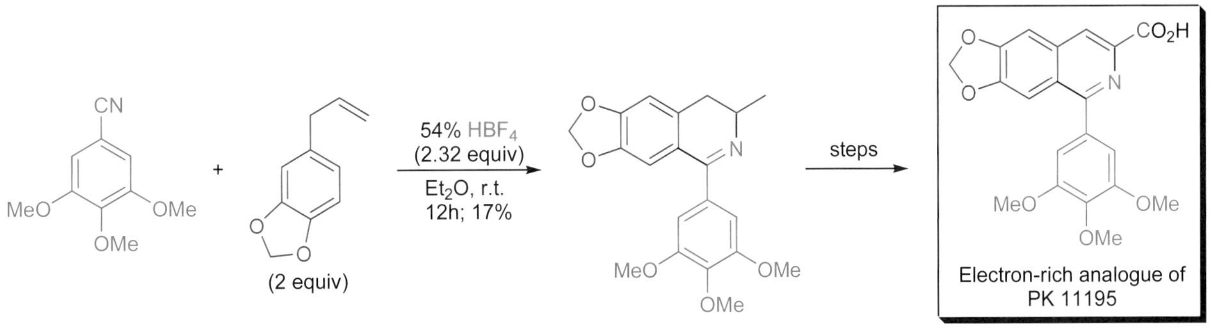

The *intramolecular Ritter reaction* was utilized by F. Compernolle and co-workers for the synthesis of a potential dopamine receptor ligand.[43] The six-membered lactam ring was formed upon treatment of the tertiary benzylic alcohol substrate with methansulfonic acid. The benzylic carbocation was captured by the nitrogen of the cyano group.

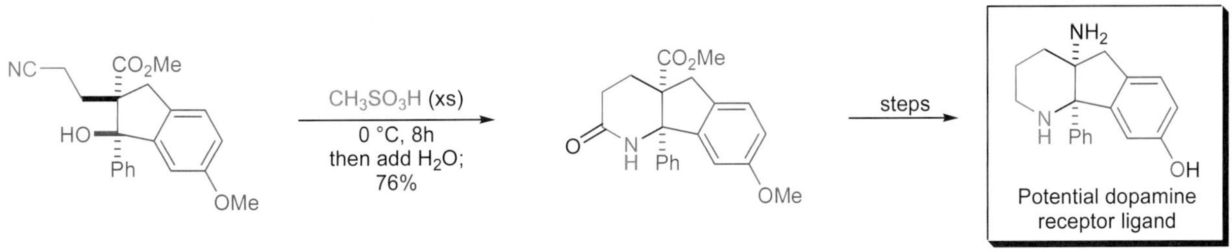

ROBINSON ANNULATION
(References are on page 665)

Importance:

[*Seminal Publications*[1,2]; *Reviews*[3-9]; *Modifications & Improvements*[10-36]]

In 1935, R. Robinson and W.S. Rapson were preparing substances related to the sterols when they found that the sodium enolate of cyclohexanone reacted with various acyclic and cyclic α,β-unsaturated ketones to afford substituted cyclohexenones.[1] Robinson recognized the generality of this transformation, which was quickly adapted by the synthetic community, and today it is widely used in the synthesis of complex natural products. The reaction of a ketone (most often a cyclic one) with an α,β-unsaturated ketone to give a substituted cyclohexenone derivative is known as the *Robinson annulation*. The general features of the reaction are: 1) it is a combination of three reactions: *Michael addition*, *intramolecular aldol reaction*, and dehydration; 2) it can be both acid- and base-catalyzed, but predominantly the reaction is conducted under basic conditions; 3) acyclic enones and cyclic ketones afford bicyclic enones, whereas cyclic enones and cyclic ketones give rise to polycyclic fused enones; 4) methyl vinyl ketone (MVK) and its various derivatives and surrogates are used most often as the enone component; 5) can be conducted as a one-pot process, but yields tend to be higher when the Michael adduct is isolated and then subjected to the *aldol reaction*; 6) the alkylation of an unsymmetrical ketone occurs regioselectively at the most substituted α-position unless severe steric interference dictates otherwise; 7) regioselective cyclization can also be achieved by using pre-formed enolates or enamines under non-equilibrium conditions; 8) the annulation can generate as many as five stereocenters, but in the dehydration step two of these chiral centers are lost; 9) the relative stereochemistry between R^3 and R^7 (*cis* or *trans*) is dependent on the reaction conditions (e.g., solvent);[11] and 10) the enantioselective version is known as the *Hajos-Parrish reaction*.[10,13]

Robinson & Rapson (1935):

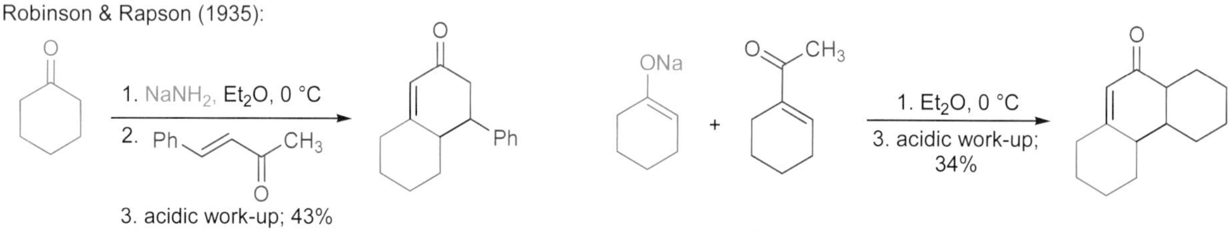

Robinson annulation:

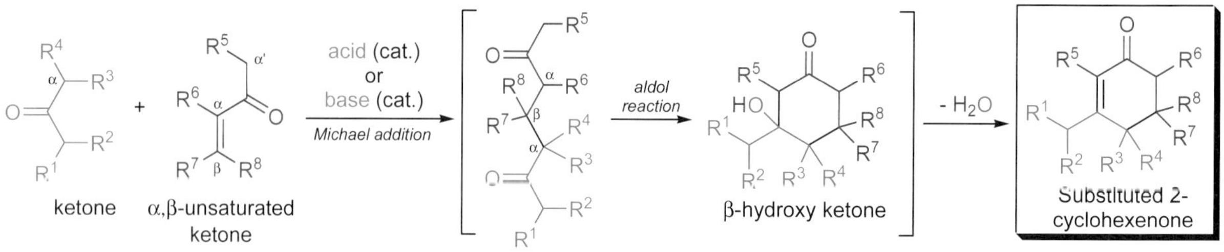

R^{1-4} = H, alkyl, aryl; R^5 = H, alkyl, aryl; R^6 = H, alkyl, aryl, SiR_3; R^{7-8} = H, alkyl, aryl

Mechanism: [11,15,4]

The *Robinson annulation* has three distinct steps: the *Michael addition* of the enol or enolate across the double bond of the α,β-unsaturated ketone to produce a 1,5-diketone (Michael adduct), followed by an *intramolecular aldol reaction*, which affords a cyclic β-hydroxy ketone (keto alcohol), and finally a base-catalyzed dehydration which gives rise to the substituted cyclohexenone. An alternative mechanism *via* disrotatory electrocyclic ring closure is possible.[11]

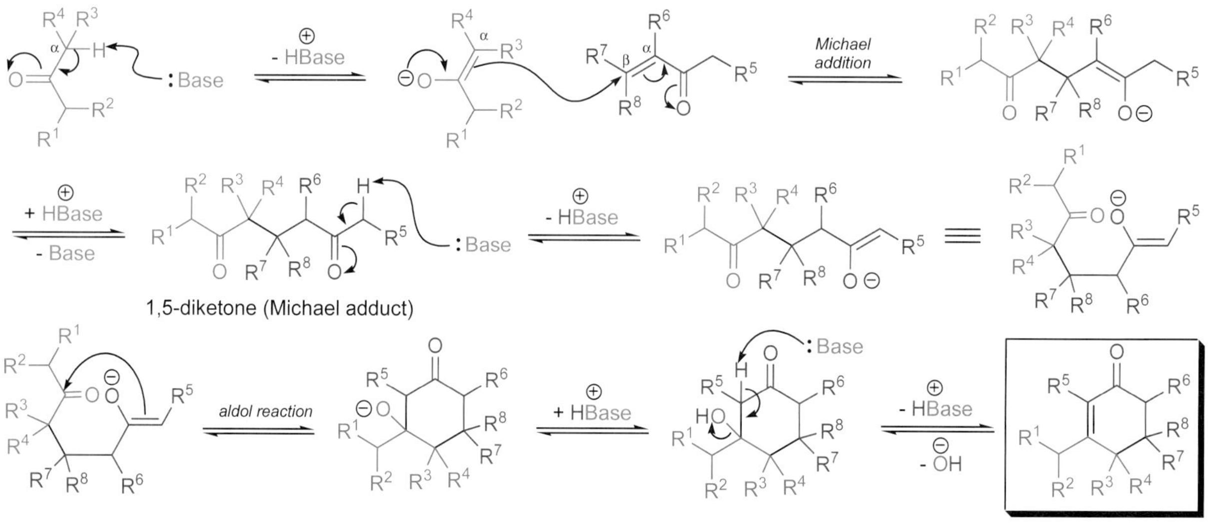

Synthetic Applications:

A *conjugate cuprate addition-Robinson annulation* sequence was utilized in the highly stereoselective total synthesis of hispidospermidin by S.J. Danishefsky et al.[37] It is a well-known fact that the MVK has a great tendency to polymerize under aprotic basic conditions that are used when the integrity of the enolate reaction partner has to be maintained. In order to avoid complications arising from the likely polymerization of MVK, α-trimethylsilyl methyl vinyl ketone (a base-stable surrogate of MVK developed by G. Stork and co-workers[12,14]) was chosen as the reaction partner. The 2-substituted cylopentenone was treated with lithium dimethyl cuprate, and the resulting enolate was trapped with α-trimethylsilyl MVK in a *Michael addition*. The crude Michael adduct was refluxed with aqueous KOH in methanol, which resulted in the desired hydrindenone as a single diastereomer.

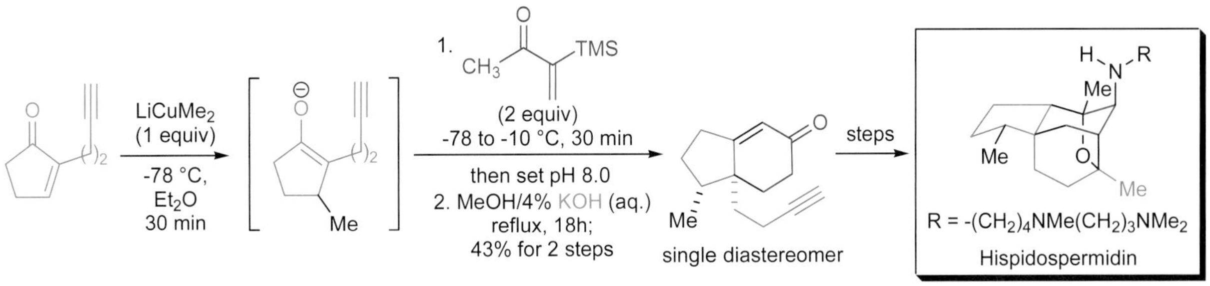

In the laboratory of J.D. White, the asymmetric total synthesis of (+)-codeine was accomplished.[38] The *Robinson annulation* was the method of choice to build a phenanthrenone precursor starting from a substituted tetralone derivative. As it is usually the case, the isolation of the Michael adduct allowed the *intramolecular aldol reaction* to proceed cleanly and to afford a higher yield of the annulated product.

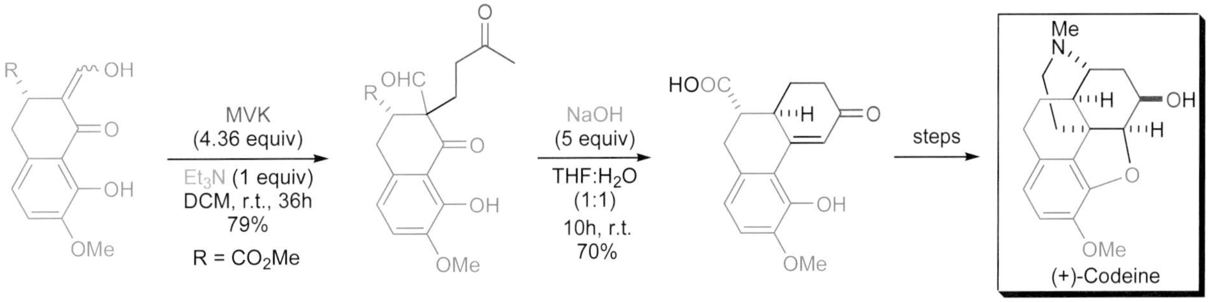

The *Hajos-Parrish reaction* can be regarded as the enantioselective version of the *Robinson annulation*. In the early stages of the synthetic effort targeting the mixed polyketide-terpenoid metabolite (–)-austalide B, L.A. Paquette and co-workers used this transformation to prepare the key bicyclic precursor in enantiopure form.[39] Ethyl vinyl ketone was reacted with 2-methyl-1,3-cyclopentanedione in the presence of catalytic amounts of L-valine to afford the bicyclic diketone with a 75% *ee*.

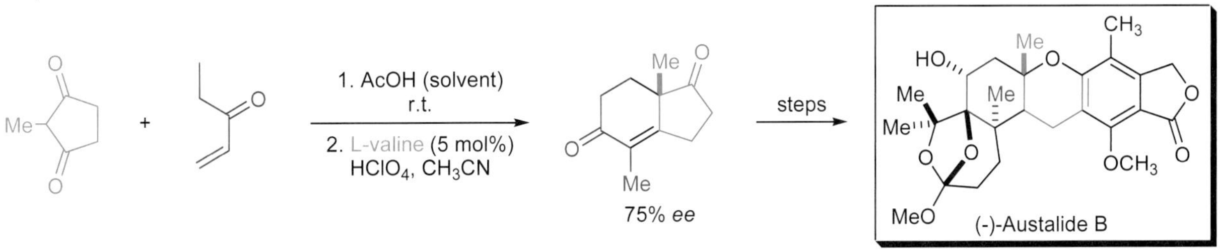

A novel variant of the *Stork-Jung modified Robinson annulation* was developed and applied to the formal total synthesis of (±)-guanacastepene A by the research group of B.B. Snider.[40] Instead of using MVK directly, they prepared the necessary 1,5-diketone by alkylating the ketone with an allylsilane and generating the ketone functionality *via* a *Fleming-Tamao oxidation*.

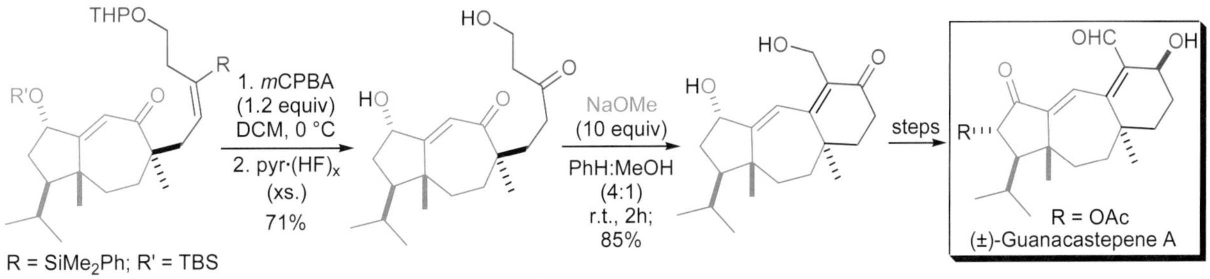

ROUSH ASYMMETRIC ALLYLATION

(References are on page 666)

Importance:

[*Seminal Publications*[1-5]; *Reviews*[6-11]; *Modifications & Improvements*[12-22]; *Theoretical studies*[23-27]]

The first example of the enantioselective synthesis of homoallylic alcohols *via* chiral nonracemic allylboronic esters was reported by R. W. Hoffmann in 1978.[1] He studied the reaction between (+)-camphor derived allylboronic ester and a series of aliphatic aldehydes. The resulting homoallylic alcohols formed with excellent yield but moderate enantioselectivity. A few years later, W.R. Roush examined the reaction of allylboronates with aldehydes and he found that diisopropyltartrate ester derived allylboronates reacted with aldehydes to give the products in good yield and enantioselectivity.[2-5] This reaction is referred to as the *Roush asymmetric allylation.* The synthesis of these allylboronates may be achieved by esterification of allylboronic acid or by transesterification of triisopropyl-allylboronate with the appropriate tartrate ester. The general features of the allylation reaction are: 1) the reaction is typically carried out in toluene, in the presence of 4Å molecular sieves at -78 °C; 2) this method provides access to both enantiomers of the homoallylic alcohol product by selecting the proper enantiomer of the diisopropyltartrate ester for the preparation of the reagent; 3) this reaction exhibits high levels of matched and mismatched diastereoselection in the case of chiral aldehydes; 4) both aliphatic and aromatic aldehydes are suitable substrates; 5) (*E*)-crotylboronate derivatives lead to the formation of the *anti* diastereomer as the major product, while (*Z*)-crotylboronates give the *syn* product; and 6) (*E*)-crotylboronates usually exhibit higher enantioselectivities than (*Z*)-crotylboronates. In addition to the *Roush asymmetric allylation*, several other methods were developed for the asymmetric synthesis of homoallylic alcohols utilizing chiral allylboranes and allylboronates: 1) H.C. Brown reported the application of *B*-allyldiisopinocampheylborane;[13,16,19,21,22] 2) E.J. Corey described the application of 1,2-diamino-1,2-diphenylethane derived allylboranes;[18,20] 3) S. Masamune developed a method where he utilized (*E*)- and (*Z*)-crotyl-2,5-dimethylborolanes;[15] and 4) chiral nonracemic allenylboronates were also utilized to form the corresponding propargyl alcohols enantioselectively.[12,14,20]

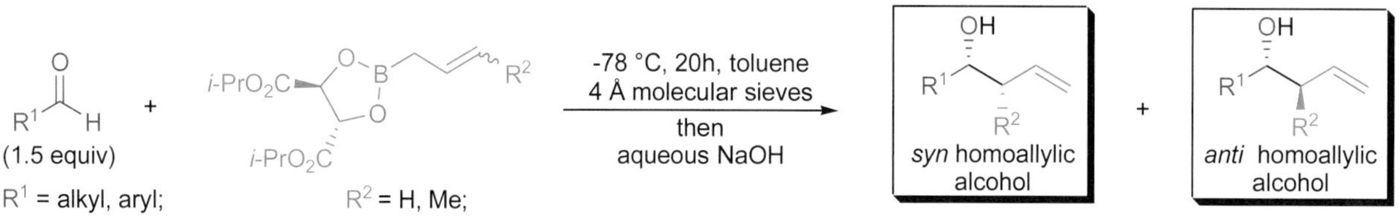

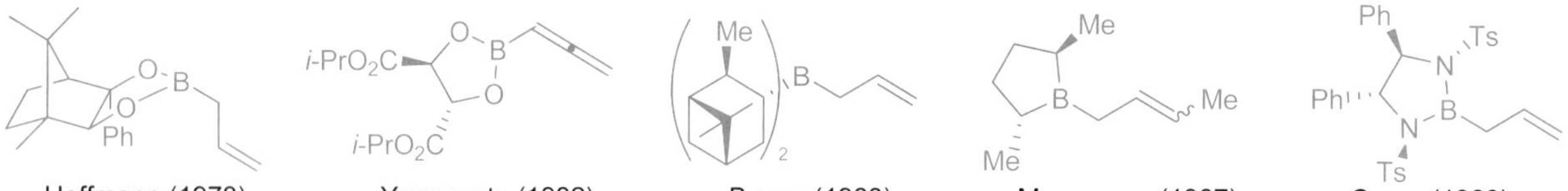

Hoffmann (1978) Yamamoto (1982) Brown (1983) Masamune (1987) Corey (1989)

Mechanism:[2]

According to Roush, the asymmetric induction can be explained by an unfavorable electronic repulsive interaction between the nonbonding electron pair of the aldehyde and ester that destabilizes transition state **B** relative to **A**.[2]

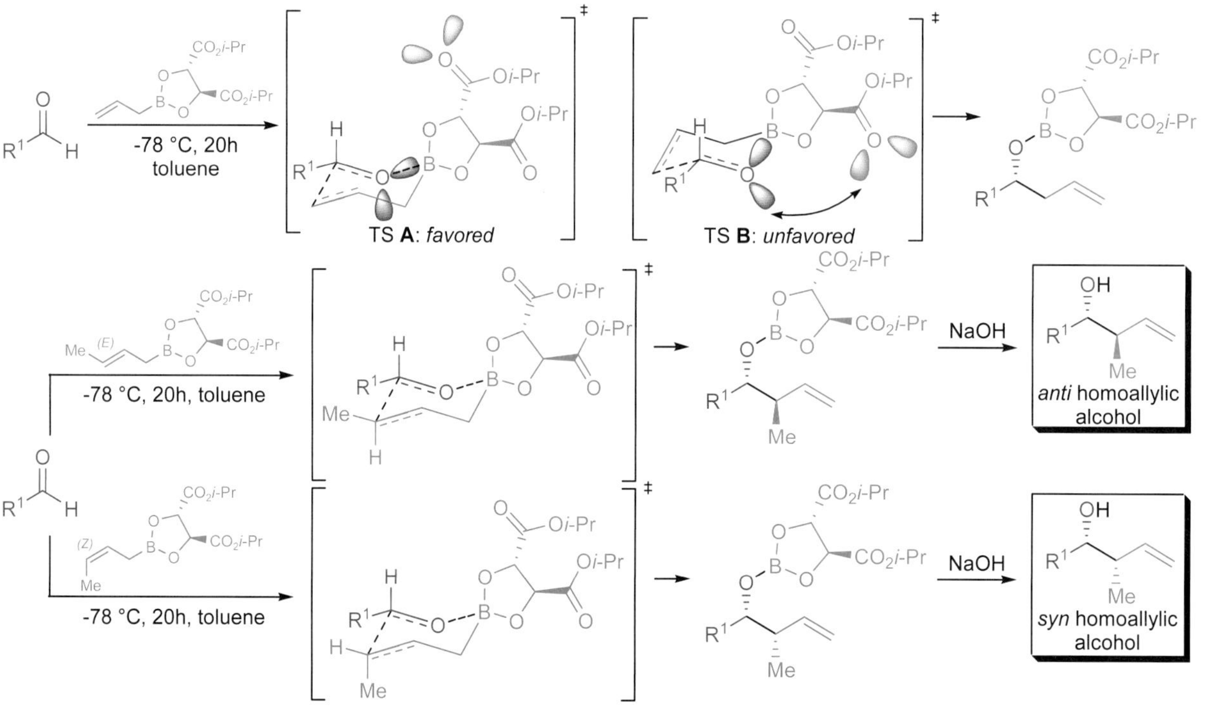

Synthetic Applications:

The total synthesis of the 20-membered macrolide *(+)-lasonolide-A* was undertaken by S.H. Kang and co-workers.[28] During the construction of the C15-C25 subunit, they utilized the *Roush asymmetric allylation* reaction to introduce the C21 and C23 stereocenters. First, (*R,R*)-diisopropyltartrate derived allylboronate was used to provide the (*S*)-homoallylic alcohol with 78% *ee*. A second asymmetric allylation was achieved utilizing the (*S,S*)-diisopropyltartrate-derived allylboronate to form the (*R*)-homoallylic alcohol with a 91% *ee*.

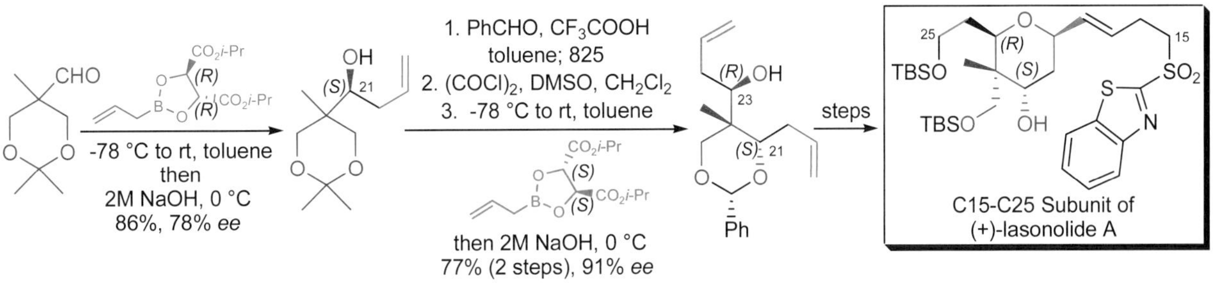

Y. Kishi and coworkers accomplished the total synthesis of *spongistatin 1*.[29] In their approach, they applied the *Roush asymmetric allylation reaction* twice during the synthesis of the C38-C51 fragment of the natural product to construct the C39, C40 and C41 stereocenters. In the first allylation, they utilized (*S,S*)-diisopropyltartrate-derived (*E*)-crotylboronate, while in the second reaction they used the (*R,R*)-diisopropyltartrate-modified allyl boronate. During their studies, they compared Roush's method with the allylation developed by H.C. Brown utilizing the corresponding crotyl- and allyldiisopinocampheylboranes. They concluded that Brown's method proceeded with higher enantioselectivity, but the ratio of the *syn* and *anti* diastereomers was higher in the *Roush asymmetric allylation*.

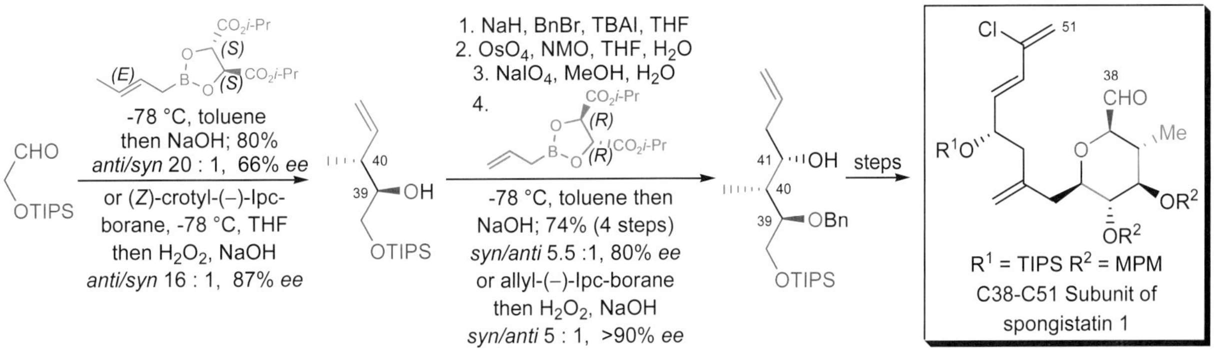

Stevastelins are depsipeptides exhibiting immunosuppressant activity. The first total synthesis of *stevastelin B* was described by Y. Yamamoto and co-workers.[30] To construct four consecutive stereocenters, the *Evans aldol reaction* and the *Roush asymmetric allylation* were utilized. In the allylation step, the authors used (*S,S*)-diisopropyltartrate-derived (*E*)-crotyl boronate. The *anti* homoallylic alcohol product formed as the only diastereomer.

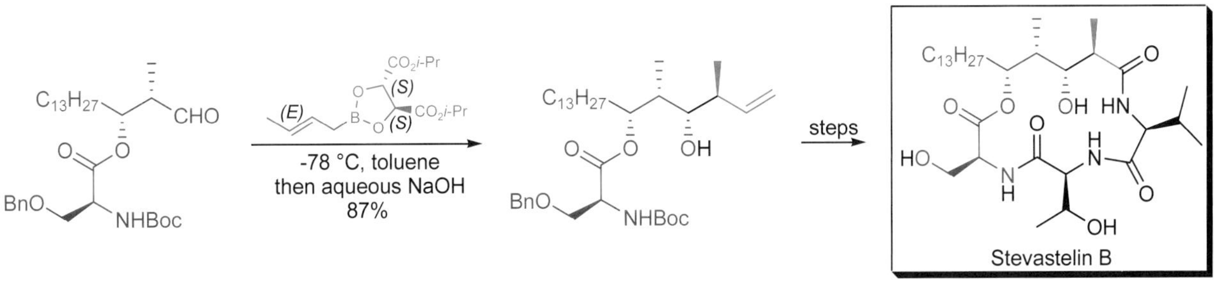

E.A. Theodorakis and co-workers reported the total synthesis of *clerocidin*, a diterpenoid antibiotic.[31] To form the C12 stereocenter and the diene moiety, they applied an asymmetric homoallenylboration method.[32] The reaction of the aldehyde and (*S,S*)-diisopropyltartrate-derived homoallenyl boronate provided the alcohol with a 6:1 diastereoselectivity and 83% yield.

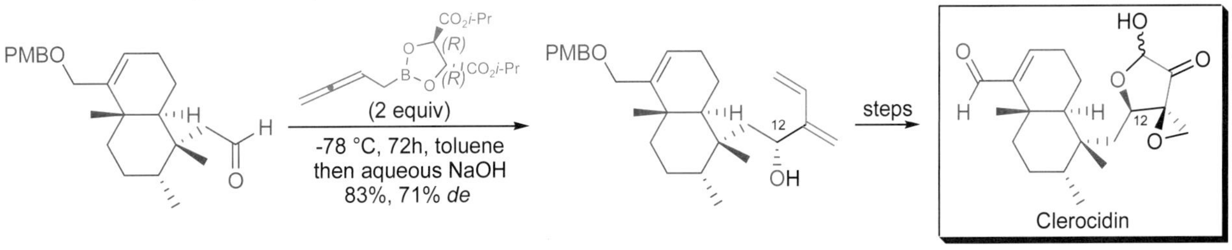

RUBOTTOM OXIDATION

(References are on page 667)

Importance:

[*Seminal Publications*[1-3]; *Modifications & Improvements*[4-14]]

In 1974, the research groups of G.M. Rubottom and A. Hassner independently developed a general and high-yielding preparation of α-hydroxy ketones (acyloins) and α-hydroxy aldehydes by the oxidation of silyl enol ethers with *m*CPBA.[2,3] The first observation of this transformation, however, was made by A.G. Brook and co-workers the same year.[1] Today the α-hydroxylation of carbonyl compounds *via* the peroxyacid oxidation of the corresponding silyl enol ethers is known as the *Rubottom oxidation*. The general features of this reaction are: 1) the silyl enol ether substrates can be prepared efficiently and regioselectively from ketones and aldehydes;[15,16] 2) both acyclic and cyclic enol ethers undergo the oxidation; 3) the oxidation readily takes place at or below room temperature (predominantly using dichloromethane as the solvent) and the reaction mixture is worked up with either acid or base to afford the α-hydroxy carbonyl compounds in good yield; 4) the silyl enol ethers derived from α,β-unsaturated ketones (2-trimethylsilyloxy-1,3-dienes) are regioselectively oxidized at the more electron-rich double bond to afford α-hydroxy or α-acyloxy enones depending on the workup conditions;[4] 5) often the initial product of the oxidation is the α-silyloxy carbonyl compound, which is readily hydrolyzed to the corresponding α-hydroxy derivative; 6) in the case of bicyclic silyl enol ethers, the reaction has to be buffered and the use of a completely non-polar solvent (e.g., pentane, toluene) is required to avoid the extensive hydrolysis of the starting material;[8] and 7) the introduction of the α-hydroxyl functionality is stereoselective in the case of bicyclic and polycyclic substrates.[8] There are a number of modifications of the *Rubottom oxidation*, and they mainly differ in the applied oxidizing agent: 1) the use of chiral oxidants such as Davis' chiral oxaziridines,[5] Shi's D-fructose-derived chiral ketone in combination with Oxone[9,12] or manganese(III)-(Salen)complexes[10] gives rise to enantiomerically enriched α-hydroxy ketones; 2) hydrogen peroxide efficiently oxidizes silyl enol ethers in the presence of MTO (methyltrioxorhenium) to give high yields of the corresponding α-hydroxy and α-silyloxy ketones;[11] and 3) HOF-acetonitrile complex (made directly from F_2 and aqueous acetonitrile) not only oxidizes silyl enol ethers but also silyl ketene acetals (derived from esters) to afford α-hydroxy ketones and esters, respectively.[17]

Rubottom & Hassner (1974):

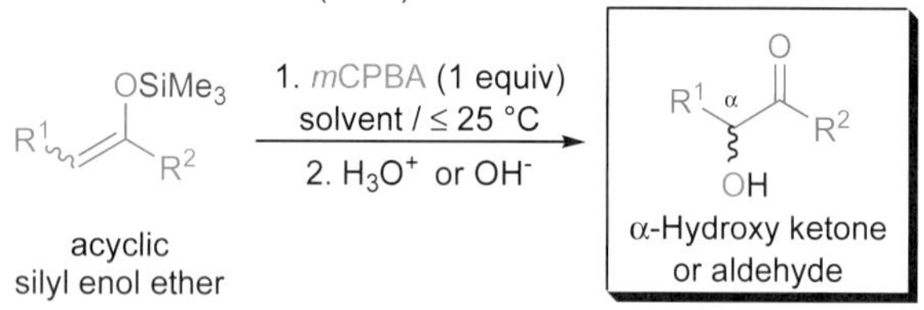

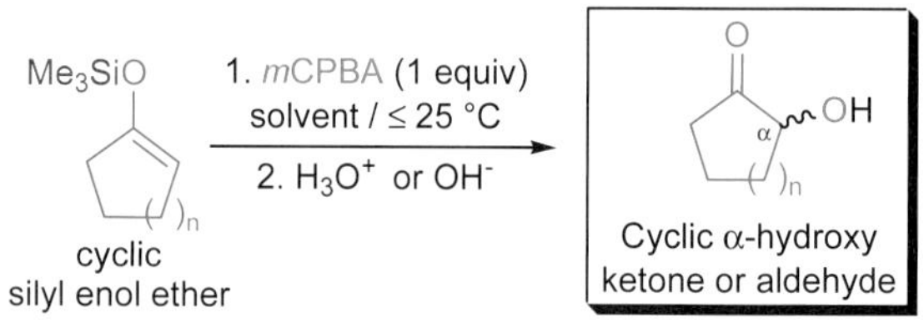

Oxidation of 2-trimethylsilyloxy-1,3-dienes:

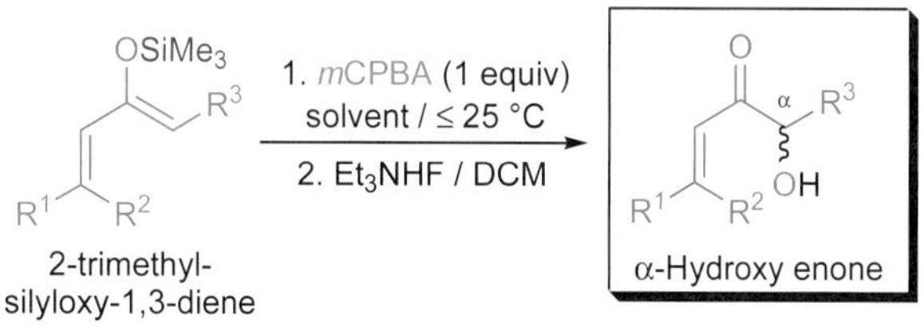

Asymmetric modification:

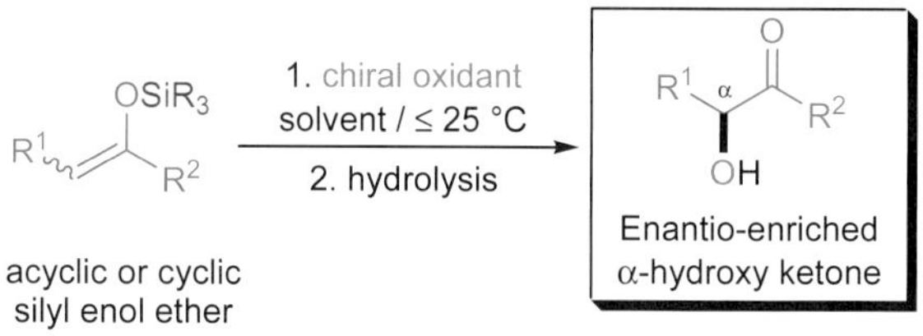

R^{1-3} = H, alkyl, aryl, substituted alkyl and aryl; SiR_3 = $SiMe_3$, $SiMe_2$(t-Bu), $SiEt_3$; solvent: CH_2Cl_2, pentane, toluene; n = 1-3; chiral oxidant: Davis' chiral oxaziridine, Shi's D-fructose derived ketone/Oxone, (Salen)manganese(III)-complexes/NaOCl or PhIO

Mechanism: [18,1,19]

The *Rubottom oxidation* proceeds through the intermediacy of a silyloxy epoxide. The epoxide ring opens under the acidic conditions to afford a stable oxocarbenium ion, which undergoes a *1,4-silyl migration* (*Brook rearrangement*)[1] to give an α-silyloxy ketone. The α-silyloxy ketone is readily hydrolyzed to the product. Until recently the silyloxy epoxide could not be isolated or observed but when the oxidation was conducted with neutral epoxidizing agents, the silyloxy epoxide intermediate could be isolated.

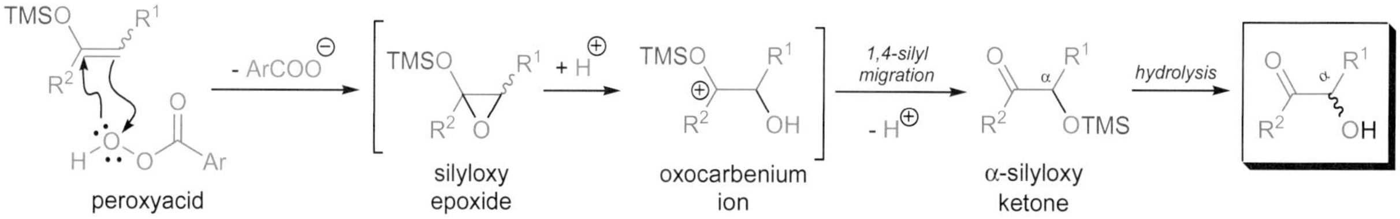

Synthetic Applications:

The highly potent antithrombotic (±)-rishirilide B was synthesized in the laboratory of S.J. Danishefsky.[20] One of the tertiary alcohol functionalities was introduced *via* the *Rubottom oxidation* of a six-membered silyl dienol ether with dimethyl dioxirane (DMDO). The oxidation was completely stereoselective, and it was guided by the proximal secondary methyl group. Subsequently, the enone was converted to the enedione, which was used as a dienophile in the key *intermolecular Diels-Alder cycloaddition* step.

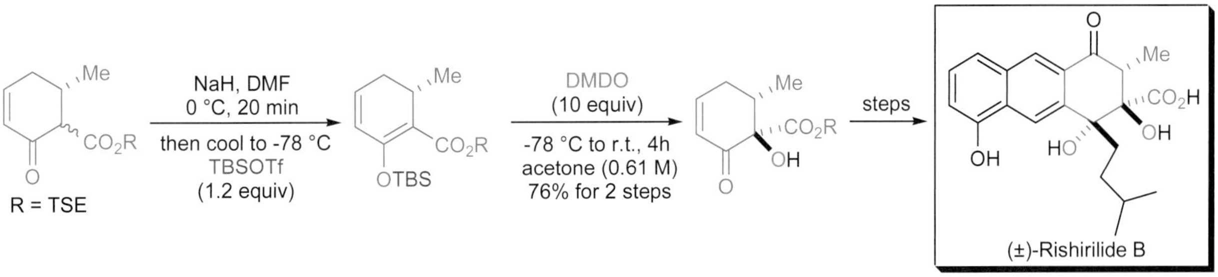

The total synthesis of the antitumor antibiotic FR901464 was accomplished by E.N. Jacobsen et al.[21] The preparation of the central six-membered fragment was achieved *via* a highly *enantioselective hetero Diels-Alder reaction* between a diene and an aldehyde. The resulting silyl enol ether was subjected to a modified *Rubottom oxidation* condition (buffer and nonpolar solvent) with mCPBA to afford the desired α-hydroxy ketone with complete diastereoselectivity.

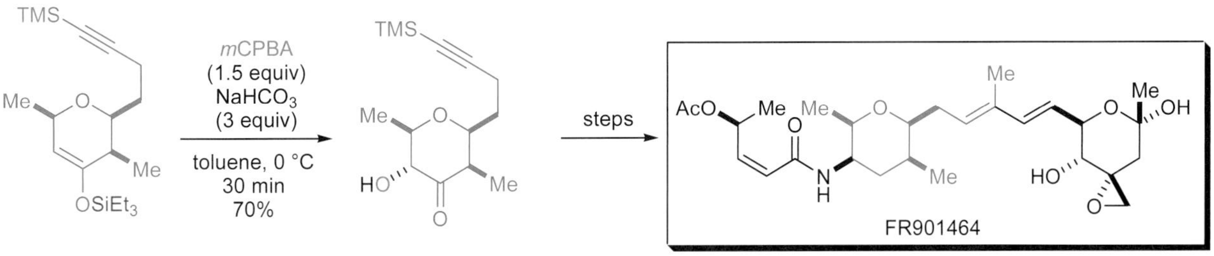

The key step in the total synthesis of the furanoditerpene *d,l*-isospongiadiol by P.A. Zoretic and co-workers was an *oxidative free-radical cyclization*, which gave rise to the tricyclic skeleton of the natural product.[22] The last stereocenter at C2 was introduced using the *Rubottom oxidation* on the fully elaborated tetracyclic intermediate. The product was a mixture of α-hydroxy and silyloxy ketone and the last step was a global deprotection with TBAF to afford the natural product.

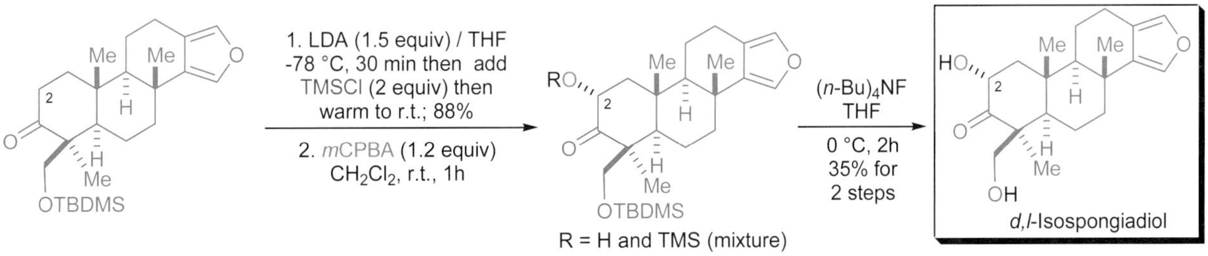

In the highly stereoselective synthesis of hispidospermidin, the oxygenation of the C10 position was achieved *via* a *Rubottom oxidation* by S.J. Danishefsky et al.[23] The tricyclic ketone was first converted to the TES enol ether, which was readily oxidized with *m*CPBA to give the corresponding α-hydroxy ketone as a single diastereomer.

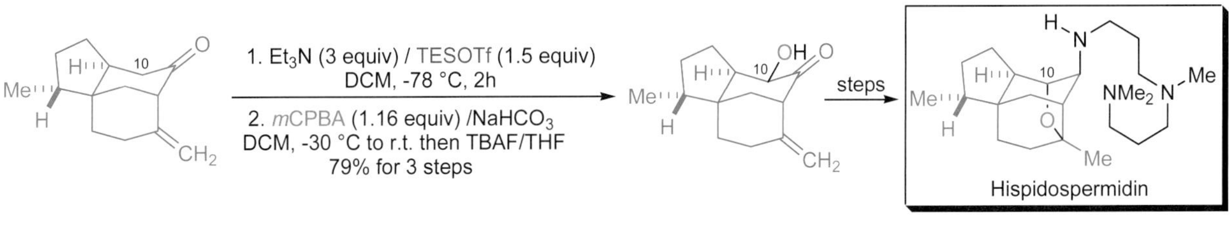

SAEGUSA OXIDATION

(References are on page 667)

Importance:

[*Seminal Publications*[1,2]; *Reviews*[3-7]; *Modifications & Improvements*[8-11]]

In 1978, T. Saegusa and co-workers reported that silyl enol ethers reacted with substoichiometric amounts of $Pd(OAc)_2$ and *p*-benzoquinone in acetonitrile at room temperature to afford the corresponding α,β-unsaturated carbonyl compounds.[2] The regioselective introduction of the α,β carbon-carbon double bond to cyclic and acyclic ketones *via* the Pd-mediated oxidation of the corresponding silyl enol ethers is known as the *Saegusa oxidation*. The general features of the transformation are: 1) the reaction is usually carried out using 0.5 equivalents of $Pd(OAc)_2$ and 0.5 equivalents of *p*-benzoquinone (co-oxidant) at room temperature; 2) when stoichiometric amounts of $Pd(OAc)_2$ are used, no co-oxidant is needed. However, less than 0.25 equivalents of $Pd(OAc)_2$ results in a substantial decrease in the reaction rate as well as isolated yield of the product; 3) the starting silyl enol ethers are easily obtained by trapping metal enolates with TMSCl (the metal enolates are either obtained by the regioselective deprotonation of ketones and aldehydes with LiHMDS or LDA or by the conjugate addition of carbon nucleophiles to α,β-unsaturated carbonyl compounds);[12,13] 4) both acyclic and cyclic silyl enol ethers undergo the transformation; 5) the oxidation proceeds with high stereoselectivity, because in acyclic systems the stereochemistry of the newly formed double bond is predominantly (*E*) even if the starting silyl enol ether was a mixture of (*E*) and (*Z*) stereoisomers; and 6) cyclic silyl enol ethers (*n*=1-7) are efficiently oxidized, and when the ring size allows, the newly introduced double bond will have the (*E*) stereochemistry. The main drawback of the *Saegusa oxidation* is the high cost of the palladium acetate. However, methods employing truly catalytic amounts of $Pd^{(II)}$- and $Pd^{(0)}$ complexes have been developed.[11] There are several modifications of the process: 1) an environmentally friendly catalytic version using only 10 mol% of $Pd(OAc)_2$ and oxygen atmosphere in DMSO (*Larock modification*);[11] 2) instead of silyl enol ethers, enol acetates can also be used when they are heated with allyl methyl carbonate, catalytic amounts of $Pd(OAc)_2$ and $MeOSnBu_3$;[10] and 3) allyl enol carbonates also undergo oxidation with catalytic amounts of $Pd(OAc)_2$/dppe.[8,7] Alternatively, silyl enol ethers can be efficiently oxidized by IBX and IBX-*N*-oxides to the corresponding enones (*Nicolaou oxidation*).[14]

Saegusa (1978):

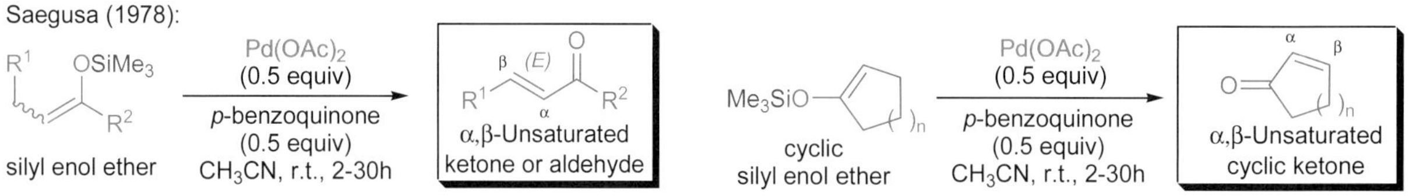

Catalytic process (*Larock modification*,1995):

Allyl enol carbonate modification:

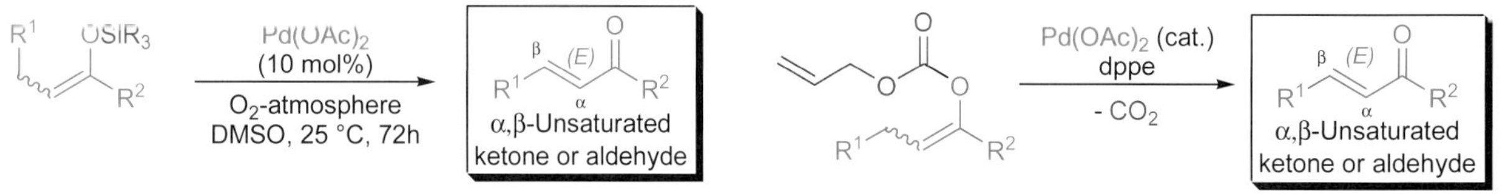

R^{1-2} = H, alkyl, aryl; SiR_3 = TMS, TBDMS; n = 1-7

Mechanism: [15,7]

When substoichiometric/stoichiometric amounts of $Pd(OAc)_2$ is used:

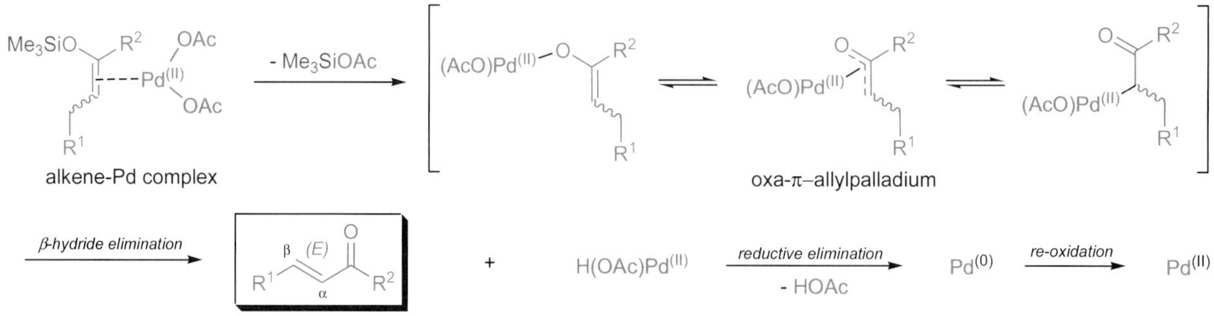

When the oxidation takes place under an oxygen atmosphere with catalytic amounts of $Pd(OAc)_2$:

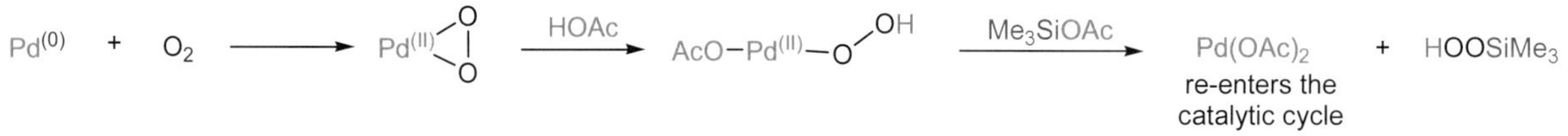

Synthetic Applications:

The first total synthesis of the marine polycyclic ether toxin (–)-gambierol was accomplished in the laboratory of M. Sasaki.[16] The introduction of the α,β-unsaturation into the seven-membered H ring of the FGH tricyclic subunit proved to be problematic, because both the conventional selenium-based method and the *Nicolaou oxidation* with IBX failed. However, when the seven-membered ketone was treated with LiHMDS in the presence of TMSCl and Et_3N, the corresponding silyl enol ether was formed, which was oxidized under Saegusa conditions to give the desired cyclic enone in high yield. Because of the small scale of the reaction, a large excess of $Pd(OAc)_2$ was used in acetonitrile so the presence of a co-oxidant was not necessary.

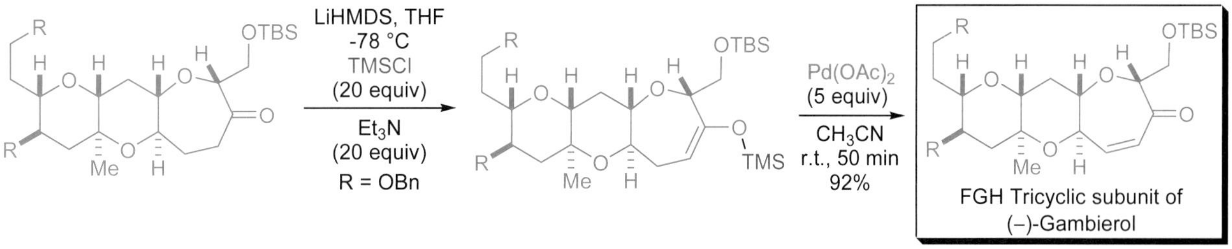

A.G.M. Barrett and co-workers reported the first total synthesis of (–)-preussomerin G.[17] In the late stages of the synthesis, the introduction of the desired cyclohexenone moiety was achieved using the *Saegusa oxidation*. The ketone was first converted to the silyl enol ether with trimethylsilyl triflate, and then it was treated with stoichiometric amounts of $Pd(OAc)_2$.

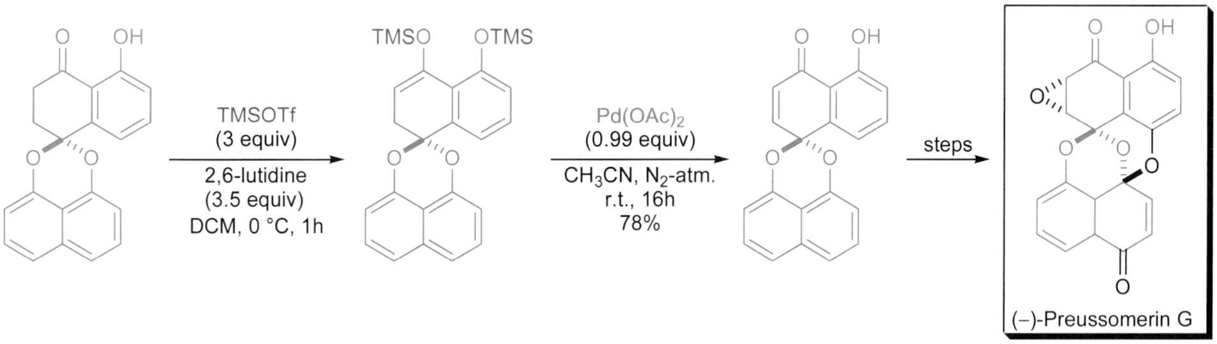

The *Larock modified Saegusa oxidation* conditions were utilized in the total synthesis of (±)-8,14-cedranoxide by M. Ihara et al.[18] The main strategy was to apply an *intramolecular double Michael addition* reaction to assemble the tricyclic cedranoid skeleton. The precursor five-membered enone was prepared in high yield from the corresponding substituted cyclopentanone in two steps.

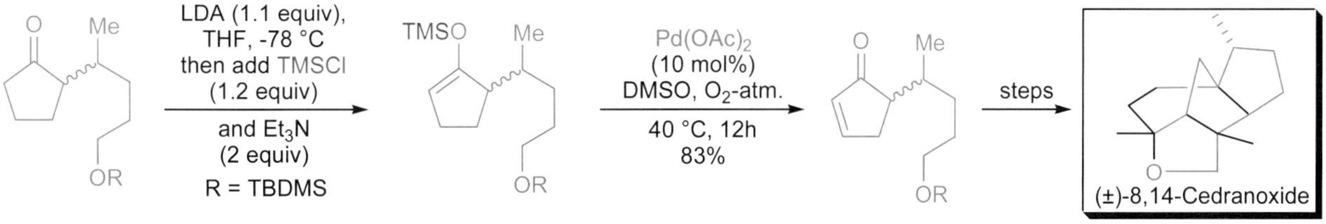

A stereodivergent synthesis was developed by H. Nemoto and co-workers for the preparation of *cis*-fused 2,5-disubstituted octahydroquinolines, which constitute the core structure of certain dendrobatid alkaloids.[19] The installation of the C5 methyl group was achieved by 1,4-cuprate addition and the resulting enolate was trapped with TMSCl. The silyl enol ether was then oxidized to the enone with $Pd(OAc)_2$.

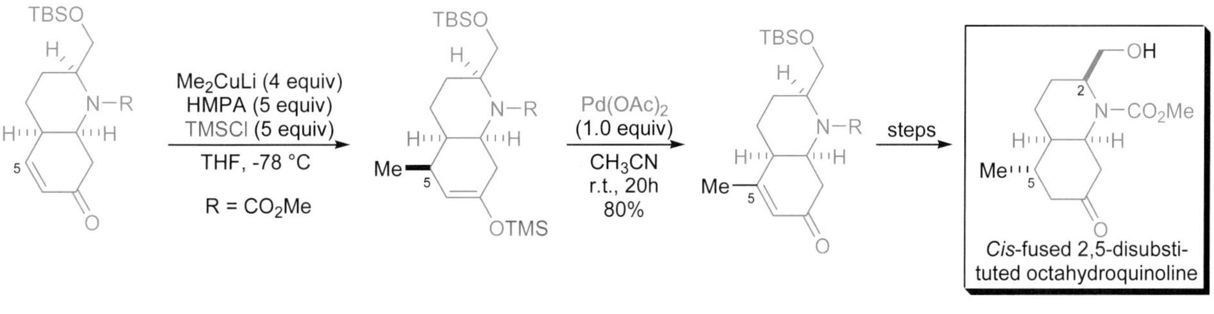

SAKURAI ALLYLATION

(References are on page 668)

Importance:

[*Seminal Publications*[1,2]; *Reviews*[3-9]; *Modifications & Improvements*[10-23]; *Theoretical Studies*[24,25]]

In 1976, H. Sakurai reported that allylsilanes react with a wide variety of aldehydes and ketones in the presence of stoichiometric quantities of $TiCl_4$ to form the corresponding homoallylic alcohols. Today, this transformation is referred to as the *Sakurai allylation*, and it is one of the most important carbon-carbon bond forming reactions. The general features of the reaction are: 1) typically, it is carried out in dichloromethane under nitrogen atmosphere at a temperature range between -78 °C and 25 °C; 2) in addition to $TiCl_4$, several other Lewis acids can be used such as $AlCl_3$, $BF_3{\cdot}OEt_2$, $SnCl_4$, $EtAlCl_2$;[1,2] 3) most commonly trimethylallylsilanes and phenyldimethylallylsilanes are utilized as the allylsilane reactant;[4,6] 4) the reaction is highly regioselective, the electrophile attacking at the C3 terminus of the allylsilane;[1,2,4] 5) C1 substituted allylsilanes give the (*E*)-alkene product;[26] 6) allenyl-,[27] propargyl-,[28] vinyl-,[29] and ethynylsilanes[29] also undergo the reaction in the presence of Lewis acids; 7) the most commonly used electrophiles are aldehydes and ketones, but acetals and ketals[30] are also often utilized; 8) dithioacetals,[31] monothioacetals,[32] alkoxymethyl-,[33] and phenylthiomethyl chlorides[34] undergo the allylation reaction; 9) α,β-unsaturated aldehydes react at the carbonyl group, while α,β-unsaturated ketones undergo conjugate addition;[35,36] 10) intramolecular reactions are also feasible;[37] and 11) C3 monosubstituted allysilanes give the *syn*-diastereomer as the major product.[38] Common side reactions in the *Sakurai allylation* are the following: 1) protodesilylation;[39] 2) allylic alcohol products, especially tertiary allylic alcohols can undergo ionization;[40] and 3) in the case of 1,1-disubstituted allylsilanes, the trisubstituted alkene product may react further.[41] Side reactions usually can be avoided by carefully controlled conditions or utilizing acetal or ketal substrates. Catalytic versions of the *Sakurai allylation* are known as well, utilizing TMSOTf,[10] TMSI,[11] Ph_3CClO_4,[12] $Cp_2Ti(CF_3SO_3)_2$,[14] TMSOMs,[19] and $InCl_3$/TMSCl[20] as catalysts. Recently, catalytic asymmetric versions were developed.[15,22,23]

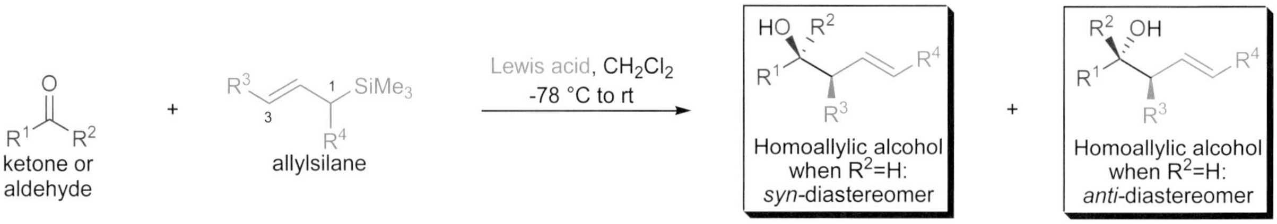

R^1 = alkyl, aryl; R^2 = H, alkyl, aryl; R^3 and R^4 = H, alkyl, aryl; Lewis acid = $TiCl_4$, $BF_3{\cdot}OEt_2$, $SnCl_4$, $EtAlCl_2$

Mechanism:[42,43,38,44-46]

The reaction starts with the activation of the carbonyl group by the Lewis acid. Subsequent carbon-carbon bond formation leads to a silyl-stabilized carbocation,[45] which after loss of the trimethylsilyl group, gives the double bond. From studies conducted on chiral allylsilanes, it was concluded that the incoming electrophile attacks the double bond on the surface opposite to the silyl group.[42] The reaction of aldehydes with C3 substituted allylsilanes leads to the *syn*-diastereomer as the major product, and (*E*)-allylsilanes give higher diastereoselectivities than (*Z*)-allylsilanes. The reaction presumably goes through an open transition state.[38] The possible transition states leading to the *syn*-diastereomer are depicted below.[43,44]

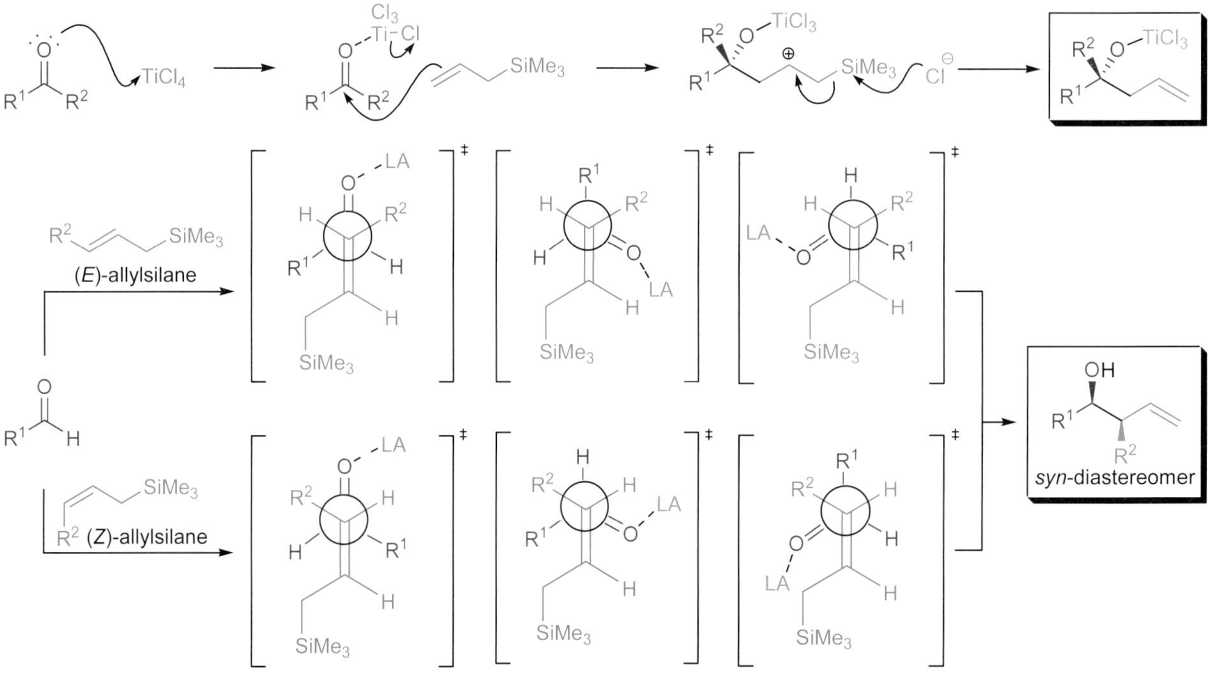

Synthetic Applications:

In the laboratory of B.M. Trost, a modular approach toward the total syntheses of furaquinocins was developed.[47] To introduce the homoallylic side chain in a diastereoselective fashion, they utilized the *Sakurai allylation reaction*. During their studies they found that the highest diastereoselectivity can be achieved using 1 equivalent of $TiCl_4$ at room temperature. Application of other Lewis acids such as $BF_3{\cdot}OEt_2$ gave the product with lower selectivity. Attempts to perform the allylation using catalytic amounts of Lewis acids such as $FeCl_3$ or $Sc(OTf)_3$ led to no conversion. The resulting homoallylic alcohol served as a common intermediate toward the syntheses of both furaquinocin A and B.

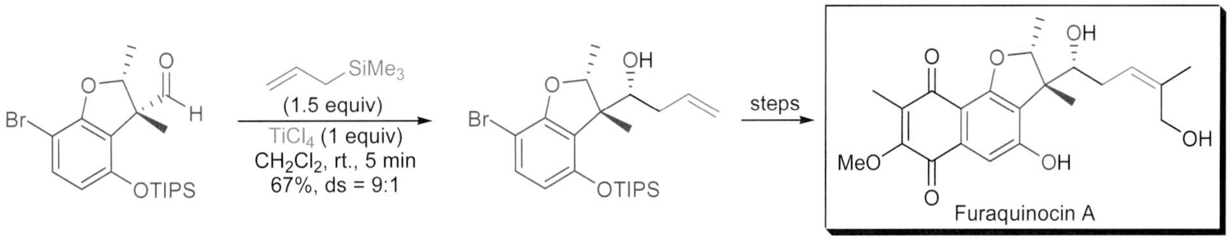

A convergent total synthesis of 15-membered macrolactone, (–)-amphidinolide P was reported by D.R. Williams and coworkers.[48] In their approach, they utilized the *Sakurai allylation* to introduce the C7 hydroxyl group and the homoallylic side chain. The transformation was effected by $BF_3{\cdot}OEt_2$ at -78 °C to provide the homoallylic alcohol as a 2:1 mixture of diastereomers. The desired alcohol proved to be the major diastereomer, as it resulted from the Felkin-Ahn controlled addition of the allylsilane to the aldehyde. The minor diastereomer was converted into the desired stereoisomer *via* a *Mitsunobu reaction*.

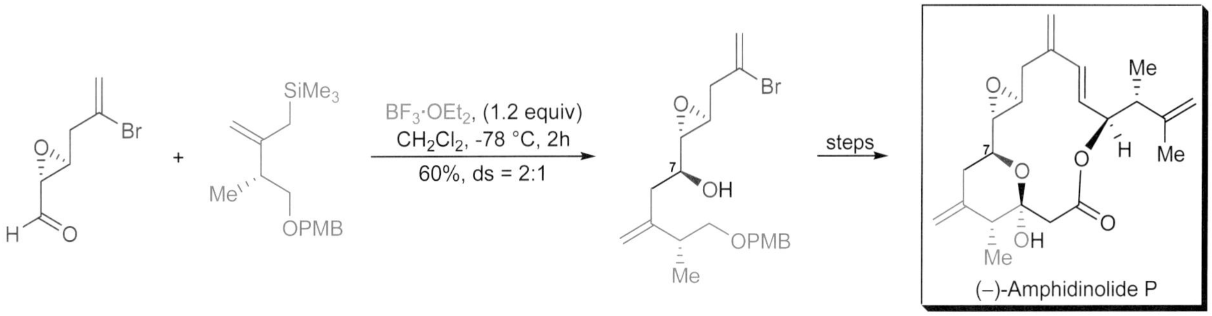

A highly convergent, enantioselective total synthesis of structurally novel, cancer therapeutic lead, (–)-laulimalide was achieved by P.A. Wender and co-workers.[49] During the synthesis, they performed an unprecedented complex asymmetric *Sakurai allylation reaction* as a key step to form the C14-C15 carbon-carbon bond. In this transformation, they utilized a chiral, nonracemic (acyloxy)borane Lewis acid that was developed by H. Yamamoto.[15] According to Yamamoto's original procedure, only a catalytic amount (10-20 mol%) of the Lewis acid was needed to bring about the desired transformation with high yield and enantioselectivity. However, in this case, one equivalent of the Lewis acid was necessary to effect the allylation. The reaction was carried out in propionitrile at -78 °C, and the product was obtained in high yield and as the only detectable diastereomer by spectroscopic methods.

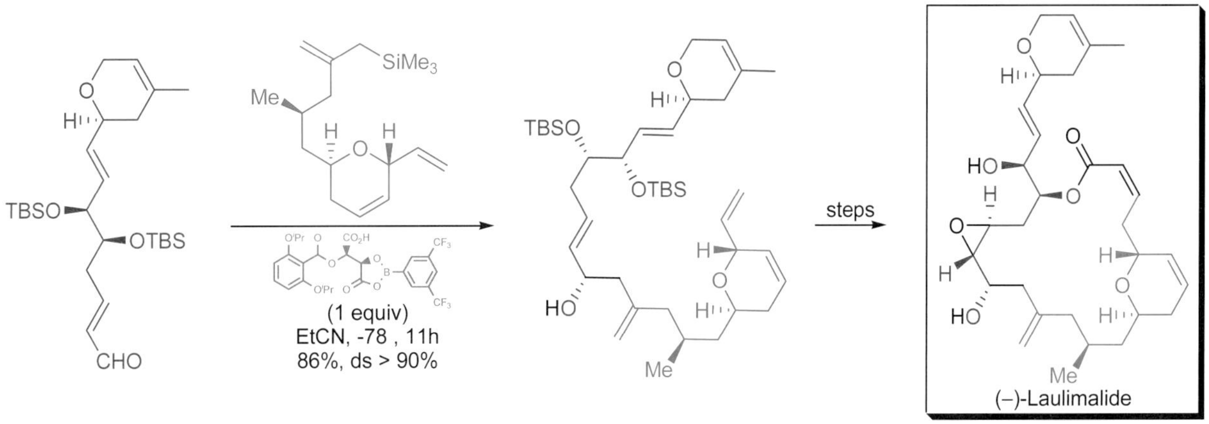

SANDMEYER REACTION

(References are on page 669)

Importance:

[*Seminal Publications*[1-4]; *Reviews*[5-11]; *Modifications & Improvements*[12-19]]

In 1884, T. Sandmeyer intended to prepare phenylacetylene by reacting benzenediazonium chloride with copper(I) acetylide, but the major product of the reaction was chlorobenzene, and no trace of the desired product was detected.[3] Careful examination of the reaction conditions revealed that copper(I) chloride was formed *in situ* and it catalyzed the replacement of the diazonium group with a chlorine atom.[4] Sandmeyer also showed that bromobenzene was formed by using copper(I) bromide, and copper(I) cyanide led to benzonitrile. The substitution of aryldiazonium salts with halides or pesudohalides is known as the *Sandmeyer reaction*. The general features of this transformation are: 1) the required aryldiazonium halides are usually prepared from arylamines *via* diazotization using either $NaNO_2$/hydrohalic acid in water or alkyl nitrites (e.g., *tert*-butyl nitrite) under anhydrous conditions; 2) the aryldiazonium halides are not isolated but reacted in the same pot with copper(I) chloride, bromide or cyanide to obtain the corresponding aryl chloride, aryl bromide, and aryl nitrile, respectively; 3) the counterion of the copper(I) salt has to match the conjugate base of the hydrohalic acid otherwise product mixtures are formed; 4) the preparation of aryl iodides does not require the use of a copper(I) salts; simply adding potassium iodide brings about the substitution accompanied by the loss of dinitrogen; and 5) the substitution pattern on the aromatic amine can be widely varied, both electron-donating and electron-withdrawing groups are tolerated. There are other useful substitution reactions of aryldiazonium salts, but these are referred to with different names (or with no specific name):[8] 1) when the aryldiazonium halides are treated with hydrogen chloride or hydrogen bromide in the presence of copper metal to afford aryl chlorides and bromides, the process is called the *Gattermann reaction*; 2) the thermal decomposition of aryldiazonium tetrafluoroborates to give aryl fluorides is known as the *Balz-Schiemann reaction*; 3) aryldiazonium tetrafluoroborates react with sodium nitrite in the presence of catalytic amounts of copper(I) salt to give nitroarenes;[20,21] and 4) aryldiazonium salts can also be converted to phenols by heating with trifluoroacetic acid, aqueous sulfuric acid, or with aqueous solution of copper salts (occasionally called the *Sandmeyer hydroxylation*).[22-24]

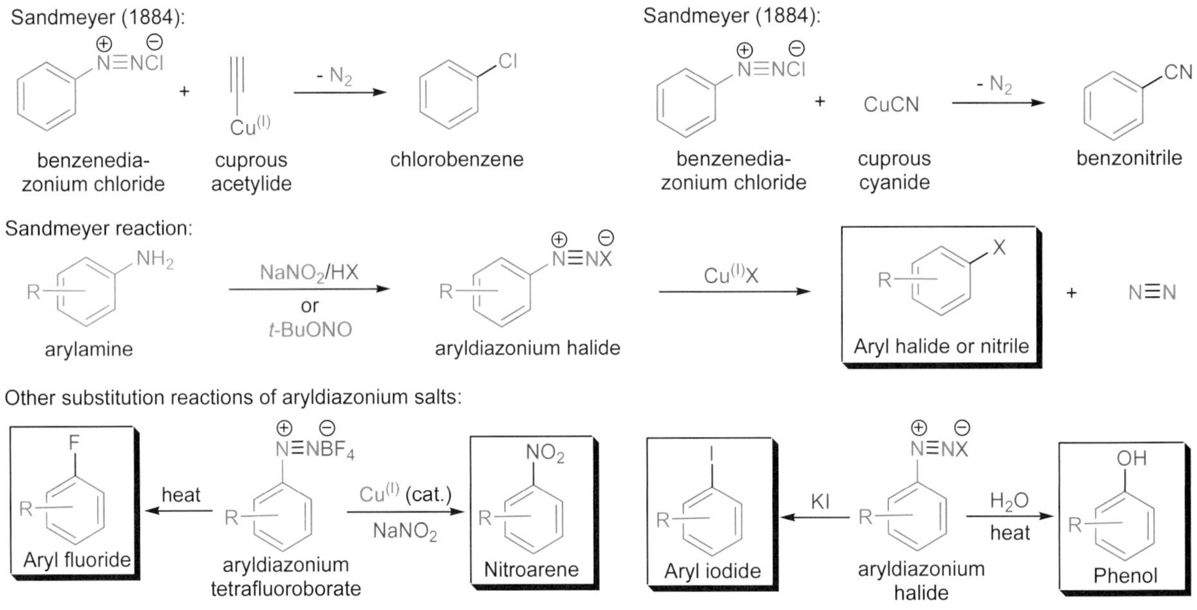

R = H, alkyl, aryl, electron-withdrawing groups (EWG) or electron-donating groups (EDG); HX: HCl, HBr; X = Cl, Br, CN

Mechanism: [25-32,9,33,34,16,35,36,19,24]

The mechanism of the *Sandmeyer reaction* is not completely understood. For a long time it was believed to proceed *via* aryl cations, but later W.A. Waters and then later J.K. Kochi proposed a radical mechanism which was catalytic for the copper(I) salt.[25,26] In a single electron-transfer event the diazonium halide is reduced to a diazonium radical which quickly loses dinitrogen to afford an aryl radical. A final ligand transfer from the copper(II) salt completes the catalytic cycle and regenerates the copper(I) species.

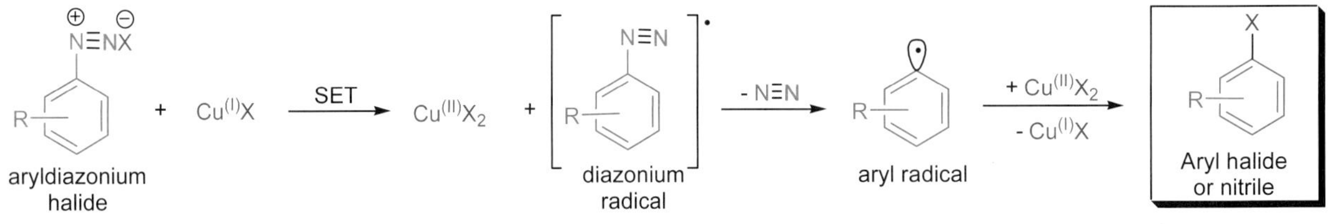

Synthetic Applications:

In the laboratory of D.A. Evans the total synthesis of the teicoplanin aglycon was accomplished.[37] In the endgame of the synthetic effort the introduction of the required chloro substituent on ring-2 under mild conditions was necessary. The authors chose the *Sandmeyer reaction* to bring about the desired transformation of the aromatic amine moiety. First the substrate was diazotized with *t*-butyl nitrite and HBF_4 in acetonitrile and then in the same pot a mixture of copper(I) chloride and copper(II) chloride in large excess was added at low temperature. The desired aryl chloride was isolated in moderate yield. To complete the synthesis, the following steps had to be carried out: 1) deprotection of the carboxy-terminal *N*-methylamide with N_2O_4 followed by a pH neutral hydrolysis; and 2) global demethylation at room temperature using $AlBr_3$/EtSH with concomitant *N*-terminal trifluoroacetamide hydrolysis.

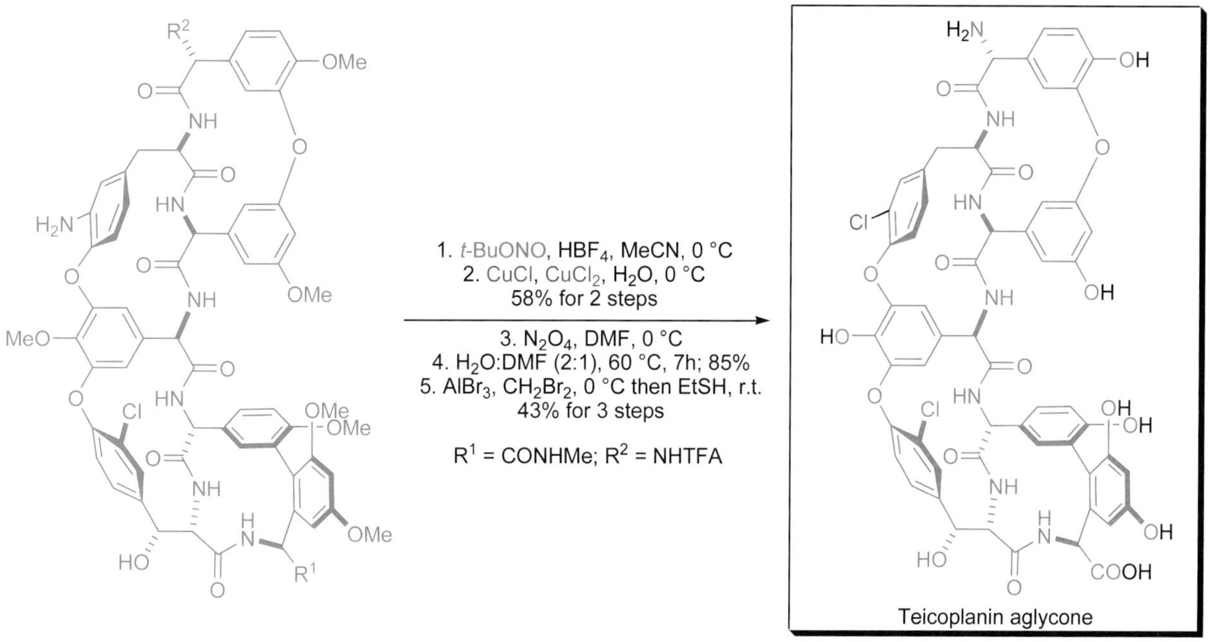

The neurotoxic quaterpyridine natural product nemertelline was successfully synthesized by S. Rault et al. using a *Suzuki cross-coupling* as the key step. The boronic acid coupling partner, required for the *Suzuki reaction*, was prepared by first subjecting 3-amino-2-chloropyridine to the conditions of the *Sandmeyer reaction* followed by a lithium-halogen exchange and trapping the lithiopyridine derivative with triisopropylborate.

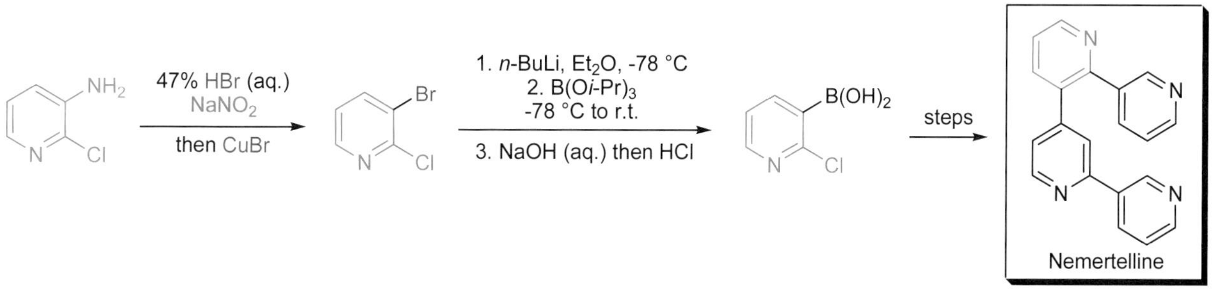

M. Nakata and co-workers completed the concise total synthesis of (±)-A80915G, which belongs to the napyradiomycin family of antibiotics.[38] There were two key carbon-carbon bond forming reactions in the synthetic sequence: a *Stille cross-coupling* between an aromatic trihalide and geranyl tributyltin and a *Diels-Alder cycloaddition* employing the Danishefsky-Brassard diene. A *Sandmeyer reaction* was used to introduce the iodine substituent to the 2-bromo-4-chloro-3,6-dimethoxy-aniline substrate in order to obtain the required trihalogenated 1,4-dimethoxy-benzene precursor.

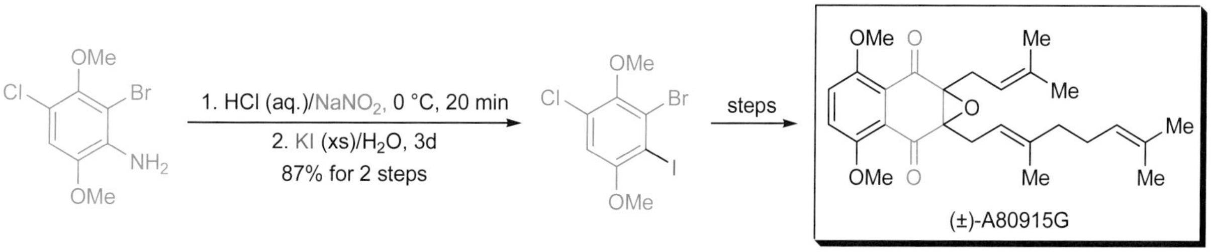

SCHMIDT REACTION

(References are on page 670)

Importance:

[*Seminal Publications*[1,2]; *Reviews*[3-10]; *Modifications & Improvements*[11-17]; *Theoretical Studies*[18-21]]

In 1923, K.F. Schmidt reported that heating hydrazoic acid (HN_3) with benzophenone in the presence of sulfuric acid, afforded benzanilide in quantitative yield.[1] Later this transformation was shown to be general for ketones, aldehydes, and carboxylic acids that underwent similar reactions with HN_3 to give amides, nitriles, and amines, respectively. The reaction of carbonyl compounds with hydrazoic acid or alkyl azides in the presence of acid catalysts is known as the *Schmidt reaction*. The general features of the *Schmidt reaction* are: 1) the transformation occurs in a single stage from carboxylic acids unlike the related *Curtius and Hoffmann rearrangements*; 2) the reaction conditions are mild, the reagents are readily available, the procedure is simple, and does not require special equipment; 3) protic acids are used as acid catalysts (e.g., H_2SO_4, PPA, trichloroacetic acid/H_2SO_4, TFA, TFAA), and sulfuric acid is by far the most widely used; 4) hydrogen azide is handled either as a solution in an inert solvent (e.g., $CHCl_3$) or generated *in situ* by adding NaN_3 to the acidic reaction mixture; 5) HN_3 is known to be toxic and explosive (especially on large scale); 6) in the case of carboxylic acids, the best results are obtained with aliphatic and sterically hindered aromatic substrates; 7) the product amines are one-carbon shorter homologs of the substrates due the loss of CO_2; 8) aromatic acids with electron-withdrawing groups require the use of very strong acid catalysts (e.g., conc. H_2SO_4 or oleum) and very electron-poor heterocyclic acids usually do not react; 9) the α-stereocenter remains unaffected and the product amine is obtained with retention of configuration; 10) carboxylic acids that are fully alkyl or aryl substituted at the α-position (have no α hydrogen atom) may undergo side reactions due to the decarboxylation of the acid to a stable carbocation; 11) 1,3-dicarboxylic acids react at only one of the carboxylic acid fuctional groups; 12) α-amino acids do not react; 13) α,β-unsaturated carboxylic acids are not good substrates, since they give rise to complex reaction mixtures; 14) aldehydes and ketones react with hydrazoic acid faster than carboxylic acids so good chemoselectivity can be achieved with keto acids; 15) aliphatic aldehydes are unstable in sulfuric acid, so mainly aromatic aldehydes are used; 16) the main product with aldehydes is the corresponding nitrile, but the formation of formamides is often a side reaction; 17) symmetrical ketones give rise to *N*-substituted amides; 18) in unsymmetrical ketones such as alkyl aryl ketones, the aryl group migrates preferentially so *N*-aryl amides are obtained; 19) cyclic ketones undergo ring-enlargement to afford cyclic amides; 20) Lewis acids are effective catalysts when alkyl azides are employed; and 21) the reaction works efficiently intramolecularly and affords *N*-substituted lactams. The disadvantages of the *Schmidt reaction* are: 1) carbonyl compounds and carboxylic acids that are unstable in aqueous acid cannot be used as substrates; 2) the reaction medium has to be fairly acidic to achieve high yields; 3) when ketones are reacted with excess HN_3, tetrazoles are formed in significant amounts; and 4) in addition to the carbonyl group, several other functional groups such as nitriles, imines, diimides, certain alkenes, and alcohols (which are dehydrated to alkenes in the acidic medium) react with HN_3.

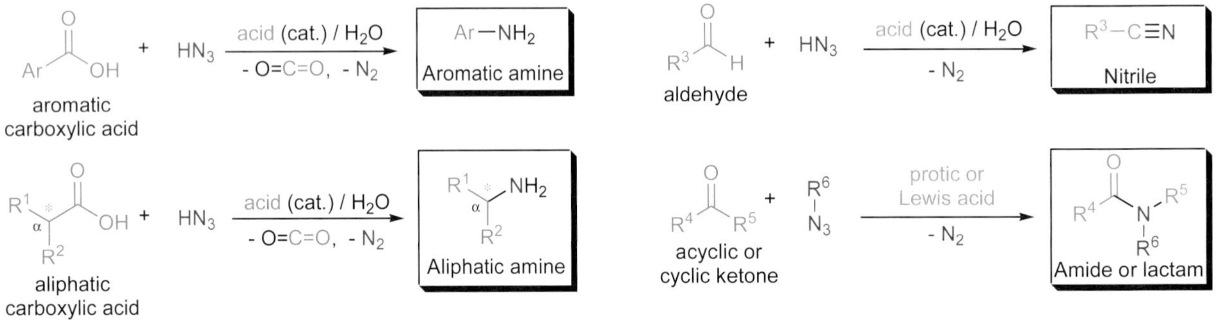

R^{1-2} = alkyl, aryl; Ar = substituted aryl or heteroaromatic; R^3 = Me, substituted aryl; R^4 = alkyl, substituted alkyl; R^5 = aryl; R^6 = H, alkyl or aryl; Lewis acid: $TiCl_4$, TFA, CH_3SO_3H; acid catalyst: H_2SO_4, PPA, Cl_3CCO_2H/H_2SO_4, TFA, TFAA

Mechanism:[22-25,7,26-28]

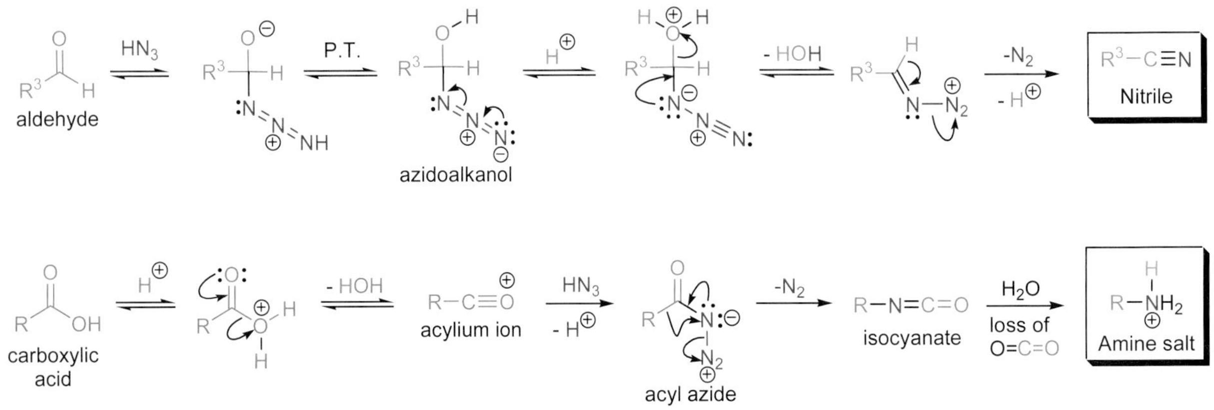

Synthetic Applications:

Research by A.G. Schultz targeted the synthesis of a new class of structural analogues of the morphine alkaloids in order to test their opioid receptor affinities.[29] A seven-membered ring homologue of 6,7-benzomorphans was prepared using the *Schmidt reaction* on a six-membered bicyclic ketone. The substrate was dissolved in TFA and the aqeous solution of NaN_3 was slowly added at room temperature. After 30 minutes, the reaction mixture was heated at 65 °C for several hours, and the desired seven-membered lactam was isolated in excellent yield. As expected, the slight excess of hydrazoic acid resulted in the formation of small amounts of a tetrazole by-product.

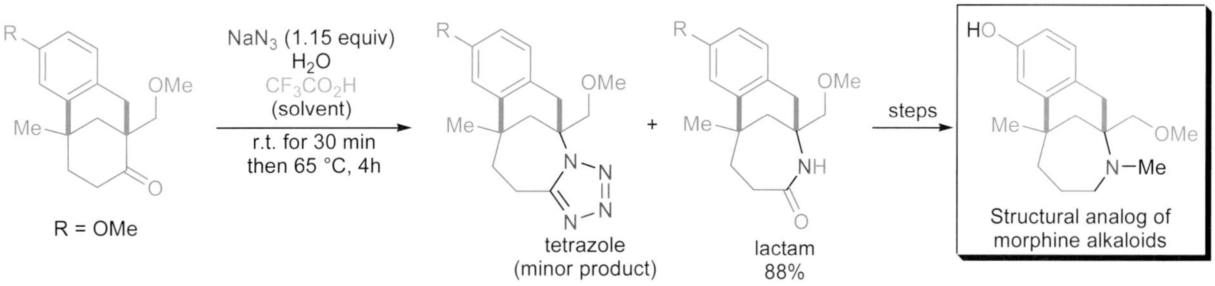

The first asymmetric total synthesis of (+)-sparteine was accomplished by J. Aubé and co-workers using two ring-expansion reactions as key steps.[30] The $TiCl_4$-mediated *intramolecular Schmidt reaction* proceeded in good yield on the bicyclic azido ketone which was derived from 2,5-norbornadione. The cyclization was accompanied by deketalization and afforded a tricyclic keto lactam. The second ring-expansion was achieved using a *photo-Beckmann rearrangement*.

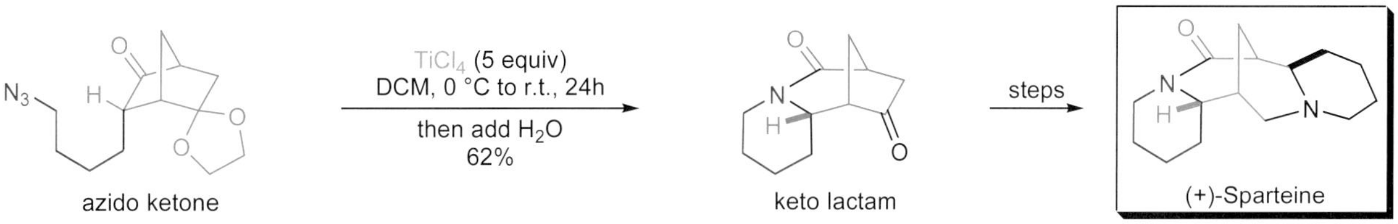

The *intramolecular Schmidt reaction* was the key step in the total synthesis of dendrobatid alkaloid 251F by J. Aubé et al.[31] The initial attempts to bring about the ring-expansion on a bicyclic substrate that contained a terminal alkene functionality failed, only degradation products were observed. However, the conversion of this alkene moiety into a primary alcohol solved the problem and the rearrangement took place smoothly to afford the full skeleton of 251F. Reduction of the six-membered lactam to the corresponding tertiary amine with LAH completed the total synthesis.

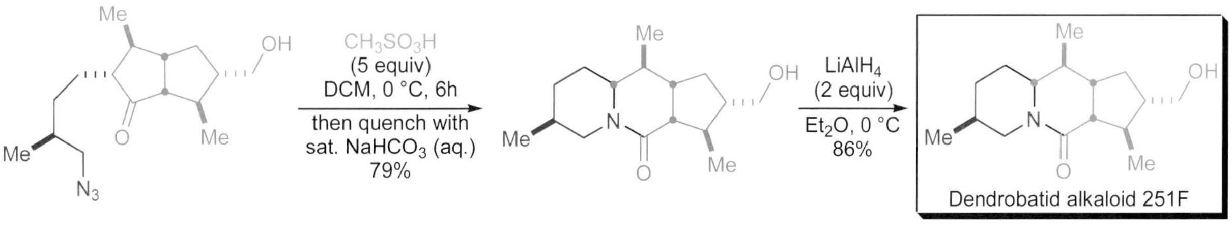

A novel asymmetric synthesis of α,α-disubstituted α-amino acids was developed in the laboratory of M. Tanaka and S. Suemune.[32] First, the diastereoselective alkylation of 2-methyl acetoacetate using (*S,S*)-cyclohexane-1,2-diol as an acetal chiral auxiliary was performed. The removal of the chiral auxiliary with $BF_3 \cdot OEt_2$ afforded enantiopure β-keto esters having a quaternary carbon atom. The β-keto esters were subjected to the *Schmidt reaction* to give the desired *N*-acetyl α-amino acids without loss of the optical activity.

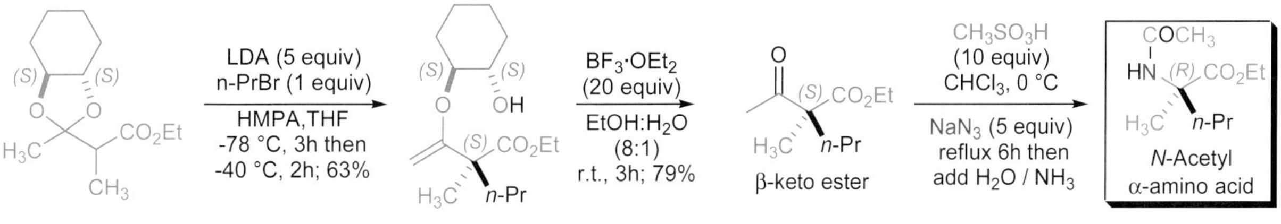

SCHOTTEN-BAUMANN REACTION

(References are on page 670)

Importance:

[*Seminal Publications*[1,2]; *Reviews*[3,4]; *Modifications & Improvements*[5-21]; *Theoretical Studies*[22]]

In 1884, C. Schotten reported an efficient method for the preparation of *N*-benzoyl piperidine from piperidine and benzoyl chloride in water and in the presence of sodium hydroxide.[1] In 1886, E. Baumann showed that the same reaction conditions were suitable for the preparation of benzoic acid esters from alcohols and benzoyl chloride.[2] The neat alcohol and benzoyl chloride were mixed in water, then the resulting mixture was treated with aqueous sodium hydroxide. The product esters were frequently crystalline and could be isolated in high yield. Baumann demonstrated the power of this method by benzoylating several polyhydroxy compounds such as glucose and glycerol. The synthesis of esters from alcohols and amides from amines with acyl halides or anhydrides in the presence of aqueous base is known as the *Schotten-Baumann reaction*. The general features of these transformations are: 1) the reaction is especially well-suited for the preparation of simple amides; 2) in the typical procedure the alcohol or ester is mixed with excess acyl halide or anhydride in the presence of aqueous sodium hydroxide or saturated aqueous sodium bicarbonate while the reaction mixture is stirred vigorously; 3) the order of reactivity for alcohols is: 1°>2°>3°, which means that sterically hindered secondary and tertiary alcohols are usually acylated sluggishly; 4) the order of reactivity of the amines is determined by their basicity and generally the more basic amine is acylated faster; 5) the success of the process depends on the reactivity of the acyl halide, and in general acyl halides that are less reactive give higher yields of the product (since less reactive acyl halides do not undergo rapid hydrolysis by water); 6) aromatic acyl halides are more stable under aqueous conditions than aliphatic acyl halides, so they are more suitable for acylation under the Schotten-Baumann conditions; 7) in the acylation of primary alcohols the presence of a base is not always necessary (but it is recommended to achieve high yields), since the by-product hydrogen halide in certain cases does not hydrolyze the product ester; 8) the use of a base is required during the acylation of secondary and tertiary alcohols; and 9) during the acylation of amines the presence of a base is crucial, since the substrate amine is rendered unreactive upon protonation by the acid by-product (the base applied must be stronger than the substrate amine). Several modifications were developed for the acylation of sterically hindered substrates. Today, the majority of acylation reactions is conducted in aprotic organic solvents in the presence of organic bases (e.g., pyridine, DMAP, etc.) and/or Lewis acids, and they can all be considered as modifications of the original Schotten-Baumann conditions.

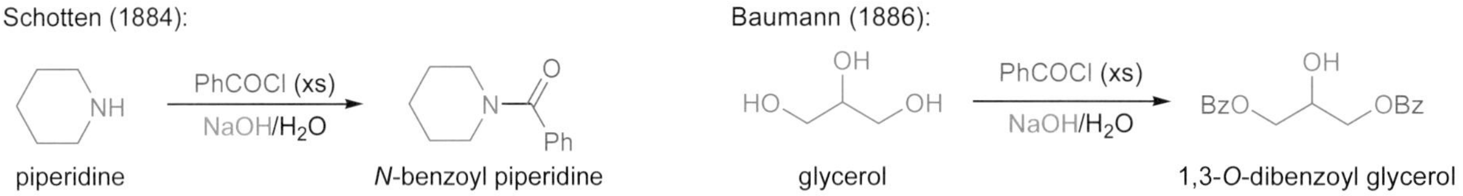

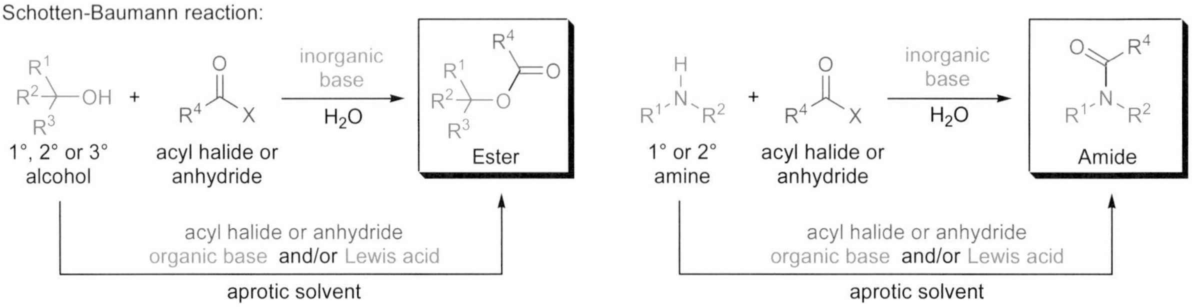

R^{1-3} = H, alkyl, aryl, heteroaryl; R^4 = alkyl, aryl; X = F, Cl, Br, $OCOR^4$; inorganic base: NaOH, KOH, Na_2CO_3, $NaHCO_3$; organic base: pyridine, DMAP, Et_3N, $(i\text{-}Pr)_2NEt$, PPh_3; Lewis acid: $MgBr_2$, $Sc(OTf)_3$, $Yb(OTf)_3$, TMSOTf

Mechanism: [23,24,4,25-27]

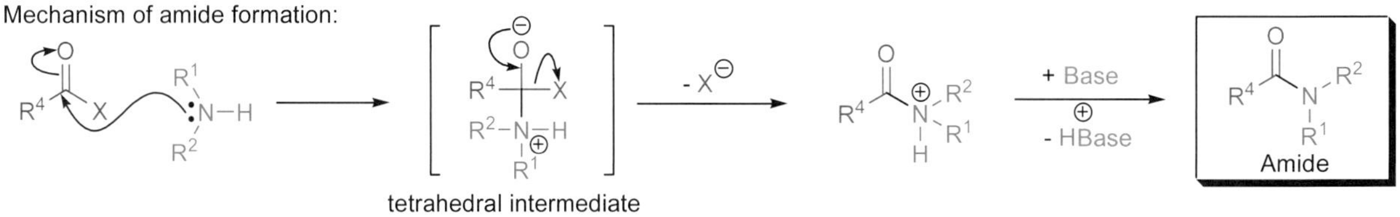

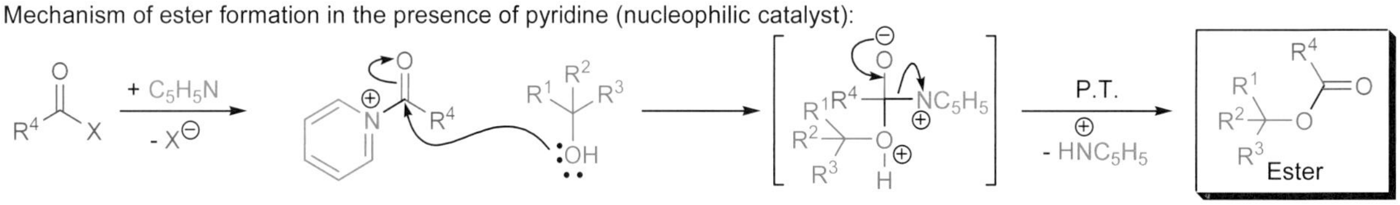

Synthetic Applications:

The first enantioselective total synthesis of (–)-tejedine was completed by P.E. Georghiou using a chiral auxiliary-assisted diastereoselective *Bischler-Napieralski cyclization* as one of the key steps.[28] The chiral auxiliary was the commercially available (*S*)-α-methylbenzylamine, which was coupled to the substrate using the original *Schotten-Baumann acylation* conditions. The acid chloride was reacted with the chiral amine in a solvent mixture containing aqueous sodium hydroxide and dichloromethane and the desired amide was isolated in excellent yield.

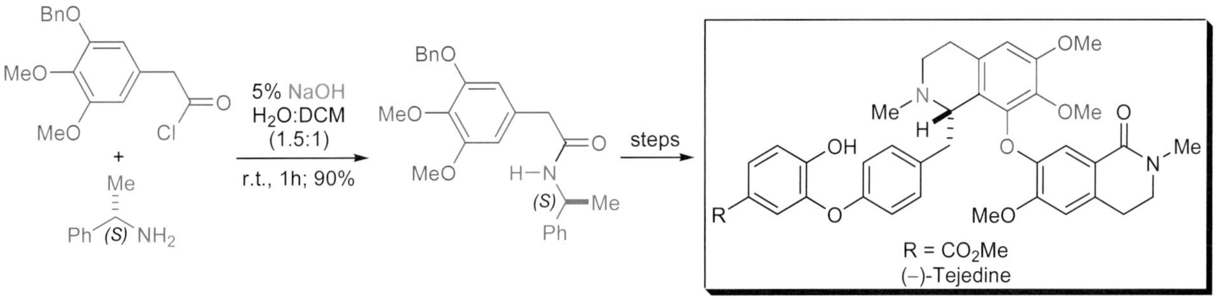

In the laboratory of A. Ganesan the short biomimetic total synthesis of the fumiquinazoline alkaloid fumiquinazoline G was accomplished.[29] The key step in the synthetic sequence was the dehydration of the anthranilamide residue in a linear tripeptide to the corresponding benzoxazine by reacting it with triphenylphosphine, iodine and Hünig's base. The authors initially prepared the linear tripeptide by condensing Fmoc-D-alanine with PyBroP as the acylating agent, but the product was formed only in a poor yield. When the peptide bond was formed under two-phase *Schotten-Baumann conditions* using sodium carbonate as the base, the desired tripeptide was isolated in high yield.

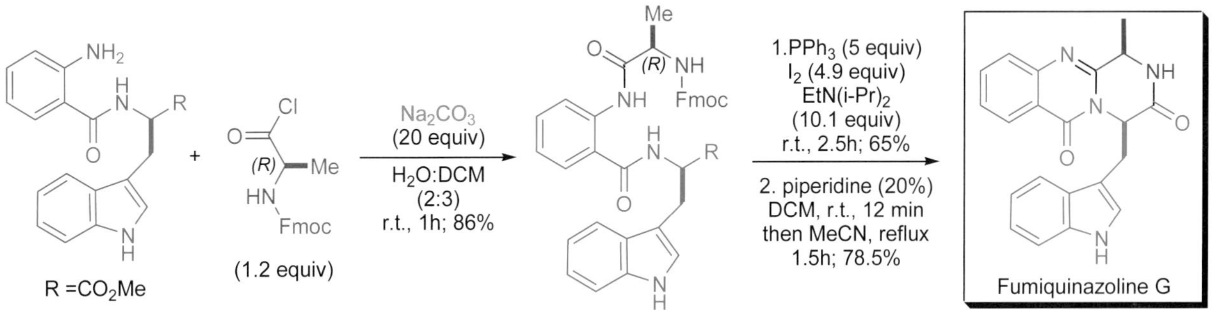

One of the intermediates in sphingolipid biosynthesis and degradation is ceramide, which influences certain cellular processes such as apoptosis and cell differentiation. The research team of P. Herdwijn prepared several ceramide analogues with substituted aromatic rings in the sphingoid moiety and evaluated their biological activity in hippocampal neurons.[30] The ceramide analogue with a thiophenyl sphingoid moiety was prepared by the *Schotten-Baumann acylation* of an amino diol with hexanoyl chloride. Since the nucleophilicity of the amino group is far greater than that of the hydroxyl groups, the acylation took place selectively to form the corresponding amide.

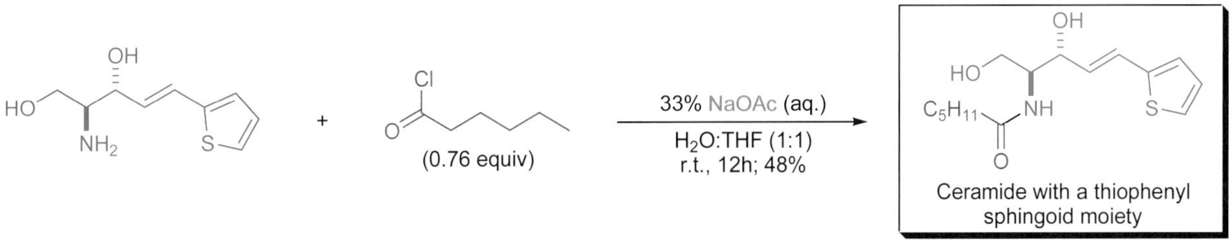

The first asymmetric synthesis of (+)-cannabisativine was achived by D.L. Comins et al. using the addition of metallo enolates to a chiral 1-acylpyridinium salt as one of the key steps.[31] The amide bond was created under the *Schotten-Baumann conditions* from a bicyclic acid chloride and a 1,4-amino alcohol.

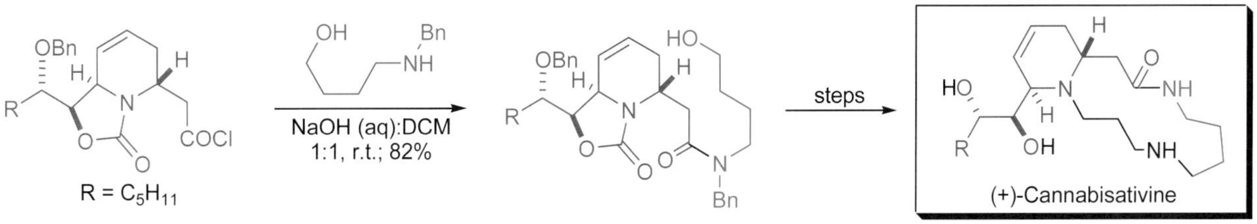

SCHWARTZ HYDROZIRCONATION

(References are on page 671)

Importance:

[*Seminal Publications*[1-4]; *Reviews*[5-18]; *Modifications & Improvements*[19-22]; *Theoretical Studies*[23-29]]

In 1970, P.C. Wailes and H. Weigold were the first to prepare zirconocene hydrochloride[2] (Cp_2ZrHCl) by the reduction of Cp_2ZrCl_2, but it was J. Schwartz who examined its reactions with a wide range of substrates and developed it as a useful reagent for organic synthesis.[3] The reaction of Cp_2ZrHCl (*Schwartz reagent*) with multiple bonds to form alkyl- and alkenylzirconium compounds is called the *Schwartz hydrozirconation*. The general features of the reaction are: 1) the hydrozirconation of alkenes and alkynes takes place at room temperature; 2) the reaction rate is orders of magnitude faster in ether solvents (e.g., THF, oxetane) than in hydrocarbon solvents such as hexanes and benzene; 3) under thermodynamic control, terminal or internal alkenes all give the terminal alkylzirconium compound because a rapid "chain walk" takes place to relieve the steric crowding;[12] 3) the order of reactivity for alkenes and alkynes are: terminal alkyne > terminal monosubstituted alkene ≈ internal alkyne > internal disubstituted alkene ≈ 2,2-disubstituted terminal alkene ≈ conjugated polyene > trisubstituted alkene; 4) tetrasubstituted alkenes generally do not react; 5) internal alkynes react regioselectively to give an alkenylzirconium compound in which the zirconium is located on the carbon closer to the smaller substituent; 6) conjugated dienes are hydrozirconated at the sterically less hindered double bond; and 7) the alkyl- and alkenylzirconium compounds are easily transmetallated to other useful organometallic compounds that can be used in various coupling reactions (e.g., *Negishi cross-coupling*) or can be trapped with small electrophiles (e.g., halogens, CO, isonitrile, H^+, etc.) with retention of configuration at carbon.

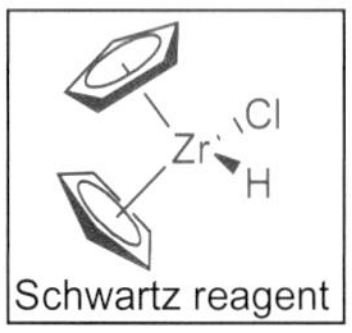

Order of reactivity in hydrozirconation reactions:

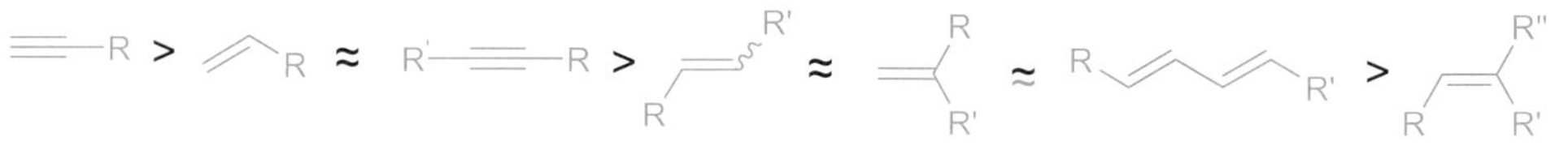

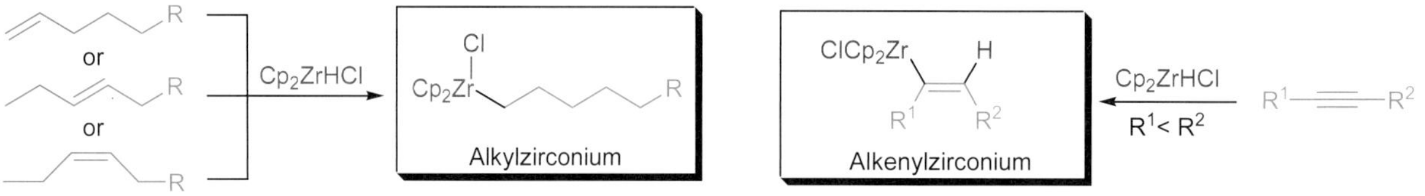

Synthetic utility of alkyl- and alkenylzirconium componds:

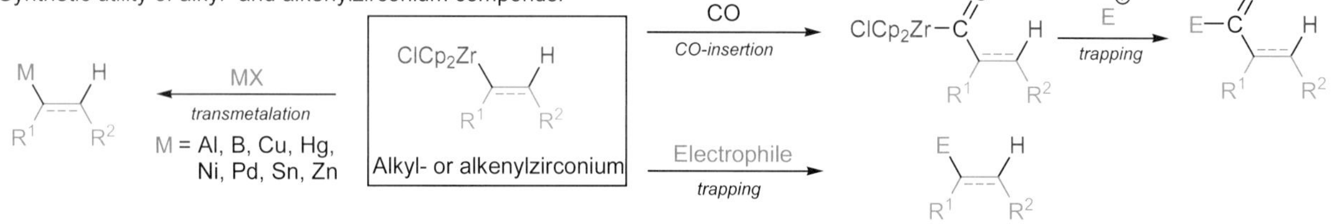

Mechanism: [30-32,25,33,34,12,35]

The *Schwartz hydrozirconation* is closely related to the *Brown hydroboration* reaction, but its mechanistic details are poorly understood mainly because of the oligomeric character of the Schwartz reagent, which makes the elucidation of the reaction kinetics very difficult. The fact that solvents with donor heteroatoms (e.g., THF, oxetane) accelerate hydrozirconations suggests that there is a rate-limiting dissociation of the oligomer before the addition to the multiple bond takes place. In THF the reaction is zero order in Schwartz reagent, while in oxetane it is first order both in the reagent and the alkene (or alkyne) substrate. The hydrozirconation proceeds *via* a four-atom concerted transition state (formally symmetry allowed due to the vacant d-orbitals on Zr), while the hydroboration is formally symmetry-forbidden.[25] The insertion into C-C multiple bonds takes place with *syn* stereochemistry. The *ab initio* study of hydrozirconation revealed that the attack of alkene at Zr is the most favorable between the Cl and H ligands. The alkene-Zr 18-electron π-complex is not stabilized by metal to olefin back-donation, because the zirconium has no d-electrons. Interestingly, the 16-electron alkylzirconium σ-complex is thermodynamically more stable (~10 kcal/mol) than the alkene-Zr complex, which is the driving force for hydrozirconation. (The alkene complexes of late-transition metals, however, are more thermodynamically stable. Therefore, they rarely undergo hydrometallation reactions.)

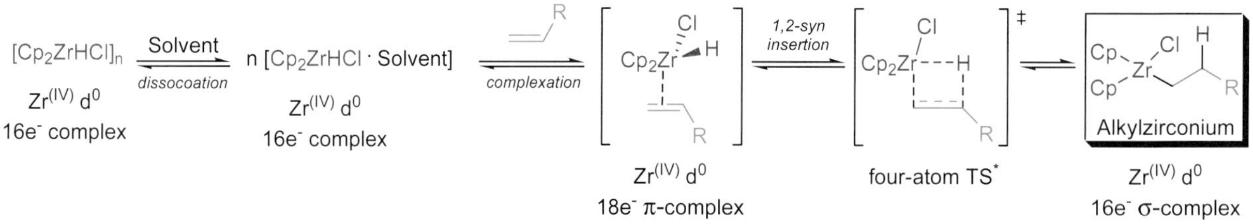

Synthetic Applications:

The *hydrozirconation* of polyunsaturated substrates can be plagued by extensive isomerization of double bonds. However, under carefully controlled reaction conditions, synthetically useful functionalization of conjugated polyenes can be realized. During the total synthesis of curacin A by P. Wipf and co-workers, a one-carbon homologation of conjugated triene substrate was achieved by *hydrozirconation* followed by an electrophile-trapping step.[36] The rate of formation of the desired terminal alkylzirconocene derivative was slow but was accomplished by heating the reaction mixture at 40 °C overnight. The treatment of the alkylzirconocene with *n*-butyl isocyanide and subsequent hydrolysis of the corresponding iminoylzirconocene with HCl gave the expected aldehyde in 54% yield.

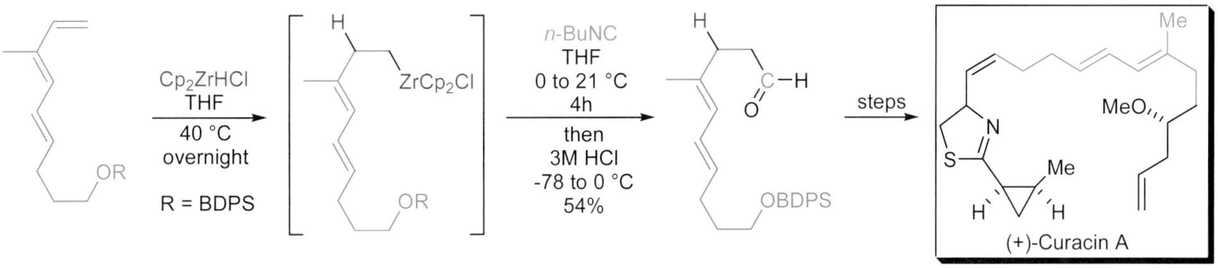

The total synthesis of apoptolidin was accomplished in the laboratory of K.C. Nicolaou.[37] The key C12-C28 vinyl iodide fragment was prepared using the *Schwartz hydrozirconation* of an internal alkyne followed by trapping of the alkenylzirconium intermediate with iodine (I_2). The vinyl iodide was formed as a 6:1 mixture of regioisomers. Under the reaction conditions, the methyl orthoester was converted to the methyl glycoside moiety at C21, which was presumably facilitated by the complexation of Zr with the pyranoside oxygen atom.

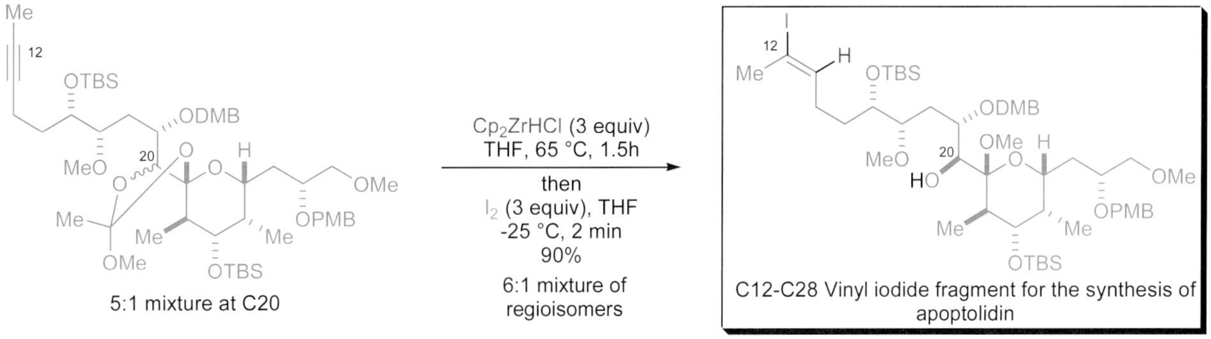

J. Montgomery and co-workers established the stereochemistry of isodomoic acid G through its first total synthesis.[38] The key step to construct the pyrrolidine ring was the nickel-catalyzed coupling of an alkynyl enone with an *in situ* formed alkenylzirconium. The terminal alkyne was then exposed to the Schwartz reagent, and subsequently the alkynyl enone was added along with catalytic amounts of $Ni(COD)_2$ and $ZnCl_2$. The initial alkenylzirconium regioselectively added across the internal alkyne and was first transmetallated to an organozinc and subsequently to an organonickel intermediate. This organonickel compound underwent an *intramolecular 1,4-addition* with the enone to form the pyrrolidine ring. This one-pot operation set all the necessary stereocenters of the natural product including the stereoselective formation of the highly substituted 1,3-diene side-chain.

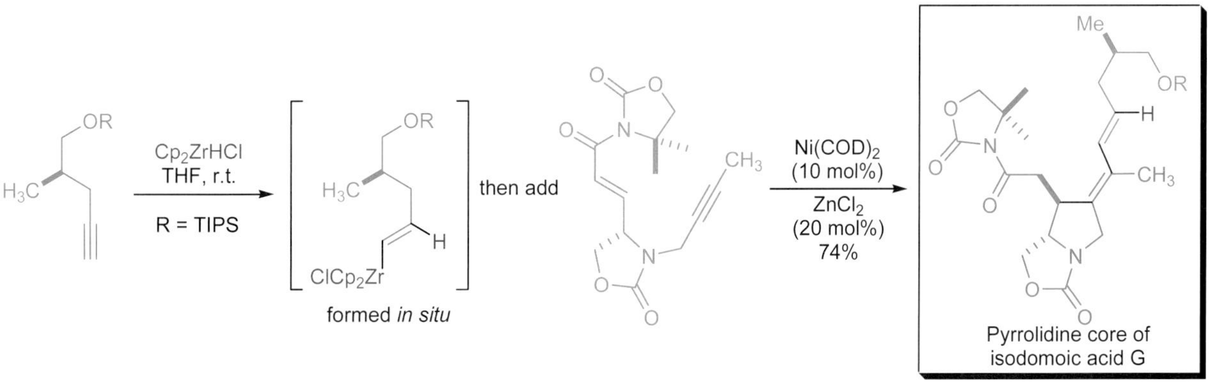

SEYFERTH-GILBERT HOMOLOGATION

(References are on page 672)

Importance:

[*Seminal Publications*[1-7]; *Reviews*[8-10]; *Modifications & Improvements*[11-17]]

In 1973, E.W. Colvin and B.J. Hamill reported a convenient one-step conversion of aldehydes and ketones to acetylenes using trimethylsilyldiazomethane or dimethylphosphonodiazomethane (a compound first synthesized by D. Seyferth[2]) under basic conditions. In a subsequent paper, the authors noted that the transformation worked well only for non-enolizable carbonyl compounds such as diaryl ketones and aromatic aldehydes with electron-withdrawing groups. In 1979, J.C. Gilbert and U. Weerasooriya disclosed an improved procedure that dramatically increased the scope of the reaction.[6] The one-pot conversion of carbonyl compounds to the corresponding terminal or internal alkynes using α-diazophosphonates under basic conditions is known as the *Seyferth-Gilbert homologation*. The general features of this transformation are: 1) the phosphonate reagents are not commercially available, but they can be prepared readily;[18] 2) in the original procedure developed by Gilbert, the dialkylphosphonodiazomethane (DAMP) was deprotonated with a strong base such as an alkyllithium or potassium *tert*-butoxide, and the carbonyl compound was added at low temperature under an inert atmosphere. The product alkyne was isolated upon a simple aqueous work-up (this procedure is only rarely used, since base-sensitive substrates do not tolerate the strongly basic conditions); 3) in the *Ohira-Bestmann modification* the dimethyl-1-diazo-2-oxopropylphosphonate is added to a solution of K_2CO_3 and the aldehyde in methanol at room temperature. After several hours of stirring, the product is isolated upon aqueous work-up in excellent yield (this modified procedure is by far the most popular). The key features of the *Ohira-Bestmann protocol* are: 1) the reaction conditions are mild, and most functional groups are tolerated; 2) highly sensitive enantiopure α-alkoxy aldehydes do not undergo racemization; 3) aliphatic, aromatic, as well as arylalkyl aldehydes are homologated to the corresponding terminal alkynes in excellent yields; 4) substrates containing highly C-H acidic bonds are homologated in high yields; and 5) α,β-unsaturated aldehydes do not undergo the transformation and the expected enynes are not formed (rather the homopropargylic methyl esters are obtained).

Colvin & Hamill (1973):

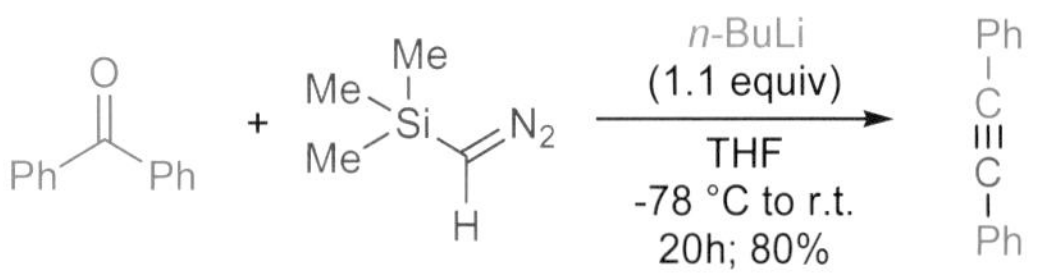

Gilbert & Weerasooriya (1979):

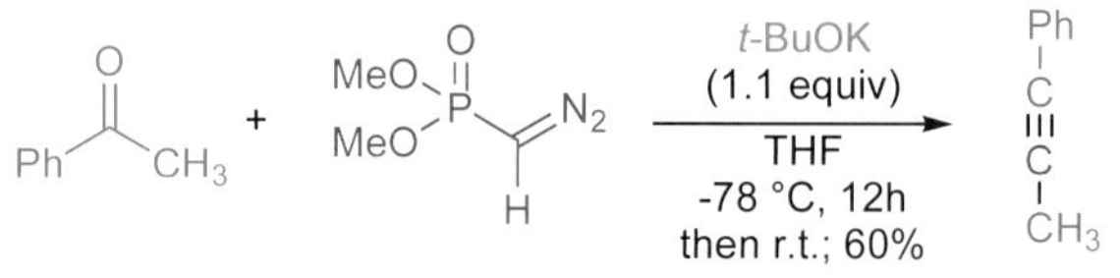

Seyferth-Gilbert homologation:

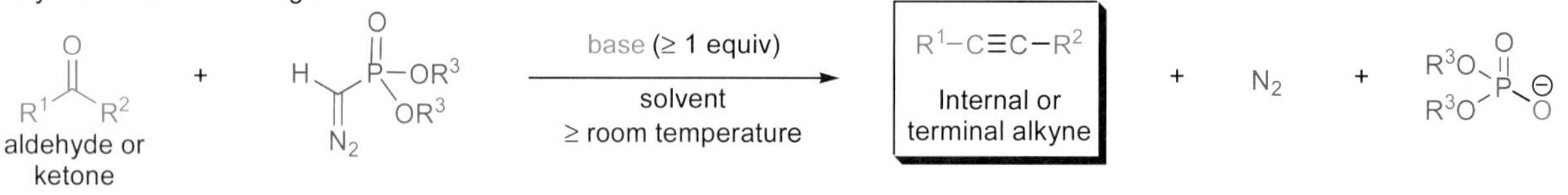

Modification for the synthesis of terminal alkynes (Ohira & Bestmann):

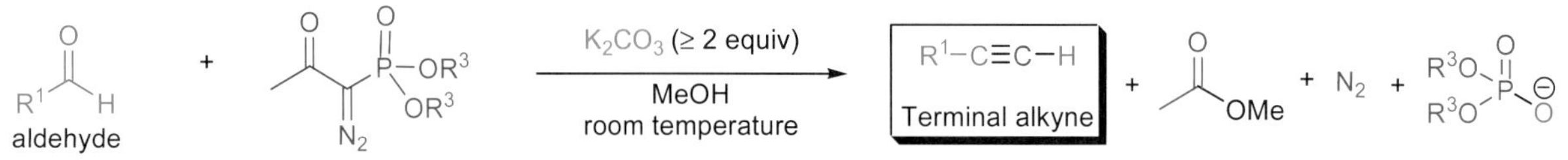

R^1 = alkyl, aryl, heteroaryl; R^2 = H, aryl, heteroaryl; R^3 = Me, Et; base: *n*-BuLi, KO-*t*Bu

Mechanism: [7]

In the *Ohira-Bestmann modified procedure* the first step is the deacylation of the reagent by a methoxide ion. The resulting carbanion (nucleophile) attacks the carbonyl group of the aldehyde or ketone and an oxaphosphetane-type intermediate is formed (just like in the *HWE olefination*), which breaks down to afford a thermally unstable diazoalkene. The diazoalkene loses dinitrogen (α-elimination) and the resulting alkylidenecarbene undergoes a 1,2-shift to give rise to the alkyne.

Formation of the dialkylphosphonodiazomethane from dialkyl-1-diazo-2-oxopropylphosphonate:

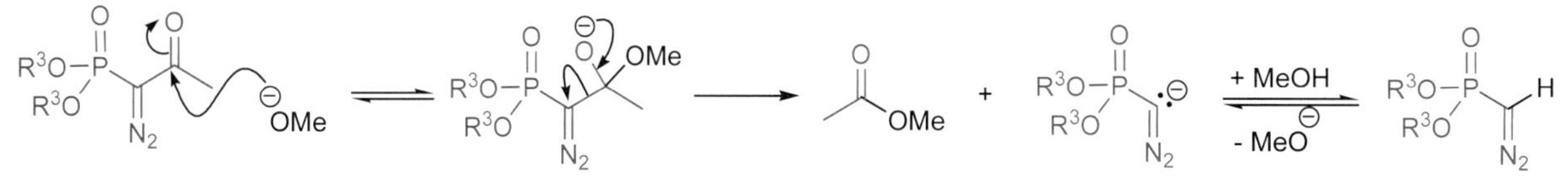

Reaction of the anion with carbonyl compounds:

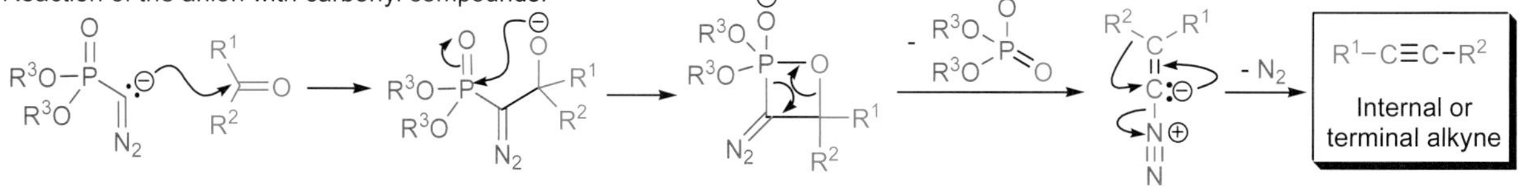

Synthetic Applications:

The total synthesis of the marine toxin polycavernoside A was achieved by J.D. White and co-workers.[19] In order to couple the central pyran moiety in a *Nozaki-Hiyama-Kishi reaction*, the aldehyde side chain had to be first homologated to the corresponding terminal alkyne and subsequently transformed into a vinyl bromide. The aldehyde substrate was treated under the *Ohira-Bestmann protocol*, and the desired alkyne product was obtained in high yield.

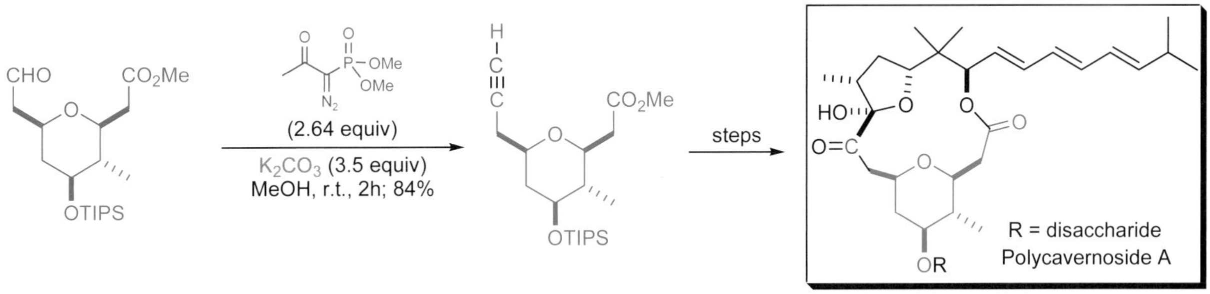

The tetraacetylenic compound (–)-minquartynoic acid was synthesized in the laboratory of B.W. Gung from commercially available azelaic acid monomethyl ester using a one-pot three-component *Cadiot-Chodkiewitz reaction* as the key step.[20] This natural product shows strong anti-cancer and anti-HIV activity. One of the alkyne components was prepared using the *modified Seyferth-Gilbert homologation*.

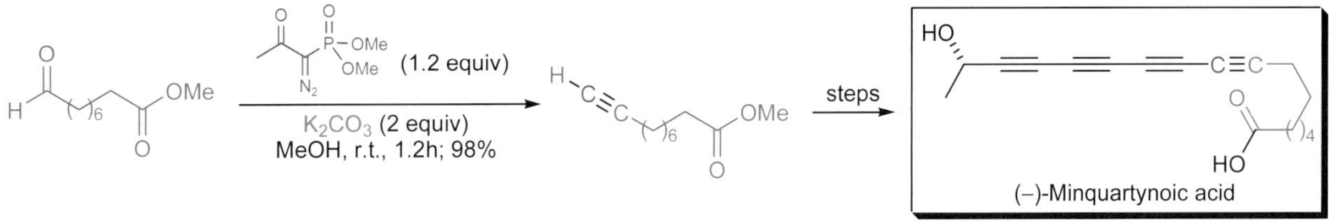

The stereoselective synthesis of the C5-C20 subunit of the aplyronine family of polyketide marine macrolides was accomplished by J.A. Marshall and co-workers.[21] The C15-C20 moiety was prepared using the original *Seyferth-Gilbert homologation* conditions. The diazophosphonate was deprotonated with potassium *tert*-butoxide at low temperature, and then the solution of the aldehyde was added slowly also at low temperature. Interestingly, the alternative *Corey-Fuchs alkyne synthesis* was unsuccessful on this substrate, since extensive decomposition was observed.

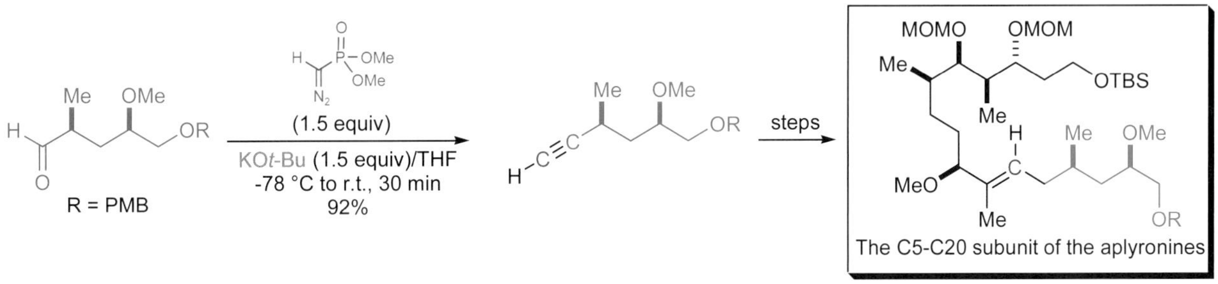

A structurally novel cancer therapeutic agent, (–)-laulimalide, was isolated from Pacific marine sponges in trace amounts, and it was shown to promote abnormal tubulin polymerization. P.A. Wender et al. applied the *modified Seyfert-Gilbert homologation* on a complex substrate in the endgame of the total synthesis to obtain the desired terminal alkyne.[22]

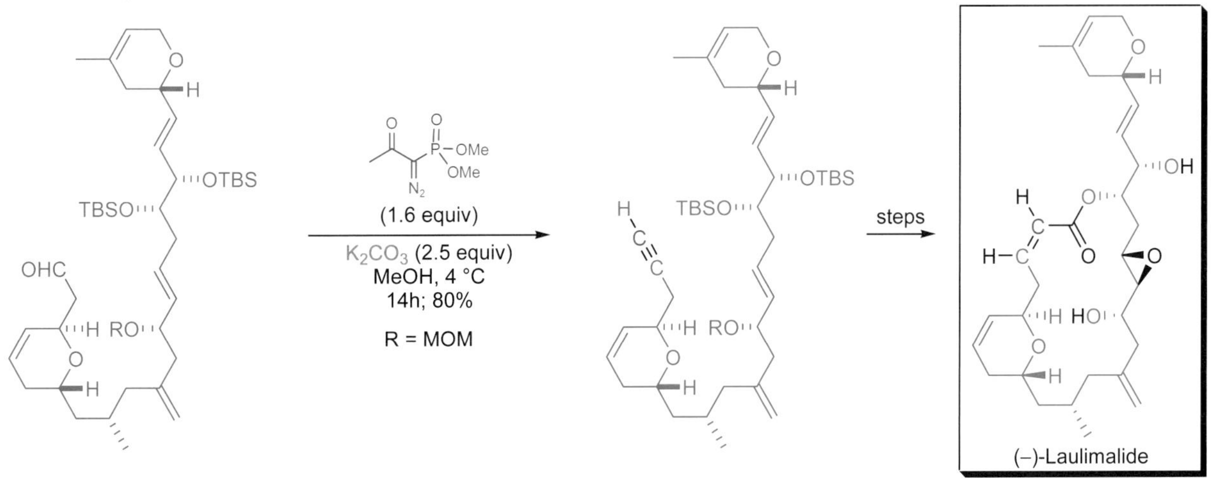

SHARPLESS ASYMMETRIC AMINOHYDROXYLATION

(References are on page 673)

Importance:

[*Seminal Publications*[1]; *Reviews*[2-8]; *Modifications & Improvements*[9-21]; *Theoretical Studies*[22]]

In 1996, K.B. Sharpless et al. reported the one-pot enantioselective synthesis of protected vicinal amino alcohols from simple alkenes.[1] This transformation is known as the *Sharpless asymmetric aminohydroxylation* (*SAA*), which complements the other asymmetric methods such as the *Sharpless asymmetric epoxidation (SAE) and dihydroxylation (SAD)* using olefins as substrates. The *SAA* is closely related to the *SAD*, since it uses the same chiral tertiary amine ligands and the factors that determine the enantioselectivity are similar. The β-amino alcohol moiety is an important pharmacophore, since it is a common structural motif in many biologically active compounds. This fact alone makes the *SAA* extremely valuable as a synthetic tool to access such compounds in good yield and with high enantioselectivity. The general features of the *SAA* are: 1) most olefins are substrates for the reaction: the best substrates have an electron-withdrawing group (e.g., CO_2R, $P(O)(OR)_2$, $CONR_2$) and tetrasubstituted alkenes do not react; 2) unlike in the *SAD*, there is no preformed reagent mixture (such as the AD-mixes) available, but the necessary components are the same except for the nitrogen source; 3) generally the nitrogen source is the alkali metal salt of an *N*-halogenated sulfonamide (X = Ms, Ns, Ts),[1,23] alkyl carbamate (X = Cbz, Boc, Teoc),[10,13,14] or amide (X = Ac);[9,24] 4) in the case of sulfonamides and acetamides, the *N*-haloamine salt is prepared from the corresponding *N*-haloamides while carbamates are prepared *in situ* by using *t*-BuOCl/NaOH; 5) the smaller the substituent (X) on the nitrogen source, the higher is the enantiopurity of the product; 6) to achieve the highest possible yield, a large excess (~3-6 equivalents) of the nitrogen source should be applied; 7) when sulfonamides are used, the substrate scope is limited to alkenes with electron-withdrawing groups, but the use of carbamates increases the substrate scope considerably; 8) just as in the *SAD,* the use of chiral bidentate tertiary amine ligands (DHQ- and DHQD-derived) give enantio-complementary results; 9) the absolute stereochemistry can be predicted with the "mnemonic device" proposed for the *SAD* and the asymmetric induction is of the same sense and similar magnitude for structurally related substrates; 10) the regioselectivity is hard to predict, since it is influenced by many factors but in the case of unsymmetrical alkenes the nitrogen generally adds to the less substituted carbon, while cinnamate esters react to give preferentially the β-amino ester product; 11) the nature of the ligand and the solvent system usually has a dramatic effect on the regioselectivity for styrene substrates;[14] 12) diols are often side-products in the *SAA* reactions, but there are several ways to reduce the extent of the dihydroxylation.[14,24,25]

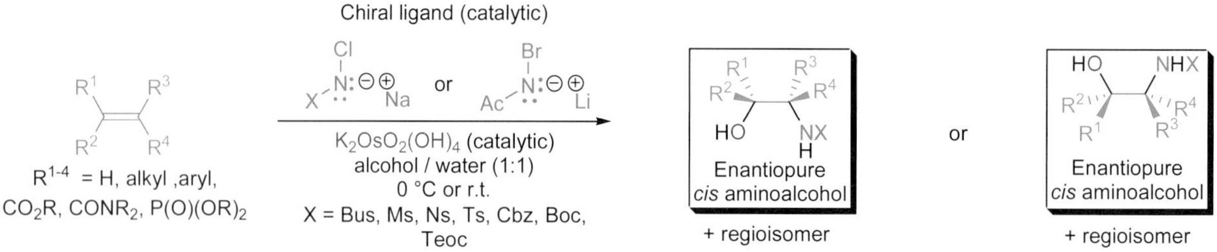

Mechanism: [23,18,24-26]

The mechanisms of the SAD and SAA are similar. The first step in mechanism of the SAA is the formal *[2+2] or [3+2] cycloaddition* of the imidotrioxoosmium(VIII) species with the olefin in a *syn*-stereospecific fashion to give eventually an osmium(VI) azaglycolate intermediate. This azaglycolate is then oxidized by the nitrogen source, while the ligand is lost and subsequent hydrolysis affords the 1,2-*cis* amino alcohol product and the imidotrioxoosmium(VIII) species which reenters the catalytic cycle.

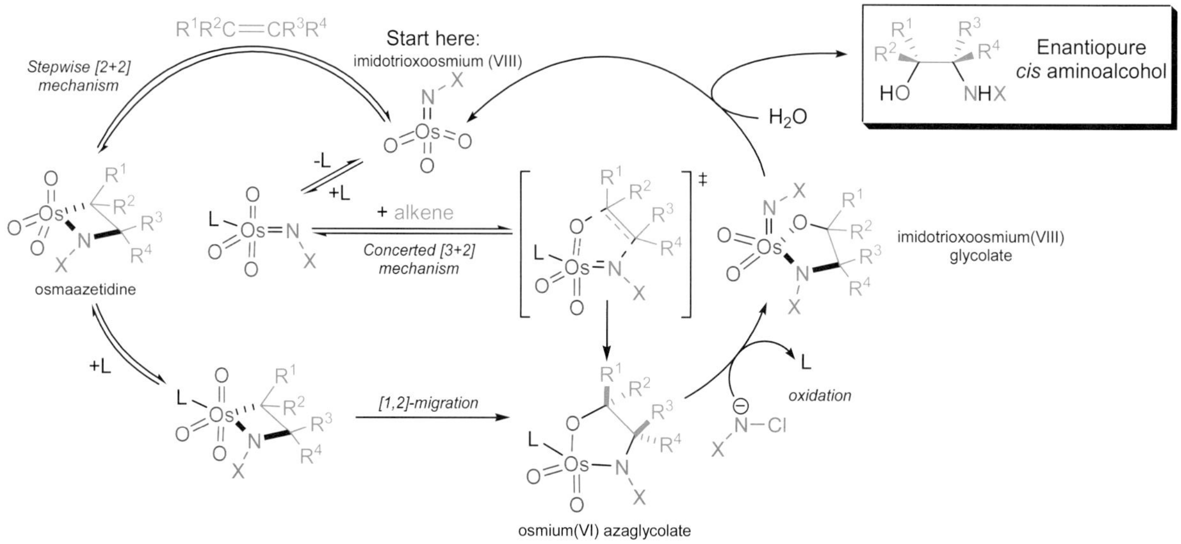

Synthetic Applications:

The *Sharpless regioreversed asymmetric aminohydroxylation* protocol was used as a key step in the total synthesis of ustiloxin D by M.M. Joullié and co-workers.[27] The (*E*)-ethyl cinnamate derivative was subjected to *in situ* generated sodium salt of the *N*-Cbz chloroamine in the presence of catalytic amounts of the anthraquinone-based chiral ligand to afford the desired *N*-Cbz protected (2*S*,3*R*)-β-hydroxy amino ester in good yield and with good diastereoselectivity.

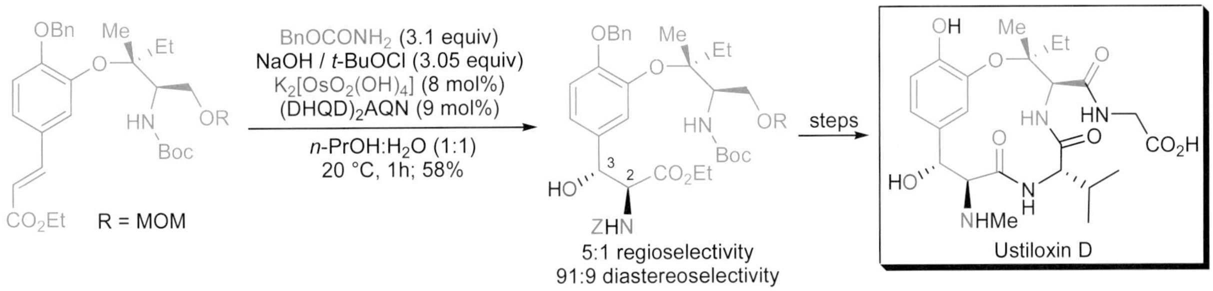

Research by B. Jiang et al. showed that the *asymmetric aminohydroxylation* of vinyl indoles can afford (*S*)-*N*-Boc protected α-indol-3-ylglycinols in moderate to good yield and with up to 94% *ee*.[28] The use of these enantiopure intermediates allowed the short enantioselective total synthesis of bisindole alkaloids, such as dragmacidin A, which contains a piperazine moiety between the indole rings.

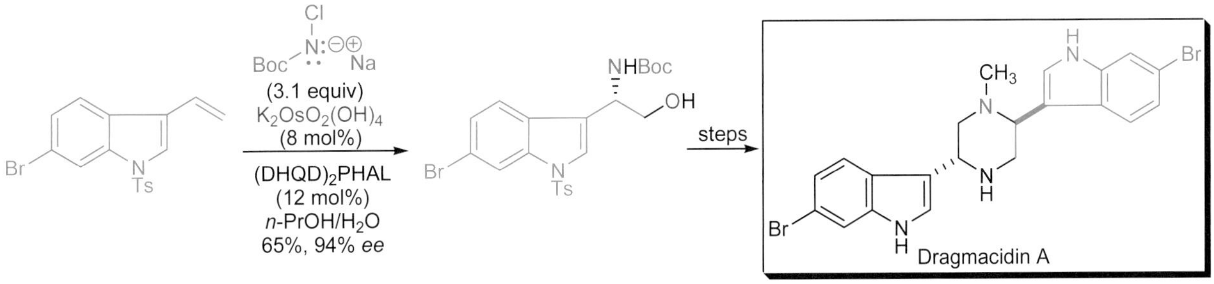

During the total synthesis of the teicoplanin aglycon, the *Sharpless asymmetric aminohydroxylation* was used twice to prepare the required G- and F-ring phenylglycine precursors by D.L. Boger and co-workers.[29] For the G-ring precursor the $(DHQD)_2PHAL$ ligand was used to obtain the *N*-Boc protected (*R*)-phenylglycinol, while the use of the pseudo enantiomer $(DHQ)_2PHAL$ ligand afforded the *N*-Cbz protected (*S*)-phenylglycinol.

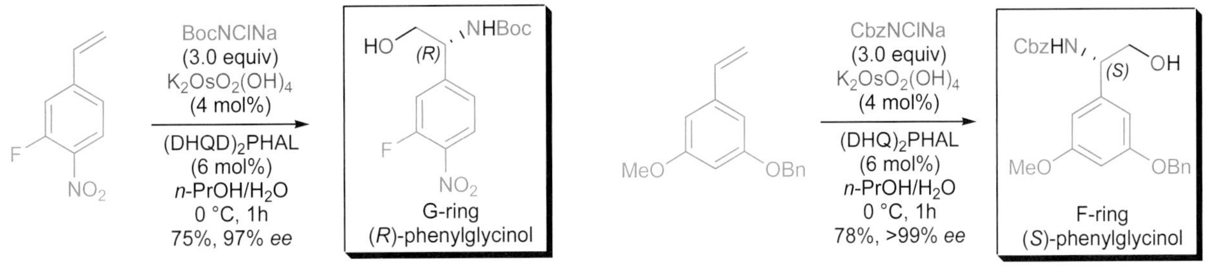

The stereocontrolled total synthesis of (–)-ephedradine A was accomplished by the research group of T. Fukuyama.[30] The highly stereoselective incorporation of the nitrogen atom at the benzylic position was achieved by using the *SAA*. Subsequently, the hydroxyl group was removed in two steps: first by conversion to the corresponding alkyl chloride, and then by subjecting the alkyl chloride to *transfer hydrogenation* to afford the β-amino ester.

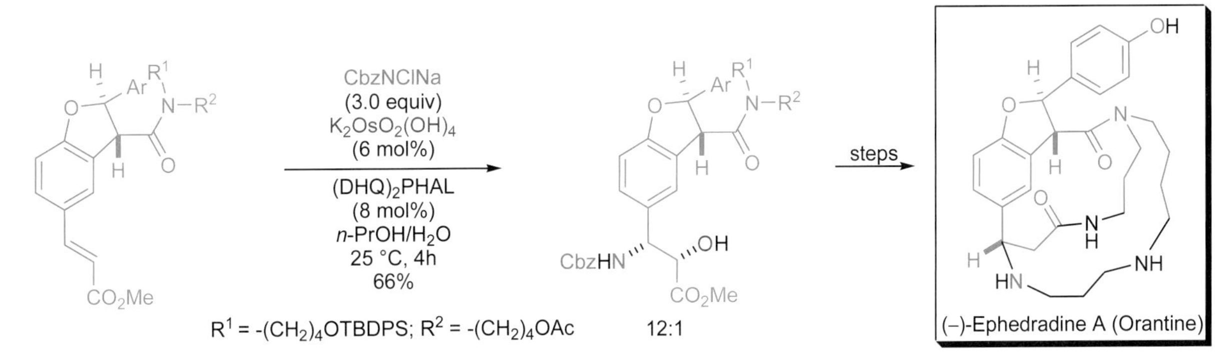

SHARPLESS ASYMMETRIC DIHYDROXYLATION

(References are on page 673)

Importance:

[*Seminal Publications*[1,2]; *Reviews*[3-21]; *Modifications & Improvements*[22-33]; *Theoretical Studies*[34-49]]

The reaction of osmium tetroxide (OsO_4) with olefins to give *cis* vicinal diols was discovered in the early 1900s[50] and since then it has undergone substantial developement.[51] At the beginning of the 1980s, the research group of K.B. Sharpless reported the first asymmetric dihydroxylation reaction of olefins with osmium tetroxide in the presence of dihydroquinine acetate, a chiral tertiary amine ligand that belongs to the family of *Chinchona* alkaloids. Today, this transformation is known as the *Sharpless asymmetric dihydroxylation* (*SAD*).[1] Sharpless's experiment was based on the observation of Criegee that certain tertiary amines (e.g., pyridine) accelerated the reaction of OsO_4 with olefins.[52] At this point the reaction was catalytic for OsO_4 but stoichiometric amount of the ligand was needed. When chiral tertiary diamines (e.g., $(DHQ)_2PHAL$ and $(DHQD)_2PHAL$) were introduced as ligands, it became feasible to use only sub-stoichiometric amounts of them, since these ligands considerably accelerated the rate of dihydroxylation compared to the monodentate chiral amines.[2] The phenomenon of rate acceleration caused by ligands is known as the *ligand accelerated catalysis* (*LAC*). The general features of the *SAD* are: 1) practically all alkenes are substrates for the reaction, but no other functional groups are affected; 2) electron-rich alkenes tend to react faster than electron-deficient ones; 3) the enantioselectivity is moderate for *cis*-disubstituted olefins having substituents that are similar in size (facial differentiation by the catalyst becomes very difficult); 4) all the reagents are solids, and they are commercially available as preformulated mixtures: AD-mix α and AD-mix β containing the necessary bidentate chiral ligand, stoichiometric oxidant, and the osmium tetroxide in the form of dipotassium osmate dihydrate ($K_2OsO_4(OH)_4$); 5) to predict the absolute configuration of the product, an empirical model (mnemonic device) was developed by Sharpless et al.[24] in which one has to examine the substrate and rank the substituents (R_S = small, R_M = medium and R_L = large) and place the large substituent in the southwestern corner (SW); to dihydroxylate from the bottom face (α-face) one should use AD-mix α and to dihydroxylate from the top face (β-face) AD-mix β should be used; and 6) the reaction is usually conducted in *tert*-butanol/water = 1:1 at room temperature and 1.4g of the necessary AD-mix is added for each mmol of the olefinic substrate.

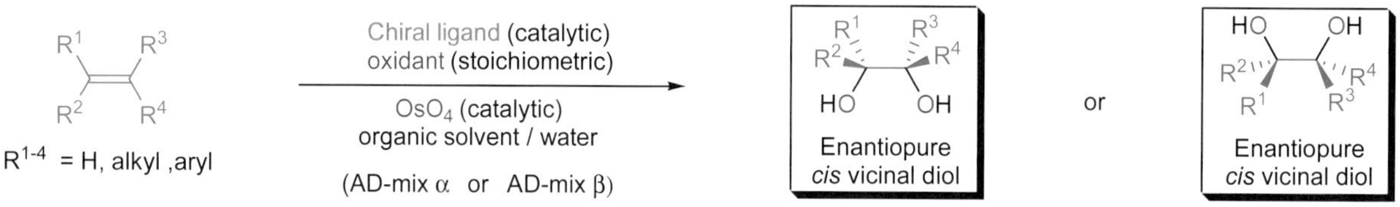

Empirical model (mnemonic device):

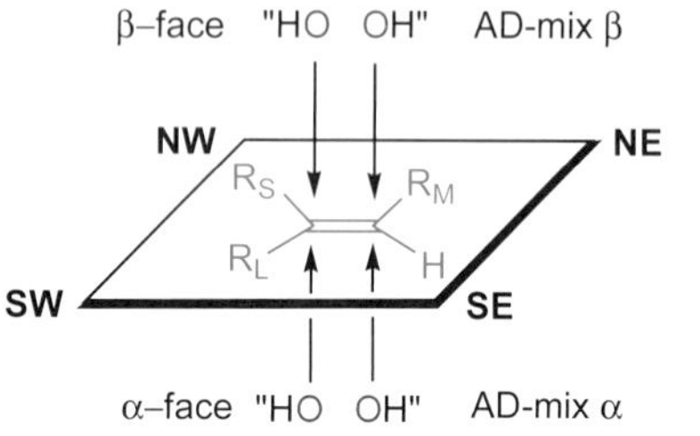

AD-mix α: $(DHQ)_2PHAL + K_2OsO_2(OH)_4 + K_3Fe(CN)_6$

AD-mix β: $(DHQD)_2PHAL + K_2OsO_2(OH)_4 + K_3Fe(CN)_6$

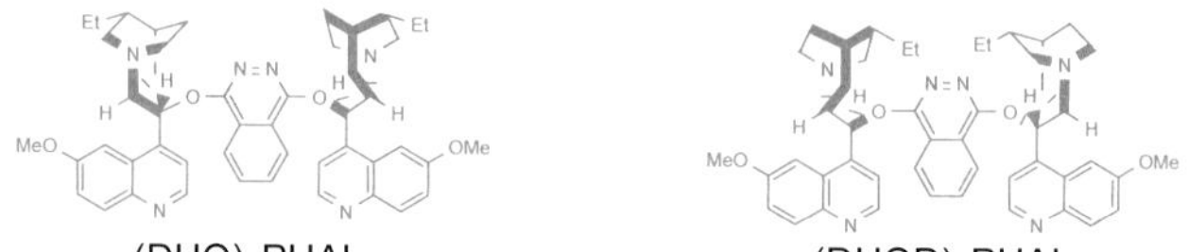

$(DHQ)_2PHAL$ $(DHQD)_2PHAL$

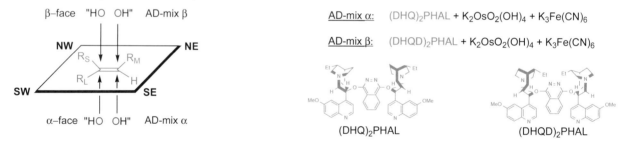

Mechanism:[53-77]

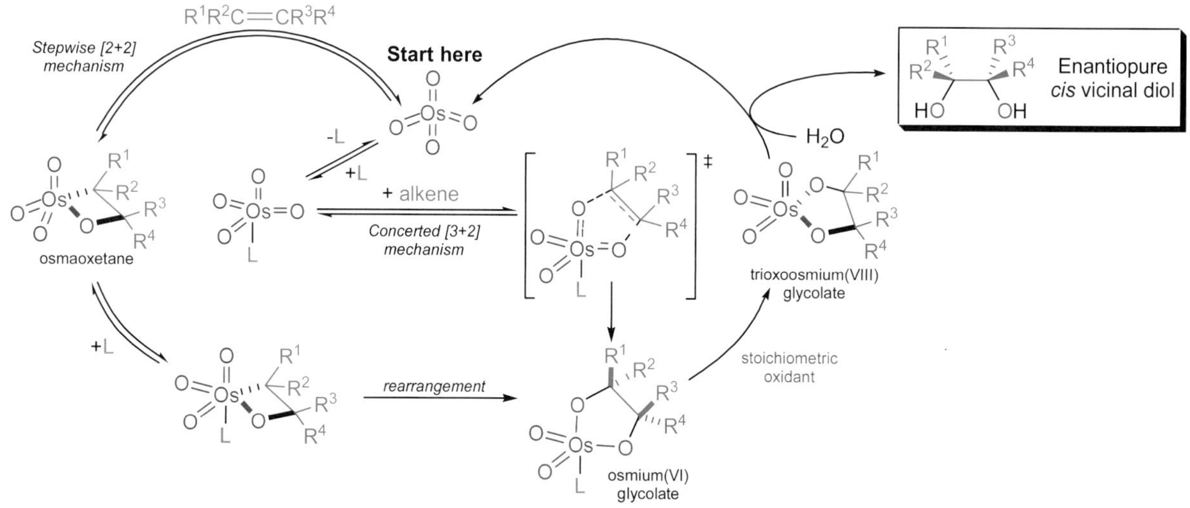

Synthetic Applications:

The total synthesis of (+)-zaragozic acid C was accomplished in the laboratory of A. Armstrong using a double *Sharpless asymmetric dihydroxylation* of a diene as the key step.[78] The stereochemistry of four contiguous stereocenters (C3 to C6) were controlled this way. Interestingly, the double dihydroxylation could not be performed efficiently (low yield, low *ee*) in one-pot, so it was conducted in two separate steps. In the first step, the diene was subjected to Super AD-mix β (commericial AD-mix supplemented with extra ligand and osmium tetroxide) for 4 days to afford regioisomeric triols in 78% yield. In the second step NMO was used as the stoichiometric oxidant, which afforded the desired pentaol with good diastereoselectivity. This two-step procedure was conducted on multigram scale, which allowed the completion of the total synthesis.

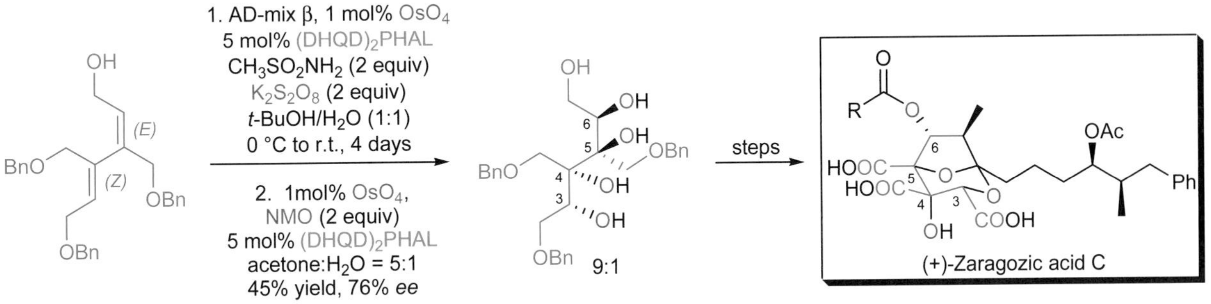

The key component of the cell wall lipopolysaccharide of Gram-negative bacteria, KDO (3-deoxy-D-manno-2-octulosonic acid), was synthesized by S.D. Burke and co-workers.[79] One of the key transformations in the synthetic sequence was a double *SAD* of a 6-vinyldihydropyran-2-carboxylate template. This 1,4-diene was cleanly converted to a mixture of two C7 epimeric tetraols in a 20:1 ratio. The endocyclic olefin had an intrinsic preference for dihydroxylation from the β-face and not from the desired α-face. This stereofacial bias was impossible to override with any ligand normally used in the *SAD*, so later in the synthesis these two stereocenters had to be inverted in order to give the required stereochemistry at C4 and C5.

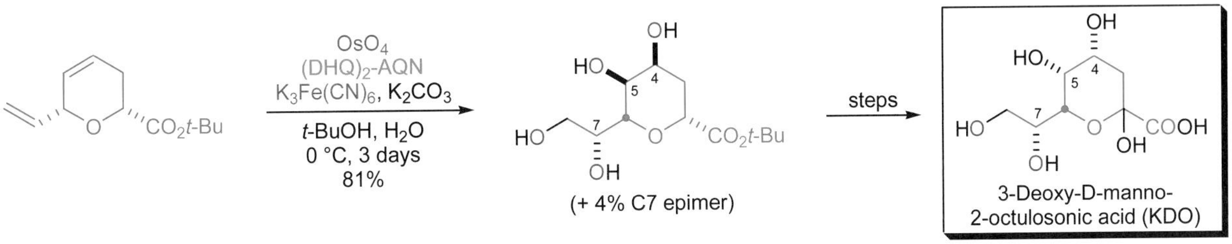

The total synthesis of (+)-1-epiaustraline, a tetrahydroxypyrrolizidine alkaloid, was achieved by S.E. Denmark et al. who used a tandem *intramolecular [4+2] / intermolecular [3+2] nitroalkene cycloaddition* as the key ring forming reaction.[80] During the endgame of the synthesis, the last stereocenter was installed by the *SAD* of the terminal olefin moiety on the tricyclic intermediate. It was found that most ligands in the dihydroxylation gave the undesired stereoisomer as the major product. Eventually, after exhaustive screening, a DHQD ligand with a phenanthracene spacer (DHQD-PHN) was found to produce the desired stereoisomer with good selectivity.

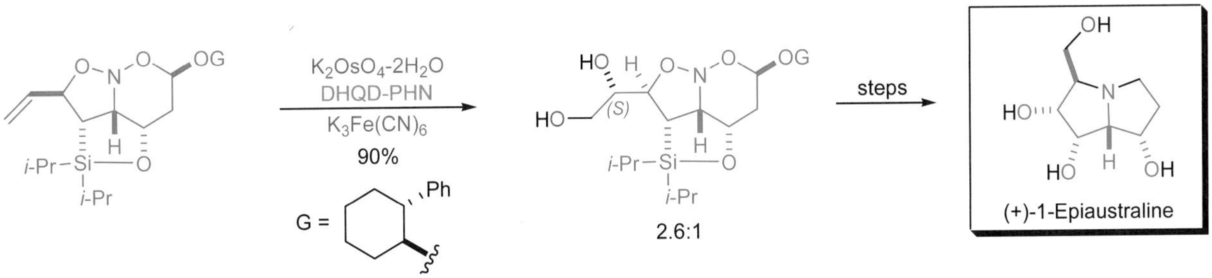

SHARPLESS ASYMMETRIC EPOXIDATION

(References are on page 675)

Importance:

[*Seminal Publications*[1]; *Reviews*[2-24]; *Modifications & Improvements*[25-27]; *Theoretical Studies*[28-33]]

In 1980, K.B. Sharpless and T. Katsuki reported the first practical method for asymmetric epoxidation.[1] They discovered that the combination of $Ti^{(IV)}$ tetraisopropoxide, optically active diethyl tartrate (DET) and *tert*-butyl hydroperoxide (TBHP) was capable of epoxidizing a wide variety of allylic alcohols in high yield and with excellent enantiomeric excess (>90% *ee*). The $Ti^{(IV)}$ alkoxide-catalyzed epoxidation of prochiral and chiral allylic alcohols in the presence of a chiral tartrate ester and an alkyl hydroperoxide to give enantiopure 2,3-epoxy alcohols is known as the *Sharpless asymmetric epoxidation* (*SAE*). The general features of this method are: 1) only allylic alcohols are good substrates for this method, since the presence of the hydroxyl group is essential; 2) allylic alcohols are epoxidized with high chemoselectivity in the presence of other olefins; 3) the epoxidation is totally reagent controlled: by using either (+)- or (-)-DET the corresponding enantiomer of the product 2,3-epoxy alcohol can be obtained; 4) the inherent diastereofacial bias of chiral allylic alcohols is overridden: in the "matched case" the reagent reinforces the inherent selectivity of the substrate and the epoxidation proceeds with extremely high stereoselectivity, while in the "mismatched case" the diastereofacial preference of the substrate and the reagent is opposite and the level stereoselectivity for the epoxidation is lower than in the matched case, but it is synthetically still useful;[34,35] 5) the enantiofacial selectivity of the *SAE* can be predicted for all prochiral allylic alcohols (no exceptions found to date!) using the scheme below; 6) if there is a chiral center at C1 (attached to OH group) the *SAE* will proceed with substantially different rates for the two enantiomers, so it can be used for the kinetic resolution of a racemic allylic alcohols; 7) the addition of catalytic amounts of molecular sieves to the reaction mixture allows the use of only catalytic amounts (5-10 mol%) of the Ti-tartrate complex; in the absence of molecular sieves, a full equivalent of this complex is needed;[25] 8) if the product is too reactive or its solubility properties make it difficult to isolate, the *in situ* derivatization (conversion to the corresponding ester) can be used to preserve the integrity of the epoxide and make the isolation easier;[26] 9) the reaction conditions tolerate most functional groups except for free amines; carboxylic acids, thiols, and phosphines; 10) in order to achieve high yield and enantiomeric excess, it is crucial to prepare the catalyst fresh by mixing the solutions of $Ti(Oi\text{-}Pr)_4$ and DET followed by the addition of TBHP at -20 °C and age the resulting mixture for 20-30 minutes prior to adding the allylic alcohol substrate; 11) the solvent of choice is alcohol-free dichloromethane; 12) most often DET is used, but occasionally DMT and DIPT are utilized; 13) titanium tetra *t*-butoxide is applied if the product epoxy alcohol (e.g., 2-substituted epoxy alcohols) is sensitive to ring-opening by the alkoxide; and 14) the molecular sieves must be activated (heat at 200 °C for 3h) and generally 3-5 Å molecular sieves are sufficient to remove any interfering amounts water.

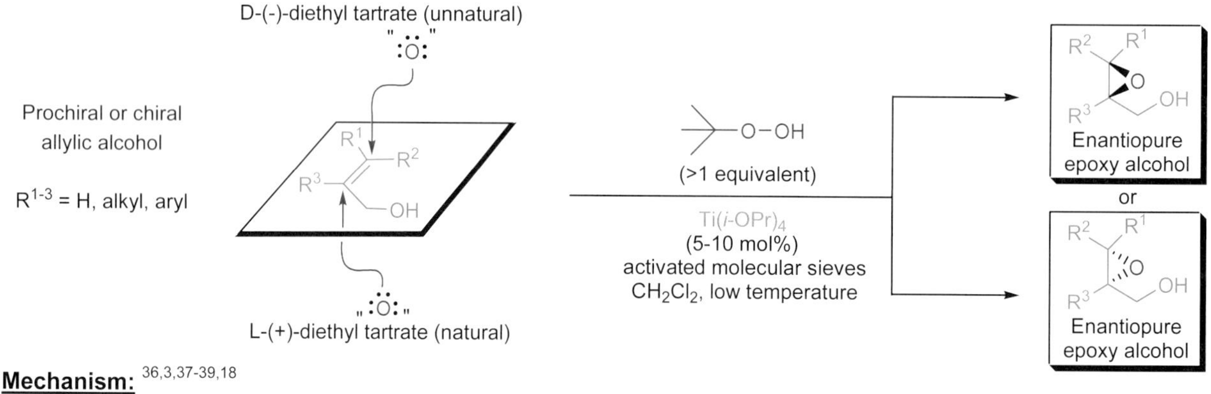

Mechanism: [36,3,37-39,18]

The first step is the rapid ligand exchange of $Ti(Oi\text{-}Pr)_4$ with DET. The resulting complex undergoes further ligand exchange with the allylic alcohol substrate and then TBHP. The exact structure of the active catalyst is difficult to determine due to the rapid ligand exchange but it is likely to have a dimeric structure. The hydroperoxide and the allylic alcohol occupy the axial coordination site on the titanium and this model accounts for the enantiofacial selectivity.

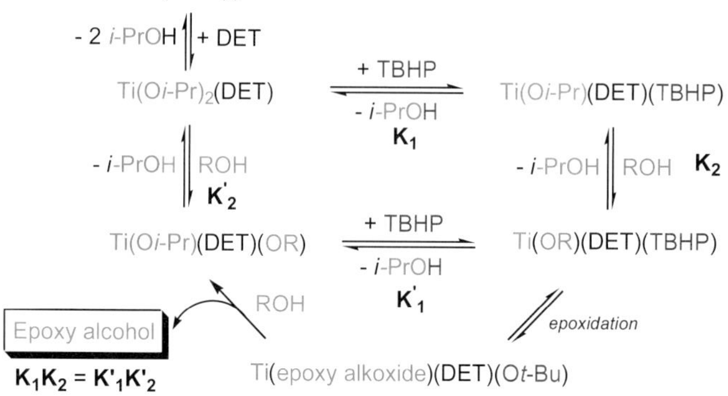

Transition state of epoxidation:

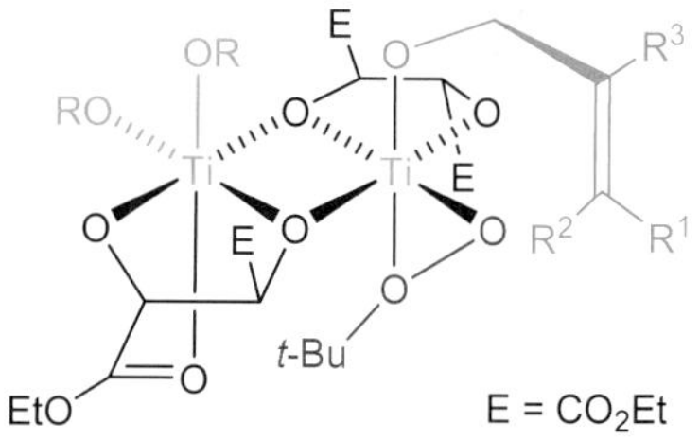

$$\text{Rate} = \frac{[Ti(Oi\text{-}Pr)_2(DET)]\,[TBHP]\,[ROH]}{[i\text{-}PrOH]^2}$$

Synthetic Applications:

The enantioselective total synthesis of the annonacenous acetogenin (+)-parviflorin was accomplished by T.R. Hoye and co-workers.[40] The *bis*-tetrahydrofuran backbone of the natural product was constructed using a sequential double *Sharpless asymmetric epoxidation* and *Sharpless asymmetric dihydroxylation*. The *bis* allylic alcohol was epoxidized using L-(+)-DET to give the essentially enantiopure *bis* epoxide in 87% yield.

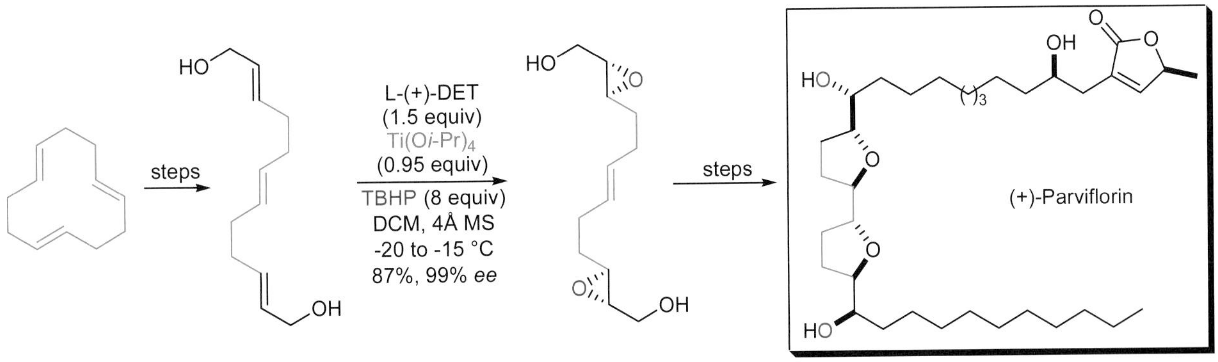

In the laboratory of D.P. Curran, the asymmetric total synthesis of (20*R*)-homocamptothecin was achieved using the *Stille coupling* and the *SAE* as key steps.[41] The *SAE* was used to install the key C20 stereocenter. The (*E*)-allylic alcohol was epoxidized rapidly in the presence of stoichiometric amounts of L-(+)-DET and TBHP at -20 °C to afford the corresponding epoxide in 93% *ee*. Interestingly, the (Z)-allylic alcohol reacted with D-(-)-DET sluggishly and gave the epoxide in very low yield and with only 31% ee.

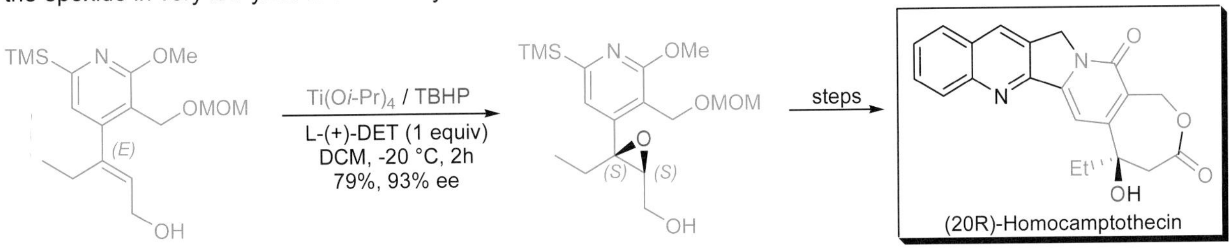

The last and key step during the total synthesis of (–)-laulimalide by I. Paterson et al. was the *Sharpless asymmetric epoxidation*.[42] The success of the total synthesis relied on the efficient kinetic differentiation of the C_{15} and C_{20} allylic alcohols during the epoxidation step. When the macrocyclic diol was oxidized in the presence of (+)-DIPT at -27 °C for 15h, only the C_{16}-C_{17} epoxide was formed.

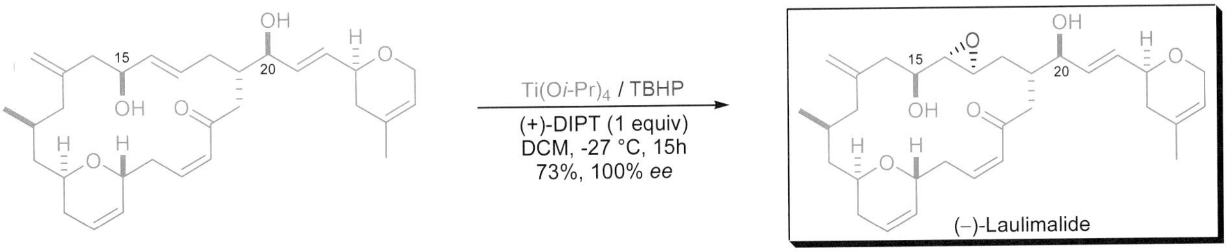

(+)-Madindoline A and (–)-madindoline B are potent and selective inhibitors of interleukin 6. The relative and absolute configuration of these natural products was determined by means of their total synthesis by A.B. Smith and S. Omura.[43] The key step was the *SAE* of the indole double bond, which led to the formation of the hydroxyfuroindole ring of both compounds.

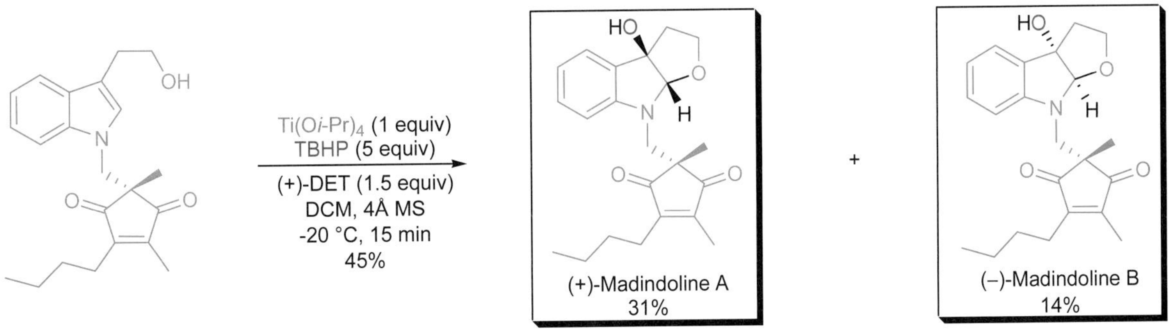

SHI ASYMMETRIC EPOXIDATION

(References are on page 676)

Importance:

[*Seminal Publications*[1-4]; *Reviews*[5-15]; *Modifications & Improvements*[16-28]; *Theoretical Studies*[29,30]]

When ketones are treated with Oxone (potassium peroxymonosulfate, $KHSO_5$), dioxiranes are formed that are capable of transferring an oxygen atom to a wide variety of substrates, and the ketones are regenerated after the oxygen-transfer.[13] For this reason dioxiranes are considered to be environmentally friendly and versatile oxidizing agents. Recently, dioxiranes have found use in asymmetric oxidation reactions such as epoxidation of alkenes. In 1984, R. Curci and co-workers reported the first chiral ketone-catalyzed asymmetric epoxidation of unfunctionalized olefins.[1] During the following decade several new chiral ketones (mainly biphenyl and binaphthyl-based ketones) were developed and tested as catalysts in asymmetric epoxidations.[8] In 1996, a fructose-derived ketone catalyst was prepared by Y. Shi and co-workers that showed very high enantioselectivities in epoxidation reactions.[2] Today, this transformation is known as the *Shi asymmetric epoxidation*. The general features of the reaction are: 1) either enantiomer of the catalyst can be prepared easily from D- or L-fructose in two steps;[31,2,4] 2) the pH of the reaction medium has a crucial effect on the outcome of the reaction: at high pH the oxidant (Oxone) decomposes rapidly, while at lower pH values the catalyst is decomposed *via* a *Baeyer-Villiger oxidation*, and this neccesitates the use of large amounts of catalyst;[3] 3) by keeping the pH at an optimum (~10.5), the epoxidation usually takes place with high enantiomeric excess at low catalyst loadings (20-30 mol%) without the need to use large excess of Oxone; 4) at the optimum pH the epoxide products are more stable than at lower pH values; 5) a wide variety of alkene substrates are epoxidized efficiently with high *ee*: homoallylic and bishomoallylic alcohols,[32] unsymmetrical dienes are epoxidized regioselectively to give vinyl epoxides,[16] conjugated enynes yield propargylic epoxides,[22] silyl enol ethers give α-hydroxy ketones;[19] and 6) *trans*-disubstituted and trisubstituted olefins give high enantioselectivities, whereas for *cis*-disubstituted and terminal olefins the *ee*'s are lower.[4]

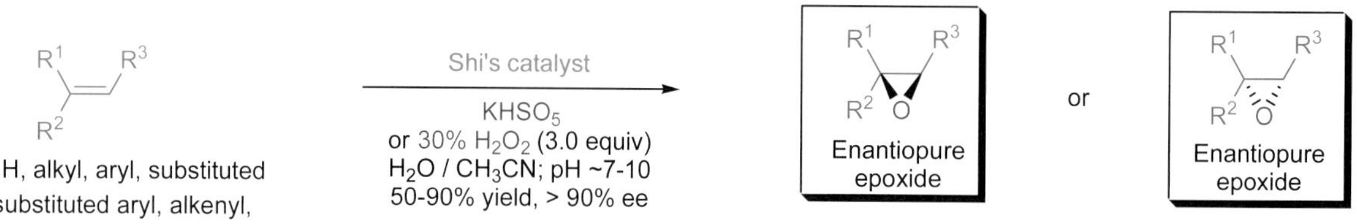

R^{1-3} = H, alkyl, aryl, substituted alkyl, substituted aryl, alkenyl, alkynyl

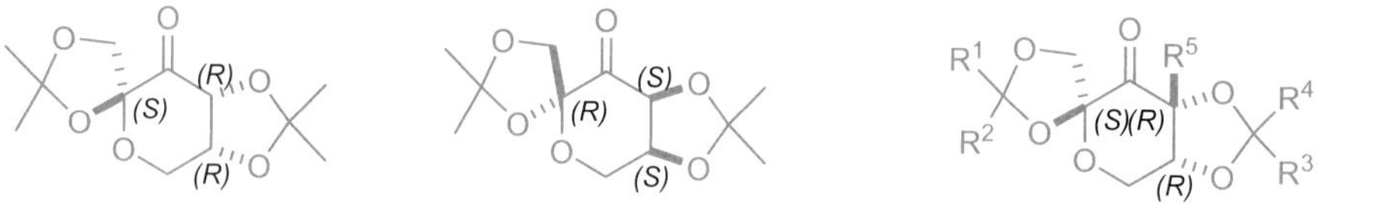

Shi's catalyst derived from D-fructose; Shi's catalyst derived from L-fructose; Shi's generalized catalysts (derived from D-fructose)

R^{1-2} = Me, Et, $-(CH_2)_4-$, $-(CH_2)_5-$, $-(CH_2)_6-$
R^{3-4} = Me, Et, i-Pr, H, F, Bn, $-(CH_2)_4-$, $-(CH_2)_5-$, $-(CH_2)_6-$
R^5 = H, F

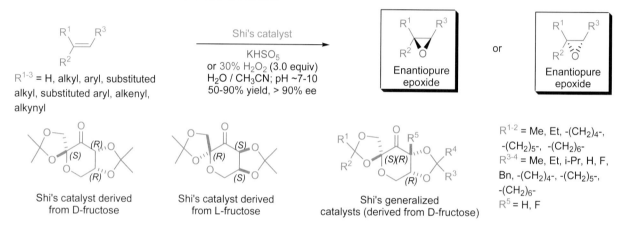

Mechanism: [2,33,4,34,20,8]

There are two possible transition states: spiro and planar. Nearly every example of *trans*-disubstituted and trisubstituted olefins which were studied with Shi's catalyst is consistent with the spiro transition state.[8] The extent of the involvement of the competing planar transition state depends on the nature of the substituents on the olefins.

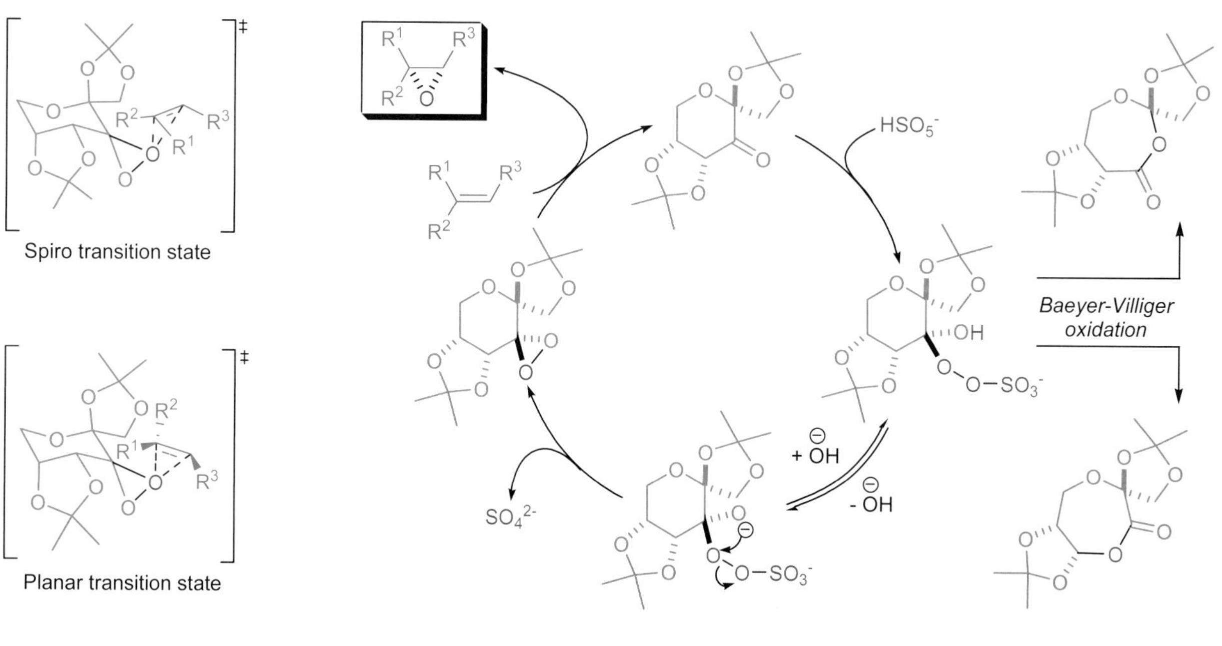

Synthetic Applications:

The synthesis of cryptophycin 52 was accomplished by E.D. Moher et al. using the *Shi epoxidation* as the key step to install the epoxide moiety diastereoselectively.[35] In the previous syntheses of this molecule, the epoxide moiety was always introduced in the last step, using common oxidants such as mCPBA or DMD, and with poor diastereoselectivity. Interestingly, the usual alkene precursor was a very poor substrate for the *Shi epoxidation*, so an earlier intermediate was subjected to the epoxidation conditions in which the pH was very carefully controlled. The desired epoxide was obtained as a 6.5:1 mixture of diastereomers.

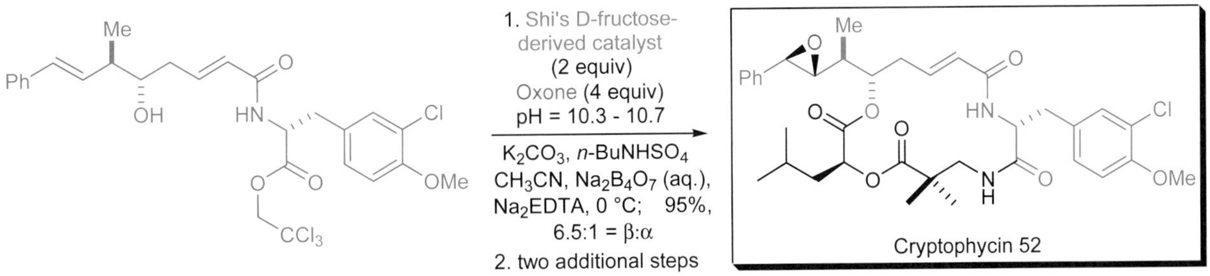

The *Shi epoxidation* employing the L-fructose derived catalyst was used during the total syntheses of (+)-murisolin and a library containing 15 of its diastereoisomers by D.P. Curran and co-workers.[36] The 4-mix/4-split strategy relied on the solution phase technique of fluorous mixture synthesis. One of the (*E*)-alkene substrates was subjected to the *Shi epoxidation* conditions to give 88% yield of the corresponding epoxide followed by ring-closure to the tetrahydrofuran by CSA. At the end of the synthesis, the four murisolin diastereomers were demixed by using FluoroFlash silica gel followed by detagging.

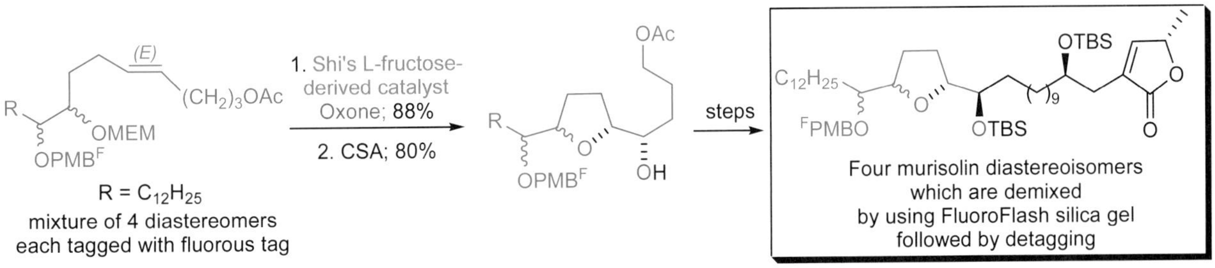

A novel *asymmetric epoxidation-ring expansion* strategy was used for the total synthesis of (+)-equilenin in the laboratory of M. Ihara.[37] This strategy involved the *Shi asymmetric epoxidation* of an aryl-substituted cyclopropylidene derivative to form a chiral oxaspiropentane followed by its enantiospecific rearrangement to the corresponding chiral cyclobutanone. The D-fructose-derived catalyst had to be used in large excess because the optimum yield and ee could be reached only at pH ~9 where the catalyst decomposed fairly rapidly. The authors also showed that by using the *Jacobsen epoxidation*, the enantiomeric excess could be slightly increased along with a slight decrease in the yield.

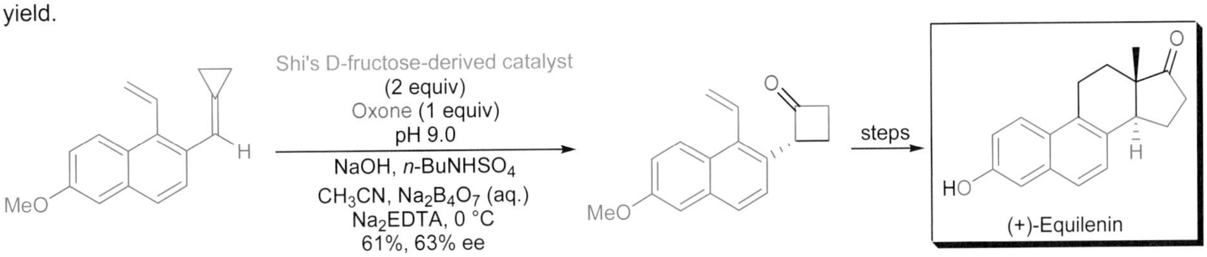

The *Shi epoxidation* was the key step in E.J. Corey's total synthesis of the chiral C_2-symmetric pentacyclic oxasqualenoid glabrescol.[38,39] Four epoxides were introduced in one step with an (*R*):(*S*) selectivity of 20:1.

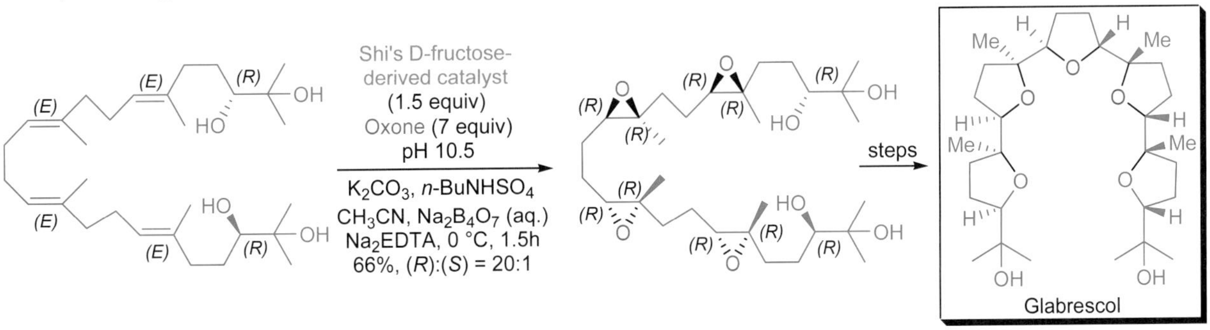

SIMMONS-SMITH CYCLOPROPANATION

(References are on page 677)

Importance:

[*Seminal Publications*[1,2]; *Reviews*[3-16]; *Modifications & Improvements*[17-33]; *Theoretical Studies*[34-45]]

In 1958, H.E. Simmons and R.D. Smith were the first to utilize diiodomethane (CH_2I_2) in the presence of zinc-copper couple (Zn-Cu) to convert unfunctionalized alkenes (e.g., cyclohexene, styrene) to cyclopropanes stereospecifically.[1] This transformation proved to be general and has become the most powerful method of cyclopropane formation: it bears the name of its discoverers and is referred to as the *Simmons-Smith cyclopropanation*. The most important features of the reaction are: 1) a wide range of alkenes can be used: simple alkenes, α,β-unsaturates ketones and aldehydes, electron rich alkenes (enol ethers, enamines, etc.); 2) due to the electrophilic nature of the reagent, the rate of cyclopropanation is faster with more electron rich alkenes. However, highly substituted alkenes may react slower due to the increased steric hindrance; 3) the cyclopropanation is stereospecific, so the stereochemical information in the alkene substrates is translated to the products; 4) when a substituted methylene group is transferred to the alkene ($R^5 \neq H$) a preference for *syn* stereochemistry is typically observed;[17] 5) in case of chiral substrates, the cyclopropanation is highly diastereoselective and occurs from the less hindered face of the double bond; 6) when the alkene has functional groups containing heteroatoms (e.g., OH, OAc, OMe, OBn, NHR), a strong directing effect is observed and the delivery of the alkylidene occurs from the face of the double bond having the closer proximity of the functional group; 7) in cycloalkenols, the stereochemical outcome depends on the ring size: 5-, 6-, and 7-membered rings give rise to high *cis*-diastereoselectivity, while large ring cycloalkenols exhibit high levels of *anti* diastereoselectivity; 8) usually no serious side reactions are observed (e.g., C-H insertion), and the reaction conditions are tolerant of most functional groups; and 9) non-coordinating solvents (e.g., DCM, DCE) are recommended, because the use of basic solvents decrease the rate of the reaction. Today the preparation of the zinc-copper couple is more convenient (treatment of zinc powder with $CuSO_4$ solution) than described in the original procedure. However, there have been several modifications to generate the active reagent: 1) zinc-silver couple tends to give higher yields and shorter reaction times;[18] 2) the use of diethylzinc with CH_2I_2 gives very reproducible results (*Furukawa modification*);[17] 3) iodo- or chloromethylsamarium iodide ($Sm/Hg/CH_2I_2$) is the reagent of choice for the chemoselective cyclopropanation of allylic alcohols in the presence of other olefins (*Molander modification*);[21] and 4) dialkyl(iodomethyl)aluminum (i-Bu_3Al/CH_2I_2) exclusively cyclopropanates unfunctionalized olefins in the presence of allylic alcohols.[46] *Asymmetric Simmons-Smith cyclopropanations* can be achieved several different ways:[11] 1) the use of cleavable chiral auxiliaries (e.g., chiral allylic ethers, acetals, boronates); 2) by the addition of stoichiometric amounts of chiral additives, such as dioxaborolane prepared from tetramethyltartaric acid diamide and butylboronic acid (*Charette asymmetric modification*). However, this method is only applicable to allylic alcohols;[25] and 3) the use of chiral catalysts, such as the chiral disulfonamide ligand derived from *trans*-cyclohexanediamine, gives high *ee*'s for allylic alcohols.[26,27]

Simmons & Smith (1958):

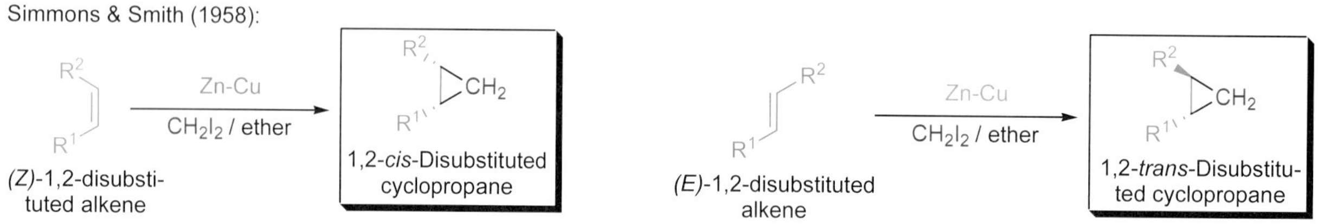

Furukawa modification (1966):

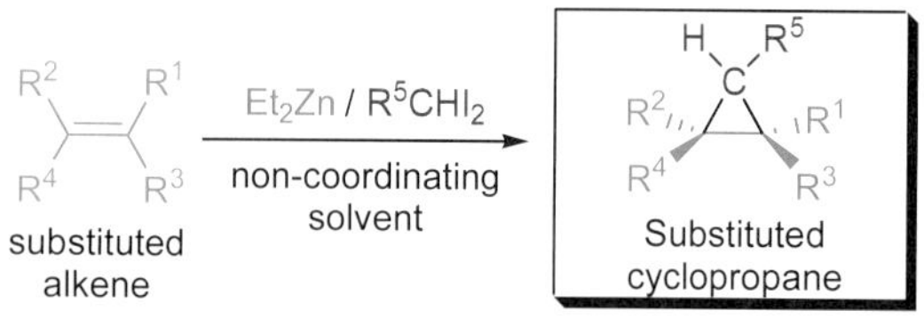

Charette asymmetric modification (1994):

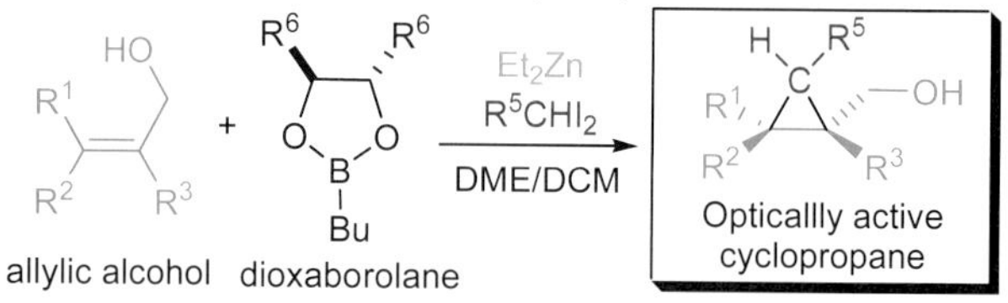

R^{1-4} = H, substituted alkyl and aryl; R^5 = H, Me, phenyl; R^6 = $CONMe_2$; non-coordinating solvent: toluene, benzene, DCM, DCE

Mechanism: [47,11,48,13,15,33]

The *Simmons-Smith cyclopropanation* is a concerted process, and it proceeds *via* a three-centered "butterfly-type" transition state. This is in agreement with the result of theoretical studies as well as the stereochemical outcome of a large number of reactions.

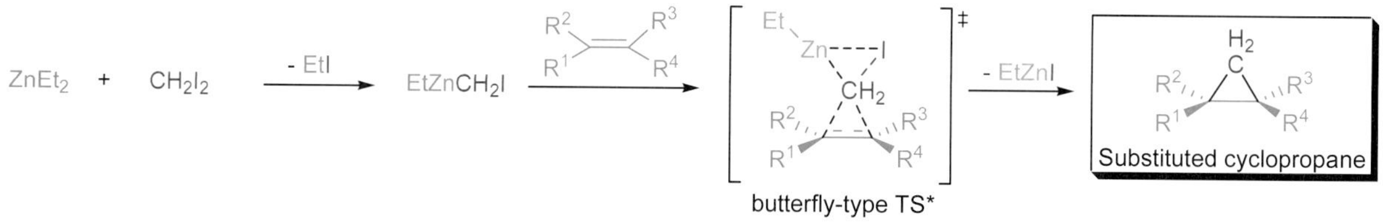

Synthetic Applications:

The highly stereocontrolled total synthesis of the antimitotic agent (+)-curacin A was achieved by S. Iwasaki and co-workers.[49] The main structural feature of this natural product is a disubstituted thiazoline ring bearing a cyclopropane ring and an aliphatic side chain. Diethyl L-tartrate was converted to a (*Z,Z*)-diene in several steps, which was subjected to a *double directed Simmons-Smith cyclopropanation reaction*. The dicyclopropane was obtained as a single diastereomer in good yield. Subsequent periodate cleavage of the diol moiety followed by oxidation led to the desired 2-methylcyclopropanecarboxylic acid, which was used to form the thiazoline portion of curacin A.

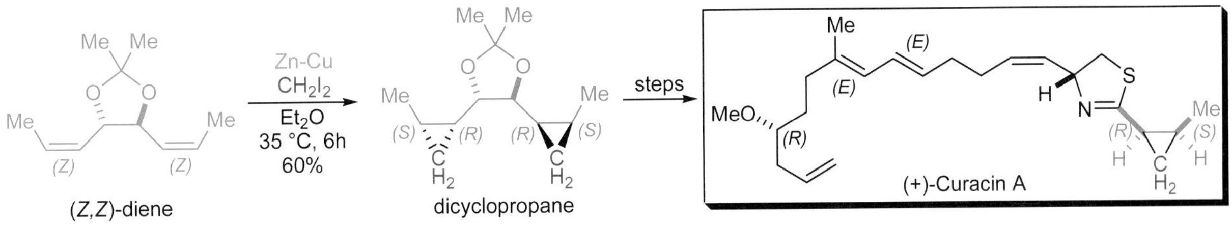

The secondary marine metabolite (+)-acetoxycrenulide has unprecedented structural features which prompted L.A. Paquette et al. to embark on its total synthesis.[50] The eight-membered carbocycle of the target was constructed *via* a *Claisen rearrangement*. The bicyclic β,γ-unsaturated lactone was subjected to Simmons-Smith conditions, that delivered the cyclopropyl ring exclusively from the β-face of the molecule as a result of the predominant ground-state conformation.

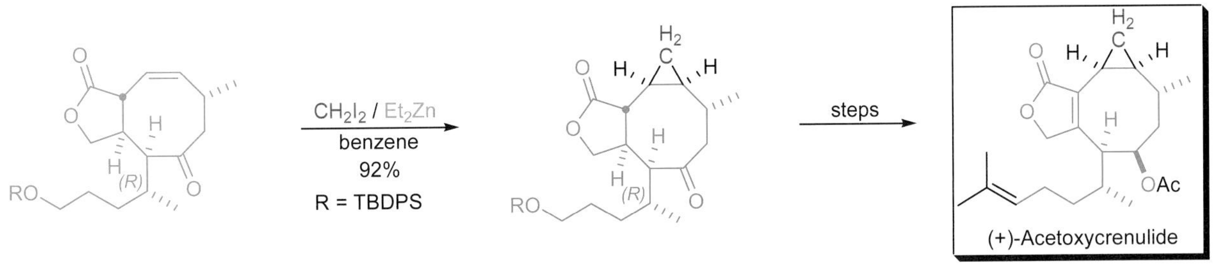

The *asymmetric Simmons-Smith cyclopropanation* (*Charette modification*) was used for the ethylidenation of an allylic alcohol moiety during the total synthesis of (+)-ambruticin in the laboratory of E.N. Jacobsen.[51] Diethylzinc was added to the solution of 1,1-diiodoethane to form the active reagent $Zn(CH_3CH_2I)_2 \cdot DME$, which was transferred to a solution of the substrate containing dioxaborolane (chiral ligand). The central cyclopropane ring was installed with high diastereoselectivity.

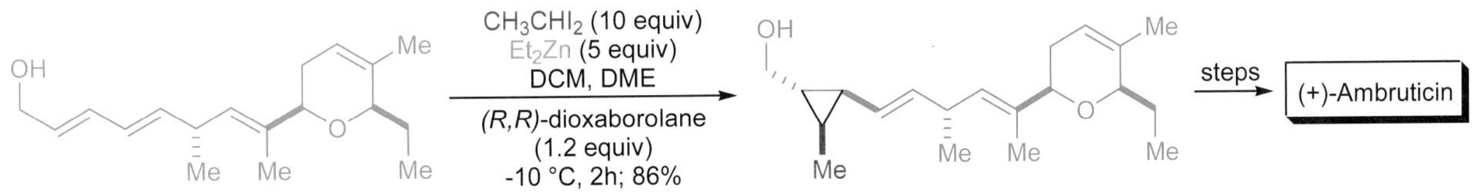

The *lactone-directed intramolecular Diels-Alder cycloaddition* was the key step in D.F. Taber's synthesis of *trans*-dihydroconfertifolin.[52] During the endgame, the *Simmons-Smith cyclopropanation* was utilized to install the *gem*-dimethyl group at C4. The trisubstituted alkene was cyclopropanated in excellent yield and the resulting cyclopropane was subjected to catalytic hydrogenation.

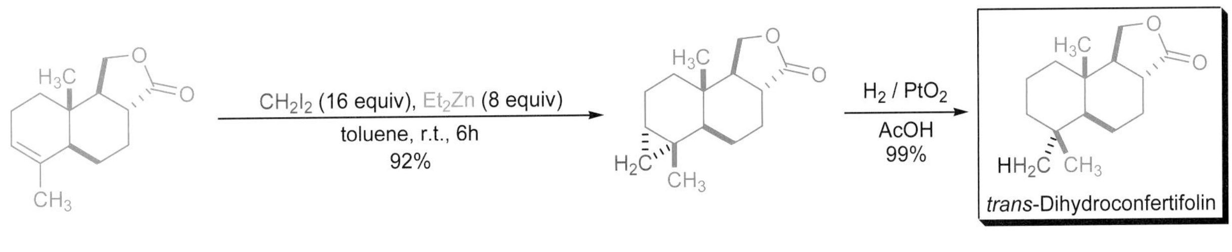

SKRAUP AND DOEBNER-MILLER QUINOLINE SYNTHESIS

(References are on page 678)

Importance:

[*Seminal Publications*[1-3]; *Reviews*[4,5]; *Modifications & Improvements*[6,7]]

In 1880, Z.H. Skraup reported the formation of quinoline by heating aniline with glycerol (1,2,3-propanetriol), sulfuric acid and an oxidizing agent (As_2O_5, $ArNO_2$, *m*-$NO_2C_6H_5SO_3H$, etc.).[2] Shortly after Skraup's discovery, O. Doebner and W. Miller successfully modified and generalized Skraup's method by using α,β-unsaturated aldehydes, ketones or 1,2-diols instead of glycerol.[3] In addition, the sulfuric acid component was replaced by HCl and zinc chloride. This modification allowed the preparation of substituted quinolines. Today these methods are known as the *Skraup* and *Doebner-Miller quinoline synthesis*. The *Skraup procedure* gives easy access to quinolines substituted on the benzene ring (containing only those substituents which were on the aniline component), while the *Doebner-Miller modification* can introduce substituents on the pyridine ring as well. A great advantage of these methods is that structurally complex quinoline derivatives can be prepared in a simple operation. However, there are a few drawbacks: 1) the carbonyl component undergoes polymerization under the strongly Lewis acidic conditions; consequently the yields are often moderate; 2) the rate of addition of the aldehyde influences the yield; 3) isolation of the product from the complex reaction mixtures is often tedious; and 4) large-scale reactions are usually impractical. A recent modification of the *Doebner-Miller synthesis* in a two-phase solvent system allows the clean preparation of the desired quinoline derivative on a large scale.[6] If the aniline substrate is unsubstituted, the oxidizing agent is usually nitrobenzene, since it is conveniently converted to aniline in the process. There are two related well-known quinoline syntheses: 1) *Friedländer synthesis*, which is the condensation of *o*-aminobenzaldehydes with α-methylene ketones to give 3-substituted quinolines;[8,9] and 2) *Combes quinoline synthesis*, which is the condensation of primary arylamines with β-diketones followed by the acid catalyzed ring-closure of the resulting Schiff base.

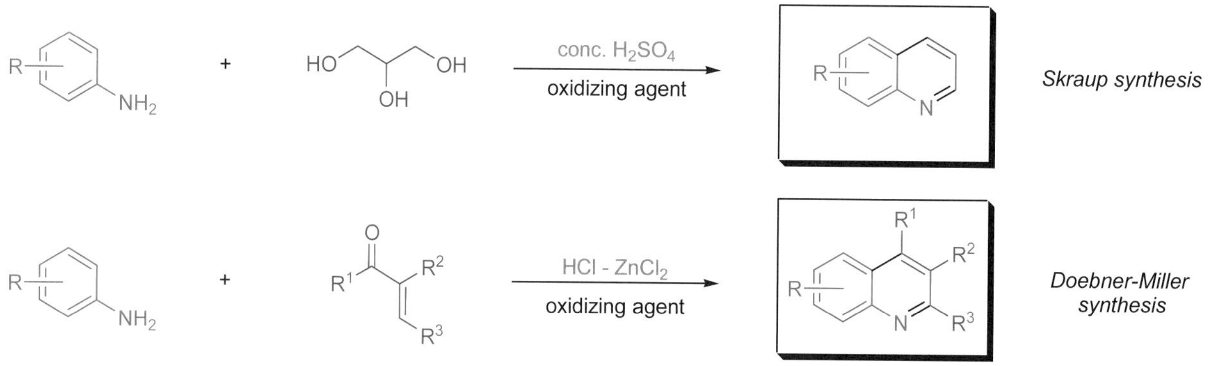

Mechanism: [10-15]

The detailed mechanism of the *Skraup* and *Doebner-Miller quinoline synthesis* has not been fully explored.[15] The two reactions are closely related, and it is assumed that the glycerol in the *Skraup procedure* is dehydrated to form acrylaldehyde (α,β-unsaturated aldehyde) or the 1,2-diol is first dehydrated to acetaldehyde, which undergoes an *aldol condensation* to afford crotonaldehyde in the *Doebner-Miller reaction*. The mechanism most likely involves the following steps: 1) condensation of the carbonyl component with the arylamine to form a Schiff-base (anil) (this step is not shown); 2) formation of a labile 1,3-diazetidinium cation intermediate from two anils; 3) ring-opening of the 1,3-diazetidinium ion to form a carbocation, which undergoes an S_EAr reaction with the aromatic ring; 4) formation of a substituted-1,2-dihydroquinoline; 5) hydride transfer (oxidation) to give a substituted quinoline.

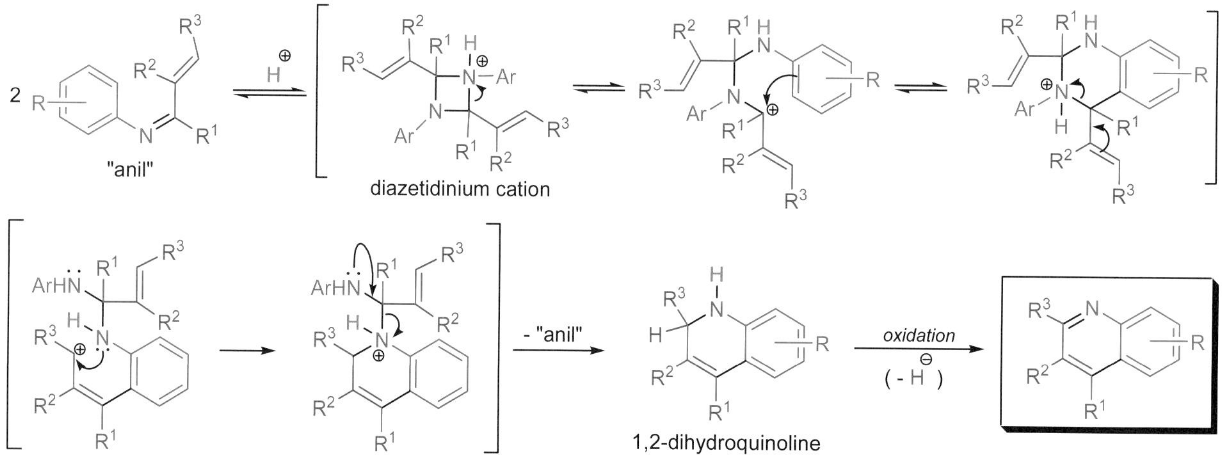

Synthetic Applications:

A new synthesis was developed by Y. Kashman et al. for the preparation of the parent pyrido[2,3,4-*kl*]acridine skeleton utilizing the *Doebner-Miller synthesis*.[16] In the first step, 3-aminoacetanilide was reacted with vinyl phenyl ketone in the presence of *m*-nitrobenzenesulfonic acid sodium salt and acetic acid to afford the corresponding 4-phenylquinolines. The acetamido group was then converted to the corresponding aryl azide, which underwent *intramolecular nitrene insertion* upon thermolysis to give the desired heterocyclic skeleton.

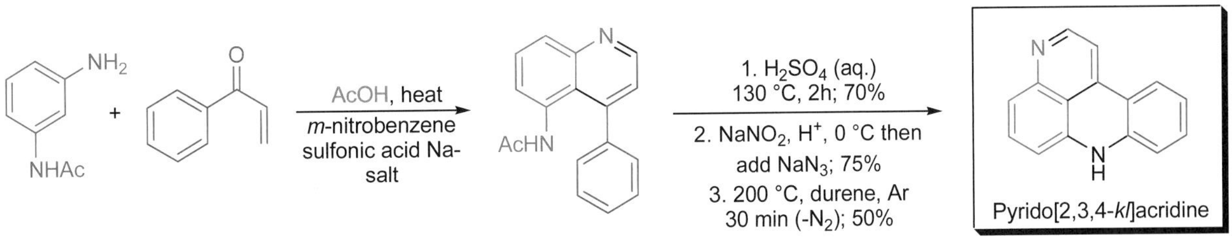

The synthesis of the antimalarial 5-fluoroprimaquine by P.M. O'Neil and co-workers involved a *Doebner-Miller reaction* of 5-fluoro-4-methoxy-2-nitroaniline with acrolein.[17] In this modified procedure 80% phosphoric acid, acrolein and arsenic acid were employed to allow a shorter reaction time and lower temperature than in the original procedure.

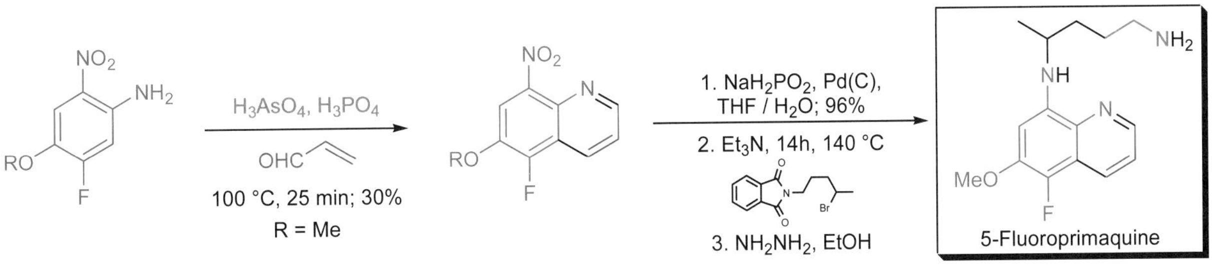

The short and convenient synthesis of novel naphthopyranoquinolines from naphthopyran chloroaldehydes *via* the *Doebner-Miller synthesis* was developed in the laboratory of J.K. Ray.[18] The chloroaldehydes were treated with 2.5 equivalents of a substituted aniline in ethanol in the presence of 2N HCl to afford enaminoimine hydrochlorides in good yield. These hydrochloride salts were exposed to heat at a temperature slightly above their melting point, resulting in ring-closure and elimination of one equivalent of arylamine hydrochloride.

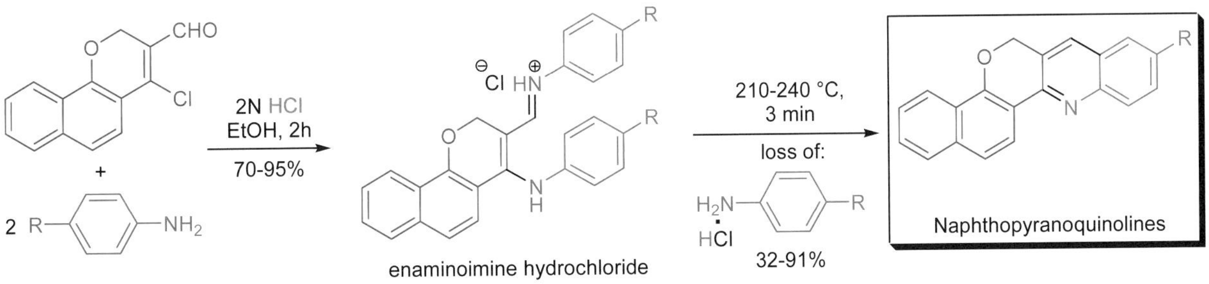

A 3,8-dialkyl phenanthroline-based asymmetric transfer hydrogenation catalyst was prepared by S. Gladiali and co-workers using two consecutive *Doebner-Miller reactions*.[19] The synthesis of the ligand commenced with the reaction between 2-nitro aniline and enantiomerically pure 2-*sec*-butylacrolein. The resulting nitroquinoline was hydrogenated to give the corresponding aminoquinoline which was subjected to the second *Doebner-Miller reaction* to afford the enantiopure phenanhroline catalyst.

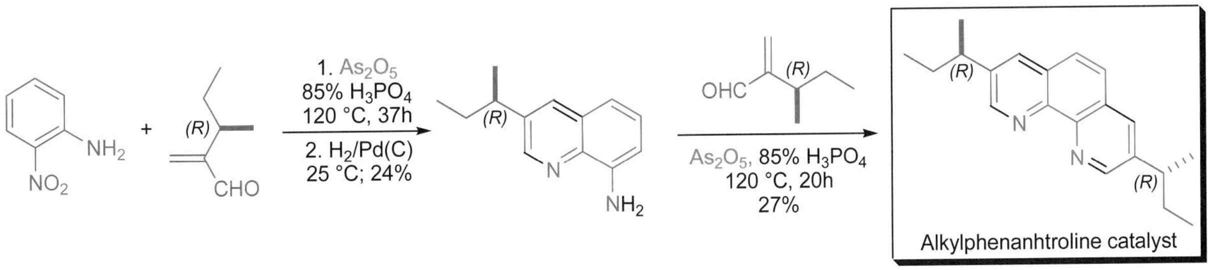

SMILES REARRANGEMENT

(References are on page 678)

Importance:

[*Seminal Publications*[1-8]; *Reviews*[9-17]; *Modifications & Improvements*[18-26]; *Theoretical Studies*[27-31]]

In 1894, R. Henriques reported that the base treatment of *bis*-(2-hydroxy-1-naphthyl) sulfide afforded an isomeric compound, 2-hydroxy-2'-mercapto-*bis*-(1-naphthyl) ether.[1] Two decades later, O. Hinsberg carried out similar experiments with the corresponding sulfones,[2,3] but it was S. Smiles and co-workers who established the structure of the products.[5-8] Smiles recognized that the transformations belonged to a new type of intramolecular nucleophilic aromatic rearrangement, which is known as the *Smiles rearrangement*. The general features of the reaction are:[11] 1) the aromatic ring needs to be activated by electron-withdrawing groups at the *ortho*- or *para* positions (e.g. NO_2, SO_2R); 2) if there is more than one activating group (when R^2=EWG), the rate of the rearrangement increases; 3) electron-withdrawing groups in the *meta* position usually do not activate the aromatic ring sufficiently; 4) in the absence of activating groups or when R^1 and R^2 are electron-donating, the rearrangement is slow or does not occur; 5) besides substituted benzene rings, the aromatic ring can also be heteroaromatic such as pyridine or pyrimidine; 6) in the presence of a strong base, when Y=SO_2 and XH=CH_3, no activating group is necessary and the process is called *Smiles-Truce rearrangement*);[18] 7) the nucleophilicity of the XH group and the ability of the Y group to function as a good leaving group are two factors that are interconnected and their combined effect have a dramatic influence on the rate of the rearrangement; 8) when XH=NH_2, usually no base is required and Y does not have to be a good leaving group for the reaction to take place; 9) the more stabilization of the negative charge is possible on Y, the faster the reaction will proceed (e.g., Y = SO_2 > SO > S); 10) when the Z groups are part of an aromatic ring (e.g., biaryl systems), electron-withdrawing substituents on this second ring tend to accelerate the reaction; 11) substituents at the 6-position of the second ring (*ortho* to Y) also accelerate the reaction because it forces the substrate to be predominantly in the reactive conformation, where the migrating ring is perpendicular to the plane of the other aromatic ring; 12) when the Y and the XH groups have very similar negative charge stabilizing abilities, the *Smiles rearrangement* becomes a reversible process. There are several modifications of the transformation: 1) the *Smiles-Truce rearrangement* utilizes a carbon-centered anion as the nucleophile and that can be generated by using a strong base (e.g., alkyllithium, KO*t*-Bu) is necessary;[18,11] 2) *photochemical Smiles rearrangement*;[21,32] and 3) rearrangement of phosphonium zwitterions, generated by the addition of an aryne to an alkylidene triarylphosphorane, affords P-substituted aromatic compounds.[19,20]

Henriques & Hinsberg (1894 & 1914):

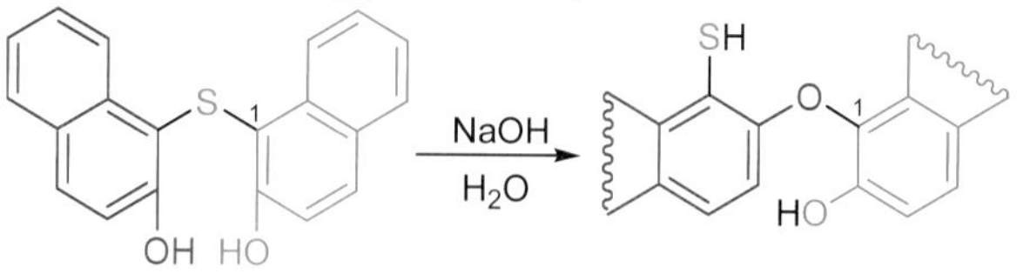

bis-(2-hydroxy-1-naphthyl) sulfide

2-hydroxy-2'-mercapto-*bis*-(1-naphthyl) ether

Smiles (1930-1936):

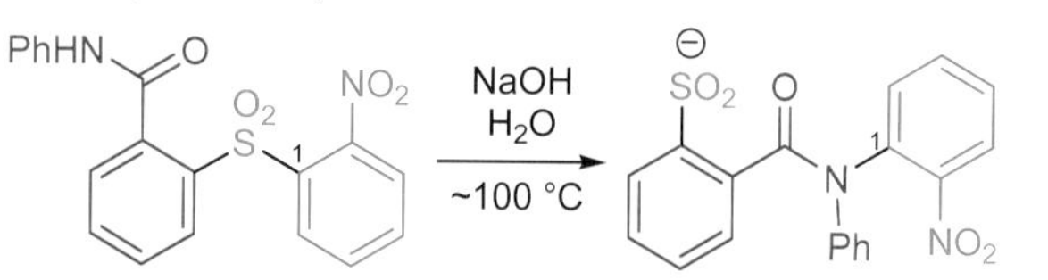

2-(2-nitro-benzenesulfonyl)-*N*-phenyl-benzamide

2-[(2-nitro-phenyl)-phenyl-carbamoyl]-benzenesulfinate

Smiles rearrangement (general equation):

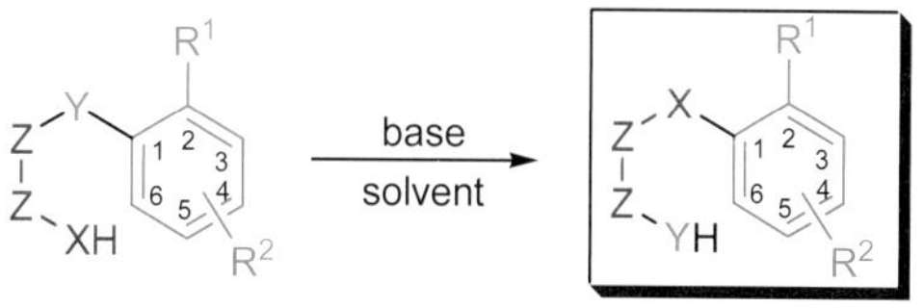

Smiles rearrangement of biaryl systems:

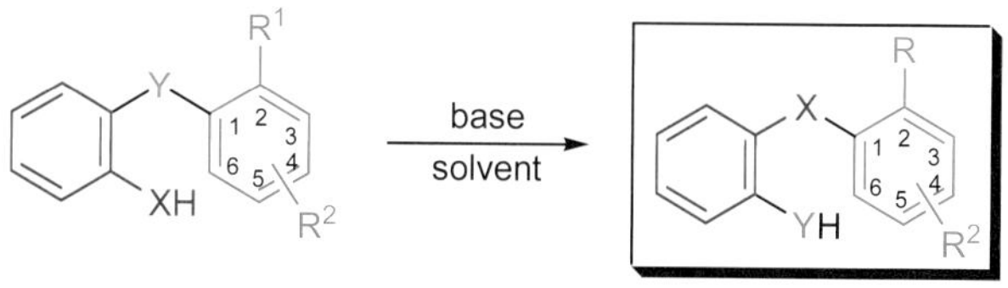

XH = NHCOR, $CONH_2$, SO_2NH_2, OH, NH_2, SH, SO_2H, CH_3 (*Smiles-Truce rearrangement*); Z = sp^2 or sp^3 hybridized substituted- or unsubstituted carbon, C=O, sp^3 nitrogen; Y = S, O, SO_2, S=O, CO_2, SO_3, I^+, P^+; R^1 = EWG = NO_2, SO_2R, Cl; R^2 = alkyl, halogen, NO_2, acyl; base: NaOH, KOH, RONa, RLi, K_2CO_3/DMSO

Mechanism: [33-47]

The first step of the reaction is the formation of the nucleophile by deprotonation. The substrate then has to adopt the reactive conformation in which the plane of the migrating ring is perpendicular to the Z-Z bond. The nucleophile attacks the ring in an *ipso* fashion to form a five-membered transition state that affords the product.

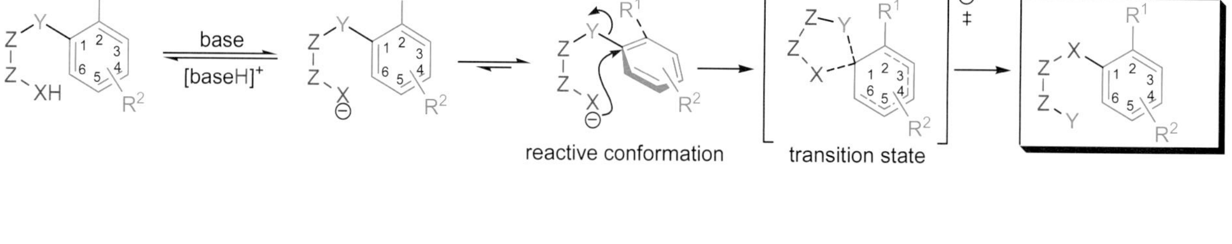

Synthetic Applications:

Frequently, the anionic product of the *Smiles rearrangement* can undergo further reaction if there are electrophilic functional groups on the aromatic ring. This approach was utilized by T. Hirota to prepare complex fused *N*-heterocyclic compounds such as the [1]benzothieno[3,2-*d*]furo[2,3-b]pyridine skeleton.[48] The substrate, cyanopropoxy-substituted benzo[*b*]thiophene, was exposed to sodium hydride in refluxing dioxane that induced the *Smiles rearrangement*. The resulting alkoxide attacked the cyano group to form an imine salt, which in turn added across the nitrile at the 2-position.

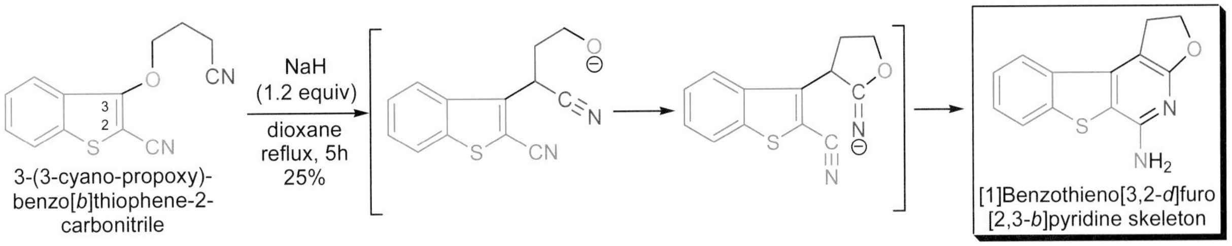

The total synthesis of the lichen diphenyl ether epiphorellic acid 1 was achieved in the laboratory of J.A. Elix using the *Smiles rearrangement* as the key step.[49] The diaryl phenolic ester substrate was heated in dry DMSO in the presence of potassium carbonate, which brought about the rearrangement. The resulting carboxylic acid was converted to the methyl ester with diazomethane and was debenzylated under catalytic hydrogenation conditions.

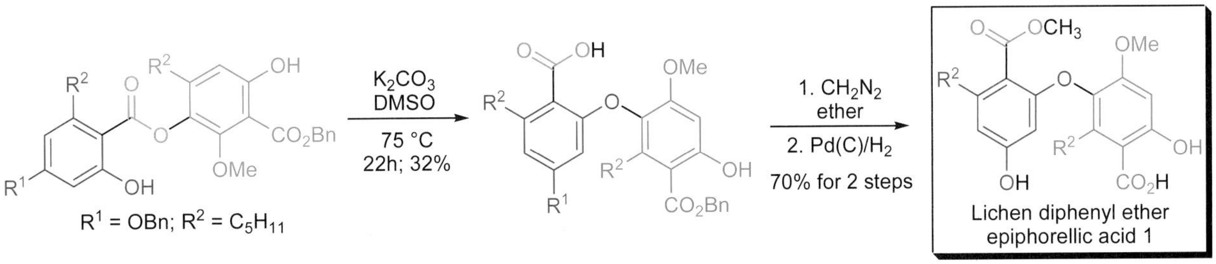

Novel non-nucleoside inhibitors of HIV-1 reverse transcriptase, dipyrido[2,3-*b*]diazepinones, were prepared by J.R. Proudfoot and co-workers.[50] These compounds are isomeric to the potent inhibitor nevirapine and available *via* the *Smiles rearrangement* of substrates that are intermediates used for the synthesis of nevarpine analogs. The deprotonated amide functionality in the rearrangement products displaces the chlorine at the 2-position to give the desired heterocycles in moderate to good yield.

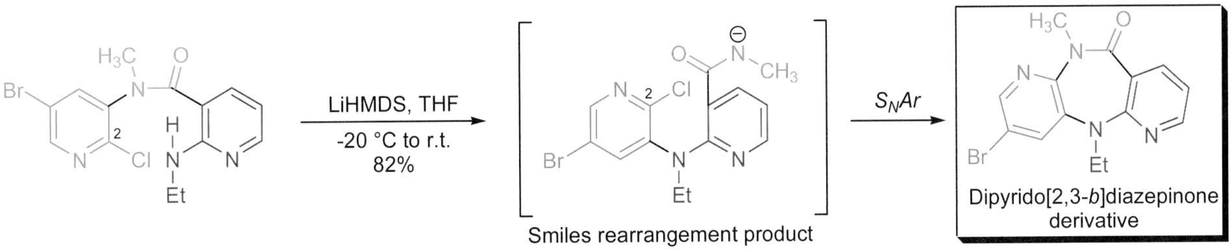

A one-pot procedure was developed for the preparation of aromatic amines from phenols *via* a one-pot *Smiles rearrangement* by N.P. Peet et al.[51] This new approach can be considered as an alternative of the *Bucherer reaction* which only works well for naphthalene derivatives and gives very poor yields for substituted benzene derivatives. In the current procedure, the phenol was reacted with 2-bromo-2-methylpropionamide to give 2-aryloxy-2-methylpropionamide which upon treatment with base underwent the *Smiles rearrangement*. The hydrolysis of the resulting *N*-aryl-2-hydroxypropionamide afforded the aromatic amine.

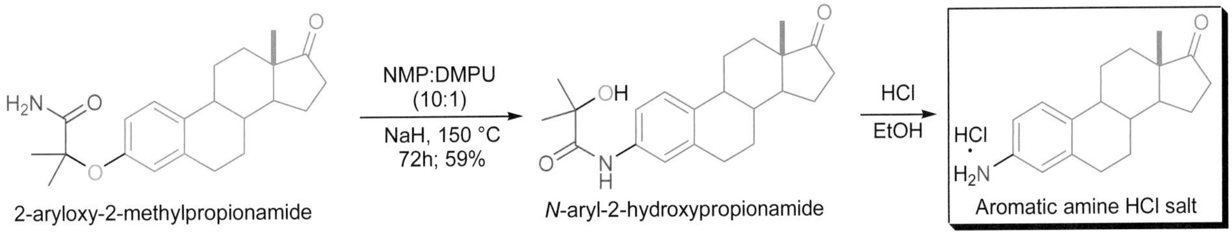

SMITH-TIETZE MULTICOMPONENT DITHIANE LINCHPIN COUPLING

(References are on page 679)

Importance:

[*Seminal Publications*[1-3]; *Review*[4]; *Modifications & Improvements*[5-9]]

The one-pot multicomponent coupling of 2-silylated-1,3-dithianes with epoxides is referred to as the *Smith-Tietze coupling*. The first application of 2-lithio-1,3-dithianes as "carbonyl anion" equivalents was described by E.J. Corey and D. Seebach in the mid-1960s.[10] In 1994, L.F. Tietze and co-workers successfully synthesized C_2-symmetrical enantiopure 1,5-diols, 3-oxo-1,5-diols and 1,3,5-triols by the symmetrical *bis*-alkylation of lithiated 2-trialkylsilyl-1,3-dithianes with epoxides.[2] Tietze's protocol began with the deprotonation of the 2-trialkylsilyl-1,3-dithiane with an alkyllithium followed by the addition of 2.2 equivalents of epoxide in the presence of one equivalent of crown ether. After the opening of the first epoxide, the resulting alkoxide intermediate underwent a spontaneous *[1,4]-Brook rearrangement*, thus generating a second dithiane anion that reacted with the remaining excess epoxide. This multicomponent coupling protocol, however, had a long reaction time, and it was unsuitable for unsymmetrical couplings. A.B. Smith et al. used HMPA or DMPU as an additive in the solvent, which significantly increased the rate of the reaction and allowed two different electrophiles (epoxides) to be coupled with the dithiane in a one-pot operation.[3] The *Smith-Tietze coupling* has the following advantages: 1) optically active terminal epoxides can be readily prepared by known methods; 2) the epoxide ring-opening is completely regioselective, the nucleophile attacks on the least substituted carbon; 3) the exact timing of the *Brook rearrangement* is possible by the addition of HMPA or DMPU to the reaction mixture (*solvent-controlled Brook rearrangement*) and the formation of symmetrical adducts can be completely avoided; 4) altering the absolute configuration of the epoxides and the stereoselective reduction of the ketone moiety after the removal of the dithiane can give rise to 1,3-polyols of any desired configuration; 5) after the second epoxide has reacted, the resulting unsymmetrical adduct has its hydroxyl groups differentiated by one of them being silylated; and 6) the use of an enantiopure *bis*-epoxide as the second epoxide component allows for a one-pot *five-component linchpin coupling*.

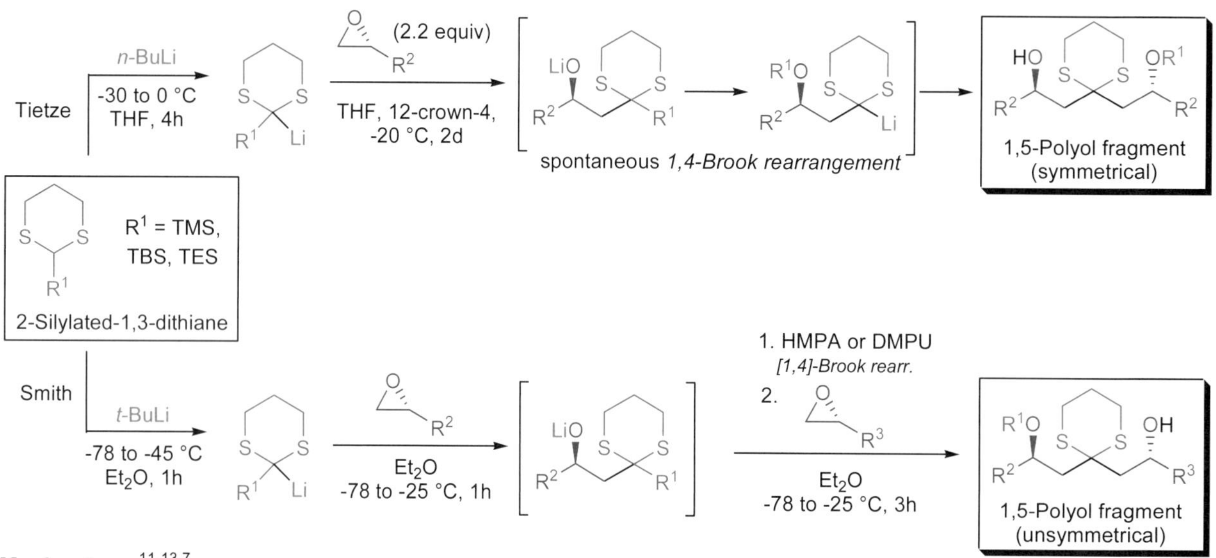

Mechanism: [11-13,7]

The key step of the mechanism is the solvent-controlled *[1,4]-Brook-rearrangement*, which proceeds through an intermediate having a pentacoordinate-silicon atom. This rearrangement does not take place until HMPA is added to the solvent. A similar solvent effect has been observed by K. Oshima, K. Utimoto and co-workers.[11,13] The rearrangement was found to be completely intramolecular based on the results of a crossover experiment by A.B. Smith et.al.[7]

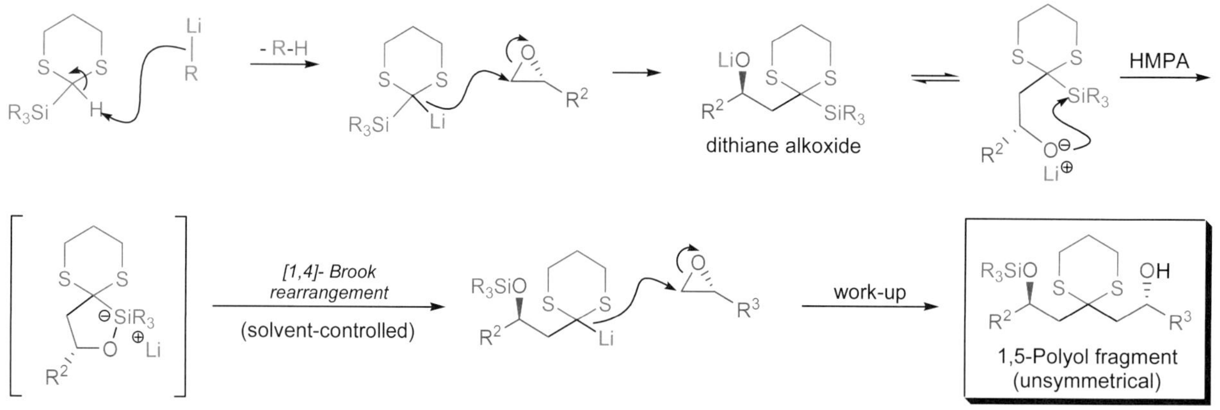

Synthetic Applications:

The stereocontrolled enantioselective synthesis of an advanced B-ring synthon of bryostatin 1 was achieved in the laboratory of K.J. Hale.[14] The key step was a *Smith-Tietze coupling* of 2-lithio-2-TBS-1,3-dithiane with a homochiral epoxide in the presence of HMPA. The resulting dithiane alkoxide was trapped with TBSCl *in situ* followed by deprotection of the dithiane moiety to give a C_2-symmetrical ketone. This ketone was then further elaborated into the target B-ring synthon.

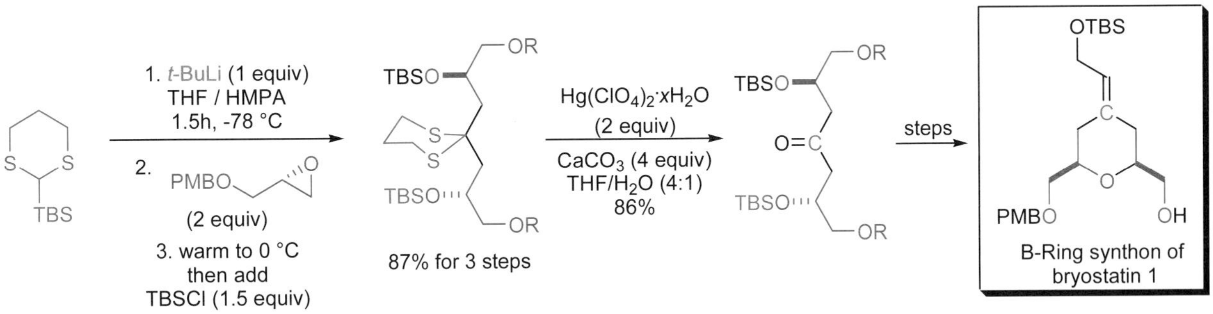

A *one-pot five-component dithiane linchpin coupling* was applied as the key synthetic transformation in A.B. Smith's approach to Schreiber's C16-C28 trisacetonide subtarget for mycoticins A and B.[7] To prevent a premature *Brook rearrangement*, ether was used instead of THF as a solvent for the initial deprotonation of 2-TBS-1,3-dithiane. The third component in the linchpin coupling was (*S*,*S*)-diepoxypentane that was added to the reaction mixture along with HMPA in THF.

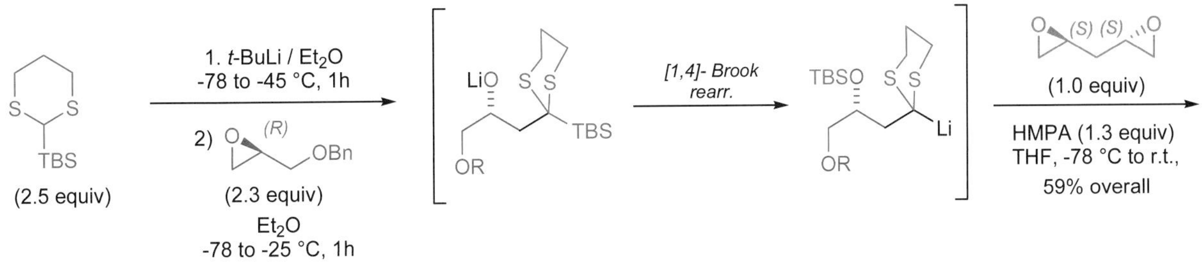

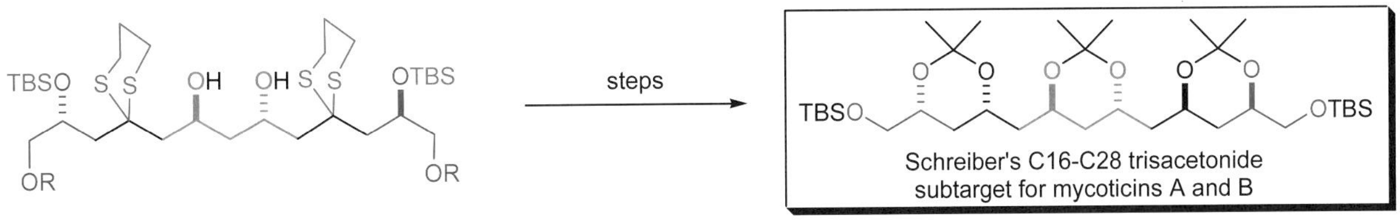

The *three-component dithiane linchpin coupling* was the key bond forming reaction during the second-generation synthesis of an advanced ABCD intermediate for spongistatins by A.B. Smith et al.[15] Both the AB and CD fragments were accessed by this multicomponent coupling. Interestingly, one of the epoxide components had to be added into the reaction mixture as its lithium alkoxide to avoid the formation of elimination products. Upon deprotection of the dithiane moiety, an *in situ* spiroketalization took place. The target AB fragment was realized in several subsequent steps.

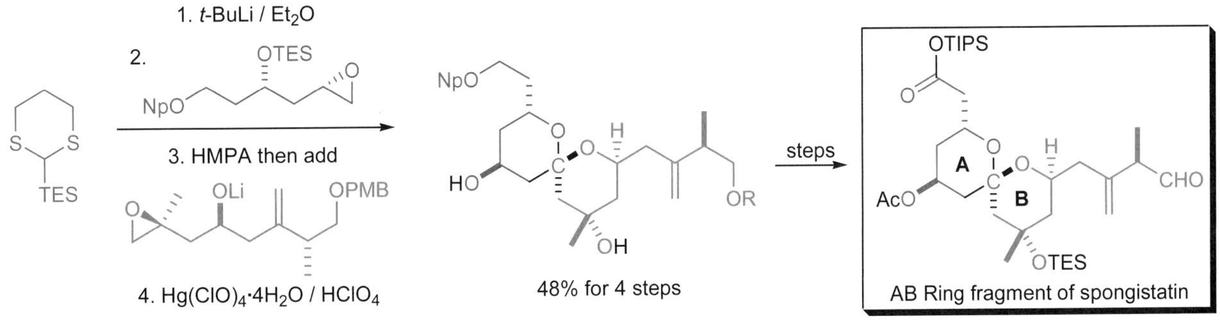

SNIECKUS DIRECTED ORTHO METALATION

(References are on page 680)

Importance:

[*Seminal Publications*[1,2]; *Reviews*[3-27]; *Modifications & Improvements*[28-33]; *Theoretical Studies*[34-36]]

In the late 1930s, the research groups of H. Gilman and G. Wittig independently discovered that the treatment of anisole (methoxybenzene) and other heteroatom-substituted aromatic compounds with *n*-BuLi resulted in the exclusive deprotonation at the *ortho* position to afford the corresponding 2-lithio derivatives.[1,2] During the 1970s alkyllithiums became commercially available, and this resulted in the widespread use of the *ortho* metallation protocol to functionalize aromatic and heteroaromatic compounds.[7] *Directed metalation* is defined as the deprotonation of an sp^2 hybridized carbon atom positioned α to a heteroatom-containing substituent on an aromatic or olefinic substrate.[8] The contributions by V. Snieckus and co-workers over the last two decades significantly expanded the scope of this method, which is often referred to as the *Snieckus directed ortho metalation (DoM)*. Before the advent of *DoM*, the preparation of contiguously substituted (e.g., 1,2-, 1,2,3- or 1,2,3,4-) aromatic compounds, using the directing effect of the various substituents in S_EAr reactions, was a major challenge and required many steps to accomplish. The general features of *DoM* reaction are: 1) the directed metalation group (Z group) must be resistant to nucleophilic attack by the metalating reagent (e.g., alkyllithiums), and it must contain at least one heteroatom, which can coordinate with the incipient *ortho* metal atom forming a 4-, 5-, or 6-membered intermediate; 2) the formation of a 5-membered intermediate is the most favorable; 3) the best Z groups are sterically demanding or charge deactivated or exhibit both of these properties at the same time; 4) the Z groups can be classified depending on the atom through which the group is attached to the aromatic ring: there are carbon linked (e.g., $CONR_2$), nitrogen linked (e.g., NHCOR), oxygen linked (e.g., $OCONR_2$), sulfur linked (e.g., SO_2R) etc. Z groups; 5) the most popular Z groups are tertiary amides and *O*-carbamates; 6) the Z groups can be ranked according to the strength of their directing effects (based on competition experiments), but the ranking changes considerably depending on the solvent, temperature and the base used to generate the metalated species: SO_2t-Bu > CON(*i*-Pr)$_2$ > OCON(*i*-Pr)$_2$ > OMOM was the hierarchy of metalation when *n*-BuLi/THF/-78 °C were used;[15] 6) in a typical procedure, the solution of the substrate is treated with the alkyllithium reagent at -78 °C under inert atmosphere followed by the addition of the electrophile; 7) substrates with Z groups having an acidic proton require the addition of at least two equivalents of the alkylithium reagent; 8) since alkyllithiums exist predominantly as aggregates in hydrocarbon solvents, the addition of basic solvents such as ethers and tertiary amines or bidentate ligands (e.g., TMEDA) is necessary to break down the aggregates to monomers and dimers to enhance their basicity; and 9) when the Z group is a carbamate ($OCONR_2$), a facile *1,3-acyl shift* occurs after the *ortho* lithiation is complete to afford a salicylamide (*anionic ortho-Fries rearrangement*). One shortcoming of the *DoM* is that the most powerful Z groups require harsh reaction conditions for their removal making it unsuitable for sensitive substrates. To address this issue, easily removable Z groups have been developed: 1) the CON(Cumyl) group is removed under mildly acidic conditions (TFA) to afford a primary amide;[31] 2) *N*-cumyl-*O*-carbamate can also be removed with mild acids.[23]

Gillman and Wittig (1939 & 1940):

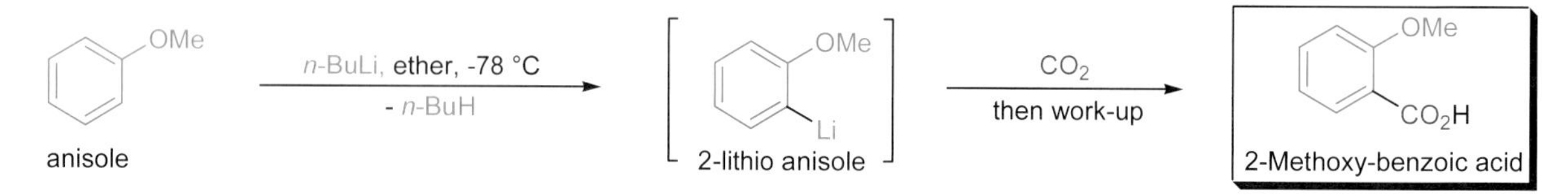

Directed ortho metalation:

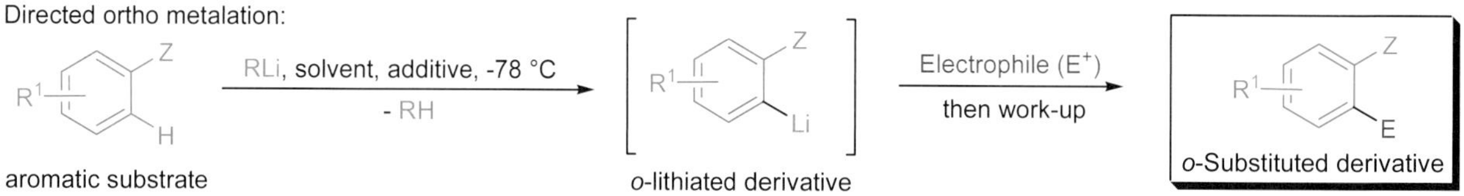

Z = directed metalation group = $CONR_2$, CONHR, CONH(Cumyl), CSNHR, 2-oxazolino, 2-imidazolino, CF_3, CH=NR, $(CH_2)_nNR_2$ where n=1 or 2, CH_2OH, NMe_2, NHCOR, $NHCO_2R$, OMe, OCH_2OMe, OCH(Me)OEt, $OCONR_2$, OSEM, $OP(O)NR_2$, SO_2NR_2, SO_2NHR, SO_2R; R = *n*-Bu, *sec*-Bu, *t*-Bu; <u>solvent</u> = THF, Et_2O, hexanes, benzene or combinations of these; <u>additive:</u> TMEDA

Mechanism: [37,27]

The *directed ortho metalation* is fundamentally a complex-induced proximity effect (CIPE) in which the formation of a pre-metalation complex brings reactive groups into proximity for directed deprotonation.

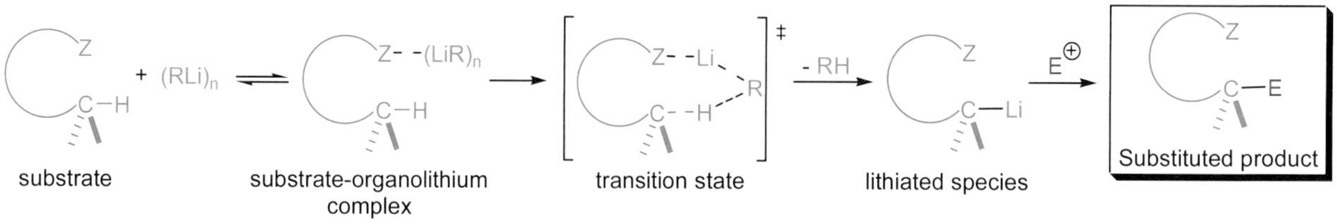

Synthetic Applications:

The synthesis of the aglycons of gilvocarcin V, M and E by V. Snieckus and co-workers involved the use of directed *o*-metalation and *remote metalation* (*anionic ortho-Fries rearrangement*).[38] The trioxygenated naphthalene ring was first *o*-metalated and the resulting lithiated species was iodinated. The 2-iodo compound was then subjected to a *Suzuki cross-coupling* to obtain a biaryl compound that was treated with excess LDA in refluxing THF to induce the remote metalation. Exposure to refluxing acetic acid gave the corresponding lactone, which was subsequently converted to the gilvocarcin M aglycone.

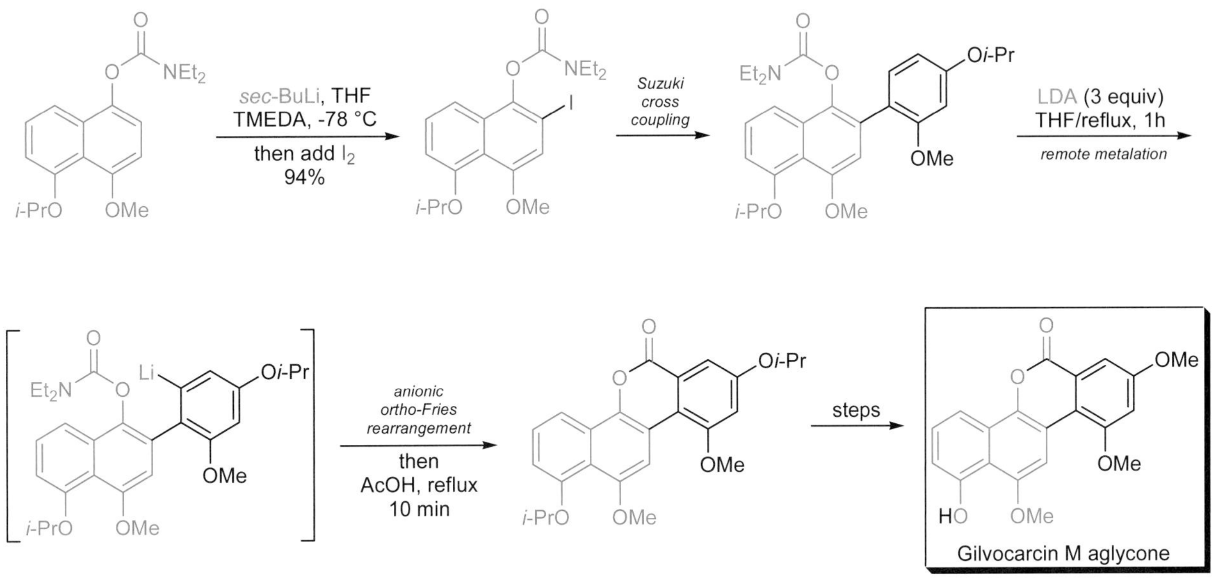

In the laboratory of M. Iwao, the first total synthesis of a new pyrroloiminoquinone marine alkaloid veiutamine was accomplished.[39] The key step was the selective functionalization of the 1,3,4,5-tetrahydropyrrolo[4,3,2-*de*]quinoline nucleus *via* an *N*-Boc *directed ortho metalation* at the C6 position. The resulting 6-lithiated compound was trapped with MOM-protected *p*-hydroxybenzaldehyde.

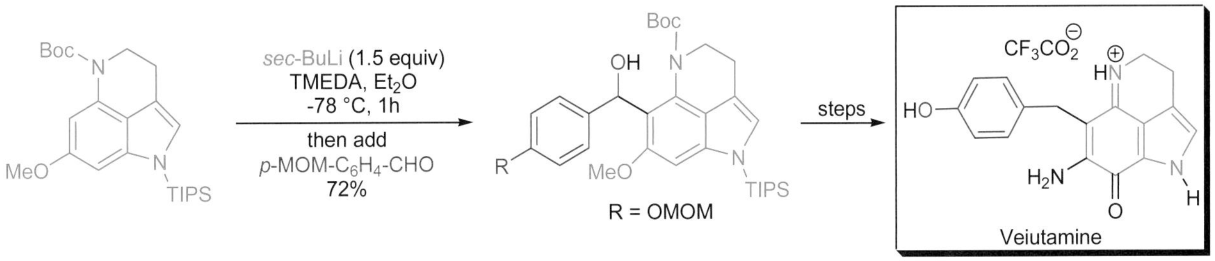

A practical six-step synthesis of (*S*)-camptothecin was developed by D.L. Comins and co-workers.[40] In order to prepare the DE ring fragment, 2-methoxypyridine was lithiated at C3 with mesityllithium and treated with *N*-formyl-*N,N',N'*-trimethyl ethylenediamine to form an α-amino alkoxide *in situ*. In the same pot, the addition of *n*-BuLi brought about a *directed lithiation* at C4 to afford a dianion, which was trapped with iodine and treated with $NaBH_4/CeCl_3$ to give the desired 4-iodo-2-methoxy-3-hydroxymethyl pyridine in 46% yield.

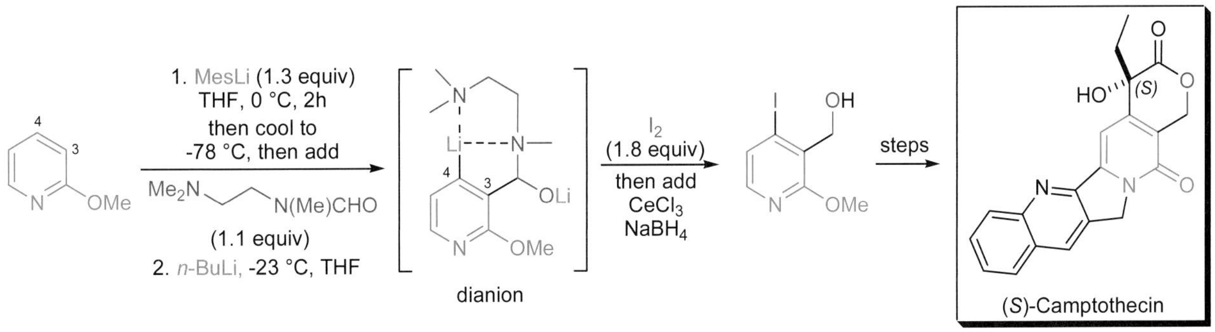

SOMMELET-HAUSER REARRANGEMENT

(References are on page 681)

Importance:

[*Seminal Publications*[1,2]; *Reviews*[3-9]; *Modifications & Improvements*[10-17]; *Theoretical Studies*[18-20]]

In 1937, M. Sommelet reported that benzhydryltrimethylammonium hydroxide rearranged to give (*o*-benzylbenzyl)dimethylamine in modest yield when kept in the desiccator over P_2O_5 exposed to sunlight.[1] The same result was obtained when the substrate was heated to 145 °C, which suggested that the sunlight only provided the heat necessary for the transformation. During the following decade several research groups reported products from similar rearrangements which accompanied the well-known *Stevens rearrangement* of quaternary ammonium salts; however, it was C.R. Hauser and co-workers who investigated this new rearrangement extensively. Hauser et al. treated benzyltrimethylammonium iodide with $NaNH_2$ in liquid ammonia and isolated dimethyl-(2-methylbenzyl)-amine as the sole product in excellent yield.[2] They also demonstrated that methyl groups could be successively introduced into the aromatic ring by exhaustively methylating the product and exposing it to $NaNH_2/NH_3$. The *[2,3]-sigmatropic rearrangement* of benzylic quaternary ammonium salts in the presence of a strong base is known as the *Sommelet-Hauser rearrangement (S.-H. rearrangement)*. The general features of this transformation are: 1) the quaternary ammonium salts are easily available by the alkylation of the corresponding tertiary amines with alkyl halides; 2) the aromatic ring can be either a substituted benzene ring or a substituted heteroaromatic ring; 3) the deprotonation of the quaternary ammonium salt to generate the reactive nitrogen ylide intermediate is most often achieved by treatment with alkali metal amides in liquid ammonia, however, there are alternative methods available for the generation of the reactive intermediate; 4) when there are two possible sites of deprotonation, usually the more stable ylide is formed (derived from the more stable carbanion); 5) when it is not possible to form the ylide by deprotonation because the initial benzylic carbanion is significantly stabilized (e.g., R^1=EWG group such as CN, NO_2, Cl, Br), the rearrangement may not occur; 6) when the alkyl groups attached to the nitrogen contain a hydrogen atom at their β-position, the *Hofmann elimination* may compete; 7) cyclic quaternary ammonium salts react by ring-expansion; 8) one major competing reaction is the *Stevens rearrangement*; 8) in systems where both the *Stevens*- and *S.-H. rearrangements* are possible, the choice of reaction conditions allow control over which of these competing processes dominate; 9) low temperatures and polar solvents (e.g., NH_3, DMSO, HMPA) usually favor the *S.-H. rearrangement*, whereas higher temperatures and nonpolar solvents (e.g., hexanes, ether) facilitate the *Stevens rearrangement*; and 10) since most quaternary ammonium salts are insoluble in nonpolar organic solvents, the use of alkyllithiums as bases is limited. There are several modifications of the *S.-H. rearrangement*: 1) when benzylsulfonium salts are deprotonated, sulfonium ylides are formed that undergo analogous rearrangement and allow an asymmetric version;[10] and 2) the generation of nitrogen ylides is possible under neutral conditions by fluoride-induced desilylation of (trimethylsilyl)methyl ammonium halides.[12,13]

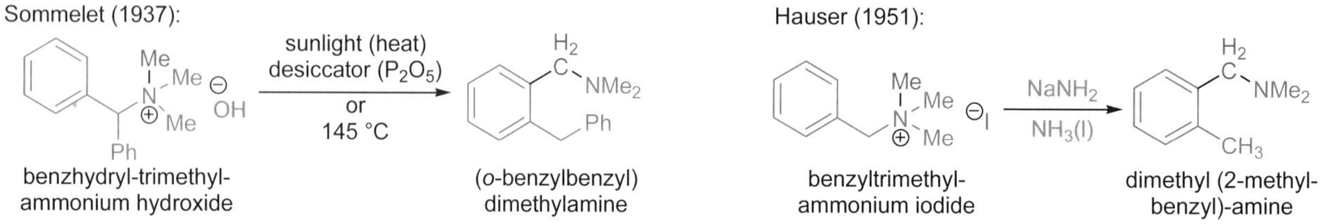

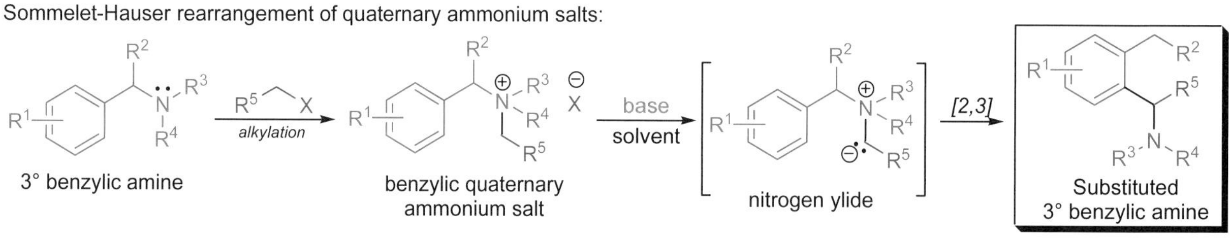

R^1 = usually EDG = H, alkyl, aryl, *O*-alkyl; R^2 = H, alkyl, aryl; R^{3-4} = CH_3, alkyl with no β-hydrogen, aryl; R^5 = most often H, 3° alkyl; X = Cl, Br, I; base = $NaNH_2$, KNH_2, alkyllithium; solvent = NH_3 (liquid), DMSO, HMPA

Mechanism:[21-25]

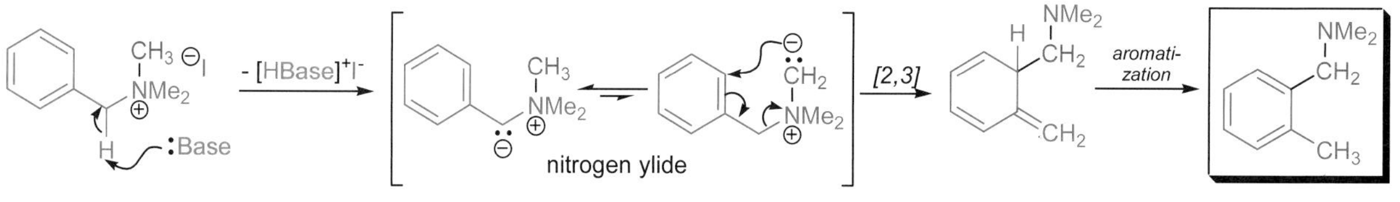

Synthetic Applications:

In the laboratory of S.M. Weinreb, the total synthesis of the potent antibiotic natural product streptonigrin was accomplished.[26] In order to obtain a fully substituted pyridine moiety under mild conditions, the *modified Sommelet-Hauser rearrangement* was utilized. The quaternary ammonium salt was derived from *N*-(cyanomethyl)pyrrolidine which could be efficiently deprotonated using KO*t*-Bu. Upon deprotonation the expected *[2,3]-sigmatropic shift* took place, and the resulting amino nitrile was immediately hydrolyzed to afford the corresponding aldehyde.

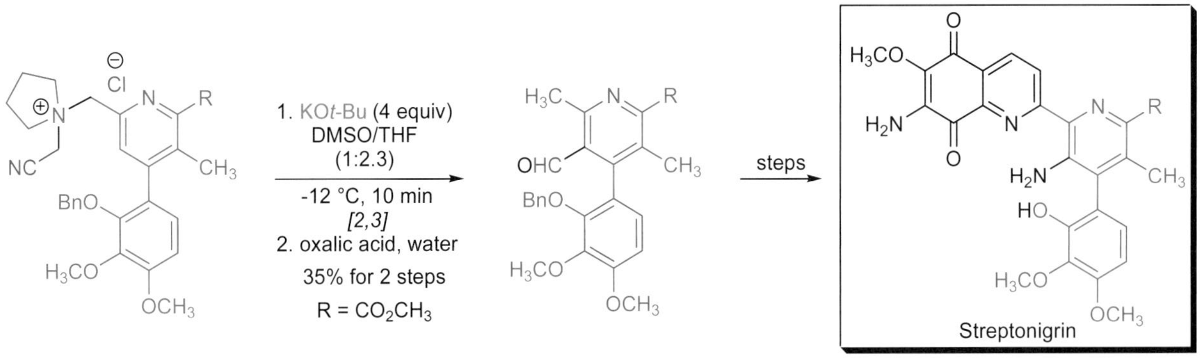

In the traditional strong base-promoted *S.-H. rearrangement*, the regioselective deprotonation of the ammonium salts is often difficult and other processes become competitive. A nonbasic modification may be accomplished when the desired nitrogen ylide is generated regiospecifically by means of fluoride ion-induced desilylation. Y. Sato and co-workers utilized this method for the ring-expansion of cyclic ammonium salts.[27] They showed that the stereochemistry of the substrate had a dramatic effect on the course of the reaction. The *cis*-stereoisomer gave predominantly the *[2,3]*-rearranged product, while the *trans*-stereoisomer gave exclusively the *Stevens rearrangement* product.

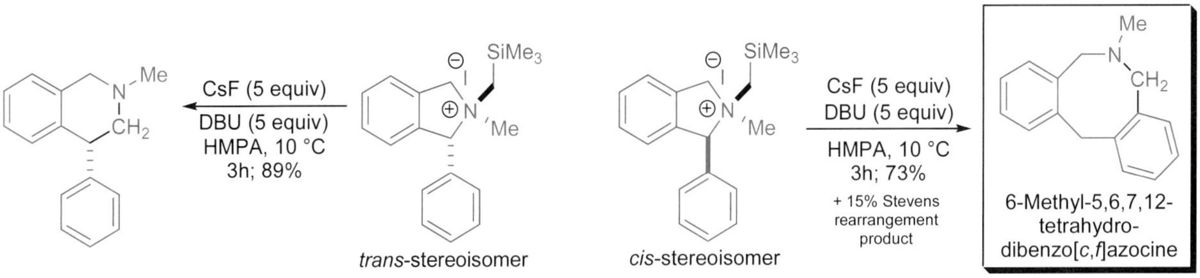

P.B. Alper and co-workers developed a practical approach for the synthesis of 4,7-disubstituted indoles based on the *Sommelet-Hauser rearrangement* of aryl sulfilimines.[28] The multihundred-gram preparation of methyl 7-chloroindole-4-carboxylate was achieved. The synthesis commenced with the activation of a sulfide precursor with $SOCl_2$ and coupling the intermediate with 3-amino-4-chlorobenzoate to afford an aromatic sulfilimine. This sulfilimine was exposed to excess triethylamine and heated to generate the sulfonium ylide that underwent the rearrangement.

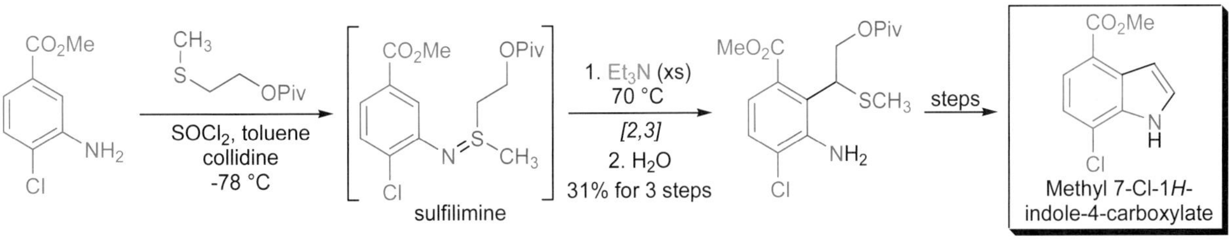

Novel regioisomeric tetrahydrophthalimide-substituted indoline-2(3*H*)-ones were prepared as potential herbicides by G.M. Karp et al. utilizing the sulfonium ylide version of the *Sommelet-Hauser rearrangement*[29] The unsymmetrical aniline substrate was treated with the chlorosulfonium salt of ethyl (methylthio)acetate and triethylamine at low temperature. The resulting regioisomeric amino esters were cyclized to the regioisomeric indoline-2(3*H*)-ones that were separated by column chromatography.

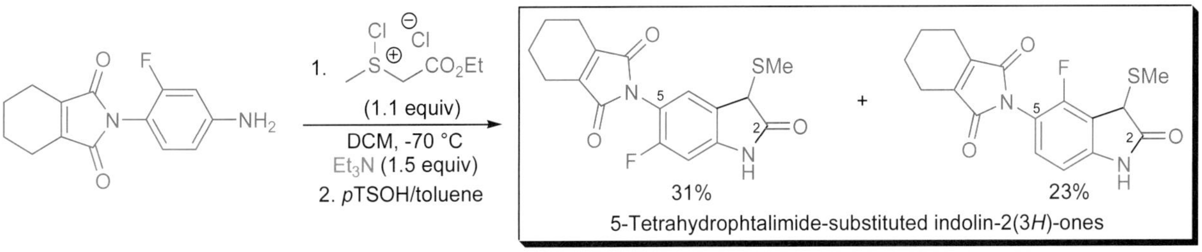

SONOGASHIRA CROSS-COUPLING

(References are on page 681)

Importance:

[*Seminal Publications*[1-3]; *Reviews*[4-30]; *Modifications & Improvements*[31-46]]

In 1975, K. Sonogashira and co-workers reported that symmetrically substituted alkynes could be prepared under mild conditions by reacting acetylene gas with aryl iodides or vinyl bromides in the presence of catalytic amounts of $Pd(PPh_3)Cl_2$ and cuprous iodide (CuI).[3] During the same year the research groups of both R.F. Heck and L. Cassar independently disclosed similar Pd-catalyzed processes, but these were not using copper co-catalysis, and the reaction conditions were harsh.[1,2] The copper-palladium catalyzed coupling of terminal alkynes with aryl and vinyl halides to give enynes is known as the *Sonogashira cross-coupling* and can be considered as the catalytic version of the *Castro-Stephens coupling*. The general features of the reaction are: 1) the coupling can usually be conducted at or slightly above room temperature, and this is a major advantage over the forcing conditions required for the alternative *Castro-Stephens coupling*; 2) the handling of the shock-sensitive/explosive copper acetylides is avoided by the use of a catalytic amounts of copper(I) salt; 3) the copper(I) salt can be the commercially available CuI or CuBr and are usually applied in 0.5-5 mol% with respect to the halide or alkyne; 4) the best palladium catalysts are $Pd(PPh_3)_2Cl_2$ or $Pd(PPh_3)_4$; 5) the solvents and the reagents do not need to be rigorously dried. However, a thorough deoxygenation is essential to maintain the activity of the Pd-catalyst; 6) often the base serves as the solvent but occasionally a co-solvent is used; 7) the reaction works well on both very small and large scale (>100g); 8) the coupling is stereospecific; the stereochemical information of the substrates is preserved in the products; 9) the order of reactivity for the aryl and vinyl halides is I ≈ OTf > Br >> Cl; 10) the difference between the reaction rates of iodides and bromides allows selective coupling with the iodides in the presence of bromides; 11) almost all functional groups are tolerated on the aromatic and vinyl halide substrates. However, alkynes with conjugated electron-withdrawing groups (R^2=CO_2Me) give *Michael addition* products and propargylic substrates with electron-withdrawing groups (R^2= CH_2CO_2Me or NH_2) tend to rearrange to allenes under the reaction conditions;[5] and 12) the exceptional functional group tolerance of the process makes it feasible to use this coupling for complex substrates in the late stages of a total synthesis. The coupling of sp^2-C halides with sp-C metal derivatives is also possible by using other reactions such as the *Negishi-, Stille-, Suzuki-, and Kumada cross-couplings*. In terms of functional group tolerance, the *Negishi cross-coupling* is the best alternative to the *Sonogashira reaction*. There are certain limitations on the *Sonogashira coupling*: 1) aryl halides and bulky substrates that are not very reactive require higher reaction temperature; and 2) at high temperatures terminal akynes undergo side reactions.

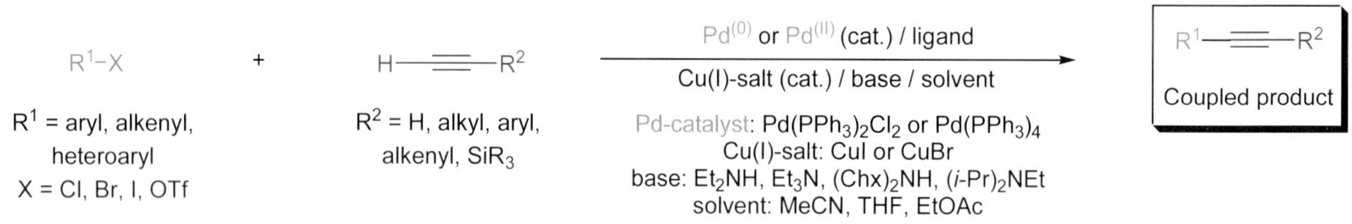

Mechanism: [47-50,27]

The mechanism of the *Sonogashira cross-coupling* follows the expected oxidative addition-reductive elimination pathway. However, the structure of the catalytically active species and the precise role of the CuI catalyst is unknown. The reaction commences with the generation of a coordinatively unsaturated $Pd^{(0)}$ species from a $Pd^{(II)}$ complex by reduction with the alkyne substrate or with an added phosphine ligand. The $Pd^{(0)}$ then undergoes oxidative addition with the aryl or vinyl halide followed by transmetallation by the copper(I)-acetylide. Reductive elimination affords the coupled product and the regeneration of the catalyst completes the catalytic cycle.

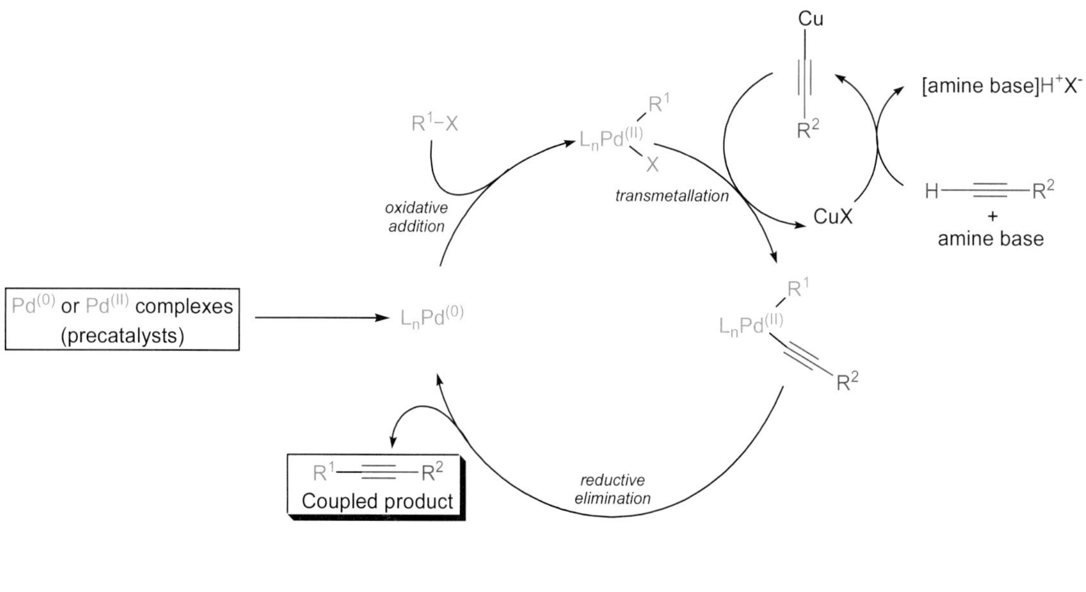

Synthetic Applications:

The novel heliannane-type sesquiterpenoid (–)-heliannuol E was synthesized in the laboratory of K. Shishido.[51] Interest in the total synthesis of this natural product was not only spurred by its irregular terpenoid structure and significant biological activity but the need to establish the absolute stereochemistry at the C2 and C4 stereocenters. The *Sonogashira reaction* was utilized to prepare the 3-arylpropargyl alcohol by coupling of a heavily substituted aryl iodide with an unprotected propargyl alcohol in quantitative yield.

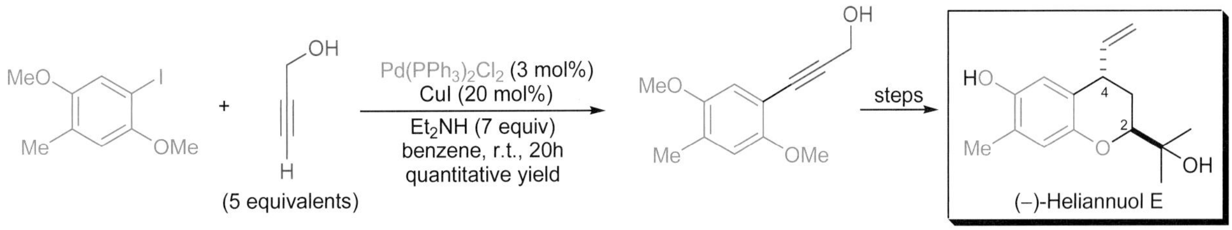

The concise formal total synthesis of mappicine was accomplished using an *intramolecular hetero Diels-Alder reaction* as the key step by M. Ihara and co-workers.[52] Introduction of the necessary acetylenic moiety at the C2 position was achieved by the *Sonogashira cross-coupling* of a 2-chloroquinoline derivative with TMS-acetylene. Several substituents at the C3 position were investigated, and it was found that the unprotected hydroxymethyl substituent gave almost quantitative yield of the desired disubstituted alkyne product.

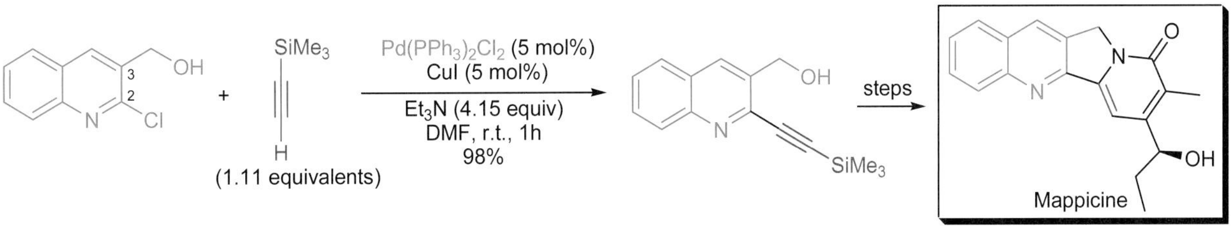

A novel member of the highly strained nine-membered enediyne antibiotic family, N1999-A2, exhibits remarkable antitumor activity against various tumor cell lines. Because the absolute configuration has not been established, the goal of the synthetic effort by M. Hirama et al. was to prove the stereochemistry unambiguously.[53] The cyclopentenyl iodide fragment was efficiently coupled with the epoxydiyne fragment under the *Sonogashira coupling* conditions. Unfortunately, the spectrum of the final product did not match the spectrum of the natural product so the proposed structure needs to be revised.

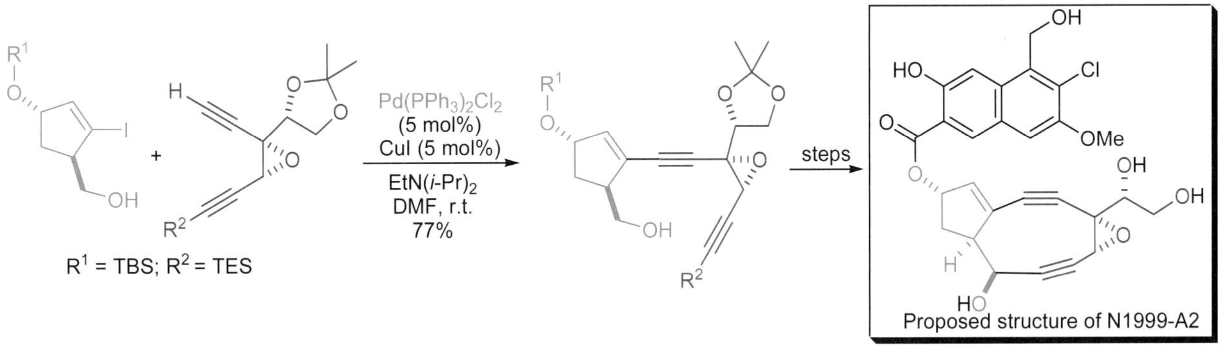

The expedient total synthesis of the callipeltoside aglycon was achieved by I. Paterson and co-workers.[54] The authors utilized a late-stage *Sonogashira coupling* between a dienyl iodide and an alkynyl cyclopropane derivative. Interestingly, the use of $Pd(PPh_3)_4$ as a catalyst did not give any of the desired coupling product. However, switching the catalyst to $Pd(PPh_3)_2Cl_2$ afforded the desired dieneyne in excellent yield.

STAUDINGER KETENE CYCLOADDITION

(References are on page 682)

Importance:

[*Seminal Publications*[1-8]; *Reviews*[9-33]; *Theoretical studies*[34-48]]

In 1908, ketene (CH_2=C=O) was independently prepared by the research groups of F. Chick and H. Staudinger.[3,6] At the same time, Staudinger exhaustively studied the reactivity of ketene and ketene derivatives, and he found that diphenylketene reacts with alkenes, ketones, and imines.[1,2,4-8,49,50] Today, the thermal [2+2] cycloaddition reaction of ketenes with carbon-carbon, carbon-oxygen, and carbon-nitrogen double bonds is referred to as the *Staudinger ketene cycloaddition*. The most common methods for the preparation of ketenes are: 1) dehydrohalogenation of acid chlorides by trialkylamines;[51] 2) dehalogenation of α-halo acid chlorides by zinc or zinc-copper alloy to form dihaloketenes;[52,53] 3) thermal[54] or photochemical[55] opening of cyclobutenones;[54] 4) *Wolff rearrangement* of α-diazoketones;[54] 5) pyrolysis of anhydrides followed by bulb to bulb distillation;[9] 6) pyrolysis of esters;[56-58] and 7) cracking commercially available diketene at atmospheric pressure leads to ketene.[9] The general features of the reaction of ketenes with alkenes are:[24] 1) the reaction leads to cyclobutanones; 2) the order of reactivity with simple alkenes is *trans* olefin < *cis* olefin < cyclic olefin< linear diene < cyclic diene; 3) the stereochemistry around the double bond is retained; 4) regiochemistry is determined by the polarization of the double bond; 5) as ketene itself is not reactive toward double bonds; usually dichloroketene is used instead, followed by dehalogenation by zinc-copper alloy; 6) in case of perfluorinated ketenes and alkoxybutadienes, the reaction may lead to the [4+2] cycloadducts; and 7) in addition to simple alkenes, allenes, enamines, and enol ethers also undergo the cycloaddition, although the yields are generally lower. The general features of the reaction of ketenes with aldehydes and ketones are:[24] 1) the reaction leads to the formation of 2-oxetanones (also called as β-lactones); 2) these reactions usually require Lewis acid activation, and the most common Lewis acids are boron trifluoride etherate, aluminum chloride, and zinc chloride; 3) amines can also be utilized as catalysts; 4) carbonyls bearing strongly electron-withdrawing substituents do not require activation; 4) a wide array of ketene substrates can be used, although aryl- and diarylketenes are generally unreactive; and 5) asymmetric versions of the cycloaddition have been developed by utilizing chiral amine bases as catalysts. The general features of the reaction of ketenes with imines are:[23,29,30,32,33] 1) the reaction is of particular importance because it leads to the formation of azetidinones (also called as β-lactams); 2) the reaction is usually carried out thermally or photochemically using acid chloride and triethylamine or α-diazoketones as the ketene precursors; 3) the diastereoselectivity of the resulting β-lactams is generally high; 4) asymmetric versions were developed by employing chiral auxiliaries attached to the imine or the ketene; 5) asymmetric catalytic methods utilizing chiral amine bases were also developed; and 6) when the reaction is carried out in sulfur dioxide, it leads to the formation of 2,3-diphenylthiazolidin-4-one-1,1-dioxide derivatives.[59] In addition to the above compounds, acetylenes, thiocarbonyls, isocyanates, carbodiimides, *N*-sulfinylamines, nitroso- and azo compounds also undergo a formal [2+2] cycloaddition with ketenes.[24]

H. Staudinger (1907):

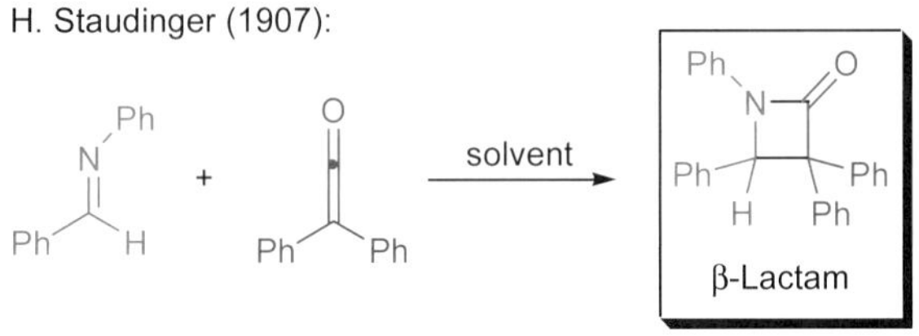

Reaction of ketenes with alkenes:

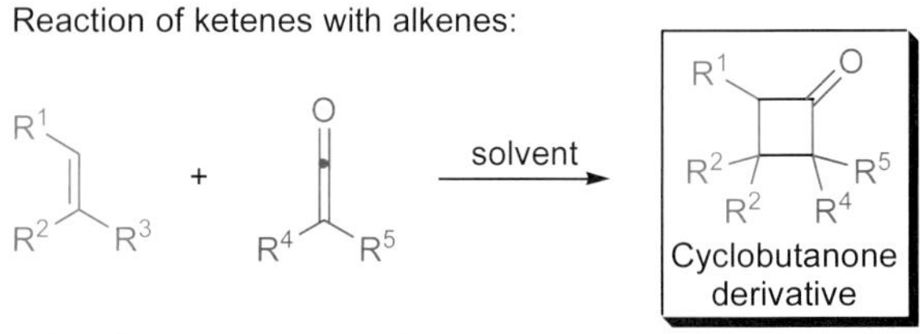

R^1 = R^2 = H, alkyl, aryl; R^3 = H, alkyl, aryl, vinyl, -OR, -NR_2; R^4 = H, alkyl, aryl, -Cl, -Br; R^5 = alkyl, aryl, -Cl, -Br, -OR;

Reaction of ketenes with aldehydes and ketones:

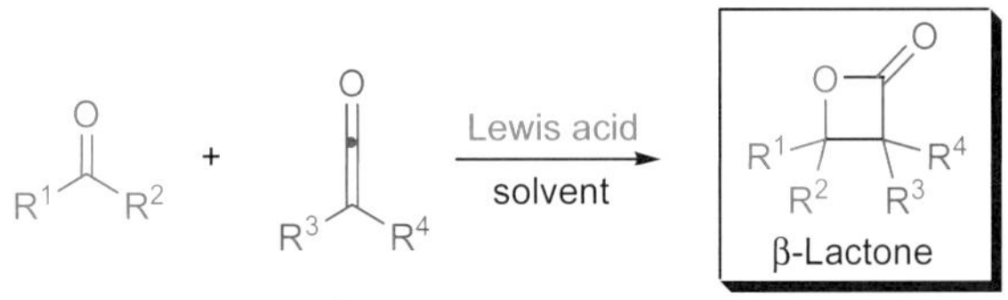

R^1 = H, alkyl, aryl ; R^2 = H, alkyl, aryl, vinyl; R^3 = H, alkyl, -Cl, -Br; R^4 = alkyl, aryl, -Cl, -Br, -OR -$SiMe_3$; Lewis acid = $BF_3{\cdot}OEt_2$, $AlCl_3$, $ZnCl_2$

Reaction of ketenes with imines:

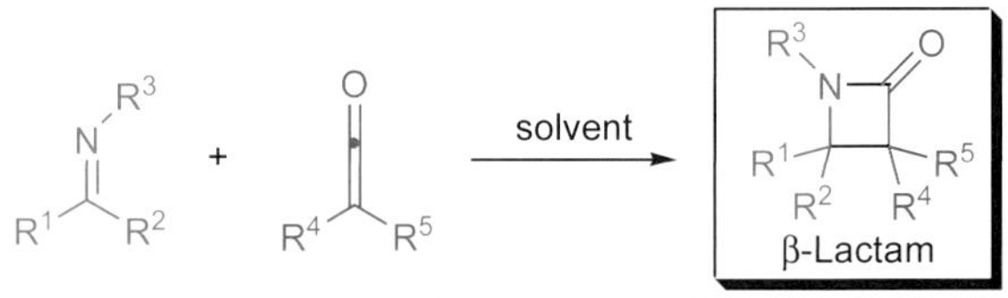

R^1 = H, alkyl, aryl; R^2 = alkyl, aryl; R^3 = alkyl, benzyl, aryl, -$SiMe_3$; R^4 = H, alkyl; R^5 = alkyl, -OR, -NR_2;

Mechanism:[60-67]

The reaction of ketenes with alkenes is assumed to occur *via* a concerted nonsynchronous mechanism, where the approach of the reacting partners is orthogonal.[60-66] As a consequence, the bulkier substituent of the ketene will end up on the sterically more crowded face of the cyclobutanone product. There are two descriptions that explain the experimental results: 1) according to the Woodward-Hoffmann rules, the LUMO of the ketene reacts antarafacially with the HOMO of the alkene that reacts suprafacially;[24] 2) the HOMO of the alkene forms a bond with the p_z orbital of the terminal carbon and the p_y orbital of the central carbon of the ketene.[67] The reaction of ketenes with carbonyls and imines follows a stepwise mechanism.

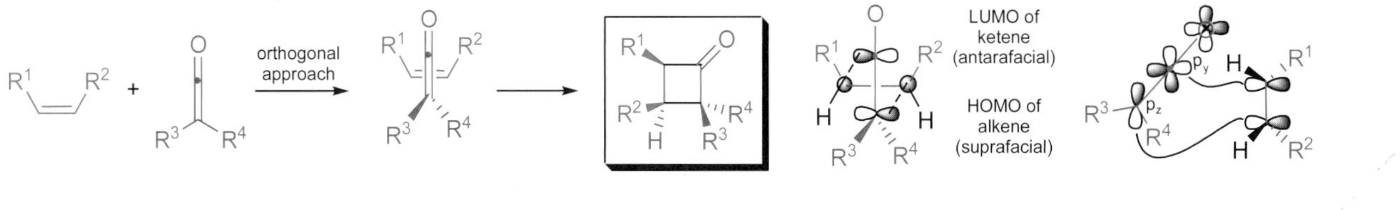

Synthetic Applications:

The *Staudinger ketene cycloaddition* was utilized as the key reaction in the synthesis of a number of bakkane natural products in the laboratory of A.E. Greene.[68] Dichloroketene was generated *in situ* from trichloroacetyl chloride by zinc-copper alloy in the presence of phosphorous oxychloride. The *[2+2] cycloaddition* between dichloroketene and 1,6-dimethylcyclohexene gave the product in high yield and excellent regio- and diastereoselectivity. The cycloadduct was successfully converted to (±)-bakkenolide A.

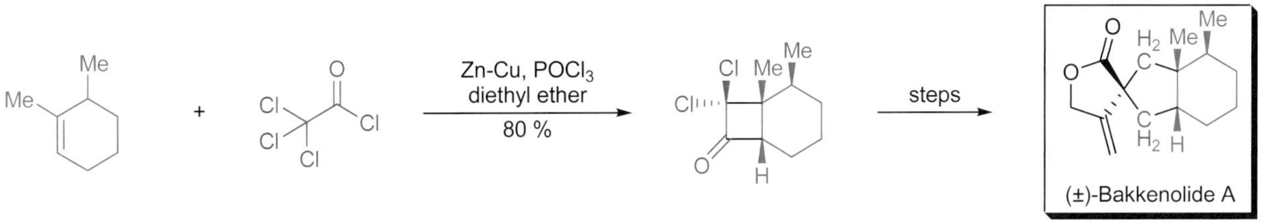

N.C. Chen and co-workers devised an efficient synthesis of the *cis*-bicyclo[3.3.0]octane ring system that was a key intermediate in the synthesis of iridoid monoterpene natural products loganin and sarracenin.[69] In their approach, they utilized a *[2+2] ketene cycloaddition* between a fulvene derivative and methylchloroketene that was generated *in situ* from 2-choropropanoyl chloride by treatment with triethylamine. The cycloaddition reaction provided the product with excellent regioselectivity and as a 8:1 mixture of diastereomers. Subsequent ring expansion and dehalogenation by zinc metal in acetic acid gave the key intermediate as a 9:1 mixture of diastereomers.

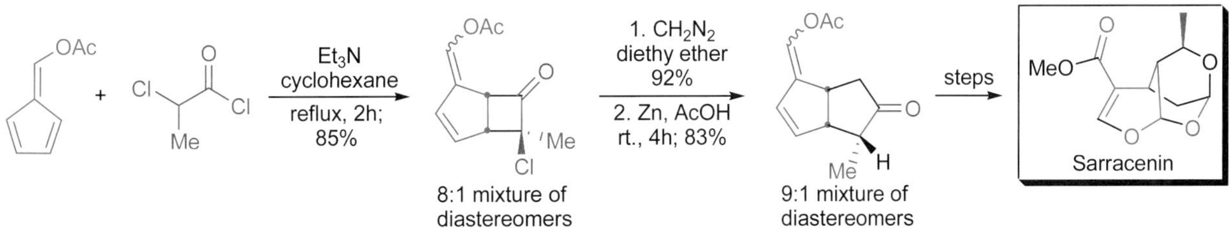

Ecteinascidin (ET)-743 is a marine natural product that exhibits potent antitumor activity. R.M. Williams and co-workers developed an approach for the synthesis of the pentacyclic framework of the molecule.[70] At an early stage in the synthesis, they used a *ketene-imine cycloaddition* utilizing a chiral *N*-protected ketene derivative to control the stereoselectivity. Subsequently, the chiral auxiliary was removed and the intermediate β-lactam was converted to the target structure.

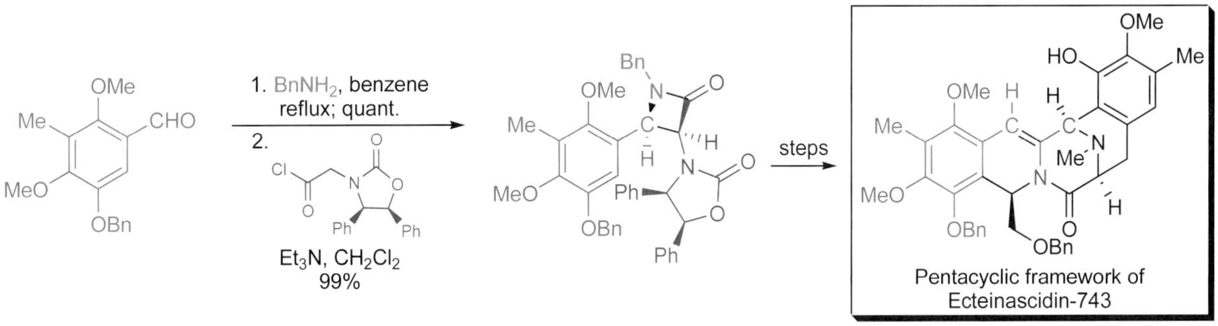

(–)-Lipstatin is a natural product that exhibits potent inhibitor activity of the pancreatic lipase, and therefore it is a potential lead for the development of antiobesity agents. P.J. Kocienski developed a synthesis for this compound that incorporates an *aldehyde-ketene cycloaddition* as the key step.[71] The reaction between the aldehyde and silylketene derivative was carried out in the presence of $EtAlCl_2$ that served as the Lewis acid activator. This transformation led to the formation of four diastereomers in 91% yield, but after desilylation, the desired stereoisomer could be isolated in 64% yield from the mixture.

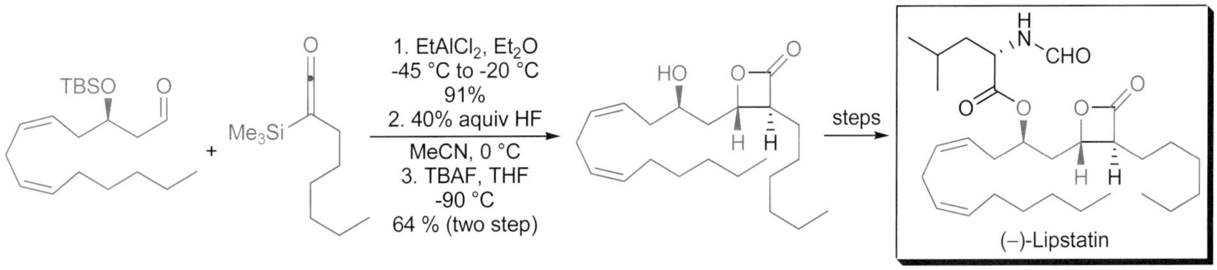

STAUDINGER REACTION

(References are on page 684)

Importance:

[*Seminal Publications*[1]; *Reviews*[2-15]; *Modifications & Improvements*[16-32]; *Theoretical Studies*[33-35]]

In 1919, H. Staudinger and J. Meyer reported the reaction between phenyl azide and triphenylphosphine, which afforded a novel compound, phosphinimine (also known as aza-ylide or iminophosphorane), in quantitative yield accompanied by the evolution of nitrogen gas.[1] It was found that benzoyl azide reacted with triphenylphosphine in an analogous fashion to afford the corresponding benzoyl aza-ylide. The authors also investigated the reactivity of phosphinimines and demonstrated that the reaction of carbon dioxide with phenyl aza-ylide gave rise to phenyl isocyanate and triphenylphosphine oxide, which is the first example of an *aza-Wittig reaction*. The reaction of organic azides with trivalent phosphorous compounds (e.g., trialkyl- or triarylphosphines) to afford the corresponding aza-ylides is known as the *Staudinger reaction*. The general features of this transformation are:[4,6,9,10] 1) the reaction is usually very fast and takes place in almost quantitative yield without the formation of side products; 2) virtually any trivalent phosphorus compound undergoes the reaction; 3) the structure of the azide component can also be widely varied; and 4) the iminophosphorane products derived from alkyl- or arylazides and trialkyl- and triarylphosphines are stable compounds that can be isolated, but alkoxy groups on the P atom tend to undergo alkyl migration. The iminophosphoranes are versatile synthetic intermediates: 1) hydrolysis with water gives rise to primary amines (this reduction of azides is highly chemo- and stereoselective); 2) inter- or intramolecular reaction with carbonyl or thiocarbonyl compounds affords imines (*aza-Wittig reaction*); 3) carboxylic acids convert iminophosphoranes to *N*-substituted amides; 4) acyl halides condense to generate imydoyl halides; and 5) ozonolysis produces nitro compounds.

Staudinger (1919):

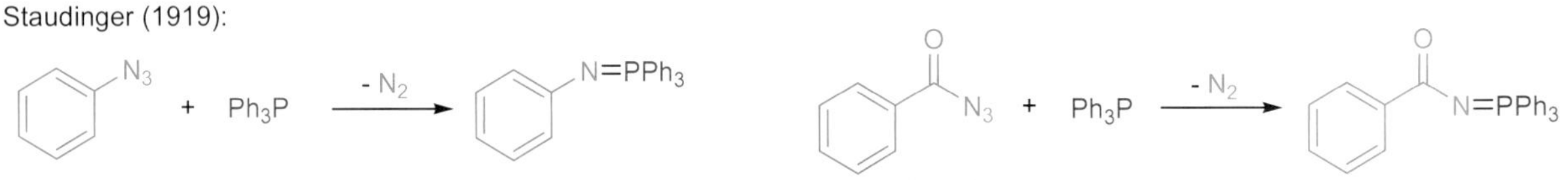

phenyl azide — benzoyl azide

Staudinger reaction:

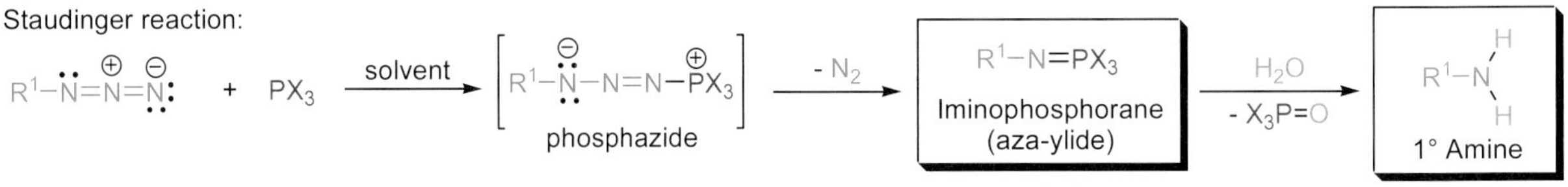

Synthetic use of iminophosphoranes:

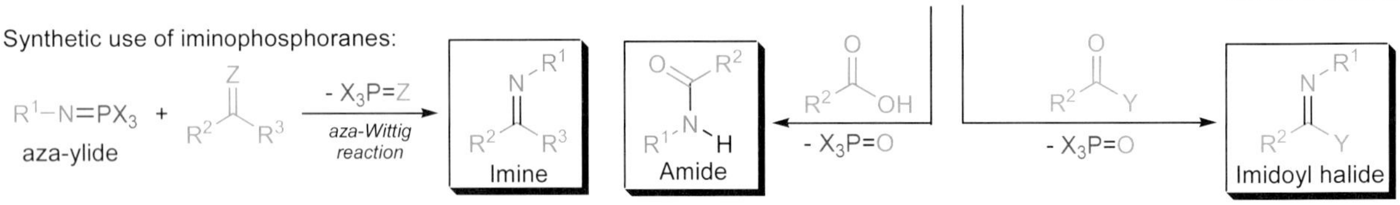

R^1 = alkyl, aryl, heteroaryl, RC(O), RSO_2, RP(O), R_2P, R_3Si, R_3Sn, R_3Ge; R^{2-3} = H, alkyl, aryl, heteroaryl; X = alkyl, aryl, O-alkyl, O-aryl, NH_2, NR_2, Cl, F, NCO, (also the combination of these ligands); Y = Cl, Br; Z = O, S; solvent: THF, Et_2O

Mechanism:[36-42]

The mechanism of the *Staudinger reaction* has been subject to a number of kinetic and theoretical studies[35] and at this point the exact mechanism remains unclear. All experimental data shows, however, that free radicals or nitrenes are not intermediates in this transformation. The first step of the mechanism is the attack of trivalent phosphorous by the unsubstituted nitrogen atom (N_α) of the azide to give the corresponding phosphazide (which occasionally can be isolated) with retention of configuration at the phosphorous atom. Next, the phosphazide goes through a four-membered transition state, which upon losing dinitrogen affords the iminophosphorane. A subtle point in the mechanism is the exact mode of attack of the phosphorous at N_α, since the $PN_\alpha N_\beta N_\gamma$ backbone is not linear. Instead, the A_{NNN} angle is approximately 170°. There are two possible trajectories of the phosphorous atom to approach N_α: 1) from the same side of the R^1 substituent on N_γ (*trans attack*); and 2) from the opposite side of the R^1 substituent on N_γ (*cis attack*). Investigations within DFT showed that the reaction prefers a *cis* TS due to the extra interaction between the P atom and N_γ.

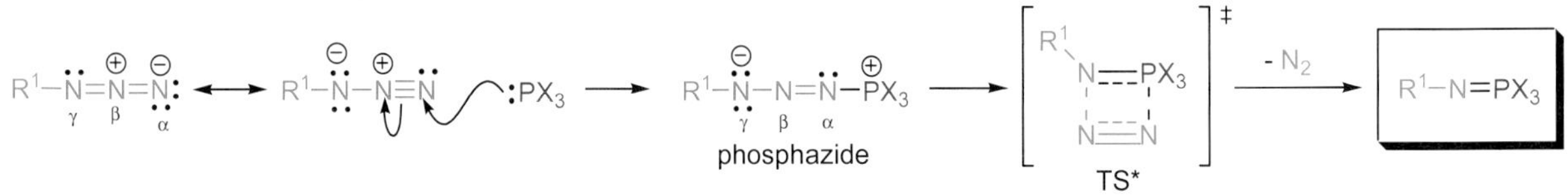

Possible modes of attack:

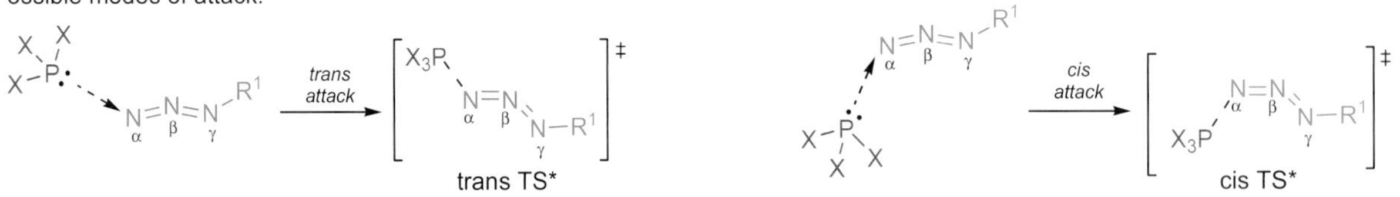

Synthetic Applications:

The total synthesis of the antiviral marine natural product (–)-hennoxazole A was accomplished by F. Yokokawa and co-workers.[43] The mild reduction of a secondary alkyl azide at C9 was carried out using triphenylphosphine in a THF/water mixture at slightly elevated temperature. The corresponding primary amine was obtained in good yield and was subsequently acylated and converted to one of the oxazole rings of the natural product.

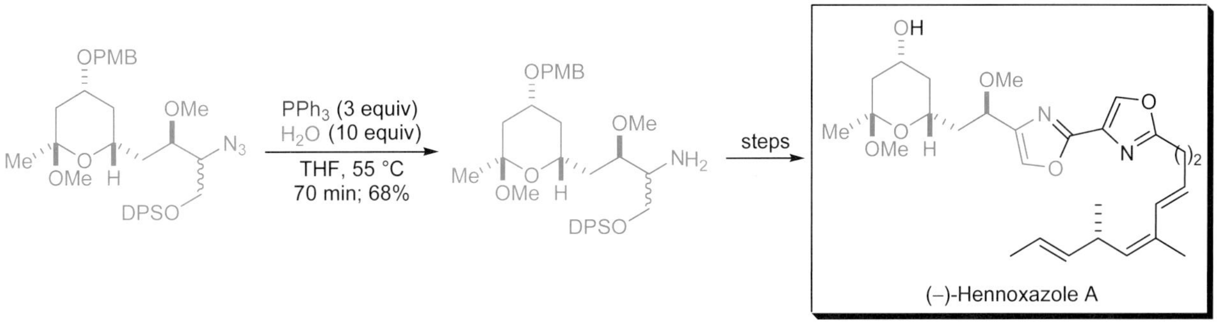

The marine indole alkaloid (+)-hamacanthin B was prepared by B. Jiang et al. using a tandem *Staudinger reaction/intramolecular aza-Wittig reaction* to convert a secondary azide to the corresponding iminophosphorane, which upon prolonged heating cyclized to the central pyrazinone ring.[44] The reduction of the azide was conducted with a slight excess of tributylphosphine in anhydrous toluene at room temperature while the *aza-Wittig cyclization* required the reflux temperature.

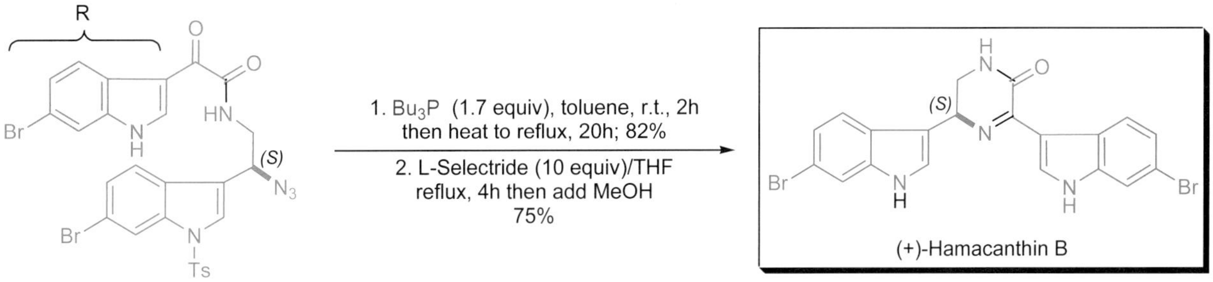

The absolute configuration of the structurally unique fungal metabolite mycosporins was determined in the laboratory of J.D. White by means of enantioselective total synthesis.[45] In the endgame of the synthetic effort, the *Staudinger reaction* was used to elaborate the side chain. The cyclic vinyl azide was first converted to a stable vinyl iminophosphorane, which was subsequently reacted with benzyl glyoxylate to afford the corresponding Schiff base. Reduction of the imine was achieved with sodium cyanoborohydride.

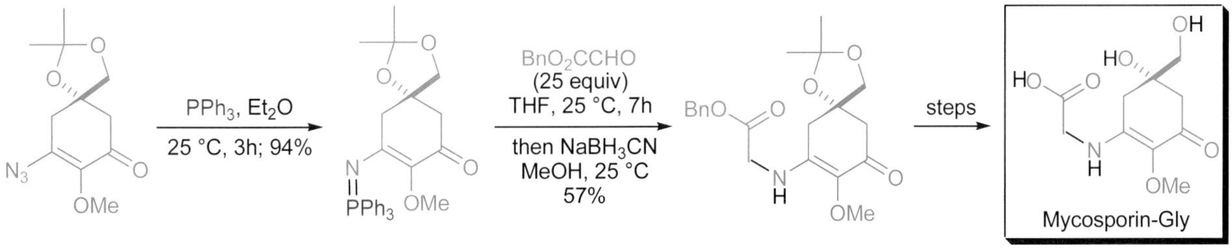

The research team of S.R. Rajski demonstrated that *o*-carboalkoxy triarylphosphines react with aryl azides to afford *Staudinger ligation* products bearing O-alkyl imidate linkages.[27] In comparison, the reaction of alkyl azides with o-carbalkoxy triarylphosphines usually gives rise to amide linkages. The importance of this technique lies in its ability to couple abiotic reagents under biocompatible conditions.

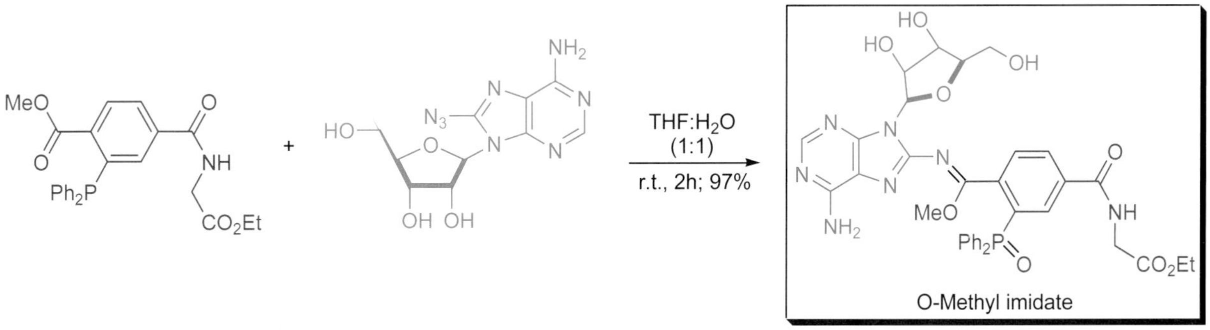

STEPHEN ALDEHYDE SYNTHESIS (STEPHEN REDUCTION)

(References are on page 685)

Importance:

[*Seminal Publications*[1]; *Reviews*[2-7]; *Modifications & Improvements*[8,9]; *Theoretical Studies*[10]]

In 1925, H. Stephen reported that when aromatic or aliphatic nitriles were added to a solution of stannous chloride ($SnCl_2$) in diethyl ether saturated with anhydrous hydrogen chloride gas, imine hydrochlorides were obtained that readily underwent hydrolysis in warm water to give the corresponding aldehydes in good yield.[1] The preparation of aldehydes by the reduction of nitriles with the combination of stannous halide/HCl in an organic solvent is known as the *Stephen aldehyde synthesis* or *Stephen reduction*. The general features of this transformation are:[4] 1) the original procedure has been modified: first the nitrile is dissolved in an inert solvent and the resulting solution is saturated with anhydrous HCl gas at 0 °C, then a solution of SnX_2/HCl in the same solvent is added; 2) if the substrate is insoluble in a given solvent, the use of a mixture of inert solvents is recommended; 3) most common solvents for the transformation are diethyl ether, dioxane, ethyl acetate, and chloroform; 4) the reduction products are aldimine hexachlorostannanes which usually precipitate from the reaction mixture as crystalline complexes and are readily hydrolyzed to the corresponding aldehydes with warm water; 5) the best substrates are aromatic nitriles that give moderate to good yields of the aldehyde; 6) aliphatic nitriles tend to give lower yields primarily due to the formation of *N,N'*-alkyliden*bis*acylamides, which are trimeric side products; 7) the yield drops sharply for aliphatic nitriles having more than six carbon atoms; 8) seldom does the *Stephen reduction* stop at the aldimine stage, but the reduction proceeds all the way to form the primary amine product; 9) yields are also strongly influenced by steric factors, so *ortho*-substituted aromatic nitriles rarely give high yield of the corresponding aldehyde; 10) the functional group tolerance is low, which renders this method only useful for robust substrates that do not have acid sensitive functional groups; and 11) if a large excess of the stannous halide is used, aromatic nitro groups also undergo reduction to yield the corresponding aromatic amines. Alternatively, nitriles can be reduced to the corresponding aldehydes by the following methods:[11-18] 1) catalytic hydrogenation with Raney nickel/H_2 in the presence of one equivalent of an acid (e.g., H_2SO_4, HCO_2H); and 2) use of metal hydride reagents (e.g., DIBAL-H, $LiAlH(OR)_3$, etc.).

Stephen (1925):

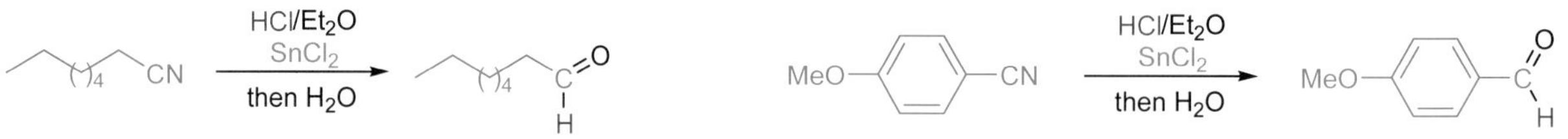

Stephen reduction:

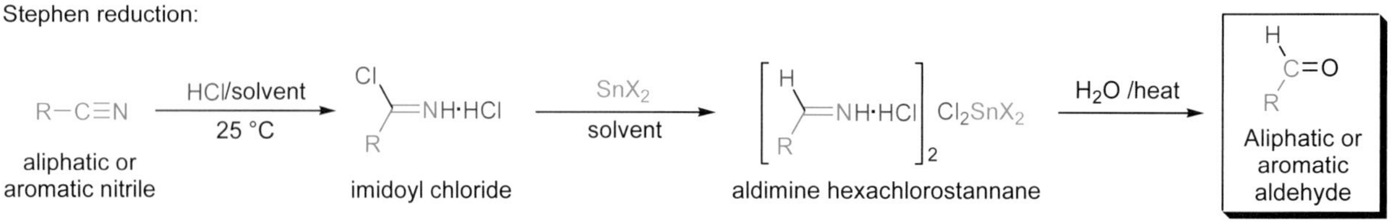

R = 1°, 2° or 3° alkyl, aryl, heteroaryl; solvent: Et_2O, dioxane, $CHCl_3$, EtOAc; X = Cl, Br

Mechanism:[4,7]

Formation of the imidoyl chloride intermediate:

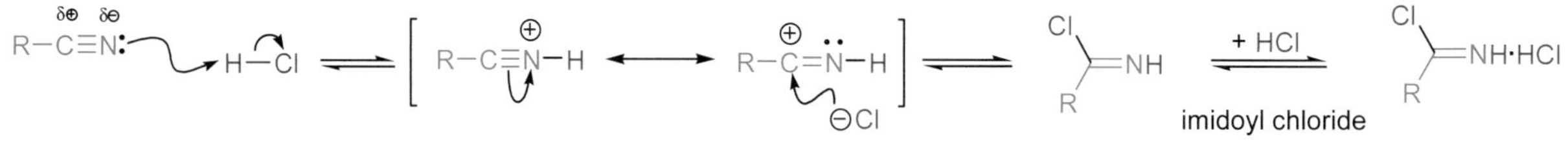

Reduction of the imidoyl chloride to the aldimine:

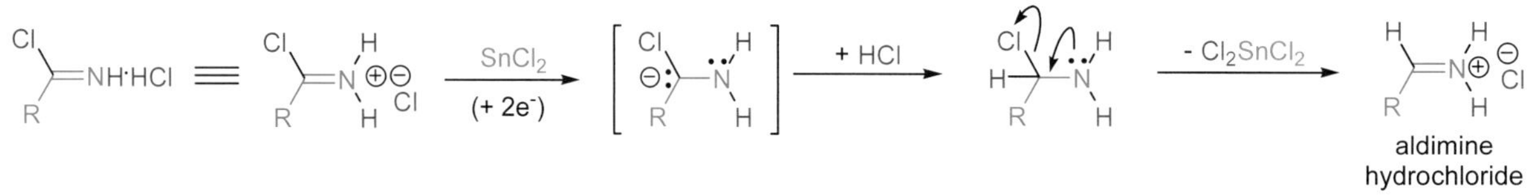

Hydrolysis of the aldimine to the corresponding aldehyde with water:

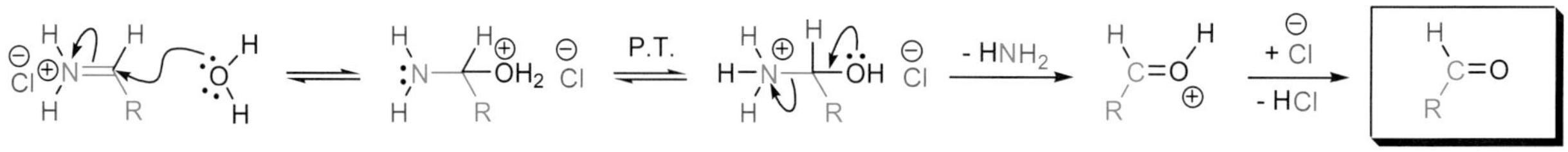

Synthetic Applications:

In the laboratory of N. Suzuki, the synthesis of several heterocyclic condensed 1,8-naphthyridine derivatives with potential antimicrobial activity was executed.[19] The preparation of pyrazolo[3,4-*b*][1,8]naphthyridines required 7-chloro-6-formyl-3-ethyl ester as the precursor that was obtained by the *Stephen reduction* of the corresponding aromatic nitrile. The solution of the aromatic nitrile in chloroform was added to the solution of $SnCl_2$/dry HCl gas in ether. After two days of stirring, the aldimine hexachlorostannane product was treated with warm water to obtain the desired aromatic aldehyde in modest yield. Heating of the aldehyde with methyl hydrazine afforded the pyrazole derivative.

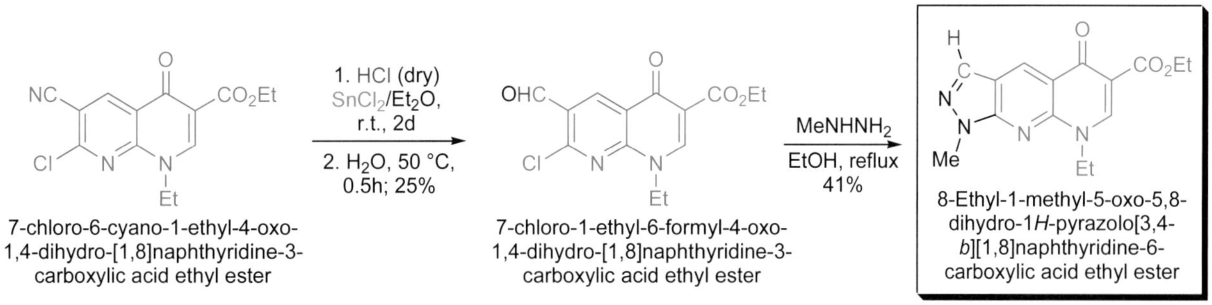

The stereoselective cyanation of [1,1']-binaphthalenyl-2,2'-diiodide was developed by M. Putala and co-workers using zinc cyanide and catalytic amounts of $Pd(dppf)_2$.[20] The resulting dinitrile was converted to the corresponding [1,1']-binaphthalenyl-2,2'-dicarbaldehyde in high yield using the *Stephen reduction*.

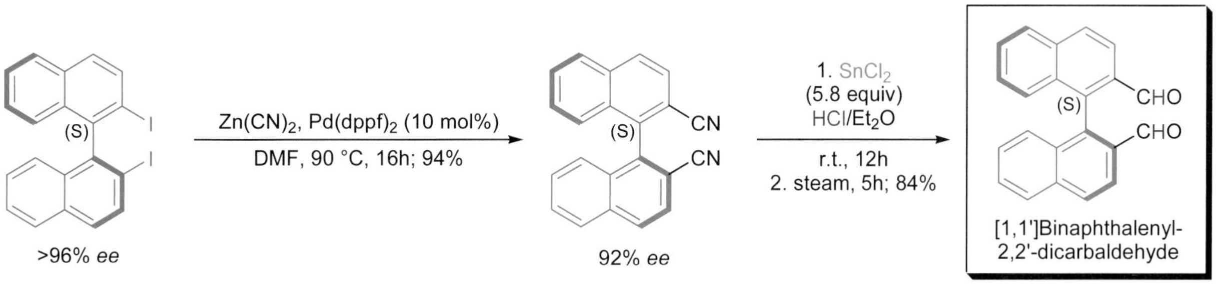

Research by P. Scrimin and U. Tonellato et al. showed that $Zn^{(II)}$ was an allosteric regulator of liposomal membrane permeability induced by synthetic template-assembled tripodal polypeptides.[21] Several copies of peptide sequences from the peptaibol family were connected to *tris*(2-aminoethyl)amine (TREN), which is a tripodal metal ion ligand. The resulting tripodal polypeptides were capable of modifying the permeability of liposomal membranes, and their activity was tunable upon metal ion coordination of the TREN subunit. The synthesis of the TREN-based template began with the *Stephen reduction* of 4-cyanomethylbenzoate followed by the reductive amination of the resulting aldehyde with TREN.

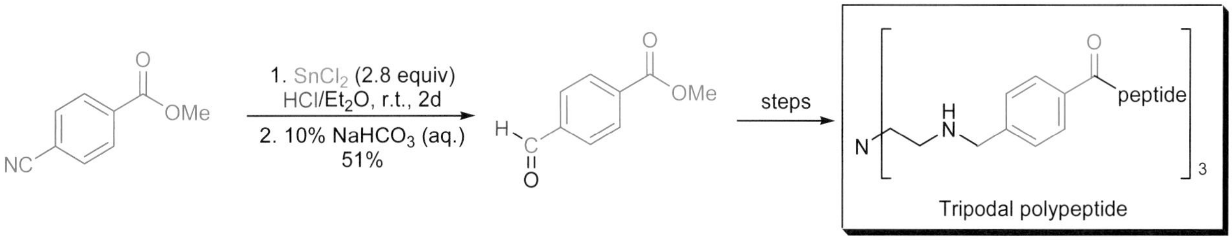

L.-M. Yang and co-workers designed and synthesized a new series of *trans*-stilbene benzenesulfonamide derivatives as potential antitumor agents.[22] A common precursor diethylphosphonate was prepared from commercially available sulfanilamide in six steps. The aromatic nitrile-to-aldehyde reduction was affected by the *modified Stephen reduction* using Raney nickel alloy in aqueous formic acid. The corresponding aldehyde was obtained in high yield.

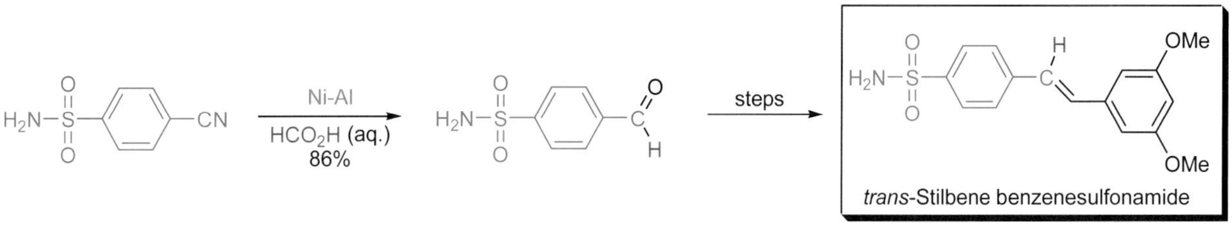

STETTER REACTION

(References are on page 685)

Importance:

[*Seminal Publications*[1]; *Reviews*[2-9]; *Modifications & Improvements*[10-24]]

In 1973, H. Stetter and M. Schreckenberg found that in the presence of catalytic amounts of sodium cyanide, aromatic aldehydes such as benzaldehyde and *p*-chlorobenzaldehyde added smoothly to α,β-unsaturated nitriles and ketones to afford the corresponding γ-oxo nitriles and and γ-diketones, respectively.[1] The method was later expanded to aliphatic aldehydes by the use of catalytic amounts of thiazolium salts in the presence of bases. The addition of aliphatic and aromatic aldehydes across activated double bonds in the presence of a nucleophilic catalyst is known as the *Stetter reaction*. The general features of this transformation are:[2,5] 1) when the reaction is catalyzed by cyanide ions, dipolar aprotic solvents (e.g., DMF, DMSO) should be used, but with thiazolium salts protic solvents (e.g. EtOH) may also be used; 2) the reaction temperature is usually above 30 °C and the reaction time is a few hours (~1-4h); 3) the cyanide catalyzed reaction is restricted to aromatic aldehydes, since aliphatic aldehydes undergo an undesired *aldol condensation*; 4) the thiazolium salts are actually precatalysts since the added base (e.g., Et_3N, NaOAc) deprotonates the highly acidic C-H bond between the nitrogen and sulfur atoms to generate an ylide structure *in situ* (this ylide behaves the same way as cyanide ions do); 5) since the mechanism involves the rapid, reversible formation of benzoins from aromatic aldehyde substrates, benzoins can be used instead of the aldehydes (aliphatic aldehydes cannot be replaced with acyloins); 6) a wide variety of activated alkene substrates can be used, and the yields are especially high with α,β-unsaturated ketones; 7) straight chain aldehydes tend to give higher yields than α-branched aldehydes; 8) the aldehyde substrates may also be α,β-unsaturated and may have isolated double or triple bonds; and 9) the reaction fails with aromatic aldehydes that have nitro substituents as well as with 2,6-disubstituted aromatic aldehydes (due to steric hindrance).

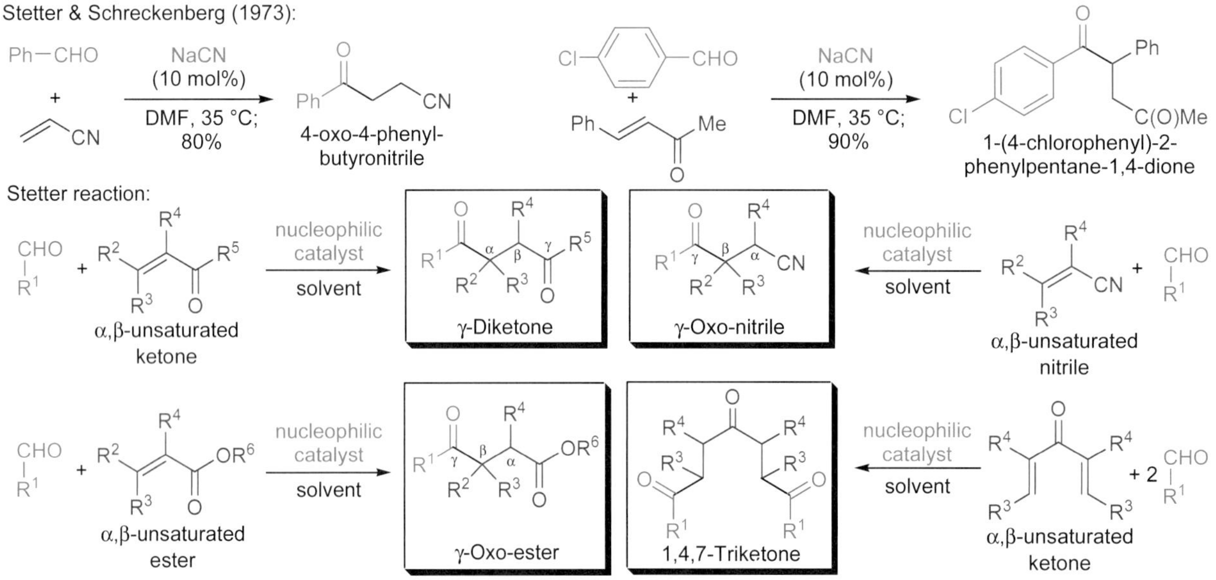

R^1 = alkyl, aryl, heteroaryl, alkenyl; R^{2-5} = H, alkyl, aryl, heteroaryl; R^6 = alkyl, aryl; nucleophilic catalyst: NaCN, KCN, thiazolium salts/base; solvent: DMF, DMSO

Mechanism: [2,5]

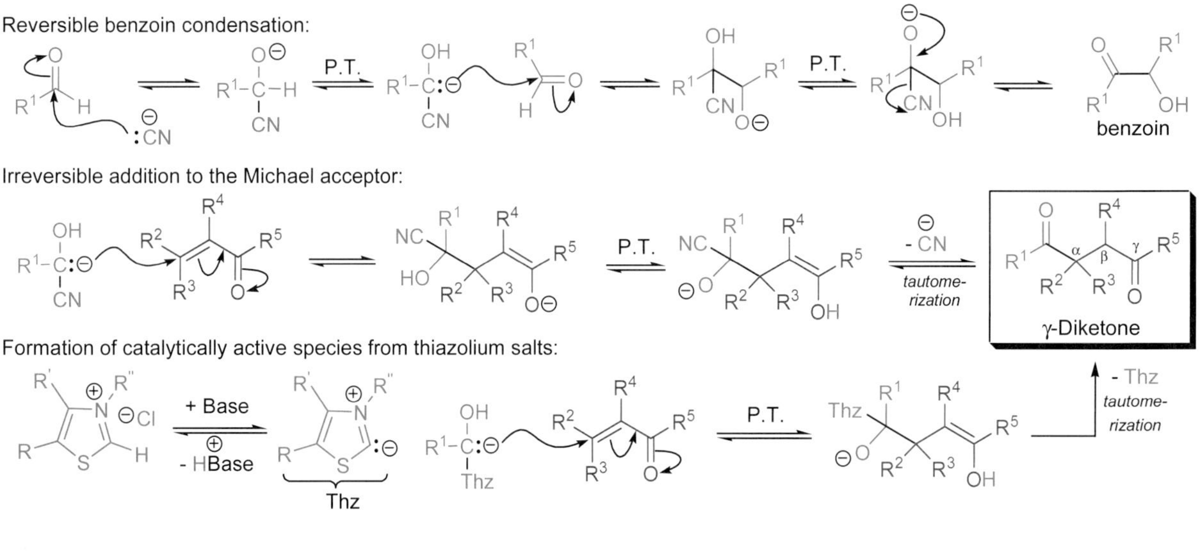

Synthetic Applications:

In the laboratory of A. Millar, the convergent enantioselective synthesis of CI-981, a potent and tissue-selective inhibitor of HMG-CoA reductase was achieved.[25] The central tetrasubstituted pyrrole ring was prepared *via* the *Paal-Knorr pyrrole synthesis*. The required 1,4-diketone precursor was efficiently prepared by the *Stetter reaction* between *p*-fluorobenzaldehyde and an unsaturated amide. Interestingly, the *N*-benzyl thiazolium chloride catalyst afforded only the *benzoin condensation* product and none of the desired diketone. However, when the *N*-ethyl thiazolium bromide catalyst was employed, under anhydrous and concentrated reaction conditions, the 1,4-diketone was formed in good yield. The authors also noted that the simple dilution of the reaction mixture resulted in a dramatic increase in the formation of the undesired benzoin condensation product.

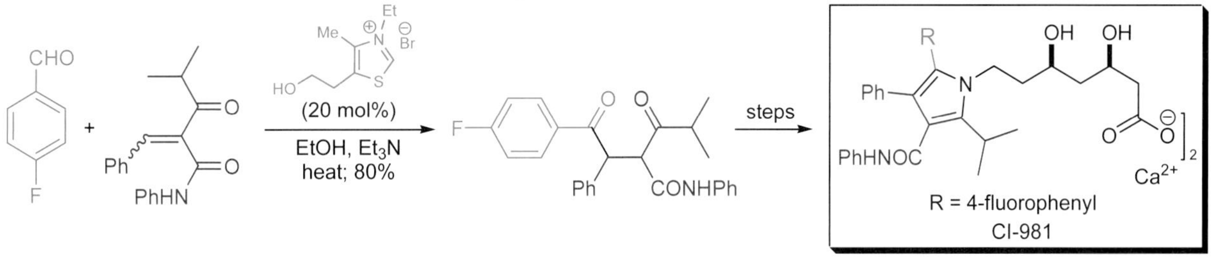

The absolute stereochemistry of natural roseophilin was determined by means of asymmetric total synthesis by M.A. Tius and co-workers.[26] The trisubstituted pyrrole moiety of the natural product was installed using the *Paal-Knorr pyrrole synthesis* starting from a macrocyclic 1,4-diketone. This diketone was prepared by reacting an exocyclic α,β-unsaturated ketone with excess 6-heptenal in the presence of 3-benzyl-5-(hydroxyethyl)-4-methylthiazolium chloride as the catalyst. The major product was the *trans* diastereomer and the macrocyclization was achieved *via alkene metathesis*. It is worth noting that when the aldehyde was tethered to the cyclopentenone, all attempts to close the macrocycle in an *intramolecular Stetter reaction* failed.

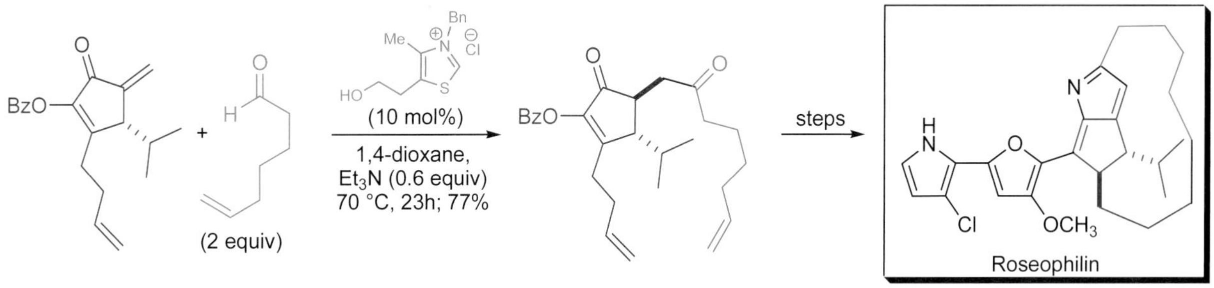

The short synthesis of (±)-*trans*-sabinene hydrate, an important flavor chemical found in a variety of essential oils from mint and herbs, was developed by C.C. Galopin.[27] The key intermediate of the synthetic sequence was 3-isopropyl-2-cyclopentenone. Initially a *Nazarov cyclization* of a dienone substrate was attempted for the synthesis of this compound, but the cyclization did not take place under a variety of conditions. For this reason, a sequential *Stetter reaction/intramolecular aldol condensation* approach was successfully implemented.

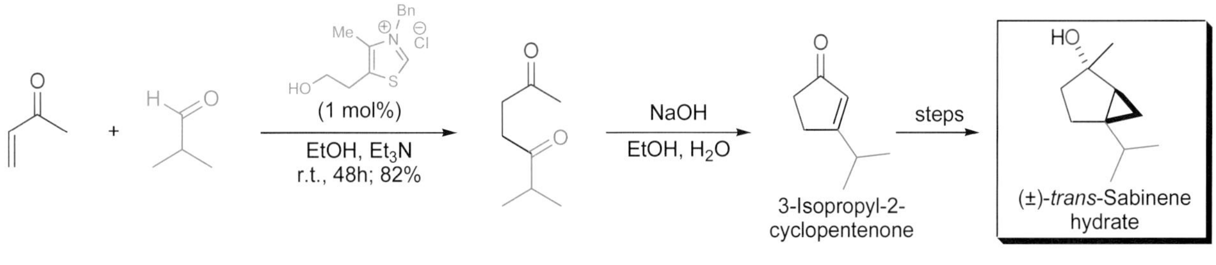

The concise enantioselective total synthesis of (+)-monomorine I, a 3,5-dialkyl-substituted indolizidine alkaloid, was completed by S. Blechert et al. using a sequential *cross-metathesis/double reductive cyclization* strategy.[28] The enedione substrate was prepared in two steps. The *Stetter reaction* between the masked equivalent of acrolein and butyl vinyl ketone was followed by a *retro Diels-Alder reaction* under flash vacuum pyrolysis (FVP) conditions.

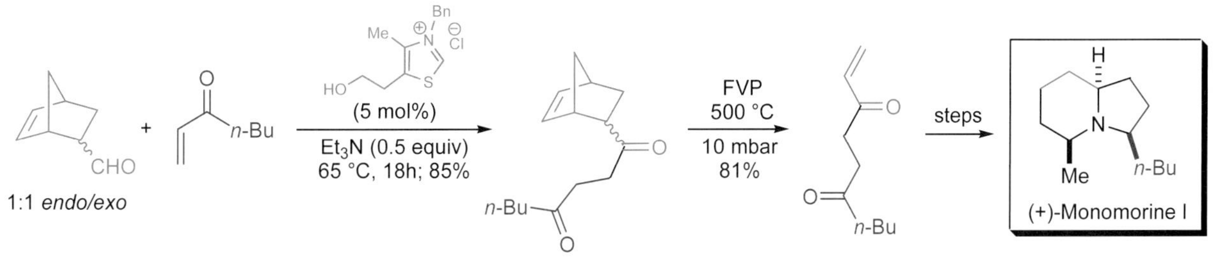

STEVENS REARRANGEMENT

(References are on page 686)

Importance:

[*Seminal Publications*[1-5]; *Reviews*[6-14]; *Modifications & Improvements*[15-25]; *Theoretical Studies*[26-34]]

In 1928, T.S. Stevens reported that phenacylbenzyldimethylammonium bromide could be converted to 1-benzoyl-2-benzyldimethylamine upon treatment with aqueous sodium hydroxide.[1] A few years later he observed an analogous transformation by exposing a sulfonium salt to sodium methoxide that rearranged to the corresponding sulfide.[4] The base-promoted transformation of sulfonium or quaternary ammonium salts to the corresponding sulfides or tertiary amines involving the [*1,2*]-*migration* of one of the groups on the nitrogen or sulfur atom is known as the *Stevens rearrangement*. The general features of this reaction are: 1) the quaternary ammonium salts are readily available by the alkylation of the corresponding tertiary amines; 2) the sulfonium salts are usually prepared by the direct alkylation of the corresponding sulfides; 3) the key intermediate of the rearrangement is the nitrogen- or sulfur ylide; 3) the R^1 group has to be able to stabilize carbanions, so it is often an electron-withdrawing group; 4) depending on the nature of R^1, the acidity of the adjacent C-H bond varies so the type of base used for the deprotonation must be chosen accordingly; 5) when R^1=aryl or heteroaryl, the *Sommelet-Hauser rearrangement* becomes competitive; 6) R^2 and R^3 groups of ammonium salts cannot contain a hydrogen at their β-position, since the *Hofmann elimination* may compete; 7) the migrating group (R^4) is usually capable of stabilizing a carbon-centered radical; 8) the migratory aptitude of benzyl groups depends on the substituents on the phenyl ring and decrease in the following order: *p*-NO_2>*p*-halogen>*p*-Me>*p*-OMe; 9) when the migrating group has a stereocenter, it is transferred with retention of configuration at the migrating terminus; 10) the degree of the retention of configuration is influenced by the nature of substituents present on the migrating group; 11) in the case of sulfonium salts, the retention of configuration at the migrating terminus occurs to a lesser extent than in the case of quaternary ammonium salts; and 12) in addition to nitrogen to carbon migrations, there are nitrogen to heteroatom migrations as well (when Y=NH).[16,35] When the regioselective deprotonation of the ammonium and sulfonium salts is problematic, the use of fluoride ion catalyzed desilylation of (trimethylsilyl)methyl ammonium- and sulfonium salts under nonbasic conditions gives the required ylides directly and with complete regioselectivity.[17,18]

Stevens (1928):

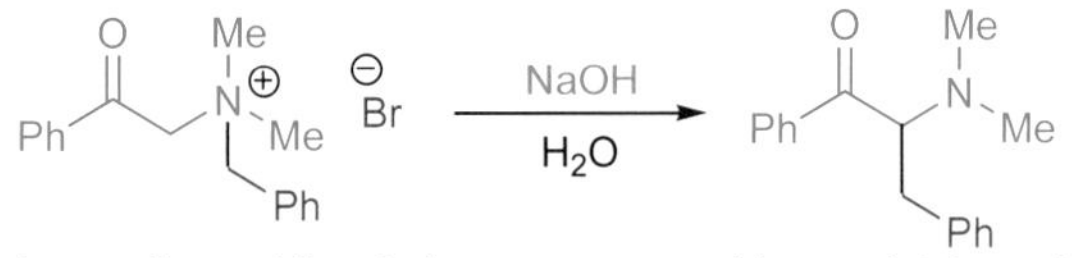

phenacylbenzyldimethyl ammonium bromide → 1-benzoyl-2-benzyl dimethylamine

Stevens (1932):

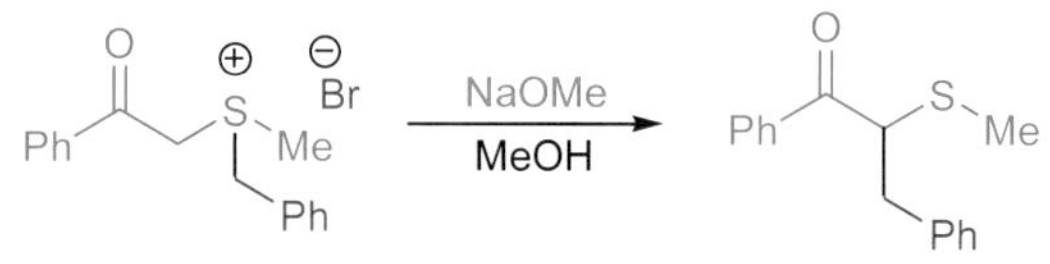

phenacylbenzylmethyl sulfonium bromide → 2-methylsulfanyl-1,3-diphenyl-propan-1-one

Stevens rearrangement of quaternary ammonium salts:

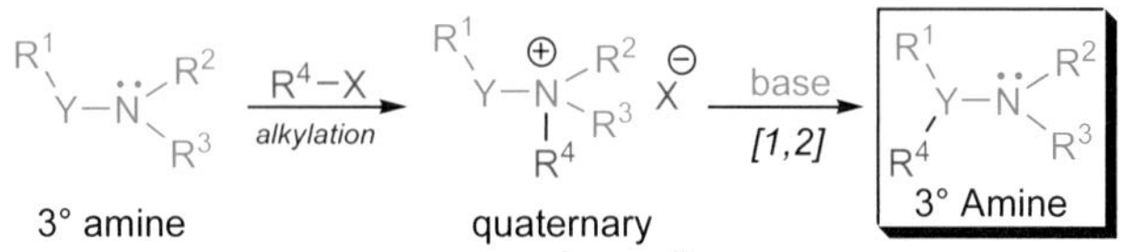

3° amine → quaternary ammonium salt

Stevens rearrangement of sulfonium salts:

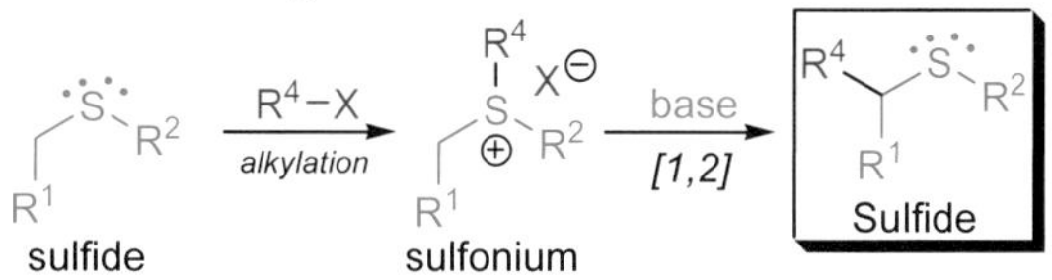

sulfide → sulfonium salt

R^1 = EWG = Ar, heteroaryl, COR, COOR, CN; Y= CH_2, CHR, NH; R^{2-3} = alkyl with no β-hydrogen, aryl; R^4 = CH_3, alkyl, allyl, benzyl, CH_2COAr; X = Cl, Br, I, OTs, OMs; base: NaH, KH, RLi, ArLi, RONa, ROK

Mechanism: [36,37,26,38-41,11,42]

If the *Stevens rearrangement* is a concerted reaction, it is a symmetry-forbidden process based on the *Woodward-Hoffmann rules*. Indeed, it was shown to occur *via* an intramolecular *homolytic cleavage-radical pair recombination* process, which explains the lack of crossover products and the observed retention of configuration at the migrating terminus.[41] The radicals are held in a solvent-cage in which there is a lack of rotation, and they recombine quickly.

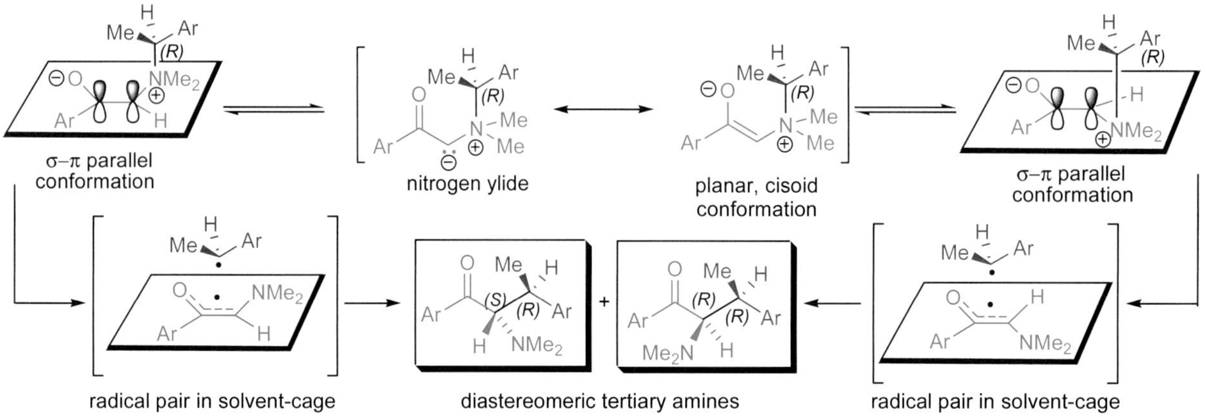

Synthetic Applications:

The nitrogen ylides required for the *Stevens rearrangement* can be accessed in a direct manner by using the transition metal catalyzed decomposition of an α-diazo carbonyl functionality tethered to tertiary amines. This tandem *ylide formation/Stevens rearrangement* strategy was used by A. Padwa et al. as a novel approach toward the preparation of isoindolo-benzazepines.[43] The diazo ester was added to a refluxing solution of rhodium(II) acetate in toluene, generating the nitrogen ylide *in situ*, which underwent a facile *[1,2]-benzyl shift* to afford the 5,7-fused heterocyclic ring system.

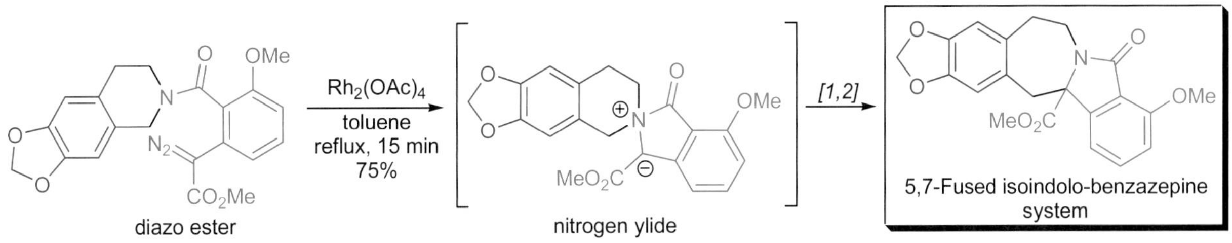

A new approach to the morphine skeleton was demonstrated by the total synthesis of (±)-desoxycodeine-D by C.-Y. Cheng and co-workers.[44] The key step was the formation of the B ring by the *Stevens rearrangement* of a tetrahydroisoquinoline-derived quaternary ammonium salt upon treatment with phenyllithium.

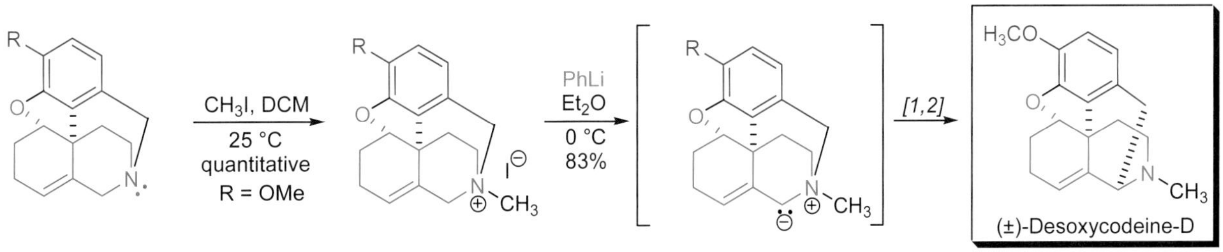

The first synthesis of 1,2-(1,1'-ferrocenediyl)ethene was accomplished in the laboratory of V.K. Aggarwal in six steps from ferrocene.[45] In order to construct the strained two-carbon bridge, several methods were tested including the *McMurry coupling* and the *Ramberg-Bäcklund rearrangement*. Unfortunately, under the McMurry conditions only intermolecularly coupled products were obtained. The α-chlorination of the sulfide or sulfone failed, therefore the α-chloro sulfone precursor for the *Ramberg-Bäcklund rearrangement* could not be prepared. Alternatively, the *Stevens rearrangement* of a sulfonium salt was successful in providing the desired ring-contracted product.

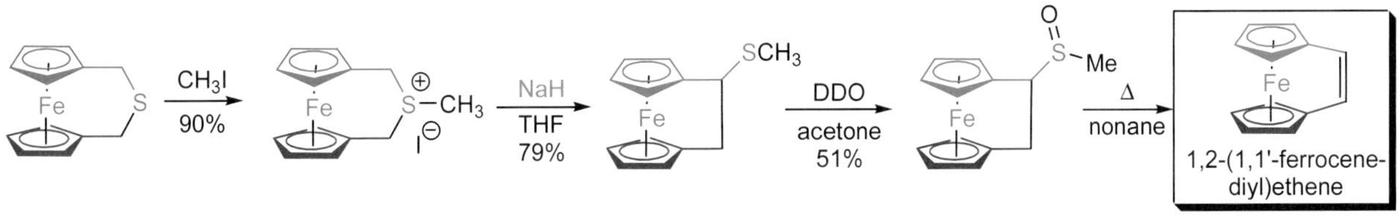

The transfer of axial chirality to central chirality during the *Stevens rearrangement* of binaphthyl compounds was investigated by I.G. Stará et al.[42] They found that the stereochemical course of the *Stevens rearrangement* of axially chiral onium salts is significantly structure-dependent. Their findings were utilized in a novel enantioselective synthesis of pentahelicene. The treatment of the optically pure binaphtyl ammonium salt with an excess of butyllithium brought about the expected *[1,2]- benzyl shift*, and the tertiary amine intermediate underwent an *in situ* base-induced 1,2-elimination to afford the optically pure pentahelicene. Interestingly, the rearrangement of analogous sulfur ylides proceeded with considerably lower stereoselectivity.

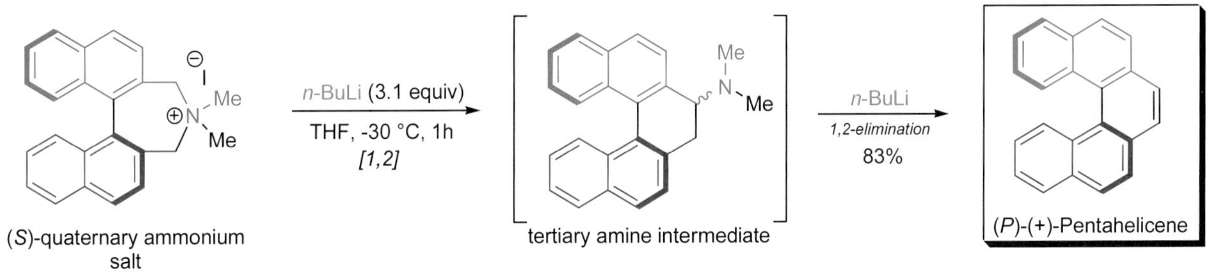

STILLE CARBONYLATIVE CROSS-COUPLING

(References are on page 687)

Importance:

[*Seminal Publications*[1-5]; *Reviews*[6,7]; *Modifications & Improvements*[8,9]]

The synthesis of ketones using the *Stille cross-coupling* initially called for the use of acid chlorides as coupling partners. However, acid chlorides are not always readily available, and their preparation is often not compatible with sensitive functional groups. To widen the scope of the synthesis of ketones, the transition metal catalyzed carbonylative cross-coupling of organic halides and pseudohalides was extensively investigated in the 1980s. The $Pd^{(0)}$-catalyzed coupling between an organostannane, carbon monoxide, and an organic electrophile to form two new C-C sigma bonds is called the Stille carbonylative cross-coupling. Advantages of this method are: 1) many organic halides are commercially available or easily prepared and indefinitely stable; 2) the coupling occurs not only with *chemo- and regioselectivity*, but also with *stereoselectivity*, generally retaining the configuration at the substituted position of both the vinyl/aryl halide and the organostannane; 3) allyl and benzyl chlorides react, and they give the corresponding ketones with inversion of configuration;[2] 4) the reaction of alkenyl iodides and alkenyltins takes place under neutral and mild conditions; and 5) the use of heterostannanes (alkoxy, thioalkoxy, and aminostannanes) allows the preparation of the corresponding carboxylic acid derivatives.[8] Disadvantages are: 1) direct coupling without CO insertion and the need to use high pressures of CO to suppress this side reaction;[10] 2) the occasional *Z/E* isomerization of alkenyl groups from both reaction components, especially with (*Z*)-alkenyl derivatives;[3] and 3) aryl chlorides react only slowly compared to aryl bromides and iodides.

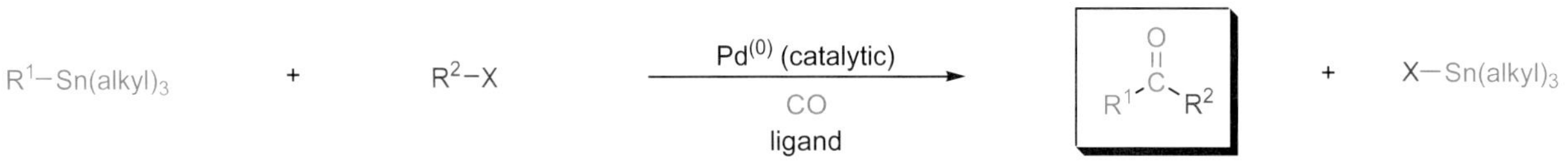

R^1 = alkyl, allyl, alkenyl, aryl; R^2 = alkenyl, aryl; X = Cl, Br, I, OTf, $OPO(OR)_2$

Mechanism: [11]

The mechanism of the *Stille carbonylative cross-coupling* is very similar to the regular *Stille cross-coupling*. The only difference between the two couplings is that a carbon-monoxide (CO) insertion takes place between the oxidative-addition step and the transmetallation step. The rate determining step is the transmetallation, so transferable groups attached to the tin atom may have β-hydrogens attached to sp^3 carbons, because the steps following the transmetallation are very fast and no β-hydride elimination is expected.

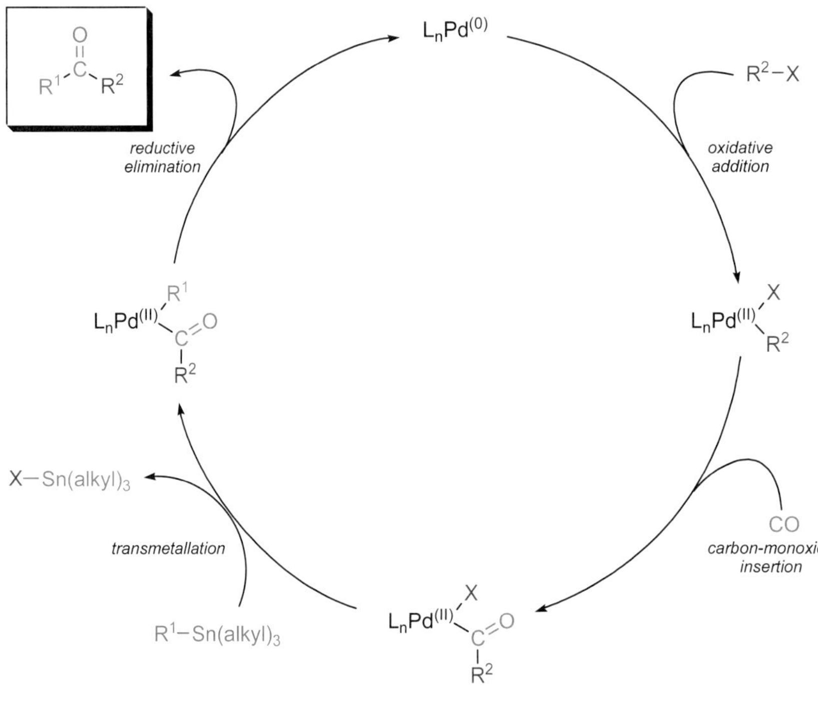

Synthetic Examples:

The first enantioselective total synhesis of (–)-strychnine was achieved by L.E. Overman and co-workers.[12] The carbon skeleton of the main precursor for the key *aza-Cope rearrangement/Mannich cyclization* was assembled by applying a Pd$^{(0)}$-catalyzed *carbonylative Stille coupling reaction*. Thus, the cyclic vinylstannane was coupled with the triazinone-protected *ortho*-iodoaniline to afford 80% yield of the aromatic enone using $Pd_2(dba)_3$ as the catalyst in the presence of carbon monoxide.

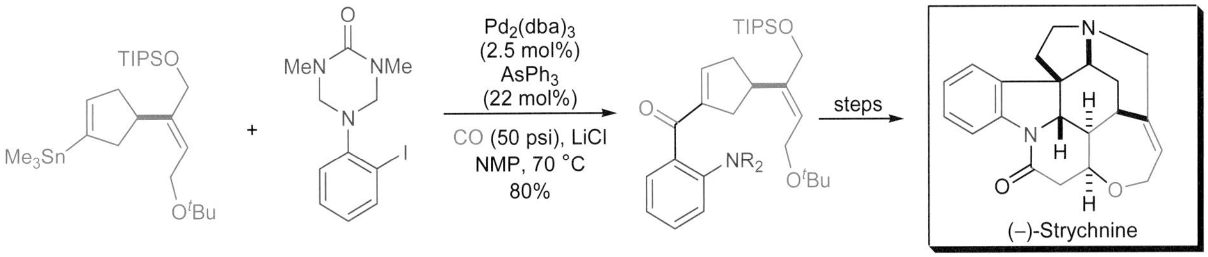

C-Disaccharides (*C*-glycosides) have an advantage over *O*-glycosides as they resist acidic and enzymatic hydrolysis. They can therefore serve as potential glycosidase inhibitors. In the laboratory of P. Vogel, a novel approach was developed for the synthesis of *C*-glycosides by a *Stille carbonylative coupling reaction* between 1-stannylglucals and 1-iodoglucals.[13]

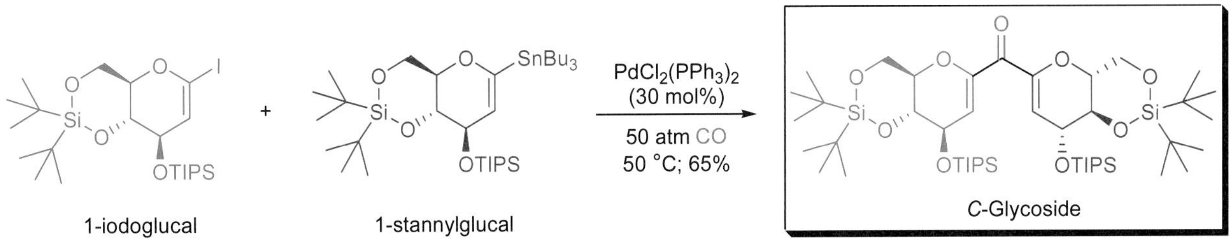

A concise synthesis of photoactivatable 4-benzoyl-L-phenylalanines and related peptides was described by G. Ortar et al. using a *carbonylative Stille cross-coupling* as the key step.[14] Surprisingly, when the coupling was attempted with tyrosine triflate derivatives, it proved to be unsuccessful. However, 4-iodo-phenylalanine derivatives reacted smoothly under standard conditions to give the corresponding 4-benzoyl derivatives.

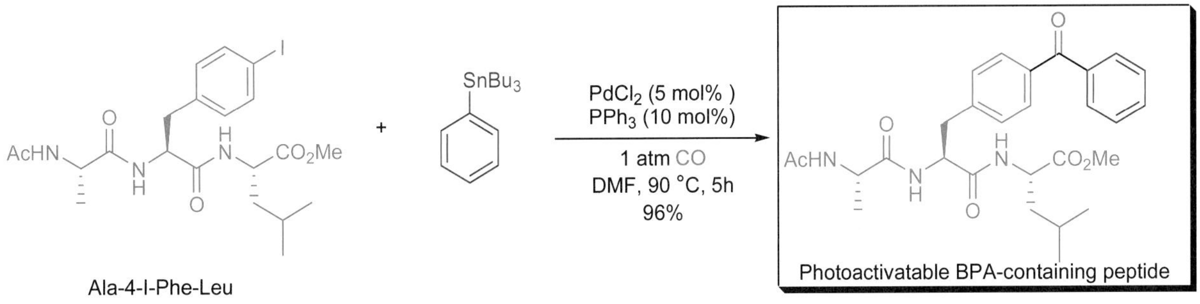

Systematic evolution of ligands by exponential enrichment (SELEX) is a procedure that generates nucleic acid ligands capable of high-affinity binding to both protein and small molecule targets. In order to synthesize a wide range of these ligands, B.E. Eaton and co-workers used the *carbonylative Stille coupling* to obtain 5-carbonyluridine analogues.[15]

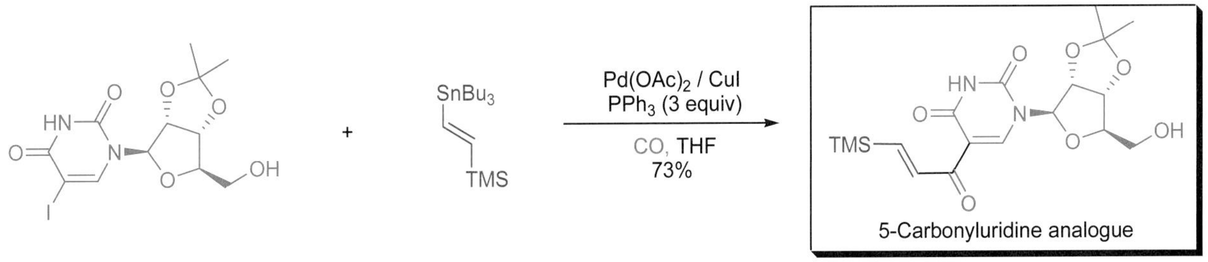

STILLE CROSS-COUPLING
(MIGITA-KOSUGI-STILLE COUPLING)

(References are on page 687)

Importance:

[*Seminal Publications*[1-8] ; *Reviews*[9-27]; *Modifications*[28-40]]

In 1976, the first palladium catalyzed reaction of organotin compounds (organostannanes) was published by C. Eaborn et al.[1] A year later in 1977, M. Kosugi and T. Migita reported transition-metal-catalyzed C-C-bond forming reactions of organotins with aryl halides[2] and acid chlorides.[3,4] In 1978, J.K. Stille used organotin compounds for the synthesis of ketones under reaction conditions much milder than Kosugi's and with significantly improved yields.[5] In the early 1980s, Stille pioneered the use of this method.[10] The $Pd^{(0)}$-catalyzed coupling reaction between an organostannane and an organic electrophile to form a new C-C sigma bond is known as the Stille cross coupling. The precursor organotin compounds have many advantages because they: 1) tolerate a wide variety of functional groups; 2) are not sensitive to moisture or oxygen unlike other reactive organometallic compounds; and 3) are easily prepared, isolated, and stored. The main disadvantages are their toxicity and the difficulty to remove the traces of tin by-products from the reaction mixture. In the past two decades, the *Stille reaction* has become one of the most powerful synthetic tools in organic chemistry, and it finds many uses in preparative chemistry. The success of the *Stille coupling* is largely attributed to the mild conditions of the method. The reaction conditions are compatible with many types of functional groups (carboxylic acid, amide, ester, nitro, ether, amine, hydroxyl, ketone, and formyl groups) and high levels of stereochemical complexity can be tolerated by both coupling partners. The only major side reaction associated with the *Stille coupling* is the oxidative homocoupling of the organostannane reagent and under harsh conditions allylic and (*Z*)-alkenyl components may undergo double bond migration and isomerization.[41,42] Metals other than palladium such as manganese,[33] nickel,[29,36] and copper[28,30-35,37] have also been found to catalyze the reaction and procedures, using only catalytic amounts of tin have been developed.[38-40]

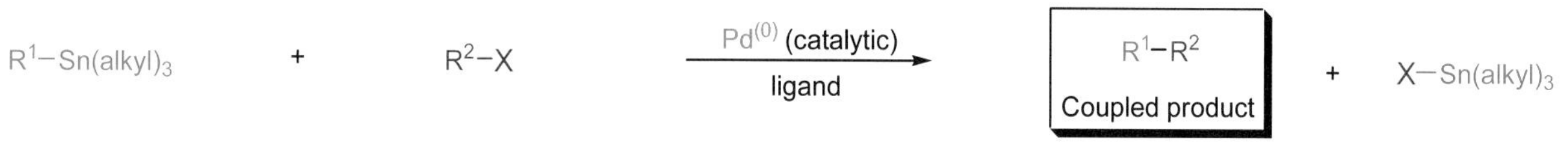

R^1 = allyl, alkenyl, aryl; R^2 = alkenyl, aryl, acyl; X = Cl, Br, I, OTf, $OPO(OR)_2$

Mechanism: [6,7,41,43-46,12,47-53,27,54]

The catalytic cycle for the Stille coupling reaction was first proposed for the reaction with benzylic and aryl halides in 1979,[6,7] although the detailed mechanism is still a matter of some debate.[12,27] The catalytic cycle has three steps: 1) oxidative addition; 2) transmetallation; and 3) reductive elimination. The active catalyst is believed to be a 14-electron $Pd^{(0)}$-complex which can be generated *in situ*. Palladium(0)-catalysts such as $Pd(PPh_3)_4$ and $Pd(dba)_2$, with or without an added ligand, are often used. Alternatively, $Pd^{(II)}$-complexes such as $Pd(OAc)_2$, $PdCl_2(MeCN)_2$, $(PdCl_2(PPh_3)_2$, $BnPdCl(PPh_3)_2$, etc. are also used as precursors for the catalytically active $Pd^{(0)}$ species, as these compounds are reduced by the organostannane[48] or by an added phosphine ligand prior to the main catalytic process. The transmetallation step is the rate-determining step in the catalytic cycle.[46,47,49,50] Different groups on the tin coupling partner transmetallate to the $Pd^{(II)}$ intermediate at different rates and the order of migration is: alkynyl > vinyl > aryl > allyl ~ benzyl »» alkyl. The very slow migration rate of the alkyl substituents allows the transfer of aryl or vinyl groups when mixed organostannanes containing three methyl or butyl groups are used.

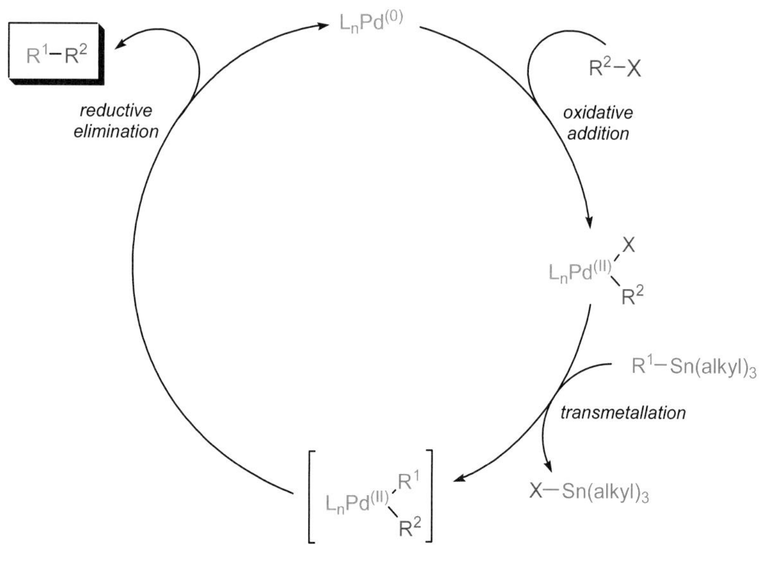

The total synthesis of the phenolic sesquiterpene (±)-parviflorine was acc[...] workers.[26] The key step in the synthetic sequence was the reaction of an [...] intermediate, which was subjected to a *Grob fragmentation* to afford the eig[...] The bicyclic ketone substrate was refluxed in benzene using a Dean-Stark tr[...] to the next step as crude material.

The biomimetic synthesis of the structurally novel bisesquiterpenoid (±)-bia[...] et al.[27] The cornerstone of the synthetic strategy was the radical dime[...] atractylolide precursor was prepared from a bicyclic ketone using the S[...] enamine was generated using large excess of pyrrolidine in refluxing benz[...] under reduced pressure). The alkylation of the crude enamine with ethyl α[...] dioxane and afforded a mixture of ethyl ester diastereomers.

In the laboratory of A.B. Smith, the synthesis of (+)-jatropholone A and B w[...] *Alder cycloaddition* between a tetrasubstituted furan and a homochiral en[...] component, the *Stork enamine synthesis* was used. The α-benzyloxy [...] corresponding morpholine enamine in quantitative yield. The enamine w[...] contrast, the corresponding piperidine or pyrrolidine enamines were obtain[...] The acylation of the enamine with O-acetoxyacetyl chloride yielded a 1,[...] desired tetrasubstituted furan component.

An intramolecular variant of the *Stork enamine synthesis* was utilized durin[...] aza-12-oxo-17-desoxoestrone by A.I. Meyers et al.[28]

Synthetic Ap[...]

A novel strate[...] catalyzed amir[...] bromo-substitu[...] aminopyridines[...] the correspond[...]

J.J. Li and [...] approaches.[12] chloropyridine[...] *coupling* cond[...]

4-chloro-3-iodo[...] pyridine

The total syn[...] Sakamoto.[13] indole precurs[...]

The cyclic *bi*[...] *coupling* as [...] conditions but [...] 9% stannylat[...] stannylated [...] removal of th[...]

Synthetic Applications:

The total synthesis of (+)-mycotrienol was accomplished by J. Panek and co-workers using a Pd$^{(0)}$-catalyzed *Stille coupling reaction* to incorporate the (*E,E,E*)-triene unit with simultaneous macrocyclization.[55] After macrocyclization, the aromatic core was oxidized with CAN and the protecting groups were removed to provide the natural product.

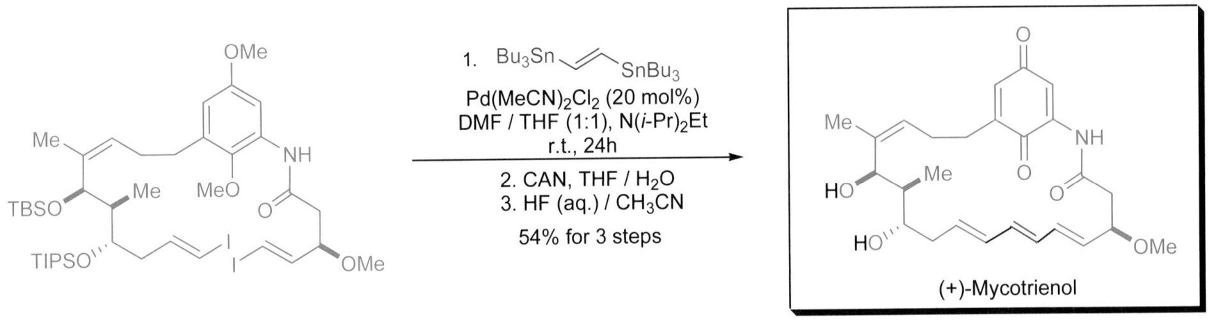

The enantioselective total synthesis of the manzamine alkaloid ircinal A was completed in the laboratory of S.F. Martin utilizing a novel strategy. A *domino Stille/Diels-Alder reaction* was used to assemble the ABC ring core of the natural product.[56] The vinyl bromide intermediate reacted with vinyl tributylstannane in the presence of Pd$^{(0)}$ to afford the 1,3-diene moiety, which cyclized *via* an *intramolecular Diels-Alder reaction* to give the ABC core.

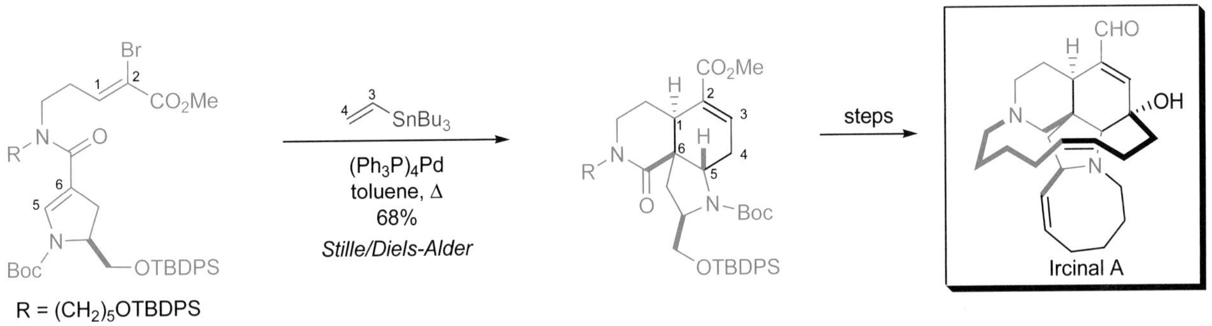

The first total synthesis of quadrigemine C, a higher-order member of the polypyrrolidinoindoline alkaloid family was published by L. Overman et al.[57] Key steps included a *double Stille cross coupling* and *catalyst-controlled double Heck cyclization*.

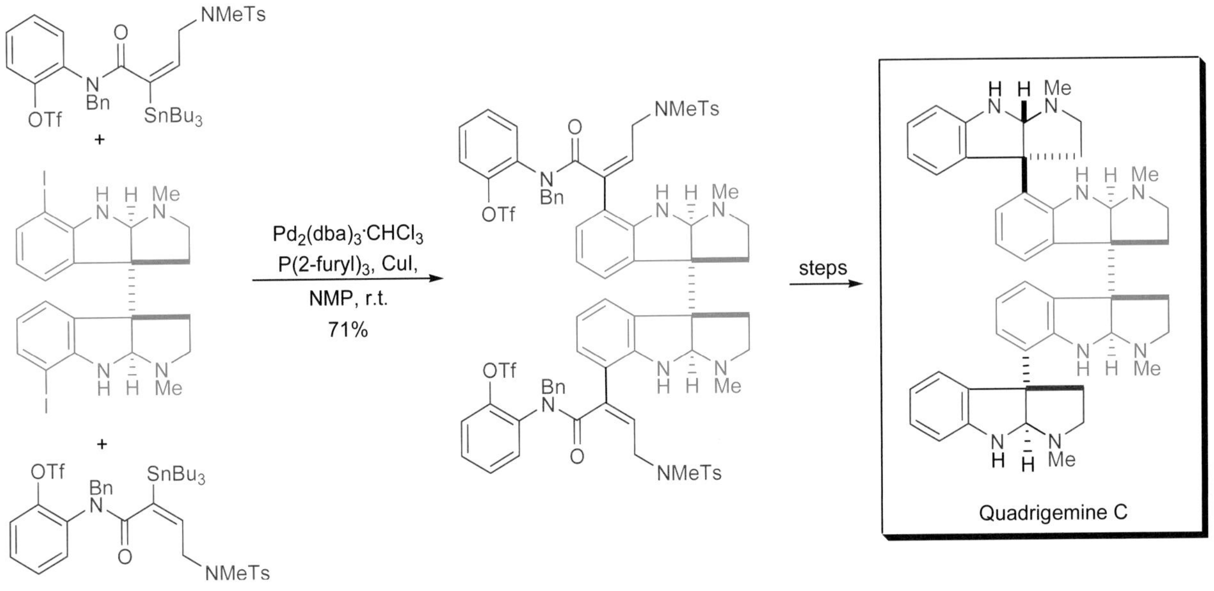

STORK ENAMINE SYN

(References are on pag

Importance:

[*Seminal Publications*[1-4]; *Reviews*[5-11]; *Modifications & Im*

In 1936, C. Mannich and H. Davidson reported that in the pres secondary amines underwent facile condensation with aldehyde enolate equivalents).[23] At that time the reaction of enamines wit established that enamines were relatively labile compounds that un aqueous acid. Two decades later, in 1954, G. Stork and co-worke alkyl- or acyl halides followed by acidic hydrolysis constituted a carbonyl compounds.[3,4] The synthesis of α-alkyl- or acyl carbonyl corresponding enamines is known as the *Stork enamine synthesis.* enamines are prepared by reacting the aldehyde or ketone with one morpholine or pyrrolidine) in the presence of a catalyst (or dehydr formation of enamine regioisomers is expected but usually the preparation of aldehyde enamines is often accompanied by the forr desired enamines by destructive distillation;[9] 4) activated alkyl and allyl-, benzyl-, propargylic-, or activated aryl halides); 5) tertiary alky undergo elimination; 6) other electrophiles such as Michael accept bulkier the ketone and the amine components, the better the yield rates tend to drop. Advantages of the *Stork enamine synthesis* are: 1 neutral conditions, which is important when the substrate is base seldom observed; 3) the alkylation takes place on the less substi version utilizing chiral enamines is also available.

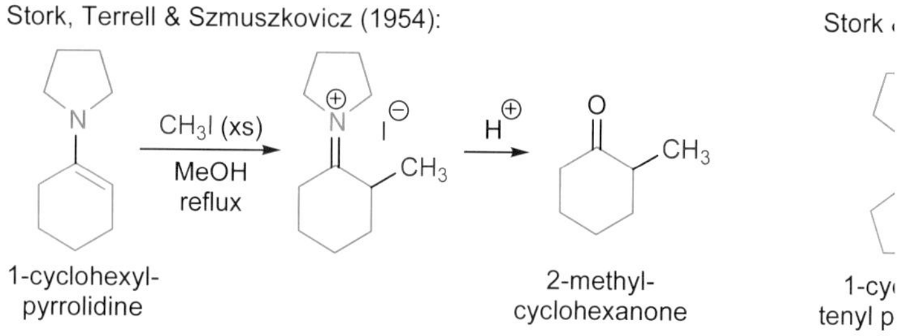

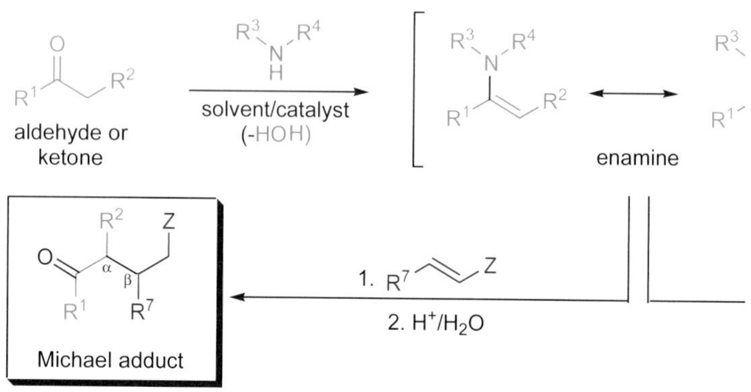

R^1 = H, alkyl, substituted alkyl; R^2 = H, alkyl, aryl; R^{3-4} = alkyl, aryl; R^5 = propargylic; R^6 = alkyl, aryl, OR, H; R^7 = H, alkyl, aryl; X = Cl, Br, I, OTs;

Mechanism: [24,25]

Formation of the enamine:

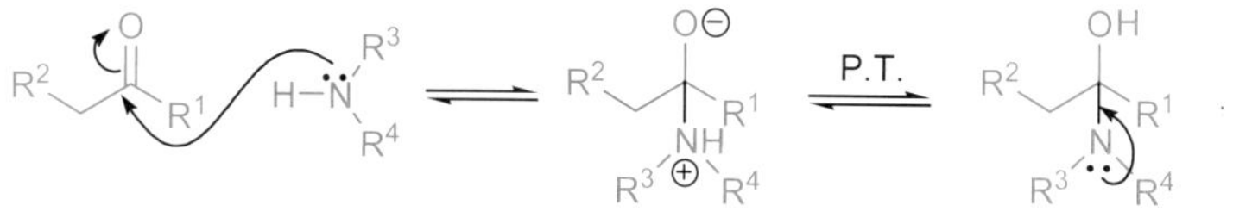

Alkylation of the enamine:

Hydrolysis with dilute ac

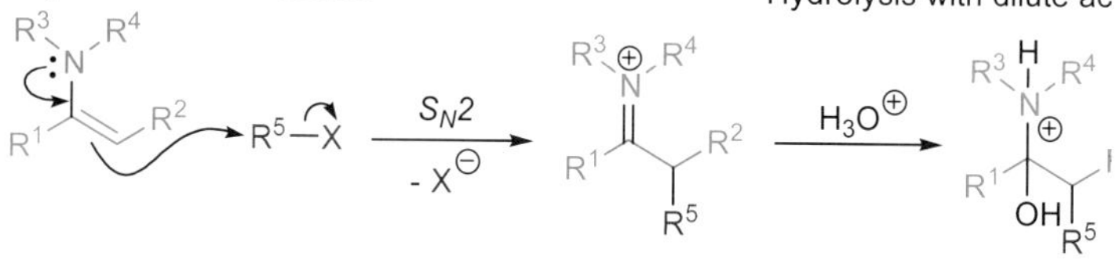

STRECKER REACTION

(References are on page 690)

Importance:

[*Seminal Publications*[1,2]; *Reviews*[3-28]; *Modifications & Improvements*[29-42]; *Theoretical Studies*[43-46]]

In 1850, A. Strecker attempted the synthesis of lactic acid by treating acetaldehyde first with aqueous ammonia followed by the addition of hydrogen cyanide and by hydrolyzing the resulting amino nitrile intermediate with aqueous acid.[1] To his surprise he did not isolate any of the desired lactic acid but instead obtained alanine. This discovery constituted the first laboratory preparation of an α-amino acid. The condensation of an aldehyde or ketone with a primary amine or ammonia and hydrogen cyanide (or their equivalents) to afford the corresponding α-amino nitrile is known as the *Strecker reaction*. The most well-known use of α-amino nitriles is their hydrolysis under acidic or basic conditions to obtain α-amino acids (*Strecker amino acid synthesis*). The general features of the *Strecker reaction* are: 1) the transformation is a one-pot three-component coupling; 2) due to the extreme toxicity of HCN, various alkali cyanides (e.g., KCN, NaCN) in buffered aqueous media are used; 3) both aldehydes and ketones are good substrates; 3) the amine component can be ammonia, primary, or secondary amine; 4) the addition of HCN to preformed aldimines and ketimines (even iminium salts) or to oximes and hydrazones leads to *N*-substituted α-amino nitriles; 5) hydrolysis of α-amino nitriles gives α-amino acids, reduction with metal hydrides affords 1,2-diamines, while strong bases can deprotonate at the α-carbon (if R^2=H) and the resulting carbanion can be trapped with a variety of electrophiles (umpolung);[22] and 6) upon treatment with heavy metal salts (e.g., $AgNO_3$), Brönsted or Lewis acids, α-amino nitriles undergo a loss of cyanide ion to form iminium ions, which can be trapped with various nucleophiles (if the nucleophile is an organometallic reagent, the transformation is called the *Bruylants reaction*). It is now possible to conduct the *Strecker reaction* asymmetrically: 1) the use of optically active amines generate chiral imines, which give rise to enantio-enriched α-amino nitriles upon the addition of cyanide ions;[8,11] and 2) asymmetric induction may be achieved by the use of organocatalysts or chiral metal catalysts.[25,27]

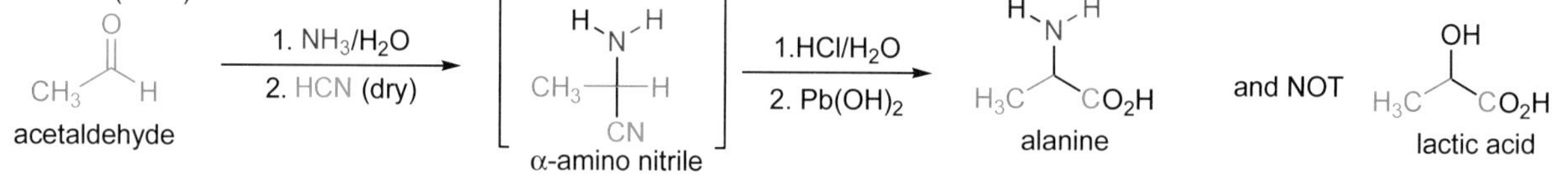

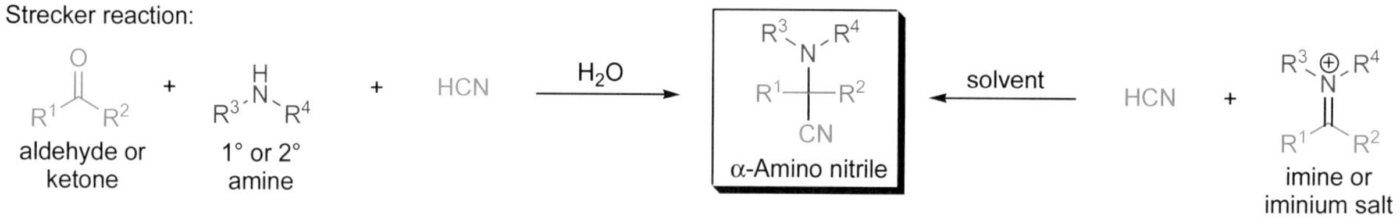

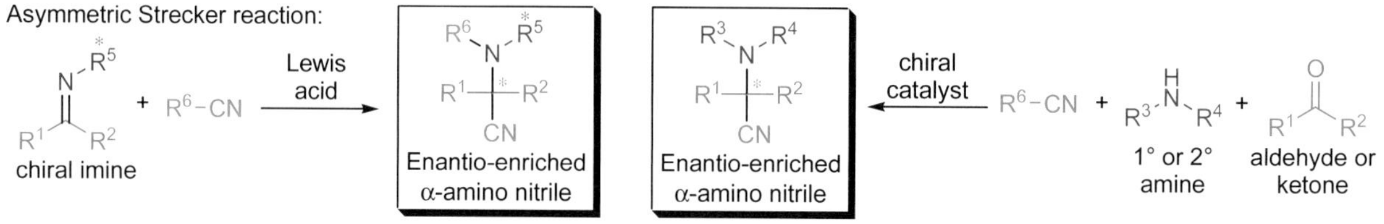

R^1 = alkyl, aryl, heteroaryl; R^2 = H, alkyl, aryl, heteroaryl; R^{3-4} = H, alkyl, aryl, heteroaryl; R^5 = group having a chiral center; R^6 = H, TMS; chiral catalyst: chiral metal catalyst or organocatalyst

Mechanism: [47-54]

Mechanism in the presence of an organocatalyst (Corey, 1999):

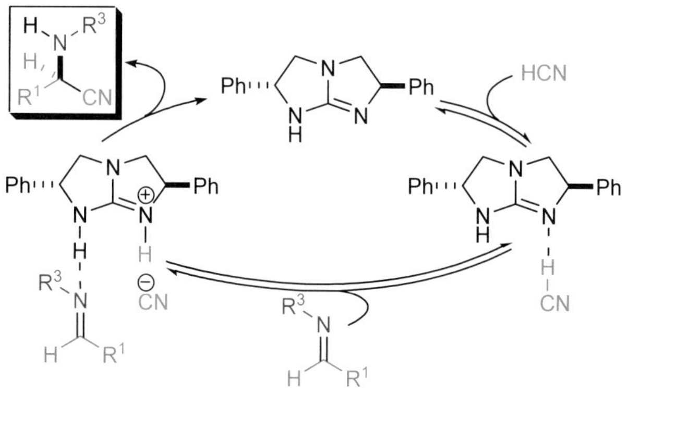

Mechanism of the classical Strecker reaction:

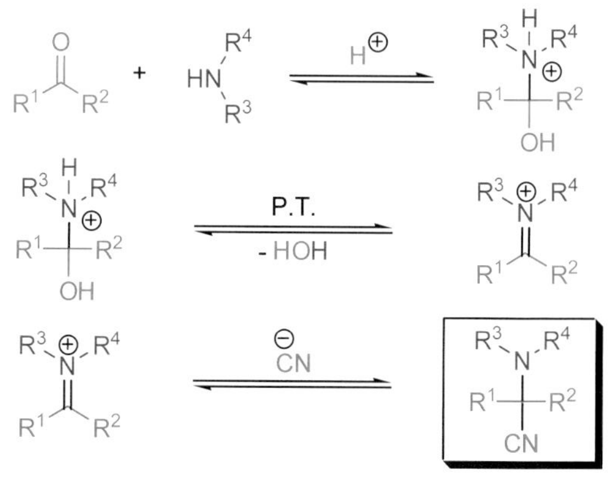

Synthetic Applications:

The enantioselective total synthesis of (–)-hemiasterlin, a marine tripeptide with cytotoxic and antimitotic activity, was achieved by E. Vedejs and co-workers.[55] The *asymmetric Strecker reaction* was used to construct the key tetramethyltryptophan subunit. The aldehyde substrate was first converted to the corresponding chiral imine with (*R*)-2-phenylglycinol under scandium triflate catalysis. The addition of tributyltin cyanide resulted in the formation of α-amino nitriles as an 8:1 mixture of diastereomers. Subsequently the cyano group was converted to a primary amide, and the chiral auxiliary was removed under catalytic hydrogenation conditions.

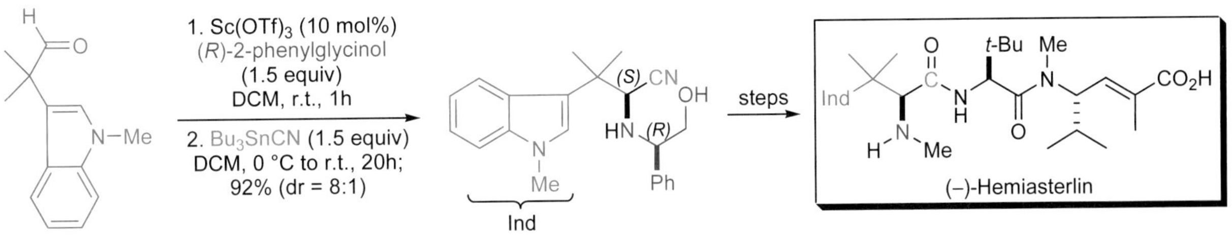

In the laboratory of B. Ganem, the asymmetric total synthesis of (–)-α-kainic acid was accomplished starting from very simple precursors. A highly stereoselective *zirconium-mediated Strecker reaction* was used to install the α-amino acid moiety of the natural product. The five-membered lactam substrate was treated with excess Schwartz reagent at low temperature which generated the corresponding cyclic imine *in situ*. This cyclic imine was not isolated but was immediately reacted with cyanotrimethylsilane to afford the all *cis* α-amino nitrile. In order to convert this intermediate to kainic acid, the cyano group was first converted by the *Pinner reaction* to a methyl ester. The resulting diester was hydrolyzed with aqueous KOH solution to give the corresponding dicarboxylic acid with complete epimerization at C2.

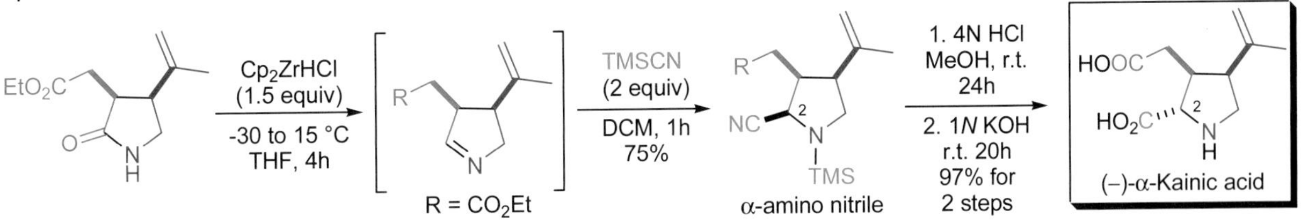

The *sulfinimine-mediated asymmetric Strecker reaction* was developed by F.A. Davis et al. This method involves the addition of ethylaluminumcyanoisopropoxide to functionalized sulfinimines and the resulting diastereomeric α-amino nitriles are easily separated. Subsequent hydrolysis directly affords the enantiopure α-amino acids. This protocol was applied for the synthesis of polyoxamic acid lactone.[56]

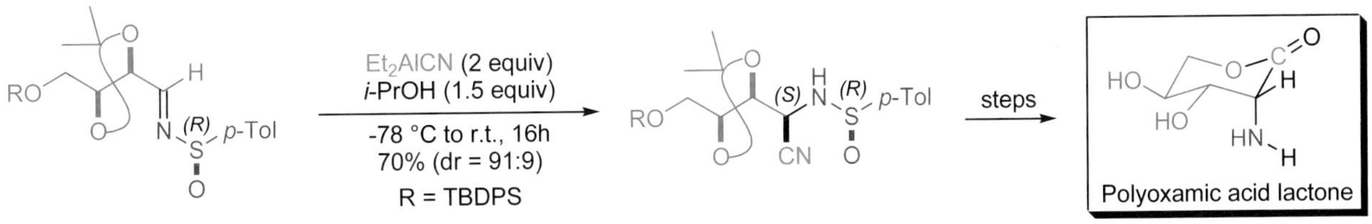

The first total synthesis of amiclenomycin, an inhibitor of biotin biosynthesis, was completed by A. Marquet and co-workers.[57] In order to prove its structure unambiguously, both the *cis* and *trans* isomers were prepared. The L-amino acid functionality was installed by a *Strecker reaction* using TMSCN in the presence of catalytic amounts of ZnI_2. The resulting O-TMS protected cyanohydrin was exposed to saturated methanolic ammonia solution, which gave rise to the corresponding α-amino nitrile. Enzymatic hydrolysis with immobilized pronase afforded the desired L-amino acid.

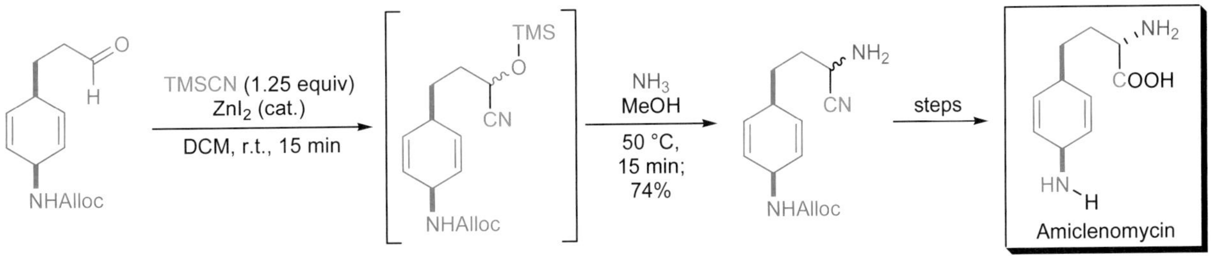

SUZUKI CROSS-COUPLING
(SUZUKI-MIYAURA CROSS-COUPLING)

(References are on page 691)

Importance:

[*Seminal Publications*[1-3] ; *Reviews*[4-38]; *Modifications & Improvements*[39-49]]

In 1979, A. Suzuki and N. Miyaura reported the stereoselective synthesis of arylated (*E*)-alkenes by the reaction of 1-alkenylboranes with aryl halides in the presence of a palladium catalyst.[1] The palladium-catalyzed cross-coupling reaction between organoboron compounds and organic halides or triflates provides a powerful and general method for the formation of carbon-carbon bonds known as the *Suzuki cross-coupling*. There are several advantages to this method: 1) mild reaction conditions; 2) commercial availability of many boronic acids; 3) the inorganic by-products are easily removed from the reaction mixture, making the reaction suitable for industrial processes; 4) boronic acids are environmentally safer and much less toxic than organostannanes (see *Stille coupling*); 5) starting materials tolerate a wide variety of functional groups, and they are unaffected by water; 6) the coupling is generally *stereo-* and *regioselective;* and 7) sp^3-hybridized alkyl boranes can also be coupled by the *B-alkyl Suzuki-Miyaura cross-coupling*. Some disadvantages are: 1) generally aryl halides react sluggishly; 2) by-products such as self-coupling products are formed because of solvent-dissolved oxygen; 3) coupling products of phosphine-bound aryls are often formed; and 4) since the reaction does not proceed in the absence of a base, side reactions such as racemization of optically active compounds or aldol condensations occur. Improvements of the *Suzuki cross-coupling* include the development of catalysts facilitating coupling of unreactive aryl halides,[39,40] the ability to react sp^3-hybridized alkyl halides,[42,44,50] and the use of alkyl, alkenyl, aryl, and alkynyl trifluoroborates in place of boronic acids.[45-47]

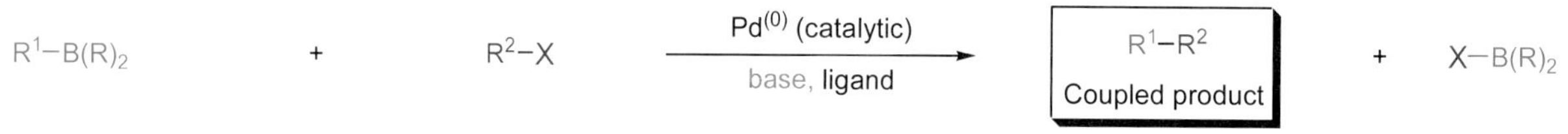

R^1 = alkyl, allyl, alkenyl, alkynyl, aryl; R = alkyl, OH, O-alkyl; R^2 = alkenyl, aryl, alkyl; X = Cl, Br,I, OTf, $OPO(OR)_2$ (enol phosphate); base = Na_2CO_3, $Ba(OH)_2$, K_3PO_4, Cs_2CO_3, K_2CO_3, TlOH, KF, CsF, Bu_4F, NaOH, $M^+(^-O$-alkyl)

Mechanism: [51-55,24,56,57,50,58-60]

The mechanism of the Suzuki cross-coupling is analogous to the catalytic cycle for the other cross-coupling reactions and has four distinct steps: 1) oxidative addition of an organic halide to the $Pd^{(0)}$-species to form $Pd^{(II)}$; 2) exchange of the anion attached to the palladium for the anion of the base (metathesis); 3) transmetallation between $Pd^{(II)}$ and the alkylborate complex; and 4) reductive elimination to form the C-C sigma bond and regeneration of $Pd^{(0)}$. Although organoboronic acids do not transmetallate to the $Pd^{(II)}$-complexes, the corresponding ate-complexes readily undergo transmetallation. The quaternization of the boron atom with an anion increases the nucleophilicity of the alkyl group and it accelerates its transfer to the palladium in the transmetallation step. Very bulky and electron-rich ligands (e.g., $P(t\text{-}Bu)_3$) increase the reactivity of otherwise unreactive aryl chlorides by accelerating the rate of the oxidative addition step.

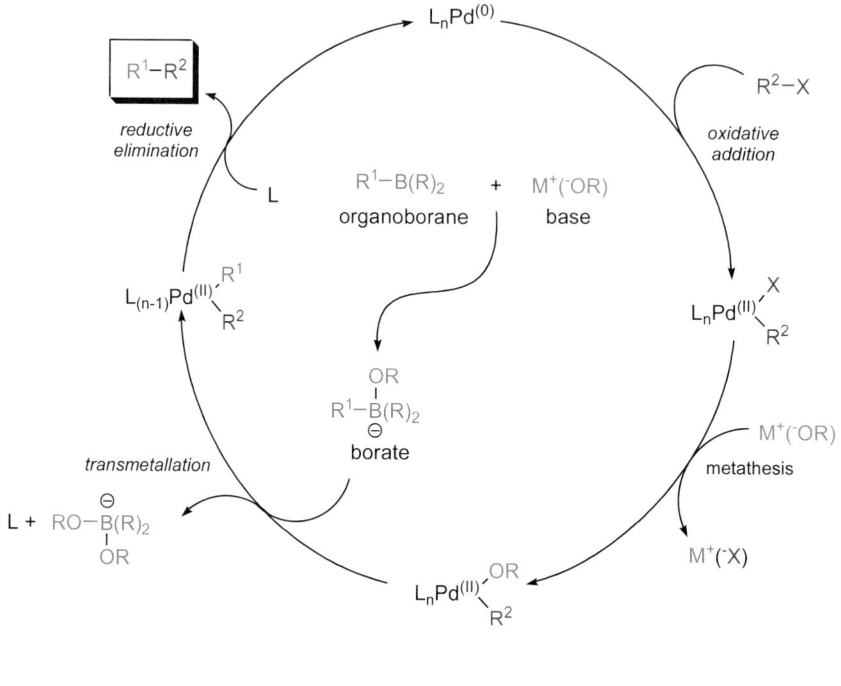

Synthetic Applications:

During the total synthesis of the proteosome inhibitor TMC-95A by S.J. Danishefsky et al., the biaryl moiety of the compound was assembled in good yield by the *Suzuki cross-coupling* of an aryl iodide and an arylboron intermediate.[61]

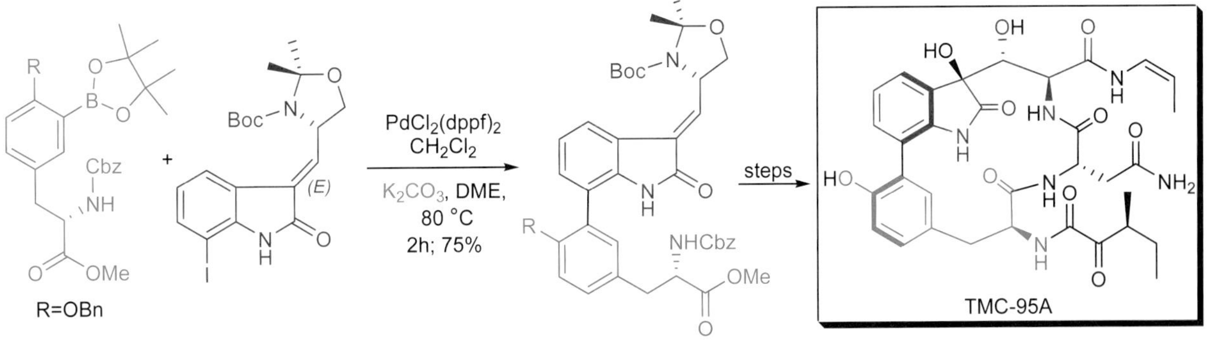

The antitumor natural product epothilone A was synthesized in the laboratory of J.S. Panek.[62] They utilized the *B-alkyl Suzuki cross-coupling* between an sp^3-hybridized alkylborane and a (*Z*)-iodoalkene for the construction of the main fragment. The alkylborane was prepared by hydroborating the terminal alkene with 9-BBN and the (*Z*)-iodoalkene was added along with the palladium catalyst and the base.

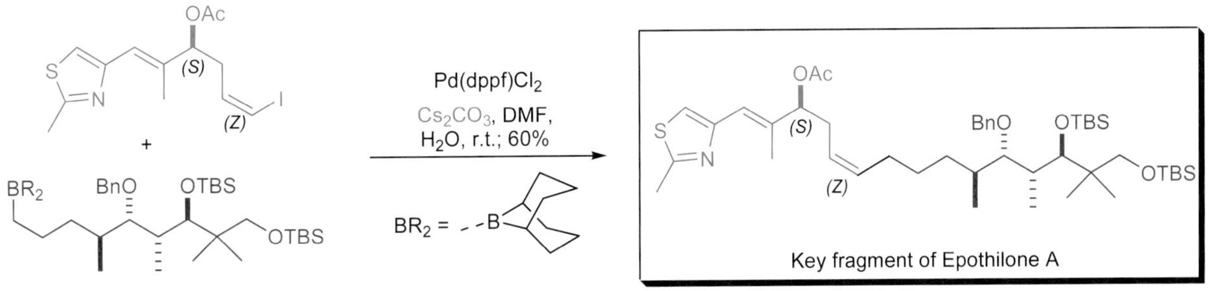

The last and key step in the total synthesis of myxalamide A by C.H. Heathcock et al. was a *Suzuki cross-coupling* between an (*E*)-vinylborane and a (*Z*)-iodotriene.[63] The (*E*)-vinylborane was prepared prior to the coupling by reacting the precursor enyne with 2 equivalents of cathecholborane. Upon completion of the hydroboration, it was combined with the (*Z*)-iodotriene and catalytic amounts of palladium acetate.

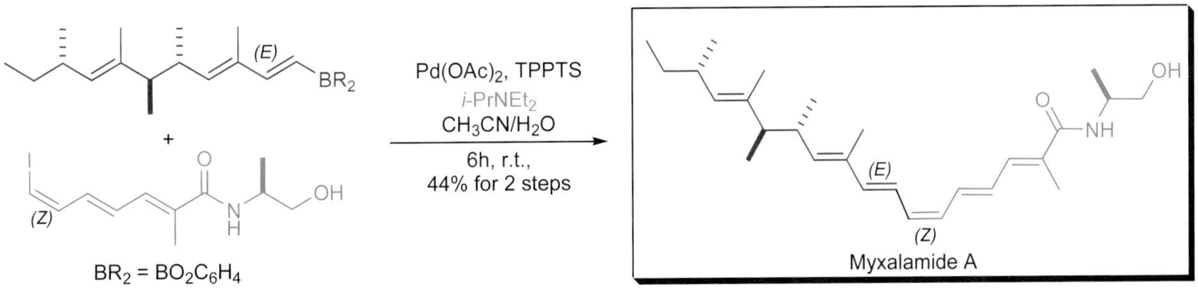

A formal total synthesis of oximidine II was achieved by G.A. Molander et al., using an *intramolecular Suzuki-type cross-coupling* between an alkenyl potassium trifluoroborate and an alkenyl bromide to construct the highly strained, polyunsaturated 12-membered macrolactone core of the natural product.[64] The stability of potassium trifluoroborates was exploited in order to establish the best conditions for the macrocyclization.

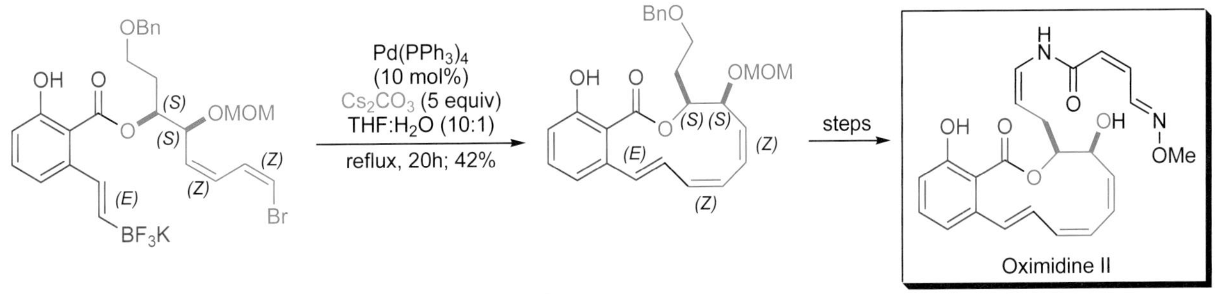

SWERN OXIDATION

(References are on page 692)

Importance:

[*Seminal Publications*[1-6]; *Reviews*[7-10]; *Modifications & Improvements*[11-16]]

In 1976, D. Swern and co-workers reported that treatment of dimethyl sulfoxide (DMSO) with trifluoroacetic anhydride (TFAA) below -50 °C in methylene chloride gave trifluoroacetoxydimethylsulfonium trifluoroacetate, which reacted rapidly with primary and secondary alcohols.[3] The resulting alkoxydimethylsulfonium trifluoroacetates, upon addition of triethylamine, afforded the corresponding aldehydes and ketones in good yield.[3] In 1978, oxalyl chloride was found to be more effective than TFAA as an activating agent for DMSO in the oxidation of alcohols.[5,6] The oxidation of primary and secondary alcohols using DMSO and TFAA or oxalyl chloride is referred to as the *Swern oxidation*. The general features of this oxidation are: 1) when no solvent is used, DMSO reacts with TFAA or oxalyl chloride violently (explosion!), so great care should be exercised while running the reaction; 2) the most common solvent is DCM; 3) when TFAA is used, the initial intermediate is unstable above -30 °C and a side product is formed *via* the *Pummerer rearrangement*; 4) when oxalyl chloride is used, the initial intermediate is unstable above -60 °C, so the oxidation is usually conducted at -78 °C; 5) the typical procedure begins with the reaction of DMSO with TFAA or oxalyl chloride at low temperature followed by the slow addition of the alcohol, then a tertiary amine; 6) the addition of a tertiary amine (e.g., DIPA, TEA) is necessary to facilitate the decomposition of the alkoxysulfonium salt; 7) the efficiency of the oxidation is not influenced by the steric hindrance of the substrate; and 8) the use of TFAA may give rise to trifluoroacetate side products, whereas in the case of oxalyl chloride side reactions are extremely rare.

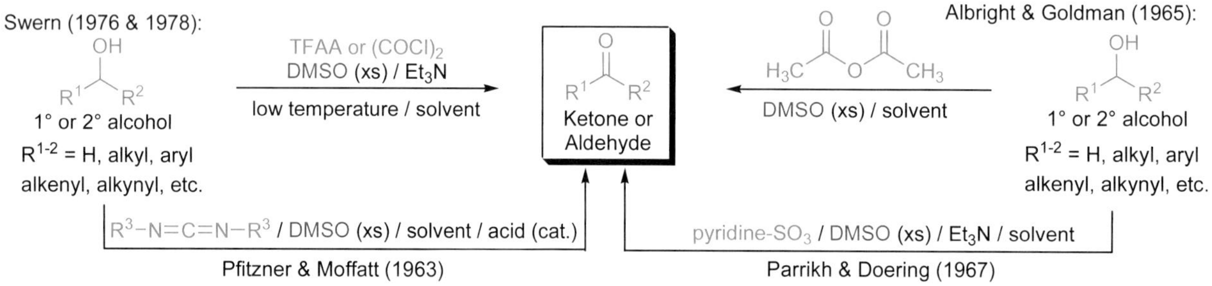

Mechanism:[6-9]

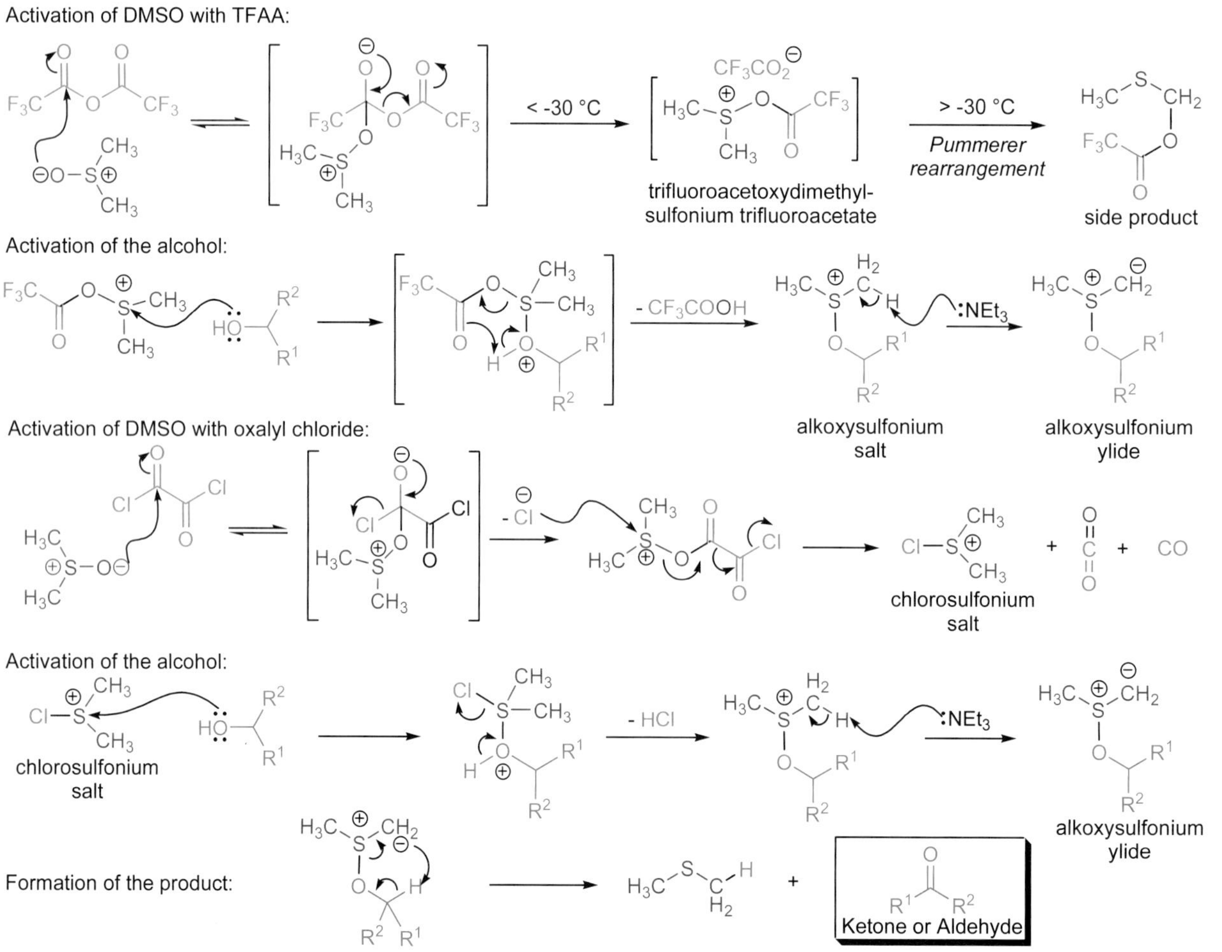

Synthetic Applications:

The first total synthesis of the marine dolabellane diterpene (+)-deoxyneodolabelline was achieved in the laboratory of D.R. Williams.[17] In the final step of the synthetic sequence, the oxidation of a secondary alcohol functionality of a 1,2-diol to the corresponding α-hydroxy ketone was required. Such 1,2-diols are known to be unstable under most oxidation conditions, and often *glycol cleavage* is observed. Indeed, when *Dess-Martin and Ley oxidations* were tried, the substrate suffered carbon-carbon bond cleavage. However, under the *Swern oxidation* conditions, the desired α-hydroxy ketone was isolated in a 65% yield. Interestingly, the substrate was a mixture of four inseparable diastereomeric diols (obtained in a *McMurry reaction*), which gave two easily separable ketone products, one of which was the natural product.

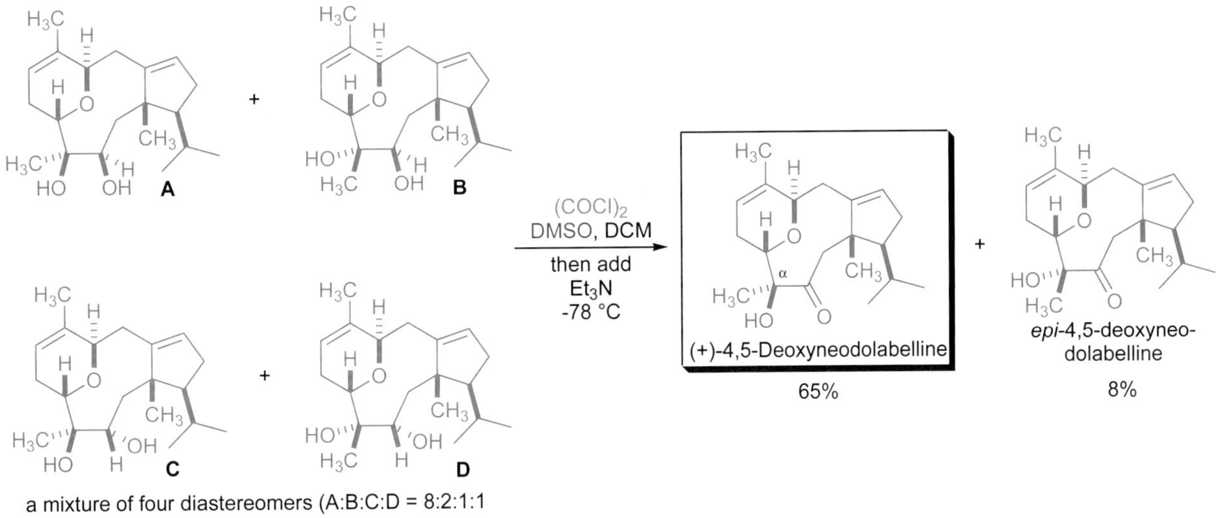

S.F. Martin and co-workers utilized a *double Swern oxidation* in their synthesis of ircinal A and related manzamine alkaloids.[18] The advanced tricyclic diol intermediate was first subjected to the *Swern oxidation* conditions at -78 °C to afford the corresponding dialdehyde in excellent yield. In the next step, the dialdehyde was exposed to excess Wittig reagent under salt-free conditions to form the two terminal alkenes.

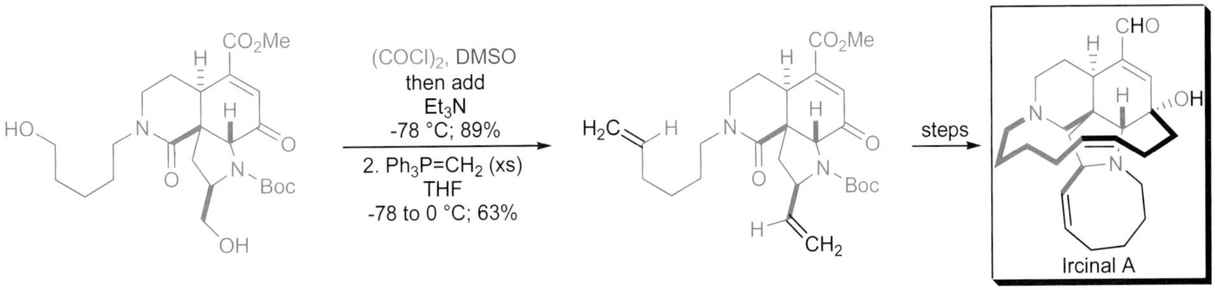

The convergent total synthesis of the mytotoxic (+)-asteltoxin was accomplished by J.K. Cha et al.[19] The coupling of the two main fragments was achieved by the *HWE olefination* of a *bis*(tetrahydrofuran) aldehyde with an α-pyrone phosphonate. The *bis*(tetrahydrofuran) aldehyde was prepared by the *Swern oxidation* of the corresponding *bis*(tetrahydrofuran) primary alcohol. Interestingly, under the oxidation conditions there was no epimerization of the α-stereocenter, but during the *HWE olefination* a small amount of C8 epimer was formed.

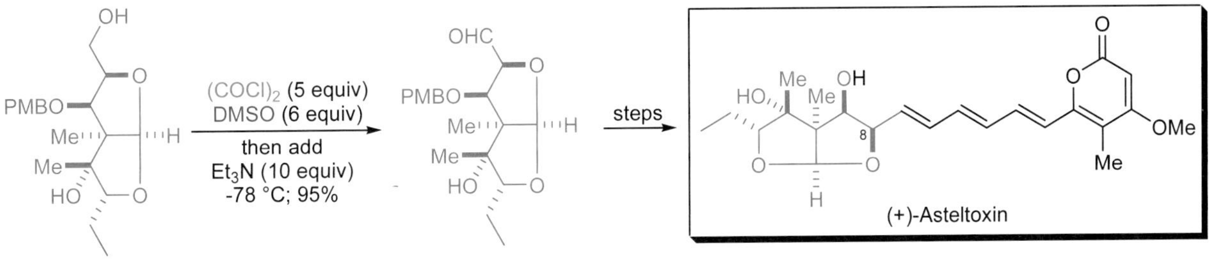

TAKAI-UTIMOTO OLEFINATION (TAKAI REACTION)

(References are on page 693)

Importance:

[*Seminal Publications*[1,2]; *Reviews*[3-8]; *Modifications & Improvements*[9-17]]

Until the second half of the 1980s there was no general method available for the stereoselective preparation of alkenyl halides from carbonyl compounds. In 1987, K. Takai and K. Utimoto introduced a simple and stereoselective method for the conversion of aldehydes to the corresponding (*E*)-alkenyl halides by treating the aldehydes with a haloform-chromium(II)-chloride (CHX_3-$CrCl_2$) system.[1] The chromium(II)-mediated one-carbon homologation of aldehydes with haloform to give the corresponding (*E*)-alkenyl halides is known as the *Takai-Utimoto olefination* (*Takai reaction*). General features of the reaction are: 1) the anhydrous $CrCl_2$ can be dissolved in the solvent just prior to the reaction or can be generated by reacting $CrCl_3$ with $LiAlH_4$; 2) aldehydes react much faster than ketones, so the chemoselective transformation of aldehydes in the presence of ketones is possible; 3) for aliphatic and aromatic aldehydes the major product is the (*E*)-alkenyl halide but for α,β-unsaturated aldehydes the stereoselectivity is usually poor; 4) the rate of the reaction is a function of the haloform used: I>Br>Cl; 5) iodoform reacts rapidly at low temperatures (~0 °C), while other haloforms require higher temperatures to react; 6) the (*E/Z*) ratio is also dependent on the haloform used (Cl>Br>I) and the best (*E*)-selectivity is observed when X=Cl; 7) when $CHBr_3/CrCl_2$ is used, a mixture of alkenyl chlorides and bromides is obtained due to a *Finkelstein reaction* of $CrCl_2$ with bromide (Br^-). However, by preparing $CrBr_2$ from $CrBr_3/LiAlH_4$ this problem is eliminated;[1,18] 8) reducing agents other than Cr(II) give unsatisfactory or no yield of the desired alkenyl halides; 9) in certain cases the applied solvent is critical to achieve good yield and stereoselectivity; 10) the reaction conditions tolerate almost any functional group; and 11) the reaction conditions are mild enough (the reagent is practically nonbasic) that even highly enolizable substrates do not racemize at their α-position. There are several important modification of the *T-U olefination*: 1) instead of haloforms, 1,1-geminal dihalides are used to afford predominantly (*E*)-olefins;[2] 2) instead of 1,1-geminal dihalides, α-acetoxy bromides can be used, which are more stable and easier to prepare and handle than 1,1-geminal dihalides;[11] and 3) one-carbon homologation of aldehydes *via* chromium enolates to the corresponding methyl ketones using $TMSCBr_3/CrBr_2$.[12] When 1,1-geminal dihalides are used, the following can be expected: 1) the (*E*)-selectivity is especially high for aliphatic substrates, and it increases with the size of R^1; 2) only 1,1-geminal diiodoalkanes are suitable; the dichlorides and dibromides undergo reduction under the reaction conditions; 3) CH_2I_2 is the most reactive. The higher homologs react slower and give lower yields; 4) aldehydes react faster than ketones; 5) the reaction can be carried out with catalytic amounts of $CrCl_3$ in the presence of samarium metal or samarium diiodide;[19] and 6) the R^2 substituent can contain heteroatoms so the preparation of alkenyl silanes,[13,15] -boronates,[14] -stannanes,[10] and sulfides[9] is possible. The use of α-acetoxy bromides has the following features: 1) the *in situ* preparation of the chromium(II) reagent and donor ligand such as DMF or TMEDA should be present; 2) high (*E*)-selectivity; and 3) exclusive reaction with aldehydes.[11]

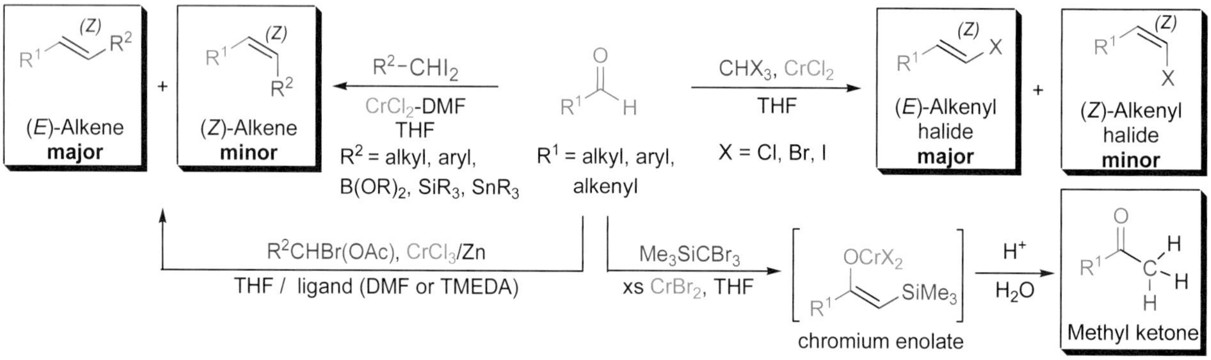

Mechanism: [20,21,2,3,15,7]

The exact mechanistic pathway is not known. However, it is believed that the *T-U olefination* proceeds *via* geminal-dichromium intermediates that are nucleophilic and attack the carbonyl compound. The (*E*)-alkene is formed from the β-oxychromium species.

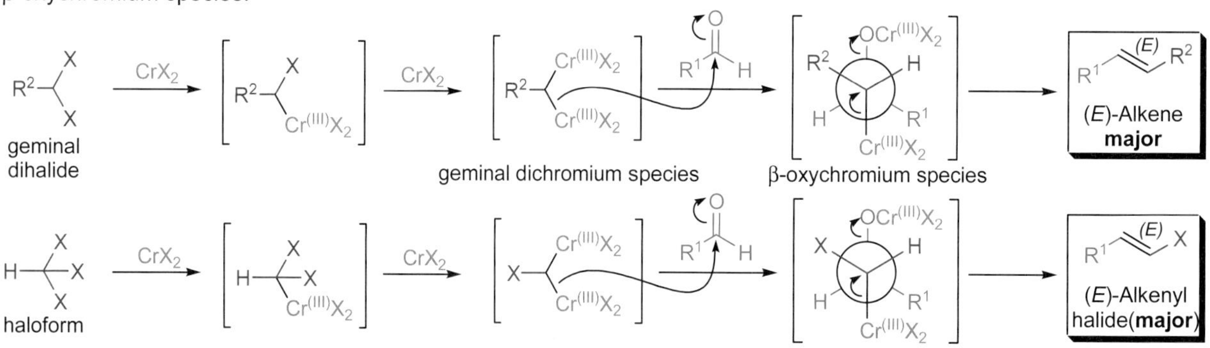

TAKAI-UTIMOTO OLEFINATION (TAKAI REACTION)

Synthetic Applications:

The first total synthesis of the cytotoxic marine natural product aplysiapyranoid C was accomplished by M.E. Jung et al.[22] The special structural feature of this natural product is the (*E*)-vinyl chloride moiety, which was introduced in high yield *via* the *Takai reaction* in the late stages of the synthetic effort. The removal of the silicon protecting group and cyclization of the dichlorodienol with TBCO (tetrabromocyclohexadienone) in nitromethane gave a mixture of four products, one of which was the desired product that was isolated in 43% yield.

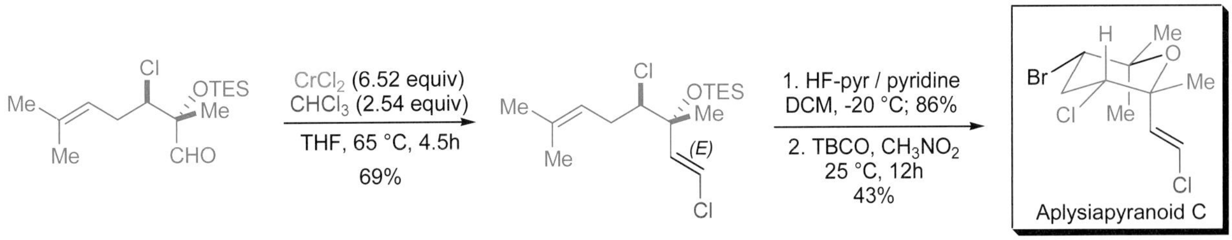

Polycephalin C is a bis(trienoyltetramic acid) linked by an unusual asymmetric cyclohexene ring. At the time of isolation and structure elucidation the absolute configuration at the C3 and C4 positions was not established. S.V. Ley and co-workers carried out the total synthesis of this natural product based on a *double Takai olefination* followed by a *double Stille cross-coupling*.[23] The dialdehyde substrate for the *Takai olefination* was prepared by the *asymmetric Diels-Alder cycloaddition* of dimenthyl fumarate with butadiene. The *double Takai olefination* proceeded with high (E)-stereoselectivity to afford the bisiodide, albeit in only 40% yield. Subsequent *double Stille coupling* proceeded in good yield and after a global deprotection the target compound was obtained.

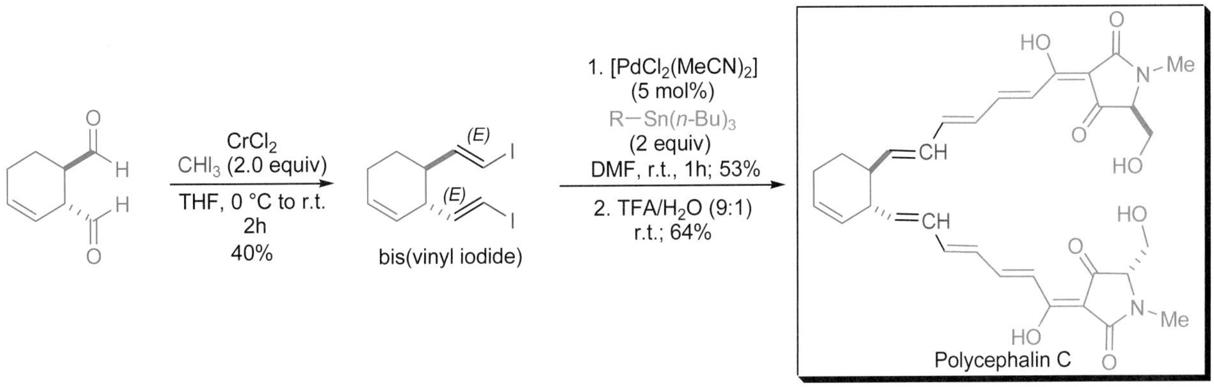

In the laboratory of F.R. Kinder Jr., the total synthesis of cytotoxic marine natural product bengamide E was completed.[24] The *Takai-Utimoto olefination* was used to introduce the (*E*)-disubstituted double bond. The aldehyde was exposed to a $CrCl_2$ solution in THF in the presence of 1,1-diiodo-2-methylpropane, and the desired olefin was obtained in 29% yield.

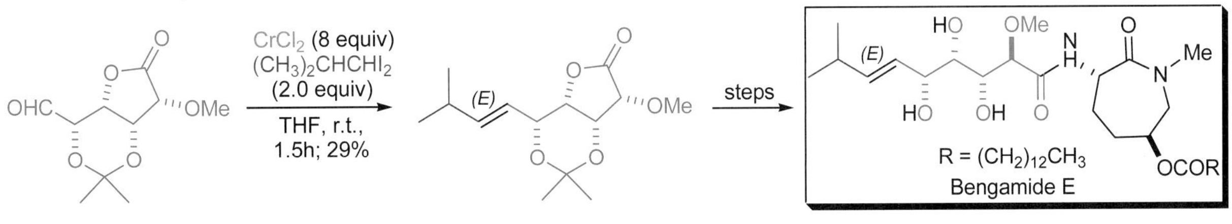

The diastereoselective Me_3Al-mediated *intramolecular Diels-Alder reaction*, a highly (*E*)-selective *Takai olefination* and a *Suzuki coupling* were the key steps in the enantioselective total synthesis of (–)-equisetin by K. Shishido et al.[25] It should be noted that the type of *T-U olefination* utilized allowed the preparation of functionalized heterosubstituted (*E*)-alkenes.

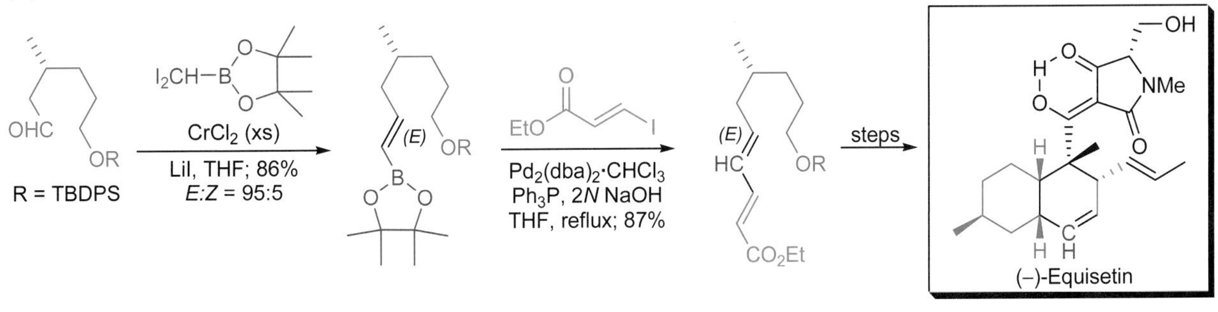

TEBBE OLEFINATION / PETASIS-TEBBE OLEFINATION

(References are on page 693)

Importance:

[*Seminal Publications*[1-3]; *Reviews*[4-15]; *Modifications & Improvements*[16-20]]

In 1976, R.R. Schrock discovered, during his studies of *alkene metathesis*, that the neopentylidene complex of tantalum was structurally analogous to phosphorous ylides (*Wittig reagents*), and it not only olefinated aldehydes and ketones but esters and amides as well.[1] In 1978, F.N. Tebbe et al. reported that titanocene dichloride reacted with two equivalents of $AlMe_3$ to produce a methylene-bridged titanium-aluminum complex (*Tebbe reagent*), which transferred a methylene group (CH_2) efficiently to various carbonyl compounds to afford olefins.[2] It was shown early on that the Tebbe reagent converted carboxylic esters, lactones, and amides to the corresponding enol ethers and enamines in high yield. The one-carbon homologation (methylenation) of carbonyl compounds using the Tebbe reagent is known as the *Tebbe olefination*. The *Tebbe reaction* has the following general features: 1) the active species (titanocene methylidene) is more nucleophilic and much less basic than the corresponding Wittig reagents. Consequently, less reactive (bulkier) and enolizable carbonyl compounds can be readily olefinated; 2) the Tebbe reagent is stable in solution and reacts at low temperature with the various carbonyl groups in the following order: aldehydes>ketones>esters>amides; 3) acid halides and anhydrides do not undergo methenylation. Instead, the corresponding titanium enolates are formed, which can be used in subsequent *aldol reactions*;[21] 4) only methenylations can be performed; higher alkenyl groups cannot be introduced with this method; 5) a wide range of functional groups are tolerated. However, the presence of the Lewis acidic aluminum may cause complications with certain substrates. The thermal decomposition of dimethyltitanocene also generates titanocene methylidene without the Lewis acidic aluminum, and it is capable of olefinating very sensitive substrates such as anhydrides, silyl esters, and acylsilanes.[18] This method is known as the *Petasis-Tebbe olefination* or *Petasis olefination*.[3]

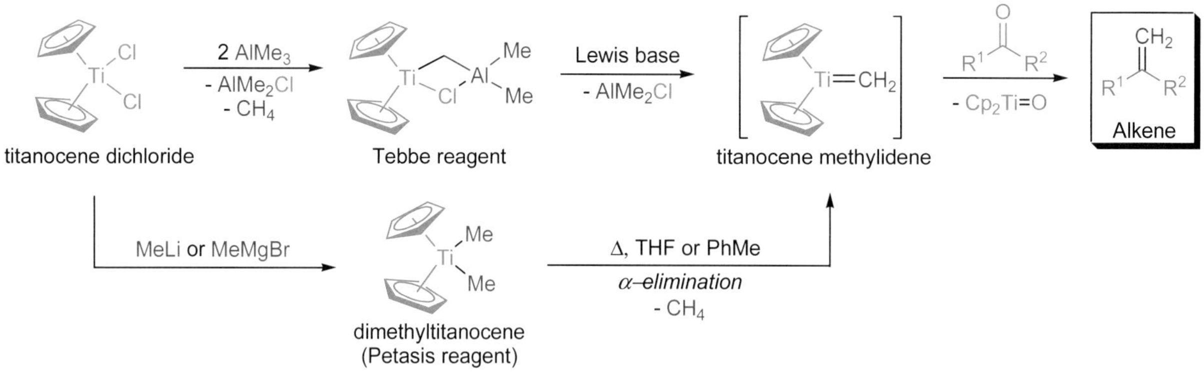

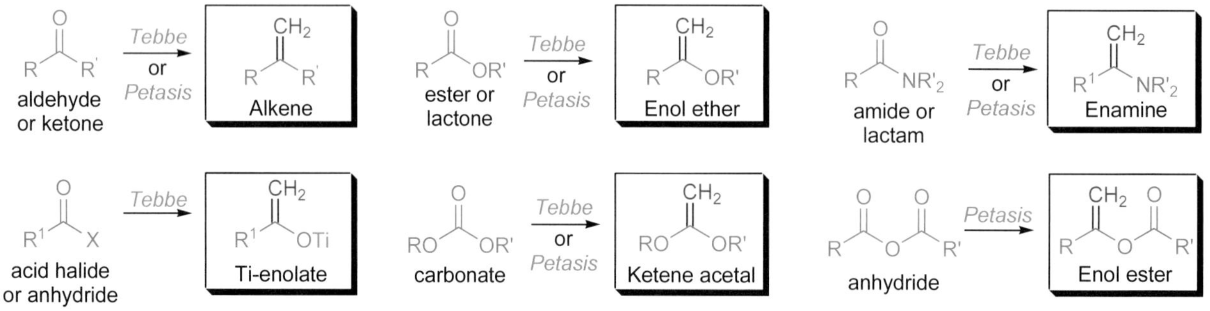

Mechanism: [4,22,7,23-30]

The active species in the *Tebbe olefination* is believed to be the nucleophilic (Schrock-type) titanocene methylidene, which is formed from the Tebbe reagent upon coordination of the aluminum with a Lewis base (e.g., pyridine). This methylidene in its uncomplexed form, however, has never been isolated or observed spectroscopically owing to its extreme reactivity. The same intermediate can also be generated by other means.[4] The titanocene methylidene reacts with the carbonyl group to form an oxatitanacyclobutane intermediate that breaks down to titanocene oxide and the desired methenylated compound (alkene). The driving force is the formation of the very strong titanium-oxygen bond.

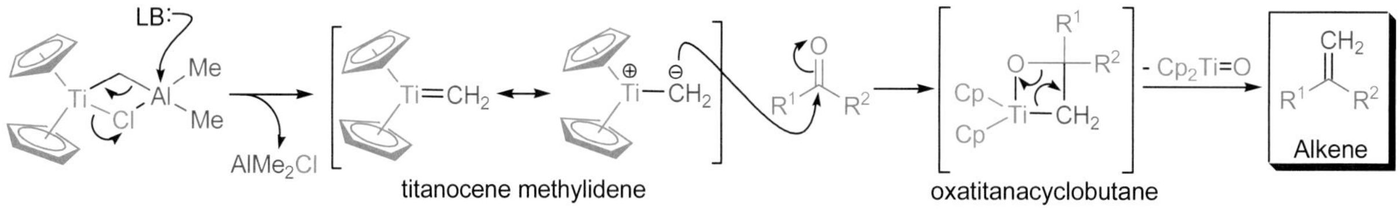

Synthetic Applications:

The enantioselective total synthesis of the cyclooctanoid natural product (+)-epoxydictymene was accomplished in the laboratory of L.A. Paquette.[31] The entire tricyclic framework was constructed by the application of a *Claisen rerrangement* via a chairlike transition state. The precursor for this *[3,3]*-sigmatropic rearrangement was obtained by treating a lactone precursor with the solution of the Tebbe reagent in the presence of pyridine. The corresponding enol ether was formed in almost quantitative yield, and immediately after isolation it was treated with triisobutylaluminum to effect the *Claisen rearrangement*.

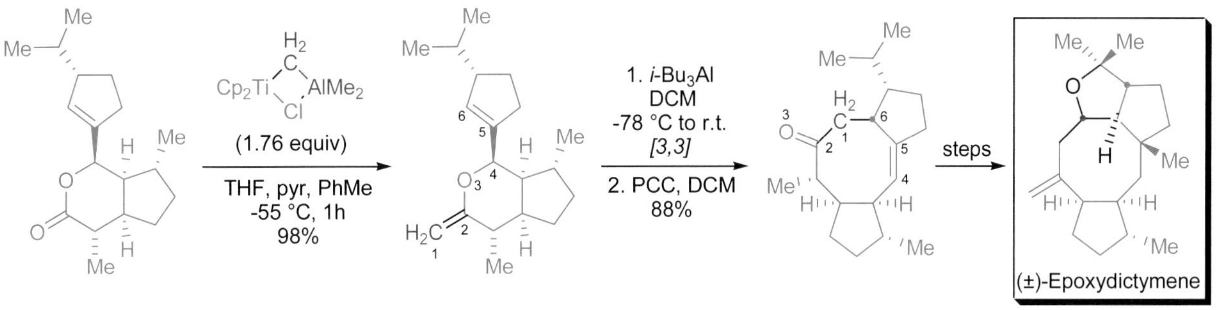

The unsaturated medium ring ether (+)-laurencin was synthesized by A.H. Holmes and co-workers.[32] Halfway into the synthetic sequence the ethyl side chain had to be introduced at C2. This task was accomplished by using sequential *Tebbe methenylation*, *diastereoselective intramolecular hydrosilation*, and displacement of a primary tosylate with dimethyl cuprate. The eight-membered lactone was exposed to the Tebbe reagent in the presence of DMAP to afford the cyclic enol ether in good yield.

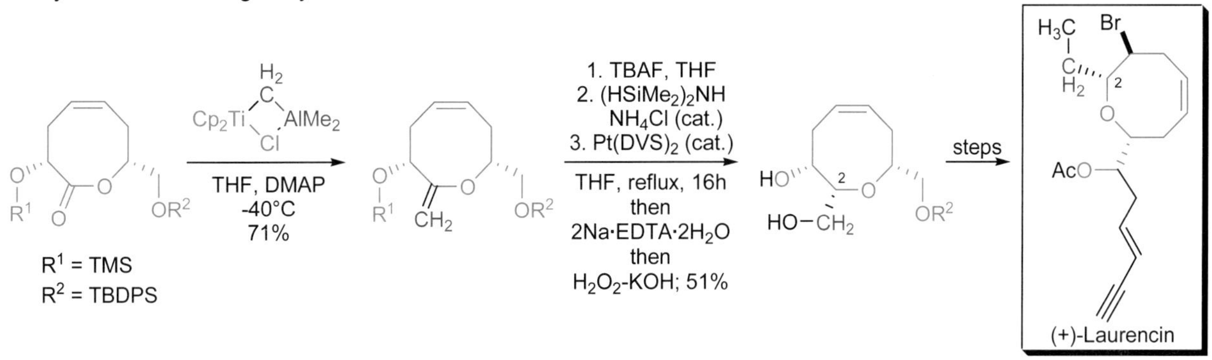

In the final step of the total synthesis of (±)-21-oxogelsemine and (±)-gelsemine by D.J. Hart et al., the introduction of the C20 vinyl group was unsuccessful when the cagelike aldehyde was treated with (methylidene)triphenylphosphorane (*Wittig reaction*).[33] This failure was attributed to two factors, namely, steric hindrance and neighboring group participation of the oxindole carbonyl group. However, when the Petasis reagent was used in refluxing tetrahydrofuran, the desired olefin was obtained in 87% yield. Since (±)-21-oxogelsemine has been converted to (±)-gelsemine before, this synthesis was also a formal total synthesis of (±)-gelsemine.

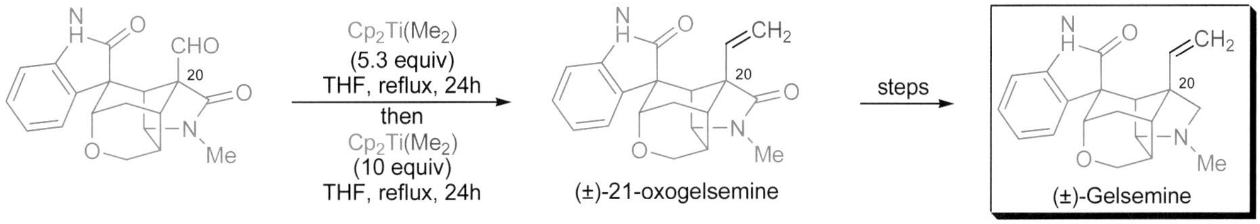

It is possible to methenylate the carbonyl group of amides and lactams provided that the nitrogen atom is substituted with an electron-withdrawing group. This was the case when A.R. Howell and co-workers successfully converted a wide range of *N*-substituted β-lactams to the corresponding 2-methyleneazetidines.[34] In the two examples it is noteworthy that the β-lactam carbonyl group reacted preferentially in the presence of the ester carbonyl group.

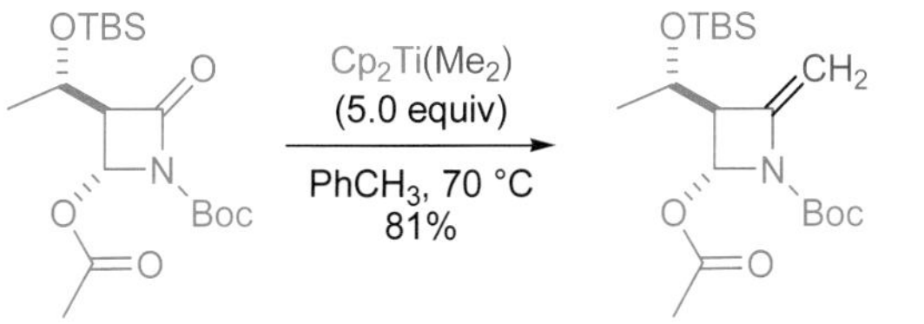

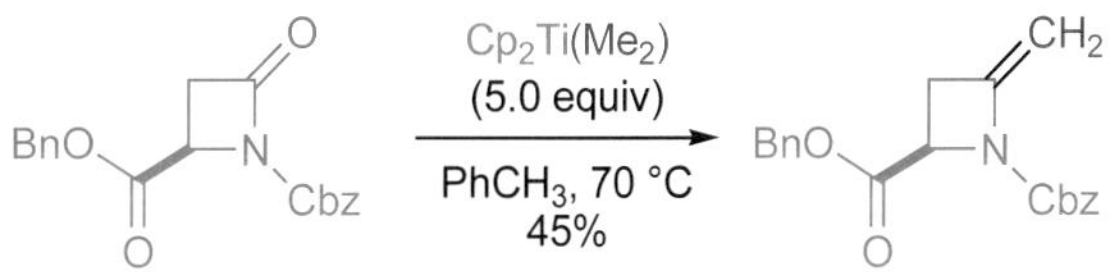

TISHCHENKO REACTION

(References are on page 694)

Importance:

[*Seminal Publications*[1-9]; *Reviews*[10-12]; *Modifications & Improvements*[13-27]; *Theoretical studies*[28]]

In 1887, L. Claisen reported the formation of benzyl benzoate from benzaldehyde in the presence of sodium alkoxides.[1] Nearly thirty years later, W.E. Tishchenko found that both enolizable and non-enolizable aldehydes can be converted to the corresponding esters in the presence of magnesium- or aluminum alkoxides.[2-9] The reaction involves a hydride shift from one aldehyde to another that leads to the formation of the ester product. This transformation is known today as the *Tishchenko reaction*. The general features of the reaction are: 1) in the traditional transformation, the reaction takes place between the same aldehydes;[11] 2) in the *crossed Tishchenko reaction*, two different aldehydes are reacted to form the ester product;[13] 3) the reaction can take place in an intramolecular fashion, yielding the corresponding lactone;[21] and 4) common side reactions are the *aldol reaction*, *Cannizzaro reaction*, *Merwein-Ponndorf-Verley reduction,* and *Oppenauer oxidation*.[11] The most general catalysts in the traditional *Tishchenko reaction* are aluminum alkoxides, but a wide-variety of catalysts can be used:[11] 1) alkali- and alkali earth metal oxides[26] and alkoxides;[18] 2) transition metal-based catalysts such as ruthenium complexes ($RuH_2(PPh_3)_4$,[22] certain rhodium-,[18] iridium-,[19] and iron complexes,[14,15] and metallocenes of group IV metals (Cp_2MH_2 M = Hf, Zr);[22] and 3) lanthanide based catalyst such as lanthanide amides ($Ln[NSiMe_2)_3]$, Ln = La, Sm, Y),[24] organolanthanoid halides (EtLnX, Ln = Pr, Nd, Sm, X = I)[16] and SmI_2.[17] A modification of the *Tishchenko reaction* is the *aldol-Tishchenko reaction* where the aldehyde first undergoes an aldol reaction followed by the *Tishchenko reaction* to form monoesters of 1,3-diols.[11,12] In the *homo aldol-Tishchenko reaction*, the same aldehyde molecules react.[29] In the *hetero aldol-Tishchenko reaction*, a ketone or aldehyde reacts with two equivalents of a different aldehyde over the catalyst.[23,25] The most widely used modification of the *Tishchenko reaction* is the *Evans-Tishchenko reaction*.[20] In this transformation, a chiral β-hydroxy ketone reacts with an aldehyde in the presence of catalytic SmI_2 to provide the *anti* 1,3-diol monoester product with excellent diastereoselectivity.

Traditional Tishchenko reaction:

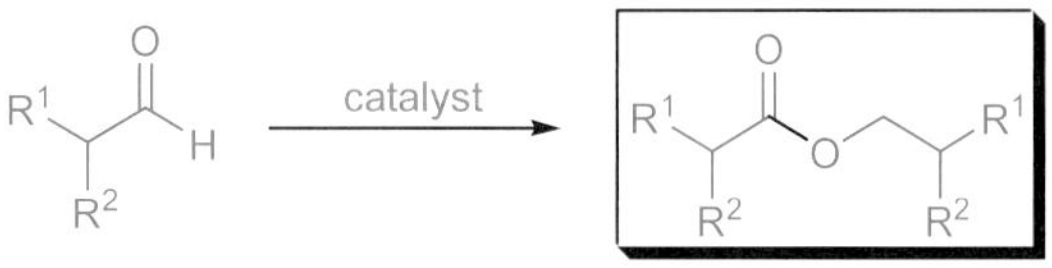

Aldol-Tishchenko reaction:

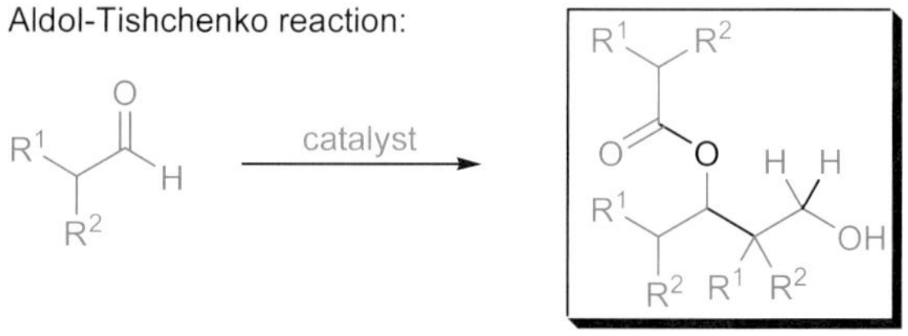

Crossed- (mixed-) Tishchenko reaction:

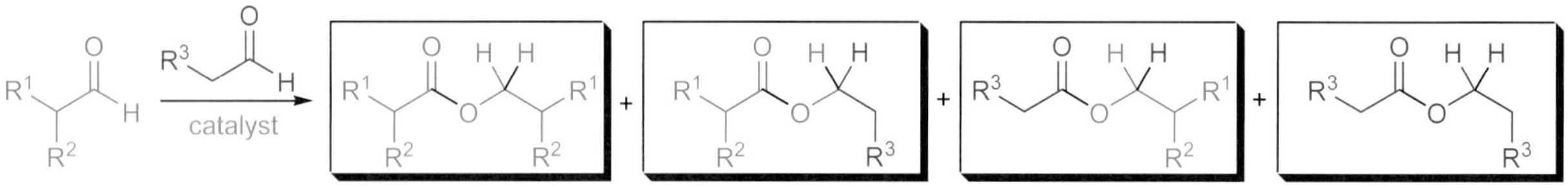

Evans-Tishchenko reaction:

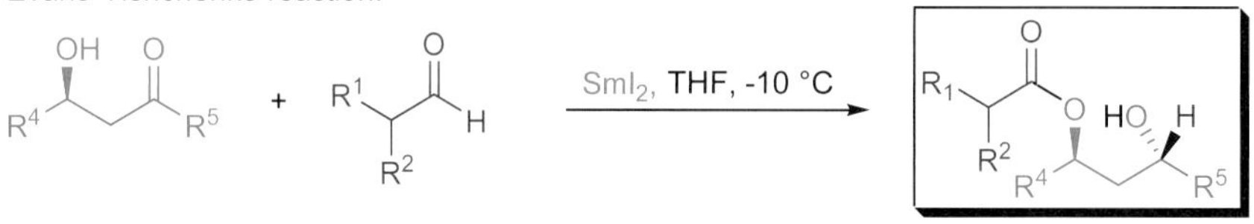

R^1, R^2, R^3 = H, alkyl, aryl; R^4, R^5 = alkyl, aryl; catalyst: $AlOR_3$; NaOR; MO, M = Ba, Sr, Mg; $RuH_2(PPh_3)_4$; $[\eta^4\text{-}C_4Ph_4\text{-}CO)Ru(CO)_3]_2$; $CpMH_2$, M = Hf, Zr; $Na_2Fe(CO)_4$; $ROIr(CO)(PPh_3)_2$; $Ln[N(SiMe_2)_3]$, Ln = La, Sm, Y; EtLnI, Ln = Pr, Nd, Sm; SmI_2

Mechanism:[13,30,31,22,11]

The mechanism of the Tishchenko reaction was extensively studied and there were three different mechanisms proposed. The most commonly accepted mechanism is depicted below.[13] According to this proposal, first the aluminum alkoxide coordinates to the aldehyde. This is followed by the attack of a second molecule of aldehyde. Subsequent hydride shift leads to the regeneration of the catalyst and formation of the product.

Traditional Tishchenko reaction:

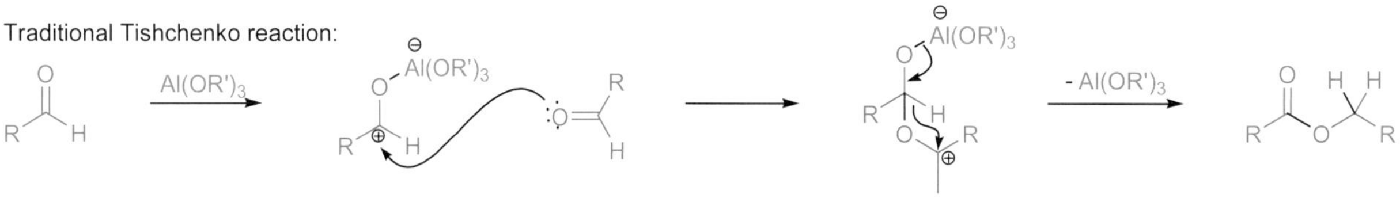

Evans-Tishchenko reaction:

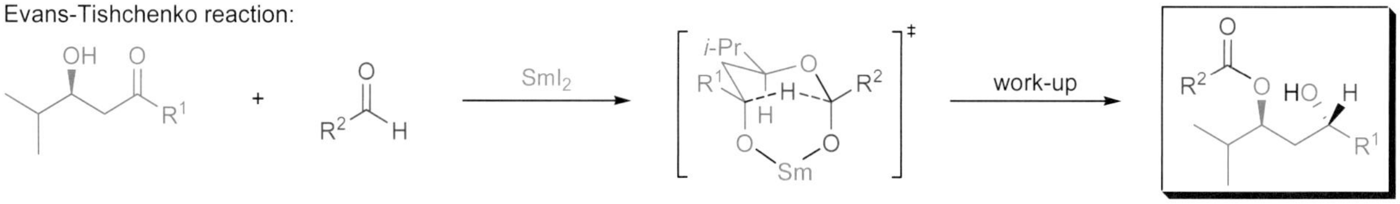

Synthetic Applications:

Sarains A-C are a family of alkaloids isolated from marine sponges. J.K. Cha and co-workers accomplished the synthesis of the western macrocyclic ring of sarain A.[32] To establish the C3 quaternary stereocenter, they treated the aldehyde substrate with formaldehyde in the presence of sodium carbonate. The aldehyde substrate underwent an *aldol reaction* followed by a *Tishchenko reaction* to provide the formate ester of the 1,3-diol product. This ester was hydrolyzed *in situ* under the reaction conditions and the 1,3-diol was isolated.

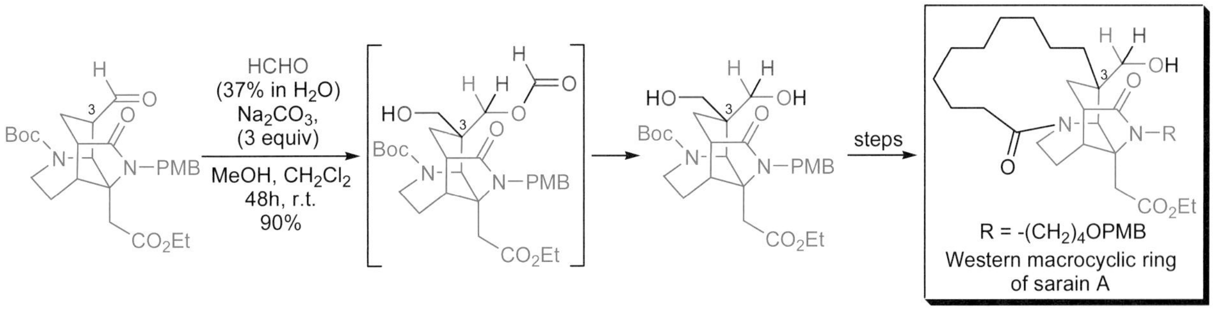

S.L. Schreiber and co-workers accomplished the total synthesis of (–)-rapamycin.[33] In their approach, they utilized an *Evans-Tishchenko reaction* of C22-C42 fragment and Boc pipecolinal. The reaction provided the product with excellent yield and as a >20:1 mixture of the *anti* and *syn* diastereomers.

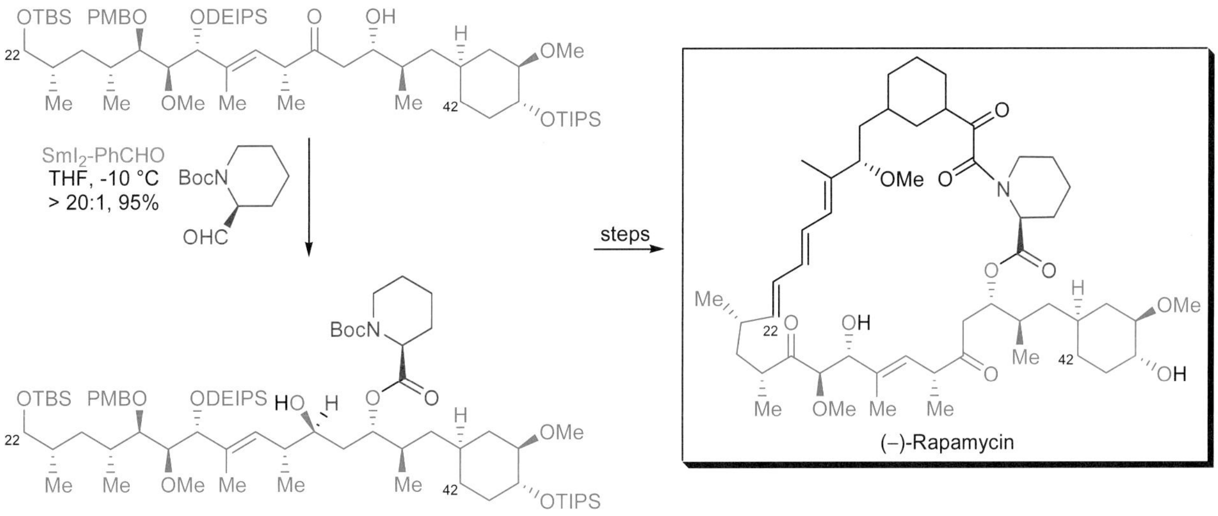

Rhizoxin D, a natural product possessing potent antitumor and antifungal activity, was synthesized by J.W. Leahy and co-workers.[34] To establish the C17 stereocenter, they utilized the *Evans-Tishchenko reaction.* To this end, the 3-hydroxyketone substrate was reacted with *p*-nitrobenzaldehyde in the presence of catalytic SmI_2. The reaction yielded the monoester of the *anti* 1,3-diol as a single product.

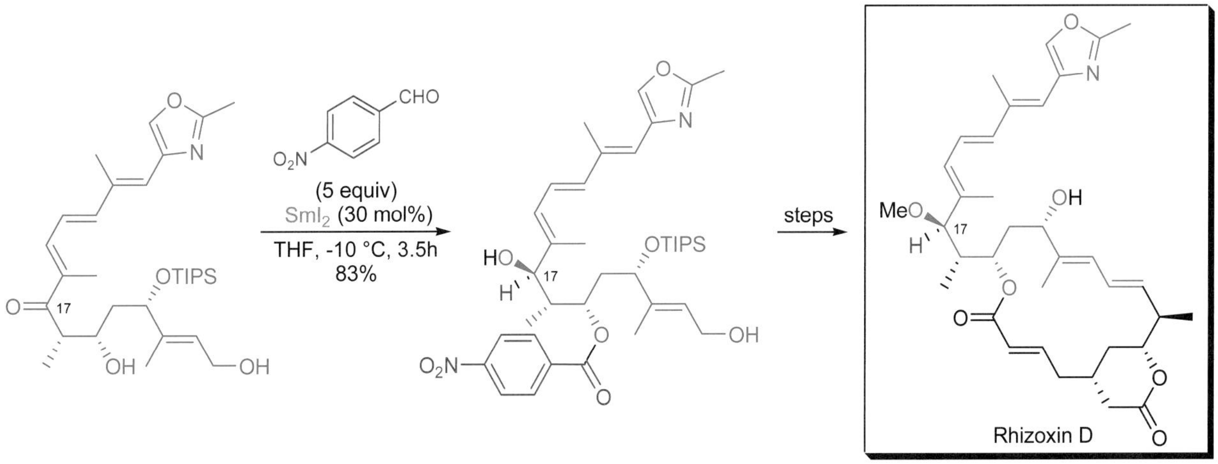

TSUJI-TROST REACTION / ALLYLATION

(References are on page 695)

Importance:

[*Seminal Publications*[1-4]; *Reviews*[5-24]; *Modifications & Improvements*[25-30]; *Theoretical Studies*[31-37]]

In 1965, J. Tsuji demonstrated that π-allylpalladium chloride could be substituted with certain nucleophiles such as enamines and the anions derived from diethyl malonate and ethyl acetoacetate.[1] Soon after this initial report, the catalytic version of this transformation was developed.[2] In 1973, B.M. Trost reported that alkyl-substituted π-allylpalladium complexes could be alkylated with soft carbon nucleophiles with complete regio- and stereoselectivity. However, hard nucleophiles (e.g., alkylithiums, alkylmagnesium halides) failed to react.[4] The Pd-catalyzed allylation of carbon nucleophiles with allylic compounds *via* π-allylpalladium complexes is called the *Tsuji-Trost reaction*. The general features of this transformation are: 1) a wide range of leaving groups (X) can be utilized to form π-allylpalladium complexes (e.g., halides, acetates, ethers, sulfones, carbonates, carbamates, epoxides, and phosphates); 2) there is a marked difference in the reactivity of the various leaving groups with the following trend: Cl > OCO_2R > OAc >> OH; 3) in the case of most substrates, the use of a stoichiometric amount of base is necessary to generate the soft nucleophiles. However, allylic carbonates undergo decarboxylation, and in the process a sufficiently basic alkoxide is formed so no extra base is needed; 4) the range of possible soft carbon nucleophiles is also wide: active methylene compounds with two electron-withdrawing groups (R^3 and R^4), enamines and enolates; 5) the catalytically active $Pd^{(0)}$ species is introduced in either the form of $Pd^{(0)}$ or by the *in situ* reduction of $Pd^{(II)}$ complexes; 6) the addition of the nucleophiles to the unsymmetrical π-allylpalladium complexes is regioselective and favors the least substituted allyl terminus regardless of the initial position of the leaving group; 7) occasionally the regioselectivity can be influenced by the nature of the ligand and the nucleophile; 8) *bis* allylic substrates having two different leaving groups can be substituted with high regioselectivity; and 9) optically active substrates are substituted by soft nucleophiles with an overall retention of configuration (double inversion), while hard nucleophiles give rise to products with an overall inversion of configuration (π-allylpalladium complexes are transmetallated). Nitrogen-, oxygen-, and sulfur-based soft nucleophiles can also be used in *Tsuji-Trost allylation* reactions.

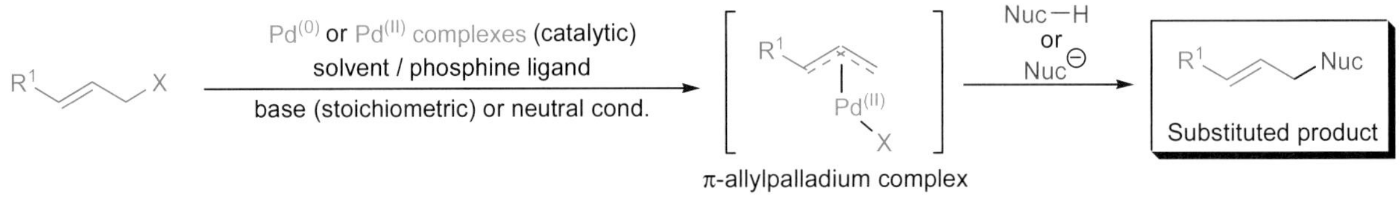

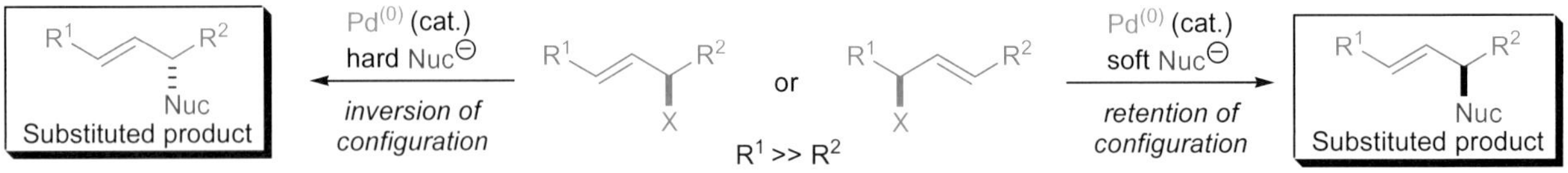

R^{1-2} = H, alkyl, aryl; X = OH, OPh, OCOR, OCONHR, OCO_2R, $OP(O)(OR)_2$, Cl, NO_2, SO_2Ph, NR_2, NR_3X, SR_2X
soft Nuc-H = $R^3R^4CH_2$, enamines, enolates; R^{3-4} = CO_2R, CN, NO_2, SO_2Ph, COR, NC, $N=(CMe_2)$, SPh, alkenyl
Pd-complexes: $Pd(PPh_3)_4$, $Pd_2(dba)_3$, $[Pd(allyl)Cl]_2$; ligands: PPh_3, dba

Mechanism: [38,5,39,40]

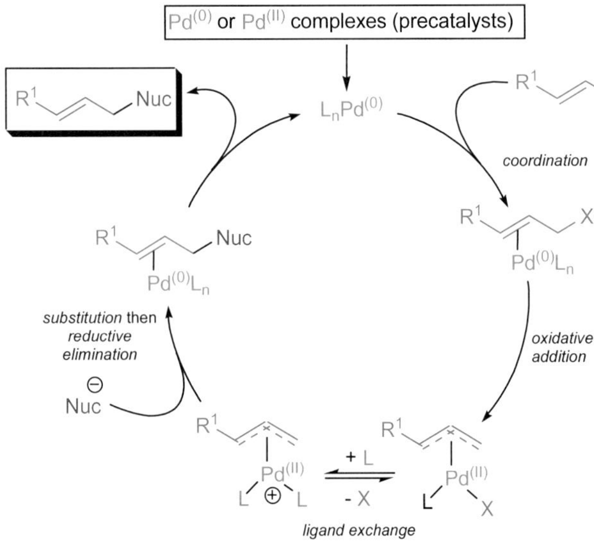

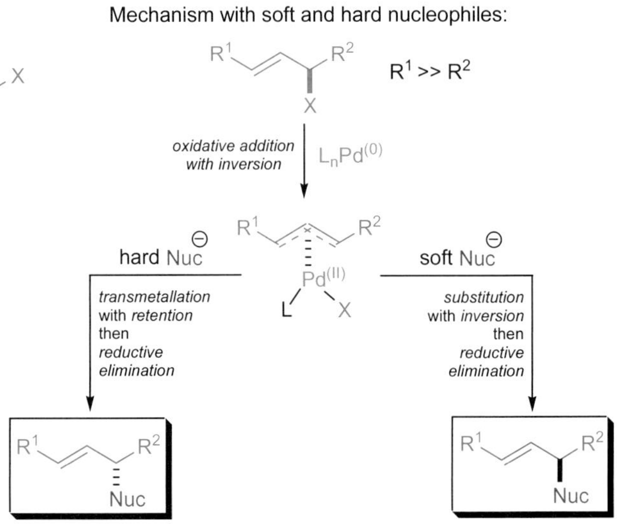

Synthetic Applications:

The scalable total synthesis of the cytotoxic natural product (+)-FR182877 was accomplished in the laboratory of E.J. Sorensen.[41] The key steps of the synthetis were an *intramolecular Tsuji-Trost allylation* to prepare the 19-membered macrocyclic pentaene followed by a double *transannular Diels-Alder cycloaddition* to obtain the desired pentacyclic structure. The allylic carbonate was exposed to 10 mol% of the Pd-catalyst under high dilution conditions in THF. The new bond between C1 and C19 was formed with complete diastereoselectivity and in good yield, although the configuration at C19 was not determined.

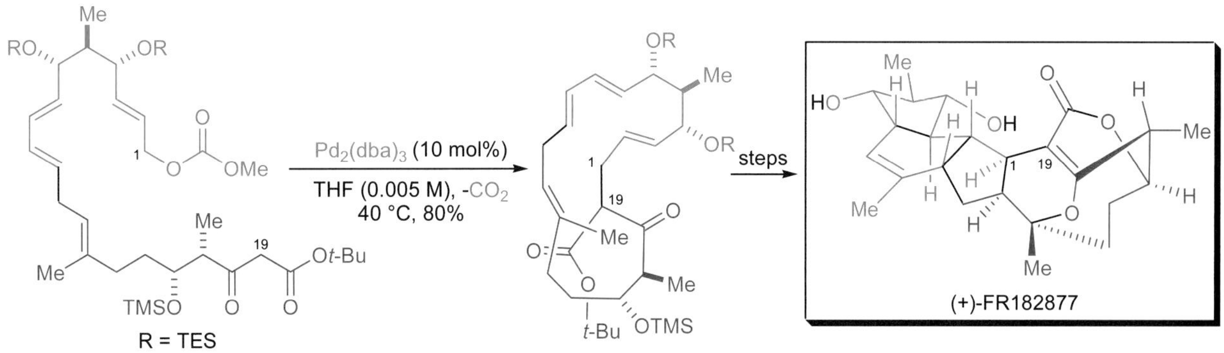

The water soluble vitamin (+)-biotin was synthesized by M. Seki and co-workers from L-cysteine in only 11 steps using inexpensive reagents and mild reaction conditions.[42] The key ring forming step was an *intramolecular allylic amination* (*Tsuji-Trost reaction* using a nitrogen nucleophile) of a *cis* allylic carbonate. As expected with a soft nucleophile, the allylation took place with an overall retention of configuration.

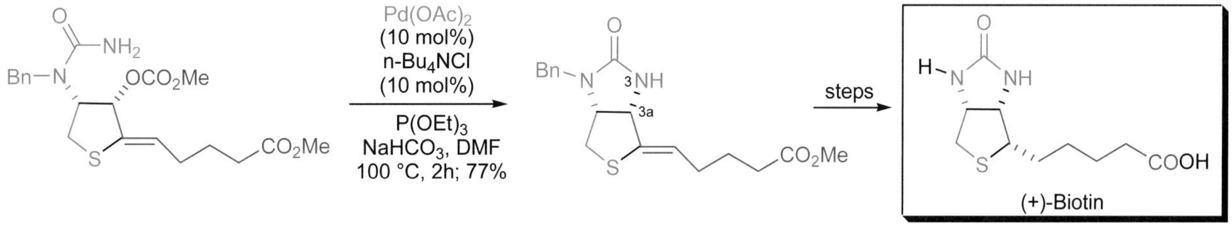

The first total synthesis of cristatic acid, a compound of considerable cytotoxic activity, was reported by A. Fürstner et al.[43] The disubstituted furan moiety was constructed by the *Tsuji-Trost allylation* of a vinyl epoxide intermediate by *bis*(phenylsulfonyl)methane. The resulting 1,4-diol was obtained in an almost quantitative yield.

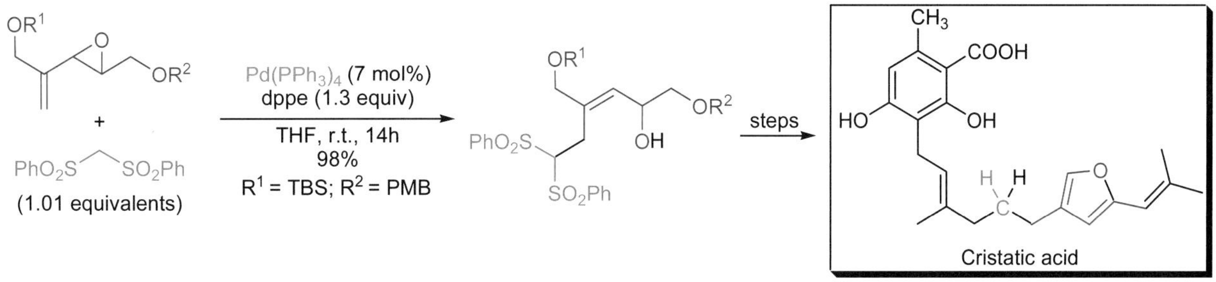

The *Tsuji-Trost reaction* using an oxygen-based soft nucleophile was applied to the synthesis of *cis*-2,5-disubstituted-3-methylenetetrahydrofurans in the laboratory of D.R. Williams.[44] This method was the basis for the preparation of the C7-C22 core of amphidinolide K. The addition of Me_3SnCl served two purposes: it accelerated the reaction and insured that the oxygen was strongly nucleophilic during the ring-closure, and it suppressed an undesired acyl migration.

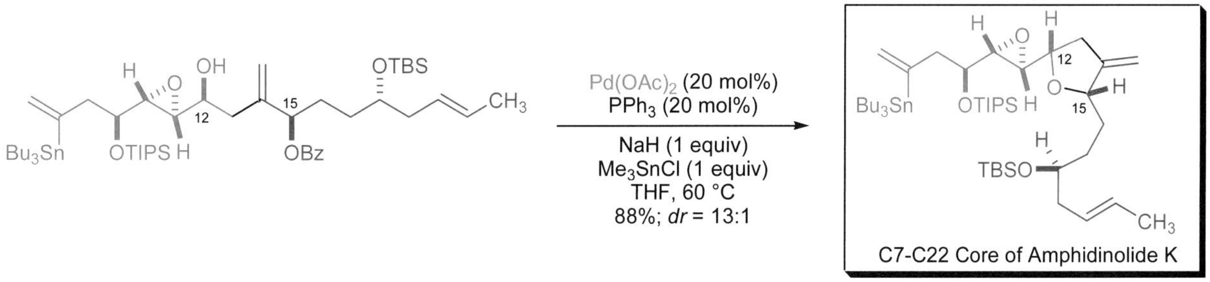

TSUJI-WILKINSON DECARBONYLATION REACTION

(References are on page 696)

Importance:

[*Seminal Publications*[1-4]; *Reviews*[5-11]; *Modifications & Improvements*[12-18]]

In 1965, J. Tsuji and K. Ono reported that aldehydes reacted with a stoichiometric amount of chloro-*tris*(triphenylphosphine)rhodium (Wilkinson's catalyst) to form chloro-carbonyl*bis*(triphenylphosphine)rhodium and the corresponding C-H compound.[1] Numerous aliphatic, aromatic, and α,β-unsaturated aldehydes were decarbonylated in good yield at or above room temperature. A few years later, the method was extended to the decarbonylation of acyl halides that afforded the corresponding halides.[3] The decarbonylation of aldehydes and acyl halides using Wilkinson's catalyst is known as the *Tsuji-Wilkinson decarbonylation reaction*. The general features of this transformation are:[7,10] 1) several transition metal complexes (e.g., Pd-complexes)[17] are capable of decarbonylating aldehydes and acyl halides, but the most efficient complex was found to be Wilkinson's catalyst; 2) the catalyst is employed in stoichiometric amounts, but the resulting carbonyl complex can be isolated and the catalyst recovered; 3) if the reaction temperature is raised above 200 °C, the reaction becomes catalytic because carbon monoxide is released from the coordination sphere of the rhodium, and the catalyst is regenerated; 4) the substrate can be an aldehyde, acyl halide, acyl cyanide,[16] or 1,2-diketone;[13] 5) for aliphatic substrates the order of reactivity is primary>secondary>tertiary; 6) in most cases the reaction takes place under mild conditions and at relatively low temperature (room temperature or at reflux temperature of the applied solvent); 7) the decarbonylation is stereospecific: the configuration of the stereocenter to which the formyl group is attached to is retained;[19] 8) the decarbonylation of α,β-unsaturated substrates proceeds without interference from the double bond; and 9) if the acyl halide contains β-hydrogen atoms the final product is an alkene rather than an alkane due to facile β-elimination.

Tsuji & Ono (1965):

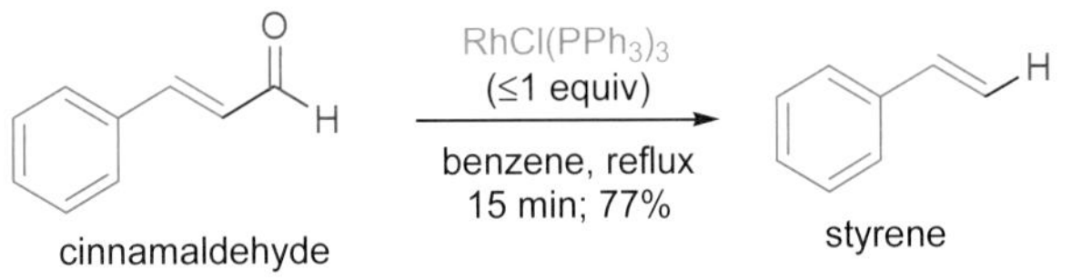

Tsuji & Ono (1968):

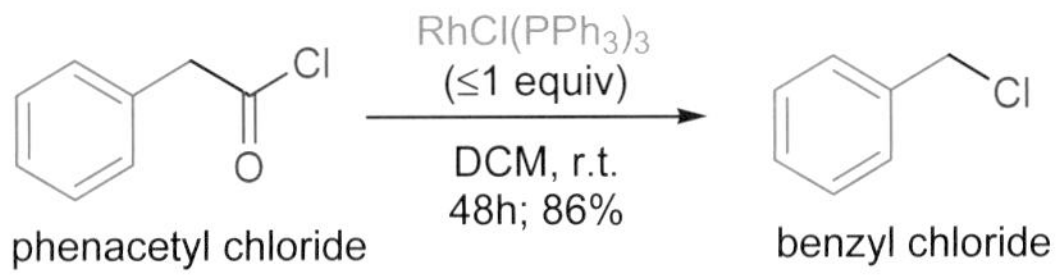

Tsuji-Wilkinson decarbonylation:

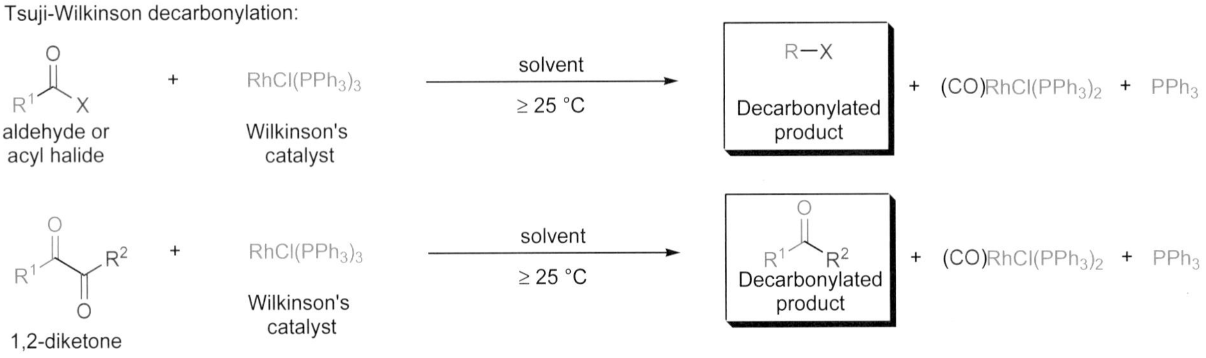

R^1 = 1°, 2° and 3° alkyl, aryl, alkenyl; X = H, Cl, Br, CN, CO-alkyl, CO-aryl, CH_2CO_2Me; R^2 = alkyl, aryl, alkenyl; solvent: C_6H_6, toluene, xylenes, acetonitrile, benzonitrile, DCM

Mechanism:[20-27,10]

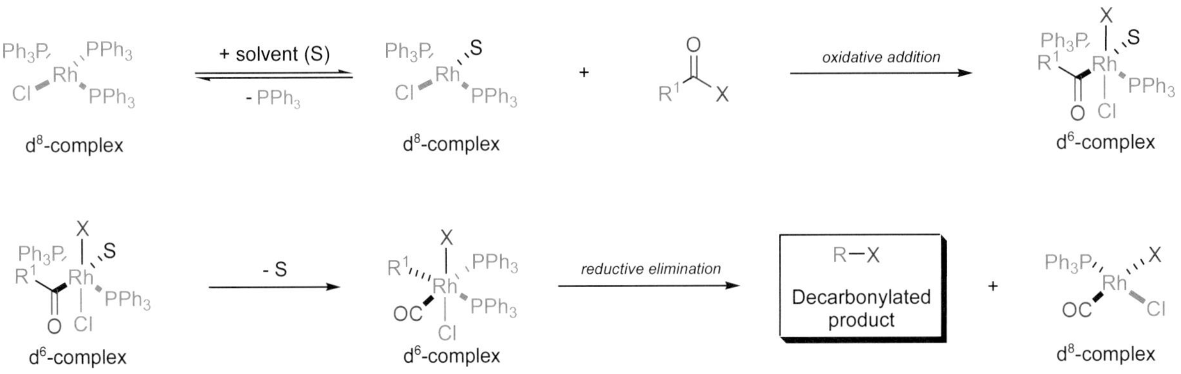

Synthetic Applications:

In the laboratory of F.E. Ziegler, the synthesis of the core nucleus of FR-900482 was accomplished.[28] In the final stages of the synthetic effort, the removal of the formyl group from the C7 quaternary center was necessary. The authors chose the *Tsuji-Wilkinson decarbonylation protocol* to effect the transformation. The 1,3-diol functionality was protected as the acetonide prior to the decarbonylation. Usually the rate of decarbonylation is slowest for aldehydes that have the formyl group attached to a quaternary carbon, so it was necessary to use more than two equivalents of the catalyst to effect the decarbonylation at the reflux temperature of xylene.

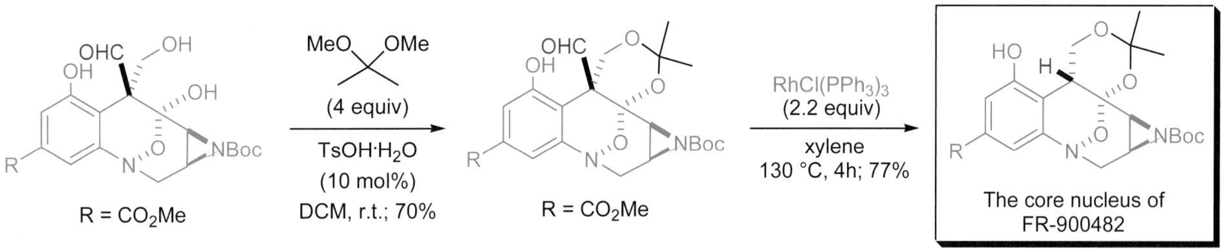

The research team of D.F. Covey developed a synthetic route to convert 5β-methyl-3-ketosteroids into 7(S)-methyl substituted analogues of neuroactive benz[*e*]indenes.[29] The synthesis began with 19-nortestosterone, in which the α,β-unsaturated cyclic ketone moiety was degraded to afford a tricyclic aldehyde. This aldehyde was unstable and could not be stored. For this reason it was immediately subjected to the *Tsuji-Wilkinson decarbonylation* to afford the decarbonylated product in high yield.

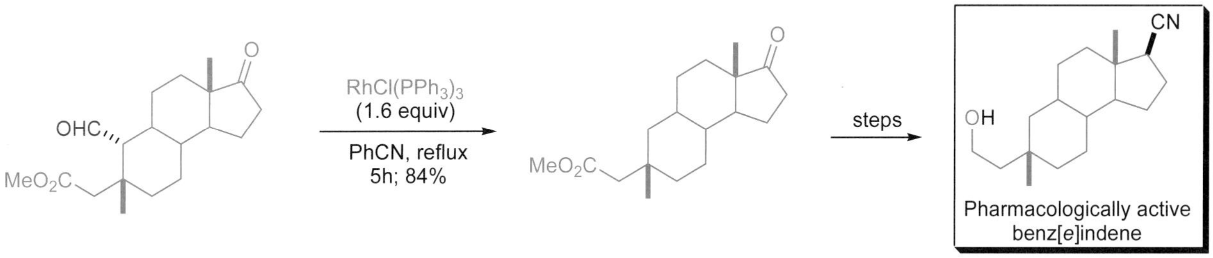

The total synthesis of (–)-gomisin J, a biologically active dibenzocyclooctane lignan, was completed by M. Tanaka and co-workers.[30] At the end of the synthesis, the removal of two aromatic formyl groups was needed. The exposure of the dialdehyde substrate to a little more than one equivalent of Wilkinson's catalyst and heating at reflux for two days afforded the deformylated product in excellent yield. The removal of the benzyl groups under catalytic hydrogenation conditions provided the natural product. Interestingly, the authors found that the decarbonylation could also be achieved *via* a *retro-Friedel-Crafts reaction*, which is a successful strategy only with electron-rich aromatic compounds.

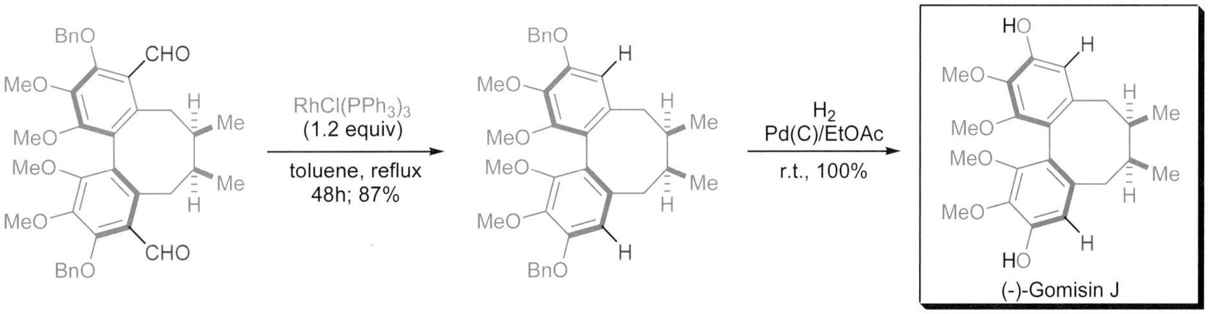

The isodaucane sesquiterpene (+)-aphanamol I was synthesized in the laboratory of B. Wickberg using the *DeMayo cycloaddition* as the key step.[31] The required starting material 3(*S*)-isopropyl-1-methylcyclopentene was prepared by the *Tsuji-Wilkinson decarbonylation* of the corresponding α,β-unsaturated aldehyde.

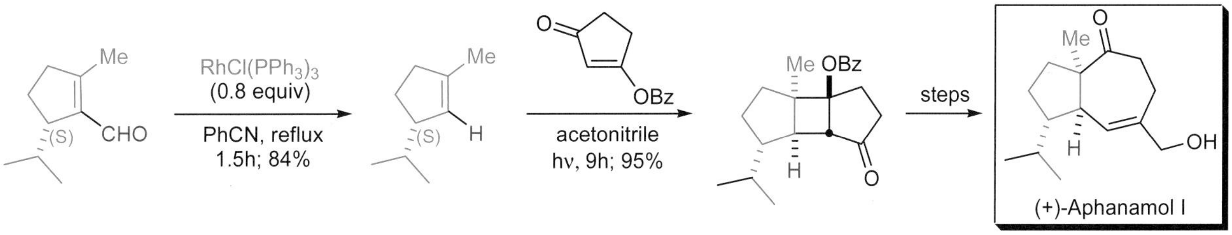

UGI MULTICOMPONENT REACTION

(References are on page 696)

Importance:

[*Seminal Publications*[1-3]; *Reviews*[4-35]; *Modifications & Improvements*[36-42,28,29,33,35]]

In 1959, I. Ugi reported that isocyanides undergo a four-component reaction (4-CR) in the presence of an amine, aldehyde or ketone and a nucleophile to provide a single condensation product.[1-3] The most commonly used nucleophiles are carboxylic acids, but hydrazoic acid, cyanates, thiocyanates, carbonic acid monoesters, salts of secondary amines, water, hydrogen sulfide, and hydrogen selenide can also be used.[1-3] Today, this transformation is referred to as the *Ugi four-component reaction* (U-4CR). The general features of the reaction are:[16] 1) it is very easy to carry out, usually, the isocyanide is added to a stirring and well cooled solution of the other three components; 2) in case of less reactive aldehydes and ketones, it is advisable to precondense the carbonyl compounds and the amine to form the imine; 3) as the reaction is very exothermic, adequate cooling is necessary; 4) methanol is generally a suitable solvent, although many other solvents can be used; 5) the reaction typically is carried out between -80 °C to 80 °C and it may take from a few minutes to a week to go to completion; 6) the amine component can be any compound with a sufficiently nucleophilic NH group such as ammonia, primary and secondary amines, hydrazine and derivatives,[36-38,40] diaziridines[42] as well as hydroxylamine;[39] 7) diarylamines are usually not nucleophilic enough to undergo the reaction; 8) with the exception of diarylketones, almost all aldehydes and ketones are suitable for the U-4CR; 9) a wide range of C-isocyanides undergo the transformation; and 10) when nonpolar solvents are used, or the reacting components are bulky, the *Passerini reaction* may occur as a side reaction leading to the formation of α-acyloxycarboxamides.[12] The Ugi reaction is a powerful synthetic transformation, where the four reaction partners are combined in one pot under mild conditions. One of the earliest and most important application of the U-4CR is peptide coupling and α-amino acid synthesis;[4,7-14,17-19,23] Several modifications of the original transformation leading to the formation of heterocyclic compounds were developed.[28,29,33,35] The *Ugi reaction* also found a widespread application in combinatorial chemistry, where the synthetic power of the reaction coupled with modern techniques allows the quick assembly of a large number of molecules from simple starting materials.[20,27,28,31-33]

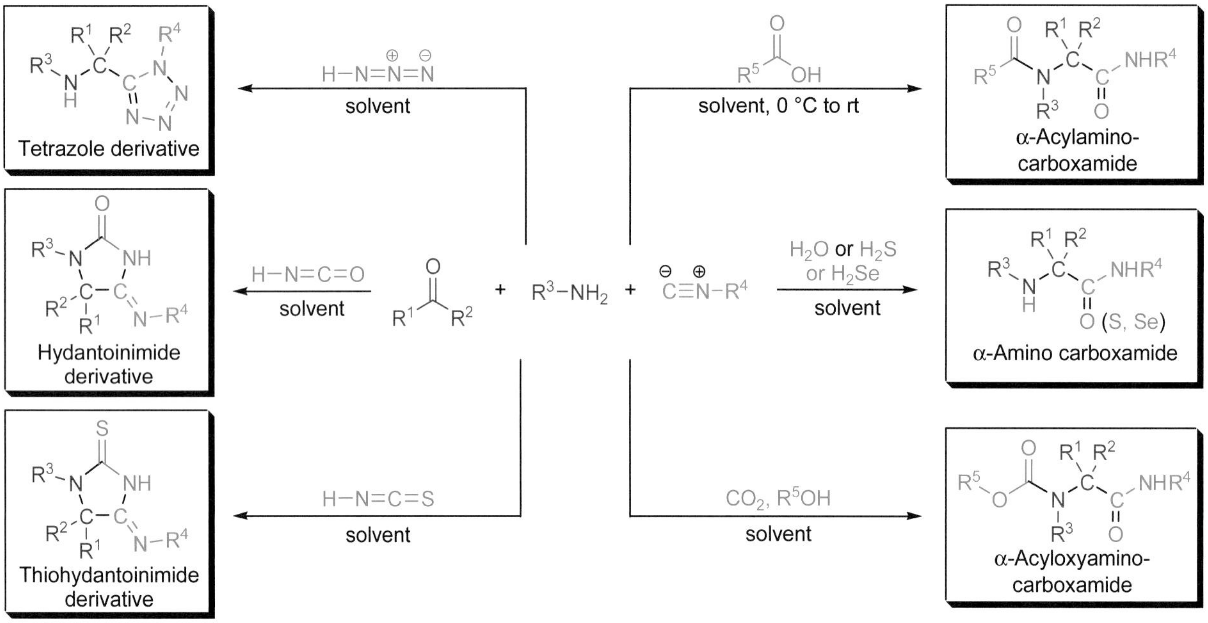

R^1 = alkyl, aryl; R^2 = H, alkyl; R^3 = alkyl, aryl; R^4 = alkyl, aryl; R^5 = alkyl, aryl; solvent: MeOH, EtOH, CF_3CH_2OH, DMF, $CHCl_3$, CH_2Cl_2, THF, dioxane, Et_2O

Mechanism:[43-45]

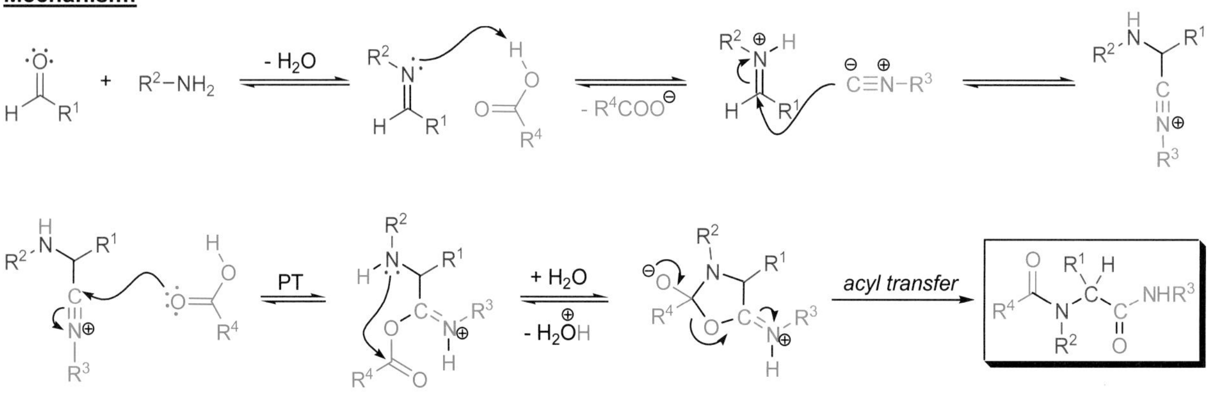

Synthetic Applications:

The potential application of the *Ugi four-component reaction* for amino acid and polypeptide natural product synthesis was recognized and utilized early on by M.M. Joullié.[46,47] A representative example is the total synthesis of (+)-furanomycin, a naturally occurring antibiotic. As the exact stereochemistry of the compound was not confirmed, total synthesis of the natural product and its stereoisomers was used to elucidate the stereochemistry.

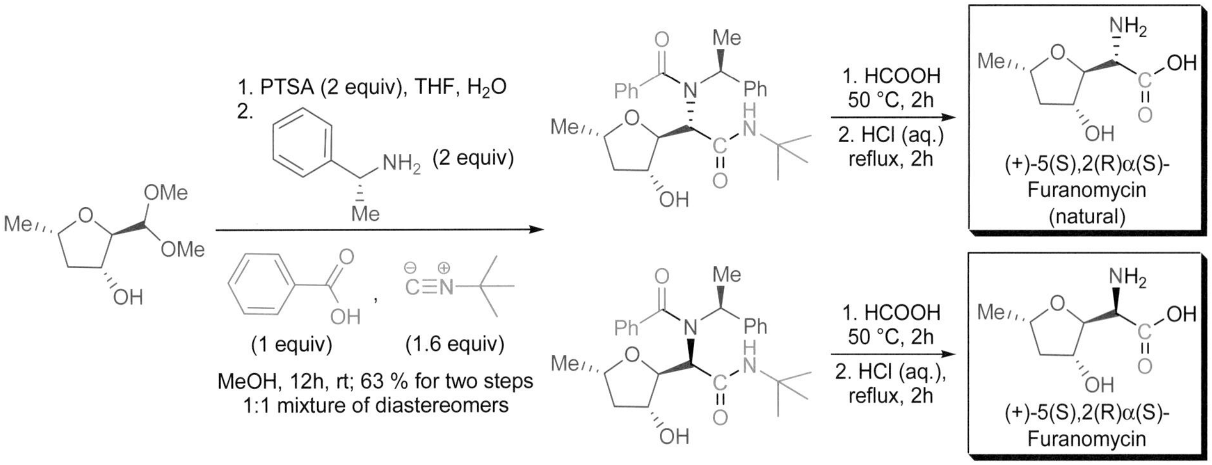

Ecteinascidin 743 is an extremely potent antitumor agent isolated from a marine tunicate. The total synthesis of this natural product was realized in the laboratory of T. Fukuyama.[48] To achieve the synthesis of the key dipeptide fragment, they utilized the *Ugi four-component reaction.* The transformation was carried out under mild conditions providing the product with excellent yield.

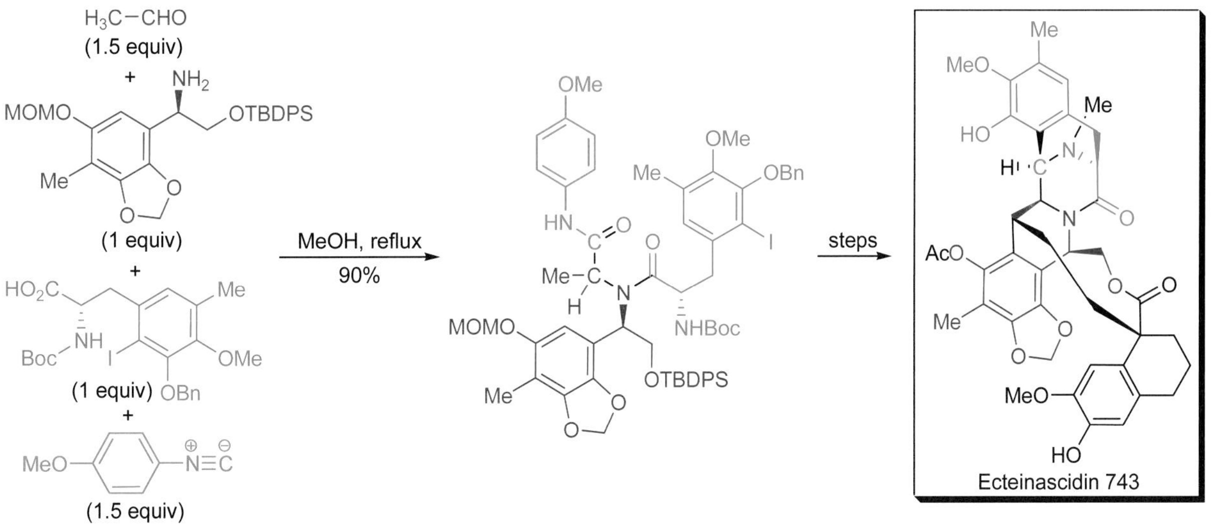

Ketopiperazines are biologically active molecules, they are antagonists of the platelet glycoprotein IIb-IIIa, and they exhibit hypocolesteremic activity.[49,50] The solution phase synthesis of ketopiperazine libraries was achieved by C. Hulme and co-workers using a *Ugi reaction/Boc-deprotection/cyclization* strategy. The four-component coupling was performed in methanol at room temperature. The deprotection and conversion of the enamide into the corresponding methyl ester was effected by acetyl chloride in methanol. Subsequent cyclization in the presence of diethylamine in dichloromethane provided the products with a 30-97% yield for the overall process. A representative ketopiperazine product is shown below.

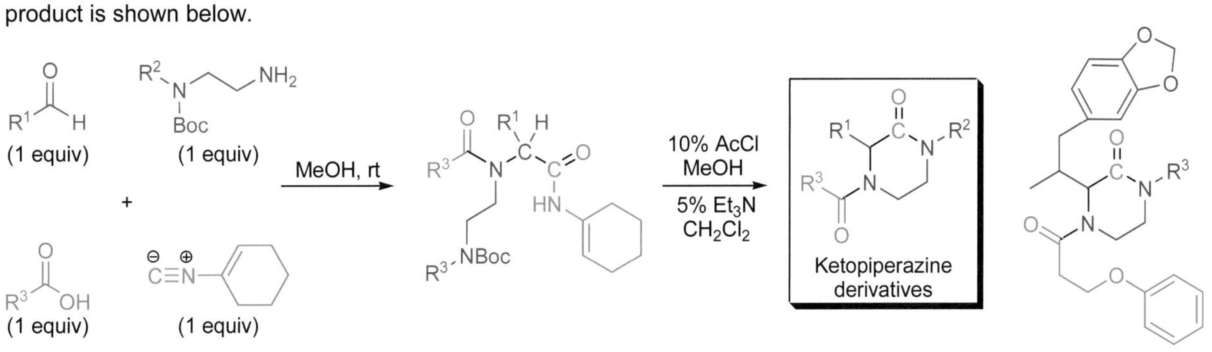

ULLMANN BIARYL ETHER AND BIARYL AMINE SYNTHESIS / CONDENSATION

(References are on page 697)

Importance:

[*Seminal Publications*[1-4]; *Reviews*[5-11]; *Modifications & Improvements*[12-46]]

In 1904, F. Ullmann observed that the reaction of aryl halides with phenols to give biaryl ethers was significantly improved in the presence of copper powder.[2] The copper mediated synthesis of biaryl ethers is known as the *Ullmann condensation* (*Ullmann biaryl ether synthesis*). In 1906, I. Goldberg disclosed the copper-mediated formation of an arylamine by reacting an aryl halide with an amide in the presence of K_2CO_3/CuI (*Goldberg reaction/Goldberg modified Ullmann condensation*). The general features of the *Ullmann condensation* are: 1) aryl iodides, bromides, and chlorides are all good substrates with the following reactivity trend: I > Br > Cl >> F (the opposite trend is observed in uncatalyzed S_NAr reactions); 2) aryl fluorides usually do not react under the reaction conditions; 3) the introduction of several aryloxy groups is possible in a stepwise manner; 4) the aromatic halide can contain many different substituents and even reactive functional groups (e.g., OH, NH_2, CHO) need not be protected unlike in the *Ullmann biaryl coupling*; 5) electron-withdrawing substituents (e.g., NO_2, CO_2R, COO^-) in the *ortho* and *para* positions have a marked activating effect and the yields for these substrates are excellent; 6) electron-donating substituents anywhere on the aromatic ring do not significantly decrease the reactivity of the aryl halide compared to the unsubstituted aryl halide; 7) the required temperature ranges from 100 to 300 °C in the presence of copper metal or a copper-derived catalyst and with or without the use of solvents; 8) the catalytic activity of the copper depends on the method of preparation; 9) a wide variety of solvents work well and most of them contain a heteroatom with a lone pair of electrons; 10) the solvent helps to solubilize the catalytically active copper species by way of complexation; 11) the phenol component can be introduced in the form of free phenols or phenolate salts; 12) when free phenols are used, a base (K_2CO_3) is added to the reaction mixture, but other salts proved to be ineffective; 13) if Cu_2O or CuO is used instead of copper, no base is required, since these substances serve as bases; and 14) since phenols and phenolates are sensitive to oxidation, the use of an inert atmosphere is often required. There are few typical side reactions of the aryl halide component: 1) *reductive dehalogenation* especially when the phenol is relatively unreactive; 2) *Ullmann biaryl homocoupling*; and 3) exchange of halogens with the Cu(I)-salt. Several modifications have been introduced to improve the somewhat harsh original reaction conditions (high temperatures, often low yields and the use of stoichiometric amounts of copper), which primarily utilize coupling partners other than aryl halides: 1) arylboronic acids in the presence of Et_3N, molecular sieves and $Cu(OAc)_2$ (*Chan-Evans-Lam modification*);[23-25] 2) potassium aryltrifluoroborates (*Batey modification*);[42,43] 3) aryl iodonium salts (*Beringer-Kang modification*);[12,29] 4) aryl lead compounds (*Barton plumbane modification*);[17] and 5) aryl bismuth compounds (*Barton modification*).[15,16,18]

Biaryl ether and amine synthesis (Ullmann 1903 & Goldberg 1906):

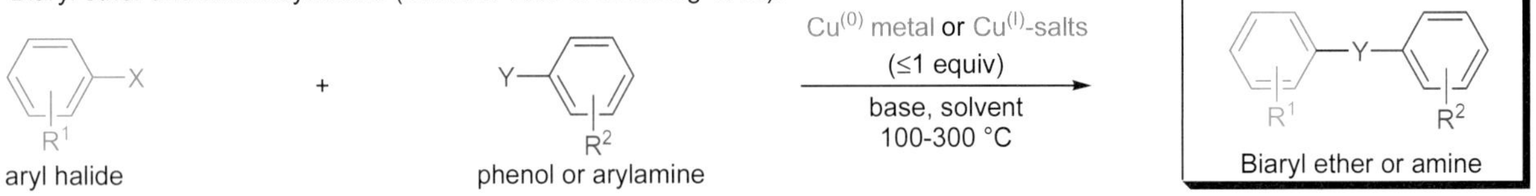

Modified Ullmann biaryl ether / thioether and biaryl amine synthesis:

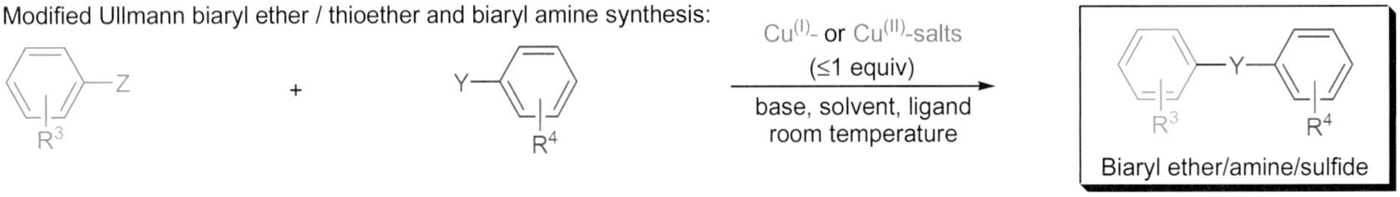

R^{1-4} = H, CN, NO_2, CO_2R, I, Br, Cl, I; X = I, Br, Cl, SCN; Y = OH, NH_2, NHR, NHCOR; <u>solvent:</u> DMF, pyridine, quinoline, DMSO, nitrobenzene, glycol, diglyme, dioxane; <u>base:</u> K_2CO_3, Et_3N, pyridine; $Cu^{(I)}$- and $Cu^{(II)}$-salts: CuI, Cu_2O, $Cu(OAc)_2$; <u>ligand</u>: diamines When Y = NH_2, OH, SH and Z = $B(OH)_2$ (*Chan-Evans-Lam modification*), Z = BF_3K (*Batey mod.*), Z = $Si(OMe)_3$ or $Sn(alkyl)_3$ (*Lam mod.*), Z = $(I\text{-aryl})^+BF_4^-$ (*Beringer-Kang mod.*), Z = $Pb(OAc)_3$ (*Barton plumbane mod.*), Z = $BiPh_2X_2$ (*Barton mod.*)

Mechanism: [47,16,48,24,49,10]

The exact nature (oxidation state) of the Cu-intermediate is not known, but radical mechanisms have been ruled out based on radical scavenger experiments. Two possible (speculated) pathways are shown.

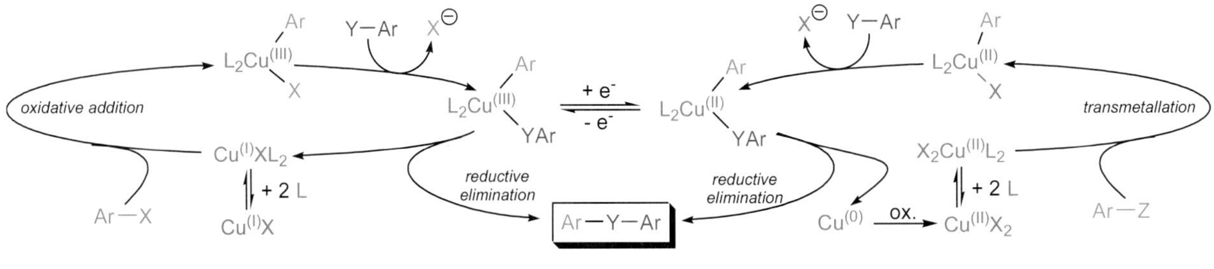

Synthetic Applications:

The *intramolecular Ullmann condensation* was used by D.L. Boger and co-workers to form the 15-membered macrocyclic ring of the cytotoxic natural product, combretastatin D-2.[50] This compound possesses unusual meta- and paracyclophane subunits, which are also found in a range of antitumor antibiotics. The first approach where the final step was a *macrolactonization* was unsuccessful, so the researchers chose to form the biaryl ether moiety as the key macrocyclization step. Methylcopper was found to mediate the cyclization and gave moderate yield of the corresponding biaryl ether. Finally *boron triiodide mediated demethylation* afforded the natural product.

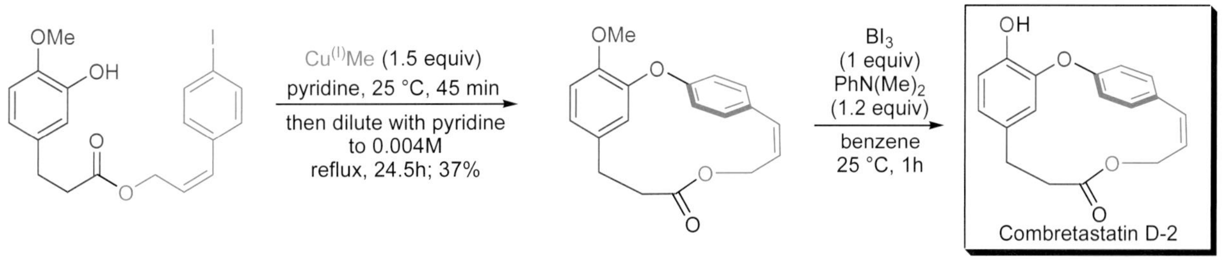

The highly oxygenated antifungal/anticancer natural product (±)-diepoxin σ was prepared in the laboratory of P. Wipf.[51] The coupling of the two substituted naphthalene rings was achieved *via* the *Ullmann condensation* of a phenolic compound with 1-iodo-8-methoxynaphthalene. The aryl iodide coupling partner was used in excess and the condensation was conducted in refluxing pyridine in the presence of a full equivalent of copper(I)-oxide.

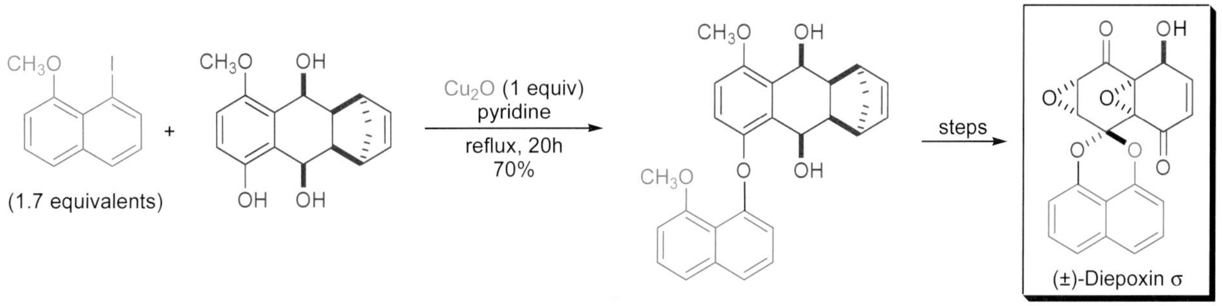

In the laboratory of K.C. Nicolaou, a novel mild method for the preparation of biaryl ethers was developed.[22] The di-*ortho*-halogenated aromatic triazenes underwent efficient coupling with phenols in the presence of CuBr. This mild *modified Ullmann condensation* was utilized in the synthesis of the DOE and COD model ring systems of vancomycin.

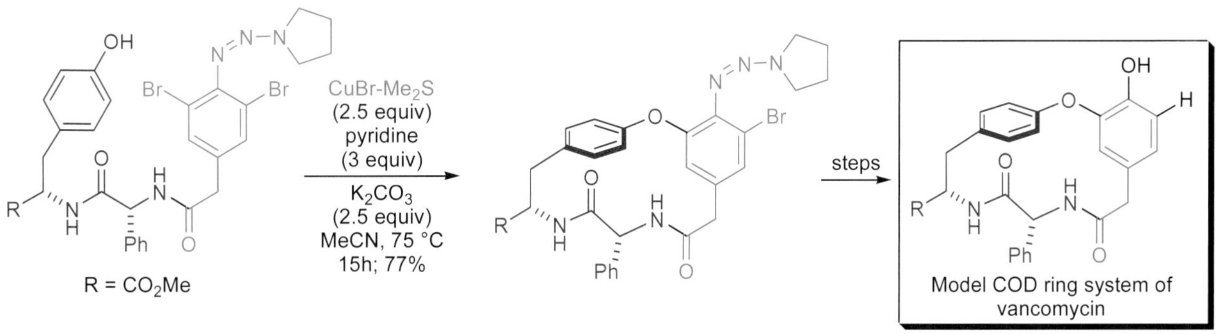

The *Ullmann biaryl amine condensation* was used in the synthesis of SB-214857, a GPIIb/IIIa receptor antagonist.[52] D. Ma and co-workers coupled aryl halides with β-amino acids and esters under relatively mild conditions using CuI as a true catalyst.

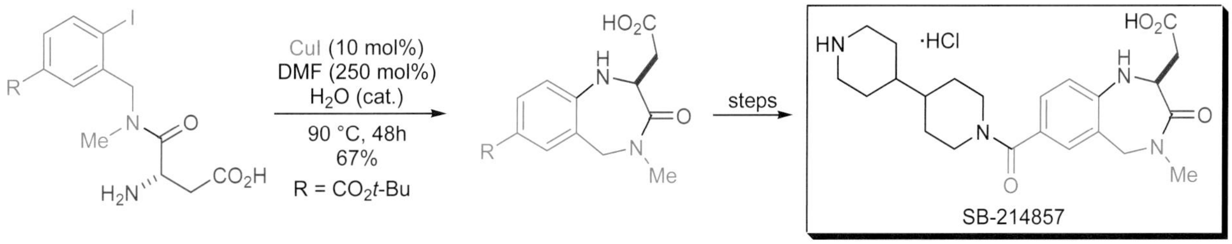

ULLMANN REACTION / COUPLING / BIARYL SYNTHESIS

(References are on page 699)

Importance:

[*Seminal Publications*[1,2]; *Reviews*[3-9]; *Modifications & Improvements*[10-21]]

In 1901, F. Ullmann reported the reaction of two equivalents of an aryl halide with one equivalent of finely divided copper at high temperature (>200 °C) to afford a symmetrical biaryl and copper halide.[1] This condensation of two aryl halides in the presence of copper to give symmetrical or unsymmetrical biaryls is now referred to as the *Ullmann reaction* (*Ullmann biaryl synthesis* or *Ullmann coupling*). Since its discovery, the *Ullmann reaction* has become a general method for the synthesis of numerous symmetrical and unsymmetrical biaryls. The general features of this reaction are: 1) halogenated benzene rings as well as halogenated heteroaromatic compounds are substrates for the coupling; 2) the order of reactivity is I > Br >> Cl, but aromatic fluorides are totally inert; 3) the reaction can take place both inter- and intramolecularly and has been used to form macrocycles (4- to 24-membered rings);[6] 4) electron-withdrawing groups (e.g., NO_2, CO_2Me, CHO) *ortho* to the halogen substituent increase the reactivity of the aryl halide; 5) generally substituents in the *ortho* position, which have a lone pair increase the reactivity regardless whether they are EWG or EDG, but these substituents have no noticeable activating effect in the *meta* or *para* positions;[22] 6) substrates that are very electron rich (e.g., multiple alkyl or alkoxy groups) tend to give lower yield of the biaryl; 7) certain unprotected functional groups (e.g., OH, NH_2, CO_2H, SO_2NH_2) open alternative reaction pathways therefore inhibit the coupling;[23] 8) bulky groups located *ortho* to the halogen tend to retard or inhibit the coupling reaction due to steric hindrance; 9) when unsymmetrical biaryls are prepared, the highest yield is obtained when one of the aryl halides is activated (more electron rich), while the other is less reactive; 10) in order to achieve good results, activated copper (preferably prepared prior to use) must be used;[17] 11) highly active copper metal can be prepared by reducing CuI with lithium naphthalenide or by reducing $CuSO_4$ with Zn powder; 12) usually temperatures over 100 °C are necessary to initiate the coupling but the use of highly active Cu-powder allows lower temperatures; 13) the most common solvent is DMF, but for higher temperatures $PhNO_2$ or *p*-$NO_2C_6H_4CH_3$ are used;[10,11] 14) sonication often improves the efficiency of the coupling;[18,19] 15) Cu(I)-salts (e.g., Cu_2O, Cu_2S) also mediate the coupling although they are less active than the activated copper metal;[12] and 16) Cu(I) thiophene 2-carboxylate (CuTC) was found to be an efficient mediator under mild conditions (usually room temperature) in NMP.[21] There are a few modifications: 1) the reaction conditions of the *Ullmann coupling* become significantly milder when $Ni^{(0)}$ complexes are used in place of copper metal;[13,9] and 2) for the preparation of highly substituted biaryls the use of preformed aryl copper species has been successful (*Ziegler modification*).[16,20]

Synthesis of symmetrical biaryls (Ullmann, 1901):

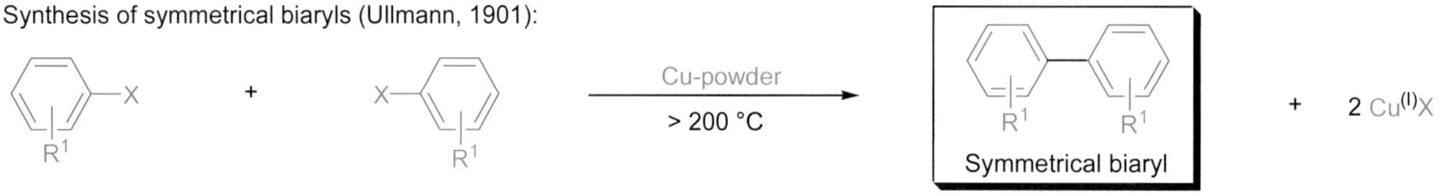

Synthesis of unsymmetrical biaryls:

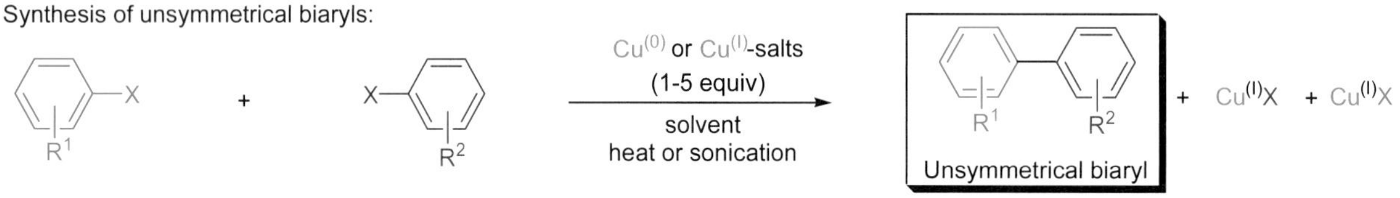

R^1, R^2 = H, CN, NO_2, CO_2R, I, Br, Cl; X = I, Br, Cl, SCN; solvent: DMF, pyridine, quinoline, nitrobenzene, *p*-nitro toluene

Mechanism: [24-26,14,27-32,9]

The exact mechanistic pathway of the *Ullmann coupling* is not known. There are two main pathways possible: 1) formation of aryl radicals or 2) the formation of aryl copper [$ArCu^{(I)}$, $ArCu^{(II)}$ and $ArCu^{(III)}$] intermediates. Currently the most widely accepted mechanism assumes the formation of aryl copper intermediates, since many of these species can be isolated and they can react with aryl halides to give biaryls.

Pathway involving aryl radicals:

Step #1: $Ar{-}X + Cu^{(0)} \longrightarrow [Ar{-}X]^{\bullet\ominus} + Cu^{(I)}$

Step #2: $[Ar{-}X]^{\bullet\ominus} + Cu^{(I)} \longrightarrow Ar\bullet + Cu^{(I)}X$

Step #3: $Ar\bullet + Ar\bullet \longrightarrow \boxed{Ar{-}Ar}$

Pathway involving arylcopper intermediates:

Step #1: $Ar{-}X + Cu^{(0)} \longrightarrow Ar{-}Cu^{(II)}X$

Step #2: $Ar{-}Cu^{(II)}X + Cu^{(0)} \longrightarrow Ar{-}Cu^{(I)} + Cu^{(I)}X$

Step #3: $Ar{-}Cu^{(I)} + Ar{-}X \longrightarrow Ar{-}Cu^{(III)}XAr$

Step #4: $Ar{-}Cu^{(III)}XAr \longrightarrow \boxed{Ar{-}Ar} + Cu^{(I)}X$

Synthetic Applications:

The *Ziegler-modified Ullmann reaction* was used for the total synthesis of pyrrolophenanthridinium alkaloid tortuosine by L.A. Flippin and co-workers.[33] First, *N*-Boc-5-methoxyindoline was lithiated at C7 with *s*-BuLi in the presence of TMEDA, and then it was transmetallated to the corresponding organocopper species that smoothly underwent the *Ullmann reaction* with a 3-iodoaryl imine. The resulting biaryl product was treated with anhydrous HCl in chloroform, which promoted the cyclization followed by dehydration to give the natural product.

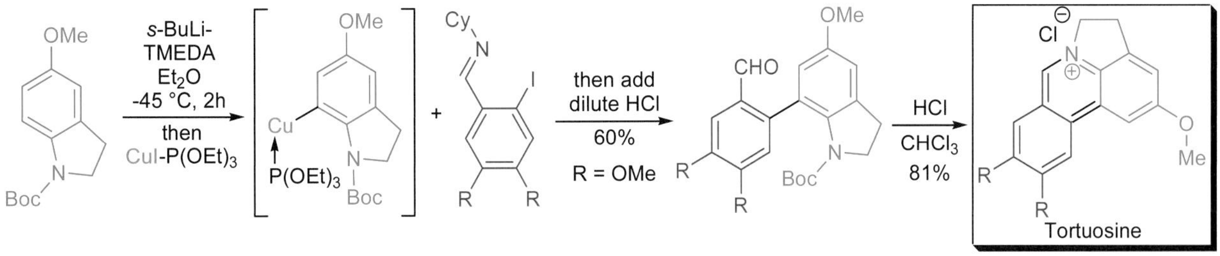

In the laboratory of A.I. Meyers, the oxazoline-mediated *asymmetric Ullmann coupling* was utilized to establish the chirality about the biaryl axis of mastigophorenes A and B.[34] The key coupling step was conducted in DMF in two stages: first the reaction mixture (0.66M) containing freshly prepared activated Cu-powder was heated at 95 °C for 8h, and then it was diluted with DMF (0.11M) and refluxed for 3 days. Interestingly, during these studies it was revealed that smaller chiral auxiliaries lead to higher atroposelection, a fact which was not previously recognized.

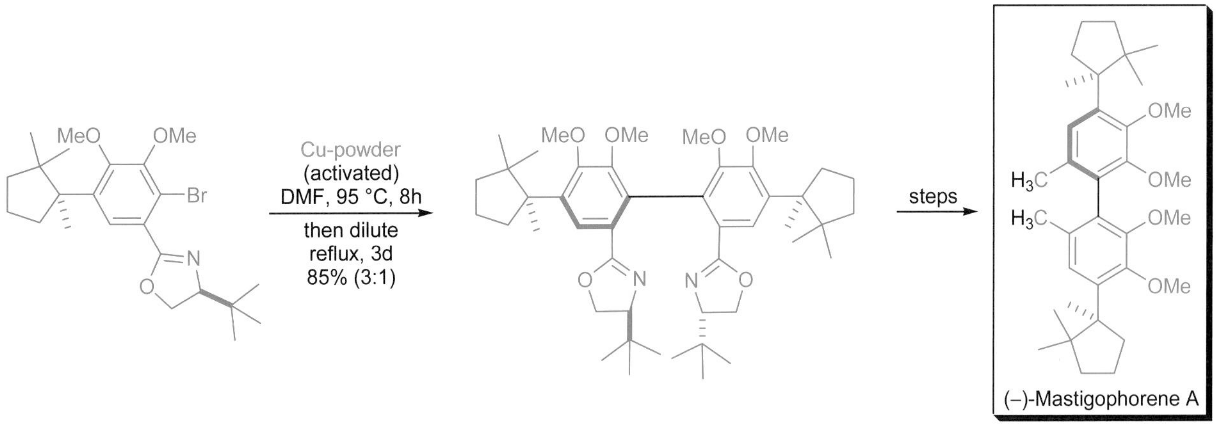

The first total synthesis of taspine was accomplished by T.R. Kelly and co-workers.[35] The central biaryl link was established by a classical *Ullmann coupling* using activated copper bronze. It is noteworthy that no other cross-coupling strategy was successful to make the C-C bond between the aromatic rings due to the severe steric hindrance.

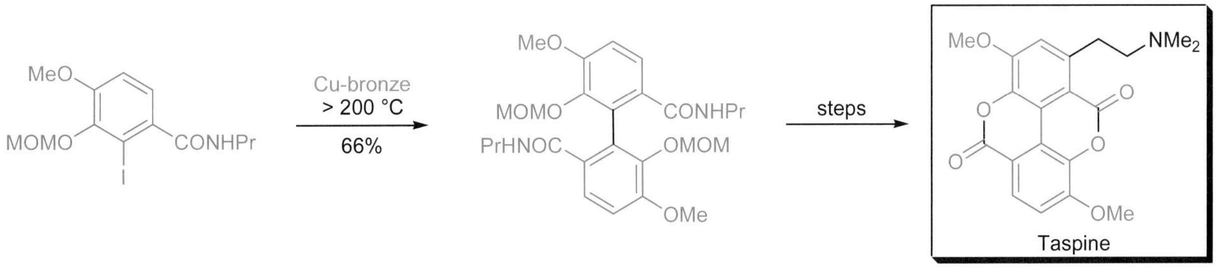

L.S. Liebeskind et al. demonstrated that CuTC could be efficiently used to mediate the *Ullmann reaction* at room temperature under very mild conditions tolerating a wide variety of functional groups.[21] One of the examples features an intramolecular process while the other demonstrates the coupling of halogenated heteroaromatics.

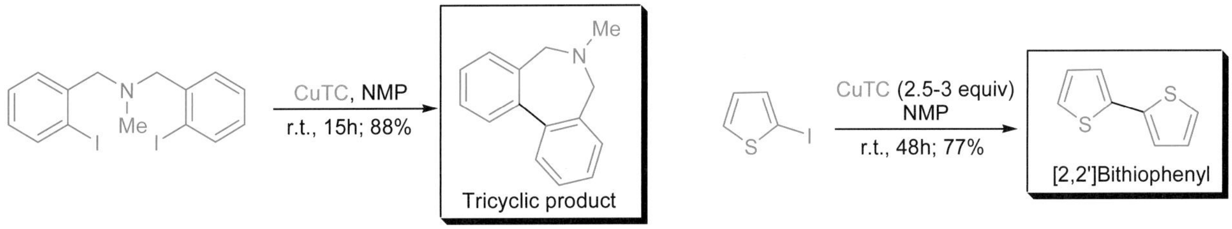

VILSMEIER-HAACK FORMYLATION

(References are on page 699)

Importance:

[*Seminal Publications*[1,2]; *Reviews*[3-16]; *Modifications & Improvements*[17-30]; *Theoretical Studies*[31-33]]

In 1925, A. Vilsmeier and co-workers reported that upon treatment with phosphoryl chloride ($POCl_3$), *N*-methylacetanilide gave rise to a mixture of products among which 4-chloro-1,2-dimethylquinolinium chloride was one of the major products.[1] Further investigation revealed that the reaction between *N*-methylformanilide and $POCl_3$ gave rise to a chloromethyliminium salt (Vilsmeier reagent), which readily reacts with electron-rich aromatic compounds to yield substituted benzaldehydes.[2] The introduction of a formyl group into electron-rich aromatic compounds using a Vilsmeier reagent is known as the *Vilsmeier-Haack formylation* (*Vilsmeier reaction*). The general features of this transformation are:[8,11] 1) the Vilsmeier reagent is prepared from any *N,N*-disubstituted formamide by reacting it with an acid chloride (e.g., $POCl_3$, $SOCl_2$, oxalyl chloride); 2) most often the combination of DMF and $POCl_3$ is used and the resulting Vilsmeier reagent is usually isolated before use; 3) mostly electron-rich aromatic or heteroaromatic compounds[8] as well as electron-rich alkenes and 1,3-dienes[11] are substrates for the transformation, since the Vilsmeier reagent is a weak electrophile; 4) the relative reactivity of five-membered heterocycles is pyrrole > furan > thiophene; 5) the solvent is usually a halogenated hydrocarbon, DMF or $POCl_3$ and the nature of the solvent has a profound effect on the electrophilicity of the reagent, so it should be carefully chosen; 6) the required reaction temperature varies widely depending on the reactivity of the substrate and it ranges from below 0 °C up to 80 °C; 7) the initial product is an iminium salt, which can be hydrolyzed with water to the corresponding aldehyde, treated with H_2S to afford thioaldehydes, reacted with hydroxylamine to afford nitriles, or reduced to give amines; 8) the transformation is regioselective favoring the less sterically hindered position (this means the *para* position on a substituted benzene ring); but electronic effects can also influence the product distribution; and 9) vinylogous chloromethyliminium salts undergo similar reaction to afford the corresponding α,β-unsaturated carbonyl compounds upon hydrolysis.

Vilsmeier (1925):

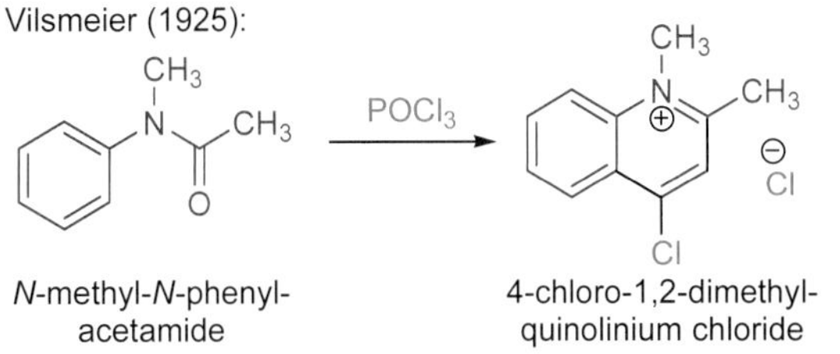

Vilsmeier (1927):

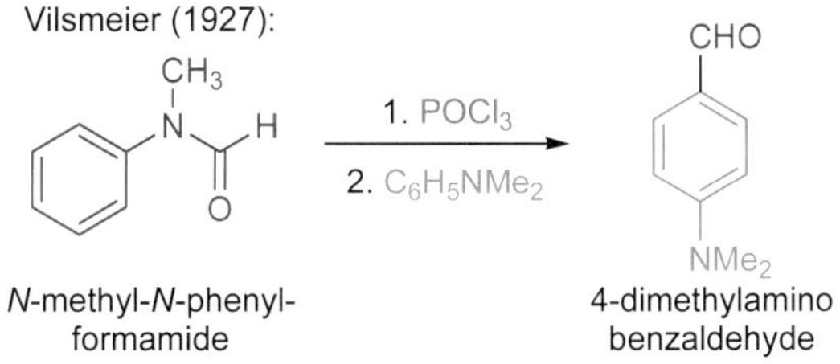

Vilsmeier-Haack formylation:

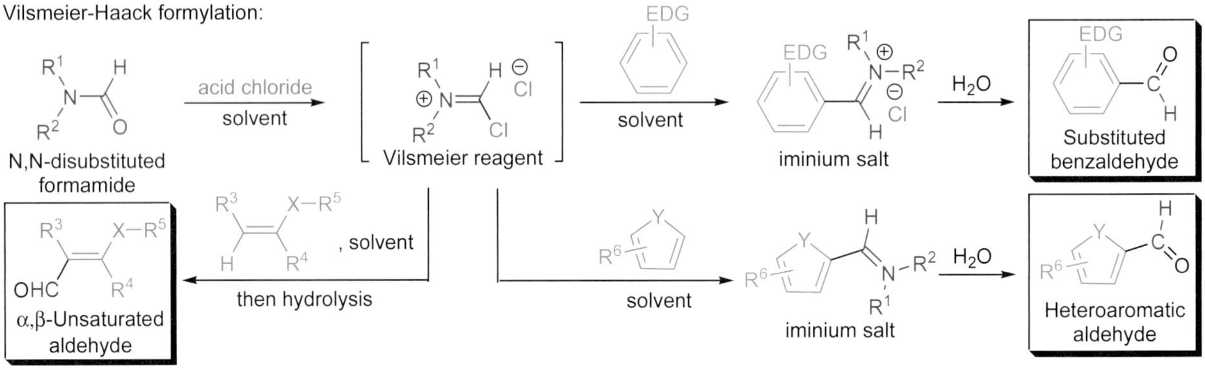

R^{1-2} = alkyl, aryl; acid chloride: $POCl_3$, $SOCl_2$, $COCl_2$, $(COCl)_2$, Ph_3PBr_2, 2,4,6-trichloro-1,3,5-triazine; solvent: DCM, DMF, $POCl_3$; EDG = OH, *O*-alkyl, *O*-aryl, NR_2; R^{3-4} = H, alkyl, aryl; R^5 = alkyl, aryl; X = O, NR, CH_2, CR_2; Y = O, S, NR, NH; R^6 = H, alkyl, aryl

Mechanism: [34-41,8,42,11]

Formation of the Vilsmeier reagent (an equilibrium mixture of iminium salts):

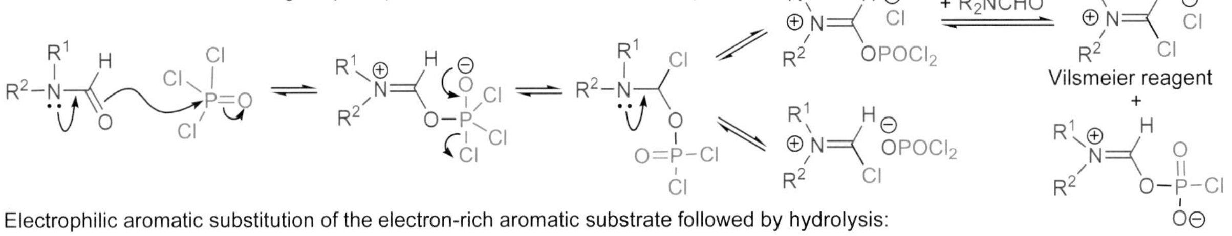

Electrophilic aromatic substitution of the electron-rich aromatic substrate followed by hydrolysis:

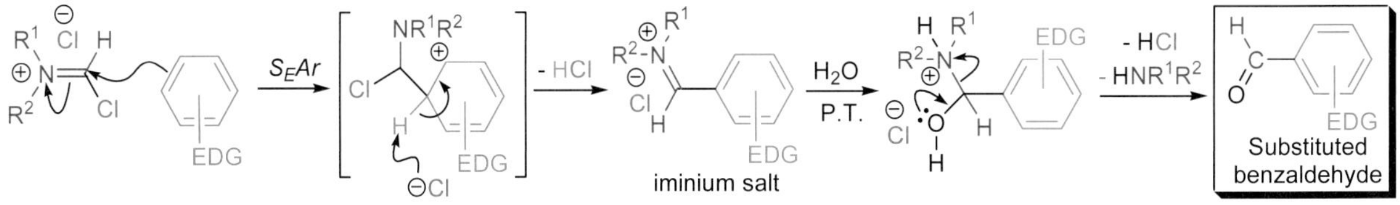

Synthetic Applications:

The total synthesis of the calophylium coumarin (–)-calanolide A was accomplished by D.C. Baker and co-workers.[43] This compound attracted considerable attention because it is a potent inhibitor of HIV-1 reverse transcriptase. In order to introduce a formyl group at C8, a regioselective *Vilsmeier reaction* was employed on a coumarin lactone substrate.

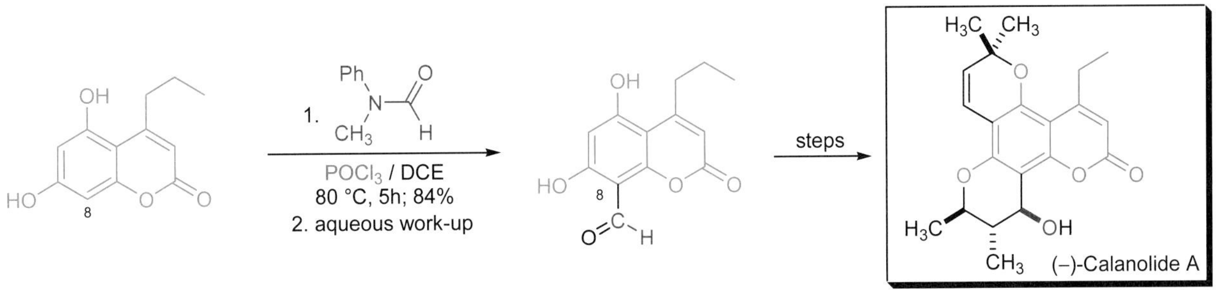

In the laboratory of F.E. Ziegler, the cyclization of a chiral aziridinyl radical into an indole nucleus was utilized to prepare the core nucleus of the potent antitumor agent FR-900482.[44] In the early stages of the synthetic effort, the *Vilsmeier-Haack formylation* was chosen to install an aldehyde functionality at the C3 position of a substituted indole substrate. The initial iminium salt was hydrolyzed under very mildly basic conditions to minimize the hydrolysis of the methyl ester moiety. Eventually the formyl group was removed from the molecule *via decarbonylation* using Wilkinson's catalyst.

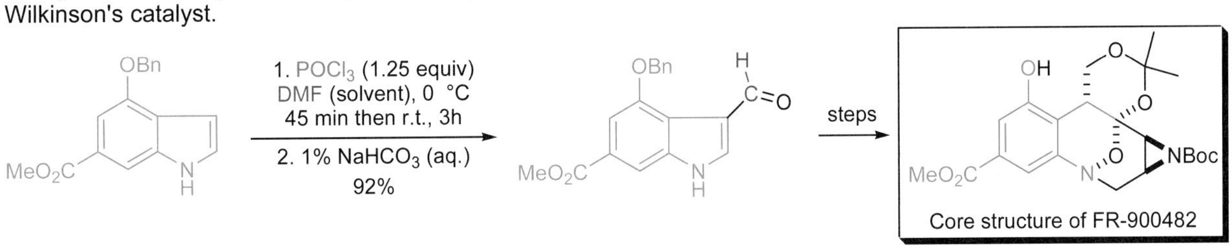

Since the *Vilsmeier-Haack formylation* is feasible on electron-rich alkenes such as enol ethers, it was a method of choice to prepare an α,β-unsaturated aldehyde during the total synthesis of (±)-illudin C by R.L. Funk et al.[45] The TES enol ether was treated with several reagent combinations (e.g., PBr_3/DMF/DCM), but unfortunately only regioisomeric product mixtures were obtained. However, the use of $POBr_3$/DMF/DCM allowed the clean preparation of the desired aldehyde regioisomer in good yield.

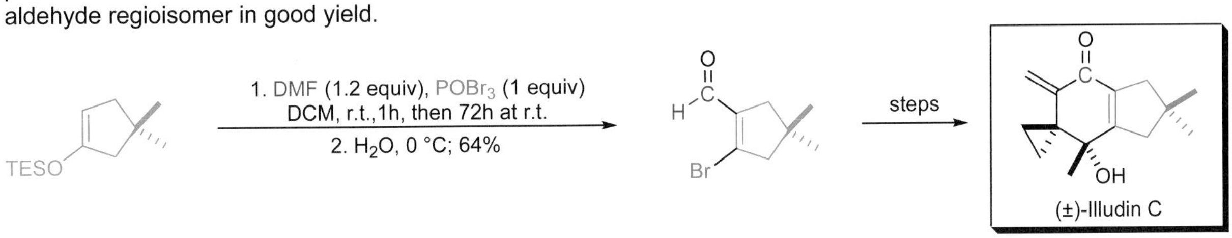

The marine sponge pigment homofascaplysin C was synthesized by the research team of G.W. Gribble.[46] The natural product had a novel 12*H*-pyrido[1,2-a:3,4-b']diindole ring system and a formyl group at the C13 position. The *Vilsmeier reaction* allowed the introduction of this substituent in excellent yield.

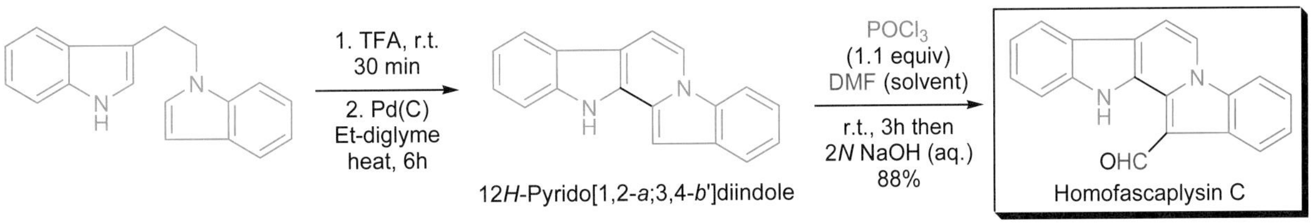

The total synthesis of (–)-(*R*)-MEM-protected arthrographol was accomplished by G.L.D. Krupadanam et al.[47] The authors used sequential *Vilsmeier reaction/Dakin oxidation* to prepare a 1,2,4-trihydroxybenzene derivative.

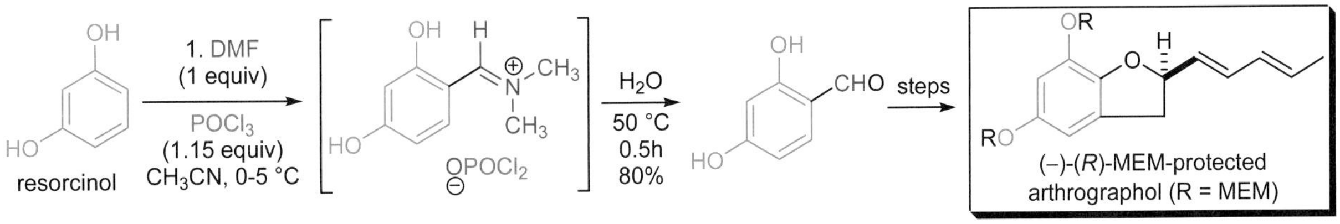

VINYLCYCLOPROPANE-CYCLOPENTENE REARRANGEMENT

(References are on page 700)

Importance:

[*Seminal Publications*[1-3]; *Reviews*[4-10]; *Modifications & Improvements*[9,11,12]; *Theoretical Studies*[13-30]]

In 1959, N.P. Neureiter investigated the reactivity of 1,1-dichloro-2-vinylcyclopropane, which he prepared by the addition of dichlorocarbene to 1,3-butadiene.[1] Surprisingly, this compound was very stable and was recovered intact after being exposed to a variety of oxidizing and reducing agents. However, under flash vacuum thermolysis conditions it cleanly underwent a rearrangement to afford a mixture of five-membered chloroolefins. A year later, C.G. Overberger and A.E. Borchert reported a novel thermal rearrangement during the acetate pyrolysis of 2-cyclopropyl ethyl acetate, which yielded cyclopentene as the major product. The transformation of substituted vinylcyclopropanes to the corresponding substituted cyclopentenes is known as the *vinylcyclopropane-cyclopentene rearrangement*. The general features of the reaction are:[4-10] 1) thermal-, photochemical-, transition metal-mediated, as well as Lewis acid-mediated conditions can be applied to affect the transformation; 2) the photochemical process works well only for a limited number and type of substrates and is mainly of mechanistic interest; 3) the rearrangement of vinylcyclopropanes under thermal conditions is the most important transformation and it may take two major pathways: conversion to cyclopentenes or formation of open-chain alkenes or dienes; 4) the pathway taken depends on many factors such as the nature of substituents on the cyclopropane ring as well as the orientation of the π-system of the vinyl group relative to the cyclopropane ring (e.g., *cis*-alkylvinylcyclopropanes tend to undergo *[1,5]-sigmatropic H-shift* (*retro-ene reaction*) rather than forming cyclopentenes); 5) the rearrangement usually requires high temperatures (often this means running the reaction in a flash vacuum pyrolysis apparatus), but the degree of substitution and the presence of extended conjugation and heteroatoms lower the activation energy and also the required temperature; 6) heteroatom substitution (e.g., *O*-alkyl, NH_2, *S*-alkyl, etc.) on the cyclopropane moiety has a dramatic activation energy-lowering effect, whereas substitution on the vinylic moiety does not have a significant influence; 7) the rearrangement can be highly regio- and stereoselective provided that the cyclopropane is opened regioselectively; 8) predictions can be made regarding which cyclopropane bond is cleaved preferentially and the prediction is based on the donor/acceptor properties of the various substituents on the cyclopropane ring; and 9) the stereochemical outcome of the rearrangement is determined by the energetics of the substituted cyclopentene product.

Neureiter (1959):

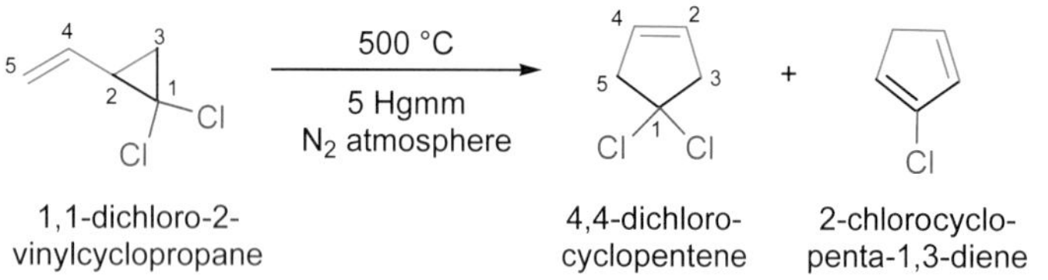

Overberger & Borchert (1960):

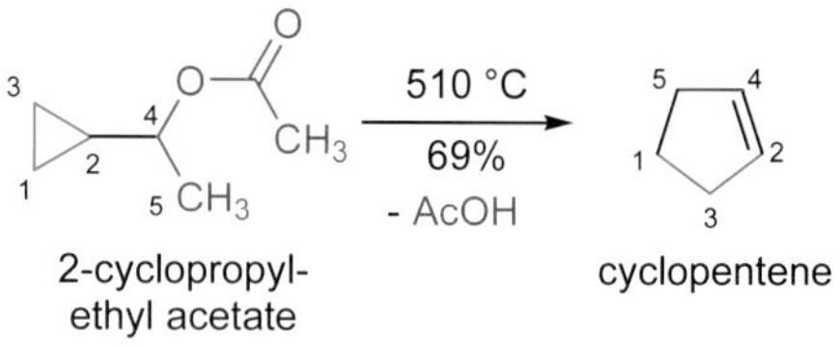

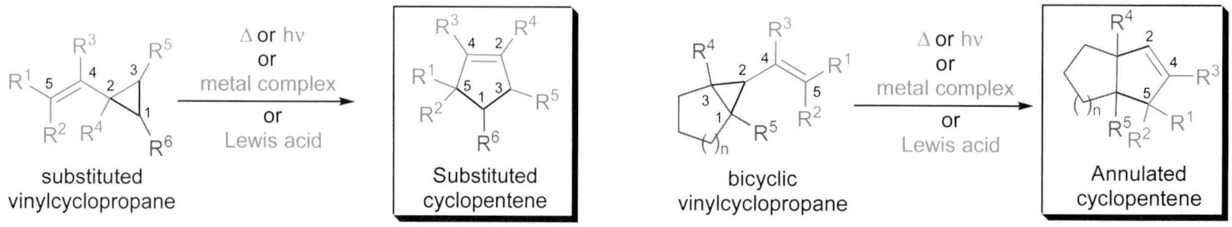

R^{1-5}, R^6 = H, alkyl, alkenyl, aryl, O-alkyl, NH_2, NH-alkyl, NR_2; n = 1-3

Mechanism:[31-61]

Cyclopentene formation *via* biradical intermediates:

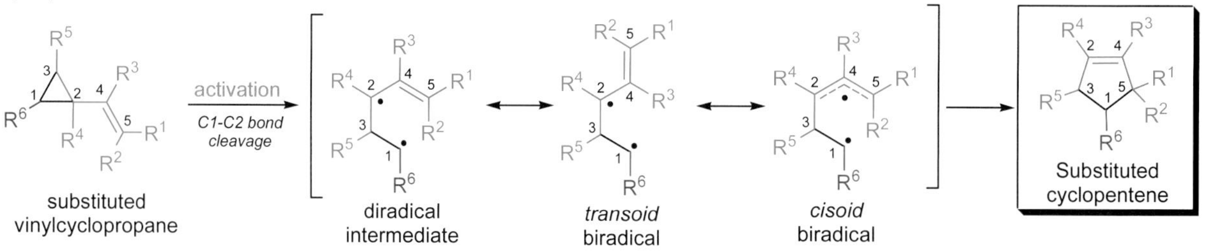

Competing retro-ene reaction (reversible process):

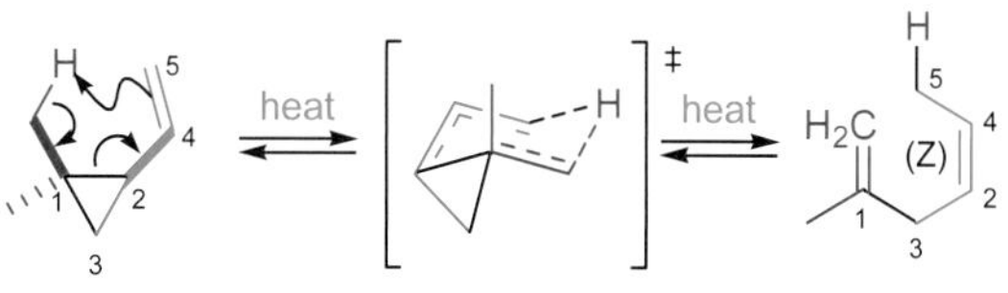

Cyclopentene formation *via* dipolar intermediates:

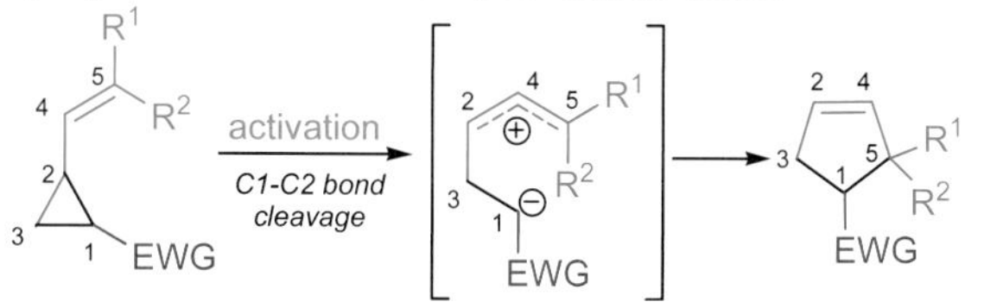

Synthetic Applications:

In the laboratory of H.R. Sonawane, both enantiomers of $\Delta^{9(12)}$-capnellene were prepared using the *photoinduced vinylcyclopropane-cyclopentene rearrangement*.[62] The conversion of (+)-Δ^3-carene to the corresponding enantiopure allylic alcohol was achieved by a two-step sequence of a *Prilezhaev reaction* and *base-induced epoxide ring-opening*. The photochemical rearrangement of the *cis*-alkyl vinylcyclopropane intermediate proceeded without the occurrence of the competing *retro-ene reaction* and gave rise to a diastereomeric mixture of cyclopentene-annulated products.

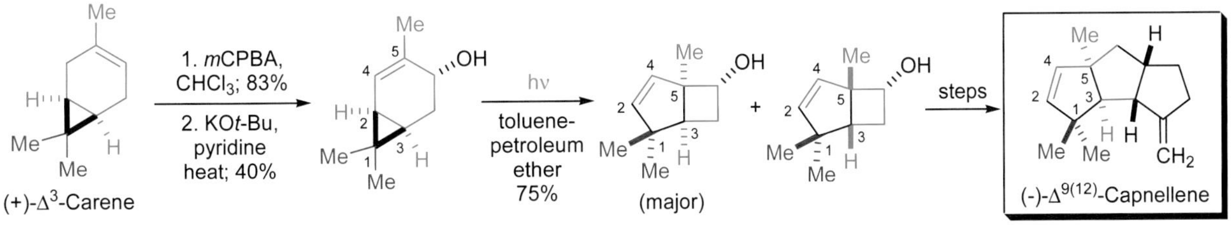

The enantioselective total synthesis of (+)-antheridic acid was accomplished by E.J. Corey and co-workers using the *Lewis-acid-mediated vinylcyclopropane-cyclopentene rearrangement* as the key step.[63] This key transformation was not possible under thermal conditions; however, the use of excess diethylaluminum chloride in DCM gave rise to the rearranged product in excellent yield.

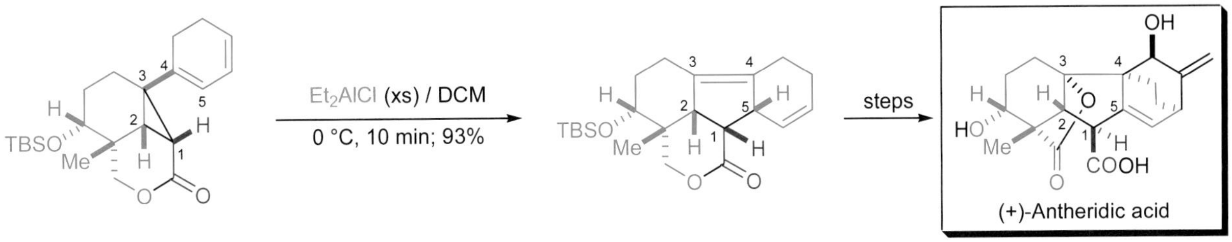

T. Hudlicky et al. achieved the short enantioselective total synthesis of (-)-retigeranic acid.[64] The C ring of the natural product was assembled *via* the *thermal vinylcyclopropane-cyclopentene rearrangement* for which the precursor was prepared by the *vinylcyclopropanation* of a bicyclic enone with a dienolate. The vinylcyclopropane was evaporated at 585 °C in high vacuum through a Vycor tube conditioned with $PbCO_3$ (flash vacuum pyrolysis) to afford the annulated product in good yield.

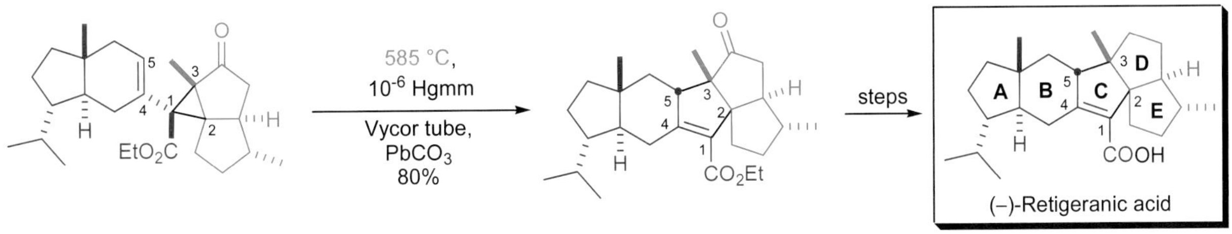

The iridoid sesquiterpene (–)-specionin, an antifeedant to the spruce budworm, was synthesized by T. Hudlicky et al. using the *low-temperature vinylcyclopropane-cyclopentene rearrangement* as the key step.[65] The substituted cyclopentenone precursor was first exposed to the lithium dienolate derived from ethyl 4-(dimethyl-*tert*-butylsilyloxy)-2-bromocrotonate at -110 °C to afford silyloxyvinylcyclopropanes as a mixture of *exo* and *endo* isomers (with respect to the vinyl group). The mixture was not separated but immediately subjected to TMSI/HMDS, and the corresponding tricyclic ketones were obtained in good yield. Similar results were obtained when TBAF in THF was used instead of TMSI.

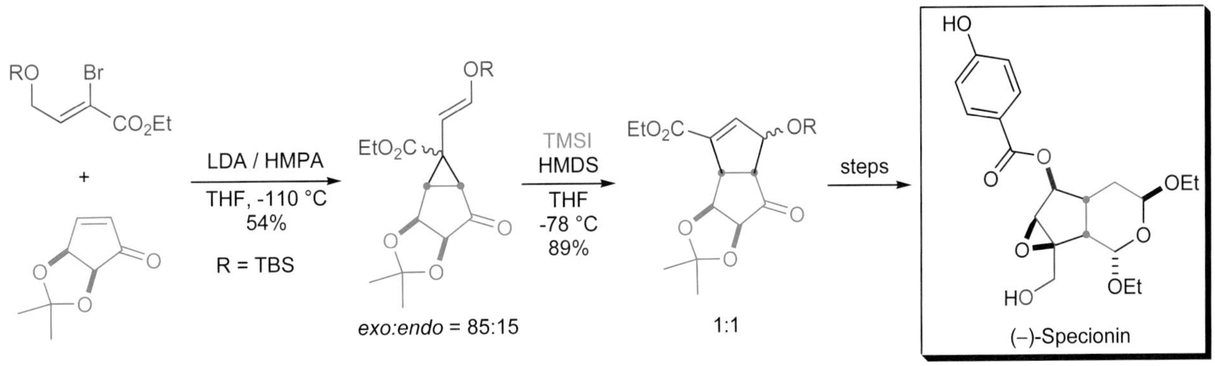

VON PECHMANN REACTION

(References are on page 702)

Importance:

[*Seminal Publications*[1,2]; *Reviews*[3,4]; *Modifications & Improvements*[5-26]]

In 1883, H. von Pechmann and C. Duisberg reported that when ethyl acetoacetate was mixed with resorcinol in the presence of concentrated sulfuric acid, 4-methyl-7-hydroxycoumarin was formed.[1] He obtained a similar result upon reacting resorcinol with malic acid and isolated 7-hydroxycoumarin as the major product.[2] The condensation of phenols with β-keto esters in the presence of protic or Lewis acids to afford substituted coumarins is known as the *von Pechmann reaction* (also as *Pechmann reaction* or *Pechmann condensation*). The general features of this transformation are: 1) the best substrates are electron-rich mono-, di-, and trihydric phenols having electron-donating substituents; 2) phenols with strongly electron-withdrawing substituents (e.g., NO_2 or CO_2H) often fail to react; 3) the position of the substituents on the phenol also has an influence on the reactivity and therefore on the rate of the condensation; 4) *ortho* substituents tend to inhibit the reaction completely, *para* substituents usually do not interfere much, and substituents in the *meta* position give the best results; 5) both cyclic and acyclic β-keto esters undergo the reaction; 6) malic acid, fumaric, and maleic acids also react, but the scope of phenolic substrates is somewhat limited with these reactants; 7) β-keto esters yield coumarins that have substituents at the C4 position while malic acid affords coumarins which are unsubstituted at C4; 8) the nature of the protic or Lewis acid catalyst has a profound effect on the outcome of the reaction: if the reaction does not take place in the presence of one particular catalyst, it may proceed in high yield in the presence of another; 9) during the 1900s the most popular catalyst was concentrated sulfuric acid, but for highly functionalized and sensitive substrates milder condensation conditions have been developed; and 10) for highly reactive phenols heating of the reaction mixture is usually not necessary, but for less reactive substrates heating is often required. There are some drawbacks of the *von Pechmann reaction*: 1) in the overwhelming majority of the cases the catalyst has to be used in excess so the process is not catalytic; and 2) extended reaction times at high temperatures can lead to side reactions such as to the formation of chromones in addition to coumarins. Numerous modifications have been developed and several of them allow the synthesis of coumarins under mild conditions and even using truly catalytic amounts of condensing agent.[27]

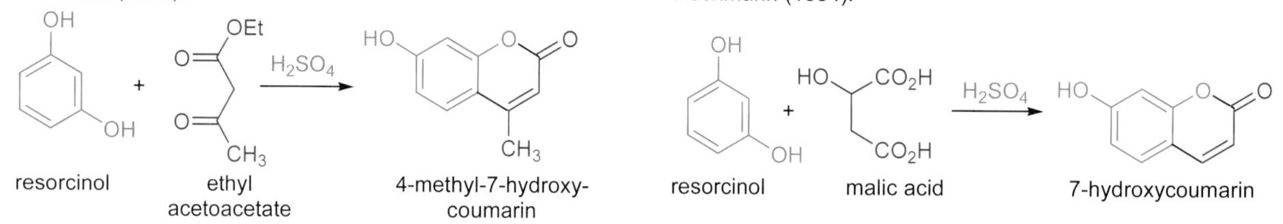

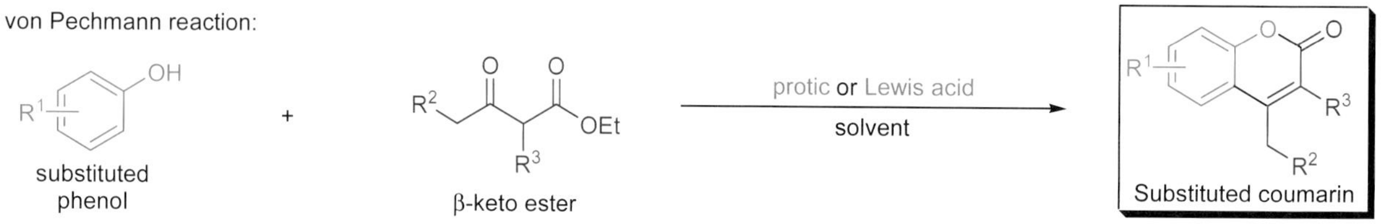

R^1 = H, OH, *O*-alkyl, NH_2, NHR, NR_2; R^2 = H, alkyl, aryl; R^3 = H, alkyl, aryl, Cl; protic acid: H_2SO_4, HCl, H_3PO_4; Lewis acid: $POCl_3$, $ZnCl_2$, $AlCl_3$, $FeCl_3$, $InCl_3$, $Yb(OTf)_3$, $SnCl_2$, $TiCl_4$, $SiCl_4$, PPA

Mechanism: [28,29]

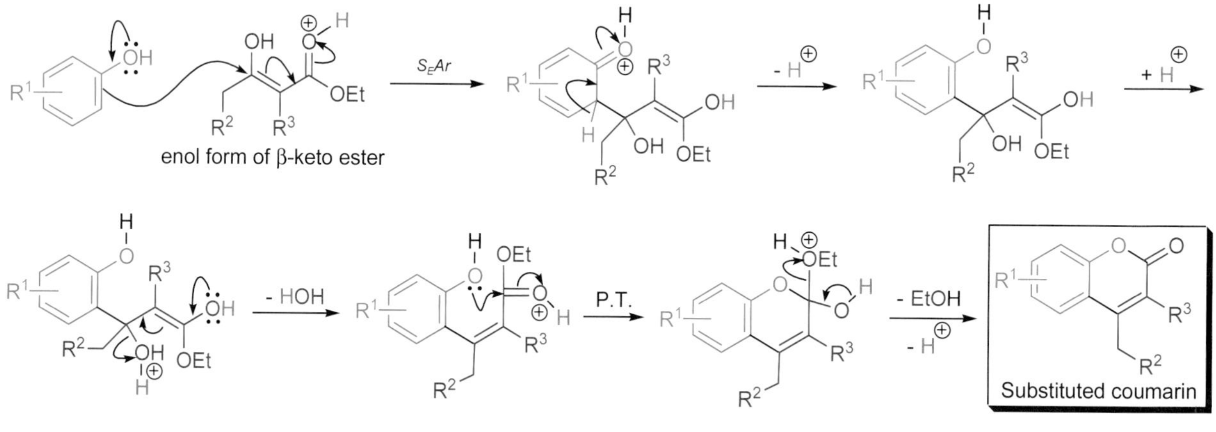

Synthetic Applications:

In the laboratory of J. Moron, the synthesis of two pyridoangelicins, the angular isomers of pyridopsoralens, was accomplished. The authors demonstrated in previous publications that pyridopsoralens exhibit high affinity toward DNA, so it was a logical next step to prepare the angular isomers and test their affinities. The skeleton of the desired compound was assembled by the *von Pechmann reaction*. The reaction between 2,3-dihydro-4-hydroxybenzofuran and 1-benzyl-3-ethoxycarbonylpiperidin-4-one was conducted in glacial acetic acid at room temperature in the presence of sulfuric acid and phosphorous oxychloride ($POCl_3$). When only hydrochloric acid was used as the condensing agent, the yield was very poor.

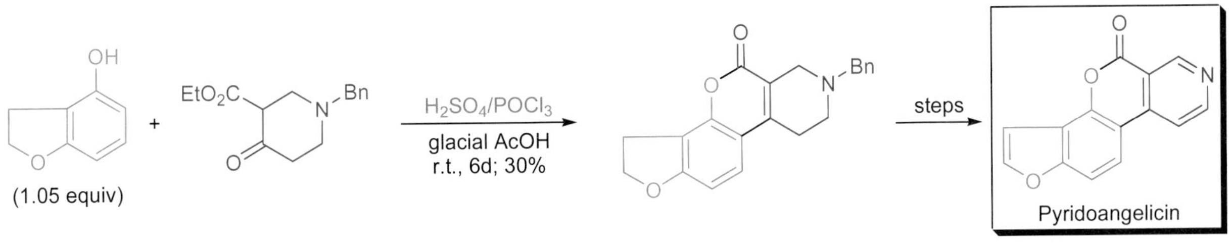

One of the mildest conditions for the *von Pechman reaction* was developed by D.S. Bose and co-workers who used indium(III) chloride as the catalyst.[27] A large number of 4-substituted coumarins were prepared in high yield by this method. Under the reaction conditions most functional groups are tolerated. In the typical procedure the substrates are heated in the presence of 10 mol% of $InCl_3$ and the reaction mixture was poured onto crushed ice which caused the product to precipitate.

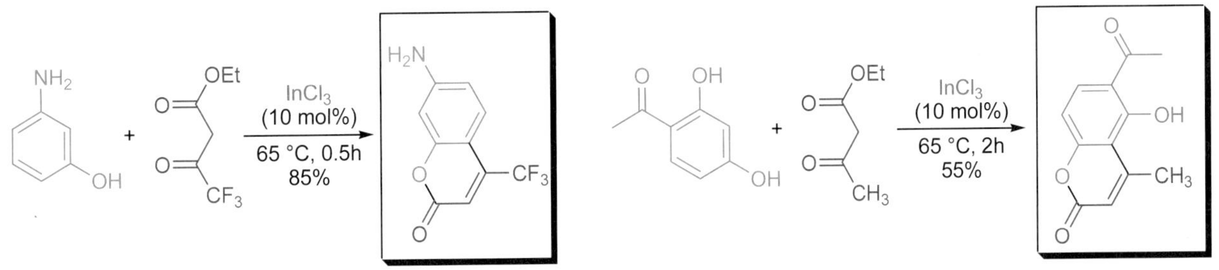

Photochemotherapy is an efficient way to treat hyperproliferative diseases. Especially the so-called PUVA therapy (psoralen + UVA light) is very common in which the psoralen is irradiated with UVA light to give rise to a covalent adduct with the pyrimidine bases of DNA by means of a photoaddition reaction. There are several undesired side effects for the patients as a result of this therapy, so the synthesis and photobiological evaluation of novel benzosporalen derivatives was undertaken by the research team of L.D. Via.[30] The key step in their synthetic sequence was the *von Pechman reaction* of 2-methoxyresorcinol with ethyl 2-oxocyclohexanecarboxylate.

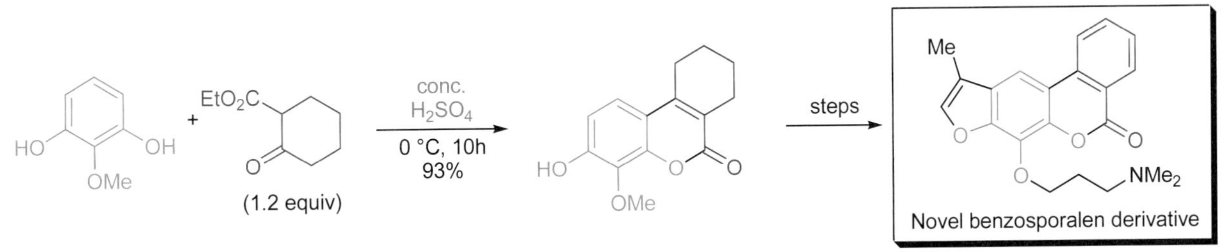

The short and efficient stereospecific synthesis of the dimer-selective retinoid X receptor modulator was carried out in the laboratory of L.G. Hamann.[31] The synthetic sequence began with the *von Pechmann reaction* between tetramethyltetrahydronaphthol and ethyl acetoacetate in 75% sulfuric acid solution. The desired coumarin was formed regioselectively and isolated in high yield.

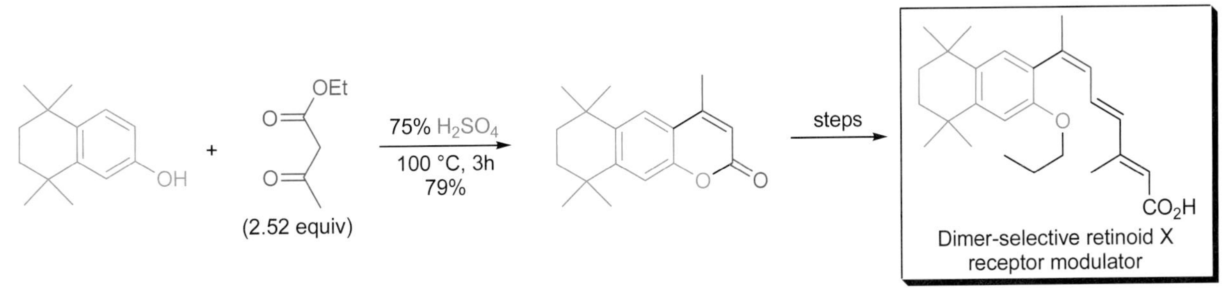

WACKER OXIDATION

(References are on page 702)

Importance:

[*Seminal Publications*[1-5]; *Reviews*[6-24]; *Modifications & Improvements*[25-45]; *Theoretical Studies*[46-56]]

The industrial oxidation of ethylene to ethanal (acetaldehyde) under an atmosphere of oxygen using $PdCl_2$ and $CuCl_2$ as catalysts is known as the *Wacker-Smidt process*. The first report of the oxidation of ethylene with stoichiometric amounts of $PdCl_2$ in an aqueous solution was made by F.C. Phillips in 1894 and later the precipitation of Pd metal from a $PdCl_2$ solution was used as a test for the presence of olefins.[1] In 1959, J. Smidt et al. (at Wacker Chemie in Germany) showed that the $Pd^{(0)}$ metal could be re-oxidized to the active $PdCl_2$ with the use of $CuCl_2$.[2,3] This discovery first turned the reaction into a commercially feasible process, and it opened the door for applications in organic synthesis.[4,5,57] The one-pot oxidation of olefins to the corresponding ketones with catalytic amounts of $Pd^{(II)}$ salts is known as the *Wacker oxidation*. The general features of this reaction are: 1) the reaction is carried out in an aqueous medium in the presence of HCl; 2) terminal alkenes react much faster than internal or 1,1-disubstituted alkenes and they are almost exclusively converted to the corresponding methyl ketones; 3) terminal alkenes can be viewed as masked ketones for synthetic purposes; 4) under the reaction conditions, internal alkenes are not oxidized to any appreciable extent; 5) α,β-unsaturated ketones and esters are oxidized regioselectively to the corresponding β-keto compounds using catalytic amounts of Na_2PdCl_4 and TBHP or H_2O_2 as co-oxidants; 6) allylic- and homoallylic ethers are regioselectively oxidized to give the corresponding β- and γ-alkoxyketones; and 7) when the oxidation is carried out in the presence of nucleophiles other than water, the process is called the *Wacker-type oxidation*, which can take place both inter- and intramolecularly.

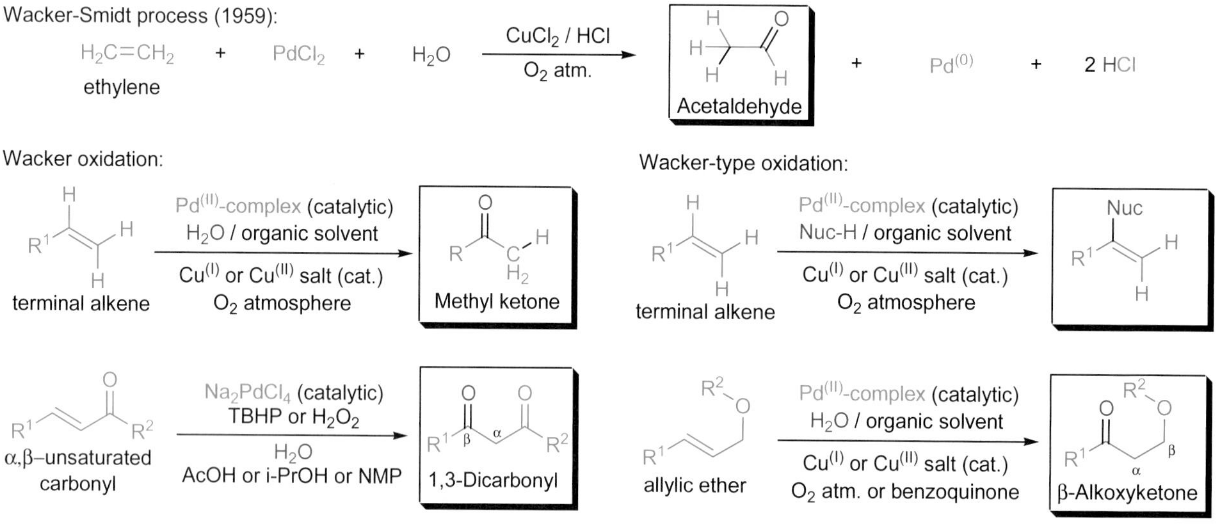

R^1 = alkyl, substituted alkyl; R^2 = alkyl, aryl, O-alkyl

Mechanism: [58-75,37,19]

Certain steps in the mechanism of the *Wacker oxidation* are still unclear despite intensive research. One of these steps, the attack of the coordinated alkene by the nucleophile (OH^- or H_2O), could be both intra- or intermolecular as the observed rate law is consistent with either possibility. One of the plausible catalytic cycles is presented.

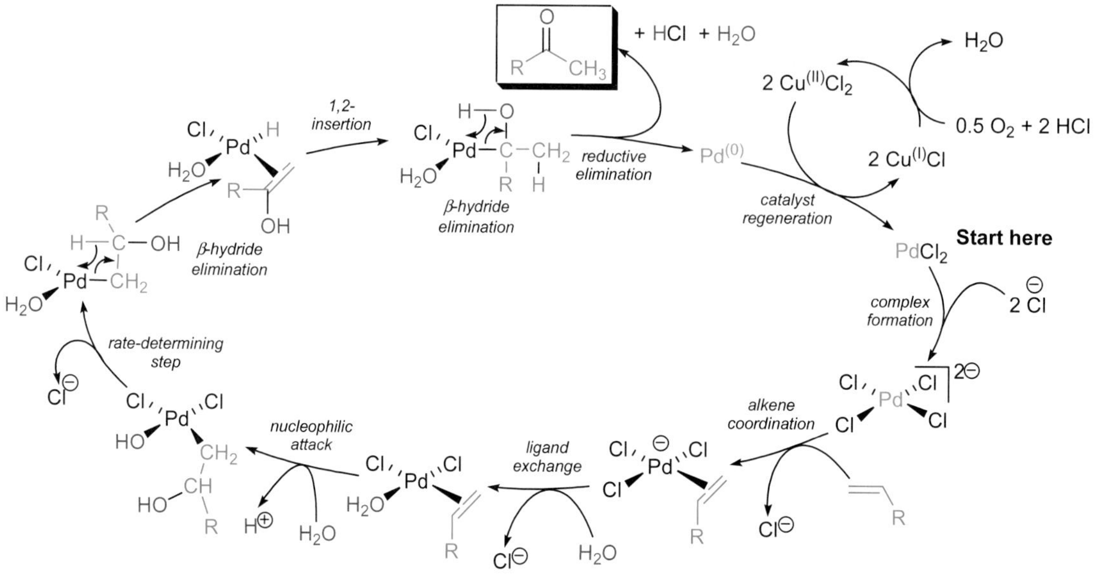

Synthetic Applications:

The asymmetric total synthesis of the putative structure of the cytotoxic diterpenoid (-)-sclerophytin A was accomplished by L.A. Paquette and co-workers.[76] At the beginning of the synthesis, a bicyclic intermediate was subjected to the *Wacker oxidation* to oxidize its terminal alkene into the corresponding methyl ketone. The oxidation took place in high yield, although the reaction time was long. The spectra obtained for the final product (proposed structure) did not match that of the natural product, consequently a structural revision was necessary.

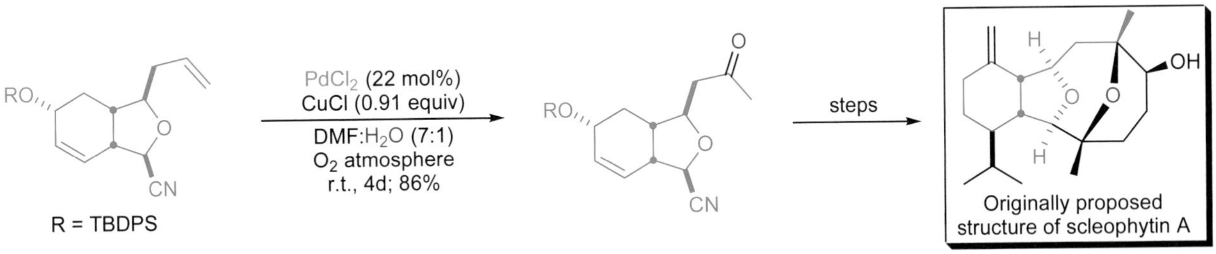

The antiviral marine natural product, (–)-hennoxazole A, was synthesized in the laboratory of F. Yokokawa.[77] The highly functionalized tetrahydropyranyl ring moiety was prepared by the sequence of a *Mukaiyama aldol reaction*, *chelation-controlled 1,3-syn reduction*, *Wacker oxidation*, and an *acid catalyzed intramolecular ketalization*. The terminal olefin functionality was oxidized by the *modified Wacker oxidation*, which utilized $Cu(OAc)_2$ as a co-oxidant.[34] Interestingly, a similar terminal alkene substrate, which had an oxazole moiety, failed to undergo oxidation to the corresponding methyl ketone under a variety of conditions.

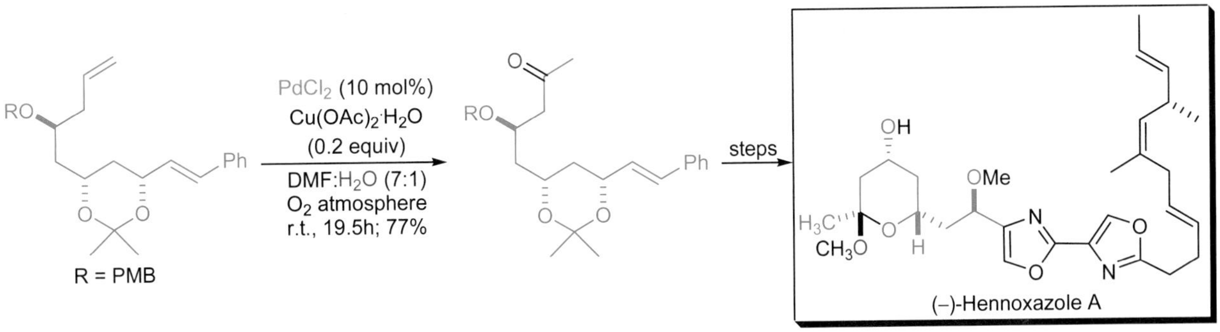

The first synthesis of the hexacyclic himandrine skeleton was achieved by L.N. Mander and co-workers.[78] The last six-membered heterocycle was formed *via* an intramolecular *Wacker-type oxidation* in which the terminal alkene side-chain reacted with the secondary amine functionality. The oxidation was conducted in anhydrous acetonitrile to insure that the Pd-alkene complex was substituted exclusively by the internal nucleophile. The resulting six-membered enamine was then hydrogenated and the MOM protecting groups removed to give the desired final product.

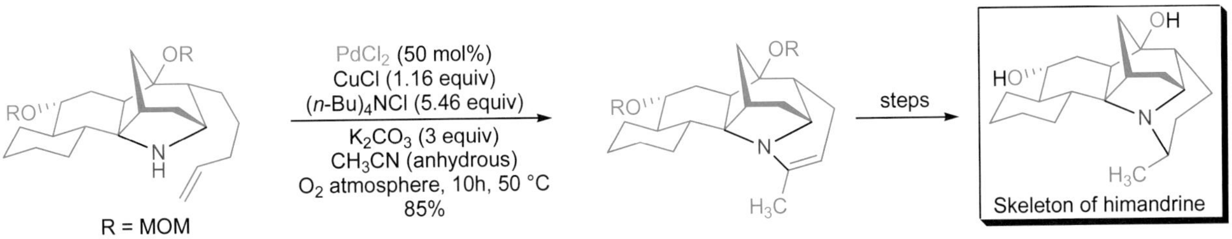

Studies in the laboratory of M. Shibasaki toward the total synthesis of garsubellin A led to the stereocontrolled synthesis of the 18-*epi*-tricyclic core of the natural product.[79] During the final stages of the synthetic sequence, the tetrahydrofuran ring was installed using a *Wacker-type process*. The reaction conditions insured that the acetonide protecting group was first removed and the C18 secondary alcohol moiety served as the internal nucleophile to form the tricyclic product.

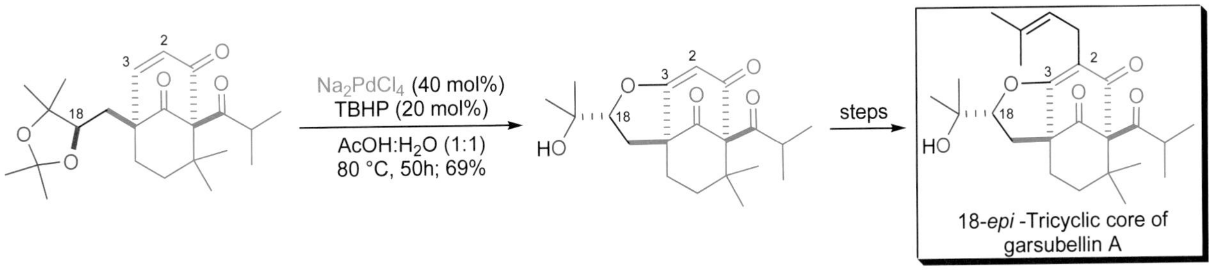

WAGNER-MEERWEIN REARRANGEMENT

(References are on page 704)

Importance:

[*Seminal Publications*[1-3]; *Reviews*[4-18]; *Modifications & Improvements*[19-25]; *Theoretical Studies*[26-30]]

In 1899, G. Wagner and W. Brickner reported the rearrangement of α-pinene to bornyl chloride in the presence of hydrogen chloride.[1] The transformation baffled chemists at the time, since it contradicted the classical structural theory that was based on the postulate of skeletal invariance.[31] It was not until 1922, when H. Meerwein and co-workers revealed the ionic nature of the rearrangement, that an explanation was offered.[3] The generation of a carbocation followed by the *[1,2]*-shift of an adjacent carbon-carbon bond to generate a new carbocation is known as the *Wagner-Meerwein rearrangement*. Originally this name referred only to skeletal rearrangements in bicyclic systems, but today it is used to describe all *[1,2]*-shifts of hydrogen, alkyl, and aryl groups. Occasionally the *[1,2]-methyl shift* in bridged bicyclic monoterpenoids and related systems is referred to as the *Nametkin rearrangement*. The general features of the *Wagner-Meerwein rearrangement* are: 1) the generation of the initial carbocation can be achieved in a variety of ways (e.g., protonation of alkenes, alcohols, epoxides or cyclopropanes, solvolysis of secondary and tertiary alkyl halides, or sulfonates in a polar protic solvent (*semipinacol rearrangement*), deamination of amines with nitrous acid (*Tiffeneau-Demjanov rearrangement*), treatment of an alkyl halide with Lewis acid, etc.; 2) the initial carbocation has a tendency to rearrange to a thermodynamically more stable structure, a change that may occur in several different ways: e.g., *[1,2]-alkyl, -aryl- or hydride shift* to afford a more stable carbocation, ring-expansion of strained small rings such as cyclopropanes and cyclobutanes to give more stable five- or six-membered products, collapse by fragmentation, etc.; 3) several consecutive *[1,2]*-shifts are possible if the substrate contains multiple structural elements that allow the formation of gradually more stable structures; 4) the various competing rearrangement pathways limit the synthetic utility of the *Wagner-Meerwein rearrangement*, since one needs to install all the structural features that will drive the rearrangement in the desired direction; 5) the final most stable carbocation's fate may be the loss of a proton to afford an alkene or capture by a nucleophile present in the reaction mixture (solvent or conjugate base of the acid used to promote the rearrangement); and 6) the stereochemistry of the migrating group is retained, which is in accordance of the *Woodward-Hofmann rules*.

Wagner & Brickner (1899):

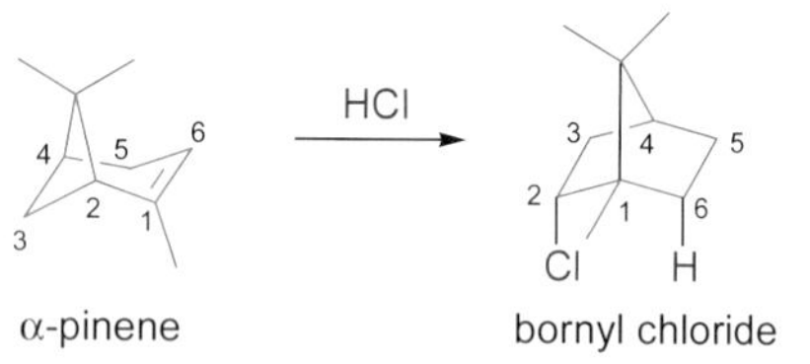

Meerwein, Ermster & Jussen (1922):

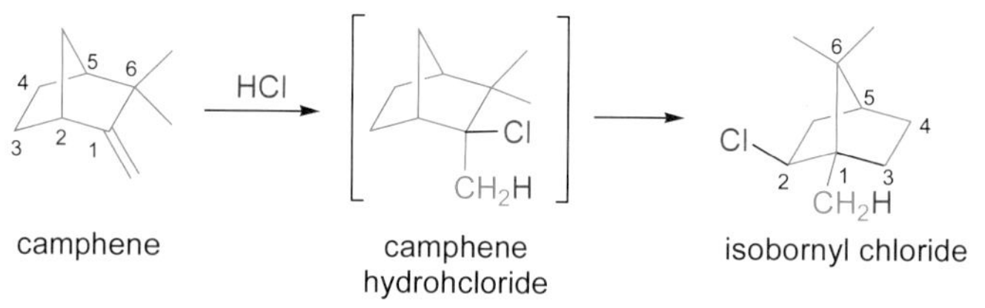

General scheme for Wagner-Meerwein rearrangement:

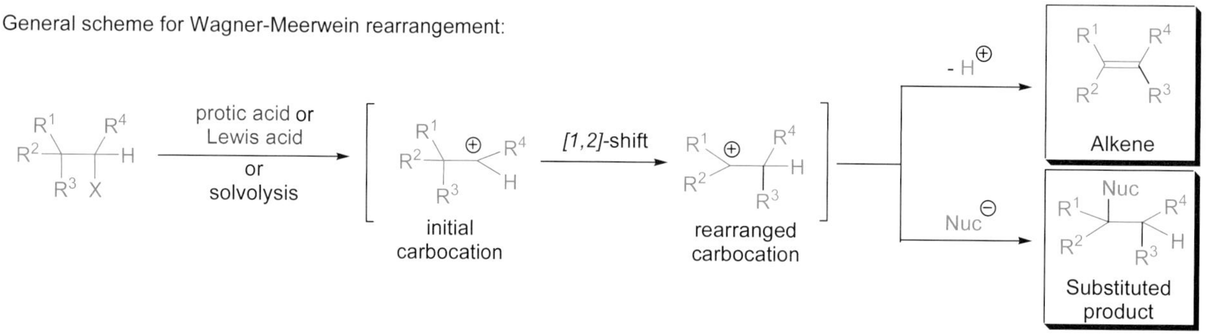

Mechanism: [32,13,15-17]

The *Wagner-Meerwein rearrangement* has been the subject of a large number of mechanistic investigations, making it probably one of the most thoroughly studied reactions in organic chemistry. Depending on the structure and stereochemistry of the substrate, the rearrangement may proceed in a concerted or stepwise fashion. When the leaving group and the migrating groups are antiperiplanar to each other, the rearrangement is concerted (especially in rigid polycyclic sytems), but in most other cases the formation of a carbocation intermediate is expected.

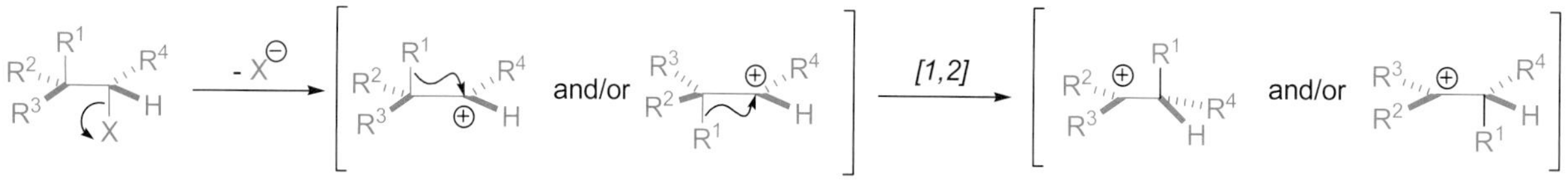

Synthetic Applications:

The large-scale synthesis of the potent antitumor agent KW-2189, derived from the antitumor antibiotic duocarmycin B2, was accomplished by T. Ogasa and co-workers who utilized the *Wagner-Meerwein rearrangement* as the key step.[33] The synthetic strategy avoided the use of protecting groups. The key rearrangement step was investigated in detail and the authors found that both protic and Lewis acids were effective. The best results were obtained with methanesulfonic acid in dichloroethane. Protonation of the 2° alcohol at C3 resulted in the loss of a water molecule and the formation of a secondary carbocation. The adjacent carboxymethyl group at C2 underwent a *[1,2]*-shift to form the more stable tertiary carbocation at C2, which was also stabilized by the lone pair of the nitrogen atom and finally the loss of proton afforded the indole nucleus.

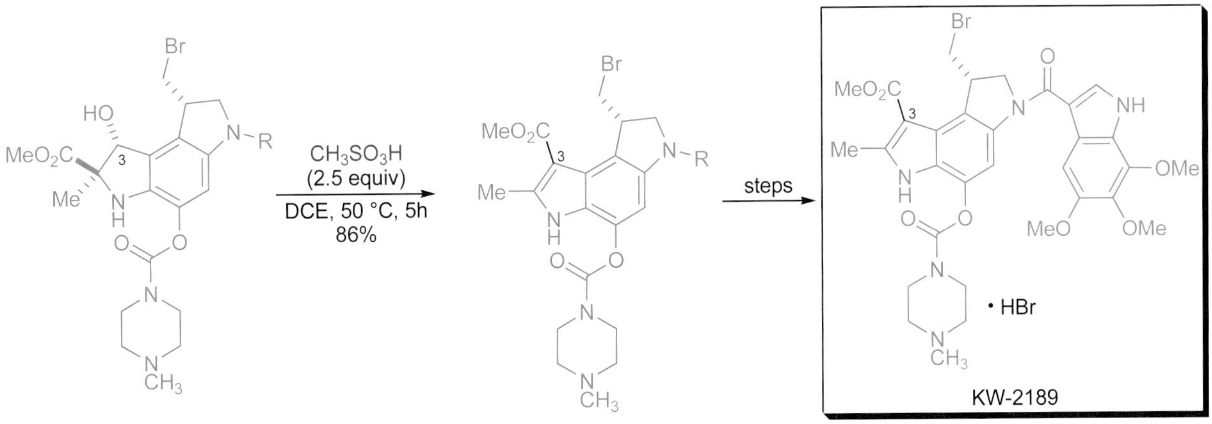

The short enantiospecific synthesis of (1*R*)-10-hydroxyfenchone from fenchone based on two consecutive *Wagner-Meerwein rearrangements* was developed in the laboratory of A.G. Martinez.[34] The preparation of this target is of great importance, since 10-hydroxyfenchone is a convenient intermediate for C10-*O*-substituted fenchones. The key intermediate in the synthetic sequence is 2-methylenenorbornan-1-ol, obtained from fenchone *via* a *Wagner-Meerwein rearrangement* (steps not shown), which was exposed to *m*CPBA at room temperature. The initially formed epoxide was protonated by *m*CPBA, generating a tertiary carbocation that underwent a facile *[1,2]-alkyl shift* to produce the more stable oxygen-stabilized carbocation.

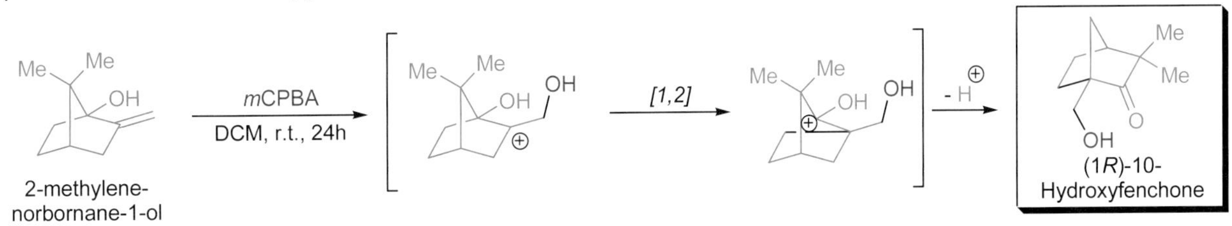

The research team of G. Fráter investigated the acid catalyzed rearrangement of β-monocyclofarnesol for the synthesis of tricyclic ketones with sesquiterpene skeleton. The substrate β-monocyclofarnesol, prepared from dihydro-β-ionone in two steps, was exposed to concentrated formic acid, which resulted in the formation of a mixture of three different formates.

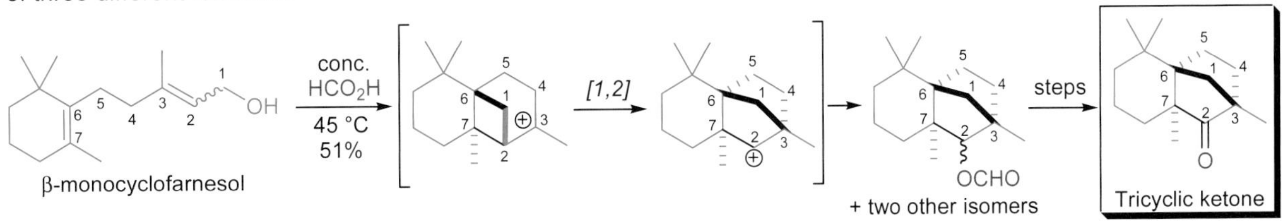

The *Wagner-Meerwein rearrangement* was one of the key steps in the total synthesis of (+)-quadrone by A.B. Smith and co-workers.[35] The propellane substrate was treated with 40% sulfuric acid, which resulted in the *[1,2]-alkyl shift* of the initially formed cyclobutylcarbinyl system.

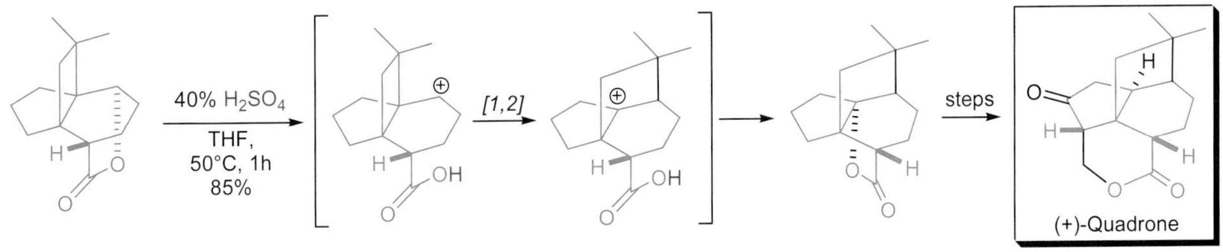

WEINREB KETONE SYNTHESIS

(References are on page 705)

Importance:

[*Seminal Publications*[1]; *Reviews*[2-5]; *Modifications & Improvements*[6-23]]

In 1981, S.M. Weinreb and S. Nahm discovered that the addition of excess Grignard reagent or organolithium species to *N*-methoxy-*N*-methylamides resulted in the formation of ketones upon acidic work-up. This observation was significant because at that time there was no general procedure available for the efficient synthesis of ketones from carboxylic acid derivatives and the then existing methods all required carefully controlled reaction conditions, and overaddition (to produce tertiary alcohols) was a major side reaction. The synthesis of ketones from *N*-methoxy-*N*-methylamides (Weinreb's amides) with organometallic reagents is known as the *Weinreb ketone synthesis*. The general features of this transformation are: 1) the Weinreb's amides can be easily prepared from activated carboxylic acid derivatives (e.g., acid chlorides or anhydrides) and *N,O*-dimethylhydroxylamine hydrochloride in the presence of a base; 2) the conversion of less active carboxylic acid derivatives such as esters and lactones to the corresponding Weinreb's amide require the use of several equivalents of trimethylaluminum (Me_3Al) or dimethylaluminum chloride (Me_2AlCl);[6,9] 3) carboxylic acids can also be converted to Weinreb's amides by the use of standard activating agents (DCC, EDCI, CBr_4/PPh_3, etc.); 4) Weinreb's amides are stable compounds; they do not require special handling, are easily purified by flash chromatography or crystallization and can be stored indefinitely; 5) the addition of at least 1.1 equivalents of Grignard reagent or organolithium species to the solution of Weinreb's amide in an ether solvent at low temperatures results in the formation of a strongly chelated metal complex, which prevents the addition of more than one equivalent of the reagent; 5) work-up with dilute aqueous acid (HCl) affords the ketone and usually does not interfere with other functional groups or protecting groups; 6) virtually any alkyl, alkenyl, alkynyl, aryl, and heteroaryl organomagnesium- or organolithium reagent can be used; 7) side reactions such as overaddition of the reagent or the epimerization of the stereocenter at the α-position are extremely rare; 8) the treatment of Weinreb's amides with excess metal hydride (e.g., LAH, DIBAL-H) results in the formation of aldehydes; and 9) the use of DIBAL-H tends to give higher yields than LAH. All the above features render the *Weinreb ketone synthesis* extraordinarily well-suited for use in the synthesis of complex molecules. One important limitation of the procedure occurs when highly basic or sterically hindered organometallic reagents are used since these are capable of removing a proton from the O-Me group resulting in the formation of *N*-methylamides.

Weinreb & Nahm (1981):

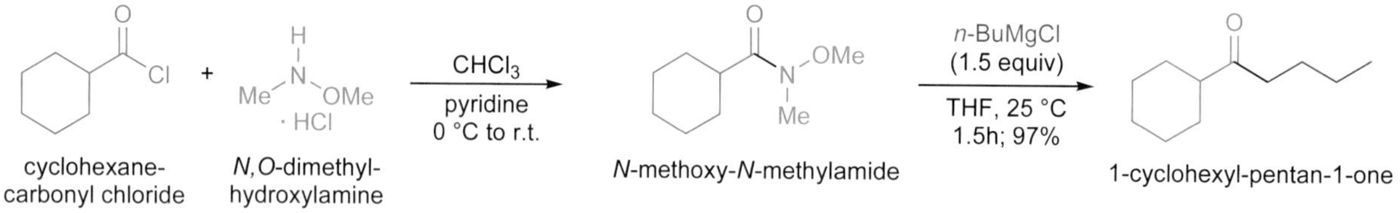

Weinreb ketone synthesis:

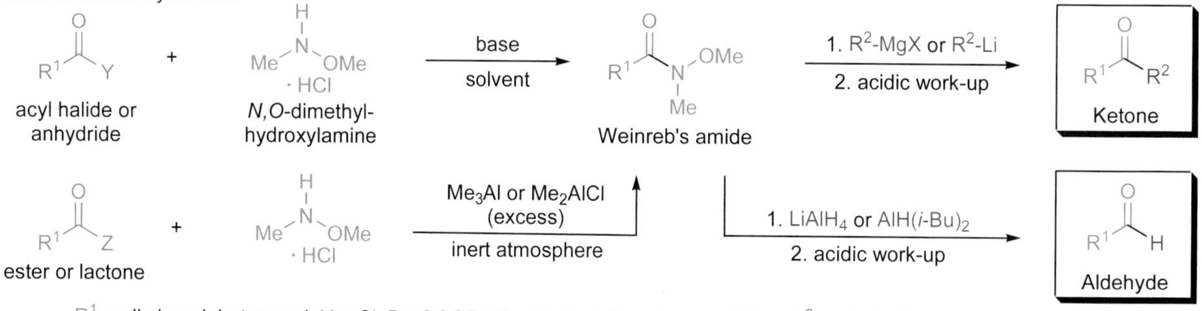

R^1 = alkyl, aryl, heteroaryl; Y = Cl, Br, OCOR'; Z = *O*-alkyl, *O*-aryl, oxazolidine; R^2 = alkyl, alkenyl, alkynyl, aryl; base: pyridine, Et_3N

Mechanism: [1]

Formation of Weinreb's amide:

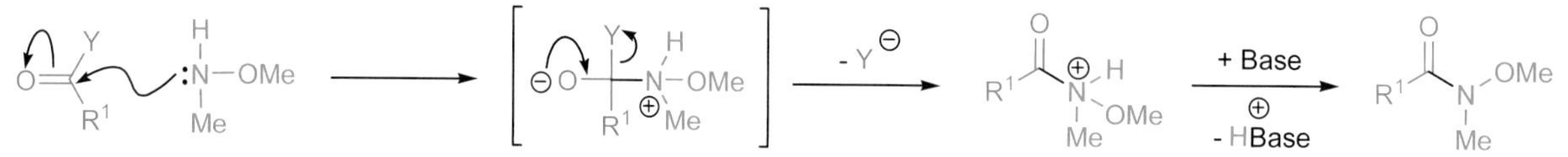

Reaction of the organometallic species with Weinreb's amide:

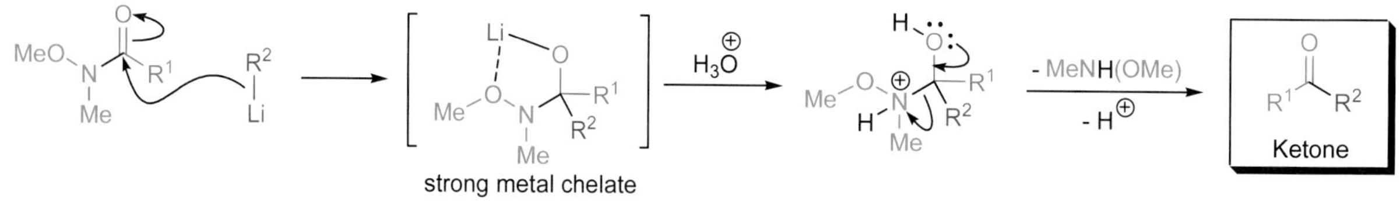

Synthetic Applications:

The first total synthesis of the *Stemona* alkaloid (–)-tuberostemonine was accomplished by P. Wipf and co-workers.[24] The installation of the butyrolactone moiety commenced with the preparation of a Weinreb's amide from a methyl ester. The tricyclic methyl ester substrate was exposed to *N,O*-dimethylhydroxylamine hydrochloride and Me_2AlCl and the tertiary amide was isolated in excellent yield. Next, the bromo ortho ester was treated with LDBB in THF to generate the corresponding primary alkyllithium species, which cleanly and efficiently added to the Weinreb's amide to afford the desired ketone.

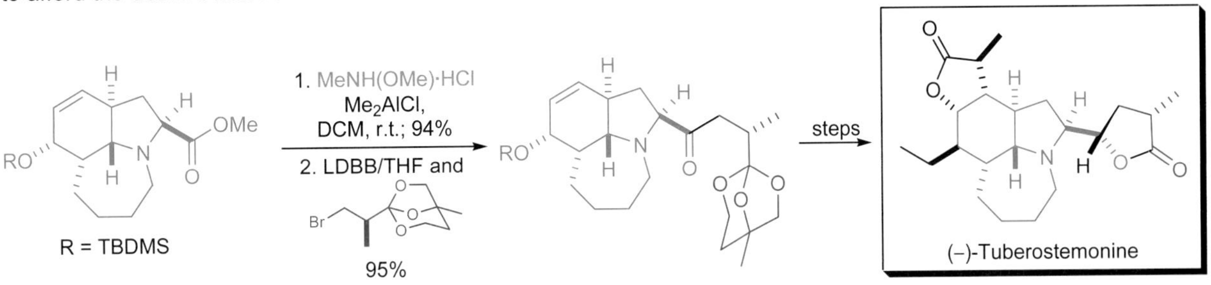

The preparation of the C1-C21 subunit of the protein phosphatase inhibitor tautomycin was completed by J.A. Marshall et al., and it constituted a formal total synthesis of the natural product.[25] The spiroketal carbon of the target was introduced by the *Weinreb ketone synthesis* between a lithioalkyne and *N*-methoxy-*N*-methylurea (a carbon monoxide equivalent). The triple bond of the resulting Weinreb's amide was first reduced under catalytic hydrogenation conditions to yield the corresponding saturated amide, which was reacted with another lithium acetylide to afford an ynone.

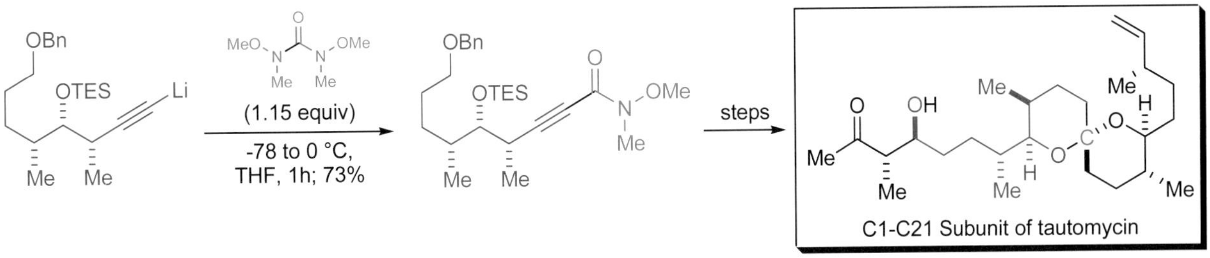

In the laboratory of E.J. Corey, the first synthesis of nicandrenones (NIC), a structurally complex steroid-derived family of natural products, was accomplished.[26] The side chain of NIC-1 was constructed from the known six-membered lactone which was converted to the Weinreb's amide by treating it with excess MeNH(OMe)·HCl and trimethyl-aluminum. The resulting primary alcohol was protected as the TBS ether. The ethynylation of this amide was carried out by reaction with two equivalents of lithium trimethylsilylacetylide to afford an ynone, which was reduced enantioselectively to the corresponding propargylic alcohol using *CBS reduction*.

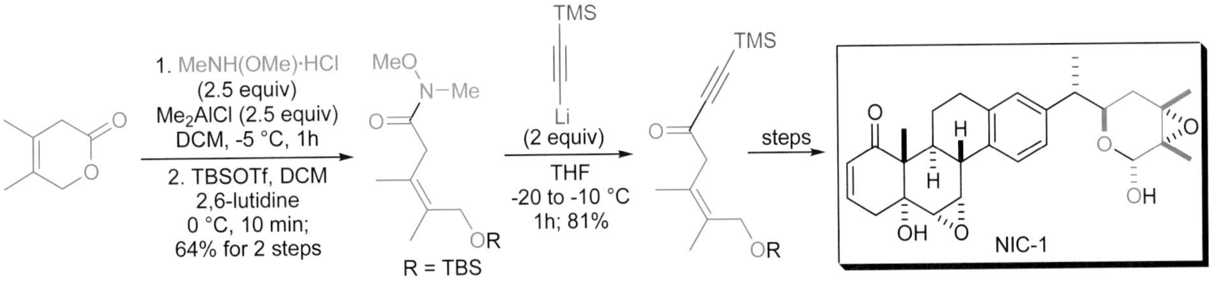

The *rhodium-catalyzed intramolecular [5+2] cycloaddition* of an allene and vinylcyclopropane was the key step in the asymmetric total synthesis of the trinorguaiane sesquiterpene (+)-dictamnol by P.A. Wender and co-workers.[27] The cyclization precursor allene-cyclopropane was assembled starting from commercially available cyclopropane-carbaldehyde. Using the *HWE olefination*, the Weinreb's amide moiety was installed and subsequently reacted with a primary alkyllithium that was generated *via* lithium-halogen exchange.

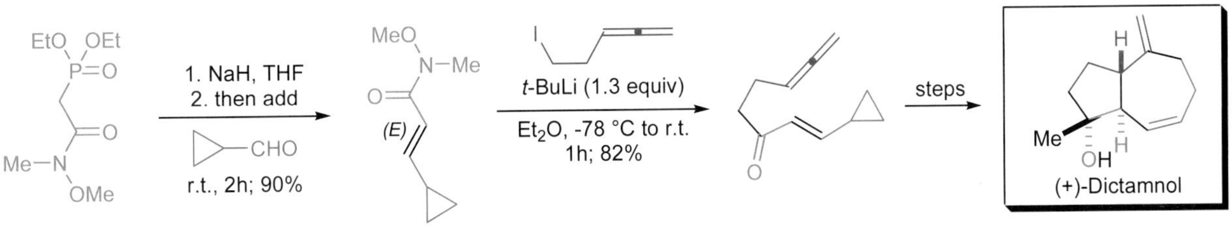

WHARTON FRAGMENTATION

(References are on page 705)

Importance:

[*Seminal Publications*[1-5]; *Reviews*;[6-11] *Modifications & Improvements*[12-17]]

In 1961, P.S. Wharton investigated the potassium-*tert*-butoxide-induced heterolytic fragmentation of a bicyclic 1,3-diol monomesylate ester (functionalized decalin system), to form a 10-membered cyclic alkene stereospecifically.[2] The base-induced stereospecific fragmentation of cyclic 1,3-diol monosulfonate esters (X=OSO_2R; Y=OH) to form medium-sized cyclic alkenes is known as the *Wharton fragmentation*. Wharton and co-workers contributed to this area extensively by uncovering the stereoelectronic requirements for the reaction as well as demonstrating its synthetic utility. This fragmentation, however, falls into the category of *Grob-type fragmentations* in which carbon chains with a variety of combinations of nucleophilic atoms (heteroatoms) and leaving groups give rise to three fragments.[18] The general features of the *Wharton fragmentation* are the following: 1) synthetically, cyclic 1,3-diol derivatives are the most useful substrates, since acyclic precursors often give rise to side-products (e.g., oxetanes, Y=O) resulting from an intramolecular displacement; 2) cyclic 1,3-hydroxy monotosylates and monomesylates are the most widely used substrates, and they are prepared by treating the unsymmetrical 1,3-diol with one equivalent of MsCl or TsCl; 3) the rate of the fragmentation depends on the concentration of the anion derived from the 1,3-diol derivative; 3) strong and less nucleophilic bases favor the fragmentation, whereas more nucleophilic bases favor intramolecular substitution and elimination of the leaving group; 4) KO*t*-Bu/*t*-BuOH and dimsylsodium/DMSO are the most often used base/solvent combination; 5) if the substrate has considerable ring strain (e.g., n=1), even weaker bases (e.g., NEt_3) will initiate successful fragmentation; 6) when the fragmentation product is labile (e.g., aldehyde), $LiAlH_4$ can serve as both a basic initiator and a reducing agent, since it instantly traps (reduces) the initial product avoiding undesired side reactions (e.g., *aldol condensation*); 7) alkenes are generated stereospecifically from cyclic substrates in high yield; 8) fragmentations leading to ketones occur more readily than those that give aldehydes; 9) more highly substituted alkenes are formed faster than less substituted ones; and 10) substrates with more ring strain generally fragment faster.

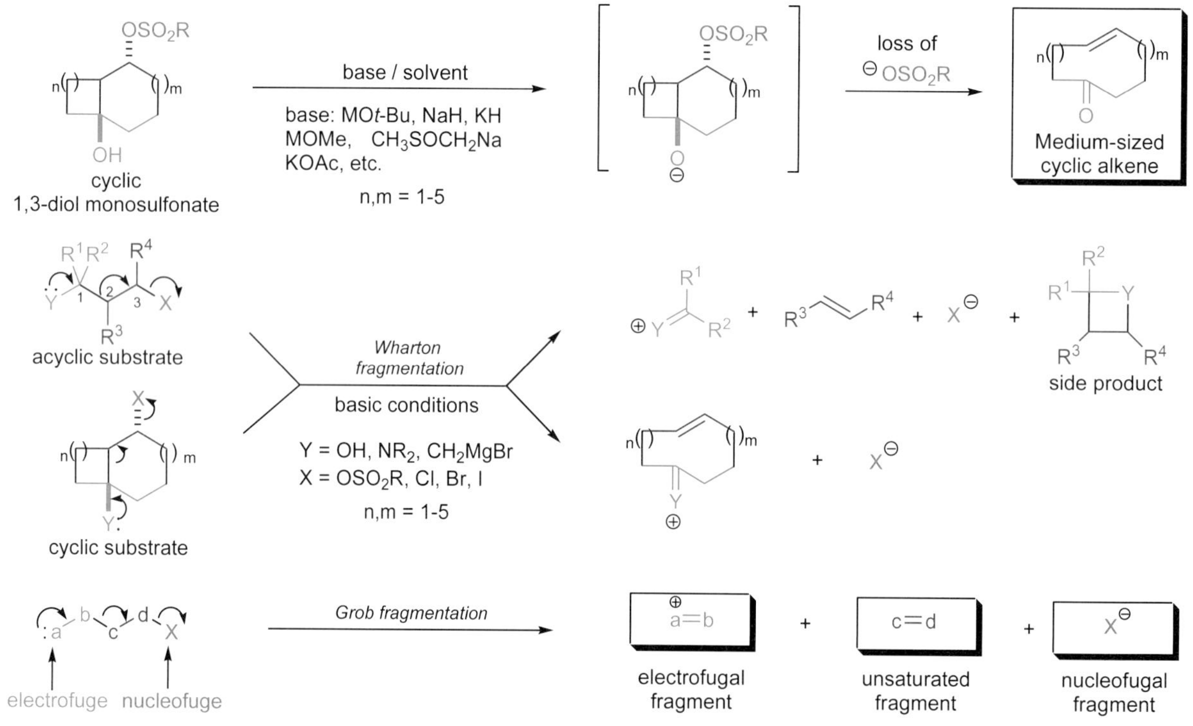

Mechanism: [4,19,10]

The *Wharton fragmentation* is a concerted reaction and the stereoelectronic requirement is that the bonds that are undergoing the cleavage must be *anti* to each other. This requirement is easily met in cyclic systems; however, acyclic systems have much larger conformational freedom, so side reactions may arise when the conformation of the bonds undergoing cleavage is *gauche*. In cyclic systems the fragmentation becomes slow and complex product mixtures are formed when the conformation of the bonds undergoing cleavage is *gauche*.

Preferred anti conformation: Side reaction:

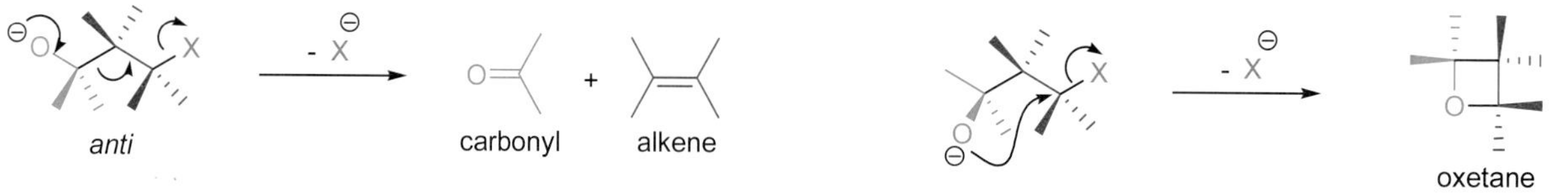

Synthetic Applications:

The *Wharton fragmentation* was used as a key step in an approach toward the total synthesis of xenicanes by H. Pfander et al.[20] Two optically active substituted *trans*-cyclononenes were synthesized starting from (-)-Hajos-Parrish ketone. First, the bicyclic 1,3-diol was protected regioselectively on the less sterically hindered hydroxyl group with *p*-toluenesulfonyl chloride in quantitative yield. Next, the monosulfonate ester was exposed to dimsylsodium in DMSO, which is a strong base, to initiate the desired heterolytic fragmentation.

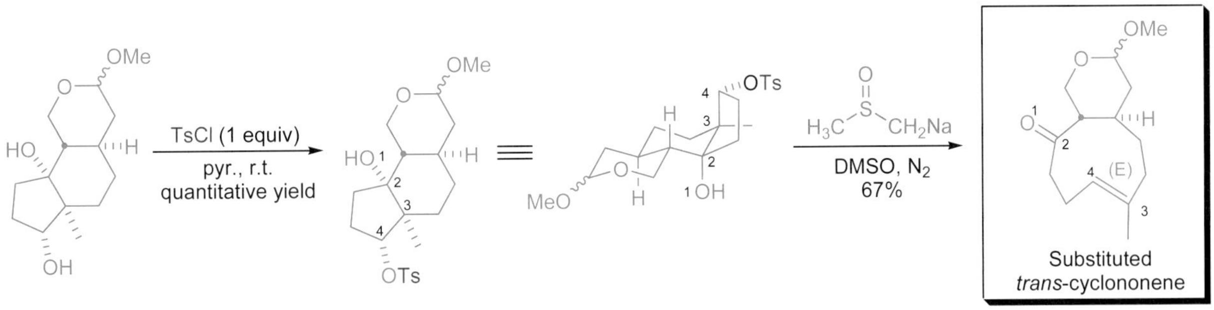

A novel synthetic approach was developed for the norbornane-based carbocyclic core of CP-263,114 in the laboratory of J.L. Wood.[21] Initial attempts to prepare the core using the *oxy-Cope rearrangement* failed even under forcing conditions, so an alternative approach utilizing the *Wharton fragmentation* was chosen. The tricyclic 1,3-diol substrate was prepared by the SmI_2-mediated 5-exo-trig *ketyl radical cyclization*. The resulting tertiary alcohol was mesylated and subjected to methanolysis, which afforded the *Wharton fragmentation* product in an almost quantitative yield.

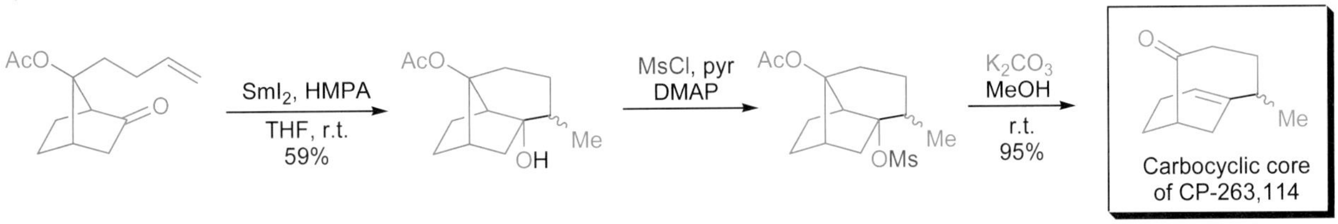

Research by S. Arseniyadis and co-workers showed that the *aldol-annelation-fragmentation* strategy could be used for the synthesis of complex structures, which are precursors of a variety of taxoid natural products.[22] This strategy allows the preparation of the twenty-carbon framework of taxanes from inexpensive and simple starting materials.

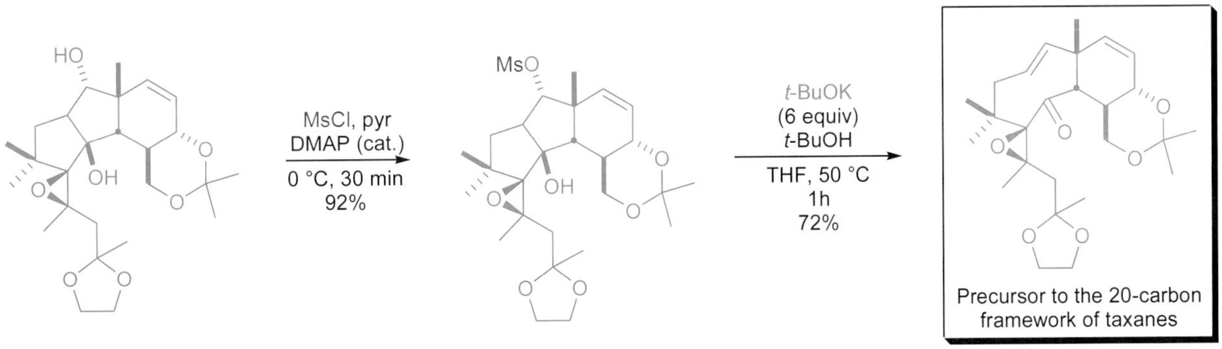

The stereocontrolled synthesis of 5β-substituted kainic acids was achieved by A. Rubio et al.[23] The C3 and C4 substituents were introduced by the *Wharton fragmentation* of a bicyclic monotosylated 1,3-diol. When this secondary alcohol was exposed to KO*t*-Bu, the corresponding fragmentation product was obtained in moderate yield. *Jones oxidation* of the aldehyde to the carboxylic acid followed by hydrolysis of the ester and removal of the Boc group resulted in the desired substituted kainic acid.

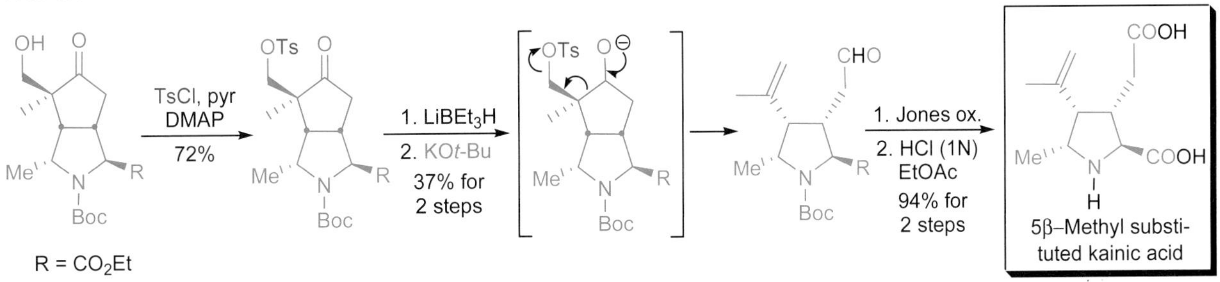

WHARTON OLEFIN SYNTHESIS (WHARTON TRANSPOSITION)

(References are on page 706)

Importance:

[*Seminal Publications*[1-4]; *Reviews*[5]; *Modifications & Improvements*[6-8]]

In 1913, N. Kishner reported that treating 2-hydroxy-2,6-dimethyloctan-3-one under standard *Wolff-Kishner reduction* conditions (N_2H_4/KOH/glycol/heat) gave the corresponding reductive elimination product 2,6-dimethyl-2-octene.[1] This transformation is known as the *Kishner eliminative reduction*. It was shown to work with a wide variety of α-substituted ketones, so it offers a convenient and regioselective introduction of a double bond into acyclic and cyclic ketones.[9,10] In 1961, an extension of this method was introduced independently by P.S. Wharton and Huang-Minlon when they described the rearrangement of α,β-epoxyketones to allylic alcohols *via* the corresponding epoxyhydrazones. Today, this transformation is referred to as the *Wharton olefin synthesis* or *Wharton transposition*.[3,4] The general features of this transformation are the following:[8] 1) the epoxidation of α,β-unsaturated ketones is achieved usually by basic hydrogen peroxide solution in high yield; 2) according to the classical Wharton conditions, the epoxyketone was treated with 2-3 equivalents of hydrazine hydrate in the presence of substoichiometric amounts of acetic acid, and the allylic alcohol product formed in a matter of minutes; 3) the classical reaction conditions are not free of water, which is unsuitable for sensitive substrates; 4) stable epoxyhydrazones can be prepared by treating the epoxyketones with hydrazyne hydrate in CH_2Cl_2, and in a separate step a strong base (e.g., KDA, KO*t*-Bu) is added at low temperature to afford the desired products; 5) unstable epoxyhydrazones can be prepared and rearranged when the corresponding epoxyketones are added to the solution of an *in situ* generated hydrazine (hydrazine salt + NEt_3), which is anhydrous; and 6) in acyclic systems there is no marked selectivity for the configuration of the new double bond.

Kishner eliminative reduction (1913):

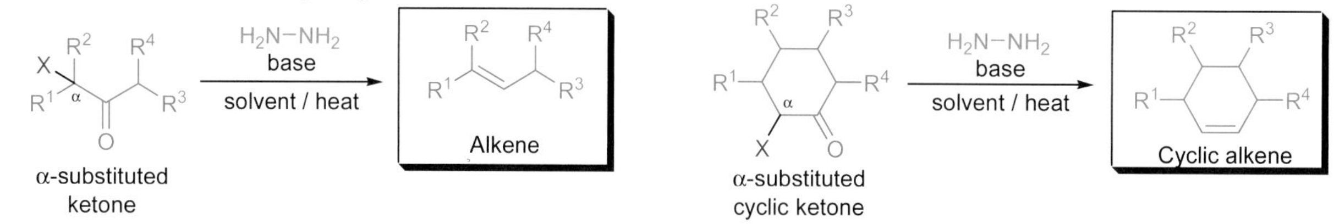

Wharton olefin synthesis (1961):

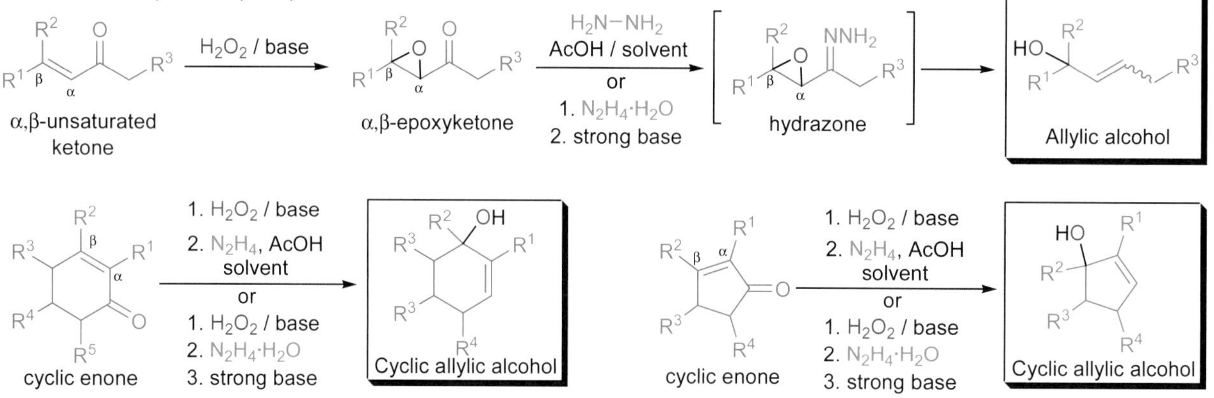

X = OH, *O*-alkyl, OPh, NR_2, *S*-alkyl, OCO-alkyl, Cl, Br, I; R^1,R^2,R^3,R^4, R^5 = H, alkyl, aryl

Mechanism:[4,6,8]

The mechanism of the *Wharton transposition* is very similar to that of the *Wolff-Kishner reaction*. The epoxyhydrazone is first deprotonated, which triggers the facile and irreversible epoxide ring-opening. The C-N bond of the resulting vinyl diazene[11,12] is broken upon another deprotonation, releasing N_2 and a vinyl anion, which in turn affords the desired allylic alcohol. Alternatively, the formation of a vinyl radical has been proposed.[6]

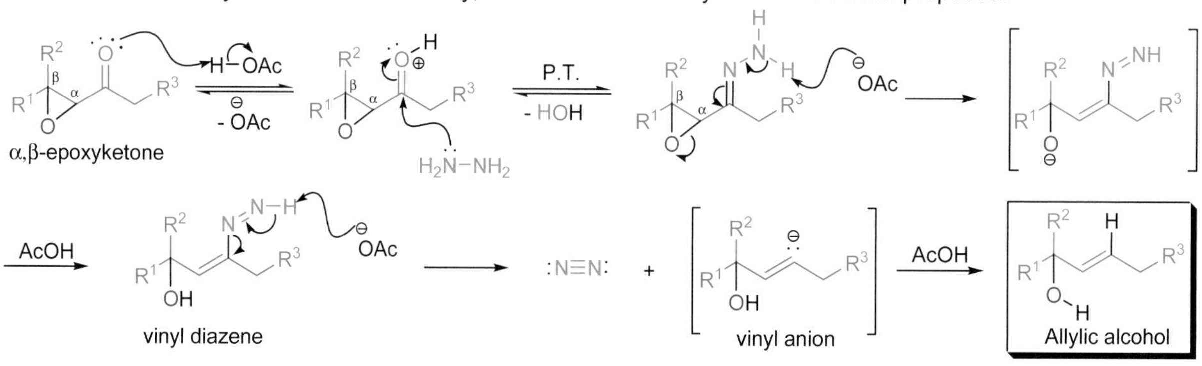

Synthetic Applications:

During the total synthesis of the anticancer natural product OSW-1, Z. Jin and co-workers explored several approaches to prepare a crucial steroid enone precursor with high stereoselectivity.[13] In one of the approaches, the commercially available 5-pergnen-16,17-epoxy-3β-ol-20-one was protected with a TBS group and was subjected to the *Wharton transposition*. The epoxyketone was treated with hydrazine hydrate in THF/MeOH under reflux to give the expected allylic alcohol in good yield. The desired enone was obtained by the *Dess-Martin oxidation* of the allylic alcohol with a slight preference for the (*Z*)-stereoisomer.

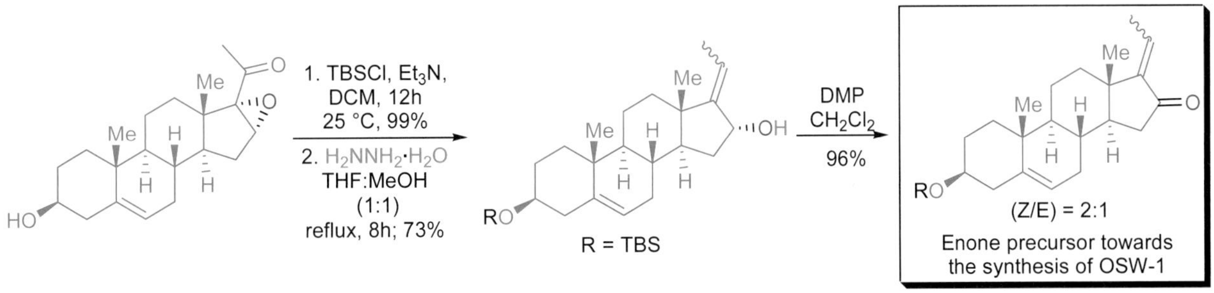

The racemic synthesis of decipienin A was accomplished in the laboratory of G.M. Massanet.[14] In the late stages of the total synthesis, the tricyclic enone lactone was converted to the corresponding α,β-epoxyketone by treatment with hydrogen peroxide in the presence of NaOH. The epoxyketone was subjected to the conditions of the *Wharton transposition* to afford the cyclic allylic alcohol in excellent yield. Several subsequent steps completed the total synthesis.

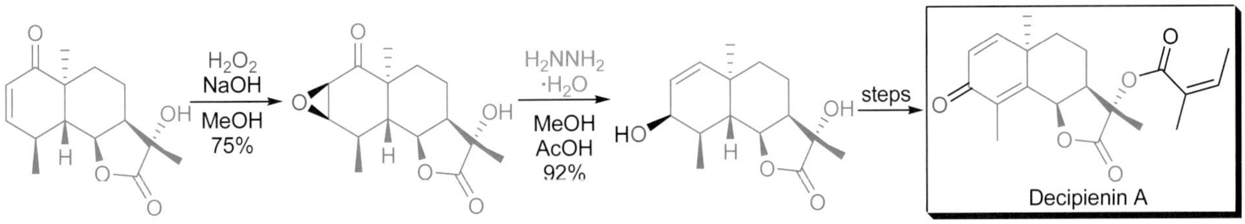

The synthesis of the bioactive natural product warburganal from (-)-sclareol was carried out by A.F. Barrero et al.[15] The bicyclic allylic acetate was epoxidized and deacetylated under basic conditions. Next, the solution of the ketoepoxide in glacial acetic acid was treated with hydrazine hydrate and the resulting mixture was heated at reflux for 30 minutes to afford the bicyclic allylic diol in excellent yield.

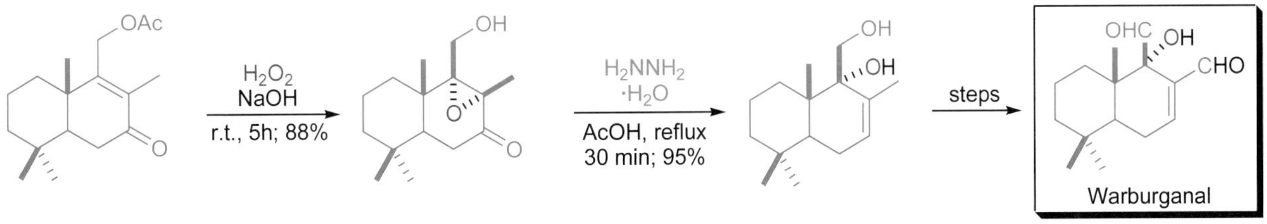

Research by M. Majewski et al. showed that the enantioselective ring opening of tropinone allowed for a novel way to synthesize tropane alkaloids such as physoperuvine.[16] The treatment of tropinone with a chiral lithium amide base resulted in an enantioslective deprotonation, which resulted in the facile opening of the five-membered ring to give a substituted cycloheptenone. This enone was subjected to the *Wharton transposition* by first epoxidation under basic conditions followed by addition of anhydrous hydrazine in MeOH in the presence of catalytic amounts of glacial acetic acid.

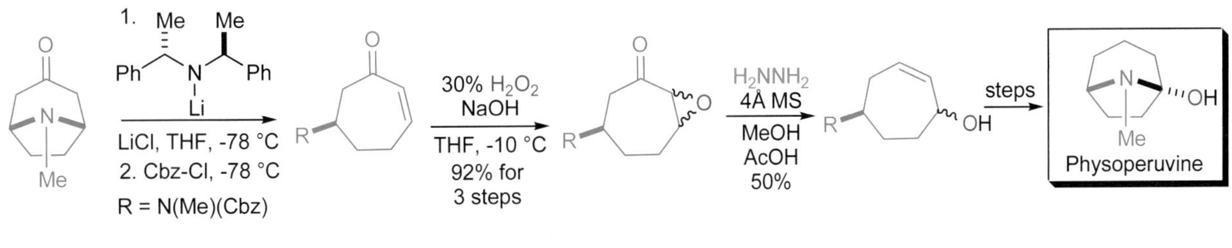

WILLIAMSON ETHER SYNTHESIS

(References are on page 706)

Importance:

[*Seminal Publications*[1,2]; *Reviews*[3-7]; *Modifications & Improvements*[8-19]; *Theoretical Studies*[20]]

In 1851, W. Williamson was the first to establish the correct formula of diethyl ether, which was first prepared by V. Cordus in 1544 by heating ethanol with sulfuric acid.[1] Williamson synthesized diethyl ether from sodium ethoxide and ethyl chloride. The reaction of aliphatic or aromatic alkoxides with alkyl, allyl, or benzyl halides to afford the corresponding ethers is known as the *Williamson ether synthesis*. The general features of this transformation are: 1) alkali metal alkoxides of simple aliphatic primary, secondary and tertiary alcohols are easily prepared by the use of strong bases such as NaH, KH, LHMDS, or LDA; 2) preparation of alkali metal salts of phenols (hydroxy-substituted aromatic or heteroaromatic compounds) are accomplished by reacting them with weak bases such as sodium- or potassium hydroxide or alkali metal carbonates such as potassium- or cesium carbonate, since phenols are more acidic than aliphatic alcohols; 3) alternatively, the alcohol can be directly reacted with alkali metals such as sodium or potassium at ambient or elevated temperatures in the neat substrate or at low temperature in liquid ammonia; the pure alkoxides are obtained by evaporating the excess alcohol or the liquid ammonia; 4) most alkali metal alkoxides and phenoxides can be obtained in crystalline form and stored indefinitely under an inert gas atmosphere and in the absence of moisture; 5) the reaction is usually carried out in a dipolar aprotic solvent such as DMF or DMSO to minimize side products as a result of dehydrohalogenation; 6) the choice of the alkyl halide component is critical to the success of the reaction: primary alkyl, methyl, allylic, and benzylic halides give the highest yields, since these undergo S_N2 type halide displacement by the alkoxide nucleophile; 7) the order of reactivity for the halides regarding the alkyl group: Me>allylic~benzylic>1° alkyl>2°alkyl while under standard conditions tertiary alkyl halides undergo E2 elimination to afford the corresponding alkenes; 8) the order of reactivity is also influenced by the nature of the leaving group: OTs~I>OMs>Br>Cl; and 9) when alkyl dihalides containing two different halogen atoms (such as Cl or I) are employed in the reaction, the chemoselective displacement of the better leaving group will occur. The preparation of diaryl ethers from phenoxides and unactivated aryl halides is not possible under the reaction conditions of the *Williamson ether synthesis*, but in the presence of copper metal or Cu(I)-salt catalysts, diaryl ethers are obtained (see *Ullmann biaryl ether synthesis*). When the aryl halide is activated (strongly electron-withdrawing substituents are present) the displacement of the halogen atom by the alkoxide is possible in the absence of catalyst (nucleophilic aromatic substitution). There are a few limitations of *Williamson ether synthesis*: 1) tertiary alkyl halides or sterically hindered primary or secondary alkyl halides tend to undergo E2 elimination in the presence of the alkoxide that in addition to being a nucleophile also acts as a base; and 2) alkali phenoxides may undergo *C*-alkylation in addition to expected *O*-alkylation.

Valerius Cordus (1544):

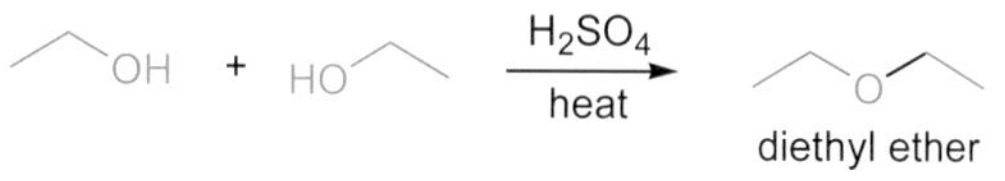

Williamson (1851):

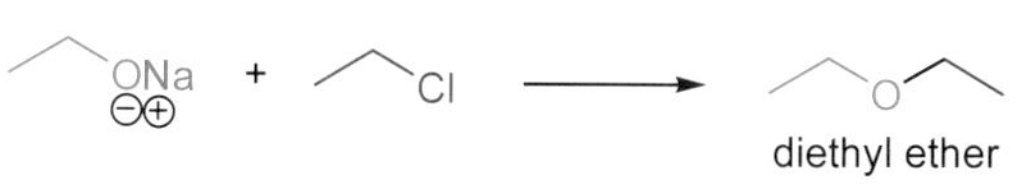

Williamson synthesis of dialkyl ethers:

Williamson synthesis of aryl ethers:

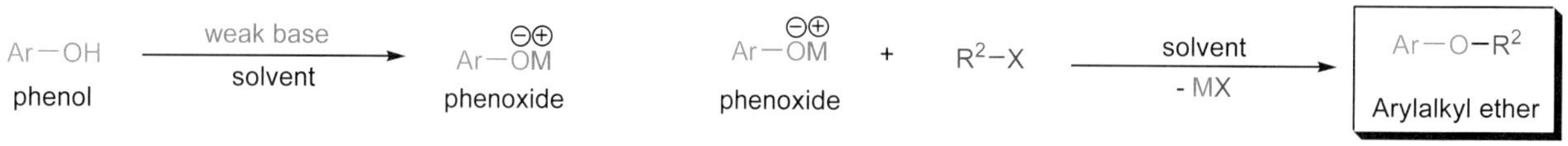

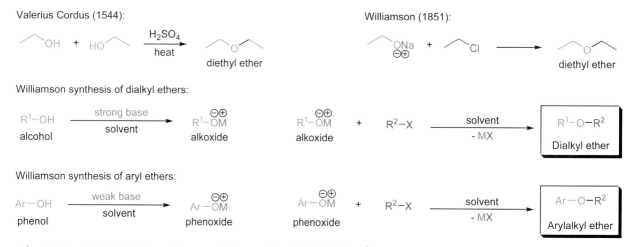

R^1 = 1°, 2° or 3° alkyl, allyl, benzyl; Ar = aryl, heteroaryl; M = Li, Na, K, Cs; R^2 = 1° or 2° alkyl, allyl, benzyl; X = Cl, Br, I, OMs, OTs; strong base: alkali metals/liquid ammonia, metal hydrides, LHMDS, LDA; weak base: NaOH, KOH, K_2CO_3, Cs_2CO_3; solvent: usually dipolar aprotic such as DMSO, DMF

Mechanism: [21-24]

In the case of most alkoxides and primary or secondary alkyl halides, the mechanism of the *Williamson ether synthesis* proceeds *via* an S_N2 process. When the alkyl halide is secondary (R''=H) with a given absolute configuration, the product ether will have a complete inversion of configuration at that particular sterereocenter. E.C. Ashby demonstrated, however, that the reaction between lithium alkoxides and alkyl iodides proceeds *via* single-electron transfer.[22]

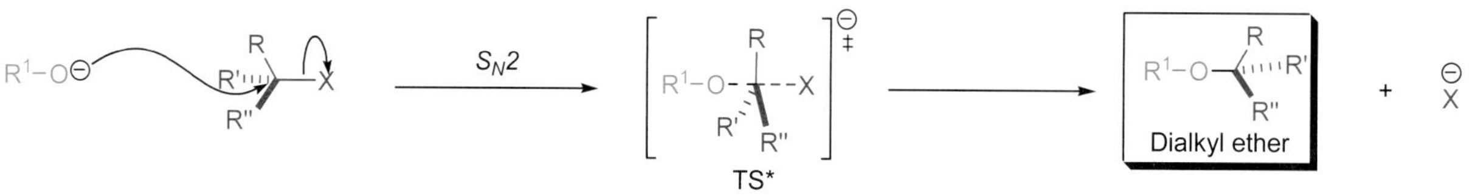

Synthetic Applications:

The redox-active natural product (±)-methanophenazine (MP) is the first phenazine to be isolated from archea. This compound is able to mediate the electron transport between membrane-bound enzymes and was characterized as the first phenazine derivative involved in the electron transport of biological systems. The research team of U. Beifuss prepared this natural product by using the *Williamson ether synthesis* in the last step of the synthetic sequence.[25] The etherification was conducted under phase-transfer conditions in a THF/water system in the presence of methyltrioctyl-ammonium chloride and using potassium hydroxide as a base.

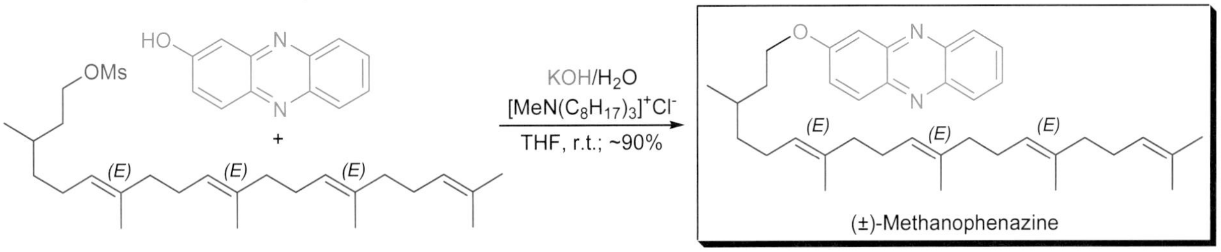

The total synthesis of (+)-asimicin, which belongs to the family of Annonaceous acetogenins, was completed by E. Keinan and co-workers.[26] In order to create one of the tetrahydrofuran rings stereospecifically, an *intramolecular Williamson ether synthesis* was performed between a secondary alcohol and a secondary mesylate using pyridine as the base.

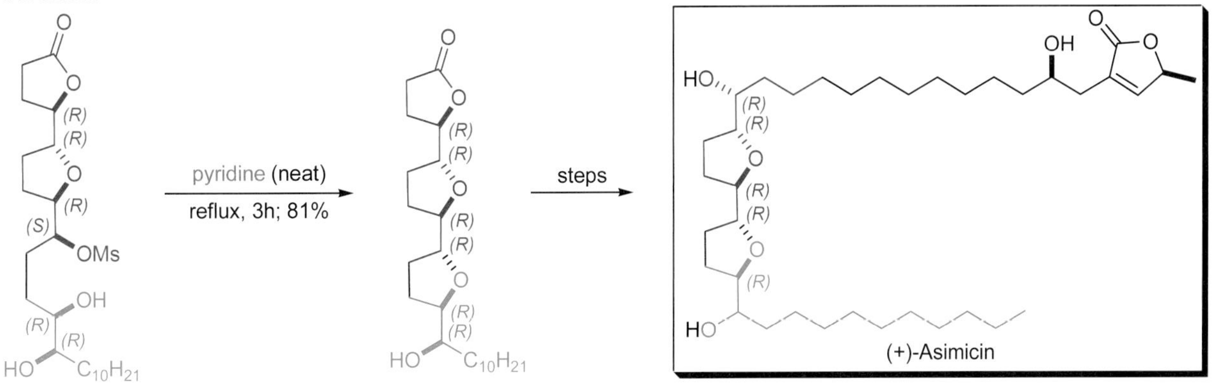

In the laboratory of D. Kim, the asymmetric total synthesis of (–)-fumagillol, the hydrolysis product of fumagillin, was accomplished.[27] The stereoselective introduction of the sensitive 1,1-disubstituted epoxide moiety took place in the final stages of the synthesis. The primary alcohol portion of the vicinal diol functionality was first selectively converted to the corresponding tosylate. Upon treatment with K_2CO_3/MeOH the epoxide formation occurred smoothly.

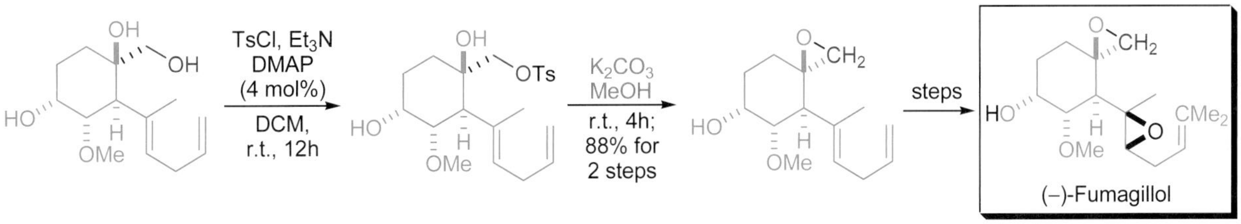

The two key ether linkages during the total synthesis of archaeal 36-membered macrocyclic diether lipid by K. Kakinuma and co-workers were formed using the *Williamson ether synthesis*.[28] Two equivalents of the enantiopure isoprenoid mesylate was added to the dialkoxide derived from 1-*O*-benzyl-glycerol and the corresponding diether was isolated in good yield. Four more steps including a *McMurry coupling* completed the synthetic sequence.

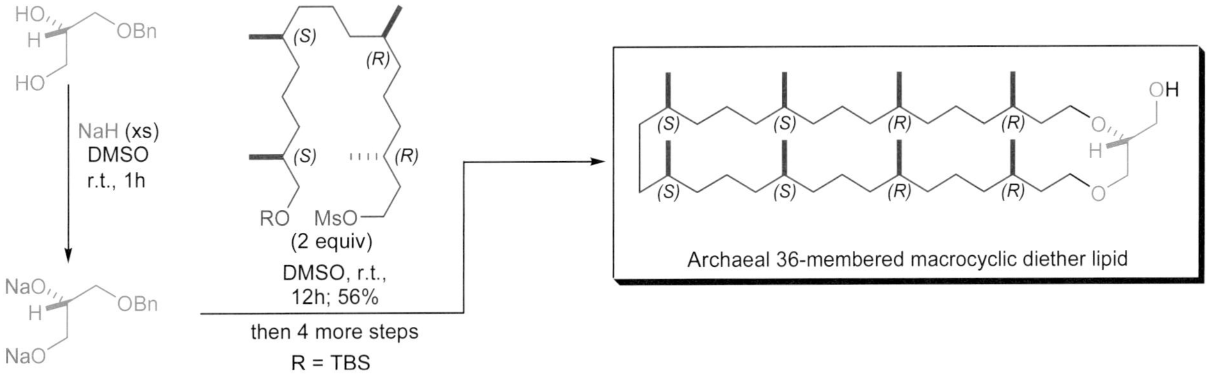

WITTIG REACTION

(References are on page 707)

Importance:

[*Seminal Publications*[1-5]; *Reviews*[6-40]; *Modifications & Improvements*[41-54]; *Theoretical Studies*[55-70]]

In the early 1950s, G. Wittig and G. Geissler investigated the chemistry of pentavalent phosphorous and described the reaction between methylenetriphenylphosphorane ($Ph_3P{=}CH_2$) and benzophenone, which gave 1,1-diphenylethene and triphenylphosphine oxide ($Ph_3P{=}O$) in quantitative yield.[3] Wittig recognized the importance of this observation and conducted a systematic study in which several phosphoranes were reacted with various aldehydes and ketones to obtain the corresponding olefins.[4,5] The formation of carbon-carbon double bonds (olefins) from carbonyl compounds and phosphoranes (phosphorous ylides) is known as the *Wittig reaction*. From a historical point of view it is important to note that Wittig was not the first to prepare a phosphorane, since Staudinger and Marvel had reported the synthesis of such compounds three decades before.[1,2] Since its discovery, the *Wittig reaction* has become one the most important and most effective method for the synthesis of alkenes. The active reagent in this transformation is the phosphorous ylide, which is usually prepared from a triaryl- or trialkylphosphine and an alkyl halide (1° or 2°) followed by deprotonation with a suitable base (e.g., RLi, NaH, NaOR, etc.). There are three different types of ylides depending on the nature of the R^2 and R^3 substituents: 1) in the "*stabilized*" ylides the alkyl halide component has at least one strong electron-withdrawing group ($-CO_2R$, $-SO_2R$, etc.), which stabilizes the formal negative charge on the carbon; 2) "semi-stabilized" ylides have at least one aryl or alkenyl substituents as the R^2 or R^3 groups, which are less stabilizing; and 3) "nonstabilized" ylides usually have only alkyl substituents, which do not stabilize the formal negative charge on the carbon. The general features of the *Wittig reaction* are: 1) the phosphonium salts are usually prepared using triphenylphosphine, and the phosphorous ylides are generated before the reaction or *in situ*; 2) the ylides are water as well as oxygen-sensitive; 3) the phosphorous ylides chemoselectively react with aldehydes (fast) and ketones (slow), other carbonyl groups (e.g., esters, amides) remain intact during the reaction; 4) the stereoselectivity, *E*-or *Z*-selectivity, is influenced by many factors: type of ylide, type of carbonyl compound, nature of solvent, and the counterion for the ylide formation; 5) "nonstabilized" ylides under salt-free conditions in a dipolar aprotic solvent with aldehydes afford olefins with high (*Z*)-selectivity; 6) "stabilized" ylides give predominantly (*E*)-olefins with aldehydes under the same salt-free conditions; 7) "semi-stabilized" ylides usually give alkenes with poorer steroselectivity; and 8) ether solvents such as THF, Et_2O, DME, MTBE, or toluene are used. The *Wittig reaction* has several important variants: 1) the *Horner-Wittig reaction* takes place when the phosphorous ylides contain phosphine oxides in place of triarylphosphines;[71] 2) the use of stabilized alkyl phosphonate carbanions is known as the *Horner-Wadsworth-Emmons reaction* in which (*E*)-α,β-unsaturated esters are formed;[72] 3) in the *Schlosser modification*, "nonstabilized" ylides can give pure (*E*)-alkenes when two equivalents of Li-halide salt is present in the reaction mixture;[73] 4) *asymmetric Wittig reaction* were also developed;[53] and 5) *Wittig reaction* on solid support allows easy separation of the products from triphenylphosphine oxide.[42]

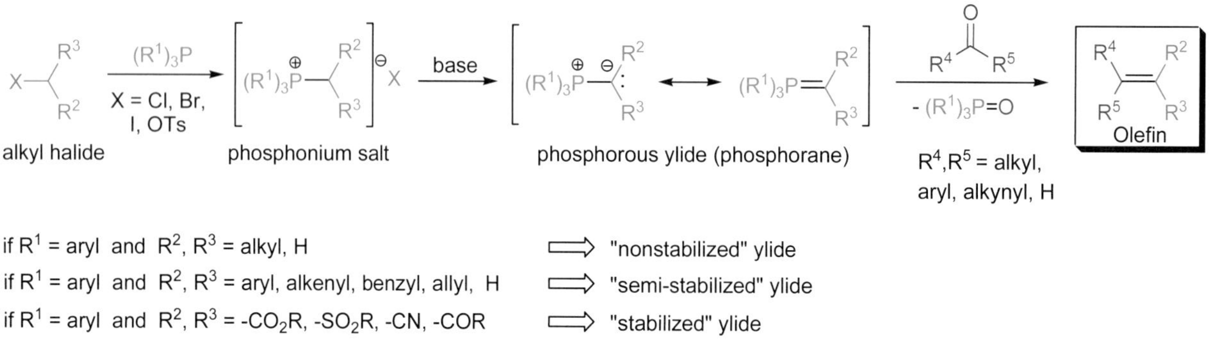

Mechanism: [9,23,74-77,28,78-82,37]

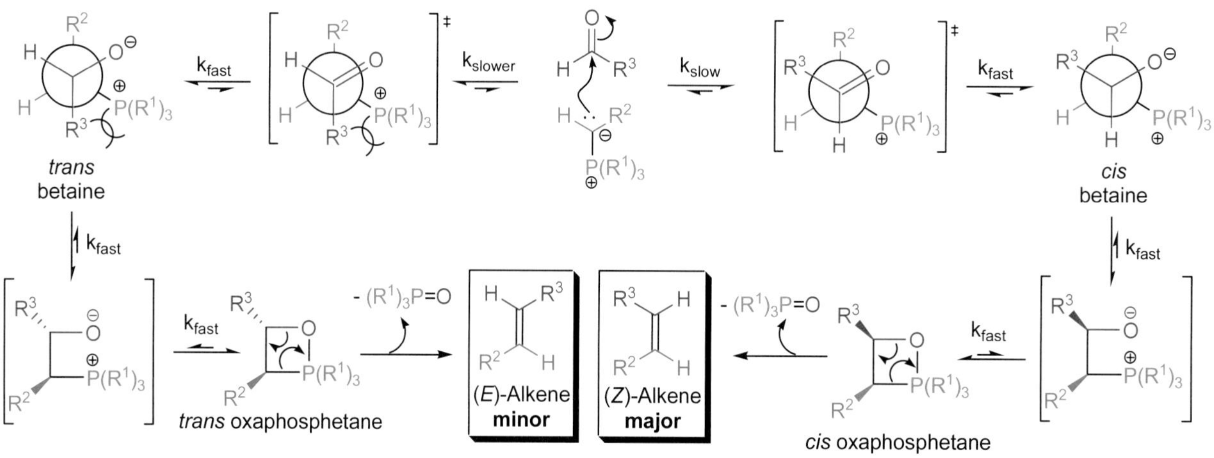

Synthetic Applications:

In the late stages of the gram-scale synthesis of (+)-discodermolide, A.B. Smith and co-workers utilized the highly *Z-selective Wittig reaction* to couple two advanced intermediates, a phosphonium salt and an aldehyde.[83] The phosphonium salt was prepared using the primary alkyl iodide, triphenylphosphine, Hünig's base, and high pressure. This procedure was necessary because the traditional methods led to the formation of substantial amounts of side-products and decomposition. The Hünig's base trapped any HI that was generated during the process and prevented the formation of decomposition products. The phosphonium salt was deprotonated with NaHMDS which, upon reacting with the aldehyde, afforded the desired C8-C9 alkene with high *Z*-selectivity.

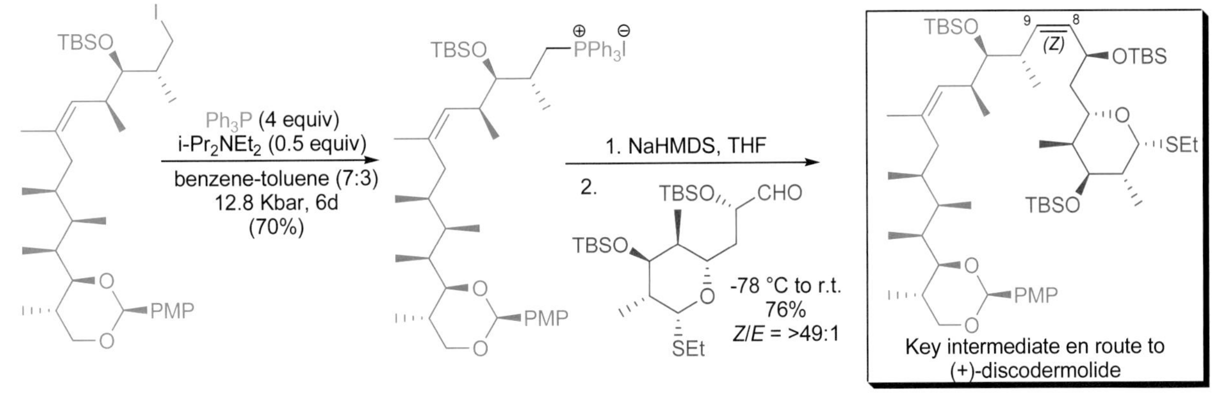

The total synthesis of amaryllidaceae alkaloid buflavin was achieved in the laboratory of A. Couture by utilizing a *Horner-Wittig reaction* between a biaryl aldehyde and a metalated carbamate.[84] The diphenyl phosphine oxide carbamate was deprotonated with *n*-BuLi. To the resulting metalated carbamate was added the solution of the biaryl aldehyde in THF. The reaction afforded the corresponding (*Z*)- and (*E*)-enecarbamates in good yield and with high *E*-selectivity.

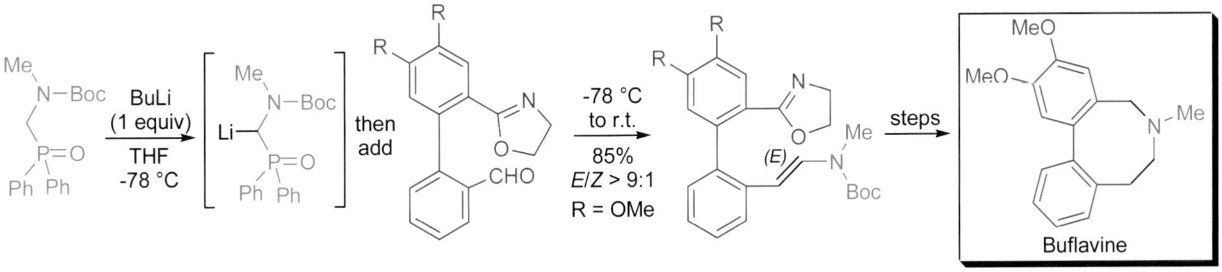

The *iterative Wittig olefination* was used to assemble β-D-C-(1,6)-linked oligoglucoses and oligogalactoses, which are connected through olefinic bridges. The strategy by A. Dondoni et al. involved the coupling of the sugar aldehyde building block with a substrate having a phosphorous ylide functionality at C6.[85] The yields were good in each step, and oligosaccharides up to pentaoses were prepared. The synthesis of a tetraose is illustrated.

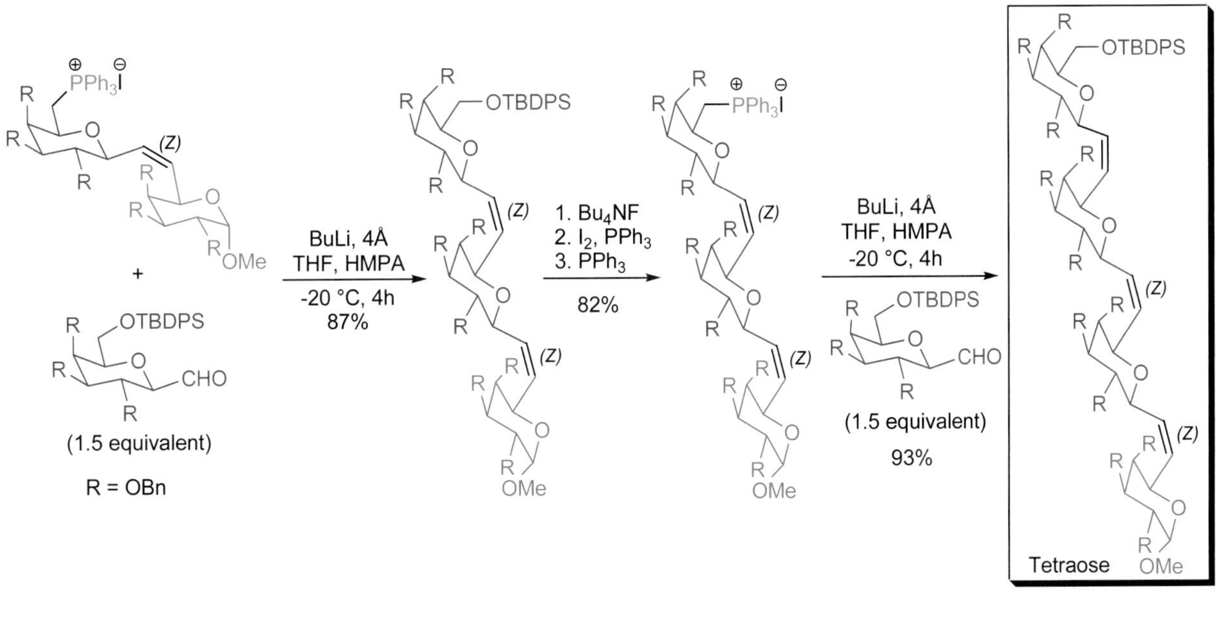

WITTIG REACTION - SCHLOSSER MODIFICATION

(References are on page 708)

Importance:

[*Seminal Publication*[1]; *Reviews*[2-6]; *Modifications & Improvements*[7-10]]

The one-pot multistep preparation of (*E*)-alkenes from "nonstabilized" phosphorous ylides and carbonyl compounds by the equilibration of the intermediate lithiobetaines is known as the *Schlosser modification of the Wittig reaction*. In the decade following the disclosure of a novel olefin synthesis using phosphorous ylides and carbonyl compounds by G. Wittig and G. Geissler,[11-13] intensive research was conducted to reveal what intermediates were involved in the reaction and what factors influenced the stereoselectivity. It was established early on that the so-called oxaphosphetanes (four-membered heterocycles containing a P-O bond) were the key intermediates, and the *cis*- and *trans* diastereomers decompose *via* cycloreversion to the corresponding *cis* and *trans* alkenes. In 1966, M. Schlosser reported that in the presence of excess lithium halide, the P-O bond of the oxaphosphetanes was rapidly cleaved and the corresponding diastereomeric lithiobetaines were formed.[1] At low temperature the lithiobetaines (pK_a = ~20) were deprotonated at their α-positions with alkyl- or aryllithiums[14] (PhLi, n-BuLi, etc.), and the resulting β-oxido phosphorous ylides rapidly equilibrated to give the thermodynamically more stable *trans* diastereomer. At this point, the diastereomerically pure *trans* β-oxido phosphorous ylide was protonated stereospecifically with one equivalent of a proton source (HCl in ether or alcohol) or an electrophile[7-10] to afford the pure *trans* lithiobetaine and the excess lithium halide was removed with KO*t*-Bu. The resulting *trans* betaine gave the corresponding (*E*)-alkene *via* the *trans* oxaphosphetane.

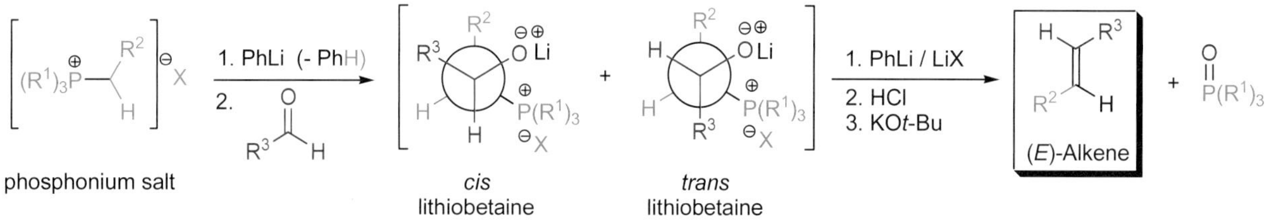

R^1 = aryl; R^2 = alkyl, H; R^3 = alkyl, aryl; X = Cl, Br, I

Mechanism:[2,15,14]

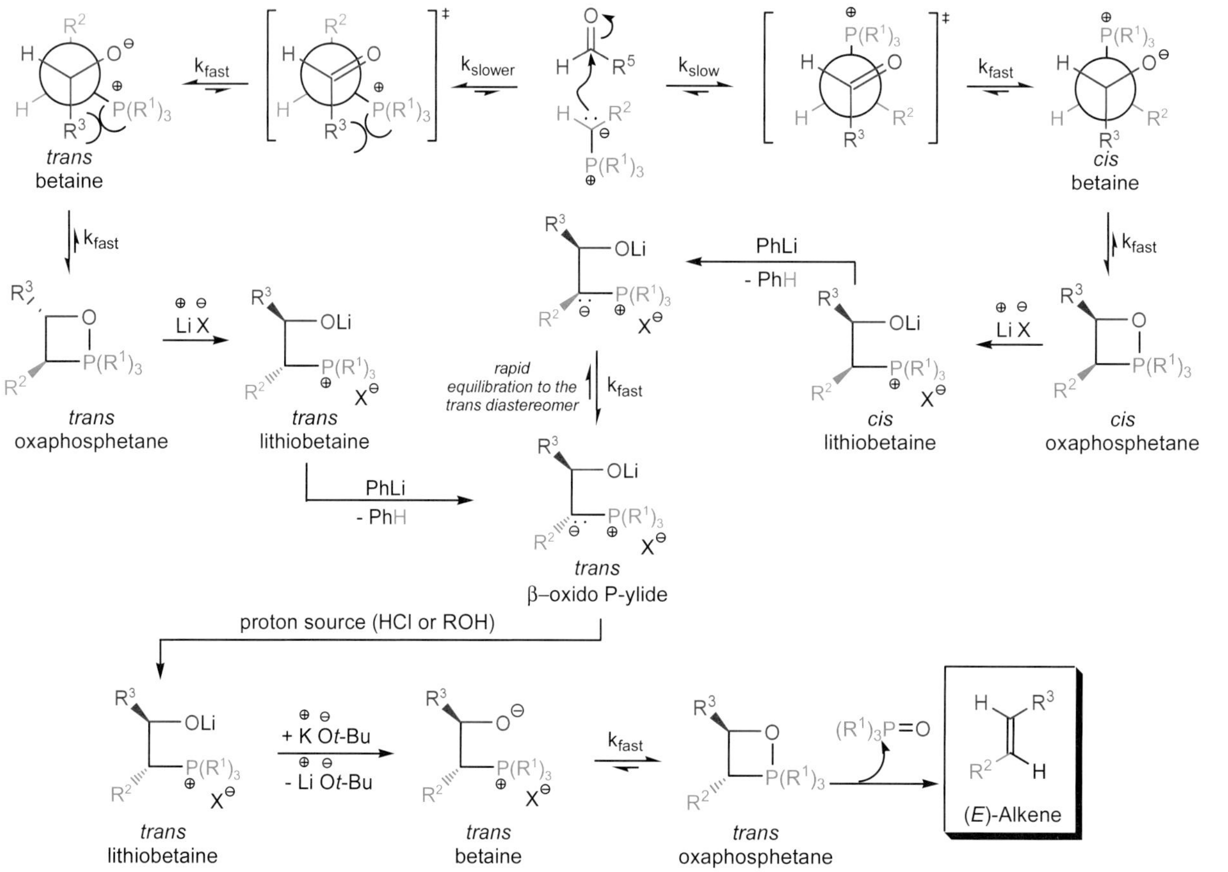

Synthetic Applications:

The asymmetric total synthesis of ISP-I (myriocin, thermozymocidin) was accomplished by utilizing the *Schlosser modified Wittig reaction* as one of the key steps in the laboratory of Y. Nagao.[16] The phosphonium bromide fragment was treated with PhLi at 0 °C to generate the phosphorous ylide which was reacted with the aldehyde at -78 °C. The resulting mixture of lithiobetaines was treated with PhLi at 0 °C to afford the desired (*E*)-alkene with excellent stereoselectivity.

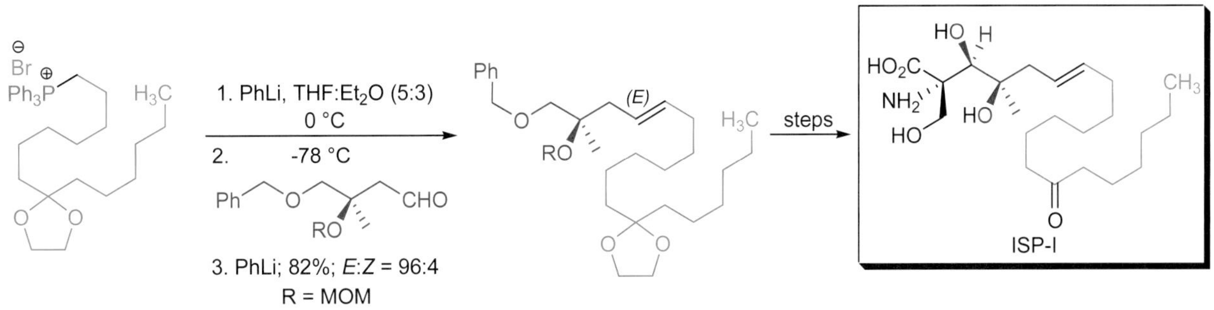

During the stereospecific total synthesis of (7*S*,15*S*) and (7*R*, 15*S*)-dolatrienoic acid by G.R. Pettit et al., the C7-C10 and C11-C16 subunits were coupled using the highly (*E*)-selective *Wittig-Schlosser reaction*.[17] The traditional Wittig conditions resulted in a mixture of alkenes in which the (*Z*)-stereoisomer was predominant. When the *Schlosser conditions* were applied, the stereoselectivity was reversed in favor of the (*E*)-alkene.

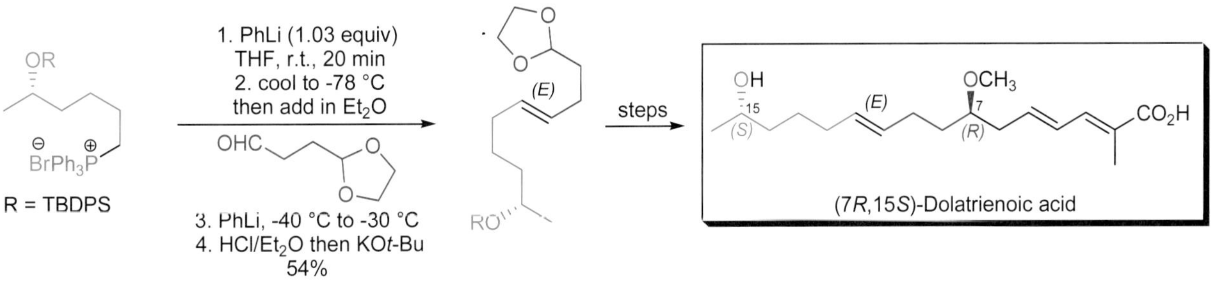

A simple and efficient method was developed by E.A. Couladouros and co-workers for the synthesis of optically pure five- or six-membered hydroxylactones.[18] The method begins from γ-butyrolactone and uses the following key transformations: reduction, *Wittig-Schlosser reaction*, *Sharpless asymmetric dihydroxylation*, oxidation, and lactonization. The preparation of antitumor agent (–)-muricatacin was achieved in 6 steps and in 43% overall yield.

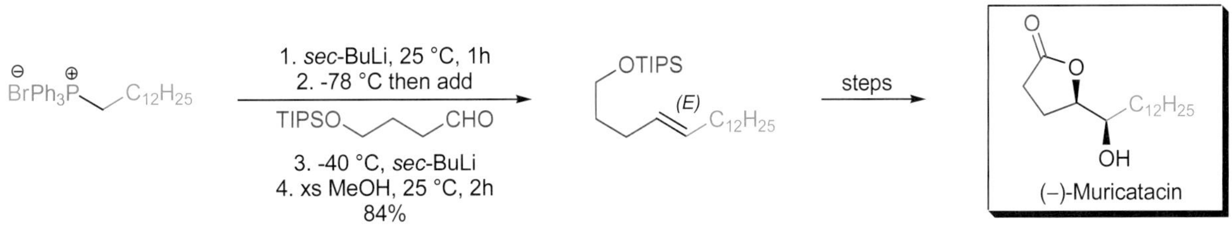

In the laboratory of M. Martin-Lomas, a short and enantiodivergent synthetic route was designed and carried out to both D-*erythro* and L-*threo*-sphingosine I and II.[19] The *trans* double bond was introduced using the *Schlosser modified Wittig reaction* by coupling tetradecyltriphenylphosphonium bromide and a chiral aldehyde. Other olefination methods proved inferior: coupling *via* the traditional *Wittig reaction* afforded mostly the *cis* olefin and the *Julia-Lythgoe olefination* gave low yield and low selectivity.

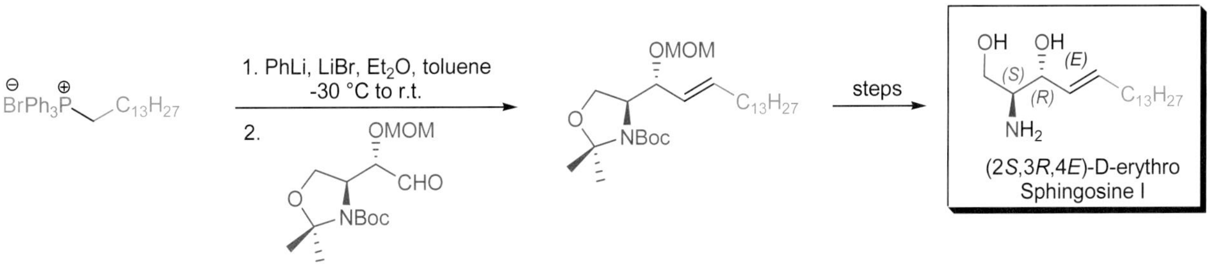

WITTIG-[1,2]- AND [2,3]-REARRANGEMENT

(References are on page 709)

Importance:

[*Seminal Publications*[1-4]; *Reviews*[5-20]; *Modifications & Improvements*[21-25]; *Theoretical Studies*[26-40]]

In 1942, G. Wittig and L. Löhmann reported that the deprotonation of benzyl methyl ether with phenyllithium afforded 1-phenylethanol upon work-up.[1] Subsequent studies showed that the transformation was general for α-lithiated aryl alkyl ethers that undergo a facile rearrangement to give lithio alkoxides in an overall *[1,2]-alkyl shift.* The rearrangement of aryl alkyl ethers to the corresponding secondary or tertiary alcohols in the presence of stoichiometric amount of a strong base is known as the *[1,2]-Wittig rearrangement*.[16,18] The most important features are: 1) the R^1 substituent has to be able to stabilize the carbanion; 2) the chiral center in the migrating group retains its configuration; 3) yields are usually moderate due to the harsh reaction conditions and the competing *[1,4]*-pathway; 4) at low temperatures, the formation of the [1,4]-product is favored, while at higher temperatures the [1,2]-product dominates. During the course of early mechanistic studies of this process, the research groups of G. Wittig and T.S. Stevens found that upon deprotonation, allylic ethers mainly underwent a *[2,3]-sigmatropic shift* to afford homoallylic alcohols, a process that is now referred to as the *[2,3]-Wittig rearrangement*.[2,4] The general features of the *[2,3]-rearrangement* are: 1) it proceeds under milder conditions and gives higher yields than the *[1,2]-rearrangement*; 2) virtually any α-(allyloxy)carbanion can udergo the rearrangement; the only limitation lies with the chemist's ability to generate a particular anion with currently available methods; 3) the R^4 substituent should be a carbanion-stabilizing group; 4) the *[1,2]*- and *[2,3]*-shifts often compete, and the amount of each product depends strongly on the structure of the substrate and the reaction temperature; 5) by carefully optimizing the reaction temperature, the formation of the *[1,2]*-rearranged product can be avoided; 6) for acyclic and cyclic substrates, the anions can be generated by a variety of different methods: with a strong base (e.g., LDA, *n*-BuLi) at -60 to -85 °C, *via* a *tin-lithium exchange reaction* (*Still variant*)[21] and by *reductive lithiation* of *O,S*-acetals; 7) because of the highly ordered cyclic transition state, the rearrangement is stereoselective with respect to the stereochemistry of the new double bond and the two new stereocenters;[15] 8) in acyclic substrates, the chirality of the C1 stereocenter of the substrate gets transferred to the product in a predictable fashion, consistent with the orbital symmetry conservation rules;[15,17] 9) the newly formed double bond generally has the (*E*)-stereochemistry, but the *Still variant* (R^4=SnR_3) gives predominantly (*Z*)-olefins; 10) the highest (*E*)-selectivity is achieved when the allylic moiety is only monosubstituted (R^5=alkyl and R^6=H); 11) the diastereoselectivity with respect to the newly created vicinal chiral centers is high: (*Z*)-substrates give *erythro* products with high levels of selectivity, while (*E*)-substrates afford *threo* products with lower selectivity, but the nature of the R^4 substituent also has a profound effect on the level of diastereoselectivity;[17] and 12) five different asymmetric versions of the rearrangement have been identified.[17]

Wittig and Löhmann (1942):

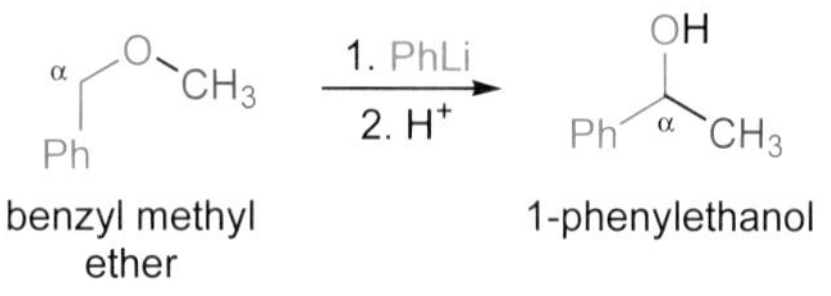

Wittig (1949) and Stevens et al. (1960):

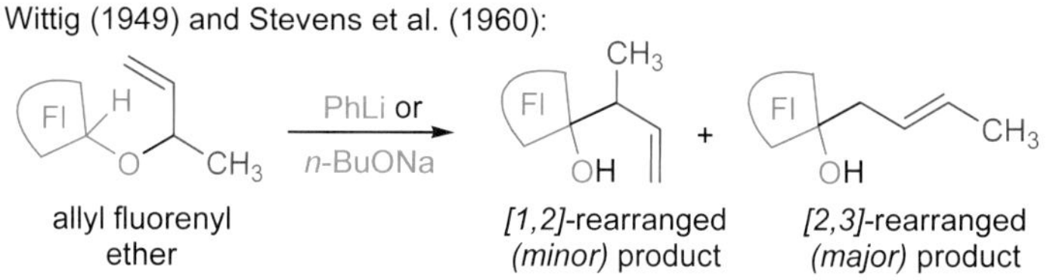

[1,2]-Wittig rearrangement:

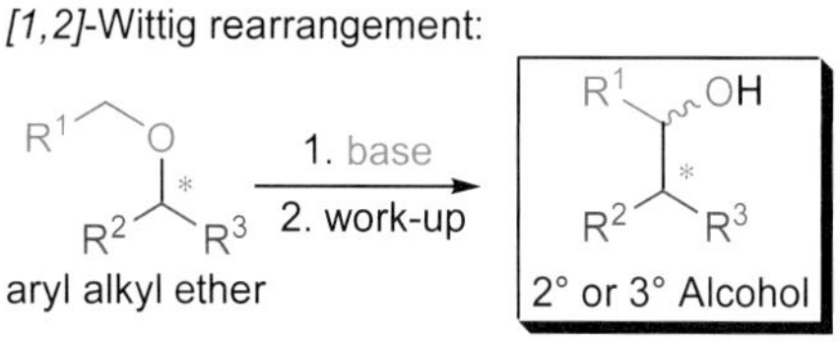

[2,3]-Wittig rearrangement:

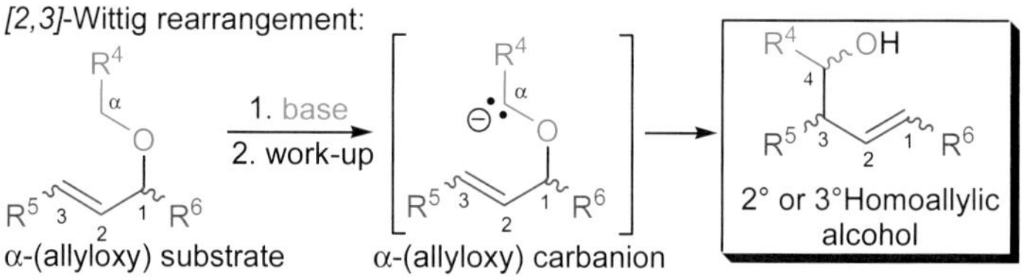

R^1 = aryl, alkenyl, alkynyl; R^{2-3} = H, alkyl; R^4 = carbanion stabilizing = aryl, alkenyl, alkynyl, COR, CN, CO_2R, $CONR_2$; when R^4 = SnR_3 (*Still variant*); R^{5-6} = H, alkyl; base: LDA, *n*-BuLi, PhLi, ROLi, $NaNH_2/NH_3$

Mechanism: [41-45,10,26,46,15]

The *[1,2]-Wittig rearrangement* proceeds *via* a radical-pair dissociation-recombination mechanism, while the *[2,3]-Wittig rearrangement* is a concerted, thermally allowed sigmatropic process proceeding *via* an envelope-like transition state in which the substituents are pseudo-equatorial.

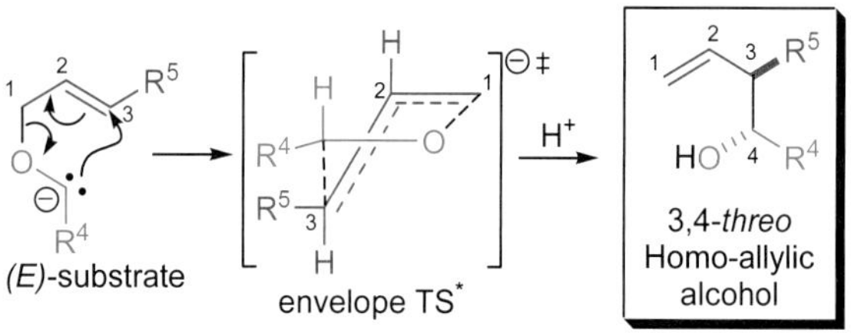

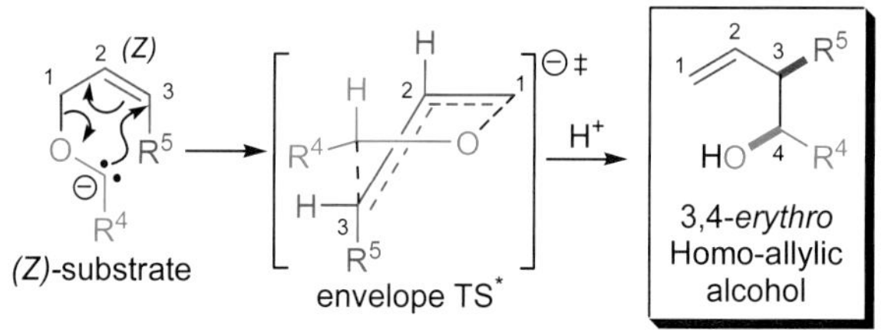

Synthetic Applications:

The *acetal version of the [1,2]-Wittig rearrangement* was utilized in the stereoselective total synthesis of zaragozic acid A by K. Tomooka and co-workers.[47] The acetal-protected *bis*(ethynyl)methanol was treated with *n*-BuLi, which brought about the sigmatropic *[1,2]-shift*. Thus, the chiral centers at C5 and C6 were established with high diastereoselectivity (95% β at C5 and 84% d.r. at C4). It is worth noting, that the intermediate anomeric radical could efficiently discriminate between the enantiotopic faces of prochiral *bis*(ethynyl)methanol radical (TMS vs. TBDPS) during the radical recombination process.

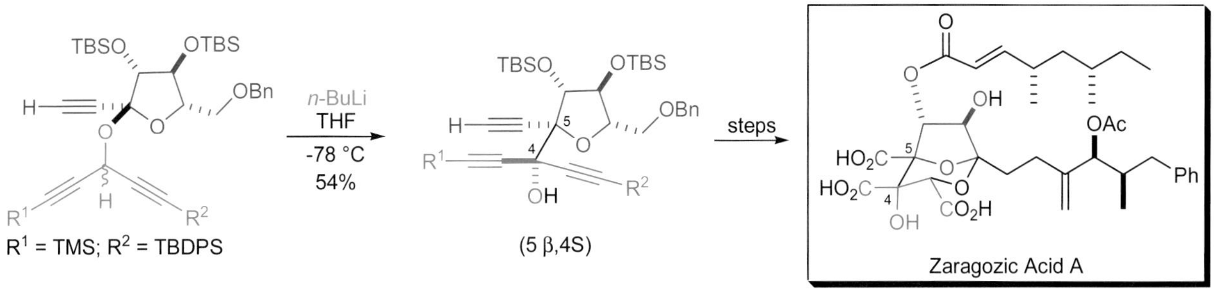

The first asymmetric total synthesis of (+)-astrophylline was accomplished in the laboratory of S. Blechert.[48] The *Still variant of the [2,3]-Wittig rearrangement* was used to generate the 1,2-*trans* relationship between the substituents of the key cyclopentene intermediate. The tributylstannylmethyl ether substrate was transmetalated with *n*-BuLi, which initiated the desired *[2,3]-sigmatropic shift* to afford the expected homoallylic alcohol as a single enantiomer.

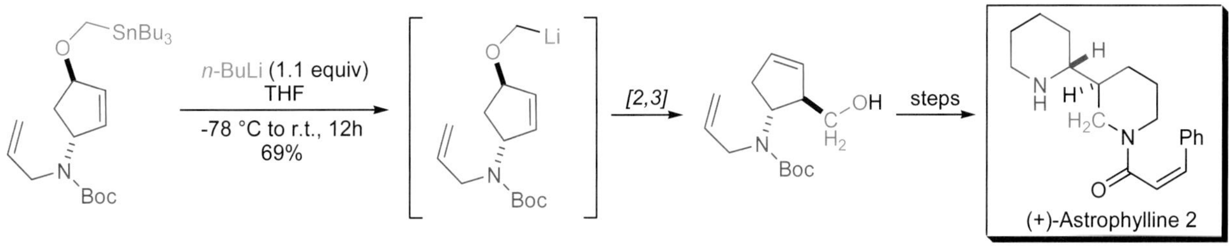

The last and key step in the total synthesis of both enantiomers of sarcophytols A and T by Y. Fukuyama et al. was a *stereospecific [2,3]-Wittig rearrangement*.[49] The deprotonation of the macrocyclic bis-allylic ether precursor occurred with complete regioselectivity at the less substituted position. The rearrangement proceeded in excellent yield and exhibited an unexpectedly high level of stereospecificity even though the substrate was highly flexible. The reaction could occur either *via* a *syn* or *anti* carbanionic intermediate, but the (*S*)-stereochemistry of the product indicated that the *anti* carbanion was operational.

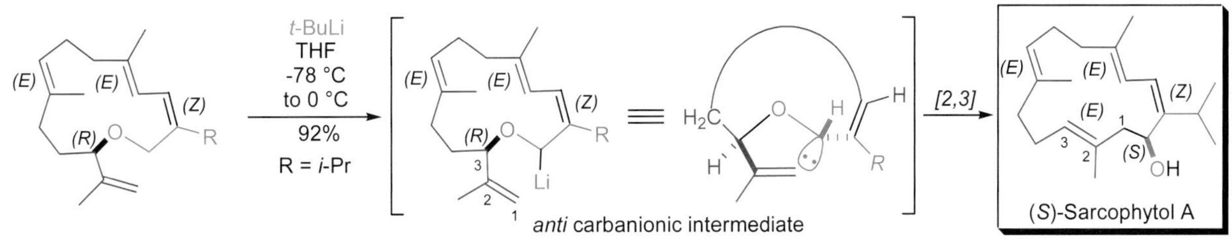

A novel approach to the asymmetric synthesis of Stork's prostaglandin intermediate was developed by T. Nakai et al.[50] This was the first example of an *asymmetric [2,3]-Wittig rearrangement*, in which three contiguous chiral centers were created in a cyclic system. Upon deprotonation, the rearrangement of the allyl propargyl ether substrate took place in excellent yield and gave rise to a single stereoisomer. Interestingly, when the TMS group was replaced with an amyl group (C_5H_{11}), the stereoselectivity diminished to only 3:1.

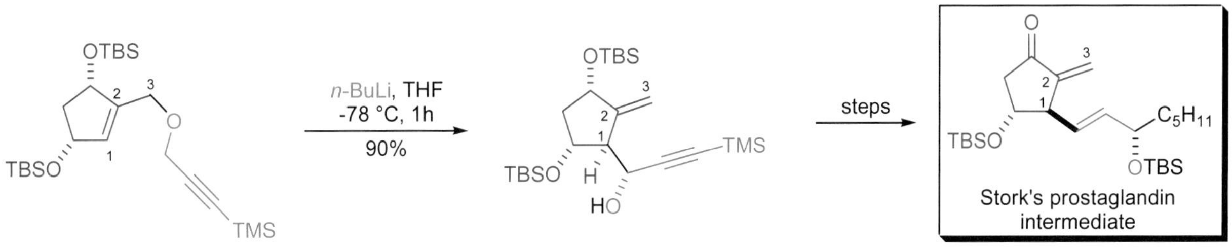

WOHL-ZIEGLER BROMINATION

(References are on page 710)

Importance:

[*Seminal Publications*[1-3]; *Reviews*[4-6]; *Modifications & Improvements*[7-12]; *Theoretical Studies*[13-15]]

In 1919, A. Wohl studied the reaction between 2,3-dimethyl-2-butene and *N*-bromoacetamide in cold diethyl ether and found that the double bond of the substrate remained intact and one of the methyl groups was substituted with a single bromine atom.[1] This observation was interesting because such a transformation was previously possible only by the reaction of alkenes with elemental bromine at high temperature, but it went unnoticed for almost two decades. In 1942, K. Ziegler and co-workers conducted a detailed study on the allylic bromination of olefins using *N*-bromosuccinimide (NBS) as a new and stable brominating agent and demonstrated the preparative value of such a halogenation process. A few years later, P. Karrer found that the addition of 5-10 mol% of dibenzoyl peroxide to the reaction mixture results in significant increase in the reaction rate and allowed the bromination of substrates that were unreactive under the original reaction conditions.[7] The introduction of a bromine substituent at the allylic position of olefins or at the benzylic position of alkylated aromatic or heteroaromatic compounds in known as the *Wohl-Ziegler bromination*. The general features of this transformation are: 1) NBS is a commercially available reagent, and it is stable when kept in the dark and away from moisture; 2) various other *N*-bromo amides and *N*-bromo imides can also be used for bromination, but NBS is by far the most effective of all, and its use is accompanied by the least amount of side products; 3) when the olefin has two allylic positions, the bromination is regioselective and favors the bromination of the more substituted position (the more stable allylic radical); 4) alkylated aromatic and heteroaromatic compounds are selectively brominated at their respective benzylic positions (on the carbon directly attached to the aromatic ring) and no halogenation on the ring takes place; 5) the best solvents are carbon tetrachloride and benzene but recent environmentally friendly modifications use ionic liquids as the reaction medium, and even solvent-free conditions have been developed;[12] 6) the reaction is usually carried out at the boiling point of the solvent in the presence of 5-20 mol% of a radical initiator (AIBN or dibenzoyl peroxide); 7) alternatively the bromination can also be conducted at lower temperatures while the reaction mixture is irradiated with UV light; and 8) when the formation of polybrominated products is a side reaction, the use of a slight excess of the olefin substrate is recommended.

Wohl (1919): Ziegler (1942):

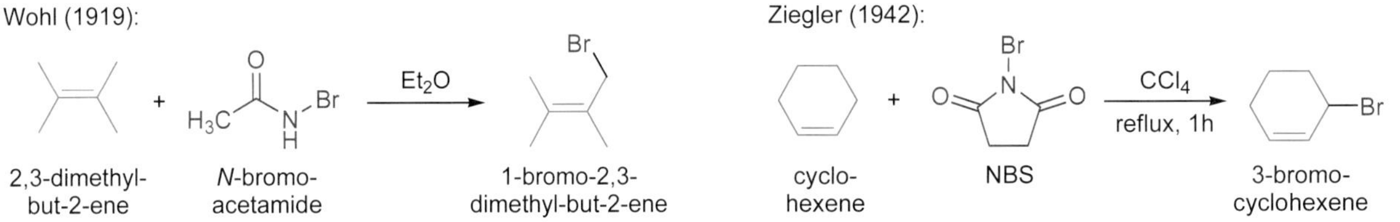

Wohl-Ziegler bromination:

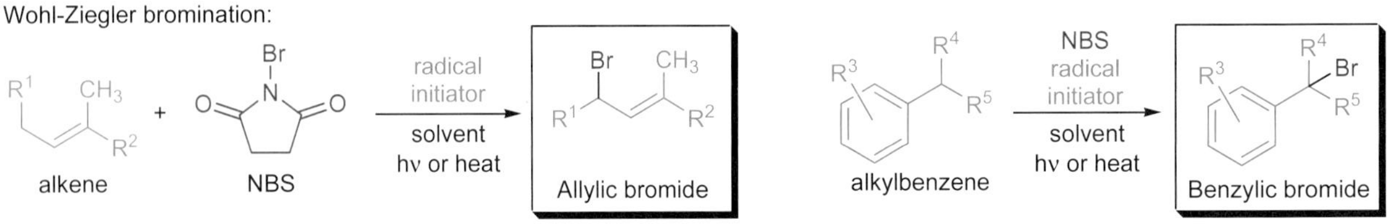

R^1 = alkyl; R^2 = H, alkyl, COR, CO_2R; R^3 = H, alkyl, aryl, O-alkyl, NR_2; R^{4-5} = H, alkyl, aryl; radical initiator: ROOR, $(Bz)_2O_2$, AIBN

Mechanism:[16-27]

The mechanism of the *Wohl-Ziegler bromination* involves bromine radicals (and not imidoyl radicals). The radical initiator is homolytically cleaved upon irradiation with heat or light, and it reacts with Br_2 (which is always present in small quantities in NBS) to generate the Br· radical, which abstracts a hydrogen atom from the allylic (or benzylic) position. The key to the success of the reaction is to maintain a low concentration of Br_2 so that the addition across the C=C double bond is avoided. The Br_2 is regenerated by the ionic reaction of NBS with the HBr by-product.

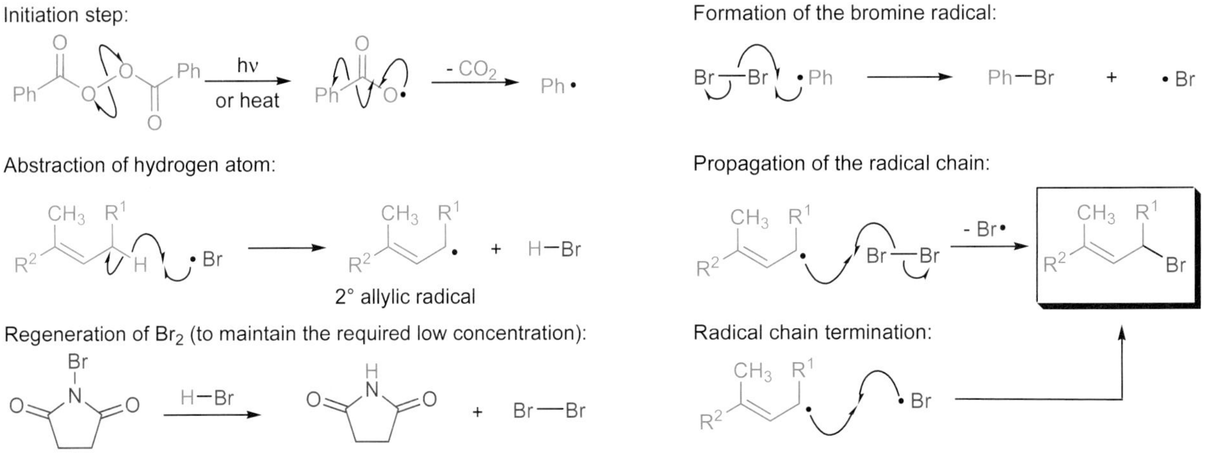

Synthetic Applications:

The first total synthesis of the novel sesquiterpene (–)-mastigophorene C was completed by G. Bringmann and co-workers.[28] This natural product has a negative effect on the growth of nerve cells. The synthetic strategy relied on the *Wohl-Ziegler bromination* to install the side-chain bromide on herbertenediol dimethyl ether. The substrate was dissolved in carbon tetrachloride; one equivalent of NBS and 20 mol% of dibenzoyl peroxide were added and the resulting mixture was heated at reflux for a few hours. The crude benzylic bromide was then hydrolyzed to the benzylic alcohol with water, which in turn was oxidized with MnO_2 to obtain the corresponding benzaldehyde derivative.

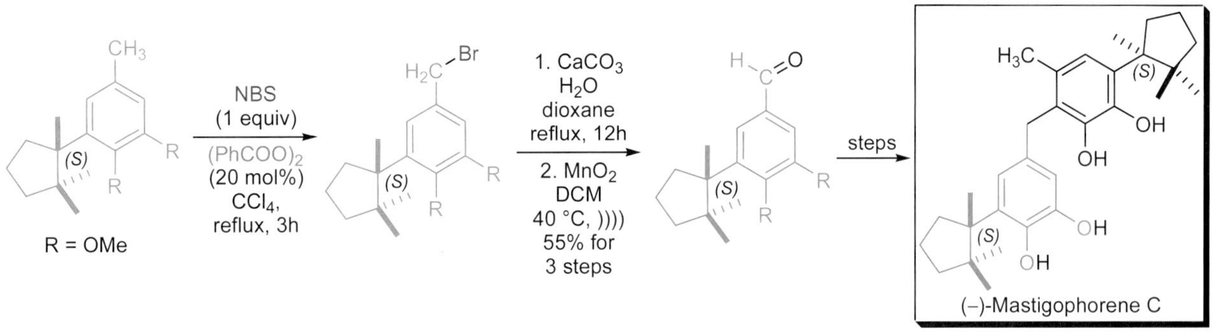

The research team of J. Tadanier prepared a series of C8-modified 3-deoxy-β-D-*manno*-2-octulosonic acid analogues as potential inhibitors of CMP-Kdo synthetase.[29] One of the derivatives was prepared from a functionalized olefinic carbohydrate substrate by means of the *Wohl-Ziegler bromination*. The stereochemistry of the double bond was (*Z*), however, under the reaction conditions a *cis-trans* isomerization took place in addition to the bromination at the allylic position (no yield was reported for this step). It is worth noting that the authors did not use a radical initiator for this transformation, the reaction mixture was simply irradiated with a 150W flood lamp. Subsequently the allylic bromide was converted to an allylic azide, which was then subjected to the *Staudinger reaction* to obtain the corresponding allylic amine.

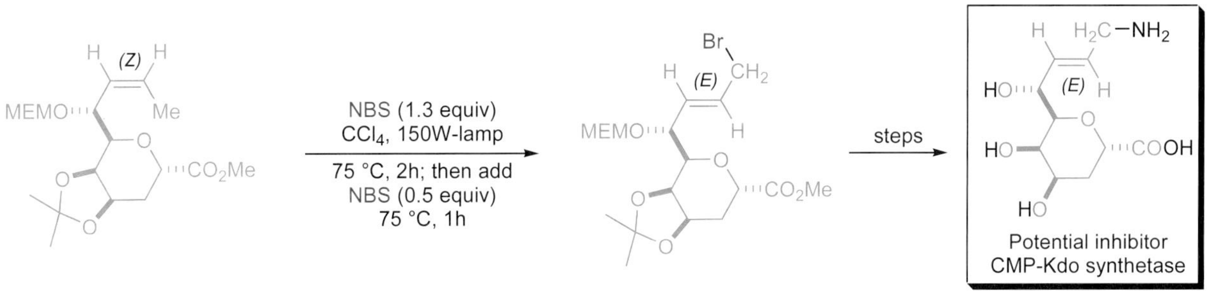

In the laboratory of J.M. Cook, the first enantioselective total synthesis of (–)-tryprostatin A was accomplished.[30] This natural product was isolated as a secondary metabolite of the marine fungal strain BM939 and was shown to inhibit cell cycle progression. The chiral center of the 2-isoprenyltryptophan moiety was introduced by the alkylation of the Schöllkopf chiral auxiliary. The alkylating agent was prepared from *N*-Boc-6-methoxy-3-methylindole using the *Wohl-Ziegler bromination*.

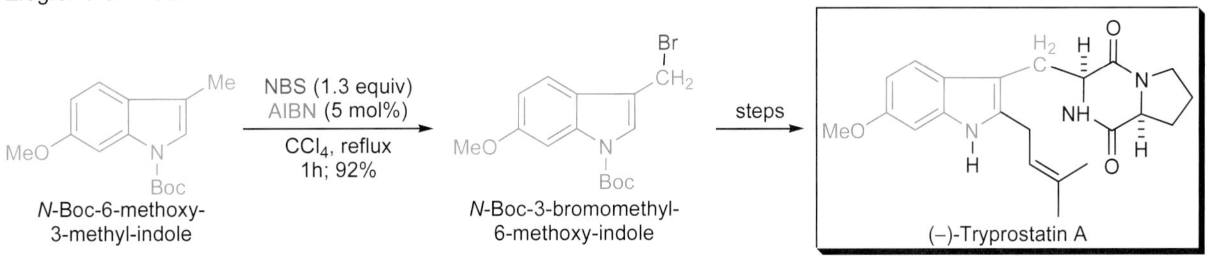

Conformationally restricted analogues of lavendustin A were prepared by M. Cushman and co-wokers as cyctotoxic inhibitors of tubulin polymerization.[31]

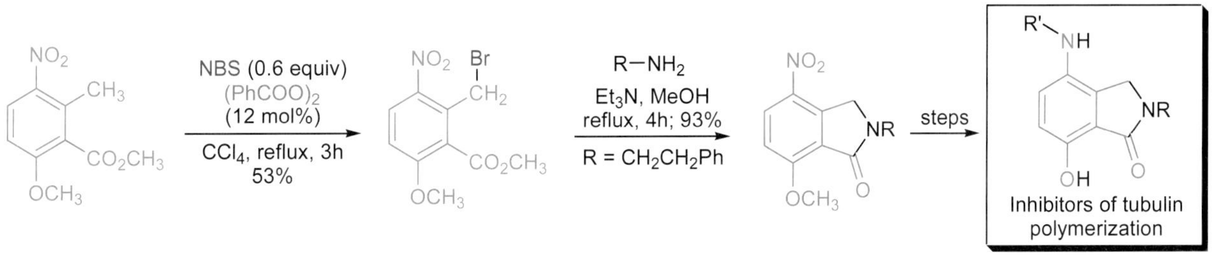

WOLFF REARRANGEMENT

(References are on page 711)

Importance:

[*Seminal Publications*[1-3]; *Reviews*[4-15]; *Modifications & Improvements*[16-23]; *Theoretical Studies*[24-56]]

In 1902, L. Wolff was studying the chemistry of α-diazo ketones when he observed that upon treatment with silver oxide and water, diazoacetophenone rearranged to give phenylacetic acid.[1] When the reaction medium contained aqueous ammonia, phenylacetamide was formed. A few years later G. Schröter published similar findings in an independent study, but the reaction remained unexplored for the next three decades due to the lack of general methods for the preparation of α-diazo ketones.[2] The conversion of α-diazo ketones into ketenes and products derived from ketenes is known as the *Wolff rearrangement*. The substrate α-diazo ketones can be prepared by various methods: 1) reaction of an acyl halide or anhydride with two equivalents of diazomethane in ether or DCM solution at room temperature or below (*Arndt-Eistert homologation*);[4] however, only one equivalent is needed of higher diazoalkanes, and low temperatures are necessary due to competing *azo coupling*; 2) sequential treatment of *N*-acyl-α-amino ketones (prepared by the *Dakin-West reaction*) with N_2O_3 and sodium methoxide in methanol affords secondary α-diazo ketones, so the cumbersome use of higher diazoalkanes is avoided; 3) transfer of the diazo group from an organic azide (e.g., tosyl azide) to a substrate containing an active methylene group (e.g., β-keto ester or β-keto nitrile) in the presence of a base (*Regitz diazo transfer*);[57-60] 4) simple diazo monoketones are synthesized from ketones by the introduction of a formyl group at the α-position *via* a *Claisen reaction* and then treatment of the resulting α-formyl derivative with tosyl azide and a tertiary amine (*deformylative diazo-transfer*);[61,62] 5) oxidation of α-ketoximes with chloramine;[63] and 6) hydroxide ion assisted decomposition of tosylhydrazones.[64] The general features of the *Wolff rearrangement* are: 1) the reaction can be initiated thermally, photolytically, or by transition metal catalysis; 2) thermal conditions are not used frequently, since delicate substrates may degrade and side reactions are frequent (e.g., direct displacement of the diazo group without rearrangement); 3) the use of transition metal complexes does not only reduce the required reaction temperature considerably compared to the thermal process, but also changes the reactivity of the α-keto carbene intermediate by the formation of less reactive metal carbene complexes (Rh- and Pd-complexes usually prevent the *Wolff rearrangement* from taking place); 4) freshly prepared silver(I)oxide or silver(I)benzoate are best suited for the reaction; 5) photochemical activation is convenient, and it takes place even at low temperatures, but it can be problematic if the product is photolabile; 6) if the migrating group has a stereocenter, the stereochemistry remains unchanged (net retention of configuration) after the migration; 7) the ketene products are electrophilic and can react with various nucleophiles as well as undergo [2+2] cycloaddition reactions with alkenes; 8) cyclic diazo ketones undergo ring-contraction, and the process is well-suited for the preparation of strained ring systems; 9) α,β-unsaturated diazo ketones undergo the *vinylogous Wolff rearrangement* to give skeletally rearranged γ,δ-unsaturated esters (alternative to *Claisen-type rearrangements*);[16] and 10) since α-diazo ketones are very reactive compounds, numerous side reactions are possible that can be avoided or minimized by the careful choice of reaction conditions.[9]

Wolff (1902):

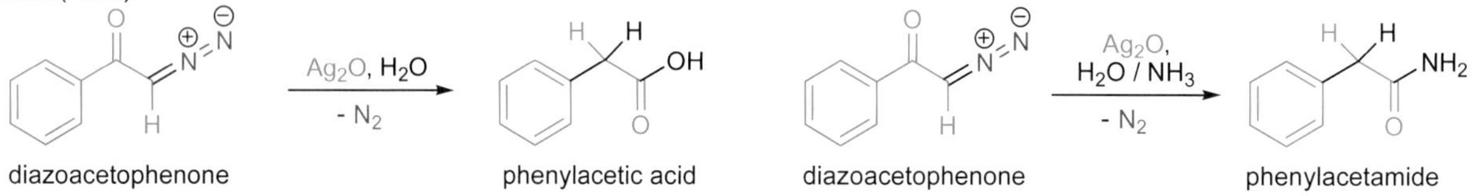

Preparation of the α-diazo ketone: Wolff rearrangement:

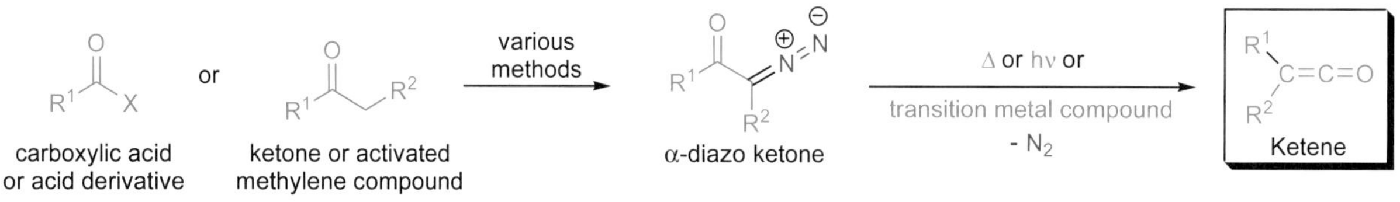

R^1 = alkyl, aryl, heteroaryl; R^2 = alkyl, aryl, H, CN, CHO, C(O)-alkyl, SO_2R, CO_2R; X = Cl, Br, OCOR

Mechanism: [65,9,13]

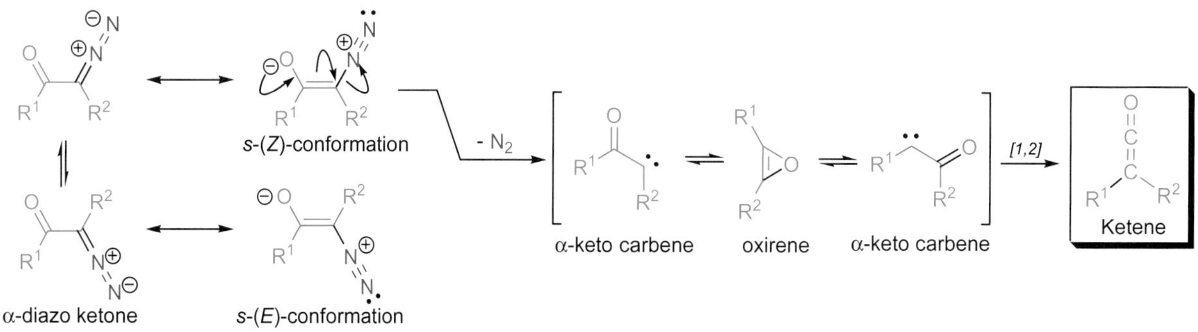

Synthetic Applications:

The stereoselective total synthesis of (±)-campherenone was accomplished by T. Uyehara and co-workers based on a *photochemical Wolff rearrangement*.[66] The bicyclic ketone was treated with 2,4,6-triisopropylbenzenesulfonyl azide (trisyl azide) under homogeneous basic conditions and the α-diazo ketone was obtained in excellent yield. The photochemical rearrangement of the diazo ketone was conducted in a THF-water mixture using a high-pressure 100 W mercury lamp. The ring-contracted acid was isolated as a 4:1 mixture of *endo* and *exo* products.

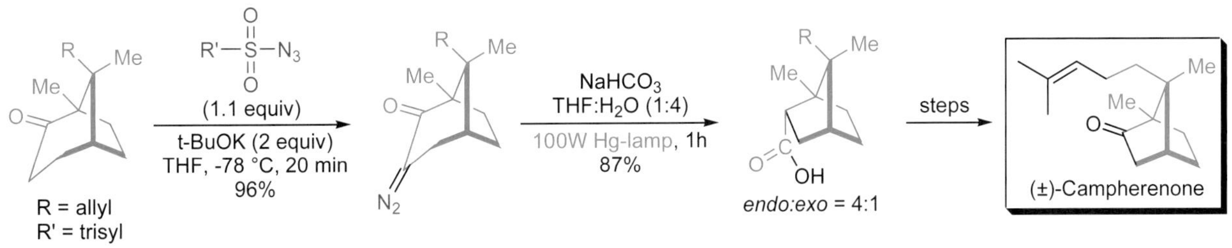

In the laboratory of K. Fukumoto, the stereoselective total synthesis of (±)-$\Delta^{9(12)}$-capnellene was carried out using an *intramolecular Diels-Alder reaction* to obtain a tricyclic 5-5-6 system.[67] Since the target molecule was a triquinane, the six-membered ring had to be converted to a five-membered one, a transformation achieved by a *Wolff rearrangement.* The required α-diazo ketone was prepared *via* a *deformylative diazo transfer reaction* and was photolyzed in methanol. The ring-contracted methyl ester was isolated as a 3:1 mixture of separable isomers favoring the α-isomer.

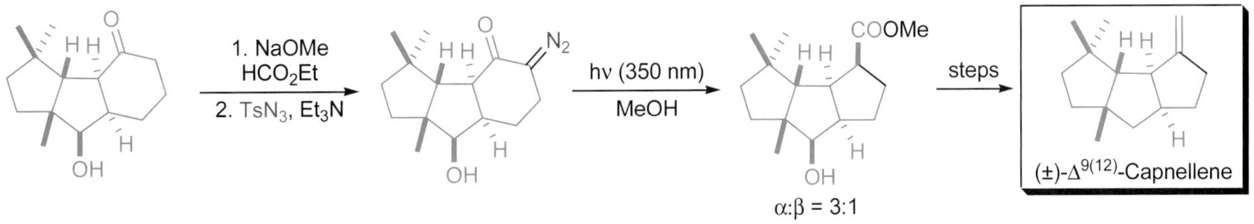

The natural product (–)-oxetanocin is an unprecedented oxetanosyl-*N*-glycoside that inhibits the *in vitro* replication of human immunodeficiency virus (HIV). In order to prepare multigram quantities of the compound, D.W. Norbeck et al. devised a short and efficient synthetic strategy.[68] The cornerstone of the strategy was the *Wolff rearrangement* of a five-membered diazo ketone. The diazo transfer was achieved by first converting the ketone to an enamino ketone followed by treatment with triflyl azide. Upon irradiation with a 450 W Pyrex filtered Hanovia lamp, the isomeric oxetanes (α:β = 2:1) were obtained in 36% yield.

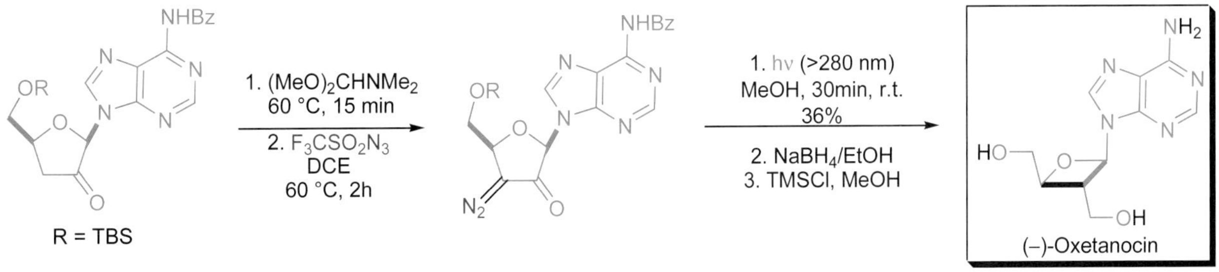

R.L. Danheiser and co-workers generated a key vinylketene intermediate *via* tandem *Wolff rearrangement*-ketene-alkyne cycloaddition to utilize it in a *photochemical aromatic annulation reaction* (*Danheiser benzannulation*) for the total synthesis of the phenalenone diterpene salvilenone.[69]

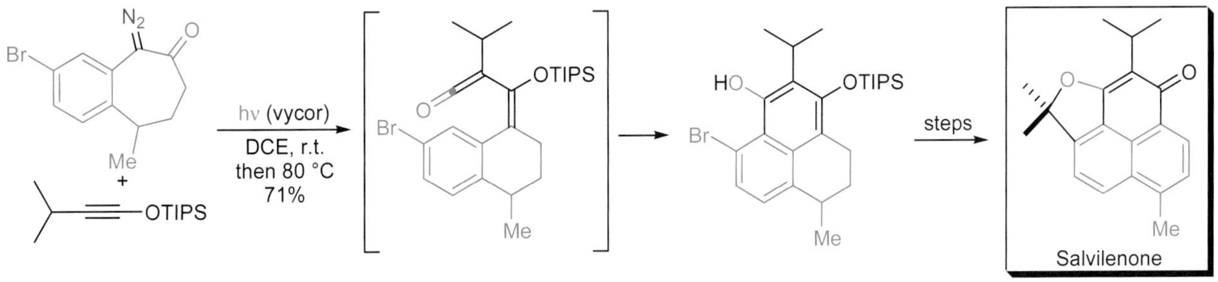

WOLFF-KISHNER REDUCTION

(References are on page 712)

Importance:

[*Seminal Publications*[1,2]; *Reviews*[3-6]; *Modifications & Improvements*[7-20]; *Theoretical Studies*[21]]

In 1911, N. Kishner reported that by adding a hydrazone dropwise onto a mixture of hot potassium hydroxide and platinized porous plate the corresponding hydrocarbon was formed.[1] A year later L. Wolff independently showed that heating the ethanol solution of semicarbazones and hydrazones in a sealed tube at ~180 °C in the presence of sodium ethoxide gives rise to the same result. The deoxygenation of aldehydes and ketones to hydrocarbons *via* the corresponding hydrazones or semicarbazones under basic conditions is known as the *Wolff-Kishner reduction* (*W-K reduction*). Since the seminal reports, the original procedure has been substantially modified to make the reaction conditions milder and improve the yields.[3,6] The standard procedure for a long time was to mix the carbonyl compound with 100% hydrazine in a high-boiling solvent (e.g., ethylene- or triethylene glycol) in the presence of excess base (sodium metal, NaOEt, etc.) and keep the reaction mixture at reflux for a couple of days. One of the main problems encountered was the temperature-lowering effect of the water generated during the formation of the hydrazone, and this resulted in long reaction times (50-100h) and the need to use an excess of the reagents and solvents. In the *Huang-Minlon modification*, the water and the excess hydrazine are removed by distillation (once the hydrazone is formed *in situ*) so the reaction temperature could rise to ~200 °C, which dramatically shortened the reaction time (3-6h), increased the yields and also allowed the use of the cheaper hydrazine hydrate along with water-soluble bases (KOH or NaOH).[9] The general features of the reaction are: 1) the reduction is usually carried out in a high boiling solvent (~180-200 °C) so that the use of a sealed tube can be avoided;[7,8,17] 2) for base-sensitive substrates better yields are achieved when the hydrazone is preformed and the base is added to the substrates at lower temperatures (e.g., 25 °C) followed by refluxing the reaction mixture; 3) esters, lactones, amides, and lactams are hydrolyzed under the reaction conditions; 4) sterically hindered carbonyl compounds are deoxygenated more slowly than unhindered ones, so higher reaction temperatures are required (*Barton modification*);[11,14] 5) the use of DMSO instead of glycols as the reaction medium containing KO*t*-Bu, followed by the slow addition of preformed hydrazones, allows the reduction to take place at room temperature (*Cram modification*). However, on small scale this method is inconvenient, and good results are very substrate dependent;[12] 6) preformed hydrazones can also be mixed with KO*t*-Bu and refluxed in toluene (~110 °C) to effect the reduction (*Henbest modification*);[13] 7) for α,β-unsaturated carbonyl compounds, the use of preformed semicarbazones is advised (hydrazine tends to give pyrazolines with these substrates), which undergo reduction under the original or most of the modified reaction conditions;[3] and 8) certain aromatic carbonyl compounds (e.g., benzophenone, benzaldehyde) do not require the use of a strong base for reduction, they are reduced when heated with excess hydrazine hydrate.[3] A powerful alternative of the *W-K reduction* is the treatment of tosylhydrazones with hydride reagents to obtain the corresponding alkanes (*Caglioti reaction*).[22] A few side reactions have been observed: 1) formation of azines; 2) reduction of ketone substrates to alcohols when the reaction is unsuccessful; 3) isomerization of double bonds especially in the case of α,β-unsaturated carbonyl compounds ; 4) elimination of the α-heteroatom substituent to afford alkenes (*Kishner-Leonard elimination*);[23,24] and 5) cleavage or rearrangement of strained rings adjacent to the carbonyl group.

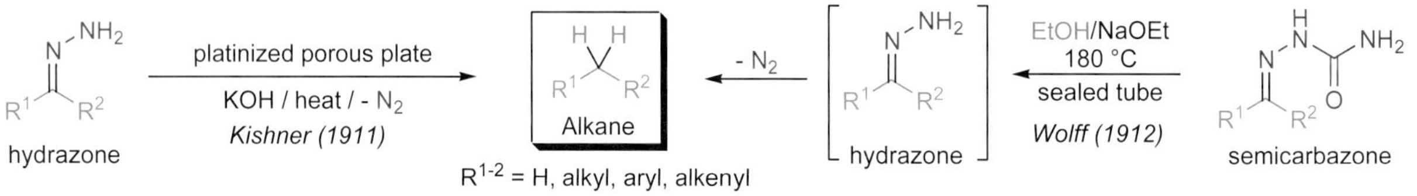

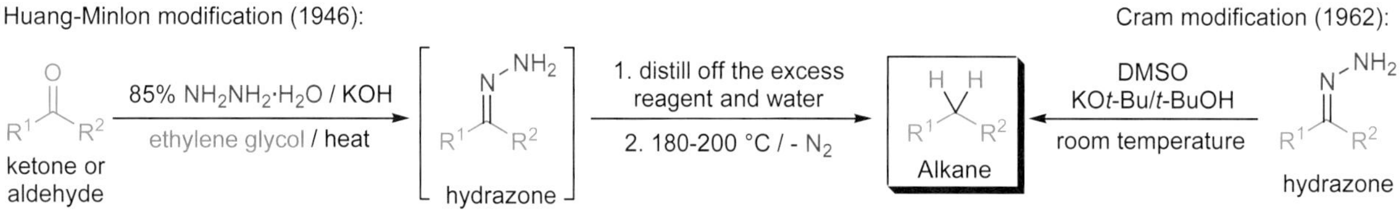

Mechanism: [25-32]

The rate-determining step is the proton capture at the carbon terminal. This process takes place in a concerted fashion with the solvent-induced proton abstraction at the nitrogen terminus to form a diimide that undergoes a loss of N_2.

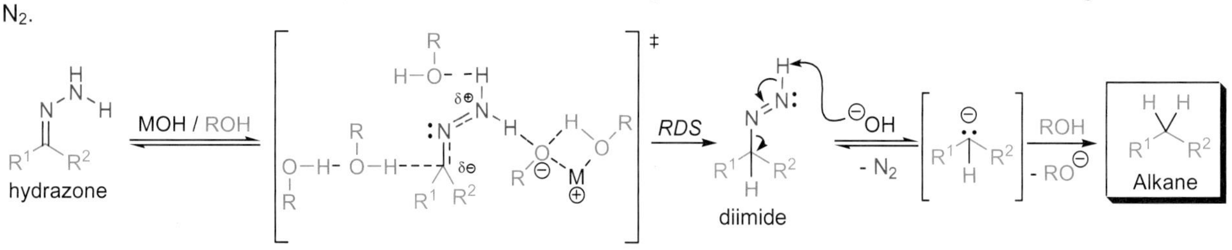

Synthetic Applications:

The asymmetric syntheses of (-)-methyl kaur-16-en-19-oate and (-)-methyl trachyloban-19-oate was achieved by M. Ihara and co-workers.[33] One of the last transformations was the deoxygenation of the ketone carbonyl group of the tetracyclic intermediate, which was effected by the *Wolff-Kishner reduction*. Under the strongly basic conditions the ester functionality was hydrolyzed, so an esterification using diazomethane was necessary as the final step. The major deoxygenated product was (-)-methyl kaur-16-en-19-oate (59%). The minor product was identified as (-)-methyl trachyloban-19-oate (16%).

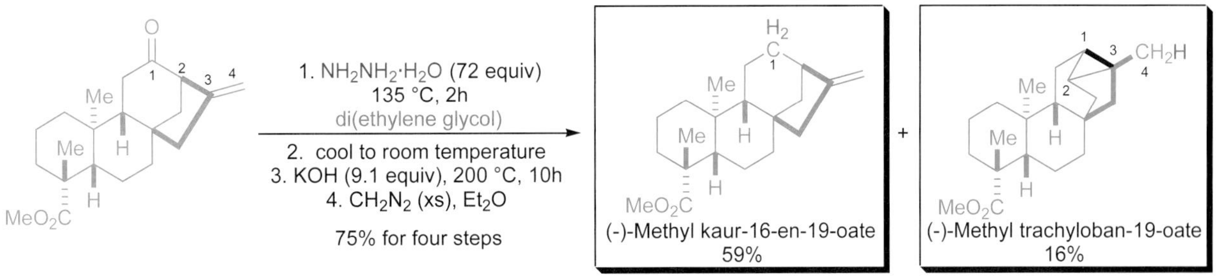

The total synthesis of (+)-aspidospermidine was accomplished in the laboratory of J.P. Marino using a novel *[3,3]-sigmatropic rearrangement* of chiral vinyl sulfoxide with a ketene as the key step.[34] During the endgame of the synthesis the pentacyclic ketone was deoxygenated using the *Wolff-Kishner reduction*. Because the ketone was sterically hindered, harsh reaction conditions had to be applied: after the formation of the hydrazone, the water and the excess hydrazine were removed and the temperature was raised to 210 °C. The final step in the synthetic sequence was the reduction of the five-membered lactam to the corresponding tertiary amine with LAH.

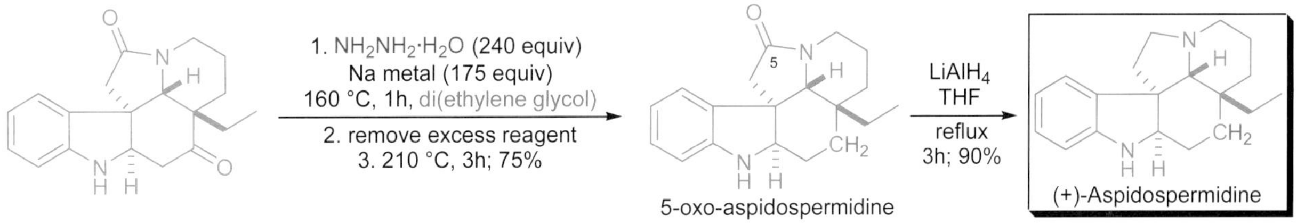

Dysidiolide is the first compound found to be a natural inhibitor of protein phosphatase cdc25A that is essential for cell proliferation. Y. Yamada et al. developed a novel total synthesis of this natural product using an *intramolecular Diels-Alder cycloaddition* as the key step.[35] Deoxygenation of the advanced bicyclic intermediate at the C24 position was achieved under *Wolff-Kishner reduction* conditions to afford the C24 methyl group.

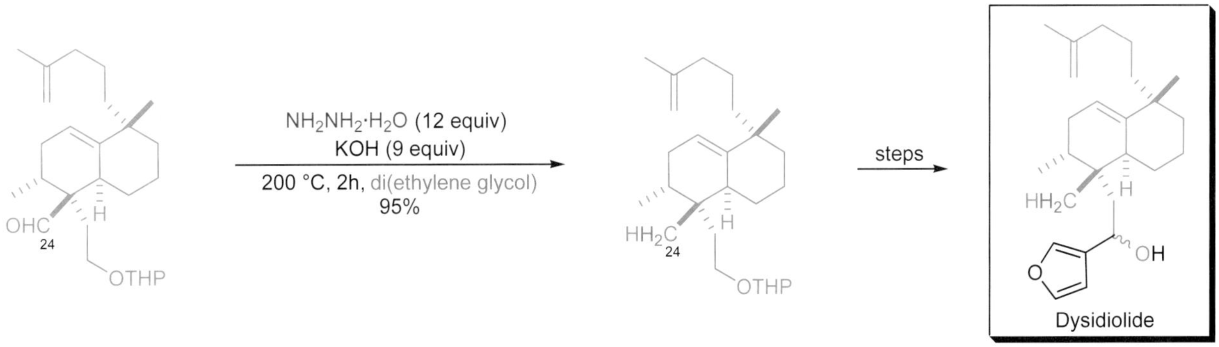

A novel two-step one-pot *modified Wolff-Kishner reduction* protocol was developed in the laboratory of A.G. Myers (*Myers modification*).[20] The carbonyl compound was first converted to the *N*-TBS-hydrazone followed by the addition of KO*t*-Bu/*t*-BuOH in DMSO at or above room temperature.

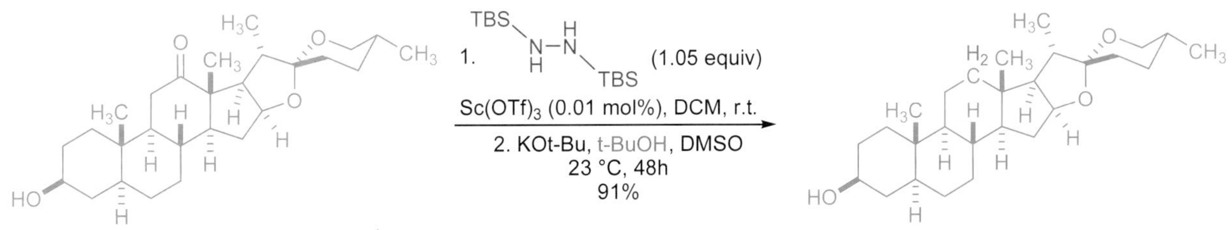

WURTZ COUPLING

(References are on page 713)

Importance:

[*Seminal Publications*[1,2]; *Reviews*[3-7]; *Modifications & Improvements*[8-16]]

In 1855, A. Wurtz treated alkyl halides with sodium metal, and he isolated the corresponding symmetrical alkane dimers.[1,2] The coupling of two sp^3-carbon centers by the treatment of alkyl or benzyl halides with sodium metal is known as the *Wurtz coupling*. When metals other than sodium are used, this transformation is referred to as a *Wurtz-type coupling*. The coupling of an alkyl and an aryl halide in the presence of sodium metal to get the corresponding alkylated aromatic compound is called the *Wurtz-Fittig reaction*. Today, the synthetic significance of the *Wurtz coupling* is fairly limited and often in widely used reactions (e.g., *Grignard reactions*) involving highly reactive organometals, such as allyl- and benzylmetals, this is the side reaction. The general features of the *Wurtz coupling* are: 1) the classical reaction is heterogeneous and relatively low-yielding, because it is plagued by side reactions such as elimination and rearrangements; 2) best results are achieved with finely dispersed sodium metal; 3) alkyl halides can be coupled both inter- and intramolecularly; 4) the order of reactivity for alkyl halides is: I >> Br >> Cl, and by far primary alkyl iodides are the best substrates; 5) secondary alkyl halides are generally poor substrates and should be avoided; 6) the method works reasonably well for intermolecular homocouplings, but the heterocoupling of two different alkyl halides often results in a statistical mixture of products in low yields; 7) intramolecularly, the coupling can give rise to strained rings as well as macrocycles (e.g., cyclopropanes, cyclobutanes, and cyclophanes) in moderate to good yield, and it has been applied extensively for the preparation of such compounds;[17,4] and 8) the *Wurtz-Fittig reaction* gives high yields of the desired product without significant side reactions mainly because aryl halides do not usually dimerize under the reaction conditions. Because of the limited synthetic value of the classical reaction conditions, several modifications were introduced: 1) the most widespread reaction condition (*Müller modification*) is to treat the alkyl halide with sodium metal in THF at -78 °C in the presence of catalytic amounts of tetraphenylethylene (TPE), which solubilizes the sodium and makes the reaction homogeneous; [8] 2) metals other than sodium[18] as well as various metal complexes have been used successfully to improve the yields and suppress side reactions: activated Cu,[13] $Mn_2(CO)_{10}$,[15] Li metal/ultrasound, Na(Hg),[10] Na-K alloy, Zn;[16] and 3) the use of sonication (ultrasound) in general improves the yield, since the metal becomes highly dispersed and as a result its reactivity increases.[11,12,14]

Wurtz homocoupling (dimerization):

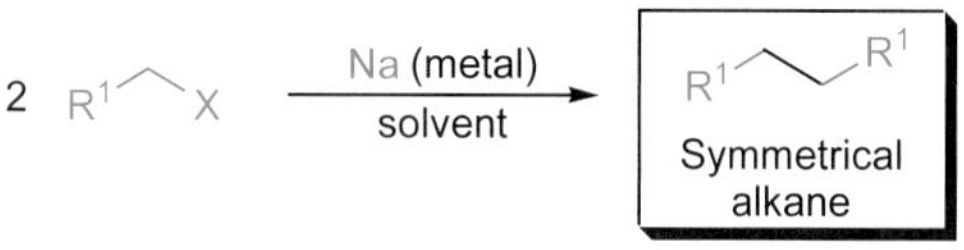

Wurtz heterocoupling (cross-coupling):

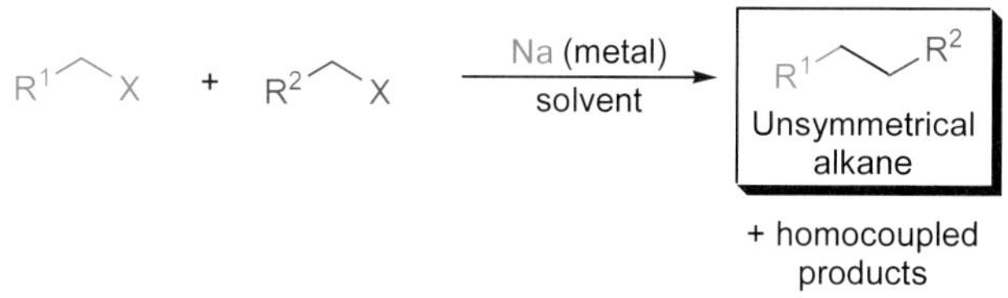

Wurtz-Fittig reaction:

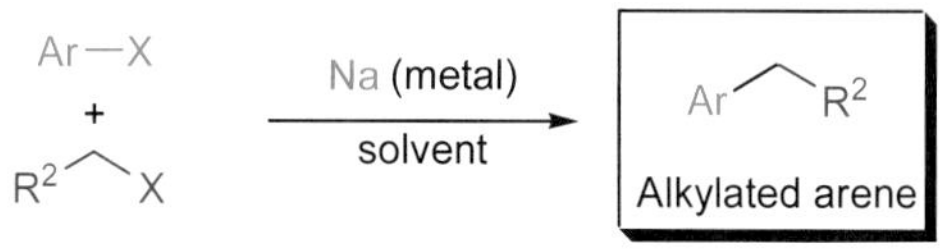

Wurtz-type coupling:

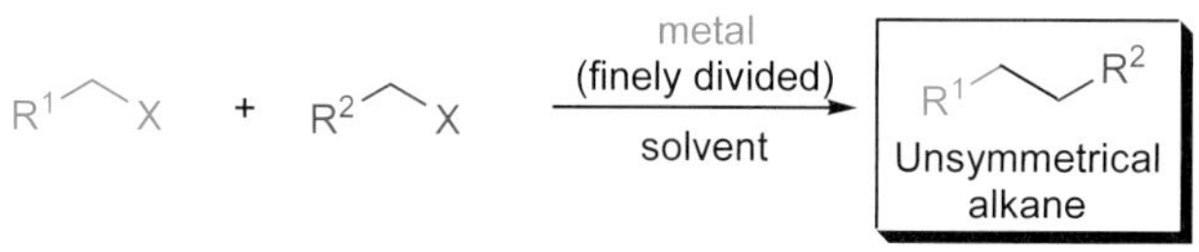

R^1 = 1° alkyl, aryl; R^2 = 1° alkyl, aryl; Ar = electron-rich and electron-poor substituted aryl; metal = K, Mg, Zn, Cu etc.; solvent = THF, Et_2O, dioxane, xylenes

Mechanism: [19-30]

The mechanism of the *Wurtz coupling* is not well understood, and the currently accepted mechanism involves two steps: 1) formation of a carbanionic organosodium compound *via* metal-halogen exchange; and 2) the displacement of the halide ion by the organosodium species in an S_N2 reaction. Alternatively, a radical process can also be envisioned, although to date there has been no experimental evidence to support this assumption.

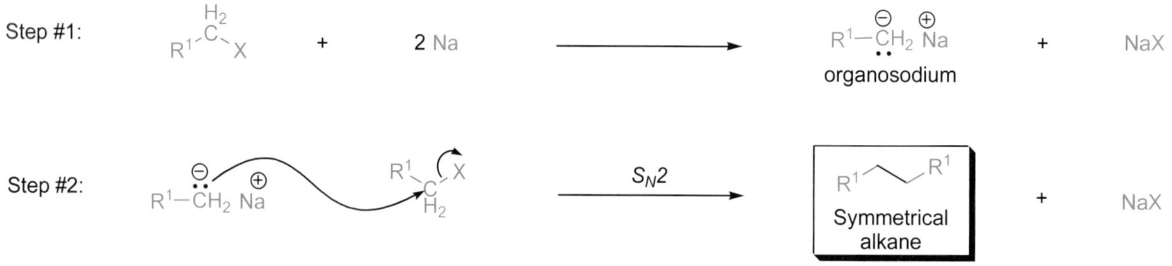

Synthetic Applications:

J.W. Morzycki and co-workers described the synthesis of dimeric steroids to be used as components of artificial lipid bilayer membranes.[31] The key coupling of two steroid derivatives was achieved by the *Wurtz reaction*. The steroid primary alkyl iodide was dissolved in anhydrous toluene and treated with an excess of sodium metal. After 20h of reflux, the desired homocoupled product was obtained in moderate yield along with a considerable amount (36%) of the reduced compound.

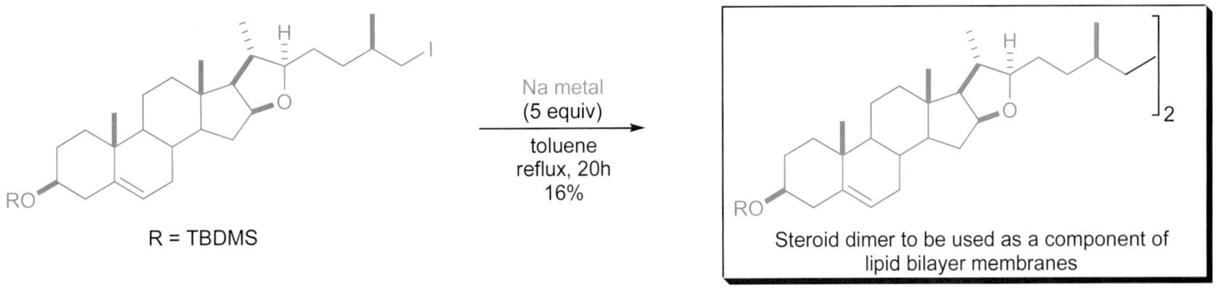

The total synthesis of the diarylheptanoid garugamblin 1 was achieved by M. Nógrádi et al. using the *modified Wurtz coupling* as the key macrocyclization step.[32] The dibromide was treated with sodium metal at room temperature in the presence of TPE to afford the desired macrocycle in moderate yield. The N-O bond of the isoxazole ring was cleaved under the reaction conditions.

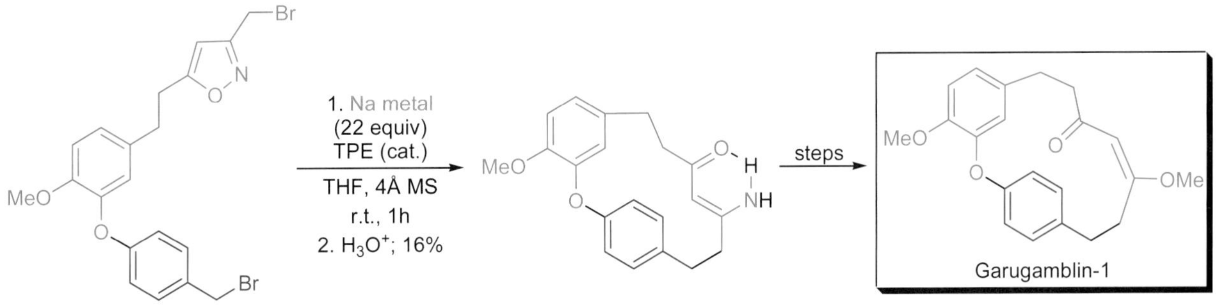

The structure of the macrocyclic bis(benzylether) natural product marchantin I was confirmed in the laboratory of M. Nógrádi.[33] The last and key step of the synthesis was the *modified Wurtz coupling* in which the 18-membered ring was formed.

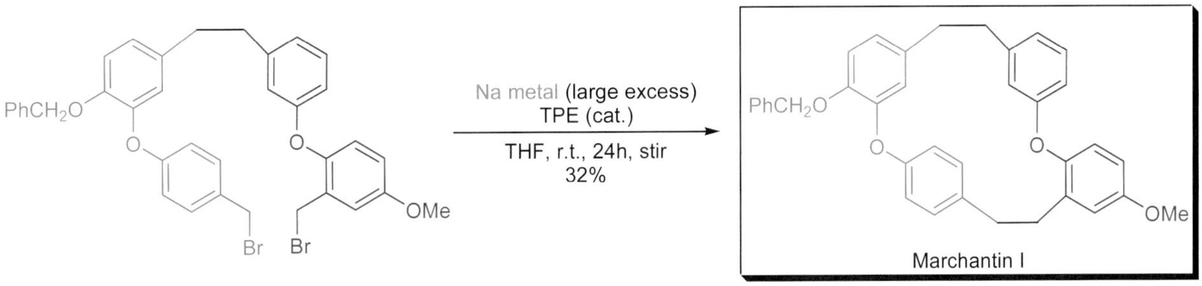

The classical preparation of cyclobutyl ketones involves the base-catalyzed reaction of 1,3-dihaloalkanes with malonate esters. However, the initial product of this reaction is a cyclobutane carboxylic acid. S.D. Van Arnum and co-workers showed that cyclobutyl ketones can be efficiently synthesized starting from acyl succinates and using the *Wurtz reaction* as the key cyclization step.[34] The cyclization was catalyzed by naphthalene.

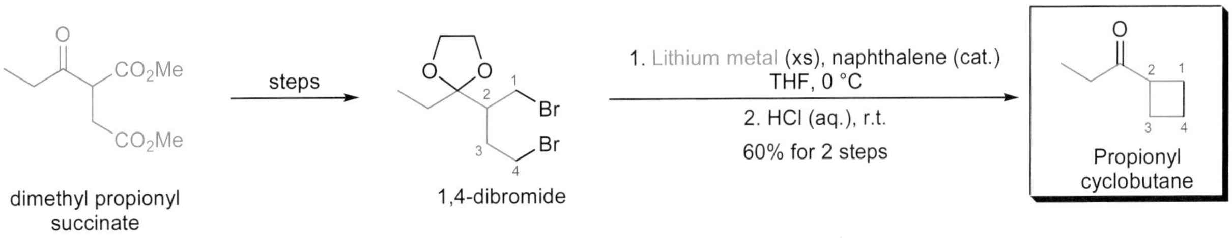

YAMAGUCHI MACROLACTONIZATION

(References are on page 714)

Importance:

[*Seminal Publications*[1]; *Reviews*[2-4]; *Modifications & Improvements*[5,6]]

In 1979, M. Yamaguchi and co-workers developed a novel procedure for the rapid preparation of esters and lactones under mild conditions *via* the alcoholysis of the corresponding mixed anhydrides.[1] As a result of their thorough study, they found that 2,4,6-trichlorobenzoyl chloride/DMAP was the best reagent combination in terms of both the high reaction rate as well as the high product yield. The procedure was put to the test and used for the lactonization of a very acid sensitive substrate that was known to rapidly decompose on contact with catalytic amounts of HCl. The substrate hydroxy acid was treated with 2,4,6-trichlorobenzoyl chloride in the presence of NEt_3, and the by-product triethylamine hydrochloride was removed. The resulting mixed anhydride was diluted with toluene and slowly added to a refluxing solution of DMAP in toluene under high dilution conditions (~0.002 M). The desired macrolactone, (±)-2,4,6-tridemethyl-3-deoxymethynolide, was obtained without the formation of any decomposition product. The formation of medium- and large-ring lactones from hydroxy acids using 2,4,6-trichlorobenzoyl chloride/DMAP is known as the *Yamaguchi macrolactonization*. The general features of this transformation are: 1) the substrate is first converted to the corresponding mixed anhydride with 2,4,6-trichlorobenzoyl chloride in the presence of a tertiary amine to activate the carboxylic acid functionality; 2) aromatic hydrocarbons such as benzene and toluene are the best solvents; 3) the reaction is conducted under high-dilution conditions to minimize intermolecular coupling; 4) the mixed anhydride is dissolved and slowly added (*via* a syringe pump) to a refluxing solution of DMAP in benzene or toluene; and 5) usually several equivalents of DMAP, a known catalyst for acyl transfer reactions, is used. The main advantages of the *Yamaguchi macrolactonization* over other existing methods are its operational simplicity, its high reaction rate and the lack of by-products.

Yamaguchi et al. (1979):

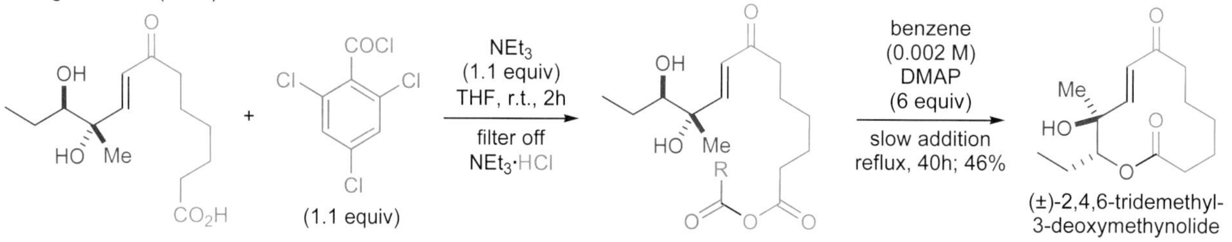

Yamaguchi macrolactonization:

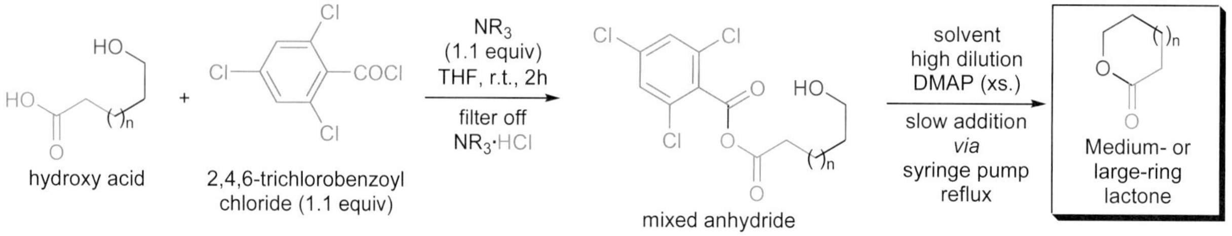

Mechanism:

Formation of the mixed anhydride (R = 2,4,6-trichlorobenzoyl):

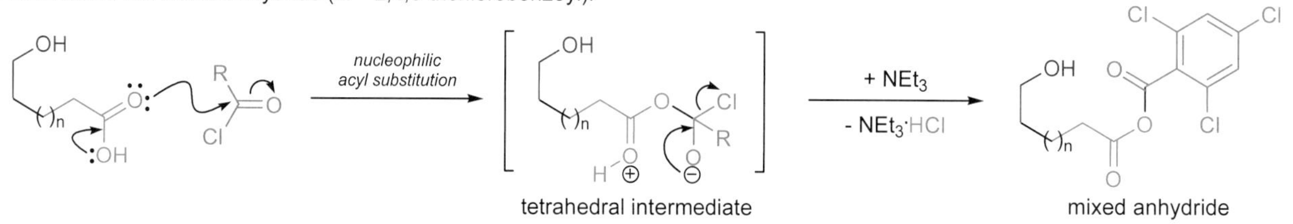

Formation of the macrolactone (R = 2,4,6-trichlorobenzoyl):

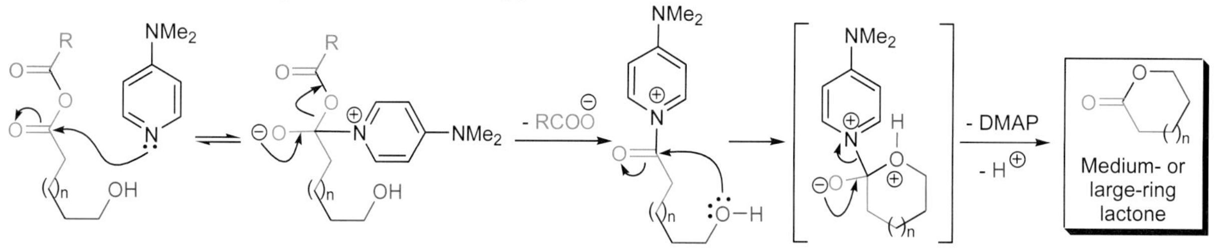

Synthetic Applications:

The stereocontrolled total synthesis of (–)-macrolactin A, a 24-membered macrolide, was achieved by J.P. Marino and co-workers.[7] The key macrocyclization step was carried out using the *Yonemitsu modification of the Yamaguchi macrolactonization*.[6] In this procedure, the mixed anhydride is added to the highly dilute solution of DMAP rapidly (in one portion) at room temperature. The final step of the total synthesis was the removal of the protecting groups under acidic condition.

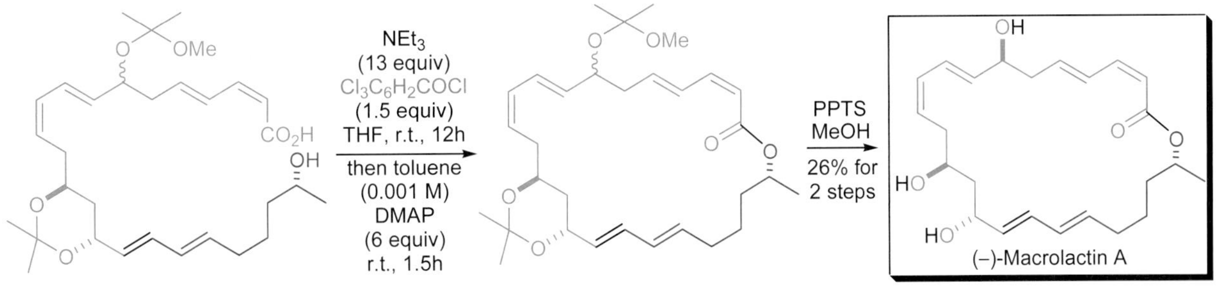

The convergent enantioselective synthesis of oleandolide, the aglycon of the macrolide antibiotic oleandomycin, was reported by J.S. Panek et al.[8] The key macrocyclization was carried out by a *modified Yamaguchi macrolactonization* protocol. The azeotropically dried dihydroxy acid was first treated with a large excess of 2,4,6-trichlorobenzoyl chloride and Hünig's base, and the resulting mixed anhydride was diluted with benzene (~0.001 M). To this dilute solution was added in one portion a large excess of DMAP. The desired 14-membered lactone was isolated in nearly quantitative yield and no trace of the undesired 12-membered lactone was detected. The unusually high efficiency of the cyclization was attributed to the strong conformational preference induced by the large substituent at C9.

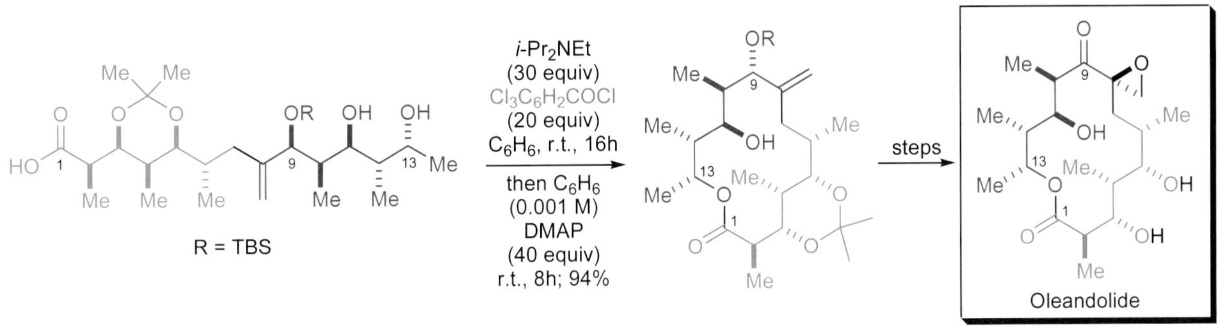

The microtubule-stabilizing and potent antitumor 18-membered macrolide, (–)-laulimalide, was synthesized in the laboratory of A.K. Ghosh.[9] The macrolactonization of the α,β-unsaturated (*Z*)-hydroxy acid under *Yamaguchi conditions* caused isomerization of the double bond. Presumably this undesired isomerization was due to the reversible *Michael addition* of the DMAP catalyst to the active acylating agent. Unfortunately, no other reaction conditions were found that could decrease the extent of the double bond isomerization, so an alternative strategy was sought. Therefore, the macrolactonization of a hydroxy alkynoic acid was performed and the triple bond was efficiently hydrogenated to the desired (*Z*)-double bond with Lindlar's catalyst. In order to complete the total synthesis, selective removal the MOM protecting group was achieved by treatment with excess PPTS in *t*-butanol at reflux. The epoxide was installed using the *Sharpless epoxidation*, which afforded the epoxide as a single diastereomer. The final step was the removal of the PMB group with DDQ.

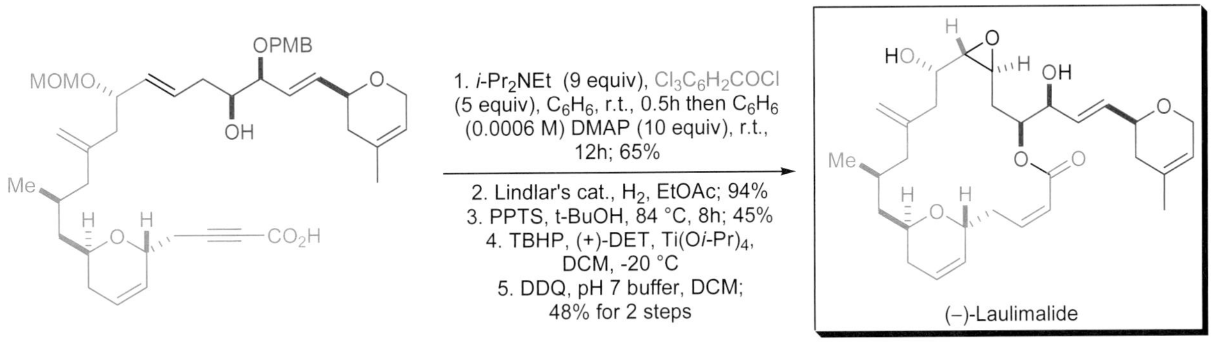

VIII. APPENDIX

8.1 Brief explanation of the organization of this section

The primary function of this section is to help advanced undergraduate students and first year graduate students in organizing the large amount of information available on various chemical transformations. It is important to note that the categorization of named reactions is a subjective one and has been addressed differently in other textbooks.

The categorization of named reactions is mainly based on the mechanism of the various processes. To make studying more friendly, we included a brief description of each named reaction and the page number for that particular transformation.

Because a large number of functional group transformations are affected by the reactions covered in the book, we felt that tables showing the interconversion of functional groups should be included.

Various functional groups are listed in alphabetical order in the first column and the functionalities that can be created from them are shown in the second column. The names of all reactions that can bring about these transformations are listed in the third column.

In the second table we listed the target functional groups in alphabetical order in the first column and showed the substrate functionalities in the second column. In the third column the names of these transformations are listed.

A note of caution: none of these tables were created with the intent to be comprehensive, since that would be beyond the scope of this book. The reader should always check the details for each reaction to find out the true scope and limitations of a given transformation. We welcome any suggestions on how to make this section more effective in future editions.

8.2 LIST OF NAMED REACTIONS IN CHRONOLOGICAL ORDER OF THEIR DISCOVERY

YEAR OF DISCOVERY	NAME OF THE TRANSFORMATION	PAGE #
1822	Lieben Haloform Reaction	264
1838	Benzilic Acid Rearrangement	52
1839	Aldol Reaction	8
1844	Dieckmann Condensation	138
1850	Strecker Reaction	446
1851	Hofmann Elimination	206
1852	Williamson Ether Synthesis	484
1853	Cannizzaro Reaction	74
1855	Wurtz Coupling	498
1860	Kolbe-Schmitt Reaction	248
1860	Pinacol and Semipinacol Rearrangement	350
1861	Acyloin Condensation	4
1861	Hunsdiecker Reaction (Borodin Reaction)	218
1868	Perkin Reaction	338
1869	Glaser Coupling Reaction	186
1869	Lossen Rearrangement	266
1876	Reimer-Tiemann Reaction	378
1877	Friedel-Crafts Acylation	176
1877	Friedel-Crafts Alkylation	178
1877	Malonic Ester Synthesis	272
1877	Pinner Reaction	352
1879	Koenigs-Knorr Glycosidation	246
1880	Skraup and Doebner-Miller Reaction	414
1881	Ciamician-Dennstedt Rearrangement	84
1881	Fries-, Photo-Fries and Anionic *Ortho*-Fries Rearrangement	180
1881	Hell-Volhard-Zelinsky Reaction	200
1881	Hofmann Rearrangement	210
1882	Hantzsch Dihydropyridine Synthesis	194
1883	Combes Quinoline Synthesis	94
1883	Fischer Indole Synthesis	172
1883	Hofmann-Löffler-Freytag Reaction	208
1883	Michael Addition	286
1883	von Pechmann Reaction	472
1884	Paal-Knorr Furan Synthesis	326
1884	Paal-Knorr Pyrrole Synthesis	328
1884	Sandmeyer Reaction	394
1884	Schotten-Baumann Reaction	398
1885	Buchner Method of Ring Enlargement (Buchner Reaction)	68
1885	Curtius Rearrangement	116
1886	Beckman Rearrangement	50
1886	Knorr Pyrrole Synthesis	244
1887	Claisen Condensation/(Claisen Reaction)	86
1887	Gabriel Synthesis	182
1887	Japp-Klingemann Reaction	224
1887	Reformatsky Reaction	374
1887	Tishchenko Reaction	456
1888	Dimroth Rearrangement	144
1891	Biginelli Reaction	58
1892	Darzens Glycidic Ester Condensation	128
1893	Bischler-Napieralski Isoquinoline Synthesis	62
1893	Dienone-Phenol Rearrangement	142

YEAR OF DISCOVERY	NAME OF THE TRANSFORMATION	PAGE #
1893	Pomeranz-Fritsch Reaction	358
1893	Stobbe Condensation	442
1894	Favorskii Rearrangement and Homo-Favorskii Rearrangement	164
1894	Knoevenagel Condensation	242
1894	Nef Reaction	308
1894	Smiles Rearrangement	416
1894	Wacker Oxidation	474
1895	Henry Reaction	202
1897	Arbuzov Reaction (Michaelis-Arbuzov Reaction)	16
1897	Gattermann and Gattermann-Koch Formylation	184
1898	Chugaev Elimination (Xanthate Ester Pyrolysis)	82
1899	Baeyer-Villiger Oxidation/Rearrangement	28
1899	Barbier Coupling Reaction	38
1899	Prins Reaction	364
1899	Wagner-Meerwein Rearrangement	476
1900	Grignard Reaction	188
1901	Demjanov Rearrangement and Tiffeneau-Demjanov Rearrangement	134
1901	Ullmann Reaction/Coupling/Biaryl Synthesis	466
1902	Feist-Bénary Furan Synthesis	166
1902	Wolff Rearrangement	494
1903	Benzoin and Retro-Benzoin Condensation	54
1903	Mannich Reaction	274
1903	Nazarov Cyclization	304
1903	Ullmann Biaryl Ether and Biaryl Amine Synthesis/Condensation	464
1905	Eschweiler-Clarke Methylation	160
1908	Staudinger Ketene Cycloaddition	426
1909	Acetoacetic Ester Synthesis	2
1909	Dakin Oxidation	118
1909	Paterno-Büchi Reaction	332
1909	Prilezhaev Reaction	362
1909	Pummerer Rearrangement	368
1910	Finkelstein Reaction	170
1910	Regitz Diazo-Transfer Reaction	376
1911	Pictet-Spengler Tetrahydroisoquinoline Synthesis	348
1911	Wolff-Kishner Reduction	496
1912	Madelung Indole Synthesis	270
1913	Claisen Rearrangement	88
1913	Clemmensen Reduction	92
1913	Wharton Olefin Synthesis (Wharton Transposition)	482
1914	Chichibabin Amination Reaction (Chichibabin Reaction)	80
1914	Ferrier Reaction/Ferrier Rearrangement	168
1915	Houben-Hoesch Reaction/Synthesis	216
1919	Aza-Wittig Reaction	24
1919	Meisenheimer Rearrangement	282
1919	Staudinger Reaction	428
1919	Wittig Reaction	486
1919	Wohl-Ziegler Bromination	492
1921	Passerini Multicomponent Reaction	330
1922	Meyer-Schuster and Rupe Rearrangement	284
1923	Schmidt Reaction	396
1925	Amadori Reaction/Rearrangement	14
1925	Meerwein-Ponndorf-Verley Reduction	280
1925	Stephen Aldehyde Synthesis	430
1926	Diels-Alder Cycloaddition	140
1926	Neber Rearrangement	306

YEAR OF DISCOVERY	NAME OF THE TRANSFORMATION	PAGE #
1927	Balz-Schiemann Reaction (Schiemann Reaction)	34
1928	Dakin-West reaction	120
1928	Stevens Rearrangement	434
1929	Nenitzescu Indole Synthesis	312
1931	Criegee Oxidation	114
1932	Riley Selenium Dioxide Oxidation	380
1933	Baker-Venkataraman Rearrangement	30
1933	Prévost Reaction	360
1935	Arndt-Eistert Homologation/Synthesis	18
1935	Payne Rearrangement	336
1935	Robinson Annulation	384
1937	Claisen-Ireland Rearrangement	90
1937	Oppenauer Oxidation	320
1937	Sommelet-Hauser Rearrangement	422
1939	Meerwein Arylation	278
1939	Quasi-Favorskii Rearrangement	370
1939	Snieckus Directed *Ortho* Metalation	420
1940	Carrol Rearrangement (Kimel-Cope Rearrangement)	76
1940	Cope Rearrangement	98
1940	Ramberg-Bäcklund Rearrangement	372
1942	Wittig-[1,2]- and [2,3]-Rearrangement	490
1943	Alder (Ene) Reaction	6
1943	Hetero Diels-Alder Reaction (HDA)	204
1944	Birch Reduction	60
1946	Jones Oxidation/Oxidation of Alcohols by Chromium Reagents	228
1947	Peterson Olefination	344
1948	Ritter Reaction	382
1949	Cope Elimination/(Cope Reaction)	96
1949	Cornforth Rearrangement	112
1952	Bamford-Stevens-Shapiro Olefination	36
1952	Grob Fragmentation	190
1952	Wharton Fragmentation	480
1954	Stork Enamine Synthesis	444
1955	Alkene (Olefin) Metathesis	10
1956	Brown Hydroboration Reaction	66
1957	Kornblum Oxidation	250
1958	Brook Rearrangement	64
1958	Doering-LaFlamme Allene Synthesis	146
1958	Horner-Wadsworth-Emmons Olefination	212
1958	Simmons-Smith Cyclopropanation	412
1959	Heine Reaction	198
1959	Ugi Multicomponent Reaction	462
1959	Vilsmeier-Haack Formylation	468
1959	Vinylcyclopropane-Cyclopentene Rearrangement	470
1960	Barton Nitrite Ester Reaction	42
1961	Kröhnke Pyridine Synthesis	254
1962	Barton Radical Decarboxylation Reaction	44
1962	Corey-Chaykovsky Epoxidation and Cyclopropanation	102
1962	DeMayo Cycloaddition (Enone-Alkene [2+2] Photocycloaddition)	132
1962	Nagata Hydrocyanation Reaction	302
1963	Castro-Stevens Coupling	78
1963	Corey-Winter Olefination	110
1963	Pfitzner-Moffatt Oxidation	346
1964	Eschenmoser-Claisen Rearrangement	156
1964	Oxy-Cope Rearrangement/Anionic Oxy-Cope Rearrangement	324

YEAR OF DISCOVERY	NAME OF THE TRANSFORMATION	PAGE #
1965	Tsuji-Trost Reaction/(Allylation)	458
1965	Tsuji-Wilkinson Decarbonylation	460
1966	Wittig Reaction-Schlosser Modification	488
1967	Aza-Claisen Rearrangement (3-Aza-Cope Rearrangement)	20
1967	Aza-Cope Rearrangement	22
1967	Eschenmoser-Tanabe Fragmentation	158
1967	Krapcho Dealkoxycarbonylation	252
1967	Mitsunobu Reaction	294
1967	Seyferth-Gilbert Homologation	402
1968	Baylis-Hillman Reaction	48
1968	Heck Reaction	196
1968	Minisci Reaction	290
1968	Mislow-Evans Rearrangement	292
1969	Prins-Pinacol Rearrangement	366
1969	Schwartz Hydrozirconation	400
1970	Burgess Dehydration Reaction	72
1970	Johnson-Claisen Rearrangement	226
1971	Aza-[2,3]-Wittig Rearrangement	26
1971	Corey-Kim Oxidation	106
1971	Eschenmoser Methenylation	154
1971	Hajos-Parrish Reaction	192
1971	Nicholas Reaction	314
1972	Bergman Cycloaromatization Reaction	56
1972	Corey-Fuchs Alkyne Synthesis	104
1972	Kumada Cross Coupling Reaction	258
1972	McMurry Coupling	276
1972	Saegusa Oxidation	390
1973	Julia-Lythgoe Olefination	230
1973	Mukaiyama Aldol Reaction	298
1973	Pauson-Khand Reaction	334
1973	Pinnick Oxidation	354
1973	Polonovski Reaction	356
1973	Stetter Reaction	432
1974	Overman Rearrangement	322
1974	Alkyne Metathesis	12
1974	Corey-Nicolau Macrolactonization	108
1974	Danishefsky's Diene Cycloaddition	126
1974	Rubottom Oxidation	388
1974	Swern Oxidation	450
1975	Barton-McCombie Radical Deoxygenation Reaction	46
1975	Dötz Benzannulation Reaction	148
1975	Sonogashira Cross-Coupling	424
1976	Enders SAMP/RAMP Hydrazone Alkylation	150
1976	Negishi Cross-Coupling	310
1976	Sakurai Allylation	392
1976	Stille Cross-Coupling (Migita-Kosugi-Stille Coupling)	438
1976	Tebbe Olefination/Petasis-Tebbe Olefination	454
1977	Davis Oxaziridine Oxidation	130
1977	Nozaki-Hiyama-Kishi Coupling	318
1978	Bartoli Indole Synthesis	40
1978	Luche Reduction	268
1978	Roush Asymmetric Allylation	386
1979	Midland Alpine Borane Reduction	288
1979	Suzuki Cross-Coupling (Suzuki-Miyaura Cross-Coupling)	448
1979	Yamaguchi Macrolactonization	500

YEAR OF DISCOVERY	NAME OF THE TRANSFORMATION	PAGE #
1980	Kagan-Molander Samarium-Diiodide Coupling	232
1980	Noyori Asymmetric Hydrogenation	316
1980	Sharpless Asymmetric Dihydroxylation Reaction	406
1980	Sharpless Asymmetric Epoxidation Reaction	408
1981	Corey-Bakshi-Shibata (CBS) Reduction	100
1981	Danheiser Cyclopentene Annulation	124
1981	Evans Aldol Reaction	162
1981	Weinreb Ketone Synthesis	478
1983	Buchwald-Hartwig Cross-Coupling	70
1983	Dess-Martin Oxidation	136
1983	Fleming-Tamao Oxidation	174
1983	Horner-Wadsworth-Emmons Olefination (Still-Gennari modification)	214
1984	Danheiser Benzannulation	122
1984	Stille Carbonylative Cross-Coupling	436
1985	Enyne Metathesis	152
1985	Keck Macrolactonization	238
1985	Ley Oxidation	262
1986	Takai-Utimoto Olefination (Takai Reaction)	452
1987	Stille-Kelly Coupling	440
1989	Kahne Glycosidation	234
1989	Kulinkovich Reaction	256
1990	Jacobsen-Katsuki Epoxidation	222
1991	Larock Indole Synthesis	260
1993	Keck Asymmetric Allylation	236
1993	Keck Radical Allylation	240
1993	Petasis Boronic Acid-Mannich reaction	340
1994	Myers Asymmetric Alkylation	300
1994	Smith-Tietze Multicomponent Dithiane Linchpin Coupling	418
1995	Jacobsen Hydrolytic Kinetic Resolution of Epoxides	220
1995	Miyaura Boration Reaction	296
1995	Petasis-Ferrier Rearrangement	342
1996	Sharpless Asymmetric Aminohydroxylation Reaction	404
1996	Shi Asymmetric Epoxidation	410

8.3 REACTION CATEGORIES

REACTION CATEGORY	NAME OF REACTIONS	BRIEF DESCRIPTION OF SYNTHETIC USE	Page#
CARBOCYCLE FORMATION			
	Acyloin condensation	Formation of cyclic α-hydroxy ketones from diesters.	4
	Alkene metathesis	Formation of cyclic alkenes from dienes.	10
	Alkyne metathesis	Formation of cyclic alkynes from diynes.	12
	Danheiser cyclopentene annulation	Formation of cyclopentenes from enones and allenes.	124
	Danishefsky's diene cycloaddition	Formation of six-membered carbocycles using 1-methoxy-3-trimethylsilyloxy-1,3-butadiene.	126
	Dieckmann condensation	Formation of cyclic β-keto esters from diesters.	138
	Diels-Alder cycloaddition	The [4+2] cycloaddition of alkenes and dienes to afford substituted cyclohexenes.	140
	Hajos-Parrish reaction	Enantio-enriched bicyclic enones from 1,5-diketones.	192
	Nazarov cyclization	Cyclopentenones and cyclopentanones from divinyl ketones.	304
	Pauson-Khand reaction	Formation of cyclopentenones from alkenes, alkynes and CO.	334
	Robinson annulation	Formation of bicyclic enones from 1,5-diketones.	384
CYCLO-AROMATIZATION			
	Bergman cycloaromatization reaction	Thermal or photochemical cycloaromatization of enediynes to form substituted benzene rings.	56
	Danheiser benzannulation	Reaction of cyclobutenones with alkynes to give highly substituted benzene rings.	122
	Dötz benzannulation	Reaction of Fischer chromium carbenes with alkynes to give substituted hydroquinone derivatives.	148
DEGRADATION			
	Hofmann rearrangement	Conversion of primary carboxamides to one-carbon shorter primary amines.	210
	Hunsdiecker reaction	Conversion of carboxylic acids to one-carbon shorter alkyl, alkenyl or aryl halides.	218
	Lieben haloform reaction	Conversion of methyl ketones to one-carbon shorter carboxylic acids.	262
ELECTROPHILIC ADDITION TO C-C MULTIPLE BONDS			
Addition to alkenes			
cyclopropanation	Simmons-Smith cyclopropanation	Formation of cyclopropanes from alkenes.	412
epoxidation	Davis' oxaziridine oxidation	Formation of epoxides from alkenes using oxaziridines.	130
epoxidation	Jacobsen-Katsuki epoxidation	Formation of epoxides from alkenes using metal salen complexes.	222
epoxidation	Prilezhaev reaction	Formation of epoxides from alkenes using peracids.	362
epoxidation	Sharpless asymmetric epoxidation	Formation of epoxy alcohols from allylic alcohols.	408
epoxidation	Shi asymmetric epoxidation	Formation of epoxides from alkenes.	410
hydrogenation	Noyori asymmetric hydrogenation	Formation of enantio-enriched carboxylic acids, alcohols and amino acids from unsaturated carboxylic acids, allylic alcohols and enamides, respectively.	316
hydrometalation	Brown hydroboration reaction	Formation of alkylboranes from alkenes.	66
hydrometalation	Schwartz hydrozirconation	Formation of alkylzirconium compounds from alkenes.	400
Addition to alkynes			
hydrometalation	Brown hydroboration reaction	Formation of alkenylboranes from alkynes.	66
hydrometalation	Schwartz hydrozirconation	Formation of alkenylzirconium compounds from alkynes.	400
ELECTROPHILIC AROMATIC SUBSTITUTION			
	Bischler-Napieralski isoquinoline synthesis	Preparation of isoquinolines from acylated phenylethylamines.	62
	Combes Quinoline synthesis	Preparation of quinolines from aryl amines and 1,3-diketones.	94
	Friedel-Crafts acylation	Synthesis of aromatic ketones using acyl halides or anhydrides.	176
	Friedel-Crafts alkylation	Synthesis of alkylbenzenes using alkyl halides.	178
	Fries rearrangement	Synthesis of acylated phenols from *O*-acyl phenols.	180

REACTION CATEGORY	NAME OF REACTIONS	BRIEF DESCRIPTION OF SYNTHETIC USE	Page#
ELECTROPHILIC AROMATIC SUBSTITUTION			
	Gattermann and Gattermann-Koch formylation	Synthesis of aromatic aldehydes using HCN or CO.	184
	Houben-Hoesch reaction	Synthesis of aromatic ketones from activated aromatic compounds (e.g. phenols) and nitriles.	216
	Kolbe-Schmitt reaction	Synthesis of salicylic acid der. from phenols and CO_2.	248
	Pictet-Spengler tetrahydro-isoquinoline synthesis	Synthesis of tetrahydroisoquinolines and isoquinolines from β-arylethylamines.	348
	Pomeranz-Fritsch reaction	Synthesis of isoquinolines from aromatic aldehydes and 2,2-dialkoxyethylamine.	358
	Reimer-Tiemann reaction	Preparation of formylated phenols from substituted phenols	378
	Vilsmeier-Haack formylation	Synthesis of substituted benzaldehydes and heteroaromatic aldehydes using chloromethyliminium salts.	468
	von Pechmann reaction	Preparation of coumarins from phenols and β-keto esters.	472
ELIMINATION REACTIONS			
	Burgess dehydration	Preparation of alkenes from 2° and 3° alcohols.	72
	Chugaev elimination	Thermal *syn* elimination of xanthate esters to form alkenes.	82
	Cope elimination	Thermal *syn* elimination of 3° amine N-oxides to form alkenes.	96
	Hofmann elimination	Formation of alkenes from quaternary ammonium salts.	206
FRAGMENTATION REACTIONS			
	Eschenmoser-Tanabe fragmentation	Formation of alkynals or alkynones from epoxy ketone hydrazones.	158
	Grob fragmentation	Regulated heterolytic cleavage of certain types of molecules to form three different fragments.	190
	Wharton fragmentation	Base-induced formation of medium-sized cyclic alkenes from 1,3-diol monosulfonates.	480
HETEROCYCLE FORMATION			
	Bartoli indole synthesis	Formation of 7-substituted indoles from ortho-substituted nitro- or nitrosoarenes.	40
	Biginelli reaction	One-pot three component formation of 3,4-dihydropyrimidin-2(1H)-ones from aromatic aldehydes, keto esters and urea.	58
	Bischler-Napieralski isoquinoline synthesis	Preparation of isoquinolines from acylated phenylethylamines.	62
	Ciamician-Dennstedt rearrangement	Synthesis of 3-halopyridines from pyrroles and 2-haloquinolines from indoles.	84
	Combes quinoline synthesis	Preparation of quinolines from aryl amines and 1,3-diketones.	94
	Dimroth rearrangement	Isomerization of heterocycles in which endocyclic or oxocyclic heteroatoms and their attached substituents are translocated via a ring-opening-ring-closure sequence.	144
	Feist-Bénary furan synthesis	Synthesis of furans from β-keto esters and α-halogenated carbonyl compounds under basic conditions.	166
	Fischer indole synthesis	Preparation of indoles from arylhydrazones of ketones and aldehydes in the presence of protic or Lewis acid catalyst.	172
	Hantzsch dihydropyridine synthesis	Preparation of dihydropyridines from 1,3-diketones, aldehydes and ammonia.	194
	Heine reaction	Intramolecular ring expansion of substituted N-acylazirdines to the corresponding substituted oxazolines.	198
	Hetero Diels-Alder reaction	The [4+2] cyclization of a diene or heterodiene and a dienophile or heterodienophile.	204
	Hofmann-Löffler-Freytag reaction	Formation of cyclic amines from N-halogenated amines via an intramolecular 1,5-hydrogen atom transfer to a nitrogen radical.	208
	Knorr pyrrole synthesis	Condensation of an α-amino ketone or an α-amino-β-ketoester with an active methylene compound to afford substituted pyrroles.	244
	Kröhnke pyridine synthesis	Condensation of an unsaturated ketone with an α-halo ketone to give highly substituted pyridines.	254
	Larock indole synthesis	Preparation of 2,3-disubstituted indoles from *ortho*-iodoanilines and disubstituted alkynes.	258

REACTION CATEGORY	NAME OF REACTIONS	BRIEF DESCRIPTION OF SYNTHETIC USE	Page#
HETEROCYCLE FORMATION			
	Madelung indole synthesis	The intramolecular cyclization of N-acylated-*ortho*-alkylanilines to afford 2,3-disubstituted indoles.	**270**
	Paal-Knorr furan synthesis	Dehydration of 1,4-diketones to the corresponding substituted furans.	**326**
	Paal-Knorr pyrrole synthesis	Condensation of primary amines with 1,4-dicarbonyl compounds to form substituted pyrroles.	**328**
	Paterno-Büchi reaction	Formation of oxetanes by the photocycloaddition of alkenes and carbonyl compounds.	**332**
	Pictet-Spengler tetrahydroisoquinoline synthesis	Condensation of a β-arylethylamine with carbonyl compounds to form tetrahydroisoquinolines.	**348**
	Pomeranz-Fritsch reaction	The acid catalyzed cyclization of benzalaminoacetals to form isoquinolines.	**358**
	Skraup and Doebner-Miller quinoline synthesis	Condensation of enones with substituted anilines to afford isoquinolines.	**414**
	von Pechman reaction	Condensation of phenols with β-keto esters to give substituted coumarins.	**472**
HOMOLOGATION			
	Arndt-Eistert homologation	One-carbon homologation of carboxylic acids.	**18**
	Corey-Fuchs alkyne synthesis	One-carbon homologation of aldehydes to form the corresponding terminal alkynes.	**104**
	Doering-LaFlamme allene synthesis	Preparation of allenes from olefins.	**146**
	Seyferth-Gilbert homologation	Synthesis of alkynes from aldehydes.	**402**
	Takai-Utimoto olefination	The chromium(II)-mediated one-carbon homologation of aldehydes to the corresponding (*E*)-alkenyl halides.	**452**
	Tebbe olefination	One-carbon homologation of carbonyl compounds to afford the corresponding 1,1-disubstituted alkenes.	**454**
METATHESIS			
	Alkene metathesis	Metal catalyzed redistribution of carbon-carbon double bonds.	**10**
	Alkyne metathesis	Metal catalyzed redistribution of carbon-carbon triple bonds.	**12**
	Enyne metathesis	Transition metal catalyzed cycloisomerization of [1,n]-enynes to the corresponding 1,3-dienes.	**152**
NUCLEOPHILIC AROMATIC SUBSTITUTION			
	Chichibabin amination reaction	Direct amination of pyridine via S_NAr reaction.	**80**
	Smiles rearrangement	Intramolecular nucleophilic aromatic rearrangement of activated aromatic substrates.	**416**
NUCLEOPHILIC SUBSTITUTION			
	Finkelstein reaction	Equilibrium exchange of the halogen atom in alkyl halides for another halogen atom.	**170**
	Gabriel synthesis	Two-step preparation of primary amines from the corresponding alkyl halides using phthalimide as the nitrogen source.	**182**
	Heine reaction	Intramolecular ring expansion of substituted N-acylaziridines by nucleophilic reagents to the corresponding substituted oxazolines.	**198**
	Kahne glycosidation	Preparation of *O*-, *S*- or *N*-glycosides via the activation of glycosyl sulfoxides.	**234**
	Koenigs-Knorr glycosidation	Synthesis of alkyl or aryl *O*-glycosides from glycosyl halides and alcohols or phenols, respectively.	**246**
	Krapcho dealkoxycarbonylation	Decarboxylation of β-keto esters using alkali metal salts.	**252**
	Mitsunobu reaction	Substitution of primary and secondary alcohols with nucleophiles in the presence of dialkyl azodicarboxylate and trialkyl- or triarylphosphine.	**294**
	Myers asymmetric alkylation	Alkylation of *N*-acylated pseudoephedrines to obtain enantio-enriched α-alkylated carbonyl compounds.	**300**
	Nicholas reaction	Trapping of dicobalt hexacarbonyl-stabilized propargylic cations with various nucleophiles.	**314**
	Payne rearrangement	Base-catalyzed intramolecular displacement of 2,3-epoxy alcohols to give isomeric 2,3-epoxy alcohols.	**336**
	Stork enamine synthesis	Alkylation of enamines with alkyl halides to afford α-alkylated aldehydes or ketones.	**444**
	Williamson ether synthesis	Alkylation of alkali alkoxides with primary or secondary alkyl halides to form ethers.	**484**

REACTION CATEGORY	NAME OF REACTIONS	BRIEF DESCRIPTION OF SYNTHETIC USE	Page#
OXIDATION			
	Baeyer-Villiger oxidation	Formation of esters from ketones upon peracid oxidation.	28
	Corey-Chaykovsky epoxidation	Preparation of epoxides from aldehydes and ketones.	102
	Corey-Kim oxidation	Oxidation of primary and secondary alcohols with NCS/DMS to afford aldehydes and ketones, respectively.	106
	Criegee oxidation	Cleavage of 1,2-diols (glycols) to the corresponding carbonyl compounds using LTA.	114
	Dakin oxidation	Conversion of aromatic aldehydes and ketones to the corresponding phenols.	118
	Davis' oxaziridine oxidation	Oxidation of electron-rich substrates (e.g. alkenes, enolates, enol ethers etc.) with oxaziridines.	130
	Dess-Martin oxidation	Oxidation of alcohols and oximes to afford the corresponding carbonyl compounds using DMP.	136
	Fleming-Tamao oxidation	Mild stereospecific oxidation of silicon-carbon bonds to the corresponding carbon-oxygen bonds	174
	Jacobsen-Katsuki epoxidation	Enantioselective epoxidation of unfunctionalized alkyl- and aryl-substituted olefins.	222
	Jones oxidation	Oxidation of primary and secondary alcohols with chromic acid to give the corresponding carboxylic acids and ketones.	228
	Kornblum oxidation	Oxidation of alkyl halides to the corresponding carbonyl compounds using DMSO as the oxidant.	250
	Ley oxidation	Oxidation of primary and secondary alcohols with TPAP/NMO to give the corresponding aldehydes and ketones.	260
	Oppenauer oxidation	Oxidation of primary and secondary alcohols with ketones in the presence of metal alkoxides to afford the corresponding aldehydes and ketones.	320
	Pfitzner-Moffatt oxidation	Oxidation of primary and secondary alcohols with DCC/DMSO to give the corresponding aldehydes and ketones.	346
	Pinnick oxidation	Mild oxidation of aldehydes directly to the corresponding carboxylic acids using $NaClO_2$ as the oxidant.	354
	Prilezhaev reaction	Oxidation of alkenes to epoxides using peroxycarboxylic acids.	362
	Riley selenium dioxide oxidation	Oxidation of the methylene group adjacent to a carbonyl group or the double bond of olefins (allylic or benzylic position) with SeO_2.	380
	Rubottom oxidation	Oxidation of silyl enol ethers with *m*CPBA to give α-hydroxy ketones or α-hydroxy aldehydes.	388
	Saegusa oxidation	Regioselective introduction of the α,β carbon-carbon double bond to cyclic and acylic ketones via Pd-mediated oxidation of the corresponding silyl enol ethers.	390
	Sharpless asymmetric aminohydroxylation	One-pot enantioselective synthesis of protected vicinal amino alcohols from simple alkenes.	404
	Sharpless asymmetric dihydroxylation	One-pot enantioselective synthesis of vicinal diols from simple alkenes.	406
	Sharpless asymmetric epoxidation	Ti-alkoxide-catalyzed epoxidation of prochiral and chiral allylic alcohols in the presence of a chiral tartrate ester and an alkyl hydroperoxide.	408
	Shi asymmetric epoxidation	Chiral-ketone catalyzed epoxidation of unfunctionalized olefins.	410
	Swern oxidation	Oxidation of primary and secondary alcohols using DMSO/TFAA or oxalyl chloride to afford the corresponding aldehydes and ketones.	450
	Tishchenko reaction	Conversion of aldehydes to the corresponding esters in the presence of metal alkoxides.	456
	Wacker oxidation	One-pot oxidation of olefins to the corresponding ketones in the presence of catalytic amounts of Pd(II)-salts	474
PERICYCLIC REACTIONS			
	Alder (ene) reaction	Activation of an allylic C-H bond and the concomitant allylic transposition of the C=C double bond of alkenes. (Formally the addition of alkenes to C=C and C=O bonds.)	6
cycloaddition	Danishefsky's diene cycloaddition	Formation of six-membered carbocycles and heterocycles using 1-methoxy-3-trimethylsilyloxy-1,3-butadiene.	126
cycloaddition	DeMayo cycloaddition	Photochemical [2+2] cycloaddition of enones and alkenes to give substituted cyclobutanes.	132
cycloaddition	Diels-Alder cycloaddition	The [4+2] cycloaddition of alkenes and dienes to afford substituted cyclohexenes.	140

REACTION CATEGORY	NAME OF REACTIONS	BRIEF DESCRIPTION OF SYNTHETIC USE	Page#
PERICYCLIC REACTIONS			
cycloaddition	Hetero Diels-Alder cycloaddition	The [4+2] cyclization of a diene or heterodiene and a dienophile or heterodienophile.	**204**
cycloaddition	Paterno-Büchi reaction	Formation of oxetanes by the photocycloaddition of alkenes and carbonyl compounds.	**332**
cycloaddition	Staudinger ketene cycloaddition	Formation of cyclobutanones from alkenes and ketenes.	**426**
electrocyclization	Cornforth rearrangement	Thermal rearrangement of 4-carbonyl substituted oxazoles to their isomeric oxazoles.	**112**
electrocyclization	Nazarov cyclization	Thermal or photochemical ring-closure of divinyl ketones.	**304**
sigmatropic rearr.	Aza-Claisen rearrangement	Thermal [3,3]-sigmatropic rearrangement of *N*-allyl enamines.	**20**
sigmatropic rearr.	Aza-Cope rearrangement	Thermal [3,3]-sigmatropic rearrangement of *N*-substituted 1,5-dienes.	**22**
sigmatropic rearr.	Aza-Wittig rearrangement	Thermal [3,3]-sigmatropic rearrangement of allylic tertiary amines to give homoallylic secondary amines.	**26**
sigmatropic rearr.	Carroll rearrangement	Thermal [3,3]-sigmatropic rearrangement of allylic β-keto esters to afford γ,δ-unsaturated ketones.	**76**
sigmatropic rearr.	Claisen rearrangement	Thermal [3,3]-sigmatropic rearrangement of allyl vinyl ethers to give γ,δ-unsaturated carbonyl compounds.	**88**
sigmatropic rearr.	Claisen-Ireland rearrangement	Thermal [3,3]-sigmatropic rearrangement of *O*-trialkylsilylketene acetals to γ,δ-unsaturated carboxylic acids.	**90**
sigmatropic rearr.	Cope rearrangement	Thermal [3,3]-sigmatropic rearrangement of 1,5-dienes to the isomeric 1,5-dienes.	**98**
sigmatropic rearr.	Eschenmoser-Claisen rearrangement	Thermal [3,3]-sigmatropic rearrangement to generate γ,δ-unsaturated amides from allylic alcohols and *N,N*-dimethylacetamide dimethyl acetal.	**156**
sigmatropic rearr.	Johnson-Claisen rearrangement	Thermal [3,3]-sigmatropic rearrangement of allyl ketene acetals to afford γ,δ-unsaturated esters.	**226**
sigmatropic rearr.	Meisenheimer rearrangement	Thermal rearrangement of certain tertiary amine N-oxides to the corresponding *O*-substituted-*N,N*-disubstituted hydroxylamines.	**282**
sigmatropic rearr.	Mislow-Evans rearrangement	Reversible 1,3-transposition of allylic sulfoxide and allylic alcohol functionalities.	**292**
sigmatropic rearr.	Overman rearrangement	The 1,3-transposition of alcohol and amine functionalities *via* the [3,3]-sigmatropic rearrangement of allylic trichloroacetimidates.	**322**
sigmatropic rearr.	Oxy-Cope rearrangement	Thermal [3,3]-sigmatropic rearrangement of 1,5-diene-3-ols to afford δ,ε-unsaturated carbonyl compounds.	**324**
sigmatropic rearr.	Sommelet-Hauser rearrangement	The thermal [2,3]-sigmatropic rearrangement of benzylic quaternary ammonium salts in the presence of a strong base.	**422**
sigmatropic rearr.	Wittig rearrangement	Thermal [1,2]-rearrangement of aryl alkyl ethers and also the thermal [2,3]-rearrangement of allyl alkyl ethers.	**490**
PHOTOCHEMICAL REACTIONS			
	Bergman cycloaromatization reaction	Thermal or photochemical cycloaromatization of enediynes to form substituted benzene rings.	**56**
	Buchner method of ring expansion	Thermal or photochemical reaction of ethyl diazoacetate with benzenes and its homologs to give the isomeric esters of cycloheptatriene carboxylic acid.	**68**
	Curtius rearrangement	Thermal or photochemical rearrangement of acyl azides to give isocyanates.	**116**
	DeMayo cycloaddition	Photochemical [2+2] cycloaddition of enones and alkenes to give substituted cyclobutanes.	**132**
	Fries rearrangement	Conversion of phenolic esters to the corresponding phenolic ketones and aldehydes.	**180**
	Nazarov cyclization	Thermal or photochemical ring-closure of divinyl ketones.	**304**
	Paterno-Büchi reaction	Formation of oxetanes by the photocycloaddition of alkenes and carbonyl compounds.	**332**
	Vinylcyclopropane-cyclopentene rearrangement	Thermal or photochemical rearrangement of substituted vinylcyclopropanes to substituted cyclopentenes.	**470**
	Wolff rearrangement	Thermal or photochemical rearrangement of α-diazo ketones to form ketenes.	**494**
RADICAL REACTIONS			
alkylation	Minisci reaction	Substitution of protonated heteroaromatic bases by nucleophilic carbon-centered radicals.	**290**
allylation	Keck radical allylation	Coupling of alkyl halides with allyltributyltin in the presence of a radical initiator (e.g. AIBN)	**240**

REACTION CATEGORY	NAME OF REACTIONS	BRIEF DESCRIPTION OF SYNTHETIC USE	Page#
RADICAL REACTIONS			
arylation	Meerwein arylation	Arylation of unsaturated carbonyl compounds using diazonium salts.	278
decarboxylation	Barton radical decarboxylation reaction	Reductive decarboxylation of thiohydroxamate esters to give alkanes.	44
decarboxylation	Hunsdiecker reaction	Halogenative decarboxylation of carboxylic acids to give one-carbon shorter alkyl halides.	218
deoxygenation	Barton-McCombie radical deoxygenation	Reductive deoxygenation of thioxoesters to give the corresponding alkanes.	46
halogenation	Sandmeyer reaction	Formation of aryl halides from the corresponding diazonium salts *via* an aryl radical.	394
halogenation	Wohl-Ziegler bromination	Bromination of alkenes and alkylbenzenes at the allylic or benzylic position.	492
remote functionalization	Barton nitrite ester reaction	Thermal or photolytic reaction of nitrite esters to afford γ-hydroxy oximes.	42
remote functionalization	Hofmann-Löffler-Freytag reaction	Thermal or photolytic reaction of *N*-halogenated amines to form cyclic amines.	208
REACTIONS INVOLVING CARBENES			
	Buchner method of ring expansion	Thermal or photochemical reaction of ethyl diazoacetate with benzenes and its homologs to give the isomeric esters of cycloheptatriene carboxylic acid.	68
	Ciamician-Dennstedt rearrangement	Synthesis of 3-halopyridines from pyrroles and 2-haloquinolines from indoles.	84
	Doering-LaFlamme allene synthesis	Preparation of allenes from olefins.	146
	Reimer-Tiemann reaction	Preparation of formylated phenols from substituted phenols	378
	Wolff rearrangement	Thermal or photochemical rearrangement of α-diazo ketones to form ketenes.	494
REACTIONS INVOLVING CARBONYL COMPOUNDS			
	Aldol reaction	Addition of an enol/enolate of a carbonyl compound to an aldehyde or ketone to form a β-hydroxycarbonyl compound.	8
	Barbier coupling reaction	Metal-mediated addition of alkyl, allyl or benzyl halides to carbonyl compounds.	38
	Baylis-Hillman reaction	Formation of a C-C single bond between the α-position of conjugated carbonyl compounds or conjugated carboxylic acid derivatives and aldehydes or ketones.	48
	Benzoin and retro-benzoin condensation	Reaction of aldehydes to form α-hydroxy ketones in the presence of a nucleophilic catalyst (e.g. cyanide ion).	54
	Corey-Chaykovsky epoxidation	Preparation of epoxides from aldehydes and ketones using sulfur ylides.	102
	Corey-Fuchs alkyne synthesis	One-carbon homologation of aldehydes to form the corresponding terminal alkynes.	104
	Dakin oxidation	Conversion of aromatic aldehydes and ketones to the corresponding phenols.	118
	Eschweiler-Clarke methylation	One-pot reductive methylation of primary and secondary amines to the corresponding tertiary amines using formaldehyde and a reducing agent.	160
	Evans aldol reaction	Reaction of boron enolates with aldehydes to afford *syn* aldol products.	162
	Grignard reaction	Addition of organomagnesium species to aldehydes and ketones to form secondary alcohols and tertiary alcohols, respectively.	188
	Hantzsch dihydropyridine synthesis	Preparation of dihydropyridines from 1,3-diketones, aldehydes and ammonia.	194
	Henry reaction	Aldol condensation between nitroalkanes and carbonyl compounds to form β-nitro alcohols.	202
	HWE olefination	Stereoselective olefination of aldehydes and ketones using phosphoryl-stabilized carbanions.	212
	HWE olefination-Still modification	Preparation of (Z)-α,β-unsaturated ketones and esters by coupling electrophilic *bis*(trifluoroalkyl) phosphonoesters with aldehydes and ketones in the presence of a strong base.	214

REACTION CATEGORY	NAME OF REACTIONS	BRIEF DESCRIPTION OF SYNTHETIC USE	Page#
REACTIONS INVOLVING CARBONYL COMPOUNDS			
	Kagan-Molander coupling	SmI_2-mediated addition of alkyl, allyl or benzyl halides to carbonyl compounds.	**232**
	Keck asymmetric allylation	The reaction of aldehydes with allyltributylstannane in the presence of Lewis acid catalysts to form homoallylic alcohols.	**236**
	Knoevenagel condensation	Condensation of aldehydes and ketones with active methylene compounds to afford α,β-unsaturated dicarbonyl or related compounds.	**242**
	Mannich reaction	The condensation of CH activated compound with a primary or secondary amine and a non-enolizable carbonyl compound to afford aminoalkylated derivatives.	**274**
	Mukaiyama aldol reaction	Lewis acid mediated addition of enol silanes to carbonyl compounds.	**298**
	Passerini multicomponent reaction	Condensation of isocyanides with carboxylic acids and carbonyl compounds to afford α-acyloxycarboxamides.	**330**
	Perkin reaction	Condensation of aromatic aldehydes with the anhydrides of aliphatic carboxylic acids to afford α,β-unsaturated carboxylic acids.	**338**
	Peterson olefination	Preparation of alkenes from α-silyl carbanions and carbonyl compounds.	**344**
	Pictet-Spengler tetrahydro-isoquinoline synthesis	Synthesis of tetrahydroisoquinolines and isoquinolines from β-arylethylamines.	**348**
	Prins reaction	Acid-catalyzed condensation of alkenes with aldehydes.	**364**
	Reformatsky reaction	Zinc-induced reaction between an α-halo ester and an aldehyde or ketone to afford a β-hydroxy ketone.	**374**
	Robinson annulation	Formation of bicyclic enones from 1,5-diketones.	**384**
	Roush asymmetric allylation	Reaction of allylboronates with aldehydes to give homoallylic alcohols.	**386**
	Sakurai allylation	Reaction of allylsilanes with a variety of aldehydes and ketones in the presence of a Lewis acid.	**392**
	Seyferth-Gilbert homologation	Preparation of alkynes from aldehydes and ketones.	**402**
	Stetter reaction	Formation of 1,4-diketones from aldehydes and α,β-unsaturated carbonyl compounds in the presence of a nucleophilic catalyst.	**432**
	Stobbe condensation	Formation of alkylidene succinic acids or their monoesters from dialkyl succinates and carbonyl compounds.	**442**
	Strecker reaction	The condensation of carbonyl compounds with amines and nitriles to afford α-amino nitriles.	**446**
	Takai-Utimoto olefination	The chromium(II)-mediated one-carbon homologation of aldehydes to the corresponding (*E*)-alkenyl halides.	**452**
	Tebbe olefination	One-carbon homologation of carbonyl compounds to afford the corresponding 1,1-disubstituted alkenes.	**454**
	Wittig reaction	Formation of carbon-carbon double bonds from carbonyl compounds and phosphorous ylides.	**486**
	Wittig reaction-Schlosser modification	One-pot multistep preparation of (*E*)-alkenes from "nonstabilized" phosphorous ylides and carbonyl compounds by the equilibration of the intermediate lithiobetaines.	**488**
REARRANGE-MENTS			
anionic	Baker-Venkataraman rearrangement	Base-catalyzed rearrangement of aromatic *ortho*-acyloxyketones to aromatic β-diketones.	**30**
anionic	Benzilic acid rearrangement	Rearrangement of 1,2-diketones to give the salts of α-hydroxy acids.	**52**
anionic	Brook rearrangement	Intramolecular anionic [1,*n*]-migration of silyl groups from a carbon to an oxygen atom.	**64**
anionic	Ciamician-Dennstedt rearrangement	Synthesis of 3-halopyridines from pyrroles and 2-haloquinolines from indoles.	**84**
anionic	Favorskii rearrangement	Skeletal rearrangement of α-halo ketones via a cyclopropanone intermediate to give carboxylic acids or carboxylic acid derivatives.	**164**
anionic	Hofmann rearrangement	Conversion of primary carboxamides to one-carbon shorter primary amines.	**210**
anionic	Lossen rearrangement	Conversion of *O*-acyl hydroxamic acids to the corresponding isocyanates.	**266**

REACTION CATEGORY	NAME OF REACTIONS	BRIEF DESCRIPTION OF SYNTHETIC USE	Page#
REARRANGE-MENTS			
anionic	Payne rearrangement	The base-catalyzed intramolecular nucleophilic displacement of 2,3-epoxy alcohols to give the isomeric 2,3-epoxy alcohols.	**336**
anionic	Quasi-Favorskii rearrangement	Skeletal rearrangement of bicyclic α-halo ketones in which the halogen is located at the bridgehead position to afford carboxylic acids or carboxylic acid derivatives.	**370**
anionic	Ramberg-Bäcklund rearrangement	Base-induced rearrangement of α-halogenated sulfones via episulfone intermediates to produce alkenes.	**372**
anionic	Smiles rearrangement	Intramolecular nucleophilic aromatic rearrangement of activated aromatic substrates.	**416**
anionic	Wittig rearrangement	Thermal [1,2]-rearrangement of aryl alkyl ethers and also the thermal [2,3]-rearrangement of allyl alkyl ethers.	**490**
ANRORC	Dimroth rearrangement	Isomerization of heterocycles in which endocyclic or oxocyclic heteroatoms and their attached substituents are translocated via a ring-opening-ring-closure sequence.	**144**
biradical or dipolar	Vinylcyclopropane-cyclopentene rearrangement	Thermal or photochemical rearrangement of substituted vinylcyclopropanes to substituted cyclopentenes.	**470**
cationic	Amadori reaction/rearrangement	The acid- or base-catalyzed isomerization of *N*-glycosides of aldoses to form 1-amino-1-deoxy ketoses.	**14**
cationic	Beckmann rearrangement	Conversion of aldoximes and ketoximes to the corresponding amides in acidic medium.	**50**
cationic	Demjanov and Tiffeneau-Demjanov rearrangement	The ring enlargement of 1-aminomethyl cycloalkanes to the corresponding cycloalkanols and the ring-enlargement of 1-aminomethyl cycloalkanols to the corresponding cycloalkanones.	**134**
cationic	Dienone-phenol rearrangement	Acid-catalyzed migration of alkyl groups in cyclohexadienones to afford substituted phenols.	**142**
cationic	Ferrier reaction	Lewis acid promoted rearrangement of unsaturated carbohydrates (glycals) in the presence of nucleophiles to the corresponding 2,3-unsaturated glycosyl compounds.	**168**
cationic	Fries rearrangement	Synthesis of acylated phenols from *O*-acyl phenols.	**180**
cationic	Meyer-Schuster and Rupe rearrangement	Acid-catalyzed isomerization of secondary and tertiary propargylic alcohols to the corresponding α,β-unsaturated aldehydes or ketones.	**284**
cationic	Petasis-Ferrier rearrangement	Lewis acid-promoted rearrangement of cyclic enol acetals to the corresponding substituted tetrahydrofurans and tetrahydropyrans.	**342**
cationic	Pinacol rearrangement	Acid-catalyzed transformation of 1,2-diols to give the corresponding rearranged ketones or aldehydes.	**350**
cationic	Prins-Pinacol rearrangement	Formation of oxacyclic and carbocyclic ring systems by terminating Prins cyclizations with the pinacol rearrangement in a tandem fashion.	**366**
cationic	Pummerer rearrangement	Formation of α-substituted sulfides from the corresponding sulfoxides.	**368**
cationic	Schmidt reaction	Reaction of carboxylic acids and carbonyl compounds with hydrazoic acid or alkyl azides to afford the corresponding amines, nitriles or amides, respectively.	**396**
cationic	Wagner-Meerwein rearrangement	Generation of a carbocation followed by the [1,2]-shift of an adjacent carbon-carbon bond to generate a new carbocation.	**476**
concerted	Baeyer-Villiger oxidation/rearrangement	Transformation of ketones to esters and cyclic ketones to lactones by peroxyacids.	**28**
dipolar	Cornforth rearrangement	Thermal rearrangement of 4-carbonyl substituted oxazoles to their isomeric oxazoles.	**112**
neutral	Curtius rearrangement	Thermal or photochemical rearrangement of acyl azides to give isocyanates.	**116**
neutral	Wolff rearrangement	Thermal or photochemical rearrangement of α-diazo ketones to form ketenes.	**494**
radical pair	Stevens rearrangement	Base-promoted transformation of sulfonium or quaternary ammonium salts to sulfides or tertiary amines.	**434**
sigmatropic (neutral)	Aza-Claisen rearrangement	Thermal [3,3]-sigmatropic rearrangement of *N*-allyl enamines.	**20**
sigmatropic (neutral)	Aza-Cope rearrangement	Thermal [3,3]-sigmatropic rearrangement of *N*-substituted 1,5-dienes.	**22**
sigmatropic (anionic)	Aza-Wittig rearrangement	Thermal [3,3]-sigmatropic rearrangement of allylic tertiary amines to give homoallylic secondary amines.	**26**
sigmatropic (neutral)	Carroll rearrangement	Thermal [3,3]-sigmatropic rearrangement of allylic β-keto esters to afford γ,δ-unsaturated ketones.	**76**

REACTION CATEGORY	NAME OF REACTIONS	BRIEF DESCRIPTION OF SYNTHETIC USE	Page#
REARRANGE-MENTS			
sigmatropic (neutral)	Claisen rearrangement	Thermal [3,3]-sigmatropic rearrangement of allyl vinyl ethers to give γ,δ-unsaturated carbonyl compounds.	**88**
sigmatropic (neutral)	Claisen-Ireland rearrangement	Thermal [3,3]-sigmatropic rearrangement of O-trialkylsilylketene acetals to γ,δ-unsaturated carboxylic acids.	**90**
sigmatropic (neutral)	Cope rearrangement	Thermal [3,3]-sigmatropic rearrangement of 1,5-dienes to the isomeric 1,5-dienes.	**98**
sigmatropic (neutral)	Eschenmoser-Claisen rearrangement	Thermal [3,3]-sigmatropic rearrangement to generate γ,δ-unsaturated amides from allylic alcohols and *N,N*-dimethylacetamide dimethyl acetal.	**156**
sigmatropic (neutral)	Johnson-Claisen rearrangement	Thermal [3,3]-sigmatropic rearrangement of allyl ketene acetals to afford γ,δ-unsaturated esters.	**226**
sigmatropic anionic for [2,3] and radical for [1,2]	Meisenheimer rearrangement	Thermal rearrangement of certain tertiary amine *N*-oxides to the corresponding *O*-substituted-*N,N*-disubstituted hydroxylamines.	**282**
sigmatropic (anionic)	Mislow-Evans rearrangement	Reversible 1,3-transposition of allylic sulfoxide and allylic alcohol functionalities.	**292**
sigmatropic (neutral)	Overman rearrangement	The 1,3-transposition of alcohol and amine functionalities *via* the [3,3]-sigmatropic rearrangement of allylic trichloroacetimidates.	**322**
sigmatropic (anionic)	Oxy-Cope rearrangement	Thermal [3,3]-sigmatropic rearrangement of 1,5-diene-3-ols to afford δ,ε-unsaturated carbonyl compounds.	**324**
sigmatropic (anionic)	Sommelet-Hauser rearrangement	The thermal [2,3]-sigmatropic rearrangement of benzylic quaternary ammonium salts in the presence of a strong base.	**422**
REDUCTION			
	Birch reduction	1,4-Reduction of aromatic rings using alkali metals dissolved in liquid ammonia as reducing agents.	**60**
	Clemmensen reduction	Conversion of a carbonyl group to the corresponding methylene group using Zn(Hg)/HCl.	**92**
	Corey-Bakshi-Shibata reduction	Enantioselective reduction of ketones with BH_3 using oxazaborolidines as catalysts.	**100**
	Eschweiler-Clarke methylation	One-pot reductive methylation of primary and secondary amines to the corresponding tertiary amines using formaldehyde and a reducing agent.	**160**
	Luche reduction	Reduction of enones to the corresponding allylic alcohols using $CeCl_3/NaBH_4$.	**268**
	Meerwein-Ponndorf-Verley reduction	The reduction of aldehydes and ketones by metal alkoxides to the corresponding alcohols	**280**
	Midland Alpine borane reduction	Enantioselective reduction of ketones using Alpine borane.	**288**
	Noyori asymmetric hydrogenation	Formation of enantio-enriched carboxylic acids, alcohols and amino acids from unsaturated carboxylic acids, allylic alcohols and enamides, respectively.	**316**
	Staudinger reduction	Reduction of azides with triphenylphosphine.	**428**
	Stephen aldehyde synthesis	Reduction of nitriles with $SnCl_2$/HCl to give the corresponding aldehydes.	**430**
	Tishchenko reaction	Conversion of aldehydes to the corresponding esters in the presence of metal alkoxides.	**456**
	Wolff-Kishner reduction	Deoxygenation of aldehydes and ketones under basic conditions to give hydrocarbons via the corresponding hydrazones or semicarbazones.	**496**
RING CONTRACTION			
	Benzilic acid rearrangement	Rearrangement of 1,2-diketones to give the salts of α-hydroxy acids.	**52**
	Favorskii rearrangement	Skeletal rearrangement of α-halo ketones via a cyclopropanone intermediate to give carboxylic acids or carboxylic acid derivatives.	**164**
	Quasi-Favorskii rearrangement	Skeletal rearrangement of bicyclic α-halo ketones in which the halogen is located at the bridgehead position to afford carboxylic acids or carboxylic acid derivatives.	**370**
RING EXPANSION			
	Buchner method of ring expansion	Thermal or photochemical reaction of ethyl diazoacetate with benzenes and its homologs to give the isomeric esters of cycloheptatriene carboxylic acid.	**68**

REACTION CATEGORY	NAME OF REACTIONS	BRIEF DESCRIPTION OF SYNTHETIC USE	Page#
RING EXPANSION			
	Ciamician-Dennstedt rearrangement	Synthesis of 3-halopyridines from pyrroles and 2-haloquinolines from indoles.	**84**
	Demjanov and Tiffeneau-Demjanov rearrangement	The ring enlargement of 1-aminomethyl cycloalkanes to the corresponding cycloalkanols and the ring-enlargement of 1-aminomethyl cycloalkanols to the corresponding cycloalkanones.	**134**
TRANSITION METAL CATALYZED COUPLINGS			
Cu-catalyzed	Castro-Stevens coupling	The copper(I)-mediated coupling of aryl or vinyl halides with aryl- or alkyl-substituted alkynes to afford disubstituted alkynes or enynes.	**78**
Cu-catalyzed	Glaser coupling	Preparation of symmetrical conjugated diynes and polyynes by the oxidative homocoupling of terminal alkynes in the presence of copper salts.	**186**
Cu-catalyzed	Ullmann biaryl ether synthesis	Cu-mediated synthesis of biaryl ethers by coupling aryl halides and phenols.	**464**
Cu-catalyzed	Ullmann reaction	Cu-mediated coupling of two aryl halides to afford symmetrical or unsymmetrical biaryls.	**466**
Pd-catalyzed	Buchwald-Hartwig cross-coupling	Direct Pd-catalyzed C-N and C-O bond formation between aryl halides and amines or alcohols.	**70**
Pd-catalyzed	Heck reaction	Pd-catalyzed arylation or alkenylation of olefins.	**196**
Pd- or Ni-catalyzed	Kumada cross-coupling	Cross-coupling of alkenyl- or aryl halides and Grignard reagents or organolithium species.	**258**
Pd-catalyzed	Larock indole synthesis	Preparation of 2,3-disubstituted indoles from *ortho*-iodoanilines and disubstituted alkynes.	**260**
Pd-catalyzed	Miyaura boration	Pd-catalyzed cross-coupling of aromatic and heteroaromatic halides or triflates with tetraalkoxydiboron compounds to give arylboronic and heteroarylboronic esters.	**296**
Pd- or Ni-catalyzed	Negishi cross-coupling	Pd- or Ni-catalyzed cross-coupling of organozincs and aryl- or alkenyl- or alkynyl halides.	**310**
Pd and Cu-catalyzed	Sonogashira cross-coupling	Cu-Pd-catalyzed coupling of terminal alkynes with aryl and vinyl halides to give enynes.	**424**
Pd-catalyzed	Stille carbonylative cross-coupling	Pd-catalyzed coupling of organostannanes and alkenyl- or aryl halides and CO to form ketones.	**436**
Pd-catalyzed	Stille cross-coupling	Pd-catalyzed coupling of organostannanes and alkenyl- or aryl halides.	**438**
Pd-catalyzed	Stille-Kelly coupling	Pd-catalyzed intramolecular biaryl coupling of aryl halides or aryl triflates in the presence of distannanes.	**440**
Pd-catalyzed	Suzuki cross-coupling	Pd-catalyzed coupling between organoboron compounds and organic halides and triflates.	**448**
Pd-catalyzed	Tsuji-Trost allylation	Pd-catalyzed allylation of carbon nucleophiles with allylic compounds via π-allylpalladium complexes.	**458**

8.4 AFFECTED FUNCTIONAL GROUPS

AFFECTED FUNCTIONAL GROUP	NEWLY FORMED FUNCTIONAL GROUP	NAME OF TRANSFORMATION
ACETAL		
	γ,δ-unsaturated amide	Eschenmoser-Claisen rearrangement
ALCOHOL		
1° alcohol	γ-hydroxy oxime	Barton nitrite ester reaction
1° alcohol	aldehyde	Corey-Kim oxidation, Dess-Martin oxidation, Ley oxidation, Oppenauer oxidation, Pfitzner-Moffatt oxidation, Swern oxidation
1° alcohol	alkane	Barton-McCombie radical deoxygenation
1° alcohol	alkene	Chugaev elimination
1° alcohol	amine	Mitsunobu reaction
1° alcohol	azide	Mitsunobu reaction
1° alcohol	carboxylic acid	Jones oxidation
1° alcohol	ester	Mitsunobu reaction
1° alcohol	ether	Mitsunobu reaction, Williamson ether synthesis
1° alcohol	lactone	Corey-Nicolaou macrolactonization, Keck macrolactonization, Yamaguchi macrolactonization
1° alcohol	nitrile	Mitsunobu reaction
1° alcohol	sulfide	Mitsunobu reaction
2° alcohol	γ-hydroxy oxime	Barton nitrite ester reaction
2° alcohol	alkane	Barton-McCombie radical deoxygenation
2° alcohol	alkene	Burgess dehydration, Chugaev elimination
2° alcohol	amine	Mitsounobu reaction
2° alcohol	azide	Mitsunobu reaction
2° alcohol	ester	Mitsunobu reaction, Schotten-Baumann reaction
2° alcohol	ether	Mitsunobu reaction, Williamson ether synthesis
2° alcohol	ketone	Corey-Kim oxidation, Dess-Martin oxidation, Jones oxidation, Ley oxidation, Oppenauer oxidation, Pfitzner-Moffatt oxidation, Swern oxidation
2° alcohol	lactone	Corey-Nicolaou macrolactonization, Keck macrolactonization, Yamaguchi macrolactonization
2° alcohol	nitrile	Mitsunobu reaction
2° alcohol	sulfide	Mitsunobu reaction
3° alcohol	γ-hydroxy oxime	Barton nitrite ester reaction
3° alcohol	alkane	Barton-McCombie radical deoxygenation
3° alcohol	alkene	Burgess dehydration, Chugaev elimination, Grob fragmentation
3° alcohol	amide	Ritter reaction
3° alcohol	ester	Schotten-Baumann reaction
3° alcohol	ether	Williamson ether synthesis
3° alcohol	lactone	Corey-Nicolaou macrolactonization, Keck macrolactonization, Yamaguchi macrolactonization
allylic alcohol	γ,δ-unsaturated amide	Eschenmoser-Claisen rearrangement
allylic alcohol	γ,δ-unsaturated ester	Johnson-Claisen rearrangement
allylic alcohol	allylic amide	Overman rearrangement
allylic alcohol	epoxy alcohol	Sharpless asymmetric epoxidation
allylic alcohol	saturated enantio-enriched alcohol	Noyori asymmetric hydrogenation
propargylic alcohol	α,β-unsaturated ketone	Meyer-Schuster and Rupe rearrangement
propargylic alcohol	propargyl-substituted compound	Nicholas reaction
ALDEHYDE		
	α,β-epoxy ester	Darzens glycidic ester condensation
	α,β-unsaturated carboxylic acid	Perkin reaction
	α-amino nitrile	Strecker reaction
	β-nitro alcohol	Henry reaction
	γ-oxo ester	Stetter reaction
	γ-oxo nitrile	Stetter reaction
	1,3-diol	Prins reaction
	1,4,7-triketone	Stetter reaction
	1,4-diketone	Stetter reaction
	alkane	Tsuji-Wilkinson decarbonylation

AFFECTED FUNCTIONAL GROUP	NEWLY FORMED FUNCTIONAL GROUP	NAME OF TRANSFORMATION
ALDEHYDE		
	alkene	McMurry coupling, Wittig reaction, Wittig reaction-Schlosser modification, Bamford-Stevens-Shapiro reaction, HWE olefination, HWE olefination-Still modification, Julia-Lythgoe olefination, Peterson olefination, Takai reaction, Tebbe olefination, Stobbe condensation, Perkin reaction, Knoevenagel condensation
	alkyne	Corey-Fuchs alkyne synthesis, Seyferth-Gilbert homologation
	allylic alcohol	Baylis-Hillman reaction
	amide	Passerini reaction, Ugi multicomponent reaction
	amine	Eschweiler-Clarke methylation, Baylis-Hillman reaction, Petasis boronic acid-Mannich reaction
	carboxylic acid	Jones oxidation, Cannizzaro reaction, Pinnick oxidation
	epoxide	Corey-Chaykovsky epoxidation
	ester	Tishchenko reaction, Dakin oxidation (aromatic aldehydes only)
	homoallylic alcohol	Sakurai allylation, Roush asymmetric allylation, Keck asymmetric allylation
	imine	Aza-Wittig reaction
	nitrile	Schmidt reaction
	nitroalkene	Henry reaction
	primary alcohol	Meerwein-Ponndorf-Verley reduction, Cannizzaro reaction
	secondary alcohol	Barbier coupling reaction, Grignard reaction, Aldol reaction, Evans aldol reaction, Nozaki-Hiyama-Kishi reaction, Sakurai allylation, Roush asymmetric allylation, Keck asymmetric allylation
	tetrahydroisoquinoline	Pictet-Spengler tetrahydroisoquinoline synthesis
ALKENE		
	1,2-diol	Sharpless asymmetric dihydroxylation, Prévost reaction
	1,3-diene	Enyne metathesis, Heck reaction
	1,3-diol	Prins reaction
	1,5-diketone	DeMayo cycloaddition
	alcohol	Brown hydroboration reaction/oxidation
	alkylborane	Brown hydroboration
	alkylzirconium	Schwartz hydrozirconation
	allene	Doering-LaFlamme allene synthesis
	allylic alcohol	Baylis-Hillman reaction
	allylic alcohol	Riley selenium dioxide oxidation, Prins reaction
	allylic bromide	Wohl-Ziegler bromination
	amide	Ritter reaction
	amino alcohol	Sharpless asymmetric aminohydroxylation
	arylated alkene	Heck reaction, Meerwein arylation
	cyclic alkene	Alkene metathesis, Diels-Alder cycloaddition
	cyclobutane	DeMayo cycloaddition
	cyclobutanone	Staudinger ketene cycloaddition
	cyclopentenone	Pauson-Khand reaction
	cyclopropane	Simmons-Smith cyclopropanation
	epoxide	Jacobsen-Katsuki epoxidation, Sharpless asymmetric epoxidation, Davis' oxaziridine oxidation, Prilezhaev reaction, Shi asymmetric epoxidation
	heteroatom-substituted alkene	Wacker oxidation
	methyl ketone	Wacker oxidation
	oxetane	Paterno-Büchi reaction
	unsymmetrically substituted alkene	Alkene metathesis
ALKYNE		
	1,3-diene	Enyne metathesis
	1,3-diyne	Glaser coupling
	2,3-disubstituted indole	Larock indole synthesis
	aldehyde	Brown hydroboration/oxidation
	aryl substituted alkyne	Castro-Stephens coupling, Sonogashira cross-coupling
	cyclopentenone	Pauson-Khand reaction
	disubstituted alkyne	Alkyne metathesis
	enyne	Sonogashira cross-coupling
	highly substituted benzene ring	Danheiser benzannulation, Dötz benzannulation

AFFECTED FUNCTIONAL GROUP	NEWLY FORMED FUNCTIONAL GROUP	NAME OF TRANSFORMATION
ALKYNE		
	ketone	Brown hydroboration/oxidation
	macrocyclic alkyne	Alkyne metathesis
	substituted 1,4-cyclohexadiene	Diels-Alder cycloaddition, Danishefsky's diene cycloaddition
	vinylborane (alkenylborane)	Brown hydroboration
ALLENE		
	substituted cyclopentene	Danheiser cyclopentene annulation
AMIDE		
1° amide	carbamate	Hofmann rearrangement
1° amide	primary amine	Hofmann rearrangement
1° amide	substituted amidine	Aza-Wittig reaction
1° amide	substituted urea	Hofmann rearrangement
2° amide	2,3-disubstituted indole	Madelung indole synthesis
2° amide	3,4-dihydro isoquinoline	Bischler-Napieralski isoquinoline synthesis
2° amide	isoquinoline	Bischler-Napieralski isoquinoline synthesis
2° amide	N-substituted enamine	Tebbe olefination
3° amide	α,β-unsaturated aldehyde	Vilsmeier-Haack formylation
3° amide	α-alkylated aldehyde	Myers asymmetric alkylation
3° amide	α-alkylated amide	Myers asymmetric alkylation
3° amide	α-alkylated carboxylic acid	Myers asymmetric alkylation
3° amide	α-diazo amide	Regitz diazo transfer
3° amide	β-hydroxy carbonyl compound	Evans aldol reaction
3° amide	ketone	Weinreb ketone synthesis
3° amide	*N,N*-disubstituted cyclopropylamine	Kulinkovich reaction
3° amide	*N,N*-disubstituted enamine	Tebbe olfination
3° amide	substituted benzaldehyde	Vilsmeier-Haack formylation
3° amide	β-alkylated primary alcohol	Myers asymmetric alkylation
AMINE		
1° amine	α-acylamino carboxamide	Ugi multicomponent reaction
1° amine	α-amino carboxamide	Ugi muticomponent reaction
1° amine	α-amino nitrile	Strecker reaction
1° amine	amide	Schotten-Baumann reaction
1° amine	cycloalkanol	Demjanov rearrangement
1° amine	cycloalkanone	Demjanov and Tiffeneau-Demjanov rearrangement
1° amine	hydantoinimide	Ugi multicomponent reaction
1° amine	tetrahydroisoquinoline	Pictet-Spengler tetrahydroisoquinoline synthesis
1° amine	Mannich base	Mannich reaction
1° amine	secondary aromatic amine	Buchwald-Hartwig cross-coupling, Chichibabin amination reaction
1° amine	tetrazole	Ugi multicomponent reaction
1° amine	thiohydantoinimide	Ugi multicomponent reaction
2° amine	α-acylamino carboxamide	Ugi multicomponent reaction
2° amine	α-amino carboxamide	Ugi multicomponent reaction
2° amine	α-amino nitrile	Strecker reaction
2° amine	allylic amine	Petasis boronic acid-Mannich reaction
2° amine	amide	Schotten-Baumann reaction
2° amine	hydroxylamine	Davis' oxaziridine oxidation
2° amine	Mannich base	Mannich reaction
2° amine	tertiary aromatic amine	Buchwald-Hartwig cross-coupling
2° amine	tetrazole	Ugi muticomponent reaction
3° amine	alkene	Cope elimination
3° amine	homoallylic secondary amine	Aza-Wittig rearrangement
3° amine	*N,N*-dialkyl hydroxylamine	Cope elimination
3° amine	*N*-oxide	Davis' oxaziridine oxidation
3° amine	rearranged tertiary amine	Stevens rearrangement
allylic amine	1,2-oxazaheterocycle	Meisenheimer rearrangement
allylic amine	homoallylic amine	Aza-[2,3]-Wittig rearrangement
allylic amine	imine	Aza-Claisen rearrangement

AFFECTED FUNCTIONAL GROUP	NEWLY FORMED FUNCTIONAL GROUP	NAME OF TRANSFORMATION
AMINE		
allylic amine	*O*-allyl-*N,N*-disubstituted hydroxylamine	Meisenheimer rearrangement
aryl amine	amide	Ugi multicomponent reaction
aryl amine	aryl bromide	Sandmeyer reaction
aryl amine	aryl chloride	Sandmeyer reaction
aryl amine	aryl fluoride	Balz-Schiemann reaction
aryl amine	aryl iodide	Sandmeyer reaction
aryl amine	aryl substituted alkene	Meerwein arylation
aryl amine	diaryl amine	Buchwald-Hartwig cross-coupling, Ullmann biaryl amine synthesis
aryl amine	*N*-aryl substituted pyrrole	Paal-Knorr pyrrole synthesis
aryl amine	*N*-methyl aryl amine	Eschweiler-Clarke methylation
aryl amine	*N*-oxide	Davis' oxaziridine oxidation
aryl amine	*ortho*-acyl aryl amine	Houben-Hoesch reaction
aryl amine	substituted quinoline	Combes quinoline synthesis, Skraup and Doebner-Miller quinoline synthesis
aryl amine	thiohydantoinimide	Ugi multicomponent reaction
N-halo amine	amine	Hofmann-Löffler-Freytag reaction
ANHYDRIDE		
	α,β-unsaturated carboxylic acid	Perkin reaction
	α-halogenated anhydride	Hell-Volhard-Zelinsky reaction
	aromatic ketone	Friedel-Crafts acylation
	enol ether	Petasis-Tebbe olefination
	tertiary amide	Polonovski reaction
	titanium enolate	Tebbe olefination
AZIDE		
acyl azide	isocyanate	Curtius rearrangement
alkyl azide	imine	Aza-Wittig reaction
alkyl azide	iminophosphorane	Staudinger reaction
aryl azide	imine	Aza-Wittig reaction
aryl azide	iminophosphorane	Staudinger reaction
CARBONATE		
	allylated products	Tsuji-Trost allylation
	ketene acetal	Tebbe olefination
CARBOXYLIC ACID		
	α-acyloxycarboxamide	Passerini multicomponent reaction
	α-bromo acid bromide	Hell-Volhard-Zelinsky reaction
	alkane	Barton radical decarboxylation reaction
	alkyl bromide	Hunsdiecker reaction
	homologated carboxylic acid	Arndt-Eistert homologation
	isocyanate	Curtius rearrangement
	lactone	Keck macrolactonization, Corey-Nicolaou macrolactonization, Yamaguchi macrolactonization
	primary amine	Curtius rearrangement, Schmidt reaction
CYCLOPROPANE		
vinylcyclopropane	cyclopentene	Vinylcyclopropane-cyclopentene rearrangement
DIENE		
1,5-diene	1,5-diene	Cope rearrangement
1,3-diene	aryl substituted diene	Heck reaction
1,3-diene	six-membered heterocycle	Hetero Diels-Alder cycloaddition
1,3-diene	substituted cyclohexene	Diels-Alder reaction, Danishefsky's diene cycloaddition
ENAMINE		
	α-alkylated aldehyde	Stork enamine synthesis
	α-alkylated ketone	Stork enamine synthesis
	β-diketone	Stork enamine synthesis
ENAMIDE		
	enantio-enriched amino acid	Noyori asymmetric hydrogenation

AFFECTED FUNCTIONAL GROUP	NEWLY FORMED FUNCTIONAL GROUP	NAME OF TRANSFORMATION
ENOL ETHER		
	α,β-unsaturated ketone	Saegusa oxidation
	β-hydroxy carbonyl compound	Mukaiyama aldol reaction
	α-hydroxy ketone	Rubottom oxidation, Davis' oxaziridine oxidation
	substituted cyclohexanone	Ferrier reaction/rearrangement
ENONE		
	1,3-dicarbonyl compound	Wacker oxidation
	1,4-diketone	Stetter reaction
	allylic alcohol	Baylis-Hillman reaction
	allylic alcohol	Luche reduction
	arylated enone	Meerwein arylation
	cyclopropane	Corey-Chaykovsky cyclopropanation
	Michael adduct	Michael addition
	phenol	Dienone-Phenol rearrangement
	quinoline	Skraup and Doebner-Miller quinoline synthesis
	substituted enone	Heck reaction
	substituted pyridine	Kröhnke pyridine synthesis
ENYNE		
	1,3-diene	Enyne metathesis
EPOXIDE		
	allylated product	Tsuji-Trost allylation
	enantiomerically pure epoxide	Jacobsen hydrolytic kinetic resolution
	polyol	Smith-Tietze multicomponent dithiane coupling
ESTER		
α,β-unsaturated ester	allylic alcohol	Baylis-Hillman reaction
α-halo ester	β-hydroxy ketone	Reformatsky reaction
β-keto ester	α-diazo-β-keto ester	Regitz diazo transfer
β-keto ester	alkylated β-keto ester	Acetoacetic ester synthesis
β-keto ester	ketone	Krapcho dealkoxycarbonylation
β-keto ester	substituted coumarin	von Pechmann reaction
β-keto ester	substituted furan	Feist-Benary furan synthesis
β-keto ester	substituted pyrrole	Knorr pyrrole synthesis
carboxylic acid ester	α,β-epoxy ester	Darzens glycidic ester condensation
carboxylic acid ester	γ,δ-unsaturated acid	Claisen-Ireland rearrangement
carboxylic acid ester	alcohol	Kagan-Molander samarium-diiodide coupling
carboxylic acid ester	cyclopropanol	Kulinkovich reaction
carboxylic acid ester	enol ether	Tebbe olefination, Petasis-Tebbe olefination
carboxylic acid ester	tertiary alcohol	Grignard reaction
diester	α-hydroxy ketone	Acyloin condensation
diester	β-keto ester	Claisen condensation, Dieckmann condensation
diester	substituted malonic ester	Malonic ester synthesis
nitrite ester	hydroxy oxime	Barton nitrite ester reaction
phenolic ester	acylated phenol	Fries rearrangement
phosphonate ester	alkene	HWE olefination, HWE olefination-Still modification
thiohydroxamate ester	alkane	Barton radical decarboxylation of thiohydroxamate esters
xanthate ester	alkene	Chugaev elimination reaction
ETHER		
	alcohol	Wittig rearrangement
allylic ether	γ,δ-unsaturated carbonyl compound	Claisen rearrangement
allylic ether	β-alkoxyketone	Wacker oxidation
allylic ether	homoallylic alcohol	Wittig-[2,3]-rearrangement
HALIDE		
α,ω-dihalide	substituted cycloalkane	Malonic ester synthesis
1,1-geminal dihalide	alkene	Takai-Utimoto olefination
1,1-geminal dihalide	allene	Doering-LaFlamme allene synthesis
acyl halide	alkyl halide (one carbon shorter)	Tsuji-Wilkinson decarbonylation

AFFECTED FUNCTIONAL GROUP	NEWLY FORMED FUNCTIONAL GROUP	NAME OF TRANSFORMATION
HALIDE		
acyl halide	amide	Schotten-Baumann reaction
acyl halide	aromatic ketone	Friedel-Crafts acylation
acyl halide	ketone	Negishi cross-coupling
alkyl halide	1° or 2° alkyl halide	Finkelstein reaction
alkyl halide	alcohol	Barbier coupling reaction, Molander-Kagan samarium-diiodide coupling
alkyl halide	aldehyde	Kornblum oxidation
alkyl halide	alkane	Wurtz coupling
alkyl halide	alkylated β-keto ester	Acetoacetic ester synthesis
alkyl halide	alkylated 1,3-diester	Malonic ester synthesis
alkyl halide	alkylated aromatic compound	Friedel-Crafts alkylation
alkyl halide	alkylated heteroaromatic compound	Minisci reaction
alkyl halide	alkylated ketone	Stork enamine synthesis
alkyl halide	amine	Gabriel synthesis
alkyl halide	ether	Williamson ether synthesis
aryl halide	ketone	Kornblum oxidation
aryl halide	phosphonate ester	Arbuzov reaction
aryl halide	rearranged carbon skeleton	Wagner-Meerwein rearrangement
aryl halide	substituted alkene	Heck reaction
aryl halide	aryl ether	Buchwald-Hartwig cross-coupling
aryl halide	aryl substituted alkene	Kumada cross-coupling, Stille cross-coupling, Suzuki cross-coupling
aryl halide	biaryl amine	Ullmann biaryl amine synthesis
aryl halide	biaryl ether	Ullmann biaryl ether synthesis
aryl halide	biaryls	Kumada cross-coupling, Stille cross-coupling, Negishi cross-coupling, Stille-Kelly coupling, Suzuki cross-coupling, Ullmann biaryl synthesis
allylic halide	allyl-substituted products	Tsuji-Trost allylation
allylic halide	*C*-allyl substituted acetoacetic ester	Acetoacetic ester synthesis
allylic halide	*C*-allyl substituted malonic ester	Malonic ester synthesis
allylic halide	homoallylic alcohol	Barbier coupling reaction, Nozaki-Hiyama-Kishi coupling
HYDRAZONE		
	α-alkylated aldehyde	Enders SAMP/RAMP hydrazone alkylation
	α-alkylated hydrazone	Enders SAMP/RAMP hydrazone alkylation
	α-alkylated ketone	Enders SAMP/RAMP hydrazone alkylation
	alkane	Wolff-Kishner reduction
	alkene	Bamford-Stevens-Shapiro reaction
	allylic alcohol	Wharton olefin synthesis
	substituted indole	Fischer indole synthesis
IMIDE		
	cyclic imine	Aza-Wittig reaction
	primary amine	Gabriel amine synthesis
IMINE		
	α-amino nitrile	Strecker reaction
	quinoline	Combes quinoline synthesis
	six-membered azaheterocycle	Hetero Diels-Alder cycloaddition
ISOCYANATE		
	carbodiimide	Aza-Wittig reaction
ISONITRILE		
	α-acyloxycarboxamide	Passerini multicomponent reaction
	α-hydroxycarboxamide	Passerini multicomponent reaction
	α-hydroxyalkyltetrazole	Passerini multicomponent reaction
KETONE		
α-halo ketone	rearranged amide	Favorskii rearrangement
α-halo ketone	rearranged ester	Favorskii rearrangement
α-halo ketone	ring-contracted ester	Favorskii rearrangement, Quasi-Favorskii rearrangement
α-halo ketone	substituted furan	Feist-Bénary furan synthesis
α-halo ketone	substituted pyridine	Kröhnke pyridine synthesis

AFFECTED FUNCTIONAL GROUP	NEWLY FORMED FUNCTIONAL GROUP	NAME OF TRANSFORMATION
KETONE		
1,2-diketone	α-hydroxy acid	Benzilic acid rearrangement
1,2-diketone	ketone	Tsuji-Wilkinson decarbonylation
1,3-diketone	α-diazo-1,3-diketone	Regitz diazo transfer
1,3-diketone	quinoline	Combes quinoline synthesis
1,5-diketone	substituted 2-cyclohexenone	Hajos-Parrish reaction, Robinson annulation
cyclic ketone	lactone	Baeyer-Villiger reaction
diazo ketone	carboxylic acid	Wolff rearrangement
diazo ketone	highly substituted aromatic ring	Danheiser benzannulation
diazo ketone	ketene	Wolff rearrangement
ketone	α,β-epoxy ester	Darzens glycidic ester condensation
ketone	β-nitro alcohol	Henry reaction
ketone	alkene	McMurry coupling, Wittig reaction, Wittig reaction-Schlosser modification, Bamford-Stevens-Shapiro reaction, HWE olefination, HWE olefination-Still modification, Julia-Lythgoe olfination, Peterson olefination, Takai-Utimoto olefination, Tebbe olefination
ketone	amide	Schmidt reaction
ketone	epoxide	Corey-Chaykovsky epoxidation
LACTONE		
	tertiary alcohol	Grignard reaction
	cyclic enol ether	Tebbe olefination
NITRILE		
aliphatic nitrile	aldehyde	Stephen aldehyde synthesis
aliphatic nitrile	aromatic ketone	Houben-Hoesch reaction
aliphatic nitrile	ester	Pinner reaction
aliphatic nitrile	imino ether	Pinner reaction
aliphatic nitrile	imino thioether	Pinner reaction
aliphatic nitrile	*N*-alkyl carboxamide	Ritter reaction
aliphatic nitrile	six-membered azaheterocycle	Hetero Diels-Alder cycloaddition
aromatic nitrile	aldehyde	Stephen aldehyde synthesis
aromatic nitrile	ester	Pinner reaction
aromatic nitrile	imino ether	Pinner reaction
aromatic nitrile	imino thioether	Pinner reaction
aromatic nitrile	*N*-alkyl carboxamide	Ritter reaction
NITRO COMPOUNDS		
aliphatic nitro cmpd.	β-nitro alcohol	Henry reaction
aliphatic nitro cmpd.	1,2-oxazaheterocycle	Hetero Diels-Alder cycloaddition
aliphatic nitro cmpd.	carbonyl compound	Nef reaction
aliphatic nitro cmpd.	carboxylic acid	Nef reaction
aliphatic nitro cmpd.	oxime	Nef reaction
aromatic nitro cmpd.	7-substituted indole	Bartoli indole synthesis
NITROALKENE		
	ketone	Nef reaction
	oxime	Nef reaction
OXIME		
	amide	Beckmann rearrangement
	α-amino ketone	Neber rearrangement
PHENOL		
	acyl-substituted phenol	Fries rearrangement
	aryl alkyl ether	Williamson ether synthesis
	biaryl ether	Ullmann biaryl ether synthesis
	ortho-formyl phenol	Reimer-Tiemann reaction
	substituted coumarin	von Pechmann reaction
	substituted salicylamide	anionic *ortho*-Fries rearrangement
	substituted salicylic acid	Kolbe-Schmitt reaction
SILANE		
acyl silane	*O*-silylated alcohol	Brook rearrangement
alkyl silane	alcohol	Fleming-Tamao oxidation

AFFECTED FUNCTIONAL GROUP	NEWLY FORMED FUNCTIONAL GROUP	NAME OF TRANSFORMATION
SILANE		
allylic silane	homoallylic alcohol	Sakurai allylation
aryl silane	alcohol	Fleming-Tamao oxidation
SULFIDE		
	sulfoxide	Davis' oxaziridine oxidation
SULFONE		
α-halo sulfone	alkene	Ramberg-Bäcklund rearrangement
aliphatic sulfone	alkene	Julia-Lythgoe olefination
SULFOXIDE		
	α-substituted sulfide	Pummerer rearrangement
	aldehyde	Pummerer rearrangement
	allylic alcohol	Mislow-Evans rearrangement
	glycoside	Kahne glycosidation
	ketone	Pummerer rearrangement
	sulfenate ester	Mislow-Evans rearrangement
allylic sulfoxide	allylic alcohol	Mislow-Evans rearrangement
allylic sulfoxide	sulfenate ester	Mislow-Evans rearrangement

8.5 PREPARATION OF FUNCTIONAL GROUPS

TARGET FUNCTIONAL GROUP	SUBSTRATE FUNCTIONAL GROUP	NAME OF TRANSFORMATION
ALCOHOL		
	α,β-epoxy alcohol	Payne rearrangement
	aldehyde	Grignard reaction, Barbier coupling reaction, Nozaki-Hiyama-Kishi reaction, Baylis-Hillman reaction, Cannizzaro reaction, Henry reaction, Keck asymmetric allylation, MPV reduction, Prins reaction, Roush asymmetric allylation, Sakurai allylation, Kagan-Molander coupling
	alkene	Sharpless asymmetric aminohydroxylation
	alkenyl halide or triflate	Nozaki-Hiyama-Kishi coupling
	aryl alkyl ether	Wittig-[1,2]-rearrangement
	enol ether and silyl enol ether	Davis' oxaziridine oxidation
	ketone	Grignard reaction, Barbier coupling reaction, Nozaki-Hiyama-Kishi reaction, Baylis-Hillman reaction, Henry reaction, Keck asymmetric allylation, MPV reduction, Prins reaction, Roush asymmetric allylation, Sakurai allylation, CBS reduction, Luche reduction, Midland Alpine borane reduction, Molander-Kagan coupling, Noyori asymmetric hydrogenation
	nitroalkane	Henry reaction
	organomagnesium species	Grignard reaction
	2° alcohol	Mitsunobu reaction
	silane	Fleming-Tamao oxidation
allylic alcohol	aldehyde	Baylis-Hillman reaction, Grignard reaction, Prins reaction, Nozaki-Hiyama-Kishi coupling
allylic alcohol	alkene	Prins reaction, Riley selenium dioxide oxidation
allylic alcohol	allylic sulfoxide	Mislow-Evans rearrangement
allylic alcohol	enone	Luche reduction, Baylis-Hillmann reaction
allylic alcohol	epoxyhydrazone	Wharton olefin synthesis
allylic alcohol	epoxyketone	Wharton olefin synthesis
allylic alcohol	ketone	Baylis-Hillman reaction, Grignard reaction, Nozaki-Hiyama-Kishi coupling, Wharton olefin synthesis
homoallylic alcohol	aldehyde	Grignard reaction, Barbier coupling reaction, Keck asymmetric allylation, Roush asymmetric allylation, Sakurai allylation
homoallylic alcohol	alkyl allyl ether	Wittig-[2,3]-rearrangement
homoallylic alcohol	ketone	Grignard reaction, Barbier coupling reaction, Keck asymmetric allylation, Roush asymmetric allylation, Sakurai allylation
propargylic alcohol	aldehyde	Barbier reaction, Grignard reaction
propargylic alcohol	ketone	Barbier reaction, Grignard reaction
ALDEHYDE		
aliphatic	aliphatic nitro compound	Nef reaction
aliphatic	cyclic epoxy hydrazone	Eschenmoser-Tanabe fragmentation
aliphatic	cyclic epoxy ketone	Eschenmoser-Tanabe fragmentation
aliphatic	3° amine *N*-oxide	Polonovski reaction
aliphatic/aromatic	1° or 2° alkyl halide	Kornblum oxidation
aliphatic/aromatic	1,2-diol	Criegee oxidation
aliphatic/aromatic	nitrile	Stephen aldehyde synthesis
aliphatic/aromatic	1° alcohol	Corey-Kim oxidation, Dess-Martin oxidation, Ley oxidation, Swern oxidation, Oppenauer oxidation, Pfitzner-Moffatt oxidation
aromatic	activated benzyl halide	Kornblum oxidation
aromatic	electron-rich heteroaromatic ring	Vilsmeier-Haack formylation
aromatic	electron-rich substituted benzene	Vilsmeier-Haack formylation, Reimer-Tiemann reation
aromatic	*N,N*-disubstituted formamide	Vilsmeier-Haack formylation
aromatic	substituted benzene	Gatterman formylation and Gatterman-Koch formylation
ALKENE		
	α-halo sulfone	Ramberg-Bäcklund rearrangement
	1,2-diol	Corey-Winter olefination
	1,3-diol monosulfonate ester	Wharton fragmentation, Grob fragmentation
	1,5-diene	Cope rearrangement
	2° or 3° alcohol	Burgess dehydration, Chugaev elimination

TARGET FUNCTIONAL GROUP	SUBSTRATE FUNCTIONAL GROUP	NAME OF TRANSFORMATION
ALKENE		
	aldehyde	HWE olefination, HWE olefination-Still modification, Wittig reaction, Wittig reaction-Schlosser modification, Tebbe olefination, Julia olefination, Peterson olefination, Takai-Utimoto olefination
	alkyl phenyl sulfone	Julia-Lythgoe olefination
	diene	Alkene metathesis
	ketone	Bamford-Stevens-Shapiro olefination, HWE olefination, HWE olefination-Still modification, Wittig reaction, Wittig reaction-Schlosser modification, Tebbe olefination, Julia-Lythgoe olefination, Peterson olefination, Takai-Utimoto olefination
	nitroalkane	Henry reaction
	phosphonate ester	HWE olefination, HWE olefination-Still modification
	quaternary ammonium salt	Hofmann elimination
	3° amine *N*-oxide	Cope elimination, Polonosvki reaction
	tosylhydrazone	Bamford-Stevens-Shapiro olefination
	xanthate ester	Chugaev elimination
ALKYNE		
	aldehyde	Corey-Fuchs alkyne synthesis, Seyferth-Gilbert homologation
	cyclic epoxy ketone	Eschenmoser-Tanabe fragmentation
	diyne	Alkyne metathesis
	ketone	Seyferth-Gilbert homologation
ALLENE		
	alkene	Doering-LaFlamme allene synthesis
	geminal dihalocyclopropane	Doering-LaFlamme allene synthesis
AMIDE		
	α-diazo ketone	Wolff rearrangement
	3° alcohol	Ritter reaction
	3° amine *N*-oxide	Polonovski reaction
	acyl halide	Schotten-Baumann reaction
	alcohol	Ugi multicomponent reaction
	aldehyde	Passerini reaction, Ugi multicomponent reaction
	alkene	Ritter reaction
	allylic alcohol	Eschenmoser-Claisen rearrangement, Overman rearrangement
	amine	Schotten-Baumann reaction, Ugi multicomponent reaction
	anhydride	Schotten-Baumann reaction
	carboxylic acid	Passerini reaction, Ugi multicomponent reaction
	ketone	Passerini reaction, Schmidt reaction, Ugi multicomponent reaction
	nitrile	Ritter reaction, Ugi multicomponent reaction
	O-aryl carbamate	Fries rearrangement
	oxime	Beckmann rearrangement
AMINE		
	1° or 2° amine	Eschweiler-Clarke methylation
	acyl azide	Curtius rearrangement
	alkyl halide	Gabriel amine synthesis
	amide	Kulinkovich reaction, Hofmann rearrangement
	aryl halide	Buchwald-Hartwig cross-coupling, Ullmann diaryl amine synthesis
	3° benzylic amine	Sommelet-Hauser rearrangement
	benzylic quarter. ammonium salt	Sommelet-Hauser rearrangement
	carboxylic acid	Schmidt reaction
	N-halogenated amine	Hofmann-Löffler-Freytag reaction
	quaternary ammonium salt	Stevens rearrangement
allylic amine	α,β-unsaturated carboxylic acid derivative	Baylis-Hillman reaction
allylic amine	2° amine	Petasis boronic acid-Mannich reaction
allylic amine	aldehyde	Petasis boronic acid-Mannich reaction
allylic amine	allylic azide	Staudinger reaction
allylic amine	imine	Baylis-Hillman reaction

TARGET FUNCTIONAL GROUP	SUBSTRATE FUNCTIONAL GROUP	NAME OF TRANSFORMATION
AMINE		
allylic amine	ketone	Petasis boronic acid-Mannich reaction
allylic amine	vinylboronic acid	Petasis boronic acid-Mannich reaction
homoallylic amine	allylic 3° amine	Aza-Wittig rearrangement
AZIDE		
alkyl azide	1° or 2° alcohol	Mitsunobu reaction
CARBOXYLIC ACID		
	α-diazo ketone	Wolff rearrangement
	aldehyde	Cannizzaro reaction, Jones oxidation, Pinnick oxidation
	anhydride	Perkin reaction
	carboxylic acid	Arndt-Eistert homologation
	methyl ketone	Lieben haloform reaction
CYCLOPROPANE		
	alkene	Simmons-Smith cyclopropanation
	amide	Kulinkovich reaction
	enone	Corey-Chaykovsky cyclopropanation
	ester	Kulinkovich reaction
DIAZO KETONE		
	β-keto ester	Regitz diazo transfer
	1,3-diketone	Regitz diazo transfer
DIENE		
1,5-diene	1,5-diene	Cope rearrangement
cyclic 1,4-diene	alkyne	Diels-Alder cycloaddition
cyclic 1,4-diene	aromatic compound	Birch reduction
α,ω-diene	cyclic alkene	Alkene metathesis
1,3-diene	enyne	Enyne metathesis
DIKETONE		
	α,β-unsaturated ester	Wacker oxidation
	α,β-unsaturated ketone	Stetter reaction, Wacker oxidation
	aldehyde	Stetter reaction
	aromatic *ortho*-acyloxyketone	Baker-Venkataraman rearrangement
	cyclic 1,2-diol	Criegee oxidation
	enamine	Stork enamine synthesis
	ketone	Riley selenium dioxide oxidation
DIOL		
	aldehyde	Prins reaction
	alkene	Prévost reaction, Prins reaction, Sharpless asymmetric dihydroxylation
	racemic epoxide	Jacobsen hydrolytic kinetic resolution
DIYNE		
	terminal alkyne	Glaser coupling
ENAMINE		
	amide	Tebbe olefination
ENYNE		
	terminal alkyne	Castro-Stephens coupling, Sonogashira cross-coupling
ENOL ETHER		
	ester	Tebbe olefination
ENONE		
	1,5-diketone	Hajos-Parrish reaction, Robinson annulation
	alkene	Pauson-Khand reaction
	alkyne	Pauson-Khand reaction
	divinyl ketone	Nazarov cyclization
	enyne	Nazarov cyclization
	propargylic alcohol	Meyer-Schuster and Rupe rearrangement
	silyl enol ether	Saegusa oxidation

TARGET FUNCTIONAL GROUP	SUBSTRATE FUNCTIONAL GROUP	NAME OF TRANSFORMATION
EPOXIDE		
	α-halo ester	Darzens glycidic ester condensation
	aldehyde	Corey-Chaykovsky epoxidation
	alkene	Prilezhaev reaction, Davis' oxaziridine oxidation, Shi asymmetric epoxidation, Jacobsen-Katsuki epoxidation
	allylic alcohol	Sharpless asymmetric epoxidation
	ketone	Corey-Chaykovsky epoxidation, Darzens glycidic ester condensation
ESTER		
	α-diazo ketone	Wolff rearrangement
	1° or 2° alcohol	Mitsunobu reaction
	1°, 2° or 3° alcohol	Schotten-Baumann reaction
	acyl halide	Schotten-Baumann reaction
	aldehyde	Stobbe condensation, Tishchenko reaction
	allylic alcohol	Johnson-Claisen rearrangement
	anhydride	Schotten-Baumann reaction
	ketone	Baeyer-Villiger oxidation
	nitrile	Pinner reaction
ETHER		
	1° or 2° alcohol	Mitsunobu reaction
	1° or 2° alkyl halide	Williamson ether synthesis
	1° or 2° or 3° alcohol	Williamson ether synthesis
	aryl halide	Ullmann biaryl ether synthesis, Buchwald-Hartwig cross-coupling
	phenol	Williamson ether synthesis, Ullmann biaryl ether synthesis
HALIDE		
alkyl halide	1° or 2° alkyl halide	Finkelstein reaction
alkyl halide	acyl chloride	Tsuji-Wilkinson decarbonylation
alkyl halide	carboxylic acid	Hunsdiecker reaction
aryl halide	aryl amine	Sandmeyer reaction
aryl halide	aryldiazonium halide	Sandmeyer reaction
aryl halide	aryldiazonium tetrafluoroborate	Balz-Schiemann reaction
HYDROXY KETONE		
	α-halo ester	Reformatsky reaction
	aldehyde	Aldol condensation, Reformatsky reaction, Benzoin condensation
	enol ether	Davis' oxaziridine oxidation, Rubottom oxidation
	ester	Acyloin condensation
	ketone	Aldol condensation, Reformatsky reaction
	metal enolate	Davis' oxaziridine oxidation
IMINE		
	aldehyde	Aza-Wittig reaction
	allyl vinyl amine	Aza-Cope rearrangement
	ketone	Aza-Wittig reaction
	phenol	Houben-Hoesch reaction
IMINE		
	nitrile	Houben-Hoesch reaction
ISOCYANATE		
	acyl azide	Curtius rearrangement
	O-acyl hydroxamate	Lossen rearrangement
KETENE		
	α-diazo ketone	Wolff rearrangement
KETONE		
	α-amino acid	Dakin-West reaction
	1,2-diol	Pinacol rearrangement
	1,2-dione	Tsuji-Wilkinson decarbonylation
	1,3-diol monosulfonate	Wharton fragmentation
	2° alcohol	Corey-Kim oxidation, Dess-Martin oxidation, Ley oxidation, Swern oxidation, Oppenauer oxidation, Pfitzner-Moffatt oxidation

TARGET FUNCTIONAL GROUP	SUBSTRATE FUNCTIONAL GROUP	NAME OF TRANSFORMATION
KETONE		
	2-hetero substituted alcohol	Semipinacol rearrangement
	alkene	Wacker oxidation
	nitroalkane	Nef reaction
	N-methoxy-*N*-methyl amide	Weinreb ketone synthesis
	substituted benzene	Friedel-Crafts acylation
	sulfoxide	Pummerer rearrangement
KETO ESTER		
	diester	Dieckmann condensation
	ester	Claisen condensation
LACTONE		
	cyclic ketone	Baeyer-Villiger oxidation
	hydroxy acid	Corey-Nicolaou macrolactonization, Keck macrolactonization, Yamaguchi macrolactonization
NITRILE		
	3-aza-1,2,5-hexatriene	Aza-Claisen rearrangement
	aldehyde	Schmidt reaction
	aldehyde	Strecker reaction
	ketone	Strecker reaction
NITROALKENE		
	aldehyde	Henry reaction
	ketone	Henry reaction
	nitroalkane	Henry reaction
OXIME		
	nitrite ester	Barton nitrite ester reaction
PHENOL		
	aromatic ketone	Dakin oxidation
	chromium carbene	Dötz benzannulation
	dienone	Dienone-phenol rearrangement
	disubstituted alkyne	Dötz benzannulation
	phenolic ester	Fries rearrangement
PHOSPHONATE ESTER		
	alkyl halide	Arbuzov reaction
	trialkyl phosphite	Arbuzov reaction
SULFIDE		
	1° or 2° alcohol	Mitsunobu reaction
	1° or 2° alkyl halide	Williamson ether synthesis
	1° or 2° or 3° thiol	Williamson ether synthesis
SULFOXIDE		
	sulfide	Davis's oxaziridine oxidation

IX. REFERENCES

1. Michael, A., Wolgast, K. Preparation of pure ketones by means of acetoacetic esters. *Ber.* **1909**, 42, 3176-3177.
2. Schroeter, G., Kesseler, H., Liesche, O., Muller, R. F. Relationships of the polymeric ketenes to cyclobutane-1,3-dione and its derivatives. *Ber.* **1916**, 49, 2697-2745.
3. Arndt, F., Nachtwey, P. Preparation of dehydroacetic acid from acetoacetic ester and the mechanism of this reaction. *Ber.* **1924**, 57B, 1489-1491.
4. Arndt, F., Eistert, B., Scholz, H., Aron, E. Synthesis of dehydracetic acid from acetoacetic ester. *Ber.* **1936**, 69B, 2373-2380.
5. Hauser, C. R., Hudson, B. E., Jr. Acetoacetic ester condensation and certain related reactions. *Org. React.* **1942**, 1, 266-302.
6. House, H. O. *Modern Synthetic Reactions (The Organic Chemistry Monograph Series). 2nd ed* (ed. Benjamin, W. A.) (Menlo Park, **1972**) 735-760.
7. Mehrotra, R. C. Reactions of metal alkoxides with β-diketones and β-keto esters. *J. Indian Chem. Soc.* **1978**, 55, 1-7.
8. Benetti, S., Romagnoli, R., De Risi, C., Spalluto, G., Zanirato, V. Mastering β-Keto Esters. *Chem. Rev.* **1995**, 95, 1065-1114.
9. Guingant, A. Asymmetric syntheses of α,α-disubstituted β-diketones and β-keto esters. *Adv. in Asymmetric Synth.* **1997**, 2, 119-188.
10. Spielman, M. A., Schmidt, M. T. Mesitylmagnesium bromide as a reagent in the acetoacetic ester condensation. *J. Am. Chem. Soc.* **1937**, 59, 2009-2010.
11. Krapcho, A. P. Synthetic applications of dealkoxycarbonylations of malonate esters, β-keto esters, α-cyano esters and related compounds in dipolar aprotic media. Part II. *Synthesis* **1982**, 893-914.
12. Krapcho, A. P. Synthetic applications of dealkoxycarbonylations of malonate esters, β-keto esters, α-cyano esters and related compounds in dipolar aprotic media - Part I. *Synthesis* **1982**, 805-822.
13. Hanamoto, T., Hiyama, T. A facile entry to β,δ-diketo and syn-β,δ-dihydroxy esters. *Tetrahedron Lett.* **1988**, 29, 6467-6470.
14. Lygo, B. Reaction of aziridines with dianions derived from β-keto esters: application to the preparation of substituted pyrrolidines. *Synlett* **1993**, 764-766.
15. Wang, K. C., Liang, C. H., Kan, W. M., Lee, S. S. Synthesis of steroid intermediates via alkylation of dianion derived from acetoacetic ester. *Bioorg. Med. Chem.* **1994**, 2, 27-34.
16. Moreno-Manas, M., Marquet, J., Vallribera, A. Transformations of β-dicarbonyl compounds by reactions of their transition metal complexes with carbon and oxygen electrophiles. *Tetrahedron* **1996**, 52, 3377-3401.
17. Moreno-Manas, M., Marquet, J., Vallribera, A. Synthetic and mechanistic aspects of α-alkylation and α-arylation of β-dicarbonyl compounds via their transition metal complexes. *Russ. Chem. Bull.* **1997**, 46, 398-406.
18. Osowska-Pacewicka, K., Zwierzak, A. Reactions of N-phosphorylated aziridines with dianions derived from ethyl acetoacetate and 1,3-diketones: new route to substituted pyrrolines and pyrrolidines. *Synth. Commun.* **1998**, 28, 1127-1137.
19. Nakada, M., Takano, M., Iwata, Y. Preparation of novel synthons, uniquely functionalized tetrahydrofuran and tetrahydropyran derivatives. *Chem. Pharm. Bull.* **2000**, 48, 1581-1585.
20. Watson, H. B. Mechanism of the addition and condensation reactions of carbonyl compounds. *Trans. Faraday Soc.* **1941**, 37, 707-713.
21. Blaauw, R. H., Briere, J.-F., de Jong, R., Benningshof, J. C. J., van Ginkel, A. E., Fraanje, J., Goubitz, K., Schenk, H., Rutjes, F. P. J. T., Hiemstra, H. Intramolecular Photochemical Dioxenone-Alkene [2 + 2] Cycloadditions as an Approach to the Bicyclo[2.1.1]hexane Moiety of Solanoeclepin A. *J. Org. Chem.* **2001**, 66, 233-242.
22. Stauffer, F., Neier, R. Synthesis of Tri- and Tetrasubstituted Furans Catalyzed by Trifluoroacetic Acid. *Org. Lett.* **2000**, 2, 3535-3537.
23. Nakada, M., Iwata, Y., Takano, M. Reaction of dianions of acetoacetic esters with epibromohydrin derivatives: a novel synthesis of tetrahydrofuran derivatives and tetrahydropyran derivatives. *Tetrahedron Lett.* **1999**, 40, 9077-9080.
24. Hinman, M. M., Heathcock, C. H. A synthetic approach to the Stemona alkaloids. *J. Org. Chem.* **2001**, 66, 7751-7756.

Related reactions: Benzoin condensation;

1. Freund, A. Oxygen containing fragments. *Liebigs Ann. Chem.* **1861**, 118, 33-43.
2. Brühl, J. W. Preparation of divaleryls. *Chem. Ber.* **1879**, 12, 315-323.
3. Bouveault, L., Blanc, G. *Compt. rend. Acad. Sci. Paris* **1903**, 136, 1676.
4. Bouveault, L., Blanc, G. Transformation of monobasic saturated acids into the corresponding alcohols. *Bull. Soc. Chim. France* **1904**, 31, 666.
5. McElvain, S. M. Acyloins. *Org. React.* **1948**, 4, 256-268.
6. Finley, K. T. Acyloin condensation as a cyclization method. *Chem. Rev.* **1964**, 64, 573-589.
7. Kwart, H., King, K. Rearrangement and cyclization reactions of carboxylic acids and esters. in *Chem. Carboxylic Acids and Esters* (ed. Patai, S.), 341-373 (Interscience-Publishers, London, New York, **1969**).
8. Bloomfield, J. J., Owsley, D. C., Nelke, J. M. The acyloin condensation. *Org. React.* **1976**, 23, 259-403.
9. Seoane, G. Enzymatic C-C bond-forming reactions in organic synthesis. *Curr. Org. Chem.* **2000**, 4, 283-304.
10. Bloomfield, J. J. The acyloin condensation. IV. Avoidance of Dieckmann condensation products in acyloin condensations. *Tetrahedron Lett.* **1968**, 591-593.
11. Ruehlmann, K. Reaction of carboxylic acid esters with sodium in the presence of trimethylchlorosilane. *Synthesis* **1971**, 236-253.
12. Tamarkin, D., Rabinovitz, M. Hyper-acyloin condensation from simple aromatic esters to phenanthrenequinones: a new reaction of C_8K. *J. Org. Chem.* **1987**, 52, 3472-3474.
13. Fadel, A., Canet, J. L., Salaun, J. Ultrasound-promoted acyloin condensation and cyclization of carboxylic esters. *Synlett* **1990**, 89-91.
14. Daynard, T. S., Eby, P. S., Hutchinson, J. H. The acyloin reaction using tethered diesters. *Can. J. Chem.* **1993**, 71, 1022-1028.
15. Yamashita, K., Osaki, T., Sasaki, K., Yokota, H., Oshima, N., Nango, M., Tsuda, K. Acyloin condensation in aqueous system by durable polymer-supported thiazolium salt catalysts. *J. Polym. Sci., Part A: Polym. Chem.* **1994**, 32, 1711-1717.
16. Cetinkaya, E., Kucukbay, H. Effective acyloin condensations catalyzed by electron-rich olefins. *Turk. J. Chem.* **1995**, 19, 24-30.
17. Kashimura, S., Murai, Y., Ishifune, M., Masuda, H., Murase, H., Shono, T. Electroorganic chemistry. 148. Cathodic coupling of aliphatic esters. Useful reaction for the synthesis of 1,2-diketone and acyloin. *Tetrahedron Lett.* **1995**, 36, 4805-4808.
18. Yamashita, K., Sasaki, S.-i., Osaki, T., Nango, M., Tsuda, K. A holoenzyme model of thiamine dependent enzyme; asymmetrical acyloin condensation using a lipid catalyst in a bilayer membrane. *Tetrahedron Lett.* **1995**, 36, 4817-4820.
19. Makosza, M., Grela, K. Convenient preparation of 'high-surface sodium' in liquid ammonia. Use in the acyloin reaction. *Synlett* **1997**, 267-268.
20. Guo, Z., Goswami, A., Mirfakhrae, K. D., Patel, R. N. Asymmetric acyloin condensation catalyzed by phenylpyruvate decarboxylase. *Tetrahedron: Asymmetry* **1999**, 10, 4667-4675.
21. Guo, Z., Goswami, A., Nanduri, V. B., Patel, R. N. Asymmetric acyloin condensation catalysed by phenylpyruvate decarboxylase. Part 2: Substrate specificity and purification of the enzyme. *Tetrahedron: Asymmetry* **2001**, 12, 571-577.
22. Heck, R., Henderson, A. P., Kohler, B., Retey, J., Golding, B. T. Crossed acyloin condensation of aliphatic aldehydes. *Eur. J. Org. Chem.* **2001**, 2623-2627.
23. Bloomfield, J. J., Owsley, D. C., Ainsworth, C., Robertson, R. E. Mechanism of the acyloin condensation. *J. Org. Chem.* **1975**, 40, 393-402.
24. Sieburth, S. M., Santos, E. D. A short synthesis of the tricyclo[3.3.2^{1,4}.0]decane ring system. *Tetrahedron Lett.* **1994**, 35, 8127-8130.

25. Blanchard, A. N., Burnell, D. J. First intramolecular geminal acylation: synthesis of bridged bicyclic diketones. *Tetrahedron Lett.* **2001**, 42, 4779-4781.

Related reactions: Prins reaction;

1. Treibs, W., Schmidt, H. Catalytic dehydrogenation of hydroaromatic compounds. *Ber.* **1927**, 60B, 2335-2341.
2. Grignard, V., Doeuvre, J. Transformation of L-isopulegol into D-citronellal. *Compt. rend.* **1930**, 190, 1164-1167.
3. Ikeda, T., Wakatsuki, K. Linaloöl. Isomerization of linaloöl by heating under pressure. I. Plinol. *J. Chem. Soc. Japan* **1936**, 57, 425-435.
4. Alder, K., Noble, T. Substituting additions. II. Addition of azodicarboxylic esters to aldehydes. *Ber.* **1943**, 76B, 54-57.
5. Alder, K., Pascher, F., Schmitz, A. Substituting additions. I. Addition of maleic anhydride and azodicarboxylic esters to singly unsaturated hydrocarbons. Substitution processes in the allyl position. *Ber.* **1943**, 76B, 27-53.
6. Alder, K., Schmidt, C.-H. Substituting additions. III. Condensation of furan and its homologs with α,β-unsaturated ketones and aldehydes. Synthesis of di-, tri-and tetraketones of the aliphatic series. *Ber.* **1943**, 76B, 183-205.
7. Hoffmann, H. M. R. Ene reaction. *Angew. Chem., Int. Ed. Engl.* **1969**, 8, 556-577.
8. Gollnick, K., Kuhn, H. J. Ene-reactions with singlet oxygen. *Org. Chem. (N. Y.)* **1979**, 40, 287-427.
9. Snider, B. B. Lewis-acid catalyzed ene reactions. *Acc. Chem. Res.* **1980**, 13, 426-432.
10. Oppolzer, W. Diastereo- and enantio-selective cycloaddition and ene reactions in organic synthesis. (*Curr. Trends Org. Synth., Proc. Int. Conf., 4th*). **1983**, 131-149
11. Oppolzer, W. Asymmetric Diels-Alder- and ene reactions in organic synthesis. *Angew. Chem.* **1984**, 96, 840-854.
12. Dubac, J., Laporterie, A. Ene and retro-ene reactions in group 14 organometallic chemistry. *Chem. Rev.* **1987**, 87, 319-334.
13. Boyd, G. V. The ene reaction. (*Chem. Double-Bonded Funct. Groups*). **1989**, 477-525
14. Trost, B. M. Palladium-catalyzed cycloisomerizations of enynes and related reactions. *Acc. Chem. Res.* **1990**, 23, 34-42.
15. Mikami, K., Shimizu, M. Asymmetric ene reactions in organic synthesis. *Chem. Rev.* **1992**, 92, 1021-1050.
16. Mikami, K., Terada, M., Narisawa, S., Nakai, T. Asymmetric catalysis for carbonyl-ene reaction. *Synlett* **1992**, 255-265.
17. Ripoll, J. L., Vallee, Y. Synthetic applications of the retro-ene reaction. *Synthesis* **1993**, 659-677.
18. Berrisford, D. J., Bolm, C. Catalytic asymmetric carbonyl-ene reactions. *Angew. Chem., Int. Ed. Engl.* **1995**, 34, 1717-1719.
19. Borzilleri, R. M., Weinreb, S. M. Imino ene reactions in organic synthesis. *Synthesis* **1995**, 347-360.
20. Davies, A. G. Hydrogen-ene and metallo-ene reactions. *Spec. Publ. - R. Soc. Chem.* **1995**, 148, 263-277.
21. Mikami, K. Supramolecular chemistry in asymmetric carbonyl-ene reactions. *Adv. in Asymmetric Synth.* **1995**, 1, 1-44.
22. Mikami, K., Terada, M., Nakai, T. Asymmetric catalysis of the glyoxylate-ene reaction and related reactions. *Adv. Catal. Processes* **1995**, 1, 123-149.
23. Mikami, K. Asymmetric catalysis of carbonyl-ene reactions and related carbon-carbon bond forming reactions. *Pure Appl. Chem.* **1996**, 68, 639-644.
24. Prein, M., Adam, W. The Schenck ene reaction: diastereoselective oxyfunctionalization with singlet oxygen in synthetic applications. *Angew. Chem., Int. Ed. Engl.* **1996**, 35, 477-494.
25. Weinreb, S. M. Synthetic applications of a novel pericyclic imino ene reaction of allenyl silanes. *J. Heterocycl. Chem.* **1996**, 33, 1429-1436.
26. Mackewitz, T. W., Regitz, M. The ene reaction in the chemistry of low-coordinate phosphorus. Part 127. *Synthesis* **1998**, 125-138.
27. Mikami, K., Terada, M. Ene-type reactions. in *Comprehensive Asymmetric Catalysis I-III* (eds. Jacobsen, E., Pfaltz, A.,Yamamoto, H.), *3*, 1143-1174 (Springer, New York, **1999**).
28. Dias, L. C. Chiral Lewis acid catalyzed ene reactions. *Curr. Org. Chem.* **2000**, 4, 305-342.
29. Mikami, K., Nakai, T. Asymmetric ene reactions. *Catal. Asymmetric Synth. (2nd Edition)* **2000**, 543-568.
30. Stratakis, M., Orfanopoulos, M. Regioselectivity in the ene reaction of singlet oxygen with alkenes. *Tetrahedron* **2000**, 56, 1595-1615.
31. Leach, A. G., Houk, K. N. Diels-Alder and ene reactions of singlet oxygen, nitroso compounds and triazolinediones: transition states and mechanisms from contemporary theory. *Chem. Commun.* **2002**, 1243-1255.
32. Adam, W., Krebs, O. The Nitroso Ene Reaction: A Regioselective and Stereoselective Allylic Nitrogen Functionalization of Mechanistic Delight and Synthetic Potential. *Chem. Rev.* **2003**, 103, 4131-4146.
33. Griesbeck, A. G., El-Idreesy, T. T., Adam, W., Krebs, O. Ene-reactions with singlet oxygen. *CRC Handbook of Organic Photochemistry and Photobiology (2nd Edition)* **2004**, 8/1-8/20.
34. Yamaguchi, K., Yabushita, S., Fueno, T., Houk, K. N. Mechanism of photooxygenation reactions. Computational evidence against the diradical mechanism of singlet oxygen ene reactions. *J. Am. Chem. Soc.* **1981**, 103, 5043-5046.
35. Tsai, T.-G., Yu, C.-H. Effect of orbital overlap in thermal reverse homo-Diels-Alder reactions and an intramolecular reverse ene reaction. *J. Chin. Chem. Soc. (Taipei)* **1994**, 41, 631-634.
36. Yliniemela, A., Konschin, H., Neagu, C., Pajunen, A., Hase, T., Brunow, G., Teleman, O. Design and Synthesis of a Transition State Analog for the Ene Reaction between Maleimide and 1-Alkenes. *J. Am. Chem. Soc.* **1995**, 117, 5120-5126.
37. Hase, T., Brunow, G., Hase, A., Kodaka, M., Neagu, C., Nevanen, T., Teeri, T., Teleman, O., Tianinen, E., et al. Search for antibody catalysts for the ene reaction. *Pure Appl. Chem.* **1996**, 68, 605-608.
38. Yliniemela, A., Teleman, O., Nevanen, T., Takkinen, K., Hemminki, A., Teeri, T. T. Towards recombinant catalytic antibodies for the ene reaction. *VTT Symp.* **1996**, 163, 277-282.
39. Chen, J. S., Houk, K. N., Foote, C. S. The Nature of the Transition Structures of Triazolinedione Ene Reactions. *J. Am. Chem. Soc.* **1997**, 119, 9852-9855.
40. Houk, K. N., Beno, B. R., Nendel, M., Black, K., Yoo, H. Y., Wilsey, S., Lee, J. Exploration of pericyclic reaction transition structures by quantum mechanical methods: competing concerted and stepwise mechanisms. *THEOCHEM* **1997**, 398-399, 169-179.
41. Pranata, J. The ene reaction: comparison of results of Hartree-Fock, Moller-Plesset, CASSCF and DFT calculations. *Int. J. Quantum Chem.* **1997**, 62, 509-514.
42. Adam, W., Bottke, N., Engels, B., Krebs, O. An Experimental and Computational Study on the Reactivity and Regioselectivity for the Nitrosoarene Ene Reaction: Comparison with Triazolinedione and Singlet Oxygen. *J. Am. Chem. Soc.* **2001**, 123, 5542-5548.
43. Musch, P. W., Engels, B. The Importance of the Ene Reaction for the C2-C6 Cyclization of Enyne-Allenes. *J. Am. Chem. Soc.* **2001**, 123, 5557-5562.
44. Morao, I., McNamara, J. P., Hillier, I. H. Carbonyl-Ene Reactions Catalyzed by Bis(oxazoline) Copper(II) Complexes Proceed by a Facile Stepwise Mechanism: DFT and ONIOM (DFT:PM3) Studies. *J. Am. Chem. Soc.* **2003**, 125, 628-629.
45. Boyd, G. V. The ene reaction. *Chem. Double-Bonded Funct. Groups* **1989**, 2, 477-525.
46. Ripoll, J. L., Vallee, Y. Synthetic Applications of the Retro-Ene Reaction. *Synthesis-Stuttgart* **1993**, 659-677.
47. Mackewitz, T. W., Regitz, M. The ene reaction in the chemistry of low-coordinate phosphorus. *Synthesis* **1998**, 125-138.
48. Huisgen, R., Pohl, H. Addition reactions of the N,N-double bond. III. The reaction of azodicarboxylic acid ester with olefins. *Chem. Ber.* **1960**, 93, 527-540.
49. Thaler, W. A., Franzus, B. The reaction of ethyl azodicarboxylate with monoolefins. *J. Org. Chem.* **1964**, 29, 2226-2235.
50. Paderes, G. D., Jorgensen, W. L. Computer-assisted mechanistic evaluation of organic reactions. 20. Ene and retro-ene chemistry. *J. Org. Chem.* **1992**, 57, 1904-1916.
51. Achmatowicz, O., Bialecka-Florjanczyk, E. Mechanism of the carbonyl-ene reaction. *Tetrahedron* **1996**, 52, 8827-8834.
52. Xia, Q., Ganem, B. Asymmetric Total Synthesis of (-)-α-Kainic Acid Using an Enantioselective, Metal-Promoted Ene Cyclization. *Org. Lett.* **2001**, 3, 485-487.
53. Snider, B. B., Ron, E. The mechanism of Lewis acid catalyzed ene reactions. *J. Am. Chem. Soc.* **1985**, 107, 8160-8164.
54. Zhang, J.-H., Wang, M.-X., Huang, Z.-T. The aza-ene reaction of heterocyclic ketene aminals with enones: an unusual and efficient formation of imidazo[1,2-*a*] pyridine and imidazo [1,2,3-*ij*] [1,8]naphthyridine derivatives. *Tetrahedron Lett.* **1998**, 39, 9237-9240.

55. Barriault, L., Deon, D. H. Total Synthesis of (+)-Arteannuin M Using the Tandem Oxy-Cope/Ene Reaction. *Org. Lett.* **2001**, 3, 1925-1927.

Related reactions: Evans aldol reaction, Mukaiyama aldol reaction, Reformatsky reaction;

1. Kane, R. About acetic acid and some of its derivatives. *J. Prakt. Chem.* **1838**, 15, 129.
2. Kane, R. About acetic acid and some of its derivatives. *Ann. Phys. Chem. Ser. 2* **1838**, 44, 475.
3. Nielsen, A. T., Houlihan, W. J. The aldol condensation. *Org. React.* **1968**, 16, 438 pp.
4. Hajos, Z. G. Aldol and related reactions. in *Carbon-Carbon Bond Formation* (ed. Augustine, R. L.), *1*, 1-84 (M. Dekker, New York, **1979**).
5. Heathcock, C. H. Acyclic stereocontrol through the aldol condensation. *Science* **1981**, 214, 395-400.
6. Evans, D. A., Nelson, J. V., Taber, T. R. Stereoselective aldol condensations. *Top. Stereochem.* **1982**, 13, 1-115.
7. Mukaiyama, T. The directed aldol reaction. *Org. React.* **1982**, 28, 203-331.
8. Heathcock, C. H. The aldol addition reaction. in *Asymmetric Synthesis 3*, 111-212 (Academic Press, New York, **1984**).
9. Heathcock, C. H. Stereoselective aldol condensations. in *Stud. Org. Chem. (Amsterdam)* (ed. Buncel, E.), *5B*, 177-237 (Elsevier, New York, **1984**).
10. Mukaiyama, T. Aldol reactions directed Ti synthetic control. in *Stud. Org. Chem. (Amsterdam)* (ed. Buncel, E.), *25*, 119-139 (Elsevier, New York, **1986**).
11. Bednarski, M. D. Applications of enzymic aldol reactions in organic synthesis. *Applied Biocatalysis* **1991**, 1, 87-116.
12. Heathcock, C. H. The Aldol Reaction: Group I and II Enolates. in *Comp. Org. Synth.* (eds. Trost, B. M.,Fleming, I.), *1*, 181-231 (Pergamon Press, Oxford, **1991**).
13. Heathcock, C. H. The Aldol Reaction: Acid and General Base Catalysis. in *Comp. Org. Synth.* (eds. Trost, B. M.,Fleming, I.), *1*, 133-179 (Pergamon Press, Oxford, **1991**).
14. Moon Kim, B., Williams, S. F., Masamune, S. The Aldol reaction: Group III Enolates. in *Comp. Org. Synth.* (eds. Trost, B. M.,Fleming, I.), *1*, 239-275 (Pergamon Press, Oxford, **1991**).
15. Paterson, I. The Aldol Reaction: Transition Metal Enolates. in *Comp. Org. Synth.* (eds. Trost, B. M.,Fleming, I.), *1*, 301-319 (Pergamon Press, Oxford, **1991**).
16. Rathke, M. W., Weipert, P. Zinc Enolates: Refortmasky and Blasie Reaction. in *Comp. Org. Synth.* (eds. Trost, B. M.,Fleming, I.), *1*, 277-299 (Pergamon Press, Oxford, **1991**).
17. Braun, M. Recent developments in stereoselective aldol reactions. *Advances in Carbanion Chemistry* **1992**, 1, 177-247.
18. Fessner, W. D. Enzyme-catalyzed aldol additions in asymmetric synthesis. Part 1. *Kontakte (Darmstadt)* **1992**, 3-9.
19. Braun, M., Sacha, H. Recent advances in stereoselective aldol reactions of ester and thioester enolates. *J. Prakt. Chem.* **1993**, 335, 653-668.
20. Fessner, W. D. Enzyme-catalyzed aldol additions in asymmetric synthesis. Part 2. *Kontakte (Darmstadt)* **1993**, 23-34.
21. Sawamura, M., Ito, Y. Asymmetric aldol reactions. in *Catal. Asymmetric Synth.* (ed. Ojima, I.), 367-388 (VCH, New York, **1993**).
22. Franklin, A. S., Paterson, I. Recent developments in asymmetric aldol methodology. *Contemp. Org. Synth.* **1994**, 1, 317-338.
23. Mukaiyama, T., Kobayashi, S. Tin(II) enolates in the aldol, Michael, and related reactions. *Org. React.* **1994**, 46, 1-103.
24. Bernardi, A., Gennari, C., Goodman, J. M., Paterson, I. The rational design and systematic analysis of asymmetric aldol reactions using enol borinates: applications of transition state computer modeling. *Tetrahedron: Asymmetry* **1995**, 6, 2613-2636.
25. Cowden, C. J., Paterson, I. Asymmetric aldol reactions using boron enolates. *Org. React.* **1997**, 51, 1-200.
26. Kiyooka, S.-I. Development of a chiral Lewis acid-promoted asymmetric aldol reaction using oxazaborolidinone. *Rev. on Heteroa. Chem.* **1997**, 17, 245-270.
27. Takayama, S., McGarvey, G. J., Wong, C.-H. Enzymes in organic synthesis: recent developments in aldol reactions and glycosylations. *Chem. Soc. Rev.* **1997**, 26, 407-415.
28. Groger, H., Vogl, E. M., Shibasaki, M. New catalytic concepts for the asymmetric aldol reaction. *Chem.-- Eur. J.* **1998**, 4, 1137-1141.
29. Mahrwald, R. Lewis acid catalysts in enantioselective aldol addition. *Rec. Res. Dev. Synt. Org. Chem.* **1998**, 1, 123-150.
30. Nelson, S. G. Catalyzed enantioselective aldol additions of latent enolate equivalents. *Tetrahedron: Asymmetry* **1998**, 9, 357-389.
31. Mahrwald, R. Diastereoselection in Lewis-Acid-Mediated Aldol Additions. *Chem. Rev.* **1999**, 99, 1095-1120.
32. Saito, S., Yamamoto, H. Directed aldol condensation. *Chem.-- Eur. J.* **1999**, 5, 1959-1962.
33. Carreira, E. M. Recent advances in asymmetric aldol addition reactions. in *Catalytic Asymmetric Synthesis (2nd Edition)* (ed. Ojima, I.), 513-541 (Wiley-VCH, New York, **2000**).
34. Casiraghi, G., Zanardi, F., Appendino, G., Rassu, G. The Vinylogous Aldol Reaction: A Valuable, Yet Understated Carbon-Carbon Bond-Forming Maneuver. *Chem. Rev.* **2000**, 100, 1929-1972.
35. Krauss, R., Koert, U. Catalytic asymmetric aldol reactions. *Organic Synthesis Highlights IV* **2000**, 144-154.
36. Machajewski, T. D., Wong, C.-H., Lerner, R. A. The catalytic asymmetric aldol reaction. *Angew. Chem., Int. Ed. Engl.* **2000**, 39, 1352-1374.
37. Sawamura, M., Ito, Y. Asymmetric aldol reactions-discovery and development. in *Catalytic Asymmetric Synthesis (2nd Edition)* (ed. Ojima, I.), 493-512 (Wiley-VCH, New York, **2000**).
38. Paterson, I., Doughty, V. A., Florence, G., Gerlach, K., McLeod, M. D., Scott, J. P., Trieselmann, T. Asymmetric aldol reactions using boron enolates: applications to polyketide synthesis. *ACS Symp. Ser.* **2001**, 783, 195-206.
39. Alcaide, B., Almendros, P. The direct catalytic asymmetric aldol reaction. *Eur. J. Org. Chem.* **2002**, 1595-1601.
40. Cauble, D. F., Jr., Krische, M. J. The first direct and enantioselective cross-Aldol reaction of aldehydes. *Chemtracts* **2002**, 15, 380-383.
41. Palomo, C., Oiarbide, M., Garcia, J. M. The aldol addition reaction: an old transformation at constant rebirth. *Chem.-- Eur. J.* **2002**, 8, 36-44.
42. Luo, Z.-B., Dai, L.-X. Novel small organic molecules for a highly enantioselective direct Aldol reaction. *Chemtracts* **2003**, 16, 843-847.
43. Abiko, A. Boron-Mediated Aldol Reaction of Carboxylic Esters. *Acc. Chem. Res.* **2004**, 37, 387-395.
44. Palomo, C., Oiarbide, M., Garcia, J. M. Current progress in the asymmetric aldol addition reaction. *Chem. Soc. Rev.* **2004**, 33, 65-75.
45. Saito, S., Yamamoto, H. Design of Acid-Base Catalysis for the Asymmetric Direct Aldol Reaction. *Acc. Chem. Res.* **2004**, 37, 570-579.
46. Shibasaki, M., Yoshikawa, N., Matsunaga, S. Direct catalytic asymmetric aldol reaction. *Comprehensive Asymmetric Catalysis, Supplement* **2004**, 1, 135-141.
47. Wuts, P. G. M., Walters, M. A. A molecular mechanics approach to determining the diastereofacial selectivity in the reaction of asymmetric aldehydes with achiral enolates. *J. Org. Chem.* **1984**, 49, 4573-4574.
48. Hoffman, R. W., Ditrich, K., Forech, S., Cremer, D. On the stereochemistry of the aldol addition. *Tetrahedron* **1985**, 41, 5517-5524.
49. Gennari, C., Todeschini, R., Beretta, M. G., Favini, G., Scolastico, C. Theoretical studies of stereoselective aldol condensations. *J. Org. Chem.* **1986**, 51, 612-616.
50. Li, Y., Paddon-Row, M. N., Houk, K. N. Transition structures of aldol reactions. *J. Am. Chem. Soc.* **1988**, 110, 3684-3686.
51. Bernardi, A., Capelli, A. M., Gennari, C., Goodman, J. M., Paterson, I. Transition-state modeling of the aldol reaction of boron enolates: a force field approach. *J. Org. Chem.* **1990**, 55, 3576-3581.
52. Goodman, J. M., Kahn, S. D., Paterson, I. Theoretical studies of aldol stereoselectivity: the development of a force field model for enol borinates and the investigation of chiral enolate π-face selectivity. *J. Org. Chem.* **1990**, 55, 3295-3303.
53. Li, Y., Paddon-Row, M. N., Houk, K. N. Transition structures for the aldol reactions of anionic, lithium, and boron enolates. *J. Org. Chem.* **1990**, 55, 481-493.
54. Bernardi, A., Capelli, A. M., Comotti, A., Gennari, C., Gardner, M., Goodman, J. M., Paterson, I. Origins of stereoselectivity in chiral boron enolate aldol reactions: a computational study using transition state modeling. *Tetrahedron* **1991**, 47, 3471-3484.
55. Bernardi, F., Robb, M. A., Suzzi-Valli, G., Tagliavini, E., Trombini, C., Umani-Ronchi, A. An MC-SCF study of the transition structures for the aldol reaction of formaldehyde with acetaldehyde boron enolate. *J. Org. Chem.* **1991**, 56, 6472-6475.

56. Guthrie, J. P. Rate-equilibrium correlations for the aldol condensation: an analysis in terms of Marcus theory. *J. Am. Chem. Soc.* **1991**, 113, 7249-7255.
57. Gennari, C., Hewkin, C. T., Molinari, F., Bernardi, A., Comotti, A., Goodman, J. M., Paterson, I. The rational design of highly stereoselective boron enolates using transition-state computer modeling: a novel, asymmetric anti aldol reaction for ketones. *J. Org. Chem.* **1992**, 57, 5173-5177.
58. Gennari, C., Vieth, S., Comotti, A., Vulpetti, A., Goodman, J. M., Paterson, I. Diastereofacial selectivity in the aldol reactions of chiral α-methyl aldehydes: a computer modelling approach. *Tetrahedron* **1992**, 48, 4439-4458.
59. Bernardi, F., Bongini, A., Cainelli, G., Robb, M. A., Valli, G. S. Theoretical study of the aldol condensation with imine-type electrophiles. *J. Org. Chem.* **1993**, 58, 750-755.
60. Vulpetti, A., Bernardi, A., Gennari, C., Goodman, J. M., Paterson, I. Origins of π-face selectivity in the aldol reactions of chiral E-enol borinates: a computational study using transition state modeling. *Tetrahedron* **1993**, 49, 685-696.
61. Bernardi, A., Comotti, A., Gennari, C., Hewkin, C. T., Goodman, J. M., Schlapbach, A., Paterson, I. Computer-assisted design of chiral boron enolates: the role of ate complexes in determining aldol stereoselectivity. *Tetrahedron* **1994**, 50, 1227-1242.
62. Henderson, K. W., Dorigo, A. E., Liu, Q.-Y., Williard, P. G., Schleyer, P. v. R., Bernstein, P. R. Structural Consequences of the Addition of Lithium Halides in Enolization and Aldol Reactions. *J. Am. Chem. Soc.* **1996**, 118, 1339-1347.
63. Bernardi, A., Gennari, C., Raimondi, L., Villa, M. B. Computational studies on the aldol-type addition of boron enolates to imines: an ab-initio approach. *Tetrahedron* **1997**, 53, 7705-7714.
64. Gennari, C. Rationally designed chiral enol borinates: powerful reagents for the stereoselective synthesis of natural products. *Pure Appl. Chem.* **1997**, 69, 507-512.
65. Bouillon, J.-P., Portella, C., Bouquant, J., Humbel, S. Theoretical Study of Intramolecular Aldol Condensation of 1,6-Diketones: Trimethylsilyl Substituent Effect. *J. Org. Chem.* **2000**, 65, 5823-5830.
66. Bahmanyar, S., Houk, K. N. Transition States of Amine-Catalyzed Aldol Reactions Involving Enamine Intermediates: Theoretical Studies of Mechanism, Reactivity, and Stereoselectivity. *J. Am. Chem. Soc.* **2001**, 123, 11273-11283.
67. Murga, J., Falomir, E., Carda, M., Marco, J. A. An ab initio study of the enolboration of 3-pentanone mediated by boron monochlorides L_2BCl. *Tetrahedron* **2001**, 57, 6239-6247.
68. Arno, M., Domingo, L. R. Density functional theory study of the mechanism of the proline-catalyzed intermolecular aldol reaction. *Theoretical Chemistry Accounts* **2002**, 108, 232-239.
69. Murga, J., Falomir, E., Gonzalez, F., Carda, M., Marco, J. A. Influence of the protecting groups on the syn/anti stereoselectivity of boron aldol additions with erythrulose derivatives. A theoretical and experimental study. *Tetrahedron* **2002**, 58, 9697-9707.
70. Ruiz, M., Ojea, V., Quintela, J. M. Computational study of the syn,anti-selective aldol additions of lithiated bis-lactim ether to 1,3-dioxolane-4-carboxaldehydes. *Tetrahedron: Asymmetry* **2002**, 13, 1863-1873.
71. Arno, M., Domingo, L. R. Theozyme for antibody aldolases. Characterization of the transition-state analogue. *Org. Biomol. Chem.* **2003**, 1, 637-643.
72. Bahmanyar, S., Houk, K. N. Origins of Opposite Absolute Stereoselectivities in Proline-Catalyzed Direct Mannich and Aldol Reactions. *Org. Lett.* **2003**, 5, 1249-1251.
73. Bahmanyar, S., Houk, K. N., Martin, H. J., List, B. Quantum Mechanical Predictions of the Stereoselectivities of Proline-Catalyzed Asymmetric Intermolecular Aldol Reactions. *J. Am. Chem. Soc.* **2003**, 125, 2475-2479.
74. Hoang, L., Bahmanyar, S., Houk, K. N., List, B. Kinetic and Stereochemical Evidence for the Involvement of Only One Proline Molecule in the Transition States of Proline-Catalyzed Intra- and Intermolecular Aldol Reactions. *J. Am. Chem. Soc.* **2003**, 125, 16-17.
75. Claisen, L., Claparede, A. The compounds formed by the reaction between benzladehydes and mesityl oxides and the constitution of acetophorones. *Ber.* **1881**, 14, 349-353.
76. Claisen, L. Introduction of acid radicals into ketones. *Ber.* **1887**, 20, 655-657.
77. Mukaiyama, T., Narasaka, K., Banno, K. New aldol type reaction. *Chem. Lett.* **1973**, 1011-1014.
78. Mukaiyama, T., Banno, K., Narasaka, K. New cross-aldol reactions. Reactions of silyl enol ethers with carbonyl compounds activated by titanium tetrachloride. *J. Am. Chem. Soc.* **1974**, 96, 7503-7509.
79. Heathcock, C. H. The aldol condensation as a tool for stereoselective organic synthesis. *Curr. Trends Org. Synth., Proc. Int. Conf., 4th* **1983**, 27-43.
80. Arya, P., Qin, H. Advances in asymmetric enolate methodology. *Tetrahedron* **2000**, 56, 917-947.
81. Petersen, M., Zannetti, M. T., Fessner, W.-D. Tandem asymmetric C-C bond formations by enzyme catalysis. *Top. Curr. Chem.* **1997**, 186, 87-117.
82. Denmark, S. E., Stavenger, R. A. Asymmetric Catalysis of Aldol Reactions with Chiral Lewis Bases. *Acc. Chem. Res.* **2000**, 33, 432-440.
83. De Brabander, J., Oppolzer, W. Enantioselective total synthesis of (-)-denticulatins A and B using a novel group-selective aldolization of a meso dialdehyde as a key step. *Tetrahedron* **1997**, 53, 9169-9202.
84. White, J. D., Blakemore, P. R., Green, N. J., Hauser, E. B., Holoboski, M. A., Keown, L. E., Kolz, C. S. N., Phillips, B. W. Total Synthesis of Rhizoxin D, a Potent Antimitotic Agent from the Fungus Rhizopus chinensis. *J. Org. Chem.* **2002**, 67, 7750-7760.
85. Yoshikawa, N., Yamada, Y. M. A., Das, J., Sasai, H., Shibasaki, M. Direct catalytic asymmetric aldol reaction. *J. Am. Chem. Soc.* **1999**, 121, 4168-4178.
86. Fujii, K., Maki, K., Kanai, M., Shibasaki, M. Formal Catalytic Asymmetric Total Synthesis of Fostriecin. *Org. Lett.* **2003**, 5, 733-736.

Alkene (Olefin) Metathesis 10

1. Anderson, A. W., Merckling, N. G. *Polymeric bicyclo[2.2.1]hept-2-ene.* US 2721189, **1955** (E. I. du Pont de Nemours & Co.).
2. Calderon, N., Chen, H. Y., Scott, K. W. Olefin metathesis, a novel reaction for skeletal transformations of unsaturated hydrocarbons. *Tetrahedron Lett.* **1967**, 3327-3329.
3. Grubbs, R. H. The olefin metathesis reaction. *J. Organomet. Chem. Libr.* **1976**, 1, 423-460.
4. Grubbs, R. H. The olefin metathesis reaction. *Prog. Inorg. Chem.* **1978**, 24, 1-50.
5. Grubbs, R. H., Tumas, W. Polymer synthesis and organotransition metal chemistry. *Science* **1989**, 243, 907-915.
6. Amass, A. J. Ring-opening metathesis polymerization of cyclic alkenes. *New Methods Polym. Synth.* **1991**, 76-106.
7. Eleuterio, H. The discovery of olefin metathesis. *Chemtech* **1991**, 21, 92-95.
8. Grubbs, R. H. The development of functional group tolerant ROMP catalysts. *J. Macromol. Sci., Pure Appl. Chem.* **1994**, A31, 1829-1833.
9. Grubbs, R. H., Miller, S. J., Fu, G. C. Ring-Closing Metathesis and Related Processes in Organic Synthesis. *Acc. Chem. Res.* **1995**, 28, 446-452.
10. Schuster, M., Blechert, S. Olefin metathesis in organic chemistry. *Angew. Chem., Int. Ed. Engl.* **1997**, 36, 2037-2056.
11. Fürstner, A. Recent advancements in ring closing olefin metathesis. *Top. in Cat.* **1998**, 4, 285-299.
12. Gibson, S. E., Keen, S. P. Cross-metathesis. *Top. Organomet. Chem.* **1998**, 1, 155-181.
13. Grubbs, R. H., Chang, S. Recent advances in olefin metathesis and its application in organic synthesis. *Tetrahedron* **1998**, 54, 4413-4450.
14. Grubbs, R. H., Lynn, D. M. *Olefin metathesis* (eds. Cornils, B.,Hermann, W. A.) (Wiley-VCH, Weinheim, New York, **1998**) 466-476.
15. Hoveyda, A. H. Catalytic ring-closing metathesis and the development of enantioselective processes. *Top. Organomet. Chem.* **1998**, 1, 105-132.
16. Ivin, K. J. Some recent applications of olefin metathesis in organic synthesis: A review. *J. Mol. Catal. A: Chemical* **1998**, 133, 1-16.
17. Nicolaou, K. C., King, N. P., He, Y. Ring-closing metathesis in the synthesis of epothilones and polyether natural products. *Top. Organomet. Chem.* **1998**, 1, 73-104.
18. Pariya, C., Jayaprakash, K. N., Sarkar, A. Alkene metathesis: new developments in catalyst design and application. *Coord. Chem. Rev.* **1998**, 168, 1-48.
19. Randall, M. L., Snapper, M. L. Selective olefin metatheses-new tools for the organic chemist: A review. *J. Mol. Catal. A: Chemical* **1998**, 133, 29-40.

20. Schrock, R. R. Olefin metathesis by well-defined complexes of molybdenum and tungsten. *Top. Organomet. Chem.* **1998**, 1, 1-36.
21. Schuster, M., Blechert, S. *Application of olefin metathesis* (eds. Beller, M.,Bolm, C.) (Wiley-VCH, Weinheim, New York, **1998**) 275-284.
22. Blechert, S. Olefin metathesis-recent applications in synthesis. *Pure Appl. Chem.* **1999**, 71, 1393-1399.
23. Grubbs, R. H., Khosravi, E. Ring-opening metathesis polymerization (ROMP) and related processes. *Mater. Sci. Technol.* **1999**, 20, 65-104.
24. Schrock, R. R. Olefin metathesis by molybdenum imido alkylidene catalysts. *Tetrahedron* **1999**, 55, 8141-8153.
25. Wright, D. L. Application of olefin metathesis to organic synthesis. *Curr. Org. Chem.* **1999**, 3, 211-240.
26. Fürstner, A. Olefin metathesis and beyond. *Angew. Chem., Int. Ed. Engl.* **2000**, 39, 3012-3043.
27. Jorgensen, M., Hadwiger, P., Madsen, R., Stutz, A. E., Wrodnigg, T. M. Olefin metathesis in carbohydrate chemistry. *Curr. Org. Chem.* **2000**, 4, 565-588.
28. Karle, M., Koert, U. Ring-closing olefin metathesis. in *Organic Synthesis Highlights IV* 91-96 (VCH, Weinheim, New York, **2000**).
29. Paddock, R. L., Nguyen, S. T., Poeppelmeier, K. R. A novel class of ruthenium catalysts for olefin metathesis. *Chemtracts* **2000**, 13, 119-123.
30. Roy, R., Das, S. K. Recent applications of olefin metathesis and related reactions in carbohydrate chemistry. *Chem. Commun.* **2000**, 519-529.
31. Brennan, P. E., Ley, S. V. New catalysts for olefin metathesis. *Chemtracts* **2001**, 14, 88-93.
32. Hoveyda, A. H., Schrock, R. R. Catalytic asymmetric olefin metathesis. *Chem.-- Eur. J.* **2001**, 7, 945-950.
33. Kotha, S., Sreenivasachary, N. Catalytic metathesis reaction in organic synthesis. *Indian J. Chem., Sect. B* **2001**, 40B, 763-780.
34. Mol, J. C. *Olefin metathesis* (eds. Sheldon, R. A.,Bekkum, H.) (Weinheim: Wiley-VCH, New York, **2001**) 562-575.
35. Trnka, T. M., Grubbs, R. H. The Development of L_2X_2Ru:CHR Olefin Metathesis Catalysts: An Organometallic Success Story. *Acc. Chem. Res.* **2001**, 34, 18-29.
36. Zaragoza, F. Olefin metathesis and related processes in combinatorial chemistry multiple bond formation. *Handbook of Combinatorial Chemistry* **2002**, 1, 585-609.
37. Anon. NHC ruthenium complexes as second generation Grubbs catalysts. *Synlett* **2003**, 423-424.
38. Buchmeiser, M. R. Well-defined transition metal catalysts for metathesis polymerization. *Late Transition Metal Polymerization Catalysis* **2003**, 155-191.
39. Connon, S. J., Blechert, S. Recent developments in olefin cross-metathesis. *Angew. Chem., Int. Ed. Engl.* **2003**, 42, 1900-1923.
40. Fuerstner, A. Ring closing metathesis. *Actualite Chimique* **2003**, 57-59.
41. Grubbs, R. H., Trnka, T. M., Sanford, M. S. Transition metal-carbene complexes in olefin metathesis and related reactions. *Current Methods in Inorganic Chemistry* **2003**, 3, 187-231.
42. Hoveyda, A. H., Schrock, R. R. Catalytic asymmetric olefin metathesis. in *Organic Synthesis Highlights V* 210-229 (VCH, Weinheim, New York, **2003**).
43. Lehman, S. E., Jr., Wagener, K. B. Catalysis in acyclic diene metathesis (ADMET) polymerization. (*Late Transition Metal Polymerization Catalysis*). **2003**, 193-229
44. Marciniec, B., Pietraszuk, C. Synthesis of unsaturated organosilicon compounds via alkene metathesis and metathesis polymerization. *Curr. Org. Chem.* **2003**, 7, 691-735.
45. Prunet, J. Recent methods for the synthesis of (E)-alkene units in macrocyclic natural products. *Angew. Chem., Int. Ed. Engl.* **2003**, 42, 2826-2830.
46. Schmidt, B. Ruthenium-catalyzed cyclizations: More than just olefin metathesis! *Angew. Chem., Int. Ed. Engl.* **2003**, 42, 4996-4999.
47. Schrock, R. R., Hoveyda, A. H. Molybdenum and tungsten imido alkylidene complexes as efficient olefin-metathesis catalysts. *Angew. Chem., Int. Ed. Engl.* **2003**, 42, 4592-4633.
48. Vernall, A. J., Abell, A. D. Cross metathesis of nitrogen-containing systems. *Aldrichimica Acta* **2003**, 36, 93-105.
49. Castarlenas, R., Fischmeister, C., Bruneau, C., Dixneuf, P. H. Allenylidene-ruthenium complexes as versatile precatalysts for alkene metathesis reactions. *J. Mol. Catal. A: Chemical* **2004**, 213, 31-37.
50. Coperet, C. Molecular design of heterogeneous catalysts: The case of olefin metathesis. *New J. Chem.* **2004**, 28, 1-10.
51. Deiters, A., Martin, S. F. Synthesis of Oxygen- and Nitrogen-Containing Heterocycles by Ring-Closing Metathesis. *Chem. Rev.* **2004**, 104, 2199-2238.
52. Grubbs, R. H. Olefin metathesis. *Tetrahedron* **2004**, 60, 7117-7140.
53. Hoveyda, A. H., Gillingham, D. G., Van Veldhuizen, J. J., Kataoka, O., Garber, S. B., Kingsbury, J. S., Harrity, J. P. A. Ru complexes bearing bidentate carbenes: from innocent curiosity to uniquely effective catalysts for olefin metathesis. *Org. Biomol. Chem.* **2004**, 2, 8-23.
54. Hoveyda, A. H., Schrock, R. R. Metathesis reactions. *Comprehensive Asymmetric Catalysis, Supplement* **2004**, 1, 207-233.
55. McReynolds, M. D., Dougherty, J. M., Hanson, P. R. Synthesis of Phosphorus and Sulfur Heterocycles via Ring-Closing Olefin Metathesis. *Chem. Rev.* **2004**, 104, 2239-2258.
56. Mol, J. C. Industrial applications of olefin metathesis. *J. Mol. Catal. A: Chemical* **2004**, 213, 39-45.
57. Mol, J. C. Catalytic Metathesis of Unsaturated Fatty Acid Esters and Oils. *Top. in Cat.* **2004**, 27, 97-104.
58. Nuyken, O., Mueller, B. Ring opening cross-metathesis of functional cyclo-olefins. *Designed Monomers and Polymers* **2004**, 7, 215-222.
59. Piscopio, A. D., Robinson, J. E. Recent applications of olefin metathesis to combinatorial chemistry. *Curr. Opin. Chem. Biol.* **2004**, 8, 245-254.
60. Schrock, R. R. Recent advances in olefin metathesis by molybdenum and tungsten imido alkylidene complexes. *J. Mol. Catal. A: Chemical* **2004**, 213, 21-30.
61. Thieuleux, C., Coperet, C., Dufaud, V., Marangelli, C., Kuntz, E., Basset, J. M. Heterogeneous well-defined catalysts for metathesis of inert and not so inert bonds. *J. Mol. Catal. A: Chemical* **2004**, 213, 47-57.
62. Chatterjee, A. K., Grubbs, R. H. Synthesis of Trisubstituted Alkenes via Olefin Cross-Metathesis. *Org. Lett.* **1999**, 1, 1751-1753.
63. Du Plessis, J. A. K., Spamer, A., Vosloo, H. C. M. A non-transition metal catalyst system for alkene metathesis. *S. African J. Chem.* **1999**, 52, 71-72.
64. Bielawski, C. W., Grubbs, R. H. Highly efficient ring-opening metathesis polymerization (ROMP) using new ruthenium catalysts containing N-heterocyclic carbene ligands. *Angew. Chem., Int. Ed. Engl.* **2000**, 39, 2903-2906.
65. Blackwell, H. E., O'Leary, D. J., Chatterjee, A. K., Washenfelder, R. A., Bussmann, D. A., Grubbs, R. H. New approaches to olefin cross-metathesis. *J. Am. Chem. Soc.* **2000**, 122, 58-71.
66. Jafarpour, L., Hillier, A. C., Nolan, S. P. Improved One-Pot Synthesis of Second-Generation Ruthenium Olefin Metathesis Catalysts. *Organometallics* **2002**, 21, 442-444.
67. Dunne, A. M., Mix, S., Blechert, S. A highly efficient olefin metathesis initiator: improved synthesis and reactivity studies. *Tetrahedron Lett.* **2003**, 44, 2733-2736.
68. Love, J. A., Sanford, M. S., Day, M. W., Grubbs, R. H. Synthesis, Structure, and Activity of Enhanced Initiators for Olefin Metathesis. *J. Am. Chem. Soc.* **2003**, 125, 10103-10109.
69. Van Veldhuizen, J. J., Gillingham, D. G., Garber, S. B., Kataoka, O., Hoveyda, A. H. Chiral Ru-Based Complexes for Asymmetric Olefin Metathesis: Enhancement of Catalyst Activity through Steric and Electronic Modifications. *J. Am. Chem. Soc.* **2003**, 125, 12502-12508.
70. Yao, Q., Zhang, Y. Olefin metathesis in the ionic liquid 1-butyl-3-methylimidazolium hexafluorophosphate using a recyclable Ru catalyst: Remarkable effect of a designer ionic tag. *Angew. Chem., Int. Ed. Engl.* **2003**, 42, 3395-3398.
71. Cundari, T. R., Gordon, M. S. Theoretical investigations of olefin metathesis catalysts. *Organometallics* **1992**, 11, 55-63.
72. Folga, E., Ziegler, T. Density functional study on molybdacyclobutane and its role in olefin metathesis. *Organometallics* **1993**, 12, 325-337.
73. Axe, F. U., Andzelm, J. W. Theoretical Characterization of Olefin Metathesis in the Bis-dicyclopentadienyltitanium(IV) System by Density Functional Theory. *J. Am. Chem. Soc.* **1999**, 121, 5396-5402.
74. Cavallo, L. Mechanism of Ruthenium-Catalyzed Olefin Metathesis Reactions from a Theoretical Perspective. *J. Am. Chem. Soc.* **2002**, 124, 8965-8973.
75. Vyboishchikov, S. F., Buhl, M., Thiel, W. Mechanism of olefin metathesis with catalysis by ruthenium carbene complexes: density functional studies on model systems. *Chem.-- Eur. J.* **2002**, 8, 3962-3975.

76. Bernardi, F., Bottoni, A., Miscione, G. P. DFT Study of the Olefin Metathesis Catalyzed by Ruthenium Complexes. *Organometallics* **2003**, 22, 940-947.
77. Bencze, L., Szilagyi, R. Molecular mechanics studies on the olefin metathesis reaction. III. Modeling of the "well-defined" carbenes. *J. Organomet. Chem.* **1994**, 475, 183-192.
78. Bencze, L., Szilagyi, R. Molecular mechanical studies on olefin metathesis reaction. Part 2: Modeling of [cyclic] {$W[C(CH_2)_3CH_2](OCH_2$-*t*-$Bu)_2Br_2\}_2$ and [cyclic] $W[C(CH_2)_3CH_2](OCH_2$-*t*-Bu_2Br_2.cntdot.$GaBr_3$. *J. Mol. Catal.* **1994**, 90, 157-170.
79. Bencze, L., Szilagyi, R. Molecular mechanical studies on the olefin metathesis reaction. I. Development and evaluation of tungsten carbene parameters: METMOD1. *J. Organomet. Chem.* **1994**, 465, 211-219.
80. Bray, M. R., Deeth, R. J., Paget, V. J. Kinetics and mechanism in transition metal chemistry: A computational perspective. *Prog. React. Kinet.* **1996**, 21, 169-214.
81. Bartlett, B., Hossain, M. M., Tysoe, W. T. Reaction pathway and stereoselectivity of olefin metathesis at high temperature. *J. Catal.* **1998**, 176, 439-447.
82. Ulman, M., Grubbs, R. H. Relative Reaction Rates of Olefin Substrates with Ruthenium(II) Carbene Metathesis Initiators. *Organometallics* **1998**, 17, 2484-2489.
83. Adlhart, C., Hinderling, C., Baumann, H., Chen, P. Mechanistic Studies of Olefin Metathesis by Ruthenium Carbene Complexes Using Electrospray Ionization Tandem Mass Spectrometry. *J. Am. Chem. Soc.* **2000**, 122, 8204-8214.
84. Adlhart, C., Volland, M. A. O., Hofmann, P., Chen, P. Comparing Grubbs-, Werner-, and Hofmann-type (carbene)ruthenium complexes: the key role of pre-equilibria for olefin metathesis. *Helv. Chim. Acta* **2000**, 83, 3306-3311.
85. Sanford, M. S., Love, J. A., Grubbs, R. H. Mechanism and Activity of Ruthenium Olefin Metathesis Catalysts. *J. Am. Chem. Soc.* **2001**, 123, 6543-6554.
86. Sanford, M. S., Ulman, M., Grubbs, R. H. New Insights into the Mechanism of Ruthenium-Catalyzed Olefin Metathesis Reactions. *J. Am. Chem. Soc.* **2001**, 123, 749-750.
87. Schwab, P., France, M. B., Ziller, J. W., Grubbs, R. H. A series of well-defined metathesis catalysts - synthesis of $[RuCl_2(:CHR')(PR_3)_2]$ and their reactions. *Angew. Chem., Int. Ed. Engl.* **1995**, 34, 2039-2041.
88. Schwab, P., Grubbs, R. H., Ziller, J. W. Synthesis and Applications of $RuCl_2(:CHR')(PR_3)_2$: The Influence of the Alkylidene Moiety on Metathesis Activity. *J. Am. Chem. Soc.* **1996**, 118, 100-110.
89. Dias, E. L., Nguyen, S. T., Grubbs, R. H. Well-Defined Ruthenium Olefin Metathesis Catalysts: Mechanism and Activity. *J. Am. Chem. Soc.* **1997**, 119, 3887-3897.
90. Smith, A. B., III, Kozmin, S. A., Adams, C. M., Paone, D. V. Assembly of (-)-Cylindrocyclophanes A and F via Remarkable Olefin Metathesis Dimerizations. *J. Am. Chem. Soc.* **2000**, 122, 4984-4985.
91. Smith, A. B., III, Adams, C. M., Kozmin, S. A., Paone, D. V. Total Synthesis of (-)-Cylindrocyclophanes A and F Exploiting the Reversible Nature of the Olefin Cross Metathesis Reaction. *J. Am. Chem. Soc.* **2001**, 123, 5925-5937.
92. Dvorak, C. A., Schmitz, W. D., Poon, D. J., Pryde, D. C., Lawson, J. P., Amos, R. A., Meyers, A. I. The synthesis of streptogramin antibiotics: (-)-griseoviridin and its C-8 epimer. *Angew. Chem., Int. Ed. Engl.* **2000**, 39, 1664-1666.
93. Crimmins, M. T., Tabet, E. A. Total Synthesis of (+)-Prelaureatin and (+)-Laurallene. *J. Am. Chem. Soc.* **2000**, 122, 5473-5476.

1. Mortreux, A., Blanchard, M. Metathesis of alkynes by a molybdenum hexacarbonyl-resorcinol catalyst. *J. Chem. Soc., Chem. Commun.* **1974**, 786-787.
2. Bencheick, A., Petit, M., Mortreux, A., Petit, F. New active and selective catalysts for homogeneous metathesis of disubstituted alkynes. *J. Mol. Catal.* **1982**, 15, 93-101.
3. Petit, M., Mortreux, A., Petit, F. Homogeneous metathesis of functionalized alkynes. *J. Chem. Soc., Chem. Commun.* **1982**, 1385-1386.
4. Weiss, K. Catalytic reactions of carbyne complexes. *Carbyne Complexes* **1988**, 205-228.
5. Szymanska-Buzar, T. Photochemical reactions of Group 6 metal carbonyls in catalytic transformation of alkenes and alkynes. *Coord. Chem. Rev.* **1997**, 159, 205-220.
6. Mori, M. Enyne metathesis. in *Top. Organomet. Chem.* (eds. Fürstner, A.,Gibson, S. E.), *1*, 133-154 (Springer, Berlin, New York, **1998**).
7. Bunz, U. H. F., Kloppenburg, L. Alkyne metathesis as a new synthetic tool: ring-closing, ring-opening, and acyclic. *Angew. Chem., Int. Ed. Engl.* **1999**, 38, 478-481.
8. Tsuji, J. Ring-closing metathesis of functionalized acetylene derivatives: a new entry into cycloalkynes. *Chemtracts* **1999**, 12, 522-525.
9. Anon. Alkyne metathesis. *Nachrichten aus der Chemie* **2000**, 48, 1242-1244.
10. Bunz, U. H. F. Poly(p-phenyleneethynylene)s by Alkyne Metathesis. *Acc. Chem. Res.* **2001**, 34, 998-1010.
11. Lindel, T. Alkyne metathesis in natural product synthesis. *Organic Synthesis Highlights V* **2003**, 27-35.
12. McCullough, L. G., Schrock, R. R. Multiple metal-carbon bonds. 34. Metathesis of acetylenes by molybdenum(VI) alkylidyne complexes. *J. Am. Chem. Soc.* **1984**, 106, 4067-4068.
13. Kaneta, N., Hikichi, K., Asaka, S.-i., Uemura, M., Mori, M. Novel synthesis of disubstituted alkyne using molybdenum catalyzed cross-alkyne metathesis. *Chem. Lett.* **1995**, 1055-1056.
14. Kaneta, N., Hirai, T., Mori, M. Reaction of alkyne having hydroxyphenyl group with $Mo(CO)_6$. *Chem. Lett.* **1995**, 627-628.
15. Fürstner, A., Seidel, G. Ring-closing metathesis of functionalized acetylene derivatives: a new entry into cycloalkynes. *Angew. Chem., Int. Ed. Engl.* **1998**, 37, 1734-1736.
16. Fürstner, A., Guth, O., Rumbo, A., Seidel, G. Ring Closing Alkyne Metathesis. Comparative Investigation of Two Different Catalyst Systems and Application to the Stereoselective Synthesis of Olfactory Lactones, Azamacrolides, and the Macrocyclic Perimeter of the Marine Alkaloid Nakadomarin A. *J. Am. Chem. Soc.* **1999**, 121, 11108-11113.
17. Fürstner, A., Grela, K. Ring-closing alkyne metathesis: application to the stereoselective total synthesis of prostaglandin E2-1,15-lactone. *Angew. Chem., Int. Ed. Engl.* **2000**, 39, 1234-1236.
18. Schleyer, D., Niessen, H. G., Bargon, J. In situ ^{1}H-PHIP-NMR studies of the stereoselective hydrogenation of alkynes to (E)-alkenes catalyzed by a homogeneous [Cp*Ru]+ catalyst. *New J. Chem.* **2001**, 25, 423-426.
19. Fürstner, A., Radkowski, K. A chemo- and stereoselective reduction of cycloalkynes to (E)-cycloalkenes. *Chem. Commun.* **2002**, 2182-2183.
20. Mortreux, A., Delgrange, J. C., Blanchard, M., Lubochinsky, B. Role of phenol in the metathesis of acetylenic hydrocarbons on catalysts based on molybdenum hexacarbonyl. *J. Mol. Catal.* **1977**, 2, 73-82.
21. Mortreux, A., Petit, F., Blanchard, M. Carbon-13 tracer studies of alkynes metathesis. *Tetrahedron Lett.* **1978**, 4967-4968.
22. Fritch, J. R., Vollhardt, K. P. C. Cyclobutadiene-metal complexes as potential intermediates of alkyne metathesis: flash thermolysis of substituted η4-cyclobutadienyl-η5-cyclopentadienylcobalt complexes. *Angew. Chem.* **1979**, 91, 439-440.
23. Leigh, G. J., Rahman, M. T., Walton, D. R. M. Carbon-carbon triple bond fission in the homogeneous catalysis of acetylene metathesis. *J. Chem. Soc., Chem. Commun.* **1982**, 541-542.
24. Freudenberger, J. H., Schrock, R. R., Churchill, M. R., Rheingold, A. L., Ziller, J. W. Metathesis of acetylenes by (fluoroalkoxy)tungstenacyclobutadiene complexes and the crystal structure of $W(C_3Et_3)[OCH(CF_3)_2]_3$. A higher order mechanism for acetylene metathesis. *Organometallics* **1984**, 3, 1563-1573.
25. Vosloo, H. C. M., du Plessis, J. A. K. Influence of phenolic compounds on the $Mo(CO)_6$ catalyzed metathetical reactions of alkynes. *J. Mol. Catal. A: Chemical* **1998**, 133, 205-211.
26. Haskel, A., Straub, T., Dash, A. K., Eisen, M. S. Oligomerization and Cross-Oligomerization of Terminal Alkynes Catalyzed by Organoactinide Complexes. *J. Am. Chem. Soc.* **1999**, 121, 3014-3024.
27. Brizius, G., Bunz, U. H. F. Increased Activity of in Situ Catalysts for Alkyne Metathesis. *Org. Lett.* **2002**, 4, 2829-2831.
28. Fürstner, A., Stelzer, F., Rumbo, A., Krause, H. Total synthesis of the turrianes and evaluation of their DNA-cleaving properties. *Chem.--Eur. J.* **2002**, 8, 1856-1871.

1. Amadori, M. Products of condensation between glucose and p-phenetidine. I. *Atti. accad. Lincci [6]* **1925**, 2, 337-342.
2. Amadori, M. The condensation product of glucose and p-anisidine. *Atti. accad. Lincci [6]* **1929**, 9, 226-230.
3. Hodge, J. E. Amadori rearrangement. in *Advances in Carbohydrate Chem. 10*, 169-205 (Academic Press Inc., New York, N.Y., **1955**).
4. Lemieux, R. U. Rearrangements and isomerizations in carbohydrate chemistry. *Mol. Rearrangements (Paul de Mayo, editor. Interscience)* **1964**, 2, 709-769.
5. Maruoka, K., Yamamoto, H. Functional group transformations via carbonyl group derivatives. in *Comp. Org. Synth.* (eds. Trost, B. M.,Fleming, I.), *6*, 789-791 (Pergamon, Oxford, **1991**).
6. Yaylayan, V. A., Huyghues-Despointes, A. Chemistry of Amadori rearrangement products: analysis, synthesis, kinetics, reactions, and spectroscopic properties. *Crit. Rev. Food Sci. Nutr.* **1994**, 34, 321-369.
7. Khalifah, R. G., Baynes, J. W., Hudson, B. G. Amadorins: Novel Post-Amadori Inhibitors of Advanced Glycation Reactions. *Biochem. Biophys. Res. Commun.* **1999**, 257, 251-258.
8. Wrodnigg, T. M., Eder, B. The Amadori and Heyns rearrangements: Landmarks in the history of carbohydrate chemistry or unrecognized synthetic opportunities? *Top. Curr. Chem.* **2001**, 215, 115-152.
9. Yaylayan, V. A. Recent advances in the chemistry of Strecker degradation and Amadori rearrangement: Implications to aroma and color formation. *Food Sci. Technol. Res.* **2003**, 9, 1-6.
10. Hollnagel, A., Kroh, L. W. Degradation of Oligosaccharides in Nonenzymatic Browning by Formation of α-Dicarbonyl Compounds via a "Peeling Off" Mechanism. *J. Agric. Food Chem.* **2000**, 48, 6219-6226.
11. Fodor, G., Sachetto, J. P. Mechanism of formation of 3-deoxygluosulose from D-glucose 3-phosphate and from difructosylglycine. *Tetrahedron Lett.* **1968**, 401-403.
12. Nursten, H. E. Key mechanistic problems posed by the Maillard reaction. *Maillard React. Food Process., Hum. Nutr. Physiol., [Proc. Int. Symp. Maillard React.], 4th* **1990**, 145-153.
13. Azema, L., Bringaud, F., Blonski, C., Perie, J. Chemical and enzymatic synthesis of fructose analogues as probes for import studies by the hexose transporter in parasites. *Bioorg. Med. Chem.* **2000**, 8, 717-722.
14. Guzi, T. J., Macdonald, T. L. A novel synthesis of piperidin-3-ones via an intramolecular Amadori-type reaction. *Tetrahedron Lett.* **1996**, 37, 2939-2942.
15. Horvat, S., Roscic, M., Varga-Defterdarovic, L., Horvat, J. Intramolecular rearrangement of the monosaccharide esters of an opioid pentapeptide: formation and identification of novel Amadori compounds related to fructose and tagatose. *J. Chem. Soc., Perkin Trans. 1* **1998**, 909-914.

1. Michaelis, A., Becker, T. The structure of phosphorous acid. *Chem. Ber.* **1897**, 30, 1003-1009.
2. Michaelis, A., Kaehne, R. The reaction of alkyl iodides with phosphites. *Chem. Ber.* **1898**, 31, 1048-1055.
3. Arbuzov, A. *J. Russ. Phys. Chem. Soc.* **1906**, 38, 687.
4. Arbuzov, A. *J. Russ. Phys. Chem. Soc.* **1910**, 42, 395.
5. Kosolapoff, G. M. Synthesis of phosphonic and phosphinic acids. *Org. React.* **1951**, 6, 273-338.
6. Freedman, L. D., Doak, G. O. The preparation and properties of phosphonic acids. *Chem. Rev.* **1957**, 57, 479-523.
7. Arbuzov, B. A. Michaelis-Arbuzov and Perkov reactions. *Pure Appl. Chem.* **1964**, 9, 307-335.
8. Marquarding, D., Ramirez, F., Ugi, I., Gillespie, P. Chemistry and logical structures. 5. Exchange reactions of phosphorus(V) compounds and their pentacoordinated intermediates. *Angew. Chem.* **1973**, 85, 99-127.
9. Bhattacharya, A. K., Thyagarajan, G. Michaelis-Arbuzov rearrangement. *Chem. Rev.* **1981**, 81, 415-430.
10. Brill, T. B., Landon, S. J. Arbuzov-like dealkylation reactions of transition-metal-phosphite complexes. *Chem. Rev.* **1984**, 84, 577-585.
11. Borowitz, G. B., Borowitz, I. J. The Perkow and related reactions. *Handb. Organophosphorus Chem.* **1992**, 115-172.
12. Waschbusch, R., Carran, J., Marinetti, A., Savignac, P. The synthesis of dialkyl α-halogenated methylphosphonates. *Synthesis* **1997**, 727-743.
13. Abalonin, B. E. Correlation of the forward and reverse Arbuzov reactions with similar transformations of derivatives of p elements with variable valence. *Russ. J. Gen. Chem. (Translation of Zhurnal Obshchei Khimii)* **1999**, 69, 26-31.
14. Iorga, B., Eymery, F., Carmichael, D., Savignac, P. Dialkyl 1-alkynylphosphonates: a range of promising reagents. *Eur. J. Org. Chem.* **2000**, 3103-3115.
15. Winum, J.-Y., Kamal, M., Agnaniet, H., Leydet, A., Montero, J.-L. Study of the Michaelis-Arbuzov reaction during ultrasonic activation. *Phosphorus, Sulfur Silicon Relat. Elem.* **1997**, 129, 83-88.
16. Kolodyazhnyi, O. I., Neda, E. V., Neda, I., Schmutzler, R. Asymmetric induction in the Arbuzov reaction. *Russ. J. Gen. Chem. (Translation of Zhurnal Obshchei Khimii)* **1998**, 68, 1159-1160.
17. Villemin, D., Simeon, F., Decreus, H., Jaffres, P.-A. Rapid and efficient Arbuzov reaction under microwave irradiation. *Phosphorus, Sulfur Silicon Relat. Elem.* **1998**, 133, 209-213.
18. Cherkasov, R. A., Polezhaeva, N. A., Galkin, V. I. Arbuzov reaction in the series of halogenocyclenes: new synthetical and mechanistical variants. *Phosphorus, Sulfur Silicon Relat. Elem.* **1999**, 144-146, 333-336.
19. Pernak, J., Kmiecik, R., Weglewski, J. Reaction of phenolic Mannich base with trialkyl phosphite. *Synth. Commun.* **2000**, 30, 1535-1541.
20. Bhanthumnavin, W., Bentrude, W. G. Photo-Arbuzov Rearrangements of 1-Arylethyl Phosphites: Stereochemical Studies and the Question of Radical-Pair Intermediates. *J. Org. Chem.* **2001**, 66, 980-990.
21. Renard, P.-Y., Vayron, P., Leclerc, E., Valleix, A., Mioskowski, C. Lewis acid catalyzed room-temperature Michaelis-Arbuzov rearrangement. *Angew. Chem., Int. Ed. Engl.* **2003**, 42, 2389-2392.
22. Renard, P.-Y., Vayron, P., Mioskowski, C. Trimethylsilyl Halide-Promoted Michaelis-Arbuzov Rearrangement. *Org. Lett.* **2003**, 5, 1661-1664.
23. Garner, A. Y., Chapin, E. C., Scanlon, P. M. Mechanism of the Michaelis-Arbuzov reaction: olefin formation. *J. Org. Chem.* **1959**, 24, 532-536.
24. Harwood, H. J., Grisley, D. W., Jr. The unexpected course of several Arbuzov-Michaelis reactions; an example of the nucleophilicity of the phosphoryl group. *J. Am. Chem. Soc.* **1960**, 82, 423-426.
25. Aksnes, G., Aksnes, D. Mechanism of the Michaelis-Arbuzov rearrangement in aceto- nitrile. *Acta Chem. Scand.* **1963**, 17, 2121-2122.
26. Benschop, H. P., Van den Berg, G. R., Platenburg, D. H. J. M. Stereochemistry of a Michaelis-Arbuzov reaction. Alkylation of optically active ethyl trimethylsilyl phenylphosphonite with retention of configuration. *J. Chem. Soc. D.* **1971**, 606-607.
27. Clemens, J., Neukomm, H., Werner, H. Reactivity of metal π-complexes. 14. Preparation and formation mechanisms of π-cyclopentadienylnickel (tert-phosphite) dialkylphosphonate complexes, an organometallic variant of the Michaelis-Arbuzov reaction. *Helv. Chim. Acta* **1974**, 57, 2000-2010.
28. Balthazor, T. M., Grabiak, R. C. Nickel-catalyzed Arbuzov reaction: mechanistic observations. *J. Org. Chem.* **1980**, 45, 5425-5426.
29. Hudson, H. R., Kow, A., Roberts, J. C. Quasiphosphonium intermediates. Part 3. Preparation, structure, and reactivity of alkoxyphosphonium halides in the reactions of neopentyl diphenylphosphinite, dineopentyl phenylphosphonite, and trineopentyl phosphite with halomethanes and the effect of phenoxy-substituents on the mechanism of alkyl-oxygen fission in Michaelis-Arbuzov reactions. *J. Chem. Soc., Perkin Trans. 2* **1983**, 1363-1368.
30. Bao, Q. B., Brill, T. B. Methyl-group transfer involving transition-metal complexes by the Michaelis-Arbuzov mechanism. *Organometallics* **1987**, 6, 2588-2589.
31. Mugrage, B., Diefenbacher, C., Somers, J., Parker, D. T., Parker, T. Phosphonic acid analogs of diclofenac: an Arbuzov reaction of trimethyl phosphite with an ortho-quinonoid intermediate. *Tetrahedron Lett.* **2000**, 41, 2047-2050.

32. Bhattacharya, A. K., Stolz, F., Schmidt, R. R. Design and synthesis of aryl/hetarylmethyl phosphonate-UMP derivatives as potential glucosyltransferase inhibitors. *Tetrahedron Lett.* **2001**, 42, 5393-5395.
33. Hansen, M. M., Bertsch, C. F., Harkness, A. R., Huff, B. E., Hutchison, D. R., Khau, V. V., LeTourneau, M. E., Martinelli, M. J., Misner, J. W., Peterson, B. C., Rieck, J. A., Sullivan, K. A., Wright, I. G. An Enantioselective Synthesis of Cis Perhydroisoquinoline LY235959. *J. Org. Chem.* **1998**, 63, 775-785.

Arndt-Eistert Homologation/Synthesis 18

Related reactions: Wolff rearrangement;

1. Arndt, F., Eistert, B. A method for conversion of carboxylic acids to higher homologs or their derivatives. *Ber.* **1935**, 68B, 200-208.
2. Bachmann, W. E., Struve, W. S. Arndt-Eistert synthesis. *Org. React.* **1942**, 1, 38-62.
3. Matthews, J. L., Braun, C., Guibourdenche, C., Overhand, M., Seebach, D. Preparation of enantiopure β-amino acids from α-amino acids using the Arndt-Eistert homologation. *Enantiosel. Synth. β-Amino Acids* **1997**, 105-126.
4. Kirmse, W. 100 years of the Wolff rearrangement. *Eur. J. Org. Chem.* **2002**, 2193-2256.
5. Winum, J.-Y., Kamal, M., Leydet, A., Roque, J.-P., Montero, J.-L. Homologation of carboxylic acids by Arndt-Eistert reaction under ultrasonic waves. *Tetrahedron Lett.* **1996**, 37, 1781-1782.
6. Katritzky, A. R., Zhang, S., Fang, Y. $BtCH_2TMS$-Assisted Homologation of Carboxylic Acids: A Safe Alternative to the Arndt-Eistert Reaction. *Org. Lett.* **2000**, 2, 3789-3791.
7. Katritzky, A. R., Zhang, S., Mostafa Hussein, A. H., Fang, Y., Steel, P. J. One-Carbon Homologation of Carboxylic Acids via $BtCH_2TMS$: A Safe Alternative to the Arndt-Eistert Reaction. *J. Org. Chem.* **2001**, 66, 5606-5612.
8. Vasanthakumar, G.-R., Patil, B. S., Suresh Babu, V. V. Homologation of α-amino acids to β-amino acids using Boc_2O. *J. Chem. Soc., Perkin Trans. 1* **2002**, 2087-2089.
9. Vasanthakumar, G. R., Babu, V. V. S. Simple and stereospecific homologation of urethane-protected α-amino acids to their higher homologs using HBTU. *J. Pept. Res.* **2003**, 61, 230-236.
10. Vasanthakumar, G. R., Babu, V. V. S. Synthesis of Fmoc-/Boc-/Z-β-amino acids via Arndt-Eistert homologation of Fmoc-/Boc-/Z-α-amino acids employing BOP and PyBOP. *Indian J. Chem., Sect. B* **2003**, 42B, 1691-1695.
11. Huggett, C., Arnold, R. T., Taylor, T. I. Mechanism of the Arndt-Eistert reaction. *J. Am. Chem. Soc.* **1942**, 64, 3043.
12. Gademann, K., Ernst, M., Hoyer, D., Seebach, D. Synthesis and biological evaluation of a cyclo-β-tetrapeptide as a somatostatin analog. *Angew. Chem., Int. Ed. Engl.* **1999**, 38, 1223-1226.
13. Nicolaou, K. C., Baran, P. S., Zhong, Y.-L., Choi, H.-S., Yoon, W. H., He, Y., Fong, K. C. Total synthesis of the CP molecules CP-263,114 and CP-225,917-part 1: synthesis of key intermediates and intelligence gathering. *Angew. Chem., Int. Ed. Engl.* **1999**, 38, 1669-1675.
14. Ancliff, R. A., Russell, A. T., Sanderson, A. J. Resolution of a citric acid derivative: synthesis of (R)-(-)-homocitric acid-γ-lactone. *Tetrahedron: Asymmetry* **1997**, 8, 3379-3382.
15. Garg, N. K., Sarpong, R., Stoltz, B. M. The First Total Synthesis of Dragmacidin D. *J. Am. Chem. Soc.* **2002**, 124, 13179-13184.

Aza-Claisen Rearrangement (3-Aza-Cope Rearrangement) 20

Related reactions: Aza-Cope rearrangement, Overman rearrangement;

1. Hill, R. K., Gilman, N. W. Nitrogen analog of the Claisen rearrangement. *Tetrahedron Lett.* **1967**, 1421-1423.
2. Ito, S., Tsunoda, T. Application of the aza-Claisen rearrangement to the total synthesis of natural products: (-)-isoiridomyrmecin. *Pure Appl. Chem.* **1994**, 66, 2071-2074.
3. Majumdar, K. C., Bhattacharyya, T. Aza-Claisen rearrangement. *J. Indian Chem. Soc.* **2002**, 79, 112-121.
4. Hill, R. K., Khatri, H. N. Titanium tetrachloride catalysis of aza-Claisen rearrangements. *Tetrahedron Lett.* **1978**, 4337-4340.
5. Padwa, A., Cohen, L. A. Aza-Claisen rearrangements in the 2-allyloxy substituted oxazole system. *Tetrahedron Lett.* **1982**, 23, 915-918.
6. Murahashi, S., Makabe, Y., Kunita, K. Palladium$^{(0)}$-catalyzed rearrangement of *N*-allyl enamines. Synthesis of δ,ε-unsaturated imines and γ,δ-unsaturated carbonyl compounds. *J. Org. Chem.* **1988**, 53, 4489-4495.
7. Welch, J. T., De Corte, B., De Kimpe, N. Regioselective aza-Cope rearrangement of α-halogenated and nonhalogenated imines. *J. Org. Chem.* **1990**, 55, 4981-4983.
8. Cook, G. R., Stille, J. R. Stereochemical consequences of the Lewis acid-promoted 3-aza-Cope rearrangement of N-alkyl-N-allyl enamines. *Tetrahedron* **1994**, 50, 4105-4124.
9. Wang, M.-X., Huang, Z.-T. Regiospecific Allylation of Benzoyl-Substituted Heterocyclic Ketene Aminals and Their Zinc Chloride-Promoted 3-Aza-Cope Rearrangement. *J. Org. Chem.* **1995**, 60, 2807-2811.
10. McComsey, D. F., Maryanoff, B. E. 3-Aza-Cope Rearrangement of Quaternary N-Allyl Enammonium Salts. Stereospecific 1,3 Allyl Migration from Nitrogen to Carbon on a Tricyclic Template. *J. Org. Chem.* **2000**, 65, 4938-4943.
11. Gomes, M. J. S., Sharma, L., Prabhakar, S., Lobo, A. M., Gloria, P. M. C. Studies in 3-oxy-assisted 3-aza Cope rearrangements. *Chem. Commun.* **2002**, 746-747.
12. Winter, R. F., Rauhut, G. Computational studies on 3-aza-Cope rearrangements: protonation-induced switch of mechanism in the reaction of vinylpropargylamine. *Chem.-- Eur. J.* **2002**, 8, 641-649.
13. Claisen, L. Rearrangement of Phenol Allyl Ethers into C-Allylphenols. *Ber.* **1912**, 45, 3157-3166.
14. Claisen, L., Eisleb, O. Rearrangement of phenol allyl ethers into the isomeric allylphenols. *Ann.* **1914**, 401, 21-119.
15. Suh, Y.-G., Kim, S.-A., Jung, J.-K., Shin, D.-Y., Min, K.-H., Koo, B.-A., Kim, H.-S. Asymmetric total synthesis of fluvirucinine A1. *Angew. Chem., Int. Ed. Engl.* **1999**, 38, 3545-3547.
16. Tsunoda, T., Nishii, T., Yoshizuka, M., Yamasaki, C., Suzuki, T., Ito, S. Total synthesis of (-)-antimycin A3b. *Tetrahedron Lett.* **2000**, 41, 7667-7671.
17. Nubbemeyer, U. Diastereoselective Zwitterionic Aza-Claisen Rearrangement: The Synthesis of Bicyclic Tetrahydrofurans and a Total Synthesis of (+)-Dihydrocanadensolide. *J. Org. Chem.* **1996**, 61, 3677-3686.

Aza-Cope Rearrangement 22

Related reactions: Aza-Claisen rearrangement;

1. Oehlschlager, A. C., Zalkow, L. H. Bridged ring compounds. X. The reaction of benzenesulfonyl azide with norbornadiene, dicyclopentadiene, and bicyclo-[2.2.2]oct-2-ene. *J. Org. Chem.* 1965, 30, 4205-4211.
2. Hill, R. K., Gilman, N. W. Nitrogen analog of the Claisen rearrangement. *Tetrahedron Lett.* 1967, 1421-1423.
3. Lipkowitz, K. B., Scarpone, S., McCullough, D., Barney, C. The synthesis of N-substituted tetrahydropyridines using the hetero-Cope rearrangement. *Tetrahedron Lett.* 1979, 2241-2244.
4. Przheval'skii, N. M., Grandberg, I. I. The aza-Cope rearrangement in organic synthesis. *Usp. Khim.* 1987, 56, 814-843.
5. Allin, S. M., Baird, R. D. Development and synthetic applications of asymmetric [3,3]-sigmatropic rearrangements. *Curr. Org. Chem.* 2001, 5, 395-415.

6. Nakamura, H., Yamamoto, Y. Rearrangement reactions catalyzed by palladium: palladium-catalyzed carbon skeletal rearrangements: Cope, Claisen, and other [3,3] rearrangements. in *Handbook of Organopalladium Chemistry for Organic Synthesis* (eds. Negishi, E.-i.,De Meijere, A.), 2, 2919-2934 (Wiley-Interscience, New York, 2002).
7. Voegtle, F., Goldschmitt, E. Dynamic stereochemistry of degenerate diaza-Cope rearrangement. *Angew. Chem.* 1974, 86, 520-521.
8. Voegtle, F., Goldschmitt, E. The diaza-Cope rearrangement. *Chem. Ber.* 1976, 109, 1-40.
9. Ent, H., De Koning, H., Speckamp, W. N. The 2-aza-Cope N-acyliminium cyclization. *Tetrahedron Lett.* 1983, 24, 2109-2112.
10. Ent, H., De Koning, H., Speckamp, W. N. N-Acyliminium cyclizations via reversible 2-aza-Cope rearrangements. *Tetrahedron Lett.* 1985, 26, 5105-5108.
11. Ent, H., De Koning, H., Speckamp, W. N. 2-Azonia-Cope rearrangement in N-acyliminium cyclizations. *J. Org. Chem.* 1986, 51, 1687-1691.
12. Beck, K., Burghard, H., Fischer, G., Huenig, S., Reinold, P. Aza-Cope rearrangement with unstabilized azo compounds. *Angew. Chem.* 1987, 99, 695-697.
13. Kawashima, T., Kihara, T., Inamoto, N. The azaphospha-Cope rearrangement of 2-aza-3-phospha-1,5-hexadiene derivatives. *Chem. Lett.* 1988, 577-580.
14. Wu, P. L., Chu, M., Fowler, F. W. The 1-aza-Cope rearrangement. *J. Org. Chem.* 1988, 53, 963-972.
15. Wu, P. L., Fowler, F. W. The 1-aza-Cope rearrangement. 2. *J. Org. Chem.* 1988, 53, 5998-6005.
16. Rousselle, D., Musick, C., Viehe, H. G., Tinant, B., Declercq, J. P. Tris-aza-Cope rearrangement of bicyclic N-cyano-N'-vinyl or N'-arylhydrazines to imidazolodiazepine derivatives. *Tetrahedron Lett.* 1991, 32, 907-910.
17. Barta, N. S., Cook, G. R., Landis, M. S., Stille, J. R. Studies of the regiospecific 3-aza-Cope rearrangement promoted by electrophilic reagents. *J. Org. Chem.* 1992, 57, 7188-7194.
18. Walters, M. A. The anionic 1-aza-Cope rearrangement. Theoretical evidence for the intramolecular reaction of imide anions with alkenes. *Tetrahedron Lett.* 1995, 36, 7055-7056.
19. Deur, C., Miller, M., Hegedus, L. S. Photochemical Reaction between Tertiary Allylic Amines and Chromium Carbene Complexes: Synthesis of Lactams via a Zwitterion Aza Cope Rearrangement. *J. Org. Chem.* 1996, 61, 2871-2876.
20. Sreekumar, R., Padmakumar, R. Aromatic 3-aza-Cope rearrangement over zeolites. *Tetrahedron Lett.* 1996, 37, 5281-5282.
21. Walters, M. A. Ab Initio Investigation of the 3-Aza-Cope Reaction. *J. Org. Chem.* 1996, 61, 978-983.
22. Ryckmans, T., Schulte, K., Viehe, H.-G. The methylation of N-cyano N-methyl hydrazones: a new access to 2-aminoimidazoles through an in situ 1,3,4-triaza Cope rearrangement. *Bull. Soc. Chim. Belg.* 1997, 106, 553-557.
23. Winter, R. F., Hornung, F. M. The Aza-Cope Rearrangement in Transition Metal Complexes. Construction of an Unsaturated C7-Ligand from Butadiyne and an Allylic Amine. *Organometallics* 1997, 16, 4248-4250.
24. Mustafin, A. G., Gimadieva, A. R., Tambovtsev, K. A., Tolstikov, G. A., Abdrakhmanov, I. B. $SnCl_4$-catalyzed Claisen and Cope rearrangements of *N*-allylanilines and *N*-allylenamines. *Russ. J. Org. Chem.* 1998, 34, 90-92.
25. Muller, P., Toujas, J.-L., Bernardinelli, G. A stereospecific "2-Aza-divinylcyclopropane" rearrangement. *Helv. Chim. Acta* 2000, 83, 1525-1534.
26. Yadav, J. S., Subba Reddy, B. V., Abdul Rasheed, M., Sampath Kumar, H. M. Zn^{2+} montmorillonite catalyzed 3-aza-Cope rearrangement under microwave irradiation. *Synlett* 2000, 487-488.
27. Allin, S. M., Baird, R. D., Lins, R. J. Synthetic applications of the amino-Cope rearrangement: enantioselective synthesis of some tetrahydropyrans. *Tetrahedron Lett.* 2002, 43, 4195-4197.
28. Gomes, M. J. S., Sharma, L., Prabhakar, S., Lobo, A. M., Gloria, P. M. C. Studies in 3-oxy-assisted 3-aza Cope rearrangements. *Chem. Commun.* 2002, 746-747.
29. Winter, R. F., Rauhut, G. Computational studies on 3-aza-Cope rearrangements: protonation-induced switch of mechanism in the reaction of vinylpropargylamine. *Chem.-- Eur. J.* 2002, 8, 641-649.
30. Marshall, J. A., Babler, J. H. Heterolytic fragmentation of 1-substituted decahydroquinolines. *J. Org. Chem.* 1969, 34, 4186-4188.
31. Hart, D. J., Tsai, Y.-M. N-Acyliminium ions: detection of a hidden 2-aza-Cope rearrangement. *Tetrahedron Lett.* 1981, 22, 1567-1570.
32. Castelhano, A. L., Krantz, A. Allenic amino acids. 1. Synthesis of γ-allenic GABA by a novel aza-Cope rearrangement. *J. Am. Chem. Soc.* 1984, 106, 1877-1879.
33. Chu, M., Wu, P. L., Givre, S., Fowler, F. W. The 1-Aza-Cope rearrangement. *Tetrahedron Lett.* 1986, 27, 461-464.
34. Jacobsen, E. J., Levin, J., Overman, L. E. Synthesis applications of cationic aza-Cope rearrangements. Part 18. Scope and mechanism of tandem cationic aza-Cope rearrangement-Mannich cyclization reactions. *J. Am. Chem. Soc.* 1988, 110, 4329-4336.
35. Welch, J. T., De Corte, B., De Kimpe, N. Regioselective aza-Cope rearrangement of α-halogenated and nonhalogenated imines. *J. Org. Chem.* 1990, 55, 4981-4983.
36. Wang, M.-X., Huang, Z.-T. Regiospecific Allylation of Benzoyl-Substituted Heterocyclic Ketene Aminals and Their Zinc Chloride-Promoted 3-Aza-Cope Rearrangement. *J. Org. Chem.* 1995, 60, 2807-2811.
37. Obrecht, D., Zumbrunn, C., Mueller, K. Formal [3+2] Cycloaddition Reaction of [1,4]Oxazin-2-ones and α-Alkynyl Ketones via a Tandem Mukaiyama-Aldol Addition/Aza-Cope Rearrangement. *J. Org. Chem.* 1999, 64, 6891-6895.
38. McComsey, D. F., Maryanoff, B. E. 3-Aza-Cope Rearrangement of Quaternary N-Allyl Enammonium Salts. Stereospecific 1,3 Allyl Migration from Nitrogen to Carbon on a Tricyclic Template. *J. Org. Chem.* 2000, 65, 4938-4943.
39. Brummond, K. M., Lu, J. Tandem Cationic aza-Cope Rearrangement-Mannich Cyclization Approach to the Core Structure of FR901483 via a Bridgehead Iminium Ion. *Org. Lett.* 2001, 3, 1347-1349.
40. Bennett, D. J., Hamilton, N. M. A facile synthesis of N-benzylallylglycine. *Tetrahedron Lett.* 2000, 41, 7961-7964.
41. Madin, A., O'Donnell, C. J., Oh, T., Old, D. W., Overman, L. E., Sharpe, M. J. Total Synthesis of (±)-Gelsemine. *Angew. Chem., Int. Ed. Engl.* 1999, 38, 2934-2936.
42. Knight, S. D., Overman, L. E., Pairaudeau, G. Asymmetric Total Syntheses of (-)- and (+)-Strychnine and the Wieland-Gumlich Aldehyde. *J. Am. Chem. Soc.* 1995, 117, 5776-5788.

1. Staudinger, H., Meyer, J. New organic compounds of phosphorus. III. Phosphinemethylene derivatives and phosphinimines. *Helv. Chim. Acta* **1919**, 2, 635-646.
2. Wittig, G. Staudinger and the history of organophosphorus-carbonyl olefination. *Pure Appl. Chem.* **1964**, 9, 245-254.
3. Eguchi, S., Matsushita, Y., Yamashita, K. The aza-Wittig reaction in heterocyclic synthesis. A review. *Org. Prep. Proced. Int.* **1992**, 24, 209-243.
4. Johnson, A. W. *Ylides and Imines of Phosphorous* (Wiley, New York, **1993**).
5. Molina, P., Alajarin, M., Lopez-Leonardo, C., Elguero, J. Four-membered heterocyclic rings from iminophosphoranes. Preparation and reactivity of 2,4-dimino-1,3-diazetidines and related compounds. *J. Prakt. Chem./Chem.-Ztg.* **1993**, 335, 305-315.
6. Nitta, M. Reaction of (vinylimino)phosphoranes and related compounds. Novel synthesis of nitrogen heterocycles. *Rev. Heteroat. Chem.* **1993**, 9, 87-121.
7. Molina, P., Vilaplana, M. J. Iminophosphoranes: useful building blocks for the preparation of nitrogen-containing heterocycles. *Synthesis* **1994**, 1197-1218.
8. Eguchi, S., Okano, T., Okawa, T. Synthesis of heterocyclic natural products and related heterocycles by the aza-Wittig methodology. *Rec. Res. Dev. Org. Chem.* **1997**, 1, 337-346.
9. Shah, S., Protasiewicz, J. D. "Phospha-variations" on the themes of Staudinger and Wittig: phosphorus analogs of Wittig reagents. *Coord. Chem. Rev.* **2000**, 210, 181-201.
10. Arques, A., Molina, P. Bis(iminophosphoranes) as useful building blocks for the preparation of complex polyaza ring systems. *Curr. Org. Chem.* **2004**, 8, 827-843.
11. Fresneda, P. M., Molina, P. Application of iminophosphorane-based methodologies for the synthesis of natural products. *Synlett* **2004**, 1-17.

12. Rzepa, H. S., Molina, P., Alajarin, M., Vidal, A. An AM1 and PM3 molecular orbital study of the pericyclic reactivity of aryl carbodiimides. *Tetrahedron* **1992**, 48, 7425-7434.
13. Koketsu, J., Ninomiya, Y., Suzuki, Y., Koga, N. Theoretical Study on the Structures of Iminopnictoranes and Their Reactions with Formaldehyde. *Inorg. Chem.* **1997**, 36, 694-702.
14. Lu, W. C., Sun, C. C., Zang, Q. J., Liu, C. B. Theoretical study of the aza-Wittig reaction X_3P=NH + O=CHCOOH -> X_3P=O + HN=CHCOOH (X=Cl, H and CH_3). *Chem. Phys. Lett.* **1999**, 311, 491-498.
15. Xue, Y., Xie, D., Yan, G. Theoretical Study of the aza-Wittig Reactions of X_3P:NH (X=H and Cl) with Formaldehyde in Gas Phase and in Solution. *J. Phys. Chem. A* **2002**, 106, 9053-9058.
16. Lu, W. C., Zhang, R. Q., Zang, Q. J., Wong, N. B. Theoretical Prediction on Efficient Formation of Imino Acid via an Aza-Wittig Reaction. *J. Phys. Chem. B* **2003**, 107, 2061-2067.
17. Xue, Y., Kim, C. K. Effects of Substituents and Solvents on the Reactions of Iminophosphorane with Formaldehyde: Ab Initio MO Calculation and Monte Carlo Simulation. *J. Phys. Chem. A* **2003**, 107, 7945-7951.
18. Kano, N., Xing, J.-H., Kawa, S., Kawashima, T. Cycloaddition reactions of an iminophosphorane bearing the Martin ligand with some double-bond compounds: syntheses, structures and thermolyses of a 1,3,2λ 5-oxazaphosphetidine and a 1,3,2λ 5-diazaphosphetidine-4-thione. *Polyhedron* **2002**, 21, 657-665.
19. Drewry, D. H., Gerritz, S. W., Linn, J. A. Solid-phase synthesis of trisubstituted guanidines. *Tetrahedron Lett.* **1997**, 38, 3377-3380.
20. Williams, D. R., Fromhold, M. G., Earley, J. D. Total Synthesis of (-)-Stemospironine. *Org. Lett.* **2001**, 3, 2721-2724.
21. Sugimori, T., Okawa, T., Eguchi, S., Kakehi, A., Yashima, E., Okamoto, Y. The first total synthesis of (-)-benzomalvin A and benzomalvin B via the intramolecular aza-Wittig reactions. *Tetrahedron* **1998**, 54, 7997-8008.
22. Neubert, B. J., Snider, B. B. Synthesis of (±)-Phloeodictine A1. *Org. Lett.* **2003**, 5, 765-768.

Aza-[2,3]-Wittig Rearrangement 26

1. Eisch, J. J., Kovacs, C. A. The π-orbital overlap requirement in 1,2-anionic rearrangements. *J. Organomet. Chem.* **1971**, 30, C97-C100.
2. Durst, T., Van den Elzen, R., LeBelle, M. J. Base-induced ring enlargements of 1-benzyl- and 1-allyl-2-azetidinones. *J. Am. Chem. Soc.* **1972**, 94, 9261-9263.
3. Vogel, C. The aza-Wittig rearrangement. *Synthesis* **1997**, 497-505.
4. Aahman, J., Somfai, P. Aza-[2,3]-Wittig Rearrangements of Vinylaziridines. *J. Am. Chem. Soc.* **1994**, 116, 9781-9782.
5. Anderson, J. C., Siddons, D. C., Smith, S. C., Swarbrick, M. E. Aza-[2,3]-Wittig sigmatropic rearrangement of crotyl amines. *J. Chem. Soc., Chem. Commun.* **1995**, 1835-1836.
6. Coldham, I., Collis, A. J., Mould, R. J., Rathmell, R. E. Ring expansion of aziridines to piperidines using the aza-Wittig rearrangement. *Tetrahedron Lett.* **1995**, 36, 3557-3560.
7. Aahman, J., Jarevaang, T., Somfai, P. Synthesis and Aza-[2,3]-Wittig Rearrangements of Vinylaziridines: Scope and Limitations. *J. Org. Chem.* **1996**, 61, 8148-8159.
8. Aahman, J., Somfai, P. A novel rearrangement of N-propargyl vinylaziridines. Mechanistic diversity in the aza-[2,3]-Wittig rearrangement. *Tetrahedron Lett.* **1996**, 37, 2495-2498.
9. Anderson, J. C., Siddons, D. C., Smith, S. C., Swarbrick, M. E. The Silicon-Assisted Aza-[2,3]-Wittig Sigmatropic Rearrangement. *J. Org. Chem.* **1996**, 61, 4820-4823.
10. Anderson, J. C., Smith, S. C., Swarbrick, M. E. Diastereoselective acyclic aza-[2,3] Wittig sigmatropic rearrangements. *J. Chem. Soc., Perkin Trans. 1* **1997**, 1517-1521.
11. Kawachi, A., Doi, N., Tamao, K. The Sila-Wittig Rearrangement. *J. Am. Chem. Soc.* **1997**, 119, 233-234.
12. Anderson, J. C., Dupau, P., Siddons, D. C., Smith, S. C., Swarbrick, M. E. The aza-[2,3]-Wittig sigmatropic rearrangement of Z(C)-alkenes. *Tetrahedron Lett.* **1998**, 39, 2649-2650.
13. Anderson, J. C., Roberts, C. A. The tri-n-butyltin group as a novel stereocontrol element and synthetic handle in the aza-[2,3]-Wittig sigmatropic rearrangement. *Tetrahedron Lett.* **1998**, 39, 159-162.
14. Anderson, J. C., Flaherty, A., Swarbrick, M. E. The Aza-[2,3]-Wittig Sigmatropic Rearrangement of Acyclic Amines: Scope and Limitations of Silicon Assistance. *J. Org. Chem.* **2000**, 65, 9152-9156.
15. Haeffner, F., Houk, K. N., Schulze, S. M., Lee, J. K. Concerted Rearrangement versus Heterolytic Cleavage in Anionic [2,3]- and [3,3]-Sigmatropic Shifts. A DFT Study of Relationships among Anion Stabilities, Mechanisms, and Rates. *J. Org. Chem.* **2003**, 68, 2310-2316.
16. Wittig, G., Lohmann, L. Cationotropic isomerization of benzyl ethers by lithium phenyl. *Ann.* **1942**, 550, 260-268.
17. Gawley, R. E., Zhang, Q., Campagna, S. Stereochemical Course of [2,3] Anionic and Ylide Rearrangements of Unstabilized α-Aminoorganolithiums. *J. Am. Chem. Soc.* **1995**, 117, 11817-11818.
18. Brückner, R. [2,3]-Sigmatropic rearrangements. in *Comp. Org. Synth.* (eds. Trost, B. M.,Fleming, I.), *6*, 873-909 (Pergamon, Oxford, **1991**).
19. Markó, I. E. The Stevens and related rearrangements. in *Comp. Org. Synth.* (eds. Trost, B. M.,Fleming, I.), *3*, 913-975 (Pergamon, Oxford, **1991**).
20. Marshall, J. A. The Wittig rearrangement. in *Comp. Org. Synth.* (eds. Trost, B. M.,Fleming, I.), *3*, 975-1015 (Pergamon, Oxford, **1991**).
21. Nakai, T., Mikami, K. The [2,3]-Wittig rearrangement. *Org. React.* **1994**, 46, 105-209.
22. Coldham, I. One or more CH and/or CC bond(s) formed by rearrangement. in *Comp. Org. Funct. Group Trans.* (eds. Katritzky, A. R., Meth-Cohn, O.,Rees, C. W.), *1*, 377-423 (Pergamon, Oxford, **1995**).
23. Anderson, J. C., Whiting, M. Total Synthesis of (±)-Kainic Acid with an Aza-[2,3]-Wittig Sigmatropic Rearrangement as the Key Stereochemical Determining Step. *J. Org. Chem.* **2003**, 68, 6160-6163.
24. Coldham, I., Collis, A. J., Mould, R. J., Rathmell, R. E. Ring expansion of aziridines to piperidines using the aza-Wittig rearrangement. *Tetrahedron Lett.* **1995**, 36, 3557-3560.
25. Coldham, I., Middleton, M. L., Taylor, P. L. Investigations into the [2,3]-aza-Wittig rearrangement of *N*-alkyl *N*-allyl α-amino esters. *J. Chem. Soc., Perkin Trans. 1* **1998**, 2817-2822.

Baeyer-Villiger Oxidation/Rearrangement 28

Related reactions: *Dakin oxidation;*

1. Baeyer, A., Villiger, V. The effect of Caro's reagent on ketones. *Ber. Dtsch. Chem. Ges.* **1899**, 32, 3625-3633.
2. Hassall, C. H. The Baeyer-Villiger oxidation of aldehydes and ketones. *Org. React.* **1957**, 9, 73-106.
3. Krow, G. R. The Baeyer-Villiger oxidation of ketones and aldehydes. *Org. React.* **1993**, 43, 251-798.
4. Battistel, E., Ricci, M. New tools for the Baeyer-Villiger oxidation of ketones. 2. Enzymic catalysis. *Chim. Ind. (Milan)* **1997**, 79, 1209-1215.
5. Ricci, M., Battistel, E. New tools for the Baeyer-Villiger oxidation of ketones. 1. Phase transfer catalysis. *Chim. Ind. (Milan)* **1997**, 79, 879-882.
6. Bolm, C., Beckmann, O., Luong, T. K. K. *Metal-catalyzed Baeyer-Villiger reactions* (eds. Beller, M.,Bolm, C.) (Wiley-VCH, Weinheim, New York, **1998**) 213-218.
7. Roberts, S. M., Wan, P. W. H. Enzyme-catalyzed Baeyer-Villiger oxidations. *J. Mol. Catal. B: Enzym.* **1998**, 4, 111-136.
8. Stewart, J. D. Cyclohexanone monooxygenase: a useful reagent for asymmetric Baeyer-Villiger reactions. *Curr. Org. Chem.* **1998**, 2, 195-216.
9. Strukul, G. Transition metal catalysis in the Baeyer-Villiger oxidation of ketones. *Angew. Chem., Int. Ed. Engl.* **1998**, 37, 1199-1209.
10. Bolm, C. Metal-catalyzed asymmetric oxidations. *Med. Res. Rev.* **1999**, 19, 348-356.

11. Bolm, C., Beckmann, O. Baeyer-Villiger reaction. in *Comprehensive Asymmetric Catalysis I-III* (eds. Jacobsen, E., Pfaltz, A.,Yamamoto, H.), *2*, 803-810 (Springer, Berlin, New York, **1999**).
12. Kayser, M., Chen, G., Stewart, J. "Designer yeast". An enantioselective oxidizing reagent for organic synthesis. *Synlett* **1999**, 153-158.
13. Renz, M., Meunier, B. 100 years of Baeyer-Villiger oxidations. *Eur. J. Org. Chem.* **1999**, 737-750.
14. Banerjee, A. Stereoselective microbial Baeyer-Villiger oxidations. (*Stereoselective Biocatalysis*). **2000**, 867-876
15. Bolm, C. Enantioselective Baeyer-Villiger reactions and sulfide oxidations. (*Peroxide Chemistry*). **2000**, 494-510
16. Kelly, D. R. Enantioselective Baeyer-Villiger reactions. Part 2. *Chimica Oggi* **2000**, 18, 52-56.
17. Kelly, D. R. Enantioselective Baeyer-Villiger reactions. Part 1. *Chimica Oggi* **2000**, 18, 33-37.
18. Alphand, V., Furstoss, R. Asymmetric Baeyer-Villiger oxidation using biocatalysis. (*Asymmetric Oxidation Reactions*). **2001**, 214-227
19. Bolm, C., Luong, T. K. K., Beckmann, O. Oxidation of carbonyl compounds: asymmetric Baeyer-Villiger oxidation. (*Asymmetric Oxidation Reactions*). **2001**, 147-151
20. Flitsch, S., Grogan, G. Baeyer-Villiger oxidations. (*Enzyme Catalysis in Organic Synthesis (2nd Edition)*). **2002**, 1202-1245
21. Mihovilovic, M. D., Muller, B., Stanetty, P. Monooxygenase-mediated Baeyer-Villiger oxidations. *Eur. J. Org. Chem.* **2002**, 3711-3730.
22. Alphand, V., Carrea, G., Wohlgemuth, R., Furstoss, R., Woodley, J. M. Towards large-scale synthetic applications of Baeyer-Villiger monooxygenases. *Trends Biotechnol.* **2003**, 21, 318-323.
23. Kamerbeek, N. M., Janssen, D. B., van Berkel, W. J. H., Fraaije, M. W. Baeyer-Villiger monooxygenases, an emerging family of flavin-dependent biocatalysts. *Adv. Syn. & Catal.* **2003**, 345, 667-678.
24. Brink, G. J. t., Arends, I. W. C. E., Sheldon, R. A. The Baeyer-Villiger Reaction: New Developments toward Greener Procedures. *Chem. Rev.* **2004**, 104, 4105-4123.
25. Mihovilovic, M. D., Rudroff, F., Groetzl, B. Enantioselective Baeyer-Villiger oxidations. *Curr. Org. Chem.* **2004**, 8, 1057-1069.
26. Camps, F., Coll, J., Messeguer, A., Pericas, M. A. Improved oxidation procedure with aromatic peroxyacids. *Tetrahedron Lett.* **1981**, 22, 3895-3896.
27. Taschner, M. J., Black, D. J. The enzymatic Baeyer-Villiger oxidation: enantioselective synthesis of lactones from mesomeric cyclohexanones. *J. Am. Chem. Soc.* **1988**, 110, 6892-6893.
28. Syper, L. The Baeyer-Villiger oxidation of aromatic aldehydes and ketones with hydrogen peroxide catalyzed by selenium compounds. A convenient method for the preparation of phenols. *Synthesis* **1989**, 167-172.
29. Baures, P. W., Eggleston, D. S., Flisak, J. R., Gombatz, K., Lantos, I., Mendelson, W., Remich, J. J. An efficient asymmetric synthesis of substituted phenyl glycidic esters. *Tetrahedron Lett.* **1990**, 31, 6501-6504.
30. Lopp, M., Paju, A., Kanger, T., Pehk, T. Asymmetric Baeyer-Villiger oxidation of cyclobutanones. *Tetrahedron Lett.* **1996**, 37, 7583-7586.
31. Ricci, M., Battistel, E. New tools for the Baeyer-Villiger oxidation of ketones. 1. Phase transfer catalysis. *Chimica e l'Industria (Milan)* **1997**, 79, 879-882.
32. Alphand, V., Furstoss, R. Asymmetric Baeyer-Villiger oxidation using biocatalysis. *Asymmetric Oxidation Reactions* **2001**, 214-227.
33. Bolm, C., Luong, T. K. K., Beckmann, O. Oxidation of carbonyl compounds: asymmetric Baeyer-Villiger oxidation. *Asymmetric Oxidation Reactions* **2001**, 147-151.
34. ten Brink, G.-J., Vis, J.-M., Arends, I. W. C. E., Sheldon, R. A. Selenium-Catalyzed Oxidations with Aqueous Hydrogen Peroxide. 2. Baeyer-Villiger Reactions in Homogeneous Solution. *J. Org. Chem.* **2001**, 66, 2429-2433.
35. Miyake, Y., Nishibayashi, Y., Uemura, S. Asymmetric Baeyer-Villiger oxidation of cyclic ketones using chiral organoselenium catalysts. *Bull. Chem. Soc. Jpn.* **2002**, 75, 2233-2237.
36. Murahashi, S.-I., Ono, S., Imada, Y. Asymmetric Baeyer-Villiger reaction with hydrogen peroxide catalyzed by a novel planar-chiral bisflavin. *Angew. Chem., Int. Ed. Engl.* **2002**, 41, 2366-2368.
37. Stoute, V. A., Winnik, M. A., Csizmadia, I. G. Theoretical model for the Baeyer-Villiger rearrangement. *J. Am. Chem. Soc.* **1974**, 96, 6388-6393.
38. Winnik, M. A., Stoute, V., Fitzgerald, P. Secondary deuterium isotope effects in the Baeyer-Villiger reaction. *J. Am. Chem. Soc.* **1974**, 96, 1977-1979.
39. Singh, R. P., Singh, V., Srivastava, J. N., Bhattacharjee, A. K. MNDO study of Baeyer-Villiger oxidation of indan-1-ones. *Indian J. Chem., Sect. B* **1996**, 35B, 1101-1103.
40. Cardenas, R., Cetina, R., Lagunez-Otero, J., Reyes, L. Ab Initio and Semiempirical Studies on the Transition Structure of the Baeyer and Villiger Rearrangement. The Reaction of Acetone with Performic Acid. *J. Phys. Chem. A* **1997**, 101, 192-200.
41. Okuno, Y. Theoretical investigation of the mechanism of the Baeyer-Villiger reaction in nonpolar solvents. *Chem.-- Eur. J.* **1997**, 3, 212-218.
42. Hannachi, H., Anoune, N., Arnaud, C., Lanteri, P., Longeray, R., Chermette, H. PM3 semi-empirical study of stereoelectronic effects in the Baeyer-Villiger reaction. *THEOCHEM* **1998**, 434, 183-191.
43. Lehtinen, C., Nevalainen, V., Brunow, G. Experimental and Computational Studies on Substituent Effects in Reactions of Peracid-Aldehyde Adducts. *Tetrahedron* **2000**, 56, 9375-9382.
44. Aoki, M., Seebach, D. Preparation of TADOOH, a hydroperoxide from TADDOL, and use in highly enantioface- and enantiomer-differentiating oxidations. *Helvetica Chimica Acta* **2001**, 84, 187-207.
45. Lehtinen, C., Nevalainen, V., Brunow, G. Experimental and computational studies on solvent effects in reactions of peracid-aldehyde adducts. *Tetrahedron* **2001**, 57, 4741-4751.
46. Calqvist, P., Eklund, R., Hult, K., Brinck, T. Rational design of a lipase to accommodate catalysis of Baeyer-Villiger oxidation with hydrogen peroxide. *J. Mol. Model.* **2003**, 9, 164-171.
47. Sever, R. R., Root, T. W. Computational Study of Tin-Catalyzed Baeyer-Villiger Reaction Pathways Using Hydrogen Peroxide as Oxidant. *J. Phys. Chem. B* **2003**, 107, 10848-10862.
48. Sever, R. R., Root, T. W. Comparison of Epoxidation and Baeyer-Villiger Reaction Pathways for Ti(IV)-H_2O_2 and Sn(IV)-H_2O_2. *J. Phys. Chem. B* **2003**, 107, 10521-10530.
49. Robertson, J. C., Swelim, A. A. M. Free radical pathway for the Baeyer-Villiger reaction. *Tetrahedron Lett.* **1967**, 2871-2874.
50. Mitsuhashi, T., Miyadera, H., Simamura, O. Mechanism of the Baeyer-Villiger reaction. *J. Chem. Soc. D* **1970**, 1301-1302.
51. Trost, B. M., Buhlmayer, P., Mao, M. An unusual effect of selenium substituents on the regiochemistry of a Baeyer-Villiger rearrangement. *Tetrahedron Lett.* **1982**, 23, 1443-1446.
52. Tobe, Y., Ohtani, M., Kakiuchi, K., Odaira, Y. The Baeyer-Villiger oxidation via carbocation. Oxidation of 7-acetyl[4.2.1]- and 7-acetyl[4.2.2]propellanes. *Tetrahedron Lett.* **1983**, 24, 3639-3642.
53. Chandrasekhar, S., Roy, C. D. Evidence for a stereoelectronic effect in the Baeyer-Villiger reaction. Introducing the intramolecular reaction. *Tetrahedron Lett.* **1987**, 28, 6371-6372.
54. Cullis, P. M., Arnold, J. R. P., Clarke, M., Howell, R., DeMira, M., Naylor, M., Nicholls, D. On the mechanism of peracid oxidation of α-diketones to acid anhydrides: an oxygen-17 and oxygen-18 isotope study. *J. Chem. Soc., Chem. Commun.* **1987**, 1088-1089.
55. Camporeale, M., Fiorani, T., Troisi, L., Adam, W., Curci, R., Edwards, J. O. On the mechanism of the Baeyer-Villiger oxidation of ketones by bis(trimethylsilyl) peroxomonosulfate. Intermediacy of dioxiranes. *J. Org. Chem.* **1990**, 55, 93-98.
56. Gordon, N. J., Evans, S. A., Jr. Acyl phosphates from acyl phosphonates. A novel Baeyer-Villiger rearrangement. *J. Org. Chem.* **1993**, 58, 4516-4519.
57. Goodman, R. M., Kishi, Y. Experimental Support for the Primary Stereoelectronic Effect Governing Baeyer-Villiger Oxidation and Criegee Rearrangement. *J. Am. Chem. Soc.* **1998**, 120, 9392-9393.
58. Cadenas, R., Reyes, L., Lagunez-Otero, J., Cetina, R. Semiempirical studies on the transition structure of the Baeyer and Villiger rearrangement. The reaction of acetone with alkyl and aryl peracids. *THEOCHEM* **2000**, 497, 211-225.
59. Crudden, C. M., Chen, A. C., Calhoun, L. A. A demonstration of the primary stereoelectronic effect in the Baeyer - Villiger oxidation of α-fluorocyclohexanones. *Angew. Chem., Int. Ed. Engl.* **2000**, 39, 2851-2855.
60. Jenner, G. High pressure mechanistic diagnosis in Baeyer-Villiger oxidation of aliphatic ketones. *Tetrahedron Lett.* **2001**, 42, 8969-8971.
61. Reiser, O. A demonstration of the primary stereoelectronic effect in the Baeyer-Villiger oxidation of α-fluorocyclohexanones. *Chemtracts* **2001**, 14, 94-99.

62. Oh, J. Synthesis of α-D-C-glucoside employing dichloroketene cycloaddition and Baeyer-Villiger oxidation. *Tetrahedron Lett.* **1997**, 38, 3249-3250.
63. Shing, T. K. M., Lee, C. M., Lo, H. Y. Synthesis of the CD ring in taxol from (S)-(+)-carvone. *Tetrahedron Lett.* **2001**, 42, 8361-8363.
64. Marchand, A. P., Kumar, V. S., Hariprakasha, H. K. Synthesis of novel cage oxaheterocycles. *J. Org. Chem.* **2001**, 66, 2072-2077.
65. Demnitz, F. W. J., Philippini, C., Raphael, R. A. Unexpected Rearrangement in the Peroxytrifluoroacetic Acid-Mediated Baeyer-Villiger Oxidation of trans-3β-Hydroxy-4,4,10β-trimethyl-9-decalone Forming a 7-Oxabicyclo[2.2.1]heptane. Structure Proof and Total Synthesis of (±)-Farnesiferol-C. *J. Org. Chem.* **1995**, 60, 5114-5120.

Baker-Venkataraman Rearrangement 30

Related reactions: Claisen condensation, Dieckmann condensation;

1. Baker, W. Molecular rearrangement of some o-acyloxyacetophenones and the mechanism of the production of 3-acylchromones. *J. Chem. Soc.* **1933**, 1381-1389.
2. Baker, W. Attempts to synthesize 5,6-dihydroxyflavone(primetin). *J. Chem. Soc.* **1934**, 1953-1954.
3. Mahal, H. S., Venkataraman, K. Synthetical experiments in the chromone group. XIV. Action of sodamide on 1-acyloxy-2-acetonaphthones. *J. Chem. Soc.* **1934**, 1767-1769.
4. Bhalla, D. C., Mahal, H. S., Venkataraman, K. Synthetical experiments in the chromone group. XVII. Further observations on the action of sodamide on o-acyloxyacetophenones. *J. Chem. Soc.* **1935**, 868-870.
5. Ollis, W. D., Weight, D. Synthesis of 3-substituted chromones by rearrangement of o-acyloxyacetophenones. *J. Chem. Soc.* **1952**, 3826-3830.
6. Hauser, C. R., Swamer, F. W., Adams, J. T. The acylation of ketones to form β-diketones or β-keto aldehydes. *Org. React.* **1954**, 59-196.
7. Gripenberg, J. Flavones. *Chem. Flavonoid Compds. (T. A. Geissman, editor. MacMillan Co., New York, N.Y.)* **1962**, 406-440.
8. Dunne, A. T. M., Gowan, J. E., Keane, J., O'Kelly, B. M., O'Sullivan, D., Roche, M. M., Ryan, P. M., Wheeler, T. S. Thermal cyclization of o-aroyloxyacetoarones. A new synthesis of flavones. *J. Chem. Soc., Abstracts* **1950**, 1252-1259.
9. Rao, A. V. S., Rao, N. V. S. Synthesis of some substituted 2-(2-furyl)chromones by the simplified Baker-Venkataraman transformation. *Curr. Sci.* **1966**, 35, 149.
10. Kraus, G. A., Fulton, B. S., Wood, S. H. Aliphatic acyl transfer in the Baker-Venkataraman reaction. *J. Org. Chem.* **1984**, 49, 3212-3214.
11. Makrandi, J. K., Kumari, V. A convenient synthesis of 2-styrylchromones by modified Baker-Venkataraman transformation using phase transfer catalysis. *Synth. Commun.* **1989**, 19, 1919-1922.
12. Dua, S., Amemiya, S., Bowie, J. H. The gas-phase Baker-Venkataraman rearrangement. *Rapid Commun. Mass Spectrom.* **1994**, 8, 475-477.
13. Song, G.-Y., Ahn, B.-Z. Synthesis of dibenzoylmethanes as intermediates for flavone synthesis by a modified Baker-Venkataraman rearrangement. *Arch. Pharm. Res.* **1994**, 17, 434-437.
14. Boers, F., Deng, B. L., Lemiere, G., Lepoivre, J., De Groot, A., Dommisse, R., Vlietinck, A. J. An improved synthesis of the anti-picornavirus flavone 3-O-methylquercetin. *Arch. Pharm. (Weinheim, Ger.)* **1997**, 330, 313-316.
15. Kalinin, A. V., Da Silva, A. J. M., Lopes, C. C., Lopes, R. S. C., Snieckus, V. Directed ortho metalation and cross coupling links. Carbamoyl rendition of the Baker-Venkataraman rearrangement. Regiospecific route to substituted 4-hydroxycoumarins. *Tetrahedron Lett.* **1998**, 39, 4995-4998.
16. Kalinin, A. V., Snieckus, V. 4,6-Dimethoxy-3,7-dimethylcoumarin from Colchicum decaisnei. Total synthesis by carbamoyl Baker-Venkataraman rearrangement and structural revision to isoeugenetin methyl ether. *Tetrahedron Lett.* **1998**, 39, 4999-5002.
17. Pinto, D. C. G. A., Silva, A. M. S., Cavaleiro, J. A. S. A convenient synthesis of new (E)-5-hydroxy-2-styrylchromones by modifications of the Baker-Venkataraman method. *New Journal of Chemistry* **2000**, 24, 85-92.
18. Szell, T., Schobel, G., Balaspiri, L. Cyclization of enol esters of o-acyloxyphenyl alkyl ketones. II. A contribution to the mechanism of the reaction. *Tetrahedron* **1969**, 25, 707-714.
19. Burrows, H. D., Topping, R. M. Base-catalyzed intramolecular nucleophilic keto-group participation in the solvolysis of the hindered ester, 2-acetylphenyl mesitoate: acetal formation under basic conditions as a mechanistic consequence of such participation. *J. Chem. Soc. B* **1970**, 1323-1329.
20. Bowden, K., Taylor, G. R. Reactions of carbonyl compounds in basic solutions. III. Mechanism of the alkaline hydrolysis of methyl 2-aroyl- and 2-acylbenzoates and related esters. *J. Chem. Soc. B* **1971**, 149-156.
21. Burrows, H. D., Topping, R. M. Intramolecular participation by enolate anions in the cleavage of aryl esters of mesitoic acid. Carbon-carbon bond formation in aqueous and alcoholic solvents. *J. Chem. Soc., Perkin Trans. 2* **1975**, 571-574.
22. Bowden, K., Chehel-Amiran, M. Reactions of carbonyl compounds in basic solutions. Part 11. The Baker-Venkataraman rearrangement. *J. Chem. Soc., Perkin Trans. 2* **1986**, 2039-2043.
23. Krohn, K., Roemer, E., Top, M. Total synthesis of aklanonic acid and derivatives by Baker-Venkataraman rearrangement. *Liebigs Ann. Chem.* **1996**, 271-277.
24. Enders, D., Geibel, G., Osborne, S. Diastereo- and enantioselective total synthesis of stigmatellin A. *Chem.-- Eur. J.* **2000**, 6, 1302-1309.
25. Thasana, N., Ruchirawat, S. The application of the Baker-Venkataraman rearrangement to the synthesis of benz[b]indeno[2,1-*e*]pyran-10,11-dione. *Tetrahedron Lett.* **2002**, 43, 4515-4517.

Baldwin's Rules/Guidelines for Ring-Closing Reactions 32

1. Baldwin, J. E. Rules for ring closure. *J. Chem. Soc., Chem. Commun.* **1976**, 734-736.
2. Juaristi, E., Cuevas, G. A mnemonics for Baldwin's Rules for ring closure. *Rev. Soc. Quim. Mex.* **1992**, 36, 48.
3. Johnson, C. D. Stereoelectronic effects in the formation of 5- and 6-membered rings: the role of Baldwin's rules. *Acc. Chem. Res.* **1993**, 26, 476-482.
4. Gregory, B., Bullock, E., Chen, T.-S. Intramolecular Michael-type additions. A 5-endo-trig ring closure? *J. Chem. Soc., Chem. Commun.* **1979**, 1070-1071.
5. Kansal, V. K., Bhaduri, A. P. Baldwin rules for ring closure - a reexamination of the concept. *Z. Naturforsch., B: Chem. Sci.* **1979**, 34B, 1567-1569.
6. Reddy, C. P., Singh, S. M., Rao, R. B. Some thoughts on the mechanism of acetal formation and related reactions: extension of Baldwin's rules for ring closure. *Tetrahedron Lett.* **1981**, 22, 973-976.
7. Baldwin, J. E., Lusch, M. J. Rules for ring closure: application to intramolecular aldol condensations in polyketonic substrates. *Tetrahedron* **1982**, 38, 2939-2947.
8. Alva Astudillo, M. E., Chokotho, N. C. J., Jarvis, T. C., Johnson, C. D., Lewis, C. C., McDonnell, P. D. Hydroxy Schiff base-oxazolidine tautomerism: apparent breakdown of Baldwin's rules. *Tetrahedron* **1985**, 41, 5919-5928.
9. Elliott, R. J. The 5-endo-dig ring closure. A "quick but late" transition state. *THEOCHEM* **1985**, 22, 79-83.
10. Wilt, J. W. Reactivity and selectivity in the cyclization of sila-5-hexen-l-yl carbon-centered radicals. *Tetrahedron* **1985**, 41, 3979-4000.
11. Clive, D. L. J., Cheshire, D. R. On Baldwin's kinetic barrier against 5-(enol-endo)-exo-trigonal closures. A comparison of ionic and analogous radical reactions, and a new synthesis of cyclopentanones. *J. Chem. Soc., Chem. Commun.* **1987**, 1520-1523.
12. Brennan, C. M., Johnson, C. D., McDonnell, P. D. Ring closure to ynone systems: 5- and 6-endo- and -exo-dig modes. *J. Chem. Soc., Perkin Trans. 2* **1989**, 957-961.
13. Piccirilli, J. A. Do enzymes obey the Baldwin rules? A mechanistic imperative in enzymic cyclization reactions. *Chem. Biol.* **1999**, 6, R59-R64.

14. Chatgilialoglu, C., Ferreri, C., Guerra, M., Timokhin, V., Froudakis, G., Gimisis, T. 5-Endo-trig Radical Cyclizations: Disfavored or Favored Processes? *J. Am. Chem. Soc.* **2002**, 124, 10765-10772.
15. Buergi, H. B., Dunitz, J. D. From crystal statics to chemical dynamics. *Acc. Chem. Res.* **1983**, 16, 153-161.
16. Boger, D. L., Hong, J. Asymmetric Total Synthesis of ent-(-)-Roseophilin: Assignment of Absolute Configuration. *J. Am. Chem. Soc.* **2001**, 123, 8515-8519.
17. Nicolaou, K. C., Bunnage, M. E., Koide, K. Total Synthesis of Balanol. *J. Am. Chem. Soc.* **1994**, 116, 8402-8403.
18. Overhand, M., Hecht, S. M. A Concise Synthesis of the Antifungal Agent (+)-Preussin. *J. Org. Chem.* **1994**, 59, 4721-4722.
19. Nacro, K., Baltas, M., Zedde, C., Gorrichon, L., Jaud, J. Lactonization and lactone ether formation of nerol geraniol compounds. Use of ^{13}C to identify the cyclization process. *Tetrahedron* **1999**, 55, 5129-5138.

Related reactions: Sandmeyer reaction;

1. Balz, G., Schiemann, G. Aromatic fluorine compounds. I. A new method for their preparation. *Ber.* **1927**, 60B, 1186-1190.
2. Roe, A. Preparation of aromatic fluorine compounds from diazonium fluoborates. The Schiemann reaction. *Org. React.* **1949**, 5, 193-228.
3. Sellers, C., Suschitzky, H. A new preparation of aryldiazonium tetrafluoroborates. *J. Chem. Soc.* **1965**, 6186-6188.
4. Suschitzky, H. The Balz-Schiemann reaction. *Advan. Fluorine Chem. (M. Stacey, J. C. Tatlow, and A. G. Sharpe, editors. Butter-worths)* **1965**, 4, 1-27.
5. Sharts, C. M. Organic fluorine chemistry. *J. Chem. Educ.* **1968**, 45, 185-192.
6. Suschitzky, H., Wakefield, B. J. Aromatic fluorine chemistry at Salford. *Fluorine Chemistry at the Millennium* **2000**, 463-473.
7. Rutherford, K. G., Redmond, W., Rigamonti, J. Use of hexafluorophosphoric acid in the Schiemann reaction. *J. Org. Chem.* **1961**, 26, 5149-5152.
8. Sellers, C., Suschitzky, H. The use of arenediazonium hexafluoro-antimonates and -arsenates in the preparation of aryl fluorides. *J. Chem. Soc., C* **1968**, 2317-2319.
9. Cohen, L. A., Kirk, K. L. Photochemical decomposition of diazonium fluoroborates. Application to the synthesis of ring-fluorinated imidazoles. *J. Am. Chem. Soc.* **1971**, 93, 3060-3061.
10. Horning, D. E., Ross, D. A., Muchowski, J. M. Synthesis of phenols from diazonium tetrafluoroborates. Useful modification. *Can. J. Chem.* **1973**, 51, 2347-2348.
11. Yoneda, N., Fukuhara, T., Mizokami, T., Suzuki, A. A facile preparation of aryl triflates. Decomposition of arenediazonium tetrafluoroborate salts in trifluoromethanesulfonic acid. *Chem. Lett.* **1991**, 459-460.
12. Dolle, F., Dolci, L., Valette, H., Hinnen, F., Vaufrey, F., Guenther, I., Fuseau, C., Coulon, C., Bottlaender, M., Crouzel, C. Synthesis and Nicotinic Acetylcholine Receptor in Vivo Binding Properties of 2-Fluoro-3-[2(S)-2-azetidinylmethoxy]pyridine: A New Positron Emission Tomography Ligand for Nicotinic Receptors. *J. Med. Chem.* **1999**, 42, 2251-2259.
13. Sawaguchi, M., Fukuhara, T., Yoneda, N. Preparation of aromatic fluorides: facile photo-induced fluorinative decomposition of arenediazonium salts and their related compounds using pyridine-nHF. *J. Fluorine Chem.* **1999**, 97, 127-133.
14. Laali, K. K., Gettwert, V. J. Fluorodediazoniation in ionic liquid solvents: new life for the Balz-Schiemann reaction. *J. Fluorine Chem.* **2001**, 107, 31-34.
15. Fukuhara, T., Sekiguchi, M., Yoneda, N. Facile preparation of aromatic fluorides by the fluoro-dediazoniation of aromatic diazonium tetrafluoroborates using HF-pyridine solution. *Chem. Lett.* **1994**, 1011-1012.
16. Makarova, L. G., Gribchenko, E. A. Decomposition of aryldiazonium fluoborates in esters of benzoic acid. *Izvest. Akad. Nauk S.S.S.R., Otdel. Khim. Nauk* **1958**, 693-697.
17. Makarova, L. G., Matveeva, M. K., Gribchenko, E. A. Decomposition of aryldiazonium fluoroborates in nitrobenzene. *Izvest. Akad. Nauk S.S.S.R., Otdel. Khim. Nauk* **1958**, 1452-1460.
18. Ishida, K., Kobori, N., Kobayashi, M., Minato, H. Decomposition of benzenediazonium tetrafluoroborate in aprotic polar solvents. *Bull. Chem. Soc. Jap.* **1970**, 43, 285-286.
19. Swain, C. G., Sheats, J. E., Gorenstein, D. G., Harbison, K. G., Rogers, R. J. Phenyl cation as an intermediate in nucleophilic displacements on benzenediazonium salts. *Tetrahedron Lett.* **1974**, 2973-2974.
20. Swain, C. G., Rogers, R. J. Mechanism of formation of aryl fluorides from arenediazonium fluoborates. *J. Am. Chem. Soc.* **1975**, 97, 799-800.
21. Becker, H. G. O., Israel, G. Ion pair effects in the photolysis and thermolysis of arenediazonium tetrafluoroborates. *J. Prakt. Chem.* **1979**, 321, 579-586.
22. Deng, Y. Study on the mechanism of Schiemann reaction by mass spectrometry. *Acta Chim. Sin.* **1989**, 422-430.
23. Suschitzky, H., Wakefield, B. J. Aromatic fluorine chemistry at Salford. (*Fluorine Chemistry at the Millennium*). **2000**, 463-473
24. Gronheid, R., Lodder, G., Okuyama, T. Photosolvolysis of (E)-Styryl(phenyl)iodonium Tetrafluoroborate. Generation and Reactivity of a Primary Vinyl Cation. *J. Org. Chem.* **2002**, 67, 693-702.
25. Holt, D. A., Levy, M. A., Ladd, D. L., Oh, H. J., Erb, J. M., Heaslip, J. I., Brandt, M., Metcalf, B. W. Steroidal A ring aryl carboxylic acids: a new class of steroid 5α-reductase inhibitors. *J. Med. Chem.* **1990**, 33, 937-942.
26. Argentini, M., Wiese, C., Weinreich, R. Syntheses of 5-fluoro-D/L-dopa and [18F]5-fluoro-L-dopa. *J. Fluorine Chem.* **1994**, 68, 141-144.
27. Mirsadeghi, S., Prasad, G. K. B., Whittaker, N., Thakker, D. R. Synthesis of the K-region monofluoro- and difluorobenzo[c]phenanthrenes. *J. Org. Chem.* **1989**, 54, 3091-3096.
28. Thompson, W. J., Anderson, P. S., Britcher, S. F., Lyle, T. A., Thies, J. E., Magill, C. A., Varga, S. L., Schwering, J. E., Lyle, P. A., et al. Synthesis and pharmacological evaluation of a series of dibenzo[a,d]cycloalkenimines as N-methyl-D-aspartate antagonists. *J. Med. Chem.* **1990**, 33, 789-808.
29. Thibault, C., L'Heureux, A., Bhide, R. S., Ruel, R. Concise and Efficient Synthesis of 4-Fluoro-1H-pyrrolo[2,3-b]pyridine. *Org. Lett.* **2003**, 5, 5023-5025.

Related reactions: Wharton olefin synthesis;

1. Bamford, W. R., Stevens, T. S. The decomposition of p-tolylsulfonylhydrazones by alkali. *J. Chem. Soc.* **1952**, 4735-4740.
2. Shapiro, R. H. Alkenes from tosylhydrazones. *Org. React.* **1976**, 23, 405-507.
3. Adlington, R. M., Barrett, A. G. M. Recent applications of the Shapiro reaction. *Acc. Chem. Res.* **1983**, 16, 55-59.
4. Chamberlin, A. R., Bloom, S. H. Lithioalkenes from arenesulfonylhydrazones. *Org. React.* **1990**, 39, 1-83.
5. Bayless, J. H., Friedman, L., Cook, F. B., Shechter, H. Effect of solvent on the course of the Bamford-Stevens reaction. *J. Am. Chem. Soc.* **1968**, 90, 531-533.
6. Chamberlin, A. R., Bond, F. T. Leaving-group variation in aprotic Bamford-Stevens carbene generation. *J. Org. Chem.* **1978**, 43, 154-155.
7. Nickon, A., Zurer, P. S. J. Isolation of a diazoalkane intermediate in the photic Bamford-Stevens reaction. *J. Org. Chem.* **1981**, 46, 4685-4694.
8. Kang, J., Kim, J. H., Jang, G. S. A Shapiro reaction with [(diethylamino)sulfonyl]hydrazones. *Bull. Korean Chem. Soc.* **1992**, 13, 192-199.
9. Sarkar, T. K., Ghorai, B. K. Silicon-directed Bamford-Stevens reaction of β-trimethylsilyl N-aziridinylimines. *J. Chem. Soc., Chem. Commun.* **1992**, 1184-1185.

10. Stern, A. G., Ilao, M. C., Nickon, A. Hydrogen migrations in a constrained cyclohexylidene. Hax/Heq shift ratios in thermal and photic Bamford-Stevens reactions. *Tetrahedron* **1993**, 49, 8107-8118.
11. Chandrasekhar, S., Mohapatra, S., Lakshman, S. Study of Bamford-Stevens reaction on α-oxy tosylhydrazones. *Chem. Lett.* **1996**, 211-212.
12. Maruoka, K., Oishi, M., Yamamoto, H. The Catalytic Shapiro Reaction. *J. Am. Chem. Soc.* **1996**, 118, 2289-2290.
13. Passafaro, M. S., Keay, B. A. A one pot in situ combined Shapiro-Suzuki reaction. *Tetrahedron Lett.* **1996**, 37, 429-432.
14. Siemeling, U., Neumann, B., Stammler, H.-G. First Example of a High-Yield Shapiro Reaction with a Substrate Containing Only Tertiary α-Carbon Atoms. *J. Org. Chem.* **1997**, 62, 3407-3408.
15. Kirmse, W. Reactive intermediates from N-aziridinyl imines. *Eur. J. Org. Chem.* **1998**, 201-212.
16. Olmstead, K. K., Nickon, A. 1,2-Hydrogen shifts in thermal and photic Bamford-Stevens reactions of cyclohexanones. Activation by an endocyclic oxygen. *Tetrahedron* **1998**, 54, 12161-12172.
17. Chandrasekhar, S., Rajaiah, G., Chandraiah, L., Swamy, D. N. Direct conversion of tosylhydrazones to tert-butyl ethers under Bamford-Stevens reaction conditions. *Synlett* **2001**, 1779-1780.
18. Kurek-Tyrlik, A., Marczak, S., Michalak, K., Wicha, J., Zarecki, A. Reaction of Arylsulfonylhydrazones of Aldehydes with α-Magnesio Sulfones. A Novel Olefin Synthesis. *J. Org. Chem.* **2001**, 66, 6994-7001.
19. Casanova, J., Waegell, B. Bamford-Stevens reaction. Various mechanisms. *Bull. Soc. Chim. Fr.* **1975**, 922-932.
20. Nickon, A., Bronfenbrenner, J. K. Migrating-group orientation in carbene rearrangements. *J. Am. Chem. Soc.* **1982**, 104, 2022-2023.
21. Doye, S., Hotopp, T., Winterfeldt, E. The enantioselective total synthesis of (-)-myltaylenol. *Chem. Commun.* **1997**, 1491-1492.
22. Yajima, A., Mori, K. Diterpenoid total synthesis. XXXII. Synthesis and absolute configuration of (-)-phytocassane D, a diterpene phytoalexin isolated from the rice plant, Oryza sativa. *Eur. J. Org. Chem.* **2000**, 4079-4091.
23. Toth, M., Somsak, L. exo-Glycals from glycosyl cyanides. First generation of C-glycosylmethylene carbenes from 2,5- and 2,6-anhydroaldose tosylhydrazones. *J. Chem. Soc., Perkin Trans. 1* **2001**, 942-943.
24. Rupert, K. C., Liu, C. C., Nguyen, T. T., Whitener, M. A., Sowa, J. R., Jr. Synthesis of Verbenindenes: A New Class of Chiral Indenyl Ligands Derived from Verbenone. *Organometallics* **2002**, 21, 144-149.
25. Trost, B. M., Higuchi, R. I. On the Diastereoselectivity of Intramolecular Pd-Catalyzed TMM Cycloadditions. An Asymmetric Synthesis of the Perhydroazulene (-)-Isoclavukerin A. *J. Am. Chem. Soc.* **1996**, 118, 10094-10105.

Barbier Coupling Reaction 38

Related reactions: Grignard reaction, Kagan-Molander samarium diiodide coupling, Nozaki-Hiyama-Kishi reaction;

1. Barbier, P. *C.R.Acad.Sci.* **1899**, 110.
2. Blomberg, C., Hartog, F. A. The Barbier reaction - a one-step alternative for syntheses via organomagnesium compounds. *Synthesis* **1977**, 18-30.
3. Li, C. J. Organic reactions in aqueous media - with a focus on carbon-carbon bond formation. *Chem. Rev.* **1993**, 93, 2023-2035.
4. Li, C.-J. Aqueous Barbier-Grignard type reaction: scope, mechanism, and synthetic applications. *Tetrahedron* **1996**, 52, 5643-5668.
5. Russo, D. A. The Barbier reaction. *Chem. Ind.* **1996**, 64, 405-439.
6. Alonso, F., Yus, M. Recent developments in Barbier-type reactions. *Rec. Res. Dev. Org. Chem.* **1997**, 1, 397-436.
7. Lubineau, A., Auge, J., Queneau, Y. Carbonyl additions and organometallic chemistry in water. (*Organic Synthesis in Water*). **1998**, 102-140
8. Krief, A., Laval, A.-M. Coupling of Organic Halides with Carbonyl Compounds Promoted by SmI_2, the Kagan Reagent. *Chem. Rev.* **1999**, 99, 745-777.
9. Li, C.-J., Chan, T.-H. Organic syntheses using indium-mediated and catalyzed reactions in aqueous media. *Tetrahedron* **1999**, 55, 11149-11176.
10. Luche, J.-L., Sarandeses, L. A. Zinc-mediated Barbier reactions. *Organozinc Reagents* **1999**, 307-323.
11. Knochel, P., Millot, N., Rodriguez, A. L., Tucker, C. E. Preparation and applications of functionalized organozinc compounds. *Org. React.* **2001**, 58, 417-731.
12. Steel, P. G. Recent developments in lanthanide mediated organic synthesis. *J. Chem. Soc., Perkin Trans. 1* **2001**, 2727-2751.
13. Banik, B. K. Samarium metal in organic synthesis. *Eur. J. Org. Chem.* **2002**, 2431-2444.
14. Pae, A. N., Cho, Y. S. Indium-mediated organic reactions in aqueous media. *Curr. Org. Chem.* **2002**, 6, 715-737.
15. Kagan, H. B. Twenty-five years of organic chemistry with diiodosamarium: an overview. *Tetrahedron* **2003**, 59, 10351-10372.
16. Podlech, J., Maier, T. C. Indium in organic synthesis. *Synthesis* **2003**, 633-655.
17. Luche, J. L., Einhorn, C., Einhorn, J., De Souza Barboza, J. C., Petrier, C., Dupuy, C., Delair, P., Allavena, C., Tuschl, T. Ultrasonic waves as promoters of radical processes in chemistry: the case of organometallic reactions. *Ultrasonics* **1990**, 28, 316-321.
18. Steurer, S., Podlech, J. Indium-catalyzed Barbier reactions of amino aldehydes. *Adv. Syn. & Catal.* **2001**, 343, 251-254.
19. Zha, Z., Wang, Y., Yang, G., Zhang, L., Wang, Z. Efficient Barbier reaction of carbonyl compounds improved by a phase transfer catalyst in water. *Green Chem.* **2002**, 4, 578-580.
20. Auge, J., Lubin-Germain, N., Marque, S., Seghrouchni, L. Indium-catalyzed Barbier allylation reaction. *J. Organomet. Chem.* **2003**, 679, 79-83.
21. Legros, J., Meyer, F., Coliboeuf, M., Crousse, B., Bonnet-Delpon, D., Begue, J.-P. Stereoselective Barbier-Type Allylation Reaction of Trifluoromethyl Aldimines. *J. Org. Chem.* **2003**, 68, 6444-6446.
22. Zha, Z., Xie, Z., Zhou, C., Chang, M., Wang, Z. High regio- and stereoselective Barbier reaction of carbonyl compounds mediated by NaBF4/Zn (Sn) in water. *New J. Chem.* **2003**, 27, 1297-1300.
23. Moyano, A., Pericas, M. A., Riera, A., Luche, J. L. A theoretical study of the Barbier reaction. *Tetrahedron Lett.* **1990**, 31, 7619-7622.
24. Bauer, P., Molle, G. The Barbier reaction: no organometallic pathway in a one-step alternative Grignard reaction. *Tetrahedron Lett.* **1978**, 4853-4856.
25. De Souza-Barboza, J. C., Luche, J. L., Petrier, C. Ultrasound in organic synthesis. 11. Retention of optical activity in Barbier reactions from (S)-(+)-2-octyl halides. Mechanistic consequences. *Tetrahedron Lett.* **1987**, 28, 2013-2016.
26. Molander, G. A., McKie, J. A. A facile synthesis of bicyclo[m.n.1]alkan-1-ols. Evidence for organosamarium intermediates in the samarium(II) iodide promoted intramolecular Barbier-type reaction. *J. Org. Chem.* **1991**, 56, 4112-4120.
27. Curran, D. P., Fevig, T. L., Jasperse, C. P., Totleben, M. J. New mechanistic insights into reductions of halides and radicals with samarium(II) iodide. *Synlett* **1992**, 943-961.
28. Danhui, Y., Einhorn, J., Einhorn, C., Aurell, M. J., Luche, J.-L. The sonochemical Barbier reaction applied to carboxylates. Study of a model case. *J. Chem. Soc., Chem. Commun.* **1994**, 1815-1816.
29. Curran, D. P., Gu, X., Zhang, W., Dowd, P. On the mechanism of the intramolecular samarium Barbier reaction. Probes for formation of radical and organosamarium intermediates. *Tetrahedron* **1997**, 53, 9023-9042.
30. Ashby, E. C. Grignard reagents. Compositions and mechanisms of reaction. *Quart. Rev., Chem. Soc.* **1967**, 21, 259-285.
31. Ashby, E. C., Laemmle, J., Neumann, H. M. Mechanisms of Grignard reagent addition to ketones. *Acc. Chem. Res.* **1974**, 7, 272-280.
32. Ashby, E. C. A detailed description of the mechanism of reaction of Grignard reagents with ketones. *Pure Appl. Chem.* **1980**, 52, 545-569.
33. Trost, B. M., Corte, J. R. Total synthesis of (+)-saponaceolide B**. *Angew. Chem., Int. Ed. Engl.* **1999**, 38, 3664-3666.
34. Trost, B. M., Corte, J. R., Gudiksen, M. S. Towards the total synthesis of saponaceolides: synthesis of cis-2,4-disubstituted 3,3-dimethylmethylenecyclohexanes. *Angew. Chem., Int. Ed. Engl.* **1999**, 38, 3662-3664.
35. Yu, P., Wang, T., Li, J., Cook, J. M. Enantiospecific Total Synthesis of the Sarpagine Related Indole Alkaloids Talpinine and Talcarpine as Well as the Improved Total Synthesis of Alstonerine and Anhydromacrosalhine-methine via the Asymmetric Pictet-Spengler Reaction. *J. Org. Chem.* **2000**, 65, 3173-3191.

36. Yanagisawa, A., Habaue, S., Yamamoto, H. Allylbarium in organic synthesis: unprecedented α-selective and stereospecific allylation of carbonyl compounds. *J. Am. Chem. Soc.* **1991**, 113, 8955-8956.
37. Yanagisawa, A., Habaue, S., Yamamoto, H. Direct insertion of alkali (alkaline earth) metals into allylic carbon-halogen bonds avoiding stereorandomization. *J. Am. Chem. Soc.* **1991**, 113, 5893-5895.
38. Abad, A., Agullo, C., Arno, M., Cunat, A. C., Meseguer, B., Zaragoza, R. J. An Efficient Stereoselective Synthesis of Stypodiol and Epistypodiol. *J. Org. Chem.* **1998**, 63, 5100-5106.

Related reactions: Fischer indole synthesis, Larock indole synthesis, Madelung indole synthesis, Nenitzescu indole synthesis;

1. Bartoli, G., Leardini, R., Medici, A., Rosini, G. Reactions of nitroarenes with Grignard reagents. General method of synthesis of alkyl-nitroso-substituted bicyclic aromatic systems. *J. Chem. Soc., Perkin Trans. 1* **1978**, 692-696.
2. Bartoli, G., Palmieri, G., Bosco, M., Dalpozzo, R. The reaction of vinyl Grignard reagents with 2-substituted nitroarenes: a new approach to the synthesis of 7-substituted indoles. *Tetrahedron Lett.* **1989**, 30, 2129-2132.
3. Bartoli, G., Bosco, M., Dalpozzo, R., Palmieri, G., Marcantoni, E. Reactivity of nitro- and nitrosoarenes with vinyl Grignard reagents: synthesis of 2-(trimethylsilyl)indoles. *J. Chem. Soc., Perkin Trans. 1* **1991**, 2757-2761.
4. Bartoli, G. Conjugate addition of alkyl Grignard reagents to mononitroarenes. *Acc. Chem. Res.* **1984**, 17, 109-115.
5. Gribble, G. W. Recent developments in indole ring synthesis-methodology and applications. *Perkin 1* **2000**, 1045-1075.
6. Joule, J. A. Product class 13: indole and its derivatives. *Science of Synthesis* **2001**, 10, 361-652.
7. Ricci, A., Fochi, M. Reactions between organomagnesium reagents and nitroarenes: Past, present, and future. *Angew. Chem., Int. Ed. Engl.* **2003**, 42, 1444-1446.
8. Dobbs, A. P., Voyle, M., Whittall, N. Synthesis of novel indole derivatives. Variations in the Bartoli reaction. *Synlett* **1999**, 1594-1596.
9. Dobbs, A. Total Synthesis of Indoles from Tricholoma Species via Bartoli/Heteroaryl Radical Methodologies. *J. Org. Chem.* **2001**, 66, 638-641.
10. Pirrung, M. C., Wedel, M., Zhao, Y. 7-Alkyl indole synthesis via a convenient formation/alkylation of lithionitrobenzenes and an improved Bartoli reaction. *Synlett* **2002**, 143-145.
11. Knepper, K., Braese, S. Bartoli Indole Synthesis on Solid Supports. *Org. Lett.* **2003**, 5, 2829-2832.
12. Bosco, M., Dalpozzo, R., Bartoli, G., Palmieri, G., Petrini, M. Mechanistic studies on the reaction of nitro- and nitrosoarenes with vinyl Grignard reagents. *J. Chem. Soc., Perkin Trans. 2* **1991**, 657-663.
13. Zhang, Z., Yang, Z., Meanwell Nicholas, A., Kadow John, F., Wang, T. A general method for the preparation of 4- and 6-azaindoles. *J. Org. Chem.* **2002**, 67, 2345-2347.
14. Harrowven, D. C., Lai, D., Lucas, M. C. A short synthesis of hippadine. *Synthesis* **1999**, 1300-1302.
15. Engler, T. A., Henry, J. R., Malhotra, S., Cunningham, B., Furness, K., Brozinick, J., Burkholder, T. P., Clay, M. P., Clayton, J., Diefenbacher, C., Hawkins, E., Iversen, P. W., Li, Y., Lindstrom, T. D., Marquart, A. L., McLean, J., Mendel, D., Misener, E., Briere, D., O'Toole, J. C., Porter, W. J., Queener, S., Reel, J. K., Owens, R. A., Brier, R. A., Eessalu, T. E., Wagner, J. R., Campbell, R. M., Vaughn, R. Substituted 3-Imidazo[1,2-a]pyridin-3-yl- 4-(1,2,3,4-tetrahydro-[1,4]diazepino- [6,7,1-hi]indol-7-yl)pyrrole-2,5-diones as Highly Selective and Potent Inhibitors of Glycogen Synthase Kinase-3. *J. Med. Chem.* **2004**, 47, 3934-3937.
16. Fonseca, T., Gigante, B., Marques, M. M., Gilchrist, T. L., De Clercq, E. Synthesis and antiviral evaluation of benzimidazoles, quinoxalines and indoles from dehydroabietic acid. *Bioorg. Med. Chem.* **2004**, 12, 103-112.

Related reactions: Hofmann-Löffler-Freytag reaction;

1. Barton, D. H. R., Beaton, J. M., Geller, L. E., Pechet, M. M. A new photochemical reaction. *J. Am. Chem. Soc.* **1960**, 82, 2640-2641.
2. Barton, D. H. R., Beaton, J. M., Geller, L. E., Pechet, M. M. A new photochemical reaction. *J. Am. Chem. Soc.* **1961**, 83, 4076-4083.
3. Nussbaum, A. L., Yuan, E. P., Robinson, C., Mitchell, A., Oliveto, E. P., Beaton, J. M., Barton, D. R. Photolysis of organic nitrites. VII. Fragmentation of the steroidal side chain. *J. Org. Chem.* **1962**, 27, 20-23.
4. Barton, D. H. R. *Photolysis of organic nitrites.* Fr 1334932, **1963** (Scherico Ltd.). 116 pp.
5. Akhtar, M., Barton, D. H. R., Sammes, P. G. Radical exchange during nitrite photolysis. *J. Am. Chem. Soc.* **1964**, 86, 3394-3395.
6. Akhtar, M., Barton, D. H. R., Sammes, P. G. Some radical exchange reactions during nitrite ester photolysis. *J. Am. Chem. Soc.* **1965**, 87, 4601-4607.
7. Robinson, C. H., Gonj, O., Mitchell, A., Oliveto, E. P., Barton, D. H. R. Photochemical rearrangement of steroidal 17-nitrites. *Tetrahedron* **1965**, 21, 743-757.
8. Hesse, R. H. Barton reaction. *Advances in Free-Radical Chemistry (London)* **1969**, 3, 83-137.
9. Majetich, G., Wheless, K. Remote intramolecular free radical functionalizations: an update. *Tetrahedron* **1995**, 51, 7095-7129.
10. Reese, P. B. Remote functionalization reactions in steroids. *Steroids* **2001**, 66, 481-497.
11. Cekovic, Z. Reactions of δ-carbon radicals generated by 1,5-hydrogen transfer to alkoxyl radicals. *Tetrahedron* **2003**, 59, 8073-8090.
12. Suginome, H. Remote functionalization by alkoxyl radicals generated by the photolysis of nitrite esters: the Barton reaction and related reactions of nitrite esters. in *CRC Handbook of Organic Photochemistry and Photobiology (2nd Edition)* 102/101-102/116 (**2004**).
13. Hornung, G., Schalley, C. A., Dieterle, M., Schroder, D., Schwarz, H. A study of the gas-phase reactivity of neutral alkoxy radicals by mass spectrometry: α-cleavages and Barton-type hydrogen migrations. *Chem.-- Eur. J.* **1997**, 3, 1866-1883.
14. Bouchoux, G., Choret, N. Intramolecular hydrogen migrations in ionized aliphatic alcohols. Barton type and related rearrangements. *Int. J. Mass Spectrom.* **2000**, 201, 161-177.
15. Barton, D. H. R., Hesse, R. H., Pechet, M. M., Smith, L. C. The mechanism of the Barton reaction. *J. Chem. Soc., Perkin Trans. 1* **1979**, 1159-1165.
16. Burke, S. D., Silks, L. A., III, Strickland, S. M. S. Remote functionalization and molecular modeling. Observations relevant to the Barton and hypoiodite reactions. *Tetrahedron Lett.* **1988**, 29, 2761-2764.
17. Konoike, T., Takahashi, K., Araki, Y., Horibe, I. Practical Partial Synthesis of Myriceric Acid A, an Endothelin Receptor Antagonist, from Oleanolic Acid. *J. Org. Chem.* **1997**, 62, 960-966.
18. Hakimelahi, G. H., Li, P.-C., Moosavi-Movahedi, A. A., Chamani, J., Khodarahmi, G. A., Ly, T. W., Valiyev, F., Leong, M. K., Hakimelahi, S., Shia, K.-S., Chao, I. Application of the Barton photochemical reaction in the synthesis of 1-dethia-3-aza-1-carba-2-oxacephem: a novel agent against resistant pathogenic microorganisms. *Org. Biomol. Chem.* **2003**, 1, 2461-2467.
19. Corey, E. J., Hahl, R. W. Synthesis of a limonoid, azadiradione. *Tetrahedron Lett.* **1989**, 30, 3023-3026.
20. Corey, E. J., Arnett, J. F., Widiger, G. N. Simple total synthesis of (±)-perhydrohistrionicotoxin. *J. Am. Chem. Soc.* **1975**, 97, 430-431.
21. Sejbal, J., Klinot, J., Vystrcil, A. Triterpenes. LXXXV. Photolysis of 19β,28-epoxy-18α-oleanan-2β-ol nitrites: functionalization of 10β- and 8β-methyl groups. *Collect. Czech. Chem. Commun.* **1988**, 53, 118-131.
22. Petrovic, G., Cekovic, Z. Free radical alkylation of the remote nonactivated δ-carbon atom. *Tetrahedron Lett.* **1997**, 38, 627-630.

Barton Radical Decarboxylation Reaction 44

Related reactions: Hunsdiecker reaction;

1. Barton, D. H. R., Serebryakov, E. P. A convenient procedure for the decarboxylation of acids. *Proc. Chem. Soc.* **1962**, 309.
2. Barton, D. H. R., Dowlatshahi, H. A., Motherwell, W. B., Villemin, D. A new radical decarboxylation reaction for the conversion of carboxylic acids into hydrocarbons. *J. Chem. Soc., Chem. Commun.* **1980**, 732-733.
3. Barton, D. H. R., Crich, D., Motherwell, W. B. A practical alternative to the Hunsdiecker reaction. *Tetrahedron Lett.* **1983**, 24, 4979-4982.
4. Barton, D. H. R., Zard, S. Z. A novel radical decarboxylation reaction. *Janssen Chim. Acta* **1986**, 4, 3-9.
5. Boivin, J., Fouquet, E., Zard, S. Z. A new and synthetically useful source of iminyl radicals. *Tetrahedron Lett.* **1991**, 32, 4299-4302.
6. Ballestri, M., Chatgilialoglu, C., Cardi, N., Sommazzi, A. The reaction of tris(trimethylsilyl)silane with acid chlorides. *Tetrahedron Lett.* **1992**, 33, 1787-1790.
7. Barton, D. H. R., Chern, C.-Y., Jaszberenyi, J. C. The invention of radical reactions. XXXIII. Homologation reactions of carboxylic acids by radical chain chemistry. *Aust. J. Chem.* **1995**, 48, 407-425.
8. Stojanovic, A., Renaud, P. Generation of 1-amidoalkyl radicals from N-protected amino acids. An alternative to the Barton decarboxylation procedure. *Synlett* **1997**, 181-182.
9. Garner, P., Anderson, J. T., Dey, S., Youngs, W. J., Galat, K. S-(1-Oxido-2-pyridinyl)-1,1,3,3-tetramethylthiouronium Hexafluorophosphate. A New Reagent for Preparing Hindered Barton Esters. *J. Org. Chem.* **1998**, 63, 5732-5733.
10. Girard, P., Guillot, N., Motherwell, W. B., Hay-Motherwell, R. S., Potier, P. The reaction of thionitrites with Barton esters: a convenient free radical chain reaction for decarboxylative nitrosation. *Tetrahedron* **1999**, 55, 3573-3584.
11. Attardi, M. E., Taddei, M. The Barton radical decarboxylation on solid phase. A versatile synthesis of peptides containing modified amino acids. *Tetrahedron Lett.* **2001**, 42, 3519-3522.
12. Kim, S., Lim, C. J., Song, S.-E., Kang, H.-Y. Decarboxylative acylation approach of thiohydroxamate esters. *Chem. Commun.* **2001**, 1410-1411.
13. Barton, D. H. R., Bridon, D., Fernandez-Picot, I., Zard, S. Z. Invention of radical reactions. Part XV. Some mechanistic aspects of the decarboxylative rearrangement of thiohydroxamic esters. *Tetrahedron* **1987**, 43, 2733-2740.
14. Ishihara, J., Nonaka, R., Terasawa, Y., Shiraki, R., Yabu, K., Kataoka, H., Ochiai, Y., Tadano, K.-i. Total Synthesis of (-)-Verrucarol. *J. Org. Chem.* **1998**, 63, 2679-2688.
15. Katoh, T., Kirihara, M., Yoshino, T., Tamura, O., Ikeuchi, F., Nakatani, K., Matsuda, F., Yamada, K., Gomi, K., Ashizawa, T., Terashima, S. Synthetic studies on quinocarcin and its related compounds. 5. Synthesis and antitumor activity of various structural types of quinocarcin congeners. *Tetrahedron* **1994**, 50, 6259-6270.
16. Zhu, J., Klunder, A. J. H., Zwanenburg, B. Synthesis of 6-functionalized tricyclodecadienones using Barton's radical decarboxylation reaction. Generation of tricyclo[5.2.1.0^{2,6}]decatrienone, a norbornene annulated cyclopentadienone. *Tetrahedron* **1995**, 51, 5099-5116.
17. Poigny, S., Guyot, M., Samadi, M. One-step Synthesis of Tyromycin A and Analogs. *J. Org. Chem.* **1998**, 63, 1342-1343.

Barton-McCombie Radical Deoxygenation Reaction 46

1. Barton, D. H. R., McCombie, S. W. New method for the deoxygenation of secondary alcohols. *J. Chem. Soc., Perkin Trans. 1* **1975**, 1574-1585.
2. Barrett, A. G. M., Prokopiou, P. A., Barton, D. H. R. Novel method for the deoxygenation of alcohols. *J. Chem. Soc., Chem. Commun.* **1979**, 1175.
3. Barton, D. H. R., Motherwell, W. B., Stange, A. Radical-induced deoxygenation of primary alcohols. *Synthesis* **1981**, 743-745.
4. Barton, D. H. R., Hartwig, W., Motherwell, R. S. H., Motherwell, W. B., Stange, A. Radical deoxygenation of tertiary alcohols. *Tetrahedron Lett.* **1982**, 23, 2019-2022.
5. Barton, D. H. R., Crich, D. A new method for the radical deoxygenation of tertiary alcohols. *J. Chem. Soc., Chem. Commun.* **1984**, 774-775.
6. Barton, D. H. R., Jaszberenyi, J. C. Improved methods for the radical deoxygenation of secondary alcohols. *Tetrahedron Lett.* **1989**, 30, 2619-2622.
7. Hartwig, W. Modern methods for the radical deoxygenation of alcohols. *Tetrahedron* **1983**, 39, 2609-2645.
8. Robins, M. J., Hansske, F., Wilson, J. S., Hawrelak, S. D., Madej, D. Selective modification and deoxygenation at C2' of nucleosides. *Nucleosides, Nucleotides, Their Biol. Appl., Proc. Int. Round Table, 5th* **1983**, 279-295.
9. Baer, H. H. Recent synthetic studies in nitrogen-containing and deoxygenated sugars. *Pure Appl. Chem.* **1989**, 61, 1217-1234.
10. Chatgilialoglu, C., Ferreri, C. Progress of the Barton-Mccombie methodology: from tin hydrides to silanes. *Res. Chem. Intermed.* **1993**, 19, 755-775.
11. David, S. Hypophosphorous acid and its salts: new reagents for radical chain deoxygenation, dehalogenation, and deamination. *Chemtracts: Org. Chem.* **1993**, 6, 55-58.
12. Hong, F.-T., Paquette, L. A. Bu_3SnH-catalyzed Barton-McCombie deoxygenation of alcohols. Single-step process for the reductive deoxygenation of unhindered alcohols. *Chemtracts* **1998**, 11, 67-72.
13. Crich, D. The use of S-alkenyl dithiocarbonates as mechanistic probes in the Barton-McCombie radical deoxygenation reaction. *Tetrahedron Lett.* **1988**, 29, 5805-5806.
14. Kirwan, J. N., Roberts, B. P., Willis, C. R. Deoxygenation of alcohols by the reactions of their xanthate esters with triethylsilane: an alternative to tributyltin hydride in the Barton-McCombie reaction. *Tetrahedron Lett.* **1990**, 31, 5093-5096.
15. Schummer, D., Hoefle, G. Tris(trimethylsilyl)silane as a reagent for the radical deoxygenation of alcohols. *Synlett* **1990**, 705-706.
16. Neumann, W. P., Peterseim, M. Elegant improvement of the deoxygenation of alcohols using a polystyrene-supported organotin hydride. *Synlett* **1992**, 801-802.
17. Crich, D., Beckwith, A. L. J., Chen, C., Yao, Q., Davison, I. G. E., Longmore, R. W., Anaya de Parrodi, C., Quintero-Cortes, L., Sandoval-Ramirez, J. Origin of the "β-Oxygen Effect" in the Barton Deoxygenation Reaction. *J. Am. Chem. Soc.* **1995**, 117, 8757-8768.
18. Lopez, R. M., Hays, D. S., Fu, G. C. Bu_3SnH-Catalyzed Barton-McCombie Deoxygenation of Alcohols. *J. Am. Chem. Soc.* **1997**, 119, 6949-6950.
19. Prudhomme, D. R., Wang, Z., Rizzo, C. J. An Improved Photosensitizer for the Photoinduced Electron-Transfer Deoxygenation of Benzoates and m-(Trifluoromethyl)benzoates. *J. Org. Chem.* **1997**, 62, 8257-8260.
20. Boussaguet, P., Delmond, B., Dumartin, G., Pereyre, M. Catalytic and supported Barton-McCombie deoxygenation of secondary alcohols: a clean reaction. *Tetrahedron Lett.* **2000**, 41, 3377-3380.
21. Siddiqui, M. A., Driscoll, J. S., Abushanab, E., Kelley, J. A., Barchi, J. J., Jr., Marquez, V. E. The "β-fluorine effect" in the non-metal hydride radical deoxygenation of fluorine-containing nucleoside xanthates. *Nucleosides, Nucleotides & Nucleic Acids* **2000**, 19, 1-12.
22. Studer, A., Amrein, S. Silylated cyclohexadienes: new alternatives to tributyltin hydride in free radical chemistry. *Angew. Chem., Int. Ed. Engl.* **2000**, 39, 3080-3082.
23. Rhee, J. U., Bliss, B. I., RajanBabu, T. V. A New Reaction Manifold for the Barton Radical Intermediates: Synthesis of N-Heterocyclic Furanosides and Pyranosides via the Formation of the C1-C2 Bond. *J. Am. Chem. Soc.* **2003**, 125, 1492-1493.
24. Studer, A., Amrein, S., Schleth, F., Schulte, T., Walton, J. C. Silylated Cyclohexadienes as New Radical Chain Reducing Reagents: Preparative and Mechanistic Aspects. *J. Am. Chem. Soc.* **2003**, 125, 5726-5733.
25. Barton, D. H. R., Crich, D., Loebberding, A., Zard, S. Z. On the mechanism of the deoxygenation of secondary alcohols by the reduction of their methyl xanthates by tin hydrides. *Tetrahedron* **1986**, 42, 2329-2338.
26. Barton, D. H. R., Jaszberenyi, J. C., Morrell, A. I. The generation and reactivity of oxygen centered radicals from the photolysis of derivatives of N-hydroxy-2-thiopyridone. *Tetrahedron Lett.* **1991**, 32, 311-314.

27. Liu, P., Panek, J. S. Studies directed toward the total synthesis of kabiramide C: asymmetric synthesis of the C1-C19 fragment. *Tetrahedron Lett.* **1998**, 39, 6147-6150.
28. Pettus, T. R. R., Inoue, M., Chen, X.-T., Danishefsky, S. J. A Fully Synthetic Route to the Neurotrophic Illicinones: Syntheses of Tricycloillicinone and Bicycloillicinone Aldehyde. *J. Am. Chem. Soc.* **2000**, 122, 6160-6168.
29. Singh, V., Prathap, S., Porinchu, M. Aromatics to Triquinanes: *p*-Cresol to (±)-Δ9(12)-Capnellene. *J. Org. Chem.* **1998**, 63, 4011-4017.
30. Luzzio, F. A., Fitch, R. W. Formal Synthesis of (+)- and (-)-Perhydrohistrionicotoxin: A "Double Henry" Enzymatic Desymmetrization Route to the Kishi Lactam. *J. Org. Chem.* **1999**, 64, 5485-5493.
31. Schlessinger, R. H., Gillman, K. W. An enantioselective solution towards synthesizing "skip" 1,3-dimethyl stereocenters. A synthesis of 4S(2E,4R*,6R*)-4,6-dimethyl-2-octenoic acid. *Tetrahedron Lett.* **1996**, 37, 1331-1334.

Baylis-Hillman Reaction 48

1. Morita, K., Suzuki, Z., Hirose, H. Tertiary phosphine-catalyzed reaction of acrylic compounds with aldehydes. *Bull. Chem. Soc. Jpn.* **1968**, 41, 2815.
2. Baylis, A. B., Hillman, M. E. D. *Acrylic compounds*. De 2155113, **1972** (Celanese Corp.). 16 pp.
3. Basavaiah, D., Rao, P. D., Hyma, R. S. The Baylis-Hillman reaction: a novel carbon-carbon bond forming reaction. *Tetrahedron* **1996**, 52, 8001-8062.
4. Ciganek, E. The catalyzed α-hydroxyalkylation and α-aminoalkylation of activated olefins (the Morita-Baylis-Hillman reaction). *Org. React.* **1997**, 51, 201-350.
5. Venkatesan, H., Liotta, D. C. The Baylis-Hillman reaction: practical improvements and asymmetric synthesis. *Chemtracts* **1998**, 11, 29-34.
6. Coelho, F., Almeida, W. P. The Baylis-Hillman reaction: a strategy for the preparation of multifunctionalized intermediates for organic synthesis. *Quim. Nova* **2000**, 23, 98-101.
7. Langer, P. New strategies for the development of an asymmetric version of the Baylis - Hillman reaction. *Angew. Chem., Int. Ed. Engl.* **2000**, 39, 3049-3052.
8. Ley, S. V., Rodriguez, F. Deracemization of Baylis-Hillman adducts. *Chemtracts* **2000**, 13, 596-601.
9. Kim, J. N., Lee, K. Y. Synthesis of cyclic compounds from the Baylis-Hillman adducts. *Curr. Org. Chem.* **2002**, 6, 627-645.
10. Basavaiah, D., Rao, A. J., Satyanarayana, T. Recent Advances in the Baylis-Hillman Reaction and Applications. *Chem. Rev.* **2003**, 103, 811-891.
11. Huddleston, R. R., Krische, M. J. Enones as latent enolates in catalytic processes: catalytic cycloreduction, cycloaddition and cycloisomerization. *Synlett* **2003**, 12-21.
12. Langer, P. The asymmetric Baylis-Hillman-reaction. in *Organic Synthesis Highlights V* 165-177 (VCH, Weinheim, New York, **2003**).
13. Murugan, R., Scriven, E. F. V. Applications of dialkylaminopyridine (DMAP) catalysts in organic synthesis. *Aldrichimica Acta* **2003**, 36, 21-27.
14. Rafel, S., Leahy, J. W. An Unexpected Rate Acceleration-Practical Improvements in the Baylis-Hillman Reaction. *J. Org. Chem.* **1997**, 62, 1521-1522.
15. Shi, M., Jiang, J. K. Amendment in Titanium(IV) Chloride and Chalcogenide-Promoted Baylis-Hillman Reaction of Aldehydes with α,β-Unsaturated Ketones. *Tetrahedron* **2000**, 56, 4793-4797.
16. Shi, M., Jiang, J.-K., Feng, Y.-S. Titanium(IV) chloride and the amine-promoted Baylis-Hillman reaction. *Org. Lett.* **2000**, 2, 2397-2400.
17. Yamada, Y. M. A., Ikegami, S. Efficient Baylis-Hillman reactions promoted by mild cooperative catalysts and their application to catalytic asymmetric synthesis. *Tetrahedron Lett.* **2000**, 41, 2165-2169.
18. Jauch, J. A new protocol for Baylis-Hillman reactions: chirality transfer in a lithium phenylselenide induced tandem-Michael-aldol-retro-Michael reaction. *J. Org. Chem.* **2001**, 66, 609-611.
19. Coelho, F., Almeida, W. P., Veronese, D., Mateus, C. R., Silva Lopes, E. C., Rossi, R. C., Silveira, G. P. C., Pavam, C. H. Ultrasound in Baylis-Hillman reactions with aliphatic and aromatic aldehydes: scope and limitations. *Tetrahedron* **2002**, 58, 7437-7447.
20. Shi, M., Xu, Y.-M. Catalytic, asymmetric Baylis-Hillman reaction of imines with methyl vinyl ketone and methyl acrylate. *Angew. Chem., Int. Ed. Engl.* **2002**, 41, 4507-4510.
21. Shi, M., Zhao, G.-L. One-pot aza-Baylis-Hillman reactions of aryl aldehydes and diphenylphosphinamide with methyl vinyl ketone in the presence of $TiCl_4$, PPh_3, and Et_3N. *Tetrahedron Lett.* **2002**, 43, 9171-9174.
22. Balan, D., Adolfsson, H. Chiral quinuclidine-based amine catalysts for the asymmetric one-pot, three-component aza-Baylis-Hillman reaction. *Tetrahedron Lett.* **2003**, 44, 2521-2524.
23. Clifford, A. A., Rose, P. M., Lee, K., Rayner, C. M. Thermodynamics of the Baylis-Hillman reaction in supercritical carbon dioxide. *ACS Symp. Ser.* **2003**, 860, 259-268.
24. Corma, A., Garcia, H., Leyva, A. Heterogeneous Baylis-Hillman using a polystyrene-bound 4-(N-benzyl-N-methylamino)pyridine as reusable catalyst. *Chem. Commun.* **2003**, 2806-2807.
25. Huang, J.-W., Shi, M. Polymer-supported lewis bases for the Baylis - Hillman reaction. *Adv. Syn. & Catal.* **2003**, 345, 953-958.
26. Kim, E. J., Ko, S. Y., Song, C. E. Acceleration of the Baylis-Hillman reaction in the presence of ionic liquids. *Helv. Chim. Acta* **2003**, 86, 894-899.
27. Matsuya, Y., Hayashi, K., Nemoto, H. A Novel Modified Baylis-Hillman Reaction of Propiolate. *J. Am. Chem. Soc.* **2003**, 125, 646-647.
28. Methot, J. L., Roush, W. R. Synthetic Studies toward FR182877. Remarkable Solvent Effect in the Vinylogous Morita-Baylis-Hillman Cyclization. *Org. Lett.* **2003**, 5, 4223-4226.
29. Octavio, R., de Souza, M. A., Vasconcellos, M. L. A. A. The use of DMAP as catalyst in the Baylis-Hillman reaction between methyl acrylate and aromatic aldehydes. *Synth. Commun.* **2003**, 33, 1383-1389.
30. Shi, M., Chen, L.-H. Chiral phosphine Lewis base catalyzed asymmetric aza-Baylis-Hillman reaction of N-sulfonated imines with methyl vinyl ketone and phenyl acrylate. *Chem. Commun.* **2003**, 1310-1311.
31. Shi, M., Xu, Y.-M. An Unexpected Highly Stereoselective Double Aza-Baylis-Hillman Reaction of Sulfonated Imines with Phenyl Vinyl Ketone. *J. Org. Chem.* **2003**, 68, 4784-4790.
32. Bode, M. L., Kaye, P. T. A kinetic and mechanistic study of the Baylis-Hillman reaction. *Tetrahedron Lett.* **1991**, 32, 5611-5614.
33. Fort, Y., Berthe, M. C., Caubere, P. The 'Baylis-Hillman reaction' mechanism and applications revisited. *Tetrahedron* **1992**, 48, 6371-6384.
34. van Rozendaal, E. L. M., Voss, B. M. W., Scheeren, H. W. Effect of solvent, pressure and catalyst on the E/Z-selectivity in the Baylis-Hillman reaction between crotononitrile and benzaldehye. *Tetrahedron* **1993**, 49, 6931-6936.
35. Shi, M., Jiang, J.-K., Cui, S.-C. Amine and titanium(IV) chloride, boron(III) chloride or zirconium(IV) chloride-promoted Baylis-Hillman reactions. *Molecules [online computer file]* **2001**, 6, 852-868.
36. Shi, M., Jiang, J.-K., Li, C.-Q. Lewis base and L-proline co-catalyzed Baylis-Hillman reaction of aryl aldehydes with methyl vinyl ketone. *Tetrahedron Lett.* **2001**, 43, 127-130.
37. Shi, M., Li, C.-Q., Jiang, J.-K. New discovery in the traditional Baylis-Hillman reaction of arylaldehydes with methyl vinyl ketone. *Chem. Commun.* **2001**, 833-834.
38. Shi, M., Li, C.-Q., Jiang, J.-K. Reexamination of the traditional Baylis-Hillman reaction. *Tetrahedron* **2003**, 59, 1181-1189.
39. Iwabuchi, Y., Sugihara, T., Esumi, T., Hatakeyama, S. An enantio- and stereocontrolled route to epopromycin B via cinchona alkaloid-catalyzed Baylis-Hillman reaction. *Tetrahedron Lett.* **2001**, 42, 7867-7871.
40. Kaye, P. T., Nocanda, X. W. A convenient general synthesis of 3-substituted 2H-chromene derivatives. *J. Chem. Soc., Perkin Trans. 1* **2002**, 1318-1323.
41. Basavaiah, D., Krishnamacharyulu, M., Hyma, R. S., Sarma, P. K. S., Kumaragurubaran, N. A Facile One-Pot Conversion of Acetates of the Baylis-Hillman Adducts to [*E*]-α-Methylcinnamic Acids. *J. Org. Chem.* **1999**, 64, 1197-1200.
42. Jauch, J. A short total synthesis of kuehneromycin A. *Angew. Chem., Int. Ed. Engl.* **2000**, 39, 2764-2765.
43. Reddy, L. R., Saravanan, P., Corey, E. J. A Simple Stereocontrolled Synthesis of Salinosporamide A. *J. Am. Chem. Soc.* **2004**, 126, 6230-6231.

Beckmann Rearrangement 50

1. Beckmann, E. Isonitroso compounds. *Ber. Dtsch. Chem. Ges.* **1886**, 19, 988-993.
2. Donaruma, L. G., Heldt, W. Z. The Beckmann rearrangement. *Org. React.* **1960**, 11, 1-156.
3. Beckwith, A. L. J. Synthesis of amides. in *Chem. Amides* (ed. Zabicky), 73-185 (Wiley, New York, **1970**).
4. Conley, R. T., Ghosh, S. "Abnormal" Beckmann rearrangements. *Mechanisms of Molecular Migrations* **1971**, 4, 197-308.
5. Tatsumi, T. *Beckmann rearrangement* (eds. Sheldon, R. A.,Bekkum, H.) (Weinheim: Wiley-VCH, New York, **2001**) 185-204.
6. Field, L., Hughmark, P. B., Shumaker, S. H., Marshall, W. S. Isomerization of aldoximes to amides under substantially neutral conditions. *J. Am. Chem. Soc.* **1961**, 83, 1983-1987.
7. Chattopadhyaya, J. B., Rao, A. V. R. Silica gel-induced isomerization of aldoximes to amides. *Tetrahedron* **1974**, 30, 2899-2900.
8. Ganboa, I., Palomo, C. Reagents and synthetic methods. 33. Improved one-step Beckmann rearrangement from ketones and hydroxylamine in formic acid solution. *Synth. Commun.* **1983**, 13, 941-944.
9. Loupy, A., Regnier, S. Solvent-free microwave-assisted Beckmann rearrangement of benzaldehyde and 2'-hydroxyacetophenone oximes. *Tetrahedron Lett.* **1999**, 40, 6221-6224.
10. Anilkumar, R., Chandrasekhar, S. Improved procedures for the Beckmann rearrangement: the reaction of ketoxime carbonates with boron trifluoride etherate. *Tetrahedron Lett.* **2000**, 41, 5427-5429.
11. Khodaei, M. M., Meybodi, F. A., Rezai, N., Salehi, P. Solvent free Beckmann rearrangement of ketoximes by anhydrous ferric chloride. *Synth. Commun.* **2001**, 31, 2047-2050.
12. Sharghi, H., Hosseini, M. Solvent-free and one-step Beckmann rearrangement of ketones and aldehydes by zinc oxide. *Synthesis* **2002**, 1057-1060.
13. Chandrasekhar, S., Gopalaiah, K. Ketones to amides via a formal Beckmann rearrangement in one pot': a solvent-free reaction promoted by anhydrous oxalic acid. Possible analogy with the Schmidt reaction. *Tetrahedron Lett.* **2003**, 44, 7437-7439.
14. Chandrasekhar, S., Gopalaiah, K. Beckmann reaction of oximes catalysed by chloral. Mild and neutral procedures. *Tetrahedron Lett.* **2003**, 44, 755-756.
15. Eshghi, H., Gordi, Z. An Easy Method for the Generation of Amides from Ketones by a Beckmann Type Rearrangement Mediated by Microwave. *Synth. Commun.* **2003**, 33, 2971-2978.
16. His, S., Meyer, C., Cossy, J., Emeric, G., Greiner, A. Solid phase synthesis of amides by the Beckmann rearrangement of ketoxime carbonates. *Tetrahedron Lett.* **2003**, 44, 8581-8584.
17. Lee, J. K., Kim, D.-C., Song, C. E., Lee, S.-g. Thermal behaviors of ionic liquids under microwave irradiation and their application on microwave-assisted catalytic Beckmann rearrangement of ketoximes. *Synth. Commun.* **2003**, 33, 2301-2307.
18. Hunt, P. A., Rzepa, H. S. A comparison of semiempirical SCF-MO and ab initio energy surfaces for the Beckmann rearrangement. *J. Chem. Soc., Chem. Commun.* **1989**, 623-625.
19. Minh Tho, N., Vanquickenborne, L. G. Mechanism of the Beckmann rearrangement of formaldehyde oxime and formaldehyde hydrazone in the gas phase. *J. Chem. Soc., Perkin Trans. 2* **1993**, 1969-1972.
20. Nguyen, M. T. Hydrogen cyanide loss from $[CH_5N_2]$+ cations: 1,2-elimination versus Beckmann rearrangement. *Int. J. Mass Spectrom. Ion Processes* **1994**, 136, 45-53.
21. Nguyen, M. T., Raspoet, G., Vanquickenborne, L. G. Important role of the Beckmann rearrangement in the gas phase chemistry of protonated formaldehyde oximes and their $[CH_4NO]$+ isomers. *J. Chem. Soc., Perkin Trans. 2* **1995**, 1791-1795.
22. Nguyen, M. T., Raspoet, G., Vanquickenborne, L. G. Mechanism of the Beckmann rearrangement: ab initio calculations suggest an active solvent catalysis. *Trends in Organic Chemistry* **1997**, 6, 169-180.
23. Nguyen, M. T., Raspoet, G., Vanquickenborne, L. G. Mechanism of the Beckmann rearrangement in sulfuric acid solution. *J. Chem. Soc., Perkin Trans. 2* **1997**, 821-825.
24. Nguyen, M. T., Raspoet, G., Vanquickenborne, L. G. A New Look at the Classical Beckmann Rearrangement: A Strong Case of Active Solvent Effect. *J. Am. Chem. Soc.* **1997**, 119, 2552-2562.
25. Mori, S., Uchiyama, K., Yayashi, Y., Narasaka, K., Nakamura, E. S_N2 substitution on sp^2 nitrogen of protonated oxime. *Chem. Lett.* **1998**, 111-112.
26. Simunic-Meznaric, V., Mihalic, Z., Vancik, H. Oxime rearrangements: ab initio calculations and reactions in the solid state. *J. Chem. Soc., Perkin Trans. 2* **2002**, 2154-2158.
27. Yamaguchi, Y., Yasutake, N., Nagaoka, M. Ab initio study of noncatalytic Beckmann rearrangement and hydrolysis of cyclohexanone-oxime in subcritical and supercritical water using the polarizable continuum model. *THEOCHEM* **2003**, 639, 137-150.
28. Butler, R. N., O'Donoghue, D. A. Direct detection of intermediates and synthetic applications of the reaction of thionyl chloride with oximes of substituted acetophenones and benzaldehydes: Beckmann rearrangements. *J. Chem. Res., Synop.* **1983**, 18-19.
29. Raspoet, G., Nguyen, M. T., Vanquickenborne, L. G. A theoretical study of the Beckmann rearrangement involving aliphatic and cyclic alkanone oximes. *Bull. Soc. Chim. Belg.* **1997**, 106, 691-697.
30. Lee, B. S., Chu, S., Lee, I. Y., Lee, B.-S., Song, C. E., Chi, D. Y. Beckmann rearrangements of 1-indanone oxime derivatives using aluminum chloride and mechanistic considerations. *Bull. Korean Chem. Soc.* **2000**, 21, 860-866.
31. Fois, G. A., Ricchiardi, G., Bordiga, S., Busco, C., Dalloro, L., Spano, G., Zecchina, A. The Beckmann rearrangement catalyzed by silicalite: a spectroscopic and computational study. *Stud. Surf. Sci. Catal.* **2001**, 135, 2477-2484.
32. Mani, N. S., Wu, M. An efficient synthetic route to chiral 4-alkyl-1,2,3,4-tetrahydroquinolines: enantioselective synthesis of (R)-4-ethyl-1,2,3,4-tetrahydroquinoline. *Tetrahedron: Asymmetry* **2000**, 11, 4687-4691.
33. White, J. D., Hrnciar, P., Stappenbeck, F. Asymmetric Total Synthesis of (+)-Codeine via Intramolecular Carbenoid Insertion. *J. Org. Chem.* **1999**, 64, 7871-7884.
34. White, J. D., Choi, Y. Catalyzed Asymmetric Diels-Alder Reaction of Benzoquinone. Total Synthesis of (-)-Ibogamine. *Org. Lett.* **2000**, 2, 2373-2376.
35. Smith, B. T., Wendt, J. A., Aube, J. First Asymmetric Total Synthesis of (+)-Sparteine. *Org. Lett.* **2002**, 4, 2577-2579.

Benzilic Acid Rearrangement 52

1. Liebig, J. *Ann.* **1838**, 27.
2. Zinin, N. Studies on benzoyl derivatives. *Ann.* **1839**, 31, 329-333.
3. Selman, S., Easthan, J. F. Benzilic acid and related rearrangements. *Quart. Rev., Chem. Soc.* **1960**, 14, 221-235.
4. Cram, D. J. *Fundamentals of Carbanion Chemistry (Organic Chemistry, Vol. 4)* (Academic, New York, **1965**) 281 pp.
5. Gill, G. B. Benzyl-benzilic acid rearrangements. in *Comp. Org. Synth.* (eds. Trost, B. M.,Fleming, I.), *3*, 821-838 (Pergamon, Oxford, **1991**).
6. Bowden, K., Fabian, W. M. F. Reactions of carbonyl compounds in basic solutions. Part 36: the base-catalysed reactions of 1,2-dicarbonyl compounds. *J. Phys. Org. Chem.* **2001**, 14, 794-796.
7. Toda, F., Tanaka, K., Kagawa, Y., Sakaino, Y. Benzilic acid rearrangement in the solid state. *Chem. Lett.* **1990**, 373-376.
8. Wasserman, H. H., Ennis, D. S., Vu, C. B., Schulte, G. K. Benzilic acid rearrangements in the reactions of aryl vicinal tricarbonyl derivatives with aldehyde Schiff bases. *Tetrahedron Lett.* **1991**, 32, 6039-6042.
9. Polackova, V., Toma, S. Effect of ultrasound on the benzil-benzilic acid rearrangement under phase-transfer conditions. *Chemical Papers* **1996**, 50, 146-147.
10. Yu, H.-M., Chen, S.-T., Tseng, M.-J., Chen, S.-T., Wang, K.-T. Microwave-assisted heterogeneous benzil-benzilic acid rearrangement. *J. Chem. Res., Synop.* **1999**, 62-63.
11. Rajyaguru, I., Rzepa, H. S. A MNDO SCF-MO study of the mechanism of the benzilic acid and related rearrangements. *J. Chem. Soc., Perkin Trans. 2* **1987**, 1819-1827.
12. O'Meara, D., Richards, G. N. Mechanism of saccharinic acid formation. IV. Influence of cations in the benzilic acid rearrangement of glyoxal. *J. Chem. Soc.* **1960**, 1944-1945.

13. S.Warren, K., Neville, O. K., Hendley, E. C. Mechanism study of a benzilic acid-type rearrangement. *J. Org. Chem.* **1963**, 28, 2152-2153.
14. Black, D. S. C., Srivastava, R. C. Metal template reactions. I. Benzilic acid rearrangement of dipyridylglyoxal compounds promoted by nickel(II) and cobalt(II) ions. *Aust. J. Chem.* **1969**, 22, 1439-1447.
15. Novelli, A., Barrio, J. R. Carbon-14 tracer studies in the benzilic acid type rearrangement of 1-phenyl-and 1-(4-methoxyphenyl)-2-(3-pyridyl) glyoxal. *Tetrahedron Lett.* **1969**, 3671-3672.
16. Screttas, C. G., Micha-Screttas, M., Cazianis, C. T. The benzilic ester rearrangement. Evidence for a set pathway in the benzilic ester and/or acid rearrangement. *Tetrahedron Lett.* **1983**, 24, 3287-3288.
17. Askin, D., Reamer, R. A., Joe, D., Volante, R. P., Shinkai, I. A mechanistic study of the FK-506 tricarbonyl system rearrangement: synthesis of C.9 labeled FK-506. *Tetrahedron Lett.* **1989**, 30, 6121-6124.
18. Robinson, J. M., Flynn, E. T., McMahan, T. L., Simpson, S. L., Trisler, J. C., Conn, K. B. Benzoin enediol dianion and hydroxide ion in DMSO: a single electron transfer reduction system driven by the irreversible benzilic acid rearrangement. *J. Org. Chem.* **1991**, 56, 6709-6712.
19. Stoltz, B. M., Wood, J. L. The stereoselective ring contraction of a pyranosylated indolecarbazole. A biosynthetic link between K252a and staurosporine? *Tetrahedron Lett.* **1996**, 37, 3929-3930.
20. Kym, P. R., Wilson, S. R., Gritton, W. H., Katzenellenbogen, J. A. Novel steroids from cetyltrimethylammonium permanganate-initiated oxidative rearrangements of 16-dehydroprogesterone. *Tetrahedron Lett.* **1994**, 35, 2833-2836.
21. Hatsui, T., Wang, J.-J., Takeshita, H. Synthetic photochemistry. LXVII. A total synthesis of (±)-hinesol and (±)-agarospirol via retro-benzilic acid rearrangement. *Bull. Chem. Soc. Jpn.* **1995**, 68, 2393-2399.
22. Grieco, P. A., Collins, J. L., Huffman, J. C. Synthetic Studies on Quassinoids: Synthesis of (±)-Shinjudilactone and (±)-13-epi-Shinjudilactone. *J. Org. Chem.* **1998**, 63, 9576-9579.

Benzoin and Retro-Benzoin Condensation 54

Related reactions: Acyloin condensation;

1. Lapworth, A. J. Reactions involving the addition of hydrogen cyanide to carbon compounds. *J. Chem. Soc.* **1903**, 995.
2. Lapworth, A. J. Reactions involving the addition of hydrogen cyanide to carbon compounds. Part II. Cyanohydrins regarded as complex acids. *J. Chem. Soc.* **1904**, 1206-1215.
3. Staudinger, H. The autooxidation of organic compounds: connection between autooxidation and benzoin formation. *Ber. Dtsch. Chem. Ges.* **1913**, 46, 3535-3538.
4. Ide, W. S., Buck, J. S. Synthesis of benzoins. *Org. React.* **1948**, 4, 269-304.
5. Imoto, M. Acyloin condensation reactions. *Setchaku* **1976**, 20, 270-271.
6. Castells, J., Lopez - Calahorra, F. Thiamine other thiazolium salts, and related compounds. Structural studies and discussion of their catalytic activity. *Trends in Heterocyclic Chemistry* **1990**, 1, 35-53.
7. Hassner, A., Rai, K. M. L. The benzoin and related acyl anion equivalent reactions. in *Comp. Org. Synth.* (eds. Trost, B. M.,Fleming, I.), *1*, 541-578 (Pergamon, Oxford, **1991**).
8. Stetter, H., Kuhlmann, H. The catalyzed nucleophilic addition of aldehydes to electrophilic double bonds. *Org. React.* **1991**, 40, 407-496.
9. Tagaki, W., Tamura, Y., Yano, Y. Asymmetric benzoin condensation catalyzed by optically active thiazolium salts in micellar two-phase media. *Bull. Chem. Soc. Jpn.* **1980**, 53, 478-480.
10. Castells, J., Dunach, E. Polymer-supported quaternary ammonium cyanides and their use as catalysts in the benzoin condensation. *Chem. Lett.* **1984**, 1859-1860.
11. Knight, R. L., Leeper, F. J. Comparison of chiral thiazolium and triazolium salts as asymmetric catalysts for the benzoin condensation. *J. Chem. Soc., Perkin Trans. 1* **1998**, 1891-1894.
12. Davis, J. H., Jr., Forrester, K. J. Thiazolium-ion based organic ionic liquids (OILs). Novel OILs which promote the benzoin condensation. *Tetrahedron Lett.* **1999**, 40, 1621-1622.
13. Duenkelmann, P., Kolter-Jung, D., Nitsche, A., Demir, A. S., Siegert, P., Lingen, B., Baumann, M., Pohl, M., Mueller, M. Development of a Donor-Acceptor Concept for Enzymatic Cross-Coupling Reactions of Aldehydes: The First Asymmetric Cross-Benzoin Condensation. *J. Am. Chem. Soc.* **2002**, 124, 12084-12085.
14. Enders, D., Kallfass, U. An efficient nucleophilic carbene catalyst for the asymmetric benzoin condensation. *Angew. Chem., Int. Ed. Engl.* **2002**, 41, 1743-1745.
15. Hachisu, Y., Bode, J. W., Suzuki, K. Catalytic Intramolecular Crossed Aldehyde-Ketone Benzoin Reactions: A Novel Synthesis of Functionalized Preanthraquinones. *J. Am. Chem. Soc.* **2003**, 125, 8432-8433.
16. Xin, L., Johnson, J. S. Kinetic control in direct α-silyloxy ketone synthesis: A new regiospecific catalyzed cross silyl benzoin reaction. *Angew. Chem., Int. Ed. Engl.* **2003**, 42, 2534-2536.
17. Castells, J., Lopez-Calahorra, F., Domingo, L. Postulation of bis(thiazolin-2-ylidene)s as the catalytic species in the benzoin condensation catalyzed by a thiazolium salt plus base. *J. Org. Chem.* **1988**, 53, 4433-4436.
18. Lopez-Celahorra, F., Castells, J., Domingo, L., Marti, J., Bofill, J. M. Use of 3,3'-polymethylene-bridged thiazolium salts plus bases as catalysts of the benzoin condensation and its mechanistic implications: proposal of a new mechanism in aprotic conditions. *Heterocycles* **1994**, 37, 1579-1597.
19. White, M. J., Leeper, F. J. Kinetics of the Thiazolium Ion-Catalyzed Benzoin Condensation. *J. Org. Chem.* **2001**, 66, 5124-5131.
20. Schowen, R. L., Kuebrich, J. P., Wang, M.-S., Lupes, M. E. Mechanism of the benzoin condensation. *J. Am. Chem. Soc.* **1971**, 93, 1214-1220.
21. Correia, J. Isolation of the intermediates in a benzoin-type condensation. *J. Org. Chem.* **1983**, 48, 3343-3344.
22. Lopez-Calahorrra, F., Castells, J. Reaction mechanism of the benzoin condensation catalyzed by a thiazolium salt plus base or by a bis(thiazolidin-2-ylidene). *Afinidad* **1993**, 50, 461-466.
23. Breslow, R., Kim, R. The thiazolium catalyzed benzoin condensation with mild base does not involve a "dimer" intermediate. *Tetrahedron Lett.* **1994**, 35, 699-702.
24. Chen, Y.-T., Barletta, G. L., Haghjoo, K., Cheng, J. T., Jordan, F. Reactions of Benzaldehyde with Thiazolium Salts in Me_2SO: Evidence for Initial Formation of 2-(α-Hydroxybenzyl)thiazolium by Nucleophilic Addition, and for Dramatic Solvent Effects on Benzoin Formation. *J. Org. Chem.* **1994**, 59, 7714-7722.
25. Marti, J., Lopez-Calahorra, F., Bofill, J. M. A theoretical study of benzoin condensation. *THEOCHEM* **1995**, 339, 179-194.
26. Breslow, R., Schmuck, C. The mechanism of thiazolium catalysis. *Tetrahedron Lett.* **1996**, 37, 8241-8242.
27. Lopez-Calahorra, F., Castro, E., Ochoa, A., Marti, J. Further evidence about the role of bis(thiazolin-2-ylidene)s as the actual catalytic species in the generalized benzoin condensation. *Tetrahedron Lett.* **1996**, 37, 5019-5022.
28. Kluger, R. Lessons from thiamin-watching. *Pure Appl. Chem.* **1997**, 69, 1957-1967.
29. Motesharei, K., Myles, D. C. Multistep Synthesis on the Surface of Self-Assembled Thiolate Monolayers on Gold: Probing the Mechanism of the Thiazolium-Promoted Acyloin Condensation. *J. Am. Chem. Soc.* **1997**, 119, 6674-6675.
30. Ikeda, H., Horimoto, Y., Nakata, M., Ueno, A. Artificial holoenzymes for benzoin condensation using thiazolio-appended β-cyclodextrin dimers. *Tetrahedron Lett.* **2000**, 41, 6483-6487.
31. Pohl, M., Lingen, B., Muller, M. Thiamin-diphosphate-dependent enzymes: new aspects of asymmetric C-C bond formation. *Chemistry--A European Journal* **2002**, 8, 5288-5295.
32. Miyashita, A., Suzuki, Y., Okumura, Y., Iwamoto, K.-I., Higashino, T. Carbon-carbon bond cleavage of α-substituted benzoins by retro-benzoin condensation; a new method of synthesizing ketones. *Chem. Pharm. Bull.* **1998**, 46, 6-11.

33. Suzuki, Y., Takemura, Y., Iwamoto, K.-i., Higashino, T., Miyashita, A. Carbon-carbon bond cleavage of α-hydroxybenzylheteroarenes catalyzed by cyanide ion: retro-benzoin condensation affords ketones and heteroarenes and benzyl migration affords benzylheteroarenes and arenecarbaldehydes. *Chem. Pharm. Bull.* **1998**, 46, 199-206.
34. Barta, T. E., Stealey, M. A., Collins, P. W., Weier, R. M. Antiinflammatory 4,5-diarylimidazoles as selective cyclooxygenase inhibitors. *Bioorg. Med. Chem. Lett.* **1998**, 8, 3443-3448.

Related reactions: Danheiser benzannulation, Dötz benzannulation reaction;

1. Jones, R. R., Bergman, R. G. p-Benzyne. Generation as an intermediate in a thermal isomerization reaction and trapping evidence for the 1,4-benzenediyl structure. *J. Am. Chem. Soc.* **1972**, 94, 660-661.
2. Bergman, R. G. Reactive 1,4-dehydroaromatics. *Acc. Chem. Res.* **1973**, 6, 25-31.
3. Nicolaou, K. C., Dai, W. M., Tsay, S. C., Estevez, V. A., Wrasidlo, W. Designed enediynes: a new class of DNA-cleaving molecules with potent and selective anticancer activity. *Science* **1992**, 256, 1172-1178.
4. Nicolaou, K. C., Smith, A. L. Molecular design, chemical synthesis, and biological action of enediynes. *Acc. Chem. Res.* **1992**, 25, 497-503.
5. Nicolaou, K. C., Smith, A. L., Yue, E. W. Chemistry and biology of natural and designed enediynes. *Proc. Natl. Acad. Sci. U. S. A.* **1993**, 90, 5881-5888.
6. Nicolaou, K. The magic of enediyne chemistry. *Chem. Br.* **1994**, 30, 33-36, 41.
7. Grissom, J. W., Gunawardena, G. U., Klingberg, D., Huang, D. The chemistry of enediynes, enyne allenes and related compounds. *Tetrahedron* **1996**, 52, 6453-6518.
8. Wang, K. K. Cascade Radical Cyclizations via Biradicals Generated from Enediynes, Enyne-Allenes, and Enyne-Ketenes. *Chem. Rev.* **1996**, 96, 207-222.
9. Aubert, C., Buisine, O., Petit, M., Slowinski, F., Malacria, M. Cobalt-mediated cyclotrimerization and cycloisomerization reactions. Synthetic applications. *Pure Appl. Chem.* **1999**, 71, 1463-1470.
10. Konig, B. Changing the reactivity of enediynes by metal-ion coordination. *Eur. J. Org. Chem.* **2000**, 381-385.
11. Basak, A., Mandal, S., Bag, S. S. Chelation-Controlled Bergman Cyclization: Synthesis and Reactivity of Enediynyl Ligands. *Chem. Rev.* **2003**, 103, 4077-4094.
12. Jones, G. B., Russell, K. C. The photo-Bergman cycloaromatization of enediynes. *CRC Handbook of Organic Photochemistry and Photobiology (2nd Edition)* **2004**, 29/21-29/21.
13. Klein, M., Walenzyk, T., Koenig, B. Electronic effects on the Bergman cyclisation of enediynes. A review. *Collect. Czech. Chem. Commun.* **2004**, 69, 945-965.
14. Rawat, D. S., Zaleski, J. M. Geometric and electronic control of thermal Bergman cyclization. *Synlett* **2004**, 393-421.
15. Alabugin, I. V., Manoharan, M., Kovalenko, S. V. Tuning Rate of the Bergman Cyclization of Benzannelated Enediynes with Ortho Substituents. *Org. Lett.* **2002**, 4, 1119-1122.
16. O'Connor, J. M., Friese, S. J., Tichenor, M. Ruthenium-Mediated Cycloaromatization of Acyclic Enediynes and Dienynes at Ambient Temperature. *J. Am. Chem. Soc.* **2002**, 124, 3506-3507.
17. Feng, L., Kumar, D., Kerwin, S. M. An Extremely Facile Aza-Bergman Rearrangement of Sterically Unencumbered Acyclic 3-Aza-3-ene-1,5-diynes. *J. Org. Chem.* **2003**, 68, 2234-2242.
18. Kraft, B. J., Coalter, N. L., Nath, M., Clark, A. E., Siedle, A. R., Huffman, J. C., Zaleski, J. M. Photothermally Induced Bergman Cyclization of Metalloenediynes via Near-Infrared Ligand-to-Metal Charge-Transfer Excitation. *Inorg. Chem.* **2003**, 42, 1663-1672.
19. Zheng, M., DiRico, K. J., Kirchhoff, M. M., Phillips, K. M., Cuff, L. M., Johnson, R. P. Photorearrangements of acyclic conjugated enynes: a photochemical analog of the Bergman rearrangement. *J. Am. Chem. Soc.* **1993**, 115, 12167-12168.
20. Lindh, R., Lee, T. J., Bernhardsson, A., Persson, B. J., Karlstroem, G. Extended ab Initio and Theoretical Thermodynamics Studies of the Bergman Reaction and the Energy Splitting of the Singlet o-, m-, and p-Benzynes. *J. Am. Chem. Soc.* **1995**, 117, 7186-7194.
21. Chen, W.-C., Chang, N.-y., Yu, C.-h. Density Functional Study of Bergman Cyclization of Enediynes. *J. Phys. Chem. A* **1998**, 102, 2584-2593.
22. Cramer, C. J. Bergman, Aza-Bergman, and Protonated Aza-Bergman Cyclizations and Intermediate 2,5-Arynes: Chemistry and Challenges to Computation. *J. Am. Chem. Soc.* **1998**, 120, 6261-6269.
23. Graefenstein, J., Hjerpe, A. M., Kraka, E., Cremer, D. An Accurate Description of the Bergman Reaction Using Restricted and Unrestricted DFT: Stability Test, Spin Density, and On-Top Pair Density. *J. Phys. Chem. A* **2000**, 104, 1748-1761.
24. Rothchild, R., Sapse, A.-M., Balkova, A., Lown, J. W. Ab initio calculations on (-)-calicheamicinone and some of its reactions involved in its activation and interaction with DNA. *J. Biomol. Struct. Dyn.* **2000**, 18, 413-421.
25. Clark, A. E., Davidson, E. R., Zaleski, J. M. UDFT and MCSCF Descriptions of the Photochemical Bergman Cyclization of Enediynes. *J. Am. Chem. Soc.* **2001**, 123, 2650-2657.
26. Jones, G. B., Warner, P. M. On the Mechanism of Quinone Formation from the Bergman Cyclization: Some Theoretical Insights. *J. Org. Chem.* **2001**, 66, 8669-8672.
27. Jones, G. B., Warner, P. M. Electronic Control of the Bergman Cyclization: The Remarkable Role of Vinyl Substitution. *J. Am. Chem. Soc.* **2001**, 123, 2134-2145.
28. Koenig, B., Pitsch, W., Klein, M., Vasold, R., Prall, M., Schreiner, P. R. Carbonyl- and Carboxyl-Substituted Enediynes: Synthesis, Computations, and Thermal Reactivity. *J. Org. Chem.* **2001**, 66, 1742-1746.
29. Rawat, D. S., Zaleski, J. M. Mg^{2+}-Induced Thermal Enediyne Cyclization at Ambient Temperature. *J. Am. Chem. Soc.* **2001**, 123, 9675-9676.
30. Sapse, A.-M., Rothchild, R., Kumar, R., Lown, J. W. Ab initio studies of the reaction of hydrogen transfer from DNA to the calicheamicinone diradical. *Molecular Medicine (Baltimore, MD, United States)* **2001**, 7, 797-802.
31. Ahlstrom, B., Kraka, E., Cremer, D. The Bergman reaction of dynemicin A - a quantum chemical investigation. *Chem. Phys. Lett.* **2002**, 361, 129-135.
32. Jones, G. B., Wright, J. M., Hynd, G., Wyatt, J. K., Warner, P. M., Huber, R. S., Li, A., Kilgore, M. W., Sticca, R. P., Pollenz, R. S. Oxa-Enediynes: Probing the Electronic and Stereoelectronic Contributions to the Bergman Cycloaromatization. *J. Org. Chem.* **2002**, 67, 5727-5732.
33. Chen, W.-C., Zou, J.-W., Yu, C.-H. Density Functional Study of the Ring Effect on the Myers-Saito Cyclization and a Comparison with the Bergman Cyclization. *J. Org. Chem.* **2003**, 68, 3663-3672.
34. Bowles, D. M., Palmer, G. J., Landis, C. A., Scott, J. L., Anthony, J. E. The Bergman reaction as a synthetic tool: advantages and restrictions. *Tetrahedron* **2001**, 57, 3753-3760.
35. Bharucha, K. N., Marsh, R. M., Minto, R. E., Bergman, R. G. Double cycloaromatization of (Z,Z)-deca-3,7-diene-1,5,9-triyne: evidence for the intermediacy and diradical character of 2,6-didehydronaphthalene. *J. Am. Chem. Soc.* **1992**, 114, 3120-3121.
36. Schmittel, M., Kiau, S. Thermal and electron-transfer induced reactions of enediynes and enyne-allenes. Part 9. Electron-transfer versus acid catalysis in enediyne cyclizations. *Liebigs Ann. Chem.* **1997**, 1391-1399.
37. Schreiner, P. R. Cyclic enediynes: relationship between ring size, alkyne carbon distance, and cyclization barrier. *Chem. Commun.* **1998**, 483-484.
38. Kaneko, T., Takahashi, M., Hirama, M. Benzannelation alters the rate limiting step in enediyne cycloaromatization. *Tetrahedron Lett.* **1999**, 40, 2015-2018.
39. Choy, N., Kim, C. S., Ballestero, C., Artigas, L., Diez, C., Lichtenberger, F., Shapiro, J., Russell, K. C. Linear free energy relationships in the Bergman cyclization of 4-substituted-1,2-diethynylbenzenes. *Tetrahedron Lett.* **2000**, 41, 6955-6958.
40. O'Connor, J. M., Lee, L. I., Gantzel, P., Rheingold, A. L., Lam, K.-C. Inhibition and Acceleration of the Bergman Cycloaromatization Reaction by the Pentamethylcyclopentadienyl Ruthenium Cation. *J. Am. Chem. Soc.* **2000**, 122, 12057-12058.

41. Choy, N., Blanco, B., Wen, J., Krishan, A., Russell, K. C. Photochemical and thermal Bergman cyclization of a pyrimidine enediynol and enediynone. *Org. Lett.* **2000**, 2, 3761-3764.
42. Aihara, H., Jaquinod, L., Nurco, D. J., Smith, K. M. Multicarbocycle formation mediated by arenoporphyrin 1,4-diradicals: Synthesis of picenoporphyrins. *Angew. Chem., Int. Ed. Engl.* **2001**, 40, 3439-3441.
43. Haseltine, J. N., Danishefsky, S. J. Installation of the allylic trisulfide functionality of the enediyne antibiotics. Thiol-induced reductive actuation of the Bergman process. *J. Org. Chem.* **1990**, 55, 2576-2578.

1. Biginelli, P. The urea-aldehyde derivatives of acetoacetic esters. *Ber.* **1891**, 24, 2962-2967.
2. Biginelli, P. The urea-aldehyde derivatives of acetoacetic esters. *Gazz. Chim. Ital.* **1893**, 23, 360-416.
3. Brown, D. J. *The Pyrimidines* (**1962**) 774 pp.
4. Brown, D. J. *The Pyrimidines, Supplement 1 (The Chemistry of Heterocyclic Compounds, Vol. 16)* (**1970**) 1127 pp.
5. Kappe, C. O. 100 Years of the Biginelli dihydropyrimidine synthesis. *Tetrahedron* **1993**, 49, 6937-6963.
6. Kappe, C. O. 4-Aryldihydropyrimidines via the Biginelli condensation: aza analogs of nifedipine-type calcium channel modulators. *Molecules [Electronic Publication]* **1998**, 3, 1-9.
7. Bannwarth, W. Multicomponent condensations (MCCs). *Methods and Principles in Medicinal Chemistry* **2000**, 9, 6-21.
8. Kappe, C. O. Biologically active dihydropyrimidones of the Biginelli-type - a literature survey. *Eur. J. Med. Chem.* **2000**, 35, 1043-1052.
9. Kappe, C. O. Recent Advances in the Biginelli Dihydropyrimidine Synthesis. New Tricks from an Old Dog. *Acc. Chem. Res.* **2000**, 33, 879-888.
10. Tuch, A., Walle, S. Multicomponent reactions. *Handbook of Combinatorial Chemistry* **2002**, 2, 685-705.
11. Dondoni, A., Massi, A. Decoration of dihydropyrimidine and dihydropyridine scaffolds with sugars via Biginelli and Hantzsch multicomponent reactions: An efficient entry to a collection of artificial nucleosides. *Mol. Divers.* **2003**, 6, 261-270.
12. Kappe, C. O. The generation of dihydropyrimidine libraries utilizing Biginelli multi-component chemistry. *QSAR & Combinatorial Science* **2003**, 22, 630-645.
13. Aron, Z. D., Overman, L. E. The tethered Biginelli condensation in natural product synthesis. *Chem. Commun.* **2004**, 253-265.
14. Kappe, C. O., Stadler, A. The Biginelli dihydropyrimidinone synthesis. *Org. React.* **2004**, 63, 1-116.
15. Atwal, K. S., O'Reilly, B. C., Gougoutas, J. Z., Malley, M. F. Synthesis of substituted 1,2,3,4-tetrahydro-6-methyl-2-thioxo-5-pyrimidinecarboxylic acid esters. *Heterocycles* **1987**, 26, 1189-1192.
16. Atwal, K. S., Rovnyak, G. C., O'Reilly, B. C., Schwartz, J. Substituted 1,4-dihydropyrimidines. 3. Synthesis of selectively functionalized 2-hetero-1,4-dihydropyrimidines. *J. Org. Chem.* **1989**, 54, 5898-5907.
17. Atwal, K. S., Rovnyak, G. C., Schwartz, J., Moreland, S., Hedberg, A., Gougoutas, J. Z., Malley, M. F., Floyd, D. M. Dihydropyrimidine calcium channel blockers: 2-heterosubstituted 4-aryl-1,4-dihydro-6-methyl-5-pyrimidinecarboxylic acid esters as potent mimics of dihydropyridines. *J. Med. Chem.* **1990**, 33, 1510-1515.
18. Studer, A., Jeger, P., Wipf, P., Curran, D. P. Fluorous Synthesis: Fluorous Protocols for the Ugi and Biginelli Multicomponent Condensations. *J. Org. Chem.* **1997**, 62, 2917-2924.
19. Kappe, C. O., Falsone, S. F. Synthesis and reactions of Biginelli compounds. Part 12. Polyphosphate ester-mediated synthesis of dihydropyrimidines. Improved conditions for the Biginelli reaction. *Synlett* **1998**, 718-720.
20. Shutalev, A. D., Kishko, E. A., Sivova, N. V., Kuznetsov, A. Y. A new convenient synthesis of 5-acyl-1,2,3,4-tetrahydropyrimidine-2-thiones/ones. *Molecules [Electronic Publication]* **1998**, 3, 100-106.
21. Lu, J., Ma, H. R. Iron(III)-catalyzed synthesis of dihydropyrimidinones. Improved conditions for the Biginelli reaction. *Synlett* **2000**, 63-64.
22. Ranu, B. C., Hajra, A., Jana, U. Indium(III) chloride-catalyzed one-pot synthesis of dihydropyrimidinones by a three-component coupling of 1,3-dicarbonyl compounds, aldehydes, and urea: an improved procedure for the Biginelli reaction. *J. Org. Chem.* **2000**, 65, 6270-6272.
23. Stefani, H. A., Gatti, P. M. 3,4-Dihydropyrimidin-2(1H)-ones: fast synthesis under microwave irradiation in solvent free conditions. *Synth. Commun.* **2000**, 30, 2165-2173.
24. Yadav, J. S., Reddy, B. V. S., Reddy, E. J., Ramalingam, T. Microwave-assisted efficient synthesis of dihydro pyrimidines: improved high yielding protocol for the Biginelli reaction. *J. Chem. Res., Synop.* **2000**, 354-355.
25. Peng, J., Deng, Y. Ionic liquids catalyzed Biginelli reaction under solvent-free conditions. *Tetrahedron Lett.* **2001**, 42, 5917-5919.
26. Ramalinga, K., Vijayalakshmi, P., Kaimal, T. N. B. Bismuth(III)-catalyzed synthesis of dihydropyrimidinones: improved protocol conditions for the Biginelli reaction. *Synlett* **2001**, 863-865.
27. Yadav, J. S., Reddy, B. V. S., Srinivas, R., Venugopal, C., Ramalingam, T. $LiClO_4$-catalyzed one-pot synthesis of dihydropyrimidinones: an improved protocol for the Biginelli reaction. *Synthesis* **2001**, 1341-1345.
28. Dondoni, A., Massi, A., Sabbatini, S. Improved synthesis and preparative scale resolution of racemic monastrol. *Tetrahedron Lett.* **2002**, 43, 5913-5916.
29. Fu, N. Y., Mei, L. P., Yuan, Y. F., Wang, J. T. Indium(III) tribromide: an excellent catalyst for Biginelli reaction. *Chin. Chem. Lett.* **2002**, 13, 921-922.
30. Fu, N.-Y., Yuan, Y.-F., Cao, Z., Wang, S.-W., Wang, J.-T., Peppe, C. Indium(III) bromide-catalyzed preparation of dihydropyrimidinones: improved protocol conditions for the Biginelli reaction. *Tetrahedron* **2002**, 58, 4801-4807.
31. Jin, T., Zhang, S., Li, T. *p*-Toluenesulfonic acid-catalyzed efficient synthesis of dihydropyrimidines: improved high yielding protocol for the Biginelli reaction. *Synth. Commun.* **2002**, 32, 1847-1851.
32. Ranu, B. C., Hajra, A., Dey, S. S. A Practical and Green Approach towards Synthesis of Dihydropyrimidinones without Any Solvent or Catalyst. *Org. Process Res. Dev.* **2002**, 6, 817-818.
33. Reddy, A. V., Reddy, V. L. N., Ravinder, K., Venkateswarlu, Y. Synthesis of dihydropyrimidinones: An improved conditions for the Biginelli reaction. *Heterocycl. Commun.* **2002**, 8, 459-464.
34. Bose, D. S., Fatima, L., Mereyala, H. B. Green Chemistry Approaches to the Synthesis of 5-Alkoxycarbonyl-4-aryl-3,4- dihydropyrimidin-2(1H)-ones by a Three-Component Coupling of One-Pot Condensation Reaction: Comparison of Ethanol, Water, and Solvent-free Conditions. *J. Org. Chem.* **2003**, 68, 587-590.
35. Maiti, G., Kundu, P., Guin, C. One-pot synthesis of dihydropyrimidinones catalyzed by lithium bromide: an improved procedure for the Biginelli reaction. *Tetrahedron Lett.* **2003**, 44, 2757-2758.
36. Sabitha, G., Reddy, G. S. K. K., Reddy, C. S., Yadav, J. S. One-pot synthesis of dihydropyrimidinones using iodotrimethylsilane. Facile and new improved protocol for the Biginelli reaction at room temperature. *Synlett* **2003**, 858-860.
37. Varala, R., Alam, M. M., Adapa, S. R. Bismuth triflate catalyzed one-pot synthesis of 3,4-dihydropyrimidin-2(1H)-ones: an improved protocol for the Biginelli reaction. *Synlett* **2003**, 67-70.
38. Xia, M., Wang, Y.-g. An efficient protocol for the liquid-phase synthesis of methyl 3,4-dihydropyrimidin-2(1H)-one-5-carboxylate derivatives. *Synthesis* **2003**, 262-266.
39. Fabian, W. M. F., Semones, M. A., Kappe, C. O. Synthesis and reactions of Biginelli compounds. Part 11. Ring conformation and ester orientation in dihydropyrimidinecarboxylates: a combined theoretical (ab initio, density functional) and x-ray crystallographic study. *THEOCHEM* **1998**, 432, 219-228.
40. Kappe, C. O., Falsone, S. F., Fabian, W. M. F., Belaj, F. Synthesis and reactions of Biginelli compounds. 13. Isolation, conformational analysis and x-ray structure determination of a trifluoromethyl-stabilized hexahydropyrimidine - an intermediate in the Biginelli reaction. *Heterocycles* **1999**, 51, 77-84.
41. Sabitha, G., Reddy, G. S. K. K., Reddy, K. B., Yadav, J. S. Vanadium(III) chloride catalyzed Biginelli condensation: solution phase library generation of dihydropyrimidin-(2H)-ones. *Tetrahedron Lett.* **2003**, 44, 6497-6499.
42. Kappe, C. O., Uray, G., Roschger, P., Lindner, W., Kratky, C., Keller, W. Synthesis and reactions of Biginelli compounds. 5. Facile preparation and resolution of a stable 5-dihydropyrimidinecarboxylic acid. *Tetrahedron* **1992**, 48, 5473-5480.

43. Wipf, P., Cunningham, A. A solid phase protocol of the Biginelli dihydropyrimidine synthesis suitable for combinatorial chemistry. *Tetrahedron Lett.* **1995**, 36, 7819-7822.
44. Xia, M., Wang, Y.-G. Soluble polymer-supported synthesis of Biginelli compounds. *Tetrahedron Lett.* **2002**, 43, 7703-7705.
45. Hinkel, L. E., Hey, D. H. The condensation of benzaldeyde and ethyl acetoacetate with urea and thiourea. *Recl. Trav. Chim. Pays-Bas* **1929**, 48, 1280-1286.
46. Folkers, K., Johnson, T. B. Pyrimidines. CXXXVI. The mechanism of formation of tetrahydropyrimidines by the Biginelli reaction. *J. Am. Chem. Soc.* **1933**, 55, 3784-3791.
47. Ehsan, A., Karimullah. Synthesis of pyrimidine. *Pak. J. Sci. Ind. Res.* **1967**, 10, 83-85.
48. Sweet, F., Fissekis, J. D. Synthesis of 3,4-dihydro-2(1H)-pyrimidinones and the mechanism of the Biginelli reaction. *J. Am. Chem. Soc.* **1973**, 95, 8741.
49. Zigeuner, G., Knopp, C., Blaschke, H. Heterocyclics, 48. Tetrahydro-6-methyl- and -6-phenyl-2-oxopyrimidine-5-carboxylic acids and derivatives. *Monatsh. Chem.* **1976**, 107, 587-603.
50. Kappe, C. O. A reexamination of the mechanism of the Biginelli dihydropyrimidine synthesis. Support for an N-acyliminium ion intermediate. *J. Org. Chem.* **1997**, 62, 7201-7204.
51. Overman, L. E., Rabinowitz, M. H. Studies toward the total synthesis of (+)-ptilomycalin A. Use of a tethered Biginelli condensation for the preparation of an advanced tricyclic intermediate. *J. Org. Chem.* **1993**, 58, 3235-3237.
52. Coffey, D. S., Overman, L. E., Stappenbeck, F. Enantioselective Total Syntheses of 13,14,15-Isocrambescidin 800 and 13,14,15-Isocrambescidin 657. *J. Am. Chem. Soc.* **2000**, 122, 4904-4914.
53. Cohen, F., Overman, L. E. Enantioselective total synthesis of batzelladine F: structural revision and stereochemical definition. *J. Am. Chem. Soc.* **2001**, 123, 10782-10783.

Birch Reduction 60

1. Birch, A. J. Reduction by dissolving metals. I. *J. Chem. Soc.* **1944**, 430-436.
2. Robinson, B. Reduction of indoles and related compounds. *Chem. Rev.* **1969**, 69, 785-797.
3. Harvey, R. G. Metal-ammonia reduction of aromatic molecules. *Synthesis* **1970**, 2, 161-172.
4. Dryden, H. L., Jr. Reductions of steroids by metal-ammonia solutions. *Org. React.* **1972**, 1, 1-60.
5. Kaiser, E. M. Comparison of methods using lithium/amine and Birch reduction systems. *Synthesis* **1972**, 391-415.
6. Hook, J. M., Mander, L. N. Recent developments in the Birch reduction of aromatic compounds. Applications to the synthesis of natural products. *Nat. Prod. Rep.* **1986**, 3, 35-85.
7. Rabideau, P. W. The metal-ammonia reduction of aromatic compounds. *Tetrahedron* **1989**, 45, 1579-1603.
8. Rabideau, P. W., Marcinow, Z. The Birch reduction of aromatic compounds. *Org. React.* **1992**, 42, 1-334.
9. Birch, A. J. The Birch reduction in organic synthesis. *Pure Appl. Chem.* **1996**, 68, 553-556.
10. Donohoe, T. J., Guyo, P. M., Raoof, A. Birch reduction of aromatic heterocycles. *Targets in Heterocyclic Systems* **1999**, 3, 117-145.
11. Schultz, A. G. The asymmetric Birch reduction and reduction-alkylation strategies for synthesis of natural products. *Chem. Commun.* **1999**, 1263-1271.
12. Zvilichovsky, G., Gurvich, V. Transformation of aromatic amino acids by reduction and ozonolysis. A novel approach to the synthesis of unnatural α-amino acids. *Rec. Res. Dev. Org. Chem.* **1999**, 3, 87-104.
13. Pellissier, H., Santelli, M. The Birch reduction of steroids. A review. *Org. Prep. Proced. Int.* **2002**, 34, 609,611-642.
14. Kraus, J. N. Enter the strange world of cold chemistry. *Chem. Eng.* **2003**, 110, 50-55.
15. Subba Rao, G. S. R. Birch reduction and its application in the total synthesis of natural products. *Pure Appl. Chem.* **2003**, 75, 1443-1451.
16. Kwart, H., Conley, R. A. Modified Birch reductions. Lithium in n-alkylamines. *J. Org. Chem.* **1973**, 38, 2011-2016.
17. Ashmore, J. W., Helmkamp, G. K. Improved procedure for the Birch reduction of indole and carbazole. *Org. Prep. Proced. Int.* **1976**, 8, 223-225.
18. Swenson, K. E., Zemach, D., Nanjundiah, C., Kariv-Miller, E. Birch reductions of methoxyaromatics in aqueous solution. *J. Org. Chem.* **1983**, 48, 1777-1779.
19. Hamilton, R. J., Mander, L. N., Sethi, S. P. Improved methods for the reductive alkylation of methoxybenzoic acids and esters. Applications to the synthesis of bicyclic ketones. *Tetrahedron* **1986**, 42, 2881-2892.
20. Radivoy, G., Alonso, F., Yus, M. The $NiCl_2.2H_2O$-Li-arene (cat.) combination as reducing system 5. Reduction of sulfonates and aromatic compounds with the $NiCl_2.2H_2O$-Li-arene (cat.) combination. *Tetrahedron* **1999**, 55, 14479-14490.
21. Donohoe, T. J., Harji, R. R., Cousins, R. P. C. The partial reduction of heterocycles: an alternative to the Birch reduction. *Tetrahedron Lett.* **2000**, 41, 1331-1334.
22. Coll Toledano, J., Olivella, S. HMO prediction of the metal-ammonia reduction products of the fluorene series. *Anales de Quimica (1968-1979)* **1979**, 75, 107-109.
23. Birch, A. J., Hinde, A. L., Radom, L. A theoretical approach to the Birch reduction. Structures and stabilities of cyclohexadienyl radicals. *J. Am. Chem. Soc.* **1980**, 102, 4074-4080.
24. Birch, A. J., Hinde, A. L., Radom, L. A theoretical approach to the Birch reduction. Structures and stabilities of cyclohexadienyl anions. *J. Am. Chem. Soc.* **1980**, 102, 6430-6437.
25. Birch, A. J., Hinde, A. L., Radom, L. A theoretical approach to the Birch reduction. Structures and stabilities of the radical anions of substituted benzenes. *J. Am. Chem. Soc.* **1980**, 102, 3370-3376.
26. Paddon-Row, M. N., Hartcher, R. Orbital interactions. 7. The Birch reduction as a tool for exploring orbital interactions through bonds. Through-four-, -five-, and -six-bond interactions. *J. Am. Chem. Soc.* **1980**, 102, 671-678.
27. Birch, A. J., Hinde, A. L., Radom, L. A theoretical approach to the Birch reduction. Structures and stabilities of cyclohexadienes. *J. Am. Chem. Soc.* **1981**, 103, 284-289.
28. Toledano, J. C., Olivella, S. HMO interpretation of the Birch reduction mechanism of aromatic carboxylates. *Anales de Quimica, Serie C: Quimica Organica y Bioquimica* **1982**, 78, 187-189.
29. Chau, D. D., Paddon-Row, M. N., Patney, H. K. Orbital interactions. XII. Product studies and competition kinetic measurements of the Birch reduction of a series of hexahydrodimethanonaphthalenes and their interpretation in terms of orbital interactions through space and through bonds. *Aust. J. Chem.* **1983**, 36, 2423-2446.
30. Zimmerman, H. E., Wang, P. A. The regioselectivity of the Birch reduction. *J. Am. Chem. Soc.* **1993**, 115, 2205-2216.
31. Sinclair, S., Jorgensen, W. L. Computer Assisted Mechanistic Evaluation of Organic Reactions. 23. Dissolving Metal Reductions with Lithium in Liquid Ammonia Including the Birch Reduction. *J. Org. Chem.* **1994**, 59, 762-772.
32. Saa, J. M., Ballester, P., Deya, P. M., Capo, M., Garcias, X. Metal-Induced Reductive Cleavage Reactions: An Experimental and Theoretical (MNDO) Study on the Stereochemical Puzzle of Birch and Vinylogous Birch Processes. *J. Org. Chem.* **1996**, 61, 1035-1046.
33. Schafer, A., Schafer, B. Diastereoselective protonation after the Birch reduction of pyrroles. *Tetrahedron* **1999**, 55, 12309-12312.
34. Vaidyanathaswamy, R., Moorthy, S. N., Devaprabhakara, D. Stereochemistry and mechanism of the Birch reduction of cyclic allenes. *Indian J. Chem., Sect. B* **1977**, 15B, 309-312.
35. Collins, C. J., Hombach, H. P., Maxwell, B., Woody, M. C., Benjamin, B. M. Carbon-carbon cleavage during Birch-Hueckel-type reductions. *J. Am. Chem. Soc.* **1980**, 102, 851-853.
36. Rabideau, P. W., Huser, D. L. Protonation of anion intermediates in metal-ammonia reduction: 1,2- vs. 1,4-dihydro aromatic products. *J. Org. Chem.* **1983**, 48, 4266-4271.
37. Zimmerman, H. E., Wang, P. A. Regioselectivity of the Birch reduction. *J. Am. Chem. Soc.* **1990**, 112, 1280-1281.
38. Moebitz, H., Boll, M. A Birch-like Mechanism in Enzymatic Benzoyl-CoA Reduction: A Kinetic Study of Substrate Analogues Combined with an ab Initio Model. *Biochemistry* **2002**, 41, 1752-1758.
39. Donohoe, T. J., Guyo, P. M. Birch Reduction of Electron-Deficient Pyrroles. *J. Org. Chem.* **1996**, 61, 7664-7665.
40. Donohoe, T. J., Guyo, P. M., Harji, R. R., Helliwell, M. The Birch reduction of 3-substituted pyrroles. *Tetrahedron Lett.* **1998**, 39, 3075-3078.

41. Donohoe, T. J., Guillermin, J.-B., Frampton, C., Walter, D. S. The synthesis of (+)-nemorensic acid. *Chem. Commun.* **2000**, 465-466.
42. Kusama, H., Hara, R., Kawahara, S., Nishimori, T., Kashima, H., Nakamura, N., Morihira, K., Kuwajima, I. Enantioselective Total Synthesis of (-)-Taxol. *J. Am. Chem. Soc.* **2000**, 122, 3811-3820.
43. Schultz, A. G., Malachowski, W. P., Pan, Y. Asymmetric Total Synthesis of (+)-Apovincamine and a Formal Synthesis of (+)-Vincamine. Demonstration of a Practical "Asymmetric Linkage" between Aromatic Carboxylic Acids and Chiral Acyclic Substrates. *J. Org. Chem.* **1997**, 62, 1223-1229.
44. Mander, L. N., McLachlan, M. M. The Total Synthesis of the Galbulimima Alkaloid GB 13. *J. Am. Chem. Soc.* **2003**, 125, 2400-2401.

Bischler-Napieralski Isoquinoline Synthesis 62

Related reactions: Pictet-Spengler tetrahydroisoquinoline synthesis, Pomeranz-Fritsch reaction;

1. Bischler, A., Napieralski, B. A new method for the synthesis of isoquinolines. *Ber.* **1893**, 26, 1903-1908.
2. Whaley, W. M., Govindachari, T. R. Preparation of 3,4-dihydroisoquinolines and related compounds by the Bischler-Napieralski reaction. *Org. React.* **1951**, 6, 74-150.
3. Rozwadowska, M. D. Recent progress in the enantioselective synthesis of isoquinoline alkaloids. *Heterocycles* **1994**, 39, 903-931.
4. Lorsbach, B. A., Kurth, M. J. Carbon-Carbon Bond Forming Solid-Phase Reactions. *Chem. Rev.* **1999**, 99, 1549-1581.
5. Fodor, G., Nagubandi, S. Correlation of the von Braun, Ritter, Bischler-Napieralski, Beckmann, and Schmidt reactions via nitrilium salt intermediates. *Tetrahedron* **1980**, 36, 1279-1300.
6. Nagubandi, S., Fodor, G. Novel condensing agents for Bischler-Napieralski type cyclodehydration. *Heterocycles* **1981**, 15, 165-177.
7. Ramesh, D., Srinivasan, M. Phosphonitrilic chloride - a reagent for Bischler-Napieralski reaction. *Synth. Commun.* **1986**, 16, 1523-1527.
8. Aguirre, J. M., Alesso, E. N., Ibanez, A. F., Tombari, D. G., Moltrasio Iglesias, G. Y. Reaction of 1,2-diarylethylamides with ethyl polyphosphate (EPP): correlation of the von Braun, Ritter and Bischler-Napieralski reactions. *J. Heterocycl. Chem.* **1989**, 26, 25-27.
9. Bkhattacharjya, A., Chattopadhyay, P., Bhaumik, M., Pakrashi, S. C. Bischler-Napieralski cyclization with triphenylphosphine-carbon tetrachloride: one-pot synthesis of dihydroisoquinolines and β-carbolines. *J. Chem. Res., Synop.* **1989**, 228-229.
10. Roblot, F., Hocquemiller, R., Cave, A. Obtention of anthranil derivatives by a modified Bischler-Napieralski reaction. *J. Chem. Res., Synop.* **1989**, 344-345.
11. Ishikawa, T., Shimooka, K., Narioka, T., Noguchi, S., Saito, T., Ishikawa, A., Yamazaki, E., Harayama, T., Seki, H., Yamaguchi, K. Anomalous Substituent Effects in the Bischler-Napieralski Reaction of 2-Aryl Aromatic Formamides. *J. Org. Chem.* **2000**, 65, 9143-9151.
12. Saito, T., Yoshida, M., Ishikawa, T. Triphosgene: A versatile reagent for Bischler-Napieralski reaction. *Heterocycles* **2001**, 54, 437-438.
13. Judeh, Z. M. A., Ching, C. B., Bu, J., McCluskey, A. The first Bischler-Napieralski cyclization in a room temperature ionic liquid. *Tetrahedron Lett.* **2002**, 43, 5089-5091.
14. Nicoletti, M., O'Hagan, D., Slawin, A. M. Z. The asymmetric Bischler-Napieralski reaction: preparation of 1,3,4-trisubstituted 1,2,3,4-tetrahydroisoquinolines. *J. Chem. Soc., Perkin Trans. 1* **2002**, 116-121.
15. Pal, B., Jaisankar, P., Giri, V. S. Microwave assisted Pictet-Spengler and Bischler-Napieralski reactions. *Synth. Commun.* **2003**, 33, 2339-2348.
16. Fodor, G., Gal, J., Phillips, B. A. Mechanism of the Bischler-Napieralski reaction. *Angew. Chem., Int. Ed. Engl.* **1972**, 11, 919-920.
17. Quirante, J., Escolano, C., Merino, A., Bonjoch, J. First Total Synthesis of (±)-Melinonine-E and (±)-Strychnoxanthine Using a Radical Cyclization Process as the Core Ring-Forming Step. *J. Org. Chem.* **1998**, 63, 968-976.
18. Pampin, M. C., Estevez, J. C., Castedo, L., Estevez, R. J. First total synthesis of the 1,2,3,4-tetrahydronaphtho[2,1-f]isoquinoline annoretine. *Tetrahedron Lett.* **2001**, 42, 2307-2308.
19. Bergmeier, S. C., Seth, P. P. Aziridine-Allylsilane-Mediated Total Synthesis of (-)-Yohimbane. *J. Org. Chem.* **1999**, 64, 3237-3243.
20. Bungard, C. J., Morris, J. C. First Total Synthesis of the 7,3'-Linked Naphthylisoquinoline Alkaloid Ancistrocladidine. *Org. Lett.* **2002**, 4, 631-633.

Brook Rearrangement 64

1. Brook, A. G. Isomerism of some α-hydroxysilanes to silyl ethers. *J. Am. Chem. Soc.* **1958**, 80, 1886-1889.
2. Brook, A. G., Warner, C. M., McGriskin, M. E. Isomerization of some α-hydroxysilanes to silyl ethers. II. *J. Am. Chem. Soc.* **1959**, 81, 981-983.
3. Brook, A. G., Schwartz, N. V. Reaction of triphenylsilyl metallics with benzophenone. I. *J. Am. Chem. Soc.* **1960**, 82, 2435-2439.
4. Brook, A. G., Iachia, B. Isomerization of α-hydroxysilanes to silyl ethers. III. Triphenylsilylcarbinol. *J. Am. Chem. Soc.* **1961**, 83, 827-831.
5. Brook, A. G. Molecular rearrangements of organosilicon compounds. *Acc. Chem. Res.* **1974**, 7, 77-84.
6. Brook, A. G., Bassindale, A. R. *Molecular rearrangements of organosilicon compounds* (Academic Press, New York, **1980**) 149-227.
7. Page, P. C. B., Klair, S. S., Rosenthal, S. Synthesis and chemistry of acyl silanes. *Chem. Soc. Rev.* **1990**, 19, 147-195.
8. Jankowski, P., Raubo, P., Wicha, J. Tandem transformations initiated by the migration of a silyl group. Some new synthetic applications of silyloxiranes. *Synlett* **1994**, 985-992.
9. Kawashima, T. Small ring compounds containing highly coordinate Group 14 elements. *J. Organomet. Chem.* **2000**, 611, 256-263.
10. Portella, C., Brigaud, T., Lefebvre, O., Plantier-Royon, R. Convergent synthesis of fluoro and gem-difluoro compounds using (trifluoromethyl)trimethylsilane. *J. Fluorine Chem.* **2000**, 101, 193-198.
11. Kira, M., Iwamoto, T. Silyl migrations. *Chemistry of Organic Silicon Compounds* **2001**, 3, 853-948.
12. Moser, W. H. The Brook rearrangement in tandem bond formation strategies. *Tetrahedron* **2001**, 57, 2065-2084.
13. Behrens, K., Kneisel, B. O., Noltemeyer, M., Brueckner, R. Preparation of α-chiral crotylsilanes by retro-[1,4]-Brook rearrangements. A stereochemical study. *Liebigs Ann. Chem.* **1995**, 385-400.
14. Honda, T., Mori, M. An Aza-Brook Rearrangement of (α-Silylallyl)amine. *J. Org. Chem.* **1996**, 61, 1196-1197.
15. Gandon, V., Bertus, P., Szymoniak, J. New transformations from a 3-silyloxy-2-aza-1,3-diene: consecutive Zr-mediated retro-Brook rearrangement and reactions with electrophiles. *Tetrahedron* **2000**, 56, 4467-4472.
16. Paredes, M. D., Alonso, R. On the Radical Brook Rearrangement. Reactivity of α-Silyl Alcohols, α-Silyl Alcohol Nitrite Esters, and β-Haloacylsilanes under Radical-Forming Conditions. *J. Org. Chem.* **2000**, 65, 2292-2304.
17. Liu, G., Sieburth, S. M. Enantioselective a-Silyl Amino Acid Synthesis by Reverse-Aza-Brook Rearrangement. *Org. Lett.* **2003**, 5, 4677-4679.
18. Paleo, M. R., Calaza, M. I., Grana, P., Sardina, F. J. Stanna-Brook Rearrangement of Carboxylic Acid Derivatives. Synthetic Utility and Mechanistic Studies. *Org. Lett.* **2004**, 6, 1061-1063.
19. Yagi, K., Tsuritani, T., Takami, K., Shinokubo, H., Oshima, K. Reaction of Silyldihalomethyllithiums with Nitriles: Formation of α-Keto Acylsilanes via Azirines and 1,3-Rearrangement of Silyl Group from C to N. *J. Am. Chem. Soc.* **2004**, 126, 8618-8619.
20. Boche, G., Opel, A., Marsch, M., Harms, K., Haller, F., Lohrenz, J. C. W., Thuemmler, C., Koch, W. α-Oxygen-substituted organolithium compounds: calculations of the configurational stability and of $LiCH_2OH$ model structures, crystal structure of diphenyl(trimethylsiloxy)methyllithium.3 THF, and the stereochemistry of the (reverse) Brook rearrangement. *Chem. Ber.* **1992**, 125, 2265-2273.
21. Schlegel, H. B., Skancke, P. N. Theoretical study of reaction pathways for $F^- + H_3SiCHO$. *J. Am. Chem. Soc.* **1993**, 115, 10916-10924.
22. Schiesser, C. H., Styles, M. L. On the radical Brook and related reactions: an ab initio study of some (1,2)-silyl, germyl and stannyl translocations. *J. Chem. Soc., Perkin Trans. 2* **1997**, 2335-2340.

23. Gossage, R. A., Munoz-Martinez, E., Frey, H., Burgath, A., Lutz, M., Spek, A. L., Van Koten, G. A novel phenol for use in convergent and divergent dendrimer synthesis: access to core functionalizable trifurcate carbosilane dendrimers-the X-ray crystal structure of [1,3,5-tris{4-(triallylsilyl)phenyl ester}benzene]. *Chem.-- Eur. J.* **1999**, 5, 2191-2197.
24. Wang, Y., Dolg, M. Theoretical confirmation of the stereoselectivity in the reverse Brook rearrangement. *Tetrahedron* **1999**, 55, 12751-12756.
25. Pezacki, J. P., Loncke, P. G., Ross, J. P., Warkentin, J., Gadosy, T. A. Silicon Migration from Oxygen to Carbon and Decarbonylation in Methoxytriphenylsiloxycarbene. *Org. Lett.* **2000**, 2, 2733-2736.
26. Speier, J. L., Jr. The preparation and properties of (hydroxyorgano) silanes and related compounds. *J. Am. Chem. Soc.* **1952**, 74, 1003-1010.
27. West, R., Lowe, R., Stewart, H. F., Wright, A. New anionic rearrangements. XII. 1,2-Anionic rearrangement of alkoxysilanes. *J. Am. Chem. Soc.* **1971**, 93, 282-283.
28. Linderman, R. J., Ghannam, A. Synthetic utility and mechanistic studies of the aliphatic reverse Brook rearrangement. *J. Am. Chem. Soc.* **1990**, 112, 2392-2398.
29. Hoffmann, R., Brueckner, R. Asymmetric induction in reductively initiated [2,3]-Wittig and retro-[1,4]-Brook rearrangements of secondary carbanions. *Chem. Ber.* **1992**, 125, 1471-1484.
30. Jiang, X.-L., Bailey, W. F. Facile Retro-[1,4]-Brook Rearrangement of a [(2-Siloxycyclopentyl)methyl]lithium Species. *Organometallics* **1995**, 14, 5704-5707.
31. Lautens, M., Delanghe, P. H. M., Goh, J. B., Zhang, C. H. Studies in the Transmetalation of Cyclopropyl, Vinyl, and Epoxy Stannanes. *J. Org. Chem.* **1995**, 60, 4213-4227.
32. Kawashima, T., Naganuma, K., Okazaki, R. Generation and Decomposition of a Pentacoordinate Spirobis[1,2-oxasiletanide]. *Organometallics* **1998**, 17, 367-372.
33. Takeda, K., Sawada, Y., Sumi, K. Stereoselective Construction of Eight-Membered Carbocycles by Brook Rearrangement-Mediated [3 + 4] Annulation. *Org. Lett.* **2002**, 4, 1031-1033.
34. Moser, W. H., Zhang, J., Lecher, C. S., Frazier, T. L., Pink, M. Stereocontrolled [3 + 2] Annulations with Arene Chromium Tricarbonyl Complexes: Construction of Spirocyclic Compounds Related to Fredericamycin A. *Org. Lett.* **2002**, 4, 1981-1984.
35. Takeda, K., Nakane, D., Takeda, M. Synthesis of the Tricyclic Skeleton of Cyathins Using Brook Rearrangement-Mediated [3 + 4] Annulation. *Org. Lett.* **2000**, 2, 1903-1905.
36. Mi, Y., Schreiber, J. V., Corey, E. J. Total Synthesis of (+)-α-Onocerin in Four Steps via Four-Component Coupling and Tetracyclization Steps. *J. Am. Chem. Soc.* **2002**, 124, 11290-11291.

Brown Hydroboration Reaction....66

Related reactions: Schwartz hydrozirconation;

1. Brown, H. C., Rao, B. C. S. A new technique for the conversion of olefins into organoboranes and related alcohols. *J. Am. Chem. Soc.* **1956**, 78, 5694-5695.
2. Zweifel, G., Brown, H. C. Hydration of olefins, dienes, and acetylenes via hydroboration. *Org. React.* **1963**, 13, 1-54.
3. Brown, H. C., Negishi, E. Cyclic hydroboration of dienes. Simple convenient route to heterocyclic organoboranes. *Pure Appl. Chem.* **1972**, 29, 527-545.
4. Brown, H. C., Jadhav, P. K. Asymmetric hydroboration. *Asymmetric Synth.* **1983**, 2, 1-43.
5. Matteson, D. S. The use of chiral organoboranes in organic synthesis. *Synthesis* **1986**, 973-985.
6. Suzuki, A., Dhillon, R. S. Selective hydroboration and synthetic utility of organoboranes thus obtained. *Top. Curr. Chem.* **1986**, 130, 23-88.
7. Brown, H. C., Prasad, J. V. N. V. Hydroboration of heterocyclic olefins - a versatile route for the synthesis of both racemic and optically active heterocyclic compounds. *Heterocycles* **1987**, 25, 641-657.
8. Brown, H. C., Singaram, B. Organoboranes for synthesis. Substitution with retention. *Pure Appl. Chem.* **1987**, 59, 879-894.
9. Brown, H. C., Singaram, B. The development of a simple general procedure for synthesis of pure enantiomers via chiral organoboranes. *Acc. Chem. Res.* **1988**, 21, 287-293.
10. Blackburn, B. K., Sharpless, K. B. Rhodium(I) catalyzed hydroboration of olefins. The documentation of regio- and stereocontrol in cyclic and acyclic systems. *Chemtracts: Org. Chem.* **1989**, 2, 33-35.
11. Brown, H. C., Rangaishenvi, M. V. Some recent applications of hydroboration/organoborane chemistry to heterocycles. *J. Heterocycl. Chem.* **1990**, 27, 13-24.
12. Burgess, K., Ohlmeyer, M. J. Transition-metal promoted hydroborations of alkenes, emerging methodology for organic transformations. *Chem. Rev.* **1991**, 91, 1179-1191.
13. Kabalka, G. W., Guindi, L. H. M. Boron: boranes in organic synthesis. Annual Survey covering the year 1989. *J. Organomet. Chem.* **1993**, 457, 1-23.
14. Kabalka, G. W., Marks, R. C. Boron: boranes in organic synthesis. Annual survey covering the year 1990. *J. Organomet. Chem.* **1993**, 457, 25-40.
15. Westcott, S. A., Nguyen, P., Blom, H. P., Taylor, N. J., Marder, T. B., Baker, R. T., Calabrese, J. C. New developments in the transition metal-catalyzed hydroboration of alkenes. *Spec. Publ. - R. Soc. Chem.* **1994**, 143, 68-71.
16. Fu, G. C., Evans, D. A., Muci, A. R. Metal-catalyzed hydroboration reactions. *Adv. Catal. Processes* **1995**, 1, 95-121.
17. Burgess, K., Van der Donk, W. A. Asymmetric hydroboration. *Adv. Asymmetric Synth.* **1996**, 181-211.
18. Beletskaya, I., Pelter, A. Hydroborations catalyzed by transition metal complexes. *Tetrahedron* **1997**, 53, 4957-5026.
19. Hayashi, T. Hydroboration of carbon-carbon double bonds. *Comprehensive Asymmetric Catalysis I-III* **1999**, 1, 351-364.
20. Itsuno, S. Hydroboration of carbonyl groups. in *Comprehensive Asymmetric Catalysis I-III* (eds. Jacobsen, E., Pfaltz, A.,Yamamoto, H.), *1*, 289-315 (Springer, New York, **1999**).
21. Kanth, J. V. B. Borane-amine complexes for hydroboration. *Aldrichimica Acta* **2002**, 35, 57-66.
22. Evans, D. A., Fu, G. C., Hoveyda, A. H. Rhodium(I)- and iridium(I)-catalyzed hydroboration reactions: scope and synthetic applications. *J. Am. Chem. Soc.* **1992**, 114, 6671-6679.
23. Dhokte, U. P., Brown, H. C. 2-Isopropylapoisopinocampheylborane, an improved reagent for the asymmetric hydroboration of representative prochiral alkenes. *Tetrahedron Lett.* **1994**, 35, 4715-4718.
24. Fu, G. C. Metal-catalyzed hydroboration reactions. in *Transition Metals for Organic Synthesis* (eds. Beller, M.,Bolm, C.), *2*, 141-146 (Wiley-VCH, Weinheim, New York, **1998**).
25. Brunel, J.-M., Buono, G. Enantioselective rhodium catalyzed hydroboration of olefins using chiral bis(aminophosphine) ligands. *Tetrahedron Lett.* **1999**, 40, 3561-3564.
26. Brinkman, J. A., Nguyen, T. T., Sowa, J. R., Jr. Trifluoromethyl-Substituted Indenyl Rhodium and Iridium Complexes Are Highly Selective Catalysts for Directed Hydroboration Reactions. *Org. Lett.* **2000**, 2, 981-983.
27. Morrill, T. C., D'Souza, C. A., Yang, L., Sampognaro, A. J. Transition-Metal-Promoted Hydroboration of Alkenes: A Unique Reversal of Regioselectivity. *J. Org. Chem.* **2002**, 67, 2481-2484.
28. Renaud, P., Ollivier, C., Weber, V. One-Pot Rhodium(I)-Catalyzed Hydroboration of Alkenes: Radical Conjugate Addition. *J. Org. Chem.* **2003**, 68, 5769-5772.
29. Brown, H. C., Zweifel, G. Hydroboration. XI. The hydroboration of acetylenes-a convenient conversion of internal acetylenes into cis-olefins and of terminal acetylenes into aldehydes. *J. Am. Chem. Soc.* **1961**, 83, 3834-3840.
30. Clark, T., Wilhelm, D., Schleyer, P. v. R. Mechanism of hydroboration in ether solvents. A model ab initio study. *J. Chem. Soc., Chem. Commun.* **1983**, 606-608.
31. Houk, K. N., Rondan, N. G., Wu, Y. D., Metz, J. T., Paddon-Row, M. N. Theoretical studies of stereoselective hydroborations. *Tetrahedron* **1984**, 40, 2257-2274.

32. Nelson, D. J., Cooper, P. J. Experimental and theoretical investigation of the influence of alkene HOMO energy levels upon the hydroboration reaction. Additional evidence supporting an early transition state which has retention of alkene character. *Tetrahedron Lett.* **1986**, 27, 4693-4696.
33. Kulkarni, S. A., Koga, N. Ab initio mechanistic investigation of samarium(III)-catalyzed olefin hydroboration reaction. *THEOCHEM* **1999**, 461-462, 297-310.
34. Widauer, C., Gruetzmacher, H., Ziegler, T. Comparative Density Functional Study of Associative and Dissociative Mechanisms in the Rhodium(I)-Catalyzed Olefin Hydroboration Reactions. *Organometallics* **2000**, 19, 2097-2107.
35. Wang, K. K., Scouten, C. G., Brown, H. C. Hydroboration kinetics. 3. Kinetics and mechanism of the hydroboration of alkynes with 9-borabicyclo[3.3.1]nonane dimer. Effect of structure on the reactivity of representative alkynes. *J. Am. Chem. Soc.* **1982**, 104, 531-536.
36. Brown, H. C., Chandrasekharan, J., Wang, K. K. Hydroboration - kinetics and mechanism. *Pure Appl. Chem.* **1983**, 55, 1387-1414.
37. Brown, H. C., Chandrasekharan, J. Mechanism of hydroboration of alkenes with borane-Lewis base complexes. Evidence that the mechanism of the hydroboration reaction proceeds through a prior dissociation of such complexes. *J. Am. Chem. Soc.* **1984**, 106, 1863-1865.
38. Brown, H. C., Chandrasekharan, J., Nelson, D. J. Hydroboration kinetics. 10. Kinetics, mechanism, and selectivity for hydroboration of representative alkenes with borinane. *J. Am. Chem. Soc.* **1984**, 106, 3768-3771.
39. Lo Sterzo, C., Ortaggi, G. Hydroboration of ferrocenylalkenes: mechanistic and synthetic aspects. *J. Chem. Soc., Perkin Trans. 2* **1984**, 345-348.
40. Evans, D. A., Fu, G. C. The rhodium-catalyzed hydroboration of olefins: a mechanistic investigation. *J. Org. Chem.* **1990**, 55, 2280-2282.
41. Evans, D. A., Fu, G. C., Anderson, B. A. Mechanistic study of the rhodium(I)-catalyzed hydroboration reaction. *J. Am. Chem. Soc.* **1992**, 114, 6679-6685.
42. Burgess, K., van der Donk, W. A. The importance of phosphine-to-rhodium ratios in enantioselective hydroborations. *Inorg. Chim. Acta* **1994**, 220, 93-98.
43. Hartwig, J. F., Muhoro, C. N. Mechanistic Studies of Titanocene-Catalyzed Alkene and Alkyne Hydroboration: Borane Complexes as Catalytic Intermediates. *Organometallics* **2000**, 19, 30-38.
44. Liu, X., Cook, J. M. General Approach for the Synthesis of Sarpagine/Macroline Indole Alkaloids. Enantiospecific Total Synthesis of the Indole Alkaloid Trinervine. *Org. Lett.* **2001**, 3, 4023-4026.
45. Van Gool, M., Vandewalle, M. Vitamin D: enantioselective synthesis of (3aR,4R,7aS)-4-hydroxy-7a-methylperhydro-1-indenone, a suitable CD-ring fragment. *Eur. J. Org. Chem.* **2000**, 3427-3431.
46. Demay, S., Volant, F., Knochel, P. New C_2-symmetrical 1,2-diphosphanes for the efficient rhodium-catalyzed asymmetric hydroboration of styrene derivatives. *Angew. Chem., Int. Ed. Engl.* **2001**, 40, 1235-1238.
47. Makabe, H., Kong, L. K., Hirota, M. Total Synthesis of (-)-Cassine. *Org. Lett.* **2003**, 5, 27-29.

Buchner Method of Ring Expansion (Buchner Reaction)........68

1. Buchner, E., Curtius, T. Synthesis of β-keto esters from aldehydes and diazoacetic acid. *Chem. Ber.* **1885**, 18, 2371-2377.
2. Buchner, E. Pseudophenyl acetic acids. *Ber.* **1896**, 29, 106-111.
3. Buchner, E., Schottenhammer, K. Action of diazoacetic ester on mesitylene. *Ber.* **1920**, 53B, 865-873.
4. Dave, V., Warnhoff, E. W. Reactions of diazoacetic esters with alkenes, alkynes, heterocyclic and aromatic compounds. *Org. React.* **1970**, 18, 217-401.
5. Kirmse, W. *Carbene Chemistry. 2nd ed* (**1971**) 622 pp.
6. Maas, G. Transition-metal catalyzed decomposition of aliphatic diazo compounds. New results and applications in organic synthesis. *Top. Curr. Chem.* **1987**, 137, 75-253.
7. Ye, T., McKervey, M. A. Organic Synthesis with α-Diazo Carbonyl Compounds. *Chem. Rev.* **1994**, 94, 1091-1160.
8. Doyle, M. P., McKervey, M. A., Ye, T. Modern Catalytic Methods for Organic Synthesis with Diazo Compounds: From Cyclopropanes to Ylides. *Chapter 6*, 289-342 (John Wiley & Sons, **1998**).
9. Lebel, H., Marcoux, J.-F., Molinaro, C., Charette, A. B. Stereoselective Cyclopropanation Reactions. *Chem. Rev.* **2003**, 103, 977-1050.
10. Anciaux, A. J., Demonceau, A., Hubert, A. J., Noels, A. F., Petiniot, N., Teyssie, P. Catalytic control of reactions of dipoles and carbenes; an easy and efficient synthesis of cycloheptatrienes from aromatic compounds by an extension of Buchner's reaction. *J. Chem. Soc., Chem. Commun.* **1980**, 765-766.
11. Anciaux, A. J., Demonceau, A., Noels, A. F., Hubert, A. J., Warin, R., Teyssie, P. Transition-metal-catalyzed reactions of diazo compounds. 2. Addition to aromatic molecules: catalysis of Buchner's synthesis of cycloheptatrienes. *J. Org. Chem.* **1981**, 46, 873-876.
12. Duddeck, H., Ferguson, G., Kaitner, B., Kennedy, M., McKervey, M. A., Maguire, A. R. The intramolecular Buchner reaction of aryl diazoketones. Synthesis and x-ray crystal structures of some polyfunctional hydroazulene lactones. *J. Chem. Soc., Perkin Trans. 1* **1990**, 1055-1063.
13. Kennedy, M., McKervey, M. A., Maguire, A. R., Tuladhar, S. M., Twohig, M. F. The intramolecular Buchner reaction of aryl diazoketones. Substituent effects and scope in synthesis. *J. Chem. Soc., Perkin Trans. 1* **1990**, 1047-1054.
14. Cordi, A. A., Lacoste, J. M., Hennig, P. A reinvestigation of the intramolecular Buchner reaction of 1-diazo-4-phenylbutan-2-ones leading to 2-tetralones. *J. Chem. Soc., Perkin Trans. 1* **1993**, 3-4.
15. Manitto, P., Monti, D., Speranza, G. Rhodium(II)-Catalyzed Decomposition of 1-Diazo-4-(2-naphthyl)butan-2-one. Direct Chemical Evidence for the Formation of the Norcaradiene System in the Intramolecular Buchner Reaction. *J. Org. Chem.* **1995**, 60, 484-485.
16. Maguire, A. R., Buckley, N. R., O'Leary, P., Ferguson, G. Excellent stereocontrol in intramolecular Buchner cyclizations and subsequent cycloadditions; stereospecific construction of polycyclic systems. *Chem. Commun.* **1996**, 2595-2596.
17. Maguire, A. R., Buckley, N. R., O'Leary, P., Ferguson, G. Stereocontrol in the intramolecular Buchner reaction of diazoketones. *J. Chem. Soc., Perkin Trans. 1* **1998**, 4077-4092.
18. Moody, C. J., Miah, S., Slawin, A. M. Z., Mansfield, D. J., Richards, I. C. Stereocontrol in the intramolecular Buchner reaction of diazoamides and diazoesters. *J. Chem. Soc., Perkin Trans. 1* **1998**, 4067-4076.
19. Doyle, M. P., Ene, D. G., Forbes, D. C., Pillow, T. H. Chemoselectivity and enantiocontrol in catalytic intramolecular metal carbene reactions of diazo acetates linked to reactive functional groups by naphthalene-1,8-dimethanol. *Chem. Commun.* **1999**, 1691-1692.
20. Merlic, C. A., Zechman, A. L., Miller, M. M. Reactivity of (η6-Arene)tricarbonylchromium Complexes with Carbenoids: Arene Activation or Protection? *J. Am. Chem. Soc.* **2001**, 123, 11101-11102.
21. Petiniot, N., Noels, A. F., Anciaux, A. J., Hubert, A. J., Teyssie, P. Novel aspects of transition metal-catalyzed reactions of carbenes. *Fundam. Res. Homogeneous Catal.* **1979**, 3, 421-432.
22. Doyle, M. P. Catalytic methods for metal carbene transformations. *Chem. Rev.* **1986**, 86, 919-940.
23. Wenkert, E., Guo, M., Pizzo, F., Ramachandran, K. Synthesis of 2-cycloalkenones (parts of 1,4-diacyl-1,3-butadiene systems) and of a heterocyclic analog by metal-catalyzed decomposition of 2-diazoacylfurans. *Helv. Chim. Acta* **1987**, 70, 1429-1438.
24. Kane, J. L., Shea, K. M., Crombie, A. L., Danheiser, R. L. A ring expansion-annulation strategy for the synthesis of substituted azulenes. Preparation and Suzuki coupling reactions of 1-azulenyl triflates. *Org. Lett.* **2001**, 3, 1081-1084.
25. Pellicciari, R., Annibali, D., Costantino, G., Marinozzi, M., Natalini, B. Dirhodium(II) tetraacetate-mediated decomposition of ethyl diazoacetate and ethyl diazomalonate in the presence of fullerene. A new procedure for the selective synthesis of [6-6]-closed methanofullerenes. *Synlett* **1997**, 1196-1198.
26. Frey, B., Wells, A. P., Rogers, D. H., Mander, L. N. Synthesis of the Unusual Diterpenoid Tropones Hainanolidol and Harringtonolide. *J. Am. Chem. Soc.* **1998**, 120, 1914-1915.

Related reactions: Ullmann biaryl ether and biaryl amine synthesis;

1. Kosugi, M., Kameyama, M., Migita, T. Palladium-catalyzed aromatic amination of aryl bromides with N,N-diethylaminotributyltin. *Chem. Lett.* **1983**, 927-928.
2. Guram, A. S., Buchwald, S. L. Palladium-Catalyzed Aromatic Aminations with in situ Generated Aminostannanes. *J. Am. Chem. Soc.* **1994**, 116, 7901-7902.
3. Paul, F., Patt, J., Hartwig, J. F. Palladium-catalyzed formation of carbon-nitrogen bonds. Reaction intermediates and catalyst improvements in the hetero cross-coupling of aryl halides and tin amides. *J. Am. Chem. Soc.* **1994**, 116, 5969-5970.
4. Wolfe, J. P., Wagaw, S., Buchwald, S. L. An Improved Catalyst System for Aromatic Carbon-Nitrogen Bond Formation: The Possible Involvement of Bis(Phosphine) Palladium Complexes as Key Intermediates. *J. Am. Chem. Soc.* **1996**, 118, 7215-7216.
5. Guram, A. S., Rennels, R. A., Buchwald, S. L., Barta, N. S., Pearson, W. H. Palladium-catalyzed amination of aryl halides with amines. *Chemtracts: Inorg. Chem.* **1996**, 8, 1-5.
6. Baranano, D., Mann, G., Hartwig, J. F. Nickel and palladium-catalyzed cross-couplings that form carbon-heteroatom and carbon-element bonds. *Curr. Org. Chem.* **1997**, 1, 287-305.
7. Hartwig, J. F. Palladium-catalyzed amination of aryl halides. Mechanism and rational catalyst design. *Synlett* **1997**, 329-340.
8. Hartwig, J. F. Carbon-Heteroatom Bond-Forming Reductive Eliminations of Amines, Ethers, and Sulfides. *Acc. Chem. Res.* **1998**, 31, 852-860.
9. Hartwig, J. F. Transition metal catalyzed synthesis of arylamines and aryl ethers from aryl halides and triflates: scope and mechanism. *Angew. Chem., Int. Ed. Engl.* **1998**, 37, 2046-2067.
10. Wolfe, J. P., Wagaw, S., Marcoux, J.-F., Buchwald, S. L. Rational Development of Practical Catalysts for Aromatic Carbon-Nitrogen Bond Formation. *Acc. Chem. Res.* **1998**, 31, 805-818.
11. Hartwig, J. F. Approaches to catalyst discovery. New carbon-heteroatom and carbon-carbon bond formation. *Pure Appl. Chem.* **1999**, 71, 1417-1423.
12. Kocovsky, P., Malkov, A. V., Vyskocil, S., Lloyd-Jones, G. C. Transition metal catalysis in organic synthesis: reflections, chirality and new vistas. *Pure Appl. Chem.* **1999**, 71, 1425-1433.
13. Yang, B. H., Buchwald, S. L. Palladium-catalyzed amination of aryl halides and sulfonates. *J. Organomet. Chem.* **1999**, 576, 125-146.
14. Muci, A. R., Buchwald, S. L. Practical palladium catalysts for C-N and C-O bond formation. *Top. Curr. Chem.* **2002**, 219, 131-209.
15. Jiang, L., Buchwald, S. L. Palladium-catalyzed aromatic carbon-nitrogen bond formation. *Metal-Catalyzed Cross-Coupling Reactions (2nd Edition)* **2004**, 2, 699-760.
16. Guram, A. S., Rennels, R. A., Buchwald, S. L. A simple catalytic method for the conversion of aryl bromides to arylamines. *Angew. Chem., Int. Ed. Engl.* **1995**, 34, 1348-1350.
17. Louie, J., Hartwig, J. F. Palladium-catalyzed synthesis of arylamines from aryl halides. Mechanistic studies lead to coupling in the absence of tin reagents. *Tetrahedron Lett.* **1995**, 36, 3609-3612.
18. Wentland, M. P., Xu, G., Cioffi, C. L., Ye, Y., Duan, W., Cohen, D. J., Colasurdo, A. M., Bidlack, J. M. 8-aminocyclazocine analogues: synthesis and structure-activity relationships. *Bioorg. Med. Chem. Lett.* **2000**, 10, 183-187.
19. Lee, S., Lee, W.-M., Sulikowski, G. A. An Enantioselective 1,2-Aziridinomitosene Synthesis via a Chemoselective Carbon-Hydrogen Insertion Reaction of a Metal Carbene. *J. Org. Chem.* **1999**, 64, 4224-4225.
20. Emoto, T., Kubosaki, N., Yamagiwa, Y., Kamikawa, T. A new route to phenazines. *Tetrahedron Lett.* **2000**, 41, 355-358.

Related reactions: Chugaev elimination, Cope elimination, Hofmann elimination;

1. Burgess, E. M., Penton, H. R., Jr., Taylor, E. A. Synthetic applications of N-carboalkoxysulfamate esters. *J. Am. Chem. Soc.* **1970**, 92, 5224-5226.
2. Burgess, E. M., Penton, H. R., Jr., Taylor, E. A. Thermal reactions of alkyl N-carbomethoxysulfamate esters. *J. Org. Chem.* **1973**, 38, 26-31.
3. Lamberth, C. Burgess reagent ([methoxycarbonylsulfamoyl]triethylammonium hydroxide, inner salt): dehydrations and more. *J. Prakt. Chem./Chem.-Ztg.* **2000**, 342, 518-522.
4. Khapli, S., Dey, S., Mal, D. Burgess reagent in organic synthesis. *J. Indian Inst. Sci.* **2001**, 81, 461-476.
5. Burgess, E. M., Penton, H. R., Jr., Taylor, E. A., Williams, W. M. Conversion of primary alcohols to urethanes via the inner salt of methyl (carboxysulfamoyl)triethylammonium hydroxide: methyl n-hexylcarbamate. *Org. Synth.* **1977**, 56, 40-43.
6. Claremon, D. A., Phillips, B. T. An efficient chemoselective synthesis of nitriles from primary amides. *Tetrahedron Lett.* **1988**, 29, 2155-2158.
7. Wipf, P., Venkatraman, S. An improved protocol for azole synthesis with PEG-supported Burgess reagent. *Tetrahedron Lett.* **1996**, 37, 4659-4662.
8. Barvian, M. R., Showalter, H. D. H., Doherty, A. M. Preparation of N,N'-bis(aryl)guanidines from electron deficient amines via masked carbodiimides. *Tetrahedron Lett.* **1997**, 38, 6799-6802.
9. Maugein, N., Wagner, A., Mioskowski, C. New conditions for the generation of nitrile oxides from primary nitroalkanes. *Tetrahedron Lett.* **1997**, 38, 1547-1550.
10. Creedon, S. M., Crowley, H. K., McCarthy, D. G. Dehydration of formamides using the Burgess reagent: a new route to isocyanides. *J. Chem. Soc., Perkin Trans. 1* **1998**, 1015-1018.
11. Brain, C. T., Paul, J. M., Loong, Y., Oakley, P. J. Novel procedure for the synthesis of 1,3,4-oxadiazoles from 1,2-diacylhydrazines using polymer-supported Burgess reagent under microwave conditions. *Tetrahedron Lett.* **1999**, 40, 3275-3278.
12. Burckhardt, S. Methyl N-(trimethylammoniumsulfonyl)carbamate: "Burgess Reagent". *Synlett* **2000**, 559.
13. Miller, C. P., Kaufman, D. H. Mild and efficient dehydration of oximes to nitriles mediated by the Burgess reagent. *Synlett* **2000**, 1169-1171.
14. Nicolaou, K. C., Huang, X., Snyder, S. A., Rao, P. B., Bella, M., Reddy, M. V. A novel regio- and stereoselective synthesis of sulfamidates from 1,2-diols using Burgess and related reagents: a facile entry into β-amino alcohols. *Angew. Chem., Int. Ed. Engl.* **2002**, 41, 834-838.
15. Nicolaou, K. C., Longbottom, D. A., Snyder, S. A., Nalbanadian, A. Z., Huang, X. A new method for the synthesis of nonsymmetrical sulfamides using Burgess-type reagents. *Angew. Chem., Int. Ed. Engl.* **2002**, 41, 3866-3870.
16. Wood, M. R., Kim, J. Y., Books, K. M. A novel, one-step method for the conversion of primary alcohols into carbamate-protected amines. *Tetrahedron Lett.* **2002**, 43, 3887-3890.
17. Rinner, U., Adams, D. R., dos Santos, M. L., Abboud, K. A., Hudlicky, T. New application of Burgess reagent in its reaction with epoxides. *Synlett* **2003**, 1247-1252.
18. Holton, R. A., Kim, H. B., Somoza, C., Liang, F., Biediger, R. J., Boatman, P. D., Shindo, M., Smith, C. C., Kim, S., et al. First total synthesis of taxol. 2. Completion of the C and D rings. *J. Am. Chem. Soc.* **1994**, 116, 1599-1600.
19. Tavares, F., Lawson, J. P., Meyers, A. I. Total Synthesis of Streptogramin Antibiotics. (-)-Madumycin II. *J. Am. Chem. Soc.* **1996**, 118, 3303-3304.
20. Ichikawa, S., Shuto, S., Matsuda, A. The First Synthesis of Herbicidin B. Stereoselective Construction of the Tricyclic Undecose Moiety by a Conformational Restriction Strategy Using Steric Repulsion between Adjacent Bulky Silyl Protecting Groups on a Pyranose Ring. *J. Am. Chem. Soc.* **1999**, 121, 10270-10280.

Cannizzaro Reaction 74

Related reactions: Meerwein-Ponndorf-Verley reduction, Oppenauer oxidation, Tishchenko reaction;

1. Cannizzaro, S. Benzyl alcohol. *Ann.* **1853**, 88, 129-130.
2. List, K., Limpricht, H. Benzoic acid and related compounds. *Ann.* **1854**, 90, 190-210.
3. Geissman, T. A. Cannizzaro reaction. *Org. React.* **1944**, 2, 94-113.
4. Kagan, J. Photo-Cannizzaro reaction of o-phthalaldehyde. *Tetrahedron Lett.* **1966**, 6097-6102.
5. Bianchi, M., Matteoli, U., Menchi, G., Frediani, P., Piacenti, F. Asymmetric synthesis by chiral ruthenium complexes. VIII. The asymmetric Cannizzaro reaction. *J. Organomet. Chem.* **1982**, 240, 65-70.
6. Polackova, V., Tomova, V., Elecko, P., Toma, S. Ultrasound-promoted Cannizzaro reaction under phase-transfer conditions. *Ultrason. Sonochem.* **1996**, 3, 15-17.
7. Thakuria, J. A., Baruah, M., Sandhu, J. S. Microwave-induced efficient synthesis of alcohols via cross-Cannizzaro reactions. *Chem. Lett.* **1999**, 995-996.
8. Entezari, M. H., Shameli, A. A. Phase-transfer catalysis and ultrasonic waves. I. Cannizzaro reaction. *Ultrason. Sonochem.* **2000**, 7, 169-172.
9. Russell, A. E., Miller, S. P., Morken, J. P. Efficient Lewis acid catalyzed intramolecular Cannizzaro reaction. *J. Org. Chem.* **2000**, 65, 8381-8383.
10. Yoshizawa, K., Toyota, S., Toda, F. Solvent-free Claisen and Cannizzaro reactions. *Tetrahedron Lett.* **2001**, 42, 7983-7985.
11. Pourjavadi, A., Soleimanzadeh, B., Marandi, G. B. Microwave-induced Cannizzaro reaction over neutral γ-alumina as a polymeric catalyst. *React. Funct. Polym.* **2002**, 51, 49-53.
12. Reddy, B. V. S., Srinivas, R., Yadav, J. S., Ramalingam, T. KF-Al_2O_3 mediated cross-Cannizzaro reaction under microwave irradiation. *Synth. Commun.* **2002**, 32, 219-223.
13. Ogawa, H., Hosoe, T., Senda, H. Na-zeolites promoted Cannizzaro reaction of p-nitrobenzaldehyde in liquid phase. *Tokyo Gakugei Daigaku Kiyo, Dai-4-bumon: Sugaku, Shizen Kagaku* **2003**, 55, 35-38.
14. Tim, B. T., Cho, C. S., Kim, T.-J., Shim, S. C. Ruthenium-catalyzed transfer hydrogenation of aromatic aldehydes with dioxane under KOH. Assistance of Cannizzaro reaction. *J. Chem. Res., Synop.* **2003**, 368-369.
15. Rzepa, H. S., Miller, J. An MNDO SCF-MO study of the mechanism of the Cannizzaro reaction. *J. Chem. Soc., Perkin Trans. 2* **1985**, 717-723.
16. Rajyaguru, I., Rzepa, H. S. A MNDO SCF-MO study of the mechanism of the benzilic acid and related rearrangements. *J. Chem. Soc., Perkin Trans. 2* **1987**, 1819-1827.
17. Jacobus, J. End group transfers. Mechanism of the Cannizzaro reaction. *J. Chem. Educ.* **1972**, 49, 349-350.
18. Swain, C. G., Powell, A. L., Sheppard, W. A., Morgan, C. R. Mechanism of the Cannizzaro reaction. *J. Am. Chem. Soc.* **1979**, 101, 3576-3583.
19. Chung, S.-K. Mechanism of the Cannizzaro reaction: possible involvement of radical intermediates. *J. Chem. Soc., Chem. Commun.* **1982**, 480-481.
20. Ashby, E. C., Coleman, D. T., III, Gamasa, M. P. Evidence supporting a single-electron-transfer path in the Cannizzaro reaction. *Tetrahedron Lett.* **1983**, 24, 851-854.
21. Fuentes, A., Sinisterra, J. V. Single electron transfer mechanism of the Cannizzaro reaction in heterogeneous phase, under ultrasonic conditions. *Tetrahedron Lett.* **1986**, 27, 2967-2970.
22. Ashby, E. C., Coleman, D., Gamasa, M. Single-electron transfer in the Cannizzaro reaction. *J. Org. Chem.* **1987**, 52, 4079-4085.
23. Bowden, K., Butt, A. M., Streater, M. Intramolecular catalysis. Part 8. The intramolecular Cannizzaro reaction of naphthalene-1,8-dicarbaldehyde and [α,α'-$2H^2$]naphthalene-1,8-dicarbaldehyde. *J. Chem. Soc., Perkin Trans. 2* **1992**, 567-571.
24. Mehta, G., Padma, S. Observation of a transannular Cannizzaro reaction in a caged [7]prismane related system. *J. Org. Chem.* **1991**, 56, 1298-1299.
25. Moore, J. A., Robello, D. R., Rebek, J., Jr., Gadwood, R. Synthesis of dibenzoheptalene bislactones via a double intramolecular Cannizzaro reaction. *Org. Prep. Proced. Int.* **1988**, 20, 87-91.
26. Albanese, D., Penso, M., Zenoni, M. A practical synthesis of 4-chloro-3-(hydroxymethyl)pyridine by regioselective one-pot lithiation/formylation/reduction of 4-chloropyridine. *Synthesis* **1999**, 1294-1296.
27. Bringmann, G., Hinrichs, J., Henschel, P., Kraus, J., Peters, K., Peters, E.-M. Novel concepts in directed biaryl synthesis, 97. Atropo-enantioselective synthesis of the natural bicoumarin (+)-isokotanin A via a configurationally stable biaryl lactone. *Eur. J. Org. Chem.* **2002**, 1096-1106.

Carroll Rearrangement (Kimel-Cope Rearrangement) 76

Related reactions: Claisen rearrangement, Claisen-Ireland rearrangement, Eschenmoser-Claisen rearrangement, Johnson-Claisen rearrangement;

1. Carroll, M. F. Addition of α,β-unsaturated alcohols to the active methylene group. I. The action of ethyl acetoacetate on linaloöl and geraniol. *J. Chem. Soc.* **1940**, 704-706.
2. Carroll, M. F. Addition of β,γ-unsaturated alcohols to the active methylene group. III. Scope and mechanism of the reaction. *J. Chem. Soc.* **1941**, 507-511.
3. Wilson, S. R., Price, M. F. The ester enolate Carroll rearrangement. *J. Org. Chem.* **1984**, 49, 722-725.
4. Castro, A. M. M. Claisen Rearrangement over the Past Nine Decades. *Chem. Rev.* **2004**, 104, 2939-3002.
5. Gilbert, J. C., Kelly, T. A. Diastereoselective formation of contiguous quaternary centers. The modified Carroll rearrangement. *Tetrahedron* **1988**, 44, 7587-7600.
6. Echavarren, A. M., De Mendoza, J., Prados, P., Zapata, A. Stereoselective synthesis of (±)-4-epiacetomycin by the ester enolate Carroll rearrangement. *Tetrahedron Lett.* **1991**, 32, 6421-6424.
7. Ouvrard, N., Rodriguez, J., Santelli, M. Stereoselective ester dienolate Carroll Rearrangement: a New Approach to the Prelog-Djerassi lactone framework. *Tetrahedron Lett.* **1993**, 34, 1149-1150.
8. Genus, J. F., Peters, D. D., Ding, J. f., Bryson, T. A. The dianion Carroll rearrangement - a cyclic application. *Synlett* **1994**, 209-210.
9. Enders, D., Knopp, M., Runsink, J., Raabe, G. Diastereo- and enantioselective synthesis of polyfunctional ketones with adjacent quaternary and tertiary centers by asymmetric Carroll rearrangement. *Angew. Chem., Int. Ed. Engl.* **1995**, 34, 2278-2280.
10. Sorgi, K. L., Scott, L., Maryanoff, C. A. The Carroll rearrangement: a facile entry into substituted arylacetones and related derivatives. *Tetrahedron Lett.* **1995**, 36, 3597-3600.
11. Enders, D., Knopp, M. Novel asymmetric syntheses of (-)-malyngolide and (+)-epi-malyngolide. *Tetrahedron* **1996**, 52, 5805-5818.
12. Enders, D., Knopp, M., Runsink, J., Raabe, G. The first asymmetric Carroll rearrangement. Diastereo- and enantioselective synthesis of polyfunctional ketones with vicinal quaternary and tertiary stereogenic centers. *Liebigs Ann. Chem.* **1996**, 1095-1116.
13. Tsuji, J. Catalytic reactions via π-allylpalladium complexes. *Pure Appl. Chem.* **1982**, 54, 197-206.
14. Sobenina, L. N., Mikhaleva, A. I., Petrova, O. V., Polovnikova, R. I., Trofimov, B. A. Unknown pathway of the Carroll reaction. *Russ. J. Org. Chem.* **1997**, 33, 1041-1042.

Related reactions: *Sonogashira coupling;*

1. Castro, C. E., Stephens, R. D. Substitutions by ligands of low valent transition metals. A preparation of tolans and heterocyclics from aryl iodides and cuprous acetylides. *J. Org. Chem.* **1963**, 28, 2163.
2. Stephens, R. D., Castro, C. E. The substitution of aryl iodides with cuprous acetylides. A synthesis of tolanes and heterocyclics. *J. Org. Chem.* **1963**, 28, 3313-3315.
3. Jukes, A. E. Organic chemistry of copper. *Adv. Organomet. Chem.* **1974**, 12, 215-322.
4. Posner, G. H. Substitution reactions using organocopper reagents. *Org. React.* **1975**, 22, 253-400.
5. Posner, G. H. *An Introduction to Synthesis Using Organocopper Reagents* (Wiley, New York, **1980**).
6. Lindley, J. Copper-assisted nucleophilic substitution of aryl halogen. *Tetrahedron* **1984**, 40, 1433-1456.
7. Sonogashira, K. Coupling Reactions Between sp^2 and sp Carbon Centers. in *Comp. Org. Synth.* (eds. Trost, B. M.,Fleming, I.), *3*, 521-549 (Pergamon, Oxford, **1991**).
8. Rossi, R., Carpita, A., Bellina, F. Palladium- and/or copper-mediated cross-coupling reactions between 1-alkynes and vinyl, aryl, 1-alkynyl, 1,2-propadienyl, propargyl and allylic halides or related compounds. A review. *Org. Prep. Proced. Int.* **1995**, 27, 127-160.
9. Ogawa, T., Kusume, K., Tanaka, M., Hayami, K., Suzuki, H. An alternative method for the stereospecific synthesis of conjugated alkenynes via the copper(I) iodide assisted cross-coupling reaction of 1-alkynes with haloalkenes. *Synth. Commun.* **1989**, 19, 2199-2207.
10. Mignani, G., Chevalier, C., Grass, F., Allmang, G., Morel, D. Synthesis of new unsaturated enynes, catalyzed by copper(I) complexes. *Tetrahedron Lett.* **1990**, 31, 5161-5164.
11. Okuro, K., Furuune, M., Miura, M., Nomura, M. Copper-catalyzed coupling reaction of aryl and vinyl halides with terminal alkynes. *Tetrahedron Lett.* **1992**, 33, 5363-5364.
12. Chowdhury, C., Kundu, N. G. Studies on copper(I) catalyzed cross-coupling reactions: a convenient and facile method for the synthesis of diversely substituted α,β-acetylenic ketones. *Tetrahedron* **1999**, 55, 7011-7016.
13. Kang, S.-K., Yoon, S.-K., Kim, Y.-M. Copper-Catalyzed Coupling Reaction of Terminal Alkynes with Aryl- and Alkenyliodonium Salts. *Org. Lett.* **2001**, 3, 2697-2699.
14. Haglund, O., Nilsson, M. A facile one-pot synthesis of isocoumestans via a novel extension of the Castro cyclization of o-iodophenols and ethyl propiolate. *Synlett* **1991**, 723-724.
15. Kinder, J. D., Tessier, C. A., Youngs, W. J. Synthesis of a para-methoxy substituted tribenzocyclotriyne. *Synlett* **1993**, 149-150.
16. Coleman, R. S., Garg, R. Stereocontrolled Synthesis of the Diene and Triene Macrolactones of Oximidines I and II: Organometallic Coupling versus Standard Macrolactonization. *Org. Lett.* **2001**, 3, 3487-3490.
17. White, J. D., Carter, R. G., Sundermann, K. F., Wartmann, M. Total Synthesis of Epothilone B, Epothilone D, and cis- and trans-9,10-Dehydroepothilone D. *J. Am. Chem. Soc.* **2001**, 123, 5407-5413.

1. Chichibabin, A. E., Zeide, O. A. New reaction for compounds containing the pyridine nucleus. *J. Russ. Phys. Chem. Soc.* **1914**, 46, 1216-1236.
2. Chichibabin, A. E. A new method of preparation of hydroxy derivatives of pyridine, quinoline and their homologs. *Ber.* **1923**, 56B, 1879-1885.
3. Leffler, M. T. Organic Reactions. I: Amination of heterocyclic bases by alkali amides. **1942**, 91-104.
4. Gibson, M. S. Introduction of the amino group. *Chem. Amino Group* **1968**, 37-77.
5. Pozharskii, A. F., Simonov, A. M., Doron'kin, V. N. Advances in the study of the Chichibabin reaction. *Usp. Khim.* **1978**, 47, 1933-1969.
6. Van der, P. H. C. Potassium permanganate in liquid ammonia. An effective reagent in the Chichibabin amination of azines (review). *Khim. Geterotsikl. Soedin.* **1987**, 1011-1027.
7. McGill, C. K., Rappa, A. Advances in the Chichibabin reaction. *Adv. Heterocycl. Chem.* **1988**, 44, 1-79.
8. Sagitullin, R. S., Shkil, G. P., Nosonova, I. I., Ferber, A. A. Chichibabin syntheses of pyridine bases. *Khim. Geterotsikl. Soedin.* **1996**, 147-161.
9. Van der Plas, H. C., Wozniak, M. Potassium permanganate in liquid ammonia. An effective reagent in the Chichibabin amination of azines. *Croat. Chem. Acta* **1986**, 59, 33-49.
10. Lawin, P. B., Sherman, A. R., Grendze, M. P. *Improved Chichibabin aminations of pyridine bases.* WO 9600216, **1996** (Reilly Industries, Inc., USA). 44 pp.
11. Ziegler, K., Zeiser, H. Alkali-organic compounds. VII. Alkali metal alkyls and pyridine (preliminary communication). *Ber.* **1930**, 63B, 1847-1851.
12. Hawes, E. M., Wibberley, G. D. 1,8-Naphthyridines. *J. Chem. Soc., Org.* **1966**, 315-321.
13. Hawes, E. M., Davis, H. L. Intramolecular nucleophilic cyclization of 3-substituted pyridylalkylamines onto the 2-position of the pyridine ring. *J. Heterocycl. Chem.* **1973**, 10, 39-42.
14. Palucki, M., Hughes, D. L., Yasuda, N., Yang, C., Reider, P. J. A highly efficient synthesis of 2-(3-aminopropyl)-5,6,7,8-tetrahydronaphthyridine via a double Suzuki reaction and a Chichibabin cyclization. *Tetrahedron Lett.* **2001**, 42, 6811-6814.
15. Abramovitch, R. A., Helmer, F., Saha, J. G. Mechanism of the Chichibabin reaction. *Tetrahedron Lett.* **1954**, 3445-3447.
16. Abramovitch, R. A., Helmer, F., Saha, J. G. Aromatic substitution. VIII. Aspects of the mechanism of the Chichibabin reaction. *Can. J. Chem.* **1965**, 43, 725-731.
17. Eckroth, D. R. An abnormal Chichibabin reaction in a lithium aluminum hydride medium. *Chem. Ind.* **1967**, 920-921.
18. Kametani, T., Nemoto, H. Syntheses of heterocyclic compounds. CCLIII. Mechanism of the formation of an abnormal product in the Chichibabin reaction of quinoline. *Chem. Pharm. Bull. (Tokyo)* **1968**, 16, 1696-1699.
19. Kessar, S. V., Nadir, U. K., Singh, M. Mechanism of the Tschitschibabin reaction. *Indian J. Chem.* **1973**, 11, 825-826.
20. Zoltewicz, J. A., Helmick, L. S., Oestreich, T. M., King, R. W., Kandetzki, P. E. Addition of amide ion to isoquinoline and quinoline in liquid ammonia. Nuclear magnetic resonance spectra of anionic σ-complexes. *J. Org. Chem.* **1973**, 38, 1947-1949.
21. Sanders, G. M., Van Dijk, M., Den Hertog, H. J. Didehydrohetarenes. XXXIV. Diversity in the course of the reactions of the 4-haloisoquinolines with potassium amide in liquid ammonia and with piperidine. *Recl. Trav. Chim. Pays-Bas* **1974**, 93, 273-277.
22. Simig, G., Van der Plas, H. C. The SN(ANRORC) mechanism. XVII. An SN(ANRORC) mechanism in the amination of phenyl-1,3,5-triazine with potassium amide in liquid ammonia. A novel mechanism for the Chichibabin reaction. *Recl. Trav. Chim. Pays-Bas* **1976**, 95, 125-126.
23. Hirota, M., Masuda, H., Hamada, Y., Takeuchi, I. A simple MO treatment on the nucleophilic substitution reactions of six-membered aza-aromatic compounds. *Bull. Chem. Soc. Jpn.* **1979**, 52, 1498-1505.
24. Breuker, J., Van der Plas, H. C. The Chichibabin amination of 4-phenyl- and 4-tert-butylpyrimidine. *Recl.: J. R. Neth. Chem. Soc.* **1983**, 102, 367-372.
25. Tondys, H., Van der Plas, H. C., Wozniak, M. σ-Adduct formation in liquid ammonia. Part 41. On the Chichibabin amination of quinoline and some nitroquinolines. *J. Heterocycl. Chem.* **1985**, 22, 353-355.
26. Breuker, K., Van der Plas, H. C., Van Veldhuizen, A. SN(ANRORC) mechanism. Part 32. Pyrimidine chemistry. Part 99. The Chichibabin amination of 5-phenylpyrimidine and phenylpyrazine. *Isr. J. Chem.* **1986**, 27, 67-72.
27. Knize, M. G., Felton, J. S. The synthesis of the cooked-beef mutagen 2-amino-1-methyl-6-phenylimidazo[4,5-b]pyridine and its 3-methyl isomer. *Heterocycles* **1986**, 24, 1815-1819.
28. Kelly, T. R., Bridger, G. J., Zhao, C. Bisubstrate reaction templates. Examination of the consequences of identical versus different binding sites. *J. Am. Chem. Soc.* **1990**, 112, 8024-8034.

29. Vedernikov, A. N., Pink, M., Caulton, K. G. Design and Synthesis of Tridentate Facially Chelating Ligands of the [2.n.1]-(2,6)-Pyridinophane Family. *J. Org. Chem.* **2003**, 68, 4806-4814.

Related reactions: *Burgess dehydration, Cope elimination, Hofmann elimination;*

1. Chugaev, L. Studies on optical activity. *Ber.* **1898**, 31, 1775-1783.
2. Chugaev, L. A new method for the preparation of unsaturated hydrocarbons. *Ber.* **1899**, 32, 3332-3335.
3. DePuy, C. H., King, R. W. Pyrolytic cis eliminations. *Chem. Rev.* **1960**, 60, 431-457.
4. Nace, H. R. The preparation of olefins by the pyrolysis of xanthates. The Chugaev reaction. *Org. React.* **1962**, 12, 57-100.
5. Bordwell, F. G., Landis, P. S. Elimination reactions. VIII. A trans Chugaev elimination. *J. Am. Chem. Soc.* **1958**, 80, 2450-2453.
6. Benkeser, R. A., Hazdra, J. J. Factors influencing the direction of elimination in the Chugaev reaction. *J. Am. Chem. Soc.* **1959**, 81, 228-231.
7. De Groot, A., Evenhuis, B., Wynberg, H. Syntheses and properties of sterically hindered butadienes. A modification of the Chugaev reaction. *J. Org. Chem.* **1968**, 33, 2214-2217.
8. Cram, D. J. Studies on stereochemistry. IV. The Chugaev reaction in the determination of configuration of certain alcohols. *J. Am. Chem. Soc.* **1949**, 71, 3883-3889.
9. Alexander, E. R., Mudrak, A. Mechanism of Chugaev and acetate thermal decompositions. II. cis- and trans-2-Methyl-1-tetralol. *J. Am. Chem. Soc.* **1950**, 72, 3194-3198.
10. O'Connor, G. L., Nace, H. R. Chemical and kinetic studies on the Chugaev reaction. *J. Am. Chem. Soc.* **1952**, 74, 5454-5459.
11. O'Connor, G. L., Nace, H. R. Further studies on the Chugaev reaction and related reactions. *J. Am. Chem. Soc.* **1953**, 75, 2118-2123.
12. Bader, R. F. W., Bourns, A. N. A kinetic isotope effect study of the Chugaev reaction. *Can. J. Chem.* **1961**, 39, 348-358.
13. Nakagawa, H., Sugahara, T., Ogasawara, K. A Concise route to (-)-kainic acid. *Org. Lett.* **2000**, 2, 3181-3183.
14. Meulemans, T. M., Stork, G. A., Macaev, F. Z., Jansen, B. J. M., de Groot, A. Total Synthesis of Dihydroclerodin from (R)-(-)-Carvone. *J. Org. Chem.* **1999**, 64, 9178-9188.
15. Hagiwara, H., Kobayashi, K., Miya, S., Hoshi, T., Suzuki, T., Ando, M. The First Total Synthesis of (-)-Solanapyrone E Based on Domino Michael Strategy. *Org. Lett.* **2001**, 3, 251-254.
16. Fu, X., Cook, J. M. General approach for the synthesis of polyquinenes via the Weiss reaction. XII. The Chugaev approach to ellacene (1,10-cyclododecanotriquinancene). *Tetrahedron Lett.* **1990**, 31, 3409-3412.
17. Kumamoto, T., Tabe, N., Yamaguchi, K., Yagishita, H., Iwasa, H., Ishikawa, T. Synthetic studies on kinamycin antibiotics: elaboration of a highly oxygenated D ring. *Tetrahedron* **2001**, 57, 2717-2728.

Related reactions: *Buchner method of ring expansion;*

1. Ciamician, G. L., Dennstedt, M. The effect of chloroform on the potassium salt of pyrroles. *Ber.* **1881**, 14, 1153-1163.
2. Ciamician, Dennstedt. Studies on pyrrole derivatives: conversion of pyrroles to pyridines. *Ber.* **1882**, 15, 1172-1181.
3. Ciamician, G. L., Silber, P. Monobromopyridine. *Ber.* **1885**, 721-725.
4. Ciamician, G. L., Silber, P. The conversion of pyrroles into pyridine derivatives. *Ber.* **1887**, 20, 191-195.
5. Castillo, R., Moliner, V., Andres, J., Oliva, M., Safont, V. S., Bohm, S. Theoretical investigation of the abnormal Reimer-Tiemann reaction. *J. Phys. Org. Chem.* **1998**, 11, 670-677.
6. Dennstedt, M., Zimmerman, J. The conversion of pyrroles into pyridines. *Ber.* **1885**, 18, 3316-3319.
7. Madnanini, P. C. The conversion of indoles into quinoline derivatives. *Ber.* **1887**, 20, 2608-2614.
8. Ellinger, A. Indoles in egg white. Oxidation of tryphtophans to β-indole aldehydes. *Ber.* **1906**, 39, 2515-2523.
9. Ellinger, A., Flamand, C. Effect of chloroform and potassium hydroxide on skatol. *Ber.* **1906**, 39, 4388-4391.
10. Rees, C. W., Smithen, C. E. The mechanism of heterocyclic ring expansions. II. Reaction of methylindoles with halocarbenes. *J. Chem. Soc.* **1964**, 938-945.
11. Rees, C. W., Smithen, C. E. The mechanism of heterocyclic ring expansions. I. Reaction of 2,3-dimethylindole with dichlorocarbene. *J. Chem. Soc.* **1964**, 928-937.
12. Nicoletti, R., Forcellese, M. L. Reactions of carbenes. I. Action of dichlorocarbene on some pyrrole derivatives. *Gazz. Chim. Ital.* **1965**, 95, 83-94.
13. Nicoletti, R., Forcellese, M. L., Germani, C. Transformation of plancher pyrrolenines by treatment with bases. II. Mechanism of the formation of 2-methyl-4-ethoxy-5-methyl-6-chloro-1-azabicyclo[3.1.0]hex-2-ene. *Gazz. Chim. Ital.* **1967**, 97, 685-693.
14. Jones, R. L., Rees, C. W. Mechanism of heterocyclic ring expansions. III. Reaction of pyrroles with dichlorocarbene. *J. Chem. Soc. C* **1969**, 2249-2251.
15. Gambacorta, A., Nicoletti, R., Forcellese, M. L. Transformation of Plancher's pyrrolenines by reaction with bases. III. Novel ring enlargement: cyclic expansion of 2-dichloromethyl-2H-pyrroles. *Tetrahedron* **1971**, 27, 985-990.
16. Gambacorta, A., Nicoletti, R., Cerrini, S., Fedeli, W., Gavuzzo, G. The reaction between 2,5-dialkylpyrroles and dichlorocarbene. *Tetrahedron* **1980**, 36, 1367-1374.
17. Botta, M., De Angelis, F., Gambacorta, A. The reaction between 2,3-dialkylindoles and dichlorocarbene. *Tetrahedron* **1982**, 38, 2315-2318.
18. Botta, M., De Angelis, F., Gambacorta, A. The reaction between 2,3-dialkylindoles and dihalocarbenes. Additional evidence for the interconversion of the reaction intermediates. *Gazz. Chim. Ital.* **1983**, 113, 129-132.
19. De Angelis, F., Inesi, A., Feroci, M., Nicoletti, R. Reaction of Electrogenerated Dichlorocarbene with Methylindoles. *J. Org. Chem.* **1995**, 60, 445-447.
20. Dhanak, D., Reese, C. B. Synthesis of [6](2,4)pyridinophanes. *J. Chem. Soc., Perkin Trans. 1* **1987**, 2829-2832.
21. Kral, V., Gale, P. A., Anzenbacher, P., Jr., Jursikova, K., Lynch, V., Sessler, J. L. Calix[4]pyridine: a new arrival in the heterocalixarene family. *Chem. Commun.* **1998**, 9-10.

Related reactions: *Dieckmann condensation, Baker-Venkataraman rearrangement;*

1. Claisen, L., Lowman, O. A new method for the synthesis of benzoyl acetic esters. *Ber.* **1887**, 20, 651-654.
2. Hauser, C. R., Hudson, B. E., Jr. Acetoacetic ester condensation and certain related reactions. *Org. React.* **1942**, 1, 266-302.
3. House, H. O. *Modern Synthetic Reactions (The Organic Chemistry Monograph Series). 2nd ed* (**1972**) 856 pp.
4. Brandaenge, S. Studies on some Claisen-type condensations. *Chem. Scr.* **1987**, 27, 553-554.
5. Masamune, S., Walsh, C. T., Sinskey, A. J., Peoples, O. P. Poly-(R)-3-hydroxybutyrate (PHB) biosynthesis: mechanistic studies on the biological Claisen condensation catalyzed by β-ketoacyl thiolase. *Pure Appl. Chem.* **1989**, 61, 303-312.
6. Tanaka, T., Hirama, M. Bio-Claisen condensation catalyzed by thiolase from Zoogloea ramigera. Active site cysteine residues. *Chemtracts: Org. Chem.* **1989**, 2, 247-251.
7. Leijonmarck, H. K. E. Studies on the intramolecular Claisen condensation and related reactions. *Chem. Commun.* **1992**, 33 pp.

8. Heath, R. J., Rock, C. O. The Claisen condensation in biology. *Nat. Prod. Rep.* **2002**, 19, 581-596.
9. Sinistierra, J. V., Garcia-Raso, A., Cabello, J. A., Marinas, J. M. An improved procedure for the Claisen-Schmidt reaction. *Synthesis* **1984**, 502-504.
10. Popic, V. V., Korneev, S. M., Nikolaev, V. A., Korobitsyna, I. K. An improved synthesis of 2-diazo-1,3-diketones. *Synthesis* **1991**, 195-198.
11. Li, J.-T., Yang, W.-Z., Wang, S.-X., Li, S.-H., Li, T.-S. Improved synthesis of chalcones under ultrasound irradiation. *Ultrason. Sonochem.* **2002**, 9, 237-239.
12. Garst, J. F. Claisen ester condensation equilibriums - model calculations. *J. Chem. Educ.* **1979**, 56, 721-722.
13. Bartmess, J. E., Hays, R. L., Caldwell, G. The addition of carbanions to the carbonyl group in the gas phase. *J. Am. Chem. Soc.* **1981**, 103, 1338-1344.
14. Gasull, E. I., Silber, J. J., Blanco, S. E., Tomas, F., Ferretti, F. H. A theoretical and experimental study of the formation mechanism of 4-X-chalcones by the Claisen-Schmidt reaction. *THEOCHEM* **2000**, 503, 131-144.
15. Burdon, J., McLoughlin, V. C. R. Sodium-prompted Claisen ester condensations of ethyl perfluoroalkane carboxylates. *Tetrahedron* **1964**, 20, 2163-2166.
16. Csuros, Z., Deak, G., Haraszthy-Papp, M., Prihradny, L. Kinetic investigation of the Claisen-Schmidt condensation of aromatic ketones and aldehydes catalyzed by anion exchange resins. *Acta Chim. Acad. Sci. Hung.* **1968**, 55, 411-436.
17. Ashby, E. C., Park, W. S. Evidence for single electron transfer in Claisen condensation. The reaction of ethyl p-nitrobenzoate with the lithium enolate of pinacolone. *Tetrahedron Lett.* **1983**, 24, 1667-1670.
18. Alcantara, A., Marinas, J. M., Sinisterra, J. V. Barium hydroxide as catalyst in organic reactions. VIII. Nature of the adsorbed species in Claisen-Schmidt reaction. *React. Kinet. Catal. Lett.* **1986**, 32, 377-385.
19. Aguilera, A., Alcantara, A. R., Marinas, J. M., Sinisterra, J. V. Barium hydroxide as the catalyst in organic reactions. Part XIV. Mechanism of Claisen-Schmidt condensation in solid-liquid conditions. *Can. J. Chem.* **1987**, 65, 1165-1171.
20. Clark, J. D., O'Keefe, S. J., Knowles, J. R. Malate synthase: proof of a stepwise Claisen condensation using the double-isotope fractionation test. *Biochemistry* **1988**, 27, 5961-5971.
21. Guida, A., Lhouty, M. H., Tichit, D., Figueras, F., Geneste, P. Hydrotalcites as base catalysts. Kinetics of Claisen-Schmidt condensation, intramolecular condensation of acetonylacetone and synthesis of chalcone. *Appl. Cat. A* **1997**, 164, 251-264.
22. Leung, S. S.-W., Streitwieser, A. The Role of Aggregates in Claisen Acylation Reactions of Two Lithium Enolates in THF. *J. Am. Chem. Soc.* **1998**, 120, 10557-10558.
23. Rahimizadeh, M., Kam, K., Jenkins, S. I., McDonald, R. S., Harrison, P. H. M. Kinetics of glycoluril template-directed Claisen condensations and mechanistic implications. *Can. J. Chem.* **2002**, 80, 517-527.
24. Heathcock, C. H., Stafford, J. A. Daphniphyllum alkaloids. 13. Asymmetric total synthesis of (-)-secodaphniphylline. *J. Org. Chem.* **1992**, 57, 2566-2574.
25. Harrowven, D. C., Bradley, M., Lois Castro, J., Flanagan, S. R. Total syntheses of justicidin B and retrojusticidin B using a tandem Horner-Emmons-Claisen condensation sequence. *Tetrahedron Lett.* **2001**, 42, 6973-6975.
26. Trotter, N. S., Takahashi, S., Nakata, T. Simple and Efficient Synthesis of (+)-Methyl 7-Benzoylpederate, a Key Intermediate toward the Mycalamides. *Org. Lett.* **1999**, 1, 957-959.

Claisen Rearrangement 88

Related reactions: Carroll rearrangement, Claisen-Ireland rearrangement, Eschenmoser-Claisen rearrangement, Johnson-Claisen rearrangement;

1. Claisen, L. Rearrangement of Phenol Allyl Ethers into C-Allylphenols. *Ber.* **1913**, 45, 3157-3166.
2. Claisen, L., Eisleb, O. Rearrangement of phenol allyl ethers into the isomeric allylphenols. *Ann.* **1914**, 401, 21-119.
3. Tarbell, D. S. Claisen rearrangement. **1944**, 1-48.
4. Rhoads, S. J., Raulins, N. R. Claisen and Cope rearrangements. *Org. React.* **1975**, 22, 1-252.
5. Bennett, G. B. The Claisen rearrangement in organic synthesis; 1967 to January 1977. *Synthesis* **1977**, 589-606.
6. Ziegler, F. E. Stereo- and regiochemistry of the Claisen rearrangement: applications to natural products synthesis. *Acc. Chem. Res.* **1977**, 10, 227-232.
7. Hill, R. K. Chirality transfer via sigmatropic rearrangements. *Asymmetric Synth.* **1984**, 3, 503-572.
8. Lutz, R. P. Catalysis of the Cope and Claisen rearrangements. *Chem. Rev.* **1984**, 84, 205-247.
9. Moody, C. J. Claisen rearrangements in heteroaromatic systems. *Adv. Heterocycl. Chem.* **1987**, 42, 203-244.
10. Murray, A. W. Molecular rearrangements. *Org. React. Mech.* **1988**, 429-526.
11. Suzuki, T., Sato, E., Unno, K. Total syntheses of biologically active natural products by Claisen rearrangement of simple chiral templates. *Akita Igaku* **1988**, 15, 759-775.
12. Ziegler, F. E. The thermal, aliphatic Claisen rearrangement. *Chem. Rev.* **1988**, 88, 1423-1452.
13. Murray, A. W. Molecular rearrangements. *Org. React. Mech.* **1989**, 457-573.
14. Altenbach, H. J. Diastereoselective Claisen rearrangements. *Org. Synth. Highlights* **1991**, 111-115.
15. Wipf, P. Claisen rearrangements. in *Comp. Org. Synth.* (eds. Trost, B. M.,Fleming, I.), *5*, 827-873 (Pergamon, Oxford, **1991**).
16. Tadano, K. Natural product synthesis starting with carbohydrates based on the Claisen rearrangement protocol. *Stud. Nat. Prod. Chem.* **1992**, 10, 405-455.
17. Enders, D., Knopp, M., Schiffers, R. Asymmetric [3.3]-sigmatropic rearrangements in organic synthesis. *Tetrahedron: Asymmetry* **1996**, 7, 1847-1882.
18. Ganem, B. The mechanism of the Claisen rearrangement: deja vu all over again. *Angew. Chem., Int. Ed. Engl.* **1996**, 35, 936-945.
19. Cambie, R. C., Milbank, J. B. J., Rutledge, P. S. Reductive Claisen rearrangements of allyloxyanthraquinones. A review. *Org. Prep. Proced. Int.* **1997**, 29, 365-407.
20. Gajewski, J. J. The Claisen Rearrangement. Response to Solvents and Substituents: The Case for Both Hydrophobic and Hydrogen Bond Acceleration in Water and for a Variable Transition State. *Acc. Chem. Res.* **1997**, 30, 219-225.
21. Gajewski, J. J. Claisen rearrangements in aqueous solution. *Organic Synthesis in Water* **1998**, 82-101.
22. Ito, H., Taguchi, T. Asymmetric Claisen rearrangement. *Chem. Soc. Rev.* **1999**, 28, 43-50.
23. Nowicki, J. Claisen, Cope and related rearrangements in the synthesis of flavor and fragrance compounds. *Molecules [online computer file]* **2000**, 5, 1033-1050.
24. Fleming, M., Rigby, J. H., Yoon, T. P., MacMillan, D. W. C. Enantioselective Claisen rearrangements: Development of a first generation asymmetric acyl-Claisen reaction. *Chemtracts* **2001**, 14, 620-624.
25. Murray, A. W. Molecular rearrangements. *Org. React. Mech.* **2001**, 473-603.
26. Werschkun, B., Thiem, J. Claisen rearrangements in carbohydrate chemistry. *Top. Curr. Chem.* **2001**, 215, 293-325.
27. Hiersemann, M., Abraham, L. Catalysis of the Claisen rearrangement of aliphatic allyl vinyl ethers. *Eur. J. Org. Chem.* **2002**, 1461-1471.
28. Lindstroem, U. M. Stereoselective Organic Reactions in Water. *Chem. Rev.* **2002**, 102, 2751-2771.
29. Nakamura, H., Yamamoto, Y. Rearrangement reactions catalyzed by palladium: palladium-catalyzed carbon skeletal rearrangements: Cope, Claisen, and other [3,3] rearrangements. *Handbook of Organopalladium Chemistry for Organic Synthesis* **2002**, 2, 2919-2934.
30. Majumdar, K. C., Ghosh, S., Ghosh, M. The thio-Claisen rearrangement 1980-2001. *Tetrahedron* **2003**, 59, 7251-7271.
31. Castro, A. M. M. Claisen Rearrangement over the Past Nine Decades. *Chem. Rev.* **2004**, 104, 2939-3002.
32. Gonda, J. The Bellus-Claisen rearrangement. *Angew. Chem., Int. Ed. Engl.* **2004**, 43, 3516-3524.
33. Carroll, M. F. Addition of α,β-unsaturated alcohols to the active methylene group. I. The action of ethyl acetoacetate on linaloöl and geraniol. *J. Chem. Soc.* **1940**, 704-706.

34. Carroll, M. F. Addition of β,γ-unsaturated alcohols to the active methylene group. II. The action of ethyl acetoacetate on cinnamyl alcohol and phenylvinylcarbinol. *J. Chem. Soc.* **1940**, 1266-1268.
35. Carroll, M. F. Addition of β,γ-unsaturated alcohols to the active methylene group. III. Scope and mechanism of the reaction. *J. Chem. Soc.* **1941**, 507-511.
36. Wick, A. E., Felix, D., Steen, K., Eschenmoser, A. Claisen rearrangement of allyl and benzyl alcohols by N,N-dimethylacetamide acetals. *Helv. Chim. Acta* **1964**, 47, 2425-2429.
37. Felix, D., Gschwend-Steen, K., Wick, A. E., Eschenmoser, A. Claisen rearrangement of allyl and benzyl alcohols with 1-dimethylamino-1-methoxyethene. *Helv. Chim. Acta* **1969**, 52, 1030-1042.
38. Johnson, W. S., Werthemann, L., Bartlett, W. R., Brocksom, T. J., Li, T.-T., Faulkner, D. J., Petersen, M. R. Simple stereoselective version of the Claisen rearrangement leading to trans-trisubstituted olefinic bonds. Synthesis of squalene. *J. Am. Chem. Soc.* **1970**, 92, 741-743.
39. Ireland, R. E., Mueller, R. H. Claisen rearrangement of allyl esters. *J. Am. Chem. Soc.* **1972**, 94, 5897-5898.
40. Baldwin, J. E., Walker, J. A. Reformatskii-Claisen reaction, new synthetically useful sigmatropic process. *J. Chem. Soc., Chem. Commun.* **1973**, 117-118.
41. Ireland, R. E., Willard, A. K. Stereoselective generation of ester enolates. *Tetrahedron Lett.* **1975**, 3975-3978.
42. Ireland, R. E., Mueller, R. H., Willard, A. K. The ester enolate Claisen rearrangement. Stereochemical control through stereoselective enolate formation. *J. Am. Chem. Soc.* **1976**, 98, 2868-2877.
43. Malherbe, R., Bellus, D. A new type of Claisen rearrangement involving 1,3-dipolar intermediates. *Helv. Chim. Acta* **1978**, 61, 3096-3099.
44. Denmark, S. E., Harmata, M. A. Carbanion-accelerated Claisen rearrangements. *J. Am. Chem. Soc.* **1982**, 104, 4972-4974.
45. Denmark, S. E., Harmata, M. A. Carbanion-accelerated Claisen rearrangements. 2. Studies on internal asymmetric induction. *J. Org. Chem.* **1983**, 48, 3369-3370.
46. Malherbe, R., Rist, G., Bellus, D. Reactions of haloketenes with allyl ethers and thioethers: a new type of Claisen rearrangement. *J. Org. Chem.* **1983**, 48, 860-869.
47. Denmark, S. E., Harmata, M. A. Carbanion-accelerated Claisen rearrangements. 3. Vicinal quaternary centers. *Tetrahedron Lett.* **1984**, 25, 1543-1546.
48. Greuter, H., Lang, R. W., Romann, A. J. Fluorine-containing organozinc reagents. V. The Reformatskii-Claisen reaction of chlorodifluoroacetic acid derivatives. *Tetrahedron Lett.* **1988**, 29, 3291-3294.
49. Burrows, C., Carpenter, B. K. Substituent effects on the aliphatic Claisen rearrangements. 2. Theoretical analysis. *J. Am. Chem. Soc.* **1981**, 103, 6984-6986.
50. Yoo, H. Y., Houk, K. N. Transition Structures and Kinetic Isotope Effects for the Claisen Rearrangement. *J. Am. Chem. Soc.* **1994**, 116, 12047-12048.
51. Sehgal, A., Shao, L., Gao, J. Transition Structure and Substituent Effects on Aqueous Acceleration of the Claisen Rearrangement. *J. Am. Chem. Soc.* **1995**, 117, 11337-11340.
52. Wiest, O., Houk, K. N., Black, K. A., Thomas, B. I. V. Secondary Kinetic Isotope Effects of Diastereotopic Protons in Pericyclic Reactions: A New Mechanistic Probe. *J. Am. Chem. Soc.* **1995**, 117, 8594-8599.
53. Guest, J. M., Craw, J. S., Vincent, M. A., Hillier, I. H. The effect of water on the Claisen rearrangement of allyl vinyl ether: theoretical methods including explicit solvent and electron correlation. *J. Chem. Soc., Perkin Trans. 2* **1997**, 71-74.
54. Meyer, M. P., DelMonte, A. J., Singleton, D. A. Reinvestigation of the Isotope Effects for the Claisen and Aromatic Claisen Rearrangements: The Nature of the Claisen Transition States. *J. Am. Chem. Soc.* **1999**, 121, 10865-10874.
55. Gozzo, F. C., Fernandes, S. A., Rodrigues, D. C., Eberlin, M. N., Marsaioli, A. J. Regioselectivity in Aromatic Claisen Rearrangements. *J. Org. Chem.* **2003**, 68, 5493-5499.
56. Watanabe, W. H., Conlon, L. E. *Vinyl transetherification.* Us 2760990, **1956** (Rohm & Haas Co.).
57. Tulshian, D. B., Tsang, R., Fraser-Reid, B. Out-of-ring Claisen rearrangements are highly stereoselective in pyranoses: routes to gem-dialkylated sugars. *J. Org. Chem.* **1984**, 49, 2347-2355.
58. Marbet, R., Saucy, G. Reaction of tertiary vinylcarbinols with vinyl ethers. New method for the preparation of γ,δ-unsaturated aldehydes. *Helv. Chim. Acta* **1967**, 50, 2095-2100.
59. Saucy, G., Marbet, R. Reaction of tertiary vinylcarbinols with isopropenyl ether. New method for the preparation of γ,δ-unsaturated ketones. *Helv. Chim. Acta* **1967**, 50, 2091-2094.
60. Mandai, T., Matsumoto, S., Kohama, M., Kawada, M., Tsuji, J., Saito, S., Moriwake, T. A new, highly efficient method for isocarbacyclin synthesis based on tandem Claisen rearrangement and ene reactions. *J. Org. Chem.* **1990**, 55, 5671-5673.
61. Mandai, T., Ueda, M., Hasegawa, S., Kawada, M., Tsuji, J., Saito, S. Preparation and rearrangement of 2-allyloxyethyl aryl sulfoxides; a mercury-free Claisen sequence. *Tetrahedron Lett.* **1990**, 31, 4041-4044.
62. Suda, M. Preparation of allyl vinyl ethers by the Wittig reaction of allyl formates. *Chem. Lett.* **1981**, 967-970.
63. Kulkarni, M. G., Pendharkar, D. S., Rasne, R. M. Wittig olefination: an efficient route for the preparation of allyl vinyl ethers - precursors for the Claisen rearrangement. *Tetrahedron Lett.* **1997**, 38, 1459-1462.
64. Pine, S. H., Zahler, R., Evans, D. A., Grubbs, R. H. Titanium-mediated methylene-transfer reactions. Direct conversion of esters into vinyl ethers. *J. Am. Chem. Soc.* **1980**, 102, 3270-3272.
65. Pine, S. H., Kim, G., Lee, V. The synthesis of enol ethers by methylenation of esters: 1-phenoxy-1-phenylethene and 3,4-dihydro-2-methylene-2H-1-benzopyran. *Org. Synth.* **1990**, 69, 72-79.
66. Claisen, L., Tietze, E. Mechanism of the rearrangement of phenyl allyl ethers. *Ber.* **1925**, 58B, 275-281.
67. Claisen, L., Tietze, E. Mechanism of the rearrangement of the phenol allyl ethers. II. *Ber.* **1926**, 59B, 2344-2351.
68. Hill, R. K., Khatri, H. N. Titanium tetrachloride catalysis of aza-Claisen rearrangements. *Tetrahedron Lett.* **1978**, 4337-4340.
69. Padwa, A., Cohen, L. A. Aza-Claisen rearrangements in the 2-allyloxy substituted oxazole system. *Tetrahedron Lett.* **1982**, 23, 915-918.
70. Curran, D. P., Suh, Y. G. Substituent effects on the Claisen rearrangement. The accelerating effect of a 6-donor substituent. *J. Am. Chem. Soc.* **1984**, 106, 5002-5004.
71. Cave, R. J., Lythgoe, B., Metcalfe, D. A., Waterhouse, I. Stereochemical aspects of some Claisen rearrangements with cyclic orthoesters. *J. Chem. Soc., Perkin Trans. 1* **1977**, 1218-1228.
72. Ireland, R. E., Wipf, P., Xiang, J. N. Stereochemical control in the ester enolate Claisen rearrangement. 2. Chairlike vs boatlike transition-state selection. *J. Org. Chem.* **1991**, 56, 3572-3582.
73. Bernardelli, P., Moradei, O. M., Friedrich, D., Yang, J., Gallou, F., Dyck, B. P., Doskotch, R. W., Lange, T., Paquette, L. A. Total Asymmetric Synthesis of the Putative Structure of the Cytotoxic Diterpenoid (-)-Sclerophytin A and of the Authentic Natural Sclerophytins A and B. *J. Am. Chem. Soc.* **2001**, 123, 9021-9032.
74. Boeckman, R. K., Jr., Rico Ferreira, M. d. R., Mitchell, L. H., Shao, P. An Enantioselective Total Synthesis of (+)- and (-)-Saudin. Determination of the Absolute Configuration. *J. Am. Chem. Soc.* **2002**, 124, 190-191.
75. Nicolaou, K. C., Li, J. "Biomimetic" cascade reactions in organic synthesis: construction of 4-oxatricyclo[4.3.1.0]decan-2-one systems and total synthesis of 1-O-methylforbesione via tandem Claisen rearrangement/Diels-Alder reactions. *Angew. Chem., Int. Ed. Engl.* **2001**, 40, 4264-4268.

Claisen-Ireland Rearrangement 90

Related reactions: Carroll rearrangement, Claisen rearrangement, Eschenmoser-Claisen rearrangement, Johnson-Claisen rearrangement;

1. Tseou, H.-F., Wang, Y.-T. Abnormal acetoacetic ester synthesis. I. The reaction of sodium with allyl, benzohydryl and cinnamyl acetates. *J. Chinese Chem. Soc.* **1937**, 5, 224-229.
2. Arnold, R. T., Parham, W. E., Dodson, R. M. Rearrangement of allyl 9-fluorenecarboxylate. *J. Am. Chem. Soc.* **1949**, 71, 2439-2440.

3. Arnold, R. T., Searles, S., Jr. A new rearrangement of allylic esters. *J. Am. Chem. Soc.* **1949**, 71, 1150-1151.
4. Brannock, K. C., Pridgen, H. S., Thompson, B. Preparation of 2,2-dialkyl-4-pentenoic acids. *J. Org. Chem.* **1960**, 25, 1815-1816.
5. Julia, S., Julia, M., Linstrumelle, G. Synthesis of (±)-cis-homocaronic acid and (±)-trans-chrysanthemic acids from substituted bicyclo[3.1.0]hexan-2-one intermediates. *Bull. Soc. Chim. France* **1964**, 2693-2694.
6. Ireland, R. E., Mueller, R. H. Claisen rearrangement of allyl esters. *J. Am. Chem. Soc.* **1972**, 94, 5897-5898.
7. Ireland, R. E., Willard, A. K. Stereoselective generation of ester enolates. *Tetrahedron Lett.* **1975**, 3975-3978.
8. Ireland, R. E., Mueller, R. H., Willard, A. K. The ester enolate Claisen rearrangement. Stereochemical control through stereoselective enolate formation. *J. Am. Chem. Soc.* **1976**, 98, 2868-2877.
9. Ziegler, F. E. The thermal, aliphatic Claisen rearrangement. *Chem. Rev.* **1988**, 88, 1423-1452.
10. Blechert, S. The hetero-Cope rearrangement in organic synthesis. *Synthesis* **1989**, 71-82.
11. Altenbach, H. J. Ester enolate Claisen rearrangements. *Org. Synth. Highlights* **1991**, 116-118.
12. Marshall, J. A. Stereochemical control in the ester enolate Claisen rearrangement. Stereoselectivity in silyl ketene acetal formation. *Chemtracts: Org. Chem.* **1991**, 4, 154-157.
13. Pereira, S., Srebnik, M. The Ireland-Claisen rearrangement. *Aldrichimica Acta* **1993**, 26, 17-29.
14. Panek, J. S., Schaus, S., Masse, C. E. Development and utility of an enantioselective Ireland-Claisen reaction. *Chemtracts: Org. Chem.* **1995**, 8, 238-241.
15. Enders, D., Knopp, M., Schiffers, R. Asymmetric [3.3]-sigmatropic rearrangements in organic synthesis. *Tetrahedron: Asymmetry* **1996**, 7, 1847-1882.
16. Kazmaier, U. Synthesis of γ,δ-unsaturated amino acids via ester enolate Claisen rearrangement of chelated allylic esters. *Amino Acids* **1996**, 11, 283-299.
17. Kazmaier, U. Application of the chelate-enolate Claisen rearrangement to the synthesis of γ,δ-unsaturated amino acids. *Liebigs Ann. Chem.* **1997**, 285-295.
18. Kazmaier, U. Reactions of chelated amino acid ester enolates and their application to natural product synthesis. *Bioorg. Chem.* **1999**, 201-206.
19. Chai, Y., Hong, S.-p., Lindsay, H. A., McFarland, C., McIntosh, M. C. New aspects of the Ireland and related Claisen rearrangements. *Tetrahedron* **2002**, 58, 2905-2928.
20. Castro, A. M. M. Claisen Rearrangement over the Past Nine Decades. *Chem. Rev.* **2004**, 104, 2939-3002.
21. Corey, E. J., Lee, D. H. Highly enantioselective and diastereoselective Ireland-Claisen rearrangement of achiral allylic esters. *J. Am. Chem. Soc.* **1991**, 113, 4026-4028.
22. Hattori, K., Yamamoto, H. Highly selective enolization method for heteroatom substituted esters; its application to the Ireland ester enolate Claisen rearrangement. *Tetrahedron* **1994**, 50, 3099-3112.
23. Corey, E. J., Roberts, B. E., Dixon, B. R. Enantioselective Total Synthesis of β-Elemene and Fuscol Based on Enantiocontrolled Ireland-Claisen Rearrangement. *J. Am. Chem. Soc.* **1995**, 117, 193-196.
24. Krafft, M. E., Dasse, O. A., Jarrett, S., Fievre, A. A Chelation-Controlled Ester Enolate Claisen Rearrangement. *J. Org. Chem.* **1995**, 60, 5093-5101.
25. Kazmaier, U., Mues, H., Krebs, A. Asymmetric chelated Claisen rearrangements in the presence of chiral ligands-scope and limitations. *Chem.-- Eur. J.* **2002**, 8, 1850-1855.
26. Khaledy, M. M., Kalani, M. Y. S., Khuong, K. S., Houk, K. N., Aviyente, V., Neier, R., Soldermann, N., Velker, J. Origins of Boat or Chair Preferences in the Ireland-Claisen Rearrangements of Cyclohexenyl Esters: A Theoretical Study. *J. Org. Chem.* **2003**, 68, 572-577.
27. Ireland, R. E., Wipf, P., Armstrong, J. D., III. Stereochemical control in the ester enolate Claisen rearrangement. 1. Stereoselectivity in silyl ketene acetal formation. *J. Org. Chem.* **1991**, 56, 650-657.
28. Ireland, R. E., Wipf, P., Xiang, J. N. Stereochemical control in the ester enolate Claisen rearrangement. 2. Chairlike vs boatlike transition-state selection. *J. Org. Chem.* **1991**, 56, 3572-3582.
29. Uchiyama, H., Kawano, M., Katsuki, T., Yamaguchi, M. Ester enolate Claisen rearrangement via boat-like transition state. *Chem. Lett.* **1987**, 351-354.
30. Paterson, I., Hulme, A. N. Total Synthesis of (-)-Ebelactone A and B. *J. Org. Chem.* **1995**, 60, 3288-3300.
31. He, F., Bo, Y., Altom, J. D., Corey, E. J. Enantioselective Total Synthesis of Aspidophytine. *J. Am. Chem. Soc.* **1999**, 121, 6771-6772.
32. Gilbert, J. C., Selliah, R. D. Enantioselective synthesis of (-)-trichodiene. *J. Org. Chem.* **1993**, 58, 6255-6265.

Clemmensen Reduction 92

Related reactions: Wolff-Kishner reduction;

1. Clemmensen, E. Reduction of ketones and aldehydes to the corresponding hydrocarbons using zinc-amalgam and hydrochloric acid. *Chem. Ber.* **1913**, 46, 1837-1843.
2. Clemmensen, E. General method for the reduction of the carbonyl group in aldehydes and ketones to the methylene group. III. *Ber.* **1914**, 47, 681-687.
3. Clemmensen, E. A general method for the reduction of the carbonyl group in aldehydes and ketones to the methylene group. II. *Ber.* **1914**, 47, 51-63.
4. Martin, E. L. Clemmensen reduction. *Org. React.* **1942**, 1, 155-209.
5. Smith, M. Dissolving metal reductions. *Reduction* **1968**, 95-170.
6. Buchanan, J. G. S. C., Woodgate, P. D. Clemmensen reduction of difunctional ketones. *Quart. Rev., Chem. Soc.* **1969**, 23, 522-536.
7. Vedejs, E. Clemmensen reduction of ketones in anhydrous organic solvents. *Org. React.* **1975**, 22, 401-422.
8. Banerjee, A. K. Molecular rearrangement of ketonic and olefinic compounds. *J. Sci. Ind. Res.* **1992**, 51, 869-876.
9. Motherwell, W. B., Nutley, C. J. The role of zinc carbenoids in organic synthesis. *Contemp. Org. Synth.* **1994**, 1, 219-241.
10. Yamamura, S., Ueda, S., Hirata, Y. Zinc reductions of oxo steroids. *Chem. Commun.* **1967**, 1049-1050.
11. Yamamura, S., Hirata, Y. Zinc reductions of keto groups to methylene groups. *J. Chem. Soc. C* **1968**, 2887-2889.
12. Toda, M., Hirata, Y., Yamamura, S. Zinc reductions of ketosteroids. *J. Chem. Soc. D* **1969**, 919-920.
13. Toda, M., Hayashi, M., Hirata, Y., Yamamura, S. Modified Clemmensen reductions of keto groups to methylene groups. *Bull. Chem. Soc. Jap.* **1972**, 45, 264-266.
14. Steinkopf, W., Wolfram, A. Reduction of the carbonyl group with zinc amalgam; theory of the reduction. *Ann.* **1923**, 430, 113-161.
15. Brewster, J. H. Mechanism of reductions at metal surfaces. II. A mechanism of the Clemmensen reduction. *J. Am. Chem. Soc.* **1954**, 76, 6364-6368.
16. Brewster, J. H., Patterson, J., Fidler, D. A. Mechanism of reductions at metal surfaces. III. Clemmensen reduction of some sterically hindered ketones. *J. Am. Chem. Soc.* **1954**, 76, 6368-6371.
17. Nakabayashi, T., Kai, K. Kinetics of Clemmensen reduction. II. Mechanism of the reduction of p-hydroxyacetophenone. *J. Chem. Soc. Jap., Pure Chem. Sect.* **1956**, 77, 657-665.
18. Staschewski, D. The mechanism of the Clemmensen reduction. *Angew. Chem.* **1959**, 71, 726-736.
19. Risinger, G. E., Eddy, C. W. Studies in the zinc reduction series: a mechanism for the zinc and alkali reduction of aromatic ketones. *Chem. Ind.* **1963**, 570-571.
20. Risinger, G. E., Mach, E. E., Barnett, K. W. Effect of dilute acid on the Clemmensen reduction. *Chem. Ind.* **1965**, 679.
21. Elphimoff-Felkin, I., Sarda, P. Reductions by zinc in the presence of acids. III. Reduction of alcohols, ethers, acetates, and allylic halides to olefins. *Tetrahedron* **1977**, 33, 511-516.
22. Burdon, J., Price, R. C. The mechanism of the Clemmensen reduction: the substrates. *J. Chem. Soc., Chem. Commun.* **1986**, 893-894.
23. Di Vona, M. L., Floris, B., Luchetti, L., Rosnati, V. Single-electron transfers in zinc-promoted reactions. The mechanisms of the Clemmensen reduction and related reactions. *Tetrahedron Lett.* **1990**, 31, 6081-6084.

24. Talapatra, S. K., Chakrabarti, S., Mallik, A. K., Talapatra, B. Some newer aspects of Clemmensen reduction of aromatic ketones. *Tetrahedron* **1990**, 46, 6047-6052.
25. Di Vona, M. L., Rosnati, V. Zinc-promoted reactions. 1. Mechanism of the Clemmensen reaction. Reduction of benzophenone in glacial acetic acid. *J. Org. Chem.* **1991**, 56, 4269-4273.
26. Davis, B. R., Hinds, M. G. Clemmensen reduction. XII. The synthesis and acidolysis of some diaryl-substituted cyclopropane-1,2-diols. The possible involvement of a cyclopropyl cation. *Aust. J. Chem.* **1997**, 50, 309-319.
27. Villiers, C., Ephritikhine, M. Reactions of aliphatic ketones R_2CO (R = Me, Et, *i*-Pr, and *t*-Bu) with the MCl_4/Li(Hg) system (M = U or Ti): mechanistic analogies between the McMurry, Wittig, and Clemmensen reactions. *Chem.--Eur. J.* **2001**, 7, 3043-3051.
28. Kappe, T., Aigner, R., Roschger, P., Schnell, B., Stadbauer, W. A simple and effective method for the reduction of acyl substituted heterocyclic 1,3-dicarbonyl compounds to alkyl derivatives by zinc-acetic acid-hydrochloric acid. *Tetrahedron* **1995**, 51, 12923-12928.
29. Naruse, M., Aoyagi, S., Kibayashi, C. Total synthesis of (-)-pumiliotoxin C by aqueous intramolecular acylnitroso Diels-Alder approach. *Tetrahedron Lett.* **1994**, 35, 9213-9216.
30. Werner, K. M., De los Santos, J. M., Weinreb, S. M., Shang, M. A Convergent Stereoselective Synthesis of the Putative Structure of the Marine Alkaloid Lepadiformine via an Intramolecular Nitrone/1,3-Diene Dipolar Cycloaddition. *J. Org. Chem.* **1999**, 64, 686-687.
31. Martins, F. J. C., Viljoen, A. M., Venter, H. J., Wessels, P. L. Synthesis of novel tetracyclo[6.3.0.02,6.03,10]undecane and tetracyclo[6.4.0.02,6.03,10]dodecane derivatives. *Tetrahedron* **1997**, 53, 14991-14996.

Combes Quinoline Synthesis 94

Related reactions: Skraup and Doebner-Miller quinoline synthesis;

1. Combes, A. Synthesis of quinoline derivatives from acetyl acetone. *Bull. Soc. Chim. France* **1883**, 49, 89.
2. Bergstrom, F. W. Heterocyclic N compounds. IIA. Hexacyclic compounds: pyridine, quinoline and isoquinoline. *Chem. Rev.* **1944**, 35, 77-277.
3. Claret, P. A. Quinolines. *Compr. Org. Chem.* **1979**, 4, 155-203.
4. Yamashkin, S. A., Yudin, L. G., Kost, A. N. Pyridine ring closure in synthesis of quinolines according to Combes (review). *Khim. Geterotsikl. Soedin.* **1992**, 1011-1024.
5. Claret, P. A., Osborne, A. G. 2,4-Diethylquinoline- an extension of the Combes synthesis. *Org. Prep. Proced.* **1970**, 2, 305-308.
6. Conrad, M., Limpach, L. The synthesis of quinoline derivatives from acetoacetic ester. *Ber.* **1887**, 20, 944-948.
7. Conrad, M., Limpach, L. Synthesis of quinoline derivatives from acetoacetic ester. *Ber.* **1891**, 24, 2990-2992.
8. Manske, R. H. F. The chemistry of quinolines. *Chem. Rev.* **1942**, 30, 113-144.
9. Born, J. L. Mechanism of formation of benzo[g]quinolones via the Combes reaction. *J. Org. Chem.* **1972**, 37, 3952-3953.
10. Gupta, S. C., Singh, D., Sadana, A., Mor, S., Saini, A., Sharma, K., Dhawan, S. N. Novel polycyclic heterocyclic ring systems: synthesis of benz[h]- and benz[f]indeno[2,1-c]quinolines. *J. Chem. Res., Synop.* **1994**, 34-35.
11. Marcos, A., Pedregal, C., Avendano, C. Reactivity of 4(7)-aminobenzimidazole as a bidentate nucleophile. *Tetrahedron* **1991**, 47, 7459-7464.
12. West, A. P., Jr., Van Engen, D., Pascal, R. A., Jr. Attempted synthesis of 1,2,3,4-tetraphenylfluoreno[1,9-gh]quinoline. *J. Org. Chem.* **1992**, 57, 784-786.

Cope Elimination / Cope Reaction 96

Related reactions: Burgess dehydration, Chugaev elimination, Hofmann elimination;

1. Cope, A. C., Foster, T. T., Towle, P. H. Thermal decomposition of amine oxides to olefins and dialkylhydroxylamines. *J. Am. Chem. Soc.* **1949**, 71, 3929-3935.
2. Cope, A. C., Towle, P. H. Rearrangement of allyldialkylamine oxides and benzyldimethylamine oxide. *J. Am. Chem. Soc.* **1949**, 71, 3423-3428.
3. Cope, A. C., Pike, R. A., Spencer, C. F. Cyclic polyolefins. XXVII. *cis*- and *trans*-Cycloöctene from *N,N*-dimethylcycloöctylamine. *J. Am. Chem. Soc.* **1953**, 75, 3212-3215.
4. Cope, A. C., Trumbull, E. R. Olefins from amines: the Hofmann elimination reaction and amine oxide pyrolysis. *Org. React.* **1960**, 11, 317-493.
5. DePuy, C. H., King, R. W. Pyrolytic cis eliminations. *Chem. Rev.* **1960**, 60, 431-457.
6. Cooper, N. J., Knight, D. W. The reverse Cope cyclization: a classical reaction goes backwards. *Tetrahedron* **2004**, 60, 243-269.
7. Ciganek, E. Reverse Cope elimination reactions. 2. Application to synthesis. *J. Org. Chem.* **1995**, 60, 5803-5807.
8. Ciganek, E., Read, J. M., Jr., Calabrese, J. C. Reverse Cope elimination reactions. 1. Mechanism and scope. *J. Org. Chem.* **1995**, 60, 5795-5802.
9. Gallagher, B. M., Pearson, W. H. Thermal cyclization of N-hydroxylamines with alkenes: the reverse Cope elimination. *Chemtracts: Org. Chem.* **1996**, 9, 126-130.
10. O'Neil, I. A., Cleator, E., Tapolczay, D. J. A convenient synthesis of secondary hydroxylamines. *Tetrahedron Lett.* **2001**, 42, 8247-8249.
11. Sammelson, R. E., Kurth, M. J. Oxidation-Cope elimination: a REM-resin cleavage protocol for the solid-phase synthesis of hydroxylamines. *Tetrahedron Lett.* **2001**, 42, 3419-3422.
12. Bach, R. D., Gonzalez, C., Andres, J. L., Schlegel, H. B. Kinetic Isotope Effects as a Guide to Transition State Geometries for the Intramolecular Cope and Ylide Elimination Reactions. An ab Initio MO Study. *J. Org. Chem.* **1995**, 60, 4653-4656.
13. Komaromi, I., Tronchet, J. M. J. Quantum Chemical Reaction Path and Transition State for a Model Cope (and Reverse Cope) Elimination. *J. Phys. Chem. A* **1997**, 101, 3554-3560.
14. Caserio, F. F., Jr., Parker, S. H., Piccolini, R., Roberts, J. D. Small-ring compounds. XX. 1,3-Dimethylenecyclobutane and related compounds. *J. Am. Chem. Soc.* **1958**, 80, 5507-5513.
15. Cope, A. C., Ciganek, E., Howell, C. F., Schweizer, E. E. Amine oxides. VIII. Medium-sized cyclic olefins from amine oxides and quaternary ammonium hydroxides. *J. Am. Chem. Soc.* **1960**, 82, 4663-4669.
16. Cope, A. C., LeBel, N. A. Amine oxides. VII. Thermal decomposition of the N-oxides of N-methylazacycloalkanes. *J. Am. Chem. Soc.* **1960**, 82, 4656-4662.
17. Cram, D. J., McCarty, J. E. Stereochemistry. XXIV. The preparation and determination of configuration of the isomers of 2-amino-3-phenylbutane, and the steric course of the amine oxide pyrolysis reaction in this system. *J. Am. Chem. Soc.* **1954**, 76, 5740-5745.
18. Bach, R. D., Andrzejewski, D., Dusold, L. R. Mechanism of the Cope elimination. *J. Org. Chem.* **1973**, 38, 1742-1743.
19. Chiao, W.-B., Saunders, W. H., Jr. Mechanisms of elimination reactions. 29. Deuterium kinetic isotope effects in eliminations from amine oxides. The consequences of nonlinear proton transfer. *J. Am. Chem. Soc.* **1978**, 100, 2802-2805.
20. Kwart, H., Brechbiel, M. Role of solvent in the mechanism of amine oxide thermolysis elucidated by the temperature dependence of a kinetic isotope effect. *J. Am. Chem. Soc.* **1981**, 103, 4650-4652.
21. Bach, R. D., Braden, M. L. Primary and secondary kinetic isotope effects in the Cope and Hofmann elimination reactions. *J. Org. Chem.* **1991**, 56, 7194-7195.
22. Remen, L., Vasella, A. Conformationally biased mimics of mannopyranosylamines: Inhibitors of β-mannosidases? *Helv. Chim. Acta* **2002**, 85, 1118-1127.
23. Garcia Martinez, A., Teso Vilar, E., Garcia Fraile, A., de la Moya Cerero, S., Lora Maroto, B. A new enantiospecific synthetic procedure to the taxoid-intermediate 10-methylenecamphor, and 10-methylenefenchone. *Tetrahedron: Asymmetry* **2002**, 13, 17-19.

Related reactions: Oxy-Cope rearrangement;

1. Cope, A. C., Hardy, E. M. Introduction of substituted vinyl groups. V. A rearrangement involving the migration of an allyl group in a three-carbon system. *J. Am. Chem. Soc.* **1940**, 62, 441-444.
2. Takeda, K. Stereospecific Cope rearrangement of the germacrene-type sesquiterpenes. *Tetrahedron* **1974**, 30, 1525-1534.
3. Rhoads, S. J., Raulins, N. R. Claisen and Cope rearrangements. *Org. React.* **1975**, 22, 1-252.
4. Lutz, R. P. Catalysis of the Cope and Claisen rearrangements. *Chem. Rev.* **1984**, 84, 205-247.
5. Murray, A. W. Molecular rearrangements. *Org. React. Mech.* **1989**, 457-573.
6. Dewar, M. J. S., Jie, C. Mechanisms of pericyclic reactions: the role of quantitative theory in the study of reaction mechanisms. *Acc. Chem. Res.* **1992**, 25, 537-543.
7. Davies, H. M. L. Tandem cyclopropanation/Cope rearrangement: a general method for the construction of seven-membered rings. *Tetrahedron* **1993**, 49, 5203-5223.
8. Wilson, S. R. Anion-assisted sigmatropic rearrangements. *Org. React.* **1993**, 43, 93-250.
9. Enders, D., Knopp, M., Schiffers, R. Asymmetric [3.3]-sigmatropic rearrangements in organic synthesis. *Tetrahedron: Asymmetry* **1996**, 7, 1847-1882.
10. Lukyanov, S. M., Koblik, A. V. Rearrangements of dienes and polyenes. *Chemistry of Dienes and Polyenes* **2000**, 2, 739-884.
11. Nowicki, J. Claisen, Cope and related rearrangements in the synthesis of flavor and fragrance compounds. *Molecules [online computer file]* **2000**, 5, 1033-1050.
12. Murray, A. W. Molecular rearrangements. *Org. React. Mech.* **2001**, 473-603.
13. Murray, A. W. Molecular rearrangements. *Org. React. Mech.* **2003**, 487-615.
14. Nubbemeyer, U. Recent advances in asymmetric [3,3]-sigmatropic rearrangements. *Synthesis* **2003**, 961-1008.
15. Delbecq, F., Nguyen Trong, A. A theoretical study of substituent effects. Influence on the rate of the Cope rearrangement. *Nouv. J. Chim.* **1983**, 7, 505-513.
16. Lalitha, S., Chandrasekhar, J., Mehta, G. Acceleration of Cope rearrangement by a remote carbenium ion center: theoretical elucidation of the electronic origin. *J. Org. Chem.* **1990**, 55, 3455-3457.
17. Houk, K. N., Gustafson, S. M., Black, K. A. Theoretical secondary kinetic isotope effects and the interpretation of transition state geometries. 1. The Cope rearrangement. *J. Am. Chem. Soc.* **1992**, 114, 8565-8572.
18. Williams, R. V., Kurtz, H. A. A theoretical investigation of through-space interactions. Part 3. A semiempirical study of the Cope rearrangement in singly annellated semibullvalenes. *J. Chem. Soc., Perkin Trans. 2* **1994**, 147-150.
19. Jiao, H., Nagelkerke, R., Kurtz, H. A., Williams, R. V., Borden, W. T., Schleyer, P. v. R. Annelated Semibullvalenes: A Theoretical Study of How They "Cope" with Strain. *J. Am. Chem. Soc.* **1997**, 119, 5921-5929.
20. Black, K. A., Wilsey, S., Houk, K. N. Alkynes, Allenes, and Alkenes in [3,3]-Sigmatropy: Functional Diversity and Kinetic Monotony. A Theoretical Analysis. *J. Am. Chem. Soc.* **1998**, 120, 5622-5627.
21. Hrovat, D. A., Chen, J., Houk, K. N., Borden, W. T. Cooperative and Competitive Substituent Effects on the Cope Rearrangements of Phenyl-Substituted 1,5-Hexadienes Elucidated by Becke3LYP/6-31G Calculations. *J. Am. Chem. Soc.* **2000**, 122, 7456-7460.
22. Sakai, S. Theoretical analysis of the Cope rearrangement of 1,5-hexadiene. *Int. J. Quantum Chem.* **2000**, 80, 1099-1106.
23. Staroverov, V. N., Davidson, E. R. Transition Regions in the Cope Rearrangement of 1,5-Hexadiene and Its Cyano Derivatives. *J. Am. Chem. Soc.* **2000**, 122, 7377-7385.
24. Karadakov, P. B. Chapter 3. Theoretical description of reaction mechanisms: reaction pathways and electronic structure rearrangements. *Annu. Rep. Prog. Chem., Sect. C, Phys. Chem.* **2001**, 97, 61-90.
25. Staroverov, V. N., Davidson, E. R. The Cope rearrangement in theoretical retrospect. *THEOCHEM* **2001**, 573, 81-89.
26. Sakai, S. Theoretical analysis of the Cope rearrangement of 1,5-hexadiene and phenyl derivatives. *THEOCHEM* **2002**, 583, 181-188.
27. Isobe, H., Yamanaka, S., Yamaguchi, K. Utility of chemical indices for transition structures of pericyclic reactions: Case study of the cope rearrangement. *Int. J. Quantum Chem.* **2003**, 95, 532-545.
28. Ozkan, I., Zora, M. Transition Structures, Energetics, and Secondary Kinetic Isotope Effects for Cope Rearrangements of cis-1,2-Divinylcyclobutane and cis-1,2-Divinylcyclopropane: A DFT Study. *J. Org. Chem.* **2003**, 68, 9635-9642.
29. Doering, W. v. E., Roth, W. R. The overlap of two allyl radicals or a four-centered transition state in the Cope rearrangement. *Tetrahedron* **1962**, 18, 67-74.
30. Wigfield, D. C., Feiner, S. Solvent effects in the Cope rearrangement. *Can. J. Chem.* **1970**, 48, 855-858.
31. Baldwin, J. E., Kaplan, M. S. Mechanistic alternative for the thermal antara-antara Cope rearrangements of bicyclo[3.2.0]hepta-2,6-dienes and bicyclo[4.2.0]octa-2,7-dienes. *J. Am. Chem. Soc.* **1971**, 93, 3969-3977.
32. Goldstein, M. J., Benzon, M. S. Boat and chair transition states of 1,5-hexadiene. *J. Am. Chem. Soc.* **1972**, 94, 7147-7149.
33. Wehrli, R., Bellus, D., Hansen, H. J., Schmid, H. The Cope rearrangement - a reaction with a manifold mechanism? *Chimia* **1976**, 30, 416-423.
34. Shea, K. J., Phillips, R. B. Diastereomeric transition states. Relative energies of the chair and boat reaction pathways in the Cope rearrangement. *J. Am. Chem. Soc.* **1980**, 102, 3156-3162.
35. Guenther, H., Runsink, J., Schmickler, H., Schmitt, P. Activation parameters for the degenerate Cope rearrangement of barbaralane and 3,7-disubstituted barbaralanes. *J. Org. Chem.* **1985**, 50, 289-293.
36. Owens, K. A., Berson, J. A. Stereochemistry of the thermal acetylenic Cope rearrangement. Experimental test for a 1,4-cyclohexenediyl as a mechanistic intermediate. *J. Am. Chem. Soc.* **1990**, 112, 5973-5985.
37. Houk, K. N., Gonzalez, J., Li, Y. Pericyclic Reaction Transition States: Passions and Punctilios, 1935-1995. *Acc. Chem. Res.* **1995**, 28, 81-90.
38. Jiao, H., Schleyer, P. v. R. The Cope rearrangement transition structure is not diradicaloid, but is it aromatic? *Angew. Chem., Int. Ed. Engl.* **1995**, 34, 334-337.
39. Wiest, O., Houk, K. N., Black, K. A., Thomas, B. I. V. Secondary Kinetic Isotope Effects of Diastereotopic Protons in Pericyclic Reactions: A New Mechanistic Probe. *J. Am. Chem. Soc.* **1995**, 117, 8594-8599.
40. Castano, O., Frutos, L.-M., Palmeiro, R., Notario, R., Andres, J.-L., Gomperts, R., Blancafort, L., Robb, M. A. The valence isomerization of cyclooctatetraene to semibullvalene. *Angew. Chem., Int. Ed. Engl.* **2000**, 39, 2095-2097.
41. von Doering, W., Birladeanu, L., Sarma, K., Blaschke, G., Scheidemantel, U., Boese, R., Benet-Bucholz, J., Klaerner, F. G., Gehrke, J.-S., Zimny, B. U., Sustmann, R., Korth, H.-G. A Non-Cope among the Cope Rearrangements of 1,3,4,6-Tetraphenylhexa-1,5-dienes. *J. Am. Chem. Soc.* **2000**, 122, 193-203.
42. Allin, S. M., Baird, R. D. Development and synthetic applications of asymmetric [3,3]-sigmatropic rearrangements. *Curr. Org. Chem.* **2001**, 5, 395-415.
43. Gajewski, J. J., Conrad, N. D., Emrani, J., Gilbert, K. E. Substituent effects on the Cope rearrangement, Neither centaurs nor chameleons can characterize them. *ARKIVOC (Gainesville, FL, United States) [online computer file]* **2002**, 18-29.
44. Limanto, J., Snapper, M. L. Sequential Intramolecular Cyclobutadiene Cycloaddition, Ring-Opening Metathesis, and Cope Rearrangement: Total Syntheses of (+)- and (-)-Asteriscanolide. *J. Am. Chem. Soc.* **2000**, 122, 8071-8072.
45. Davies, H. M. L., Doan, B. D. Total Synthesis of (±)-Tremulenolide A and (±)-Tremulenediol A via a Stereoselective Cyclopropanation/Cope Rearrangement Annulation Strategy. *J. Org. Chem.* **1998**, 63, 657-660.
46. Hudlicky, T., Boros, C. H., Boros, E. E. A model study directed towards a practical enantioselective total synthesis of (-)-morphine. *Synthesis* **1992**, 174-178.

Corey-Bakshi-Shibata Reduction (CBS Reduction) 100

Related reactions: *Luche reduction, Midland alpine borane reduction, Noyori asymmetric reduction;*

1. Hirao, A., Itsuno, S., Nakahama, S., Yamazaki, N. Asymmetric reduction of aromatic ketones with chiral alkoxyamine-borane complexes. *J. Chem. Soc., Chem. Commun.* **1981**, 315-317.
2. Corey, E. J., Bakshi, R. K., Shibata, S. Highly enantioselective borane reduction of ketones catalyzed by chiral oxazaborolidines. Mechanism and synthetic implications. *J. Am. Chem. Soc.* **1987**, 109, 5551-5553.
3. Corey, E. J., Bakshi, R. K., Shibata, S., Chen, C. P., Singh, V. K. A stable and easily prepared catalyst for the enantioselective reduction of ketones. Applications to multistep syntheses. *J. Am. Chem. Soc.* **1987**, 109, 7925-7926.
4. Corey, E. J., Shibata, S., Bakshi, R. K. An efficient and catalytically enantioselective route to (S)-(-)-phenyloxirane. *J. Org. Chem.* **1988**, 53, 2861-2863.
5. Ganem, B. Highly enantioselective borane reduction of ketones catalyzed by chiral oxazaborolidines: mechanism and synthetic implications. *Chemtracts: Org. Chem.* **1988**, 1, 40-42.
6. Singh, V. K. Practical and useful methods for the enantioselective reduction of unsymmetrical ketones. *Synthesis* **1992**, 607-617.
7. Wallbaum, S., Martens, J. Asymmetric syntheses with chiral oxazaborolidines. *Tetrahedron: Asymmetry* **1992**, 3, 1475-1504.
8. Deloux, L., Srebnik, M. Asymmetric boron-catalyzed reactions. *Chem. Rev.* **1993**, 93, 763-784.
9. Corey, E. J., Helal, C. J. Reduction of carbonyl compounds with chiral oxazaborolidine catalysts: A new paradigm for enantioselective catalysis and a powerful new synthetic method. *Angew. Chem., Int. Ed. Engl.* **1998**, 37, 1986-2012.
10. Brandt, P., Andersson, P. G. Exploring the chemistry of 3-substituted 2-azanorbornyls in asymmetric catalysis. *Synlett* **2000**, 1092-1106.
11. Kadyrov, R., Selke, R. Highly enantioselective catalytic reduction of ketones paying particular attention to aliphatic derivatives. in *Organic Synthesis Highlights IV* 194-206 (VCH, Weinheim, New York, **2000**).
12. Woodward, S. Going soft (or hard) in asymmetric catalysis? *Curr. Sci.* **2000**, 78, 1314-1317.
13. Molt, O., Schrader, T. Asymmetric synthesis with chiral cyclic phosphorus auxiliaries. *Synthesis* **2002**, 2633-2670.
14. Price, M. D., Sui, J. K., Kurth, M. J., Schore, N. E. Oxazaborolidines as Functional Monomers: Ketone Reduction Using Polymer-Supported Corey, Bakshi, and Shibata Catalysts. *J. Org. Chem.* **2002**, 67, 8086-8089.
15. Nevalainen, V. Quantum chemical modeling of chiral catalysis. Part 2. On the origin of enantioselection in the coordination of carbonyl compounds to borane adducts of chiral oxazaborolidines. *Tetrahedron: Asymmetry* **1991**, 2, 429-435.
16. Nevalainen, V. Quantum chemical modeling of chiral catalysis. On the mechanism of catalytic enantioselective reduction of carbonyl compounds by chiral oxazaborolidines. *Tetrahedron: Asymmetry* **1991**, 2, 63-74.
17. Linney, L. P., Self, C. R., Williams, I. H. Computational elucidation of the catalytic mechanism for ketone reduction by an oxazaborolidine-borane adduct. *J. Chem. Soc., Chem. Commun.* **1994**, 1651-1652.
18. Linney, L. P., Self, C. R., Williams, I. H. A theoretical investigation of hydride bridging in chiral oxazaborolidine-borane adducts: the importance of electron correlation. *Tetrahedron: Asymmetry* **1994**, 5, 813-816.
19. Nevalainen, V. Quantum-chemical modeling of chiral catalysis. Part 15. On the role of hydride-bridged borane-alkoxyborane complexes in the catalytic enantioselective reduction of ketones promoted by chiral oxazaborolidines. *Tetrahedron: Asymmetry* **1994**, 5, 289-296.
20. Nevalainen, V. Quantum chemical modeling of chiral catalysis. Part 17. On the diborane derivatives of chiral oxazaborolidines used as catalysts in the enantioselective reduction of ketones. *Tetrahedron: Asymmetry* **1994**, 5, 395-402.
21. Quallich, G. J., Blake, J. F., Woodall, T. M. A combined synthetic and ab initio study of chiral oxazaborolidines structure and enantioselectivity relationships. *J. Am. Chem. Soc.* **1994**, 116, 8516-8525.
22. Li, M., Tian, A. Enantioselective reduction of 3,3-dimethyl-butanone-2 with borane catalyzed by oxazaborolidine. Part 1. Quantum chemical computations on the structures and properties of catalyst and catalyst-borane-ketone adducts. *THEOCHEM* **2001**, 544, 25-35.
23. Alagona, G., Ghio, C., Persico, M., Tomasi, S. Quantum Mechanical Study of Stereoselectivity in the Oxazaborolidine-Catalyzed Reduction of Acetophenone. *J. Am. Chem. Soc.* **2003**, 125, 10027-10039.
24. Evans, D. A. Stereoselective organic reactions: catalysts for carbonyl addition processes. *Science* **1988**, 240, 420-426.
25. Corey, E. J. New enantioselective routes to biologically interesting compounds. *Pure Appl. Chem.* **1990**, 62, 1209-1216.
26. Jones, D. K., Liotta, D. C., Shinkai, I., Mathre, D. J. Origins of the enantioselectivity observed in oxazaborolidine-catalyzed reductions of ketones. *J. Org. Chem.* **1993**, 58, 799-801.
27. Mathre, D. J., Thompson, A. S., Douglas, A. W., Hoogsteen, K., Carroll, J. D., Corley, E. G., Grabowski, E. J. J. A practical process for the preparation of tetrahydro-1-methyl-3,3-diphenyl-1H,3H-pyrrolo[1,2-c][1,3,2]oxazaborole-borane. A highly enantioselective stoichiometric and catalytic reducing agent. *J. Org. Chem.* **1993**, 58, 2880-2888.
28. Rodriguez, A., Nomen, M., Spur, B. W., Godfroid, J.-J. An efficient asymmetric synthesis of prostaglandin E1. *Eur. J. Org. Chem.* **1999**, 2655-2662.
29. Corey, E. J., Roberts, B. E. Total Synthesis of Dysidiolide. *J. Am. Chem. Soc.* **1997**, 119, 12425-12431.
30. Sabes, S. F., Urbanek, R. A., Forsyth, C. J. Efficient Synthesis of Okadaic Acid. 2. Synthesis of the C1-C14 Domain and Completion of the Total Synthesis. *J. Am. Chem. Soc.* **1998**, 120, 2534-2542.

Corey-Chaykovsky Epoxidation and Cyclopropanation 102

Related reactions: *Simmons-Smith cyclopropanation, Darzens glycidic ester condensation;*

1. Corey, E. J., Chaykovsky, M. Dimethylsulfoxonium methylide. *J. Am. Chem. Soc.* **1962**, 84, 867-868.
2. Corey, E. J., Chaykovsky, M. Dimethyloxosulfonium methylide and dimethylsulfonium methylide. Formation and application to organic synthesis. *J. Am. Chem. Soc.* **1965**, 87, 1353-1364.
3. Trost, B. M. Sulfur as a key element in synthesis and biosynthesis. *Org. Sulphur Chem., [Proc. Int. Conf.], 6th* **1975**, 237-263.
4. Trost, B. M., Melvin, L. S., Jr. *Organic Chemistry, Vol. 31: Sulfur Ylides, Emerging Synthetic Intermediates* (Academic Press, New York, **1975**) 346 pp.
5. Block, E. *Reactions of Organosulfur Compounds* (Academic Press, New York, **1978**) 336 pp.
6. Li, A.-H., Dai, L.-X., Aggarwal, V. K. Asymmetric Ylide Reactions: Epoxidation, Cyclopropanation, Aziridination, Olefination, and Rearrangement. *Chem. Rev.* **1997**, 97, 2341-2372.
7. Aggarwal, V. K. Catalytic asymmetric epoxidation and aziridination mediated by sulfur ylides. Evolution of a project. *Synlett* **1998**, 329-336.
8. Aggarwal, V. K. Epoxide formation of enones and aldehydes. in *Comprehensive Asymmetric Catalysis I-III* (eds. Jacobsen, E., Pfaltz, A.,Yamamoto, H.), *2*, 679-696 (Springer-Verlag, Berlin Heidelberg, **1999**).
9. Aggarwal, V. K. in *Comprehensive Asymmetric Catalysis* (eds. Jacobsen, E., Pfaltz, N.,Yamamoto, H.), *II*, 679-693 (Springer-Verlag, Heildelberg, **1999**).
10. Herrera, F. J. L., Garcia, F. R. S., Gonzalez, M. S. P. The chemistry of sulfur ylides and diazo compounds in the carbohydrate field: Reactivity and synthetic applications. *Rec. Res. Dev. Org. Chem.* **2000**, 4, 465-490.
11. Lakeev, S. N., Maydanova, I. O., Galin, F. Z., Tolstikov, G. A. Sulfur ylides in the synthesis of heterocyclic and carbocyclic compounds. *Russ. Chem. Rev.* **2001**, 70, 655-672.
12. Johnson, C. R. Utilization of sulfoximines and derivatives as reagents for organic synthesis. *Acc. Chem. Res.* **1973**, 6, 341-347.
13. Ng, J. S. Epoxide formation from aldehydes and ketones - a modified method for preparing the Corey-Chaykovsky reagents. *Synth. Commun.* **1990**, 20, 1193-1202.
14. Saito, T., Sakairi, M., Akiba, D. Enantioselective synthesis of aziridines from imines and alkyl halides using a camphor-derived chiral sulfide mediator via the imino Corey-Chaykovsky reaction. *Tetrahedron Lett.* **2001**, 42, 5451-5454.

15. Volatron, F., Eisenstein, O. Theoretical study of the reactivity of phosphonium and sulfonium ylides with carbonyl groups. *J. Am. Chem. Soc.* **1984**, 106, 6117-6119.
16. Volatron, F., Eisenstein, O. Wittig versus Corey-Chaykovsky Reaction. Theoretical study of the reactivity of phosphonium methylide and sulfonium methylide with formaldehyde. *J. Am. Chem. Soc.* **1987**, 109, 1-4.
17. Das, G. K. Substituent effect on transition structure of Corey-Chaykovsky reaction: a semiempirical study. *Indian J. Chem., Sect. A: Inorg., Bio-inorg., Phys., Theor. Anal. Chem.* **2001**, 40A, 23-29.
18. Rocquet, F., Sevin, A. Mechanism of the addition of trimethylsulphoxonium ylid to α,β-ethylenic ketones. *Bull. Soc. Chim. Fr.* **1974**, 881-887.
19. Aggarwal, V. K., Abdel-Rahman, H., Fan, L., Jones, R. V. H., Standen, M. C. H. A novel catalytic cycle for the synthesis of epoxides using sulfur ylides. *Chem.-- Eur. J.* **1996**, 2, 1024-1030.
20. Ohno, F., Kawashima, T., Okazaki, R. Synthesis, Crystal Structure, and Thermolysis of a Pentacoordinate 1,2λ6-Oxathietane: An Intermediate of the Corey-Chaykovsky Reaction of Oxosulfonium Ylides? *J. Am. Chem. Soc.* **1996**, 118, 697-698.
21. Aggarwal, V. K., Bell, L., Coogan, M. P., Jubault, P. Bifunctional catalysts for asymmetric sulfur ylide epoxidation of carbonyl compounds. *J. Chem. Soc., Perkin Trans. 1* **1998**, 2037-2042.
22. Aggarwal, V. K., Ford, J. G., Fonquerna, S., Adams, H., Jones, R. V. H., Fieldhouse, R. Catalytic Asymmetric Epoxidation of Aldehydes. Optimization, Mechanism, and Discovery of Stereoelectronic Control Involving a Combination of Anomeric and Cieplak Effects in Sulfur Ylide Epoxidations with Chiral 1,3-Oxathianes. *J. Am. Chem. Soc.* **1998**, 120, 8328-8339.
23. Lindvall, M. K., Koskinen, A. M. P. Origins of Stereoselectivity in the Corey-Chaykovsky Reaction. Insights from Quantum Chemistry. *J. Org. Chem.* **1999**, 64, 4596-4606.
24. Zanardi, J., Leriverend, C., Aubert, D., Julienne, K., Metzner, P. A catalytic cycle for the asymmetric synthesis of epoxides using sulfur ylides. *J. Org. Chem.* **2001**, 66, 5620-5623.
25. Aggarwal, V. K., Harvey, J. N., Richardson, J. Unraveling the Mechanism of Epoxide Formation from Sulfur Ylides and Aldehydes. *J. Am. Chem. Soc.* **2002**, 124, 5747-5756.
26. Smith, A. B., III, Fukui, M., Vaccaro, H. A., Empfield, J. R. Phyllanthoside-phyllanthostatin synthetic studies. 7. Total synthesis of (+)-phyllanthocin and (+)-phyllanthocindiol. *J. Am. Chem. Soc.* **1991**, 113, 2071-2092.
27. Hansson, T., Wickberg, B. A short enantiospecific route to isodaucane sesquiterpenes from limonene. On the absolute configuration of (+)-aphanamol I and II. *J. Org. Chem.* **1992**, 57, 5370-5376.
28. Hsu, L. F., Chang, C. P., Li, M. C., Chang, N. C. Bicyclo[3.2.1]octenones as building blocks in natural products synthesis. 1. Formal synthesis of (±)-mussaenoside and (±)-8-epiloganin aglycons. *J. Org. Chem.* **1993**, 58, 4756-4757.
29. Thompson, S. K., Heathcock, C. H. Total synthesis of some marasmane and lactarane sesquiterpenes. *J. Org. Chem.* **1992**, 57, 5979-5989.

Corey-Fuchs Alkyne Synthesis 104

Related reactions: Seyferth-Gilbert homologation;

1. Corey, E. J., Fuchs, P. L. Synthetic method for conversion of formyl groups into ethynyl groups (RCHO --> RCCH or RCCR1). *Tetrahedron Lett.* **1972**, 3769-3772.
2. Eymery, F., Iorga, B., Savignac, P. The usefulness of phosphorus compounds in alkyne synthesis. *Synthesis* **2000**, 185-213.
3. Bestmann, H. J., Frey, H. Reactions with phosphinealkylenes. XXXIX. New methods for the preparation of 1-bromoacetylenes and aromatic and conjugated enynes. *Liebigs Ann. Chem.* **1980**, 2061-2071.
4. Bestmann, H. J., Li, K. Reactions with phosphine alkylenes. XL. Sequence for the preparation of acetylenes from aldehydes. *Chem. Ber.* **1982**, 115, 828-831.
5. Michel, P., Gennet, D., Rassat, A. A one-pot procedure for the synthesis of alkynes and bromoalkynes from aldehydes. *Tetrahedron Lett.* **1999**, 40, 8575-8578.
6. Ramirez, F., Desai, N. B., McKelvie, N. New synthesis of 1,1-dibromoolefins via phosphinedi-bromomethylenes. The reaction of triphenylphosphine with carbon tetrabromide. *J. Am. Chem. Soc.* **1962**, 84, 1745-1747.
7. Leeuwenburgh, M. A., Litjens, R. E. J. N., Codee, J. D. C., Overkleeft, H. S., Van der Marel, G. A., VanBoom, J. H. Radical Cyclization of Sugar-Derived β-(Alkynyloxy)acrylates: Synthesis of Novel Fused Ethers. *Org. Lett.* **2000**, 2, 1275-1277.
8. Wong, L. S. M., Sharp, L. A., Xavier, N. M. C., Turner, P., Sherburn, M. S. The Bromopentadienyl Acrylate Approach to Himbacine. *Org. Lett.* **2002**, 4, 1955-1957.
9. Donkervoort, J. G., Gordon, A. R., Johnstone, C., Kerr, W. J., Lange, U. Development of modified Pauson-Khand reactions with ethylene and utilization in the total synthesis of (+)-taylorione. *Tetrahedron* **1996**, 52, 7391-7420.
10. Oppolzer, W., Robyr, C. Synthesis of (±)-hirsutene by a catalytic allylpalladium-alkyne cyclization/carbonylation cascade. *Tetrahedron* **1994**, 50, 415-424.

Corey-Kim Oxidation 106

Related reactions: Dess-Martin oxidation, Jones oxidation, Ley oxidation, Oppenauer oxidation, Pfitzner-Moffatt oxidation, Swern oxidation;

1. Vilsmaier, E., Spruegel, W. Halo thioethers. I. Reaction of thioethers with N-chlorosuccinimide. *Liebigs Ann. Chem.* **1971**, 747, 151-157.
2. Corey, E. J., Kim, C. U. New and highly effective method for the oxidation of primary and secondary alcohols to carbonyl compounds. *J. Am. Chem. Soc.* **1972**, 94, 7586-7587.
3. Johnson, C. R., Bacon, C. C., Kingsbury, W. D. Chemistry of sulfoxides and related compounds. XXXVIII. Oxidation of sulfides with 1-chlorobenzotriazole. Preparation of amino- and alkoxysulfonium salts. *Tetrahedron Lett.* **1972**, 501-503.
4. Crich, D., Neelamkavil, S. The fluorous Swern and Corey-Kim reactions: scope and mechanism. *Tetrahedron* **2002**, 58, 3865-3870.
5. Nishide, K., Ohsugi, S.-i., Fudesaka, M., Kodama, S., Node, M. New odorless protocols for the Swern and Corey-Kim oxidations. *Tetrahedron Lett.* **2002**, 43, 5177-5179.
6. Ho, T.-L., Wong, C. M. Nitrile synthesis. Use of Corey-Kim system as dehydrant of aldoximes. *Synth. Commun.* **1975**, 5, 423-425.
7. Katayama, S., Fukuda, K., Watanabe, T., Yamauchi, M. Synthesis of 1,3-dicarbonyl compounds by the oxidation of 3-hydroxycarbonyl compounds with Corey-Kim reagent. *Synthesis* **1988**, 178-183.
8. Katayama, S., Watanabe, T., Yamauchi, M. Convenient synthesis of stable sulfur ylides by reaction of active methylene compounds with Corey-Kim reagent. *Chem. Pharm. Bull.* **1990**, 38, 3314-3316.
9. Katayama, S., Watanabe, T., Yamauchi, M. Convenient synthesis of 3H-indoles (indolenines) by reaction of 1H-indoles with Corey-Kim reagent. *Chem. Pharm. Bull.* **1992**, 40, 2836-2838.
10. Tanino, K., Onuki, K., Asano, K., Miyashita, M., Nakamura, T., Takahashi, Y., Kuwajima, I. Total Synthesis of Ingenol. *J. Am. Chem. Soc.* **2003**, 125, 1498-1500.
11. Gyorkos, A. C., Stille, J. K., Hegedus, L. S. The total synthesis of (±)-epi-jatrophone and (±)-jatrophone using palladium-catalyzed carbonylative coupling of vinyl triflates with vinyl stannanes as the macrocycle-forming step. *J. Am. Chem. Soc.* **1990**, 112, 8465-8472.
12. Kuehne, M. E., Bornmann, W. G., Parsons, W. H., Spitzer, T. D., Blount, J. F., Zubieta, J. Total syntheses of (±)-cephalotaxine and (±)-8-oxocephalotaxine. *J. Org. Chem.* **1988**, 53, 3439-3450.
13. Denmark, S. E., Fu, J. Asymmetric Construction of Quaternary Centers by Enantioselective Allylation: Application to the Synthesis of the Serotonin Antagonist LY426965. *Org. Lett.* **2002**, 4, 1951-1953.

Related reactions: ***Keck macrolactonization, Yamaguchi macrolactonization;***

1. Corey, E. J., Nicolaou, K. C. Efficient and mild lactonization method for the synthesis of macrolides. *J. Am. Chem. Soc.* **1974**, 96, 5614-5616.
2. Corey, E. J., Nicolaou, K. C., Melvin, L. S., Jr. Synthesis of novel macrocyclic lactones in the prostaglandin and polyether antibiotic series. *J. Am. Chem. Soc.* **1975**, 97, 653-654.
3. Back, T. G. The synthesis of macrocyclic lactones. Approaches to complex macrolide antibiotics. *Tetrahedron* **1977**, 33, 3041-3059.
4. Masamune, S., Bates, G. S., Corcoran, J. W. Macrolides. Recent progress in chemistry and biochemistry. *Angew. Chem., Int. Ed. Engl.* **1977**, 16, 585-607.
5. Nicolaou, K. C. Synthesis of macrolides. *Tetrahedron* **1977**, 33, 683-710.
6. Haslam, E. Recent developments in methods for the esterification and protection of the carboxyl group. *Tetrahedron* **1980**, 36, 2409-2433.
7. Rossa, L., Voegtle, F. Synthesis of medio- and macrocyclic compounds by high dilution principle techniques. *Top. Curr. Chem.* **1983**, 113, 1-86.
8. Paterson, I., Mansuri, M. M. Recent developments in the total synthesis of macrolide antibiotics. *Tetrahedron* **1985**, 41, 3569-3624.
9. Mulzer, J. Synthesis of Esters, Activated Esters and Lactones. in *Comp. Org. Synth.* (eds. Trost, B. M.,Fleming, I.), *6*, 323-380 (Pergamon, Oxford, **1991**).
10. Meng, Q., Hesse, M. Ring-closure methods in the synthesis of macrocyclic natural products. *Top. Curr. Chem.* **1992**, 161, 107-176.
11. Rousseau, G. Medium ring lactones. *Tetrahedron* **1995**, 51, 2777-2849.
12. Roxburgh, C. J. The syntheses of large-ring compounds. *Tetrahedron* **1995**, 51, 9767-9822.
13. Nakata, T. Total synthesis of macrolides. *Macrolide Antibiotics (2nd Edition)* **2002**, 181-284.
14. Gerlach, H., Thalmann, A. Formation of esters and lactones by silver ion catalysis. *Helv. Chim. Acta* **1974**, 57, 2661-2663.
15. Corey, E. J., Brunelle, D. J. New Reagents for Conversion of Hydroxy-Acids to Macrolactones by Double Activation Method. *Tetrahedron Lett.* **1976**, 17, 3409-3412.
16. Corey, E. J., Clark, D. A. New Method for the Synthesis of 2-Pyridinethiol Carboxylic Esters. *Tetrahedron Lett.* **1979**, 2875-2878.
17. Deretey, E. Computational support for the 'double activation' mechanism of macrolide ring closure. *J. Mol. Struct.-Theochem* **1999**, 459, 273-286.
18. Mukaiyama, T., Matsueda, R., Suzuki, M. Peptide synthesis via the oxidation-reduction condensation by the use of 2,2'-dipyridyldisulfide as an oxidant. *Tetrahedron Lett.* **1970**, 1901-1904.
19. Corey, E. J., Brunelle, D. J., Stork, P. J. Mechanistic Studies on Double Activation Method for Synthesis of Macrocyclic Lactones. *Tetrahedron Lett.* **1976**, 17, 3405-3408.
20. Behinpour, K., Hopkins, A., Williams, A. Macrolide Ring-Closure - Double Activation Mechanism. *Tetrahedron Lett.* **1981**, 22, 275-278.
21. Nicolaou, K. C., Bunnage, M. E., McGarry, D. G., Shi, S., Somers, P. K., Wallace, P. A., Chu, X.-J., Agrios, K. A., Gunzner, J. L., Yang, Z. Total synthesis of brevetoxin A: Part 1: first generation strategy and construction of BCD ring system. *Chem.-- Eur. J.* **1999**, 5, 599-617.
22. Sasaki, T., Inoue, M., Hirama, M. Synthetic studies toward C-1027 chromophore: construction of a highly unsaturated macrocycle. *Tetrahedron Lett.* **2001**, 42, 5299-5303.
23. Hansen, T. V., Stenstrom, Y. First total synthesis of (-)-aplyolide A. *Tetrahedron: Asymmetry* **2001**, 12, 1407-1409.
24. Andrus, M. B., Shih, T.-L. Synthesis of Tuckolide, a New Cholesterol Biosynthesis Inhibitor. *J. Org. Chem.* **1996**, 61, 8780-8785.

1. Corey, E. J., Winter, A. E. New stereospecific olefin synthesis from 1,2-diols. *J. Am. Chem. Soc.* **1963**, 85, 2677-2678.
2. Corey, E. J., Carey, F. A., Winter, R. A. E. Stereospecific syntheses of olefins from 1,2-thionocarbonates and 1,2-trithiocarbonates. trans-Cycloheptene. *J. Am. Chem. Soc.* **1965**, 87, 934-935.
3. Block, E. Olefin synthesis by deoxygenation of vicinal diols. *Org. React.* **1984**, 30, 457-566.
4. Corey, E. J., Markl, G. Generation of phosphite ylides from trithiocarbonates and trimethyl phosphite and their application to the extension of carbon chains. *Tetrahedron Lett.* **1967**, 3201-3204.
5. Corey, E. J., Winter, R. A. E. Structure of the product $C_{21}H_{12}O_6$ from *o*-phenylene thiono-carbonate and trimethyl phosphite. *Chem. Commun.* **1965**, 208-209.
6. Horton, D., Tindall, C. G., Jr. Synthesis and reactions of unsaturated sugars. XI. Evidence for a carbenoid intermediate in the Corey-Winter alkene synthesis. *J. Org. Chem.* **1970**, 35, 3558-3559.
7. Borden, W. T., Concannon, P. W., Phillips, D. I. Synthesis and pyrolysis of carbonate tosylhydrazone salts derived from vicinal glycols. *Tetrahedron Lett.* **1973**, 3161-3164.
8. Shing, T. K. M., Tam, E. K. W. Enantiospecific Syntheses of (+)-Crotepoxide, (+)-Boesenoxide, (+)-β-Senepoxide, (+)-Pipoxide Acetate, (-)-iso-Crotepoxide, (-)-Senepoxide, and (-)-Tingtanoxide from (-)-Quinic Acid. *J. Org. Chem.* **1998**, 63, 1547-1554.
9. Noguchi, H., Aoyama, T., Shioiri, T. Determination of the absolute configuration and total synthesis of radiosumin, a trypsin inhibitor from a freshwater blue-green alga. *Tetrahedron Lett.* **1997**, 38, 2883-2886.
10. Rigby, J. H., Bazin, B., Meyer, J. H., Mohammadi, F. Synthetic Studies on the Ingenane Diterpenes. An Improved Entry into a trans-Intrabridgehead System. *Org. Lett.* **2002**, 4, 799-801.
11. Davis, B., Bell, A. A., Nash, R. J., Watson, A. A., Griffiths, R. C., Jones, M. G., Smith, C., Fleet, G. W. J. L-(+)-Swainsonine and other pyrrolidine inhibitors of naringinase: through an enzymic looking glass from D-mannosidase to L-rhamnosidase? *Tetrahedron Lett.* **1996**, 37, 8565-8568.

1. Cornforth, J. W., Clarke, H. T., et al. *Oxazoles and oxazolones* (Princeton University Press, Princeton, **1949**) 688-848.
2. Turchi, I. J., Dewar, M. J. S. Chemistry of oxazoles. *Chem. Rev.* **1975**, 75, 389-437.
3. Taylor, E. C., Turchi, I. J. 1,5-Dipolar cyclizations. *Chem. Rev.* **1979**, 79, 181-231.
4. Turchi, I. J. Oxazole chemistry. A review of recent advances. *Ind. Eng. Chem. Prod. Res. Dev.* **1981**, 20, 32-76.
5. L'Abbe, G. Molecular rearrangements of five-membered ring heteromonocycles. *J. Heterocycl. Chem.* **1984**, 21, 627-638.
6. Turchi, I. J. *Oxazoles* (Wiley, New York, **1986**) 1-341.
7. Hartner, F. W., Jr. Oxazoles. in *Comprehensive Heterocyclic Chemistry II* (eds. Katritzky, A. R., Rees, C. W.,Scriven, E. F. V.), *3*, 261-318 (Pergamon, New York, **1996**).
8. Williams, D. R., McClymont, E. L. Carbanion methodology for alkylations and acylations in the synthesis of substituted oxazoles. The formation of Cornforth rearrangement products. *Tetrahedron Lett.* **1993**, 34, 7705-7708.
9. Dewar, M. J. S., Turchi, I. J. Ground states of molecules. Part 35. MINDO/3 study of the Cornforth rearrangement. *J. Chem. Soc., Perkin Trans. 2* **1977**, 724-729.
10. Fabian, W. M. F., Kollenz, G. Iminofurandione-pyrroledione rearrangement: a semi-empirical molecular orbital study. *ECHET98: Electronic Conference on Heterocyclic Chemistry, June 29-July 24, 1998* **1998**, 106-116.
11. Fabian, W. M. F., Kappe, C. O., Bakulev, V. A. Ab Initio and Density Functional Calculations on the Pericyclic vs Pseudopericyclic Mode of Conjugated Nitrile Ylide 1,5-Electrocyclizations. *J. Org. Chem.* **2000**, 65, 47-53.
12. Dewar, M. J. S., Turchi, I. J. Scope and limitations of the Cornforth rearrangement. *J. Org. Chem.* **1975**, 40, 1521-1523.
13. Dewar, M. J. S., Turchi, I. J. Cornforth rearrangement. *J. Am. Chem. Soc.* **1974**, 96, 6148-6152.

14. Korte, F., Storiko, K. Acyl-lactone rearrangement. XV. The rearrangement of 4-acyl-5-oxazolones. *Chem. Ber.* **1960**, 93, 1033-1042.
15. Dewar, M. J. S., Spanninger, P. A., Turchi, I. J. Nature of the intermediate in the Cornforth rearrangement. *J. Chem. Soc., Chem. Commun.* **1973**, 925-926.
16. Hoefle, G., Steglich, W. Conversion of 4-acyl-2- and 2-acyl-3-oxazolin-5-ones into trisubstituted oxazoles. A simple access to 3,3,3-trifluoroalanine and 2-amino-3,3,3-trifluoropropionyl derivatives. *Chem. Ber.* **1971**, 104, 1408-1419.
17. Corrao, S. L., Macielag, M. J., Turchi, I. J. Rearrangement of 4-(aminothiocarbonyl)oxazoles to 5-aminothiazoles. Synthetic and MINDO/3 MO studies. *J. Org. Chem.* **1990**, 55, 4484-4487.

Criegee Oxidation 114

1. Criegee, R. Oxidation with quadrivalent lead salts. II. Oxidative cleavage of glycols. *Ber.* **1931**, 64B, 260-266.
2. Criegee, R., et al. *Newer Methods of Preparative Organic Chemistry* (Interscience Publishers, New York, **1948**) 657 pp.
3. Perlin, A. S. Action of lead tetraacetate on the sugars. *Adv. Carbohydr. Chem.* **1959**, 14, 9-61.
4. Bunton, C. A. Glycol cleavage and relayed reactions. *Oxidation Org. Chem.* **1965**, 367-407.
5. Bentley, K. W., Kirby, G. W. *Elucidation of Organic Structures by Physical and Chemical Methods, Pt. 2. 2nd Ed* (Wiley, Chichester, Engl., **1973**) 561 pp.
6. Arseniyadis, S., Yashunsky, D. V., de Freitas, R. P., Dorado, M. M., Potier, P., Toupet, L. Left and right-half taxoid building blocks from (S)-(+)-Hajos-Parrish ketone. *Tetrahedron* **1996**, 52, 12443-12458.
7. Arseniyadis, S., Brondi-Alves, R., Del Moral, J., Yashunsky, D. V., Potier, P. Lead tetraacetate mediated "one-pot" multistage transformations on selected unsaturated 1,2-diols: the Wieland-Miescher series. *Tetrahedron* **1998**, 54, 5949-5958.
8. Unaleroglu, C., Aviyente, V., Arseniyadis, S. Lead Tetraacetate Mediated One-Pot Multistage Transformations: Theoretical Studies on the Diverging Behavior in the Hajos-Parrish and Wieland-Miescher Series. *J. Org. Chem.* **2002**, 67, 2447-2452.
9. Reeves, R. E. Direct titration of cis-glycols with lead tetraacetate. *Anal. Chem.* **1949**, 21, 751.
10. Gillet, A. The Criegee reaction and the Grignard reaction. *Bull. Soc. Chim. Belg.* **1937**, 46, 171-172.
11. Criegee, R., Buchner, E. The velocity of glycol cleavage with lead tetraacetate in relation to the solvent. *Ber.* **1940**, 73B, 563-571.
12. Criegee, R., Buchner, E., Walther, W. The velocity of glycol cleavage with lead tetraacetate in relation to the constitution of the glycol. *Ber.* **1940**, 73B, 571-575.
13. Criegee, R., Hoger, E., Huber, G., Kruck, P., Marktscheffel, F., Schellenberger, H. The velocity of cleavage with lead tetraacetate in relation to the constitution and configuration of the glycol. III. *Ann.* **1956**, 599, 81-125.
14. Bell, R. P., Rivlin, V. G., Waters, W. A. Acid catalysis of glycol fission by lead tetraacetate. *J. Chem. Soc.* **1958**, 1696-1697.
15. Moriconi, E. J., Wallenberger, F. T., O'Connor, W. F. Lead tetraacetate oxidation of cis- and trans-9,10-diaryl-9,10-dihydro-9,10-phenanthrenediols. A kinetic study. *J. Am. Chem. Soc.* **1958**, 80, 656-661.
16. Moriconi, E. J., O'Connor, W. F., Keneally, E. A., Wallenberger, F. T. Oxidation kinetics of vic-diols in cyclic systems. II. Lead tetraacetate oxidation of cis- and trans-1,2-diaryl-1,2-acenaphthenediols. *J. Am. Chem. Soc.* **1960**, 82, 3122-3126.
17. Perlin, A. S., Suzuki, S. Spectrophotometric observations on the cleavage of vic-diols by lead tetraacetate. *Can. J. Chem.* **1962**, 40, 1226-1229.
18. Trahanovsky, W. S., Gilmore, J. R., Heaton, P. C. Oxidation of Organic Compounds with Cerium(IV). XV. Electronic and Steric Effects on the Oxidative Cleavage of 1,2-Glycols by Cerium(IV) and Lead(IV). *J. Org. Chem.* **1973**, 38, 760-763.
19. Ferreira, M. d. R. R., Hernando, J. I. M., Lena, J. I. C., Toupet, L., Birlirakis, N., Arseniyadis, S. $Pb(OAc)_4$ mediated oxidative cleavage of steroidal unsaturated 1,2-diols: influence of the angular substitution. *Tetrahedron Lett.* **1999**, 40, 7679-7682.
20. Hernando, J. I. M., Ferreira, M. d. R. R., Lena, J. I. C., Toupet, L., Birlirakis, N., Arseniyadis, S. Influence of the substitution pattern on the $Pb(OAc)_4$ mediated oxidative cleavage of steroidal 1,2-diols. *Tetrahedron: Asymmetry* **1999**, 10, 3977-3989.
21. Biju, P. J., Rao, G. S. R. S. Aromatics to polyquinanes: a general method for the construction of tricyclo[6.3.0.04,8]-and tricyclo[6.3.0.02,6]undecanes. *Tetrahedron Lett.* **1999**, 40, 9379-9382.
22. Baba, Y., Saha, G., Nakao, S., Iwata, C., Tanaka, T., Ibuka, T., Ohishi, H., Takemoto, Y. Asymmetric Total Synthesis of Halicholactone. *J. Org. Chem.* **2001**, 66, 81-88.
23. Moricz, A., Gassman, E., Bienz, S., Hesse, M. Synthesis of (±)-pyrenolide B. *Helv. Chim. Acta* **1995**, 78, 663-669.
24. Shizuri, Y., Ohkubo, M., Yamamura, S. Synthesis of (±)-silphinene using electrochemical method as a key step. *Tetrahedron Lett.* **1989**, 30, 3797-3798.

Curtius Rearrangement 116

*Related reactions: **Hofmann rearrangement, Lossen rearrangement, Schmidt reaction;***

1. Buchner, E., Curtius, T. Synthesis of β-keto esters from aldehydes and diazoacetic acid. *Chem. Ber.* **1885**, 18, 2371-2377.
2. Curtius, T. Hydrazoic acid. *Ber.* **1890**, 23, 3023-3041.
3. Curtius, T. Hydrazides and azides of organic acids. *J. Prakt. Chem.* **1894**, 50, 275.
4. Curtius, T. New Observations on Acid Azides. *Chem. Ztg.* **1912**, 35, 249.
5. Curtius, T. Rearrangement of sulfonazides. *J. Prakt. Chem.* **1930**, 125, 303-424.
6. Smith, P. A. S. Curtius reaction. *Org. React.* **1946**, 337-349.
7. Saunders, J. H., Slocombe, R. J. The chemistry of the organic isocyanates. *Chem. Rev.* **1948**, 43, 203-218.
8. Smith, P. A. S. Carbon-to-nitrogen migrations; what the last decade has brought. *Trans. N. Y. Acad. Sci.* **1969**, 31, 504-515.
9. Banthorpe, D. V. Rearrangements involving azido groups. in *The Chemistry of the Azido Group* (ed. Patai, S.), 397-340 (Wiley, New York, **1971**).
10. Majoral, J. P., Bertrand, G., Ocando-Mavarez, E., Baceiredo, A. Phosphorus azides, powerful reagents in heterocyclic chemistry. *Bull. Soc. Chim. Belg.* **1986**, 95, 945-957.
11. Weinstock, J. Modified Curtius reaction. *J. Org. Chem.* **1961**, 26, 3511.
12. Shioiri, T., Ninomiya, K., Yamada, S. Diphenylphosphoryl azide. New convenient reagent for a modified Curtius reaction and for peptide synthesis. *J. Am. Chem. Soc.* **1972**, 94, 6203-6205.
13. Warren, J. D., Press, J. B. Trimethylsilylazide/KN_3/18-crown-6. Formation and Curtius rearrangement of acyl azides from unreactive acid chlorides. *Synth. Commun.* **1980**, 10, 107-110.
14. Capson, T. L., Poulter, C. D. A facile synthesis of primary amines from carboxylic acids by the Curtius rearrangement. *Tetrahedron Lett.* **1984**, 25, 3515-3518.
15. Rauk, A., Alewood, P. F. A theoretical study of the Curtius rearrangement. The electronic structures and interconversions of the CHNO species. *Can. J. Chem.* **1977**, 55, 1498-1510.
16. Nguyen Minh, T. Mechanism of the Curtius-type rearrangement in the boron series. An ab initio study of the borylnitrene (H_2B-N)-iminoborane (HB=NH) isomerization. *J. Chem. Soc., Chem. Commun.* **1987**, 342-344.
17. Nguyen Minh, T., Fitzpatrick, N. J. Intermediacy of nitrene in the Curtius-type rearrangement of phosphinic azides. Insights from ab initio study of the H_2P(:O)N .dblharw. HP(:O):NH interconversion. *Polyhedron* **1988**, 7, 223-227.
18. Yukawa, Y., Tsuno, Y. The Curtius rearrangement. III. The decomposition of substituted benzazides in acidic solvents, the acid catalysis. *J. Am. Chem. Soc.* **1959**, 81, 2007-2012.
19. Fahr, E., Neumann, L. Curtius reaction with boron trihalides. *Angew. Chem.* **1965**, 77, 591.
20. Prakash, G. K. S., Iyer, P. S., Arvanaghi, M., Olah, G. A. Synthetic methods and reactions. 121. Zinc iodide catalyzed preparation of aroyl azides from aroyl chlorides and trimethylsilyl azide. *J. Org. Chem.* **1983**, 48, 3358-3359.

21. Pozsgay, V., Jennings, H. J. Azide synthesis with stable nitrosyl salts. *Tetrahedron Lett.* **1987**, 28, 5091-5092.
22. Eibler, E., Sauer, J. Contribution to isocyanate formation in the photolysis of acyl azides. *Tetrahedron Lett.* **1974**, 2569-2572.
23. Harger, M. J. P., Westlake, S. Photolysis of some unsymmetrical phosphinic azides in methanol. Relative migratory aptitudes of alkyl groups and phenyl in the Curtius-like rearrangement. *Tetrahedron* **1982**, 38, 3073-3078.
24. Denmark, S. E., Dorow, R. L. The stereochemical course of migration from phosphorus to nitrogen in the photo-Curtius rearrangement of phosphinic azides (Harger reaction). *J. Org. Chem.* **1989**, 54, 5-6.
25. Denmark, S. E., Dorow, R. L. Stereospecific cleavage of carbon-phosphorus bonds: stereochemical course of the phosphinoyl curtius (Harger) reaction. *Chirality* **2002**, 14, 241-257.
26. Newman, M. S., Gildenhorn, H. L. Mechanism of the Schmidt reactions and observations on the Curtius rearrangement. *J. Am. Chem. Soc.* **1948**, 70, 317-319.
27. Linke, S., Tisue, G. T., Lwowski, W. Curtius and Lossen rearrangements. II. Pivaloyl azide. *J. Am. Chem. Soc.* **1967**, 89, 6308-6310.
28. L'Abbe, G. Decomposition and addition reactions of organic azides. *Chem. Rev.* **1969**, 69, 345-363.
29. Benecke, H. P., Wikel, J. H. Curtius rearrangement in aminimides. *Tetrahedron Lett.* **1972**, 289-292.
30. Batori, S., Messmer, A., Timpe, H. J. Condensed as-triazines. Part XI. Photoinduced fragmentation of pyrido[2,1-f]-as-triazinium-4-olate and its benzolog. Mechanism of Curtius rearrangement. *Heterocycles* **1991**, 32, 649-654.
31. Carda, M., Gonzalez, F., Sanchez, R., Marco, J. A. Stereoselective synthesis of (-)-cytoxazone. *Tetrahedron: Asymmetry* **2002**, 13, 1005-1010.
32. Boger, D. L., Cassidy, K. C., Nakahara, S. Total synthesis of streptonigrone. *J. Am. Chem. Soc.* **1993**, 115, 10733-10741.
33. Nagumo, S., Nishida, A., Yamazaki, C., Matoba, A., Murashige, K., Kawahara, N. Total synthesis of antimuscarinic alkaloid, (±)-TAN1251A. *Tetrahedron* **2002**, 58, 4917-4924.
34. Kim, S., Ko, H., Kim, E., Kim, D. Stereocontrolled Total Synthesis of Pancratistatin. *Org. Lett.* **2002**, 4, 1343-1345.

Dakin Oxidation 118

Related reactions: Baeyer-Villiger oxidation;

1. Dakin, H. D. The oxidation of hydroxy derivatives of benzaldehyde, acetophenone and related substances. *Am. Chem. J.* **1909**, 42, 477-498.
2. Dakin, H. D. Oxidation of Hydroxy Derivatives of Benzaldehyde and Acetophenone. *Proc. Chem. Soc.* **1910**, 25, 194.
3. Dakin, H. D. Catechol (Pyrocatechol). in *Org. Synth.* (ed. Gilman, H.), *1*, 149-154 (John Wiley and Sons, New York, **1941**).
4. Leffler, J. E. Cleavages and rearrangements involving oxygen radicals and cations. *Chem. Rev.* **1949**, 45, 385-417.
5. Hassall, C. H. The Baeyer-Villiger oxidation of aldehydes and ketones. *Org. React.* **1957**, 9, 73-106.
6. Schubert, W. M., Kintner, R. R. Decarbonylation. *Chem. Carbonyl Group. 1966* **1966**, 695-760.
7. Lee, J. B., Uff, B. C. Organic reactions involving electrophilic oxygen. *Quart. Rev., Chem. Soc.* **1967**, 21, 429-457.
8. Syper, L. The Baeyer-Villiger oxidation of aromatic aldehydes and ketones with hydrogen peroxide catalyzed by selenium compounds. A convenient method for the preparation of phenols. *Synthesis* **1989**, 167-172.
9. Varma, R. S., Naicker, K. P. The Urea-Hydrogen Peroxide Complex: Solid-State Oxidative Protocols for Hydroxylated Aldehydes and Ketones (Dakin Reaction), Nitriles, Sulfides, and Nitrogen Heterocycles. *Org. Lett.* **1999**, 1, 189-191.
10. Hocking, M. B. Dakin oxidation of o-hydroxyacetophenone and some benzophenones. Rate enhancement and mechanistic aspects. *Can. J. Chem.* **1973**, 51, 2384-2392.
11. Matsumoto, M., Kobayashi, K., Hotta, Y. Acid-catalyzed oxidation of benzaldehydes to phenols by hydrogen peroxide. *J. Org. Chem.* **1984**, 49, 4740-4741.
12. Ogata, Y., Sawaki, Y. Kinetics of the Baeyer-Villiger reaction of benzaldehydes with perbenzoic acid in aquo-organic solvents. *J. Org. Chem.* **1969**, 34, 3985-3991.
13. Boeseken, J., Coden, W. D., Kip, C. J. The synthesis of sesamol and of its β-glucoside. The Baudouin reaction. *Rec. trav. chim.* **1936**, 55, 815-820.
14. Kabalka, G. W., Reddy, N. K., Narayana, C. Sodium percarbonate: a convenient reagent for the Dakin reaction. *Tetrahedron Lett.* **1992**, 33, 865-866.
15. Hocking, M. B., Ong, J. H. Kinetic studies of Dakin oxidation of o- and p-hydroxyacetophenones. *Can. J. Chem.* **1977**, 55, 102-110.
16. Hocking, M. B., Ko, M., Smyth, T. A. Detection of intermediates and isolation of hydroquinone monoacetate in the Dakin oxidation of p-hydroxyacetophenone. *Can. J. Chem.* **1978**, 56, 2646-2649.
17. Hocking, M. B., Bhandari, K., Shell, B., Smyth, T. A. Steric and pH effects on the rate of Dakin oxidation of acylphenols. *J. Org. Chem.* **1982**, 47, 4208-4215.
18. Bolitt, V., Mioskowski, C., Kollah, R. O., Manna, S., Rajapaksa, D., Falck, J. R. Total synthesis of vineomycinone B2 methyl ester via double Bradsher cyclization. *J. Am. Chem. Soc.* **1991**, 113, 6320-6321.
19. Jung, M. E., Lazarova, T. I. Efficient Synthesis of Selectively Protected L-Dopa Derivatives from L-Tyrosine via Reimer-Tiemann and Dakin Reactions. *J. Org. Chem.* **1997**, 62, 1553-1555.
20. Lyttle, M. H., Carter, T. G., Cook, R. M. Improved synthetic procedures for 4,7,2',7'-tetrachloro- and 4',5'-dichloro-2',7'-dimethoxy-5(and 6)-carboxyfluoresceins. *Org. Process Res. Dev.* **2001**, 5, 45-49.

Dakin-West Reaction 120

Related reactions: Neber rearrangement;

1. Perkin, W. H., Sr. The formation of acids from aldehydes by the action of anhydrides and salts and the formation of ketones from the compounds resulting from the union of anhydrides and salts. *J. Chem. Soc.* **1886**, 41, 317-328.
2. Dakin, H. D., West, R. Some aromatic derivatives of substituted acetylaminoacetones. *J. Biol. Chem.* **1928**, 78, 757-764.
3. Dakin, H. D., West, R. A general reaction of amino acids. II. *J. Biol. Chem.* **1928**, 78, 745-756.
4. Buchanan, G. L. The Dakin-West reaction. *Chem. Soc. Rev.* **1988**, 17, 91-109.
5. Kawase, M., Hirabayashi, M., Saito, S. Anomalous Dakin-West reactions of secondary α-amino acids with trifluoroacetic anhydride. *Rec. Res. Dev. Org. Chem.* **2000**, 4, 283-293.
6. Nicholson, J. W., Wilson, A. D. The conversion of carboxylic acids to ketones: A repeated discovery. *J. Chem. Educ.* **2004**, 81, 1362-1366.
7. Bullerwell, R. A. F., Lawson, A., Morley, H. V. 2-Mercaptoglyoxalines. VIII. Preparation of 2-mercaptoglyoxalines from glutaric acid. *J. Chem. Soc., Abstracts* **1954**, 3283-3287.
8. Steglich, W., Hoefle, G. N,N-Dimethyl-4-pyridinamine, a very effective acylation catalyst. *Angew. Chem., Int. Ed. Engl.* **1969**, 8, 981.
9. Hoefle, G., Steglich, W., Vorbrueggen, H. New synthetic methods. 25. 4-Dialkylaminopyridines as acylation catalysts. 4. Pyridine syntheses. 1. 4-Dialkylaminopyridines as highly active acylation catalysts. *Angew. Chem.* **1978**, 90, 602-615.
10. Orain, D., Canova, R., Dattilo, M., Kloppner, E., Denay, R., Koch, G., Giger, R. Efficient solution and solid-phase synthesis of a 3,9-diazabicyclo[3.3.1]non-6-en-2-one scaffold. *Synlett* **2002**, 1443-1446.
11. Cornforth, J. W., Elliott, D. F. Mechanism of the Dakin and West reaction. *Science* **1950**, 112, 534-535.
12. Otvos, L., Marton, J., Meisel-Agoston, J. Investigations with radioactive acetic anhydride. I. On the mechanism of the Dakin-West reaction. II. The mechanism of the reaction between aromatic isocyanates and acid anhydrides. *Acta Chim. Acad. Sci. Hung.* **1960**, 24, 327-331.
13. Otvos, L., Marton, J., Meisel-Agoston, J. Investigations with radioactive acetic anhydride. I. On the mechanism of the Dakin-West reaction. *Acta Chim. Acad. Sci. Hung.* **1960**, 24, 321-325.

14. Singh, G., Singh, S. Synthesis of a new mesoionic aromatic system and the mechanism of the Dakin-West reaction. *Tetrahedron Lett.* **1964**, 3789-3793.
15. Iwakura, Y., Toda, F., Suzuki, H. Synthesis of N-[1-(1-substituted 2-oxopropyl)]acrylamides and -methylacrylamides. Isolation and some reactions of intermediates of the Dakin-West reaction. *J. Org. Chem.* **1967**, 32, 440-443.
16. Steglich, W., Hoefle, G. Mechanism of the Dakin-West reaction. *Tetrahedron Lett.* **1968**, 1619-1624.
17. Gerencevic, N., Prostenik, M. n-Acyl amino acids in the Dakin-West reaction: replacements of the acyl groups. *Bulletin Scientifique, Section A: Sciences Naturelles, Techniques et Medicales (Zagreb)* **1970**, 15, 158.
18. Knorr, R., Huisgen, R. Mechanism of the Dakin-West reaction. I. Reaction of secondary N-acylamino acids with acetic anhydride. *Chem. Ber.* **1970**, 103, 2598-2610.
19. Knorr, R. Mechanism of the Dakin-West reaction. III. Course of ring opening during the Dakin-West reaction of an oxazolium 5-olate. *Chem. Ber.* **1971**, 104, 3633-3643.
20. Knorr, R., Staudinger, G. K. Mechanism of the Dakin-West reaction. II. Kinetics and mechanisms of the Dakin-West reaction of N-acylated secondary amino acids. *Chem. Ber.* **1971**, 104, 3621-3632.
21. Steglich, W., Hoefle, G. Mechanism of the Dakin-West reaction. II. Acylation of D2-oxazolin-5-ones by carboxylic anhydrides/pyridine. *Chem. Ber.* **1971**, 104, 3644-3652.
22. Hoefle, G., Prox, A., Steglich, W. Mechanism of the Dakin-West reaction. III. Ring opening of 4-acyl-2-oxazolin-5-ones by carboxylic acids. *Chem. Ber.* **1972**, 105, 1718-1725.
23. Allinger, N. L., Wang, G. L., Dewhurst, B. B. Kinetic and mechanistic studies of the Dakin-West reaction. *J. Org. Chem.* **1974**, 39, 1730-1735.
24. Kawase, M. Unusual ring expansion observed during the Dakin-West reaction of tetrahydroisoquinoline-1-carboxylic acids using trifluoroacetic anhydride: an expedient synthesis of 3-benzazepine derivatives bearing a trifluoromethyl group. *J. Chem. Soc., Chem. Commun.* **1992**, 1076-1077.
25. Kawasi, M., Miyamae, H., Narita, M., Kurihara, T. Unexpected product from the Dakin-West reaction of N-acylprolines using trifluoroacetic anhydride: a novel access to 5-trifluoromethyloxazoles. *Tetrahedron Lett.* **1993**, 34, 859-862.
26. Devulapalli, G. K., Rajanna, K. C., Sai Prakash, P. K. Pyridine catalyzed kinetic and mechanistic studies of Dakin-West reaction. *Book of Abstracts, 214th ACS National Meeting, Las Vegas, NV, September 7-11* **1997**, PHYS-054.
27. Loksha, Y. M., El-Badawi, M. A., El-Barbary, A. A., Pedersen, E. B., Nielsen, C. Synthesis of 2-methylsulfanyl-1H-imidazoles as novel non-nucleoside reverse transcriptase inhibitors (NNRTIs). *Arch. Pharm. (Weinheim, Ger.)* **2003**, 336, 175-180.
28. Cheng, L., Goodwin, C. A., Schully, M. F., Kakkar, V. V., Claeson, G. Synthesis and biological activity of ketomethylene pseudopeptide analogs as thrombin inhibitors. *J. Med. Chem.* **1992**, 35, 3364-3369.
29. Godfrey, A. G., Brooks, D. A., Hay, L. A., Peters, M., McCarthy, J. R., Mitchell, D. Application of the Dakin-West Reaction for the Synthesis of Oxazole-Containing Dual PPARα/γ Agonists. *J. Org. Chem.* **2003**, 68, 2623-2632.

Related reactions: Bergman cycloaromatization reaction, Dötz benzannulation reaction;

1. Danheiser, R. L., Gee, S. K. A regiocontrolled annulation approach to highly substituted aromatic compounds. *J. Org. Chem.* **1984**, 49, 1672-1674.
2. Danheiser, R. L., Brisbois, R. G., Kowalczyk, J. J., Miller, R. F. An annulation method for the synthesis of highly substituted polycyclic aromatic and heteroaromatic compounds. *J. Am. Chem. Soc.* **1990**, 112, 3093-3100.
3. Danheiser, R. L., Gee, S. K., Perez, J. J. Total synthesis of mycophenolic acid. *J. Am. Chem. Soc.* **1986**, 108, 806-810.
4. Danheiser, R. L., Nishida, A., Savariar, S., Trova, M. P. Trialkylsilyloxyalkynes: synthesis and aromatic annulation reactions. *Tetrahedron Lett.* **1988**, 29, 4917-4920.
5. Kowalski, C. J., Lal, G. S. Cycloaddition reactions of silyloxyacetylenes with ketenes: synthesis of cyclobutenones, resorcinols, and Δ-6-tetrahydrocannabinol. *J. Am. Chem. Soc.* **1988**, 110, 3693-3695.
6. Marvell, E. N. *Thermal Electrocyclic Reactions* (Academic Press, New York, **1980**) 422 pp.
7. Meier, H., Zeller, K. P. Wolff rearrangement of α-diazo carbonyl compounds. *Angew. Chem.* **1975**, 87, 52-63.
8. Kowalski, C. J., Lal, G. S., Haque, M. S. Ynol silyl ethers via O-silylation of ester-derived ynolate anions. *J. Am. Chem. Soc.* **1986**, 108, 7127-7128.
9. Smith, A. B., III, Adams, C. M., Kozmin, S. A., Paone, D. V. Total Synthesis of (-)-Cylindrocyclophanes A and F Exploiting the Reversible Nature of the Olefin Cross Metathesis Reaction. *J. Am. Chem. Soc.* **2001**, 123, 5925-5937.

1. Danheiser, R. L., Carini, D. J., Basak, A. (Trimethylsilyl)cyclopentene annulation: a regiocontrolled approach to the synthesis of five-membered rings. *J. Am. Chem. Soc.* **1981**, 103, 1604-1606.
2. Danheiser, R. L., Carini, D. J., Fink, D. M., Basak, A. Scope and stereochemical course of the (trimethylsilyl)cyclopentene annulation. *Tetrahedron* **1983**, 39, 935-947.
3. Danheiser, R. L., Fink, D. M. The reaction of allenylsilanes with α,β-unsaturated acylsilanes: new annulation approaches to five and six-membered carbocyclic compounds. *Tetrahedron Lett.* **1985**, 26, 2513-2516.
4. Danheiser, R. L., Kwasigroch, C. A., Tsai, Y. M. Application of allenylsilanes in [3+2] annulation approaches to oxygen and nitrogen heterocycles. *J. Am. Chem. Soc.* **1985**, 107, 7233-7235.
5. Danheiser, R. L., Becker, D. A. Application of allenylsilanes in a regiocontrolled [3 + 2] annulation route to substituted isoxazoles. *Heterocycles* **1987**, 25, 277-281.
6. Danheiser, R. L., Fink, D. M., Tsai, Y. M. A general [3 + 2] annulation: cis-4-exo-isopropenyl-1,9-dimethyl-8-(trimethylsilyl)bicyclo[4.3.0]non-8-en-2-one. *Org. Synth.* **1988**, 66, 8-13.
7. Becker, D. A., Danheiser, R. L. A new synthesis of substituted azulenes. *J. Am. Chem. Soc.* **1989**, 111, 389-391.
8. Danheiser, R. L., Stoner, E. J., Koyama, H., Yamashita, D. S., Klade, C. A. A new synthesis of substituted furans. *J. Am. Chem. Soc.* **1989**, 111, 4407-4413.
9. Chan, T. H., Fleming, I. Electrophilic substitution of organosilicon compounds - applications to organic synthesis. *Synthesis* **1979**, 761-786.
10. Stang, P. J., Rappoport, Z., Hanack, M., Subramanian, L. R. *Vinyl Cations* (Academic Press, New York, **1979**) 513 pp.
11. Brook, A. G., Bassindale, A. R. *Molecular rearrangements of organosilicon compounds* (Academic Press, New York, **1980**) 149-227.
12. Suginome, M., Matsunaga, S.-i., Ito, Y. Disilanyl group as a synthetic equivalent of the hydroxyl group. *Synlett* **1995**, 941-942.
13. Friese, J. C., Krause, S., Schafer, H. J. Formal total synthesis of the trinorguaiane sesquiterpenes (±)-clavukerin A and (±)-isoclavukerin. *Tetrahedron Lett.* **2002**, 43, 2683-2685.

Related reactions: Diels-Alder cycloaddition; Hetero Diels-Alder reaction;

1. Danishefsky, S., Kitahara, T. Useful diene for the Diels-Alder reaction. *J. Am. Chem. Soc.* **1974**, 96, 7807.
2. Danishefsky, S. Siloxy dienes in total synthesis. *Acc. Chem. Res.* **1981**, 14, 400-406.

3. Danishefsky, S. J., Deninno, M. P. Totally Synthetic Routes to the Higher Monosaccharides. *Angew. Chem., Int. Ed. Engl.* **1987**, 26, 15-23.
4. Danishefsky, S. Cycloaddition and cyclocondensation reactions of highly functionalized dienes: applications to organic synthesis. *Chemtracts: Org. Chem.* **1989**, 2, 273-297.
5. Herczzegh, P., Kovacs, I., Erdosi, G., Varga, T., Agocs, A., Szilagyi, L., Sztaricskai, F., Berecibar, A., Lukacs, G., Olesker, A. Stereoselective cycloaddition reactions of carbohydrate derivatives. *Pure Appl. Chem.* **1997**, 69, 519-524.
6. Kerwin, J. F., Jr., Danishefsky, S. On the Lewis acid catalyzed cyclocondensation of imines with a siloxydiene. *Tetrahedron Lett.* **1982**, 23, 3739-3742.
7. Kozmin, S. A., Rawal, V. H. Preparation and Diels-Alder reactivity of 1-amino-3-siloxy-1,3- butadienes. *J. Org. Chem.* **1997**, 62, 5252-5253.
8. Wang, Y., Wilson, S. R. Solid phase synthesis of 2,3-dihydro-4-pyridones: reaction of Danishefsky's diene with polymer-bound imines. *Tetrahedron Lett.* **1997**, 38, 4021-4024.
9. Kozmin, S. A., Green, M. T., Rawal, V. H. On the reactivity of 1-amino-3-siloxy-1,3-dienes: Kinetics investigation and theoretical interpretation. *J. Org. Chem.* **1999**, 64, 8045-8047.
10. Kozmin, S. A., Janey, J. M., Rawal, V. H. 1-amino-3-siloxy-1,3-butadienes: Highly reactive dienes for the Diels-Alder reaction. *J. Org. Chem.* **1999**, 64, 3039-3052.
11. Simonsen, K. B., Svenstrup, N., Roberson, M., Jorgensen, K. A. Development of an unusually highly enantioselective hetero-Diels - Alder reaction of benzaldehyde with activated dienes catalyzed by hyper-coordinating chiral aluminum complexes. *Chem.-- Eur. J.* **2000**, 6, 123-128.
12. Amil, H., Kobayashi, T., Terasawa, H., Uneyama, K. Difluorinated Danishefsky's diene: A versatile C-4 building block for the fluorinated six-membered rings. *Org. Lett.* **2001**, 3, 3103-3105.
13. Inokuchi, T., Okano, M., Miyamoto, T. Catalyzed Diels-Alder Reaction of Alkylidene- or Arylideneacetoacetates and Danishefsky's Dienes with Lanthanide Salts Aimed at Selective Synthesis of cis-4,5-Dimethyl-2-cyclohexenone Derivatives. *J. Org. Chem.* **2001**, 66, 8059-8063.
14. Motoyama, Y., Koga, Y., Nishiyama, H. Asymmetric hetero Diels-Alder reaction of Danishefsky's dienes and glyoxylates with chiral bis(oxazolinyl)phenylrhodium(III) aqua complexes, and its mechanistic studies. *Tetrahedron* **2001**, 57, 853-860.
15. Kuethe, J. T., Davies, I. W., Dormer, P. G., Reamer, R. A., Mathre, D. J., Reider, P. J. Asymmetric aza-Diels-Alder reactions of indole 2-carboxaldehydes. *Tetrahedron Lett.* **2002**, 43, 29-32.
16. Josephsohn, N. S., Snapper, M. L., Hoveyda, A. H. Efficient and Practical Ag-Catalyzed Cycloadditions between Arylimines and the Danishefsky Diene. *J. Am. Chem. Soc.* **2003**, 125, 4018-4019.
17. Paredes, E., Biolatto, B., Kneeteman, M., Mancini, P. M. Nitronaphthalenes as Diels-Alder dienophiles. *Tetrahedron Lett.* **2000**, 41, 8079-8082.
18. Petrzilka, M., Grayson, J. I. Preparation and Diels-Alder Reactions of Hetero-Substituted 1,3-Dienes. *Synthesis* **1981**, 753-786.
19. Danishefsky, S., Larson, E., Askin, D., Kato, N. On the scope, mechanism and stereochemistry of the Lewis acid catalyzed cyclocondensation of activated dienes with aldehydes: an application to the erythronolide problem. *J. Am. Chem. Soc.* **1985**, 107, 1246-1255.
20. Baldoli, C., Maiorana, S., Licandro, E., Zinzalla, G., Lanfranchi, M., Tiripicchio, A. Stereoselective hetero Diels-Alder reactions of chiral tricarbonyl (η6-benzaldehyde)chromium complexes. *Tetrahedron: Asymmetry* **2001**, 12, 2159-2167.
21. Keck, G. E., Li, X.-Y., Krishnamurthy, D. Catalytic Enantioselective Synthesis of Dihydropyrones via Formal Hetero Diels-Alder Reactions of "Danishefsky's Diene" with Aldehydes. *J. Org. Chem.* **1995**, 60, 5998-5999.
22. Corey, E. J., Cywin, C. L., Roper, T. D. Enantioselective Mukaiyama-aldol and aldol-dihydropyrone annulation reactions catalyzed by a tryptophan-derived oxazaborolidine. *Tetrahedron Lett.* **1992**, 33, 6907-6910.
23. Mujica, M. T., Afonso, M. M., Galindo, A., Palenzuela, J. A. Hetero Diels-Alder vs Mukaiyama aldol pathways in the reaction of monoactivated dienes and aldehydes. A Lewis acid study. *Tetrahedron* **1996**, 52, 2167-2176.
24. Schaus, S. E., Brnalt, J., Jacobsen, E. N. Asymmetric Hetero-Diels-Alder Reactions Catalyzed by Chiral (Salen)Chromium(III) Complexes. *J. Org. Chem.* **1998**, 63, 403-405.
25. Bednarski, M., Maring, C., Danishefsky, S. Chiral induction in the cyclocondensation of aldehydes with siloxydienes. *Tetrahedron Lett.* **1983**, 24, 3451-3454.
26. Han, G., LaPorte, M. G., Folmer, J. J., Werner, K. M., Weinreb, S. M. Total Syntheses of the Securinega Alkaloids (+)-14,15-Dihydronorsecurinine, (-)-Norsecurinine, and Phyllanthine. *J. Org. Chem.* **2000**, 65, 6293-6306.
27. Ojida, A., Tanoue, F., Kanematsu, K. Total Syntheses of Marine Furanosesquiterpenoids, Tubipofurans. *J. Org. Chem.* **1994**, 59, 5970-5976.
28. Takemura, S., Hirayama, A., Tokunaga, J., Kawamura, F., Inagaki, K., Hashimoto, K., Nakata, M. A concise total synthesis of (±)-A80915G, a member of the napyradiomycin family of antibiotics. *Tetrahedron Lett.* **1999**, 40, 7501-7505.

Darzens Glycidic Ester Condensation 128

Related reactions: Corey-Chaykovsky epoxidation and cyclopropanation;

1. Erlenmeyer, E. Phenyl-α-oxypropionoic acid and phenyl-α,β-propionic acid. *Liebigs Ann. Chem.* **1892**, 271, 137-163.
2. Claisen, L. Application of sodium amide in a few transfomations. *Ber.* **1905**, 38, 693-694.
3. Darzens, G. New Method of Preparing Glycidic Esters. *Compt. rend.* **1911**, 151, 883-884.
4. Newman, M. S., Magerlein, B. J. Darzens glycidic ester condensation. *Org. React.* **1949**, 5, 413-440.
5. Ballester, M. Mechanisms of the Darzens and related condensations. *Chem. Rev.* **1955**, 55, 283-300.
6. Bachelor, F. W., Bansal, R. K. Darzens glycidic ester condensation. *J. Org. Chem.* **1969**, 34, 3600-3604.
7. Rosen, T. Darzens glycidic ester condensation. in *Comp. Org. Synth.* (eds. Trost, B. M.,Fleming, I.), *2*, 409-441 (Pergamon, Oxford, **1991**).
8. Deyrup, J. A. Darzens aziridine synthesis. *J. Org. Chem.* **1969**, 34, 2724-2727.
9. Takahashi, T., Muraoki, M., Capo, M., Koga, K. Enantioselective Darzens reaction: asymmetric synthesis of trans-glycidic esters mediated by chiral lithium amides. *Chem. Pharm. Bull.* **1995**, 43, 1821-1823.
10. Arai, S., Shirai, Y., Ishida, T., Shioiri, T. Phase-transfer-catalyzed asymmetric Darzens reaction. *Tetrahedron* **1999**, 55, 6375-6386.
11. Davis, F. A., Liu, H., Zhou, P., Fang, T., Reddy, G. V., Zhang, Y. Aza-Darzens asymmetric synthesis of N-(p-toluenesulfinyl)aziridine 2-carboxylate esters from sulfinimines (N-sulfinyl imines). *J. Org. Chem.* **1999**, 64, 7559-7567.
12. McLaren, A. B., Sweeney, J. B. Inverted Diastereoselectivity in Asymmetric Aziridine Synthesis via Aza-Darzens Reaction of (2S)-N-Bromoacyl Camphorsultam. *Org. Lett.* **1999**, 1, 1339-1341.
13. Tanaka, K., Shiraishi, R. Darzens condensation reaction in water. *Green Chem.* **2001**, 3, 135-136.
14. Aggarwal, V. K., Hynd, G., Picoul, W., Vasse, J.-L. Highly Enantioselective Darzens Reaction of a Camphor-Derived Sulfonium Amide to Give Glycidic Amides and Their Applications in Synthesis. *J. Am. Chem. Soc.* **2002**, 124, 9964-9965.
15. Arai, S., Shioiri, T. Asymmetric Darzens reaction utilizing chloromethyl phenyl sulfone under phase-transfer catalyzed conditions. *Tetrahedron* **2002**, 58, 1407-1413.
16. Arai, S., Suzuki, Y., Tokumaru, K., Shioiri, T. Diastereoselective Darzens reactions of α-chloro esters, amides and nitriles with aromatic aldehydes under phase-transfer catalyzed conditions. *Tetrahedron Lett.* **2002**, 43, 833-836.
17. Shibata, I., Yamasaki, H., Baba, A., Matsuda, H. Stereoselective synthesis of α,β-epoxy ketones by the Darzen's reaction with methyl N-ethyl-N-(tributylstannyl)carbamate. *Synlett* **1990**, 490-492.
18. Hirashita, T., Kinoshita, K., Yamamura, H., Kawai, M., Araki, S. A facile preparation of indium enolates and their Reformatskii- and Darzens-type reactions. *Perkin 1* **2000**, 825-828.
19. Vogt, P. F., Tavares, D. F. α,β-Epoxy sulfones. Darzens condensation with α-halosulfones. *Can. J. Chem.* **1969**, 47, 2875-2881.
20. Stork, G., Worrall, W. S., Pappas, J. J. Synthesis and reactions of glycidonitriles. Transformation into a-haloacyl compounds and aminoalcohols. *J. Am. Chem. Soc.* **1960**, 82, 4315-4323.

21. Sulmon, P., De Kimpe, N., Schamp, N., Declercq, J. P., Tinant, B. A novel Darzens-type condensation using α-chloro ketimines. *J. Org. Chem.* **1988**, 53, 4457-4462.
22. Dagli, D. J., Yu, P.-S., Wemple, J. Darzens synthesis of glycidic thiol esters. Formation of a β-lactone by-product. *J. Org. Chem.* **1975**, 40, 3173-3178.
23. Blanchard, E. P., Jr., Buechi, G. The conversion of glycidic esters to aldehydes and ketones. *J. Am. Chem. Soc.* **1963**, 85, 955-958.
24. Munch-Petersen, J. Darzens' glycidic ester condensation. The isolation of an aldol intermediate. *Acta Chem. Scand.* **1953**, 7, 1041-1044.
25. Kwart, H., Kirk, L. G. Steric considerations in base catalyzed condensation; the Darzens reaction. *J. Org. Chem.* **1957**, 22, 116-120.
26. Ballester, M., Perez-Blanco, D. Mechanism of the Darzens condensation. Isolation of two aldol intermediates. *J. Org. Chem.* **1958**, 23, 652.
27. Deschamps, B., Seyden-Penne, J. Mechanism of the Darzens reaction. Formation of glycidonitriles. *C. R. Seances Acad. Sci. C.* **1970**, 271, 1097-1099.
28. Deschamps, B., Seyden-Penne, J. Solvent effects on the stereochemistry of the Darzens reaction. III. Condensation of chloroacetonitrile and aromatic aldehydes in a basic medium. *Tetrahedron* **1971**, 27, 3959-3964.
29. Yliniemela, A., Brunow, G., Flugge, J., Teleman, O. A Cyclic Transition State for the Darzens Reaction. *J. Org. Chem.* **1996**, 61, 6723-6726.
30. Mizuno, H., Domon, K., Masuya, K., Tanino, K., Kuwajima, I. Total Synthesis of (-)-Coriolin. *J. Org. Chem.* **1999**, 64, 2648-2656.
31. Aldous, D. J., Dalencon, A. J., Steel, P. G. A Short Synthesis of (±)-Epiasarinin. *Org. Lett.* **2002**, 4, 1159-1162.
32. Schwartz, A., Madan, P. B., Mohacsi, E., O'Brien, J. P., Todaro, L. J., Coffen, D. L. Enantioselective synthesis of calcium channel blockers of the diltiazem group. *J. Org. Chem.* **1992**, 57, 851-856.

Davis' Oxaziridine Oxidations........130

Related reactions: Jacobsen-Katsuki epoxidation, Prilezhaev oxidation, Rubottom oxidation, Sharpless asymmetric epoxidation, Shi epoxidation;

1. Davis, F. A., Nadir, U. K. Photolysis of 2-arenesulfonyl-3-phenyloxaziridines. *Tetrahedron Lett.* **1977**, 1721-1724.
2. Davis, F. A., Nadir, U. K., Kluger, E. W. 2-Arylsulfonyl-3-phenyloxaziridines: a new class of stable oxaziridine derivatives. *J. Chem. Soc., Chem. Commun.* **1977**, 25-26.
3. Davis, F. A., Jenkins, R., Jr., Yocklovich, S. G. 2-Arenesulfonyl-3-aryloxaziridines: a new class of aprotic oxidizing agents (oxidation of organic sulfur compounds). *Tetrahedron Lett.* **1978**, 5171-5174.
4. Davis, F. A., Abdul-Malik, N. F., Awad, S. B., Harakal, M. E. Epoxidation of olefins by oxaziridines. *Tetrahedron Lett.* **1981**, 22, 917-920.
5. Davis, F. A., Billmers, J. M. Chemistry of oxaziridines. 5. Kinetic resolution of sulfoxides using chiral 2-sulfonyloxaziridines. *J. Org. Chem.* **1983**, 48, 2672-2675.
6. Davis, F. A., Harakal, M. E., Awad, S. B. Chemistry of oxaziridines. 4. Asymmetric epoxidation of unfunctionalized alkenes using chiral 2-sulfonyloxaziridines: evidence for a planar transition state geometry. *J. Am. Chem. Soc.* **1983**, 105, 3123-3126.
7. Davis, F. A., Stringer, O. D., Billmers, J. M. Oxidation of selenides to selenoxides using 2-sulfonyloxaziridines. *Tetrahedron Lett.* **1983**, 24, 1213-1216.
8. Davis, F. A., Vishwakarma, L. C., Billmers, J. G., Finn, J. Synthesis of α-hydroxycarbonyl compounds (acyloins): direct oxidation of enolates using 2-sulfonyloxaziridines. *J. Org. Chem.* **1984**, 49, 3241-3243.
9. Davis, F. A., Billmers, J. M., Gosciniak, D. J., Towson, J. C., Bach, R. D. Chemistry of oxaziridines. 7. Kinetics and mechanism of the oxidation of sulfoxides and alkenes by 2-sulfonyloxaziridines. Relationship to the oxygen-transfer reactions of metal peroxides. *J. Org. Chem.* **1986**, 51, 4240-4245.
10. Davis, F. A., McCauley, J. P., Jr., Chattopadhyay, S., Harakal, M. E., Watson, W. H., Tavanaiepour, I. New synthetic applications of chiral sulfamides. Optically active sulfamyloxaziridines in the oxidation of nonfunctionalized substrates with high enantioselectivity. *Stud. Org. Chem. (Amsterdam)* **1987**, 28, 153-165.
11. Davis, F. A., Sheppard, A. C. Applications of oxaziridines in organic synthesis. *Tetrahedron* **1989**, 45, 5703-5742.
12. Davis, F. A., Haque, M. S. Oxygen-transfer reactions of oxaziridines. *Adv. Oxygenated Processes* **1990**, 2, 61-116.
13. Davis, F. A., Chen, B. C. Asymmetric hydroxylation of enolates with N-sulfonyloxaziridines. *Chem. Rev.* **1992**, 92, 919-934.
14. Davis, F. A., Reddy, R. T., Han, W., Reddy, R. E. Asymmetric synthesis using N-sulfonyloxaziridines. *Pure Appl. Chem.* **1993**, 65, 633-640.
15. Davis, F. A., Reddy, R. T. Oxaziridines and Oxazirines. in *Comprehensive Heterocyclic Chemistry II* (eds. Katritzky, A. R., Rees, C. W.,Scriven, E. F. V.), *1A*, 365-413 (Pergamon, New York, **1996**).
16. McCoull, W., Davis, F. A. Recent synthetic applications of chiral aziridines. *Synthesis* **2000**, 1347-1365.
17. Zhou, P., Chen, B. C., Davis, F. A. Asymmetric hydroxylations of enolates and enol derivatives. *Asymmetric Oxidation Reactions* **2001**, 128-145.
18. Davis, F. A., ThimmaReddy, R., Weismiller, M. C. (-)-α,α-Dichlorocamphorsulfonyloxaziridine: a superior reagent for the asymmetric oxidation of sulfides to sulfoxides. *J. Am. Chem. Soc.* **1989**, 111, 5964-5965.
19. Petrov, V. A., Resnati, G. Polyfluorinated Oxaziridines: Synthesis and Reactivity. *Chem. Rev.* **1996**, 96, 1809-1823.
20. Davis, F. A., Reddy, R. E., Kasu, P. V. N., Portonovo, P. S., Carroll, P. J. Synthesis and Reactions of exo-(Camphorylsulfonyl)oxaziridine. *J. Org. Chem.* **1997**, 62, 3625-3630.
21. Wolfe, M. S., Dutta, D., Aube, J. Stereoselective Synthesis of Freidinger Lactams Using Oxaziridines Derived from Amino Acids. *J. Org. Chem.* **1997**, 62, 654-663.
22. Bohe, L., Lusinchi, M., Lusinchi, X. Oxygen atom transfer from a chiral oxaziridinium salt. Asymmetric epoxidation of unfunctionalized olefins. *Tetrahedron* **1999**, 55, 141-154.
23. Bach, R. D., Wolber, G. J. Mechanism of oxygen transfer from oxaziridine to ethylene: the consequences of HOMO-HOMO interactions on frontier orbital narrowing. *J. Am. Chem. Soc.* **1984**, 106, 1410-1415.
24. Bach, R. D., Coddens, B. A., McDouall, J. J. W., Schlegel, H. B., Davis, F. A. The mechanism of oxygen transfer from an oxaziridine to a sulfide and a sulfoxide: a theoretical study. *J. Org. Chem.* **1990**, 55, 3325-3330.
25. Bach, R. D., Andres, J. L., Davis, F. A. Mechanism of oxygen atom transfer from oxaziridine to a lithium enolate. A theoretical study. *J. Org. Chem.* **1992**, 57, 613-618.
26. Emmons, W. D. Synthesis of oxaziranes. *J. Am. Chem. Soc.* **1956**, 78, 6208-6209.
27. Horner, L., Jurgens, E. Preparation and properties of some isonitrones (oxaziranes). *Chem. Ber.* **1957**, 90, 2184-2189.
28. Mata, E. G. Recent advances in the synthesis of sulfoxides from sulfides. *Phosphorus, Sulfur Silicon Relat. Elem.* **1996**, 117, 231-286.
29. Bohe, L., Lusinchi, M., Lusinchi, X. Oxygen atom transfer from a chiral N-alkyloxaziridine promoted by acid. The asymmetric oxidation of sulfides to sulfoxides. *Tetrahedron* **1999**, 55, 155-166.
30. Zajac, W. W., Jr., Walters, T. R., Darcy, M. G. Oxidation of amines with 2-sulfonyloxaziridines (Davis' reagents). *J. Org. Chem.* **1988**, 53, 5856-5860.
31. Davis, F. A., Mancinelli, P. A., Balasubramanian, K., Nadir, U. K. Coupling and hydroxylation of lithium and Grignard reagents by oxaziridines. *J. Am. Chem. Soc.* **1979**, 101, 1044-1045.
32. Maccagnani, G., Innocenti, A., Zani, P., Battaglia, A. Oxidation of thiones to sulfines (thione S-oxides) with N-sulfonyloxaziridines: synthetic and mechanistic aspects. *J. Chem. Soc., Perkin Trans. 2* **1987**, 1113-1116.
33. Davis, F. A., Sheppard, A. C., Chen, B. C., Haque, M. S. Chemistry of oxaziridines. 14. Asymmetric oxidation of ketone enolates using enantiomerically pure (camphorylsulfonyl)oxaziridine. *J. Am. Chem. Soc.* **1990**, 112, 6679-6690.
34. Leslie, D. R., Beaudry, W. T., Szafraniec, L. L., Rohrbaugh, D. K. Mechanistic implications of pyrophosphate formation in the oxidation of O,S-dimethyl phosphoramidothioate. *J. Org. Chem.* **1991**, 56, 3459-3462.
35. Beak, P., Anderson, D. R., Jarboe, S. G., Kurtzweil, M. L., Woods, K. W. Mechanisms and consequences of oxygen transfer reactions. *Pure Appl. Chem.* **2000**, 72, 2259-2264.

36. Davis, F. A., Weismiller, M. C., Murphy, C. K., Reddy, R. T., Chen, B. C. Chemistry of oxaziridines. 18. Synthesis and enantioselective oxidations of the [(8,8-dihalocamphoryl)sulfonyl]oxaziridines. *J. Org. Chem.* **1992**, 57, 7274-7285.
37. White, J. D., Carter, R. G., Sundermann, K. F. A Highly Stereoselective Synthesis of Epothilone B. *J. Org. Chem.* **1999**, 64, 684-685.
38. Dounay, A. B., Forsyth, C. J. Abbreviated Synthesis of the C3-C14 (Substituted 1,7-Dioxaspiro[5.5]undec-3-ene) System of Okadaic Acid. *Org. Lett.* **1999**, 1, 451-453.
39. Snider, B. B., Zeng, H. Total Syntheses of (-)-Fumiquinazolines A, B, and I. *Org. Lett.* **2000**, 2, 4103-4106.

De Mayo Cycloaddition (Enone-Alkene [2+2] Photocycloaddition) 132

1. De Mayo, P., Takeshita, H., Sattar, A. B. M. A. Photochemical synthesis of 1,5-diketones and their cyclization-new annulation process. *Proc. Chem. Soc.* **1962**, 119.
2. Eaton, P. E. On the mechanism of the photodimerization of cyclopentenone. *J. Am. Chem. Soc.* **1962**, 84, 2454-2455.
3. De Mayo, P., Takeshita, H. Photochemical syntheses. VI. The formation of heptanediones from acetylacetone and alkenes. *Can. J. Chem.* **1963**, 41, 440-449.
4. Corey, E. J., Bass, J. D., Le Mahieu, R., Mitra, R. B. Photochemical reactions of 2-cyclohexenones with substituted olefins. *J. Am. Chem. Soc.* **1964**, 86, 5570-5583.
5. Eaton, P. E. Photochemical reactions of simple alicyclic enones. *Acc. Chem. Res.* **1968**, 1, 50-57.
6. Bauslaugh, P. G. Photochemical cycloaddition reactions of enones to alkenes; synthetic applications. *Synthesis* **1970**, 2, 287-300.
7. De Mayo, P. Photochemical syntheses. 37. Enone photoannelation. *Acc. Chem. Res.* **1971**, 4, 41-48.
8. Lenz, G. R. Photocycloaddition reactions of conjugated enones. *Rev. Chem. Intermed.* **1980**, 4, 369-404.
9. Oppolzer, W. The intramolecular [2 + 2] photoaddition/cyclobutane-fragmentation sequence in organic synthesis. *Acc. Chem. Res.* **1982**, 15, 135-141.
10. Crimmins, M. T. Synthetic applications of intramolecular enone-olefin photocycloadditions. *Chem. Rev.* **1988**, 88, 1453-1473.
11. Crimmins, M. T., Reinhold, T. L. Enone olefin [2 + 2] photochemical cycloadditions. *Org. React.* **1993**, 44, 297-588.
12. Schuster, D. I., Lem, G., Kaprinidis, N. A. New insights into an old mechanism: [2 + 2] photocycloaddition of enones to alkenes. *Chem. Rev.* **1993**, 93, 3-22.
13. Horspool, W. M. Enone cycloadditions and rearrangements: Photoreactions of dienones and quinones. *Photochemistry* **2001**, 32, 74-116.
14. Schuster, D. I. Mechanistic issues in [2 + 2]- photocycloadditions of cyclic enones to alkenes. *CRC Handbook of Organic Photochemistry and Photobiology (2nd Edition)* **2004**, 72/71-72/24.
15. Sato, M., Takayama, K., Sekiguchi, K., Abe, Y., Furuya, T., Inukai, N., Kaneko, C. Synthesis of optically active cyclopenta[c]pyran-4-carboxylic acid derivatives, building blocks for iridoids. An attractive alternative to the asymmetric de Mayo reaction. *Chem. Lett.* **1989**, 1925-1928.
16. Sato, M., Abe, Y., Takayama, K., Sekiguchi, K., Kaneko, C., Inoue, N., Furuya, T., Inukai, N. Use of 1,3-dioxin-4-ones and their related compounds in synthesis. Part 28. Asymmetric de Mayo reactions using chiral spirocyclic dioxinones. *J. Heterocycl. Chem.* **1991**, 28, 241-252.
17. Galatsis, P., Ashbourne, K. J., Manwell, J. J., Wendling, P., Dufault, R., Hatt, K. L., Ferguson, G., Gallagher, J. F. Synthesis of fused-ring cyclobutenones via a tandem [2 + 2] cycloaddition-β-elimination sequence. *J. Org. Chem.* **1993**, 58, 1491-1495.
18. Sato, M., Sunami, S., Kogawa, T., Kaneko, C. An efficient synthesis of cis-hydroindan-5-ones by novel modified de Mayo reaction using 2,3-dihydro-4-pyrones as the enone chromophore. *Chem. Lett.* **1994**, 2191-2194.
19. Ciamician, G., Silber, P. Chemical Action of Light (XIII). *Ber.* **1908**, 41, 1928-1935.
20. Buchi, G., Goldman, I. M. Photochemical reactions. VII. The intramolecular cyclization of carvone to carvonecamphor. *J. Am. Chem. Soc.* **1957**, 79, 4741-4748.
21. Nozaki, H., Kurita, M., Mori, T., Noyori, R. Photochemical behavior of enolic β-diketones towards cycloolefins. *Tetrahedron* **1968**, 24, 1821-1828.
22. Loutfy, R. O., De Mayo, P. Photochemical synthesis. 67. Mechanism of enone photoannelation: activation energies and the role of exciplexes. *J. Am. Chem. Soc.* **1977**, 99, 3559-3565.
23. Burshtein, K. Y., Serebryakov, E. P. The regioselectivity of α,β-enone photoannelation with monosubstituted acetylenes: a possible effect of dipole-dipole interactions. *Tetrahedron* **1978**, 34, 3233-3238.
24. Schuster, D. I., Brown, P. B., Capponi, L. J., Rhodes, C. A., Scaiano, J. C., Tucker, P. C., Weir, D. Photochemistry of ketones in solution. Part 79. Mechanistic alternatives in photocycloaddition of cyclohexenones to alkenes. *J. Am. Chem. Soc.* **1987**, 109, 2533-2534.
25. Swapna, G. V. T., Lakshmi, A. B., Rao, J. M., Kunwar, A. C. Mechanistic implications of photoannelation reaction of 4,4-dimethylcyclohex-2-en-1-one and acrylonitrile - regio- and stereochemistry of the major photoadduct by proton and carbon-13 NMR spectroscopy. *Tetrahedron* **1989**, 45, 1777-1782.
26. Kumar, M. S., Rao, J. M. Effect of micellar medium on photoannelation vs. energy transfer in the excited system cyclohex-2-en-1-one-cyclopentadiene. *Tetrahedron* **1990**, 46, 5383-5388.
27. Andrew, D., Hastings, D. J., Weedon, A. C. The Mechanism of the Photochemical Cycloaddition Reaction between 2-Cyclopentenone and Polar Alkenes: Trapping of Triplet 1,4-Biradical Intermediates with Hydrogen Selenide. *J. Am. Chem. Soc.* **1994**, 116, 10870-10882.
28. Broeker, J. L., Eksterowicz, J. E., Belk, A. J., Houk, K. N. On the Regioselectivity of Photocycloadditions of Triplet Cyclohexenones to Alkenes. *J. Am. Chem. Soc.* **1995**, 117, 1847-1848.
29. Benchikh le-Hocine, M., Do Khac, D., Fetizon, M. Model studies in taxane diterpene synthesis. Part III. *Synth. Commun.* **1992**, 22, 245-255.
30. Shipe, W. D., Sorensen, E. J. A Convergent Synthesis of the Tricyclic Architecture of the Guanacastepenes Featuring a Selective Ring Fragmentation. *Org. Lett.* **2002**, 4, 2063-2066.
31. Winkler, J. D., Rouse, M. B., Greaney, M. F., Harrison, S. J., Jeon, Y. T. The First Total Synthesis of (±)-Ingenol. *J. Am. Chem. Soc.* **2002**, 124, 9726-9728.
32. Lange, G. L., Gottardo, C., Merica, A. Synthesis of Terpenoids Using a Free Radical Fragmentation/Elimination Sequence. *J. Org. Chem.* **1999**, 64, 6738-6744.

Demjanov Rearrangement and Tiffeneau-Demjanov Rearrangement 134

Related reactions: Pinacol and semipinacol rearrangement;

1. Demjanov, N. J., Luschnikov, M. On the treatment of aminomethyl cyclobutane with nitrous acid about bromomethyl cyclobutane. *J.Russ. Phys.-Chem. Soc.* **1901**, 33, 279-283.
2. Demjanov, N. J., Luschnikov, M. Products of the reaction of nitrous acid with aminomethyl cyclobutane. *J. Russ. Phys. Chem. Soc.* **1903**, 35, 26-42.
3. Tiffeneau, M., Tchoubar, B. Argentic dehalogenation of α-cyclanediol iodohydrins. *Compt. rend.* **1937**, 205, 1411-1413.
4. Smith, P. A. S., Baer, D. R. The Demjanov and Tiffeneau-Demjanov ring expansions. *Org. React.* **1960**, 11, 157-188.
5. Krow, G. R. One carbon ring expansions of bridged bicyclic ketones. *Tetrahedron* **1987**, 43, 3-38.
6. Fattori, D., Henry, S., Vogel, P. The Demjanov and Tiffeneau-Demjanov one-carbon ring enlargements of 2-aminomethyl-7-oxabicyclo[2.2.1]heptane derivatives. The stereo- and regioselective additions of 8-oxabicyclo[3.2.1]oct-6-en-2-one to soft electrophiles. *Tetrahedron* **1993**, 49, 1649-1664.
7. Ou, Z., Chen, Z., Jiang, O. Demjanov rearrangement on synthetic zeolites. *Kexue Tongbao (Foreign Language Edition)* **1987**, 32, 462-464.
8. Tchoubar, B. Extension of alicylic rings of 1-(aminomethyl)cycloalkanols by nitrous deamination. II. Nitrous deamination of (aminomethyl)cycloalkanols. *Bull. soc. chim. France* **1949**, 164-169.

9. Roberts, J. D., Gorham, W. F. Syntheses of some bicyclo[3.3.0]octane derivatives. *J. Am. Chem. Soc.* **1952**, 74, 2278-2282.
10. Tchoubar, B. Extension of alicylic rings of 1-(aminomethyl)cycloalkanols by nitrous deamination. III. Theoretical consideration of the reaction mechanism of nitrous deamination of amino alcohols. *Bull. soc. chim. France* **1949**, 169-172.
11. Dave, V., Stothers, J. B., Warnhoff, E. W. Ring expansion of cyclic ketones: the reliable determination of migration ratios for 3-keto steroids by carbon-13 nuclear magnetic resonance and the general implications thereof. *Can. J. Chem.* **1979**, 57, 1557-1568.
12. Cooper, C. N., Jenner, P. J., Perry, N. B., Russell-King, J., Storesund, H. J., Whiting, M. C. Classical carbonium ions. Part 13. Rearrangements from secondary to primary alkyl groups during reactions involving carbonium ions. *J. Chem. Soc., Perkin Trans. 2* **1982**, 605-611.
13. Stern, A. G., Nickon, A. Synthesis of brexan-2-one and ring-expanded congeners. *J. Org. Chem.* **1992**, 57, 5342-5352.
14. Thomas, R. C., Fritzen, E. L. Spectinomycin modification. V. The synthesis and biological activity of spectinomycin analogs with ring-expanded sugars. *J. Antibiot.* **1988**, 41, 1445-1451.

Dess-Martin Oxidation 136

Related reactions: *Corey-Kim oxidation, Jones oxidation, Ley oxidation, Oppenauer oxidation, Pfitzner-Moffatt oxidation, Swern oxidation;*

1. Dess, D. B., Martin, J. C. Readily accessible 12-I-5 oxidant for the conversion of primary and secondary alcohols to aldehydes and ketones. *J. Org. Chem.* **1983**, 48, 4155-4156.
2. Speicher, A., Bomm, V., Eicher, T. Dess-Martin periodinane (DMP). *J. Prakt. Chem./Chem.-Ztg.* **1996**, 338, 588-590.
3. Kitamura, T., Fujiwara, Y. Recent progress in the use of hypervalent iodine reagents in organic synthesis. A review. *Org. Prep. Proced. Int.* **1997**, 29, 409-458.
4. Akiba, K.-Y. Structure and reactivity of hypervalent organic compounds: general aspects. *Chemistry of Hypervalent Compounds* **1999**, 9-47.
5. Wirth, T., Hirt, U. H. Hypervalent iodine compounds. Recent advances in synthetic applications. *Synthesis* **1999**, 1271-1287.
6. Schilling, G. Dess-Martin periodinane. *GIT Labor-Fachzeitschrift* **2002**, 46, 84.
7. Zhdankin, V. V., Stang, P. J. Recent Developments in the Chemistry of Polyvalent Iodine Compounds. *Chem. Rev.* **2002**, 102, 2523-2584.
8. Tohma, H., Kita, Y. Hypervalent iodine reagents for the oxidation of alcohols and their application to complex molecule synthesis. *Adv. Syn. & Catal.* **2004**, 346, 111-124.
9. Dess, D. B., Martin, J. C. A useful 12-I-5 triacetoxyperiodinane (the Dess-Martin periodinane) for the selective oxidation of primary or secondary alcohols and a variety of related 12-I-5 species. *J. Am. Chem. Soc.* **1991**, 113, 7277-7287.
10. Ireland, R. E., Liu, L. An improved procedure for the preparation of the Dess-Martin periodinane. *J. Org. Chem.* **1993**, 58, 2899.
11. Meyer, S. D., Schreiber, S. L. Acceleration of the Dess-Martin Oxidation by Water. *J. Org. Chem.* **1994**, 59, 7549-7552.
12. Stevenson, P. J., Treacy, A. B., Nieuwenhuyzen, M. Preparation of Dess-Martin periodinane - the role of the morphology of 1-hydroxy-1,2-benziodoxol-3(1H)-one 1-oxide precursor. *J. Chem. Soc., Perkin Trans. 2* **1997**, 589-591.
13. Frigerio, M., Santagostino, M., Sputore, S. A user-friendly entry to 2-iodoxybenzoic acid (IBX). *J. Org. Chem.* **1999**, 64, 4537-4538.
14. Boeckman, R. K., Jr., Shao, P., Mullins, J. J. The Dess-Martin periodinane: 1,1,1-triacetoxy-1,1-dihydro-1,2-benziodoxol-3(1H)-one. *Org. Synth.* **2000**, 77, 141-152.
15. Depernet, D., Francois, B. *Stabilized o-iodoxybenzoic acid compositions.* Wo 0257210, **2002** (Simafex, Fr.).
16. Amey, R. L., Martin, J. C. An alkoxyaryltrifluoroperiodinane. A stable heterocyclic derivative of pentacoordinated organoiodine(V). *J. Am. Chem. Soc.* **1978**, 100, 300-301.
17. Amey, R. L., Martin, J. C. Synthesis and reactions of stable alkoxyaryltrifluoroperiodinanes. A "tamed" analog of iodine pentafluoride for use in oxidations of amines, alcohols, and other species. *J. Am. Chem. Soc.* **1979**, 101, 5294-5299.
18. Wirth, T. IBX-new reactions with an old reagent. *Angew. Chem., Int. Ed. Engl.* **2001**, 40, 2812-2814.
19. Hartmann, C., Meyer, V. Iodobenzoic acids. *Chem.Ber.* **1893**, 26, 1727-1732.
20. Plumb, J. B., Harper, D. J. 2-Iodoxybenzoic acid. *Chem. Eng. News* **1990**, 68, 3.
21. Lawrence, N. J., Crump, J. P., McGown, A. T., Hadfield, J. A. Reaction of Baylis-Hillman products with Swern and Dess-Martin oxidants. *Tetrahedron Lett.* **2001**, 42, 3939-3941.
22. Chaudhari, S. S., Akamanchi, K. G. A mild, chemoselective, oxidative method for deoximation using Dess-Martin periodinane. *Synthesis* **1999**, 760-764.
23. Jenkins, N. E., Ware, R. W., Jr., Atkinson, R. N., King, S. B. Generation of acyl nitroso compounds by the oxidation of N-acyl hydroxylamines with the Dess-Martin periodinane. *Synth. Commun.* **2000**, 30, 947-953.
24. Nicolaou, K. C., Sugita, K., Baran, P. S., Zhong, Y.-L. New synthetic technology for the construction of N-containing quinones and derivatives thereof: total synthesis of epoxyquinomycin B. *Angew. Chem., Int. Ed. Engl.* **2001**, 40, 207-210.
25. Myers, A. G., Zhong, B., Movassaghi, M., Kung, D. W., Lanman, B. A., Kwon, S. Synthesis of highly epimerizable *N*-protected α-amino aldehydes of high enantiomeric excess. *Tetrahedron Lett.* **2000**, 41, 1359-1362.
26. Nicolaou, K. C., Baran, P. S., Zhong, Y. L., Sugita, K. Iodine(V) Reagents in Organic Synthesis. Part 1. Synthesis of Polycyclic Heterocycles via Dess-Martin Periodinane-Mediated Cascade Cyclization: Generality, Scope, and Mechanism of the Reaction. *J. Am. Chem. Soc.* **2002**, 124, 2212-2220.
27. De Munari, S., Frigerio, M., Santagostino, M. Hypervalent Iodine Oxidants: Structure and Kinetics of the Reactive Intermediates in the Oxidation of Alcohols and 1,2-Diols by o-Iodoxybenzoic Acid and Dess-Martin Periodinane. A Comparative 1H-NMR Study. *J. Org. Chem.* **1996**, 61, 9272-9279.
28. Nicolaou, K. C., Baran, P. S., Kranich, R., Zhong, Y.-L., Sugita, K., Zou, N. Mechanistic studies of periodinane-mediated reactions of anilides and related systems. *Angew. Chem., Int. Ed. Engl.* **2001**, 40, 202-206.
29. Cao, B., Park, H., Joullie, M. M. Total Synthesis of Ustiloxin D. *J. Am. Chem. Soc.* **2002**, 124, 520-521.
30. Nicolaou, K. C., He, Y., Fong, K. C., Yoon, W. H., Choi, H.-S., Zhong, Y.-L., Baran, P. S. Novel Strategies to Construct the γ-Hydroxy Lactone Moiety of the CP Molecules. Synthesis of the CP-225,917 Core Skeleton. *Org. Lett.* **1999**, 1, 63-66.
31. Smith, A. B., III, Lin, Q., Doughty, V. A., Zhuang, L., McBriar, M. D., Kerns, J. K., Brook, C. S., Murase, N., Nakayama, K. The spongistatins: architecturally complex natural products. Part two. Synthesis of the C(29-51) subunit, fragment assembly, and final elaboration to (+)-spongistatin 2. *Angew. Chem., Int. Ed. Engl.* **2001**, 40, 196-199.

Dieckmann Condensation 138

Related reactions: *Claisen condensation, Baker-Venkataraman rearrangement;*

1. Fehling. Succinic acid and its derivatives. *Ann.* **1844**, 49, 154-212.
2. Dieckmann, W. Formation of rings from chains. *Ber.* **1894**, 27, 102.
3. Hauser, C. R., Hudson, B. E., Jr. Acetoacetic ester condensation and certain related reactions. *Org. React.* **1942**, 1, 266-302.
4. Schaefer, J. P., Bloomfield, J. J. Dieckmann condensation. (Including the Thorpe-Ziegler condensation). *Org. React.* **1967**, 15, 1-203.
5. Kwart, H., King, K. Rearrangement and cyclization reactions of carboxylic acids and esters. in *Chem. Carboxylic Acids and Esters* (ed. Patai, S.), 341-373 (Interscience-Publishers, London, New York, **1969**).
6. Banerjee, D. K. Dieckmann cyclization and utilization of the products in the synthesis of steroids. *Proc. - Indian Acad. Sci. - Section A* **1974**, 79, 282-309.

7. Davis, D. R., Garratt, P. J. Dieckmann Condensation. in *Comp. Org. Synth.* (eds. Trost, B. M.,Fleming, I.), *2*, 795-863 (Pergamon, Oxford, **1991**).
8. Gorobets, E. V., Miftakhov, M. S., Valeev, F. A. Tandem transformations initiated and determined by the Michael reaction. *Russ. Chem. Rev.* **2000**, 69, 1001-1019.
9. Liu, H.-J., Lai, H. K. A dithiol ester version of Dieckmann condensation. *Tetrahedron Lett.* **1979**, 1193-1196.
10. Nee, G., Tchoubar, B. Extension of the Dieckmann cyclization to an α,α'-dialkyl diester: methyl α,α'-dimethylpimelate. *Tetrahedron Lett.* **1979**, 3717-3720.
11. Crowley, J. I., Rapoport, H. Unidirectional Dieckmann cyclizations on a solid phase and in solution. *J. Org. Chem.* **1980**, 45, 3215-3227.
12. Kodpinid, M., Thebtaranonth, Y. Vinylogous Dieckmann condensation: an application of Baldwin's rules. *Tetrahedron Lett.* **1984**, 25, 2509-2512.
13. Thebtaranonth, Y. Synthesis and chemistry of cyclopentenoid antibiotics. *Pure Appl. Chem.* **1986**, 58, 781-788.
14. Tanabe, Y. The selective Claisen and Dieckmann ester condensations promoted by dichlorobis(trifluoromethanesulfonato)titanium(IV). *Bull. Chem. Soc. Jpn.* **1989**, 62, 1917-1924.
15. Bunce, R. A., Harris, C. R. Six-membered cyclic β-keto esters by tandem conjugate addition-Dieckmann condensation reactions. *J. Org. Chem.* **1992**, 57, 6981-6985.
16. Toda, F., Suzuki, T., Higa, S. Solvent-free Dieckmann condensation reactions of diethyl adipate and pimelate. *J. Chem. Soc., Perkin Trans. 1* **1998**, 3521-3522.
17. Tanabe, Y., Makita, A., Funakoshi, S., Hamasaki, R., Kawakusu, T. Practical synthesis of (Z)-civetone utilizing Ti-Dieckmann condensation. *Adv. Syn. & Catal.* **2002**, 344, 507-510.
18. Granik, V. G., Kadushkin, A. V., Liebscher, J. Synthesis of amino derivatives of five-membered heterocycles by Thorpe-Ziegler cyclization. *Adv. Heterocycl. Chem.* **1998**, 72, 79-125.
19. Fleming, F. F., Shook, B. C. Nitrile anion cyclizations. *Tetrahedron* **2002**, 58, 1-23.
20. Brown, C. A. Saline hydrides and superbases in organic reactions. X. Rapid, high yield condensations of esters and nitriles via kaliation. *Synthesis* **1975**, 326-327.
21. Leonard, N. J., Schimelpfenig, C. W., Jr. Synthesis of medium- and large-ring ketones via the Dieckmann condensation. *J. Org. Chem.* **1958**, 23, 1708-1710.
22. Schimelpfenig, C. W. Synthesis of oxometacyclophanes with the Dieckmann condensation. *J. Org. Chem.* **1975**, 40, 1493-1494.
23. Schimelpfenig, C. W. Synthesis of a macrocyclic triketone by the Dieckmann condensation. *Tex. J. Sci.* **1981**, 33, 73-76.
24. Yoshida, Y., Hayashi, R., Sumibara, H., Tanabe, Y. $TiCl_4/Bu_3N$/(catalytic TMSOTf): efficient agent for direct aldol addition and Claisen condensation. *Tetrahedron Lett.* **1997**, 38, 8727-8730.
25. Reed, R. I., Thornley, M. B. Pyrogenesis of ketones. II. The formation of some substituted cyclopentanones by the Dieckmann reaction. *J. Chem. Soc.* **1954**, 2148-2150.
26. Carrick, W. L., Fry, A. A carbon-14 isotope effect study of the Dieckmann condensation of diethyl phenylenediacetate. *J. Am. Chem. Soc.* **1955**, 77, 4381-4387.
27. Thaoker, M. R., Bagavant, G. Dieckmann cyclization of triethyl α-oxalylglutarate. *Indian J. Chem.* **1969**, 7, 232-233.
28. Hromatka, O., Binder, D., Eichinger, K. Mechanism of the Dieckmann-reaction of methyl-3-(methoxycarbonylmethylthio)propionic acid methyl ester. *Monatsh. Chem.* **1973**, 104, 1520-1525.
29. Paranjpe, P. P., Bagavant, G. Hinsberg's reaction. Condensation of thiodiacetic esters with pyruvic esters. *Indian J. Chem.* **1973**, 11, 313-314.
30. Bagavant, G., Thaoker, M. R., Raich, N. K. Mechanism of the Dieckmann-Komppa reaction of diethyl oxalate with diethyl oxydiacetate. *Curr. Sci.* **1974**, 43, 248-249.
31. Burinsky, D. J., Cooks, R. G. Gas-phase Dieckmann ester condensation characterized by mass spectrometry/mass spectrometry. *J. Org. Chem.* **1982**, 47, 4864-4869.
32. Covarrubias-Zuniga, A., Gonzalez-Lucas, A., Dominguez, M. M. Total synthesis of mycophenolic acid. *Tetrahedron* **2003**, 59, 1989-1994.
33. Tse, B. Total Synthesis of (-)-Galbonolide B and the Determination of Its Absolute Stereochemistry. *J. Am. Chem. Soc.* **1996**, 118, 7094-7100.
34. Grossman, R. B., Rasne, R. M. Short Total Syntheses of Both the Putative and Actual Structures of the Clerodane Diterpenoid (±)-Sacacarin by Double Annulation. *Org. Lett.* **2001**, 3, 4027-4030.

Diels-Alder Cycloaddition 140

Related reactions: Danishefsky's diene cycloaddition, Hetero Diels-Alder reaction;

1. Diels, O., Alder, K. Cause of the "azo ester" reaction. *Ann.* **1926**, 450, 237-254.
2. Diels, O., Adler, K. Syntheses in the hydroaromatic series. I. Addition of "diene" hydrocarbons. *Ann.* **1928**, 460, 98-122.
3. Diels, O., Alder, K. Syntheses in the hydroaromatic series. V. δ–4-Tetrahydro-*o*-phthalic acid (reply to the communication of E. H. Farmer and F. L. Warren: Properties of conjugated double bonds. (7). *Ber.* **1929**, 62B, 2087-2090.
4. Diels, O., Alder, K., Pries, P. Syntheses in the hydroaromatic series. IV. Addition of maleic anhydride to arylated dienes, trienes and fulvenes. *Ber.* **1929**, 62B, 2081-2087.
5. Holmes, H. L. Diels-Alder reaction: ethylenic and acetylenic dienophiles. *Org. React.* **1948**, 4, 60-173.
6. Kloetzel, M. C. Diels-Alder reaction with maleic anhydride. *Org. React.* **1948**, 4, 1-59.
7. Butz, L. W. Diels-Alder reaction: quinones and other cyclenones. *Org. React.* **1949**, 5, 136-192.
8. Kwart, H., King, K. The reverse Diels-Alder or retrodiene reaction. *Chem. Rev.* **1968**, 68, 415-447.
9. Gompper, R. Cycloadditions with polar intermediates. *Angew. Chem., Int. Ed. Engl.* **1969**, 8, 312-327.
10. McCabe, J. R., Eckert, C. A. Role of high-pressure kinetics in studies of the transition states of Diels-Alder reactions. *Acc. Chem. Res.* **1974**, 7, 251-257.
11. Houk, K. N. Frontier molecular orbital theory of cycloaddition reactions. *Acc. Chem. Res.* **1975**, 8, 361-369.
12. Brieger, G., Bennett, J. N. The intramolecular Diels-Alder reaction. *Chem. Rev.* **1980**, 80, 63-97.
13. Ciganek, E. The intramolecular Diels-Alder reaction. *Org. React.* **1984**, 32, 1-374.
14. Ichihara, A. Retro-Diels-Alder strategy in natural product synthesis. *Synthesis* **1987**, 207-222.
15. Weinreb, S. M. Synthetic methodology based upon N-sulfinyl dienophile [4 + 2]-cycloaddition reactions. *Acc. Chem. Res.* **1988**, 21, 313-318.
16. Bauld, N. L. Cation radical cycloadditions and related sigmatropic reactions. *Tetrahedron* **1989**, 45, 5307-5363.
17. Oppolzer, W. Intermolecular Diels-Alder reactions. in *Comp. Org. Synth.* (eds. Trost, B. M.,Fleming, I.), *5*, 315-401 (Pergamon Press, Oxford, **1991**).
18. Roush, W. R. Intramolecular Diels-Alder reactions. in *Comp. Org. Synth.* (eds. Trost, B. M.,Fleming, I.), *5*, 513-551 (Pergamon Press, Oxford, **1991**).
19. Sweger, R. W., Czarnik, A. W. Retrograde Diels-Alder reactions. in *Comp. Org. Synth.* (eds. Trost, B. M.,Fleming, I.), *5*, 551-593 (Pergamon Press, Oxford, **1991**).
20. Thomas, E. J. Cytochalasan synthesis: macrocycle formation via intramolecular Diels-Alder reactions. *Acc. Chem. Res.* **1991**, 24, 229-235.
21. Kagan, H. B., Riant, O. Catalytic asymmetric Diels Alder reactions. *Chem. Rev.* **1992**, 92, 1007-1019.
22. Pindur, U., Lutz, G., Otto, C. Acceleration and selectivity enhancement of Diels-Alder reactions by special and catalytic methods. *Chem. Rev.* **1993**, 93, 741-761.
23. Coxon, J. M., McDonald, D. Q., Steel, P. J. Diastereofacial selectivity in the Diels-Alder reaction. *Adv. Detailed React. Mech.* **1994**, 3, 131-166.

24. Bols, M., Skrydstrup, T. Silicon-Tethered Reactions. *Chem. Rev.* **1995**, 95, 1253-1277.
25. Winkler, J. D. Tandem Diels-Alder cycloadditions in organic synthesis. *Chem. Rev.* **1996**, 96, 167-176.
26. Dias, L. C. Chiral Lewis acid catalysts in Diels-Alder cycloadditions: mechanistic aspects and synthetic applications of recent systems. *J. Braz. Chem. Soc.* **1997**, 8, 289-332.
27. Neuschuetz, K., Velker, J., Neier, R. Tandem reactions combining Diels-Alder reactions with sigmatropic rearrangement processes and their use in synthesis. *Synthesis* **1998**, 227-255.
28. Barluenga, J., Suarez-Sobrino, A., Lopez, L. A. Chiral heterosubstituted 1,3-butadienes: synthesis and [4+2] cycloaddition reactions. *Aldrichimica Acta* **1999**, 32, 4-15.
29. Coxon, J. M., Froese, R. D. J., Ganguly, B., Marchand, A. P., Morokuma, K. On the origins of diastereofacial selectivity in Diels-Alder cycloadditions. *Synlett* **1999**, 1681-1703.
30. Fallis, A. G. Harvesting Diels and Alder's Garden: Synthetic Investigations of Intramolecular [4 + 2] Cycloadditions. *Acc. Chem. Res.* **1999**, 32, 464-474.
31. Klunder, A. J. H., Zhu, J., Zwanenburg, B. The Concept of Transient Chirality in the Stereoselective Synthesis of Functionalized Cycloalkenes Applying the Retro-Diels-Alder Methodology. *Chem. Rev.* **1999**, 99, 1163-1190.
32. Lee, L., Snyder, J. K. Indole as a dienophile in inverse electron demand Diels-Alder and related reactions. *Adv. Cycloadd.* **1999**, 6, 119-171.
33. Ruano, J. L. G., De la Plata, B. C. Asymmetric [4+2] cycloadditions mediated by sulfoxides. *Top. Curr. Chem.* **1999**, 204, 1-126.
34. Woodard, B. T., Posner, G. H. Recent advances in Diels-Alder cycloadditions of 2-pyrones. *Adv. Cycloadd.* **1999**, 5, 47-83.
35. Mehta, G., Uma, R. Stereoelectronic Control in Diels-Alder Reaction of Dissymmetric 1,3-Dienes. *Acc. Chem. Res.* **2000**, 33, 278-286.
36. Bear, B. R., Sparks, S. M., Shea, K. J. The type 2 intramolecular Diels-Alder reaction: synthesis and chemistry of bridgehead alkenes. *Angew. Chem., Int. Ed. Engl.* **2001**, 40, 820-849.
37. Fringuelli, F., Piermatti, O., Pizzo, F., Vaccaro, L. Recent advances in Lewis acid catalyzed Diels-Alder reactions in aqueous media. *Eur. J. Org. Chem.* **2001**, 439-455.
38. Kumar, A. Salt Effects on Diels-Alder Reaction Kinetics. *Chem. Rev.* **2001**, 101, 1-19.
39. Marsault, E., Toro, A., Nowak, P., Deslongchamps, P. The transannular Diels-Alder strategy: applications to total synthesis. *Tetrahedron* **2001**, 57, 4243-4260.
40. Yli-Kauhaluoma, J. Diels-Alder reactions on solid supports. *Tetrahedron* **2001**, 57, 7053-7071.
41. Corey, E. J. Catalytic enantioselective Diels-Alder reactions: Methods, mechanistic fundamentals, pathways, and applications. *Angew. Chem., Int. Ed. Engl.* **2002**, 41, 1650-1667.
42. Hayashi, Y. Catalytic asymmetric Diels-Alder reactions. *Cycloaddition Reactions in Organic Synthesis* **2002**, 5-55.
43. Liao, C.-C., Peddinti, R. K. Masked o-Benzoquinones in Organic Synthesis. *Acc. Chem. Res.* **2002**, 35, 856-866.
44. Marchand, A. P., Coxon, J. M. On the Origins of Diastereofacial Selectivity of [4 + 2] Cycloadditions in Cage-Annulated and Polycarbocyclic Diene/Dienophile Systems. *Acc. Chem. Res.* **2002**, 35, 271-277.
45. Nicolaou, K. C., Snyder, S. A., Montagnon, T., Vassilikogiannakis, G. The Diels-Alder reaction in total synthesis. *Angew. Chem., Int. Ed. Engl.* **2002**, 41, 1668-1698.
46. Konovalov, A. I., Kiselev, V. D. Diels-Alder reaction. Effect of internal and external factors on the reactivity of diene-dienophile systems. *Russ. Chem. Bull.* **2003**, 52, 293-311.
47. Stocking, E. M., Williams, R. M. Chemistry and biology of biosynthetic Diels-Alder reactions. *Angew. Chem., Int. Ed. Engl.* **2003**, 42, 3078-3115.
48. Houk, K. N. Ab initio and empirical computations of mechanism and stereoselectivity. *Pure Appl. Chem.* **1989**, 61, 643-650.
49. Bauld, N. L. An ab initio theoretical reaction path study of the cation radical Diels-Alder reaction. *J. Am. Chem. Soc.* **1992**, 114, 5800-5804.
50. Li, Y., Houk, K. N. Diels-Alder dimerization of 1,3-butadiene: an ab initio CASSCF study of the concerted and stepwise mechanisms and butadiene-ethylene revisited. *J. Am. Chem. Soc.* **1993**, 115, 7478-7485.
51. de Pascual-Teresa, B., Gonzalez, J., Asensio, A., Houk, K. N. Ab Initio-Based Force Field Modeling of the Transition States And Stereoselectivities of Lewis Acid-Catalyzed Asymmetric Diels-Alder Reactions. *J. Am. Chem. Soc.* **1995**, 117, 4347-4356.
52. Jursic, B., Zdravkovski, Z. DFT study of the Diels-Alder reactions between ethylene with buta-1,3-diene and cyclopentadiene. *J. Chem. Soc., Perkin Trans. 2* **1995**, 1223-1226.
53. Beno, B. R., Houk, K. N., Singleton, D. A. Synchronous or Asynchronous? An "Experimental" Transition State from a Direct Comparison of Experimental and Theoretical Kinetic Isotope Effects for a Diels-Alder Reaction. *J. Am. Chem. Soc.* **1996**, 118, 9984-9985.
54. de Pascual-Teresa, B., Houk, K. N. The ionic Diels-Alder reaction of the allyl cation and butadiene: theoretical investigation of the mechanism. *Tetrahedron Lett.* **1996**, 37, 1759-1762.
55. Domingo, L. R., Arno, M., Andres, J. The tandem Diels-Alder reaction of dimethyl acetylenedicarboxylate to bicyclopentadiene. A theoretical study of the molecular mechanisms. *Tetrahedron Lett.* **1996**, 37, 7573-7576.
56. Branchadell, V. Density functional study of Diels-Alder reactions between cyclopentadiene and substituted derivatives of ethylene. *Int. J. Quantum Chem.* **1997**, 61, 381-388.
57. Froese, R. D. J., Coxon, J. M., West, S. C., Morokuma, K. Theoretical Studies of Diels-Alder Reactions of Acetylenic Compounds. *J. Org. Chem.* **1997**, 62, 6991-6996.
58. Manoharan, M., Venuvanalingam, P. Gain or loss of aromaticity in Diels-Alder transition states and adducts: a theoretical investigation. *J. Phys. Org. Chem.* **1998**, 11, 133-140.
59. Marchand, A. P., Ganguly, B., Shukla, R. Stereoselectivities of Diels-Alder cycloadditions of p-facially nonequivalent dienes to MTAD, PTAD, and N-methylmaleimide: a theoretical study. *Tetrahedron* **1998**, 54, 4477-4484.
60. Silvero, G., Lucero, M. J., Winterfeldt, E., Houk, K. N. Theoretical study of the facial selectivity in Diels-Alder reactions of 4,4-disubstituted cyclohexadienones. *Tetrahedron* **1998**, 54, 7293-7300.
61. Tian, J., Houk, K. N., Klaerner, F. G. Substituent effect on stereospecificity and energy of concert of the retro-Diels-Alder reaction of isopropylidenenorbornene. *J. Phys. Chem. A* **1998**, 102, 7662-7667.
62. Bachmann, C., Boeker, N., Mondon, M., Gesson, J.-P. Density Functional Study of π-Facial Selectivity in Diels-Alder Reactions. *J. Org. Chem.* **2000**, 65, 8089-8092.
63. Manoharan, M., De Proft, F., Geerlings, P. Aromaticity Interplay between Quinodimethanes and C60 in Diels-Alder Reactions: Insights from a Theoretical Study. *J. Org. Chem.* **2000**, 65, 6132-6137.
64. Marchand, A. P., Chong, H.-S., Ganguly, B., Coxon, J. M. π-Facial selectivity in Diels-Alder cycloadditions. *Croat. Chem. Acta* **2000**, 73, 1027-1038.
65. Cannizzaro, C. E., Ashley, J. A., Janda, K. D., Houk, K. N. Experimental Determination of the Absolute Enantioselectivity of an Antibody-Catalyzed Diels-Alder Reaction and Theoretical Explorations of the Origins of Stereoselectivity. *J. Am. Chem. Soc.* **2003**, 125, 2489-2506.
66. Isobe, H., Takano, Y., Kitagawa, Y., Kawakami, T., Yamanaka, S., Yamaguchi, K., Houk, K. N. Systematic Comparisons between Broken Symmetry and Symmetry-Adapted Approaches to Transition States by Chemical Indices: A Case Study of the Diels-Alder Reactions. *J. Phys. Chem. A* **2003**, 107, 682-694.
67. Wieland. Studies on dicyclopentadiene. *Ber.* **1906**, 39, 1492-1499.
68. Albrecht, W. Addition products of cyclopentadiene and quinones. *Ann.* **1906**, 31-49.
69. Euler, H. V., Josephson, K. O. Condensations at double bonds. I. Condensation of isoprene with benzoquinone. *Ber.* **1920**, 53B, 822-826.
70. Tietze, L. F., Kettschau, G. Hetero Diels-Alder reactions in organic chemistry. *Top. Curr. Chem.* **1997**, 189, 1-120.
71. Waldmann, H. Asymmetric hetero Diels-Alder reactions. *Synthesis* **1994**, 535-551.
72. Jorgensen, K. A. Catalytic asymmetric hetero-Diels-Alder reactions of carbonyl compounds and imines. *Angew. Chem., Int. Ed. Engl.* **2000**, 39, 3558-3588.
73. Motoyama, Y., Koga, Y., Nishiyama, H. Asymmetric hetero Diels-Alder reaction of Danishefsky's dienes and glyoxylates with chiral bis(oxazolinyl)phenylrhodium(III) aqua complexes, and its mechanistic studies. *Tetrahedron* **2001**, 57, 853-860.
74. Gugelchuk, M. M., Wisner, J. Density Functional Study of Intermediates in the Nickel-Catalyzed Homo-Diels-Alder Reaction of Norbornadiene with Alkenes. *Organometallics* **1995**, 14, 1834-1839.

75. Yueh, W., Bauld, N. L. Mechanistic aspects of aminium salt-catalyzed Diels-Alder reactions: the substrate ionization step. *J. Phys. Org. Chem.* **1996**, 9, 529-538.
76. Branchadell, V., Font, J., Moglioni, A. G., Ochoa de Echagueen, C., Oliva, A., Ortuno, R. M., Veciana, J., Vidal-Gancedo, J. A Biradical Mechanism in the Diels-Alder Reactions of 5-Methylene-2(5H)-furanones: Experimental Evidence and Theoretical Rationalization. *J. Am. Chem. Soc.* **1997**, 119, 9992-10003.
77. Corey, E. J., Barnes-Seeman, D., Lee, T. W. The formyl C-H--O hydrogen bond as a critical factor in enantioselective reactions of aldehydes. Part 4. Aldol, ethylation, hydrocyanation and Diels-Alder reactions catalyzed by chiral B, Ti and Al Lewis acids. *Tetrahedron Lett.* **1997**, 38, 4351-4354.
78. Wender, P. A., Smith, T. E. Transition metal-catalyzed intramolecular [4 + 2] cycloadditions: mechanistic and synthetic investigations. *Tetrahedron* **1998**, 54, 1255-1275.
79. Telan, L. A., Firestone, R. A. Heavy atom effects reveal diradical intermediates. I. An aqueous Diels-Alder reaction. *Tetrahedron* **1999**, 55, 14269-14280.
80. Garcia, J. I., Mayoral, J. A., Salvatella, L. Do Secondary Orbital Interactions Really Exist? *Acc. Chem. Res.* **2000**, 33, 658-664.
81. Marchand, A. P., Chong, H.-S., Ganguly, B., Coxon, J. M. p-Facial selectivity in Diels-Alder cycloadditions. *Croatica Chemica Acta* **2000**, 73, 1027-1038.
82. Sakai, S. Theoretical Analysis of Concerted and Stepwise Mechanisms of Diels-Alder Reaction between Butadiene and Ethylene. *J. Phys. Chem. A* **2000**, 104, 922-927.
83. Christian Atherton, J. C., Jones, S. Mechanistic investigations in diastereoselective Diels-Alder additions of chiral 9-anthrylethanol derivatives. *J. Chem. Soc., Perkin Trans. 1* **2002**, 2166-2173.
84. Hermitage, S., Jay, D. A., Whiting, A. Evidence for the non-concerted [4+2]-cycloaddition of N-aryl imines when acting as both dienophiles and dienes under Lewis acid-catalyzed conditions. *Tetrahedron Lett.* **2002**, 43, 9633-9636.
85. Lightfoot, A. P., Pritchard, R. G., Wan, H., Warren, J. E., Whiting, A. A novel scandium ortho-methoxynitrosobenzene-dimer complex: mechanistic implications for the nitroso-Diels-Alder reaction. *Chem. Commun.* **2002**, 2072-2073.
86. Rodriguez, D., Navarro-Vazquez, A., Castedo, L., Dominguez, D., Saa, C. Cyclic Allene Intermediates in Intramolecular Dehydro Diels-Alder Reactions: Labeling and Theoretical Cycloaromatization Studies. *J. Org. Chem.* **2003**, 68, 1938-1946.
87. Kürti, L., Szilagyi, L., Antus, S., Nógrádi, M. Oxidation of 2-methoxyphenols with a hypervalent iodine reagent. Improved synthesis of asatone and demethoxyasatone. *Eur. J. Org. Chem.* **1999**, 2579-2581.
88. Boger, D. L., Ichikawa, S., Jiang, H. Total Synthesis of the Rubrolone Aglycon. *J. Am. Chem. Soc.* **2000**, 122, 12169-12173.
89. Lee, T. W., Corey, E. J. Enantioselective Total Synthesis of Eunicenone A. *J. Am. Chem. Soc.* **2001**, 123, 1872-1877.
90. Allen, J. G., Danishefsky, S. J. The Total Synthesis of (±)-Rishirilide B. *J. Am. Chem. Soc.* **2001**, 123, 351-352.

Dienone-Phenol Rearrangement 142

1. Andreocci, A. About two new isomers of santonin and santonin acid. *Gazz. Chim. Ital.* **1893**, 23, 468-476.
2. Auwers, K. V., Ziegler, K. Hydrocarbons of the semibenzene group. *Ann.* **1921**, 425, 217-280.
3. Clemo, G. R., Haworth, R. D., Walton, E. Constitution of santonin. II. Synthesis of racemic desmotroposantonin. *J. Chem. Soc.* **1930**, 1110-1115.
4. Selman, S., Easthan, J. F. Benzilic acid and related rearrangements. *Quart. Rev., Chem. Soc.* **1960**, 14, 221-235.
5. Collins, C. J., Eastham, J. F. Rearrangements involving the carbonyl group. in *Chem. Carbonyl Group. 1966* (ed. Patai, S.), 761-821 (Interscience Publishres, New York, **1966**).
6. Miller, B. Rearrangements of cyclohexadienones. *Mech. Mol. Migr.* **1968**, 1, 247-313.
7. Miller, B. Too many rearrangements of cyclohexadienones. *Acc. Chem. Res.* **1975**, 8, 245-256.
8. Whiting, D. A. Dienone-Phenol rearrangements and related reactions. in *Comp. Org. Synth.* (eds. Trost, B. M.,Fleming, I.), *3*, 803-821 (Pergamon Press, Oxford, **1991**).
9. Chalais, S., Laszlo, P., Mathy, A. Catalysis of the cyclohexadienone-phenol rearrangement by a Lewis-acidic clay system. *Tetrahedron Lett.* **1986**, 27, 2627-2630.
10. Reymond, J. L., Chen, Y., Lerner, R. A. *Antibody catalysis of cyclohexadienone rearrangements.* US 5500358, **1996** (Scripps Research Institute, USA).
11. Wijsman, G. W., Boesveld, W. M., Beekman, M. C., Goedheijt, M. S., Van Baar, B. L. M., De Kanter, F. J. J., De Wolf, W. H., Bickelhaupt, F. Unusual reactions of halo[5]metacyclophanes. *Eur. J. Org. Chem.* **2002**, 614-629.
12. Hemetsberger, H. Kinetic investigations and LCAO-MO calculations of the dienone-phenol rearrangement. *Monatsh. Chem.* **1968**, 99, 1724-1732.
13. Wilds, A. L., Djerassi, C. Dienone-phenol rearrangement applied to chrysene derivatives. The synthesis of 3-hydroxy-1-methylchrysene and related compounds. *J. Am. Chem. Soc.* **1946**, 68, 1715-1719.
14. Futaki, R. Tracer studies on the mechanism of the dienone-phenol rearrangement. *Tetrahedron Lett.* **1964**, 3059-3064.
15. Futaki, R. Tracer studies on the mechanism of the dienonephenol rearrangement with mineral acids. *Tetrahedron Lett.* **1967**, 2455-2458.
16. Vitullo, V. P. Cyclohexadienyl cations. II. Evidence for a protonated cyclohexadienone during the dienone-phenol rearrangement. *J. Org. Chem.* **1970**, 35, 3976-3978.
17. Vitullo, V. P., Grossman, N. Nature of rate-determining step in the dienone-phenol rearrangement. *Tetrahedron Lett.* **1970**, 1559-1562.
18. Shine, H. J., Schoening, C. E. Dienone-phenol rearrangement. So-called medium effect. *J. Org. Chem.* **1972**, 37, 2899-2901.
19. Cook, K. L., Waring, A. J. Kinetics of the dienone-phenol rearrangement of 4,4-dimethyl-2,5-cyclohexadienones. *J. Chem. Soc., Perkin Trans. 2* **1973**, 88-92.
20. Vitullo, V. P., Logue, E. A. Cyclohexadienyl cation. V. Acidity dependence of the dienone-phenol rearrangement. *J. Org. Chem.* **1973**, 38, 2265-2267.
21. Hughes, M. J., Waring, A. J. Kinetics of the dienone-phenol rearrangement and basicity studies of cyclohexa-2,5-dienones. *J. Chem. Soc., Perkin Trans. 2* **1974**, 1043-1051.
22. Jacquesy, J. C., Jacquesy, R., Ly, U. H. Hyperacid media. Dienone-phenol and phenol-phenol rearrangements. *Tetrahedron Lett.* **1974**, 2199-2202.
23. Vitullo, V. P., Logue, E. A. Methyl-trideuteriomethyl isotope effects in the acid catalyzed dienone-phenol rearrangement. *J. Chem. Soc., Chem. Commun.* **1974**, 228-229.
24. Suehiro, T., Yamazaki, S. Establishing the migration of ethoxycarbonyl residues in dienone-phenol rearrangement. *Bull. Chem. Soc. Jpn.* **1975**, 48, 3655-3659.
25. Pilkington, J. W., Waring, A. J. Cyclohexadienones. Use of the dienone-phenol rearrangement in measuring migratory aptitudes of alkyl groups. *J. Chem. Soc., Perkin Trans. 2* **1976**, 1349-1359.
26. Vitullo, V. P., Logue, E. A. Cyclohexadienyl cations. 6. Methyl group isotope effects in the dienone-phenol rearrangement. *J. Am. Chem. Soc.* **1976**, 98, 5906-5909.
27. Palmer, J. D., Waring, A. J. The migratory aptitude of the sec-butyl group in a cationic rearrangement. *J. Chem. Soc., Perkin Trans. 2* **1979**, 1089-1092.
28. Waring, A. J., Zaidi, J. H., Pilkington, J. W. Dienone-phenol rearrangements of bicyclic cyclohexa-2,5-dien-1-ones; kinetic studies of the importance of a multistage mechanism. *J. Chem. Soc., Perkin Trans. 2* **1981**, 935-939.
29. Guo, Z., Schultz, A. G. Preparation and Photochemical Rearrangements of 2-Phenyl-2,5-cyclohexadien-1-ones. An Efficient Route to Highly Substituted Phenols. *Org. Lett.* **2001**, 3, 1177-1180.
30. Parker, K. A., Koh, Y.-h. Methodology for the Regiospecific Synthesis of Bis C-Aryl Glycosides. Models for Kidamycins. *J. Am. Chem. Soc.* **1994**, 116, 11149-11150.
31. Hart, D. J., Kim, A., Krishnamurthy, R., Merriman, G. H., Waltos, A. M. Synthesis of 6H-dibenzo[b,d]pyran-6-ones via dienone-phenol rearrangements of spiro[2,5-cyclohexadiene-1,1'(3'H)-isobenzofuran]-3'-ones. *Tetrahedron* **1992**, 48, 8179-8188.

1. Rathke, B. Monophenyl isocyanuric acid. *Ber.Dtsch.Chem.Ges.* **1888**, 21, 867-877.
2. Dimroth, O. Intramolecular Rearrangements. *Ann.* **1909**, 364, 183-226.
3. Dimroth, O., Michaelis, W. Intramolecular rearrangement of 5-amino-1,2,3-triazole. *Ann.* **1927**, 459, 39-46.
4. Brown, D. J., Harper, J. S. Ease of rearrangement of aminopteridines and aminopyrimidines alkylated on the ring nitrogen. *Pteridine Chem., Proc. Intern. Symp., 3rd, Stuttgart* **1964**, 1962, 219-231,discussion 231-212.
5. Brown, D. J. Amidine rearrangements (the Dimroth rearrangements). *Mech. Mol. Migr.* **1968**, 1, 209-245.
6. Brown, D. J. *The Pyrimidines, Supplement 1 (The Chemistry of Heterocyclic Compounds, Vol. 16)* (**1970**) 1127 pp.
7. L'Abbe, G. Dimroth reaction. *Ind. Chim. Belge* **1971**, 36, 3-10.
8. Fujii, T., Itaya, T., Saito, T. Base-catalyzed ring opening and reclosure of the adenine ring: mechanism, substituent effects, and synthetic utility. *Symp. Heterocycl., [Pap.]* **1977**, 129-134.
9. Fujii, T., Itaya, T. The Dimroth rearrangement in the adenine series: a review updated. *Heterocycles* **1998**, 48, 359-390.
10. El Ashry, E. S. H., El Kilany, Y., Rashed, N., Assafir, H. Dimroth rearrangement: Translocation of heteroatoms in heterocyclic rings and its role in ring transformations of heterocycles. *Adv. Heterocycl. Chem.* **1999**, 75, 79-167.
11. Mint Tho, N., Leroy, G., Sana, M., Elguero, J. Reaction mechanism of the Dimroth rearrangement. Ab initio study. *J. Heterocycl. Chem.* **1982**, 19, 943-944.
12. Brown, D. J., Hoerger, E., Mason, S. F. Simple pyrimidines. III. Methylation and structure of the aminopyrimidines. *J. Chem. Soc.* **1955**, 4035-4040.
13. Carrington, H. C., Curd, F. H. S., Richardson, D. N. The synthesis of trypanocides. V. Rearrangement of some 6-amino-1-methylpyrimidinium salts and synthesis of 4-amino-1,2-dimethyl-6-(1,2-dimethyl-6-methylaminopyrimidinium-4-amino)quinolinium diiodide. *J. Chem. Soc.* **1955**, 1858-1862.
14. Brown, D. J., Harper, J. S. The Dimroth rearrangement. I. Alkylated 2-iminopyrimidines. *J. Chem. Soc.* **1963**, 1276-1284.
15. L'Abbe, G., Vanderstede, E. Dimroth rearrangement of 5-hydrazino-1,2,3-thiadiazoles. *J. Heterocycl. Chem.* **1989**, 26, 1811-1814.
16. Nagamatsu, T., Fujita, T. The first reliable, general synthesis of the 5-oxo derivatives of 5,6-dihydro-1,2,4-triazolo[4,3-c]pyrimidine and the rates of isomerization of the [4,3-c] compounds into their [1,5-c] isomers. *Heterocycles* **2002**, 57, 631-636.
17. Loakes, D., Brown, D. M., Salisbury, S. A. Cyclization and rearrangement of N4-acylaminodeoxycytidines. *Tetrahedron Lett.* **1998**, 39, 3865-3868.
18. Loakes, D., Brown, D. M., Salisbury, S. A. A Dimroth rearrangement of pyrimidine nucleosides. *J. Chem. Soc., Perkin Trans. 1* **1999**, 1333-1338.
19. Ogata, Y., Takagi, K., Hayashi, E. Photochemical Dimroth rearrangement of 1,4-diphenyl-5-amino- and 4-phenyl-5-anilino-1,2,3-triazoles. *Bull. Chem. Soc. Jpn.* **1977**, 50, 2505-2506.
20. Fanghaenel, E., Kordts, B., Richter, A. M., Dutschmann, K. Lewis-acid and photochemically induced Dimroth rearrangement of 3H,6H-2,5-bis(p-N,N-dimethylaminophenyl)1,2-thiazolino[5,4-d]1,2-thiazoline-3,6-dithione. *J. Prakt. Chem.* **1990**, 332, 387-393.
21. Guerret, P., Jacquier, R., Maury, G. Minimal structural conditions for the Dimroth-type rearrangement in the polyazaindolizine series. *J. Heterocycl. Chem.* **1971**, 8, 643-650.
22. Brown, D. J., Nagamatsu, T. Isomerizations akin to the Dimroth rearrangement. III. The conversion of simple s-triazolo[4,3-a]pyrimidines into their [1,5-a] isomers. *Aust. J. Chem.* **1977**, 30, 2515-2525.
23. Perrin, D. D. The Dimroth rearrangement. II. Kinetic studies. *J. Chem. Soc.* **1963**, 1284-1290.
24. Perrin, D. D., Pitman, I. H. The Dimroth rearrangement. V. The mechanism of the rearrangement of 1-alkyl-1,2-dihydro-2-iminopyrimidines in aqueous solution. *J. Chem. Soc., Abstracts* **1965**, 7071-7082.
25. Itaya, T., Hozumi, Y., Kanai, T., Ohta, T. Syntheses of the marine ascidian purine aplidiamine and its 9-β-D-ribofuranoside. *Tetrahedron Lett.* **1998**, 39, 4695-4696.
26. Pagano, A. R., Zhao, H., Shallop, A., Jones, R. A. Syntheses of [1,7-15N_2]- and [1,7,NH_2-15N_3]Adenosine and 2'-Deoxyadenosine via an N^1-Alkoxy-Mediated Dimroth Rearrangement. *J. Org. Chem.* **1998**, 63, 3213-3217.
27. Volkova, N. N., Tarasov, E. V., Van Meervelt, L., Toppet, S., Dehaen, W., Bakulev, V. A. Reaction of 5-halo-1,2,3-thiadiazoles with arylenediamines as a new approach to tricyclic 1,3,6-thiadiazepines. *J. Chem. Soc., Perkin Trans. 1* **2002**, 1574-1580.

1. Doering, W. v. E., Hoffmann, A. K. The addition of dichlorocarbene to olefins. *J. Am. Chem. Soc.* **1954**, 76, 6162-6165.
2. Doering, W. v. E., LaFlamme, P. The cis addition of dibromocarbene and methylene to cis- and trans-butene. *J. Am. Chem. Soc.* **1956**, 78, 5447-5448.
3. v. E. Doering, W., LaFlamme, P. M. A two-step synthesis of Allenes from olefins. *Tetrahedron* **1958**, 2, 75-79.
4. Hopf, H. The preparation of allenes and cumulenes. in *The Chemistry of Ketenes, Allenes and Related Compounds* (ed. Patai, S.), *2*, 779-901 (John Wiley & Sons, New York, **1980**).
5. Kostikov, R. R., Molchanov, A. P., Hopf, H. Gem-dihalocyclopropanes in organic synthesis. *Top. Curr. Chem.* **1990**, 155, 41-80.
6. Banwell, M. G., Reum, M. E. gem-Dihalocyclopropanes in chemical synthesis. *Adv. Strain Org. Chem.* **1991**, 1, 19-64.
7. Fedorynski, M. Syntheses of gem-Dihalocyclopropanes and Their Use in Organic Synthesis. *Chem. Rev.* **2003**, 103, 1099-1132.
8. Moore, W. R., Ward, H. R. Reactions of gem-dibromocyclopropanes with alkyllithium reagents. Formation of allenes, spiropentanes, and a derivative of bicyclopropylidene. *J. Org. Chem.* **1960**, 25, 2073.
9. Moore, W. R., Ward, H. R. Formation of allenes from gem-dihalocyclopropanes by reaction with alkyllithium reagents. *J. Org. Chem.* **1962**, 27, 4179-4181.
10. Baird, M. S., Nizovtsev, A. V., Bolesov, I. G. Bromine-magnesium exchange in gem-dibromocyclopropanes using Grignard reagents. *Tetrahedron* **2002**, 58, 1581-1593.
11. Moore, W. R., Hill, J. B. Competitive bicyclobutane and allene formation from phenyl-substituted gem-dibromocyclopropanes. *Tetrahedron Lett.* **1970**, 4553-4556.
12. Lilje, K. C., Macomber, R. S. tert-Butylallene. Reversibility of carbenoid formation. *J. Org. Chem.* **1974**, 39, 3600-3601.
13. Creary, X., Jang, Z., Butchko, M., McLean, K. Silyl-substituted cyclopropyl carbenoids. *Tetrahedron Letters* **1996**, 37, 579-582.
14. De Meijere, A., Faber, D., Heinecke, U., Walsh, R., Muller, T., Apeloig, Y. On the question of cyclopropylidene intermediates in cyclopropene-to-allene rearrangements - tetrakis(trimethylsilyl)cyclopropene, 3-alkenyl-1,2,3-tris(trimethylsilyl)cyclopropenes, and related model compounds. *Eur. J. Org. Chem.* **2001**, 663-680.
15. Ekhato, I. V., Robinson, C. H. Synthesis of novel 4a-substituted sterols. *J. Org. Chem.* **1989**, 54, 1327-1331.
16. de Meijere, A., von Seebach, M., Zollner, S., Kozhushkov, S. I., Belov, V. N., Boese, R., Haumann, T., Benet-Buchholz, J., Yufit, D. S., Howard, J. A. K. Spirocyclopropanated bicyclopropylidenes: straightforward preparation, physical properties, and chemical transformations. *Chem.-- Eur. J.* **2001**, 7, 4021-4034.
17. Lahrech, M., Hacini, S., Parrain, J.-L., Santelli, M. A general synthesis of β-silylallenes from allylsilanes. *Tetrahedron Lett.* **1997**, 38, 3395-3398.
18. Braverman, S., Duar, Y. Thermal rearrangements of allenes. Synthesis and mechanisms of cycloaromatization of π and heteroatom bridged diallenes. *J. Am. Chem. Soc.* **1990**, 112, 5830-5837.

Dötz Benzannulation Reaction 148

Related reactions: Bergman cycloaromatization reaction, Danheiser benzannulation;

1. Dötz, K. H. Synthesis of the naphthol skeleton from pentacarbonyl[methoxy(phenyl)carbene]chromium(0) and tolan. *Angew. Chem.* **1975**, 87, 672-673.
2. Dötz, K. H. Synthesis of Naphthol Skeleton from Pentacarbonyl Methoxy(Phenyl)Carbene Chromium(0) and Tolan. *Angew. Chem., Int. Ed. Engl.* **1975**, 14, 644-645.
3. Semmelhack, M. F., Tamura, R., Schnatter, W., Park, J., Steigerwald, M., Ho, S. Carbene-metal complexes. New processes and applications in organic synthesis. *Stud. Org. Chem. (Amsterdam)* **1986**, 25, 21-42.
4. Schore, N. E. Transition metal-mediated cycloaddition reactions of alkynes in organic synthesis. *Chem. Rev.* **1988**, 88, 1081-1119.
5. Dötz, K. H. Carbene complexes in stereoselective cycloaddition reactions. *New J. Chem.* **1990**, 14, 433-445.
6. Hua, D. H., Saha, S. Gilvocarcins. *Recl. Trav. Chim. Pays-Bas* **1995**, 114, 341-355.
7. de Meijere, A. β-Aminosubstituted α,β-unsaturated Fischer carbene complexes as chemical multitalents. *Pure Appl. Chem.* **1996**, 68, 61-72.
8. Harvey, D. F., Sigano, D. M. Carbene-Alkyne-Alkene Cyclization Reactions. *Chem. Rev.* **1996**, 96, 271-288.
9. Bernasconi, C. F. Developing the physical organic chemistry of Fischer carbene complexes. *Chem. Soc. Rev.* **1997**, 26, 299-308.
10. Frenking, G., Pidun, U. Ab initio studies of transition-metal compounds: the nature of the chemical bond to a transition metal. *J. Chem. Soc., Dalton Trans.* **1997**, 1653-1662.
11. Alcaide, B., Casarrubios, L., Dominguez, G., Sierra, M. A. Reactions of group 6 metal carbene complexes with ylides and related dipolar species. *Curr. Org. Chem.* **1998**, 2, 551-574.
12. Wulff, W. D. Asymmetric Synthesis with Fischer Carbene Complexes: The Development of Imidazolidinone and Oxazolidinone Complexes. *Organometallics* **1998**, 17, 3116-3134.
13. Barluenga, J. Recent advances in selective organic synthesis mediated by transition metal complexes. *Pure Appl. Chem.* **1999**, 71, 1385-1391.
14. Dötz, K. H., Tomuschat, P. Annulation reactions of chromium carbene complexes: scope, selectivity and recent developments. *Chem. Soc. Rev.* **1999**, 28, 187-198.
15. Aumann, R. 1-Metalla-1,3,5-hexatrienes and related compounds. *Eur. J. Org. Chem.* **2000**, 17-31.
16. Barluenga, J., Fananas, F. J. Metalloxy Fischer Carbene Complexes: An Efficient Strategy to Modulate Their Reactivity. *Tetrahedron* **2000**, 56, 4597-4628.
17. De Meijere, A., Schirmer, H., Duetsch, M. Fischer carbene complexes as chemical multitalents: the incredible range of products from carbenepentacarbonylmetal α,β-unsaturated complexes. *Angew. Chem., Int. Ed. Engl.* **2000**, 39, 3964-4002.
18. Sierra, M. A. Di- and Polymetallic Heteroatom Stabilized (Fischer) Metal Carbene Complexes. *Chem. Rev.* **2000**, 100, 3591-3637.
19. Barluenga, J., Florez, J., Fananas, F. J. Carbon nucleophile addition to sp^2-unsaturated Fischer carbene complexes. *J. Organomet. Chem.* **2001**, 624, 5-17.
20. Dötz, K. H., Stendel, J., Jr. The chromium-templated carbene benzannulation approach to densely functionalized arenes (Dötz reaction). *Modern Arene Chemistry* **2002**, 250-296.
21. Merlic, C. A., Burns, E. E., Xu, D., Chen, S. Y. Aminobenzannulation via metathesis of isonitriles using chromium carbene complexes. *J. Am. Chem. Soc.* **1992**, 114, 8722-8724.
22. Merlic, C. A., Xu, D., Gladstone, B. G. Aminobenzannulation via photocyclization reactions of chromium dienyl(amino)carbene complexes. Synthesis of o-amino aromatic alcohols. *J. Org. Chem.* **1993**, 58, 538-545.
23. Barluenga, J., Aznar, F., Palomero, M. A. Eight-membered carbocycles from a Dötz-like reaction. *Angew. Chem., Int. Ed. Engl.* **2000**, 39, 4346-4348.
24. Caldwell, J. J., Colman, R., Kerr, W. J., Magennis, E. J. Novel use of a selenoalkyne within untraditionally mild Dötz benzannulation processes; total synthesis of a Calceolaria andina L. natural hydroxylated naphthoquinone. *Synlett* **2001**, 1428-1430.
25. Torrent, M. Novel mechanistic proposal for the Dötz reaction derived from a density functional study: the chromahexatriene route. *Chem. Commun.* **1998**, 999-1000.
26. Torrent, M., Duran, M., Sola, M. Density Functional Study on the Preactivation Scenario of the Dötz Reaction: Carbon Monoxide Dissociation versus Alkyne Addition as the First Reaction Step. *Organometallics* **1998**, 17, 1492-1501.
27. Torrent, M., Duran, M., Sola, M. Weighing Different Mechanistic Proposals for the Dötz Reaction: A Density Functional Study. *J. Am. Chem. Soc.* **1999**, 121, 1309-1316.
28. Torrent, M., Sola, M., Frenking, G. Theoretical Studies of Some Transition-Metal-Mediated Reactions of Industrial and Synthetic Importance. *Chem. Rev.* **2000**, 100, 439-493.
29. Chan, K. S., Peterson, G. A., Brandvold, T. A., Faron, K. L., Challener, C. A., Hyldahl, C., Wulff, W. D. Solvent, chelation and concentration effects on the benzannulation reaction of chromium carbene complexes and acetylenes. *J. Organomet. Chem.* **1987**, 334, 9-56.
30. Bos, M. E., Wulff, W. D., Miller, R. A., Chamberlin, S., Brandvold, T. A. Substrate regulation of product distribution in the reactions of arylchromium carbene complexes with alkynes. *J. Am. Chem. Soc.* **1991**, 113, 9293-9319.
31. Hofmann, P., Hammerle, M. The Mechanism of the Dötz Reaction - Chromacyclobutenes by Alkyne-Carbene Coupling. *Angew. Chem., Int. Ed. Engl.* **1989**, 28, 908-910.
32. Knorr, J. R., Brown, T. L. Low-Temperature Infrared Spectral Study of the Photochemical Reaction of $(CO)_5Cr$:C(OMe)Ph with Solvent and Alkynes. *Organometallics* **1994**, 13, 2178-2185.
33. Gleichmann, M. M., Dötz, K. H., Hess, B. A. Intermediates and Transition Structures of the Benzannulation of Heteroatom-Stabilized Chromium Carbene Complexes with Ethyne: A Density Functional Study. *J. Am. Chem. Soc.* **1996**, 118, 10551-10560.
34. Waters, M. L., Bos, M. E., Wulff, W. D. Mechanistic Studies on the Reaction of Fischer Carbene Complex with Alkynes: Does the Alkyne Insertion Intermediate Form Irreversibly? *J. Am. Chem. Soc.* **1999**, 121, 6403-6413.
35. Sola, M., Duran, M., Torrent, M. The Dötz reaction: a chromium fischer carbene-mediated benzannulation reaction. *Catal. Metal Compl.* **2002**, 25, 269-287.
36. Dötz, K. H. Vinyl ketenes. 1. Stable vinyl ketenes by metal complex-induced olefination and carbonylation of alkynes. *Angew. Chem.* **1979**, 91, 1021-1022.
37. Dötz, K. H., Fuegen-Koester, B. Vinylketenes. II. Stable silyl-substituted vinylketenes. *Chem. Ber.* **1980**, 113, 1449-1457.
38. Casey, C. P. Metal-carbene complexes [as reactive intermediates]. *Reactive Intermediates (Wiley)* **1981**, 2, 135-174.
39. Tang, P. C., Wulff, W. D. Cyclohexadienone annulation via α,β-unsaturated Fischer carbene complexes. *J. Am. Chem. Soc.* **1984**, 106, 1132-1133.
40. Merlic, C. A., Aldrich, C. C., Albaneze-Walker, J., Saghatelian, A., Mammen, J. Total Synthesis of the Calphostins: Application of Fischer Carbene Complexes and Thermodynamic Control of Atropisomers. *J. Org. Chem.* **2001**, 66, 1297-1309.
41. King, J., Quayle, P., Malone, J. F. A new synthesis of 12-O-methylroyleanone. *Tetrahedron Lett.* **1990**, 31, 5221-5224.
42. Pulley, S. R., Carey, J. P. C-Aryl Glycosides via a Benzannulation Mediated by Fischer Chromium Carbene Complexes. *J. Org. Chem.* **1998**, 63, 5275-5279.

Enders SAMP/RAMP Hydrazone Alkylation 150

Related reactions: Myers' asymmetric alkylation,

1. Enders, D., Eichenauer, H. Asymmetric synthesis of α-substituted ketones by metalation and alkylation of chiral hydrazones. *Angew. Chem. Int. Ed. Engl.* **1976**, 15, 549-550.

2. Enders, D., Eichenauer, H. Enantioselective alkylation of aldehydes via metalated chiral hydrazones. *Tetrahedron Lett.* **1977**, 191-194.
3. Davenport, K. G., Eichenauer, H., Enders, D., Newcomb, M., Bergbreiter, D. E. Stereoselective formation and electrophilic substitution of aldehyde hydrazone lithio anions. *J. Am. Chem. Soc.* **1979**, 101, 5654-5659.
4. Enders, D., Eichenauer, H. Asymmetric synthesis of ant alarm pheromones - α-alkylation of acyclic ketones with practically complete asymmetric induction. *Angew. Chem. Int. Ed. Engl.* **1979**, 18, 397.
5. Enders, D., Eichenauer, H., Baus, U., Schubert, H., Kremer, K. A. M. Asymmetric syntheses via metalated chiral hydrazones. Overall enantioselective α-alkylation of acyclic ketones. *Tetrahedron* **1984**, 40, 1345-1359.
6. Enders, D. Alkylation of chiral hydrazones. *Asymmetric Synth.* **1984**, 3, 275-339.
7. Enders, D., Kipphardt, H. Asymmetric synthesis. (S)-2-methoxymethylpyrrolidine - a chiral auxiliary. *Nachrichten aus Chemie, Technik und Laboratorium* **1985**, 33, 882-888.
8. Enders, D. SADP and SAEP. Novel chiral hydrazine auxiliaries for asymmetric synthesis. *Acros Organics Acta* **1995**, 1, 35-36.
9. Enders, D., Bettray, W. Recent advances in the development of highly enantioselective synthetic methods. *Pure Appl. Chem.* **1996**, 68, 569-580.
10. Enders, D., Klatt, M. Asymmetric synthesis with (S)-2-methoxymethylpyrrolidine (SMP). A pioneer auxiliary. *Synthesis* **1996**, 1403-1418.
11. Enders, D., Bettray, W., Schankat, J., Wiedemann, J. Diastereo- and enantioselective synthesis of β-amino acids via SAMP hydrazones and hetero Michael addition using TMS-SAMP as a chiral equivalent of ammonia. *Enantioselective Synthesis of β-Amino Acids* **1997**, 187-210.
12. Enders, D., Bolkenius, M., Vazquez, J., Lassaletta, J. M., Fernandez, R. Formaldehyde SAMP-hydrazone. A neutral chiral formyl anion and cyanide equivalent. *J. Prakt. Chem.* **1998**, 340, 281-285.
13. Enders, D., Wortmann, L., Peters, R. Recovery of Carbonyl Compounds from N,N-Dialkylhydrazones. *Acc. Chem. Res.* **2000**, 33, 157-169.
14. Job, A., Janeck, C. F., Bettray, W., Peters, R., Enders, D. The SAMP/RAMP-hydrazone methodology in asymmetric synthesis. *Tetrahedron* **2002**, 58, 2253-2329.
15. Alam, M. M. (S)-(-)-1-amino-2-methoxypyrrolidine (SAMP) and (R)-(+)-1-Amino-2-methoxypyrrolidine (RAMP) as versatile chiral auxiliaries. *Synlett* **2003**, 1755-1756.
16. Enders, D., Kipphardt, H., Gerdes, P., Brena-Valle, L. J., Bhushan, V. Large-scale preparation of versatile chiral auxiliaries derived from (S)-proline. *Bull. Soc. Chim. Belg.* **1988**, 97, 691-704.
17. Martens, J., Luebben, S. (1S,3S,5S)-2-amino-3-methoxymethyl-2-azabicyclo[3.3.0]octane: SAMBO, a new chiral auxiliary. *Liebigs Ann. Chem.* **1990**, 949-952.
18. Wilken, J., Thorey, C., Groger, H., Haase, D., Saak, W., Pohl, S., Muzart, J., Martens, J. Utilization of industrial waste materials. Part 11. Synthesis of new, chiral β-sec-amino alcohols. Diastereodivergent addition of Grignard reagents to α-amino aldehydes based on the (all-R)-2-azabicyclo[3.3.0]octane system. *Liebigs Ann. Chem.* **1997**, 2133-2146.
19. Enders, D., Schubert, H. Enantioselective synthesis of β-substituted primary amines, α-alkylation/reductive amination of aldehydes via SAMP hydrazones. *Angew. Chem., Int. Ed. Engl.* **1984**, 23, 365-366.
20. Enders, D., Plant, A. Enantioselective synthesis of α-substituted nitriles by oxidative cleavage of aldehyde SAMP-hydrazones with magnesium monoperoxyphthalate. *Synlett* **1994**, 1054-1056.
21. Diez, E., Lopez, A. M., Pareja, C., Martin, E., Fernandez, R., Lassaletta, J. M. Direct synthesis of dithioketals from N,N-dialkylhydrazones. *Tetrahedron Lett.* **1998**, 39, 7955-7958.
22. Enders, D., Eichenauer, H. Asymmetric syntheses via metalated chiral hydrazones. Enantioselective alkylation of cyclic ketones and aldehydes. *Chem. Ber.* **1979**, 112, 2933-2960.
23. Enders, D., Eichenauer, H., Pieter, R. Enantioselective synthesis of (-)-R- and (+)-S-[6]-gingerol - aromatic principle of ginger. *Chem. Ber.* **1979**, 112, 3703-3714.
24. Enders, D., Fey, P., Kipphardt, H. Efficient preparation of the chiral auxiliaries SAMP and RAMP. N-Amination via Hofmann degradation. *Org. Prep. Proced. Int.* **1985**, 17, 1-9.
25. Enders, D., Eichenauer, H., Brauer, S., Baus, U., Andrade, J., Schleyer, P. V. R. University of Bonn. Unpublished results.
26. Ahlbrecht, H., Dueber, E. O., Enders, D., Eichenauer, H., Weuster, P. NMR spectroscopic investigation of deprotonation of imines and hydrazones. *Tetrahedron Lett.* **1978**, 3691-3694.
27. Enders, D., Baus, U. Asymmetric synthesis of both enantiomers of (E)-4,6-dimethyl-6-octen-3-one, the defensive substance of daddy longlegs, Leiobunum vittatum and L. calcar (Opiliones). *Liebigs Ann. Chem.* **1983**, 1439-1445.
28. Enders, D., Bachstaedter, G., Kremer, K. A. M., Marsch, M., Harms, K., Boche, G. The structure of a chiral lithium azaenolate: monomeric intramolecular chelated lithio-2-acetylnaphthalene-SAMP-hydrazone. *Angew. Chem., Int. Ed. Engl.* **1988**, 27, 1522-1524.
29. Bauer, W., Seebach, D. Determination of the degree of aggregation of organolithium compounds by cryoscopy in tetrahydrofuran. *Helv. Chim. Acta* **1984**, 67, 1972-1988.
30. Schwaebe, M., Little, R. D. Asymmetric Reductive Cyclization. Total Synthesis of (-)-C10-Desmethy-Arteannuin B. *J. Org. Chem.* **1996**, 61, 3240-3244.
31. Ziegler, F. E., Becker, M. R. Total synthesis of (-)-denticulatins A and B: marine polypropionates from Siphonaria denticulata. *J. Org. Chem.* **1990**, 55, 2800-2805.
32. Toro, A., Nowak, P., Deslongchamps, P. Transannular Diels-Alder Entry into Stemodanes: First Asymmetric Total Synthesis of (+)-Maritimol. *J. Am. Chem. Soc.* **2000**, 122, 4526-4527.
33. Enders, D., Hundertmark, T., Lampe, C., Jegelka, U., Scharfbillig, I. Highly diastereo- and enantioselective synthesis of protected anti-1,3-diols. *Eur. J. Org. Chem.* **1998**, 2839-2849.
34. Enders, D., Hundertmark, T. Asymmetric synthesis of (+)- and (-)-streptenol A. *Eur. J. Org. Chem.* **1999**, 751-756.

Enyne Metathesis..152

1. Katz, T. J., Sivavec, T. M. Metal-catalyzed rearrangement of alkene-alkynes and the stereochemistry of metallacyclobutene ring opening. *J. Am. Chem. Soc.* **1985**, 107, 737-738.
2. Korkowski, P. F., Hoye, T. R., Rydberg, D. B. Fischer carbene-mediated conversions of enynes to bi- and tricyclic cyclopropane-containing carbon skeletons. *J. Am. Chem. Soc.* **1988**, 110, 2676-2678.
3. Hoye, T. R., Suriano, J. A. Reactions of pentacarbonyl(1-methoxyethylidene)molybdenum and -tungsten with α,ω-enynes: comparison with the chromium analog and resulting mechanistic ramifications. *Organometallics* **1992**, 11, 2044-2050.
4. Kim, S.-H., Bowden, N., Grubbs, R. H. Catalytic Ring Closing Metathesis of Dienynes: Construction of Fused Bicyclic Rings. *J. Am. Chem. Soc.* **1994**, 116, 10801-10802.
5. Kinoshita, A., Mori, M. Ruthenium catalyzed enyne metathesis. *Synlett* **1994**, 1020-1022.
6. Katz, T. J. Reactions of acetylenes and alkenes induced by catalysts of olefin metathesis. *NATO ASI Ser., Ser. C* **1989**, 269, 293-304.
7. Trost, B. M. Palladium-catalyzed cycloisomerizations of enynes and related reactions. *Acc. Chem. Res.* **1990**, 23, 34-42.
8. Trost, B. M. Transition metal-catalyzed cycloisomerizations of enynes. *Janssen Chimica Acta* **1991**, 9, 3-9.
9. Grubbs, R. H., Miller, S. J., Fu, G. C. Ring-Closing Metathesis and Related Processes in Organic Synthesis. *Acc. Chem. Res.* **1995**, 28, 446-452.
10. Ivin, K. J. Some recent applications of olefin metathesis in organic synthesis: A review. *J. Mol. Catal. A: Chemical* **1998**, 133, 1-16.
11. Mori, M. Enyne metathesis. in *Top. Organomet. Chem.* (eds. Fürstner, A.,Gibson, S. E.), *1*, 133-154 (Springer, Berlin, New York, **1998**).
12. Mori, M., Kitamura, T., Sato, Y. Synthesis of medium-sized ring compounds using enyne metathesis. *Synthesis* **2001**, 654-664.
13. Aubert, C., Buisine, O., Malacria, M. The Behavior of 1,*n*-Enynes in the Presence of Transition Metals. *Chem. Rev.* **2002**, 102, 813-834.
14. Semeril, D., Bruneau, C., Dixneuf, P. H. Imidazolium and imidazolinium salts as carbene precursors or solvents for ruthenium-catalyzed diene and enyne metathesis. *Adv. Syn. & Catal.* **2002**, 344, 585-595.
15. Poulsen, C. S., Madsen, R. Enyne metathesis catalyzed by ruthenium carbene complexes. *Synthesis* **2003**, 1-18.

16. Diver, S. T., Giessert, A. J. Enyne metathesis (enyne bond reorganization). *Chem. Rev.* **2004**, 104, 1317-1382.
17. Kim, S.-H., Zuercher, W. J., Bowden, N. B., Grubbs, R. H. Catalytic Ring Closing Metathesis of Dienynes: Construction of Fused Bicyclic [n.m.0] Rings. *J. Org. Chem.* **1996**, 61, 1073-1081.
18. Yao, Q. Rapid Assembly of Structurally Defined and Highly Functionalized Conjugated Dienes via Tethered Enyne Metathesis. *Org. Lett.* **2001**, 3, 2069-2072.
19. Hansen, E. C., Lee, D. Enyne Metathesis for the Formation of Macrocyclic 1,3-Dienes. *J. Am. Chem. Soc.* **2003**, 125, 9582-9583.
20. Kang, B., Kim, D.-H., Do, Y., Chang, S. Conjugated Enynes as a New Type of Substrates for Olefin Metathesis. *Org. Lett.* **2003**, 5, 3041-3043.
21. Kulkarni, A. A., Diver, S. T. Cycloheptadiene Ring Synthesis by Tandem Intermolecular Enyne Metathesis. *Org. Lett.* **2003**, 5, 3463-3466.
22. Lee, H.-Y., Kim, B. G., Snapper, M. L. A Stereoselective Enyne Cross Metathesis. *Org. Lett.* **2003**, 5, 1855-1858.
23. Royer, F., Vilain, C., Elkaiem, L., Grimaud, L. Selective Domino Ring-Closing Metathesis-Cross-Metathesis Reactions between Enynes and Electron-Deficient Alkenes. *Org. Lett.* **2003**, 5, 2007-2009.
24. Cavallo, L. Mechanism of Ruthenium-Catalyzed Olefin Metathesis Reactions from a Theoretical Perspective. *J. Am. Chem. Soc.* **2002**, 124, 8965-8973.
25. Vyboishchikov, S. F., Buhl, M., Thiel, W. Mechanism of olefin metathesis with catalysis by ruthenium carbene complexes: density functional studies on model systems. *Chem.-- Eur. J.* **2002**, 8, 3962-3975.
26. Stragies, R., Schuster, M., Blechert, S. A crossed yne-ene metathesis showing atom economy. *Angew. Chem., Int. Ed. Engl.* **1997**, 36, 2518-2520.
27. Fürstner, A., Stelzer, F., Szillat, H. Platinum-Catalyzed Cycloisomerization Reactions of Enynes. *J. Am. Chem. Soc.* **2001**, 123, 11863-11869.
28. Chatani, N., Morimoto, T., Muto, T., Murai, S. Highly Selective Skeletal Reorganization of 1,6- and 1,7-Enynes to 1-Vinylcycloalkenes Catalyzed by $[RuCl_2(CO)_3]_2$. *J. Am. Chem. Soc.* **1994**, 116, 6049-6050.
29. Chatani, N., Inoue, H., Morimoto, T., Muto, T., Murai, S. Iridium(I)-Catalyzed Cycloisomerization of Enynes. *J. Org. Chem.* **2001**, 66, 4433-4436.
30. Fürstner, A., Ackermann, L., Gabor, B., Goddard, R., Lehmann, C. W., Mynott, R., Stelzer, F., Thiel, O. R. Comparative investigation of ruthenium-based metathesis catalysts bearing N-heterocyclic carbene (NHC) ligands. *Chem.-- Eur. J.* **2001**, 7, 3236-3253.
31. Kitamura, T., Sato, Y., Mori, M. Effects of substituents on the multiple bonds on ring-closing metathesis of enynes. *Adv. Syn. & Catal.* **2002**, 344, 678-693.
32. Mori, M., Sakakibara, N., Kinoshita, A. Remarkable effect of ethylene gas in the intramolecular enyne metathesis of terminal alkynes. *J. Org. Chem.* **1998**, 63, 6082-6083.
33. Kitamura, T., Mori, M. Ruthenium-catalyzed ring-opening and ring-closing enyne metathesis. *Org. Lett.* **2001**, 3, 1161-1163.
34. Trost, B. M., Trost, M. K. Mechanistic dichotomies in palladium catalyzed enyne metathesis of cyclic olefins. *Tetrahedron Lett.* **1991**, 32, 3647-3650.
35. Trost, B. M., Chang, V. K. An approach to botrydianes: on the steric demands of a metal catalyzed enyne metathesis. *Synthesis* **1993**, 824-832.
36. Trost, B. M., Yanai, M., Hoogsteen, K. A palladium-catalyzed [2 + 2] cycloaddition. Mechanism of a Pd-catalyzed enyne metathesis. *J. Am. Chem. Soc.* **1993**, 115, 5294-5295.
37. Fürstner, A., Szillat, H., Gabor, B., Mynott, R. Platinum- and Acid-Catalyzed Enyne Metathesis Reactions: Mechanistic Studies and Applications to the Syntheses of Streptorubin B and Metacycloprodigiosin. *J. Am. Chem. Soc.* **1998**, 120, 8305-8314.
38. Zuercher, W. J., Scholl, M., Grubbs, R. H. Ruthenium-Catalyzed Polycyclization Reactions. *J. Org. Chem.* **1998**, 63, 4291-4298.
39. Fürstner, A., Szillat, H., Stelzer, F. Novel Rearrangements of Enynes Catalyzed by $PtCl_2$. *J. Am. Chem. Soc.* **2000**, 122, 6785-6786.
40. Sanford, M. S., Love, J. A., Grubbs, R. H. Mechanism and Activity of Ruthenium Olefin Metathesis Catalysts. *J. Am. Chem. Soc.* **2001**, 123, 6543-6554.
41. Randl, S., Lucas, N., Connon, S. J., Blechert, S. A mechanism switch in enyne metathesis reactions involving rearrangement: influence of heteroatoms in the propargylic position. *Adv. Syn. & Catal.* **2002**, 344, 631-633.
42. Hoye, T. R., Donaldson, S. M., Vos, T. J. An Enyne Metathesis/(4 + 2)-Dimerization Route to (±)-Differolide. *Org. Lett.* **1999**, 1, 277-279.
43. Kinoshita, A., Mori, M. Total Synthesis of (-)-Stemoamide Using Ruthenium-Catalyzed Enyne Metathesis Reaction. *J. Org. Chem.* **1996**, 61, 8356-8357.
44. Layton, M. E., Morales, C. A., Shair, M. D. Biomimetic Synthesis of (-)-Longithorone A. *J. Am. Chem. Soc.* **2002**, 124, 773-775.

Eschenmoser Methenylation 154

Related reactions: Mannich reaction;

1. Schreiber, J., Maag, H., Hashimoto, N., Eschenmoser, A. Dimethyl(methylene)ammonium iodide. *Angew. Chem., Int. Ed. Engl.* **1971**, 10, 330-331.
2. Winterfeldt, E. Eschenmoser's salt: $H_2C:N^+I^-$ [sic]. *J. Prakt. Chem.* **1994**, 336, 91-92.
3. Jasor, Y., Luche, M. J., Gaudry, M., Marquet, A. Regioselective synthesis of Mannich bases from unsymmetrical ketones and immonium salts. *J. Chem. Soc., Chem. Commun.* **1974**, 253-254.
4. Jasor, Y., Gaudry, M., Luche, M. J., Marquet, A. Regioselective synthesis of Mannich bases from disymmetric ketones. *Tetrahedron* **1977**, 33, 295-303.
5. Cravotto, G., Giovenzana, G. B., Pilati, T., Sisti, M., Palmisano, G. Azomethine Ylide Cycloaddition/Reductive Heterocyclization Approach to Oxindole Alkaloids: Asymmetric Synthesis of (-)-Horsfiline. *J. Org. Chem.* **2001**, 66, 8447-8453.
6. Dudley, G. B., Tan, D. S., Kim, G., Tanski, J. M., Danishefsky, S. J. Remarkable stereoselectivity in the alkylation of a hydroazulenone: progress towards the total synthesis of guanacastepene. *Tetrahedron Lett.* **2001**, 42, 6789-6791.
7. Njardarson, J. T., McDonald, I. M., Spiegel, D. A., Inoue, M., Wood, J. L. An Expeditious Approach toward the Total Synthesis of CP-263,114. *Org. Lett.* **2001**, 3, 2435-2438.
8. Dauben, W. G., Wang, T. Z., Stephens, R. W. Total synthesis of (±)-crassin acetate methyl ether. *Tetrahedron Lett.* **1990**, 31, 2393-2396.
9. Ng, F. W., Lin, H., Danishefsky, S. J. Explorations in Organic Chemistry Leading to the Total Synthesis of (±)-Gelsemine. *J. Am. Chem. Soc.* **2002**, 124, 9812-9824.

Eschenmoser-Claisen Rearrangement 156

Related reactions: Carroll rearrangement, Claisen rearrangement, Claisen-Ireland rearrangement, Johnson-Claisen rearrangement;

1. Wick, A. E., Felix, D., Steen, K., Eschenmoser, A. Claisen rearrangement of allyl and benzyl alcohols by N,N-dimethylacetamide acetals. *Helv. Chim. Acta* **1964**, 47, 2425-2429.
2. Felix, D., Gschwend-Steen, K., Wick, A. E., Eschenmoser, A. Claisen rearrangement of allyl and benzyl alcohols with 1-dimethylamino-1-methoxyethene. *Helv. Chim. Acta* **1969**, 52, 1030-1042.
3. Ziegler, F. E. The thermal, aliphatic Claisen rearrangement. *Chem. Rev.* **1988**, 88, 1423-1452.
4. Castro, A. M. M. Claisen Rearrangement over the Past Nine Decades. *Chem. Rev.* **2004**, 104, 2939-3002.
5. Gradl, S. N., Kennedy-Smith, J. J., Kim, J., Trauner, D. A practical variant of the Claisen-Eschenmoser rearrangement: synthesis of unsaturated morpholine amides. *Synlett* **2002**, 411-414.

6. Daub, G. W., Edwards, J. P., Okada, C. R., Allen, J. W., Maxey, C. T., Wells, M. S., Goldstein, A. S., Dibley, M. J., Wang, C. J., Ostercamp, D. P., Chung, S., Cunningham, P. S., Berliner, M. A. Acyclic Stereoselection in the Ortho Ester Claisen Rearrangement. *J. Org. Chem.* **1997**, 62, 1976-1985.
7. Chen, C. Y., Hart, D. J. A Diels-Alder approach to Stemona alkaloids: total synthesis of stenine. *J. Org. Chem.* **1993**, 58, 3840-3849.
8. Daniewski, A. R., Wovkulich, P. M., Uskokovic, M. R. Remote diastereoselection in the asymmetric synthesis of pravastatin. *J. Org. Chem.* **1992**, 57, 7133-7139.
9. Loh, T.-P., Hu, Q.-Y. Synthetic Studies toward Anisatin: A Formal Synthesis of (±)-8-Deoxyanisatin. *Org. Lett.* **2001**, 3, 279-281.
10. Williams, D. R., Brugel, T. A. Intramolecular Diels-Alder Cyclizations of (E)-1-Nitro-1,7,9-decatrienes: Synthesis of the AB Ring System of Norzoanthamine. *Org. Lett.* **2000**, 2, 1023-1026.

Eschenmoser-Tanabe Fragmentation 158

Related reactions: *Grob fragmentation, Wharton fragmentation;*

1. Eschenmoser, A., Felix, D., Ohloff, G. New fragmentation reaction of α,β-unsaturated carbonyls. Synthesis of exaltone and rac-muscone from cyclododecanone. *Helv. Chim. Acta* **1967**, 50, 708-713.
2. Schreiber, J., et al. Synthesis of acetylenic carbonyl compounds by fragmentation of α,β-epoxy ketones with *p*-tolylsulfonylhydrazine. *Helv. Chim. Acta* **1967**, 50, 2101-2108.
3. Tanabe, M., Crowe, D. F., Dehn, R. L. Novel fragmentation reaction of α,β-epoxyketones. Synthesis of acetylenic ketones. *Tetrahedron Lett.* **1967**, 3943-3946.
4. Tanabe, M., Crowe, D. F., Dehn, R. L., Detre, G. Synthesis of secosteroid acetylenic ketones. *Tetrahedron Lett.* **1967**, 3739-3743.
5. Felix, D., Schreiber, J., Ohloff, G., Eschenmoser, A. Synthetic methods. 3. α,β-Epoxy ketone->alkynone fragmentation. I. Synthesis of exaltone and (±)-muscone from cyclododecanone. *Helv. Chim. Acta* **1971**, 54, 2896-2912.
6. Weyerstahl, P., Marschall, H. Fragmentation Reactions. in *Comp. Org. Synth.* (eds. Trost, B. M.,Fleming, I.), *6*, 1041-1070 (Pergamon, Oxford, **1991**).
7. Borrevang, P., Hjort, J., Rapala, R. T., Edie, R. Novel ring fragmentation products via diazirines and its conversion to A-nor steroids. *Tetrahedron Lett.* **1968**, 4905-4907.
8. Felix, D., Schreiber, J., Piers, K., Horn, U., Eschenmoser, A. New version of epoxy ketone-> alkynone fragmentation. Thermal decomposition of hydrazones from α,β-epoxycarbonyl compounds and *N*-aminoaziridines. *Helv. Chim. Acta* **1968**, 51, 1461-1465.
9. Felix, D., Mueller, R. K., Horn, U., Joos, R., Schreiber, J., Eschenmoser, A. Synthetic methods. 4. α,β-Epoxyketone.far. alkynone fragmentation. II. Pyrolytic decomposition of hydrazones from α,β-epoxyketones and *N*-aminoaziridines. *Helv. Chim. Acta* **1972**, 55, 1276-1319.
10. Corey, E. J., Sachdev, H. S. 2,4-Dinitrobenzenesulfonylhydrazine, a useful reagent for the Eschenmoser α,β cleavage of α,β-epoxy ketones. Conformational control of halolactonization. *J. Org. Chem.* **1975**, 40, 579-581.
11. MacAlpine, G. A., Warkentin, J. Thermolysis of D3-1,3,4-oxadiazolin-2-ones and 2-phenylimino-D3-1,3,4-oxadiazolines derived from α,β-epoxyketones. An alternative method for the conversion of α,β-epoxyketones to alkynones and alkynals. *Can. J. Chem.* **1978**, 56, 308-315.
12. Fehr, C., Ohloff, G., Büchi, G. A new α,β-enone -> alkynone fragmentation. Synthesis of Exaltone and (±)-muscone. *Helv. Chim. Acta* **1979**, 62, 2655-2660.
13. Felix, D., Wintner, C., Eschenmoser, A. Fragmentation of α,β-epoxyketones to acetylenic aldehydes and ketones: preparation of 2,3-epoxycyclohexanone and its fragmentation to 5-hexynal. *Org. Synth.* **1976**, 55, 52-56.
14. Wieland, P., Kaufmann, H., Eschenmoser, A. Fragmentation of α,β-epoxy ketoximes to acetylenic ketones. *Helv. Chim. Acta* **1967**, 50, 2108-2110.
15. Coates, R. M., Freidinger, R. M. Total synthesis of sesquicarene. *J. Chem. Soc., Chem. Commun.* **1969**, 15, 871-872.
16. Zbiral, E., Nestler, G., Kischa, K. Transfer reactions with lead(IV) acetate. IV. General single stage synthesis of seco-oxonitrile forms of steroids with lead(IV) acetate-trimethylsilyl azide. New type of fragmentation principle. *Tetrahedron* **1970**, 26, 1427-1434.
17. Morioka, M., Kato, M., Yoshida, H., Ogata, T. Anomalous Bamford-Stevens reaction of cis-N-alkyl-3-phenyl-2-aziridinyl phenyl ketones. Preparation of 1,6-dihydro-1,2,3-triazine derivatives. *Heterocycles* **1996**, 43, 1759-1765.
18. Herges, R. Ordering principle of complex reactions and theory of contracted transition states. *Angew. Chem.* **1994**, 106, 261-283 (See also Angew. Chem., Int. Ed. Engl., 1994, 1933(1993), 1255-1976).
19. Mueck-Lichtenfeld, C. Theoretical Prediction of the Stability and Intramolecular Rearrangement Reactions of Heteroanalogues of Cyclopropylcarbene: 2-Oxiranyl-, 2-Aziridinyl-, and 1-Aziridinylcarbene. *J. Org. Chem.* **2000**, 65, 1366-1375.
20. Mander, L. N., McLachlan, M. M. The Total Synthesis of the Galbulimima Alkaloid GB 13. *J. Am. Chem. Soc.* **2003**, 125, 2400-2401.
21. Dai, W., Katzenellenbogen, J. A. New approaches to the synthesis of alkyl-substituted enol lactone systems, inhibitors of the serine protease elastase. *J. Org. Chem.* **1993**, 58, 1900-1908.
22. Trost, B. M., Chang, V. K. An approach to botrydianes: on the steric demands of a metal catalyzed enyne metathesis. *Synthesis* **1993**, 824-832.
23. Gordon, D. M., Danishefsky, S. J., Schulte, G. K. Studies in the benzannulation of a cycloalkynone: an approach to the synthesis of antibiotics containing the benz[a]anthracene core structure. *J. Org. Chem.* **1992**, 57, 7052-7055.

Eschweiler-Clarke Methylation (Reductive Alkylation) 160

1. Eschweiler, W. Substitution of hydrogen atoms bound to nitrogen for a methyl group with formaldehyde. *Ber.* **1905**, 38, 880-887.
2. Clarke, H. T., Gillespie, H. B., Weisshaus, S. Z. Action of formaldehyde on amines and amino acids. *J. Am. Chem. Soc.* **1933**, 55, 4571-4587.
3. Cope, A. C., Burrows, W. D. Cyclization in the course of Clarke-Eschweiler methylation. *J. Org. Chem.* **1965**, 30, 2163-2165.
4. Cope, A. C., Burrows, W. D. Clarke-Eschweiler cyclization. Scope and mechanism. *J. Org. Chem.* **1966**, 31, 3099-3103.
5. Emerson, W. S. Preparation of amines by reductive alkylation. *Org. React.* **1948**, 4, 174-255.
6. Deno, N. C., Peterson, H. J., Saines, G. S. The hydride-transfer reaction. *Chem. Rev.* **1960**, 60, 7-14.
7. Gibson, H. W. Chemistry of formic acid and its simple derivatives. *Chem. Rev.* **1969**, 69, 673-692.
8. Borch, R. F., Hassid, A. I. New method for the methylation of amines. *J. Org. Chem.* **1972**, 37, 1673-1674.
9. Fache, F., Jacquot, L., Lemaire, M. Extension of the Eschweiler-Clarke procedure to the N-alkylation of amides. *Tetrahedron Lett.* **1994**, 35, 3313-3314.
10. Bulman Page, P. C., Heaney, H., Rassias, G. A., Reignier, S., Sampler, E. P., Talib, S. The reductive cleavage of cyclic aminol ethers to N,N-dialkylamino-derivatives. Modifications to the Eschweiler-Clarke procedure. *Synlett* **2000**, 104-106.
11. Torchy, S., Barbry, D. N-alkylation of amines under microwave irradiation: modified Eschweiler-Clarke reaction. *J. Chem. Res., Synop.* **2001**, 292-293.
12. Harding, J. R., Jones, J. R., Lu, S.-Y., Wood, R. Development of a microwave-enhanced isotopic labeling procedure based on the Eschweiler-Clarke methylation reaction. *Tetrahedron Lett.* **2002**, 43, 9487-9488.
13. Rosenau, T., Potthast, A., Rohrling, J., Hofinger, A., Sixta, H., Kosma, P. A solvent-free and formalin-free Eschweiler-Clarke methylation of amines. *Synth. Commun.* **2002**, 32, 457-465.
14. Moore, M. L. Leuckart reaction. *Org. React.* **1949**, 5, 301-330.
15. Leuckart, R. A new synthesis of tribenzylamines. *Ber.* **1885**, 18, 2341-2344.
16. Wallach. Menthylamine. *Ber.* **1891**, 24, 3992-3993.

17. Mattson, R. J., Pham, K. M., Leuck, D. J., Cowen, K. A. An improved method for reductive alkylation of amines using titanium(IV) isopropoxide and sodium cyanoborohydride. *J. Org. Chem.* **1990**, 55, 2552-2554.
18. Ramanjulu, J. M., Joullie, M. M. N-alkylation of amino acid esters using sodium triacetoxyborohydride. *Synth. Commun.* **1996**, 26, 1379-1384.
19. Yoon, N. M., Kim, E. G., Son, H. S., Choi, J. Borohydride exchange resin, a new reducing agent for reductive amination. *Synth. Commun.* **1993**, 23, 1595-1599.
20. Felpin, F.-X., Girard, S., Vo-Thanh, G., Robins, R. J., Villieras, J., Lebreton, J. Efficient Enantiomeric Synthesis of Pyrrolidine and Piperidine Alkaloids from Tobacco. *J. Org. Chem.* **2001**, 66, 6305-6312.
21. Pollard, C. B., Young, D. C., Jr. The mechanism of the Leuckart reaction. *J. Org. Chem.* **1951**, 16, 661-672.
22. Lukasiewiez, A. Mechanism of chemical reactions. I. Mechanism of the Leuckart-Wallach reaction and of the reduction of Schiff bases by formic acid. *Tetrahedron* **1963**, 19, 1789-1799.
23. Ito, K., Oba, H., Sekiya, M. Studies on Leuckart-Wallach reaction paths. *Bull. Chem. Soc. Jpn.* **1976**, 49, 2485-2490.
24. Subbaiah, G., Sethuram, B., Mahadevan, E. G., Rao, T. N. Kinetics of methylation of primary alkyl amine hydrochlorides with formaldehyde formic acid. *Indian J. Chem., Sect. B* **1978**, 16B, 1009-1011.
25. Awachie, P. I., Agwada, V. C. Evidence for rate limiting carbon-hydrogen bond cleavage in the Leuckart reaction. *Tetrahedron* **1990**, 46, 1899-1910.
26. Martinez, A. G., Vilar, E. T., Fraile, A. G., Ruiz, P. M., San Antonio, R. M., Alcazar, M. P. M. On the mechanism of the Leuckart reaction. Enantiospecific preparation of (1R,2R)- and (1S,2S)-N-(3,3-dimethyl-2-formylamino-1-norbornyl)acetamide. *Tetrahedron: Asymmetry* **1999**, 10, 1499-1505.
27. Smith, A. B., III, Friestad, G. K., Barbosa, J., Bertounesque, E., Duan, J. J. W., Hull, K. G., Iwashima, M., Qiu, Y., Spoors, P. G., Salvatore, B. A. Total Synthesis of (+)-Calyculin A and (-)-Calyculin B: Cyanotetraene Construction, Asymmetric Synthesis of the C(26-37) Oxazole, Fragment Assembly, and Final Elaboration. *J. Am. Chem. Soc.* **1999**, 121, 10478-10486.
28. Lakshmaiah, G., Kawabata, T., Shang, M., Fuji, K. Total Synthesis of (-)-Horsfiline via Asymmetric Nitroolefination. *J. Org. Chem.* **1999**, 64, 1699-1704.
29. Argouarch, G., Gibson, C. L., Stones, G., Sherrington, D. C. The synthesis of chiral annulet 1,4,7-triazacyclononanes. *Tetrahedron Lett.* **2002**, 43, 3795-3798.

Evans Aldol Reaction 162

Related reactions: Aldol reaction, Mukaiyama aldol reaction, Reformatsky reaction;

1. Evans, D. A., Bartroli, J., Shih, T. L. Enantioselective aldol condensations. 2. Erythro-selective chiral aldol condensations via boron enolates. *J. Am. Chem. Soc.* **1981**, 103, 2127-2109.
2. Evans, D. A., Takacs, J. M., McGee, L. R., Ennis, M. D., Mathre, D. J., Bartroli, J. Chiral enolate design. *Pure Appl. Chem.* **1981**, 53, 1109-1127.
3. Evans, D. A. Studies in asymmetric synthesis. The development of practical chiral enolate synthons. *Aldrichimica Acta* **1982**, 15, 23-32.
4. Kim, B. M., Williams, S. F., Masamune, S. The Aldol Reaction: Group III Enolates. in *Comp. Org. Synth.* (ed. Trost, B. M.), 2, 239-275 (Pergamon Press, Oxford, **1991**).
5. Hoveyda, A. H., Evans, D. A., Fu, G. C. Substrate-directable chemical reactions. *Chem. Rev.* **1993**, 93, 1307-1370.
6. Franklin, A. S., Paterson, I. Recent developments in asymmetric aldol methodology. *Contemp. Org. Synth.* **1994**, 1, 317-338.
7. Cowden, C. J., Paterson, I. Asymmetric aldol reactions using boron enolates. *Org. React.* **1997**, 51, 1-200.
8. Saito, S., Yamamoto, H. Directed aldol condensation. *Chem.-- Eur. J.* **1999**, 5, 1959-1962.
9. Arya, P., Qin, H. Advances in asymmetric enolate methodology. *Tetrahedron* **2000**, 56, 917-947.
10. Evans, D. A., Shaw, J. T. Recent advances in asymmetric synthesis with chiral imide auxiliaries. *Actualite Chimique* **2003**, 35-38.
11. Hsiao, C. N., Liu, L., Miller, M. J. Cysteine- and serine-derived thiazolidinethiones and oxazolidinethiones as efficient chiral auxiliaries in aldol condensations. *J. Org. Chem.* **1987**, 52, 2201-2206.
12. Bermejo Gonzalez, F., Perez Baz, J., Santinelli, F., Mayer Real, F. Synthesis and aldol stereoselectivity of 2-oxazolidinones derived from L-histidine. *Bull. Chem. Soc. Jpn.* **1991**, 64, 674-681.
13. Davies, S. G., Mortlock, A. A. Bifunctional chiral auxiliaries. 1. The aldol reaction between dialkylboron enolates of 1,3-dipropionyl-trans-4,5-tetramethyleneimidazolidin-2-one and aldehydes. *Tetrahedron Lett.* **1991**, 32, 4787-4790.
14. Yan, T. H., Chu, V. V., Lin, C., Tseng, W. H., Cleng, T. W. A superior chiral auxiliary in aldol condensations: camphor-based oxazolidone. *Tetrahedron Lett.* **1991**, 32, 5563-5566.
15. Ghosh, A. K., Duong, T. T., McKee, S. P. Highly enantioselective aldol reaction: development of a new chiral auxiliary from cis-1-amino-2-hydroxyindan. *J. Chem. Soc., Chem. Commun.* **1992**, 1673-1674.
16. Yan, T. H., Tan, C. W., Lee, H. C., Lo, H. C., Huang, T. Y. Asymmetric aldol reactions: a novel model for switching between chelation- and non-chelation-controlled aldol reactions. *J. Am. Chem. Soc.* **1993**, 115, 2613-2621.
17. Banks, M. R., Cadogan, J. I. G., Gosney, I., Grant, K. J., Hodgson, P. K. G., Thorburn, P. Synthesis of enantiomerically pure (5S)-4-aza-2-oxa-6,6-dimethyl-7,10-methylene-5-spiro[4.5]decan-3-one, a novel chiral oxazolidin-2-one from (-)-camphene for use as a recyclable chiral auxiliary in asymmetric transformations. *Heterocycles* **1994**, 37, 199-206.
18. Davies, S. G., Doisneau, G. J. M., Prodger, J. C., Sanganee, H. J. Synthesis of 5-substituted-3,3-dimethyl-2-pyrrolidinones: "quat" chiral auxiliaries. *Tetrahedron Lett.* **1994**, 35, 2369-2372.
19. Davies, S. G., Doisneau, G. J. M., Prodger, J. C., Sanganee, H. J. Asymmetric aldol and alkylation reactions mediated by the "quat" chiral auxiliary (R)-(-)-5-methyl-3,3-dimethyl-2-pyrrolidinone. *Tetrahedron Lett.* **1994**, 35, 2373-2376.
20. Davies, S. G., Edwards, A. J., Evans, G. B., Mortlock, A. A. Bifunctional chiral auxiliaries. 7. Aldol reactions of enolates derived from 1,3-diacylimidazolidine-2-thiones and 1,3-diacylimidazolidin-2-ones. *Tetrahedron* **1994**, 50, 6621-6642.
21. Boeckman, R. K., Jr., Connell, B. T. Toward the Development of a General Chiral Auxiliary. 3. Design and Evaluation of a Novel Chiral Bicyclic Lactam for Asymmetric Aldol Condensations: Evidence for the Importance of Dipole Alignment in the Transition State. *J. Am. Chem. Soc.* **1995**, 117, 12368-12369.
22. Hintermann, T., Seebach, D. A useful modification of the Evans auxiliary. 4-Isopropyl-5,5-diphenyloxazolidin-2-one. *Helv. Chim. Acta* **1998**, 81, 2093-2126.
23. Bernardi, A., Capelli, A. M., Gennari, C., Goodman, J. M., Paterson, I. Transition-state modeling of the aldol reaction of boron enolates: a force field approach. *J. Org. Chem.* **1990**, 55, 3576-3581.
24. Li, Y., Paddon-Row, M. N., Houk, K. N. Transition structures for the aldol reactions of anionic, lithium, and boron enolates. *J. Org. Chem.* **1990**, 55, 481-493.
25. Bernardi, F., Robb, M. A., Suzzi-Valli, G., Tagliavini, E., Trombini, C., Umani-Ronchi, A. An MC-SCF study of the transition structures for the aldol reaction of formaldehyde with acetaldehyde boron enolate. *J. Org. Chem.* **1991**, 56, 6472-6475.
26. Gennari, C., Vieth, S., Comotti, A., Vulpetti, A., Goodman, J. M., Paterson, I. Diastereofacial selectivity in the aldol reactions of chiral α-methyl aldehydes: a computer modelling approach. *Tetrahedron* **1992**, 48, 4439-4458.
27. Vulpetti, A., Bernardi, A., Gennari, C., Goodman, J. M., Paterson, I. Origins of π-face selectivity in the aldol reactions of chiral E-enol borinates: a computational study using transition state modeling. *Tetrahedron* **1993**, 49, 685-696.
28. Makino, Y., Iseki, K., Fujii, K., Oishi, S., Hirano, T., Kobayashi, Y. Reversal of π-face selectivity in the Evans aldol reaction with fluoral: a computational study on the transition states using semiempirical calculations. *Tetrahedron Lett.* **1995**, 36, 6527-6530.
29. Zachariasen, W. H. The crystal structure of monoclinic metaboric acid. *Acta Cryst.* **1963**, 16, 385-389.
30. Evans, D. A., Vogel, E., Nelson, J. V. Stereoselective aldol condensations via boron enolates. *J. Am. Chem. Soc.* **1979**, 101, 6120-6123.
31. Masamune, S., Mori, S., Van Horn, D., Brooks, D. W. E- and Z-Vinyloxyboranes (alkenyl borinates): stereoselective formation and aldol condensation. *Tetrahedron Lett.* **1979**, 1665-1668.

32. Evans, D. A., Bartroli, J. Stereoselective reactions of chiral enolates. Application to the synthesis of (+)-Prelog-Djerassi lactonic acid. *Tetrahedron Lett.* **1982**, 23, 807-810.
33. Evans, D. A., Ennis, M. D., Mathre, D. J. Asymmetric alkylation reactions of chiral imide enolates. A practical approach to the enantioselective synthesis of α-substituted carboxylic acid derivatives. *J. Am. Chem. Soc.* **1982**, 104, 1737-1739.
34. Evans, D. A., Britton, T. C., Ellman, J. A. Contrasteric carboximide hydrolysis with lithium hydroperoxide. *Tetrahedron Lett.* **1987**, 28, 6141-6144.
35. Damon, R. E., Coppola, G. M. Cleavage of N-acyloxazolidones. *Tetrahedron Lett.* **1990**, 31, 2849-2852.
36. Thaisrivongs, S., Pals, D. T., Kroll, L. T., Turner, S. R., Han, F. S. Renin inhibitors. Design of angiotensinogen transition-state analogs containing novel (2R,3R,4R,5S)-5-amino-3,4-dihydroxy-2-isopropyl-7-methyloctanoic acid. *J. Med. Chem.* **1987**, 30, 976-982.
37. Evans, D. A., Bender, S. L. Total synthesis of the ionophore antibiotic X-206. Studies relevant to the stereoselective synthesis of the C(17)-C(26) synthon. *Tetrahedron Lett.* **1986**, 27, 799-802.
38. Evans, D. A., Britton, T. C., Dorow, R. L., Dellaria, J. F. Stereoselective amination of chiral enolates. A new approach to the asymmetric synthesis of α-hydrazino and α-amino acid derivatives. *J. Am. Chem. Soc.* **1986**, 108, 6395-6397.
39. Evans, D. A., Britton, T. C. Electrophilic azide transfer to chiral enolates. A general approach to the asymmetric synthesis of α-amino acids. *J. Am. Chem. Soc.* **1987**, 109, 6881-6883.
40. Evans, D. A., Britton, T. C., Dellaria, J. F., Jr. The asymmetric synthesis of α-amino and α-hydrazino acid derivatives via the stereoselective amination of chiral enolates with azodicarboxylate esters. *Tetrahedron* **1988**, 44, 5525-5540.
41. Evans, D. A., Britton, T. C., Ellman, J. A., Dorow, R. L. The asymmetric synthesis of α-amino acids. Electrophilic azidation of chiral imide enolates, a practical approach to the synthesis of (R)- and (S)–α-azido carboxylic acids. *J. Am. Chem. Soc.* **1990**, 112, 4011-4030.
42. Evans, D. A., Morrissey, M. M., Dorow, R. L. Asymmetric oxygenation of chiral imide enolates. A general approach to the synthesis of enantiomerically pure α-hydroxy carboxylic acid synthons. *J. Am. Chem. Soc.* **1985**, 107, 4346-4348.
43. Fuerstner, A., Ruiz-Caro, J., Prinz, H., Waldmann, H. Structure Assignment, Total Synthesis, and Evaluation of the Phosphatase Modulating Activity of Glucolipsin A. *J. Org. Chem.* **2004**, 69, 459-467.
44. Boger, D. L., Menezes, R. F. Synthesis of tri- and tetrapeptide S: the extended C-terminus of bleomycin A2. *J. Org. Chem.* **1992**, 57, 4331-4333.
45. Boger, D. L., Colletti, S. L., Honda, T., Menezes, R. F. Total Synthesis of Bleomycin A2 and Related Agents. 1. Synthesis and DNA Binding Properties of the Extended C-Terminus: Tripeptide S, Tetrapeptide S, Pentapeptide S, and Related Agents. *J. Am. Chem. Soc.* **1994**, 116, 5607-5618.
46. Evans, D. A., Starr, J. T. A cascade cycloaddition strategy leading to the total synthesis of (-)-FR182877. *Angew. Chem., Int. Ed. Engl.* **2002**, 41, 1787-1790.
47. Evans, D. A., Starr, J. T. A Cycloaddition Cascade Approach to the Total Synthesis of (-)-FR182877. *J. Am. Chem. Soc.* **2003**, 125, 13531-13540.
48. Gage, J. R., Evans, D. A. (S)-4-(Phenylmethyl)-2-oxazolidinone [preparation]. *Org. Synth.* **1990**, 68, 77-82.

1. Favorskii, A. *J. Russ. Phys. Chem. Soc.* **1894**, 26, 559.
2. Favorskii, A. *J. Prakt. Chem./Chem.-Ztg.* **1895**, 51, 533-563.
3. Kende, A. S. The Favorski rearrangement of haloketones. *Org. React.* **1960**, 11, 261-316.
4. Wasserman, H. H., Clark, G. M., Turley, P. C. Recent aspects of cyclopropanone chemistry. *Top. Curr. Chem.* **1974**, 47, 73-156.
5. Chenier, P. J. The Favorskii rearrangement in bridged polycyclic compounds. *J. Chem. Educ.* **1978**, 55, 286-291.
6. Baretta, A., Waegell, B. A survey of Favorskii rearrangement mechanisms: influence of the nature and strain of the skeleton. *React. Intermed. (Plenum)* **1982**, 2, 527-585.
7. Mann, J. The Favorskii Rearrangement. in *Comp. Org. Synth.* (eds. Trost, B. M.,Fleming, I.), *3*, 839-861 (Pergamon Press, Oxford, **1991**).
8. Moulay, S. The most well-known rearrangements in organic chemistry at hand. *Chem. Ed.: Res. Pract. Eur.* **2002**, 3, 33-64.
9. Cymerman Craig, J., Dinner, A., Mulligan, P. J. Novel variant of the Favorskii reaction. *J. Org. Chem.* **1972**, 37, 3539-3541.
10. De Kimpe, N., Sulmon, P., Moens, L., Schamp, N., Declercq, J. P., Van Meerssche, M. The Favorskii rearrangement of α-chloro ketimines. *J. Org. Chem.* **1986**, 51, 3839-3848.
11. De Kimpe, N., Stanoeva, E., Schamp, N. Intramolecular trapping of a cyclopropylidenamine during the Favorskii rearrangement of α-chloro ketimines. *Tetrahedron Lett.* **1988**, 29, 589-592.
12. Satoh, T., Motohashi, S., Kimura, S., Tokutake, N., Yamakawa, K. The asymmetric Favorskii rearrangement: a synthesis of optically active α-alkyl amides from aldehydes and (-)-1-chloroalkyl p-tolyl sulfoxide. *Tetrahedron Lett.* **1993**, 34, 4823-4826.
13. Moliner, V., Castillo, R., Safont, V. S., Oliva, M., Bohn, S., Tunon, I., Andres, J. A theoretical study of the Favorskii rearrangement. calculation of gas-phase reaction paths and solvation effects on the molecular mechanism for the transposition of the α-chlorocyclobutanone. *J. Am. Chem. Soc.* **1997**, 119, 1941-1947.
14. Castillo, R., Andres, J., Moliner, V. Quantum Mechanical/Molecular Mechanical Study on the Favorskii Rearrangement in Aqueous Media. *J. Phys. Chem. B* **2001**, 105, 2453-2460.
15. House, H. O., Gilmore, W. F. The stereochemistry of the Favorskii rearrangement. *J. Am. Chem. Soc.* **1961**, 83, 3980-3985.
16. Abad, A., Arno, M., Pedro, J. R., Seoane, E. Selective Favorskii rearrangement in macrocyclic rings. *Tetrahedron Lett.* **1981**, 22, 1733-1736.
17. Eaton, P. E., Or, Y. S., Branca, S. J., Shankar, B. K. R. The synthesis of pentaprismane. *Tetrahedron* **1986**, 42, 1621-1631.
18. Mouk, R. W., Patel, K. M., Reusch, W. Favorskii rearrangement of α,β-epoxy ketones. *Tetrahedron* **1975**, 31, 13-19.
19. Bhat, K. L., Trivedi, G. Base catalyzed reaction of 1β,2β-epoxy-γ-tetrahydrosantonin. *Synth. Commun.* **1982**, 12, 585-593.
20. Wenkert, E., Bakuzis, P., Baumgarten, R. J., Leicht, C. L., Schenk, H. P. Homo-Favorskii rearrangement. *J. Am. Chem. Soc.* **1971**, 93, 3208-3216.
21. Wong, H. N. C., Lau, K. L., Tam, K. F. The application of cyclobutane derivatives in organic synthesis. *Top. Curr. Chem.* **1986**, 133, 83-157.
22. Loftfield, R. B. The alkaline rearrangement of α-haloketones. II. The mechanism of the Favorskii reaction. *J. Am. Chem. Soc.* **1951**, 73, 4707-4714.
23. Turro, N. J., Hammond, W. B. Tetramethylcyclopropanone. II. Mechanism of the Favorskii rearrangement. *J. Am. Chem. Soc.* **1965**, 87, 3258-3259.
24. Bordwell, F. G., Frame, R. R., Scamehorn, R. G., Strong, J. G., Meyerson, S. Favorskii reactions. I. Nature of the rate-determining step. *J. Am. Chem. Soc.* **1967**, 89, 6704-6711.
25. Warnhoff, E. W., Wong, C. M., Tai, W.-T. Mechanistic changes in a Favorskii reaction. *J. Am. Chem. Soc.* **1968**, 90, 514-515.
26. Rappe, C., Knutsson, L., Turro, N. J., Gagosian, R. B. Favorskii rearrangements. Evidence for steric control in the fission of crowded cyclopropanone intermediates. *J. Am. Chem. Soc.* **1970**, 92, 2032-2035.
27. Knutsson, L. Favorsky rearrangements. XVII. Studies on the mechanism of the rearrangement of 1,1,3-tribromoacetone using carbon-13-NMR spectroscopy. Large intramolecular secondary deuterium isotope effect. *Chem. Scr.* **1972**, 2, 227-229.
28. Rappe, C., Knutsson, L. Cyclopropanones and the Favorskii rearrangement. An unexpectedly large secondary isotope effect. *Angew. Chem., Int. Ed. Engl.* **1972**, 11, 329-330.
29. Wolff, S., Agosta, W. C. Evidence against a ketene intermediate in the Homo-Favorskii Reaction. *J. Chem. Soc., Chem. Commun.* **1973**, 771.
30. Schamp, N., De Kimpe, N., Coppens, W. Favorskii rearrangement of dichlorinated methyl ketones. *Tetrahedron* **1975**, 31, 2081-2087.
31. McGrath, M. J. A. Favorskii rearrangements. I. One electron transfer from α'-enolate intermediates to triplet oxygen in aprotic, polar protic, and mixed media. *Tetrahedron* **1976**, 32, 377-387.

32. Takeshita, H., Kawakami, H., Ikeda, Y., Mori, A. Synthetic Photochemistry. 65. Synthesis of Hexacyclo[6.4.2.$0^{2,7}$.$0^{3,11}$.$0^{6,10}$.$0^{9,12}$]tetradecane. *J. Org. Chem.* **1994**, 59, 6490-6492.
33. Lee, E., Yoon, C. H. Stereoselective Favorskii rearrangement of carvone chlorohydrin; expedient synthesis of (+)-dihydronepetalactone and (+)-iridomyrmecin. *J. Chem. Soc., Chem. Commun.* **1994**, 479-481.
34. Zhang, L., Koreeda, M. Stereocontrolled Synthesis of Kelsoene by the Homo-Favorskii Rearrangement. *Org. Lett.* **2002**, 4, 3755-3758.

Related reactions: Paal-Knorr furan synthesis;

1. Feist, F. Studies on the furan and pyrrole group: Condensation of β-keto esters with chloroacetone and ammonia. *Chem.Ber.* **1902**, 35, 1545.
2. Benary, E. Synthesis of Pyridine Derivatives from Dichloroether and β-Aminocrotonic Ester. *Ber.* **1911**, 44, 489-493.
3. Friedrichsen, W. Furans and their Benzo Derivatives: Synthesis. in *Comprehensive Heterocyclic Chemistry II.* (eds. Katritzky, A. R.,Scriven, E. F. V.), *2*, 359 (Pergamon: Elsevier Science Ltd., Oxford, **1996**).
4. Bisagni, E., Marquet, J. P., Andre-Louisfert, J., Cheutin, A., Feinte, F. 2,3-Disubstituted furans and pyrroles. I. Extension of the Feist-Benary reaction to β-diketones. New synthesis of 3-acylated furans and pyrroles. *Bull. Soc. Chim. Fr.* **1967**, 2796-2780.
5. Cambie, R. C., Moratti, S. C., Rutledge, P. S., Woodgate, P. D. 1,2-Dibromoethyl acetate, a reagent for Feist-Benary condensations. *Synth. Commun.* **1990**, 20, 1923-1929.
6. Lavoisier-Gallo, T., Rodriguez, J. Facile One-Pot Preparation of Functionalized 2-Allenylidenehydrofurans by Tandem C-O-Cycloalkylation of Stabilized Carbanions. *J. Org. Chem.* **1997**, 62, 3787-3788.
7. Calter, M. A., Zhu, C. Scope and Diastereoselectivity of the "Interrupted" Feist-Benary Reaction. *Org. Lett.* **2002**, 4, 205-208.
8. Bambury, R. E., Yaktin, H. K., Wyckoff, K. K. Trifluoromethylfurans. *J. Heterocycl. Chem.* **1968**, 5, 95-100.
9. Couffignal, R. Synthesis of ethyl and tert-butyl 2,4-dialkyl-5-methylene-4,5-dihydrofuran-3-carboxylates. *Synthesis* **1978**, 581-583.
10. Yuste, F., Vergel, H., Barrios, H., Ortiz, B., Sanchez-Obregon, R. Preparation of 4-(carbethoxy)-5-alkyl- and -5-phenyl-2-furanacetic acids and their methyl esters. *Org. Prep. Proced. Int.* **1988**, 20, 173-177.
11. Dunlop, A. P., Hurd, C. D. Base-catalyzed condensation of α–halogenated ketones with β–keto esters. *J. Org. Chem.* **1950**, 15, 1160-1164.
12. Bisagni, E., Marquet, J. P., Bourzat, J. D., Pepin, J. J., Andre-Louisfert, J. 2,3-Disubstituted furans and pyrroles. XI. Reaction of chloroacetaldehyde with β-keto esters. Formation of the expected furans and pyrroles and of 1,4-diacyl-1,4-cyclohexadienes. *Bull. Soc. Chim. Fr.* **1971**, 4041-4047.
13. Calter, M. A., Zhu, C., Lachicotte, R. J. Rapid Synthesis of the 7-Deoxy Zaragozic Acid Core. *Org. Lett.* **2002**, 4, 209-212.
14. Smith, J. O., Mandal, B. K., Filler, R., Beery, J. W. Reaction of ethyl 4,4,4-trifluoroacetoacetate enolate with 3-bromo,1,1,1-trifluoroacetone: synthesis of 2,4-bis(trifluoromethyl)furan. *J. Fluorine Chem.* **1997**, 81, 123-128.
15. Anxin, W., Mingyi, W., Yonghong, G., Xinfu, P. An Expeditious Synthetic Route to Furolignans having Two Different Aryl Groups. *J. Chem. Res., Synop.* **1998**, 136-137.
16. Tada, M., Ohtsu, K., Chiba, K. Synthesis of patulin and its cyclohexane analog from furan derivatives. *Chem. Pharm. Bull.* **1994**, 42, 2167-2169.

Related reactions: Petasis-Ferrier rearrangement;

1. Fischer, E. New reduction products of glucose: glucal and hydroglucal. *Ber.* **1914**, 47, 196-210.
2. Ferrier, R. J., Overend, W. G., Ryan, A. E. The reaction between 3,4,6-tri-O-acetyl-D-glucal and *p*-nitrophenol. *J. Chem. Soc., Abstracts* **1962**, 3667-3670.
3. Ferrier, R. J. Unsaturated carbohydrates. II. Three reactions leading to unsaturated glycopyranosides. *J. Chem. Soc., Abstracts* **1964**, 5443-5449.
4. Ferrier, R. J., Prasad, N., Sankey, G. H. Unsaturated carbohydrates. VIII. Intramolecular allylic isomerizations of 1-deoxyald-1-enopyranose (2-hydroxyglycal) esters. *J. Chem. Soc. C* **1968**, 974-977.
5. Ferrier, R. J. Unsaturated carbohydrates. Part 21. A carbocyclic ring closure of a hex-5-enopyranoside derivative. *J. Chem. Soc., Perkin Trans. 1* **1979**, 1455-1458.
6. Williams, N. R., Davison, B. E., Ferrier, R. J., Furneaux, R. H. Synthesis of enantiomerically pure noncarbohydrate compounds. *Carbohydr. Chem.* **1985**, 17, 244-255.
7. Ferrier, R. J., Middleton, S. The conversion of carbohydrate derivatives into functionalized cyclohexanes and cyclopentanes. *Chem. Rev.* **1993**, 93, 2779-2831.
8. Ferrier, R. J. Synthesis of enantiomerically pure non-carbohydrate compounds. *Carbohydr. Chem.* **1995**, 27, 312-360.
9. Ferrier, R. J., Blattner, R., Clinch, K., Furneaux, R. H., Gardiner, J. M., Tyler, P. C., Wightman, R. H., Williams, N. R. Synthesis of enantiomerically pure non-carbohydrate compounds. *Carbohydr. Chem.* **1996**, 28, 345-379.
10. Fraser-Reid, B. Some Progeny of 2,3-Unsaturated Sugars-They Little Resemble Grandfather Glucose: Twenty Years Later. *Acc. Chem. Res.* **1996**, 29, 57-66.
11. Ferrier, R. J. The conversion of carbohydrates to cyclohexane derivatives. *Prep. Carbohydr. Chem.* **1997**, 569-594.
12. Paquette, L. A. Oxonium ion-initiated pinacolic ring expansion reactions. *Rec. Res. Dev. Chem. Sci.* **1997**, 1, 1-16.
13. Sinay, P. Recent advances in the synthesis of carbohydrate mimics. *Pure Appl. Chem.* **1998**, 70, 1495-1499.
14. Ferrier, R. J. Direct conversion of 5,6-unsaturated hexopyranosyl compounds to functionalized cyclohexanones. *Top. Curr. Chem.* **2001**, 215, 277-291.
15. Ferrier, R. J. Substitution-with-allylic-rearrangement reactions of glycal derivatives. *Top. Curr. Chem.* **2001**, 215, 153-175.
16. Jarosz, S. C=C bond formation. *Glycoscience* **2001**, 1, 365-383.
17. Kozikowski, A. P., Park, P. U. Synthesis of streptazolin: use of the aza-Ferrier reaction in conjunction with the INOC process to deliver a unique but sensitive natural product. *J. Org. Chem.* **1990**, 55, 4668-4682.
18. Chida, N., Ohtsuka, M., Ogura, K., Ogawa, S. Synthesis of optically active substituted cyclohexenones from carbohydrates by catalytic Ferrier rearrangement. *Bull. Chem. Soc. Jpn.* **1991**, 64, 2118-2121.
19. Lopez, J. C., Fraser-Reid, B. n-Pentenyl esters facilitate an oxidative alternative to the Ferrier rearrangement. An expeditious route to sucrose. *J. Chem. Soc., Chem. Commun.* **1992**, 94-96.
20. Sobti, A., Sulikowski, G. A. Mitsunobu reactions of glycals with phenoxide nucleophiles are S_N2'-selective. *Tetrahedron Lett.* **1994**, 35, 3661-3664.
21. Koreeda, M., Houston, T. A., Shull, B. K., Klemke, E., Tuinman, R. J. Iodine-catalyzed Ferrier reaction. 1. A mild and highly versatile glycosidation of hydroxyl and phenolic groups. *Synlett* **1995**, 90-92.
22. Pelyvas, I. F., Madi-Puskas, M., Toth, Z. G., Varga, Z., Hornyak, M., Batta, G., Sztaricskai, F. Synthesis of new pseudo-disaccharide amino glycoside antibiotics from carbohydrates. *J. Antibiot.* **1995**, 48, 683-695.
23. Toshima, K., Matsuo, G., Ishizuka, T., Ushiki, Y., Nakata, M., Matsumura, S. Aryl and Allyl C-Glycosidation Methods Using Unprotected Sugars. *J. Org. Chem.* **1998**, 63, 2307-2313.
24. Masson, C., Soto, J., Bessodes, M. Ferric chloride: a new and very efficient catalyst for the Ferrier glycosylation reaction. *Synlett* **2000**, 1281-1282.

25. Bergmann, M. Unsaturated reduction products of the sugars and their derivatives. X. Pseudoglucal and dihydropseudoglucal. *Ann.* **1925**, 443, 223-242.
26. Laszlo, P., Pelyvas, I. F., Sztaricskai, F., Szilagyi, L., Somogyi, A. Novel aspects of the Ferrier carbocyclic ring-transformation reaction. *Carbohydr. Res.* **1988**, 175, 227-239.
27. Ferrier, R. J., Prasad, N. Unsaturated carbohydrates. IX. Synthesis of 2,3-dideoxy-a-D-erythro-hex-2-enopyranosides from tri-O-acetyl-D-glucal. *J. Chem. Soc. C.* **1969**, 570-575.
28. Guthrie, R. D., Irvine, R. W. Allylic nucleophilic substitution reactions in sugars. I. Tri-O-acetylglycals and related compounds. *Carbohydr. Res.* **1980**, 82, 207-224.
29. Machado, A. S., Dubreuil, D., Cleophax, J., Gero, S. D., Thomas, N. F. Expedient syntheses of inososes from carbohydrates: conformational and stereoelectronic aspects of the Ferrier reaction. *Carbohydr. Res.* **1992**, 233, C5-C8.
30. Yamauchi, N., Terachi, T., Eguchi, T., Kakinuma, K. Mechanistic and stereochemical studies on Ferrier reaction by means of chirally deuterated glucose. *Tetrahedron* **1994**, 50, 4125-4136.
31. Abada, P. B., Shull, B. K., Koreeda, M. On the mechanism of the iodine-catalyzed ferrier glycosylation reaction. *Book of Abstracts, 213th ACS National Meeting, San Francisco, April 13-17* **1997**, CARB-091.
32. Dubreuil, D., Cleophax, J., De Almeida, M. V., Verre-Sebrie, C., Liaigre, J., Vass, G., Gero, S. D. Stereoselective synthesis of 6-deoxy and 3,6-dideoxy-D-myoinositol precursors of deoxy myoinositol phosphate analogs from D-galactose. *Tetrahedron* **1997**, 53, 16747-16766.
33. Paquette, L. A., Kinney, M. J., Dullweber, U. Practical Synthesis of Spirocyclic Bis-C,C-glycosides. Mechanistic Models in Explanation of Rearrangement Stereoselectivity and the Bifurcation of Reaction Pathways. *J. Org. Chem.* **1997**, 62, 1713-1722.
34. Gemmell, N., Meo, P., Osborn, H. M. I. Stereoselective Entry to β-Linked C-Disaccharides Using a Carbon-Ferrier Reaction. *Org. Lett.* **2003**, 5, 1649-1652.
35. Williams, D. R., Heidebrecht, R. W., Jr. Total Synthesis of (+)-4,5-Deoxyneodolabelline. *J. Am. Chem. Soc.* **2003**, 125, 1843-1850.
36. Amano, S., Takemura, N., Ohtsuka, M., Ogawa, S., Chida, N. Total synthesis of paniculide A from D-glucose. *Tetrahedron* **1999**, 55, 3855-3870.
37. Chida, N., Ohtsuka, M., Ogawa, S. Total synthesis of (+)-lycoricidine and its 2-epimer from D-glucose. *J. Org. Chem.* **1993**, 58, 4441-4447.

Finkelstein Reaction 170

1. Finkelstein, H. Preparation of Organic Iodides from the Corresponding Bromides and Chlorides. *Ber.* **1910**, 43, 1528-1532.
2. Sharts, C. M., Sheppard, W. A. Modern methods to prepare monofluoroaliphatic compounds. *Org. React.* **1974**, 21, 125-406.
3. Miller, J. A., Nunn, M. J. Synthesis of alkyl iodides. *J. Chem. Soc., Perkin Trans. 1* **1976**, 416-420.
4. Olah, G. A., Narang, S. C., Field, L. D. Synthetic methods and reactions. 103. Preparation of alkyl iodides from alkyl fluorides and chlorides with iodotrimethylsilane or its in situ analogs. *J. Org. Chem.* **1981**, 46, 3727-3728.
5. Clark, J. H., Jones, C. W. The preparation of alkyl iodides from alkyl chlorides and bromides using potassium iodide supported on alumina. *J. Chem. Res., Synop.* **1990**, 39.
6. Majetich, G., Hicks, R. Applications of microwave accelerated organic chemistry. *Res. Chem. Intermed.* **1994**, 20, 61-77.
7. Albanese, D., Landini, D., Penso, M. Hydrated Tetrabutylammonium Fluoride as a Powerful Nucleophilic Fluorinating Agent. *J. Org. Chem.* **1998**, 63, 9587-9589.
8. Williams, R., Kennedy, A., Hijji, Y., Tadesse, S. Finkelstein halogen exchange reaction using microwave energy and a binary solvent system. *Proceedings - NOBCChE* **2001**, 28, 58-63.
9. Klapars, A., Buchwald, S. L. Copper-Catalyzed Halogen Exchange in Aryl Halides: An Aromatic Finkelstein Reaction. *J. Am. Chem. Soc.* **2002**, 124, 14844-14845.
10. McLennan, D. J. Semi-empirical calculation of rates of S_N2 Finkelstein reactions in solution by a quasi-thermodynamic cycle. *Aust. J. Chem.* **1978**, 31, 1897-1909.
11. Chalk, C. D., McKenna, J., Williams, I. H. NPE effects in bimolecular nucleophilic substitution. *J. Am. Chem. Soc.* **1981**, 103, 272-281.
12. Tucker, S. C., Truhlar, D. G. Ab initio calculations of the transition-state geometry and vibrational frequencies of the S_N2 reaction of chloride with chloromethane. *J. Phys. Chem.* **1989**, 93, 8138-8142.
13. Yamataka, H. Theoretical calculations of organic reactions in solution. *Rev. on Heteroa. Chem.* **1999**, 21, 277-291.
14. Jaworski, J. S. Looking for a contribution of the non-equilibrium solvent polarization to the activation barrier of the S_N2 reaction. *J. Phys. Org. Chem.* **2002**, 15, 319-323.
15. Perkin, W. H., Duppa, B. F. Iodoacetic acid. *Liebigs Ann. Chem.* **1859**, 112, 125-127.
16. Lange, U., Senning, A. Improved synthesis of fluoromethanesulfonyl chloride. *Chem. Ber.* **1991**, 124, 1879-1880.
17. Hayami, J., Tanaka, N., Hihara, N., Kaji, A. S_N2 reactions in dipolar aprotic solvents. V. Nucleophile-substrate complex in solution. Detection of chloride-organic chloride association and the potential role of the complexes in the S_N2 reaction. *Tetrahedron Lett.* **1973**, 385-388.
18. Holman, J. The Finkelstein reaction. *Sch. Sci. Rev.* **1977**, 58, 476-477.
19. Hayami, J., Koyanagi, T., Hihara, N., Kaji, A. Substrate-nucleophile association in the Finkelstein reaction system in a dipolar aprotic solvent. Formation of complex between substituted chloromethanes and halide ion in acetonitrile. *Bull. Chem. Soc. Jpn.* **1978**, 51, 891-896.
20. Hayami, J., Hihara, N., Kaji, A. S_N2 reactions in dipolar aprotic solvents. IX. An estimation of nucleophilicities and nucleofugicities of anionic nucleophiles studied in the reversible Finkelstein reactions of benzyl derivatives in acetonitrile - dissociative character of the reaction as studied by the nucleofugicity approach. *Chem. Lett.* **1979**, 413-414.
21. Hayami, J., Koyanagi, T., Kaji, A. S_N2 reactions in dipolar aprotic solvents. VIII. Chlorine isotopic exchange reaction of (arylsulfonyl)chloromethanes, (arylsulfinyl)chloromethanes, and 2-chloro-1-arylethanones in acetonitrile. A role of the nucleophile-substrate interaction in the Finkelstein reaction. *Bull. Chem. Soc. Jpn.* **1979**, 52, 1441-1446.
22. Smith, W. B., Branum, G. D. The abnormal Finkelstein reaction. A sequential ionic-free radical reaction mechanism. *Tetrahedron Lett.* **1981**, 22, 2055-2058.
23. Maartmann-Moe, K., Sanderud, K. A., Songstad, J. Reactions of benzylic compounds. Nucleophilicity, leaving group ability and carbon basicity of some ionic nucleophiles in acetonitrile. Comments on the utility of the Finkelstein reaction in synthesis. *Acta Chem. Scand.* **1982**, B36, 211-223.
24. Hayami, J., Otani, S., Hashimoto, S. Solute-solvent interactions in the Finkelstein reaction system. Characterization of chloride and perchlorate anion. *Stud. Org. Chem. (Amsterdam)* **1987**, 31, 561-566.
25. Lin, S. N., Jwo, J. J. Kinetic study of the substitution reactions of benzyl halides and halide ions in acetone. *J. Chin. Chem. Soc.* **1988**, 35, 85-103.
26. Hsu, M. C., Jwo, J. J. Kinetic study of the catalyzed substitution reaction of benzal chloride and sodium iodide in acetone. *J. Chin. Chem. Soc.* **1989**, 36, 403-412.
27. Landini, D., Albanese, D., Mottadelli, S., Penso, M. Finkelstein reaction with aqueous hydrogen halides efficiently catalyzed by lipophilic quaternary onium salts. *J. Chem. Soc., Perkin Trans. 1* **1992**, 2309-2311.
28. Morimoto, Y., Iwahashi, M., Kinoshita, T., Nishida, K. Stereocontrolled total synthesis of the Stemona alkaloid (-)-stenine. *Chem.-- Eur. J.* **2001**, 7, 4107-4116.
29. Boisnard, S., Carbonnelle, A.-C., Zhu, J. Studies on the Total Synthesis of RP 66453: Synthesis of Fully Functionalized 15-Membered Biaryl-Containing Macrocycle. *Org. Lett.* **2001**, 3, 2061-2064.
30. Stahl, P., Kissau, L., Mazitschek, R., Huwe, A., Furet, P., Giannis, A., Waldmann, H. Total synthesis and biological evaluation of the nakijiquinones. *J. Am. Chem. Soc.* **2001**, 123, 11586-11593.
31. Kim, D., Lee, J., Shim, P. J., Lim, J. I., Jo, H., Kim, S. Asymmetric Total Synthesis of (+)-Brefeldin A from (S)-Lactate by Triple Chirality Transfer Process and Nitrile Oxide Cycloaddition. *J. Org. Chem.* **2002**, 67, 764-771.

Related reactions: *Bartoli indole synthesis, Larock indole synthesis, Madelung indole synthesis, Nenitzescu indole synthesis;*

1. Fischer, E., Jourdan, F. The hydrazone of pyruvic acid. *Ber.* **1883**, 16, 2241-2245.
2. Fischer, E., Hess, O. The synthesis of indole derivatives. *Ber.* **1884**, 17.
3. Robinson, B. Studies on the Fischer indole synthesis. *Chem. Rev.* **1969**, 69, 227-250.
4. Robinson, B. *The Fischer Indole Synthesis* (Wiley, New York, N. Y., **1982**) 923 pp.
5. Ambekar, S. Y. Recent developments in the Fischer indole synthesis. *Curr. Sci.* **1983**, 52, 578-582.
6. Thummel, R. P. The application of Friedlaender and Fischer methodologies to the synthesis of organized polyaza cavities. *Synlett* **1992**, 1-12.
7. Hughes, D. L. Progress in the Fischer indole reaction. A review. *Org. Prep. Proced. Int.* **1993**, 25, 607-632.
8. Martin, M. J., Dorn, L. J., Cook, J. M. Novel pyridodiindoles, azadiindoles, and indolopyridoimidazoles via the Fischer-indole cyclization. *Heterocycles* **1993**, 36, 157-189.
9. Downing, R. S., Kunkeler, P. J. *The Fischer indole synthesis* (eds. Sheldon, R. A.,Bekkum, H.) (Weinheim: Wiley-VCH, New York, **2001**) 178-183.
10. Katritzky, A. R., Rachwal, S., Bayyuk, S. An improved Fischer synthesis of nitroindoles. 1,3-Dimethyl-4,5- and 6-nitroindoles. *Org. Prep. Proced. Int.* **1991**, 23, 357-363.
11. Chen, C.-y., Senanayake, C. H., Bill, T. J., Larsen, R. D., Verhoeven, T. R., Reider, P. J. Improved Fischer Indole Reaction for the Preparation of N,N-Dimethyltryptamines: Synthesis of L-695,894, a Potent 5-HT1D Receptor Agonist. *J. Org. Chem.* **1994**, 59, 3738-3741.
12. Zimmermann, T. A facile synthesis of 3H-indolium perchlorates by one-pot hydrazone formation/Fischer indolization. *J. Heterocycl. Chem.* **2000**, 37, 1571-1574.
13. Lipin'ska, T. 1,2,4-Triazines in organic synthesis. 9. Synthesis of 3-(3-ethylindol-2-yl)-5,6,7,8-tetrahydroisoquinoline using the Fischer reaction under the usual conditions and with microwave irradiation. *Chem. Het. Comp. (New York) (Translation of Khim. Geterot. Soed.)* **2001**, 37, 231-236.
14. Lacoume, B., Milcent, G., Olivier, A. Regioselectivity in the Fischer indole synthesis using 3-substituted cyclanones. *Tetrahedron* **1972**, 28, 667-674.
15. Pulici, M., Sello, G. Studies toward a model for the prediction of the regioselectivity in the Fischer indole synthesis. Part 2. Derivatives of cyclic ketones. *THEOCHEM* **1993**, 107, 245-254.
16. Pulici, M., Sello, G. Studies toward a model for the prediction of the regioselectivity in the Fischer indole synthesis. *THEOCHEM* **1993**, 100, 195-206.
17. Keresclidze, J., Raevski, N. Quantum-chemical study of conformation of phenylhydrazone ethyl pyruvate. *Bull. of the Georgian Acad. Sci.* **1996**, 153, 380-381.
18. Kereselidze, J., Raevski, K. The quantum-chemical study of N-N bond cleavage in phenylhydrazones. *Izv. Akad. Nauk Gruz. SSR, Ser. Khim.* **1996**, 22, 170-172.
19. Rosas-Garcia, V. M., Quintanilla-Licea, R., Longoria R, F. E. The Fischer indole synthesis: a semiempirical study. *ECHET98: Electronic Conference on Heterocyclic Chemistry, June 29-July 24, 1998* **1998**, 237-243.
20. Kereselidze, J. A. New views on hydrazone-enehydrazine tautomerism. *Chem. Het. Comp. (New York) (Translation of Khim. Geterot. Soed.)* **1999**, 35, 666-670.
21. Gverdtsiteli, M., Samsonia, N. Investigation of Fischer's reaction within the scope of quasi-ANB-matrices method. *Bull. of the Georgian Acad. Sci.* **2001**, 164, 68-69.
22. Robinson, G. M., Robinson, R. Mechanism of E. Fischer's synthesis of indoles. Application of the method to the preparation of a pyrindole derivative. *J. Chem. Soc., Abstracts* **1924**, 125, 827-840.
23. Owellen, R. J., Fitzgerald, J. A., Fitzgerald, B. M., Welsh, D. A., Walker, D. M., Southwick, P. L. Cyclization phase of the Fischer indole synthesis. The structure and significance of Plieninger's intermediate. *Tetrahedron Lett.* **1967**, 1741-1746.
24. Elgersma, R. H. C., Havinga, E. Fischer indole synthesis. I. Structure of a supposed intermediate. *Tetrahedron Lett.* **1969**, 1735-1736.
25. Palmer, M. H., McIntyre, P. S. Fischer indole synthesis on unsymmetrical ketones. Effect of the acid catalyst. *J. Chem. Soc. B.* **1969**, 446-449.
26. Ishii, H., Murakami, Y., Suzuki, Y., Ikeda, N. Substitution and migration of methoxyl group in the Fischer indolization of ethyl pyruvate 2-methoxyphenylhydrazone. *Tetrahedron Lett.* **1970**, 1181-1184.
27. Forrest, T. P., Chen, F. M. F. Isolation of a 2-aminoindoline derivative. Suggested intermediate in the Fischer indole synthesis. *J. Chem. Soc., Chem. Commun.* **1972**, 1067.
28. Fusco, R., Sannicolo, F. Fischer indole synthesis. III. Evidence for a double 1,2-shift of a methyl group. *Gazz. Chim. Ital.* **1975**, 105, 465-472.
29. Fusco, R., Sannicolo, F. Studies on the Fischer Indole synthesis. V. Shift and elimination of substituent. *Gazz. Chim. Ital.* **1976**, 106, 85-94.
30. Miller, B., Matjeka, E. R. The mechanism of 1,4-methyl migration is the Fischer Indole reaction. *Tetrahedron Lett.* **1977**, 131-134.
31. Miller, F. M., Schinske, W. N. Direction of cyclization in the Fischer indole synthesis. Mechanistic considerations. *J. Org. Chem.* **1978**, 43, 3384-3388.
32. Douglas, A. W. In situ nitrogen-15 nuclear magnetic resonance observation of the Fischer indolization reaction. Nitrogen-15 NMR characterization of amide-imine intermediates. *J. Am. Chem. Soc.* **1979**, 101, 5676-5678.
33. Ishii, H., Sugiura, T., Akiyama, Y., Ichikawa, Y., Watanabe, T., Murakami, Y. Fischer indolization and its related compounds. XXIII. Fischer indolization of ethyl pyruvate 2-(2,6-dimethoxyphenyl)phenylhydrazone. *Chem. Pharm. Bull.* **1990**, 38, 2118-2126.
34. Ishii, H., Sugiura, T., Kogusuri, K., Watanabe, T., Murakami, Y. Fischer indolization and its related compounds. XXIV. Fischer indolization of ethyl pyruvate 2-(2-methoxyphenyl)phenylhydrazone. *Chem. Pharm. Bull.* **1991**, 39, 572-578.
35. Hughes, D. L., Zhao, D. Mechanistic studies of the Fischer indole reaction. *J. Org. Chem.* **1993**, 58, 228-233.
36. Hughes, D. L. An unusual parabolic dependence of rate on acidity in the Fischer indole reaction. *J. Phys. Org. Chem.* **1994**, 7, 625-628.
37. Murakami, Y., Watanabe, T., Otsuka, T., Iwata, T., Yamada, Y., Yokoyama, Y. Fischer indolization of ethyl pyruvate 2-bis(2-methoxyphenyl)hydrazone and new insight into the mechanism of Fischer indolization. Fischer indolization and its related compounds. XXVII. *Chem. Pharm. Bull.* **1995**, 43, 1287-1293.
38. Fujii, H., Mizusuna, A., Tanimura, R., Nagase, H. A novel abnormal rearrangement in the Fischer indole synthesis. *Heterocycles* **1997**, 45, 2109-2112.
39. Bast, K., Durst, T., Huisgen, R., Lindner, K., Temme, R. 1,3-Dipolar cycloadditions. 104. Can the progress of Fischer's indole synthesis be stopped? *Tetrahedron* **1998**, 54, 3745-3764.
40. Bonjoch, J., Catena, J., Valls, N. Total Synthesis of (±)-Deethylibophyllidine: Studies of a Fischer Indolization Route and a Successful Approach via a Pummerer Rearrangement/Thionium Ion-Mediated Indole Cyclization. *J. Org. Chem.* **1996**, 61, 7106-7115.
41. Iyengar, R., Schildknegt, K., Aube, J. Regiocontrol in an Intramolecular Schmidt Reaction: Total Synthesis of (+)-Aspidospermidine. *Org. Lett.* **2000**, 2, 1625-1627.
42. Roberson, C. W., Woerpel, K. A. Development of the [3 + 2] Annulations of Cyclohexenylsilanes and Chlorosulfonyl Isocyanate: Application to the Total Synthesis of (±)-Peduncularine. *J. Am. Chem. Soc.* **2002**, 124, 11342-11348.
43. Gan, T., Liu, R., Yu, P., Zhao, S., Cook, J. M. Enantiospecific Synthesis of Optically Active 6-Methoxytryptophan Derivatives and Total Synthesis of Tryprostatin A. *J. Org. Chem.* **1997**, 62, 9298-9304.

1. Tamao, K., Akita, M., Kumada, M. Silafunctional compounds in organic synthesis. XVIII. Oxidative cleavage of the silicon-carbon bond in alkenylfluorosilanes to carbonyl compounds: synthetic and mechanistic aspects. *J. Organomet. Chem.* **1983**, 254, 13-22.
2. Tamao, K., Ishida, N., Kumada, M. (Diisopropoxymethylsilyl)methyl Grignard reagent: a new, practically useful nucleophilic hydroxymethylating agent. *J. Org. Chem.* **1983**, 48, 2120-2122.
3. Tamao, K., Ishida, N., Tanaka, T., Kumada, M. Silafunctional compounds in organic synthesis. Part 20. Hydrogen peroxide oxidation of the silicon-carbon bond in organoalkoxysilanes. *Organometallics* **1983**, 2, 1694-1696.
4. Tamao, K., Kakui, T., Akita, M., Iwahara, T., Kanatani, R., Yoshida, J., Kumada, M. Organofluorosilicates in organic synthesis. Part 17. Oxidative cleavage of silicon-carbon bonds in organosilicon fluorides to alcohols. *Tetrahedron* **1983**, 39, 983-990.
5. Fleming, I., Henning, R., Plaut, H. The phenyldimethylsilyl group as a masked form of the hydroxy group. *J. Chem. Soc., Chem. Commun.* **1984**, 29-31.
6. Tamao, K., Ishida, N. Silafunctional compounds in organic synthesis. XXVI. Silyl groups synthetically equivalent to the hydroxy group. *J. Organomet. Chem.* **1984**, 269, C37-C39.
7. Fleming, I., Sanderson, P. E. J. A one-pot conversion of the phenyldimethylsilyl group into an hydroxyl group. *Tetrahedron Lett.* **1987**, 28, 4229-4232.
8. Colvin, E. W. *Silicon Reagents in Organic Synthesis* (Academic Press, London, San Diego, **1988**) 147 pp.
9. Colvin, E. W. Oxidation of Carbon-Silicon Bonds. in *Comp. Org. Synth.* (eds. Trost, B. M.,Fleming, I.), *7*, 641-653 (Pergamon Press, Oxford, **1991**).
10. Fleming, I. Silyl-to-hydroxy conversion in organic synthesis. *Chemtracts: Org. Chem.* **1996**, 9, 1-64.
11. Jones, G. R., Landais, Y. The oxidation of the carbon-silicon bond. *Tetrahedron* **1996**, 52, 7599-7662.
12. Tamao, K. Oxidative cleavage of the silicon-carbon bond: Development, mechanism, scope, and limitations. *Adv. Silicon Chem.* **1996**, 3, 1-62.
13. Tamao, K., Hayashi, T., Ito, Y. Oxidative cleavage of carbon-silicon bonds by dioxygen: catalysis by a flavin-dihydronicotinamide redox system. *J. Chem. Soc., Chem. Commun.* **1988**, 795-797.
14. Magar, S. S., Fuchs, P. L. Synthesis of tertiary alcohols via the use of the allyldimethylsilyl moiety as a latent hydroxyl group in the Kumada-Fleming-Tamao reaction. *Tetrahedron Lett.* **1991**, 32, 7513-7516.
15. Suginome, M., Matsunaga, S.-i., Ito, Y. Disilanyl group as a synthetic equivalent of the hydroxyl group. *Synlett* **1995**, 941-942.
16. Itami, K., Mitsudo, K., Yoshida, J.-i. Oxidation of 2-Pyridyldimethylsilyl Group to Hydroxyl Group by H_2O_2/KF. Implication of Fluoride Ion Accelerated 2-Pyridyl-Silyl Bond Cleavage. *J. Org. Chem.* **1999**, 64, 8709-8714.
17. Denmark, S. E., Hurd, A. R. Synthesis of (+)-Casuarine. *J. Org. Chem.* **2000**, 65, 2875-2886.
18. Mader, M. M., Norrby, P.-O. Computational investigation of the role of fluoride in Tamao oxidations. *Chem.-- Eur. J.* **2002**, 8, 5043-5048.
19. Vanecko, J. A., West, F. G. A Novel, Stereoselective Silyl-Directed Stevens [1,2]-Shift of Ammonium Ylides. *Org. Lett.* **2002**, 4, 2813-2816.
20. Sun, P., Sun, C., Weinreb, S. M. Stereoselective Total Syntheses of the Racemic Form and the Natural Enantiomer of the Marine Alkaloid Lepadiformine via a Novel *N*-Acyliminium Ion/Allylsilane Spirocyclization Strategy. *J. Org. Chem.* **2002**, 67, 4337-4345.
21. Usuda, H., Kanai, M., Shibasaki, M. Studies toward the Total Synthesis of Garsubellin A: A Concise Synthesis of the 18-epi-Tricyclic Core. *Org. Lett.* **2002**, 4, 859-862.
22. Marshall, J. A., Yanik, M. M. Synthesis of a C1-C21 Subunit of the Protein Phosphatase Inhibitor Tautomycin: A Formal Total Synthesis. *J. Org. Chem.* **2001**, 66, 1373-1379.

Related reactions: Fries-, Photo-Fries and Anionic Ortho-Fries rearrangement, Houben-Hoesch reaction, Minisci reaction;

1. Crafts, J. M., Ador, E. The reaction of phosgene with toluene in the presence of aluminum chloride. *Ber.* **1877**, 10, 2173-2176.
2. Crafts, J. M., Ador, E. Effect of phthalic anhydride on naphthalin in the presence of aluminum trichloride. *Bull. Soc. Chim. France* **1880**, 531-532.
3. Calloway, N. O. The Friedel-Crafts syntheses. *Chem. Rev.* **1935**, 17, 327-392.
4. Berliner, E. Friedel and Crafts reaction with aliphatic dibasic acid anhydrides. *Org. React.* **1949**, 5, 229-289.
5. Gore, P. H. The Friedel-Crafts acylation reaction and its application to polycyclic aromatic hydrocarbons. *Chem. Rev.* **1955**, 55, 229-281.
6. Olah, G. A. Miscellaneous Reactions, Cumulative Indexes. in *Friedel-Crafts and Related Reactions 4*, 1191 pp. (Interscience Publishers, New York, **1965**).
7. Groves, J. K. Friedel-Crafts acylation of alkenes. *Chem. Soc. Rev.* **1972**, 1, 73-97.
8. Pearson, D. E., Buehler, C. A. Friedel-Crafts acylations with little or no catalyst. *Synthesis* **1972**, 533-542.
9. Olah, G. A. *Interscience Monographs on Organic Chemistry: Friedel-Crafts Chemistry* (Wiley-Interscience, New York, N. Y., **1973**) 581 pp.
10. Gore, P. H. Friedel-Crafts acylations. Unusual aspects of selectivity. *Chem. Ind. (London)* **1974**, 727-731.
11. Yakobson, G. G., Furin, G. G. Antimony pentahalides as catalysts of Friedel-Crafts type reactions. *Synthesis* **1980**, 345-364.
12. Ashforth, R., Desmurs, J.-R. Friedel-Crafts acylation: Interactions between Lewis acids-acyl chlorides and Lewis acids-aryl ketones. *Ind. Chem. Library* **1996**, 8, 3-14.
13. Desmurs, J.-R., Labrouillere, M., Dubac, J., Laporterie, A., Gaspard, H., Metz, F. Bismuth(III) salts in Friedel-Crafts acylation. *Ind. Chem. Library* **1996**, 8, 15-28.
14. Hasumoto, I., Takatoshi, K., Badea, F. D., Sawada, T., Mataka, S., Tashiro, M. Regioselectivity of Friedel-Crafts acylation of aromatic compounds with several cyclic anhydrides. *Res. Chem. Intermed.* **1996**, 22, 855-869.
15. Spagnol, M., Gilbert, L., Alby, D. Friedel-Crafts acylation of aromatics using zeolites. *Ind. Chem. Library* **1996**, 8, 29-38.
16. Mahato, S. B. Advances in the chemistry of Friedel-Crafts acylation. *J. Indian Chem. Soc.* **2000**, 77, 175-191.
17. Metivier, P. *Friedel-Crafts acylation* (eds. Sheldon, R. A.,Bekkum, H.) (Weinheim: Wiley-VCH, New York, **2001**) 161-172.
18. Kozhevnikov, I. V. Friedel-Crafts acylation and related reactions catalyzed by heteropoly acids. *Appl. Cat. A* **2003**, 256, 3-18.
19. Galatsis, P., Manwell, J. J., Blackwell, J. M. Indenone synthesis. Improved synthetic protocol and effect of substitution on the intramolecular Friedel-Crafts acylation. *Can. J. Chem.* **1994**, 72, 1656-1659.
20. Hachiya, I., Moriwaki, M., Kobayashi, S. Hafnium(IV) trifluoromethanesulfonate, an efficient catalyst for the Friedel-Crafts acylation and alkylation reactions. *Bull. Chem. Soc. Jpn.* **1995**, 68, 2053-2060.
21. Hachiya, I., Moriwaki, M., Kobayashi, S. Catalytic Friedel-Crafts acylation reactions using hafnium triflate as a catalyst in lithium perchlorate-nitromethane. *Tetrahedron Lett.* **1995**, 36, 409-412.
22. Ranu, B. C., Ghosh, K., Jana, U. Simple and Improved Procedure for Regioselective Acylation of Aromatic Ethers with Carboxylic Acids on the Solid Surface of Alumina in the Presence of Trifluoroacetic Anhydride. *J. Org. Chem.* **1996**, 61, 9546-9547.
23. Smyth, T. P., Corby, B. W. Toward a Clean Alternative to Friedel-Crafts Acylation: In Situ Formation, Observation, and Reaction of an Acyl Bis(trifluoroacetyl)phosphate and Related Structures. *J. Org. Chem.* **1998**, 63, 8946-8951.
24. Nakano, H., Kitazume, T. Friedel-Crafts reaction in fluorous fluids. *Green Chem.* **1999**, 1, 179-181.
25. Le Roux, C., Dubac, J. Bismuth(III) chloride and triflate: novel catalysts for acylation and sulfonylation reactions. Survey and mechanistic aspects. *Synlett* **2002**, 181-200.
26. McMills, M. C., Wright, D. L., Weekly, R. M. Synthesis of highly functionalized arene systems. Novel selectivities of intra- and intermolecular Friedel-Crafts reactions. *Synth. Commun.* **2002**, 32, 2417-2425.
27. Ross, J., Xiao, J. Friedel-Crafts acylation reactions using metal triflates in ionic liquid. *Green Chem.* **2002**, 4, 129-133.
28. Gmouh, S., Yang, H., Vaultier, M. Activation of Bismuth(III) Derivatives in Ionic Liquids: Novel and Recyclable Catalytic Systems for Friedel-Crafts Acylation of Aromatic Compounds. *Org. Lett.* **2003**, 5, 2219-2222.

29. Jorgensen, K. A. Asymmetric Friedel-Crafts reactions: Catalytic enantioselective addition of aromatic and heteroaromatic C-H bonds to activated alkenes, carbonyl compounds, and imines. *Synthesis* **2003**, 1117-1125.
30. Morrill, T. C., Opitz, R., Replogle, L. L., Katsumoto, K., Schroeder, W., Hess, B. A., Jr. Correspondence between theoretically predicted and experimentally observed sites of electrophilic substitution on a fused tricyclic heteroaromatic (azulene) system. *Tetrahedron Lett.* **1975**, 2077-2080.
31. Branchadell, V., Oliva, A., Bertran, J. A theoretical insight into the catalytic action in Friedel-Crafts reactions. *J. Mol. Catal.* **1988**, 44, 285-294.
32. Branchadell, V., Oliva, A., Bertran, J. Theoretical study of the acid-catalyzed Friedel-Crafts reaction between methyl fluoride and methane. *J. Chem. Soc., Perkin Trans. 2* **1989**, 1091-1096.
33. Ertel, T. S., Bertagnolli, H. EXAFS spectroscopy and MNDO/AM1/PM3 calculations: a structural study of a model system for Friedel-Crafts alkylation. *J. Mol. Struct.* **1993**, 301, 143-154.
34. Xu, T., Barich, D. H., Torres, P. D., Haw, J. F. Benzenium Ion Chemistry on Solid Metal Halide Superacids: In Situ ^{13}C NMR Experiments and Theoretical Calculations. *J. Am. Chem. Soc.* **1997**, 119, 406-414.
35. Xu, T., Barich, D. H., Torres, P. D., Nicholas, J. B., Haw, J. F. Carbon-13 Chemical Shift Tensors for Acylium Ions: A Combined Solid State NMR and Ab Initio Molecular Orbital Study. *J. Am. Chem. Soc.* **1997**, 119, 396-405.
36. Tarakeshwar, P., Lee, J. Y., Kim, K. S. Role of Lewis Acid ($AlCl_3$)-Aromatic Ring Interactions in Friedel-Craft's Reaction: An Ab Initio Study. *J. Phys. Chem. A* **1998**, 102, 2253-2255.
37. Gothelf, A. S., Hansen, T., Jorgensen, K. A. Studies on aluminum mediated asymmetric Friedel-Crafts hydroxyalkylation reactions of pyridinecarbaldehydes. *J. Chem. Soc., Perkin Trans. 1* **2001**, 854-860.
38. Csihony, S., Mehdi, H., Homonnay, Z., Vertes, A., Farkas, O., Horvath, I. T. In situ spectroscopic studies related to the mechanism of the Friedel-Crafts acetylation of benzene in ionic liquids using $AlCl_3$ and $FeCl_3$. *J. Chem. Soc., Dalton Trans.* **2002**, 680-685.
39. Meric, P., Finiels, A., Moreau, P. Kinetics of 2-methoxynaphthalene acetylation with acetic anhydride over dealuminated HY zeolites. *J. Mol. Catal. A: Chemical* **2002**, 189, 251-262.
40. Olah, G. A., Toeroek, B., Joschek, J. P., Bucsi, I., Esteves, P. M., Rasul, G., Prakash, G. K. S. Efficient Chemoselective Carboxylation of Aromatics to Arylcarboxylic Acids with a Superelectrophilically Activated Carbon Dioxide-Al_2Cl_6/Al System. *J. Am. Chem. Soc.* **2002**, 124, 11379-11391.
41. Meima, G. R., Lee, G. S., Garces, J. M. *Friedel-Crafts acylation* (eds. Sheldon, R. A.,Bekkum, H.) (Weinheim: Wiley-VCH, New York, **2001**) 161-172.
42. Pines, S. H., Douglas, A. W. Friedel-Crafts chemistry. A mechanistic study of the reaction of 3-chloro-4'-fluoro-2-methylpropiophenone with aluminum chloride and aluminum chloride-nitromethane. *J. Org. Chem.* **1978**, 43, 3126-3131.
43. Beak, P., Berger, K. R. Scope and mechanism of the reaction of olefins with anhydrides and zinc chloride to give β,γ-unsaturated ketones. *J. Am. Chem. Soc.* **1980**, 102, 3848-3856.
44. Lee, C. C., Zohdi, H. F., Sallam, M. M. M. Hydrogen-deuterium exchanges in a Friedel-Crafts reaction. *J. Org. Chem.* **1985**, 50, 705-707.
45. Selvin, R., Sivasankar, B., Rengaraj, K. Kinetic studies on Friedel-Crafts acylation of anisole by clayzic. *React. Kinet. Catal. Lett.* **1999**, 67, 319-324.
46. Csihony, S., Mehdi, H., Horvath, I. T. In situ infrared spectroscopic studies of the Friedel-Crafts acetylation of benzene in ionic liquids using $AlCl_3$ and $FeCl_3$. *Green Chem.* **2001**, 3, 307-309.
47. Effenberger, F., Maier, A. H. Changing the Ortho/Para Ratio in Aromatic Acylation Reactions by Changing Reaction Conditions: A Mechanistic Explanation from Kinetic Measurements. *J. Am. Chem. Soc.* **2001**, 123, 3429-3433.
48. Overman, L. E., Tomasi, A. L. Enantioselective Total Synthesis of Hispidospermidin. *J. Am. Chem. Soc.* **1998**, 120, 4039-4040.
49. Boger, D. L., Hong, J., Hikota, M., Ishida, M. Total Synthesis of Phomazarin. *J. Am. Chem. Soc.* **1999**, 121, 2471-2477.
50. Krohn, K., Zimmermann, G. Transition-Metal-Catalyzed Oxidations. 11. Total Synthesis of (±)-Lacinilene C Methyl Ether by β-Naphthol to α-Ketol Oxidation. *J. Org. Chem.* **1998**, 63, 4140-4142.
51. Fürstner, A., Weintritt, H. Total Synthesis of the Potent Antitumor Agent Roseophilin: A Concise Approach to the Macrotricyclic Core. *J. Am. Chem. Soc.* **1997**, 119, 2944-2945.

Friedel-Crafts Alkylation 178

Related reactions: Minisci reaction;

1. Friedel, C., Crafts, J. M. A new general synthetical method of producing hydrocarbons. *J. Chem.Soc.* **1877**, 32, 725.
2. Friedel, C., Crafts, J. M. *Bull. Soc. Chim. France* **1877**, 27, 530.
3. Calloway, N. O. The Friedel-Crafts syntheses. *Chem. Rev.* **1935**, 17, 327-392.
4. Price, C. C. Alkylation of aromatic compounds by the Friedel-Crafts method. *Org. React.* **1946**, 1-82.
5. Olah, G. A. Miscellaneous Reactions, Cumulative Indexes. in *Friedel-Crafts and Related Reactions 4*, 1191 pp. (Interscience Publishers, New York, **1965**).
6. Olah, G. A. *Interscience Monographs on Organic Chemistry: Friedel-Crafts Chemistry* (**1973**) 581 pp.
7. Olah, G. A., Meidar, D. *Friedel-Crafts reactions* (Wiley, New York, **1980**) 269-300.
8. Yakobson, G. G., Furin, G. G. Antimony pentahalides as catalysts of Friedel-Crafts type reactions. *Synthesis* **1980**, 345-364.
9. Olah, G. A., Krishnamurti, R., Surya, G. K. Friedel-Crafts alkylations. in *Comp. Org. Synth.* (eds. Trost, B. M.,Fleming, I.), *3*, 293-339 (Pergamon, Oxford, **1991**).
10. Jung, I. N., Yoo, B. R. Friedel-Crafts alkylations with silicon compounds. *Adv. Organomet. Chem.* **2000**, 46, 145-180.
11. Meima, G. R., Lee, G. S., Garces, J. M. *Friedel-Crafts alkylation* (eds. Sheldon, R. A.,Bekkum, H.) (Weinheim: Wiley-VCH, New York, **2001**) 151-160.
12. Wan, Y., Ding, K., Dai, L., Ishii, A., Soloshonok, V. A., Mikami, K., Gathergood, N., Zhuang, W., Jorgensen, K. A., Jesen, K. B., Thorhauge, J., Hazell, R. G. Enantioselective Friedel-Crafts reaction: from stoichiometric to catalytic. *Chemtracts* **2001**, 14, 610-615.
13. Bandini, M., Melloni, A., Umani-Ronchi, A. New catalytic approaches in the stereoselective Friedel-Crafts alkylation reaction. *Angew. Chem., Int. Ed. Engl.* **2004**, 43, 550-556.
14. Roberts, R. M., Anderson, G. P., Jr., Khalaf, A. A., Low, C.-E. New Friedel-Crafts chemistry. XXV. Friedel-Crafts cycloalkylations and bicyclialkylations with diphenylalkyl chlorides. *J. Org. Chem.* **1971**, 36, 3342-3345.
15. Mayr, H., Striepe, W. Scope and limitations of aliphatic Friedel-Crafts alkylations. Lewis acid catalyzed addition reactions of alkyl chlorides to carbon-carbon double bonds. *J. Org. Chem.* **1983**, 48, 1159-1165.
16. Mine, N., Fujiwara, Y., Taniguchi, H. Trichlorolanthanoid($LnCl_3$)-catalyzed Friedel-Crafts alkylation reactions. *Chem. Lett.* **1986**, 357-360.
17. Moodie, R. B. Electrophilic aromatic substitution. *Org. React. Mech.* **1986**, 269-281.
18. Clark, J. H., Kybett, A. P., Macquarrie, D. J., Barlow, S. J., Landon, P. Montmorillonite supported transition metal salts as Friedel-Crafts alkylation catalysts. *J. Chem. Soc., Chem. Commun.* **1989**, 1353-1354.
19. Cativiela, C., Garcia, J. I., Garcia-Matres, M., Mayoral, J. A., Figueras, F., Fraile, J. M., Cseri, T., Chiche, B. Clay-catalyzed Friedel-Crafts alkylation of anisole with dienes. *Appl. Cat. A* **1995**, 123, 273-287.
20. Hachiya, I., Moriwaki, M., Kobayashi, S. Hafnium(IV) trifluoromethanesulfonate, an efficient catalyst for the Friedel-Crafts acylation and alkylation reactions. *Bull. Chem. Soc. Jpn.* **1995**, 68, 2053-2060.
21. Desmurs, J.-R., Labrouillere, M., Dubac, J., Laporterie, A., Gaspard, H., Metz, F. Bismuth(III) salts in Friedel-Crafts acylation. *Ind. Chem. Library* **1996**, 8, 15-28.
22. Retey, J. Enzymic catalysis by Friedel-Crafts-type reactions. *Naturwissenschaften* **1996**, 83, 439-447.
23. Spagnol, M., Gilbert, L., Alby, D. Friedel-Crafts acylation of aromatics using zeolites. *Ind. Chem. Library* **1996**, 8, 29-38.
24. Ghorpade, S. P., Darshane, V. S., Dixit, S. G. Liquid-phase Friedel-Crafts alkylation using $CuCr_2$-$xFexO_4$ spinel catalysts. *Appl. Cat. A* **1998**, 166, 135-142.

25. Sukumar, R., Sabu, K. R., Bindu, L. V., Lalithambika, M. Kaolinite supported metal chlorides as Friedel-Crafts alkylation catalysts. *Stud. Surf. Sci. Catal.* **1998**, 113, 557-562.
26. Yonezawa, N., Hino, T., Ikeda, T. New approaches in Friedel-Crafts type carbon-carbon bond formation using novel types of Friedel-Crafts mediators. *Rec. Res. Dev. Synt. Org. Chem.* **1998**, 1, 213-223.
27. Nakano, H., Kitazume, T. Friedel-Crafts reaction in fluorous fluids. *Green Chem.* **1999**, 1, 179-181.
28. Fleming, I. Improving the Friedel-Crafts reaction. *Chemtracts* **2001**, 14, 405-406.
29. Paras, N. A., MacMillan, D. W. C. New Strategies in Organic Catalysis: The First Enantioselective Organocatalytic Friedel-Crafts Alkylation. *J. Am. Chem. Soc.* **2001**, 123, 4370-4371.
30. Corma, A., Garcia, H., Moussaif, A., Sabater, M. J., Zniber, R., Redouane, A. Chiral copper(II) bisoxazoline covalently anchored to silica and mesoporous MCM-41 as a heterogeneous catalyst for the enantioselective Friedel-Crafts hydroxyalkylation. *Chem. Commun.* **2002**, 1058-1059.
31. McMills, M. C., Wright, D. L., Weekly, R. M. Synthesis of highly functionalized arene systems. Novel selectivities of intra- and intermolecular Friedel-Crafts reactions. *Synth. Commun.* **2002**, 32, 2417-2425.
32. Wasserscheid, P., Sesing, M., Korth, W. Hydrogen sulfate and tetrakis(hydrogen sulfato)borate ionic liquids: synthesis and catalytic application in highly Bronsted-acidic systems for Friedel-Crafts alkylation. *Green Chem.* **2002**, 4, 134-138.
33. Evans, D. A., Scheidt, K. A., Fandrick, K. R., Lam, H. W., Wu, J. Enantioselective Indole Friedel-Crafts Alkylations Catalyzed by Bis(oxazolinyl)pyridine-Scandium(III) Triflate Complexes. *J. Am. Chem. Soc.* **2003**, 125, 10780-10781.
34. Jorgensen, K. A. Asymmetric Friedel-Crafts reactions: Catalytic enantioselective addition of aromatic and heteroaromatic C-H bonds to activated alkenes, carbonyl compounds, and imines. *Synthesis* **2003**, 1117-1125.
35. Kumarraja, M., Pitchumani, K. Divalent transition metal ion-exchanged faujasites as mild, efficient, heterogeneous Friedel-Crafts benzylation catalysts. *Synth. Commun.* **2003**, 33, 105-111.
36. Saber, A., Smahi, A., Solhy, A., Nazih, R., Elaabar, B., Maizi, M., Sebti, S. Heterogeneous catalysis of Friedel-Crafts alkylation by the fluorapatite alone and doped with metal halides. *J. Mol. Catal. A: Chemical* **2003**, 202, 229-237.
37. Morrill, T. C., Opitz, R., Replogle, L. L., Katsumoto, K., Schroeder, W., Hess, B. A., Jr. Correspondence between theoretically predicted and experimentally observed sites of electrophilic substitution on a fused tricyclic heteroaromatic (azulene) system. *Tetrahedron Lett.* **1975**, 2077-2080.
38. Branchadell, V., Oliva, A., Bertran, J. A theoretical insight into the catalytic action in Friedel-Crafts reactions. *J. Mol. Catal.* **1988**, 44, 285-294.
39. Branchadell, V., Oliva, A., Bertran, J. Theoretical study of the acid-catalyzed Friedel-Crafts reaction between methyl fluoride and methane. *J. Chem. Soc., Perkin Trans. 2* **1989**, 1091-1096.
40. Ertel, T. S., Bertagnolli, H. EXAFS spectroscopy and MNDO/AM1/PM3 calculations: a structural study of a model system for Friedel-Crafts alkylation. *J. Mol. Struct.* **1993**, 301, 143-154.
41. Xu, T., Barich, D. H., Torres, P. D., Haw, J. F. Benzenium Ion Chemistry on Solid Metal Halide Superacids: In Situ ^{13}C NMR Experiments and Theoretical Calculations. *J. Am. Chem. Soc.* **1997**, 119, 406-414.
42. Xu, T., Barich, D. H., Torres, P. D., Nicholas, J. B., Haw, J. F. Carbon-13 Chemical Shift Tensors for Acylium Ions: A Combined Solid State NMR and Ab Initio Molecular Orbital Study. *J. Am. Chem. Soc.* **1997**, 119, 396-405.
43. Tarakeshwar, P., Lee, J. Y., Kim, K. S. Role of Lewis Acid ($AlCl_3$)-Aromatic Ring Interactions in Friedel-Craft's Reaction: An Ab Initio Study. *J. Phys. Chem. A* **1998**, 102, 2253-2255.
44. Gothelf, A. S., Hansen, T., Jorgensen, K. A. Studies on aluminum mediated asymmetric Friedel-Crafts hydroxyalkylation reactions of pyridinecarbaldehydes. *J. Chem. Soc., Perkin Trans. 1* **2001**, 854-860.
45. Olah, G. A., Toeroek, B., Joschek, J. P., Bucsi, I., Esteves, P. M., Rasul, G., Prakash, G. K. S. Efficient Chemoselective Carboxylation of Aromatics to Arylcarboxylic Acids with a Superelectrophilically Activated Carbon Dioxide-Al_2Cl_6/Al System. *J. Am. Chem. Soc.* **2002**, 124, 11379-11391.
46. Brown, H. C., Grayson, M. The catalytic halides. IX. Kinetics of the reaction of representative benzyl halides with aromatic compounds; evidence for a displacement mechanism in the Friedel-Crafts reactions of primary halides. *J. Am. Chem. Soc.* **1953**, 75, 6285-6292.
47. Brown, H. C., Jungk, H. The catalytic halides. XII. The reaction of benzene and toluene with methyl bromide and iodide in the presence of aluminum bromide; evidence for a displacement mechanism in the methylation of aromatic compounds. *J. Am. Chem. Soc.* **1955**, 77, 5584-5589.
48. Brown, H. C., Jungk, H. The catalytic halides. XI. The isomerization of *o*- and *p*-xylenes and some related alkylbenzenes under the influence of hydrogen bromide and aluminum bromide; the relative isomerization aptitudes of alkyl groups. *J. Am. Chem. Soc.* **1955**, 77, 5579-5584.
49. DeHaan, F. P., Delker, G. L., Covey, W. D., Ahn, J., Anisman, M. S., Brehm, E. C., Chang, J., Chicz, R. M., Cowan, R. L., et al. Electrophilic aromatic substitution. 8. A kinetic study of the Friedel-Crafts benzylation reaction in nitromethane, nitrobenzene, and sulfolane. Substituent effects in Friedel-Crafts benzylation. *J. Am. Chem. Soc.* **1984**, 106, 7038-7046.
50. Lee, C. C., Zohdi, H. F., Sallam, M. M. M. Hydrogen-deuterium exchanges in a Friedel-Crafts reaction. *J. Org. Chem.* **1985**, 50, 705-707.
51. Aschi, M., Attina, M., Cacace, F. An Alternative Route To Electrophilic Substitution. 2. Aromatic Alkylation in the Ion Neutral Complexes Formed Upon Addition of Gaseous Arenium Ions to Olefins. *J. Am. Chem. Soc.* **1995**, 117, 12832-12839.
52. Macknight, E., McClelland, R. A. A photochemical retro-Friedel-Crafts alkylation. Rapid rearrangement of cyclohexadienyl cations. *Can. J. Chem.* **1996**, 74, 2518-2527.
53. Janssens, B., Catry, P., Claessens, R., Baron, G., Jacobs, P. A. Zeolite catalysts for the Friedel-Crafts alkylation of methyl benzoate, a strongly deactivated aromatic substrate. *Stud. Surf. Sci. Catal.* **1997**, 105B, 1211-1218.
54. Molnar, A., Torok, B., Bucsi, I., Foldvari, A. Alkylation of aromatics with diols in superacidic media. *Top. in Cat.* **1998**, 6, 9-16.
55. Taunton, J., Wood, J. L., Schreiber, S. L. Total syntheses of di- and tri-O-methyl dynemicin A methyl esters. *J. Am. Chem. Soc.* **1993**, 115, 10378-10379.
56. Posner, G. H., Parker, M. H., Northrop, J., Elias, J. S., Ploypradith, P., Xie, S., Shapiro, T. A. Orally Active, Hydrolytically Stable, Semisynthetic, Antimalarial Trioxanes in the Artemisinin Family. *J. Med. Chem.* **1999**, 42, 300-304.
57. Patil, M. L., Borate, H. B., Ponde, D. E., Bhawal, B. M., Deshpande, V. H. First total synthesis of (±)-brasiliquinone B. *Tetrahedron Lett.* **1999**, 40, 4437-4438.
58. Ruell, J. A., De Clercq, E., Pannecouque, C., Witvrouw, M., Stup, T. L., Turpin, J. A., Buckheit, R. W., Jr., Cushman, M. Synthesis and Anti-HIV Activity of Cosalane Analogues with Substituted Benzoic Acid Rings Attached to the Pharmacophore through Methylene and Amide Linkers. *J. Org. Chem.* **1999**, 64, 5858-5866.

Fries-, Photo-Fries, and Anionic Ortho-Fries Rearrangement..180

Related reactions: ***Friedel-Crafts acylation, Houben-Hoesh reaction, Minisci reaction;***

1. Döbner, O. Benzoyl derivatives. *Ann.* **1881**, 210, 246-284.
2. Bialobrezeski, M., Nencki, N. About acetylsalicylic acid. *Ber.* **1897**, 30, 1776-1779.
3. Fries, K., Finck, G. Homologues of Cumaranone and their Derivatives. *Ber.* **1909**, 41, 4271-4284.
4. Fries, k., Pfaffendorf, W. Condensation Product of Cumaranone and Its Transformation into Oxindirubin. *Ber.* **1910**, 43, 212-219.
5. Blatt, A. H. Fries reaction. *Org. React.* **1942**, 1, 342-369.
6. Gerecs, A. The Fries Reaction. in *Friedel-Crafts and Related Reactions* (ed. Olah, G. A.), *3*, 499-533 (Interscience Publishers, New York, **1964**).
7. Finnegan, R. A., Mattice, J. J. Photochemical studies. II. Photorearrangement of aryl esters. *Tetrahedron* **1965**, 21, 1015-1026.
8. Bellus, D., Hrdlovic, P. Photochemical rearrangement of aryl, vinyl, and substituted vinyl esters and amides of carboxylic acids. *Chem. Rev.* **1967**, 67, 599-609.

9. Kwart, H., King, K. Rearrangement and cyclization reactions of carboxylic acids and esters. in *Chem. Carboxylic Acids and Esters* (ed. Patai, S.), 341-373 (Interscience-Publishers, London, New York, **1969**).
10. Martin, R. Uses of the Fries rearrangement for the preparation of hydroxyaryl ketones. A review. *Org. Prep. Proced. Int.* **1992**, 24, 369-435.
11. Rusu, E., Comanita, E., Onciu, M. Photo-Fries rearrangement. *Roumanian Chem. Quart. Rev.* **2000**, 7, 241-250.
12. Guisnet, M., Perot, G. *The Fries Rearrangement* (eds. Sheldon, R. A.,Bekkum, H.) (W, New York, **2001**) 211-215.
13. Miranda, M. A., Galindo, F. The photo-Fries rearrangement. *Molecular and Supramolecular Photochemistry* **2003**, 9, 43-131.
14. Taylor, C. M., Watson, A. J. The anionic phospho-Fries rearrangement. *Curr. Org. Chem.* **2004**, 8, 623-636.
15. Pappas, S. P., Alexander, J. E., Long, G. L., Zehr, R. D. Vinylogous Fries and photo-Fries rearrangements. *J. Org. Chem.* **1972**, 37, 1258-1259.
16. Olah, G. A., Arvanaghi, M., Krishnamurthy, V. V. Heterogeneous catalysis by solid superacids. 17. Polymeric perfluorinated resin sulfonic acid (Nafion-H) catalyzed Fries rearrangement of aryl esters. *J. Org. Chem.* **1983**, 48, 3359-3360.
17. Sibi, M. P., Snieckus, V. The directed ortho lithiation of O-aryl carbamates. An anionic equivalent of the Fries rearrangement. *J. Org. Chem.* **1983**, 48, 1935-1937.
18. Garcia, H., Miranda, M. A., Primo, J. The photo-Fries rearrangement of acetoxyacetophenones using cyclic acetals as carbonyl blocking groups, in the presence of potassium carbonate. An improved procedure for the synthesis of diacylphenols. *J. Chem. Res., Synop.* **1986**, 100-101.
19. Hallberg, A., Svensson, A., Martin, A. R. An intramolecular anionic Fries rearrangement of N-acylphenothiazines. *Tetrahedron Lett.* **1986**, 27, 1959-1962.
20. Horne, S., Rodrigo, R. A complex induced proximity effect in the anionic Fries rearrangement of o-iodophenyl benzoates: synthesis of dihydro-O-methylsterigmatocystin and other xanthones. *J. Org. Chem.* **1990**, 55, 4520-4522.
21. Lassila, K. R., Ford, M. E. Solid acid catalysis of the Fries rearrangement: thermodynamic limitations based on solvent polarity. *Chem. Ind.* **1992**, 47, 169-180.
22. Tabuchi, H., Hamamoto, T., Ichihara, A. Modification of the Fries type rearrangement of the O-enol acyl group using *N,N*-dicyclohexylcarbodiimide and 4-dimethylaminopyridine. *Synlett* **1993**, 651-652.
23. Vogt, A., Kouwenhoven, H. W., Prins, R. Fries rearrangement over zeolitic catalysts. *Appl. Cat. A* **1995**, 123, 37-49.
24. Venkatachalapathy, C., Pitchumani, K. Fries rearrangement of esters in montmorillonite clays:steric control on selectivity. *Tetrahedron* **1997**, 53, 17171-17176.
25. Balkus, K. J., Jr., Khanmamedova, A. K., Woo, R. Fries rearrangement of acetanilide over zeolite catalysts. *J. Mol. Catal. A: Chemical* **1998**, 134, 137-143.
26. Cambie, R. C., Mitchell, L. H., Rutledge, P. S. Acid-promoted Fries rearrangements of benzannulated lactones. *Aust. J. Chem.* **1998**, 51, 1167-1174.
27. Middel, O., Greff, Z., Taylor, N. J., Verboom, W., Reinhoudt, D. N., Snieckus, V. The First Lateral Functionalization of Calix[4]arenes by a Homologous Anionic Ortho-Fries Rearrangement. *J. Org. Chem.* **2000**, 65, 667-675.
28. Clark, J. H., Dekamin, M. G., Moghaddam, F. M. Genuinely catalytic Fries rearrangement using sulfated zirconia. *Green Chem.* **2002**, 4, 366-368.
29. Charmant, J. P. H., Dyke, A. M., Lloyd-Jones, G. C. The anionic thia-Fries rearrangement of aryl triflates. *Chem. Commun.* **2003**, 380-381.
30. Mouhtady, O., Gaspard-Iloughmane, H., Roques, N., Le Roux, C. Metal triflates-methanesulfonic acid as new catalytic systems: application to the Fries rearrangement. *Tetrahedron Lett.* **2003**, 44, 6379-6382.
31. Wang, H., Zou, Y. Modified β–Zeolite as Catalyst for Fries Rearrangement Reaction. *Catal. Lett.* **2003**, 86, 163-167.
32. Tsutsumi, K., Matsui, K., Shizuka, H. Substituent effects on the photo-Fries rearrangement of aryloxy-s-triazines: Norrish type I dissociation. *Mol. Photochem.* **1976**, 7, 325-342.
33. Mehlhorn, A., Schwenzer, B., Schwetlick, K. MO-calculations of the energy transfer activities of organic *p*-structures in the photo-Fries rearrangement. II. Selection of sensitizers and inhibitors of the photo-Fries reaction based on theoretical absorption and fluorescence data. *Tetrahedron* **1977**, 33, 1489-1491.
34. Mehlhorn, A., Schwenzer, B., Brueckner, H. J., Schwetlick, K. MO-calculations of the energy transfer-activities of organic *p*-structures in the photo-Fries rearrangement. III. A theoretical index for the energy transfer efficiency of organic *p*-systems in the photo-Fries reaction. *Tetrahedron* **1978**, 34, 481-486.
35. Katagi, T. Theoretical studies on the photo-Fries rearrangement of O-aryl N-methylcarbamates. *Nippon Noyaku Gakkaishi* **1991**, 16, 57-62.
36. Grimme, S. MO theoretical investigation on the photodissociation of carbon-oxygen bonds in aromatic compounds. *Chem. Phys.* **1992**, 163, 313-330.
37. Cui, C., Wang, X., Weiss, R. G. Investigation of the Photo-Fries Rearrangements of Two 2-Naphthyl Alkanoates by Experiment and Theory. Comparison with the Acid-Catalyzed Reactions. *J. Org. Chem.* **1996**, 61, 1962-1974.
38. Shizuka, H., Tobita, S. Tunneling effects on the sigmatropic hydrogen shifts in the photorearranged intermediates of phenyl acetate and N-acetylpyrrole studied by laser photolysis. *JAERI-Conf* **1998**, 98-002, 76-84.
39. Gerecs, A., Windholz, M. The role of hydrochloric acid in the Fries reaction. III. *Acta Chim. Acad. Sci. Hung.* **1955**, 8, 295-302.
40. Bisanz, T. Fries rearrangement of β-naphthol esters and of their derivatives. I. Evidence for the mechanism of the reaction. *Roczniki Chem.* **1956**, 30, 87-102.
41. Dewar, M. J. S., Hart, L. S. Aromatic rearrangements in the benzene series. I. Fries rearrangement of phenyl benzoate: the benzoylation of phenol. *Tetrahedron* **1970**, 26, 973-1000.
42. Munavalli, S. Mechanism of Fries rearrangement. Intermolecular versus intramolecular acylation. *Chem. Ind.* **1972**, 293-294.
43. Cohen, N., Lopresti, R. J., Williams, T. H. Fries rearrangement of trimethylhydroquinone diacetate. A novel hydroquinone to resorcinol transformation. *J. Org. Chem.* **1978**, 43, 3723-3726.
44. Warshawsky, A., Kalir, R., Patchornik, A. Interpolymeric reactions. The Fries rearrangement of acetoxy and benzyloxy derivatives of 4-hydroxy-3-nitrobenzylated polystyrene and 5-polystyrylmethyl-8-quinolinol. *J. Am. Chem. Soc.* **1978**, 100, 4544-4550.
45. Banks, M. R. Fries rearrangement of some 3-acetoxy- and 3-(propionyloxy)thiophenes. *J. Chem. Soc., Perkin Trans. 1* **1986**, 507-513.
46. Dawson, I. M., Gibson, J. L., Hart, L. S., Waddington, C. R. Aromatic rearrangements in the benzene series. Part 5. The Fries rearrangement of phenyl benzoate: the rearranging species. The effect of tetrabromoaluminate ion on the ortho/para ratio: the noninvolvement of the proton as a cocatalyst. *J. Chem. Soc., Perkin Trans. 2* **1989**, 2133-2139.
47. Gibson, J. L., Hart, L. S. Aromatic rearrangements in the benzene series. Part 6. The Fries rearrangement of phenyl benzoate: the role of tetrabromoaluminate ion as an aluminum bromide transfer agent. *J. Chem. Soc., Perkin Trans. 2* **1991**, 1343-1348.
48. Sharghi, H., Eshghi, H. The mechanism of the Fries rearrangement and acylation reaction in polyphosphoric acid. *Bull. Chem. Soc. Jpn.* **1993**, 66, 135-139.
49. Boyer, J. L., Krum, J. E., Myers, M. C., Fazal, A. N., Wigal, C. T. Synthetic Utility and Mechanistic Implications of the Fries Rearrangement of Hydroquinone Diesters in Boron Trifluoride Complexes. *J. Org. Chem.* **2000**, 65, 4712-4714.
50. Yoon, H. J., Ko, S. H., Ko, M. K., Chae, W. K. The trivial mechanism for the photo-Fries reaction of phenyl acetate and biphenylyl acetates. *Bull. Korean Chem. Soc.* **2000**, 21, 901-904.
51. Bringmann, G., Menche, D., Kraus, J., Muehlbacher, J., Peters, K., Peters, E.-M., Brun, R., Bezabih, M., Abegaz, B. M. Atropo-Enantioselective Total Synthesis of Knipholone and Related Antiplasmodial Phenylanthraquinones. *J. Org. Chem.* **2002**, 67, 5595-5610.
52. Lampe, J. W., Hughes, P. F., Biggers, C. K., Smith, S. H., Hu, H. Total Synthesis of (-)- and (+)-Balanol. *J. Org. Chem.* **1996**, 61, 4572-4581.
53. Magnus, P., Lescop, C. Photo-Fries rearrangement for the synthesis of the diazonamide macrocycle. *Tetrahedron Lett.* **2001**, 42, 7193-7196.

1. Gabriel, S. Synthesis of primary amines from the corresponding alkyl halides. *Ber.* **1887**, 20, 2224-2236.
2. Gibson, M. S., Bradshaw, R. W. Gabriel synthesis of primary amines. *Angew. Chem., Int. Ed. Engl.* **1968**, 7, 919-930.
3. Mitsunobu, O. Synthesis of Amines and Ammonium Salts. in *Comp. Org. Synth.* (eds. Trost, B. M.,Fleming, I.), *6*, 65-101 (Pergamon, Oxford, **1991**).
4. Ragnarsson, U., Grehn, L. Novel Gabriel reagents. *Acc. Chem. Res.* **1991**, 24, 285-289.
5. Ing, H. R., Manske, R. H. F. Modification of the Gabriel synthesis of amines. *J. Chem. Soc., Abstracts* **1926**, 2348-2351.
6. Bacon, R. G. R., Karim, A. Copper-catalyzed substitution of aryl halides by potassium phthalimide: an extension of the Gabriel reaction. *J. Chem. Soc., Chem. Commun.* **1969**, 578.
7. Landini, D., Rolla, F. A convenient synthesis of *N*-alkylphthalimides in a solid-liquid two-phase system in the presence of phase-transfer catalysts. *Synthesis* **1976**, 6, 389-391.
8. Pasquini, M. A., Le Goaller, R., Pierre, J. L. Effects of cryptands and activation of bases. V. Action of alkali hydrides on weak acids. II. Alkylation of anions obtained. *Tetrahedron* **1980**, 36, 1223-1226.
9. Sato, M., Ebine, S., Akabori, S. Condensation of halobenzenes and haloferrocenes with phthalimide in the presence of copper(I) oxide; a simplified Gabriel reaction. *Synthesis* **1981**, 472-473.
10. Soai, K., Ookawa, A., Kato, K. A facile one-pot synthesis of *N*-substituted phthalimides using a catalytic amount of crown ether. *Bull. Chem. Soc. Jpn.* **1982**, 55, 1671-1672.
11. Osby, J. O., Martin, M. G., Ganem, B. An exceptionally mild deprotection of phthalimides. *Tetrahedron Lett.* **1984**, 25, 2093-2096.
12. Slusarska, E., Zwierzak, A. Conversion of alcohols into primary amines, a new Mitsunobu version of Gabriel-type synthesis. *Liebigs Ann. Chem.* **1986**, 402-405.
13. Grehn, L., Ragnarsson, U. A convenient one-flask preparation of di-*tert*-butyl iminodicarbonate: a versatile Gabriel reagent. *Synthesis* **1987**, 275-276.
14. Han, Y., Hu, H. A convenient synthesis of primary amines using sodium diformylamide as a modified Gabriel reagent. *Synthesis* **1990**, 122-124.
15. Khan, M. N. Suggested Improvement in the Ing-Manske Procedure and Gabriel Synthesis of Primary Amines: Kinetic Study on Alkaline Hydrolysis of N-Phthaloylglycine and Acid Hydrolysis of N-(o-Carboxybenzoyl)glycine in Aqueous Organic Solvents. *J. Org. Chem.* **1996**, 61, 8063-8068.
16. Zwierzak, A. An optimized version of Gabriel-type nucleophilic amination. *Synth. Commun.* **2000**, 30, 2287-2293.
17. Graebe, C., Pictet, A. Methyl phtalimide. *Ber.* **1884**, 17, 1173-1175.
18. Sheehan, J. C., Bolhofer, W. A. An improved procedure for the condensation of potassium phthalimide with organic halides. *J. Am. Chem. Soc.* **1950**, 72, 2786-2788.
19. Sen, A. K., Sarma, S. Alkylations in dimethylformamide. *J. Indian Chem. Soc.* **1967**, 44, 644-645.
20. Inoue, Y., Taguchi, M., Hashimoto, H. *N*-Alkylation of imides with *O*-alkylisourea under neutral conditions. *Synthesis* **1986**, 332-334.
21. Krafft, G. A., Siddall, T. L. Stereospecific displacement of sulfur from chiral centers. Activation via thiaphosphonium salts. *Tetrahedron Lett.* **1985**, 26, 4867-4870.
22. Rancurel, A., Grenier, G. *Aminopropanols.* De 2606106, **1976** (Laboratoires Pharmascience, Fr.). 24 pp.
23. Coxon, B., Reynolds, R. C. Synthesis of nitrogen-15-labeled amino sugar derivatives by addition of phthalimide-15N to a carbohydrate epoxide. *Carbohydr. Res.* **1982**, 110, 43-54.
24. Moe, O. A., Warner, D. T. 1,4-Addition reactions. III. The addition of cyclic imides to α,β-unsaturated aldehydes. A synthesis of balanine hydrochloride. *J. Am. Chem. Soc.* **1949**, 71, 1251-1253.
25. Dumas, D. J. Total synthesis of peramine. *J. Org. Chem.* **1988**, 53, 4650-4653.
26. Burgess, K., Henderson, I. A new approach to swainsonine and castanospermine analogs. *Tetrahedron Lett.* **1990**, 31, 6949-6952.
27. Kubo, A., Kubota, H., Takahashi, M., Nunami, K.-i. Dynamic kinetic resolution utilizing 2-oxoimidazolidine-4-carboxylate as a chiral auxiliary: stereoselective synthesis of α-amino acids by Gabriel reaction. *Tetrahedron Lett.* **1996**, 37, 4957-4960.
28. Rossi, F. M., Powers, E. T., Yoon, R., Rosenberg, L., Meinwald, J. Preparation of 2,3-diamino acids: stereocontrolled synthesis of an aminated analog of the taxol side chain. *Tetrahedron* **1996**, 52, 10279-10286.

Related reactions: Reimer-Tiemann reaction, Vilsmeier-Haack formylation;

1. Gattermann, L., Koch, J. A. Synthesis of aromatic aldehydes. *Ber.* **1897**, 30, 1622-1624.
2. Gattermann, L. Syntheses of Aromatic Aldehydes. First Paper. *Ann.* **1906**, 347, 347-386.
3. Gattermann, L. Syntheses of Aromatic Aldehydes. Second Paper. *Ann.* **1908**, 357, 313-383.
4. Crounse, N. N. Gattermann-Koch reaction. *Org. React.* **1949**, 5, 290-301.
5. Gore, P. H. The Friedel-Crafts acylation reaction and its application to polycyclic aromatic hydrocarbons. *Chem. Rev.* **1955**, 55, 229-281.
6. Eyley, S. C., Rainey, D. K. Aldehydes and ketones. *General and Synthetic Methods* **1981**, 4, 26-86.
7. Hollingworth, G. J. Aldehydes: Aryl and heteroaryl aldehydes. in *Comp. Org. Funct. Group Trans.* (eds. Katritzky, A. R., Meth-Cohn, O.,Rees, C. W.), *3*, 81-109, 733-856 (Pergamon, Oxford, New York, **1995**).
8. Tanaka, M. A new aspect in electrophilic aromatic substitutions: intracomplex and conventional electrophilic aromatic substitutions in Gattermann-Koch formylation. *Trends in Organic Chemistry* **1998**, 7, 45-61.
9. Karrer, P. Hydroxycarbonyl compounds. I. A new synthesis of hydroxyaldehydes. *Helv. Chim. Acta* **1919**, 2, 89-94.
10. Adams, R., Levine, I. Simplification of the Gattermann synthesis of hydroxy aldehydes. *J. Am. Chem. Soc.* **1923**, 45, 2373-2377.
11. Gresham, W. F., Tabet, G. E. *Aromatic aldehydes.* US 2485237, **1949** (E. I. du Pont de Nemours & Co.).
12. Toniolo, L., Graziani, M. Metals in organic syntheses. V. The Gattermann-Koch synthesis of aromatic aldehydes promoted by $CuCl(PPh_3)_n$ (n = 0,1 or 3). Is the cuprous complex necessary in the synthesis of tolualdehyde? *J. Organomet. Chem.* **1980**, 194, 221-228.
13. Olah, G. A., Ohannesian, L., Arvanaghi, M. Formylating agents. *Chem. Rev.* **1987**, 87, 671-686.
14. Alagona, G., Tomasi, J. The mechanism of addition to a C-N triple bond. An ab initio study of the first stages of the Stephen, Gattermann and Houben-Hoesch reactions. *THEOCHEM* **1983**, 8, 263-281.
15. Tanaka, M., Fujiwara, M., Xu, Q., Ando, H., Raeker, T. J. Influence of Conformation and Proton-Transfer Dynamics in the Dibenzyl s-Complex on Regioselectivity in Gattermann-Koch Formylation via Intracomplex Reaction. *J. Org. Chem.* **1998**, 63, 4408-4412.
16. Niedzielski, E. L., Nord, F. F. Mechanism of the Gattermann reaction. II. *J. Org. Chem.* **1943**, 8, 147-152.
17. Dilke, M. H., Eley, D. D. The Gattermann-Koch reaction. II. Reaction kinetics. *J. Chem. Soc., Abstracts* **1949**, 2613-2620.
18. Dilke, M. H., Eley, D. D. The Gattermann-Koch reaction. I. Thermodynamics. *J. Chem. Soc., Abstracts* **1949**, 2601-2612.
19. Yato, M., Ohwada, T., Shudo, K. Requirements for Houben-Hoesch and Gattermann reactions. Involvement of diprotonated cyanides in the reactions with benzene. *J. Am. Chem. Soc.* **1991**, 113, 691-692.
20. Tanaka, M., Fujiwara, M., Ando, H. Dual Reactivity of the Formyl Cation as an Electrophile and a Broensted Acid in Superacids. *J. Org. Chem.* **1995**, 60, 3846-3850.
21. Tanaka, M., Fujiwara, M., Ando, H., Souma, Y. The influence of aromatic compound protonation on the regioselectivity of Gattermann-Koch formylation. *Chem. Commun.* **1996**, 159-160.
22. Clingenpeel, T. H., Wessel, T. E., Biaglow, A. I. ^{13}C NMR Study of the Carbonylation of Benzene with CO in Sulfated Zirconia. *J. Am. Chem. Soc.* **1997**, 119, 5469-5470.
23. Tanaka, M., Fujiwara, M., Xu, Q., Souma, Y., Ando, H., Laali, K. K. Evidence for the Intracomplex Reaction in Gattermann-Koch Formylation in Superacids: Kinetic and Regioselectivity Studies. *J. Am. Chem. Soc.* **1997**, 119, 5100-5105.

24. Burke, J. M., Stevenson, R. Synthesis of calebertin and caleprunin A. *J. Nat. Prod.* **1986**, 49, 522-523.
25. Wong, H. N. C., Niu, C. R., Yang, Z., Hon, P. M., Chang, H. M., Lee, C. M. Compounds from danshen. 8. Regiospecific introduction of carbon-3 formyl group to 2,5-dialkyl-7-methoxybenzo[b]furans: synthesis of potential ligands for adenosine A1 receptors. *Tetrahedron* **1992**, 48, 10339-10344.
26. Shanmugasundaram, K., Prasad, K. J. R. Synthesis of 2-hydroxypyrido[2,3-a]carbazoles and 2-hydroxypyrimido[4,5-a]carbazoles from 1-hydroxycarbazoles. *Heterocycles* **1999**, 51, 2163-2169.
27. Vela, M. A., Fronczek, F. R., Horn, G. W., McLaughlin, M. L. Syntheses of 1- and 2-naphthol analogs of DL-tyrosine. Potential fluorescent probes of peptide structure and dynamics in complex environments. *J. Org. Chem.* **1990**, 55, 2913-2918.

Glaser Coupling 186

1. Glaser, C. Contribution to the chemistry of phenylacetylenes. *Ber.* **1869**, 2, 422-424.
2. Eglington, G., McCrae, W. in *Advances in Organic Chemistry, Method and Results* (eds. Raphael, R. A., Taylor, C.,Wynberg, H.), *4*, 225 (Wiley, New York, **1963**).
3. Rutledge, T. F. *Acetylenic Compounds* (Reinhold, New York, **1968**) Chapter 6.
4. Cadiot, P., Chodkiewicz, W. *Couplings of Acetylenes* (ed. Viehe, H. G.) (Dekker, New York, **1969**) 597-647.
5. Nigh, W. G. *Oxidation in Organic Chemistry* (ed. Trahanovsky, W. S.) (Academic Press, New York, **1973**).
6. Simándi, L. I. The chemistry of triple-bonded functional groups. in *The chemistry of functional groups* (eds. Patai, S.,Rappoport, Z.), *Supplement C*, 529-534 (Wiley, New York, **1983**).
7. Brandsma, L. *Preparative Acetylenic Chemistry* (Elsevier, Amsterdam, **1988**) Chapter 10.
8. Sonogashira, K. Coupling Reactions Between sp Carbon Centers. in *Compr.Org.Synth.* (eds. Trost, B. M.,Fleming, I.), *3*, 551-561 (Pergamon, Oxford, **1991**).
9. Siemsen, P., Livingston, R. C., Diederich, F. Acetylenic coupling: a powerful tool in molecular construction. *Angew. Chem., Int. Ed. Engl.* **2000**, 39, 2632-2657.
10. Eglinton, G., Galbraith, A. R. Cyclic diynes. *Chem. Ind.* **1956**, 737-738.
11. Hay, A. S. Oxidative coupling of acetylenes. *J. Org. Chem.* **1960**, 25, 1275-1276.
12. Hay, A. S. Oxidative couplings of acetylenes. II. *J. Org. Chem.* **1962**, 27, 3320-3321.
13. Li, J., Jiang, H. Glaser coupling reaction in supercritical carbon dioxide. *Chem. Commun.* **1999**, 2369-2370.
14. Kabalka, G. W., Wang, L., Pagni, R. M. Microwave enhanced Glaser coupling under solvent-free conditions. *Synlett* **2001**, 108-110.
15. Krafft, M. E., Hirosawa, C., Dalal, N., Ramsey, C., Stiegman, A. Cobalt-catalyzed homocoupling of terminal alkynes: synthesis of 1,3-diynes. *Tetrahedron Lett.* **2001**, 42, 7733-7736.
16. Yadav, J. S., Reddy, B. V. S., Reddy, K. B., Gayathri, K. U., Prasad, A. R. Glaser oxidative coupling in ionic liquids: an improved synthesis of conjugated 1,3-diynes. *Tetrahedron Lett.* **2003**, 44, 6493-6496.
17. Chodkiewicz, W., Cadiot, P. New synthesis of symmetrical and asymmetrical conjugated polyacetylenes. *Compt. rend.* **1955**, 241, 1055-1057.
18. Chodkiewicz, W. Synthesis of acetylenic compounds. *Ann. chim. (Paris) [13]* **1957**, 2, 819-869.
19. Chodkiewicz, W., Alhuwalia, J. S., Cadiot, P., Willemart, A. Preparation of aliphatic bifuncfional compounds. *Compt. rend.* **1957**, 245, 322-324.
20. Wityak, J., Chan, J. B. Synthesis of 1,3-diynes using palladium-copper catalysis. *Synth. Commun.* **1991**, 21, 977-979.
21. Klebanskii, A. L., Grachev, I. V., Kuznetsova, O. M. Reaction of formation of diacetylenic compounds, from monosubstituted derivatives of acetylene. I. Mechanism of formation of diacetylenic compounds. *Zh. Obshch. Khim.* **1957**, 27, 2977-2983.
22. Clifford, A. A., Waters, W. A. Oxidations of organic compounds by cupric salts. III. The oxidation of propargyl alcohol. *J. Chem. Soc., Abstracts* **1963**, 3056-3062.
23. Bohlmann, F., Schoenowsky, H., Inhoffen, E., Grau, G. Polyacetylenic compounds. LII. The mechanism of oxidative dimerization of acetylene compounds. *Ber.* **1964**, 97, 794-800.
24. Kevelam, H. J., De Jong, K. P., Meinders, H. C., Challa, G. Kinetics of oxidative polymerization of 1,8-nonadiyne. *Makromol. Chem.* **1975**, 176, 1369-1381.
25. Challa, G., Meinders, H. C. Copper-polymer complexes as catalysts for oxidative coupling reactions. *J. Mol. Catal.* **1977**, 3, 185-190.
26. Brailovskii, S. M., Man'Khoan, K., Temkin, O. N. Kinetics and mechanism of additive oxidative chlorination of acetylene in solutions of Cu(I) and Cu(II) chlorides. *Kinet. Katal.* **1994**, 35, 734-740.
27. Huynh Manh, H., Brailovskii, S. M., Temkin, O. N. Kinetics of dialkyne synthesis in aqueous solutions of Cu(I) and Cu(II) chlorides. *Kinet. Katal.* **1994**, 35, 266-270.
28. Kennedy, J. C., MacCallum, J. R., MacKerron, D. H. Synthesis and characterization of a series of poly(α,ω-alkyldiynes) and copoly(α,ω-alkyldiynes). *Can. J. Chem.* **1995**, 73, 1914-1923.
29. Mykhalichko, B. M. Copper(I) acetylenide complexes. synthesis and structure of cluster p compound $(AnH)_2[Cu_4Cl_5(CCCH_2OH)]$ (AnH^+ = anilinium cation). *Russ. J. Coord. Chem. (Translation of Koordinatsionnaya Khimiya)* **1999**, 25, 336-341.
30. Hebert, N., Beck, A., Lennox, R. B., Just, G. A new reagent for the removal of the 4-methoxybenzyl ether: application to the synthesis of unusual macrocyclic and bolaform phosphatidylcholines. *J. Org. Chem.* **1992**, 57, 1777-1783.
31. Nicolaou, K. C., Petasis, N. A., Uenishi, J., Zipkin, R. E. The endiandric acid cascade. Electrocyclizations in organic synthesis. 2. Stepwise, stereocontrolled total synthesis of endiandric acids C-G. *J. Am. Chem. Soc.* **1982**, 104, 5557-5558.
32. Nicolaou, K. C., Petasis, N. A., Zipkin, R. E. The endiandric acid cascade. Electrocyclizations in organic synthesis. 4. Biomimetic approach to endiandric acids A-G. Total synthesis and thermal studies. *J. Am. Chem. Soc.* **1982**, 104, 5560-5562.
33. Nicolaou, K. C., Petasis, N. A., Zipkin, R. E., Uenishi, J. The endiandric acid cascade. Electrocyclizations in organic synthesis. I. Stepwise, stereocontrolled total synthesis of endiandric acids A and B. *J. Am. Chem. Soc.* **1982**, 104, 5555-5557.
34. Nicolaou, K. C., Zipkin, R. E., Petasis, N. A. The endiandric acid cascade. Electrocyclizations in organic synthesis. 3. "Biomimetic" approach to endiandric acids A-G. Synthesis of precursors. *J. Am. Chem. Soc.* **1982**, 104, 5558-5560.
35. Bukownik, R. R., Wilcox, C. S. Synthetic receptors. 3,6-anhydro-7-benzenesulfonamido-1,7-dideoxy-4,5-O-isopropylidene-D-altro-hept-1-ynitol: a useful component for the preparation of chiral water-soluble cyclophanes based on carbohydrate precursors. *J. Org. Chem.* **1988**, 53, 463-467.
36. Jung, F., Burger, A., Biellmann, J.-F. Synthesis of Nucleoside Dimers Bridged on Ribose with a Butadiynyl Group. *Org. Lett.* **2003**, 5, 383-385.

Grignard Reaction 188

Related reactions: Barbier coupling reaction, Kagan-Molander samarium diiodide coupling, Nozaki-Hiyama-Kishi reaction;

1. Grignard, V. Some new organometallic combinations of magnesium and their application to the synthesis of alcohols and hydrocarbons. *C.R.Acad.Sci.* **1900**, 1322-1324.
2. Grignard, V. Mixed organomagnesium combinations and their application in acid, alcohol and hydrocarbon synthesis. *Ann.Chim.* **1901**, 7, 433-490.
3. Kharasch, M. S., Reinmuth, O. *Grignard Reactions of Nonmetallic Substances* (Prentice-Hall, New York, **1954**) 1267 pp.
4. Shirley, D. A. The synthesis of ketones from acid halides and organometallic compounds of magnesium, zinc, and cadmium. *Org. React.* **1954**, 8, 28-58.
5. Ashby, E. C. Grignard reagents. Compositions and mechanisms of reaction. *Quart. Rev., Chem. Soc.* **1967**, 21, 259-285.
6. Felkin, H., Swierczewski, G. Activation of Grignard reagents by transition metal compounds. *Tetrahedron* **1975**, 31, 2735-2748.

7. Erdik, E. Copper(I)-catalyzed reactions of organolithiums and Grignard reagents. *Tetrahedron* **1984**, 40, 641-657.
8. Sato, F. The preparation of Grignard reagents via the hydromagnesation reaction and their uses in organic synthesis. *J. Organomet. Chem.* **1985**, 285, 53-64.
9. Blomberg, C. Structure-reactivity relationships [of Grignard reagents]. *Chem. Ind.* **1996**, 64, 249-269.
10. Cannon, K. C., Krow, G. R. Dihalide-derived di-Grignard reagents: Preparation and reactions. *Chem. Ind.* **1996**, 64, 497-526.
11. Umeno, M., Suzuki, A. Alkynyl Grignard reagents and their uses. *Chem. Ind.* **1996**, 64, 645-666.
12. Urabe, H., Sato, F. Metal-catalyzed [Grignard] reactions. *Chem. Ind.* **1996**, 64, 577-632.
13. Franzen, R. G. Utilization of Grignard reagents in solid-phase synthesis: a review of the literature. *Tetrahedron* **2000**, 56, 685-691.
14. Hill, E. A. Nucleophilic displacements at carbon by Grignard reagents. in *Grignard Reagents* (ed. Richey, H. G.), 27-64 (Wiley, Chichester, **2000**).
15. Raston, C. L. Applications of magnesium anthracene in forming Grignard reagents. in *Grignard Reagents* (ed. Richey, H. G.), 277-298 (Wiley, Chichester, **2000**).
16. Hoffmann, R. W. The quest for chiral Grignard reagents. *Chem. Soc. Rev.* **2003**, 32, 225-230.
17. Garst, J. F., Soriaga, M. P. Grignard reagent formation. *Coord. Chem. Rev.* **2004**, 248, 623-652.
18. Gawley, R. E. Stereoselective additions of chiral Grignard reagents to aldehydes: stereochemical and mechanistic principles, with examples using α-amino Grignard reagents. in *Grignard Reagents* (ed. Richey, H. G.), 139-164 (Wiley, Chichester, **2000**).
19. Fleming, I. An enantiometrically enriched Grignard reagent. *Chemtracts* **2001**, 14, 505-508.
20. Oestreich, M., Hoppe, D. Stereospecific preparation of highly enantiomerically enriched organomagnesium reagents. *Chemtracts* **2001**, 14, 100-105.
21. Davis, S. R. Ab initio study of the insertion reaction of magnesium into the carbon-halogen bond of fluoro- and chloromethane. *J. Am. Chem. Soc.* **1991**, 113, 4145-4150.
22. Liu, L., Davis, S. R. Ab initio study of the Grignard reaction between magnesium atoms and fluoroethylene and chloroethylene. *J. Phys. Chem.* **1991**, 95, 8619-8625.
23. Peralez, E., Negrel, J.-C., Goursot, A., Chanon, M. Mechanism of the Grignard reagent formation - Part 1. Theoretical investigations of the Mgn and RMgn participation in the mechanism. *Main Group Metal Chem.* **1999**, 22, 185-200.
24. Yoo, S.-E., Gong, Y.-D., Kim, S.-K. Theoretical study on the regioselectivity of tetrazolylimines with alkyl Grignard reagents. *Bull. Korean Chem. Soc.* **1999**, 20, 441-444.
25. Aitken, D. J., Beaufort, V., Chalard, P., Cladiere, J.-L., Dufour, M., Pereira, E., Thery, V. Theoretical and model studies on the chemoselectivity of a Grignard reagent's reaction with a combined aminonitrile-oxazolidine system. *Tetrahedron* **2002**, 58, 5933-5940.
26. Benhallam, R., Zair, T., Jarid, A., Ibrahim-Ouali, M. Theoretical study of the stereochemistry of crotylmagnesium chloride's addition on a series of cyclic and acyclic enones. *THEOCHEM* **2003**, 626, 1-17.
27. Ashby, E. C., Laemmle, J., Neumann, H. M. Mechanisms of Grignard reagent addition to ketones. *Acc. Chem. Res.* **1974**, 7, 272-280.
28. Bickelhaupt, F. Free radicals in Grignard reactions. *Angew. Chem., Int. Ed. Engl.* **1974**, 13, 419-420.
29. Ashby, E. C. A detailed description of the mechanism of reaction of Grignard reagents with ketones. *Pure Appl. Chem.* **1980**, 52, 545-569.
30. Ashby, E. C. Single-electron transfer, a major reaction pathway in organic chemistry. An answer to recent criticisms. *Acc. Chem. Res.* **1988**, 21, 414-421.
31. Walborsky, H. M. Mechanism of Grignard reagent formation. The surface nature of the reaction. *Acc. Chem. Res.* **1990**, 23, 286-293.
32. Blomberg, C. Mechanisms of reactions of Grignard reagents. *Chem. Ind.* **1996**, 64, 219-248.
33. Garst, J. F., Ungvary, F. Mechanisms of Grignard reagent formation. in *Grignard Reagents* (ed. Richey, H. G.), 185-275 (Wiley, Chichester, **2000**).
34. Holm, T., Crossland, I. Mechanistic features of the reactions of organomagnesium compounds. in *Grignard Reagents* (ed. Richey, H. G.), 1-26 (Wiley, Chichester, **2000**).
35. Sun, P., Sun, C., Weinreb, S. M. Stereoselective Total Syntheses of the Racemic Form and the Natural Enantiomer of the Marine Alkaloid Lepadiformine via a Novel *N*-Acyliminium Ion/Allylsilane Spirocyclization Strategy. *J. Org. Chem.* **2002**, 67, 4337-4345.
36. White, J. D., Shin, H., Kim, T.-S., Cutshall, N. S. Total Synthesis of the Sesquiterpenoid Polyols (±)-Euonyminol and (±)-3,4-Dideoxymaytol, Core Constituents of Esters of the Celastraceae. *J. Am. Chem. Soc.* **1997**, 119, 2404-2419.
37. Kuehne, M. E., Xu, F. Syntheses of Strychnan- and Aspidospermatan-Type Alkaloids. 11. Total Syntheses of (-)-Lochneridine and (-)- and Racemic 20-epi-Lochneridine. *J. Org. Chem.* **1998**, 63, 9434-9439.
38. Taunton, J., Collins, J. L., Schreiber, S. L. Synthesis of Natural and Modified Trapoxins, Useful Reagents for Exploring Histone Deacetylase Function. *J. Am. Chem. Soc.* **1996**, 118, 10412-10422.

Grob Fragmentation 190

Related reactions: Eschenmoser-Tanabe fragmentation, Wharton fragmentation;

1. Eschenmoser, A., Frey, A. Cleavage of methanesulfonyl esters of 2-methyl-2-(hydroxymethyl)cyclopentanone with bases. *Helv. Chim. Acta* **1952**, 35, 1660-1666.
2. Grob, C. A., Baumann, W. 1,4-Elimination reaction with simultaneous fragmentation. *Helv. Chim. Acta* **1955**, 38, 594-610.
3. Grob, C. A. The principle of ethylogy in organic chemistry. *Experientia* **1957**, 13, 126-129.
4. Grob, C. A., Schiess, P. W. Heterolytic fragmentation. A class of organic reactions. *Angew. Chem., Int. Ed. Engl.* **1967**, 6, 1-15.
5. Grob, C. A. Mechanisms and stereochemistry of heterolytic fragmentation. *Angew. Chem., Int. Ed. Engl.* **1969**, 8, 535-546.
6. Weyerstahl, P., Marschall, H. Fragmentation Reactions. in *Comp. Org. Synth.* (eds. Trost, B. M.,Fleming, I.), *6*, 1041-1070 (Pergamon, Oxford, **1991**).
7. Jones, P. G., Edwards, M. R., Kirby, A. J. Bond length and reactivity: structure of a Grob fragmentation substrate, 4aα,5β,8aβ-1-methyldecahydroquinolin-5-yl 3,5-dinitrobenzoate. *Acta Crystallogr., Sect. C: Cryst. Struct. Commun.* **1986**, C42, 1372-1374.
8. Zimmerman, H. E., Weinhold, F. Use of Hueckel Methodology with ab Initio Molecular Orbitals: Polarizabilities and Prediction of Organic Reactions. *J. Am. Chem. Soc.* **1994**, 116, 1579-1580.
9. Alder, R. W., Harvey, J. N., Oakley, M. T. Aromatic 4-Tetrahydropyranyl and 4-Quinuclidinyl Cations. Linking Prins with Cope and Grob. *J. Am. Chem. Soc.* **2002**, 124, 4960-4961.
10. Nagata, W., Hirai, S., Aoki, T., Takeda, K. Angular substituted polycyclic compounds. III. Alkaline degradation of 3α-alkoxy-3β-amino-5β-cholestane-5-carboxylic acid γ-lactam. *Chem. Pharm. Bull.* **1961**, 9, 845-854.
11. Armesto, X. L., Canle L, M., Losada, M., Santaballa, J. A. Concerted Grob Fragmentation in *N*-Halo-α-amino Acid Decomposition. *J. Org. Chem.* **1994**, 59, 4659-4664.
12. Hu, W.-P., Wang, J.-J., Tsai, P.-C. Novel Examples of 3-Aza-Grob Fragmentation. *J. Org. Chem.* **2000**, 65, 4208-4209.
13. Queralt, J. J., Andres, J., Canle L, M., Cobas, J. H., Santaballa, J. A., Sambrano, J. R. A joint theoretical and kinetic investigation on the fragmentation of (N-halo)-2-amino cycloalkanecarboxylates. *Chem. Phys.* **2002**, 280, 1-14.
14. Paquette, L. A., Yang, J., Long, Y. O. Concerning the Antileukemic Agent Jatrophatrione: The First Total Synthesis of a [5.9.5] Tricyclic Diterpene. *J. Am. Chem. Soc.* **2002**, 124, 6542-6543.
15. Winkler, J. D., Quinn, K. J., MacKinnon, C. H., Hiscock, S. D., McLaughlin, E. C. Tandem Diels-Alder/Fragmentation Approach to the Synthesis of Eleutherobin. *Org. Lett.* **2003**, 5, 1805-1808.
16. Molander, G. A., Le Huerou, Y., Brown, G. A. Sequenced Reactions with Samarium(II) Iodide. Sequential Intramolecular Barbier Cyclization/Grob Fragmentation for the Synthesis of Medium-Sized Carbocycles. *J. Org. Chem.* **2001**, 66, 4511-4516.

Related reactions: Robinson annulation;

1. Eder, U., Sauer, G., Wiechert, R. Total synthesis of optically active steroids. 6. New type of asymmetric cyclization to optically active steroid CD partial structures. *Angew. Chem., Int. Ed. Engl.* **1971**, 10, 496-497.
2. Eder, U., Wiechert, R., Sauer, G. *Optically active 1,5-indandione and 1,6-naphthalenedione derivatives.* Application: DE 70-2014757 **1971** (Schering A.-G.)
3. Hajos, Z. G., Parrish, D. R. *Asymmetric synthesis of optically active polycyclic organic compounds.* Application: DE DE 71-2102623 **1971** (Hoffmann-La Roche, F., und Co., A.-G.).
4. Hajos, Z. G., Parrish, D. R. Asymmetric synthesis of bicyclic intermediates of natural product chemistry. *J. Org. Chem.* **1974**, 39, 1615-1621.
5. Cohen, N. Asymmetric induction in 19-norsteroid total synthesis. *Acc. Chem. Res.* **1976**, 9, 412-417.
6. Dalko, P. I., Moisan, L. Enantioselective organocatalysis. *Angew. Chem., Int. Ed. Engl.* **2001**, 40, 3726-3748.
7. List, B. Asymmetric aminocatalysis. *Synlett* **2001**, 1675-1686.
8. Gathergood, N. Asymmetric organocatalysis: Proline an essential amino acid? *Aust. J. Chem.* **2002**, 55, 615.
9. Jarvo, E. R., Miller, S. J. Amino acids and peptides as asymmetric organocatalysts. *Tetrahedron* **2002**, 58, 2481-2495.
10. List, B. Proline-catalyzed asymmetric reactions. *Tetrahedron* **2002**, 58, 5573-5590.
11. Anon. Pyrrolidine-2-carboxylic acid (L-proline). *Synlett* **2003**, 582-583.
12. Allemann, C., Gordillo, R., Clemente, F. R., Cheong, P. H.-Y., Houk, K. N. Theory of Asymmetric Organocatalysis of Aldol and Related Reactions: Rationalizations and Predictions. *Acc. Chem. Res.* **2004**, 37, 558-569.
13. Takano, S., Kasahara, C., Ogasawara, K. Enantioselective synthesis of the gibbane framework. *J. Chem. Soc., Chem. Commun.* **1981**, 635-637.
14. Agami, C., Meynier, F., Puchot, C., Guilhem, J., Pascard, C. Stereochemistry - 59. New insights into the mechanism of the proline-catalyzed asymmetric Robinson cyclization; structure of two intermediates. Asymmetric dehydration. *Tetrahedron* **1984**, 40, 1031-1038.
15. Kondo, K., Yamano, T., Takemoto, K. Functional monomers and polymers, 129. Asymmetric Robinson cyclization reaction catalyzed by polymer-bound L-proline. *Makromol. Chem.* **1985**, 186, 1781-1785.
16. Kwiatkowski, S., Syed, A., Brock, C. P., Watt, D. S. Enantioselective synthesis of (-)-(7aS)-2,3,7,7a-tetrahydro-7a-phenylthio-1H-indene-1,5(6H)-dione and (+)-(8aS)-3,4,8,8a-tetrahydro-8a-phenylthio-1,6(2H,7H)-naphthalenedione. *Synthesis* **1989**, 818-820.
17. Blazejewski, J. C. The angular trifluoromethyl group. Part 2. Synthesis of (+)-2,3,7,7a-tetrahydro-7a-(trifluoromethyl)-1H-indene-1,5-(6H)-dione. *J. Fluorine Chem.* **1990**, 46, 515-519.
18. Przezdziecka, A., Stepanenko, W., Wicha, J. Catalytic enantioselective annulation using phenylsulfanylmethyl vinyl ketone. An approach to trans-hydrindane building blocks for ent-vitamin D3 synthesis. *Tetrahedron: Asymmetry* **1999**, 10, 1589-1598.
19. Bahmanyar, S., Houk, K. N. Proline-catalyzed direct asymmetric aldol reactions. Catalytic asymmetric synthesis of anti-1,2-diols. *Chemtracts* **2000**, 13, 904-911.
20. Bahmanyar, S., Houk, K. N. The Origin of Stereoselectivity in Proline-Catalyzed Intramolecular Aldol Reactions. *J. Am. Chem. Soc.* **2001**, 123, 12911-12912.
21. Bahmanyar, S., Houk, K. N. Transition States of Amine-Catalyzed Aldol Reactions Involving Enamine Intermediates: Theoretical Studies of Mechanism, Reactivity, and Stereoselectivity. *J. Am. Chem. Soc.* **2001**, 123, 11273-11283.
22. Hoang, L., Bahmanyar, S., Houk, K. N., List, B. Kinetic and Stereochemical Evidence for the Involvement of Only One Proline Molecule in the Transition States of Proline-Catalyzed Intra- and Intermolecular Aldol Reactions. *J. Am. Chem. Soc.* **2003**, 125, 16-17.
23. Gutzwiller, J., Buchschacher, P., Fuerst, A. A procedure for the preparation of (S)-8a-methyl-3,4,8,8a-tetrahydro-1,6-(2H,7H)naphthalenedione. *Synthesis* **1977**, 167-168.
24. Danishefsky, S., Cain, P. Optically specific synthesis of estrone and 19-norsteroids from 2,6-lutidine. *J. Am. Chem. Soc.* **1976**, 98, 4975-4983.
25. Jung, M. E. A review of annulation. *Tetrahedron* **1976**, 32, 3-31.
26. Brown, K. L., Damm, L., Dunitz, J. D., Eschenmoser, A., Hobi, R., Kratky, C. Structural studies on crystalline enamines. *Helv. Chim. Acta* **1978**, 61, 3108-3135.
27. Agami, C., Puchot, C., Sevestre, H. Is the mechanism of the proline-catalyzed enantioselective aldol reaction related to biochemical processes? *Tetrahedron Lett.* **1986**, 27, 1501-1504.
28. Puchot, C., Samuel, O., Dunach, E., Zhao, S., Agami, C., Kagan, H. B. Nonlinear effects in asymmetric synthesis. Examples in asymmetric oxidations and aldolization reactions. *J. Am. Chem. Soc.* **1986**, 108, 2353-2357.
29. Agami, C. Mechanism of the proline-catalyzed enantioselective aldol reaction. Recent advances. *Bull. Soc. Chim. Fr.* **1988**, 499-507.
30. Corey, E. J., Huang, A. X. A Short Enantioselective Total Synthesis of the Third-Generation Oral Contraceptive Desogestrel. *J. Am. Chem. Soc.* **1999**, 121, 710-714.
31. Sakai, H., Hagiwara, H., Ito, Y., Hoshi, T., Suzuki, T., Ando, M. Total synthesis of (+)-cyclomyltaylan-5a-ol isolated from the Taiwanese liverwort Reboulia hemisphaerica. *Tetrahedron Lett.* **1999**, 40, 2965-2968.
32. Takahashi, S., Oritani, T., Yamashita, K. Total synthesis of (+)-methyl trisporate B, fungal sex hormone. *Tetrahedron* **1988**, 44, 7081-7088.
33. Patin, A., Kanazawa, A., Philouze, C., Greene, A. E., Muri, E., Barreiro, E., Costa, P. C. C. Highly Stereocontrolled Synthesis of Natural Barbacenic Acid, Novel Bisnorditerpene from Barbacenia flava. *J. Org. Chem.* **2003**, 68, 3831-3837.

Related reactions: Kröhnke pyridine synthesis;

1. Hantzsch, A. Synthesis of pyridine derivatives from acetoacetic ester and aldehydeammoniak. *Liebigs Ann. Chem.* **1882**, 215, 1-82.
2. Bergstrom, F. W. Heterocyclic N compounds. IIA. Hexacyclic compounds: pyridine, quinoline and isoquinoline. *Chem. Rev.* **1944**, 35, 77-277.
3. Phillips, A. P. Hantzsch's pyridine synthesis. *J. Am. Chem. Soc.* **1949**, 71, 4003-4007.
4. Barnes, R. A., Brody, F., Ruby, P. R. *Pyridine and its Derivatives* (ed. Klingsberg, E.) (Interscience Publishers, New York, **1960**) 613 pp.
5. Eisner, U., Kuthan, J. Chemistry of dihydropyridines. *Chem. Rev.* **1972**, 72, 1-42.
6. Lyle, R. E. *Partially reduced pyridines* (Wiley, Chichester, United Kingdom, **1974**) 137-182.
7. Bossert, F., Meyer, H., Wehinger, E. 4-Aryldihydropyridine, a new class of highly active calcium antagonists. *Angew. Chem.* **1981**, 93, 755-763.
8. Kuthan, J., Kurfurst, A. Development in dihydropyridine chemistry. *Ind. Eng. Chem. Prod. Res. Dev.* **1982**, 21, 191-261.
9. Stout, D. M., Meyers, A. I. Recent advances in the chemistry of dihydropyridines. *Chem. Rev.* **1982**, 82, 223-243.
10. Sausins, A., Duburs, G. Synthesis of 1,4-dihydropyridines by cyclocondensation reactions. *Heterocycles* **1988**, 27, 269-289.
11. Goldmann, S., Stoltefuss, J. 1,4-Dihydropyridines: effect of chirality and conformation on the calcium-antagonistic and -agonistic effects. *Angew. Chem.* **1991**, 103, 1587-1605 (See also *Angew. Chem., Int. Ed. Engl.*, **1991**, 1530(1512), 1559-1578).
12. Bannwarth, W. Multicomponent condensations (MCCs). *Methods and Principles in Medicinal Chemistry* **2000**, 9, 6-21.
13. Horton, D. A., Bourne, G. T., Smythe, M. L. The Combinatorial Synthesis of Bicyclic Privileged Structures or Privileged Substructures. *Chem. Rev.* **2003**, 103, 893-930.
14. Berson, J. A., Brown, E. Dihydropyridines. I. The preparation of unsymmetrical 4-aryl-1,4-dihydropyridines by the Hantzsch-Beyer synthesis. *J. Am. Chem. Soc.* **1955**, 77, 444-447.

15. Snyder, C. A., Thorn, M. A., Klijanowicz, J. E., Southwick, P. L. Preparation of compounds in the new dipyrrolo[3,4-b:3',4'-e]pyridine series from 1-benzylidene-2,3-dioxopyrrolidines. A variation of the Hantzsch synthesis. *J. Heterocycl. Chem.* **1982**, 19, 603-607.
16. Watanabe, Y., Shiota, K., Hoshiko, T., Ozaki, S. An efficient procedure for the Hantzsch dihydropyridine synthesis. *Synthesis* **1983**, 761.
17. Enders, D., Mueller, S., Demir, A. S. Enantioselective Hantzsch dihydropyridine synthesis via metalated chiral alkyl acetoacetate hydrazones. *Tetrahedron Lett.* **1988**, 29, 6437-6440.
18. Penieres, G., Garcia, O., Franco, K., Hernandez, O., Alvarez, C. A modification to the Hantzsch method to obtain pyridines in a one pot reaction: use of a bentonitic clay in a dry medium. *Heterocycl. Commun.* **1996**, 2, 359-360.
19. Anniyappan, M., Muralidharan, D., Perumal, P. T. Synthesis of hantzsch 1,4-dihydropyridines under microwave irradiation. *Synth. Commun.* **2002**, 32, 659-663.
20. Litvic, M., Cepanec, I., Vinkovic, V. A convenient Hantzsch synthesis of 1,4-dihydropyridines using tetraethyl orthosilicate. *Heterocycl. Commun.* **2003**, 9, 385-390.
21. Sabitha, G., Reddy, G. S. K. K., Reddy, C. S., Yadav, J. S. A novel TMSI-mediated synthesis of Hantzsch 1,4-dihydropyridines at ambient temperature. *Tetrahedron Lett.* **2003**, 44, 4129-4131.
22. Vanden Eynde, J. J., Mayence, A. Synthesis and aromatization of Hantzsch 1,4-dihydropyridines under microwave irradiation. An overview. *Molecules* **2003**, 8, 381-391.
23. Pfister, J. R. Rapid, high-yield oxidation of Hantzsch-type 1,4-dihydropyridines with ceric ammonium nitrate. *Synthesis* **1990**, 689-690.
24. Mashraqui, S. H., Karnik, M. A. Catalytic oxidation of Hantzsch 1,4-dihydropyridines by $RuCl_3$ under oxygen atmosphere. *Tetrahedron Lett.* **1998**, 39, 4895-4898.
25. Zolfigol, M. A., Kiany-Borazjani, M., Sadeghi, M. M., Mohammadpoor-Baltork, I., Memarian, H. R. Aromatization of 1,4-dihydropyridines under mild and heterogeneous conditions. *Synth. Commun.* **2000**, 30, 3919-3923.
26. Memarian, H. R., Sadeghi, M. M., Momeni, A. R. Aromatization of Hantzsch 1,4-dihydropyridines using barium manganate. *Synth. Commun.* **2001**, 31, 2241-2244.
27. Cheng, D.-P., Chen, Z.-C. Hypervalent iodine in synthesis. Part 76. An efficient oxidation of 1,4-dihydropyridines to pyridines using iodobenzene diacetate. *Synth. Commun.* **2002**, 32, 793-798.
28. Zolfigol, M. A., Shirini, F., Choghamarani, A. G., Mohammadpoor-Baltork, I. Silica modified sulfuric acid/NaNO2 as a novel heterogeneous system for the oxidation of 1,4-dihydropyridines under mild conditions. *Green Chem.* **2002**, 4, 562-564.
29. Hashemi, M. M., Ahmadibeni, Y. Cobalt and Manganese Salts of *p*-Aminobenzoic Acid Supported on Silica Gel: A Versatile Catalyst for Oxidation by Molecular Oxygen. *Monatsh. Chem.* **2003**, 134, 411-418.
30. Sabitha, G., Reddy, G. S. K. K., Reddy, C. S., Fatima, N., Yadav, J. S. $Zr(NO_3)_4$: A versatile oxidizing agent for aromatization of Hantzsch 1,4-dihydropyridines and 1,3,5-trisubstituted pyrazolines. *Synthesis* **2003**, 1267-1271.
31. Chatterjea, J. N. 1,3-Dimethyl-2-azafluorenone Meyer's pyridine synthesis. *J. Indian Chem. Soc.* **1952**, 29, 323-326.
32. Nemes, P., Balazs, B., Toth, G., Scheiber, P. Synthesis of fused heterocycles from β-enamino nitrile and carbonyl compounds. *Synlett* **1999**, 222-224.
33. Goerlitzer, K., Bartke, U. 3-(Nitrobenzylidene)-2,4(3H,5H)-furandiones in the Hantzsch pyridine synthesis: Part 1. A new approach to furo[3,4-b]pyridines. *Pharmazie* **2002**, 57, 672-678.
34. Knoevenagel, E. 1,5-Diketones. *Liebigs Ann. Chem.* **1894**, 281, 25-126.
35. Micheel, F., Moeller, W. Pyridine syntheses. II. Pyridine derivatives from penta- O-acetyl-aldehydo-D-glucose. *Ann.* **1963**, 670, 63-68.
36. Katritzky, A. R., Ostercamp, D. L., Yousaf, T. I. Mechanism of heterocyclic ring closures. 3. Mechanism of the Hantzsch pyridine synthesis: a study by nitrogen-15 and carbon-13 NMR spectroscopy. *Tetrahedron* **1986**, 42, 5729-5738.
37. Katritzky, A. R., Ostercamp, D. L., Yousaf, T. I. The mechanisms of heterocyclic ring closures. *Tetrahedron* **1987**, 43, 5171-5186.
38. Bredenkamp, M. W., Holzapfel, C. W., Synman, R. M., Van Zyl, W. J. Observations on the Hantzsch reaction: synthesis of N-Boc-S-dolaphenine. *Synth. Commun.* **1992**, 22, 3029-3039.
39. Dolle, F., Hinnen, F., Valette, H., Fuseau, C., Duval, R., Peglion, J.-L., Crouzel, C. Synthesis of two optically active calcium channel antagonists labeled with carbon-11 for in vivo cardiac PET imaging. *Bioorg. Med. Chem.* **1997**, 5, 749-764.
40. Dondoni, A., Massi, A., Minghini, E., Sabbatini, S., Bertolasi, V. Model Studies toward the Synthesis of Dihydropyrimidinyl and Pyridyl α-Amino Acids via Three-Component Biginelli and Hantzsch Cyclocondensations. *J. Org. Chem.* **2003**, 68, 6172-6183.
41. Natale, N. R., Rogers, M. E., Staples, R., Triggle, D. J., Rutledge, A. Lipophilic 4-Isoxazolyl-1,4-dihydropyridines: Synthesis and Structure-Activity Relationships. *J. Med. Chem.* **1999**, 42, 3087-3093.
42. Raboin, J.-C., Kirsch, G., Beley, M. On the way to unsymmetrical terpyridines carrying carboxylic acids. *J. Heterocycl. Chem.* **2000**, 37, 1077-1080.

Heck Reaction 196

Related reactions: Meerwein arylation;

1. Heck, R. F. Acylation, methylation, and carboxyalkylation of olefins by Group VIII metal derivatives. *J. Am. Chem. Soc.* **1968**, 90, 5518-5526.
2. Mizoroki, T., Mori, K., Ozaki, A. Arylation of olefin with aryl iodide catalyzed by palladium. *Bull. Chem. Soc. Jpn.* **1971**, 44, 581.
3. Heck, R. F., Nolley, J. P., Jr. Palladium-catalyzed vinylic hydrogen substitution reactions with aryl, benzyl, and styryl halides. *J. Org. Chem.* **1972**, 37, 2320-2322.
4. Dieck, H. A., Heck, R. F. Organophosphinepalladium complexes as catalysts for vinylic hydrogen substitution reactions. *J. Am. Chem. Soc.* **1974**, 96, 1133-1136.
5. Heck, R. F. Palladium-catalyzed vinylation of organic halides. *Org. React.* **1982**, 27, 345-390.
6. Daves, G. D., Jr., Hallberg, A. 1,2-Additions to heteroatom-substituted olefins by organopalladium reagents. *Chem. Rev.* **1989**, 89, 1433-1445.
7. Heck, R. F. Vinyl substitution and organopalladium intermediates. in *Comp. Org. Synth.* (eds. Trost, B. M.,Fleming, I.), *4*, 833-865 (Pergamon Press, Oxford, **1991**).
8. Ley, S. V., Marsden, S. P. Diastereoselective and enantioselective formation of quaternary carbon centers via the intramolecular Heck reaction: the influence of the coordination state of the palladium catalyst. *Chemtracts: Org. Chem.* **1993**, 6, 23-26.
9. de Meijere, A., Meyer, F. E. Clothes make the people: the Heck reaction in new clothing. *Angew. Chem.* **1994**, 106, 2473-2506 (See also Angew. Chem., Int. Ed. Engl., 1994, 2433(2423/2424), 2379-2411).
10. Overman, L. E. Application of intramolecular Heck reactions for forming congested quaternary carbon centers in complex molecule total synthesis. *Pure Appl. Chem.* **1994**, 66, 1423-1430.
11. Cabri, W., Candiani, I. Recent Developments and New Perspectives in the Heck Reaction. *Acc. Chem. Res.* **1995**, 28, 2-7.
12. Gibson, S. E., Middleton, R. J. The intramolecular Heck reaction. *Contemp. Org. Synth.* **1996**, 3, 447-471.
13. Heumann, A., Reglier, M. The stereochemistry of palladium-catalyzed cyclization reactions. Part C: Cascade reactions. *Tetrahedron* **1996**, 52, 9289-9346.
14. Jeffery, T. Recent improvements and developments in Heck-type reactions and their potential in organic synthesis. *Advances in Metal-Organic Chemistry* **1996**, 5, 153-260.
15. Shibasaki, M., Boden, C. D. J., Kojima, A. The asymmetric Heck reaction. *Tetrahedron* **1997**, 53, 7371-7395.
16. Brase, S., De Meijere, A. Palladium-catalyzed coupling of organyl halides to alkenes - the Heck reaction. in *Metal-Catalyzed Cross-Coupling Reactions* (eds. Diederich, F.,Stang, P. J.), 99-166 (Wiley-VCH, Weinhem, New York, **1998**).
17. Link, J. T., Overman, L. E. Intramolecular Heck reactions in natural product chemistry. in *Metal-Catalyzed Cross-Coupling Reactions* (eds. Diederich, F.,Stang, P. J.), 231-269 (Wiley-VCH, Weinheim, New York, **1998**).
18. Loiseleur, O., Hayashi, M., Keenan, M., Schmees, N., Pfaltz, A. Enantioselective Heck reactions using chiral P,N-ligands. *J. Organomet. Chem.* **1999**, 576, 16-22.

19. Shibasaki, M., Vogl, E. M. Heck reaction. in *Comprehensive Asymmetric Catalysis I-III* (eds. Jacobsen, E., Pfaltz, A.,Yamamoto, H.), *1*, 457-487 (Springer, Berlin, New York, **1999**).
20. Shibasaki, M., Vogl, E. M. The palladium-catalyzed arylation and vinylation of alkenes-enantioselective fashion. *J. Organomet. Chem.* **1999**, 576, 1-15.
21. Amatore, C., Jutand, A. Anionic $Pd^{(0)}$ and $Pd^{(II)}$ Intermediates in Palladium-Catalyzed Heck and Cross-Coupling Reactions. *Acc. Chem. Res.* **2000**, 33, 314-321.
22. Beletskaya, I. P., Cheprakov, A. V. The Heck Reaction as a Sharpening Stone of Palladium Catalysis. *Chem. Rev.* **2000**, 100, 3009-3066.
23. Donde, Y., Overman, L. E. Asymmetric intramolecular Heck reactions. in *Catal. Asymmetric Synth. (2nd Edition)* (ed. Ojima, I.), 675-697 (Wiley-VCH, New York, **2000**).
24. Jachmann, M., Schmalz, H.-G. Enantioselective Heck reactions. in *Organic Synthesis Highlights IV* 136-143 (VCH, Weinheim, New York, **2000**).
25. Wu, T.-C., Ramachandran, V. Pharmaceutical applications of Heck chemistry. *Innovations in Pharmaceutical Technology* **2000**, 97-101.
26. Biffis, A., Zecca, M., Basato, M. Palladium metal catalysts in Heck C-C coupling reactions. *J. Mol. Catal. A: Chemical* **2001**, 173, 249-274.
27. de Vries, J. G. The Heck reaction in the production of fine chemicals. *Can. J. Chem.* **2001**, 79, 1086-1092.
28. Eisenstadt, A., Ager, D. J. *Heck coupling* (eds. Sheldon, R. A.,Bekkum, H.) (Weinheim: Wiley-VCH, New York, **2001**) 576-587.
29. Frost, C. G. Palladium catalysed coupling reactions. in *Rodd's Chemistry of Carbon Compounds (2nd Edition) 5*, 315-350 (Elsevier, Amsterdam, New York, **2001**).
30. Stephan, M. S., De Vries, J. G. Homogeneous catalysis for fine chemicals: The Heck reaction as a clean alternative for Friedel-Crafts chemistry. *Chem. Ind.* **2001**, 82, 379-390.
31. Whitcombe, N. J., Hii, K. K., Gibson, S. E. Advances in the Heck chemistry of aryl bromides and chlorides. *Tetrahedron* **2001**, 57, 7449-7476.
32. Herrmann, W. A. Catalytic carbon-carbon coupling by palladium complexes: heck reactions. *Applied Homogeneous Catalysis with Organometallic Compounds (2nd Edition)* **2002**, 2, 775-793.
33. Larhed, M., Hallberg, A. The Heck reaction (alkene substitution via carbopalladation-dehydropalladation) and related carbopalladation reactions. in *Handbook of Organopalladium Chemistry for Organic Synthesis* (eds. Negishi, E.-i.,De Meijere, A.), *1*, 1133-1178 (Wiley-Interscience, New York, **2002**).
34. Link, J. T. The intramolecular Heck reaction. *Org. React.* **2002**, 60, 157-534.
35. Shibasaki, M., Miyazaki, F. Asymmetric Heck reactions. in *Handbook of Organopalladium Chemistry for Organic Synthesis* (eds. Negishi, E.-i.,De Meijere, A.), *1*, 1283-1315 (Wiley-Interscience, New York, **2002**).
36. Dounay, A. B., Overman, L. E. The Asymmetric Intramolecular Heck Reaction in Natural Product Total Synthesis. *Chem. Rev.* **2003**, 103, 2945-2963.
37. Braese, S., de Meijere, A. Cross-coupling of organic halides with alkenes: The Heck reaction. *Metal-Catalyzed Cross-Coupling Reactions (2nd Edition)* **2004**, 1, 217-315.
38. Guiry, P. J., Kiely, D. The development of the intramolecular asymmetric heck reaction. *Curr. Org. Chem.* **2004**, 8, 781-794.
39. Shibasaki, M., Vogl, E. M., Ohshima, T. Heck reaction. *Comprehensive Asymmetric Catalysis, Supplement* **2004**, 1, 73-81.
40. Eisenstadt, A. Utilization of the heterogeneous palladium-on-carbon catalyzed Heck reaction in applied synthesis. *Chem. Ind.* **1998**, 75, 415-427.
41. Guiry, P. J., Hennessy, A. J., Cahill, J. P. The asymmetric Heck reaction: recent developments and applications of new palladium diphenylphosphinopyrrolidine complexes. *Top. in Cat.* **1998**, 4, 311-326.
42. Walters, M. A. Macrocyclization on solid support using Heck reaction. *Chemtracts* **1998**, 11, 291-296.
43. Herrmann, W. A., Bohm, V. P. W., Reisinger, C.-P. Application of palladacycles in Heck type reactions. *J. Organomet. Chem.* **1999**, 576, 23-41.
44. Poli, G., Scolastico, C. Phosphapalladacycles: new efficient catalysts. *Chemtracts* **1999**, 12, 643-655.
45. Wang, J.-X., Liu, Z., Hu, Y., Wei, B., Bai, L. Microwave-promoted palladium-catalyzed Heck cross-coupling reaction in water. *J. Chem. Res., Synop.* **2000**, 484-485.
46. Farrington, E. J., Brown, J. M., Barnard, C. F. J., Roswell, E. Ruthenium-catalyzed oxidative Heck reactions. *Angew. Chem., Int. Ed. Engl.* **2002**, 41, 169-171.
47. Solabannavar, S. B., Desai, U. V., Mane, R. B. Heck reaction in aqueous medium using Amberlite IRA-400 (basic). *Green Chem.* **2002**, 4, 347-348.
48. Albert, K., Gisdakis, P., Roesch, N. On C-C Coupling by Carbene-Stabilized Palladium Catalysts: A Density Functional Study of the Heck Reaction. *Organometallics* **1998**, 17, 1608-1616.
49. Deeth, R. J., Smith, A., Hii, K. K., Brown, J. M. The Heck olefination reaction; a DFT study of the elimination pathway. *Tetrahedron Lett.* **1998**, 39, 3229-3232.
50. Shmidt, A. F., Khalaika, A., Nindakova, L. O., Shmidt, E. Y. Mechanism of alkene insertion into the Pd-Ar bond in the Heck reaction. *Kinetics and Catalysis (Translation of Kinetika i Kataliz)* **1998**, 39, 200-206.
51. Hii, K. K., Claridge, T. D. W., Brown, J. M., Smith, A., Deeth, R. J. The intermolecular asymmetric Heck reaction: mechanistic and computational studies. *Helv. Chim. Acta* **2001**, 84, 3043-3056.
52. Rosner, T., Le Bars, J., Pfaltz, A., Blackmond, D. G. Kinetic Studies of Heck Coupling Reactions Using Palladacycle Catalysts: Experimental and Kinetic Modeling of the Role of Dimer Species. *J. Am. Chem. Soc.* **2001**, 123, 1848-1855.
53. Sundermann, A., Uzan, O., Martin, J. M. L. Computational study of a new Heck reaction mechanism catalyzed by palladium(II/IV) species. *Chem.-- Eur. J.* **2001**, 7, 1703-1711.
54. Yates, B. Computational organometallic chemistry. *Chem. Austr.* **2001**, 68, 16-18.
55. Iserloh, U., Curran, D. P. Catalytic asymmetric synthesis of quaternary carbon centers: investigation of intramolecular Heck reactions and the application to calabar alkaloid synthesis. *Chemtracts* **1999**, 12, 289-296.
56. Herrmann, W. A., Reisinger, C.-P. *Carbon-carbon coupling by Heck-type reactions* (eds. Cornils, B.,Hermann, W. A.) (Wiley-VCH, Weinheim, New York, **1998**) 383-392.
57. Pierre Genet, J., Savignac, M. Recent developments of palladium(0) catalyzed reactions in aqueous medium. *J. Organomet. Chem.* **1999**, 576, 305-317.
58. Crisp, G. T. Variations on a theme: recent developments on the mechanism of the Heck reaction and their implications for synthesis. *Chem. Soc. Rev.* **1998**, 27, 427-436.
59. Amatore, C., Jutand, A. Mechanistic and kinetic studies of palladium catalytic systems. *J. Organomet. Chem.* **1999**, 576, 254-278.
60. Endo, A., Yanagisawa, A., Abe, M., Tohma, S., Kan, T., Fukuyama, T. Total Synthesis of Ecteinascidin 743. *J. Am. Chem. Soc.* **2002**, 124, 6552-6554.
61. Govek, S. P., Overman, L. E. Total Synthesis of Asperazine. *J. Am. Chem. Soc.* **2001**, 123, 9468-9469.
62. Fürstner, A., Thiel, O. R., Kindler, N., Bartkowska, B. Total Syntheses of (S)-(-)-Zearalenone and Lasiodiplodin Reveal Superior Metathesis Activity of Ruthenium Carbene Complexes with Imidazol-2-ylidene Ligands. *J. Org. Chem.* **2000**, 65, 7990-7995.

Heine Reaction 198

1. Heine, H. W., Fetter, M. E., Nicholson, E. M. Isomerization of some 1-aroylaziridines. II. *J. Am. Chem. Soc.* **1959**, 81, 2202-2204.
2. Heine, H. W., Kenyon, W. G., Johnson, E. M. The isomerization of aziridine derivatives. IV. *J. Am. Chem. Soc.* **1961**, 83, 2570-2574.
3. Heine, H. W., King, D. C., Portland, L. A. Aziridines. XII. The isomerization of some cis- and trans-1-(p-nitrobenzoyl)-2,3-substituted aziridines. *J. Org. Chem.* **1966**, 31, 2662-2665.
4. Heine, H. W., Kaplan, M. S. Aziridines. XVI. Isomerization of some 1-aroyl-aziridines. *J. Org. Chem.* **1967**, 32, 3069-3073.
5. Heine, H. W. Rearrangements of aziridine derivatives. *Angew. Chem.* **1962**, 74, 772-776.
6. Heine, H. W. The isomerization of aziridine derivatives. *Angew. Chem., Int. Ed. Engl.* **1962**, 1, 528-532.

7. Frump, J. A. Oxazolines. Their preparation, reactions, and applications. *Chem. Rev.* **1971**, 71, 483-506.
8. McCoull, W., Davis, F. A. Recent synthetic applications of chiral aziridines. *Synthesis* **2000**, 1347-1365.
9. Zwanenburg, B., ten Holte, P. The synthetic potential of three-membered ring aza-heterocycles. *Top. Curr. Chem.* **2001**, 216, 93-124.
10. Bonini, B. F., Fochi, M., Comes-Franchini, M., Ricci, A., Thijs, L., Zwanenburg, B. Synthesis of ferrocenyl-oxazolines by ring expansion of N-ferrocenoyl-aziridine-2-carboxylic esters. *Tetrahedron: Asymmetry* **2003**, 14, 3321-3327.
11. Tronchet, J. M. J., Massoud, M. A. M. Glycosylaziridine derivatives. *Heterocycles* **1989**, 29, 419-426.
12. Prabhakaran, E. N., Nandy, J. P., Shukla, S., Tewari, A., Kumar Das, S., Iqbal, J. Synthesis and conformation of proline containing tripeptides constrained with phenylalanine-like aziridine and dehydrophenylalanine residues. *Tetrahedron Lett.* **2002**, 43, 6461-6466.

Hell-Volhard-Zelinsky Reaction 200

1. Hell, C. A new method for the bromination of organic acids. *Ber.* **1881**, 14, 891-893.
2. Volhard, J. Preparation of α–brominated acids. *Ann.* **1887**, 242, 141-163.
3. Zelinsky, N. A convenient preparation of α-bromopropionic acid ester. *Ber.* **1887**, 20, 2026.
4. Watson, H. B. Reactions of halogens with compounds containing the carbonyl group. *Chem. Rev.* **1930**, 7, 173-201.
5. Sonntag, N. O. V. The reactions of aliphatic acid chlorides. *Chem. Rev.* **1953**, 52, 237-416.
6. Harwood, H. J. Reactions of the hydrocarbon chain of fatty acids. *Chem. Rev.* **1962**, 62, 99-154.
7. Ogata, Y., Harada, T., Matsuyama, K., Ikejiri, T. α-Chlorination of aliphatic acids by molecular chlorine. *J. Org. Chem.* **1975**, 40, 2960-2962.
8. Ogata, Y., Sugimoto, T., Inaishi, M. α-Chlorination of long-chain aliphatic acids. *Bull. Chem. Soc. Jpn.* **1979**, 52, 255-256.
9. Crawford, R. J. An improved α-chlorination of carboxylic acids. *J. Org. Chem.* **1983**, 48, 1364-1366.
10. Stevens, C., De Buyck, L., De Kimpe, N. The acylphosphonate function as an activating and masking moiety for the α-chlorination of fatty acids. *Tetrahedron Lett.* **1998**, 39, 8739-8742.
11. Stevens, C. V., Vanderhoydonck, B. Use of acylphosphonates for the synthesis of α-chlorinated carboxylic and α,α'-dichloro dicarboxylic acids and their derivatives. *Tetrahedron* **2001**, 57, 4793-4800.
12. Little, J. C., Sexton, A. R., Tong, Y.-L. C., Zurawic, T. E. Chlorination. II. Free radical vs. Hell-Volhard-Zelinsky chlorination of cyclohexanecarboxylic acid. *J. Am. Chem. Soc.* **1969**, 91, 7098-7103.
13. Little, J. C., Tong, Y.-L. C., Heeschen, J. P. Chlorination. I. Physical evidence for polar effects in the products of the chlorination of cyclohexanecarboxylic acid. *J. Am. Chem. Soc.* **1969**, 91, 7090-7097.
14. Harpp, D. N., Bao, L. Q., Black, C. J., Gleason, J. G., Smith, R. A. Efficient α-halogenation of acyl chlorides by N-bromosuccinimide, N-chlorosuccinimide, and molecular iodine. *J. Org. Chem.* **1975**, 40, 3420-3427.
15. Lapworth, A. J. *J. Chem.Soc.* **1904**, 85, 30.
16. Aschan, O. Mechanism of the Hell-Volhard Reaction. *Ber.* **1912**, 45, 1913-1919.
17. Aschan, O. Mechanism of the Hell-Volhard Reaction. II. *Ber.* **1913**, 46, 2162-2171.
18. Kwart, H., Scalzi, F. V. Observations regarding the mechanism and steric course of the a-bromination of carboxylic acid derivatives. An electrophilic substitution reaction in nonpolar media. *J. Am. Chem. Soc.* **1964**, 86, 5496-5503.
19. Turner, J. A., Kubler, D. G. Mechanism of the Hell-Volhard-Zelinsky reaction. *Furman Univ. Bull., Furman Studies* **1965**, 12, 45-52.
20. Liu, H. J., Luo, W. A convenient procedure for the conversion of carboxylic acids to α-bromo thiolesters. *Synth. Commun.* **1991**, 21, 2097-2102.
21. Watson, H. A., Jr., O'Neill, B. T. A reinvestigation and improvement in the synthesis of meso-2,5-dibromoadipates by application of Le Chatelier's principle. *J. Org. Chem.* **1990**, 55, 2950-2952.
22. Gibson, T. Thermal rearrangement of a 2-methylenebicyclo[2.1.1]hexane. *J. Org. Chem.* **1981**, 46, 1073-1076.

Henry Reaction 202

1. Henry, L. *C.R.Acad.Sci. Ser. C.* **1895**, 120, 1265.
2. Henry, L. *Bull. Soc. Chim. France* **1895**, 13, 999.
3. Hass, H. B., Riley, E. F. The nitro paraffins. *Chem. Rev.* **1943**, 32, 373-430.
4. Baer, H. H., Urbas, L. Activating and directing effects of the nitro group in aliphatic systems. in *Chem. Nitro Nitroso Groups* (ed. Feuer, H.), *2*, 75-200 (Interscience, New York, **1970**).
5. Fischer, R. H., Weitz, H. M. Preparation and reactions of cyclic α-nitroketones. *Synthesis* **1980**, 261-282.
6. Seebach, D., Beck, A. K., Mukhopadhyay, T., Thomas, E. Diastereoselective synthesis of nitroaldol derivatives. *Helv. Chim. Acta* **1982**, 65, 1101-1133.
7. Yoshikoshi, A., Miyashita, M. Oxoalkylation of carbonyl compounds with conjugated nitro olefins. *Acc. Chem. Res.* **1985**, 18, 284-290.
8. Barrett, A. G. M., Graboski, G. G. Conjugated nitroalkenes: versatile intermediates in organic synthesis. *Chem. Rev.* **1986**, 86, 751-762.
9. Varma, R. S., Kabalka, G. W. Nitroalkenes in the synthesis of heterocyclic compounds. *Heterocycles* **1986**, 24, 2645-2677.
10. Kabalka, G. W., Varma, R. S. Syntheses and selected reductions of conjugated nitroalkenes. A review. *Org. Prep. Proced. Int.* **1987**, 19, 283-328.
11. Rosini, G., Ballini, R. Functionalized nitroalkanes as useful reagents for alkyl anion synthons. *Synthesis* **1988**, 833-847.
12. Rosini, G. The Henry (Nitroaldol) Reaction. in *Comp. Org. Synth.* (eds. Trost, B. M.,Fleming, I.), *2*, 321-340 (Pergamon, Oxford, **1991**).
13. Bianchini, C., Glendenning, L. Efficient diastereoselective and enantioselective nitro aldol reactions from prochiral starting materials: utilization of La-Li-6,6'-distributed BINOL complexes as asymmetric catalysts. *Chemtracts: Org. Chem.* **1996**, 9, 327-330.
14. Ballini, R., Bosica, G. Formation of carbon-carbon bonds via nitroalkanes with heterogeneous catalysts. *Rec. Res. Dev. Org. Chem.* **1997**, 1, 11-24.
15. Shibasaki, M., Groger, H. Nitro aldol reaction. in *Comprehensive Asymmetric Catalysis I-III* (eds. Jacobsen, E., Pfaltz, A.,Yamamoto, H.), *3*, 1075-1090 (Springer, Berlin, New York, **1999**).
16. Iseki, K. Catalytic asymmetric synthesis of fluoro-organic compounds: Mukaiyama-aldol and Henry reactions. *ACS Symp. Ser.* **2000**, 746, 38-51.
17. Jacobsen, E. The nitro-aldol (Henry) reaction. in *The Nitro Group in Organic Synthesis* (ed. Ono, N.), 30-69 (Wiley, New York, **2001**).
18. Luzzio, F. A. The Henry reaction: recent examples. *Tetrahedron* **2001**, 57, 915-945.
19. Seebach, D., Lehr, F. α,α-Doubly deprotonated nitroalkanes. Increase in nitronate carbon nucleophilicity. *Angew. Chem.* **1976**, 88, 540-541.
20. Yamada, K., Tanaka, S., Kohmoto, S., Yamamoto, M. Novel regioselective generation of nitroalkane dianions. *J. Chem. Soc., Chem. Commun.* **1989**, 110-111.
21. Sasai, H., Itoh, N., Suzuki, T., Shibasaki, M. Catalytic asymmetric nitroaldol reaction: an efficient synthesis of (S)-propranolol using the lanthanum binaphthol complex. *Tetrahedron Lett.* **1993**, 34, 855-858.
22. Chinchilla, R., Najera, C., Sanchez-Agullo, P. Enantiomerically pure guanidine-catalyzed asymmetric nitro aldol reaction. *Tetrahedron: Asymmetry* **1994**, 5, 1393-1402.
23. Ballini, R., Bosica, G. Nitroaldol Reaction in Aqueous Media: An Important Improvement of the Henry Reaction. *J. Org. Chem.* **1997**, 62, 425-427.
24. Morao, I., Cossio, F. P. Dendritic catalysts for the nitroaldol (Henry) reaction. *Tetrahedron Lett.* **1997**, 38, 6461-6464.
25. Niyazmbetova, Z. I., Evans, D. H. Electrochemical version of the Henry reaction. *Electrochemical Processing Technologies, International Forum, Electrolysis in the Chemical Industry, 11th, Clearwater Beach, Fla., Nov. 2-6, 1997* **1997**, 465-469.
26. Bulbule, V. J., Deshpande, V. H., Velu, S., Sudalai, A., Sivasankar, S., Sathe, V. T. Heterogeneous Henry reaction of aldehydes: diastereoselective synthesis of nitro alcohol derivatives over Mg-Al hydrotalcites. *Tetrahedron* **1999**, 55, 9325-9332.

27. Kisanga, P. B., Verkade, J. G. P(RNCH2CH2)$_3$N: An efficient promoter for the nitroaldol (Henry) reaction. *J. Org. Chem.* **1999**, 64, 4298-4303.
28. Knudsen, K. R., Risgaard, T., Nishiwaki, N., Gothelf, K. V., Jorgensen, K. A. The First Catalytic Asymmetric Aza-Henry Reaction of Nitronates with Imines: A Novel Approach to Optically Active β-Nitro-α-Amino Acid- and α,β-Diamino Acid Derivatives. *J. Am. Chem. Soc.* **2001**, 123, 5843-5844.
29. Lin, W.-W., Jang, Y.-J., Wang, Y., Liu, J.-T., Hu, S.-R., Wang, L.-Y., Yao, C.-F. An Improved and Easy Method for the Preparation of 2,2-Disubstituted 1-Nitroalkenes. *J. Org. Chem.* **2001**, 66, 1984-1991.
30. Christensen, C., Juhl, K., Hazell, R. G., Jorgensen, K. A. Copper-Catalyzed Enantioselective Henry Reactions of α-Keto Esters: An Easy Entry to Optically Active β-Nitro-α-hydroxy Esters and β-Amino-α-hydroxy Esters. *J. Org. Chem.* **2002**, 67, 4875-4881.
31. Klein, G., Pandiaraju, S., Reiser, O. Activation of nitroaldol reactions by diethylzinc and amino alcohols or diamines as promoters. *Tetrahedron Lett.* **2002**, 43, 7503-7506.
32. Rajasekhar, C. V., Maheswaran, H. Enantioselective Michael addition and Henry reaction catalyzed by a new heterobimetallic aluminum-lithium complex derived from (+)-2,3-O-isopropylidine threitol. *Indian J. Chem., Sect. A: Inorg., Bio-inorg., Phys., Theor. Anal. Chem.* **2002**, 41A, 2503-2506.
33. Trost, B. M., Yeh, V. S. C., Ito, H., Bremeyer, N. Effect of Ligand Structure on the Zinc-Catalyzed Henry Reaction. Asymmetric Syntheses of (-)-Denopamine and (-)-Arbutamine. *Org. Lett.* **2002**, 4, 2621-2623.
34. Evans, D. A., Seidel, D., Rueping, M., Lam, H. W., Shaw, J. T., Downey, C. W. A New Copper Acetate-Bis(oxazoline)-Catalyzed, Enantioselective Henry Reaction. *J. Am. Chem. Soc.* **2003**, 125, 12692-12693.
35. Gao, J., Martell, A. E. Novel chiral N_4S_2- and N_6S_3-donor macrocyclic ligands: synthesis, protonation constants, metal-ion binding and asymmetric catalysis in the Henry reaction. *Org. Biomol. Chem.* **2003**, 1, 2801-2806.
36. Risgaard, T., Gothelf, K. V., Jorgensen, K. A. Catalytic asymmetric Henry reactions of silyl nitronates with aldehydes. *Org. Biomol. Chem.* **2003**, 1, 153-156.
37. Westermann, B. Asymmetric catalytic aza-Henry reactions leading to 1,2-diamines and 1,2-diaminocarboxylic acids. *Angew. Chem., Int. Ed. Engl.* **2003**, 42, 151-153.
38. Zhou, C. L., Zhou, Y. Q., Wang, Z. Y. Henry reaction in aqueous media: Chemoselective addition of aldehydes. *Chin. Chem. Lett.* **2003**, 14, 355-358.
39. Lecea, B., Arrieta, A., Morao, I., Cossio, F. P. Ab initio models for the nitroaldol (Henry) reaction. *Chem.-- Eur. J.* **1997**, 3, 20-28.
40. Serrano, J. A., Caceres, L. E., Roman, E. Asymmetric Diels-Alder reactions of a chiral sugar nitroalkene: diastereofacial selectivity and regioselectivity. *J. Chem. Soc., Perkin Trans. 1* **1992**, 941-942.
41. Avalos, M., Babiano, R., Cintas, P., Higes, F. J., Jimenez, J. L., Palacios, J. C., Silva, M. A. Substrate-Controlled Stereo-differentiation of Tandem [4+2]/[3+2] Cycloadditions by a Vicinal Carbohydrate-Based Template. *J. Org. Chem.* **1996**, 61, 1880-1882.
42. Fringuelli, F., Matteucci, M., Piermatti, O., Pizzo, F., Burla, M. C. [4 + 2] Cycloadditions of Nitroalkenes in Water. Highly Asymmetric Synthesis of Functionalized Nitronates. *J. Org. Chem.* **2001**, 66, 4661-4666.
43. Areces, P., Gil, M. V., Higes, F. J., Roman, E., Serrano, J. A. Stereoselective Michael addition reactions of 5-glyco-4-nitro-1-cyclohexenes. *Tetrahedron Lett.* **1998**, 39, 8557-8560.
44. Oehrlein, R., Jaeger, V. 3-Nitrobutanal. A new building block for the synthesis of methyl-branched sugars. *Tetrahedron Lett.* **1988**, 29, 6083-6086.
45. Wollenberg, R. H., Miller, S. J. Nitroalkane synthesis. A convenient method for aldehyde reductive nitromethylation. *Tetrahedron Lett.* **1978**, 3219-3222.
46. Melot, J. M., Texier-Boullet, F., Foucaud, A. Preparation and oxidation of α-nitro alcohols with supported reagents. *Tetrahedron Lett.* **1986**, 27, 493-496.
47. Akutu, K., Kabashima, H., Seki, T., Hattori, H. Nitroaldol reaction over solid base catalysts. *Appl. Cat. A* **2003**, 247, 65-74.
48. Sasai, H., Arai, S., Shibasaki, M. Catalytic Aldol Reaction with Sm(HMDS)$_3$ and Its Application for the Introduction of a Carbon-Carbon Triple Bond at C-13 in Prostaglandin Synthesis. *J. Org. Chem.* **1994**, 59, 2661-2664.
49. Ballini, R., Bosica, G., Livi, D., Palmieri, A., Maggi, R., Sartori, G. Use of heterogeneous catalyst KG-60-NEt2 in Michael and Henry reactions involving nitroalkanes. *Tetrahedron Lett.* **2003**, 44, 2271-2273.
50. Simoni, D., Invidiata, F. P., Manfredini, S., Ferroni, R., Lampronti, I., Roberti, M., Pollini, G. P. Facile synthesis of 2-nitroalkanols by tetramethylguanidine (TMG)-catalyzed addition of primary nitroalkanes to aldehydes and alicyclic ketones. *Tetrahedron Lett.* **1997**, 38, 2749-2752.
51. Phiasivongsa, P., Samoshin, V. V., Gross, P. H. Henry condensations with 4,6-O-benzylidenylated and non-protected D-glucose and L-fucose via DBU-catalysis. *Tetrahedron Lett.* **2003**, 44, 5495-5498.
52. Ono, N., Katayama, H., Nisyiyama, S., Ogawa, T. Regioselective synthesis of 5-unsubstituted benzyl pyrrole-2-carboxylates from benzyl isocyanoacetate. *J. Heterocycl. Chem.* **1994**, 31, 707-710.
53. Jenner, G. Effect of high pressure on Michael and Henry reactions between ketones and nitroalkanes. *New J. Chem.* **1999**, 23, 525-529.
54. Soengas, R. G., Estevez, J. C., Estevez, R. J. Transformation of D-Glucose into 1D-3-Deoxy-3-hydroxymethyl-myo-inositol by Stereocontrolled Intramolecular Henry Reaction. *Org. Lett.* **2003**, 5, 4457-4459.
55. Heffner, R. J., Jiang, J., Joullie, M. M. Total synthesis of (-)-nummularine F. *J. Am. Chem. Soc.* **1992**, 114, 10181-10189.
56. Adamczyk, M., Johnson, D. D., Reddy, R. E. Bone collagen cross-links: a convergent synthesis of (+)-deoxypyrrololine. *J. Org. Chem.* **2001**, 66, 11-19.
57. Ishikawa, T., Shimizu, Y., Kudoh, T., Saito, S. Conversion of D-Glucose to Cyclitol with Hydroxymethyl Substituent via Intramolecular Silyl Nitronate Cycloaddition Reaction: Application to Total Synthesis of (+)-Cyclophellitol. *Org. Lett.* **2003**, 5, 3879-3882.

Hetero Diels-Alder Cycloaddition (HDA)....................204

Related reactions: Danishefsky's diene cycloaddition;

1. Alder, K. *Recent methods of preparative organic chemistry* (Verlag Chemie, Weinheim, **1943**).
2. Kresze, G., Albrecht, R. Heterocycles by diene synthesis. Dienophilic azomethines and their diene adducts. *Ber.* **1964**, 97, 490-493.
3. Albrecht, R., Kresze, G. Heterocycles by diene synthesis. II. N-Carbobutyloxymethylene-p-toluenesulfonamide, a new dienophile for the preparation of pyridine, piperideine, and piperidine derivatives by the Diels-Alder synthesis. *Chem. Ber.* **1965**, 98, 1431-1434.
4. Weinreb, S. M., Levin, J. I. Synthesis of nitrogen-containing heterocycles by the imino Diels-Alder reaction. *Heterocycles* **1979**, 12, 949-975.
5. Weinreb, S. M., Staib, R. R. Synthetic aspects of Diels-Alder cycloadditions with heterodienophiles. *Tetrahedron* **1982**, 38, 3087-3128.
6. Boger, D. L. Diels-Alder reactions of heterocyclic aza dienes. Scope and applications. *Chem. Rev.* **1986**, 86, 781-794.
7. Schmidt, R. R. Hetero-Diels-Alder reaction in highly functionalized natural product synthesis. *Acc. Chem. Res.* **1986**, 19, 250-259.
8. Boger, D. L., Weinreb, S. M. Hetero Diels-Alder Methodology in Organic Synthesis. in *Organic Chemistry (N.Y.) 47*, 366 pp. (Academic Press, San Diego, **1987**).
9. Kametani, T., Hibino, S. Synthesis of natural heterocyclic products by hetero Diels-Alder cycloaddition reactions. *Adv. Heterocycl. Chem.* **1987**, 42, 245-333.
10. Taschner, M. J. Asymmetric Diels-Alder reactions. *Org. Synth.: Theory Appl.* **1989**, 1, 1-101.
11. Weinreb, S. M., Scola, P. M. N-acyl imines and related hetero dienes in [4+2]-cycloaddition reactions. *Chem. Rev.* **1989**, 89, 1525-1534.
12. Marcelis, A. T. M., van der Plas, H. C. Diels-Alder reactions of diazines and pyridines. *Trends Heterocycl. Chem.* **1990**, 1, 111-123.
13. Boger, D. L. Heterodiene Additions. in *Comp. Org. Synth.* (eds. Trost, B. M.,Fleming, I.), *5*, 451-512 (Pergamon, Oxford, **1991**).
14. Ibata, T. Diels-Alder reactions of acyclic and alicyclic dienes. in *Org. Synth. High Pressures* (eds. Matsumoto, K.,Acheson, R. M.), 213-285 (Wiley, New York, **1991**).

15. Matsumoto, K., Toda, M., Uchida, T. Diels-Alder reactions of heterocyclic dienes. in *Org. Synth. High Pressures* (eds. Matsumoto, K.,Acheson, R. M.), 287-326 (Wiley, New York, **1991**).
16. Weinreb, S. M. Heterodienophile Additions to Dienes. in *Comp. Org. Synth.* (eds. Trost, B. M.,Fleming, I.), *5*, 401-449 (Pergamon, Oxford, **1991**).
17. Giuliano, R. M. Cycloaddition reactions in carbohydrate chemistry. An overview. *ACS Symp. Ser.* **1992**, 494, 1-23.
18. Lopez, J. C., Lukacs, G. Pyranose-derived dienes and conjugated enals. Preparation and Diels-Alder cycloaddition reactions. *ACS Symp. Ser.* **1992**, 494, 33-49.
19. Martin, S. F. Strategies for the synthesis of heterocyclic natural products. *J. Heterocycl. Chem.* **1994**, 31, 679-686.
20. Oh, T., Reilly, M. Reagent-controlled asymmetric Diels-Alder reactions. A review. *Org. Prep. Proced. Int.* **1994**, 26, 129-158.
21. Streith, J., Defoin, A. Hetero Diels-Alder reactions with nitroso dienophiles: application to the synthesis of natural product derivatives. *Synthesis* **1994**, 1107-1117.
22. Waldmann, H. Asymmetric hetero Diels-Alder reactions. *Synthesis* **1994**, 535-551.
23. Zamojski, A. Asymmetric hetero-Diels-Alder reaction catalyzed by stable and easily prepared CAB catalysts. *Chemtracts: Org. Chem.* **1994**, 7, 337-339.
24. Waldmann, H. Amino acid esters: versatile chiral auxiliary groups for the asymmetric synthesis of nitrogen heterocycles. *Synlett* **1995**, 133-141.
25. Zamojski, A. De novo synthesis of enantiopure carbohydrates: preparation of ethyl β-D- and β-L-mannopyranosides by an asymmetrically induced hetero Diels-Alder reaction. *Chemtracts: Org. Chem.* **1995**, 8, 69-72.
26. Enders, D., Meyer, O. Diastereo- and enantioselective Diels-Alder reaction of 2-amino-1,3-dienes. *Liebigs Annalen* **1996**, 1023-1035.
27. Streith, J., Defoin, A. Aza sugar syntheses and multi-step cascade rearrangements via hetero Diels-Alder cycloadditions with nitroso dienophiles. *Synlett* **1996**, 189-200.
28. Fringuelli, F., Piermatti, O., Pizzo, F. Hetero Diels-Alder reactions in aqueous medium. *Targets in Heterocyclic Systems* **1997**, 1, 57-73.
29. Gonzalez, J., Sordo, J. A. Theoretical results on hetero-Diels-Alder reactions. *Recent Research Developments in Organic Chemistry* **1997**, 1, 239-257.
30. Laschat, S. New synthetic pathways to nitrogen heterocycles. *Liebigs Annalen/Recueil* **1997**, 1-11.
31. Tietze, L. F., Kettschau, G. Hetero Diels-Alder reactions in organic chemistry. *Top. Curr. Chem.* **1997**, 189, 1-120.
32. Johannsen, M., Yao, S., Graven, A., Jorgensen, K. A. Metal-catalyzed asymmetric hetero-Diels-Alder reactions of unactivated dienes with glyoxylates. *Pure Appl. Chem.* **1998**, 70, 1117-1122.
33. Parker, D. T. Hetero Diels-Alder reactions. *Organic Synthesis in Water* **1998**, 47-81.
34. Tietze, L. F., Kettschau, G., Gewert, J. A., Schuffenhauer, A. Hetero-Diels-Alder reactions of 1-oxa-1,3-butadienes. *Curr. Org. Chem.* **1998**, 2, 19-62.
35. Ooi, T., Maruoka, K. Hetero Diels-Alder and related reactions. in *Comprehensive Asymmetric Catalysis I-III* (eds. Jacobsen, E., Pfaltz, A.,Yamamoto, H.), *3*, 1237-1254 (Springer, New York, **1999**).
36. Ruano, J. L. G., De la Plata, B. C. Asymmetric [4+2] cycloadditions mediated by sulfoxides. *Topics in Current Chemistry* **1999**, 204, 1-126.
37. Johnson, J. S., Evans, D. A. Chiral Bis(oxazoline) Copper(II) Complexes: Versatile Catalysts for Enantioselective Cycloaddition, Aldol, Michael, and Carbonyl Ene Reactions. *Acc. Chem. Res.* **2000**, 33, 325-335.
38. Jorgensen, K. A. Catalytic asymmetric hetero-Diels-Alder reactions of carbonyl compounds and imines. *Angew. Chem., Int. Ed. Engl.* **2000**, 39, 3558-3588.
39. Stevenson, P. J. Synthetic methods part (ii) pericyclic methods. *Annu. Rep. Prog. Chem., Sect. B, Org. Chem.* **2000**, 96, 23-38.
40. Bianchini, C., Giambastiani, G. Discovery of exceptionally efficient catalysts for solvent-free enantioselective hetero-Diels-Alder reaction. *Chemtracts* **2002**, 15, 672-676.
41. Kobayashi, S. Catalytic enantioselective aza Diels-Alder reactions. *Cycloaddition Reactions in Organic Synthesis* **2002**, 187-209.
42. Jorgensen, K. A. Hetero-Diels-Alder reactions of ketones - a challenge for chemists. *Eur. J. Org. Chem.* **2004**, 2093-2102.
43. Osborn, H. M. I., Coisson, D. Application of the asymmetric hetero diels-alder reaction for synthesising carbohydrate derivatives and glycosidase inhibitors. *Mini-Reviews in Organic Chemistry* **2004**, 1, 41-54.
44. Mertes, J., Mattay, J. Thermal reactions of donor-acceptor systems. Part 3. Captodative olefins in normal and inverse Diels-Alder reactions. *Helv. Chim. Acta* **1988**, 71, 742-748.
45. Tietze, L. F., Geissler, H., Fennen, J., Brumby, T., Brand, S., Schulz, G. Intra- and intermolecular hetero-Diels-Alder reactions. 45. Simple and induced diastereoselectivity in intramolecular hetero-Diels-Alder reactions of 1-oxa-1,3-butadienes. Experimental data and calculations. *J. Org. Chem.* **1994**, 59, 182-191.
46. Fan, B. T., Barbu, A., Doucet, J.-P. Quantum chemical AM1 study of dimerization by hetero-Diels-Alder reaction of methyl 4,6-O-benzylidene-3-deoxy-3-C-methylene-a-D-hexopyranoside-2-ulose. *J. Chem. Soc., Perkin Trans. 2* **1997**, 1937-1941.
47. Venturini, A., Joglar, J., Fustero, S., Gonzalez, J. Diels-Alder Reactions of 2-Azabutadienes with Aldehydes: Ab Initio and Density Functional Theoretical Study of the Reaction Mechanism, Regioselectivity, Acid Catalysis, and Stereoselectivity. *J. Org. Chem.* **1997**, 62, 3919-3926.
48. Margetic, D., Johnston, M. R., Warrener, R. N. High-level computational study of the site-, facial- and stereoselectivities for the Diels-Alder reaction between o-benzoquinone and norbornadiene. *Molecules [online computer file]* **2000**, 5, 1417-1428.
49. Elliott, M. C., Kruiswijk, E., Willock, D. J. Asymmetric hetero-Diels-Alder reactions. Reactions of oxazolo[3,2-c]pyrimidines. *Tetrahedron* **2001**, 57, 10139-10146.
50. Orlova, G., Goddard, J. D. Competition between Diradical Stepwise and Concerted Mechanisms in Chalcogeno-Diels-Alder Reactions: A Density Functional Study. *J. Org. Chem.* **2001**, 66, 4026-4035.
51. Roberson, M., Jepsen, A. S., Jorgensen, K. A. On the mechanism of catalytic enantioselective hetero-Diels-Alder reactions of carbonyl compounds catalyzed by chiral aluminum complexes-a concerted, step-wise or Mukaiyama-aldol pathway. *Tetrahedron* **2001**, 57, 907-913.
52. Monnat, F., Vogel, P., Rayon, V. M., Sordo, J. A. Ab Initio and Experimental Studies on the Hetero-Diels-Alder and Cheletropic Additions of Sulfur Dioxide to (E)-1-Methoxybutadiene: A Mechanism Involving Three Molecules of SO2. *J. Org. Chem.* **2002**, 67, 1882-1889.
53. Monnat, F., Vogel, P., Sordo, J. A. Hetero-Diels-Alder and cheletropic additions of sulfur dioxide to 1,2-dimethylidenecycloalkanes. Determination of thermochemical and kinetics parameters for reactions in solution and comparison with estimates from quantum calculations. *Helv. Chim. Acta* **2002**, 85, 712-732.
54. Ujaque, G., Lee, P. S., Houk, K. N., Hentemann, M. F., Danishefsky, S. J. The origin of *endo* stereoselectivity in the hetero-Diels-Alder reactions of aldehydes with *ortho*-xylylenes: CH - π, π - π, and steric effects on stereoselectivity. *Chem.-- Eur. J.* **2002**, 8, 3423-3430.
55. Wang, Y., Zeng, X.-I. Ab initio theoretical study of the hetero Diels-Alder reactions of 1,3-butadiene and furan with thioformaldehyde. *Xinyang Shifan Xueyuan Xuebao, Ziran Kexueban* **2002**, 15, 177-180.
56. Alves, C. N., Romero, O. A. S., Da Silva, A. B. F. Theoretical study on the stereochemistry of intramolecular hetero Diels-Alder cycloaddition reactions of azoalkenes. *Int. J. Quantum Chem.* **2003**, 95, 133-136.
57. Ding, Y.-Q., Fang, D.-C. Theoretical Studies on Cycloaddition Reactions between the 2-Aza-1,3-butadiene Cation and Olefins. *J. Org. Chem.* **2003**, 68, 4382-4387.
58. Domingo, L. R., Andres, J. Enhancing Reactivity of Carbonyl Compounds via Hydrogen-Bond Formation. A DFT Study of the Hetero-Diels-Alder Reaction between Butadiene Derivative and Acetone in Chloroform. *J. Org. Chem.* **2003**, 68, 8662-8668.
59. Sakai, S. Theoretical analysis of concerted and stepwise mechanisms of the hetero-Diels-Alder reaction of butadiene with formaldehyde and thioformaldehyde. *THEOCHEM* **2003**, 630, 177-185.
60. Birkinshaw, T. N., Tabor, A. B., Holmes, A. B., Kaye, P., Mayne, P. M., Raithby, P. R. The products of an imino Diels-Alder reaction with 2-(trimethylsiloxy)cyclohexadiene: synthesis, x-ray crystal structures, and mechanistic implications. *J. Chem. Soc., Chem. Commun.* **1988**, 1599-1601.
61. Matsuoka, T., Harano, K., Uemura, T., Hisano, T. Hetero Diels-Alder reaction of N-acyl imines. I. The reaction of N'-thiobenzoyl-N,N-dimethylformamidine with electron-deficient dienophiles. Stereochemical and mechanistic aspects. *Chem. Pharm. Bull.* **1993**, 41, 50-54.
62. Mikami, K., Motoyama, Y., Terada, M. Asymmetric Catalysis of Diels-Alder Cycloadditions by an MS-Free Binaphthol-Titanium Complex: Dramatic Effect of MS, Linear vs Positive Nonlinear Relationship, and Synthetic Applications. *J. Am. Chem. Soc.* **1994**, 116, 2812-2820.

63. Linkert, F., Laschat, S., Kotila, S., Fox, T. Evidence for a stepwise mechanism in formal hetero-Diels-Alder reactions of N-arylimines. *Tetrahedron* **1996**, 52, 955-970.
64. Wijnen, J. W., Zavarise, S., Engberts, J. B. F. N., Charton, M. Substituent Effects on an Inverse Electron Demand Hetero Diels-Alder Reaction in Aqueous Solution and Organic Solvents: Cycloaddition of Substituted Styrenes to Di(2-pyridyl)-1,2,4,5-tetrazine. *J. Org. Chem.* **1996**, 61, 2001-2005.
65. Evans, D. A., Johnson, J. S., Burgey, C. S., Campos, K. R. Reversal in enantioselectivity of *tert*-butyl versus phenyl-substituted bis(oxazoline) copper(II) catalyzed hetero Diels-alder and ene reactions. Crystallographic and mechanistic studies. *Tetrahedron Lett.* **1999**, 40, 2879-2882.
66. Zhuo, J.-C., Wyler, H. Hetero-Diels-Alder cycloadditions of α,β-unsaturated acyl cyanides. Part 4. Substituent effects in reactions with *p*-substituted styrenes. *Helv. Chim. Acta* **1999**, 82, 1122-1134.
67. Evans, D. A., Johnson, J. S., Olhava, E. J. Enantioselective synthesis of dihydropyrans. Catalysis of hetero Diels-Alder reactions by bis(oxazoline) copper(II) complexes. *J. Am. Chem. Soc.* **2000**, 122, 1635-1649.
68. Mayr, H., Ofial, A. R., Sauer, J., Schmied, B. [2+4] cycloadditions of iminium ions - concerted or stepwise mechanism of aza Diels-Alder reactions? *Eur. J. Org. Chem.* **2000**, 2013-2020.
69. Simonsen, K. B., Svenstrup, N., Roberson, M., Jorgensen, K. A. Development of an unusually highly enantioselective hetero-Diels-Alder reaction of benzaldehyde with activated dienes catalyzed by hyper-coordinating chiral aluminum complexes. *Chemistry--A European Journal* **2000**, 6, 123-128.
70. Cuzzupe, A. N., Hutton, C. A., Lilly, M. J., Mann, R. K., McRae, K. J., Zammit, S. C., Rizzacasa, M. A. Total Synthesis of the Epidermal Growth Factor Inhibitor (-)-Reveromycin B. *J. Org. Chem.* **2001**, 66, 2382-2393.
71. Ito, M., Clark, C. W., Mortimore, M., Goh, J. B., Martin, S. F. Biogenetically inspired approach to the Strychnos alkaloids. Concise syntheses of (±)-akuammicine and (±)-strychnine. *J. Am. Chem. Soc.* **2001**, 123, 8003-8010.
72. Ozawa, T., Aoyagi, S., Kibayashi, C. Total Synthesis of the Marine Alkaloids (-)-Lepadins A, B, and C Based on Stereocontrolled Intramolecular Acylnitroso-Diels-Alder Reaction. *J. Org. Chem.* **2001**, 66, 3338-3347.
73. Swindell, C. S., Tao, M. Chiral auxiliary-mediated asymmetric induction in a thermal inverse electron demand hetero-Diels-Alder reaction. Enantioselective synthesis of the taxol A-ring side chain. *J. Org. Chem.* **1993**, 58, 5889-5891.

Hofmann Elimination 206

*Related reactions: **Burgess dehydration, Chugaev elimination, Cope elimination;***

1. Hofmann, A. W. Volatile organic bases. *Ann.* **1851**, 78, 253-286.
2. Hofmann, A. W. Volatile organic bases. *Ann.* **1851**, 79, 11-37.
3. Hofmann, A. W. The effect of heat on ammonium bases. *Ber.* **1881**, 14, 659-669.
4. Hofmann, A. W. The effect of heat on ammonium bases. *Ber.* **1881**, 14, 494-496.
5. Brewster, J. H., Eliel, E. L. Carbon-carbon alkylations with amines and ammonium salts. *Org. React.* **1953**, 7, 99-197.
6. Cope, A. C., Trumbull, E. R. Olefins from amines: the Hofmann elimination reaction and amine oxide pyrolysis. *Org. React.* **1960**, 11, 317-493.
7. Sunderwirth, S. G., Wood, J. K. Stereochemistry and orientation in bimolecular elimination reactions. *Trans. Kans. Acad. Sci.* **1967**, 70, 17-32.
8. Parker, A. J. Generation of olefins via elimination promoted by weak bases. *Chem. Technol.* **1971**, 297-303.
9. Coke, J. L. Stereochemistry of Hofmann eliminations. *Selec. Org. Transform.* **1972**, 2, 269-307.
10. Archer, D. A. Behavior of quaternary salts under reduced pressure. I. Improvement in yield during hofmann elimination. *J. Chem. Soc. C.* **1971**, 1327-1329.
11. Bach, R. D., Bair, K. W., Andrzejewski, D. Wittig modification of the Hofmann elimination reaction. Evidence for an α',β mechanism. *J. Am. Chem. Soc.* **1972**, 94, 8608-8610.
12. Lewis, D. E., Sims, L. B., Yamataka, H., McKenna, J. Calculations of kinetic isotope effects in the Hofmann eliminations of substituted (2-phenylethyl)trimethylammonium ions. *J. Am. Chem. Soc.* **1980**, 102, 7411-7419.
13. Baumgarten, R. J. Substitution products in the Hofmann elimination. *J. Chem. Educ.* **1968**, 45, 122.
14. Paquette, L. A., Wise, L. D. Unsaturated heterocyclic systems. XI. The Hofmann elimination of 9-methyl-3,9-diazabicyclo[4.2.1]nonan-4-one methiodide. Nature of the product and mechanism. *J. Org. Chem.* **1965**, 30, 228-231.
15. Coke, J. L., Cooke, M. P., Jr. Elimination reactions. II. Hofmann elimination in bicyclic compounds. *J. Am. Chem. Soc.* **1967**, 89, 6701-6704.
16. Coke, J. L., Cooke, M. P., Jr. Hofmann elimination. I. An example of a cis E2 mechanism. *J. Am. Chem. Soc.* **1967**, 89, 2779-2780.
17. Pankova, M., Sicher, J., Zavada, J. Syn-anti elimination dichotomy: a common feature in Hofmann elimination. *Chem. Commun.* **1967**, 394-396.
18. Banthorpe, D. V., Davies, H. f. S. Elimination reactions. II. Pyrolytic and base-promoted decompositions of thujyl compounds. *J. Chem. Soc. B.* **1968**, 1339-1346.
19. Coke, J. L., Cooke, M. P., Jr. Hofmann elimination of N,N,N-trimethyl-3,3-dimethcyclopentylammonium hydroxide. *Tetrahedron Lett.* **1968**, 18, 2253-2256.
20. Coke, J. L., Cooke, M. P., Jr., Mourning, M. C. Hofmann elimination in cyclic compounds. *Tetrahedron Lett.* **1968**, 2247-2251.
21. Coke, J. L., Mourning, M. C. Elimination reactions. IV. Hofmann elimination of N,N,N-trimethylcyclooctylammonium hydroxide. *J. Am. Chem. Soc.* **1968**, 90, 5561-5563.
22. Cooke, M. P., Jr., Coke, J. L. Elimination reactions. III. Hofmann elimination in cyclic compounds. *J. Am. Chem. Soc.* **1968**, 90, 5556-5561.
23. Zavada, J., Svoboda, M., Sicher, J. Stereochemical studies. LIII. Steric course of cycloalkyl onium base eliminations. Direct evidence for the syn-anti elimination dichotomy using deuterium-labeled cyclodecyl derivatives. *Collect. Czech. Chem. Commun.* **1968**, 33, 4027-4038.
24. Molnar, A., Bartok, M., Kovacs, K. Chemistry of 1,3-bifunctional compounds. II. Decomposition of the quaternary salts of 1,3-aminoalcohols. *Acta Chim. Acad. Sci. Hung.* **1969**, 59, 133-156.
25. Saunders, W. H., Jr., Ashe, T. A. Mechanisms of elimination reactions. XII. Hydrogen isotope effects and the nature of the transition state in eliminations from alicyclic quaternary ammonium salts. *J. Am. Chem. Soc.* **1969**, 91, 4473-4478.
26. Brown, K. C., Saunders, W. H., Jr. Mechanisms of elimination reactions. XIV. Stereochemistry and isotope effects in elimination from cyclopentyl- and 3,3-dimethylcyclopentyltrimethylammonium salts. *J. Am. Chem. Soc.* **1970**, 92, 4292-4295.
27. Sicher, J., Svoboda, M., Pankova, M., Zavada, J. Stereochemistry. LXII. Hofmann-Saytzeff and the syn-anti elimination dichotomy. Relation between the two phenomena. *Collect. Czech. Chem. Commun.* **1971**, 36, 3633-3649.
28. Coke, J. L. Stereochemistry of Hofmann eliminations. *Selective Organic Transformations* **1972**, 2, 269-307.
29. Coke, J. L., Smith, G. D., Britton, G. H., Jr. Elimination reactions. V. Steric effects in Hofmann elimination. *J. Am. Chem. Soc.* **1975**, 97, 4323-4327.
30. Kirby, A. J., Logan, C. J. Addition of amine nitrogen to an unactivated double bond. The mechanisms of the reverse Hofmann elimination. *J. Chem. Soc., Perkin Trans. 2* **1978**, 642-648.
31. Wu, S. L., Tao, Y. T., Saunders, W. H., Jr. Mechanisms of elimination reactions. 38. Why is the effect of successive β-alkyl substitution on the rates of elimination from quaternary ammonium salts nonadditive? *J. Am. Chem. Soc.* **1984**, 106, 7583-7588.
32. Bach, R. D., Braden, M. L. Primary and secondary kinetic isotope effects in the Cope and Hofmann elimination reactions. *J. Org. Chem.* **1991**, 56, 7194-7195.
33. Burch, R. R., Manring, L. E. N-Alkylation and Hofmann elimination from thermal decomposition of R_4N^+ salts of aromatic polyamide polyanions: synthesis and stereochemistry of N-alkylated aromatic polyamides. *Macromolecules* **1991**, 24, 1731-1735.
34. Eubanks, J. R. I., Sims, L. B., Fry, A. Carbon isotope effect studies of the mechanism of the Hofmann elimination reaction of para-substituted (2-phenylethyl-1-^{14}C)- and (2-phenylethyl-2-^{14}C)-trimethylammonium bromides. *J. Am. Chem. Soc.* **1991**, 113, 8821-8829.

35. Sekine, A., Ohshima, T., Shibasaki, M. An enantioselective formal synthesis of 4-demethoxydaunomycin using the catalytic asymmetric ring opening reaction of meso-epoxide with p-anisidine. *Tetrahedron* **2002**, 58, 75-82.
36. Gupta, R. B., Franck, R. W. The total synthesis of (-)-cryptosporin. *J. Am. Chem. Soc.* **1989**, 111, 7668-7670.
37. Kametani, T., Honda, T., Fukumoto, K., Toyota, M., Ihara, M. Synthetic approach to diterpene alkaloids - a simple and novel synthesis of the A,B,C and D ring part from 1-benzyl-1,2,3,4-tetrahydroisoquinoline. *Heterocycles* **1981**, 16, 1673-1676.
38. Kawada, K., Kim, M., Watt, D. S. Synthesis of quassinoids. 13. An enantioselective total synthesis of (+)-picrasin B. *Tetrahedron Lett.* **1989**, 30, 5989-5992.

Related reactions: Barton nitrite ester reaction;

1. Hofmann, A. W. Effect of basic bromine solution on amines. *Ber.* **1883**, 16, 558-560.
2. Hofmann, A. W. Study of the coniin group. *Ber.* **1885**, 18, 5-23.
3. Hofmann, A. W. Study of the coniin group. *Ber.* **1885**, 18, 109-131.
4. Löffler, K., Freytag, C. A new synthesis of N-alkyl pyrrolidines. *Ber.* **1909**, 42, 3427-3431.
5. Wolff, M. E. Cyclization of N-halogenated amines (the Hoffmann-Löffler reaction). *Chem. Rev.* **1963**, 63, 55-64.
6. Davidson, R. S. Hydrogen abstraction in the liquid phase by free radicals. *Quart. Rev., Chem. Soc.* **1967**, 21, 249-258.
7. Schonberg, A. *Preparative Organic Photochemistry* (Springer-Verlag, West Berlin, **1968**) p242.
8. Neale, R. S. Nitrogen radicals as synthesis intermediates. N-Haloamide rearrangements and additions to unsaturated hydrocarbons. *Synthesis* **1971**, 1-15.
9. Mackiewicz, P., Furstoss, R. Amidyl radicals: structure and reactivity. *Tetrahedron* **1978**, 34, 3241-3260.
10. Majetich, G., Wheless, K. Remote intramolecular free radical functionalizations: an update. *Tetrahedron* **1995**, 51, 7095-7129.
11. Feray, L., Kuznetsov, N., Renaud, P. in *Radicals in Organic Synthesis* (ed. Renaud, P.), 2, 254-256 (Wiley-VCH, Weinheim, **2001**).
12. Pellissier, H., Santelli, M. Functionalization of the 18-methyl group of steroids. A review. *Org. Prep. Proced. Int.* **2001**, 33, 455-476.
13. Stella, L. in *Radicals in Organic Synthesis* (ed. Renaud, P.), *2*, 409-426 (Wiley-VCH, Weinheim, **2001**).
14. Togo, H., Katohgi, M. Synthetic uses of organohypervalent iodine compounds through radical pathways. *Synlett* **2001**, 565-581.
15. Kimura, M., Ban, Y. A synthesis of 1,3-diaza heterocycles. A Hofmann-Löffler type of photocyclization in the absence of strong acid. *Synthesis* **1976**, 201-202.
16. Betancor, C., Concepcion, J. I., Hernandez, R., Salazar, J. A., Suarez, E. Intramolecular functionalization of nonactivated carbons by amidylphosphate radicals. Synthesis of 1,4-epimine compounds. *J. Org. Chem.* **1983**, 48, 4430-4432.
17. De Armas, P., Carrau, R., Concepcion, J. I., Francisco, C. G., Hernandez, R., Suarez, E. Synthesis of 1,4-epimine compounds. Iodosobenzene diacetate, an efficient reagent for neutral nitrogen radical generation. *Tetrahedron Lett.* **1985**, 26, 2493-2496.
18. Carrau, R., Hernandez, R., Suarez, E., Betancor, C. Intramolecular functionalization of N-cyanamide radicals. Synthesis of 1,4- and 1,5-N-cyanoepimino compounds. *J. Chem. Soc., Perkin Trans. 1* **1987**, 937-943.
19. Hernandez, R., Medina, M. C., Salazar, J. A., Suarez, E. Intramolecular functionalization of amides leading to lactams. *Tetrahedron Lett.* **1987**, 28, 2533-2536.
20. De Armas, P., Francisco, C. G., Hernandez, R., Salazar, J. A., Suarez, E. Steroidal N-nitroamines. Part 4. Intramolecular functionalization of N-nitroamine radicals: synthesis of 1,4-nitroimine compounds. *J. Chem. Soc., Perkin Trans. 1* **1988**, 3255-3265.
21. Dorta, R. L., Francisco, C. G., Suarez, E. Hypervalent organoiodine reagents in the transannular functionalization of medium-sized lactams: synthesis of 1-azabicyclo compounds. *J. Chem. Soc., Chem. Commun.* **1989**, 1168-1169.
22. Francisco, C. G., Herrera, A. J., Suarez, E. Intramolecular Hydrogen Abstraction Reaction Promoted by N-Radicals in Carbohydrates. Synthesis of Chiral 7-Oxa-2-azabicyclo[2.2.1]heptane and 8-Oxa-6-azabicyclo[3.2.1]octane Ring Systems. *J. Org. Chem.* **2003**, 68, 1012-1017.
23. Yates, B. F., Radom, L. Intramolecular hydrogen migration in ionized amines: a theoretical study of the gas-phase analogs of the Hofmann-Löffler and related rearrangements. *J. Am. Chem. Soc.* **1987**, 109, 2910-2915.
24. Ban, Y., Kimura, M., Oishi, T. A synthesis of (±)-dihydrodeoxyepiallocernuine by application of a facile Hofmann-Löffler type of photocyclization. *Chem. Pharm. Bull.* **1976**, 24, 1490-1496.
25. Baldwin, S. W., Doll, R. J. Synthesis of the 2-aza-7-oxatricyclo[4.3.2.04,8]undecane: nucleus of some gelsemium alkaloids. *Tetrahedron Lett.* **1979**, 3275-3278.
26. Chow, Y. L., Mojelsky, T. W., Magdzinski, L. J., Tichy, M. Chemistry of amidyl radicals. Intramolecular hydrogen abstraction as related to amidyl radical configurations. *Can. J. Chem.* **1985**, 63, 2197-2202.
27. Nikishin, G. I., Troyanskii, E. I., Lazareva, M. I. Regioselective one-step γ-chlorination of alkanesulfonamides. Preponderance of 1,5-H migration from sulfonyl versus amide moiety in sulfonylamidyl radicals. *Tetrahedron Lett.* **1985**, 26, 3743-3744.
28. Corey, E. J., Hertler, W. R. A study of the formation of halo amines and cyclic amines by the free radical chain decomposition of N-haloammonium ions (Hofmann-Löffler reaction). *J. Am. Chem. Soc.* **1960**, 82, 1657-1668.
29. Neale, R. S., Walsh, M. R., Marcus, N. L. The influence of solvent and chloramine structure on the free-radical rearrangement products of N-chlorodialkylamines. *J. Org. Chem.* **1965**, 30, 3683-3688.
30. Hammerum, S. Rearrangement and hydrogen abstraction reactions of amine cation radicals; a gas-phase analogy to the Hofmann-Löffler-Freytag reaction. *Tetrahedron Lett.* **1981**, 22, 157-160.
31. Green, M. M., Boyle, B. A., Vairamani, M., Mukhopadhyay, T., Saunders, W. H., Jr., Bowen, P., Allinger, N. L. Temperature-dependent stereoselectivity and hydrogen deuterium kinetic isotope effect for γ-hydrogen transfer to 2-hexyloxy radical. The transition state for the Barton reaction. *J. Am. Chem. Soc.* **1986**, 108, 2381-2387.
32. Shibanuma, Y., Okamoto, T. Synthetic approach to diterpene alkaloids: construction of the bridged azabicyclic ring system of kobusine. *Chem. Pharm. Bull.* **1985**, 33, 3187-3194.
33. Katohgi, M., Togo, H., Yamaguchi, K., Yokoyama, M. New synthetic method to 1,2-benzisothiazoline-3-one-1,1-dioxides and 1,2-benzisothiazoline-3-one-1-oxides from N-alkyl(o-methyl)arenesulfonamides. *Tetrahedron* **1999**, 55, 14885-14900.

Related reactions: Curtius rearrangement, Lossen rearrangement, Schmidt reaction;

1. Hofmann, A. W. The effect of bromine on amides in basic solutions. *Ber.* **1881**, 14, 2725-2736.
2. Hofmann, A. W. The effect of bromine on amides in basic solutions. *Ber.* **1882**, 15, 407-416.
3. Hofmann, A. W. The effect of bromine on amides in basic solutions. *Ber.* **1882**, 15, 762-775.
4. Hofmann, A. W. The effect of bromine on amides in basic solutions. *Ber.* **1884**, 17, 1406-1412.
5. Hofmann, A. W. The effect of bromine on amides in basic solutions. *Ber.* **1885**, 18, 2734-2741.
6. Wallis, E. S., Lane, J. F. Hofmann reaction. *Org. React.* **1946**, 267-306.
7. Applequist, D. E., Roberts, J. D. Displacement reactions at bridgeheads of bridged polycarbocyclic systems. *Chem. Rev.* **1954**, 54, 1065-1089.
8. Smith, P. A. S. Carbon-to-nitrogen migrations; what the last decade has brought. *Trans. N. Y. Acad. Sci.* **1969**, 31, 504-515.
9. Kovacic, P., Lowery, M. K., Field, K. W. Chemistry of N-bromamines and N-chloramines. *Chem. Rev.* **1970**, 70, 639-665.
10. Grillot, G. F. Hofmann-Martius rearrangement. *Mechanisms of Molecular Migrations* **1971**, 3, 237-270.

11. Jew, S. S., Park, H. G., Park, H. J., Park, M. S., Cho, Y. S. New methods for Hofmann rearrangement. *Ind. Chem. Library* **1991**, 3, 147-153.
12. Shioiri, T. Degradation Reactions. in *Comp. Org. Synth.* (eds. Trost, B. M.,Fleming, I.), *6*, 795-828 (Pergamon, Oxford, **1991**).
13. Kajigaeshi, S., Kakinami, T. Bromination and oxidation with benzyltrimethylammonium tribromide. *Ind. Chem. Library* **1995**, 7, 29-48.
14. Waldmann, H. Hypervalent iodine reagents. in *Organic Synthesis Highlights II* (ed. Waldmann, H.), 223-230 (VCH, Weinheim, **1995**).
15. Zhdankin, V. V., Stang, P. J. Recent Developments in the Chemistry of Polyvalent Iodine Compounds. *Chem. Rev.* **2002**, 102, 2523-2584.
16. Simons, S. S., Jr. Lead tetraacetate and pyridine. New, mild conditions for a Hofmann-like rearrangement. New synthesis of 2-oxazolidinones. *J. Org. Chem.* **1973**, 38, 414-416.
17. Baumgarten, H. E., Smith, H. L., Staklis, A. Reactions of amines. XVIII. Oxidative rearrangement of amides with lead tetraacetate. *J. Org. Chem.* **1975**, 40, 3554-3561.
18. Shono, T., Matsumura, Y., Yamane, S., Kashimura, S. The Hofmann rearrangement induced by an electroorganic method. *Chem. Lett.* **1982**, 565-568.
19. Loudon, G. M., Radhakrishna, A. S., Almond, M. R., Blodgett, J. K., Boutin, R. H. Conversion of aliphatic amides into amines with [I,I-bis(trifluoroacetoxy)iodo]benzene. 1. Scope of the reaction. *J. Org. Chem.* **1984**, 49, 4272-4276.
20. Vasudevan, A., Koser, G. F. Direct conversion of long-chain carboxamides to alkylammonium tosylates with hydroxy(tosyloxy)iodobenzene, a notable improvement over the classical Hofmann reaction. *J. Org. Chem.* **1988**, 53, 5158-5160.
21. Kajigaeshi, S., Asano, K., Fujisaki, S., Kakinami, T., Okamoto, T. Oxidation using quaternary ammonium polyhalides. I. An efficient method for the Hofmann degradation of amides by use of benzyltrimethylammonium tribromide. *Chem. Lett.* **1989**, 463-464.
22. Jew, S. S., Park, H. G., Kang, M. H., Lee, T. H., Cho, Y. S. Practical Hofmann rearrangement. *Arch. Pharm. Res.* **1992**, 15, 333-335.
23. Moriarty, R. M., Chany, C. J., II, Vaid, R. K., Prakash, O., Tuladhar, S. M. Preparation of methyl carbamates from primary alkyl- and arylcarboxamides using hypervalent iodine. *J. Org. Chem.* **1993**, 58, 2478-2482.
24. Rane, D. S., Sharma, M. M. New strategies for the Hofmann reaction. *J. Chem. Technol. Biotechnol.* **1994**, 59, 271-277.
25. Raynor, R. J., Knowles, T. A. *Isocyanates and their preparation via N-chlorination of amides with hypochlorous acid followed by phase-transfer-catalyzed Hofmann rearrangement.* US 92-997376, **1994** (Olin Corp., USA). 5 pp
26. Matsumura, Y., Maki, T., Satoh, Y. Electrochemically induced Hofmann rearrangement. *Tetrahedron Lett.* **1997**, 38, 8879-8882.
27. Varvoglis, A. Chemical transformations induced by hypervalent iodine reagents. *Tetrahedron* **1997**, 53, 1179-1255.
28. Matsumura, Y., Satoh, Y., Maki, T., Onomura, O. The electrochemically induced Hofmann rearrangement and its comparison with the classic Hofmann rearrangement. *Electrochim. Acta* **2000**, 45, 3011-3020.
29. Togo, H., Katohgi, M. Synthetic uses of organohypervalent iodine compounds through radical pathways. *Synlett* **2001**, 565-581.
30. Keillor, J. W., Huang, X. Methyl carbamate formation via modified Hofmann rearrangement reactions: Methyl N-(p-methoxyphenyl)carbamate. *Org. Synth.* **2002**, 78, 234-238.
31. Sy, A. O., Raksis, J. W. Synthesis of aliphatic isocyanates via a two-phase Hofmann reaction. *Tetrahedron Lett.* **1980**, 21, 2223-2226.
32. Gandhi, M. L., Chopra, S. L., Bhatia, I. S. Hofmann reaction of α,β-unsaturated amides. *Indian J. Chem.* **1968**, 6, 121-122.
33. Jew, S.-s., Kang, M.-h. Hofmann rearrangement of α-hydroxyamides. *Arch. Pharm. Res.* **1994**, 17, 490-491.
34. Joshi, K. M., Shah, K. K. Kinetics of the Hofmann bromoamide reaction. *J. Indian Chem. Soc.* **1966**, 43, 481-484.
35. Judd, W. P., Swedlund, B. E. Rearrangements of N-bromoamides. *Chem. Commun.* **1966**, 43-44.
36. Imamoto, T., Kim, S.-G., Tsuno, Y., Yukawa, Y. Hofmann rearrangement. IV. Kinetic isotope effect of N-chlorobenzamide. *Bull. Chem. Soc. Jpn.* **1971**, 44, 2776-2779.
37. Imamoto, T., Tsuno, Y., Yukawa, Y. Hofmann rearrangement. II. Kinetic substituent effects of ortho-, meta-, and para-substituted N-chlorobenzamides. *Bull. Chem. Soc. Jpn.* **1971**, 44, 1639-1643.
38. Imamoto, T., Tsuno, Y., Yukawa, Y. Hofmann rearrangement. III. Kinetic substituent effects of 4- and 5-substituted 2,N-dichlorobenzamides. *Bull. Chem. Soc. Jpn.* **1971**, 44, 1644-1648.
39. Imamoto, T., Tsuno, Y., Yukawa, Y. Hofmann rearrangement. I. Kinetic substituent effects of ortho-, meta-, and para-substituted N-bromobenzamides. *Bull. Chem. Soc. Jpn.* **1971**, 44, 1632-1638.
40. Boutin, R. H., Loudon, G. M. Conversion of aliphatic amides into amines with [I,I-bis(trifluoroacetoxy)iodo]benzene. 2. Kinetics and mechanism. *J. Org. Chem.* **1984**, 49, 4277-4284.
41. Senanayake, C. H., Fredenburgh, L. E., Reamer, R. A., Larsen, R. D., Verhoeven, T. R., Reider, P. J. Nature of N-Bromosuccinimide in Basic Media: The True Oxidizing Species in the Hofmann Rearrangement. *J. Am. Chem. Soc.* **1994**, 116, 7947-7948.
42. Evans, D. A., Scheidt, K. A., Downey, C. W. Synthesis of (-)-epibatidine. *Org. Lett.* **2001**, 3, 3009-3012.
43. Schultz, A. G., Wang, A. First Asymmetric Synthesis of a Hasubanan Alkaloid. Total Synthesis of (+)-Cepharamine. *J. Am. Chem. Soc.* **1998**, 120, 8259-8260.
44. Verma, R., Ghosh, S. K. A silicon controlled total synthesis of the antifungal agent (+)-preussin. *Chem. Commun.* **1997**, 1601-1602.
45. DeMong, D. E., Williams, R. M. Asymmetric Synthesis of (2S,3R)-Capreomycidine and the Total Synthesis of Capreomycin IB. *J. Am. Chem. Soc.* **2003**, 125, 8561-8565.

Horner-Wadsworth-Emmons Olefination 212

Related reactions:, Horner-Wadsworth-Emmons olefination - Still-Gennari modification, Julia-Lithgoe olefination, Peterson olefination, Takai-Utimoto olefination, Tebbe olefination, Wittig reaction, Wittig reaction – Schlosser modification;

1. Horner, L., Hoffmann, H., Wippel, H. G. Phosphorus organic compounds. XII. Phosphine oxides as reagents for the olefin formation. *Chem. Ber.* **1958**, 91, 61-63.
2. Horner, L., Hoffman, H., Wippel, H. G., Klahre, G. Phosphorus organic compounds. XX. Phosphine oxides as reagents for olefin formation. *Chem. Ber.* **1959**, 92, 2499-2505.
3. Wadsworth, W. S., Jr., Emmons, W. D. The utility of phosphonate carbanions in olefin synthesis. *J. Am. Chem. Soc.* **1961**, 83, 1733-1738.
4. Wadsworth, D. H., Schupp, I. O. E., Sous, E. J., Ford, J. J. A. The stereochemistry of the phosphonate modification of the Wittig reaction. *J. Org. Chem.* **1965**, 30, 680-685.
5. Boutagy, J., Thomas, R. Olefin synthesis with organic phosphonate carbanions. *Chemical Reviews (Washington, DC, United States)* **1974**, 74, 87-99.
6. Maier, L., Kunz, W. Preparation of triazolylmethylphosphonates and of triazolylmethylphosphonium salts and their application in the Wittig-Horner reaction. *Phosphorus and Sulfur and the Related Elements* **1987**, 30, 201-204.
7. Maryanoff, B. E., Reitz, A. B. The Wittig olefination reaction and modifications involving phosphoryl-stabilized carbanions. Stereochemistry, mechanism, and selected synthetic aspects. *Chem. Rev.* **1989**, 89, 863-927.
8. Kulkarni, Y. S. Carboxyolefination. *Aldrichimica Acta* **1990**, 23, 39-42.
9. Kelly, S. E. Alkene Synthesis. in *Comp. Org. Synth.* (eds. Trost, B. M.,Fleming, I.), *1*, 729-818 (Pergamon, Oxford, **1991**).
10. Yamaguchi, M., Hirama, M. Kinetic resolution of racemic aldehydes and ketones by the asymmetric Horner-Wadsworth-Emmons reaction. *Chemtracts: Org. Chem.* **1994**, 7, 401-405.
11. Gosney, I., Lloyd, D. One or more C=C bond(s) formed by condensation: Condensation of P, As, Sb, Bi, Si or metal functions. in *Comp. Org. Funct. Group Trans.* *1*, 719-770 (Pergamon, Cambridge, UK, **1995**).
12. Heron, B. M. Heterocycles from intramolecular Wittig, Horner and Wadsworth-Emmons reactions. *Heterocycles* **1995**, 41, 2357-2386.
13. Ernst, H., Muenster, P. Carotenoid synthesis. Wittig and Horner-Emmons reaction. *Carotenoids* **1996**, 2, 307-310.
14. Lawrence, N. J. The Wittig reaction and related methods. *Preparation of Alkenes* **1996**, 19-58.
15. Nicolaou, K. C., Harter, M. W., Gunzner, J. L., Nadin, A. The Wittig and related reactions in natural product synthesis. *Liebigs Annalen/Recueil* **1997**, 1283-1301.

16. Iorga, B., Eymery, F., Mouries, V., Savignac, P. Phosphorylated aldehydes: preparations and synthetic uses. *Tetrahedron* **1998**, 54, 14637-14677.
17. Motoyoshiya, J. Recent developments in Z-selective Horner-Wadsworth-Emmons reactions. *Trends in Organic Chemistry* **1998**, 7, 63-73.
18. Lorsbach, B. A., Kurth, M. J. Carbon-Carbon Bond Forming Solid-Phase Reactions. *Chemical Reviews (Washington, D. C.)* **1999**, 99, 1549-1581.
19. Rein, T., Vares, L., Kawasaki, I., Pedersen, T. M., Norrby, P.-O., Brandt, P., Tanner, D. Asymmetric Horner-Wadsworth-Emmons reactions with meso-dialdehydes: scope, mechanism, and synthetic applications. *Phosphorus, Sulfur Silicon Relat. Elem.* **1999**, 144-146, 169-172.
20. Jarosz, S. Synthesis of higher carbon sugars via coupling of simple monosaccharides-Wittig, Horner-Emmons, and related methods. *Journal of Carbohydrate Chemistry* **2001**, 20, 93-107.
21. Minami, T., Okauchi, T., Kouno, R. α-Phosphonovinyl carbanions in organic synthesis. *Synthesis* **2001**, 349-357.
22. Kellogg, R. M. Enantioconvergent synthesis by sequential asymmetric Horner-Wadsworth-Emmons and palladium-catalyzed allylic substitution reactions. *Chemtracts* **2002**, 15, 69-73.
23. Rein, T., Pedersen, T. M. Asymmetric Wittig type reactions. *Synthesis* **2002**, 579-594.
24. Prunet, J. Recent methods for the synthesis of (E)-alkene units in macrocyclic natural products. *Angewandte Chemie, International Edition* **2003**, 42, 2826-2830.
25. Corey, E. J., Kwiatkowski, G. T. Synthesis of olefins from carbonyl compounds and phosphonic acid bisamides. *J. Am. Chem. Soc.* **1966**, 88, 5652-5653.
26. Corey, E. J., Kwiatkowski, G. T. Synthesis of olefins from carbonyl compounds and phosphonic acid bis amides. *J. Am. Chem. Soc.* **1968**, 90, 6816-6821.
27. Still, W. C., Gennari, C. Direct synthesis of Z-unsaturated esters. A useful modification of the Horner-Emmons olefination. *Tetrahedron Lett.* **1983**, 24, 4405-4408.
28. Blanchette, M. A., Choy, W., Davis, J. T., Essenfeld, A. P., Masamune, S., Roush, W. R., Sakai, T. Horner-Wadsworth-Emmons reaction: use of lithium chloride and an amine for base-sensitive compounds. *Tetrahedron Lett.* **1984**, 25, 2183-2186.
29. Hanessian, S., Delorme, D., Beaudoin, S., Leblanc, Y. Design and reactivity of topologically unique, chiral phosphonamides. Remarkable diastereofacial selectivity in asymmetric olefination and alkylation. *J. Am. Chem. Soc.* **1984**, 106, 5754-5756.
30. Heathcock, C. H., Von Geldern, T. W. Total synthesis of (±)-norsecurinine. *Heterocycles* **1987**, 25, 75-78.
31. Ando, K. Practical synthesis of Z-unsaturated esters by using a new Horner-Emmons reagent, ethyl diphenylphosphonoacetate. *Tetrahedron Lett.* **1995**, 36, 4105-4108.
32. Ruebsam, F., Evers, A. M., Michel, C., Giannis, A. Z-Selective olefination of base-sensitive chiral β-hydroxy-α-aminoaldehydes using a modified Horner-Wadsworth-Emmons reaction. *Tetrahedron* **1997**, 53, 1707-1714.
33. Sano, s., Yokoyama, K., Fukushima, M., Yagi, T., Nagao, Y. New reaction mode of the Horner-Wadsworth-Emmons reaction using $Sn(OSO_2CF_3)_2$ and N-ethylpiperidine. *Chem. Commun.* **1997**, 559-560.
34. Salvino, J. M., Kiesow, T. J., Darnbrough, S., Labaudiniere, R. Solid-phase Horner-Emmons synthesis of olefins. *J. Comb. Chem.* **1999**, 1, 134-139.
35. Ando, K., Oishi, T., Hirama, M., Ohno, H., Ibuka, T. Z-Selective Horner-Wadsworth-Emmons Reaction of Ethyl (Diarylphosphono)acetates Using Sodium Iodide and DBU. *J. Org. Chem.* **2000**, 65, 4745-4749.
36. Molt, O., Schrader, T. Asymmetric synthesis with chiral cyclic phosphorus auxiliaries. *Synthesis* **2002**, 2633-2670.
37. Pihko, P. M., Salo, T. M. Excess sodium ions improve Z selectivity in Horner-Wadsworth-Emmons olefinations with the Ando phosphonate. *Tetrahedron Lett.* **2003**, 44, 4361-4364.
38. Reichwein, J. F., Pagenkopf, B. L. New Mixed Phosphonate Esters by Transesterification of Pinacol Phosphonates and Their Use in Aldehyde and Ketone Coupling Reactions with Nonstabilized Phosphonates. *J. Org. Chem.* **2003**, 68, 1459-1463.
39. Reichwein, J. F., Pagenkopf, B. L. A New Horner-Wadsworth-Emmons Type Coupling Reaction between Nonstabilized β-Hydroxy Phosphonates and Aldehydes or Ketones. *J. Am. Chem. Soc.* **2003**, 125, 1821-1824.
40. Gushurst, A. J., Jorgensen, W. L. Computer-assisted mechanistic evaluation of organic reactions. 14. Reactions of sulfur and phosphorus ylides, iminophosphoranes, and P=X-activated anions. *Journal of Organic Chemistry* **1988**, 53, 3397-3408.
41. Brandt, P., Norrby, P.-O., Martin, I., Rein, T. A Quantum Chemical Exploration of the Horner-Wadsworth-Emmons Reaction. *J. Org. Chem.* **1998**, 63, 1280-1289.
42. Kokin, K., Iitake, K.-I., Takaguchi, Y., Aoyama, H., Hayashi, S., Motoyoshiya, J. A study on the Z-selective Horner-Wadsworth-Emmons (HWE) reaction of methyl diarylphosphonoacetates. *Phosphorus, Sulfur Silicon Relat. Elem.* **1998**, 133, 21-40.
43. Ando, K. A Mechanistic Study of the Horner-Wadsworth-Emmons Reaction: Computational Investigation on the Reaction Pass and the Stereochemistry in the Reaction of Lithium Enolate Derived from Trimethyl Phosphonoacetate with Acetaldehyde. *J. Org. Chem.* **1999**, 64, 6815-6821.
44. Norrby, P.-O., Brandt, P., Rein, T. Rationalization of Product Selectivities in Asymmetric Horner-Wadsworth-Emmons Reactions by Use of a New Method for Transition-State Modeling. *J. Org. Chem.* **1999**, 64, 5845-5852.
45. Norrby, P. O. Selectivity in asymmetric synthesis from QM-guided molecular mechanics. *THEOCHEM* **2000**, 506, 9-16.
46. Motoyoshiya, J., Kusaura, T., Kokin, K., Yokoya, S. i., Takaguchi, Y., Narita, S., Aoyama, H. The Horner-Wadsworth-Emmons reaction of mixed phosphonoacetates and aromatic aldehydes: geometrical selectivity and computational investigation. *Tetrahedron* **2001**, 57, 1715-1721.
47. Maryanoff, B. E., Reitz, A. B. The Wittig olefination reaction and modifications involving phosphoryl-stabilized carbanions. Stereochemistry, mechanism, and selected synthetic aspects. *Chem. Rev.* **1989**, 89, 863-927.
48. Vedejs, E., Peterson, M. J. Stereochemistry and mechanism in the Wittig reaction. *Top. Stereochem.* **1994**, 21, 1-157.
49. Hoye, T. R., Humpal, P. E., Moon, B. Total Synthesis of (-)-Cylindrocyclophane A via a Double Horner-Emmons Macrocyclic Dimerization Event. *J. Am. Chem. Soc.* **2000**, 122, 4982-4983.
50. Stocksdale, M. G., Ramurthy, S., Miller, M. J. Asymmetric Total Synthesis of an Important 3-(Hydroxymethyl)carbacephalosporin. *J. Org. Chem.* **1998**, 63, 1221-1225.
51. Trost, B. M., Dirat, O., Gunzner, J. L. Callipeltoside A: assignment of absolute and relative configuration by total synthesis. *Angew. Chem., Int. Ed. Engl.* **2002**, 41, 841-843.

Horner-Wadsworth-Emmons Olefination – Still-Gennari Modification 214

Related reactions: Horner-Wadsworth-Emmons olefination, Julia olefination, Peterson olefination, Takai-Utimoto olefination, Tebbe olefination, Wittig reaction, Wittig reaction – Schlosser modification;

1. Still, W. C., Gennari, C. Direct synthesis of Z-unsaturated esters. A useful modification of the Horner-Emmons olefination. *Tetrahedron Lett.* **1983**, 24, 4405-4408.
2. Kelly, S. E. Alkene Synthesis. in *Comp. Org. Synth.* (eds. Trost, B. M.,Fleming, I.), *1*, 729-818 (Pergamon, Oxford, **1991**).
3. Motoyoshiya, J. Recent developments in Z-selective Horner-Wadsworth-Emmons reactions. *Trends in Organic Chemistry* **1998**, 7, 63-73.
4. Davis, A. A., Rosen, J. J., Kiddle, J. J. A new bisphosphonate reagent for the synthesis of (Z)-olefins and bis(trifluoroethyl)phosphonates. *Tetrahedron Lett.* **1998**, 39, 6263-6266.
5. Yu, W., Su, M., Jin, Z. A highly selective synthesis of (Z)-α,β-unsaturated ketones. *Tetrahedron Lett.* **1999**, 40, 6725-6728.
6. Tago, K., Kogen, H. Bis(2,2,2-trifluoroethyl) bromophosphonoacetate, a Novel HWE Reagent for the Preparation of (E)-α-Bromoacrylates: A General and Stereoselective Method for the Synthesis of Trisubstituted Alkenes. *Org. Lett.* **2000**, 2, 1975-1978.
7. Sano, S., Takehisa, T., Ogawa, S., Yokoyama, K., Nagao, Y. Stereoselective synthesis of tetrasubstituted (Z)-alkenes from aryl alkyl ketones utilizing the Horner-Wadsworth-Emmons reaction. *Chem. Pharm. Bull.* **2002**, 50, 1300-1302.

8. Sano, S., Yokoyama, K., Shiro, M., Nagao, Y. A facile method for the stereoselective Horner-Wadsworth-Emmons reaction of aryl alkyl ketones. *Chem. Pharm. Bull.* **2002**, 50, 706-709.
9. Franci, X., Martina, S. L. X., McGrady, J. E., Webb, M. R., Donald, C., Taylor, R. J. K. A comparison of the Still-Gennari and Ando HWE-methodologies with α,β-unsaturated aldehydes; unexpected results with stannyl substituted systems. *Tetrahedron Lett.* **2003**, 44, 7735-7740.
10. Sano, S., Takemoto, Y., Nagao, Y. (E)-Selective Horner-Wadsworth-Emmons reaction of aryl alkyl ketones with bis(2,2,2-trifluoroethyl)phosphonoacetic acid. *Tetrahedron Lett.* **2003**, 44, 8853-8855.
11. Motoyoshiya, J., Kusaura, T., Kokin, K., Yokoya, S. i., Takaguchi, Y., Narita, S., Aoyama, H. The Horner-Wadsworth-Emmons reaction of mixed phosphonoacetates and aromatic aldehydes: geometrical selectivity and computational investigation. *Tetrahedron* **2001**, 57, 1715-1721.
12. Maryanoff, B. E., Reitz, A. B. The Wittig olefination reaction and modifications involving phosphoryl-stabilized carbanions. Stereochemistry, mechanism, and selected synthetic aspects. *Chem. Rev.* **1989**, 89, 863-927.
13. Forsyth, C. J., Ahmed, F., Cink, R. D., Lee, C. S. Total Synthesis of Phorboxazole A. *J. Am. Chem. Soc.* **1998**, 120, 5597-5598.
14. Bates, R. W., Fernandez-Megia, E., Ley, S. V., Ruck-Braun, K., Tilbrook, D. M. G. Total synthesis of the cholesterol biosynthesis inhibitor 1233A via a (π-allyl)tricarbonyliron lactone complex. *J. Chem. Soc., Perkin Trans. 1* **1999**, 1917-1925.
15. Broady, S. D., Rexhausen, J. E., Thomas, E. J. Total synthesis of AI-77-B: stereoselective hydroxylation of 4-alkenylazetidinones. *J. Chem. Soc., Perkin Trans. 1* **1999**, 1083-1094.
16. Mergott, D. J., Frank, S. A., Roush, W. R. Application of the Intramolecular Vinylogous Morita-Baylis-Hillman Reaction toward the Synthesis of the Spinosyn A Tricyclic Nucleus. *Org. Lett.* **2002**, 4, 3157-3160.

Houben-Hoesch Reaction/Synthesis 216

Related reactions: Friedel-Crafts acylation, Fries-, photo-Fries and anionic ortho-Fries rearrangement, Minisci reaction;

1. Hoesch, K. A new synthesis of aromatic ketones. I. Preparation of some phenol ketones. *Ber.* **1915**, 48, 1122-1133.
2. Hoesch, K., von Zarzecki, T. A new synthesis of aromatic ketones. II. Artificial production of maclurin and related ketones. *Ber.* **1917**, 50, 462-468,660.
3. Houben, J. Nucleus condensation of phenols and phenol ethers with nitriles to phenol and phenol ether ketimides and ketones. I. *Ber.* **1926**, 59B, 2878-2891.
4. Hoesch, K. Final reply to J. Houben (nucleus condensation of phenols, etc., with nitriles, etc.). *Ber.* **1927**, 60B, 2537.
5. Hoesch, K. Reply to Houben (nucleus condensation of phenols, etc., with nitriles etc.). *Ber.* **1927**, 60B, 389.
6. Houben, J., Fischer, W. Nucleus-condensation of phenols and phenol ethers with nitriles to phenol and phenol ether ketimides and ketones. III. Syntheses of cotogenin, protocotoin, isorpotocotoin and methylprotocotoin. *J. Prakt. Chem.* **1929**, 123, 89-109.
7. Calloway, N. O. The Friedel-Crafts syntheses. *Chem. Rev.* **1935**, 17, 327-392.
8. Migrdichian, V. in *The Chemistry of Organic Cyanogen Compounds* pp 235 (Reinhold Publ. Corp., New York, **1947**).
9. Spoerri, P. E., DuBois, A. S. Hoesch synthesis. *Org. React.* **1949**, 5, 387-412.
10. Ruske, W. Houben-Hoesch and Related Syntheses. in *Friedel-Crafts and Related Reactions* (ed. Olah, G. A.), *3*, 383-497 (Interscience Publishers, New York, **1964**).
11. Zil'berman, E. N., Rybakova, N. A. Hoesch synthesis. Preparation of benzoresorcinol. *Zh. Obshch. Khim.* **1960**, 30, 1992-1996.
12. Sato, K., Amakasu, T. Coumarins. V. Acid-catalyzed reaction of phenols with β-oxonitriles. *J. Org. Chem.* **1968**, 33, 2446-2450.
13. Sugasawa, T., Adachi, M., Sasakura, K., Kitagawa, A. Aminohaloborane in organic synthesis. 2. Simple synthesis of indoles and 1-acyl-3-indolinones using specific ortho a-chloroacetylation of anilines. *J. Org. Chem.* **1979**, 44, 578-586.
14. Sanchez-Viesca, F., Gomez, M. R. Synthetic applications of the anomalous Hoesch reaction. *Revista Latinoamericana de Quimica* **1982**, 13, 24-26.
15. Amer, M. I., Booth, B. L., Noori, G. F. M., Proenca, M. F. J. R. P. The chemistry of nitrilium salts. Part 3. The importance of triazinium salts in Houben-Hoesch reactions catalyzed by trifluoromethanesulfonic acid. *J. Chem. Soc., Perkin Trans. 1* **1983**, 1075-1081.
16. Bigi, F., Maggi, R., Sartori, G., Casnati, G., Bocelli, G. Template Houben-Hoesch reaction on metal phenolates. Synthesis of aromatic ketones, nitriles and amides. Crystal structure of dichloro[2-(1-imino-2,2,2-trichloroethyl)-4-methoxyphenoxido-O,N]boron. *Gazz. Chim. Ital.* **1992**, 122, 283-289.
17. Udwary, D. W., Casillas, L. K., Townsend, C. A. Synthesis of 11-hydroxyl O-methylsterigmatocystin and the role of a cytochrome P-450 in the final step of aflatoxin biosynthesis. *J. Am. Chem. Soc.* **2002**, 124, 5294-5303.
18. Alagona, G., Tomasi, J. The mechanism of addition to a CN triple bond. An ab initio study of the first stages of the Stephen, Gattermann and Houben-Hoesch reactions. *THEOCHEM* **1983**, 8, 263-281.
19. Jeffery, E. A., Satchell, D. P. N. A kinetic study of the formation of ketimine hydrochlorides. The mechanism of the Houben-Hoesch reaction. *J. Chem. Soc. B.* **1966**, 579-586.
20. Yato, M., Ohwada, T., Shudo, K. Requirements for Houben-Hoesch and Gattermann reactions. Involvement of diprotonated cyanides in the reactions with benzene. *Journal of the American Chemical Society* **1991**, 113, 691-692.
21. Sato, Y., Yato, M., Ohwada, T., Saito, S., Shudo, K. Involvement of Dicationic Species as the Reactive Intermediates in Gattermann, Houben-Hoesch, and Friedel-Crafts Reactions of Nonactivated Benzenes. *J. Am. Chem. Soc.* **1995**, 117, 3037-3043.
22. Cameron, D. W., Deutscher, K. R., Feutrill, G. I., Hunt, D. E. Synthesis of azaanthraquinones: homolytic substitution of pyridines. *Aust. J. Chem.* **1982**, 35, 1451-1468.
23. Dixon, R. A., Ferreira, D. Genistein. *Phytochemistry* **2002**, 60, 205-211.
24. Balasubramanian, S., Nair, M. G. An efficient "one pot" synthesis of isoflavones. **2000**, 30, 469-484.
25. Kawecki, R., Mazurek, A. P., Kozerski, L., Maurin, J. K. Synthesis of benzofuro[2,3-b]benzofuran derivatives under Hoesch reaction conditions. *Synthesis* **1999**, 751-753.

Hunsdiecker Reaction 218

Related reactions: Barton radical decarboxylation reaction;

1. Borodine, A. Bromovaleric acid and bromobutyric acid. *Ann.* **1861**, 119, 121-123.
2. Hunsdiecker, H., Hunsdiecker, C., Vogt, E. *Halogen-containing organic compounds.* US 2176181, **1939**,
3. Hunsdiecker, H., Hunsdiecker, C. Degradation of the salts of aliphatic acids by bromine. *Ber.* **1942**, 75B, 291-297.
4. Johnson, R. G., Ingham, R. K. The degradation of carboxylic acid salts by means of halogen. The Hunsdiecker reaction. *Chem. Rev.* **1956**, 56, 219-269.
5. Wilson, C. V. The reaction of halogens with silver salts of carboxylic acids. *Org. React.* **1957**, 332-387.
6. Sheldon, R. A., Kochi, J. K. Oxidative decarboxylation of acids by lead tetraacetate. *Org. React.* **1972**, 19, 279-421.
7. Crich, D. The Hunsdiecker and Related Reactions. in *Comp. Org. Synth.* (eds. Trost, B. M.,Fleming, I.), *7*, 717-734 (Pergamon Press, Oxford, **1991**).
8. Rice, F. A. H. Decarboxylation via the acid chloride of penta-O-acetyl-D-gluconic acid. *J. Am. Chem. Soc.* **1956**, 78, 3173-3175.
9. Rice, F. A. H., Morganroth, W. Reaction of the acid chlorides of aromatic acids with bromine and silver oxide. *J. Org. Chem.* **1956**, 21, 1388-1389.
10. Cristol, S. J., Firth, W. C., Jr. Convenient synthesis of alkyl halides from carboxylic acids. *J. Org. Chem.* **1961**, 26, 280.

11. Barton, D. H. R., Faro, H. P., Serebryakov, E. P., Woolsey, N. F. Photochemical transformations. XVII. Improved methods for the decarboxylation of acids. *J. Chem. Soc., Abstracts* **1965**, 2438-2444.
12. Davis, J. A., Herynk, J., Carroll, S., Bunds, J., Johnson, D. Modifications of the Hunsdiecker reaction. *J. Org. Chem.* **1965**, 30, 415-417.
13. McKillop, A., Bromley, D. Thallium in organic synthesis. VIII. Preparation of aromatic bromides. *Tetrahedron Lett.* **1969**, 1623-1626.
14. Lampman, G. M., Aumiller, J. C. Mercuric oxide-modified Hunsdiecker reaction. 1-Bromo-3-chlorocyclobutane. *Org. Synth.* **1971**, 51, 106-108.
15. Meyers, A. I., Fleming, M. P. Photoassisted Cristol-Firth-Hunsdiecker reaction. *J. Org. Chem.* **1979**, 44, 3405-3406.
16. Cambie, R. C., Hayward, R. C., Jurlina, J. L., Rutledge, P. S., Woodgate, P. D. Thallium(I) carboxylate modification of the Hunsdiecker reaction. *J. Chem. Soc., Perkin Trans. 1* **1981**, 2608-2614.
17. Barton, D. H. R., Crich, D., Motherwell, W. B. A practical alternative to the Hunsdiecker reaction. *Tetrahedron Lett.* **1983**, 24, 4979-4982.
18. Patrick, T. B., Johri, K. K., White, D. H. Fluoro-decarboxylation with xenon difluoride. *J. Org. Chem.* **1983**, 48, 4158-4159.
19. Barton, D. H. R., Crich, D., Motherwell, W. B. The invention of new radical chain reactions. Part VIII. Radical chemistry of thiohydroxamic esters; a new method for the generation of carbon radicals from carboxylic acids. *Tetrahedron* **1985**, 41, 3901-3924.
20. Concepcion, J. I., Francisco, C. G., Freire, R., Hernandez, R., Salazar, J. A., Suarez, E. Iodosobenzene diacetate, an efficient reagent for the oxidative decarboxylation of carboxylic acids. *J. Org. Chem.* **1986**, 51, 402-404.
21. Patrick, T. B., Johri, K. K., White, D. H., Bertrand, W. S., Mokhtar, R., Kilbourn, M. R., Welch, M. J. Replacement of the carboxylic acid function with fluorine. *Can. J. Chem.* **1986**, 64, 138-141.
22. Chowdhury, S., Roy, S. Manganese(II)-catalyzed Hunsdiecker reaction: a facile entry to α-(dibromomethyl)benzenemethanol. *Tetrahedron Lett.* **1996**, 37, 2623-2624.
23. Chowdhury, S., Roy, S. The First Example of a Catalytic Hunsdiecker Reaction: Synthesis of β-Halostyrenes. *J. Org. Chem.* **1997**, 62, 199-200.
24. Naskar, D., Chowdhury, S., Roy, S. Is metal necessary in the Hunsdiecker-Borodin reaction? *Tetrahedron Lett.* **1998**, 39, 699-702.
25. Camps, P., Lukach, A. E., Pujol, X., Vazquez, S. Hunsdiecker-type bromodecarboxylation of carboxylic acids with iodosobenzene diacetate-bromine. *Tetrahedron* **2000**, 56, 2703-2707.
26. Kuang, C., Senboku, H., Tokuda, M. Stereoselective synthesis of (E)-β-arylvinyl halides by microwave-induced Hunsdiecker reaction. *Synlett* **2000**, 1439-1442.
27. Naskar, D., Roy, S. Catalytic Hunsdiecker reaction and one-pot catalytic Hunsdiecker-heck strategy: synthesis of α,β-unsaturated aromatic halides, α-(dihalomethyl)benzenemethanols, 5-aryl-2,4-pentadienoic acids, dienoates and dienamides. *Tetrahedron* **2000**, 56, 1369-1377.
28. Sinha, J., Layek, S., Mandal, G. C., Bhattacharjee, M. A green Hunsdiecker reaction: synthesis of β-bromostyrenes from the reaction of α,β-unsaturated aromatic carboxylic acids with KBr and H_2O_2 catalyzed by $Na_2MoO_4.2H_2O$ in aqueous medium. *Chem. Commun.* **2001**, 1916-1917.
29. Das, J. P., Roy, S. Catalytic Hunsdiecker Reaction of α,β-Unsaturated Carboxylic Acids: How Efficient Is the Catalyst? *J. Org. Chem.* **2002**, 67, 7861-7864.
30. Das, J. P., Sinha, P., Roy, S. A Nitro-Hunsdiecker Reaction: From Unsaturated Carboxylic Acids to Nitrostyrenes and Nitroarenes. *Org. Lett.* **2002**, 4, 3055-3058.
31. Barton, D. H. R., Lacher, B., Zard, S. Z. The invention of radical reactions. Part XVI. Radical decarboxylative bromination and iodination of aromatic acids. *Tetrahedron* **1987**, 43, 4321-4328.
32. Cristol, S. J., Douglass, J. R., Firth, W. C., Jr., Krall, R. E. Bridged polycyclic compounds. XII. A mechanism for the Hunsdiecker reaction. *J. Am. Chem. Soc.* **1960**, 82, 1829-1830.
33. Jennings, P. W., Ziebarth, T. D. Mechanism of the modified Hunsdiecker reaction. *J. Org. Chem.* **1969**, 34, 3216-3217.
34. Bunce, N. J., Urban, L. O. Decomposition of benzoyl hypochlorite in the presence of metal ions. *Can. J. Chem.* **1971**, 49, 821-827.
35. Cason, J., Walba, D. M. Reaction pathway in the modified Hunsdiecker reaction. *J. Org. Chem.* **1972**, 37, 669-671.
36. Britten-Kelley, M. R., Goosen, A., Scheffer, A. Kinetic studies on the photodecarboxylation reactions of acyl hypoiodites. *J. S.African Chem. Inst.* **1975**, 28, 224-234.
37. Norula, J. L. Mechanism of the reaction of bromine with the silver salt of a carboxylic acid. *Chemical Era* **1975**, 11, 40-42.
38. Chenier, P. J., Southard, D. A., Jr. Tricyclo[3.2.2.0^{2,4}]non-2(4)-ene: synthesis and trapping of a strained cyclopropene. *J. Org. Chem.* **1990**, 55, 4333-4337.
39. Sebahar, P. R., Williams, R. M. The Asymmetric Total Synthesis of (+)- and (-)-Spirotryprostatin B. *J. Am. Chem. Soc.* **2000**, 122, 5666-5667.

Jacobsen Hydrolytic Kinetic Resolution 220

1. Emziane, M., Sutowardoyo, K. I., Sinou, D. Asymmetric ring-opening of cyclohexene oxide with trimethylsilyl azide in the presence of titanium isopropoxide/chiral ligand. *J. Organomet. Chem.* **1988**, 346, C7-C10.
2. Nugent, W. A. Chiral Lewis acid catalysis. Enantioselective addition of azide to meso epoxides. *J. Am. Chem. Soc.* **1992**, 114, 2768-2769.
3. Martinez, L. E., Leighton, J. L., Carsten, D. H., Jacobsen, E. N. Highly Enantioselective Ring Opening of Epoxides Catalyzed by (salen)Cr(III) Complexes. *J. Am. Chem. Soc.* **1995**, 117, 5897-5898.
4. Jacobsen, E. N., Kakiuchi, F., Konsler, R. G., Larrow, J. F., Tokunaga, M. Enantioselective catalytic ring opening of epoxides with carboxylic acids. *Tetrahedron Lett.* **1997**, 38, 773-776.
5. Tokunaga, M., Larrow, J. F., Kakiuchi, F., Jacobsen, E. N. Asymmetric catalysis with water: efficient kinetic resolution of terminal epoxides by means of catalytic hydrolysis. *Science* **1997**, 277, 936-938.
6. Finney, N. S. Enantioselective epoxide hydrolysis: catalysis involving microbes, mammals and metals. *Chem. Biol.* **1998**, 5, R73-R79.
7. Hoveyda, A. H., Didiuk, M. T. Metal-catalyzed kinetic resolution processes. *Curr. Org. Chem.* **1998**, 2, 489-526.
8. Tonks, L., Williams, J. M. J. Catalytic applications of transition metals in organic synthesis. *J. Chem. Soc., Perkin Trans. 1* **1998**, 3637-3652.
9. Canali, L., Sherrington, D. C. Utilization of homogeneous and supported chiral metal(salen) complexes in asymmetric catalysis. *Chem. Soc. Rev.* **1999**, 28, 85-93.
10. Jacobsen, E. N., Wu, M. H. Ring opening of epoxides and related reactions. *Comprehensive Asymmetric Catalysis I-III* **1999**, 3, 1309-1326.
11. Orru, R. V. A., Archelas, A., Furstoss, R., Faber, K. Epoxide hydrolases and their synthetic applications. *Adv. Biochem. Eng. Biotechnol.* **1999**, 63, 145-167.
12. Willis, M. C. Enantioselective desymmetrization. *J. Chem. Soc., Perkin Trans. 1* **1999**, 1765-1784.
13. Cook, G. R. Transition metal-mediated kinetic resolution. *Curr. Org. Chem.* **2000**, 4, 869-885.
14. Jacobsen, E. N. Asymmetric Catalysis of Epoxide Ring-Opening Reactions. *Acc. Chem. Res.* **2000**, 33, 421-431.
15. Keith, J. M., Larrow, J. F., Jacobsen, E. N. Practical considerations in kinetic resolution reactions. *Adv. Syn. & Catal.* **2001**, 343, 5-26.
16. Furrow, M. E., Schaus, S. E., Jacobsen, E. N. Practical access to highly enantioenriched C-3 building blocks via hydrolytic kinetic resolution. *J. Org. Chem.* **1998**, 63, 6776-6777.
17. Breinbauer, R., Jacobsen, E. N. Cooperative asymmetric catalysis with dendrimeric [Co(salen)] complexes. *Angew. Chem., Int. Ed. Engl.* **2000**, 39, 3604-3607.
18. Schaus, S. E., Brandes, B. D., Larrow, J. F., Tokunaga, M., Hansen, K. B., Gould, A. E., Furrow, M. E., Jacobsen, E. N. Highly Selective Hydrolytic Kinetic Resolution of Terminal Epoxides Catalyzed by Chiral (salen)CoIII Complexes. Practical Synthesis of Enantioenriched Terminal Epoxides and 1,2-Diols. *J. Am. Chem. Soc.* **2002**, 124, 1307-1315.
19. Song, Y., Yao, X., Chen, H., Bai, C., Hu, X., Zheng, Z. Highly enantioselective resolution of terminal epoxides using polymeric catalysts. *Tetrahedron Lett.* **2002**, 43, 6625-6627.
20. Kim, G.-J., Lee, H., Kim, S.-J. Catalytic activity and recyclability of new enantioselective chiral Co-salen complexes in the hydrolytic kinetic resolution of epichlorohydrin. *Tetrahedron Lett.* **2003**, 44, 5005-5008.

21. Liu, Y., Dimare, M., Marchese, S. A., Jacobsen, E. N., Jasmin, S. *Hydrolytic kinetic resolution of epoxides using chiral cobalt catalysts.* WO 2002-US26729 (2003018520), **2003** (Rhodia/Chirex, Inc., USA). 47 pp
22. Oh, C. R., Choo, D. J., Shim, W. H., Lee, D. H., Roh, E. J., Lee, S.-g., Song, C. E. Chiral Co(III)(salen)-catalyzed hydrolytic kinetic resolution of racemic epoxides in ionic liquids. *Chem. Commun.* **2003**, 1100-1101.
23. Song, Y., Chen, H., Hu, X., Bai, C., Zheng, Z. Highly enantioselective resolution of terminal epoxides with crosslinked polymeric salen-Co(III) complexes. *Tetrahedron Lett.* **2003**, 44, 7081-7085.
24. White, D. E., Jacobsen, E. N. New oligomeric catalyst for the hydrolytic kinetic resolution of terminal epoxides under solvent-free conditions. *Tetrahedron: Asymmetry* **2003**, 14, 3633-3638.
25. Hansen, K. B., Leighton, J. L., Jacobsen, E. N. On the Mechanism of Asymmetric Nucleophilic Ring-Opening of Epoxides Catalyzed by (Salen)CrIII Complexes. *J. Am. Chem. Soc.* **1996**, 118, 10924-10925.
26. Annis, D. A., Jacobsen, E. N. Polymer-supported chiral Co(salen) complexes: synthetic applications and mechanistic investigations in the hydrolytic kinetic resolution of terminal epoxides. *J. Am. Chem. Soc.* **1999**, 121, 4147-4154.
27. Blackmond, D. G. Kinetic Resolution Using Enantioimpure Catalysts: Mechanistic Considerations of Complex Rate Laws. *J. Am. Chem. Soc.* **2001**, 123, 545-553.
28. Nielsen, L. P. C., Stevenson, C. P., Blackmond, D. G., Jacobsen, E. N. Mechanistic Investigation Leads to a Synthetic Improvement in the Hydrolytic Kinetic Resolution of Terminal Epoxides. *J. Am. Chem. Soc.* **2004**, 126, 1360-1362.
29. Ahmed, A., Hoegenauer, E. K., Enev, V. S., Hanbauer, M., Kaehlig, H., Oehler, E., Mulzer, J. Total Synthesis of the Microtubule Stabilizing Antitumor Agent Laulimalide and Some Nonnatural Analogues: The Power of Sharpless' Asymmetric Epoxidation. *J. Org. Chem.* **2003**, 68, 3026-3042.
30. Chavez, D. E., Jacobsen, E. N. Total synthesis of fostriecin (CI-920). *Angew. Chem., Int. Ed. Engl.* **2001**, 40, 3667-3670.
31. Jiang, S., Liu, Z.-H., Sheng, G., Zeng, B.-B., Cheng, X.-G., Wu, Y.-L., Yao, Z.-J. Mimicry of Annonaceous Acetogenins: Enantioselective Synthesis of a (4R)-Hydroxy Analogue Having Potent Antitumor Activity. *J. Org. Chem.* **2002**, 67, 3404-3408.

Jacobsen-Katsuki Epoxidation..........222

Related reactions: Davis oxaziridine oxidation, Prilezhaev reaction, Sharpless asymmetric epoxidation, Shi asymmetric epoxidation;

1. Srinivasan, K., Michaud, P., Kochi, J. K. Epoxidation of olefins with cationic (salen)manganese(III) complexes. The modulation of catalytic activity by substituents. *J. Am. Chem. Soc.* **1986**, 108, 2309-2320.
2. Zhang, W., Loebach, J. L., Wilson, S. R., Jacobsen, E. N. Enantioselective epoxidation of unfunctionalized olefins catalyzed by salen manganese complexes. *J. Am. Chem. Soc.* **1990**, 112, 2801-2803.
3. Irie, R., Noda, K., Ito, Y., Katsuki, T. Enantioselective epoxidation of unfunctionalized olefins using chiral (salen)manganese(III) complexes. *Tetrahedron Lett.* **1991**, 32, 1055-1058.
4. Irie, R., Noda, K., Ito, Y., Matsumoto, N., Katsuki, T. Catalytic asymmetric epoxidation of unfunctionalized olefins using chiral (salen)manganese(III) complexes. *Tetrahedron: Asymmetry* **1991**, 2, 481-494.
5. Jacobsen, E. N., Zhang, W., Muci, A. R., Ecker, J. R., Deng, L. Highly enantioselective epoxidation catalysts derived from 1,2-diaminocyclohexane. *J. Am. Chem. Soc.* **1991**, 113, 7063-7064.
6. Schurig, V., Betschinger, F. Metal-mediated enantioselective access to unfunctionalized aliphatic oxiranes: prochiral and chiral recognition. *Chem. Rev.* **1992**, 92, 873-888.
7. Jacobsen, E. N. Asymmetric catalytic epoxidation of unfunctionalized olefins. in *Catal. Asymmetric Synth.* (ed. Ojima, I.), 159-202 (VCH, New York, **1993**).
8. Katsuki, T. Mn-salen catalyst, competitor of enzymes, for asymmetric epoxidation. *J. Mol. Catal. A: Chemical* **1996**, 113, 87-107.
9. Katsuki, T. Asymmetric reactions using metallosalen complexes as catalysts. *Recent Research Developments in Pure & Applied Chemistry* **1997**, 1, 35-44.
10. Linker, T. The Jacobsen-Katsuki epoxidation and its controversial mechanism. *Angew. Chem., Int. Ed. Engl.* **1997**, 36, 2060-2062.
11. Woodward, S. Transition metal-promoted oxidations. *Transition Metals in Organic Synthesis* **1997**, 1-34.
12. Dalton, C. T., Ryan, K. M., Wall, V. M., Bousquet, C., Gilheany, D. G. Recent progress towards the understanding of metal-salen catalyzed asymmetric alkene epoxidation. *Top. in Cat.* **1998**, 5, 75-91.
13. Muniz-Fernandez, K., Bolm, C. Manganese-catalyzed epoxidations. *Transition Metals for Organic Synthesis* **1998**, 2, 271-282.
14. Flessner, T., Doye, S. N,N'-bis(3,5-di-t-butylsalicylidene)-1,2-cyclohexanediaminomanganese(III) chloride. The Jacobsen catalyst. *J. Prakt. Chem.* **1999**, 341, 436-444.
15. Houk, K. N., Liu, J., Strassner, T. Transition state modeling of asymmetric epoxidation catalysts. *ACS Symp. Ser.* **1999**, 721, 33-48.
16. Ito, Y. N., Katsuki, T. Asymmetric Catalysis of New Generation Chiral Metallosalen Complexes. *Bull. Chem. Soc. Jpn.* **1999**, 72, 603-619.
17. Jacobsen, E. N., Wu, M. H. Epoxidation of alkenes other than allylic alcohols. *Comprehensive Asymmetric Catalysis I-III* **1999**, 2, 649-677.
18. Katsuki, T. Metallosalen-catalyzed asymmetric oxygen-transfer reaction: Dynamics of salen ligand conformation. *Peroxide Chemistry* **2000**, 303-319.
19. Dhal, P. K., De, B. B., Sivaram, S. Polymeric metal complex catalyzed enantioselective epoxidation of olefins. *J. Mol. Catal. A: Chemical* **2001**, 177, 71-87.
20. Ito, Y. N., Katsuki, T. Oxidation of the C:C bond: metal catalyzed epoxidation of simple olefins. *Asymmetric Oxidation Reactions* **2001**, 19-37.
21. Adam, W., Malisch, W., Roschmann, K. J., Saha-Moller, C. R., Schenk, W. A. Catalytic oxidations by peroxy, peroxo and oxo metal complexes: an interdisciplinary account with a personal view. *J. Organomet. Chem.* **2002**, 661, 3-16.
22. Corsi, M. Jacobsen's catalyst. *Synlett* **2002**, 2127-2128.
23. Katsuki, T. Chiral metallosalen complexes: structures and catalyst tuning for asymmetric epoxidation and cyclopropanation. *Adv. Syn. & Catal.* **2002**, 344, 131-147.
24. Noyori, R., Hashiguchi, S., Yamano, T. Asymmetric synthesis. *Applied Homogeneous Catalysis with Organometallic Compounds (2nd Edition)* **2002**, 1, 557-585.
25. Katsuki, T. Some recent advances in metallosalen chemistry. *Synlett* **2003**, 281-297.
26. Yoon, T. P., Jacobsen, E. N. Privileged chiral catalysts. *Science* **2003**, 299, 1691-1693.
27. Nishikori, H., Ohta, C., Katsuki, T. Enantioselective epoxidation of conjugated trans-olefins with (salen)manganese(III) complexes as catalysts. *Synlett* **2000**, 1557-1560.
28. Nakata, K., Takeda, T., Mihara, J., Hamada, T., Irie, R., Katsuki, T. Asymmetric epoxidation with a photoactivated [Ru(salen)] complex. *Chem.-- Eur. J.* **2001**, 7, 3776-3782.
29. Jitsukawa, K., Shiozaki, H., Masuda, H. Epoxidation activities of mononuclear ruthenium-oxo complexes with a square planar 6,6'-bis(benzoylamino)-2,2'-bipyridine and axial ligands. *Tetrahedron Lett.* **2002**, 43, 1491-1494.
30. Mirza-Aghayan, M., Ghassemzadeh, M., Hoseini, M., Bolourtchian, M. Microwave-assisted synthesis of the tetradentate Schiff-bases under solvent-free and catalyst-free condition. *Synth. Commun.* **2003**, 33, 521-525.
31. Rose, E., Ren, Q.-z., Andrioletti, B. A unique binaphthyl strapped iron-porphyrin catalyst for the enantioselective epoxidation of terminal olefins. *Chem.-- Eur. J.* **2004**, 10, 224-230.
32. Linde, C., Aakermark, B., Norrby, P.-O., Svensson, M. Timing Is Critical: Effect of Spin Changes on the Diastereoselectivity in Mn(salen)-Catalyzed Epoxidation. *J. Am. Chem. Soc.* **1999**, 121, 5083-5084.
33. Strassner, T., Houk, K. N. Predictions of Geometries and Multiplicities of the Manganese-Oxo Intermediates in the Jacobsen Epoxidation. *Org. Lett.* **1999**, 1, 419-421.
34. El-Bahraoui, J., Wiest, O., Feichtinger, D., Plattner, D. A. Rate enhancement and enantioselectivity of the Jacobsen-Katsuki epoxidation: the significance of the sixth coordination site. *Angew. Chem., Int. Ed. Engl.* **2001**, 40, 2073-2076.

35. Feichtinger, D., Plattner, D. A. Probing the reactivity of oxomanganese-salen complexes: an electrospray tandem mass spectrometric study of highly reactive intermediates. *Chem.-- Eur. J.* **2001**, 7, 591-599.
36. Adam, W., Roschmann, K. J., Saha-Moeller, C. R., Seebach, D. cis-Stilbene and (1a,2b,3a)-(2-Ethenyl-3-methoxycyclopropyl)benzene as Mechanistic Probes in the MnIII(salen)-Catalyzed Epoxidation: Influence of the Oxygen Source and the Counterion on the Diastereoselectivity of the Competitive Concerted and Radical-Type Oxygen Transfer. *J. Am. Chem. Soc.* **2002**, 124, 5068-5073.
37. Cavallo, L., Jacobsen, H. Transition Metal Mediated Epoxidation as Test Case for the Performance of Different Density Functionals: A Computational Study. *J. Phys. Chem. A* **2003**, 107, 5466-5471.
38. Cavallo, L., Jacobsen, H. Manganese-salen complexes as oxygen-transfer agents in catalytic epoxidations - a density functional study of mechanistic aspects. *Eur. J. Inorg. Chem.* **2003**, 892-902.
39. Cavallo, L., Jacobsen, H. Electronic Effects in (salen)Mn-Based Epoxidation Catalysts. *J. Org. Chem.* **2003**, 68, 6202-6207.
40. Abashkin, Y. G., Burt, S. K. (Salen)Mn-Catalyzed Epoxidation of Alkenes: A Two-Zone Process with Different Spin-State Channels as Suggested by DFT Study. *Org. Lett.* **2004**, 6, 59-62.
41. Feichtinger, D., Plattner, D. A. Direct proof for O:MnV(salen) complexes. *Angew. Chem., Int. Ed. Engl.* **1997**, 36, 1718-1719.
42. Hughes, D. L., Smith, G. B., Liu, J., Dezeny, G. C., Senanayake, C. H., Larsen, R. D., Verhoeven, T. R., Reider, P. J. Mechanistic Study of the Jacobsen Asymmetric Epoxidation of Indene. *J. Org. Chem.* **1997**, 62, 2222-2229.
43. Palucki, M., Finney, N. S., Pospisil, P. J., Gueler, M. L., Ishida, T., Jacobsen, E. N. The mechanistic basis for electronic effects on enantioselectivity in the (salen)$Mn^{(III)}$-catalyzed epoxidation reaction. *J. Am. Chem. Soc.* **1998**, 120, 948-954.
44. Meou, A., Garcia, M. A., Brun, P. Oxygen transfer mechanism in the Mn-salen catalyzed epoxidation of olefins. *J. Mol. Catal. A: Chemical* **1999**, 138, 221-226.
45. Adam, W., Mock-Knoblauch, C., Saha-Moeller, C. R., Herderich, M. Are $Mn^{(IV)}$ Species Involved in Mn(Salen)-Catalyzed Jacobsen-Katsuki Epoxidations? A Mechanistic Elucidation of Their Formation and Reaction Modes by EPR Spectroscopy, Mass-Spectral Analysis, and Product Studies: Chlorination versus Oxygen Transfer. *J. Am. Chem. Soc.* **2000**, 122, 9685-9691.
46. Cavallo, L., Jacobsen, H. Radical intermediates in the Jacobsen-Katsuki epoxidation. *Angew. Chem., Int. Ed. Engl.* **2000**, 39, 589-592.
47. Campbell, K. A., Lashley, M. R., Wyatt, J. K., Nantz, M. H., Britt, R. D. Dual-Mode EPR Study of $Mn^{(III)}$ Salen and the $Mn^{(III)}$ Salen-Catalyzed Epoxidation of cis-β-Methylstyrene. *J. Am. Chem. Soc.* **2001**, 123, 5710-5719.
48. Jacobsen, H., Cavallo, L. A possible mechanism for enantioselectivity in the chiral epoxidation of olefins with [Mn(salen)] catalysts. *Chem.-- Eur. J.* **2001**, 7, 800-807.
49. Chellamani, A., Harikengaram, S. Kinetics and mechanism of (salen)Mn^{III}-catalysed oxidation of organic sulfides with sodium hypochlorite. *J. Phys. Org. Chem.* **2003**, 16, 589-597.
50. Kureshy, R. I., Khan, N.-u. H., Abdi, S. H. R., Singh, S., Ahmed, I., Shukla, R. S., Jasra, R. V. Chiral $Mn^{(III)}$ salen complex-catalyzed enantioselective epoxidation of nonfunctionalized alkenes using urea-H_2O_2 adduct as oxidant. *J. Catal.* **2003**, 219, 1-7.
51. Higashibayashi, S., Mori, T., Shinko, K., Hashimoto, K., Nakata, M. Synthetic studies on thiostrepton family of peptide antibiotics: synthesis of the tetrasubstituted dihydroquinoline portion of siomycin D_1. *Heterocycles* **2002**, 57, 111-122.
52. Boger, D. L., McKie, J. A., Boyce, C. W. Asymmetric synthesis of the 1,2,9,9a-tetrahydrocyclopropa[c]benzo[e]indol-4-one (CBI) alkylation subunit of the CC-1065 and duocarmycin analogs. *Synlett* **1997**, 515-517.
53. Lee, J., Hoang, T., Lewis, S., Weissman, S. A., Askin, D., Volante, R. P., Reider, P. J. Asymmetric synthesis of (2S,3S)-3-hydroxy-2-phenylpiperidine via ring expansion. *Tetrahedron Lett.* **2001**, 42, 6223-6225.
54. Lynch, J. E., Choi, W. B., Churchill, H. R. O., Volante, R. P., Reamer, R. A., Ball, R. G. Asymmetric Synthesis of CDP840 by Jacobsen Epoxidation. An Unusual Syn Selective Reduction of an Epoxide. *J. Org. Chem.* **1997**, 62, 9223-9228.

Japp-Klingemann Reaction 224

1. Japp, F. R., Klingemann, F. Studies on aromatic azo and hydrazo fatty acids. *Ber.* **1887**, 20, 2942-2944.
2. Japp, F. R., Klingemann, F. Studies of aromatic azo and hydrazo propionic acids. **1887**, 20, 3284-3286.
3. Japp, F. R., Klingemann, F. Mixed azo compounds. *Ber.* **1887**, 20, 3398-3401.
4. Japp, F. R., Klingemann, F. The structure of some mixed azo compounds. *Ann.* **1888**, 247, 190-225.
5. Phillips, R. R. The Japp-Klingeman reaction. *Organic Reactions (Roger Adams, editor, John Wiley & Sons, Inc.)* **1959**, 10, 143-178.
6. Robinson, B. Studies on the Fischer indole synthesis. *Chem. Rev.* **1969**, 69, 227-250.
7. Heckendorn, R. Novel heterocycles by the malonic ester variation of the Japp-Klingemann reaction. *Bull. Soc. Chim. Belg.* **1986**, 95, 921-943.
8. Neplyuev, V. M., Bazavova, I. M., Lozinskii, M. O. Japp-Klingemann reaction: nontraditional substrates and leaving groups. *Zh. Org. Khim.* **1989**, 25, 2225-2236.
9. Buzykin, B. I., Sokolov, M. P., Pavlov, V. A., Ivanova, V. N., Chertanova, L. F., Zyablikova, T. A. Disubstituted acetaldehydes containing a phosphoryl group in the Japp-Klingemann reaction. *Zh. Obshch. Khim.* **1990**, 60, 546-555.
10. Bazavova, I. M., Esipenko, A. N., Neplyuev, V. M., Lozinskii, M. O. Unusual reaction of 1,3-dicyano-2-thiapropane 2,2-dioxide with aryldiazonium salts. Sulfonyl as a leaving group in the Japp-Klingemann reaction. *Zh. Org. Khim.* **1996**, 32, 1278.
11. Atlan, V., Kaim, L. E., Supiot, C. New versatile approach to α-hydrazonoesters and amino acid derivatives through a modified Japp-Klingemann reaction. *Chem. Commun.* **2000**, 1385-1386.
12. Dimroth, O., Hartmann, M. Rearrangement of Azo Compounds into Hydrazones. *Ber.* **1908**, 40, 4460-4465.
13. Doree, C., Petrow, V. A. Hydrogenation under the action of selenium. I. The action of selenium on cholesterol at 230 Deg. *J. Chem. Soc., Abstracts* **1935**, 1391-1393.
14. and, H.-C., Resnick, P. Azo-hydrazone conversion. I. The Japp-Klingemann reaction. *J. Am. Chem. Soc.* **1962**, 84, 3514-3517.
15. Eistert, B., Regitz, M. Japp-Klingemann cleavages. I. Cleavage of the coupling products of p-nitrobenzenediazonium chloride with carboxylic acid esters of tetrahydrothiopyran-3-one and thiophan-3-one. *Ann.* **1963**, 666, 97-112.
16. Hamana, M., Kumadaki, I. Reaction of aromatic N-oxides with indoles in the presence of an acylating agent. *Chem. Pharm. Bull.* **1967**, 15, 363-366.
17. Genkina, N. K., Gordeev, E. N., Suvorov, N. N. Kinetic study of the Japp-Klingemann reaction. *Trudy Instituta - Moskovskii Khimiko-Tekhnologicheskii Institut imeni D. I. Mendeleeva* **1975**, 86, 35-37.
18. Genkina, N. K., Gordeev, E. N., Suvorov, N. N. Kinetic study of the Japp-Klingemann reaction. *Khimiya I Farmakol. Indol'n. Soedinenii* **1975**, 15-18.
19. Genkina, N. K., Gordeev, E. N., Suvorov, N. N. Kinetic principles of the Japp-Klingemann reaction. *Zh. Org. Khim.* **1976**, 12, 1462-1466.
20. Reichardt, C., Wuerthwein, E. U. Mechanism of the Japp-Klingemann reaction of 1,3-dialdehydes. *Chem. Ber.* **1976**, 109, 3735-3737.
21. Genkina, N. K., Gordeev, E. N., Suvorov, N. N. Kinetics of the splitting of azo compounds - intermediate products of the Japp-Klingemann reaction. *Zh. Org. Khim.* **1978**, 14, 1501-1506.
22. Jiricek, J., Blechert, S. Enantioselective Synthesis of (-)-Gilbertine via a Cationic Cascade Cyclization. *J. Am. Chem. Soc.* **2004**, 126, 3534-3538.
23. Delfourne, E., Roubin, C., Bastide, J. The First Synthesis of the Pentacyclic Pyridoacridine Marine Alkaloids: Arnoamines A and B. *J. Org. Chem.* **2000**, 65, 5476-5479.
24. Loubinoux, B., Sinnes, J.-L., O'Sullivan, A. C., Winkler, T. Synthesis of Southern-Part Models of Soraphen A. *J. Org. Chem.* **1995**, 60, 953-959.
25. Chetoni, F., Da Settimo, F., Marini, A. M., Primofiore, G. Synthesis of some 5H,12H-[1]benzoxepino[4,3-b]indol-6-ones. A new heterocyclic ring system. *J. Heterocycl. Chem.* **1993**, 30, 1481-1484.

Related reactions: *Carroll rearrangement, Claisen rearrangement, Claisen-Ireland rearrangement, Eschenmoser-Claisen rearrangement;*

1. Johnson, W. S., Werthemann, L., Bartlett, W. R., Brocksom, T. J., Li, T.-T., Faulkner, D. J., Petersen, M. R. Simple stereoselective version of the Claisen rearrangement leading to trans-trisubstituted olefinic bonds. Synthesis of squalene. *J. Am. Chem. Soc.* **1970**, 92, 741-743.
2. Bennett, G. B. The Claisen rearrangement in organic synthesis; 1967 to January 1977. *Synthesis* **1977**, 589-606.
3. Ziegler, F. E. Stereo- and regiochemistry of the Claisen rearrangement: applications to natural products synthesis. *Acc. Chem. Res.* **1977**, 10, 227-232.
4. Ziegler, F. E. The thermal, aliphatic Claisen rearrangement. *Chem. Rev.* **1988**, 88, 1423-1452.
5. Lounasmaa, M. Synthetic studies in the field of indole alkaloids. Part 3. *Curr. Org. Chem.* **1998**, 2, 63-90.
6. Castro, A. M. M. Claisen Rearrangement over the Past Nine Decades. *Chem. Rev.* **2004**, 104, 2939-3002.
7. Jones, G. B., Huber, R. S., Chau, S. The Claisen rearrangement in synthesis: acceleration of the Johnson Orthoester Protocol en route to bicyclic lactones. *Tetrahedron* **1992**, 49, 369-380.
8. Daub, G. W., Edwards, J. P., Okada, C. R., Allen, J. W., Maxey, C. T., Wells, M. S., Goldstein, A. S., Dibley, M. J., Wang, C. J., Ostercamp, D. P., Chung, S., Cunningham, P. S., Berliner, M. A. Acyclic Stereoselection in the Ortho Ester Claisen Rearrangement. *J. Org. Chem.* **1997**, 62, 1976-1985.
9. Schlama, T., Baati, R., Gouveneur, V., Valleix, A., Flack, J. R., Mioskowski, C. Total synthesis of (±)-halomon by a Johnson-Claisen rearrangement. *Angew. Chem., Int. Ed. Engl.* **1998**, 37, 2085-2087.
10. Birman, V. B., Danishefsky, S. J. The total synthesis of (±)-merrilactone A. *J. Am. Chem. Soc.* **2002**, 124, 2080-2081.
11. Kim, D., Lee, J., Shim, P. J., Lim, J. I., Jo, H., Kim, S. Asymmetric Total Synthesis of (+)-Brefeldin A from (S)-Lactate by Triple Chirality Transfer Process and Nitrile Oxide Cycloaddition. *J. Org. Chem.* **2002**, 67, 764-771.
12. Ng, F. W., Lin, H., Danishefsky, S. J. Explorations in Organic Chemistry Leading to the Total Synthesis of (±)-Gelsemine. *J. Am. Chem. Soc.* **2002**, 124, 9812-9824.

Related reactions: *Corey-Kim oxidation, Dess-Martin oxidation, Ley oxidation, Oppenauer oxidation, Pfitzner-Moffatt oxidation, Pinnick oxidation;*

1. Bowden, K., Heilbron, I. M., Jones, E. R. H., Weedon, B. C. L. Acetylenic compounds. I. Preparation of acetylenic ketones by oxidation of acetylenic carbinols and glycols. *J. Chem. Soc.* **1946**, 39-45.
2. Bowers, A., Halsall, T. G., Jones, E. R. H., Lemin, A. J. Chemistry of the triterpenes and related compounds. XVIII. Elucidation of the structure of polyporenic acid C. *J. Chem. Soc., Abstracts* **1953**, 2548-2560.
3. Wiberg, K. B. Oxidation by chromic acid and chromyl compounds. in *Oxid. Org. Chem.* (ed. Wiberg, K. B.), *5A*, 69-184 (Academic Press, New York, **1965**).
4. Freeman, F. Oxidation by oxochromium(VI) compounds. in *Org. Synth. Oxid. Met. Compd.* (eds. Mijs, W. J.,de Jonge, C. R. H. I.), 41-118 (Plenum Press, New York, **1986**).
5. Luzzio, F. A., Guziec, F. S., Jr. Recent applications of oxochromiumamine complexes as oxidants in organic synthesis. A review. *Org. Prep. Proced. Int.* **1988**, 20, 533-584.
6. Ley, S. V., Madin, A. Oxidation Adjacent to Oxygen of Alcohols by Chromium Reagents. in *Comp. Org. Synth.* (eds. Trost, B. M.,Fleming, I.), *7*, 251-289 (Pergamon Press, Oxford, **1991**).
7. Luzzio, F. A. The oxidation of alcohols by modified oxochromium(VI)-amine reagents. *Org. React.* **1998**, 53, 1-221.
8. Poos, G. I., Arth, G. E., Beyler, R. E., Sarett, H. Approaches to the total synthesis of adrenal steroids. V. 4b-Methyl-7-ethylenedioxy-1,2,3,4,4aα,4b,5,6,7,8,10,10a-β-dodecahydrophenanthren-4β-ol-1-one and related tricyclic derivatives. *J. Am. Chem. Soc.* **1953**, 75, 422-429.
9. Hampton, J., Leo, A., Westheimer, F. H. Mechanism of the cleavage of phenyl-tert-butylcarbinol by chromic acid. *J. Am. Chem. Soc.* **1956**, 78, 306-312.
10. Walker, B. H. Effect of manganese on the chromic acid oxidation of secondary-tertiaryvicinal glycols. *J. Org. Chem.* **1967**, 32, 1098-1103.
11. Collins, J. C., Hess, W. W., Frank, F. J. Dipyridine-chromium(VI) oxide oxidation of alcohols in dichloromethane. *Tetrahedron Lett.* **1968**, 3363-3366.
12. Corey, E. J., Suggs, J. W. Pyridinium chlorochromate. Efficient reagent for oxidation of primary and secondary alcohols to carbonyl compounds. *Tetrahedron Lett.* **1975**, 2647-2650.
13. Harding, K. E., May, L. M., Dick, K. F. Selective oxidation of allylic alcohols with chromic acid. *J. Org. Chem.* **1975**, 40, 1664-1665.
14. Rogers, H. R., McDermott, J. X., Whitesides, G. M. Oxidation of terminal olefins to methyl ketones by Jones reagent is catalyzed by mercury(II). *J. Org. Chem.* **1975**, 40, 3577-3580.
15. Cainelli, G., Cardillo, G., Orena, M., Sandri, S. Polymer supported reagents. Chromic acid on anion exchange resins. A simple and practical oxidation of alcohols to aldehydes and ketones. *J. Am. Chem. Soc.* **1976**, 98, 6737-6738.
16. Corey, E. J., Schmidt, G. Useful procedures for the oxidation of alcohols involving pyridinium dichromate in aprotic media. *Tetrahedron Lett.* **1979**, 399-402.
17. Henry, J. R., Weinreb, S. M. A convenient, mild method for oxidative cleavage of alkenes with Jones reagent/osmium tetraoxide. *J. Org. Chem.* **1993**, 58, 4745.
18. Allanson, N. M., Llu, D., Chi, F., Jain, R. K., Chen, A., Ghosh, M., Hong, L., Sofia, M. J. Synthesis of phenyl 1-thioglycopyranosiduronic acids using a sonicated Jones oxidation. *Tetrahedron Lett.* **1998**, 39, 1889-1892.
19. Ali, M. H., Wiggin, C. J. Silica gel supported Jones reagent (SJR): a simple, versatile, and efficient reagent for oxidation of alcohols in non-aqueous media. *Synth. Commun.* **2001**, 31, 3383-3393.
20. Ali, M. H., Wiggin, C. J. Silica gel supported Jones reagent (SJR): a simple and efficient reagent for oxidation of benzyl alcohols to benzaldehydes. *Synth. Commun.* **2001**, 31, 1389-1397.
21. Westheimer, F. H. The mechanisms of chromic acid oxidations. *Chem. Rev.* **1949**, 45, 419-451.
22. Farrell, R. P., Lay, P. A. New insights into the structures and reactions of chromium(V) complexes: implications for chromium(VI) and chromium(V) oxidations of organic substrates and the mechanisms of chromium-induced cancers. *Comments on Inorganic Chemistry* **1992**, 13, 133-175.
23. Scott, S. L., Bakac, A., Espenson, J. H. Oxidation of alcohols, aldehydes, and carboxylates by the aquachromium(IV) ion. *J. Am. Chem. Soc.* **1992**, 114, 4205-4213.
24. Das, A. K. Kinetics and mechanistic aspects of catalysis by different chelating agents in chromium(VI) oxidation. *Oxidation Communications* **2001**, 24, 321-334.
25. Waizumi, N., Itoh, T., Fukuyama, T. Total Synthesis of (-)-CP-263,114 (Phomoidride B). *J. Am. Chem. Soc.* **2000**, 122, 7825-7826.
26. Crimmins, M. T., Jung, D. K., Gray, J. L. Synthetic studies on the ginkgolides: total synthesis of (±)-bilobalide. *J. Am. Chem. Soc.* **1993**, 115, 3146-3155.
27. Sha, C.-K., Chiu, R.-T., Yang, C.-F., Yao, N.-T., Tseng, W.-H., Liao, F.-L., Wang, S.-L. Total Synthesis of (-)-Dendrobine via α-Carbonyl Radical Cyclization. *J. Am. Chem. Soc.* **1997**, 119, 4130-4135.
28. Hagiwara, H., Kobayashi, K., Miya, S., Hoshi, T., Suzuki, T., Ando, M. The First Total Synthesis of (-)-Solanapyrone E Based on Domino Michael Strategy. *Org. Lett.* **2001**, 3, 251-254.

Related reactions: *Horner-Wadsworth-Emmons olefination, Horner-Wadsworth-Emmons olefination - Still-Gennari modification, Peterson olefination, Takai-Utimoto olefination, Tebbe olefination, Wittig reaction, Wittig reaction – Schlosser modification;*

1. Julia, M., Paris, J. M. Syntheses with the help of sulfones. V. General method of synthesis of double bonds. *Tetrahedron Lett.* **1973**, 4833-4836.
2. Julia, M. Recent advances in double bond formation. *Pure Appl. Chem.* **1985**, 57, 763-768.
3. Kocienski, P. Recent sulfone-based olefination reactions. *Phosphorus Sulfur* **1985**, 24, 477-507.
4. Trost, B. M. Chemical chameleons. Organosulfones as synthetic building blocks. *Bull. Chem. Soc. Jpn.* **1988**, 61, 107-124.
5. Kelly, S. E. Alkene Synthesis. in *Comp. Org. Synth.* (eds. Trost, B. M.,Fleming, I.), *1*, 729-818 (Pergamon, Oxford, **1991**).
6. Breit, B. Dithioacetals as an entry to titanium-alkylidene chemistry: a new and efficient carbonyl olefination. *Angew. Chem., Int. Ed. Engl.* **1998**, 37, 453-456.
7. Prilezhaeva, E. N. Sulfones and sulfoxides in the total synthesis of biologically active natural compounds. *Russ. Chem. Rev.* **2000**, 69, 367-408.
8. Blakemore, P. R. The modified Julia olefination: alkene synthesis via the condensation of metallated heteroarylalkylsulfones with carbonyl compounds. *J. Chem. Soc., Perkin Trans. 1* **2002**, 2563-2585.
9. Dumeunier, R., Marko, I. E. The Julia reaction. *Modern Carbonyl Olefination* **2004**, 104-150.
10. Kocienski, P. J., Lythgoe, B., Roberts, D. A. Calciferol and its relatives. Part 23. An alternative synthesis of Windaus and Grundmann's C19 ketone. *J. Chem. Soc., Perkin Trans. 1* **1978**, 834-837.
11. Kocienski, P. J., Lythgoe, B., Ruston, S. Scope and stereochemistry of an olefin synthesis from β-hydroxy-sulfones. *J. Chem. Soc., Perkin Trans. 1* **1978**, 829-834.
12. Kocienski, P. J., Lythgoe, B., Ruston, S. Calciferol and its relatives. Part 24. A synthesis of vitamin D4. *J. Chem. Soc., Perkin Trans. 1* **1979**, 1290-1293.
13. Kocienski, P. J., Lythgoe, B., Waterhouse, I. The influence of chain-branching on the steric outcome of some olefin-forming reactions. *J. Chem. Soc., Perkin Trans. 1* **1980**, 1045-1050.
14. Kocienski, P. J. A new and useful olefin synthesis based on sulfones. *Chem. Ind.* **1981**, 548-551.
15. Baudin, J. B., Hareau, G., Julia, S. A., Ruel, O. A direct synthesis of olefins by the reaction of carbonyl compounds with lithio derivatives of 2-[alkyl- or 2'-alkenyl- or benzylsulfonyl]benzothiazoles. *Tetrahedron Lett.* **1991**, 32, 1175-1178.
16. Keck, G. E., Savin, K. A., Weglarz, M. A. Use of Samarium Diiodide as an Alternative to Sodium/Mercury Amalgam in the Julia-Lythgoe Olefination. *J. Org. Chem.* **1995**, 60, 3194-3204.
17. Blakemore, P. R., Cole, W. J., Kocienski, P. J., Morley, A. A stereoselective synthesis of trans-1,2-disubstituted alkenes based on the condensation of aldehydes with metalated 1-phenyl-1H-tetrazol-5-yl sulfones. *Synlett* **1998**, 26-28.
18. Satoh, T., Yamada, N., Asano, T. Ligand exchange reaction of sulfoxides in organic synthesis: sulfoxide version of the Julia-Lythgoe olefination. *Tetrahedron Lett.* **1998**, 39, 6935-6938.
19. Kocienski, P. J., Bell, A., Blakemore, P. R. 1-tert-Butyl-1H-tetrazol-5-yl sulfones in the modified Julia olefination. *Synlett* **2000**, 365-366.
20. Kurek-Tyrlik, A., Marczak, S., Michalak, K., Wicha, J. Synthesis of alkenes by the reaction of magnesium sulfone derivatives with arylsulfonylhydrazones of aldehydes. *Synlett* **2000**, 547-549.
21. Satoh, T., Hanaki, N., Yamada, N., Asano, T. A Sulfoxide Version of the Julia-Lythgoe Olefination: A New Method for the Synthesis of Olefins from Carbonyl Compounds and Sulfoxides with Carbon-Carbon Coupling. *Tetrahedron* **2000**, 56, 6223-6234.
22. Marko, I. E., Murphy, F., Kumps, L., Ates, A., Touillaux, R., Craig, D., Carballares, S., Dolan, S. Efficient preparation of trisubstituted alkenes using the SmI_2 modification of the Julia-Lythgoe olefination of ketones and aldehydes. *Tetrahedron* **2001**, 57, 2609-2619.
23. Kim, G., Chu-Moyer, M. Y., Danishefsky, S. J., Schulte, G. K. The total synthesis of indolizomycin. *J. Am. Chem. Soc.* **1993**, 115, 30-39.
24. Smith, A. B., III, Brandt, B. M. Total Synthesis of (-)-Callystatin A. *Org. Lett.* **2001**, 3, 1685-1688.
25. Liu, P., Jacobsen, E. N. Total Synthesis of (+)-Ambruticin. *J. Am. Chem. Soc.* **2001**, 123, 10772-10773.

Related reactions: *Barbier reaction, Grignard reaction, Nozaki-Hiyama-Kishi reaction;*

1. Namy, J. L., Girard, P., Kagan, H. B. A new preparation of some divalent lanthanide iodides and their usefulness in organic synthesis. *Nouv. J. Chim.* **1977**, 1, 5-7.
2. Girard, P., Namy, J. L., Kagan, H. B. Divalent lanthanide derivatives in organic synthesis. 1. Mild preparation of samarium iodide and ytterbium iodide and their use as reducing or coupling agents. *J. Am. Chem. Soc.* **1980**, 102, 2693-2698.
3. Molander, G. A., Etter, J. B. Lanthanides in organic synthesis. Synthesis of bicyclic alcohols. *Tetrahedron Lett.* **1984**, 25, 3281-3284.
4. Kagan, H. B. Divalent samarium compounds: perspectives for organic chemistry. *New J. Chem.* **1990**, 14, 453-460.
5. Wang, S. H. Samarium diiodide: a most useful material. *Reviews in Inorganic Chemistry* **1990**, 11, 1-20.
6. Soderquist, J. A. Samarium(II) iodide in organic synthesis. *Aldrichimica Acta* **1991**, 24, 15-23.
7. Curran, D. P., Fevig, T. L., Jasperse, C. P., Totleben, M. J. New mechanistic insights into reductions of halides and radicals with samarium(II) iodide. *Synlett* **1992**, 943-961.
8. Curran, D. P. Synthesis of substituted cyclooctanols by a samarium(II) iodide promoted 8-endo radical cyclization process. On the stereoselectivity of radical 4-exo-trig-cyclization of optically active ethyl (2E)-6-oxohex-2-enoates with samarium(II) iodide. *Chemtracts: Org. Chem.* **1994**, 7, 351-354.
9. Kamochi, Y., Kudo, T. Rapid reduction of organic functionalities using samarium diiodide. *Rev. on Heteroa. Chem.* **1994**, 11, 165-190.
10. Molander, G. A. Reductions with samarium(II) iodide. *Org. React.* **1994**, 46, 211-367.
11. Krief, A., Laval, A.-M. Kagan's reagent (SmI_2). A mild yet powerful reagent for the hydrogenolysis of organic halides. *Acros Organics Acta* **1996**, 2, 17-19.
12. Molander, G. A., Harris, C. R. Sequencing reactions with samarium(II) iodide. *Chem. Rev.* **1996**, 96, 307-338.
13. Khan, F. A., Zimmer, R. Samarium diiodide. A mild and selective reagent in organic synthesis. *J. Prakt. Chem.* **1997**, 339, 101-104.
14. Lobben, P. C., Paquette, L. A. Sequenced reactions with samarium(II) iodide. Tandem nucleophilic acyl/ketyl-olefin coupling reactions. *Chemtracts* **1997**, 10, 284-288.
15. Skrydstrup, T. New sequential reactions with single-electron-donating agents. *Angew. Chem., Int. Ed. Engl.* **1997**, 36, 345-347.
16. Chiara, J. L. New reductive carbocyclizations of carbohydrate derivatives promoted by samarium diiodide. *Carbohydrate Mimics* **1998**, 123-156.
17. Utimoto, K., Matsubara, S. Samarium diiodide-mediated reaction of organic halides with carbonyl compounds. *Yuki Gosei Kagaku Kyokaishi* **1998**, 56, 908-918.
18. Kagan, H. B., Namy, J.-L. Influence of solvents or additives on the organic chemistry mediated by diiodosamarium. *Top. Organomet. Chem.* **1999**, 2, 155-198.
19. Krief, A., Laval, A.-M. Coupling of Organic Halides with Carbonyl Compounds Promoted by SmI_2, the Kagan Reagent. *Chem. Rev.* **1999**, 99, 745-777.
20. Kunishima, M. Generation and application of organosamariums mediated by samarium diiodide. *Rev. on Heteroa. Chem.* **1999**, 21, 117-137.
21. Yu, M.-X., Zhang, Y.-M., Bao, W.-L. Application of samarium reagents in organic synthesis. *Chin. J. Chem.* **1999**, 17, 4-15.
22. Bradley, D., Williams, G., Blann, K., Caddy, J. Fragmentation and cleavage reactions mediated by SmI_2. Part 1: X-Y, X-X and C-C substrates. *Org. Prep. Proced. Int.* **2001**, 33, 565-602.

23. Banik, B. K. Samarium metal in organic synthesis. *Eur. J. Org. Chem.* **2002**, 2431-2444.
24. Kagan, H. B. Twenty-five years of organic chemistry with diiodosamarium: an overview. *Tetrahedron* **2003**, 59, 10351-10372.
25. Dahlen, A., Hilmersson, G. Samarium(II) iodide-mediated reductions. Influence of various additives. *Eur. J. Inorg. Chem.* **2004**, 3393-3403.
26. Edmonds, D. J., Johnston, D., Procter, D. J. Samarium(II)-Iodide-Mediated Cyclizations in Natural Product Synthesis. *Chem. Rev.* **2004**, 104, 3371-3403.
27. Molander, G. A., Etter, J. B. Lanthanides in organic synthesis. 3. A general procedure for five- and six-membered ring annulation. *J. Org. Chem.* **1986**, 51, 1778-1786.
28. Molander, G. A., McKie, J. A. Intramolecular nucleophilic acyl substitution reactions of halo-substituted esters and lactones. New applications of organosamarium reagents. *J. Org. Chem.* **1993**, 58, 7216-7227.
29. Molander, G. A., Harris, C. R. Sequenced Reactions with Samarium(II) Iodide. Tandem Intramolecular Nucleophilic Acyl Substitution/Intramolecular Barbier Cyclizations. *J. Am. Chem. Soc.* **1995**, 117, 3705-3716.
30. Kagan, H. B., Namy, J. L. Preparation of divalent ytterbium and samarium derivatives and their use in organic chemistry. *Handb. Phys. Chem. Rare Earths* **1984**, 6, 525-565.
31. Shabangi, M., Flowers, R. A., II. Electrochemical investigation of the reducing power of SmI_2 in THF and the effect of HMPA cosolvent. *Tetrahedron Lett.* **1997**, 38, 1137-1140.
32. Kagan, H. B., Namy, J. L., Girard, P. Divalent lanthanide derivatives in organic synthesis. II. Mechanism of SmI_2 reactions in presence of ketones and organic halides. *Tetrahedron, Supplement* **1981**, 175-180.
33. Curran, D. P., Fevig, T. L., Totleben, M. J. Sequential radical cyclization-organometallic addition. The mechanism of the samarium(II) mediated Barbier reaction in the presence of hexamethylphosphoric triamide. *Synlett* **1990**, 773-774.
34. Molander, G. A., McKie, J. A. A facile synthesis of bicyclo[m.n.1]alkan-1-ols. Evidence for organosamarium intermediates in the samarium(II) iodide promoted intramolecular Barbier-type reaction. *J. Org. Chem.* **1991**, 56, 4112-4120.
35. Curran, D. P., Gu, X., Zhang, W., Dowd, P. On the mechanism of the intramolecular samarium Barbier reaction. Probes for formation of radical and organosamarium intermediates. *Tetrahedron* **1997**, 53, 9023-9042.
36. Ha, D.-C., Yun, C.-S. Mechanistic studies on the samarium diiodide-promoted cyclization of N-iodoalkyl cyclic imides. *Bull. Korean Chem. Soc.* **1997**, 18, 1039-1041.
37. Hou, Z., Zhang, Y., Wakatsuki, Y. Molecular structures of HMPA-coordinated samarium(II) and ytterbium(II) iodide complexes. A structural basis for the HMPA effects in SmI_2-promoted reactions. *Bull. Chem. Soc. Jpn.* **1997**, 70, 149-153.
38. Enemaerke, R. J., Daasbjerg, K., Skrydstrup, T. Is samarium diiodide an inner- or outer-sphere electron donating agent? *Chem. Commun.* **1999**, 343-344.
39. Shabangi, M., Kuhlman, M. L., Flowers, R. A., II. Mechanism of Reduction of Primary Alkyl Radicals by SmI_2-HMPA. *Org. Lett.* **1999**, 1, 2133-2135.
40. Miller, R. S., Sealy, J. M., Shabangi, M., Kuhlman, M. L., Fuchs, J. R., Flowers, R. A., II. Reactions of SmI_2 with Alkyl Halides and Ketones: Inner-Sphere vs Outer-Sphere Electron Transfer in Reactions of Sm(II) Reductants. *J. Am. Chem. Soc.* **2000**, 122, 7718-7722.
41. Lin, T.-Y., Fuh, M.-R., Chen, Y.-Y. Rate study of haloadamantane reduction by samarium diiodide. *J. Chin. Chem. Soc.* **2002**, 49, 969-973.
42. Prasad, E., Flowers, R. A., II. Reduction of Ketones and Alkyl Iodides by SmI_2 and Sm(II)-HMPA Complexes. Rate and Mechanistic Studies. *J. Am. Chem. Soc.* **2002**, 124, 6895-6899.
43. Prasad, E., Flowers, R. A., II. Mechanistic Study of β-Substituent Effects on the Mechanism of Ketone Reduction by SmI_2. *J. Am. Chem. Soc.* **2002**, 124, 6357-6361.
44. Villar, H., Guibe, F., Aroulanda, C., Lesot, P. Investigation of SmI_2-mediated cyclization of δ-iodo-α,β-unsaturated esters by deuterium 2D NMR in oriented solvents. *Tetrahedron: Asymmetry* **2002**, 13, 1465-1475.
45. Chopade, P. R., Prasad, E., Flowers, R. A. The Role of Proton Donors in SmI_2-Mediated Ketone Reduction: New Mechanistic Insights. *J. Am. Chem. Soc.* **2004**, 126, 44-45.
46. Molander, G. A., Quirmbach, M. S., Silva, L. F., Jr., Spencer, K. C., Balsells, J. Toward the Total Synthesis of Variecolin. *Org. Lett.* **2001**, 3, 2257-2260.
47. Matsuda, F., Kito, M., Sakai, T., Okada, N., Miyashita, M., Shirahama, H. Efficient construction of 8-membered ring framework of vinigrol through SmI_2-induced coupling cyclization. *Tetrahedron* **1999**, 55, 14369-14380.
48. Kawamura, K., Hinou, H., Matsuo, G., Nakata, T. Efficient strategy for convergent synthesis of trans-fused polycyclic ethers based on an intramolecular SmI_2-promoted cyclization of iodo ester. *Tetrahedron Lett.* **2003**, 44, 5259-5261.
49. Takemura, T., Nishii, Y., Takahashi, S., Kobayashi, J. i., Nakata, T. Total synthesis of pederin, a potent insect toxin: the efficient synthesis of the right half, (+)-benzoylpedamide. *Tetrahedron* **2002**, 58, 6359-6365.

Kahne Glycosidation 234

Related reactions: Koenigs-Knorr glycosidation;

1. Kahne, D., Walker, S., Cheng, Y., Van Engen, D. Glycosylation of unreactive substrates. *J. Am. Chem. Soc.* **1989**, 111, 6881-6882.
2. Norberg, T. Glycosylation properties and reactivity of thioglycosides, sulfoxides, and other S-glycosides: current scope and future prospects. *Frontiers in Natural Product Research* **1996**, 1, 82-106.
3. Boons, G. J. Strategies and tactics in oligosaccharide synthesis. *Carbohydr. Chem.* **1998**, 175-222.
4. Nicolaou, K. C., Bockovich, N. J. Chemical synthesis of complex carbohydrates. *Bioorg.Chem.: Carbohydrates* **1999**, 134-173, 565-567.
5. Nicolaou, K. C., Mitchell, H. J. Adventures in carbohydrate chemistry: new synthetic technologies, chemical synthesis, molecular design, and chemical biology. *Angew. Chem., Int. Ed. Engl.* **2001**, 40, 1576-1624.
6. Taylor, C. M. The sulfoxide glycosylation method and its application to solid-phase oligosaccharide synthesis and the generation of combinatorial libraries. *Solid Support Oligosaccharide Synthesis and Combinatorial Carbohydrate Libraries* **2001**, 41-65.
7. Kartha, K. P. R., Field, R. A. Synthesis and activation of carbohydrate donors: thioglycosides and sulfoxides. *Carbohydrates* **2003**, 121-145.
8. Crich, D., Lim, L. B. L. Glycosylation with sulfoxides and sulfinates as donors or promoters. *Org. React.* **2004**, 64, 115-251.
9. Pellissier, H. The glycosylation of steroids. *Tetrahedron* **2004**, 60, 5123-5162.
10. Yan, L., Taylor, C. M., Goodnow, R., Jr., Kahne, D. Glycosylation on the Merrifield Resin Using Anomeric Sulfoxides. *J. Am. Chem. Soc.* **1994**, 116, 6953-6954.
11. Tingoli, M., Temperini, A., Testaferri, L., Tiecco, M., Resnati, G. Glycosylation reaction using anomeric selenoxides. *Carbohydr. Lett.* **1998**, 3, 39-46.
12. Berkowitz, D. B., Choi, S., Bhuniya, D., Shoemaker, R. K. Novel "Reverse Kahne-Type Glycosylation": Access to O-, N-, and C-Linked Epipodophyllotoxin Conjugates. *Org. Lett.* **2000**, 2, 1149-1152.
13. Crich, D., Li, H. Direct Stereoselective Synthesis of β-Thiomannosides. *J. Org. Chem.* **2000**, 65, 801-805.
14. Nagai, H., Matsumura, S., Toshima, K. A novel promoter, heteropoly acid, mediated chemo- and stereoselective sulfoxide glycosidation reactions. *Tetrahedron Lett.* **2000**, 41, 10233-10237.
15. Nagai, H., Kawahara, K., Matsumura, S., Toshima, K. Novel stereocontrolled α- and β-glycosidations of mannopyranosyl sulfoxides using environmentally benign heterogeneous solid acids. *Tetrahedron Lett.* **2001**, 42, 4159-4162.
16. Wipf, P., Reeves, J. T. Glycosylation via $Cp_2ZrCl_2/AgClO_4$-Mediated Activation of Anomeric Sulfoxides. *J. Org. Chem.* **2001**, 66, 7910-7914.
17. Crich, D. Chemistry of glycosyl triflates: Synthesis of β-mannopyranosides. *J. Carbohydr. Chem.* **2002**, 21, 667-690.
18. Marsh, S. J., Kartha, K. P. R., Field, R. A. Observations on iodine-promoted β-mannosylation. *Synlett* **2003**, 1376-1378.
19. Chambers, D. J., Evans, G. R., Fairbanks, A. J. Elimination reactions of glycosyl selenoxides. *Tetrahedron* **2004**, 60, 8411-8419.
20. Gildersleeve, J., Pascal, R. A., Jr., Kahne, D. Sulfenate Intermediates in the Sulfoxide Glycosylation Reaction. *J. Am. Chem. Soc.* **1998**, 120, 5961-5969.

21. Callam, C. S., Gadikota, R. R., Krein, D. M., Lowary, T. L. 2,3-Anhydrosugars in Glycoside Bond Synthesis. NMR and Computational Investigations into the Mechanism of Glycosylations with 2,3-Anhydrofuranosyl Glycosyl Sulfoxides. *J. Am. Chem. Soc.* **2003**, 125, 13112-13119.
22. Crich, D., Mataka, J., Sun, S., Wink, D. J., Lam, K. C., Rheingold, A. L. Stereoselective sulfoxidation of α-mannopyranosyl thioglycosides: the exo-anomeric effect in action. *Chem. Commun.* **1998**, 2763-2764.
23. Chen, M.-Y., Patkar, L. N., Chen, H.-T., Lin, C.-C. An efficient and selective method for preparing glycosyl sulfoxides by oxidizing glycosyl sulfides with OXONE or t-BuOOH on SiO_2. *Carbohydr. Res.* **2003**, 338, 1327-1332.
24. Chen, M.-Y., Patkar, L. N., Lin, C.-C. Selective Oxidation of Glycosyl Sulfides to Sulfoxides Using Magnesium Monoperoxyphthalate and Microwave Irradiation. *J. Org. Chem.* **2004**, 69, 2884-2887.
25. Raghavan, S., Kahne, D. A one step synthesis of the ciclamycin trisaccharide. *J. Am. Chem. Soc.* **1993**, 115, 1580-1581.
26. Crich, D., Sun, S. Are Glycosyl Triflates Intermediates in the Sulfoxide Glycosylation Method? A Chemical and ^{1}H, ^{13}C, and ^{19}F NMR Spectroscopic Investigation. *J. Am. Chem. Soc.* **1997**, 119, 11217-11223.
27. Nukada, T., Berces, A., Whitfield, D. M. Can the stereochemical outcome of glycosylation reactions be controlled by the conformational preferences of the glycosyl donor? *Carbohydr. Res.* **2002**, 337, 765-774.
28. Boeckman, R. K., Jr., Liu, Y. Toward the Development of a General Chiral Auxiliary. 5. High Diastereofacial Selectivity in Cycloadditions with Trienol Silyl Ethers: An Application to an Enantioselective Synthesis of (-)-Cassioside. *J. Org. Chem.* **1996**, 61, 7984-7985.
29. Gildersleeve, J., Smith, A., Sakurai, K., Raghavan, S., Kahne, D. Scavenging Byproducts in the Sulfoxide Glycosylation Reaction: Application to the Synthesis of Ciclamycin 0. *J. Am. Chem. Soc.* **1999**, 121, 6176-6182.

Keck Asymmetric Allylation..236

Related reactions: Roush asymmetric allylation, Sakurai allylation;

1. Keck, G. E., Geraci, L. S. Catalytic asymmetric allylation (CAA) reactions. II. A new enantioselective allylation procedure. *Tetrahedron Lett.* **1993**, 34, 7827-7828.
2. Keck, G. E., Krishnamurthy, D., Grier, M. C. Catalytic asymmetric allylation reactions. 3. Extension to methallylstannane, comparison of procedures, and observation of a nonlinear effect. *J. Org. Chem.* **1993**, 58, 6543-6544.
3. Keck, G. E., Tarbet, K. H., Geraci, L. S. Catalytic asymmetric allylation of aldehydes. *J. Am. Chem. Soc.* **1993**, 115, 8467-8468.
4. Cozzi, P. G., Tagliavini, E., Umani-Ronchi, A. Enantioselective addition of allylic silanes and stannanes to aldehydes mediated by chiral Lewis acids. *Gazz. Chim. Ital.* **1997**, 127, 247-254.
5. Ramon, D. J., Yus, M. Recent developments in enantioselective reactions promoted by titanium(IV) reagents bearing a chiral ligand. *Rec. Res. Dev. Org. Chem.* **1998**, 2, 489-523.
6. Cozzi, P. G., Tagliavini, E., Umani-Ronchi, A. Enantio- and diastereoselective addition of organometallic reagents to aldehydes and imines. *Current Trends in Organic Synthesis, [Proceedings of the International Conference on Organic Synthesis], 12th, Venezia, June 28-July 2, 1998* **1999**, 239-246.
7. Mikami, K., Terada, M. Chiral titanium complexes for enantioselective catalysis. *Lewis Acid Reagents* **1999**, 93-136.
8. Yanagisawa, A. Allylation of carbonyl groups. *Comprehensive Asymmetric Catalysis I-III* **1999**, 2, 965-979.
9. Marshall, J. A. Preparation and addition reactions of allylic and allenic tin and indium reagents. *Lewis Acids in Organic Synthesis* **2000**, 1, 453-522.
10. Mikami, K., Terada, M. Chiral Titanium(IV) Lewis acids. *Lewis Acids in Organic Synthesis - 2 vols.* **2000**, 2, 799-847.
11. Denmark, S. E., Fu, J. Catalytic Enantioselective Addition of Allylic Organometallic Reagents to Aldehydes and Ketones. *Chem. Rev.* **2003**, 103, 2763-2793.
12. Faller, J. W., Sams, D. W. I., Liu, X. Catalytic Asymmetric Synthesis of Homoallylic Alcohols: Chiral Amplification and Chiral Poisoning in a Titanium/BINOL Catalyst System. *J. Am. Chem. Soc.* **1996**, 118, 1217-1218.
13. Yu, C.-M., Choi, H.-S., Jung, W.-H., Lee, S.-S. Catalytic asymmetric allylation of aldehydes with BINOL-Ti(IV) complex accelerated by i-$PrSSiMe_3$. *Tetrahedron Lett.* **1996**, 37, 7095-7098.
14. Yu, C.-M., Choi, H.-S., Jung, W.-H., Kim, H.-J., Shin, J. Bifunctional molecular accelerator for catalytic asymmetric allylation: R_2MSR' (M = B, Al) as a useful synergetic reagent. *Chem. Commun.* **1997**, 761-762.
15. Yamago, S., Furukawa, M., Azuma, A., Yoshida, J.-i. Synthesis of optically active dendritic binaphthols and their metal complexes for asymmetric catalysis. *Tetrahedron Lett.* **1998**, 39, 3783-3786.
16. Bandin, M., Casolari, S., Cozzi, P. G., Proni, G., Schmohel, E., Spada, G. P., Tagliavini, E., Umani-Ronchi, A. Synthesis and characterization of new enantiopure 7,7'-disubstituted 2,2'-dihydroxy-1,1'-binaphthyls: useful ligands for the asymmetric allylation reaction of aldehydes. *Eur. J. Org. Chem.* **2000**, 491-497.
17. Brenna, E., Scaramelli, L., Serra, S. An efficient atropisomeric chiral biaryl ligand for catalytic stereoselective allylation of aldehydes: a novel approach to 2,2'-binol analogs. *Synlett* **2000**, 357-358.
18. Kii, S., Maruoka, K. Practical approach for catalytic asymmetric allylation of aldehydes with a chiral bidentate titanium(IV) complex. *Tetrahedron Lett.* **2001**, 42, 1935-1939.
19. Kii, S., Maruoka, K. Catalytic enantioselective allylation of ketones with novel chiral bis-titanium(IV) catalyst. *Chirality* **2003**, 15, 68-70.
20. Aoki, S., Mikami, K., Terada, M., Nakai, T. Enantio- and diastereoselective catalysis of addition reaction of allylic silanes and stannanes to glyoxylates by binaphthol-derived titanium complex. *Tetrahedron* **1993**, 49, 1783-1792.
21. Costa, A. L., Piazza, M. G., Tagliavini, E., Trombini, C., Umani-Ronchi, A. Catalytic asymmetric synthesis of homoallylic alcohols. *J. Am. Chem. Soc.* **1993**, 115, 7001-7002.
22. Weigand, S., Bruckner, R. Ti(IV)-BINOLate-catalyzed highly enantioselective additions of β-substituted allylstannanes to aldehydes. *Chem.-- Eur. J.* **1996**, 2, 1077-1084.
23. Almendros, P., Gruttadauria, M., Helliwell, M., Thomas, E. J. Stereoselective synthesis of 4-alkoxy-3-methylidenealkanols using reactions between 2-(1-alkoxyalkyl)propenylstannanes and aldehydes: X-ray crystal structure of (1R,4R)-3-methylidene-1-(4-nitrophenyl)-pentane-1,4-diol. *J. Chem. Soc., Perkin Trans. 1* **1997**, 2549-2560.
24. Keck, G. E., Yu, T. Catalytic Asymmetric Allylation Reactions Using BITIP Catalysis and 2-Substituted Allylstannanes as Surrogates for β-Keto Ester Dianions. *Org. Lett.* **1999**, 1, 289-291.
25. Keck, G. E., Covel, J. A., Schiff, T., Yu, T. Pyran Annulation: Asymmetric Synthesis of 2,6-Disubstituted-4-methylene Tetrahydropyrans. *Org. Lett.* **2002**, 4, 1189-1192.
26. Denmark, S. E., Hosoi, S. Stereochemical Studies on the Addition of Allylstannanes to Aldehydes. The SE' Component. *J. Org. Chem.* **1994**, 59, 5133-5135.
27. Corey, E. J., Lee, T. W. The formyl C-H...O hydrogen bond as a critical factor in enantioselective Lewis-acid catalyzed reactions of aldehydes. *Chem. Commun.* **2001**, 1321-1329.
28. Fuerstner, A., Langemann, K. Total Syntheses of (+)-Ricinelaidic Acid Lactone and of (-)-Gloeosporone Based on Transition-Metal-Catalyzed C-C Bond Formations. *J. Am. Chem. Soc.* **1997**, 119, 9130-9136.
29. Meng, D., Bertinato, P., Balog, A., Su, D.-S., Kamenecka, T., Sorensen, E., Danishefsky, S. J. Total Syntheses of Epothilones A and B. *J. Am. Chem. Soc.* **1997**, 119, 10073-10092.
30. Smith, A. B., III, Doughty, V. A., Sfouggatakis, C., Bennett, C. S., Koyanagi, J., Takeuchi, M. Spongistatin Synthetic Studies. An Efficient, Second-Generation Construction of an Advanced ABCD Intermediate. *Org. Lett.* **2002**, 4, 783-786.
31. Keck, G. E., Wager, C. A., Wager, T. T., Savin, K. A., Covel, J. A., McLaws, M. D., Krishnamurthy, D., Cee, V. J. Asymmetric total synthesis of Rhizoxin D. *Angew. Chem., Int. Ed. Engl.* **2001**, 40, 231-234.

Related reactions: Corey-Nicolaou macrolactonization, Yamaguchi macrolactonization;

1. Boden, E. P., Keck, G. E. Proton-transfer steps in Steglich esterification: a very practical new method for macrolactonization. *J. Org. Chem.* **1985**, 50, 2394-2395.
2. Meng, Q., Hesse, M. Ring-closure methods in the synthesis of macrocyclic natural products. *Top. Curr. Chem.* **1992**, 161, 107-176.
3. Nakata, T. Total synthesis of macrolides. *Macrolide Antibiotics (2nd Edition)* **2002**, 181-284.
4. Keck, G. E., Sanchez, C., Wager, C. A. Macrolactonization of hydroxy acids using a polymer-bound carbodiimide. *Tetrahedron Lett.* **2000**, 41, 8673-8676.
5. Neises, B., Steglich, W. 4-Dialkylaminopyridines as acylation catalysts. 5. Simple method for the esterification of carboxylic acids. *Angew. Chem.* **1978**, 90, 556-557.
6. Cetusic Jeannie, R. P., Green Frederick, R., 3rd, Graupner Paul, R., Oliver, M. P. Total synthesis of hectochlorin. *Org. Lett.* **2002**, 4, 1307-1310.
7. Hanessian, S., Ma, J., Wang, W. Total Synthesis of Bafilomycin A1 Relying on Iterative 1,2-Induction in Acyclic Precursors. *J. Am. Chem. Soc.* **2001**, 123, 10200-10206.
8. Lewis, A., Stefanuti, I., Swain, S. A., Smith, S. A., Taylor, R. J. K. A formal total synthesis of (+)-apicularen A: Base-induced conversion of apicularen-derived intermediates into salicylihalamide-like products. *Org. Biomol. Chem.* **2003**, 1, 104-116.
9. Mulzer, J., Mantoulidis, A., Oehler, E. Total Syntheses of Epothilones B and D. *J. Org. Chem.* **2000**, 65, 7456-7467.

1. Grignon, J., Pereyre, M. Mechanism of the substitution of halogen derivatives by allylic organotin compounds. *J. Organomet. Chem.* **1973**, 61, C33-C35.
2. Kosugi, M., Kurino, K., Takayama, K., Migita, T. Reaction of organic halides with allyltrimethyltin. *J. Organomet. Chem.* **1973**, 56, C11-C13.
3. Grignon, J., Servens, C., Pereyre, M. Reactivity of allylic organotin compounds with halogen derivatives. Synthetic aspects and mechanism. *J. Organomet. Chem.* **1975**, 96, 225-235.
4. Keck, G. E., Yates, J. B. Carbon-carbon bond formation via the reaction of trialkylallylstannanes with organic halides. *J. Am. Chem. Soc.* **1982**, 104, 5829-5831.
5. Jarosz, S., Kozlowska, E. Synthesis and application of allyltin derivatives in organic chemistry. *Pol. J. Chem.* **1998**, 72, 815-831.
6. Walton, J. C. Homolytic substitution: a molecular menage a trois. *Acc. Chem. Res.* **1998**, 31, 99-107.
7. Marshall, R. L. Product subclass 28: allylstannanes. *Science of Synthesis* **2003**, 5, 573-605.
8. Thomas, E. J. Tin compounds. *Science of Synthesis* **2003**, 5, 195-204.
9. Keck, G. E., Yates, J. B. A novel synthesis of (±)-perhydrohistrionicotoxin. *J. Org. Chem.* **1982**, 47, 3590-3591.
10. Birman, V. B., Danishefsky, S. J. The total synthesis of (±)-merrilactone A. *J. Am. Chem. Soc.* **2002**, 124, 2080-2081.
11. Wipf, P., Rector, S. R., Takahashi, H. Total Synthesis of (-)-Tuberostemonine. *J. Am. Chem. Soc.* **2002**, 124, 14848-14849.
12. Roe, B. A., Boojamra, C. G., Griggs, J. L., Bertozzi, C. R. Synthesis of β-C-Glycosides of N-Acetylglucosamine via Keck Allylation Directed by Neighboring Phthalimide Groups. *J. Org. Chem.* **1996**, 61, 6442-6445.
13. Campbell, J. A., Hart, D. J. Synthesis of a tetracyclic substructure of manzamine A. *Tetrahedron Lett.* **1992**, 33, 6247-6250.

1. Knoevenagel, E. Method for the synthesis of glutaric acid. *Ber.* **1894**, 27, 2345-2346.
2. Knoevenagel, E. The preparation of benzylidene acetoacetic esters. *Ber.* **1896**, 29, 172-174.
3. Johnson, J. R. Perkin reaction and related reactions. *Org. React.* **1942**, 1, pp 210-265.
4. Jones, G. Knoevenagel condensation. *Org. React.* **1967**, 15, 204-599.
5. Bhagade, S. S., Haran, N. P., Nageshwar, G. D. Catalysis by ion exchange resins. Knoevenagel condensation. *Chemicals & Petro-Chemicals Journal* **1977**, 8, 21-22.
6. Tietze, L. F. Domino-reactions: the tandem-Knoevenagel-hetero-Diels-Alder reaction and its application in natural product synthesis. *J. Heterocycl. Chem.* **1990**, 27, 47-69.
7. Tietze, L. F., Beifuss, U. The Knoevenagel Reaction. in *Comp. Org. Synth.* (eds. Trost, B. M.,Fleming, I.), *2*, 341-394 (Pergamon, Oxford, **1991**).
8. Macquarrie, D. J., Clark, J. H., Jackson, D. B., Lambert, A., Mdoe, J. E. G., Priest, A. New solid bases derived from organically modified silicas and their use in the Knoevenagel reaction. *Spec. Publ. - R. Soc. Chem.* **1998**, 216, 174-181.
9. Tietze, L. F., Modi, A. Multicomponent domino reactions for the synthesis of biologically active natural products and drugs. *Med. Res. Rev.* **2000**, 20, 304-322.
10. Corma, A., Iborra, S. Zeolites and related materials in Knoevenagel condensations and Michael additions. *Fine Chemicals through Heterogeneous Catalysis* **2001**, 309-326.
11. Prout, F. S. Amino acid catalysis of the Knoevenagel reaction. *J. Org. Chem.* **1953**, 18, 928-933.
12. Lelean, P. M., Morris, J. A. Fluoride catalysis of the Knoevenagel reaction. *Chem. Commun.* **1968**, 239.
13. ApSimon, J. W., Hooper, J. W., Laishes, B. A. Potassium fluoride catalyzed reactions between malononitrile and α,β–unsaturated ketones. *Can. J. Chem.* **1970**, 48, 3064-3075.
14. Lehnert, W. Improved varients of the Knoevenagel condensation with titanium tetrachloride-THF-pyridine. I. Alkylidene- and arylideremalonic esters at 0-25.deg. *Tetrahedron Lett.* **1970**, 4723-4724.
15. Lehnert, W. Knoevenagel condensation with a titanium(IV) chloride base. V. 3-Alkylidene- and 3-arylidene-2,4-pentanedione from aldehydes and acetylacetones. *Synthesis* **1974**, 667-669.
16. Texier-Boullet, F., Foucaud, A. Knoevenagel condensation catalyzed by aluminum oxide. *Tetrahedron Lett.* **1982**, 23, 4927-4928.
17. Cabello, J. A., Campelo, J. M., Garcia, A., Luna, D., Marinas, J. M. Aluminum phosphate-aluminum oxide catalytic system as a reagent in the Knoevenagel condensation of aromatic aldehydes. *Actas Simp. Iberoam. Catal., 9th* **1984**, 2, 1543-1544.
18. Prajapati, D., Sandhu, J. S. Bismuth(III)chloride as a new catalyst for Knoevenagel condensation in the absence of solvent. *Chem. Lett.* **1992**, 1945-1946.
19. Sebti, S., Saber, A., Rhihil, A. Natural phosphate and trisodium phosphate: novel solid catalysts for the Knoevenagel condensation in heterogeneous media. *Tetrahedron Lett.* **1994**, 35, 9399-9400.
20. Torres, D. A., Ferraris, J. P. Knoevenagel-like condensations with highly stabilized active methylene compounds. *Tetrahedron Lett.* **1994**, 35, 7589-7592.
21. Kantam, M. L., Bharathi, B. Mn(III) salen catalyst for Knoevenagel condensation - a novel heterogeneous system. *Catal. Lett.* **1998**, 55, 235-237.
22. McNulty, J., Steere, J. A., Wolf, S. The ultrasound promoted Knoevenagel condensation of aromatic aldehydes. *Tetrahedron Lett.* **1998**, 39, 8013-8016.
23. Betts, R. L., Diebes, A., Frenette, C. R., Langler, R. F. Novel knoevenagel condensations of a β-keto sulfone and a β-carboalkoxy sulfone. *Sulfur Letters* **1999**, 23, 11-31.
24. Rodriguez, I., Sastre, G., Corma, A., Iborra, S. Catalytic Activity of Proton Sponge: Application to Knoevenagel Condensation Reactions. *J. Catal.* **1999**, 183, 14-23.
25. Simpson, J., Rathbone, D. L., Billington, D. C. New solid phase Knoevenagel catalyst. *Tetrahedron Lett.* **1999**, 40, 7031-7033.

26. Balalaie, S., Nemati, N. Ammonium acetate-basic alumina catalyzed Knoevenagel condensation under microwave irradiation under solvent-free condition. *Synth. Commun.* **2000**, 30, 869-875.
27. Li, Y.-Q. Potassium phosphate as a catalyst for the Knoevenagel condensation. *J. Chem. Res., Synop.* **2000**, 524-525.
28. Shi, D., Wang, Y., Lu, Z., Dai, G. Condensation of aromatic aldehydes with acidic methylene compounds without catalyst. *Synth. Commun.* **2000**, 30, 713-726.
29. Jenner, G. Steric effects in high pressure Knoevenagel reactions. *Tetrahedron Lett.* **2001**, 42, 243-245.
30. Siebenhaar, B., Casagrande, B., Studer, M., Blaser, H.-U. An easy-to-use heterogeneous catalyst for the Knoevenagel condensation. *Can. J. Chem.* **2001**, 79, 566-569.
31. Hayashi, Y., Miyamoto, Y., Shoji, M. β-Ketothioester as a reactive Knoevenagel donor. *Tetrahedron Lett.* **2002**, 43, 4079-4082.
32. Ren, Z., Cao, W., Tong, W. The Knoevenagel condensation reaction of aromatic aldehydes with malononitrile by grinding in the absence of solvents and catalysts. *Synth. Commun.* **2002**, 32, 3475-3479.
33. Shi, D.-q., Wang, X.-s., Yao, C.-s., Mu, L. Knoevenagel condensation in the heterogeneous phase using KF-montmorillonite as a new catalyst. *J. Chem. Res., Synop.* **2002**, 344-345.
34. Li, Y. Q., Xu, X. M., Zhou, M. Y. n-Butyl pyridinium nitrate as a reusable ionic liquid medium for Knoevenagel condensation. *Chin. Chem. Lett.* **2003**, 14, 448-450.
35. Narsaiah, A. V., Nagaiah, K. An efficient Knoevenagel condensation catalyzed by $LaCl_3.7H_2O$ in heterogeneous medium. *Synth. Commun.* **2003**, 33, 3825-3832.
36. Su, C., Chen, Z.-C., Zheng, Q.-G. Organic reactions in ionic liquids: Knoevenagel condensation catalyzed by ethylenediammonium diacetate. *Synthesis* **2003**, 555-559.
37. Wang, S.-G. Amino groups immobilized on MCM-48: an efficient heterogeneous catalyst for the Knoevenagel reaction. *Catal. Commun.* **2003**, 4, 469-470.
38. Hayashi, M., Nakamura, N., Yamashita, K. Novel Knoevenagel-type reaction via titanium enolate derived from $Ti(O\text{-}i\text{-}Pr)_4$ and diketene. *Tetrahedron* **2004**, 60, 6777-6783.
39. Khan, F. A., Dash, J., Satapathy, R., Upadhyay, S. K. Hydrotalcite catalysis in ionic liquid medium: a recyclable reaction system for heterogeneous Knoevenagel and nitroaldol condensation. *Tetrahedron Lett.* **2004**, 45, 3055-3058.
40. Kubota, Y., Nishizaki, Y., Ikeya, H., Saeki, M., Hida, T., Kawazu, S., Yoshida, M., Fujii, H., Sugi, Y. Organic-silicate hybrid catalysts based on various defined structures for Knoevenagel condensation. *Microporous and Mesoporous Materials* **2004**, 70, 135-149.
41. Yadav, J. S., Reddy, B. V. S., Basak, A. K., Visali, B., Narsaiah, A. V., Nagaiah, K. Phosphane-catalyzed Knoevenagel condensation: A facile synthesis of α-cyanoacrylates and α-cyanoacrylonitriles. *Eur. J. Org. Chem.* **2004**, 546-551.
42. Hann, A. C. O., Lapworth, A. Optically active esters of β-ketonic and β-aldehydic acid. Part IV. Condensation of aldehydes with menthyl acetoacetate. *J. Chem. Soc.* **1904**, 85, 46-56.
43. Van der Baan, J. L., Bickelhaupt, F. Knoevenagel reaction of malononitrile with cyclic β-keto esters. II. Mechanism of formation of heterocyclic reaction products. *Tetrahedron* **1974**, 30, 2447-2453.
44. Cabello, J. A., Campelo, J. M., Garcia, A., Luna, D., Marinas, J. M. Knoevenagel condensation in the heterogeneous phase using aluminum phosphate-aluminum oxide as a new catalyst. *J. Org. Chem.* **1984**, 49, 5195-5197.
45. Kinastowski, S., Mroczyk, W. Kinetic investigations on aldolic stage of Knoevenagel's reaction. *Pol. J. Chem.* **1984**, 58, 179-184.
46. Tanikaga, R., Konya, N., Kaji, A. Stereochemistry of amine-catalyzed Knoevenagel reactions. *Chem. Lett.* **1985**, 1583-1586.
47. Tanikaga, R., Konya, N., Tamura, T., Kaji, A. Stereochemistry in the Knoevenagel reaction of methyl (arylsulfinyl)acetate and aldehydes. *J. Chem. Soc., Perkin Trans. 1* **1987**, 825-830.
48. Tanaka, M., Oota, O., Hiramatsu, H., Fujiwara, K. The Knoevenagel reactions of aldehydes with carboxy compounds. I. Reactions of p-nitrobenzaldehyde with active methine compounds. *Bull. Chem. Soc. Jpn.* **1988**, 61, 2473-2479.
49. Kinastowski, S., Mroczyk, W. The mechanism of the Knoevenagel reaction of malonic ester with benzaldehyde, catalyzed by pyrrolidine. *Bull. Pol. Acad. Sci., Chem.* **1989**, 37, 109-116.
50. Mroczyk, W., Grabarkiewicz-Szczesna, J., Kinastowski, S. Mechanism of Knoevenagel reaction of malonic ester with benzaldehyde catalyzed by pyrrolidine: kinetic studies of the deamination stage. *Roczniki Akademii Rolniczej w Poznaniu* **1995**, 281, 57-64.
51. Bojilova, A., Nikolova, R., Ivanov, C., Rodios, N. A., Terzis, A., Raptopoulou, C. P. A comparative study of the interaction of salicylaldehydes with phosphonoacetates under Knoevenagel reaction conditions. Synthesis of 1,2-benzoxaphosphorins and their dimers. *Tetrahedron* **1996**, 52, 12597-12612.
52. Bogdal, D. Influence of microwave irradiation on the rate of coumarin synthesis by the Knoevenagel condensation. *ECHET98: Electronic Conference on Heterocyclic Chemistry, June 29-July 24, 1998* **1998**, 387-390.
53. Boucard, V. Kinetic Study of the Knoevenagel Condensation Applied to the Synthesis of Poly[bicarbazolylene-alt-phenylenebis(cyanovinylene)]s. *Macromolecules* **2001**, 34, 4308-4313.
54. Medien, H. A. A. Kinetic studies of condensation of aromatic aldehydes with Meldrum's acid. *Z. Naturforsch., B: Chem. Sci.* **2002**, 57, 1320-1326.
55. Pivonka Don, E., Empfield James, R. Real-time in situ Raman analysis of microwave-assisted organic reactions. *Appl. Spectrosc.* **2004**, 58, 41-46.
56. Nicolaou, K. C., Xu, J. Y., Kim, S., Ohshima, T., Hosokawa, S., Pfefferkorn, J. Synthesis of the Tricyclic Core of Eleutherobin and Sarcodictyins and Total Synthesis of Sarcodictyin A. *J. Am. Chem. Soc.* **1997**, 119, 11353-11354.
57. Tietze, L. F., Zhou, Y. Highly efficient, enantioselective total synthesis of the active anti-influenza A virus indole alkaloid hirsutine and related compounds by domino reactions. *Angew. Chem., Int. Ed. Engl.* **1999**, 38, 2045-2047.
58. Snider, B. B., Lu, Q. Total Synthesis of (±)-Leporin A. *J. Org. Chem.* **1996**, 61, 2839-2844.
59. Fukuyama, T., Liu, G. Stereocontrolled Total Synthesis of (±)-Gelsemine. *J. Am. Chem. Soc.* **1996**, 118, 7426-7427.

Knorr Pyrrole Synthesis 244

Related reactions: Paal-Knorr pyrrole synthesis;

1. Knorr, L. *Ann.* **1886**, 236, 290.
2. Knorr, L., Lange, H. The synthesis of pyrrole derivatives from isonitroso ketones. *Ber.* **1902**, 35, 2998-3008.
3. Jones, R. A., Bean, G. P. The Chemistry of Pyrroles. in *Organic Chemistry* (eds. Blomquist, A. T.,Wasserman, H. H.), *34*, 525 pp (Academic Press, New York, **1977**).
4. Hort, E. V., Anderson, L. R. Pyrrole and pyrrole derivatives. *Kirk-Othmer Encycl. Chem. Technol., 3rd Ed.* **1982**, 19, 499-520.
5. Sundberg, R. J. Pyrroles and their Benzo Derivatives: Synthesis. in *Comprehensive Organic Functional Group Transformations II* (eds. Katritzky, A. R., Rees, C. W.,Scriven, E. F. V.), *2*, 119-200 (Pergamon, Oxford, New York, **1995**).
6. Ferreira, V. F., De Souza, M. C. B. V., Cunha, A. C., Pereira, L. O. R., Ferreira, M. L. G. Recent advances in the synthesis of pyrroles. *Organic Preparations and Procedures International* **2001**, 33, 411-454.
7. Black, D. S. Product class 13: 1H-pyrroles. *Science of Synthesis* **2002**, 9, 441-552.
8. Kel'in, A. V., Maioli, A. Recent advances in the chemistry of 1,3-diketones: Structural modifications and synthetic applications. *Curr. Org. Chem.* **2003**, 7, 1855-1886.
9. Bondietti, G., Lions, F. Extension of Knorr's pyrrole synthesis. *Journal and Proceedings of the Royal Society of New South Wales* **1933**, 66, 477-485.
10. Davidson, D. An extension of Knorr's pyrrole synthesis. *J. Org. Chem.* **1938**, 3, 361-364.
11. MacDonald, S. F., Stedman, R. J. A modified Knorr pyrrole synthesis. *Can. J. Chem.* **1954**, 32, 812-813.
12. Tamura, Y., Kato, S., Ikeda, M. One-Step Knorr pyrrole synthesis with hydroxylamine O-sulfonic acid. *Chem. Ind.* **1971**, 767.

13. Hamby, J. M., Hodges, J. C. α-Amino ketones from amino acids as precursors for the Knorr pyrrole synthesis. *Heterocycles* **1993**, 35, 843-850.
14. Aoyagi, Y., Mizusaki, T., Ohta, A. Facile and efficient synthesis of pyrroles and indoles via palladium-catalyzed oxidation of hydroxy-enamines and -amines. *Tetrahedron Lett.* **1996**, 37, 9203-9206.
15. Nagafuji, P., Cushman, M. A General Synthesis of Pyrroles and Fused Pyrrole Systems from Ketones and Amino Acids. *J. Org. Chem.* **1996**, 61, 4999-5003.
16. Zhang, Y., Jiang, Y.-Z., Liang, X.-T. A new variant of Knorr's pyrrole synthesis. *Chinese Journal of Chemistry* **1997**, 15, 371-378.
17. Alberola, A., Ortega, A. G., Sadaba, M. L., Sanudo, C. Versatility of Weinreb amides in the Knorr pyrrole synthesis. *Tetrahedron* **1999**, 55, 6555-6566.
18. Katritzky, A. R., Ostercamp, D. L., Yousaf, T. I. The mechanisms of heterocyclic ring closures. *Tetrahedron* **1987**, 43, 5171-5186.
19. Fabiano, E., Golding, B. T. On the mechanism of pyrrole formation in the Knorr pyrrole synthesis and by porphobilinogen synthase. *J. Chem. Soc., Perkin Trans. 1* **1991**, 3371-3375.
20. Yaylayan, V. A., Keyhani, A. Elucidation of the mechanism of pyrrole formation during thermal degradation of 13C-labeled L-serines. *Food Chem.* **2001**, 74, 1-9.
21. Goudie, A. C., Rosenberg, H. E., Ward, R. W. 4,5,8,9-Tetrahydro-8-methyl-9-oxothieno[3',2':5,6]cyclohepta[1,2-b]pyrrole-7-acetic acid. A new anti-inflammatory/analgesic agent. *J. Heterocycl. Chem.* **1983**, 20, 1027-1030.
22. Bellingham, R. K., Carey, J. S., Hussain, N., Morgan, D. O., Oxley, P., Powling, L. C. A Practical Synthesis of a Potent δ-Opioid Antagonist: Use of a Modified Knorr Pyrrole Synthesis. *Org. Process Res. Dev.* **2004**, 8, 279-282.
23. Coffen, D. L., Hengartner, U., Katonak, D. A., Mulligan, M. E., Burdick, D. C., Olson, G. L., Todaro, L. J. Syntheses of an antipsychotic pyrrolo[2,3-g]isoquinoline from areca alkaloids. *J. Org. Chem.* **1984**, 49, 5109-5113.

Koenigs-Knorr Glycosidation 246

Related reactions: Kahne gycosidation;

1. Michael, A. On the synthesis of Helicin and Phenolglucoside. *Am. Chem. J.* **1879**, 1, 305.
2. Koenigs, W., Knorr, E. Some derivatives of grape sugars and galactose. *Ber.* **1901**, 34, 957-981.
3. Igarashi, K. The Koenigs-Knorr reaction. *Adv. Carbohydr. Chem. Biochem.* **1977**, 34, 243-283.
4. Pigman, W., Horton, D., Editors. *The Carbohydrates, Vol. 1B: Chemistry and Biochemistry. 2nd Ed* (Academic Press, New York, N. Y., **1980**) 984 pp.
5. Paulsen, H. Progress in the selective chemical synthesis of complex oligosaccharides. *Angew. Chem., Int. Ed. Engl.* **1982**, 21, 155-173.
6. Paulsen, H. Synthesis of complex oligosaccharide chains of glycoproteins. *Chem. Soc. Rev.* **1984**, 13, 15-45.
7. Schmidt, R. R. New methods of glycoside and oligosaccharide syntheses - are there alternatives to the Koenigs-Knorr method? *Angew. Chem.* **1986**, 98, 213-236.
8. Krohn, K. Synthesis of O-glycosides. *Org. Synth. Highlights* **1991**, 277-285.
9. Schmidt, R. R. Synthesis of Glycosides. in *Comp. Org. Synth.* (eds. Trost, B. M.,Fleming, I.), *6*, 33-64 (Pergamon, Oxford, **1991**).
10. Toshima, K., Tatsuta, K. Recent progress in O-glycosylation methods and its application to natural products synthesis. *Chem. Rev.* **1993**, 93, 1503-1531.
11. Paulsen, H. Twenty five years of carbohydrate chemistry; an overview of oligosaccharide synthesis. *Frontiers in Natural Product Research* **1996**, 1, 1-19.
12. Veeneman, G. H. Chemical synthesis of O-glycosides. *Carbohydr. Chem.* **1998**, 98-174.
13. Schmidt, R. R., Castro-Palomino, J. C., Retz, O. New aspects of glycoside bond formation. *Pure Appl. Chem.* **1999**, 71, 729-744.
14. Capozzi, G., Menichetti, S., Nativi, C. Selective glycosidation reactions and their use in medicinal chemistry. *Methods and Principles in Medicinal Chemistry* **2000**, 7, 221-259.
15. Nicolaou, K. C., Mitchell, H. J. Adventures in carbohydrate chemistry: new synthetic technologies, chemical synthesis, molecular design, and chemical biology. *Angew. Chem., Int. Ed. Engl.* **2001**, 40, 1576-1624.
16. Nitz, M., Bundle, D. R. Glycosyl halides in oligosaccharides synthesis. *Glycoscience* **2001**, 2, 1497-1542.
17. Ferrier, R. J., Blattner, R., Field, R. A., Furneaux, R. H., Gardiner, J. M., Hoberg, J. O., Kartha, K. P. R., Tilbrook, D. M. G., Tyler, P. C., Wightman, R. H. Glycosides and disaccharides. *Carbohydr. Chem.* **2002**, 33, 16-61.
18. Demchenko, A. V. Stereoselective chemical 1,2-cis O-glycosylation: From "sugar ray" to modern techniques of the 21st century. *Synlett* **2003**, 1225-1240.
19. Demchenko, A. V. 1,2-cis O-glycosylation: methods, strategies, principles. *Curr. Org. Chem.* **2003**, 7, 35-79.
20. Pietruszka, J. Modern glycosidation methods: tuning of reactivity. *Carbohydrates* **2003**, 195-218.
21. Pellissier, H. The glycosylation of steroids. *Tetrahedron* **2004**, 60, 5123-5162.
22. Bernstein, S., Conrow, R. B. Steroid conjugates. VI. Improved Koenigs--Knorr synthesis of aryl glucuronides using cadmium carbonate, a new and effective catalyst. *J. Org. Chem.* **1971**, 36, 863-870.
23. Lemieux, R. U., Hendriks, K. B., Stick, R. V., James, K. Halide ion catalyzed glycosidation reactions. Syntheses of α-linked disaccharides. *J. Am. Chem. Soc.* **1975**, 97, 4056-4062.
24. Ackermann, I. E., Banthorpe, D. V., Fordham, W. D., Kinder, J. P., Poots, I. Preparation of new terpenyl β-D-glucopyranosides by a modified Koenigs-Knorr procedure. *Liebigs Ann. Chem.* **1989**, 79-81.
25. Li, Z., Xiao, G., Cai, M. Studies on carbohydrates. XII. An improved Koenigs-Knorr method for highly stereoselective synthesis of 1-O-acyl-β-D-galactopyranose tetraacetates. *Chin. Chem. Lett.* **1992**, 3, 711-712.
26. Hadd, M. J., Gervay, J. Glycosyl iodides are highly efficient donors under neutral conditions. *Carbohydr. Res.* **1999**, 320, 61-69.
27. Hanessian, S., Lou, B. Stereocontrolled Glycosyl Transfer Reactions with Unprotected Glycosyl Donors. *Chem. Rev.* **2000**, 100, 4443-4463.
28. Krepinsky, J. J., Douglas, S. P. Polymer-supported synthesis of oligosaccharides. *Carbohydrates in Chemistry and Biology* **2000**, 1, 239-265.
29. Desmares, G., Lefebvre, D., Renevret, G., Le Drian, C. Selective formation of β-D-glucosides of hindered alcohols. *Helv. Chim. Acta* **2001**, 84, 880-889.
30. Crich, D. Chemistry of glycosyl triflates: Synthesis of β-mannopyranosides. *J. Carbohydr. Chem.* **2002**, 21, 667-690.
31. Shingu, Y., Nishida, Y., Dohi, H., Matsuda, K., Kobayashi, K. Convenient access to halide ion-catalyzed α-glycosylation free from noxious fumes at the donor synthesis. *J. Carbohydr. Chem.* **2002**, 21, 605-611.
32. Fairbanks, A. J. Intramolecular aglycon delivery (IAD): The solution to 1,2-cis stereocontrol for oligosaccharide synthesis? *Synlett* **2003**, 1945-1958.
33. Shingu, Y., Nishida, Y., Dohi, H., Kobayashi, K. An easy access to halide ion-catalytic α-glycosylation using carbon tetrabromide and triphenylphosphine as multifunctional reagents. *Org. Biomol. Chem.* **2003**, 1, 2518-2521.
34. Mukaiyama, T., Kobashi, Y. Highly α-selective synthesis of disaccharide using glycosyl bromide by the promotion of phosphine oxide. *Chem. Lett.* **2004**, 33, 10-11.
35. Nukada, T., Berces, A., Zgierski, M. Z., Whitfield, D. M. Exploring the Mechanism of Neighboring Group Assisted Glycosylation Reactions. *J. Am. Chem. Soc.* **1998**, 120, 13291-13295.
36. Berces, A., Enright, G., Nukada, T., Whitfield, D. M. The Conformational Origin of the Barrier to the Formation of Neighboring Group Assistance in Glycosylation Reactions: A Dynamical Density Functional Theory Study. *J. Am. Chem. Soc.* **2001**, 123, 5460-5464.
37. Nukada, T., Berces, A., Whitfield, D. M. Can the stereochemical outcome of glycosylation reactions be controlled by the conformational preferences of the glycosyl donor? *Carbohydr. Res.* **2002**, 337, 765-774.

38. Wulff, G., Roehle, G. Glycoside synthesis. VI. Kinetic investigations on the mechanism of Koenigs-Knorr reaction. *Chem. Ber.* **1972**, 105, 1122-1132.
39. Wallace, J. E., Schroeder, L. R. Koenigs-Knorr reactions. Part II. A mechanistic study of mercury(II) cyanide-promoted reactions of 2,3,4,6-tetra-O-methyl-a-D-glucopyranosyl bromide with cyclohexanol in benzene-nitromethane. *J. Chem. Soc., Perkin Trans. 2* **1976**, 1632-1636.
40. Wallace, J. E., Schroeder, L. R. Koenigs-Knorr reactions. Part 3. Mechanistic study of mercury(II) cyanide promoted reactions of 2-O-acetyl-3,4,6-tri-O-methyl-a-D-glucopyranosyl bromide with cyclohexanol in benzene-nitromethane. *J. Chem. Soc., Perkin Trans. 2* **1977**, 795-802.
41. Banoub, J., Bundle, D. R. 1,2-Orthoacetate intermediates in silver trifluoromethanesulfonate promoted Koenigs-Knorr synthesis of disaccharide glycosides. *Can. J. Chem.* **1979**, 57, 2091-2097.
42. Bowden, T., Garegg, P. J., Maloisel, J.-L., Konradsson, P. A mechanistic study: nucleophile dependence in glucosylation with glucosyl bromides. *Isr. J. Chem.* **2000**, 40, 271-277.
43. Fürstner, A., Radkowski, K., Grabowski, J., Wirtz, C., Mynott, R. Ring-Closing Alkyne Metathesis. Application to the Total Synthesis of Sophorolipid Lactone. *J. Org. Chem.* **2000**, 65, 8758-8762.
44. Josien-Lefebvre, D., Le Drian, C. Total synthesis of (-)-lithospermoside. *Helv. Chim. Acta* **2003**, 86, 661-672.

Kolbe-Schmitt Reaction 248

1. Kolbe, H., Lautemann, E. *Liebigs Ann. Chem.* **1860**, 113, 125-127.
2. Kolbe, H., Lautemann, E. Constitution of salicylic acid and its basicity. *Liebigs Ann. Chem.* **1860**, 115, 157-206.
3. Kolbe, H., Lautemann, E. *Liebigs Ann. Chem.* **1860**, 115, 178.
4. Kolbe, H., Lautemann, E. *Liebigs Ann. Chem.* **1860**, 115, 201.
5. Lindsey, A. S., Jeskey, H. The Kolbe-Schmitt reaction. *Chem. Rev.* **1957**, 57, 583-620.
6. Aresta, M., Quaranta, E., Tommasi, I., Giannoccaro, P., Ciccarese, A. Enzymic versus chemical carbon dioxide utilization. Part I. The role of metal centers in carboxylation reactions. *Gazz. Chim. Ital.* **1995**, 125, 509-538.
7. Schmitt, R. German Patent 29939, **1884**, 233
8. Schmitt, R. *J. Prakt. Chem./Chem.-Ztg.* **1885**, 31, 397.
9. Cate, L. A. An efficient carboxylation of 1-naphthols using magnesium methyl carbonate. *Synthesis* **1983**, 385-386.
10. Baxter, J., Yamaguchi, T. Effect of cation capture by crown ether and polar solvent in the carboxylation with CO_2 of alkali metal 2-naphtholates under ordinary conditions. *J. Chem. Res., Synop.* **1997**, 374-375.
11. Sikkema, D. J., Reichwein, A. M. *Process for dicarboxylating dihydric phenols.* Application: WO 96-EP4611 9717315, **1997** (Akzo Nobel N.V., Neth.). 15 pp
12. Kosugi, Y., Takahashi, K., Imaoka, Y. Solvent-assisted carboxylation of alkali metal phenoxide with carbon dioxide. *J. Chem. Res., Synop.* **1999**, 114-115.
13. Samuels, M. R., Yabroff, R. M. *Drying of alkali metal aryloxides and carboxylation in preparation of aromatic hydroxycarboxylic acids.* Application: US 98-189599 5977405, **1999** (E. I. Du Pont de Nemours & Co., USA). 10 pp , Division of U S Ser No 832,738
14. Rahim, M. A., Matsui, Y., Kosugi, Y. Effects of alkali and alkaline earth metals on the Kolbe-Schmitt reaction. *Bull. Chem. Soc. Jpn.* **2002**, 75, 619-622.
15. Rahim, M. A., Matsui, Y., Matsuyama, T., Kosugi, Y. Regioselective carboxylation of phenols with carbon dioxide. *Bull. Chem. Soc. Jpn.* **2003**, 76, 2191-2195.
16. Markovic, Z., Engelbrecht, J. P., Markovic, S. Theoretical study of the Kolbe-Schmitt reaction mechanism. *Z. Naturforsch., A: Phys. Sci.* **2002**, 57, 812-818.
17. Kosugi, Y., Imaoka, Y., Gotoh, F., Rahim Mohammad, A., Matsui, Y., Sakanishi, K. Carboxylations of alkali metal phenoxides with carbon dioxide. *Org. Biomol. Chem.* **2003**, 1, 817-821.
18. Baumann, E. *Ber.* **1907**, 11, 1878.
19. Bhagwanth, M. R. R., Rao, A. V. R., Venkataraman, K. Applications of NMR and mass spectroscopy to some problems concerning synthetic dyes. V. Kolbe-Schmitt reaction products from 2-hydroxycarbazole and the structure of naphthol AS-LB. *Indian J. Chem.* **1969**, 7, 1065-1069.
20. Phadtare, P. G., Doraiswamy, L. K. Kolbe-Schmitt carbonation of 2-naphthol. Confirmation of the mass transfer model and process optimization. *Industrial & Engineering Chemistry Process Design and Development* **1969**, 8, 165-173.
21. Pizzi, A., Garcia, R., Wang, S. On the networking mechanisms of additives-accelerated phenol-formaldehyde polycondensates. *J. Appl. Polym. Sci.* **1997**, 66, 255-266.
22. Kosugi, Y., Takahashi, K. Carboxylation reaction with carbon dioxide. Mechanistic studies on the Kolbe-Schmitt reaction. *Stud. Surf. Sci. Catal.* **1998**, 114, 487-490.
23. Kosugi, Y., Rahim, M. A., Takahashi, K., Imaoka, Y., Kitayama, M. Carboxylation of alkali metal phenoxide with carbon dioxide at ambient temperature. *Applied Organometallic Chemistry* **2000**, 14, 841-843.
24. Kosugi, Y., Imaoka, Y., Gotoh, F., Rahim, M. A., Matsui, Y., Sakanishi, K. Carboxylations of alkali metal phenoxides with carbon dioxide. *Org. Biomol. Chem.* **2003**, 1, 817-821.
25. Dunne, A. M., Mix, S., Blechert, S. A highly efficient olefin metathesis initiator: improved synthesis and reactivity studies. *Tetrahedron Lett.* **2003**, 44, 2733-2736.
26. Chu, Y., Gao, Y.-C., Shen, F.-J., Liu, X., Wang, Y.-Y., Shi, Q.-Z. Synthesis, crystal structure and characterization of ternary complexes of lanthanide(III) 3,5-di-tert-butyl-γ-resorcylate with substituted pyridine-N-oxide. *Polyhedron* **1999**, 18, 3723-3727.
27. Gnaim, J. M., Green, B. S., Arad-Yellin, R., Keehn, P. M. Improved preparation of the clathrate host compound tri-o-thymotide and related trisalicylide derivatives. *J. Org. Chem.* **1991**, 56, 4525-4529.
28. Roedel, T., Gerlach, H. Enantioselective synthesis of (R)-(+)-pulvilloric acid. *Liebigs Ann. Chem.* **1997**, 213-216.

Kornblum Oxidation 250

1. Kornblum, N., Powers, J. W., Anderson, G. J., Jones, W. J., Larson, H. O., Levand, O., Weaver, W. M. New and selective method of oxidation. *J. Am. Chem. Soc.* **1957**, 79, 6562.
2. Kornblum, N., Jones, W. J., Anderson, G. J. A new and selective method of oxidation. Conversion of alkyl halides and alkyl tosylates to aldehydes. *J. Am. Chem. Soc.* **1959**, 81, 4113-4114.
3. Carnduff, J. Recent advances in aldehyde synthesis. *Quart. Rev. (London)* **1966**, 20, 169-189.
4. Epstein, W. W., Sweat, F. W. Dimethyl sulfoxide oxidations. *Chem. Rev.* **1967**, 67, 247-260.
5. Mancuso, A. J., Swern, D. Activated dimethyl sulfoxide: useful reagents for synthesis. *Synthesis* **1981**, 165-185.
6. Tidwell, T. T. Oxidation of alcohols by activated dimethyl sulfoxide and related reactions: an update. *Synthesis* **1990**, 857-870.
7. Tidwell, T. T. Oxidation of alcohols to carbonyl compounds via alkoxysulfonium ylides: the Moffat, Swern, and related oxidations. *Org. React.* **1990**, 39, 297-572.
8. Kilenyi, S. N. Oxidation of Carbon-Halogen Bonds. in *Comp. Org. Synth.* (eds. Trost, B. M.,Fleming, I.), *7*, 653-670 (Pergamon, Oxford, **1991**).
9. Kornblum, N., Frazier, H. W. New and convenient synthesis of glyoxals, glyoxalate esters, and α-diketones. *J. Am. Chem. Soc.* **1966**, 88, 865-866.
10. Epstein, W. W., Ollinger, J. Silver ion assisted dimethyl sulfoxide oxidations of organic halides. *J. Chem. Soc., Chem. Commun.* **1970**, 1338-1339.
11. Ganem, B., Boeckman, R. K., Jr. Silver-assisted dimethyl sulfoxide oxidations. Improved synthesis of aldehydes and ketones. *Tetrahedron Lett.* **1974**, 917-920.

12. Dave, P., Byun, H. S., Engel, R. Improved direct oxidation of alkyl halides to aldehydes. *Synth. Commun.* **1986**, 16, 1343-1346.
13. Godfrey, A. G., Ganem, B. Ready oxidation of halides to aldehydes using trimethylamine N-oxide in dimethyl sulfoxide. *Tetrahedron Lett.* **1990**, 31, 4825-4826.
14. Kitagawa, H., Matsuo, J.-I., Iida, D., Mukaiyama, T. Oxidation of primary alkyl triflates to the corresponding aldehydes via alkoxy(N-tert-butylamino)(methyl)sulfonium triflates. *Chem. Lett.* **2001**, 580-581.
15. Bettadaiah, B. K., Gurudutt, K. N., Srinivas, P. Direct Conversion of tert-β-Bromo Alcohols to Ketones with Zinc Sulfide and DMSO. *J. Org. Chem.* **2003**, 68, 2460-2462.
16. Bhat, K. S., Srinivas, S., Srinivas, P., Gurudutt, K. N. Oxidation of benzylic bromides by DMSO in the presence of zinc salts: A new variant of Kornblum's method. *Indian J. Chem., Sect. B* **2004**, 43B, 426-429.
17. Feely, W., Lehn, W. L., Boekelheide, V. Alkaline decomposition of quaternary salts of amine oxides. *J. Org. Chem.* **1957**, 22, 1135.
18. Manning, R. E., Schaefer, F. M. Mechanism of the base-catalyzed conversion of N-alkoxypyridinium salts to aldehydes. *Tetrahedron Lett.* **1975**, 213-214.
19. Kornblum, N. The synthesis of aliphatic and alicyclic nitro compounds. *Org. React.* **1962**, 12, 101-156.
20. Kornblum, N., Wade, P. A. Mild, nonacidic, method for converting secondary nitro compounds into ketones. *J. Org. Chem.* **1973**, 38, 1418-1420.
21. McKillop, A., Ford, M. E. Improved procedure for the conversion of benzyl halides into benzaldehydes. *Synth. Commun.* **1974**, 4, 45-50.
22. Angyal, S. J. The Sommelet reaction. *Org. React.* **1954**, 197-217.
23. Kroehnke, F. Syntheses with the aid of pyridinium salts. *Angew. Chem.* **1963**, 75, 317-329.
24. Fischer, B., Kabha, E., Gendron, F.-P., Beaudoin, A. R. Synthesis, mechanism and fluorescence properties of 8-(aryl)-3-β-D-ribofuranosylimidazo[2,1-i]purine 5'-phosphate derivatives. *Nucleosides, Nucleotides & Nucleic Acids* **2000**, 19, 1033-1054.
25. Ly, T. W., Liao, J.-H., Shia, K.-S., Liu, H.-J. A highly effective Diels-Alder approach to cis-clerodane natural products: First total synthesis of solidago alcohol. *Synthesis* **2004**, 271-275.
26. Mueller, K., Prinz, H., Gawlik, I., Ziereis, K., Huang, H.-S. Simple Analogs of Anthralin: Unusual Specificity of Structure and Antiproliferative Activity. *J. Med. Chem.* **1997**, 40, 3773-3780.
27. Paquette, W. D., Taylor, R. E. Enantioselective Preparation of the C1-C11 Fragment of Apoptolidin. *Org. Lett.* **2004**, 6, 103-106.

Krapcho Decarboxylation 252

1. Bernhard, A. Benzoylation of acetoacetic esters. *Liebigs Ann. Chem.* **1894**, 282, 153-191.
2. Meerwein, H. New Method of Ketone Decomposition of β-Ketonic Esters. *Ann.* **1913**, 398, 242-250.
3. Meerwein, H., Schurmann, W. Synthesis of Derivatives of Bicyclo(1,3,3)nonane. I. *Ann.* **1913**, 398, 196-242.
4. MacDonald, S. F., Stedman, R. J. Synthesis of hemopyrroledicarboxylic acid and of some dipyrromethenes. *Can. J. Chem.* **1955**, 33, 458-467.
5. Krapcho, A. P., Glynn, G. A., Grenon, B. J. Decarbethoxylation of geminal dicarbethoxy compounds. *Tetrahedron Lett.* **1967**, 215-217.
6. Krapcho, A. P., Mundy, B. P. Stereospecific synthesis of 2-isopropylidene-cis,cis-4,8-dimethyl-6-keto-cis-decahydroazulene. *Tetrahedron* **1970**, 26, 5437-5446.
7. McMurry, J. Ester cleavages via S_N2-type dealkylation. *Org. React.* **1976**, 24, 187-224.
8. Krapcho, A. P. Synthetic applications of dealkoxycarbonylations of malonate esters, β-keto esters, α-cyano esters and related compounds in dipolar aprotic media. Part II. *Synthesis* **1982**, 893-914.
9. Krapcho, A. P. Synthetic applications of dealkoxycarbonylations of malonate esters, β-keto esters, α-cyano esters and related compounds in dipolar aprotic media - Part I. *Synthesis* **1982**, 805-822.
10. Krapcho, A. P., Lovey, A. J. Decarbalkoxylations of geminal diesters, β-keto esters, and α-cyano esters effected by sodium chloride in dimethyl sulfoxide. *Tetrahedron Lett.* **1973**, 957-960.
11. Krapcho, A. P., Jahngen, E. G. E., Jr., Lovey, A. J., Short, F. W. Decarbalkoxylations of germinal diesters and β-keto esters in wet dimethyl sulfoxide. Effect of added sodium chloride on the decarbalkoxylation rates of mono- and disubstituted malonate esters. *Tetrahedron Lett.* **1974**, 1091-1094.
12. Krapcho, A. P., Gadamasetti, G. Facile de-tert-butoxycarbonylations of β-keto esters and mixed malonate esters using water in dimethyl sulfoxide. *J. Org. Chem.* **1987**, 52, 1880-1881.
13. Loupy, A., Pigeon, P., Ramdani, M., Jacquault, P. A new solvent-free procedure using microwave technology as an alternative to the Krapcho reaction. *J. Chem. Res., Synop.* **1993**, 36-37.
14. Melo, J. O. F., Pereira, E. H. T., Donnici, C. L., Wladislaw, B., Marzorati, L. A novel and easy de-ethoxycarbonylation of α-substituted malonic esters. *Synth. Commun.* **1998**, 28, 4179-4185.
15. Lynn Zara, C., Jin, T., Giguere, R. J. Microwave heating in organic synthesis: decarboxylation of malonic acid derivatives in water. *Synth. Commun.* **2000**, 30, 2099-2104.
16. Krapcho, A. P., Weimaster, J. F., Eldridge, J. M., Jahngen, E. G. E., Jr., Lovey, A. J., Stephens, W. P. Synthetic applications and mechanism studies of the decarbalkoxylations of geminal diesters and related systems effected in dimethyl sulfoxide by water and/or by water with added salts. *J. Org. Chem.* **1978**, 43, 138-147.
17. Krapcho, A. P., Weimaster, J. F. Stereochemistry of decarbalkoxylation of cyclic geminal diesters effected by water and lithium chloride in dimethyl sulfoxide. *J. Org. Chem.* **1980**, 45, 4105-4111.
18. Bernard, A., Cerioni, G., Piras, P. P. Mechanism of decarbalkoxylation of arylmethylenepropanedioic acid dimethyl esters. *Tetrahedron* **1990**, 46, 3929-3940.
19. Gilligan, P. J., Krenitsky, P. J. Divergent mechanisms for the dealkoxycarbonylation of a 2-(3-azetidinyl)malonate by chloride and cyanide. *Tetrahedron Lett.* **1994**, 35, 3441-3444.
20. Fürstner, A., Krause, H. Flexible Synthesis of Metacycloprodigiosin and Functional Derivatives Thereof. *J. Org. Chem.* **1999**, 64, 8281-8286.
21. Evans, D. A., Scheidt, K. A., Downey, C. W. Synthesis of (-)-epibatidine. *Org. Lett.* **2001**, 3, 3009-3012.
22. Molander, G. A., St. Jean, D. J., Jr., Haas, J. Toward a General Route to the Eunicellin Diterpenes: The Asymmetric Total Synthesis of Deacetoxyalcyonin Acetate. *J. Am. Chem. Soc.* **2004**, 126, 1642-1643.
23. Jonasson, C., Roenn, M., Baeckvall, J.-E. An Enantioselective Route to Paeonilactone A via Palladium- and Copper-Catalyzed Reactions. *J. Org. Chem.* **2000**, 65, 2122-2126.

Kröhnke Pyridine Synthesis 254

Related reactions: Hantzsch dihydropyridine synthesis;

1. Zecher, W., Kröhnke, F. A new synthesis of substituted pyridines. II. Some variations and special cases. *Chem. Ber.* **1961**, 94, 698-706.
2. Zecher, W., Kröhnke, F. A new synthesis of substituted pyridines. I. Principles of the synthesis. *Chem. Ber.* **1961**, 94, 690-697.
3. Kröhnke, F. Syntheses using pyridinium salts. 5. The specific synthesis of pyridines and oligopyridines. *Synthesis* **1976**, 1-24.
4. Grosche, P., Holtzel, A., Walk, T. B., Trautwein, A. W., Jung, G. Pyrazole, pyridine, and pyridone synthesis on solid support. *Synthesis* **1999**, 1961-1970.
5. Fujimori, T., Wirsching, P., Janda, K. D. Preparation of a Kröhnke Pyridine Combinatorial Library Suitable for Solution-Phase Biological Screening. *J. Comb. Chem.* **2003**, 5, 625-631.
6. Ahlbrecht, H., Kröhnke, F. N-Vinylpyridinium salts. X. Action of methyl ketones on substituted N-vinylpyridinium salts. *Liebigs Ann. Chem.* **1967**, 704, 133-139.

7. Alvarez-Builla, J., Novella, J. L., Galvez, E., Smith, P., Florencio, F., Garcia-Blanco, S., Bellanato, J., Santos, M. Synthesis and structural study on α-substituted-1-styrylpyridinium salts. Reinvestigation of the Kröhnke condensation. *Tetrahedron* **1986**, 42, 699-708.
8. Litvinov, V. P., Shestopalov, A. M. Pyridinium ylides in organic synthesis. Part 4. Pyridinium ylides in nucleophilic addition-elimination (AdN-E) reactions. *Russ. J. Org. Chem.* **1997**, 33, 903-940.
9. Osmialowski, B., Janota, H., Gawinecki, R. Stability of 1-phenacylpyridinium and 1-(2-hydroxy-2-phenylvinyl)pyridinium cations. *Pol. J. Chem.* **2003**, 77, 169-177.
10. Malkov, A. V., Bella, M., Stara, I. G., Kocovsky, P. Modular pyridine-type P,N-ligands derived from monoterpenes: application in asymmetric Heck addition. *Tetrahedron Lett.* **2001**, 42, 3045-3048.
11. Kelly, T. R., Lee, Y.-J., Mears, R. J. Synthesis of Cyclo-2,2':4',4'':2'',2''':4''',4'''':2'''',2''''':4''''',4-sexipyridine. *J. Org. Chem.* **1997**, 62, 2774-2781.
12. Zhao, L.-X., Moon, Y.-S., Basnet, A., Kim, E.-k., Jahng, Y., Park, J. G., Jeong, T. C., Cho, W.-J., Choi, S.-U., Lee, C. O., Lee, S.-Y., Lee, C.-S., Lee, E.-S. Synthesis, topoisomerase I inhibition and structure-activity relationship study of 2,4,6-trisubstituted pyridine derivatives. *Bioorg. Med. Chem. Lett.* **2004**, 14, 1333-1337.
13. Chelucci, G., Muroni, D., Pinna, G. A., Saba, A., Vignola, D. Chiral 2-(2-phenylthiophenyl)-5,6,7,8-tetrahydroquinolines: new N-S ligands for asymmetric catalysis. Palladium-catalyzed allylic alkylation and copper-catalyzed cyclopropanation reactions. *J. Mol. Catal. A: Chemical* **2003**, 191, 1-8.

Kulinkovich Reaction 256

1. Kulinkovich, O. G., Sviridov, S. V., Vasilevskii, D. A., Pritytskaya, T. S. Reaction of ethylmagnesium bromide with carboxylic acid esters in the presence of tetraisopropoxytitanium. *Zh. Org. Khim.* **1989**, 25, 2244-2245.
2. Kulinkovich, O. G., Sviridov, S. V., Vasilevskii, D. A. Titanium(IV) isopropoxide-catalyzed formation of 1-substituted cyclopropanols in the reaction of ethylmagnesium bromide with methyl alkanecarboxylates. *Synthesis* **1991**, 234.
3. Kulinkovich, O. G. Alkylation of carbonyl compounds through conversion into oxycyclopropane intermediates. *Pol. J. Chem.* **1997**, 71, 849-882.
4. Pfaltz, A. Cyclopropanation. *Transition Metals for Organic Synthesis* **1998**, 1, 100-113.
5. Sato, F., Urabe, H., Okamoto, S. Bicyclization of dienes, enynes, and diynes with $Ti^{(II)}$ reagent. New developments towards asymmetric synthesis. *Pure Appl. Chem.* **1999**, 71, 1511-1519.
6. Breit, B. Bis(alkoxy)titanacyclopropanes and -propenes (Kulinkovich reagents): Versatile reagents for carbon-carbon bond formation. *J. Prakt. Chem.* **2000**, 342, 211-214.
7. Kulinkovich, O. G. Titanacyclopropanes as versatile intermediates for carbon-carbon bond formation in reactions with unsaturated compounds. *Pure Appl. Chem.* **2000**, 72, 1715-1719.
8. Kulinkovich, O. G., de Meijere, A. 1,n-Dicarbanionic titanium intermediates from monocarbanionic organometallics and their application in organic synthesis. *Chem. Rev.* **2000**, 100, 2789-2834.
9. Sato, F., Urabe, H., Okamoto, S. Synthesis of Organotitanium Complexes from Alkenes and Alkynes and Their Synthetic Applications. *Chem. Rev.* **2000**, 100, 2835-2886.
10. Sato, F., Urabe, H., Okamoto, S. Synthetic reactions mediated by a $Ti(OiPr)_4/2$ *i*PrMgX reagent. *Synlett* **2000**, 753-775.
11. Sato, F., Okamoto, S. The divalent titanium complex $Ti(O\text{-}i\text{-}Pr)_4/2$ *i*-PrMgX as an efficient and practical reagent for fine chemical synthesis. *Adv. Syn. & Catal.* **2001**, 343, 759-784.
12. de Meijere, A., Kozhushkov, S. I., Savchenko, A. I. Titanium-mediated syntheses of cyclopropanols and cyclopropylamines. *Titanium and Zirconium in Organic Synthesis* **2002**, 390-434.
13. Sato, F., Urabe, H. Titanium(II) alkoxides in organic synthesis. *Titanium and Zirconium in Organic Synthesis* **2002**, 319-354.
14. Kulinkovich, O. G. The Chemistry of Cyclopropanols. *Chem. Rev.* **2003**, 103, 2597-2632.
15. de Meijere, A., Kozhushkov, S. I., Savchenko, A. I. Titanium-mediated syntheses of cyclopropylamines. *J. Organomet. Chem.* **2004**, 689, 2033-2055.
16. Takeda, K., Nakatani, J., Nakamura, H., Sako, K., Yoshii, E., Yamaguchi, K. Formation of 1,2-cyclopropanediols in the reaction of acylsilanes with ketone enolates. *Synlett* **1993**, 841-843.
17. Corey, E. J., Rao, S. A., Noe, M. C. Catalytic Diastereoselective Synthesis of Cis-1,2-Disubstituted Cyclopropanols from Esters Using a Vicinal Dicarbanion Equivalent. *J. Am. Chem. Soc.* **1994**, 116, 9345-9346.
18. Chaplinski, V., de Meijere, A. Cyclopropyl building blocks for organic synthesis. 33. A versatile new preparation of cyclopropylamines from acid dialkylamides. *Angew. Chem., Int. Ed. Engl.* **1996**, 35, 413-414.
19. Lee, J., Kim, H., Cha, J. K. A New Variant of the Kulinkovich Hydroxycyclopropanation. Reductive Coupling of Carboxylic Esters with Terminal Olefins. *J. Am. Chem. Soc.* **1996**, 118, 4198-4199.
20. Lee, J., Cha, J. K. Facile Preparation of Cyclopropylamines from Carboxamides. *J. Org. Chem.* **1997**, 62, 1584-1585.
21. Takeda, K., Ubayama, H., Sano, A., Yoshii, E., Koizumi, T. Comparing α-carbanion-stabilizing ability of substituents using the Brook rearrangement. *Tetrahedron Lett.* **1998**, 39, 5243-5246.
22. Hamada, T., Suzuki, D., Urabe, H., Sato, F. Titanium Alkoxide-Based Method for Stereoselective Synthesis of Functionalized Conjugated Dienes. *J. Am. Chem. Soc.* **1999**, 121, 7342-7344.
23. Ollero, L., Mentink, G., Rutjes, F. P. J. T., Speckamp, W. N., Hiemstra, H. A Kulinkovich Entry into Tertiary N-Acyliminium Ion Chemistry. *Org. Lett.* **1999**, 1, 1331-1334.
24. Wu, Y.-D., Yu, Z.-X. A Theoretical Study on the Mechanism and Diastereoselectivity of the Kulinkovich Hydroxycyclopropanation Reaction. *J. Am. Chem. Soc.* **2001**, 123, 5777-5786.
25. Epstein, O. L., Savchenko, A. I., Kulinkovich, O. G. $Titanium^{(IV)}$ isopropoxide-catalyzed reaction of alkylmagnesium halides with ethyl acetate in the presence of styrene. Non-hydride mechanism of ligand exchange in the titanacyclopropanes. *Tetrahedron Lett.* **1999**, 40, 5935-5938.
26. Epstein, O. L., Savchenko, A. I., Kulinkovich, O. G. On the mechanism of titanium-catalyzed cyclopropanation of esters with aliphatic organomagnesium compounds. Deuterium distribution in the reaction products of $(CD_3)_2CHMgBr$ with ethyl 3-chloropropionate in the presence of titanium tetraisopropoxide. *Russ. Chem. Bull.* **2000**, 49, 378-380.
27. Eisch, J. J., Gitua, J. N., Otieno, P. O., Shi, X. Carbon-carbon bond formation via oxidative-addition processes of titanium(II) reagents with π-bonded organic substrates. Reactivity modifications by Lewis acids and Lewis bases Part 22. Organic chemistry of subvalent transition metal complexes. *J. Organomet. Chem.* **2001**, 624, 229-238.
28. Eisch, J. J., Adeosun, A. A., Gitua, J. N. Organic chemistry of subvalent transition metal complexes, 27. Mechanism of the kulinkovich cyclopropanol synthesis: Transfer-epititanation of the alkene in generating the key titanacyclopropane intermediate. *Eur. J. Org. Chem.* **2003**, 4721-4727.
29. Esposito, A., Piras, P. P., Ramazzotti, D., Taddei, M. First Stereocontrolled Synthesis of (S)-Cleonin and Related Cyclopropyl-Substituted Amino Acids. *Org. Lett.* **2001**, 3, 3273-3275.
30. Cao, B., Xiao, D., Joullie, M. M. Synthesis of Bicyclic Cyclopropylamines by Intramolecular Cyclopropanation of N-Allylamino Acid Dimethylamides. *Org. Lett.* **1999**, 1, 1799-1801.
31. Lee, H. B., Sung, M. J., Blackstock, S. C., Cha, J. K. Radical Cation-Mediated Annulation. Stereoselective Construction of Bicyclo[5.3.0]decan-3-ones by Aerobic Oxidation of Cyclopropylamines. *J. Am. Chem. Soc.* **2001**, 123, 11322-11324.
32. Lee, J., Kim, H., Cha, J. K. Diastereoselective Synthesis of cis-1,2-Dialkenylcyclopropanols and Subsequent Oxy-Cope Rearrangement. *J. Am. Chem. Soc.* **1995**, 117, 9919-9920.

Related reactions: Negishi cross-coupling, Stille cross-coupling, Suzuki cross-coupling;

1. Corriu, R. J. P., Masse, J. P. Activation of Grignard reagents by transition-metal complexes. New and simple synthesis of trans-stilbenes and polyphenyls. *J. Chem. Soc., Chem. Commun.* **1972**, 144.
2. Tamao, K., Kiso, Y., Sumitani, K., Kumada, M. Alkyl group isomerization in the cross-coupling reaction of secondary alkyl Grignard reagents with organic halides in the presence of nickel-phosphine complexes as catalysts. *J. Am. Chem. Soc.* **1972**, 94, 9268-9269.
3. Tamao, K., Sumitani, K., Kumada, M. Selective carbon-carbon bond formation by cross-coupling of Grignard reagents with organic halides. Catalysis by nickel-phosphine complexes. *J. Am. Chem. Soc.* **1972**, 94, 4374-4376.
4. Tamao, K., Sumitani, K., Kiso, Y., Zembayashi, M., Fujioka, A., Kodama, S., Nakajima, I., Minato, A., Kumada, M. Nickel-phosphine complex-catalyzed Grignard coupling. I. Cross-coupling of alkyl, aryl, and alkenyl Grignard reagents with aryl and alkenyl halides: general scope and limitations. *Bull. Chem. Soc. Jpn.* **1976**, 49, 1958-1969.
5. Luh, T.-Y. Transition metal-catalyzed cross-coupling reactions of unactivated aliphatic C-X bonds. *Rev. on Heteroa. Chem.* **1996**, 15, 61-82.
6. Negishi, E.-I., Liu, F. Palladium- or nickel-catalyzed cross-coupling with organometals containing zinc, magnesium, aluminum, and zirconium. in *Metal-Catalyzed Cross-Coupling Reactions* (eds. Diederich, F.,Stang, P. J.), 1-47 (Wiley-VCH, Weinheim, Germany, **1998**).
7. De Vries, J. G., De Vries, A. H. M., Tucker, C. E., Miller, J. A. Palladium catalysis in the production of pharmaceuticals. *Innovations in Pharmaceutical Technology* **2001**, 01, 125-126, 128, 130.
8. Anastasia, L., Negishi, E.-i. Palladium-catalyzed aryl-aryl coupling. *Handbook of Organopalladium Chemistry for Organic Synthesis* **2002**, 1, 311-334.
9. Anctil, E. J. G., Snieckus, V. The directed ortho metalation-cross coupling symbiosis. Regioselective methodologies for biaryls and heterobiaryls. Deployment in aromatic and heteroaromatic natural product synthesis. *J. Organomet. Chem.* **2002**, 653, 150-160.
10. Banno, T., Hayakawa, Y., Umeno, M. Some applications of the Grignard cross-coupling reaction in the industrial field. *J. Organomet. Chem.* **2002**, 653, 288-291.
11. Hayashi, T. Palladium-catalyzed asymmetric cross-coupling. *Handbook of Organopalladium Chemistry for Organic Synthesis* **2002**, 1, 791-806.
12. Hillier, A. C., Grasa, G. A., Viciu, M. S., Lee, H. M., Yang, C., Nolan, S. P. Catalytic cross-coupling reactions mediated by palladium/nucleophilic carbene systems. *J. Organomet. Chem.* **2002**, 653, 69-82.
13. Huo, S., Negishi, E.-i. Palladium-catalyzed alkenyl-aryl, aryl-alkenyl, and alkenyl-alkenyl coupling reactions. *Handbook of Organopalladium Chemistry for Organic Synthesis* **2002**, 1, 335-408.
14. Murahashi, S.-I. Palladium-catalyzed cross-coupling reaction of organic halides with Grignard reagents, organolithium compounds and heteroatom nucleophiles. *J. Organomet. Chem.* **2002**, 653, 27-33.
15. Negishi, E.-i. A genealogy of Pd-catalyzed cross-coupling. *J. Organomet. Chem.* **2002**, 653, 34-40.
16. Tamao, K. Discovery of the cross-coupling reaction between Grignard reagents and C(sp2) halides catalyzed by nickel-phosphine complexes. *J. Organomet. Chem.* **2002**, 653, 23-26.
17. Tamao, K., Miyaura, N. Introduction to cross-coupling reactions. *Top. Curr. Chem.* **2002**, 219, 1-9.
18. Herrmann, W. A., Ofele, K., von Preysing, D., Schneider, S. K. Phospha-palladacycles and N-heterocyclic carbene palladium complexes: efficient catalysts for CC-coupling reactions. *J. Organomet. Chem.* **2003**, 687, 229-248.
19. Cassar, L. Synthesis of aryl- and vinyl-substituted acetylene derivatives by the use of nickel and palladium complexes. *J. Organomet. Chem.* **1975**, 93, 253-257.
20. Yamamura, M., Moritani, I., Murahashi, S.-I. Reaction of s-vinylpalladium complexes with alkyllithiums. Stereospecific synthesis of olefins from vinyl halides and alkyllithiums. *J. Organomet. Chem.* **1975**, 91, C39-C42.
21. Fauvarque, J. F., Jutand, A. Reaction of various nucleophiles with organopalladium compounds. *Bull. Soc. Chim. Fr.* **1976**, 765-770.
22. Sekiya, A., Ishikawa, N. Palladium metal-catalyzed cross-coupling of aryl iodides with arylmagnesium bromides. Synthesis of fluorobiphenyls. *J. Organomet. Chem.* **1977**, 125, 281-290.
23. Dang, H. P., Linstrumelle, G. An efficient stereospecific synthesis of olefins by the palladium-catalyzed reaction of Grignard reagents with alkenyl iodides. *Tetrahedron Lett.* **1978**, 191-194.
24. Hayashi, T., Konishi, M., Kumada, M. Dichloro[1,1'-bis(diphenylphosphino)ferrocene]palladium(II): an effective catalyst for cross-coupling reaction of a secondary alkyl Grignard reagent with organic halides. *Tetrahedron Lett.* **1979**, 1871-1874.
25. Kondo, K., Murahashi, S. Selective transformation of organoboranes to Grignard reagents by using pentane-1,5-di(magnesium bromide). Synthesis of the pheromones of Southern armyworm moth and Douglas fir tussock moth. *Tetrahedron Lett.* **1979**, 1237-1240.
26. Murahashi, S., Yamamura, M., Yanagisawa, K., Mita, N., Kondo, K. Stereoselective synthesis of alkenes and alkenyl sulfides from alkenyl halides using palladium and ruthenium catalysts. *J. Org. Chem.* **1979**, 44, 2408-2417.
27. Huang, J., Nolan, S. P. Efficient Cross-Coupling of Aryl Chlorides with Aryl Grignard Reagents (Kumada Reaction) Mediated by a Palladium/Imidazolium Chloride System. *J. Am. Chem. Soc.* **1999**, 121, 9889-9890.
28. Fuerstner, A., Leitner, A., Mendez, M., Krause, H. Iron-Catalyzed Cross-Coupling Reactions. *Journal of the American Chemical Society* **2002**, 124, 13856-13863.
29. Furstner, A., Leitner, A. Iron-catalyzed cross-coupling reactions of alkyl-Grignard reagents with aryl chlorides, tosylates, and triflates. *Angewandte Chemie, International Edition* **2002**, 41, 609-612.
30. Korn, T. J., Cahiez, G., Knochel, P. New cobalt-catalyzed cross-coupling reactions of heterocyclic chlorides with aryl and heteroaryl magnesium halides. *Synlett* **2003**, 1892-1894.
31. Tamura, M., Kochi, J. Coupling of Grignard reagents with organic halides. *Synthesis* **1971**, 303-305.
32. Tamura, M., Kochi, J. K. Vinylation of Grignard reagents. Catalysis by iron. *J. Am. Chem. Soc.* **1971**, 93, 1487-1489.
33. Negishi, E., Takahashi, T., Baba, S., Van Horn, D. E., Okukado, N. Nickel- or palladium-catalyzed cross coupling. 31. Palladium- or nickel-catalyzed reactions of alkenylmetals with unsaturated organic halides as a selective route to arylated alkenes and conjugated dienes: scope, limitations, and mechanism. *J. Am. Chem. Soc.* **1987**, 109, 2393-2401.
34. Loar, M. K., Stille, J. K. Mechanisms of 1,1-reductive elimination from palladium: coupling of styrylmethylpalladium complexes. *J. Am. Chem. Soc.* **1981**, 103, 4174-4181.
35. Negishi, E., Takahashi, T., Akiyoshi, K. Bis(triphenylphosphine)palladium: its generation, characterization, and reactions. *J. Chem. Soc., Chem. Commun.* **1986**, 1338-1339.
36. Amatore, C., Azzabi, M., Jutand, A. Role and effects of halide ions on the rates and mechanisms of oxidative addition of iodobenzene to low-ligated zerovalent palladium complexes Pd0(PPh3)2. *J. Am. Chem. Soc.* **1991**, 113, 8375-8384.
37. Amatore, C., Jutand, A., Suarez, A. Intimate mechanism of oxidative addition to zerovalent palladium complexes in the presence of halide ions and its relevance to the mechanism of palladium-catalyzed nucleophilic substitutions. *J. Am. Chem. Soc.* **1993**, 115, 9531-9541.
38. Hoelzer, B., Hoffmann, R. W. Kumada-Corriu coupling of Grignard reagents, probed with a chiral Grignard reagent. *Chem. Commun.* **2003**, 732-733.
39. Liu, P., Jacobsen, E. N. Total Synthesis of (+)-Ambruticin. *J. Am. Chem. Soc.* **2001**, 123, 10772-10773.
40. Rivkin, A., Njardarson, J. T., Biswas, K., Chou, T.-C., Danishefsky, S. J. Total Syntheses of [17]- and [18]Dehydrodesoxyepothilones B via a Concise Ring-Closing Metathesis-Based Strategy: Correlation of Ring Size with Biological Activity in the Epothilone Series. *J. Org. Chem.* **2002**, 67, 7737-7740.
41. William, A. D., Kobayashi, Y. Synthesis of Tetrahydrocannabinols Based on an Indirect 1,4-Addition Strategy. *J. Org. Chem.* **2002**, 67, 8771-8782.
42. Ikunaka, M., Maruoka, K., Okuda, Y., Ooi, T. A Scalable Synthesis of (R)-3,5-Dihydro-4H-dinaphth[2,1-c:1'2'-e]azepine. *Org. Process Res. Dev.* **2003**, 7, 644-648.

Larock Indole Synthesis 260

Related reactions: Bartoli indole synthesis, Fischer indole synthesis, Madelung indole synthesis, Nenitzescu indole synthesis;

1. Larock, R. C., Yum, E. K. Synthesis of indoles via palladium-catalyzed heteroannulation of internal alkynes. *J. Am. Chem. Soc.* **1991**, 113, 6689-6690.
2. Cacchi, S. Heterocycles via cyclization of alkynes promoted by organopalladium complexes. *J. Organomet. Chem.* **1999**, 576, 42-64.
3. Larock, R. C. Palladium-catalyzed annulation. *Pure Appl. Chem.* **1999**, 71, 1435-1442.
4. Larock, R. C. Palladium-catalyzed annulation. *J. Organomet. Chem.* **1999**, 576, 111-124.
5. Gribble, G. W. Recent developments in indole ring synthesis-methodology and applications. *Perkin 1* **2000**, 1045-1075.
6. Poli, G., Giambastiani, G., Heumann, A. Palladium in Organic Synthesis: Fundamental Transformations and Domino Processes. *Tetrahedron* **2000**, 56, 5959-5989.
7. Battistuzzi, G., Cacchi, S., Fabrizi, G. The aminopalladation/reductive elimination domino reaction in the construction of functionalized indole rings. *Eur. J. Org. Chem.* **2002**, 2671-2681.
8. Cacchi, S., Fabrizi, G., Parisi, L. M. Nitrogen-containing heterocycles via palladium-catalyzed reaction of alkynes with organic halides or triflates. *Heterocycles* **2002**, 58, 667-682.
9. Undheim, K. Heteroaromatics via palladium-catalyzed cross-coupling. *Handbook of Organopalladium Chemistry for Organic Synthesis* **2002**, 1, 409-492.
10. Alonso, F., Beletskaya, I. P., Yus, M. Transition-Metal-Catalyzed Addition of Heteroatom-Hydrogen Bonds to Alkynes. *Chem. Rev.* **2004**, 104, 3079-3159.
11. Kirsch, G., Hesse, S., Comel, A. Synthesis of five- and six-membered heterocycles through palladium-catalyzed reactions. *Current Organic Synthesis* **2004**, 1, 47-63.
12. Larock, R. C., Zenner, J. M. Enantioselective, Palladium-Catalyzed Hetero- and Carboannulation of Allenes Using Functionally-Substituted Aryl and Vinylic Iodides. *J. Org. Chem.* **1995**, 60, 482-483.
13. Xu, L., Lewis, I. R., Davidsen, S. K., Summers, J. B. Transition metal catalyzed synthesis of 5-azaindoles. *Tetrahedron Lett.* **1998**, 39, 5159-5162.
14. Roesch, K. R., Larock, R. C. Synthesis of Isoindolo[2,1-a]indoles by the Palladium-Catalyzed Annulation of Internal Alkynes. *Org. Lett.* **1999**, 1, 1551-1553.
15. Roesch, K. R., Larock, R. C. Synthesis of isoindolo[2,1-a]indoles by the palladium-catalyzed annulation of internal acetylenes. *J. Org. Chem.* **2001**, 66, 412-420.
16. Lee, M. S., Yum, E. K. Synthesis of trisubstituted 6-azaindoles via palladium-catalyzed heteroannulation. *Bull. Korean Chem. Soc.* **2002**, 23, 535-536.
17. Nishikawa, T., Wada, K., Isobe, M. Synthesis of novel α-C-glycosylamino acids and reverse regioselectivity in Larock's heteroannulation for the synthesis of the indole nucleus. *Biosci. Biotechnol. Biochem.* **2002**, 66, 2273-2278.
18. Huang, Q., Larock, R. C. Synthesis of Substituted Naphthalenes and Carbazoles by the Palladium-Catalyzed Annulation of Internal Alkynes. *J. Org. Chem.* **2003**, 68, 7342-7349.
19. Kang, S. S., Yum, E. K., Sung, N.-d. Synthesis of pyridopyrrolo[2,1-a]isoindoles by palladium-catalyzed annulation. *Heterocycles* **2003**, 60, 2727-2736.
20. Chae, J., Konno, T., Ishihara, T., Yamanaka, H. A facile synthesis of various fluorine-containing indole derivatives via palladium-catalyzed annulation of internal alkynes. *Chem. Lett.* **2004**, 33, 314-315.
21. Larock, R. C., Yum, E. K., Refvik, M. D. Synthesis of 2,3-Disubstituted Indoles via Palladium-Catalyzed Annulation of Internal Alkynes. *J. Org. Chem.* **1998**, 63, 7652-7662.
22. Zhou, H., Liao, X., Cook, J. M. Regiospecific, Enantiospecific Total Synthesis of the 12-Alkoxy-Substituted Indole Alkaloids, (+)-12-Methoxy-Na-methylvellosimine, (+)-12-Methoxyaffinisine, and (-)-Fuchsiaefoline. *Org. Lett.* **2004**, 6, 249-252.

Ley Oxidation 262

Related reactions: Corey-Kim oxidation, Dess-Martin oxidation, Ley oxidation, Oppenauer oxidation, Pfitzner-Moffatt oxidation, Jones oxidation;

1. Dengel, A. C., Hudson, R. A., Griffith, W. P. Tetrabutylammonium perruthenate: a new mild oxidant for alcohols. *Transition Metal Chemistry (Dordrecht, Netherlands)* **1985**, 10, 98-99.
2. Griffith, W. P., Ley, S. V., Whitcombe, G. P., White, A. D. Preparation and use of tetrabutylammonium perruthenate (TBAP reagent) and tetrapropylammonium perruthenate (TPAP reagent) as new catalytic oxidants for alcohols. *J. Chem. Soc., Chem. Commun.* **1987**, 1625-1627.
3. Griffith, W. P., Ley, S. V. TPAP: tetra-n-propylammonium perruthenate, a mild and convenient oxidant for alcohols. *Aldrichimica Acta* **1990**, 23, 13-19.
4. Ley, S. V. Oxidation Adjacent to Oxygen of Alcohols by Other Methods. in *Comp. Org. Synth.* (eds. Trost, B. M.,Fleming, I.), *7*, 305-327 (Pergamon, Oxford, **1991**).
5. Che, C. M., Yam, V. W. W. High-valent complexes of ruthenium and osmium. *Advances in Inorganic Chemistry* **1992**, 39, 233-325.
6. Griffith, W. P. Ruthenium oxo complexes as organic oxidants. *Chem. Soc. Rev.* **1992**, 21, 179-185.
7. Ley, S. V., Norman, J., Griffith, W. P., Marsden, S. P. Tetrapropylammonium perruthenate, $Pr_4N^+RuO_4^-$, TPAP: a catalytic oxidant for organic synthesis. *Synthesis* **1994**, 639-666.
8. Friedrich, H. B. The oxidation of alcohols to aldehydes or ketones high oxidation state ruthenium compounds as catalysts. *Platinum Metals Review* **1999**, 43, 94-102.
9. Langer, P. Tetra-n-propyl ammonium perruthenate (TPAP) - an efficient and selective reagent for oxidation reactions in solution and on the solid phase. *J. Prakt. Chem.* **2000**, 342, 728-730.
10. Sheldon, R. A., Arends, I. W. C. E. Catalytic oxidations of alcohols. *Catal. Metal Compl.* **2003**, 26, 123-155.
11. Che, C.-M., Lau, T.-C. Ruthenium and osmium: high oxidation states. *Comprehensive Coordination Chemistry II* **2004**, 5, 733-847.
12. Zhan, B.-Z., Thompson, A. Recent developments in the aerobic oxidation of alcohols. *Tetrahedron* **2004**, 60, 2917-2935.
13. Bailey, A. J., Griffith, W. P., Mostafa, S. I., Sherwood, P. A. Studies on transition-metal oxo and nitrido complexes. 13. Perruthenate and ruthenate anions as catalytic organic oxidants. *Inorg. Chem.* **1993**, 32, 268-271.
14. Hinzen, B., Ley, S. V. Polymer supported perruthenate: a new oxidant for clean organic synthesis. *J. Chem. Soc., Perkin Trans. 1* **1997**, 1907-1908.
15. Lenz, R., Ley, S. V. Tetra-n-propylammonium perruthenate (TPAP)-catalyzed oxidations of alcohols using molecular oxygen as a co-oxidant. *J. Chem. Soc., Perkin Trans. 1* **1997**, 3291-3292.
16. Marko, I. E., Giles, P. R., Tsukazaki, M., Chelle-Regnaut, I., Urch, C. J., Brown, S. M. Efficient, Aerobic, Ruthenium-Catalyzed Oxidation of Alcohols into Aldehydes and Ketones. *J. Am. Chem. Soc.* **1997**, 119, 12661-12662.
17. Yates, M. H. One-pot conversion of olefins to carbonyl compounds by hydroboration/NMO-TPAP oxidation. *Tetrahedron Lett.* **1997**, 38, 2813-2816.
18. Hinzen, B., Lenz, R., Ley, S. V. Polymer supported perruthenate (PSP). Clean oxidation of primary alcohols to carbonyl compounds using oxygen as cooxidant. *Synthesis* **1998**, 977-979.
19. Lee, D. G., Congson, L. N. Kinetics and mechanism of the oxidation of alcohols by ruthenate and perruthenate ions. *Can. J. Chem.* **1990**, 68, 1774-1779.

20. Lee, D. G., Wang, Z., Chandler, W. D. Autocatalysis during the reduction of tetra-n-propylammonium perruthenate by 2-propanol. *J. Org. Chem.* **1992**, 57, 3276-3277.
21. Tony, K. J., Mahadevan, V., Rajaram, J., Swamy, C. S. Oxidation of secondary alcohols by N-methylmorpholine-N-oxide (NMO) catalyzed by a trans-dioxo ruthenium(VI) complex or perruthenate complex: a kinetic study. *React. Kinet. Catal. Lett.* **1997**, 62, 105-116.
22. Wang, Z., Chandler, W. D., Lee, D. G. Mechanisms for the oxidation of secondary alcohols by dioxoruthenium(VI) complexes. *Can. J. Chem.* **1998**, 76, 919-928.
23. Hasan, M., Musawir, M., Davey, P. N., Kozhevnikov, I. V. Oxidation of primary alcohols to aldehydes with oxygen catalyzed by tetrapropylammonium perruthenate. *J. Mol. Catal. A: Chemical* **2002**, 180, 77-84.
24. Mucientes, A. E., Santiago, F., Almena, M. C., Poblete, F. J., Rodriguez-Cervantes, A. M. Kinetic study of the ruthenium(VI)-catalyzed oxidation of benzyl alcohol by alkaline hexacyanoferrate(III). *Int. J. Chem. Kinet.* **2002**, 34, 421-429.
25. Davis, B. G., Emmerson, D. P. G., Williams, J. A. G. Oxidation and reduction. *Org. React. Mech.* **2003**, 217-252.
26. Keck, G. E., Knutson, C. E., Wiles, S. A. Total Synthesis of the Immunosuppressant (-)-Pironetin (PA48153C). *Org. Lett.* **2001**, 3, 707-710.
27. Ward, D. E., Gai, Y., Qiao, Q. A General Approach to Cyathin Diterpenes. Total Synthesis of Allocyathin B3. *Org. Lett.* **2000**, 2, 2125-2127.
28. Nielsen, T. E., Tanner, D. Stereoselective Synthesis of (E)-β-Tributylstannyl-α,β-unsaturated Ketones: Construction of a Key Intermediate for the Total Synthesis of Zoanthamine. *J. Org. Chem.* **2002**, 67, 6366-6371.
29. Hu, T., Panek, J. S. Total synthesis of (-)-Motuporin. *J. Org. Chem.* **1999**, 64, 3000-3001.

Lieben Haloform Reaction 264

1. Serullas. Effect of iodine on the basic solution of alcohols. *Ann. chim. (Paris) [13]* **1822**, 20, 165.
2. von Liebig, J. Reaction of alcohols with chlorine gas. *Ann. Phys. Chem. Ser. 2* **1831**, 23, 444.
3. Lieben, A. The formation of iodoform and the application of this reaction in the chemical analysis. *Liebigs Ann. Chem.* **1870**, Supp. 7, 218-236.
4. Fuson, R. C., Bull, B. A. The haloform reaction. *Chem. Rev.* **1934**, 15, 275-309.
5. Turney, T. A., Seelye, R. N. The Iodoform Reaction. *J. Chem. Educ.* **1959**, 36, 572-574.
6. Chakrabartty, S. K. Alkaline Hypohalite Oxidations. in *Oxidation in Organic Chemistry, Part C* (ed. Trahanovsky, W. S.), 343-370 (Academic Press, New York, **1978**).
7. Olofson, R. A. New, useful reactions of novel haloformates and related reagents. *Pure Appl. Chem.* **1988**, 60, 1715-1724.
8. Hrutfiord, B. F., Negri, A. R. Chemistry of chloroform formation in pulp bleaching: a review. *Tappi J.* **1990**, 73, 219-225.
9. Hashmi, M. H., Mahmood ul, H., Lateef, A. B. A modification of the haloform reaction. *Pak. J. Sci. Res.* **1963**, 15, 7-10.
10. Del Buttero, P., Maiorana, S. Haloform reaction on β-sulfonyl methyl ketones. *Gazz. Chim. Ital.* **1973**, 103, 809-812.
11. Kajigaeshi, S., Kakinami, T. Bromination and oxidation with benzyltrimethylammonium tribromide. *Ind. Chem. Library* **1995**, 7, 29-48.
12. Rothenberg, G., Sasson, Y. Extending the haloform reaction to non-methyl ketones: oxidative cleavage of cycloalkanones to dicarboxylic acids using sodium hypochlorite under phase transfer catalysis conditions. *Tetrahedron* **1996**, 52, 13641-13648.
13. Madler, M. M., Klucik, J., Soell, P. S., Brown, C. W., Liu, S., Berlin, K. D., Benbrook, D. M., Birckbichler, P. J., Nelson, E. C. Lithium hypochlorite-clorox as a novel oxidative mixture for methyl ketones and methyl carbinols. *Org. Prep. Proced. Int.* **1998**, 30, 230-234.
14. Gillis, B. T. Iodoform test. *J. Org. Chem.* **1959**, 24, 1027-1029.
15. Bartlett, P. D. Enolization as directed by acid and basic catalysts. II. Enolic mechanism of the haloform reaction. *J. Am. Chem. Soc.* **1934**, 56, 967-969.
16. Aston, J. G., Newkirk, J. D., Dorsky, J., Jenkins, D. M. Mechanism of the haloform reaction. Preparation of mixed haloforms. *J. Am. Chem. Soc.* **1942**, 64, 1413-1416.
17. Guthrie, J. P., Cossar, J. The chlorination of acetone: a complete kinetic analysis. *Can. J. Chem.* **1986**, 64, 1250-1266.
18. Hailes, H. C., Isaac, B., Hashim Javaid, M. Synthesis of methyl epijasmonate and cis-3-(2-oxopropyl)-2-(pent-2Z-enyl)-cyclopentan-1-one. *Tetrahedron* **2001**, 57, 10329-10333.
19. Storm, J. P., Andersson, C.-M. Iron-Mediated Synthetic Routes to Unsymmetrically Substituted, Sterically Congested Benzophenones. *J. Org. Chem.* **2000**, 65, 5264-5274.
20. Bennasar, M. L., Vidal, B., Bosch, J. Total Synthesis of Indole Alkaloids of the Ervatamine Group. A Biomimetic Approach. *J. Org. Chem.* **1996**, 61, 1916-1917.
21. Ihara, M., Taniguchi, T., Tokunaga, Y., Fukumoto, K. Ring Contraction of Cyclobutanes and a Novel Cascade Reaction: Application to Synthesis of (±)-Anthoplalone and (±)-Lepidozene. *J. Org. Chem.* **1994**, 59, 8092-8100.

Lossen Rearrangement 266

*Related reactions: **Curtius rearrangement, Hofmann rearrangement, Schmidt reaction;***

1. Lossen, W. A method for the conversion of aromatic carboxylic acids to the corresponding amides. *Liebigs Ann. Chem.* **1869**, 150, 313-325.
2. Lossen, W. The structure of hydroxylamines and their amide derivatives. *Liebigs Ann. Chem.* **1872**, 175, 271.
3. Lossen, W. Benzoyl derivatives of hydroxylamines. *Liebigs Ann. Chem.* **1872**, 161, 347-362.
4. Yale, H. L. The hydroxamic acids. *Chem. Rev.* **1943**, 33, 209-256.
5. Bauer, L., Exner, O. Chemistry of hydroxamic acids and N-hydroxyimides. *Angew. Chem.* **1974**, 86, 419-428.
6. Lipczynska-Kochany, E. Some new aspects of hydroxamic acid chemistry. *Pr. Nauk. - Politech. Warsz., Chem.* **1988**, 46, 3-98.
7. Shioiri, T. Degradation Reactions. in *Comp. Org. Synth.* (eds. Trost, B. M.,Fleming, I.), *6*, 795-828 (Pergamon, Oxford, **1991**).
8. Romine, J. L. Bis-protected hydroxylamines as reagents in organic synthesis. A review. *Org. Prep. Proced. Int.* **1996**, 28, 249-288.
9. Boche, G., Lohrenz, J. C. W. The Electrophilic Nature of Carbenoids, Nitrenoids, and Oxenoids. *Chemical Reviews (Washington, D. C.)* **2001**, 101, 697-756.
10. Hoare, D. G., Olson, A., Koshland, D. E., Jr. The reaction of hydroxamic acids with water-soluble carbodiimides. A lossen rearrangement. *J. Am. Chem. Soc.* **1968**, 90, 1638-1643.
11. Daniher, F. A. Sulfation of hydroxamic acids. *J. Org. Chem.* **1969**, 34, 2908-2911.
12. Bittner, S., Grinberg, S., Kartoon, I. Novel variation of the Lossen rearrangement. *Tetrahedron Lett.* **1974**, 1965-1968.
13. King, F. D., Pike, S., Walton, D. R. M. Silylhydroxylamines as reagents for high-yield RCOCl->RNCO conversions. *J. Chem. Soc., Chem. Commun.* **1978**, 351-352.
14. Sheradsky, T., Avramovici-Grisaru, S. 1,2-Aryl migration onto acylnitrenium ions. *Tetrahedron Lett.* **1978**, 2325-2326.
15. Aly, N. F., Abd-El Aleem, A. E. A. H., Eltamany, S. H., Abou-Hadeed, K. Base-catalyzed Lossen rearrangement and acid-catalyzed Beckmann rearrangement with N-(arylsulfonyloxy)naphthalene-2,3-dicarboximides. *Egypt. J. Chem.* **1990**, 31, 133-140.
16. Salomon, C. J., Breuer, E. Spontaneous Lossen Rearrangement of (Phosphonoformyl)hydroxamates. The Migratory Aptitude of the Phosphonyl Group. *J. Org. Chem.* **1997**, 62, 3858-3861.
17. Stafford, J. A., Gonzales, S. S., Barrett, D. G., Suh, E. M., Feldman, P. L. Degradative Rearrangements of N-(t-Butyloxycarbonyl)-O-methanesulfonylhydroxamic Acids: A Novel, Reagent-Based Alternative to the Lossen Rearrangement. *J. Org. Chem.* **1998**, 63, 10040-10044.
18. Zalipsky, S. Alkyl succinimidyl carbonates undergo Lossen rearrangement in basic buffers. *Chem. Commun.* **1998**, 69-70.
19. Anilkumar, R., Chandrasekhar, S., Sridhar, M. N,O-Bis(ethoxycarbonyl)hydroxylamine: a convenient reagent for the Lossen transformation. *Tetrahedron Lett.* **2000**, 41, 5291-5293.

20. Linke, S., Tisue, G. T., Lwowski, W. Curtius and Lossen rearrangements. II. Pivaloyl azide. *Journal of the American Chemical Society* **1967**, 89, 6308-6310.
21. Joensson, N. A., Moses, P. 3,3-dialkylindolin-2-ones and 3,3-dialkylisoindolin-1-ones. 2. Hofmann and Lossen degradation of 4,4-dialkyl-1,3-dioxo-1,2,3,4-tetrahydroisoquinolines (4,4-dialkylhomophthalimides). Mechanistic study. *Acta Chem. Scand.* **1974**, 28, 441-448.
22. Groutas, W. C., Stanga, M. A., Brubaker, M. J. ^{13}C NMR evidence for an enzyme-induced Lossen rearrangement in the mechanism-based inactivation of α-chymotrypsin by 3-benzyl-N-((methylsulfonyl)oxy)succinimide. *J. Am. Chem. Soc.* **1989**, 111, 1931-1932.
23. Adams, G. W., Bowie, J. H., Hayes, R. N. Does the Lossen rearrangement occur in the gas phase? *J. Chem. Soc., Perkin Trans. 2* **1991**, 689-693.
24. Ohmoto, K., Yamamoto, T., Horiuchi, T., Kojima, T., Hachiya, K., Hashimoto, S., Kawamura, M., Nakai, H., Toda, M. Improved synthesis of a new nonpeptidic inhibitor of human neutrophil elastase. *Synlett* **2001**, 299-301.
25. Marzoni, G., Varney, M. D. An Improved Large-Scale Synthesis of Benz[cd]indol-2(1H)-one and 5-Methylbenz[cd]indol-2(1H)-one. *Org. Process Res. Dev.* **1997**, 1, 81-84.
26. Needs, P. W., Rigby, N. M., Ring, S. G., MacDougall, A. J. Specific degradation of pectins via a carbodiimide-mediated Lossen rearrangement of methyl esterified galacturonic acid residues. *Carbohydr. Res.* **2001**, 333, 47-58.

Luche Reduction 268

Related reactions: Corey-Bakshi-Shibata (CBS) reduction, Noyori asymmetric hydrogenation;

1. Luche, J. L. Lanthanides in organic chemistry. 1. Selective 1,2 reductions of conjugated ketones. *J. Am. Chem. Soc.* **1978**, 100, 2226-2227.
2. Luche, J. L., Rodriguez-Hahn, L., Crabbe, P. Reduction of natural enones in the presence of cerium trichloride. *J. Chem. Soc., Chem. Commun.* **1978**, 601-602.
3. Luche, J. L., Gemal, A. L. Lanthanoids in organic synthesis. 5. Selective reductions of ketones in the presence of aldehydes. *J. Am. Chem. Soc.* **1979**, 101, 5848-5849.
4. Gemal, A. L., Luche, J. L. Lanthanoids in organic synthesis. 6. Reduction of α-enones by sodium borohydride in the presence of lanthanoid chlorides: synthetic and mechanistic aspects. *J. Am. Chem. Soc.* **1981**, 103, 5454-5459.
5. Wade, R. C. Catalyzed reductions of organofunctional groups with sodium borohydride. *J. Mol. Catal.* **1983**, 18, 273-297.
6. Kagan, H. B., Namy, J. L. Lanthanides in organic synthesis. *Tetrahedron* **1986**, 42, 6573-6614.
7. Molander, G. A. Application of lanthanide reagents in organic synthesis. *Chem. Rev.* **1992**, 92, 29-68.
8. Seyden-Penne, J. Electrophilic assistance in the reduction of six-membered cyclic ketones by alumino- and borohydrides. *ACS Symp. Ser.* **1996**, 641, 70-83.
9. Nakai, T., Tomooka, K. Lanthanide(III) reagents. *Lewis Acid Reagents* **1999**, 203-223.
10. Periasamy, M., Thirumalaikumar, M. Methods of enhancement of reactivity and selectivity of sodium borohydride for applications in organic synthesis. *J. Organomet. Chem.* **2000**, 609, 137-151.
11. Sumino, Y., Tomisaka, Y., Ogawa, A. Cerium reagents in organic synthesis. *Materials Integration* **2003**, 16, 37-41.
12. Fukuzawa, S., Fujinami, T., Yamauchi, S., Sakai, S. 1,2-Regioselective reduction of α,β-unsaturated carbonyl compounds with lithium aluminum hydride in the presence of lanthanoid salts. *J. Chem. Soc., Perkin Trans. 1* **1986**, 1929-1932.
13. Komiya, S., Tsutsumi, O. Selective 1,2-reduction of α,β-unsaturated carbonyl compounds with $LnCpCl_2(THF)_3/NaBH_4$. *Bull. Chem. Soc. Jpn.* **1987**, 60, 3423-3424.
14. Singh, J., Kaur, I., Kaur, J., Bhalla, A., Kad, G. L. Speedy and regioselective 1,2-reduction of conjugated α,β-unsaturated aldehydes and ketones using $NaBH_4/I_2$. *Synth. Commun.* **2003**, 33, 191-197.
15. Zeynizadeh, B., Shirini, F. Mild and efficient reduction of α,β-unsaturated carbonyl compounds, α-diketones and acyloins with the sodium borohydride/Dowex1-x8 system. *Bull. Korean Chem. Soc.* **2003**, 24, 295-298.
16. Dewar, M. J. S., McKee, M. L. Ground states of molecules. 50. MNDO study of hydroboration and borohydride reduction. Implications concerning cyclic conjugation and pericyclic reactions. *J. Am. Chem. Soc.* **1978**, 100, 7499-7505.
17. Lefour, J. M., Loupy, A. The effect of cations on nucleophilic additions to carbonyl compounds: carbonyl complexation control versus ionic association control. Application to the regioselectivity of addition to α-enones. *Tetrahedron* **1978**, 34, 2597-2605.
18. Ohwada, T. Orbital-Controlled Stereoselections in Sterically Unbiased Cyclic Systems. *Chem. Rev.* **1999**, 99, 1337-1375.
19. Marks, T. J., Kolb, J. R. Covalent transition metal, lanthanide, and actinide tetrahydroborate complexes. *Chem. Rev.* **1977**, 77, 263-293.
20. Cockerill, A. F., Davies, G. L. O., Harden, R. C., Rackham, D. M. Lanthanide shift reagents for nuclear magnetic resonance spectroscopy. *Chem. Rev.* **1973**, 73, 553-588.
21. Hudlicky, T., Rinner, U., Gonzalez, D., Akgun, H., Schilling, S., Siengalewicz, P., Martinot, T. A., Pettit, G. R. Total Synthesis and Biological Evaluation of Amaryllidaceae Alkaloids: Narciclasine, ent-7-Deoxypancratistatin, Regioisomer of 7-Deoxypancratistatin, 10β-epi-Deoxypancratistatin, and Truncated Derivatives. *J. Org. Chem.* **2002**, 67, 8726-8743.
22. Paquette, L. A., Meister, P. G., Friedrich, D., Sauer, D. R. Enantioselective total synthesis of (-)-subergorgic acid. *J. Am. Chem. Soc.* **1993**, 115, 49-56.
23. White, J. D., Shin, H., Kim, T.-S., Cutshall, N. S. Total Synthesis of the Sesquiterpenoid Polyols (±)-Euonyminol and (±)-3,4-Dideoxymaytol, Core Constituents of Esters of the Celastraceae. *J. Am. Chem. Soc.* **1997**, 119, 2404-2419.
24. Kurosu, M., Marcin, L. R., Grinsteiner, T. J., Kishi, Y. Total Synthesis of (±)-Batrachotoxinin A. *J. Am. Chem. Soc.* **1998**, 120, 6627-6628.

Madelung Indole Synthesis 270

Related reactions: Bartoli indole synthesis, Fischer indole synthesis, Larock indole synthesis, Nenitzescu indole synthesis;

1. Madelung, W. *Indole substituted in the α-position.* DE 262327, **1912**,
2. Madelung, W. A new synthesis of substituted indoles. *Ber.* **1912**, 45, 1128-1134.
3. Willette, R. E. Monoazaindoles: the pyrrolopyridines. *Adv. Heterocycl. Chem.* **1968**, 9, 27-105.
4. Brown, R. K. Synthesis of the indole nucleus. in *Chemistry of Heterocyclic Compounds: Indoles Part One* (ed. Houlihan, W. J.), *25*, 385-396 (Wiley, Chichester, **1972**).
5. Cheeseman, G. W. H., Bird, C. W. Synthesis of Five-membered Rings with One Heteroatom. in *Comprehensive Heterocyclic Chemistry* (eds. Katritzky, A. R.,Rees, C. W.), *4*, 89-147 (Pergamon Press, Oxford, **1984**).
6. Gribble, G. W. Recent developments in indole ring synthesis-methodology and applications. *Perkin 1* **2000**, 1045-1075.
7. Joule, J. A. Product class 13: indole and its derivatives. *Science of Synthesis* **2001**, 10, 361-652.
8. Verley, A. Sodamide as a dehydrating agent. Preparation of indole, methylindole and their homologs. *Bull. soc. chim.* **1924**, 35, 1039-1040.
9. Verley, A., Beduwe, J. A general method for the preparation of substituted indole derivatives. *Bull. soc. chim.* **1925**, 37, 189-191.
10. Houlihan, W. J., Parrino, V. A., Uike, Y. Lithiation of N-(2-alkylphenyl)alkanamides and related compounds. A modified Madelung indole synthesis. *J. Org. Chem.* **1981**, 46, 4511-4515.
11. Smith, A. B., III, Visnick, M., Haseltine, J. N., Sprengeler, P. A. Organometallic reagents in synthesis. A new protocol for construction of the indole nucleus. *Tetrahedron* **1986**, 42, 2957-2969.
12. Verboom, W., Orlemans, E. O. M., Berga, H. J., Scheltinga, M. W., Reinhoudt, D. N. Synthesis of dihydro-1H-pyrrolo- and tetrahydropyrido[1,2-a]indoles via a modified Madelung reaction. *Tetrahedron* **1986**, 42, 5053-5064.
13. Orlemans, E. O. M., Schreuder, A. H., Conti, P. G. M., Verboom, W., Reinhoudt, D. N. Synthesis of 3-substituted indoles via a modified Madelung reaction. *Tetrahedron* **1987**, 43, 3817-3826.

14. Hands, D., Bishop, B., Cameron, M., Edwards, J. S., Cottrell, I. F., Wright, S. H. B. A convenient method for the preparation of 5-, 6-, and 7-azaindoles and their derivatives. *Synthesis* **1996**, 877-882.
15. Hughes, I. Application of polymer-bound phosphonium salts as traceless supports for solid phase synthesis. *Tetrahedron Lett.* **1996**, 37, 7595-7598.
16. Miyashita, K., Kondoh, K., Tsuchiya, K., Miyabe, H., Imanishi, T. Novel indole-ring formation by thermolysis of 2-(N-acylamino)benzylphosphonium salts. Effective synthesis of 2-trifluoromethylindoles. *J. Chem. Soc., Perkin Trans. 1* **1996**, 1261-1268.
17. Miyashita, K., Tsuchiya, K., Kondoh, K., Miyabe, H., Imanishi, T. Novel indole-ring construction method for the synthesis of 2-trifluoromethylindoles. *Heterocycles* **1996**, 42, 513-516.
18. Kim, G., Keum, G. A new route to quinolone and indole skeletons via ketone and ester-imide cyclodehydration reactions. *Heterocycles* **1997**, 45, 1979-1988.
19. Takahashi, M., Suga, D. Synthesis of 2-aryl-3-(arylsulfonyl)indoles and 2-anilino-3-(arylsulfonyl)indoles from 2-[(arylsulfonyl)methyl]anilines using the aza-Wittig reaction of iminophosphoranes. *Synthesis* **1998**, 986-990.
20. Wacker, D. A., Kasireddy, P. Efficient solid-phase synthesis of 2,3-substituted indoles. *Tetrahedron Lett.* **2002**, 43, 5189-5191.
21. Smith, A. B., Kanoh, N., Ishiyama, H., Minakawa, N., Rainier, J. D., Hartz, R. A., Cho, Y. S., Cui, H., Moser, W. H. Tremorgenic Indole Alkaloids. The Total Synthesis of (-)-Penitrem D. *J. Am. Chem. Soc.* **2003**, 125, 8228-8237.
22. Chen, H. Z., Jin, Y. D., Xu, R. S., Peng, B. X., Desseyn, H., Janssens, J., Heremans, P., Borghs, G., Geise, H. J. Synthesis, optical and electroluminescent properties of a novel Indacene. *Synth. Met.* **2003**, 139, 529-534.
23. Kouznetsov, V., Zubkov, F., Palma, A., Restrepo, G. A simple synthesis of spiro-C6-annulated hydrocyclopenta[g]indole derivatives. *Tetrahedron Lett.* **2002**, 43, 4707-4709.

Malonic Ester Synthesis 272

1. Wislicenus, J. About acetoacetic ester synthesis. *Liebigs Ann. Chem.* **1877**, 186, 161-228.
2. Wislicenus, J. The reaction of organic halides with the sodium salts of organic compounds. *Liebigs Ann. Chem.* **1882**, 212, 239-250.
3. Hauser, C. R., Hudson, B. E., Jr. Acetoacetic ester condensation and certain related reactions. *Org. React.* **1942**, 1, 266-302.
4. Cope, A. C., Holmes, H. L., House, H. O. The alkylation of esters and nitriles. *Org. React.* **1957**, 107-331.
5. House, H. O. *Modern Synthetic Reactions (The Organic Chemistry Monograph Series). 2nd ed* (**1972**) 856 pp.
6. Carruthers, W. *Some Modern Methods of Organic Synthesis. 3rd Ed* (Cambridge University Press, Cambridge, **1986**) 526 pp.
7. Russell, R. R., VanderWerf, C. A. Malonic ester synthesis with styrene oxide and with butadiene oxide. *J. Am. Chem. Soc.* **1947**, 69, 11-13.
8. Mizuno, Y., Adachi, K., Ikeda, K. Condensed systems of aromatic nitrogenous series. XIII. Extension of malonic ester synthesis to the heterocyclic series. *Pharmaceutical Bulletin* **1954**, 2, 225-234.
9. Sommer, L. H., Goldberg, G. M., Barnes, G. H., Stone, L. S., Jr. Malonic ester syntheses with organosilicon compounds. New silicon-containing malonic esters, mono- and dicarboxylic acids, barbituric acids, and a disiloxanetetracarboxylic acid. *J. Am. Chem. Soc.* **1954**, 76, 1609-1612.
10. Mamedov, S., Khydyrov, D. N., Gevorkyan, A. N., Rustamov, V. R., Ismailov, R. G. Malonic ester synthesis based on aromatic γ-chloro ethers. *Doklady - Akademiya Nauk Azerbaidzhanskoi SSR* **1969**, 25, 10-12.
11. McMurry, J. E., Musser, J. H. Simple one-step alternative to the malonic ester synthesis. *J. Org. Chem.* **1975**, 40, 2556-2557.
12. Hunter, D. H., Perry, R. A. Synthetic applications of crown ethers; the malonic ester synthesis. *Synthesis* **1977**, 37-39.
13. Verbrugge, P. A., Tieleman, J. H., Van der Jagt, P. J., Bickelhaupt, F. Bridged ring compounds by malonic ester synthesis. *Synth. Commun.* **1977**, 7, 1-11.
14. Padgett, H. C., Csendes, I. G., Rapoport, H. The alkoxycarbonyl moiety as blocking group. A generally useful variation of the malonic ester synthesis. *J. Org. Chem.* **1979**, 44, 3492-3496.
15. Obaza, J., Smith, F. X. A malonic ester synthesis with acid chlorides. The homologation of dioic acids. *Synth. Commun.* **1982**, 12, 19-23.
16. Sato, T., Otera, J. CsF in Organic Synthesis. Malonic Ester Synthesis Revisited for Stereoselective Carbon-Carbon Bond Formation. *J. Org. Chem.* **1995**, 60, 2627-2629.
17. Geuther. *Jahresber. Schweiz. Akad. Med. Wiss.* **1863**, 16, 324.
18. Grigsby, W. E., Hind, J., Chanley, J., Westheimer, F. H. Malonic ester synthesis and Walden inversion. *J. Am. Chem. Soc.* **1942**, 64, 2606-2610.
19. Baker, R., Castro, J. L. Total synthesis of (+)-macbecin I. *J. Chem. Soc., Perkin Trans. 1* **1990**, 47-65.
20. Fukuyama, Y., Hirono, M., Kodama, M. Total synthesis of (+)-bicyclohumulenone. *Chem. Lett.* **1992**, 167-170.
21. Takaya, Y., Kikuchi, H., Terui, Y., Komiya, J., Furukawa, K.-i., Seya, K., Motomura, S., Ito, A., Oshima, Y. Novel acyl α-pyronoids, dictyopyrone A, B, and C, from Dictyostelium cellular slime molds. *J. Org. Chem.* **2000**, 65, 985-989.
22. Bergner, E. J., Helmchen, G. Enantioselective Synthesis of (+)-Juvabione. *J. Org. Chem.* **2000**, 65, 5072-5074.

Mannich Reaction 274

Related reactions: Eschenmoser methenylation;

1. Tollens, B., Marle, v. The formaldehyde derivatives of acetophenones. *Ber.* **1903**, 36, 1351-1357.
2. Mannich, C. Synthesis of β-ketonic bases. *J. Chem. Soc., Abstracts* **1917**, 112, 634.
3. Mannich, C. Synthesis of β-ketonic bases. *Arch. Pharm.* **1917**, 255, 261-276.
4. Blicke, F. F. Organic Reactions. I: Mannich reaction. **1942**, pp 303-341.
5. Reichert, B. *Die Mannich-Reaktion* (Springer-Verlag, **1959**) 195 pp.
6. Thompson, B. B. The Mannich reaction. Mechanistic and technological considerations. *J. Pharm. Sci.* **1968**, 57, 715-733.
7. Tramontini, M. Advances in the chemistry of Mannich bases. *Synthesis* **1973**, 703-775.
8. Varma, R. S. The application of Mannich reaction in the field of drug research. *Labdev, Part B* **1974**, 12, 126-133.
9. Gevorgyan, G. A., Agababyan, A. G., Mndzhoyan, O. L. Advances in the chemistry of β-amino ketones. *Usp. Khim.* **1984**, 53, 971-1013.
10. Tramontini, M., Angiolini, L., Ghedini, N. Mannich bases in polymer chemistry. *Polymer* **1988**, 29, 771-788.
11. Tramontini, M., Angiolini, L. Further advances in the chemistry of Mannich bases. *Tetrahedron* **1990**, 46, 1791-1837.
12. Heaney, H. The Bimolecular Aromatic Mannich Reaction. in *Comp. Org. Synth.* (eds. Trost, B. M.,Fleming, I.), *2*, 953-973 (Pergamon, Oxford, **1991**).
13. Kleinman, E. F. The Bimolecular Aliphatic Mannich and Related Reactions. in *Comp. Org. Synth.* (eds. Trost, B. M.,Fleming, I.), *2*, 893-951 (Pergamon, Oxford, **1991**).
14. Overman, L. E., Ricca, D. J. The Intramolecular Mannich and Related Reactions. in *Comprehensive Organic Synthesis* (eds. Trost, B. M.,Fleming, I.), *2*, 1007-1046 (Pergamon, Oxford, **1991**).
15. Overman, L. E. Charge as a key component in reaction design. The invention of cationic cyclization reactions of importance in synthesis. *Acc. Chem. Res.* **1992**, 25, 352-359.
16. Tramontini, M., Angiolini, L. *Mannich Bases - Chemistry and Uses* (CRC, Boca Raton, Fla., **1994**) 304 pp.
17. Bordunov, A. V., Bradshaw, J. S., Pastushok, V. N., Izatt, R. M. Application of the Mannich reaction for the synthesis of azamacroheterocycles. *Synlett* **1996**, 933-948.
18. Arend, M., Westermann, B., Risch, N. Modern variants of the Mannich reaction. *Angew. Chem., Int. Ed. Engl.* **1998**, 37, 1045-1070.
19. Costisor, O., Linert, W. Metal directed Mannich synthesis of ligands. *Reviews in Inorganic Chemistry* **2000**, 20, 63-127.
20. Habata, Y., Akabori, S., Bradshaw, J. S., Izatt, R. M. Synthesis of Armed and Double-Armed Macrocyclic Ligands by the Mannich Reaction: A Short Review. *Ind. & Eng. Chem. Res.* **2000**, 39, 3465-3470.

21. Sheehan, S. M. Proline-catalyzed asymmetric Mannich reactions: The highly enantioselective synthesis of amino acid derivatives and 1,2-amino alcohols. *Chemtracts* **2002**, 15, 384-390.
22. Cordova, A. The Direct Catalytic Asymmetric Mannich Reaction. *Acc. Chem. Res.* **2004**, 37, 102-112.
23. Kobayashi, S., Ueno, M. Mannich reaction. *Comprehensive Asymmetric Catalysis, Supplement* **2004**, 1, 143-150.
24. Seebach, D., Betschart, C., Schiess, M. Diastereoselective synthesis of novel Mannich bases through titanium derivatives. *Helv. Chim. Acta* **1984**, 67, 1593-1597.
25. Ishitani, H., Ueno, M., Kobayashi, S. Catalytic Enantioselective Mannich-Type Reactions Using a Novel Chiral Zirconium Catalyst. *J. Am. Chem. Soc.* **1997**, 119, 7153-7154.
26. Atlan, V., Bienayme, H., El Kaim, L., Majee, A. The use of hydrazones for efficient Mannich type coupling with aldehydes and secondary amines. *Chem. Commun.* **2000**, 1585-1586.
27. List, B. The Direct Catalytic Asymmetric Three-Component Mannich Reaction. *J. Am. Chem. Soc.* **2000**, 122, 9336-9337.
28. Muller, R., Rottele, H., Henke, H., Waldmann, H. Asymmetric steering of the Mannich reaction with phthaloyl amino acids. *Chem.-- Eur. J.* **2000**, 6, 2032-2043.
29. Bur, S. K., Martin, S. F. Vinylogous Mannich reactions: selectivity and synthetic utility. *Tetrahedron* **2001**, 57, 3221-3242.
30. Juhl, K., Gathergood, N., Jorgensen, K. A. Catalytic asymmetric direct Mannich reactions of carbonyl compounds with α-imino esters. *Angew. Chem., Int. Ed. Engl.* **2001**, 40, 2995-2997.
31. McReynolds, M. D., Hanson, P. R. The three-component boronic acid Mannich reaction: structural diversity and stereoselectivity. *Chemtracts* **2001**, 14, 796-801.
32. Enders, D., Adam, J., Oberborsch, S., Ward, D. Asymmetric Mannich reactions by a-silyl controlled aminomethylation of ketones. *Synthesis* **2002**, 2737-2748.
33. List, B., Pojarliev, P., Biller, W. T., Martin, H. J. The Proline-Catalyzed Direct Asymmetric Three-Component Mannich Reaction: Scope, Optimization, and Application to the Highly Enantioselective Synthesis of 1,2-Amino Alcohols. *J. Am. Chem. Soc.* **2002**, 124, 827-833.
34. Martin, S. F. Evolution of the Vinylogous Mannich Reaction as a Key Construction for Alkaloid Synthesis. *Acc. Chem. Res.* **2002**, 35, 895-904.
35. Cordova, A., Barbas, C. F. Direct organocatalytic asymmetric Mannich-type reactions in aqueous media: one-pot Mannich-allylation reactions. *Tetrahedron Lett.* **2003**, 44, 1923-1926.
36. Notz, W., Tanaka, F., Watanabe, S., Chowdari, N. S., Turner, J. M., Thayumanavan, R., Barbas, C. F. The Direct Organocatalytic Asymmetric Mannich Reaction: Unmodified Aldehydes as Nucleophiles. *J. Org. Chem.* **2003**, 68, 9624-9634.
37. Xiao, H., Tang, Z. A quantum-chemical study on the Mannich reaction with polynitromethanes. *Acta Chim. Sin.* **1989**, 289-294.
38. Zhang, Y., Dong, W., Shi, J., Li, W. Relationships between electron structures of N-methylolamines and their reactivity in the Mannich reaction. *Propellants, Explosives, Pyrotechnics* **1994**, 19, 103-106.
39. Li, Y., Xiao, H., Wu, J. The study on the mechanism of iminium salts as potential Mannich reagents. Part 2. Ethylene as pseudo-acid component. *THEOCHEM* **1995**, 333, 165-170.
40. Li, Y. M., Xiao, H. M. Studies on the mechanism of Mannich reaction involving iminium salt as potential Mannich reagent. III. Furan as pseudo acid component. *Int. J. Quantum Chem.* **1995**, 54, 293-297.
41. Stankovicova, H., Fabian, W. M. F., Lacova, M. Synthesis and theoretical study of Mannich type reaction products of 3-formylchromones with triazoles and amides and nucleophilic formation of 2,3-disubstituted-4-chromanones. *Molecules [Electronic Publication]* **1996**, 1, 223-235.
42. Xiao, H.-M., Ling, Y., Zhai, Y.-F., Li, Y.-M. Theoretical studies on the mechanism of Mannich reaction involving iminium salt as potential Mannich reagent. Use of acetaldehyde as pseudo-acid component. *Chemical Research in Chinese Universities* **1997**, 13, 324-329.
43. Zhang, Y. J., Yu, M. H., Li, W. M. Relationship between electron structures of substituted ureas and their reactivity in Mannich reaction. *Propellants, Explosives, Pyrotechnics* **1997**, 22, 279-283.
44. Maran, U., Katritzky, A. R., Karelson, M. Theoretical study of aminoalkylation in the Mannich reaction of furan with methyleneimminium salt. *Int. J. Quantum Chem.* **1998**, 67, 359-366.
45. Murata, K., Kitazume, T. Study of the effect of fluorine atom(s) on fluoromethylated imines in the tandem Mannich-Michael-type reaction. *Isr. J. Chem.* **1999**, 39, 163-166.
46. Bur, S. K., Martin, S. F. Vinylogous Mannich Reactions: Some Theoretical Studies on the Origins of Diastereoselectivity. *Org. Lett.* **2000**, 2, 3445-3447.
47. Tomazela, D. M., Moraes, L. A. B., Pilli, R. A., Eberlin, M. N., D'Oca, M. G. M. Mannich-Type Reactions in the Gas-Phase: The Addition of Enol Silanes to Cyclic N-Acyliminium Ions. *J. Org. Chem.* **2002**, 67, 4652-4658.
48. Mondal, N., Mandal, S. C., Das, G. K., Mukherjee, S. Theoretical study on the mechanism of Robinson's synthesis of tropinone. *J. Chem. Res., Synop.* **2003**, 580-583.
49. Allemann, C., Gordillo, R., Clemente, F. R., Cheong, P. H.-Y., Houk, K. N. Theory of Asymmetric Organocatalysis of Aldol and Related Reactions: Rationalizations and Predictions. *Acc. Chem. Res.* **2004**, 37, 558-569.
50. Benkovic, S. J., Benkovic, P. A., Comfort, D. R. Kinetic detection of the imminium cation in formaldehyde-amine condensations in neutral aqueous solution. *J. Am. Chem. Soc.* **1969**, 91, 1860-1861.
51. Toczko, M. A., Heathcock, C. H. Total Synthesis of (±)-Aspidospermidine. *J. Org. Chem.* **2000**, 65, 2642-2645.
52. Paquette, L. A., Backhaus, D., Braun, R. Direct asymmetric entry into the cytotoxic 8,9-secokaurene diterpenoids. Total synthesis of (-)-O-methylshikoccin and (+)-O-(methylepoxy)shikoccin. *J. Am. Chem. Soc.* **1996**, 118, 11990-11991.
53. Brueggemann, M., McDonald, A. I., Overman, L. E., Rosen, M. D., Schwink, L., Scott, J. P. Total Synthesis of (±)-Didehydrostemofoline (Asparagamine A) and (±)-Isodidehydrostemofoline. *J. Am. Chem. Soc.* **2003**, 125, 15284-15285.
54. Martin, S. F., Barr, K. J. Vinylogous Mannich Reactions. The Asymmetric Total Synthesis of (+)-Croomine. *J. Am. Chem. Soc.* **1996**, 118, 3299-3300.
55. Martin, S. F., Barr, K. J., Smith, D. W., Bur, S. K. Applications of Vinylogous Mannich Reactions. Concise Enantiospecific Total Syntheses of (+)-Croomine. *J. Am. Chem. Soc.* **1999**, 121, 6990-6997.

McMurry Coupling 276

1. Sharpless, K. B., Flood, T. C. Direct deoxygenation of vicinal diols with tungsten(IV). New olefin synthesis. *J. Chem. Soc., Chem. Commun.* **1972**, 370-371.
2. Sharpless, K. B., Umbreit, M. A., Nieh, M. T., Flood, T. C. Lower valent tungsten halides. New class of reagents for deoxygenation of organic molecules. *J. Am. Chem. Soc.* **1972**, 94, 6538-6540.
3. Mukaiyama, T., Sato, T., Hanna, J. Reductive coupling of carbonyl compounds to pinacols and olefins using titanium(IV) chloride and zinc. *Chem. Lett.* **1973**, 1041-1044.
4. Tyrlik, S., Wolochowicz, I. Application of transition metal complexes with low oxidation states in organic synthesis. I. New synthesis of olefins from carbonyl compounds. *Bull. Soc. Chim. Fr.* **1973**, 2147-2148.
5. McMurry, J. E., Fleming, M. P. New method for the reductive coupling of carbonyls to olefins. Synthesis of β-carotene. *J. Am. Chem. Soc.* **1974**, 96, 4708-4709.
6. McMurry, J. E. Organic chemistry of low-valent titanium. *Acc. Chem. Res.* **1974**, 7, 281-286.
7. Lai, Y.-H. Organic reductive coupling with titanium and vanadium chlorides. A review. *Org. Prep. Proced. Int.* **1980**, 12, 361-391.
8. McMurry, J. E. Titanium-induced dicarbonyl-coupling reactions. *Acc. Chem. Res.* **1983**, 16, 405-411.
9. Welzel, P. McMurry reaction for alkene syntheses. *Nachrichten aus Chemie, Technik und Laboratorium* **1983**, 31, 814-816.
10. McMurry, J. E. Carbonyl-coupling reactions using low-valent titanium. *Chem. Rev.* **1989**, 89, 1513-1524.
11. Fürstner, A., Bogdanovic, B. New developments in the chemistry of low-valent titanium. *Angew. Chem., Int. Ed. Engl.* **1996**, 35, 2443-2469.
12. Lectka, T. The McMurry reaction. *Active Metals* **1996**, 85-131.
13. Ephritikhine, M. A new look at the McMurry reaction. *Chem. Commun.* **1998**, 2549-2554.

14. Fürstner, A. The McMurry reaction and related transformations. *Transition Metals for Organic Synthesis* **1998**, 1, 381-401.
15. Hirao, T. A catalytic system for reductive transformations via one-electron transfer. *Synlett* **1999**, 175-181.
16. Gansaeuer, A., Bluhm, H. Reagent-Controlled Transition-Metal-Catalyzed Radical Reactions. *Chem. Rev.* **2000**, 100, 2771-2788.
17. Griesbeck, A. G., Schieffer, S. Electron-transfer reactions of carbonyl compounds. *Electron Transfer in Chemistry* **2001**, 2, 457-493.
18. Herrmann, W. A., Schneider, H. Catalytic McMurry coupling: olefins from keto compounds. *Applied Homogeneous Catalysis with Organometallic Compounds (2nd Edition)* **2002**, 3, 1093-1099.
19. Ephritikhine, M., Villiers, C. The McMurry coupling and related reactions. *Modern Carbonyl Olefination* **2004**, 223-285.
20. McMurry, J. E., Fleming, M. P., Kees, K. L., Krepski, L. R. Titanium-induced reductive coupling of carbonyls to olefins. *J. Org. Chem.* **1978**, 43, 3255-3266.
21. McMurry, J. E., Miller, D. D. Titanium-induced cyclization of keto esters: a new method of cycloalkanone synthesis. *J. Am. Chem. Soc.* **1983**, 105, 1660-1661.
22. McMurry, J. E., Lectka, T., Rico, J. G. An optimized procedure for titanium-induced carbonyl coupling. *J. Org. Chem.* **1989**, 54, 3748-3749.
23. Fürstner, A., Jumbam, D. N. Titanium-induced syntheses of furans, benzofurans and indoles. *Tetrahedron* **1992**, 48, 5991-6010.
24. Fürstner, A., Hupperts, A., Ptock, A., Janssen, E. "Site Selective" Formation of Low-Valent Titanium Reagents: An "Instant" Procedure for the Reductive Coupling of Oxo Amides to Indoles. *J. Org. Chem.* **1994**, 59, 5215-5229.
25. Fürstner, A., Hupperts, A. Carbonyl Coupling Reactions Catalytic in Titanium and the Use of Commercial Titanium Powder for Organic Synthesis. *J. Am. Chem. Soc.* **1995**, 117, 4468-4475.
26. Fürstner, A., Seidel, G. High-surface sodium as a reducing agent for $TiCl_3$. *Synthesis* **1995**, 63-68.
27. Fürstner, A., Seidel, G., Gabor, B., Kopiske, C., Krueger, C., Mynott, R. Unprecedented McMurry reactions with acylsilanes: enedisilane formation versus Brook rearrangement. *Tetrahedron* **1995**, 51, 8875-8888.
28. Lipski, T. A., Hilfiker, M. A., Nelson, S. G. Ligand-Modified Catalysts for the McMurry Pinacol Reaction. *J. Org. Chem.* **1997**, 62, 4566-4567.
29. Stahl, M., Pidun, U., Frenking, G. Theoretical studies of organometallic compounds. XXVII. On the mechanism of the McMurry reaction. *Angew. Chem., Int. Ed. Engl.* **1997**, 36, 2234-2237.
30. Fujiwara, Y., Ishikawa, R., Akiyama, F., Teranishi, S. Reductive coupling of carbonyl compounds to olefins by tungsten hexachloride-lithium aluminum hydride and some tungsten and molybdenum carbonyls. *J. Org. Chem.* **1978**, 43, 2477-2480.
31. Dams, R., Malinowski, M., Geise, H. J. Reductive couplings of ketones by low-valent titanium, prepared from titanium tetrachloride and reducing agents. *Bull. Soc. Chim. Belg.* **1981**, 90, 1141-1152.
32. Dams, R., Malinowski, M., Geise, H. J. Reductive couplings with low-valent titanium compounds (McMurry reaction). An ESR investigation into the $TiCl_3/LiAlH_4$/ROH system. *Transition Metal Chemistry (Dordrecht, Netherlands)* **1982**, 7, 37-40.
33. Dams, R., Malinowski, M., Westdorp, I., Geise, H. On the mechanism of the titanium-induced reductive coupling of ketones to olefins. *J. Org. Chem.* **1982**, 47, 248-259.
34. Bryan, J. C., Mayer, J. M. Oxidative addition of carbon-oxygen and carbon-nitrogen double bonds to $WCl_2(PMePh_2)_4$. Synthesis of tungsten metallaoxirane and tungsten oxo- and imido-alkylidene complexes. *J. Am. Chem. Soc.* **1990**, 112, 2298-2308.
35. Chisholm, M. H., Folting, K., Klang, J. A. Reaction between benzophenone and ditungsten hexaalkoxides. Molecular structure and reactivity of W(OCH$_2$-*tert*-Bu)$_4$(py)(η2-OCPh$_2$). *Organometallics* **1990**, 9, 607-613.
36. Chisholm, M. H., Folting, K., Klang, J. A. Reductive cleavage of ketonic carbon-oxygen bonds in the reactions between ketones and ditungsten hexaalkoxides. Structural characterization of a ditungsten μ-propylidene derivative. *Organometallics* **1990**, 9, 602-606.
37. Pierce, K. G., Barteau, M. A. Ketone Coupling on Reduced TiO_2 (001) Surfaces: Evidence of Pinacol Formation. *J. Org. Chem.* **1995**, 60, 2405-2410.
38. Villiers, C., Ephritikhine, M. New insights into the mechanism of the McMurry reaction. *Angew. Chem., Int. Ed. Engl.* **1997**, 36, 2380-2382.
39. Sherrill, A. B., Lusvardi, V. S., Eng, J., Chen, J. G., Barteau, M. A. NEXAFS investigation of benzaldehyde reductive coupling to form stilbene on reduced surfaces of TiO_2(0 0 1). *Catal. Today* **2000**, 63, 43-51.
40. Villiers, C., Ephritikhine, M. Reactions of aliphatic ketones R_2CO (R = Me, Et, *i*-Pr, and *t*-Bu) with the MCl_4/Li(Hg) system (M = U or Ti): mechanistic analogies between the McMurry, Wittig, and Clemmensen reactions. *Chem.--Eur. J.* **2001**, 7, 3043-3051.
41. Liu, Z., Zhang, T., Li, Y. First enantioselective total synthesis of (-)-13-hydroxyneocembrene. *Tetrahedron Lett.* **2001**, 42, 275-277.
42. Mikami, K., Takahashi, K., Nakai, T. Asymmetric tandem Claisen-ene strategy for steroid total synthesis: an efficient access to (+)-9(11)-dehydroestrone methyl ether. *J. Am. Chem. Soc.* **1990**, 112, 4035-4037.
43. Casimiro-Garcia, A., Micklatcher, M., Turpin, J. A., Stup, T. L., Watson, K., Buckheit, R. W., Cushman, M. Novel Modifications in the Alkenyldiarylmethane (ADAM) Series of Non-Nucleoside Reverse Transcriptase Inhibitors. *J. Med. Chem.* **1999**, 42, 4861-4874.
44. Eguchi, T., Ibaragi, K., Kakinuma, K. Total Synthesis of Archaeal 72-Membered Macrocyclic Tetraether Lipids. *J. Org. Chem.* **1998**, 63, 2689-2698.

Meerwein Arylation 278

Related reactions: Heck reaction;

1. Meerwein, H., Buchner, E., van Emster, K. Reaction of aromatic diazo compounds upon α,β-unsaturated carbonyl compounds. *J. Prakt. Chem.* **1939**, 152, 237-266.
2. Rondestvedt, C. S., Jr. Arylation of unsaturated compounds by diazonium salts (the Meerwein arylation reaction). *Org. React.* **1960**, 11, 189-260.
3. Rondestvedt, C. S., Jr. Arylation of unsaturated compounds by diazonium salts (the Meerwein arylation reaction). *Org. React.* **1976**, 24, 225-259.
4. Galli, C. Radical reactions of arenediazonium ions: An easy entry into the chemistry of the aryl radical. *Chem. Rev.* **1988**, 88, 765-792.
5. Weis, C. D. Meerwein arylation reactions of olefins with anthraquinone diazonium hydrogen sulfates: formation of new carbon bonds at the carbon atoms C-1 and at C-1,5 of the anthraquinone system. *Dyes Pigm.* **1988**, 9, 1-20.
6. Curran, D. P. Radical Addition Reactions. in *Comp. Org. Synth.* (eds. Trost, B. M.,Fleming, I.), *4*, 715-777 (Pergamon, Oxford, **1991**).
7. Minisci, F., Fontana, F., Vismara, E. Polar and enthalpic effects in free-radical reactions. Free-radical diazo coupling and reactivity of carbohydrate radicals. *Gazz. Chim. Ital.* **1993**, 123, 9-18.
8. Truce, W. E., Breiter, J. J., Tracy, J. E. The Meerwein arylation of vinyl sulfones. *J. Org. Chem.* **1964**, 29, 3009-3014.
9. Filler, R., White, A. B., Taqui-Khan, B., Gorelic, L. Synthesis of aromatic α-amino acids via Meerwein arylation. *Can. J. Chem.* **1967**, 45, 329-331.
10. Doyle, M. P., Siegfried, B., Elliott, R. C., Dellaria, J. F., Jr. Alkyl nitrite-metal halide deamination reactions. 3. Arylation of olefinic compounds in the deamination of arylamines by alkyl nitrites and copper(II) halides. A convenient and effective variation of the Meerwein arylation reaction. *J. Org. Chem.* **1977**, 42, 2431-2436.
11. Rondestvedt, C. S., Jr. Meerwein arylation of fluorinated olefins. *J. Org. Chem.* **1977**, 42, 2618-2620.
12. Ganushchak, N. I., Obushak, N. D., Luka, G. Y. Catalytic action of iron(II) chloride in a Meerwein reaction. *Zh. Org. Khim.* **1981**, 17, 870-872.
13. Obushak, N. D., Ganushchak, N. I., Lyakhovich, M. B. 1-Naphthyldiazonium tetrachlorocuprate. New arylating reagent. *Zh. Org. Khim.* **1991**, 27, 1757-1762.
14. Nock, H., Schottenberger, H. Direct arylation of ferrocenylacetylenes and ferrocenylethenes under autocatalytic Meerwein conditions. *J. Org. Chem.* **1993**, 58, 7045-7048.
15. Brunner, H., Bluchel, C., Doyle, M. P. Asymmetric catalysis. Part 108. Copper catalysts with optically active ligands in the enantioselective Meerwein arylation of activated olefins. *J. Organomet. Chem.* **1997**, 541, 89-95.
16. Obushak, M. D., Lyakhovych, M. B., Ganushchak, M. I. Arenediazonium tetrachlorocuprates(II). Modification of the Meerwein and Sandmeyer reactions. *Tetrahedron Lett.* **1998**, 39, 9567-9570.

17. Navon, N., Cohen, H., Paoletti, P., Valtancoli, B., Bencini, A., Meyerstein, D. Design of Ligands Which Improve Cu(I) Catalysis. *Ind. & Eng. Chem. Res.* **2000**, 39, 3536-3540.
18. Obushak, N. D., Lyakhovich, M. B., Bilaya, E. E. Arenediazonium tetrachlorocuprates(II). Modified versions of the Meerwein and Sandmeyer reactions. *Russ. J. Org. Chem.* **2002**, 38, 38-46.
19. Takahashi, I., Takeyama, N., Morita, T., Mori, H., Yamamoto, M., Nishimura, H., Kitajima, H. MNDO calculation-based examination on the product distribution in Meerwein arylation of naphthalene-1,4-diones. *Chem. Express* **1993**, 8, 289-292.
20. Takahashi, I., Muramatsu, O., Fukuhara, J., Hosokawa, Y., Takeyama, N., Morita, T., Kitajima, H. Studies on the Meerwein arylation-based preparation of 2,3-diarylbenzene-1,4-diones and its theoretical interpretation. *Chem. Lett.* **1994**, 465-468.
21. Mella, M., Coppo, P., Guizzardi, B., Fagnoni, M., Freccero, M., Albini, A. Photoinduced, Ionic Meerwein Arylation of Olefins. *J. Org. Chem.* **2001**, 66, 6344-6352.
22. Kochi, J. K. The Meerwein reaction. Catalysis by cuprous chloride. *J. Am. Chem. Soc.* **1955**, 77, 5090-5092.
23. Kochi, J. K. The reduction of cupric chloride by carbonyl compounds. *J. Am. Chem. Soc.* **1955**, 77, 5274-5278.
24. Epstein, J. W., McKenzie, T. C., Lovell, M. F., Perkinson, N. A. Diels-Alder dimerization of 2-arylmaleimides. X-ray crystal structure of the dimer of 2-para-tolylmaleimide. *J. Chem. Soc., Chem. Commun.* **1980**, 314-315.
25. Wells, G. J., Tao, M., Josef, K. A., Bihovsky, R. 1,2-Benzothiazine 1,1-Dioxide P2-P3 Peptide Mimetic Aldehyde Calpain I Inhibitors. *J. Med. Chem.* **2001**, 44, 3488-3503.
26. Baldwin, J. E., Forrest, A. K., Monaco, S., Young, R. J. Synthetic entry into N(5)-ergolines. *J. Chem. Soc., Chem. Commun.* **1985**, 1586-1587.
27. McKenzie, T. C., Hassen, W., Macdonald, S. J. F. Synthesis of the Gilvocarcin-M aglycone. *Tetrahedron Lett.* **1987**, 28, 5435-5436.
28. Sohda, T., Mizuno, K., Momose, Y., Ikeda, H., Fujita, T., Meguro, K. Studies on antidiabetic agents. 11. Novel thiazolidinedione derivatives as potent hypoglycemic and hypolipidemic agents. *J. Med. Chem.* **1992**, 35, 2617-2626.

Meerwein-Ponndorf-Verley Reduction..280

Related reactions: Cannizzaro reaction, Tishchenko reaction;

1. Meerwein, H., Schmidt, R. New method for the reduction of aldehydes and ketones. *Ann.* **1925**, 444, 221-238.
2. Verley, A. The exchange of functional groups between two molecules. The passage of ketones to alcohols and the reverse. *Bull. soc. chim.* **1925**, 37, 871-874.
3. Ponndorf, W. The reversible exchange of oxygen between aldehydes or ketones on the one hand and primary or secondary alcohols on the other hand. *Z. angew. Chem.* **1926**, 39, 138-143.
4. Wilds, A. L. Reduction with Al alkoxides (Meerwein-Ponndorf-Verley reduction). *Org. React.* **1944**, 2, 178-223.
5. Inch, T. D. Asymmetric synthesis. *Synthesis* **1970**, 2, 466-473.
6. Kellogg, R. M. Reduction of C=X to CHXH by hydride delivery from carbon. in *Comp. Org. Synth.* (eds. Trost, B. M.,Fleming, I.), *8*, 79-106 (Pergamon, Oxford, **1991**).
7. Mehrotra, R. C., Rai, A. K. Aluminum alkoxides, β-diketonates and carboxylates. *Polyhedron* **1991**, 10, 1967-1994.
8. de Graauw, C. F., Peters, J. A., van Bekkum, H., Huskens, J. Meerwein-Ponndorf-Verley reductions and Oppenauer oxidations: an integrated approach. *Synthesis* **1994**, 1007-1017.
9. Creyghton, E. J., Downing, R. S. Shape-selective hydrogenation and hydrogen transfer reactions over zeolite catalysts. *J. Mol. Catal. A: Chemical* **1998**, 134, 47-61.
10. Ohkuma, T., Noyori, R. Hydrogenation of carbonyl groups. *Comprehensive Asymmetric Catalysis I-III* **1999**, 1, 199-246.
11. Palmer, M. J., Wills, M. Asymmetric transfer hydrogenation of C:O and C:N bonds. *Tetrahedron: Asymmetry* **1999**, 10, 2045-2061.
12. Ooi, T., Maruoka, K. Achiral Al(III) Lewis acids. *Lewis Acids in Organic Synthesis* **2000**, 1, 191-281.
13. Creyghton, E. J., Van der Waal, J. C. Meerwein-Ponndorf-Verley reduction, Oppenauer oxidation, and related reactions. *Fine Chemicals through Heterogeneous Catalysis* **2001**, 438-448.
14. Maruoka, K. Bidentate Lewis acids for organic synthesis. *Catal. Today* **2001**, 66, 33-45.
15. Nishide, K., Node, M. Recent development of asymmetric syntheses based on the Meerwein-Ponndorf-Verley reduction. *Chirality* **2002**, 14, 759-767.
16. Ooi, T., Miura, T., Itagaki, Y., Ichikawa, H., Maruoka, K. Catalytic Meerwein-Ponndorf-Verley (MPV) and Oppenauer (OPP) reactions: remarkable acceleration of the hydride transfer by powerful bidentate aluminum alkoxides. *Synthesis* **2002**, 279-291.
17. Budzelaar, P. H. M., Talarico, G. Insertion and β-hydrogen transfer at aluminum. in *Structure and Bonding* (ed. Mingos, D. M. P.), *105*, (Springer-Verlag, Berlin, Heidelberg, **2003**).
18. Jerome, J. E., Sergent, R. H. Catalytic applications of aluminum isopropoxide in organic synthesis. *Chem. Ind.* **2003**, 89, 97-114.
19. Kow, R., Nygren, R., Rathke, M. W. Rate enhancement of the Meerwein-Ponndorf-Verley-Oppenauer reaction in the presence of proton acids. *J. Org. Chem.* **1977**, 42, 826-827.
20. Hutton, J. Diisobornyloxyaluminum isopropoxide - a new highly selective Meerwein-Ponndorf-Verley reagent. *Synth. Commun.* **1979**, 9, 483-486.
21. Namy, J. L., Souppe, J., Collin, J., Kagan, H. B. New preparations of lanthanide alkoxides and their catalytical activity in Meerwein-Ponndorf-Verley-Oppenauer reactions. *J. Org. Chem.* **1984**, 49, 2045-2049.
22. Okano, T., Matsuoka, M., Konishi, H., Kiji, J. Meerwein-Ponndorf-Verley reduction of ketones and aldehydes catalyzed by lanthanide tri-2-propoxides. *Chem. Lett.* **1987**, 181-184.
23. Evans, D. A., Nelson, S. G., Gagne, M. R., Muci, A. R. A chiral samarium-based catalyst for the asymmetric Meerwein-Ponndorf-Verley reduction. *J. Am. Chem. Soc.* **1993**, 115, 9800-9801.
24. Akamanchi, K. G., Varalakshmy, N. R. Aluminum isopropoxide - TFA, a modified catalyst for highly accelerated Meerwein - Ponndorf - Verley (MPV) reduction. *Tetrahedron Lett.* **1995**, 36, 3571-3572.
25. Akamanchi, K. G., Varalakshmy, N. R., Chaudari, B. A. Diisopropoxyaluminum trifluoroacetate. A new off-the-shelf metal alkoxide-type reducing agent for reduction of aldehydes and ketones. *Synlett* **1997**, 371-372.
26. Anwander, R., Palm, C. Meerwein-Ponndorf-Verley reductions mediated by lanthanide-alkoxide-functionalized mesoporous silicates. *Stud. Surf. Sci. Catal.* **1998**, 117, 413-420.
27. Node, M., Nishide, K., Shigeta, Y., Shiraki, H., Obata, K. A Novel Tandem Michael Addition/Meerwein-Ponndorf-Verley Reduction: Asymmetric Reduction of Acyclic α,β-Unsaturated Ketones Using A Chiral Mercapto Alcohol. *J. Am. Chem. Soc.* **2000**, 122, 1927-1936.
28. Warner, B. P., D'Alessio, J. A., Morgan, A. N., III, Burns, C. J., Schake, A. R., Watkin, J. G. Plutonium(III)-catalyzed Meerwein-Ponndorf-Verley reactions. *Inorg. Chim. Acta* **2000**, 309, 45-48.
29. Ooi, T., Ichikawa, H., Maruoka, K. Practical approach to the Meerwein-Ponndorf-Verley reduction of carbonyl substrates with new aluminum catalysts. *Angew. Chem., Int. Ed. Engl.* **2001**, 40, 3610-3612.
30. Cha, J. S., Park, J. H. Reaction of aldehydes and ketones with boron triisopropoxide. The Meerwein-Ponndorf-Verley type reduction of boron alkoxides. 1. *Bull. Korean Chem. Soc.* **2002**, 23, 1051-1052.
31. Cha, J. S., Park, J. H. Reaction of epoxides with boron triisopropoxide. The Meerwein-Ponndorf-Verley type reduction of boron alkoxides. 2. *Bull. Korean Chem. Soc.* **2002**, 23, 1377-1378.
32. Sastre, G., Corma, A. Relation between structure and Lewis acidity of Ti-β and TS-1 zeolites A quantum-chemical study. *Chem. Phys. Lett.* **1999**, 302, 447-453.
33. Sominsky, L., Rozental, E., Gottlieb, H., Gedanken, A., Hoz, S. Uncatalyzed Meerwein-Ponndorf-Oppenauer-Verley reduction of aldehydes and ketones under supercritical conditions. *J. Org. Chem.* **2004**, 69, 1492-1496.
34. Lund, H. Aluminum isopropylate as a reducing agent. A general method for carbonyl reduction. *Ber.* **1937**, 70B, 1520-1525.
35. Woodward, R. B., Wendler, N. L., Brutschy, F. J. Quininone. *J. Am. Chem. Soc.* **1945**, 67, 1425-1429.

36. Doering, W. v. E., Aschner, T. C. Mechanism of the alkoxide-catalyzed carbinol-carbonyl equilibrium. *J. Am. Chem. Soc.* **1953**, 75, 393-397.
37. Moulton, W. N., Atta, R. E. V., Ruch, R. R. Mechanism of the Meerwein-Ponndorf-Verley reduction. *J. Org. Chem.* **1961**, 26, 290-292.
38. Otvos, L., Gruber, L., Meisel-Agoston, J. The Meerwein-Ponndorf-Verley-Oppenauer reaction. I. Investigation of the reaction mechanism with radiocarbon. Racemization of secondary alcohols. *Acta Chim. Acad. Sci. Hung.* **1965**, 43, 149-153.
39. Yager, B. J., Hancock, C. K. Equilibrium and kinetic studies of the Meerwein-Ponndorf-Verley-Oppenauer (MPVO) reaction. *J. Org. Chem.* **1965**, 30, 1174-1179.
40. Snyder, C. H., Micklus, M. J. Meerwein-Ponndorf-Verley reduction of 1,2-cyclopentanedione. Stereochemical evidence for dual reductive paths. *J. Org. Chem.* **1970**, 35, 264-267.
41. Screttas, C. G., Cazianis, C. T. Mechanism of Meerwein-Pondorf-Verley type reductions. *Tetrahedron* **1978**, 34, 993-940.
42. Ashby, E. C., Argyropoulos, J. N. Single electron transfer in the Meerwein-Ponndorf-Verley reduction of benzophenone by lithium alkoxides. *J. Org. Chem.* **1986**, 51, 3593-3597.
43. Nugent, W. A., Zubyk, R. M. Catalytic hydrogen-deuterium exchange in deuteriated alcohols promoted by early-transition-metal alkoxides. Insight into a mechanistic puzzle. *Inorg. Chem.* **1986**, 25, 4604-4606.
44. Ashby, E. C. Single-electron transfer, a major reaction pathway in organic chemistry. An answer to recent criticisms. *Acc. Chem. Res.* **1988**, 21, 414-421.
45. Brunne, J., Hoffmann, N., Scharf, H.-D. The temperature dependence of the diastereoselective reduction of 2-t-butylcyclohexanone with diisobutylaluminum-2,6-di-t-butyl-4-methylphenoxide. *Tetrahedron* **1994**, 50, 6819-6824.
46. Ivanov, V. A., Bachelier, J., Audry, F., Lavalley, J. C. Study of the Meerwein-Pondorff-Verley reaction between ethanol and acetone on various metal oxides. *J. Mol. Catal.* **1994**, 91, 45-59.
47. Liu, Y.-C., Ko, B.-T., Huang, B.-H., Lin, C.-C. Reduction of Aldehydes and Ketones Catalyzed by a Novel Aluminum Alkoxide: Mechanistic Studies of Meerwein-Ponndorf-Verley Reaction. *Organometallics* **2002**, 21, 2066-2069.
48. Klomp, D., Maschmeyer, T., Hanefeld, U., Peters Joop, A. Mechanism of homogeneously and heterogeneously catalysed meerwein-ponndorf-verley-oppenauer reactions for the racemisation of secondary alcohols. *Chemistry (Weinheim an der Bergstrasse, Germany)* **2004**, 10, 2088-2093.
49. Toyota, M., Odashima, T., Wada, T., Ihara, M. Application of Palladium-Catalyzed Cycloalkenylation Reaction to C20 Gibberellin Synthesis: Formal Syntheses of GA_{12}, GA_{111}, and GA_{112}. *J. Am. Chem. Soc.* **2000**, 122, 9036-9037.
50. Sano, T., Toda, J., Maehara, N., Tsuda, Y. Synthesis of erythrina and related alkaloids. 17. Total synthesis of dl-coccuvinine and dl-coccolinine. *Can. J. Chem.* **1987**, 65, 94-98.
51. Evans, D. A., Rieger, D. L., Jones, T. K., Kaldor, S. W. Assignment of stereochemistry in the oligomycin/rutamycin/cytovaricin family of antibiotics. Asymmetric synthesis of the rutamycin spiroketal synthon. *J. Org. Chem.* **1990**, 55, 6260-6268.
52. Gammill, R. B. The synthesis and chemistry of functionalized furochromones. 2. The synthesis, Sommelet-Hauser rearrangement, and conversion of 4,9-dimethoxy-7-[(methylthio)methyl]-5H-furo(3,2-g)benzopyran-5-one to ammiol. *J. Org. Chem.* **1984**, 49, 5035-5041.

Meisenheimer Rearrangement 282

Related reactions: Mislow-Evans rearrangement;

1. Meisenheimer, J. A peculiar rearrangement of methylallylaniline N-oxide. *Ber.* **1919**, 52B, 1667-1677.
2. Meisenheimer, J., Greeske, H., Willmersdorf, A. Behavior of allyl- and benzylamine oxides towards sodium hydroxide. **1922**, 55B, 512-532.
3. Johnstone, R. A. W. Meisenheimer rearrangement of tertiary amine oxides. *Mechanisms of Molecular Migrations* **1969**, 2, 249-266.
4. Oae, S., Ogino, K. Rearrangements of tert-amine oxides. *Heterocycles* **1977**, 6, 583-675.
5. Albini, A. Synthetic utility of amine N-oxides. *Synthesis* **1993**, 263-277.
6. Khuthier, A. H., Ahmed, T. Y., Jallo, L. I. Aryl migration in the Meisenheimer rearrangement. *J. Chem. Soc., Chem. Commun.* **1976**, 1001-1002.
7. Enders, D., Kempen, H. Enantioselective synthesis of allylic alcohols via asymmetric [2,3]-sigmatropic Meisenheimer rearrangement. *Synlett* **1994**, 969-971.
8. Buston, J. E. H., Coldham, I., Mulholland, K. R. Chirality transfer from nitrogen to carbon in the [2,3]-Meisenheimer rearrangement. *Synlett* **1997**, 322-324.
9. Bergbreiter, D. E., Walchuk, B. Meisenheimer Rearrangement of Allyl N-Oxides as a Route to Initiators for Nitroxide-Mediated "Living" Free Radical Polymerizations. *Macromolecules* **1998**, 31, 6380-6382.
10. Buston, J. E. H., Coldham, I., Mulholland, K. R. Studies into the asymmetric meisenheimer rearrangement. *Tetrahedron: Asymmetry* **1998**, 9, 1995-2009.
11. Blanchet, J., Bonin, M., Micouin, L., Husson, H.-P. [2,3]-Meisenheimer rearrangement of N-allyl phenylglycinol derivatives. N-C versus C-C chirality transfer. *Tetrahedron Lett.* **2000**, 41, 8279-8283.
12. Guarna, A., Occhiato, E. G., Pizzetti, M., Scarpi, D., Sisi, S., van Sterkenburg, M. Stereoselective Meisenheimer rearrangement using BTAa's as chiral auxiliaries. *Tetrahedron: Asymmetry* **2000**, 11, 4227-4238.
13. Szabo, A., Galambos-Farago, A., Mucsi, Z., Timari, G., Vasvari-Debreczy, L., Hermecz, I. Solvent-dependent competitive rearrangements of cyclic tertiary propargylamine N-oxides. *Eur. J. Org. Chem.* **2004**, 687-694.
14. Kurihara, T., Sakamoto, Y., Takai, M., Ohishi, H., Harusawa, S., Yoneda, R. Meisenheimer rearrangement of azetopyridoindoles. VII. Ring expansion of 2-phenylhexahydroazeto[1',2':1,2]pyrido[3,4-b]indoles by oxidation with m-chloroperbenzoic acid. *Chem. Pharm. Bull.* **1995**, 43, 1089-1095.
15. Molina, J. M., El-Bergmi, R., Dobado, J. A., Portal, D. On the Aromaticity and Meisenheimer Rearrangement of Strained Heterocyclic Amine, Phosphine, and Arsine Oxides. *J. Org. Chem.* **2000**, 65, 8574-8581.
16. Greenberg, A., DuBois, T. D. Amide N-oxides: an ab initio molecular orbital study. *J. Mol. Struct.* **2001**, 567-568, 303-317.
17. Mucsi, Z., Szabo, A., Hermecz, I. Ab-initio study on the competitive rearrangements of tertiary N-propargylamine-N-oxides. *THEOCHEM* **2003**, 666-667, 547-556.
18. Kleinschmidt, R. F., Cope, A. C. Rearrangement of allyl groups in dyad systems. Amine oxides. *J. Am. Chem. Soc.* **1944**, 66, 1929-1933.
19. Castagnoli, N., Jr., Cymerman Craig, J., Melikian, A. P., Roy, S. K. Amine-N-oxide rearrangements. Mechanism and products of thermolysis. *Tetrahedron* **1970**, 26, 4319-4327.
20. Lorand, J. P., Grant, R. W., Samuel, P. A., Sister Elizabeth, M. O. C., Zaro, J., Pilotte, J., Wallace, R. W. Radicals and scavengers. II. Scavengers, viscosity, and the cage effect in a Meisenheimer rearrangement1,2. *J. Org. Chem.* **1973**, 38, 1813-1821.
21. Davies, S. G., Smyth, G. D. Asymmetric synthesis of (R)-sulcatol. *Tetrahedron: Asymmetry* **1996**, 7, 1005-1006.
22. Kurihara, T., Doi, M., Hamaura, K., Ohishi, H., Harusawa, S., Yoneda, R. Meisenheimer rearrangement of 2-ethenyl-1,4,5,10b-tetrahydro-2H-azetopyrido[3,4-b]indole N-oxides: new route to the 12(S)carba-eudistomin skeleton. *Chem. Pharm. Bull.* **1991**, 39, 811-813.
23. Kondo, H., Sakamoto, F., Uno, T., Kawahata, Y., Tsukamoto, G. Studies on prodrugs. 11. Synthesis and antimicrobial activity of N-[(4-methyl-5-methylene-2-oxo-1,3-dioxolan-4-yl)oxy]norfloxacin. *J. Med. Chem.* **1989**, 32, 671-674.
24. Yoneda, R., Sakamoto, Y., Oketo, Y., Harusawa, S., Kurihara, T. An efficient synthesis of magallanesine using [1,2]-Meisenheimer rearrangement and Heck cyclization. *Tetrahedron* **1996**, 52, 14563-14576.

Meyer-Schuster and Rupe Rearrangement 284

1. Meyer, K. H., Schuster, K. Rearrangement of tertiary ethynylcarbinols into unsaturated ketones. *Ber.* **1922**, 55B, 819-823.
2. Rupe, H., Glenz, K. Influence of constitution upon the rotatory power of optically active substances. XVI. Acetylene derivatives, ketones and isonitriles. *Ann.* **1924**, 436, 184-204.

3. Rupe, H., Kambli, E. Unsaturated aldehydes from acetylene alcohols. *Helv. Chim. Acta* **1926**, 9, 672.
4. Rupe, H., Kambli, E. Influence of the constitution upon the optical activity of optically active substances. XX. Influence of the triple bond. *Ann.* **1927**, 459, 195-217.
5. Rupe, H., Giesler, L. Aldehydes and acetylenecarbinols. II. Dimethyloctenaldehyde, tert-butylmethylacrolein and experiments with the acetylenecarbinol prepared from acetophenone. *Basel. Helv. Chim. Acta* **1928**, 11, 656-669.
6. Rupe, H., Messner, W., Kambli, E. Aldehydes from acetylenecarbinols. I. Cyclohexylideneacetaldehyde. *Helv. Chim. Acta* **1928**, 11, 449-462.
7. Rupe, H., Wirz, A., Lotter, P. Aldehydes from acetylenecarbinols. III. Preparation of two dimethylhexenaldehydes. *Anstalt fur organ. Chemie, Basel. Helv. Chim. Acta* **1928**, 11, 965-971.
8. Swaminathan, S., Narayanan, K. V. Rupe and Meyer-Schuster rearrangements. *Chem. Rev.* **1971**, 71, 429-438.
9. Newman, M. S. Reactions of acetylenic compounds catalyzed by sulfonated polystyrene resins. *J. Am. Chem. Soc.* **1953**, 75, 4740-4742.
10. Olah, G. A., Fung, A. P. Synthetic methods and reactions; 98. Improved solid super acid (Nafion-H) catalyzed Rupe rearrangement of α-ethynyl alcohols to α,β-unsaturated carbonyl compounds. *Synthesis* **1981**, 473-474.
11. Barre, V., Massias, F., Uguen, D. Sulfone-mediated Rupe and Raphael rearrangements. *Tetrahedron Lett.* **1989**, 30, 7389-7392.
12. Erman, M. B., Gulyi, S. E., Aulchenko, I. S. A new efficient catalytic system for the Meyer-Schuster rearrangement. *Mendeleev Commun.* **1994**, 89.
13. Yoshimatsu, M., Naito, M., Kawahigashi, M., Shimizu, H., Kataoka, T. Meyer-Schuster Rearrangement of γ-Sulfur-Substituted Propargyl Alcohols: A Convenient Synthesis of α,β-Unsaturated Thioesters. *J. Org. Chem.* **1995**, 60, 4798-4802.
14. Lorber, C. Y., Osborn, J. A. Cis-dioxomolybdenum(VI) complexes as new catalysts for the Meyer-Schuster rearrangement. *Tetrahedron Lett.* **1996**, 37, 853-856.
15. Weinmann, H., Harre, M., Neh, H., Nickisch, K., Skoetsch, C., Tilstam, U. The Rupe Rearrangement: A New Efficient Method for Large-Scale Synthesis of Unsaturated Ketones in the Pilot Plant. *Org. Process Res. Dev.* **2002**, 6, 216-219.
16. Andres, J., Arnau, A., Silla, E., Bertran, J., Tapia, O. A theoretical study of the intramolecular solvolytic mechanism of the Meyer-Schuster reaction. MINDO/3 and CNDO/2 calculations of minimum energy paths. *THEOCHEM* **1983**, 14, 49-54.
17. Andres, J., Silla, E., Tapia, O. A quantum chemical study of protonated intermediates in Rupe and Meyer-Schuster rearrangement mechanisms. *THEOCHEM* **1983**, 14, 307-314.
18. Andres, J., Silla, E., Tapia, O. Quantum-chemical studies of the energy hypersurface for the Meyer-Schuster rearrangement. STO-3G calculation of minimum-energy paths. Intermolecular mechanism. *Chem. Phys. Lett.* **1983**, 94, 193-197.
19. Tapia, O., Andres, J. A simple protocol to help calculate saddle points. Transition-state structures for the Meyer-Schuster reaction in nonaqueous media: an ab initio MO study. *Chem. Phys. Lett.* **1984**, 109, 471-477.
20. Andres, J., Cardenas, R., Silla, E., Tapia, O. A theoretical study of the Meyer-Schuster reaction mechanism: minimum-energy profile and properties of transition-state structure. *J. Am. Chem. Soc.* **1988**, 110, 666-674.
21. Andres, J., Pascual-Ahuir, J. L., Silla, E., Bertran, J. Theoretical study of the intermolecular mechanism of the Meyer-Schuster reaction. MINDO/3 and CNDO/2 minimum energy pathways. *Anales de Quimica, Serie A: Quimica Fisica e Ingenieria Quimica* **1988**, 84, 159-162.
22. Hennion, G. F., Davis, R. B., Maloney, D. E. The mechanism of the Rupe reaction. *J. Am. Chem. Soc.* **1949**, 71, 2813-2814.
23. Edens, M., Boerner, D., Chase, C. R., Nass, D., Schiavelli, M. D. The mechanism of the Meyer-Schuster rearrangement. *J. Org. Chem.* **1977**, 42, 3403-3408.
24. Stevens, K. E., Paquette, L. A. Stereocontrolled total synthesis of (±)-$\Delta^{9(12)}$-capnellene. *Tetrahedron Lett.* **1981**, 22, 4393-4396.
25. Stark, H., Sadek, B., Krause, M., Huels, A., Ligneau, X., Ganellin, C. R., Arrang, J.-M., Schwartz, J.-C., Schunack, W. Novel Histamine H3-Receptor Antagonists with Carbonyl-Substituted 4-(3-(Phenoxy)propyl)-1H-imidazole Structures like Ciproxifan and Related Compounds. *J. Med. Chem.* **2000**, 43, 3987-3994.
26. Welch, S. C., Hagan, C. P., White, D. H., Fleming, W. P., Trotter, J. W. A stereoselective total synthesis of the antifungal mold metabolite 7a-methoxy-3a,10b-dimethyl-1,2,3,3aa,5aa,7,10bb,10ca-octahydro-4H,9H-furo[2',3',4':4,5]naphtho[2,1-c]pyran-4,10-dione. *J. Am. Chem. Soc.* **1977**, 99, 549-556.

Michael Addition Reaction 286

Related reactions: Nagata hydrocyanation, Stetter reaction;

1. Komnenos, T. The reaction of unsaturated aldehydes with malonoic acid and ethyl malonate. *Liebigs Ann. Chem.* **1883**, 218, 145-169.
2. Michael, A. *J. Prakt. Chem./Chem.-Ztg.* **1887**, 35, 349.
3. Michael, A. *Am. Chem. J.* **1887**, 9, 115.
4. Michael, A. Addition of sodium acetoacetate and sodium diethyl malonate to unsaturated acids. *J. Prakt. Chem./Chem.-Ztg.* **1894**, 49, 20.
5. Bergmann, E. D., Ginsburg, D., Pappo, R. The Michael reaction. *Org. React.* **1959**, 10, 179-563.
6. Hunt, D. A. Michael addition of organolithium compounds. A review. *Org. Prep. Proced. Int.* **1989**, 21, 705-749.
7. Oare, D. A., Heathcock, C. H. Stereochemistry of the base-promoted Michael addition reaction. *Top. Stereochem.* **1989**, 19, 227-407.
8. Hulce, M. Nucleophilic Addition-Electrophilic Coupling with a Carbanion Intermediate. in *Comp. Org. Synth.* (eds. Trost, B. M.,Fleming, I.), *4*, 237-268 (Pergamon, Oxford, **1991**).
9. Jung, M. E. Stabilized Nucleophiles with Electron Deficient Alkenes and Alkynes. in *Comp. Org. Synth.* (eds. Trost, B. M.,Fleming, I.), *4*, 1-67 (Pergamon, Oxford, **1991**).
10. Kozlowski, J. A. Organocuprates in the Conjugate Addition Reaction. in *Comp. Org. Synth.* (eds. Trost, B. M.,Fleming, I.), *4*, 169-198 (Pergamon, Oxford, **1991**).
11. Lee, V. J. Conjugate Additions of Reactive Carbanions to Activated Alkenes and Alkynes. in *Comp. Org. Synth.* (eds. Trost, B. M.,Fleming, I.), *4*, 69-137 (Pergamon, Oxford, **1991**).
12. Lee, V. J. Conjugate Additions of Carbon Ligands to Activated Alkenes and Alkynes Mediated by Lewis Acids. in *Comp. Org. Synth.* (eds. Trost, B. M.,Fleming, I.), *4*, 139-168 (Pergamon, Oxford, **1991**).
13. Oare, D. A., Heathcock, C. H. Acyclic stereocontrol in Michael addition reactions of enamines and enol ethers. *Top. Stereochem.* **1991**, 20, 87-170.
14. Schmalz, H.-G. Asymmetric Nucleophilic Addition to Electron Deficient Alkenes. in *Comp. Org. Synth.* (eds. Trost, B. M.,Fleming, I.), *4*, 199-236 (Pergamon, Oxford, **1991**).
15. Rele, D., Trivedi, G. K. Recent developments in Michael reaction: a convenient tool for annelation and annulation. *J. Sci. Ind. Res.* **1993**, 52, 13-28.
16. Bernardi, A. Stereoselective conjugate addition of enolates to α,β-unsaturated carbonyl compounds. *Gazz. Chim. Ital.* **1995**, 125, 539-547.
17. Little, R. D., Masjedizadeh, M. R., Wallquist, O., McLoughlin, J. I. The intramolecular Michael reaction. *Org. React.* **1995**, 47, 315-552.
18. Guingant, A. Asymmetric syntheses of α,α-disubstituted β-diketones and β-keto esters. *Advances in Asymmetric Synthesis* **1997**, 2, 119-188.
19. Juaristi, E., Garcia-Barradas, O. Asymmetric addition of amines to α,β-unsaturated esters and nitriles in the enantioselective synthesis of β-amino acids. *Enantioselective Synthesis of .beta.-Amino Acids* **1997**, 139-149.
20. Geirsson, J. K. F. The use of 1-aza-1,3-butadienes as Michael acceptors in the preparation of biologically interesting compounds. *Rec. Res. Dev. Org. Chem.* **1998**, 2, 609-622.
21. Katritzky, A. R., Qi, M. Michael additions of benzotriazole-stabilized carbanions. A review. *Collect. Czech. Chem. Commun.* **1998**, 63, 599-613.
22. Leonard, J., Diez-Barra, E., Merino, S. Control of asymmetry through conjugate addition reactions. *Eur. J. Org. Chem.* **1998**, 2051-2061.
23. Yamaguchi, M. Conjugate addition of stabilized carbanions. *Comprehensive Asymmetric Catalysis I-III* **1999**, 3, 1121-1139.
24. Gil, M. V., Roman, E., Serrano, J. A. Nitro compounds in asymmetric Michael reactions. *Trends in Organic Chemistry* **2001**, 9, 17-28.

25. Berner, O. M., Tedeschi, L., Enders, D. Asymmetric Michael additions to nitroalkenes. *Eur. J. Org. Chem.* **2002**, 1877-1894.
26. Notz, W., Tanaka, F., Barbas, C. F., III. Enamine-Based Organocatalysis with Proline and Diamines: The Development of Direct Catalytic Asymmetric Aldol, Mannich, Michael, and Diels-Alder Reactions. *Acc. Chem. Res.* **2004**, 37, 580-591.
27. Posner, G. H. Asymmetric Michael and Diels-Alder reactions using sulfoxides and sulfones. *Stud. Org. Chem. (Amsterdam)* **1987**, 28, 145-152.
28. Roush, W. R. The catalytic asymmetric Michael reaction of tin(II) enethiolates. *Chemtracts: Org. Chem.* **1988**, 1, 439-442.
29. Roush, W. R. An asymmetric intramolecular Michael reaction. Construction of chiral building blocks for the synthesis of several alkaloids. *Chemtracts: Org. Chem.* **1988**, 1, 233-235.
30. D'Angelo, J., Desmaele, D., Dumas, F., Guingant, A. The asymmetric Michael addition reactions using chiral imines. *Tetrahedron: Asymmetry* **1992**, 3, 459-505.
31. Angelo, J. d., Cave, C., Desmaele, D., Dumas, F. The asymmetric Michael addition reactions using chiral imines: Application to the synthesis of compounds of biological interest. *Trends in Organic Chemistry* **1993**, 4, 555-616.
32. Rudorf, W. D., Schwarz, R. Intramolecular Michael and anti-Michael additions to carbon-carbon triple bonds. *Synlett* **1993**, 369-374.
33. Enders, D. TMS-SAMP. Novel chiral hydrazine auxiliary for hetero-Michael additions and aza-Peterson olefinations. *Acros Organics Acta* **1995**, 1, 37-38.
34. Enders, D., Bettray, W., Schankat, J., Wiedemann, J. Diastereo- and enantioselective synthesis of β-amino acids via SAMP hydrazones and hetero Michael addition using TMS-SAMP as a chiral equivalent of ammonia. *Enantioselective Synthesis of β-Amino Acids* **1997**, 187-210.
35. Ruck-Braun, K., Kunz, H. A new multifunctional heterobimetallic asymmetric catalyst for Michael additions and tandem Michael-aldol reactions. *Chemtracts* **1997**, 10, 519-521.
36. Christoffers, J. Transition-metal catalysis of the Michael reaction of 1,3-dicarbonyl compounds and acceptor-activated alkenes. *Eur. J. Org. Chem.* **1998**, 1259-1266.
37. Fringuelli, F., Piermatti, O., Pizzo, F. Base-catalyzed aldol- and Michael-type condensations in aqueous media. *Organic Synthesis in Water* **1998**, 250-261.
38. Krause, N. Copper-catalyzed enantioselective Michael additions: recent progress with new phosphorus ligands. *Angew. Chem., Int. Ed. Engl.* **1998**, 37, 283-285.
39. Krause, N., Thorand, S. Copper-mediated 1,6-, 1,8-, 1,10- and 1,12-addition and 1,5-substitution reactions in organic synthesis. *Inorg. Chim. Acta* **1999**, 296, 1-11.
40. Yamazaki, T. Asymmetric Michael addition reactions of fluoro carbons. *Enantiocontrolled Synthesis of Fluoro-Organic Compounds* **1999**, 263-286.
41. Kanai, M., Shibasaki, M. Asymmetric Michael reactions. *Catal. Asymmetric Synth. (2nd Edition)* **2000**, 569-592.
42. Krause, N. Copper-catalyzed enantioselective Michael additions: recent progress with new phosphorus ligands. *Organic Synthesis Highlights IV* **2000**, 182-186.
43. Christoffers, J. Catalysis of the Michael reaction and the vinylogous Michael reaction by ferric chloride hexahydrate. *Synlett* **2001**, 723-732.
44. Krause, N., Hoffmann-Roder, A. Recent advances in catalytic enantioselective Michael additions. *Synthesis* **2001**, 171-196.
45. Christoffers, J. Formation of quaternary stereocenters by copper-catalyzed Michael reactions with L-valine amides as auxiliaries. *Chem.--Eur. J.* **2003**, 9, 4862-4867.
46. Christoffers, J., Baro, A. Construction of quaternary stereocenters: New perspectives through enantioselective Michael reactions. *Angew. Chem., Int. Ed. Engl.* **2003**, 42, 1688-1690.
47. Ellis, G. W. L., Tavares, D. F., Rauk, A. The mechanism of an intramolecular Michael addition: a MNDO study. *Can. J. Chem.* **1985**, 63, 3510-3515.
48. Lavallee, J. F., Berthiaume, G., Deslongchamps, P. Intramolecular Michael addition of cyclic β-keto esters onto conjugated acetylenic ketones. *Tetrahedron Lett.* **1986**, 27, 5455-5458.
49. Bayly, C. I., Grein, F. Comparison of an intramolecular Michael-type addition with its intermolecular counterpart: an ab initio theoretical study. *Can. J. Chem.* **1989**, 67, 2173-2177.
50. Bernardi, A., Capelli, A. M., Gennari, C., Scolastico, C. 1,4-Addition to α,β-unsaturated carbonyl compounds bearing a γ-stereocenter: a molecular mechanics model for steric interactions in the transition state. *Tetrahedron: Asymmetry* **1990**, 1, 21-32.
51. Hori, K., Higuchi, S., Kamimura, A. Theoretical and experimental study on the stereoselectivity of Michael addition of alkoxide anion to nitro olefin. *J. Org. Chem.* **1990**, 55, 5900-5905.
52. Sevin, A., Masure, D., Giessner-Prettre, C., Pfau, M. A theoretical investigation of enantioselectivity: Michael reaction of secondary enamines with enones. *Helv. Chim. Acta* **1990**, 73, 552-573.
53. Bernardi, A., Capelli, A. M., Cassinari, A., Comotti, A., Gennari, C., Scolastico, C. A computational study of the 1,4-addition of lithium enolates to conjugated carbonyl compounds. *J. Org. Chem.* **1992**, 57, 7029-7034.
54. Pardo, L., Osman, R., Weinstein, H., Rabinowitz, J. R. Mechanisms of nucleophilic addition to activated double bonds: 1,2- and 1,4-Michael addition of ammonia. *J. Am. Chem. Soc.* **1993**, 115, 8263-8269.
55. Thomas, B. E., Kollman, P. A. An ab initio molecular orbital study of the first step of the catalytic mechanism of thymidylate synthase: the Michael addition of sulfur and oxygen nucleophiles. *J. Org. Chem.* **1995**, 60, 8375-8381.
56. Lucero, M. J., Houk, K. N. Conformational Transmission of Chirality: The Origin of 1,4-Asymmetric Induction in Michael Reactions of Chiral Imines. *J. Am. Chem. Soc.* **1997**, 119, 826-827.
57. Ramirez, M. A., Padron, J. M., Palazon, J. M., Martin, V. S. Stereocontrolled Synthesis of Cyclic Ethers by Intramolecular Hetero-Michael Addition. 6. A Computational Study of the Annelation to 2,3-Disubstituted Tetrahydropyrans. *J. Org. Chem.* **1997**, 62, 4584-4590.
58. Dau, M. E. T. H., Riche, C., Dumas, F., d'Angelo, J. The origin of the stereoselectivity in the asymmetric Michael reaction using chiral imines/secondary enamines under neutral conditions: a computational investigation. *Tetrahedron: Asymmetry* **1998**, 9, 1059-1064.
59. Yonemitsu, O., Yamazaki, T., Uenishi, J.-I. On the stereoselective construction of the B and A rings of halichondrin B. A PM3 study. *Heterocycles* **1998**, 49, 89-92.
60. Dumas, F., Fressigne, C., Langlet, J., Giessner-Prettre, C. Theoretical Investigations of the Influence of Pressure on the Selectivity of the Michael Addition of Diphenylmethanamine to Stereogenic Crotonates. *J. Org. Chem.* **1999**, 64, 4725-4732.
61. Okumoto, S., Yamabe, S. A theoretical study of curing reactions of maleimide resins through Michael additions of amines. *J. Org. Chem.* **2000**, 65, 1544-1548.
62. Poon, T., Mundy, B. P., Shattuck, T. W. The Michael reaction. *J. Chem. Educ.* **2002**, 79, 264-267.
63. Pelzer, S., Kauf, T., Vvan Wuellen, C., Christoffers, J. Catalysis of the Michael reaction by iron(III): calculations, mechanistic insights and experimental consequences. *J. Organomet. Chem.* **2003**, 684, 308-314.
64. Yasuda, M., Chiba, K., Ohigashi, N., Katoh, Y., Baba, A. Michael Addition of Stannyl Ketone Enolate to α,β-Unsaturated Esters Catalyzed by Tetrabutylammonium Bromide and an ab Initio Theoretical Study of the Reaction Course. *J. Am. Chem. Soc.* **2003**, 125, 7291-7300.
65. Yasuda, M., Chiba, K., Ohigashi, N., Katoh, Y., Baba, A. Michael addition of stannyl ketone enolate to α,β-unsaturated esters catalyzed by tetrabutylammonium bromide and an ab initio theoretical study of the reaction course. *J. Am. Chem. Soc.* **2003**, 125, 7291-7300.
66. Chatfield, D. C., Augsten, A., D'Cunha, C., Lewandowska, E., Wnuk, S. F. Theoretical and experimental study of the regioselectivity of Michael additions. *Eur. J. Org. Chem.* **2004**, 313-322.
67. Hoz, S. Is the transition state indeed intermediate between reactants and products? The Michael addition reaction as a case study. *Acc. Chem. Res.* **1993**, 26, 69-74.
68. Kurosu, M., Marcin, L. R., Grinsteiner, T. J., Kishi, Y. Total Synthesis of (±)-Batrachotoxinin A. *J. Am. Chem. Soc.* **1998**, 120, 6627-6628.
69. Boger, D. L., Hueter, O., Mbiya, K., Zhang, M. Total Synthesis of Natural and ent-Fredericamycin A. *J. Am. Chem. Soc.* **1995**, 117, 11839-11849.
70. Takasu, K., Mizutani, S., Noguchi, M., Makita, K., Ihara, M. Total Synthesis of (±)-Culmorin and (±)-Longiborneol: An Efficient Construction of Tricyclo[6.3.0.03,9]undecan-10-one by Intramolecular Double Michael Addition. *J. Org. Chem.* **2000**, 65, 4112-4119.
71. Masaki, H., Maeyama, J., Kamada, K., Esumi, T., Iwabuchi, Y., Hatakeyama, S. Total Synthesis of (-)-Dysiherbaine. *J. Am. Chem. Soc.* **2000**, 122, 5216-5217.

Related reactions: Corey-Bakshi-Shibata (CBS) reduction, Noyori asymmetric hydrogenation;

1. Mikhailov, B. M., Bubnov, Y. N., Kiselev, V. G. Organoboron compounds. CLVIII. Comparative ability of trialkylborons to eliminate olefins. *Zh. Obshch. Khim.* **1966**, 36, 62-66.
2. Midland, M. M., Tramontano, A., Zderic, S. A. Preparation of optically active benzyl-α-d alcohol via reduction by B-3α-pinanyl-9-borabicyclo[3.3.1]nonane. A new highly effective chiral reducing agent. *J. Am. Chem. Soc.* **1977**, 99, 5211-5213.
3. Midland, M. M., Tramontano, A., Zderic, S. A. The facile reaction of B-alkyl-9-borabicyclo[3.3.1]nonanes with benzaldehyde. *J. Organomet. Chem.* **1977**, 134, C17-C19.
4. Midland, M. M., Tramontano, A. B-Alkyl-9-borabicyclo[3.3.1]nonanes as mild, chemoselective reducing agents for aldehydes. *J. Org. Chem.* **1978**, 43, 1470-1471.
5. Midland, M. M., Tramontano, A., Zderic, S. A. The reaction of B-alkyl-9-borabicyclo[3.3.1]nonanes with aldehydes and ketones. A facile elimination of the alkyl group by aldehydes. *J. Organomet. Chem.* **1978**, 156, 203-211.
6. Midland, M. M., Greer, S., Tramontano, A., Zderic, S. A. Chiral trialkylborane reducing agents. Preparation of 1-deuterio primary alcohols of high enantiomeric purity. *J. Am. Chem. Soc.* **1979**, 101, 2352-2355.
7. Midland, M. M., McDowell, D. C., Hatch, R. L., Tramontano, A. Reduction of α,β-acetylenic ketones with B-3-pinanyl-9-borabicyclo[3.3.1]nonane. High asymmetric induction in aliphatic systems. *J. Am. Chem. Soc.* **1980**, 102, 867-869.
8. Brown, H. C., Jadhav, P. K., Mandal, A. K. Asymmetric syntheses via chiral organoborane reagents. *Tetrahedron* **1981**, 37, 3547-3587.
9. Midland, M. M. Asymmetric reductions with organoborane reagents. *Chem. Rev.* **1989**, 89, 1553-1561.
10. Singh, V. K. Practical and useful methods for the enantioselective reduction of unsymmetrical ketones. *Synthesis* **1992**, 607-617.
11. Brown, H. C., Ramachandran, P. V. Versatile α-pinene-based borane reagents for asymmetric syntheses. *J. Organomet. Chem.* **1995**, 500, 1-19.
12. Itsuno, S. Enantioselective reduction of ketones. *Org. React.* **1998**, 52, 395-576.
13. Farina, V., Roth, G. P. Asymmetric synthesis with chiral reagents derived from α-pinene. *Chimica Oggi* **1999**, 17, 39-47.
14. Cho, B. T., Chun, Y. S. Asymmetric reduction of α-functionalized ketones with organoboron-based chiral reducing agents. *ACS Symp. Ser.* **2001**, 783, 122-135.
15. Midland, M. M., Tramontano, A., Kazubski, A., Graham, R. S., Tsai, D. J. S., Cardin, D. B. Asymmetric reductions of propargyl ketones. An effective approach to the synthesis of optically active compounds. *Tetrahedron* **1984**, 40, 1371-1380.
16. Midland, M. M., Lee, P. E. Efficient asymmetric reduction of acyl cyanides with B-3-pinanyl 9-BBN (Alpine-borane). *J. Org. Chem.* **1985**, 50, 3237-3239.
17. Brown, H. C., Chandrasekharan, J., Ramachandran, P. V. Chiral synthesis via organoboranes. 14. Selective reductions. 41. Diisopinocampheylchloroborane, an exceptionally efficient chiral reducing agent. *J. Am. Chem. Soc.* **1988**, 110, 1539-1546.
18. Brown, H. C., Ramachandran, P. V. Selective reductions. 45. Asymmetric reduction of prochiral ketones by iso-2-methyl-, iso-2-ethyl-, and [iso-2-[2-(benzyloxy)ethyl]apopinocampheyl]-tert-butylchloroboranes. Evidence for a major influence of the steric requirements of the 2-substituent on the efficiency of asymmetric reduction. *J. Org. Chem.* **1989**, 54, 4504-4511.
19. Brown, H. C., Srebnik, M., Ramachandran, P. V. Chiral synthesis via organoboranes. 22. Selective reductions. 44. The effect of the steric requirements of the alkyl substituent in isopinocampheylalkylchloroboranes for the asymmetric reduction of representative ketones. *J. Org. Chem.* **1989**, 54, 1577-1583.
20. Midland, M. M., McLoughlin, J. I., Gabriel, J. Asymmetric reductions of prochiral ketones with B-3-pinanyl-9-borabicyclo[3.3.1]nonane (Alpine-Borane) at elevated pressures. *J. Org. Chem.* **1989**, 54, 159-165.
21. Rogic, M. M. Conformational Analysis and the Transition State in Asymmetric Reductions with Boranes Based on (+)-α-Pinene. 1. Benzaldehyde Reduction with Alpine Borane and Other B-Alkyl-9-borabicyclo[3.3.1]nonanes. A Semiempirical Study. *J. Org. Chem.* **1996**, 61, 1341-1346.
22. Midland, M. M., Petre, J. E., Zderic, S. A., Kazubski, A. Thermal reactions of B-alkyl-9-borabicyclo[3.3.1]nonane (9-BBN). Evidence for unusually facile dehydroboration with B-pinanyl-9-BBN. *J. Am. Chem. Soc.* **1982**, 104, 528-531.
23. Midland, M. M., Zderic, S. A. Kinetics of reductions of substituted benzaldehydes with B-alkyl-9-borabicyclo[3.3.1]nonane (9-BBN). *J. Am. Chem. Soc.* **1982**, 104, 525-528.
24. Rogic, M. M., Ramachandran, P. V., Zinnen, H., Brown, L. D., Zheng, M. The origins of stereoselectivity in asymmetric reductions with boranes based on (+)-α-pinene. II. The geometries of competing transition-states and the nature of the reaction. A semiempirical study. *Tetrahedron: Asymmetry* **1997**, 8, 1287-1303.
25. Bland, L., Panigot, M. J. Reaction of Alpine-Borane with aldehydes: reactivity rate assessment by observation of the disappearance of the carbonyl n - P* peak by UV-visible spectroscopy. *Journal of the Arkansas Academy of Science* **2000**, 54, 24-32.
26. Murakami, N., Nakajima, T., Kobayashi, M. Total synthesis of lembehyne A, a neuritogenic spongean polyacetylene. *Tetrahedron Lett.* **2001**, 42, 1941-1943.
27. Xu, L., Price, N. P. J. Stereoselective synthesis of chirally deuterated (S)-D-(6-^{2}H1)glucose. *Carbohydr. Res.* **2004**, 339, 1173-1178.
28. Dussault, P. H., Eary, C. T., Woller, K. R. Total Synthesis of the Alkoxydioxines (+)- and (-)-Chondrillin and (+)- and (-)-Plakorin via Singlet Oxygenation/Radical Rearrangement. *J. Org. Chem.* **1999**, 64, 1789-1797.
29. Walker, J. R., Curley, J. R. W. Improved synthesis of (R)-glycine-d-^{15}N. *Tetrahedron* **2001**, 57, 6695-6701.

Related reactions: Friedel-Crafts alkylation, Friedel-Crafts acylation;

1. Minisci, F., Galli, R., Cecere, M., Malatesta, V., Caronna, T. Nucleophilic character of alkyl radicals: new syntheses by alkyl radicals generated in redox processes. *Tetrahedron Lett.* **1968**, 5609-5612.
2. Caronna, T., Quilico, A., Minisci, F. Free radical reactivity of nitriloxides. 1,3-Addition. *Tetrahedron Lett.* **1970**, 3633-3636.
3. Gardini, G. P., Minisci, F. Nucleophilic character of acyl radicals. Homolytic acylation of quinoxaline. *J. Chem. Soc., C* **1970**, 929.
4. Gardini, G. P., Minisci, F. Nucleophilic character of alkyl radicals. IV. Reactivity with quinoline and quinoxaline. *Ann. Chim. (Rome)* **1970**, 60, 746-752.
5. Minisci, F., Galli, R., Malatesta, V., Caronna, T. Nucleophilic character of alkyl radicals. II. Selective alkylation of pyridine, quinoline, and acridine by hydroperoxides and oxaziranes. *Tetrahedron* **1970**, 26, 4083-4091.
6. Minisci, F., Gardini, G. P., Bertini, F. Metal ion initiated halogenation reaction of N-haloamines. *Can. J. Chem.* **1970**, 48, 544-545.
7. Minisci, F., Gardini, G. P., Galli, R., Bertini, F. New selective type of aromatic substitution: homolytic amidation. *Tetrahedron Lett.* **1970**, 15-16.
8. Minisci, F., Zammori, P., Bernardi, R., Cecere, M., Galli, R. Nucleophilic character of alkyl radicals generated in redox processes. III. Reactivity of alkyl radicals towards conjugated olefins. *Tetrahedron* **1970**, 26, 4153-4166.
9. Buratti, W., Gardini, G. P., Minisci, F., Bertini, F., Galli, R., Perchinunno, M. Nucleophilic character of alkyl radicals. V. Selective homolytic α-oxyalkylation of heteroaromatic bases. *Tetrahedron* **1971**, 27, 3655-3668.
10. Gardini, G. P., Minisci, F., Palla, G. Polar character of th methyl radical. *Chim. Ind. (Milan)* **1971**, 53, 263-264.
11. Gardini, G. P., Minisci, F., Palla, G., Arnone, A., Galli, R. Homolytic amidation of heteroaromatic bases: a new selective process. *Tetrahedron Lett.* **1971**, 59-62.
12. Minisci, F. Novel applications of free-radical reactions in preparative organic chemistry. *Synthesis* **1973**, 1-24.
13. Heinisch, G. Advances in the synthesis of substituted pyridazines via introduction of carbon functional groups into the parent heterocycles. *Heterocycles* **1987**, 26, 481-496.

14. Minisci, F. Selective syntheses via radical reactions. *Chim. Ind. (Milan)* **1988**, 70, 82-94.
15. Minisci, F., Vismara, E., Fontana, F. Recent developments of free-radical substitutions of heteroaromatic bases. *Heterocycles* **1989**, 28, 489-519.
16. Vismara, E., Fontana, F., Minisci, F. Alkyl iodides as source of alkyl radicals, useful for selective syntheses. *NATO ASI Ser., Ser. C* **1989**, 260, 53-69.
17. Minisci, F., Bertini, F., Galli, R., Quilico, A. *Alkylation of pyridines.* De 2153234, **1972** (Montecatini Edison S.p.A.). 12 pp.
18. Minisci, F., Bertini, F., Galli, R., Quilico, A. *Alkylation of pyridine derivatives.* It 906418, **1972** (Montedison S.p.A., Italy). 14 pp.
19. Caronna, T., Fronza, G., Minisci, F., Porta, O. Homolytic acylation of protonated pyridine and pyrazine derivatives. *J. Chem. Soc., Perkin Trans. 2* **1972**, 2035-2038.
20. Gebauer, M., Heinisch, G., Lotsch, G. Pyridazines. XXXIX. N-Substituted 1,2-dihydro-1,2-diazines formed in homolytic alkoxycarbonylation reactions of pyridazines. *Tetrahedron* **1988**, 44, 2449-2455.
21. Bertini, F., Galli, R., Minisci, F., Porta, O. Free radical reactivity of the imidazole ring. *Chim. Ind. (Milan)* **1972**, 54, 223.
22. Minisci, F., Caronna, T., Galli, R., Malatesta, V. Homolytic acylation of benzothiazole. Diagnostic criterion for the presence of acyl radicals. *J. Chem. Soc. C* **1971**, 1747-1750.
23. Malatesta, V., Minisci, F. Steric effects in homolytic substitution of benzene by dialkylamine radicals. Effect of the alkyl group nature. *Chim. Ind. (Milan)* **1971**, 53, 1154-1155.
24. Citterio, A., Gentile, A., Minisci, F., Serravalle, M., Ventura, S. Polar effects in free-radical reactions. Carbamoylation and α-N-amidoalkylation of heteroaromatic bases by amides and hydroxylamine-O-sulfonic acid. *J. Org. Chem.* **1984**, 49, 3364-3367.
25. Fontana, F., Minisci, F., Vismara, E. New general and convenient sources of alkyl radicals, useful for selective syntheses. *Tetrahedron Lett.* **1988**, 29, 1975-1978.
26. Fontana, F., Minisci, F., Barbosa, M. C. N., Vismara, E. New general processes of homolytic alkylation of heteroaromatic bases by tert-butyl peroxide or di-tert-butyl peroxide and alkyl iodides. *Acta Chem. Scand.* **1989**, 43, 995-999.
27. Minisci, F., Vismara, E., Fontana, F. Redox catalysis and electron-transfer processes in selective organic syntheses. *NATO ASI Ser., Ser. C* **1989**, 257, 29-60.
28. Biyouki, M. A. A., Smith, R. A. J., Bedford, J. J., Leader, J. P. Hydroxymethylation and carbamoylation of di- and tetramethylpyridines using radical substitution (Minisci) reactions. *Synth. Commun.* **1998**, 28, 3817-3825.
29. Rothenberg, G., Feldberg, L., Wiener, H., Sasson, Y. Copper-catalyzed homolytic and heterolytic benzylic and allylic oxidation using tert-butyl hydroperoxide. *J. Chem. Soc., Perkin Trans. 2* **1998**, 2429-2434.
30. Bertini, F., Caronna, T., Galli, R., Minisci, F., Porta, O. New processes for the homolytic alkylation of protonated heteroaromatic bases. *Chim. Ind. (Milan)* **1972**, 54, 425-426.
31. Minisci, F., Kintzinger, J. P., Porta, O., Barilli, P., Gardini, G. P. Nucleophilic character of alkyl radicals. VIII. Kinetics and mechanism of induced decomposition of decanoyl peroxide in the homolytic alkylation of protonated quinoline. *Tetrahedron* **1972**, 28, 2415-2427.
32. minisci, F., Mondelli, R., Gardini, G. p., Porta, O. Nucleophilic character of alkyl radicals. VII. Substituent effects on the homolytic alkylation of protonated heteroaromatic bases with methyl, primary, secondary, and tertiary alkyl radicals. *Tetrahedron* **1972**, 28, 2403-2413.
33. Clerici, A., Minisci, F., Porta, O. Nucleophilic character of alkyl radicals. X. Polar and steric effects in the alkylation of 3-substituted pyridines by tert-butyl radical. *Tetrahedron* **1974**, 30, 4201-4203.
34. Minisci, F., Giordano, C., Vismara, E., Levi, S., Tortelli, V. Polar effects in free radical reactions. Induced decompositions of peroxo compounds in the substitution of heteroaromatic bases by nucleophilic radicals. *J. Am. Chem. Soc.* **1984**, 106, 7146-7150.
35. Minisci, F., Vismara, E., Fontana, F., Platone, E., Faraci, G. Chlorinations by hypochlorous acid: free-radical versus electrophilic reactions. *Chim. Ind. (Milan)* **1988**, 70, 52-55.
36. Fontana, F., Minisci, F., Vismara, E., Faraci, G., Platone, E. Chlorination by hypochlorous acid. Free-radical versus electrophilic reactions. *NATO ASI Ser., Ser. C* **1989**, 260, 269-282.
37. Vismara, E., Donna, A., Minisci, F., Naggi, A., Pastori, N., Torri, G. Reactivity of carbohydrate radicals derived from iodo sugars and dibenzoyl peroxide. Homolytic heteroaromatic and aromatic substitution, reduction, and oxidation. *J. Org. Chem.* **1993**, 58, 959-963.
38. Doll, M. K. H. A Short Synthesis of the 8-Azaergoline Ring System by Intramolecular Tandem Decarboxylation-Cyclization of the Minisci-Type Reaction. *J. Org. Chem.* **1999**, 64, 1372-1374.
39. Cowden, C. J. Use of *N*-Protected Amino Acids in the Minisci Radical Alkylation. *Org. Lett.* **2003**, 5, 4497-4499.

Mislow-Evans Rearrangement 292

Related reactions: Meisenheimer rearrangement;

1. Bickart, P., Carson, F. W., Jacobus, J., Miller, E. G., Mislow, K. Thermal racemization of allylic sulfoxides and interconversion of allylic sulfoxides and sulfenates. Mechanism and stereochemistry. *J. Am. Chem. Soc.* **1968**, 90, 4869-4876.
2. Tang, R., Mislow, K. Rates and equilibria in the interconversion of allylic sulfoxides and sulfenates. *J. Am. Chem. Soc.* **1970**, 92, 2100-2104.
3. Evans, D. A., Andrews, G. C., Sims, C. L. Reversible 1,3 transposition of sulfoxide and alcohol functions. Potential synthetic utility. *J. Am. Chem. Soc.* **1971**, 93, 4956-4957.
4. Evans, D. A., Andrews, G. C. Allylic sulfoxides. Useful intermediates in organic synthesis. *Acc. Chem. Res.* **1974**, 7, 147-155.
5. Hoffmann, R. W. Stereochemistry of [2,3]sigmatropic rearrangements. *Angew. Chem., Int. Ed. Engl.* **1979**, 18, 563-572.
6. Altenbach, H. J. Functional group transformations via allyl rearrangement. in *Comp. Org. Synth.* (eds. Trost, B. M.,Fleming, I.), *6*, 829-871 (Pergamon, Oxford, **1991**).
7. Brückner, R. [2,3]-Sigmatropic rearrangements. in *Comp. Org. Synth.* (eds. Trost, B. M.,Fleming, I.), *6*, 873-909 (Pergamon, Oxford, **1991**).
8. Prilezhaeva, E. N. Rearrangements of sulfoxides and sulfones in the total synthesis of natural compounds. *Russian Chemical Reviews* **2001**, 70, 897-920.
9. Andrews, G., Evans, D. A. Stereochemistry of the rearrangement of allylic sulfonium ylides. New method for the stereoselective formation of asymmetry at quaternary carbon. *Tetrahedron Lett.* **1972**, 5121-5124.
10. Evans, D. A., Andrews, G. C., Fujimoto, T. T., Wells, D. Application of allylic sulfoxide anions as vinyl anion equivalents. General synthesis of allylic alcohols. *Tetrahedron Lett.* **1973**, 1385-1388.
11. Masaki, Y., Sakuma, K., Kaji, K. Facile synthesis of (E)-allylic alcohols by acid-catalyzed modification of the Mislow-Evans rearrangement of allylic sulfoxides. *Chem. Pharm. Bull.* **1985**, 33, 2531-2534.
12. Zhou, Z. S., Flohr, A., Hilvert, D. An Antibody-Catalyzed Allylic Sulfoxide-Sulfenate Rearrangement. *J. Org. Chem.* **1999**, 64, 8334-8341.
13. Jones-Hertzog, D. K., Jorgensen, W. L. Regioselective Synthesis of Allylic Alcohols Using the Mislow-Evans Rearrangement: A Theoretical Rationalization. *J. Org. Chem.* **1995**, 60, 6682-6683.
14. Jones-Hertzog, D. K., Jorgensen, W. L. Elucidation of Transition Structures and Solvent Effects for the Mislow-Evans Rearrangement of Allylic Sulfoxides. *J. Am. Chem. Soc.* **1995**, 117, 9077-9078.
15. Evans, D. A., Andrews, G. C. Nucleophilic cleavage of allylic sulfenate esters. Mechanistic observations. *J. Am. Chem. Soc.* **1972**, 94, 3672-3674.
16. Amaudrut, J., Wiest, O. The Thermal Sulfenate-Sulfoxide Rearrangement: A Radical Pair Mechanism. *J. Am. Chem. Soc.* **2000**, 122, 3367-3374.
17. Taber, D. F., Teng, D. Total Synthesis of the Ethyl Ester of the Major Urinary Metabolite of Prostaglandin E2. *J. Org. Chem.* **2002**, 67, 1607-1612.
18. Engstrom, K. M., Mendoza, M. R., Navarro-Villalobos, M., Gin, D. Y. Total synthesis of (+)-pyrenolide D. *Angew. Chem., Int. Ed. Engl.* **2001**, 40, 1128-1130.
19. Majetich, G., Song, J. S., Ringold, C., Nemeth, G. A., Newton, M. G. Intramolecular additions of allylsilanes to conjugated dienones. A direct stereoselective synthesis of (±)-14-deoxyisoamijiol. *J. Org. Chem.* **1991**, 56, 3973-3988.

20. Baba, Y., Saha, G., Nakao, S., Iwata, C., Tanaka, T., Ibuka, T., Ohishi, H., Takemoto, Y. Asymmetric Total Synthesis of Halicholactone. *J. Org. Chem.* **2001**, 66, 81-88.

1. Mitsunobu, O., Yamada, M. Preparation of esters of carboxylic and phosphoric acid via quaternary phosphonium salts. *Bull. Chem. Soc. Jpn.* **1967**, 40, 2380-2382.
2. Mitsunobu, O., Yamada, M., Mukaiyama, T. Preparation of esters of phosphoric acid by the reaction of trivalent phosphorus compounds with diethyl azodicarboxylate in the presence of alcohols. *Bull. Chem. Soc. Jpn.* **1967**, 40, 935-939.
3. Mitsunobu, O. The use of diethyl azodicarboxylate and triphenylphosphine in synthesis and transformation of natural products. *Synthesis* **1981**, 1-28.
4. Castro, B. R. Replacement of alcoholic hydroxyl groups by halogens and other nucleophiles via oxyphosphonium intermediates. *Org. React.* **1983**, 29, 1-162.
5. Hughes, D. L. The Mitsunobu reaction. *Org. React.* **1992**, 42, 335-656.
6. Hughes, D. L. Progress in the Mitsunobu reaction. A review. *Org. Prep. Proced. Int.* **1996**, 28, 127-164.
7. Dodge, J. A., Jones, S. A. Advances in the Mitsunobu reaction for the stereochemical inversion of hindered secondary alcohols. *Rec. Res. Dev. Org. Chem.* **1997**, 1, 273-283.
8. Simon, C., Hosztafi, S., Makleit, S. Application of the Mitsunobu reaction in the field of alkaloids. *J. Heterocycl. Chem.* **1997**, 34, 349-365.
9. Wisniewski, K., Koldziejczyk, A. S., Falkiewicz, B. Applications of the Mitsunobu reaction in peptide chemistry. *Journal of Peptide Science* **1998**, 4, 1-14.
10. Lawrence, S. The Mitsunobu reaction. *PharmaChem* **2002**, 1, 12-14.
11. Dandapani, S., Curran, D. P. Separation-friendly Mitsunobu reactions: A microcosm of recent developments in separation strategies. *Chem.-- Eur. J.* **2004**, 10, 3130-3138.
12. Dembinski, R. Recent advances in the Mitsunobu reaction: Modified reagents and the quest for chromatography-free separation. *Eur. J. Org. Chem.* **2004**, 2763-2772.
13. Martin, S. F., Dodge, J. A. Efficacious modification of the Mitsunobu reaction for inversions of sterically hindered secondary alcohols. *Tetrahedron Lett.* **1991**, 32, 3017-3020.
14. Charette, A. B., Cote, B., Monroc, S., Prescott, S. Synthesis of Monoprotected 2-Alkylidene-1,3-propanediols by an Unusual SN2' Mitsunobu Reaction. *J. Org. Chem.* **1995**, 60, 6888-6894.
15. Anderson, N. G., Lust, D. A., Colapret, K. A., Simpson, J. H., Malley, M. F., Gougoutas, J. Z. Sulfonation with Inversion by Mitsunobu Reaction: An Improvement on the Original Conditions. *J. Org. Chem.* **1996**, 61, 7955-7958.
16. Harvey, P. J., von Itzstein, M., Jenkins, I. D. The formation of anhydrides in the Mitsunobu reaction. *Tetrahedron* **1997**, 53, 3933-3942.
17. Kiankarimi, M., Lowe, R., McCarthy, J. R., Whitten, J. P. Diphenyl 2-pyridylphosphine and di-tert-butyl azodicarboxylate: convenient reagents for the Mitsunobu reaction. *Tetrahedron Lett.* **1999**, 40, 4497-4500.
18. Paul, N. M., Gabriel, C. J., Parquette, J. R. Developments in fluorous Mitsunobu chemistry. *Chemtracts* **2002**, 15, 617-622.
19. Curran, D. P., Dandapani, S. *Fluorous nucleophilic substitution of alcohols and reagents for use therein, specifically, perfluoroalkyl-containing phosphines and azodicarboxylates as polyfluorinated reagents for the Mitsunobu reaction.* 2002-US26045 2003016246, **2003** (University of Pittsburgh, USA).
20. Mukaiyama, T., Shintou, T., Fukumoto, K. A Convenient Method for the Preparation of Inverted tert-Alkyl Carboxylates from Chiral tert-Alcohols by a New Type of Oxidation-Reduction Condensation Using 2,6-Dimethyl-1,4-benzoquinone. *J. Am. Chem. Soc.* **2003**, 125, 10538-10539.
21. Shintou, T., Mukaiyama, T. Efficient method for the preparation of primary, inverted secondary and tertiary alkyl carboxylates from alcohols and carboxylic acids by a new type of oxidation-reduction condensation using simple 1,4-benzoquinone. *Chem. Lett.* **2003**, 32, 1100-1101.
22. Mukaiyama, T., Masutani, K., Hagiwara, Y. Preparation of nitriles from primary alcohols by a new type of oxidation-reduction condensation using 2,6-dimethyl-1,4-benzoquinone and diethyl cyanophosphonate. *Chem. Lett.* **2004**, 33, 1192-1193.
23. Shintou, T., Fukumoto, K., Mukaiyama, T. Efficient method for the preparation of inverted alkyl carboxylates and phenyl carboxylates via oxidation-reduction condensation using 2,6-dimethyl-1,4-benzoquinone or simple 1,4-benzoquinone. *Bull. Chem. Soc. Jpn.* **2004**, 77, 1569-1579.
24. Shintou, T., Mukaiyama, T. Efficient Methods for the Preparation of Alkyl-Aryl and Symmetrical or Unsymmetrical Dialkyl Ethers between Alcohols and Phenols or Two Alcohols by Oxidation-Reduction Condensation. *J. Am. Chem. Soc.* **2004**, 126, 7359-7367.
25. Grochowski, E., Hilton, B. D., Kupper, R. J., Michejda, C. J. Mechanism of the triphenylphosphine and diethyl azodicarboxylate induced dehydration reactions (Mitsunobu reaction). The central role of pentavalent phosphorus intermediates. *J. Am. Chem. Soc.* **1982**, 104, 6876-6877.
26. Guthrie, R. D. G., Jenkins, I. D. The mechanism of the Mitsunobu reaction. A phosphorus-31 NMR study. *Aust. J. Chem.* **1982**, 35, 767-774.
27. Townsend, C. A., Nguyen, L. T. Improved asymmetric synthesis of (-)-3-aminonocardicinic acid and further observations of the Mitsunobu reaction for β-lactam formation in seryl peptides. *Tetrahedron Lett.* **1982**, 23, 4859-4862.
28. Von Itzstein, M., Jenkins, I. D. The mechanism of the Mitsunobu reaction. II. Dialkoxytriphenylphosphoranes. *Aust. J. Chem.* **1983**, 36, 557-563.
29. Adam, W., Narita, N., Nishizawa, Y. On the mechanism of the triphenylphosphine-azodicarboxylate (Mitsunobu reaction) esterification. *J. Am. Chem. Soc.* **1984**, 106, 1843-1845.
30. Varasi, M., Walker, K. A. M., Maddox, M. L. A revised mechanism for the Mitsunobu reaction. *J. Org. Chem.* **1987**, 52, 4235-4238.
31. Hughes, D. L., Reamer, R. A., Bergan, J. J., Grabowski, E. J. J. A mechanistic study of the Mitsunobu esterification reaction. *J. Am. Chem. Soc.* **1988**, 110, 6487-6491.
32. Camp, D., Jenkins, I. D. The mechanism of the Mitsunobu esterification reaction. Part I. The involvement of phosphoranes and oxyphosphonium salts. *J. Org. Chem.* **1989**, 54, 3045-3049.
33. Camp, D., Jenkins, I. D. The mechanism of the Mitsunobu esterification reaction. Part II. The involvement of (acyloxy)alkoxyphosphoranes. *J. Org. Chem.* **1989**, 54, 3049-3054.
34. Crich, D., Dyker, H., Harris, R. J. Some observations on the mechanism of the Mitsunobu reaction. *J. Org. Chem.* **1989**, 54, 257-259.
35. Camp, D., Jenkins, I. D. The mechanism of the Mitsunobu reaction. III. The use of tributylphosphine. *Aust. J. Chem.* **1992**, 45, 47-55.
36. Kodaka, M., Tomohiro, T., Okuno, H. The mechanism of the Mitsunobu reaction and its application to carbon dioxide fixation. *J. Chem. Soc., Chem. Commun.* **1993**, 81-82.
37. Macor, J. E., Wehner, J. M. The use of (o-nitroaryl)acetonitriles in the Mitsunobu reaction: mechanistic implications and synthetic applications. *Heterocycles* **1993**, 35, 349-365.
38. Afonso, C. M., Barros, M. T., Godinho, L. S., Maycock, C. D. The mechanism of the Mitsunobu azide modification and the effect of additives on the rate of hydroxyl group activation. *Tetrahedron* **1994**, 50, 9671-9678.
39. Camp, D., Hanson, G. R., Jenkins, I. D. Formation of Radicals in the Mitsunobu Reaction. *J. Org. Chem.* **1995**, 60, 2977-2980.
40. Moravcova, J., Rollin, P., Lorin, C., Gardon, V., Capkova, J., Mazac, J. Mechanism of regioselective Mitsunobu thio-functionalization of pentofuranoses. *J. Carbohydr. Chem.* **1997**, 16, 113-127.
41. Eberson, L., Persson, O., Svensson, J. O. Structure of the radicals formed in the Mitsunobu reaction. *Acta Chem. Scand.* **1998**, 52, 1293-1300.
42. Sung, D. D., Choi, M. J., Ha, K. M., Uhm, T. S. Reactivity and reaction mechanism for reactions of 1,1'-(azodicarbonyl)dipiperidine with triphenylphosphines. *Bull. Korean Chem. Soc.* **1999**, 20, 935-938.
43. Watanabe, T., Gridnev, I. D., Imamoto, T. Synthesis of a new enantiomerically pure P-chiral phosphine and its use in probing the mechanism of the Mitsunobu reaction. *Chirality* **2000**, 12, 346-351.
44. Ahn, C., Correia, R., DeShong, P. Mechanistic study of the Mitsunobu reaction. *J. Org. Chem.* **2002**, 67, 1751-1753.

45. Elson, K. E., Jenkins, I. D., Loughlin, W. A. The Hendrickson reagent and the Mitsunobu reaction: a mechanistic study. *Org. Biomol. Chem.* **2003**, 1, 2958-2965.
46. Smith, A. B., III, Safonov, I. G., Corbett, R. M. Total Synthesis of (+)-Zampanolide. *J. Am. Chem. Soc.* **2001**, 123, 12426-12427.
47. Overman, L. E., Paone, D. V. Enantioselective Total Syntheses of Ditryptophenaline and ent-WIN 64821. *J. Am. Chem. Soc.* **2001**, 123, 9465-9467.
48. Boger, D. L., McKie, J. A., Nishi, T., Ogiku, T. Enantioselective Total Synthesis of (+)-Duocarmycin A, epi-(+)-Duocarmycin A, and Their Unnatural Enantiomers. *J. Am. Chem. Soc.* **1996**, 118, 2301-2302.
49. Abe, H., Aoyagi, S., Kibayashi, C. First Total Synthesis of the Marine Alkaloids (±)-Fasicularine and (±)-Lepadiformine Based on Stereocontrolled Intramolecular Acylnitroso-Diels-Alder Reaction. *J. Am. Chem. Soc.* **2000**, 122, 4583-4592.

Miyaura Boration 296

1. Ishiyama, T., Matsuda, N., Miyaura, N., Suzuki, A. Platinum(0)-catalyzed diboration of alkynes. *J. Am. Chem. Soc.* **1993**, 115, 11018-11019.
2. Ishiyama, T., Murata, M., Miyaura, N. Palladium(0)-Catalyzed Cross-Coupling Reaction of Alkoxydiboron with Haloarenes: A Direct Procedure for Arylboronic Esters. *J. Org. Chem.* **1995**, 60, 7508-7510.
3. Ishiyama, T., Miyaura, N. Synthesis of arylboronates via palladium-catalyzed cross-coupling reaction of alkoxydiboron with aryl halides or triflates. *Spec. Publ. - R. Soc. Chem.* **1997**, 201, 92-95.
4. Marder, T. B., Norman, N. C. Transition metal catalyzed diboration. *Top. in Cat.* **1998**, 5, 63-73.
5. Suzuki, A. Cross-coupling reactions of organoboron compounds with organic halides. *Metal-Catalyzed Cross-Coupling Reactions* **1998**, 49-97.
6. Suzuki, A. Recent advances in the cross-coupling reactions of organoboron derivatives with organic electrophiles, 1995-1998. *J. Organomet. Chem.* **1999**, 576, 147-168.
7. Chemler, S. R., Trauner, D., Danishefsky, S. J. The B-alkyl Suzuki-Miyaura cross-coupling reaction: development, mechanistic study, and applications in natural product synthesis. *Angewandte Chemie, International Edition* **2001**, 40, 4544-4568.
8. Miyaura, N. Organoboron compounds. *Top. Curr. Chem.* **2002**, 219, 11-59.
9. Prim, D., Campagne, J.-M., Joseph, D., Andrioletti, B. Palladium-catalyzed reactions of aryl halides with soft, non-organometallic nucleophiles. *Tetrahedron* **2002**, 58, 2041-2075.
10. Liu, X. Bis(pinacolato)diboron. *Synlett* **2003**, 2442-2443.
11. Dembitsky, V. M., Ali, H. A., Srebnik, M. Recent chemistry of the diboron compounds. *Adv. Organomet. Chem.* **2004**, 51, 193-250.
12. Ishiyama, T., Miyaura, N. Metal-catalyzed reactions of diborons for synthesis of organoboron compounds. *Chemical Record* **2004**, 3, 271-280.
13. Ahiko, T.-a., Ishiyama, T., Miyaura, N. A sequence of palladium-catalyzed borylation of allyl acetates with bis(pinacolato)diboron and intramolecular allylboration for the cyclization of oxo-2-alkenyl acetates. *Chem. Lett.* **1997**, 811-812.
14. Giroux, A., Han, Y., Prasit, P. One pot biaryl synthesis via in situ boronate formation. *Tetrahedron Lett.* **1997**, 38, 3841-3844.
15. Ishiyama, T., Itoh, Y., Kitano, T., Miyaura, N. Synthesis of arylboronates via the palladium(0)-catalyzed cross-coupling reaction of tetra(alkoxo)diborons with aryl triflates. *Tetrahedron Lett.* **1997**, 38, 3447-3450.
16. Murata, M., Watanabe, S., Masuda, Y. Novel Palladium(0)-Catalyzed Coupling Reaction of Dialkoxyborane with Aryl Halides: Convenient Synthetic Route to Arylboronates. *J. Org. Chem.* **1997**, 62, 6458-6459.
17. Murata, M., Oyama, T., Watanabe, S., Masuda, Y. Synthesis of alkenylboronates via palladium-catalyzed borylation of alkenyl triflates (or iodides) with pinacolborane. *Synthesis* **2000**, 778-780.
18. Murata, M., Oyama, T., Watanabe, S., Masuda, Y. Palladium-Catalyzed Borylation of Aryl Halides or Triflates with Dialkoxyborane: A Novel and Facile Synthetic Route to Arylboronates. *J. Org. Chem.* **2000**, 65, 164-168.
19. Murata, M., Watanabe, S., Masuda, Y. Regio- and stereoselective synthesis of allylboranes via platinum(0)-catalyzed borylation of allyl halides with pinacolborane. *Tetrahedron Lett.* **2000**, 41, 5877-5880.
20. Takahashi, K., Takagi, J., Ishiyama, T., Miyaura, N. Synthesis of 1-alkenylboronic esters via palladium-catalyzed cross-coupling reaction of bis(pinacolato)diboron with 1-alkenyl halides and triflates. *Chem. Lett.* **2000**, 126-127.
21. Willis, D. M., Strongin, R. M. Palladium-catalyzed borylation of aryldiazonium tetrafluoroborate salts. A new synthesis of arylboronic esters. *Tetrahedron Lett.* **2000**, 41, 8683-8686.
22. Yang, F.-Y., Wu, M.-Y., Cheng, C.-H. Highly Regio- and Stereoselective Acylboration of Allenes Catalyzed by Palladium Complexes: An Efficient Route to a New Class of 2-Acylallylboronates. *J. Am. Chem. Soc.* **2000**, 122, 7122-7123.
23. Ishiyama, T., Ishida, K., Miyaura, N. Synthesis of pinacol arylboronates via cross-coupling reaction of bis(pinacolato)diboron with chloroarenes catalyzed by palladium(0)-tricyclohexylphosphine complexes. *Tetrahedron* **2001**, 57, 9813-9816.
24. Ishiyama, T., Oohashi, Z., Ahiko, T.-A., Miyaura, N. Nucleophilic borylation of benzyl halides with bis(pinacolato)diboron catalyzed by palladium(0) complexes. *Chem. Lett.* **2002**, 780-781.
25. Takagi, J., Sato, K., Hartwig, J. F., Ishiyama, T., Miyaura, N. Iridium-catalyzed C-H coupling reaction of heteroaromatic compounds with bis(pinacolato)diboron: regioselective synthesis of heteroarylboronates. *Tetrahedron Lett.* **2002**, 43, 5649-5651.
26. Appukkuttan, P., Van der Eycken, E., Dehaen, W. Microwave enhanced formation of electron rich arylboronates. *Synlett* **2003**, 1204-1206.
27. Ishiyama, T., Takagi, J., Kamon, A., Miyaura, N. Palladium-catalyzed cross-coupling reaction of bis(pinacolato)diboron with vinyl triflates β-substituted by a carbonyl group: efficient synthesis of β-boryl-α,β-unsaturated carbonyl compounds and their synthetic utility. *J. Organomet. Chem.* **2003**, 687, 284-290.
28. Ma, Y., Song, C., Jiang, W., Xue, G., Cannon, J. F., Wang, X., Andrus, M. B. Borylation of Aryldiazonium Ions with N-Heterocyclic Carbene-Palladium Catalysts Formed without Added Base. *Org. Lett.* **2003**, 5, 4635-4638.
29. Wolan, A., Zaidlewicz, M. Synthesis of arylboronates by the palladium catalyzed cross-coupling reaction in ionic liquids. *Org. Biomol. Chem.* **2003**, 1, 3274-3276.
30. Broutin, P.-E., Cerna, I., Campaniello, M., Leroux, F., Colobert, F. Palladium-Catalyzed Borylation of Phenyl Bromides and Application in One-Pot Suzuki-Miyaura Biphenyl Synthesis. *Org. Lett.* **2004**, 6, 4419-4422.
31. Melaimi, M., Thoumazet, C., Ricard, L., Floch, P. L. Syntheses of a 2,6-bis-(methylphospholyl)pyridine ligand and its cationic Pd(II) and Ni(II) complexes - application in the palladium-catalyzed synthesis of arylboronic esters. *J. Organomet. Chem.* **2004**, 689, 2988-2994.
32. Sumimoto, M., Iwane, N., Takahama, T., Sakaki, S. Theoretical Study of Trans-metalation Process in Palladium-Catalyzed Borylation of Iodobenzene with Diboron. *J. Am. Chem. Soc.* **2004**, 126, 10457-10471.
33. Lin, S., Danishefsky, S. J. The total synthesis of proteasome inhibitors TMC-95A and TMC-95B: discovery of a new method to generate cis-propenyl amides. *Angew. Chem., Int. Ed. Engl.* **2002**, 41, 512-515.
34. Carbonnelle, A.-C., Zhu, J. A Novel Synthesis of Biaryl-Containing Macrocycles by a Domino Miyaura Arylboronate Formation: Intramolecular Suzuki Reaction. *Org. Lett.* **2000**, 2, 3477-3480.
35. Miyashita, K., Sakai, T., Imanishi, T. Total Synthesis of (±)-Spiroxin C. *Org. Lett.* **2003**, 5, 2683-2686.
36. Wang, B. B., Smith, P. J. Synthesis of a terbenzimidazole topoisomerase I poison via iterative borinate ester couplings. *Tetrahedron Lett.* **2003**, 44, 8967-8969.

Mukaiyama Aldol Reaction 298

Related reactions: Aldol reaction, Evans aldol reaction, Reformatsky reaction;

1. Mukaiyama, T., Narasaka, K., Banno, K. New aldol type reaction. *Chem. Lett.* **1973**, 1011-1014.

2. Mukaiyama, T., Banno, K., Narasaka, K. New cross-aldol reactions. Reactions of silyl enol ethers with carbonyl compounds activated by titanium tetrachloride. *J. Am. Chem. Soc.* **1974**, 96, 7503-7509.
3. Heathcock, C. H. Acyclic stereocontrol through the aldol condensation. *Science* **1981**, 214, 395-400.
4. Mukaiyama, T. The directed aldol reaction. *Org. React.* **1982**, 28, 203-331.
5. Suzuki, T., Hirama, M. Asymmetric aldol reaction of silyl enol ethers with aldehydes promoted by the combined use of chiral diamine coordinated tin(II) triflate and tributyltin fluoride. *Chemtracts: Org. Chem.* **1989**, 2, 268-270.
6. Altenbach, H. J. Chiral Lewis acids. *Org. Synth. Highlights* **1991**, 66-70.
7. Bianchini, C., Glendenning, L. Homogeneous catalysis. Mechanisms of the catalytic Mukaiyama aldol and Sakurai allylation reactions. *Chemtracts: Inorg. Chem.* **1995**, 7, 107-111.
8. Bianchini, C., Glendenning, L. Homogeneous catalysis. Mechanisms of the catalytic Mukaiyama aldol and Sakurai allylation reactions. *Chemtracts: Org. Chem.* **1996**, 9, 331-335.
9. Ellis, W. W., Bosnich, B. Mechanisms of the catalyzed Mukaiyama cross-aldol reaction. *Organic Synthesis via Organometallics, Proceedings of the Symposium, 5th, Heidelberg, Sept. 26-28, 1996* **1997**, 209-227.
10. Groger, H., Vogl, E. M., Shibasaki, M. New catalytic concepts for the asymmetric aldol reaction. *Chem.-- Eur. J.* **1998**, 4, 1137-1141.
11. Mahrwald, R. Lewis acid catalysts in enantioselective aldol addition. *Rec. Res. Dev. Synt. Org. Chem.* **1998**, 1, 123-150.
12. Bellassoued, M., Chelain, E. Silylketene acetals: Preparation and selective aldol reactions. *Rec. Res. Dev. Org. Chem.* **1999**, 3, 357-383.
13. Carreira, E. M. Mukaiyama aldol reaction. *Comprehensive Asymmetric Catalysis I-III* **1999**, 3, 997-1065.
14. Mahrwald, R. Diastereoselection in Lewis-Acid-Mediated Aldol Additions. *Chem. Rev.* **1999**, 99, 1095-1120.
15. Casiraghi, G., Zanardi, F., Appendino, G., Rassu, G. The Vinylogous Aldol Reaction: A Valuable, Yet Understated Carbon-Carbon Bond-Forming Maneuver. *Chem. Rev.* **2000**, 100, 1929-1972.
16. Machajewski, T. D., Wong, C.-H., Lerner, R. A. The catalytic asymmetric aldol reaction. *Angew. Chem., Int. Ed. Engl.* **2000**, 39, 1352-1374.
17. Shibasaki, M., Yamada, K.-i., Yoshikawa, N. Lanthanide Lewis acid catalysis. *Lewis Acids in Organic Synthesis - 2 vols.* **2000**, 2, 911-944.
18. Kobayashi, S., Manabe, K., Ishitani, H., Matsuo, J. I. Product subclass 16: silyl enol ethers. *Science of Synthesis* **2002**, 4, 317-369.
19. Palomo, C., Oiarbide, M., Garcia, J. M. The aldol addition reaction: an old transformation at constant rebirth. *Chem.-- Eur. J.* **2002**, 8, 36-44.
20. Rechavi, D., Lemaire, M. Enantioselective Catalysis Using Heterogeneous Bis(oxazoline) Ligands: Which Factors Influence the Enantioselectivity? *Chem. Rev.* **2002**, 102, 3467-3493.
21. Murray, B. A. Reactions of aldehydes and ketones and their derivatives. *Org. React. Mech.* **2003**, 1-33.
22. Palomo, C., Oiarbide, M., Garcia, J. M. Current progress in the asymmetric aldol addition reaction. *Chem. Soc. Rev.* **2004**, 33, 65-75.
23. Gung, B. W., Zhu, Z., Fouch, R. A. Transition State of the Silicon-Directed Aldol Reaction: An ab Initio Molecular Orbital Study. *J. Org. Chem.* **1995**, 60, 2860-2864.
24. Noyori, R., Yokoyama, K., Sakata, J., Kuwajima, I., Nakamura, E., Shimizu, M. Fluoride ion catalyzed aldol reaction between enol silyl ethers and carbonyl compounds. *J. Am. Chem. Soc.* **1977**, 99, 1265-1267.
25. Denmark, S. E., Winter, S. B. D., Su, X., Wong, K.-T. Chemistry of Trichlorosilyl Enolates. 1. New Reagents for Catalytic, Asymmetric Aldol Additions. *J. Am. Chem. Soc.* **1996**, 118, 7404-7405.
26. Denmark, S. E., Stavenger, R. A., Wong, K.-T. Lewis Base-Catalyzed, Asymmetric Aldol Additions of Methyl Ketone Enolates. *J. Org. Chem.* **1998**, 63, 918-919.
27. Denmark, S. E., Stavenger, R. A. Asymmetric Catalysis of Aldol Reactions with Chiral Lewis Bases. *Acc. Chem. Res.* **2000**, 33, 432-440.
28. Heathcock, C. H., Hug, K. T., Flippin, L. A. Acyclic stereoselection. 27. Simple diastereoselection in the Lewis acid mediated reactions of enolsilanes with aldehydes. *Tetrahedron Lett.* **1984**, 25, 5973-5976.
29. Reetz, M. T., Kesseler, K., Jung, A. Aldol-additions to α- and β-alkoxy aldehydes: the effect of chelation on simple diastereoselectivity. *Tetrahedron* **1984**, 40, 4327-4336.
30. Chan, T. H., Brook, M. A. INEPT silicon-29 NMR study of a titanium tetrachloride-mediated reaction of an enol silyl ether. *Tetrahedron Lett.* **1985**, 26, 2943-2946.
31. Kuwajima, I., Nakamura, E. Reactive enolates from enol silyl ethers. *Acc. Chem. Res.* **1985**, 18, 181-187.
32. Reetz, M. T. Selective reactions of organotitanium reagents. *Pure Appl. Chem.* **1985**, 57, 1781-1788.
33. Heathcock, C. H., Davidsen, S. K., Hug, K. T., Flippin, L. A. Acyclic stereoselection. 36. Simple diastereoselection in the Lewis acid mediated reactions of enol silanes with aldehydes. *J. Org. Chem.* **1986**, 51, 3027-3037.
34. Mori, I., Ishihara, K., Heathcock, C. H. Acyclic stereoselection. 50. New stereoselective propanal/propanoic acid synthons for aldol reactions. *J. Org. Chem.* **1990**, 55, 1114-1117.
35. Reetz, M. T., Raguse, B., Marth, C. F., Huegel, H. M., Bach, T., Fox, D. N. A. A rapid injection NMR study of the chelation controlled Mukaiyama aldol addition: $TiCl_4$ versus $LiClO_4$ as the Lewis acid. *Tetrahedron* **1992**, 48, 5731-5742.
36. Reetz, M. T. Structural, mechanistic, and theoretical aspects of chelation-controlled carbonyl addition reactions. *Acc. Chem. Res.* **1993**, 26, 462-468.
37. Denmark, S. E., Lee, W. Investigations on Transition-State Geometry in the Lewis Acid- (Mukaiyama) and Fluoride-Promoted Aldol Reactions. *J. Org. Chem.* **1994**, 59, 707-709.
38. Hollis, T. K., Bosnich, B. Homogeneous Catalysis. Mechanisms of the Catalytic Mukaiyama Aldol and Sakurai Allylation Reactions. *J. Am. Chem. Soc.* **1995**, 117, 4570-4581.
39. Evans, D. A., Dart, M. J., Duffy, J. L., Yang, M. G. A Stereochemical Model for Merged 1,2- and 1,3-Asymmetric Induction in Diastereoselective Mukaiyama Aldol Addition Reactions and Related Processes. *J. Am. Chem. Soc.* **1996**, 118, 4322-4343.
40. Panek, J. S., Jain, N. F. Total Synthesis of Rutamycin B and Oligomycin C. *J. Org. Chem.* **2001**, 66, 2747-2756.
41. Kobayashi, S., Horibe, M. Highly Enantioselective Synthesis of Enantiomeric 2,3-Dihydroxy Thioesters by Using Similar Types of Chiral Sources Derived from L-Proline. *J. Am. Chem. Soc.* **1994**, 116, 9805-9806.
42. Kobayashi, S., Horibe, M., Saito, Y. Enantioselective synthesis of both diastereomers, including the α-alkoxy-β-hydroxy-β-methyl(phenyl) units, by chiral tin(II) Lewis acid-mediated asymmetric aldol reactions. *Tetrahedron* **1994**, 50, 9629-9642.
43. Kobayashi, S., Furuta, T., Hayashi, T., Nishijima, M., Hanada, K. Catalytic Asymmetric Syntheses of Antifungal Sphingofungins and Their Biological Activity as Potent Inhibitors of Serine Palmitoyltransferase (SPT). *J. Am. Chem. Soc.* **1998**, 120, 908-919.
44. Rychnovsky, S. D., Khire, U. R., Yang, G. Total Synthesis of the Polyene Macrolide Roflamycoin. *J. Am. Chem. Soc.* **1997**, 119, 2058-2059.
45. Carreira, E. M., Singer, R. A., Lee, W. Catalytic, Enantioselective Aldol Additions with Methyl and Ethyl Acetate O-Silyl Enolates: A Chiral Tridentate Chelate as a Ligand for Titanium(IV). *J. Am. Chem. Soc.* **1994**, 116, 8837-8838.

Myers' Asymmetric Alkylation 300

Related reactions: Enders SAMP/RAMP hydrazone alkylation;

1. Larcheveque, M., Ignatova, E., Cuvigny, T. Asymmetric synthesis of α-substituted ketones and acids via chiral N,N-substituted amides. *Tetrahedron Lett.* **1978**, 3961-3964.
2. Larcheveque, M., Ignatova, E., Cuvigny, T. Asymmetric alkylation of chiral N,N-disubstituted amides. *J. Organomet. Chem.* **1979**, 177, 5-15.
3. Myers, A. G., Yang, B. H., Chen, H., Gleason, J. L. Use of Pseudoephedrine as a Practical Chiral Auxiliary for Asymmetric Synthesis. *J. Am. Chem. Soc.* **1994**, 116, 9361-9362.
4. Myers, A. G., Yang, B. H., Chen, H., McKinstry, L., Kopecky, D. J., Gleason, J. L. Pseudoephedrine as a Practical Chiral Auxiliary for the Synthesis of Highly Enantiomerically Enriched Carboxylic Acids, Alcohols, Aldehydes, and Ketones. *J. Am. Chem. Soc.* **1997**, 119, 6496-6511.

5. Ager, D. J., Prakash, I., Schaad, D. R. 1,2-Amino Alcohols and Their Heterocyclic Derivatives as Chiral Auxiliaries in Asymmetric Synthesis. *Chem. Rev.* **1996**, 96, 835-875.
6. Anakabe, E., Badia, D., Carrillo, L., Rodriguez, M., Vicario, J. L. Stereocontrolled electrophilic additions on amide enolates employing (S,S)-(+)-pseudoephedrine as chiral auxiliary. *Trends in Organic Chemistry* **2001**, 9, 29-52.
7. Mikami, K., Shimizu, M., Zhang, H. C., Maryanoff, B. E. Acyclic stereocontrol between remote atom centers via intramolecular and intermolecular stereo-communication. *Tetrahedron* **2001**, 57, 2917-2951.
8. Myers, A. G., Yang, B. H., Chen, H., Kopecky, D. J. Asymmetric synthesis of 1,3-dialkyl-substituted carbon chains of any stereochemical configuration by an iterable process. *Synlett* **1997**, 457-459.
9. Duffey, M. O., LeTiran, A., Morken, J. P. Enantioselective Total Synthesis of Borrelidin. *J. Am. Chem. Soc.* **2003**, 125, 1458-1459.
10. Colby, E. A., O'Brien, K. C., Jamison, T. F. Synthesis of Amphidinolide T1 via Catalytic, Stereoselective Macrocyclization. *J. Am. Chem. Soc.* **2004**, 126, 998-999.
11. Paterson, I., Britton, R., Delgado, O., Meyer, A., Poullennec, K. G. Total synthesis and configurational assignment of (-)-dictyostatin, a microtubule-stabilizing macrolide of marine sponge origin. *Angew. Chem., Int. Ed. Engl.* **2004**, 43, 4629-4633.
12. White, J. D., Lee, C.-S., Xu, Q. Total synthesis of (+)-kalkitoxin. *Chem. Commun.* **2003**, 2012-2013.

Nagata Hydrocyanation Reaction 302

Related reactions: Michael addition;

1. Claus, A. Reactions of dichloroglycids. *Ann. Chem., Justus Liebigs* **1873**, 170, 125-136.
2. Bredt, J., Kallen, J. The addition of HCN to unsaturated carboxylic acids. *Ann. Chem., Justus Liebigs* **1896**, 293, 338-371.
3. Nagata, W., Yoshioka, M., Hirai, S. Angular substituted polycyclic compounds. IX. A new hydrocyanation method. *Tetrahedron Lett.* **1962**, 461-466.
4. Nagata, W., Yoshioka, M. Hydrocyanation. III. Alkylaluminum cyanides as potent reagents for hydrocyanation. *Tetrahedron Lett.* **1966**, 1913-1918.
5. Nagata, W., Yoshioka, M. Hydrocyanation and its application to steroid syntheses. *Proc. Int. Congr. Horm. Steroids, 2nd* **1967**, 327-335.
6. Nagata, W., Okumura, T., Yoshioka, M. Hydrocyanation. VIII. Conjugate hydrocyanation of steroidal α,β-unsaturated carboxylic acid derivatives. *J. Chem. Soc. C.* **1970**, 2347-2355.
7. Nagata, W., Yoshioka, M. Hydrocyanation of conjugated carbonyl compounds. *Org. React.* **1977**, 25, 255-476.
8. Nagata, W., Yoshioka, M. Preparation of cyano compounds using alkylaluminum intermediates. I. Diethylaluminum cyanide. *Org. Synth.* **1972**, 52, 90-95.
9. Nagata, W., Yoshioka, M., Hirai, S. Hydrocyanation. IV. New hydrocyanation methods using hydrogen cyanide and an alkylaluminum, and an alkylaluminum cyanide. *J. Am. Chem. Soc.* **1972**, 94, 4635-4643.
10. Overman, L. E., Ricca, D. J., Tran, V. D. Total Synthesis of (±)-Scopadulcic Acid B. *J. Am. Chem. Soc.* **1997**, 119, 12031-12040.
11. Hirukawa, T., Shudo, T., Kato, T. Synthesis of secotrinervitanes, unique bicyclic diterpenes from termites. *J. Chem. Soc., Perkin Trans. 1* **1993**, 217-225.
12. Ihara, M., Katsumata, A., Egashira, M., Suzuki, S., Tokunaga, Y., Fukumoto, K. Stereoselective Construction of the Diterpene Part of Indole Alkaloids, Radarins, by Way of Intramolecular Diels-Alder Reaction. *J. Org. Chem.* **1995**, 60, 5560-5566.

Nazarov Cyclization 304

1. Vorlander, D., Schroeter, G. The effect of sulfuric acid and acetic acid anhydride on dibenzylideneacetone. *Ber.* **1903**, 36, 1490-1497.
2. Blomquist, A. T., Marvel, C. S. Reactions of some substituted divinylacetylenes. *J. Am. Chem. Soc.* **1933**, 55, 1655-1662.
3. Mitchell, D. T., Marvel, C. S. Cyclization of substituted divinylacetylenes. *J. Am. Chem. Soc.* **1933**, 55, 4276-4279.
4. Nazarov, I. N., Zaretskaya, I. I. Derivatives of acetylene. XXVII. Hydration of divinylacetylene. *Bull. acad. sci. U.R.S.S., Classe sci. chim.* **1942**, 200-209.
5. Nazarov, I. N., Zaretskaya, I. I. Structure of products of hydration of divinylethynyl hydrocarbons. *Zh. Obshch. Khim.* **1957**, 27, 693-713.
6. Nazarov, I. N., Zaretskaya, I. I., Sorkina, T. I. Cyclopentanolones from the cyclization of divinyl ketones. *Zh. Obshch. Khim.* **1960**, 30, 746-754.
7. Santelli-Rouvier, C., Santelli, M. The Nazarov cyclization. *Synthesis* **1983**, 429-442.
8. Denmark, S. E. Nazarov and Related Cationic Cyclizations. in *Comp. Org. Synth.* (eds. Trost, B. M.,Fleming, I.), *5*, 751-784 (Pergamon, Oxford, **1991**).
9. Krohn, K. Nazarov and Pauson-Khand reactions. *Org. Synth. Highlights* **1991**, 137-144.
10. Habermas, K. L., Denmark, S. E., Jones, T. K. The Nazarov cyclization. *Org. React.* **1994**, 45, 1-158.
11. Tius, M. A. Cationic Cyclopentannelation of Allene Ethers. *Acc. Chem. Res.* **2003**, 36, 284-290.
12. Hirano, S., Takagi, S., Hiyama, T., Nozaki, H. Abnormal Nazarov reaction. A new synthetic approach to 2,3-disubstituted 2-cyclopentenones. *Bull. Chem. Soc. Jpn.* **1980**, 53, 169-173.
13. Denmark, S. E., Jones, T. K. Silicon-directed Nazarov cyclization. *J. Am. Chem. Soc.* **1982**, 104, 2642-2645.
14. Peel, M. R., Johnson, C. R. Tin-directed Nazarov cyclizations: a versatile route to cyclopentenoids. *Tetrahedron Lett.* **1986**, 27, 5947-5950.
15. Leitich, J., Heise, I., Werner, S., Krueger, C., Schaffner, K. The photo-Nazarov cyclization of 1-cyclohexenyl phenyl ketone revisited. Observation of intermediates. *J. Photochem. Photobiol., A* **1991**, 57, 127-151.
16. Kang, H. T., Kim, S. S., Lee, J. C., U, J. S. Synthesis of α-methylenecyclopentanones via silicon-directed Nazarov reaction of α-trimethylsilylmethyl-substituted divinyl ketones. *Tetrahedron Lett.* **1992**, 33, 3495-3498.
17. Ichikawa, J., Miyazaki, S., Fujiwara, M., Minami, T. Fluorine-Directed Nazarov Cyclizations: A Controlled Synthesis of Cross-Conjugated 2-Cyclopenten-1-ones. *J. Org. Chem.* **1995**, 60, 2320-2321.
18. Bender, J. A., Blize, A. E., Browder, C. C., Giese, S., West, F. G. Highly diastereoselective cycloisomerization of acyclic trienones. The interrupted Nazarov reaction. *J. Org. Chem.* **1998**, 63, 2430-2431.
19. Giese, S., West, F. G. The reductive Nazarov cyclization. *Tetrahedron Lett.* **1998**, 39, 8393-8396.
20. Ichikawa, J., Fujiwara, M., Okauchi, T., Minami, T. Fluorine-directed Nazarov cyclizations. Part 2. Regioselective synthesis of 5-trifluoromethyl-2-cyclopentenones. *Synlett* **1998**, 927-929.
21. Zuev, D., Paquette, L. A. First examples of the interrupted Nazarov reaction. *Chemtracts* **1999**, 12, 1019-1025.
22. Giese, S., West, F. G. Ionic hydrogenation of oxyallyl intermediates: the reductive Nazarov cyclization. *Tetrahedron* **2000**, 56, 10221-10228.
23. Tius, M. A., Chu, C. C., Nieves-Colberg, R. An imino Nazarov cyclization. *Tetrahedron Lett.* **2001**, 42, 2419-2422.
24. Harmata, M., Lee, D. R. The Retro-Nazarov Reaction. *J. Am. Chem. Soc.* **2002**, 124, 14328-14329.
25. Aggarwal, V. K., Belfield, A. J. Catalytic Asymmetric Nazarov Reactions Promoted by Chiral Lewis Acid Complexes. *Org. Lett.* **2003**, 5, 5075-5078.
26. He, W., Sun, X., Frontier, A. J. Polarizing the Nazarov Cyclization: Efficient Catalysis under Mild Conditions. *J. Am. Chem. Soc.* **2003**, 125, 14278-14279.
27. Janka, M., He, W., Frontier, A. J., Eisenberg, R. Efficient Catalysis of Nazarov Cyclization Using a Cationic Iridium Complex Possessing Adjacent Labile Coordination Sites. *J. Am. Chem. Soc.* **2004**, 126, 6864-6865.
28. Liang, G., Trauner, D. Enantioselective Nazarov Reactions through Catalytic Asymmetric Proton Transfer. *J. Am. Chem. Soc.* **2004**, 126, 9544-9545.
29. Smith, D. A., Ulmer, C. W., II. Theoretical studies of the Nazarov cyclization. 1. 1,4-Pentadien-3-one. *Tetrahedron Lett.* **1991**, 32, 725-728.

30. Smith, D. A., Ulmer, C. W., II. Theoretical studies of the Nazarov cyclization. 2. The effect of β-silyl and β-methyl groups. *J. Org. Chem.* **1991**, 56, 4444-4447.
31. Smith, D. A., Ulmer, C. W., II. Theoretical studies of the Nazarov cyclization 3. Torquoselectivity and hyperconjugation in the Nazarov cyclization. The effects of inner versus outer β-methyl and β-silyl groups. *J. Org. Chem.* **1993**, 58, 4118-4121.
32. Braude, E. A., Coles, J. A. Syntheses of polycyclic systems. III. Some hydroindanones and hydrofluorenones. The mechanism of the Nazarov cyclization reaction. *J. Chem. Soc., Abstracts* **1952**, 1430-1433.
33. Kursanov, D. N., Parnes, Z. N., Zaretskaya, I. I., Nazarov, I. N. Reaction mechanism of the cyclization by means of deuterium. I. Cyclization of isopropenyl allyl ketone. *Bull. Acad.Sci. USSR, Chem. Sci. (English Translation)* **1953**, 103-107.
34. Nazarov, I. N., Zaretskaya, I. I., Parnes, Z. N., Kursanov, D. N. The mechanism of the cyclization reaction by means of deuterium. II. *Bull. Acad.Sci. USSR, Chem. Sci. (English Translation)* **1953**, 467-470.
35. Kursanov, D. N., Parnes, Z. N., Zaretskaya, I. I., Nazarov, I. N. Reaction mechanism of the cyclization by means of deuterium. III. *Bull. Acad.Sci. USSR, Chem. Sci. (English Translation)* **1954**, 743-746.
36. Jones, T. K., Denmark, S. E. Silicon-directed Nazarov reactions. III. Stereochemical and mechanistic considerations. *Helv. Chim. Acta* **1983**, 66, 2397-2411.
37. Denmark, S. E., Hite, G. A. Silicon-directed Nazarov cyclizations. Part VI. The anomalous cyclization of vinyl dienyl ketones. *Helv. Chim. Acta* **1988**, 71, 195-208.
38. Harding, K. E., Clement, K. S., Tseng, C. Y. Stereoselective synthesis of (±)-trichodiene. *J. Org. Chem.* **1990**, 55, 4403-4410.
39. Miesch, M., Miesch-Gross, L., Franck-Neumann, M. Total synthesis of (±)-silphinene: non photochemical cyclobutenic route to a crucial intermediate. *Tetrahedron* **1997**, 53, 2103-2110.
40. Balczewski, P., Mikolajczyk, M. An Expeditious Synthesis of (±)-Desepoxy-4,5-didehydromethylenomycin A Methyl Ester. *Org. Lett.* **2000**, 2, 1153-1155.
41. Cheng, K.-F., Cheung, M.-K. Synthesis of inverto-yuehchukene and its 10-(indol-3'-yl) isomer. X-ray structures of (4aRS,10aRS)-1,1,3-trimethyl-1,2,4a,5,10,10a-hexahydroindeno[1,2-b]indol-10-one. *J. Chem. Soc., Perkin Trans. 1* **1996**, 1213-1218.

Neber Rearrangement 306

Related reactions: Dakin-West reaction;

1. Neber, P. W., v. Friedolsheim, A. New kind of rearrangement of oximes. *Ann* **1926**, 449, 109-134.
2. Neber, P. W., Uber, A. New kind of rearrangement of oximes. II. *Ann.* **1928**, 467, 52-72.
3. Neber, P. W., Burgard, A. Course of the reaction in a new type of rearrangement of ketoximes. III. *Ann.* **1932**, 493, 281-294.
4. Neber, P. W., Huh, G. New general method for preparation of α-amino ketones. I. *Ann.* **1935**, 515, 283-296.
5. Neber, P. W., Burgard, A., Thier, W. New general method for the preparation of α-amino- and α,γ-diamino keto compounds. *Ann.* **1936**, 526, 277-294.
6. O'Brien, C. The rearrangement of ketoxime O-sulfonates to amino ketones (The Neber rearrangement). *Chem. Rev.* **1964**, 64, 81-90.
7. McCarty, C. G. *syn-anti* Isomerizations and rearrangements. in *Chem. Carbon-Nitrogen Double Bond* (ed. Patai, S.), 363-464 (Interscience Publishers, **1970**).
8. Conley, R. T., Ghosh, S. "Abnormal" Beckmann rearrangements. *Mechanisms of Molecular Migrations* **1971**, 4, 197-308.
9. Maruoka, K., Yamamoto, H. Functional group transformations via Carbonyl derivatives. in *Comp. Org. Synth.* (eds. Trost, B. M.,Fleming, I.), *6*, 763-795 (Pergamon, Oxford, **1991**).
10. Palacios, F., Ochoa de Retana, A. M., Martinez de Marigorta, E., Manuel De los Santos, J. 2H-azirines as synthetic tools in organic chemistry. *Eur. J. Org. Chem.* **2001**, 2401-2414.
11. Palacios, F., Ochoa de Retana, A. M., Martinez de Marigorta, E., Manuel de los Santos, J. Preparation, properties and synthetic applications of 2H-Azirines: A review. *Org. Prep. Proced. Int.* **2002**, 34, 219-269.
12. Baumgarten, H. E., Dirks, J. E., Petersen, J. M., Zey, R. L. Reactions of amines. XV. Synthesis of α-amino acids from imino esters. *J. Org. Chem.* **1966**, 31, 3708-3711.
13. Graham, W. H. General synthesis of α-amino acid orthoesters from nitriles via N-chloroimidates. *Tetrahedron Lett.* **1969**, 2223-2225.
14. Hyatt, J. A. Neber rearrangement of amidoxime sulfonates. Synthesis of 2-amino-1-azirines. *J. Org. Chem.* **1981**, 46, 3953-3955.
15. Eremeev, A. V., Piskunova, I. P., El'kinson, R. S. Synthesis of 2-amino-1-azirines and their reactions with carboxylic acids. *Khim. Geterotsikl. Soedin.* **1985**, 1202-1206.
16. Piskunova, I. P., Eremeev, A. V., Mishnev, A. F., Vosekalna, I. A. Synthesis and structure of optically active 3-amino-2H-azirines. *Tetrahedron* **1993**, 49, 4671-4676.
17. Verstappen, M. M. H., Ariaans, G. J. A., Zwanenburg, B. Asymmetric Synthesis of 2H-Azirine Carboxylic Esters by an Alkaloid-Mediated Neber Reaction. *J. Am. Chem. Soc.* **1996**, 118, 8491-8492.
18. Palacios, F., Ochoa de Retana, A. M., Gil, J. I., Ezpeleta, J. M. Simple asymmetric synthesis of 2H-azirines derived from phosphine oxides. *J. Org. Chem.* **2000**, 65, 3213-3217.
19. Ooi, T., Takahashi, M., Doda, K., Maruoka, K. Asymmetric Induction in the Neber Rearrangement of Simple Ketoxime Sulfonates under Phase-Transfer Conditions: Experimental Evidence for the Participation of an Anionic Pathway. *J. Am. Chem. Soc.* **2002**, 124, 7640-7641.
20. Smith, P. A. S., Most, E. E., Jr. Quaternary hydrazones and their rearrangement. *J. Org. Chem.* **1957**, 22, 358-362.
21. Baumgarten, H. E., Bower, F. A. Reactions of amines. I. A novel rearrangement of N,N-dichloro-sec-alkylamines. *J. Am. Chem. Soc.* **1954**, 76, 4561-4564.
22. Alt, G. H., Knowles, W. S. Mechanism of the N,N-dichloro-sec-alkylamine rearrangement. *J. Org. Chem.* **1960**, 25, 2047-2048.
23. Hallinan, K. O., Crout, D. H. G., Errington, W. Simple synthesis of L- and D-vinylglycine (2-aminobut-3-enoic acid) and related amino acids. *J. Chem. Soc., Perkin Trans. 1* **1994**, 3537-3543.
24. Cram, D. J., Hatch, M. J. The problem of the unsaturated three-membered ring containing nitrogen. *J. Am. Chem. Soc.* **1953**, 75, 33-38.
25. Hatch, M. J., Cram, D. J. The mechanism and scope of the Neber rearrangement. *J. Am. Chem. Soc.* **1953**, 75, 38-44.
26. House, H. O., Berkowitz, W. F. The stereochemistry of the Neber rearrangement. *J. Org. Chem.* **1963**, 28, 2271-2276.
27. Morrow, D. F., Butler, M. E. Stereoselectivity in the Neber rearrangement-synthesis of a steroidal spiroazirine. *J. Heterocycl. Chem.* **1964**, 1, 53-54.
28. Adams, G. W., Bowie, J. H., Hayes, R. N. The complex anionic rearrangements of deprotonated α-oximino carbonyl derivatives in the gas phase. *J. Chem. Soc., Perkin Trans. 2* **1991**, 1809-1818.
29. Chung, J. Y. L., Ho, G.-J., Chartrain, M., Roberge, C., Zhao, D., Leazer, J., Farr, R., Robbins, M., Emerson, K., Mathre, D. J., McNamara, J. M., Hughes, D. L., Grabowski, E. J. J., Reider, P. J. Practical chemoenzymatic synthesis of a 3-pyridylethanolamino β_3 adrenergic receptor agonist. *Tetrahedron Lett.* **1999**, 40, 6739-6743.
30. Diez, A., Voldoire, A., Lopez, I., Rubiralta, M., Segarra, V., Pages, L., Palacios, J. M. Synthetic applications of 2-aryl-4-piperidones. X. Synthesis of 3-aminopiperidines, potential substances P antagonists. *Tetrahedron* **1995**, 51, 5143-5156.

Nef Reaction 308

1. Konovalov, M. *J. Russ. Phys. Chem. Soc.* **1893**, 25, 509.
2. Nef, J. U. Nitroparaffin salt constitution. *Liebigs Ann. Chem.* **1894**, 280, 263-342.
3. Salomaa, P. Formation of carbonyl groups in hydrolytic reactions. *Chem. Carbonyl Group. 1966* **1966**, 177-210.
4. Pinnick, H. W. The Nef reaction. *Org. React.* **1990**, 38, 655-792.

5. Grierson, D. S., Husson, H.-P. Polonovski- and Pummerer-type reactions and the Nef reaction. in *Comp. Org. Synth.* (eds. Trost, B. M.,Fleming, I.), *6*, 909-947 (Pergamon, Oxford, **1991**).
6. Lawrence, N. J. Aldehydes and ketones. *J. Chem. Soc., Perkin Trans. 1* **1998**, 1739-1749.
7. Adams, J. P., Box, D. S. Nitro and related compounds. *J. Chem. Soc., Perkin Trans. 1* **1999**, 749-764.
8. Petrus, L., Petrusova, M., Pham-Huu, D.-P., Lattova, E., Pribulova, B., Turjan, J. Conversions of nitroalkyl to carbonyl groups in carbohydrates. *Monatsh. Chem.* **2002**, 133, 383-392.
9. Ballini, R., Petrini, M. Recent synthetic developments in the nitro to carbonyl conversion (Nef reaction). *Tetrahedron* **2004**, 60, 1017-1047.
10. McMurry, J. E., Melton, J. New method for the conversion of nitro groups into carbonyls. *J. Org. Chem.* **1973**, 38, 4367.
11. Steliou, K., Poupart, M. A. Reagents for organic synthesis. 5. Synthesis of aldehydes and ketones from nitro paraffins. *J. Org. Chem.* **1985**, 50, 4971-4973.
12. Urpi, F., Vilarrasa, J. New synthetic tricks. A novel one-pot procedure for the conversion of primary nitro groups into aldehydes. *Tetrahedron Lett.* **1990**, 31, 7499-7500.
13. Saville-Stones, E. A., Lindell, S. D. Direct transformation of nitroalkanes to alkanoic acids under mild conditions. *Synlett* **1991**, 591-592.
14. Bordoloi, M. Cadmium chloride-magnesium-water: a new system for regioselective transformation of conjugated nitroalkenes to ketocompounds; conversion of 6-nitro-D5-steroids to 6-ketosteroids. *J. Chem. Soc., Chem. Commun.* **1993**, 922-923.
15. Kumaran, G., Kulkarni, G. H. A facile conversion of nitro olefins to functionalized hydroximoyl chlorides as nitrile oxide precursors. *Tetrahedron Lett.* **1994**, 35, 5517-5518.
16. DeHaan, F. P., Allen, M. W., Ang, M. C., Balacuit, D. B., Bentley, C. A., Bergstrand, R. J., Cho, E. R., DeHaan, D. O., Farris, T. E., et al. Scope and Mechanism of "Double-Agent" Halogenation. *J. Org. Chem.* **1995**, 60, 8320-8323.
17. Das, N. B., Sarangi, C., Nanda, B., Nayak, A., Sharma, R. P. $SnCl_2.2H_2O$-Mg-H_2O: a mild reagent system for the regioselective transformation of conjugated nitroalkenes to carbonyl compounds. *J. Chem. Res., Synop.* **1996**, 28-29.
18. Saikia, A. K., Barua, N. C., Sharma, R. P., Ghosh, A. C. The zinc-trifluoroacetic acid reaction in organic solvents: a facile procedure for the conversion of nitroolefins into carbonyl compounds under mild conditions. *J. Chem. Res., Synop.* **1996**, 124-125.
19. Matt, C., Wagner, A., Mioskowski, C. Novel Transformation of Primary Nitroalkanes and Primary Alkyl Bromides to the Corresponding Carboxylic Acids. *J. Org. Chem.* **1997**, 62, 234-235.
20. Tokunaga, Y., Ihara, M., Fukumoto, K. A mild oxidative transformation of nitro compounds into ketones by tetrapropylammonium perruthenate. *J. Chem. Soc., Perkin Trans. 1* **1997**, 207-209.
21. Adam, W., Makosza, M., Saha-Moeller, C. R., Zhao, C.-G. A mild and efficient Nef reaction for the conversion of nitro to carbonyl group by dimethyldioxirane (DMD) oxidation of nitronate anions. *Synlett* **1998**, 1335-1336.
22. Adam, W., Makosza, M., Stalinski, K., Zhao, C.-G. DMD Oxidation of in-Situ-Generated sH Adducts Derived from Nitroarenes and the Carbanion of 2-Phenylpropionitrile to Phenols: The First Direct Substitution of a Nitro by a Hydroxy Group. *J. Org. Chem.* **1998**, 63, 4390-4391.
23. Ballini, R., Curini, M., Epifano, F., Marcotullio, M. C., Rosati, O. A new, modulated, oxidative ring-cleavage of α-nitro cycloalkanones by Oxone. Synthesis of α,ω-dicarboxylic acids and α,ω-dicarboxylic monomethyl esters. *Synlett* **1998**, 1049-1050.
24. Ceccherelli, P., Curini, M., Marcotullio, M. C., Epifano, F., Rosati, O. Oxone promoted Nef reaction. Simple conversion of nitro group into carbonyl. *Synth. Commun.* **1998**, 28, 3057-3064.
25. Bezbarua, M. S., Bez, G., Barua, N. C. A facile procedure for the conversion of nitroolefins to carbonyl compounds using the Al-$NiCl_2.6H_2O$-THF system. *Chem. Lett.* **1999**, 325-326.
26. Hwu, J. R., Tseng, W. N., Patel, H. V., Wong, F. F., Horng, D.-N., Liaw, B. R., Lin, L. C. Mono-deoxygenation of Nitroalkanes, Nitrones, and Heterocyclic N-Oxides by Hexamethyldisilane through 1,2-Elimination: Concept of "Counterattack Reagent". *J. Org. Chem.* **1999**, 64, 2211-2218.
27. Shahi, S. P., Vankar, Y. D. Chemistry of nitro compounds. Part 10. Nef reaction of benzylic and secondary nitro compounds using bis(trimethylsilyl) peroxide. *Synth. Commun.* **1999**, 29, 4321-4325.
28. Nikalje, M. D., Ali, I. S., Dewkar, G. K., Sudalai, A. Synthesis of aryl α-keto acids via the Cu-catalyzed conversion of aryl nitro-aldol products. *Tetrahedron Lett.* **2000**, 41, 959-961.
29. Ballini, R., Bosica, G., Fiorini, D., Petrini, M. Unprecedented, selective Nef reaction of secondary nitroalkanes promoted by DBU under basic homogeneous conditions. *Tetrahedron Lett.* **2002**, 43, 5233-5235.
30. Ogibin, Y. N., Ilovaiskii, A. I., Merkulova, V. M., Nikishin, G. I. Electrolysis of Nitro Compound Salts: Application in Ketone Synthesis. *Russ. J. Electrochem.* **2003**, 39, 1220-1227.
31. Van Tamelen, E. E., Thiede, R. J. Synthetic application and mechanism of the Nef reaction. *J. Am. Chem. Soc.* **1952**, 74, 2615-2618.
32. Leitch, L. C. Synthesis of organic deuterium compounds. XIII. The mechanism of the Nef reaction. Synthesis of ethanal-1-d. *Can. J. Chem.* **1955**, 33, 400-404.
33. Hawthorne, M. F. aci-Nitroalkanes. II. The mechanism of the Nef reaction. *J. Am. Chem. Soc.* **1957**, 79, 2510-2515.
34. Feuer, H., Nielsen, A. T. Direct Nef reaction by acid-catalyzed hydrolysis of 2-nitrooctane to 2-octanone. *J. Am. Chem. Soc.* **1962**, 84, 688.
35. Kornblum, N., Brown, R. A. The action of acids on nitronic esters and nitroparaffin salts. Concerning the mechanisms of the Nef and the hydroxamic acid forming reactions of nitroparaffins. *J. Am. Chem. Soc.* **1965**, 87, 1742-1747.
36. Sun, S.-F., Folliard, J. T. Participation of water in the Nef reaction of aci-nitro compounds. *Tetrahedron* **1971**, 27, 323-330.
37. Wilson, H., Lewis, E. S. Neighboring group participation in proton transfers. *J. Am. Chem. Soc.* **1972**, 94, 2283-2285.
38. Chapas, R. B., Knudsen, R. D., Nystrom, R. F., Snyder, H. R. Nef-type transformation in basic solution. *J. Org. Chem.* **1975**, 40, 3746-3748.
39. Gorvin, J. H. Hydroxydenitration by a Nef-type process; the more general case. *J. Chem. Soc., Chem. Commun.* **1976**, 972-973.
40. Balogh-Hergovich, E., Greczi, Z., Kaizer, J., Speier, G., Reglier, M., Giorgi, M., Parkanyi, L. Kinetics and mechanism of the copper-catalysed oxygenation of 2-nitropropane. *Eur. J. Inorg. Chem.* **2002**, 1687-1696.
41. Kaizer, J., Greczi, Z., Speier, G. Ferroxime(II)-catalyzed oxygenation of nitroalkanes by tert-butyl hydroperoxide and its relevance to 2-nitropropane dioxygenase. *J. Mol. Catal. A: Chemical* **2002**, 179, 35-39.
42. Capecchi, T., de Koning, C. B., Michael, J. P. Synthesis of the bisbenzannelated spiroketal core of the γ-rubromycins. The use of a novel Nef-type reaction mediated by Pearlman's catalyst. *Perkin 1* **2000**, 2681-2688.
43. Bagul, T. D., Lakshmaiah, G., Kawabata, T., Fuji, K. Total synthesis of spirotryprostatin B via asymmetric nitroolefination. *Org. Lett.* **2002**, 4, 249-251.
44. Trost, B. M., Patterson, D. E., Hembre, E. J. AAA in KAT/DYKAT processes: first- and second-generation asymmetric syntheses of (+)- and (-)-cyclophellitol. *Chem.-- Eur. J.* **2001**, 7, 3768-3775.
45. Mineno, T., Miller, M. J. Stereoselective Total Synthesis of Racemic BCX-1812 (RWJ-270201) for the Development of Neuraminidase Inhibitors as Anti-influenza Agents. *J. Org. Chem.* **2003**, 68, 6591-6596.

Negishi Cross-Coupling 310

Related reactions: Kumada cross-coupling, Stille cross-coupling, Suzuki cross-coupling;

1. Baba, S., Negishi, E. A novel stereospecific alkenyl-alkenyl cross-coupling by a palladium- or nickel-catalyzed reaction of alkenylalanes with alkenyl halides. *J. Am. Chem. Soc.* **1976**, 98, 6729-6731.
2. Negishi, E., Baba, S. Novel stereoselective alkenyl-aryl coupling via nickel-catalyzed reaction of alkenylalanes with aryl halides. *J. Chem. Soc., Chem. Commun.* **1976**, 596-597.
3. King, A. O., Okukado, N., Negishi, E. Highly general stereo-, regio-, and chemo-selective synthesis of terminal and internal conjugated enynes by the palladium-catalyzed reaction of alkynylzinc reagents with alkenyl halides. *J. Chem. Soc., Chem. Commun.* **1977**, 683-684.

4. Negishi, E., King, A. O., Okukado, N. Selective carbon-carbon bond formation via transition metal catalysis. 3. A highly selective synthesis of unsymmetrical biaryls and diarylmethanes by the nickel- or palladium-catalyzed reaction of aryl- and benzylzinc derivatives with aryl halides. *J. Org. Chem.* **1977**, 42, 1821-1823.
5. Negishi, E., Van Horn, D. E. Selective carbon-carbon bond formation via transition metal catalysis. 4. A novel approach to cross-coupling exemplified by the nickel-catalyzed reaction of alkenylzirconium derivatives with aryl halides. *J. Am. Chem. Soc.* **1977**, 99, 3168-3170.
6. King, A. O., Negishi, E., Villani, F. J., Jr., Silveira, A., Jr. A general synthesis of terminal and internal arylalkynes by the palladium-catalyzed reaction of alkynylzinc reagents with aryl halides. *J. Org. Chem.* **1978**, 43, 358-360.
7. Negishi, E. Selective carbon-carbon bond formation via transition metal catalysis: is nickel or palladium better than copper? in *Aspects Mech. Organomet. Chem., [Proc. Symp.]* (ed. Brewster, J. H.), 285-317 (Plenum, New York, **1978**).
8. Negishi, E. Palladium- or nickel-catalyzed cross coupling. A new selective method for carbon-carbon bond formation. *Acc. Chem. Res.* **1982**, 15, 340-348.
9. Negishi, E. Palladium- or nickel-catalyzed cross coupling involving proximally heterofunctional reagents. *Curr. Trends Org. Synth., Proc. Int. Conf., 4th* **1983**, 269-280.
10. Negishi, E., Takahashi, T., Akiyoshi, K. Aspects of cross-coupling reactions catalyzed by palladium and nickel complexes. *Chem. Ind.* **1988**, 33, 381-407.
11. Erdik, E. Transition metal catalyzed reactions of organozinc reagents. *Tetrahedron* **1992**, 48, 9577-9648.
12. Erdik, E., Editor. *Organozinc Reagents in Organic Synthesis* (**1996**) 464 pp.
13. Negishi, E.-I., Liu, F. Palladium- or nickel-catalyzed cross-coupling with organometals containing zinc, magnesium, aluminum, and zirconium. in *Metal-Catalyzed Cross-Coupling Reactions* (eds. Diederich, F.,Stang, P. J.), 1-47 (Wiley-VCH, Weinheim, Germany, **1998**).
14. Stanforth, S. P. Catalytic cross-coupling reactions in biaryl synthesis. *Tetrahedron* **1998**, 54, 263-303.
15. Green, L., Chauder, B., Snieckus, V. The directed ortho metalation-cross-coupling symbiosis in heteroaromatic synthesis. *J. Heterocycl. Chem.* **1999**, 36, 1453-1468.
16. Knochel, P., Jones, P., Langer, F. Transition metal catalyzed reactions of zinc organometallics. *Organozinc Reagents* **1999**, 179-212.
17. De Vries, J. G., De Vries, A. H. M., Tucker, C. E., Miller, J. A. Palladium catalysis in the production of pharmaceuticals. *Innovations in Pharmaceutical Technology* **2001**, 01, 125-126, 128, 130.
18. Anctil, E. J. G., Snieckus, V. The directed ortho metalation-cross coupling symbiosis. Regioselective methodologies for biaryls and heterobiaryls. Deployment in aromatic and heteroaromatic natural product synthesis. *J. Organomet. Chem.* **2002**, 653, 150-160.
19. Negishi, E.-i. A genealogy of Pd-catalyzed cross-coupling. *J. Organomet. Chem.* **2002**, 653, 34-40.
20. Negishi, E.-i. Palladium-catalyzed carbon-carbon cross-coupling. Overview of the Negishi protocol with Zn, Al, Zr, and related metals. in *Handbook of Organopalladium Chemistry for Organic Synthesis* (ed. Negishi, E.-i.), *1*, 229-247 (John Wiley & Sons Inc., New York, **2002**).
21. Negishi, E.-i., Dumond, Y. Palladium-catalyzed cross-coupling substitution. *Handbook of Organopalladium Chemistry for Organic Synthesis* **2002**, 1, 767-789.
22. Woltermann, C. J. Recent advances in phosphine ligands for palladium-catalyzed carbon-carbon bond forming reactions. *PharmaChem* **2002**, 1, 11-14.
23. Herrmann, W. A., Ofele, K., von Preysing, D., Schneider, S. K. Phospha-palladacycles and N-heterocyclic carbene palladium complexes: efficient catalysts for CC-coupling reactions. *J. Organomet. Chem.* **2003**, 687, 229-248.
24. Lessene, G. Advances in the Negishi Coupling. *Aust. J. Chem.* **2004**, 57, 107.
25. Negishi, E., Ay, M., Gulevich, Y. V., Noda, Y. Highly stereoselective and general synthesis of (Z)-3-methyl-2-alken-1-ols via palladium-catalyzed cross coupling of (Z)-3-iodo-2-buten-1-ol with organozincs and other organometals. *Tetrahedron Lett.* **1993**, 34, 1437-1440.
26. Weichert, A., Bauer, M., Wirsig, P. Palladium(0) catalyzed cross coupling reactions of hindered, double activated aryl halides with organozinc reagents - the effect of copper(I) cocatalysis. *Synlett* **1996**, 473-474.
27. Dai, C., Fu, G. C. The first general method for palladium-catalyzed Negishi cross-coupling of aryl and vinyl chlorides: use of commercially available $Pd(P(t\text{-}Bu)_3)_2$ as a catalyst. *J. Am. Chem. Soc.* **2001**, 123, 2719-2724.
28. Yus, M., Gomis, J. Negishi cross-coupling with functionalized organozinc compounds prepared by lithium-zinc transmetallation. *Tetrahedron Lett.* **2001**, 42, 5721-5724.
29. Peyrat, J.-F., Thomas, E., L'Hermite, N., Alami, M., Brion, J.-D. Versatile palladium(II)-catalyzed Negishi coupling reactions with functionalized conjugated alkenyl chlorides. *Tetrahedron Lett.* **2003**, 44, 6703-6707.
30. Zhou, J., Fu, G. C. Cross-Couplings of Unactivated Secondary Alkyl Halides: Room-Temperature Nickel-Catalyzed Negishi lkyl Bromides and Iodides. *J. Am. Chem. Soc.* **2003**, 125, 14726-14727.
31. Zhou, J., Fu, G. C. Palladium-Catalyzed Negishi Cross-Coupling Reactions of Unactivated Alkyl Iodides, Bromides, Chlorides, and Tosylates. *J. Am. Chem. Soc.* **2003**, 125, 12527-12530.
32. Walla, P., Kappe, C. O. Microwave-assisted Negishi and Kumada cross-coupling reactions of aryl chlorides. *Chem. Commun.* **2004**, 564-565.
33. Erdik, E. Use of activation methods for organozinc reagents. *Tetrahedron* **1987**, 43, 2203-2212.
34. Zhu, L., Wehmeyer, R. M., Rieke, R. D. The direct formation of functionalized alkyl(aryl)zinc halides by oxidative addition of highly reactive zinc with organic halides and their reactions with acid chlorides, α,β-unsaturated ketones, and allylic, aryl, and vinyl halides. *J. Org. Chem.* **1991**, 56, 1445-1453.
35. Miyaura, N., Suzuki, A. Palladium-Catalyzed Cross-Coupling Reactions of Organoboron Compounds. *Chem. Rev.* **1995**, 95, 2457-2483.
36. Sammakia, T., Stangeland, E. L., Whitcomb, M. C. Total Synthesis of Caerulomycin C via the Halogen Dance Reaction. *Org. Lett.* **2002**, 4, 2385-2388.
37. Hu, T., Panek, J. S. Total synthesis of (-)-Motuporin. *J. Org. Chem.* **1999**, 64, 3000-3001.
38. Williams, D. R., Kissel, W. S. Total Synthesis of (+)-Amphidinolide J. *J. Am. Chem. Soc.* **1998**, 120, 11198-11199.

Nenitzescu Indole Synthesis 312

Related reactions: ***Bartoli indole synthesis, Fischer indole synthesis, Larock indole synthesis, Madelung indole synthesis;***

1. Nenitzescu, C. D. Derivatives of 2-methyl-5-hydroxyindole. *Bull soc. chim. Romania* **1929**, 11, 37-43.
2. Brown, R. K. Synthesis of the indole nucleus. in *Chemistry of Heterocyclic Compounds: Indoles Part One* (ed. Houlihan, W. J.), *25*, 413-436 (Wiley, Chichester, **1972**).
3. Allen, G. R., Jr. Synthesis of 5-hydroxyindoles by the Nenitzescu reaction. *Org. React.* **1973**, 20, 337-454.
4. Gribble, G. W. Recent developments in indole ring synthesis-methodology and applications. *Perkin 1* **2000**, 1045-1075.
5. Joule, J. A. Product class 13: indole and its derivatives. *Science of Synthesis* **2001**, 10, 361-652.
6. Brase, S., Gil, C., Knepper, K. The recent impact of solid-phase synthesis on medicinally relevant benzannelated nitrogen heterocycles. *Bioorg. Med. Chem.* **2002**, 10, 2415-2437.
7. Adams, R., Samuels, W. P., Jr. Quinone imides. XXXVII. Conversion of p-quinone diimides to indoles. *J. Am. Chem. Soc.* **1955**, 77, 5375-5382.
8. Adams, R., Werbel, L. M., Nair, M. D. Quinone imides. XLVI. The addition of heterocyclic active methylene compounds to p-benzoquinone diimides. *J. Am. Chem. Soc.* **1958**, 80, 3291-3293.
9. Parker, K. A., Kang, S.-K. Convergent approaches to indoloquinones: additions to quinone monoimides. *J. Org. Chem.* **1979**, 44, 1536-1540.
10. Bernier, J. L., Henichart, J. P., Vaccher, C., Houssin, R. Condensation of p-benzoquinone with 4-cyano- and 4-nitroanilines. An extension of the Nenitzescu reaction. *J. Org. Chem.* **1980**, 45, 1493-1496.
11. Aggarwal, V., Kumar, A., Ila, H., Junjappa, H. Polarized ketene N,N- and S,N-acetals as novel enamine components for the Nenitzescu indole synthesis. XV. *Synthesis* **1981**, 157-158.

12. Lyubchanskaya, V. M., Alekseeva, L. M., Granik, V. G. The first example of dienediamine utilization in the Nenitzescu reaction. *Mendeleev Commun.* **1995**, 68-69.
13. Lyubchanskaya, V. M., Alekseeva, L. M., Granik, V. G. The first example of aza-Nenitzescu reaction. A new approach to heterocyclic quinone synthesis. *Tetrahedron* **1997**, 53, 15005-15010.
14. Alekseeva, L. M., Mukhanova, T. I., Panisheva, E. K., Anisimova, O. S., Turchin, K. F., Komkov, A. V., Dorokhov, V. A., Granik, V. G. Acetyl ketene aminals in the Nenitzescu reaction. *Russ. Chem. Bull.* **1999**, 48, 160-165.
15. Lyubchanskaya, V. M., Alekseeva, L. M., Granik, V. G. The aza-Nenitzescu reaction. Synthesis of indazole derivatives by condensation of quinones with hydrazones. *Chem. Het. Comp. (New York) (Translation of Khim. Geterot. Soed.)* **1999**, 35, 570-574.
16. Lyubchanskaya, V. M., Alekseeva, L. M., Savina, S. A., Granik, V. G. Indazolequinones in the Nenitzescu reaction. Synthesis of pyrrolo[2,3-e]- and furo[2,3-e]indazoles. *Chem. Het. Comp. (New York) (Translation of Khim. Geterot. Soed.)* **2000**, 36, 1276-1283.
17. Lyubchanskaya, V. M., Alekseeva, L. M., Savina, S. A., Shashkov, A. S., Granik, V. G. Novel synthesis of chromene and benzofuran derivatives via the Nenitzescu reaction. *Mendeleev Commun.* **2002**, 15-17.
18. Katkevica, D., Trapencieris, P., Boman, A., Kalvins, I., Lundstedt, T. The Nenitzescu reaction: An initial screening of experimental conditions for improvment of the yield of a model reaction. *Journal of Chemometrics* **2004**, 18, 183-187.
19. Lyubchanskaya, V. M., Savina, S. A., Alekseeva, L. M., Shashkov, A. S., Granik, V. G. The use of enehydrazines in the Nenitzescu reaction. *Mendeleev Commun.* **2004**, 73-75.
20. Mukhanova, T. I., Alekseeva, L. M., Shashkov, A. S., Granik, V. G. Heterocyclic quinones in the nenitzescu reaction. Synthesis of furo- and pyrroloquinolines from 2-methoxycarbonyl-4-oxo-5,8-quinolinequinone. *Chem. Het. Comp. (New York) (Translation of Khim. Geterot. Soed.)* **2004**, 40, 16-21.
21. Kucklaender, U. Mechanism of the Nenitzescu reaction. *Tetrahedron* **1972**, 28, 5251-5259.
22. Kucklaender, U. Mechanism of the Nenitzescu reaction. II. *Tetrahedron* **1973**, 29, 921-927.
23. Kucklaender, U. Mechanism of the nenitzescu reaction. III. Acyl migrations. I. *Tetrahedron* **1975**, 31, 1631-1639.
24. Kucklaender, U. Mechanism of the Nenitzescu reaction, IV. Synthesis of benzindole derivatives. *Liebigs Ann. Chem.* **1978**, 129-139.
25. Kucklaender, U. Mechanism of the Nenitzescu reaction, V. Synthesis of naphthofuran derivatives. *Liebigs Ann. Chem.* **1978**, 140-149.
26. Kucklaender, U., Huehnermann, W. Studies on the mechanism of the Nenitzescu reaction. Synthesis of 6-hydroxyindole derivatives. *Arch. Pharm. (Weinheim, Ger.)* **1979**, 312, 515-526.
27. Patrick, J. B., Saunders, E. K. Studies on the Nenitzescu synthesis of 5-hydroxyindoles. *Tetrahedron Lett.* **1979**, 4009-4012.
28. Kinugawa, M., Arai, H., Nishikawa, H., Sakaguchi, A., Ogasa, T., Tomioka, S., Kasai, M. Facile synthesis of the key intermediate of EO 9 via the formation of the indole skeleton using the Nenitzescu reaction. *J. Chem. Soc., Perkin Trans. 1* **1995**, 2677-2678.
29. Pawlak, J. M., Khau, V. V., Hutchison, D. R., Martinelli, M. J. A Practical, Nenitzescu-Based Synthesis of LY311727, the First Potent and Selective s-PLA2 Inhibitor. *J. Org. Chem.* **1996**, 61, 9055-9059.
30. Ketcha, D. M., Wilson, L. J., Portlock, D. E. The solid-phase Nenitzescu indole synthesis. *Tetrahedron Lett.* **2000**, 41, 6253-6257.
31. Engler, T. A., Wanner, J. Lewis acid-directed reactions of benzoquinone mono-/bis-imines: application to syntheses of substituted β- and γ-tetrahydrocarbolines. *Tetrahedron Lett.* **1997**, 38, 6135-6138.

1. Nicholas, K. M., Pettit, R. Alkyne protecting group. *Tetrahedron Lett.* **1971**, 37, 3475-3478.
2. Nicholas, K. M., Pettit, R. Stability of α-(alkynyl)dicobalt hexacarbonyl carbonium ions. *J. Organomet. Chem.* **1972**, 44, C21-C24.
3. Connor, R. E., Nicholas, K. M. Isolation, characterization, and stability of α-[(ethynyl)dicobalt hexacarbonyl] carbonium ions. *J. Organomet. Chem.* **1977**, 125, C45-C48.
4. Lockwood, R. F., Nicholas, K. M. Transition metal-stabilized carbenium ions as synthetic intermediates. I. a-[(Alkynyl)dicobalt hexacarbonyl] carbenium ions as propargylating agents. *Tetrahedron Lett.* **1977**, 4163-4166.
5. Nicholas, K. M. Chemistry and synthetic utility of cobalt-complexed propargyl cations. *Acc. Chem. Res.* **1987**, 20, 207-214.
6. Iqbal, J., Bhatia, B., Khanna, V. Cobalt carbonyls: a versatile reagent and catalyst in organic synthesis. *J. Indian Inst. Sci.* **1994**, 74, 411-471.
7. Nicholas, K. M., Caffyn, A. J. M. Transition metal alkyne complexes: Transition metal-stabilized propargyl systems. in *Comprehensive Organometallic Chemistry II.* (eds. Abel, E. W., Stone, F. G. A.,Wilkinson, F.), *12*, 685-702 (Oxford, **1995**).
8. Jacobi, P. A., Zheng, W. Enantioselective synthesis of β-amino acids using the Nicholas reaction. *Enantioselective Synthesis of β-Amino Acids* **1997**, 359-372.
9. Went, M. J. Synthesis and reactions of polynuclear cobalt-alkyne complexes. *Adv. Organomet. Chem.* **1997**, 41, 69-125.
10. Fletcher, A. J., Christie, S. D. R. Cobalt mediated cyclizations. *Perkin 1* **2000**, 1657-1668.
11. Green, J. R. Chemistry of propargyldicobalt cations: recent developments in the Nicholas and related reactions. *Curr. Org. Chem.* **2001**, 5, 809-826.
12. Muller, T. J. J. Stereoselective propargylations with transition-metal-stabilized propargyl cations. *Eur. J. Org. Chem.* **2001**, 2021-2033.
13. Malacria, M., Aubert, C., Renaud, J. L. Product class 4: organometallic complexes of cobalt. *Science of Synthesis* **2002**, 1, 439-530.
14. Teobald, B. J. The Nicholas reaction: the use of dicobalt hexacarbonyl-stabilized propargylic cations in synthesis. *Tetrahedron* **2002**, 58, 4133-4170.
15. Roth, K. D. Reaction of dicobalt hexacarbonyl propargyl cations with aldehydic N,N-dibenzylenamines. *Synlett* **1992**, 435-438.
16. Tyrrell, E., Skinner, G. A., Bashir, T. The synthesis of bridged and fused ring carbocycles using a novel variation of an intramolecular Nicholas reaction. *Synlett* **2001**, 1929-1931.
17. Cassel, J. A., Leue, S., Gachkova, N. I., Kann, N. C. Solid-Phase Synthesis of Substituted Alkynes Using the Nicholas Reaction. *J. Org. Chem.* **2002**, 67, 9460-9463.
18. Betancort, J. M., Martin, T., Palazon, J. M., Martin, V. S. Stereoselective Synthesis of Cyclic Ethers by Intramolecular Trapping of Dicobalt Hexacarbonyl-Stabilized Propargylic Cations. *J. Org. Chem.* **2003**, 68, 3216-3224.
19. Crisostomo, F. R. P., Martin, T., Martin, V. S. Stereoselective intramolecular Nicholas reaction using epoxides as nucleophiles. *Org. Lett.* **2004**, 6, 565-568.
20. Kuhn, O., Rau, D., Mayr, H. How Electrophilic Are Cobalt Carbonyl Stabilized Propargylium Ions? *J. Am. Chem. Soc.* **1998**, 120, 900-907.
21. Soleilhavoup, M., Saccavini, C., Lepetit, C., Lavigne, G., Maurette, L., Donnadieu, B., Chauvin, R. Parallel Approaches to Mono- and Bis-Propargylic Activation via Co2(CO)8 and [Ru3(m-Cl)(CO)10]. *Organometallics* **2002**, 21, 871-883.
22. Padmanabhan, S., Nicholas, K. M. Carbon-13 NMR study of (propargyl)dicobalt hexacarbonyl cations: a structurally unique class of metal-stabilized carbenium ions. *J. Organomet. Chem.* **1984**, 268, C23-C27.
23. Melikyan, G. G., Bright, S., Monroe, T., Hardcastle, K. I., Ciurash, J. Overcoming a longstanding challenge: X-ray structure of a [$Co_2(CO)_6$]-complexed propargyl cation. *Angew. Chem., Int. Ed. Engl.* **1998**, 37, 161-164.
24. Jacobi, P. A., Murphree, S., Rupprecht, F., Zheng, W. Formal Total Syntheses of the β-Lactam Antibiotics Thienamycin and PS-5. *J. Org. Chem.* **1996**, 61, 2413-2427.
25. Mukai, C., Moharram, S. M., Azukizawa, S., Hanaoka, M. Total Syntheses of (+)-Secosyrins and (+)-Syributins. *J. Org. Chem.* **1997**, 62, 8095-8103.
26. Jamison, T. F., Shambayati, S., Crowe, W. E., Schreiber, S. L. Tandem Use of Cobalt-Mediated Reactions to Synthesize (+)-Epoxydictymene, a Diterpene Containing a Trans-Fused 5-5 Ring System. *J. Am. Chem. Soc.* **1997**, 119, 4353-4363.
27. Mukai, C., Yamashita, H., Ichiryu, T., Hanaoka, M. A new procedure for construction of oxocane and oxonane derivatives based on alkyne-Co2(CO)6 complexes. *Tetrahedron* **2000**, 56, 2203-2209.

Related reactions: Luche readuction, Corey-Bakshi-Shibata (CBS) reduction, Midland alpine borane reduction;

1. Miyashita, A., Yasuda, A., Takaya, H., Toriumi, K., Ito, T., Souchi, T., Noyori, R. Synthesis of 2,2'-bis(diphenylphosphino)-1,1'-binaphthyl (BINAP), an atropisomeric chiral bis(triaryl)phosphine, and its use in the rhodium(I)-catalyzed asymmetric hydrogenation of α-(acylamino)acrylic acids. *J. Am. Chem. Soc.* **1980**, 102, 7932-7934.
2. Noyori, R., Ohta, M., Hsiao, Y., Kitamura, M., Ohta, T., Takaya, H. Asymmetric synthesis of isoquinoline alkaloids by homogeneous catalysis. *J. Am. Chem. Soc.* **1986**, 108, 7117-7119.
3. Kitamura, M., Ohkuma, T., Inoue, S., Sayo, N., Kumobayashi, H., Akutagawa, S., Ohta, T., Takaya, H., Noyori, R. Homogeneous asymmetric hydrogenation of functionalized ketones. *J. Am. Chem. Soc.* **1988**, 110, 629-631.
4. Kitamura, M., Tokunaga, M., Ohkuma, T., Noyori, R. Convenient preparation of BINAP-ruthenium(II) complexes for catalyzing the asymmetric hydrogenation of functionalized ketones. *Tetrahedron Lett.* **1991**, 32, 4163-4166.
5. Matteoli, U., Frediani, P., Bianchi, M., Botteghi, C., Gladiali, S. Asymmetric homogeneous catalysis by ruthenium complexes. *J. Mol. Catal.* **1981**, 12, 265-319.
6. James, B. R., Pacheco, A., Rettig, S. J., Thorburn, I. S., Ball, R. G., Ibers, J. A. Activation of dihydrogen by ruthenium(II)-chelating phosphine complexes, and activation of dioxygen by ruthenium(II) porphyrin complexes: an update. *J. Mol. Catal.* **1987**, 41, 147-161.
7. Akutagawa, S. Asymmetric hydrogenation with Ru-BINAP catalysts. *Chirality Ind.* **1992**, 325-339.
8. Genet, J. P. General synthesis of chiral RuII catalysts (P*P)RuX_2 using CODRu-(2-methylallyl)$_2$. Efficient catalysts for asymmetric hydrogenations. *Acros Organics Acta* **1995**, 1, 4-9.
9. Genet, J. P. New developments in chiral ruthenium (II) catalysts for asymmetric hydrogenation and synthetic applications. *ACS Symp. Ser.* **1996**, 641, 31-51.
10. Kagan, H. B. Development of asymmetric catalysis by chiral metal complexes: the example of asymmetric hydrogenation. *C. R. l'Academie. Sci., Ser. IIb Univers* **1996**, 322, 131-143.
11. Noyori, R. Asymmetric hydrogenation. *Acta Chem. Scand.* **1996**, 50, 380-390.
12. Ager, D. J., Laneman, S. A. Reductions of 1,3-dicarbonyl systems with ruthenium-biarylbisphosphine catalysts. *Tetrahedron: Asymmetry* **1997**, 8, 3327-3355.
13. Bianchini, C., Glendenning, L. Ruthenium(II)-catalyzed asymmetric transfer hydrogenation of ketones using a formic acid-triethylamine mixture. Asymmetric transfer hydrogenation of imines. *Chemtracts* **1997**, 10, 333-338.
14. Noyori, R., Hashiguchi, S. Asymmetric Transfer Hydrogenation Catalyzed by Chiral Ruthenium Complexes. *Acc. Chem. Res.* **1997**, 30, 97-102.
15. Sun, Y., Wang, J., LeBlond, C., Landau, R. N., Joseph, L., Sowa, J. R., Jr., Blackmond, D. G. Kinetic influences on enantioselectivity in asymmetric catalytic hydrogenation. *J. Mol. Catal. A: Chemical* **1997**, 115, 495-502.
16. Sun, Y., Wang, J., LeBlond, C., Landau, R. N., Laquidara, J., Sowa, J. R., Jr., Blackmond, D. G. Kinetic influences on enantioselectivity in asymmetric catalytic hydrogenation. *J. Mol. Catal. A: Chemical* **1997**, 115, 495-502.
17. Nagel, U., Albrecht, J. The enantioselective hydrogenation of N-acyl dehydroamino acids. *Top. in Cat.* **1998**, 5, 3-23.
18. Naota, T., Takaya, H., Murahashi, S.-I. Ruthenium-Catalyzed Reactions for Organic Synthesis. *Chem. Rev.* **1998**, 98, 2599-2660.
19. Ratovelomanana-Vidal, V., Genet, J.-P. Enantioselective ruthenium-mediated hydrogenation: developments and applications. *J. Organomet. Chem.* **1998**, 567, 163-172.
20. Brown, J. M. Hydrogenation of functionalized carbon-carbon double bonds. *Comprehensive Asymmetric Catalysis I-III* **1999**, 1, 121-182.
21. Genet, J. P. Recent developments in asymmetric hydrogenation with chiral Ru(II) catalysts and synthetic applications to biologically active molecules. *Current Trends in Organic Synthesis, [Proceedings of the International Conference on Organic Synthesis], 12th, Venezia, June 28-July 2, 1998* **1999**, 229-237.
22. Ratovelomanana-Vidal, V., Genet, J.-P. Synthetic applications of the ruthenium-catalyzed hydrogenation via dynamic kinetic resolution. *Can. J. Chem.* **2000**, 78, 846-851.
23. Kumobayashi, H., Miura, T., Sayo, N., Saito, T., Zhang, X. Recent advances of BINAP chemistry in the industrial aspects. *Synlett* **2001**, 1055-1064.
24. Rossen, K. Ru- and Rh-catalyzed asymmetric hydrogenations: recent surprises from an old reaction. *Angew. Chem., Int. Ed. Engl.* **2001**, 40, 4611-4613.
25. Genet, J. P. Recent studies on asymmetric hydrogenation. New catalysts and synthetic applications in organic synthesis. *Pure Appl. Chem.* **2002**, 74, 77-83.
26. McCague, R. Can asymmetric hydrogenation chemocatalysis be predicted? *Speciality Chemicals Magazine* **2002**, 22, 26, 28-29.
27. Noyori, R. Asymmetric catalysis: science and opportunities (Nobel Lecture). *Angew. Chem., Int. Ed. Engl.* **2002**, 41, 2008-2022.
28. Stibbs, B. The 2001 nobel prize in chemistry: Prize awarded for the development of catalytic asymmetric synthesis. *Can. Chem. News* **2002**, 54, 26-27.
29. Genet, J.-P. Asymmetric Catalytic Hydrogenation. Design of New Ru Catalysts and Chiral Ligands: From Laboratory to Industrial Applications. *Acc. Chem. Res.* **2003**, 36, 908-918.
30. Tang, W., Zhang, X. New Chiral Phosphorus Ligands for Enantioselective Hydrogenation. *Chem. Rev.* **2003**, 103, 3029-3069.
31. Ohkuma, T., Noyori, R. Hydrogenation of carbonyl groups. *Comprehensive Asymmetric Catalysis, Supplement* **2004**, 1, 1-41.
32. Pettinari, C., Marchetti, F., Martini, D. Metal complexes as hydrogenation catalysts. *Comprehensive Coordination Chemistry II* **2004**, 9, 75-139.
33. Xiao, J., Nefkens, S. C. A., Jessop, P. G., Ikariya, T., Noyori, R. Asymmetric hydrogenation of α,β-unsaturated carboxylic acids in supercritical carbon dioxide. *Tetrahedron Lett.* **1996**, 37, 2813-2816.
34. Dijkstra, H. P., van Klink, G. P. M., van Koten, G. The Use of Ultra- and Nanofiltration Techniques in Homogeneous Catalyst Recycling. *Acc. Chem. Res.* **2002**, 35, 798-810.
35. Wu, J., Ji, J.-X., Guo, R., Yeung, C.-H., Chan, A. S. C. Chiral [RuCl_2(dipyridylphosphane)(1,2-diamine)] catalysts: Applications in asymmetric hydrogenation of a wide range of simple ketones. *Chem.-- Eur. J.* **2003**, 9, 2963-2968.
36. Makino, K., Goto, T., Hiroki, Y., Hamada, Y. Stereoselective synthesis of anti-β-hydroxy-α-amino acids through dynamic kinetic resolution. *Angew. Chem., Int. Ed. Engl.* **2004**, 43, 882-884.
37. Valenrod, Y., Myung, J., Ben, R. N. Dynamic kinetic resolution (DKR) using immobilized amine nucleophiles. *Tetrahedron Lett.* **2004**, 45, 2545-2549.
38. Kless, A., Boerner, A., Heller, D., Selke, R. Ab Initio Studies of Rhodium(I)-N-Alkenylamide Complexes with cis- and trans-Coordinating Phosphines: Relevance for the Mechanism of Catalytic Asymmetric Hydrogenation of Prochiral Dehydroamino Acids. *Organometallics* **1997**, 16, 2096-2100.
39. Landis, C. R., Hilfenhaus, P., Feldgus, S. Structures and Reaction Pathways in Rhodium(I)-Catalyzed Hydrogenation of Enamides: A Model DFT Study. *J. Am. Chem. Soc.* **1999**, 121, 8741-8754.
40. Takaya, H., Mashima, K., Koyano, K., Yagi, M., Kumobayashi, H., Taketomi, T., Akutagawa, S., Noyori, R. Practical synthesis of (R)- or (S)-2,2'-bis(diarylphosphino)-1,1'-binaphthyls (BINAPs). *J. Org. Chem.* **1986**, 51, 629-635.
41. Takaya, H., Akutagawa, S., Noyori, R. (R)-(+)- and (S)-(-)-2,2'-Bis(diphenylphosphino)-1,1'-binaphthyl (BINAP). *Org. Synth.* **1989**, 67, 20-32.
42. Ohta, T., Takaya, H., Kitamura, M., Nagai, K., Noyori, R. Asymmetric hydrogenation of unsaturated carboxylic acids catalyzed by BINAP-ruthenium(II) complexes. *J. Org. Chem.* **1987**, 52, 3174-3176.
43. Takaya, H., Ohta, T., Sayo, N., Kumobayashi, H., Akutagawa, S., Inoue, S., Kasahara, I., Noyori, R. Enantioselective hydrogenation of allylic and homoallylic alcohols. *J. Am. Chem. Soc.* **1987**, 109, 1596-1597.
44. Lubell, W. D., Kitamura, M., Noyori, R. Enantioselective synthesis of β-amino acids based on BINAP-ruthenium(II) catalyzed hydrogenation. *Tetrahedron: Asymmetry* **1991**, 2, 543-554.

45. Kitamura, M., Yoshimura, M., Tsukamoto, M., Noyori, R. Synthesis of α-amino phosphonic acids by asymmetric hydrogenation. *Enantiomer* **1996**, 1, 281-303.
46. Noyori, R., Ohkuma, T., Kitamura, M., Takaya, H., Sayo, N., Kumobayashi, H., Akutagawa, S. Asymmetric hydrogenation of β-keto carboxylic esters. A practical, purely chemical access to β-hydroxy esters in high enantiomeric purity. *J. Am. Chem. Soc.* **1987**, 109, 5856-5858.
47. Noyori, R., Ikeda, T., Ohkuma, T., Widhalm, M., Kitamura, M., Takaya, H., Akutagawa, S., Sayo, N., Saito, T., et al. Stereoselective hydrogenation via dynamic kinetic resolution. *J. Am. Chem. Soc.* **1989**, 111, 9134-9135.
48. Kitamura, M., Tokunaga, M., Noyori, R. Quantitative expression of dynamic kinetic resolution of chirally labile enantiomers: stereoselective hydrogenation of 2-substituted 3-oxo carboxylic esters catalyzed by BINAP-ruthenium(II) complexes. *J. Am. Chem. Soc.* **1993**, 115, 144-152.
49. Kawano, H., Ikariya, T., Ishii, Y., Saburi, M., Yoshikawa, S., Uchida, Y., Kumobayashi, H. Asymmetric hydrogenation of prochiral alkenes catalyzed by ruthenium complexes of (R)-(+)-2,2'-bis(diphenylphosphino)-1,1'-binaphthyl. *J. Chem. Soc., Perkin Trans. 1* **1989**, 1571-1575.
50. Ohta, T., Takaya, H., Noyori, R. Stereochemistry and mechanism of the asymmetric hydrogenation of unsaturated carboxylic acids catalyzed by BINAP-ruthenium(II) dicarboxylate complexes. *Tetrahedron Lett.* **1990**, 31, 7189-7192.
51. Ashby, M. T., Halpern, J. Kinetics and mechanism of catalysis of the asymmetric hydrogenation of α,β-unsaturated carboxylic acids by bis(carboxylato) {2,2'-bis(diphenylphosphino)-1,1'-binaphthyl}ruthenium(II), [RuII(BINAP) $(O_2CR)_2$]. *J. Am. Chem. Soc.* **1991**, 113, 589-594.
52. Yoshikawa, K., Murata, M., Yamamoto, N., Inoguchi, K., Achiwa, K. Asymmetric reactions catalyzed by chiral metal complexes. III. The origin of the enantioselection in the ruthenium(II)-catalyzed asymmetric hydrogenation of α,β-unsaturated carboxylic acid. *Chem. Pharm. Bull.* **1992**, 40, 1072-1074.
53. Chan, A. S. C., Chen, C. C., Yang, T. K., Huang, J. H., Lin, Y. C. Mechanistic aspects of Ru(BINAP)-catalyzed asymmetric hydrogenation of vinyl carboxylic acid derivatives. *Inorg. Chim. Acta* **1995**, 234, 95-100.
54. Girard, C., Genet, J.-P., Bulliard, M. Non-linear effects in ruthenium-catalyzed asymmetric hydrogenation with atropisomeric diphosphines. *Eur. J. Org. Chem.* **1999**, 2937-2942.
55. Brown, J. M., Giernoth, R. New mechanistic aspects of the asymmetric homogeneous hydrogenation of alkenes. *Current Opinion in Drug Discovery & Development* **2000**, 3, 825-832.
56. Petra, D. G. I., Reek, J. N. H., Handgraaf, J.-W., Meijer, E. J., Dierkes, P., Kamer, P. C. J., Brussee, J., Schoemaker, H. E., Van Leeuwen, P. W. N. M. Chiral induction effects in ruthenium(II) amino alcohol catalyzed asymmetric transfer hydrogenation of ketones: an experimental and theoretical approach. *Chem.-- Eur. J.* **2000**, 6, 2818-2829.
57. Gridnev, I. D., Yasutake, M., Higashi, N., Imamoto, T. Asymmetric Hydrogenation of Enamides with Rh-BisP and Rh-MiniPHOS Catalysts. Scope, Limitations, and Mechanism. *J. Am. Chem. Soc.* **2001**, 123, 5268-5276.
58. Noyori, R., Yamakawa, M., Hashiguchi, S. Metal-Ligand Bifunctional Catalysis: A Nonclassical Mechanism for Asymmetric Hydrogen Transfer between Alcohols and Carbonyl Compounds. *J. Org. Chem.* **2001**, 66, 7931-7944.
59. Abdur-Rashid, K., Clapham, S. E., Hadzovic, A., Harvey, J. N., Lough, A. J., Morris, R. H. Mechanism of the Hydrogenation of Ketones Catalyzed by trans-Dihydrido(diamine)ruthenium(II) Complexes. *J. Am. Chem. Soc.* **2002**, 124, 15104-15118.
60. Daley, C. J. A., Bergens, S. H. The First Complete Identification of a Diastereomeric Catalyst-Substrate (Alkoxide) Species in an Enantioselective Ketone Hydrogenation. Mechanistic Investigations. *J. Am. Chem. Soc.* **2002**, 124, 3680-3691.
61. Kitamura, M., Tsukamoto, M., Bessho, Y., Yoshimura, M., Kobs, U., Widhalm, M., Noyori, R. Mechanism of asymmetric hydrogenation of α-(acylamino)acrylic esters catalyzed by BINAP-ruthenium(II) diacetate. *J. Am. Chem. Soc.* **2002**, 124, 6649-6667.
62. Andraos, J. Quantification and Optimization of Dynamic Kinetic Resolution. *J. Phys. Chem. A* **2003**, 107, 2374-2387.
63. Sandoval Christian, A., Ohkuma, T., Muniz, K., Noyori, R. Mechanism of asymmetric hydrogenation of ketones catalyzed by BINAP/1,2-diamine-rutheniumII complexes. *J. Am. Chem. Soc.* **2003**, 125, 13490-13503.
64. Noyori, R., Kitamura, M., Ohkuma, T. Toward efficient asymmetric hydrogenation: architectural and functional engineering of chiral molecular catalysts. *Proc. Natl. Acad. Sci. U. S. A.* **2004**, 101, 5356-5362.
65. Taber, D. F., Wang, Y. Synthesis of (-)-Haliclonadiamine. *J. Am. Chem. Soc.* **1997**, 119, 22-26.
66. Holson, E. B., Roush, W. R. Diastereoselective Synthesis of the C(17)-C(28) Fragment (The C-D Spiroketal Unit) of Spongistatin 1 (Altohyrtin A) via a Kinetically Controlled Iodo-Spiroketalization Reaction. *Org. Lett.* **2002**, 4, 3719-3722.
67. Shimizu, H., Shimada, Y., Tomita, A., Mitsunobu, O. Pronounced enhancement of stereoselectivity in asymmetric hydrogenation of 2-substituted 2-propen-1-ols by transient acylation. *Tetrahedron Lett.* **1997**, 38, 849-852.
68. Meuzelaar, G. J., Van Vliet, M. C. A., Maat, L., Sheldon, R. A. Chemistry of opium alkaloids. Part 45. Improvements in the total synthesis of morphine. *Eur. J. Org. Chem.* **1999**, 2315-2321.

Nozaki-Hiyama-Kishi Reaction 318

Related reactions: Barbier reaction, Grignard reaction, Kagan-Molander samarium diiodide coupling;

1. Okude, Y., Hirano, S., Hiyama, T., Nozaki, H. Grignard-type carbonyl addition of allyl halides by means of chromous salt. A chemospecific synthesis of homoallyl alcohols. *J. Am. Chem. Soc.* **1977**, 99, 3179-3181.
2. Okude, Y., Hiyama, T., Nozaki, H. Reduction of organic halides by means of chromium(III) chloride-lithium aluminum hydride reagent in anhydrous media. *Tetrahedron Lett.* **1977**, 3829-3830.
3. Hiyama, T., Okude, Y., Kimura, K., Nozaki, H. Highly selective carbon-carbon bond forming reactions mediated by chromium(II) reagents. *Bull. Chem. Soc. Jpn.* **1982**, 55, 561-568.
4. Nozaki, H., Hiyama, T., Oshima, K., Takai, K. Highly selective synthesis with novel metallic reagents. *ACS Symp. Ser.* **1982**, 185, 99-108.
5. Takai, K., Kimura, K., Kuroda, T., Hiyama, T., Nozaki, H. Selective Grignard-type carbonyl addition of alkenyl halides mediated by chromium(II) chloride. *Tetrahedron Lett.* **1983**, 24, 5281-5284.
6. Jin, H., Uenishi, J., Christ, W. J., Kishi, Y. Catalytic effect of nickel(II) chloride and palladium(II) acetate on chromium(II)-mediated coupling reaction of iodo olefins with aldehydes. *J. Am. Chem. Soc.* **1986**, 108, 5644-5646.
7. Takai, K., Tagashira, M., Kuroda, T., Oshima, K., Utimoto, K., Nozaki, H. Reactions of alkenylchromium reagents prepared from alkenyl trifluoromethanesulfonates (triflates) with chromium(II) chloride under nickel catalysis. *J. Am. Chem. Soc.* **1986**, 108, 6048-6050.
8. Cintas, P. Addition of organochromium compounds to aldehydes: the Nozaki-Hiyama reaction. *Synthesis* **1992**, 248-257.
9. Fürstner, A. Low-valent transition metal induced C-C bond formations: stoichiometric reactions evolving into catalytic processes. *Pure Appl. Chem.* **1998**, 70, 1071-1076.
10. Fürstner, A. Multicomponent catalysis for reductive bond formations. *Chem.-- Eur. J.* **1998**, 4, 567-570.
11. Hodgson, D. M., Comina, P. J. *Chromium(II)-mediated C-C coupling reactions* (eds. Beller, M.,Bolm, C.) (Wiley-VCH, Weinheim, New York, **1998**) 418-424.
12. Avalos, M., Babiano, R., Cintas, P., Jimenez, J. L., Palacios, J. C. Synthetic variations based on low-valent chromium: new developments. *Chem. Soc. Rev.* **1999**, 28, 169-177.
13. Fürstner, A. Carbon-Carbon Bond Formation Involving Organochromium(III) Reagents. *Chem. Rev.* **1999**, 99, 991-1045.
14. Hirao, T. A catalytic system for reductive transformations via one-electron transfer. *Synlett* **1999**, 175-181.
15. Wessjohann, L. A., Scheid, G. Recent advances in chromium(II)- and chromium(III)-mediated organic synthesis. *Synthesis* **1999**, 1-36.
16. Takai, K., Nozaki, H. Nucleophilic addition of organochromium reagents to carbonyl compounds. *Proceedings of the Japan Academy, Series B: Physical and Biological Sciences* **2000**, 76B, 123-131.
17. Wipf, P., Lim, S. Addition of organochromium reagents to aldehydes, ketones and enones: a low-temperature version of the Nozaki-Hiyama reaction. *J. Chem. Soc., Chem. Commun.* **1993**, 1654-1656.
18. Fürstner, A., Shi, N. Nozaki-Hiyama-Kishi Reactions Catalytic in Chromium. *J. Am. Chem. Soc.* **1996**, 118, 12349-12357.

19. Fürstner, A., Shi, N. A Multicomponent Redox System Accounts for the First Nozaki-Hiyama-Kishi Reactions Catalytic in Chromium. *J. Am. Chem. Soc.* **1996**, 118, 2533-2534.
20. Maguire, R. J., Mulzer, J., Bats, J. W. 1,4-Asymmetric Induction in the Chromium(II)- and Indium-Mediated Coupling of Allyl Bromides to Aldehydes. *J. Org. Chem.* **1996**, 61, 6936-6940.
21. Marshall, J. A., McNulty, L. M. A multicomponent redox system accounts for the first Nozaki-Hiyama-Kishi reactions catalytic in chromium. *Chemtracts* **1997**, 10, 50-52.
22. Bandini, M., Cozzi, P. G., Melchiorre, P., Umani-Ronchi, A. The first catalytic enantioselective Nozaki-Hiyama reaction. *Angew. Chem., Int. Ed. Engl.* **1999**, 38, 3357-3359.
23. Bandini, M., Cozzi, P. G., Umani-Ronchi, A. The first catalytic enantioselective Nozaki-Hiyama-Kishi reaction. *Polyhedron* **2000**, 19, 537-539.
24. Bandini, M., Cozzi, P. G., Melchiorre, P., Morganti, S., Umani-Ronchi, A. Cr(Salen)-Catalyzed Addition of 1,3-Dichloropropene to Aromatic Aldehydes. A Simple Access to Optically Active Vinyl Epoxides. *Org. Lett.* **2001**, 3, 1153-1155.
25. Durandetti, M., Nedelec, J.-Y., Perichon, J. An electrochemical coupling of organic halide with aldehydes, catalytic in chromium and nickel salts. The Nozaki-Hiyama-Kishi reaction. *Org. Lett.* **2001**, 3, 2073-2076.
26. Micskei, K., Kiss-Szikszai, A., Gyarmati, J., Hajdu, C. Carbon-carbon bond formation in neutral aqueous medium by modification of the Nozaki-Hiyama reaction. *Tetrahedron Lett.* **2001**, 42, 7711-7713.
27. Berkessel, A., Menche, D., Sklorz, C. A., Schroder, M., Paterson, I. A highly enantioselective catalyst for the asymmetric Nozaki-Hiyama-Kishi reaction of allylic and vinylic halides. *Angew. Chem., Int. Ed. Engl.* **2003**, 42, 1032-1035.
28. Inoue, M., Suzuki, T., Nakada, M. Asymmetric Catalysis of Nozaki-Hiyama Allylation and Methallylation with A New Tridentate Bis(oxazolinyl)carbazole Ligand. *J. Am. Chem. Soc.* **2003**, 125, 1140-1141.
29. Lombardo, M., Licciulli, S., Morganti, S., Trombini, C. 3-Chloropropenyl pivaloate in organic synthesis: the first asymmetric catalytic entry to syn-alk-1-ene-3,4-diols. *Chem. Commun.* **2003**, 1762-1763.
30. Suzuki, T., Kinoshita, A., Kawada, H., Nakada, M. A new asymmetric tridentate carbazole ligand: Its preparation and application to Nozaki-Hiyama allylation. *Synlett* **2003**, 570-572.
31. Molander, G. A., St. Jean, D. J., Jr., Haas, J. Toward a General Route to the Eunicellin Diterpenes: The Asymmetric Total Synthesis of Deacetoxyalcyonin Acetate. *J. Am. Chem. Soc.* **2004**, 126, 1642-1643.
32. Panek, J. S., Liu, P. Total Synthesis of the Actin-Depolymerizing Agent (-)-Mycalolide A: Application of Chiral Silane-Based Bond Construction Methodology. *J. Am. Chem. Soc.* **2000**, 122, 11090-11097.
33. Pilli, R. A., Victor, M. M., De Meijere, A. First Total Synthesis of Aspinolide B, a New Pentaketide Produced by Aspergillus ochraceus. *J. Org. Chem.* **2000**, 65, 5910-5916.
34. Taylor, R. E., Chen, Y. Total Synthesis of Epothilones B and D. *Org. Lett.* **2001**, 3, 2221-2224.

Oppenauer Oxidation 320

Related reactions: Cannizzaro reaction, Tishchenko reaction;

1. Oppenauer, R. V. Dehydration of secondary alcohols to ketones. I. Preparation of sterol ketones and sex hormones. *Recl. Trav. Chim. Pays-Bas* **1937**, 56, 137-144.
2. Djerassi, C. Oppenauer oxidation. *Org. React.* **1951**, 6, 207-272.
3. Procter, G. Oxidation Adjacent to Oxygen of Alcohols by Othr Methods. in *Comp. Org. Synth.* (eds. Trost, B. M.,Fleming, I.), *7*, 305-327 (Pergamon, Oxford, **1991**).
4. de Graauw, C. F., Peters, J. A., van Bekkum, H., Huskens, J. Meerwein-Ponndorf-Verley reductions and Oppenauer oxidations: an integrated approach. *Synthesis* **1994**, 1007-1017.
5. Creyghton, E. J., Van der Waal, J. C. Meerwein-Ponndorf-Verley reduction, Oppenauer oxidation, and related reactions. *Fine Chemicals through Heterogeneous Catalysis* **2001**, 438-448.
6. Budzelaar, P. H. M., Talarico, G. Insertion and β-hydrogen transfer at aluminum. in *Structure and Bonding* (ed. Mingos, D. M. P.), *105*, (Springer-Verlag, Berlin, Heidelberg, **2003**).
7. Jerome, J. E., Sergent, R. H. Catalytic applications of aluminum isopropoxide in organic synthesis. *Chem. Ind.* **2003**, 89, 97-114.
8. Reich, R., Keana, J. F. W. Oppenauer oxidations using 1-methyl-4-piperidone as the hydride acceptor. *Synth. Commun.* **1972**, 2, 323-325.
9. Kow, R., Nygren, R., Rathke, M. W. Rate enhancement of the Meerwein-Ponndorf-Verley-Oppenauer reaction in the presence of proton acids. *J. Org. Chem.* **1977**, 42, 826-827.
10. Namy, J. L., Souppe, J., Collin, J., Kagan, H. B. New preparations of lanthanide alkoxides and their catalytical activity in Meerwein-Ponndorf-Verley-Oppenauer reactions. *J. Org. Chem.* **1984**, 49, 2045-2049.
11. Akamanchi, K. G., Chaudhari, B. A. Diisopropoxyaluminum trifluoroacetate/4-nitrobenzaldehyde - a new Oppenauer oxidation system for accelerated oxidation of secondary alcohols to the corresponding ketones. *Tetrahedron Lett.* **1997**, 38, 6925-6928.
12. Creyghton, E. J., Ganeshie, S. D., Downing, R. S., van Bekkum, H. Stereoselective Meerwein-Ponndorf-Verley and Oppenauer reactions catalyzed by zeolite BEA. *J. Mol. Catal. A: Chemical* **1997**, 115, 457-472.
13. Ishihara, K., Kurihara, H., Yamamoto, H. Bis(pentafluorophenyl)borinic Acid as a Highly Effective Oppenauer Oxidation Catalyst for Allylic and Benzylic Alcohols. *J. Org. Chem.* **1997**, 62, 5664-5665.
14. Ooi, T., Miura, T., Itagaki, Y., Ichikawa, H., Maruoka, K. Catalytic Meerwein-Ponndorf-Verley (MPV) and Oppenauer (OPP) reactions: remarkable acceleration of the hydride transfer by powerful bidentate aluminum alkoxides. *Synthesis* **2002**, 279-291.
15. Ooi, T., Otsuka, H., Miura, T., Ichikawa, H., Maruoka, K. Practical Oppenauer (OPP) oxidation of alcohols with a modified aluminum catalyst. *Org. Lett.* **2002**, 4, 2669-2672.
16. Suzuki, T., Morita, K., Tsuchida, M., Hiroi, K. Iridium-Catalyzed Oppenauer Oxidations of Primary Alcohols Using Acetone or 2-Butanone as Oxidant. *J. Org. Chem.* **2003**, 68, 1601-1602.
17. Sominsky, L., Rozental, E., Gottlieb, H., Gedanken, A., Hoz, S. Uncatalyzed Meerwein-Ponndorf-Oppenauer-Verley reduction of aldehydes and ketones under supercritical conditions. *J. Org. Chem.* **2004**, 69, 1492-1496.
18. Sastre, G., Corma, A. Relation between structure and Lewis acidity of Ti-β and TS-1 zeolites A quantum-chemical study. *Chem. Phys. Lett.* **1999**, 302, 447-453.
19. Meerwein, H., Schmidt, R. New method for the reduction of aldehydes and ketones. *Ann.* **1925**, 444, 221-238.
20. Verley, A. The exchange of functional groups between two molecules. The passage of ketones to alcohols and the reverse. *Bull. soc. chim.* **1925**, 37, 871-874.
21. Ponndorf, W. The reversible exchange of oxygen between aldehydes or ketones on the one hand and primary or secondary alcohols on the other hand. *Z. angew. Chem.* **1926**, 39, 138-143.
22. Bersin, T. New methods in organic synthesis. II. Reduction according to Meerwein-Ponndorf and oxidation according to Oppenauer. *Angew. Chem.* **1940**, 53, 266-271,299.
23. Woodward, R. B., Wendler, N. L., Brutschy, F. J. Quininone. *J. Am. Chem. Soc.* **1945**, 67, 1425-1429.
24. Otvos, L., Gruber, L., Meisel-Agoston, J. The Meerwein-Ponndorf-Verley-Oppenauer reaction. I. Investigation of the reaction mechanism with radiocarbon. Racemization of secondary alcohols. *Acta Chim. Acad. Sci. Hung.* **1965**, 43, 149-153.
25. Yager, B. J., Hancock, C. K. Equilibrium and kinetic studies of the Meerwein-Ponndorf-Verley-Oppenauer (MPVO) reaction. *J. Org. Chem.* **1965**, 30, 1174-1179.
26. Ashby, E. C. Single-electron transfer, a major reaction pathway in organic chemistry. An answer to recent criticisms. *Acc. Chem. Res.* **1988**, 21, 414-421.
27. Laxmi, Y. R. S., Backvall, J.-E. Mechanistic studies on ruthenium-catalyzed hydrogen transfer reactions. *Chem. Commun.* **2000**, 611-612.
28. Pamies, O., Backvall, J.-E. Studies on the mechanism of metal-catalyzed hydrogen transfer from alcohols to ketones. *Chem.-- Eur. J.* **2001**, 7, 5052-5058.

29. Klomp, D., Maschmeyer, T., Hanefeld, U., Peters Joop, A. Mechanism of homogeneously and heterogeneously catalysed meerwein-ponndorf-verley-oppenauer reactions for the racemisation of secondary alcohols. *Chemistry (Weinheim an der Bergstrasse, Germany)* **2004**, 10, 2088-2093.
30. Kocovsky, P., Baines, R. S. Synthesis of estrone via a thallium(III)-mediated fragmentation of a 19-hydroxyandrost-5-ene precursor. *Tetrahedron Lett.* **1993**, 34, 6139-6140.
31. Sternbach, D. D., Ensinger, C. L. Synthesis of polyquinanes. 3. The total synthesis of (±)-hirsutene: the intramolecular Diels-Alder approach. *J. Org. Chem.* **1990**, 55, 2725-2736.
32. Heathcock, C. H., Kleinman, E. F., Binkley, E. S. Total synthesis of lycopodium alkaloids: (±)-lycopodine, (±)-lycodine, and (±)-lycodoline. *J. Am. Chem. Soc.* **1982**, 104, 1054-1068.
33. Shing, T. K. M., Lee, C. M., Lo, H. Y. Synthesis of the CD ring in taxol from (S)-(+)-carvone. *Tetrahedron Lett.* **2001**, 42, 8361-8363.

Overman Rearrangement 322

Related reactions: Claisen rearrangement;

1. Mumm, O., Möller, F. Experiments on the theory of the allyl rearrangement. *Ber.* **1937**, 70B, 2214-2227.
2. Overman, L. E. Thermal and mercuric ion catalyzed [3,3]-sigmatropic rearrangement of allylic trichloroacetimidates. 1,3 Transposition of alcohol and amine functions. *J. Am. Chem. Soc.* **1974**, 96, 597-599.
3. Overman, L. E. A general method for the synthesis of amines by the rearrangement of allylic trichloroacetimidates. 1,3 Transposition of alcohol and amine functions. *J. Am. Chem. Soc.* **1976**, 98, 2901-2910.
4. McCarty, C. G., Garner, L. A. Rearrangements involving imidic acid derivatives. in *Chem. Amidines Imidates* (ed. Patai, S.), 189-240 (Wiley, New York, **1975**).
5. Overman, L. E. Allylic and propargylic imidic esters in organic synthesis. *Acc. Chem. Res.* **1980**, 13, 218-224.
6. Overman, L. E. New synthetic methods. (46). Mercury(II)- and palladium(II)-catalyzed [3.3]-sigmatropic rearrangements. *Angew. Chem.* **1984**, 96, 565-573.
7. Altenbach, H. J. Functional group transformations via allyl rearrangement. in *Comp. Org. Synth.* (eds. Trost, B. M.,Fleming, I.), *6*, 829-871 (Pergamon, Oxford, **1991**).
8. Ritter, K. Formation of C-N bonds by sigmatropic rearrangements. in *Houben-Weyl. Stereoselective Synthesis* (eds. Hoffmann, R. W., Mulzer, J.,Schaumann, E.), *9*, 5677-5699 (Thieme, Stuttgart, **1995**).
9. Overman, L. E., Clizbe, L. A. Synthesis of trichloroacetamido-1,3-dienes. Useful aminobutadiene equivalents for the Diels-Alder reaction. *J. Am. Chem. Soc.* **1976**, 98, 2352-2354.
10. Overman, L. E., Kakimoto, M. Preparation of rearranged allylic isocyanates from the reaction of allylic alkoxides with cyanogen chloride. *J. Org. Chem.* **1978**, 43, 4564-4567.
11. Savage, I., Thomas, E. J. Asymmetric α-amino acid synthesis: synthesis of (+)-polyoxamic acid using a [3,3]allylic trifluoroacetimidate rearrangement. *J. Chem. Soc., Chem. Commun.* **1989**, 717-719.
12. Chen, A., Savage, I., Thomas, E. J., Wilson, P. D. Asymmetric α-amino acid synthesis using [3.3] rearrangement of allylic trifluoroacetimidates: synthesis of thymine polyoxin C. *Tetrahedron Lett.* **1993**, 34, 6769-6772.
13. Calter, M., Hollis, T. K., Overman, L. E., Ziller, J., Zipp, G. G. First Enantioselective Catalyst for the Rearrangement of Allylic Imidates to Allylic Amides. *J. Org. Chem.* **1997**, 62, 1449-1456.
14. Toshio, N., Masanori, A., Norio, O., Minoru, I. Improved Conditions for Facile Overman Rearrangement. *J. Org. Chem.* **1998**, 63, 188-192.
15. Donde, Y., Overman, L. E. High Enantioselection in the Rearrangement of Allylic Imidates with Ferrocenyl Oxazoline Catalysts. *J. Am. Chem. Soc.* **1999**, 121, 2933-2934.
16. Savage, I., Thomas, E. J., Wilson, P. D. Stereoselective synthesis of allylic amines by rearrangement of allylic trifluoroacetimidates: stereoselective synthesis of polyoxamic acid and derivatives of other α-amino acids. *J. Chem. Soc., Perkin Trans. 1* **1999**, 3291-3303.
17. Banert, K., Melzer, A. The first direct observation of an allylic [3,3] sigmatropic cyanate-isocyanate rearrangement. *Tetrahedron Lett.* **2001**, 42, 6133-6135.
18. Anderson, C. E., Overman, L. E. Catalytic Asymmetric Rearrangement of Allylic Trichloroacetimidates. A Practical Method for Preparing Allylic Amines and Congeners of High Enantiomeric Purity. *J. Am. Chem. Soc.* **2003**, 125, 12412-12413.
19. Overman, L. E., Owen, C. E., Pavan, M. M., Richards, C. J. Catalytic asymmetric rearrangement of allylic N-aryl trifluoroacetimidates. A useful method for transforming prochiral allylic alcohols to chiral allylic amines. *Org. Lett.* **2003**, 5, 1809-1812.
20. Lee, E. E., Batey, R. A. Palladium-catalyzed [3,3] sigmatropic rearrangement of (allyloxy)iminodiazaphospholidines: Allylic transposition of C-O and C-N functionality. *Angew. Chem., Int. Ed. Engl.* **2004**, 43, 1865-18687.
21. Eguchi, T., Koudate, T., Kakinuma, K. Diacetone glucose architecture as a chirality template. III. The Overman rearrangement on a diacetone-D-glucose template: kinetic and theoretical studies on the chirality transcription. *Tetrahedron* **1993**, 49, 4527-4540.
22. Cramer, F., Pawelzik, K., Baldauf, H. J. Imido esters. I. Preparation of trichloroacetimidic acid esters. *Chem. Ber.* **1958**, 91, 1049-1054.
23. Clizbe, L. A., Overman, L. E. Allylically transposed amines from allylic alcohols: 3,7-dimethyl-1,6-octadien-3-amine. *Org. Synth.* **1978**, 58, 4-11.
24. Nagashima, H., Wakamatsu, H., Ozaki, N., Ishii, T., Watanabe, M., Tajima, T., Itoh, K. Transition metal catalyzed radical cyclization: new preparative route to γ-lactams from allylic alcohols via the [3.3]-sigmatropic rearrangement of allylic trichloroacetimidates and the subsequent ruthenium-catalyzed cyclization of N-allyltrichloroacetamides. *J. Org. Chem.* **1992**, 57, 1682-1689.
25. Yamamoto, N., Isobe, M. Direct preparation of guanidine from trichloroacetamide. A potentially important method to (-)-tetrodotoxin. *Chem. Lett.* **1994**, 2299-2302.
26. Overman, L. E., Campbell, C. B. Mercury(II)-catalyzed 3,3-sigmatropic rearrangements of allylic N,N-dimethylcarbamates. A mild method for allylic equilibrations and contrathermodynamic allylic isomer enrichments. *J. Org. Chem.* **1976**, 41, 3338-3340.
27. Doherty, A. M., Kornberg, B. E., Reily, M. D. A study of the 3,3-sigmatropic rearrangement of chiral trichloroacetamidic esters. *J. Org. Chem.* **1993**, 58, 795-798.
28. Oishi, T., Ando, K., Inomiya, K., Sato, H., Iida, M., Chida, N. Total Synthesis of Sphingofungin E from D-Glucose. *Org. Lett.* **2002**, 4, 151-154.
29. Danishefsky, S., Lee, J. Y. Total synthesis of (±)-pancratistatin. *J. Am. Chem. Soc.* **1989**, 111, 4829-4837.
30. Mehmandoust, M., Petit, Y., Larcheveque, M. Synthesis of (E)-β,γ-unsaturated a-amino acids by rearrangement of allyltrichloracetimidates. *Tetrahedron Lett.* **1992**, 33, 4313-4316.
31. Kim, S., Lee, T., Lee, E., Lee, J., Fan, G.-J., Lee, S. K., Kim, D. Asymmetric Total Syntheses of (-)-Antofine and (-)-Cryptopleurine Using (R)-(E)-4-(Tributylstannyl)but-3-en-2-ol. *J. Org. Chem.* **2004**, 69, 3144-3149.

Oxy-Cope Rearrangement and Anionic Oxy-Cope Rearrangement...324

Related reactions: Cope rearrangement;

1. Berson, J. A., Jones, M., Jr. Stepwise mechanisms in the oxy-Cope rearrangement. *J. Am. Chem. Soc.* **1964**, 86, 5017-5018.
2. Berson, J. A., Jones, M., Jr. Synthesis of ketones by the thermal isomerization of 3-hydroxy-1,5-hexadienes. The oxy-Cope rearrangement. *J. Am. Chem. Soc.* **1964**, 86, 5019-5020.
3. Lutz, R. P. Catalysis of the Cope and Claisen rearrangements. *Chem. Rev.* **1984**, 84, 205-247.
4. Swaminathan, S. Base catalyzed rearrangements of oxy-Cope systems. *J. Indian Chem. Soc.* **1984**, 61, 99-107.
5. Paquette, L. A. Stereocontrolled synthesis of complex cyclic ketones by oxy-Cope rearrangement. *Angew. Chem.* **1990**, 102, 642-660.

6. Paquette, L. A. Carbonyl group regeneration with substantive enhancement of structural complexity. *Synlett* **1990**, 67-73.
7. Hill, R. K. Cope, oxy-Cope and anionic oxy-Cope rearrangements. in *Comp. Org. Synth.* (eds. Trost, B. M.,Fleming, I.), *5*, 785-827 (Pergamon Press, Oxford, **1991**).
8. Wilson, S. R. Anion-assisted sigmatropic rearrangements. *Org. React.* **1993**, 43, 93-250.
9. Durairaj, K. Complex cyclic ketones via oxy-Cope rearrangement -- studies relevant to stereocontrolled synthesis. *Curr. Sci.* **1994**, 66, 917-922.
10. Eichinger, P. C. H., Dua, S., Bowie, J. H. A comparison of skeletal rearrangement reactions of even-electron anions in solution and in the gas phase. *Int. J. Mass Spectrom. Ion Processes* **1994**, 133, 1-12.
11. Paquette, L. A. Bridgehead unsaturation in compounds of nature: a proper forum for unleashing the potential of organic synthesis. *Chem. Soc. Rev.* **1995**, 24, 9-17.
12. Paquette, L. A. Recent applications of anionic oxy-Cope rearrangements. *Tetrahedron* **1997**, 53, 13971-14020.
13. Paquette, L. A. Cascade rearrangements following twofold addition of alkenyl anions to squarate esters. *Eur. J. Org. Chem.* **1998**, 1709-1728.
14. Schneider, C. The silyloxy-Cope rearrangement of syn-aldol products: evolution of a powerful synthetic strategy. *Synlett* **2001**, 1079-1091.
15. Evans, D. A., Golob, A. M. [3,3]Sigmatropic rearrangements of 1,5-diene alkoxides. Powerful accelerating effects of the alkoxide substituent. *J. Am. Chem. Soc.* **1975**, 97, 4765-4766.
16. Evans, D. A., Baillargeon, D. J., Nelson, J. V. A general approach to the synthesis of 1,6-dicarbonyl substrates. New applications of base-accelerated oxy-Cope rearrangements. *J. Am. Chem. Soc.* **1978**, 100, 2242-2244.
17. Baumann, H., Chen, P. Density functional study of the oxy-Cope rearrangement. *Helv. Chim. Acta* **2001**, 84, 124-140.
18. Schulze, S. M., Santella, N., Grabowski, J. J., Lee, J. K. The Anionic Oxy-Cope Rearrangement: Using Chemical Reactivity to Reveal the Facile Isomerization of the Parent Substrates in the Gas Phase. *Journal of Organic Chemistry* **2001**, 66, 7247-7253.
19. Viola, A., Iorio, E. J., Chen, K. K. N., Glover, G. M., Nayak, U., Kocienski, P. J. Vapor-phase thermolyses of 3-hydroxy-1,5-hexadienes. II. Effects of methyl substitution. *J. Am. Chem. Soc.* **1967**, 89, 3462-3470.
20. Haeffner, F., Houk, K. N., Reddy, Y. R., Paquette, L. A. Mechanistic Variations and Rate Effects of Alkoxy and Thioalkoxy Substituents on Anionic Oxy-Cope Rearrangements. *J. Am. Chem. Soc.* **1999**, 121, 11880-11884.
21. Lee, E., Lee, Y. R., Moon, B., Kwon, O., Shim, M. S., Yun, J. S. Oxyanion Orientation in Anionic Oxy-Cope Rearrangements. *J. Org. Chem.* **1994**, 59, 1444-1456.
22. Paquette, L. A., Gao, Z., Ni, Z., Smith, G. F. Total Synthesis of Spinosyn A. 1. Enantioselective Construction of a Key Tricyclic Intermediate by a Multiple Configurational Inversion Scheme. *J. Am. Chem. Soc.* **1998**, 120, 2543-2552.
23. MacDougall, J. M., Santora, V. J., Verma, S. K., Turnbull, P., Hernandez, C. R., Moore, H. W. Cyclobutenone-Based Syntheses of Polyquinanes and Bicyclo[6.3.0]undecanes by Tandem Anionic Oxy-Cope Reactions. Total Synthesis of (±)-Precapnelladiene. *J. Org. Chem.* **1998**, 63, 6905-6913.
24. Ogawa, Y., Ueno, T., Karikomi, M., Seki, K., Haga, K., Uyehara, T. Synthesis of 2-acetoxy[5]helicene by sequential double aromatic oxy-Cope rearrangement. *Tetrahedron Lett.* **2002**, 43, 7827-7829.

Paal-Knorr Furan Synthesis 326

Related reactions: Feist-Benary furan synthesis;

1. Knorr, L. Synthesis of furan derivatives from succinic acid esters. *Ber.* **1884**, 17, 2863-2870.
2. Paal, C. Derivatives of acetophenoneacetoacetic ester. *Ber.* **1884**, 17, 2756-2767.
3. Paal, C. Synthesis of thiophene and pyrrole derivatives. *Ber.* **1885**, 367-371.
4. Cheeseman, G. W. H., Bird, C. W. Synthesis of Five-membered Rings with One Heteroatom. in *Comprehensive Heterocyclic Chemistry* (eds. Katritzky, A. R.,Rees, C. W.), *4*, 89-147 (Pergamon Press, Oxford, **1984**).
5. Friedrichsen, W. Furans and their Benzo Derivatives: Synthesis. in *Comprehensive Heterocyclic Chemistry II.* (eds. Katritzky, A. R.,Scriven, E. F. V.), *2*, 359 (Pergamon: Elsevier Science Ltd., Oxford, **1996**).
6. Koenig, B. Product class 9: furans. *Science of Synthesis* **2002**, 9, 183-285.
7. Raghavan, S., Anuradha, K. Solid-phase synthesis of heterocycles from 1,4-diketone synthons. *Synlett* **2003**, 711-713.
8. Minetto, G., Raveglia, L. F., Taddei, M. Microwave-assisted Paal-Knorr reaction. A rapid approach to substituted pyrroles and furans. *Org. Lett.* **2004**, 6, 389-392.
9. Cormier, R. A., Francis, M. D. The epoxyketone-furan rearrangement. *Synth. Commun.* **1981**, 11, 365-369.
10. Lie Ken Jie, M. S. F., Zheng, Y. F. A convenient route to a linear C18 carboxylic acid derivative containing a thiophene ring in the chain via a 9,10-epithio-12-oxo intermediate. *Synthesis* **1988**, 467-468.
11. Ji, J., Lu, X. Facile synthesis of 2,5-disubstituted furans via palladium complex and perfluorinated resin sulfonic acid catalyzed isomerization-dehydration of alkynediols. *J. Chem. Soc., Chem. Commun.* **1993**, 764-765.
12. Foglia, T. A., Sonnet, P. E., Nunez, A., Dudley, R. L. Selective oxidations of methyl ricinoleate: diastereoselective epoxidation with titanium(IV) catalyst. *J. Am. Oil Chem. Soc.* **1998**, 75, 601-607.
13. Amarnath, V., Amarnath, K. Intermediates in the Paal-Knorr Synthesis of Furans. *J. Org. Chem.* **1995**, 60, 301-307.
14. Hart, H., Takehira, Y. Adducts derived from furan macrocycles and benzyne. *J. Org. Chem.* **1982**, 47, 4370-4372.
15. Christopfel, W. C., Miller, L. L. Synthesis of a soluble nonacenetriquinone via a bisisobenzofuran. *J. Org. Chem.* **1986**, 51, 4169-4175.
16. Lai, Y. H., Chen, P. 2,5B,10b,11-Tetramethyldihydropyreno[5,6-c]furan: the first furan-isoannulated [14]annulene that sustains as strong a diamagnetic ring current as the parent system. *Tetrahedron Lett.* **1988**, 29, 3483-3486.
17. Cooper, C. S., Klock, P. L., Chu, D. T. W., Fernandes, P. B. The synthesis and antibacterial activities of quinolones containing five- and six-membered heterocyclic substituents at the 7-position. *J. Med. Chem.* **1990**, 33, 1246-1252.

Paal-Knorr Pyrrole Synthesis 328

Related reactions: Knorr pyrrole synthesis;

1. Knorr, L. Synthesis of furan derivatives from succinic acid esters. *Ber.* **1884**, 17, 2863-2870.
2. Paal, C. Derivatives of acetophenoneacetoacetic ester. *Ber.* **1884**, 17, 2756-2767.
3. Jones, R. A., Bean, G. P. The Chemistry of Pyrroles. in *Organic Chemistry* (eds. Blomquist, A. T.,Wasserman, H. H.), *34*, 525 pp (Academic Press, New York, **1977**).
4. Hort, E. V., Anderson, L. R. Pyrrole and pyrrole derivatives. *Kirk-Othmer Encycl. Chem. Technol., 3rd Ed.* **1982**, 19, 499-520.
5. Cheeseman, G. W. H., Bird, C. W. Synthesis of Five-membered Rings with One Heteroatom. in *Comprehensive Heterocyclic Chemistry* (eds. Katritzky, A. R.,Rees, C. W.), *4*, 89-147 (Pergamon Press, Oxford, **1984**).
6. Sundberg, R. J. Pyrroles and their Benzo Derivatives: Synthesis. in *Comprehensive Organic Functional Group Transformations II* (eds. Katritzky, A. R., Rees, C. W.,Scriven, E. F. V.), *2*, 119-200 (Pergamon, Oxford, New York, **1995**).
7. Korostova, S. E., Mikhaleva, A. I., Vasil'tsov, A. M., Trofimov, B. A. Arylpyrroles: development of classical and modern methods of synthesis. Part II. *Russ. J. Org. Chem.* **1998**, 34, 1691-1714.
8. Ferreira, V. F., De Souza, M. C. B. V., Cunha, A. C., Pereira, L. O. R., Ferreira, M. L. G. Recent advances in the synthesis of pyrroles. *Org. Prep. Proced. Int.* **2001**, 33, 411-454.
9. Kostyanovsky, R. G., Kadorkina, G. K., Mkhitaryan, A. G., Chervin, I. I., Aliev, A. E. New scope and limitations in the Knorr-Paal synthesis of pyrroles. *Mendeleev Commun.* **1993**, 21-23.

10. Yu, S.-X., Le Quesne, P. W. Quararibea metabolites. 3. Total synthesis of (±)-funebral, a rotationally restricted pyrrole alkaloid, using a novel Paal-Knorr reaction. *Tetrahedron Lett.* **1995**, 36, 6205-6208.
11. Dong, Y., Pai, N. N., Ablaza, S. L., Yu, S.-X., Bolvig, S., Forsyth, D. A., Le Quesne, P. W. Quararibea Metabolites. 4.Total Synthesis and Conformational Studies of (±)-Funebrine and (±)-Funebral. *J. Org. Chem.* **1999**, 64, 2657-2666.
12. Braun, R. U., Zeitler, K., Mueller, T. J. J. A Novel One-Pot Pyrrole Synthesis via a Coupling-Isomerization-Stetter-Paal-Knorr Sequence. *Org. Lett.* **2001**, 3, 3297-3300.
13. Surya Prakash Rao, H., Jothilingam, S. One-pot synthesis of pyrrole derivatives from (E)-1,4-diaryl-2-butene-1,4-diones. *Tetrahedron Lett.* **2001**, 42, 6595-6597.
14. Quiclet-Sire, B., Quintero, L., Sanchez-Jimenez, G., Zard, S. Z. A practical variation on the Paal-Knorr pyrrole synthesis. *Synlett* **2003**, 75-78.
15. Banik, B. K., Samajdar, S., Banik, I. Simple Synthesis of Substituted Pyrroles. *J. Org. Chem.* **2004**, 69, 213-216.
16. Bharadwaj, A. R., Scheidt, K. A. Catalytic Multicomponent Synthesis of Highly Substituted Pyrroles Utilizing a One-Pot Sila-Stetter/Paal-Knorr Strategy. *Org. Lett.* **2004**, 6, 2465-2468.
17. Minetto, G., Raveglia, L. F., Taddei, M. Microwave-assisted Paal-Knorr reaction. A rapid approach to substituted pyrroles and furans. *Org. Lett.* **2004**, 6, 389-392.
18. Tracey, M. R., Hsung, R. P., Lambeth, R. H. Allylated β-keto esters as precursors in Paal-Knorr-type pyrrole synthesis: Preparations of chiral and bispyrroles. *Synthesis* **2004**, 918-922.
19. Wang, B., Gu, Y., Luo, C., Yang, T., Yang, L., Suo, J. Pyrrole synthesis in ionic liquids by Paal-Knorr condensation under mild conditions. *Tetrahedron Lett.* **2004**, 45, 3417-3419.
20. Yuguchi, M., Tokuda, M., Orito, K. Pd(0)-Catalyzed conjugate addition of benzylzinc chlorides to α,β-enones in an atmosphere of carbon monoxide: Preparation of 1,4-diketones. *J. Org. Chem.* **2004**, 69, 908-914.
21. Katritzky, A. R., Ostercamp, D. L., Yousaf, T. I. The mechanisms of heterocyclic ring closures. *Tetrahedron* **1987**, 43, 5171-5186.
22. Amarnath, V., Anthony, D. C., Amarnath, K., Valentine, W. M., Wetterau, L. A., Graham, D. G. Intermediates in the Paal-Knorr synthesis of pyrroles. *J. Org. Chem.* **1991**, 56, 6924-6931.
23. Amarnath, V., Amarnath, K. Intermediates in the Paal-Knorr Synthesis of Furans. *J. Org. Chem.* **1995**, 60, 301-307.
24. Cafeo, G., Garozzo, D., Kohnke, F. H., Pappalardo, S., Parisi, M. F., Pistone Nascone, R., Williams, D. J. From calixfurans to heterocyclophanes containing isopyrazole units. *Tetrahedron* **2004**, 60, 1895-1902.
25. Cafeo, G., Kohnke, F. H., La Torre, G. L., White, A. J. P., Williams, D. J. From large furan-based calixarenes to calixpyrroles and calix[n]furan[m]pyrroles: syntheses and structures. *Angew. Chem., Int. Ed. Engl.* **2000**, 39, 1496-1498.
26. Cafeo, G., Kohnke, F. H., Parisi, M. F., Nascone, R. P., La Torre, G. L., Williams, D. J. The Elusive β-Unsubstituted Calix[5]pyrrole Finally Captured. *Org. Lett.* **2002**, 4, 2695-2697.
27. Trost, B. M., Doherty, G. A. An Asymmetric Synthesis of the Tricyclic Core and a Formal Total Synthesis of Roseophilin via an Enyne Metathesis. *J. Am. Chem. Soc.* **2000**, 122, 3801-3810.
28. Taber, D. F., Nakajima, K. Unsymmetrical ozonolysis of a Diels-Alder adduct: practical preparation of a key intermediate for heme total synthesis. *J. Org. Chem.* **2001**, 66, 2515-2517.
29. Dong, Y., Le Quesne, P. W. Total synthesis of magnolamide. *Heterocycles* **2002**, 56, 221-225.

Passerini Multicomponent Reaction 330

1. Passerini, M. Isonitriles. II. Compounds with aldehydes or with ketones and monobasic organic acids. *Gazz. Chim. Ital.* **1921**, 51, 181-189.
2. Ugi, I. The α-addition of immonium ions and anions to isonitriles coupled with secondary reactions. *Angew. Chem. Int. Ed. Engl.* **1962**, 1, 8-21.
3. Ferosie, I. Isonitriles. *Aldrichimica Acta* **1971**, 4, 21-23.
4. Marquarding, D., Gokel, G., Hoffmann, P., Ugi, I. Passerini reaction and related reactions. *Isonitrile Chem.* **1971**, 133-143.
5. Ugi, I., Lohberger, S., Karl, R. The Passerini and Ugi Reactions. in *Comp. Org. Synth.* (eds. Trost, B. M.,Fleming, I.), *2*, 1083-1109 (Pergamon Press, Oxford, **1991**).
6. Ugi, I. K. MCR.XXIII. The highly variable multidisciplinary preparative and theoretical possibilities of the Ugi multicomponent reactions in the past, now, and in the future. *Proc. Est. Acad. of Sci. Chem.* **1998**, 47, 107-127.
7. Bienayme, H., Hulme, C., Oddon, G., Schmitt, P. Maximizing synthetic efficiency: multi-component transformations lead the way. *Chem.--Eur. J.* **2000**, 6, 3321-3329.
8. Basso, A., Wrubl, F. MCR approach to synthesis of peptidomimetic libraries. *Speciality Chemicals Magazine* **2003**, 23, 28-30.
9. Hulme, C., Gore, V. Multi-component reactions: emerging chemistry in drug discovery from xylocain to crixivan. *Curr. Med. Chem.* **2003**, 10, 51-80.
10. Hulme, C., Nixey, T. Rapid assembly of molecular diversity via exploitation of isocyanide-based multi-component reactions. *Current Opinion in Drug Discovery & Development* **2003**, 6, 921-929.
11. Nerdinger, S., Beck, B. New heterocycle synthesis by using bifunctional reactants in multicomponent reaction chemistry: the use of arylglyoxals and cinnamaldehyde in the Ugi-4CR and Passerini-3CR. *Chemtracts* **2003**, 16, 233-237.
12. Ostaszewski, R., Portlock, D. E., Fryszkowska, A., Jeziorska, K. Combination of enzymic procedures with multicomponent condensations. *Pure Appl. Chem.* **2003**, 75, 413-419.
13. Passerini, M. The isonitriles. III. Reaction with halogen aldehyde hydrates. *Gazz. Chim. Ital.* **1922**, 52, 432-435.
14. Passerini, M. The isonitriles. V. Reaction with levulinic acid. *Gazz. Chim. Ital.* **1923**, 53, 331-333.
15. McFarland, J. W. Reactions of cyclohexyl isonitrile and isobutyraldehyde with various nucleophiles and catalysts. *J. Org. Chem.* **1963**, 28, 2179-2181.
16. Mueller, E., Zeeh, B. Lewis acid-catalyzed reaction of carbonyl compounds with tertbutyl isonitrile. *Liebigs Ann. Chem.* **1966**, 696, 72-80.
17. Zeeh, B., Mueller, E. Lewis acid-catalyzed reaction of aliphatic ketones and tert-butyl isonitrile. *Liebigs Ann. Chem.* **1968**, 715, 47-51.
18. Grunewald, G. L., Brouillette, W. J., Finney, J. A. Synthesis of α-hydroxy amides via the cyanosilylation of aromatic ketones. *Tetrahedron Lett.* **1980**, 21, 1219-1220.
19. Sebti, S., Foucaud, A. A convenient conversion of 2-acyloxy-3-chlorocarboxamides to 3-acyloxy-2-azetidinones in heterogeneous media. *Synthesis* **1983**, 546-549.
20. Bossio, R., Marcaccini, S., Pepino, R. Studies on isocyanides and related compounds. Synthesis of oxazole derivatives via the Passerini reaction. *Liebigs Ann. Chem.* **1991**, 1107-1108.
21. Bossio, R., Marcaccini, S., Pepino, R., Torroba, T. Studies on isocyanides and related compounds: a novel synthetic route to furan derivatives. *Synthesis* **1993**, 783-785.
22. Bossio, R., Marcaccini, S., Pepino, R., Torroba, T. Studies on isocyanides and related compounds: synthesis of benzo[c]thiophenes by way of acid-induced three-component reactions. *J. Chem. Soc., Perkin Trans. 1* **1996**, 229-230.
23. Bossio, R., Marcos, C. F., Marcaccini, S., Pepino, R. A facile synthesis of β-lactams based on the isocyanide chemistry. *Tetrahedron Lett.* **1997**, 38, 2519-2520.
24. Kobayashi, K., Matoba, T., Irisawa, S., Matsumoto, T., Morikawa, O., Konishi, H. Synthesis of pyrrolo[1,2-a]quinoxaline and its 4-(1-hydroxyalkyl) derivatives by Lewis acid-catalyzed reactions of 1-(2-isocyanophenyl)pyrrole. *Chem. Lett.* **1998**, 551-552.
25. Xia, Q., Ganem, B. Metal-Promoted Variants of the Passerini Reaction Leading to Functionalized Heterocycles. *Org. Lett.* **2002**, 4, 1631-1634.
26. Denmark, S. E., Fan, Y. The First Catalytic, Asymmetric α-Additions of Isocyanides. Lewis-Base-Catalyzed, Enantioselective Passerini-Type Reactions. *J. Am. Chem. Soc.* **2003**, 125, 7825-7827.
27. Frey, R., Galbraith, S. G., Guelfi, S., Lamberth, C., Zeller, M. First examples of a highly stereoselective Passerini reaction: A new access to enantiopure mandelamides. *Synlett* **2003**, 1536-1538.

28. Kusebauch, U., Beck, B., Messer, K., Herdtweck, E., Doemling, A. Massive Parallel Catalyst Screening: Toward Asymmetric MCRs. *Org. Lett.* **2003**, 5, 4021-4024.
29. Baker, R. H., Schlesinger, A. H. Application of the Passerini reaction to steroid ketones. *J. Am. Chem. Soc.* **1945**, 67, 1499-1500.
30. Baker, R. H., Stanonis, D. The Passerini reaction. III. Stereochemistry and mechanism. *J. Am. Chem. Soc.* **1951**, 73, 699-702.
31. Ugi, I., Meyr, R. Isonitriles. V. Extended scope of the Passerini reaction. *Chem. Ber.* **1961**, 94, 2229-2233.
32. Hagedorn, I., Eholzer, U. Isonitrfles. VII. Single-step synthesis of α-hydroxy acid amides by a modification of the Passerini reaction. *Chem. Ber.* **1965**, 98, 936-940.
33. Carofiglio, T., Floriani, C., Chiesi-Villa, A., Guastini, C. Isocyanide complexes of titanium(IV) and vanadium(V): concerning the nonexistent insertion of isocyanides into a metal-chloride bond. *Inorg. Chem.* **1989**, 28, 4417-4419.
34. Carofiglio, T., Floriani, C., Chiesi-Villa, A., Rizzoli, C. Nonorganometallic pathway of the Passerini reaction assisted by titanium tetrachloride. *Organometallics* **1991**, 10, 1659-1660.
35. Carofiglio, T., Cozzi, P. G., Floriani, C., Chiesi-Villa, A., Rizzoli, C. Nonorganometallic pathway of the Passerini reaction assisted by titanium tetrachloride. *Organometallics* **1993**, 12, 2726-2736.
36. Bergemann, M., Neidlein, R. Studies on the reactivity of α-cyano α-isocyano alkanoates. Versatile synthons for the assembly of imidazoles. *Helv. Chim. Acta* **1999**, 82, 909-918.
37. Mandair, G. S., Light, M., Russell, A., Hursthouse, M., Bradley, M. Re-evaluation of the outcome of a multiple component reaction-2- and 3-amino-imidazo[1,2-a]pyrimidines? *Tetrahedron Lett.* **2002**, 43, 4267-4269.
38. Owens, T. D., Araldi, G. L., Nutt, R. F., Semple, J. E. Concise total synthesis of the prolyl endopeptidase inhibitor eurystatin A via a novel Passerini reaction-deprotection-acyl migration strategy. *Tetrahedron Lett.* **2001**, 42, 6271-6274.
39. Banfi, L., Guanti, G., Riva, R. Passerini multicomponent reaction of protected α-amino aldehydes as a tool for combinatorial synthesis of enzyme inhibitors. *Chem. Commun.* **2000**, 985-986.
40. Tadesse, S., Balan, C., Jones, W., Viswanadhan, V., Hulme, C. A facile three-step one-pot synthesis of norstatines using the Passerini reaction. *Abstracts of Papers, 223rd ACS National Meeting, Orlando, FL, United States, April 7-11, 2002* **2002**, ORGN-241.

Paterno-Büchi Reaction 332

1. Paterno, E. Organic Syntheses Induced by Light. Introductory Note. *Gazz. Chim. Ital.* **1909**, 39, 237-250.
2. Paterno, E., Chieffi, G. Synthesis in organic chemistry using light. Note II. Compounds of unsaturated hydrocarbons with aldehydes and ketones. *Gazz. Chim. Ital.* **1909**, 39, 341-361.
3. Paterno, E., Chieffi, G. Synthesis in Organic Chemistry by Means of Light. II. Compounds of the Unsaturated Hydrocarbons with Aldehydes and Ketones. *Gazz. Chim. Ital.* **1911**, 39, 341-361.
4. Büchi, G., Inman, C. G., Lipinsky, E. S. Light-catalyzed organic reactions. I. The reaction of carbonyl compounds with 2-methyl-2-butene in the presence of ultraviolet light. *J. Am. Chem. Soc.* **1954**, 76, 4327-4331.
5. Arnold, D. R. Photocycloaddition of carbonyl compounds to unsaturated systems: syntheses of oxetanes. *Adv. Photochem.* **1968**, 6, 301-423.
6. Jones, G., II. Synthetic applications of the Paterno-Büchi reaction. *Organic Photochemistry* **1981**, 5, 1-122.
7. Porco, J. A., Jr., Schreiber, S. L. The Paterno-Büchi Reaction. in *Comp. Org. Synth.* (eds. Trost, B. M.,Fleming, I.), *5*, 151-192 (Pergamon, Oxford, **1991**).
8. Antoulinakis, E. G., Schultz, A. G. The intramolecular Paterno-Büchi photoaddition-fragmentation route to polyquinane natural products. *Chemtracts: Org. Chem.* **1996**, 9, 224-227.
9. Bach, T. The Paterno-Büchi reaction of 3-heteroatom-substituted alkenes as a stereoselective entry to polyfunctional cyclic and acyclic molecules. *Liebigs Ann. Chem.* **1997**, 1627-1634.
10. Bach, T. Stereoselective intermolecular [2+2] photocycloaddition reactions and their application in synthesis. *Synthesis* **1998**, 683-703.
11. Bach, T. The Paterno-Büchi reaction of N-acyl enamines and aldehydes - the development of a new synthetic method and its application to total synthesis and molecular recognition studies. *Synlett* **2000**, 1699-1707.
12. Vargas, F., Rivas, C. Photochemical formation of oxetanes derived from aromatic ketones and substituted thiophenes and selenophenes. *International Journal of Photoenergy* **2000**, 2, 97-101.
13. D'Auria, M., Emanuele, L., Racioppi, R., Romaniello, G. The Paterno-Büchi reaction on furan derivatives. *Curr. Org. Chem.* **2003**, 7, 1443-1459.
14. Griesbeck, A. G. Spin-selectivity in photochemistry: A tool for organic synthesis. *Synlett* **2003**, 451-472.
15. Abe, M. Photochemical oxetane formation: addition to heterocycles. *CRC Handbook of Organic Photochemistry and Photobiology (2nd Edition)* **2004**, 62/61-62/10.
16. D'Auria, M., Emanuele, L., Racioppi, R. Regio- and diastereoselectivity in the Paterno-Büchi reaction on furan derivatives. *Spectrum (Bowling Green, OH, United States)* **2004**, 17, 22-27.
17. Griesbeck, A. G., Bondock, S. Oxetane formation: intermolecular additions. *CRC Handbook of Organic Photochemistry and Photobiology (2nd Edition)* **2004**, 60/61-60/21.
18. Griesbeck, A. G., Bondock, S. Oxetane formation: stereocontrol. *CRC Handbook of Organic Photochemistry and Photobiology (2nd Edition)* **2004**, 59/51-59/19.
19. Palmer, I. J., Ragazos, I. N., Bernardi, F., Olivucci, M., Robb, M. A. An MC-SCF Study of the (Photochemical) Paterno-Büchi Reaction. *J. Am. Chem. Soc.* **1994**, 116, 2121-2132.
20. Buhr, S., Griesbeck, A. G., Lex, J., Mattay, J., Schroeer, J. Electronic control of stereoselectivity in photocycloaddition reactions. 7. Stereoselectivity in the Paterno-Büchi reaction of 2,2-diisopropyl-1,3-dioxole with methyl trimethylpyruvate. *Tetrahedron Lett.* **1996**, 37, 1195-1196.
21. Minaev, B. F., Agren, H. Spin-orbit coupling in oxygen containing diradicals. *THEOCHEM* **1998**, 434, 193-206.
22. Abe, M., Fujimoto, K., Nojima, M. Notable Sulfur Atom Effects on the Regio- and Stereoselective Formation of Oxetanes in Paterno-Büchi Photocycloaddition of Aromatic Aldehydes with Silyl O,S-Ketene Acetals. *J. Am. Chem. Soc.* **2000**, 122, 4005-4010.
23. D'Auria, M., Racioppi, R. Paterno-Büchi reaction on 5-methyl-2-furylmethanol derivatives. *ARKIVOC [online computer file]* **2000**, 1, 145-152.
24. Rochat, S., Minardi, C., De Saint Laumer, J.-Y., Herrmann, A. Controlled release of perfumery aldehydes and ketones by Norrish type-II photofragmentation of α–keto esters in undegassed solution. *Helv. Chim. Acta* **2000**, 83, 1645-1671.
25. Abe, M., Tachibana, K., Fujimoto, K., Nojima, M. Regioselective formation of 3-selanyl-3-siloxyoxetanes in the Paterno-Büchi reaction of silyl O,Se-ketene acetals (O,Se-SKA). *Synthesis* **2001**, 1243-1247.
26. D'Auria, M., Emanuele, L., Poggi, G., Racioppi, R., Romaniello, G. On the stereoselectivity of the Paterno-Büchi reaction between carbonyl compounds and 2-furylmethanol derivatives. The case of aliphatic aldehydes and ketones. *Tetrahedron* **2002**, 58, 5045-5051.
27. Morris, T. H., Smith, E. H., Walsh, R. Oxetane synthesis. Methyl vinyl sulfides as new traps of excited benzophenone in a stereoselective and regiospecific Paterno-Büchi reaction. *J. Chem. Soc., Chem. Commun.* **1987**, 964-965.
28. Yang, N.-C., Loeschen, R. L., Mitchell, D. Mechanism of the Paterno-Büchi reaction. *J. Am. Chem. Soc.* **1967**, 89, 5465-5466.
29. Yang, N. C. Photochemical reactions of carbonyl compounds in solution. II. Paterno-Büchi reaction and energy transfer. *Photochem. Photobiol.* **1968**, 7, 767-773.
30. Turro, N. J., Wriede, P. A. Molecular photochemistry. XXI. Photocycloaddition of acetone to 1-methoxy-1-butene. Comparison of singlet and triplet mechanisms and singlet and triplet biradical intermediates. *J. Am. Chem. Soc.* **1970**, 92, 320-329.
31. Yang, N.-C., Eisenhardt, W. Mechanism of Paterno-Büchi reaction of alkanals. *J. Am. Chem. Soc.* **1971**, 93, 1277-1279.
32. Funke, C. W., Cerfontain, H. Photochemical oxetane formation: the Paterno-Büchi reaction of aliphatic aldehydes and ketones with alkenes and dienes. *J. Chem. Soc., Perkin Trans. 2* **1976**, 1902-1908.
33. Freilich, S. C., Peters, K. S. Observation of the 1,4 biradical in the Paterno-Büchi reaction. *J. Am. Chem. Soc.* **1981**, 103, 6255-6257.

34. Bolivar, R. A., Rivas, C. Quencher effect of thiophene and its monomethyl derivatives on photoreduction and photocycloaddition reactions of ketones. *J. Photochem.* **1982**, 19, 95-99.
35. Farneth, W. E., Johnson, D. G. Chemiluminescence in the infrared photochemistry of oxetanes: the formal reverse of ketone photocycloaddition. *J. Am. Chem. Soc.* **1984**, 106, 1875-1876.
36. Mazzocchi, P. H., Klingler, L. Photoreduction of N-methylphthalimide with 2,3-dimethyl-2-butene. Evidence for reaction through an electron transfer generated ion pair. *J. Am. Chem. Soc.* **1984**, 106, 7567-7572.
37. Freilich, S. C., Peters, K. S. Picosecond dynamics of the Paterno-Büchi reaction. *J. Am. Chem. Soc.* **1985**, 107, 3819-3822.
38. Farneth, W. E., Johnson, D. G. Infrared photochemistry of oxetanes: mechanism of chemiluminescence. *J. Am. Chem. Soc.* **1986**, 108, 773-780.
39. Adam, W., Kliem, U., Peters, E. M., Peters, K., Von Schnering, H. G. Preparative vis-laser photochemistry. Qinghaosu-type 1,2,4-trioxanes by molecular oxygen trapping of Paterno-Büchi triplet 1,4-diradicals derived from the bicyclic enol lactones D1,6- and D1,10-2-oxabicyclo[4.4.0]decen-2-one and p-benzoquinone. *J. Prakt. Chem.* **1988**, 330, 391-405.
40. Buschmann, H., Scharf, H. D., Hoffmann, N., Plath, M. W., Runsink, J. Chiral induction in photochemical reactions. 10. The principle of isoinversion: a model of stereoselection developed from the diastereoselectivity of the Paterno-Büchi reaction. *J. Am. Chem. Soc.* **1989**, 111, 5367-5373.
41. Yuan, H., Yan, B. Photochemical [2+2] cycloaddition via radical ion intermediates. A CIDNP evidence. *Chin. Chem. Lett.* **1992**, 3, 25-28.
42. Eckert, G., Goez, M. Paterno-Büchi reactions via correlated radical ion pairs-a CIDNP investigation. *Journal of Information Recording* **1996**, 22, 561-565.
43. Goez, M., Eckert, G. Photoinduced Electron Transfer Reactions of Aryl Olefins. 2. Cis-Trans Isomerization and Cycloadduct Formation in Anethole-Fumaronitrile Systems in Polar Solvents. *J. Am. Chem. Soc.* **1996**, 118, 140-154.
44. Sun, D., Hubig, S. M., Kochi, J. K. Oxetanes from [2+2] cycloaddition of stilbenes to quinone via photoinduced electron transfer. *J. Org. Chem.* **1999**, 64, 2250-2258.
45. Griesbeck, A. G., Fiege, M., Bondock, S., Gudipati, M. S. Spin-Directed Stereoselectivity of Carbonyl-Alkene Photocycloadditions. *Org. Lett.* **2000**, 2, 3623-3625.
46. Xue, J., Zhang, Y., Wu, T., Fun, H.-K., Xu, J.-H. Photoinduced [2 + 2] cycloadditions (the Paterno-Büchi reaction) of 1H-1-acetylindole-2,3-dione with alkenes. *J. Chem. Soc., Perkin Trans. 1* **2001**, 183-191.
47. Adam, W., Stegmann, V. R. Unusual Temperature Dependence in the cis/trans-Oxetane Formation Discloses Competitive Syn versus Anti Attack for the Paterno-Büchi Reaction of Triplet-Excited Ketones with cis- and trans-Cyclooctenes. Conformational Control of Diastereoselectivity in the Cyclization and Cleavage of Preoxetane Diradicals. *J. Am. Chem. Soc.* **2002**, 124, 3600-3607.
48. Goez, M., Frisch, I. Activation Energy of a Biradical Rearrangement Measured by Photo-CIDNP. *J. Phys. Chem. A* **2002**, 106, 8079-8084.
49. Griesbeck, A. G. The link between stereoselectivity and spin selectivity in intermolecular and intramolecular photochemical reactions. *J. Photosci.* **2003**, 10, 49-60.
50. Wang, X. Y., Yan, B. Z., Wang, T. CIDNP study of photoinduced [2+2] cycloadditions (the Paterno-Büchi coupling) of arylacetylenes and quinone. *Chin. Chem. Lett.* **2003**, 14, 270-273.
51. Abe, M., Kawakami, T., Ohata, S., Nozaki, K., Nojima, M. Mechanism of Stereo- and Regioselectivity in the Paterno-Büchi Reaction of Furan Derivatives with Aromatic Carbonyl Compounds: Importance of the Conformational Distribution in the Intermediary Triplet 1,4-Diradicals. *J. Am. Chem. Soc.* **2004**, 126, 2838-2846.
52. Bach, T., Brummerhop, H. Unprecedented facial diastereoselectivity in the Paterno - Büchi reaction of a chiral dihydropyrrole-a short total synthesis of (+)-preussin. *Angew. Chem., Int. Ed. Engl.* **1999**, 37, 3400-3402.
53. Bach, T., Brummerhop, H., Harms, K. The synthesis of (+)-preussin and related pyrrolidinols by diastereoselective Paterno-Büchi reactions of chiral 2-substituted 2,3-dihydropyrroles. *Chem.-- Eur. J.* **2000**, 6, 3838-3848.
54. Reddy, T. J., Rawal, V. H. Expeditious Syntheses of (±)-5-Oxosilphiperfol-6-ene and (±)-Silphiperfol-6-ene. *Org. Lett.* **2000**, 2, 2711-2712.
55. Aungst, R. A., Jr., Funk, R. L. Synthesis of (Z)-2-Acyl-2-enals via Retrocycloadditions of 5-Acyl-4-alkyl-4H-1,3-dioxins: Application in the Total Synthesis of the Cytotoxin (±)-Euplotin A. *J. Am. Chem. Soc.* **2001**, 123, 9455-9456.
56. Schreiber, S. L., Hoveyda, A. H., Wu, H. J. A photochemical route to the formation of threo aldols. *J. Am. Chem. Soc.* **1983**, 105, 660-661.
57. Schreiber, S. L., Hoveyda, A. H. Synthetic studies of the furan-carbonyl photocycloaddition reaction. A total synthesis of (±)-avenaciolide. *J. Am. Chem. Soc.* **1984**, 106, 7200-7202.

Pauson-Khand Reaction 334

1. Khand, I. U., Knox, G. R., Pauson, P. L., Watts, W. E., Foreman, M. I. Organocobalt complexes. II. Reaction of acetylenehexacarbonyl dicobalt complexes, $(RC_2R_1)Co_2(CO)_6$, with norbornene and its derivatives. *J. Chem. Soc., Perkin Trans. 1* **1973**, 977-981.
2. Pauson, P. L. The Khand reaction. A convenient and general route to a wide range of cyclopentenone derivatives. *Tetrahedron* **1985**, 41, 5855-5860.
3. Brunner, H. Right or left - this is the question (enantioselective catalysis with transition metal compounds). *Pure Appl. Chem.* **1990**, 62, 589-594.
4. Ganem, B. Pauson-Khand cycloadditions of polymer-linked substrates. *Chemtracts: Org. Chem.* **1990**, 3, 211-212.
5. Krohn, K. Nazarov and Pauson-Khand reactions. *Org. Synth. Highlights* **1991**, 137-144.
6. Schore, N. E. The Pauson-Khand cycloaddition reaction for synthesis of cyclopentenones. *Org. React.* **1991**, 40, 1-90.
7. Schore, N. E. The Pauson-Khand Reaction. in *Comp. Org. Synth.* (eds. Trost, B. M.,Fleming, I.), *5*, 1037-1064 (Pergamon, Oxford, **1991**).
8. Castro, J., Green, A. E., Moyano, A., Pericas, M. A., Poch, M., Riera, A., Sola, L., Verdaguer, X. Enantioselective Pauson-Khand reactions. *An. Quim.* **1993**, 89, 135-136.
9. Blagg, J. Stoichiometric applications of organotransition metal complexes in organic synthesis. *Contemp. Org. Synth.* **1994**, 1, 125-143.
10. Rautenstrauch, V. Pauson-Khand reaction. *Applied Homogeneous Catalysis with Organometallic Compounds* **1996**, 2, 1092-1101.
11. Geis, O., Schmalz, H.-G. New developments in the Pauson-Khand reaction. *Angew. Chem., Int. Ed. Engl.* **1998**, 37, 911-914.
12. Ingate, S. T., Marco-Contelles, J. The asymmetric Pauson-Khand reaction. A review. *Org. Prep. Proced. Int.* **1998**, 30, 121-143.
13. Jeong, N. Pauson-Khand reactions. *Transition Metals for Organic Synthesis* **1998**, 1, 560-577.
14. Buchwald, S. L., Hicks, F. A. Pauson-Khand type reactions. *Comprehensive Asymmetric Catalysis I-III* **1999**, 2, 491-510.
15. Keun Chung, Y. Transition metal alkyne complexes: the Pauson-Khand reaction. *Coord. Chem. Rev.* **1999**, 188, 297-341.
16. Sugihara, T., Yamaguchi, M., Nishizawa, M. Lewis base promoted reactions of alkyne-dicobalt hexacarbonyls. *Rev. on Heteroa. Chem.* **1999**, 21, 179-194.
17. Brummond, K. M., Kent, J. L. Recent advances in the Pauson-Khand reaction and related [2+2+1] cycloadditions. *Tetrahedron* **2000**, 56, 3263-3283.
18. Geis, O., Schmalz, H.-G. New developments in the Pauson-Khand reaction. *Organic Synthesis Highlights IV* **2000**, 116-122.
19. Kellogg, R. M. Rhodium(I)-catalyzed asymmetric intramolecular Pauson-Khand-type reaction. *Chemtracts* **2000**, 13, 708-710.
20. Sugihara, T., Yamaguchi, M., Nishizawa, M. Advances in the Pauson-Khand reaction: development of reactive cobalt complexes. *Chem.-- Eur. J.* **2001**, 7, 1589-1595.
21. Welker, M. E. Organocobalt complexes in organic synthesis. *Curr. Org. Chem.* **2001**, 5, 785-807.
22. Fryatt, R., Christie, S. D. R. Applications of stoichiometric transition metal complexes in organic synthesis. *J. Chem. Soc., Perkin Trans. 1* **2002**, 447-458.
23. Hanson, B. E. Catalytic Pauson Khand and related cycloaddition reactions. *Comments on Inorganic Chemistry* **2002**, 23, 289-318.
24. Herrmann, W. A. Pauson-Khand reaction. *Applied Homogeneous Catalysis with Organometallic Compounds (2nd Edition)* **2002**, 3, 1241-1252.
25. Jeong, N., Sung, B. K., Kim, J. S., Park, S. B., Seo, S. D., Shin, J. Y., In, K. Y., Choi, Y. K. Pauson-Khand-type reaction mediated by Rh(I) catalysts. *Pure Appl. Chem.* **2002**, 74, 85-91.

26. Kubota, H., Lim, J., Depew, K. M., Schreiber, S. L. Pathway development and pilot library realization in diversity-oriented synthesis. Exploring Ferrier and Pauson-Khand reactions on a glycal template. *Chem. Biol.* **2002**, 9, 265-276.
27. Rivero, M. R., Adrio, J., Carretero, J. C. Pauson-Khand reactions of electron-deficient alkenes. *Eur. J. Org. Chem.* **2002**, 2881-2889.
28. Gibson, S. E., Stevenazzi, A. The Pauson-Khand reaction: The catalytic age is here! *Angew. Chem., Int. Ed. Engl.* **2003**, 42, 1800-1810.
29. Preston, A. J., Parquette, J. R. A pyridylsilyl group expands the scope of the intermolecular Pauson-Khand reactions. *Chemtracts* **2003**, 16, 435-438.
30. Alcaide, B., Almendros, P. The allenic Pauson-Khand reaction in synthesis. *Eur. J. Org. Chem.* **2004**, 3377-3383.
31. Blanco-Urgoiti, J., Anorbe, L., Perez-Serrano, L., Dominguez, G., Perez-Castells, J. The Pauson-Khand reaction, a powerful synthetic tool for the synthesis of complex molecules. *Chem. Soc. Rev.* **2004**, 33, 32-42.
32. Becker, D. P., Flynn, D. L. Studies of the solid-phase Pauson-Khand reaction: selective in-situ enone reduction to 3-azabicyclo[3.3.0]octanones. *Tetrahedron Lett.* **1993**, 34, 2087-2090.
33. Spitzer, J. L., Kurth, M. J., Schore, N. E., Najdi, S. D. Polymer-supported synthesis as a tool for improving chemoselectivity: Pauson-Khand reaction. *Tetrahedron* **1997**, 53, 6791-6808.
34. Kerr, W. J., Lindsay, D. M. Preparation of an amine N-oxide on solid phase: an efficient promoter of the Pauson-Khand reaction. *Chem. Commun.* **1999**, 2551-2552.
35. Comely, A. C., Gibson, S. E., Hales, N. J. Polymer-supported cobalt carbonyl complexes as novel solid-phase catalysts of the Pauson-Khand reaction. *Chem. Commun.* **2000**, 305-306.
36. Castro, J., Moyano, A., Pericas, M. A., Riera, A. A qualitative molecular mechanics approach to the stereoselectivity of intramolecular Pauson-Khand reactions. *Tetrahedron* **1995**, 51, 6541-6556.
37. Verdaguer, X., Vazquez, J., Fuster, G., Bernardes-Genisson, V., Greene, A. E., Moyano, A., Pericas, M. A., Riera, A. Camphor-Derived, Chelating Auxiliaries for the Highly Diastereoselective Intermolecular Pauson-Khand Reaction: Experimental and Computational Studies. *J. Org. Chem.* **1998**, 63, 7037-7052.
38. Verdaguer, X., Moyano, A., Pericas, M. A., Riera, A., Alvarez-Larena, A., Piniella, J.-F. Alkyne Dicobalt Carbonyl Complexes with Sulfide Ligands. Synthesis, Crystal Structure, and Dynamic Behavior. *Organometallics* **1999**, 18, 4275-4285.
39. Balsells, J., Vazquez, J., Moyano, A., Pericas, M. A., Riera, A. Low-Energy Pathway for Pauson-Khand Reactions: Synthesis and Reactivity of Dicobalt Hexacarbonyl Complexes of Chiral Ynamines. *J. Org. Chem.* **2000**, 65, 7291-7302.
40. de Bruin, T. J., Milet, A., Robert, F., Gimbert, Y., Greene, A. E. Theoretical study of the regiochemistry-determining step of the Pauson-Khand reaction. *J. Am. Chem. Soc.* **2001**, 123, 7184-7185.
41. Robert, F., Milet, A., Gimbert, Y., Konya, D., Greene, A. E. Regiochemistry in the Pauson-Khand Reaction: Has a Trans Effect Been Overlooked? *J. Am. Chem. Soc.* **2001**, 123, 5396-5400.
42. Yamanaka, M., Nakamura, E. Density Functional Studies on the Pauson-Khand Reaction. *J. Am. Chem. Soc.* **2001**, 123, 1703-1708.
43. Pericas, M. A., Balsells, J., Castro, J., Marchueta, I., Moyano, A., Riera, A., Vazquez, J., Verdaguer, X. Toward the understanding of the mechanism and enantioselectivity of the Pauson-Khand reaction. Theoretical and experimental studies. *Pure Appl. Chem.* **2002**, 74, 167-174.
44. Schulte, J. H., Gleiter, R., Rominger, F. Regiochemistry of S-Alkyl-Substituted Alkynes in Pauson-Khand Reactions. Is a Correlation with X-ray Data and Charge Distribution Calculations of the $Co_2(CO)_6$-Alkyne Complexes Possible? *Org. Lett.* **2002**, 4, 3301-3304.
45. Imhof, W., Anders, E., Gobel, A., Gorls, H. A theoretical study on the complete catalytic cycle of the hetero-Pauson-Khand-type [2+2+1] cycloaddition reaction of ketimines, carbon monoxide and ethylene catalyzed by iron carbonyl complexes. *Chem.-- Eur. J.* **2003**, 9, 1166-1181.
46. Imhof, W., Anders, E., Gobel, A., Gorls, H. A theoretical study on the complete catalytic cycle of the hetero-Pauson-Khand-type [2+2+1] cycloaddition reaction of ketimines, carbon monoxide and ethylene catalyzed by iron carbonyl complexes. *Chemistry (Weinheim an der Bergstrasse, Germany)* **2003**, 9, 1166-1181.
47. de Bruin Theodorus, J. M., Milet, A., Greene Andrew, E., Gimbert, Y. Insight into the reactivity of olefins in the Pauson-Khand reaction. *J. Org. Chem.* **2004**, 69, 1075-1080.
48. Krafft, M. E. Steric control in the Pauson cycloaddition: further support for the proposed mechanism. *Tetrahedron Lett.* **1988**, 29, 999-1002.
49. Krafft, M. E., Scott, I. L., Romero, R. H., Feibelmann, S., Van Pelt, C. E. Effect of coordinating ligands on the Pauson-Khand cycloaddition: trapping of an intermediate. *J. Am. Chem. Soc.* **1993**, 115, 7199-7207.
50. Krafft, M. E., Wilson, A. M., Dasse, O. A., Shao, B., Cheung, Y. Y., Fu, Z., Bonaga, L. V. R., Mollman, M. K. The Interrupted Pauson-Khand Reaction. *J. Am. Chem. Soc.* **1996**, 118, 6080-6081.
51. Gordon, C. M., Kiszka, M., Dunkin, I. R., Kerr, W. J., Scott, J. S., Gebicki, J. Elucidating the mechanism of the photochemical Pauson-Khand reaction: matrix photochemistry of (phenylacetylene)hexacarbonyldicobalt. *J. Organomet. Chem.* **1998**, 554, 147-154.
52. Breczinski, P. M., Stumpf, A., Hope, H., Krafft, M. E., Casalnuovo, J. A., Schore, N. E. Stereoselectivity in the intramolecular Pauson-Khand reaction: towards a simple predictive model. *Tetrahedron* **1999**, 55, 6797-6812.
53. Derdau, V., Laschat, S., Dix, I., Jones, P. G. Cobalt-Alkyne Complexes with Diphosphine Ligands as Mechanistic Probes for the Pauson-Khand Reaction. *Organometallics* **1999**, 18, 3859-3864.
54. Bitterwolf, T. E., Scallorn, W. B., Weiss, C. A. Photochemical studies of $Co_2(CO)_6$(acetylene) complexes and their phosphine derivatives in frozen Nujol matrices. *J. Organomet. Chem.* **2000**, 605, 7-14.
55. Castro, J., Moyano, A., Pericas, M. A., Riera, A., Alvarez-Larena, A., Piniella, J. F. Acetylene-Dicobaltcarbonyl Complexes with Chiral Phosphinooxazoline Ligands: Synthesis, Structural Characterization, and Application to Enantioselective Intermolecular Pauson-Khand Reactions. *J. Am. Chem. Soc.* **2000**, 122, 7944-7952.
56. Derdau, V., Laschat, S., Jones, P. G. Evaluation of the regioselectivity in Pauson-Khand reactions of substituted norbornenes and diazabicyclo[2.2.1]heptanes with terminal alkynes. *Eur. J. Org. Chem.* **2000**, 681-689.
57. Kennedy, A. R., Kerr, W. J., Lindsay, D. M., Scott, J. S., Watson, S. P. Stereochemical and mechanistic features of asymmetric Pauson-Khand processes. *Perkin 1* **2000**, 4366-4372.
58. Krafft, M. E., Bonaga, L. V., Hirosawa, C. Practical cobalt carbonyl catalysis in the thermal Pauson--Khand reaction: efficiency enhancement using Lewis bases. *J. Org. Chem.* **2001**, 66, 3004-3020.
59. Sugihara, T., Yamaguchi, M., Nishizawa, M. Advances in the Pauson-Khand reaction: development of reactive cobalt complexes. *Chemistry (Weinheim an der Bergstrasse, Germany)* **2001**, 7, 1589-1595.
60. Vazquez, J., Fonquerna, S., Moyano, A., Pericas, M. A., Riera, A. Bornane-2,10-sultam: a highly efficient chiral controller and mechanistic probe for the intermolecular Pauson-Khand reaction. *Tetrahedron: Asymmetry* **2001**, 12, 1837-1850.
61. Gimbert, Y., Lesage, D., Milet, A., Fournier, F., Greene, A. E., Tabet, J.-C. On Early Events in the Pauson-Khand Reaction. *Org. Lett.* **2003**, 5, 4073-4075.
62. de Bruin, T. J. M., Milet, A., Greene, A. E., Gimbert, Y. Insight into the Reactivity of Olefins in the Pauson-Khand Reaction. *J. Org. Chem.* **2004**, 69, 1075-1080.
63. Donkervoort, J. G., Gordon, A. R., Johnstone, C., Kerr, W. J., Lange, U. Development of modified Pauson-Khand reactions with ethylene and utilization in the total synthesis of (+)-taylorione. *Tetrahedron* **1996**, 52, 7391-7420.
64. Paquette, L. A., Borrelly, S. Studies Directed toward the Total Synthesis of Kalmanol. An Approach to Construction of the C/D Diquinane Substructure. *J. Org. Chem.* **1995**, 60, 6912-6921.
65. Jamison, T. F., Shambayati, S., Crowe, W. E., Schreiber, S. L. Tandem Use of Cobalt-Mediated Reactions to Synthesize (+)-Epoxydictymene, a Diterpene Containing a Trans-Fused 5-5 Ring System. *J. Am. Chem. Soc.* **1997**, 119, 4353-4363.
66. Cassayre, J., Gagosz, F., Zard, S. Z. A short synthesis of (±)-13-deoxyserratine. *Angew. Chem., Int. Ed. Engl.* **2002**, 41, 1783-1785.

1. Kohler, E. P., Bickel, C. L. Properties of certain β-oxanols. *J. Am. Chem. Soc.* **1935**, 57, 1099-1101.
2. Lake, W. H. G., Peat, S. Conversion of D-glucose into D-idose. *J. Chem. Soc., Abstracts* **1939**, 1069-1074.
3. Angyal, S. J., Gilham, P. T. Cyclitols. VII. Anhydroinositols and the "Epoxide migration." *J. Chem. Soc., Abstracts* **1957**, 3691-3699.
4. Payne, G. B. Epoxide migrations with α,β-epoxy alcohols. *J. Org. Chem.* **1962**, 27, 3819-3822.
5. Ibuka, T. The aza-Payne rearrangement: a synthetically valuable equilibration. *Chem. Soc. Rev.* **1998**, 27, 145-154.
6. Lundt, I., Madsen, R. Synthetically useful base-induced rearrangements of aldonolactones. *Top. Curr. Chem.* **2001**, 215, 177-191.
7. Hanson, R. M. Epoxide migration (Payne rearrangement) and related reactions. *Org. React.* **2002**, 60, 1-156.
8. Pena, P. C. A., Roberts, S. M. The chemistry of epoxy alcohols. *Curr. Org. Chem.* **2003**, 7, 555-571.
9. Ibuka, T., Nakai, K., Habashita, H., Hotta, Y., Otaka, A., Tamamura, H., Fujii, N., Mimura, N., Miwa, Y., et al. Aza-Payne Rearrangement of Activated 2-Aziridinemethanols and 2,3-Epoxy Amines under Basic Conditions. *J. Org. Chem.* **1995**, 60, 2044-2058.
10. Brnalt, J., Kvarnstroem, I., Classon, B., Samuelsson, B. Synthesis of [4,5-Bis(hydroxymethyl)-1,3-oxathiolan-2-yl]nucleosides as Potential Inhibitors of HIV via Stereospecific Base-Induced Rearrangement of a 2,3-Epoxy Thioacetate. *J. Org. Chem.* **1996**, 61, 3604-3610.
11. Ibuka, T., Nakai, K., Akaji, M., Tamamura, H., Fujii, N., Yamamoto, Y. An aza-Payne rearrangement-epoxide ring opening reaction of 2-aziridinemethanols in a one-pot manner: a regio- and stereoselective synthetic route to diastereomerically pure N-protected 1,2-amino alcohols. *Tetrahedron* **1996**, 52, 11739-11752.
12. Wu, M. H., Hansen, K. B., Jacobsen, E. N. Regio- and enantioselective cyclization of epoxy alcohols catalyzed by a Co-III(salen) complex. *Angew. Chem., Int. Ed. Engl.* **1999**, 38, 2012-2014.
13. Dua, S., Bowie, J. H., Taylor, M. S., Buntine, M. A. The gas phase Payne rearrangement. Part 2. Methyl substitution: a joint ab initio and experimental study. *Int. J. Mass Spectrom. Ion Processes* **1997**, 165/166, 139-153.
14. Dua, S., Taylor, M. S., Buntine, M. A., Bowie, J. H. The degenerate Payne rearrangement of the 2,3-epoxypropoxide anion in the gas phase. A joint theoretical and experimental study. *J. Chem. Soc., Perkin Trans. 2* **1997**, 1991-1997.
15. Bouyacoub, A., Volatron, F. The aza-Payne rearrangement: a theoretical DFT study of the counter-ion and solvent effects. *Eur. J. Org. Chem.* **2002**, 4143-4150.
16. Rinner, U., Siengalewicz, P., Hudlicky, T. Total synthesis of epi-7-deoxypancratistatin via aza-Payne rearrangement and intramolecular cyclization. *Org. Lett.* **2002**, 4, 115-117.
17. Sasaki, M., Koike, T., Sakai, R., Tachibana, K. Total synthesis of (-)-dysiherbaine, a novel neuroexcitotoxic amino acid. *Tetrahedron Lett.* **2000**, 41, 3923-3926.
18. Birman, V. B., Danishefsky, S. J. The total synthesis of (±)-merrilactone A. *J. Am. Chem. Soc.* **2002**, 124, 2080-2081.

1. Perkin, W. H. On the hydride of aceto-salicyl. *J. Chem. Soc.* **1868**, 21, 181-186.
2. Perkin, W. H. *J. Chem. Soc.* **1877**, 31, 388-427.
3. Johnson, J. R. Perkin reaction and related reactions. *Org. React.* **1942**, 1, pp 210-265.
4. Rosen, T. The Perkin reaction. in *Comp. Org. Synth.* (eds. Trost, B. M.,Fleming, I.), 2, 395-408 (Pergamon, Oxford, **1991**).
5. Oglialoro. Synthesis of phenylcinnamic acid. *Gazz. Chim. Ital.* **1878**, 8, 429-434.
6. Plöchl, J. About phenylglycidic acid. *Ber.* **1883**, 16, 2815-2825.
7. Erlenmeyer, E. The condensation of hippuric acid with phthalic anhydride and benzaldehyde. *Liebigs Ann. Chem.* **1893**, 275, 1-3.
8. Carter, H. E. Azlactones. *Org. React.* **1946**, 198-239.
9. Baltazzi, E. The chemistry of 5-oxazolones. *Quart. Revs. (London)* **1955**, 9, 150-173.
10. Obretenov, T., Kratchanov, C., Kurtev, B. The low-temperature Perkin reaction: synthesis and stereochemistry of the diastereomeric 3-hydroxy-2-phenylbutanoic acids. *Izvestiya po Khimiya* **1975**, 8, 44-50.
11. Rai, M., Krishan, K., Singh, A. Perkin reaction of azlactone with vanillin Schiff bases. *Indian J. Chem., Sect. B* **1977**, 15B, 847-848.
12. Gaset, A., Gorrichon, J. P. The use of ion-exchange resins in the Perkin reaction for the synthesis of azlactones from aldehydes of plant origin. *Synth. Commun.* **1982**, 12, 71-79.
13. Jayamani, M., Pillai, C. N. Reaction of carboxylic acids with carbonyl compounds over alumina. *J. Catal.* **1984**, 87, 93-97.
14. Koepp, E., Voegtle, F. Perkin syntheses with cesium acetate. *Synthesis* **1987**, 177-179.
15. Mukerjee, A. K. Azlactones: retrospect and prospect. *Heterocycles* **1987**, 26, 1077-1097.
16. Ivanova, G. A modification of the Ploechl-Erlenmeyer reaction. I. Synthesis of 2-phenyl-4-diphenylmethylene-5(4H)-oxazolone. *Tetrahedron* **1992**, 48, 177-186.
17. Limaye, P. A., Huddar, P. H., Ghate, S. M. Application of Perkin's reaction to terpene aldehyde β-cyclocitral. *Asian J. Chem.* **1993**, 5, 230-231.
18. Bellassoued, M., Lensen, N., Bakasse, M., Mouelhi, S. Two-Carbon Homologation of Aldehydes via Silyl Ketene Acetals: A New Stereoselective Approach to (E)-Alkenoic Acids. *J. Org. Chem.* **1998**, 63, 8785-8789.
19. Veverkova, E., Pacherova, E., Toma, S. Examination of the Perkin reaction under microwave irradiation. *Chemical Papers* **1999**, 53, 257-259.
20. Bautista, F. M., Campelo, J. M., Garcia, A., Luna, D., Marinas, J. M., Romero, A. A. Study on dry-media microwave azalactone synthesis on different supported KF catalysts: influence of textural and acid-base properties of supports. *J. Chem. Soc., Perkin Trans. 2* **2002**, 227-234.
21. Mogilaiah, K., Prashanthi, M., Reddy, C. S. Solid support Erlenmeyer synthesis of azlactones using microwaves. *Indian J. Chem., Sect. B* **2003**, 42B, 2126-2128.
22. Cativiela, C., Diaz-de-Villegas, M. D. 5(2H)-oxazolones and 5(4H)-oxazolones. *Chemistry of Heterocyclic Compounds (Hoboken, NJ, United States)* **2004**, 60, 129-330.
23. Kalnin, P. The mechanism of the Perkin synthesis. *Helv. Chim. Acta* **1928**, 11, 977-1003.
24. Dippy, J. F. J., Evans, R. M. The nature of the catalyst in the Perkin condensation. *J. Org. Chem.* **1950**, 15, 451-456.
25. Buckles, R. E., Bremer, K. G. A kinetic study of the Perkin condensation. *J. Am. Chem. Soc.* **1953**, 75, 1487-1489.
26. Crawford, M., Moore, G. W. Stereospecificity in the Perkin-Oglialoro reaction. The stereochemical configurations of some substituted α-phenylcinnamic acids. *J. Chem. Soc., Abstracts* **1955**, 3445-3448.
27. Kinastowski, S., Kasprzyk, H., Grabarkiewicz, J. Perkin reaction mechanism. *Bulletin de l'Academie Polonaise des Sciences, Serie des Sciences Chimiques* **1975**, 23, 211-214.
28. Pohjala, E. Indolizine derivatives. IV. Evidence for a disproportionation-dehydrogenation mechanism in the Perkin reaction of 2-pyridinecarbaldehyde in the presence of α,β-unsaturated carbonyl compounds to give 1-acylpyrrolo[2,1,5-cd]indolizines. *Heterocycles* **1975**, 3, 615-618.
29. Kinastowski, S., Kasprzyk, H. Ketene intermediates in the Perkin reaction catalyzed by tertiary amines. *Bulletin de l'Academie Polonaise des Sciences, Serie des Sciences Chimiques* **1978**, 26, 907-915.
30. Poonia, N. S., Sen, S., Porwal, P. K., Jayakumar, A. Coordinative role of alkali cations in organic reactions. V. The Perkin reaction. *Bull. Chem. Soc. Jpn.* **1980**, 53, 3338-3343.
31. Kinastowski, S., Nowacki, A. β-Lactone as intermediate in the Perkin reaction catalyzed by tertiary amines. *Tetrahedron Lett.* **1982**, 23, 3723-3724.
32. Bowden, K., Battah, S. Reactions of carbonyl compounds in basic solutions. Part 32. The Perkin rearrangement. *J. Chem. Soc., Perkin Trans. 2* **1998**, 1603-1606.
33. Palinko, I., Kukovecz, A., Torok, B., Kortvelyesi, T. On the mechanism of a modified Perkin condensation leading to α-phenylcinnamic acid stereoisomers - experiments and molecular modelling. *Monatsh. Chem.* **2000**, 131, 1097-1104.

34. Kasprzyk, H., Kinastowski, S. Kinetic investigations on the Perkin reaction catalyzed by tertiary amines. *React. Kinet. Catal. Lett.* **2002**, 77, 3-12.
35. Marton, G. I., Marton, A. L., Badea, F. Is the Perkin condensation of benzaldehyde reversible? *Scientific Bulletin - University "Politehnica" of Bucharest, Series B: Chemistry and Materials Science* **2002**, 64, 21-26.
36. Gaukroger, K., Hadfield, J. A., Hepworth, L. A., Lawrence, N. J., McGown, A. T. Novel Syntheses of Cis and Trans Isomers of Combretastatin A-4. *J. Org. Chem.* **2001**, 66, 8135-8138.
37. Ma, D., Tian, H., Zou, G. Asymmetric Strecker-type reaction of α-aryl ketones. Synthesis of (S)-αM4CPG, (S)-MPPG, (S)-AIDA, and (S)-APICA, the antagonists of metabotropic glutamate receptors. *J. Org. Chem.* **1999**, 64, 120-125.
38. Federsel, H. J. Development of a process for a chiral aminochroman antidepressant: A case story. *Org. Process Res. Dev.* **2000**, 4, 362-369.
39. Konkel, J. T., Fan, J., Jayachandran, B., Kirk, K. L. Syntheses of 6-fluoro-meta-tyrosine and its metabolites. *J. Fluorine Chem.* **2002**, 115, 27-32.

Petasis Boronic Acid-Mannich Reaction 340

1. Petasis, N. A., Akritopoulou, I. The boronic acid Mannich reaction: a new method for the synthesis of geometrically pure allylamines. *Tetrahedron Lett.* **1993**, 34, 583-586.
2. Dyker, G. Amino acid derivatives by multicomponent reactions. *Angew. Chem., Int. Ed. Engl.* **1997**, 36, 1700-1702.
3. Petasis, N. A., Zavialov, I. A. New reactions of alkenylboronic acids. *Spec. Publ. - R. Soc. Chem.* **1997**, 201, 179-182.
4. Dyker, G. Amino acid derivatives by multicomponent reactions. *Organic Synthesis Highlights IV* **2000**, 53-57.
5. McReynolds, M. D., Hanson, P. R. The three-component boronic acid Mannich reaction: structural diversity and stereoselectivity. *Chemtracts* **2001**, 14, 796-801.
6. Meester, W. J. N., van Maarseveen, J. H., Schoemaker, H. E., Hiemstra, H., Rutjes, F. P. J. T. Glyoxylates as versatile building blocks for the synthesis of α-amino acid and α-alkoxy acid derivatives via cationic intermediates. *Eur. J. Org. Chem.* **2003**, 2519-2529.
7. Orru, R. V. A., de Greef, M. Recent advances in solution-phase multicomponent methodology for the synthesis of heterocyclic compounds. *Synthesis* **2003**, 1471-1499.
8. Petasis, N. A., Goodman, A., Zavialov, I. A. A new synthesis of α-arylglycines from aryl boronic acids. *Tetrahedron* **1997**, 53, 16463-16470.
9. Petasis, N. A., Zavialov, I. A. A New and Practical Synthesis of α-Amino Acids from Alkenyl Boronic Acids. *J. Am. Chem. Soc.* **1997**, 119, 445-446.
10. Batey, R. A., MacKay, D. B., Santhakumar, V. Alkenyl and aryl boronates-mild nucleophiles for the stereoselective formation of functionalized N-heterocycles. *J. Am. Chem. Soc.* **1999**, 121, 5075-5076.
11. Schlienger, N., Bryce, M. R., Hansen, T. K. The Boronic Mannich Reaction in a Solid-Phase Approach. *Tetrahedron* **2000**, 56, 10023-10030.
12. Portlock, D. E., Naskar, D., West, L., Li, M. Petasis boronic acid-Mannich reactions of substituted hydrazines: synthesis of α-hydrazino carboxylic acids. *Tetrahedron Lett.* **2002**, 43, 6845-6847.
13. Naskar, D., Roy, A., Seibel, W. L., Portlock, D. E. Hydroxylamines and sulfinamide as amine components in the Petasis boronic acid-Mannich reaction: synthesis of N-hydroxy or alkoxy-α-aminocarboxylic acids and N-(tert-butyl sulfinyl)-α-amino carboxylic acids. *Tetrahedron Lett.* **2003**, 44, 8865-8868.
14. Naskar, D., Roy, A., Seibel, W. L., Portlock, D. E. Novel Petasis boronic acid-Mannich reactions with tertiary aromatic amines. *Tetrahedron Lett.* **2003**, 44, 5819-5821.
15. Kabalka, G. W., Venkataiah, B., Dong, G. The use of potassium alkynyltrifluoroborates in Mannich reactions. *Tetrahedron Lett.* **2004**, 45, 729-731.
16. Tremblay-Morin, J.-P., Raeppel, S., Gaudette, F. Lewis acid-catalyzed Mannich type reactions with potassium organotrifluoroborates. *Tetrahedron Lett.* **2004**, 45, 3471-3474.
17. Sugiyama, S., Arai, S., Ishii, K. Short synthesis of both enantiomers of cytoxazone using the Petasis reaction. *Tetrahedron: Asymmetry* **2004**, 15, 3149-3153.
18. Golebiowski, A., Klopfenstein, S. R., Chen, J. J., Shao, X. Solid supported high-throughput organic synthesis of peptide β-turn mimetics via tandem Petasis reaction/diketopiperazine formation. *Tetrahedron Lett.* **2000**, 41, 4841-4844.
19. Wang, Q., Finn, M. G. 2H-Chromenes from Salicylaldehydes by a Catalytic Petasis Reaction. *Org. Lett.* **2000**, 2, 4063-4065.
20. Batey, R. A., MacKay, D. B. Total synthesis of (±)-6-deoxycastanospermine: an application of the addition of organoboronates to N-acyliminium ions. *Tetrahedron Lett.* **2000**, 41, 9935-9938.

Petasis-Ferrier Rearrangement 342

Related reactions: *Ferrier reaction;*

1. Petasis, N. A., Lu, S.-P. New Stereocontrolled Synthesis of Substituted Tetrahydrofurans from 1,3-Dioxolan-4-ones. *J. Am. Chem. Soc.* **1995**, 117, 6394-6395.
2. Petasis, N. A., Lu, S.-P. Stereocontrolled synthesis of substituted tetrahydropyrans from 1,3-dioxan-4-ones. *Tetrahedron Lett.* **1996**, 37, 141-144.
3. Smith, A. B., III, Minbiole, K. P., Verhoest, P. R., Beauchamp, T. J. Phorboxazole Synthetic Studies. 2. Construction of a C(20-28) Subtarget, a Further Extension of the Petasis-Ferrier Rearrangement. *Org. Lett.* **1999**, 1, 913-916.
4. Smith, A. B., III, Verhoest, P. R., Minbiole, K. P., Lim, J. J. Phorboxazole Synthetic Studies. 1. Construction of a C(3-19) Subtarget Exploiting an Extension of the Petasis-Ferrier Rearrangement. *Org. Lett.* **1999**, 1, 909-912.
5. Smith, A. B., III, Minbiole, K. P., Verhoest, P. R., Schelhaas, M. Total synthesis of (+)-phorboxazole A exploiting the Petasis-Ferrier rearrangement. *J. Am. Chem. Soc.* **2001**, 123, 10942-10953.
6. Smith, A. B., III, Safonov, I. G., Corbett, R. M. Total Synthesis of (+)-Zampanolide. *J. Am. Chem. Soc.* **2001**, 123, 12426-12427.
7. Smith, A. B., Safonov, I. G., Corbett, R. M. Total Syntheses of (+)-Zampanolide and (+)-Dactylolide Exploiting a Unified Strategy. *J. Am. Chem. Soc.* **2002**, 124, 11102-11113.
8. Baldwin, J. E. Rules for ring closure. *J. Chem. Soc., Chem. Commun.* **1976**, 734-736.
9. Baldwin, J. E., Lusch, M. J. Rules for ring closure: application to intramolecular aldol condensations in polyketonic substrates. *Tetrahedron* **1982**, 38, 2939-2947.

Peterson Olefination 344

Related reactions: *Horner-Wadsworth-Emmons olefination, Horner-Wadsworth-Emmons olefination - Still-Gennari modification, Julia-Lithgoe olefination, Takai-Utimoto olefination, Tebbe olefination, Wittig reaction, Wittig reaction – Schlosser modification;*

1. Whitmore, F. C., Sommer, L. H., Gold, J., Strien, R. E. V. Fisson of β-oxygenated organosilicon compounds. *J. Am. Chem. Soc.* **1947**, 69, 1551.
2. Gilman, H., Tomasi, R. A. α-Silyl-substituted ylides. Tetraphenylallene via the Wittig reaction. *J. Org. Chem.* **1962**, 27, 3647-3650.
3. Peterson, D. J. Carbonyl olefination reaction using silyl-substituted organometallic compounds. *J. Org. Chem.* **1968**, 33, 780-784.

4. Colvin, E. W. Silicon in organic synthesis. *Chem. Soc. Rev.* **1978**, 7, 15-64.
5. Birkofer, L., Stuhl, O. Silylated synthons. Facile organic reagents of great applicability. *Top. Curr. Chem.* **1980**, 88, 33-88.
6. Colvin, E. W. *Silicon in Organic Synthesis* (Butterworths, Boston, London, **1981**) 288 pp.
7. Ager, D. J. The Peterson reaction. *Synthesis* **1984**, 384-398.
8. Colvin, E. W. Preparation and use of organosilicon compounds in organic synthesis. *Chem. Met.-Carbon Bond* **1987**, 4, 539-621.
9. Ager, D. J. The Peterson olefination reaction. *Org. React.* **1990**, 38, 1-223.
10. Barrett, A. G. M., Hill, J. M., Wallace, E. M., Flygare, J. A. Recent studies on the Peterson olefination reaction. *Synlett* **1991**, 764-770.
11. Kelly, S. E. Alkene Synthesis. in *Comp. Org. Synth.* (eds. Trost, B. M.,Fleming, I.), *1*, 729-818 (Pergamon, Oxford, **1991**).
12. Luh, T. Y., Wong, K. T. Silyl-substituted conjugated dienes: versatile building blocks of organic synthesis. *Synthesis* **1993**, 349-370.
13. Gosney, I., Lloyd, D. One or more C=C bond(s) formed by condensation: Condensation of P, As, Sb, Bi, Si or metal functions. in *Comp. Org. Funct. Group Trans. 1*, 719-770 (Pergamon, Cambridge, UK, **1995**).
14. Kawashima, T., Okazaki, R. Synthesis and reactions of the intermediates of the Wittig, Peterson, and their related reactions. *Synlett* **1996**, 600-608.
15. Krempner, C., Reinke, H., Oehme, H. The synthesis of transient silenes using the principle of the Peterson reaction. *Organosilicon Chemistry II: From Molecules to Materials, [Muenchner Silicontage], 2nd, Munich, 1994* **1996**, 389-398.
16. Colvin, E. W. Recent synthetic applications of organosilicon reagents. *Chemistry of Organic Silicon Compounds* **1998**, 2, 1667-1685.
17. Ager, D. J. Product subclass 37:β-silyl alcohols and the Peterson reaction. *Science of Synthesis* **2002**, 4, 789-809.
18. Baines, K. M., Samuel, M. S. Product subclass 2: silenes. *Science of Synthesis* **2002**, 4, 125-134.
19. Lawrence, N. J. Product subclass 27: α-haloalkylsilanes. *Science of Synthesis* **2002**, 4, 579-594.
20. Sarkar, T. K. Product subclass 40: allylsilanes. *Science of Synthesis* **2002**, 4, 837-925.
21. van Staden, L. F., Gravestock, D., Ager, D. J. New developments in the Peterson olefination reaction. *Chem. Soc. Rev.* **2002**, 31, 195-200.
22. Whitham, G. H. Product subclass 29: α,β-epoxysilanes. *Science of Synthesis* **2002**, 4, 633-646.
23. Kano, N., Kawashima, T. The Peterson and related reactions. *Modern Carbonyl Olefination* **2004**, 18-103.
24. Fleming, I., Floyd, C. D. The reactions of tris(trimethylsilyl)methyllithium with some carbon electrophiles. *J. Chem. Soc., Perkin Trans. 1* **1981**, 969-976.
25. Johnson, C. R., Tait, B. D. A cerium(III) modification of the Peterson reaction: methylenation of readily enolizable carbonyl compounds. *J. Org. Chem.* **1987**, 52, 281-283.
26. Savignac, P., Teulade, M. P., Collignon, N. Preparation and properties of α-silyl phosphonates, $(RO)_2P(O)CR^1R_2SiR^3R^4R^5$, and α,α-disilyl phosphonates, $(RO)_2P(O)CR^1(SiMe_3)_2$. *J. Organomet. Chem.* **1987**, 323, 135-144.
27. Fleming, I., Morgan, I. T., Sarkar, A. K. The stereochemistry of the vinylogous Peterson elimination. *J. Chem. Soc., Chem. Commun.* **1990**, 1575-1577.
28. Olah, G. A., Reddy, V. P., Prakash, G. K. S. Catalysis by solid superacids. 26. Peterson (silyl-Wittig) methylenation of carbonyl compounds using Nafion-H catalyzed hydroxy-trimethylsilane elimination of β-hydroxysilanes. *Synthesis* **1991**, 29-30.
29. Chen, F., Mudryk, B., Cohen, T. Generation, rearrangements and some synthetic use of bishomoallyllithiums. *Tetrahedron* **1994**, 50, 12793-12810.
30. Suzuki, T., Oriyama, T. New olefination of acetals with $TMSCH_2Cu(PBu_3)$.LiI under the influence of $BF_3.OEt_2$. *Synlett* **2000**, 859-861.
31. Trindle, C., Hwang, J.-T., Carey, F. A. CNDO-MO [complete neglect of differential overlap-molecular orbital] exploration of concerted and stepwise pathways for the Wittig and Peterson olefination reactions. *J. Org. Chem.* **1973**, 38, 2664-2669.
32. Gushurst, A. J., Jorgensen, W. L. Computer-assisted mechanistic evaluation of organic reactions. 14. Reactions of sulfur and phosphorus ylides, iminophosphoranes, and P=X-activated anions. *J. Org. Chem.* **1988**, 53, 3397-3408.
33. Apeloig, Y., Bendikov, M., Yuzefovich, M., Nakash, M., Bravo-Zhivotovskii, D., Blaser, D., Boese, R. Novel Stable Silenes via a Sila-Peterson-type Reaction. Molecular Structure and Reactivity. *J. Am. Chem. Soc.* **1996**, 118, 12228-12229.
34. Gillies, M. B., Tonder, J. E., Tanner, D., Norrby, P.-O. Quantum Chemical Calculations on the Peterson Olefination with α-Silyl Ester Enolates. *J. Org. Chem.* **2002**, 67, 7378-7388.
35. Hernandez, D., Larson, G. L. Chemistry of α-silyl carbonyl compounds. 9. Synthesis of tri- and tetrasubstituted olefins from α-silyl esters. *J. Org. Chem.* **1984**, 49, 4285-4287.
36. Brown, P. A., Bonnert, R. V., Jenkins, P. R., Lawrence, N. J., Selim, M. R. Silicon-directed diene synthesis. *J. Chem. Soc., Perkin Trans. 1* **1991**, 1893-1900.
37. Cuadrado, P., Gonzalez-Nogal, A. M. Regio- and stereospecific cleavage of α,β-epoxysilanes with lithium phenylsulfide. *Tetrahedron Lett.* **2000**, 41, 1111-1114.
38. Fürstner, A., Brehm, C., Cancho-Grande, Y. Stereoselective Synthesis of Enamides by a Peterson Reaction Manifold. *Org. Lett.* **2001**, 3, 3955-3957.
39. Matsuda, I., Okada, H., Sato, S., Izumi, Y. A regioselective enolate formation of trimethylsilylmethyl ketones. Application to the (E)-selective synthesis of α,β-unsaturated ketones. *Tetrahedron Lett.* **1984**, 25, 3879-3882.
40. Mallya, M. N., Nagendrappa, G. cis-Hydroxylation of cyclic vinylsilanes using cetyltrimethylammonium permanganate. *Synthesis* **1999**, 37-39.
41. Yamamoto, K., Kimura, T., Tomo, Y. Novel competing reaction of 1-methoxy-2-(trimethylsilyl)-3-hydroxy moiety in base-induced Peterson olefination; mechanistic rationale of the reaction. *Tetrahedron Lett.* **1984**, 25, 2155-2158.
42. Boeckman, R. K., Jr., Chinn, R. L. Counterion effects on geometric control in the Peterson reaction of bistrimethylsilyl esters: synthetic scope and mechanistic implications. *Tetrahedron Lett.* **1985**, 26, 5005-5008.
43. Bassindale, A. R., Ellis, R. J., Lau, J. C. Y., Taylor, P. G. The mechanism of the Peterson reaction. Part 2. The effect of reaction conditions, and a new model for the addition of carbanions to carbonyl derivatives in the absence of chelation control. *J. Chem. Soc., Perkin Trans. 2* **1986**, 593-597.
44. Hudrlik, P. F., Agwaramgbo, E. L. O., Hudrlik, A. M. Concerning the mechanism of the Peterson olefination reaction. *J. Org. Chem.* **1989**, 54, 5613-5618.
45. Kang, K. T., Sung, T. M., Lee, K. R., Lee, J. G., Jyung, K. K. Reactions of acylsilanes with phenylthio(trimethylsilyl)methyllithium. Competitive Peterson and Brook rearrangement-elimination reactions in the β-thiophenyl-α,β-disilylalkoxides. *Bull. Korean Chem. Soc.* **1993**, 14, 757-759.
46. Hoffmann, D., Reinke, H., Oehme, H. The reaction of tris(trimethylsilyl)silyllithium with dibenzosuberenone. *J. Organomet. Chem.* **1996**, 526, 185-189.
47. Van Staden, L. F., Bartels-Rahm, B., Field, J. S., Emslie, N. D. Stereoselective Peterson olefinations of silylated benzyl carbamates. *Tetrahedron* **1998**, 54, 3255-3278.
48. Kawashima, T., Okazaki, R. Diheteracyclobutanes containing highly coordinate main group elements: syntheses, structures, and thermolyses. *Advances in Strained and Interesting Organic Molecules* **1999**, 7, 1-41.
49. Naganuma, K., Kawashima, T., Okazaki, R. Control factors of two reaction modes of pentacoordinate 1,2-oxasiletanides, the Peterson reaction and homo-Brook rearrangement. *Chem. Lett.* **1999**, 1139-1140.
50. Tonder, J. E., Begtrup, M., Hansen, J. B., Olesen, P. H. Exploring the stereoselectivity in the Peterson reaction of several 2-substituted 1-azabicyclo[2.2.2]octan-3-ones. *Tetrahedron* **2000**, 56, 1139-1146.
51. Toro, A., Nowak, P., Deslongchamps, P. Transannular Diels-Alder Entry into Stemodanes: First Asymmetric Total Synthesis of (+)-Maritimol. *J. Am. Chem. Soc.* **2000**, 122, 4526-4527.
52. Harrington, P. E., Tius, M. A. A Formal Total Synthesis of Roseophilin: Cyclopentannelation Approach to the Macrocyclic Core. *Org. Lett.* **1999**, 1, 649-651.
53. Denmark, S. E., Yang, S.-M. Intramolecular Silicon-Assisted Cross-Coupling: Total Synthesis of (+)-Brasilenyne. *J. Am. Chem. Soc.* **2002**, 124, 15196-15197.
54. Galano, J. M., Audran, G., Monti, H. First enantioselective total synthesis of both enantiomers of lancifolol. Correlation: absolute configuration/specific rotation. *Tetrahedron Lett.* **2001**, 42, 6125-6128.

Pfitzner-Moffatt Oxidation ... 346

Related reactions: Corey-Kim oxidation, Dess-Martin oxidation, Jones oxidation, Ley oxidation, Oppenauer oxidation, Swern oxidation;

1. Pfitzner, K. E., Moffatt, J. G. A new and selective oxidation of alcohols. *J. Am. Chem. Soc.* **1963**, 85, 3027-3028.
2. Pfitzner, K. E., Moffatt, J. G. Sulfoxide-carbodiimide reactions. I. A facile oxidation of alcohols. *J. Am. Chem. Soc.* **1965**, 87, 5661-5670.
3. Pfitzner, K. E., Moffatt, J. G. Sulfoxide-carbodiimide reactions. II. Scope of the oxidation reaction. *J. Am. Chem. Soc.* **1965**, 87, 5670-5678.
4. Epstein, W. W., Sweat, F. W. Dimethyl sulfoxide oxidations. *Chem. Rev.* **1967**, 67, 247-260.
5. Hanessian, S., Butterworth, R. F. Selected methods of oxidation in carbohydrate chemistry. *Synthesis* **1971**, 70-88.
6. Moffatt, J. G. Sulfoxide-carbodiimide and related oxidations. in *Oxidation* (eds. Augustine, R. L.,Trecker, D. J.), *2*, 1-64 (Dekker, New Yprk, **1971**).
7. Mancuso, A. J., Swern, D. Activated dimethyl sulfoxide: useful reagents for synthesis. *Synthesis* **1981**, 165-185.
8. Tidwell, T. T. Oxidation of alcohols by activated dimethyl sulfoxide and related reactions: an update. *Synthesis* **1990**, 857-870.
9. Tidwell, T. T. Oxidation of alcohols to carbonyl compounds via alkoxysulfonium ylides: the Moffat, Swern, and related oxidations. *Org. React.* **1990**, 39, 297-572.
10. Albright, J. D., Goldman, L. Dimethyl sulfoxide-acid anhydride mixtures. New reagent for oxidation of alcohols. *J. Am. Chem. Soc.* **1965**, 87, 4214-4216.
11. Onodera, K., Hirano, S., Kashimura, N. Oxidation of carbohydrates with dimethyl sulfoxide containing phosphorus pentaoxide. *J. Am. Chem. Soc.* **1965**, 87, 4651-4652.
12. Onodera, K., Hirano, S., Kashimura, N., Yajima, T. Reaction of dimethyl sulfoxide with organic compounds in the presence of phosphorus pentoxide. *Tetrahedron Lett.* **1965**, 4327-4331.
13. Albright, J. D., Goldman, L. Dimethyl sulfoxide-acid anhydride mixtures for the oxidation of alcohols. *J. Am. Chem. Soc.* **1967**, 89, 2416-2423.
14. Parikh, J. R., Doering, W. v. E. Sulfur trioxide in the oxidation of alcohols by dimethyl sulfoxide. *J. Am. Chem. Soc.* **1967**, 89, 5505-5507.
15. Weinshenker, N. M., Shen, C. M. Polymeric reagents. I. Synthesis of an insoluble polymeric carbodiimide. *Tetrahedron Lett.* **1972**, 3281-3284.
16. Omura, K., Sharma, A. K., Swern, D. Dimethyl sulfoxide-trifluoroacetic anhydride. New reagent for oxidation of alcohols to carbonyls. *J. Org. Chem.* **1976**, 41, 957-962.
17. Omura, K., Swern, D. Oxidation of alcohols by "activated" dimethyl sulfoxide. A preparative steric and mechanistic study. *Tetrahedron* **1978**, 34, 1651-1660.
18. Roa-Gutierrez, F., Liu, H.-J. Use of silyl chlorides as dimethyl sulfoxide activators for the oxidation of alcohols. *Bulletin of the Institute of Chemistry, Academia Sinica* **2000**, 47, 19-26.
19. Fenselau, A. H., Moffatt, J. G. Sulfoxide-carbodiimide reactions. III. Mechanism of the oxidation reaction. *J. Am. Chem. Soc.* **1966**, 88, 1762-1765.
20. Torssell, K. Mechanisms of dimethylsulfoxide oxidations. *Tetrahedron Lett.* **1966**, 4445-4451.
21. Moffatt, J. G. Sulfoxide-carbodiimide reactions. X. Mechanism of the oxidation reaction. *J. Org. Chem.* **1971**, 36, 1909-1913.
22. Ichikawa, S., Shuto, S., Matsuda, A. The First Synthesis of Herbicidin B. Stereoselective Construction of the Tricyclic Undecose Moiety by a Conformational Restriction Strategy Using Steric Repulsion between Adjacent Bulky Silyl Protecting Groups on a Pyranose Ring. *J. Am. Chem. Soc.* **1999**, 121, 10270-10280.
23. Smith, A. B., III, Kingery-Wood, J., Leenay, T. L., Nolen, E. G., Sunazuka, T. Indole diterpene synthetic studies. 8. The total synthesis of (+)-paspalicine and (+)-paspalinine. *J. Am. Chem. Soc.* **1992**, 114, 1438-1449.
24. Begley, M. J., Bowden, M. C., Patel, P., Pattenden, G. New stereoselective approach to hydroxy-substituted tetrahydrofurans. Total synthesis of (±)-citreoviral. *J. Chem. Soc., Perkin Trans. 1* **1991**, 1951-1958.
25. Mori, K., Takaishi, H. Synthesis of mono- and sesquiterpenoids. XVI. Synthesis of (-)-pereniporins A and B, sesquiterpene antibiotics from a basidiomycete. *Liebigs Ann. Chem.* **1989**, 939-943.

Pictet-Spengler Tetrahydroisoquinoline Synthesis ... 348

Related reactions: Bischler-Napieralski isoquinoline synthesis, Pomeranz-Fritsch reaction;

1. Pictet, A., Spengler, T. Formation of Isoquinoline Derivatives by the Action of Methylal on Phenylethylamine, Phenylalanine and Tyrosine. *Ber.* **1911**, 44, 2030-2036.
2. Whaley, W. M., Govindachari, T. R. The Pictet-Spengler synthesis of tetrahydroisoquinolines and related compounds. *Org. React.* **1951**, 6, 151-190.
3. Abramovitch, R. A., Spenser, I. D. The carbolines. *Advan. Heterocyclic Chem. (A. R. Katritzky, editor. Academic)* **1964**, 3, 79-207.
4. Farrar, W. V. Formaldehyde-amine reactions. *Rec. Chem. Progr.* **1968**, 29, 85-101.
5. Batra, H. R. Aminoalkyl chain in medicinal chemistry. *Pharmacos* **1970**, 15, 57-61.
6. Stuart, K., Woo-Ming, R. The β-carboline alkaloids. *Heterocycles* **1975**, 3, 223-264.
7. Ungemach, F., Cook, J. M. The spiroindolenine intermediate. A review. *Heterocycles* **1978**, 9, 1089-1119.
8. Overman, L. E., Ricca, D. J. The Intramolecular Mannich and Related Reactions. in *Comp. Org. Synth.* (eds. Trost, B. M.,Fleming, I.), *2*, 1007-1046 (Pergamon, Oxford, **1991**).
9. Cox, E. D., Cook, J. M. The Pictet-Spengler condensation: a new direction for an old reaction. *Chem. Rev.* **1995**, 95, 1797-1842.
10. Czerwinski, K. M., Cook, J. M. Stereochemical control of the Pictet-Spengler reaction in the synthesis of natural products. *Advances in Heterocyclic Natural Product Synthesis* **1996**, 3, 217-277.
11. Hino, T., Nakagawa, M. Pictet-Spengler reactions of Nβ-hydroxytryptamines and their application to the synthesis of eudistomins. *Heterocycles* **1998**, 49, 499-530.
12. Chrzanowska, M., Rozwadowska, M. D. Asymmetric Synthesis of Isoquinoline Alkaloids. *Chem. Rev.* **2004**, 104, 3341-3370.
13. Royer, J., Bonin, M., Micouin, L. Chiral Heterocycles by Iminium Ion Cyclization. *Chem. Rev.* **2004**, 104, 2311-2352.
14. Cesati, R. R., III, Katzenellenbogen, J. A. Preparation of hexahydrobenzo[f]isoquinolines using a vinylogous Pictet-Spengler cyclization. *Org. Lett.* **2000**, 2, 3635-3638.
15. Gremmen, C., Wanner, M. J., Koomen, G.-J. Enantiopure tetrahydroisoquinolines via N-sulfinyl Pictet-Spengler reactions. *Tetrahedron Lett.* **2001**, 42, 8885-8888.
16. Connors, R. V., Zhang, A. J., Shuttleworth, S. J. Pictet-Spengler synthesis of tetrahydro-β-carbolines using vinylsulfonylmethyl resin. *Tetrahedron Lett.* **2002**, 43, 6661-6663.
17. Cutter, P. S., Miller, R. B., Schore, N. E. Synthesis of protoberberines using a silyl-directed Pictet-Spengler cyclization. *Tetrahedron* **2002**, 58, 1471-1478.
18. Horiguchi, Y., Kodama, H., Nakamura, M., Yoshimura, T., Hanezi, K., Hamada, H., Saitoh, T., Sano, T. A convenient synthesis of 1,1-disubstituted 1,2,3,4-tetrahydroisoquinolines via Pictet-Spengler reaction using titanium(IV) isopropoxide and acetic-formic anhydride. *Chem. Pharm. Bull.* **2002**, 50, 253-257.
19. Miles, W. H., Heinsohn, S. K., Brennan, M. K., Swarr, D. T., Eidam, P. M., Gelato, K. A. The oxa-Pictet-Spengler reaction of 1-(3-furyl)alkan-2-ols. *Synthesis* **2002**, 1541-1545.
20. Nielsen, T. E., Diness, F., Meldal, M. The Pictet-Spengler reaction in solid-phase combinatorial chemistry. *Current Opinion in Drug Discovery & Development* **2003**, 6, 801-814.

21. Pal, B., Jaisankar, P., Giri, V. S. Microwave assisted Pictet-Spengler and Bischler-Napieralski reactions. *Synth. Commun.* **2003**, 33, 2339-2348.
22. Srinivasan, N., Ganesan, A. Highly efficient Lewis acid-catalysed Pictet-Spengler reactions discovered by parallel screening. *Chem. Commun.* **2003**, 916-917.
23. Tsuji, R., Nakagawa, M., Nishida, A. An efficient synthetic approach to optically active β-carboline derivatives via Pictet-Spengler reaction promoted by trimethylchlorosilane. *Tetrahedron: Asymmetry* **2003**, 14, 177-180.
24. Alberch, L., Bailey, P. D., Clingan, P. D., Mills, T. J., Price, R. A., Pritchard, R. G. The cis-specific Pictet-Spengler reaction. *Eur. J. Org. Chem.* **2004**, 1887-1890.
25. Hegedues, A., Hell, Z. One-step preparation of 1-substituted tetrahydroisoquinolines via the Pictet-Spengler reaction using zeolite catalysts. *Tetrahedron Lett.* **2004**, 45, 8553-8555.
26. Nielsen, T. E., Meldal, M. Solid-Phase Intramolecular N-Acyliminium Pictet-Spengler Reactions as Crossroads to Scaffold Diversity. *J. Org. Chem.* **2004**, 69, 3765-3773.
27. Taylor, M. S., Jacobsen, E. N. Highly enantioselective catalytic acyl-Pictet-Spengler reactions. *J. Am. Chem. Soc.* **2004**, 126, 10558-10559.
28. Kowalski, P., Mokrosz, J. L. Structure and spectral properties of β-carbolines. Part 9. New arguments against direct rearrangement of the spiroindolenine intermediate into β-carboline system in the Pictet-Spengler cyclization. An MNDO approach. *Bull. Soc. Chim. Belg.* **1997**, 106, 147-149.
29. Bailey, P. D., Morgan, K. M. The total synthesis of (-)-suaveoline. *Perkin 1* **2000**, 3578-3583.
30. Xu, Y.-C., Kohlman, D. T., Liang, S. X., Erikkson, C. Stereoselective, Oxidative C-C Bond Coupling of Naphthopyran Induced by DDQ: Stereocontrolled Total Synthesis of Deoxyfrenolicin. *Org. Lett.* **1999**, 1, 1599-1602.
31. Pearson, W. H., Lian, B. W. Application of the 2-azaallyl anion cyclo-addition method to an enantioselective total synthesis of (+)-coccinine. *Angew. Chem., Int. Ed. Engl.* **1998**, 37, 1724-1726.
32. Zhou, B., Guo, J., Danishefsky, S. J. Studies Directed to the Total Synthesis of ET 743 and Analogues Thereof: An Expeditious Route to the ABFGH Subunit. *Org. Lett.* **2002**, 4, 43-46.

Pinacol and Semipinacol Rearrangement 350

Related reactions: Demjanov and Tiffeneau-Demjanov rearrangement;

1. Fittig, R. Certain derivatives of acetone. *Liebigs Ann. Chem.* **1860**, 114, 54-63.
2. Tiffeneau, M., Levy, J. Pinacolic and semi-pinacolic transpositions. Comparative migratory tendencies of different radicals. *Compt. rend.* **1923**, 176, 312-314.
3. Bayer, O. in *Methoden der Organischen Chemie (Houben-Weyl) 7*, 230-236 (Thieme, Stuttgart, **1954**).
4. Collins, C. J. Pinacol rearrangement. *Quarterly Revs.* **1960**, 14, 357-377.
5. Pocker, Y. Wagner-Meerwein and pinacolic rearrangements in acyclic and cyclic systems. in *Molecular Rearrangements* (ed. De Mayo, P.), *1*, 1-25 (Wiley, New York, **1963**).
6. Collins, C. J., Eastham, J. F. Rearrangements involving the carbonyl group. *Chemistry of the Carbonyl Group* **1966**, 761-821.
7. Bartok, M., Molnar, A. Dehydration of 1,2-diols. in *The Chemistry of Functional Groups. Supplement E: The Chemistry of Hydroxyl, Ether and Peroxide Groups* (ed. Patai, S.), 761-821 (Wiley, New York, **1980**).
8. Krow, G. R. One carbon ring expansions of bridged bicyclic ketones. *Tetrahedron* **1987**, 43, 3-38.
9. Coveney, D. J. The Semipinacol and Other Rearrangements. in *Comp. Org. Synth.* (eds. Trost, B. M.,Fleming, I.), *3*, 777-801 (Pergamon, Oxford, **1991**).
10. Rickborn, B. The Pinacol Rearrangement. in *Comp. Org. Synth.* (eds. Trost, B. M.,Fleming, I.), *3*, 721-732 (Pergamon, Oxford, **1991**).
11. Toda, F. Solid-to-solid organic reactions. *React. Mol. Cryst.* **1993**, 177-201.
12. Eichinger, P. C. H., Dua, S., Bowie, J. H. A comparison of skeletal rearrangement reactions of even-electron anions in solution and in the gas phase. *Int. J. Mass Spectrom. Ion Processes* **1994**, 133, 1-12.
13. Coldham, I. One or more CH and/or CC bond(s) formed by rearrangement. in *Comp. Org. Funct. Group Trans.* (eds. Katritzky, A. R., Meth-Cohn, O.,Rees, C. W.), *1*, 377-423 (Pergamon, Oxford, **1995**).
14. Paquette, L. A. Oxonium ion-initiated pinacolic ring expansion reactions. *Recent Research Developments in Chemical Sciences* **1997**, 1, 1-16.
15. Molnar, A. The pinacol rearrangement. *Fine Chemicals through Heterogeneous Catalysis* **2001**, 232-241.
16. Blanco, F. E., Harris, F. L. Semipinacol rearrangements involving trifluoromethylphenyl groups. *J. Org. Chem.* **1977**, 42, 868-871.
17. Arce de Sanabia, J., Carrion, A. E. Radical cation catalyzed pinacol-pinacolone rearrangement. *Tetrahedron Lett.* **1993**, 34, 7837-7840.
18. Lopez, L., Mele, G., Mazzeo, C. Pinacol-pinacolone rearrangement induced by aminium salts. *J. Chem. Soc., Perkin Trans. 1* **1994**, 779-781.
19. Paquette, L. A., Dullweber, U., Branan, B. M. Thionium ion-activated pinacol rearrangements. Generality and scope. *Heterocycles* **1994**, 37, 187-191.
20. Kimura, M., Kobayashi, K., Yamamoto, Y., Sawaki, Y. Electrooxidative pinacol-type rearrangement of β-hydroxy sulfides. Efficient C-S cleavage mediated by chloride ion oxidation. *Tetrahedron* **1996**, 52, 4303-4310.
21. Hornyak, G., Fetter, J., Nemeth, G., Poszavacz, L., Simig, G. A trifluoromethyl group directed semipinacol rearrangement: synthesis of trifluoroacetyldiarylmethanes. *J. Fluorine Chem.* **1997**, 84, 49-51.
22. Hoang, M., Gadosy, T., Ghazi, H., Hou, D.-F., Hopkinson, A. C., Johnston, L. J., Lee-Ruff, E. Photochemical Pinacol Rearrangement. *J. Org. Chem.* **1998**, 63, 7168-7171.
23. Bickley, J. F., Hauer, B., Pena, P. C. A., Roberts, S. M., Skidmore, J. The semi-pinacol rearrangement of homochiral epoxy alcohols catalyzed by rare earth triflates. *J. Chem. Soc., Perkin Trans. 1* **2001**, 1253-1255.
24. Fan, C.-A., Wang, B.-M., Tu, Y.-Q., Song, Z.-L. Samarium-catalyzed tandem semipinacol rearrangement/Tishchenko reaction of α-hydroxy epoxides: a novel approach to highly stereoselective construction of 2-quaternary 1,3-diol units. *Angew. Chem., Int. Ed. Engl.* **2001**, 40, 3877-3880.
25. Fenster, M. D. B., Patrick, B. O., Dake, G. R. Construction of Azaspirocyclic Ketones through a-Hydroxyiminium Ion or α-Siloxy Epoxide Semipinacol Rearrangements. *Org. Lett.* **2001**, 3, 2109-2112.
26. Sugihara, Y., Iimura, S., Nakayama, J. Aza-pinacol rearrangement: acid-catalyzed rearrangement of aziridines to imines. *Chem. Commun.* **2002**, 134-135.
27. Li, X., Wu, B., Zhao, X. Z., Jia, Y. X., Tu, Y. Q., Li, D. R. An interesting $AlEt_3$-promoted stereoselective tandem rearrangement/reduction of α-hydroxy (or amino) heterocyclopropane. *Synlett* **2003**, 623-626.
28. Shionhara, T., Suzuki, K. Facile one-pot procedure for Et_3Al-promoted asymmetric pinacol-type rearrangement. *Synthesis* **2003**, 141-146.
29. Wang, B. M., Song, Z. L., Fan, C. A., Tu, Y. Q., Chen, W. M. Halogen cation induced stereoselective semipinacol-type rearrangement of allylic alcohols. A highly efficient approach to α-quaternary β-haloketo compounds. *Synlett* **2003**, 1497-1499.
30. Hu, X.-D., Fan, C.-A., Zhang, F.-M., Tu, Y. Q. A tandem semipinacol rearrangement/alkylation of α-epoxy alcohols: An efficient and stereoselective approach to multifunctional 1,3-diols. *Angew. Chem., Int. Ed. Engl.* **2004**, 43, 1702-1705.
31. Mladenova, G., Singh, G., Acton, A., Chen, L., Rinco, O., Johnston, L. J., Lee-Ruff, E. Photochemical Pinacol Rearrangements of Unsymmetrical Diols. *J. Org. Chem.* **2004**, 69, 2017-2023.
32. Shinde, A. B., Shrigadi, N. B., Bhat, R. P., Samant, S. D. Pinacol-Pinacolone Rearrangement on $FeCl_3$ Modified Montmorillonite K10. *Synth. Commun.* **2004**, 34, 309-314.
33. Nakamura, K., Osamura, Y. MO study of the possibility of a concerted mechanism in the pinacol rearrangement. *J. Phys. Org. Chem.* **1990**, 3, 737-745.

34. Nakamura, K., Osamura, Y. Theoretical study of the reaction mechanism and migratory aptitude of the pinacol rearrangement. *J. Am. Chem. Soc.* **1993**, 115, 9112-9120.
35. Bouchoux, G., Choret, N., Flammang, R. Unimolecular Chemistry of Protonated Diols in the Gas Phase: Internal Cyclization and Hydride Ion Transfer. *J. Phys. Chem. A* **1997**, 101, 4271-4282.
36. Haque, A., Ghatak, A., Ghosh, S., Ghoshal, N. A Facile Access to Densely Functionalized Substituted Cyclopentanes and Spiro Cyclopentanes. Carbocation Stabilization Directed Bond Migration in Rearrangement of Cyclobutanes. *J. Org. Chem.* **1997**, 62, 5211-5214.
37. Smith, W. B. Hydrogen as a migrating group in some pinacol rearrangements: a DFT study. *J. Phys. Org. Chem.* **1999**, 12, 741-746.
38. Smith, W. B. Ethylene glycol to acetaldehyde-dehydration or a concerted mechanism. *Tetrahedron* **2002**, 58, 2091-2094.
39. Berson, J. A. What is a discovery? Carbon skeletal rearrangements as counter-examples to the rule of minimal structural change. *Angew. Chem., Int. Ed. Engl.* **2002**, 41, 4655-4660.
40. Kursanov, D. N., Parnes, Z. N. Mechanism of pinacolone rearrangement by deuterium exchange study. *Zh. Obshch. Khim.* **1957**, 27, 737-739.
41. Matsumoto, K. Pinacol rearrangement. V. Rearrangements of cis- and trans-1,2-diphenyl-1,2-ditolylethylene oxides. *Tetrahedron* **1968**, 24, 6851-6862.
42. Matsumoto, K. Pinacol rearrangement. IV. The kinetics and mechanism of the rearrangement of meso- and (±)-2,2'-dimethoxybenzopinacol. *Bull. Chem. Soc. Jpn.* **1968**, 41, 1356-1360.
43. Bhatia, K., Fry, A. Mechanism of the sulfuric acid-catalyzed rearrangement of methyl and carbonyl carbon-14-labeled 3,3-dimethyl-2-butanone. *J. Org. Chem.* **1969**, 34, 806-811.
44. Moriyoshi, T., Tamura, K. Effects of pressure on organic reactions. II. Acid-catalyzed rearrangement of pinacol. *Rev. Phys. Chem. Japan* **1970**, 40, 48-58.
45. Pocker, Y., Ronald, B. P. Kinetics and mechanism of vic-diol dehydration. II. p-Anisyl group in pinacolic rearrangement. *J. Org. Chem.* **1970**, 35, 3362-3367.
46. Pocker, Y., Ronald, B. P. Kinetics and mechanism of vic-diol dehydration. I. Origin of epoxide intermediates in certain pinacolic rearrangements. *J. Am. Chem. Soc.* **1970**, 92, 3385-3392.
47. Dubois, J. E., Bauer, P. Metathetical transposition of bis-tert-alkyl ketones. 1. A model for a study of group migration. *J. Am. Chem. Soc.* **1976**, 98, 6993-6999.
48. Herlihy, K. P. Rearrangement of diols. II. Kinetics of the pinacol rearrangement of propane-1,2-diol. *Aust. J. Chem.* **1981**, 34, 107-114.
49. Wistuba, E., Ruechardt, C. Intrinsic migration aptitudes of alkyl groups in a pinacol rearrangement. *Tetrahedron Lett.* **1981**, 22, 4069-4072.
50. Kaupp, G., Haak, M., Toda, F. Atomic force microscopy and solid-state rearrangement of benzopinacol. *J. Phys. Org. Chem.* **1995**, 8, 545-551.
51. Clericuzio, M., Cobianco, S., Fabbi, M., Lezzi, A., Montanari, L. The cationic ring-opening polymerization of 7-tetradecene oxide with methyl trifluoromethansulfonate. an investigation of the mechanism and the kinetics by means of ^{1}H, ^{13}C and ^{19}F NMR. *Polymer* **1998**, 40, 1839-1851.
52. Rathore, R., Kochi, J. K. Acid catalysis vs. electron-transfer catalysis via organic cations or cation-radicals as the reactive intermediate. Are these distinctive mechanisms? *Acta Chem. Scand.* **1998**, 52, 114-130.
53. De Lezaeta, M., Sattar, W., Svoronos, P., Karimi, S., Subramaniam, G. Effect of various acids at different concentrations on the pinacol rearrangement. *Tetrahedron Lett.* **2002**, 43, 9307-9309.
54. Dai, Z., Hatano, B., Tagaya, H. Catalytic dehydration of propylene glycol with salts in near-critical water. *Appl. Cat. A* **2004**, 258, 189-193.
55. Seki, M., Sakamoto, T., Suemune, H., Kanematsu, K. Total synthesis of (±)-furoscrobiculin B. *J. Chem. Soc., Perkin Trans. 1* **1997**, 1707-1714.
56. Pettit, G. R., Lippert, J. W., III, Herald, D. L. A Pinacol Rearrangement/Oxidation Synthetic Route to Hydroxyphenstatin. *J. Org. Chem.* **2000**, 65, 7438-7444.
57. Wendt, J. A., Gauvreau, P. J., Bach, R. D. Synthesis of (±)-Fredericamycin A. *J. Am. Chem. Soc.* **1994**, 116, 9921-9926.
58. Suzuki, K., Tomooka, K., Katayama, E., Matsumoto, T., Tsuchihashi, G. Stereocontrolled asymmetric total synthesis of protomycinolide IV. *J. Am. Chem. Soc.* **1986**, 108, 5221-5229.

Pinner Reaction..352

1. Pinner, A., Klein, F. The conversion of nitriles to imides. *Ber.* **1877**, 10, 1889-1897.
2. Pinner, A., Klein, F. The conversion of nitriles to imides. *Ber.* **1878**, 11, 1475-1487.
3. Pinner, A. The conversion of nitriles to imides. *Ber.* **1883**, 16, 1643-1655.
4. Brotherton, T. K., Lynn, J. W. The synthesis and chemistry of cyanogen. *Chem. Rev.* **1959**, 59, 841-883.
5. Roger, R., Neilson, D. G. The chemistry of imidates. *Chem. Rev.* **1961**, 61, 179-211.
6. Zil'berman, E. N. Reactions of nitriles with hydrogen halides and nucleophilic reagents. *Russ. Chem. Rev.* **1962**, 31, 615-633.
7. *The Chemistry of Functional Groups: the Chemistry of Amidines and Imidates* (ed. Patai, S.) (**1975**) 679 pp.
8. *The Chemistry of Amidines and Imidates, Vol. 2* (eds. Patai, S.,Rappoport, Z.) (**1991**) 918 pp.
9. Schaefer, F. C., Peters, G. A. Base-catalyzed reaction of nitriles with alcohols. A convenient route to imidates and amidine salts. *J. Org. Chem.* **1961**, 26, 412-418.
10. Poupaert, J., Bruylants, A., Crooy, P. N-acyl-α-aminonitriles in the Pinner reaction. *Synthesis* **1972**, 622-624.
11. Lee, Y. B., Goo, Y. M., Lee, Y. Y., Lee, J. K. Conversion of α-amino nitriles to amides by a new Pinner-type reaction. *Tetrahedron Lett.* **1990**, 31, 1169-1170.
12. Magedov, I. V., Usorov, M. I., Smushkevich, Y. I. Extending the application of the Pinner reaction by a new modification. *Zh. Org. Khim.* **1991**, 27, 282-284.
13. Shishkin, V. E., Mednikov, E. V., Anishchenko, O. V., No, B. I. A two-stage version of the Pinner reaction as a route to dialkoxyphosphorylalkyl imidate hydrochlorides. *Russ. J. Gen. Chem. (Translation of Zhurnal Obshchei Khimii)* **1999**, 69, 1673.
14. Luzyanin, K. V., Kukushkin, V. Y., Kuznetsov, M. L., Garnovskii, D. A., Haukka, M., Pombeiro, A. J. L. Novel Reactivity Mode of Hydroxamic Acids: A Metalla-Pinner Reaction. *Inorg. Chem.* **2002**, 41, 2981-2986.
15. Luzyanin, K. V., Kukushkin, V. Y., Haukka, M., Frausto da Silva, J. J. R., Pombeiro, A. J. L. The metalla-Pinner reaction between Pt(IV)-bound nitriles and alkylated oxamic and oximic forms of hydroxamic acids. *Dalton Transactions* **2004**, 2728-2732.
16. Hartigan, R. H., Cloke, J. B. Thermal and hydrolytic behavior of imido and thioimido ester salts. *J. Am. Chem. Soc.* **1945**, 67, 709-715.
17. Cramer, F., Pawelzik, K., Lichtenthaler, F. W. Imido esters. II. The reaction of imido esters with acids (Pinner cleavage). *Chem. Ber.* **1958**, 91, 1555-1562.
18. Lee, Y. B., Goo, Y. M., Lee, Y. Y. Another evidence for the formation of 2-amino-1,3-oxathiolane tetrahedral intermediate in the Pinner type reaction of nitriles with 2-mercaptoethanol; formation of 2-chlorothio esters and 2-mercaptoethyl esters from nitriles. *Bull. Korean Chem. Soc.* **1992**, 13, 9-10.
19. Hamada, Y., Hara, O., Kawai, A., Kohno, Y., Shioiri, T. New methods and reagents in organic synthesis. 97. Efficient total synthesis of AI-77-B, a gastroprotective substance from Bacillus pumilus AI-77. *Tetrahedron* **1991**, 47, 8635-8652.
20. Grossman, R. B., Rasne, R. M. Short Total Syntheses of Both the Putative and Actual Structures of the Clerodane Diterpenoid (±)-Sacacarin by Double Annulation. *Org. Lett.* **2001**, 3, 4027-4030.
21. Schaerer, K., Morgenthaler, M., Seiler, P., Diederich, F., Banner, D. W., Tschopp, T., Obst-Sander, U. Enantiomerically pure thrombin inhibitors for exploring the molecular-recognition features of the oxyanion hole. *Helv. Chim. Acta* **2004**, 87, 2517-2538.

Related reactions: *Jones oxidation;*

1. Lindgren, B. O., Nilsson, T. Preparation of carboxylic acids from aldehydes (including hydroxylated benzaldehydes) by oxidation with chlorite. *Acta Chemica Scandinavica (1947-1973)* **1973**, 27, 888-890.
2. Kraus, G. A., Roth, B. Synthetic studies toward verrucarol. 2. Synthesis of the AB ring system. *J. Org. Chem.* **1980**, 45, 4825-4830.
3. Kraus, G. A., Taschner, M. J. Model studies for the synthesis of quassinoids. 1. Construction of the BCE ring system. *J. Org. Chem.* **1980**, 45, 1175-1176.
4. Bal, B. S., Childers, W. E., Jr., Pinnick, H. W. Oxidation of α,β-unsaturated aldehydes. *Tetrahedron* **1981**, 37, 2091-2096.
5. Raach, A., Reiser, O. Sodium chlorite-hydrogen peroxide, a mild and selective reagent for the oxidation of aldehydes to carboxylic acids. *J. Prakt. Chem.* **2000**, 342, 605-608.
6. Dalcanale, E., Montanari, F. Selective oxidation of aldehydes to carboxylic acids with sodium chlorite-hydrogen peroxide. *J. Org. Chem.* **1986**, 51, 567-569.
7. Takemoto, T., Yasuda, K., Ley, S. V. Solid-supported reagents for the oxidation of aldehydes to carboxylic acids. *Synlett* **2001**, 1555-1556.
8. Fabian, I. The reactions of transition metal ions with chlorine(III). *Coord. Chem. Rev.* **2001**, 216-217, 449-472.
9. Smith, A. B., Kürti, L. *Unpublished results from the laboratory of Prof. A.B. Smith.*
10. Kudesia, V. P. Mechanism of chlorite oxidations. I. Kinetics of the oxidation of formaldehyde by chlorite ion. *Bull. Soc. Chim. Belg.* **1972**, 81, 623-628.
11. Overman, L. E., Paone, D. V. Enantioselective Total Syntheses of Ditryptophenaline and ent-WIN 64821. *J. Am. Chem. Soc.* **2001**, 123, 9465-9467.
12. Armstrong, A., Barsanti, P. A., Jones, L. H., Ahmed, G. Total Synthesis of (+)-Zaragozic Acid C. *J. Org. Chem.* **2000**, 65, 7020-7032.
13. Wong, L. S. M., Sherburn, M. S. IMDA-Radical Cyclization Approach to (+)-Himbacine. *Org. Lett.* **2003**, 5, 3603-3606.

1. Polonovski, M., Polonovski, M. Amine oxides of the alkaloids. III. Action of organic acid chlorides and anhydrides. Preparation of the nor bases. *Bull. soc. chim.* **1927**, 1190-1208.
2. Polonovski, M. The aminoxide function and its transpositions in the alkaloid group. *Bull. Soc. Chim. Belg.* **1930**, 39, 1-39.
3. Katritzky, A. R., Lagowski, J. M. *Chemistry of the Heterocyclic N-Oxides (Organic Chemistry, a Series of Monographs)* (Academic Press, New York, N. Y., **1971**) 599 pp.
4. Potier, P. Is the modified Polonovski reaction biomimetic? *Annu. Proc. Phytochem. Soc. Eur.* **1980**, 17, 159-169.
5. Koskinen, A. Regiospecific functionalization of carbon atoms α to heterocyclic nitrogen. *Ann. Acad. Sci. Fenn., Ser. A2* **1983**, 198, 20 pp.
6. Lounasmaa, M., Koskinen, A. Modified Polonovski reaction, a versatile synthetic tool. *Heterocycles* **1984**, 22, 1591-1612.
7. Potier, P. Chemistry of N-oxides. Further developments. *Lect. Heterocycl. Chem.* **1984**, 7, 59-62.
8. Grierson, D. The Polonovski reaction. *Org. React.* **1990**, 39, 85-295.
9. Grierson, D. S., Husson, H.-P. Polonovski- and Pummerer-type reactions and the Nef reaction. in *Comp. Org. Synth.* (eds. Trost, B. M.,Fleming, I.), *6*, 909-947 (Pergamon, Oxford, **1991**).
10. Cave, A., Kan-Fan, C., Potier, P., Le Men, J. Modification of the Polonovski reaction. Reaction of trifluoroacetic anhydride with an amine oxide. *Tetrahedron* **1967**, 23, 4681-4689.
11. Ferris, J. P., Gerwe, R. D., Gapski, G. R. Detoxication mechanisms. II. Iron-catalyzed dealkylation of trimethylamine oxide. *J. Am. Chem. Soc.* **1967**, 89, 5270-5275.
12. Edward, J. T., Whiting, J. Reactions of amine oxides and hydroxylamines with sulfur dioxide. *Can. J. Chem.* **1971**, 49, 3502-3514.
13. Grierson, D. S., Harris, M., Husson, H. P. Synthesis and chemistry of 5,6-dihydropyridinium salt adducts. Synthons for general electrophilic and nucleophilic substitution of the piperidine ring system. *J. Am. Chem. Soc.* **1980**, 102, 1064-1082.
14. Okazaki, R., Itoh, Y. Selenium Polonovski reaction using benzeneselenenyl triflate. *Chem. Lett.* **1987**, 1575-1578.
15. Tokito, N., Okazaki, R. Silicon Polonovskii reaction. Formation and synthetic application of α-siloxy amines. *Bull. Chem. Soc. Jpn.* **1987**, 60, 3291-3297.
16. Bonjoch, J., Casamitjana, N., Bosch, J. Functionalized 2-azabicyclo[3.3.1]nonanes. VIII. New synthesis of 5-phenylmorphans. *Tetrahedron* **1988**, 44, 1735-1741.
17. Tokitoh, N., Okazaki, R. A new method for deoxygenation of tertiary amine N-oxides with acetic formic anhydride. *Chem. Lett.* **1985**, 1517-1520.
18. Rosenau, T., Potthast, A., Ebner, G., Kosma, P. Deoxygenation of amine oxides by in-situ-generated formic pivalic anhydride. *Synlett* **1999**, 623-625.
19. Renaud, R. N., Leitch, L. C. Reinvestigation of the Polonovski reaction. Synthesis of deuterated dimethylamine and formaldehyde. *Can. J. Chem.* **1968**, 46, 385-390.
20. Hayashi, Y., Nagano, Y., Hongyo, S., Teramura, K. Trapping an intermediate of the Polonovski reaction. *Tetrahedron Lett.* **1974**, 1299-1302.
21. Jessop, R. A., Smith, J. R. L. Amine oxidation. Part XII. Reactions of some N,N-dimethylbenzylamine N-oxides with acetic anhydride and of some N-acetoxy-N,N-dimethylbenzylammonium perchlorates with acetate ion. The Polonovski reaction. *J. Chem. Soc., Perkin Trans. 1* **1976**, 1801-1805.
22. Manninen, K., Hakala, E. Trapping an intermediate with styrene and 2-phenylbicyclo[2.2.1]hept-2-ene in the Polonovski demethylation reaction. *Acta Chem. Scand.* **1986**, B40, 598-600.
23. Hunt, P. J. 125 pp (1992).
24. Sundberg, R. J., Gadamasetti, K. G., Hunt, P. J. Mechanistic aspects of the formation of anhydrovinblastine by Potier-Polonovski oxidative coupling of catharanthine and vindoline. Spectroscopic observation and chemical reactions of intermediates. *Tetrahedron* **1992**, 48, 277-296.
25. Morita, H., Kobayashi, J. i. A Biomimetic Transformation of Serratinine into Serratezomine A through a Modified Polonovski Reaction. *J. Org. Chem.* **2002**, 67, 5378-5381.
26. Shair, M. D., Yoon, T. Y., Mosny, K. K., Chou, T. C., Danishefsky, S. J. The Total Synthesis of Dynemicin A Leading to Development of a Fully Contained Bioreductively Activated Enediyne Prodrug. *J. Am. Chem. Soc.* **1996**, 118, 9509-9525.
27. Kende, A. S., Liu, K., Jos Brands, K. M. Total Synthesis of (-)-Altemicidin: A Novel Exploitation of the Potier-Polonovski Rearrangement. *J. Am. Chem. Soc.* **1995**, 117, 10597-10598.
28. Ziegler, F. E., Belema, M. Chiral Aziridinyl Radicals: An Application to the Synthesis of the Core Nucleus of FR-900482. *J. Org. Chem.* **1997**, 62, 1083-1094.

Related reactions: *Bischler-Napieralski isoquinoline synthesis, Pictet-Spengler tetrahydroisoquinoline synthesis;*

1. Fritsch, P. Syntheses in the isocoumarin and isoquinoline series. *Ber.* **1893**, 26, 419-422.
2. Pomeranz, C. A new isoquinoline synthesis. *Monatsh. Chem.* **1893**, 14, 116-119.
3. Pomeranz, C. The synthesis of isoquinoline and its derivatives. *Monatsh. Chem.* **1894**, 15, 299-306.
4. Fritsch, P. Synthesis of isoquinoline derivatives. *Ann* **1895**, 286, 1-17.

5. Pomeranz, C. The synthesis of isoquinolines and its derivatives. *Monatsh. Chem.* **1897**, 18, 1-5.
6. Gensler, W. J. Synthesis of isoquinolines by the Pomeranz-Fritsch Reaction. *Org. React.* **1951**, VI, 191-206.
7. Jones, G. Pyridines and their Benzo Derivatives: (v) Synthesis. in *Comprehensive Heterocyclic Chemistry* (eds. Katritzky, A. R.,Rees, C. W.), *2*, 395-510 (Pergamon Press, Oxford, **1984**).
8. Bobbitt, J. M., Bourque, A. J. Synthesis of heterocycles using aminoacetals. *Heterocycles* **1987**, 25, 601-616.
9. Rozwadowska, M. D. Recent progress in the enantioselective synthesis of isoquinoline alkaloids. *Heterocycles* **1994**, 39, 903-931.
10. Jones, G. Pyridines and their Benzo Derivatives: Synthesis. in *Comprehensive Organic Functional Group Transformations II* (eds. Katritzky, A. R., Rees, C. W.,Scriven, E. F. V.), *5*, 167-243 (Pergamon, Oxford, New York, **1995**).
11. Chrzanowska, M., Rozwadowska, M. D. Asymmetric Synthesis of Isoquinoline Alkaloids. *Chem. Rev.* **2004**, 104, 3341-3370.
12. Schlittler, E., Muller, J. A new modification of the isoquinoline synthesis according to Pomeranz-Fritsch. *Helv. Chim. Acta* **1948**, 31, 914-924.
13. Bobbitt, J. M., Kiely, J. M., Khanna, K. L., Ebermann, R. Synthesis of isoquinolines. III. A new synthesis of 1,2,3,4- tetrahydroisoquinolines. *J. Org. Chem.* **1965**, 30, 2247-2250.
14. Bevis, M. J., Forbes, E. J., Uff, B. C. Use of polyphosphoric acid in the Pomeranz-Fritsch synthesis of isoquinolines. *Tetrahedron* **1969**, 25, 1585-1589.
15. Bevis, M. J., Forbes, E. J., Naik, N. N., Uff, B. C. Synthesis of isoquinolines, indoles, and benzothiophene by an improved Pomeranz-Fritsch reaction, using boron trifluoride in trifluoroacetic anhydride. *Tetrahedron* **1971**, 27, 1253-1259.
16. Birch, A. J., Jackson, A. H., Shannon, P. V. R. New modification of the Pomeranz-Fritsch isoquinoline synthesis. *J. Chem. Soc., Perkin Trans. 1* **1974**, 2185-2190.
17. Katritzky, A. R., Yang, Z., Cundy, D. J. A mild and efficient synthesis of intermediates for the Pomeranz-Fritsch reaction. *Heteroatom Chem.* **1994**, 5, 103-106.
18. Kamochi, Y., Kudo, T. Cyclization of iminoacetals with lanthanide triflates as acid catalyst. *Kidorui* **1999**, 34, 304-305.
19. Gluszynska, A., Rozwadowska, M. D. Enantioselective addition of methyllithium to a prochiral imine-the substrate in the Pomeranz-Fritsch-Bobbitt synthesis of tetrahydroisoquinoline derivatives mediated by chiral monooxazolines. *Tetrahedron: Asymmetry* **2004**, 15, 3289-3295.
20. Brown, E. V. The Pomeranz-Fritsch reaction, isoquinoline vs. oxazoles. *J. Org. Chem.* **1977**, 42, 3208-3209.
21. Gluszynska, A., Rozwadowska, M. D. Enantioselective modification of the Pomeranz-Fritsch-Bobbitt synthesis of tetrahydroisoquinoline alkaloids synthesis of (-)-salsolidine and (-)-carnegine. *Tetrahedron: Asymmetry* **2000**, 11, 2359-2366.
22. Zhou, B., Guo, J., Danishefsky, S. J. Studies Directed to the Total Synthesis of ET 743 and Analogues Thereof: An Expeditious Route to the ABFGH Subunit. *Org. Lett.* **2002**, 4, 43-46.
23. Kunitomo, J., Miyata, Y., Oshikata, M. Studies on the alkaloids of menispermaceous plants. Part 285. Synthesis of dl-4-hydroxycrebanine. *Chem. Pharm. Bull.* **1985**, 33, 5245-5249.
24. Hirsenkorn, R. Short-cut in the Pomeranz-Fritsch synthesis of 1-benzylisoquinolines; short and efficient syntheses of norreticuline derivatives and of papaverine. *Tetrahedron Lett.* **1991**, 32, 1775-1778.

Prévost Reaction 360

Related reactions: Sharpless asymmetric dihydroxylation;

1. Prevost, C. Iodo-silver benzoate and its use in the oxidation of ethylene derivatives into α-glycols. *Compt. rend.* **1933**, 196, 1129-1131.
2. Prevost, C. Silver halide complexes of the carboxylic acids. *Compt. rend.* **1933**, 197, 1661-1663.
3. Prevost, C., Wiemann, J. The iodinating properties of a complex iodo-silver benzoate. *Compt. rend.* **1937**, 204, 700-701.
4. Wilson, C. V. The reaction of halogens with silver salts of carboxylic acids. *Org. React.* **1957**, 332-387.
5. Grewal, G. S. Wet and dry Prevost reactions of cyclopentene. *Journal of Research (Punjab Agricultural University)* **1977**, 14, 468-472.
6. Rodriguez, J., Dulcere, J. P. Cohalogenation in organic synthesis. *Synthesis* **1993**, 1177-1205.
7. Vaino, A. R., Szarek, W. A. Iodine in carbohydrate chemistry. *Adv. Carbohydr. Chem. Biochem.* **2001**, 56, 9-63.
8. Woodward, R. B., Brutcher, F. V., Jr. Cis hydroxylation of a synthetic steroid intermediate with iodine, silver acetate, and wet acetic acid. *J. Am. Chem. Soc.* **1958**, 80, 209-211.
9. Cambie, R. C., Hayward, R. C., Roberts, J. L., Rutledge, P. S. Iodolactonizations using thallium(I) carboxylates. *J. Chem. Soc., Perkin Trans. 1* **1974**, 1864-1867.
10. Cambie, R. C., Hayward, R. C., Roberts, J. L., Rutledge, P. S. Reactions of thallium(I) carboxylates and iodine with alkenes. *J. Chem. Soc., Perkin Trans. 1* **1974**, 1858-1864.
11. Cambie, R. C., Rutledge, P. S. Stereoselective hydroxylation with thallium(I) acetate and iodine: trans- and cis-1,2-cyclohexanediols. *Org. Synth.* **1980**, 59, 169-176.
12. Campi, E. M., Deacon, G. B., Edwards, G. L., Fitzroy, M. D., Giunta, N., Jackson, W. R., Trainor, R. Bismuth(III) acetate: a cheap, efficient, and environmentally acceptable reagent for wet and dry Prevost reactions. *J. Chem. Soc., Chem. Commun.* **1989**, 407-408.
13. Trainor, R. W., Deacon, G. B., Jackson, W. R., Giunta, N. The use of bismuth(III) acetate in 'wet' and 'dry' Prevost reactions. *Aust. J. Chem.* **1992**, 45, 1265-1280.
14. Hamm, S., Hennig, L., Findeisen, M., Muller, D., Welzel, P. Submission of some iodoformates to Woodward-Prevost conditions. *Tetrahedron* **2000**, 56, 1345-1348.
15. Iranpoor, N., Shekarriz, M. Regioselective 1,2-alkoxy, hydroxy, and acetoxy iodination of alkenes with I_2 catalyzed by $Ce(SO_3CF_3)_4$. *Tetrahedron* **2000**, 56, 5209-5211.
16. Myint, Y. Y., Pasha, M. A. Preparation of α-iodoacetates from alkenes by $Co(OAc)_2$ catalysed Woodward-Prevost reaction. *Indian J. Chem., Sect. B* **2004**, 43B, 590-592.
17. Winstein, S., Buckles, R. E. Role of neighboring groups in replacement reactions. II. The effects of small amounts of water on the reaction of silver acetate in acetic acid with some butene and cyclohexene derivatives. *J. Am. Chem. Soc.* **1942**, 64, 2787-2790.
18. Winstein, S., Buckles, R. E. Role of neighboring groups in replacement reactions. I. Retention of configuration in the reaction of some dihalides and acetoxyhalides with silver acetate. *J. Am. Chem. Soc.* **1942**, 64, 2780-2786.
19. Wiberg, K. B., Saegebarth, K. A. An oxygen-18 tracer study of the "wet" and "dry" Prevost reactions. *J. Am. Chem. Soc.* **1957**, 79, 6256-6261.
20. Briggs, L. H., Cain, B. F., Davis, B. R. Novel Prevost reaction. *Tetrahedron Letters* **1960**, 9-11.
21. Kumar, S., Kole, P. L., Sehgal, R. K. Synthesis of the phenolic derivatives of highly tumorigenic trans-7,8-dihydroxy-7,8-dihydrobenzo[a]pyrene. *J. Org. Chem.* **1989**, 54, 5272-5277.
22. Sabat, M., Johnson, C. R. Synthesis of (2R,4R)- and (2S,4S)-4-hydroxypipecolic acid derivatives and (2S,4S)-(-)-SS20846A. *Tetrahedron Lett.* **2001**, 42, 1209-1212.
23. Germain, J., Deslongchamps, P. Total Synthesis of (±)-Momilactone A. *J. Org. Chem.* **2002**, 67, 5269-5278.
24. Lansbury, P. T., Nickson, T. E., Vacca, J. P., Sindelar, R. D., Messinger, J. M., II. Total synthesis of pseudoguaianolides. V. Stereocontrolled approaches to the fastigilins: (±)-2,3-dihydrofastigilin C. *Tetrahedron* **1987**, 43, 5583-5592.

Prilezhaev Reaction 362

Related reactions: Davis oxaziridine oxidation, Jacobsen-Katsuki epoxidation, Sharpless asymmetric epoxidation, Shi asymmetric epoxidation;

1. Prilezhaev, N. Oxidation of Unsaturated Compounds by Means of Organic Peroxides. *Ber.* **1909**, 42, 4811-4815.

2. Prilezhaev, N. Oxidation of Unsaturated Compounds by Organic Peroxides. III. *Zhurnal Russkago Fiziko-Khimicheskago Obshchestva* **1912**, 44, 613-647.
3. Prilezhaev, N. Oxidation of Unsaturated Compounds with Organic Peroxides. II. Oxidation of Derivatives of Unsaturated Hydrocarbons with One Double Union. *Zhurnal Russkago Fiziko-Khimicheskago Obshchestva* **1912**, 43, 609-620.
4. Swern, D. Organic peracids. *Chem. Rev.* **1949**, 45, 1-68.
5. Swern, D. Epoxidation and hydroxylation of ethylenic compounds with organic peracids. *Org. React.* **1953**, VII, 378-433.
6. Lewis, S. N. Peracid and peroxide oxidations. *Oxidation* **1969**, 1, 213-258.
7. Berti, G. Stereochemical aspects of the synthesis of 1,2-epoxides. *Top. Stereochem.* **1973**, 7, 93-251.
8. Plesnicar, B. Oxidations with peroxy acids and other peroxides. in *Oxidation in Organic Chemistry* (ed. Trahanovsky, W. S.), *5*, 211-294 (Academic Press, New York, **1978**).
9. Plesnicar, B. Polar reaction mechanisms involving peroxides in solution. in *The Chemistry of Peroxides* (ed. Patai, S.), 521-584 (Wiley, New York, **1983**).
10. Dryuk, V. G. Advances in the development of methods for the epoxidation of olefins. *Russ. Chem. Rev.* **1985**, 54, 1674-1705.
11. Rao, A. S. Addition Reactions with Formation of Carbon-Oxygen Bonds: General Methods of Epoxidation. in *Comp. Org. Synth.* (eds. Trost, B. M.,Fleming, I.), *7*, 357-387 (Pergamon, Oxford, **1991**).
12. Dryuk, V. G., Kartsev, V. G. Mechanism of the directing influence of functional groups and the geometry of reactant molecules on peroxide epoxidation of alkenes. *Russian Chemical Reviews* **1999**, 68, 183-201.
13. Payne, G. B. Simplified procedure for epoxidation by benzonitrile-hydrogen peroxide. Selective oxidation of 2-allylcyclo-hexanone. *Tetrahedron* **1962**, 18, 763-765.
14. Rebek, J., Jr. Progress in the development of new epoxidation reagents. *Heterocycles* **1981**, 15, 517-545.
15. Sawaki, Y., Ogata, Y. Photoepoxidation of olefins with benzoins and oxygen. Epoxidation with acylperoxy radical. *J. Am. Chem. Soc.* **1981**, 103, 2049-2053.
16. Brougham, P., Cooper, M. S., Cummerson, D. A., Heaney, H., Thompson, N. Oxidation reactions using magnesium monoperphthalate: a comparison with m-chloroperoxybenzoic acid. *Synthesis* **1987**, 1015-1017.
17. Adam, W., Curci, R., Edwards, J. O. Dioxiranes: a new class of powerful oxidants. *Acc. Chem. Res.* **1989**, 22, 205-211.
18. Majetich, G., Hicks, R., Sun, G.-r., McGill, P. Carbodiimide-promoted olefin epoxidation with aqueous hydrogen peroxide. *J. Org. Chem.* **1998**, 63, 2564-2573.
19. Ueno, S., Yamaguchi, K., Yoshida, K., Ebitani, K., Kaneda, K. Hydrotalcite catalysis: heterogeneous epoxidation of olefins using hydrogen peroxide in the presence of nitriles. *Chem. Commun.* **1998**, 295-296.
20. Iwahama, T., Sakaguchi, S., Ishii, Y. Epoxidation of alkenes using dioxygen in the presence of an alcohol catalyzed by N-hydroxyphthalimide and hexafluoroacetone without any metal catalyst. *Chem. Commun.* **1999**, 727-728.
21. Klaas, M. R., Warwel, S. Chemoenzymatic Epoxidation of Alkenes by Dimethyl Carbonate and Hydrogen Peroxide. *Org. Lett.* **1999**, 1, 1025-1026.
22. Mohajer, D., Tayebee, R., Goudarziafshar, H. Sodium Periodate Epoxidation of Alkenes Catalyzed by Manganese Porphyrins. *J. Chem. Res., Synop.* **1999**, 168-169.
23. Adam, W., Saha-Moeller, C. R., Zhao, C.-G. Dioxirane epoxidation of alkenes. *Org. React.* **2002**, 61, 219-516.
24. Yonezawa, T., Kato, H., Yamamoto, O. A molecular orbital investigation of the electronic structures of alkyl peroxides. II. Peracids and peresters. *Bull. Chem. Soc. Jpn.* **1967**, 40, 307-311.
25. Bach, R. D., Owensby, A. L., Gonzalez, C., Schlegel, H. B., McDouall, J. J. W. Transition structure for the epoxidation of alkenes with peroxy acids. A theoretical study. *J. Am. Chem. Soc.* **1991**, 113, 2338-2339.
26. Warmerdam, E. G. J. C., van den Nieuwendijk, A. M. C. H., Brussee, J., Kruse, C. G., van der Gen, A. Asymmetric epoxidation of chiral allylic alcohols. *Tetrahedron: Asymmetry* **1996**, 7, 2539-2550.
27. Yamabe, S., Kondou, C., Minato, T. A Theoretical Study of the Epoxidation of Olefins by Peracids. *J. Org. Chem.* **1996**, 61, 616-620.
28. Marchand, A. P., Ganguly, B., Shukla, R., Krishnudu, K., Kumar, V. S., Watson, W. H., Bodige, S. G. Experimental and theoretical investigations into the stereoselectivities of peracid promoted epoxidations of substituted norbornenes and norbornadienes. *Tetrahedron* **1999**, 55, 8313-8322.
29. Adam, W., Bach, R. D., Dmitrenko, O., Saha-Moeller, C. R. A Computational Study of the Hydroxy-Group Directivity in the Peroxyformic Acid Epoxidation of the Chiral Allylic Alcohol (Z)-3-Methyl-3-penten-2-ol: Control of Threo Diastereoselectivity through Allylic Strain and Hydrogen Bonding. *J. Org. Chem.* **2000**, 65, 6715-6728.
30. Freccero, M., Gandolfi, R., Sarzi-Amade, M., Rastelli, A. Facial Selectivity in Epoxidation of 2-Cyclohexen-1-ol with Peroxy Acids. A Computational DFT Study. *J. Org. Chem.* **2000**, 65, 8948-8959.
31. Washington, I., Houk, K. N. CH...O hydrogen bonding influences π-facial stereoselective epoxidations. *Angew. Chem., Int. Ed. Engl.* **2001**, 40, 4485-4488.
32. Bach, R. D., Dmitrenko, O. Spiro versus Planar Transition Structures in the Epoxidation of Simple Alkenes. A Reassessment of the Level of Theory Required. *J. Phys. Chem. A* **2003**, 107, 4300-4306.
33. Bach, R. D., Dmitrenko, O., Adam, W., Schambony, S. Relative Reactivity of Peracids versus Dioxiranes (DMDO and TFDO) in the Epoxidation of Alkenes. A Combined Experimental and Theoretical Analysis. *J. Am. Chem. Soc.* **2003**, 125, 924-934.
34. Hoveyda, A. H., Evans, D. A., Fu, G. C. Substrate-directable chemical reactions. *Chem. Rev.* **1993**, 93, 1307-1370.
35. Joergensen, K. A. Transition-metal-catalyzed epoxidations. *Chem. Rev.* **1989**, 89, 431-458.
36. Kwart, H., Starcher, P. S., Tinsley, S. W. Mechanism of epoxidation of olefins and further possible applications of the 1,3-dipolar mechanism of oxidations with peroxy acids. *Chem. Commun.* **1967**, 335-337.
37. Hanzlik, R. P., Shearer, G. O. Transition state structure for peracid epoxidation. Secondary deuterium isotope effects. *J. Am. Chem. Soc.* **1975**, 97, 5231-5233.
38. Dryuk, V. G. The mechanism of epoxidation of olefins by peracids. *Tetrahedron* **1976**, 32, 2855-2866.
39. Roof, A. A. M., Winter, W. J., Bartlett, P. D. syn- and anti-Sesquinorbornenes as mechanistic probes in reactions of the carbon-carbon double bond. *J. Org. Chem.* **1985**, 50, 4093-4098.
40. Cieplak, A. S., Tait, B. D., Johnson, C. R. Reversal of π-facial diastereoselection upon electronegative substitution of the substrate and the reagent. *J. Am. Chem. Soc.* **1989**, 111, 8447-8462.
41. Woods, K. W., Beak, P. The endocyclic restriction test: an experimental evaluation of the geometry at oxygen in the transition structure for epoxidation of an alkene by a peroxy acid. *J. Am. Chem. Soc.* **1991**, 113, 6281-6283.
42. Shea, K. J., Kim, J. S. Influence of strain on chemical reactivity. Relative reactivity of torsionally distorted double bonds in MCPBA epoxidations. *J. Am. Chem. Soc.* **1992**, 114, 3044-3051.
43. Angelis, Y. S., Orfanopoulos, M. A Reinvestigation of the Structure of the Transition State in Peracid Epoxidations. α- and β-Secondary Isotope Effects. *J. Org. Chem.* **1997**, 62, 6083-6085.
44. Fehr, C. Diastereoface-selective epoxidations: dependency on the reagent electrophilicity. *Angew. Chem., Int. Ed. Engl.* **1998**, 37, 2407-2409.
45. Adam, W., Mitchell, C. M., Saha-Moeller, C. R. Regio- and Diastereoselective Catalytic Epoxidation of Acyclic Allylic Alcohols with Methyltrioxorhenium: A Mechanistic Comparison with Metal (Peroxy and Peroxo Complexes) and Nonmetal (Peracids and Dioxirane) Oxidants. *J. Org. Chem.* **1999**, 64, 3699-3707.
46. Beak, P., Anderson, D. R., Jarboe, S. G., Kurtzweil, M. L., Woods, K. W. Mechanisms and consequences of oxygen transfer reactions. *Pure Appl. Chem.* **2000**, 72, 2259-2264.
47. Kas'yan, L. I., Okovityi, S. I., Bomushkar, M. F., Driuk, V. G. Epoxidation of stereoisomeric substituted norbornenes. Kinetic and theoretical investigation. *Russ. J. Org. Chem.* **2000**, 36, 195-202.
48. Smith, A. B., Cui, H. Total Synthesis of (-)-21-Isopentenylpaxilline. *Org. Lett.* **2003**, 5, 587-590.
49. Corminboeuf, O., Overman, L. E., Pennington, L. D. Enantioselective Total Synthesis of Briarellins E and F: The First Total Syntheses of Briarellin Diterpenes. *J. Am. Chem. Soc.* **2003**, 125, 6650-6652.

50. Nishikawa, T., Asai, M., Isobe, M. Asymmetric Total Synthesis of 11-Deoxytetrodotoxin, a Naturally Occurring Congener. *J. Am. Chem. Soc.* **2002**, 124, 7847-7852.
51. Mulzer, J., Mantoulidis, A., Oehler, E. Total Syntheses of Epothilones B and D. *J. Org. Chem.* **2000**, 65, 7456-7467.
52. Taylor, R. E., Chen, Y. Total Synthesis of Epothilones B and D. *Org. Lett.* **2001**, 3, 2221-2224.

Related reactions: Alder (ene) reaction;

1. Kriewitz, O. Additive compounds of formaldehyde with terpenes. *Ber.* **1899**, 32, 57-60.
2. Kriewitz, O. Additive compounds of formaldehyde with terpenes. *J. Chem. Soc.* **1899**, 76, 298.
3. Prins, H. J. The reciprocal condensation of unsaturated organic compounds. *Chem. Weekblad* **1919**, 16, 1510-1526.
4. Prins, H. J. Condensation of formaldehyde with some unsaturated compounds. *Chem. Weekblad* **1919**, 16, 1072-1073.
5. Arundale, E., Mikeska, L. A. The olefin-aldehyde condensation. The Prins reaction. *Chem. Rev.* **1952**, 51, 505-555.
6. Adams, D. R., Bhatnagar, S. P. The Prins reaction. *Synthesis* **1977**, 661-672.
7. Delmas, M., Gaset, A. Supported acid catalysis with ion-exchange resins. II. Mechanism of the condensation reaction between aqueous formaldehyde and aromatic alkenes. *J. Mol. Catal.* **1982**, 14, 269-282.
8. Snider, B. B. The Prins and Carbonyl Ene Reactions. in *Comp. Org. Synth.* (eds. Trost, B. M.,Fleming, I.), *2*, 527-561 (Pergamon, Oxford, **1991**).
9. Maruoka, K., Hoshino, Y., Shirasaka, T., Yamamoto, H. Asymmetric ene reaction catalyzed by chiral organoaluminum reagent. *Tetrahedron Lett.* **1988**, 29, 3967-3970.
10. Yang, J., Viswanathan, G. S., Li, C.-J. Highly effective synthesis of 4-halotetrahydropyrans via a highly diastereoselective in situ Prins-type cyclization reaction. *Tetrahedron Lett.* **1999**, 40, 1627-1630.
11. Zhang, W.-C., Viswanathan, G. S., Li, C.-J. Scandium triflate catalyzed in situ Prins-type cyclization: formations of 4-tetrahydropyranols and ethers. *Chem. Commun.* **1999**, 291-292.
12. Bach, T., Lobel, J. Selective Prins reaction of styrenes and formaldehyde catalyzed by 2,6-di-tert-butylphenoxy(difluoro)borane. *Synthesis* **2002**, 2521-2526.
13. Keh, C. C. K., Li, C.-J. Direct formation of 2,4-disubstituted tetrahydropyranols in water mediated by an acidic solid resin. *Green Chem.* **2003**, 5, 80-81.
14. Yadav, J. S., Reddy, B. V. S., Bhaishya, G. $InBr_3$-[bmim]PF_6: a novel and recyclable catalytic system for the synthesis of 1,3-dioxane derivatives. *Green Chem.* **2003**, 5, 264-266.
15. Meresz, O., Leung, K. P., Denes, A. S. Intermediacy of oxetanes in the Prins reaction. *Tetrahedron Lett.* **1972**, 2797-2800.
16. Dolby, L. J., Schwarz, M. J. The mechanism of the Prins reaction. IV. Evidence against acetoxonium ion intermediates. *J. Org. Chem.* **1965**, 30, 3581-3586.
17. Smissman, E. E., Schnettler, R. A., Portoghese, P. S. Mechanism of the Prins reaction. Stereoaspects of the formation of 1,3-dioxanes. *J. Org. Chem.* **1965**, 30, 797-801.
18. Dolby, L. J., Wilkins, C., Frey, T. G. The mechanism of the Prins reaction. V. The Prins reaction of styrenes. *J. Org. Chem.* **1966**, 31, 1110-1116.
19. Kovacs, O., Kovari, I. Chemistry of 1,3-diols. IV. Stereochemistry of the Prins reaction of 4-tert-butylcyclohexene. *Acta Chim. Acad. Sci. Hung.* **1966**, 48, 147-160.
20. Watanabe, S. Thermal Prins reaction. II. The mechanism of the thermal Prins reaction. *Kogakubu Kenkyu Hokoku (Chiba Daigaku)* **1966**, 17, 17-21.
21. Dolby, L. J., Meneghini, F. A., Koizumi, T. The mechanism of the Prins reaction. VI. The solvolysis of optically active trans-2-hydroxymethylcyclohexyl brosylate and related arenesulfonates. *J. Org. Chem.* **1968**, 33, 3060-3066.
22. Dolby, L. J., Wilkins, C. L., Rodia, R. M. Mechanism of the Prins reaction. VII. Kinetic studies of the Prins reaction of styrenes. *J. Org. Chem.* **1968**, 33, 4155-4158.
23. Schowen, K. B., Smissman, E. E., Schowen, R. L. Mechanism of the Prins reaction. Kinetics and product composition in acetic acid. *J. Org. Chem.* **1968**, 33, 1873-1876.
24. Wilkins, C. L., Marianelli, R. S., Pickett, C. S. Stereochemical applications of deuterium magnetic resonance. I. Prins reaction of trans-styrene-β-d. *Tetrahedron Lett.* **1968**, 5109-5112.
25. Wilkins, C. L., Marianelli, R. S. Mechanism of the Prins reaction of styrenes. Prins reaction of trans-β-deuterostyrene. *Tetrahedron* **1970**, 26, 4131-4138.
26. Dai, Q., Liu, R., Li, Y. Research of stereochemistry on the Prins reaction of cyclohexene. *Chin. Sci. Bull.* **1989**, 34, 2045-2049.
27. Dumitriu, E., On, D. T., Kaliaguine, S. Isoprene by Prins condensation over acidic molecular sieves. *J. Catal.* **1997**, 170, 150-160.
28. Dumitriu, E., Hulea, V., Fechete, I., Catrinescu, C., Auroux, A., Lacaze, J.-F., Guimon, C. Prins condensation of isobutylene and formaldehyde over Fe-silicates of MFI structure. *Appl. Cat. A* **1999**, 181, 15-28.
29. Kocovsky, P., Ahmed, G., Srogl, J., Malkov, A. V., Steele, J. New Lewis-Acidic Molybdenum(II) and Tungsten(II) Catalysts for Intramolecular Carbonyl Ene and Prins Reactions. Reversal of the Stereoselectivity of Cyclization of Citronellal. *J. Org. Chem.* **1999**, 64, 2765-2775.
30. Toro, A., L'Heureux, A., Deslongchamps, P. Transannular Diels-Alder Studies on the Asymmetric Total Synthesis of Chatancin: The Pyranophane Approach. *Org. Lett.* **2000**, 2, 2737-2740.
31. Kopecky, D. J., Rychnovsky, S. D. Mukaiyama Aldol-Prins Cyclization Cascade Reaction: A Formal Total Synthesis of Leucascandrolide A. *J. Am. Chem. Soc.* **2001**, 123, 8420-8421.
32. Marumoto, S., Jaber, J. J., Vitale, J. P., Rychnovsky, S. D. Synthesis of (-)-Centrolobine by Prins Cyclizations that Avoid Racemization. *Org. Lett.* **2002**, 4, 3919-3922.
33. Welch, S. C., Chou, C., Gruber, J. M., Assercq, J. M. Total syntheses of (±)-seychellene, (±)-isocycloseychellene, and (±)-isoseychellene. *J. Org. Chem.* **1985**, 50, 2668-2676.

1. Martinet, P., Mousset, G., Colineau, M. Activated montmorillonite as a catalyst in the synthesis of cyclic acetals. Evidence of side reactions. *C. R. Seances Acad. Sci. C.* **1969**, 268, 1303-1306.
2. Martinet, P., Mousset, G. Isomerization of cyclic acetals. I. Stereochemical influences on the participation of ethylenic systems. *Bull. Soc. Chim. Fr.* **1970**, 1071-1076.
3. Hopkins, M. H., Overman, L. E. Stereocontrolled preparation of tetrahydrofurans by acid-catalyzed rearrangement of allylic acetals. *J. Am. Chem. Soc.* **1987**, 109, 4748-4749.
4. Brown, M. J., Harrison, T., Herrinton, P. M., Hopkins, M. H., Hutchinson, K. D., Overman, L. E., Mishra, P. Acid-promoted reaction of cyclic allylic diols with carbonyl compounds. Stereoselective ring-enlarging tetrahydrofuran annulations. *J. Am. Chem. Soc.* **1991**, 113, 5365-5378.
5. Brown, M. J., Harrison, T., Overman, L. E. General approach to halogenated tetrahydrofuran natural products from red algae of the genus Laurencia. Total synthesis of (±)-trans-kumausyne and demonstration of an asymmetric synthesis strategy. *J. Am. Chem. Soc.* **1991**, 113, 5378-5384.
6. Hopkins, M. H., Overman, L. E., Rishton, G. M. Stereocontrolled preparation of tetrahydrofurans from acid-promoted rearrangements of allylic acetals. *J. Am. Chem. Soc.* **1991**, 113, 5354-5365.

7. Overman, L. E. Charge as a key component in reaction design. The invention of cationic cyclization reactions of importance in synthesis. *Acc. Chem. Res.* **1992**, 25, 352-359.
8. Overman, L. E. New reactions for forming heterocycles and their use in natural products synthesis. *Aldrichimica Acta* **1995**, 28, 107-120.
9. Overman, L. E., Pennington, L. D. Strategic Use of Pinacol-Terminated Prins Cyclizations in Target-Oriented Total Synthesis. *J. Org. Chem.* **2003**, 68, 7143-7157.
10. Ando, S., Minor, K. P., Overman, L. E. Ring-Enlarging Cyclohexane Annulations. *J. Org. Chem.* **1997**, 62, 6379-6387.
11. Minor, K. P., Overman, L. E. Prins-pinacol spiroannulations. *Tetrahedron* **1997**, 53, 8927-8940.
12. Cloninger, M. J., Overman, L. E. Stereocontrolled Synthesis of Trisubstituted Tetrahydropyrans. *J. Am. Chem. Soc.* **1999**, 121, 1092-1093.
13. Gahman, T. C., Overman, L. E. Stereoselective synthesis of carbocyclic ring systems by pinacol-terminated Prins cyclizations. *Tetrahedron* **2002**, 58, 6473-6483.
14. Burke, B. J., Lebsack, A. D., Overman, L. E. Scope and limitations of the thionium ion-initiated prins-pinacol synthesis of carbocycles. *Synlett* **2004**, 1387-1393.
15. Hanaki, N., Link, J. T., MacMillan, D. W. C., Overman, L. E., Trankle, W. G., Wurster, J. A. Stereoselection in the Prins-Pinacol Synthesis of 2,2-Disubstituted 4-Acyltetrahydrofurans. Enantioselective Synthesis of (-)-Citreoviral. *Org. Lett.* **2000**, 2, 223-226.
16. Cohen, F., MacMillan, D. W. C., Overman, L. E., Romero, A. Stereoselection in the Prins-Pinacol Synthesis of Acyltetrahydrofurans. *Org. Lett.* **2001**, 3, 1225-1228.
17. Corminboeuf, O., Overman, L. E., Pennington, L. D. Enantioselective Total Synthesis of Briarellins E and F: The First Total Syntheses of Briarellin Diterpenes. *J. Am. Chem. Soc.* **2003**, 125, 6650-6652.
18. Hirst, G. C., Johnson, T. O., Jr., Overman, L. E. First total synthesis of Lycopodium alkaloids of the magellanane group. Enantioselective total syntheses of (-)-magellanine and (+)-magellaninone. *J. Am. Chem. Soc.* **1993**, 115, 2992-2993.
19. Lebsack, A. D., Overman, L. E., Valentekovich, R. J. Enantioselective Total Synthesis of Shahamin K. *J. Am. Chem. Soc.* **2001**, 123, 4851-4852.

Pummerer Rearrangement......368

1. Pummerer, R. Phenylsulphoxylacetic Acid. *Ber.* **1909**, 42, 2282-2291.
2. Pummerer, R. Phenylsulphoxyacetic Acid. II. *Ber.* **1910**, 43, 1401-1412.
3. Johnson, C. R. Current chemistry of sulfoxides. *Quart. Rep. Sulfur Chem.* **1968**, 3, 91-94.
4. Russell, G. A., Mikol, G. J. Acid-catalyzed rearrangements of sulfoxides and amine oxides. The Pummerer and Polonovski reactions. *Mech. Mol. Migr.* **1968**, 1, 157-207.
5. Schneller, S. W. Name reactions in sulfur chemistry. II. *International Journal of Sulfur Chemistry* **1973**, 8, 485-503.
6. Schneller, S. W. Name reactions in sulfur chemistry. Part II. *International Journal of Sulfur Chemistry* **1976**, 8, 579-597.
7. Tillett, J. G. Nucleophilic substitution at tricoordinate sulfur. *Chem. Rev.* **1976**, 76, 747-772.
8. Oae, S. The Pummerer rearrangements. *Top Org. Sulphur Chem., Plenary Lect. Int. Symp., 8th* **1978**, 289-336.
9. Oae, S., Numata, T. The Pummerer and Pummerer type of reactions. *Isotopes in Organic Chemistry* **1980**, 5, 45-102.
10. Warren, S. Phosphorus and sulfur reagents in synthesis. *Chem. Ind. (London)* **1980**, 824-828.
11. Oae, S. Several unsolved problems of chemical behavior of the sulfur atom in organic sulfur chemistry. *THEOCHEM* **1989**, 55, 321-345.
12. De Lucchi, O., Miotti, U., Modena, G. The Pummerer reaction of sulfinyl compounds. *Org. React.* **1991**, 40, 157-405.
13. Grierson, D. S., Husson, H.-P. Polonovski- and Pummerer-type reactions and the Nef reaction. in *Comp. Org. Synth.* (eds. Trost, B. M.,Fleming, I.), *6*, 909-947 (Pergamon, Oxford, **1991**).
14. Carreno, M. C. Applications of Sulfoxides to Asymmetric Synthesis of Biologically Active Compounds. *Chem. Rev.* **1995**, 95, 1717-1760.
15. Kita, Y., Shibata, N. Asymmetric Pummerer-type reactions induced by O-silylated ketene acetals. *Synlett* **1996**, 289-296.
16. Kita, Y. Some recent advances in Pummerer-type reactions. *Phosphorus, Sulfur and Silicon and the Related Elements* **1997**, 120 & 121, 145-164.
17. Padwa, A., Gunn, D. E., Jr., Osterhout, M. H. Application of the Pummerer reaction toward the synthesis of complex carbocycles and heterocycles. *Synthesis* **1997**, 1353-1377.
18. Furukawa, N. Creation of organo-sulfur, -selenium- and - tellurium multi-cation species. *Phosphorus, Sulfur and Silicon and the Related Elements* **1998**, 136,137&138, 43-58.
19. Padwa, A. Heterocyclic synthesis using the Pummerer reaction. *Phosphorus, Sulfur Silicon Relat. Elem.* **1999**, 153-154, 23-40.
20. Padwa, A., Waterson, A. G. Synthesis of nitrogen heterocycles using the intramolecular Pummerer reaction. *Curr. Org. Chem.* **2000**, 4, 175-203.
21. Murray, A. W. Molecular rearrangements. *Organic Reaction Mechanisms* **2001**, 473-603.
22. Prilezhaeva, E. N. Rearrangements of sulfoxides and sulfones in the total synthesis of natural compounds. *Russian Chemical Reviews* **2001**, 70, 897-920.
23. Padwa, A., Bur, S. K., Danca, D. M., Ginn, J. D., Lynch, S. M. Linked Pummerer-Mannich ion cyclizations for heterocyclic chemistry. *Synlett* **2002**, 851-862.
24. Wang, C.-C., Huang, H.-C., Reitz, D. B. New developments in the use of enantiomerically enriched sulfoxides in stereoselective syntheses. *Org. Prep. Proced. Int.* **2002**, 34, 271-319.
25. Bur, S. K., Padwa, A. The Pummerer Reaction: Methodology and Strategy for the Synthesis of Heterocyclic Compounds. *Chem. Rev.* **2004**, 104, 2401-2432.
26. Sharma, A. K., Swern, D. Trifluoroacetic anhydride. New activating agent for dimethyl sulfoxide in the synthesis of iminosulfuranes. *Tetrahedron Lett.* **1974**, 15, 1503-1506.
27. Numata, T., Ito, O., Oae, S. Unusually high asymmetric induction in the Pummerer reaction of optically active sulfoxides. *Tetrahedron Lett.* **1979**, 1869-1870.
28. Kita, Y., Shibata, N., Yoshida, N., Fujita, S. First highly asymmetric Pummerer-type reaction in chiral, non-racemic acyclic sulfoxides induced by O-silylated ketene acetal. *J. Chem. Soc., Perkin Trans. 1* **1994**, 3335-3341.
29. Kita, Y., Shibata, N., Kawano, N., Tohjo, T., Fujimori, C., Matsumoto, K., Fujita, S. Highly asymmetric Pummerer-type cyclization of chiral, non-racemic β-amido sulfoxides. *J. Chem. Soc., Perkin Trans. 1* **1995**, 2405-2410.
30. Shibata, N., Matsugi, M., Kawano, N., Fukui, S., Fujimori, C., Gotanda, K., Murata, K., Kita, Y. Highly asymmetric Pummerer-type reaction induced by ethoxyvinyl esters. *Tetrahedron: Asymmetry* **1997**, 8, 303-310.
31. Padwa, A., Kuethe, J. T. Additive and Vinylogous Pummerer Reactions of Amido Sulfoxides and Their Use in the Preparation of Nitrogen Containing Heterocycles. *J. Org. Chem.* **1998**, 63, 4256-4268.
32. Ruano, J. L. G., Paredes, C. G. Intramolecular asymmetric Pummerer reactions as a key step in the synthesis of bicyclic precursors of anthracyclinones. *Tetrahedron Lett.* **1999**, 41, 261-265.
33. Solladie, G., Wilb, N., Bauder, C. Highly stereoselective synthesis of enantiomerically pure β-hydroxy-γ-sulfenyl-γ-butyrolactone by asymmetric Pummerer type cyclization. *Tetrahedron Lett.* **2000**, 41, 4189-4192.
34. McAllister, L. A., Brand, S., de Gentile, R., Procter, D. J. The first Pummerer cyclizations on solid phase. Convenient construction of oxindoles enabled by a sulfur-link to resin. *Chem. Commun.* **2003**, 2380-2381.
35. Feldman, K. S., Vidulova, D. B. Extending Pummerer Reaction Chemistry. Application to the Oxidative Cyclization of Indole Derivatives. *Org. Lett.* **2004**, 6, 1869-1871.
36. Ruano, J. L. G., Aleman, J., Padwa, A. A Novel Asymmetric Vinylogous Tin-Pummerer Rearrangement. *Org. Lett.* **2004**, 6, 1757-1760.
37. Ginsburg, J. L., Darvesh, K. V., Axworthy, P., Langler, R. F. An ab initio molecular orbital study of sulfur-substituted carbanions: toward an understanding of regiochemistry in the chlorination of unsymmetrical sulfides. *Aust. J. Chem.* **1997**, 50, 517-521.
38. Horner, L., Kaiser, P. The course of substitution. XVIII. The reaction of carboxylic acid anhydrides with sulfoxides. *Ann.* **1959**, 626, 19-25.
39. Brook, A. G. Molecular rearrangements of organosilicon compounds. *Acc. Chem. Res.* **1974**, 7, 77-84.

40. Colobert, F., Tito, A., Khiar, N., Denni, D., Medina, M. A., Martin-Lomas, M., Ruano, J.-L. G., Solladie, G. Enantioselective Approach to Polyhydroxylated Compounds Using Chiral Sulfoxides: Synthesis of Enantiomerically Pure myo-Inositol and Pyrrolidine Derivatives. *J. Org. Chem.* **1998**, 63, 8918-8921.
41. Roush, W. R., Limberakis, C., Kunz, R. K., Barda, D. A. Diastereoselective Synthesis of the endo- and exo-Spirotetronate Subunits of the Quartromicins. The First Enantioselective Diels-Alder Reaction of an Acyclic (Z)-1,3-Diene. *Org. Lett.* **2002**, 4, 1543-1546.
42. Bonjoch, J., Catena, J., Valls, N. Total Synthesis of (±)-Deethylibophyllidine: Studies of a Fischer Indolization Route and a Successful Approach via a Pummerer Rearrangement/Thionium Ion-Mediated Indole Cyclization. *J. Org. Chem.* **1996**, 61, 7106-7115.
43. Hagiwara, H., Kobayashi, K., Miya, S., Hoshi, T., Suzuki, T., Ando, M., Okamoto, T., Kobayashi, M., Yamamoto, I., Ohtsubo, S., Kato, M., Uda, H. First Total Syntheses of the Phytotoxins Solanapyrones D and E via the Domino Michael Protocol. *J. Org. Chem.* **2002**, 67, 5969-5976.

Quasi-Favorskii Rearrangement.....370

1. Tchoubar, B., Sackur, O. Alkaline dehalogenation of 1-chlorocyclohexyl methyl ketone and 1-chlorocyclohexyl phenyl ketone. Transposition into a-substituted cyclohexanecarboxylic acids. *Compt. rend.* **1939**, 208, 1020-1022.
2. Cope, A. C., Graham, E. S. Reactions of 1-bromobicyclo[3.3.1]nonan-9-one. *J. Am. Chem. Soc.* **1951**, 73, 4702-4706.
3. Stevens, C. L., Farkas, E. The formation of 1-phenylcyclohexanecarboxylic acid from α-halocyclohexyl phenyl ketones. *J. Am. Chem. Soc.* **1952**, 74, 5352-5355.
4. Kende, A. S. The Favorski rearrangement of haloketones. *Org. React.* **1960**, 11, 261-316.
5. Eaton, P. E., Cole, T. W., Jr. Cubane. *J. Am. Chem. Soc.* **1964**, 86, 3157-3158.
6. Eaton, P. E., Cole, T. W., Jr. Cubane system. *J. Am. Chem. Soc.* **1964**, 86, 962-964.
7. Warnhoff, E. W., Wong, C. M., Tai, W.-T. Mechanistic changes in a Favorskii reaction. *J. Am. Chem. Soc.* **1968**, 90, 514-515.
8. Kraus, G. A., Shi, J. Rearrangements of bridgehead bromides. A direct synthesis of epi-modhephene. *J. Org. Chem.* **1990**, 55, 5423-5424.
9. Kraus, G. A., Shi, J. Reactions of bridgehead halides. A synthesis of modhephene, isomodhephene, and epi-modhephene. *J. Org. Chem.* **1991**, 56, 4147-4151.
10. Harmata, M., Shao, L., Kürti, L., Abeywardane, A. 4+3 Cycloaddition reactions of halogen-substituted cyclohexenyl oxyallylic cations. *Tetrahedron Lett.* **1999**, 40, 1075-1078.
11. Harmata, M., Bohnert, G., Kürti, L., Barnes, C. L. Intramolecular 4+3 cycloadditions. A cyclohexenyl cation, its halogenated congener and a quasi-Favorskii rearrangement. *Tetrahedron Lett.* **2002**, 43, 2347-2349.
12. Smissman, E. E., Hite, G. Quasi-Favorskii rearrangement. I. Preparation of Demerol and β-pethidine. *J. Am. Chem. Soc.* **1959**, 81, 1201-1203.
13. Smissman, E. E., Hite, G. Quasi-Favorski rearrangement. II. Stereochemistry and mechanism. *J. Am. Chem. Soc.* **1960**, 82, 3375-3381.
14. Baudry, D., Begue, J. P., Charpentier-Morize, M. Favorsky-like rearrangement mechanism. *Tetrahedron Lett.* **1970**, 2147-2150.
15. Baudry, D., Begue, J. P., Charpentier-Morize, M. Stereochemistry and mechanism of the dehalogenation of α-bromo ketones without α'-hydrogen atoms under quasi-Favorsky rearrangement conditions. *Bull. Soc. Chim. Fr.* **1971**, 1416-1424.
16. Harmata, M., Rashatasakhon, P. A 4 + 3 Cycloaddition Approach to the Synthesis of Spatol. A Formal Total Synthesis of Racemic Spatol. *Org. Lett.* **2001**, 3, 2533-2535.
17. Harmata, M., Bohnert, G. J. A 4+3 Cycloaddition Approach to the Synthesis of (±)-Sterpurene. *Org. Lett.* **2003**, 5, 59-61.

Ramberg-Bäcklund Rearrangement.....372

1. Ramberg, L., Bäcklund, B. The reactions of some monohalogen derivatives of diethyl sulfone. *Arkiv Kemi, Minerat. Geol.* **1940**, 13A, 50 pp.
2. Bordwell, F. G. Ramberg-Bäcklund reaction. *Organosulfur Chem.* **1967**, 271-284.
3. Paquette, L. A. The base-induced rearrangement of α-halo sulfones. *Acc. Chem. Res.* **1968**, 1, 209-216.
4. Magnus, P. D. Recent developments in sulfone chemistry. *Tetrahedron* **1977**, 33, 2019-2045.
5. Paquette, L. A. The Ramberg-Bäcklund rearrangement. *Org. React.* **1977**, 25, 1-71.
6. Clough, J. M. The Ramberg-Bäcklund rearrangement. in *Comp. Org. Synth.* (eds. Trost, B. M.,Fleming, I.), *3*, 861-886 (Pergamon, Oxford, **1991**).
7. Braverman, S., Cherkinsky, M., Raj, P. Recent progress on rearrangements of sulfones. *Sulfur Reports* **1999**, 22, 49-84.
8. Taylor, R. J. K. Recent developments in Ramberg-Bäcklund and episulfone chemistry. *Chem. Commun.* **1999**, 217-227.
9. Prilezhaeva, E. N. Rearrangements of sulfoxides and sulfones in the total synthesis of natural compounds. *Russ. Chem. Rev.* **2001**, 70, 897-920.
10. Meyers, C. Y., Malte, A. M., Matthews, W. S. Ionic reactions of carbon tetrachloride. Survey of reactions with ketones, alcohols, and sulfones. *J. Am. Chem. Soc.* **1969**, 91, 7510-7512.
11. Chen, T. B. R. A., Burger, J. J., De Waard, E. R. The Michael induced Ramberg-Bäcklund olefin synthesis. *Tetrahedron Lett.* **1977**, 4527-4530.
12. Hartman, G. D., Hartman, R. D. The phase-transfer catalyzed Ramberg-Bäcklund reaction. *Synthesis* **1982**, 504-506.
13. Becker, K. B., Labhart, M. P. The intramolecular Ramberg-Bäcklund reaction: a convenient method for the synthesis of strained bridgehead olefins. *Helv. Chim. Acta* **1983**, 66, 1090-1100.
14. Chan, T.-L., Fong, S., Li, Y., Man, T.-O., Poon, C.-D. A new one-flask Ramberg-Bäcklund reaction. *J. Chem. Soc., Chem. Commun.* **1994**, 1771-1772.
15. Lawrence, N. J., Muhammad, F. Ramberg-Bäcklund type reactions of phosphonium salts. *Tetrahedron Lett.* **1994**, 35, 5903-5906.
16. Wladislaw, B., Marzorati, L., Russo, V. F. T., Zim, M. H., Di Vitta, C. Novel reaction: decarboxylative Ramberg-Bäcklund rearrangement in some α-isopropylsulfonyl carboxylic esters. *Tetrahedron Lett.* **1995**, 36, 8367-8370.
17. Evans, P., Taylor, R. J. The epoxy-Ramberg-Bäcklund reaction: a new route to allylic alcohols. *Tetrahedron Lett.* **1997**, 38, 3055-3058.
18. Evans, P., Taylor, R. J. K. Novel tandem conjugate addition/Ramberg-Bäcklund rearrangements. *Synlett* **1997**, 1043-1044.
19. Cao, X., Yang, Y., Wang, X. A direct route to conjugated enediynes from dipropargylic sulfones by a modified one-flask Ramberg-Bäcklund reaction. *J. Chem. Soc., Perkin Trans. 1* **2002**, 2485-2489.
20. Cao, X.-P. Stereoselective synthesis of substituted all-trans 1,3,5,7-octatetraenes by a modified Ramberg-Bäcklund reaction. *Tetrahedron* **2002**, 58, 1301-1307.
21. Bordwell, F. G., Cooper, G. D. The mechanism of formation of olefins by the reaction of sodium hydroxide with α-halo sulfones. *J. Am. Chem. Soc.* **1951**, 73, 5187-5190.
22. Bordwell, F. G., Cooper, G. D. The effect of the sulfonyl group on the nucleophilic displacement of halogen of α-halo sulfones and related substances. *J. Am. Chem. Soc.* **1951**, 73, 5184-5186.
23. Bordwell, F. G., Doomes, E., Corfield, P. W. R. Structure and stereochemical behavior of asymmetric α-sulfonyl carbanions. *J. Am. Chem. Soc.* **1970**, 92, 2581-2583.
24. Isaac, P. A. H. 223 pp (1973).
25. Bordwell, F. G., Doomes, E. Driving forces for 1,3-elimination reactions. Dehydrohalogenation of 1-halo-2-thia-2,3-dihydrophenalene 2,2-dioxides in a Ramberg-Bäcklund reaction. *J. Org. Chem.* **1974**, 39, 2531-2534.
26. Bordwell, F. G., Wolfinger, M. D. Solvent and substituent effects in the Ramberg-Bäcklund reaction. *J. Org. Chem.* **1974**, 39, 2521-2525.
27. Langler, R. F., Mantle, W. S., Newman, M. J. Investigation of the Ramberg-Bäcklund process. *Org. Mass Spectrom.* **1975**, 10, 1135-1140.
28. King, J. F., Hillhouse, J. H., Khemani, K. C. Organic sulfur mechanisms. 27. A reexamination of the reaction of thiirane 1,1-dioxide with aqueous hydroxide. *Can. J. Chem.* **1985**, 63, 1-5.

29. Sutherland, A. G., Taylor, R. J. K. The first isolation of an episulfone intermediate from a Ramberg-Bäcklund reaction. *Tetrahedron Lett.* **1989**, 30, 3267-3270.
30. Ewin, R. A., Loughlin, W. A., Pyke, S. M., Morales, J. C., Taylor, R. J. K. The isolation of episulfones from the Ramberg-Bäcklund rearrangement; part 3. *Synlett* **1993**, 660-662.
31. Jeffery, S. M., Sutherland, A. G., Pyke, S. M., Powell, A. K., Taylor, R. J. K. Isolation of episulfones from the Ramberg-Bäcklund rearrangement. Part 2. X-ray molecular structure of 2,3-epithio-8,8-dimethyl-6,10-dioxaspiro[4.5]decane S,S-dioxide and of r-6-benzyl-t-7,t-8-epithio-1,4-dioxaspiro[4.4]nonane S,S-dioxide. *J. Chem. Soc., Perkin Trans. 1* **1993**, 2317-2327.
32. Trost, B. M., Shi, Z. A Concise Convergent Strategy to Acetogenins. (+)-Solamin and Analogs. *J. Am. Chem. Soc.* **1994**, 116, 7459-7460.
33. Boeckman, R. K., Jr., Yoon, S. K., Heckendorn, D. K. Synthetic studies directed toward the eremantholides. 2. A novel application of the Ramberg-Bäcklund rearrangement to a highly stereoselective synthesis of (+)-eremantholide A. *J. Am. Chem. Soc.* **1991**, 113, 9682-9684.
34. Rigby, J. H., Warshakoon, N. C., Payen, A. J. Studies on Chromium(0)-Promoted Higher-Order Cycloaddition-Based Benzannulation. Total Synthesis of (+)-Estradiol. *J. Am. Chem. Soc.* **1999**, 121, 8237-8245.
35. MaGee, D. I., Beck, E. J. The use of the Ramberg-Bäcklund rearrangement for the formation of aza-macrocycles: a total synthesis of manzamine C. *Can. J. Chem.* **2000**, 78, 1060-1066.

Reformatsky Reaction 374

Related reactions: *Aldol reaction;*

1. Reformatsky, S. New synthesis of monobasic acids from ketones. *Ber.* **1887**, 20, 1210-1211.
2. Shriner, R. L. Reformatskii reaction. *Org. React.* **1942**, pp 1-37.
3. Diaper, D. G. M., Kuksis, A. Synthesis of alkylated alkanedioic acids. *Chem. Rev.* **1959**, 59, 89-178.
4. Gaudemar, M. Reformatskii reaction during the last thirty years. *Organometallic Chemistry Reviews, Section A: Subject Reviews* **1972**, 8, 183-233.
5. Rathke, M. W. Reformatskii reaction. *Org. React.* **1975**, 22, 423-460.
6. Heathcock, C. H. The aldol addition reaction. *Asymmetric Synth.* **1984**, 3, 111-212.
7. Fuerstner, A. Recent advancements in the Reformatskii reaction. *Synthesis* **1989**, 571-590.
8. Inanaga, J. Carbon-carbon bond formation via samarium iodide (SmI_2)-promoted electron transfer process. *Trends in Organic Chemistry* **1990**, 1, 23-30.
9. Rathke, M. W. Zinc enolates: the Reformatsky and Blaise reactions. in *Comp. Org. Synth.* (eds. Trost, B. M.,Fleming, I.), 2, 277-299 (Pergamon, Oxford, **1991**).
10. Erdik, E. Transition metal catalyzed reactions of organozinc reagents. *Tetrahedron* **1992**, 48, 9577-9648.
11. Fuerstner, A. Carbon-Carbon Bond Formation Involving Organochromium(III) Reagents. *Chem. Rev.* **1999**, 99, 991-1045.
12. Furstner, A. The Reformatskii reaction. *Organozinc Reagents* **1999**, 287-305.
13. Wessjohann, L. A., Scheid, G. Recent advances in chromium(II)- and chromium(III)-mediated organic synthesis. *Synthesis* **1999**, 1-36.
14. Marshall, J. A. Rhodium-catalyzed Reformatskii reaction. *Chemtracts* **2000**, 13, 705-707.
15. Banik, B. K. Samarium metal in organic synthesis. *Eur. J. Org. Chem.* **2002**, 2431-2444.
16. Podlech, J., Maier, T. C. Indium in organic synthesis. *Synthesis* **2003**, 633-655.
17. Nair, V., Ros, S., Jayan, C. N., Pillai, B. S. Indium- and gallium-mediated carbon-carbon bond-forming reactions in organic synthesis. *Tetrahedron* **2004**, 60, 1959-1982.
18. Ocampo, R., Dolbier, W. R., Jr. The Reformatsky reaction in organic synthesis. Recent advances. *Tetrahedron* **2004**, 60, 9325-9374.
19. Orsini, F., Sello, G. Transition metals-mediated Reformatsky reactions. *Current Organic Synthesis* **2004**, 1, 111-135.
20. Moriwake, T. Reformatskii reaction. I. Condensation of ketones and tert-butyl bromoacetate by magnesium. *J. Org. Chem.* **1966**, 31, 983-985.
21. Chao, L.-C., Rieke, R. D. Activated metals. IX. New reformatsky reagent involving activated indium for the preparation of β-hydroxy esters. *J. Org. Chem.* **1975**, 40, 2253-2255.
22. Villieras, J., Perriot, P., Bourgain, M., Normant, J. F. Enolates of esters. V. Preparation of the lithium analogs of Reformatsky reagents from α,α-dichloro and α-monohalo esters. Reactivity. *J. Organomet. Chem.* **1975**, 102, 129-140.
23. Kagan, H. B., Namy, J. L., Girard, P. Divalent lanthanide derivatives in organic synthesis. II. Mechanism of SmI_2 reactions in presence of ketones and organic halides. *Tetrahedron, Supplement* **1981**, 175-180.
24. Imamoto, T., Kusumoto, T., Tawarayama, Y., Sugiura, Y., Mita, T., Hatanaka, Y., Yokoyama, M. Carbon-carbon bond-forming reactions using cerium metal or organocerium(III) reagents. *J. Org. Chem.* **1984**, 49, 3904-3912.
25. Ishihara, T., Yamanaka, T., Ando, T. New low-valent titanium catalyzed reaction of chlorodifluoromethyl ketones leading to α,α-difluorinated β-hydroxy ketones. *Chem. Lett.* **1984**, 1165-1168.
26. Matsubara, S., Tsuboniwa, N., Morizawa, Y., Oshima, K., Nozaki, H. Reformatskii type reaction with new aluminum reagents containing an aluminum-tin or aluminum-lead linkage. *Bull. Chem. Soc. Jpn.* **1984**, 57, 3242-3246.
27. Tsuboniwa, N., Matsubara, S., Morizawa, Y., Oshima, K., Nozaki, H. Reformatskii type reaction by means of (tributyltin)diethylaluminum or (tributyllead)diethylaluminum. *Tetrahedron Lett.* **1984**, 25, 2569-2572.
28. Burkhardt, E. R., Rieke, R. D. The direct preparation of organocadmium compounds from highly reactive cadmium metal powders. *J. Org. Chem.* **1985**, 50, 416-417.
29. Dubois, J. E., Axiotis, G., Bertounesque, E. Chromium(II) chloride: a new reagent for cross-aldol reactions. *Tetrahedron Lett.* **1985**, 26, 4371-4372.
30. Fukuzawa, S., Fujinami, T., Sakai, S. Carbon-carbon bond formation between α-halo ketones and aldehydes promoted by cerium(III) iodide or cerium(III) chloride-sodium iodide. *J. Chem. Soc., Chem. Commun.* **1985**, 777-778.
31. Inaba, S., Rieke, R. D. Reformatskii type additions of haloacetonitriles to aldehydes mediated by metallic nickel. *Tetrahedron Lett.* **1985**, 26, 155-156.
32. Tabuchi, T., Kawamura, K., Inanaga, J., Yamaguchi, M. Preparation of medium- and large-ring lactones. SmI_2-Induced cyclization of w-(α-bromoacyloxy) aldehydes. *Tetrahedron Lett.* **1986**, 27, 3889-3890.
33. Molander, G. A., Etter, J. B. Lanthanides in organic synthesis. 8. 1.3-Asymmetric induction in intramolecular Reformatskii-type reactions promoted by samarium diiodide. *J. Am. Chem. Soc.* **1987**, 109, 6556-6558.
34. Araki, S., Ito, H., Butsugan, Y. Synthesis of β-hydroxyesters by Reformatsky reaction using indium metal. *Synth. Commun.* **1988**, 18, 453-458.
35. Orsini, F., Pelizzoni, F., Pulici, M., Vallarino, L. M. A cobalt-phosphine complex as mediator in the formation of carbon-carbon bonds. *J. Org. Chem.* **1994**, 59, 1-3.
36. Kagoshima, H., Hashimoto, Y., Oguro, D., Saigo, K. An Activated Germanium Metal-Promoted, Highly Diastereoselective Reformatsky Reaction. *J. Org. Chem.* **1998**, 63, 691-697.
37. Yanagisawa, A., Takahashi, H., Arai, T. Reactive barium-promoted Reformatsky-type reaction of α-chloro ketones with aldehydes. *Chem. Commun.* **2004**, 580-581.
38. Dewar, M. J. S., Merz, K. M., Jr. The Reformatskii reaction. *J. Am. Chem. Soc.* **1987**, 109, 6553-6554.
39. Orsini, F., Pelizzoni, F., Shillady, D. D., Vallarino, L. M. Theoretical cluster model of a zinc-carbon Reformatskii intermediate. *NATO ASI Ser., Ser. B* **1987**, 158, 457-461.
40. Mainz, J., Arrieta, A., Lopez, X., Ugalde, J. M., Cossio, F. P., Lecea, B. Transition structures for the Reformatskii reaction. A theoretical (MNDO-PM3) study. *Tetrahedron Lett.* **1993**, 34, 6111-6114.
41. Loeffler, A., Pratt, R. D., Pucknat, J., Gelbart, G., Dreiding, A. S. Preparation of α-methylenebutyrolactones by the Reformatskii reaction; synthesis of protolichesterinic acid. *Chimia* **1969**, 23, 413-416.

42. Rice, L. E., Boston, M. C., Finklea, H. O., Suder, B. J., Frazier, J. O., Hudlicky, T. Regioselectivity in the Reformatskii reaction of 4-bromocrotonate. Role of the catalyst and the solvent in the normal vs. abnormal modes of addition to carbonyl substrates. *J. Org. Chem.* **1984**, 49, 1845-1848.
43. Gedge, D. R., Pattenden, G., Smith, A. G. New syntheses of pulvinic acids via Reformatsky-type reactions with aryl methoxymaleic anhydrides. *J. Chem. Soc., Perkin Trans. 1* **1986**, 2127-2131.
44. Sato, T., Itoh, T., Fujisawa, T. Facile synthesis of β-oxo esters by a coupling reaction of the Reformatskii reagent with acyl chlorides catalyzed by a palladium complex. *Chem. Lett.* **1982**, 1559-1560.
45. Stamm, H., Steudle, H. Nitrones. XI. Isoxazolidine compounds. VIII. N-substituted 5-isoxazolidinones by Reformatskii reaction with nitrones. *Tetrahedron* **1979**, 35, 647-650.
46. Alvernhe, G., Lacombe, S., Laurent, A., Marquet, B. Addition of the Reformatskii reagent to azirines. Synthesis of 4-amino lactones. *J. Chem. Res., Synop.* **1980**, 54-55.
47. Gilman, H., Speeter, M. Reformatskii reaction with benzalaniline. *J. Am. Chem. Soc.* **1943**, 65, 2255-2256.
48. Blaise, E. E. *C. R. Hebd. Seances Acad. Sci.* **1901**, 478.
49. Rieke, R. D., Uhm, S. J. Activated metals. XI. Improved procedure for the preparation of β-hydroxy esters using activated zinc. *Synthesis* **1975**, 452-453.
50. Arnold, R. T., Kulenovic, S. T. Activated metals. A procedure for the preparation of activated magnesium and zinc. *Synth. Commun.* **1977**, 7, 223-232.
51. Rieke, R. D., Li, P. T.-J., Burns, T. P., Uhm, S. T. Preparation of highly reactive metal powders. New procedure for the preparation of highly reactive zinc and magnesium metal powders. *J. Org. Chem.* **1981**, 46, 4323-4324.
52. Boldrini, G. P., Savoia, D., Tagliavini, E., Trombini, C., Umani-Ronchi, A. Active metals from potassium-graphite. Zinc-graphite promoted synthesis of β-hydroxy esters, homoallylic alcohols and α-methylene-γ-butyrolactones. *J. Org. Chem.* **1983**, 48, 4108-4111.
53. Orsini, F., Pelizzoni, F., Ricca, G. Reformatskii intermediate. A C-metalated species. *Tetrahedron Lett.* **1982**, 23, 3945-3948.
54. Dekker, J., Boersma, J., Van der Kerk, G. J. M. The structure of the Reformatskii reagent. *J. Chem. Soc., Chem. Commun.* **1983**, 553-555.
55. Dekker, J., Budzelaar, P. H. M., Boersma, J., Van der Kerk, G. J. M., Spek, A. J. The nature of the Reformatsky reagent. Crystal structure of $(BrZnCH_2COO\text{-t-Bu.THF})_2$. *Organometallics* **1984**, 3, 1403-1407.
56. Orsini, F., Pelizzoni, F., Ricca, G. C-Metallated Reformatsky intermediates. Structure and reactivity. *Tetrahedron* **1984**, 40, 2781-2787.
57. Hansen, M. M., Bartlett, P. A., Heathcock, C. H. Preparation and reactions of an alkylzinc enolate. *Organometallics* **1987**, 6, 2069-2074.
58. Vedejs, E., Duncan, S. M. A Synthesis of C(16),C(18)-Bis-epi-cytochalasin D via Reformatsky Cyclization. *J. Org. Chem.* **2000**, 65, 6073-6081.
59. Inoue, M., Sasaki, M., Tachibana, K. A convergent synthesis of the decacyclic ciguatoxin model containing the F-M ring framework. *J. Org. Chem.* **1999**, 64, 9416-9429.
60. Gabriel, T., Wessjohann, L. The chromium-Reformatskii reaction: asymmetric synthesis of the aldol fragment of the cytotoxic epothilons from 3-(2-bromoacyl)-2-oxazolidinones. *Tetrahedron Lett.* **1997**, 38, 1363-1366.
61. Pettit, G. R., Grealish, M. P. A Cobalt-Phosphine Complex Directed Reformatskii Approach to a Stereospecific Synthesis of the Dolastatin 10 Unit Dolaproine (Dap). *J. Org. Chem.* **2001**, 66, 8640-8642.

Regitz Diazo Transfer 376

1. Dimroth, O., et al. Intramolecular Rearrangements. IV. Hydroxytriazoles and Diazocarboxylic Acid Amides. *Ann.* **1910**, 373, 336-370.
2. Regitz, M. Reaction of active methylene compounds with azides. I. New synthesis of α-diazo-β-dicarbonyl compounds from benzenesulfonyl azides and β-diketones. *Ann.* **1964**, 676, 101-109.
3. Regitz, M. Reactions of active methylene compounds with azides. IV. A new synthesis of α-diazocarbonyl compounds. *Tetrahedron Lett.* **1964**, 1403-1407.
4. Regitz, M. Reactions of active methylene compounds with azides. III. Diazo, azino, and triphenylphosphazino derivatives of anthrone and thioxanthene S,S-dioxide. *Ber.* **1964**, 97, 2742-2754.
5. Regitz, M., Heck, G. Syntheses and some reactions of 2-diazo- and 2-hydroxyindan-1,3-dione. *Ber.* **1964**, 97, 1482-1501.
6. Regitz, M., Anschuetz, W., Bartz, W., Liedhegener, A. Reactions of CH-active compounds with azides. XXII. Synthesis and some properties of α-diazophosphine oxides and α-diazophosphonates. *Tetrahedron Lett.* **1968**, 3171-3174.
7. Regitz, M. Transfer of diazo groups. *Angew. Chem., Int. Ed. Engl.* **1967**, 6, 733-749.
8. Regitz, M. Reactions of carbon-hydrogen active compounds with azides. XIII. Diazo group transfer. *Neuere Method. Praep. Org. Chem.* **1970**, 6, 76-118.
9. Regitz, M. Recent synthetic methods in diazo chemistry. *Synthesis* **1972**, 351-373.
10. Regitz, M., Korobitsyna, I. K., Rodina, L. L. Aliphatic diazo compounds. *Method. Chim.* **1975**, 6, 205-299.
11. Askani, R., Taber, D. F. Synthesis of Nitroso, Nitro and Related Compounds. in *Comp. Org. Synth.* (eds. Trost, B. M.,Fleming, I.), *6*, 103-132 (Pergamon, Oxford, **1991**).
12. Ye, T., McKervey, M. A. Organic Synthesis with α-Diazo Carbonyl Compounds. *Chem. Rev.* **1994**, 94, 1091-1160.
13. Heydt, H. Product class 21: diazo compounds. *Science of Synthesis* **2004**, 27, 843-935.
14. Hendrickson, J. B., Wolf, W. A. Direct introduction of the diazo function in organic synthesis. *J. Org. Chem.* **1968**, 33, 3610-3618.
15. Regitz, M., Rueter, J. Reactions of CH-active compounds with azides. XVIII. Synthesis of 2-oxo-1-diazo cycloalkanes by deformylative diazo-group transfer. *Chem. Ber.* **1968**, 101, 1263-1270.
16. Taber, D. F., Ruckle, R. E., Jr., Hennessy, M. J. Mesyl azide: a superior reagent for diazo transfer. *J. Org. Chem.* **1986**, 51, 4077-4078.
17. Baum, J. S., Shook, D. A., Davies, H. M. L., Smith, H. D. Diazo transfer reactions with p-acetamidobenzenesulfonyl azide. *Synth. Commun.* **1987**, 17, 1709-1716.
18. Ben Alloum, A., Villemin, D. Potassium fluoride on alumina: an easy preparation of diazo carbonyl compounds. *Synth. Commun.* **1989**, 19, 2567-2571.
19. Danheiser, R. L., Miller, R. F., Brisbois, R. G., Park, S. Z. An improved method for the synthesis of α-diazo ketones. *J. Org. Chem.* **1990**, 55, 1959-1964.
20. Koskinen, A. M. P., Munoz, L. Diazo transfer reactions under mildly basic conditions. *J. Chem. Soc., Chem. Commun.* **1990**, 652-653.
21. McGuiness, M., Schechter, H. Azidotris(diethylamino)phosphonium bromide: a self-catalyzing diazo transfer reagent. *Tetrahedron Lett.* **1990**, 31, 4987-4990.
22. Ghosh, S., Datta, I. Diazo transfer reaction in solid state. *Synth. Commun.* **1991**, 21, 191-200.
23. Lee, J. C., Yuk, J. Y. An improved and efficient method for diazo transfer reaction of active methylene compounds. *Synth. Commun.* **1995**, 25, 1511-1515.
24. Taber, D. F., Gleave, D. M., Herr, R. J., Moody, K., Hennessy, M. J. A New Method For the Construction of α-Diazo ketones. *J. Org. Chem.* **1995**, 60, 2283-2285.
25. Taber, D. F., You, K., Song, Y. A Simple Preparation of α-Diazo Esters. *J. Org. Chem.* **1995**, 60, 1093-1094.
26. Danheiser, R. L., Miller, R. F., Brisbois, R. G. Detrifluoroacetylative diazo group transfer: (E)-1-diazo-4-phenyl-3-butene-2-one (3-buten-2-one, 1-diazo-4-phenyl-). *Org. Synth.* **1996**, 73, 134-143.
27. Benati, L., Calestani, G., Nanni, D., Spagnolo, P., Volta, M. Diazo transfer reaction of 2-(trimethylsilyl)-1,3-dithiane with tosyl azide. Carbenic reactivity of transient 2-diazo-1,3-dithiane. *Tetrahedron* **1997**, 53, 9269-9278.
28. Benati, L., Calestani, G., Nanni, D., Spagnolo, P. Reactions of Benzocyclic β-Keto Esters with Tosyl and 4-Nitrophenyl Azide. Structural Influence of Dicarbonyl Substrate and Azide Reagent on Distribution of Diazo, Azide and Ring-Contraction Products. *J. Org. Chem.* **1998**, 63, 4679-4684.
29. Charette, A. B., Wurz, R. P., Ollevier, T. Trifluoromethanesulfonyl Azide: A Powerful Reagent for the Preparation of α-Nitro-α-diazocarbonyl Derivatives. *J. Org. Chem.* **2000**, 65, 9252-9254.

30. Xu, Y., Wang, Y., Zhu, S. Reactions of fluoroalkanesulfonyl azides with carbocyclic β-keto esters: structural influence of dicarbonyl substrate on distribution of diazo and ring-contraction products. *J. Fluorine Chem.* **2000**, 105, 25-30.
31. Green, G. M., Peet, N. P., Metz, W. A. Polystyrene-Supported Benzenesulfonyl Azide: A Diazo Transfer Reagent That Is Both Efficient and Safe. *J. Org. Chem.* **2001**, 66, 2509-2511.
32. de S. Rianelli, R., de Souza, M. C., Ferreira, V. F. Mild diazo transfer reaction catalyzed by modified clays. *Synth. Commun.* **2004**, 34, 951-959.
33. Evans, D. A., Britton, T. C., Ellman, J. A., Dorow, R. L. The asymmetric synthesis of α-amino acids. Electrophilic azidation of chiral imide enolates, a practical approach to the synthesis of (R)- and (S)-α-azido carboxylic acids. *J. Am. Chem. Soc.* **1990**, 112, 4011-4030.
34. Marino, J. P., Jr., Osterhout, M. H., Padwa, A. An Approach to Lysergic Acid Utilizing an Intramolecular Isomünchone Cycloaddition Pathway. *J. Org. Chem.* **1995**, 60, 2704-2713.
35. Swain, N. A., Brown, R. C. D., Bruton, G. A Versatile Stereoselective Synthesis of endo,exo-Furofuranones: Application to the Enantioselective Synthesis of Furofuran Lignans. *J. Org. Chem.* **2004**, 69, 122-129.
36. Hughes, C. C., Kennedy-Smith, J. J., Trauner, D. Synthetic studies toward the guanacastepenes. *Org. Lett.* **2003**, 5, 4113-4115.
37. Padwa, A., Sheehan, S. M., Straub, C. S. An Isomünchone-Based Method for the Synthesis of Highly Substituted 2(1H)-Pyridones. *J. Org. Chem.* **1999**, 64, 8648-8659.

Reimer-Tiemann Reaction..378

Related reactions: Gattermann and Gattermann-Koch formylation, Vilsmeier-Haack formylation;

1. Reimer, K. A new synthesis of aromatic aldehydes. *Ber. Dtsch. Chem. Ges.* **1876**, 9, 423-424.
2. Reimer, K., Tiemann, F. The effect of chloroform on phenolates. *Ber. Dtsch. Chem. Ges.* **1876**, 9, 824-828.
3. Reimer, K., Tiemann, F. The effect of chloroform on phenol and especially on the alkaline solution of aromatic oxyacids. *Ber. Dtsch. Chem. Ges.* **1876**, 9, 1268-1278.
4. Wynberg, H. The Reimer-Tiemann reaction. *Chem. Rev.* **1960**, 60, 169-184.
5. Mullins, R. M. Hydroxybenzaldehydes. *Kirk-Othmer Encycl. Chem. Technol., 3rd Ed.* **1981**, 13, 70-79.
6. Wynberg, H., Meijer, E. W. The Reimer-Tiemann reaction. *Org. React.* **1982**, 28, 1-36.
7. Wynberg, H. The Reimer-Tiemann Reaction. in *Comp. Org. Synth.* (eds. Trost, B. M.,Fleming, I.), *2*, 769-775 (Pergamon, Oxford, **1991**).
8. Hirao, K., Ikegame, M., Yonemitsu, O. Photochemical Reimer-Tiemann reaction of phenols, anilines, and indolines. *Tetrahedron* **1974**, 30, 2301-2305.
9. Sasson, Y., Yonovich, M. The effect of phase transfer catalysts on the Reimer-Tiemann reaction. *Tetrahedron Lett.* **1979**, 3753-3756.
10. Fahmy, A. M., Mahgoub, S. A., Aly, M. M., Badr, M. Z. A. The photo Reimer-Tiemann reaction. *Bull. Facult. Sci. Assiut University* **1982**, 11, 17-23.
11. Bird, C. W., Brown, A. L., Chan, C. C. A new type of abnormal Reimer-Tiemann reaction. *Tetrahedron* **1985**, 41, 4685-4690.
12. Smith, K. M., Bobe, F. W., Minnetian, O. M., Hope, H., Yanuck, M. D. Novel substituent orientation in Reimer-Tiemann reactions of pyrrole-2-carboxylates. *J. Org. Chem.* **1985**, 50, 790-792.
13. Neumann, R., Sasson, Y. Increased para selectivity in the Reimer-Tiemann reaction by use of polyethylene glycol as complexing agent. *Synthesis* **1986**, 569-570.
14. Thoer, A., Denis, G., Delmas, M., Gaset, A. The Reimer-Tiemann reaction in slightly hydrated solid-liquid medium: a new method for the synthesis of formyl and diformyl phenols. *Synth. Commun.* **1988**, 18, 2095-2101.
15. Cochran, J. C., Melville, M. G. The Reimer-Tiemann reaction, enhanced by ultrasound. *Synth. Commun.* **1990**, 20, 609-616.
16. Gaonkar, A. V., Kirtany, J. K. Reimer-Tiemann reaction using carbon tetrachloride. *Indian J. Chem., Sect. B* **1991**, 30B, 800-801.
17. Langlois, B. R. Anomalous Reimer-Tiemann reaction from phenol, chloroform and potassium fluoride in sulfolane. *Tetrahedron Lett.* **1991**, 32, 3691-3694.
18. Divakar, S., Maheswaran, M. M., Narayan, M. S. Reimer-Tiemann reactions of guaiacol and catechol in the presence of β-cyclodextrin. *Indian J. Chem., Sect. B* **1992**, 31B, 543-546.
19. Jimenez, M. C., Miranda, M. A., Tormos, R. Formation of dichloromethyl phenyl ethers as major products in the photo-Reimer-Tiemann reaction without base. *Tetrahedron* **1995**, 51, 5825-5830.
20. Ravichandran, R. β-Cyclodextrin mediated regioselective photo-Reimer-Tiemann reaction of phenols. *J. Mol. Catal. A: Chemical* **1998**, 130, L205-L207.
21. Castillo, R., Moliner, V., Andres, J., Oliva, M., Safont, V. S., Bohm, S. Theoretical investigation of the abnormal Reimer-Tiemann reaction. *J. Phys. Org. Chem.* **1998**, 11, 670-677.
22. Castillo, R., Moliner, V., Andres, J. A theoretical study on the molecular mechanism for the normal Reimer-Tiemann reaction. *Chem. Phys. Lett.* **2000**, 318, 270-275.
23. Auwers, K. *Ber. Dtsch. Chem. Ges.* **1884**, 17, 2976.
24. Auwers, K. Tribromo pseudocumenol bromide and its analogs. *Ber. Dtsch. Chem. Ges.* **1896**, 29, 1109-1110.
25. Kemp, D. S. Relative ease of 1,2-proton shifts. Origin of the formyl proton of salicylaldehyde obtained by the Reimer-Tiemann reaction. *J. Org. Chem.* **1971**, 36, 202-204.
26. Gu, X.-H., Yu, H., Jacobson, A. E., Rothman, R. B., Dersch, C. M., George, C., Flippen-Anderson, J. L., Rice, K. C. Design, Synthesis, and Monoamine Transporter Binding Site Affinities of Methoxy Derivatives of Indatraline. *J. Med. Chem.* **2000**, 43, 4868-4876.
27. Makela, T., Matikainen, J., Wahala, K., Hase, T. Development of a novel hapten for radioimmunoassay of the lignan, enterolactone in plasma (serum). Total synthesis of (±)-trans-5-carboxymethoxyenterolactone and several analogues. *Tetrahedron* **2000**, 56, 1873-1882.

Riley Selenium Dioxide Oxidation..380

1. Riley, H. L., Morley, J. F., Friend, N. A. C. Selenium dioxide, a new oxidizing agent. I. Its reactions with aldehydes and ketones. *J. Chem. Soc.* **1932**, 1875-1883.
2. Guillemonat, A. Oxidation of ethylenic hydrocarbons with selenium dioxide. *Annali di Chimica Applicata* **1939**, 11, 143-211.
3. Waitkins, G. R., Clark, C. W. Selenium dioxide: preparation, properties, and use as oxidizing agent. *Chem. Rev.* **1945**, 36, 235-289.
4. Rabjohn, N. Selenium dioxide oxidation. *Org. React.* **1949**, 5, 331-386.
5. Trachtenberg, E. N. Selenium dioxide oxidation. in *Oxidation* (ed. Augustine, R. L.), *1*, 119-187 (Marcel Dekker, New York, **1969**).
6. Jerussi, R. A. Selective oxidations with selenium dioxide. *Selective Organic Transformations* **1970**, 1, 301-326.
7. Rabjohn, N. Selenium dioxide oxidation. *Org. React.* **1976**, 24, 261-415.
8. Laitalainen, T. Selenium dioxide oxidation of cyclohexanone derivatives. Preparation of cyclopentane-1,2-dione derivatives and characterization of organic selenium compounds. *Annales Academiae Scientiarum Fennicae, Series A2: Chemica* **1982**, 195, 51 pp.
9. Bulman Page, P. C., McCarthy, T. J. Oxidation Adjacent to C=C Bonds. in *Comp. Org. Synth.* (eds. Trost, B. M.,Fleming, I.), *7*, 83-117 (Pergamon, Oxford, **1991**).
10. Huguet, J. L. Oxidation of olefins catalyzed by selenium. *Advances in Chemistry Series* **1968**, No. 76, 345-351.
11. Schaefer, J. P., Horvath, B., Klein, H. P. Selenium dioxide oxidations. III. Oxidation of olefins. *J. Org. Chem.* **1968**, 33, 2647-2655.
12. Kariyone, K., Yazawa, H. Oxidative cleavage of β,γ- unsaturated ether. *Tetrahedron Lett.* **1970**, 2885-2888.
13. Trachtenberg, E. N., Carver, J. R. Stereochemistry of selenium dioxide oxidation of cyclohexenyl systems. *J. Org. Chem.* **1970**, 35, 1646-1653.
14. Rapoport, H., Bhalerao, U. T. Stereochemistry of allylic oxidation with selenium dioxide. Stereospecific oxidation of gem-dimethyl olefins. *J. Am. Chem. Soc.* **1971**, 93, 4835-4840.

15. Hellman, H. M., Jerussi, R. A., Rosegay, A. Oxidative rearrangement of ketones to carboxylic acids. *Ann. N. Y. Acad. Sci.* **1972**, 193, 44-48.
16. Howe, R., Johnson, D. Oxidative fission of α-substitutted β-diketones by selenium dioxide. *J. Chem. Soc., Perkin Trans. 1* **1972**, 977-981.
17. Francis, M. J., Grant, P. K., Low, K. S., Weavers, R. T. Diterpene chemistry. VI. Selenium dioxide-hydrogen peroxide oxidations of exocyclic olefins. *Tetrahedron* **1976**, 32, 95-101.
18. Ishii, Y., Murai, S., Sonoda, N. Oxidation of aldehydes by hydrogen peroxide in the presence of selenium dioxide catalyst. *Technol. Rep. Osaka Univ.* **1976**, 26, 623-626.
19. Cain, M., Campos, O., Guzman, F., Cook, J. M. Selenium dioxide oxidations in the indole area. Synthesis of β–carboline alkaloids. *J. Am. Chem. Soc.* **1983**, 105, 907-913.
20. San Feliciano, A., Medarde, M., Lopez, J. L., Pereira, J. A. P., Caballero, E., Perales, A. Reaction of selenium dioxide with dienes. 1. Linalyl acetate. *Tetrahedron* **1989**, 45, 5073-5080.
21. Lee, J. C., Park, H.-J., Park, J. Y. Rapid microwave-promoted solvent-free oxidation of α-methylene ketones to α–diketones. *Tetrahedron Lett.* **2002**, 43, 5661-5663.
22. Tagawa, Y., Yamashita, K., Higuchi, Y., Goto, Y. Improved oxidation of an active methyl group of N-heteroaromatic compounds by selenium dioxide in the presence of tert-butyl hydroperoxide. *Heterocycles* **2003**, 60, 953-957.
23. Ra, C. S., Park, G. Ab initio studies of the allylic hydroxylation: DFT calculation on the reaction of 2-methyl-2-butene with selenium dioxide. *Tetrahedron Lett.* **2003**, 44, 1099-1102.
24. Trachtenberg, E. N., Nelson, C. H., Carver, J. R. Mechanism of selenium dioxide oxidation of olefins. *J. Org. Chem.* **1970**, 35, 1653-1658.
25. Sharpless, K. B., Lauer, R. F. Selenium dioxide oxidation of olefins. Evidence for the intermediacy of allylseleninic acids. *J. Am. Chem. Soc.* **1972**, 94, 7154-7155.
26. Arigoni, D., Vasella, A., Sharpless, K. B., Jensen, H. P. Selenium dioxide oxidations of olefins. Trapping of the allylic seleninic acid intermediate as a seleninolactone. *J. Am. Chem. Soc.* **1973**, 95, 7917-7919.
27. Sharpless, K. B., Gordon, K. M. Selenium dioxide oxidation of ketones and aldehydes. Evidence for the intermediacy of β-ketoseleninic. *J. Am. Chem. Soc.* **1976**, 98, 300-301.
28. Stephenson, L. M., Speth, D. R. Mechanism of allylic hydroxylation by selenium dioxide. *J. Org. Chem.* **1979**, 44, 4683-4689.
29. Woggon, W. D., Ruther, F., Egli, H. The mechanism of allylic oxidation by selenium dioxide. *J. Chem. Soc., Chem. Commun.* **1980**, 706-708.
30. Pati, S. C., Mishra, M. M. Kinetics and mechanism of the oxidation of aryl aliphatic ketones by selenium dioxide. *Indian Journal of Physical and Natural Sciences* **1981**, 1, 54-63.
31. Warpehoski, M. A., Chabaud, B., Sharpless, K. B. Selenium dioxide oxidation of endocyclic olefins. Evidence for a dissociation-recombination pathway. *J. Org. Chem.* **1982**, 47, 2897-2900.
32. Valechha, N. D., Pandey, A. K. Kinetics of oxidation of allyl, crotyl and cinnamic alcohols by selenium dioxide. *J. Indian Chem. Soc.* **1986**, 63, 670-673.
33. Valechha, N. D., Sewanee, J. P. Kinetics of oxidation of acetophenones by selenium dioxide. *J. Indian Chem. Soc.* **1986**, 63, 970-973.
34. Hassan, R. M. Kinetics and mechanism of selenium(IV) oxidation of ascorbic acid in aqueous perchlorate solutions. *Croat. Chem. Acta* **1991**, 64, 229-236.
35. Hassan, R. M., El-Gaiar, S. A., El-Hady, A., El-Summan, M. Kinetics and mechanism of oxidation of selenium(IV) by permanganate ion in aqueous perchlorate solutions. *Collect. Czech. Chem. Commun.* **1993**, 58, 538-546.
36. Valeccha, N. D., Khan, M. U., Verma, J. K., Singh, V. R. Oxidation of some aliphatic aldehydes by selenium dioxide in acetic acid-water and sulfuric acid medium. A kinetic study. *Oxidation Communications* **1995**, 18, 312-320.
37. Shafer, C. M., Molinski, T. F. Oxidative Rearrangement of 2-Substituted Oxazolines. A Novel Entry to 5,6-Dihydro-2H-1,4-oxazin-2-ones and Morpholin-2-ones. *J. Org. Chem.* **1996**, 61, 2044-2050.
38. Shafer, C. M., Morse, D. I., Molinski, T. F. Mechanism of SeO_2 promoted oxidative rearrangement of 2-substituted oxazolines to dihydrooxazinones: isotopic labeling and kinetic studies. *Tetrahedron* **1996**, 52, 14475-14486.
39. Aziz, S., Khan, A. U. Oxidation of Di Ethyl: Ethyl Aceto Acetate by selenium-di-oxide in acidic medium: A KINETIC STUDY. *Ultra Scientist of Physical Sciences* **1998**, 10, 240-244.
40. Tiwari, S., Khan, M. U., Tiwari, B. M. L., Tiwari, K. S., Valechha, N. D. Kinetics and mechanism of oxidation of some 2-alkanones by selenium dioxide in aqueous acetic acid and perchloric acid media. *Oxidation Communications* **1999**, 22, 416-423.
41. Singleton, D. A., Hang, C. Isotope effects and the mechanism of allylic hydroxylation of alkenes with selenium dioxide. *J. Org. Chem.* **2000**, 65, 7554-7560.
42. Mehta, G., Shinde, H. M. Enantiospecific total synthesis of 6-epi-(-)-hamigeran B. Intramolecular Heck reaction in a sterically constrained environment. *Tetrahedron Lett.* **2003**, 44, 7049-7053.
43. Xu, P.-F., Chen, Y.-S., Lin, S.-I., Lu, T.-J. Chiral Tricyclic Iminolactone Derived from (1R)-(+)-Camphor as a Glycine Equivalent for the Asymmetric Synthesis of α-Amino Acids. *J. Org. Chem.* **2002**, 67, 2309-2314.
44. Fürstner, A., Gastner, T. Total Synthesis of Cristatic Acid. *Org. Lett.* **2000**, 2, 2467-2470.
45. Corey, E. J., Wu, L. I. Enantioselective total synthesis of miroestrol. *J. Am. Chem. Soc.* **1993**, 115, 9327-9328.

Ritter Reaction 382

1. Ritter, J. J., Kalish, J. New reaction of nitriles. II. Synthesis of t-carbinamines. *J. Am. Chem. Soc.* **1948**, 70, 4048-4050.
2. Ritter, J. J., Minieri, P. P. New reaction of nitriles. I. Amides from alkenes and mononitriles. *J. Am. Chem. Soc.* **1948**, 70, 4045-4048.
3. Zil'berman, E. N. Some reactions of nitriles which lead to formation of new nitrogen-carbon bonds. *Russ. Chem. Rev.* **1960**, 26, 331-344.
4. Johnson, F., Madronero, R. Heterocyclic syntheses involving nitrilium salts and nitriles under acidic conditions. *Adv. Heterocycl. Chem.* **1966**, 6, 95-146.
5. Krimen, L. I., Cota, D. J. Ritter reaction. *Org. React.* **1969**, 17, 213-325.
6. Beckwith, A. L. J. Synthesis of amides. *Chem. Amides* **1970**, 73-185.
7. Meyers, A. I., Sircar, J. C. in *The Chemistry of the Cyano Group* (ed. Rappoport, Z.), 341 (Wiley Interscience, New York, **1970**).
8. Bishop, R. Ritter-type reactions. in *Comp. Org. Synth.* (eds. Trost, B. M.,Fleming, I.), *6*, 261-300 (Pergamon, Oxford, **1991**).
9. Larock, R. C., Leong, W. W. Addition of H-X reagents to alkenes and alkynes. in *Comp. Org. Synth.* (eds. Trost, B. M.,Fleming, I.), *4*, 269-327 (Pergamon, Oxford, **1991**).
10. Biermann, U., Fuermeier, S., Metzger, J. O. Some carbon-nitrogen and carbon-oxygen bond forming additions to unsaturated fatty compounds. *Fett/Lipid* **1998**, 100, 236-246.
11. Olah, G. A., Gupta, B. G. B., Narang, S. C. Synthetic methods and reactions; 66. Nitrosonium ion induced preparation of amides from alkyl (arylalkyl) halides with nitriles, a mild and selective Ritter-type reaction. *Synthesis* **1979**, 274-276.
12. Sharghi, H., Niknam, K. Conversion of alcohols to amides using alumina-methanesulfonic acid (AMA) in nitrile solvents. *Iranian Journal of Chemistry & Chemical Engineering* **1999**, 18, 36-39.
13. Jirgensons, A., Kauss, V., Kalvinsh, I., Gold, M. R. A practical synthesis of tert-alkylamines via the Ritter reaction with chloroacetonitrile. *Synthesis* **2000**, 1709-1712.
14. Salehi, P., Motlagh, A. R. Silica gel supported ferric perchlorate: a new and efficient reagent for one pot synthesis of amides from benzylic alcohols. *Synth. Commun.* **2000**, 30, 671-675.
15. Okuhara, T. Ritter-type reactions catalyzed by H-ZSM-5 zeolites. *Zeoraito* **2001**, 18, 100-106.
16. Okuhara, T., Chen, X. Ritter-type reactions catalyzed by high-silica MFI zeolites. *Microporous and Mesoporous Materials* **2001**, 48, 293-299.
17. Salehi, P., Khodaei, M. M., Zolfigol, M. A., Keyvan, A. Facile conversion of alcohols into N-substituted amides by magnesium hydrogensulfate under heterogeneous conditions. *Synth. Commun.* **2001**, 31, 1947-1951.

18. Darbeau, R. W., Pease, R. S., Perez, E. V., Gibble, R. E., Ayo, F. A., Sweeney, A. W. N-Nitrosamide-mediated Ritter-type reactions.Part II. The operation of "persistent steric" and "π*-acceptor agostic-type" effects. *J. Chem. Soc., Perkin Trans. 2* **2002**, 2146-2153.
19. Eastgate, M. D., Fox, D. J., Morley, T. J., Warren, S. Sulfur mediated Ritter reactions: the synthesis of cyclic amides. *Synthesis* **2002**, 2124-2128.
20. Lebedev, M. Y., Erman, M. B. Lower primary alkanols and their esters in a Ritter-type reaction with nitriles. An efficient method for obtaining N-primary-alkyl amides. *Tetrahedron Lett.* **2002**, 43, 1397-1399.
21. Sakaguchi, S., Hirabayashi, T., Ishii, Y. First Ritter-type reaction of alkylbenzenes using N-hydroxyphthalimide as a key catalyst. *Chem. Commun.* **2002**, 516-517.
22. Welniak, M. The reactions of selected terpene alcohols with acetonitrile in the presence of boron trifluoride etherate. *Pol. J. Chem.* **2002**, 76, 1405-1411.
23. Booker-Milburn, K. I., Guly, D. J., Cox, B., Procopiou, P. A. Ritter-Type Reactions of N-Chlorosaccharin: A Method for the Electrophilic Diamination of Alkenes. *Org. Lett.* **2003**, 5, 3313-3315.
24. Reddy, K. L. An efficient method for the conversion of aromatic and aliphatic nitriles to the corresponding N-tert-butyl amides. A modified Ritter reaction. *Tetrahedron Lett.* **2003**, 44, 1453-1455.
25. Ho, T.-L., Chein, R.-J. Intervention of Phenonium Ion in Ritter Reactions. *J. Org. Chem.* **2004**, 69, 591-592.
26. Janin, Y. L., Decaudin, D., Monneret, C., Poupon, M.-F. Synthesis of methylenedioxy-bearing 1-aryl-3-carboxylisoquinolines using a modified Ritter reaction procedure. *Tetrahedron* **2004**, 60, 5481-5485.
27. Colominas, C., Orozco, M., Luque, F. J., Borrell, J. I., Teixido, J. A Priori Prediction of Substituent and Solvent Effects in the Basicity of Nitriles. *J. Org. Chem.* **1998**, 63, 4947-4953.
28. Hessley, R. K. Computational investigations for undergraduate organic chemistry: predicting the mechanism of the Ritter reaction. *J. Chem. Educ.* **2000**, 77, 202-205.
29. Benson, F. R., Ritter, J. J. A new reaction of nitriles. III. Amides from dinitriles. *J. Am. Chem. Soc.* **1949**, 71, 4128-4129.
30. Roe, E. T., Swern, D. Fatty acid amides. VI. Preparation of substituted amidostearic acids by addition of nitriles to oleic acid. *J. Am. Chem. Soc.* **1953**, 75, 5479-5481.
31. Deno, N. C., Edwards, T., Perizzolo, C. Carbonium ions. V. The nature of the tert-butyl cation as indicated by a study of the formation of N-tert-butylacrylamide. *J. Am. Chem. Soc.* **1957**, 79, 2108-2112.
32. Deno, N. C., Gaugler, R. W., Wisotsky, M. J. Base strengths and chemical behavior of nitriles in sulfuric acid and oleum systems. *J. Org. Chem.* **1966**, 31, 1967-1968.
33. Glikmans, G., Torck, B., Hellin, M., Coussemant, F. Ritter reaction between isobutene and acrylonitrile. II. Kinetic study. *Bull. Soc. Chim. Fr.* **1966**, 1383-1388.
34. Glikmans, G., Torck, B., Hellin, M., Coussemant, F. Ritter reaction between isobutene and acrylonitrile. I. De-scriptive study. *Bull. Soc. Chim. Fr.* **1966**, 1376-1383.
35. Norell, J. R. Organic reactions in liquid hydrogen fluoride. I. Synthetic aspects of the Ritter reaction in hydrogen fluoride. *J. Org. Chem.* **1970**, 35, 1611-1618.
36. Janout, V., Cefelin, P. The Ritter reaction of polyacrylonitrile with tert-butyl alcohol. *Eur. Polym. J.* **1980**, 16, 1075-1078.
37. Cacace, F., Ciranni, G., Giacomello, P. Alkylation of nitriles with gaseous carbenium ions. The Ritter reaction in the dilute gas state. *J. Am. Chem. Soc.* **1982**, 104, 2258-2261.
38. Thibault-Starzyk, F., Payen, R., Lavalley, J.-C. IR evidence of zeolitic hydroxy insertion in amide formation by the Ritter reaction. *Chem. Commun.* **1996**, 2667-2668.
39. Colombo, M. I., Bohn, M. L., Ruveda, E. A. The mechanism of the Ritter reaction in combination with Wagner-Meerwin rearrangements: A cooperative learning experience. *J. Chem. Educ.* **2002**, 79, 484-485.
40. Gerasimova, N. P., Nozhnin, N. A., Ermolaeva, V. V., Ovchinnikova, A. V., Moskvichev, Y. A., Alov, E. M., Danilova, A. S. The Ritter reaction mechanism: New corroboration in the synthesis of arylsulfonyl(thio)propionic acid N-(1-adamantyl)amides. *Mendeleev Commun.* **2003**, 82-84.
41. Stoermer, D., Heathcock, C. H. Total synthesis of (-)-alloaristoteline, (-)-serratoline, and (+)-aristotelone. *J. Org. Chem.* **1993**, 58, 564-568.
42. Ho, T.-L., Kung, L.-R., Chein, R.-J. Total Synthesis of (±)-2-Isocyanoallopupukeanane. *J. Org. Chem.* **2000**, 65, 5774-5779.
43. Van Emelen, K., De Wit, T., Hoornaert, G. J., Compernolle, F. Synthesis of cis-fused hexahydro-4aH-indeno[1,2-b]pyridines via intramolecular Ritter reaction and their conversion to tricyclic analogs of NK-1 and dopamine receptor ligands. *Tetrahedron* **2002**, 58, 4225-4236.

Robinson Annulation..384

Related reactions: Hajos-Parrish reaction;

1. Rapson, W. S., Robinson, R. Synthesis of substances related to the sterols. II. New general method for the synthesis of substituted cyclohexenones. *J. Chem. Soc.* **1935**, 1285-1288.
2. du Feu, E. C., McQuillin, F. J., Robinson, R. Synthesis of substances related to the sterols. XIV. A simple synthesis of certain octalones and ketotetrahydrohydrindenes which may be of angle-methyl-substituted type. A theory of the biogenesis of the sterols. *J. Chem. Soc., Abstracts* **1937**, 53-60.
3. Bergmann, E. D., Ginsburg, D., Pappo, R. The Michael reaction. *Org. React.* **1959**, 10, 179-563.
4. Gawley, R. E. The Robinson annelation and related reactions. *Synthesis* **1976**, 777-794.
5. Jung, M. E. A review of annulation. *Tetrahedron* **1976**, 32, 3-31.
6. Heathcock, C. H. The Aldol Reaction: Acid and General Base Catalysis. in *Comp. Org. Synth.* (eds. Trost, B. M.,Fleming, I.), *1*, 133-179 (Pergamon Press, Oxford, **1991**).
7. Sera, A. Addition reactions to conjugated compounds. *Org. Synth. High Pressures* **1991**, 179-200.
8. Varner, M. A., Grossman, R. B. Annulation routes to trans-decalins. *Tetrahedron* **1999**, 55, 13867-13886.
9. Jarvo, E. R., Miller, S. J. Amino acids and peptides as asymmetric organocatalysts. *Tetrahedron* **2002**, 58, 2481-2495.
10. Hajos, Z. G., Parrish, D. R. *Asymmetric synthesis of optically active polycyclic organic compounds.* De 2102623, **1971** (Hoffmann-La Roche, F., und Co., A.-G.). 42 pp.
11. Scanio, C. J. V., Starrett, R. M. Remarkably stereoselective Robinson annulation reaction. *J. Am. Chem. Soc.* **1971**, 93, 1539-1540.
12. Stork, G., Ganem, B. α-Silylated vinyl ketones. New class of reagents for the annelation of ketones. *J. Am. Chem. Soc.* **1973**, 95, 6152-6153.
13. Hajos, Z. G., Parrish, D. R. Asymmetric synthesis of bicyclic intermediates of natural product chemistry. *J. Org. Chem.* **1974**, 39, 1615-1621.
14. Stork, G., Jung, M. E. Vinylsilanes as carbonyl precursors. Use in annelation reactions. *J. Am. Chem. Soc.* **1974**, 96, 3682-3684.
15. Telschow, J. E., Reusch, W. Enamino ketone variant of the Robinson annelation. *J. Org. Chem.* **1975**, 40, 862-865.
16. Zoretic, P. A., Branchaud, B., Maestrone, T. Robinson annelations with a β-chloro ketone in the presence of an acid. *Tetrahedron Lett.* **1975**, 527-528.
17. Kamat, P. L., Shaligram, A. M. Acid-catalyzed Robinson annulation of 4,4-ethylenedioxy-2-methylcyclohexanone with ethyl vinyl ketone. *Indian J. Chem., Sect. B* **1980**, 19B, 904-905.
18. Ziegler, F. E., Hwang, K. J. On the aprotic Robinson annelation of dihydrocarvone and 2-methylcyclohexanone with methyl and ethyl vinyl ketone. *J. Org. Chem.* **1983**, 48, 3349-3351.
19. Huffman, J. W., Potnis, S. M., Satish, A. V. A silyl enol ether variation of the Robinson annulation. *J. Org. Chem.* **1985**, 50, 4266-4270.
20. Kuo, F., Fuchs, P. L. Bruceantin support studies. II. Use of 1-penten-3-one-4-phosphonate as a kinetic ethyl vinyl ketone equivalent in the Robinson annulation reaction. *Synth. Commun.* **1986**, 16, 1745-1759.

21. Olsen, R. S., Fataftah, Z. A., Rathke, M. W. Convenient procedure for carboxylation and Robinson annulations of ketones using triethylamine in the presence of magnesium chloride. *Synth. Commun.* **1986**, 16, 1133-1139.
22. Haynes, R. K., Vonwiller, S. C., Hambley, T. W. Use of β-sulfonyl vinyl ketones as equivalents to vinyl ketones in the Robinson annelation. Convergent, highly stereoselective preparation of a hydrindanol related to vitamin D from 2-methylcyclopent-2-enone and lithiated (E)-but-2-enyldiphenylphosphine oxide. *J. Org. Chem.* **1989**, 54, 5162-5170.
23. Sato, T., Wakahara, Y., Otera, J., Nozaki, H. Organotin triflates as functional Lewis acids. A new entry to simple and efficient Robinson annulation. *Tetrahedron Lett.* **1990**, 31, 1581-1584.
24. Kim, S., Emeric, G., Fuchs, P. L. Use of β-silylethyl vinyl ketone as a β-hydroxyethyl vinyl ketone synthon in the Robinson annulation reaction. *J. Org. Chem.* **1992**, 57, 7362-7364.
25. Hudlicky, M. The Wichterle reaction. *Collect. Czech. Chem. Commun.* **1993**, 58, 2229-2244.
26. Okano, T., Tamura, M., Kiji, J. Asymmetric Robinson annelation catalyzed by lanthanoid alkoxides. *Kidorui* **1997**, 30, 300-301.
27. Saito, S., Shimada, I., Takamori, Y., Tanaka, M., Maruoka, K., Yamamoto, H. Regioselective Robinson annulation realized by the combined use of lithium enolates and aluminum tris(2,6-diphenylphenoxide) (ATPH). *Bull. Chem. Soc. Jpn.* **1997**, 70, 1671-1681.
28. Zhong, G., Hoffmann, T., Lerner, R. A., Danishefsky, S., Barbas, C. F., III. Antibody-Catalyzed Enantioselective Robinson Annulation. *J. Am. Chem. Soc.* **1997**, 119, 8131-8132.
29. Bui, T., Barbas, C. F. A proline-catalyzed asymmetric Robinson annulation reaction. *Tetrahedron Lett.* **2000**, 41, 6951-6954.
30. Miyamoto, H., Kanetaka, S., Tanaka, K., Yoshizawa, K., Toyota, S., Toda, F. Solvent-free Robinson annelation reaction. *Chem. Lett.* **2000**, 888-889.
31. Rajagopal, D., Narayanan, R., Swaminathan, S. Enantioselective solvent-free Robinson annulation reactions. *Proc. - Indian Acad. Sci., Chem. Sci.* **2001**, 113, 197-213.
32. Rajagopal, D., Narayanan, R., Swaminathan, S. Asymmetric one-pot Robinson annulations. *Tetrahedron Lett.* **2001**, 42, 4887-4890.
33. Snider, B. B., Shi, B. A novel extension of the Stork-Jung vinylsilane Robinson annelation procedure for the introduction of the cyclohexene of guanacastepene. *Tetrahedron Lett.* **2001**, 42, 9123-9126.
34. Liu, H.-J., Ly, T. W., Tai, C.-L., Wu, J.-D., Liang, J.-K., Guo, J.-C., Tseng, N.-W., Shia, K.-S. A modified Robinson annulation process to α,α-disubstituted-β,γ-unsaturated cyclohexanone system. Application to the total synthesis of nanaimoal. *Tetrahedron* **2003**, 59, 1209-1226.
35. Takatori, K., Nakayama, M., Futaishi, N., Yamada, S., Hirayama, S., Kajiwara, M. Solid-supported Robinson annulation under microwave irradiation. *Chem. Pharm. Bull.* **2003**, 51, 455-457.
36. Kawanami, H., Ikushima, Y. Promotion of one-pot Robinson annelation achieved by gradual pressure and temperature manipulation under supercritical conditions. *Tetrahedron Lett.* **2004**, 45, 5147-5150.
37. Frontier, A. J., Raghavan, S., Danishefsky, S. J. A Highly Stereoselective Total Synthesis of Hispidospermidin: Derivation of a Pharmacophore Model. *J. Am. Chem. Soc.* **2000**, 122, 6151-6159.
38. White, J. D., Hrnciar, P., Stappenbeck, F. Asymmetric Total Synthesis of (+)-Codeine via Intramolecular Carbenoid Insertion. *J. Org. Chem.* **1999**, 64, 7871-7884.
39. Paquette, L. A., Wang, T.-Z., Sivik, M. R. Total Synthesis of (-)-Austalide B. A Generic Solution to Elaboration of the Pyran/p-Cresol/Butenolide Triad. *J. Am. Chem. Soc.* **1994**, 116, 11323-11334.
40. Shi, B., Hawryluk, N. A., Snider, B. B. Formal Synthesis of (±)-Guanacastepene A. *J. Org. Chem.* **2003**, 68, 1030-1042.

Roush Asymmetric Allylation 386

Related reactions: Keck asymmetric allylation, Sakurai allylation;

1. Hoffmann, R. W., Herold, T. Enantioselective synthesis of homoallyl alcohols via chiral allylboronic esters. *Angew. Chem. Int. Ed.* **1978**, 17, 768-769.
2. Roush, W. R., Walts, A. E., Hoong, L. K. Diastereo- and enantioselective aldehyde addition reactions of 2-allyl-1,3,2-dioxaborolane-4,5-dicarboxylic esters, a useful class of tartrate ester modified allylboronates. *J. Am. Chem. Soc.* **1985**, 107, 8186-8190.
3. Roush, W. R., Ando, K., Powers, D. B., Halterman, R. L., Palkowitz, A. D. Enantioselective synthesis using diisopropyl tartrate-modified (E)- and (Z)-crotylboronates: reactions with achiral aldehydes. *Tetrahedron Lett.* **1988**, 29, 5579-5582.
4. Roush, W. R., Grover, P. T. Diisopropyl tartrate (E)-γ-(dimethylphenylsilyl)allylboronate, a chiral allylic alcohol β-carbanion equivalent for the enantioselective synthesis of 2-butene-1,4-diols from aldehydes. *Tetrahedron Lett.* **1990**, 31, 7567-7570.
5. Roush, W. R., Grover, P. T., Lin, X. Diisopropyl tartrate modified (E)-γ-[(cyclohexyloxy)dimethylsilyl]allylboronate, a chiral reagent for the stereoselective synthesis of anti 1,2-diols via the formal α-hydroxyallylation of aldehydes. *Tetrahedron Lett.* **1990**, 31, 7563-7566.
6. Kabalka, G. W. The reactions of organoboranes with carbonyl compounds and their derivatives. *Spec. Publ. - R. Soc. Chem.* **1997**, 201, 139-150.
7. Inomata, K., Ukaji, Y. Development of new asymmetric reactions utilizing tartaric acid ester as a chiral auxiliary: design of an efficient chiral dinucleating system. *Rev. on Heteroa. Chem.* **1998**, 18, 119-140.
8. Williams, D. R., Brooks, D. A., Meyer, K. G., Clark, M. P. Asymmetric allylation. An effective strategy for the convergent synthesis of highly functionalized homoallylic alcohols. *Tetrahedron Lett.* **1998**, 39, 7251-7254.
9. Mulzer, J. Basic principles of asymmetric synthesis. *Comprehensive Asymmetric Catalysis I-III* **1999**, 1, 33-97.
10. Ramachandran, P. V. Pinane-based versatile "allyl" boranes. *Aldrichimica Acta* **2002**, 35, 23-35.
11. Persichini, P. J., III. Carbon-carbon bond formation via boron mediated transfer. *Curr. Org. Chem.* **2003**, 7, 1725-1736.
12. Haruta, R., Ishiguro, M., Ikeda, N., Yamamoto, H. Chiral allenylboronic esters: a practical reagent for enantioselective carbon-carbon bond formation. *J. Am. Chem. Soc.* **1982**, 104, 7667-7669.
13. Brown, H. C., Jadhav, P. K. Asymmetric carbon-carbon bond formation via β-allyldiisopinocampheylborane. Simple synthesis of secondary homoallylic alcohols with excellent enantiomeric purities. *J. Am. Chem. Soc.* **1983**, 105, 2092-2093.
14. Ikeda, N., Omori, K., Yamamoto, H. Complete 1,3-asymmetric induction in the reactions of allenylboronic acid with β-hydroxy ketones. *Tetrahedron Lett.* **1986**, 27, 1175-1178.
15. Garcia, J., Kim, B. M., Masamune, S. Asymmetric addition of (E)- and (Z)-crotyl-trans-2,5-dimethylborolanes to aldehydes. *J. Org. Chem.* **1987**, 52, 4831-4832.
16. Brown, H. C., Jadhav, P. K., Bhat, K. S. Chiral synthesis via organoboranes. 13. A highly diastereoselective and enantioselective addition of [(Z)-γ-alkoxyallyl]diisopinocampheylboranes to aldehydes. *J. Am. Chem. Soc.* **1988**, 110, 1535-1538.
17. Roush, W. R., Banfi, L. N,N'-dibenzyl-N,N'-ethylenetartramide: a rationally designed chiral auxiliary for the allylboration reaction. *J. Am. Chem. Soc.* **1988**, 110, 3979-3982.
18. Corey, E. J., Yu, C. M., Kim, S. S. A practical and efficient method for enantioselective allylation of aldehydes. *J. Am. Chem. Soc.* **1989**, 111, 5495-5496.
19. Brown, H. C., Randad, R. S. B-2'-Isoprenyldiisopinocampheylborane: an efficient reagent for the chiral isoprenylation of aldehydes. A convenient route to both enantiomers of ipsenol and ipsdienol. *Tetrahedron Lett.* **1990**, 31, 455-458.
20. Corey, E. J., Yu, C. M., Lee, D. H. A practical and general enantioselective synthesis of chiral propa-1,2-dienyl and propargyl carbinols. *J. Am. Chem. Soc.* **1990**, 112, 878-879.
21. Brown, H. C., Bhat, K. S., Jadhav, P. K. Chiral synthesis via organoboranes. Part 32. Synthesis of B-(cycloalk-2-enyl)diisopinocampheylboranes of high enantiomeric purity via the asymmetric hydroboration of cycloalka-1,3-dienes. Successful asymmetric allylborations of aldehydes with B-(cycloalk-2-enyl)diisopinocampheylboranes. *J. Chem. Soc., Perkin Trans. 1* **1991**, 2633-2638.
22. Racherla, U. S., Brown, H. C. Chiral synthesis via organoboranes. 27. Remarkably rapid and exceptionally enantioselective (approaching 100% ee) allylboration of representative aldehydes at -100 °C under new, salt-free conditions. *J. Org. Chem.* **1991**, 56, 401-404.

23. Omoto, K., Fujimoto, H. Theoretical Study of the Effects of Structure and Substituents on Reactivity in Allylboration. *J. Org. Chem.* **1998**, 63, 8331-8336.
24. Gung, B. W., Xue, X. Asymmetric transition states of allylation reaction: an ab initio molecular orbital study. *Tetrahedron: Asymmetry* **2001**, 12, 2955-2959.
25. Ruiz, M., Ojea, V., Quintela, J. M. Computational study of the syn,anti-selective aldol additions of lithiated bis-lactim ether to 1,3-dioxolane-4-carboxaldehydes. *Tetrahedron: Asymmetry* **2002**, 13, 1863-1873.
26. Kozlowski, M. C., Panda, M. Computer-Aided Design of Chiral Ligands. Part 2. Functionality Mapping as a Method To Identify Stereocontrol Elements for Asymmetric Reactions. *J. Org. Chem.* **2003**, 68, 2061-2076.
27. Lipkowitz, K. B., Kozlowski, M. C. Understanding stereoinduction in catalysis via computer: New tools for asymmetric synthesis. *Synlett* **2003**, 1547-1565.
28. Kang, S. H., Kang, S. Y., Kim, C. M., Choi, H.-w., Jun, H.-S., Lee, B. M., Park, C. M., Jeong, J. W. Total synthesis of natural (+)-lasonolide A. *Angew. Chem., Int. Ed. Engl.* **2003**, 42, 4779-4782.
29. Hayward, M. M., Roth, R. M., Duffy, K. J., Dalko, P. I., Stevens, K. L., Guo, J., Kishi, Y. Total synthesis of altohyrtin A (spongistatin 1): Part 2. *Angew. Chem., Int. Ed. Engl.* **1998**, 37, 192-196.
30. Kohyama, N., Yamamoto, Y. Total synthesis of stevastelin B, a novel immunosuppressant. *Synlett* **2001**, 694-696.
31. Xiang, A. X., Watson, D. A., Ling, T., Theodorakis, E. A. Total synthesis of clerocidin via a novel, enantioselective homoallenylboration methodology. *J. Org. Chem.* **1998**, 63, 6774-6775.
32. Soundararajan, R., Li, G., Brown, H. C. Chiral Syntheses via Organoboranes. 44. Racemic and Diastereo- and Enantioselective Homoallenylboration Using Dialkyl 2,3-Butadien-1-ylboronate Reagents. Another Novel Application of the Tandem Homologation-Allylboration Strategy. *J. Org. Chem.* **1996**, 61, 100-104.

Rubottom Oxidation 388

Related reactions: Davis oxaziridine oxidation;

1. Brook, A. G., Macrae, D. M. 1,4-Silyl rearrangements of siloxyalkenes to siloxyketones during peroxidation. *J. Organomet. Chem.* **1974**, 77, C19-C21.
2. Rubottom, G. M., Vazquez, M. A., Pelegrina, D. R. Peracid oxidation of trimethylsilyl enol ethers. Facile α-hydroxylation procedure. *Tetrahedron Lett.* **1974**, 4319-4322.
3. Hassner, A., Reuss, R. H., Pinnick, H. W. Synthetic methods. VIII. Hydroxylation of carbonyl compounds via silyl enol ethers. *J. Org. Chem.* **1975**, 40, 3427-3429.
4. Rubottom, G. M., Gruber, J. M. m-Chloroperbenzoic acid oxidation of 2-trimethylsilyloxy-1,3-dienes. Synthesis of α-hydroxy and α-acetoxy enones. *J. Org. Chem.* **1978**, 43, 1599-1602.
5. Davis, F. A., Sheppard, A. C. Oxidation of silyl enol ethers using 2-sulfonyloxaziridines. Synthesis of α-siloxy epoxides and α-hydroxy carbonyl compounds. *J. Org. Chem.* **1987**, 52, 954-955.
6. Kaye, P. T., Learmonth, R. A. Chiral organosilicon compounds. Part 3. Peracid oxidation of chiral silyl enol ethers. *Synth. Commun.* **1990**, 20, 1333-1338.
7. Reddy, D. R., Thornton, E. R. A very mild, catalytic and versatile procedure for α-oxidation of ketone silyl enol ethers using (salen)manganese(III) complexes; a new, chiral complex giving asymmetric induction. A possible model for selective biochemical oxidative reactions through enol formation. *J. Chem. Soc., Chem. Commun.* **1992**, 172-173.
8. Jauch, J. Stereochemistry of the Rubottom oxidation with bicyclic silyl enol ethers; synthesis and dimerization reactions of bicyclic α-hydroxy ketones. *Tetrahedron* **1994**, 50, 12903-12912.
9. Adam, W., Fell, R. T., Saha-Moller, C. R., Zhao, C.-G. Synthesis of optically active α-hydroxy ketones by enantioselective oxidation of silyl enol ethers with a fructose-derived dioxirane. *Tetrahedron: Asymmetry* **1998**, 9, 397-401.
10. Adam, W., Fell, R. T., Stegmann, V. R., Saha-Moeller, C. R. Synthesis of Optically Active α-Hydroxy Carbonyl Compounds by the Catalytic, Enantioselective Oxidation of Silyl Enol Ethers and Ketene Acetals with (Salen)manganese(III) Complexes. *J. Am. Chem. Soc.* **1998**, 120, 708-714.
11. Stankovic, S., Espenson, J. H. Facile Oxidation of Silyl Enol Ethers with Hydrogen Peroxide Catalyzed by Methyltrioxorhenium. *J. Org. Chem.* **1998**, 63, 4129-4130.
12. Zhu, Y., Tu, Y., Yu, H., Shi, Y. Highly enantioselective epoxidation of esters and enol silyl ethers. *Tetrahedron Lett.* **1998**, 39, 7819-7822.
13. Adam, W., Saha-Moller, C. R., Ganeshpure, P. A. Synthetic applications of nonmetal catalysts for homogeneous oxidations. *Chem. Rev.* **2001**, 101, 3499-3548.
14. Solladie-Cavallo, A., Lupattelli, P., Jierry, L., Bovicelli, P., Angeli, F., Antonioletti, R., Klein, A. Asymmetric oxidation of silyl enol ethers using chiral dioxiranes derived from α-fluoro cyclohexanones. *Tetrahedron Lett.* **2003**, 44, 6523-6526.
15. Brownbridge, P. Silyl enol ethers in synthesis - part II. *Synthesis* **1983**, 85-104.
16. Brownbridge, P. Silyl enol ethers in synthesis - part I. *Synthesis* **1983**, 1-28.
17. Dayan, S., Bareket, Y., Rozen, S. An efficient α-hydroxylation of carbonyls using the HOF.CH3CN complex. *Tetrahedron* **1999**, 55, 3657-3664.
18. Brook, A. G. Molecular rearrangements of organosilicon compounds. *Acc. Chem. Res.* **1974**, 7, 77-84.
19. Rubottom, G. M., Gruber, J. M., Boeckman, R. K., Jr., Ramaiah, M., Medwid, J. B. Clarification of the mechanism of rearrangement of enol silyl ether epoxides. *Tetrahedron Lett.* **1978**, 19, 4603-4606.
20. Allen, J. G., Danishefsky, S. J. The Total Synthesis of (±)-Rishirilide B. *J. Am. Chem. Soc.* **2001**, 123, 351-352.
21. Thompson, C. F., Jamison, T. F., Jacobsen, E. N. Total Synthesis of FR901464. Convergent Assembly of Chiral Components Prepared by Asymmetric Catalysis. *J. Am. Chem. Soc.* **2000**, 122, 10482-10483.
22. Zoretic, P. A., Wang, M., Zhang, Y., Shen, Z., Ribeiro, A. A. Total Synthesis of d,l-Isospongiadiol: An Intramolecular Radical Cascade Approach to Furanoditerpenes. *J. Org. Chem.* **1996**, 61, 1806-1813.
23. Frontier, A. J., Raghavan, S., Danishefsky, S. J. A Highly Stereoselective Total Synthesis of Hispidospermidin: Derivation of a Pharmacophore Model. *J. Am. Chem. Soc.* **2000**, 122, 6151-6159.

Saegusa Oxidation 390

1. Bierling, B., Kirschke, K., Oberender, H., Schulz, M. Dehydrogenation of ketones with palladium(II) compounds. *J. Prakt. Chem.* **1972**, 314, 170-180.
2. Ito, Y., Hirao, T., Saegusa, T. Synthesis of α,β-unsaturated carbonyl compounds by palladium(II)-catalyzed dehydrosilylation of silyl enol ethers. *J. Org. Chem.* **1978**, 43, 1011-1013.
3. Buckle, D. R., Pinto, I. L. Oxidation Adjacent to C=X Bonds by Dehydrogenation. in *Comp. Org. Synth.* (eds. Trost, B. M.,Fleming, I.), 7, 119-149 (Pergamon, Oxford, **1991**).
4. Heumann, A., Jens, K.-J., Reglier, M. Palladium complex catalyzed oxidation reactions. *Prog. Inorg. Chem.* **1994**, 42, 483-576.
5. Friesen, R. W. Product subclass 2: palladium-allyl complexes. *Science of Synthesis* **2002**, 1, 113-264.
6. Ito, Y., Suginome, M. Palladium-catalyzed or -promoted oxidation via 1,2- or 1,4-elimination: oxidation of silyl enol ethers and related enol derivatives to α,β-unsaturated enones and other carbonyl compounds. *Handbook of Organopalladium Chemistry for Organic Synthesis* **2002**, 2, 2873-2879.
7. Toyota, M., Ihara, M. Development of palladium-catalyzed cycloalkenylation and its application to natural product synthesis. *Synlett* **2002**, 1211-1222.

8. Shimizu, I., Minami, I., Tsuji, J. Palladium-catalyzed synthesis of α,β-unsaturated ketones from ketones via allyl enol carbonates. *Tetrahedron Lett.* **1983**, 24, 1797-1800.
9. Tsuji, J., Minami, I., Shimizu, I. A novel palladium-catalyzed method for preparation of α,β-unsaturated ketones and aldehydes from saturated ketones and aldehydes via their silyl enol ethers. *Tetrahedron Lett.* **1983**, 24, 5635-5638.
10. Tsuji, J., Minami, I., Shimizu, I. One-step synthesis of α,β-unsaturated ketones by the reaction of enol acetates with allyl methyl carbonate catalyzed by palladium and tin compounds. *Tetrahedron Lett.* **1983**, 24, 5639-5640.
11. Larock, R. C., Hightower, T. R., Kraus, G. A., Hahn, P., Zheng, D. A simple effective new palladium-catalyzed conversion of enol silanes to enones and enals. *Tetrahedron Lett.* **1995**, 36, 2423-2426.
12. Brownbridge, P. Silyl enol ethers in synthesis - part II. *Synthesis* **1983**, 85-104.
13. Brownbridge, P. Silyl enol ethers in synthesis - part I. *Synthesis* **1983**, 1-28.
14. Nicolaou, K. C., Gray, D. L. F., Montagnon, T., Harrison, S. T. Oxidation of silyl enol ethers by using IBX and IBX.N-oxide complexes: A mild and selective reaction for the synthesis of enones. *Angew. Chem., Int. Ed. Engl.* **2002**, 41, 996-1000.
15. Porth, S., Bats, J. W., Trauner, D., Giester, G., Mulzer, J. Insight into the mechanism of the Saegusa oxidation: isolation of a novel palladium(0)-tetraolefin complex. *Angew. Chem., Int. Ed. Engl.* **1999**, 38, 2015-2016.
16. Fuwa, H., Kainuma, N., Tachibana, K., Sasaki, M. Total Synthesis of (-)-Gambierol. *J. Am. Chem. Soc.* **2002**, 124, 14983-14992.
17. Barrett, A. G. M., Blaney, F., Campbell, A. D., Hamprecht, D., Meyer, T., White, A. J. P., Witty, D., Williams, D. J. Unified Route to the Palmarumycin and Preussomerin Natural Products. Enantioselective Synthesis of (-)-Preussomerin G. *J. Org. Chem.* **2002**, 67, 2735-2750.
18. Ihara, M., Makita, K., Takasu, K. Facile construction of the tricyclo 5.2.1.0(1,5) decane ring system by intramolecular double Michael reaction: Highly stereocontrolled total synthesis of (±)-8,14-cedranediol and (±)-8,14-cedranoxide. *J. Org. Chem.* **1999**, 64, 1259-1264.
19. Toyooka, N., Okumura, M., Nemoto, H. Stereodivergent process for the synthesis of the decahydroquinoline type of dendrobatid alkaloids. *J. Org. Chem.* **2002**, 67, 6078-6081.

Sakurai Allylation 392

Related reactions: ***Keck asymmetric allylation, Roush asymmetric allylation;***

1. Hosomi, A., Endo, M., Sakurai, H. Chemistry of organosilicon compounds. 91. Allylsilanes as synthetic intermediates. II. Syntheses of homoallyl ethers from allylsilanes and acetals promoted by titanium tetrachloride. *Chem. Lett.* **1976**, 941-942.
2. Hosomi, A., Sakurai, H. Chemistry of organosilicon compounds. 89. Syntheses of γ,δ-unsaturated alcohols from allylsilanes and carbonyl compounds in the presence of titanium tetrachloride. *Tetrahedron Lett.* **1976**, 1295-1298.
3. Schinzer, D. Intramolecular addition reactions of allylic and propargylic silanes. *Synthesis* **1988**, 263-273.
4. Fleming, I., Dunogues, J., Smithers, R. The electrophilic substitution of allylsilanes and vinylsilanes. *Org. React.* **1989**, 37, 57-575.
5. Yamamoto, Y., Sasaki, N. The stereochemistry of the Sakurai reaction. *Stereochemistry of Organometallic and Inorganic Compounds* **1989**, 3, 363-441.
6. Fleming, I. Allylsilanes, allylstannanes and related systems. in *Comp. Org. Synth.* (eds. Trost, B. M.,Fleming, I.), *6*, 563-593 (Pergamon Press, Oxford, **1991**).
7. Bianchini, C., Glendenning, L. Homogeneous catalysis. Mechanisms of the catalytic Mukaiyama aldol and Sakurai allylation reactions. *Chemtracts: Inorg. Chem.* **1995**, 7, 107-111.
8. Bianchini, C., Glendenning, L. Homogeneous catalysis. Mechanisms of the catalytic Mukaiyama aldol and Sakurai allylation reactions. *Chemtracts: Org. Chem.* **1996**, 9, 331-335.
9. Dai, L.-X., Lin, Y.-R., Hou, X.-L., Zhou, Y.-G. Stereoselective reactions with imines. *Pure Appl. Chem.* **1999**, 71, 1033-1040.
10. Tsunoda, T., Suzuki, M., Noyori, R. Trialkylsilyl triflates. III. Trimethylsilyl trifluoromethanesulfonate as a catalyst of the reaction of allyltrimethylsilane and acetals. *Tetrahedron Lett.* **1980**, 21, 71-74.
11. Sakurai, H., Sasaki, K., Hosomi, A. Chemistry of organosilicon compounds. 143. Regiospecific allylation of acetals with allylsilanes catalyzed by iodotrimethylsilane. Synthesis of homoallylethers. *Tetrahedron Lett.* **1981**, 22, 745-748.
12. Mukaiyama, T., Nagaoka, H., Murakami, M., Ohshima, M. A facile synthesis of homoallyl ethers. The reaction of acetals with allyltrimethylsilanes promoted by trityl perchlorate or diphenylboryl triflate. *Chem. Lett.* **1985**, 977-980.
13. Davis, A. P., Jaspars, M. Superacid catalysis of the addition of allysilanes to carbonyl compounds. *J. Chem. Soc., Chem. Commun.* **1990**, 1176-1178.
14. Hollis, T. K., Robinson, N. p., Whelan, J., Bosnich, B. Homogeneous catalysis. Use of the [$TiCp_2(CF_3SO_3)_2$] catalyst for the Sakurai reaction of allylic silanes with orthoesters, acetals, ketals and carbonyl compounds. *Tetrahedron Lett.* **1993**, 34, 4309-4312.
15. Ishihara, K., Mouri, M., Gao, Q., Maruyama, T., Furuta, K., Yamamoto, H. Catalytic asymmetric allylation using a chiral (acyloxy)borane complex as a versatile Lewis acid catalyst. *J. Am. Chem. Soc.* **1993**, 115, 11490-11495.
16. Polla, M., Frejd, T. Lewis Acid-induced alkoxyalkylation of allylsilanes with acetals (the Sakurai reaction): regio- and stereochemical aspects. *Acta Chem. Scand.* **1993**, 47, 716-720.
17. Hollis, T. K., Bosnich, B. Homogeneous Catalysis. Mechanisms of the Catalytic Mukaiyama Aldol and Sakurai Allylation Reactions. *J. Am. Chem. Soc.* **1995**, 117, 4570-4581.
18. Kumagai, T., Itsuno, S. Asymmetric Allylation Polymerization: Novel Polyaddition of Bis(allylsilane) and Dialdehyde Using Chiral (Acyloxy)borane Catalyst. *Macromolecules* **2000**, 33, 4995-4996.
19. Wang, M. W., Chen, Y. J., Wang, D. Catalytic allylation of aldehydes with allyltrimethylsilane using in situ-generated trimethylsilyl methanesulfonate (TMSOMs) as a catalyst. *Synlett* **2000**, 385-387.
20. Lee, P. H., Lee, K., Sung, S.-y., Chang, S. The Catalytic Sakurai Reaction. *J. Org. Chem.* **2001**, 66, 8646-8649.
21. Onishi, Y., Ito, T., Yasuda, M., Baba, A. Indium(III) chloride/chlorotrimethylsilane as a highly active Lewis acid catalyst system for the Sakurai-Hosomi reaction. *Eur. J. Org. Chem.* **2002**, 1578-1581.
22. Cesarotti, E., Araneo, S., Rimoldi, I., Tassi, S. Enantioselective Mukaiyama aldol and Sakurai allylation reactions catalysed by silver(I) complexes with chiral atropisomeric chelating ligands. *J. Mol. Catal. A: Chemical* **2003**, 204-205, 221-226.
23. Wadamoto, M., Ozasa, N., Yanagisawa, A., Yamamoto, H. BINAP/AgOTf/KF/18-Crown-6 as New Bifunctional Catalysts for Asymmetric Sakurai-Hosomi Allylation and Mukaiyama Aldol Reaction. *J. Org. Chem.* **2003**, 68, 5593-5601.
24. Bottoni, A., Costa, A. L., Di Tommaso, D., Rossi, I., Tagliavini, E. New Computational and Experimental Evidence for the Mechanism of the Sakurai Reaction. *J. Am. Chem. Soc.* **1997**, 119, 12131-12135.
25. Organ, M. G., Dragan, V., Miller, M., Froese, R. D. J., Goddard, J. D. Sakurai Addition and Ring Annulation of Allylsilanes with α,β-Unsaturated Esters. Experimental Results and ab Initio Theoretical Predictions Examining Allylsilane Reactivity. *J. Org. Chem.* **2000**, 65, 3666-3678.
26. Hayashi, T., Konishi, M., Ito, H., Kumada, M. Optically active allylsilanes. 1. Preparation by palladium-catalyzed asymmetric Grignard cross-coupling and anti stereochemistry in electrophilic substitution reactions. *J. Am. Chem. Soc.* **1982**, 104, 4962-4963.
27. Danheiser, R. L., Carini, D. J., Kwasigroch, C. A. Scope and stereochemical course of the addition of (trimethylsilyl)allenes to ketones and aldehydes. A regiocontrolled synthesis of homopropargylic alcohols. *J. Org. Chem.* **1986**, 51, 3870-3878.
28. Pornet, J., Randrianoelina, B. Action of 1-trimethylsilyl-2-butyne on carbonyl derivatives in the presence of catalysts: synthesis of α-allenic alcohols or chloroprenic derivatives. *Tetrahedron Lett.* **1981**, 22, 1327-1328.
29. Deleris, G., Dunogues, J., Calas, R. Addition of silylated hydrocarbons with an activated silicon-carbon bond to some carbonyl compounds. I. Addition to chloral. *J. Organomet. Chem.* **1975**, 93, 43-50.
30. Mukaiyama, T., Murakami, M. Cross-coupling reactions based on acetals. *Synthesis* **1987**, 1043-1054.
31. Mori, I., Bartlett, P. A., Heathcock, C. H. High diastereofacial selectivity in nucleophilic additions to chiral thionium ions. *J. Am. Chem. Soc.* **1987**, 109, 7199-7200.

32. Nishiyama, H., Narimatsu, S., Sakuta, K., Itoh, K. Reaction of allylsilanes and monothioacetals in the presence of Lewis acids: regioselectivity in the cleavage of the acetals. *J. Chem. Soc., Chem. Commun.* **1982**, 459-460.
33. Shikhmamedbekova, A. Z., Sultanov, R. A. Addition of α-chlorodimethyl ether to trialkylalkenylsilanes. *Zh. Obshch. Khim.* **1970**, 40, 77-84.
34. Ishibashi, H., Nakatani, H., Umei, Y., Yamamoto, W., Ikeda, M. Reaction of [arylthio(chloro)methyl]trimethylsilanes with arenes and 1-alkenes in the presence of Lewis acid: syntheses of [aryl(arylthio)methyl]- and [1-(arylthio)-3-alkenyl]trimethylsilanes. *J. Chem. Soc., Perkin Trans. 1* **1987**, 589-593.
35. Hosomi, A., Sakurai, H. Chemistry of organosilicon compounds. 99. Conjugate addition of allylsilanes to α,β-enones. A New method of stereoselective introduction of the angular allyl group in fused cyclic α,β-enones. *J. Am. Chem. Soc.* **1977**, 99, 1673-1675.
36. Sakurai, H., Hosomi, A., Hayashi, J. Conjugate allylation of α,β-unsaturated ketones with allylsilanes: 4-phenyl-6-hepten-2-one (6-hepten-2-one, 4-phenyl-). *Org. Synth.* **1984**, 62, 86-94.
37. Kuwajima, I., Tanaka, T., Atsumi, K. Preparation and reactions of 2-substituted 3-trimethylsilyl 4-en-1-one system. *Chem. Lett.* **1979**, 779-782.
38. Hayashi, T., Kabeta, K., Hamachi, I., Kumada, M. Erythroselectivity in addition of γ-substituted allylsilanes to aldehydes in the presence of titanium chloride. *Tetrahedron Lett.* **1983**, 24, 2865-2868.
39. Fleming, I., Langley, J. A. The mechanism of the protodesilylation of allylsilanes and vinylsilanes. *J. Chem. Soc., Perkin Trans. 1* **1981**, 1421-1423.
40. Cella, J. A. Reductive alkylation/arylation of arylcarbinols and ketones with organosilicon compounds. *J. Org. Chem.* **1982**, 47, 2125-2130.
41. Hosomi, A., Sakurai, H. Chemistry of organosilicon compounds. 112. Synthesis of α,α-dimethylallylsilanes, a reagent of regiospecific prenylation of acetals and carbonyl compounds. *Tetrahedron Lett.* **1978**, 2589-2592.
42. Hayashi, T., Konishi, M., Kumada, M. Optically active allylsilanes. 2. High stereoselectivity in asymmetric reaction with aldehydes producing homoallylic alcohols. *J. Am. Chem. Soc.* **1982**, 104, 4963-4965.
43. Denmark, S. E., Weber, E. J. On the stereochemistry of allylmetal-aldehyde condensations. Preliminary communication. *Helv. Chim. Acta* **1983**, 66, 1655-1660.
44. Denmark, S. E., Henke, B. R., Weber, E. $SnCl_4(4\text{-tert-}BuC_6H_4CHO)_2$. X-ray crystal structure, solution NMR, and implications for reactions at complexed carbonyls. *J. Am. Chem. Soc.* **1987**, 109, 2512-2514.
45. Sugimura, H., Uematsu, M. Unusual [2+2]cycloaddition reaction of allylsilanes with 2,3-O-isopropylidene derivatives of aldehydo-aldose catalyzed by boron trifluoride etherate. *Tetrahedron Lett.* **1988**, 29, 4953-4956.
46. Reetz, M. T., Raguse, B., Marth, C. F., Huegel, H. M., Bach, T., Fox, D. N. A. A rapid injection NMR study of the chelation controlled Mukaiyama aldol addition: $TiCl_4$ versus $LiClO_4$ as the Lewis acid. *Tetrahedron* **1992**, 48, 5731-5742.
47. Trost, B. M., Thiel, O. R., Tsui, H.-C. Total Syntheses of Furaquinocin A, B, and E. *J. Am. Chem. Soc.* **2003**, 125, 13155-13164.
48. Williams, D. R., Myers, B. J., Mi, L. Total Synthesis of (-)-Amphidinolide P. *Org. Lett.* **2000**, 2, 945-948.
49. Wender, P. A., Hegde, S. G., Hubbard, R. D., Zhang, L. Total Synthesis of (-)-Laulimalide. *J. Am. Chem. Soc.* **2002**, 124, 4956-4957.

Sandmeyer Reaction 394

Related reactions: Balz-Schiemann reaction;

1. Griess, J. P. *Philos. Trans. R. Soc. London* **1864**, 164, 693.
2. Griess, J. P. A new group of organic compounds in which the hydrogen is replaced by nitrogen. *Ann. Chem., Justus Liebigs* **1866**, 137, 39-91.
3. Sandmeyer, T. The substitution of the amine group with chlorine atom in aromatic systems. *Ber. Dtsch. Chem. Ges.* **1884**, 17, 1633-1635.
4. Sandmeyer, T. The substitution of the amine group with chloride, bromide or cyanide in aromatic systems. *Ber. Dtsch. Chem. Ges.* **1884**, 17, 2650-2653.
5. Hodgson, H. H. The Sandmeyer reaction. *Chem. Rev.* **1947**, 40, 251-277.
6. Cowdrey, W. A., Davies, D. S. Sandmeyer and related reactions. *Quart. Revs. (London)* **1952**, 6, 358-379.
7. Nonhebel, D. C. Copper-catalyzed single-electron oxidations and reductions. *Special Publication - Chemical Society* **1970**, No. 24, 409-437.
8. Wulfman, D. S. Synthetic applications of diazonium ions. in *The Chemistry of Diazonium and Diazo Groups* (ed. Patai, S.), *Part 1*, 247-339 (Wiley, **1978**).
9. Galli, C. Radical reactions of arenediazonium ions: An easy entry into the chemistry of the aryl radical. *Chem. Rev.* **1988**, 88, 765-792.
10. Merkushev, E. B. Advances in the synthesis of iodoaromatic compounds. *Synthesis* **1988**, 923-937.
11. Bohlmann, R. Synthesis of Halides. in *Comp. Org. Synth.* (eds. Trost, B. M.,Fleming, I.), *6*, 203-223 (Pergamon, Oxford, **1991**).
12. Doyle, M. P., Siegfried, B., Dellaria, J. F., Jr. Alkyl nitrite-metal halide deamination reactions. 2. Substitutive deamination of arylamines by alkyl nitrites and copper(II) halides. A direct and remarkably efficient conversion of arylamines to aryl halides. *J. Org. Chem.* **1977**, 42, 2426-2431.
13. Oae, S., Shinhama, K., Kim, Y. H. Direct conversion of arylamines to the corresponding halides, biphenyls, and sulfides with tert-butyl thionitrate. *Bull. Chem. Soc. Jpn.* **1980**, 53, 2023-2026.
14. Oae, S., Shinhama, K., Kim, Y. H. Direct conversion of arylamines to the halides by deamination with thionitrite or related compounds and anhydrous copper(II) halides. *Bull. Chem. Soc. Jpn.* **1980**, 53, 1065-1069.
15. Lee, J. G., Cha, H. T. One step conversion of anilines to aryl halides using sodium nitrite and halotrimethylsilane. *Tetrahedron Lett.* **1992**, 33, 3167-3168.
16. Obushak, M. D., Lyakhovych, M. B., Ganushchak, M. I. Arenediazonium tetrachlorocuprates(II). Modification of the Meerwein and Sandmeyer reactions. *Tetrahedron Lett.* **1998**, 39, 9567-9570.
17. Suzuki, H., Nonoyama, N. Nitrogen dioxide-sodium iodide as an efficient reagent for the one-pot conversion of aryl amines to aryl iodides under nonaqueous conditions. *Tetrahedron Lett.* **1998**, 39, 4533-4536.
18. Ozeki, N., Shimomura, N., Harada, H. A new Sandmeyer iodination of 2-aminopurines in non-aqueous conditions: combination of alkali metal iodide and iodine as iodine sources. *Heterocycles* **2001**, 55, 461-464.
19. Obushak, N. D., Lyakhovich, M. B., Bilaya, E. E. Arenediazonium tetrachlorocuprates(II). Modified versions of the Meerwein and Sandmeyer reactions. *Russ. J. Org. Chem.* **2002**, 38, 38-46.
20. Bagal, L. I., Pevzner, M. S., Frolov, A. N. Reaction of diazonium salts of the benzene series with sodium nitrite in the absence of a catalyst. *Zh. Org. Khim.* **1969**, 5, 1820-1828.
21. Opgenorth, H. J., Ruechardt, C. Aromatic diazo compounds. V. Reaction of aromatic diazonium salts with nitrite ions. *Liebigs Ann. Chem.* **1974**, 1333-1347.
22. Horning, D. E., Ross, D. A., Muchowski, J. M. Synthesis of phenols from diazonium tetrafluoroborates. Useful modification. *Can. J. Chem.* **1973**, 51, 2347-2348.
23. Cohen, T., Dietz, A. G., Jr., Miser, J. R. A simple preparation of phenols from diazonium ions via the generation and oxidation of aryl radicals by copper salts. *J. Org. Chem.* **1977**, 42, 2053-2058.
24. Hanson, P., Rowell, S. C., Walton, P. H., Timms, A. W. Promotion of Sandmeyer hydroxylation (homolytic hydroxydediazoniation) and hydrodediazoniation by chelation of the copper catalyst: bidentate ligands. *Org. Biomol. Chem.* **2004**, 2, 1838-1855.
25. Waters, W. A. Decomposition reactions of the aromatic diazo compounds. X. Mechanism of the Sandmeyer reaction. *J. Chem. Soc., Abstracts* **1942**, 266-270.
26. Kochi, J. K. The mechanism of the Sandmeyer and Meerwein reactions. *J. Am. Chem. Soc.* **1957**, 79, 2942-2948.
27. Dickerman, S. C., DeSouza, D. J., Jacobson, N. Role of copper chlorides in the Sandmeyer and Meerwein reactions. *J. Org. Chem.* **1969**, 34, 710-713.
28. Nakatani, Y. Sandmeyer reaction with ferrous chloride. *Tetrahedron Lett.* **1970**, 4455-4458.

29. Galli, C. An investigation of the two step nature of Sandmeyer reaction. *J. Chem. Soc., Perkin Trans. 2* **1981**, 1459-1461.
30. Galli, C. Evidence for the intermediacy of the aryl radical in the Sandmeyer reaction. *J. Chem. Soc., Perkin Trans. 2* **1982**, 1139-1141.
31. Singh, P. R., Kumar, R., Khanna, R. K. Radical nucleophilic substitution mechanism in the reactions of arenediazonium cations with nitrite ion. *Tetrahedron Lett.* **1982**, 23, 5191-5194.
32. Galli, C. Substituent effects on the Sandmeyer reaction. Quantitative evidence for rate-determining electron transfer. *J. Chem. Soc., Perkin Trans. 2* **1984**, 897-902.
33. Hanson, P., Jones, J. R., Gilbert, B. C., Timms, A. W. Sandmeyer reactions. Part 1. A comparative study of the transfer of halide and water ligands from complexes of copper(II) to aryl radicals. *J. Chem. Soc., Perkin Trans. 2* **1991**, 1009-1017.
34. Hanson, P., Hammond, R. C., Gilbert, B. C., Timms, A. W. Sandmeyer reactions. Part 3. Estimation of absolute rate constants for the transfer of chloride ligands from CuII to 2-benzoylphenyl radical (Pschorr radical clock) and further investigations of the relative rates of transfer of chloride and water ligands to other substituted phenyl radicals. *J. Chem. Soc., Perkin Trans. 2* **1995**, 2195-2202.
35. Hanson, P., Jones, J. R., Taylor, A. B., Walton, P. H., Timms, A. W. Sandmeyer reactions. Part 7.1 An investigation into the reduction steps of Sandmeyer hydroxylation and chlorination reactions. *J. Chem. Soc., Perkin Trans. 2* **2002**, 1135-1150.
36. Hanson, P., Rowell, S. C., Taylor, A. B., Walton, P. H., Timms, A. W. Sandmeyer reactions. Part 6.1 A mechanistic investigation into the reduction and ligand transfer steps of Sandmeyer cyanation. *J. Chem. Soc., Perkin Trans. 2* **2002**, 1126-1134.
37. Evans, D. A., Katz, J. L., Peterson, G. S., Hintermann, T. Total Synthesis of Teicoplanin Aglycon. *J. Am. Chem. Soc.* **2001**, 123, 12411-12413.
38. Takemura, S., Hirayama, A., Tokunaga, J., Kawamura, F., Inagaki, K., Hashimoto, K., Nakata, M. A concise total synthesis of (±)-A80915G, a member of the napyradiomycin family of antibiotics. *Tetrahedron Lett.* **1999**, 40, 7501-7505.

Schmidt Reaction 396

Related reactions: Curtius rearrangement, Hofmann rearrangement, Lossen rearrangement;

1. Schmidt, K. F. *Z. angew. Chem.* **1923**, 36, 511.
2. Schmidt, K. F. The imine residue. *Ber.* **1924**, 57B, 704-706.
3. Wolff, H. Schmidt reaction. *Org. React.* **1946**, 307-336.
4. Smith, P. A. S. Rearrangements involving migration to an electron-deficient nitrogen or oxygen. in *Molecular Rearrangements* (ed. Mayo, P.), *1*, 457-591 (Wiley, New York, **1963**).
5. Beckwith, A. L. J. Synthesis of amides. in *Chem. Amides* (ed. Zabicky), 73-185 (Wiley, New York, **1970**).
6. Koldobskii, G. I., Tereshchenko, G. F., Gerasimova, E. S., Bagal, L. I. Schmidt reaction with ketones. *Russ. Chem. Rev.* **1971**, 40, 835.
7. Koldobskii, G. I., Ostrovskii, V. A., Gidaspov, B. V. Schmidt reaction with aldehydes and carboxylic acids. *Russ. Chem. Rev.* **1978**, 47, 1084-1094.
8. Krow, G. R. Nitrogen insertion reactions of bridged bicyclic ketones. Regioselective lactam formation. *Tetrahedron* **1981**, 37, 1283-1307.
9. Shioiri, T. Degradation Reactions. in *Comp. Org. Synth.* (eds. Trost, B. M.,Fleming, I.), *6*, 795-828 (Pergamon, Oxford, **1991**).
10. Pearson, W. H. Aliphatic azides as Lewis bases. Application to the synthesis of heterocyclic compounds. *J. Heterocycl. Chem.* **1996**, 33, 1489-1496.
11. Trost, B. M., Vaultier, M., Santiago, M. L. Thionium ions as carbonyl substitutes. Synthesis of cyclic imino thioethers and lactams. *J. Am. Chem. Soc.* **1980**, 102, 7929-7932.
12. Ohkata, K., Mase, M., Akiba, K. Reaction of silyl enol ethers with N-chlorosuccinimide: trapping of the siloxycarbinyl cation by an azide anion. *J. Chem. Soc., Chem. Commun.* **1987**, 1727-1728.
13. Aube, J., Milligan, G. L. Intramolecular Schmidt reaction of alkyl azides. *J. Am. Chem. Soc.* **1991**, 113, 8965-8966.
14. Aube, J., Milligan, G. L., Mossman, C. J. Titanium tetrachloride-mediated reactions of alkyl azides with cyclic ketones. *J. Org. Chem.* **1992**, 57, 1635-1637.
15. Gracias, V., Milligan, G. L., Aube, J. Efficient Nitrogen Ring-Expansion Process Facilitated by in Situ Hemiketal Formation. An Asymmetric Schmidt Reaction. *J. Am. Chem. Soc.* **1995**, 117, 8047-8048.
16. Mossman, C. J., Aube, J. Intramolecular Schmidt reactions of alkyl azides with ketals and enol ethers. *Tetrahedron* **1996**, 52, 3403-3408.
17. Sahasrabudhe, K., Gracias, V., Furness, K., Smith, B. T., Katz, C. E., Reddy, D. S., Aube, J. Asymmetric Schmidt Reaction of Hydroxyalkyl Azides with Ketones. *J. Am. Chem. Soc.* **2003**, 125, 7914-7922.
18. Bach, R. D., Wolber, G. J. Theoretical study of the barrier to nitrogen inversion in N-cyano- and N-diazoformimine. Mechanism of the Schmidt reaction. *J. Org. Chem.* **1982**, 47, 239-245.
19. Amyes, T. L., Richard, J. P. Kinetic and thermodynamic stability of α-azidobenzyl carbocations: putative intermediates in the Schmidt reaction. *J. Am. Chem. Soc.* **1991**, 113, 1867-1869.
20. Pearson, W. H., Walavalkar, R., Schkeryantz, J. M., Fang, W. K., Blickensdorf, J. D. Intramolecular Schmidt reactions of azides with carbocations: synthesis of bridged-bicyclic and fused-bicyclic tertiary amines. *J. Am. Chem. Soc.* **1993**, 115, 10183-10194.
21. Hewlett, N. D., Aube, J., Radkiewicz-Poutsma, J. L. Ab Initio Approach to Understanding the Stereoselectivity of Reactions between Hydroxyalkyl Azides and Ketones. *J. Org. Chem.* **2004**, 69, 3439-3446.
22. Mirek, J. Mechanism of the Schmidt reaction. *Bulletin de l'Academie Polonaise des Sciences, Serie des Sciences Chimiques* **1962**, 10, 421-426.
23. Pyun, H. C. Preequilibrium in the Schmidt reaction of benzhydrols. *Taehan Hwahakhoe Chi* **1964**, 8, 25-29.
24. Lansbury, P. T., Mancuso, N. R. Nonstereospecificity in the Beckmann and Schmidt reactions. *Tetrahedron Lett.* **1965**, 2445-2450.
25. Rutherford, K. G., Ing, S. Y.-S., Thibert, R. J. The reaction of some aromatic acids with sodium azide in a trifluoroacetic acid-trifluoroacetic anhydride medium. *Can. J. Chem.* **1965**, 43, 541-546.
26. Vogler, E. A., Hayes, J. M. Carbon isotopic fractionation in the Schmidt decarboxylation: evidence for two pathways to products. *J. Org. Chem.* **1979**, 44, 3682-3686.
27. Richard, J. P., Amyes, T. L., Lee, Y.-G., Jagannadham, V. Demonstration of the Chemical Competence of an Iminodiazonium Ion to Serve as the Reactive Intermediate of a Schmidt Reaction. *J. Am. Chem. Soc.* **1994**, 116, 10833-10834.
28. Kaye, P. T., Mphahlele, M. J., Brown, M. E. Benzodiazepine analogs. Part 9. Kinetics and mechanism of the azidotrimethylsilane-mediated Schmidt reaction of flavanones. *J. Chem. Soc., Perkin Trans. 2* **1995**, 835-838.
29. Schultz, A. G., Wang, A., Alva, C., Sebastian, A., Glick, S., Deecher, D. C., Bidlack, J. M. Asymmetric Syntheses, Opioid Receptor Affinities, and Antinociceptive Effects of 8-Amino-5,9-methanobenzocyclooctenes, a New Class of Structural Analogs of the Morphine Alkaloids. *J. Med. Chem.* **1996**, 39, 1956-1966.
30. Smith, B. T., Wendt, J. A., Aube, J. First Asymmetric Total Synthesis of (+)-Sparteine. *Org. Lett.* **2002**, 4, 2577-2579.
31. Wrobleski, A., Sahasrabudhe, K., Aube, J. Asymmetric Total Synthesis of Dendrobatid Alkaloid 251F. *J. Am. Chem. Soc.* **2002**, 124, 9974-9975.
32. Tanaka, M., Oba, M., Tamai, K., Suemune, H. Asymmetric Synthesis of α,α-Disubstituted α-Amino Acids Using (S,S)-Cyclohexane-1,2-diol as a Chiral Auxiliary. *J. Org. Chem.* **2001**, 66, 2667-2673.

Schotten-Baumann Reaction 398

1. Schotten, C. The oxidation of piperidines. *Chem. Ber.* **1884**, 21, 2544-2547.
2. Baumann, E. A method for the synthesis of benzoyl esters. *Chem.Ber.* **1886**, 19, 3218-3222.
3. Sonntag, N. O. V. The reactions of aliphatic acid chlorides. *Chem. Rev.* **1953**, 52, 237-416.
4. Challis, B. C., Butler, A. R. Substitution at an amino nitrogen. *Chem. Amino Group* **1968**, 277-347.

5. Yamada, M., Yashiro, S., Yamano, T., Nakatani, Y., Ourisson, G. Efficient acylation of alcohols with acylthiazolidine-2-thiones and cesium fluoride. *Bull. Soc. Chim. Fr.* **1990**, 824-829.
6. Ishihara, K., Kubota, M., Kurihara, H., Yamamoto, H. Scandium Trifluoromethanesulfonate as an Extremely Active Lewis Acid Catalyst in Acylation of Alcohols with Acid Anhydrides and Mixed Anhydrides. *J. Org. Chem.* **1996**, 61, 4560-4567.
7. Fitt, J., Prasad, K., Repic, O., Blacklock, T. J. Sodium 2-ethylhexanoate: a mild acid scavenger useful in acylation of amines. *Tetrahedron Lett.* **1998**, 39, 6991-6992.
8. Gopi, H. N., Babu, V. V. S. Synthesis of peptides employing Fmoc-amino acid chlorides and commercial zinc dust. *Tetrahedron Lett.* **1998**, 39, 9769-9772.
9. Sano, T., Ohashi, K., Oriyama, T. Remarkably fast acylation of alcohols with benzoyl chloride promoted by TMEDA. *Synthesis* **1999**, 1141-1144.
10. Orita, A., Tanahashi, C., Kakuda, A., Otera, J. Highly efficient and versatile acylation of alcohols with $Bi(OTf)_3$ as catalyst. *Angew. Chem., Int. Ed. Engl.* **2000**, 39, 2877-2879.
11. Jursic, B. S., Neumann, D. Preparation of N-acyl derivatives of amino acids from acyl chlorides and amino acids in the presence of cationic surfactants. A variation of the Schotten-Baumann method of benzoylation of amino acids. *Synth. Commun.* **2001**, 31, 555-564.
12. Orita, A., Tanahashi, C., Kakuda, A., Otera, J. Highly Powerful and Practical Acylation of Alcohols with Acid Anhydride Catalyzed by $Bi(OTf)_3$. *J. Org. Chem.* **2001**, 66, 8926-8934.
13. Bartoli, G., Bosco, M., Dalpozzo, R., Marcantoni, E., Massaccesi, M., Rinaldi, S., Sambri, L. $Mg(ClO_4)_2$ as a powerful catalyst for the acylation of alcohols under solvent-free conditions. *Synlett* **2003**, 39-42.
14. Bartoli, G., Bosco, M., Dalpozzo, R., Marcantoni, E., Massaccesi, M., Sambri, L. $Zn(ClO_4)_2.6H_2O$ as a powerful catalyst for a practical acylation of alcohols with acid anhydrides. *Eur. J. Org. Chem.* **2003**, 4611-4617.
15. Cho, D. H., Kim, J. G., Jang, D. O. Indium-promoted convenient method for acylation of alcohols with acid chlorides. *Bull. Korean Chem. Soc.* **2003**, 24, 155-156.
16. Jin, T.-S., Xiao, J.-C., Wang, Z.-H., Li, T.-S. Silica gel-supported phosphotungstic acid (PTA) catalyzed acylation of alcohols and phenols with acetic anhydride under mild reaction conditions. *J. Chem. Res., Synop.* **2003**, 412-414.
17. Ranu, B. C., Dey, S. S., Hajra, A. Highly efficient acylation of alcohols, amines and thiols under solvent-free and catalyst-free conditions. *Green Chem.* **2003**, 5, 44-46.
18. Cho, D. H., Jang, D. O. Indium-mediated mild and facile method for the synthesis of amides. *Tetrahedron Lett.* **2004**, 45, 2285-2287.
19. Constantinou-Kokotou, V., Peristeraki, A. Microwave-Assisted $NiCl_2$ Promoted Acylation of Alcohols. *Synth. Commun.* **2004**, 34, 4227-4232.
20. Naik, S., Bhattacharjya, G., Kavala, V. R., Patel, B. K. Mild and eco-friendly chemoselective acylation of amines in aqueous medium. *ARKIVOC (Gainesville, FL, United States)* **2004**, 55-63.
21. Naik, S., Bhattacharjya, G., Talukdar, B., Patel, B. K. Chemoselective acylation of amines in aqueous media. *Eur. J. Org. Chem.* **2004**, 1254-1260.
22. Fox, J. M., Dmitrenko, O., Liao, L.-A., Bach, R. D. Computational Studies of Nucleophilic Substitution at Carbonyl Carbon: the S_N2 Mechanism versus the Tetrahedral Intermediate in Organic Synthesis. *J. Org. Chem.* **2004**, 69, 7317-7328.
23. Williams, E. G., Hinshelwood, C. N. The factors determining the velocity of reactions in solution. Molecular statistics of the benzoylation of amines. *J. Chem. Soc., Abstracts* **1934**, 1079-1084.
24. Stubbs, F. J., Hinshelwood, C. N. Benzoylation of substituted anilines - investigation into the additive effects of substituents. *J. Chem. Soc., Abstracts* **1949**, S71-77.
25. Kluger, R., Hunt, J. C. Circumventive catalysis: contrasting reaction patterns of tertiary and primary amines with cyclic anhydrides and the avoidance of intermediates. *J. Am. Chem. Soc.* **1989**, 111, 3325-3328.
26. Bentley, T. W., Llewellyn, G., McAlister, J. A. S_N2 Mechanism for Alcoholysis, Aminolysis, and Hydrolysis of Acetyl Chloride. *J. Org. Chem.* **1996**, 61, 7927-7932.
27. King, J. A., Jr. Nucleophilic versus general base catalysis in carbonyl substitution reactions: the influence of acylpyridinium/acylammonium salt formation on the observed reaction rate. *Trends in Organic Chemistry* **1997**, 6, 67-89.
28. Wang, Y.-C., Georghiou, P. E. First Enantioselective Total Synthesis of (-)-Tejedine. *Org. Lett.* **2002**, 4, 2675-2678.
29. Wang, H., Ganesan, A. Total Synthesis of the Fumiquinazoline Alkaloids: Solution-Phase Studies. *J. Org. Chem.* **2000**, 65, 1022-1030.
30. Van Overmeire, I., Boldin, S. A., Venkataraman, K., Zisling, R., De Jonghe, S., Van Calenbergh, S., De Keukeleire, D., Futerman, A. H., Herdewijn, P. Synthesis and Biological Evaluation of Ceramide Analogues with Substituted Aromatic Rings or an Allylic Fluoride in the Sphingoid Moiety. *J. Med. Chem.* **2000**, 43, 4189-4199.
31. Kuethe, J. T., Comins, D. L. Addition of Metallo Enolates to Chiral 1-Acylpyridinium Salts: Total Synthesis of (+)-Cannabisativine. *Org. Lett.* **2000**, 2, 855-857.

Schwartz Hydrozirconation 400

Related reactions: Brown hydroboration reaction;

1. Kautzner, B., Wailes, P. C., Weigold, H. Hydrides of bis(cyclopentadienyl)zirconium. *J. Chem. Soc., Chem. Commun.* **1969**, 1105.
2. Wailes, P. C., Weigold, H. Hydrido complexes of zirconium. I. Preparation. *J. Organomet. Chem.* **1970**, 24, 405-411.
3. Hart, D. W., Schwartz, J. Hydrozirconation. Organic synthesis via organozirconium intermediates. Synthesis and rearrangement of alkylzirconium(IV) complexes and their reaction with electrophiles. *J. Am. Chem. Soc.* **1974**, 96, 8115.
4. Schwartz, J., Labinger, J. A. New synthetic methods. 16. Hydrozirconation: organic syntheses with a new transition metal reagent. *Angew. Chem.* **1976**, 88, 402-409.
5. Ganem, B. Hydrozirconation-transmetalation: a mild, direct route to higher-order vinylic cuprates from monosubstituted alkenes. One-pot synthesis of protected prostaglandins from alkynes and cyclopentenones: in situ generation of higher-order cyanocuprates from alkenylzirconium intermediates. *Chemtracts: Org. Chem.* **1991**, 4, 44-47.
6. Annby, U., Karlsson, S., Gronowitz, S., Hallberg, A., Alvhaell, J., Svenson, R. Hydrozirconation-isomerization. Reactions of terminally functionalized olefins with zirconocene hydrides and general aspects. *Acta Chem. Scand.* **1993**, 47, 425-433.
7. Kalesse, M. Hydrozirconation with Schwartz's reagent. A convenient approach for C-C bond formation. *Acros Organics Acta* **1995**, 1, 29-31.
8. Majoral, J. P., Zablocka, M., Igau, A., Cenac, N. Zirconium species as tools in phosphorus chemistry. Part 1. $[Cp_2ZrHCl]_n$, a versatile reagent. *Chem. Ber.* **1996**, 129, 879-886.
9. Wipf, P., Jahn, H. Synthetic applications of organochlorozirconocene complexes. *Tetrahedron* **1996**, 52, 12853-12910.
10. Wipf, P., Xu, W., Takahashi, H., Jahn, H., Coish, P. D. G. Synthetic applications of organozirconocenes. *Pure Appl. Chem.* **1997**, 69, 639-644.
11. Negishi, E.-i., Takahashi, T. Alkene and alkyne complexes of zirconocene. Their preparation, structure, and novel transformations. *Bull. Chem. Soc. Jpn.* **1998**, 71, 755-769.
12. Wipf, P., Takahashi, H., Zhuang, N. Kinetic vs. thermodynamic control in hydrozirconation reactions. *Pure Appl. Chem.* **1998**, 70, 1077-1082.
13. Fernandez-Megia, E. Zirconocene hydrochloride, "Schwartz reagent". *Synlett* **1999**, 1179.
14. Whtie, J. M., Tunoori, A. R., Georg, G. I. Enabling science: Selective reduction with Cp_2ZrHCl. *Chemical Innovation* **2000**, 30, 23-28.
15. Lipshutz, B. H., Pfeiffer, S. S., Noson, K., Tomioka, T. Hydrozirconation and further transmetalation reactions. *Titanium and Zirconium in Organic Synthesis* **2002**, 110-148.
16. Negishi, E.-I., Huo, S. Synthesis and reactivity zirconocene derivatives. *Titanium and Zirconium in Organic Synthesis* **2002**, 1-49.
17. Wipf, P., Kendall, C. Novel applications of alkenyl zirconocenes. *Chem.-- Eur. J.* **2002**, 8, 1778-1784.
18. Negishi, E. I., Takahashi, T. Product class 11: organometallic complexes of zirconium and hafnium. *Science of Synthesis* **2003**, 2, 681-848.

19. Chang, B. H., Grubbs, R. H., Brubaker, C. H., Jr. The preparation and catalytic applications of supported zirconocene and hafnocene complexes. *J. Organomet. Chem.* **1985**, 280, 365-376.
20. Swanson, D. R., Nguyen, T., Noda, Y., Negishi, E. A convenient procedure for hydrozirconation of alkynes with iso-BuZrCp2Cl generated in situ by treatment of Cp2ZrCl2 with tert-butylmagnesium chloride. *J. Org. Chem.* **1991**, 56, 2590-2591.
21. Luinstra, G. A., Rief, U., Prosenc, M. H. Synthesis and Reactivity of $Cp_2ZrH(OSO_2CF_3)$, a Soluble Monomeric Alternative for Schwartz's Reagent, and the Solid-State Structure of Its Dimer, $[Cp_2Zr(OSO_2CF_3)(\mu\text{-}H)]_2$.0.5THF. *Organometallics* **1995**, 14, 1551-1552.
22. Zhang, Y., Keaton, R. J., Sita, L. R. A Case for Asymmetric Hydrozirconation. *J. Am. Chem. Soc.* **2003**, 125, 8746-8747.
23. Erker, G., Zwettler, R., Krueger, C., Hyla-Kryspin, I., Gleiter, R. Reactions of β-CH agostic alkenylzirconocene complexes. *Organometallics* **1990**, 9, 524-530.
24. Hyla-Kryspin, I., Gleiter, R., Krueger, C., Zwettler, R., Erker, G. Formation of β-CH agostic alkenylzirconocene complexes. *Organometallics* **1990**, 9, 517-523.
25. Koga, N., Morokuma, K. Ab initio molecular orbital studies of catalytic elementary reactions and catalytic cycles of transition-metal complexes. *Chem. Rev.* **1991**, 91, 823-842.
26. Endo, J., Koga, N., Morokuma, K. Theoretical study on hydrozirconation. *Organometallics* **1993**, 12, 2777-2787.
27. Watson, L. A., Yandulov, D. V., Caulton, K. G. C-D0 (D0 = p-donor, F) Cleavage in H_2C:CH(D0) by $(Cp_2ZrHCl)_n$: Mechanism, Agostic Fluorines, and a Carbene of Zr(IV). *J. Am. Chem. Soc.* **2001**, 123, 603-611.
28. Zhang, L., Borysenko, C. W., Albright, T. A., Bittner, E. R., Lee, T. R. The Cis-Trans Isomerization of 1,2,5,6-Tetrasilacycloocta-3,7-dienes: Analysis by Mechanistic Probes and Density Functional Theory. *J. Org. Chem.* **2001**, 66, 5275-5283.
29. Diaz, J. L., Villacampa, B., Lopez-Calahorra, F., Velasco, D. Synthesis of polyconjugated carbazolyl-oxazolones by a tandem hydrozirconation-Erlenmeyer reaction. Study of their hyperpolarizability values. *Tetrahedron Lett.* **2002**, 43, 4333-4337.
30. Carr, D. B., Schwartz, J. Transmetalation: preparation of organometallic reagents for organic synthesis by transfer of organic groups from one metal to another. Transmetalation from zirconium to aluminum. *J. Am. Chem. Soc.* **1977**, 99, 638-640.
31. Czisch, P., Erker, G., Korth, H. G., Sustmann, R. Transfer and coupling of zirconocene-bound alkenyl ligands. An alternative route to (s-trans-η^4-conjugated diene)$ZrCp_2$ complexes. *Organometallics* **1984**, 3, 945-947.
32. Nelson, J. E., Bercaw, J. E., Labinger, J. A. An unexpected isotope scrambling process accompanies hydrozirconation of styrene. *Organometallics* **1989**, 8, 2484-2486.
33. Mashima, K., Yamakawa, M., Takaya, H. Preparation, characterization and thermal reactions of 1-oxa-2-zirconacyclopentanes and 1-oxa-2-zirconacyclohexane: crystal structures of $[\{Zr(C_5H_5)_2(OCH_2CH_2CHMe)\}_2]$ and $[Zr(C_5Me_5)_2\{OCH_2(CH_2)_2CH_2\}]$. *Journal of the Chemical Society, Dalton Transactions: Inorganic Chemistry (1972-1999)* **1991**, 2851-2858.
34. Negishi, E.-i., Kondakov, D. Y., Choueiry, D., Kasai, K., Takahashi, T. Multiple Mechanistic Pathways for Zirconium-Catalyzed Carboalumination of Alkynes. Requirements for Cyclic Carbometalation Processes Involving C-H Activation. *J. Am. Chem. Soc.* **1996**, 118, 9577-9588.
35. Huang, X., Duan, D., Zheng, W. Studies on hydrozirconation of 1-alkynyl sulfoxides or sulfones and the application for the synthesis of stereodefined vinyl sulfoxides or sulfones. *J. Org. Chem.* **2003**, 68, 1958-1963.
36. Wipf, P., Xu, W. Total Synthesis of the Antimitotic Marine Natural Product (+)-Curacin A. *J. Org. Chem.* **1996**, 61, 6556-6562.
37. Nicolaou, K. C., Li, Y., Fylaktakidou, K. C., Mitchell, H. J., Sugita, K. Total synthesis of apoptolidin: part 2. Coupling of key building blocks and completion of the synthesis. *Angew. Chem., Int. Ed. Engl.* **2001**, 40, 3854-3857.
38. Ni, Y., Amarasinghe, K. K. D., Ksebati, B., Montgomery, J. First Total Synthesis and Stereochemical Definition of Isodomoic Acid G. *Org. Lett.* **2003**, 5, 3771-3773.

Seyferth-Gilbert Homologation 402

Related reactions: Corey-Fuchs alkyne synthesis;

1. Seyferth, D., Hilbert, P., Marmor, R. S. Novel diazo alkanes and the first carbene containing the dimethyl phosphite group. *J. Am. Chem. Soc.* **1967**, 89, 4811-4812.
2. Seyferth, D., Marmor, R. S. Dimethyl diazomethylphosphonate: its preparation and reactions. *Tetrahedron Lett.* **1970**, 2493-2496.
3. Seyferth, D., Marmor, R. S., Hilbert, P. Reactions of dimethylphosphono-substituted diazoalkanes. $(MeO)_2P(O)CR$ transfer to olefins and 1,3-dipolar additions of $(MeO)_2P(O)C(N_2)R$. *J. Org. Chem.* **1971**, 36, 1379-1386.
4. Colvin, E. W., Hamill, B. J. One-step conversion of carbonyl compounds into acetylenes. *J. Chem. Soc., Chem. Commun.* **1973**, 151-152.
5. Colvin, E. W., Hamill, B. J. A simple procedure for the elaboration of carbonyl compounds into homologous alkynes. *J. Chem. Soc., Perkin Trans. 1* **1977**, 869-874.
6. Gilbert, J. C., Weerasooriya, U. Elaboration of aldehydes and ketones to alkynes: improved methodology. *J. Org. Chem.* **1979**, 44, 4997.
7. Gilbert, J. C., Weerasooriya, U. Diazoethenes: their attempted synthesis from aldehydes and aromatic ketones by way of the Horner-Emmons modification of the Wittig reaction. A facile synthesis of alkynes. *J. Org. Chem.* **1982**, 47, 1837-1845.
8. Kirmse, W. Alkenylidenes in organic synthesis. *Angew. Chem., Int. Ed. Engl.* **1997**, 36, 1164-1170.
9. Walker, B. J. Ylides and related compounds. *Organophosphorus Chem.* **1997**, 28, 237-284.
10. Heydt, H. Product class 21: diazo compounds. *Science of Synthesis* **2004**, 27, 843-935.
11. Ohira, S. Methanolysis of dimethyl (1-diazo-2-oxopropyl)phosphonate: generation of dimethyl (diazomethyl)phosphonate and reaction with carbonyl compounds. *Synth. Commun.* **1989**, 19, 561-564.
12. Brown, D. G., Velthuisen, E. J., Commerford, J. R., Brisbois, R. G., Hoye, T. R. A Convenient Synthesis of Dimethyl (Diazomethyl)phosphonate (Seyferth/Gilbert Reagent). *J. Org. Chem.* **1996**, 61, 2540-2541.
13. Mueller, S., Liepold, B., Roth, G. J., Bestmann, H. J. An improved one-pot procedure for the synthesis of alkynes from aldehydes. *Synlett* **1996**, 521-522.
14. Marinetti, A., Savignac, P. Diethyl (dichloromethyl)phosphonate. Preparation and use in the synthesis of alkynes: (4-methoxyphenyl)ethyne. *Org. Synth.* **1997**, 74, 108-114.
15. Barrett, A. G. M., Hopkins, B. T., Love, A. C., Tedeschi, L. Parallel Synthesis of Terminal Alkynes Using a ROMPgel-Supported Ethyl 1-Diazo-2-oxopropylphosphonate. *Org. Lett.* **2004**, 6, 835-837.
16. Dickson, H. D., Smith, S. C., Hinkle, K. W. A convenient scalable one-pot conversion of esters and Weinreb amides to terminal alkynes. *Tetrahedron Lett.* **2004**, 45, 5597-5599.
17. Roth, G. J., Liepold, B., Mueller, S. G., Bestmann, H. J. Further improvements of the synthesis of alkynes from aldehydes. *Synthesis* **2004**, 59-62.
18. Callant, P., D'Haenens, L., Vandewalle, M. An efficient preparation and the intramolecular cyclopropanation of α-diazo-β-ketophosphonates and α-diazophosphonoacetates. *Synth. Commun.* **1984**, 14, 155-161.
19. White, J. D., Blakemore, P. R., Browder, C. C., Hong, J., Lincoln, C. M., Nagornyy, P. A., Robarge, L. A., Wardrop, D. J. Total Synthesis of the Marine Toxin Polycavernoside A via Selective Macrolactonization of a Trihydroxy Carboxylic Acid. *J. Am. Chem. Soc.* **2001**, 123, 8593-8595.
20. Gung, B. W., Dickson, H. Total Synthesis of (-)-Minquartynoic Acid: An Anti-Cancer, Anti-HIV Natural Product. *Org. Lett.* **2002**, 4, 2517-2519.
21. Marshall, J. A., Johns, B. A. Stereoselective Synthesis of C5-C20 and C21-C34 Subunits of the Core Structure of the Aplyronines. Applications of Enantioselective Additions of Chiral Allenylindium Reagents to Chiral Aldehydes. *J. Org. Chem.* **2000**, 65, 1501-1510.
22. Wender, P. A., Hegde, S. G., Hubbard, R. D., Zhang, L. Total Synthesis of (-)-Laulimalide. *J. Am. Chem. Soc.* **2002**, 124, 4956-4957.

1. Li, G., Chang, H.-T., Sharpless, K. B. Catalytic asymmetric aminohydroxylation (AA) of olefins. *Angew. Chem., Int. Ed. Engl.* **1996**, 35, 451-454.
2. Reiser, O. The Sharpless asymmetric aminohydroxylation of olefins. *Angew. Chem., Int. Ed. Engl.* **1996**, 35, 1308-1309.
3. O'Brien, P. Sharpless asymmetric aminohydroxylation: scope, limitations, and use in synthesis. *Angew. Chem., Int. Ed. Engl.* **1999**, 38, 326-329.
4. Bodkin, J. A., McLeod, M. D. The Sharpless asymmetric aminohydroxylation. *J. Chem. Soc., Perkin Trans. 1* **2002**, 2733-2746.
5. Nilov, D., Reiser, O. The Sharpless asymmetric aminohydroxylation - scope and limitation. *Adv. Syn. & Catal.* **2002**, 344, 1169-1173.
6. Donohoe, T. J., Johnson, P. D., Pye, R. J. The tethered aminohydroxylation (TA) reaction. *Org. Biomol. Chem.* **2003**, 1, 2025-2028.
7. Nilov, D., Reiser, O. Recent advances on the sharpless asymmetric aminohydroxylation. *Organic Synthesis Highlights V* **2003**, 118-124.
8. Muniz, K. Imido-osmium(VIII) compounds in organic synthesis: aminohydroxylation and diamination reactions. *Chem. Soc. Rev.* **2004**, 33, 166-174.
9. Bruncko, M., Schlingloff, G., Sharpless, K. B. N-Bromoacetamide - a new nitrogen source for the catalytic asymmetric aminohydroxylation of olefins. *Angew. Chem., Int. Ed. Engl.* **1997**, 36, 1483-1486.
10. Li, G., Angert, H. H., Sharpless, K. B. N-Halocarbamate salts lead to more efficient catalytic asymmetric aminohydroxylation. *Angew. Chem., Int. Ed. Engl.* **1997**, 35, 2813-2817.
11. Rubin, A. E., Sharpless, K. B. A highly efficient aminohydroxylation process. *Angew. Chem., Int. Ed. Engl.* **1997**, 36, 2637-2640.
12. Pilcher, A. S., Yagi, H., Jerina, D. M. A Novel Synthetic Method for Cis-Opened Benzo[a]pyrene 7,8-Diol 9,10-Epoxide Adducts at the Exocyclic N6-Amino Group of Deoxyadenosine. *J. Am. Chem. Soc.* **1998**, 120, 3520-3521.
13. Reddy, K. L., Dress, K. R., Sharpless, K. B. N-Chloro-N-sodio-2-trimethylsilyl ethyl carbamate: a new nitrogen source for the catalytic asymmetric aminohydroxylation. *Tetrahedron Lett.* **1998**, 39, 3667-3670.
14. Reddy, K. L., Sharpless, K. B. From Styrenes to Enantiopure α-Arylglycines in Two Steps. *J. Am. Chem. Soc.* **1998**, 120, 1207-1217.
15. Tao, B., Schlingloff, G., Sharpless, K. B. Reversal of regioselection in the asymmetric aminohydroxylation of cinnamates. *Tetrahedron Lett.* **1998**, 39, 2507-2510.
16. Gontcharov, A. V., Liu, H., Sharpless, K. B. tert-Butylsulfonamide. A New Nitrogen Source for Catalytic Aminohydroxylation and Aziridination of Olefins. *Org. Lett.* **1999**, 1, 783-786.
17. Goossen, L. J., Liu, H., Dress, R., Sharpless, K. B. Catalytic asymmetric aminohydroxylation with amino-substituted heterocycles as nitrogen sources. *Angew. Chem., Int. Ed. Engl.* **1999**, 38, 1080-1083.
18. Han, H., Cho, C.-W., Janda, K. D. A substrate-based methodology that allows the regioselective control of the catalytic aminohydroxylation reaction. *Chem.-- Eur. J.* **1999**, 5, 1565-1569.
19. Pringle, W., Sharpless, K. B. The osmium-catalyzed aminohydroxylation of Baylis-Hillman olefins. *Tetrahedron Lett.* **1999**, 40, 5151-5154.
20. Thomas, A. A., Sharpless, K. B. The Catalytic Asymmetric Aminohydroxylation of Unsaturated Phosphonates. *J. Org. Chem.* **1999**, 64, 8379-8385.
21. Fokin, V. V., Sharpless, K. B. A practical and highly efficient aminohydroxylation of unsaturated carboxylic acids. *Angew. Chem., Int. Ed. Engl.* **2001**, 40, 3455-3457.
22. DelMonte, A. J., Haller, J., Houk, K. N., Sharpless, K. B., Singleton, D. A., Strassner, T., Thomas, A. A. Experimental and Theoretical Kinetic Isotope Effects for Asymmetric Dihydroxylation. Evidence Supporting a Rate-Limiting "(3 + 2)" Cycloaddition. *J. Am. Chem. Soc.* **1997**, 119, 9907-9908.
23. Rudolph, J., Sennhenn, P. C., Vlaar, C. P., Sharpless, K. B. Smaller substituents on nitrogen facilitate the osmium-catalyzed asymmetric aminohydroxylation. *Angew. Chem., Int. Ed. Engl.* **1997**, 35, 2810-2813.
24. Demko, Z. P., Bartsch, M., Sharpless, K. B. Primary Amides. A General Nitrogen Source for Catalytic Asymmetric Aminohydroxylation of Olefins. *Org. Lett.* **2000**, 2, 2221-2223.
25. Wuts, P. G. M., Anderson, A. M., Goble, M. P., Mancini, S. E., VanderRoest, R. J. Concentration Dependence of the Sharpless Asymmetric Amidohydroxylation of Isopropyl Cinnamate. *Org. Lett.* **2000**, 2, 2667-2669.
26. Lohray, B. B., Bhushan, V., Reddy, G. J., Reddy, A. S. Mechanistic investigation of asymmetric aminohydroxylation of alkenes. *Indian J. Chem., Sect. B* **2002**, 41B, 161-168.
27. Cao, B., Park, H., Joullie, M. M. Total Synthesis of Ustiloxin D. *J. Am. Chem. Soc.* **2002**, 124, 520-521.
28. Yang, C.-G., Wang, J., Tang, X.-X., Jiang, B. Asymmetric aminohydroxylation of vinyl indoles: a short enantioselective synthesis of the bisindole alkaloids dihydrohamacanthin A and dragmacidin A. *Tetrahedron: Asymmetry* **2002**, 13, 383-394.
29. Boger, D. L., Kim, S. H., Mori, Y., Weng, J.-H., Rogel, O., Castle, S. L., McAtee, J. J. First and Second Generation Total Synthesis of the Teicoplanin Aglycon. *J. Am. Chem. Soc.* **2001**, 123, 1862-1871.
30. Kurosawa, W., Kan, T., Fukuyama, T. Stereocontrolled total synthesis of (-)-ephedradine A (orantine). *J. Am. Chem. Soc.* **2003**, 125, 8112-8113.

Related reactions: Prevost reaction;

1. Hentges, S. G., Sharpless, K. B. Asymmetric induction in the reaction of osmium tetroxide with olefins. *J. Am. Chem. Soc.* **1980**, 102, 4263-4265.
2. Jacobsen, E. N., Marko, I., Mungall, W. S., Schroeder, G., Sharpless, K. B. Asymmetric dihydroxylation via ligand-accelerated catalysis. *J. Am. Chem. Soc.* **1988**, 110, 1968-1970.
3. Lohray, B. B. Recent advances in the asymmetric dihydroxylation of alkenes. *Tetrahedron: Asymmetry* **1992**, 3, 1317-1349.
4. Johnson, R. A., Sharpless, K. B. Catalytic asymmetric dihydroxylation. *Catal. Asymmetric Synth.* **1993**, 227-222.
5. Dovletoglou, A., Thorp, H. H. Toward an understanding of the high enantioselectivity in the osmium-catalyzed asymmetric dihydroxylation (AD) kinetics. *Chemtracts: Inorg. Chem.* **1994**, 6, 136-141.
6. Jayamma, Y., Nandanan, E., Lohray, B. B. Emerging applications of asymmetric dihydroxylation chemistry. *J. Indian Inst. Sci.* **1994**, 74, 309-328.
7. Kolb, H. C., VanNieuwenhze, M. S., Sharpless, K. B. Catalytic Asymmetric Dihydroxylation. *Chem. Rev.* **1994**, 94, 2483-2547.
8. Cha, J. K., Kim, N.-S. Acyclic Stereocontrol Induced by Allylic Alkoxy Groups. Synthetic Applications of Stereoselective Dihydroxylation in Natural Product Synthesis. *Chem. Rev.* **1995**, 95, 1761-1795.
9. Pini, D., Petri, A., Mastantuono, A., Salvadori, P. Heterogeneous enantioselective hydrogenation and dihydroxylation of carbon carbon double bond mediated by transition metal asymmetric catalysts. *Chiral Reactions in Heterogeneous Catalysis, [Proceedings of the European Symposium on Chiral Reactions in Heterogeneous Catalysis], 1st, Brussels, Oct. 25-26, 1993* **1995**, 155-176.
10. Sanchez-Delgado, R. A., Rosales, M., Esteruelas, M. A., Oro, L. A. Homogeneous catalysis by osmium complexes. A review. *J. Mol. Catal. A: Chemical* **1995**, 96, 231-243.
11. Waldmann, H. Enantioselective cis-dihydroxylation. *Organic Synthesis Highlights II* **1995**, 9-18.
12. Beller, M., Sharpless, K. B. Diols via catalytic dihydroxylation. *Applied Homogeneous Catalysis with Organometallic Compounds* **1996**, 2, 1009-1024.
13. Bolm, C., Gerlach, A. Polymer-supported catalytic asymmetric Sharpless dihydroxylations of olefins. *Eur. J. Org. Chem.* **1998**, 21-27.
14. Kolb, H. C., Sharpless, K. B. Asymmetric dihydroxylation. *Transition Metals for Organic Synthesis* **1998**, 2, 219-242.
15. Marko, I. E., Svendsen, J. S. Dihydroxylation of carbon-carbon double bonds. *Comprehensive Asymmetric Catalysis I-III* **1999**, 2, 713-787.
16. Bolm, C., Hildebrand, J. P., Muniz, K. Recent advances in asymmetric dihydroxylation and aminohydroxylation. *Catal. Asymmetric Synth. (2nd Edition)* **2000**, 399-428.

17. Johnson, R. A., Sharpless, K. B. Catalytic asymmetric dihydroxylation-discovery and development. *Catal. Asymmetric Synth. (2nd Edition)* **2000**, 357-398.
18. Salvadori, P., Pini, D., Petri, A., Mandoli, A. Catalytic heterogeneous enantioselective dihydroxylation and epoxidation. *Chiral Catalyst Immobilization and Recycling* **2000**, 235-259.
19. Takahata, H. Organic synthesis with enantiomeric enhancement by dual asymmetric dihydroxylation. *Trends in Organic Chemistry* **2000**, 8, 101-119.
20. Becker, H., Sharpless, K. B. Asymmetric dihydroxylation. *Asymmetric Oxidation Reactions* **2001**, 81-104.
21. Beller, M., Sharpless, K. B. Diols via catalytic dihydroxylation. *Applied Homogeneous Catalysis with Organometallic Compounds (2nd Edition)* **2002**, 3, 1149-1164.
22. Shibata, T., Gilheany, D. G., Blackburn, B. K., Sharpless, K. B. Ligand-based improvement of enantioselectivity in the catalytic asymmetric dihydroxylation of dialkyl-substituted olefins. *Tetrahedron Lett.* **1990**, 31, 3817-3820.
23. Sharpless, K. B., Amberg, W., Beller, M., Chen, H., Hartung, J., Kawanami, Y., Lubben, D., Manoury, E., Ogino, Y., et al. New ligands double the scope of the catalytic asymmetric dihydroxylation of olefins. *J. Org. Chem.* **1991**, 56, 4585-4588.
24. Sharpless, K. B., Amberg, W., Bennani, Y. L., Crispino, G. A., Hartung, J., Jeong, K. S., Kwong, H. L., Morikawa, K., Wang, Z. M., et al. The osmium-catalyzed asymmetric dihydroxylation: a new ligand class and a process improvement. *J. Org. Chem.* **1992**, 57, 2768-2771.
25. Arrington, M. P., Bennani, Y. L., Gobel, T., Walsh, P., Zhao, S. H., Sharpless, K. B. Modified cinchona alkaloid ligands: improved selectivities in the osmium tetroxide catalyzed asymmetric dihydroxylation (AD) of terminal olefins. *Tetrahedron Lett.* **1993**, 34, 7375-7378.
26. Crispino, G. A., Jeong, K. S., Kolb, H. C., Wang, Z. M., Xu, D., Sharpless, K. B. Improved enantioselectivity in asymmetric dihydroxylations of terminal olefins using pyrimidine ligands. *J. Org. Chem.* **1993**, 58, 3785-3786.
27. Pini, D., Petri, A., Salvadori, P. Heterogeneous catalytic asymmetric dihydroxylation of olefins: a new polymeric support and a process improvement. *Tetrahedron* **1994**, 50, 11321-11328.
28. Becker, H., King, S. B., Taniguchi, M., Vanhessche, K. P. M., Sharpless, K. B. New Ligands and Improved Enantioselectivities for the Asymmetric Dihydroxylation of Olefins. *J. Org. Chem.* **1995**, 60, 3940-3941.
29. Riedl, R., Tappe, R., Berkessel, A. Probing the Scope of the Asymmetric Dihydroxylation of Polymer-Bound Olefins. Monitoring by HRMAS NMR Allows for Reaction Control and On-Bead Measurement of Enantiomeric Excess. *J. Am. Chem. Soc.* **1998**, 120, 8994-9000.
30. Salvadori, P., Pini, D., Petri, A. Insoluble polymer-bound (IPB) approach to the catalytic asymmetric dihydroxylation of alkenes. *Synlett* **1999**, 1181-1190.
31. Mehltretter, G. M., Dobler, C., Sundermeier, U., Beller, M. An improved version of the Sharpless asymmetric dihydroxylation. *Tetrahedron Lett.* **2000**, 41, 8083-8087.
32. Huang, J., Corey, E. J. A Mechanistically Guided Design Leads to the Synthesis of an Efficient and Practical New Reagent for the Highly Enantioselective, Catalytic Dihydroxylation of Olefins. *Org. Lett.* **2003**, 5, 3455-3458.
33. Jiang, R., Kuang, Y., Sun, X., Zhang, S. An improved catalytic system for recycling OsO4 and chiral ligands in the asymmetric dihydroxylation of olefins. *Tetrahedron: Asymmetry* **2004**, 15, 743-746.
34. Wu, Y. D., Wang, Y., Houk, K. N. A new model for the stereoselectivities of dihydroxylations of alkenes by chiral diamine complexes of osmium tetroxide. *J. Org. Chem.* **1992**, 57, 1362-1369.
35. Becker, H., Ho, P. T., Kolb, H. C., Loren, S., Norrby, P.-O., Sharpless, K. B. Comparing two models for the selectivity in the asymmetric dihydroxylation reaction (AD). *Tetrahedron Lett.* **1994**, 35, 7315-7318.
36. Norrby, P. O., Kolb, H. C., Sharpless, K. B. Calculations on the reaction of ruthenium tetroxide with olefins using density functional theory (DFT). Implications for the possibility of intermediates in osmium-catalyzed asymmetric dihydroxylation. *Organometallics* **1994**, 13, 344-347.
37. Norrby, P.-O., Becker, H., Sharpless, K. B. Toward an Understanding of the High Enantioselectivity in the Osmium-Catalyzed Asymmetric Dihydroxylation. 3. New Insights into Isomeric Forms of the Putative Osmaoxetane Intermediate. *J. Am. Chem. Soc.* **1996**, 118, 35-42.
38. Ujaque, G., Maseras, F., Lledos, A. A theoretical evaluation of steric and electronic effects on the structure of [OsO4(NR3)] (NR3 = bulky chiral alkaloid derivative) complexes. *Theor. Chim. Acta* **1996**, 94, 67-73.
39. Haller, J., Strassner, T., Houk, K. N. Models for Stereoselective Additions to Chiral Allylic Ethers: Osmium Tetraoxide Bis-hydroxylations. *J. Am. Chem. Soc.* **1997**, 119, 8031-8034.
40. Torrent, M., Deng, L., Duran, M., Sola, M., Ziegler, T. Density Functional Study of the [2+2]- and [2+3]-Cycloaddition Mechanisms for the Osmium-Catalyzed Dihydroxylation of Olefins. *Organometallics* **1997**, 16, 13-19.
41. Ujaque, G., Maseras, F., Lledos, A. Theoretical Characterization of an Intermediate for the [3 + 2] Cycloaddition Mechanism in the Bis(dihydroxy- quinidine)-3,6-Pyridazine.Osmium Tetroxide-Catalyzed Dihydroxylation of Styrene. *J. Org. Chem.* **1997**, 62, 7892-7894.
42. Deubel, D. V., Frenking, G. Are There Metal Oxides That Prefer a [2 + 2] Addition over a [3 + 2] Addition to Olefins? Theoretical Study of the Reaction Mechanism of LReO3 Addition (L = O-, Cl, Cp) to Ethylene. *J. Am. Chem. Soc.* **1999**, 121, 2021-2031.
43. Houk, K. N., Strassner, T. Establishing the (3 + 2) mechanism for the permanganate oxidation of alkenes by theory and kinetic isotope effects. *J. Org. Chem.* **1999**, 64, 800-802.
44. Maseras, F. Hybrid quantum mechanics/molecular mechanics methods in transition metal chemistry. *Top. Organomet. Chem.* **1999**, 4, 165-191.
45. Norrby, P.-O., Rasmussen, T., Haller, J., Strassner, T., Houk, K. N. Rationalizing the Stereoselectivity of Osmium Tetroxide Asymmetric Dihydroxylations with Transition State Modeling Using Quantum Mechanics-Guided Molecular Mechanics. *J. Am. Chem. Soc.* **1999**, 121, 10186-10192.
46. Ujaque, G., Maseras, F., Lledos, A. Theoretical Study on the Origin of Enantioselectivity in the Bis(dihydroquinidine)-3,6-pyridazine.Osmium Tetroxide-Catalyzed Dihydroxylation of Styrene. *J. Am. Chem. Soc.* **1999**, 121, 1317-1323.
47. Moitessier, N., Maigret, B., Chretien, F., Chapleur, Y. Molecular dynamics-based models explain the unexpected diastereoselectivity of the Sharpless asymmetric dihydroxylation of allyl D-xylosides. *Eur. J. Org. Chem.* **2000**, 995-1005.
48. Moitessier, N., Henry, C., Len, C., Chapleur, Y. Toward a Computational Tool Predicting the Stereochemical Outcome of Asymmetric Reactions. 1. Application to Sharpless Asymmetric Dihydroxylation. *J. Org. Chem.* **2002**, 67, 7275-7282.
49. Fristrup, P., Tanner, D., Norrby, P.-O. Updating the asymmetric osmium-catalyzed dihydroxylation (AD) mnemonic: Q2MM modeling and new kinetic measurements. *Chirality* **2003**, 15, 360-368.
50. Makowka, O. Contribution to the Knowledge of Osmium. *Ber.* **1908**, 41, 943-944.
51. Schroeder, M. Osmium tetraoxide cis hydroxylation of unsaturated substrates. *Chem. Rev.* **1980**, 80, 187-213.
52. Criegee, R., Marchand, B., Wannowius, H. Organic osmium compounds. II. *Ann.* **1942**, 550, 99-133.
53. Wai, J. S. M., Marko, I., Svendsen, J. S., Finn, M. G., Jacobsen, E. N., Sharpless, K. B. A mechanistic insight leads to a greatly improved osmium-catalyzed asymmetric dihydroxylation process. *J. Am. Chem. Soc.* **1989**, 111, 1123-1125.
54. Joergensen, K. A. A mechanistic approach to the asymmetric epoxidation of allylic alcohols and osmylation of alkenes. *Tetrahedron: Asymmetry* **1991**, 2, 515-532.
55. Ogino, Y., Chen, H., Kwong, H. L., Sharpless, K. B. The timing of hydrolysis-reoxidation in the osmium-catalyzed asymmetric dihydroxylation of olefins using potassium ferricyanide as the reoxidant. *Tetrahedron Lett.* **1991**, 32, 3965-3968.
56. Lohay, B. B., Bhushan, V. Mechanism of osmium-catalyzed asymmetric dihydroxylation (ADH) of alkenes. *Tetrahedron Lett.* **1992**, 33, 5113-5116.
57. Bruckner, C., Dolphin, D. Temperature effects in asymmetric dihydroxylation: evidence for a stepwise mechanism. *Chemtracts: Org. Chem.* **1993**, 6, 364-367.
58. Corey, E. J., Noe, M. C., Sarshar, S. The origin of high enantioselectivity in the dihydroxylation of olefins using osmium tetraoxide and cinchona alkaloid catalysts. *J. Am. Chem. Soc.* **1993**, 115, 3828-3829.
59. Kolb, H. C., Andersson, P. G., Bennani, Y. L., Crispino, G. A., Jeong, K. S., Kwong, H. L., Sharpless, K. B. On "The origin of high enantioselectivity in the dihydroxylation of olefins using osmium tetraoxide and cinchona alkaloid catalysts". *J. Am. Chem. Soc.* **1993**, 115, 12226-12227.
60. Corey, E. J., Noe, M. C., Grogan, M. J. A mechanistically designed mono-cinchona alkaloid is an excellent catalyst for the enantioselective dihydroxylation of olefins. *Tetrahedron Lett.* **1994**, 35, 6427-6430.

61. Lohray, B. B., Bhushan, V., Nandanan, E. On the mechanism of asymmetric dihydroxylation (AD) of alkenes. *Tetrahedron Lett.* **1994**, 35, 4209-4210.
62. Norrby, P.-O., Kolb, H. C., Sharpless, K. B. Toward an Understanding of the High Enantioselectivity in the Osmium-Catalyzed Asymmetric Dihydroxylation. 2. A Qualitative Molecular Mechanics Approach. *J. Am. Chem. Soc.* **1994**, 116, 8470-8478.
63. Veldkamp, A., Frenking, G. Mechanism of the Enantioselective Dihydroxylation of Olefins by OsO4 in the Presence of Chiral Bases. *J. Am. Chem. Soc.* **1994**, 116, 4937-4946.
64. Corey, E. J., Guzman-Perez, A., Noe, M. C. The application of a mechanistic model leads to the extension of the Sharpless asymmetric dihydroxylation to allylic 4-methoxybenzoates and conformationally related amine and homoallylic alcohol derivatives. *J. Am. Chem. Soc.* **1995**, 117, 10805-10816.
65. Corey, E. J., Noe, M. C., Guzman-Perez, A. Kinetic Resolution by Enantioselective Dihydroxylation of Secondary Allylic 4-Methoxybenzoate Esters Using a Mechanistically Designed Cinchona Alkaloid Catalyst. *J. Am. Chem. Soc.* **1995**, 117, 10817-10824.
66. Corey, E. J., Noe, M. C., Lin, S. A mechanistically designed bis-cinchona alkaloid ligand allows position- and enantioselective dihydroxylation of farnesol and other oligoprenyl derivatives at the terminal isopropylidene unit. *Tetrahedron Lett.* **1995**, 36, 8741-8744.
67. Corey, E. J., Noe, M. C. A Critical Analysis of the Mechanistic Basis of Enantioselectivity in the Bis-Cinchona Alkaloid Catalyzed Dihydroxylation of Olefins. *J. Am. Chem. Soc.* **1996**, 118, 11038-11053.
68. Corey, E. J., Noe, M. C. Kinetic Investigations Provide Additional Evidence That an Enzyme-like Binding Pocket Is Crucial for High Enantioselectivity in the Bis-Cinchona Alkaloid Catalyzed Asymmetric Dihydroxylation of Olefins. *J. Am. Chem. Soc.* **1996**, 118, 319-329.
69. Corey, E. J., Noe, M. C., Ting, A. Y. Improved Enantioselective Dihydroxylation of Bishomoallylic Alcohol Derivatives Using a Mechanistically Inspired Bis[cinchona] Alkaloid Catalyst. *Tetrahedron Lett.* **1996**, 37, 1735-1738.
70. Dapprich, S., Ujaque, G., Maseras, F., Lledos, A., Musaev, D. G., Morokuma, K. Theory Does Not Support an Osmaoxetane Intermediate in the Osmium-Catalyzed Dihydroxylation of Olefins. *J. Am. Chem. Soc.* **1996**, 118, 11660-11661.
71. Lohray, B. B., Bhushan, V., Nandanan, E. The mechanism of catalytic asymmetric dihydroxylation (AD) of alkenes. *Indian J. Chem., Sect. B* **1996**, 35B, 1119-1122.
72. Norrby, P.-O., Gable, K. P. Kinetic constraints on possible reaction pathways for osmium-catalyzed asymmetric dihydroxylation. *J. Chem. Soc., Perkin Trans. 2* **1996**, 171-178.
73. DelMonte, A. J., Haller, J., Houk, K. N., Sharpless, K. B., Singleton, D. A., Strassner, T., Thomas, A. A. Experimental and Theoretical Kinetic Isotope Effects for Asymmetric Dihydroxylation. Evidence Supporting a Rate-Limiting "(3 + 2)" Cycloaddition. *J. Am. Chem. Soc.* **1997**, 119, 9907-9908.
74. Nelson, D. W., Gypser, A., Ho, P. T., Kolb, H. C., Kondo, T., Kwong, H.-L., McGrath, D. V., Rubin, A. E., Norrby, P.-O., Gable, K. P., Sharpless, K. B. Toward an understanding of the high enantioselectivity in the osmium-catalyzed asymmetric dihydroxylation. 4. Electronic effects in amine-accelerated osmylations. *J. Am. Chem. Soc.* **1997**, 119, 1840-1858.
75. Bayer, A., Svendsen, J. S. Substrate binding in the asymmetric dihydroxylation reaction - investigation of the stereoselectivity in the dihydroxylation of Cs-symmetric divinylcarbinol derivatives. *Eur. J. Org. Chem.* **2001**, 1769-1780.
76. Corey, E. J., Zhang, J. Highly Effective Transition Structure Designed Catalyst for the Enantio- and Position-Selective Dihydroxylation of Polyisoprenoids. *Org. Lett.* **2001**, 3, 3211-3214.
77. Lohray, B. B., Singh, S. K., Bhushan, V. A mechanistically designed cinchona alkaloid ligand in the osmium catalyzed asymmetric dihydroxylation of alkenes. *Indian J. Chem., Sect. B* **2002**, 41B, 1226-1233.
78. Armstrong, A., Barsanti, P. A., Jones, L. H., Ahmed, G. Total Synthesis of (+)-Zaragozic Acid C. *J. Org. Chem.* **2000**, 65, 7020-7032.
79. Burke, S. D., Sametz, G. M. Total Synthesis of 3-Deoxy-D-manno-2-octulosonic Acid (KDO) and 2-Deoxy-β-KDO. *Org. Lett.* **1999**, 1, 71-74.
80. Denmark, S. E., Cottell, J. J. Synthesis of (+)-1-Epiaustraline. *J. Org. Chem.* **2001**, 66, 4276-4284.

Sharpless Asymmetric Epoxidation 408

Related reactions: *Davis oxaziridine oxidation, Jacobsen-Katsuki epoxidation, Prilezhaev reaction, Shi asymmetric epoxidation;*

1. Katsuki, T., Sharpless, K. B. The first practical method for asymmetric epoxidation. *J. Am. Chem. Soc.* **1980**, 102, 5974-5976.
2. Rossiter, B. E. Asymmetric epoxidation. *Chem. Ind.* **1985**, 22, 295-308.
3. Sharpless, K. B., Finn, M. G. in *Asymmetric Synthesis* (ed. Morrison, J. D.), *5*, 193 (Academic Press, Orlando, FL, **1985**).
4. Pfenninger, A. Asymmetric epoxidation of allylic alcohols: the Sharpless epoxidation. *Synthesis* **1986**, 89-116.
5. Johnson, R. A., Sharpless, K. B. Addition Reactions with Formation of Carbon-Oxygen Bonds: Asymmetric Methods of Epoxidation. in *Comp. Org. Synth.* (eds. Trost, B. M.,Fleming, I.), *7*, 389-436 (Pergamon, Oxford, **1991**).
6. Hoeft, E. Enantioselective epoxidation with peroxidic oxygen. *Top. Curr. Chem.* **1993**, 164, 63-77.
7. Schinzer, D. Asymmetric synthesis. The Sharpless epoxidation. *Organic Synthesis Highlights II* **1995**, 3-8.
8. Katsuki, T., Martin, V. S. Asymmetric epoxidation of allylic alcohols: The Katsuki-Sharpless epoxidation reaction. *Org. React.* **1996**, 48, 1-299.
9. Stephenson, G. R. Asymmetric oxidation. *Advanced Asymmetric Synthesis* **1996**, 367-391.
10. Shum, W. P., Cannarsa, M. J. Sharpless asymmetric epoxidation: scale-up and industrial production. *Chirality in Industry II* **1997**, 363-380.
11. Inomata, K., Ukaji, Y. Development of new asymmetric reactions utilizing tartaric acid ester as a chiral auxiliary: design of an efficient chiral dinucleating system. *Rev. on Heteroa. Chem.* **1998**, 18, 119-140.
12. Katsuki, T. Titanium-catalyzed epoxidation. *Transition Metals for Organic Synthesis* **1998**, 2, 261-271.
13. Sherrington, D. C. Polymer-supported transition metal complex alkene epoxidation catalysts. *Spec. Publ. - R. Soc. Chem.* **1998**, 216, 220-228.
14. Kagan, H. B. Historical perspective. *Comprehensive Asymmetric Catalysis I-III* **1999**, 1, 9-30.
15. Katsuki, T. Epoxidation of allylic alcohols. *Comprehensive Asymmetric Catalysis I-III* **1999**, 2, 621-648.
16. Mahrwald, R. Ti(OiPr)4 in stereoselective synthesis. *J. Prakt. Chem.* **1999**, 341, 191-194.
17. Dusi, M., Mallat, T., Baiker, A. Epoxidation of functionalized olefins over solid catalysts. *Catalysis Reviews - Science and Engineering* **2000**, 42, 213-278.
18. Johnson, R. A., Sharpless, K. B. Catalytic asymmetric epoxidation of allylic alcohols. *Catal. Asymmetric Synth. (2nd Edition)* **2000**, 231-280.
19. Sherrington, D. C. Polymer-supported metal complex alkene epoxidation catalysts. *Catal. Today* **2000**, 57, 87-104.
20. Liu, M. Epoxidation of alkenes. *Rodd's Chemistry of Carbon Compounds (2nd Edition)* **2001**, 5, 1-32.
21. Martin, V. S. Asymmetric epoxidation of olefins bearing precoordinating functional groups. *Asymmetric Oxidation Reactions* **2001**, 50-69.
22. Sheldon, R. A., Van Vliet, M. C. A. Epoxidation. *Fine Chemicals through Heterogeneous Catalysis* **2001**, 473-490.
23. Bonini, C., Righi, G. A critical outlook and comparison of enantioselective oxidation methodologies of olefins. *Tetrahedron* **2002**, 58, 4981-5021.
24. Sharpless, K. B. Searching for new reactivity (Nobel Lecture). *Angew. Chem., Int. Ed. Engl.* **2002**, 41, 2024-2032.
25. Hanson, R. M., Sharpless, K. B. Procedure for the catalytic asymmetric epoxidation of allylic alcohols in the presence of molecular sieves. *J. Org. Chem.* **1986**, 51, 1922-1925.
26. Gao, Y., Klunder, J. M., Hanson, R. M., Masamune, H., Ko, S. Y., Sharpless, K. B. Catalytic asymmetric epoxidation and kinetic resolution: modified procedures including in situ derivatization. *J. Am. Chem. Soc.* **1987**, 109, 5765-5780.
27. Suresh, P. S., Srinivasan, M., Pillai, V. N. R. Copolymers of polystyrene-new polymer supports for asymmetric epoxidation of allylic alcohols. *J. Polym. Sci., Part A: Polym. Chem.* **2000**, 38, 161-169.
28. Joergensen, K. A., Wheeler, R. A., Hoffmann, R. Electronic and steric factors determining the asymmetric epoxidation of allylic alcohols by titanium-tartrate complexes (the Sharpless epoxidation). *J. Am. Chem. Soc.* **1987**, 109, 3240-3246.
29. Potvin, P. G., Bianchet, S. The nature of the Katsuki-Sharpless asymmetric epoxidation catalyst. *J. Org. Chem.* **1992**, 57, 6629-6635.
30. Wu, Y.-D., Lai, D. K. W. Transition Structure for the Epoxidation Mediated by Titanium(IV) Peroxide. A Density Functional Study. *J. Org. Chem.* **1995**, 60, 673-680.

31. Wu, Y.-D., Lai, D. K. W. A Density Functional Study on the Stereocontrol of the Sharpless Epoxidation. *J. Am. Chem. Soc.* **1995**, 117, 11327-11336.
32. Yudanov, I. V., Gisdakis, P., Di Valentin, C., Rosch, N. Activity of peroxo and hydroperoxo complexes of Ti(IV) in olefin epoxidation. A density functional model study of energetics and mechanism. *Eur. J. Inorg. Chem.* **1999**, 2135-2145.
33. Cui, M., Adam, W., Shen, J. H., Luo, X. M., Tan, X. J., Chen, K. X., Ji, R. Y., Jiang, H. L. A Density-Functional Study of the Mechanism for the Diastereoselective Epoxidation of Chiral Allylic Alcohols by the Titanium Peroxy Complexes. *J. Org. Chem.* **2002**, 67, 1427-1435.
34. Masamune, S., Choy, W., Petersen, J. S., Sita, L. R. Double stereodifferentiation and a new strategy for stereocontrol in organic syntheses. *Angew. Chem.* **1985**, 97, 1-31.
35. Kolodiazhnyi, O. I. Multiple stereoselectivity and its application in organic synthesis. *Tetrahedron* **2003**, 59, 5953-6018.
36. Sharpless, K. B., Woodard, S. S., Finn, M. G. On the mechanism of titanium-tartrate catalyzed asymmetric epoxidation. *Pure Appl. Chem.* **1983**, 55, 1823-1836.
37. Corey, E. J. On the origin of enantioselectivity in the Katsuki-Sharpless epoxidation procedure. *J. Org. Chem.* **1990**, 55, 1693-1694.
38. Finn, M. G., Sharpless, K. B. Mechanism of asymmetric epoxidation. 2. Catalyst structure. *J. Am. Chem. Soc.* **1991**, 113, 113-126.
39. Woodard, S. S., Finn, M. G., Sharpless, K. B. Mechanism of asymmetric epoxidation. 1. Kinetics. *J. Am. Chem. Soc.* **1991**, 113, 106-113.
40. Hoye, T. R., Ye, Z. Highly Efficient Synthesis of the Potent Antitumor Annonaceous Acetogenin (+)-Parviflorin. *J. Am. Chem. Soc.* **1996**, 118, 1801-1802.
41. Gabarda, A. E., Du, W., Isarno, T., Tangirala, R. S., Curran, D. P. Asymmetric total synthesis of (20R)-homocamptothecin, substituted homocamptothecins and homosilatecans. *Tetrahedron* **2002**, 58, 6329-6341.
42. Paterson, I., De Savi, C., Tudge, M. Total Synthesis of the Microtubule-Stabilizing Agent (-)-Laulimalide. *Org. Lett.* **2001**, 3, 3149-3152.
43. Sunazuka, T., Hirose, T., Shirahata, T., Harigaya, Y., Hayashi, M., Komiyama, K., Omura, S., Smith, A. B., III. Total Synthesis of (+)-Madindoline A and (-)-Madindoline B, Potent, Selective Inhibitors of Interleukin 6. Determination of the Relative and Absolute Configurations. *J. Am. Chem. Soc.* **2000**, 122, 2122-2123.

Shi Asymmetric Epoxidation 410

Related reactions: Davis oxaziridine oxidation, Jacobsen-Katsuki epoxidation, Prilezhaev reaction, Sharpless asymmetric epoxidation;

1. Curci, R., Fiorentino, M., Serio, M. R. Asymmetric epoxidation of unfunctionalized alkenes by dioxirane intermediates generated from potassium peroxomonosulfate and chiral ketones. *J. Chem. Soc., Chem. Commun.* **1984**, 155-156.
2. Tu, Y., Wang, Z.-X., Shi, Y. An Efficient Asymmetric Epoxidation Method for trans-Olefins Mediated by a Fructose-Derived Ketone. *J. Am. Chem. Soc.* **1996**, 118, 9806-9807.
3. Wang, Z.-X., Tu, Y., Frohn, M., Shi, Y. A Dramatic pH Effect Leads to a Catalytic Asymmetric Epoxidation. *J. Org. Chem.* **1997**, 62, 2328-2329.
4. Wang, Z.-X., Tu, Y., Frohn, M., Zhang, J.-R., Shi, Y. An Efficient Catalytic Asymmetric Epoxidation Method. *J. Am. Chem. Soc.* **1997**, 119, 11224-11235.
5. Denmark, S. E., Wu, Z. The development of chiral, nonracemic dioxiranes for the catalytic, enantioselective epoxidation of alkenes. *Synlett* **1999**, 847-859.
6. Jacobsen, E. N., Wu, M. H. Epoxidation of alkenes other than allylic alcohols. *Comprehensive Asymmetric Catalysis I-III* **1999**, 2, 649-677.
7. Adam, W., Mitchell, C. M., Saha-Moller, C. R., Weichold, O. Structure, reactivity, and selectivity of metal-peroxo complexes versus dioxiranes. *Struct. Bonding (Berlin)* **2000**, 97, 237-285.
8. Frohn, M., Shi, Y. Chiral ketone-catalyzed asymmetric epoxidation of olefins. *Synthesis* **2000**, 1979-2000.
9. Ojima, I., Editor. *Catalytic Asymmetric Synthesis, Second Edition* (**2000**) 864 pp.
10. Chen, B. C., Zhou, P., Davis, F. A. Asymmetric epoxidation using peroxides and related reagents. *Asymmetric Oxidation Reactions* **2001**, 37-50.
11. Dalko, P. I., Moisan, L. Enantioselective organocatalysis. *Angewandte Chemie, International Edition* **2001**, 40, 3726-3748.
12. Davis, B. G., Williams, J. A. G. Oxidation and reduction. *Org. React. Mech.* **2001**, 179-219.
13. Adam, W., Saha-Moeller, C. R., Zhao, C.-G. Dioxirane epoxidation of alkenes. *Org. React.* **2002**, 61, 219-516.
14. Ge, H. Q. Chiral ketone catalysts derived from D-fructose. *Synlett* **2004**, 2046-2047.
15. Shi, Y. Organocatalytic Asymmetric Epoxidation of Olefins by Chiral Ketones. *Acc. Chem. Res.* **2004**, 37, 488-496.
16. Frohn, M., Dalkiewicz, M., Tu, Y., Wang, Z.-X., Shi, Y. Highly Regio- and Enantioselective Monoepoxidation of Conjugated Dienes. *J. Org. Chem.* **1998**, 63, 2948-2953.
17. Frohn, M., Wang, Z.-X., Shi, Y. A Mild and Efficient Epoxidation of Olefins Using in Situ Generated Dimethyldioxirane at High pH. *J. Org. Chem.* **1998**, 63, 6425-6426.
18. Tu, Y., Wang, Z.-X., Frohn, M., He, M., Yu, H., Tang, Y., Shi, Y. Structural Probing of Ketone Catalysts for Asymmetric Epoxidation. *J. Org. Chem.* **1998**, 63, 8475-8485.
19. Zhu, Y., Tu, Y., Yu, H., Shi, Y. Highly enantioselective epoxidation of esters and enol silyl ethers. *Tetrahedron Lett.* **1998**, 39, 7819-7822.
20. Frohn, M., Zhou, X., Zhang, J.-R., Tang, Y., Shi, Y. Kinetic resolution of racemic cyclic olefins via chiral dioxirane. *J. Am. Chem. Soc.* **1999**, 121, 7718-7719.
21. Shu, L., Shi, Y. Asymmetric epoxidation using hydrogen peroxide (H_2O_2) as primary oxidant. *Tetrahedron Lett.* **1999**, 40, 8721-8724.
22. Wang, Z.-X., Cao, G.-A., Shi, Y. Chiral Ketone Catalyzed Highly Chemo- and Enantioselective Epoxidation of Conjugated Enynes. *J. Org. Chem.* **1999**, 64, 7646-7650.
23. Wang, Z.-X., Miller, S. M., Anderson, O. P., Shi, Y. A Class of C2 and Pseudo C2 Symmetric Ketone Catalysts for Asymmetric Epoxidation. Conformational Effect on Catalysis. *J. Org. Chem.* **1999**, 64, 6443-6458.
24. Tian, H., She, X., Shi, Y. Electronic probing of ketone catalysts for asymmetric epoxidation. Search for more robust catalysts. *Organic Letters* **2001**, 3, 715-718.
25. Wang, Z.-X., Miller, S. M., Anderson, O. P., Shi, Y. Asymmetric Epoxidation by Chiral Ketones Derived from Carbocyclic Analogues of Fructose. *Journal of Organic Chemistry* **2001**, 66, 521-530.
26. Zhu, Y., Shu, L., Tu, Y., Shi, Y. Enantioselective synthesis and stereoselective rearrangements of enol ester epoxides. *J. Org. Chem.* **2001**, 66, 1818-1826.
27. Tian, H., She, X., Yu, H., Shu, L., Shi, Y. Designing New Chiral Ketone Catalysts. Asymmetric Epoxidation of cis-Olefins and Terminal Olefins. *Journal of Organic Chemistry* **2002**, 67, 2435-2446.
28. Bez, G., Zhao, C.-G. First highly enantioselective epoxidation of alkenes with aldehyde/Oxone. *Tetrahedron Lett.* **2003**, 44, 7403-7406.
29. Houk, K. N., Liu, J., DeMello, N. C., Condroski, K. R. Transition States of Epoxidations: Diradical Character, Spiro Geometries, Transition State Flexibility, and the Origins of Stereoselectivity. *J. Am. Chem. Soc.* **1997**, 119, 10147-10152.
30. Jenson, C., Liu, J., Houk, K. N., Jorgensen, W. L. Elucidation of Transition Structures and Solvent Effects for Epoxidation by Dimethyldioxirane. *J. Am. Chem. Soc.* **1997**, 119, 12982-12983.
31. Chen, C. C., Whistler, R. L. Synthesis of L-fructose. *Carbohydr. Res.* **1988**, 175, 265-271.
32. Wang, Z.-X., Shi, Y. A pH Study on the Chiral Ketone Catalyzed Asymmetric Epoxidation of Hydroxyalkenes. *J. Org. Chem.* **1998**, 63, 3099-3104.
33. Yang, D., Wang, X.-C., Wong, M.-K., Yip, Y.-C., Tang, M.-W. Highly Enantioselective Epoxidation of trans-Stilbenes Catalyzed by Chiral Ketones. *J. Am. Chem. Soc.* **1996**, 118, 11311-11312.
34. Yang, D., Wong, M.-K., Yip, Y.-C., Wang, X.-C., Tang, M.-W., Zheng, J.-H., Cheung, K.-K. Design and Synthesis of Chiral Ketones for Catalytic Asymmetric Epoxidation of Unfunctionalized Olefins. *J. Am. Chem. Soc.* **1998**, 120, 5943-5952.
35. Hoard, D. W., Moher, E. D., Martinelli, M. J., Norman, B. H. Synthesis of Cryptophycin 52 Using the Shi Epoxidation. *Org. Lett.* **2002**, 4, 1813-1815.

36. Zhang, Q., Lu, H., Richard, C., Curran, D. P. Fluorous Mixture Synthesis of Stereoisomer Libraries: Total Syntheses of (+)-Murisolin and Fifteen Diastereoisomers. *J. Am. Chem. Soc.* **2004**, 126, 36-37.
37. Yoshida, M., Abdel-Hamid Ismail, M., Nemoto, H., Ihara, M. Asymmetric total synthesis of (+)-equilenin utilizing two types of cascade ring expansion reactions of small ring systems. *Perkin 1* **2000**, 2629-2635.
38. Xiong, Z., Corey, E. J. Simple Enantioselective Total Synthesis of Glabrescol, a Chiral C2-Symmetric Pentacyclic Oxasqualenoid. *J. Am. Chem. Soc.* **2000**, 122, 9328-9329.
39. Xiong, Z., Corey, E. J. Simple Total Synthesis of the Pentacyclic Cs-Symmetric Structure Attributed to the Squalenoid Glabrescol and Three Cs-Symmetric Diastereomers Compel Structural Revision. *J. Am. Chem. Soc.* **2000**, 122, 4831-4832.

Simmons-Smith Cyclopropanation 412

Related reactions: Corey-Chaykovsky epoxidation and cyclopropanation;

1. Simmons, H. E., Smith, R. D. A new synthesis of cyclopropanes from olefins. *J. Am. Chem. Soc.* **1958**, 80, 5323-5324.
2. Simmons, H. E., Smith, R. D. A new synthesis of cyclopropanes. *J. Am. Chem. Soc.* **1959**, 81, 4256-4264.
3. Sawada, S. Simmons-Smith reaction. *Bulletin of the Institute for Chemical Research, Kyoto University* **1969**, 47, 451-479.
4. Simmons, H. E., Cairns, T. L., Vladuchick, S. A., Hoiness, C. M. Cyclopropanes from unsaturated compounds, methylene iodide, and zinc-copper couple. *Org. React.* **1973**, 20, 1-131.
5. Conia, J. M. The cyclopropanation of silyl enol ethers. A powerful synthetic tool. *Pure Appl. Chem.* **1975**, 43, 317-326.
6. Helquist, P. M. Methylene and Nonfunctionalized Alkylidene Transfer to Form Cyclopropanes. in *Comp. Org. Synth.* (eds. Trost, B. M.,Fleming, I.), *4*, 951-998 (Pergamon, Oxford, **1991**).
7. Kasdorf, K., Liotta, D. C. Development of a complexes chiral auxiliary for the asymmetric cyclopropanation of allylic alcohols. *Chemtracts* **1997**, 10, 533-535.
8. Charette, A. B. Cyclopropanation mediated by zinc organometallics. *Organozinc Reagents* **1999**, 263-285.
9. Schuppan, J., Koert, U. Stereocontrolled Simmons-Smith cyclopropanation. *Organic Synthesis Highlights IV* **2000**, 3-10.
10. Boche, G., Lohrenz, J. C. W. The Electrophilic Nature of Carbenoids, Nitrenoids, and Oxenoids. *Chem. Rev.* **2001**, 101, 697-756.
11. Charette, A. B., Beauchemin, A. Simmons-Smith cyclopropanation reaction. *Org. React.* **2001**, 58, 1-415.
12. Donaldson, W. A. Synthesis of cyclopropane containing natural products. *Tetrahedron* **2001**, 57, 8589-8627.
13. Denmark, S. E., Beutner, G. Enantioselective [2+1] cycloaddition: cyclopropanation with zinc carbenoids. *Cycloaddition Reactions in Organic Synthesis* **2002**, 85-150.
14. Noels, A. F., Demonceau, A. Catalytic cyclopropanation. *Applied Homogeneous Catalysis with Organometallic Compounds (2nd Edition)* **2002**, 2, 793-808.
15. Lebel, H., Marcoux, J.-F., Molinaro, C., Charette, A. B. Stereoselective Cyclopropanation Reactions. *Chem. Rev.* **2003**, 103, 977-1050.
16. Matsubara, S., Oshima, K. Bis(iodozincio)methane as a synthetic tool. *Proceedings of the Japan Academy, Series B: Physical and Biological Sciences* **2003**, 79B, 71-77.
17. Furukawa, J., Kawabata, N., Nishimura, J. Novel route to cyclopropanes from olefins. *Tetrahedron Lett.* **1966**, 3353-3354.
18. Denis, J. M., Girard, C., Conia, J. M. Improved Simmons-Smith reactions. *Synthesis* **1972**, 549-551.
19. Arai, I., Mori, A., Yamamoto, H. An asymmetric Simmons-Smith reaction. *J. Am. Chem. Soc.* **1985**, 107, 8254-8256.
20. Mori, A., Arai, I., Yamamoto, H., Nakai, H., Arai, Y. Asymmetric Simmons-Smith reactions using homochiral protecting groups. *Tetrahedron* **1986**, 42, 6447-6458.
21. Molander, G. A., Etter, J. B. Lanthanides in organic synthesis. 7. Samarium-promoted, stereocontrolled cyclopropanation reactions. *J. Org. Chem.* **1987**, 52, 3942-3944.
22. Takahashi, H., Yoshioka, M., Ohno, M., Kobayashi, S. A catalytic enantioselective reaction using a C2-symmetric disulfonamide as a chiral ligand: cyclopropanation of allylic alcohols by the diethylzinc-diiodomethane-disulfonamide system. *Tetrahedron Lett.* **1992**, 33, 2575-2578.
23. Ukaji, Y., Nishimura, M., Fujisawa, T. Enantioselective construction of cyclopropane rings via asymmetric Simmons-Smith reaction of allylic alcohols. *Chem. Lett.* **1992**, 61-64.
24. Charette, A. B., Marcoux, J. F. The use of 1,2-trans-cyclohexanediol as an efficient chiral auxiliary for the asymmetric cyclopropanation of allylic ethers. *Tetrahedron Lett.* **1993**, 34, 7157-7160.
25. Charette, A. B., Juteau, H. Design of Amphoteric Bifunctional Ligands: Application to the Enantioselective Simmons-Smith Cyclopropanation of Allylic Alcohols. *J. Am. Chem. Soc.* **1994**, 116, 2651-2652.
26. Denmark, S. E., Christenson, B. L., Coe, D. M., O'Connor, S. P. Catalytic enantioselective cyclopropanation with bis(halomethyl)zinc reagents. I. Optimization of reaction protocol. *Tetrahedron Lett.* **1995**, 36, 2215-2218.
27. Denmark, S. E., Christenson, B. L., O'Connor, S. P. Catalytic enantioselective cyclopropanation with bis(halomethyl)zinc reagents. II. The effect of promoter structure on selectivity. *Tetrahedron Lett.* **1995**, 36, 2219-2222.
28. Kitajima, H., Aoki, Y., Ito, K., Katsuki, T. Asymmetric Simmons-Smith cyclopropanation of E-allylic alcohols using 1,1'-bi-w-naphthol-3,3'-dicarboxamide as a chiral auxiliary. *Chem. Lett.* **1995**, 1113-1114.
29. Takahashi, H., Yoshioka, M., Shibasaki, M., Ohno, M., Imai, N., Kobayashi, S. A catalytic enantioselective reaction using a C2-symmetric disulfonamide as a chiral ligand: Simmons-Smith cyclopropanation of allylic alcohols by the Et_2Zn-CH_2I_2-disulfonamide system. *Tetrahedron* **1995**, 51, 12013-12026.
30. Kitajima, H., Ito, K., Aoki, Y., Katsuki, T. N,N,N',N'-Tetraalkyl-2,2'-dihydroxy-1,1'-binaphthalene-3,3'-dicarboxamides: novel chiral auxiliaries for asymmetric Simmons-Smith cyclopropanation of allylic alcohols and for asymmetric diethylzinc addition to aldehydes. *Bull. Chem. Soc. Jpn.* **1997**, 70, 207-217.
31. Aggarwal, V. K., Fang, G. Y., Meek, G. Highly Diastereoselective Simmons-Smith Cyclopropanation of Allylic Amines. *Org. Lett.* **2003**, 5, 4417-4420.
32. Long, J., Yuan, Y., Shi, Y. Asymmetric Simmons-Smith Cyclopropanation of Unfunctionalized Olefins. *J. Am. Chem. Soc.* **2003**, 125, 13632-13633.
33. Lorenz, J. C., Long, J., Yang, Z., Xue, S., Xie, Y., Shi, Y. A Novel Class of Tunable Zinc Reagents ($RXZnCH_2Y$) for Efficient Cyclopropanation of Olefins. *J. Org. Chem.* **2004**, 69, 327-334.
34. Dargel, T. K., Koch, W. Density functional study on the mechanism of the Simmons-Smith reaction. *J. Chem. Soc., Perkin Trans. 2* **1996**, 877-881.
35. Bernardi, F., Bottoni, A., Miscione, G. P. A DFT study of the Simmons-Smith cyclopropanation reaction. *J. Am. Chem. Soc.* **1997**, 119, 12300-12305.
36. Hirai, A., Nakamura, M., Nakamura, E. Theoretical studies on cyclopropanation reaction with lithium and zinc carbenoids. *Chem. Lett.* **1998**, 927-928.
37. Nakamura, E., Hirai, A., Nakamura, M. Theoretical Studies on Lewis Acid Acceleration in Simmons-Smith Reaction. *J. Am. Chem. Soc.* **1998**, 120, 5844-5845.
38. Hermann, H., Lohrenz, J. C. W., Kuhn, A., Boche, G. The influence of the leaving group X (X = F, Cl, Br, I, OH) on the carbenoid nature of the carbenoids $LiCH_2X$ and $XZnCH_2X$- a theoretical study. *Tetrahedron* **2000**, 56, 4109-4115.
39. Bernardi, F., Bottoni, A., Miscione, G. P. DFT Study of the Palladium-Catalyzed Cyclopropanation Reaction. *Organometallics* **2001**, 20, 2751-2758.
40. Bottoni, A., Bernardi, F., Miscione, G. P. Applications of density functional theory in the study of organometallic reactivity problems. *Journal of Computational Methods in Sciences and Engineering* **2002**, 2, 319-333.
41. Fang, W.-H., Phillips David, L., Wang, D.-q., Li, Y.-L. A density functional theory investigation of the Simmons-Smith cyclopropanation reaction: examination of the insertion reaction of zinc into the C-I bond of CH(2)I(2) and subsequent cyclopropanation reactions. *J. Org. Chem.* **2002**, 67, 154-160.

42. Fang, W.-H., Phillips, D. L., Wang, D., Li, Y.-L. A Density Functional Theory Investigation of the Simmons-Smith Cyclopropanation Reaction: Examination of the Insertion Reaction of Zinc into the C-I Bond of CH_2I_2 and Subsequent Cyclopropanation Reactions. *J. Org. Chem.* **2002**, 67, 154-160.
43. Nakamura, M., Hirai, A., Nakamura, E. Reaction Pathways of the Simmons-Smith Reaction. *J. Am. Chem. Soc.* **2003**, 125, 2341-2350.
44. Zhao, C., Wang, D., Phillips David, L. Theoretical study of samarium (II) carbenoid ($ISmCH_2I$) promoted cyclopropanation reactions with ethylene and the effect of THF solvent on the reaction pathways. *J. Am. Chem. Soc.* **2003**, 125, 15200-15209.
45. Zhao, C., Wang, D.-Q., Phillips, D. L. Density functional study of selected mono-zinc and gem-dizinc radical carbenoid cyclopropanation reactions: observation of an efficient radical zinc carbenoid cyclopropanation reaction and the influence of the leaving group on ring closure. *Journal of Theoretical & Computational Chemistry* **2003**, 2, 357-369.
46. Maruoka, K., Fukutani, Y., Yamamoto, H. Trialkylaluminum-alkylidene iodide. A powerful cyclopropanation agent with unique selectivity. *J. Org. Chem.* **1985**, 50, 4412-4414.
47. Motoyama, Y. Mechanism of Lewis acid-promoted Simmons-Smith reaction. *Organometallic News* **1995**, 128.
48. Li, Y.-L., Leung, K. H., Phillips, D. L. Time-Resolved Resonance Raman Study of the Reaction of Isodiiodomethane with Cyclohexene: Implications for the Mechanism of Photocyclopropanation of Olefins Using Ultraviolet Photolysis of Diiodomethane. *J. Phys. Chem. A* **2001**, 105, 10621-10625.
49. Onoda, T., Shirai, R., Koiso, Y., Iwasaki, S. Asymmetric total synthesis of curacin A. *Tetrahedron Lett.* **1996**, 37, 4397-4400.
50. Paquette, L. A., Wang, T.-Z., Pinard, E. Total Synthesis of Natural (+)-Acetoxycrenulide. *J. Am. Chem. Soc.* **1995**, 117, 1455-1456.
51. Liu, P., Jacobsen, E. N. Total Synthesis of (+)-Ambruticin. *J. Am. Chem. Soc.* **2001**, 123, 10772-10773.
52. Taber, D. F., Nakajima, K., Xu, M., Rheingold, A. L. Lactone-Directed Intramolecular Diels-Alder Cyclization: Synthesis of trans-Dihydroconfertifolin. *J. Org. Chem.* **2002**, 67, 4501-4504.

Skraup and Doebner-Miller Quinoline Synthesis 414

Related reactions: Combes quinoline synthesis;

1. Koenigs. Synthesis of quinolines. *Ber.Dtsch.Chem.Ges.* **1880**, 911-913.
2. Skraup, Z. H. synthesis of quinolines. *Ber.Dtsch.Chem.Ges.* **1880**, 13, 2086-2087.
3. Doebner, O., von Miller, W. Homologes of quinoline. *Ber. Dtsch. Chem. Ges.* **1881**, 13, 2812-2817.
4. Bergstrom, F. W. Heterocyclic N compounds. IIA. Hexacyclic compounds: pyridine, quinoline and isoquinoline. *Chem. Rev.* **1944**, 35, 77-277.
5. Manske, R. H. F., Kulka, M. The Skraup synthesis of quinolines. *Org. React.* **1953**, 7, 59-98.
6. Matsugi, M., Tabusa, F., Minamikawa, J.-I. Doebner-Miller synthesis in a two-phase system: practical preparation of quinolines. *Tetrahedron Lett.* **2000**, 41, 8523-8525.
7. Theoclitou, M.-E., Robinson, L. A. Novel facile synthesis of 2,2,4-substituted-1,2-dihydroquinolines via a modified Skraup reaction. *Tetrahedron Lett.* **2002**, 43, 3907-3910.
8. Manske, R. H. F. The chemistry of quinolines. *Chem. Rev.* **1942**, 30, 113-144.
9. Cheng, C. C., Yan, S. J. The Friedlander synthesis of quinolines. *Org. React.* **1982**, 28, 37-201.
10. Blaise, E. E., Maire, M. Alkyl-4-quinolines. Mechanism of the Reactions of Skraup and of Doebner and Miller. *Bull. soc. chim. [4]* **1908**, 3, 667-674.
11. Badger, G. M., Crocker, H. P., Ennis, B. C., Gayler, J. A., Matthews, W. E., Raper, W. O. C., Samuel, E. L., Spotswood, T. M. Doebner-Miller, Skraup, and related reactions. I. Isolation of intermediates in the formation of quinolines. *Aust. J. Chem.* **1963**, 16, 814-827.
12. Dauphinee, G. A., Forrest, T. P. Sequential intermediates in the Doebner-Miller reaction. *J. Chem. Soc., Chem. Commun.* **1969**, 327.
13. Forrest, T. P., Dauphinee, G. A., Miles, W. F. Mechanism of the Doebner-Miller reaction. *Can. J. Chem.* **1969**, 47, 2121-2122.
14. Schindler, O., Michaelis, W. Stereochemistry of the intermediates (aldol bases) of the Doebner-von Miller quinoline synthesis. *Helv. Chim. Acta* **1970**, 53, 776-779.
15. Eisch, J. J., Dluzniewski, T. Mechanism of the Skraup and Doebner-von Miller quinoline syntheses. Cyclization of α,β-unsaturated *N*-aryliminium salts via 1,3-diazetidinium ion intermediates. *J. Org. Chem.* **1989**, 54, 1269-1274.
16. Gellerman, G., Rudi, A., Kashman, Y. Synthesis of pyrido[2,3,4-kl]acridines. A building block for the synthesis of pyridoacridine alkaloids. *Tetrahedron Lett.* **1992**, 33, 5577-5580.
17. O'Neill, P. M., Storr, R. C., Park, B. K. Synthesis of the 8-aminoquinoline antimalarial 5-fluoroprimaquine. *Tetrahedron* **1998**, 54, 4615-4622.
18. Sami, I., Kar, G., Ray, J. K. Synthesis of polycyclic oxacoumarins, potential antitumor agents and a short and convenient synthesis of naphthopyranoquinolines from naphthopyran chloroaldehydes. *Tetrahedron* **1992**, 48, 5199-5208.
19. Gladiali, S., Pinna, L., Delogu, G., De Martin, S., Zassinovich, G., Mestroni, G. Optically active phenanthrolines in asymmetric catalysis. III. Highly efficient enantioselective transfer hydrogenation of acetophenone by chiral rhodium/3-alkyl phenanthroline catalysts. *Tetrahedron: Asymmetry* **1990**, 1, 635-648.

Smiles Rearrangement 416

1. Henriques, R. The thio-derivatives of β-naphthols. *Ber.* **1894**, 27, 2993-3005.
2. Hinsberg, O. β-Naphthol sulfide and iso-β-naphthol sulfide. *J. Prakt. Chem.* **1914**, 90, 345-353.
3. Hinsberg, O. β-Naphthol sulfide and iso-β-naphthol sulfide. II. *J. Prakt. Chem.* **1915**, 91, 307-324.
4. Hinsberg, O. β-Naphthol sulfide and iso-β-naphthol sulfide. III. *J. Prakt. Chem.* **1916**, 93, 277-301.
5. Warren, L. A., Smiles, S. Dehydro-2-naphthol sulfone. *J. Chem. Soc., Abstracts* **1930**, 1327-1331.
6. Levy, A. A., Rains, H. C., Smiles, S. Rearrangement of hydroxy sulfones. I. *J. Chem. Soc.* **1931**, 3264-3269.
7. Evans, W. J., Smiles, S. Rearrangement of o-acetamido sulfones and sulfides. *J. Chem. Soc.* **1935**, 181-188.
8. Evans, W. J., Smiles, S. Rearrangement of carbamyl sulfones and sulfides. *J. Chem. Soc.* **1936**, 329-331.
9. Bunnett, J. F., Zahler, R. E. Aromatic nucleophilic substitution reactions. *Chem. Revs.* **1951**, 49, 273-412.
10. Shine, H. J. *Aromatic Rearrangements (Reaction Mechanisms in Organic Chemistry. Monograph 6)* (**1967**) 380 pp.
11. Truce, W. E., Kreider, E. M., Brand, W. W. The smiles and related rearrangements of aromatic systems. *Org. React.* **1970**, 18, 99-215.
12. Drozd, V. N. Carbanion rearrangement of o-methyldiaryl sulfones (the Truce rearrangement). *International Journal of Sulfur Chemistry* **1973**, 8, 443-467.
13. Schneller, S. W. Name reactions in sulfur chemistry. II. *International Journal of Sulfur Chemistry* **1973**, 8, 485-503.
14. Schneller, S. W. Name reactions in sulfur chemistry. Part II. *International Journal of Sulfur Chemistry* **1976**, 8, 579-597.
15. Braverman, S. The chemistry of sulfinic acids, esters and their derivatives. Rearrangements. *Chem. Sulphinic Acids, Esters Their Deriv.* **1990**, 297-349.
16. Truce, W. E. Forty years in organosulfur chemistry. *Sulfur Reports* **1990**, 9, 351-357.
17. Gerasimova, T. N., Kolchina, E. F. The Smiles rearrangement in the polyfluoroaromatic series. *J. Fluorine Chem.* **1994**, 66, 69-74.
18. Truce, W. E., Ray, W. J., Jr., Norman, O. L., Eickemeyer, D. B. Rearrangements of aryl sulfones. I. The metalation and rearrangement of mesityl phenyl sulfone. *J. Am. Chem. Soc.* **1958**, 80, 3625-3629.
19. Zbiral, E. Arynes as electrophilic reagents. I. Reaction with alkyl phosphoranes. *Monatsh. Chem.* **1964**, 95, 1759-1780.
20. Zbiral, E. Organophosphorous compounds. III. Mechanism of the rearrangement of alkylene phosphoranes brought about by arynes. *Tetrahedron Lett.* **1964**, 3963-3967.
21. Matsui, K., Maeno, N., Suzuki, S., Shizuka, H., Morita, T. Photo-Smiles rearrangements. *Tetrahedron Lett.* **1970**, 1467-1469.

22. Bayne, D. W., Nicol, A. J., Tennant, G. Synthesis of 2-acyl-3-hydroxyquinolines embodying a novel variant of the Smiles rearrangement. *J. Chem. Soc., Chem. Commun.* **1975**, 782-783.
23. Bayles, R., Johnson, M. C., Maisey, R. F., Turner, R. W. A Smiles rearrangement involving non-activated aromatic systems; the facile conversion of phenols to anilines. *Synthesis* **1977**, 33-34.
24. Mettey, Y., Vierfond, J. M. A novel synthetic route to cyanophenothiazines. First example of Smiles rearrangement from halobenzonitriles. *Heterocycles* **1993**, 36, 987-993.
25. Selvakumar, N., Srinivas, D., Azhagan, A. M. Observation of O->N type Smiles rearrangement in certain alkyl aryl nitro compounds. *Synthesis* **2002**, 2421-2425.
26. Tada, M., Shijima, H., Nakamura, M. Smiles-type free radical rearrangement of aromatic sulfonates and sulfonamides: syntheses of arylethanols and arylethylamines. *Org. Biomol. Chem.* **2003**, 1, 2499-2505.
27. Shizuka, H., Maeno, N., Matsui, K. Photo-Smiles rearrangements of o-aminophenoxy-s-triazines. *Mol. Photochem.* **1972**, 4, 335-351.
28. Mutai, K., Nakagaki, R. A rationalization of orientation in nucleophilic aromatic photosubstitution. *Chem. Lett.* **1984**, 1537-1540.
29. Kim, C. K., Lee, I., Lee, B. S. Determination of reactivity by MO theory. 71. Theoretical studies on the gas-phase Smiles rearrangement. *J. Phys. Org. Chem.* **1991**, 4, 315-329.
30. Mulholland, J. A., Akki, U., Yang, Y., Ryu, J.-Y. Temperature dependence of DCDD/F isomer distributions from chlorophenol precursors. *Chemosphere* **2001**, 42, 719-727.
31. Musaev, D. G., Galloway, A. L., Menger, F. M. The roles of steric and electronic effects in the 2-hydroxy-2'-nitrodiphenyl sulfones to 2-(o-nitrophenoxy)-benzene-sulfinic acids rearrangement (Smiles). Computational study. *THEOCHEM* **2004**, 679, 45-52.
32. Mizuno, K., Maeda, H., Sugimoto, A., Chiyonobu, K. Photocycloaddition and photoaddition reactions of aromatic compounds. *Molecular and Supramolecular Photochemistry* **2001**, 8, 127-241.
33. Okada, K., Sekiguchi, S. Aromatic nucleophilic substitution. 9. Kinetics of the formation and decomposition of anionic s complexes in the Smiles rearrangements of N-acetyl-β-aminoethyl 2-X-4-nitro-1-phenyl or N-acetyl-β-aminoethyl 5-nitro-2-pyridyl ethers in aqueous dimethyl sulfoxide. *J. Org. Chem.* **1978**, 43, 441-447.
34. Knipe, A. C., Sridhar, N. Role of intramolecular catalysis in the kinetics of Smiles' rearrangement of N-[2-(p-nitrophenoxy)ethyl]ethylenediamine. *J. Chem. Soc., Chem. Commun.* **1979**, 791-792.
35. Sunamoto, J., Kondo, H., Yanase, F., Okamoto, H. Kinetic studies on the N,N-type Smiles rearrangement. *Bull. Chem. Soc. Jpn.* **1980**, 53, 1361-1365.
36. Knipe, A. C., Lound-Keast, J., Sridhar, N. Interpretation of the kinetics of general-base-catalyzed Smiles rearrangement of 2-(p-nitrophenoxy)ethylamine into 2-(p-nitroanilino)ethanol: rate-limiting deprotonation of a spiro-Meisenheimer intermediate. *J. Chem. Soc., Perkin Trans. 2* **1984**, 1885-1891.
37. Knipe, A. C., Sridhar, N., Lound-Keast, J. Effects of N-alkyl substitution on the formation and rate-limiting deprotonation of the spiro-Meisenheimer intermediate of Smiles rearrangement of 2-(p-nitrophenoxy)ethylamine, in aqueous solution. *J. Chem. Soc., Perkin Trans. 2* **1984**, 1893-1899.
38. Nakagaki, R., Hiramatsu, M., Mutai, K., Nakakura, S. Photo-Smiles rearrangement (IV). Electron-transfer mechanism of an intramolecular aromatic nucleophilic substitution. *Mol. Cryst. Liq. Cryst.* **1985**, 126, 69-75.
39. Wubbels, G. G., Sevetson, B. R., Kaganove, S. N. Effect of α-cyclodextrin complexation on a general base catalyzed photo-Smiles rearrangement. *Tetrahedron Lett.* **1986**, 27, 3103-3106.
40. Machacek, V., Hassanien, M. M. H., Sterba, V. Kinetics and mechanism of spiro adduct formation from and Smiles rearrangement of N-methyl-N-(2,4,6-trinitrophenyl)aminoacetanilide. Base-catalyzed transformation of N-(2,4,6-trinitrophenylamino)acetanilide into 2-nitroso-4,6-dinitroaniline. *Collect. Czech. Chem. Commun.* **1987**, 52, 2225-2240.
41. Eichinger, P. C. H., Bowie, J. H., Hayes, R. N. The gas-phase Smiles rearrangement: a heavy atom labeling study. *J. Am. Chem. Soc.* **1989**, 111, 4224-4227.
42. Knyazev, V. N., Drozd, V. N. A reversible double Smiles rearrangement through intermediate formation of two tautomeric Meisenheimer spiro complexes. *Tetrahedron Lett.* **1989**, 30, 2273-2276.
43. Wubbels, G. G., Sevetson, B. R., Sanders, H. Competitive catalysis and quenching by amines of photo-Smiles rearrangement as evidence for a zwitterionic triplet as the proton-donating intermediate. *J. Am. Chem. Soc.* **1989**, 111, 1018-1022.
44. Yilmaz, I., Shine, H. J. Heavy-atom kinetic isotope effects in the base-catalyzed Smiles rearrangement of N-methyl-2-(4-nitrophenoxy)ethanamine. *Gazz. Chim. Ital.* **1989**, 119, 603-607.
45. Wubbels, G. G., Cotter, W. D., Sanders, H., Pope, C. Broensted Catalysis Law Plots for Heterolytic, General Base-Catalyzed Smiles Photorearrangement. *J. Org. Chem.* **1995**, 60, 2960-2961.
46. Bezsoudnova, K. Y., Yatsimirsky, A. K. Cyclodextrin catalysis of the Smiles rearrangement of 4-nitrophenyl salicylate. *React. Kinet. Catal. Lett.* **1997**, 62, 63-69.
47. Izod, K., O'Shaughnessy, P., Clegg, W. Unusual Solvent-Promoted Smiles Rearrangement of Two Different Phosphorus-Containing Organolithium Compounds to the Same Lithium Phosphide. Crystal Structure of $MeP\{C_6H_4\text{-}2\text{-}CH(C_6H_4\text{-}2\text{-}CH_2NMe_2)NMe_2\}Li(THF)_2$. *Organometallics* **2002**, 21, 641-646.
48. Hirota, T., Tomita, K.-I., Sasaki, K., Okuda, K., Yoshida, M., Kashino, S. Polycyclic N-heterocyclic compounds. 57. Syntheses of fused furo(or thieno)[2,3-b]pyridine derivatives via Smiles rearrangement and cyclization. *Heterocycles* **2001**, 55, 741-752.
49. Elix, J. A., Jenie, U. A. A synthesis of the lichen diphenyl ether epiphorellic acid 1. *Aust. J. Chem.* **1989**, 42, 987-994.
50. Hargrave, K. D., Proudfoot, J. R., Grozinger, K. G., Cullen, E., Kapadia, S. R., Patel, U. R., Fuchs, V. U., Mauldin, S. C., Vitous, J., et al. Novel non-nucleoside inhibitors of HIV-1 reverse transcriptase. 1. Tricyclic pyridobenzo- and dipyridodiazepinones. *J. Med. Chem.* **1991**, 34, 2231-2241.
51. Weidner, J. J., Peet, N. P. Direct conversion of hydroxy aromatic compounds to heteroarylamines via a one-pot Smiles rearrangement procedure. *J. Heterocycl. Chem.* **1997**, 34, 1857-1860.

1. Jones, P. F., Lappert, M. F., Szary, A. C. Wittig-type reactions of 2-lithio-2-(trimethylsilyl)-1,3-dithiane and related reactions. *J. Chem. Soc., Perkin Trans. 1* **1973**, 2272-2277.
2. Tietze, L. F., Geissler, H., Gewert, J. A., Jakobi, U. Tandem-bisalkylation of 2-trialkylsilyl-1,3-dithiane: a new sequential transformation for the synthesis of C2-symmetrical enantiopure 1,5-diols and β,β'-dihydroxyketones as well as of enantiopure 1,3,5-triols. *Synlett* **1994**, 511-512.
3. Smith, A. B., III, Boldi, A. M. Multicomponent Linchpin Couplings of Silyl Dithianes via Solvent-Controlled Brook Rearrangement. *J. Am. Chem. Soc.* **1997**, 119, 6925-6926.
4. Yus, M., Najera, C., Foubelo, F. The role of 1,3-dithianes in natural product synthesis. *Tetrahedron* **2003**, 59, 6147-6212.
5. Osborn, H. M. I., Sweeney, J. B., Howson, B. Ring-opening of *N*-tosyl aziridines by sulfur-stabilized nucleophiles. *Synlett* **1993**, 675-676.
6. Howson, W., Osborn, H. M. I., Sweeney, J. Ring-opening of *N*-tosyl aziridines by 2-lithiodithianes. *J. Chem. Soc., Perkin Trans. 1* **1995**, 2439-2445.
7. Smith, A. B., III, Pitram, S. M. Multicomponent Linchpin Couplings of Silyl Dithianes: Synthesis of the Schreiber C(16-28) Trisacetonide Subtarget for Mycoticins A and B. *Org. Lett.* **1999**, 1, 2001-2004.
8. Smith, A. B., III, Pitram, S. M., Gaunt, M. J., Kozmin, S. A. Dithiane Additions to Vinyl Epoxides: Steric Control over the S_N2 and S_N2' Manifolds. *J. Am. Chem. Soc.* **2002**, 124, 14516-14517.
9. Smith, A. B., III, Pitram, S. M., Boldi, A. M., Gaunt, M. J., Sfouggatakis, C., Moser, W. H. Multicomponent Linchpin Couplings. Reaction of Dithiane Anions with Terminal Epoxides, Epichlorohydrin, and Vinyl Epoxides: Efficient, Rapid, and Stereocontrolled Assembly of Advanced Fragments for Complex Molecule Synthesis. *J. Am. Chem. Soc.* **2003**, 125, 14435-14445.
10. Corey, E. J., Seebach, D. Carbanions of 1,3-dithianes. Reagents for C-C bond formation by nucleophilic displacement and carbonyl addition. *Angew. Chem., Int. Ed. Engl.* **1965**, 4, 1075-1077.

11. Shinokubo, H., Miura, K., Oshima, K., Utimoto, K. *tert*-Butyldimethylsilyldichloromethyllithium as a dichloromethylene dianion synthon. 1,3-Rearrangement of silyl group from carbon to oxide. *Tetrahedron Lett.* **1993**, 34, 1951-1954.
12. Reich, H. J., Borst, J. P., Dykstra, R. R. Solution ion pair structure of 2-lithio-1,3-dithianes in THF and THF-HMPA. *Tetrahedron* **1994**, 50, 5869-5880.
13. Shinokubo, H., Miura, K., Oshima, K., Utimoto, K. *tert*-Butyldimethylsilyldihalomethyllithium as a dihalomethylene dianion synthon. 1,3-Rearrangement and 1,4-rearrangement of silyl group from carbon to oxide. *Tetrahedron* **1996**, 52, 503-514.
14. Hale, K. J., Hummersone, M. G., Bhatia, G. S. Control of Olefin Geometry in the Bryostatin B-Ring through Exploitation of a C_2-Symmetry Breaking Tactic and a Smith-Tietze Coupling Reaction. *Org. Lett.* **2000**, 2, 2189-2192.
15. Smith, A. B., III, Doughty, V. A., Sfouggatakis, C., Bennett, C. S., Koyanagi, J., Takeuchi, M. Spongistatin Synthetic Studies. An Efficient, Second-Generation Construction of an Advanced ABCD Intermediate. *Org. Lett.* **2002**, 4, 783-786.

Snieckus Directed Ortho Metalation 420

Related reactions: Friedel-Crafts acylation, Friedel-Crafts alkylation;

1. Gilman, H., Bebb, R. L. Relative Reactivities of Organometallic Compounds. XX. Metalation. *J. Am. Chem. Soc.* **1939**, 61, 109-112.
2. Wittig, G., Fuhrmann, G. Exchange reactions with phenyllithium. V. Behavior of the halogenated anisoles toward phenyllithium. *Ber.* **1940**, 73B, 1197-1218.
3. Gilman, H., Morton, J. W., Jr. The metalation reaction with organolithium compounds. *Org. React.* **1954**, 258-304.
4. Mallan, J. M., Bebb, R. L. Metalations by organolithium compounds. *Chem. Rev.* **1969**, 69, 693-755.
5. Kaiser, E. M., Slocum, D. W. Carbanions. in *Org. Reactive Intermed.* (ed. McManus, S. P.), 337-422 (Academic Press, New York, **1973**).
6. Slocum, D. W., Sugarman, D. I. Directed metalation. *Advan. Chem. Ser.* **1974**, 130, 222-247.
7. Gschwend, H. W., Rodriguez, H. R. Heteroatom-facilitated lithiations. *Org. React.* **1979**, 26, 1-360.
8. Snieckus, V. Heterocycles via ortho-lithiated benzamides. *Heterocycles* **1980**, 14, 1649-1676.
9. Beak, P., Snieckus, V. Directed lithiation of aromatic tertiary amides: an evolving synthetic methodology for polysubstituted aromatics. *Acc. Chem. Res.* **1982**, 15, 306-312.
10. Narasimhan, N. S., Mali, R. S. Syntheses of heterocyclic compounds involving aromatic lithiation reactions in the key step. *Synthesis* **1983**, 957-986.
11. Snieckus, V. New directions in heterocyclic synthesis using metalated benzamides. *Lect. Heterocycl. Chem.* **1984**, 7, 95-106.
12. Narasimhan, N. S., Mali, R. S. Heteroatom directed aromatic lithiation reactions for the synthesis of condensed heterocyclic compounds. *Top. Curr. Chem.* **1987**, 138, 63-147.
13. Snieckus, V. Directed ortho-lithiation of aromatic compounds. New methodologies and applications in organic synthesis. *Bull. Soc. Chim. Fr.* **1988**, 67-78.
14. Snieckus, V. The directed ortho metalation reaction. Methodology, applications, synthetic links, and a non-aromatic ramification. *Pure Appl. Chem.* **1990**, 62, 2047-2056.
15. Snieckus, V. Directed ortho metalation. Tertiary amide and O-carbamate directors in synthetic strategies for polysubstituted aromatics. *Chem. Rev.* **1990**, 90, 879-933.
16. Snieckus, V. Regioselective synthetic processes based on the aromatic directed metalation strategy. *Pure Appl. Chem.* **1990**, 62, 671-680.
17. Queguiner, G., Marsais, F., Snieckus, V., Epsztajn, J. Directed metalation of π-deficient azaaromatics: strategies of functionalization of pyridines, quinolines, and diazines. *Adv. Heterocycl. Chem.* **1991**, 52, 187-304.
18. Snieckus, V., Editor. *Advances in Carbanion Chemistry, Vol. 1* (**1992**) 291 pp.
19. Snieckus, V. Combined directed ortho metalation-cross coupling strategies. Design for natural product synthesis. *Pure Appl. Chem.* **1994**, 66, 2155-2158.
20. Snieckus, V. Directed aromatic metalation. A continuing education in flatland chemistry. *NATO ASI Ser., Ser. E* **1996**, 320, 191-221.
21. Snieckus, V., Editor. *Advances in Carbanion Chemistry, Volume 2* (**1996**) 272 pp.
22. Chauder, B., Green, L., Snieckus, V. The directed ortho metalation-transition metal-catalyzed reaction symbiosis in heteroaromatic synthesis. *Pure Appl. Chem.* **1999**, 71, 1521-1529.
23. Green, L., Chauder, B., Snieckus, V. The directed ortho metalation-cross-coupling symbiosis in heteroaromatic synthesis. *J. Heterocycl. Chem.* **1999**, 36, 1453-1468.
24. Mongin, F., Queguiner, G. Advances in the directed metallation of azines and diazines (pyridines, pyrimidines, pyrazines, pyridazines, quinolines, benzodiazines and carbolines). Part 1: Metallation of pyridines, quinolines and carbolines. *Tetrahedron* **2001**, 57, 4059-4090.
25. Turck, A., Ple, N., Mongin, F., Queguiner, G. Advances in the directed metalation of azines and diazines (pyridines, pyrimidines, pyrazines, pyridazines, quinolines, benzodiazines and carbolines). Part 2. Metalation of pyrimidines, pyrazines, pyridazines and benzodiazines. *Tetrahedron* **2001**, 57, 4489-4505.
26. Hartung, C. G., Snieckus, V. The directed ortho metalation reaction - a point of departure for new synthetic aromatic chemistry. *Modern Arene Chemistry* **2002**, 330-367.
27. Whisler, M. C., MacNeil, S., Snieckus, V., Beak, P. Beyond thermodynamic acidity: A perspective on the complex-induced proximity effect (CIPE) in deprotonation reactions. *Angew. Chem., Int. Ed. Engl.* **2004**, 43, 2206-2225.
28. Quesnelle, C., Iihama, T., Aubert, T., Perrier, H., Snieckus, V. The tert-butyl sulfoxide directed ortho metalation group. New synthetic methodology for substituted aromatics and pyridines and comparison with other metalation directors. *Tetrahedron Lett.* **1992**, 33, 2625-2628.
29. Gray, M., Chapell, B. J., Felding, J., Taylor, N. J., Snieckus, V. The di-tert-butylphosphinyl directed ortho metalation group. Synthesis of hindered dialkylarylphosphines. *Synlett* **1998**, 422-424.
30. Kalinin, A. V., Bower, J. F., Riebel, P., Snieckus, V. The Directed Ortho Metalation-Ullmann Connection. A New Cu(I)-Catalyzed Variant for the Synthesis of Substituted Diaryl Ethers. *J. Org. Chem.* **1999**, 64, 2986-2987.
31. Metallinos, C., Nerdinger, S., Snieckus, V. N-Cumyl Benzamide, Sulfonamide, and Aryl O-Carbamate Directed Metalation Groups. Mild Hydrolytic Lability for Facile Manipulation of Directed Ortho Metalation Derived Aromatics. *Org. Lett.* **1999**, 1, 1183-1186.
32. Metallinos, C., Snieckus, V. (-)-Sparteine-Mediated Metalation of Ferrocenesulfonates. The First Case of Double Asymmetric Induction of Ferrocene Planar Chirality. *Org. Lett.* **2002**, 4, 1935-1938.
33. Milburn, R. R., Snieckus, V. The tertiary sulfonamide as a latent directed-metalation group: $Ni^{(0)}$-catalyzed reductive cleavage and cross-coupling reactions of aryl sulfonamides with Grignard reagents. *Angew. Chem., Int. Ed. Engl.* **2004**, 43, 888-891.
34. Bauer, W., Schleyer, P. v. R. Mechanistic evidence for ortho-directed lithiations from one- and two-dimensional NMR spectroscopy and MNDO calculations. *J. Am. Chem. Soc.* **1989**, 111, 7191-7198.
35. Kremer, T., Junge, M., Schleyer, P. v. R. Mechanisms of Competitive Ring-Directed and Side-Chain-Directed Metalations in Ortho-Substituted Toluenes. *Organometallics* **1996**, 15, 3345-3359.
36. Wheatley, A. E. H. The directed lithiation of benzenoid aromatic systems. *Eur. J. Inorg. Chem.* **2003**, 3291-3303.
37. Stratakis, M. On the Mechanism of the Ortho-Directed Metalation of Anisole by n-Butyllithium. *J. Org. Chem.* **1997**, 62, 3024-3025.
38. James, C. A., Snieckus, V. Combined directed metalation - cross coupling strategies. Total synthesis of the aglycons of gilvocarcin V, M and E. *Tetrahedron Lett.* **1997**, 38, 8149-8152.
39. Moro-Oka, Y., Fukuda, T., Iwao, M. The first total synthesis of veiutamine, a new type of pyrroloiminoquinone marine alkaloid. *Tetrahedron Lett.* **1999**, 40, 1713-1716.
40. Comins, D. L., Nolan, J. M. A Practical Six-Step Synthesis of (S)-Camptothecin. *Org. Lett.* **2001**, 3, 4255-4257.

Sommelet-Hauser Rearrangement 422

1. Sommelet, M. A special kind of molecular rearrangement. *Compt. rend.* **1937**, 205, 56-58.
2. Kantor, S. W., Hauser, C. R. Isomerizations of carbanions. II. Rearrangements of benzyltrimethylammonium ion and related quaternary ammonium ions by sodium amide involving migration into the ring. *J. Am. Chem. Soc.* **1951**, 73, 4122-4131.
3. Brewster, J. H., Eliel, E. L. Carbon-carbon alkylations with amines and ammonium salts. *Org. React.* **1953**, 7, 99-197.
4. Zimmerman, H. E. Base-catalyzed rearrangements. *Mol. Rearrangements* **1963**, 1, 345-406.
5. Pine, S. H. Base-promoted rearrangements of quaternary ammonium salts. *Org. React.* **1970**, 18, 403-464.
6. Marko, I. E. The Stevens and related rearrangement. in *Comp. Org. Synth.* (eds. Trost, B. M.,Fleming, I.), *3*, 913-974 (Pergamon, Oxford, **1991**).
7. Li, A.-H., Dai, L.-X., Aggarwal, V. K. Asymmetric Ylide Reactions: Epoxidation, Cyclopropanation, Aziridination, Olefination, and Rearrangement. *Chem. Rev.* **1997**, 97, 2341-2372.
8. Beall, L. S., Padwa, A. Application of nitrogen ylide cyclizations for organic synthesis. *Advances in Nitrogen Heterocycles* **1998**, 3, 117-158.
9. Clark, J. S. Nitrogen, oxygen and sulfur ylides: an overview. *Nitrogen, Oxygen and Sulfur Ylide Chemistry* **2002**, 1-113.
10. Campbell, S. J., Darwish, D. Asymmetric induction in the Sommelet rearrangement of chiral benzylsulfonium salts. *Can. J. Chem.* **1976**, 54, 193-201.
11. Nakano, M., Sato, Y. A convenient synthesis of o-methylbenzylamine derivatives from benzyl halides: the improved Sommelet-Hauser rearrangement. *J. Chem. Soc., Chem. Commun.* **1985**, 1684-1685.
12. Nakano, M., Sato, Y. Rearrangement of (substituted benzyl)trimethylammonium ylides in a nonbasic medium: the improved Sommelet-Hauser rearrangement. *J. Org. Chem.* **1987**, 52, 1844-1847.
13. Shirai, N., Sato, Y. Ylide rearrangement of benzyltrialkylammonium salts: the improved Sommelet-Hauser rearrangement. *J. Org. Chem.* **1988**, 53, 194-196.
14. Shirai, N., Sumiya, F., Sato, Y., Hori, M. A stable intermediate in the Sommelet-Hauser rearrangement of 1-methyl-2-phenylpiperidinium 1-methylylides. The improved Sommelet-Hauser rearrangement. *J. Chem. Soc., Chem. Commun.* **1988**, 370.
15. Yamamoto, M., Kakinuma, M., Kohmoto, S., Yamada, K. Sommelet-Hauser rearrangement of oxygen- and sulfur-containing heteroaromatic sulfonium ylides. *Bull. Chem. Soc. Jpn.* **1989**, 62, 958-960.
16. Berger, R., Ziller, J. W., Van Vranken, D. L. Stereoselectivity of the Thia-Sommelet [2,3]-Dearomatization. *J. Am. Chem. Soc.* **1998**, 120, 841-842.
17. McComas, C. C., Van Vranken, D. L. Application of chiral lithium amide bases to the thia-Sommelet dearomatization reaction. *Tetrahedron Lett.* **2003**, 44, 8203-8205.
18. Heard, G. L., Yates, B. F. Competing Rearrangements of Ammonium Ylides: A Quantum Theoretical Study. *J. Org. Chem.* **1996**, 61, 7276-7284.
19. Okada, K., Tanaka, M. Reinvestigation of base-induced skeletal conversion via a spirocyclic intermediate of dibenzodithiocinium derivatives and a computational study using the HF/6-31G* basis set. *J. Chem. Soc., Perkin Trans. 1* **2002**, 2704-2711.
20. Okada, K., Tanaka, M., Takagi, R. Computational study of base-induced skeletal conversion via a spirocyclic intermediate in dibenzodithiocinium derivatives by ab initio MO calculations. *J. Phys. Org. Chem.* **2003**, 16, 271-278.
21. Pine, S. H. Para-Sommelet-Hauser rearrangement. *Tetrahedron Lett.* **1967**, 3393-3397.
22. Archer, D. A. Behavior of quaternary salts under reduced pressure. II. Decomposition of the N-methylammonium hydroxides of benzyl-, dibenzyl-, and diphenylmethylamines. *J. Chem. Soc. C.* **1971**, 1329-1331.
23. Pine, S. H., Munemo, E. M., Phillips, T. R., Bartolini, G., Cotton, W. D., Andrews, G. C. Base-promoted rearrangements of α-arylneopentylammonium salts. *J. Org. Chem.* **1971**, 36, 984-991.
24. Giumanini, A. G., Trombini, C., Lercker, G., Lepley, A. R. Heterobenzyl quaternary ammonium salts. IV. 2-Thenyl group as terminus and migrating moiety in the Stevens and Sommelet rearrangements of a quaternary ammonium ion. *J. Org. Chem.* **1976**, 41, 2187-2193.
25. Sumiya, F., Shiral, N., Sato, Y. Conjugated-triene intermediates in the Sommelet-Hauser rearrangement of cyclic 1-methyl-2-phenylammonium 1-methylides. *Chem. Pharm. Bull.* **1991**, 39, 36-40.
26. Weinreb, S. M., Basha, F. Z., Hibino, S., Khatri, N. A., Kim, D., Pye, W. E., Wu, T. T. Total synthesis of the antitumor antibiotic streptonigrin. *J. Am. Chem. Soc.* **1982**, 104, 536-544.
27. Sakuragi, A., Shirai, N., Sato, Y., Kurono, Y., Hatano, K. Rearrangement of cis and trans-2-methyl-1-(substituted phenyl)isoindolinium 2-methylides. *J. Org. Chem.* **1994**, 59, 148-153.
28. Alper, P. B., Nguyen, K. T. Practical Synthesis and Elaboration of Methyl 7-Chloroindole-4-carboxylate. *J. Org. Chem.* **2003**, 68, 2051-2053.
29. Karp, G. M., Condon, M. E. Preparation and alkylation of regioisomeric tetrahydrophthalamide-substituted indolin-2(3H)-ones. *J. Heterocycl. Chem.* **1994**, 31, 1513-1520.

Sonogashira Cross-Coupling 424

Related reactions: Castro-Stevens coupling;

1. Cassar, L. Synthesis of aryl- and vinyl-substituted acetylene derivatives by the use of nickel and palladium complexes. *J. Organomet. Chem.* **1975**, 93, 253-257.
2. Dieck, H. A., Heck, F. R. Palladium catalyzed synthesis of aryl, heterocyclic, and vinylic acetylene derivatives. *J. Organomet. Chem.* **1975**, 93, 259-263.
3. Sonogashira, K., Tohda, Y., Hagihara, N. Convenient synthesis of acetylenes. Catalytic substitutions of acetylenic hydrogen with bromo alkenes, iodo arenes, and bromopyridines. *Tetrahedron Lett.* **1975**, 4467-4470.
4. Sonogashira, K. Coupling Reactions Between sp^2 and sp Carbon Centers. in *Comp. Org. Synth.* (eds. Trost, B. M.,Fleming, I.), *3*, 521-549 (Pergamon, Oxford, **1991**).
5. Campbell, I. B. The Sonogashira Cu-Pd-catalyzed alkyne coupling reaction. *Organocopper Reagents* **1994**, 217-235.
6. Geissler, H. Transition metal-catalyzed cross coupling reactions. *Transition Metals for Organic Synthesis* **1998**, 1, 158-183.
7. Sonogashira, K. Cross-coupling reactions to sp carbon atoms. *Metal-Catalyzed Cross-Coupling Reactions* **1998**, 203-229.
8. Osakada, K., Yamamoto, T. Alkynylcopper(I) complexes. The structure and chemical properties relevant to synthetic organic reactions. *Trends in Organometallic Chemistry* **1999**, 3, 219-225.
9. Pierre Genet, J., Savignac, M. Recent developments of palladium(0) catalyzed reactions in aqueous medium. *Journal of Organometallic Chemistry* **1999**, 576, 305-317.
10. Blaser, H.-U., Indolese, A., Schnyder, A. Applied homogeneous catalysis by organometallic complexes. *Curr. Sci.* **2000**, 78, 1336-1344.
11. De Vries, J. G., De Vries, A. H. M., Tucker, C. E., Miller, J. A. Palladium catalysis in the production of pharmaceuticals. *Innovations in Pharmaceutical Technology* **2001**, 01, 125-126, 128, 130.
12. Beller, M., Zapf, A. Palladium-catalyzed coupling reactions for industrial fine chemical syntheses. *Handbook of Organopalladium Chemistry for Organic Synthesis* **2002**, 1, 1209-1222.
13. Hillier, A. C., Grasa, G. A., Viciu, M. S., Lee, H. M., Yang, C., Nolan, S. P. Catalytic cross-coupling reactions mediated by palladium/nucleophilic carbene systems. *Journal of Organometallic Chemistry* **2002**, 653, 69-82.
14. Miura, M., Nomura, M. Direct arylation via cleavage of activated and unactivated C-H bonds. *Top. Curr. Chem.* **2002**, 219, 211-241.
15. Negishi, E.-i. A genealogy of Pd-catalyzed cross-coupling. *J. Organomet. Chem.* **2002**, 653, 34-40.
16. Negishi, E.-i., Xu, C. Palladium-catalyzed alkynylation with alkynylmetals and alkynyl electrophiles. *Handbook of Organopalladium Chemistry for Organic Synthesis* **2002**, 1, 531-549.
17. Schiedel, M.-S., Briehn, C. A., Bauerle, P. C-C Cross-coupling reactions for the combinatorial synthesis of novel organic materials. *J. Organomet. Chem.* **2002**, 653, 200-208.

18. Sonogashira, K. Palladium-catalyzed alkynylation. *Handbook of Organopalladium Chemistry for Organic Synthesis* **2002**, 1, 493-529.
19. Sonogashira, K. Development of Pd-Cu catalyzed cross-coupling of terminal acetylenes with sp^2-carbon halides. *J. Organomet. Chem.* **2002**, 653, 46-49.
20. Tamao, K., Miyaura, N. Introduction to cross-coupling reactions. *Topics in Current Chemistry* **2002**, 219, 1-9.
21. Tucker, C. E., De Vries, J. G. Homogeneous catalysis for the production of fine chemicals. Palladium- and nickel-catalyzed aromatic carbon-carbon bond formation. *Topics in Catalysis* **2002**, 19, 111-118.
22. Uozumi, Y., Hayashi, T. Solid-phase palladium catalysis for high-throughput organic synthesis. *Handbook of Combinatorial Chemistry* **2002**, 1, 531-584.
23. Zapf, A., Beller, M. Fine chemical synthesis with homogeneous palladium catalysts: examples, status and trends. *Topics in Catalysis* **2002**, 19, 101-109.
24. Bionchini, C., Giambastiani, G. Fluorous biphasic catalysis without perfluorinated solvents: application to Pd-mediated Suzuki and Sonogashira couplings. *Chemtracts* **2003**, 16, 485-490.
25. Brase, S., Kirchhoff, J. H., Kobberling, J. Palladium-catalyzed reactions in solid phase organic synthesis. *Tetrahedron* **2003**, 59, 885-939.
26. Fairlamb, I. J. S. Transition metals in organic synthesis. Part 1. Catalytic applications. *Annu. Rep. Prog. Chem., Sect. B, Org. Chem.* **2003**, 99, 104-137.
27. Herrmann, W. A., Ofele, K., von Preysing, D., Schneider, S. K. Phospha-palladacycles and N-heterocyclic carbene palladium complexes: efficient catalysts for CC-coupling reactions. *J. Organomet. Chem.* **2003**, 687, 229-248.
28. Negishi, E.-I., Anastasia, L. Palladium-Catalyzed Alkynylation. *Chem. Rev.* **2003**, 103, 1979-2017.
29. Tykwinski, R. R. Evolution in the palladium-catalyzed cross-coupling of sp- and sp^2-hybridized carbon atoms. *Angew. Chem., Int. Ed. Engl.* **2003**, 42, 1566-1568.
30. Jutand, A. Dual role of nucleophiles in palladium-catalyzed Heck, Stille, and Sonogashira reactions. *Pure Appl. Chem.* **2004**, 76, 565-576.
31. Li, C.-J., Slaven, W. T., John, V. T., Banerjee, S. Palladium catalyzed polymerization of aryl diiodides with acetylene gas in aqueous medium: a novel synthesis of areneethynylene polymers and oligomers. *Chem. Commun.* **1997**, 1569-1570.
32. Dibowski, H., Schmidtchen, F. P. Sonogashira cross-couplings using biocompatible conditions in water. *Tetrahedron Lett.* **1998**, 39, 525-528.
33. Kiji, J., Okano, T., Kimura, H., Saiki, K. Palladium-catalyzed carbonylative coupling of iodobenzene and 2-methyl-3-butyn-2-ol under biphasic conditions: Formation of furanones. *J. Mol. Catal. A: Chemical* **1998**, 130, 95-100.
34. Kingsbury, C. L., Mehrman, S. J., Takacs, J. M. A comprehensive review of the applications of transition metal-catalyzed reactions to solid phase synthesis. *Curr. Org. Chem.* **1999**, 3, 497-555.
35. Liao, Y., Fathi, R., Reitman, M., Zhang, Y., Yang, Z. Optimization study of Sonogashira cross-coupling reaction on high-loading macrobeads using a silyl linker. *Tetrahedron Lett.* **2001**, 42, 1815-1818.
36. Lopez-Deber, M. P., Castedo, L., Granja, J. R. Synthesis of N-(3-Arylpropyl)amino Acid Derivatives by Sonogashira Types of Reaction in Aqueous Media. *Org. Lett.* **2001**, 3, 2823-2826.
37. Mori, A., Ahmed, M. S. M., Sekiguchi, A., Masui, K., Koike, T. Sonogashira coupling with aqueous ammonia. *Chem. Lett.* **2002**, 756-757.
38. Pal, M., Parasuraman, K., Gupta, S., Yeleswarapu, K. R. Regioselective synthesis of 4-substituted-1-aryl-1-butanones using a Sonogashira-hydration strategy: copper-free palladium-catalyzed reaction of terminal alkynes with aryl bromides. *Synlett* **2002**, 1976-1982.
39. Quignard, F., Larbot, S., Goutodier, S., Choplin, A. Sonogashira coupling: silica supported aqueous phase palladium catalysts versus their homogeneous analogs. *J. Chem. Soc., Dalton Trans.* **2002**, 1147-1152.
40. Xia, M., Wang, Y.-G. A novel microwave-activated Sonogashira coupling reaction and cleavage using polyethylene glycol as phase-transfer catalyst and polymer support. *J. Chem. Res., Synop.* **2002**, 173-175.
41. Ahmed, M. S. M., Mori, A. Carbonylative Sonogashira Coupling of Terminal Alkynes with Aqueous Ammonia. *Org. Lett.* **2003**, 5, 3057-3060.
42. Beletskaya, I. P., Latyshev, G. V., Tsvetkov, A. V., Lukashev, N. V. The nickel-catalyzed Sonogashira-Hagihara reaction. *Tetrahedron Lett.* **2003**, 44, 5011-5013.
43. Erdelyi, M., Gogoll, A. Rapid Microwave Promoted Sonogashira Coupling Reactions on Solid Phase. *J. Org. Chem.* **2003**, 68, 6431-6434.
44. Najera, C., Gil-Molto, J., Karlstroem, S., Falvello, L. R. Di-2-pyridylmethylamine-Based Palladium Complexes as New Catalysts for Heck, Suzuki, and Sonogashira Reactions in Organic and Aqueous Solvents. *Org. Lett.* **2003**, 5, 1451-1454.
45. Utesch, N. F., Diederich, F. Acetylenic scaffolding on solid support: Poly(triacetylene)-derived oligomers by Sonogashira and Cadiot-Chodkiewicz-type cross-coupling reactions. *Org. Biomol. Chem.* **2003**, 1, 237-239.
46. Wang, L., Li, P. Sonogashira coupling reaction with palladium powder, potassium fluoride in aqueous media. *Synth. Commun.* **2003**, 33, 3679-3685.
47. Pirguliyev, N. S., Brel, V. K., Zefirov, N. S., Stang, P. J. Stereoselective synthesis of conjugated alkenynes via palladium-catalyzed coupling of alkenyl iodonium salts with terminal alkynes. *Tetrahedron* **1999**, 55, 12377-12386.
48. Bohm, V. P. W., Herrmann, W. A. Coordination chemistry and mechanisms of metal-catalyzed C-C coupling reactions, 13: a copper-free procedure for the palladium-catalyzed Sonogashira reaction of aryl bromides with terminal alkynes at room temperature. *Eur. J. Org. Chem.* **2000**, 3679-3681.
49. Osakada, K., Yamamoto, T. Transmetalation of alkynyl and aryl complexes of group 10 transition metals. *Coord. Chem. Rev.* **2000**, 198, 379-399.
50. Amatore, C., Jutand, A. Structural and mechanistic aspects of palladium-catalyzed cross-coupling. *Handbook of Organopalladium Chemistry for Organic Synthesis* **2002**, 1, 943-972.
51. Sato, K., Yoshimura, T., Shindo, M., Shishido, K. Total Synthesis of (-)-Heliannuol E. *J. Org. Chem.* **2001**, 66, 309-314.
52. Toyota, M., Komori, C., Ihara, M. A Concise Formal Total Synthesis of Mappicine and Nothapodytine B via an Intramolecular Hetero Diels-Alder Reaction. *J. Org. Chem.* **2000**, 65, 7110-7113.
53. Kobayashi, S., Reddy, R. S., Sugiura, Y., Sasaki, D., Miyagawa, N., Hirama, M. Investigation of the Total Synthesis of N1999-A2: Implication of Stereochemistry. *J. Am. Chem. Soc.* **2001**, 123, 2887-2888.
54. Paterson, I., Davies, R. D. M., Marquez, R. Total synthesis of the callipeltoside aglycon. *Angew. Chem., Int. Ed. Engl.* **2001**, 40, 603-607.

Staudinger Ketene Cycloaddition 426

1. Staudinger, H. Ketenes. IV. Communication: Reactions of Diphenylketene. *Ber.* **1907**, 40, 1145-1148.
2. Staudinger, H., Klever, H. W. Ketenes. V. Communication. Reactions of Dimethylketene. *Ber.* **1907**, 40, 1149-1153.
3. Chick, F., Wilsmore, N. T. M. Acetylketen: a Polymeride of Keten. *J. Chem. Soc., Abstracts* **1908**, 93-4, 946-950.
4. Staudinger, H. Ketenes. VIII. Preparation of Quinoidal Hydrocarbons from Diphenyl Ketene. *Ber.* **1908**, 41, 1355-1363.
5. Staudinger, H. Contribution to our Knowledge of the Ketenes. First Paper. Diphenylketene. *Ann.* **1908**, 356, 51-123.
6. Staudinger, H., Klever, H. W. Ketenes. VI. Communication. Ketene. *Ber.* **1908**, 41, 594-600.
7. Staudinger, H., Klever, H. W. Ketene. Remarks on the Article of V. T. Wilsmore and A. W. Stewart. *Ber.* **1908**, 41, 1516-1517.
8. Staudinger, H., Klever, H. W. Ketene. VII. Classification of the Ketenes. *Ber.* **1908**, 41, 906-909.
9. Quadbeck, G. Ketenes in synthetic organic chemistry. *Angew. Chem.* **1956**, 68, 361-370.
10. Roberts, J. D., Sharts, C. M. Cyclobutane derivatives from thermal cycloaddition reactions. *Org. React.* **1962**, 12, 1-56.
11. Luknitskii, F. I., Vovsi, B. A. Ketenes in situ and cycloaddition reactions with them. *Usp. Khim.* **1969**, 38, 1072-1088.
12. Ziegler, E. Ketene chemistry. *Chimia* **1970**, 24, 62-68.
13. Houk, K. N. Frontier molecular orbital theory of cycloaddition reactions. *Acc. Chem. Res.* **1975**, 8, 361-369.
14. Holder, R. W. Ketene cycloadditions. *J. Chem. Educ.* **1976**, 53, 81-85.
15. Hasek, R. H. Ketenes and related substances. *Kirk-Othmer Encycl. Chem. Technol., 3rd Ed.* **1981**, 13, 873-893.
16. Scheeren, J. W. Synthetic and mechanistic aspects of thermal [2 + 2] cycloadditions of ketene acetals with electron-poor alkenes and carbonyl compounds. *Rec. Trav. Chim. Pays-Bas* **1986**, 105, 71-84.

17. Seikaly, H. R., Tidwell, T. T. Addition reactions of ketenes. *Tetrahedron* **1986**, 42, 2587-2613.
18. Cooper, R. D. G., Daugherty, B. W., Boyd, D. B. Chiral control of the Staudinger reaction. *Pure Appl. Chem.* **1987**, 59, 485-492.
19. Snider, B. B. Intramolecular cycloaddition reactions of ketenes and keteniminium salts with alkenes. *Chem. Rev.* **1988**, 88, 793-811.
20. Svendsen, J. S., Sharpless, K. B. Type I intramolecular cycloadditions of vinylketenes. *Chemtracts: Org. Chem.* **1988**, 1, 399-402.
21. Cossio, F. P., Lecea, B., Cuevas, C., Mielgo, A., Palomo, C. A novel entry for the asymmetric Staudinger reaction: experimental and computational studies on the formation of β-lactams through [2+2] cycloaddition reaction of ketenes to imines. *An. Quim.* **1993**, 89, 119-122.
22. Georg, G. I., Ravikumar, V. T. Stereocontrolled ketene-imine cycloaddition reactions. *Org. Chem. β-Lactams* **1993**, 295-368.
23. Palonao, C., Aizpurua, J. M. Azetidinone frameworks as chiral templates via asymmetric Staudinger reaction. *Trends in Organic Chemistry* **1993**, 4, 637-659.
24. Hyatt, J. A., Raynolds, P. W. Ketene cycloadditions. *Org. React.* **1994**, 45, 159-646.
25. Ojima, I. Asymmetric syntheses by means of the β-lactam synthon method. *Adv. in Asymmetric Synth.* **1995**, 1, 95-146.
26. Parvulescu, L. Cycloaddition reactions of ketenes with bicyclic systems. *Roumanian Chem. Quart. Rev.* **1995**, 3, 187-200.
27. Lopez, R., Del Rio, E., Diaz, N., Suarez, D., Menendez, M. I., Sordo, T. L. Theoretical studies of the formation of β-lactams. *Recent Research Developments in Physical Chemistry* **1998**, 2, 245-257.
28. Hayashi, Y., Narasaka, K. [2+2] Cycloaddition reactions. *Comprehensive Asymmetric Catalysis I-III* **1999**, 3, 1255-1269.
29. Palomo, C., Aizpurua, J. M. Asymmetric synthesis of 3-amino-β-lactams via Staudinger ketene-imine cycloaddition reaction. *Chem. Het. Comp. (New York) (Translation of Khim. Geterot. Soed.)* **1999**, 34, 1222-1236.
30. Palomo, C., Aizpurua, J. M., Ganboa, I., Oiarbide, M. Asymmetric synthesis of β-lactams by Staudinger ketene-imine cycloaddition reaction. *Eur. J. Org. Chem.* **1999**, 3223-3235.
31. Orr, R. K., Calter, M. A. Asymmetric synthesis using ketenes. *Tetrahedron* **2003**, 59, 3545-3565.
32. Singh, G. S. Recent progress in the synthesis and chemistry of azetidinones. *Tetrahedron* **2003**, 59, 7631-7649.
33. Palomo, C., Aizpurua, J. M., Ganboa, I., Oiarbide, M. Asymmetric synthesis of β-lactams through the Staudinger reaction and their use as building blocks of natural and non-natural products. *Curr. Med. Chem.* **2004**, 11, 1837-1872.
34. Burke, L. A. Theoretical study of [2 + 2] cycloadditions. Ketene with ethylene. *J. Org. Chem.* **1985**, 50, 3149-3155.
35. Sordo, J. A., Gonzalez, J., Sordo, T. L. An ab initio study on the mechanism of the ketene-imine cycloaddition reaction. *J. Am. Chem. Soc.* **1992**, 114, 6249-6251.
36. Cossio, F. P., Ugalde, J. M., Lopez, X., Lecea, B., Palomo, C. A semiempirical theoretical study on the formation of β-lactams from ketenes and imines. *J. Am. Chem. Soc.* **1993**, 115, 995-1004.
37. Arrastia, I., Arrieta, A., Ugalde, J. M., Cossio, F. P., Lecea, B. Theoretical and experimental studies on the periselectivity of the cycloaddition reaction between activated ketenes and conjugated imines. *Tetrahedron Lett.* **1994**, 35, 7825-7828.
38. Arrieta, A., Ugalde, J. M., Cossio, F. P., Lecea, B. Role of the isomerization pathways in the Staudinger reaction. A theoretical study on the interaction between activated ketenes and imidates. *Tetrahedron Lett.* **1994**, 35, 4465-4468.
39. Cossio, F. P., Arrieta, A., Lecea, B., Ugalde, J. M. Chiral Control in the Staudinger Reaction between Ketenes and Imines. A Theoretical SCF-MO Study on Asymmetric Torquoselectivity. *J. Am. Chem. Soc.* **1994**, 116, 2085-2093.
40. Lopez, R., Suarez, D., Ruiz-Lopez, M. F., Gonzalez, J., Sordo, J. A., Sordo, T. L. Solvent effects on the stereoselectivity of ketene-imine cycloaddition reactions. *J. Chem. Soc., Chem. Commun.* **1995**, 1677-1678.
41. Lecea, B., Arrastia, I., Arrieta, A., Roa, G., Lopez, X., Arriortua, M. I., Ugalde, J. M., Cossio, F. P. Solvent and Substituent Effects in the Periselectivity of the Staudinger Reaction between Ketenes and α,β-Unsaturated Imines. A Theoretical and Experimental Study. *J. Org. Chem.* **1996**, 61, 3070-3079.
42. Salzner, U., Bachrach, S. M. Cycloaddition Reactions between Cyclopentadiene and Ketene. Ab Initio Examination of [2 + 2] and [4 + 2] Pathways. *J. Org. Chem.* **1996**, 61, 237-242.
43. Alajarin, M., Vidal, A., Tovar, F., Arrieta, A., Lecea, B., Cossio, F. P. Surpassing torquoelectronic effects in conrotatory ring closures: origins of stereocontrol in intramolecular ketenimine - imine [2+2] cycloadditions. *Chem.-- Eur. J.* **1999**, 5, 1106-1117.
44. Arrieta, A., Cossio, F. P., Lecea, B. New Insights on the Origins of the Stereocontrol of the Staudinger Reaction: [2 + 2] Cycloaddition between Ketenes and N-Silylimines. *J. Org. Chem.* **2000**, 65, 8458-8464.
45. Bongini, A., Panunzio, M., Piersanti, G., Bandini, E., Martelli, G., Spunta, G., Venturini, A. Stereochemical aspects of a two-step Staudinger reaction - asymmetric synthesis of chiral azetidine-2-ones. *Eur. J. Org. Chem.* **2000**, 2379-2390.
46. Alonso, E., del Pozo, C., Gonzalez, J. Staudinger reactions of unsymmetrical cyclic ketenes: a synthetically useful approach to spiro β-lactams and derivatives. Reaction mechanism and theoretical studies. *J. Chem. Soc., Perkin Trans. 1* **2002**, 571-576.
47. Venturini, A., Gonzalez, J. A CASPT2 and CASSCF Approach to the Cycloaddition of Ketene and Imine: A New Mechanistic Scheme of the Staudinger Reaction. *J. Org. Chem.* **2002**, 67, 9089-9092.
48. Macias, A., Alonso, E., Del Pozo, C., Venturini, A., Gonzalez, J. Diastereoselective [2+2]-Cycloaddition Reactions of Unsymmetrical Cyclic Ketenes with Imines: Synthesis of Modified Prolines and Theoretical Study of the Reaction Mechanism. *J. Org. Chem.* **2004**, 69, 7004-7012.
49. Staudinger, H. Ketenes. XXXI. Cyclobutanedione derivatives and the polymeric ketenes. *Ber.* **1920**, 53B, 1085-1092.
50. Staudinger, H., Suter, E. Ketenes. XXXII. Cyclobutane derivatives from diphenylketene and ethylene compounds. *Ber.* **1920**, 53B, 1092-1105.
51. Smith, C. W., Norton, D. G. Dimethylketene. *Org. Synth.* **1953**, 33, 29-31.
52. Brady, W. T. Halogenated ketenes. II. Dibromoketene. *J. Org. Chem.* **1966**, 31, 2676-2678.
53. Brady, W. T., Liddell, H. G., Vaughn, W. L. Halogenated ketenes. I. Dichloroketene. *J. Org. Chem.* **1966**, 31, 626-627.
54. Danheiser, R. L., Gee, S. K., Sard, H. A [4 + 4] annulation approach to eight-membered carbocyclic compounds. *J. Am. Chem. Soc.* **1982**, 104, 7670-7672.
55. Hassner, A., Naidorf, S. Cycloadditions. 31. Photochemical generation of vinylketenes by electrocyclic opening of cyclobutenones. *Tetrahedron Lett.* **1986**, 27, 6389-6392.
56. Leyendecker, F., Bloch, R., Conia, J. M. Flash thermolysis. Generation and reactivity of ketoketenes in intramolecular reactions towards unpolarized double bonds. *Tetrahedron Lett.* **1972**, 3703-3706.
57. Leyendecker, F. Reactivity of acylaldoketenes in intramolecular cycloaddition reactions. *Tetrahedron* **1976**, 32, 349-353.
58. Arya, F., Bouquant, J., Chuche, J. Preparation of 3-azabicyclo[3.2.0]heptenones by intramolecular [2+2] cycloaddition. *Tetrahedron Lett.* **1986**, 27, 1913-1914.
59. Joullié, M. M., Gomes, A. S. The chemistry of a ketene-sufur dioxide adduct. *J. Heterocycl. Chem.* **1969**, 6, 729.
60. Katz, T. J., Dessau, R. Hydrogen isotope effects and the mechanism of cycloaddition. *J. Am. Chem. Soc.* **1963**, 85, 2172-2173.
61. Baldwin, J. E., Kapecki, J. A. Kinetic isotope effects on the [2+2] cycloadditions of diphenylketene with α- and β-deuteriostyrene. *J. Am. Chem. Soc.* **1969**, 91, 3106-3107.
62. Baldwin, J. E., Kapecki, J. A. Stereochemistry and secondary deuterium kinetic isotope effects in the cycloadditions of diphenylketene with styrene and deuteriostyrenes. *J. Am. Chem. Soc.* **1970**, 92, 4874-4879.
63. Baldwin, J. E., Kapecki, J. A. Kinetics of the cycloadditions of diphenylketene with 1,1-diarylethylenes and styrenes. *J. Am. Chem. Soc.* **1970**, 92, 4868-4873.
64. Isaacs, N. S., Stanbury, P. F. Steric factors in [2 + 2]-cycloadditions of ketenes to olefins. *J. Chem. Soc., Chem. Commun.* **1970**, 1061-1062.
65. Rey, M., Roberts, S. M., Dieffenbacher, A., Dreiding, A. S. Stereochemical aspects of ketene addition to cyclopentadiene. *Helv. Chim. Acta* **1970**, 53, 417-432.
66. Huisgen, R., Mayr, H. Steric course, kinetics, and mechanism of the [2 + 2] cycloadditions of alkylphenylketenes to ethyl cis and trans-propenyl ether. *Tetrahedron Lett.* **1975**, 2969-2972.
67. Pasto, D. J. Reinterpretation of the mechanisms of concerted cycloaddition and cyclodimerization of allenes. *J. Am. Chem. Soc.* **1979**, 101, 37-46.

68. Brocksom, T. J., Coelho, F., Depres, J.-P., Greene, A. E., Freire de Lima, M. E., Hamelin, O., Hartmann, B., Kanazawa, A. M., Wang, Y. First Comprehensive Bakkane Approach: Stereoselective and Efficient Dichloroketene-Based Total Syntheses of (±)- and (-)-9-Acetoxyfukinanolide, (±)- and (+)-Bakkenolide A, (-)-Bakkenolides III, B, C, H, L, V, and X, (±)- and (-)-Homogynolide A, (±)-Homogynolide B, and (±)-Palmosalide C. *J. Am. Chem. Soc.* **2002**, 124, 15313-15325.
69. Tai, H.-M., Chang, M.-Y., Lee, A.-Y., Chang, N.-C. A Versatile Diquinane from Fulvene as a Building Block in Natural Product Synthesis. 1. A Facile Synthesis of the Iridoids Loganin and Sarracenin. *J. Org. Chem.* **1999**, 64, 659-662.
70. Jin, W., Metobo, S., Williams, R. M. Synthetic Studies on Ecteinascidin-743: Constructing a Versatile Pentacyclic Intermediate for the Synthesis of Ecteinascidins and Saframycins. *Org. Lett.* **2003**, 5, 2095-2098.
71. Pommier, A., Pons, J.-M., Kocienski, P. J. The First Total Synthesis of (-)-Lipstatin. *J. Org. Chem.* **1995**, 60, 7334-7339.

Staudinger Reaction 428

1. Staudinger, H., Meyer, J. New organic compounds of phosphorus. III. Phosphinemethylene derivatives and phosphinimines. *Helv. Chim. Acta* **1919**, 2, 635-646.
2. Stuckwisch, C. G. Azomethine ylides, azomethine imines, and iminophosphoranes in organic syntheses. *Synthesis* **1973**, 469-483.
3. Abel, E. W., Mucklejohn, S. A. The chemistry of phosphinimines. *Phosphorus and Sulfur and the Related Elements* **1981**, 9, 235-266.
4. Gololobov, Y. G., Zhmurova, I. N., Kasukhin, L. F. Sixty years of Staudinger reaction. *Tetrahedron* **1981**, 37, 437-472.
5. Cooper, R. D. G., Daugherty, B. W., Boyd, D. B. Chiral control of the Staudinger reaction. *Pure Appl. Chem.* **1987**, 59, 485-492.
6. Gololobov, Y. G., Kasukhin, L. F., Petrenko, V. S. New aspects of the Staudinger reaction. *Phosphorus and Sulfur and the Related Elements* **1987**, 30, 393-396.
7. Zwierzak, A., Koziara, A. Novel synthetic aspects of the Staudinger reaction. *Phosphorus and Sulfur and the Related Elements* **1987**, 30, 331-334.
8. Scriven, E. F. V., Turnbull, K. Azides: their preparation and synthetic uses. *Chem. Rev.* **1988**, 88, 297-368.
9. Gololobov, Y. G., Kasukhin, L. F. Recent advances in the Staudinger reaction. *Tetrahedron* **1992**, 48, 1353-1406.
10. Molina, P., Vilaplana, M. J. Iminophosphoranes: useful building blocks for the preparation of nitrogen-containing heterocycles. *Synthesis* **1994**, 1197-1218.
11. Dehnicke, K., Weller, F. Phosphorane iminato complexes of main group elements. *Coord. Chem. Rev.* **1997**, 158, 103-169.
12. Taillefer, M., Cristau, H.-J., Jouanin, I., Inguimbert, N., Jager, L. Phosphonium diylides in organic synthesis. *Phosphorus, Sulfur Silicon Relat. Elem.* **1999**, 144-146, 401-404.
13. Gilbertson, S. High-yielding Staudinger ligation of phosphinoesters and azides to form amides. *Chemtracts* **2001**, 14, 524-528.
14. Hartung, R., Paquette, L. Recent synthetic applications of the tandem Staudinger/Aza-Wittig reaction. *Chemtracts* **2004**, 17, 72-82.
15. Koehn, M., Breinbauer, R. The Staudinger ligation - A gift to chemical biology. *Angew. Chem., Int. Ed. Engl.* **2004**, 43, 3106-3116.
16. Koziara, A., Zwierzak, A. Optimized procedures for one-pot conversion of alkyl bromides into amines via the Staudinger reaction. *Synthesis* **1992**, 1063-1065.
17. Lopusinski, A., Luczak, L., Michalski, J. Stereospecific synthesis of P-epimeric (RP1, RP2)-bis-[O-l-menthylphenylphosphonothionyl] diselenide. A new variant of the stereoselective Staudinger reaction. *Heteroatom Chem.* **1995**, 6, 365-370.
18. Afonso, C. A. M. Transformation of an azido group to an N-(tert-butoxycarbonyl)amino group via the Staudinger reaction. *Synth. Commun.* **1998**, 28, 261-276.
19. O'Neil, I. A., Thompson, S., Murray, C. L., Kalindjian, S. B. DPPE: a convenient replacement for triphenylphosphine in the Staudinger and Mitsunobu reactions. *Tetrahedron Lett.* **1998**, 39, 7787-7790.
20. Malkinson, J. P., Falconer, R. A., Toth, I. Synthesis of C-terminal glycopeptides from resin-bound glycosyl azides via a modified Staudinger reaction. *J. Org. Chem.* **2000**, 65, 5249-5252.
21. Nilsson, B. L., Kiessling, L. L., Raines, R. T. Staudinger ligation: a peptide from a thioester and azide. *Org. Lett.* **2000**, 2, 1939-1941.
22. Vanek, P., Klan, P. An efficient one-pot conversion of alkyl bromides into imines via the Staudinger reaction. *Synth. Commun.* **2000**, 30, 1503-1507.
23. Andersen, N. G., Ramsden, P. D., Che, D., Parvez, M., Keay, B. A. A Simple Resolution Procedure Using the Staudinger Reaction for the Preparation of P-Stereogenic Phosphine Oxides. *J. Org. Chem.* **2001**, 66, 7478-7486.
24. Bonini, C., D'Auria, M., Funicello, M., Romaniello, G. Novel N-(2-benzo[b]thienyl)iminophosphoranes and their use in the synthesis of benzo[b]thieno[2,3-b]pyridines. *Tetrahedron* **2002**, 58, 3507-3512.
25. David, O., Messter, W. J. N., Bieraeugel, H., Schoemaker, H. E., Hiemstra, H., van Maarseveen, J. H. Intramolecular Staudinger ligation: A powerful ring-closure method to form medium-sized lactams. *Angew. Chem., Int. Ed. Engl.* **2003**, 42, 4373-4375.
26. Koehn, M., Wacker, R., Peters, C., Schroeder, H., Soulere, L., Breinbauer, R., Neimeyer, C. M., Waldmann, H. Staudinger ligation: A new immobilization strategy for the preparation of small-molecule arrays. *Angew. Chem., Int. Ed. Engl.* **2003**, 42, 5830-5834.
27. Restituyo, J. A., Comstock, L. R., Petersen, S. G., Stringfellow, T., Rajski, S. R. Conversion of Aryl Azides to O-Alkyl Imidates via Modified Staudinger Ligation. *Org. Lett.* **2003**, 5, 4357-4360.
28. Sauers, R. R., Van Arnum, S. D. A Thio-Staudinger Reaction: Thermolysis of a Vinyl Azide in the Presence of t-Butyl Mercaptan. *Phosphorus, Sulfur Silicon Relat. Elem.* **2003**, 178, 2169-2181.
29. Wang, C. C. Y., Seo, T. S., Li, Z., Ruparel, H., Ju, J. Site-specific fluorescent labeling of DNA using Staudinger ligation. *Bioconjug. Chem.* **2003**, 14, 697-701.
30. Basato, M., Benetollo, F., Facchin, G., Michelin, R. A., Mozzon, M., Pugliese, S., Sgarbossa, P., Sbovata, S. M., Tassan, A. The Staudinger reaction of platinum(II)- and palladium(II)-coordinated 2-(azidomethyl)phenyl isocyanide. X-ray structure of [cyclic] trans-[PtCl{CN(H)C_6H_4-2-CH_2N(H)}$(PPh_3)_2$][BF_4] $CDCl_3$ H_2O. *J. Organomet. Chem.* **2004**, 689, 454-462.
31. Bianchi, A., Bernardi, A. Selective synthesis of anomeric α-glycosyl acetamides via intramolecular Staudinger ligation of the α-azides. *Tetrahedron Lett.* **2004**, 45, 2231-2234.
32. He, Y., Hinklin, R. J., Chang, J., Kiessling, L. L. Stereoselective N-Glycosylation by Staudinger Ligation. *Org. Lett.* **2004**, 6, 4479-4482.
33. Alajarin, M., Conesa, C., Rzepa, H. S. Ab initio SCF-MO study of the Staudinger phosphorylation reaction between a phosphane and an azide to form a phosphazene. *J. Chem. Soc., Perkin Trans. 2* **1999**, 1811-1814.
34. Widauer, C., Grutzmacher, H., Shevchenko, I., Gramlich, V. Insights into the Staudinger reaction. Experimental and theoretical studies on the stabilization of cis-phosphazides. *Eur. J. Inorg. Chem.* **1999**, 1659-1664.
35. Tian, W. Q., Wang, Y. A. Mechanisms of Staudinger Reactions within Density Functional Theory. *J. Org. Chem.* **2004**, 69, 4299-4308.
36. Leffler, J. E., Temple, R. D. Staudinger reaction between triarylphosphines and azides. Mechanism. *J. Am. Chem. Soc.* **1967**, 89, 5235-5246.
37. Sasaki, T., Kanematsu, K., Murata, M. Tetrazolo-azido isomerization in heteroaromatics. III. Staudinger reaction of tetrazolopyridines with triphenylphosphine. *Tetrahedron* **1971**, 27, 5359-5366.
38. Goldwhite, H., Gysegem, P., Schow, S., Swyke, C. Structure and decomposition of a Staudinger reaction intermediate [1-methyl(or phenyl)-3-tris(dimethylamino)phosphoranylidenetriazene]. *Journal of the Chemical Society, Dalton Transactions: Inorganic Chemistry (1972-1999)* **1975**, 16-18.
39. Kroshefsky, R. D., Verkade, J. G. Staudinger reactions of aminophosphines. Influence of phosphorus basicity. *Inorg. Chem.* **1975**, 14, 3090-3095.
40. Lei, G., Xu, H., Xu, X. Staudinger reaction between bicyclic phosphites and azides. *Phosphorus, Sulfur Silicon Relat. Elem.* **1992**, 66, 101-106.
41. Shalev, D. E., Chiacchiera, S. M., Radkowsky, A. E., Kosower, E. M. Sequence of Reactant Combination Alters the Course of the Staudinger Reaction of Azides with Acyl Derivatives. Bimanes. 30. *J. Org. Chem.* **1996**, 61, 1689-1701.
42. Thirupathi, N., Liu, X., Verkade John, G. Reactions of tris(amino)phosphines with arylsulfonyl azides: product dependency on tris(amino)phosphine structure. *Inorg. Chem.* **2003**, 42, 389-397.

43. Yokokawa, F., Asano, T., Shioiri, T. Total Synthesis of the Antiviral Marine Natural Product (-)-Hennoxazole A. *Org. Lett.* **2000**, 2, 4169-4172.
44. Jiang, B., Yang, C.-G., Wang, J. Enantioselective Synthesis of Marine Indole Alkaloid Hamacanthin B. *J. Org. Chem.* **2002**, 67, 1396-1398.
45. White, J. D., Cammack, J. H., Sakuma, K. The synthesis and absolute configuration of mycosporins. A novel application of the Staudinger reaction. *J. Am. Chem. Soc.* **1989**, 111, 8970-8972.

Stephen Aldehyde Synthesis (Stephen Reduction) 430

1. Stephen, H. New synthesis of aldehydes. *J. Chem. Soc., Abstracts* **1925**, 127, 1874-1877.
2. Ferguson, L. N. The synthesis of aromatic aldehydes. *Chem. Rev.* **1946**, 38, 227-254.
3. Mosettig, E. The synthesis of aldehydes from carboxylic acids. *Org. React.* **1954**, 218-257.
4. Zil'berman, E. N. Reactions of nitriles with hydrogen halides and nucleophilic reagents. *Russ. Chem. Rev.* **1962**, 31, 615-633.
5. Fuson, R. C. Formation of aldehydes and ketones from carboxylic acids and their derivatives. in *The Chemistry of the Carbonyl Group* (ed. Patai, S.), (Interscience Publishers, New York, **1966**).
6. Rabinovitz, M. in *The Chemistry of the Cyano Group* (ed. Rappoport, Z.), 307 (Wiley Interscience, New York, **1970**).
7. Davis, A. P. Reduction of Carboxylic Acids to Aldehydes by Other Methods. in *Comp. Org. Synth.* (eds. Trost, B. M.,Fleming, I.), *8*, 283-305 (Pergamon, Oxford, **1991**).
8. Stephen, T., Stephen, H. Modification of the procedure for converting nitriles to aldehydes. *J. Chem. Soc., Abstracts* **1956**, 4695-4696.
9. Tolbert, T. L., Houston, B. The preparation of aldimines through the Stephen reaction. *J. Org. Chem.* **1963**, 28, 695-697.
10. Alagona, G., Tomasi, J. The mechanism of addition to a C-N triple bond. An ab initio study of the first stages of the Stephen, Gattermann and Houben-Hoesch reactions. *THEOCHEM* **1983**, 8, 263-281.
11. Pietra, S., Trinchera, C. The reduction of nitriles to aldehydes by means of hydrazine and Raney nickel. *Gazz. Chim. Ital.* **1955**, 85, 1705-1709.
12. Backeberg, O. G., Staskun, B. A novel reduction of nitriles to aldehydes. *J. Chem. Soc., Abstracts* **1962**, 3961-3963.
13. Staskun, B., Backeberg, O. G. Reduction of hindered nitriles to aldehydes. *J. Chem. Soc., Suppl.* **1964**, No. 1, 5880-5881.
14. Ferris, J. P., Antonucci, F. R. Hydrated electron in organic synthesis. Reduction of nitriles to aldehydes. *J. Chem. Soc., Chem. Commun.* **1971**, 1294-1295.
15. Ferris, J. P., Antonucci, F. R. Hydrated electron in organic synthesis. Reduction of nitriles to aldehydes. *J. Am. Chem. Soc.* **1972**, 94, 8091-8095.
16. Cha, J. S., Oh, S. Y., Kim, J. E. Partial reduction of nitriles to aldehydes by thexylbromoborane-methyl sulfide. *Bull. Korean Chem. Soc.* **1987**, 8, 301-304.
17. Cha, J. S., Chang, S. W., Kwon, O. O., Kim, J. M. Partial reduction of nitriles to aldehydes by catecholalane (1,3,2-benzodioxaluminole). *Synlett* **1996**, 165-166.
18. Cha, J. S., Jang, S. H., Kwon, S. Y. Selective conversion of aromatic nitriles to aldehydes by lithium N,N'-dimethylethylenediaminoaluminum hydride. *Bull. Korean Chem. Soc.* **2002**, 23, 1697-1698.
19. Suzuki, N. Synthesis of antimicrobial agents. V. Synthesis and antimicrobial activities of some heterocyclic condensed 1,8-naphthyridine derivatives. *Chem. Pharm. Bull.* **1980**, 28, 761-768.
20. Kasak, P., Putala, M. Stereoconservative cyanation of [1,1'-binaphthalene]-2,2'-dielectrophiles. An alternative approach to homochiral C_2-symmetric [1,1'-binaphthalene]-2,2'-dicarbonitrile and its transformations. *Collect. Czech. Chem. Commun.* **2000**, 65, 729-740.
21. Scrimin, P., Tecilla, P., Tonellato, U., Veronese, A., Crisma, M., Formaggio, F., Toniolo, C. Zinc(II) as an allosteric regulator of liposomal membrane permeability induced by synthetic template-assembled tripodal polypeptides. *Chem.-- Eur. J.* **2002**, 8, 2753-2763.
22. Yang, L.-M., Lin, S.-J., Hsu, F.-L., Yang, T.-H. Antitumor agents. Part 3: synthesis and cytotoxicity of new trans-stilbene benzenesulfonamide derivatives. *Bioorg. Med. Chem. Lett.* **2002**, 12, 1013-1015.

Stetter Reaction 432

1. Stetter, H., Schreckenberg, M. Addition of aldehydes to activated double bonds. *Angew. Chem.* **1973**, 85, 89.
2. Stetter, H. New synthetic methods. 17. The catalyzed addition of aldehydes to activated double bonds - a new synthesis principle. *Angew. Chem.* **1976**, 88, 695-704.
3. Kuhlmann, H., Stetter, H. The addition of aldehydes to activated double bonds and its application in the synthesis of jasmine fragrances. *Fragrance Flavor Subst., Proc. Int. Haarmann Reimer Symp., 2nd* **1980**, 99-110.
4. Anon. 1,4-Dicarbonyl compounds using the Stetter reaction. *Nachrichten aus Chemie, Technik und Laboratorium* **1981**, 29, 172-173.
5. Stetter, H., Kuhlmann, H. The catalyzed nucleophilic addition of aldehydes to electrophilic double bonds. *Org. React.* **1991**, 40, 407-496.
6. Enders, D., Breuer, K. Addition of acyl carbanion equivalents to carbonyl groups and enones. *Comprehensive Asymmetric Catalysis I-III* **1999**, 3, 1093-1102.
7. Dalko, P. I., Moisan, L. Asymmetric catalysis: In the Golden Age of Organocatalysis. *Angew. Chem., Int. Ed. Engl.* **2004**, 43, 5138-5175.
8. Enders, D., Balensiefer, T. Nucleophilic Carbenes in Asymmetric Organocatalysis. *Acc. Chem. Res.* **2004**, 37, 534-541.
9. Johnson, J. S. Catalyzed reactions of acyl anion equivalents. *Angew. Chem., Int. Ed. Engl.* **2004**, 43, 1326-1328.
10. Ho, T. L., Liu, S. H. Stetter condensation catalyzed by a polymer-bound thiazolium ylide. *Synth. Commun.* **1983**, 13, 1125-1127.
11. Enders, D., Breuer, K., Runsink, J., Teles, J. H. The first asymmetric intramolecular Stetter reaction. *Helv. Chim. Acta* **1996**, 79, 1899-1902.
12. Murry, J. A., Frantz, D. E., Soheili, A., Tillyer, R., Grabowski, E. J. J., Reider, P. J. Synthesis of α-Amido Ketones via Organic Catalysis: Thiazolium-Catalyzed Cross-Coupling of Aldehydes with Acylimines. *J. Am. Chem. Soc.* **2001**, 123, 9696-9697.
13. Gong, J. H., Im, Y. J., Lee, K. Y., Kim, J. N. Tributylphosphine-catalyzed Stetter reaction of N,N-dimethylacrylamide: synthesis of N,N-dimethyl-3-aroylpropionamides. *Tetrahedron Lett.* **2002**, 43, 1247-1251.
14. Kerr, M. S., Read de Alaniz, J., Rovis, T. A Highly Enantioselective Catalytic Intramolecular Stetter Reaction. *J. Am. Chem. Soc.* **2002**, 124, 10298-10299.
15. Kerr, M. S., Rovis, T. Effect of the Michael acceptor in the asymmetric intramolecular Stetter reaction. *Synlett* **2003**, 1934-1936.
16. Murry, J. A., Frantz, D., Soheili, A., Tillyer, R., Grabowski, E. J. J., Reider, P. J. Thiazolium-catalyzed cross-coupling of aldehydes with acylimines: a new method for the synthesis of α-amidoketones. *Chemtracts* **2003**, 16, 579-586.
17. Rovis, T. Metal and nonmetal catalysts for carbon-carbon bond-forming reactions leading to desymmetrized 1,4-dicarbonyl compounds. *Chemtracts* **2003**, 16, 542-553.
18. Yadav, J. S., Anuradha, K., Reddy, B. V. S., Eeshwaraiah, B. Microwave-accelerated conjugate addition of aldehydes to α,β-unsaturated ketones. *Tetrahedron Lett.* **2003**, 44, 8959-8962.
19. Anjaiah, S., Chandrasekhar, S., Gree, R. Stetter reaction in room temperature ionic liquids and application to the synthesis of haloperidol. *Adv. Syn. & Catal.* **2004**, 346, 1329-1334.
20. Barrett, A. G. M., Love, A. C., Tedeschi, L. ROMPgel-Supported Thiazolium Iodide: An Efficient Supported Organic Catalyst for Parallel Stetter Reactions. *Org. Lett.* **2004**, 6, 3377-3380.
21. Gacem, B., Jenner, G. Effect of pressure on Stetter reactions: synthesis of hindered aliphatic acyloins and γ-ketonitriles. *High Pressure Research* **2004**, 24, 233-236.
22. Kerr, M. S., Rovis, T. Enantioselective synthesis of quaternary stereocenters via a catalytic asymmetric Stetter reaction. *J. Am. Chem. Soc.* **2004**, 126, 8876-8877.
23. Mattson, A. E., Bharadwaj, A. R., Scheidt, K. A. The Thiazolium-Catalyzed Sila-Stetter Reaction: Conjugate Addition of Acylsilanes to Unsaturated Esters and Ketones. *J. Am. Chem. Soc.* **2004**, 126, 2314-2315.
24. Pesch, J., Harms, K., Bach, T. Preparation of axially chiral N,N'-diarylimidazolium and N-arylthiazolium salts and evaluation of their catalytic potential in the benzoin and in the intramolecular Stetter reactions. *Eur. J. Org. Chem.* **2004**, 2025-2035.

25. Baumann, K. L., Butler, D. E., Deering, C. F., Mennen, K. E., Millar, A., Nanninga, T. N., Palmer, C. W., Roth, B. D. The convergent synthesis of CI-981, an optically active, highly potent, tissue-selective inhibitor of HMG-CoA reductase. *Tetrahedron Lett.* **1992**, 33, 2283-2284.
26. Harrington, P. E., Tius, M. A. Synthesis and Absolute Stereochemistry of Roseophilin. *J. Am. Chem. Soc.* **2001**, 123, 8509-8514.
27. Galopin, C. C. A short and efficient synthesis of (±)-trans-sabinene hydrate. *Tetrahedron Lett.* **2001**, 42, 5589-5591.
28. Randl, S., Blechert, S. Concise Enantioselective Synthesis of 3,5-Dialkyl-Substituted Indolizidine Alkaloids via Sequential Cross-Metathesis-Double-Reductive Cyclization. *J. Org. Chem.* **2003**, 68, 8879-8882.

Stevens Rearrangement..434

1. Stevens, T. S., Creigton, E. M., Gordon, A. B., MacNicol, M. The degradation of quaternary ammonium salts. I. *J. Chem. Soc., Abstracts* **1928**, 3193-3197.
2. Dunn, J. L., Stevens, T. S. Degradation of quaternary ammonium salts. VI. Effect of substitution on velocity of intramolecular rearrangement. *J. Chem. Soc., Abstracts* **1932**, 1926-1931.
3. Thomson, T., Stevens, T. S. Degradation of quaternary ammonium salts. VII. New cases of radical migration. *J. Chem. Soc., Abstracts* **1932**, 1932-1940.
4. Thomson, T., Stevens, T. S. Degradation of quaternary ammonium salts. V. Molecular rearrangements in related sulfur compounds. *J. Chem. Soc., Abstracts* **1932**, 69-73.
5. Thomson, T., Stevens, T. S. Degradation of quaternary ammonium salts. IV. Relative migratory velocities of substituted benzyl radicals. *J. Chem. Soc., Abstracts* **1932**, 55-69.
6. Zimmerman, H. E. Base-catalyzed rearrangements. in *Molecular Rearrangements* (ed. De Mayo, P.), *1*, 345-406 (Wiley, New York, **1963**).
7. Pine, S. H. Base-promoted rearrangements of quaternary ammonium salts. *Org. React.* **1970**, 18, 403-464.
8. Hudson, R. F. Ylid chemistry. *Chemistry in Britain* **1971**, 7, 287-294.
9. Lepey, A. R., Giumanini, A. G. Stevens and Sommelet rearrangements. *Mechanisms of Molecular Migrations* **1971**, 3, 297-440.
10. Pant, J., Joshi, B. C. Stevens rearrangement. *Indian Journal of Chemical Education* **1980**, 7, 11-16.
11. Marko, I. E. The Stevens and related rearrangement. in *Comp. Org. Synth.* (eds. Trost, B. M.,Fleming, I.), *3*, 913-974 (Pergamon, Oxford, **1991**).
12. Li, A.-H., Dai, L.-X., Aggarwal, V. K. Asymmetric Ylide Reactions: Epoxidation, Cyclopropanation, Aziridination, Olefination, and Rearrangement. *Chemical Reviews (Washington, D. C.)* **1997**, 97, 2341-2372.
13. Beall, L. S., Padwa, A. Application of nitrogen ylide cyclizations for organic synthesis. *Advances in Nitrogen Heterocycles* **1998**, 3, 117-158.
14. Clark, J. S. Nitrogen, oxygen and sulfur ylides: an overview. *Nitrogen, Oxygen and Sulfur Ylide Chemistry* **2002**, 1-113.
15. Hill, R. K., Chan, T. H. Transfer of asymmetry from nitrogen to carbon in the Stevens rearrangement. *J. Am. Chem. Soc.* **1966**, 88, 866-867.
16. Benecke, H. P., Wikel, J. H. Stevens rearrangement in aminimides. *Tetrahedron Lett.* **1971**, 37, 3479-3482.
17. Sato, Y., Sakakibara, H. Formation of ammonium ylides by the cleavage of silicon-carbon bonds of triphenylsilylmethylammonium salts. *J. Organomet. Chem.* **1979**, 166, 303-307.
18. Sato, Y., Yagi, Y., Koto, M. Ylide reactions of benzyldimethyl[(triorganosilyl)methyl]ammonium halides. *J. Org. Chem.* **1980**, 45, 613-617.
19. Zhang, J. J., Schuster, G. B. Photo-Stevens rearrangement of 9-dimethylsulfonium fluorenylide. *J. Org. Chem.* **1988**, 53, 716-719.
20. Eberlein, T. H., West, F. G., Tester, R. W. The Stevens [1,2]-shift of oxonium ylides: a route to substituted tetrahydrofuranones. *J. Org. Chem.* **1992**, 57, 3479-3482.
21. West, F. G., Naidu, B. N. Applications of Stevens [1,2]-Shifts of Cyclic Ammonium Ylides. A Route to Morpholin-2-ones. *J. Org. Chem.* **1994**, 59, 6051-6056.
22. Feldman, K. S., Wrobleski, M. L. Alkynyliodonium salts in organic synthesis. Dihydrofuran formation via a formal stevens shift of a carbon substituent within a disubstituted-carbon oxonium ylide. *J. Org. Chem.* **2000**, 65, 8659-8668.
23. Vanecko, J. A., West, F. G. A Novel, Stereoselective Silyl-Directed Stevens [1,2]-Shift of Ammonium Ylides. *Org. Lett.* **2002**, 4, 2813-2816.
24. Marmsaeter, F. P., Murphy, G. K., West, F. G. Cyclooctanoid Ring Systems from Mixed Acetals via Heteroatom-Assisted [1,2]-Shift of Oxonium Ylides. *J. Am. Chem. Soc.* **2003**, 125, 14724-14725.
25. Harada, M., Nakai, T., Tomooka, K. Stevens rearrangement of a cyclic hemiacetal system: Diastereoselective approach to chiral α-amino ketone. *Synlett* **2004**, 365-367.
26. Dewar, M. J. S., Ramsden, C. A. Stevens rearrangement. Antiaromatic pericyclic reaction. *J. Chem. Soc., Perkin Trans. 1* **1974**, 1839-1844.
27. Heard, G. L., Frankcombe, K. E., Yates, B. F. A theoretical study of the Stevens rearrangement of methylammonium methylide and methylammonium formylmethylide. *Aust. J. Chem.* **1993**, 46, 1375-1388.
28. Heard, G. L., Yates, B. F. Theoretical studies of the Stevens' rearrangement of alkylammonium ylides. *THEOCHEM* **1994**, 116, 197-204.
29. Heard, G. L., Yates, B. F. Steric and electronic effects on the mechanism of the Stevens rearrangement - large organic ylides of unusually high symmetry. *Aust. J. Chem.* **1994**, 47, 1685-1694.
30. Heard, G. L., Yates, B. F. Theoretical evaluation of alternative pathways in the Stevens rearrangement. *Aust. J. Chem.* **1995**, 48, 1413-1423.
31. Heard, G. L., Yates, B. F. Competing Rearrangements of Ammonium Ylides: A Quantum Theoretical Study. *Journal of Organic Chemistry* **1996**, 61, 7276-7284.
32. Heard, G. L., Yates, B. F. Hybrid supermolecule-polarizable continuum approach to solvation: application to the mechanism of the Stevens rearrangement. *J. Comput. Chem.* **1996**, 17, 1444-1452.
33. Makita, K., Koketsu, J., Ando, F., Ninomiya, Y., Koga, N. Theoretical Investigation of Stevens Rearrangement of P and As Ylides. Migration of H, CH_3, $CH{:}CH_2$, SiH_3, and GeH_3 Groups on P and As Atoms. *J. Am. Chem. Soc.* **1998**, 120, 5764-5770.
34. Okada, K., Tanaka, M. Reinvestigation of base-induced skeletal conversion via a spirocyclic intermediate of dibenzodithiocinium derivatives and a computational study using the HF/6-31G* basis set. *Journal of the Chemical Society, Perkin Transactions 1* **2002**, 2704-2711.
35. Chantrapromma, K., Ollis, W. D., Sutherland, I. O. Base catalyzed rearrangements involving ylide intermediates. Part 16. The preparation and thermal rearrangement of allylammonioamidates. *J. Chem. Soc., Perkin Trans. 1* **1983**, 1029-1039.
36. Jemison, R. W., Morris, D. G. Mechanistic implications of nuclear polarization in the Stevens rearrangement of N,N-dimethyl p-nitrobenzylamine acetimide. *J. Chem. Soc., Chem. Commun.* **1969**, 1226-1227.
37. Baldwin, J. E., Erickson, W. F., Hackler, R. E., Scott, R. M. Simultaneous observation of a radical pathway and retention in a Stevens rearrangement of a sulfonium ylide: significance for a general theory of ylide rearrangements. *J. Chem. Soc., Chem. Commun.* **1970**, 576-578.
38. Ollis, W. D., Rey, M., Sutherland, I. O., Closs, G. L. Mechanism of the Stevens rearrangement. *J. Chem. Soc., Chem. Commun.* **1975**, 543-545.
39. Pine, S. H., Cheney, J. Allowed and forbidden sigmatropic pathways in the Stevens rearrangement of a phenacylammonium ylide. *J. Org. Chem.* **1975**, 40, 870-872.
40. Giumanini, A. G., Trombini, C., Lercker, G., Lepley, A. R. Heterobenzyl quaternary ammonium salts. IV. 2-Thenyl group as terminus and migrating moiety in the Stevens and Sommelet rearrangements of a quaternary ammonium ion. *Journal of Organic Chemistry* **1976**, 41, 2187-2193.
41. Ollis, W. D., Rey, M., Sutherland, I. O. Base catalyzed rearrangements involving ylide intermediates. Part 15. The mechanism of the Stevens [1,2] rearrangement. *J. Chem. Soc., Perkin Trans. 1* **1983**, 1009-1027.
42. Stara, I. G., Stary, I., Tichy, M., Zavada, J., Hanus, V. Stereochemical Dichotomy in the Stevens Rearrangement of Axially Twisted Dihydroazepinium and Dihydrothiepinium Salts. A Novel Enantioselective Synthesis of Pentahelicene. *J. Am. Chem. Soc.* **1994**, 116, 5084-5088.

43. Padwa, A., Beall, L. S., Eidell, C. K., Worsencroft, K. J. An Approach toward Isoindolobenzazepines Using the Ammonium Ylide/Stevens [1,2]-Rearrangement Sequence. *J. Org. Chem.* **2001**, 66, 2414-2421.
44. Liou, J.-P., Cheng, C.-Y. Total synthesis of (±)-desoxycodeine-D: a novel route to the morphine skeleton. *Tetrahedron Lett.* **2000**, 41, 915-918.
45. Aggarwal, V. K., Jones, D., Turner, M. L., Adams, H. First synthesis and X-ray crystal structure of 1,2-(1,1'-ferrocenediyl)ethene. *J. Organomet. Chem.* **1996**, 524, 263-266.

Stille Carbonylative Cross-Coupling 436

1. Meerifield, J. H., Godschal, J. P., Stille, J. K. Synthesis of unsymmetrical diallyl ketones: the palladium-catalyzed coupling of allyl halides with allyltin reagents in the presence of carbon monoxide. *Organometallics* **1984**, 3, 1108-1112.
2. Sheffy, F. K., Godschalx, J. P., Stille, J. K. Palladium-catalyzed cross coupling of allyl halides with organotin reagents: a method of joining highly functionalized partners regioselectively and stereospecifically. *J. Am. Chem. Soc.* **1984**, 106, 4833-4840.
3. Baillargeon, V. P., Stille, J. K. Palladium-catalyzed formylation of organic halides with carbon monoxide and tin hydride. *J. Am. Chem. Soc.* **1986**, 108, 452-461.
4. Stille, J. K. Palladium-catalyzed coupling reactions of organic electrophiles with organic tin compounds. *Angew. Chem.* **1986**, 98, 504-519.
5. Echavarren, A. M., Stille, J. K. Palladium-catalyzed carbonylative coupling of aryl triflates with organostannanes. *J. Am. Chem. Soc.* **1988**, 110, 1557-1565.
6. Farina, V., Krishnamurthy, V., Scott, W. J. The Stille reaction. *Org. React.* **1997**, 50, 1-652.
7. Miyaura, N., Editor. Cross-Coupling Reactions. A Practical Guide. *Top. Curr. Chem.* **2002**, 219, 248 pp.
8. Bumagin, N. A., Gulevich, Y. V., Beletskaya, I. P. Palladium-catalyzed synthesis of aromatic acid derivatives by carbonylation of aryl iodides and Alk_3SnNu (Nu = MeO, Et_2N, PhS, EtS). *J. Organomet. Chem.* **1985**, 285, 415-418.
9. Caldirola, P., Chowdhury, R., Johansson, A. M., Hacksell, U. Intramolecular Transfer of CO from (η6-arene)$Cr(CO)_3$ Complexes in Stille-Type Palladium-Catalyzed Cross-Coupling Reactions. *Organometallics* **1995**, 14, 3897-3900.
10. Ceccarelli, S., Piarulli, U., Gennari, C. Effect of ligands and additives on the palladium-promoted carbonylative coupling of vinyl stannanes and electron-poor enol triflates. *J. Org. Chem.* **2000**, 65, 6254-6256.
11. Skoda-Foldes, R., Horvath, J., Tuba, Z., Kollar, L. Homogeneous coupling and carbonylation reactions of steroids possessing iodoalkene moieties. Catalytic and mechanistic aspects. *J. Organomet. Chem.* **1999**, 586, 94-100.
12. Knight, S. D., Overman, L. E., Pairaudeau, G. Synthesis applications of cationic aza-Cope rearrangements. 26. Enantioselective total synthesis of (-)-strychnine. *J. Am. Chem. Soc.* **1993**, 115, 9293-9294.
13. Jeanneret, V., Meerpoel, L., Vogel, P. C-Glycosides and C-disaccharide precursors through carbonylative Stille coupling reactions. *Tetrahedron Lett.* **1997**, 38, 543-546.
14. Morera, E., Ortar, G. A concise synthesis of photoactivatable 4-aroyl-L-phenylalanines. *Bioorg. Med. Chem. Lett.* **2000**, 10, 1815-1818.
15. Dewey, T. M., Mundt, A., Crouch, G. J., Zyzniewski, M. C., Eaton, B. E. New Uridine Derivatives for Systematic Evolution of RNA Ligands by Exponential Enrichment. *J. Am. Chem. Soc.* **1995**, 117, 8474-8475.

Stille Cross-Coupling (Migita-Kosugi-Stille Coupling) 438

Related reactions: *Kumada cross-coupling, Negishi cross-coupling, Suzuki cross-coupling;*

1. Azarian, D., Dua, S. S., Eaborn, C., Walton, D. R. M. Reactions of organic halides with R_3MMR_3 compounds (M = silicon, germanium, tin) in the presence of tetrakis(triarylphosphine)palladium. *J. Organomet. Chem.* **1976**, 117, C55-C57.
2. Kosugi, M., Sasazawa, K., Shimizu, Y., Migita, T. Reactions of allyltin compounds. III. Allylation of aromatic halides with allyltributyltin in the presence of tetrakis(triphenylphosphine)palladium$^{(0)}$. *Chem. Lett.* **1977**, 301-302.
3. Kosugi, M., Shimizu, Y., Migita, T. Alkylation, arylation, and vinylation of acyl chlorides by means of organotin compounds in the presence of catalytic amounts of tetrakis(triphenylphosphine)palladium$^{(0)}$. *Chem. Lett.* **1977**, 1423-1424.
4. Kosugi, M., Shimizu, Y., Migita, T. Reaction of allyltin compounds. II. Facile preparation of allyl ketones via allyltins. *J. Organomet. Chem.* **1977**, 129, C36-C38.
5. Milstein, D., Stille, J. K. A general, selective, and facile method for ketone synthesis from acid chlorides and organotin compounds catalyzed by palladium. *J. Am. Chem. Soc.* **1978**, 100, 3636-3638.
6. Milstein, D., Stille, J. K. Mechanism of reductive elimination. Reaction of alkylpalladium(II) complexes with tetraorganotin, organolithium, and Grignard reagents. Evidence for palladium(IV) intermediacy. *J. Am. Chem. Soc.* **1979**, 101, 4981-4991.
7. Milstein, D., Stille, J. K. Palladium-catalyzed coupling of tetraorganotin compounds with aryl and benzyl halides. Synthetic utility and mechanism. *J. Am. Chem. Soc.* **1979**, 101, 4992-4998.
8. Milstein, D., Stille, J. K. Mild, selective, general method of ketone synthesis from acid chlorides and organotin compounds catalyzed by palladium. *J. Org. Chem.* **1979**, 44, 1613-1618.
9. Mitchell, T. N. Transition-metal catalysis in organotin chemistry. *J. Organomet. Chem.* **1986**, 304, 1-16.
10. Stille, J. K. Palladium-catalyzed coupling reactions of organic electrophiles with organic tin compounds. *Angew. Chem.* **1986**, 98, 504-519.
11. Mitchell, T. N. Palladium-catalyzed reactions of organotin compounds. *Synthesis* **1992**, 803-815.
12. Farina, V. New perspectives in the cross-coupling reactions of organostannanes. *Pure Appl. Chem.* **1996**, 68, 73-78.
13. Luh, T.-Y. Transition metal-catalyzed cross-coupling reactions of unactivated aliphatic C-X bonds. *Rev. on Heteroa. Chem.* **1996**, 15, 61-82.
14. Stephenson, G. R. *Asymmetric palladium-catalyzed coupling reactions* (ed. Stephenson, G. R.) (Chapman & Hall, London, **1996**) 275-298.
15. Browning, A. F., Greeves, N. Palladium-catalyzed carbon-carbon bond formation. *Transition Metals in Organic Synthesis* **1997**, 35-64.
16. Farina, V., Krishnamurthy, V., Scott, W. J. The Stille reaction. *Org. React.* **1997**, 50, 1-652.
17. Mitchell, T. N. Organotin reagents in cross-coupling. in *Metal-Catalyzed Cross-Coupling Reactions* (eds. Diederich, F.,Stang, P. J.), 167-202 (Wiley-VCH, Weinheim, New York, **1998**).
18. Stanforth, S. P. Catalytic cross-coupling reactions in biaryl synthesis. *Tetrahedron* **1998**, 54, 263-303.
19. Duncton, M. A. J., Pattenden, G. The intramolecular Stille reaction. *J. Chem. Soc., Perkin Trans. 1* **1999**, 1235-1246.
20. Pierre Genet, J., Savignac, M. Recent developments of palladium(0) catalyzed reactions in aqueous medium. *J. Organomet. Chem.* **1999**, 576, 305-317.
21. Jafarpour, L., Grasa, G. A., Viciu, M. S., Hillier, A. C., Nolan, S. P. Convenient and efficient cross-coupling of aryl halides mediated by palladium/bulky nucleophilic carbenes and related ligands. *Chimica Oggi* **2001**, 19, 10-16.
22. Kosugi, M., Fugami, K. A historical note of the Stille reaction. *J. Organomet. Chem.* **2002**, 653, 50-53.
23. Kosugi, M., Fugami, K. Overview of the Stille protocol with Sn. *Handbook of Organopalladium Chemistry for Organic Synthesis* **2002**, 1, 263-283.
24. Miyaura, N., Editor. Cross-Coupling Reactions. A Practical Guide. *Top. Curr. Chem.* **2002**, 219, 248 pp.
25. Pattenden, G., Sinclair, D. J. The intramolecular Stille reaction in some target natural product syntheses. *J. Organomet. Chem.* **2002**, 653, 261-268.
26. Ricci, A., Lo Sterzo, C. A new frontier in the metal-catalyzed cross-coupling reaction field. The palladium-promoted metal-carbon bond formation. Scope and mechanism of a new tool in organometallic synthesis. *Journal of Organometallic Chemistry* **2002**, 653, 177-194.
27. Espinet, P., Echavarren, A. M. C-C coupling: The mechanisms of the Stille reaction. *Angew. Chem., Int. Ed. Engl.* **2004**, 43, 4704-4734.
28. Falck, J. R., Bhatt, R. K., Ye, J. Tin-Copper Transmetalation: Cross-Coupling of α-Heteroatom-Substituted Alkyltributylstannanes with Organohalides. *J. Am. Chem. Soc.* **1995**, 117, 5973-5982.

29. Percec, V., Bae, J. Y., Hill, D. H. Aryl Mesylates in Metal-Catalyzed Homo-Coupling and Cross- Coupling Reactions .4. Scope and Limitations of Aryl Mesylates in Nickel-Catalyzed Cross-Coupling Reactions. *J. Org. Chem.* **1995**, 60, 6895-6903.
30. Takeda, T., Matsunaga, K.-i., Kabasawa, Y., Fujiwara, T. The copper(I) iodide-promoted allylation of vinylstannanes with allylic halides. *Chem. Lett.* **1995**, 771-772.
31. Allred, G. D., Liebeskind, L. S. Copper-Mediated Cross-Coupling of Organostannanes with Organic Iodides at or below Room Temperature. *J. Am. Chem. Soc.* **1996**, 118, 2748-2749.
32. Piers, E., Romero, M. A. Intramolecular CuCl-Mediated Oxidative Coupling of Alkenyltrimethylstannane Functions: An Effective Method for the Construction of Carbocyclic 1,3-Diene Systems. *J. Am. Chem. Soc.* **1996**, 118, 1215-1216.
33. Kang, S. K., Kim, J. S., Choi, S. C. Copper- and manganese-catalyzed cross-coupling of organostannanes with organic iodides in the presence of sodium chloride. *J. Org. Chem.* **1997**, 62, 4208-4209.
34. Kang, S. K., Kim, J. S., Yoon, S. K., Lim, K. H., Yoon, S. S. Copper-catalyzed coupling of polymer bound iodide with organostannanes. *Tetrahedron Lett.* **1998**, 39, 3011-3012.
35. Kang, S. K., Kim, W. Y., Jiao, X. G. Copper-catalyzed cross-coupling of 1-iodoalkynes with organostannanes. *Synthesis-Stuttgart* **1998**, 1252-1254.
36. Shirakawa, E., Yamasaki, K., Hiyama, T. Cross-coupling reaction of organostannanes with aryl halides catalyzed by nickel-triphenylphosphine or nickel-lithium halide complex. *Synthesis-Stuttgart* **1998**, 1544-1549.
37. Naso, F., Babudri, F., Farinola, G. M. Organometallic chemistry directed towards the synthesis of electroactive materials: stereoselective routes to extended polyconjugated systems. *Pure Appl. Chem.* **1999**, 71, 1485-1492.
38. Maleczka, R. E., Jr., Gallagher, W. P., Terstiege, I. Stille Couplings Catalytic in Tin: Beyond Proof-of-Principle. *J. Am. Chem. Soc.* **2000**, 122, 384-385.
39. Gallagher, W. P., Terstiege, I., Maleczka, R. E., Jr. Stille Couplings Catalytic in Tin: The "Sn-O" Approach. *J. Am. Chem. Soc.* **2001**, 123, 3194-3204.
40. Maleczka, R. E., Jr., Gallagher, W. P. Stille Couplings Catalytic in Tin: A "Sn-F" Approach. *Org. Lett.* **2001**, 3, 4173-4176.
41. Labadie, J. W., Tueting, D., Stille, J. K. Synthetic utility of the palladium-catalyzed coupling reaction of acid chlorides with organotins. *J. Org. Chem.* **1983**, 48, 4634-4642.
42. Echavarren, A. M., Stille, J. K. Palladium-catalyzed coupling of aryl triflates with organostannanes. *J. Am. Chem. Soc.* **1987**, 109, 5478-5486.
43. Farina, V., Krishnan, B. Large rate accelerations in the stille reaction with tri-2-furylphosphine and triphenylarsine as palladium ligands: mechanistic and synthetic implications. *J. Am. Chem. Soc.* **1991**, 113, 9585-9595.
44. Farina, V., Krishnan, B., Marshall, D. R., Roth, G. P. Palladium-catalyzed coupling of arylstannanes with organic sulfonates: a comprehensive study. *J. Org. Chem.* **1993**, 58, 5434-5444.
45. Farina, V., Kapadia, S., Krishnan, B., Wang, C., Liebeskind, L. S. On the Nature of the "Copper Effect" in the Stille Cross-Coupling. *J. Org. Chem.* **1994**, 59, 5905-5911.
46. Louie, J., Hartwig, J. F. Transmetalation, Involving Organotin Aryl, Thiolate, and Amide Compounds. An Unusual Type of Dissociative Ligand Substitution Reaction. *J. Am. Chem. Soc.* **1995**, 117, 11598-11599.
47. Casado, A. L., Espinet, P. Mechanism of the Stille Reaction. 1. The Transmetalation Step. Coupling of R1I and R_2SnBu_3 Catalyzed by trans-[PdR1IL$_2$] (R1 = $C_6Cl_2F_3$; R2 = Vinyl, 4-Methoxyphenyl; L = $AsPh_3$). *J. Am. Chem. Soc.* **1998**, 120, 8978-8985.
48. Shirakawa, E., Hiyama, T. The palladium-iminophosphine catalyst for the reactions of organostannanes. *J. Organomet. Chem.* **1999**, 576, 169-178.
49. Casado, A. L., Espinet, P., Gallego, A. M. Mechanism of the Stille Reaction. 2. Couplings of Aryl Triflates with Vinyltributyltin. Observation of Intermediates. A More Comprehensive Scheme. *J. Am. Chem. Soc.* **2000**, 122, 11771-11782.
50. Casado, A. L., Espinet, P., Gallego, A. M., Martinez-Ilarduya, J. M. Snapshots of a Stille reaction. *Chem. Commun.* **2001**, 339-340.
51. Casares, J. A., Espinet, P., Salas, G. 14-electron t-shaped [PdRXL] complexes: Evidence or illusion? Mechanistic consequences for the stille reaction and related processes. *Chem.-- Eur. J.* **2002**, 8, 4843-4853.
52. Dupont, J., de Souza, R. F., Suarez, P. A. Z. Ionic Liquid (Molten Salt) Phase Organometallic Catalysis. *Chem. Rev.* **2002**, 102, 3667-3691.
53. Amatore, C., Bahsoun, A. A., Jutand, A., Meyer, G., Ntepe, A. N., Ricard, L. Mechanism of the Stille Reaction Catalyzed by Palladium Ligated to Arsine Ligand: PhPdI($AsPh_3$)(DMF) Is the Species Reacting with Vinylstannane in DMF. *J. Am. Chem. Soc.* **2003**, 125, 4212-4222.
54. Jutand, A. Mechanism of palladium-catalyzed reactions: Role of chloride ions. *Applied Organometallic Chemistry* **2004**, 18, 574-582.
55. Masse, C. E., Yang, M., Solomon, J., Panek, J. S. Total Synthesis of (+)-Mycotrienol and (+)-Mycotrienin I: Application of Asymmetric Crotylsilane Bond Constructions. *J. Am. Chem. Soc.* **1998**, 120, 4123-4134.
56. Martin, S. F., Humphrey, J. M., Ali, A., Hillier, M. C. Enantioselective Total Syntheses of Ircinal A and Related Manzamine Alkaloids. *J. Am. Chem. Soc.* **1999**, 121, 866-867.
57. Lebsack, A. D., Link, J. T., Overman, L. E., Stearns, B. A. Enantioselective Total Synthesis of Quadrigemine C and Psycholeine. *J. Am. Chem. Soc.* **2002**, 124, 9008-9009.

Stille-Kelly Coupling 440

1. Echavarren, A. M., Stille, J. K. Palladium-catalyzed coupling of aryl triflates with organostannanes. *J. Am. Chem. Soc.* **1987**, 109, 5478-5486.
2. Kelly, T. R., Li, Q., Bhushan, V. Intramolecular biaryl coupling: asymmetric synthesis of the chiral B-ring diol unit of pradimicinone. *Tetrahedron Lett.* **1990**, 31, 161-164.
3. Stanforth, S. P. Catalytic cross-coupling reactions in biaryl synthesis. *Tetrahedron* **1998**, 54, 263-303.
4. Marshall, J. A. Synthesis and Reactions of Allylic, Allenic, Vinylic, and Arylmetal Reagents from Halides and Esters via Transient Organopalladium Intermediates. *Chem. Rev.* **2000**, 100, 3163-3185.
5. Fouquet, E. Synthetic applications of organic germanium, tin and lead compounds (excluding R_3MH). *Chemistry of Organic Germanium, Tin and Lead Compounds* **2002**, 2, 1333-1399.
6. Hosomi, A., Miura, K. Palladium-catalyzed carbon-metal bond formation via reductive elimination. *Handbook of Organopalladium Chemistry for Organic Synthesis* **2002**, 1, 1107-1119.
7. Mori, M., Kaneta, N., Shibasaki, M. The use of silyltriorganostannanes, $R_3SiSnR'_3$, in organic synthesis. A novel palladium-catalyzed tandem transmetalation-cyclization reaction. *J. Org. Chem.* **1991**, 56, 3486-3493.
8. Nishiyama, Y., Tokunaga, K., Sonoda, N. New Synthetic Method of Diorganyl Selenides: Palladium-Catalyzed Reaction of $PhSeSnBu_3$ with Aryl and Alkyl Halides. *Org. Lett.* **1999**, 1, 1725-1727.
9. Fukuyama, Y., Yaso, H., Mori, T., Takahashi, H., Minami, H., Kodama, M. Total syntheses of plagiochins A and D, macrocyclic bis(bibenzyls), by Pd(0) catalyzed intramolecular Stille-Kelly reaction. *Heterocycles* **2001**, 54, 259-274.
10. Olivera, R., SanMartin, R., Tellitu, I., Dominguez, E. The amine exchange/biaryl coupling sequence: a direct entry to the phenanthro[9,10-d]heterocyclic framework. *Tetrahedron* **2002**, 58, 3021-3037.
11. Iwaki, T., Yasuhara, A., Sakamoto, T. Novel synthetic strategy of carbolines via palladium-catalyzed amination and arylation reaction. *J. Chem. Soc., Perkin Trans. 1* **1999**, 1505-1510.
12. Yue, W. S., Li, J. J. A Concise Synthesis of All Four Possible Benzo[4,5]furopyridines via Palladium-Mediated Reactions. *Org. Lett.* **2002**, 4, 2201-2203.
13. Sakamoto, T., Yasuhara, A., Kondo, Y., Yamanaka, H. Condensed heteroaromatic ring systems. XXIII. A concise synthesis of hippadine. *Heterocycles* **1993**, 36, 2597-2600.

1. Stobbe, H. A new synthesis of tetraconic acid. *Ber.* **1893**, 26, 2312-2319.
2. Johnson, W. S., Daub, G. H. The Stobbe condensation. *Org. React.* **1951**, VI, 1-73.
3. Arason, K. M., Bergmeier, S. C. The synthesis of succinic acids and derivatives. A review. *Org. Prep. Proced. Int.* **2002**, 34, 337, 339-366.
4. Reutrakul, V., Kusamran, K., Wattanasin, S. Reaction of diethyl lithiosuccinate with carbonyl compounds: Stobbe condensation of α-keto esters. *Heterocycles* **1977**, 6, 715-719.
5. Salem, M. R., Enayat, E. I., Kandil, A. A. Stobbe condensation of aromatic aldehydes with diethyl diglycolate. *J. Prakt. Chem.* **1979**, 321, 741-746.
6. El-Rayyes, N., Al-Johary, A. J. The Stobbe condensation. 6. Reaction of aryl aldehydes with dimethyl adipate. *J. Chem. Eng. Data* **1987**, 32, 123-125.
7. Rizzacasa, M. A., Sargent, M. V. The Wittig reaction of 2-(tert-butoxycarbonyl)-1-(methoxycarbonyl)ethylidenetriphenylphosphorane: a surrogate for the Stobbe reaction. *Aust. J. Chem.* **1987**, 40, 1737-1743.
8. El-Bassiouny, F. A., Habashy, M. M., El-Nagdy, S. I., Mahmoud, M. R. Stobbe condensation involving diethyl β,β-dimethylglutarate. II. *Egypt. J. Chem.* **1991**, 32, 467-474.
9. Jabbar, S., Gupta, G. Migration of double bond in Stobbe reactions. *Curr. Sci.* **1991**, 60, 493-494.
10. Mezger, F., Simchen, G., Fischer, P. The homologous "silyl-Stobbe-reaction". *Synthesis* **1991**, 375-378.
11. El-Bassiouny, F. A., Habashy, M. M., El-Nagdy, S. I., Mahmoud, M. R. Stobbe condensation involving diethyl β,β-dimethylglutarate. II. *Communications de la Faculte des Sciences de l'Universite d'Ankara, Series B: Chemistry and Chemical Engineering* **1992**, 35, 115-121.
12. Zerrer, R., Simchen, G. The silyl-Stobbe condensation - synthesis of fulgimides. *Synthesis* **1992**, 922-924.
13. Owton, W. M., Gallagher, P. T., Juan-Montesinos, A. tert-Butyl 3-carboxyethyl-3-phosphonodiethylpropionate. A novel reagent for Stobbe-like condensations. *Synth. Commun.* **1993**, 23, 2119-2125.
14. Jabbar, S., Banerjee, S. Photochromism in Stobbe condensation and cyclized products. *Curr. Sci.* **1996**, 70, 959.
15. Tanaka, K., Sugino, T., Toda, F. Selective Stobbe condensation under solvent-free conditions. *Green Chem.* **2000**, 2, 303-304.
16. Yvon, B. L., Datta, P. K., Le, T. N., Charlton, J. L. Synthesis of magnoshinin and cyclogalgravin: modified Stobbe condensation reaction. *Synthesis* **2001**, 1556-1560.
17. Newman, M. S., Linsk, J. Steric hindrance in the Stobbe condensation. *J. Am. Chem. Soc.* **1949**, 71, 936-937.
18. Jeffery, D., Fry, A. An oxygen-18 tracer study of the Stobbe condensation. *J. Org. Chem.* **1957**, 22, 735-739.
19. Ganeshpure, P. A. Cyclolignans through Stobbe condensation: Part III. Stereochemistry and mechanism of diarylidenesuccinic acid formation in the Stobbe condensation: a synthesis of justicidin-E. *Indian J. Chem., Sect. B* **1979**, 17B, 202-206.
20. Anjaneyulu, A. S. R., Raghu, P., Rao, K. V. R., Sastry, C. V. M., Umasundari, P., Satyanarayana, P. On the stereoselectivity of Stobbe condensation with ortho-substituted aromatic aldehydes: the (E, Z) configuration of monobenzylidenesuccinates and dibenzylidenesuccinic anhydrides. *Curr. Sci.* **1984**, 53, 239-243.
21. Gupta, G., Banerjee, S. Stereochemistry of α-alkylidene-β-arylidenesuccinates formed in the Stobbe condensation. *Indian J. Chem., Sect. B* **1990**, 29B, 787-790.
22. Jabbar, S., Gupta, G. Study of dissociation constants of Stobbe condensation and cyclization products. *J. Indian Chem. Soc.* **1991**, 68, 60-61.
23. White, J. D., Hrnciar, P., Stappenbeck, F. Asymmetric Total Synthesis of (+)-Codeine via Intramolecular Carbenoid Insertion. *J. Org. Chem.* **1999**, 64, 7871-7884.
24. Huffman, J. W., Yu, S. Synthesis of a tetracyclic, conformationally constrained analog of D8-THC. *Bioorg. Med. Chem.* **1998**, 6, 2281-2288.
25. Liu, J., Brooks, N. R. Stereoselective Stobbe Condensation of Ethyl Methyl Diphenylmethylenesuccinate with Aromatic Aldehydes. *Org. Lett.* **2002**, 4, 3521-3524.

1. Benary, E. Acylation of α-Aminocrotonic Ester and Related Compounds. *Ber.* **1910**, 42, 3912-3925.
2. Robinson, R. Extension of the theory of addition to conjugated unsaturated systems. II. The C-alkylation of certain derivatives of β-aminocrotonic acid and the mechanism of the alkylation of ethyl acetoacetate and similar substances. *J. Chem. Soc., Abstracts* **1916**, 109, 1038-1046.
3. Stork, G., Terrell, R., Szmuszkovicz, J. Synthesis of 2-alkyl and 2-acyl ketones. *J. Am. Chem. Soc.* **1954**, 76, 2029-2030.
4. Stork, G., Landesman, H. K. A new alkylation of carbonyl compounds. II. *J. Am. Chem. Soc.* **1956**, 78, 5128-5129.
5. *Enamines: Synthesis, Structure, and Reactions* (ed. Cook, A. G.) (**1969**) 515 pp.
6. Dyke, S. F. *The Chemistry of Enamines* (**1973**) 93 pp.
7. Hickmott, P. W. Enamines: recent advances in synthetic, spectroscopic, mechanistic, and stereochemical aspects. II. *Tetrahedron* **1982**, 38, 3363-3446.
8. Hickmott, P. W. Enamines: recent advances in synthetic, spectroscopic, mechanistic, and stereochemical aspects. I. *Tetrahedron* **1982**, 38, 1975-2050.
9. Whitesell, J. K., Whitesell, M. A. Alkylation of ketones and aldehydes via their nitrogen derivatives. *Synthesis* **1983**, 517-536.
10. *Enamines: Synthesis, Structure, and Reactions* (ed. Cook, A. G.) (Marcel Dekker Inc., New York, **1988**) 717 pp.
11. Caine, D. Alkylations of Enols and Enolates. in *Comp. Org. Synth.* (eds. Trost, B. M.,Fleming, I.), *1*, 1-63 (Pergamon, Oxford, **1991**).
12. Stork, G., Birnbaum, G. Alkylation of enamines from α,β-unsaturated ketones. *Tetrahedron Lett.* **1961**, 313-316.
13. Von Strandtmann, M., Cohen, M. P., Shavel, J., Jr. Carbon-carbon alkylations of enamines with Mannich bases. *J. Org. Chem.* **1965**, 30, 3240-3242.
14. Curphey, T. J., Hung, J. C. Y. C-Alkylation of aldehyde enamines. *Chem. Commun.* **1967**, 510.
15. Yamada, S., Hiroi, K., Achiwa, K. Asymmetric synthesis with amino acids. I. Asymmetric induction in the alkylation of ketone enamines. *Tetrahedron Lett.* **1969**, 4233-4236.
16. Curphey, T. J., Hung, J. C. Y., Chu, C. C. C. Alkylation of enamines derived from sterically hindered amines. *J. Org. Chem.* **1975**, 40, 607-614.
17. Doutheau, A., Gore, J. Alkylation of enamines by 5-halo-3-en-1-ynes. *Tetrahedron* **1976**, 32, 2705-2711.
18. Nilsson, L., Rappe, C. Alkylation of enamines. A convenient route to 1,4-dicarbonyl compounds. *Acta Chem. Scand.* **1976**, B30, 1000-1002.
19. Schubert, S., Renaud, P., Carrupt, P. A., Schenk, K. Stereoselectivity of the radical reductive alkylation of enamines: importance of the allylic 1,3-strain model. *Helv. Chim. Acta* **1993**, 76, 2473-2489.
20. Katritzky, A. R., Fang, Y., Silina, A. Preparation of β-Amido Ketones and Aldehydes via Amidoalkylation of Enamines, Enol Silyl Ethers, and Vinyl Ethers. *J. Org. Chem.* **1999**, 64, 7622-7624.
21. Vignola, N., List, B. Catalytic asymmetric intramolecular α-alkylation of aldehydes. *J. Am. Chem. Soc.* **2004**, 126, 450-451.
22. Sevin, A., Tortajada, J., Pfau, M. Toward a transition-state model in the asymmetric alkylation of chiral ketone secondary enamines by electron-deficient alkenes. A theoretical MO study. *J. Org. Chem.* **1986**, 51, 2671-2675.
23. Mannich, C., Davidsen, H. Simple enamines with nitrogen in tertiary combination. *Ber.* **1936**, 69B, 2106-2112.
24. Kirrmann, A., Elkik, E. Mechanisms of the alkylation of enamines. *C. R. Seances Acad. Sci. C.* **1968**, 267, 623-625.
25. Kempf, B., Hampel, N., Ofial, A. R., Mayr, H. Structure - nucleophilicity relationships for enamines. *Chem.-- Eur. J.* **2003**, 9, 2209-2218.
26. Covarrubias-Zuniga, A., Cantu, F., Maldonado, L. A. A Total Synthesis of the Racemic Sesquiterpene Parvifoline. *J. Org. Chem.* **1998**, 63, 2918-2921.
27. Bagal, S. K., Adlington, R. M., Baldwin, J. E., Marquez, R., Cowley, A. Biomimetic Synthesis of Biatractylolide and Biepiasterolide. *Org. Lett.* **2003**, 5, 3049-3052.

28. Meyers, A. I., Elworthy, T. R. Chiral formamidines. The total asymmetric synthesis of (-)-8-azaestrone and related (-)-8-aza-12-oxo-17-desoxoestrone. *J. Org. Chem.* **1992**, 57, 4732-4740.

Strecker Reaction 446

1. Strecker, A. The artificial synthesis of lactic acid and a new homologue of glycine. *Liebigs Ann. Chem.* **1850**, 75, 27-45.
2. Strecker, A. The preparation of a new material by the reaction of acetaldehyde-ammonia imine and hydrogen cyanide. *Liebigs Ann. Chem.* **1854**, 91, 349-351.
3. Block, R. J. The isolation and synthesis of the naturally occurring α-amino acids. *Chem. Rev.* **1946**, 38, 501-571.
4. Greenstein, J. P., Winitz, M. Synthesis of α-amino acids. in *Chemistry of the Amino Acids 1*, 597-714 (John Wiley & Sons, Inc., New York, **1961**).
5. Miller, S. L., Van Trump, J. E. The Strecker synthesis in the primitive ocean. *Origin Life, Proc. ISSOL Meet., 3rd* **1981**, 135-141.
6. Barrett, G. C., Editor. *Chemistry and Biochemistry of the Amino Acids* (**1985**) 683 pp.
7. Miller, S. L. Current status of the prebiotic synthesis of small molecules. *Chem Scr* **1986**, 26B, 5-11.
8. Williams, R. M. in *Synthesis of optically active α-amino acids Chapter 5*, 208 (Pergamon Press, Oxford, **1989**).
9. Kunz, H., Sager, W., Pfrengle, W., Laschat, S., Schanzenbach, D. Stereoselective synthesis of amino acid derivatives using carbohydrates as templates. *Chemistry of Peptides and Proteins* **1993**, 5/6, 91-98.
10. Davis, F. A., Reddy, R. E., Portonovo, P. S. Asymmetric Strecker synthesis using enantiopure sulfinimines: a convenient synthesis of α-amino acids. *Tetrahedron Lett.* **1994**, 35, 9351-9354.
11. Duthaler, R. O. Recent developments in the stereoselective synthesis of α-amino acids. *Tetrahedron* **1994**, 50, 1539-1650.
12. Tolman, V. Syntheses of fluorine-containing amino acids by methods of classical amino acid chemistry. *Fluorine-Containing Amino Acids* **1995**, 1-70.
13. Iyer, M. S., Gigstad, K. M., Namdev, N. D., Lipton, M. Asymmetric catalysis of the Strecker amino acid synthesis by a cyclic dipeptide. *Amino Acids* **1996**, 11, 259-268.
14. Tolmann, V. Syntheses of fluorinated amino acids. From the classical to the modern concept. *Amino Acids* **1996**, 11, 15-36.
15. Dyker, G. Amino acid derivatives by multicomponent reactions. *Angew. Chem., Int. Ed. Engl.* **1997**, 36, 1700-1702.
16. Ohfune, Y., Horikawa, M. Asymmetric synthesis of α,α-disubstituted α-amino acids via an intramolecular Strecker synthesis. *Yuki Gosei Kagaku Kyokaishi* **1997**, 55, 982-993.
17. Zubay, G. Did carbohydrates provide carbon skeletons for the first amino acids to be synthesized on planet Earth? *Chemtracts* **1997**, 10, 407-413.
18. Kunz, H., Hofmeister, A., Glaser, B. Stereoselective syntheses using carbohydrates as carriers of chiral information. *Polysaccharides* **1998**, 539-567.
19. Calmes, M., Daunis, J. How to build optically active α-amino acids. *Amino acids* **1999**, 16, 215-250.
20. Kobayashi, S., Ishitani, H. Catalytic Enantioselective Addition to Imines. *Chem. Rev.* **1999**, 99, 1069-1094.
21. Dyker, G. Amino acid derivatives by multicomponent reactions. *Organic Synthesis Highlights IV* **2000**, 53-57.
22. Enders, D., Shilvock, J. P. Some recent applications of α-amino nitrile chemistry. *Chem. Soc. Rev.* **2000**, 29, 359-373.
23. Ager, D. J., Fotheringham, I. G. Methods for the synthesis of unnatural amino acids. *Current Opinion in Drug Discovery & Development* **2001**, 4, 800.
24. Yet, L. Recent developments in catalytic asymmetric Strecker-type reactions. *Angew. Chem., Int. Ed. Engl.* **2001**, 40, 875-877.
25. Gröger, H. Catalytic Enantioselective Strecker Reactions and Analogous Syntheses. *Chem. Rev.* **2003**, 103, 2795-2827.
26. Yet, L. Recent developments in catalytic asymmetric Strecker-type reactions. *Organic Synthesis Highlights V* **2003**, 187-193.
27. Spino, C. Recent developments in the catalytic asymmetric cyanation of ketimines. *Angew. Chem., Int. Ed. Engl.* **2004**, 43, 1764-1766.
28. Vachal, P., Jacobsen, E. N. Cyanation of carbonyl and imino groups. *Comprehensive Asymmetric Catalysis, Supplement* **2004**, 1, 117-130.
29. Bucherer, H. T., Libe, V. A. Syntheses of hydantoins. II. Formation of substituted hydantoins from aldehydes and ketones. *J. Prakt. Chem.* **1934**, 141, 5-43.
30. Bucherer, H. T., Steiner, W. Syntheses of hydantoins. I. Reactions of α-hydroxy and α-amino nitriles. *J. Prakt. Chem.* **1934**, 140, 291-316.
31. Bousquet, C., Tadros, Z., Tonnel, J., Mion, L., Taillades, J. Auxiliary chiral ketones in the asymmetric synthesis of α-amino acids by Strecker reaction. *Bull. Soc. Chim. Fr.* **1993**, 130, 513-520.
32. Davis, F. A., Portonovo, P. S., Reddy, R. E., Chiu, Y.-h. Asymmetric Strecker Synthesis Using Enantiopure Sulfinimines and Diethylaluminum Cyanide: The Alcohol Effect. *J. Org. Chem.* **1996**, 61, 440-441.
33. Byrne, J. J., Chavarot, M., Chavant, P.-Y., Vallee, Y. Asymmetric Strecker reactions of ketimines catalysed by titanium-based complexes. *Tetrahedron Lett.* **2000**, 41, 873-876.
34. Ishitani, H., Komiyama, S., Hasegawa, Y., Kobayashi, S. Catalytic Asymmetric Strecker Synthesis. Preparation of Enantiomerically Pure α-Amino Acid Derivatives from Aldimines and Tributyltin Cyanide or Achiral Aldehydes, Amines, and Hydrogen Cyanide Using a Chiral Zirconium Catalyst. *J. Am. Chem. Soc.* **2000**, 122, 762-766.
35. Sigman, M. S., Vachal, P., Jacobsen, E. N. A general catalyst for the asymmetric Strecker reaction. *Angew. Chem., Int. Ed. Engl.* **2000**, 39, 1279-1281.
36. Chavarot, M., Byrne, J. J., Chavant, P. Y., Vallee, Y. Sc(BINOL)$_2$Li: a new heterobimetallic catalyst for the asymmetric Strecker reaction. *Tetrahedron: Asymmetry* **2001**, 12, 1147-1150.
37. Mabic, S., Cordi, A. A. Synthesis of enantiomerically pure ethylenediamines from chiral sulfinimines: a new twist to the Strecker reaction. *Tetrahedron* **2001**, 57, 8861-8866.
38. Nogami, H., Matsunaga, S., Kanai, M., Shibasaki, M. Enantioselective Strecker-type reaction promoted by polymer-supported bifunctional catalyst. *Tetrahedron Lett.* **2001**, 42, 279-283.
39. Enders, D., Moser, M. Asymmetric Strecker synthesis by addition of trimethylsilyl cyanide to aldehyde SAMP-hydrazones. *Tetrahedron Lett.* **2003**, 44, 8479-8481.
40. Kato, N., Suzuki, M., Kanai, M., Shibasaki, M. Catalytic enantioselective Strecker reaction of ketimines using catalytic amount of TMSCN and stoichiometric amount of HCN. *Tetrahedron Lett.* **2004**, 45, 3153-3155.
41. Kato, N., Suzuki, M., Kanai, M., Shibasaki, M. General and practical catalytic enantioselective Strecker reaction of keto-imines: significant improvement through catalyst tuning by protic additives. *Tetrahedron Lett.* **2004**, 45, 3147-3151.
42. Nakamura, S., Sato, N., Sugimoto, M., Toru, T. A new approach to enantioselective cyanation of imines with Et_2AlCN. *Tetrahedron: Asymmetry* **2004**, 15, 1513-1516.
43. Inaba, T., Fujita, M., Ogura, K. Thermodynamically controlled 1,3-asymmetric induction in an acyclic system: equilibration of α-amino nitriles derived from α-alkylbenzylamines and aldehydes. *J. Org. Chem.* **1991**, 56, 1274-1279.
44. Cativiela, C., Diaz-de-Villegas, M. D., Galvez, J. A., Garcia, J. L. Diastereoselective Strecker reaction of D-glyceraldehyde derivatives. A novel route to (2S,3S)- and (2R,3S)-2-amino-3,4-dihydroxybutyric acid. *Tetrahedron* **1996**, 52, 9563-9574.
45. Kitayama, T., Watanabe, T., Takahashi, O., Morihashi, K., Kikuchi, O. Parity-violating energy for the chirality-producing step in Strecker synthesis of L-alanine. *THEOCHEM* **2002**, 584, 89-94.
46. Li, J., Jiang, W.-Y., Han, K.-L., He, G.-Z., Li, C. Density Functional Study on the Mechanism of Bicyclic Guanidine-Catalyzed Strecker Reaction. *J. Org. Chem.* **2003**, 68, 8786-8789.
47. Ogata, Y., Kawasaki, A. Mechanistic aspects of the Strecker aminonitrile synthesis. *J. Chem. Soc. B.* **1971**, 325-329.
48. Taillades, J., Commeyras, A. Strecker and related systems. II. Mechanism of formation in aqueous solution of α-alkylaminoisobutyronitriles from acetone, hydrocyanic acid, ammonia, and methyl- or dimethylamine. *Tetrahedron* **1974**, 30, 2493-2501.
49. Stout, D. M., Black, L. A., Matier, W. L. Asymmetric Strecker synthesis: isolation of pure enantiomers and mechanistic implications. *J. Org. Chem.* **1983**, 48, 5369-5373.

50. Corey, E. J., Grogan, M. J. Enantioselective Synthesis of α-Amino Nitriles from N-Benzhydryl Imines and HCN with a Chiral Bicyclic Guanidine as Catalyst. *Org. Lett.* **1999**, 1, 157-160.
51. Takamura, M., Hamashima, Y., Usuda, H., Kanai, M., Shibasaki, M. A catalytic asymmetric Strecker-type reaction promoted by Lewis acid-Lewis base bifunctional catalyst. *Chem. Pharm. Bull.* **2000**, 48, 1586-1592.
52. Josephsohn, N. S., Kuntz, K. W., Snapper, M. L., Hoveyda, A. H. Mechanism of Enantioselective Ti-Catalyzed Strecker Reaction: Peptide-Based Metal Complexes as Bifunctional Catalysts. *J. Am. Chem. Soc.* **2001**, 123, 11594-11599.
53. Vachal, P., Jacobsen, E. N. Structure-Based Analysis and Optimization of a Highly Enantioselective Catalyst for the Strecker Reaction. *J. Am. Chem. Soc.* **2002**, 124, 10012-10014.
54. Atherton, J. H., Blacker, J., Crampton, M. R., Grosjean, C. The Strecker reaction: kinetic and equilibrium studies of cyanide addition to iminium ions. *Org. Biomol. Chem.* **2004**, 2, 2567-2571.
55. Vedejs, E., Kongkittingam, C. A Total Synthesis of (-)-Hemiasterlin Using N-Bts Methodology. *J. Org. Chem.* **2001**, 66, 7355-7364.
56. Davis, F. A., Prasad, K. R., Carroll, P. J. Asymmetric Synthesis of Polyhydroxy α-Amino Acids with the Sulfinimine-Mediated Asymmetric Strecker Reaction: 2-Amino 2-Deoxy L-Xylono-1,5-lactone (Polyoxamic Acid Lactone). *J. Org. Chem.* **2002**, 67, 7802-7806.
57. Mann, S., Carillon, S., Breyne, O., Marquet, A. Total synthesis of amiclenomycin, an inhibitor of biotin biosynthesis. *Chem.-- Eur. J.* **2002**, 8, 439-450.

Suzuki Cross-Coupling (Suzuki-Miyaura Cross-Coupling)..448

Related reactions: *Kumada cross-coupling, Negishi cross-coupling, Stille cross-coupling;*

1. Miyaura, N., Suzuki, A. Stereoselective synthesis of arylated (E)-alkenes by the reaction of alk-1-enylboranes with aryl halides in the presence of palladium catalyst. *J. Chem. Soc., Chem. Commun.* **1979**, 866-867.
2. Miyaura, N., Yamada, K., Suzuki, A. A new stereospecific cross-coupling by the palladium-catalyzed reaction of 1-alkenylboranes with 1-alkenyl or 1-alkynyl halides. *Tetrahedron Lett.* **1979**, 3437-3440.
3. Miyaura, N., Yanagi, T., Suzuki, A. The palladium-catalyzed cross-coupling reaction of phenylboronic acid with haloarenes in the presence of bases. *Synth. Commun.* **1981**, 11, 513-519.
4. Suzuki, A. Organoboron compounds in new synthetic reactions. *Pure Appl. Chem.* **1985**, 57, 1749-1758.
5. Suzuki, A. Synthetic studies via the cross-coupling reaction of organoboron derivatives with organic halides. *Pure Appl. Chem.* **1991**, 63, 419-422.
6. Martin, A. R., Yang, Y. Palladium-catalyzed cross-coupling reactions of organoboronic acids with organic electrophiles. *Acta Chem. Scand.* **1993**, 47, 221-230.
7. Suzuki, A. New synthetic transformations via organoboron compounds. *Pure Appl. Chem.* **1994**, 66, 213-222.
8. Miyaura, N., Suzuki, A. Palladium-Catalyzed Cross-Coupling Reactions of Organoboron Compounds. *Chem. Rev.* **1995**, 95, 2457-2483.
9. Stephenson, G. R. *Asymmetric palladium-catalyzed coupling reactions* (ed. Stephenson, G. R.) (Chapman & Hall, London, **1996**) 275-298.
10. Browning, A. F., Greeves, N. *Palladium-catalyzed carbon-carbon bond formation* (eds. Beller, M.,Bolm, C.) (Wiley-VCH, Weinheim, New York, **1997**) 35-64.
11. Herrmann, W. A., Reisinger, C.-P. *Carbon-carbon coupling by Heck-type reactions* (eds. Cornils, B.,Hermann, W. A.) (Wiley-VCH, Weinheim, New York, **1998**) 383-392.
12. Miyaura, N. Synthesis of biaryls via the cross-coupling reaction of arylboronic acids. *Advances in Metal-Organic Chemistry* **1998**, 6, 187-243.
13. Stanforth, S. P. Catalytic cross-coupling reactions in biaryl synthesis. *Tetrahedron* **1998**, 54, 263-303.
14. Kocovsky, P., Malkov, A. V., Vyskocil, S., Lloyd-Jones, G. C. Transition metal catalysis in organic synthesis: reflections, chirality and new vistas. *Pure Appl. Chem.* **1999**, 71, 1425-1433.
15. Li, J. J. Applications of palladium chemistry to the total syntheses of naturally occurring indole alkaloids. *Alkaloids: Chemical and Biological Perspectives* **1999**, 14, 437-503.
16. Oehme, G., Grassert, I., Paetzold, E., Meisel, R., Drexler, K., Fuhrmann, H. Complex catalyzed hydrogenation and carbon-carbon bond formation in aqueous micelles. *Coord. Chem. Rev.* **1999**, 185-186, 585-600.
17. Pierre Genet, J., Savignac, M. Recent developments of palladium(0) catalyzed reactions in aqueous medium. *J. Organomet. Chem.* **1999**, 576, 305-317.
18. Suzuki, A. Recent advances in the cross-coupling reactions of organoboron derivatives with organic electrophiles, 1995-1998. *J. Organomet. Chem.* **1999**, 576, 147-168.
19. Franzen, R. The Suzuki, the Heck, and the Stille reaction; three versatile methods for the introduction of new C-C bonds on solid support. *Can. J. Chem.* **2000**, 78, 957-962.
20. Groziak, M. P. Boron heterocycles as platforms for building new bioactive agents. *Progress in Heterocyclic Chemistry* **2000**, 12, 1-21.
21. Marshall, J. A. Pd-catalyzed borylation of aryl halides. *Chemtracts* **2000**, 13, 219-222.
22. Sharman, W. M., Van Lier, J. E. Use of palladium catalysis in the synthesis of novel porphyrins and phthalocyanines. *Journal of Porphyrins and Phthalocyanines* **2000**, 4, 441-453.
23. Shen, W. The versatility of 1,1-dibromo-1-alkenes in palladium-catalyzed coupling reactions. *Frontiers of Biotechnology & Pharmaceuticals* **2000**, 1, 349-372.
24. Chemler, S. R., Trauner, D., Danishefsky, S. J. The B-alkyl Suzuki-Miyaura cross-coupling reaction: development, mechanistic study, and applications in natural product synthesis. *Angew. Chem., Int. Ed. Engl.* **2001**, 40, 4544-4568.
25. De Vries, J. G., De Vries, A. H. M., Tucker, C. E., Miller, J. A. Palladium catalysis in the production of pharmaceuticals. *Innovations in Pharmaceutical Technology* **2001**, 01, 125-126, 128, 130.
26. Lloyd-Williams, P., Giralt, E. Atropisomerism, biphenyls and the Suzuki coupling: peptide antibiotics. *Chem. Soc. Rev.* **2001**, 30, 145-157.
27. Dupont, J., de Souza, R. F., Suarez, P. A. Z. Ionic Liquid (Molten Salt) Phase Organometallic Catalysis. *Chem. Rev.* **2002**, 102, 3667-3691.
28. Hassan, J., Sevignon, M., Gozzi, C., Schulz, E., Lemaire, M. Aryl-Aryl Bond Formation One Century after the Discovery of the Ullmann Reaction. *Chem. Rev.* **2002**, 102, 1359-1469.
29. Herrmann, W. A. The Suzuki cross-coupling. *Applied Homogeneous Catalysis with Organometallic Compounds (2nd Edition)* **2002**, 1, 591-598.
30. Kotha, S., Lahiri, K., Kashinath, D. Recent applications of the Suzuki-Miyaura cross-coupling reaction in organic synthesis. *Tetrahedron* **2002**, 58, 9633-9695.
31. Lakshman, M. K. Palladium-catalyzed C-N and C-C cross-couplings as versatile, new avenues for modifications of purine 2'-deoxynucleosides. *J. Organomet. Chem.* **2002**, 653, 234-251.
32. Miyaura, N., Editor. Cross-Coupling Reactions. A Practical Guide. *Top. Curr. Chem.* **2002**, 219, 248 pp.
33. Nakamura, I., Yamamoto, Y. Room-temperature alkyl-alkyl Suzuki cross-coupling of alkyl bromides that possess β-hydrogens. *Chemtracts* **2002**, 15, 102-105.
34. Suzuki, A. The Suzuki reaction with arylboron compounds in arene chemistry. *Modern Arene Chemistry* **2002**, 53-106.
35. Tucker, C. E., De Vries, J. G. Homogeneous catalysis for the production of fine chemicals. Palladium- and nickel-catalyzed aromatic carbon-carbon bond formation. *Top. in Cat.* **2002**, 19, 111-118.
36. Yasuda, N. Application of cross-coupling reactions in Merck. *J. Organomet. Chem.* **2002**, 653, 279-287.
37. Miura, M. Rational ligand design in constructing efficient catalyst systems for Suzuki-Miyaura coupling. *Angew. Chem., Int. Ed. Engl.* **2004**, 43, 2201-2203.
38. Sasaki, M., Fuwa, H. Total synthesis of polycyclic ether natural products based on Suzuki-Miyaura cross-coupling. *Synlett* **2004**, 1851-1874.
39. Littke, A. F., Fu, G. C. A convenient and general method for Pd-catalyzed Suzuki cross-couplings of aryl chlorides and arylboronic acids. *Angew. Chem., Int. Ed. Engl.* **1999**, 37, 3387-3388.

40. Wolfe, J. P., Singer, R. A., Yang, B. H., Buchwald, S. L. Highly Active Palladium Catalysts for Suzuki Coupling Reactions. *J. Am. Chem. Soc.* **1999**, 121, 9550-9561.
41. Liu, S.-Y., Choi, M. J., Fu, G. C. A surprisingly mild and versatile method for palladium-catalyzed Suzuki cross-couplings of aryl chlorides in the presence of a triarylphosphine. *Chem. Commun.* **2001**, 2408-2409.
42. Netherton, M. R., Dai, C., Neuschuetz, K., Fu, G. C. Room-Temperature Alkyl-Alkyl Suzuki Cross-Coupling of Alkyl Bromides that Possess β Hydrogens. *J. Am. Chem. Soc.* **2001**, 123, 10099-10100.
43. Jones, W. D. Synthetic chemistry: The key to successful organic synthesis is. *Science* **2002**, 295, 289-290.
44. Kirchhoff, J. H., Dai, C., Fu, G. C. A method for palladium-catalyzed cross-couplings of simple alkyl chlorides: Suzuki reactions catalyzed by $[Pd_2(dba)_3]/PCy_3$. *Angew. Chem., Int. Ed. Engl.* **2002**, 41, 1945-1947.
45. Molander, G. A., Bernardi, C. R. Suzuki-Miyaura Cross-Coupling Reactions of Potassium Alkenyltrifluoroborates. *J. Org. Chem.* **2002**, 67, 8424-8429.
46. Molander, G. A., Biolatto, B. Efficient Ligandless Palladium-Catalyzed Suzuki Reactions of Potassium Aryltrifluoroborates. *Org. Lett.* **2002**, 4, 1867-1870.
47. Molander, G. A., Katona, B. W., Machrouhi, F. Development of the Suzuki-Miyaura Cross-Coupling Reaction: Use of Air-Stable Potassium Alkynyltrifluoroborates in Aryl Alkynylations. *J. Org. Chem.* **2002**, 67, 8416-8423.
48. Molander, G. A., Rivero, M. R. Suzuki Cross-Coupling Reactions of Potassium Alkenyltrifluoroborates. *Org. Lett.* **2002**, 4, 107-109.
49. Molander, G. A., Yun, C.-S. Cross-coupling reactions of primary alkylboronic acids with aryl triflates and aryl halides. *Tetrahedron* **2002**, 58, 1465-1470.
50. Kirchhoff, J. H., Netherton, M. R., Hills, I. D., Fu, G. C. Boronic Acids: New Coupling Partners in Room-Temperature Suzuki Reactions of Alkyl Bromides. Crystallographic Characterization of an Oxidative-Addition Adduct Generated under Remarkably Mild Conditions. *J. Am. Chem. Soc.* **2002**, 124, 13662-13663.
51. Smith, G. B., Dezeny, G. C., Hughes, D. L., King, A. O., Verhoeven, T. R. Mechanistic Studies of the Suzuki Cross-Coupling Reaction. *J. Org. Chem.* **1994**, 59, 8151-8156.
52. Moreno-Manas, M., Perez, M., Pleixats, R. Palladium-Catalyzed Suzuki-Type Self-Coupling of Arylboronic Acids. A Mechanistic Study. *J. Org. Chem.* **1996**, 61, 2346-2351.
53. Matos, K., Soderquist, J. A. Alkylboranes in the Suzuki-Miyaura Coupling: Stereochemical and Mechanistic Studies. *J. Org. Chem.* **1998**, 63, 461-470.
54. Littke, A. F., Dai, C., Fu, G. C. Versatile Catalysts for the Suzuki Cross-Coupling of Arylboronic Acids with Aryl and Vinyl Halides and Triflates under Mild Conditions. *J. Am. Chem. Soc.* **2000**, 122, 4020-4028.
55. Bedford, R. B., Cazin, C. S. J., Hursthouse, M. B., Light, M. E., Pike, K. J., Wimperis, S. Silica-supported imine palladacycles---recyclable catalysts for the Suzuki reaction? *J. Organomet. Chem.* **2001**, 633, 173-181.
56. Choudary, B. M., Madhi, S., Chowdari, N. S., Kantam, M. L., Sreedhar, B. Layered Double Hydroxide Supported Nanopalladium Catalyst for Heck-, Suzuki-, Sonogashira-, and Stille-Type Coupling Reactions of Chloroarenes. *J. Am. Chem. Soc.* **2002**, 124, 14127-14136.
57. Grasa, G. A., Viciu, M. S., Huang, J., Zhang, C., Trudell, M. L., Nolan, S. P. Suzuki-Miyaura Cross-Coupling Reactions Mediated by Palladium/Imidazolium Salt Systems. *Organometallics* **2002**, 21, 2866-2873.
58. Li, G. Y. Highly Active, Air-Stable Palladium Catalysts for the C-C and C-S Bond-Forming Reactions of Vinyl and Aryl Chlorides: Use of Commercially Available $[(t\text{-}Bu)_2P(OH)]2PdCl_2$, $[(t\text{-}Bu)_2P(OH)PdCl_2]_2$, and $[[(t\text{-}Bu)_2PO\cdots H\cdots OP(t\text{-}Bu)_2]PdCl]_2$ as Catalysts. *J. Org. Chem.* **2002**, 67, 3643-3650.
59. Miyaura, N. Cross-coupling reaction of organoboron compounds via base-assisted transmetalation to palladium$^{(II)}$ complexes. *J. Organomet. Chem.* **2002**, 653, 54-57.
60. Organ, M. G., Arvanitis, E. A., Dixon, C. E., Cooper, J. T. Controlling Chemoselectivity in Vinyl and Allylic C-X Bond Activation with Palladium Catalysis: A pKa-Based Electronic Switch. *J. Am. Chem. Soc.* **2002**, 124, 1288-1294.
61. Lin, S., Danishefsky, S. J. The total synthesis of proteasome inhibitors TMC-95A and TMC-95B: discovery of a new method to generate cis-propenyl amides. *Angew. Chem., Int. Ed. Engl.* **2002**, 41, 512-515.
62. Zhu, B., Panek, J. S. Total Synthesis of Epothilone A. *Org. Lett.* **2000**, 2, 2575-2578.
63. Mapp, A. K., Heathcock, C. H. Total Synthesis of Myxalamide A. *J. Org. Chem.* **1999**, 64, 23-27.
64. Molander, G. A., Dehmel, F. Formal Total Synthesis of Oximidine II via a Suzuki-Type Cross-Coupling Macrocyclization Employing Potassium Organotrifluoroborates. *J. Am. Chem. Soc.* **2004**, 126, 10313-10318.

Swern Oxidation........450

Related reactions: Corey-Kim oxidation, Dess-Martin oxidation, Jones oxidation, Ley oxidation, Oppenauer oxidation, Pfitzner-Moffatt oxidation;

1. Sharma, A. K., Swern, D. Trifluoroacetic anhydride. New activating agent for dimethyl sulfoxide in the synthesis of iminosulfuranes. *Tetrahedron Lett.* **1974**, 15, 1503-1506.
2. Sharma, A. K., Ku, T., Dawson, A. D., Swern, D. Iminosulfuranes. XV. Dimethyl sulfoxide-trifluoroacetic anhydride. New and efficient reagent for the preparation of iminosulfuranes. *J. Org. Chem.* **1975**, 40, 2758-2764.
3. Omura, K., Sharma, A. K., Swern, D. Dimethyl sulfoxide-trifluoroacetic anhydride. New reagent for oxidation of alcohols to carbonyls. *J. Org. Chem.* **1976**, 41, 957-962.
4. Huang, S. L., Omura, K., Swern, D. Further studies on the oxidation of alcohols to carbonyl compounds by dimethyl sulfoxide/trifluoroacetic anhydride. *Synthesis* **1978**, 297-299.
5. Huang, S. L., Swern, D. Preparation of iminosulfuranes utilizing the dimethyl sulfoxide-oxalyl chloride reagent. *J. Org. Chem.* **1978**, 43, 4537-4538.
6. Omura, K., Swern, D. Oxidation of alcohols by "activated" dimethyl sulfoxide. A preparative steric and mechanistic study. *Tetrahedron* **1978**, 34, 1651-1660.
7. Mancuso, A. J., Swern, D. Activated dimethyl sulfoxide: useful reagents for synthesis. *Synthesis* **1981**, 165-185.
8. Tidwell, T. T. Oxidation of alcohols by activated dimethyl sulfoxide and related reactions: an update. *Synthesis* **1990**, 857-870.
9. Tidwell, T. T. Oxidation of alcohols to carbonyl compounds via alkoxysulfonium ylides: the Moffat, Swern, and related oxidations. *Org. React.* **1990**, 39, 297-572.
10. Arterburn, J. B. Selective oxidation of secondary alcohols. *Tetrahedron* **2001**, 57, 9765-9788.
11. Harris, J. M., Liu, Y., Chai, S., Andrews, M. D., Vederas, J. C. Modification of the Swern oxidation: use of a soluble polymer-bound, recyclable, and odorless sulfoxide. *J. Org. Chem.* **1998**, 63, 2407-2409.
12. Bisai, A., Chandrasekhar, M., Singh, V. K. An alternative to the Swern oxidation. *Tetrahedron Lett.* **2002**, 43, 8355-8357.
13. Crich, D., Neelamkavil, S. The fluorous Swern and Corey-Kim reactions: scope and mechanism. *Tetrahedron* **2002**, 58, 3865-3870.
14. Matsuo, J.-I., Iida, D., Tatani, K., Mukaiyama, T. A new method for oxidation of various alcohols to the corresponding carbonyl compounds by using N-t-butylbenzenesulfinimidoyl chloride. *Bull. Chem. Soc. Jpn.* **2002**, 75, 223-234.
15. Crich, D., Neelamkavil, S. *Improved method of oxidizing primary and secondary alcohols by Swern or Corey-Kim oxidation using a recyclable fluorous sulfoxide as the oxidizing agent.* WO 2002-US19274 2003002526, **2003** (The Board of Trustees of the University of Illinois, USA).
16. Firouzabadi, H., Hassani, H., Hazarkhani, H. Heterogeneous Swern Oxidation. Selective Oxidation of Alcohols by DMSO/SiO_2-Cl System. *Phosphorus, Sulfur Silicon Relat. Elem.* **2003**, 178, 149-153.
17. Williams, D. R., Heidebrecht, R. W., Jr. Total Synthesis of (+)-4,5-Deoxyneodolabelline. *J. Am. Chem. Soc.* **2003**, 125, 1843-1850.
18. Martin, S. F., Humphrey, J. M., Ali, A., Hillier, M. C. Enantioselective Total Syntheses of Ircinal A and Related Manzamine Alkaloids. *J. Am. Chem. Soc.* **1999**, 121, 866-867.
19. Eom, K. D., Raman, J. V., Kim, H., Cha, J. K. Total Synthesis of (+)-Asteltoxin. *J. Am. Chem. Soc.* **2003**, 125, 5415-5421.

Related reactions: Horner-Wadsworth-Emmons olefination, Horner-Wadsworth-Emmons olefination - Still-Gennari modification, Julia-Lithgoe olefination, Peterson olefination, Tebbe olefination, Wittig reaction, Wittig reaction – Schlosser modification;

1. Takai, K., Nitta, K., Utimoto, K. Simple and selective method for aldehydes (RCHO) -> (E)-haloalkenes (RCH=CHX) conversion by means of a haloform-chromous chloride system. *J. Am. Chem. Soc.* **1986**, 108, 7408-7410.
2. Okazoe, T., Takai, K., Utimoto, K. (E)-Selective olefination of aldehydes by means of gem-dichromium reagents derived by reduction of gem-diiodoalkanes with chromium(II) chloride. *J. Am. Chem. Soc.* **1987**, 109, 951-953.
3. Saccomano, N. A. Organochromium reagents. in *Comp. Org. Synth.* (eds. Trost, B. M.,Fleming, I.), 173-209 (Pergamon Press, Oxford, **1991**).
4. Hodgson, D. M., Boulton, L. T. Chromium- and titanium-mediated synthesis of alkenes from carbonyl compounds. *Preparation of Alkenes* **1996**, 81-93.
5. Hodgson, D. M., Comina, P. J. *Chromium(II)-mediated C-C coupling reactions* (eds. Beller, M.,Bolm, C.) (Wiley-VCH, Weinheim, New York, **1998**) 418-424.
6. Avalos, M., Babiano, R., Cintas, P., Jimenez, J. L., Palacios, J. C. Synthetic variations based on low-valent chromium: new developments. *Chem. Soc. Rev.* **1999**, 28, 169-177.
7. Fürstner, A. Carbon-Carbon Bond Formation Involving Organochromium(III) Reagents. *Chem. Rev.* **1999**, 99, 991-1045.
8. Wessjohann, L. A., Scheid, G. Recent advances in chromium(II)- and chromium(III)-mediated organic synthesis. *Synthesis* **1999**, 1-36.
9. Takai, K., Kataoka, Y., Okazoe, T., Utimoto, K. Stereoselective synthesis of (E)-alkenylsilanes from aldehydes with a reagent prepared by chromium(II) reduction of trimethyl(dibromomethyl)silane. *Tetrahedron Lett.* **1987**, 28, 1443-1446.
10. Hodgson, D. M. Chromium(II)-mediated synthesis of (E)-alkenylstannanes from aldehydes and $Bu_3SnCHBr_2$. *Tetrahedron Lett.* **1992**, 33, 5603-5604.
11. Knecht, M., Boland, W. (E)-Selective alkylidenation of aldehydes with reagents derived from α-acetoxy bromides, zinc and chromium trichloride. *Synlett* **1993**, 837-838.
12. Hodgson, D. M., Comina, P. J. One-step chromium(II)-mediated homologation of aldehydes to methyl ketones using Me_3SiCBr_3. *Synlett* **1994**, 663-664.
13. Hodgson, D. M., Comina, P. J. Chromium(II)-mediated synthesis of 1,1-bis(trimethylsilyl)alkenes from aldehydes and $(Me_3Si)_2CBr_2$. *Tetrahedron Lett.* **1994**, 35, 9469-9470.
14. Takai, K., Shinomiya, N., Kaihara, H., Yoshida, N., Moriwake, T., Utimoto, K. Transformation of aldehydes into (E)-1-alkenylboronic esters with a geminal dichromium reagent derived from a dichloromethylboronic ester and $CrCl_2$. *Synlett* **1995**, 963-964.
15. Hodgson, D. M., Comina, P. J., Drew, M. G. B. Chromium(II)-mediated synthesis of vinylbis(silanes) from aldehydes and a study of acid- and base-induced reactions of the derived epoxybis(silanes): a synthesis of acylsilanes. *J. Chem. Soc., Perkin Trans. 1* **1997**, 2279-2289.
16. Boeckman, R. K., Jr., Hudack, R. A. A Variant of the Takai-Utimoto Reaction of Acrolein Acetals with Aldehydes Catalytic in Chromium: A Highly Stereoselective Route to Anti Diol Derivatives. *J. Org. Chem.* **1998**, 63, 3524-3525.
17. Auge, J., Boucard, V., Gil, R., Lubin-Germain, N., Picard, J., Uziel, J. An alternative procedure in the Takai reaction using chromium(III) chloride hexahydrate as a convenient source of chromium(II). *Synth. Commun.* **2003**, 33, 3733-3739.
18. Trost, B. M., Dumas, J., Villa, M. New strategies for the synthesis of vitamin D metabolites via palladium-catalyzed reactions. *J. Am. Chem. Soc.* **1992**, 114, 9836-9845.
19. Matsubara, S., Horiuchi, M., Takai, K., Utimoto, K. Alkylidenation of ketones by gem-dibromoalkane, SmI_2, and Sm in the presence of catalytic amount of $CrCl_3$. *Chem. Lett.* **1995**, 259-260.
20. Dodd, D., Johnson, M. D. σ-Bonded organotransition-metal ions. V. Formation of mono- and dihalomethylchromium(III) ions and their reaction with mercuric nitrate. *Journal of the Chemical Society [Section] A: Inorganic, Physical, Theoretical* **1968**, 34-38.
21. Bertini, F., Grasselli, P., Zubiani, G., Cainelli, G. Geminal dimetallic compounds. Reactivity of methylene magnesium halides and related compounds. General carbonyl olefination reaction. *Tetrahedron* **1970**, 26, 1281-1290.
22. Jung, M. E., Fahr, B. T., D'Amico, D. C. Total Syntheses of the Cytotoxic Marine Natural Product, Aplysiapyranoid C. *J. Org. Chem.* **1998**, 63, 2982-2987.
23. Longbottom, D. A., Morrison, A. J., Dixon, D. J., Ley, S. V. Total synthesis of polycephalin C and determination of the absolute configurations at the 3",4" ring junction. *Angew. Chem., Int. Ed. Engl.* **2002**, 41, 2786-2790.
24. Kinder, F. R., Jr., Wattanasin, S., Versace, R. W., Bair, K. W., Bontempo, J., Green, M. A., Lu, Y. J., Marepalli, H. R., Phillips, P. E., Roche, D., Tran, L. D., Wang, R., Waykole, L., Xu, D. D., Zabludoff, S. Total Syntheses of Bengamides B and E. *J. Org. Chem.* **2001**, 66, 2118-2122.
25. Yuki, K., Shindo, M., Shishido, K. Enantioselective total synthesis of (-)-equisetin using a Me_3Al-mediated intramolecular Diels-Alder reaction. *Tetrahedron Lett.* **2001**, 42, 2517-2519.

Related reactions: Horner-Wadsworth-Emmons olefination, Horner-Wadsworth-Emmons olefination - Still-Gennari modification, Julia-Lithgoe olefination, Peterson olefination, Takai-Utimoto olefination, Wittig reaction, Wittig reaction – Schlosser modification;

1. Schrock, R. R. Multiple metal-carbon bonds. 5. The reaction of niobium and tantalum neopentylidene complexes with the carbonyl function. *J. Am. Chem. Soc.* **1976**, 98, 5399-5400.
2. Tebbe, F. N., Parshall, G. W., Reddy, G. S. Olefin homologation with titanium methylene compounds. *J. Am. Chem. Soc.* **1978**, 100, 3611-3613.
3. Petasis, N. A., Bzowej, E. I. Titanium-mediated carbonyl olefinations. 1. Methylenations of carbonyl compounds with dimethyltitanocene. *J. Am. Chem. Soc.* **1990**, 112, 6392-6394.
4. Brown-Wensley, K. A., Buchwald, S. L., Cannizzo, L., Clawson, L., Ho, S., Meinhardt, D., Stille, J. R., Straus, D., Grubbs, R. H. Cp_2TiCH_2 complexes in synthetic applications. *Pure Appl. Chem.* **1983**, 55, 1733-1744.
5. Anon. Methylenations with Tebbe-Grubbs reagents. *Nachrichten aus Chemie, Technik und Laboratorium* **1986**, 34, 562-565.
6. Kelly, S. E. Alkene Synthesis. in *Comp. Org. Synth.* (eds. Trost, B. M.,Fleming, I.), *1*, 729-818 (Pergamon, Oxford, **1991**).
7. Pine, S. H. Carbonyl methylenation and alkylidenation using titanium-based reagents. *Org. React.* **1993**, 43, 1-91.
8. Paquette, L. A. Stereocontrolled construction of cyclooctanoid natural products by Claisen-based ring expansion. *Stereocontrolled Organic Synthesis* **1994**, 313-335.
9. Hong, F.-T., Paquette, L. A. Olefin metathesis in cyclic ether formation. Direct conversion of olefinic esters to cyclic enol ethers with Tebbe-type reagents. Copper(I)-promoted Stille cross-coupling of stannyl enol ethers with enol triflates: construction of complex polyether frameworks. *Chemtracts* **1997**, 10, 14-19.
10. Breit, B. Dithioacetals as an entry to titanium-alkylidene chemistry: a new and efficient carbonyl olefination. *Angewandte Chemie, International Edition* **1998**, 37, 453-456.
11. Walters, M. A. Chameleon catches in combinatorial chemistry: Tebbe olefination of polymer supported esters and the synthesis of amines, cyclohexanones, enones, methyl ketones and thiazoles. *Chemtracts* **1999**, 12, 679-683.
12. Kulinkovich, O. G., de Meijere, A. 1,n-Dicarbanionic titanium intermediates from monocarbanionic organometallics and their application in organic synthesis. *Chem. Rev.* **2000**, 100, 2789-2834.
13. Beckhaus, R., Santamaria, C. Carbene complexes of titanium group metals - formation and reactivity. *J. Organomet. Chem.* **2001**, 617-618, 81-97.

14. Hartley, R. C., McKiernan, G. J. Titanium reagents for the alkylidenation of carboxylic acid and carbonic acid derivatives. *J. Chem. Soc., Perkin Trans. 1* **2002**, 2763-2793.
15. Takeda, T. Titanium-based olefin metathesis and related reactions. *Titanium and Zirconium in Organic Synthesis* **2002**, 475-500.
16. Howard, T. R., Lee, J. B., Grubbs, R. H. Titanium metallacarbene-metallacyclobutane reactions: stepwise metathesis. *J. Am. Chem. Soc.* **1980**, 102, 6876-6878.
17. Pine, S. H., Zahler, R., Evans, D. A., Grubbs, R. H. Titanium-mediated methylene-transfer reactions. Direct conversion of esters into vinyl ethers. *J. Am. Chem. Soc.* **1980**, 102, 3270-3272.
18. Petasis, N. A., Lu, S.-P. Methylenations of heteroatom-substituted carbonyls with dimethyl titanocene. *Tetrahedron Lett.* **1995**, 36, 2393-2396.
19. Petasis, N. A., Staszewski, J. P., Fu, D.-K. Tris(trimethylsilyl)titanacyclobutene: a new mild reagent for the conversion of carbonyls to alkenyl silanes. *Tetrahedron Lett.* **1995**, 36, 3619-3622.
20. Breit, B. Dithioacetals as an entry to titanium-alkylidene chemistry: new and efficient carbonyl olefination. *Organic Synthesis Highlights IV* **2000**, 110-115.
21. Stille, J. R., Grubbs, R. H. Synthetic applications of titanocene methylene complexes: selective formation of ketone enolates and their reactions. *J. Am. Chem. Soc.* **1983**, 105, 1664-1665.
22. Anslyn, E. V., Grubbs, R. H. Mechanism of titanocene metallacyclobutane cleavage and the nature of the reactive intermediate. *J. Am. Chem. Soc.* **1987**, 109, 4880-4890.
23. Schioett, B., Joergensen, K. A. Addition of a carbonyl functionality to titanium carbenes. A study of the mechanism and intermediates in the Tebbe reaction. *Journal of the Chemical Society, Dalton Transactions: Inorganic Chemistry (1972-1999)* **1993**, 337-344.
24. Hughes, D. L., Payack, J. F., Cai, D., Verhoeven, T. R., Reider, P. J. A Mechanistic Study of Ester Olefinations Using Dimethyltitanocene. *Organometallics* **1996**, 15, 663-667.
25. Beckhaus, R. C2 building blocks in the co-ordination sphere of electron-poor transition metals. Aspects of the chemistry of early-transition-metal carbenoid complexes. *J. Chem. Soc., Dalton Trans.* **1997**, 1991-2001.
26. Beckhaus, R. Carbenoid complexes of electron-deficient transition metals-syntheses of and with short-lived building blocks. *Angew. Chem., Int. Ed. Engl.* **1997**, 36, 687-713.
27. Hart, S. L., McCamley, A., Taylor, P. C. Ti(η5-C_5H_5)(η5-C_5H_4tBu)$(CH_2Ph)_2$. A probe of the course of the Petasis benzylidenation reaction. *Synlett* **1999**, 90-92.
28. Takeda, T., Fujiwara, T. Titanocene(II)-promoted reactions of thioacetals with organic molecules having a multiple bond. *Rev. on Heteroa. Chem.* **1999**, 21, 93-115.
29. Siebeneicher, H., Doye, S. Dimethyltitanocene Cp_2TiMe_2: a useful reagent for C-C and C-N bond formation. *J. Prakt. Chem.* **2000**, 342, 102-106.
30. Meurer, E. C., Santos, L. S., Pilli, R. A., Eberlin, M. N. Probing the Mechanism of the Petasis Olefination Reaction by Atmospheric Pressure Chemical Ionization Mass and Tandem Mass Spectrometry. *Org. Lett.* **2003**, 5, 1391-1394.
31. Paquette, L. A., Sun, L.-Q., Friedrich, D., Savage, P. B. Total Synthesis of (+)-Epoxydictymene. Application of Alkoxy-Directed Cyclization to Diterpenoid Construction. *J. Am. Chem. Soc.* **1997**, 119, 8438-8450.
32. Robinson, R. A., Clark, J. S., Holmes, A. B. Synthesis of (+)-laurencin. *J. Am. Chem. Soc.* **1993**, 115, 10400-10401.
33. Atarashi, S., Choi, J.-K., Ha, D.-C., Hart, D. J., Kuzmich, D., Lee, C.-S., Ramesh, S., Wu, S. C. Free Radical Cyclizations in Alkaloid Total Synthesis: (±)-21-Oxogelsemine and (±)-Gelsemine. *J. Am. Chem. Soc.* **1997**, 119, 6226-6241.
34. Martinez, I., Howell, A. R. The reaction of dimethyltitanocene with N-substituted-β-lactams. *Tetrahedron Lett.* **2000**, 41, 5607-5611.

Tishchenko Reaction 456

Related reactions: Cannizzaro reaction, Meerwein-Ponndorf-Verley reduction, Oppenauer oxidation;

1. Claisen, L. The effect of sodium alkoxide onto benzaldehyde. *Ber.* **1887**, 20, 646.
2. Tishchenko, W. *J. Russ. Phys. Chem. Soc.* **1906**, 38, 355.
3. Tishchenko, W. The effect of aluminum alcoholates on aldehydes. The ester condensation as a new condesationform of aldehydes. *Chem. Zentr.* **1906**, II, 1309-1311.
4. Tishchenko, W. The effect of aluminum alcoholates on aldehydes. The ester condensation as a new condesationform of aldehydes. *Chem. Zentr.* **1906**, II, 1552-1555.
5. Tishchenko, W. The effect of magnesium amalgam on isobyraldehyde. *Chem. Zentr.* **1906**, II, 1555-1556.
6. Tishchenko, W. The effect of magnesium amalgam on isobyraldehyde. *Chem. Zentr.* **1906**, II, 1556.
7. Tishchenko, W. *J. Russ. Phys. Chem. Soc.* **1906**, 38, 482.
8. Tishchenko, W. *J. Russ. Phys. Chem. Soc.* **1906**, 38, 540.
9. Tishchenko, W. *J. Russ. Phys. Chem. Soc.* **1906**, 38, 547.
10. Hattori, H. Solid base catalysts: generation of basic sites and application to organic synthesis. *Appl. Cat. A* **2001**, 222, 247-259.
11. Tormakangas, O. P., Koskinen, A. M. P. The Tishchenko reaction and its modifications in organic synthesis. *Recent Research Developments in Organic Chemistry* **2001**, 5, 225-255.
12. Mahrwald, R. The aldol-Tishchenko reaction: A tool in stereoselective synthesis. *Curr. Org. Chem.* **2003**, 7, 1713-1723.
13. Lin, I., Day, A. R. The mixed Tishchenko reaction. *J. Am. Chem. Soc.* **1952**, 74, 5133-5135.
14. Yamashita, M., Watanabe, Y., Mitsudo, T.-a., Takegami, Y. The reaction of disodium tetracarbonylferrate(-II) with aldehydes. *Bull. Chem. Soc. Jpn.* **1976**, 49, 3597-3600.
15. Komiya, S., Taneichi, S., Yamamoto, A., Yamamoto, T. Transition metal alkoxides. Preparation and properties of bis(aryloxy)iron(II) and bis(alkoxy)iron(II) complexes having 2,2'-bipyridine ligands. *Bull. Chem. Soc. Jpn.* **1980**, 53, 673-679.
16. Yokoo, K., Mine, N., Taniguchi, H., Fujiwara, Y. Chemistry of organolanthanoids: lanthanoid-catalyzed Tishchenko condensation of aldehydes to esters. *J. Organomet. Chem.* **1985**, 279, C19-C21.
17. Collin, J., Namy, J. L., Kagan, H. B. Samarium diiodide, an efficient catalyst precursor in some Oppenauer oxidations. *Nouv. J. Chim.* **1986**, 10, 229-232.
18. Bunce, R. A., Shellhammer, A. J., Jr. Formate esters by Cannizzaro-Tishchenko reaction of Grignard and sodium alkoxides with formaldehyde. *Org. Prep. Proced. Int.* **1987**, 19, 161-166.
19. Bernard, K. A., Atwood, J. D. Evidence for carbon-oxygen bond formation, aldehyde decarbonylation, and dimerization by reaction of formaldehyde and acetaldehyde with trans-ROIr(CO)$(PPh_3)_2$. *Organometallics* **1988**, 7, 235-236.
20. Evans, D. A., Hoveyda, A. H. Samarium-catalyzed intramolecular Tishchenko reduction of β-hydroxy ketones. A stereoselective approach to the synthesis of differentiated anti 1,3-diol monoesters. *J. Am. Chem. Soc.* **1990**, 112, 6447-6449.
21. Uenishi, J., Masuda, S., Wakabayashi, S. Intramolecular Sm^{2+} and Sm^{3+} promoted reaction of γ-oxy-δ-keto aldehyde: stereocontrolled formation of pinacol and lactone. *Tetrahedron Lett.* **1991**, 32, 5097-5100.
22. Morita, K., Nishiyama, Y., Ishii, Y. Selective dimerization of aldehydes to esters catalyzed by zirconocene and hafnocene complexes. *Organometallics* **1993**, 12, 3748-3752.
23. Mahrwald, R., Costisella, B. Titanium-mediated aldol-Tishchenko reaction. A stereoselective synthesis of differentiated anti 1,3-diol monoesters. *Synthesis* **1996**, 1087-1089.
24. Onozawa, S.-y., Sakakura, T., Tanaka, M., Shiro, M. Lanthanoid-catalyzed Tishchenko reaction of mono- or di-aldehydes. *Tetrahedron* **1996**, 52, 4291-4302.
25. Bodnar, P. M., Shaw, J. T., Woerpel, K. A. Tandem Aldol-Tishchenko Reactions of Lithium Enolates: A Highly Stereoselective Method for Diol and Triol Synthesis. *J. Org. Chem.* **1997**, 62, 5674-5675.
26. Idriss, H., Seebauer, E. G. Effect of oxygen electronic polarizability on catalytic reactions over oxides. *Catal. Lett.* **2000**, 66, 139-145.

27. Smith, A., B., 3rd, Lee, D., Adams, C., M., Kozlowski, M., C. SmI_2-promoted oxidation of aldehydes in the presence of electron-rich heteroatoms. *Org. Lett.* **2002**, 4, 4539-4541.
28. Abu-Hasanayn, F., Streitwieser, A. Kinetics and Isotope Effects of the Aldol-Tishchenko Reaction between Lithium Enolates and Aldehydes. *J. Org. Chem.* **1998**, 63, 2954-2960.
29. Villani, F. J., Nord, F. F. Glycol esters from aldehydes. *J. Am. Chem. Soc.* **1946**, 68, 1674-1675.
30. Ogata, Y., Kawasaki, A. Alkoxide transfer from aluminum alkoxide to aldehyde in the Tishchenko reaction. *Tetrahedron* **1969**, 25, 929-935.
31. Horino, H., Ito, T., Yamamoto, A. A new Tishchenko-type ester formation catalyzed by ruthenium complexes. *Chem. Lett.* **1978**, 17-20.
32. Sung, M. J., Lee, H. I., Lee, H. B., Cha, J. K. Synthetic Studies toward Sarain A. Formation of the Western Macrocyclic Ring. *J. Org. Chem.* **2003**, 68, 2205-2208.
33. Romo, D., Meyer, S. D., Johnson, D. D., Schreiber, S. L. Total synthesis of (-)-rapamycin using an Evans-Tishchenko fragment coupling. *J. Am. Chem. Soc.* **1993**, 115, 7906-7907.
34. Lafontaine, J. A., Provencal, D. P., Gardelli, C., Leahy, J. W. Enantioselective Total Synthesis of the Antitumor Macrolide Rhizoxin D. *J. Org. Chem.* **2003**, 68, 4215-4234.

Tsuji-Trost Reaction/Allylation 458

1. Tsuji, J., Takahashi, H., Morikawa, M. Organic syntheses by means of noble metal and compounds. XVII. Reaction of π-allylpalladium chloride with nucleophiles. *Tetrahedron Lett.* **1965**, 4387-4388.
2. Atkins, K. E., Walker, W. E., Manyik, R. M. Palladium catalyzed transfer of allylic groups. *Tetrahedron Lett.* **1970**, 3821-3824.
3. Hata, G., Takahashi, K., Miyake, A. Palladium-catalyzed exchange of allylic groups of ethers and esters with active-hydrogen compounds. *J. Chem. Soc., Chem. Commun.* **1970**, 1392-1393.
4. Trost, B. M., Fullerton, T. J. New synthetic reactions. Allylic alkylation. *J. Am. Chem. Soc.* **1973**, 95, 292-294.
5. Fiaud, J. C. Mechanisms in stereo-differentiating metal-catalyzed reactions. Enantioselective palladium-catalyzed allylation. *Catal. Metal Compl.* **1991**, 12, 107-131.
6. Godleski, S. A. Nucleophiles with Allyl-Metal Complexes. in *Comp. Org. Synth.* (eds. Trost, B. M.,Fleming, I.), *4*, 585-662 (Pergamon, Oxford, **1991**).
7. Frost, C. G., Howart, J., Williams, J. M. J. Selectivity in palladium catalyzed allylic substitution. *Tetrahedron: Asymmetry* **1992**, 3, 1089-1122.
8. Trost, B. M. Cyclizations made easy by transition metal catalysts. *Advances in Chemistry Series* **1992**, 230, 463-478.
9. Moreno-Manas, M., Pleixats, R. Palladium$^{(0)}$-catalyzed allylation of ambident nucleophilic aromatic heterocycles. *Adv. Heterocycl. Chem.* **1996**, 66, 73-129.
10. Heumann, A. Palladium-catalyzed allylic substitutions. *Transition Metals for Organic Synthesis* **1998**, 1, 251-264.
11. Sesay, S. J., Williams, J. M. J. Palladium-catalyzed enantioselective allylic substitution reactions. *Adv. in Asymmetric Synth.* **1998**, 3, 235-271.
12. Helmchen, G. Enantioselective palladium-catalyzed allylic substitutions with asymmetric chiral ligands. *J. Organomet. Chem.* **1999**, 576, 203-214.
13. Moberg, C., Bremberg, U., Hallman, K., Svensson, M., Norrby, P.-O., Hallberg, A., Larhed, M., Csoregh, I. Selectivity and reactivity in asymmetric allylic alkylation. *Pure Appl. Chem.* **1999**, 71, 1477-1483.
14. Poli, G., Scolastico, C. New modes of regiocontrol in palladium-catalyzed allylic alkylations. *Chemtracts* **1999**, 12, 822-836.
15. Tsuji, J. Recollections of organopalladium chemistry. *Pure Appl. Chem.* **1999**, 71, 1539-1547.
16. Tsuji, J. Organopalladium chemistry in the '60s and '70s. *New J. Chem.* **2000**, 24, 127-135.
17. van Leeuwen, P. W. N. M., Kamer, P. C. J., Reek, J. N. H., Dierkes, P. Ligand Bite Angle Effects in Metal-catalyzed C-C Bond Formation. *Chem. Rev.* **2000**, 100, 2741-2769.
18. Frost, C. G. Palladium catalyzed coupling reactions. *Rodd's Chemistry of Carbon Compounds (2nd Edition)* **2001**, 5, 315-350.
19. Acemoglu, L., Williams, J. M. J. Synthetic scope of the Tsuji-Trost reaction with allylic halides, carboxylates, ethers, and related oxygen nucleophiles as starting compounds. in *Handbook of Organopalladium Chemistry for Organic Synthesis* (ed. Negishi, E.-i.), *2*, 1689-1705 (John Wiley & Sons, New York, **2002**).
20. Acemoglu, L., Williams, J. M. J. Palladium-catalyzed asymmetric allylation and related reactions. *Handbook of Organopalladium Chemistry for Organic Synthesis* **2002**, 2, 1945-1979.
21. Mandai, T. Palladium-catalyzed allylic, propargylic, and allenic substitution with nitrogen, oxygen, and other groups 15-17 heteroatom nucleophiles: Palladium-catalyzed substitution reactions of allylic, propargylic, and related electrophiles with heteroatom nucleophiles. *Handbook of Organopalladium Chemistry for Organic Synthesis* **2002**, 2, 1845-1858.
22. Trost, B. M. Pd asymmetric allylic alkylation (AAA). A powerful synthetic tool. *Chem. Pharm. Bull.* **2002**, 50, 1-14.
23. Tsuji, J. Palladium-catalyzed nucleophilic substitution involving allylpalladium, propargylpalladium, and related derivatives: the Tsuji-Trost reaction and related carbon-carbon bond formation reactions: overview of the palladium-catalyzed carbon-carbon bond formation via π-allylpalladium and propargylpalladium intermediates. in *Handbook of Organopalladium Chemistry for Organic Synthesis* (ed. Negishi, E.-i.), *2*, 1669-1687 (John Wiley & Sons, New York, **2002**).
24. Graening, T., Schmalz, H.-G. Pd-catalyzed enantioselective allylic substitution: New strategic options for the total synthesis of natural products. *Angew. Chem., Int. Ed. Engl.* **2003**, 42, 2580-2584.
25. Sinou, D. Allylic substitution. *Aqueous-Phase Organometallic Catalysis* **1998**, 401-407.
26. Dos Santos, S., Quignard, F., Sinou, D., Choplin, A. Allylic substitution catalyzed by silica-supported aqueous phase palladium(0) catalysts. *Top. in Cat.* **2000**, 13, 311-318.
27. Kaiser, N. F. K., Bremberg, U., Larhed, M., Moberg, C., Hallberg, A. Microwave-mediated palladium-catalyzed asymmetric allylic alkylation; an example of highly selective fast chemistry. *J. Organomet. Chem.* **2000**, 603, 2-5.
28. Dos Santos, S., Moineau, J., Pozzi, G., Quignard, F., Sinou, D., Choplin, A. Immobilization of palladium catalysts for Trost-Tsuji C-C and C-N bond formation. Which method? *Chem. Ind.* **2001**, 82, 509-520.
29. Negishi, E.-i. Palladium-catalyzed cross-coupling involving β-hetero-substituted compounds. Palladium-catalyzed α-substitution reactions of enolates and related derivatives other than the Tsuji-Trost allylation reaction. *Handbook of Organopalladium Chemistry for Organic Synthesis* **2002**, 1, 693-719.
30. Sato, Y., Yoshino, T., Mori, M. Pd-Catalyzed Allylic Substitution Using Nucleophilic N-Heterocyclic Carbene as a Ligand. *Org. Lett.* **2003**, 5, 31-33.
31. Sakaki, S., Nishikawa, M., Ohyoshi, A. A palladium-catalyzed reaction of a π-allyl ligand with a nucleophile. An MO study about a feature of the reaction and a ligand effect on the reactivity. *J. Am. Chem. Soc.* **1980**, 102, 4062-4069.
32. Trost, B. M., Hung, M. H. On the regiochemistry of metal-catalyzed allylic alkylation: a model. *J. Am. Chem. Soc.* **1984**, 106, 6837-6839.
33. Pregosin, P. S., Ruegger, H., Salzmann, R., Albinati, A., Lianza, F., Kunz, R. W. X-ray diffraction, multidimensional NMR spectroscopy, and MM2* calculations on chiral allyl complexes of palladium(II). *Organometallics* **1994**, 13, 83-90.
34. Szabo, K. J. Effects of β-Substituents and Ancillary Ligands on the Structure and Stability of (η^3-Allyl)palladium Complexes. Implications for the Regioselectivity in Nucleophilic Addition Reactions. *J. Am. Chem. Soc.* **1996**, 118, 7818-7826.
35. Van Leeuwen, P. W. N. M., Kamer, P. C. J., Reek, J. N. H. The bite angle makes the catalyst. *Pure Appl. Chem.* **1999**, 71, 1443-1452.
36. Kamer, P. C. J., van Leeuwen, P. W. N. M., Reek, J. N. H. Wide Bite Angle Diphosphines: Xantphos Ligands in Transition Metal Complexes and Catalysis. *Acc. Chem. Res.* **2001**, 34, 895-904.
37. Szabo, K. J. Nature of the interaction between β-substituents and the allyl moiety in (η^3-allyl)palladium complexes. *Chem. Soc. Rev.* **2001**, 30, 136-143.
38. Kurosawa, H. Molecular basis of catalytic reactions involving η^3-allyl complexes of group 10 metals as key intermediates. *J. Organomet. Chem.* **1987**, 334, 243-253.

39. Saitoh, A., Achiwa, K., Tanaka, K., Morimoto, T. Versatile Chiral Bidentate Ligands Derived from α-Amino Acids: Synthetic Applications and Mechanistic Considerations in the Palladium-Mediated Asymmetric Allylic Substitutions. *J. Org. Chem.* **2000**, 65, 4227-4240.
40. Nomura, N., Tsurugi, K., Okada, M. Mechanistic rationale of a palladium-catalyzed allylic substitution polymerization-carbon-carbon bond-forming polycondensation out of stoichiometric control by cascade bidirectional allylation. *Angew. Chem., Int. Ed. Engl.* **2001**, 40, 1932-1935.
41. Vanderwal, C. D., Vosburg, D. A., Weiler, S., Sorensen, E. J. An Enantioselective Synthesis of FR182877 Provides a Chemical Rationalization of Its Structure and Affords Multigram Quantities of Its Direct Precursor. *J. Am. Chem. Soc.* **2003**, 125, 5393-5407.
42. Seki, M., Mori, Y., Hatsuda, M., Yamada, S. A Novel Synthesis of (+)-Biotin from L-Cysteine. *J. Org. Chem.* **2002**, 67, 5527-5536.
43. Fuerstner, A., Gastner, T. Total Synthesis of Cristatic Acid. *Organic Letters* **2000**, 2, 2467-2470.
44. Williams, D. R., Meyer, K. G. Palladium-Induced Cyclizations for the Synthesis of cis-2,5-Disubstituted-3-methylenetetrahydrofurans: Studies of the C7-C22 Core of Amphidinolide K. *Org. Lett.* **1999**, 1, 1303-1305.

Tsuji-Wilkinson Decarbonylation Reaction 460

1. Tsuji, J., Ono, K. Organic syntheses with noble metal compounds. XXI. Decarbonylation of aldehydes using rhodium complex. *Tetrahedron Lett.* **1965**, 3969-3971.
2. Tsuji, J., Ono, K., Kajimoto, T. Organic syntheses with noble metal compounds. XX. Decarbonylation of acyl chloride and aldehyde catalyzed by palladium and its relation with the Rosenmund reduction. *Tetrahedron Lett.* **1965**, 4565-4568.
3. Ohno, K., Tsuji, J. Organic synthesis by means of noble metal compounds. XXXV. Novel decarbonylation reactions of aldehydes and acyl halides using rhodium complexes. *J. Am. Chem. Soc.* **1968**, 90, 99-107.
4. Tsuji, J., Ohno, K. Organic syntheses by means of noble metal compounds. XXXIV. Carbonylation and decarbonylation reactions catalyzed by palladium. *J. Am. Chem. Soc.* **1968**, 90, 94-98.
5. Tsuji, J., Ohno, K. Organic syntheses by means of noble metal compounds. XXXI. Carbonylation of olefins and decarbonylation of acyl halides and aldehydes. *Advances in Chemistry Series* **1968**, No. 70, 155-167.
6. Tsuji, J., Ohno, K. Decarbonylation reactions using transition metal compounds. *Synthesis* **1969**, 157-169.
7. Kozikowski, A. P., Wetter, H. F. Transition metals in organic synthesis. *Synthesis* **1976**, 561-590.
8. Tsuji, J. Decarbonylation reactions using transition metal compounds. *Org. Synth. Met. Carbonyls* **1977**, 2, 595-654.
9. Jardine, F. H. Chlorotris(triphenylphosphine)rhodium(I): its chemical and catalytic reactions. *Prog. Inorg. Chem.* **1981**, 28, 63-202.
10. Thompson, D. J. Carbonylation and Decarbonylation Reactions. in *Comp. Org. Synth.* (eds. Trost, B. M.,Fleming, I.), *3*, 1015-1043 (Pergamon, Oxford, **1991**).
11. Murakami, M., Ito, Y. Cleavage of carbon-carbon single bonds by transition metals. *Top. Organomet. Chem.* **1999**, 3, 97-129.
12. Tsuji, J., Ohno, K. Organic syntheses by noble metal compounds. XXXII. Selective decarbonylation of α,β-unsaturated aldehydes using rhodium complexes. *Tetrahedron Lett.* **1967**, 2173-2176.
13. Kaneda, K., Azuma, H., Wayaku, M., Teranishi, S. Decarbonylation of α- and β-diketones catalyzed by rhodium compounds. *Chem. Lett.* **1974**, 215-216.
14. Ehrenkaufer, R. E., MacGregor, R. R., Wolf, A. P. Decarbonylation of aroyl fluorides using Wilkinson's catalyst: a reevaluation. *J. Org. Chem.* **1982**, 47, 2489-2491.
15. Hori, K., Ando, M., Takaishi, N., Inamoto, Y. Palladium-catalyzed decarbonylation of tricyclic bridgehead acid chlorides. *Tetrahedron Lett.* **1986**, 27, 4615-4618.
16. Murahashi, S., Naota, T., Nakajima, N. Palladium-catalyzed decarbonylation of acyl cyanides. *J. Org. Chem.* **1986**, 51, 898-901.
17. Tsuji, J. Other reactions of acylpalladium derivatives: palladium-catalyzed decarbonylation of acyl halides and aldehydes. *Handbook of Organopalladium Chemistry for Organic Synthesis* **2002**, 2, 2643-2653.
18. Daugulis, O., Brookhart, M. Decarbonylation of Aryl Ketones Mediated by Bulky Cyclopentadienylrhodium Bis(ethylene) Complexes. *Organometallics* **2004**, 23, 527-534.
19. Walborsky, H. M., Allen, L. E. Stereochemistry of tris(triphenylphosphine)rhodium chloride decarbonylation of aldehydes. *J. Am. Chem. Soc.* **1971**, 93, 5465-5468.
20. Stille, J. K., Fries, R. W. Mechanism of decarbonylation of acid chlorides by chlorotris(triphenylphosphine)rhodium(I). Stereochemistry. *J. Am. Chem. Soc.* **1974**, 96, 1514-1518.
21. Stille, J. K., Huang, F., Regan, M. T. Mechanism of acid chloride decarbonylation with chlorotris(triphenylphosphine)rhodium(I). Stereochemistry and direction of elimination. *J. Am. Chem. Soc.* **1974**, 96, 1518-1522.
22. Stille, J. K., Regan, M. T. Mechanism and kinetics of the decarbonylation of para-substituted benzoyl and phenylacetyl chlorides by chlorotris(triphenylphosphine)rhodium(I). *J. Am. Chem. Soc.* **1974**, 96, 1508-1514.
23. Stille, J. K., Regan, M. T., Fries, R. W., Huang, F., McCarley, T. Rhodium catalyzed decarbonylations. *Advances in Chemistry Series* **1974**, 132, 181-191.
24. Delgado, F., Cabrera, A., Gomez-Lara, J. Steric and electronic influences on the reaction mechanism of the catalytic decarbonylation of acid halides in homogeneous phase using rhodium carbonyl complexes. *J. Mol. Catal.* **1983**, 22, 83-87.
25. Kampmeier, J. A., Harris, S. H., Mergelsberg, I. Intramolecular trapping of alkyl- and arylrhodium hydride intermediates in the decarbonylation of aldehydes by chlorotris(triphenylphosphine)rhodium. *J. Org. Chem.* **1984**, 49, 621-625.
26. Gassman, P. G., Macomber, D. W., Willging, S. M. Isolation and characterization of reactive intermediates and active catalysts in homogeneous catalysis. *J. Am. Chem. Soc.* **1985**, 107, 2380-2388.
27. Baldwin, J. E., Barden, T. C., Pugh, R. L., Widdison, W. C. Partial loss of deuterium label in Wilkinson's catalyst promoted decarbonylations of deuterioaldehydes. *J. Org. Chem.* **1987**, 52, 3303-3307.
28. Ziegler, F. E., Belema, M. Chiral Aziridinyl Radicals: An Application to the Synthesis of the Core Nucleus of FR-900482. *J. Org. Chem.* **1997**, 62, 1083-1094.
29. Zeng, C.-m., Han, M., Covey, D. F. Neurosteroid Analogues. 7. A Synthetic Route for the Conversion of 5β-Methyl-3-ketosteroids into 7(S)-Methyl-Substituted Analogues of Neuroactive Benz[e]indenes. *J. Org. Chem.* **2000**, 65, 2264-2266.
30. Tanaka, M., Ohshima, T., Mitsuhashi, H., Maruno, M., Wakamatsu, T. Total syntheses of the lignans isolated from Schisandra chinensis. *Tetrahedron* **1995**, 51, 11693-11702.
31. Hansson, T., Wickberg, B. A short enantiospecific route to isodaucane sesquiterpenes from limonene. On the absolute configuration of (+)-aphanamol I and II. *J. Org. Chem.* **1992**, 57, 5370-5376.

Ugi Multicomponent Reaction 462

1. Ugi, I., Meyr, R., Fetzer, U., Steinbruckner, C. Studies on isonitriles. *Angew. Chem.* **1959**, 71, 386.
2. Ugi, I., Steinbruckner, C. Concerning a new condensation principle. *Angew. Chem.* **1960**, 72, 267-268.
3. Ugi, I. The α-addition of immonium ions and anions to isonitriles coupled with secondary reactions. *Angew. Chem.* **1962**, 74, 9-22.
4. Ugi, I. Novel synthetic approach to peptides by computer planned stereoselective four component condensations of α-ferrocenyl alkylamines and related reactions. *Rec. Chem. Prog.* **1969**, 30, 289-311.
5. Gokel, G., Luedke, G., Ugi, I. Four-component condensations and related reactions. *Isonitrile Chem.* **1971**, 145-199.
6. Ugi, I. Potential of four component condensations for peptide syntheses. Study in isonitrile and ferrocene chemistry as well as stereochemistry and logics of syntheses. *Intra-Science Chemistry Reports* **1971**, 5, 229-261.
7. Ugi, I., Arora, A., Burghard, H., Eberle, G., Eckert, H., George, G., Gokel, G., Herlinger, H., Von Hinrichs, E., et al. Four component condensations (4 CC), a potential alternative to conventional peptide synthesis. Solution of the stereoselectivity and auxiliary group removal problems. *Pept., Proc. Eur. Pept. Symp., 13th* **1975**, 71-92.

8. Ugi, I., Aigner, H., Beijer, B., Ben-Efraim, D., Burghard, H., Bukall, P., Eberle, G., Eckert, H., Marquarding, D., et al. New methods for peptide synthesis with organometallic reagents and isocyanides. *Pept., Proc. Eur. Pept. Symp., 14th* **1976**, 159-181.
9. Ugi, I., Eberle, G., Eckert, H., Lagerlund, I., Marquarding, D., Skorna, G., Urban, R., Wackerle, L., Von Zychlinski, H. The present status of peptide synthesis by four-component condensation and related chemistry. *Pept., Proc. Am. Pept. Symp., 5th* **1977**, 484-487.
10. Ugi, I. The four component synthesis. *Peptides (New York, 1979-1987)* **1980**, 2, 365-381.
11. Ugi, I., Breuer, W., Bukall, P., Falou, S., Giesemann, G., Herrmann, R., Huebener, G., Marquarding, D., Seidel, P., Urban, R. New aspects of peptide synthesis by four component condensations. *Chem. Pept. Proteins, Proc. USSR-FRG Symp., 3rd* **1982**, 203-208.
12. Ugi, I., Marquarding, D., Urban, R. Synthesis of peptides by four-component condensation. in *Chemistry and Biochemistry of Amino Acids, Peptides, and Proteins 6*, 245-289 (**1982**).
13. Ugi, I., Baumeister, M., Fleck, C., Herrmann, R., Obrecht, R., Siglmueller, F., Youn, J. H. Is there hope that four component condensations will become useful for peptide and protein chemistry? *Pept., Proc. Eur. Pept. Symp., 19th* **1987**, 103-106.
14. Totah, N. I., Schreiber, S. L. Asymmetric synthesis on carbohydrate templates: stereoselective Ugi synthesis of α-amino acid derivatives. *Chemtracts: Org. Chem.* **1988**, 1, 302-305.
15. Ugi, I. Four component condensations, a versatile principle in synthesis. *Eesti Teaduste Akadeemia Toimetised, Keemia* **1991**, 40, 1-13.
16. Ugi, I., Lohberger, S., Karl, R. The Passerini and Ugi reactions. in *Comp. Org. Synth.* (eds. Trost, B.,Fleming, I.), 2, 1083-1109 (Pergamon Press, Oxford, **1991**).
17. Ugi, I., Goebel, M., Bachmeyer, N., Demharter, A., Fleck, C., Gleixner, R., Lehnhoff, S. Peptide syntheses by stereoselective four component condensations with O-alkyl 1-β-glucopyranosylamines and other chiral amines. *Chemistry of Peptides and Proteins* **1993**, 5/6, 67-72.
18. Westinger, B., Fleck, C., Goebel, M., Herrmann, R., Karl, R., Lohberger, S., Reil, S., Siglmueller, F., Ugi, I. New chiral templates for peptide synthesis by four component condensations as well as related methods and reagents. *Chemistry of Peptides and Proteins* **1993**, 5/6, 59-65.
19. Dyker, G. Amino acid derivatives by multicomponent reactions. *Angew. Chem., Int. Ed. Engl.* **1997**, 36, 1700-1702.
20. Domling, A. isocyanide based multi component reactions in combinatorial chemistry. *Combinatorial Chemistry and High Throughput Screening* **1998**, 1, 1-22.
21. Ugi, I., Almstetter, M., Bock, H., Domling, A., Ebert, B., Gruber, B., Hanusch-Kompa, C., Heck, S., Kehagia-Drikos, K., Lorenz, K., Papathoma, S., Raditschnig, R., Schmid, T., Werner, B., Von Zychlinski, A. MCR XVII. Three types of MCRs and the libraries - their chemistry of natural events and preparative chemistry. *Croat. Chem. Acta* **1998**, 71, 527-547.
22. Ugi, I. K. MCR.XXIII. The highly variable multidisciplinary preparative and theoretical possibilities of the Ugi multicomponent reactions in the past, now, and in the future. *Proceedings of the Estonian Academy of Sciences, Chemistry* **1998**, 47, 107-127.
23. Ugi, I., Domling, A., Gruber, B., Heck, S., Heilingbrunner, M. From liquid-phase multicomponent reactions to solid phase libraries. *Innovation and Perspectives in Solid Phase Synthesis & Combinatorial Libraries: Peptides, Proteins and Nucleic Acids--Small Molecule Organic Chemical Diversity, Collected Papers, International Symposium, 5th, London, Sept. 2-6, 1997* **1999**, 201-204.
24. Bienayme, H., Hulme, C., Oddon, G., Schmitt, P. Maximizing synthetic efficiency: multi-component transformations lead the way. *Chem.--Eur. J.* **2000**, 6, 3321-3329.
25. Domling, A., Ugi, I. Multicomponent reactions with isocyanides. *Angew. Chem., Int. Ed. Engl.* **2000**, 39, 3168-3210.
26. Dyker, G. Amino acid derivatives by multicomponent reactions. *Organic Synthesis Highlights IV* **2000**, 53-57.
27. Ugi, I., Domling, A. Multi-component reactions (MCRs) of isocyanides and their chemical libraries. *Combinatorial Chemistry* **2000**, 287-302.
28. Ugi, I., Domling, A., Werner, B. Since 1995 the new chemistry of multicomponent reactions and their libraries, including their heterocyclic chemistry. *J. Heterocycl. Chem.* **2000**, 37, 647-658.
29. Ugi, I., Werner, B., Domling, A. Multicomponent reactions of isocyanides and the formation of heterocycles. *Targets in Heterocyclic Systems* **2000**, 4, 1-23.
30. Ugi, I. Recent progress in the chemistry of multicomponent reactions. *Pure Appl. Chem.* **2001**, 73, 187-191.
31. Hulme, C., Gore, V. Multi-component reactions: emerging chemistry in drug discovery from xylocain to crixivan. *Curr. Med. Chem.* **2003**, 10, 51-80.
32. Hulme, C., Nixey, T. Rapid assembly of molecular diversity via exploitation of isocyanide-based multi-component reactions. *Current Opinion in Drug Discovery & Development* **2003**, 6, 921-929.
33. Nerdinger, S., Beck, B. New heterocycle synthesis by using bifunctional reactants in multicomponent reaction chemistry: the use of arylglyoxals and cinnamaldehyde in the Ugi-4CR and Passerini-3CR. *Chemtracts* **2003**, 16, 233-237.
34. Ostaszewski, R., Portlock, D. E., Fryszkowska, A., Jeziorska, K. Combination of enzymic procedures with multicomponent condensations. *Pure Appl. Chem.* **2003**, 75, 413-419.
35. Zhu, J. Recent developments in the isonitrile-based multicomponent synthesis of heterocycles. *Eur. J. Org. Chem.* **2003**, 1133-1144.
36. Ugi, I., Bodesheim, F. Isonitriles. VIII. Reaction of isonitriles with hydrazones and hydrazoic acid. *Chem. Ber.* **1961**, 94, 2797-2801.
37. Ugi, I., Bodesheim, F. Isonitriles. XIV. Reactions of isonitriles with hydrazones and carboxylic acids. *Ann.* **1963**, 666, 61-64.
38. Zinner, G., Kliegel, W. Ugi reaction with hydrazines. I. *Arch. Pharm. (Weinheim, Ger.)* **1966**, 299, 746-756.
39. Zinner, G., Moderhack, D., Kliegel, W. Hydroxylamine derivatives. XXXVII. Hydroxylamines in the Ugi four-component condensation. *Chem. Ber.* **1969**, 102, 2536-2546.
40. Zinner, G., Bock, W. Ugi reaction with hydrazines. II. *Arch. Pharm. Ber. Dtsch. Pharm. Ges.* **1971**, 304, 933-943.
41. Failli, A., Nelson, V., Immer, H., Goetz, M. Model experiments directed towards the synthesis of N-aminopeptides. *Can. J. Chem.* **1973**, 51, 2769-2775.
42. Zinner, G., Bock, W. Ugi-reactions with hydrazines. III. Ugi-reaction with diaziridines. *Arch. Pharm. (Weinheim, Ger.)* **1973**, 306, 94-96.
43. McFarland, J. W. Reactions of cyclohexyl isonitrile and isobutyraldehyde with various nucleophiles and catalysts. *J. Org. Chem.* **1963**, 28, 2179-2181.
44. Ugi, I., Kaufhold, G. Isonitriles. XXIII. Stereoselective syntheses. 4. Reaction mechanism of stereoselective four-component condensations. *Liebigs Ann. Chem.* **1967**, 709, 11-28.
45. Lohberger, S., Fontain, E., Ugi, I., Mueller, G., Lachmann, J. Malonamide derivatives as by-products of four-component condensations. The computer-assisted investigation of a reaction mechanism. *New J. Chem.* **1991**, 15, 913-917.
46. Joullie, M. M., Wang, P. C., Semple, J. E. Total synthesis and revised structural assignment of (+)-furanomycin. *J. Am. Chem. Soc.* **1980**, 102, 887-889.
47. Semple, J. E., Wang, P. C., Lysenko, Z., Joullie, M. M. Total synthesis of (+)-furanomycin and stereoisomers. *J. Am. Chem. Soc.* **1980**, 102, 7505-7510.
48. Endo, A., Yanagisawa, A., Abe, M., Tohma, S., Kan, T., Fukuyama, T. Total Synthesis of Ecteinascidin 743. *J. Am. Chem. Soc.* **2002**, 124, 6552-6554.
49. Hulme, C., Morrissette, M. M., Volz, F. A., Burns, C. J. The solution phase synthesis of diketopiperazine libraries via the Ugi reaction: novel application of Armstrong's convertible isonitrile. *Tetrahedron Lett.* **1998**, 39, 1113-1116.
50. Hulme, C., Peng, J., Louridas, B., Menard, P., Krolikowski, P., Kumar, N. V. Applications of N-BOC-diamines for the solution phase synthesis of ketopiperazine libraries utilizing a Ugi/De-BOC/cyclization (UDC) strategy. *Tetrahedron Lett.* **1998**, 39, 8047-8050.

Ullmann Biaryl Ether and Biaryl Amine Synthesis/Condensation 464

Related reactions: Buchwald-Hartwig cross coupling;

1. Ullmann, F. A new path for preparing diphenylamine derivatives. *Ber.* **1903**, 36, 2382-2384.
2. Ullmann, F. A new path for preparing phenyl ether salicylic acid. *Ber.* **1904**, 37, 853-854.
3. Ullmann, F., Sponagel, P. Phenylation of phenols. *Ber.* **1905**, 38, 2211-2212.
4. Goldberg, I. Phenylation in the presence of copper as catalyst. *Ber.* **1906**, 39, 1691-1692.

5. Bunnett, J. F., Zahler, R. E. Aromatic nucleophilic substitution reactions. *Chem. Rev.* **1951**, 49, 273-412.
6. Moroz, A. A., Shvartsberg, M. S. The Ullmann Ether Condensation. *Russ. Chem. Rev.* **1974**, 43, 1443-1461.
7. Lindley, J. Copper-assisted nucleophilic substitution of aryl halogen. *Tetrahedron* **1984**, 40, 1433-1456.
8. Scott Sawyer, J. Recent Advances in Diaryl Ether Synthesis. *Tetrahedron* **2000**, 56, 5045-5065.
9. Kunz, K., Scholz, U., Ganzer, D. Renaissance of Ullmann and Goldberg reactions - progress in copper catalyzed C-N-, C-O- and C-S-coupling. *Synlett* **2003**, 2428-2439.
10. Ley, S. V., Thomas, A. W. Modern synthetic methods for copper-mediated C(aryl)-O, C(aryl)-N, and C(aryl)-S bond formation. *Angew. Chem., Int. Ed. Engl.* **2003**, 42, 5400-5449.
11. Ley, S. V., Thomas, A. W. Modern synthetic methods for copper-mediated C(aryl)-O, C(aryl)-N, and C(aryl)-S bond formation. *Angew. Chem., Int. Ed. Engl.* **2004**, 43, 1043.
12. Beringer, F. M., Brierley, A., Drexler, M., Gindler, E. M., Lumpkin, C. C. Diaryliodonium salts. II. The phenylation of organic and inorganic bases. *J. Am. Chem. Soc.* **1953**, 75, 2708-2712.
13. Tomita, M., Fumitani, K., Aoyagi, Y. Cupric oxide as an efficient catalyst in Ullmann condensation reaction. *Chem. Pharm. Bull.* **1965**, 13, 1341-1345.
14. Bacon, R. G. R., Karim, A. Copper-catalyzed substitution of aryl halides by potassium phthalimide: an extension of the Gabriel reaction. *J. Chem. Soc., Chem. Commun.* **1969**, 578.
15. Barton, D. H. R., Finet, J. P. Bismuth(V) reagents in organic synthesis. *Pure Appl. Chem.* **1987**, 59, 937-946.
16. Barton, D. H. R., Finet, J. P., Khamsi, J. Copper salt catalysis of N-phenylation of amines by trivalent organobismuth compounds. *Tetrahedron Lett.* **1987**, 28, 887-890.
17. Barton, D. H. R., Yadav-Bhatnagar, N., Finet, J. P., Khamsi, J. Phenylation of aromatic and aliphatic amines by phenyllead triacetate using copper catalysis. *Tetrahedron Lett.* **1987**, 28, 3111-3114.
18. Barton, D. H. R., Finet, J. P., Khamsi, J. N-Phenylation of amino acid derivatives. *Tetrahedron Lett.* **1989**, 30, 937-940.
19. Schmittling, E. A., Sawyer, J. S. Synthesis of diaryl ethers, diaryl thioethers, and diarylamines mediated by potassium fluoride-alumina and 18-crown-6. *J. Org. Chem.* **1993**, 58, 3229-3230.
20. Marcoux, J.-F., Doye, S., Buchwald, S. L. A General Copper-Catalyzed Synthesis of Diaryl Ethers. *J. Am. Chem. Soc.* **1997**, 119, 10539-10540.
21. Nicolaou, K. C., Boddy, C. N. C., Natarajan, S., Yue, T. Y., Li, H., Braese, S., Ramanjulu, J. M. New Synthetic Technology for the Synthesis of Aryl Ethers: Construction of C-O-D and D-O-E Ring Model Systems of Vancomycin. *J. Am. Chem. Soc.* **1997**, 119, 3421-3422.
22. Nicolaou, K. C., Chu, X.-J., Ramanjulu, J. M., Natarajan, S., Brase, S., Rubsam, F., Boddy, C. N. C. New technology for the synthesis of vancomycin-type biaryl ring systems. *Angew. Chem., Int. Ed. Engl.* **1997**, 36, 1539-1540.
23. Chan, D. M. T., Monaco, K. L., Wang, R.-P., Winters, M. P. New N- and O-arylation with phenylboronic acids and cupric acetate. *Tetrahedron Lett.* **1998**, 39, 2933-2936.
24. Evans, D. A., Katz, J. L., West, T. R. Synthesis of diaryl ethers through the copper-promoted arylation of phenols with arylboronic acids. An expedient synthesis of thyroxine. *Tetrahedron Lett.* **1998**, 39, 2937-2940.
25. Lam, P. Y. S., Clark, C. G., Saubern, S., Adams, J., Winters, M. P., Chan, D. M. T., Combs, A. New aryl/heteroaryl C-N bond cross-coupling reactions via arylboronic acid/cupric acetate arylation. *Tetrahedron Lett.* **1998**, 39, 2941-2944.
26. Sawyer, J. S., Schmittling, E. A., Palkowitz, J. A., Smith, W. J., III. Synthesis of diaryl ethers, diaryl thioethers, and diaryl amines mediated by potassium fluoride-alumina and 18-crown-6: Expansion of scope and utility. *J. Org. Chem.* **1998**, 63, 6338-6343.
27. Kalinin, A. V., Bower, J. F., Riebel, P., Snieckus, V. The Directed Ortho Metalation-Ullmann Connection. A New Cu(I)-Catalyzed Variant for the Synthesis of Substituted Diaryl Ethers. *J. Org. Chem.* **1999**, 64, 2986-2987.
28. Herradura, P. S., Pendola, K. A., Guy, R. K. Copper-Mediated Cross-Coupling of Aryl Boronic Acids and Alkyl Thiols. *Org. Lett.* **2000**, 2, 2019-2022.
29. Kang, S.-K., Lee, S.-H., Lee, D. Copper-catalyzed N-arylation of amines with hypervalent iodonium salts. *Synlett* **2000**, 1022-1024.
30. Olivera, R., SanMartin, R., Dominguez, E. A novel palladium-catalyzed intramolecular diaryl ether formation. *Tetrahedron Lett.* **2000**, 41, 4357-4360.
31. Palomo, C., Oiarbide, M., Lopez, R., Gomez-Bengoa, E. Phosphazene bases for the preparation of biaryl thioethers from aryl iodides and arenethiols. *Tetrahedron Lett.* **2000**, 41, 1283-1286.
32. Yadav, J. S., Subba Reddy, B. V. CsF-Al_2O_3 mediated rapid condensation of phenols with aryl halides: comparative study of conventional heating vs. microwave irradiation. *New J. Chem.* **2000**, 24, 489-491.
33. Antilla, J. C., Buchwald, S. L. Copper-Catalyzed Coupling of Arylboronic Acids and Amines. *Org. Lett.* **2001**, 3, 2077-2079.
34. Klapars, A., Antilla, J. C., Huang, X., Buchwald, S. L. A general and efficient copper catalyst for the amidation of aryl halides and the N-arylation of nitrogen heterocycles. *J. Am. Chem. Soc.* **2001**, 123, 7727-7729.
35. Lam, P. Y. S., Vincent, G., Clark, C. G., Deudon, S., Jadhav, P. K. Copper-catalyzed general C-N and C-O bond cross-coupling with arylboronic acid. *Tetrahedron Lett.* **2001**, 42, 3415-3418.
36. Petrassi, H. M., Sharpless, K. B., Kelly, J. W. The copper-mediated cross-coupling of phenylboronic acids and N-hydroxyphthalimide at room temperature: synthesis of aryloxyamines. *Org. Lett.* **2001**, 3, 139-142.
37. Kwong, F. Y., Buchwald, S. L. A General, Efficient, and Inexpensive Catalyst System for the Coupling of Aryl Iodides and Thiols. *Org. Lett.* **2002**, 4, 3517-3520.
38. Lam, P. Y. S., Vincent, G., Bonne, D., Clark, C. G. Copper-promoted C-N bond cross-coupling with phenylstannane. *Tetrahedron Lett.* **2002**, 43, 3091-3094.
39. Savarin, C., Srogl, J., Liebeskind, L. S. A Mild, Nonbasic Synthesis of Thioethers. The Copper-Catalyzed Coupling of Boronic Acids with N-Thio(alkyl, aryl, heteroaryl)imides. *Org. Lett.* **2002**, 4, 4309-4312.
40. Li, F., Wang, Q., Ding, Z., Tao, F. Microwave-Assisted Synthesis of Diaryl Ethers without Catalyst. *Org. Lett.* **2003**, 5, 2169-2171.
41. Ma, D., Cai, Q. N,N-Dimethyl Glycine-Promoted Ullmann Coupling Reaction of Phenols and Aryl Halides. *Org. Lett.* **2003**, 5, 3799-3802.
42. Quach, T. D., Batey, R. A. Ligand- and Base-Free Copper(II)-Catalyzed C-N Bond Formation: Cross-Coupling Reactions of Organoboron Compounds with Aliphatic Amines and Anilines. *Org. Lett.* **2003**, 5, 4397-4400.
43. Quach, T. D., Batey, R. A. Copper(II)-Catalyzed Ether Synthesis from Aliphatic Alcohols and Potassium Organotrifluoroborate Salts. *Org. Lett.* **2003**, 5, 1381-1384.
44. Rebeiro, G. L., Khadilkar, B. M. Microwave assisted aromatic nucleophilic substitution reaction under solventless condition. *Synth. Commun.* **2003**, 33, 1405-1410.
45. Zhao, J. K., Wang, Y. G. An efficient synthesis of diaryl ethers by coupling aryl halides with substituted phenoxytrimethylsilane in the presence of TBAF. *Chin. Chem. Lett.* **2003**, 14, 1012-1014.
46. Cristau, H.-J., Cellier, P. P., Hamada, S., Spindler, J.-F., Taillefer, M. A General and Mild Ullmann-Type Synthesis of Diaryl Ethers. *Org. Lett.* **2004**, 6, 913-916.
47. Litvak, V. V., Shein, S. M. Nucleophilic substitution in an aromatic series. LI. Kinetics of the reaction of bromobenzene with phenols, catalyzed by copper salts, in the presence of alkali metal carbonates. *Zh. Org. Khim.* **1975**, 11, 92-96.
48. Paine, A. J. Mechanisms and models for copper mediated nucleophilic aromatic substitution. 2. Single catalytic species from three different oxidation states of copper in an Ullmann synthesis of triarylamines. *J. Am. Chem. Soc.* **1987**, 109, 1496-1502.
49. Collman, J. P., Zhong, M. An Efficient Diamine-Copper Complex-Catalyzed Coupling of Arylboronic Acids with Imidazoles. *Org. Lett.* **2000**, 2, 1233-1236.
50. Boger, D. L., Sakya, S. M., Yohannes, D. Total synthesis of combretastatin D-2: intramolecular Ullmann macrocyclization reaction. *J. Org. Chem.* **1991**, 56, 4204-4207.
51. Wipf, P., Jung, J.-K. Formal Total Synthesis of (+)-Diepoxin σ. *J. Org. Chem.* **2000**, 65, 6319-6337.
52. Ma, D., Xia, C. CuI-catalyzed coupling reaction of β-amino acids or esters with aryl halides at temperature lower than that employed in the normal Ullmann reaction. Facile synthesis of SB-214857. *Org. Lett.* **2001**, 3, 2583-2586.

Related reactions: *Kumada cross-coupling, Negishi cross-coupling, Stille coupling, Stille-Kelly coupling, Suzuki cross-coupling;*

1. Ullmann, F., Bielecki, J. Synthesis in the biphenyl series. *Ber.* **1901**, 34, 2174-2185.
2. Ullmann, F. Symmetrical biphenyl derivative. *Ann.* **1904**, 332, 38-81.
3. Fanta, P. E. The Ullmann synthesis of biaryls. *Chem. Rev.* **1946**, 38, 139-196.
4. Fanta, P. E. Ullmann synthesis of biaryls. *Synthesis* **1974**, 9-21.
5. Sainsbury, M. Modern methods of aryl-aryl bond formation. *Tetrahedron* **1980**, 36, 3327-3359.
6. Knight, D. W. Coupling Reactions Between sp^2 Carbon Centers. in *Comp. Org. Synth.* (eds. Trost, B. M.,Fleming, I.), *3*, 481-520 (Pergamon, Oxford, **1991**).
7. Hassan, J., Sevignon, M., Gozzi, C., Schulz, E., Lemaire, M. Aryl-Aryl Bond Formation One Century after the Discovery of the Ullmann Reaction. *Chem. Rev.* **2002**, 102, 1359-1469.
8. Moreno, I., Tellitu, I., Herrero, M. T., SanMartin, R., Dominguez, E. New perspectives for iodine (III) reagents in (hetero)biaryl coupling reactions. *Curr. Org. Chem.* **2002**, 6, 1433-1452.
9. Nelson, T. D., Crouch, R. D. Cu, Ni, and Pd mediated homocoupling reactions in biaryl syntheses: the Ullmann reaction. *Org. React.* **2004**, 63, 265-555.
10. Kornblum, N., Kendall, D. L. The use of dimethylformamide in the Ullmann reaction. *J. Am. Chem. Soc.* **1952**, 74, 5782.
11. Forrest, J. Ullmann biaryl synthesis. III. The influence of diluents on the reaction between iodobenzene and copper. *J. Chem. Soc., Abstracts* **1960**, 581-588.
12. Bacon, R. G. R., Pande, S. G. Metal ions and complexes in organic reactions. XI. Reactions in pyridine between copper species and aryl halides, in particular between copper(I) oxide and 2-bromonitrobenzene. *J. Chem. Soc. C.* **1970**, 1967-1973.
13. Semmelhack, M. F., Helquist, P. M., Jones, L. D. Synthesis with zerovalent nickel. Coupling of aryl halides with bis(1,5-cyclooctadiene)nickel(0). *J. Am. Chem. Soc.* **1971**, 93, 5908-5910.
14. Cohen, T., Cristea, I. Copper(I)-induced reductive dehalogenation, hydrolysis, or coupling of some aryl and vinyl halides at room temperature. *J. Org. Chem.* **1975**, 40, 3649-3651.
15. Cohen, T., Tirpak, J. G. Rapid, room-temperature Ullmann-type couplings and ammonolyses of activated aryl halides in homogeneous solutions containing copper(I) ions. *Tetrahedron Lett.* **1975**, 143-146.
16. Ziegler, F. E., Fowler, K. W., Kanfer, S. The chemospecific, homogeneous, ambient temperature Ullmann coupling of o-haloarylimines. *J. Am. Chem. Soc.* **1976**, 98, 8282-8283.
17. Rieke, R. D., Rhyne, L. D. Preparation of highly reactive metal powders. Activated copper and uranium. The Ullmann coupling and preparation of organometallic species. *J. Org. Chem.* **1979**, 44, 3445-3446.
18. Lindley, J., Lorimer, J. P., Mason, T. J. Enhancement of an Ullmann coupling reaction induced by ultrasound. *Ultrasonics* **1986**, 24, 292-293.
19. Lindley, J., Mason, T. J., Lorimer, J. P. Sonochemically enhanced Ullmann reactions. *Ultrasonics* **1987**, 25, 45-48.
20. Ziegler, F. E., Fowler, K. W., Rodgers, W. B., Wester, R. T. Ambient temperature Ullmann reaction: 4,5,4',5'-tetramethoxy-1,1'-biphenyl-2,2'-dicarboxaldehyde ([1,1'-biphenyl]-2,2'-dicarboxaldehyde, 4,4',5,5'-tetramethoxy-). *Org. Synth.* **1987**, 65, 108-118.
21. Zhang, S., Zhang, D., Liebeskind, L. S. Ambient Temperature, Ullmann-like Reductive Coupling of Aryl, Heteroaryl, and Alkenyl Halides. *J. Org. Chem.* **1997**, 62, 2312-2313.
22. Forrest, J. Ullmann biaryl synthesis. V. The influence of ring substituents on the rate of self-condensation of an aryl halide. *J. Chem. Soc., Abstracts* **1960**, 592-594.
23. Nilsson, M. A new biaryl synthesis illustrating a connection between the Ullmann biaryl synthesis and copper-catalyzed decarboxylation. *Acta Chem. Scand.* **1966**, 20, 423-426.
24. Rapson, W. S., Shuttleworth, R. G. Free aryl radicals in the Fittig and Ullmann reactions. *Nature (London, United Kingdom)* **1941**, 147, 675.
25. Lewin, A. H., Cohen, T. Mechanism of the Ullmann reaction. Detection of an organocopper intermediate. *Tetrahedron Lett.* **1965**, 4531-4536.
26. Cohen, T., Poeth, T. Copper-induced coupling of vinyl halides. Stereochemistry of the Ullmann reaction. *J. Am. Chem. Soc.* **1972**, 94, 4363-4364.
27. Cohen, T., Cristea, I. Kinetics and mechanism of the copper(I)-induced homogeneous Ullmann coupling of o-bromonitrobenzene. *J. Am. Chem. Soc.* **1976**, 98, 748-753.
28. Ebert, G. W., Rieke, R. D. Preparation of aryl, alkynyl, and vinyl organocopper compounds by the oxidative addition of zerovalent copper to carbon-halogen bonds. *J. Org. Chem.* **1988**, 53, 4482-4488.
29. Douglass, S. E., Massey, S. T., Woolard, S. G., Zoellner, R. W. Reductive versus coupling pathways in the reactions of nickel and copper vapors with the monohalobenzenes. *Transition Metal Chemistry (Dordrecht, Netherlands)* **1990**, 15, 317-324.
30. Negrel, J. C., Gony, M., Chanon, M., Lai, R. Reactivity of copper metal vapors with substituted bromobenzenes. Formation and molecular structure of Cu(PMe3)3Br. *Inorg. Chim. Acta* **1993**, 207, 59-63.
31. Xi, M., Bent, B. E. Mechanisms of the Ullmann coupling reaction in adsorbed monolayers. *J. Am. Chem. Soc.* **1993**, 115, 7426-7433.
32. Meyers, J. M., Gellman, A. J. Effect of substituents on the phenyl coupling reaction on Cu(111). *Surf. Sci.* **1995**, 337, 40-50.
33. Stark, L. M., Lin, X.-F., Flippin, L. A. Total Synthesis of Amaryllidaceae Pyrrolophenanthridinium Alkaloids via the Ziegler-Ullmann Reaction: Tortuosine, Criasbetaine, and Ungeremine. *J. Org. Chem.* **2000**, 65, 3227-3230.
34. Degnan, A. P., Meyers, A. I. Total Syntheses of (-)-Herbertenediol, (-)-Mastigophorene A, and (+)-Mastigophorene B. Combined Utility of Chiral Bicyclic Lactams and Chiral Aryl Oxazolines. *J. Am. Chem. Soc.* **1999**, 121, 2762-2769.
35. Kelly, T. R., Xie, R. L. Total Synthesis of Taspine. *J. Org. Chem.* **1998**, 63, 8045-8048.

Related reactions: *Gattermann and Gattermann-Koch formylation, Reimer-Tiemann formylation;*

1. Fischer, O., Muller, A., Vilsmeier, A. Action of phosphorus oxychloride upon methyl- and ethylacetanilide. Synthesis of γ-chloroisoquinocyanines. *J. Prakt. Chem.* **1925**, 109, 69-87.
2. Vilsmeier, A., Haack, A. Action of phosphorus halides on alkylformanilides. A new method for the preparation of secondary and tertiary p-alkylaminonobenzaldehydes. *Ber.* **1927**, 60B, 119-122.
3. Burn, D. Alkylation with the Vilsmeier reagent. *Chem. Ind.* **1973**, 870-873.
4. Seshadri, S. Vilsmeier-Haack reaction and its synthetic applications. *J. Sci. Ind. Res.* **1973**, 32, 128-149.
5. Jutz, C. The Vilsmeier-Haack-Arnold acylations. Carbon-carbon bond-forming reactions of chloromethyleniminium ions. *Advances in Organic Chemistry* **1976**, 9, Pt. 1, 225-342.
6. Meth-Cohn, O., Tarnowski, B. Cyclizations under Vilsmeier conditions. *Adv. Heterocycl. Chem.* **1982**, 31, 207-236.
7. Meth-Cohn, O. The Vilsmeier-Haack Reaction. in *Comp. Org. Synth.* (eds. Trost, B. M.,Fleming, I.), *2*, 777-794 (Pergamon, Oxford, **1991**).
8. Jones, G., Stanforth, S. P. The Vilsmeier reaction of fully conjugated carbocycles and heterocycles. *Org. React.* **1997**, 49, 1-330.
9. Sharma, S. D., Kanwar, S. Phosphorous oxychloride ($POCl_3$): a key molecule in organic synthesis. *Indian J. Chem., Sect. B* **1998**, 37B, 965-978.
10. Reichardt, C. Vilsmeier-Haack-Arnold formylations of aliphatic substrates with N-chloromethylene-N,N-dimethylammonium salts. *J. Prakt. Chem.* **1999**, 341, 609-615.
11. Jones, G., Stanforth, S. P. The Vilsmeier reaction of non-aromatic compounds. *Org. React.* **2000**, 56, 355-659.
12. Perumal, P. T. Synthesis of heterocyclic compounds using Vilsmeier reagent. *Indian J. Heterocycl. Chem.* **2001**, 11, 1-8.

13. Kantlehner, W. New methods for the preparation of aromatic aldehydes. *Eur. J. Org. Chem.* **2003**, 2530-2546.
14. Ramesh, N. G., Balasubramanian, K. K. 2-C-formyl glycals: Emerging chiral synthons in organic synthesis. *Eur. J. Org. Chem.* **2003**, 4477-4487.
15. Tasneem. Vilsmeier-Haack reagent (halomethyleneiminium salt). *Synlett* **2003**, 138-139.
16. Lellouche, J.-P., Kotlyar, V. Vilsmeier-Haack reagents. Novel electrophiles for the one-step formylation of O-silylated ethers to O-formates. *Synlett* **2004**, 564-571.
17. Katritzky, A. R., Shcherbakova, I. V., Tack, R. D., Steel, P. J. Reactions of unactivated olefins with Vilsmeier reagents. *Can. J. Chem.* **1992**, 70, 2040-2045.
18. Marson, C. M. Reactions of carbonyl compounds with (monohalo) methyleniminium salts (Vilsmeier reagents). *Tetrahedron* **1992**, 48, 3659-3726.
19. Balser, D., Calmes, M., Daunis, J., Natt, F., Tardy-Delassus, A., Jacquier, R. Improvement of the Vilsmeier-Haack reaction. *Org. Prep. Proced. Int.* **1993**, 25, 338-341.
20. Koeller, S., Lellouche, J.-P. Preparation of formate esters from O-TBDMS/O-TES protected alcohols. A one-step conversion using the Vilsmeier-Haack complex $POCl_3$/DMF. *Tetrahedron Lett.* **1999**, 40, 7043-7046.
21. Selvi, S., Perumal, P. T. Synthetic utility of the Vilsmeier reaction: more vinamidinium salts. *Synth. Commun.* **1999**, 29, 73-77.
22. Katritzky, A. R., Huang, T.-B., Voronkov, M. V. Direct and efficient synthesis of dimethylformamidrazones using benzotriazole Vilsmeier reagent. *J. Org. Chem.* **2000**, 65, 2246-2248.
23. Paul, S., Gupta, M., Gupta, R. Vilsmeier reagent for formylation in solvent-free conditions using microwaves. *Synlett* **2000**, 1115-1118.
24. Cohen, Y., Kotlyar, V., Koeller, S., Lellouche, J.-P. Reaction of C_2-symmetrical dialkoxysilanes $R^1O-Si(R^2)_2-OR^1$ with the two Vilsmeier-Haack complexes $POCl_3$.DMF and $(CF_3SO_2)_2O$.DMF: an efficient one-step conversion to the corresponding formates R1-OCHO. *Synlett* **2001**, 1543-1546.
25. Thomas, A. D., Asokan, C. V. Vilsmeier-Haack reaction of tertiary alcohols: formation of functionalised pyridines and naphthyridines. *J. Chem. Soc., Perkin Trans. 1* **2001**, 2583-2587.
26. Ali, M. M., Sana, S., Tasneem, Rajanna, K. C., Saiprakash, P. K. Ultrasonically accelerated Vilsmeier Haack cyclization and formylation reactions. *Synth. Commun.* **2002**, 32, 1351-1356.
27. Sridhar, R., Perumal, P. T. Synthesis of acyl azides using the Vilsmeier complex. *Synth. Commun.* **2003**, 33, 607-611.
28. Liu, Y., Dong, D., Liu, Q., Qi, Y., Wang, Z. A novel and facile synthesis of dienals and substituted 2H-pyrans via the Vilsmeier reaction of a-oxo-ketene dithioacetals. *Org. Biomol. Chem.* **2004**, 2, 28-30.
29. Rajanna, K. C., Moazzam Ali, M., Sana, S., Tasneem, Saiprakash, P. K. Vilsmeier Haack acetylation in micellar media. An efficient one-pot synthesis of 2-chloro-3-acetyl quinolines. *J. Dispersion Sci. and Tech.* **2004**, 25, 17-21.
30. Thomas, A. D., Josemin, K. N. N., Asokan, C. V. Vilsmeier-Haack reactions of carbonyl compounds: synthesis of substituted pyrones and pyridines. *Tetrahedron* **2004**, 60, 5069-5076.
31. Das, G. K., Choudhury, B., Das, K., Das, B. P. Vilsmeier reaction on carbazole: theoretical and experimental aspects. *J. Chem. Res., Synop.* **1999**, 244-245.
32. Patsenker, L. D. Theoretical study of the pathway for diazine ring formation in a series of 4-dimethylaminonaphthalic acid derivatives under Vilsmeier-Haack reaction conditions. *Theoretical and Experimental Chemistry (Translation of Teoreticheskaya i Eksperimental'naya Khimiya)* **2001**, 36, 183-186.
33. Semenova, O. N., Galkina, O. S., Patsenker, L. D., Yermolenko, I. G., Fedyunyayeva, I. A. Experimental and theoretical investigation of the reaction of 2,5-diphenyl-1,3-oxazole and 2,5-diphenyl-1,3,4-oxadiazole dimethylamino derivatives with the Vilsmeier reagent. *Functional Materials* **2004**, 11, 67-75.
34. Linda, P., Marino, G., Santini, S. Electrophilic substitutions in five-membered heteroaromatic systems. XIII. Kinetics and mechanism of the Vilsmeier formylation of thiophene derivatives. *Tetrahedron Lett.* **1970**, 4223-4224.
35. Alunni, S., Linda, P., Marino, G., Santini, S., Savelli, G. Mechanism of the Vilsmeier-Haack reaction. II. Kinetic study of the formylation of thiophene derivatives with dimethylformamide and phosphorus oxychloride or carbonyl chloride in 1,2-dichloroethane. *J. Chem. Soc., Perkin Trans. 2* **1972**, 2070-2073.
36. Martin, G. J., Poignant, S. Nuclear magnetic resonance investigations of carbonium ion intermediates. I. Kinetic and mechanism of formation of the Vilsmeier-Haack reagent. *J. Chem. Soc., Perkin Trans. 2* **1972**, 1964-1966.
37. Linda, P., Lucarelli, A., Marino, G., Savelli, G. Mechanism of the Vilsmeier-Haack reaction. III. Structural and solvent effects. *J. Chem. Soc., Perkin Trans. 2* **1974**, 1610-1612.
38. Martin, G. J., Poignant, S. Nuclear magnetic resonance investigations of carbonium ion intermediates. II. Exchange reactions in chloro iminium salts (Vilsmeier-Haack reagents). *J. Chem. Soc., Perkin Trans. 2* **1974**, 642-646.
39. Downie, I. M., Earle, M. J., Heaney, H., Shuhaibar, K. F. Vilsmeier formylation and glyoxylation reactions of nucleophilic aromatic compounds using pyrophosphoryl chloride. *Tetrahedron* **1993**, 49, 4015-4034.
40. Rajanna, K. C., Solomon, F., Ali, M. M., Prakash, P. K. S. Vilsmeier-Haack formylation of coumarin derivatives. A solvent dependent kinetic study. *Int. J. Chem. Kinet.* **1996**, 28, 865-872.
41. Rajanna, K. C., Solomon, F., Ali, M. M., Saiprakash, P. K. Kinetics and mechanism of Vilsmeier-Haack synthesis of 3-formyl chromones derived from o-hydroxy aryl alkyl ketones: a structure reactivity study. *Tetrahedron* **1996**, 52, 3669-3682.
42. Mayr, H., Ofial, A. R. Electrophilicities of iminium ions. *Tetrahedron Lett.* **1997**, 38, 3503-3506.
43. Deshpande, P. P., Tagliaferri, F., Victory, S. F., Yan, S., Baker, D. C. Synthesis of Optically Active Calanolides A and B. *J. Org. Chem.* **1995**, 60, 2964-2965.
44. Ziegler, F. E., Belema, M. Chiral Aziridinyl Radicals: An Application to the Synthesis of the Core Nucleus of FR-900482. *J. Org. Chem.* **1997**, 62, 1083-1094.
45. Aungst, R. A., Jr., Chan, C., Funk, R. L. Total Synthesis of the Sesquiterpene (±)-Illudin C via an Intramolecular Nitrile Oxide Cycloaddition. *Org. Lett.* **2001**, 3, 2611-2613.
46. Gribble, G. W., Pelcman, B. Total syntheses of the marine sponge pigments fascaplysin and homofascaplysin B and C. *J. Org. Chem.* **1992**, 57, 3636-3642.
47. Ramadas, S., Krupadanam, G. L. D. Enantioselective acylation of 2-hydroxymethyl-2,3-dihydrobenzofurans catalyzed by lipase from Pseudomonas cepacia (Amano PS) and total stereoselective synthesis of (-)-(R)-MEM-protected arthrographol. *Tetrahedron: Asymmetry* **2000**, 11, 3375-3393.

Vinylcyclopropane-Cyclopentene Rearrangement 470

1. Neureiter, N. P. Pyrolysis of 1,1-dichloro-2-vinylcyclopropane. Synthesis of 2-chlorocyclopentadiene. *J. Org. Chem.* **1959**, 24, 2044-2046.
2. Overberger, C. G., Borchert, A. E. Novel thermal rearrangements accompanying acetate pyrolysis in small ring systems. *J. Am. Chem. Soc.* **1960**, 82, 1007-1008.
3. Vogel, E. Small carbon rings. *Angew. Chem.* **1960**, 72, 4-26.
4. Frey, H. M., Walsh, R. Thermal unimolecular reactions of hydrocarbons. *Chem. Rev.* **1969**, 69, 103-123.
5. Hudlicky, T., Kutchan, T. M., Naqvi, S. M. The vinylcyclopropane-cyclopentene rearrangement. *Org. React.* **1985**, 33, 247-335.
6. Goldschmidt, Z., Crammer, B. Vinylcyclopropane rearrangements. *Chem. Soc. Rev.* **1988**, 17, 229-267.
7. Hudlicky, T., Reed, J. W. Rearrangements of Vinylcyclopropanes and Related Systems. in *Comp. Org. Synth.* (eds. Trost, B. M.,Fleming, I.), *5*, 899-970 (Pergamon, Oxford, **1991**).
8. Dolbier, W. R., Jr. Thermal rearrangement of fluorine-containing cyclopropanes. *Adv. Strain Org. Chem.* **1993**, 3, 1-58.
9. Sonawane, H. R., Bellur, N. S., Kulkarni, D. G., Ahuja, J. R. Photoinduced vinylcyclopropane-cyclopentene rearrangement (photo-VCP-CP): a methodology for chiral bicyclo[3.2.0]heptenes and their application in natural product syntheses. *Synlett* **1993**, 875-884.
10. Baldwin, J. E. Thermal Rearrangements of Vinylcyclopropanes to Cyclopentenes. *Chem. Rev.* **2003**, 103, 1197-1212.

11. Van Eis, M. J., Nijbacker, T., De Kanter, F. J. J., De Wolf, W. H., Lammertsma, K., Bickelhaupt, F. The 2-Vinylphosphirane 3-Phospholene Rearrangement: Biradicaloid and Concerted Features. *J. Am. Chem. Soc.* **2000**, 122, 3033-3036.
12. Lin, Y.-L., Turos, E. Studies of Silyl-Accelerated 1,5-Hydrogen Migrations in Vinylcyclopropanes. *J. Org. Chem.* **2001**, 66, 8751-8759.
13. Dewar, M. J. S., Fonken, G. J., Kirschner, S., Minter, D. E. Mechanism of the vinylcyclopropane rearrangement. Rearrangement of cyclopropylallene and MINDO/3 calculations. *J. Am. Chem. Soc.* **1975**, 97, 6750-6753.
14. Yin, T. K., Radziszewski, J. G., Renzoni, G. E., Downing, J. W., Michl, J., Borden, W. T. Thermal reorganization of two pyramidalized alkenes by reverse vinylcyclopropane rearrangements. *J. Am. Chem. Soc.* **1987**, 109, 820-822.
15. Quirante, J. J., Enriquez, F., Hernando, J. M. The vinylcyclopropane rearrangement: an AM1 study. *THEOCHEM* **1990**, 63, 193-200.
16. Quirante, J. J., Enriquez, F., Hernando, J. M. AM1 study of the cycloaddition of singlet methylene to butadiene and the vinylcyclopropane rearrangement. *THEOCHEM* **1992**, 86, 493-504.
17. Davidson, E. R., Gajewski, J. J. Calculational Evidence for Lack of Intermediates in the Thermal Unimolecular Vinylcyclopropane to Cyclopentene 1,3-Sigmatropic Shift. *J. Am. Chem. Soc.* **1997**, 119, 10543-10544.
18. Houk, K. N., Nendel, M., Wiest, O., Storer, J. W. The Vinylcyclopropane-Cyclopentene Rearrangement: A Prototype Thermal Rearrangement Involving Competing Diradical Concerted and Stepwise Mechanisms. *J. Am. Chem. Soc.* **1997**, 119, 10545-10546.
19. Roth, H. D., Weng, H., Herbertz, T. CIDNP study and ab-initio calculations of rigid vinylcyclopropane systems: evidence for delocalized "ring-closed" radical cations. *Tetrahedron* **1997**, 53, 10051-10070.
20. Baldwin, J. E. Thermal isomerizations of vinylcyclopropanes to cyclopentenes. *J. Comput. Chem.* **1998**, 19, 222-231.
21. Doubleday, C., Nendel, M., Houk, K. N., Thweatt, D., Page, M. Direct Dynamics Quasiclassical Trajectory Study of the Stereochemistry of the Vinylcyclopropane-Cyclopentene Rearrangement. *J. Am. Chem. Soc.* **1999**, 121, 4720-4721.
22. Oxgaard, J., Wiest, O. The Vinylcyclopropane Radical Cation Rearrangement and Related Reactions on the C5H8.bul.+ Hypersurface. *J. Am. Chem. Soc.* **1999**, 121, 11531-11537.
23. Sperling, D., Fabian, J. Substituent effects on the vinylcyclopropane-cyclopentene rearrangement. A theoretical study by restricted and unrestricted density functional theory. *Eur. J. Org. Chem.* **1999**, 215-220.
24. Sperling, D., Reissig, H.-U., Fabian, J. [1,3]-sigmatropic rearrangements of divinylcyclopropane derivatives and hetero analogs in competition with Cope-type rearrangements. A DFT study. *Eur. J. Org. Chem.* **1999**, 1107-1114.
25. Tian, F., Bartberger, M. D., Dolbier, W. R., Jr. Density Functional Theory Calculations of the Effect of Fluorine Substitution on the Kinetics of Cyclopropylcarbinyl Radical Ring Openings. *J. Org. Chem.* **1999**, 64, 540-546.
26. Nendel, M., Sperling, D., Wiest, O., Houk, K. N. Computational Explorations of Vinylcyclopropane-Cyclopentene Rearrangements and Competing Diradical Stereoisomerizations. *J. Org. Chem.* **2000**, 65, 3259-3268.
27. Doubleday, C. Mechanism of the Vinylcyclopropane-Cyclopentene Rearrangement Studied by Quasiclassical Direct Dynamics. *J. Phys. Chem. A* **2001**, 105, 6333-6341.
28. Doubleday, C., Li, G., Hase, W. L. Dynamics of the biradical mediating vinylcyclopropane-cyclopentene rearrangement. *Physical Chemistry Chemical Physics* **2002**, 4, 304-312.
29. Vysotskii, Y. B., Bryantsev, V. S. Quantum-Chemical Treatment of Cyclization and Recyclization Reactions: XXV. Skeletal Rearrangements of Radicals, and Lowest Triplet-State Systems. *Russ. J. Org. Chem.* **2002**, 38, 1244-1251.
30. Oxgaard, J., Wiest, O. Substituent effects in the vinylcyclopropane radical cation rearrangement: A computational road to a new synthetic tool. *Eur. J. Org. Chem.* **2003**, 1454-1462.
31. Ketley, A. D., McClanahan, J. L. Rearrangement of 1-(p-substituted)-phenyl-1-cyclopropylethylenes. *J. Org. Chem.* **1965**, 30, 942-943.
32. Crawford, R. J., Cameron, D. M. Pyrolysis of 3-vinyl-1-pyrazoline and 3-vinyl-1-pyrazoline-5,5-d2, and its relation to the vinylcyclopropane to cyclopentene rearrangement. *Can. J. Chem.* **1967**, 45, 691-696.
33. Willcott, M. R., III, Cargle, V. H. Stereochemical fate of the cyclopropyl ring in the vinylcyclopropane rearrangement. *J. Am. Chem. Soc.* **1967**, 89, 723-724.
34. O'Neal, H. E., Benson, S. W. Biradical mechanism in small ring compound reactions. *J. Phys. Chem.* **1968**, 72, 1866-1887.
35. Willcott, M. R., III, Cargle, V. H. Structural evidence for the existence of a symmetrical high energy species in the degenerate vinylcyclopropane rearrangement. *J. Am. Chem. Soc.* **1969**, 91, 4310-4311.
36. Cooke, R. S. Photochemical vinylcyclopropane to cyclopentene rearrangement. *J. Chem. Soc., Chem. Commun.* **1970**, 454-455.
37. Mazzocchi, P. H., Tamburin, H. J. Mechanism of the vinylcyclopropane-cyclopentene rearrangement. Evidence against a concerted process. *J. Am. Chem. Soc.* **1970**, 92, 7220-7221.
38. Retzloff, D. G., Coull, B. M., Coull, J. Microchemical study of gas-phase kinetics for three irreversible reactions. *J. Phys. Chem.* **1970**, 74, 2455-2459.
39. Zimmerman, H. E., Pratt, A. C. Organic photochemistry. LIII. Directionality of the singlet di-p-methane. Rearrangement and alkyl migration in a unique photochemical vinylcyclopropane transformation. *J. Am. Chem. Soc.* **1970**, 92, 1407-1409.
40. Swenton, J. S., Wexler, A. Stereospecific vinylcyclopropane rearrangement due to hindered rotation in the biradical. *J. Am. Chem. Soc.* **1971**, 93, 3066-3068.
41. Kende, A. S., Riecke, E. E. Allylidenecyclopropane-methylenecyclopentene energy surface. Evidence for a conducted tour mechanism. *J. Am. Chem. Soc.* **1972**, 94, 1397-1399.
42. Pickenhagen, W., Naef, F., Ohloff, G., Mueller, P., Perlberger, J. C. Thermal and photochemical rearrangements of divinylcyclopropanes to cycloheptadienes. Model for the biosynthesis of the cycloheptadiene derivatives found in a seaweed (Dictyopteris). *Helv. Chim. Acta* **1973**, 56, 1868-1874.
43. Andrews, G. D., Baldwin, J. E. Thermal isomerization of (+)-(1S,2S)-trans,trans-2-methyl-1-propenylcyclopropane: quantification of four stereochemical paths in a vinylcyclopropane rearrangement. *J. Am. Chem. Soc.* **1976**, 98, 6705-6706.
44. Meyer, L. U., De Meijere, A. A modern flow reactor for kinetic measurements of gas phase thermal rearrangements: thermolysis kinetics of some new small ring compounds. *Chem. Ber.* **1977**, 110, 2545-2560.
45. Dolbier, W. R., Jr., Al-Sader, B. H., Sellers, S. F., Koroniak, H. Thermal rearrangement of 2,2-difluorovinylcyclopropane. A concerted pathway? *J. Am. Chem. Soc.* **1981**, 103, 2138-2139.
46. Trost, B. M., Scudder, P. H. Kinetics of the siloxyvinylcyclopropane rearrangement using a micro stirred flow reactor. *J. Org. Chem.* **1981**, 46, 506-509.
47. Klumpp, G. W., Schakel, M. Diradicals in the vinylcyclopropane rearrangement of bicyclo[3.1.0]hexane derivatives. *Tetrahedron Lett.* **1983**, 24, 4595-4598.
48. Dinnocenzo, J. P., Conlon, D. A. Accelerating symmetry forbidden reactions: the vinylcyclopropane ---> cyclopentene cation radical rearrangement. *J. Am. Chem. Soc.* **1988**, 110, 2324-2326.
49. Gajewski, J. J., Squicciarini, M. P. Evidence for concert in the vinylcyclopropane rearrangement. A reinvestigation of the pyrolysis of trans-1-methyl-2-(1-tert-butylethenyl)cyclopropane. *J. Am. Chem. Soc.* **1989**, 111, 6717-6728.
50. Zimmerman, H. E., Oaks, F. L., Campos, P. An assortment of highly unusual rearrangements in the photochemistry of vinylcyclopropanes. Mechanistic and exploratory organic photochemistry. *J. Am. Chem. Soc.* **1989**, 111, 1007-1018.
51. Baldwin, J. E., Ghatlia, N. D. Stereochemistry of the thermal isomerization of trans-1-ethenyl-2-methylcyclopropane to 4-methylcyclopentene. *J. Am. Chem. Soc.* **1991**, 113, 6273-6274.
52. Gajewski, J. J., Olson, L. P. Evidence for a dominant suprafacial-inversion pathway in the thermal unimolecular vinylcyclopropane to cyclopentene 1,3-sigmatropic shift. *J. Am. Chem. Soc.* **1991**, 113, 7432-7433.
53. Tanko, J. M., Drumright, R. E., Suleman, N. K., Brammer, L. E., Jr. Radical Ion Probes. 3. The Importance of Resonance vs. Strain Energy in the Design of SET Probes Based upon the Cyclopropylcarbinyl-to-Homoallyl Radical Rearrangement. *J. Am. Chem. Soc.* **1994**, 116, 1785-1791.
54. Dinnocenzo, J. P., Conlon, D. A. The cation radical vinylcyclopropane ----> cyclopentene rearrangement: reaction mechanism and periselectivity. *Tetrahedron Lett.* **1995**, 36, 7415-7418.
55. Dolbier, W. R., Jr., McClinton, M. A. A kinetic study of the thermal rearrangement of 2-(trifluoromethyl)-1-vinylcyclopropane. *J. Fluorine Chem.* **1995**, 70, 249-253.

56. Baldwin, J. E., Bonacorsi, S. J., Jr. Stereochemistry of the Thermal Isomerizations of trans-1-Ethenyl-2-phenylcyclopropane to 4-Phenylcyclopentene. *J. Am. Chem. Soc.* **1996**, 118, 8258-8265.
57. Buchert, M., Reissig, H. U. Rearrangement of donor-acceptor-substituted vinylcyclopropanes to functionalized cyclopentene derivatives. Evidence for zwitterionic intermediates. *Liebigs Ann. Chem.* **1996**, 2007-2013.
58. Kohmoto, S., Nakayama, N., Takami, J.-i., Kishikawa, K., Yamamoto, M., Yamada, K. On the mechanism of the rearrangement of 7-vinylnorcaradienes. *Tetrahedron Lett.* **1996**, 37, 7761-7764.
59. Wang, B., Lake, C. H., Lammertsma, K. Epimerization of Cyclic Vinylphosphirane Complexes: The Intermediacy of Biradicals. *J. Am. Chem. Soc.* **1996**, 118, 1690-1695.
60. Belevskii, V. N., Shchapin, I. Y. Rearrangement and ion-molecular reactions of C5H8+.-related radical cations as studied by EPR spectroscopy in the solid and liquid phase. *Acta Chem. Scand.* **1997**, 51, 1085-1091.
61. Baldwin, J. E., Shukla, R. Thermal Stereomutations and Vinylcyclopropane-to- Cyclopentene Rearrangement of 2-Methylene-3-spiro-cyclopropanebicyclo[2.2.1]heptane. *J. Am. Chem. Soc.* **1999**, 121, 11018-11019.
62. Sonawane, H. R., Nanjundiah, B. S., Shah, V. G., Kulkarni, D. G., Ahuja, J. R. Synthesis of naturally-occurring (-)-Δ9(12)-capnellene and its antipode: an application of the photoinduced vinylcyclopropane-cyclopentene rearrangement. *Tetrahedron Lett.* **1991**, 32, 1107-1108.
63. Corey, E. J., Kigoshi, H. A route for the enantioselective total synthesis of antheridic acid, the antheridium-inducing factor from Anemia phyllitidis. *Tetrahedron Lett.* **1991**, 32, 5025-5028.
64. Hudlicky, T., Radesca-Kwart, L., Li, L. Q., Bryant, T. Short, enantioselective synthesis of (-)-retigeranic acid via [2 + 3] annulation. *Tetrahedron Lett.* **1988**, 29, 3283-3286.
65. Hudlicky, T., Natchus, M. Chemoenzymic enantiocontrolled synthesis of (-)-specionin. *J. Org. Chem.* **1992**, 57, 4740-4746.

von Pechman Reaction 472

1. v. Pechmann, H., Duisberg, C. Compounds derived from phenols and acetoacetic ester. *Ber. Dtsch. Chem. Ges.* **1883**, 16, 2119-2128.
2. v. Pechmann, H. Novel synthesis of coumarins. *Ber. Dtsch. Chem. Ges.* **1884**, 17, 929-936.
3. Sethna, S., Phadke, R. The Pechmann reaction. *Org. React.* **1953**, VII, 1-58.
4. Holden, M. S., Crouch, R. D. The Pechmann reaction. *J. Chem. Educ.* **1998**, 75, 1631.
5. Petschek, E., Simonis, H. New Chromone Synthesis. *Ber.* **1913**, 46, 2014-2020.
6. Simonis, H., Lehmann, C. B. A. Alkylated chromones and their cleavage products. *Ber.* **1914**, 47, 692-699.
7. Simonis, H., Remmert, P. New flavone synthesis. *Ber.* **1914**, 47, 2229-2233.
8. Dixit, V. M., Kanakudati, A. M. Aluminum chloride as a condensing agent for the Pechmann reaction. *J. Indian Chem. Soc.* **1951**, 28, 323-327.
9. Kapil, R. S., Joshi, S. S. Polyphosphoric acid as a condensing agent in the Pechmann reaction. *J. Indian Chem. Soc.* **1959**, 36, 596.
10. Trivedi, K. N. Silicon tetrachloride, a new condensing agent for the Pechmann reaction. *Curr. Sci.* **1959**, 28, 67.
11. Trivedi, K. N. Use of cation-exchange resins as catalyst in the Pechmann reaction. *J. Indian Chem. Soc.* **1965**, 42, 273-274.
12. Kappe, T., Ziegler, E. Modification of the Pechmann reaction. *Organic Preparations and Procedures* **1969**, 1, 61-62.
13. Ruwet, A., Janne, D., Renson, M. Formation of chromones (oxygen, sulfur, selenium) and other derivatives by the action of diketene on phenol, thiophenol, and selenophenol. Pechmann and Simonis reactions. *Bull. Soc. Chim. Belg.* **1970**, 79, 81-88.
14. Kappe, T., Mayer, C. A new modification of the Pechmann reaction. *Synthesis* **1981**, 524-526.
15. Li, T.-S., Zhang, Z.-H., Yang, F., Fu, C.-G. Montmorillonite clay catalysis. Part 7. An environmentally friendly procedure for the synthesis of coumarins via Pechmann condensation of phenols with ethyl acetoacetate. *J. Chem. Res., Synop.* **1998**, 38-39.
16. Frere, S., Thiery, V., Besson, T. Microwave acceleration of the Pechmann reaction on graphite/montmorillonite K10: application to the preparation of 4-substituted 7-aminocoumarins. *Tetrahedron Lett.* **2001**, 42, 2791-2794.
17. Potdar, M. K., Mohile, S. S., Salunkhe, M. M. Coumarin syntheses via Pechmann condensation in Lewis acidic chloroaluminate ionic liquid. *Tetrahedron Lett.* **2001**, 42, 9285-9287.
18. Reddy, B. M., Reddy, V. R., Giridhar, D. Synthesis of coumarins catalyzed by eco-friendly W/ZrO_2 solid acid catalyst. *Synth. Commun.* **2001**, 31, 3603-3607.
19. Khandekar, A. C., Khadilkar, B. M. Pechmann reaction in chloroaluminate ionic liquid. *Synlett* **2002**, 152-154.
20. Helavi, V. B., Solabannavar, S. B., Salunkhe, R. S., Mane, R. B. Microwave-assisted solventless Pechmann condensation. *J. Chem. Res., Synop.* **2003**, 279-280.
21. Laufer, M. C., Hausmann, H., Holderich, W. F. Synthesis of 7-hydroxycoumarins by Pechmann reaction using Nafion resin/silica nanocomposites as catalysts. *J. Catal.* **2003**, 218, 315-320.
22. Shockravi, A., Heravi, M. M., Valizadeh, H. An Efficient and Convenient Synthesis of Furocoumarins via Pechmann Reaction on $ZnCl_2/Al_2O_3$ Under Microwave Irradiation. *Phosphorus, Sulfur Silicon Relat. Elem.* **2003**, 178, 143-147.
23. Wang, L., Xia, J., Tian, H., Qian, C., Ma, Y. Synthesis of coumarin by $Yb(OTf)_3$ catalyzed Pechmann reaction under the solvent-free conditions. *Indian J. Chem., Sect. B* **2003**, 42B, 2097-2099.
24. Gangadasu, B., Narender, P., Raju, B. C., Rao, V. J. $ZrCl_4$ catalysed solvent free synthesis of coumarins. *J. Chem. Res.* **2004**, 480-481.
25. Singh, P. R., Singh, D. U., Samant, S. D. Sulphamic acid - an efficient and cost-effective solid acid catalyst for the pechmann reaction. *Synlett* **2004**, 1909-1912.
26. Smitha, G., Sanjeeva Reddy, C. $ZrCl_4$-Catalyzed Pechmann Reaction: Synthesis of Coumarins Under Solvent-Free Conditions. *Synth. Commun.* **2004**, 34, 3997-4003.
27. Subhas Bose, D., Rudradas, A. P., Hari Babu, M. The indium(III) chloride-catalyzed von Pechmann reaction: a simple and effective procedure for the synthesis of 4-substituted coumarins. *Tetrahedron Lett.* **2002**, 43, 9195-9197.
28. Barris, B. E., Israelstam, S. S. Use of small quantities of sulfuric acid and polyphosphoric acid in the von Pechmann reaction. *J. S.African Chem. Inst.* **1960**, 13, 125-128.
29. John, E. V. O., Israelstam, S. S. Use of cation exchange resins in organic reactions. I. The von Pechmann reaction. *J. Org. Chem.* **1961**, 26, 240-242.
30. Dalla Via, L., Uriarte, E., Quezada, E., Dolmella, A., Ferlin Maria, G., Gia, O. Novel pyrone side tetracyclic psoralen derivatives: synthesis and photobiological evaluation. *J. Med. Chem.* **2003**, 46, 3800-3810.
31. Hamann, L. G. An Efficient, Stereospecific Synthesis of the Dimer-Selective Retinoid X Receptor Modulator (2E,4E,6Z)-7-[5,6,7,8-Tetrahydro-5,5,8,8-tetramethyl-2- (n-propyloxy)naphthalen-3-yl]-3- methylocta-2,4,6-trienoic Acid. *J. Org. Chem.* **2000**, 65, 3233-3235.

Wacker Oxidation 474

1. Phillips, F. C. Researches upon the phenomena of oxidation and chemical properties of gases. *Am. Chem. J.* **1894**, 16, 255-277.
2. Smidt, J., Hafner, W., Jira, R., Sedlmeier, J., Sieber, R., Ruttinger, R., Kojer, H. Catalytic reactions of olefins on compounds of the platinum group. *Angew. Chem.* **1959**, 71, 176-182.
3. Smidt, J., Sieber, R. Reactions of palladium dichloride with olefinic double bonds. *Angew. Chem.* **1959**, 71, 626.
4. Clement, W. H., Selwitz, C. M. Improved procedures for converting higher α-olefins into methyl ketones with palladium chloride. *J. Org. Chem.* **1964**, 29, 241-243.
5. Lloyd, W. G., Luberoff, B. J. Oxidations of olefins with alcoholic palladium(II) salts. *J. Org. Chem.* **1969**, 34, 3949-3952.
6. Henry, P. M. *Catalysis by Metal Complexes, Vol. 2: Palladium Catalyzed Oxidation of Hydrocarbons* (ed. Reidel, D.) (Dordrecht, Holland, **1980**) 41.
7. Tsuji, J. Synthetic applications of the palladium-catalyzed oxidation of olefins to ketones. *Synthesis* **1984**, 369-384.
8. Lyons, J. E. Selective oxidation of hydrocarbons via carbon-hydrogen bond activation by soluble and supported palladium catalysts. *Catal. Today* **1988**, 3, 245-258.

9. Sherrington, D. C. Polymer-supported metal complex oxidation catalysts. *Pure Appl. Chem.* **1988**, 60, 401-414.
10. Hegedus, L. S. Heteroatom Nucleophiles with Metal-Activated Alkenes and Alkynes. in *Comp. Org. Synth.* (eds. Trost, B. M.,Fleming, I.), *4*, 551-570 (Pergamon, Oxford, **1991**).
11. Tsuji, J. Addition Reactions with Formation of Carbon-Oxygen Bonds: The Wacker Oxidation and Related Reactions. in *Comprehensive Organic Synthesis* (eds. Trost, B. M.,Fleming, I.), *7*, 449-468 (Pergamon, Oxford, **1991**).
12. Warwel, S., Sojka, M., Ruesch Klaas, M. Synthesis of dicarboxylic acids by transition-metal catalyzed oxidative cleavage of terminal-unsaturated fatty acids. *Top. Curr. Chem.* **1993**, 164, 49-98.
13. Dedieu, A. Wacker reactions. *Catal. Metal Compl.* **1995**, 18, 167-195.
14. Jira, R. Oxidation of olefins to carbonyl compounds (Wacker process). *Applied Homogeneous Catalysis with Organometallic Compounds* **1996**, 1, 374-393.
15. Feringa, B. L. Wacker oxidation. *Transition Metals for Organic Synthesis* **1998**, 2, 307-315.
16. Monflier, E., Mortreux, A. Wacker-type oxidations. *Aqueous-Phase Organometallic Catalysis* **1998**, 513-518.
17. Balme, G., Bouyssi, D., Monteiro, N. Palladium-catalyzed reactions involving attack on palladium-alkene, palladium-alkyne, and related π-complexes by carbon nucleophiles. *Handbook of Organopalladium Chemistry for Organic Synthesis* **2002**, 2, 2245-2265.
18. Braese, S., Kobberling, J., Griebenow, N. Organopalladium reactions in combinatorial chemistry. *Handbook of Organopalladium Chemistry for Organic Synthesis* **2002**, 2, 3031-3126.
19. Henry, P. M. Palladium-catalyzed reactions involving nucleophilic attack on π-ligands of palladium-alkene, palladium-alkyne, and related derivatives: the Wacker oxidation and related intermolecular reactions involving oxygen and other group 16 atom nucleophiles: the Wacker oxidation and related asymmetric syntheses. *Handbook of Organopalladium Chemistry for Organic Synthesis* **2002**, 2, 2119-2139.
20. Hosokawa, T., Murahashi, S.-I. Other intermolecular oxypalladation-dehydropalladation reactions. *Handbook of Organopalladium Chemistry for Organic Synthesis* **2002**, 2, 2141-2159.
21. Jira, R. Oxidations: oxidation of olefins to carbonyl compounds (Wacker process). *Applied Homogeneous Catalysis with Organometallic Compounds (2nd Edition)* **2002**, 1, 386-405.
22. Xu, C., Negishi, E.-I. Synthesis of natural products via nucleophilic attack on v-ligands of palladium-alkene, palladium-alkyne, and related v-complexes. *Handbook of Organopalladium Chemistry for Organic Synthesis* **2002**, 2, 2289-2305.
23. Takacs, J. M., Jiang, X.-t. The Wacker reaction and related alkene oxidation reactions. *Curr. Org. Chem.* **2003**, 7, 369-396.
24. Yokota, T., Sakaguchi, S., Ishii, Y. Oxidation of unsaturated hydrocarbons by $Pd(OAc)_2$ using a molybdovanadophosphate/dioxygen system. *Journal of the Japan Petroleum Institute* **2003**, 46, 15-27.
25. Roussel, M., Mimoun, H. Palladium-catalyzed oxidation of terminal olefins to methyl ketones by hydrogen peroxide. *J. Org. Chem.* **1980**, 45, 5387-5390.
26. Sage, J. M., Gore, J., Guilmet, E. Oxidation of olefins by oxygen with a mixed palladium/silver nitrite catalyst in alcohols. *Tetrahedron Lett.* **1989**, 30, 6319-6322.
27. Grate, J. H., Hamm, D. R., Mahajan, S. New technology for olefin oxidation to carbonyls using a palladium and polyoxoanion catalyst system. *Chem. Ind.* **1994**, 53, 213-264.
28. Centi, G., Stella, G. Synthesis of 2-butanone by selective oxidation on solid Wacker-type catalysts. *Chem. Ind.* **1995**, 62, 319-329.
29. Kim, Y., Kim, H., Lee, J., Sim, K., Han, Y., Paik, H. A modified Wacker catalysis using hetero polyacid: Interaction of heteropoly anion with Cu(II) in cyclohexene oxidation. *Appl. Cat. A* **1997**, 155, 15-26.
30. Nowinska, K., Dudko, D. $Mn_{(5-n)}/2HnPMOV_2$ as re-oxidant of palladium(II) in solid Wacker catalysts. *React. Kinet. Catal. Lett.* **1997**, 61, 187-192.
31. Betzemeier, B., Lhermitte, F., Knochel, P. Wacker oxidation of alkenes using a fluorous biphasic system. A mild preparation of polyfunctional ketones. *Tetrahedron Lett.* **1998**, 39, 6667-6670.
32. Monflier, E. Versatile inverse phase transfer catalysts for the functionalization of substrates in aqueous-organic two-phase systems: the chemically modified β-cyclodextrins. *Rec. Res. Dev. Org. Chem.* **1998**, 2, 623-635.
33. Reilly, C. R., Lerou, J. J. Supported liquid phase catalysis in selective oxidation. *Catal. Today* **1998**, 41, 433-441.
34. Smith, A. B., III, Cho, Y. S., Friestad, G. K. Convenient Wacker oxidations with substoichiometric cupric acetate. *Tetrahedron Lett.* **1998**, 39, 8765-8768.
35. Jiang, H., Jia, L., Li, J. Wacker reaction in supercritical carbon dioxide. *Green Chem.* **2000**, 2, 161-164.
36. Gaunt, M. J., Spencer, J. B. Derailing the Wacker Oxidation: Development of a Palladium-Catalyzed Amidation Reaction. *Org. Lett.* **2001**, 3, 25-28.
37. El-Qisairi, A. K., Qaseer, H. A., Henry, P. M. Oxidation of olefins by palladium(II). 18. Effect of reaction conditions, substrate structure and chiral ligand on the bimetallic palladium(II) catalyzed asymmetric chlorohydrin synthesis. *J. Organomet. Chem.* **2002**, 656, 168-176.
38. Lambert, A., Derouane, E. G., Kozhevnikov, I. V. Kinetics of One-Stage Wacker-Type Oxidation of C2-C4 Olefins Catalysed by an Aqueous $PdCl_2$-Heteropoly-Anion System. *J. Catal.* **2002**, 211, 445-450.
39. Namboodiri, V. V., Varma, R. S., Sahle-Demessie, E., Pillai, U. R. Selective oxidation of styrene to acetophenone in the presence of ionic liquids. *Green Chem.* **2002**, 4, 170-173.
40. Athawale, A. A., Bhagwat, S. V. Synthesis and characterization of novel copper/polyaniline nanocomposite and application as a catalyst in the Wacker oxidation reaction. *J. Appl. Polym. Sci.* **2003**, 89, 2412-2417.
41. Choi, K.-M., Mizugaki, T., Ebitani, K., Kaneda, K. Nanoscale palladium cluster immobilized on a TiO_2 surface as an efficient catalyst for liquid-phase Wacker oxidation of higher terminal olefins. *Chem. Lett.* **2003**, 32, 180-181.
42. Ho, T.-L., Chang, M. H., Chen, C. Abnormal and regioselective Wacker oxidation of 1,5-dienes. *Tetrahedron Lett.* **2003**, 44, 6955-6957.
43. Karakhanov, E. A., Zhuchkova, A. Y., Filippova, T. Y., Maksimov, A. L. Supramolecular cyclodextrin-based catalyst systems in Wacker oxidation. *Neftekhimiya* **2003**, 43, 302-307.
44. Trend, R. M., Ramtohul, Y. K., Ferreira, E. M., Stoltz, B. M. Palladium-catalyzed oxidative Wacker cyclizations in nonpolar organic solvents with molecular oxygen: A stepping stone to asymmetric aerobic cyclizations. *Angew. Chem., Int. Ed. Engl.* **2003**, 42, 2892-2895.
45. Chen, M. S., White, M. C. A sulfoxide-promoted, catalytic method for the regioselective synthesis of allylic acetates from monosubstituted olefins via C-H oxidation. *J. Am. Chem. Soc.* **2004**, 126, 1346-1347.
46. Armstrong, D. R., Fortune, R., Perkins, P. G. A molecular orbital investigation of the Wacker process for the oxidation of ethylene to acetaldehyde. *J. Catal.* **1976**, 45, 339-348.
47. Shinoda, S., Saito, Y. Quantum-chemical characterization of metal ions for the Wacker-type reactions. *J. Mol. Catal.* **1977**, 2, 369-377.
48. Ding, Y., Fu, X. Theoretical study on the configuration of ethanol dichloro palladium $[PdCl_2(C_2H_4OH)]$. *Chin. Chem. Lett.* **1990**, 1, 183-184.
49. Ding, Y., Fu, X. Theoretical study on trans-influence of palladium(II)-ethene complexes. *Chin. Sci. Bull.* **1990**, 35, 789-790.
50. Siegbahn, P. E. M. A theoretical study of some steps in the Wacker process. *Structural Chemistry* **1995**, 6, 271-279.
51. Siegbahn, P. E. M. Two, Three, and Four Water Chain Models for the Nucleophilic Addition Step in the Wacker Process. *J. Phys. Chem.* **1996**, 100, 14672-14680.
52. Dedieu, A. Theoretical treatment of organometallic reaction mechanisms and catalysis. *Top. Organomet. Chem.* **1999**, 4, 69-107.
53. Kragten, D. D., van Santen, R. A., Lerou, J. J. Density Functional Study of the Palladium Acetate Catalyzed Wacker Reaction in Acetic Acid. *J. Phys. Chem. A* **1999**, 103, 80-88.
54. Dedieu, A. Theoretical Studies in Palladium and Platinum Molecular Chemistry. *Chem. Rev.* **2000**, 100, 543-600.
55. Niu, S., Hall, M. B. Theoretical Studies on Reactions of Transition-Metal Complexes. *Chem. Rev.* **2000**, 100, 353-405.
56. Nelson, D. J., Li, R., Brammer, C. Correlation of Relative Rates of $PdCl_2$ Oxidation of Functionalized Acyclic Alkenes versus Alkene Ionization Potentials, HOMOs, and LUMOs. *J. Am. Chem. Soc.* **2001**, 123, 1564-1568.
57. Tsuji, J., Shimizu, I., Yamamoto, K. Convenient general synthetic method for 1,4- and 1,5-diketones by palladium catalyzed oxidation of a-allyl and a-3-butenyl ketones. *Tetrahedron Lett.* **1976**, 2975-2976.
58. Henry, P. M. Oxidation of olefins by palladium(II). II. Effect of structure on rate in aqueous solution. *J. Am. Chem. Soc.* **1966**, 88, 1595-1597.
59. Okada, H., Noma, T., Katsuyama, Y., Hashimoto, H. Reactions of transition-metal-olefin complexes. III. Kinetics of the oxidation of substituted styrenes catalyzed by palladium salts in aqueous tetrahydrofuran. *Bull. Chem. Soc. Jpn.* **1968**, 41, 1395-1400.

60. Schwartz, A., Holbrook, L. L., Wise, H. Catalytic oxidation studies with platinum and palladium. *J. Catal.* **1971**, 21, 199-207.
61. Kosaki, M., Isemura, M., Kitaura, Y., Shinoda, S., Saito, Y. Comparative study on catalytic oxidation of ethylene by palladium(II) in aqueous and acetic acid solutions. *J. Mol. Catal.* **1977**, 2, 351-359.
62. Baeckvall, J. E., Akermark, B., Ljunggren, S. O. Stereochemistry and mechanism for the palladium(II)-catalyzed oxidation of ethene in water (the Wacker process). *J. Am. Chem. Soc.* **1979**, 101, 2411-2416.
63. Stille, J. K., Divakaruni, R. Mechanism of the Wacker process. Stereochemistry of the hydroxypalladation. *J. Organomet. Chem.* **1979**, 169, 239-248.
64. Wan, W. K., Zaw, K., Henry, P. M. Evidence for the rate determining step in the Wacker reaction. *J. Mol. Catal.* **1982**, 16, 81-87.
65. Shinoda, S., Koie, Y., Saito, Y. Mechanism of the Wacker reaction from the viewpoint of the trans influence of ligands (OH->Cl->H2O). *Bull. Chem. Soc. Jpn.* **1986**, 59, 2938-2940.
66. Akermark, B., Soederberg, B. C., Hall, S. S. The mechanism of the Wacker process. Corroborative evidence for distal addition of water and palladium. *Organometallics* **1987**, 6, 2608-2610.
67. Van der Heide, E., Ammerlaan, J. A. M., Gerritsen, A. W., Scholten, J. J. F. Kinetics and mechanism of the gas-phase oxidation of 1-butene to butanone over a new heterogenized surface vanadate Wacker catalyst. *J. Mol. Catal.* **1989**, 55, 320-329.
68. Zaw, K., Henry, P. M. Oxidation of olefins by palladium(II). 12. Product distributions and kinetics of the oxidation of 3-buten-2-ol and 2-buten-1-ol by tetrachloropalladate ($PdCl_4$-) in aqueous solution. *J. Org. Chem.* **1990**, 55, 1842-1847.
69. Espeel, P. H., De Peuter, G., Tielen, M. C., Jacobs, P. A. Mechanism of the Wacker Oxidation of Alkenes over Cu-Pd-Exchanged Y Zeolites. *J. Phys. Chem.* **1994**, 98, 11588-11596.
70. Hronec, M., Cvengrosova, Z., Holotik, S. Is metallic palladium formed in Wacker oxidation of alkenes? *J. Mol. Catal.* **1994**, 91, 343-352.
71. Francis, J. W., Henry, P. M. Oxidation of olefins by palladium(II). Part XIV. Product distribution and kinetics of the oxidation of ethene by $PdCl_3$(pyridine)- in aqueous solution in the presence and absence of $CuCl_2$: a modified Wacker catalyst with altered reactivity. *J. Mol. Catal. A: Chemical* **1995**, 99, 77-86.
72. Monflier, E., Tilloy, S., Blouet, E., Barbaux, Y., Mortreux, A. Wacker oxidation of various olefins in the presence of per(2,6-di-O-methyl)-β-cyclodextrin: mechanistic investigations of a multistep catalysis in a solvent-free two-phase system. *J. Mol. Catal. A: Chemical* **1996**, 109, 27-35.
73. Noronha, G., Henry, P. M. Heterogenized polymetallic catalysts: Part III. Catalytic air oxidation of alcohols by Pd(II) complexed to a polyphenylene polymer containing β-di- and tri-ketone surface ligands. *J. Mol. Catal. A: Chemical* **1997**, 120, 75-87.
74. Pellissier, H., Michellys, P.-Y., Santelli, M. Regiochemistry of Wacker-type oxidation of vinyl group in the presence of neighboring oxygen functions. Part 2. *Tetrahedron* **1997**, 53, 10733-10742.
75. Pellissier, H., Michellys, P.-Y., Santelli, M. Regiochemistry of Wacker-type oxidation of vinyl group in the presence of neighboring oxygen functions. Part 1. *Tetrahedron* **1997**, 53, 7577-7586.
76. Bernardelli, P., Moradei, O. M., Friedrich, D., Yang, J., Gallou, F., Dyck, B. P., Doskotch, R. W., Lange, T., Paquette, L. A. Total Asymmetric Synthesis of the Putative Structure of the Cytotoxic Diterpenoid (-)-Sclerophytin A and of the Authentic Natural Sclerophytins A and B. *J. Am. Chem. Soc.* **2001**, 123, 9021-9032.
77. Yokokawa, F., Asano, T., Shioiri, T. Total synthesis of (-)-hennoxazole A. *Tetrahedron* **2001**, 57, 6311-6327.
78. O'Connor, P. D., Mander, L. N., McLachlan, M. M. W. Synthesis of the Himandrine Skeleton. *Org. Lett.* **2004**, 6, 703-706.
79. Usuda, H., Kanai, M., Shibasaki, M. Studies toward the Total Synthesis of Garsubellin A: A Concise Synthesis of the 18-epi-Tricyclic Core. *Org. Lett.* **2002**, 4, 859-862.

Wagner-Meerwein Rearrangement 476

1. Wagner, G., Brickner, W. The conversion of pinene halohydrates to the haloanhydrides of borneol. *Ber.* **1899**, 32, 2302-2325.
2. Meerwein, H. Pinacolin rearrangement. III. Mechanism of the transformation of borneol into camphene. *Ann.* **1914**, 405, 129-175.
3. Meerwein, H., Van Emster, K., Joussen, J. The equilibrium isomerism between bornyl chloride, isobornyl chloride and camphene hydrochloride. *Ber.* **1922**, 55B, 2500-2528.
4. Streitwieser, A., Jr. Solvolytic displacement reactions at saturated carbon atoms. *Chem. Rev.* **1956**, 56, 571-752.
5. Berson, J. A. Carbonium ion rearrangements in bridged bicyclic systems. in *Molecular Rearrangements* (ed. De Mayo, P.), *1*, 111-231 (Wiley, New York, **1963**).
6. Pocker, Y. Wagner-Meerwein and pinacolic rearrangements in acyclic and cyclic systems. in *Molecular Rearrangements* (ed. De Mayo, P.), *1*, 1-25 (Wiley, New York, **1963**).
7. Poeker, Y., de Mayo, P. Wagner-Meerwein and pinacolic rearrangements in acyclic and cyclic systems. *Molecular Rearrangements* **1963**, 1, 1-25.
8. Prilezhaeva, E. N. Thiylation of multiple bonds. Wagner-Meerwein type of rearrangement of sulfur- and chlorine-containing norbornenyl radicals. *Organosulfur Chem.* **1967**, 57-74.
9. Finley, K. T. Development of the carbonium ion hypothesis. *Org. Chem. Bull.* **1969**, 41, 5 pp.
10. Schreiber, P. 70 years of the Wagner rearrangement. Critical review on the history of chemistry. *Chemiker-Zeitung, Chemische Apparatur* **1969**, 93, 957-964.
11. Cargill, R. L., Jackson, T. E., Peet, N. P., Pond, D. M. Acid-catalyzed rearrangements of β,γ-unsaturated ketones. *Acc. Chem. Res.* **1974**, 7, 106-113.
12. Olah, G. A. Stable carbocations, 189. The s-bridged 2-norbornyl cation and its significance to chemistry. *Acc. Chem. Res.* **1976**, 9, 41-52.
13. Sorensen, T. S. Terpene rearrangements from a superacid perspective. *Acc. Chem. Res.* **1976**, 9, 257-265.
14. Hogeveen, H., Van Kruchten, E. M. G. A. Wagner-Meerwein rearrangements in long-lived polymethyl substituted bicyclo[3.2.0]heptadienyl cations. *Top. Curr. Chem.* **1979**, 80, 89-124.
15. Creary, X. Electronegatively substituted carbocations. *Chem. Rev.* **1991**, 91, 1625-1678.
16. Hanson, J. R. Wagner-Meerwein rearrangements. in *Comp. Org. Synth.* (eds. Trost, B. M.,Fleming, I.), *3*, 705-719 (Pergamon, Oxford, **1991**).
17. Saunders, M., Jimenez-Vazquez, H. A. Recent studies of carbocations. *Chem. Rev.* **1991**, 91, 375-397.
18. Greci, L., Carloni, P., Stipa, P. Acid catalyzed rearrangements on indole systems. Wagner-Meerwein type transpositions. *Topics in Heterocyclic Systems: Synthesis, Reactions and Properties* **1996**, 1, 53-61.
19. Plieninger, H., Kraemer, H. P. Enantioselective Wagner-Meerwein rearrangement in chiral solvents under high pressure. *Angew. Chem.* **1976**, 88, 230-231.
20. Kraemer, H. P., Plieninger, H. High pressure experiments. X. High pressure enantioselective Wagner-Meerwein rearrangement in chiral solvents. *Tetrahedron* **1978**, 34, 891-896.
21. Berner, D., Cox, D. P., Dahn, H. 1,2-Shift of a carboxyl group in a Wagner-Meerwein rearrangement. *Helv. Chim. Acta* **1982**, 65, 2061-2070.
22. Cristol, S. J., Opitz, R. J. Photochemical transformations. 40. syn and anti Migration in photo-Wagner-Meerwein rearrangements. *J. Org. Chem.* **1985**, 50, 4558-4563.
23. Cristol, S. J., Ali, M. B., Sankar, I. V. Photochemical transformations. 48. The nonconcertedness of nucleofuge loss and anti-aryl migration in photochemical Wagner-Meerwein rearrangements. *J. Am. Chem. Soc.* **1989**, 111, 8207-8211.
24. Taskesenligil, Y., Balci, M. An unusual zinc-promoted reductive retro-Wagner-Meerwein rearrangement. *Turk. J. Chem.* **1996**, 20, 335-340.
25. Trost, B. M., Yasukata, T. A catalytic asymmetric Wagner-Meerwein shift. *J. Am. Chem. Soc.* **2001**, 123, 7162-7163.
26. Carrupt, P. A., Vogel, P. Ab initio MO calculations on the rearrangements of 7-oxa-2-bicyclo[2.2.1]heptyl cations. The facile migration of acyl groups in Wagner-Meerwein rearrangements. *J. Phys. Org. Chem.* **1988**, 1, 287-298.
27. Shaler, T. A., Morton, T. H. Fluorine Shifts in Gaseous Cations. Analogs of Wagner-Meerwein Rearrangements. *J. Am. Chem. Soc.* **1994**, 116, 9222-9226.
28. Smith, W. B. A DFT Study of the Camphene Hydrochloride Rearrangement. *J. Org. Chem.* **1999**, 64, 60-64.

29. Smith, W. B. Nature of the 2-Bicyclo[3.2.1]octanyl and 2-Bicyclo[3.2.2]nonanyl Cations. *J. Org. Chem.* **2001**, 66, 376-380.
30. Pachuau, Z., Lyngdoh, R. H. D. Molecular orbital studies on the Wagner-Meerwein migration in some acyclic pinacol-pinacolone rearrangements. *J. Chem. Sci. (Bangalore, India)* **2004**, 116, 83-91.
31. Weininger, S. J. "What's in a name?" from designation to denunciation - the nonclassical cation controversy. *Bulletin for the History of Chemistry* **2000**, 25, 123-131.
32. Shono, T., Fujita, K., Kumai, S. Stereochemistry of migrating carbon in Wagner-Meerwein rearrangement. *Tetrahedron Lett.* **1973**, 3123-3126.
33. Kinugawa, M., Nagamura, S., Sakaguchi, A., Masuda, Y., Saito, H., Ogasa, T., Kasai, M. Practical Synthesis of the High-Quality Antitumor Agent KW-2189 from Duocarmycin B2 Using a Facile One-Pot Synthesis of an Intermediate. *Org. Process Res. Dev.* **1998**, 2, 344-350.
34. Garcia Martinez, A., Teso Vilar, E., Garcia Fraile, A., de la Moya Cerero, S., Lora Maroto, B. A novel enantiospecific route to 10-hydroxyfenchone: a convenient intermediate for C(10)-O-substituted fenchones. *Tetrahedron: Asymmetry* **2002**, 12, 3325-3327.
35. Smith, A. B., III, Konopelski, J. P. Total synthesis of (+)-quadrone: assignment of absolute stereochemistry. *J. Org. Chem.* **1984**, 49, 4094-4095.

Weinreb Ketone Synthesis 478

1. Nahm, S., Weinreb, S. M. N-Methoxy-N-methylamides as effective acylating agents. *Tetrahedron Lett.* **1981**, 22, 3815-3818.
2. Sibi, M. P. Chemistry of N-methoxy-N-methylamides. Applications in synthesis. A review. *Org. Prep. Proced. Int.* **1993**, 25, 15-40.
3. Mentzel, M., Hoffmann, H. M. R. N-Methoxy N-methyl amides (Weinreb amides) in modern organic synthesis. *J. Prakt. Chem.* **1997**, 339, 517-524.
4. Singh, J., Satyamurthi, N., Aidhen, I. S. The growing synthetic utility of Weinreb's amide. *J. Prakt. Chem.* **2000**, 342, 340-347.
5. Khlestkin, V. K., Mazhukin, D. G. Recent advances in the application of N,O-dialkylhydroxylamines in organic chemistry. *Curr. Org. Chem.* **2003**, 7, 967-993.
6. Levin, J. I., Turos, E., Weinreb, S. M. An alternative procedure for the aluminum-mediated conversion of esters to amides. *Synth. Commun.* **1982**, 12, 989-993.
7. Smith, L. A., Wang, W. B., Barnell-Curty, C., Roskamp, E. J. Conversion of esters to amides with amino halo stannylenes. *Synlett* **1993**, 850-852.
8. Williams, J. M., Jobson, R. B., Yasuda, N., Marchesini, G., Dolling, U.-H., Grabowski, E. J. J. A new general method for preparation of N-methoxy-N-methylamides. Application in direct conversion of an ester to a ketone. *Tetrahedron Lett.* **1995**, 36, 5461-5464.
9. Shimizu, T., Osako, K., Nakata, T. Efficient method for preparation of N-methoxy-N-methyl amides by reaction of lactones or esters with Me_2AlCl-MeONHMe·HCl. *Tetrahedron Lett.* **1997**, 38, 2685-2688.
10. Wallace, O. B. Solid phase synthesis of ketones from esters. *Tetrahedron Lett.* **1997**, 38, 4939-4942.
11. Lipshutz, B. H., Pfeiffer, S. S., Chrisman, W. Formylations of anions with a "Weinreb" formamide: N-methoxy-N-methylformamide. *Tetrahedron Lett.* **1999**, 40, 7889-7892.
12. Raghuram, T., Vijaysaradhi, S., Singh, I., Singh, J. Convenient conversion of acids to Weinreb's amides. *Synth. Commun.* **1999**, 29, 3215-3219.
13. Tunoori, A. R., White, J. M., Georg, G. I. A One-Flask Synthesis of Weinreb Amides from Chiral and Achiral Carboxylic Acids Using the Deoxo-Fluor Fluorinating Reagent. *Org. Lett.* **2000**, 2, 4091-4093.
14. Banwell, M., Smith, J. A mild, one-pot method for the conversion of carboxylic acids into the corresponding Weinreb amides. *Synth. Commun.* **2001**, 31, 2011-2019.
15. de Luca, L., Giacomelli, G., Taddei, M. An easy and convenient synthesis of Weinreb amides and hydroxamates. *J. Org. Chem.* **2001**, 66, 2534-2537.
16. Guo, Z., Dowdy, E. D., Li, W. S., Polniaszek, R., Delaney, E. A novel method for the mild and selective amidation of diesters and the amidation of monoesters. *Tetrahedron Lett.* **2001**, 42, 1843-1845.
17. Huang, P.-Q., Zheng, X., Deng, X.-M. DIBAL-H-H_2NR and DIBAL-H-HNR^1R^2·HCl complexes for efficient conversion of lactones and esters to amides. *Tetrahedron Lett.* **2001**, 42, 9039-9041.
18. Lee, J. I., Park, H. A convenient synthesis of N-methoxy-N-methylamides from carboxylic acids using S,S-di-2-pyridyl dithiocarbonate. *Bull. Korean Chem. Soc.* **2001**, 22, 421-423.
19. Katritzky, A. R., Yang, H., Zhang, S., Wang, M. An efficient conversion of carboxylic acids into Weinreb amides. *ARKIVOC (Gainesville, FL, United States) [online computer file]* **2002**, 39-44.
20. Sibi, M. P., Hasegawa, H., Ghorpade, S. R. A Convenient Method for the Conversion of N-Acyloxazolidinones to Hydroxamic Acids. *Org. Lett.* **2002**, 4, 3343-3346.
21. Kummer, D. A., Brenneman, J. B., Martin, S. F. An efficient synthesis of α-branched enones. *Synlett* **2004**, 1431-1433.
22. Labeeuw, O., Phansavath, P., Genet, J.-P. Synthesis of modified Weinreb amides: N-tert-butoxy-N-methylamides as effective acylating agents. *Tetrahedron Lett.* **2004**, 45, 7107-7110.
23. Woo, J. C. S., Fenster, E., Dake, G. R. A Convenient Method for the Conversion of Hindered Carboxylic Acids to N-Methoxy-N-methyl (Weinreb) Amides. *J. Org. Chem.* **2004**, 69, 8984-8986.
24. Wipf, P., Rector, S. R., Takahashi, H. Total Synthesis of (-)-Tuberostemonine. *J. Am. Chem. Soc.* **2002**, 124, 14848-14849.
25. Marshall, J. A., Yanik, M. M. Synthesis of a C1-C21 Subunit of the Protein Phosphatase Inhibitor Tautomycin: A Formal Total Synthesis. *J. Org. Chem.* **2001**, 66, 1373-1379.
26. Stoltz, B. M., Kano, T., Corey, E. J. Enantioselective Total Synthesis of Nicandrenones. *J. Am. Chem. Soc.* **2000**, 122, 9044-9045.
27. Wender, P. A., Fuji, M., Husfeld, C. O., Love, J. A. Rhodium-Catalyzed [5+2] Cycloadditions of Allenes and Vinylcyclopropanes: Asymmetric Total Synthesis of (+)-Dictamnol. *Org. Lett.* **1999**, 1, 137-139.

Wharton Fragmentation 480

Related reactions: Eschenmoser-Tanabe fragmentation, Grob fragmentation;

1. Eschenmoser, A., Frey, A. Cleavage of methanesulfonyl esters of 2-methyl-2-(hydroxymethyl)cyclopentanone with bases. *Helv. Chim. Acta* **1952**, 35, 1660-1666.
2. Wharton, P. S. Stereospecific synthesis of 6-methyl-trans-5-cyclodecenone. *J. Org. Chem.* **1961**, 26, 4781-4782.
3. Wharton, P. S., Hiegel, G. A., Coombs, R. V. trans-5-Cyclodecenone. *J. Org. Chem.* **1963**, 28, 3217-3219.
4. Wharton, P. S., Hiegel, G. A. Fragmentation of 1,10-decalindiol monotosylates. *J. Org. Chem.* **1965**, 30, 3254-3257.
5. Wharton, P. S., Baird, M. D. Conformation and reactivity in the cis,trans-2,6-cycloodecadienyl system. *J. Org. Chem.* **1971**, 36, 2932-2937.
6. Grob, C. A., Schiess, P. W. Heterolytic fragmentation. A class of organic reactions. *Angew. Chem., Int. Ed. Engl.* **1967**, 6, 1-15.
7. Faulkner, D. J. Stereoselective synthesis of trisubstituted olefins. *Synthesis* **1971**, 175-189.
8. Reucroft, J., Sammes, P. G. Stereoselective and stereospecific olefin synthesis. *Quart. Rev., Chem. Soc.* **1971**, 25, 135-169.
9. Stirling, C. J. M. Nucleophilic eliminative ring fission. *Chem. Rev.* **1978**, 78, 517-567.
10. Caine, D. Wharton fragmentations of cyclic 1,3-diol derivatives. A review. *Org. Prep. Proced. Int.* **1988**, 20, 1-51.
11. Wijnberg, J. B. P. A., De Groot, A. Induced ionization in 1,4-diol monosulfonate esters and its application in the synthesis of natural products. *Curr. Org. Chem.* **2003**, 7, 257-274.
12. Trahanovsky, W. S., Himstedt, A. L. Oxidation of organic compounds with Cerium(IV). XX. Abnormally rapid rate of oxidative cleavage of (β-trimethylsilylethyl)phenylmethanol. *J. Am. Chem. Soc.* **1974**, 96, 7974-7976.

13. Tietze, L. F., Kinast, G., Uzar, H. C. Fragmentation of γ-hydroxyammonium compounds to unsaturated aldehydes by short-time thermolysis. *Angew. Chem.* **1979**, 91, 576.
14. Gibbons, E. G. Total synthesis of (±)-pleuromutilin. *J. Am. Chem. Soc.* **1982**, 104, 1767-1769.
15. Yadav, J. S., Patil, D. G., Krishna, R. R., Chawla, H. P. S., Dev, S. Heterolytic cleavage of homoallylic alcohols. I. Fragmentation of 6-hydroxycamphene derivatives. *Tetrahedron* **1982**, 38, 1003-1007.
16. Nakatani, K., Isoe, S. Oxidative fragmentation of γ-hydroxyalkylstannanes. Stereospecific formation of (E)- and (Z)-keto olefins. *Tetrahedron Lett.* **1984**, 25, 5335-5338.
17. Moelm, D., Floerke, U., Risch, N. Fragmentation reactions of quaternized γ-amino alcohols. Diastereoselective synthesis of highly functionalized oxetanes and unsaturated aldehydes and ketones with a (Z)-C:C double bond. *Eur. J. Org. Chem.* **1998**, 2185-2191.
18. Grob, C. A., Baumann, W. 1,4-Elimination reaction with simultaneous fragmentation. *Helv. Chim. Acta* **1955**, 38, 594-610.
19. Zurflueh, R., Wall, E. N., Siddall, J. B., Edwards, J. A. Synthetic studies on insect hormones. VII. An approach to stereospecific synthesis of juvenile hormones. *J. Am. Chem. Soc.* **1968**, 90, 6224-6225.
20. Liu, G., Smith, T. C., Pfander, H. Synthesis of optically active trans-cyclononenes. A possible approach to xenicanes. *Tetrahedron Lett.* **1995**, 36, 4979-4982.
21. Njardarson, J. T., Wood, J. L. Evolution of a Synthetic Approach to CP-263,114. *Org. Lett.* **2001**, 3, 2431-2434.
22. Arseniyadis, S., Ferreira, M. D. R. R., Quilez del Moral, J., Hernando, J. I. M., Potier, P., Toupet, L. Studies towards the total synthesis of taxoids: a rapid entry into bicyclo[6.4.0]dodecane ring system. Part 1. *Tetrahedron: Asymmetry* **1998**, 9, 4055-4071.
23. Collado, I., Ezquerra, J., Mateo, A. I., Pedregal, C., Rubio, A. Stereocontrolled Synthesis of 5α- and 5β–Substituted Kainic Acids. *J. Org. Chem.* **1999**, 64, 4304-4314.

Wharton Olefin Synthesis (Wharton Transposition) 482

Related reactions: Bamford-Stevens-Shapiro olefination;

1. Kishner, N. On the composition of acylhydrazones. *J. Russ. Phys. Chem. Soc.* **1913**, 45, 973-993.
2. Ames, D. E., Bowman, R. E. Synthetic long-chain aliphatic compounds. VI. Some anomalous reductions of 9-keto-10-methoxyoctadecanoic acid. *J. Chem. Soc., Abstracts* **1951**, 2752-2753.
3. Huang, M., Chung, T. S. Conversion of 16α,17α-epoxypregnenolone into 3β,16α-dihydroxy-5,17(20)-pregnadiene by the modified Wolff-Kishner reduction. *Tetrahedron Lett.* **1961**, 666-668.
4. Wharton, P. S., Bohlen, D. H. Hydrazine reduction of α,β-epoxy ketones to allylic alcohols. *J. Org. Chem.* **1961**, 26, 3615-3616.
5. Chamberlin, A. R., Sall, D., J. Reduction of Ketones to Alkenes. in *Comp. Org. Synth.* (eds. Trost, B. M.,Fleming, I.), *8*, 923-953 (Pergamon, Oxford, **1991**).
6. Stork, G., Williard, P. G. Five- and six-membered-ring formation from olefinic α,β-epoxy ketones and hydrazine. *J. Am. Chem. Soc.* **1977**, 99, 7067-7068.
7. Barton, D. H. R., Motherwell, R. S. H., Motherwell, W. B. Radical-induced ring opening of epoxides. A convenient alternative to the Wharton rearrangement. *J. Chem. Soc., Perkin Trans. 1* **1981**, 2363-2367.
8. Dupuy, C., Luche, J. L. New developments in the Wharton transposition. *Tetrahedron* **1989**, 45, 3437-3444.
9. Leonard, N. J., Gelfand, S. The Kishner reduction-elimination. II. α-Substituted pinacolones. *J. Am. Chem. Soc.* **1955**, 77, 3272-3278.
10. Leonard, N. J., Gelfand, S. The Kishner reduction-elimination. I. Cyclic and open chain α-amino ketones. *J. Am. Chem. Soc.* **1955**, 77, 3269-3271.
11. Tsuji, T., Kosower, E. M. Alkyldiazenes. *J. Am. Chem. Soc.* **1970**, 92, 1429-1430.
12. Kosower, E. M. Monosubstituted diazenes (diimides). Surprising intermediates. *Acc. Chem. Res.* **1971**, 4, 193-198.
13. Yu, W., Jin, Z. Total Synthesis of the Anticancer Natural Product OSW-1. *J. Am. Chem. Soc.* **2002**, 124, 6576-6583.
14. Moreno-Dorado, F. J., Guerra, F. M., Aladro, F. J., Bustamante, J. M., Jorge, Z. D., Massanet, G. M. An easy route to 11-hydroxy-eudesmanolides. Synthesis of (±)-decipienin A. *Tetrahedron* **1999**, 55, 6997-7010.
15. Barrero, A. F., Cortes, M., Manzaneda, E. A., Cabrera, E., Chahboun, R., Lara, M., Rivas, A. R. Synthesis of 11,12-Epoxydrim-8,12-en-11-ol, 11,12-Diacetoxydrimane, and Warburganal from (-)-Sclareol. *J. Nat. Prod.* **1999**, 62, 1488-1491.
16. Majewski, M., Lazny, R. Stereoselective synthesis of tropane alkaloids. Physoperuvine and dihydroxytropanes. *Synlett* **1996**, 785-786.

Williamson Ether Synthesis 484

1. Williamson, W. About the theory of the ether bond. *Liebigs Ann. Chem.* **1851**, 77, 37-49.
2. Williamson, W. *J. Chem. Soc.* **1852**, 106, 229.
3. Dermer, O. C. Metallic salts of alcohols and alcohol analogs. *Chem. Rev.* **1934**, 14, 385-430.
4. Feuer, H., Hooz, J. The Chemistry of the Ether Linkage: Methods of formation of the ether linkage. in *The Chemistry of Functional Groups* (ed. Patai, S.), 445-498 (Wiley, New York, **1967**).
5. Patai, S., Editor. in *The Chemistry of the Hydroxyl Group* (ed. Patai, S.), *1*, 454 (Wiley, New York, **1971**).
6. Koert, U. Stereoselective synthesis of oligo-tetrahydrofurans. *Synthesis* **1995**, 115-132.
7. Hill, M., Dronsfield, A. Williamson's 1852 pioneering synthesis. *Educ. Chem.* **2002**, 39, 47-49.
8. Masada, H., Sakajiri, T. Synthesis of hindered tert-alkyl ethers. *Bull. Chem. Soc. Jpn.* **1978**, 51, 866-868.
9. Benedict, D. R., Bianchi, T. A., Cate, L. A. Synthesis of simple unsymmetrical ethers from alcohols and alkyl halides or sulfates: the potassium hydroxide/dimethyl sulfoxide system. *Synthesis* **1979**, 428-429.
10. Pasquini, M. A., Le Goaller, R., Pierre, J. L. Effects of cryptands and activation of bases. V. Action of alkali hydrides on weak acids. II. Alkylation of anions obtained. *Tetrahedron* **1980**, 36, 1223-1226.
11. Lee, J. C., Yuk, J. Y., Cho, S. H. Facile synthesis of alkyl phenyl ethers using cesium carbonate. *Synth. Commun.* **1995**, 25, 1367-1370.
12. Basak, A., Nayak, M. K., Chakraborti, A. K. Chemoselective O-methylation of phenols under nonaqueous condition. *Tetrahedron Lett.* **1998**, 39, 4883-4886.
13. Bogdal, D., Pielichowski, J., Jaskot, K. A rapid Williamson synthesis under microwave irradiation in dry medium. *Org. Prep. Proced. Int.* **1998**, 30, 427-432.
14. Parrish, J. P., Sudaresan, B., Jung, K. W. Improved Cs_2CO_3 promoted O-alkylation of phenols. *Synth. Commun.* **1999**, 29, 4423-4431.
15. Rao, H. S. P., Senthilkumar, S. P. A convenient procedure for the synthesis of allyl and benzyl ethers from alcohols and phenols. *Proc. - Indian Acad. Sci., Chem. Sci.* **2001**, 113, 191-196.
16. Peng, Y., Song, G. Combined microwave and ultrasound assisted Williamson ether synthesis in the absence of phase-transfer catalysts. *Green Chem.* **2002**, 4, 349-351.
17. Paul, S., Gupta, M. Zinc-catalyzed Williamson ether synthesis in the absence of base. *Tetrahedron Lett.* **2004**, 45, 8825-8829.
18. Sarju, J., Danks, T. N., Wagner, G. Rapid microwave-assisted synthesis of phenyl ethers under mildly basic and nonaqueous conditions. *Tetrahedron Lett.* **2004**, 45, 7675-7677.
19. Yadav, G. D., Bisht, P. M. Novelties of microwave assisted liquid-liquid phase transfer catalysis in enhancement of rates and selectivities in alkylation of phenols under mild conditions. *Catal. Commun.* **2004**, 5, 259-263.
20. Wright, A. R. Application of modeling and computer simulation to pharmaceutical processes: the Williamson synthesis. *Chem. Eng. Res. Design* **1984**, 62, 391-397.
21. Norula, J. L. Mechanism of Williamson synthesis. *Chemical Era* **1975**, 11, 20-22.
22. Ashby, E. C., Bae, D. H., Park, W. S., Depriest, R. N., Yang Su, W. Evidence for single electron transfer in the reaction of alkoxides with alkyl halides. *Tetrahedron Lett.* **1984**, 25, 5107-5110.

23. Takeuchi, H., Miwa, Y., Morita, S., Okada, J. Kinetic studies on an improved Williamson ether synthesis using a polymer-supported phase-transfer catalyst. *Chem. Pharm. Bull.* **1985**, 33, 3101-3106.
24. Tan, S. N., Dryfe, R. A., Girault, H. H. Electrochemical study of phase-transfer catalysis reactions: the Williamson ether synthesis. *Helv. Chim. Acta* **1994**, 77, 231-242.
25. Beifuss, U., Tietze, M., Baumer, S., Deppenmeier, U. Methanophenazine: structure, total synthesis, and function of a new cofactor from methanogenic Archaea. *Angew. Chem., Int. Ed. Engl.* **2000**, 39, 2470-2472.
26. Avedissian, H., Sinha, S. C., Yazbak, A., Sinha, A., Neogi, P., Sinha, S. C., Keinan, E. Total Synthesis of Asimicin and Bullatacin. *J. Org. Chem.* **2000**, 65, 6035-6051.
27. Kim, D., Ahn, S. K., Bae, H., Choi, W. J., Kim, H. S. An asymmetric total synthesis of (-)-fumagillol. *Tetrahedron Lett.* **1997**, 38, 4437-4440.
28. Eguchi, T., Arakawa, K., Terachi, T., Kakinuma, K. Total Synthesis of Archaeal 36-Membered Macrocyclic Diether Lipid. *J. Org. Chem.* **1997**, 62, 1924-1933.

Related reactions: Horner-Wadsworth-Emmons olefination, Horner-Wadsworth-Emmons olefination - Still-Gennari modification, Julia-Lithgoe olefination, Peterson olefination, Takai-Utimoto olefination, Tebbe olefination, Wittig reaction – Schlosser modification;

1. Staudinger, H., Meyer, J. New organic compounds of phosphorus. III. Phosphinemethylene derivatives and phosphinimines. *Helv. Chim. Acta* **1919**, 2, 635-646.
2. Coffman, D. D., Marvel, C. S. Reaction between alkali metal alkyls and quaternary phosophonium halides. *J. Am. Chem. Soc.* **1929**, 51, 3496-3501.
3. Wittig, G., Geissler, G. Course of reactions of pentaphenylphosphorus and certain derivatives. *Ann.* **1953**, 580, 44-57.
4. Wittig, G., Schollkopf, U. Triphenylphosphinemethylene as an olefin-forming reagent. I. *Chem. Ber.* **1954**, 97, 1318-1330.
5. Wittig, G., Haag, W. Triphenylphosphinemethylenes as olefin-forming reagents. II. *Chem. Ber.* **1955**, 88, 1654-1666.
6. Schollkopf, U. Carbonyl-olefin transformation with $(C_6H_5)_3P=CH_2$.-Wittig reaction. *Angew. Chem.* **1959**, 71, 260-273.
7. Wittig, G. Staudinger and the history of organophosphorus-carbonyl olefination. *Pure Appl. Chem.* **1964**, 9, 245-254.
8. Maercker, A. The Wittig reaction. *Org. React.* **1965**, 14, 270-490.
9. Reeves, R. L. Condensations of carbonyl groups leading to double bonds. *Chemistry of the Carbonyl Group* **1966**, 567-619.
10. Hopps, H. B., Biel, J. H. Wittig reaction. *Aldrichimica Acta* **1969**, 2, 3-6.
11. Hurd, C. D. Wittig reaction. *Quarterly Reports on Sulfur Chemistry* **1969**, 4, 159-227.
12. Schlosser, M. Stereochemistry of the Wittig reaction. *Top. Stereochem.* **1970**, 5, 1-30.
13. Hudson, R. F. Ylid chemistry. *Chem. Br.* **1971**, 7, 287-294.
14. Zhdanov, Y. A., Alekseev, Y. E., Alekseeva, V. G. Wittig reaction in carbohydrate chemistry. *Advances in Carbohydrate Chemistry* **1972**, 27, 227-299.
15. Boutagy, J., Thomas, R. Olefin synthesis with organic phosphonate carbanions. *Chem. Rev.* **1974**, 74, 87-99.
16. Vollhardt, K. P. C. Bis-Wittig reactions in the synthesis of nonbenzenoid aromatic ring systems. *Synthesis* **1975**, 765-780.
17. Bajaj, K. L. New advance in the use of the Wittig reaction for synthesis (of natural materials). *Riechstoffe, Aromen, Koerperpflegemittel* **1976**, 26, 224, 226-228.
18. Wadsworth, W. S., Jr. Synthetic applications of phosphoryl-stabilized anions. *Org. React.* **1977**, 25, 73-253.
19. Gosney, I., Rowley, A. G. Transformations via phosphorus-stabilized anions. 1: Stereoselective syntheses of alkenes via the Wittig reaction. *Organophosphorus Reagents Org. Synth.* **1979**, 17-153.
20. Becker, K. B. Cycloalkenes by intramolecular Wittig reaction. *Tetrahedron* **1980**, 36, 1717-1745.
21. Bestmann, H. J., Hellwinkel, D., Krebs, A., Pommer, H., Schoellkopf, U., Thieme, P. C., Vostrowsky, O., Wilke, J. Wittig Chemistry. in *Top. Curr. Chem. 109*, 236 pp (**1983**).
22. Bestmann, H. J., Vostrowsky, O. Selected topics of the Wittig reaction in the synthesis of natural products. *Top. Curr. Chem.* **1983**, 109, 85-163.
23. Schlosser, M., Oi, R., Schaub, B. The Wittig reaction: 30 years later. *Phosphorus and Sulfur and the Related Elements* **1983**, 18, 171-174.
24. Julia, M. Recent advances in double bond formation. *Pure Appl. Chem.* **1985**, 57, 763-768.
25. Cristau, H. J., Ribeill, Y., Plenat, F., Chiche, L. Use of phosphonium diylides in organic synthesis. *Phosphorus and Sulfur and the Related Elements* **1987**, 30, 135-138.
26. Murphy, P. J., Brennan, J. The Wittig olefination reaction with carbonyl compounds other than aldehydes and ketones. *Chem. Soc. Rev.* **1988**, 17, 1-30.
27. Seyden-Penne, J. Lithium coordination by Wittig-Horner reagents formed by β carbonyl substituted phosphonates and phosphine oxide: a review. *Bull. Soc. Chim. Fr.* **1988**, 238-242.
28. Maryanoff, B. E., Reitz, A. B. The Wittig olefination reaction and modifications involving phosphoryl-stabilized carbanions. Stereochemistry, mechanism, and selected synthetic aspects. *Chem. Rev.* **1989**, 89, 863-927.
29. Kelly, S. E. Alkene Synthesis. in *Comp. Org. Synth.* (eds. Trost, B. M.,Fleming, I.), *1*, 729-818 (Pergamon, Oxford, **1991**).
30. Walker, B. J. Ylides and related compounds. *Organophosphorus Chem.* **1991**, 22, 252-297.
31. Gosney, I., Lloyd, D. One or more C=C bond(s) formed by condensation: Condensation of P, As, Sb, Bi, Si or metal functions. in *Comp. Org. Funct. Group Trans. 1*, 719-770 (Pergamon, Cambridge, UK, **1995**).
32. Heron, B. M. Heterocycles from intramolecular Wittig, Horner and Wadsworth-Emmons reactions. *Heterocycles* **1995**, 41, 2357-2386.
33. Clayden, J., Warren, S. Stereocontrol in organic synthesis using the diphenylphosphoryl group. *Angew. Chem., Int. Ed. Engl.* **1996**, 35, 241-270.
34. Kawashima, T., Okazaki, R. Synthesis and reactions of the intermediates of the Wittig, Peterson, and their related reactions. *Synlett* **1996**, 600-608.
35. Lawrence, N. J. The Wittig reaction and related methods. *Preparation of Alkenes* **1996**, 19-58.
36. Vedejs, E., Peterson, M. J. The Wittig reaction: stereoselectivity and a history of mechanistic ideas (1953-1995). in *Advances in Carbanion Chemistry* (ed. Snieckus, V.), *2*, 1-85 (JAI Press Inc., Greenwich, London, **1996**).
37. Walker, B. J. Ylides and related compounds. *Organophosphorus Chem.* **1996**, 27, 264-307.
38. Nicolaou, K. C., Harter, M. W., Gunzner, J. L., Nadin, A. The Wittig and related reactions in natural product synthesis. *Liebigs Ann. Chem.* **1997**, 1283-1301.
39. Murphy, P. J., Lee, S. E. Recent synthetic applications of the non-classical Wittig reaction. *J. Chem. Soc., Perkin Trans. 1* **1999**, 3049-3066.
40. Edmonds, M., Abell, A. The Wittig reaction. *Modern Carbonyl Olefination* **2004**, 1-17.
41. Schlosser, M., Christmann, K. F. Olefination with phosphorus ylides. I. Mechanism and stereochemistry of the Wittig reaction. *Liebigs Ann. Chem.* **1967**, 708, 1-35.
42. Ford, W. T. Wittig reactions on polymer supports. *ACS Symp. Ser.* **1986**, 308, 155-185.
43. Moreno-Manas, M., Ortuno, R. M., Prat, M., Galan, M. A. The one-pot palladium catalyzed Wittig reaction with allylic alcohols. Scope and limitations. *Synth. Commun.* **1986**, 16, 1003-1013.
44. Maier, L., Kunz, W. Preparation of triazolylmethylphosphonates and of triazolylmethylphosphonium salts and their application in the Wittig-Horner reaction. *Phosphorus and Sulfur and the Related Elements* **1987**, 30, 201-204.
45. Ganem, B. The first example of a catalytic Wittig-type reaction. Tributylarsine-catalyzed olefination in the presence of triphenyl phosphite. *Chemtracts: Org. Chem.* **1989**, 2, 300-301.
46. Mathey, F., Marinetti, A., Bauer, S., Le Floch, P. Chemistry of phosphorus-carbon double bonds in the coordination sphere of transition metals. *Pure Appl. Chem.* **1991**, 63, 855-858.
47. Toda, F. Solid-to-solid organic reactions. *React. Mol. Cryst.* **1993**, 177-201.

48. Rein, T., Reiser, O. Recent advances in asymmetric Wittig-type reactions. *Acta Chem. Scand.* **1996**, 50, 369-379.
49. Erker, G., Hock, R., Wilker, S., Laurent, C., Puke, C., Wurthwein, E.-U., Aust, N. C., Frohlich, R. New aspects of the thio-Wittig reaction. *Phosphorus, Sulfur Silicon Relat. Elem.* **1999**, 153-154, 79-97.
50. Lorsbach, B. A., Kurth, M. J. Carbon-Carbon Bond Forming Solid-Phase Reactions. *Chemical Reviews (Washington, D. C.)* **1999**, 99, 1549-1581.
51. Shah, S., Protasiewicz, J. D. "Phospha-variations" on the themes of Staudinger and Wittig: phosphorus analogs of Wittig reagents. *Coord. Chem. Rev.* **2000**, 210, 181-201.
52. Habermann, J., Ley, S. V. The Bestmann ylide as a multi-purpose wittig reagent. *Chemtracts* **2001**, 14, 386-390.
53. Rein, T., Pedersen, T. M. Asymmetric Wittig type reactions. *Synthesis* **2002**, 579-594.
54. Valentine, D. H., Jr., Hillhouse, J. H. Alkyl phosphines as reagents and catalysts in organic synthesis. *Synthesis* **2003**, 317-334.
55. Bestmann, H. J. Old and new ylide chemistry. *Pure Appl. Chem.* **1980**, 52, 771-788.
56. Hoeller, R., Lischka, H. A theoretical investigation on the model Wittig reaction $PH_3CH_2 + CH_2O \dashrightarrow PH_3O + C_2H_4$. *J. Am. Chem. Soc.* **1980**, 102, 4632-4635.
57. Volatron, F., Eisenstein, O. Theoretical study of the reactivity of phosphonium and sulfonium ylides with carbonyl groups. *Journal of the American Chemical Society* **1984**, 106, 6117-6119.
58. Volatron, F., Eisenstein, O. Wittig versus Corey-Chaykovsky Reaction. Theoretical study of the reactivity of phosphonium methylide and sulfonium methylide with formaldehyde. *Journal of the American Chemical Society* **1987**, 109, 1-4.
59. Mari, F., Lahti, P. M., McEwen, W. E. Molecular modeling of oxaphosphetane intermediates of Wittig olefination reactions. *Heteroatom Chem.* **1990**, 1, 255-259.
60. Mari, F., Lahti, P. M., McEwen, W. E. Molecular modeling of the Wittig olefination reaction: Part 2: A molecular orbital approach at the MNDO-PM3 level. *Heteroatom Chem.* **1991**, 2, 265-276.
61. Mari, F., Lahti, P. M., McEwen, W. E. Molecular modeling of the Wittig reaction. 3. A theoretical study of the Wittig olefination reaction: MNDO-PM3 treatment of the Wittig half-reaction of unstabilized ylides with aldehydes. *J. Am. Chem. Soc.* **1992**, 114, 813-821.
62. Naito, T., Nagase, S., Yamataka, H. Theoretical Study of the Structure and Reactivity of Ylides of N, P, As, Sb, and Bi. *J. Am. Chem. Soc.* **1994**, 116, 10080-10088.
63. Restrepo, A. A., Gonzalez, C. A., Mari, F. Theoretical study of the Wittig olefination reaction: Ab initio treatment of the mythical wittig half-reaction. *Book of Abstracts, 212th ACS National Meeting, Orlando, FL, August 25-29* **1996**, ORGN-402.
64. Armstrong, D. R., Barr, D., Davidson, M. G., Hutton, G., O'Brien, P., Snaith, R., Warren, S. Experimental and molecular orbital calculational study of the stereoselective Horner-Wittig reaction with phosphine oxides: control of stereoselectivity by lithium. *J. Organomet. Chem.* **1997**, 529, 29-33.
65. Restrepo-Cossio, A. A., Cano, H., Mari, F., Gonzalez, C. A. Molecular modeling of the Wittig reaction. 6. Theoretical study of the mechanism of the Wittig reaction: ab initio and MNDO-PM3 treatment of the reaction of unstabilized, semistabilized and stabilized ylides with acetaldehyde. *Heteroatom Chem.* **1997**, 8, 557-569.
66. Restrepo-Cossio, A. A., Gonzalez, C. A., Mari, F. Comparative ab Initio Treatment (Hartree-Fock, Density Functional Theory, MP2, and Quadratic Configuration Interactions) of the Cycloaddition of Phosphorus Ylides with Formaldehyde in the Gas Phase. *J. Phys. Chem. A* **1998**, 102, 6993-7000.
67. Yamataka, H., Nagase, S. Theoretical Calculations on the Wittig Reaction Revisited. *J. Am. Chem. Soc.* **1998**, 120, 7530-7536.
68. Yamataka, H. Theoretical calculations of organic reactions in solution. *Reviews on Heteroatom Chemistry* **1999**, 21, 277-291.
69. Lu, W. C., Wong, N. B., Zhang, R. Q. Theoretical study on the substituent effect of a Wittig reaction. *Theoretical Chemistry Accounts* **2002**, 107, 206-210.
70. Grabarnick, M., Zamir, S. Thorough Examination of a Wittig-Horner Reaction Using Reaction Calorimetry (RC-1), LabMax, and ReactIR. *Org. Process Res. Dev.* **2003**, 7, 237-243.
71. Horner, L., Hoffman, H., Wippel, H. G., Klahre, G. Phosphorus organic compounds. XX. Phosphine oxides as reagents for olefin formation. *Chem. Ber.* **1959**, 92, 2499-2505.
72. Wadsworth, W. S., Jr., Emmons, W. D. The utility of phosphonate carbanions in olefin synthesis. *J. Am. Chem. Soc.* **1961**, 83, 1733-1738.
73. Schlosser, M., Christmann, K. F. Trans-selective olefin synthesis. *Angew. Chem., Int. Ed. Engl.* **1966**, 5, 126.
74. McEwen, W. E., Beaver, B. D., Cooney, J. V. Mechanisms of Wittig reactions; a new possibility for salt-free reactions. *Phosphorus and Sulfur and the Related Elements* **1985**, 25, 255-271.
75. Maryanoff, B. E., Reitz, A. B. Delving into the Wittig reaction stereochemistry and mechanism. Stereochemical idiosyncrasies and mechanistic implications. *Phosphorus and Sulfur and the Related Elements* **1986**, 27, 167-189.
76. Vedejs, E., Marth, C. F. Mechanism of the Wittig reaction: the role of substituents at phosphorus. *J. Am. Chem. Soc.* **1988**, 110, 3948-3958.
77. Vedejs, E., Marth, C. F., Ruggeri, R. Substituent effects and the Wittig mechanism: the case of stereospecific oxaphosphetane decomposition. *J. Am. Chem. Soc.* **1988**, 110, 3940-3948.
78. McKenna, E. G., Walker, B. J. The mechanism and stereochemistry of Wittig reactions of phosphonium ylide-anions. *Phosphorus, Sulfur Silicon Relat. Elem.* **1990**, 49-50, 445-448.
79. McEwen, W. E., Mari, F., Lahti, P. M., Baughman, L. L., Ward, W. J., Jr. Mechanisms of the Wittig reaction. *ACS Symp. Ser.* **1992**, 486, 149-161.
80. Vedejs, E., Marth, C. F. 31P NMR detection and analysis of Wittig intermediates. *Phosphorus-31 NMR Spectral Prop. Compd. Charact. Struct. Anal.* **1994**, 297-313.
81. Vedejs, E., Peterson, M. J. Stereochemistry and mechanism in the Wittig reaction. *Top. Stereochem.* **1994**, 21, 1-157.
82. Vedejs, E., Peterson, M. J. The Wittig reaction: stereoselectivity and a history of mechanistic ideas (1953-1995). *Advances in Carbanion Chemistry* **1996**, 2, 1-85.
83. Smith, A. B., III, Beauchamp, T. J., LaMarche, M. J., Kaufman, M. D., Qiu, Y., Arimoto, H., Jones, D. R., Kobayashi, K. Evolution of a Gram-Scale Synthesis of (+)-Discodermolide. *J. Am. Chem. Soc.* **2000**, 122, 8654-8664.
84. Hoarau, C., Couture, A., Deniau, E., Grandclaudon, P. Total Synthesis of Amaryllidaceae Alkaloid Buflavine. *J. Org. Chem.* **2002**, 67, 5846-5849.
85. Dondoni, A., Marra, A., Mizuno, M., Giovannini, P. P. Linear Total Synthetic Routes to β-D-C-(1,6)-Linked Oligoglucoses and Oligogalactoses up to Pentaoses by Iterative Wittig Olefination Assembly. *J. Org. Chem.* **2002**, 67, 4186-4199.

Wittig Reaction - Schlosser Modification 488

Related reactions: ***Horner-Wadsworth-Emmons olefination, Horner-Wadsworth-Emmons olefination - Still-Gennari modification, Julia-Lithgoe olefination, Peterson olefination, Takai-Utimoto olefination, Tebbe olefination, Wittig reaction;***

1. Schlosser, M., Christmann, K. F. Trans-selective olefin synthesis. *Angew. Chem., Int. Ed. Engl.* **1966**, 5, 126.
2. Schlosser, M., Christmann, K. F. Olefination with phosphorus ylides. I. Mechanism and stereochemistry of the Wittig reaction. *Liebigs Ann. Chem.* **1967**, 708, 1-35.
3. Schlosser, M. Stereochemistry of the Wittig reaction. *Top. Stereochem.* **1970**, 5, 1-30.
4. Maryanoff, B. E., Reitz, A. B. Delving into the Wittig reaction stereochemistry and mechanism. Stereochemical idiosyncrasies and mechanistic implications. *Phosphorus and Sulfur and the Related Elements* **1986**, 27, 167-189.
5. Maryanoff, B. E., Reitz, A. B. The Wittig olefination reaction and modifications involving phosphoryl-stabilized carbanions. Stereochemistry, mechanism, and selected synthetic aspects. *Chem. Rev.* **1989**, 89, 863-927.
6. Kelly, S. E. Alkene Synthesis. in *Comp. Org. Synth.* (eds. Trost, B. M.,Fleming, I.), *1*, 729-818 (Pergamon, Oxford, **1991**).
7. Schlosser, M., Christmann, K. F. Carbonyl olefination with α-substitution. *Synthesis* **1969**, 1, 38-39.

8. Schlosser, M., Christmann, K. F., Piskala, A., Coffinet, D. α-Substitution plus carbonyl olefination via β-oxido phosphorus ylids (S.C.O.O.P.Y.-reactions) scope and stereoselectivity. *Synthesis* **1971**, 29-31.
9. Schlosser, M., Coffinet, D. α-Substitution plus carbonyl olefination via β-oxido phosphorus ylides (SCOOPY) reactions. Stereoselectivity of allyl alcohol synthesis via betaine ylides. *Synthesis* **1971**, 380-381.
10. Schlosser, M., Coffinet, D. SCOOPY [α-substitution plus carbonyl olefination via β-oxido phosphorus ylides] reactions. Regioselectivity of alkenol synthesis. *Synthesis* **1972**, 575-576.
11. Wittig, G., Geissler, G. Course of reactions of pentaphenylphosphorus and certain derivatives. *Ann.* **1953**, 580, 44-57.
12. Wittig, G., Schollkopf, U. Triphenylphosphinemethylene as an olefin-forming reagent. I. *Chem. Ber.* **1954**, 97, 1318-1330.
13. Wittig, G., Haag, W. Triphenylphosphinemethylenes as olefin-forming reagents. II. *Chem. Ber.* **1955**, 88, 1654-1666.
14. Wang, Q., Deredas, D., Huynh, C., Schlosser, M. Sequestered alkyllithiums: why phenyllithium alone is suitable for betaine-ylide generation. *Chem.-- Eur. J.* **2003**, 9, 570-574.
15. Schlosser, M., Christmann, K. F., Piskala, A. Olefination reactions with phosphorus ylides. II. β-Oxido phosphorus ylides in the presence and absence of soluble alkaline metal salts. *Chem. Ber.* **1970**, 103, 2814-2820.
16. Sano, S., Kobayashi, Y., Kondo, T., Takebayashi, M., Maruyama, S., Fujita, T., Nagao, Y. Asymmetric total synthesis of ISP-I (myriocin, thermozymocidin), a potent immunosuppressive principle in the Isaria sinclairii metabolite. *Tetrahedron Lett.* **1995**, 36, 2097-2100.
17. Duffield, J. J., Pettit, G. R. Synthesis of (7S,15S)- and (7R,15S)-Dolatrienoic Acid. *J. Nat. Prod.* **2001**, 64, 472-479.
18. Couladouros, E. A., Mihou, A. P. A general synthetic route towards γ- and δ-lactones Total asymmetric synthesis of (-)-muricatacin and the mosquito oviposition pheromone (5R,6S)-6-acetoxy-hexadecanolide. *Tetrahedron Lett.* **1999**, 40, 4861-4862.
19. Khiar, N., Singh, K., Garcia, M., Martin-Lomas, M. A short enantiodivergent synthesis of D-erythro and L-threo sphingosine. *Tetrahedron Lett.* **1999**, 40, 5779-5782.

Wittig-[1,2]- and [2,3]- Rearrangement 490

1. Wittig, G., Lohmann, L. Cationotropic isomerization of benzyl ethers by lithium phenyl. *Ann.* **1942**, 550, 260-268.
2. Wittig, G., Doser, H., Lorenz, I. Isomerization of metalated fluorenyl ether. *Liebigs Ann. Chem.* **1949**, 562, 192-205.
3. Wittig, G. Progress and problems in organic anion chemistry. *Experientia* **1958**, 14, 389-395.
4. Cast, J., Stevens, T. S., Holmes, J. Molecular rearrangement and fission of ethers by alkaline reagents. *J. Chem. Soc., Abstracts* **1960**, 3521-3527.
5. Zimmerman, H. E. Base-catalyzed rearrangements. in *Molecular Rearrangements* (ed. De Mayo, P.), *1*, 345-406 (Wiley, New York, **1963**).
6. Jefferson, A., Scheinmann, F. Molecular rearrangements related to the Claisen rearrangement. *Quart. Rev., Chem. Soc.* **1968**, 22, 390-420.
7. Schoellkopf, U. Recent results in carbanion chemistry. *Angew. Chem., Int. Ed. Engl.* **1970**, 9, 763-773.
8. Tennant, G. Molecular rearrangements [in organic chemistry]. *Annu. Rep. Prog. Chem., Sect. B, Org. Chem.* **1972**, 68, 241-272.
9. Hoffmann, R. W. Stereochemistry of [2,3]sigmatropic rearrangements. *Angew. Chem.* **1979**, 91, 625-634.
10. Nakai, T., Mikami, K. [2,3]-Wittig sigmatropic rearrangements in organic synthesis. *Chem. Rev.* **1986**, 86, 885-902.
11. Altenbach, H. J. Functional group transformations via allyl rearrangement. in *Comp. Org. Synth.* (eds. Trost, B. M.,Fleming, I.), *6*, 829-871 (Pergamon, Oxford, **1991**).
12. Brückner, R. [2,3]-Sigmatropic rearrangements. in *Comp. Org. Synth.* (eds. Trost, B. M.,Fleming, I.), *6*, 873-909 (Pergamon, Oxford, **1991**).
13. Marshall, J. A. The Wittig rearrangement. in *Comp. Org. Synth.* (eds. Trost, B. M.,Fleming, I.), *3*, 975-1015 (Pergamon, Oxford, **1991**).
14. Mikami, K., Nakai, T. Acyclic stereocontrol via [2,3]-Wittig sigmatropic rearrangement. *Synthesis* **1991**, 594-604.
15. Nakai, T., Mikami, K. The [2,3]-Wittig rearrangement. *Org. React.* **1994**, 46, 105-209.
16. Tomooka, K., Nakai, T. [1,2]-Wittig rearrangement. Stereochemical features and synthetic utilities. *Yuki Gosei Kagaku Kyokaishi* **1996**, 54, 1000-1008.
17. Nakai, T., Tomooka, K. Asymmetric [2,3]-Wittig rearrangement as a general tool for asymmetric synthesis. *Pure Appl. Chem.* **1997**, 69, 595-600.
18. Tomooka, K., Yamamoto, H., Nakai, T. Recent developments in the [1,2]-Wittig rearrangement. *Liebigs Ann. Chem.* **1997**, 1275-1281.
19. McGowan, G. Applications of the [2,3]-Wittig rearrangement to total synthesis. *Aust. J. Chem.* **2002**, 55, 799.
20. Hodgson, D. M., Tomooka, K., Gras, E. Enantioselective synthesis by lithiation adjacent to oxygen and subsequent rearrangement. *Top. Organomet. Chem.* **2003**, 5, 217-250.
21. Still, W. C., Mitra, A. A highly stereoselective synthesis of Z-trisubstituted olefins via [2,3]-sigmatropic rearrangement. Preference for a pseudoaxially substituted transition state. *J. Am. Chem. Soc.* **1978**, 100, 1927-1928.
22. Fujii, K., Hara, O., Sakagami, Y. Highly stereoselective synthesis of (E)-substituted allylsilanes via the Still-Wittig rearrangement. *Biosci. Biotechnol. Biochem.* **1997**, 61, 1394-1396.
23. Maleczka, R. E., Jr., Geng, F. Methyllithium-Promoted Wittig Rearrangements of α-Alkoxysilanes. *Org. Lett.* **1999**, 1, 1115-1118.
24. Tomooka, K., Igarashi, T., Kishi, N., Nakai, T. Olefinic stereoselection in the [2,3]-Wittig rearrangement of α,β-disubstituted allylic ethers forming trisubstituted olefins. *Tetrahedron Lett.* **1999**, 40, 6257-6260.
25. Hiersemann, M., Abraham, L., Pollex, A. The ester dienolate [2,3]-Wittig rearrangement - development, opportunities, and limitations. *Synlett* **2003**, 1088-1095.
26. Mikami, K., Nakai, T. Transition-state model for diastereoselection in [2,3]Wittig sigmatropic rearrangement. *Stud. Org. Chem. (Amsterdam)* **1987**, 31, 153-160.
27. Wu, Y. D., Houk, K. N., Marshall, J. A. Transition structure for the [2,3]-Wittig rearrangement and analysis of stereoselectivities. *J. Org. Chem.* **1990**, 55, 1421-1423.
28. Kim, C. K., Lee, I., Lee, H. W., Lee, B. S. Determination of reactivity by MO theory. Part 75. Theoretical studies on the gas-phase Wittig-oxy-Cope rearrangement of deprotonated diallyl ether. *Bull. Korean Chem. Soc.* **1991**, 12, 678-681.
29. Wu, Y. D., Houk, K. N. Theoretical studies of transition structures and stereoselectivities of the [2,3]-Wittig rearrangement of sulfur ylides. *J. Org. Chem.* **1991**, 56, 5657-5661.
30. Hoffmann, R., Brueckner, R. Asymmetric induction in reductively initiated [2,3]-Wittig and retro-[1,4]-Brook rearrangements of secondary carbanions. *Chemische Berichte* **1992**, 125, 1471-1484.
31. Antoniotti, P., Tonachini, G. Ab initio theoretical investigation on the Wright-West and Wittig anionic migration reactions. *J. Org. Chem.* **1993**, 58, 3622-3632.
32. Antoniotti, P., Canepa, C., Tonachini, G. Theoretical studies on Wittig-type anionic migrations of alkyl, silyl and germyl groups. *Trends in Organic Chemistry* **1995**, 5, 189-201.
33. Okajima, T., Fukazawa, Y. Transition structures of [2,3]-Wittig rearrangement of 2-Oxa-5-methylhexene-1-carboxylic acid. *Chem. Lett.* **1997**, 81-82.
34. Antoniotti, P., Tonachini, G. Mechanism of the Anionic Wittig Rearrangement. An ab Initio Theoretical Study. *J. Org. Chem.* **1998**, 63, 9756-9762.
35. Chia, C. S. B., Taylor, M. S., Dua, S., Blanksby, S. J., Bowie, J. H. The collision induced loss of carbon monoxide from deprotonated benzyl benzoate in the gas phase. An anionic 1,2-Wittig type rearrangement. *J. Chem. Soc., Perkin Trans. 2* **1998**, 1435-1441.
36. Jursic, B. S. High level of ab initio and density functional theory evaluation of the C-O bond dissociation energies in the dimethyl ether anion. *Int. J. Quantum Chem.* **1999**, 73, 299-306.
37. Sheldon, J. C., Taylor, M. S., Bowie, J. H., Dua, S., Chia, C. S. B., Eichinger, P. C. H. The gas phase 1,2-Wittig rearrangement is an anion reaction. A joint experimental and theoretical study. *J. Chem. Soc., Perkin Trans. 2* **1999**, 333-340.
38. Fokin, A. A., Kushko, A. O., Kirij, A. V., Yurchenko, A. G., Schleyer, P. v. R. Direct Transformations of Ketones to γ-Unsaturated Thiols via [2,3]-Sigmatropic Rearrangement of Allyl Sulfinyl Carbanions: A Combined Experimental and Computational Study. *J. Org. Chem.* **2000**, 65, 2984-2995.

39. Hart, S. A., Trindle, C. O., Etzkorn, F. A. Solvent-dependent stereoselectivity in a Still-Wittig rearrangement: an experimental and ab initio study. *Org. Lett.* **2001**, 3, 1789-1791.
40. Haeffner, F., Houk, K. N., Schulze, S. M., Lee, J. K. Concerted Rearrangement versus Heterolytic Cleavage in Anionic [2,3]- and [3,3]-Sigmatropic Shifts. A DFT Study of Relationships among Anion Stabilities, Mechanisms, and Rates. *Journal of Organic Chemistry* **2003**, 68, 2310-2316.
41. Lansbury, P. T., Pattison, V. A., Sidler, J. D., Bieber, J. B. Mechanistic aspects of the rearrangement and elimination reactions of α-metalated benzyl alkyl ethers. *J. Am. Chem. Soc.* **1966**, 88, 78-84.
42. Makisumi, Y., Notzumoto, S. Wittig rearrangement of allyl ethers of 2-quinolinemethanol and 9-fluorenol. SNi' mechanism. *Tetrahedron Lett.* **1966**, 6393-6397.
43. Gaspar, P. P., Carpenter, T. C. Mechanism of the Wittig rearrangement of N-methyl-N-phenylisoindolinium iodide. *Angew. Chem., Int. Ed. Engl.* **1967**, 6, 559-560.
44. Garst, J. F., Smith, C. D. Mechanisms of Wittig rearrangements and ketyl-alkyl iodide reactions. *J. Am. Chem. Soc.* **1973**, 95, 6870-6871.
45. Eisch, J. J., Kovacs, C. A., Rhee, S.-G. Rearrangements of organometallic compounds. X. Mechanism of 1,2-aryl migration in the Wittig rearrangement of a-metallated benzyl aryl ethers. *J. Organomet. Chem.* **1974**, 65, 289-301.
46. Eichinger, P. C. H., Bowie, J. H. Gas-phase carbanion rearrangements. Does the Wittig rearrangement occur for deprotonated vinyl ethers? *J. Chem. Soc., Perkin Trans. 2* **1990**, 1763-1768.
47. Tomooka, K., Kikuchi, M., Igawa, K., Suzuki, M., Keong, P.-H., Nakai, T. Stereoselective total synthesis of zaragozic acid A based on an acetal [1,2] Wittig rearrangement. *Angew. Chem., Int. Ed. Engl.* **2000**, 39, 4502-4505.
48. Schaudt, M., Blechert, S. Total Synthesis of (+)-Astrophylline. *J. Org. Chem.* **2003**, 68, 2913-2920.
49. Kodama, M., Yoshio, S., Yamaguchi, S., Fukuyama, Y., Takayanagi, H., Morinaka, Y., Usui, S., Fukazawa, Y. Total syntheses of both enantiomers of sarcophytols A and T based on stereospecific [2,3]Wittig rearrangement. *Tetrahedron Lett.* **1993**, 34, 8453-8456.
50. Nakazawa, M., Sakamoto, Y., Takahashi, T., Tomooka, K., Ishikawa, K., Nakai, T. A new approach to asymmetric synthesis of Stork's prostaglandin intermediate. *Tetrahedron Lett.* **1993**, 34, 5923-5926.

Wohl-Ziegler Bromination 492

1. Wohl, A. Bromination of unsaturated compounds with N-bromoacetamide, a contribution to the study of the course of chemical processes. *Ber.* **1919**, 52B, 51-63.
2. Wohl, A., Jaschinowski, K. Further experiments on the bromination of unsaturated compounds with N-bromoacetamide. *Ber.* **1921**, 54B, 476-484.
3. Ziegler, K., Spath, A., Schaaf, E., Schumann, W., Winkelmann, E. Halogenation of unsaturated substances in the allyl position. *Ann.* **1942**, 551, 80-119.
4. Djerassi, C. Brominations with N-bromosuccinimide and related compounds. The Wohl-Ziegler reaction. *Chem. Rev.* **1948**, 43, 271-317.
5. Horner, L., Winkelmann, E. H. Course of substitution. XV. N-Bromosuccinimide-properties and reactions. *Angew. Chem.* **1959**, 71, 349-365.
6. Nechvatal, A. Allylic halogenation. *Advances in Free-Radical Chemistry (London)* **1972**, 4, 175-201.
7. Schmid, H., Karrer, P. Improvement and extension of bromination with bromosuccinimide. *Helv. Chim. Acta* **1946**, 29, 573-581.
8. Liu, P., Chen, Y., Deng, J., Tu, Y. An efficient method for the preparation of benzylic bromides. *Synthesis* **2001**, 2078-2080.
9. Khazaei, A., Vaghei, R. G., Karkhanei, E. Bromination of organic allylic compounds by using N,N'-dibromo-N,N'-1,2-ethane diyl bis(2,5-dimethyl benzene sulfonyl)amine. *Synth. Commun.* **2002**, 32, 2107-2113.
10. Amijs, C. H. M., van Klink, G. P. M., van Koten, G. Carbon tetrachloride free benzylic brominations of methyl aryl halides. *Green Chem.* **2003**, 5, 470-474.
11. Baag, M. M., Kar, A., Argade, N. P. N-Bromosuccinimide-dibenzoyl peroxide/azobisisobutyronitrile: a reagent for Z- to E-alkene isomerization. *Tetrahedron* **2003**, 59, 6489-6492.
12. Togo, H., Hirai, T. Environmentally-friendly Wohl-Ziegler bromination: Ionic-liquid reaction and solvent-free reaction. *Synlett* **2003**, 702-704.
13. Greenwood, J. R., Vaccarella, G., Capper, H. R., Mewett, K. N., Allan, R. D., Johnston, G. A. R. Theoretical studies on the free-radical bromination of methylpyridazines in the synthesis of novel heterocyclic analogs of neutro-transmitters. *THEOCHEM* **1996**, 368, 235-243.
14. Gainsforth, J. L., Klobukowski, M., Tanner, D. D. Structure and Reactions of the Succinimidyl Radical: A Density Functional Study. *J. Am. Chem. Soc.* **1997**, 119, 3339-3346.
15. Rothenberg, G., Sasson, Y. Cyclic vs. acyclic allylic hydrogen abstraction: an entropy motivated process? *Tetrahedron* **1998**, 54, 5417-5422.
16. Dauben, H. J., Jr., McCoy, L. L. N-Bromosuccinimide. I. Allylic bromination, a general survey of reaction variables. *J. Am. Chem. Soc.* **1959**, 81, 4863-4873.
17. Dauben, H. J., Jr., McCoy, L. L. N-Bromosuccinimide. III. Stereochemical course of benzylic bromination. *J. Am. Chem. Soc.* **1959**, 81, 5404-5409.
18. Dauben, H. J., Jr., McCoy, L. L. N-Bromosuccinimide. II. Allylic bromination of tertiary hydrogens. *J. Org. Chem.* **1959**, 24, 1577-1579.
19. Russell, G. A., DeBoer, C. Directive effects in aliphatic substitutions. XVIII. Substitutions at saturated carbon-hydrogen bonds utilizing molecular bromine or bromotrichloromethane. *J. Am. Chem. Soc.* **1963**, 85, 3136-3139.
20. Russell, G. A., Desmond, K. M. Directive effects in aliphatic substitutions. XIX. Photobromination with N-bromosuccinimide. *J. Am. Chem. Soc.* **1963**, 85, 3139-3141.
21. Day, J. C., Lindstrom, M. J., Skell, P. S. Succinimidyl radical as a chain carrier. Mechanism of allylic bromination. *J. Am. Chem. Soc.* **1974**, 96, 5616-5617.
22. Tanner, D. D., Ruo, T. C. S., Takiguchi, H., Guillaume, A., Reed, D. W., Setiloane, B. P., Tan, S. L., Meintzer, C. P. On the mechanism of N-bromosuccinimide brominations. Bromination of cyclohexane and cyclopentane with N-bromosuccinimide. *J. Org. Chem.* **1983**, 48, 2743-2747.
23. Walling, C., El-Taliawi, G. M., Zhao, C. Radical chain carriers in N-bromosuccinimide brominations. *J. Am. Chem. Soc.* **1983**, 105, 5119-5124.
24. McMillen, D. W., Grutzner, J. B. Radical Bromination of Cyclohexene in CCl_4 by Bromine: Addition versus Substitution. *J. Org. Chem.* **1994**, 59, 4516-4528.
25. Dneprovskii, A. S., Eliseenkov, E. V., Osmonov, T. A. Mechanism of radical chlorination of hydrocarbons with N-halosulfonamides. Effect of structural factors on the reaction selectivity. *Russ. J. Org. Chem.* **1998**, 34, 27-30.
26. Lind, J., Merenyi, G. Imidyl radicals. *N-Centered Radicals* **1998**, 563-575.
27. Kim, S. S., Kim, C. S. Photobrominations of substituted cumenes by N-bromosuccinimide: charge delocalizations, inductive effects, and spin dispersions triggered by substituents. *J. Org. Chem.* **1999**, 64, 9261-9264.
28. Bringmann, G., Pabst, T., Henschel, P., Michel, M. First total synthesis of the mastigophorenes C and D and of simplified unnatural analogs. *Tetrahedron* **2001**, 57, 1269-1275.
29. Tadanier, J., Lee, C. M., Whittern, D., Wideburg, N. Synthesis of some C-8-modified 3-deoxy-β-D-manno-2-octulosonic acid analogs as inhibitors of CMP-Kdo synthetase. *Carbohydr. Res.* **1990**, 201, 185-207.
30. Gan, T., Liu, R., Yu, P., Zhao, S., Cook, J. M. Enantiospecific Synthesis of Optically Active 6-Methoxytryptophan Derivatives and Total Synthesis of Tryprostatin A. *J. Org. Chem.* **1997**, 62, 9298-9304.
31. Mu, F., Lee, D. J., Pryor, D. E., Hamel, E., Cushman, M. Synthesis and Investigation of Conformationally Restricted Analogues of Lavendustin A as Cytotoxic Inhibitors of Tubulin Polymerization. *J. Med. Chem.* **2002**, 45, 4774-4785.

Related reactions: Arndt-Eistert homologation;

1. Wolff, L. Diazo anhydrides. *Liebigs Ann. Chem.* **1902**, 325, 129-195.
2. Schroeter, G. Hofmann-Curtius', Beckmann's and the Benzilic Acid Rearrangements. *Ber.* **1909**, 42, 2336-2349.
3. Wolff, L. Diazo Anhydrides (1,2,3-Oxydiazoles or Diazo Oxides) and Diazo Ketones. *Ann.* **1913**, 394, 23-59.
4. Bachmann, W. E., Struve, W. S. Arndt-Eistert synthesis. *Org. React.* **1942**, 1, 38-62.
5. Smith, P. A. S. Rearrangements involving migration to an electron-deficient nitrogen or oxygen. *Mol. Rearrangements* **1963**, 1, 457-591.
6. Meier, H., Zeller, K. P. Wolff rearrangement of .alpha.-diazo carbonyl compounds. *Angewandte Chemie* **1975**, 87, 52-63.
7. Torres, M., Lown, E. M., Gunning, H. E., Strausz, O. P. 4n π Electron antiaromatic heterocycles. *Pure Appl. Chem.* **1980**, 52, 1623-1643.
8. Ludescher, H., Mak, C. P., Schulz, G., Fliri, H. Chemistry of penicillin diazo ketones. Part II. From β-lactam to β-lactone. *Heterocycles* **1987**, 26, 885-894.
9. Gill, G. B. The Wolff rearrangement. in *Comp. Org. Synth.* (eds. Trost, B. M.,Fleming, I.), *3*, 887-912 (Pergamon, Oxford, **1991**).
10. Ye, T., McKervey, M. A. Organic Synthesis with α-Diazo Carbonyl Compounds. *Chem. Rev.* **1994**, 94, 1091-1160.
11. Tidwell, T. T. in *Ketenes* 77-100 (Wiley, New York, **1995**).
12. Toscano, J. P. Laser flash photolysis studies of carbonyl carbenes. *Advances in Carbene Chemistry* **1998**, 2, 215-244.
13. Kirmse, W. 100 years of the Wolff rearrangement. *Eur. J. Org. Chem.* **2002**, 2193-2256.
14. Zeller, K. P. Product class 1: oxirenes. *Science of Synthesis* **2002**, 9, 19-42.
15. Celius, T. C., Wang, Y., Toscano, J. P. Photochemical reactivity of α-diazocarbonyl compounds. *CRC Handbook of Organic Photochemistry and Photobiology (2nd Edition)* **2004**, 90/91-90/16.
16. Smith, A. B., III. Vinylogous Wolff rearrangement. Copper sulfate-catalyzed decomposition of unsaturated diazomethyl ketones. *J. Chem. Soc., Chem. Commun.* **1974**, 695-696.
17. Smith, A. B., III, Toder, B. H., Branca, S. J. Stereochemical consequences of the vinylogous Wolff rearrangement. *J. Am. Chem. Soc.* **1976**, 98, 7456-7458.
18. Motallebi, S., Mueller, P. The vinylogous Wolff rearrangement catalyzed with rhodium(II) complexes. *Chimia* **1992**, 46, 119-122.
19. Ceccherelli, P., Curini, M., Marcotullio, M. C., Rosati, O. Dirhodium tetraacetate-catalyzed decomposition of β,γ-unsaturated diazo ketones: a new entry to vinylogous Wolff rearrangement. *Gazz. Chim. Ital.* **1994**, 124, 177-179.
20. Ceccherelli, P., Curini, M., Epifano, F., Marcotullio, M. C., Rosati, O. Vinylogous Wolff rearrangement of β,γ-unsaturated α-diazo-β-keto esters: a novel method for the preparation of substituted malonates. *Synth. Commun.* **1995**, 25, 301-308.
21. Marsden, S. P., Pang, W.-K. Efficient, general synthesis of silylketenes via an unusual rhodium mediated Wolff rearrangement. *Chem. Commun.* **1999**, 1199-1200.
22. Lawlor, M. D., Lee, T. W., Danheiser, R. L. Rhodium-Catalyzed Rearrangement of α-Diazo Thiol Esters to Thio-Substituted Ketenes. Application in the Synthesis of Cyclobutanones, Cyclobutenones, and β-Lactams. *J. Org. Chem.* **2000**, 65, 4375-4384.
23. Bucher, G., Strehl, A., Sander, W. Laser flash photolysis of disulfonyldiazomethanes: Partitioning between hetero-Wolff rearrangement and intramolecular carbene oxidation by a sulfonyl group. *Eur. J. Org. Chem.* **2003**, 2153-2158.
24. Thornton, D. E., Gosavi, R. K., Strausz, O. P. Mechanism of the Wolff rearrangement. II. *J. Am. Chem. Soc.* **1970**, 92, 1768-1769.
25. Csizmadia, I. G., Gunning, H. E., Gosavi, R. K., Strausz, O. P. Mechanism of the Wolff rearrangement. V. Semiempirical molecular orbital calculations on α-diazo ketones, oxirenes, and related reaction intermediates. *J. Am. Chem. Soc.* **1973**, 95, 133-137.
26. Hopkinson, A. C. Nonempirical molecular orbital study of the Wolff rearrangement. *J. Chem. Soc., Perkin Trans. 2* **1973**, 794-795.
27. Strausz, O. P., Gosavi, R. K., Denes, A. S., Csizmadia, I. G. Mechanism of the Wolff rearrangement. 6. Ab initio molecular orbital calculations on the thermodynamic and kinetic stability of the oxirene molecule. *J. Am. Chem. Soc.* **1976**, 98, 4784-4786.
28. Hopkinson, A. C., Lien, M., Yates, K., Csizmadia, I. G. A non-empirical molecular orbital study of oxirene and its valence tautomers. *Progress in Theoretical Organic Chemistry* **1977**, 2, 230-247.
29. Strausz, O. P., Gosavi, R. K., Gunning, H. E. Ab initio molecular orbital calculations on the reaction path of the ketocarbene-ketene rearrangement. *J. Chem. Phys.* **1977**, 67, 3057-3060.
30. Bargon, J., Tanaka, K., Yoshimine, M. Computer chemistry studies of organic reactions: the Wolff rearrangement. *Comput. Methods Chem., [Proc. Int. Symp.]* **1980**, 239-274.
31. Tanaka, K., Yoshimine, M. An ab initio study on ketene, hydroxyacetylene, formylmethylene, oxirene, and their rearrangement paths. *J. Am. Chem. Soc.* **1980**, 102, 7655-7662.
32. Novoa, J. J., McDouall, J. J. W., Robb, M. A. The diradical nature of keto carbenes occurring in the Wolff rearrangement. *J. Chem. Soc., Fraday Trans. 2: Mol. Chem. Phys.* **1987**, 83, 1629-1636.
33. Tsuda, M., Oikawa, S., Nagayama, K. Reactive intermediates produced in the decomposition of 2-diazo ketones: mechanism of the Wolff rearrangement. *Chem. Pharm. Bull.* **1987**, 35, 1-8.
34. Tsuda, M., Oikawa, S. Elementary reactions in photochemistry of 2-diazo quinones and 2-diazo ketones. *J. Photopolymer Sci. Tech.* **1989**, 2, 325-339.
35. Tsuda, M., Oikawa, S. Mechanism of the Wolff rearrangement. *Chem. Pharm. Bull.* **1989**, 37, 573-575.
36. Bachmann, C., N'Guessan, T. Y., Debu, F., Monnier, M., Pourcin, J., Aycard, J. P., Bodot, H. Oxirenes and ketocarbenes from α-diazoketone photolysis: experiments in rare gas matrices. Relative stabilities and isomerization barriers from MNDOC-BWEN calculations. *J. Am. Chem. Soc.* **1990**, 112, 7488-7497.
37. Torres, M., Gosavi, R. K., Lown, E. M., Piotrkowski, E. J., Kim, B., Bourdelande, J. L., Font, J., Strausz, O. P. The Wolff rearrangement of α-diazo ketones: the role of oxirene and its isomers. *Stud. Phys. Theor. Chem.* **1992**, 77, 184-211.
38. Bachmann, C., N'Guessan, T. Y. Photoactivated α-ketocarbenes: formation and isomerization reactions. RRKM calculations with semi-empirical parameters (AM1, MNDOC). *Int. J. Chem. Kinet.* **1994**, 26, 643-664.
39. Nguyen, T. M., Hajnal, M. R., Vanquickenborne, L. G. Theoretical evidence of a singlet α-oxocarbene intermediate in the retro-Wolff rearrangement of azafulvenone. *J. Chem. Soc., Perkin Trans. 2* **1994**, 169-170.
40. Scott, A. P., Nobes, R. H., Schaefer, H. F., III, Radom, L. The Wolff Rearrangement: The Relevant Portion of the Oxirene-Ketene Potential Energy Hypersurface. *J. Am. Chem. Soc.* **1994**, 116, 10159-10164.
41. Termath, V., Tozer, D. J., Handy, N. C. Density functional theory studies of 4-π-electron systems. *Chem. Phys. Lett.* **1994**, 228, 239-245.
42. Calvo-Losada, S., Quirante, J. J. DFT study of competitive Wolff rearrangement and [1,2]-hydrogen shift of β-oxy-α-diazo carbonyl compounds. *THEOCHEM* **1997**, 398-399, 435-443.
43. Kim, C. K., Lee, I. Theoretical studies on the gas-phase Wolff rearrangement of α-ketocarbenes. *Bull. Korean Chem. Soc.* **1997**, 18, 395-401.
44. Borisov, Y. A., Garrett, B. C., Feller, D. Ab initio study of the Wolff rearrangement of C_6H_4O intermediate in the gas phase. *Russ. Chem. Bull.* **1999**, 48, 1642-1646.
45. Calvo-Losada, S., Suarez, D., Sordo, T. L., Quirante, J. J. Competition between Wolff Rearrangement and 1,2-Hydrogen Shift in β-oxy-α-ketocarbenes: Electrostatic and Specific Solvent Effects. *J. Phys. Chem. B* **1999**, 103, 7145-7150.
46. Calvo-Losada, S., Sordo, T. L., Lopez-Herrera, F. J., Quirante, J. J. The influence of protecting the hydroxyl group of β-oxy-α-diazo carbonyl compounds in the competition between Wolff rearrangement and [1,2]-hydrogen shift. Density functional theory study and topological analysis of the charge density. *Theoretical Chemistry Accounts* **2000**, 103, 423-430.
47. Likhotvorik, I., Zhu, Z., Tae, E. L., Tippmann, E., Hill, B. T., Platz, M. S. Carbomethoxychlorocarbene: Spectroscopy, Theory, Chemistry and Kinetics. *J. Am. Chem. Soc.* **2001**, 123, 6061-6068.
48. Platz, M. S., Tippmann, E. M. Carbomethoxyhalocarbenes: Spectroscopy, theory, chemistry, and kinetics. *Abstracts of Papers, 222nd ACS National Meeting, Chicago, IL, United States, August 26-30, 2001* **2001**, ORGN-112.

49. Acton, A. W., Allen, A. D., Antunes, L. M., Fedorov, A. V., Najafian, K., Tidwell, T. T., Wagner, B. D. Amination of Pyridylketenes: Experimental and Computational Studies of Strong Amide Enol Stabilization by the 2-Pyridyl Group. *J. Am. Chem. Soc.* **2002**, 124, 13790-13794.
50. Sudrik, S. G., Chavan, S. P., Chandrakumar, K. R. S., Pal, S., Date, S. K., Chavan, S. P., Sonawane, H. R. Microwave Specific Wolff Rearrangement of α-Diazoketones and Its Relevance to the Nonthermal and Thermal Effect. *J. Org. Chem.* **2002**, 67, 1574-1579.
51. Wilson, P. J., Tozer, D. J. A Kohn-Sham study of the oxirene-ketene potential energy surface. *Chem. Phys. Lett.* **2002**, 352, 540-544.
52. Bogdanova, A., Popik, V. V. Experimental and Theoretical Investigation of Reversible Interconversion, Thermal Reactions, and Wavelength-Dependent Photochemistry of Diazo Meldrum's Acid and its Diazirine Isomer, 6,6-Dimethyl-5,7-dioxa-1,2-diaza-spiro[2,5]oct-1-ene-4,8-dione. *J. Am. Chem. Soc.* **2003**, 125, 14153-14162.
53. Chiang, Y., Gaplovsky, M., Kresge, A. J., Leung, K. H., Ley, C., Mac, M., Persy, G., Phillips, D. L., Popik, V. V., Roedig, C., Wirz, J., Zhu, Y. Photoreactions of 3-Diazo-3H-benzofuran-2-one; Dimerization and Hydrolysis of Its Primary Photoproduct, A Quinonoid Cumulenone: A Study by Time-Resolved Optical and Infrared Spectroscopy. *J. Am. Chem. Soc.* **2003**, 125, 12872-12880.
54. Julian, R. R., May, J. A., Stoltz, B. M., Beauchamp, J. L. Gas-Phase Synthesis of Charged Copper and Silver Fischer Carbenes from Diazomalonates: Mechanistic and Conformational Considerations in Metal-Mediated Wolff Rearrangements. *J. Am. Chem. Soc.* **2003**, 125, 4478-4486.
55. Sato, T., Niino, H., Yabe, A. Ketene Formation in Benzdiyne Chemistry: Ring Cleavage versus Wolff Rearrangement. *J. Am. Chem. Soc.* **2003**, 125, 11936-11941.
56. Zimmerman, H. E., Wang, P. An unusual abnormal Wolff rearrangement. *Can. J. Chem.* **2003**, 81, 517-524.
57. Regitz, M. Transfer of diazo groups. *Angew. Chem., Int. Ed. Engl.* **1967**, 6, 733-749.
58. Regitz, M. Reactions of carbon-hydrogen active compounds with azides. XIII. Diazo group transfer. *Neuere Method. Praep. Org. Chem.* **1970**, 6, 76-118.
59. Regitz, M. Recent synthetic methods in diazo chemistry. *Synthesis* **1972**, 351-373.
60. Regitz, M., Korobitsyna, I. K., Rodina, L. L. Aliphatic diazo compounds. *Method. Chim.* **1975**, 6, 205-299.
61. Regitz, M., Rueter, J. Reactions of CH-active compounds with azides. XVIII. Synthesis of 2-oxo-1-diazo cycloalkanes by deformylative diazo-group transfer. *Chem. Ber.* **1968**, 101, 1263-1270.
62. Regitz, M., Menz, F., Liedhegener, A. Reactions of CH-active compounds with azides. XXVIII. Synthesis of α,β-unsaturated diazoketones by deformylating diazo group transfer. *Liebigs Ann. Chem.* **1970**, 739, 174-184.
63. Cava, M. P., Litle, R. L., Napier, D. R. Condensed cyclobutane aromatic systems. V. The synthesis of some α-diazoindanones: ring contraction in the indan series. *J. Am. Chem. Soc.* **1958**, 80, 2257-2263.
64. Cava, M. P., Litle, R. L. New synthesis of α-diazo ketones. *Chem. Ind.* **1957**, 367.
65. Lewars, E. G. Oxirenes. *Chem. Rev.* **1983**, 83, 519-534.
66. Uyehara, T., Takehara, N., Ueno, M., Sato, T. Rearrangement approaches to cyclic skeletons. IX. Stereoselective total synthesis of (±)-campherenone based on a ring-contraction of bicyclo[3.2.1]oct-6-en-2-one. Reliable one-step diazo transfer followed by a Wolff rearrangement. *Bull. Chem. Soc. Jpn.* **1995**, 68, 2687-2694.
67. Ihara, M., Suzuki, T., Katogi, M., Taniguchi, N., Fukumoto, K. A stereoselective total synthesis of (±)-$\Delta^{9(12)}$-capnellene via the intramolecular Diels-Alder approach. *J. Chem. Soc., Chem. Commun.* **1991**, 646-647.
68. Norbeck, D. W., Kramer, J. B. Synthesis of (-)-oxetanocin. *J. Am. Chem. Soc.* **1988**, 110, 7217-7218.
69. Danheiser, R. L., Helgason, A. L. Total Synthesis of the Phenalenone Diterpene Salvilenone. *J. Am. Chem. Soc.* **1994**, 116, 9471-9479.

Wolff-Kishner Reduction 496

Related reactions: Clemmensen reduction;

1. Kishner, N. *J. Russ. Phys. Chem. Soc.* **1911**, 43, 582.
2. Wolff, L. Diazo anhydride (1,2,3-oxydiazoles or diazooxides) and diazo ketones. *Liebigs Ann. Chem.* **1912**, 394, 23-108.
3. Todd, D. Wolff-Kishner reduction. *Org. React.* **1948**, 4, 378-422.
4. Buu-Hoi, N. P., Hoan, N., Xuong, N. D. Limitations of the Wolff-Kishner reaction. *Recl. Trav. Chim. Pays-Bas* **1952**, 71, 285-291.
5. Reusch, W. Deoxygenation of carbonyl compounds. in *Reduction* (ed. Augustine, R. L.), 171-211 (Dekker, New York, **1968**).
6. Hutchins, R. O. Reduction of C=X to CH_2 by Wolff-Kishner and other hydrazone methods. in *Comp. Org. Synth.* (eds. Trost, B. M.,Fleming, I.), *8*, 327-362 (Pergamon, Oxford, **1991**).
7. Herr, C. H., Whitmore, F. C., Schiessler, R. W. Wolff-Kishner reaction at atmospheric pressure. *J. Am. Chem. Soc.* **1945**, 67, 2061-2063.
8. Soffer, M. D., Soffer, M. B., Sherk, K. W. Low-pressure method for Wolff-Kishner reduction. *J. Am. Chem. Soc.* **1945**, 67, 1435-1436.
9. Huang, M. Simple modification of the Wolff-Kishner reduction. *J. Am. Chem. Soc.* **1946**, 68, 2487-2488.
10. Huang, M. Reduction of steroid ketones and other carbonyl compounds by modified Wolff-Kishner method. *J. Am. Chem. Soc.* **1949**, 71, 3301-3303.
11. Barton, D. H. R., Ives, D. A. J., Thomas, B. R. A Wolff-Kishner reduction procedure for sterically hindered carbonyl groups. *J. Chem. Soc., Abstracts* **1955**, 2056.
12. Cram, D. J., Sahyun, M. R. V., Knox, G. R. Room temperature Wolff-Kishner and Cope eliminations. *J. Am. Chem. Soc.* **1962**, 84, 1734-1735.
13. Grundon, M. F., Henbest, H. B., Scott, M. D. Reactions of hydrazones and related compounds with strong bases. I. Modified Wolff-Kishner procedure. *J. Chem. Soc., Abstracts* **1963**, 1855-1858.
14. Nagata, W., Itazaki, H. Simplified modification of Wolff-Kisimer reduction for hindered or masked carbonyl groups. *Chem. Ind.* **1964**, 1194-1195.
15. Parquet, E., Lin, Q. Microwave-assisted Wolff-Kishner reduction reaction. *J. Chem. Educ.* **1997**, 74, 1225.
16. Gadhwal, S., Baruah, M., Sandhu, J. S. Microwave induced synthesis of hydrazones and Wolff-Kishner reduction of carbonyl compounds. *Synlett* **1999**, 1573-1574.
17. Eisenbraun, E. J., Payne, K. W., Bymaster, J. S. Multiple-Batch, Wolff-Kishner Reduction Based on Azeotropic Distillation Using Diethylene Glycol. *Ind. & Eng. Chem. Res.* **2000**, 39, 1119-1123.
18. Chattopadhyay, S., Banerjee, S. K., Mitra, A. K. The Huang-Minlon modification of Wolff-Kishner reduction in rapid and simple way using microwave technology. *J. Indian Chem. Soc.* **2002**, 79, 906-907.
19. Jaisankar, P., Pal, B., Giri, V. S. Microwave assisted McFadyen-Stevens and Huang-Minlon reactions. *Synth. Commun.* **2002**, 32, 2569-2573.
20. Furrow, M. E., Myers, A. G. Practical Procedures for the Preparation of N-tert-Butyldimethylsilylhydrazones and Their Use in Modified Wolff-Kishner Reductions and in the Synthesis of Vinyl Halides and gem-Dihalides. *J. Am. Chem. Soc.* **2004**, 126, 5436-5445.
21. Szendi, Z., Forgo, P., Tasi, G., Bocskei, Z., Nyerges, L., Sweet, F. 1,5-Hydride shift in Wolff-Kishner reduction of (20R)-3b,20,26-trihydroxy-27-norcholest-5-en-22-one: synthetic, quantum chemical, and NMR studies. *Steroids* **2002**, 67, 31-38.
22. Caglioti, L., Magi, M. Reaction of tolylsulfonylhydrazones with lithium aluminum hydride. *Tetrahedron* **1963**, 19, 1127-1131.
23. Leonard, N. J., Gelfand, S. The Kishner reduction-elimination. II. α-Substituted pinacolones. *J. Am. Chem. Soc.* **1955**, 77, 3272-3278.
24. Leonard, N. J., Gelfand, S. The Kishner reduction-elimination. I. Cyclic and open chain α-amino ketones. *J. Am. Chem. Soc.* **1955**, 77, 3269-3271.
25. Morris Kupchan, S., Abushanab, E. Synthetic approach to the 9(10 -> 19)-abeo-pregnane system, involving carbocyclic ring cleavage during Wolff-Kishner reduction. *Tetrahedron Lett.* **1965**, 3075-3081.
26. Szmant, H. H. Mechanism of the Wolff-Kishner reduction, elimination, and isomerization reactions. *Angew. Chem., Int. Ed. Engl.* **1968**, 7, 120-128.

27. Szmant, H. H., Alciaturi, C. E. Mechanistic aspects of the Wolff-Kishner reaction. 6. Comparison of the hydrazones of benzophenone, fluorenone, dibenzotropone, and dibenzosuberone. *J. Org. Chem.* **1977**, 42, 1081-1082.
28. Szmant, H. H., Birke, A., Lau, M. P. Mechanistic aspects of the Wolff-Kishner reaction. 7. The W-K reaction of benzophenone hydrazone in dimethyl sulfoxide. *J. Am. Chem. Soc.* **1977**, 99, 1863-1871.
29. Szmant, H. H., Alciaturi, C. E. Mechanistic aspects of the Wolff-Kishner reaction. V. The cation effect on the reaction of benzophenone hydrazone in butyl carbitol and decyl alcohol. *J. Solution Chem.* **1978**, 7, 269-281.
30. Brecknell, D. J., Carman, R. M., Schumann, R. C. Kinetic versus thermodynamic effects during a Wolff-Kishner reduction. *Aust. J. Chem.* **1989**, 42, 527-539.
31. Taber, D. F., Stachel, S. J. On the mechanism of the Wolff-Kishner reduction. *Tetrahedron Lett.* **1992**, 33, 903-906.
32. Szendi, Z., Forgo, P., Tasi, G., Bocskei, Z., Nyerges, L., Sweet, F. 1,5-Hydride shift in Wolff-Kishner reduction of (20R)-3β,20, 26-trihydroxy-27-norcholest-5-en-22-one: synthetic, quantum chemical, and NMR studies. *Steroids* **2002**, 67, 31-38.
33. Toyota, M., Wada, T., Ihara, M. Total Syntheses of (-)-Methyl Atis-16-en-19-oate, (-)-Methyl Kaur-16-en-19-oate, and (-)-Methyl Trachyloban-19-oate by a Combination of Palladium-Catalyzed Cycloalkenylation and Homoallyl-Homoallyl Radical Rearrangement. *J. Org. Chem.* **2000**, 65, 4565-4570.
34. Marino, J. P., Rubio, M. B., Cao, G., de Dios, A. Total Synthesis of (+)-Aspidospermidine: A New Strategy for the Enantiospecific Synthesis of Aspidosperma Alkaloids. *J. Am. Chem. Soc.* **2002**, 124, 13398-13399.
35. Miyaoka, H., Kajiwara, Y., Hara, Y., Yamada, Y. Total Synthesis of Natural Dysidiolide. *J. Org. Chem.* **2001**, 66, 1429-1435.

Wurtz Coupling 498

1. Wurtz, A. *Ann. Chim. Phys.* **1855**, 44, 275.
2. Wurtz, A. A new class of organic groups. *Ann.* **1855**, 96, 364-375.
3. Asinger, F., Vogel, H. H. The preparation of alkanes and cycloalkanes. in *Houben-Weyl, Methoden der Organischen Chemie* (ed. Müller, E.), *5*, 347 (Georg Thieme Verlag, Stuttgart, **1970**).
4. Billington, D. C. Coupling Reactions Between sp^3 Carbon Centers. in *Comp. Org. Synth.* (eds. Trost, B. M.,Fleming, I.), *3*, 413-434 (Pergamon, Oxford, **1991**).
5. Jones, R. G., Holder, S. J. Synthesis of polysilanes by the Wurtz reductive-coupling reaction. *Silicon-Containing Polymers* **2000**, 353-373.
6. Banno, T., Hayakawa, Y., Umeno, M. Some applications of the Grignard cross-coupling reaction in the industrial field. *Journal of Organometallic Chemistry* **2002**, 653, 288-291.
7. Herrmann, W. A. Arene coupling reactions. *Applied Homogeneous Catalysis with Organometallic Compounds (2nd Edition)* **2002**, 2, 822-828.
8. Muller, E., Roscheisen, G. A variation of the Wurtz synthesis. I. Catalyzed reactions of benzyl and allyl halides with alkali metals. *Chem. Ber.* **1957**, 90, 543-553.
9. Muller, E., Roscheisen, G. The reactive behavior of disodiotetraphenylethane toward aromatic halides. *Chem. Ber.* **1958**, 91, 1106-1114.
10. Craig, A. D., MacDiarmid, A. G. Application of the Wurtz reaction to the synthesis of disilane and 1,2-dimethyldisilane. *Journal of Inorganic and Nuclear Chemistry* **1962**, 24, 161-164.
11. Lash, T. D., Berry, D. Promotion of organic reactions by ultrasound: coupling of alkyl and aryl halides in the presence of lithium metal and ultrasound. *J. Chem. Educ.* **1985**, 62, 85.
12. Osborne, A. G., Glass, K. J., Staley, M. L. Ultrasound-promoted coupling of heteroaryl halides in the presence of lithium wire. Novel formation of isomeric bipyridines in a Wurtz-type reaction. *Tetrahedron Lett.* **1989**, 30, 3567-3568.
13. Ginah, F. O., Donovan, T. A., Jr., Suchan, S. D., Pfennig, D. R., Ebert, G. W. Homocoupling of alkyl halides and cyclization of α,ω-dihaloalkanes via activated copper. *J. Org. Chem.* **1990**, 55, 584-589.
14. Mistryukov, E. A. Ultrasound in organic synthesis. Electron-transfer catalysis in Li-TMSCl reductive benzene silylation and TMSCl Wurtz coupling. *Mendeleev Commun.* **1993**, 251.
15. Gilbert, B. C., Lindsay, C. I., McGrail, P. T., Parsons, A. F., Whittaker, D. T. E. Efficient radical coupling of organobromides using dimanganese decacarbonyl. *Synth. Commun.* **1999**, 29, 2711-2718.
16. Marton, D., Tari, M. Wurtz-type reductive coupling reaction of primary alkyl iodides and haloorganotins in cosolvent/$H_2O(NH_4Cl)$/Zn media as a route to mixed alkylstannanes and hexaalkyldistannanes. *J. Organomet. Chem.* **2000**, 612, 78-84.
17. Voegtle, F., Neumann, P. Synthesis of [2.2]phanes. *Synthesis* **1973**, 85-103.
18. Meszaros, L., Soos, K., Sirokman, F. Reactivity of various metal powders and their applicability primarily in Wurtz syntheses. *Acta Physica et Chemica* **1970**, 16, 51-55.
19. Gilman, H., Wright, G. F. Mechanism of the Wurtz-Fittig reaction. The direct preparation of an organo-sodium (potassium) compound from an RX compound. *J. Am. Chem. Soc.* **1933**, 55, 2893-2896.
20. Richards, R. B. Mechanism of the Wurtz reaction. *Transactions of the Faraday Society* **1940**, 36, 956-960.
21. Saffer, A., Davis, T. W. Products from the Wurtz reaction and the mechanism of their formation. *J. Am. Chem. Soc.* **1942**, 64, 2039-2043.
22. Sirks, J. F. The Wurtz-Fittig reaction. II. *Recl. Trav. Chim. Pays-Bas* **1946**, 65, 850-858.
23. Emblem, H. G., Ridge, D., Todd, M. Mechanism of the Wurtz-Fittig reaction between organic halides, tetrachlorosilane, and sodium. *Chem. Ind.* **1955**, 905-906.
24. Malinovskii, M. S., Yavorovskii, A. A. Mechanism of the Grignard-Wurtz reaction. I. Synthesis of some alkaromatic hydrocarbons from benzyl chloride, α-bromoethylbenzene, and α-bromo-α-methylethylbenzene. *Zh. Obshch. Khim.* **1955**, 25, 2169-2173.
25. LeGoff, E., Ulrich, S. E., Denney, D. B. Mechanism of the Wurtz reaction. The configurations of 2-bromoöctane, 3-methylnonane, and 7,8-dimethyltetradecane. *J. Am. Chem. Soc.* **1958**, 80, 622-625.
26. Skell, P. S., Krapcho, A. P. Carbene intermediates in the Wurtz reaction. α-Elimination of HCl from neopentyl chloride. *J. Am. Chem. Soc.* **1961**, 83, 754-755.
27. Anteunis, M., van Schoote, J. Grignard reaction. VII. Mechanism of the Grignard reagent formation and the Wurtz side reaction in ether. *Bull. Soc. Chim. Beiges* **1963**, 72, 787-796.
28. Garst, J. F., Cox, R. H. Wurtz reaction. Chemically induced nuclear spin polarization in reactions of alkyl iodides with sodium mirrors. *J. Am. Chem. Soc.* **1970**, 92, 6389-6391.
29. Garst, J. F., Hart, P. W. Evidence against alkyl dimer formation through S_N2 processes in Wurtz reactions of alkyl iodides with sodium in 1,2-dimethoxyethane. Bineopentyl from neopentyl iodide. *J. Chem. Soc., Chem. Commun.* **1975**, 215-216.
30. Forou, M. A., Reynolds, J. L. The Wurtz cross-coupling reaction revisited. *Main Group Metal Chem.* **1994**, 17, 399-402.
31. Morzycki, J. W., Kalinowski, S., Lotowski, Z., Rabiczko, J. Synthesis of dimeric steroids as components of lipid membranes. *Tetrahedron* **1997**, 53, 10579-10590.
32. Vermes, B., Keseru, G. M., Mezey-Vandor, G., Nogradi, M., Toth, G. Synthesis of garugamblin-1. *Tetrahedron* **1993**, 49, 4893-4900.
33. Dienes, Z., Nogradi, M., Vermes, B., Kajtar-Peredy, M. Synthesis of marchantin I, a macrocyclic bis(bibenzyl ether) from Riccardia multifida. *Liebigs Ann. Chem.* **1989**, 1141-1143.
34. Ramig, K., Dong, Y., Van Arnum, S. D. A convenient preparation of cyclobutyl ketones: naphthalene-catalyzed reductive cyclization of substituted 1,4-dihalobutanes. *Tetrahedron Lett.* **1996**, 37, 443-446.

Related reactions: Corey-Nicolaou macrolactonization, Keck macrolactonization;

1. Inanaga, J., Hirata, K., Saeki, H., Katsuki, T., Yamaguchi, M. A rapid esterification by mixed anhydride and its application to large-ring lactonization. *Bull. Chem. Soc. Jpn.* **1979**, 52, 1989-1993.
2. Haslam, E. Recent developments in methods for the esterification and protection of the carboxyl group. *Tetrahedron* **1980**, 36, 2409-2433.
3. Mulzer, J. Synthesis of Esters, Activated Esters and Lactones. in *Comp. Org. Synth.* (eds. Trost, B. M.,Fleming, I.), *6*, 323-380 (Pergamon, Oxford, **1991**).
4. Meng, Q., Hesse, M. Ring-closure methods in the synthesis of macrocyclic natural products. *Top. Curr. Chem.* **1992**, 161, 107-176.
5. Thijs, L., Egenberger, D. M., Zwanenburg, B. An enantioselective total synthesis of the macrolide patulolide C. *Tetrahedron Lett.* **1989**, 30, 2153-2156.
6. Hikota, M., Tone, H., Horita, K., Yonemitsu, O. Chiral synthesis of polyketide-derived natural products. 31. Stereoselective synthesis of erythronolide A by extremely efficient lactonization based on conformational adjustment and high activation of seco-acid. *Tetrahedron* **1990**, 46, 4613-4628.
7. Marino, J. P., McClure, M. S., Holub, D. P., Comasseto, J. V., Tucci, F. C. Stereocontrolled Synthesis of (-)-Macrolactin A. *J. Am. Chem. Soc.* **2002**, 124, 1664-1668.
8. Hu, T., Takenaka, N., Panek, J. S. Asymmetric Crotylation Reactions in Synthesis of Polypropionate-Derived Macrolides: Application to Total Synthesis of Oleandolide. *J. Am. Chem. Soc.* **2002**, 124, 12806-12815.
9. Ghosh, A. K., Wang, Y., Kim, J. T. Total Synthesis of Microtubule-Stabilizing Agent (-)-Laulimalide. *J. Org. Chem.* **2001**, 66, 8973-8982.

X. INDEX

C

E

G

H

I

L

N

O

Q

R

U

V

W

X

Y

Z